Prentice Hall

COURSE 2

MATHEMATICS

Common Core

Charles
Illingworth
McNemar
Mills
Ramirez
Reeves

PEARSON

Boston, Massachusetts • Chandler, Arizona • Glenview, Illinois • Upper Saddle River, New Jersey

Acknowledgments appear on p. T738, which constitutes an extension of this copyright page.

DK Dorling Kindersley (DK) is an international publishing company that specializes in the creation of high-quality, illustrated information books for children and adults. Dorling Kindersley's unique graphic presentation style is used in this program to motivate students in learning about real-world applications of mathematics. DK is part of the Pearson family of companies.

ISBN-13: 978-0-13-319671-9
ISBN-10: 0-13-319671-2
2 3 4 5 6 7 8 9 10 V063 15 14 13 12 11

Teacher's Edition Contents

Mathematics Teacher Handbook

Student Edition with Teacher Notes

Authors

Series Author

Randall I. Charles, Ph.D., is Professor Emeritus in the Department of Mathematics and Computer Science at San Jose State University, San Jose, California. He began his career as a high school mathematics teacher, and he was a mathematics supervisor for five years. Dr. Charles has been a member of several NCTM committees and is the former Vice President of the National Council of Supervisors of Mathematics. Much of his writing and research has been in the area of problem solving. He has authored more than 75 mathematics textbooks for kindergarten through college. *Scott Foresman-Prentice Hall Mathematics Series Author Kindergarten through Algebra 2*

Program Authors

Mark Illingworth has taught in both elementary and high school math programs for more than twenty years. During this time, he received the Christa McAuliffe sabbatical to develop problem solving materials and projects for middle grades math students, and he was granted the Presidential Award for Excellence in Mathematics Teaching. Mr. Illingworth's specialty is in teaching mathematics through applications and problem solving. He has written two books on these subjects and has contributed to math and science textbooks at Prentice Hall.

Bonnie McNemar is a mathematics educator with more than 30 years' experience in Texas schools as a teacher, administrator, and consultant. She began her career as a middle school mathematics teacher and served as a supervisor at the district, county, and state levels. Ms. McNemar was the director of the Texas Mathematics Staff Development Program, now known as TEXTEAMS, for five years, and she was the first director of the Teachers Teaching with Technology (T^3) Program. She remains active in both of these organizations as well as in several local, state, and national mathematics organizations, including NCTM.

Darwin Mills, an administrator for the public school system in Newport News, Virginia, has been involved in secondary level mathematics education for more than fourteen years. Mr. Mills has served as a high school teacher, a community college adjunct professor, a department chair, and a district level mathematics supervisor. He has received numerous teaching awards, including teacher of the year for 1999–2000, and an Excellence in Teaching award from the College of Wooster, Ohio, in 2002. He is a frequent presenter at workshops and conferences. He believes that all students can learn mathematics if given the proper instruction.

Alma Ramirez is co-director of the Mathematics Case Project at WestEd a nonprofit educational institute in Oakland, California. A former bilingual elementary and middle school teacher, Ms. Ramirez has considerable expertise in mathematics teaching and learning, second language acquisition, and professional development. She has served as a consultant on a variety of projects and has extensive experience as an author for elementary and middle grades texts. In addition, her work has appeared in the 2004 NCTM Yearbook. Ms. Ramirez is a frequent presenter at professional meetings and conferences.

Andy Reeves, Ph.D., teaches at the University of South Florida in St. Petersburg. His career in education spans 30 years and includes seven years as a middle grades teacher. He subsequently served as Florida's K–12 mathematics supervisor, and more recently he supervised the publication of The Mathematics Teacher, Mathematics Teaching in the Middle School, and Teaching Children Mathematics for NCTM. Prior to entering education, he worked as an engineer for Douglas Aircraft.

Contributing Author

Denisse R. Thompson, Ph.D., is a Professor of Mathematics Education at the University of South Florida. She has particular interests in the connections between literature and mathematics and in the teaching and learning of mathematics in the middle grades. Dr. Thompson contributed to the Guided Problem Solving features.

Reviewers

Course 2 Reviewers

Cami Craig
Prince William County Public Schools
Marsteller Middle School
Bristow, Virginia

Donald O. Cram
Lincoln Middle School
Rio Rancho, New Mexico

Pat A. Davidson
Jacksonville Junior High School
Jacksonville, Arkansas

Yvette Drew
DeKalb County School System
Open Campus High School
Atlanta, Georgia

Robert S. Fair
K–12 District Mathematics Coordinator
Cherry Creek School District
Greenwood Village, Colorado

Michael A. Landry
Glastonbury Public Schools
Glastonbury, Connecticut

Nancy Ochoa
Weeden Middle School
Florence, Alabama

Charlotte J. Phillips
Wichita USD 259
Wichita, Kansas

Mary Lynn Raith
Mathematics Curriculum Specialist
Pittsburgh Public Schools
Pittsburgh, Pennsylvania

Tammy Rush
Consultant, Middle School Mathematics
Hillsborough County Schools
Tampa, Florida

Judith R. Russ
Prince George's County Public Schools
Capitol Heights, Maryland

Tim Tate
Math/Science Supervisor
Lafayette Parish School System
Lafayette, Louisiana

Dondi J. Thompson
Alcott Middle School
Norman, Oklahoma

Candace Yamagata
Hyde Park Middle School
Las Vegas, Nevada

Content Consultants

Ann Bell
Mathematics
Prentice Hall Consultant
Franklin, Tennessee

Blanche Brownley
Mathematics
Prentice Hall Consultant
Olney, Maryland

Joe Brumfield
Mathematics
Prentice Hall Consultant
Altadena, California

Linda Buckhalt
Mathematics
Prentice Hall Consultant
Derwood, Maryland

Andrea Gordon
Mathematics
Prentice Hall Consultant
Atlanta, Georgia

Eleanor Lopes
Mathematics
Prentice Hall Consultant
New Castle, Delaware

Sally Marsh
Mathematics
Prentice Hall Consultant
Baltimore, Maryland

Bob Pacyga
Mathematics
Prentice Hall Consultant
Darien, Illinois

Judy Porter
Mathematics
Prentice Hall Consultant
Fuquay Varina, North Carolina

Rose Primiani
Mathematics
Prentice Hall Consultant
Harbor City, New Jersey

Jayne Radu
Mathematics
Prentice Hall Consultant
Scottsdale, Arizona

Pam Revels
Mathematics
Prentice Hall Consultant
Sarasota, Florida

Barbara Rogers
Mathematics
Prentice Hall Consultant
Raleigh, North Carolina

Michael Seals
Mathematics
Prentice Hall Consultant
Edmond, Oklahoma

Margaret Thomas
Mathematics
Prentice Hall Consultant
Indianapolis, Indiana

Common Core State Standards for Mathematics

ABOUT THE STANDARDS

The Common Core State Standards Initiative is a state-led initiative coordinated by the National Governors Association Center for Best Practices (NGA Center) and the Council of Chief Stat School Officers (CCSSO), with a goal of developing a set of standards in math and in English language arts that would be implemented in many, if not most states in the United States.

The **Common Core State Standards for Mathematics** were developed by mathematicians and math educators and were reviewed by state department of education representatives of the 48 participating states. The members of the writing committee looked at state standards from high-performing states in the United States and from high-performing countries around the world and developed standards that reflect the intent and content of these exemplars.

The final draft was released in June 2010 after nearly 12 months of intense development, review, and revision. To date, over 40 states have adopted these new standards and are currently working to develop model curricula or curriculum frameworks based on these standards.

These standards identify the knowledge and skills students should gain throughout their K–12 careers so that upon graduation from high school. they will be college- or career-ready. The standards include rigorous content and application of knowledge through higher-order thinking skills.

The CCSSM consist of two interrelated sets of standards, the Standards for Mathematical Practice and the Standards for Mathematical Content. The **Standards for Mathematical Practice** describe the processes, practices, and dispositions that lead to mathematical proficiency. The eight Standards are common to all grade levels, K–12, highlighting that these processes, practices, and dispositions are developed throughout one's school career. A discussion of these standards is found on pages T26 to T31.

The **Standards for Mathematical Content** are grade-specific for Kindergarten through Grade 8. They are organized by domains that are similar to the strands in the *Principles and Standards for School Mathematics* (2000) and *Curriculum Focal Points for Prekindergarten through Grade 8 Mathematics* (2006) from the National Council for Teachers of Mathematics (NCTM). Within each domain are clusters, which each consist of one or more standards. An overview of these standards for Grade 7 is found on pages T32 to T35.

Assessing the Common Core State Standards

Your students will not only be learning concepts and skills based on a new set of standards, they will also likely be taking a new test to measure how well they are meeting these new standards. Two different groups are developing new assessment that they will soon be taking. Here is some information about the two groups.

Partnership for Assessment of Readiness for College and Careers (PARCC)

The PARCC assessment system will be made up of three Through-Course Assessments and one End-of-Year Comprehensive Assessment.

- The Through-Course Assessments will be given at the end of the first, second, and third quarters and will focus on the critical areas for each grade. The first and second Through-Course Assessment will require one class period to complete; the third Through-Course Assessment may require more than one class period to complete. Your students will take these assessments primarily on computers or other digital devices. They will have a range of item types, including performance tasks and computer-enhanced items.
- The End-of-Year Assessment will sample all of the standards at the grade level. Your students will take this assessment online some time during the last month of the school year. Each test will have 40 to 65 items, with a range of item types (i.e., selected-response, constructed-response, performance tasks).

Final scores will be based on the scores on the three Through-Course Assessments and the End-of-Year Assessment.

SMARTER Balanced Assessment Consortium (SBAC)

The SBAC summative assessment system consists of performance tasks and one End-of-Year Adaptive Assessment.

- Performance tasks: Students will complete two performance tasks during the last 12 weeks of the school year. These tasks will measure their ability to integrate knowledge and skills from the CCSSM. Students will take these assessment primarily on computers or other digital devices.
- The End-of-Year Assessment will also be administered during the last 12 weeks of the school year. It will be made up of 40 to 65 items, with a range of item types (i.e., selected-response, constructed-response, performance tasks). Some items will be computer-scored while others will be human-scored.

Summative scores will be based on the scores on the Performance Tasks and the End-of-Year Adaptive Assessment.

Supplemental Common Core Lessons

Throughout *Prentice Hall Course 2*, you will find many opportunities to develop and build on the Standards for Mathematical Practice and the Standards for Mathematical Content. In the **Problem Solving Handbook**, found on pages xxxii through xlix (Teacher's Edition pages T54–T71), your students will find strategies to help them make sense of problems, persevere in solving them, reason abstractly and quantitatively, and use tools appropriately. With the guided problem solving exercises, they can strengthen these same practices. The **More than One Way** activities offer students the opportunity to look at and critique the reasoning and thinking of others and to defend their solutions. The **Take Note** boxes model mathematically correct language to help students be more precise in speaking, writing, and thinking. Many of the exercises also help students develop mathematical models to represent real-life situations. With this program, your students are well on their way to becoming proficient students of mathematics!

You will also find in the program that the lessons align well to the Standards for Mathematical Content. To ensure in-depth and comprehensive coverage of all of the Standards for Mathematical Content, Pearson developed some supplemental Common Core lessons. Each lesson addresses a particular content standard, shown on the first page of the lesson and was developed to be studied with existing lessons. In the left margin of the lesson, you will find that lesson listed.

You will notice that some exercises have a small Ⓒ logo next to them. This logo indicates that these exercises are *particularly* focused on helping students become more proficient with Standards for Mathematical Practice. All of the exercises help develop mastery of the Standards for Mathematical Content.

Listing of Supplemental Common Core Lessons

Addition and Subtraction of Rational Numbers

Use this activity after Lesson 3-3.
Common Core State Standards: 7.NS.1, 7.NS.1.c, 7.NS.1.d

Students apply and extend the rules for adding and subtracting integers to all rational numbers. They represent sums and differences of rational numbers on horizontal and vertical number lines.

Guided Instruction

Teaching Tip
Before beginning this activity, review the rules for adding and subtracting integers. Review the addition and subtraction of integers on number lines in Lesson 1-7.

Activity 1

Students must attend to precision when adding and subtracting signed numbers, carefully tracking the symbols they use. If students have difficulty subtracting signed numbers, remind them that subtracting a number is the same as adding the number's opposite.

$-(-b) = +b$ and $-b = +(-b)$.

Because of this, the subtraction of two numbers with different signs can be written as the addition of two numbers with the same sign, and the subtraction of two numbers with the same sign can be rewritten as the addition of two numbers with different signs.

Answers

Activity 1

1. The yellow arrow is labeled $\frac{1}{2}$, and the blue arrow is labeled $+\left(-\frac{3}{4}\right)$.
2. $-\frac{1}{4}$
3. Sample: yes because $+\left(-\frac{3}{4}\right)$ equals $-\frac{3}{4}$
4. Check students' number lines.
   a. $-\frac{1}{8}$ b. 1 c. $\frac{1}{4}$ d. $-\frac{7}{8}$

5.

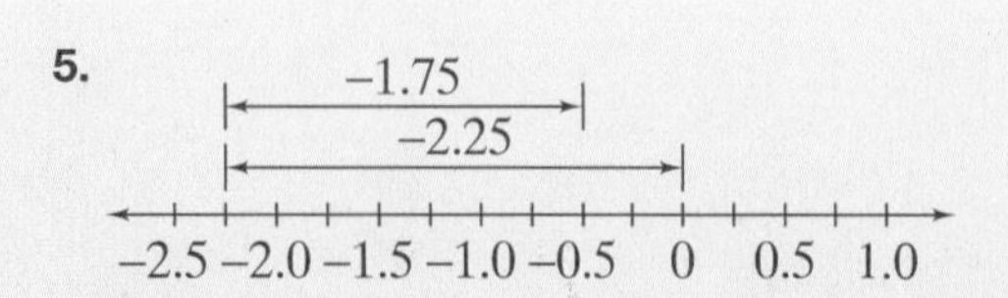

CC-1 Addition and Subtraction of Rational Numbers

CONTENT STANDARDS

7.NS.1 Apply and extend previous understandings of addition and subtraction to add and subtract rational numbers; represent addition and subtraction on a horizontal or vertical number line diagram.

7.NS.1.c Understand subtraction of rational numbers as adding the additive inverse, $p - q = p + (-q)$. Show that the distance between two rational numbers on the number line is the absolute value of their difference, and apply this principle in real-world contexts.

7.NS.1.d Apply properties of operations as strategies to add and subtract rational numbers.

GO for Help to Lesson 1-7

Use after Lesson 3-3.

You already know how to add and subtract integers. The table shows the rules, where a, b, c, and d are positive integers and $a > b$.

	Same Sign	Different Sign
Add: The sum has the sign of the addend with greater absolute value.	$a + b = c$ $-a + (-b) = c$	$a + (-b) = d$ $-a + b = -d$
Subtract: Rewrite as adding the additive inverse.	$a - b = a + (-b) = d$ $-a - (-b) = -a + b = -d$	$a - (-b) = a + b = c$ $-a - b = -a + (-b) = -c$

You also know how to add positive decimals, fractions, and mixed numbers. Use these skills to add and subtract any rational numbers.

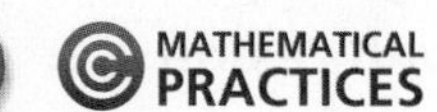

MATHEMATICAL PRACTICES

1. The sum $\frac{1}{2} + \left(-\frac{3}{4}\right)$ can be represented on a horizontal number line diagram.

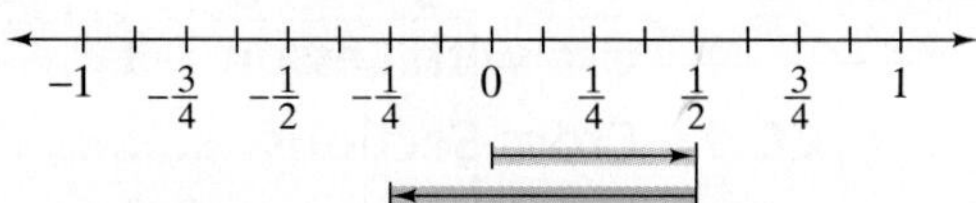

Copy the number line diagram and label the parts that represent $\frac{1}{2}$ and $\left(-\frac{3}{4}\right)$.

2. What is the sum of these two fractions?
3. Can you use the same number line diagram to represent $\frac{1}{2} - \frac{3}{4}$?
4. Represent each sum or difference on a horizontal number line. Then find each sum or difference.
   a. $-\frac{3}{4} + \frac{5}{8}$
   b. $\frac{1}{4} - \left(-\frac{3}{4}\right)$
   c. $\frac{3}{4} + \left(-\frac{1}{2}\right)$
   d. $-\frac{1}{4} - \frac{5}{8}$
5. Use a horizontal number line like the one below to represent the sum of $-2.25 + 1.75$.

−2.5 −2.0 −1.5 −1.0 −0.5 0 0.5 1.0

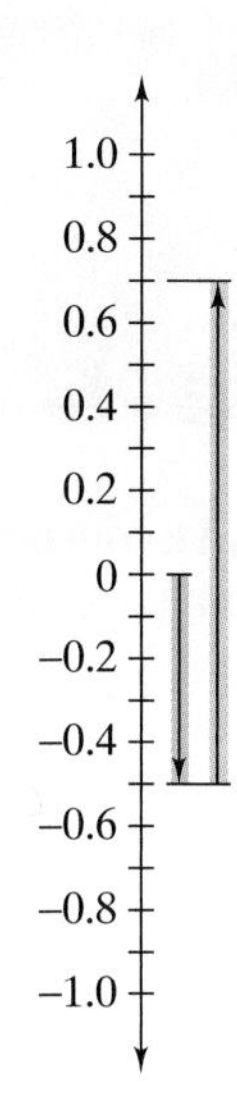

6. The vertical number line at the left represents $-0.5 - (-1.2)$.
   a. How else could you write this expression?
   b. What is the value of this expression?
7. Represent $-1\frac{1}{4} - 5\frac{1}{2}$ on a vertical number line diagram and find the difference.

MATHEMATICAL PRACTICES

The altitude at sea level is 0 meters. A scuba diver is standing on a platform 3.1 meters above sea level on a boat in the ocean.

1. Draw a diagram. At what altitude are the diver's feet?
2. The diver jumps into the water and stops at the top of a kelp plant 8.25 meters below sea level. Add this information to your diagram.
3. Find the absolute value of the difference between 3.1 and –8.25 to determine the distance the diver descended.
4. The diver follows the kelp plant down to its base at 21.3 meters below sea level. Complete your diagram.
5. Simplify the expression 3.1 – (–21.3) to find the total distance the diver descended.

Exercises

Represent each sum or difference on a horizontal or vertical number line diagram. Then find each sum or difference.

1. $3.5 + (-2.8)$
2. $-\frac{3}{4} + \left(-\frac{7}{8}\right)$
3. $-2.8 + 3.5$
4. $2.1 - (-1.7)$
5. $-1\frac{5}{8} - \left(-4\frac{3}{8}\right)$
6. $1\frac{1}{4} - 2\frac{7}{8}$

Add or subtract the following rational numbers.

7. $-16 - 26.6$
8. $-15.2 + 15.2$
9. $-9\frac{4}{5} - 4\frac{3}{5}$
10. $15\frac{1}{5} - \left(-15\frac{1}{5}\right)$
11. $-9.7 - (-8.8)$
12. $8\frac{3}{8} + \left(-6\frac{1}{4}\right)$
13. What is the temperature difference between 120°C and −50°C?
14. It was 75.5°F at 2 P.M. and then the temperature dropped 15.1°F in an hour. What was the temperature at 3 P.M.?
15. It was −20.5°F at 6 A.M. and 22°F at noon. How much did the temperature change?
16. Writing in Math Explain why $p - q = p + (-q)$ is true for all rational numbers.

Activity 2

Remind students that when finding the distance between two points, the order of subtraction is not important because distance is always reported as a positive value.

Resources

- graph paper
- colored pencils or markers

Answers

Activity 1

6a. Sample: −0.5 + 1.2
b. 0.7
7.

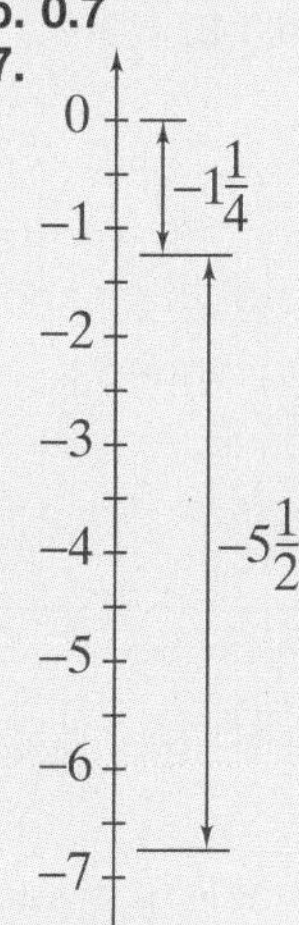

Check students' work; $-6\frac{3}{4}$.

Answers

Activity 2

1. 3.1 meters.
2. Check students' work.
3. 11.35 meters
4. Check students' work.
5. 24.4 meters

Exercises

1. 0.7
2. $-1\frac{5}{8}$
3. 0.7
4. 3.8
5. $2\frac{3}{4}$
6. $-1\frac{5}{8}$
7. −42.6
8. 0
9. $-14\frac{2}{5}$
10. $30\frac{2}{5}$
11. −0.9
12. $2\frac{1}{8}$
13. 170°F
14. 60.4°F
15. 42.5°F
16. This is true because $+(-q)$ equals $-q$.

Multiplication of Rational Numbers

Use this activity after Lesson 3-4.
Common Core State Standards: 7.NS.2, 7.NS.2.a, 7.NS.2.c

Students apply and extend their understanding of how to multiply integers to operations with rational numbers.

Guided Instruction

Teaching Tip
Review the use of number lines to show multiplication as repeated addition in Lesson 1-8.

Activity 1
Students need to reason abstractly to make sense of these quantities. They should recognize that $2\frac{1}{2}$ means two wholes plus one-half, or $2 + \frac{1}{2}$.

Answers

Activity 1

1a. Sample: The Distributive Property states that $\left(2 + \frac{1}{2}\right)$ times $-1\frac{1}{2}$ is equal to $2\left(-1\frac{1}{2}\right) + \frac{1}{2}\left(-1\frac{1}{2}\right)$.

1b. Sample: The red sections show two groups of $-1\frac{1}{2}$, and the blue section shows $\frac{1}{2}$ group of $-1\frac{1}{2}$, or $-\frac{3}{4}$.

1c. $-3\frac{3}{4}$

1d. $-3\frac{3}{4}$

2. Sample: The red section shows 1 group of $-2\frac{1}{2}$; the blue section shows $\frac{1}{2}$ of $-2\frac{1}{2}$, or $-1\frac{1}{4}$.

3a. The expressions are equivalent.

3b. Yes. Sample: You can write $2\frac{1}{2}$ as $1 \cdot 2\frac{1}{2}$ and you can write $-1\frac{1}{2}$ as $(-1) \cdot 1\frac{1}{2}$.

3c. the Commutative Property of Multiplication

3d. negative

3e. Sample: Because you can write the signs of the factors as 1 and -1, and $(1)(-1)$ is -1.

4a. $(-1)(-1)(2.2)(0.45)$

4b. 0.99

4c. Sample: The product of two positive or two negative numbers is positive; the product of one positive number and one negative number is negative.

CC-2 Multiplication of Rational Numbers

CONTENT STANDARDS

7.NS.2 Apply and extend previous understandings of multiplication and division and of fractions to multiply and divide rational numbers.

7.NS.2.a Understand that multiplication is extended from fractions to rational numbers by requiring that operations continue to satisfy the properties of operations, particularly the distributive property, leading to products such as $(-1)(-1) = 1$ and the rules for multiplying signed numbers. Interpret products of rational numbers by describing real-world contexts.

7.NS.2.c Apply properties of operations as strategies to multiply and divide rational numbers.

GO for Help to Lesson 1-8.

Use after Lesson 3-4.

You already know how to multiply integers. You also know how to multiply positive decimals, fractions, and mixed numbers. Use these skills to multiply rational numbers.

ACTIVITY MATHEMATICAL PRACTICES

1. The model below shows the product $\left(2\frac{1}{2}\right) \cdot \left(-1\frac{1}{2}\right)$.

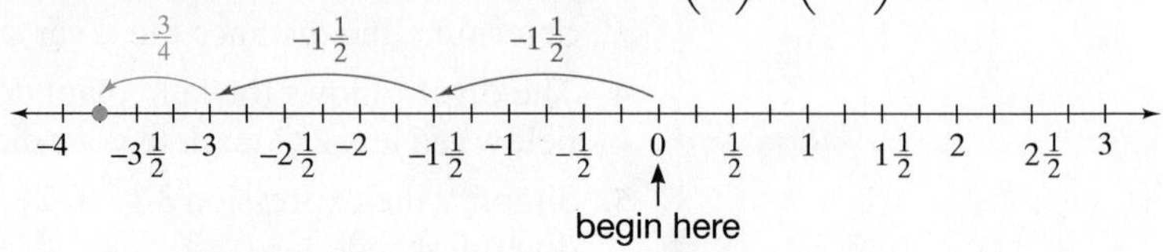

a. Explain which property justifies why $2\frac{1}{2} \cdot \left(-1\frac{1}{2}\right)$ is equal to $2 \cdot \left(-1\frac{1}{2}\right) + \frac{1}{2} \cdot \left(-1\frac{1}{2}\right)$.

b. Explain how the number line above illustrates $2 \cdot \left(-1\frac{1}{2}\right) + \frac{1}{2} \cdot \left(-1\frac{1}{2}\right)$.

c. What is the sum of $2\frac{1}{2}$ groups of $-1\frac{1}{2}$?

d. What is the product $\left(2\frac{1}{2}\right) \cdot \left(-1\frac{1}{2}\right)$?

2. How does this model show $\left(1\frac{1}{2}\right) \cdot \left(-2\frac{1}{2}\right)$?

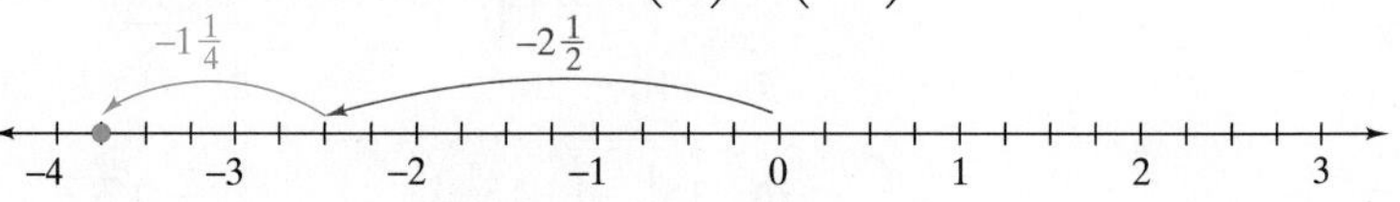

3. Compare $2\frac{1}{2} \cdot \left(-1\frac{1}{2}\right)$ and $1\frac{1}{2} \cdot \left(-2\frac{1}{2}\right)$.

a. What is true about the product of these expressions?

b. Is $2\frac{1}{2} \cdot \left(-1\frac{1}{2}\right)$ the same as $(1) \cdot \left(2\frac{1}{2}\right) \cdot (-1) \cdot \left(1\frac{1}{2}\right)$? Explain.

c. Why does each expression equal $(1) \cdot (-1) \cdot \left(1\frac{1}{2}\right) \cdot \left(2\frac{1}{2}\right)$?

d. What is the sign of $(1)(-1)$?

e. Why is the product of a positive rational number and a negative rational number negative?

4. a. Write the expression $(-2.2)(-0.45)$ as a product of four factors that includes $(-1)(-1)$.
   b. What is the product of $(-2.2)(-0.45)$?
   c. Explain how finding the sign of the product of two rational numbers is similar to finding the sign of the product of two integers.

Jolene graduated from college last year and is repaying her student loan. Every month she makes a payment of \$124.18 from her checking account.

1. Does each payment increase or decrease her checking account balance?
2. Jolene has paid 7 months of her loan so far. Simplify the expression $(-124.18) \times 7$ to determine the change to her checking account balance caused by these seven loan payments.
3. Jolene gets a promotion and decides to pay 4.5 months of loan payments in advance. Write an expression to represent the change this will make to her checking account balance.
4. How much will she pay in advance toward the loan?
5. What is the sign of your answer? Explain why.

Exercises

Describe the product each number line models.

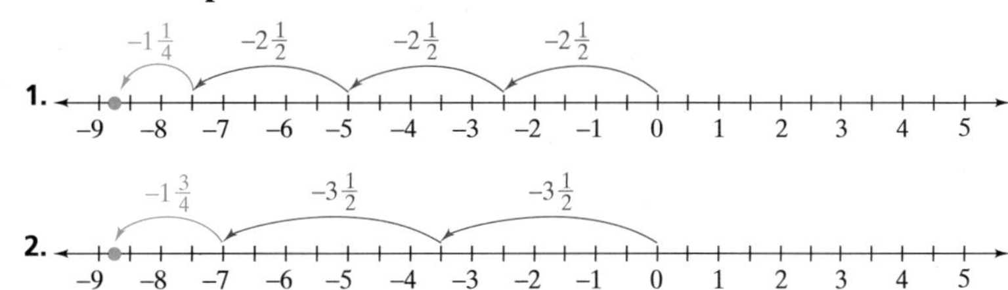

3. **Writing in Math** Use properties of mathematics to explain why the products in Exercises 1 and 2 have the same value.
4. For the past three months, Arturo's phone bill was \$58.93 each month. He made each payment from the same checking account. How much did these payments change his account balance?

Phone Bill

Total: \$58.93

Find each product.

5. $3.5 \cdot (-2.4)$
6. $-16 \cdot 6.6$
7. $-1\frac{5}{8} \cdot \left(-3\frac{1}{7}\right)$
8. $-2\frac{1}{3} \cdot -1\frac{3}{4}$
9. $-5.2 \cdot (-5.2)$
10. $3\frac{2}{5} \cdot -3\frac{1}{3}$

Activity 2

If students do not understand why Jolene's payment is negative, point out that any payment made by a check or a debit card from an account must be subtracted from that account. Remind students that subtracting a number is the same as adding the number's opposite. In this case, subtracting 124.18 is the same as adding −124.18. A payment made seven times equals the repeated addition of seven groups of −124.18, or $(-124.18) \cdot 7$.

Answers

Activity 2

1. decreases her balance
2. −\$869.26
3. $(124.18) \cdot 4.5$
4. \$558.81
5. Positive; the amount paid is positive. The amount applied to the loan balance is negative, since the loan amount is reduced.

Homework Exercises

1. $\left(3\frac{1}{2}\right) \cdot \left(-2\frac{1}{2}\right)$
2. $\left(2\frac{1}{2}\right) \cdot \left(-3\frac{1}{2}\right)$
3. Sample: You can write $\left(3\frac{1}{2}\right) \cdot \left(-2\frac{1}{2}\right)$ as the factors $\left(3\frac{1}{2}\right) \cdot (-1) \cdot \left(2\frac{1}{2}\right)$. Use the Commutative Property of Multiplication to rewrite as $\left(2\frac{1}{2}\right) \cdot (-1) \cdot \left(3\frac{1}{2}\right)$ which equals $\left(2\frac{1}{2}\right) \cdot \left(-3\frac{1}{2}\right)$.
4. −\$176.79
5. −8.4
6. −105.6
7. $5\frac{3}{28}$
8. $4\frac{1}{12}$
9. 27.04
10. $-11\frac{1}{3}$

Division of Rational Numbers

Use this activity after Lesson 3-5.
Common Core State Standards: 7.NS.2, 7.NS.2.b, 7.NS.2.c

Students apply their knowledge of operating with integers, fractions, and decimals to divide rational numbers. Students divide rational numbers in real-world contexts and interpret the quotient.

Guided Instruction

Activity 1

Review the rules for multiplying mixed numbers, fractions, and decimals. Then review how to write a division problem as multiplication using the multiplicative inverse (reciprocal).

Ask: *Since* $-2 \div \frac{1}{4} = -2 \times \frac{4}{1} = -2 \times 4$, *how do you determine the sign of the quotient?*

Sample: Because multiplying two numbers with opposite signs results in a negative product, this quotient is negative.

Make sure students attend to precision by using the correct sign when they divide signed rational numbers. Ask: *How does the decision-making process compare for determining the sign of a quotient or a product?*

Sample: The process is the same because every division problem can be written as a multiplication problem.

Answers

Activity 1

1. **Negative; the signs are opposite.**
2. $-\frac{10}{3} \div \frac{7}{3}$
3. $-\frac{10}{3} \times \frac{3}{7}$
4. $-1\frac{3}{7}$
5. **Answers may vary. Sample: No, the quotient is always negative. For example,** $-\frac{10}{3} \div \frac{3}{7} = (-1) \times \frac{10}{3} \times \frac{7}{3} = \frac{10}{3} \times (-1) \times \frac{7}{3} = \frac{10}{3} \times -\frac{7}{3} = \frac{10}{3} \div -\frac{3}{7}$**, by the Commutative Property of Multiplication.**

CC-3 Division of Rational Numbers

CONTENT STANDARDS

7.NS.2 Apply and extend previous understandings of multiplication and division and of fractions to multiply and divide rational numbers.

7.NS.2.b ...Interpret quotients of rational numbers by describing real-world contexts.

7.NS.2.c Apply properties of operations as strategies to multiply and divide rational numbers.

You already know how to divide integers.

Two Numbers	Sign of Quotient	Examples
Same Sign	Quotient is positive.	$16 \div 2 = 8$ $-16 \div (-2) = 8$
Opposite Signs	Quotient is negative.	$-16 \div 2 = (-8)$ $16 \div (-2) = (-8)$

You also know how to divide positive decimals, fractions, and mixed numbers. Use these skills to divide rational numbers.

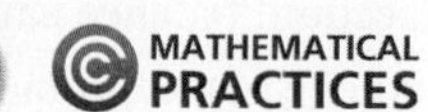

ACTIVITY — **MATHEMATICAL PRACTICES**

1. Write the division problem $-3\frac{1}{3} \div 2\frac{1}{3}$. Will the sign of the quotient be positive or negative? Why?
2. Rewrite each mixed number as an improper fraction.
3. Rewrite the division problem as a multiplication problem. To divide by a fraction, multiply by its reciprocal.
4. What is the product? Write your answer as a mixed number.
5. When dividing two numbers with opposite signs, does it matter which number is negative and which number is positive? Use properties of operations to justify your answer.

ACTIVITY — **MATHEMATICAL PRACTICES**

Denise takes her father's old bicycle to a mechanic to have it restored. Afterwards, the mechanic sends her a bill with the charges shown.

Repair Bill

Tune-up	\$50.00
Brake Shoes	\$15.42
Tires	\$34.19
Drive Train	\$105.50

1. Find the sum of the charges on the bill.
2. From Denise's point of view, money that she receives is a positive number, and money that she owes is a negative number. To Denise, is the sum of the charges on the bill positive or negative? Explain.

Use after Lesson 3-5.

3. The mechanic wants full payment of the bill in four and a half months. Write an expression for how Denise's checking account will change each month.
4. What does the sign of the quotient represent?
5. What does the quotient tell Denise about paying her bill?
6. Next year, Denise has two more parts replaced. She uses a $25 gift card to pay for part of the work. Her expenses are shown below.

   −$54.50
   −$62.75
   +$25.00

   What is the sum?
7. If Denise pays off the bill in 15 weeks, or 3.75 months, how much will her checking account change per month?

Exercises

Find each quotient.

1. $-14.28 \div 4.2$
2. $14.28 \div 4.2$
3. $14.28 \div (-4.2)$
4. $-2\frac{3}{4} \div 11$
5. $2\frac{3}{4} \div 11$
6. $2\frac{3}{4} \div (-11)$
7. $-\frac{3}{8} \div \left(-1\frac{1}{10}\right)$
8. $-10.5 \div (-0.5)$
9. $-12.96 \div (-10.8)$
10. $3\frac{1}{2} \div -2\frac{2}{3}$
11. How does the quotient $10.8 \div (-2.4)$ compare to the quotient $-10.8 \div 2.4$?
12. A spreadsheet program uses rational numbers to keep track of income and expenses for a business.

Business Expenses and Income

	Phone	Supplies	Wages	Income
Monday	−$4.32	$0.00	−$135.50	$0.00
Tuesday	−$4.32	−$17.25	−$135.50	$782.00
Wednesday	−$4.32	$0.00	−$135.50	$525.00
Thursday	−$4.32	−$25.00	−$135.50	$782.00
Friday	−$4.32	−$9.25	−$135.50	$0.00

a. What is the total balance for Monday?
b. Find the total balance for each other day of the week.
c. What is the average daily balance?

Guided Instruction

Activity 2

Make sure students understand what a positive value and a negative value represent in terms of the problem and the solution. They need to understand that in this activity, negative numbers mean amounts that Denise owes. Ask: *How might the signs of your answers change if you take the mechanic's point of view?*

Amounts would be positive.

Answers

Activity 2

1. $205.11
2. Negative; the sum of the charges is money that she owes.
3. −$45.58
4. Negative; it represents money that Denise owes.
5. She needs to save an extra $45.58 each month to pay her bill on time.
6. −$92.25
7. −$24.60

Homework Exercises

1. −3.4
2. 3.4
3. −3.4
4. $-\frac{1}{4}$
5. $\frac{1}{4}$
6. $-\frac{1}{4}$
7. $\frac{15}{44}$
8. 21
9. 1.2
10. $-\frac{21}{16}$
11. The quotients are the same.
12a. −$139.82
b. Tuesday $624.93; Wednesday $385.18; Thursday $617.18; Friday −$149.07
c. $267.68

1. Plan

Use this lesson after Lesson 4-1. Common Core State Standard: 7.EE.1

Math Background

Algebraic expressions sometimes appear in forms that can be simplified. The same variable may appear more than once in a sum. Numbers may be added or subtracted multiple times. The Commutative Property of Addition, the Associative Property of Addition, and the Distributive Property can all be used to simplify algebraic expressions.

Properties of operations used in this lesson include:

Commutative Property of Addition:

$a + b = b + a$

Associative Property of Addition:

$a + (b + c) = (a + b) + c$

Distributive Property:

$a(b + c) = ab + ac$

2. Teach

Guided Instruction

Example 1

Have students draw diagrams as they simplify expressions, updating the position of variables and the value of numbers at each step of the process.

Quick Check 1

Make sure students are aware that the term x has a coefficient of 1.

Example 2

Students can check answers by substituting the same number into both the original and simplified expressions. When the expressions are evaluated, the values should match.

Error Prevention!

Make sure students are distributing negatives correctly over parentheses.

CC-4 Simplifying Expressions

7.EE.1 Apply properties of operations as strategies to add, subtract, factor, and expand linear expressions with rational coefficients.

You evaluate algebraic expressions by substituting values for variables. To simplify algebraic expressions, use properties of operations.

Like terms are terms that have the same variable factors. For example, $12x$ and $3x$ are like terms, but $4a$ and $5b$ are not like terms. You can use the properties of operations to order, group, and combine like terms.

EXAMPLE Using Properties to Add and Subtract

1 Simplify $5x + 9 + 2x - 4$.

$5x + 9 + 2x - 4$ ← **Identify which parts of the expression are like terms.**
$= 5x + 2x + 9 - 4$ ← **Commutative Property of Addition**
$= (5 + 2)x + 9 - 4$ ← **Distributive Property**
$= 7x + 9 - 4$ ← **Simplify the coefficient.**
$= 7x + 5$ ← **Simplify.**

The simplified expression is $7x + 5$.

Test Prep Tip

The model illustrates the Distributive Property.

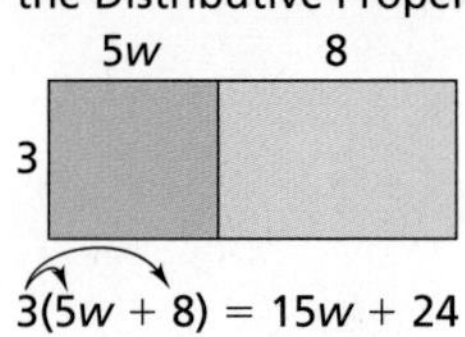

$3(5w + 8) = 15w + 24$

Quick Check

1. Simplify each expression.
   a. $2x + 8 + 4x - 5$ **b.** $6 + 7y - 4y + 1$ **c.** $10r - 5 + 3 + r$

Sometimes an expression should be expanded before it can be simplified.

EXAMPLE Expanding Expressions

2 Simplify $3(5w + 8) - 6$.

$3(5w + 8) - 6$
$= (15w + 24) - 6$ ← **Distributive Property**
$= 5w + (24 + (-6))$ ← **Associative Property of Addition**
$= 15w + 18$ ← **Simplify.**

The simplified expression is $15w + 18$.

Quick Check

2. Simplify each expression.
   a. $6(2x + 3) - 4$ **b.** $2(1 - 8v) + 5$ **c.** $9 - 4(3z + 2)$

You can use the Distributive Property to rewrite an addition expression as a product of two factors. Use the greatest common factor (GCF) so the expression is factored completely.

Use after Lesson 4-1.

Answers

Quick Check

1a. $6x + 3$
1b. $3y + 7$
1c. $11r - 2$
2a. $12x + 14$
2b. $-16v + 7$
2c. $-12z + 1$
3a. $3(3x + 5)$
3b. $12(3 + 2t)$
3c. $4(2c - 5)$

EXAMPLE Factoring Expressions

3 Factor $4x + 14$.

GCF of 4 and 14 is 2. ← **Identify the GCF.**

$4x + 14 = 2 \cdot 2x + 2 \cdot 7$ ← **Factor each term by the GCF.**

$= 2(2x + 7)$ ← **Distributive Property**

The factored expression is $2(2x + 7)$.

Quick Check

3. Factor each expression completely.
a. $9x + 15$ **b.** $36 + 24t$ **c.** $8c - 20$

Homework Exercises

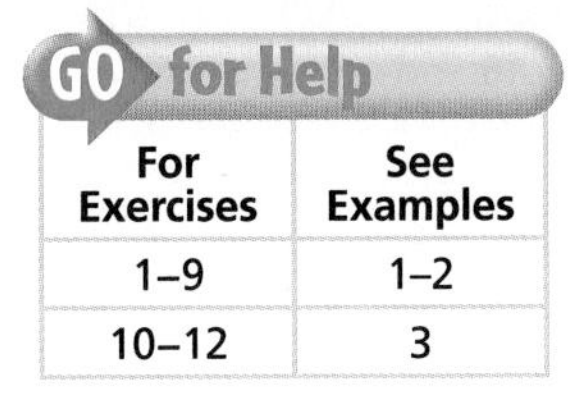

GO for Help

For Exercises	See Examples
1–9	1–2
10–12	3

Simplify each expression.

1. $9x + 3 + 2x - 2$ **2.** $6y + 7 - 3y + 3$ **3.** $12w - 7 + 1 + 4w$

4. $4 - 6x + 8 + 3x$ **5.** $3a + 8 + a - 9 - 2a$ **6.** $4(5x + 2) - 6$

7. $10 + 3(2v - 3)$ **8.** $4 - 5(3t + 3)$ **9.** $6(y + 2) - 6 + 5y$

Factor each expression completely.

10. $6x + 10$ **11.** $30 + 20y$ **12.** $12x - 28$

13. Guided Problem Solving Simplify $\frac{6x + 4}{2}$.

- **Make a Plan** Rewrite the quotient as a product. Apply the Distributive Property. Write all fractions in simplest form.
- **Carry Out the Plan**

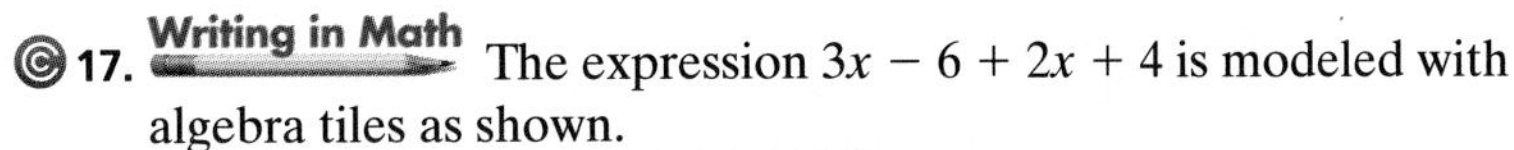

$$\frac{6x + 4}{2} = \frac{\blacksquare}{\blacksquare}(6x + 4) = \frac{\blacksquare}{\blacksquare}x + \frac{\blacksquare}{\blacksquare} = \blacksquare x + \blacksquare$$

Simplify each expression.

14. $\frac{24x + 16}{8}$ **15.** $\frac{15 + 36y}{3}$ **16.** $\frac{96b - 24}{12}$

17. Writing in Math The expression $3x - 6 + 2x + 4$ is modeled with algebra tiles as shown.

Explain how to use the model to simplify the expression.

Simplify each expression.

18. $8.4x + 10.2 + 4.3x - 2.9$ **19.** $5y + 4.7 + 2.08 - 0.6y$

20. $\frac{2}{3}(8x + 12) - \frac{4}{3}$ **21.** $1 + \frac{1}{2}x - \frac{3}{4}\left(\frac{1}{6} - \frac{2}{9}x\right)$

22. Error Analysis Jane did the work shown. Explain her error.

$9y - 2 + 4y$
$9y - 4y + 2$
$5y + 2$

Example 3

To apply the Distributive Property correctly when factoring, it may help students to think $ab + ac = a(b + c)$, where a is the greatest common factor.

Closure

- *When an algebraic expression contains like terms, how can you simplify the expression?* Use properties of operations to group like terms and combine them.

3. Practice

Assignment Guide

Homework Exercises

A Practice by Example 1–12
B Apply Your Skills 13–22

Homework Quick Check

To check students' understanding of key skills and concepts, go over Exercises 2, 7, 12, 17, and 22.

Answers

Homework Exercises

1. $11x + 1$
2. $3y + 10$
3. $16w - 6$
4. $-3x + 12$
5. $2a - 1$
6. $20x + 2$
7. $6v + 1$
8. $-15t - 11$
9. $11y + 6$
10. $2(3x + 5)$
11. $10(3 + 2y)$
12. $4(3x - 7)$
13. $3x + 2$
14. $3x + 2$
15. $5 + 12y$
16. $8b - 2$
17. Sample: Rearrange so x tiles are grouped and unit tiles are grouped. There are five x tiles. There are four zero pairs and two -1 tiles left over. So, the expression is $5x - 2$.
18. $12.7x + 7.3$
19. $4.4y + 6.78$
20. $\frac{16}{3}x + \frac{20}{3}$
21. $\frac{2}{3}x + \frac{7}{8}$
22. Sample: Jane applied the Commutative Property of Addition to 2 and $4y$ as if both were added. Since 2 was subtracted, the resulting expression should have been $9y + 4y - 2$. The simplified expression is $13y - 2$.

1. Plan

Use this lesson after Lesson 4-6.
Common Core State Standard:
7.EE.4.a

Math Background

When an equation is given in the form $p(x + q) = r$, students can use the Distributive Property to put the equation in a form they are familiar with, $px + q = r$. Then they can solve the problem using the rules for a two-step equation.

2. Teach

Guided Instruction

Example 1

When using the Distributive Property, remind students to multiply every term inside the parentheses by the factor outside the parentheses. Once the factor in front is distributed, the equation is solved like a two-step equation. Rational numbers follow the rules for properties of equality.

More than One Way

Refer students to Lesson 4-6. Ask them to compare Sarah's and Ryan's methods for solving a two-step equation. Then guide students through each method. With Ryan's method, help students model the situation with a formula by identifying important quantities and the relationships among them. Then verify that students use the correct steps to solve this equation accurately.

Closure

- *How do you solve the equation $1.5(2.2 - x) = 76.5$?* **First, multiply 1.5 by 2.2 and by $-x$, and then you can solve it like a two-step equation. The rational numbers do not change the way this is solved.**

CC-5 Solving Equations of the Form $p(x + q) = r$

CONTENT STANDARDS

7.EE.4.a Solve word problems leading to equations of the form $px + q = r$ and $p(x + q) = r$, where p, q, and r are specific rational numbers. Solve equations of these forms fluently. Compare an algebraic solution to an arithmetic solution, identifying the sequence of the operations used in each approach.

You can use the Distributive Property to solve equations in the form $p(x + q) = r$.

EXAMPLE Solve Using the Distributive Property

Solve $10(a - 6) = -25$.

$10(a) + 10(-6) = -25$ ← **Use the Distributive Property.**
$10a - 60 = -25$ ← **Simplify.**
$10a = 35$ ← **Add 60 to both sides.**
$a = 3.5$ ← **Divide both sides by 10.**

Quick Check

Solve these equations.

1. $-4.5 = -3(b + 15)$ **2.** $8\frac{1}{2}(c - 16) = 340$

More Than One Way

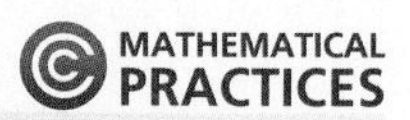

Each of 4 workers in a gourmet bakery makes $2\frac{1}{2}$ pounds of cranberry granola every day. The workers also make almond granola. Together, they make 24 pounds of granola every day. How many pounds of almond granola does each worker make daily?

Sarah's Method

I can use number sense. The amount of cranberry granola made daily is $4 \times 2\frac{1}{2}$ or 10 pounds. Since $24 - 10 = 14$ and $14 \div 4 = 3\frac{1}{2}$, each worker makes $3\frac{1}{2}$ pounds of almond granola daily.

Ryan's Method

I can write an equation. Let b represent the pounds of almond granola that each worker makes. Together, the workers make $4(2\frac{1}{2} + b)$ pounds of granola.

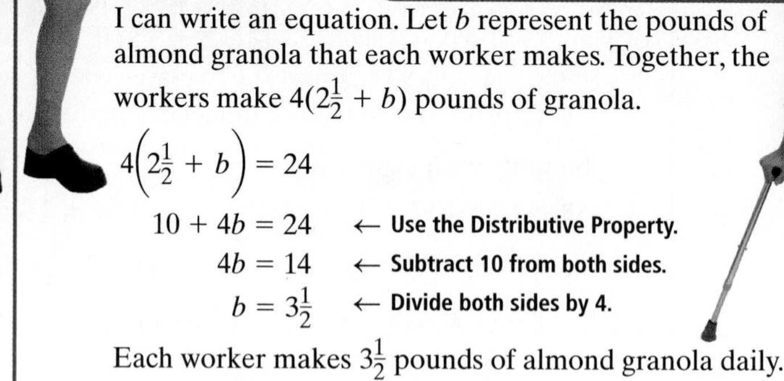

$4\left(2\frac{1}{2} + b\right) = 24$

$10 + 4b = 24$ ← **Use the Distributive Property.**
$4b = 14$ ← **Subtract 10 from both sides.**
$b = 3\frac{1}{2}$ ← **Divide both sides by 4.**

Each worker makes $3\frac{1}{2}$ pounds of almond granola daily.

Use after Lesson 4-6.

Answers

Quick Check

1a. $b = -13.5$
b. $c = 56$

Compare the Methods

1. Compare the sequence of operations in Ryan's and Sarah's methods. Explain how they are alike and how they differ.
2. Which method do you prefer? Can you improve on that method?

Homework Exercises

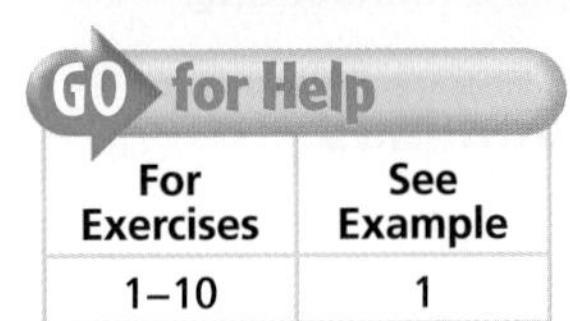

For Exercises	See Example
1–10	1

Solve.

1. $-8(x + 2) = -10$
2. $10(2.2 - b) = 15.2$
3. $10\left(c + 6\frac{1}{5}\right) = -102$
4. $3\frac{2}{3}(12 - a) = 43$
5. $17\frac{5}{8} = -2(x + 15)$
6. $4(m + 2.5) = 7.5$
7. $14(0.5 + k) = -14$
8. $3(0.2 + y) = 9.6$
9. $100(a - 4.5) = 350$
10. $138.75 = 9.25(-6 + t)$

11. **Guided Problem Solving** Sandra buys a kit that has exactly enough material to replace the seats and backs of two antique chairs. The seats and backs are woven from cane. Each seat uses $83\frac{1}{3}$ yards of cane. If the kit has $416\frac{2}{3}$ yards of cane, how much is provided for each back?
    - Each chair seat is made from $83\frac{1}{3}$ yards of cane. Write an expression that represents the amount of cane for one chair.
    - Write an expression that represents the cane for both chairs.
    - Write an equation that represents the situation and solve it.
12. **Food Preparation** Annie made fruit punch for 12 people. The punch contains sparkling water and $\frac{2}{3}$ pint of fruit juice per person. If there are $10\frac{2}{5}$ pints of fruit punch, how many pints of sparkling water did Annie add per person?
13. **Algebra** Write and solve an equation using the distributive property modeled by the algebra tiles below.

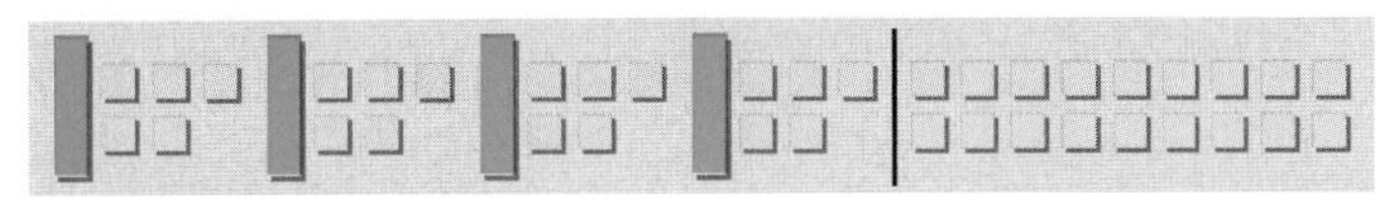

14. **Cell Phone** Paulo pays \$45.99 per month for unlimited calls, with additional charges for text messages. His bill for 4 months is \$207.96 If he sends 100 texts each month, how much is he charged per text?
15. **Writing in Math** Is it possible to solve the equation $5(b - 2) = 20$ by first dividing each side by 5? Explain.

Answers

Compare Methods

1. Sample: The Distributive Property is used for both methods, but the steps are not done in the same order. Sarah eliminates the cranberry granola from the equation completely first. Ryan writes a formal equation and uses the Distributive Property.
2. Check students' answers. Sample: I would improve Ryan's method by dividing both sides of the equation by 4 first, then solving $2\frac{1}{2} + b = 6$.

3. Practice

Assignment Guide

Homework Exercises

A Practice by Example	1–10
B Apply Your Skills	11–15

Homework Quick Check

To check students' understanding of key skills and concepts, go over Exercises 3, 5, 10, 12, and 15.

Answers

Homework Exercises

1. $x = -0.75$
2. $b = 0.68$
3. $c = -16.4$
4. $a = \frac{3}{11}$
5. $x = -23\frac{13}{16}$
6. $m = -0.625$
7. $k = -1.5$
8. $y = 3$
9. $a = 8$
10. $t = 21$
11. 125 yards
12. $\frac{1}{5}$ pint
13. $4(x + 5) = 18$; $x = -0.5$
14. \$0.06 per call
15. Yes;the Division Property of Equality allows you to divide each side of an equation by the same non-zero number.

1. Plan

Use this lesson after Lesson 4-9.
Common Core State Standard:
7.EE.4.b

Math Background

Strategies used to solve one-step inequalities and two-step equations can be applied to solving word problems represented by inequalities of the form $px + q > r$ or $px + q < r$, where p, q, and r are rational numbers.

2. Teach

Guided Instruction

Example 1

As students graph solutions to these inequalities, make sure they attend to precision by accurately representing the solution on a number line. They should clearly indicate an open or closed circle and use an appropriate scale on the number line.

Error Prevention!

When learning to solve a two-step inequality with p positive and q negative, students may reverse the direction of the inequality symbol. Point out that the direction of the symbol only changes when multiplying or dividing by a negative number, not when adding or subtracting.

Example 2

Help students attend to precision by identifying points of difficulty in writing an inequality so that it coherently represents the problem. For example, make sure students differentiate among the inequality symbols and use them correctly when setting up an inequality.

Have students interpret the real-world problem solutions in the context of the situation and reflect on whether the results make sense.

CC-6 Solving Inequalities

CONTENT STANDARDS

7.EE.4.b Solve word problems leading to inequalities of the form $px + q > r$ or $px + q < r$, where p, q, and r are specific rational numbers.

Test Prep Tip
Reverse the direction of the inequality symbol when you multiply or divide each side of an inequality by a negative number.

To solve two-step inequalities you follow the same steps as solving two-step equations. The solution can be graphed on a number line.

EXAMPLE Solving Two-Step Inequalities

1 Solve $-3.5x + 6 > 10.2$. Graph the solution.

$-3.5x + 6 > 10.2$	← **Write the inequality.**
$-3.5x + 6 - 6 > 10.2 - 6$	← **Subtract 6 from each side.**
$-3.5x > 4.2$	← **Simplify.**
$\frac{-3.5x}{-3.5} < \frac{4.2}{-3.5}$	← **Divide each side by −3.5.**
$x < -1.2$	← **Simplify.**

−3 −2 −1 0 1 2 3

Quick Check

1. Solve $-\frac{1}{3}a + \frac{1}{2} \leq \frac{1}{5}$. Graph the solution on a number line.

EXAMPLE Application: Music Downloads

2 A music club charges \$0.75 per song download plus a membership fee of \$5.70. Diego can spend at most \$15. Write and graph the inequality for the number of songs Diego can download.

Words \$0.75 times number of songs plus monthly fee is at most \$15

Let s = the number of songs

Expression $0.75 \cdot s + 5.7 \leq 15$

$0.75s + 5.7 \leq 15$	← **Write the inequality.**
$0.75s + 5.7 - 5.7 \leq 15 - 5.7$	← **Subtract 5.7 from each side.**
$0.75s \leq 9.3$	← **Simplify.**
$\frac{0.75s}{0.75} \leq \frac{9.3}{0.75}$	← **Divide each side by 0.75.**
$s \leq 12.4$	← **Simplify.**

10 11 12 13 14

Only whole-number solutions are reasonable in this context, so Diego can download at least 0 songs and no more than 12 songs.

Use after Lesson 4-9.

Closure

- *Describe the steps to solve an inequality of the form $px + q > r$.* **Subtract q from both sides, then divide sides by p. If p is negative, reverse the direction of the inequality symbol.**

Answers

Quick Check

1. $a \geq 0.9$

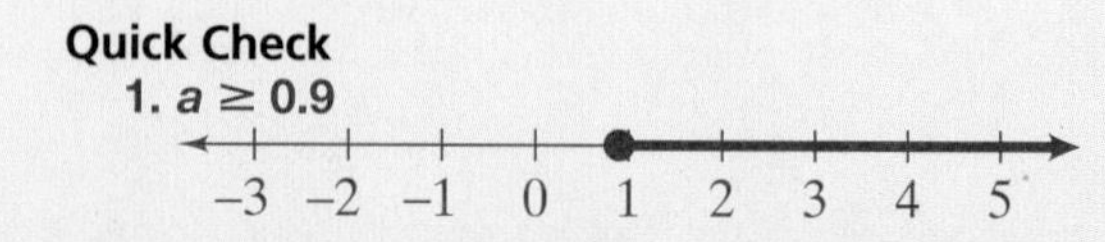

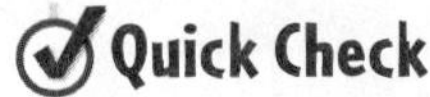

Quick Check

2. A phone plan charges \$0.20 per text message plus a monthly fee of \$42.50. Lin can spend at most \$50. Write an inequality for the number of text messages Lin can send. Graph and describe the solutions.

Homework Exercises

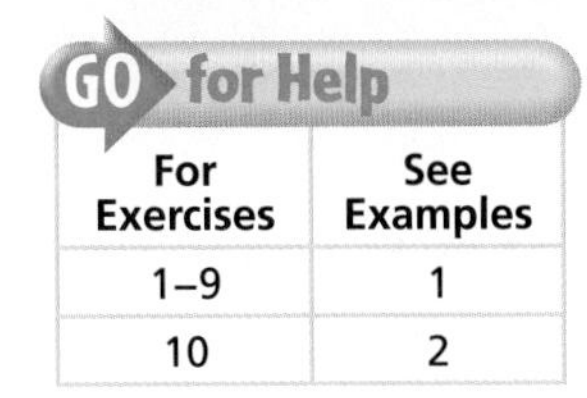

For Exercises	See Examples
1–9	1
10	2

Solve each inequality. Graph the solution.

1. $-1.6x + 5 > 7.4$
2. $3.7y + 2.8 \geq 25$
3. $-5.6c - 7.2 \leq 6.8$
4. $2.4w - 7.1 < 8.5$
5. $-\frac{5}{7} - \frac{6}{7}z < -\frac{2}{7}$
6. $-9.4x + 6 > -8.1$
7. $\frac{x}{5} + \frac{1}{2} \geq \frac{1}{10}$
8. $-\frac{3}{4}y - \frac{3}{8} < \frac{1}{8}$
9. $\frac{d}{3} - \frac{1}{6} > \frac{1}{12}$

10. Tricia receives a \$5 allowance every week. She also earns \$6.50 for every hour that she baby-sits. Next week she wants to earn at least \$21.25 to buy a present. Write an inequality to find the number of hours she needs to baby-sit. Graph and describe the solutions.

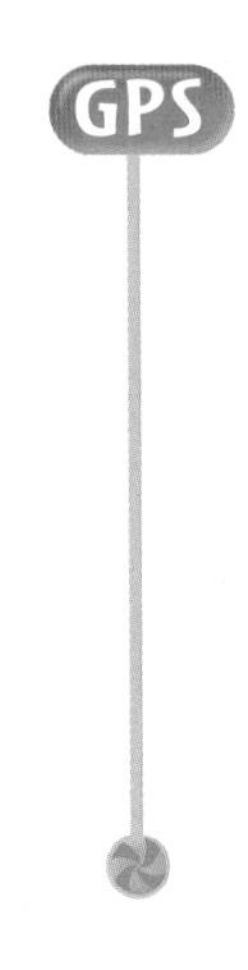

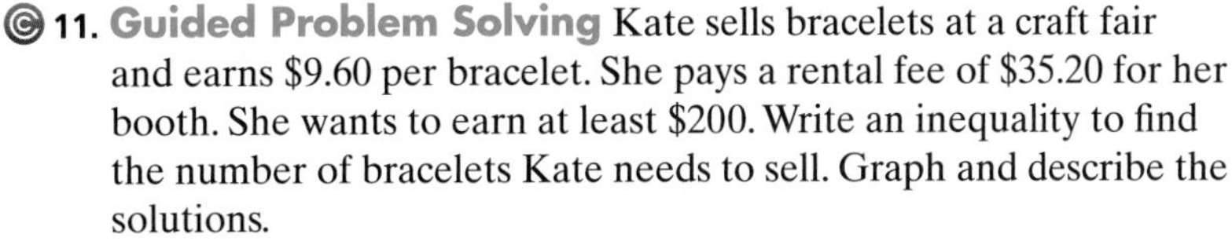

11. **Guided Problem Solving** Kate sells bracelets at a craft fair and earns \$9.60 per bracelet. She pays a rental fee of \$35.20 for her booth. She wants to earn at least \$200. Write an inequality to find the number of bracelets Kate needs to sell. Graph and describe the solutions.
    - **Make a Plan** Complete the chart below:

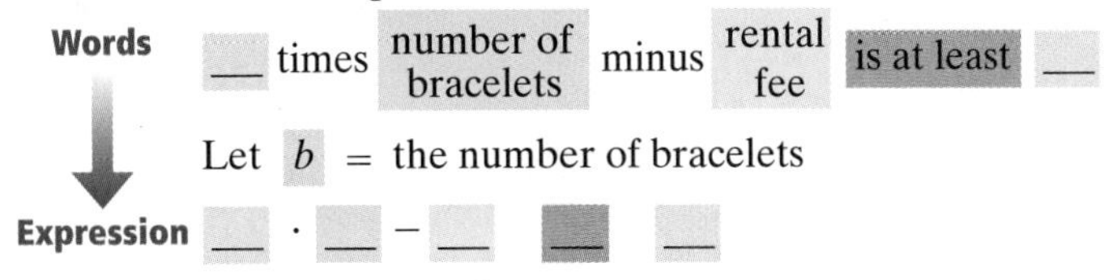

    - **Carry Out the Plan** Solve the inequality. Graph and describe the solutions.

12. Darrell wants to make a pennant with the pattern shown.

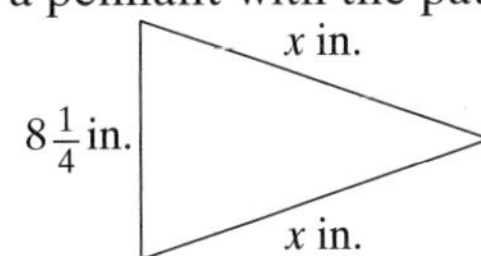

He has 4 feet of gold trim. Write an inequality for the value of x so that there is enough gold trim to go around all three edges of the pennant. Graph and describe the solutions.

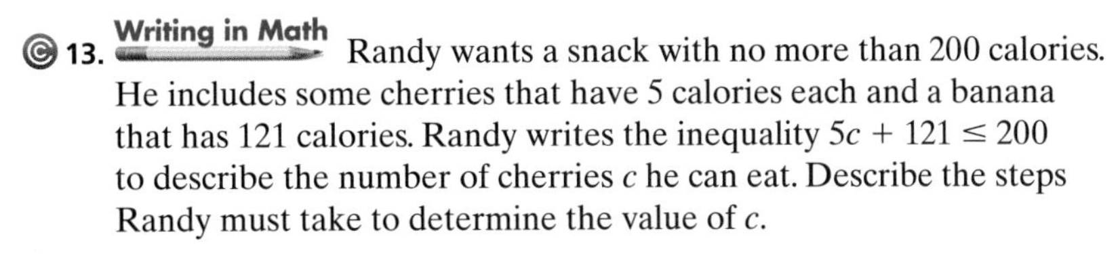

13. **Writing in Math** Randy wants a snack with no more than 200 calories. He includes some cherries that have 5 calories each and a banana that has 121 calories. Randy writes the inequality $5c + 121 \leq 200$ to describe the number of cherries c he can eat. Describe the steps Randy must take to determine the value of c.

Answers

Quick Check

2. $0.2m + 42.5 \leq 50$; $m \leq 37.5$;

35 36 37 38 39

Only whole number solutions are reasonable. Lin can send at most 37 messages.

3. Practice

Assignment Guide

Homework Exercises

A Practice by Example 1–10
B Apply Your Skills 11–13

Homework Quick Check

To check students' understanding of key skills and concepts, go over Exercises 6, 9, 10, 12, and 13.

Answers

Homework Exercises

Exercises 1–10. Check students' graphs.

1. $x < -1.5$
2. $y \geq 6$
3. $c \geq -2.5$
4. $w < 6.5$
5. $z > -\frac{1}{2}$
6. $x < 1.5$
7. $x \geq -2$
8. $y > -\frac{2}{3}$
9. $d > \frac{3}{4}$
10. $x > 2.5$; Tricia must babysit at least 2.5 hours.
11. $9.6b - 35.2 \geq 200$; $b \geq 24.5$

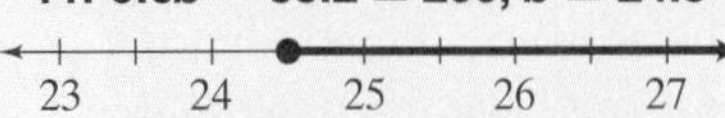

Since only whole number solutions are reasonable in this context, Kate must sell at least 25 bracelets.

12. $2x + 8\frac{1}{4} \leq 48$; $x \leq 19\frac{7}{8}$

17 18 19 20 21

Each unknown side length can be no more than $19\frac{7}{8}$ inches long.

13. Randy subtracts 121 from both sides of the inequality, then divides by 5 to find that $c \leq 15.8$. Since only whole number solutions are reasonable, Randy can eat at most 15 cherries.

1. Plan

Use this lesson after Lesson 5-2.
Common Core State Standard: 7.RP.1

Math Background

The rates and unit rates in Lesson 5-2 compare whole-number quantities. In this lesson, unit rates and scale factors are calculated using fractional and mixed-number values. When dividing such numbers, write any mixed numbers as improper fractions.

2. Teach

Guided Instruction

Example 1
Have students reason abstractly and consider whether the answer makes sense in this context. Prompt students to recognize that if Cindy walks a certain distance in $\frac{1}{4}$ hour and she continues to walk at the same speed, she will travel 4 times that distance in 1 hour. Some students may benefit from using the diagram, which shows that there are $2\frac{2}{5}$ of the $\frac{1}{4}$-unit lengths in $\frac{6}{10}$.

Example 2
Students may informally check the answer by comparing the ratio $\frac{1}{4} : \frac{5}{8}$ with $\frac{2}{5} : 1$. Note that $\frac{5}{8}$ is more than double $\frac{1}{4}$. The same is true for 1 and $\frac{2}{5}$.

Closure

- *How would you label the unit rates that compare the speeds of two bicycle racers over a short course?*

  Sample: miles per hour or miles per minute

Answers

Quick Check

1a. $\frac{2}{5}$ mile per hour

b. $1\frac{3}{4}$ containers per minute

2. $\frac{5}{8}$ inch

CC-7 Unit Rates and Ratios of Fractions

CONTENT STANDARDS

7.RP.1 Compute unit rates associated with ratios of fractions, including ratios of lengths, areas and other quantities measured in like or different units.

You know how to find unit rates using whole numbers and decimals. You can also find unit rates from data expressed as fractions.

EXAMPLE Determining Unit Rates

1 Cindy walks $\frac{6}{10}$ mile in $\frac{1}{4}$ hour. Over that distance, what is her speed in miles per hour?

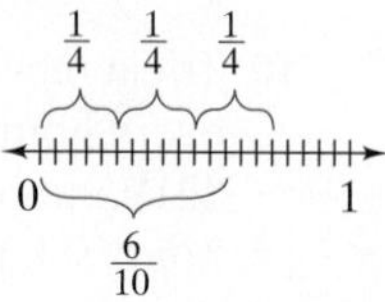

$$\text{miles} \div \text{hour} = \frac{6}{10} \div \frac{1}{4} \quad \leftarrow \text{Divide miles by hours.}$$
$$= \frac{6}{10} \cdot \frac{4}{1} \quad \leftarrow \text{Multiply by } \tfrac{4}{1}\text{, the reciprocal of } \tfrac{1}{4}.$$
$$= \frac{6}{\cancel{10}_5} \cdot \frac{\cancel{4}^2}{1} \quad \leftarrow \text{Divide 10 and 4 by their GCF, 2.}$$
$$= \frac{12}{5} \quad \leftarrow \text{Multiply.}$$
$$= 2\frac{2}{5} \quad \leftarrow \text{Write as a mixed number.}$$

Cindy walks $2\frac{2}{5}$ miles per hour.

Quick Check

1. Find the unit rate.
   - **a.** $\frac{3}{10}$ mile in $\frac{3}{4}$ hour
   - **b.** $\frac{7}{8}$ container in $\frac{1}{2}$ minute

Unit rates can be used for many comparisons.

EXAMPLE Scale Factors

2 How many inches on the map equal 1 mile?
Think: How many inches per mile? Set up a division expression.

$$\text{inches} \div \text{mile} = \frac{1}{4} \div \frac{5}{8} \quad \leftarrow \text{Divide inches by miles.}$$
$$= \frac{1}{4} \cdot \frac{8}{5} \quad \leftarrow \text{Multiply by } \tfrac{8}{5}\text{, the reciprocal of } \tfrac{5}{8}.$$
$$= \frac{1}{\cancel{4}_1} \cdot \frac{\cancel{8}^2}{5} \quad \leftarrow \text{Divide 4 and 8 by their GCF, 4.}$$
$$= \frac{2}{5} \quad \leftarrow \text{Multiply.}$$

So, $\frac{2}{5}$ inch on the map equals 1 mile.

Use after Lesson 5-2.

2. A map scale is $\frac{1}{4}$ inch = $\frac{2}{5}$ mile. How many inches represent 1 mile?

Homework Exercises

GO for Help

For Exercises	See Examples
1–4	1
5–8	2

Find the unit rate.

1. $\frac{1}{2}$ dozen pencils in $\frac{1}{3}$ box
2. $\frac{4}{5}$ chapter in $\frac{1}{4}$ hour
3. $\frac{3}{5}$ page in $\frac{3}{4}$ minute
4. $\frac{7}{12}$ gallon in $\frac{3}{10}$ kilometer

Convert the scale to a unit rate. Label your answer.

5. $\frac{1}{4}$ in. = $\frac{3}{4}$ mi
6. $\frac{3}{8}$ in. = $\frac{3}{4}$ yd
7. $\frac{7}{10}$ cm = $\frac{5}{6}$ km
8. $\frac{3}{10}$ cm = $\frac{2}{5}$ km

9. **Guided Problem Solving** In a science experiment, $16\frac{1}{2}$ grams of a powdered substance must be added to a liquid uniformly during exactly 1 minute 45 seconds. What is this rate in grams per minute?
   - Write 1 minute 45 seconds as a mixed number of minutes.
   - Write both numbers as improper fractions.
10. **Equestrian** Rayelle's horse can run $2\frac{1}{2}$ laps in 3 minutes 6 seconds. What is this rate in laps per minute?
11. How many centimeters equal 1 kilometer on the map?

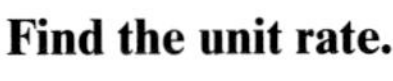

Find the unit rate.

12. $2\frac{1}{2}$ miles in $11\frac{1}{2}$ minutes
13. $2\frac{1}{5}$ sandwiches in $4\frac{2}{5}$ minutes
14. $3\frac{1}{4}$ baskets in $2\frac{1}{2}$ days
15. $3\frac{3}{8}$ cups in $1\frac{1}{2}$ servings
16. Oren is tiling a bathroom. He uses $72\frac{1}{4}$ tiles in $8\frac{1}{2}$ rows to complete the floor. What is the unit rate in tiles per row?
17. **Landscaping** A landscaper used $\frac{1}{10}$ pound of fertilizer in the soil for every $22\frac{1}{3}$ square feet of lawn. What is the unit rate in pounds per square foot?
18. Maxine can peel, core, and cut $\frac{3}{4}$ pound of apples in $2\frac{1}{2}$ minutes. What is Maxine's unit rate in pounds per hour?
19. **Writing in Math** Arturo can paint $\frac{1}{2}$ of a room in $2\frac{1}{2}$ hours. Explain how to calculate two different unit rates using this data.
20. **Open-Ended** Write a scenario and find the unit rate:

$$1\frac{2}{3} \text{ pint in } \frac{1}{6} \text{ square meter}$$

3. Practice

Assignment Guide

Homework Exercises
A Practice by Example 1–8
B Apply Your Skills 9–20

Homework Quick Check
To check students' understanding of key skills and concepts, go over Exercises 4, 8, 10, and 15.

Answers

Homework Exercises

1. $1\frac{1}{2}$ dozen pencils per box
2. $3\frac{1}{5}$ chapters per hour
3. $\frac{4}{5}$ page per minute
4. $1\frac{17}{18}$ gallons per kilometer
5. $\frac{1}{3}$ inch per mile
6. $\frac{1}{2}$ inch per yard
7. $\frac{21}{25}$ centimeter per kilometer
8. $\frac{3}{4}$ centimeter per kilometer
9. $9\frac{3}{7}$ grams per minute
10. $\frac{25}{31}$ lap per minute
11. $1\frac{1}{4}$ centimeters
12. $\frac{5}{23}$ mile per minute
13. $\frac{1}{2}$ sandwich per minute
14. $1\frac{3}{10}$ baskets per day
15. $2\frac{1}{4}$ cups per serving
16. $8\frac{1}{2}$ tiles per row
17. $\frac{3}{670}$ pound per square foot
18. 18 pounds per hour
19. Sample: Find how much of the room Arturo can paint in one hour by dividing $\frac{1}{2}$ by $2\frac{1}{2}$; or find the number of hours it takes Arturo to paint one room by dividing $2\frac{1}{2}$ by $\frac{1}{2}$.
20. Check students' work; unit rate is 10 pints per square meter.

Drawing Triangles

Use this lesson after Lesson 7-3.
Common Core State Standard: 7.G.2

Given different combinations of side and angle measures, students construct triangles where possible. Students generalize about which conditions determine a unique triangle, more than one triangle, or no triangle.

Guided Instruction

Activity 1

Observe students' strategic use of tools including protractors and rulers when constructing triangles. Demonstrate correct techniques as needed. Help students anticipate the introduction of the compass to the construction process by pointing out steps in the construction where such a tool would be useful.

Activity 2

If available, have students use technological tools, such as computer software for drawing geometric shapes, to explore and deepen their understanding of this topic.

Answers

Activity 1

1a. Check students' drawings.
1b. same shape, different sizes
1c. No; because side lengths can vary.
2a. Check students' drawings. Answers may vary. Sample: Draw a 2-inch line segment with a ruler. Use a protractor to measure a 30° angle at one end and a 40° angle at the other. Extend the sides of the angles until they intersect.
2b. Answers may vary. Sample: They are the same shape and size when the 2-inch side is between the same two angles; they are different when the 2-inch side is between two other angles.

CC-8 Drawing Triangles

7.G.2 Draw (freehand, with ruler and protractor, and with technology) geometric shapes with given conditions. Focus on constructing triangles from three measures of angles or sides, noticing when the conditions determine a unique triangle, more than one triangle, or no triangle.

Every triangle has three measures of angles and three measures of sides. A triangle is unique if there is exactly one triangle that can be determined from given measures.

1. a. Use a ruler to draw a line segment. Measure and draw a 25° angle at one end of the line segment such as the one below.

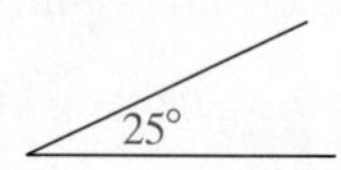

Complete your triangle by drawing a line segment that intersects the other two line segments at 50° and 105°.

b. Compare triangles with a classmate. Are your triangles the same?

c. Do three angle measures determine a unique triangle? Explain.

2. a. Construct a triangle with angle measures of 30°, 40°, and 110° and a 2-inch side length. Describe the steps you followed.

b. Compare your triangle with the triangle below. Are the triangles the same or different? Explain.

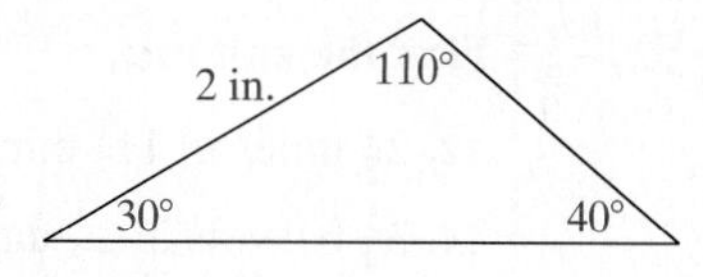

c. Do three angle measures and a side measure determine a unique triangle? Explain.

3. a. One student started constructing a triangle, as shown below, with side lengths of 7.5 cm, 10 cm, and 12.5 cm. What information is not given? Why is this construction challenging?

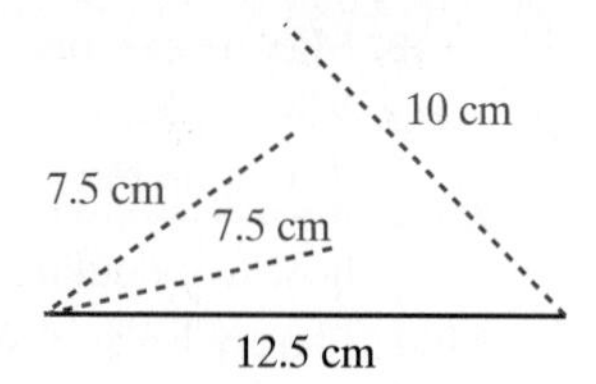

b. Choose available tools and try to construct the triangle.

c. Describe the strategy you used, explaining why you chose the tools that you did for the construction.

Use after Lesson 7-3.

2c. No, unless the given side is between two specified angles.
3a. Check students' drawings. The angles are unknown.
3b. Sample: straws and a ruler
3c. Answers may vary. Sample: I cut straws to 7.5, 10, and 12.5 cm lengths, then rotated them until they formed a triangle.
3d. No; if three side lengths are given, then triangle size and shape are determined.
3e. yes

d. Can triangles with the same three side measures have different shapes? Explain your reasoning.
e. Is a triangle unique when only three side measures are given?

MATHEMATICAL PRACTICES

1. Cut straws to lengths of 2 cm, 3 cm, 4 cm, 5 cm, and 6 cm. For each combination of 3 straws, identify the type of a triangle you can make using three straws: unique, more than one, or none. Make a table to list the combinations.

Lengths	Type of Triangle
2 cm, 3 cm, and 4 cm	Unique
2 cm, 3 cm, and 5 cm	None

2. Is it possible to construct two different triangles using the same set of three straws? Explain.
3. Do you notice any similarities about the lengths of the straws when you cannot make a triangle?

Exercises

1. The table shows three measures for five triangles. Tell whether each determines a unique triangle, more than one triangle, or no triangle.

Triangle	Measure 1	Measure 2	Measure 3
A	75°	70°	35°
B	10 cm	6 cm	3 cm
C	7 cm	7 cm	9 cm
D	6 in.	5 in.	40°
E	36°	49°	4 in.

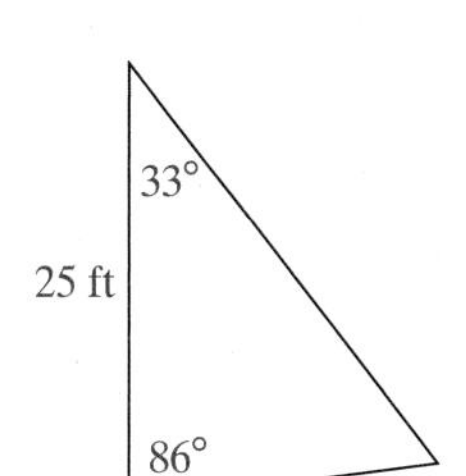

2. The triangle at the left has 25-foot side and angle measures of 33° and 86°. How many different triangles also have a side measuring 25 feet and angle measures of 33° and 86°? Explain.
3. Draw each shape.
   a. a parallelogram with angle measures of 45° and 135°
   b. a hexagon with sides that all have a measure of 1 inch
   c. a rhombus with a side length of 3 cm and an 80° angle
   d. a trapezoid with one side 2 inches long and two 90° angles
4. **Writing in Math** Write directions telling how to use a ruler and a protractor to construct a unique triangle that has a side measuring 4 centimeters that connects two angles with measures of 50° and 95°.

Guided Instruction

Activity 2

Have students justify their conclusions about the conditions that lead to either a unique triangle or no triangle.

Resources

- protractors
- plastic straws
- safety scissors
- rulers
- string
- clay

Answers

Activity 2

1. 2, 3, 4: unique; 2, 3, 5: none; 2, 3, 6: none; 2, 4, 5: unique; 2, 4, 6: none; 2, 5, 6: unique; 3, 4, 5: unique; 3, 4, 6: unique; 3, 5, 6: unique; 4, 5, 6: unique.
2. No, three side lengths determine a unique triangle.
3. If one given side is longer than or equals the sum of the other two sides, no triangle is determined. Otherwise, the triangle is unique.

Exercises

1. Check students' drawings. A: More than one; triangles can be different sizes. B: None; one side is longer than sum of other two. C: Unique; three sides determine a unique triangle. D: More than one; shape will depend upon location of 40° angle compared to sides. E: More than one; shape will depend upon location of 4 in. side compared to angles.
2. Two other: 25-ft side between 33° and 61° angles; 25-ft side between 61° and 86° angles.

3a. Size of parallelograms may vary. Sample:

135° 45° 45° 135°

3b.

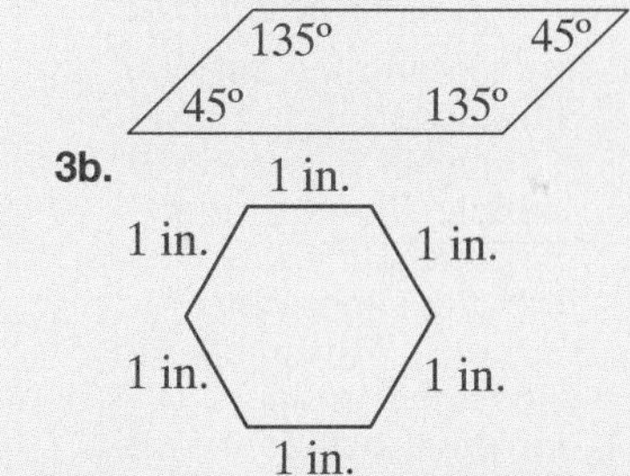

3c.

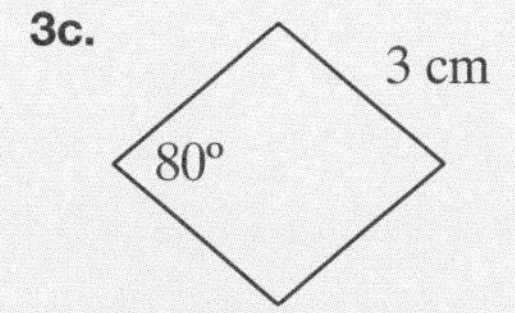

3d. Answers will vary. Sample:

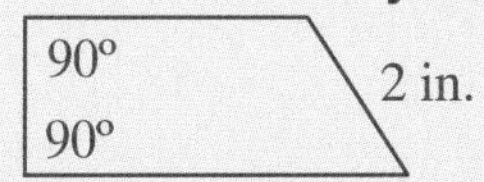

4. Use a ruler to draw a 4 cm line segment. Use a protractor to draw a 95° angle at one end and a 50° angle at the other. Extend the sides of both angles. They will intersect at a 35° angle.

Cross Sections

Use this lesson after Lesson 8-8.
Common Core State Standard: 7.G.3

When a slice is made through a three-dimensional figure, the result is a two-dimensional figure called a *cross-section*. Cross-sections of complicated figures may result in different shapes depending on the way the slice is made.

Guided Instruction

Activity

Discuss how cross sections model real-world situations. For example, professionals use cross-sections to find perimeters and areas of spaces for architectural models and scientific imaging.

Be sure students attend to precision by using the correct geometric terms when describing the shapes of the cross-sections.

Differentiated Instruction

Tactile Learner
Have students use modeling clay and string to model the three-dimensional shapes, and then make slices through their models to see the cross sections.

Answers

Activity

1a. Triangle; the cross section for the vertical slice has three sides, so it is a triangle.
b. Square; the horizontal slice is parallel to the base, so it is a square.
2a. Both cross sections are rectangles.
b. Yes; a second rip cut can be made so that the cross section is in the shape of the rectangular faces from the top and the bottom.
3a. a triangle
b. a triangle
c. Sample: The shape of the slice will always be a triangle; slices closer to a corner will be smaller than slices near a side.
4. Jorge's cross-section is a circle; Patti's cross section is a rectangle.
5. All the faces are circles.

CC-9 Cross Sections

7.G.3 Describe the two-dimensional figures that result from slicing three-dimensional figures, as in plane sections of right rectangular prisms and right rectangular pyramids.

A **cross section** is the two-dimensional shape that you see after slicing through a three-dimensional object.

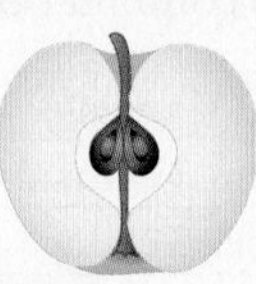
cross section of an apple

cross section of a tree trunk

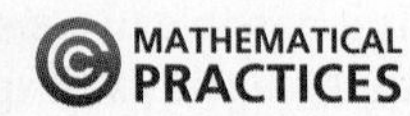

MATHEMATICAL PRACTICES

1. Jim made a clay model of a square pyramid. He shows the cross section of the pyramid by slicing the pyramid with a string.

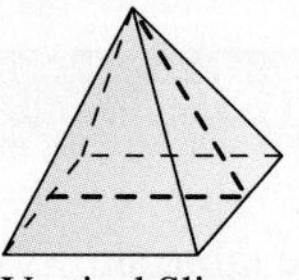
Vertical Slice

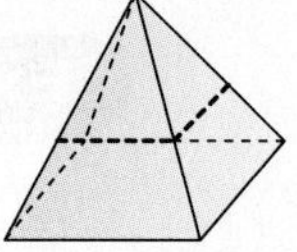
Horizontal Slice

   a. Sketch the two-dimensional shape that will result if Jim slices the pyramid vertically. Tell how you determined the shape.
   b. Sketch the two-dimensional figure that will result if Jim slices the pyramid horizontally. Tell how you determined the shape.

2. Ripping refers to cutting wood with the grain of the wood, and crosscutting refers to cutting wood across the grain of the wood.

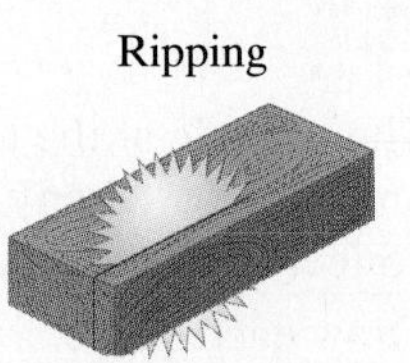
Ripping

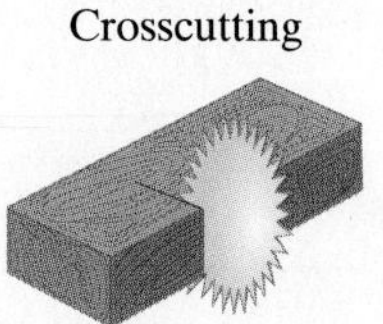
Crosscutting

   a. Identify the two-dimensional shape of the cross section for each type of cut and compare them.
   b. Would any other cuts produce a cross section in the same geometric shape? If so, describe how the cut would be made.

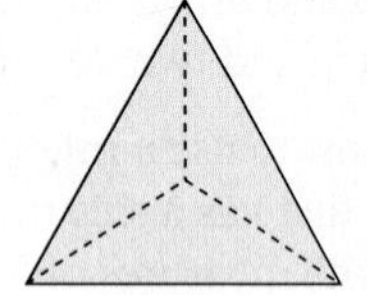

3. You have a block of cheese in the shape shown at the left.
   a. What shape will the slice of cheese be if the cheese is sliced horizontally?
   b. What shape will the slice of cheese be if the cheese is sliced vertically?
   c. Does it matter where the cuts are made? Explain your reasoning.

Use after Lesson 8-8.

4. Jorge and Patti both bought sushi rolls for lunch. The sushi rolls were shaped like cylinders. Jorge cut his sushi roll vertically, and Patti cut her sushi roll horizontally as shown below.

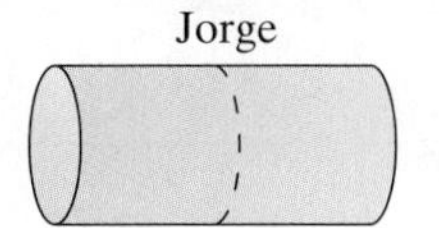

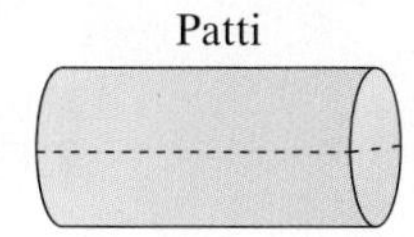

They compared the shapes of the cross sections that resulted from the cuts. Describe what they saw.

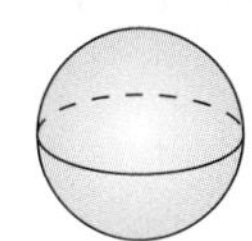

5. Sasha cut an orange shaped like a sphere through the center. What shape is the cross section that she sees?

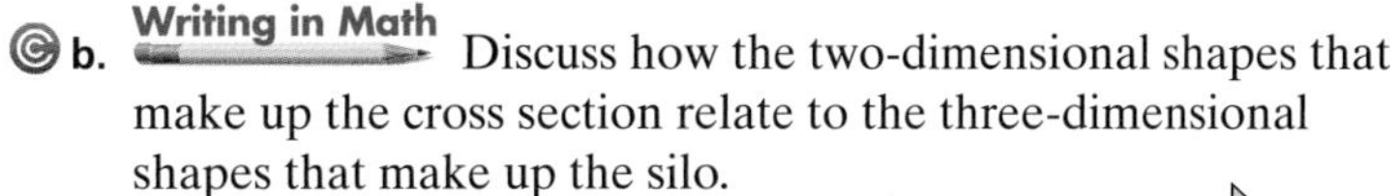

1. Parallel vertical slices are made through a rectangular pyramid as shown below.

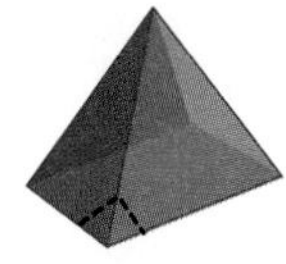
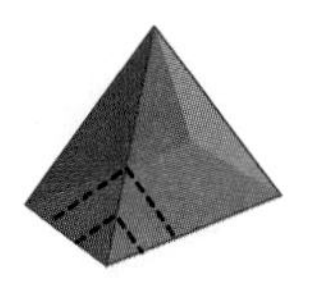
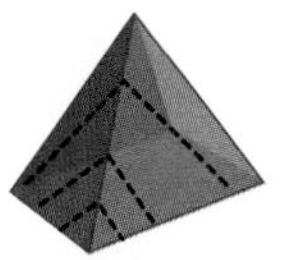

What are the shapes of the cross sections? Describe how the cross sections will change as additional cuts are made.

2. A barn silo has the shape of the figure shown at the left.
   a. If it is sliced in half with a vertical cut, what geometric shapes make up the cross section?
   b. **Writing in Math** Discuss how the two-dimensional shapes that make up the cross section relate to the three-dimensional shapes that make up the silo.
3. What three-dimensional figure can have a cross section in the same shape as the triangle shown at the right? Explain your reasoning.
4. A three-dimensional figure has a rectangular vertical cross section and a horizontal cross section in the shape of a hexagon.

vertical cross section

horizontal cross section

What is the three-dimensional figure?

5. **Reasoning** A cross section of a rectangular prism is a rectangle with sides measuring 5 cm and 7 cm. Can the exact dimensions of the prism be determined? Explain your reasoning.

Resources

- clay
- string

Answers

Exercises

1. Triangles; the cross sections will be the same shape, but will increase in size until the slice is through the vertex at the top, then they will decrease in size.

2a. triangle and rectangle

2b. The silo is made up of a cone and a cylinder; the triangle is a cross section of the cone, the rectangle is a cross section of the cylinder

3. Sample: The shape can be a cross section of more than one three-dimensional figure; for example, from slicing a triangular prism or from cutting a corner from a rectangular pyramid or rectangular prism.

4. a hexagonal prism

5. No; the dimensions of the rectangle only determine 2 of the 3 measurements for a rectangular prism—the third measurement is unknown.

Graphs and Proportional Relationships

Use this activity after Lesson 10-3.
Common Core State Standard: 7.RP.2.a, 7.RP.2.d

In a table, each pair of values representing quantities in a proportional relationship have equivalent ratios. In the coordinate plane, the graph representing a proportional relationship is a straight line through the origin and the slope is the constant of proportionality.

Guided Instruction

Activity 1

Linear relationships may not be proportional. Point out that proportional graphs contain (0, 0); non-proportional graphs do not.

Answers

Activity 1

1. **Bank A: all equal $\frac{1}{2}$; Bank B: all are different.**
2. **Bank A**

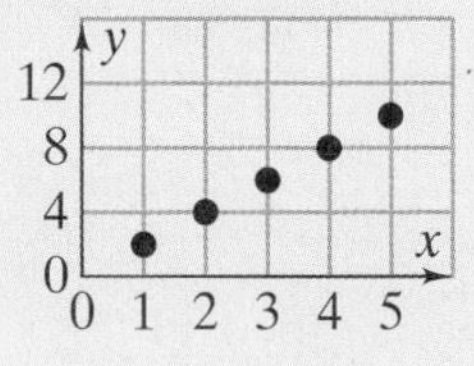

Bank B

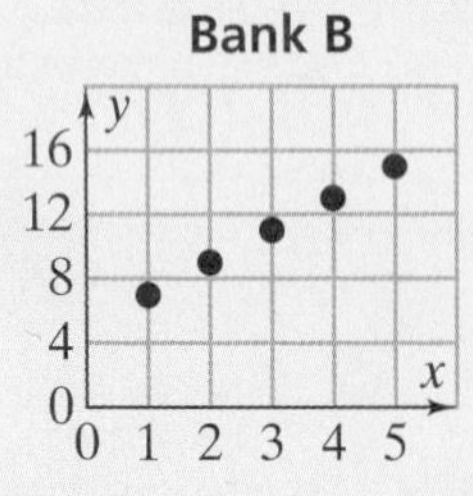

3. **Bank A: $0; Bank B: $5**
4. **Sample: Only the Bank A graph contains (0, 0).**
5. **Bank A: proportional; Bank B: not proportional. See students' answers.**
6. **$4 bonus**
7. **Bank C: $\frac{1}{3}$; Bank D: ratios are all different.**

CC-10 Graphs and Proportional Relationships

7.RP.2.a Decide whether two quantities are in a proportional relationship, e.g., by testing for equivalent ratios in a table or graphing on a coordinate plane and observing whether the graph is a straight line through the origin.

7.RP.2.d Explain what a point (*x*, *y*) on the graph of a proportional relationship means in terms of the situation, with special attention to the points (0, 0) and (1, *r*) where *r* is the unit rate.

You can use tables and graphs to decide whether or not two quantities have a proportional relationship.

ACTIVITY

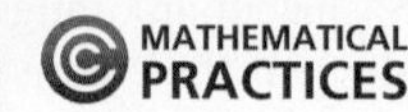

The tables below show a person's earnings at the end of each year at two different banks on a deposit of $100 over a five-year period. For each bank, earnings include simple interest paid annually at the same rate each year. In addition, Bank B gives depositors a $5 bonus when an account is first opened.

Bank A

Years	1	2	3	4	5
Earnings	$2	$4	$6	$8	$10

Bank B

Years	1	2	3	4	5
Earnings	$7	$9	$11	$13	$15

1. For each bank, find the ratio comparing years to earnings for all pairs of values in the table. What do you notice about the ratios?
2. Make a graph for each bank, plotting all the values in each table on a coordinate grid.
3. Extend the lines. What are the earnings at each bank for year 0?
4. How are the graphs different?
5. For each bank, determine if years and earnings are proportional. Explain your answer using the tables and the graphs from Step 2.
6. Information for earnings at Banks C and D is given below.

Bank C

Years	1	2	3	4	5
Earnings	$3	$6	$9	$12	$15

Bank D

Years	1	2	3	4	5
Earnings	$7	$10	$13	$16	$19

How much does Bank D give as a bonus for opening an account?

7. Find the ratio comparing years to earnings for all pairs of values in each table.
8. Make a graph for Bank C and Bank D.
9. Look for patterns between the tables for all four banks. Give a general rule for using a table to decide if two quantities are proportional.
10. Look for patterns between the graphs for all four banks. Give a general rule for using a graph to decide if two quantities are proportional.

Use after Lesson 10-3.

8. **Bank C**

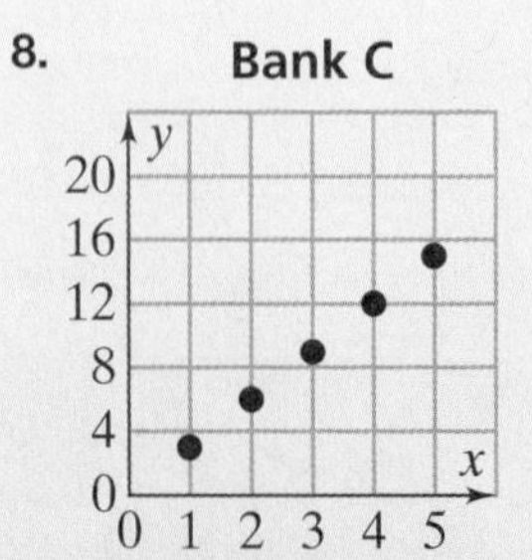

Bank D

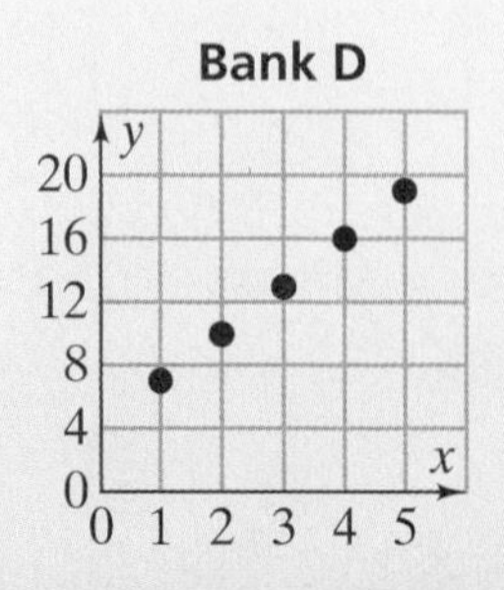

9. **If all pairs of values have equivalent ratios, the quantities are proportional.**
10. **If graph is a line through (0, 0), quantities are proportional.**

Keisha and Dave are riding in a bike-a-thon. The tables below show distances they traveled.

Keisha

Hours	0	2	4	5	7
Miles	0	13	26	32.5	45.5

Dave

Hours	0	3	6	8	9
Miles	0	18.6	37.2	49.6	55.8

1. For each cyclist, graph the relationship between distance and time.
2. Is there a proportional relationship between time and distance for either or both cyclists? Explain.
3. What does the point (4, 26) represent?
4. What is the meaning of the point (0, 0) in this situation?
5. Where $x = 1$ on each graph, what does the y value represent?
6. In the graph of any proportional relationship, what is represented by r at the point $(1, r)$?
7. How does r compare with the unit rate for each cyclist?

Exercises

For Exercises 1–6, determine whether there is a proportional relationship. Explain your reasoning.

1.

x	1	2	4	7	9
y	5	9	17	29	37

2.

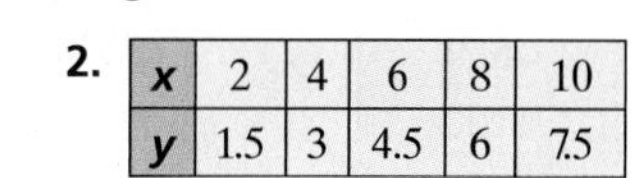

x	2	4	6	8	10
y	1.5	3	4.5	6	7.5

3.

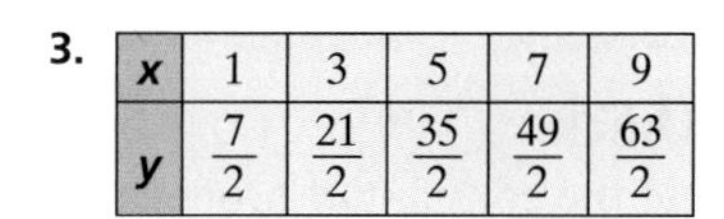

x	1	3	5	7	9
y	$\frac{7}{2}$	$\frac{21}{2}$	$\frac{35}{2}$	$\frac{49}{2}$	$\frac{63}{2}$

4.

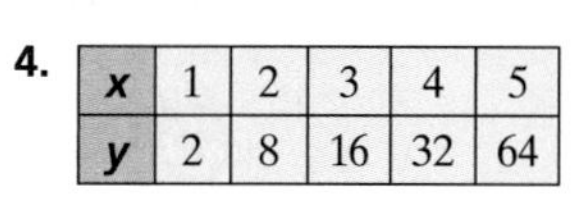

x	1	2	3	4	5
y	2	8	16	32	64

5.

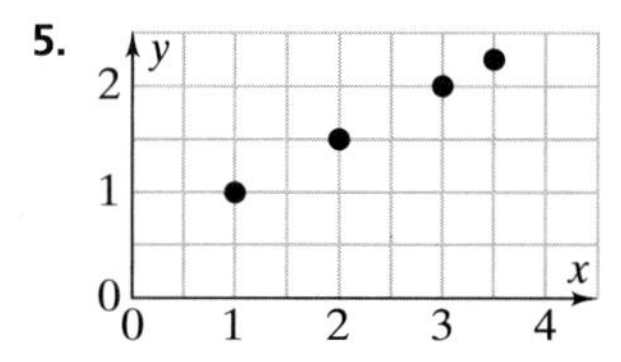

6.

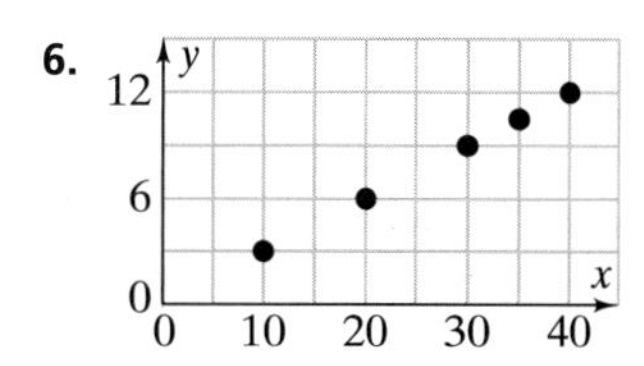

For Exercises 7–8, explain what the point with *x*-coordinate 3 represents. Then find the unit rate, *r*.

7. **Walking**

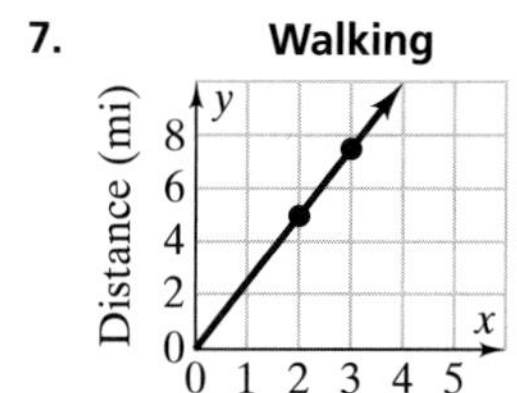

8. **Hot Air Balloon**

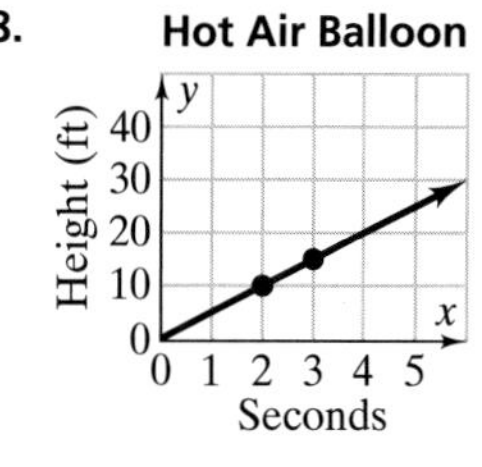

Guided Instruction

Activity 2

Ask students to think of other real-world applications where graphing a proportional relationship would be useful to find a unit rate.

Resources

- graph paper

Answers

Activity 2

1.

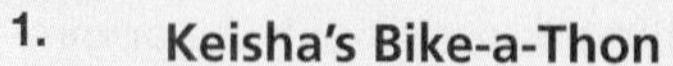

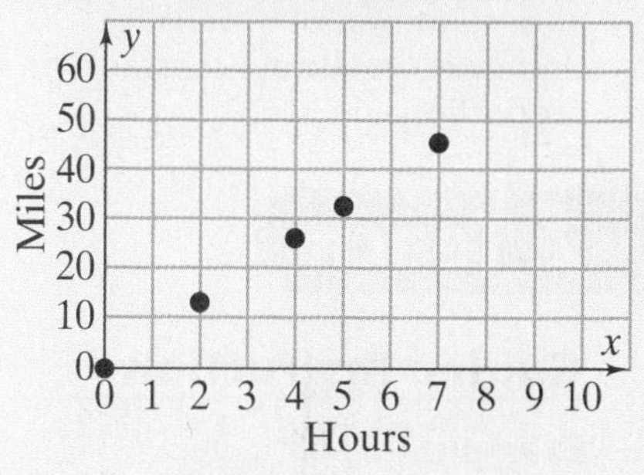

Dave's Bike-a-Thon

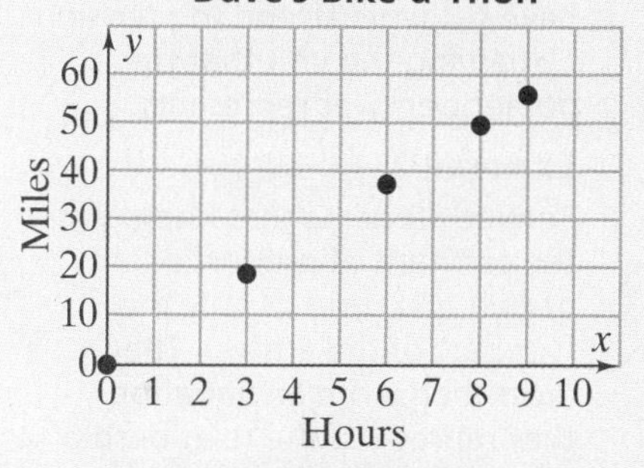

2. Both relationships are proportional because all the ratios in each table are equivalent and both graphs are straight lines that include the point (0, 0).
3. Keisha traveled 26 miles in 4 hours.
4. This point represents the start of the bike-a-thon, when no time has passed and no distance has been traveled.
5. distance traveled in 1 hour: 6.5 mi and 6.2 mi
6. the unit rate
7. They are equivalent.

Exercises

1. Not proportional; the ratios for each pair are not equal.
2. Proportional; ratios for each pair are equivalent.
3. Proportional; ratios for each pair are equivalent.
4. Not proportional; the ratios for each pair are not equivalent.
5. Not proportional; graph does not pass through the origin.
6. Proportional; graph is a straight line that passes through the origin.
7. Point (3, 7.5) represents 7.5 miles walked in 3 hours; 2.5 miles per hour
8. Point (3, 15) represents a height of 15 feet after 3 seconds; 5 feet per second

1. Plan

Use this lesson after Lesson 10-3.
Common Core State Standard: 7.RP.2.b, 7.RP.2.c

Math Background

If two quantities, such as minutes and seconds or circumference and diameter, always have the same ratio, that ratio is called the constant of proportionality. For example, the slope of a graph of a proportional relationship—a line that passes through the origin—is a constant of proportionality. Unit rates are always constants of proportionality.

2. Teach

Guided Instruction

Example 1
In each Quick Check exercise, have students attend to precision by naming the unit that the numeric answer represents.

Example 2
Remind students that finding the constant of proportionality of two quantities is only the first part of the answer. They must then write an equation that represents the relationship between the quantities.

Teaching Tip
Each example in this lesson shows a proportional relationship.
Ask: *Can every proportional relationship be represented by a linear equation? Explain.*
Yes; the graph of a proportional relationship is a straight line that contains the point (0, 0).
Does every linear equation show a proportional relationship? Explain. **No; the graph of a linear equation does not have to go through the origin.**

Closure

- *Write an equation using the constant of proportionality for converting days to hours.*
  h = 24*d*

CC-11 Constant of Proportionality

CONTENT STANDARDS

7.RP.2.b Identify the constant of proportionality (unit rate) in tables, graphs, equations, ... and verbal descriptions of proportional relationships.

7.RP.2.c Represent proportional relationships by equations.

When the ratio of two quantities is always the same, the quantities are proportional. The value of the ratio is called the **constant of proportionality.** This value is also equivalent to the unit rate.

The graph of a proportional relationship is a straight line through the origin with a slope equal to the constant of proportionality.

EXAMPLE Identifying Unit Rate

1 The table at the left shows a proportional relationship between the number of minutes and the amount the customer pays for cell phone service. Identify the constant of proportionality.

Minutes, *m*	Price, *p* (dollars)
100	\$10
500	\$50
1,000	\$100
1,500	\$150

Step 1: Use one data point to find the constant of proportionality *c*.

$\frac{\text{price}}{\text{minutes}} = \frac{10}{100}$ ← **Find the price per minute by dividing the price by the number of minutes.**

$= 0.1$ ← **Simplify.**

Step 2: Check by multiplying *c* times the first quantity.

$100 \times 0.1 = 10$ ✓ $500 \times 0.1 = 50$ ✓
$1{,}000 \times 0.1 = 100$ ✓ $1{,}500 \times 0.1 = 150$ ✓

The constant of proportionality is 0.1.
This unit rate represents a payment of \$0.10 per minute.

Quick Check

1. Find the constant of proportionality for each table of values.
   a. yards of cloth per blanket b. pay per hour

Yards (*y*)	16	32	40
Blankets (*b*)	8	16	20

Hours (*h*)	2	10	16
Pay (*p*)	\$11	\$55	\$88

EXAMPLE Representing Proportional Relationships

2 A ruler shows 12 inches on one edge and 30.48 centimeters on the other. The graph shows the relationship of inches to centimeters. Write a formula to find the number of centimeters *c* in *n* inches.

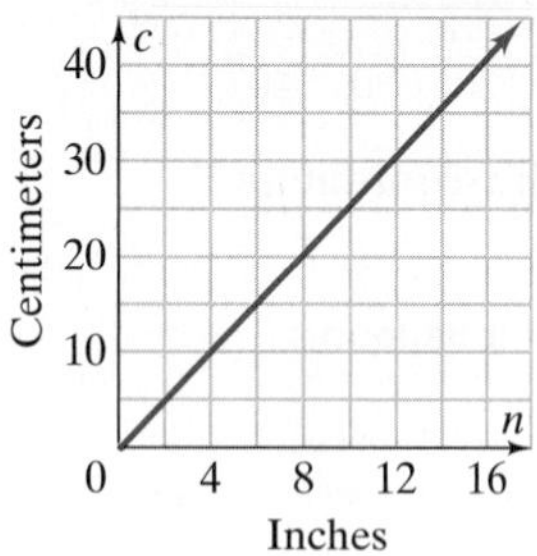

Step 1: Use one data point to find the constant of proportionality.

$\frac{\text{centimeters}}{\text{inches}} = \frac{30.48}{12}$ ← **To find the number of centimeters per inch, write a rate with inches in the denominator.**

$= 2.54$ ← **Divide.**

Step 2: Write an equation for *c* in terms of *n*.

$$c = 2.54n$$

Use after Lesson 10-3.

Answers

Quick Check
1a. *c* = 2
b. *c* = \$5.50
2a. *g* = 500*d*
b. *w* = \$7.25*h*

Quick Check

2. Write an equation to describe the relationship.
   a. An inn uses 3,500 gallons of water each week. Predict the number of gallons used for d days.
   b. Tim is paid $58 for 8 hours. Find his wage for h hours.

Homework Exercises

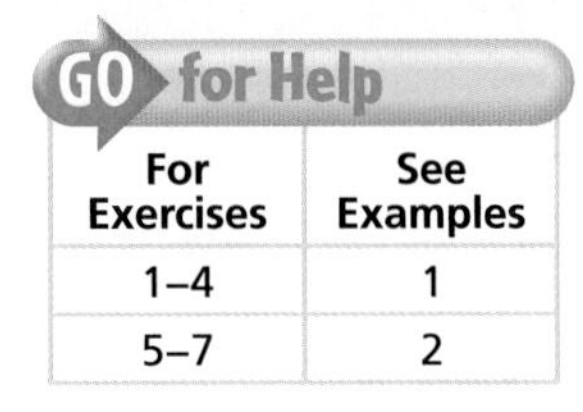

GO for Help

For Exercises	See Examples
1–4	1
5–7	2

Find the constant of proportionality for each table of values.

1. profit per shirt sold

Shirts	5	10	15
Profit	$7.50	$15.00	$22.50

2. wages per day

Days	5	10	15
Wages	$51.25	$102.50	$153.75

3. price per pound

Apples (lb)	4	5	6
Price	$7.96	$9.95	$11.94

4. pounds per bag

Bags	3	8	11
Dog Food (lb)	7.5	20	27.5

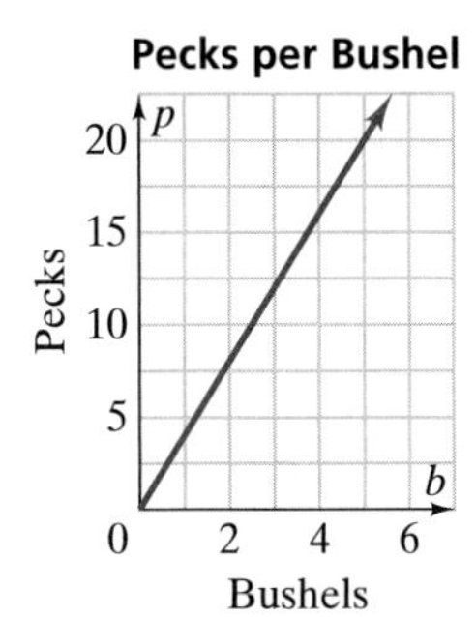

Write an equation using the constant of proportionality to describe the relationship.

5. The graph at left shows the relationship between bushels and pecks. Find the number of pecks p in b bushels.
6. A horse that is 16 hands tall is 64 inches tall. Find the number of hands h in n inches.
7. One day, 16 U.S. dollars was worth 10 British pounds. Find the number of dollars d in p pounds.

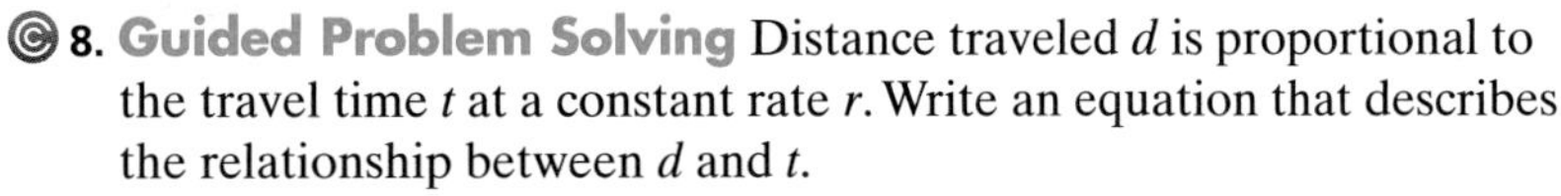

8. **Guided Problem Solving** Distance traveled d is proportional to the travel time t at a constant rate r. Write an equation that describes the relationship between d and t.
   - What is r in terms of d and t?
   - What equation tells how to find d given r and t?
9. The ratio of the circumference C of a circle to its diameter d is π. Write an equation that describes the relationship between C and d.

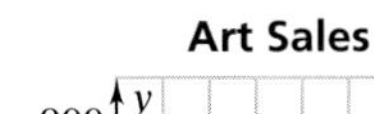

Orange Prices

$	$8	$10	$20
lbs	4	6	10

10. **Error Analysis** A salesperson showed the table on the left while explaining that oranges are the same price per pound, no matter what size bag they come in. Why is the salesperson wrong?

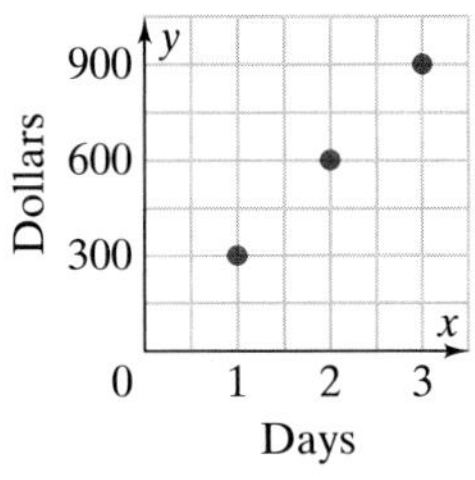

Use the graph for questions 11–13.

11. What is the constant of proportionality, dollars per day?
12. Make a table of values to show the data in the graph.
13. Write an equation to find the amount for any number of days.
14. **Writing in Math** Explain how to identify the constant of proportionality from a graph of any proportional relationship.

3. Practice

Assignment Guide

Homework Exercises
A Practice by Example 1–7
B Apply Your Skills 8–14

Homework Quick Check
To check students' understanding of key skills and concepts, go over Exercises 4, 7, 9, and 14.

Answers

Homework Exercises

1. $c = 1.50$
2. $c = 10.25$
3. $c = 1.99$
4. $c = 2.5$
5. $p = 4b$
6. $h = \frac{1}{4}n$
7. $p = 1.6d$
8. $r = \frac{d}{t}$; $d = rt$
9. $C = \pi d$
10. The price per pound is $2 for the 4-pound bag and the 10-pound bag, but about $1.67 for the 6-pound bag.
11. $c = 300$
12.

Days (d)	1	2	3
Dollars (m)	300	600	900

13. $m = 300d$
14. The constant of proportionality is the slope of the line.

Data Variability

Use this lesson after Lesson 11-3.
Common Core State Standard: 7.SP.3

Students will informally assess the visual overlap of two data sets. They will express the difference between the centers of these data sets as a multiple of either the interquartile range (IQR) or the mean absolute deviation (MAD). They will compare and analyze the relationship between the amount of visual overlap and the size of the multiples.

Guided Instruction

Activity 1

Students attend to precision by reading each box-and-whisker plot correctly and accurately calculating each IQR, median, distance between medians, and the number that, multiplied by the IQR, equals that distance.

At the end of the activity, help students summarize what they have learned. The amount of visual overlap is greater when the difference between medians is a large multiple of the IQR and less when it is a small multiple of the IQR.

Answers

Activity 1

1. Yes, they overlap between 15 and 16 lbs.
2. 3 for both breeds
3. 2
4. More than half of the upper Australian terrier data overlaps with the lower three-quarters of the border terrier data.
5. $1 = \frac{1}{3} \times 3$
6. Yes, a lower multiple shows a higher amount of overlap. In the first example, the difference between medians is larger than the IQR; in the second, the difference is smaller than the IQR.
7. $\frac{1}{5}$

CC-12 Data Variability

CONTENT STANDARDS

7.SP.3 Informally assess the degree of visual overlap of two numerical data distributions with similar variabilities, measuring the difference between the centers by expressing it as a multiple of a measure of variability.

GO for Help Activity 1-10b

Data displays can be used to assess the visual overlap of two data sets. You can compare their centers, such as mean or median, and their **variability**, the way data is spread out.

A veterinarian collects data about the weights of dogs she treats.

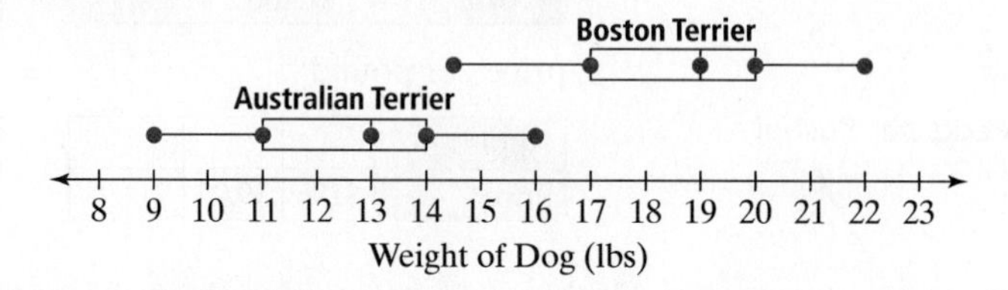

1. Do the two data sets overlap? Explain your reasoning.
2. The interquartile range (IQR) measures variability. The IQR is the difference between the upper and lower quartiles. Determine the IQR for each breed.
3. The difference between the median weights of these two breeds is 6 pounds. What number multiplied by the IQR equals 6?
4. These plots show data for Australian terriers and border terriers.

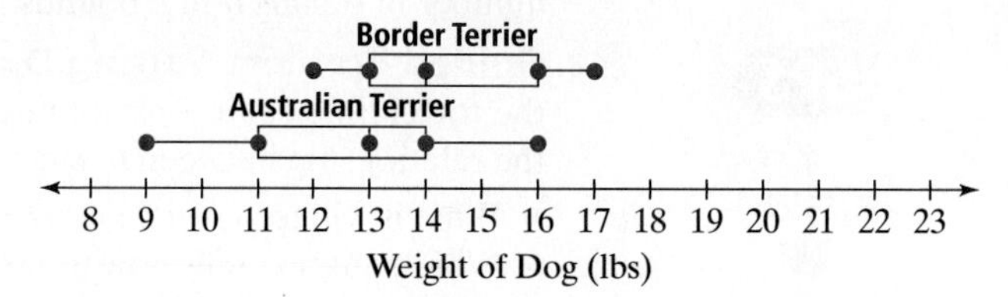

   Describe the visual overlap between the two data sets.
5. Express the difference between medians as a multiple of the IQR.
6. Can you use this multiple to assess the amount of visual overlap between two data sets? Explain.
7. Box-and-whisker plots for two other breeds are shown below.

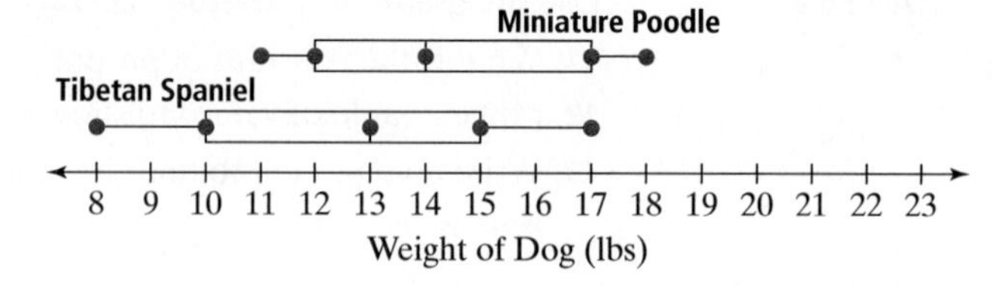

   What multiple of the IQR is the difference of medians?

Use after Lesson 11-3.

1. Describe the overlap of the two data sets below.

Dog Weights (lbs)

Pug		Dachshund
✗ ✗	13	✗
✗ ✗	14	
✗ ✗ ✗	15	
✗	16	✗ ✗
✗	17	✗ ✗ ✗
✗	18	✗ ✗
	19	✗ ✗

2. Calculate the mean of each data set.
3. You can use the mean absolute deviation (MAD) to measure variability of a data set. The MAD measures the average distance from the mean to each data point. The MAD of pug weights is calculated in this table.

Weight (lbs)	13	13	14	14	15	15	15	16	17	18
Mean	15	15	15	15	15	15	15	15	15	15
Distances	2	2	1	1	0	0	0	1	2	3
MAD = **Total Distances/# Weights**	$\frac{12}{10} = \frac{6}{5} = 1.2$									

Copy the table below to calculate the MAD of dachshund weights.

Weight (lbs)										
Mean										
Distances										
MAD = **Total Distances/# Weights**										

Dog Weights (lbs)

Pug		Miniature Dachshund
	6	✗
	7	✗
	8	✗
	9	✗ ✗ ✗
	10	✗ ✗ ✗
	11	
	12	✗
✗ ✗	13	
✗ ✗	14	
✗ ✗ ✗	15	
✗	16	
✗	17	
✗	18	

4. Can you verify that the variability for both sets is the same by looking at the shape of the data? Explain your reasoning.
5. What number multiplied by the MAD equals the difference between the means?
6. The line plots at the left show the weights of pugs and miniature dachshunds. Describe the overlap of these two sets.
7. Calculate the mean of the miniature dachshund weights. Then find the MAD.
8. How does the MAD of the miniature dachshund weights compare with the MAD of pug weights?
9. What number multiplied by the MAD equals the difference between these two means?
10. Can you use this multiple to assess the amount of visual overlap between two data sets? Explain.

Guided Instruction

Activity 2

Discuss the information about data distributions that can be observed from the visual appearance of these data plots.

Teaching Tip

When examining the data on the back-to-back line plots, students may find it easier to see the spread and overlap if they rotate the line plot 90° degrees or redraw it as two separate horizontal line plots, one directly above the other.

Answers

Activity 2

1. **All but the highest dachshund weights fall within the range of the pug weights.**
2. **15 lbs and 17 lbs**
3.

Weight (lbs)	**13**	**16**	**16**	**17**	**17**	**17**	**18**	**18**	**19**	**19**
Mean	**17**	**17**	**17**	**17**	**17**	**17**	**17**	**17**	**17**	**17**
Distances	**4**	**1**	**1**	**0**	**0**	**0**	**1**	**1**	**2**	**2**
MAD = $\frac{\text{Total Distances}}{\text{\# of Weights}}$	$\frac{12}{10} = \frac{6}{5} = 1.2$									

4. **Yes: you can count the number of data points and the total distances from the means. Since both total distances are 12, and both sets have 10 data points, the variability is the same.**
5. $\frac{5}{3}$
6. **There is no overlap.**
7. **9 lbs**
8. **Both equal $\frac{6}{5}$, or 1.2.**
9. **5**
10. **Yes; in general, the larger the multiple, the less visual overlap between the sets.**

Random Samples

Use this lesson after Lesson 11-5.
Common Core State Standard: 7.SP.2

Students will reason inductively about sample data to make and justify conclusions. They should communicate precisely when making claims about the population and justify their reasoning.

Guided Instruction

Activity 1

Remind students of their work with random samples used to make predictions about populations. Ask: *How can you use random samples to predict population size?*
Mark some members of the population, and mix them back into the population. Next, randomly sample the entire population. The ratio of marked to unmarked members in the sample should equal the ratio of all marked to all members of the population.

Teaching Tip
Inferences become more reliable if they are made from multiple same-size samples. If these samples are wildly inconsistent, adjusting the sample size or randomness may help.

Answers

Activity 1

1. **Sample: Solve the proportion for *x*. This will give the prediction based on the sample.**
2. **60 cards**
3. **80 cards**
4. **Sample 3 prediction is 40 red cards; Sample 4 prediction is 80 red cards; Sample 5 prediction is 80 red cards; Sample 6 prediction is 60 red cards.**
5. **Lowest prediction is 40 red cards; highest prediction is 80 red cards.**
6. **Only one sample shows a prediction of 40 red cards, one prediction is 60 red cards, and three predict 80 red cards.**

CC-13 Random Samples

CONTENT STANDARDS

7.SP.2 Use data from a random sample to draw inferences about a population with an unknown characteristic of interest. Generate multiple samples (or simulated samples) of the same size to gauge the variation in estimates or predictions.

A random sample of a population is used to make predictions about an entire population. These predictions are called **inferences**.

ACTIVITY — **MATHEMATICAL PRACTICES**

1. A deck of 100 cards has either a circle or a square on it, and the shape on the card is shaded either red or blue. Kris chose this sample of five cards:

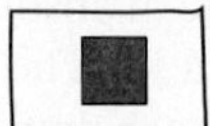 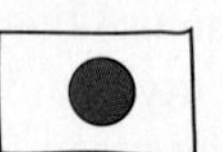 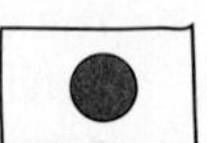 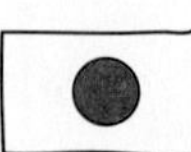 

   Explain why you can use the proportion $\frac{3}{5} = \frac{x}{100}$ to predict the number of red cards in the deck.
2. Make an inference about the number of red cards in the whole deck.
3. Kris returned the cards, shuffled the deck, and chose five new cards.

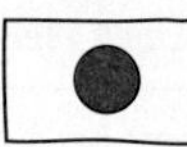 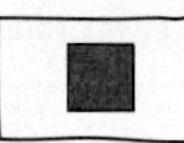 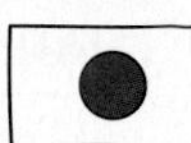 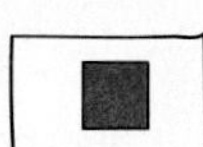 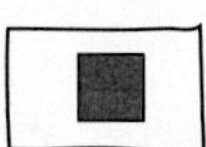

   Based on the results of this sample, use a proportion to make an inference about the number of red cards in the deck.
4. Kris repeated this three more times and recorded the results.

Sample	1	2	3	4	5
Red Cards	3	4	2	4	4
Blue Cards	2	1	3	1	1

   Make separate predictions for the number of red cards in the whole deck based on Samples 3, 4, and 5.
5. What are the highest and lowest predictions?
6. Describe the variation of all your predictions.
7. Which of your predictions do you think is most accurate?
8. Make an inference on the number of red cards that are in the whole deck based on all six samples combined.
9. Kris also recorded the shapes on the cards.

Sample	1	2	3	4	5
Squares	2	3	4	4	3
Circles	3	2	1	1	2

   Make an inference about the number of circles in the deck.

Use after Lesson 11-5.

7. **Answers may vary. Sample: I think that 80 is the best prediction because 80 occured the most often and is the median.**
8. **Answers will vary. Sample: I think there are about 70 red cards in the whole deck, because there were some predictions lower than 80.**
9. **Answers will vary. Sample: There are about 35 circles in the deck.**

1. Make a deck of 20 cards. Each card should have a blue circle, a blue square, a red circle, or a red square. Trade decks with a partner, shuffle, and choose five cards. How many circles did you choose?
2. Return the cards to the deck, shuffle, pick five new cards, and record the results four more times. Record the results in a table.

Sample	1	2	3	4	5
Squares					
Circles					

3. For each sample, estimate the number of cards in the full deck that are circles.
4. Describe the variations in your estimates.
5. Make an inference about the number of squares and the number of circles that are in the deck after looking at all five samples.
6. How do your final results compare to your first estimate?
7. Repeat the experiment and record the color of each card. Predict the number of red and blue cards in the deck.
8. Sort the cards. How do your inferences compare to the population?

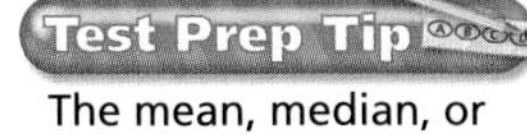

The mean, median, or mode of the predictions can be used to make an inference that is in the center of the predictions.

The principal asked the four student council officers to survey samples of the student body about after-school activities.

Favorite After-School Activity

	Sports	Band	Clubs	Tutoring	No Activity
Bo	15	10	13	8	4
Mel	15	8	12	7	8
Lea	14	7	12	8	9
Zoe	19	8	13	2	8

1. How many observations are in each student's sample?
2. Five hundred students are enrolled are in the whole school. Make four separate predictions for the number of students who favor tutoring based on the results of each survey.
3. Gauge the variation in the predictions.
4. Make an inference about the number of students who favor tutoring based on the median of your predictions.
5. *Writing in Math* Is it advantageous to make inferences based on multiple samples instead of just one sample? Explain.

Guided Instruction

Activity 2

Students use an appropriate tool to conduct an experiment, taking random samples of a population and inferring characteristics of the entire population.

Teaching Tip

Prepare a set of 20 index cards. Draw a circle on each of twelve cards and a square on each of eight cards. Color half of the cards red and half blue. Use this deck to demonstrate how to shuffle, sample five cards, and make inferences from the data set. Have students work in pairs, trading decks so that they do not know the characteristics of the deck they are sampling.

Alternative Method

If you have centimeter cubes of various colors, give each student a bag with 20 cubes. Have them analyze the characteristics of their sets of cubes, by taking random samples to infer the characteristics of the set of cubes.

Answers

Activity 2

1–7. Answers will vary. Check students' work.

Activity 3

1. **50**
2. **Based on Bo's survey, 80 students favor tutoring; based on Mel's survey, 70 students favor tutoring; based on Lea's survey, 80 students favor tutoring; based on Zoe's survey, 20 students favor tutoring.**
3. **The predications range from 20 to 80, with more results much closer to 80 than to 20.**
4. **Sample: About 75 students favor tutoring as an after-school activity.**
5. **Answers will vary. Sample: Yes; the more samples you have the more accurate the prediction; each sample reveals different results, and by combining them you get a more accurate prediction.**

Resources

- index cards (20 per student)
- blue and red markers
- centimeter cubes (optional)

Simulating Compound Events

Use this lesson after Lesson 12-4.
Common Core State Standard: 7.SP.8.c

Simulations using the proper tools can be used to find the experimental probability of a single event or compound events. For example, flipping a coin can be used to simulate a single event with two equally likely results because the probability of heads is 50% and the probability of tails is 50%. When compound events are simulated, appropriate tools must be used to generate frequencies for each single event that is involved.

Guided Instruction

Activity 1

Help students strategically select appropriate tools to simulate a compound event. Ask: *How are some other ways to simulate an event that occurs 50% of the time?* **Sample: flipping heads on a coin; rolling an odd number on a number cube**

Answers

Activity 1

1a. **Jemal could flip a coin: 1 side represents the event that a student enters the science fair and the other the event that a student does not. He could also roll a number cube: 3 sides could represent entering the science fair and the other 3 digits represent not entering.**

b. **He can spin the spinner: 2 of the 5 digits represent the event that a student competes while the other 3 digits represent the event that a student does not.**

2a. **the event that a student does not enter the science fair and does not compete in the spelling bee**

b. **T1 and T2**

c. $\frac{5}{24}$**, or about 20.8%**

CC-14 Simulating Compound Events

CONTENT STANDARDS

7.SP.8.c Design and use a simulation to generate frequencies for compound events.

GO for Help Lesson 12-2

Vocabulary Tip

A *compound event* consists of two or more events.

Use after Lesson 12-4.

A **simulation** is a model used to calculate probabilities for an experimental situation. A compound event consists of two or more simple events. These events may have different probabilities of occurring.

ACTIVITY **MATHEMATICAL PRACTICES**

One half of the students enter projects in the science fair and two fifths of the students compete in the annual spelling bee.

1. Jemal wants to find the experimental probability that a student enters the science fair and also participates in the spelling bee. The available tools are shown below.

Coins

Number Cubes

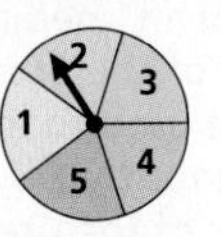

Spinner

a. Which tools can Jemal use to simulate a student entering the science fair? Explain how he can use them.

b. Which tools can Jemal use to simulate a student participating in the spelling bee? Explain how he can use them.

2. Jemal decides to use a coin and a five-section spinner.

a. Jemal lets tails (T) represent a student entering the science fair, and sections 1 and 2 on the spinner represent a student participating in the spelling bee. What does the result below mean?

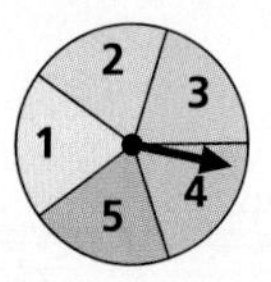

b. Jemal records the result as H4 and repeats the simulation 23 more times. He records the results in a table.

H4	T4	T3	H2	T3	H5	H2	T4
T2	H2	H1	H4	H1	T1	T4	T5
H3	H1	H4	T1	T1	H2	T2	H5

Which compound events represent a student participating in the science fair and competing in the spelling bee?

c. What is the experimental probability that a student enters the science fair and participates in the spelling bee?

At an art school, 30% of the students are left-handed. Denise wants to know the probability that in a group of 4 students, at least 1 is left-handed.

GO for Help
Activity 12-2b

1. Denise generates random digits from 0 to 9 and lets the digits 0, 1, and 2 represent a left-handed student. Is this a good tool to simulate the event that a student is left-handed?
2. The table below shows randomly generated 4-digit numbers.

7982	5839	4965	8814	3900
3933	6042	9397	4856	8373
3890	2305	3601	8174	4919
6022	6107	7903	9409	8271

   Use the first 4-digit number in the table to simulate the results of asking one group of 4 students. How many are left-handed?
3. Identify all of the 4-digit numbers in the table that represent the event *at least 1 of the 4 students is left-handed.*
4. What is the experimental probability that in a group of 4 students, at least 1 is left-handed?
5. If you did not have this table of random numbers, would a coin or a number cube be a good tool to simulate this event? Explain.
6. How can you simulate asking 30 groups of 5 students if they are left-handed?

Exercises

1. a. At Perlina's school, 30% of the students prefer folk music, 10% prefer country, 20% prefer rock, and 40% prefer hip-hop. Describe how to use a random number table with digits 0-9 to simulate finding the probability that in a group of 3 students, at least 2 prefer hip-hop.
   b. Use the random number table below to find the probability that in a group of 3 students, at least 2 like hip-hop.

165	108	952	944	542
661	827	647	333	457
950	593	087	169	813
614	869	738	027	284

2. **Writing in Math** One third of students walk to school, and 80% buy hot lunch. You want to know the probability that a student walks to school but does not buy hot lunch. Describe how you could simulate 25 trials to determine how many students walk to school but do not buy hot lunch.

Guided Instruction

Activity 2

In this activity, students demonstrate the ability to contextualize by recognizing the meaning of each 4-digit number in a table of randomly generated numbers.

Technology Tip
Random number generation using technology is a useful tool that can simplify the process of designing a simulation for compound events. If students have access to a graphing calculator or computer, have them generate random values.

Resources

- number cubes
- 5-section spinners
- coins

Answers

Activity 2

1. **Yes. Since 30% of the students are left-handed, 3 of the 10 digits can represent the event that a student is left-handed.**
2. **The digit 2 in the four-digit number 7982 simulates the result that one of the four students is left-handed.**
3. **7982, 3890, 6022, 6042, 2305, 6107, 3601, 7903, 8814, 8174, 9409, 3900, 4919, 8271**
4. $\frac{7}{10}$
5. **No: one toss of a coin or roll of a number cube cannot be used to simulate a 30% probability.**
6. **Make a table with 30, randomly generated, 5-digit numbers.**

Exercises

1. **Answers may vary. Sample: $\frac{7}{20}$; folk: 0, 1, 2; country: 3; rock: 4, 5; hip-hop: 6, 7, 8, 9**
2. **Make 2 of the 6 faces of the number cube represent the event that a student walks to school and make 1 section of a 5-section spinner represent the event that a student does not buy a hot lunch. Roll the cube and spin the spinner twenty-five times.**

Decimals and Integers

Student Support

Vocabulary

Vocabulary Review 4, 8, 14, 20, 26, 31, 38, 44, 48, 53
New Vocabulary 4, 8, 14, 31, 38, 48, 53
Vocabulary Builder 35
Vocabulary Tip 15, 20, 48
Exercises 6, 10, 16, 22, 28, 33, 41, 50, 55

Go Online

Video Tutor Help 8, 53
Active Math 32, 40
Homework Video Tutor 7, 11, 17, 22, 30, 34, 42, 47, 51, 57
Lesson Quiz 7, 11, 17, 23, 29, 33, 41, 47, 51, 57
Vocabulary Quiz 60
Chapter Test 62
Math at Work 18

GPS Guided Problem Solving

Exercises 6, 10, 16, 22, 29, 33, 41, 46, 50, 56
Choosing Operations 24
DK Applications: Applying Integers, 64–65

Assessment and Test Prep

Exponents, Factors, and Fractions

Student Support

Vocabulary

Vocabulary Review 68, 74, 82, 87, 91, 96, 102, 106
New Vocabulary 68, 74, 82, 87, 91, 96, 102, 106
Vocabulary Tip 69, 83, 97
Exercises 70, 77, 84, 89, 93, 99, 104, 108

GO Online

Video Tutor Help 68, 87
Active Math 70, 92
Homework Video Tutor 71, 78, 85, 90, 94, 100, 105, 108
Lesson Quiz 71, 77, 85, 89, 93, 99, 105, 109
Vocabulary Quiz 112
Chapter Test 114

GPS Guided Problem Solving

Exercises 70, 77, 84, 89, 93, 99, 104, 108
Using LCM and GCF 80
DK Applications: Applying Fractions, 116–117

Assessment and Test Prep

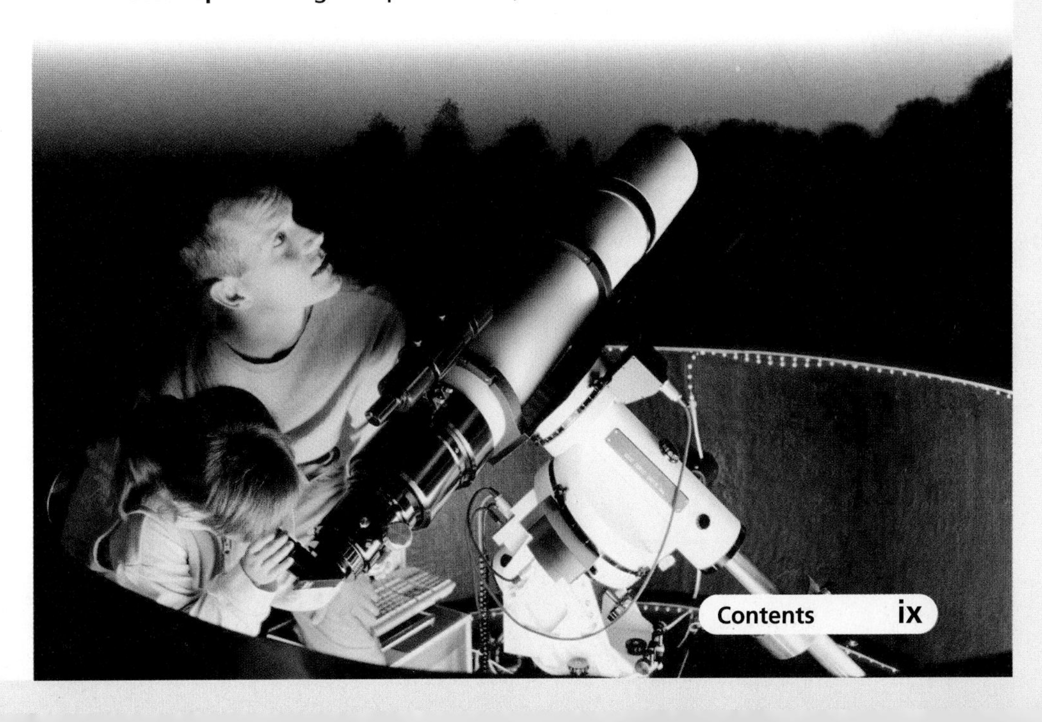

Operations With Fractions

Student Support

Vocabulary

GO Online

GPS Guided Problem Solving

Assessment and Test Prep

Algebra

Equations and Inequalities

Student Support

Vocabulary

Vocabulary Review 169, 174, 180, 186, 194, 200, 205, 210, 214
New Vocabulary 169, 174, 180, 186, 205, 210, 214
Vocabulary Builder 185
Vocabulary Tip 182, 188, 205, 208
Exercises 171, 176, 182, 188, 196, 207, 212, 216

Go Online

Video Tutor Help 180, 187
Active Math 170, 216
Homework Video Tutor 172, 177, 184, 190, 198, 204, 208, 213, 217
Lesson Quiz 171, 177, 183, 189, 197, 203, 207, 213, 217
Vocabulary Quiz 220
Chapter Test 222

GPS Guided Problem Solving

Exercises 171, 176, 183, 189, 197, 203, 207, 212, 217
Practice Solving Problems 192
DK Applications: Applying Inequalities, 224–225

Ratios, Rates, and Proportions

Student Support

Vocabulary

Vocabulary Review 228, 232, 238, 244, 252, 259
New Vocabulary 228, 232, 238, 252, 259
Vocabulary Builder 257
Vocabulary Tip 259
Exercises 230, 234, 240, 254, 261

GO Online

Video Tutor Help 228, 238
Active Math 229, 253
Homework Video Tutor 231, 235, 241, 248, 254, 262
Lesson Quiz 231, 235, 241, 247, 255, 263
Vocabulary Quiz 266
Chapter Test 268
Math at Work 237

GPS Guided Problem Solving

Exercises 230, 234, 240, 247, 254, 262
Proportions and Equations 249
DK Applications: Applying Ratios, 270–271

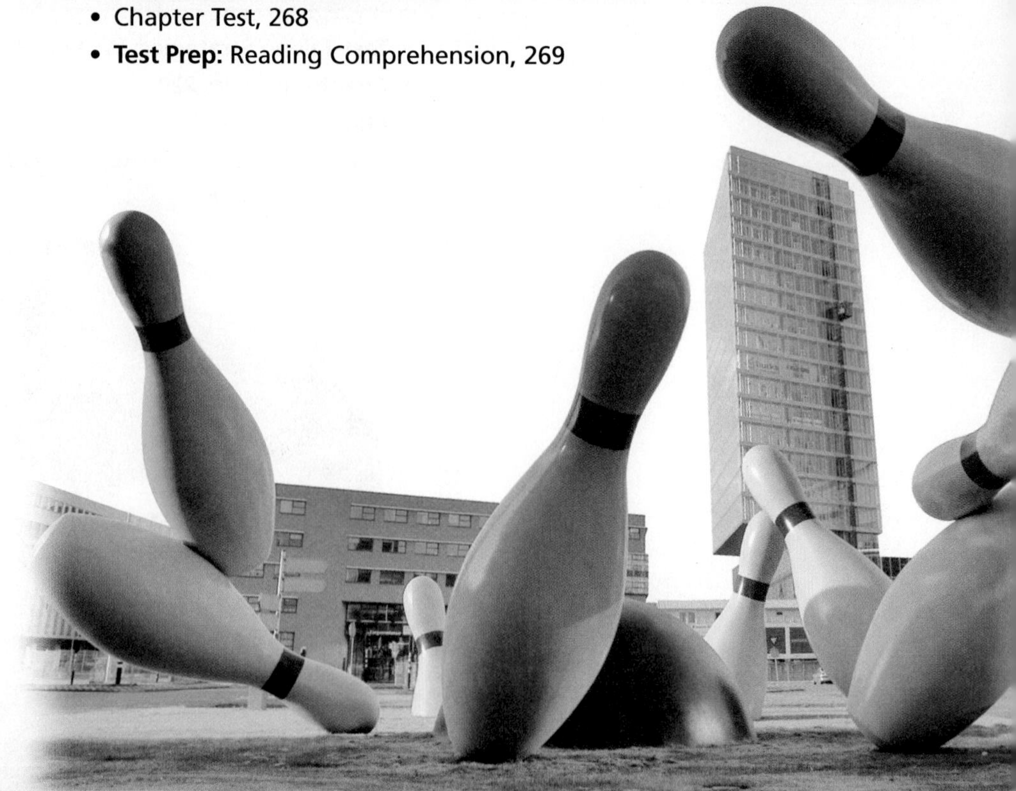

Percents

Student Support

Vocabulary

Vocabulary Review 274, 279, 284, 290, 294, 298, 304, 310
New Vocabulary 274, 304, 310
Vocabulary Tip 274
Exercises 306, 312

GO Online

Video Tutor Help 298, 305
Active Math 280, 294
Homework Video Tutor 277, 283, 287, 293, 297, 301, 306, 314
Lesson Quiz 277, 283, 287, 293, 297, 301, 307, 313
Vocabulary Quiz 316
Chapter Test 318

GPS Guided Problem Solving

Exercises 276, 282, 286, 292, 296, 300, 306, 313
Solving Percent Problems 302
DK Applications: Applying Percents, 320–321

Assessment and Test Prep

Table of Contents

Geometry

Student Support

Vocabulary

Vocabulary Review 324, 330, 336, 340, 346, 350, 354, 361
New Vocabulary 324, 330, 336, 340, 346, 350, 354, 361
Vocabulary Builder 328
Vocabulary Tip 325, 330, 341, 351, 363
Exercises 326, 333, 338, 342, 348, 352, 356, 363

Go Online

Video Tutor Help 331, 354
Active Math 337, 342
Homework Video Tutor 327, 334, 339, 344, 349, 352, 357, 364
Lesson Quiz 327, 333, 339, 343, 349, 353, 357, 363
Vocabulary Quiz 366
Chapter Test 368
Math at Work 345

GPS Guided Problem Solving

Exercises 326, 333, 338, 343, 348, 352, 356, 363
Percents and Circle Graphs 359
DK Applications: Applying Geometry, 370–371

Assessment and Test Prep

Measurement

Student Support

Vocabulary

Go Online

GPS Guided Problem Solving

Assessment and Test Prep

Table of Contents

Algebra

Patterns and Rules

Student Support

Vocabulary

Vocabulary Review 437, 442, 446, 452, 456, 461, 468, 472
New Vocabulary 442, 452, 468, 472
Vocabulary Tip 442, 447
Exercises 439, 444, 448, 454, 457, 470

GO Online

Video Tutor Help 468, 473
Active Math 462, 469
Homework Video Tutor 440, 445, 449, 455, 458, 464, 471, 474
Lesson Quiz 439, 445, 449, 455, 459, 463, 471, 475
Vocabulary Quiz 478
Chapter Test 480
Math at Work 450

GPS Guided Problem Solving

Exercises 439, 444, 448, 454, 458, 463, 470, 474
Solving Pattern Problems 466
DK Applications: Applying Graphs, 482–483

Assessment and Test Prep

Algebra

Graphing in the Coordinate Plane

Student Support

Vocabulary

Vocabulary Review 486, 491, 498, 504, 510, 514, 519
New Vocabulary 486, 491, 498, 504, 510, 514, 519
Vocabulary Builder 508
Vocabulary Tip 492, 499, 511, 519
Exercises 488, 493, 500, 506, 512, 516, 521

Video Tutor Help 492, 505
Active Math 498, 520
Homework Video Tutor 489, 494, 501, 507, 513, 517, 522
Lesson Quiz 489, 493, 501, 507, 513, 517, 521
Vocabulary Quiz 524
Chapter Test 526

GPS Guided Problem Solving

Exercises 488, 493, 500, 506, 512, 516, 521
Practice Solving Problems 496
DK Applications: Applying Coordinates, 528–529

Assessment and Test Prep

Displaying and Analyzing Data

Student Support

Vocabulary

Vocabulary Review 532, 538, 544, 550, 554, 560, 567
New Vocabulary 532, 538, 544, 550, 567
Vocabulary Tip 533, 551
Exercises 534, 540, 546, 551, 555, 568

GO Online

Video Tutor Help 567
Active Math 554
Homework Video Tutor 536, 541, 547, 552, 556, 564, 570
Lesson Quiz 535, 541, 547, 553, 555, 563, 569
Vocabulary Quiz 572
Chapter Test 574
Math at Work 565

GPS Guided Problem Solving

Exercises 535, 541, 546, 552, 555, 563, 569
Describing Data 558
DK Applications: Applying Data Analysis, 576–577

Assessment and Test Prep

Using Probability

Student Support

Vocabulary

Vocabulary Review 580, 586, 591, 598, 606, 610
New Vocabulary 580, 586, 591, 598, 606, 610
Vocabulary Builder 603
Vocabulary Tip 580, 610
Exercises 582, 588, 593, 600, 608, 611

GO Online

Video Tutor Help 581, 586
Active Math 599, 606
Homework Video Tutor 582, 589, 594, 602, 609, 612
Lesson Quiz 583, 589, 595, 601, 609, 613
Vocabulary Quiz 616
Chapter Test 618

GPS Guided Problem Solving

Exercises 582, 588, 594, 601, 608, 612
Practice With Probability 604
DK Applications: Applying Probability, 622–623

Assessment and Test Prep

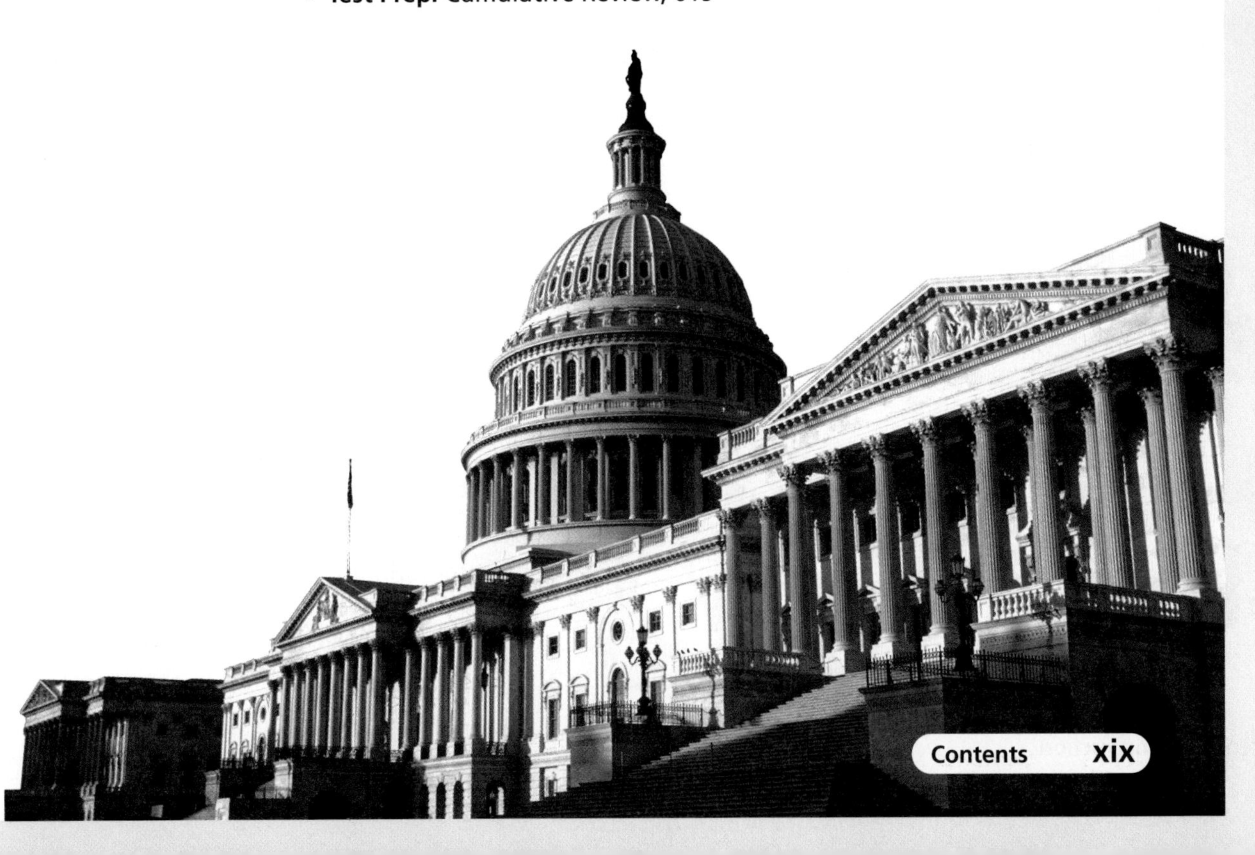

Table of Contents

Differentiate Instruction with Ease

Students develop and learn in different ways at different paces. Accessible content, presented in a variety of formats, acknowledges these unique differences while providing options for learning. The goal of *Prentice Hall Mathematics* is for all students to be successful and for you to have the tools you need to accomplish this. *Prentice Hall Mathematics* Grade 6 through Algebra 2 provides better solutions for meeting the needs of every student in the classroom by achieving two goals:

- Providing superior teacher support materials for planning how to effectively differentiate instruction
- Providing unique resources for the various populations of students.

Adapted Resources for Differentiating Instruction

In addition to the support provided in the Teacher's Editions, Prentice Hall has created resources developed uniquely for Below Level and Special Needs students.

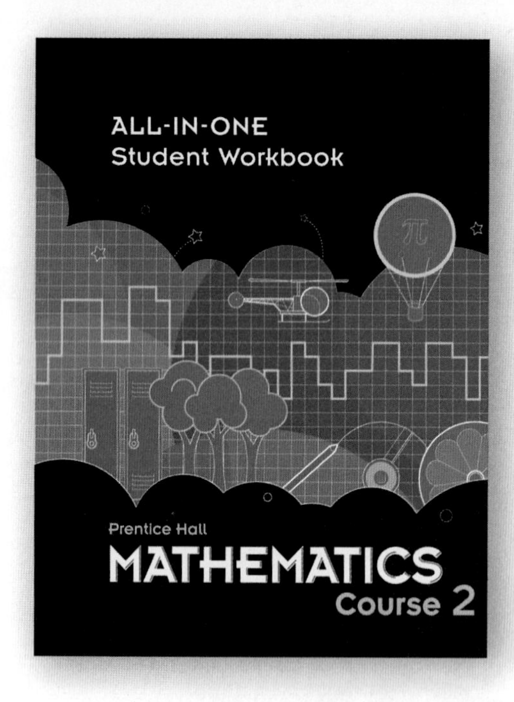

ALL-IN-ONE
Student Workbook
Prentice Hall
MATHEMATICS
Course 2

All-in-One Student Workbook Adapted Version

This resource includes adapted practice and adapted daily notetaking worksheets to support special needs students. By providing these critical resources ready for you to use, you can cover the same mathematical content with the students, but provide a more appropriate resource for them to take notes and practice the lesson's mathematics.

Differentiated Assessments

Prentice Hall Mathematics also recognizes the importance of not only differentiating instruction, but also differentiating the assessments used to monitor student progress and inform future instruction. To achieve this, three versions of each chapter test are provided: L2 for Below Level L3 for All Students and L4 for Advanced Learners.

Exam*View* 5.0 Assessment Suite

To provide the ultimate flexibility in creating assessments and practice worksheets for all students, the *Prentice Hall Mathematics ExamView* Test Banks contain adapted items written exclusively for your Special Needs and Below Level students.

Prentice Hall Mathematics Teacher's Editions offer comprehensive support in differentiating instruction.

- **L1** Special Needs
- **L2** Below Level
- **L3** All Students
- **L4** Advanced Learners
- **ELL** English Language Learners

Prentice Hall Mathematics uses a consistent method for identifying resources for differentiating instruction. This consistency helps you to easily identify and choose the appropriate resources for your students.

Planning and Using Differentiated Resources

These chapter level support pages provide you an easy-to-read overview of the resources available and suggested ways in the instructional lesson to use the resources.

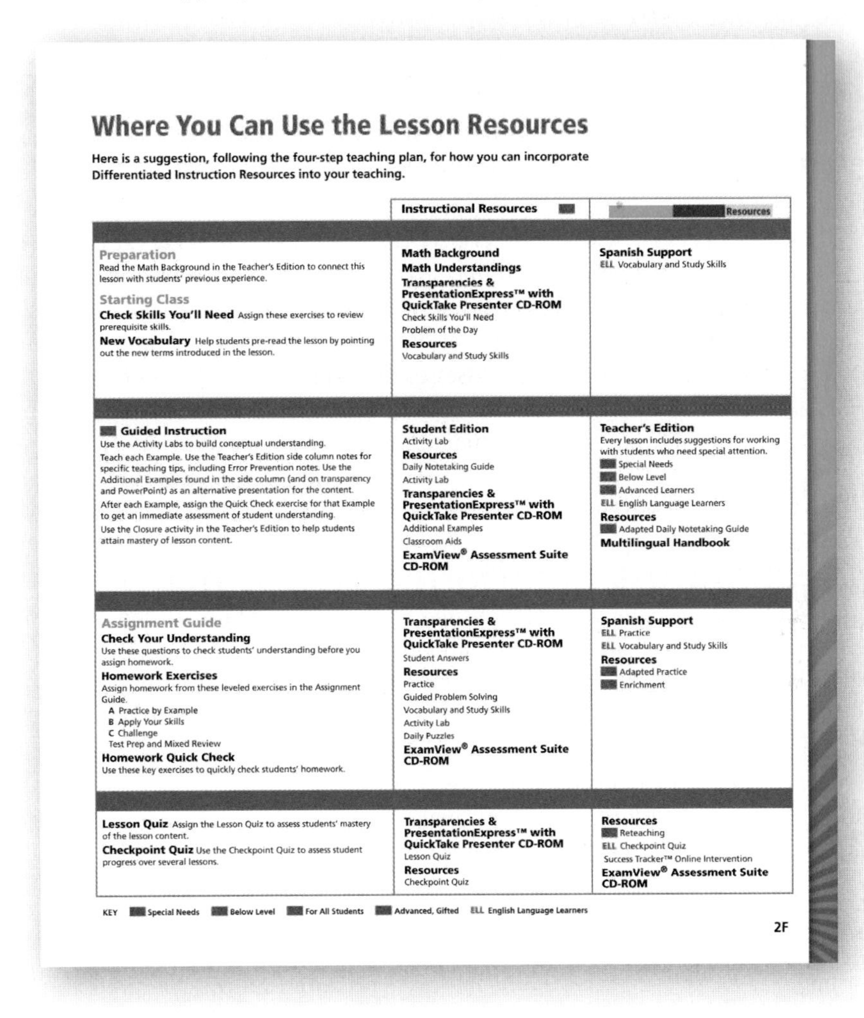

Where You Can Use the Lesson Resources

Here is a suggestion, following the four-step teaching plan, for how you can incorporate Differentiated Instruction Resources into your teaching.

	Instructional Resources	Resources
Preparation Read the Math Background in the Teacher's Edition to connect this lesson with students' previous experience. **Starting Class** **Check Skills You'll Need** Assign these exercises to review prerequisite skills. **New Vocabulary** Help students pre-read the lesson by pointing out the new terms introduced in the lesson.	**Math Background** **Math Understandings** **Transparencies & PresentationExpress™ with QuickTake Presenter CD-ROM** Check Skills You'll Need Problem of the Day **Resources** Vocabulary and Study Skills	**Spanish Support** ELL Vocabulary and Study Skills
Guided Instruction Use the Activity Labs to build conceptual understanding. Teach each Example. Use the Teacher's Edition side column notes for specific teaching tips, including Error Prevention notes. Use the Additional Examples found in the side column (and on transparency and PowerPoint) as an alternative presentation for the content. After each Example, assign the Quick Check exercise for that Example to get an immediate assessment of student understanding. Use the Closure activity in the Teacher's Edition to help students attain mastery of lesson content.	**Student Edition** Activity Lab **Resources** Daily Notetaking Guide Activity Lab **Transparencies & PresentationExpress™ with QuickTake Presenter CD-ROM** Additional Examples Classroom Aids **ExamView® Assessment Suite CD-ROM**	**Teacher's Edition** Every lesson includes suggestions for working with students who need special attention. Special Needs Below Level Advanced Learners ELL English Language Learners **Resources** Adapted Daily Notetaking Guide **Multilingual Handbook**
Assignment Guide **Check Your Understanding** Use these questions to check students' understanding before you assign homework. **Homework Exercises** Assign homework from these leveled exercises in the Assignment Guide. A Practice by Example B Apply Your Skills C Challenge Test Prep and Mixed Review **Homework Quick Check** Use these key exercises to quickly check students' homework.	**Transparencies & PresentationExpress™ with QuickTake Presenter CD-ROM** Student Answers **Resources** Practice Guided Problem Solving Vocabulary and Study Skills Activity Lab Daily Puzzles **ExamView® Assessment Suite CD-ROM**	**Spanish Support** ELL Practice ELL Vocabulary and Study Skills **Resources** Adapted Practice Enrichment
Lesson Quiz Assign the Lesson Quiz to assess students' mastery of the lesson content. **Checkpoint Quiz** Use the Checkpoint Quiz to assess student progress over several lessons.	**Transparencies & PresentationExpress™ with QuickTake Presenter CD-ROM** Lesson Quiz **Resources** Checkpoint Quiz	**Resources** Reteaching ELL Checkpoint Quiz Success Tracker™ Online Intervention **ExamView® Assessment Suite CD-ROM**

KEY Special Needs Below Level For All Students Advanced, Gifted ELL English Language Learners

2F

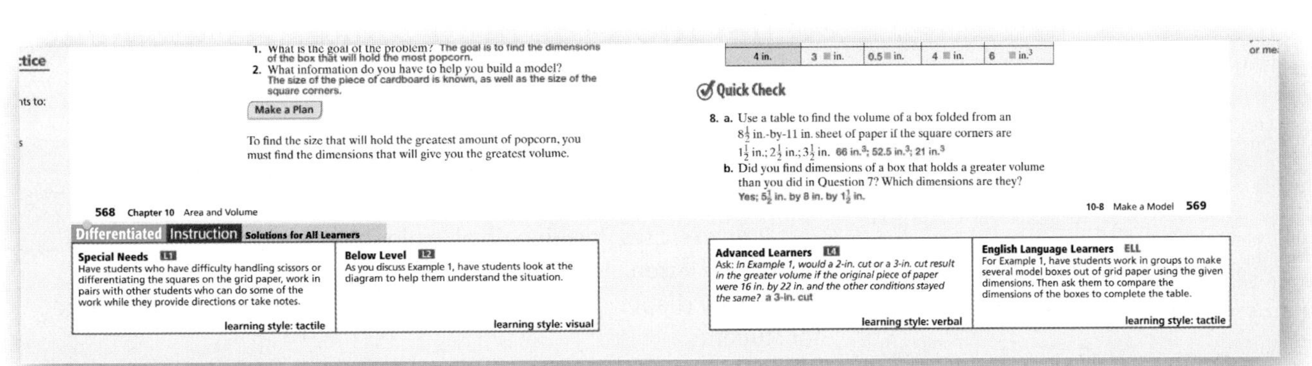

1. What is the goal of the problem? The goal is to find the dimensions of the box that will hold the most popcorn.
2. What information do you have to help you build a model? The size of the piece of cardboard is known, as well as the size of the square corners.

Make a Plan

To find the size that will hold the greatest amount of popcorn, you must find the dimensions that will give you the greatest volume.

568 Chapter 10 Area and Volume

4 in.	3 in.	0.5 in.	4 in.	6 in.^{3}

Quick Check

8. a. Use a table to find the volume of a box folded from an $8\frac{1}{2}$ in.-by-11 in. sheet of paper if the square corners are $1\frac{1}{2}$ in.; $2\frac{1}{2}$ in.; $3\frac{1}{2}$ in. 66 in.^{3}; 52.5 in.^{3}; 21 in.^{3}
   b. Did you find dimensions of a box that holds a greater volume than you did in Question 7? Which dimensions are they? Yes; $5\frac{1}{2}$ in. by 8 in. by $1\frac{1}{2}$ in.

10-8 Make a Model 569

Differentiated Instruction Solutions for All Learners

Special Needs L1
Have students who have difficulty handling scissors or differentiating the squares on the grid paper, work in pairs with other students who can do some of the work while they provide directions or take notes.
learning style: tactile

Below Level L2
As you discuss Example 1, have students look at the diagram to help them understand the situation.
learning style: visual

Advanced Learners L4
Ask: *In Example 1, would a 2-in. cut or a 3-in. cut result in the greater volume if the original piece of paper were 16 in. by 22 in. and the other conditions stayed the same?* a 3-in. cut
learning style: verbal

English Language Learners ELL
For Example 1, have students work in groups to make several model boxes out of grid paper using the given dimensions. Then ask them to compare the dimensions of the boxes to complete the table.
learning style: tactile

Differentiated Instruction teaching notes

These useful notes help you differentiate the lesson for all learners, including Special Needs, Below Level, Advanced, and English Language Learners.

Assessment to Inform Instruction

Assessment is integral to mathematics instruction. Student assessment should occur often and with a variety of different measures. *Prentice Hall Mathematics* provides an ongoing assessment strand that addresses assessment *for* learning and assessment *of* learning. The formative assessment features (before and during instruction) offer a variety of methods for teachers to assess student understanding and inform future instruction. The summative assessment features (after instruction) document student mastery of mathematical concepts and skills and further prepare students for success in today's tests.

Instant Check System™ for Ongoing Assessment

This unique feature of *Prentice Hall Mathematics* ensures that students make progress every day, in every lesson. It helps teachers assess necessary prerequisite skills and monitor student understanding. The Instant Check System™ assessments include:

- Check Your Readiness – Assesses prerequisite skills for each chapter
- Check Skills You'll Need – Assesses prerequisite skills for each lesson
- Quick Check – Assesses student understanding after every example in the book
- Check Your Understanding – Assesses understanding before the homework exercises
- Check Point – Assesses understanding after completing lessons

Progress Monitoring Assessments

This comprehensive teacher support resource contains all the program assessments needed to evaluate student understanding, monitor student progress, and inform future instruction. The following assessments are included:

- Screening Test
- Benchmark Tests
- Test-Taking Strategy Practice
- Quarter, Mid-Course, and Final Tests – regular and below level versions
- Comprehensive Report Forms
- Answers to all of the tests

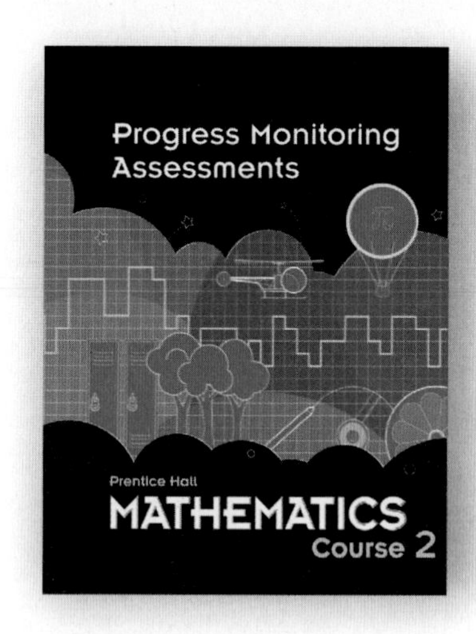

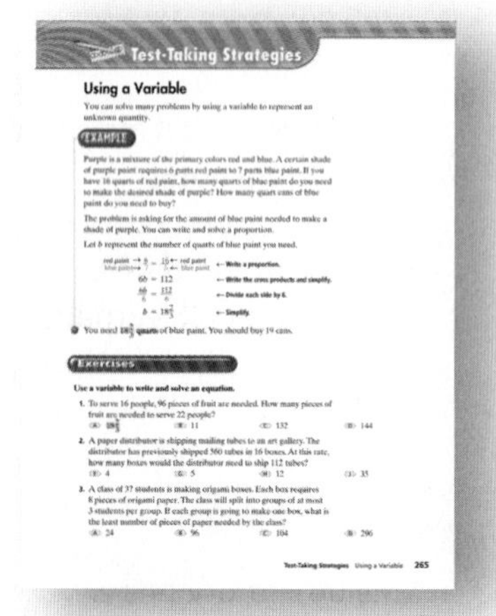

Test-Taking Strategies

Using a Variable

You can solve many problems by using a variable to represent an unknown quantity.

EXAMPLE

Exercises

Use a variable to write and solve an equation.

Test-Taking Strategies Using a Variable 265

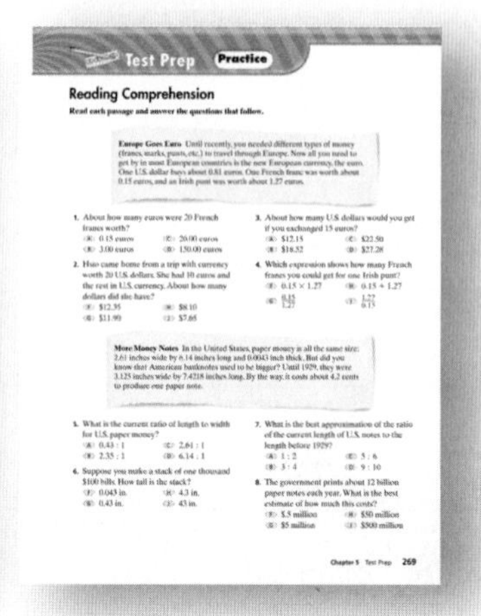

Test Prep Practice

Reading Comprehension

Read each passage and answer the questions that follow.

Chapter 5 Test Prep 269

Preparation for High-Stakes Assessment

Prentice Hall Mathematics recognizes how critical it is for teachers to prepare every student for test success.

The following features help achieve this:

- Separate Test-Taking Strategy lessons focus on specific strategies necessary for test success.
- Test Prep exercises, focusing on all major question types, are included after every lesson.
- After every chapter, Test Prep review pages provide additional practice for students.

Help All Students Become Problem Solvers

One of the major goals of a mathematics program is to develop students' ability to solve problems in class, on assessments, in the context of real-world situations, and outside the classroom. *Prentice Hall Mathematics* helps support this goal by embedding problem-solving instruction in every lesson, providing targeted support for problem-solving strategies throughout the Student Edition, and including sufficient problems to help students practice and reinforce problem solving skills.

Guided Problem Solving Features

These features throughout the Student Edition provide scaffolded support in solving problems. The student walks through how to solve one representative problem, focusing on both the reasoning and the computation that must be done.

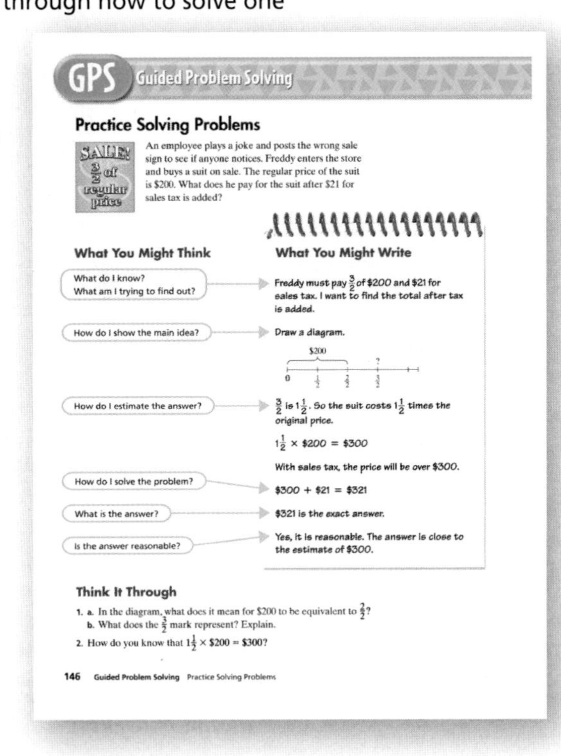

GPS Guided Problem Solving

Practice Solving Problems

An employee plays a joke and posts the wrong sale sign to see if anyone notices. Freddy enters the store and buys a suit on sale. The regular price of the suit is $200. What does he pay for the suit after $21 for sales tax is added?

What You Might Think	What You Might Write
What do I know? What am I trying to find out?	Freddy must pay $\frac{3}{2}$ of \$200 and \$21 for sales tax. I want to find the total after tax is added.
How do I show the main idea?	Draw a diagram.
How do I estimate the answer?	$\frac{3}{2}$ is $1\frac{1}{2}$. So the suit costs $1\frac{1}{2}$ times the original price. $1\frac{1}{2} \times \$200 = \$300$ With sales tax, the price will be over \$300.
How do I solve the problem?	$\$300 + \$21 = \$321$
What is the answer?	\$321 is the exact answer.
Is the answer reasonable?	Yes, it is reasonable. The answer is close to the estimate of \$300.

Think It Through

1. a. In the diagram, what does it mean for $200 to be equivalent to $\frac{2}{2}$?
   b. What does the $\frac{3}{2}$ mark represent? Explain.
2. How do you know that $1\frac{1}{2} \times \$200 = \$300$?

146 Guided Problem Solving Practice Solving Problems

Activities and Activity Labs

Throughout the Student Edition are in-lesson activities and full feature Activity Labs. These provide students an opportunity to complete more in-depth problems and applications of the mathematical content they're learning.

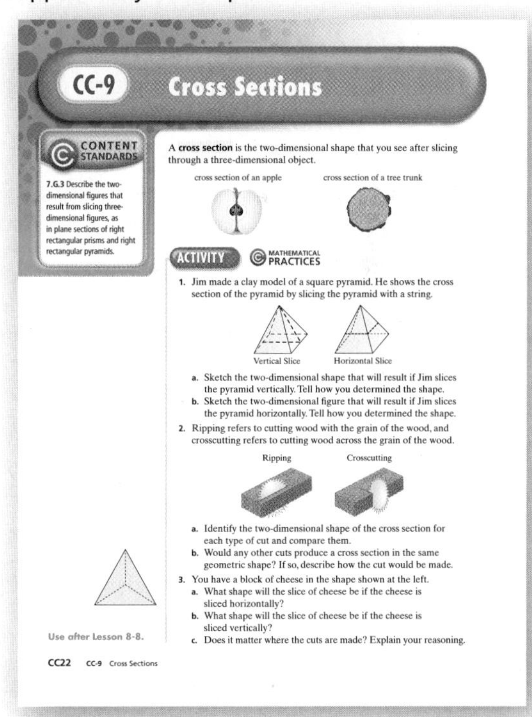

CC-9 Cross Sections

CONTENT STANDARDS

7.G.3 Describe the two-dimensional figures that result from slicing three-dimensional figures, as in plane sections of right rectangular prisms and right rectangular pyramids.

A **cross section** is the two-dimensional shape that you see after slicing through a three-dimensional object.

ACTIVITY MATHEMATICAL PRACTICES

1. Jim made a clay model of a square pyramid. He shows the cross section of the pyramid by slicing the pyramid with a string.

   a. Sketch the two-dimensional shape that will result if Jim slices the pyramid vertically. Tell how you determined the shape.
   b. Sketch the two-dimensional figure that will result if Jim slices the pyramid horizontally. Tell how you determined the shape.
2. Ripping refers to cutting wood with the grain of the wood, and crosscutting refers to cutting wood across the grain of the wood.

   a. Identify the two-dimensional shape of the cross section for each type of cut and compare them.
   b. Would any other cuts produce a cross section in the same geometric shape? If so, describe how the cut would be made.
3. You have a block of cheese in the shape shown at the left.
   a. What shape will the slice of cheese be if the cheese is sliced horizontally?
   b. What shape will the slice of cheese be if the cheese is sliced vertically?
   c. Does it matter where the cuts are made? Explain your reasoning.

Use after Lesson 8-8.

CC22 CC-9 Cross Sections

Problem Solving Practice

Every lesson in the Student Edition includes a comprehensive, leveled exercise set. This provides students the opportunity to solve a variety of problems and apply problem solving strategies on a daily basis. Furthermore, additional problems have been added to the end of each book in the Extra Skill and Word Problem Practice section.

Technology – not simply added, but added value

Take Learning to a New Level

StudentEXPRESS™ CD-ROM

A suite of learning tools to help students study, learn, and succeed in class. An interactive textbook with instructional videos, built-in activities, vocabulary support, and instant feedback assessments make this the most powerful student study tool available. Interactive Text also available online.

Homework Video Tutor and Online Active Math
Homework Video Tutors provide built-in homework help for every lesson. Narrated by real teachers, these engaging interactive tutorials cover the key concepts of each day's lesson. Additionally, Online Active Math interactivities provide an opportunity to explore math concepts.

Use Assessment Technology to Inform Instruction

Exam*View*® 5.0 Assessment Suite

The most powerful test generator available—with the most comprehensive test banks. Create and modify custom-made tests with ease. Access the Math Art Gallery to instantly add math images to your questions. Also-instantly translate any test into Spanish.

MindPoint® Quiz Show

This creative product allows teachers to involve the entire class in a fun, end-of-chapter review game.

Success Tracker™ Online

Personalized for each student with individual assessment, diagnosis, and remediation. Color-coded reports make it easy for teachers to monitor progress.

Superior Planning and Teaching Tools

PRENTICE HALL
TeacherEXPRESS™
Plan · Teach · Assess

TeacherEXPRESS™ CD-ROM—powered by LessonView® planning software

An Interactive Teacher's Edition, LessonView planning software, correlations to national and state standards, instructional tools, plus professional development to help teachers plan, teach, and assess.

PresentationEXPRESS™ CD-ROM with ExamView® QuickTake Presenter

This innovative component includes all the transparencies in interactive, PowerPoint format, making it easier for you to teach and to customize based on your teaching preferences. QuickTake assessments are embedded in every lesson, allowing teachers to quickly and easily monitor student progress.

Worksheets Online

All program worksheets are also conveniently posted online so you and your students can access them anywhere - one more way to make your planning and assignment management easier.

Professional Development that meets your needs

In-Service On Demand 

Now, you have the freedom to access your Prentice Hall in-service training online, anytime, anywhere, at PHSchool.com. This online tutorial library offers in-service training specifically for the Prentice Hall textbook and technology you're teaching with right now.

In-service designed around you!
In-Service On Demand is Prentice Hall's new Web site of in-service training tutorials for your Prentice Hall products. Now you can access the same training for using your Prentice Hall textbook and technology that you would learn in a traditional in-service—from the convenience of your computer!

In-Service for your Prentice Hall program!
The In-Service On Demand library features video-based tutorials to help you maximize the effectiveness of the Prentice Hall program you use. This extensive library is continually being updated with tutorials for Prentice Hall's newest products! Visit the site as often as you like!

In-Service is just a click away!

1. Go to PHSchool.com and click on In-Service On Demand.
2. Select mathematics and then select your *Prentice Hall Mathematics* book.
3. Watch the video-based tutorials and get the in-service training you need when and where it's convenient for you.
4. Download and print PDF tutorial guides on what you've just seen.

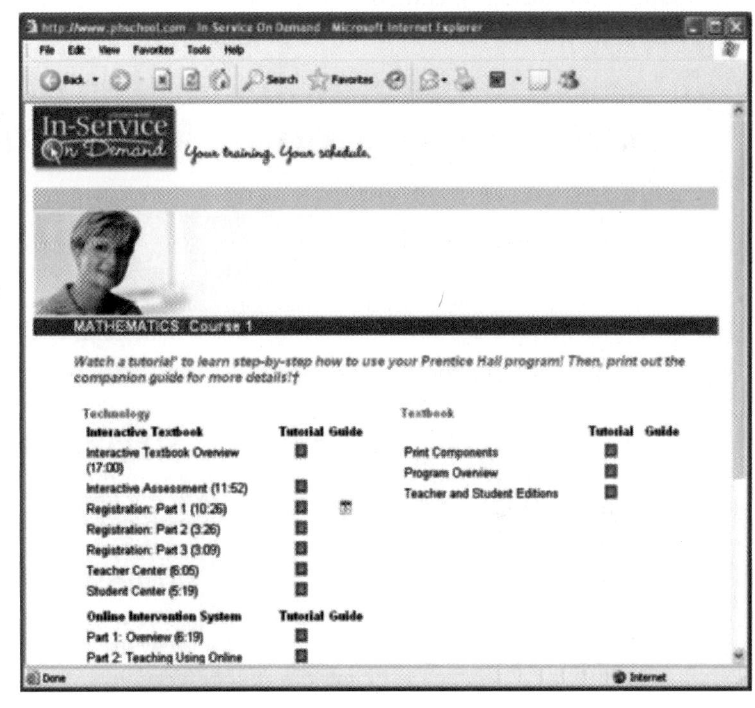

Day-to-Day Professional Development

Math Background in Teacher's Edition
Support instruction at the chapter and lesson level as every Chapter of Prentice Hall Mathematics begins with Math Background related to the content of the chapter in both middle school and high school.

Research-Based and Proven Effective

The stakes for mathematics educators are high. You are expected to raise student achievement. Prentice Hall Mathematics programs are backed by efficacy research to give you the confidence to meet this challenge.

In developing Prentice Hall programs, the use of research is a guiding, central construct. This research was conducted in three phases:

1 Exploratory Needs Assessment

(Quantitative and Qualitative)

Key research events include —

- Teacher interviews
- Classroom observations
- Mail surveys
- Reviews of educational research

2 Formative, Prototype Development and Field Testing

(Quantitative and Qualitative)

Key research events include —

- Field testing of prototypes
- Classroom observations
- Teacher reviews
- Supervisor reviews
- Educator advisory panels

3 Summative, Validation Research

(Experimental and Quasi-Experimental Study Designs & Qualitative Research)

Key research events include —

- Pre-publication learner verification research
- Post-publication efficacy studies
- Classroom observations
- Evaluation of in-market results on standardized tests
- Effect size studies

The facing page contains an example of the latest research carried out for Prentice Hall Middle School Mathematics.

2005 Research Results: Prentice Hall Middle School Mathematics

Independent research confirms Middle School students using Prentice Hall Mathematics achieve greater success in mathematics

Results of independent research indicate that students using Prentice Hall Mathematics Course 2 showed significant improvement, outperforming students using other mathematics programs.The randomized control trial, conducted by PRES Associates, Inc., a national educational evaluation firm with central offices in Jackson,WY, confirmed students using the Prentice Hall curricula showed greater improvement from pre- to post-tests than their counterparts using other programs as measured by two different standardized assessments. Improvement was evident on all mathematics objectives measured. Additionally, the program was especially effective with low-performing students.

The study is part of a multi-year research effort commissioned by the publisher and is one of many slated to evaluate the effectiveness of Prentice Hall's educational materials across disciplines and grade levels. Participants represented a mix of urban and urban-fringe districts with diverse socio-economic, ethnic, and academic backgrounds.

Among the key findings, PRES Associates reported:

- Student performance significantly improved from the beginning of the school year to its end as measured by the TerraNova Basic Multiple Assessment, chosen because it is aligned to national NCTM standards and has proven reliability and validity.
- Students using Prentice Hall Mathematics Course 2 improved to a greater extent in pre- to post-test scores than those using other programs.
- Assessment results suggest that Prentice Hall Mathematics Course 2 is particularly effective with low-performing students, as evidenced by the significant gains in low-performing student test scores.
- Prentice Hall Mathematics Course 2 students reported feeling significantly more comfortable with math than students using other programs.They also reported higher math aspirations (i.e., plans to take advanced math in high school).
- Teachers participating in the study reported that Prentice Hall Mathematics Course 2 provided significantly better assistance than other programs in the following areas (1) individualizing instruction, (2) reinforcing previously taught concepts, (3) providing test preparation, and (4) making connections to real-life.
- Teachers identified many aspects of the Prentice Hall program as effective, including the Guided Problem Solving workbook, Check Skills You'll Need exercises, and Check Understanding exercises.

The study was designed to fully meet the evidence criteria put forth by the What Works Clearinghouse, the Federal agency established in 2002 to provide the educators and the public with a trusted source of scientific evidence of what works in education.This study was designed so that accurate and appropriate inferences could be made regarding the effectiveness of the Prentice Hall Mathematics Course 2 program.

Visit PHSchool.com/MathResearch for the full report and additional research in support of Prentice Hall Mathematics programs.

STANDARDS FOR MATHEMATICAL PRACTICES

The Standards for Mathematical Practice are an important part of the Common Core State Standards. They describe varieties of proficiency that teachers should focus on developing in their students. These practices draw from the NCTM process standards of problem solving, reasoning and proof, communication, representation, and connections, and the strands of mathematical proficiency specified in the National Research Council's report *Adding It Up:* adaptive reasoning, strategic competence, conceptual understanding, procedural fluency, and productive disposition.

For each of the Standards for Mathematical Practice presented in the text that follows, is a explanation of the different features and elements of *Pearson's Prentice Hall Course 2* that help students develop mathematical proficiency.

1 MAKE SENSE OF PROBLEMS AND PERSEVERE IN SOLVING THEM.

Mathematically proficient students start by explaining to themselves the meaning of a problem and looking for entry points to its solution. They analyze givens, constraints, relationships, and goals. They make conjectures about the form and meaning of the solution and plan a solution pathway rather than simply jumping into a solution attempt. They consider analogous problems, and try special cases and simpler forms of the original problem in order to gain insight into its solution. They monitor and evaluate their progress and change course if necessary. Older students might, depending on the context of the problem, transform algebraic expressions or change the viewing window on their graphing calculator to get the information they need. Mathematically proficient students can explain correspondences between equations, verbal descriptions, tables, and graphs or draw diagrams of important features and relationships, graph data, and search for regularity or trends. Younger students might rely on using concrete objects or pictures to help conceptualize and solve a problem. Mathematically proficient students check their answers to problems using a different method, and they continually ask themselves, "Does this make sense?" They can understand the approaches of others to solving complex problems and identify correspondences between different approaches.

The structure of the Pearson's *Prentice Hall Middle Grade Mathematics* program supports students in making sense of problems and in persevering in solving them. The program was designed around a 4-step problem-solving approach, the first step being understand. Students are reminded of the problem-solving plan in the **Problem Solving Handbook** that is found in the front matter of the Student Edition. In every lesson, students have opportunities to make sense of problems with the two **Guided Problem Solving exercises** found within the homework exercises. One of the Guided Problem Solving exercises has an accompanying student workbook page that models the questions students can ask themselves to analyze the givens of the problems, determine a solution plan, and persevere to a solution. Students are also encouraged to check whether the answers they found are reasonable, that is whether they make sense within the context of the problem. Each chapter includes a **Guided Problem Solving activity** and a **DK Problem Solving Application activity,** both of which require that students apply their sense-making and perseverance skills to solve real-world problems.

For examples, see *Prentice Hall Course 2,* pages xxxii–xlix, 24–25, 80–81, 146–147, 192–193, 249–250, 302–303, 359–360, 398–399, 466–467, 496–497, 558–559, 604–605

2 REASON ABSTRACTLY AND QUANTITATIVELY.

Mathematically proficient students make sense of quantities and their relationships in problem situations. They bring two complementary abilities to bear on problems involving quantitative relationships: the ability to *decontextualize*—to abstract a given situation and represent it symbolically and manipulate the representing symbols as if they have a life of their own, without necessarily attending to their referents—and the ability to *contextualize*, to pause as needed during the manipulation process in order to probe into the referents for the symbols involved. Quantitative reasoning entails habits of creating a coherent representation of the problem at hand; considering the units involved; attending to the meaning of quantities, not just how to compute them; and knowing and flexibly using different properties of operations and objects.

Reasoning is another important theme of the Pearson's *Prentice Hall Middle Grades Mathematics Program*. Many of the examples in lessons are application examples in which students are guided to represent the situation symbolically, either numerically or algebraically. Through the solving process, as they manipulate expressions, students are reminded to check back to the problem situation with the **Check for Reasonableness** features. Each lesson ends with a **Check Your Understanding** feature in which students explain their thinking related to the concepts studied in the lesson. Throughout the exercise sets are **Reasoning exercises** that focus students' attention on the structure or meaning of an operation rather than the solution.

For examples, see *Prentice Hall Course 2,* pages xx–xxii, 8, 14, 27, 127, 130–131, 137, 142, 233, 280, 285, 299, 302, 329, 384, 423, 446, 462

3 CONSTRUCT VIABLE ARGUMENTS AND CRITIQUE THE REASONING OF OTHERS.

Mathematically proficient students understand and use stated assumptions, definitions, and previously established results in constructing arguments. They make conjectures and build a logical progression of statements to explore the truth of their conjectures. They are able to analyze situations by breaking them into cases, and can recognize and use counterexamples. They justify their conclusions, communicate them to others, and respond to the arguments of others. They reason inductively about data, making plausible arguments that take into account the context from which the data arose. Mathematically proficient students are also able to compare the effectiveness of two plausible arguments, distinguish correct logic or reasoning from that which is flawed, and—if there is a flaw in an argument—explain what it is. Elementary students can construct arguments using concrete referents such as objects, drawings, diagrams, and actions. Such arguments can make sense and be correct, even though they are not generalized or made formal until later grades. Later, students learn to determine domains to which an argument applies. Students at all grades can listen or read the arguments of others, decide whether they make sense, and ask useful questions to clarify or improve the arguments.

> Consistent with a focus on reasoning and sense-making is a focus on critical reasoning – argumentation and critique of arguments. In *Pearson's Prentice Hall Middle Grades Mathematics* program, students are frequently asked to explain their solutions and the thinking that led them to these solutions. The many **Reasoning exercises** found throughout the program specifically call for students to justify or explain their solutions. In the **More Than One Way** features, students analyze and critique the solution plans and reasoning of two students, each of whom presents a different solution plan for the same problem. The **Error Analysis** exercises provide students additional opportunities to analyze and critique the solution presented to a problem.
>
> For examples, see *Prentice Hall Course 2,* pages 6, 28, 50, 55, 76, 78, 84, 129, 133, 139, 143-144, 151, 182, 190, 202, 218, 231, 240, 246, 248, 261, 312-313, 327, 332, 334, 390, 418, 473, 494, 501, 505, 534, 556, 600

4 MODEL WITH MATHEMATICS.

Mathematically proficient students can apply the mathematics they know to solve problems arising in everyday life, society, and the workplace. In early grades, this might be as simple as writing an addition equation to describe a situation. In middle grades, a student might apply proportional reasoning to plan a school event or analyze a problem in the community. By high school, a student might use geometry to solve a design problem or use a function to describe how one quantity of interest depends on another. Mathematically proficient students who can apply what they know are comfortable making assumptions and approximations to simplify a complicated situation, realizing that these may need revision later. They are able to identify important quantities in a practical situation and map their relationships using such tools as diagrams, two-way tables, graphs, flowcharts and formulas. They can analyze those relationships mathematically to draw conclusions. They routinely interpret their mathematical results in the context of the situation and reflect on whether the results make sense, possibly improving the model if it has not served its purpose.

Students in Pearson's *Prentice Hall Middle Grades Mathematics* program are guided to build mathematical models using equations, graphs, tables, and technology. In many lessons, one of the **Examples** is an application example for which students are shown how the mathematical concept under study can be applied as a model for a real-world problem situation. The stepped-out process makes explicit for students the thinking that can help them apply models to the problem situations presented. The **chapter projects** found at the back of the Student Edition, also provide students with opportunities to apply the mathematics they are learning to solve meaningful, real-life situations.

For examples, see *Prentice Hall Course 2,* pages xx–xxii, 49, 69, 148, 196, 253, 285, 347, 389, 473, 499, 554–555, 587, 624–629

5 USE APPROPRIATE TOOLS STRATEGICALLY.

Mathematically proficient students consider the available tools when solving a mathematical problem. These tools might include pencil and paper, concrete models, a ruler, a protractor, a calculator, a spreadsheet, a computer algebra system, a statistical package, or dynamic geometry software. Proficient students are sufficiently familiar with tools appropriate for their grade or course to make sound decisions about when each of these tools might be helpful, recognizing both the insight to be gained and their limitations. For example, mathematically proficient high school students analyze graphs of functions and solutions generated using a graphing calculator. They detect possible errors by strategically using estimation and other mathematical knowledge. When making mathematical models, they know that technology can enable them to visualize the results of varying assumptions, explore consequences, and compare predictions with data. Mathematically proficient students at various grade levels are able to identify relevant external mathematical resources, such as digital content located on a website, and use them to pose or solve problems. They are able to use technological tools to explore and deepen their understanding of concepts.

Students become fluent in the use of a wide assortment of tools ranging from physical devices, such as rulers, protractors, and even pencil and paper, to technological tools, such as calculators, graphing calculators, and computers. They use various manipulatives and technology tools in the **Activity Labs.** By developing fluency in the use of different tools, students are able to select the appropriate tool(s) to solve a given problem. The **Choose a Method** exercises strengthen students' ability to articulate the difference in use of various tools.

Technology and technology tools, such as graphing calculators, are an integral part of Pearson's *Prentice Hall Middle Grades Mathematics* program and are used in these ways:

- to develop understanding of mathematical concepts;
- to solve problems that would be unapproachable without the use of technology; and
- to build models based on real-world data.

For examples, see *Prentice Hall Course 2,* pages 28–29, 55, 57, 72, 76, 78, 92, 143, 144, 173, 202, 246, 248, 256, 311, 314, 332, 390-391, 460, 473–474, 502, 505, 507, 534, 536, 543, 590, 600, 602

6 ATTEND TO PRECISION.

Mathematically proficient students try to communicate precisely to others. They try to use clear definitions in discussion with others and in their own reasoning. They state the meaning of the symbols they choose, including using the equal sign consistently and appropriately. They are careful about specifying units of measure, and labeling axes to clarify the correspondence with quantities in a problem. They calculate accurately and efficiently, express numerical answers with a degree of precision appropriate for the problem context. In the elementary grades, students give carefully formulated explanations to each other. By the time they reach high school they have learned to examine claims and make explicit use of definitions.

Students are expected to use mathematical terms and symbols with precision. Key terms are highlighted in each lesson and important concepts explained in the **Key Concepts** features. In the **Check Your Understanding** feature, students revisit these key terms and provide explicit definitions or explanations of the terms. For the **Writing in Math** exercises, students are once again expected to provide clear, concise explanations of terms, concepts, or processes or to use specific terminology accurately and precisely. Students are reminded to use appropriate units of measure when working through solutions and accurate labels on axes when making graphs to represent solutions.

For examples, see *Prentice Hall Course 2,* pages 7, 23, 26–30, 42, 47, 58, 61, 78, 94, 108, 139, 148, 150–151, 153, 158, 172, 177, 197, 209, 213, 231, 246, 256, 277, 297, 313, 334, 349, 352, 368, 374–375, 387, 402, 413, 430, 436–437, 449, 462–463, 470, 501, 513, 526, 541, 547, 553, 560–564, 579, 582, 590, 609, 618, 668

7 LOOK FOR AND MAKE USE OF STRUCTURE.

Mathematically proficient students look closely to discern a pattern or structure. Young students, for example, might notice that three and seven more is the same amount as seven and three more, or they may sort a collection of shapes according to how many sides the shapes have. Later, students will see 7×8 equals the well remembered $7 \times 5 + 7 \times 3$, in preparation for learning about the distributive property. In the expression $x^2 + 9x + 14$, older students can see the 14 as 2×7 and the 9 as $2 + 7$. They recognize the significance of an existing line in a geometric figure and can use the strategy of drawing an auxiliary line for solving problems. They also can step back for an overview and shift perspective. They can see complicated things, such as some algebraic expressions, as single objects or as being composed of several objects. For example, they can see $5 - 3(x - y)^2$ as 5 minus a positive number times a square and use that to realize that its value cannot be more than 5 for any real numbers x and y.

Throughout the program, students are encouraged to discern patterns and structure as they look to formulate solution pathways. The **Pattern/Look for a Pattern** exercises explicitly ask students to look for patterns in operations or graphic displays.

For examples, see *Prentice Hall Course 2,* pages xxxv, 21–22, 43–44, 71, 90, 168, 363, 404, 419, 436–440, 441–445, 446–449

8 LOOK FOR AND EXPRESS REGULARITY IN REPEATED REASONING.

Mathematically proficient students notice if calculations are repeated, and look both for general methods and for shortcuts. Upper elementary students might notice when dividing 25 by 11 that they are repeating the same calculations over and over again, and conclude they have a repeating decimal. By paying attention to the calculation of slope as they repeatedly check whether points are on the line through (1, 2) with slope 3, middle school students might abstract the equation $(y - 2)/(x - 1) = 3$. Noticing the regularity in the way terms cancel when expanding $(x - 1)(x + 1)$, $(x - 1)(x^2 + x + 1)$, and $(x - 1)(x^3 + x^2 + x + 1)$ might lead them to the general formula for the sum of a geometric series. As they work to solve a problem, mathematically proficient students maintain oversight of the process, while attending to the details. They continually evaluate the reasonableness of their intermediate results.

Once again, throughout each course and the program as a whole, students are prompted to look for repetition in calculations to devise general methods or shortcuts that can make the problem-solving process more efficient. Students are prompted to look for similar problems they have previously encountered or to generalize results to other problem situations. The **Online Active Math** activities offer students opportunities to notice regularity in the way operations or functions behave by easily inputting different values.

For examples, see *Prentice Hall Course 2,* pages xxxiv–xlix, 32, 40, 70, 92, 137, 142, 170, 216, 229, 253, 280, 294, 337, 342, 405, 423, 462, 469, 498, 520, 554, 559, 606

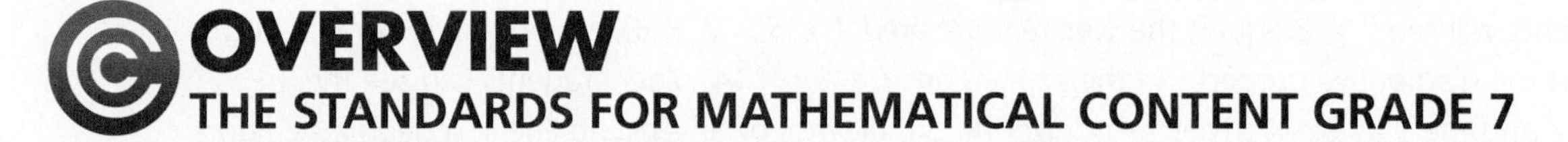

OVERVIEW
THE STANDARDS FOR MATHEMATICAL CONTENT GRADE 7

Two guiding principles framed the development of the Common Core State Standards: that they be **based on evidence and research and build on the strengths of current state standards** and that at each grade level, **the standards be fewer in number, but clearer and more rigorous.** The Overview that follows presents the main areas of emphasis in Grade 7, new concepts or approaches that these standards advance, and the progression of the concepts through Grade 8.

MAIN AREAS OF EMPHASIS

The four main areas of emphasis in Grade 7 are:

- Proportional relationships and applying those relationships to solve problems
- Operations with rational numbers; expressions and linear equations
- Scale drawings and informal geometric constructions; attributes of circles
- Drawing inferences about populations based on samples; concepts of chance

In Grade 7, students extend their study of ratios and rates to develop an understanding of proportionality. They solve problems related to proportional relationships, including a range of percent problems (e.g., discounts, interest, taxes, tips, percent of increase/decrease). They apply the concept of proportionality to scale drawings and solve problems of scale. They graph proportional relationships and develop an informal understanding of slope. They differentiate proportional relationships from other relationships.

Students develop a holistic understanding of number as they express rational numbers in different representations (e.g., fractions, decimals, and percents) and interpret negative numbers in everyday context (e.g., temperature change). Students extend their knowledge of operations and of properties of operations to solve problems with any rational numbers, including negative rational numbers. They write expressions and equations in one variable to solve problems.

Students expand their reasoning about shapes to include circles. They build on their knowledge of geometric attributes and formulas to determine the circumference and area of a circle and the surface area and volume of any polyhedron. They reason about relationships among two-dimensional figures using scale drawings and informal geometric constructions. They solve problems involving angles formed by intersecting lines, and they examine cross-sections of three-dimensional figures.

Students continue their study of statistical thinking by exploring the concept of random sampling. By comparing two data distributions and describing differences between the populations, students begin to understand the importance of representative samples for drawing valid inferences about populations.

NEW APPROACHES TO CONTENT IN GRADE 7

The Common Core State Standards consistently promote a more conceptual and analytical approach to mathematics instruction. In the elementary years, the focus is on interpreting operations to help students develop strategic competence, allowing them to make sense of operations conceptually and to apply this conceptual understanding in different contexts with different forms of numbers. In the middle school years, the focus is on strengthening students' multiplicative reasoning skills and on building a solid foundation of algebraic concepts and skills.

Grade 7 students represent and analyze proportional relationships using different strategies and models. They look to identify the constant of proportionality, using different representations and descriptions of the quantities, and write equations to represent proportional relationships. They represent these relationships graphically and can explain the meaning of any given point on the graph. Students also use these relationships to solve multi-step ratio and percent problems.

The development of algebraic concepts and skills grows from students' understanding of arithmetic operations. Students consistently apply their knowledge of place value, properties of operations, and the inverse relationships between operations (addition and subtraction; multiplication and division) to write and solve algebraic equations. Students manipulate parts of the expression and explore the meaning of the expression when it is rewritten in different forms. This analytic focus helps students to look more fully at the equations and expressions so that they begin to see patterns in the structure. These patterns will be useful when students explore more complex algebraic concepts.

In geometry, students in Grade 7 synthesize their understanding of geometric attributes and properties as they draw geometric shapes with given attributes or condition. They draw from their knowledge base to solve problems involving scale drawings while exploring concepts of similarity.

Grade 7 students continue their study of statistics. After working with measures of center and measures of variability (interquartile range and deviation) in Grade 6, students in Grade 7 analyze the data in light of the sample population from which the data were collected. Students consider the effect of the sampling on the data collected. They understand the importance of random sampling when drawing inferences or making generalizations about a population based on the data collected.

Students undertake a comprehensive study of probability in Grade 7, students' one and only encounter with probability concepts in their elementary and middle years. Students develop probability models from which they find theoretical probability of events. They compare the theoretical probabilities to observed frequencies (experimental probabilities). They investigate both simple and compound events and represent theoretical probabilities using different models.

IMPORTANT PROGRESSIONS ACROSS GRADES

In Grade 6, students expanded their working knowledge of numbers to the system of rational numbers, including positive and negative rational numbers. In Grade 7, students gain an understanding of terminating and repeating decimals as well as complex fractions, leading to an exploration of irrational numbers in Grade 8. They perform operations with any rational numbers, including negative rational numbers (e.g., integers). They can explain the additive inverse and its value when subtracting negative rational numbers. As they carry out operations, students interpret the solutions by describing real-world contexts.

Students began a study of algebraic expressions in Grade 6. They evaluated expressions and manipulated the parts of the expressions, describing each part in mathematical terms and the relationship of each part within the expression. In Grade 7, students extend their analysis of expressions to explain the relationships among the quantities that are revealed when an expression is rewritten. They expand their understanding of expressions to solving multi-step algebraic equations. In Grade 8, students solve linear equations with more than two steps as well as pairs of simultaneous linear equations. Students continue their exploration of inequalities, solving word problems that lead to inequalities and graphing the solutions on number lines.

In Grade 7, students build on their knowledge of ratio and rates from Grade 6 to explore proportional relationships. They solve a wide variety of percent problems, including those involving discounts, interest, taxes, tips, and percent of increase or decrease. They also graph proportional relationships and understand unit rate as a measure of the steepness (slope) of the related line. In Grade 8, students compare two different proportional relationships and formally study slope of linear equations.

In Grade 6, students explored statistical variability, by findings measures of center (mean, median, mode) and measures of variability (interquartile range and mean absolute deviation) to describe a data set. Grade 7 students expand their study of statistics to work with samples to draw inferences about populations and to compare two populations. In Grade 8, students investigate patterns with bivariate data. They look at graphical representations of the data (scatter plots) to describe the pattern shown (if any).

Students began their study of angles and angle measures in Grade 4. In Grade 7, students solve a variety of problems involving supplementary, complementary, vertical, and adjacent angles and in Grade 8, students examine the angle sum and exterior angle of triangles, the angles created when parallel lines are cut by a transversal, and angle relationships in congruent and similar figures.

Students in Grade 7 extend previous work with perimeter and area to include circumference and area of circles. Grade 6 students find the surface area of pyramids and prisms as well as the volume of right rectangular prisms. This work is expanded in Grade 7 to include other polyhedra. In Grade 8, students find the volume of cones, cylinders, and spheres. In preparation for their work with congruence, similarity, and the Pythagorean Theorem in Grade 8, students in Grade 7 work with scale drawings and informal geometric constructions.

WHAT'S DIFFERENT?

The Common Core State Standards identify a limited number of topics at each grade level, allowing enough time for students to achieve fluency, if not mastery of these concepts. The subsequent year of study builds on the concepts of the previous year. While some review of topics from earlier grades is appropriate and encouraged, the CCSS writers argue that re-teaching of these topics should not be needed.

The Common Core State Standards in Grades K–7 are developed to ensure that students have a strong foundation to being study of algebra in Grade 8. Much of the work with number, which begins with counting and cardinality at the kindergarten level, is completed in Grade 7. Students are expected to achieve fluency with operations with most rational numbers by the end of Grade 7.

Certain topics that have often been part of the Grade 7 curriculum are not included in the CCSS. Among the most noticeable are a pared-down set of geometry, and a different focus in the data analysis standards.

Number and Operations Students are expected to have achieved fluency with operations involving whole numbers, decimals, and fractions, so these are not part of the standards.

Geometry The CCSS introduce the study of congruence and transformations in Grade 8.

CORRELATION OF STANDARDS FOR MATHEMATICAL CONTENT

PRENTICE HALL COURSE 2

The following shows the alignment of *Prentice Hall Course 2 Common Core Edition* ©2012 to the Grade 7 Common Core State Standards for Mathematics. The Lessons and Activity Labs in this book provide complete coverage of the Grade 7 Common Core State Standards for Mathematics.

Standards for Mathematical Content		PH Course 2 Common Core Edition ©2012
RATIOS AND PROPORTIONAL RELATIONSHIPS		
Analyze proportional relationships and use them to solve real-world and mathematical problems.		
7.RP.1	Compute unit rates associated with ratios of fractions, including ratios of lengths, areas and other quantities measured in like or different units.	5-1, 5-2, 5-4, 5-5, CC-7, Activity Labs 5-4a, 5-5a
7.RP.2	Recognize and represent proportional relationships between quantities.	5-3, 5-4, 5-5 Activity Lab 5-4a
7.RP.2.a	Decide whether two quantities are in a proportional relationship, e.g., by testing for equivalent ratios in a table or graphing on a coordinate plane and observing whether the graph is a straight line through the origin.	5-3, CC-10
7.RP.2.b	Identify the constant of proportionality (unit rate) in tables, graphs, equations, diagrams, and verbal descriptions of proportional relationships.	5-3, CC-11
7.RP.2.c	Represent proportional relationships by equations.	5-4, CC-11, GPS p. 249
7.RP.2.d	Explain what a point (x, y) on the graph of a proportional relationship means in terms of the situation, with special attention to the points (0, 0) and $(1, r)$ where r is the unit rate.	10-2, 10-3, CC-10
7.RP.3	Use proportional relationships to solve multistep ratio and percent problems. Examples: simple interest, tax, markups and markdowns, gratuities and commissions, fees, percent increase and decrease, percent error.	6-7, 6-8. 9-7

	Standards for Mathematical Content	PH Course 2 Common Core Edition ©2012
THE NUMBER SYSTEM		
Apply and extend previous understandings of operations with fractions to add, subtract, multiply, and divide rational numbers.		
7.NS.1	Apply and extend previous understandings of addition and subtraction to add and subtract rational numbers; represent addition and subtraction on a horizontal or vertical number line diagram.	1-7, 3-2, 3-3, CC-1
7.NS.1.a	Describe situations in which opposite quantities combine to make 0.	1-7, Activity Lab 1-7a
7.NS.1.b	Understand $p + q$ as the number located a distance $\lvert q \rvert$ from p, in the positive or negative direction depending on whether q is positive or negative. Show that a number and its opposite have a sum of 0 (are additive inverses). Interpret sums of rational numbers by describing real-world contexts.	1-6, 1-7
7.NS.1.c	Understand subtraction of rational numbers as adding the additive inverse, $p - q = p + (-q)$. Show that the distance between two rational numbers on the number line is the absolute value of their difference, and apply this principle in real-world contexts.	1-7, CC-1
7.NS.1.d	Apply properties of operations as strategies to add and subtract rational numbers.	1-7, CC-1, Activity Lab 1-7a
7.NS.2	Apply and extend previous understandings of multiplication and division and of fractions to multiply and divide rational numbers.	1-3, 1-4, 1-8, 2-7, 3-4, 3-5, CC-2, CC-3
7.NS.2.a	Understand that multiplication is extended from fractions to rational numbers by requiring that operations continue to satisfy the properties of operations, particularly the distributive property, leading to products such as $(-1)(-1) = 1$ and the rules for multiplying signed numbers. Interpret products of rational numbers by describing real-world contexts.	1-8, CC-2
7.NS.2.b	Understand that integers can be divided, provided that the divisor is not zero, and every quotient of integers (with non-zero divisor) is a rational number. If p and q are integers, then $-(p/q) = (-p)/q = p/(-q)$. Interpret quotients of rational numbers by describing real-world contexts.	1-8, CC-3
7.NS.2.c	Apply properties of operations as strategies to multiply and divide rational numbers.	CC-2, CC-3
7.NS.2.d	Convert a rational number to a decimal using long division; know that the decimal form of a rational number terminates in 0s or eventually repeats.	2-6, 2-7
7.NS.3	Solve real-world and mathematical problems involving the four operations with rational numbers. NOTE: Computations with rational numbers extend the rules for manipulating fractions to complex fractions.	1-3, 1-4, 1-8, 3-4, 3-5

	Standards for Mathematical Content	PH Course 2 Common Core Edition ©2012
EXPRESSIONS AND EQUATIONS		
Use properties of operations to generate equivalent expressions.		
7.EE.1	Apply properties of operations as strategies to add, subtract, factor, and expand linear expressions with rational coefficients.	CC-4
7.EE.2	Understand that rewriting an expression in different forms in a problem context can shed light on the problem and how the quantities in it are related.	6-4, 6-6, 6-7, 9-8 , Activity Lab 9-8b
7.EE.3	Solve multi-step real-life and mathematical problems posed with positive and negative rational numbers in any form (whole numbers, fractions, and decimals), using tools strategically. Apply properties of operations to calculate with numbers in any form; convert between forms as appropriate; and assess the reasonableness of answers using mental computation and estimation strategies.	1-2, 1-3, 1-4, 2-6, 3-1, 3-2, 3-3, 3-4, 3-5, 6-2, 6-3, 6-4, 6-5, 6-6, 6-7, 6-8
7.EE.4	Use variables to represent quantities in a real-world or mathematical problem, and construct simple equations and inequalities to solve problems by reasoning about the quantities.	4-1, 4-2
7.EE.4 .a	Solve word problems leading to equations of the form $px - q = r$ and $p(x + q) = r$, where p, q, and r are specific rational numbers. Solve equations of these forms fluently. Compare an algebraic solution to an arithmetic solution, identifying the sequence of the operations used in each approach.	4-5, 4-6, CC-5
7.EE.4.b	Solve word problems leading to inequalities of the form $px + q > r$ or $px + q < r$, where p, q, and r are specific rational numbers. Graph the solution set of the inequality and interpret it in the context of the problem.	CC-6

	Standards for Mathematical Content	PH Course 2 Common Core Edition ©2012
GEOMETRY		
Draw, construct, and describe geometrical figures and describe the relationships between them.		
7.G.1	Solve problems involving scale drawings of geometric figures, including computing actual lengths and areas from a scale drawing and reproducing a scale drawing at a different scale.	5-5, 5-6, Activity Lab 5-6a
7.G.2	Draw (freehand, with ruler and protractor, and with technology) geometric shapes with given conditions. Focus on constructing triangles from three measures of angles or sides, noticing when the conditions determine a unique triangle, more than one triangle, or no triangle.	CC-8
7.G.3	Describe the two-dimensional figures that result from slicing three dimensional figures, as in plane sections of right rectangular prisms and right rectangular pyramids.	CC-9
Solve real-life and mathematical problems involving angle measure, area, surface area, and volume.		
7.G.4	Know the formulas for the area and circumference of a circle and use them to solve problems; give an informal derivation of the relationship between the circumference and area of a circle.	8-5, Activity Lab 8-5a
7.G.5	Use facts about supplementary, complementary, vertical, and adjacent angles in a multi-step problem to write and solve simple equations for an unknown angle in a figure.	7-2
7.G.6	Solve real-world and mathematical problems involving area, volume and surface area of two- and three-dimensional objects composed of triangles, quadrilaterals, polygons, cubes, and right prisms.	8-2, 8-3, 8-4, 8-9, 8-10

Standards for Mathematical Content		PH Course 2 Common Core Edition ©2012
STATISTICS AND PROBABILITY		
Use random sampling to draw inferences about a population.		
7.SP.1	Understand that statistics can be used to gain information about a population by examining a sample of the population; generalizations about a population from a sample are valid only if the sample is representative of that population. Understand that random sampling tends to produce representative samples and support valid inferences.	11-4
7.SP.2	Use data from a random sample to draw inferences about a population with an unknown characteristic of interest. Generate multiple samples (or simulated samples) of the same size to gauge the variation in estimates or predictions.	11-5, CC-13
Draw informal comparative inferences about two populations.		
7.SP.3	Informally assess the degree of visual overlap of two numerical data distributions with similar variabilities, measuring the difference between the centers by expressing it as a multiple of a measure of variability.	CC-12
7.SP.4	Use measures of center and measures of variability for numerical data from random samples to draw informal comparative inferences about two populations.	Activity Lab 1-10b

	Standards for Mathematical Content	PH Course 2 Common Core Edition ©2012
Investigate chance processes and develop, use, and evaluate probability models.		
7.SP.5	Understand that the probability of a chance event is a number between 0 and 1 that expresses the likelihood of the event occurring. Larger numbers indicate greater likelihood. A probability near 0 indicates an unlikely event, a probability around 1/2 indicates an event that is neither unlikely nor likely, and a probability near 1 indicates a likely event.	12-1
7.SP.6	Approximate the probability of a chance event by collecting data on the chance process that produces it and observing its long-run relative frequency, and predict the approximate relative frequency given the probability.	12-1, 12-2, Activity Lab 12-2a
7.SP.7	Develop a probability model and use it to find probabilities of events. Compare probabilities from a model to observed frequencies; if the agreement is not good, explain possible sources of the discrepancy.	12-2, Activity Lab 12-2a,
7.SP.7.a	Develop a uniform probability model by assigning equal probability to all outcomes, and use the model to determine probabilities of events.	12-1
7.SP.7.b	Develop a probability model (which may not be uniform) by observing frequencies in data generated from a chance process.	12-2, Activity Lab 12-2a
7.SP.8	Find probabilities of compound events using organized lists, tables, tree diagrams, and simulation.	12-3, 12-4
7.SP.8.a	Understand that, just as with simple events, the probability of a compound event is the fraction of outcomes in the sample space for which the compound event occurs.	12-4
7.SP.8.b	Represent sample spaces for compound events using methods such as organized lists, tables and tree diagrams. For an event described in everyday language (e.g., "rolling double sixes"), identify the outcomes in the sample space which compose the event.	12-3, 12-4
7.SP.8.c	Design and use a simulation to generate frequencies for compound events.	12-4, CC-14

PACING
FOR A COMMON CORE CURRICULUM WITH

PRENTICE HALL COURSE 2

This pacing chart can help you plan your course as you transition to a curriculum based on the Common Core State Standards (CCSS). The Chart indicates the Standard(s) for Mathematical Content that each lesson addresses and proposes pacing for each chapter. *Prentice Hall Course* 2 provides comprehensive coverage of all of the Common Core State Standards for Grade 7.

The suggested number of days for each chapter is based on a traditional 45-minute class period and on a 90-minute block period. The total of 146 days of instruction leaves time for review and enrichment lessons, additional activity labs, assessments, and projects.

✔ **Content to meet the Grade 7 Common Core State Standards**
✔ **Review and Intervention for mastery of prior standards**
✔ **Content for Enrichment to prepare for future study**

			Differentiated Instruction	
		Standards of Mathematical Content	**Core**	**Advanced**
	Chapter 1 Decimals and Integers	**Traditional 12 days Block 6 days**		
1-1	Using Estimation Strategies	Reviews 4.OA.3	✔	
1-2	Adding and Subtracting Decimals	7.EE.3	✔	✔
1-3	Multiplying Decimals	7.NS.2, 7.NS.3, 7.EE.3	✔	✔
1-4	Dividing Decimals	7.NS.2, 7.NS.3, 7.EE.3	✔	✔
1-5	Measuring in Metric Units	Reviews 5.MD.1	✔	
1-6	Comparing and Ordering Integers	7.NS.1.b	✔	✔
1-7a	Activity Lab: Modeling Integer Addition and Subtraction	7.NS.1.a, 7.NS.1.d	✔	✔
1-7	Adding and Subtracting Integers	7.NS.1, 7.NS.1.a, 7.NS.1.b, 7.NS.1.c, 7.NS.1.d	✔	✔
1-8	Multiplying and Dividing Integers	7.NS.2, 7.NS.2.a, 7.NS.2.b, 7.NS.3	✔	✔
1-9	Order of Operations and the Distributive Property	Reviews 6.EE.3	✔	
1-10	Mean, Median, Mode, and Range	Reviews 6.SP.5	✔	✔
1-10b	Mean, Median, Mode, and Range	7.SP.4	✔	✔
	Chapter 2 Exponents, Factors, and Fractions	**Traditional 20 days Block 10 days**		
2-1	Exponents and Order of Operations	Reviews 6.EE.1	✔	
2-2	Prime Factorization	Reviews 6.NS.4	✔	
2-3	Simplifying Fractions	Reviews 5.NF.1	✔	
2-4	Comparing and Ordering Fractions	Reviews 4.NF.2	✔	
2-5	Mixed Numbers and Improper Fractions	Reviews 5.NF.1, 5.NF. 2	✔	
2-6	Fractions and Decimals	7.NS.2.d, 7.EE.3	✔	✔
2-7	Rational Numbers	7.NS.2, 7.NS.2.d	✔	✔
2-8	Scientific Notation	Prepares for 8.EE.4		✔

✔ Content to meet the Grade 7 Common Core State Standards.
✔ Review and Intervention for mastery of prior standards
✔ Content for Enrichment to prepare for future study

		Standards of Mathematical Content	Differentiated Instruction: Core	Differentiated Instruction: Advanced
	Chapter 3 Operations with Fractions	**Traditional 12 days Block 6 days**		
3-1	Estimating with Fractions and Mixed Numbers	7.EE.3	✔	✔
3-2	Adding and Subtracting Fractions	7.NS.1, 7.EE.3	✔	✔
CC-1	Addition and Subtraction of Rational Numbers	7.NS.1, 7.NS.1.c, 7.NS.1.d	✔	✔
3-3	Adding and Subtracting Mixed Numbers	7.NS.1, 7.EE.3	✔	✔
CC-2	Multiplication of Rational Numbers	7.NS.2, 7.NS.2.a, 7.NS.2.b, 7.NS.2.c	✔	✔
3-4	Multiplying Fractions and Mixed Numbers	7.NS.2, 7.NS.3, 7.EE.3	✔	✔
3-5	Dividing Fractions and Mixed Numbers	7.NS.2, 7.NS.3, 7.EE.3	✔	✔
CC-3	Division of Rational Numbers	7.NS.2, 7.NS.2.a, 7.NS.2.b, 7.NS.2.c	✔	✔
3-6	Changing Units in the Customary System	Reviews 5.MD.1	✔	
3-7	Precision			✔
3-8	The Distributive Property	Reviews 6.EE.2, 6.EE.2.b, 6.EE.3	✔	✔
	Chapter 4 Number Theory and Fractions	**Traditional 12 days Block 6 days**		
4-1	Evaluating and Writing Algebraic Expressions	7.EE.4	✔	✔
CC-4	Simplifying Expressions	7.EE.1	✔	✔
4-2	Using Number Sense to Solve Equations	7.EE.4	✔	✔
4-3	Solving Equations by Adding or Subtracting	Reviews 6.EE.7	✔	✔
4-4	Solving Equations by Multiplying or Dividing	Reviews 6.EE.7	✔	✔
4-5	Exploring Two-Step Problems	7.EE.4.a	✔	✔
4-6	Solving Two-Step Equations	7.EE.4.a	✔	✔
CC-5	Solving Equations of the Form $p(x + q) = r$	7.EE.4.a	✔	✔
4-7	Graphing and Writing Inequalities	Reviews 6.EE.8	✔	
4-8	Solving Inequalities by Adding and Subtracting	Prepares for 7.EE.4.b	✔	✔
4-9	Solving Inequalities by Multiplying and Dividing	Prepares for 7.EE.4.b	✔	✔
CC-6	Solving Inequalities	7.EE.4.b	✔	✔
	Chapter 5 Ratios, Rates, and Proportions	**Traditional 16 days Block 8 days**		
5-1	Ratios	7.RP.1	✔	✔
5-2	Unit Rates and Proportional Reasoning	7.RP.1	✔	✔
5-2	Extension Using Conversion Factors	7.RP.1	✔	✔
CC-7	Unit Rates and Ratios of Fractions	7.RP.1	✔	✔
5-3	Proportions	7.RP.2, 7.RP.2.a, 7.RP.2.b	✔	✔
5-4a	Activity Lab Exploring Similar Figures	7.RP.1	✔	✔
5-4	Solving Proportions	7.RP.1, 7.RP.2, 7.RP.2.c	✔	✔
5-5a	Activity Lab: Exploring Similar Figures	7.RP.1	✔	✔
5-5	Using Similar Figures	7.RP.1, 7.RP.2, 7.G.1	✔	✔
5-6	Maps and Scale Drawings	7.G.1	✔	✔

- ✔ Content to meet the Grade 6 Common Core State Standards
- ✔ Review and Intervention for mastery of prior standards
- ✔ Content for Enrichment to prepare for future study

		Standards of Mathematical Content	Differentiated Instruction: Core	Differentiated Instruction: Advanced
	Chapter 6 Percents	**Traditional 14 days Block 7 days**		
6-1	Understanding Percents	Reviews 6.RP.3.c	✔	
6-2	Percents, Fractions, and Decimals	7.EE.3	✔	✔
6-3	Percents Greater Than 100% or Less Than 1%	7.EE.3	✔	✔
6-4	Finding a Percent of a Number	7.EE.3	✔	✔
6-5	Solving Percent Problems Using Proportions	7.EE.3	✔	✔
6-6	Solving Percent Problems Using Equations	7.EE.3	✔	✔
6-7	Applications of Percent	7.RP.3, 7.EE.3	✔	✔
6-8	Finding Percent of Change	7.RP.3, 7.EE.3	✔	✔
	Chapter 7 Geometry	**Traditional 4 days Block 2 days**		
7-1	Lines and Planes	Reviews 4.G.1	✔	
7-2	Identifying and Classifying Angles	7.G.5	✔	✔
7-3	Triangles	Reviews 5.G.4	✔	
CC-8	Drawing Triangles	7.G.2	✔	✔
7-4	Quadrilaterals and Other Polygons	Reviews 5.G.4	✔	
7-5	Congruent Figures	Prepares for 8.G.2		✔
7-6	Circles	Prepares for 7.G.4	✔	
7-7	Circle Graphs		✔	✔
7-8	Constructions		✔	✔
	Chapter 8 Measurement	**Traditional 15 days Block 8 days**		
8-1	Estimating Perimeter and Area	Reviews 6.G.1	✔	
8-2	Area of a Parallelogram	7.G.6	✔	✔
8-3	Perimeter and Area of a Triangle	7.G.6	✔	✔
8-4	Areas of Other Figures	7.G.6	✔	✔
8-5a	Activity Lab: Modeling a Circle	7.G.4	✔	✔
8-5	Circumference and Area of a Circle	7.G.4	✔	✔
8-6	Square Roots and Irrational Numbers	Prepares for 8.NS.1		✔
8-7	The Pythagorean Theorem	Prepares for 8.G.6		✔
8-8	Three-Dimensional Figures	Reviews 5.MD.5.a	✔	
CC-9	Cross Sections	7.G.3	✔	✔
8-9	Surface Areas of Prisms and Cylinders	7.G.6	✔	✔
8-10	Volumes of Prisms and Cylinders	7.G.6	✔	✔
	Chapter 9 Patterns and Rules	**Traditional 7 days Block 4 days**		
9-1	Patterns and Graphs	Reviews 5.OA.3	✔	
9-2	Number Sequences	Reviews 5.OA.3	✔	
9-3	Patterns and Tables	Prepares for 8.SP.4	✔	✔
9-4	Function Rules	Prepares for 8.F.1		✔

✔ Content to meet the Grade 6 Common Core State Standards.
✔ Review and Intervention for mastery of prior standards
✔ Content for Enrichment to prepare for future study

Lesson	Title	Standards of Mathematical Content	Differentiated Instruction	
			Core	Advanced
9-5	Using Tables, Rules, and Graphs	Prepares for 8.F.2		✔
9-6	Interpreting Graphs	Prepares for 8.F.5		✔
9-7	Simple and Compound Interest	7.RP.3	✔	✔
9-8	Transforming Formulas	7.EE.2	✔	✔
9-8b	Activity Lab: More About Formulas	7.EE.2	✔	✔
	Chapter 10 Graphing on the Coordinate Plane	**Traditional 8 days Block 4 days**		
10-1	Graphing Points in Four Quadrants	Reviews 6.NS.8	✔	
10-2	Graphing Linear Equations	7.RP.2.d	✔	✔
10-3	Finding the Slope of a Line	7.RP.2.d	✔	✔
CC-10	Graphs and Proportional Relationships	7.RP.2.a, 7.RP.2.d	✔	✔
CC-11	Constant of Proportionality	7.RP.2.b, 7.RP.2.c	✔	✔
10-4	Graphing Nonlinear Relationships	Prepares for 8.F.3, 8.F.4		✔
10-5	Translations	Prepares for 8.G.1		✔
10-6	Line Symmetry and Reflections	Prepares for 8.G.1		✔
10-7	Rotational Symmetry and Rotations	Prepares for 8.G.1		✔
	Chapter 11 Displaying and Analyzing Data	**Traditional 12 days Block 6 days**		
11-1	Reporting Frequency	Reviews 6.SP.4	✔	
11-2	Spreadsheets and Data Displays	Reviews 6.SP.4	✔	
11-3	Stem-and-Leaf Plots		✔	✔
CC-12	Data Variability	7.SP.3	✔	✔
11-4	Random Samples and Surveys	7.SP.1	✔	✔
11-5	Estimating Population Size	7.SP.2	✔	✔
CC-13	Random Samples	7.SP.2	✔	✔
11-6	Using Data to Persuade		✔	✔
11-7	Exploring Scatter Plots	Prepares for 8.SP.1		✔
	Chapter 12 Using Probability	**Traditional 14 days Block 7 days**		
12-1	Probability	7.SP.5, 7.SP.6, 7.SP.7.a	✔	✔
12-2a	Activity Lab: Exploring Experimental Probability	7.SP.7, 7.SP.7.b	✔	✔
12-2	Experimental Probability	7.SP.6, 7.SP.7, 7.SP.7.b	✔	✔
12-2b	Activity Lab: Random Numbers	7.SP.7, 7.SP.7.b	✔	✔
12-3	Sample Spaces	7.SP.8, 7.SP.8.b	✔	✔
12-4	Compound Events	7.SP.8, 7, SP.8.a, 7.SP.8.b, 7.SP.8.c	✔	✔
CC-14	Simulating Compound Events	7.SP.8.c	✔	✔
12-5	Permutations			✔
12-6	Combinations			✔

Using Your Book for Success

Welcome to *Prentice Hall Course 2*. There are many features built into the daily lessons of this text that will help you learn the important skills and concepts you will need to be successful in this course. Look through the following pages for some study tips that you will find useful as you complete each lesson.

Getting Ready to Learn

Check Your Readiness

Complete the *Check Your Readiness* exercises to see what topics you may need to review before you begin the chapter.

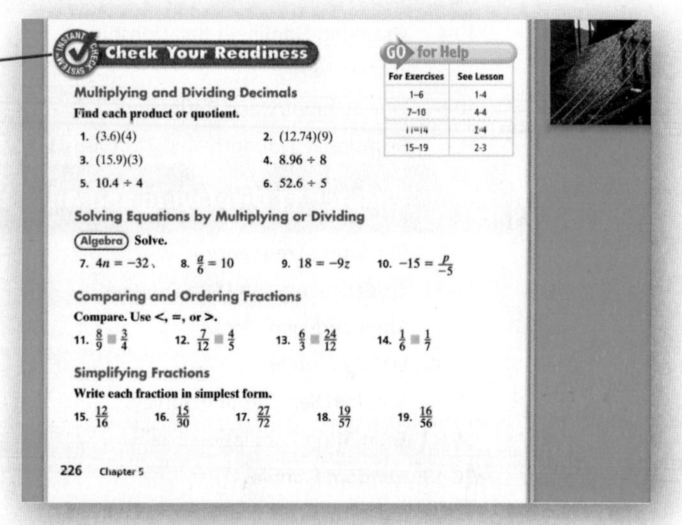

Check Your Readiness

For Exercises	See Lesson
1–6	1-4
7–10	4-4
11–14	2-4
15–19	2-3

Multiplying and Dividing Decimals

Find each product or quotient.

1. (3.6)(4)
2. (12.74)(9)
3. (15.9)(3)
4. $8.96 \div 8$
5. $10.4 \div 4$
6. $52.6 \div 5$

Solving Equations by Multiplying or Dividing

Algebra Solve.

7. $4n = -32$
8. $\frac{a}{6} = 10$
9. $18 = -9z$
10. $-15 = \frac{p}{-5}$

Comparing and Ordering Fractions

Compare. Use <, =, or >.

11. $\frac{8}{9}$ ■ $\frac{3}{4}$
12. $\frac{7}{12}$ ■ $\frac{4}{5}$
13. $\frac{6}{3}$ ■ $\frac{24}{12}$
14. $\frac{1}{6}$ ■ $\frac{1}{7}$

Simplifying Fractions

Write each fraction in simplest form.

15. $\frac{12}{16}$
16. $\frac{15}{30}$
17. $\frac{27}{72}$
18. $\frac{19}{57}$
19. $\frac{16}{56}$

226 Chapter 5

Check Skills You'll Need

Complete the *Check Skills You'll Need* exercises to make sure you have the skills needed to successfully learn the concepts in the lesson.

New Vocabulary

New Vocabulary is listed for each lesson, so you can pre-read the text. As each term is introduced, it is highlighted in yellow.

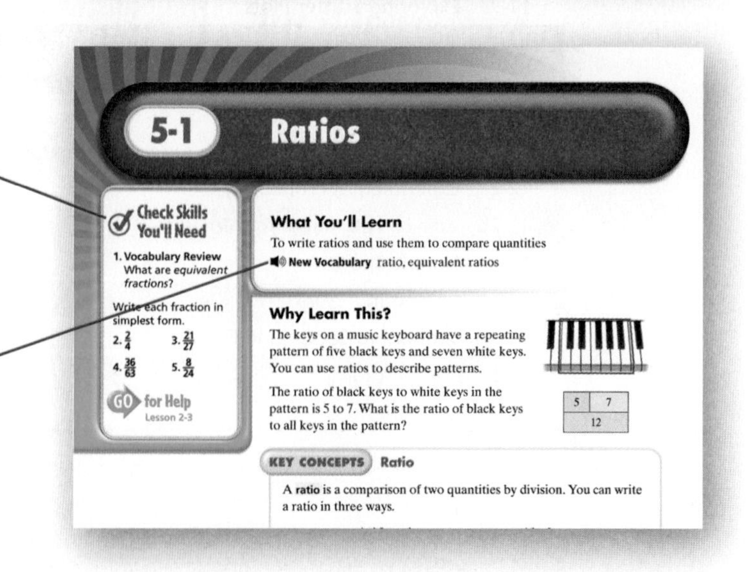

5-1 Ratios

Check Skills You'll Need

1. Vocabulary Review What are *equivalent fractions*?

Write each fraction in simplest form.

2. $\frac{2}{4}$
3. $\frac{21}{27}$
4. $\frac{36}{63}$
5. $\frac{8}{24}$

GO for Help Lesson 2-3

What You'll Learn

To write ratios and use them to compare quantities

New Vocabulary ratio, equivalent ratios

Why Learn This?

The keys on a music keyboard have a repeating pattern of five black keys and seven white keys. You can use ratios to describe patterns.

The ratio of black keys to white keys in the pattern is 5 to 7. What is the ratio of black keys to all keys in the pattern?

KEY CONCEPTS Ratio

A **ratio** is a comparison of two quantities by division. You can write a ratio in three ways.

Built-In Help

Go for Help

Look for the green labels throughout your book that tell you where to "Go" for help. You'll see this built-in help in the lessons and in the homework exercises.

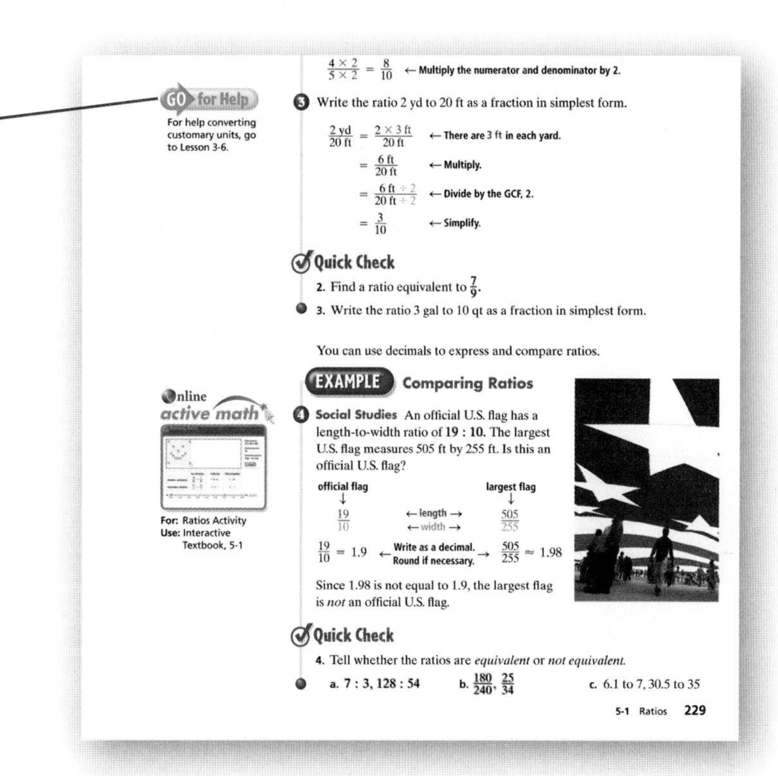

Video Tutor Help

Go online to see engaging videos to help you better understand important math concepts.

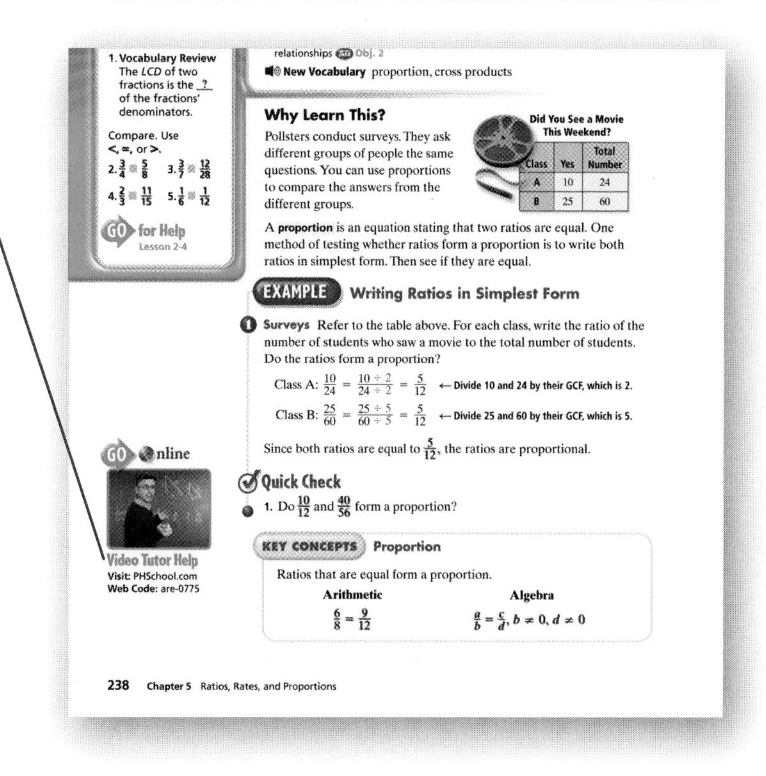

Using Your Book for Success

Understanding the Mathematics

Quick Check

Every lesson includes numerous examples, each followed by a *Quick Check* question that you can do on your own to see if you understand the skill being introduced. Check your progress with the answers at the back of the book.

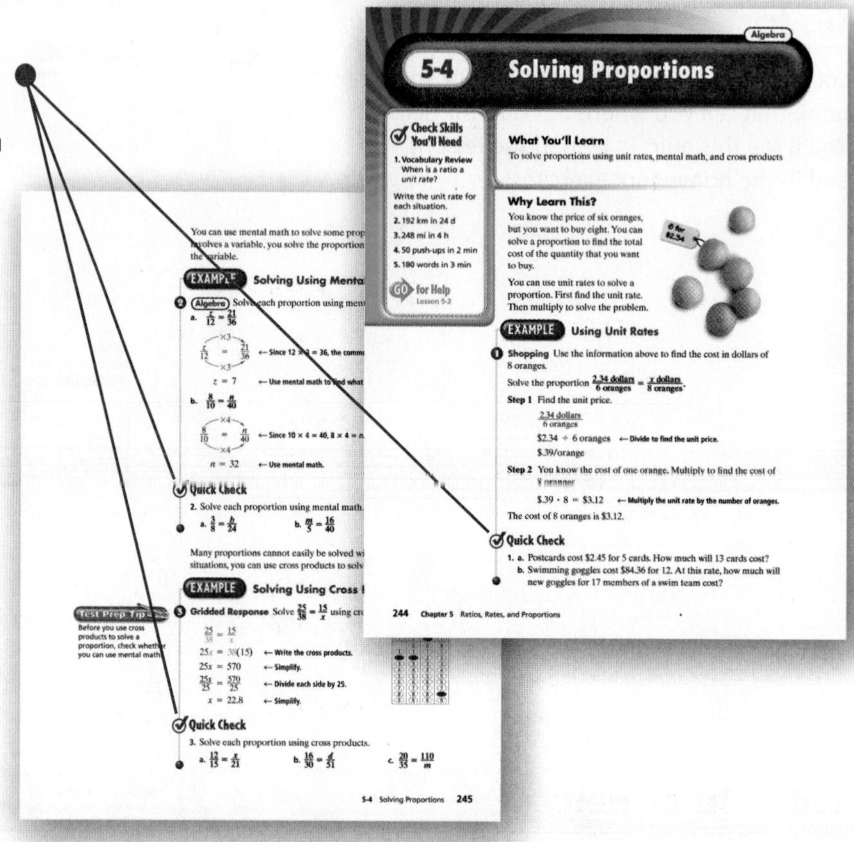

Understanding Key Concepts

Frequent *Key Concept* boxes summarize important definitions, formulas, and properties. Use these to review what you've learned.

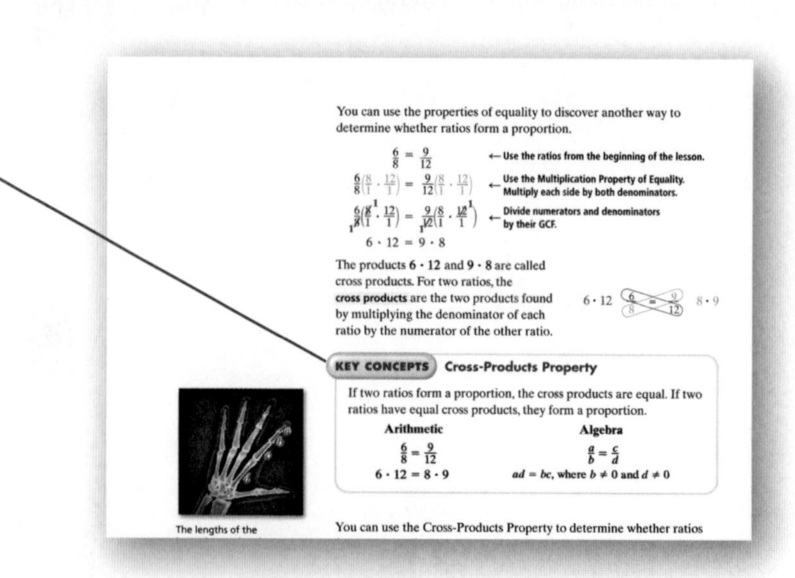

Online Active Math

Make math come alive with these online activities. Review and practice important math concepts with these engaging online tutorials.

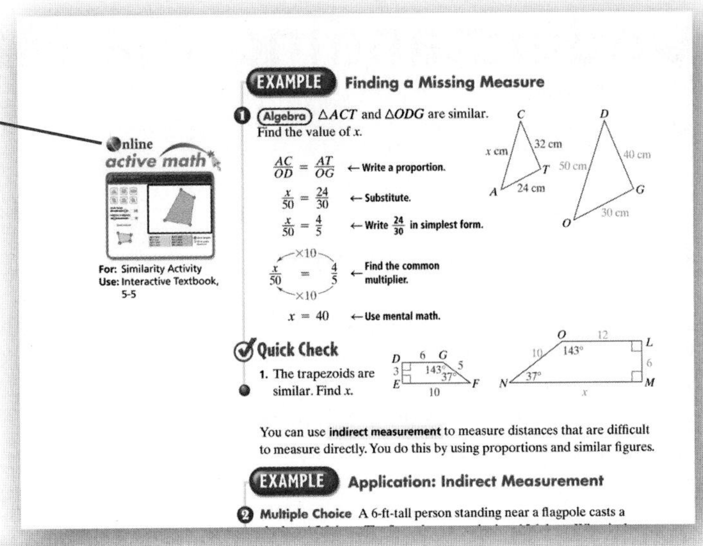

EXAMPLE Finding a Missing Measure

1 Algebra $\triangle ACT$ and $\triangle ODG$ are similar. Find the value of x.

$\frac{AC}{OD} = \frac{AT}{OG}$ ← Write a proportion.

$\frac{x}{50} = \frac{24}{30}$ ← Substitute.

$\frac{x}{50} = \frac{4}{5}$ ← Write $\frac{24}{30}$ in simplest form.

$\frac{x}{50} = \frac{4}{5}$ (×10) ← Find the common multiplier.

$x = 40$ ← Use mental math.

Online active math
For: Similarity Activity
Use: Interactive Textbook, 5-5

Quick Check

1. The trapezoids are similar. Find x.

You can use **indirect measurement** to measure distances that are difficult to measure directly. You do this by using proportions and similar figures.

EXAMPLE Application: Indirect Measurement

2 **Multiple Choice** A 6-ft-tall person standing near a flagpole casts a

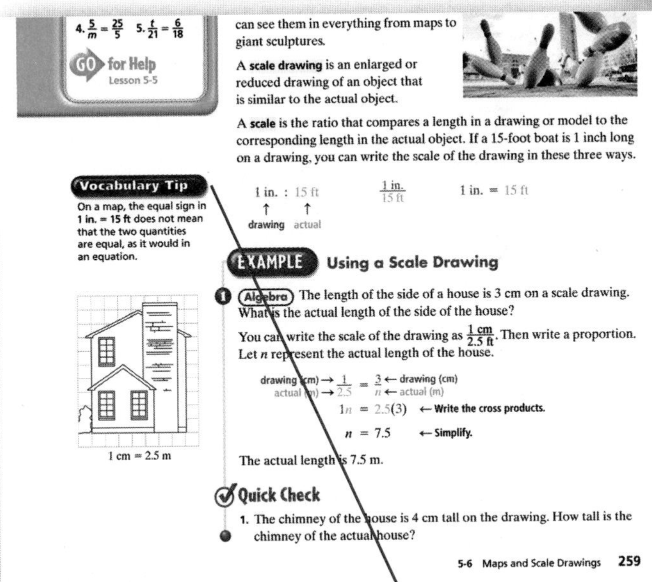

4. $\frac{5}{m} = \frac{25}{5}$ 5. $\frac{t}{21} = \frac{6}{18}$

GO for Help Lesson 5-5

can see them in everything from maps to giant sculptures.

A **scale drawing** is an enlarged or reduced drawing of an object that is similar to the actual object.

A **scale** is the ratio that compares a length in a drawing or model to the corresponding length in the actual object. If a 15-foot boat is 1 inch long on a drawing, you can write the scale of the drawing in these three ways.

1 in. : 15 ft $\quad \frac{1 \text{ in.}}{15 \text{ ft}} \quad$ 1 in. = 15 ft

drawing actual

Vocabulary Tip

On a map, the equal sign in 1 in. = 15 ft does not mean that the two quantities are equal, as it would in an equation.

EXAMPLE Using a Scale Drawing

1 Algebra The length of the side of a house is 3 cm on a scale drawing. What is the actual length of the side of the house?

You can write the scale of the drawing as $\frac{1 \text{ cm}}{2.5 \text{ ft}}$. Then write a proportion. Let n represent the actual length of the house.

drawing (cm) → $\frac{1}{2.5} = \frac{3}{n}$ ← drawing (cm)
actual (m) → ← actual (m)

$1n = 2.5(3)$ ← Write the cross products.

$n = 7.5$ ← Simplify.

The actual length is 7.5 m.

1 cm = 2.5 m

Quick Check

1. The chimney of the house is 4 cm tall on the drawing. How tall is the chimney of the actual house?

5-6 Maps and Scale Drawings 259

Vocabulary Support

Understanding mathematical vocabulary is an important part of studying mathematics. *Vocabulary Tips* and *Vocabulary Builders* throughout the book help focus on the language of math.

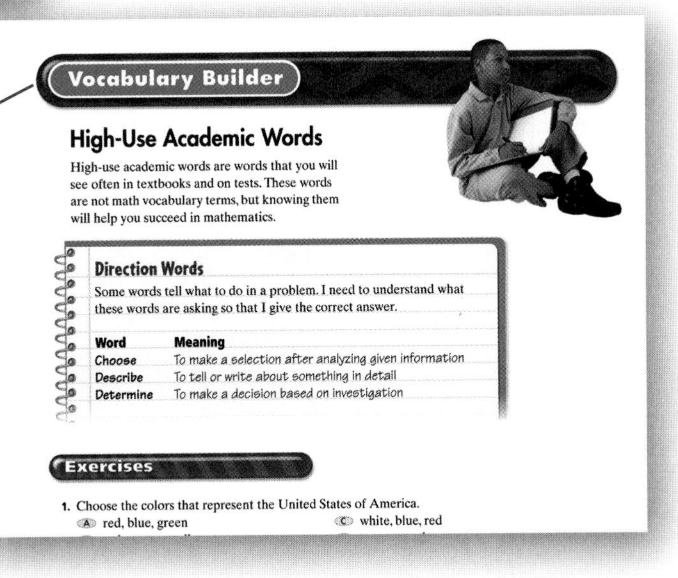

Vocabulary Builder

High-Use Academic Words

High-use academic words are words that you will see often in textbooks and on tests. These words are not math vocabulary terms, but knowing them will help you succeed in mathematics.

Direction Words

Some words tell what to do in a problem. I need to understand what these words are asking so that I give the correct answer.

Word	Meaning
Choose	To make a selection after analyzing given information
Describe	To tell or write about something in detail
Determine	To make a decision based on investigation

Exercises

1. Choose the colors that represent the United States of America.
A red, blue, green C white, blue, red

Understanding the Mathematics

Guided Problem Solving

These features throughout your Student Edition provide practice in problem solving. Solved from a student's point of view, this feature focuses on the thinking and reasoning that goes into solving a problem.

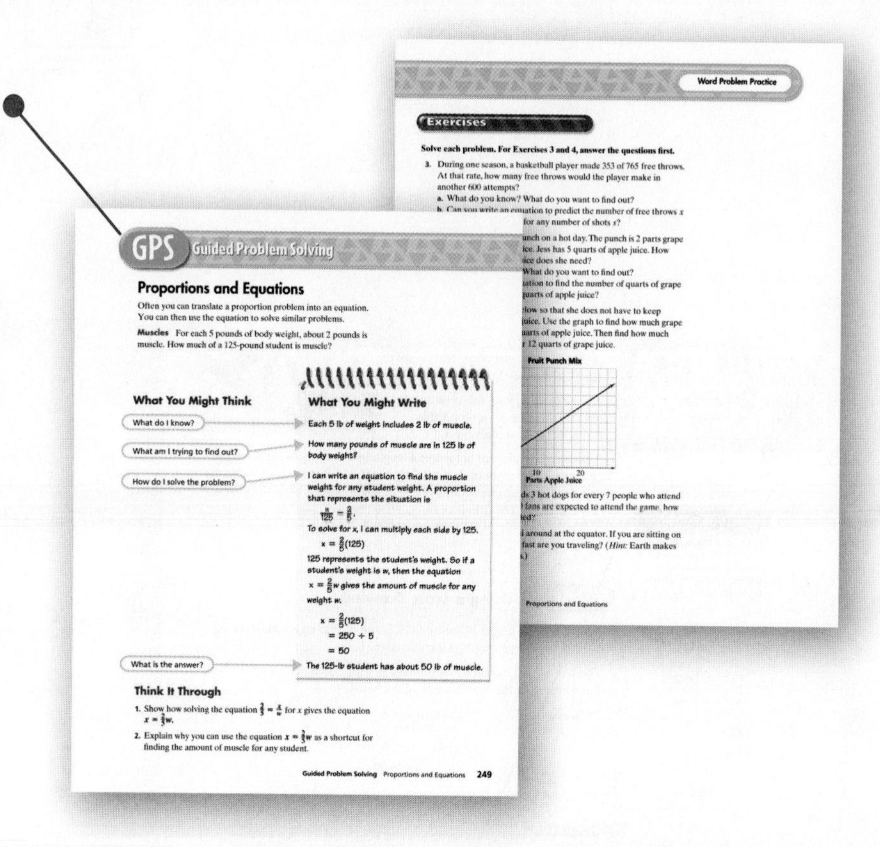

Word Problem Practice

Exercises

Solve each problem. For Exercises 3 and 4, answer the questions first.

3. During one season, a basketball player made 353 of 765 free throws. At that rate, how many free throws would the player make in another 600 attempts?
   a. What do you know? What do you want to find out?

GPS Guided Problem Solving

Proportions and Equations

Often you can translate a proportion problem into an equation. You can then use the equation to solve similar problems.

Muscles For each 5 pounds of body weight, about 2 pounds is muscle. How much of a 125-pound student is muscle?

What You Might Think	What You Might Write
What do I know?	Each 5 lb of weight includes 2 lb of muscle.
What am I trying to find out?	How many pounds of muscle are in 125 lb of body weight?
How do I solve the problem?	I can write an equation to find the muscle weight for any student weight. A proportion that represents the situation is $\frac{x}{125} = \frac{2}{5}$. To solve for x, I can multiply each side by 125. $x = \frac{2}{5}(125)$ 125 represents the student's weight. So if a student's weight is w, then the equation $x = \frac{2}{5}w$ gives the amount of muscle for any weight w. $x = \frac{2}{5}(125) = 250 \div 5 = 50$
What is the answer?	The 125-lb student has about 50 lb of muscle.

Think It Through

1. Show how solving the equation $\frac{2}{5} = \frac{x}{w}$ for x gives the equation $x = \frac{2}{5}w$.
2. Explain why you can use the equation $x = \frac{2}{5}w$ as a shortcut for finding the amount of muscle for any student.

Guided Problem Solving Proportions and Equations 249

Activity Labs

Activity Labs throughout the book give you an opportunity to explore a concept. Apply the skills you've learned in these engaging activities.

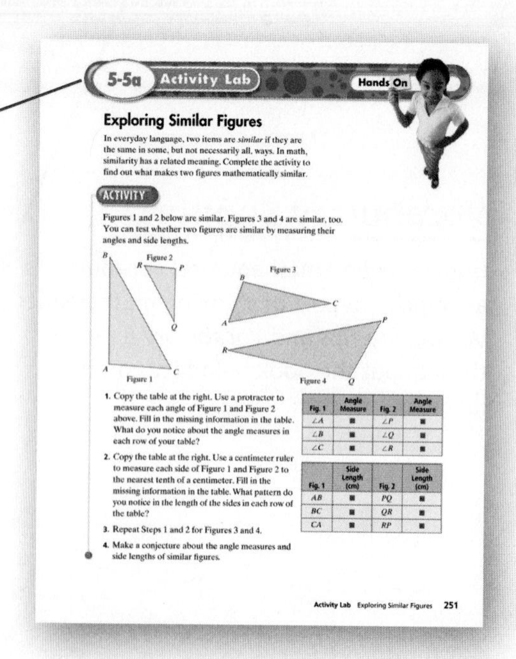

5-5a Activity Lab Hands On

Exploring Similar Figures

In everyday language, two items are *similar* if they are the same in some, but not necessarily all, ways. In math, similarity has a related meaning. Complete the activity to find out what makes two figures mathematically similar.

ACTIVITY

Figures 1 and 2 below are similar. Figures 3 and 4 are similar, too. You can test whether two figures are similar by measuring their angles and side lengths.

1. Copy the table at the right. Use a protractor to measure each angle of Figure 1 and Figure 2 above. Fill in the missing information in the table. What do you notice about the angle measures in each row of your table?
2. Copy the table at the right. Use a centimeter ruler to measure each side of Figure 1 and Figure 2 to the nearest tenth of a centimeter. Fill in the missing information in the table. What pattern do you notice in the length of the sides in each row of the table?
3. Repeat Steps 1 and 2 for Figures 3 and 4.
4. Make a conjecture about the angle measures and side lengths of similar figures.

Fig. 1	Angle Measure	Fig. 2	Angle Measure
∠A	■	∠P	■
∠B	■	∠Q	■
∠C	■	∠R	■

Fig. 1	Side Length (cm)	Fig. 2	Side Length (cm)
AB	■	PQ	■
BC	■	QR	■
CA	■	RP	■

Activity Lab Exploring Similar Figures 251

Practice What You've Learned

There are numerous exercises in each lesson that give you the practice you need to master the concepts in the lesson. The following exercises are included in each lesson.

Check Your Understanding

These exercises help you prepare for the Homework Exercises.

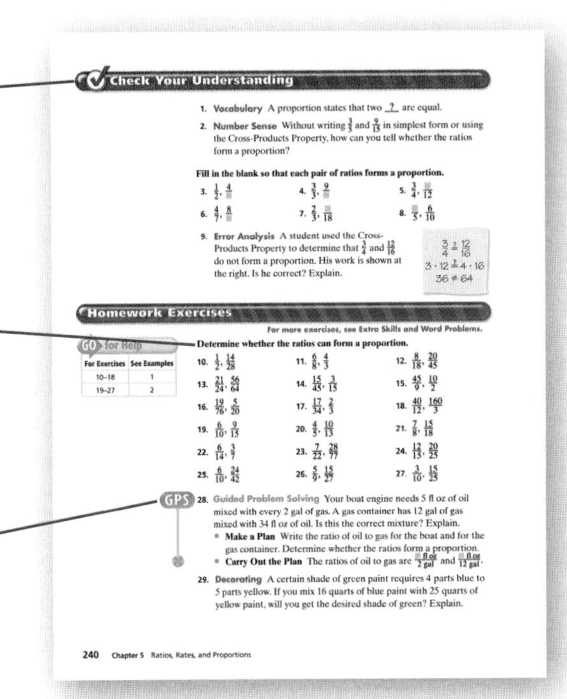

Practice by example

These exercises refer you back to the Examples in the lesson, in case you need help with completing these exercises.

Apply your skills

These exercises combine skills from earlier lessons to offer you richer skill exercises and multi-step application problems.

Homework Video Tutor

These interactive tutorials provide you with homework help for *every lesson.*

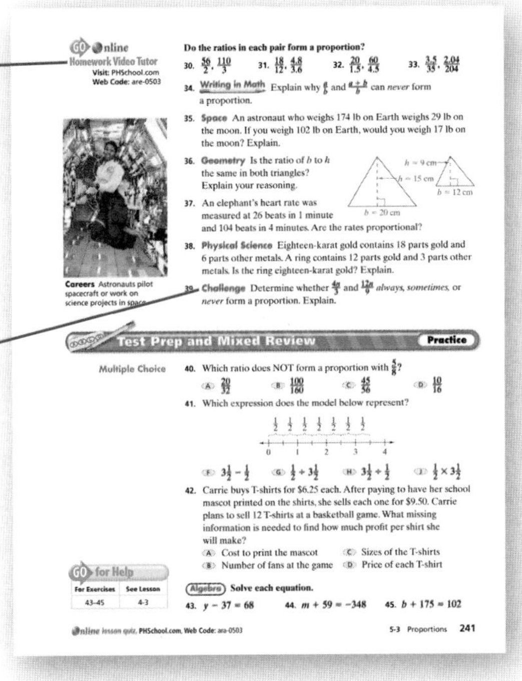

Challenge

This exercise gives you an opportunity to extend and stretch your thinking.

Using Your Book for Success

Intervention This Diagnostic Test covers pre-course skills that students need to succeed in this math program. For intervention, direct students to the following pages from the Skills Handbook in the back of their textbooks.

Exercise	Page
1,2	658
3, 4	659
5–7	654
8	655
9–11	660
12–14	661
15–18	662
19–22	663
23–27	664
28–33	665
34–37	666

Beginning-of-Course Diagnostic Test

1. Write the value of the underlined digit in 523.6$\underline{5}$4.

2. Write the value of the underlined digit in 402.$\underline{6}$59.

3. Write a number for fifty-one and six thousandths.

4. Write 7.325 in words.

Use < or > to compare the whole numbers.

5. 2,648 ■ 264

6. 625 ■ 6,250

7. 42,509 ■ 42,709

8. Round 75,845 to the nearest thousand.

9. Round 256.24 to the nearest whole number.

10. Round 546.256 to the nearest tenth.

11. Round 2.5879 to the nearest hundredth.

Multiply.

12. 4.6×0.7

13. 0.421×5.6

14. 3.08×12.4

15. 0.7×0.02

16. 0.032×0.06

17. 0.28×0.07

18. 0.06×0.2

1. 5 hundreths
2. 6 tenths
3. 51.006
4. seven and three hundred twenty-five thousandths
5. >
6. <

Divide.

19. $5\overline{)10.16}$

20. $13\overline{)34.918}$

21. $27.05 \div 2$

22. $0.036 \div 24$

Multiply.

23. $0.07 \times 1{,}000$

24. 478.24×0.01

25. 0.001×0.04

26. $0.9 \times 1{,}000$

27. 6.04×0.01

Divide.

28. $0.832 \div 0.26$

29. $0.5031 \div 0.039$

30. $0.42\overline{)0.273}$

31. $0.03\overline{)0.144}$

32. $0.00027 \div 0.18$

33. $0.018 \div 0.9$

Add or subtract. Write the answer in simplest form.

34. $\frac{5}{9} + \frac{2}{9}$

35. $\frac{7}{12} - \frac{3}{12}$

36. $5\frac{4}{8} + 4\frac{6}{8}$

37. $4\frac{8}{10} - 2\frac{6}{10}$

Beginning-of-Course Diag...

7. <
8. 76,000
9. 256
10. 546.3
11. 2.59
12. 3.22
13. 2.3576
14. 38.192
15. 0.014
16. 0.00192
17. 0.0196
18. 0.012
19. 2.032
20. 2.686
21. 13.525
22. 0.0015
23. 70
24. 4.7824
25. 0.00004
26. 900
27. 0.0604
28. 3.2
29. 12.9
30. 0.65
31. 4.8
32. 0.0015
33. 0.02
34. $\frac{7}{9}$
35. $\frac{1}{3}$
36. $10\frac{1}{4}$
37. $2\frac{1}{5}$

XXX
#1, 3, 4, 7, 8, 10, 12, 17, 19, 30, 36

Using the Problem Solving Plan

Throughout this text, students will be encouraged to use the four-step problem-solving plan that is outlined in this lesson. This approach gives students a simple yet effective framework for organizing their work in the process of solving a problem. Rather than having students haphazardly approach the task of problem solving, this four-step plan gives them an organized procedure to follow for a wide range of problems.

Guided Instruction

Students should remember these key phrases: Understand the Problem; Make a Plan; Carry out the Plan; Check for Reasonableness.

Call attention to the list of problem-solving strategies in the text. Ask for an example of each.

Have students brainstorm strategies that they can use to solve real-world problems such as Make a Table. Write up their ideas on poster board and display them in the room for students' reference.

Error Prevention!

Students often focus on one condition of a problem and forget another. Stress the importance of checking that a proposed solution satisfies all the conditions of the problem.

Using the Problem Solving Plan

One of the most important skills you can have is the ability to solve problems. An integral part of learning mathematics is how adept you become at unraveling problems and looking back to see how you found the solution. Maybe you don't realize it, but you solve problems every day—some problems are easy to solve, and others are challenging and require a good plan of action. In this Problem Solving Handbook you will learn how to work though mathematical problems using a simple four-step plan:

THE 4-STEP PLAN

1. **Understand** **Understand the problem.**
   Read the problem. Ask yourself, "What information is given? What is missing? What am I being asked to find or to do?"

2. **Plan** **Make a plan to solve the problem.**
   Choose a strategy. As you use problem solving strategies throughout this book, you will decide which one is best for the problem you are trying to solve.

3. **Carry Out** **Carry out the plan.**
   Solve the problem using your plan. Organize your work.

4. **Check** **Check the answer to be sure it is reasonable.**
   Look back at your work and compare it against the information and question(s) in the problem. Ask yourself, "Is my answer reasonable? Did I check my work?"

Problem Solving Strategies

Creating a good plan to solve a problem means that you will need to choose a strategy. What is the best way to solve that challenging problem? Perhaps drawing a diagram or making a table will lead to a solution. A problem may seem to have too many steps. Maybe working a simpler problem is the key. There are a number of strategies to choose from. You will decide which strategy is most effective.

As you work through this book, you will encounter many opportunities to improve your problem solving and reasoning skills. Working through mathematical problems using this four-step process will help you to organize your thoughts, develop your reasoning skills, and explain how you arrived at a particular solution.

Putting this problem solving plan to use will allow you to work through mathematical problems with confidence. Getting in the habit of planning and strategizing for problem solving will result in success in future math courses and high scores on those really important tests!

Good Luck!

THE STRATEGIES

Here are some examples of problem solving strategies. Which one will work best for the problem you are trying to solve?

- **Draw a Picture**
- **Look for a Pattern**
- **Systematic Guess and Check**
- **Act It Out**
- **Make a Table**
- **Work a Simpler Problem**
- **Work Backward**
- **Write an Equation**

Data Analysis Connection
Point out that problem-solving strategies are used to solve many types of problems, mathematical and non-mathematical.
Ask questions such as:

- *What are some real-world problems that scientists, mathematicians, professional athletes and coaches, or other professionals might need to analyze?* **Accept all reasonable answers.**

Differentiated Instruction

Visual Learners
To help students remember the four steps of the Problem-Solving Plan, display these phrases on a poster and have students suggest graphics to illustrate each. For instance, they might suggest a telescope to signal Check for Reasonableness.

Draw a Picture

The *Draw a Picture* strategy often helps students to clarify relationships among the numbers in a problem and to see the conditions of the problem more clearly.

Guided Instruction

Explain to students that a picture could be a table, diagram, or sketch. Different pictures are better for different situations, so students will have to choose the most appropriate picture to use to help them solve the problem.

Have volunteers read the text. Call attention to the problem solving steps:

- Understand the Problem
- Make a Plan
- Carry Out the Plan
- Check the Answer

Ask students questions such as:

- *What other kinds of pictures could you use to help you solve this problem?*
- *Could you solve this problem without a picture? How does the picture help you?*

Differentiated Instruction

Special Needs L1

Copy the table onto a transparency or onto the chalkboard. Use one color to highlight each "end of day" row of the table. Use a different color to highlight each "end of night" row. This should help students better see the patterns in the worm's positions.

Draw a Picture

When to Use This Strategy You can *Draw a Picture* to show a problem visually. A picture often helps you understand a problem better.

A worm is trying to escape from a well 10 ft deep. The worm climbs up 2 ft per day, but each night it slides back 1 ft. How many days will the worm take to climb out of the well?

Understand

The total distance to travel is 10 ft. The worm gains 2 ft during the day, but loses 1 ft each night. The goal is to find out how many days the worm will take to get out of the well.

Plan

Draw a picture to track the worm's position from day to day.

Carry Out

The worm reaches 10 ft and climbs out of the well at the end of the ninth day.

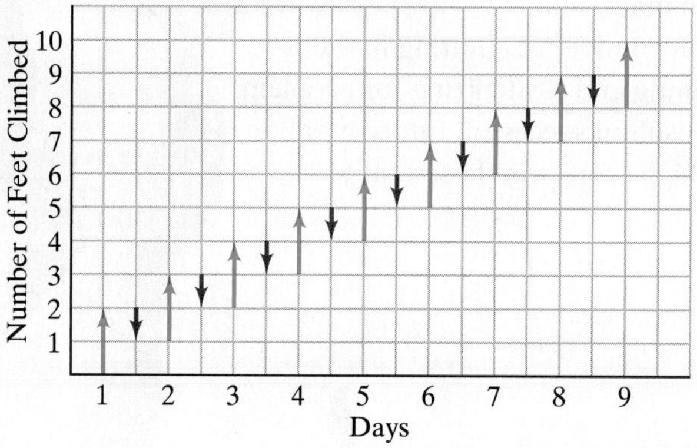

Worm's Progress

Number of Feet Climbed: 1 2 3 4 5 6 7 8 9 10

Days: 1 2 3 4 5 6 7 8 9

Check

You might think that the worm progresses 1 ft each day and so needs 10 days to escape. The worm does move a total of 1 ft each day, except on the ninth day. On the ninth day, it climbs 2 ft to the edge of the well.

Practice

1. Suppose the worm in the example above climbs up 3 ft per day and slides back 2 ft per night. How many days will it take for the worm to climb out of the 10-ft well?

2. **Multiple Choice** You schedule the games for your basketball league's tournament. If a team loses a game, it is eliminated. There are 32 teams. How many games do you need to schedule to determine the league champion?

   A. 30 games
   B. 31 games
   C. 32 games
   D. 64 games

3. There are 10 girlcs and 8 boys in a club. The club advisor can send one boy and one girl to a conference. How many different pairs of students can go to the conference?

4. A bricklayer is removing a square section of a rectangular patio. The patio is 12 feet long by 20 feet wide. She needs to remove a section 5 ft long × 5 ft wide What is the area of the patio without the square section?

5. Use the pattern below:

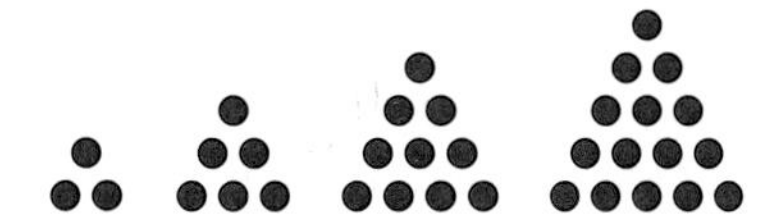

   How many dots will make up the 12th pattern?

6. A pizza party is having pizzas with pepperoni, pineapple chunks and green pepper slices. How many different pizzas can be made with these toppings? *Hint: A pizza with all the toppings is shown in the picture.*

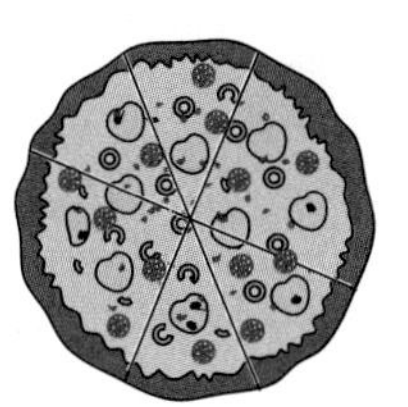

Differentiated Instruction

Visual Learners

Provide students with red, yellow, and green crayons or markers to help them draw the different possible pizzas in Practice 6.

Advanced Learners

Ask students to come up with a rule for each Practice problem, based on a variable amount.

- Problem 1: The worm climbs up *u* feet per day.
- Problem 2: There are *t* teams.
- Problem 3: There are *g* girls.
- Problem 4: The square section has side length *s*.
- Problem 5: The *n*-th pattern has how many dots?
- Problem 6: How many pizzas can you make with *t* toppings?

Students will be able to test their rules later in their study of Course 2.

Practice Answers

1. 8 days
2. B.
3. 80 pairs
4. 215 ft^2
5. 45
6. 7 pizzas

Problem Solving Handbook

Look For a Pattern

Patterns are fundamental to all of mathematics. *Look for a Pattern* is a good problem-solving strategy to use when teaching about number patterns, sequences, and similarities among geometric figures.

Guided Instruction

Tell students that when a problem asks them to analyze numerical or geometric relationships, they can use the strategy of looking for a pattern.

Have students read the problem and list the problem-solving steps they will use:

- Understand the Problem
- Make a Plan
- Carry Out the Plan
- Check the Answer

Some students may find making a table helpful.

Error Prevention!

Students may divide 12 by 3 and multiply this result of 4 by 180° to find the sum of the angles for a 12-sided polygon. This falsely assumes that only 4 triangles form the 12-sided polygon. This further emphasizes the need to draw a diagram.

Look for a Pattern

When to Use This Strategy Certain problems allow you to look at similar cases. You can *Look for a Pattern* in the solutions of these cases to solve the original problem.

Geometry What is the sum of the measures of the angles of a 12-sided polygon?

Understand

The goal is to find the sum of the measures of the angles of a 12-sided polygon. The sum of the measures in a triangle is 180°

Plan

Draw polygons with 3, 4, 5, and 6 sides. Divide each polygon into triangles by drawing diagonals from one vertex.
Look for a pattern.

Carry Out

The diagrams below show polygons divided into triangles.

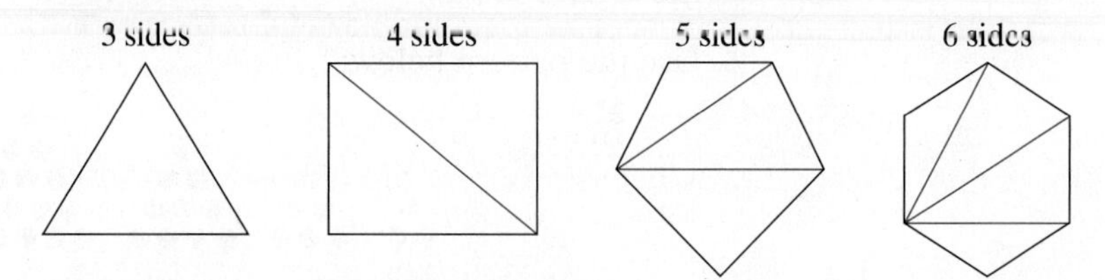

The number of triangles formed is two less than the number of sides of the polygon. This means that the sum of the measures of the angles of each polygon is the number of triangles times 180° For a 12-sided polygon, the number of triangles is $12 - 2 = 10$. The sum of the measures of the angles is $10 \times 180° = 1{,}800°$.

Check

Draw a diagram to check that exactly ten triangles are formed when you draw diagonals from one vertex of a 12-sided polygon.

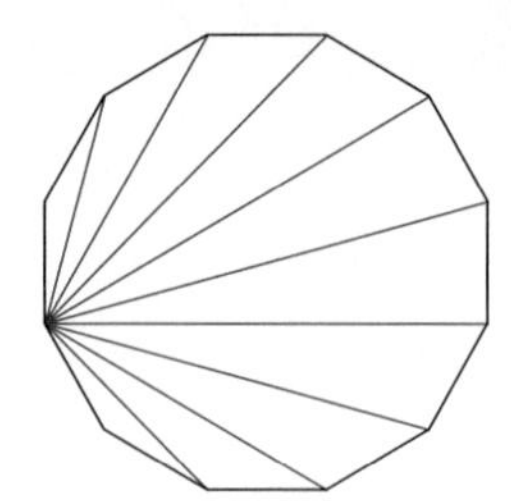

Practice

1. The figure below shows a pattern of black and white tiles. How many black tiles will you need for nine rows?

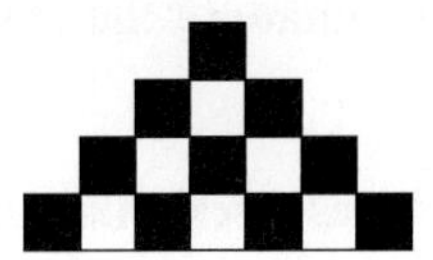

2. The figure below has four rows of small triangles. How many small triangles will you need for eight rows?

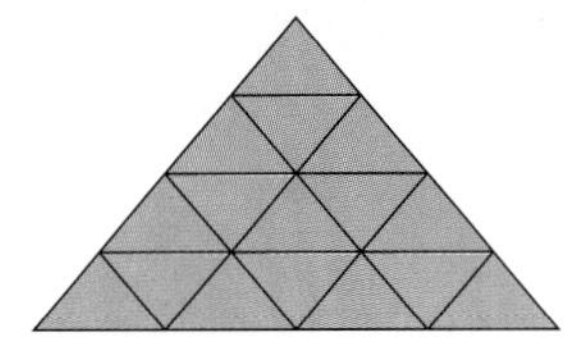

3. In a 3×3 grid, there are 14 squares of different sizes. There are nine 1×1 squares, four 2×2 squares, and one 3×3 square. How many squares of different sizes are in a 5×5 grid?

4. Your sister's new jobs pays $150 per week. After the first week, she decides to put $37.50 in a new savings account. After the second week, she puts $45.00 in the savings account. After the third week, she puts $52.50 in savings.

   a. If the pattern continues, how much will she put in the account after the fourth and fifth weeks?

   b. What will be the total amount in the savings account after the fifth week?

5. 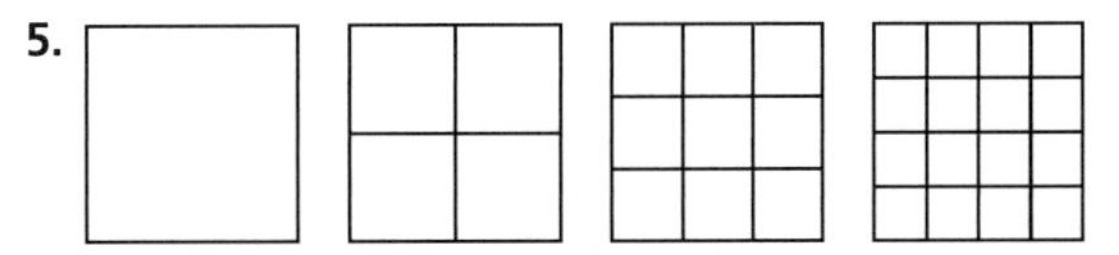

   Which series of numbers best describe the next two shapes in the pattern?

Differentiated Instruction

Tactile Learners
Provide students with a set of algebra tiles or squares to make the pattern shown in Practice 1. Ask students to describe the change from each figure to the next.

Below Level
Provide students with play money to realize the pattern in Practice 4.

Practice Answers

1. **45 black tiles**
2. **64 small triangles**
3. **55 squares**
4. **a. week 5 –$60; week 6 –$67.50**
   b. $262.50
5. **$5 \times 5, 6 \times 6$**

Problem Solving Handbook

Systematic Guess and Check

For the *Systematic Guess and Check* strategy, students make a reasonable estimate of the answer, check if the estimate satisfies the conditions of the problem, and change the number, if necessary. Students arrive at the correct solution by making increasingly more reasonable guesses.

Guided Instruction

Discuss with students how to complete a jigsaw puzzle. Talk about what to do if the piece does not fit properly in the jigsaw puzzle, that is, going back and trying another piece until one is found that does fit.

Elicit the problem solving steps:
- Understand the Problem
- Make a Plan
- Carry Out the Plan
- Check the Answer

Example

Discuss the triangular shape of the sails. Ask questions, such as:
- What other shape sails have you seen?
- Where did you see them?
- What have you noticed about the size of the sail and the size of the boat?
- Why do you suppose there is a relationship between them?

Systematic Guess and Check

When to Use This Strategy The strategy *Systematic Guess and Check* works well when you can start by making a reasonable estimate of the answer.

Construction A group of students is building a sailboat. The students have 48 ft^2 of material to make a sail. They design the sail in the shape of a right triangle as shown below. Find the length of the base and the height.

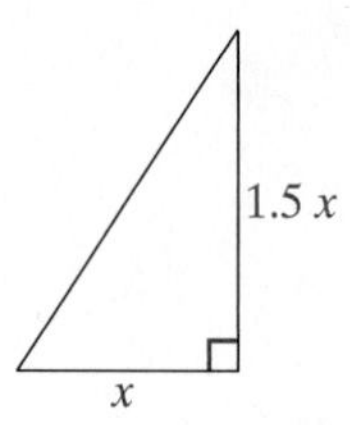

Understand

The diagram shows that the height is 1.5 times the length of the base.

Plan

Test possible dimensions of the triangle formed by the boom (base), the mast (height), and the sail. Check to see if they produce the desired area. Organize your results in a table.

Carry Out

Boom	Mast	Area	Conclusion
6	9	$\frac{1}{2} \cdot 6 \cdot 9 = 27$	Too low
10	15	$\frac{1}{2} \cdot 10 \cdot 15 = 75$	Too high
8	12	$\frac{1}{2} \cdot 8 \cdot 12 = 48$	✔

Check

A triangle with a base length of 8 ft and a height of 12 ft has an area of 48 ft^2.

Practice

1. The width of a rectangle is 4 cm less than its length. The area of the rectangle is 96 cm^2. Find the length and width of the rectangle.

2. **Multiple Choice** A dance floor is a square with an area of 1,444 ft^2. What are the dimensions of the dance floor?

   A. 38 ft × 38 ft
   B. 38 ft^2 × 38 ft^2
   C. 361 ft × 361 ft
   D. 361 ft^2 × 361 ft^2

3. You are building a rectangular tabletop for a workbench. The perimeter of the tabletop is 22 ft. The area of the tabletop is 24 ft^2. What are the dimensions of the tabletop?

4. You want to put up a fence around a rectangular garden. The length of the garden is twice its width. If the garden has an area of 2,450 ft^2, how much fencing material do you need?

5. The owner of the hot dog stand made $64.50 from selling 18 items from the menu. How many hot dogs did he sell?

Practice

Remind students to use the result of the first time they check an answer to improve subsequent guesses.

Teaching Tip

Encourage students to use a table to organize their work.

Differentiated Instruction

Advanced Learners

Ask students to find the minimum amount of fencing needed to section off 2,500 square feet of area.

Practice Answers

1. 12 cm; 8 cm
2. A.
3. 8 ft × 3 ft
4. 210 ft
5. 12 hot dogs

Problem Solving Handbook

Act It Out

Simulations use mathematical models to find experimental probabilities for events for which it is difficult to collect data directly. To design a simulation, model the possible number of outcomes by selecting a device such as coins, number cubes, or spinners. Then collect data. The more trials that are performed, the more reliable the results of the simulation will be.

Guided Instruction

When simulating or acting out a problem, students should be sure to choose a device that matches the problem. In this case, a coin makes sense since it has two outcomes, heads or tails, and each is equally likely. Have students note how each of the following steps is used to solve the problem:

- Understand the Problem
- Make a Plan
- Carry Out the Plan
- Check the Answer

Differentiated Instruction

Special Needs L1

Some students may be uncomfortable with the idea that finding experimental probability may not yield the same right answer for everyone. Discuss the meaning of the word "experiment" and how the number of favorable outcomes may differ each time.

Act It Out

When to Use This Strategy You can use the strategy *Act It Out* to simulate a problem.

A cat is expecting a litter of four kittens. The probabilities of having male and female kittens are equal. What is the probability that the litter contains three females and one male?

Understand

Your goal is to find the experimental probability that the litter of four kittens contains three females and 1 male.

Plan

Act out the problem by tossing a coin. Let heads represent a male and tails represent a female. Toss the coin 100 times. Separate the results into groups of 4 to represent the four kittens in a litter.

Carry Out

The table below shows 25 "litters" of 4.

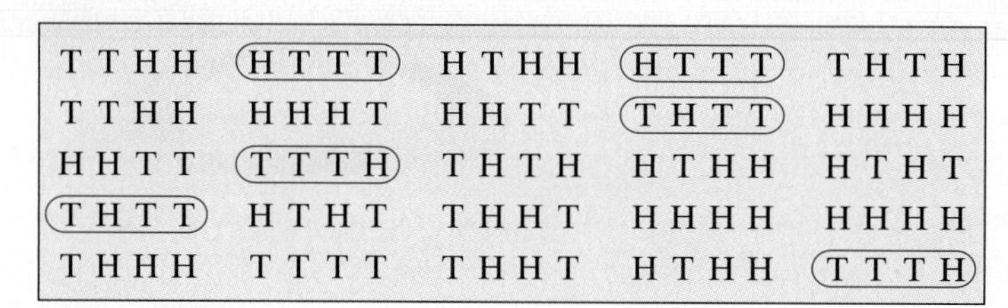

T T H H	H T T T	H T H H	H T T T	T H T H
T T H H	H H H T	H H T T	T H T T	H H H H
H H T T	T T T H	T H T H	H T H H	H T H T
T H T T	H T H T	T H H T	H H H H	H H H H
T H H H	T T T T	T H H T	H T H H	T T T H

Six groups out of twenty-five contain three females and one male. So $P(3 \text{ females and } 1 \text{ male}) = \frac{6}{25}$. The experimental probability is 0.24.

Check

Make a list of all 16 possible outcomes for having male and female kittens. Of these outcomes, only 4 have three females and one male. So the theoretical probability is $\frac{4}{16}$, or 0.25. This is close to the experimental value.

M M M M	F M M M	F M M F	M M F F
M M M F	F F M M	M F F M	F M F F
M M F M	F F F M	M F M F	F F M F
M F M M	F F F F	F M F M	M F F F

Practice

1. A sports jersey number has two digits. Even and odd digits are equally likely. Use a simulation to find the probability that both digits are even.
2. You are taking a 4-question true-or-false quiz. You do not know any of the answers. Use a simulation to find the probability that you guess exactly 3 out of 4 answers correctly.
3. A restaurant gives away a model car with each meal. You are equally likely to get any of the five cars. Use a simulation to find the probability that you get two of the same car after two meals.
4. Three friends are going bowling. Use a simulation to determine how many bowling orders are possible.
5. Conduct a simulation using a game spinner divided into 10 equal sections and numbered 1 to 10 to determine the probability of the spinner landing on an odd number three times in a row.

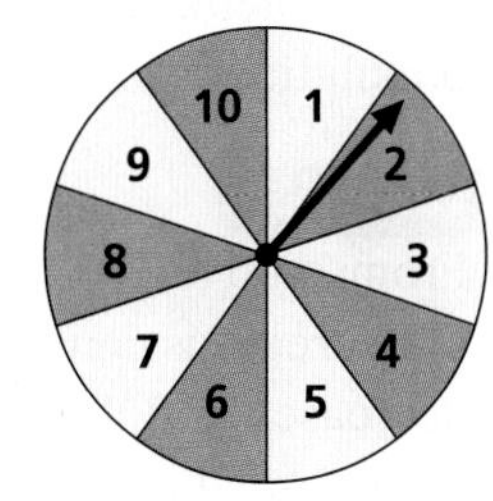

6. A bag of marbles contains 10 red marbles and 8 blue marbles. Some marbles spill out as shown.

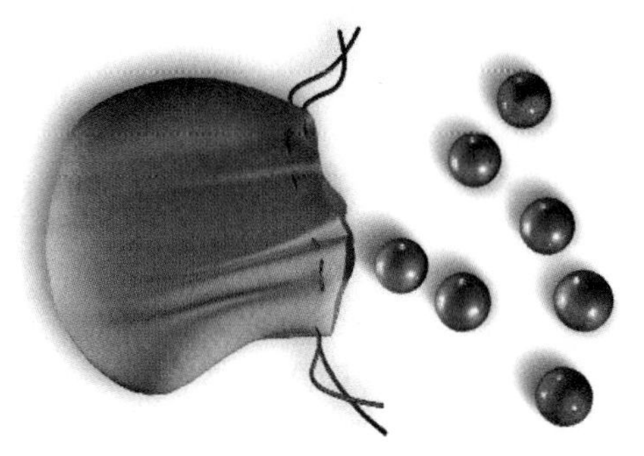

You randomly select two marbles from the marbles remaining in the bag. Use a simulation to determine the probability of selecting one red marble and one blue marble.

Teaching Tip
Divide students into six groups, with at least four students in each group. Make stations around the classroom for the groups to rotate through as they Act Out each of the practice problems. Give each group three minutes to compete each station, then one minute to rotate groups to the next station.
- Station 1: calculator with random number generator
- Station 2: a coin for each student in the largest group
- Station 3: two bags with five different colored beads in each bag
- Station 4: three cards with a different name written on each card
- Station 5: a spinner with 10 numbered sections
- Station 5: a spinner with 10 numbered sections
- Station 6: a bag with 8 red marbles and 3 blue marbles

Differentiated Instruction
Tactile Learners
Throughout the year, keep pennies, cards, tokens, and other objects available in the classroom so that students can use them when they use the problem solving strategy Act It Out.

Practice Answers
1–3. Check students' work
4. 6
5. $\frac{1}{8}$
6. 120% chance

Make a Table

Making a table helps you organize the information given in a problem. Once the information is organized, you can use logical reasoning to observe patterns and extend the table.

Guided Instruction

Be sure students understand what the rate of change or decrease is and how to find it. Call attention to the use of problem-solving steps:
- Understand the Problem
- Make a Plan
- Carry Out the Plan
- Check the Answer

Have students identify where in the problem each is used.

Emphasize the fact that students should think about making a table when they are asked to compare two or more things, or when the numbers they are using change according to a pattern.

Differentiated Instruction

Special Needs L1

Some students may prefer to do their problem solving mentally rather than making tables. Assure students that they are practicing strategies that allow them to organize their thinking, which becomes essential with more complicated problems. Encourage students to continue to use mental math to predict the solution to a problem.

Make a Table

When to Use This Strategy A real-world problem may ask you to examine a set of data and draw a conclusion. In such a problem, you can *Make a Table* to organize the data.

Biology A wildlife preserve surveyed its wolf population in 1996 and counted 56 wolves. In 2000, there were 40 wolves. In 2002, there were 32 wolves. If the wolf population changes at a constant rate, in what year will there be fewer than 15 wolves?

Understand

Given the wolf population in 1996, 2000, and 2002, you want to find the year in which there will be fewer than 15 wolves.

Plan

Find the rate of change. *Make a Table* to organize the information in the problem. Use the rate of change to extend the table until the wolf population is less than 15.

Carry Out

From 2000 to 2002, the wolf population decreased by 8. Since the rate of decrease is constant, you can say that every 2 years, the population decreases by 8. In the beginning of 2006, there will be about 16 wolves. The population will be less than 15 later that year.

Year	Wolves
2000	40
2002	$40 - 8 = 32$
2004	$32 - 8 = 24$
2006	$24 - 8 = 16$

Check

Step 1→ For there to be 15 wolves, the population must decrease by $56 - 15$ wolves, or 41 wolves.
Step 2→ The population decreases by 4 wolves per year. Let x represent the number of years until there are 15 wolves.
Step 3→ Solve $4x = 41$. The value of x is about 10 years.
So $1996 + 10 = 2006$.

Practice

1. **Multiple Choice** You are starting a business selling lemonade. You know that it costs $6 to make 20 c of lemonade and $7 to make 30 c of lemonade. How much will it cost to make 50 c of lemonade?
   A. $8
   B. $9
   C. $10
   D. $12
2. You have $10 saved and plan to save an additional $2 each week. How much will you have after 7 weeks?
3. A driver of a car slows to a stop. The decrease in speed is constant. When the driver first applies the brakes, the car is going 50 mi/h. After 5 s, the car is traveling 30 mi/h. About how long does it take the car to stop?
4. A family drives 127 miles on their first day of vacation. They drive an additional 35 miles each day after that.
   a. On what day will they have driven a total of 372 miles?
   b. How many total miles did they drive after the first 5 days?
5. A house plant starts at 13 in. and grows 4 in. each week. How tall will it be at the end of 7 weeks?
6. How many different groups of letters can be made with the following letter tiles?

Error Prevention!

Remind students that they will often need to start their table with a row zero. This row will look different than the other rows in the table.

Teaching Tip
Students can add additional rows to their tables to show their calculations. This will make it easier for them to check their work. For example, in Practice 2, students may have columns labeled Week Number, Savings at Start of Week, Additional Savings, Savings at End of Week.

Practice Answers

1. **B.**
2. **$24**
3. **$12.5 s**
4. **a. On the 8th day**
   b. 985 miles
5. **41 inches**
6. **24**

Work a Simpler Problem

When reading a problem, students sometimes see only a tangle of numbers and mysterious operations. Solving a *simpler problem* may reduce anxiety over large numbers or operations involved and help to focus attention on the nature of the operations.

Guided Instruction

Tell students that this strategy will help them when they face a complex problem. Often, it will be used in conjunction with another strategy, such as *Make a Table*. Explain that this strategy also is helpful when finding numerical or geometric patterns.

Ask volunteers to read the parts of the problem identifying each of the problem-solving steps:

- Understand the Problem
- Make a Plan
- Carry Out the Plan
- Check the Answer

Teaching Tip
Remind students that the simpler problem they set up needs to be similar to the original problem.

Work a Simpler Problem

When to Use This Strategy If a problem seems to have many steps, you may be able to *Work a Simpler Problem* first. The result may give you a clue about the solution of the original problem.

When you simplify 3^{50}, what number is in the ones place?

Understand

You know that 3^{50} is a large number to calculate. You need to find the number in the ones place.

Plan

It is not easy to simplify 3^{50} with paper and pencil. Simplify easier expressions such as 3^2, 3^3, and 3^4, to see what number is in the ones place.

Carry Out

The table shows the values of the first 10 powers of 3.

Power	Value
3^1	3
3^2	9
3^3	27
3^4	81
3^5	243
3^6	729
3^7	2,187
3^8	6,561
3^9	19,683
3^{10}	59,049

Notice that the ones digits in the value column repeat in the pattern 3, 9, 7, and 1. Every fourth power of 3 has a ones digit of 1. Since 4 is divisible by 4, 3^{48} has a ones digit of 1. Then the ones digit of 3^{49} is 3 and the ones digit of 3^{50} is 9.

Check

When the exponent of 3 is 2, 6, or 10, the ones digit is 9. The numbers 2, 6, and 10 are divisible by 2 but not by 4. Since 50 is divisible by 2 but not by 4, the ones digit of 3^{50} is 9.

Practice

1. a. What is the pattern for the ones digit of any power of 8?
   b. When you simplify 8^{63}, what number is in the ones place?
2. a. What is the pattern for the ones digit of any power of 7?
   b. When you simplify 7^{21}, what number is in the ones place?
3. The table shows the values of powers of 2 with even exponents from 10 to 20.

Power	Value
2^{10}	1,024
2^{12}	4,096
2^{14}	16,384
2^{16}	65,536
2^{18}	262,144
2^{20}	1,048,576

What is the ones digit of 2^{80}?

4. What is the value of $(-1)^{427}$? Explain.
5. When you simplify 10^{347}, what digit is in the ones place?
6. Find the sum of the first 10 powers of 10, that is, the sum of $10^1 + 10^2 + 10^3 + \ldots 10^{10}$.
7. How many minutes are there in one week?

June						
1	2	3	4	5	6	7
8	9	10	11	12	13	14
15	16	17	18	19	20	21
22	23	24	25	26	27	28
29	30					

Differentiated Instruction

Below Level L2

The problem in this lesson might frustrate students who have a brief attention span, especially if they cannot immediately perceive the correct pattern. Have each student work with a partner with whom they can discuss any patterns they see.

Visual Learners

You can present similar problems for students who recognize visual patterns. For example, you can use the repeating pattern square, triangle, circle, square, triangle, circle, ... and ask students to predict which shape will come much later in the pattern. You can add colors to the pattern, alternating red and blue, for added challenge.

Teaching Tip

For Practice 4, have students identify how many minutes are in one hour, how many hours are in one day, how many days in one week.

Practice Answers

1a. 8, 4, 2, 6, 8, 4, 2, 6, . . .

1b. 2

2a. 7, 9, 3, 1, 7, 9, 3, 1, . . .

2b. 7

3. 6

4. −1; all odd powers of −1 are negative.

5. 10,080

6. 11,111,111,110

Work Backward

In some problem-solving situations, it may be helpful to work backward. These situations often involve a time of day or a known end result of a process. To solve these problems, begin with the final result and work backward to the beginning.

Guided Instruction

Give students exercises like these to practice converting between minutes and fractional hours:

30 min $\frac{1}{2}$ h	45 min $\frac{3}{4}$ h
20 min $\frac{1}{3}$ h	$\frac{2}{3}$ h 40 min
$\frac{1}{4}$ h 15 min	$1\frac{1}{2}$ h 90 min

Then elicit problem-solving steps:

- Understand the Problem
- Make a Plan
- Carry Out the Plan
- Check the Answer

Have students explain how each one is used in the example.

Differentiated Instruction

Visual Learners

Have students sketch a series of clocks to help them keep track of changing times with each step as they test their estimates.

Work Backward

When to Use This Strategy You can use the strategy *Work Backward* when a problem asks you to find an initial value.

You and your friends are going to dinner and then to a concert that starts at 8:00 P.M., It will take $\frac{3}{4}$ h to drive to the restaurant and $1\frac{1}{4}$ h to eat and drive to the theater. You want to arrive at the theater 15 min before the concert starts. At what time should you leave?

Understand

Your goal is to find out what time you should leave home to arrive 15 min early for the 8:00 P.M. concert. It takes $\frac{3}{4}$ h to drive to the restaurant and $1\frac{1}{4}$ h to eat and drive to the theater.

You know that the series of events must end at 8:00 P.M. It makes sense to *Work Backward* to find when you must leave your house.

Carry Out

Concert starts.
8:00 P.M.

Arrive at theater.
7:45 P.M.

Arrive at dinner.
6:30 P.M.

Leave home.
5:45 P.M.

Working backward shows that when you leave at 5:45 P.M., you will get to the concert 15 min early.

Check

Find the total time it takes for the series of events to happen.

$\frac{3}{4} + 1\frac{1}{4} + \frac{1}{4} = 2\frac{1}{4}$.

Subtract $2\frac{1}{4}$ h from 8:00 P.M., and you get 5:45 P.M.

Practice

1. After school today, you spent $1\frac{3}{4}$ h at band practice and then a half hour in the library. It took you 15 min to get home at 6:00 P.M. What time did you start practice?

2. Your friend spends one third of her money on lunch. You then give her $2.50 to repay a loan. After school, your friend spends $4.00 for a movie and $2.50 for a snack. She has $4.90 left. How much money did she have before lunch?

3. If you start with a number, add 5, and then multiply by 7, the result is 133. What was the original number?

4. A bakery uses 12 bags of flour on Monday, 8 bags on Tuesday, 14 bags on Wednesday, and half of the remaining bags on Thursday. After Thursday there are 3 bags of flour left. How many bags of flour did the bakery start with on Monday?

5. A concert begins at 7:00 P.M. The walk to the bus station will take 20 minutes. The bus ride to a friend's house is 30 minutes. From there it is a 17-minute walk to the concert. The opening act is set to perform for 45 minutes. What time should you leave if you want to make it to the concert after the opening act is finished?

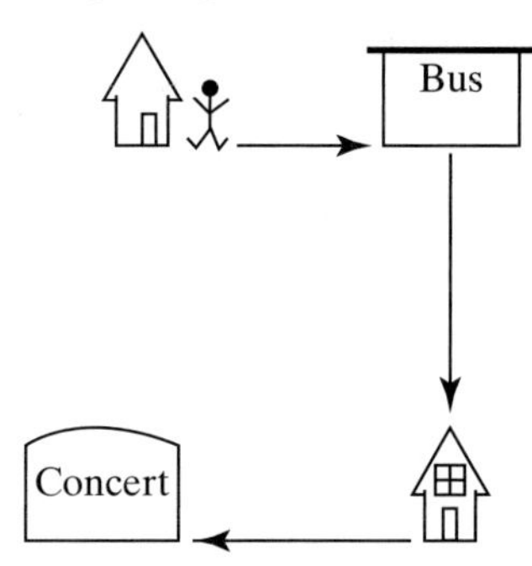

6. A delivery driver used $\frac{1}{3}$ of a tank of gas on his way to his first stop. He used $\frac{1}{8}$ of a tank from there to his second stop. He got $\frac{1}{2}$ a tank at the gas station and used another $\frac{1}{8}$ of a tank on the drive to the warehouse. His final gas gauge is shown.

How much gas did the driver start with?

Error Prevention!

Remind students to check their solution by using it to Work Forward through the problems and verify that their answer is correct.

Differentiated Instruction

Special Needs

Give students manipulatives such as play money or blocks to check their answers to Practice problems 2-4.

Practice Answers

1. 3:30 P.M.
2. $26.70
3. 14
4. 40 bags of flour
5. 6:18
6. $\frac{5}{6}$

Write an Equation

Many real-world problems describe relationships between numbers. You can often find a solution by using the strategy *Write an Equation.* Equations use variables to represent values that are unknown or subject to change.

Guided Instruction

Tell students that writing an equation is a good way to organize the information needed to solve the problem.

Point out the translation of words to equation shown in the problem. Have students identify where each of the problem-solving steps below is used.

- Understand the Problem
- Make a Plan
- Carry Out the Plan
- Check the Answer

Error Prevention!

When choosing variables for equations, suggest that it helps to use variables that have meaning in the problem, such as *d* for days.

Write an Equation

When to Use This Strategy You can *Write an Equation* to represent a real-world situation that involves two variables.

The cost of materials needed to make one toboggan is \$8. A craftsman has a budget of \$2,000. How many toboggans can he make?

Understand

Your goal is to find the number of toboggans the craftsman can make given his budget and the cost of materials.

Plan

Write an Equation to model the situation. Then solve the equation.

Carry Out

Write an equation to represent the number of toboggans.

Words	amount budgeted	divided by	cost of materials	is the	number of toboggans
	Let b = amount budgeted. Let n = number of toboggans.				
Equation	b	$\div$	\$8	$=$	n

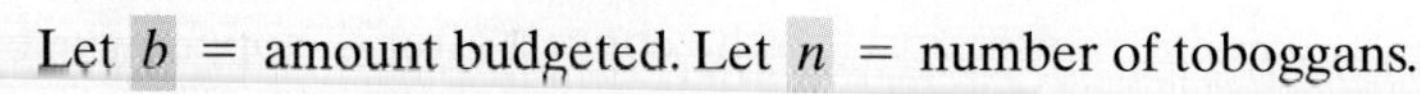

$\frac{b}{8} = n$

$\frac{2,000}{8} = n$ ← **Substitute *b* for 2,000.**

$250 = n$ ← **Simplify.**

The craftsman can make 250 toboggans.

Check

The total cost of making 250 toboggans is 250 · \$8, or \$2,000. This equals the craftsman's budget. The answer checks.

Practice

1. Family membership at a science museum costs $89 per year. The shows at the museum theater cost $22.50 per family each visit. How many shows can a family see if its yearly budget is $300?

2. **Multiple Choice** A pair of boots costs $10 more than twice the cost of a pair of shoes. The boots cost $76.50. How much do the shoes cost?
   A. $33.25
   B. $38.25
   C. $44.25
   D. $70.50

3. Your family's car can travel 23 mi using 1 gal of gas. To the nearest gallon, how many gallons of gas will your car need for a 540-mi trip?

4. An oven preheats at 15 degrees per minute. How long will it take for the oven to heat up to 375 degrees?

5. A courier service ships packages for $50 for the first 40 pounds. It costs an additional $.79 for each additional pound. The weight of a package is shown below.

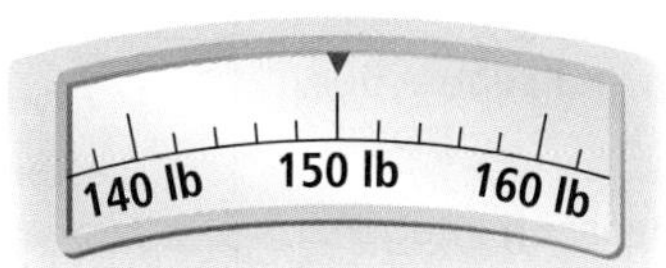

How much will it cost to ship the package?

Alternative Method
Students can use the problem solving strategies Make a Table to solve these problems. Demonstrate to students that writing and solving an equation is a more efficient way to solve these problems than making a talbe.

Differentiated Instruction

Tactile Learners
Provide algebra tiles for those students who are more comfortable solving equations using manipulatives.

Practice Answers

1. **9 shows**
2. **A.**
3. **23 gal**
4. **25 minutes**
5. **$136.90**

1 Decimals and Integers

Chapter at a Glance

Lesson Titles, Objectives, and Features	Assessment	NCTM Standards	Local Standards
1-1 Using Estimation Strategies • To estimate using rounding, front-end estimation, and compatible numbers	Lesson Quiz	1, 4, 5, 6, 7, 8, 9, 10	
1-2 Adding and Subtracting Decimals • To add and subtract decimals and to do mental math using the properties of addition **Extension:** Using Mental Math	Lesson Quiz	1, 2, 4, 6, 7, 8, 9, 10	
1-3a Activity Lab: Modeling Decimal Multiplication **1-3 Multiplying Decimals** • To multiply decimals and to do mental math using the properties of multiplication	Lesson Quiz Checkpoint Quiz 1	1, 2, 6, 7, 8, 9, 10	
1-4a Activity Lab, Hands On: Modeling Decimal Division **1-4 Dividing Decimals** • To divide decimals and to solve problems by dividing decimals **Guided Problem Solving:** Choosing Operations	Lesson Quiz	1, 2, 6, 7, 8, 9, 10	
1-5 Measuring in Metric Units • To use and convert metric units of measure	Lesson Quiz	1, 4, 5, 6, 7, 8, 9, 10	
1-6 Comparing and Ordering Integers • To compare and order integers and to find absolute values **Vocabulary Builder:** Learning New Math Terms	Lesson Quiz Checkpoint Quiz 2	1, 2, 4, 5, 6, 7, 8, 9, 10	
1-7a Activity Lab, Hands On: Modeling Integer Addition and Subtraction **1-7 Adding and Subtracting Integers** • To add and subtract integers and to solve problems involving integers	Lesson Quiz	1, 5, 6, 7, 8, 9, 10	
1-8a Activity Lab, Hands On: Modeling Integer Multiplication **1-8 Multiplying and Dividing Integers** • To multiply and divide integers and to solve problems involving integers	Lesson Quiz	1, 4, 5, 6, 7, 8, 9, 10	
1-9 Order of Operations and the Distributive Property • To use the order of operations and the Distributive Property **1-9b Activity Lab, Algebra Thinking:** Properties and Equality	Lesson Quiz	1, 2, 6, 7, 8, 9, 10	
1-10 Mean, Median, Mode, and Range • To describe data using mean, median, mode, and range **1-10b Activity Lab, Data Analysis:** Box-and-Whisker Plots	Lesson Quiz	1, 4, 5, 6, 7, 8, 9, 10	
Problem Solving Application: Applying Integers			

NCTM Standards 2000
1 Number and Operations **2** Algebra **3** Geometry **4** Measurement **5** Data Analysis and Probability
6 Problem Solving **7** Reasoning and Proof **8** Communication **9** Connections **10** Representation

Correlations to Standardized Tests

All content for these tests is contained in *Prentice Hall Math,* Course 2. This chart reflects coverage in this chapter only.

	1-1	1-2	1-3	1-4	1-5	1-6	1-7	1-8	1-9	1-10
Terra Nova CAT6 (Level 17)										
Number and Number Relations	✔	✔	✔	✔	✔	✔	✔	✔	✔	✔
Computation and Numerical Estimation	✔	✔	✔	✔	✔	✔	✔	✔	✔	✔
Operation Concepts	✔	✔	✔	✔	✔	✔	✔	✔	✔	✔
Measurement	✔	✔			✔	✔	✔	✔		
Geometry and Spatial Sense										
Data Analysis, Statistics, and Probability	✔				✔	✔	✔	✔		✔
Patterns, Functions, Algebra	✔	✔	✔	✔		✔	✔	✔	✔	
Problem Solving and Reasoning	✔	✔	✔	✔	✔	✔	✔	✔	✔	✔
Communication	✔	✔	✔	✔	✔	✔	✔	✔	✔	✔
Decimals, Fractions, Integers, Percent									✔	
Order of Operations	✔	✔	✔	✔	✔	✔	✔	✔	✔	✔
Terra Nova CTBS (Level 17)										
Decimals, Fractions, Integers, Percents	✔	✔	✔	✔	✔	✔	✔	✔	✔	
Order of Operations, Numeration, Number Theory	✔	✔	✔	✔	✔	✔	✔	✔	✔	✔
Data Interpretation	✔	✔		✔	✔	✔	✔	✔	✔	✔
Pre-algebra		✔	✔	✔		✔			✔	
Measurement	✔	✔	✔	✔	✔	✔	✔	✔		
Geometry										
ITBS (Level 13)										
Number Properties and Operations	✔	✔	✔	✔	✔	✔	✔	✔	✔	✔
Algebra		✔	✔	✔		✔			✔	
Geometry										
Measurement	✔	✔			✔	✔	✔	✔		
Probability and Statistics	✔	✔		✔	✔	✔	✔	✔	✔	✔
Estimation	✔	✔	✔	✔	✔	✔				
SAT 10 (Int 3 Level)										
Number Sense and Operations	✔	✔	✔	✔	✔	✔	✔	✔	✔	✔
Patterns, Relationships, and Algebra	✔	✔	✔	✔	✔	✔	✔	✔	✔	✔
Data, Statistics, and Probability	✔	✔		✔	✔	✔	✔	✔	✔	✔
Geometry and Measurement	✔	✔	✔	✔	✔	✔	✔	✔		
NAEP										
Number Sense, Properties, and Operations		✔	✔	✔			✔	✔		
Measurement					✔					
Geometry and Spatial Sense										
Data Analysis, Statistics, and Probability										✔
Algebra and Functions										

CAT6 California Achievement Test, 6th Ed.
SAT10 Stanford Achievement Test, 10th Ed.
CTBS Comprehensive Test of Basic Skills
NAEP National Assessment of Educational Progress 2005 Mathematics Objectives
ITBS Iowa Test of Basic Skills, Form M

Math Background

Skills Trace

BEFORE Chapter 1

Course 1 or Grade 6 reviewed decimal operations and introduced integer operations.

DURING Chapter 1

Chapter 1 reviews decimals and integers and uses these numbers to solve real-world applications.

AFTER Chapter 1

This course uses operations with decimals and integers throughout algebra, geometry, measurement, statistics, and probability.

1-1 Using Estimation Strategies

Math Understandings

- Estimation is used when an exact answer is not needed.
- Estimation is used with mental math to check answers.
- Most estimation techniques involve replacing numbers with ones that are close and easy to compute with mentally.

The symbol $\approx$ means "is approximately equal to" and is used with estimated answers.

Estimation strategies in addition include **estimation by rounding, front-end estimation,** and **using compatible numbers.**

Estimation strategies in multiplication and division include **estimation by rounding** and **using compatible numbers.**

Example: To estimate 37×54, students can use compatible numbers to change 37 to 40 and 54 to 50, and use the basic fact 4×5 to estimate 40×50 as about 2,000.

1-2 1-3 1-4 Adding, Subtracting, Multiplying, and Dividing Decimals

Math Understandings

- The properties of addition and multiplication can help you compute with decimals mentally.

Properties of Addition and Multiplication	
Arithmetic	**Algebra**
Commutative Property	
$2 + 3 = 3 + 2$	$a + b = b + a$
$2 \cdot 3 = 3 \cdot 2$	$a \cdot b = b \cdot a$
Associative Property	
$(2 + 3) + 4 = 2 + (3 + 4)$	$(a + b) + c = a + (b + c)$
$(2 \cdot 3) \cdot 4 = 2 \cdot (3 \cdot 4)$	$(a \cdot b) \cdot c = a \cdot (b \cdot c)$
Identity Property	
$6 + 0 = 6$	$a + 0 = a$
$6 \cdot 1 = 6$	$a \cdot 1 = a$
Zero Property	
$5 \cdot 0 = 0$	$a \cdot 0 = 0$
$0 \cdot 5 = 0$	$0 \cdot a = 0$

To add or subtract decimals, align the decimal points.

To multiply decimals, first multiply as if the factors were whole numbers. Then, mark off from right to left in the product, the sum of the number of decimal places in both factors.

To divide decimals, rewrite the divisor as a whole number by multiplying both the divisor and the dividend by a power of 10. Annex zeros in the dividend as needed. Place the decimal point in the quotient above the decimal point in the dividend.

1-5 Measuring in Metric Units

Math Understandings

- The metric system of measurement is a decimal system based on powers of ten.
- The **meter** is the basic unit for length; **gram** is the basic unit of mass; and **liter** is the basic unit of capacity.
- The prefixes *deci, centi,* and *milli* describe measures that are less than a basic unit. The prefixes *deca, hecto,* and *kilo* describe measures that are more than a basic unit.

1-6 Comparing and Ordering Integers

Math Understandings

- Integers are the set of positive whole numbers (positive integers), their opposites (negative integers), and zero (neither positive nor negative).

When numbers are compared on a number line, the numbers increase from left to right. The **absolute value** of a number is its distance from zero on a number line.

1-7 1-8 Adding, Subtracting, Multiplying, and Dividing Integers

Math Understandings

- Absolute value, properties, and the number line can be used to show why the rules for operations with integers make sense.

The product or quotient of two numbers with the same sign is positive; for two numbers with different signs, it is negative.

1-9 Order of Operations and the Distributive Property

Math Understandings

- Mathematicians have agreed upon an order of operations to avoid confusion.
- The Distributive Property allows you to evaluate expressions that have a number multiplying a sum or a difference.

The Order of Operations

- Work inside grouping symbols.
- Multiply and divide in order from left to right.
- Add and subtract in order from left to right.

The Distributive Property	
Arithmetic	Algebra
$9(4 + 5) = 9(4) + 9(5)$	$a(b + c) = a(b) + a(c)$
$5(8 - 2) = 5(8) - 5(2)$	$a(b - c) = a(b) - a(c)$

1-10 Mean, Median, Mode, and Range

Math Understandings

- There are different numerical methods for describing the "center" of a numerical data set.
- The mean, median, and mode are common measures for the central tendency of a data set.
- The range of a set of data is the difference between the greatest and least values.

The **mean** of a data set is the sum of the data values divided by the number of data items. The **median** of a data set is the middle value when the data are arranged in numerical order. The median for an even number of data items is the mean of the two middle items. The **mode** of a data set is the data item that occurs most often. There can be more than one mode for a data set. The **range** of a data set describes the spread of the data, the difference between the greatest and least values.

Additional Professional Development Opportunities

Math Background Notes for Chapter 1: Every lesson has a Math Background in the PLAN section.

Research Overview, Mathematics Strands
Additional support for these topics and more is in the front of the Teacher's Edition.

LessonLab
LessonLab, a Pearson Education company, offers comprehensive, facilitated professional development designed to help teachers to improve student achievement. To learn more, please visit lessonlab.com.

Chapter 1 Resources

	1-1	1-2	1-3	1-4	1-5	1-6	1-7	1-8	1-9	1-10	For the Chapter
Print Resources											
L3 Practice	●	●	●	●	●	●	●	●	●	●	
L1 Adapted Practice	●	●	●	●	●	●	●	●	●	●	
L3 Guided Problem Solving	●	●	●	●	●	●	●	●	●	●	
L2 Reteaching	●	●	●	●	●	●	●	●	●	●	
L4 Enrichment	●	●	●	●	●	●	●	●	●	●	
L3 Daily Notetaking Guide	●	●	●	●	●	●	●	●	●	●	
L1 Adapted Daily Notetaking Guide	●	●	●	●	●	●	●	●	●	●	
L3 Vocabulary and Study Skills Worksheet	●		●			●		●		●	●
L3 Daily Puzzles	●	●	●	●	●	●	●	●	●	●	
L3 Activity Labs	●	●	●	●	●	●	●	●	●	●	
L3 Checkpoint Quiz			●			●					
L3 Chapter Project											●
L2 Below Level Chapter Test											●
L3 Chapter Test											●
L4 Alternative Assessment											●
L3 Cumulative Review											●
Spanish Resources ELL											
L3 Practice	●	●	●	●	●	●	●	●	●	●	
L3 Vocabulary and Study Skills Worksheet	●		●			●		●		●	●
L3 Checkpoint Quiz			●			●					
L2 Below Level Chapter Test											●
L3 Chapter Test											●
L4 Alternative Assessment											●
L3 Cumulative Review											●
Transparencies											
Check Skills You'll Need	●	●	●	●	●	●	●	●	●	●	
Additional Examples	●	●	●	●	●	●	●	●	●	●	
Problem of the Day	●	●	●	●	●	●	●	●	●	●	
Classroom Aid		●	●	●				●	●	●	
Student Edition Answers	●	●	●	●	●	●	●	●	●	●	●
Lesson Quiz	●	●	●	●	●	●	●	●	●	●	
Test-Taking Strategies											●
Technology											
Interactive Textbook Online	●	●	●	●	●	●	●	●	●	●	●
StudentExpress™ CD-ROM	●	●	●	●	●	●	●	●	●	●	●
Success Tracker™ Online Intervention	●	●	●	●	●	●	●	●	●	●	●
TeacherExpress™ CD-ROM	●	●	●	●	●	●	●	●	●	●	●
PresentationExpress™ with QuickTake Presenter CD-ROM	●	●	●	●	●	●	●	●	●	●	●
ExamView® Assessment Suite CD-ROM	●	●	●	●	●	●	●	●	●	●	●
MindPoint® Quiz Show CD-ROM											●
Prentice Hall Web Site: PHSchool.com	●	●	●	●	●	●	●	●	●	●	●

Also available: **Prentice Hall Assessment System**
- Progress Monitoring Assessments
- Skills and Concepts Review
- Test Prep Workbook

Other Resources
Algebra Readiness Tests
All-in-One Student Workbook
All-in-One Student Workbook, Adapted Version
Multilingual Handbook
Solution Key
Math Notes Study Folder
Spanish Cumulative Assessment

Where You Can Use the Lesson Resources

Here is a suggestion, following the four-step teaching plan, for how you can incorporate Differentiated Instruction Resources into your teaching.

	Instructional Resources L3	**Differentiated Instruction Resources**
1. Plan		
Preparation Read the Math Background in the Teacher's Edition to connect this lesson with students' previous experience. **Starting Class** **Check Skills You'll Need** Assign these exercises to review prerequisite skills. **New Vocabulary** Help students pre-read the lesson by pointing out the new terms introduced in the lesson.	**Math Background** **Math Understandings** **Transparencies & PresentationExpress™ with QuickTake Presenter CD-ROM** Check Skills You'll Need Problem of the Day **Resources** Vocabulary and Study Skills	**Spanish Support** ELL Vocabulary and Study Skills
2. Teach		
L3 **Guided Instruction** Use the Activity Labs to build conceptual understanding. Teach each Example. Use the Teacher's Edition side column notes for specific teaching tips, including Error Prevention notes. Use the Additional Examples found in the side column (and on transparency and PowerPoint) as an alternative presentation for the content. After each Example, assign the Quick Check exercise for that Example to get an immediate assessment of student understanding. Use the Closure activity in the Teacher's Edition to help students attain mastery of lesson content.	**Student Edition** Activity Lab **Resources** Daily Notetaking Guide Activity Lab **Transparencies & PresentationExpress™ with QuickTake Presenter CD-ROM** Additional Examples Classroom Aids **ExamView® Assessment Suite CD-ROM**	**Teacher's Edition** Every lesson includes suggestions for working with students who need special attention. L1 Special Needs L2 Below Level L4 Advanced Learners ELL English Language Learners **Resources** L1 Adapted Daily Notetaking Guide **Multilingual Handbook**
3. Practice		
Assignment Guide **Check Your Understanding** Use these questions to check students' understanding before you assign homework. **Homework Exercises** Assign homework from these leveled exercises in the Assignment Guide. A Practice by Example B Apply Your Skills C Challenge Test Prep and Mixed Review **Homework Quick Check** Use these key exercises to quickly check students' homework.	**Transparencies & PresentationExpress™ with QuickTake Presenter CD-ROM** Student Answers **Resources** Practice Guided Problem Solving Vocabulary and Study Skills Activity Lab Daily Puzzles **ExamView® Assessment Suite CD-ROM**	**Spanish Support** ELL Practice ELL Vocabulary and Study Skills **Resources** L1 Adapted Practice L4 Enrichment
4. Assess & Reteach		
Lesson Quiz Assign the Lesson Quiz to assess students' mastery of the lesson content. **Checkpoint Quiz** Use the Checkpoint Quiz to assess student progress over several lessons.	**Transparencies & PresentationExpress™ with QuickTake Presenter CD-ROM** Lesson Quiz **Resources** Checkpoint Quiz	**Resources** L2 Reteaching ELL Checkpoint Quiz Success Tracker™ Online Intervention **ExamView® Assessment Suite CD-ROM**

KEY L1 Special Needs L2 Below Level L3 For All Students L4 Advanced, Gifted ELL English Language Learners

CHAPTER 1 Decimals and Integers

What You've Learned

- In a previous course, you compared and ordered whole numbers.
- You used the properties of addition and multiplication to add, subtract, multiply, and divide whole numbers.
- You used the order of operations and the Distributive Property to simplify expressions with whole numbers.

Check Your Readiness

GO for Help

For Exercises	See Skills Handbook
1–2	p. 654
3–6	p. 657
7–11	p. 658
12–16	p. 659
17–21	p. 660

Comparing and Ordering Whole Numbers

Use > or < to compare the whole numbers.

1. 72 ■ 720 < 2. 3,972 ■ 3,927 >

Dividing Whole Numbers

Find each quotient.

3. $8\overline{)296}$ 37 4. $9\overline{)684}$ 76 5. $11\overline{)2{,}376}$ 216 6. $68\overline{)14{,}552}$ 214

Place Value and Decimals

Write the value of the underlined digit.

7. 24.<u>3</u>5 3 tenths 8. 4.08<u>6</u> 6 thousandths 9. 17<u>9</u>.8 9 ones 10. 59.0<u>3</u> 3 hundredths 11. 1.046<u>7</u> 7 ten-thousandths

Reading and Writing Decimals

Write each number in words. Use *tenths, hundredths,* or *thousandths.* 12–16. See margin.

12. 421.5 13. 5,006.25 14. 15.004 15. 0.329 16. 710.413

Rounding Decimals

Round to the nearest hundredth.

17. 34.124 34.12 18. 278.786 278.79 19. 3.602 3.60 20. 81.796 81.80 21. 16.999 17.00

12. four hundred twenty-one and five tenths
13. five thousand six and twenty-five hundredths
14. fifteen and four thousandths
15. three hundred twenty-nine thousandths
16. seven hundred ten and four hundred thirteen thousandths

Decimals and Integers

Answers are in the back of the textbook.

For intervention, direct students to:

Comparing and Ordering Whole Numbers
Skills Handbook, p. 654

Dividing Whole Numbers
Skills Handbook, p. 657

Place Value and Decimals
Skills Handbook, p. 658

Reading and Writing Decimals
Skills Handbook, p. 659

Rounding Decimals
Skills Handbook, p. 660

Spanish Vocabulary/Study Skills ELL

Vocabulary/Study Skills L3

1A: Graphic Organizer For use before Lesson 1-1

Study Skill: As you begin a new textbook, look through the table of contents to see what kind of information you will be learning during the year. Notice that some of the topics were introduced last year. Get a head start by reviewing your old notes and problems.

Write your answers.

1. What is the chapter title? Decimals and Integers
2. How many lessons are there in this chapter? 10
3. What is the topic of the Math at Work page? Reading Your Textbook
4. What is the topic of the Test-Taking Strategies page? Writing Gridded Responses
5. Complete the graphic organizer below as you work through the chapter.
   - In the center, write the title of the chapter.
   - When you begin a lesson, write the lesson name in a rectangle.
   - When you complete a lesson, write a skill or key concept in a circle linked to that lesson block.
   - When you complete the chapter, use this graphic organizer to help you review.

Check students' diagrams.

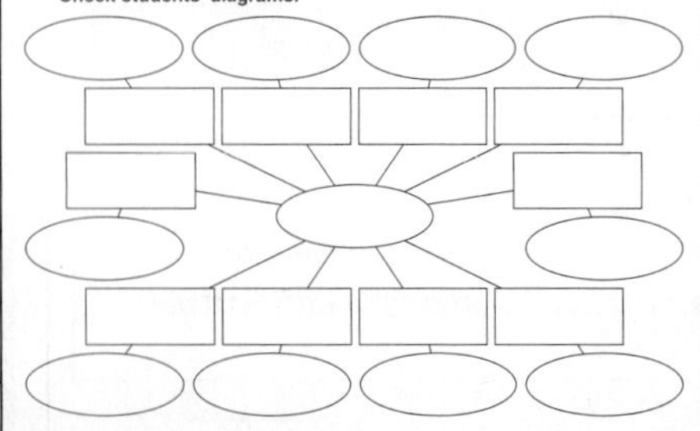

Chapter 1 Overview

In this chapter, students will build on their knowledge of mental math and estimation strategies as they compute with decimals and measures in metric units. They apply their prior knowledge of algebraic properties and number sense as they learn to compare and order integers, and to add, subtract, multiply, and divide them. They also apply skills learned in the chapter to find measures of central tendency and range.

Activating Prior Knowledge

In this chapter, students will build on their knowledge of decimal concepts and of estimation strategies to compute with decimals and metric measurements. They also draw upon their understanding of integers to compute with them. Ask questions such as:

- *How can you estimate sums of whole numbers?* **One way: round each addend to the greatest place and add.**
- *How can you estimate the product of two whole number factors?* **One way: round each factor to a number that makes the computation easier.**

What You'll Learn Next

- In this chapter, you will compare and order integers.
- You will use the properties of addition and multiplication to add, subtract, multiply, and divide decimals and integers.
- You will use the order of operations and the Distributive Property to simplify expressions with decimals and integers.

Key Vocabulary

- absolute value (p. 31)
- additive inverses (p. 38)
- compatible numbers (p. 5)
- integers (p. 31)
- mean (p. 53)
- median (p. 54)
- mode (p. 54)
- opposites (p. 31)
- order of operations (p. 48)
- outlier (p. 53)
- range (p. 55)

Problem Solving Application On pages 64 and 65, you will work an extended activity on integers.

1-1

1. Plan

Objective
To estimate using rounding, front-end estimation, and compatible numbers

Examples
1 Estimating by Rounding
2 Estimating by Front-End Estimation
3 Estimating Using Compatible Numbers

Math Understandings: p. 2C

Math Background

Learning to estimate the result of a computation is as important as learning to calculate the exact result. *Rounding* is probably the most commonly used estimation technique. When estimating sums and differences, however, *front-end estimation* often yields a better estimate than rounding. Also, when estimating quotients, it is frequently easier to work with *compatible numbers* instead of rounded numbers. In this lesson, students have an opportunity to practice all three techniques.

More Math Background: p. 2C

Lesson Planning and Resources

See p. 2E for a list of the resources that support this lesson.

Check Skills You'll Need
Use student page, transparency, or PowerPoint. For intervention, direct students to:
Rounding Whole Numbers
Skills Handbook, p. 655

1-1 Using Estimation Strategies

Check Skills You'll Need

1. **Vocabulary Review** How is an estimate different from a guess? **See below.**

Round to the place of the underlined digit.

2. 82,729 **83,000**
3. 449 **400**
4. 24,106 **24,110**
5. 3,528 **3,500**

for Help
Skills Handbook p. 655

Check Skills You'll Need

1. **An estimate is an answer with a calculation; a guess is an answer without a calculation.**

What You'll Learn

To estimate using rounding, front-end estimation, and compatible numbers

New Vocabulary compatible numbers

Why Learn This?

You can estimate an answer before you calculate it. Sometimes an estimate is all you need.

You can use rounding to estimate sums, differences, and products. You can use a number line to help you round decimals.

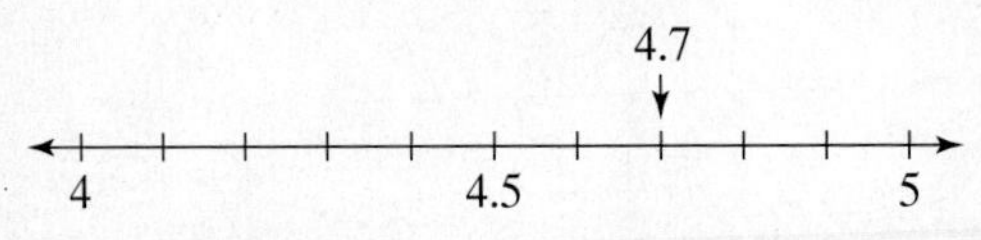

4.7 is between 4.0 and 5.0. You can round 4.7 to 5.0.

When you estimate, use the symbol ≈, which means "is approximately equal to."

EXAMPLE Estimating by Rounding

1 **Biology** The span of an eagle ray's fins is 3.27 m. The span of her baby's fins is 0.88 m. Estimate the difference between the spans. Round to the nearest whole number before you find the difference.

$3.27 - 0.88 \approx 3 - 1$ ← **Round to the nearest whole number.**

$= 2$ ← **Subtract.**

The difference in spans is about 2 m.

For help with rounding decimals, go to Skills Handbook p. 660.

Quick Check

1. Estimate. First round to the nearest whole number.
   a. $1.75 + 0.92$ **3**
   b. $14.34 - 7.8$ **6**
   c. 4.90×6.25 **30**

Differentiated Instruction Solutions for All Learners

Special Needs L1	**Below Level** L2
In Example 2, write the extra step of adding \$7 and \$2 for students so that they can see where the \$9 estimate came from. **learning style: visual**	Have students visualize rounding on a number line. For instance, this diagram shows that, to the nearest whole number, 9.25 is rounded to 9. (number line: 9.00, **9.25**, 9.50, 10.00) **learning style: visual**

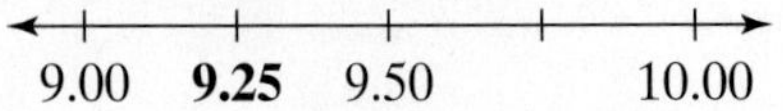

You can use front-end estimation when you want to find the sum of several numbers. First add the front numbers. Then estimate the sum of the lesser numbers and adjust the estimate.

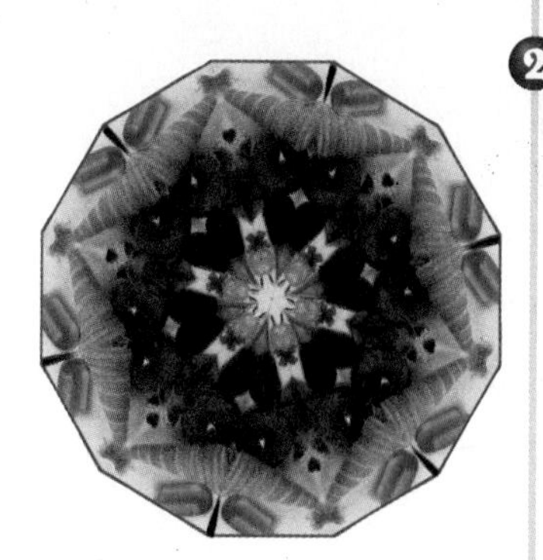

EXAMPLE Estimating by Front-End Estimation

2 **Shopping** You have \$10 and want to purchase several gift items. You select a kaleidoscope for \$4.39, gift wrap for \$1.49, and a card for \$2.95. Estimate the total cost to determine whether you have enough money.

Step 1 Add the front-end digits.

$$\begin{array}{r} \$4.39 \\ \$1.49 \\ +\ \$2.95 \\ \hline \$7 \end{array}$$

Step 2 Estimate the sum of the cents to the nearest dollar.

$$\left.\begin{array}{r} \$4.39 \\ \$1.49 \end{array}\right\} \leftarrow \text{about } \$1$$
$$\$2.95 \quad \leftarrow \text{about } \$1$$

$$\$7 + \$2 = \$9$$

The total cost is about \$9. You have enough money.

Quick Check

2. Estimate to the nearest dollar the total cost of a dog collar for \$5.79, a dog toy for \$2.48, and a dog dish for \$5.99. **about \$14**

Compatible numbers are numbers that are easy to compute mentally. You can use compatible numbers to estimate. Simply adjust the numbers in the problem to ones that are close to make the calculation easier. Compatible numbers are particularly useful for finding quotients.

EXAMPLE Estimating Using Compatible Numbers

3 **Movies** Suppose you have saved \$50.25. About how many DVDs can you buy from category D in the table?

DVD Price List

Category	Price
A	\$23.95
B	\$15.95
C	\$12.95
D	\$7.95

$50.25 \div 7.95$ ← **Use division.**

$48 \div 8$ ← **Choose compatible numbers such as 48 and 8.**

6 ← **Simplify.**

You can buy about six DVDs from category D.

Quick Check

3. Your friend says you can buy about twice as many DVDs from category D as from category B. Is your friend correct? Explain.

3. Yes; you can buy about 48 ÷ 16, or 3, DVDs from category B. This is half the number of DVDs you can buy from category D.

2. Teach

Activity Lab

Use before the lesson.

All in One Teaching Resources

Activity Lab 1-1: Using Estimates

Guided Instruction

Example 1

Some students might have difficulty distinguishing the symbol for "is approximately equal to" from the equal sign. Discuss the symbol ≈; ask: *How would you compare this symbol to an equal sign?* **Sample: It has two identical segments, like an equal sign, but they are "wavy."**

Example 3

Make sure students understand how to identify compatible numbers. Ask: *Why do you use 48 in the division instead of 50?* **Sample: It is easy to compute 48 ÷ 8 mentally, so 48 is compatible with 8.**

PowerPoint

Additional Examples

1 The length of a Colombian black spider monkey's body is 58.31 cm. The length of its tail is 78.96 cm. To the nearest tenth of a centimeter, estimate the total length of the spider monkey. **about 137.3 cm**

2 You order a taco that costs \$3.79, a juice that costs \$.89, and a yogurt that costs \$1.39. Estimate the total cost of your order. **about \$6**

3 Suppose you have saved \$61.80. About how many CDs can you buy if each costs \$15.95? **about 4 CDs**

All in One Teaching Resources

- Daily Notetaking Guide 1-1 L3
- Adapted Notetaking 1-1 L1

Closure

- Describe three methods for estimating the results of decimal computations. **rounding, front-end estimation, compatible numbers**

Advanced Learners L4

Your shopping cart contains food items with prices \$1.65, \$4.97, \$0.79, \$2.29, \$0.39, and \$2.09. Estimate the amount of change you will receive if you give the cashier \$20. **about \$8**

learning style: visual

English Language Learners ELL

Ask students to work in pairs. Have one third of the pairs focus on explaining how to estimate by rounding using examples, words, and symbols. The other two thirds should do the same with estimation using compatible numbers and front-end estimation. Have pairs explain their strategies.

learning style: verbal

3. Practice

Assignment Guide

Check Your Understanding

Go over Exercises 1–5 in class before assigning the Homework Exercises.

Homework Exercises

A	Practice by Example	6–24
B	Apply Your Skills	25–32
C	Challenge	33
Test Prep and Mixed Review		34–37

Homework Quick Check

To check students' understanding of key skills and concepts, go over Exercises 14, 22, 27, 31, and 32.

Differentiated Instruction Resources

Adapted Practice 1-1 L1

Practice 1-1 Using Estimation Strategies L3

Use rounding to estimate the nearest half-dollar.

1. $4.85 + 1.47 — $6.50
2. $6.79 − 3.95 — $3.00
3. $14.19 + 5.59 — $19.50
4. $25.43 − 21.20 — $4.00

Use front-end estimation to estimate each sum.

5. 4.76 + 6.15 — 11
6. 1.409 + 3.512 — 5
7. 2.479 + 6.518 — 9
8. 3.17 + 2.72 — 6
9. 9.87 + 2.16 — 12
10. 5.89 + 7.21 — 13

Use compatible numbers to estimate each quotient.

11. 76.32 ÷ 24.98 — 3
12. 42.693 ÷ 4.7 — 10
13. 54.36 ÷ 11.001 — 5

Use any estimation strategy to calculate. Tell which strategy you used.
Sample answers are given.

14. $66.93 + $72.18 + $69.18 + $71.94 + $65.75 — $346; rounding
15. 93.26 − 69.78 — 20; rounding
16. 51.12 × 87.906 — 4,500; rounding
17. 457.03 + 592.8 — 1,100; rounding
18. 702 ÷ 61 — 12; compatible numbers
19. 12.87 + 14.31 + 15.09 — 42; front-end estimation
20. 536 ÷ 41 — 13; compatible numbers

Find each estimate.

21. A rare truffle once sold for $13.20 for a 0.44 oz can. Approximately how much would 1 lb of this truffle cost?
about $480
22. The length of the longest loaf of bread measured 1,405 ft $1\frac{3}{4}$ in. It was cut into slices $\frac{1}{2}$ in. thick. How many slices were there?
about 33,600 slices

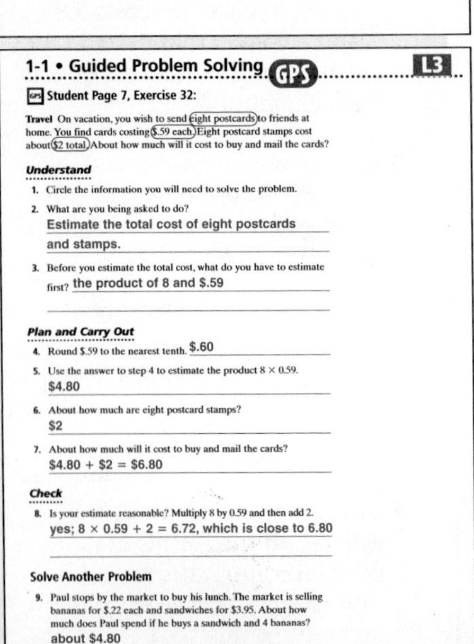

1-1 • Guided Problem Solving GPS L3

Student Page 7, Exercise 32:

Travel On vacation, you wish to send eight postcards to friends at home. You find cards costing $.59 each. Eight postcard stamps cost about $2 total. About how much will it cost to buy and mail the cards?

Understand

1. Circle the information you will need to solve the problem.
2. What are you being asked to do?
Estimate the total cost of eight postcards and stamps.
3. Before you estimate the total cost, what do you have to estimate first? the product of 8 and $.59

Plan and Carry Out

4. Round $.59 to the nearest tenth. $.60
5. Use the answer to step 4 to estimate the product 8 × 0.59.
$4.80
6. About how much are eight postcard stamps?
$2
7. About how much will it cost to buy and mail the cards?
$4.80 + $2 = $6.80

Check

8. Is your estimate reasonable? Multiply 8 by 0.59 and then add 2.
yes; 8 × 0.59 + 2 = 6.72, which is close to 6.80

Solve Another Problem

9. Paul stops by the market to buy his lunch. The market is selling bananas for $.22 each and sandwiches for $3.95. About how much does Paul spend if he buys a sandwich and 4 bananas?
about $4.80

Check Your Understanding

1. **Vocabulary** Explain the difference between rounded numbers and compatible numbers.

1. Rounded numbers are numbers rounded to the nearest whole number, whereas compatible numbers are chosen because they are easy to compute mentally.

2. **Error Analysis** A student estimated the cost of three shirts. What error was made?

$$\$17.99 + \$23.45 + \$20.15 \approx 18 + 23 + 2 = 43$$

The student rounded 20.15 to 2 instead of 20.

Match each problem with the method you would use to estimate.

3. 22.4 ÷ 3.21 **C**
4. 7.3 + 22.4 + 6.5 + 13.6 **A**
5. 21.8 − 17.4 **B**

A. front-end estimation
B. rounding
C. compatible numbers

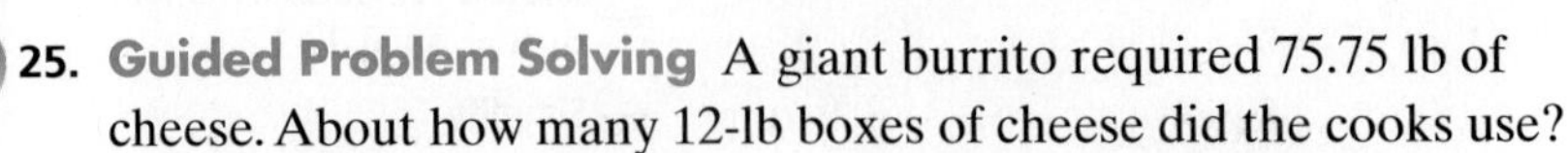

Homework Exercises

For more exercises, see Extra Skills and Word Problems.

GO for Help

For Exercises	See Examples
6–11	1
12–17	2
18–24	3

A

Estimate. First round to the nearest whole number.

6. 5.82 − 1.76 **4**
7. 10.13 + 1.46 **11**
8. 6.07 × 3.29 **18**
9. 9.86 × 9.13 **90**
10. 21.18 − 17.92 **3**
11. 11.53 + 7.23 **19**

Use front-end estimation to estimate each sum.

12. 5.43 + 2.67 **8**
13. 8.09 + 11.24 **19**
14. 7.18 + 5.89 **13**
15. 4.39 + 9.57 **14**
16. 24.21 + 16.03 **40**
17. 3.62 + 2.31 **6**

Use compatible numbers to estimate each quotient.

18. 76.5 ÷ 8.8 **9** or 8
19. 19.45 ÷ 4.92 **4**
20. 27.36 ÷ 3.14 **9**
21. 103.6 ÷ 9.72 **10**
22. 32.2 ÷ 7.56 **4**
23. 3.963 ÷ 1.79 **2**

24. **Trains** The world's fastest train travels about 162 mi/h. Estimate how far the train travels in 4.75 h. **about 800 mi**

GPS **B**

25. **Guided Problem Solving** A giant burrito required 75.75 lb of cheese. About how many 12-lb boxes of cheese did the cooks use?
   - Which operation will you use to solve the problem?
   - Which numbers will you use to make your estimate?

   about 6 boxes

26. You can buy 12 magnets for $34.68, or 3 magnets for $9.57. Estimate to decide which is the better buy. Explain your reasoning.
26–27. See left.
27. **Reasoning** When you estimate to determine if you have enough money, why should you underestimate the amount you have?

26. The first choice; the first choice is less than $3 per magnet, while the second choice is more than $3 per magnet.

27. Answers may vary. Sample: Underestimates ensure you have enough money.

Homework Video Tutor
Visit: PHSchool.com
Web Code: are-0101

Estimate the cost of each group of items.

28–30. Answers may vary. Samples are given.

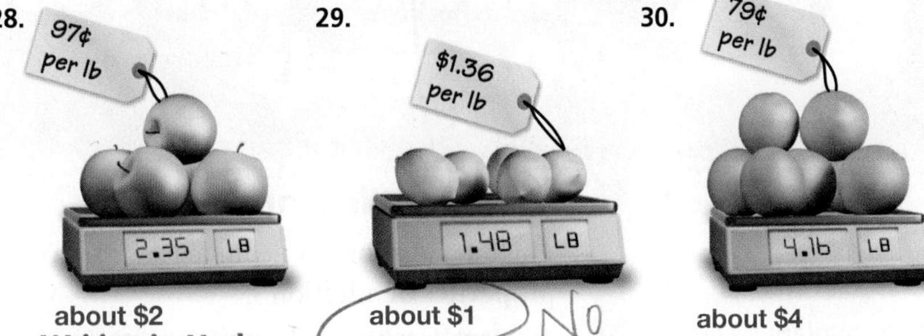

28. about $2 **29.** about $1 NO **30.** about $4

31. **Writing in Math** When is an estimate *not* as useful as an exact answer? Give an example.

31. Answers may vary. Sample: Your GPA is an exact answer and is not an estimate.

32. On vacation, you wish to send eight postcards to friends at home. You find cards costing $.59 each. Eight postcard stamps cost about $2 total. About how much will it cost to buy and mail the cards? about $7

33. **Challenge** Your family is going to visit a relative who lives 389.2 mi from your home. The family car gets 29.6 miles per gallon of gasoline. If the price of a gallon of gas is $2.16, about how much will the gasoline for the trip cost? about $26; $(390 \div 30) \times \$2$

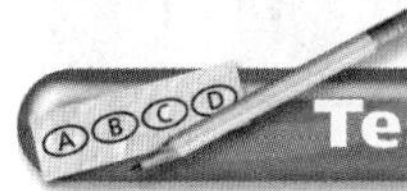

Test Prep and Mixed Review — Practice

Multiple Choice

34. The table shows the number of students at Stewart Middle School who belong to several groups. Which conclusion is reasonable? B

Stewart Middle School Groups

Group	Number of Students
Pep Club	123
Band	68
Drama Club	32
Student Council	54

(A) Student Council has twice as many members as Drama Club.
(B) Pep Club has about 4 times as many members as Drama Club.
(C) Band has about twice as many members as Pep Club.
(D) Student Council has the fewest members.

35. Carrie earned $612 in a year at a part-time job. Which is the best estimate of her weekly earnings? J

(F) $18 (G) $16 (H) $14 (J) $12

GO for Help

For Exercises	See Skills Handbook
36–37	p. 654

Order the numbers from least to greatest.

36. 4.0 4,004 40 403 — 4.0, 40, 403, 4,004

37. 761.8 768.1 768.0 706.8 — 706.8, 761.8, 768.0, 768.1

4. Assess & Reteach

Lesson Quiz

Use rounding to estimate to the nearest tenth.

1. 5.75 + 1.34 about 7.1
2. 64.54 − 30.19 about 34.3

Use front-end estimation or compatible numbers to estimate.

3. 31.28 + 3.05 about 34
4. 59.63 ÷ 9.84 about 6

Alternative Assessment

Have students work in pairs. Each student writes a decimal problem for each operation: addition, subtraction, multiplication, and division. Then have partners exchange papers and estimate the answers. Have partners discuss the technique they used for each type of problem.

Test Prep

Resources

For additional practice with a variety of test item formats:

- Test-Taking Strategies, p. 59
- Test Prep, p. 63
- Test-Taking Strategies with Transparencies

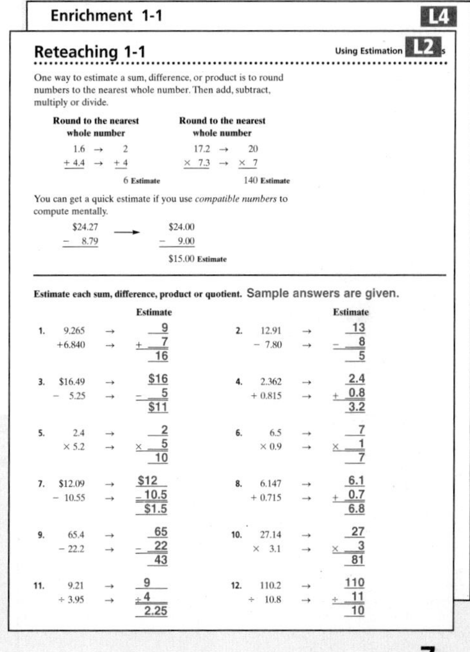

Enrichment 1-1 L4

Reteaching 1-1 Using Estimation L2

One way to estimate a sum, difference, or product is to round numbers to the nearest whole number. Then add, subtract, multiply or divide.

Round to the nearest whole number: 1.6 → 2; + 4.4 → + 4; 6 Estimate

Round to the nearest whole number: 17.2 → 20; × 7.3 → × 7; 140 Estimate

You can get a quick estimate if you use *compatible numbers* to compute mentally.

$24.27 − 8.79 → $24.00 − 9.00 = $15.00 Estimate

Estimate each sum, difference, product or quotient. Sample answers are given.

	Problem	Estimate		Problem	Estimate
1.	9.265 + 6.840	9 + 7 = 16	2.	12.91 − 7.80	13 − 8 = 5
3.	$16.49 − 5.25	$16 − 5 = $11	4.	2.362 + 0.815	2.4 + 0.8 = 3.2
5.	2.4 × 5.2	2 × 5 = 10	6.	6.5 × 0.9	7 × 1 = 7
7.	$12.09 − 10.55	$12 − 10.5 = $1.5	8.	6.147 + 0.715	6.1 + 0.7 = 6.8
9.	65.4 − 22.2	65 − 22 = 43	10.	27.14 × 3.1	27 × 3 = 81
11.	9.21 ÷ 3.95	9 ÷ 4 = 2.25	12.	110.2 ÷ 10.8	110 ÷ 11 = 10

1-2

1. Plan

Objective
To add and subtract decimals and to do mental math using the properties of addition

Examples
1 Adding Decimals
2 Subtracting Decimals
3 Using Properties of Addition

Math Understandings: p. 2C

Professional Development

Math Background

The standard methods for adding and subtracting decimals require that the numbers be matched place for place. That is, tenths must be added to or subtracted from tenths, hundredths must be added to or subtracted from hundredths, and so on. For this reason, the first step in an addition or subtraction is a vertical alignment of the decimal points. Then digits with the same place value are matched and the addition or subtraction can proceed.

More Math Background: p. 2C

Lesson Planning and Resources

See p. 2E for a list of the resources that support this lesson.

Bell Ringer Practice

Check Skills You'll Need
Use student page, transparency, or PowerPoint. For intervention, direct students to:
Place Value and Decimals
Skills Handbook, p. 658

1-2 Adding and Subtracting Decimals

Check Skills You'll Need

1. **Vocabulary Review** What is *place value?* **1–5. See below.**

Write the value of the digit 8 in each number.

2. 0.385 3. 1.879
4. 812.4 5. 0.238

for Help Skills Handbook p. 658

What You'll Learn

To add and subtract decimals and to do mental math using the properties of addition

New Vocabulary Identity Property of Addition, Commutative Property of Addition, Associative Property of Addition

Why Learn This?

Sometimes an estimate is not good enough. You can add and subtract decimals to find an exact answer.

When you add decimals, you must align the decimal points so that corresponding place values are added. You may need to insert, or "annex," zeros so you have the same number of decimal places in each number.

EXAMPLE Adding Decimals

1 Music A marching band has 6.5 minutes to perform at a football game. Does the band have enough time to play "America the Beautiful" and "Yankee Doodle Dandy"?

Band Selections

Title	Min
"76 Trombones"	3.12
"Stars and Stripes Forever"	3.52
"Born in the U.S.A."	4.65
"America the Beautiful"	3.63
"Yankee Doodle Dandy"	2.5
"The Star-Spangled Banner"	4.5

Estimate $3.63 + 2.5 \approx 4 + 3$, or 7

Align the decimal points.

$$\begin{array}{r} 3.63 \\ +\ 2.50 \\ \hline 6.13 \end{array}$$

← **Insert a zero so both addends have the same number of decimal places.**

The band will play for 6.13 minutes. The band will have enough time.

Check for Reasonableness Since 6.13 is close to 7, the answer is reasonable.

Quick Check

1. How long will the band take to play "Stars and Stripes Forever" and "The Star-Spangled Banner"? **8.02 min**

GO Online

Video Tutor Help
Visit: PHSchool.com
Web Code: are-0775

Check Skills You'll Need

1. **Place value is the position and value of a digit in a decimal.**
2. **8 hundredths**
3. **8 tenths**
4. **8 hundreds**
5. **8 thousandths**

Differentiated Instruction Solutions for All Learners

Special Needs L1
Review number pairs that add up to multiples of 10. For example, numbers ending in 7 and 3, 6 and 4, or 2 and 8 can always be grouped together to make calculations easier.
learning style: verbal

Below Level L2
Give students simple additions such as the following to focus on the need to line up the decimal points.
0.4 + 0.01 **0.41** 0.12 + 8 **8.12**
0.5 + 16 + 0.03 **16.53**
learning style: visual

When you subtract decimals, you may need to regroup.

EXAMPLE **Subtracting Decimals**

2 Find 89.9 − 46.78.

Align the decimal points.	Regroup.	Subtract.
89.90 ← Insert a zero.	89.9̸0 (8 10)	89.9̸0 (8 10)
− 46.78	− 46.78	− 46.78
		43.12

Quick Check

2. Find 26.7 − 14.81. **11.89**

You can use the properties of addition to add mentally.

KEY CONCEPTS **Properties of Addition**

Identity Property of Addition The sum of 0 and a is a.

Arithmetic $5.6 + 0 = 5.6$ **Algebra** $a + 0 = a$

Commutative Property of Addition Changing the order of the addends does not change the sum.

Arithmetic $1.2 + 3.4 = 3.4 + 1.2$ **Algebra** $a + b = b + a$

Associative Property of Addition Changing the grouping of the addends does not change the sum.

Arithmetic $(2.5 + 6) + 4 = 2.5 + (6 + 4)$

Algebra $(a + b) + c = a + (b + c)$

EXAMPLE **Using Properties of Addition**

3 **Mental Math** Find 0.7 + 12.5 + 1.3.

What You Think

I should look for compatible numbers. The sum of 0.7 and 1.3 is 2. Then I can add 2 and 12.5 for a total of 14.5.

Why It Works

$0.7 + 12.5 + 1.3 = 0.7 + 1.3 + 12.5$ ← **Commutative Property of Addition**

$= (0.7 + 1.3) + 12.5$ ← **Associative Property of Addition**

$= 2 + 12.5 = 14.5$

Test Prep Tip

Look for compatible numbers when you add several numbers.

Quick Check

3. Use mental math to find 4.4 + 5.3 + 0.6. **10.3**

2. Teach

Activity Lab

Use before the lesson.

All in One Teaching Resources

Activity Lab 1-2: Adding Decimals on the Road

Guided Instruction

Example 3

Have students write "0.7"; "12.5"; "1.3"; "("; ")"; and two "+" signs on seven small cards, one number or symbol per card. Tell them to use the cards to form 0.7 + 12.5 + 1.3. Students can then change the order and insert the parentheses to act out the properties.

PowerPoint

Additional Examples

1 Find 2.15 + 7.632 + 16.5. **26.282**

2 You have 5.08 min left on a CD. How much time will you have left after you record a 2.5 minute song? **2.58 min**

3 Use mental math to find 0.2 + 15.7 + 3.8. **19.7**

All in One Teaching Resources

- Daily Notetaking Guide 1-2 **L3**
- Adapted Notetaking 1-2 **L1**

Closure

- *How do you add or subtract decimals?* **Sample: Align the decimal points; insert zeros so each number has the same number of decimal places; then add or subtract as with whole numbers.**

Advanced Learners L4

Fill in each box with one of the digits 1, 2, 3, 4, 5 to make a true statement. Each digit is used exactly once. **5.3 − 2.14**

□.□ − □.□□ = 3.16

learning style: visual

English Language Learners ELL

Make sure students understand the difference between annexing zeros and just writing zeros anywhere in a decimal number. Tell students that the word *annex* means to add on at the end.

learning style: verbal

3. Practice

Assignment Guide

Check Your Understanding

Go over Exercises 1–7 in class before assigning the Homework Exercises.

Homework Exercises

A Practice by Example	8–32
B Apply Your Skills	33–40
C Challenge	41
Test Prep and Mixed Review	42–47

Homework Quick Check

To check students' understanding of key skills and concepts, go over Exercises 13, 21, 34, 38, and 39.

Differentiated Instruction Resources

Adapted Practice 1-2 L1

Practice 1-2 — Adding and Subtracting Decimals L3

Identify each property shown.

1. (8.7 + 6.3) + 3.7 = 8.7 + (6.3 + 3.7) Associative Property
2. 9.06 + 0 = 9.06 Identity Property
3. 4.06 + 8.92 = 8.92 + 4.06 Commutative Property
4. 0 + 7.13 = 7.13 + 0 Commutative Property

Find each sum.

5. 4.6 + 8.79 13.39
6. 14.8 + 29.07 43.87
7. 20.16 + 15.703 35.863
8. 36.12 + 5.793 41.913
9. 8.9 + 2.14 + 7.1 18.14
10. 3.6 + 5.27 + 8.93 17.8
11. 107.5 + 6 113.5
12. 15.26 + 13.29 + 38.96 67.51
13. 46.21 + 53.942 100.152

Find each difference.

14. 8.7 − 2.03 6.67
15. 53.86 − 4.02 49.84
16. 14.59 − 8.3 6.29
17. 42.75 − 26.36 16.39
18. 53.86 − 16.47 37.39
19. 56.89 − 48.91 7.98
20. 5.06 − 3.297 1.763
21. 3.4 − 2.768 0.632
22. 5.002 − 4.3 0.702

Use the advertisement at the right. Find each cost.

23. 1 egg $.55
24. toast $.40
25. bacon $.75
26. milk $.50
27. 1 egg and milk $1.05
28. 1 egg and bacon $1.30

2 eggs, toast, bacon, milk	$2.75
1 egg, toast, bacon, milk	$2.20
toast, milk	$0.90
toast, bacon, milk	$1.65
1 egg, toast	$0.95

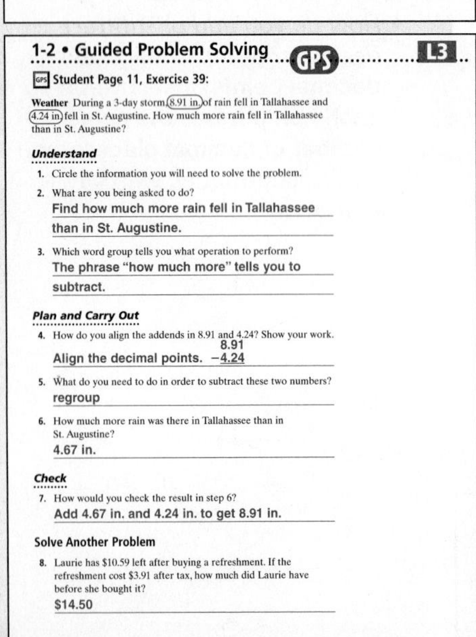

1-2 • Guided Problem Solving GPS L3

Student Page 11, Exercise 39:

Weather During a 3-day storm 8.91 in. of rain fell in Tallahassee and 4.24 in. fell in St. Augustine. How much more rain fell in Tallahassee than in St. Augustine?

Understand

1. Circle the information you will need to solve the problem.
2. What are you being asked to do? Find how much more rain fell in Tallahassee than in St. Augustine.
3. Which word group tells you what operation to perform? The phrase "how much more" tells you to subtract.

Plan and Carry Out

4. How do you align the addends in 8.91 and 4.24? Show your work. Align the decimal points. 8.91 −4.24
5. What do you need to do in order to subtract these two numbers? regroup
6. How much more rain was there in Tallahassee than in St. Augustine? 4.67 in.

Check

7. How would you check the result in step 6? Add 4.67 in. and 4.24 in. to get 8.91 in.

Solve Another Problem

8. Laurie has $10.59 left after buying a refreshment. If the refreshment cost $3.91 after tax, how much did Laurie have before she bought it? $14.50

Check Your Understanding

1. **Vocabulary** Which property allows you to regroup addends? **Assoc. Prop. of Add.**
2. **Number Sense** Which pair of addends forms compatible numbers in the expression 6.1 + 8.4 + 1.6? **8.4 and 1.6**

Mental Math Find each sum.

3. 23 + 17 **40**
4. 0.8 + 1.2 **2**
5. 16.9 + 2.1 **19**

Find each missing number.

6. 2.7 + ■ = 1.5 + 2.7 **1.5**
7. 144.98 + ■ = 144.98 **0**

Homework Exercises

For more exercises, see Extra Skills and Word Problems.

GO for Help

For Exercises	See Examples
8–17	1
18–26	2
27–32	3

A **Find each sum.**

8. 4.56 + 2.9 **7.46**
9. 102.8 + 3 **105.8**
10. 3.061 + 1.8 **4.861**
11. 0.582 + 7 **7.582**
12. 3.29 + 2 + 6.71 **12**
13. 1.913 + 0.08 + 3 **4.993**
14. 1.4 + 3.75 + 6 **11.15**
15. 7.58 + 2.4 + 0.101 **10.081**
16. 0.005 + 0.5 + 5 **5.505**
17. The average length of a king cobra is 3.7 m. The record length is 1.88 m longer than the average. How long is the record holder? **5.58 m**

Find each difference.

18. 5.3 − 0.12 **5.18**
19. 12.46 − 7.2 **5.26**
20. 3.102 − 0.89 **2.212**
21. 0.1305 − 0.066 **0.0645**
22. 0.08 − 0.002 **0.078**
23. 100 − 31.93 **68.07**
24. 2.101 − 1.22 **0.881**
25. 46.2 − 38.25 **7.95**
26. 0.15 − 0.015 **0.135**

Use mental math to find each sum.

27. 16.2 + 23.5 + 3.8 **43.5**
28. 24.4 + (5.6 + 11) **41.0**
29. 27.4 + 0 + 12.1 **39.5**
30. 9.2 + 1.8 + 0 **11.0**
31. (4.7 + 10.6) + 0.3 **15.6**
32. 8.5 + 6.3 + 1.5 **16.3**

B  

33. **Guided Problem Solving** A band can play as much as 8 minutes of music in a competition. If the band plays a 2.33-minute song and a 4.25-minute song, how many minutes does it have left? **1.42 min**
    - **Make a Plan** First find the total number of minutes in the two songs. Then subtract the total from the maximum time allowed.
    - **Carry Out the Plan** The total time is ■. The time remaining is ■.
34. **Writing in Math** How can you use the Identity Property of Addition to find 51.23 − 51.23 + 97.9? **See left.**

34. Since 51.23 − 51.23 = 0, you use the Ident. Prop. of Add. to say 0 + 97.9 = 97.9.

Visit: PHSchool.com
Web Code: are-0102

Use <, =, or > to complete each statement.

35. $3.45 + 2.9 \;\blacksquare\; 8.9 - 2.75$ >

36. $15 - 6.82 \;\blacksquare\; 32.18 - 24$ =

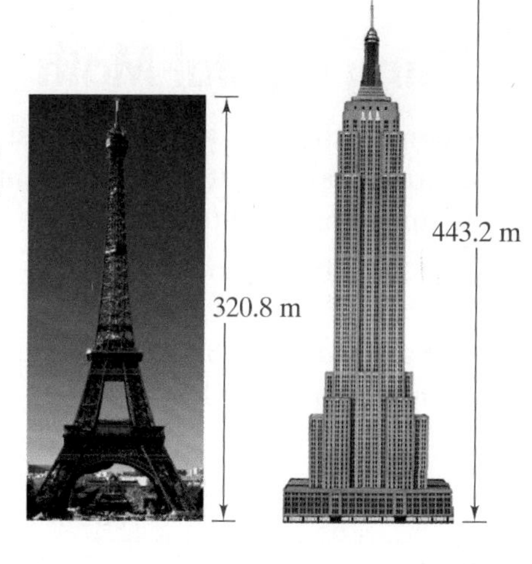

37. **Architecture** Find the difference in the heights of the Empire State Building and the Eiffel Tower. **122.4 m**

38. You decide to save some money. In week 1 you save \$4.20, in week 2 you save \$3.85, and in week 3 you save \$2.50. Estimate your total savings. Then find the exact amount you saved. **about \$11; \$10.55**

39. **Weather** During a 3-day storm, 8.91 in. of rain fell in Tallahassee and 4.24 in. fell in St. Augustine. How much more rain fell in Tallahassee than in St. Augustine? **4.67 in.**

GPS

40. The original price for a jacket is \$79.95. How much do you save if you buy the jacket on sale for \$62.79? **\$17.16**

Ⓒ **41.** **Challenge** The fastest mammal, the cheetah, can run as fast as 67.912 mi/h. The fastest fish, the sailfish, can swim as fast as 1.132 mi/min. Which animal is faster? How much faster is it? **sailfish; 0.008 mi/h**

Test Prep and Mixed Review

Practice

Multiple Choice

42. You receive \$50 for your birthday. You buy a book for \$14.95 and a baseball cap for \$24.95. How much money do you have left? **D**

Ⓐ \$60.10 Ⓑ \$39.90 Ⓒ \$29.90 Ⓓ \$10.10

43. The average rainfall for Houston, Texas, is 4.5 inches in October, 4.2 inches in November, and 3.7 inches in December. Find the total average rainfall for the last three months of the year. **H**

Ⓕ 11.4 in. Ⓖ 11.9 in. Ⓗ 12.4 in. Ⓙ 12.9 in.

44. Carla bought a CD for \$12.99 and a book for \$7.29. She also rented a movie for \$3.99. Which expression can be used to find the best estimate of the total amount Carla spent, not including tax? **D**

Ⓐ $12 + 7 + 3$ Ⓒ $13 + 7 + 3$

Ⓑ $12 + 7 + 4$ Ⓓ $13 + 7 + 4$

GO for Help

For Exercises	See Skills Handbook
45–47	p. 656

Find each product.

45. 17×8 **136** **46.** 15×0 **0** **47.** 4×56 **224**

4. Assess & Reteach

Lesson Quiz

Find each sum or difference.

1. $15 - 8.72$ **6.28**
2. $3.91 + 11.8$ **15.71**
3. $0.048 + 4.65$ **4.698**
4. $35.9 - 18.33$ **17.57**

Reteaching 1-2 — Adding and Subtracting **L2**

Add 3.19 + 6.098 + 26.7.

① Round to estimate.
3.19 → 3
6.098 → 6
+ 26.7 → + 27
36

② Line up the decimal points.
3.19
6.098
+ 26.700

③ Write zeros. Then add.
3.190
6.098
+ 26.700
35.988

Compare to make sure your answer is reasonable: 35.988 is close to 36.

Subtract 8.7 − 4.97.

① Round to estimate.
8.7 → 9
− 4.97 → − 5
4

② Line up the decimal points.
8.7
− 4.97

③ Write zeros. Then subtract.
8.70
− 4.97
3.73

Compare to make sure your answer is reasonable: 3.73 is close to 4.

Find each sum or difference.

1. 46.2 − 34.09 **12.11**
2. 3.31 + 9.075 **12.385**
3. 9.06 − 7.2 **1.86**
4. 84.32 + 6.94 **91.26**
5. 8.037 + 1.9 **9.937**
6. 10.6 − 4.59 **6.01**
7. 4.102 + 7.7 **11.802**
8. 5.4 − 1.6 **3.8**
9. 7.09 + 4.3 + 20.1 **31.49**
10. 0.392 − 0.26 **0.132**
11. 15.64 − 8.5 **7.14**
12. 8.709 + 3.2 **11.909**

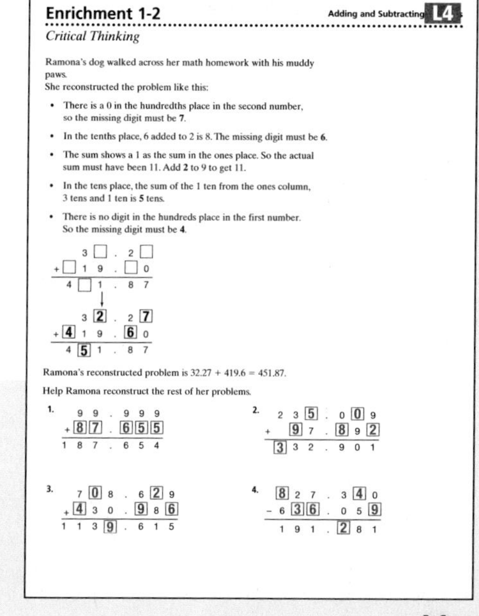

Enrichment 1-2 — Adding and Subtracting **L4**

Critical Thinking

Ramona's dog walked across her math homework with his muddy paws.
She reconstructed the problem like this:

- There is a 0 in the hundredths place in the second number, so the missing digit must be **7**.
- In the tenths place, 6 added to 2 is 8. The missing digit must be **6**.
- The sum shows a 1 as the sum in the ones place. So the actual sum must have been 11. Add **2** to 9 to get 11.
- In the tens place, the sum of the 1 ten from the ones column, 3 tens and 1 ten is **5** tens.
- There is no digit in the hundreds place in the first number. So the missing digit must be **4**.

3 □ . 2 □
+ □ 1 9 . □ 0
4 □ 1 . 8 7

↓

3 [2] . 2 [7]
+ [4] 1 9 . [6] 0
4 [5] 1 . 8 7

Ramona's reconstructed problem is 32.27 + 419.6 = 451.87.

Help Ramona reconstruct the rest of her problems.

1. 9 9 . 9 9 9
+ [8] [7] . [6] [5] [5]
1 8 7 . 6 5 4

2. 2 3 [5] . 0 [0] 9
+ [9] 7 . [8] 9 [2]
[3] 3 2 . 9 0 1

3. 7 [0] 8 . 6 [2] 9
+ [4] 3 0 . [9] 8 [6]
1 1 3 [9] . 6 1 5

4. [8] 2 7 . 3 [4] 0
− 6 [3] [6] . 0 5 [9]
1 9 1 . [2] 8 1

Alternative Assessment

Have students write and solve Exercises 8–16 and 18–26 on graph paper using vertical format. Stress that all the decimal points and all digits with the same place value must be aligned.

Test Prep

Resources

For additional practice with a variety of test item formats:

- Test-Taking Strategies, p. 59
- Test Prep, p. 63
- Test-Taking Strategies with Transparencies

E tension

Using Mental Math

Students can use the mental math strategy of compensation to find sums and differences without pencil or paper.

Guided Instruction

Sometimes when people go shopping, they have to make a quick estimate, in their heads, of how much they have spent in order to make sure they have enough money for all they intend to buy. Discuss that to do so, they can round some prices up and others down to get a rough estimate. Then point out that when people use this *estimation* strategy, they are using an approach similar to the *mental math* strategy of compensation.

Error Prevention!

Watch for students who confuse compensation with rounding. Compensation provides an exact answer, not an estimate.

Work through the Example parts (a) and (b) with students. Ask: *What aspect of the compensation process is similar in these two examples?* **Sample: numbers were added to form a whole number in each case—$5 in one and 2 in the other.**

Differentiated Instruction

Special Needs L1

Elicit from students that forming numbers that are easy to compute with mentally is effective. Provide additional examples, such as $19 + 34 \rightarrow \mathbf{20} + \mathbf{33} = 53$.

Resources

- Classroom Aid 7

Extension

Using Mental Math

Compensation allows you to adjust the numbers and make the expressions easier to calculate mentally. You can use compensation to find sums and differences.

The sum of two numbers remains the same if you add a number to one addend and subtract the same number from the other addend.

Subtract 3: 2 ← 5; Add 3: 7 → 10

$7 + 5 \rightarrow 10 + 2 = 12$

The difference between two numbers remains the same if you add (or subtract) the same number from both numbers.

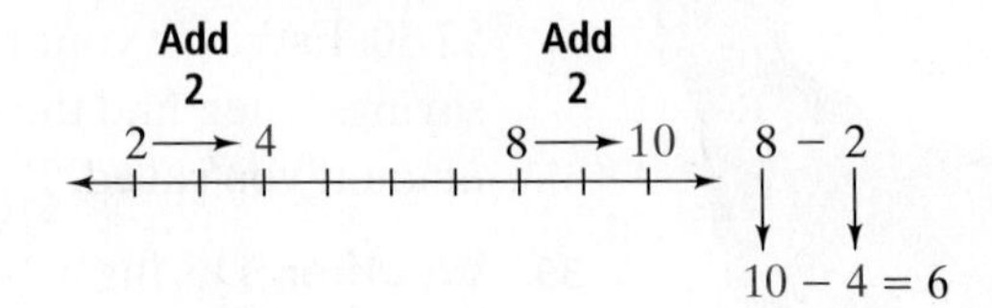

EXAMPLE Using Compensation

Find each sum or difference using compensation.

a. $4.96 + $3.79

$\$4.96 + \$3.79$

$+\,0.04 \quad -\,0.04$ ← **Add 0.04 to one addend and subtract 0.04 from the other addend.**

$\$5.00 + \$3.75 = \$8.75$

b. $6.1 - 1.3$

$6.1 - 1.3$

$+\,0.7 \quad +\,0.7$ ← **Add 0.7 to both numbers so that you subtract a whole number.**

$6.8 - 2 = 4.8$

Exercises

Find each sum or difference using compensation.

1. $0.95 + 1.45$ **2.4**
2. $2.54 + 8.16$ **10.7**
3. $3.89 + 1.73$ **5.62**
4. $72.2 + 14.9$ **87.1**
5. $38.0 - 11.1$ **26.9**
6. $9.3 - 6.8$ **2.5**
7. $102 - 77$ **25**
8. $41.6 - 0.7$ **40.9**
9. $0.4 + 7.8$ **8.2**
10. $117 + 96$ **213**
11. $74.6 - 35.8$ **38.8**
12. $12.4 - 8.3$ **4.1**
13. $11.7 - 6.9$ **4.8**
14. $2.9 + 10.5$ **13.4**
15. $984 - 852$ **132**
16. $72.3 + 8.1$ **80.4**
17. **Money** You have five items to purchase at the grocery store. The prices are $.98, $3.95, $2.08, $4.99, and $1. Use compensation to determine the amount you owe at the checkout. **$13**

1-3a Activity Lab

Modeling Decimal Multiplication

You can use models to multiply decimals. The grid below is divided into 10 columns and 10 rows. Each column or row represents one tenth of the whole, or 0.1. Each square represents one hundredth of the whole, or 0.01.

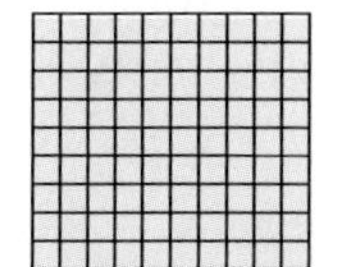 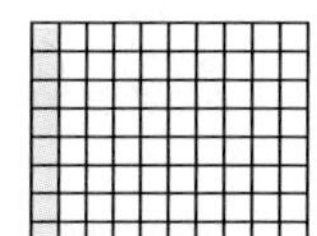 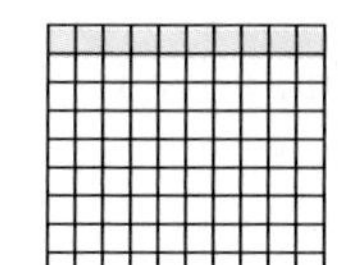 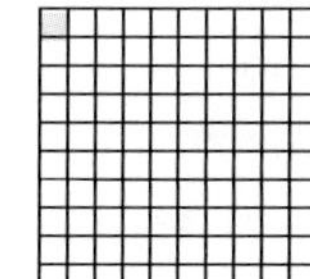

ACTIVITY

You can use three grids to construct a model to find 0.7×0.4.

Step 1 Model 0.7 by shading 7 columns in one color. Model 0.4 by shading 4 rows in a different color.

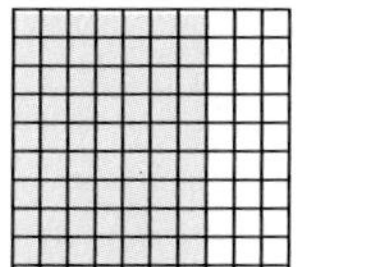 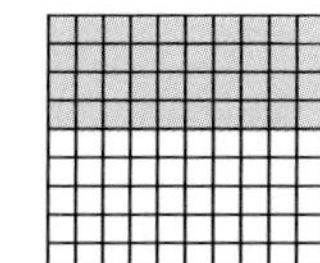

Step 2 Shade 0.7 and 0.4 on the same grid. Count the number of squares that are shaded in both colors.

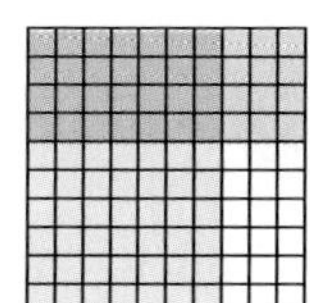

There are 28 squares shaded in both colors, so $0.7 \times 0.4 = 0.28$.

Exercises

Draw a model to find each product. 1–3. See margin.

1. 0.2×0.9
2. 0.5×0.3
3. 0.4×0.4

Write the product represented by each decimal grid.

4. $0.1 \times 0.7 = 0.07$
5. $0.6 \times 0.3 = 0.18$

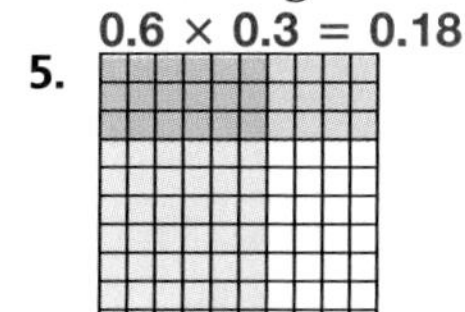

6. $0.9 \times 0.5 = 0.45$

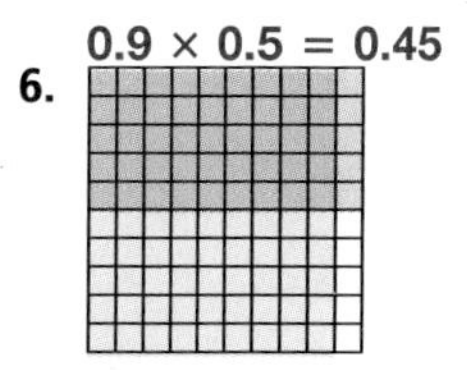

7. Use Exercises 1–3. Multiply each factor by 10. Find the new products. How do the new products compare to the original products? The product becomes a whole number.

1\.

2\. 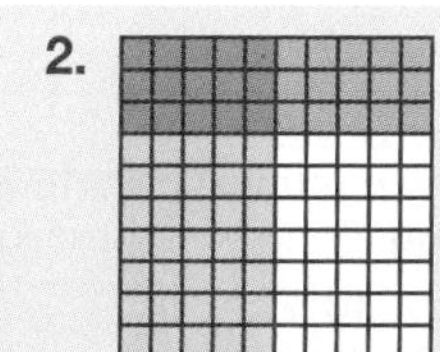

3\. 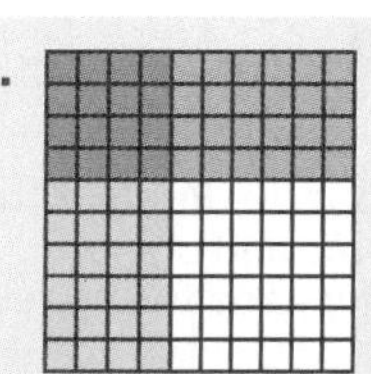

Modeling Decimal Multiplication

Students use grids to model multiplication with decimals.

Guided Instruction

Activity

Be sure students understand that Step 1 shows both factors and Step 2 shows the product in the overlap area.

Exercises

Have students work independently on Exercises 1–3 and with partners on Exercises 4–7. Suggest that they write the factors for each grid in Exercises 4–6.

Alternative Method

Students can also represent decimal multiplication with place-value models. They can use one hundred squares to represent ones, tens squares to represent tenths, and ones squares for hundredths.

Differentiated Instruction

Visual Learners

For Exercise 7, ask students to represent each problem visually as an array.

Advanced Learners L4

Ask students whether the products found in Exercises 1–6 are greater or less than the factors. Have them explain their answer. Less; the product of two factors less than 1 is less than either factor.

Resources

- Activity Lab 1-3: Taxing Problems
- grid models
- Classroom Aid 2
- Student Manipulatives Kit

1-3

1. Plan

Objective
To multiply decimals and to do mental math using the properties of multiplication

Examples
1 Multiplying Decimals
2 Using Multiplication Properties

Math Understandings: p. 2C

Professional Development

Math Background

In mathematics, a *postulate* is a statement that is assumed to be true. An important set of postulates governs the basic arithmetic operations. The properties of multiplication presented in this lesson are among these postulates.

Students sometimes wonder why there are lists of properties for addition and multiplication but no lists of properties for subtraction. The reason is that, algebraically, subtraction is defined in terms of addition. Specifically, for real numbers a and b: $a - b = a + (-b)$

More Math Background: p. 2C

Lesson Planning and Resources

See p. 2E for a list of the resources that support this lesson.

Bell Ringer Practice

Check Skills You'll Need
Use student page, transparency, or PowerPoint. For intervention, direct students to:
Using Estimation Strategies
Lesson 1-1
Extra Skills and Word Problems Practice, Ch. 1

1-3 Multiplying Decimals

Check Skills You'll Need

1. **Vocabulary Review** How is estimating different from finding an exact answer? **See below.**

Estimate.

2. 2.7 × 5.3 **15**
3. 4.09 × 6.8 **28**
4. 1.134 × 9.76 **10**

GO for Help Lesson 1-1

What You'll Learn

To multiply decimals and to do mental math using the properties of multiplication

New Vocabulary Properties of Multiplication: Identity Property, Zero Property, Commutative Property, Associative Property

Why Learn This?

Multiplying decimals can help you find how much money you will earn working at a part-time job.

To multiply decimals, treat the factors as whole numbers. Then multiply. Count the decimal places in both factors. Use the total to locate the decimal point in the product.

Check Skills You'll Need

1. **Estimating gives an approximate answer.**

EXAMPLE Multiplying Decimals

1 At your part-time job, you earn $7.30 per hour. You work for two and a half hours. How much money do you earn?

Estimate $7.30 \times 2.5 \approx 7 \times 3 = 21$.

Step 1 Multiply as if the numbers are whole numbers.

```
  730
×  25
 3650
1460
18250
```

Step 2 Locate the decimal point in the product by adding the decimal places of the factors.

```
  7.30   ← two decimal places (hundredths)
 × 2.5   ← one decimal place (tenths)
  3650
 1460
18.250   ← hundredths × tenths = thousandths
           Use three decimal places.
```

You earn $18.25.

Check for Reasonableness The product 18.25 is close to the estimate of 21. The answer is reasonable.

Quick Check

1. Estimate 14.3×0.81. Then find the product. **11.583**

Differentiated Instruction Solutions for All Learners

Special Needs L1
To show students that in the Associative Property the order of the addends stays the same and only the grouping changes, have them color-code the addends on each side of the equal sign.

learning style: visual

Below Level L2
Review multiplying decimals by 10, 100, and 1,000 by reading aloud exercises like these.

2.7 × 10 **27** 2.7 × 1,000 **2,700**
4.93 × 100 **493** 49.3 × 100 **4,930**

learning style: verbal

Four ways to write "3 times 5" using different symbols are shown below.

3×5 $3 \cdot 5$ $3(5)$ $(3)(5)$

KEY CONCEPTS **Properties of Multiplication**

Identity Property of Multiplication The product of 1 and a is a.

Arithmetic	Algebra
$5 \cdot 1 = 5$	$a \cdot 1 = a$
$1 \cdot 5 = 5$	$1 \cdot a = a$

Zero Property The product of 0 and any number is 0.

Arithmetic	Algebra
$5 \cdot 0 = 0$	$a \cdot 0 = 0$
$0 \cdot 5 = 0$	$0 \cdot a = 0$

Commutative Property of Multiplication Changing the order of factors does not change the product.

Arithmetic	Algebra
$5 \cdot 2 = 2 \cdot 5$	$a \cdot 2 = 2 \cdot a$

Associative Property of Multiplication Changing the grouping of factors does not change the product.

Arithmetic	Algebra
$(3 \cdot 2) \cdot 5 = 3 \cdot (2 \cdot 5)$	$(a \cdot b) \cdot c = a \cdot (b \cdot c)$

Vocabulary Tip

To *commute* means "to change places." A commutative property lets numbers change places.

To *associate* means "to gather in groups." An associative property forms groups of numbers.

You can use the properties of multiplication to multiply mentally.

EXAMPLE **Using Multiplication Properties**

2 **Mental Math** Use mental math to find $0.25 \cdot 3.58 \cdot 4$.

What you think

I should look for compatible numbers. The product of 0.25 and 4 is 1. Then the product of 1 and 3.58 is 3.58.

Why it works

$0.25 \cdot 3.58 \cdot 4 = 0.25 \cdot 4 \cdot 3.58$ ← Commutative Property of Multiplication

$= (0.25 \cdot 4) \cdot 3.58$ ← Associative Property of Multiplication

$= 1 \cdot 3.58$ ← Simplify.

$= 3.58$ ← Identity Property of Multiplication

Quick Check

2. Use mental math to find $2.5 \cdot 6.3 \cdot 4$. **63**

2. Teach

Activity Lab

Use before the lesson.
Student Edition Activity Lab 1-3a, Modeling Decimal Multiplication, p. 13

All in One Teaching Resources
Activity Lab 1-3: Taxing Problems

Guided Instruction

Example 1
Make sure students understand that the term "decimal places" refers to the places to the right of the decimal point—tenths, hundredths, thousandths, and so on.

Example 2
Encourage students to use the complete name of each property. For instance, they should say "the Identity Property of Multiplication" and "the Identity Property of Addition" instead of simply "the Identity Property." The properties of addition and multiplication are related to each other, but distinct.

PowerPoint

Additional Examples

1. Find the product of 11.4 and 3.6. **41.04**
2. Use mental math to find $2 \cdot 4.097 \cdot 0.5$. **4.097**

All in One Teaching Resources
- Daily Notetaking Guide 1-3 L3
- Adapted Notetaking 1-3 L1

Closure

- *How do you multiply decimals?* **First multiply as if the factors were whole numbers. Then add the number of decimal places in both factors to find the number of decimal places in the product.**

Advanced Learners L4
Show three ways to place the decimal points in the factors so that the product of 543×21 is 114.03.
Samples: 543×0.21; 54.3×2.1; 5.43×21

learning style: visual

English Language Learners ELL
In Example 2, help students to understand why they reordered and regrouped the factors. Explicitly state that they were reordered for compatibility and regrouped to make the multiplication easier.

learning style: verbal

3. Practice

Assignment Guide

Check Your Understanding
Go over Exercises 1–10 in class before assigning the Homework Exercises.

Homework Exercises

A Practice by Example	11–29
B Apply Your Skills	30–40
C Challenge	41
Test Prep and Mixed Review	42–47

Homework Quick Check
To check students' understanding of key skills and concepts, go over Exercises 13, 27, 37, 39, and 40.

Differentiated Instruction Resources

Adapted Practice 1-3 L1

Practice 1-3 Multiplying Decimals L3

Find each product.

1. 28 × 6 **168**
2. 7.3 · 0.9 **6.57**
3. 58 · 2.1 **121.8**
4. 15(187) **2,805**
5. 6.6 × 25 **165**
6. (1.8)(0.7) **1.26**
7. 0.91 · 2.7 **2.457**
8. 4.6(3.9) **17.94**
9. 17.3 × 15.23 **263.479**
10. 2.33(3.56) **8.2948**
11. 12.15 × 19 **230.85**
12. 481.51 × 623.42 **300,182.9642**

Rewrite each equation with the decimal point in the correct place in the product.

13. 5.6 × 1.2 = 672 **5.6 × 1.2 = 6.72**
14. 3.7 × 2.4 = 888 **3.7 × 2.4 = 8.88**
15. 6.5 × 2.5 = 1625 **6.5 × 2.5 = 16.25**
16. 1.02 × 6.9 = 7038 **1.02 × 6.9 = 7.038**
17. 4.4 × 6.51 = 28644 **4.4 × 6.51 = 28.644**
18. 0.6 × 9.312 = 55872 **0.6 × 9.312 = 5.5872**

Name the property of multiplication shown.

19. 3 × 4 = 4 × 3 **Commutative Property**
20. 9 × (6 × 3) = (9 × 6) × 3 **Associative Property**
21. 2 × 0 = 0 **Zero Property**
22. 10 × 1 = 10 **Identity Property**

Solve.

23. Each trip on a ride at the carnival costs $1.25. If Tara goes on 4 rides, how much will it cost her? **$5.00**
24. Postage stamps cost $0.37 each. How much does a book of 50 stamps cost? **$18.50**

1-3 • Guided Problem Solving GPS L3

GPS **Student Page 17, Exercise 39:**

In 2003, Texas had about 2.6 times the number of head of cattle as Oklahoma. Oklahoma had 5.2 million. How many head of cattle did Texas have?

Understand

1. Circle the information you will need to solve the problem.
2. What are you being asked to do? **Find how many head of cattle Texas had.**

Plan and Carry Out

3. How many million head of cattle were in Oklahoma in 2003? **5.2**
4. How many times more cattle did Texas have than Oklahoma? **2.6 times more**
5. Write an expression for the number of cattle in Texas (in millions). **2.6 × 5.2**
6. How many decimal places are in both factors? **2 decimal places**
7. Multiply as if the numbers are whole numbers. **1,352**
8. Place the decimal in the product using the total number of decimal places from step 6. **13.52 million**

Check

9. Estimate the product of 2.6 × 5.2, and compare it to your answer. Is your answer reasonable? **15; yes, the estimate is slightly high, but it is reasonable**

Solve Another Problem

10. You buy 6.5 yards of fabric that costs $7.95 per yard. How much money does it cost? **$51.68**

Check Your Understanding

1. **Vocabulary** Which property of multiplication allows you to switch the order of the factors? **Comm. Prop. of Mult.**
2. What product does the model at the left represent? **0.5 × 0.2 = 0.1**
3. **Number Sense** The product of 4 and 25 is 100. What happens to the product when you change one factor from 25 to 2.5? **The product is divided by 10.**

Multiply.

4. 0.5×6 **3**
5. $0.25 \cdot 8$ **2**
6. 1.5×2 **3**

Find the missing numbers. Name the property of multiplication shown.

7. $3.6 \cdot ■ = 0$ **0; Zero Property**
8. $■ \cdot 1 = 25.5$ **25.5; Identity Property**
9. $■ \cdot 4 = 4 \cdot 3$ **3; Commutative Property**
10. $(5 \cdot ■) \cdot 2.3 = 5 \cdot (1.4 \cdot 2.3)$ **1.4; Associative Property**

Homework Exercises

For more exercises, see Extra Skills and Word Problems.

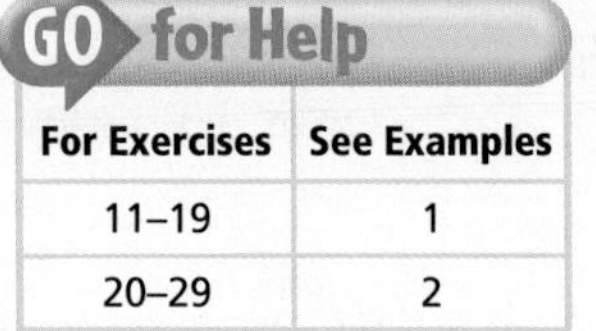
GO for Help

For Exercises	See Examples
11–19	1
20–29	2

A Estimate. Then find each product.

11. 0.2×0.7 **0.14**
12. 0.4×0.6 **0.24**
13. 0.3×0.5 **0.15**
14. 1.02×3.6 **3.672**
15. 8.7×0.45 **3.915**
16. 1.45×2.6 **3.77**
17. 41×7.5 **307.5**
18. 1.3×0.05 **0.065**
19. 1.1×1.1 **1.21**

Mental Math Find each product.

20. $0.2 \cdot 3.41 \cdot 5$ **3.41**
21. $1.09 \cdot 23.6 \cdot 0$ **0**
22. $(2.3 \cdot 0.5) \cdot 4$ **4.6**
23. $5 \cdot (4.3 \cdot 1)$ **21.5**
24. $0 \cdot 2.78 \cdot 1$ **0**
25. $0.4 \cdot 3.29 \cdot 25$ **32.9**
26. $0.4 \cdot (0.5 \cdot 0.2)$ **0.04**
27. $3.6 \cdot 2.5 \cdot 2$ **18**
28. $7.1 \cdot 6.2 \cdot 3.5 \cdot 0$ **0**

29. **Biking** If your speed is 3.5 mi/h, how far will you bike in 1.2 h? **4.2 mi**

B GPS

30. **Guided Problem Solving** You have 17 pennies, 31 nickels, 22 dimes, and 14 quarters. How much money will you have if you lend $6.50 to a friend? **$.92**
    - How much money do you have in each type of coin?
    - What is the total amount of money you have?
31. **Money** A penny weighs about 0.1 oz. How much is a pound of pennies worth? (1 lb = 16 oz) **$1.60**
32. A cubic foot of water weighs 62.4 lb. A water storage tank can hold 89.5 cubic feet of water. How much will the water in the storage tank weigh when the tank is full? **5,584.8 lb**

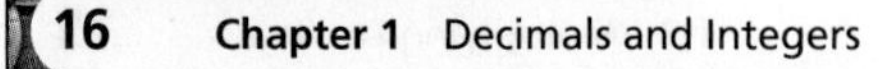

37. **They both give you the same number you started with, but the identity is zero for addition and one for multiplication.**

Homework Video Tutor
Visit: PHSchool.com
Web Code: are-0103

Find each product.

33. $0.36 \cdot 1.2$ **0.432** **34.** $3.6 \cdot 0.12$ **0.432** **35.** $0.36 \cdot 0.012$ **0.00432** **36.** $0.036 \cdot 0.12$ **0.00432**

37. **Writing in Math** How are the identity properties of multiplication and addition similar? How are they different? **See margin.**

38. New tennis balls must bounce back to no less than 0.53 and no more than 0.58 of the starting height. A ball is dropped from 200 cm. Within what range of heights should it bounce? **between 106 cm and 116 cm**

39. One year, Texas had about 2.6 times as many head of cattle as Oklahoma. Oklahoma had 5.2 million. How many head of cattle did Texas have? **13.52 million**

40. **Reasoning** You multiply two decimals that are both less than 1. How does the size of the product compare to each factor? Explain.

40. The product of any two numbers between 0 and 1 will always be less than either of the numbers.

41. **Challenge** White flour costs \$1.30/kg, and whole wheat flour costs \$1.10/kg. Flour is sold in bags of 1.2 kg and 3.4 kg. A chef buys 4 bags of flour. If each bag of flour the chef buys is different, how much must the chef pay? **\$11.04**

Test Prep and Mixed Review

Practice

Multiple Choice

42. Which model represents the expression 0.3×0.6? **C**

A
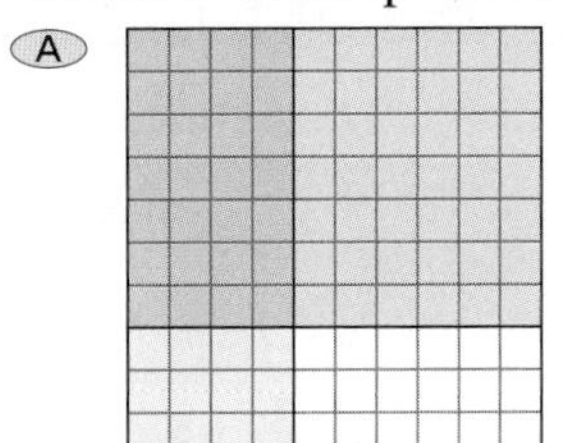

C
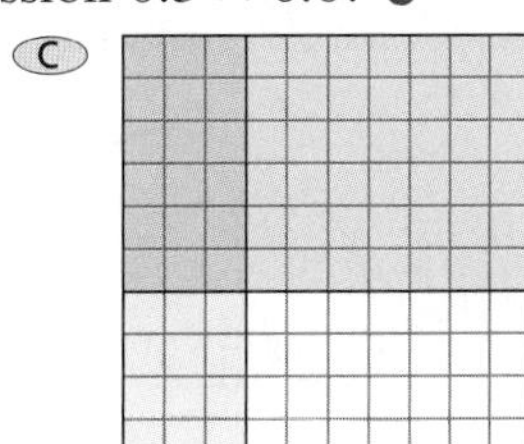

B
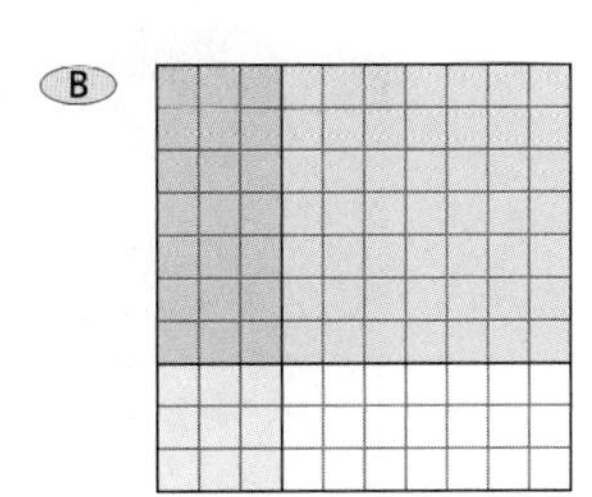

D
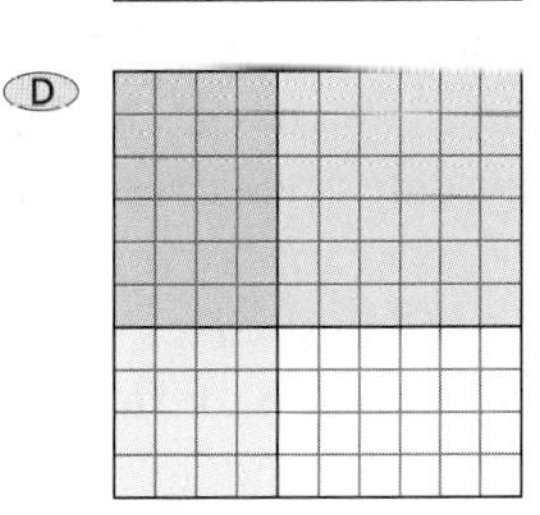

43. Mr. Porter bought 24 bagels at 6 for \$2.49 and 12 cartons of juice at 6 for \$1.98. What was the total cost of the bagels and juice Mr. Porter bought, not including tax? **J**

(F) \$8.84 (G) \$9.96 (H) \$11.88 (J) \$13.92

GO for Help

For Exercises	See Lesson
44–47	1-2

Find each sum or difference.

44. $8.56 + 3.11$ **11.67** **45.** $9.843 - 8.2$ **1.643** **46.** $9.4 - 7.024$ **2.376** **47.** $17.1 + 3.09$ **20.19**

Online lesson quiz, PHSchool.com, Web Code: ara-0103

4. Assess & Reteach

Lesson Quiz

Find each product.

1. 0.9×4.25 **3.825**
2. 6.7×8.08 **54.136**
3. 0.44×2.5 **1.1**
4. 6.3×0.7 **4.41**

Exercises

Have students use 10×10 grids or place-value models to help them multiply decimals.

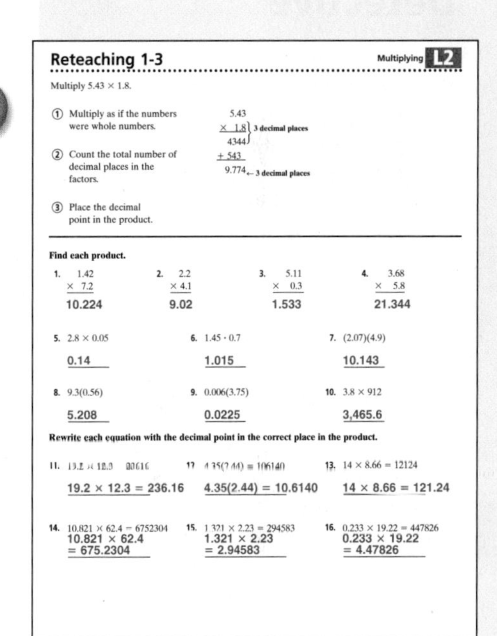

Reteaching 1-3 Multiplying L2

Multiply 5.43×1.8.

① Multiply as if the numbers were whole numbers.

② Count the total number of decimal places in the factors.

③ Place the decimal point in the product.

$$\begin{array}{r} 5.43 \\ \times\ 1.8 \\ \hline 4344 \\ +\ 543 \\ \hline 9.774 \end{array}$$
3 decimal places (factors) — 3 decimal places (product)

Find each product.

1. 1.42×7.2 **10.224**
2. 2.2×4.1 **9.02**
3. 5.11×0.3 **1.533**
4. 3.68×5.8 **21.344**
5. 2.8×0.05 **0.14**
6. $1.45 \cdot 0.7$ **1.015**
7. $(2.07)(4.9)$ **10.143**
8. $9.3(0.56)$ **5.208**
9. $0.006(3.75)$ **0.0225**
10. 3.8×912 **3,465.6**

Rewrite each equation with the decimal point in the correct place in the product.

11. $19.2 \times 12.3 = 23616$ **$19.2 \times 12.3 = 236.16$**
12. $4.35(2.44) = 106140$ **$4.35(2.44) = 10.6140$**
13. $14 \times 8.66 = 12124$ **$14 \times 8.66 = 121.24$**
14. $10.821 \times 62.4 = 6752304$ **$10.821 \times 62.4 = 675.2304$**
15. $1.321 \times 2.23 = 294583$ **$1.321 \times 2.23 = 2.94583$**
16. $0.233 \times 19.22 = 447826$ **$0.233 \times 19.22 = 4.47826$**

Enrichment 1-3 Multiplying L4

Patterns in Numbers

Find each product.

1. 26×3 **78**; 26×0.3 **7.8**; 26×0.03 **0.78**; 26×0.003 **0.078**
2. 26×3 **78**; 2.6×3 **7.8**; 0.26×3 **0.78**; 0.026×3 **0.078**
3. What pattern do you notice in the products? **Products where the factors have the same number of decimal places are the same.**
4. Use the digits 3 and 26 as factors. Find three different pairs of numbers using any combination of decimal places whose product is 0.0000078. **Sample answer: (3)0.0000026; (0.3)0.000026; (0.03)0.00026**

Rewrite each equation, adding one or more decimal points to make the equation true. You may also need to add zeros. Each exercise may have more than one correct answer.

Sample answers given.

5. $16 \times 1{,}175 = 188$ **$16 \times 11.75 = 188$**
6. $16 \times 1{,}175 = 18.8$ **$16 \times 1.175 = 18.8$**
7. $16 \times 1{,}175 = 1.88$ **$16 \times 0.1175 = 1.88$**
8. $16 \times 1{,}175 = 0.188$ **$16 \times 0.01175 = 0.188$**
9. $16 \times 1{,}175 = 0.0188$ **$16 \times 0.001175 = 0.0188$**
10. $16 \times 1{,}175 = 0.00188$ **$16 \times 0.0001175 = 0.00188$**
11. $16 \times 1{,}175 = 0.000188$ **$16 \times 0.00001175 = 0.000188$**
12. $16 \times 1{,}175 = 0.0000188$ **$16 \times 0.000001175 = 0.0000188$**
13. $16 \times 1{,}175 = 0.0000188$ **$0.000000016 \times 1175 = 0.0000188$**

Alternative Assessment

Have students work in pairs. Each student writes a decimal for a factor. Partners show their decimals and together determine the number of decimal places in their product before they multiply the decimal factors.

Test Prep

Resources

For additional practice with a variety of test item formats:

- Test-Taking Strategies, p. 59
- Test Prep, p. 63
- Test-Taking Strategies with Transparencies

Checkpoint Quiz

Use this Checkpoint Quiz to check students' understanding of the skills and concepts of Lessons 1-1 through 1-3.

Resources

- All in One Teaching Resources
- Checkpoint Quiz 1
- ExamView CD-ROM
- Success Tracker™ Online Intervention

MATH AT WORK

Detective

Students may not be aware of all the mathematics a detective uses. Some may appreciate that detectives often deal with specific proof that involves mathematics in analyzing crime scenes, working with forensics, and profiling.

Guided Instruction

Have students give examples of how detectives use mathematics from some of their favorite books, television shows, or movies. Then ask questions, such as:

- *For how much time do you think a detective would use mathematics on an ordinary workday? Support your conclusion.* **Sample: two hours doing paperwork and figuring out times and distances for a case**
- *How would a detective use math to figure out when a crime had been committed?* **Sample: checking times with witnesses or recording devices**
- *What financial information might be useful to a detective?* **Sample: how, where, and when a victim or criminal spent money or used credit cards**

Checkpoint Quiz 1

Lessons 1-1 through 1-3

Use each strategy to estimate $3.07 + $3.48 + $4.24.

1. rounding to the nearest dollar first **$10**
2. front-end estimation **$11**

Use compatible numbers to estimate each answer.

3–4. Answers may vary. Samples are given.

3. $129.4 \div 23$ **about 6**
4. $37.6 \div 3.05$ **about 13**

Find each sum or difference.

5. $2.99 + 3.08 + 18.5$ **24.57**
6. $9.718 + 4.603$ **14.321**
7. $11.64 - 8.72$ **2.92**
8. $22.4 - 0.54$ **21.86**

Find each product.

9. $1.36 \cdot 8.94$ **12.1584**
10. 2.4×0.04 **0.096**
11. 12.16×4.2 **51.072**
12. $5.45 \cdot 2.04$ **11.118**
13. Use $<$, $=$, or $>$ to complete $61.25 - 30.17$ ■ 14.8×2.1. **=**
14. Eduardo can type 65 words per minute. How many words can he type in 7.5 minutes? **487.5 words**
15. **Crafts** A box of supplies contains 0.8 lb of red clay, 1.3 lb of green clay, and 2.1 lb of white clay, as well as three cans of paint that weigh 0.75 lb each. What is the total weight of the supplies? **6.45 lb**

MATH AT WORK

Detective

Most people think of a detective as a person in a trench coat, looking for clues. In reality, detectives dress like anyone else. They can work for lawyers, government agencies, and businesses. Detectives may gather information to trace debtors or conduct background investigations.

Detectives use mathematics to locate stolen funds, develop financial profiles, or monitor expense accounts.

Go Online PHSchool.com **For:** Information about detectives **Web Code:** arb-2031

1-4a Activity Lab

Hands On

Modeling Decimal Division

You can use models to divide decimals. Follow the steps in the activity to model the quotient $0.6 \div 0.2$.

ACTIVITY

Step 1 Cut out a rectangular strip of paper that is 10 in. long. This strip represents 1 whole. Mark each inch of the strip. Notice that you now have 10 equal parts, each representing 0.1, or one tenth, of the whole.

1

Step 2 Cut the strip so that you have 2 segments. One segment should represent 0.6 of the whole.

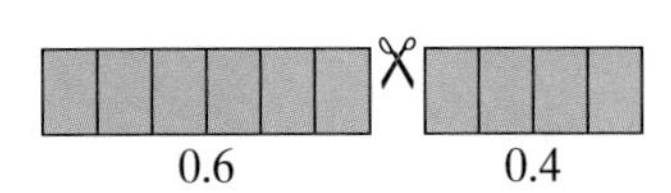

Step 3 Use another piece of paper and repeat Step 1.

Step 4 Cut this strip into 5 equal segments, so that each segment represents 0.2 of the whole.

0.2 0.2 0.2 0.2 0.2

Step 5 Align the segments that represent 0.2 under the segment that represents 0.6 of the whole.

0.6

Step 6 Count the number of 0.2 segments needed to equal the length of the 0.6 segment.

0.2 0.2 0.2

You used 3 segments, each representing 0.2 unit, so $0.6 \div 0.2 = 3$.

Exercises

Use a model to find each quotient. 1–3. See margin.

1. $0.3 \div 0.1$
2. $0.4 \div 0.2$
3. $0.8 \div 0.4$

Write the quotient represented by each model.

4. $0.8 \div 0.2 = 4$

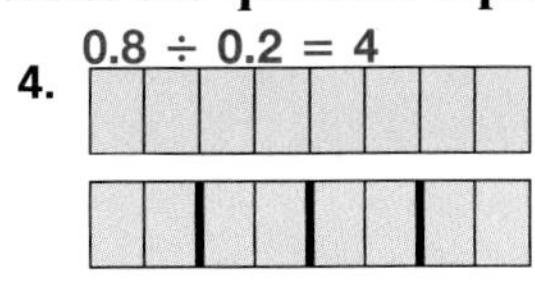

5. $0.9 \div 0.3 = 3$

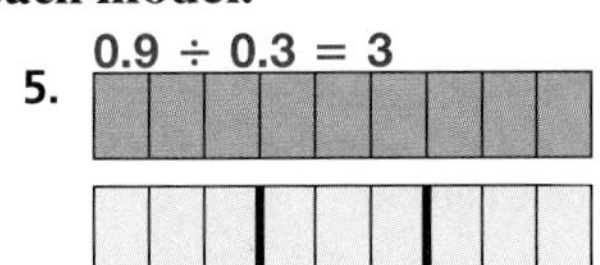

6. $0.5 \div 0.1 = 5$

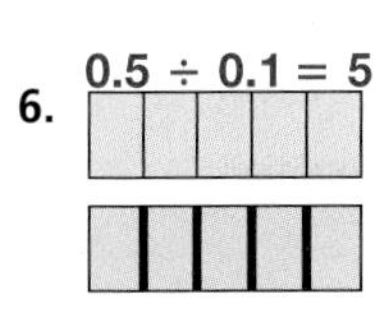

7. **Reasoning** The quotient of 0.6 and 0.2 is 3. What happens to the quotient when you multiply the dividend and divisor by 10? Explain. See margin.

1. 3;
2. 2;
3. 2;

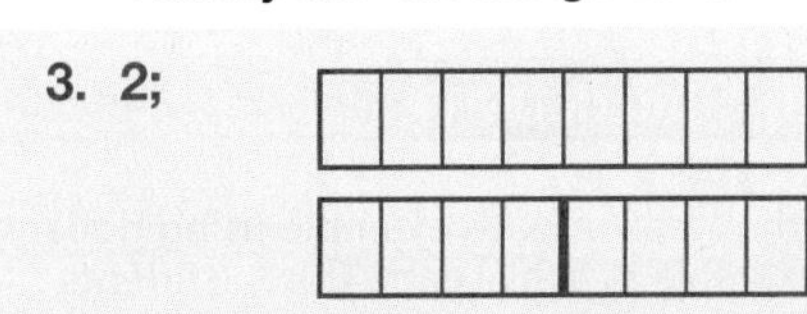

7. The quotient is the same. You are still dividing the same number of pieces.

Activity Lab

Modeling Decimal Division

Students use paper strips to model dividing decimals.

Guided Instruction

Activity
In Step 2, ask: *What does the second segment represent?* 0.4 of the whole

Exercises
Have students work in pairs to do Exercises 1–6. Have them discuss their work and share any questions with the class. Use Exercise 7 for class discussion.

Alternative Method
Students can also represent decimal division using 10 × 10 grids. To model $0.6 \div 0.2$, shade 60 squares and cut the shaded squares into three groups of 20 squares.

Differentiated Instruction

Advanced Learners L4
Ask students to explain how the model used in the activity could be used to represent the problem in Exercise 7, $6 \div 2$. Relabel the segments as 6 and 2.

Resources

- Activity Lab 1-4: Patterns in Numbers
- rectangular paper strips
- Classroom Aid 16, 17
- Student Manipulatives Kit

1-4

1. Plan

Objective
To divide decimals and to solve problems by dividing decimals

Examples
1 Dividing a Decimal by a Decimal
2 Annexing Zeros to Divide

Math Understandings: p. 2C

Math Background

Students may wonder why there are no lists of properties for division. The reason is similar to that for subtraction. Algebraically, division is defined in terms of multiplication. Specifically, for real numbers *a* and *b*:

$\frac{a}{b} = a \cdot \frac{1}{b}$, where $b \neq 0$

Since any division problem can be rewritten as a multiplication equivalent, there is no need for a separate set of properties for division.

More Math Background: p. 2C

Lesson Planning and Resources

See p. 2E for a list of the resources that support this lesson.

Bell Ringer Practice

Check Skills You'll Need
Use student page, transparency, or PowerPoint. For intervention, direct students to:
Using Estimation Strategies
Lesson 1-1
Extra Skills and Word Problems Practice, Ch. 1

1-4 Dividing Decimals

Check Skills You'll Need

1. **Vocabulary Review** What are *compatible numbers?*

1–4. See below.
Estimate each quotient using compatible numbers.

2. 103.8 ÷ 13.2
3. 128.62 ÷ 9.86
4. 41.77 ÷ 6.07

GO for Help Lesson 1-1

Check Skills You'll Need
1. numbers that are easy to compute mentally
2–4. Answers may vary. Samples are given.
2. 8
3. 13
4. 7

What You'll Learn

To divide decimals and to solve problems by dividing decimals

Why Learn This?

You may need to divide decimals to plan how many items you can buy with the money you have.

Suppose you have \$1.20 and you want to buy pencils that cost \$.30 each. How many pencils can you buy? Three ways to write "1.2 divided by 0.3" appear below.

$1.2 \div 0.3 \qquad 0.3\overline{)1.2} \qquad \frac{1.2}{0.3}$

To divide decimals, multiply both the divisor and the dividend by the power of 10 that makes the divisor a whole number. Then divide.

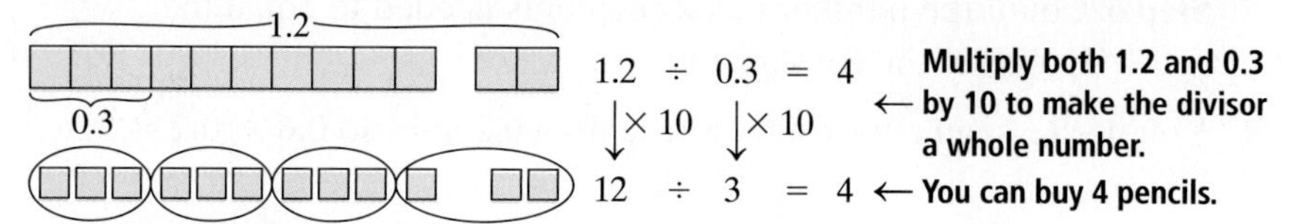

$1.2 \div 0.3 = 4$ ← Multiply both 1.2 and 0.3 by 10 to make the divisor a whole number.
↓× 10 ↓× 10
$12 \div 3 = 4$ ← You can buy 4 pencils.

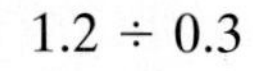

EXAMPLE Dividing a Decimal by a Decimal

1 Find 2.064 ÷ 0.24.

$0.24\overline{)2.064} \rightarrow 24\overline{)206.4}$ with quotient 8.6 ← Place the decimal point in the quotient above the decimal point in the dividend.

```
     8.6
24)206.4
   192
    144
    144
      0
```

↑ Multiply the divisor and the dividend by 100 to make the divisor a whole number.

Vocabulary Tip
Recall the names of the parts of a division problem.

divisor → $6\overline{)18}$ with 3 ← quotient
↑ dividend

Quick Check

1. Find each quotient.
   a. 12.42 ÷ 5.4 **2.3** b. 67.84 ÷ 0.64 **106** c. 144.06 ÷ 9.8 **14.7**

Differentiated Instruction Solutions for All Learners

Special Needs L1
Slowly go over the steps of each example using dollar amounts. For example in Why Learn This?, students may miss why \$.30 is written as 0.3 as the divisor.
learning style: visual

Below Level L2
Review dividing decimals by 10, 100, and 1,000 by reading aloud exercises like these.
35 ÷ 10 **3.5** 3.5 ÷ 100 **0.035**
776 ÷ 100 **7.76** 776 ÷ 1,000 **0.776**
learning style: verbal

When you divide by a decimal, sometimes you need to annex extra zeros in the dividend.

EXAMPLE Annexing Zeros to Divide

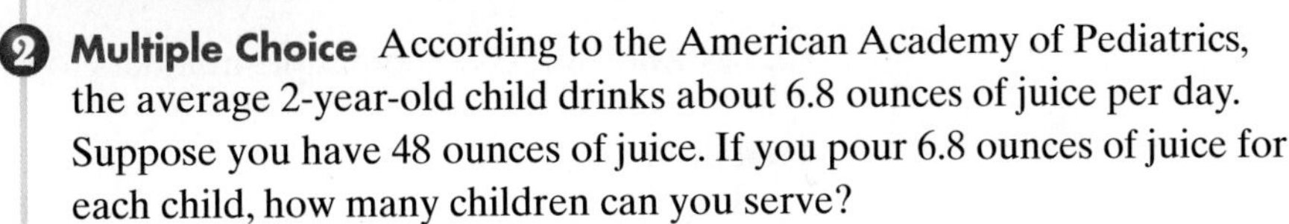

2 Multiple Choice According to the American Academy of Pediatrics, the average 2-year-old child drinks about 6.8 ounces of juice per day. Suppose you have 48 ounces of juice. If you pour 6.8 ounces of juice for each child, how many children can you serve?

Ⓐ 2 Ⓑ 7 Ⓒ 8 Ⓓ 41

Step 1 Estimate to eliminate unreasonable answers.

$48 \div 6.8 \approx 49 \div 7 = 7$ ← **Use compatible numbers 49 and 7.**

Since the estimate is 7, only choices B and C are reasonable. You can eliminate choices A and D.

Step 2 Calculate to decide which answer is correct.

$6.8\overline{)48.0} \rightarrow 68\overline{)480.}$ ← **Annex the zero in the dividend. Multiply the divisor and dividend by 10 to make the divisor a whole number.**

$$\begin{array}{r} 7. \\ 68\overline{)480.} \\ \underline{476} \\ 4 \end{array} \rightarrow \begin{array}{r} 7.05 \\ 68\overline{)480.00} \\ \underline{476} \\ 400 \\ \underline{340} \\ 60 \end{array}$$

← **Annex two zeros and divide. The quotient is about 7.05, which is only slightly more than 7. Only 7 children can be served.**

You can serve 7 children with 48 oz of juice. The answer is B.

Test Prep Tip

To eliminate unreasonable choices, estimate the answer to a multiple-choice question before calculating.

Quick Check

2. You use 0.6 lb of bananas in each smoothie. How many smoothies can you make with 3.12 lb of bananas? **5.2 smoothies**

Notice the patterns in the divisors and the quotients at the right. As the divisors decrease by a factor of 10, the quotients increase by a factor of 10.

Dividend		Divisor		Quotient
50	÷	100	=	0.5
50	÷	10	=	5
50	÷	1	=	50
50	÷	0.1	=	500
50	÷	0.01	=	5,000
50	÷	0.001	=	50,000
50	÷	0.0001	=	500,000

What happens when you try to divide by zero? Consider these related problems.

$3 \cdot 2 = 6 \rightarrow 6 \div 3 = 2$

$0 \cdot ■ = 12 \rightarrow 12 \div 0 = ■$

No value for ■ makes sense! So, division by zero is undefined.

Advanced Learners L4
Show three ways to place the decimal points in the divisor and dividend of this expression so that the quotient is 8.3.
3 4 8 6 ÷ 4 2
Samples: 34.86 ÷ 4.2; 3.486 ÷ 0.42; 348.6 ÷ 42.0
learning style: visual

English Language Learners ELL
Help students understand the purpose of making the divisor a whole number when dividing decimals. Stress that students are not simply adding a zero when they multiply by 10, but they are moving the decimal point one place to the right.
learning style: visual

2. Teach

Activity Lab

Use before the lesson.
Student Edition Activity Lab, Hands On 1-4a, Modeling Decimal Division, p. 19

All in One Teaching Resources
Activity Lab 1-4: Patterns in Numbers

Guided Instruction

Example 1
Students sometimes have difficulty translating a division from the horizontal form to the "division house" form. Suggest that they make a display like the one below in their notebooks. Have them write each of the words *divisor, dividend,* and *quotient* in a different color, using the same color for the word in each form of the division.

dividend ÷ divisor = quotient

$$\begin{array}{r} \text{quotient} \\ \text{divisor}\overline{)\text{dividend}} \end{array}$$

Example 2
Ask: *How many zeros would you need to annex to divide 48 by 0.68?* 2

Additional Examples

1. Find 40.2616 ÷ 6.824. **5.9**
2. In a restaurant, one serving of cole slaw weighs 3.5 oz. The owner buys cole slaw in cans that weigh 98 oz each. How many servings are in each can? **28 servings**

All in One Teaching Resources
- Daily Notetaking Guide 1-4 L3
- Adapted Notetaking 1-4 L1

Closure

- *How do you divide decimals?* **Rewrite the divisor as a whole number by multiplying both the divisor and the dividend by 10, 100, or 1,000. Then put the decimal point in the quotient and divide. Annex extra zeros as needed.**

3. Practice

Assignment Guide

Check Your Understanding

Go over Exercises 1–6 in class before assigning the Homework Exercises.

Homework Exercises

A	Practice by Example	7–22
B	Apply Your Skills	23–37
C	Challenge	38
Test Prep and Mixed Review		39–43

Homework Quick Check

To check students' understanding of key skills and concepts, go over Exercises 11, 18, 33, 35, and 36.

Differentiated Instruction Resources

Adapted Practice 1-4 L1

Practice 1-4 Dividing L3

Find each quotient.

1. 0.7 ÷ 100 **0.007**
2. 4.85 ÷ 0.1 **48.5**
3. 7.08 ÷ 10 **0.708**
4. 3.5 ÷ 0.1 **35**
5. 847 ÷ 0.01 **84,700**
6. 0.3 ÷ 0.1 **3**
7. 32.6 ÷ 0.01 **3,260**
8. 5.02 ÷ 0.1 **50.2**
9. $2.1\overline{)12.6}$ **6**
10. 29.75 ÷ 0.7 **42.5**
11. 37 ÷ 0.2 **185**
12. 4.74 ÷ 0.06 **79**
13. 1.414 ÷ 1.4 **1.01**
14. $0.78\overline{)0.16614}$ **0.213**
15. 0.154 ÷ 5.5 **0.028**
16. $0.85\overline{)0.0527}$ **0.062**

Annex zeros to find each quotient.

17. 29.2 ÷ 2.3 **12.69**
18. $9.24\overline{)42.1}$ **4.55**
19. 48.24 ÷ 2.22 **21.72**
20. 36.78 ÷ 2.4 **15.325**
21. $\frac{126.24}{12.6}$ **10.019**
22. $4.2\overline{)188.6}$ **44.90**

Solve.

23. Alicia paid $1.32 for a bag of pinto beans. The beans cost $.55 per lb. How much did the bag of pinto beans weigh? **2.4 lb**
24. Nina and 3 friends ate lunch at a cafe. They decided to split the bill evenly. The total bill was $17.84. How much was each person's share? **$4.46**

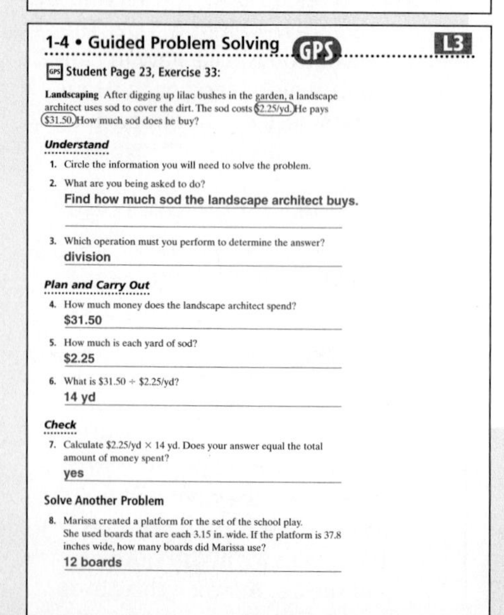

1-4 • Guided Problem Solving GPS L3

GPS Student Page 23, Exercise 33:

Landscaping After digging up lilac bushes in the garden, a landscape architect uses sod to cover the dirt. The sod costs $2.25/yd. He pays $31.50. How much sod does he buy?

Understand

1. Circle the information you will need to solve the problem.
2. What are you being asked to do? **Find how much sod the landscape architect buys.**
3. Which operation must you perform to determine the answer? **division**

Plan and Carry Out

4. How much money does the landscape architect spend? **$31.50**
5. How much is each yard of sod? **$2.25**
6. What is $31.50 ÷ $2.25/yd? **14 yd**

Check

7. Calculate $2.25/yd × 14 yd. Does your answer equal the total amount of money spent? **yes**

Solve Another Problem

8. Marissa created a platform for the set of the school play. She used boards that are each 3.15 in. wide. If the platform is 37.8 inches wide, how many boards did Marissa use? **12 boards**

Check Your Understanding

1. **Vocabulary** The number being divided in a division problem is called the ? . **dividend**
2. **Estimation** Use compatible numbers to estimate 22.54 ÷ 3.99. **6** **5 or**

Find each quotient.

3. $75\overline{)300}$ **4**
4. $7.5\overline{)300}$ **40**
5. $0.75\overline{)300}$ **400**
6. **Patterns** Look at the divisors in Exercises 3–5. Notice that the decimal point moves one place to the left from one exercise to the next. Describe what happens to the quotients.

6. **Answers may vary. Sample: The decimal point in the quotient moves one place to the right.**

Homework Exercises

For more exercises, see Extra Skills and Word Problems.

GO for Help

For Exercises	See Examples
7–15	1
16–22	2

A **Find each quotient.**

7. 19.2 ÷ 3.2 **6**
8. $1.8\overline{)7.74}$ **4.3**
9. $\frac{56.4}{4.7}$ **12**
10. $\frac{17.8}{8.9}$ **2**
11. 83.7 ÷ 2.7 **31**
12. $5.4\overline{)43.2}$ **8**
13. $9\overline{)641.7}$ **71.3**
14. $\frac{0.0882}{6}$ **0.0147**
15. 325.28 ÷ 30.4 **10.7**

Annex zeros to find each quotient.

16. $0.04\overline{)10}$ **250**
17. 5.4 ÷ 7.2 **0.75**
18. $\frac{126}{1.2}$ **105**
19. 592 ÷ 0.8 **740**
20. 0.21 ÷ 0.6 **0.35**
21. $\frac{0.003}{0.5}$ **0.006**

22. **Food** Nuts cost $1.75 per jar. How many jars can you buy with $14? **8 jars**

B GPS 23. **Guided Problem Solving** A store buys 12 pens for $11.28. The store sells each pen for $1.99. What is the store's profit per pen? **$1.05**
- **Make a Plan** Divide to find the cost of one pen. Then subtract the store's cost from the selling price to find the profit.
- **Carry Out the Plan** A single pen costs ■. The profit per pen is ■.

24. You spend $13.92 for fabric. Each yard costs $4.35. How many yards of fabric do you buy? **3.2 yd**
25. **Movies** You buy five movie tickets for a total of $23.75. Your friend gives you $5 for one ticket. How much change should you give your friend? **$.25**
26. A car travels 360.25 mi. It uses 13.1 gal of gas. How many miles per gallon of gas does the car travel? **27.5 mi**

GO Online
Homework Video Tutor
Visit: PHSchool.com
Web Code: are-0104

Find each quotient.

27. $224.5 \div 0.05$ **4,490** **28.** $1.25\overline{)0.21}$ **0.168** **29.** $4.5\overline{)13.59}$ **3.02**

30. $654 \div 0.12$ **5,450** **31.** $1.25\overline{)3.85}$ **3.08** **32.** $5.95\overline{)7.3423}$ **1.234**

Careers Landscape architects plan and develop landscape projects.

35. **The quotient is greater than the divisor; dividing by a number less than one is the same as multiplying by a number greater than one.**

36. **No; the quotient will not be the same if you switch the dividend and divisor; $6 \div 2 \neq 2 \div 6$.**

33. **Landscaping** (GPS) After digging up lilac bushes in a garden, a landscape architect uses sod to cover the ground. The sod costs $2.25/yd. He pays $31.50. How much sod does he buy? **14 yd**

34. You have a dog-walking business. Last week you worked 7.5 hours and you earned $41.25. How much do you earn per hour? **$5.50**

35. **Reasoning** When you divide a whole number by a decimal divisor less than 1, how does the size of the quotient relate to the divisor? Explain. **See left.**

36. **Writing in Math** Do you think there is a commutative property of division? Explain why or why not. Give examples. **See left.**

37. You are making a homework planner that is 8.5 in. wide. The first column is 1.25 in. wide and lists the subjects. The next five columns are equally wide and represent the five school days. How many inches wide is the column for Monday? **1.45 in.**

Ⓒ **38.** **Challenge** You want to cover a square floor that measures 127.2 in. on a side with square tiles. Each tile measures 2.4 in. on a side. How many tiles do you need? **2,809 tiles**

Test Prep and Mixed Review

Practice

Multiple Choice

39. Which expression does the model represent? **B**

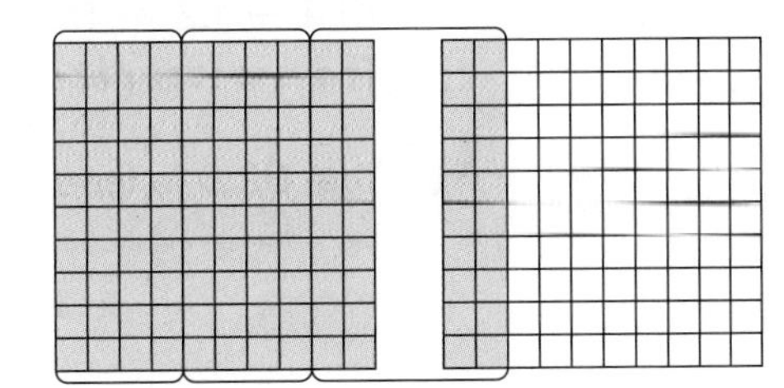

Ⓐ $1.2 \div 4$ Ⓑ $1.2 \div 3$ Ⓒ $0.2 \div 4$ Ⓓ $0.2 \div 3$

40. At Tony's Shoppe, the turkey sandwich contains 5.5 ounces of sliced turkey. Tony has a boneless turkey breast weighing 154 ounces. How many turkey sandwiches can Tony make? **H**

Ⓕ 2 Ⓖ 3 Ⓗ 28 Ⓙ 29

GO for Help

For Exercises	See Lesson
41–43	1-3

Mental Math **Find each product.**

41. $0.5 \cdot 6.7 \cdot 2$ **6.7** **42.** $5.3 \cdot 4.9 \cdot 0$ **0** **43.** $8.2 \cdot 2.5 \cdot 4$ **82**

Online lesson quiz, PHSchool.com, Web Code: ara-0104

4. Assess & Reteach

Lesson Quiz

(PowerPoint)

1. $38.164 \div 4.7$ **8.12**
2. $0.7 \div 0.5$ **1.4**
3. $14 \div 0.28$ **50**
4. $336 \div 0.006$ **56,000**

Alternative Assessment

Have students work in pairs. Each student writes a decimal. Partners can then use each decimal as a divisor and dividend. Have partners determine whether they will need to annex zeros to complete each division. Then have partners find both quotients.

Exercises

Have students use 10 × 10 grids or strips of paper to model decimal division for exercises such as Exercise 16.

Test Prep

Resources

For additional practice with a variety of test item formats:

- Test-Taking Strategies, p. 59
- Test Prep, p. 63
- Test-Taking Strategies with Transparencies

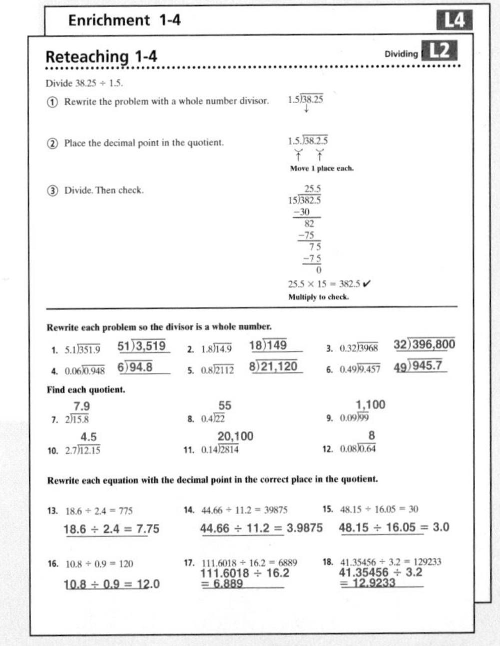

Enrichment 1-4 L4

Reteaching 1-4 Dividing L2

Divide $38.25 \div 1.5$.

① Rewrite the problem with a whole number divisor. $1.5\overline{)38.25}$

② Place the decimal point in the quotient. $1.5.\overline{)38.2.5}$

Move 1 place each.

③ Divide. Then check. $15\overline{)382.5}$ = 25.5; −30, 82, −75, 75, −75, 0

$25.5 \times 15 = 382.5$ ✔

Multiply to check.

Rewrite each problem so the divisor is a whole number.

1. $5.1\overline{)351.9}$ **$51\overline{)3,519}$**
2. $1.8\overline{)14.9}$ **$18\overline{)149}$**
3. $0.32\overline{)3968}$ **$32\overline{)396,800}$**
4. $0.06\overline{)0.948}$ **$6\overline{)94.8}$**
5. $0.8\overline{)2112}$ **$8\overline{)21,120}$**
6. $0.49\overline{)9.457}$ **$49\overline{)945.7}$**

Find each quotient.

7. $2\overline{)15.8}$ **7.9**
8. $0.4\overline{)22}$ **55**
9. $0.09\overline{)99}$ **1,100**
10. $2.7\overline{)12.15}$ **4.5**
11. $0.14\overline{)2814}$ **20,100**
12. $0.08\overline{)0.64}$ **8**

Rewrite each equation with the decimal point in the correct place in the quotient.

13. $18.6 \div 2.4 = 775$ **$18.6 \div 2.4 = 7.75$**
14. $44.66 \div 11.2 = 39875$ **$44.66 \div 11.2 = 3.9875$**
15. $48.15 \div 16.05 = 30$ **$48.15 \div 16.05 = 3.0$**
16. $10.8 \div 0.9 = 120$ **$10.8 \div 0.9 = 12.0$**
17. $111.6018 \div 16.2 = 6889$ **$111.6018 \div 16.2 = 6.889$**
18. $41.35456 \div 3.2 = 129233$ **$41.35456 \div 3.2 = 12.9233$**

GPS Guided Problem Solving

Choosing Operations

In this feature, students first identify what they know and what they want to find out. They show the main idea by drawing a diagram, choosing an operation, estimating the answer, solving the problem, and checking that their answer is reasonable.

Guided Instruction

Discuss with students the need to read the problem carefully. Students need to decide how information is related in the problem.

Have a volunteer read the problem aloud.

Ask:

- *How are the two odometer readings related?* **The odometer was at 15 when the trip began and at 126.4 when the gas tank was refilled.**
- *How can you find the distance the scooter traveled?* **Subtract. Find 126.4 – 15.**
- *After you find the distance, what do you do next and why?* **Divide the distance by 1.2 gallons to find miles traveled per gallon.**

Choosing Operations

Gas Up! The odometer of a motor scooter with a full tank of gas read 15 miles. It took 1.2 gallons of gas to refill the tank when the odometer read 126.4 miles. How many miles per gallon did the scooter get?

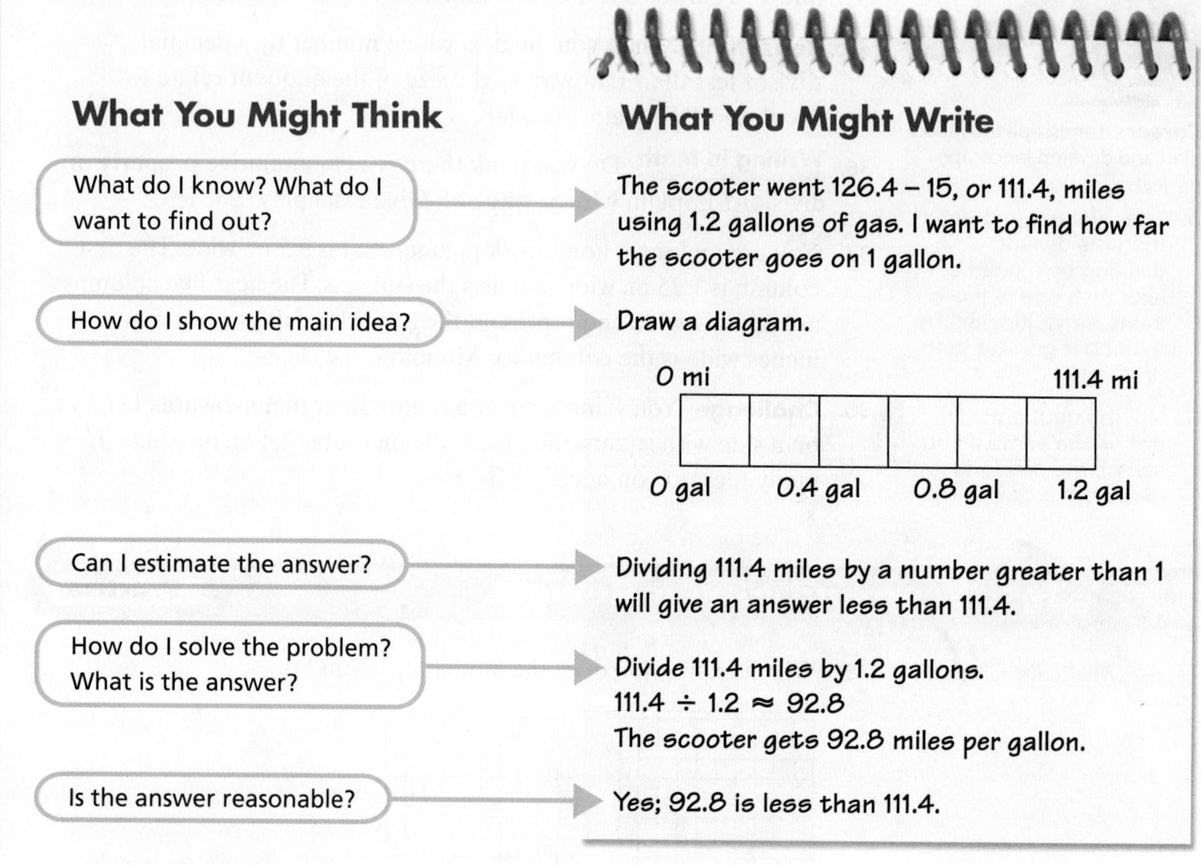

What You Might Think
What You Might Write
What do I know? What do I want to find out?
The scooter went 126.4 – 15, or 111.4, miles using 1.2 gallons of gas. I want to find how far the scooter goes on 1 gallon.
How do I show the main idea?
Draw a diagram.
0 mi
111.4 mi
0 gal
0.4 gal
0.8 gal
1.2 gal
Can I estimate the answer?
Dividing 111.4 miles by a number greater than 1 will give an answer less than 111.4.
How do I solve the problem? What is the answer?
Divide 111.4 miles by 1.2 gallons.
111.4 ÷ 1.2 ≈ 92.8
The scooter gets 92.8 miles per gallon.
Is the answer reasonable?
Yes; 92.8 is less than 111.4.

Think It Through

1–3. See right.

1. The diagram above shows the number of gallons used on the same vertical line as the number of miles traveled. Explain why.
2. **Reasoning** Explain how you know that dividing 111.4 by a number greater than 1 will give an answer less than 111.4.
3. **Number Sense** Suppose the scooter had used 0.95 gallons of gas instead of 1.2 gallons to travel the same distance. Would the mileage have been better or worse? Explain.

1. **It shows how many gallons of gas were used as the scooter traveled a certain number of miles.**
2. **Dividing by a number greater than 1 is the same as finding a fraction of that number.**
3. **The mileage would be better because it used fewer gallons for the same number of miles.**

Exercises

Solve each problem. For Exercises 4 and 5, answer the questions first.

4. At an amusement park, Tanya finds a poster for $3.75 and a shirt for $14.95. She has $20. Can she buy both items and still have enough money for a bus ticket home that costs $1.75? Explain.
   a. What do you know? What do you want to find out?
   b. How can you use the diagram below to help find the answer?

$20		
$14.95	$3.75	■

5. In May, 1860, the longest run in the history of the Pony Express was made using four horses. The table at the right shows the distances run by the horses. What was the average distance? **40.5 mi**
   a. What do you know? What do you want to find out?
   b. How can you use the diagram below to help find the answer?

Horse	Distance (miles)
1	60
2	35
3	37
4	30

60	35	37	30

6. Early settlers often sold their furniture to lighten their wagons. Suppose a settler sold furniture for 0.2 times the amount he paid. If he sold a chair for $3.80, what did he pay for it? **$19**

7. In the 1850s, a wind wagon was invented that was half sailboat and half wagon. The wind wagon took about 133 days to travel 1,968 mi. About how many miles did the wind wagon travel each day? **about 14.8 mi/day**

8. The Mississippi River is 3.2 times longer than the Platte-South Platte River. The Mississippi River is 2,340 mi long. How long is the Platte-South Platte River? **731.25 mi**

4. No, she would be short $.45.

Error Prevention!

If students are not sure they used the correct operation, have them restate the problem using their answer. **Sample: On average, the scooter traveled 92.8 miles for each gallon of gas it used. When it traveled 111.4 miles, it used 1.2 gallons of gas.**

Exercises

Have students do Exercises 1–8 with a partner. Then have them form groups to discuss, share, evaluate, and adjust their answers. Ask:

- *Will your answer for Exercise 6 be less than or greater than $3.80?* **greater**

Differentiated Instruction

English Language Learners ELL

Ask:

- *How could you restate the problem on page 24 so that it might be easier to solve?* **Sample: The odometer or mileage gauge on a motor scooter read 15 miles. The scooter had a full tank of gas. The scooter traveled until the gauge read 126.4 miles. It used 1.2 gallons of gas to travel that far. How far did the scooter travel? How many miles per gallon did the scooter get?**

1-5

1. Plan

Objective

To use and convert metric units of measure

Examples

1 Choosing a Reasonable Estimate
2 Multiplying to Change Units

Math Understandings: p. 2C

Math Background

The basic units of measurement in the metric system are the *meter* (length), the *gram* (mass), and the *liter* (liquid capacity). All other units of length, mass, and capacity in the system are related to these basic units by powers of ten. So conversions from larger to smaller units can be accomplished by multiplying by 10, 100, 1,000, and so on. Similarly, conversions from smaller to larger units are done by multiplying by 0.1, 0.01, 0.001, and so on.

More Math Background: p. 2C

Lesson Planning and Resources

See p. 2E for a list of the resources that support this lesson.

Bell Ringer Practice

Check Skills You'll Need
Use student page, transparency, or PowerPoint. For intervention, direct students to:
Multiplying Decimals
Lesson 1-3
Extra Skills and Word Problems Practice, Ch. 1

1-5 Measuring in Metric Units

Check Skills You'll Need

1. **Vocabulary Review** 4 · 8 = 8 · 4 is an example of the ? Property of Multiplication. **Commutative**

Simplify.

2. 0.25 · 10 **2.5**
3. 4.567 · 1,000 **4,567**
4. 0.03 · 100 **3**
5. 0.07 · 1,000 **70**

Lesson 1-3

What You'll Learn

To use and convert metric units of measure

Why Learn This?

Countries around the world use the metric system for measurement. Measurements are easy to convert in the metric system because it is a decimal system.

The table below is a guide for choosing an appropriate metric unit.

Type	Unit	Reference Example
Length	millimeter (mm)	about the thickness of a dime
	centimeter (cm)	about the width of your little finger
	meter (m)	about the distance from a doorknob to the floor
	kilometer (km)	about the length of 11 football fields
Capacity	milliliter (mL)	a small spoon holds about 5 mL
	liter (L)	a little more than 1 quart
Mass	milligram (mg)	about the mass of a mosquito
	gram (g)	about the mass of a paper clip
	kilogram (kg)	about the mass of a bunch of bananas

EXAMPLE Choosing a Reasonable Estimate

1 Choose a reasonable estimate. Explain your choice.

a. height of a classroom: 3 cm 3 m 3 km

3 m; a classroom is about 3 times as high as the distance from a doorknob to the floor.

b. mass of a bag of flour: 2.3 mg 2.3 g 2.3 kg

2.3 kg; a bag of flour is much heavier than a few paper clips.

Quick Check

1. Choose a reasonable estimate. Explain your choice. **See left.**

a. capacity of a soup bowl: 180 mL 180 L 180 kL

b. mass of a butterfly: 500 mg 500 g 500 kg

1a. 180 mL
b. 500 mg

Differentiated Instruction Solutions for All Learners

Special Needs L1

Show students a meter stick and the centimeter marks. Have them hold a gram cube in their hands, or something that has a mass of about a gram, so that they can make more informed decisions about what units to use.

learning style: tactile

Below Level L2

Give students several pairs of exercises like these.

4 m = ◆ cm **400** 9 g = ◆ kg **0.009**
4 cm = ◆ m **0.04** 9 kg = ◆ g **9,000**

learning style: visual

The basic unit for length in the metric system is the meter (m). The prefixes *deci-*, *centi-*, and *milli-* describe measures that are less than one basic unit. The prefixes *deca-*, *hecto-*, and *kilo-* describe measures that are greater than one basic unit.

Each unit in the table is 10 times the value of the unit to its right.

Unit	kilo-meter	hecto-meter	deca-meter	meter	deci-meter	centi-meter	milli-meter
Symbol	km	hm	dam	m	dm	cm	mm
Value	1,000 m	100 m	10 m	1 m	0.1 m	0.01 m	0.001 m

The basic unit of mass is the gram. The basic unit of capacity is the liter.

You can change a measurement from one unit to another by finding the relationship between the two units and multiplying.

EXAMPLES Multiplying to Change Units

2 Change 245 milliliters to liters.

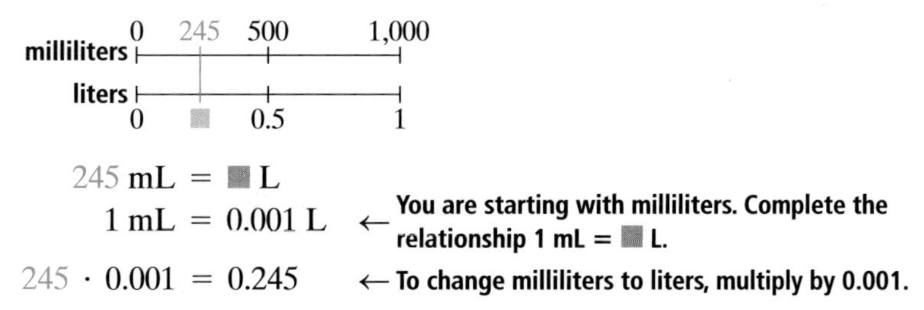

$245 \text{ mL} = \blacksquare \text{ L}$

$1 \text{ mL} = 0.001 \text{ L}$ ← You are starting with milliliters. Complete the relationship 1 mL = ■ L.

$245 \cdot 0.001 = 0.245$ ← To change milliliters to liters, multiply by 0.001.

245 milliliters equals 0.245 liters.

3 Change 1.4 grams to milligrams.

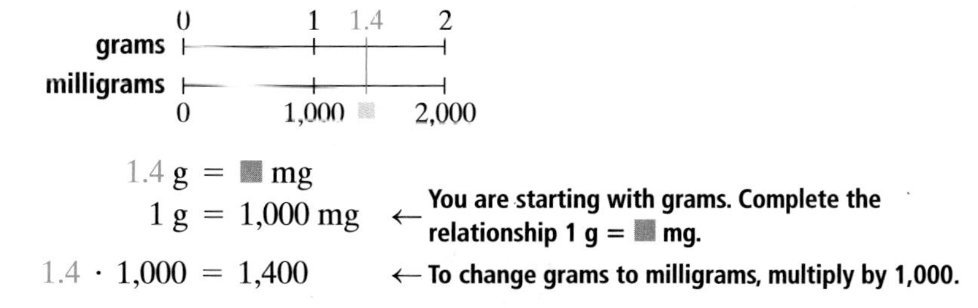

$1.4 \text{ g} = \blacksquare \text{ mg}$

$1 \text{ g} = 1{,}000 \text{ mg}$ ← You are starting with grams. Complete the relationship 1 g = ■ mg.

$1.4 \cdot 1{,}000 = 1{,}400$ ← To change grams to milligrams, multiply by 1,000.

1.4 grams equals 1,400 milligrams.

Check for Reasonableness A gram is greater than a milligram, so the number of grams should be less than the number of milligrams. Since $1.4 < 1{,}400$, the answer is reasonable.

Test Prep Tip
Pay attention to the prefixes when you convert metric units.

Quick Check

2. Change 34 liters to milliliters. **34,000 mL**
3. Change 4,690 grams to kilograms. **4.690 kg**

2. Teach

Activity Lab

Use before the lesson.

All in One Teaching Resources

Activity Lab 1-5: Fractions and Rulers

Guided Instruction

Example 1
Students might wonder about the difference between mass and weight. *Mass* is a measure of the quantity of material that makes up an object. *Weight* is a measure of the force an object experiences due to the pull of gravity.

Example 2
After discussing the solution, ask: *How would the problem be different if you had to change 245 liters to milliliters?* **You would multiply 245 L by 1,000. The answer would be 245,000 mL.**

PowerPoint

Additional Examples

1 Choose a reasonable estimate. Explain your choice. **Accept all reasonable explanations.**

a. length of a pencil: **19 cm**
   19 mm 19 cm 19 m

b. mass of a grape: **9.2 g**
   9.2 mg 9.2 g 9.2 kg

c. capacity of a bucket: **12 L**
   12 mL 12 L 12 kL

2 Change 871 centimeters to meters. **8.71 m**

3 a. Change 6 m 1 cm to meters. **6.01 m**

b. Change 6 m 1 cm to centimeters. **601 cm**

All in One Teaching Resources

- Daily Notetaking Guide 1-5 L3
- Adapted Notetaking 1-5 L1

Advanced Learners L4
One micrometer (1 μm) is one millionth of a meter. Write a number to make each statement true.

$2 \ \mu\text{m} = \blacklozenge$ m **0.000002**
$9.41 \text{ m} = \blacklozenge \ \mu\text{m}$ **9,410,000**

learning style: visual

English Language Learners ELL
Many students from other countries have a good understanding of the metric system. Take this opportunity to tap into their knowledge and have them explain the system to other students.

learning style: verbal

Closure

- *How do you choose a reasonable metric unit for measuring an object?* Sample: Decide what type of measure is involved: length, mass, or capacity. Then compare the size of the object you are measuring to the size of a reference object for each unit.
- *How do you change a measure from one unit to another?* Find the relationship between the units and multiply.

More Than One Way

A fruit punch recipe calls for 1 L of orange juice, 400 mL of pineapple juice, 60 mL of lemon juice, 2 L of apple juice, and 840 mL of water. Can you make a batch of this punch in a 5-L punch bowl?

Anna's Method

I can start by subtracting 1 L of orange juice and 2 L of apple juice from the 5 L available. That leaves 2 L of capacity in the punch bowl.

400 mL + 60 mL + 840 mL = 1,300 mL ← **Add the ingredients in milliliters.**

$2 \cdot 1{,}000 = 2{,}000$; 2 L = 2,000 mL ← **Change remaining capacity to milliliters.**

Since 1,300 mL is less than 2,000 mL, I can make the punch in the bowl.

Ryan's Method

I can convert all the measures in milliliters to liters by multiplying each measure by 0.001.

$400 \cdot 0.001 = 0.4$; 400 mL = 0.4 L

$60 \cdot 0.001 = 0.06$; 60 mL = 0.06 L ← **Change milliliters to liters.**

$840 \cdot 0.001 = 0.84$; 840 mL = 0.84 L

1 L + 0.4 L + 0.06 L + 2 L + 0.84 L = 4.3 L ← **Add all ingredients.**

The capacity of the punch bowl is 5 L. I can make a full batch of the punch in the punch bowl.

Choose a Method

For a craft project, you need ribbon in lengths of 3 m, 25 cm, 4 m, 58 cm, 1.5 m, and 70 cm. You have 10 m of ribbon. Is that enough? Explain why you chose the method you used. **See left.**

Answers may vary. Sample: No; by Ryan's method all measurements are changed to meters. The sum is 10.03 m. You have 10 m of ribbon.

Check Your Understanding

1. **Vocabulary** In the metric system, the prefix *kilo-* means that the unit is ■ times the basic unit of measure. **1,000**

What number should you multiply by to change each unit?

2. grams to kilograms **0.001**
3. centiliters to liters **0.01**
4. liters to decaliters **10**

Write the number that makes each statement true.

5. 64 g = ■ kg **0.064**
6. ■ L = 302 mL **0.302**
7. 8,490 mm = ■ km **0.00849**

Homework Exercises

For more exercises, see Extra Skills and Word Problems.

GO for Help

For Exercises	See Examples
8–10	1
11–17	2–3

A Choose a reasonable estimate.

8. capacity of a small bottle 250 mL 250 L 250 kL
   250 mL
9. height of an oak tree **22 m** 22 cm 22 m 22 km
10. mass of an adult bullfrog **0.5 kg** 0.5 mg 0.5 g 0.5 kg

Complete each statement. You may find a number line helpful.

11. 0.9 kg = ■ g **900**
12. ■ L = 90 mL **0.090**
13. 58 m = ■ mm **58,000**
14. 7,800 g = ■ kg **7.8**
15. 7 m = ■ km **0.007**
16. ■ L = 240 kL **240,000**
17. The capacity of a plastic cup is 350 mL. How many cups can you fill from a 2.1-L bottle of juice? **6 cups**

B GPS

18. **Guided Problem Solving** You are making a drawing of a family crest from a book. You have a piece of paper that is 21.5 cm wide and want to leave a 35-mm margin on each side. How wide can you draw the crest? **145 mm or 14.5 cm**
    - What is the width in centimeters of the margin on one side?
    - What is the combined width of the margins on the two sides?
    - How can you find the maximum width of the crest?

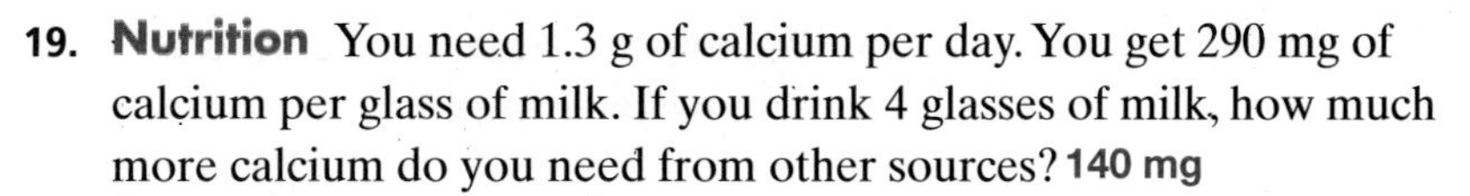

19. **Nutrition** You need 1.3 g of calcium per day. You get 290 mg of calcium per glass of milk. If you drink 4 glasses of milk, how much more calcium do you need from other sources? **140 mg**

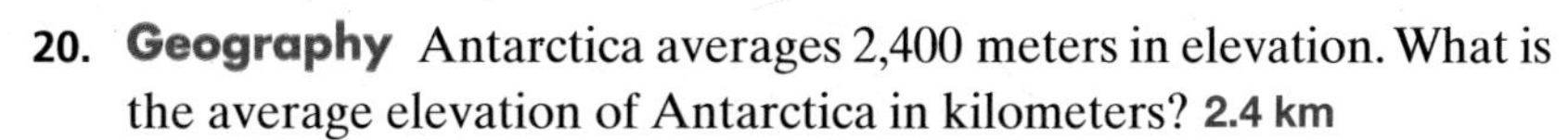

20. **Geography** Antarctica averages 2,400 meters in elevation. What is the average elevation of Antarctica in kilometers? **2.4 km**
21. The world's tallest man was 272 cm tall. Find his height in meters. **2.72 m**
22. **Choose a Method** A bag of birdseed mix contains 400 g of sunflower seed, 300 g of thistle seed, and 0.5 kg of mixed seeds. You order six bags of birdseed mix. How many grams of birdseed is your order? Explain why you chose the method you used. **7,200 g; answers will vary.**

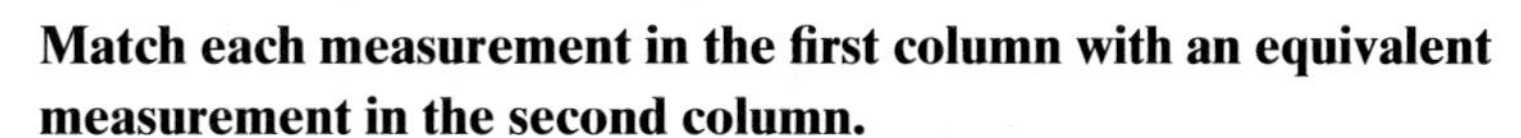

Match each measurement in the first column with an equivalent measurement in the second column.

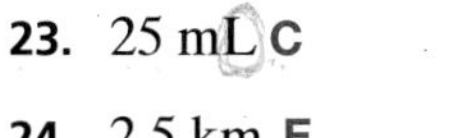

23. 25 mL **C**
24. 2.5 km **E**
25. 0.25 L **D**
26. 250 mm **B**
27. 25,000 mg **A**

A. 0.025 kg
B. 25 cm
C. 0.025 L
D. 25 cL
E. 2,500 m

3. Practice

Assignment Guide

Check Your Understanding
Go over Exercises 1–7 in class before assigning the Homework Exercises.

Homework Exercises

A Practice by Example	8–17
B Apply Your Skills	18–36
C Challenge	37
Test Prep and Mixed Review	38–43

Homework Quick Check
To check students' understanding of key skills and concepts, go over Exercises 9, 14, 32, 35, and 36.

Differentiated Instruction Resources

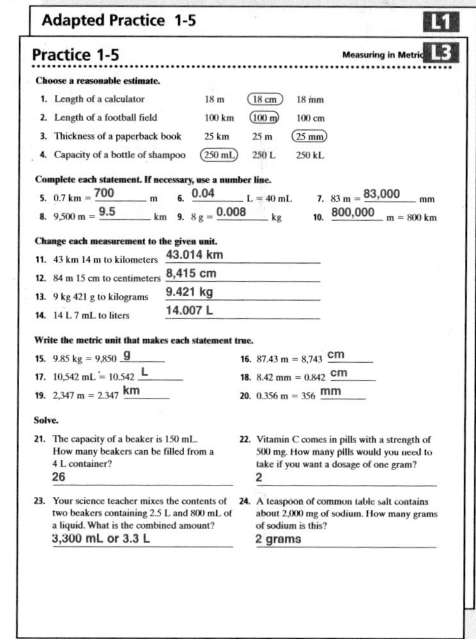

Adapted Practice 1-5 L1

Practice 1-5 Measuring in Metric L3

Choose a reasonable estimate.

1. Length of a calculator 18 m (18 cm) 18 mm
2. Length of a football field 100 km (100 m) 100 cm
3. Thickness of a paperback book 25 km 25 m (25 mm)
4. Capacity of a bottle of shampoo (250 mL) 250 L 250 kL

Complete each statement. If necessary, use a number line.

5. 0.7 km = 700 m
6. 0.04 L = 40 mL
7. 83 m = 83,000 mm
8. 9,500 m = 9.5 km
9. 8 g = 0.008 kg
10. 800,000 m = 800 km

Change each measurement to the given unit.

11. 43 km 14 m to kilometers 43.014 km
12. 84 m 15 cm to centimeters 8,415 cm
13. 9 kg 421 g to kilograms 9.421 kg
14. 14 L 7 mL to liters 14.007 L

Write the metric unit that makes each statement true.

15. 9.85 kg = 9,850 g
16. 87.43 m = 8,743 cm
17. 10,542 mL = 10.542 L
18. 8.42 mm = 0.842 cm
19. 2,347 m = 2.347 km
20. 0.356 m = 356 mm

Solve.

21. The capacity of a beaker is 150 mL. How many beakers can be filled from a 4 L container? 26
22. Vitamin C comes in pills with a strength of 500 mg. How many pills would you need to take if you want a dosage of one gram? 2
23. Your science teacher mixes the contents of two beakers containing 2.5 L and 800 mL of a liquid. What is the combined amount? 3,300 mL or 3.3 L
24. A teaspoon of common table salt contains about 2,000 mg of sodium. How many grams of sodium is this? 2 grams

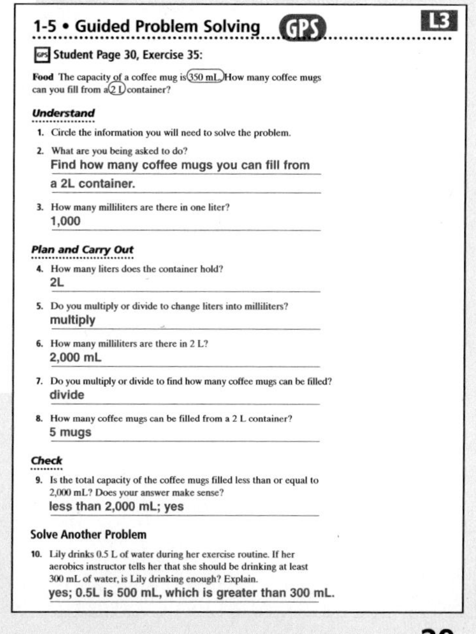

1-5 • Guided Problem Solving GPS L3

GPS Student Page 30, Exercise 35:

Food The capacity of a coffee mug is (350 mL). How many coffee mugs can you fill from a (2 L) container?

Understand

1. Circle the information you will need to solve the problem.
2. What are you being asked to do?
   Find how many coffee mugs you can fill from a 2L container.
3. How many milliliters are there in one liter?
   1,000

Plan and Carry Out

4. How many liters does the container hold?
   2L
5. Do you multiply or divide to change liters into milliliters?
   multiply
6. How many milliliters are there in 2 L?
   2,000 mL
7. Do you multiply or divide to find how many coffee mugs can be filled?
   divide
8. How many coffee mugs can be filled from a 2 L container?
   5 mugs

Check

9. Is the total capacity of the coffee mugs filled less than or equal to 2,000 mL? Does your answer make sense?
   less than 2,000 mL; yes

Solve Another Problem

10. Lily drinks 0.5 L of water during her exercise routine. If her aerobics instructor tells her that she should be drinking at least 300 mL of water, is Lily drinking enough? Explain.
    yes; 0.5L is 500 mL, which is greater than 300 mL.

4. Assess & Reteach

Lesson Quiz

Choose a reasonable estimate.

1. length of desk
   0.12 m   1.2 m   12 m **1.2 m**

Write a number that makes each statement true.

2. 0.8 km = ■ m **800**
3. 640 mL = ■ L **0.64**
4. 7 kg 300 g = ■ kg **7.3**

Alternative Assessment

Provide a balance scale. On one side place objects with known masses, such as a dollar bill with a mass of 1 g. Have small groups work together to record estimates for several classroom objects. Using the balance, have them find and record the actual masses of the objects. Similarly, direct students to estimate, record, and use metric rulers to measure the lengths of classroom objects. Have groups present their estimates to the class.

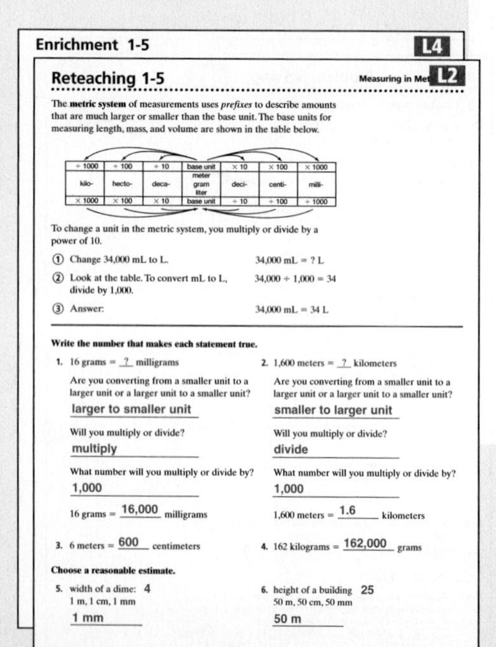

Enrichment 1-5 L4

Reteaching 1-5 Measuring in Met L2

The **metric system** of measurements uses *prefixes* to describe amounts that are much larger or smaller than the base unit. The base units for measuring length, mass, and volume are shown in the table below.

÷ 1000	÷ 100	÷ 10	base unit	× 10	× 100	× 1000
kilo-	hecto-	deca-	meter gram liter	deci-	centi-	milli-
× 1000	× 100	× 10	base unit	÷ 10	÷ 100	÷ 1000

To change a unit in the metric system, you multiply or divide by a power of 10.

① Change 34,000 mL to L. 34,000 mL = ? L

② Look at the table. To convert mL to L, divide by 1,000. 34,000 ÷ 1,000 = 34

③ Answer: 34,000 mL = 34 L

Write the number that makes each statement true.

1. 16 grams = ? milligrams
   Are you converting from a smaller unit to a larger unit or a larger unit to a smaller unit? larger to smaller unit
   Will you multiply or divide? multiply
   What number will you multiply or divide by? 1,000
   16 grams = 16,000 milligrams

2. 1,600 meters = ? kilometers
   Are you converting from a smaller unit to a larger unit or a larger unit to a smaller unit? smaller to larger unit
   Will you multiply or divide? divide
   What number will you multiply or divide by? 1,000
   1,600 meters = 1.6 kilometers

3. 6 meters = 600 centimeters

4. 162 kilograms = 162,000 grams

Choose a reasonable estimate.

5. width of a dime: 4
   1 m, 1 cm, 1 mm
   1 mm

6. height of a building 25
   50 m, 50 cm, 50 mm
   50 m

GO Online
Homework Video Tutor
Visit: PHSchool.com
Web Code: are-0105

Write the metric unit that makes each statement true.

28. 2,034 mg = 2.034 ? **grams**
29. 3.456 cm = 34.56 ? **millimeters**
30. 9,023 mL = 90.23 ? **deciliters**
31. 0.1347 m = 134.7 ? **millimeters**
32. **Writing in Math** Explain how to change units in the metric system.
32. You can change a measure from one unit to another by finding the relationship between the two units and multiplying.
33. **Money** A roll of 50 pennies has a mass of 125 g. Find the mass of $5 in pennies. **1,250 g**
34. The mass of a basketball is 620 g. The mass of a soccer ball is 0.45 kg. How much greater is the mass of a basketball in grams? **170 g**
35. **Food** The capacity of a coffee mug is 350 mL. How many coffee mugs can you fill from a 2 L container? **5 mugs**
    GPS
36. **Reasoning** Suppose you want to change 125 kg 84 g to a single unit. Would you choose kilograms or grams? Explain your choice. **Answers may vary. Sample: 125.084 kg**
37. Ⓒ **Challenge** A recipe for modeling clay requires 470 mL of baking soda, 240 mL of cornstarch, and 300 mL of water. Can you make a double batch in a 1.9-L pan? Explain. **No; a double batch needs 2.02 L.**

Test Prep and Mixed Review

Practice

Multiple Choice

38. A company is manufacturing a sports water jug that holds 1,350 mL. What is the capacity of the jug in liters? **B**
    Ⓐ 1,350,000 L   Ⓑ 1.35 L   Ⓒ 1.305 L   Ⓓ 1.035 L
39. Tina lives 350 m from school. She walks to school and home again each day. During a 5-day school week, how far does she walk? **H**
    Ⓕ 1750 km   Ⓖ 70 km   Ⓗ 3.5 km   Ⓙ 1.75 km
40. A diagram of a bike path is shown below. Shawn biked all the way around the path once. How far did Shawn ride? **C**

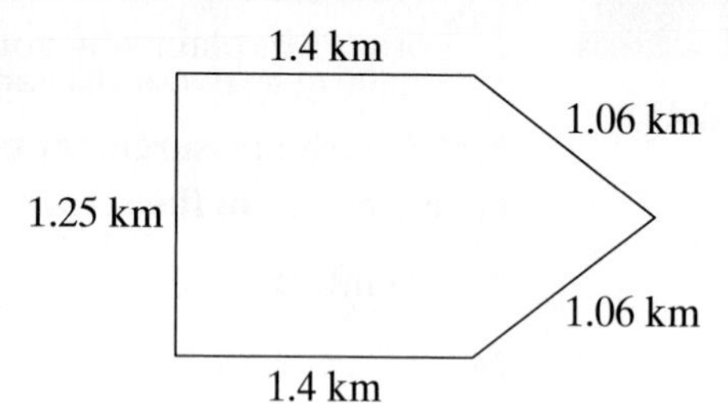

Ⓐ 3.71 km   Ⓑ 5.6 km   Ⓒ 6.17 km   Ⓓ 6.27 km

GO for Help

For Exercises	See Lesson
41–43	1-4

Find each quotient.

41. 15.621 ÷ 2.46 **6.35**
42. 0.17595 ÷ 1.035 **0.17**
43. 5.58 ÷ 9.3 **0.6**

Test Prep

Resources

For additional practice with a variety of test item formats:

- Test-Taking Strategies, p. 59
- Test Prep, p. 63
- Test-Taking Strategies with Transparencies

Algebra

1-6 Comparing and Ordering Integers

Check Skills You'll Need

1. **Vocabulary Review** You know that $5 \cdot (b \cdot 2) = (5 \cdot b) \cdot 2$, because of the _?_ Property of Multiplication. **Associative**

Find each product.

2. $530.6 \cdot 8$ **4,244.8**
3. $0.0771 \cdot 7$ **0.5397**
4. $214.17 \cdot 30$ **6,425.1**

for Help Lesson 1-3

What You'll Learn

To compare and order integers and to find absolute values

New Vocabulary integers, opposites, absolute value

Why Learn This?

Most shipwrecks lie under water. You can use integers to describe distances above and below sea level.

Integers are the set of positive whole numbers, their opposites, and zero. The wreck of *La Belle,* a ship from the 1600s, lies 12 feet below sea level off the coast of Texas. You can use −12 to describe the wreck's depth.

Two numbers that are the same distance from 0 on a number line, but in opposite directions, are **opposites.** You can use integers to find opposites.

EXAMPLE 1 Finding an Opposite

Find the opposite of −12.

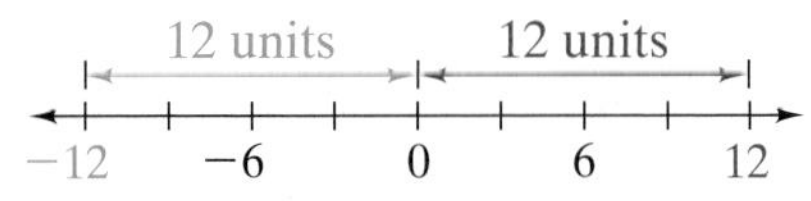

The opposite of −12 is 12, because −12 and +12 are both twelve units from 0, but in opposite directions.

GO for Help

For help with ordering whole numbers, see Skills Handbook p. 654.

Quick Check

1. Find the opposite of each number.
   a. −8 **8**
   b. 13 **−13**
   c. −22 **22**

The **absolute value** of a number is its distance from 0 on a number line. You write "the absolute value of −3" as $|-3|$.

Differentiated Instruction Solutions for All Learners

Special Needs L1
Help students visualize positive and negative integers by showing them both vertical and horizontal number lines. The vertical ones work well to show distance above or below sea level or temperature changes.

learning style: visual

Below Level L2
Review the proper use of the symbols > (is greater than) and < (is less than). Remind students that the greater number is placed at the larger, or open, side of the symbol.

learning style: visual

1-6

1. Plan

Objective

To compare and order integers and to find absolute values

Examples

1 Finding an Opposite
2 Finding Absolute Value
3 Comparing Integers
4 Ordering Integers

Math Understandings: p. 2D

Math Background

The *integers* are the numbers . . ., −3, −2, −1, 0, 1, 2, 3, . . . They can be pictured as points on a number line, with 0 as the *origin.* To the right of 0 are the *positive integers* 1, 2, 3, . . . To the left of 0 are −1, −2, −3, . . . called the *negative integers.* Positive and negative numbers are often called *signed numbers.* Zero is neither positive nor negative.

More Math Background: p. 2D

Lesson Planning and Resources

See p. 2E for a list of the resources that support this lesson.

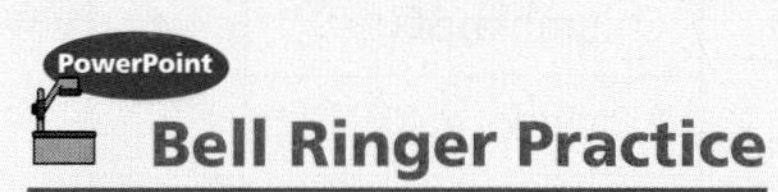

Bell Ringer Practice

Check Skills You'll Need
Use student page, transparency, or PowerPoint. For intervention, direct students to:
Multiplying Decimals
Lesson 1-3
Extra Skills and Word Problems Practice, Ch. 1

2. Teach

Activity Lab

Use before the lesson.

All in One Teaching Resources

Activity Lab 1-6: Integers and Sports

Guided Instruction

Example 1

To emphasize the distinction between positive and negative integers, have students use the word *positive* when naming integers to the right of 0. They read −7 as "negative seven," and they read 1 as "positive one."

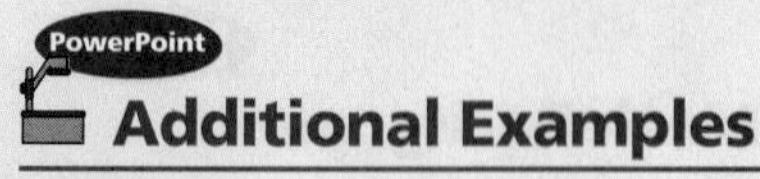

Additional Examples

1. Find the opposite of 2. **−2**
2. Find $|-5|$ and $|5|$. **5; 5**
3. Compare 3 and −8 using <, >, or =. **3 > −8 or −8 < 3**
4. Arrange the cities by temperature, coldest to warmest.

Low Temperatures for March (°F)

City	Low
Albuquerque, NM	8
Chicago, IL	−8
Cleveland, OH	−5
Columbia, SC	4
Providence, RI	1
Reno, NV	−2

Chicago; Cleveland; Reno; Providence; Columbia; Albuquerque

All in One Teaching Resources

- Daily Notetaking Guide 1-6 **L3**
- Adapted Notetaking 1-6 **L1**

Closure

- Explain how to compare two integers. **Locate the integers on a number line. The integer farther to the left is the lesser integer.**

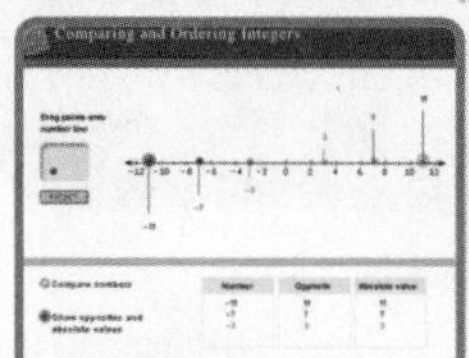

For: Integers Activity
Use: Interactive Textbook, 1-6

EXAMPLE Finding Absolute Value

2 Find $|-3|$ and $|3|$.

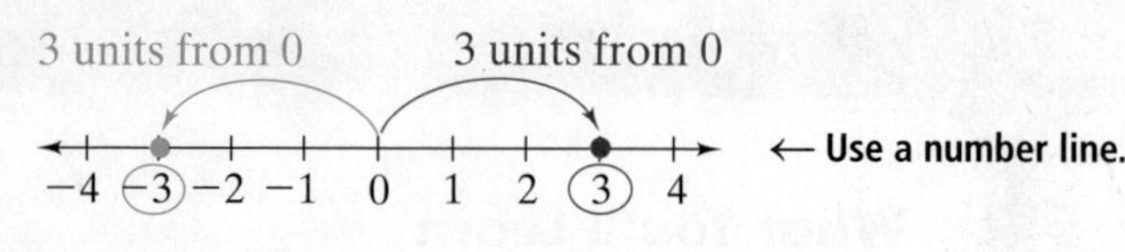

$|-3|$ and $|3| = 3$.

Quick Check

2. Find $|-8|$. **8**

You can compare and order integers by graphing.

EXAMPLE Comparing Integers

Test Prep Tip

Of two integers on a number line, the one farther to the right is greater.

3 Compare −7 and 1 using <, =, or >.

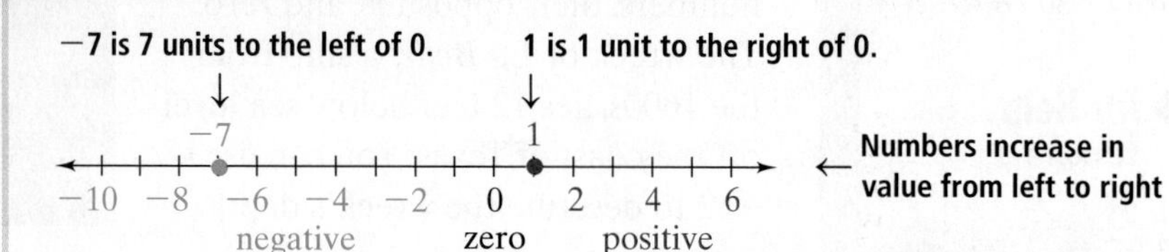

Since −7 is to the left of 1 on the number line, $-7 < 1$.

Quick Check

3. Compare −8 and −2 using <, =, or >. **−8 < −2**

EXAMPLE Ordering Integers

4 Climate Order the cities on the map from coldest to warmest by graphing.

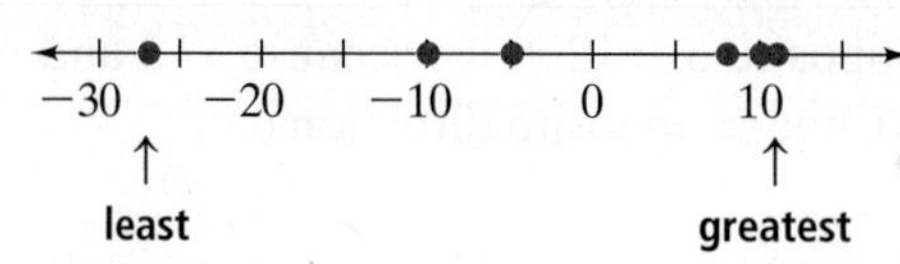

Coldest to warmest:
Fairbanks, Nome, Anchorage, Valdez, Kodiak, Juneau.

Quick Check

4. Order the numbers 3, −1, −4, and 2 from least to greatest. **−4, −1, 2, 3**

Differentiated Instruction Solutions for All Learners

Advanced Learners L4

Find the value of each expression.

−(−8) **8**	−|−1| **−1**
|−(−3)| **3**	(−|−6|) **−6**
−(−(−9)) **−9**	−|−(−2)| **−2**

learning style: visual

English Language Learners ELL

To help students understand absolute value, have them count the units from 0 to −3, and then count the units from 0 to 3. This way they can see that it is the distance that determines the absolute value.

learning style: verbal

Check Your Understanding

1. Whole numbers do not include negative numbers.

3a. sometimes true
b. sometimes true
c. sometimes true
d. sometimes true

1. **Vocabulary** How are integers different from whole numbers?
2. **Number Sense** Which two numbers have an absolute value of 1? −1 and 1
3. **Reasoning** Decide if each of the following is *always true*, *sometimes true*, or *never true* for all integer values of x. See left.
   a. $|x| = x$ b. $|-x| = x$ c. $-|x| = x$ d. $|x| = -x$

Find the opposite of each number.

4. 2 −2   5. 4 −4   6. 3 −3   7. −2 2

Which number in each pair is farther away from zero?

8. 4, −5 −5   9. 2, 5 5   10. −1, −3 −3   11. −12, 11 −12

Homework Exercises

For more exercises, see Extra Skills and Word Problems.

GO for Help

For Exercises	See Examples
12–21	1
22–31	2
32–39	3
40–42	4

A **Find the opposite of each number. You may find a number line helpful.**

12. −1 1   13. −8 8   14. 15 −15   15. 11 −11   16. 90 −90
17. −45 45   18. 20 −20   19. −20 20   20. −123 123   21. 160 −160

Find each absolute value.

22. $|10|$ 10   23. $|-11|$ 11   24. $|-16|$ 16   25. $|-1|$ 1   26. $|4|$ 4
27. $|7|$ 7   28. $|-3|$ 3   29. $|-5|$ 5   30. $|6|$ 6   31. $|-10|$ 10

Compare using <, =, or >.

32. 0 ■ −2 (>)   33. −6 ■ −3 (<)   34. −14 ■ 14 (<)   35. −23 ■ 0 (<)
36. −4 ■ −5 (>)   37. 17 ■ −18 (>)   38. 7 ■ −12 (>)   39. 5 ■ −1 (>)

Order the numbers from least to greatest.

40. −4, 8, −2, −6, 3 −6, −4, −2, 3, 8
41. −2, 0, 7, −1, −5 −5, −2, −1, 0, 7
42. 2, −3, −7, 1, 10 −7, −3, 1, 2, 10

43. −12, −7, −3, +2, +4

44. Start by ordering negative numbers by decreasing absolute value, and then order positive numbers by increasing absolute value. In this case, −5, −4, and 12.

B 43. **Guided Problem Solving** Scores in a golf tournament are reported by the number of strokes each player is above or below par. The scores for five players are −12, +2, −7, +4, and −3. Order the scores from the lowest under par to the greatest over par.
- Which score is farthest to the left on a number line?
- Which score is the next-farthest to the left? See left.

44. **Writing in Math** A friend does not know how to order integers. Explain how to order 12, −4, and −5 from least to greatest. See left.

Online lesson quiz, PHSchool.com, Web Code: ara-0106

3. Practice

Assignment Guide

Check Your Understanding
Go over Exercises 1–11 in class before assigning the Homework Exercises.

Homework Exercises

A Practice by Example	12–42
B Apply Your Skills	43–49
C Challenge	50
Test Prep and Mixed Review	51–55

Homework Quick Check
To check students' understanding of key skills and concepts, go over Exercises 24, 36, 44, 48, and 49.

Differentiated Instruction Resources

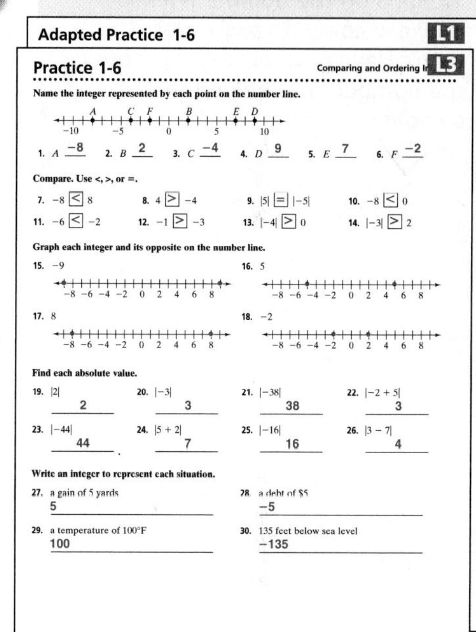

Adapted Practice 1-6 L1

Practice 1-6 Comparing and Ordering In[tegers] L3

Name the integer represented by each point on the number line.

1. A −8 2. B 2 3. C −4 4. D 9 5. E 7 6. F −2

Compare. Use <, >, or =.

7. −8 < 8 8. 4 > −4 9. |5| = |−5| 10. −8 < 0
11. −6 < −2 12. −1 > −3 13. |−4| > 0 14. |−3| > 2

Graph each integer and its opposite on the number line.

15. −9 16. 5
17. 8 18. −2

Find each absolute value.

19. |2| 2 20. |−3| 3 21. |−38| 38 22. |−2 + 5| 3
23. |−44| 44 24. |5 + 2| 7 25. |−16| 16 26. |3 − 7| 4

Write an integer to represent each situation.

27. a gain of 5 yards 5
28. a debt of $5 −5
29. a temperature of 100°F 100
30. 135 feet below sea level −135

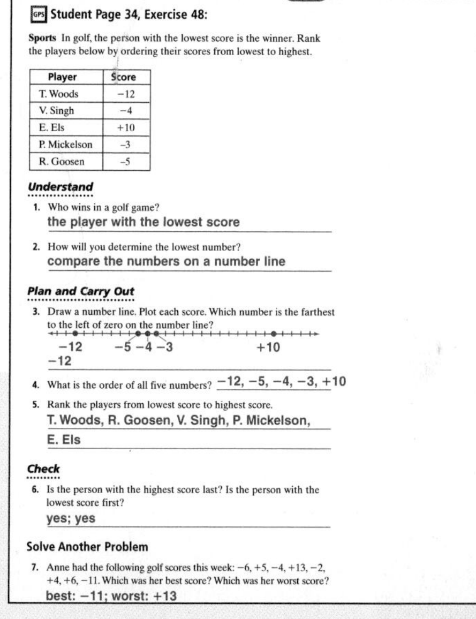

1-6 • Guided Problem Solving GPS L3

Student Page 34, Exercise 48:

Sports In golf, the person with the lowest score is the winner. Rank the players below by ordering their scores from lowest to highest.

Player	Score
T. Woods	−12
V. Singh	−4
E. Els	+10
P. Mickelson	−3
R. Goosen	−5

Understand

1. Who wins in a golf game? the player with the lowest score
2. How will you determine the lowest number? compare the numbers on a number line

Plan and Carry Out

3. Draw a number line. Plot each score. Which number is the farthest to the left of zero on the number line? −12
4. What is the order of all five numbers? −12, −5, −4, −3, +10
5. Rank the players from lowest score to highest score. T. Woods, R. Goosen, V. Singh, P. Mickelson, E. Els

Check

6. Is the person with the highest score last? Is the person with the lowest score first? yes; yes

Solve Another Problem

7. Anne had the following golf scores this week: −6, +5, −4, +13, −2, +4, +6, −11. Which was her best score? Which was her worst score? best: −11; worst: +13

4. Assess & Reteach

Lesson Quiz

Compare using <, >, or =.

1. 2 ■ −9 >   2. −7 ■ −4 <
3. Find |3| and |−1|. 3; 1

Order from least to greatest.

4. 0, −4, 7, −5, 2 −5, −4, 0, 2, 7
5. 6, 8, −3, −7, 1 −7, −3, 1, 6, 8

Alternative Assessment

Distribute number lines to pairs of students. Refer them to Exercises 32–42. Have partners work together to graph the integers on the number lines so they can compare and order them. Have partners explain how the number lines helped them compare integers.

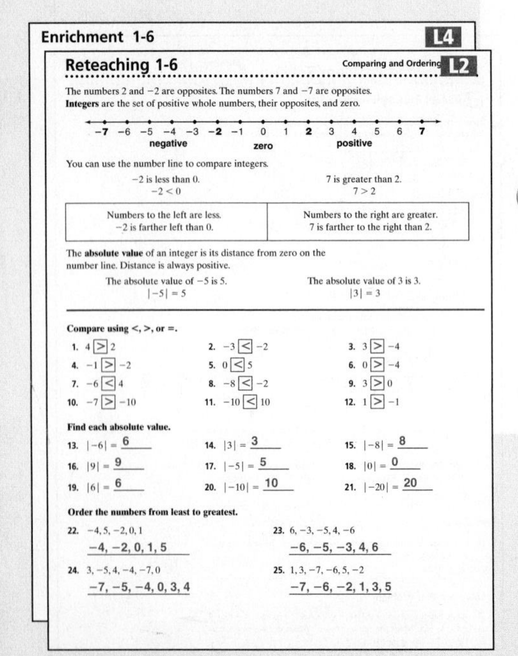

Enrichment 1-6 L4

Reteaching 1-6 Comparing and Ordering L2

The numbers 2 and −2 are opposites. The numbers 7 and −7 are opposites.
Integers are the set of positive whole numbers, their opposites, and zero.

−7 −6 −5 −4 −3 −2 −1 0 1 2 3 4 5 6 7
negative zero positive

You can use the number line to compare integers.

−2 is less than 0.
$-2 < 0$

7 is greater than 2.
$7 > 2$

Numbers to the left are less. −2 is farther left than 0.	Numbers to the right are greater. 7 is farther to the right than 2.

The **absolute value** of an integer is its distance from zero on the number line. Distance is always positive.

The absolute value of −5 is 5.
$|-5| = 5$

The absolute value of 3 is 3.
$|3| = 3$

Compare using <, >, or =.

1. 4 > 2   2. −3 < −2   3. 3 > −4
4. −1 > −2   5. 0 < 5   6. 0 > −4
7. −6 < 4   8. −8 < −2   9. 3 > 0
10. −7 > −10   11. −10 < 10   12. 1 > −1

Find each absolute value.

13. $|-6|$ = 6   14. $|3|$ = 3   15. $|-8|$ = 8
16. $|9|$ = 9   17. $|-5|$ = 5   18. $|0|$ = 0
19. $|6|$ = 6   20. $|-10|$ = 10   21. $|-20|$ = 20

Order the numbers from least to greatest.

22. −4, 5, −2, 0, 1
−4, −2, 0, 1, 5

23. 6, −3, −5, 4, −6
−6, −5, −3, 4, 6

24. 3, −5, 4, −4, −7, 0
−7, −5, −4, 0, 3, 4

25. 1, 3, −7, −6, 5, −2
−7, −6, −2, 1, 3, 5

Visit: PHSchool.com
Web Code: are-0106

45. a. Which city has the highest normal temperature? **Omaha, Nebr.**
b. Which city has the greatest difference between its normal high and normal low temperatures? **Bismarck, N. Dak.**

Normal Temperatures for January (°F)

City	High	Low
Barrow, Alaska	−8	−20
Bismarck, N. Dak.	21	−1
Caribou, Maine	19	0
Duluth, Minn.	18	−1
Omaha, Nebr.	32	13

SOURCE: National Climatic Data Center.
Go to **PHSchool.com** for a data update.
Web Code arg-9041

Order from least to greatest.

46. −14, −15, |−14|, 12, |−16| **−15, −14, 12, |−14|, |−16|**

47. −3551, −3155, −3151, −3515 **−3551, −3515, −3155, −3151**

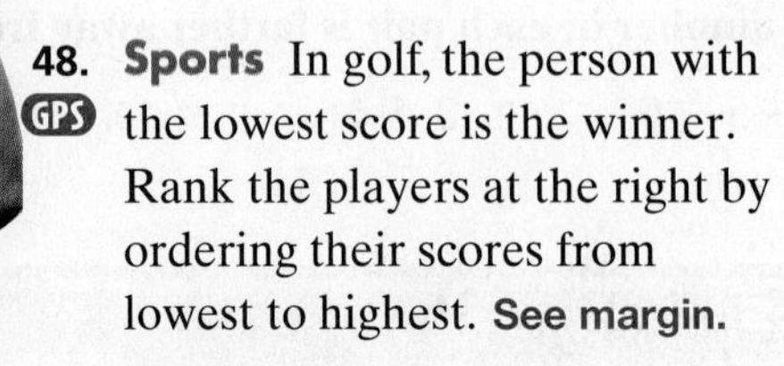

48. **Sports** In golf, the person with the lowest score is the winner. Rank the players at the right by ordering their scores from lowest to highest. **See margin.**

Player	Score
T. Woods	−12
V. Singh	−4
E. Els	+10
P. Mickelson	−3
R. Goosen	−5

49. **Reasoning** Write three numbers that are between −3 and −4. Are the numbers you wrote integers? Explain. **See margin.**

Ⓒ 50. **Challenge** The number −5 is 5 units away from 0. This means that $|-5| = 5$. How far away is −3 from 2? What is $|-3 - 2|$? **5; 5**

Test Prep and Mixed Review — Practice

Multiple Choice

51. The table shows the lowest altitudes on four continents. Which continent has the lowest altitude? **B**

Ⓐ Africa
Ⓑ Asia
Ⓒ Europe
Ⓓ North America

Lowest Altitudes

Continent	Altitude (ft below sea level)
Africa	−512
Asia	−1,348
Europe	−92
N. America	−282

52. Three friends have a 2.79-liter bottle of water to share. About how much water will each person receive? **F**

Ⓕ 950 mL Ⓖ 95 mL Ⓗ 760 mL Ⓙ 76 mL

GO for Help

For Exercises	See Lesson
53–55	1-5

Write a number that makes each statement true.

53. 45.3 cm = ■ mm **453**   54. 26.78 mL = ■ L **0.02678**   55. 256 mg = ■ g **0.256**

Test Prep

Resources

For additional practice with a variety of test item formats:

- Test-Taking Strategies, p. 59
- Test Prep, p. 63
- Test-Taking Strategies with Transparencies

48. T. Woods, R. Goosen, V. Singh, P. Mickelson, E. Els

49. No; there are no integers between the integers −3 and −4.

Checkpoint Quiz 2

Lessons 1-4 through 1-6

Find each quotient.

1. $4.2 \div 3.5$ **1.2**
2. $6.93 \div 2.2$ **3.15**
3. $3.1\overline{)5.27}$ **1.7**

Write the number that makes each statement true.

4. 5,000 mL = ■ cL **500**
5. 410 cm = ■ m **4.1**
6. 1.7 kg = ■ g **1,700**

Compare using $<$, $=$, or $>$.

7. -4 ■ -5 **$>$**
8. -2 ■ 0 **$<$**
9. $|-7|$ ■ $|7|$ **$=$**

10. **Baking** A bread recipe calls for 0.24 L of milk. Your measuring cup is marked in milliliters. How many milliliters of milk do you need? **240 mL**

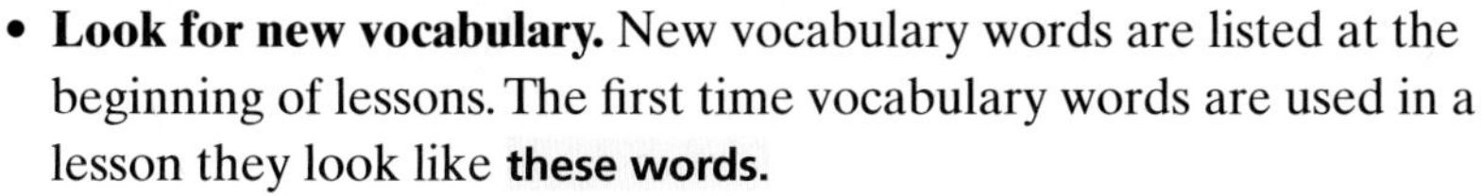

Vocabulary Builder

Learning New Math Terms

Your textbook has many features designed to help you as you read. When you aren't sure what a word means, keep these hints in mind.

- **Look for new vocabulary.** New vocabulary words are listed at the beginning of lessons. The first time vocabulary words are used in a lesson they look like **these words.**
- **Review the key concepts.** Important mathematical terms are explained in Key Concepts boxes.
- **Look for vocabulary tips.** They help you remember what a word means.
- **Use the glossary.** This book contains a glossary, which defines words and refers you to the page where the word is explained.
- **Read carefully.** If necessary, reread a section with new vocabulary until you understand all the information.

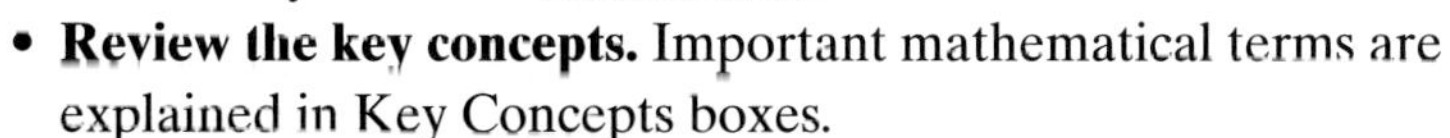

Exercises

Look through Lessons 1-1 to 1-3. Write down the page numbers where these items appear.

1. Vocabulary Tip **p. 15**
2. Key Concepts **pp. 9, 15**
3. New Vocabulary **pp. 4, 8, 14**

Checkpoint Quiz

Use this Checkpoint Quiz to check students' understanding of the skills and concepts of Lessons 1-4 through 1-6.

Resources

- All in One Teaching Resources
- Checkpoint Quiz 2
- ExamView CD-ROM
- Success Tracker™ Online Intervention

ocabu ary ui der

Learning New Math Terms

Students learn strategies for remembering and using new math terms.

Guided Instruction

Call students' attention to the bulleted suggestions. Ask students:

- *What words are new in Lesson 1-6?* **Integers, opposites, absolute value**
- *In Lessons 1-4 to 1-6, are there any Key Concept boxes?* **No**
- *Where is the glossary?* **on pp. 676–713**

Exercises

Have students work in pairs to do Exercises 1–3.

Activity Lab

Modeling Integer Addition and Subtraction

Students use chips of different colors to model positive and negative integers and addition and subtraction of integers with the same signs. They make "zero pairs" to model addition and subtraction of integers with different signs.

Guided Instruction

Activity

Students note that all four parts of both activities use integers with the same absolute values: 5 and 2. However, answers differ depending on the signs of the integers.

Before students do Exercises 1–12, you may want them to work through the activities again with a different pair of numbers.

Exercises

Students work independently on Exercises 1–12. Then form three groups to share, evaluate, and adjust their answers. For Exercise 13, have each group write one of the three rules (a), (b), (c). Have groups share their rules.

Differentiated Instruction

Special Needs L1

Students who have trouble with chips and zero pairs may prefer the number line models in Lesson 1-7.

Resources

- Activity Lab 1-7: Integers
- chips of two different colors
- Student Manipulatives Kit

1-7a Activity Lab

Modeling Integer Addition and Subtraction

You can use models to add and subtract integers. Use chips of two different colors. Let one color represent positive integers and the other color represent negative integers.

ACTIVITY

1. Find $5 + 2$.

Show 5 "+" chips. Then add 2 "+" chips.

There are 7 "+" chips. So $5 + 2 = 7$.

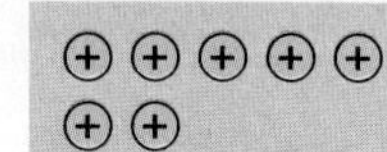

2. Find $-5 + (-2)$.

Show 5 "−" chips. Then add 2 "−" chips.

There are 7 "−" chips. So $-5 + (-2) = -7$.

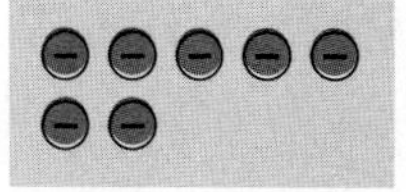

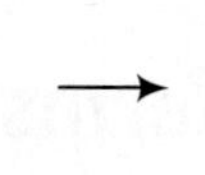

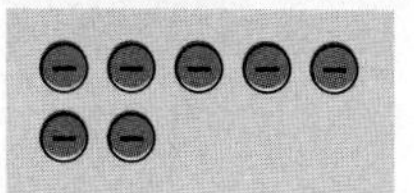

To add integers with different signs, use zero pairs. These chips ⊕ ⊖ are a *zero pair* because ⊕ ⊖ = 0. Removing a zero pair does not change the sum.

3. Find $5 + (-2)$.

Show 5 "+" chips. Then add 2 "−" chips.

Pair the "+" and "−" chips. Remove the pairs.

There are 3 "+" chips left. So $5 + (-2) = 3$.

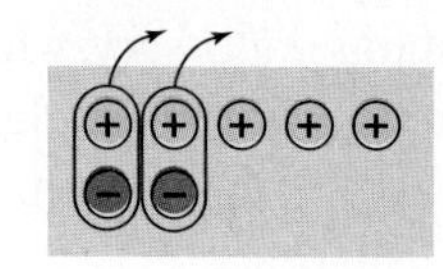

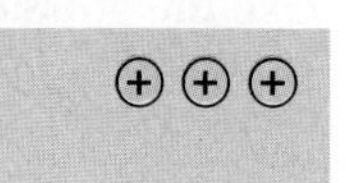

4. Find $-5 + 2$.

Show 5 "−" chips. Then add 2 "+" chips.

Pair the "+" and "−" chips. Remove the pairs.

There are 3 "−" chips left. So $-5 + 2 = -3$.

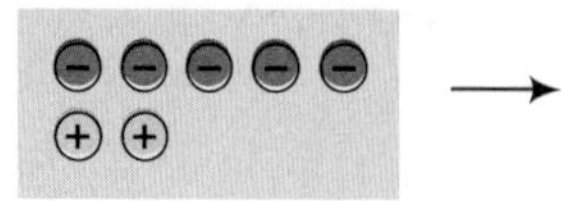

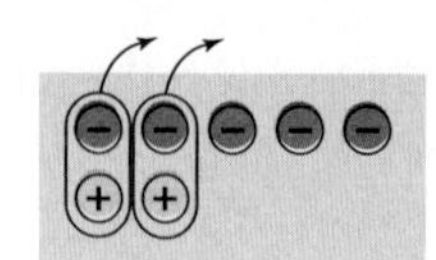

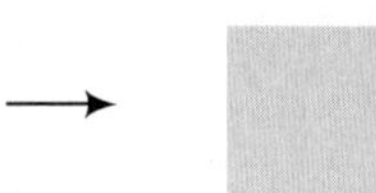

ACTIVITY

1. Find $5 - 2$.

Show 5 "+" chips. → Take away 2 "+" chips. → There are 3 "+" chips left. So $5 - 2 = 3$.

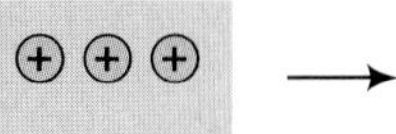
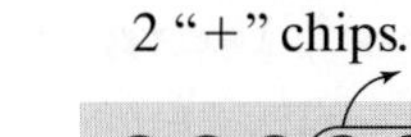
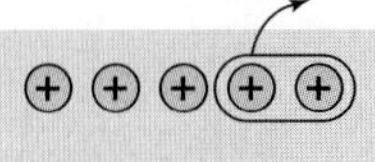
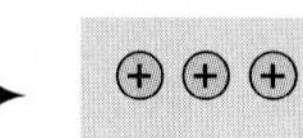

2. Find $-5 - (-2)$.

Show 5 "−" chips. → Take away 2 "−" chips. → There are 3 "−" chips left. So $-5 - (-2) = -3$.

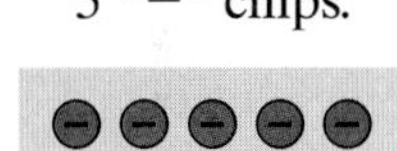
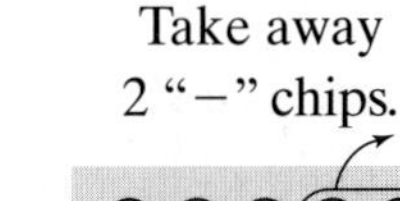
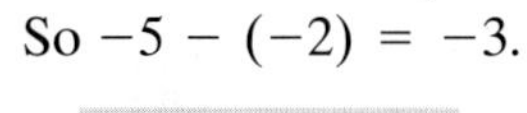

Sometimes you need to insert zero pairs in order to subtract.

3. Find $5 - (-2)$.

Show 5 "+" chips. → Insert two zero pairs. Then take away 2 "−" chips. → There are 7 "+" chips left. So $5 - (-2) = 7$.

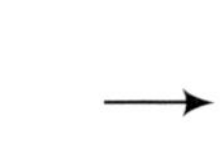
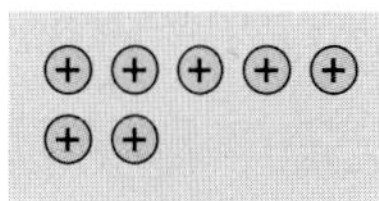

4. Find $-5 - 2$.

Show 5 "−" chips. → Insert two zero pairs. Then take away 2 "+" chips. → There are 7 "−" chips left. So $-5 - 2 = -7$.

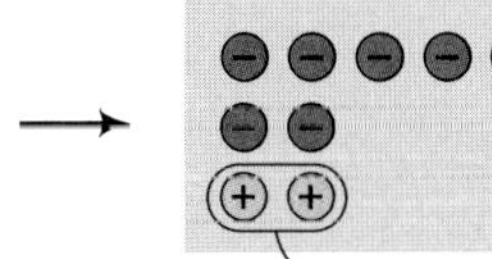
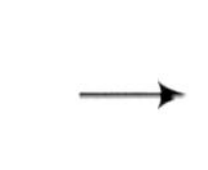
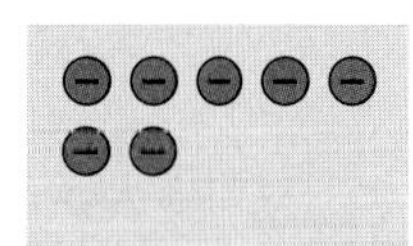

Exercises

Use chips or mental math to add or subtract the following integers.

1. $4 + 9$ **13**
2. $9 + (-3)$ **6**
3. $13 + (-8)$ **5**
4. $-14 + 6$ **−8**
5. $-7 + (-12)$ **−19**
6. $8 + (-11)$ **−3**
7. $11 - 3$ **8**
8. $-4 - (-6)$ **2**
9. $5 - 12$ **−7**
10. $-13 - 7$ **−20**
11. $5 - (-9)$ **14**
12. $-8 - (-13)$ **5**
13. Write a rule for adding: (a) two positive integers, (b) two negative integers, and (c) two integers with different signs. **See margin.**

13. Answers may vary. Sample:

a. To add two positive integers, add their absolute values. The result is the desired sum.

b. To add two negative integers, add their absolute values. The opposite of the result is the desired sum.

c. To add two integers with different signs, find the difference of the absolute values. Use the sign of the number with the greater absolute value.

1-7

1. Plan

Objective
To add and subtract integers and to solve problems involving integers

Examples
1 Adding Integers With a Number Line
2 Adding Integers
3, 4 Subtracting Integers
5 Application: Weather

Math Understandings: p. 2D

Math Background
You can model the addition of integers as a series of moves on a number line. Zero is the starting point. Positive numbers are shown as moves to the right. Negative numbers are shown as moves to the left. The ending point of the last move indicates the sum. Subtracting an integer is the same as adding its opposite.

More Math Background: p. 2D

Lesson Planning and Resources
See p. 2E for a list of the resources that support this lesson.

Bell Ringer Practice
Check Skills You'll Need
Use student page, transparency, or PowerPoint. For intervention, direct students to:
Comparing and Ordering Integers
Lesson 1-6
Extra Skills and Word Problems Practice, Ch. 1

Algebra

1-7 Adding and Subtracting Integers

Check Skills You'll Need
1. **Vocabulary Review** On a number line, how far from zero is the opposite of a number? **See below.**

Find the opposite of each number.

2. 73 **−73**
3. −49 **49**
4. 22 **−22**
5. 13 **−13**
6. −424 **424**
7. −13 **13**

Lesson 1-6

What You'll Learn
To add and subtract integers and to solve problems involving integers

New Vocabulary additive inverses

Why Learn This?
You can add and subtract integers to keep track of money.

Suppose you start the week with no money. You borrow \$10, and then you earn \$10 babysitting to pay back the money you borrowed. You can add integers on a number line to see how much money you have.

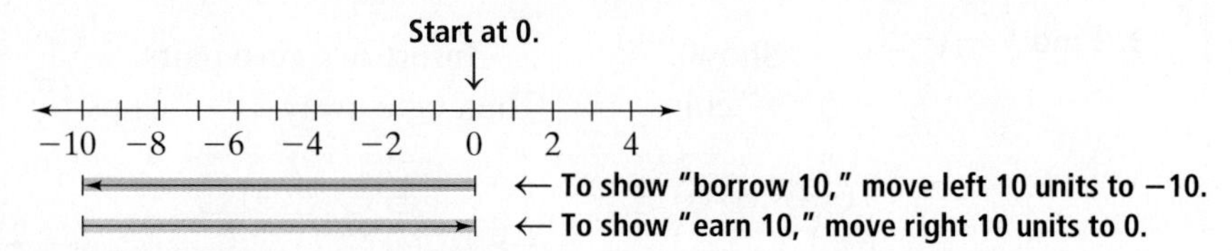

The number line shows that the sum of −10 and 10 is 0. You are back at zero where you started the week. Two numbers whose sum is 0 are **additive inverses.** You can use the following rules to add integers.

Check Skills You'll Need
1. the same distance from zero as the number

KEY CONCEPTS Adding Integers

Same Sign The sum of two positive numbers is positive. The sum of two negative numbers is negative.

Examples $3 + 5 = 8$ $\quad -3 + (-5) = -8$

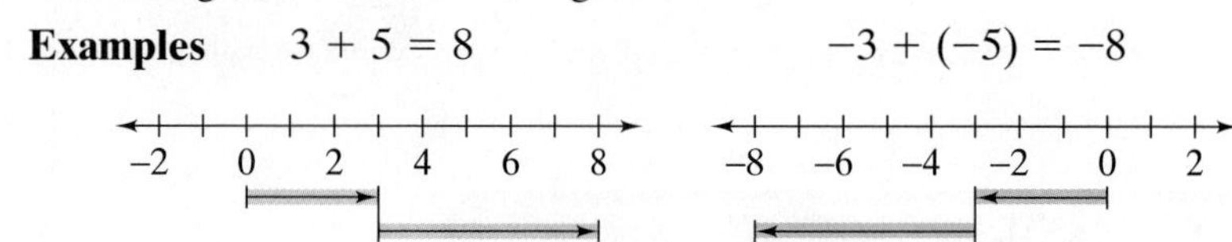

Different Signs Find the absolute value of each number. Subtract the lesser absolute value from the greater. The sum has the sign of the integer with the greater absolute value.

Examples $-3 + 5 = 2$ $\quad 3 + (-5) = -2$

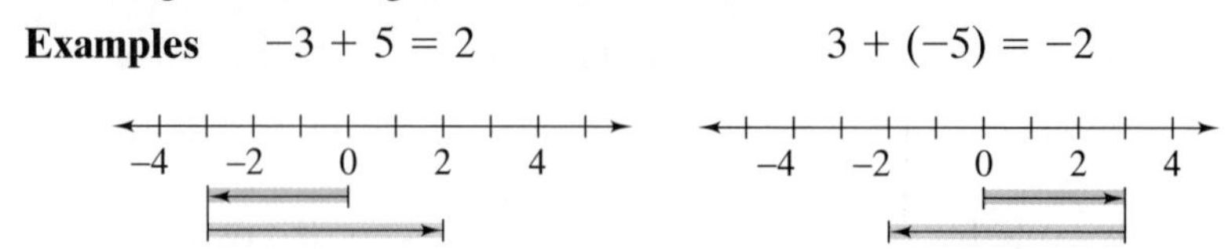

Differentiated Instruction Solutions for All Learners

Special Needs L1
Students can use several number lines to highlight their sums as they work through addition problems. They can use one color to show a negative movement and a different color to show a positive movement.

learning style: visual

Below Level L2
Have students make number line models for a set of additions like these.

$2 + 8$ **10** $\quad -2 + 8$ **6**
$-2 + (-8)$ **−10** $\quad 2 + (-8)$ **−6**

learning style: visual

EXAMPLE Adding Integers With a Number Line

1 Use a number line to find each sum.

a. $5 + (-4)$

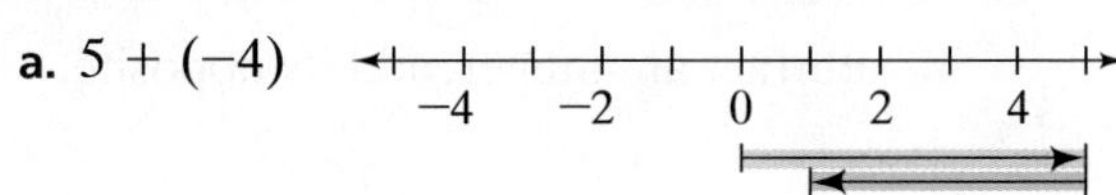

The sum is 1.

b. $-5 + (-2)$

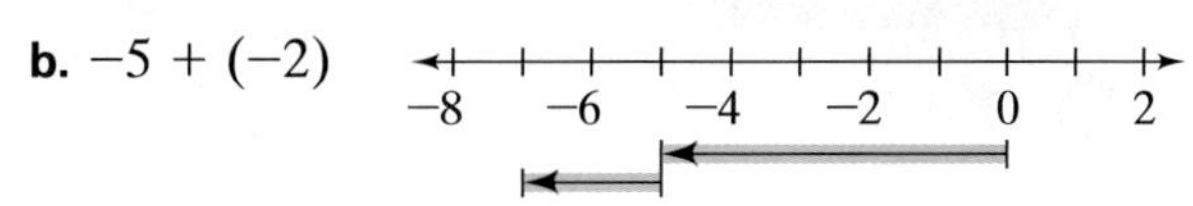

The sum is −7.

Quick Check

1. Use a number line to find each sum.

a. $-8 + 1$ **−7** **b.** $-1 + (-7)$ **−8** **c.** $-6 + 6$ **0**

You can also add integers by using the absolute value of an integer.

EXAMPLE Adding Integers

Check your answer by sketching a number line.

2 Find each sum.

a. $-18 + (-16) = -34$ ← **Both integers are negative. The sum is negative.**

b. $-23 + 8$

$|-23| = 23$ and $|8| = 8$ ← **Find the absolute value of each integer.**

$23 - 8 = 15$ ← **Subtract 8 from 23 because $|8| < |-23|$.**

$-23 + 8 = -15$ ← **The sum has the same sign as −23.**

Quick Check

2. Find each sum.

a. $-97 + (-65)$ **−162** **b.** $21 + (-39)$ **−18** **c.** $22 + (-22)$ **0**

You can subtract integers, too. The number line shows that $9 - 5 = 4$ and $9 + (-5) = 4$. Subtracting 5 is the same as adding −5.

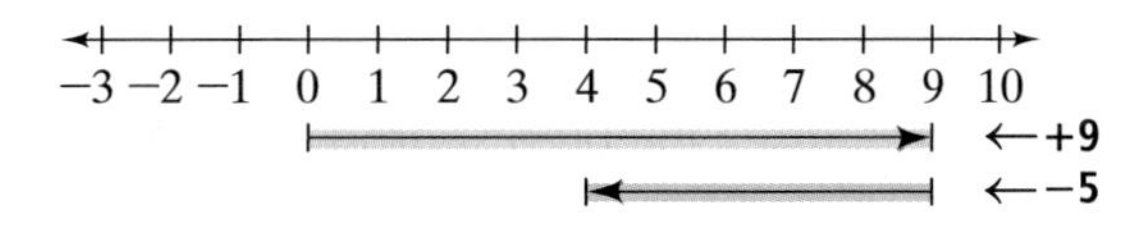

Subtract 5. Add the opposite of 5.

$9 - 5 = 4$ $9 + (-5) = 4$

The answer is 4.

2. Teach

Activity Lab

Use before the lesson.
Student Edition Activity Lab, Hands On 1-7a, Modeling Integer Addition and Subraction, pp. 36-37

All in One Teaching Resources
Activity Lab 1-7: Integers

Guided Instruction

Example 2
Point out that students can perform any integer addition by drawing a number line and counting units. However, the farther the numbers are from 0, the less practical this method becomes. That is why it is important to learn the rules.

Example 3
Write 4 + (−6) directly below 4 − 6. Use arrows, as shown below, to help students visualize the two important changes: The plus sign is changed to a minus sign, and the 6 is changed to its opposite, which is −6.

4 − 6
↓ ↓
4 + (−6)

PowerPoint Additional Examples

1 Use a number line to find each sum.

a. (−3) + 1 **The sum is −2.**

−4 −3 −2 −1 0 1 2 3

b. 1 + (−3) **The sum is −2.**

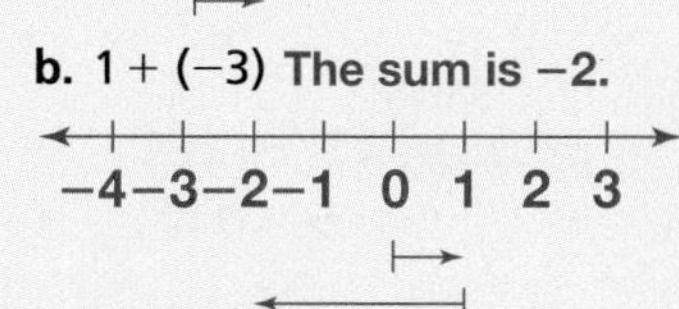

2 Find each sum.

a. 24 + (−6) **18**

b. −12 + (−19) **−31**

Advanced Learners L4
Suppose *a* and *b* are any numbers. Is each statement *always, sometimes,* or *never* true?

$|a + b| = |a| + |b|$ **sometimes**
$|a - b| = |a| - |b|$ **sometimes**

learning style: visual

English Language Learners ELL
To enable students to understand the concept of adding integers, continue to provide borrowing and earning money situations for the problems in Key Concepts.

learning style: visual

PowerPoint

Additional Examples

3 Use a number line to find each difference.

a. 5 − (−3) 8

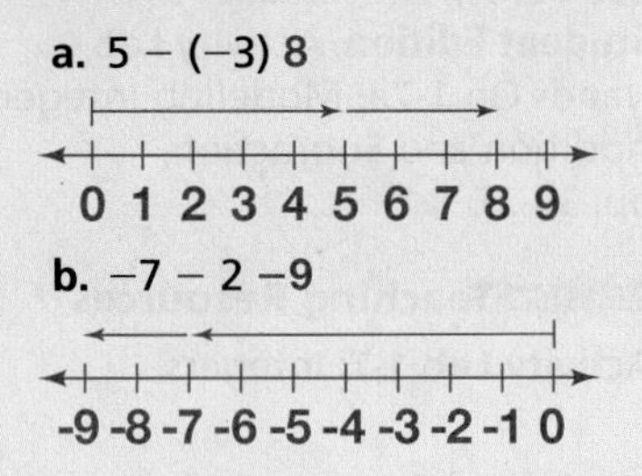

b. −7 − 2 −9

4 Find each difference.

a. – – (–) 5

b. −1 − 5 −6

5 Recorded temperatures at Amundsen-Scott Station in Antarctica have ranged from a low of −89°F to a high of −13°F. Find the difference. 76°F

All in One Teaching Resources

- Daily Notetaking Guide 1-7 L3
- Adapted Notetaking 1-7 L1

Closure

- *What are the rules for adding two integers?* Sample: The sum of two positive integers is positive. The sum of two negative integers is negative. If two integers have different signs, subtract the lesser absolute value from the greater; give the sum the sign of the integer with the greater absolute value.
- *What is the rule for subtracting integers?* Sample: To subtract an integer, add its opposite.

This result suggests a rule for subtracting integers.

KEY CONCEPTS **Subtracting Integers**

To subtract an integer, add its opposite.

EXAMPLES Subtracting Integers

Online active math

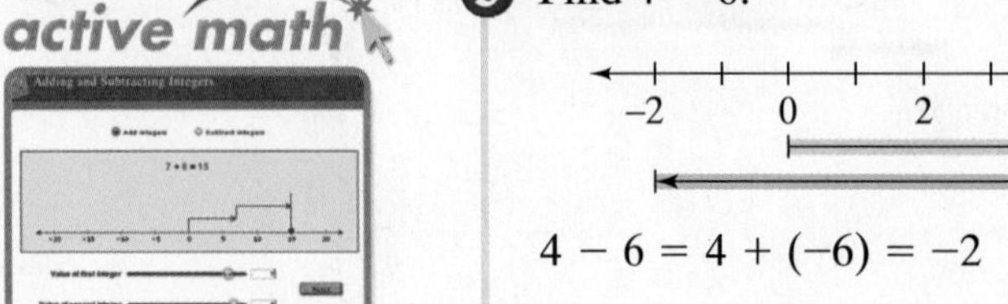

For: Integer Operations Activity
Use: Interactive Textbook, 1-7

3 Find 4 − 6.

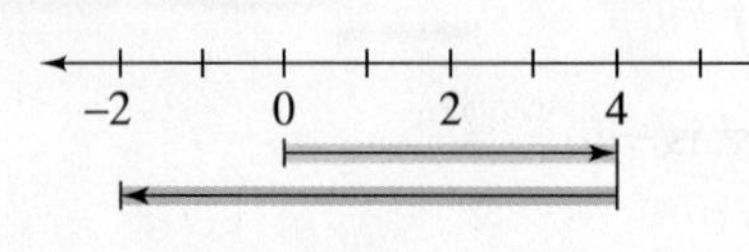

Start at 0. Move 4 units right. Then add the opposite of 6, which is −6.

$4 - 6 = 4 + (-6) = -2$

4 Find −2 − (−5).

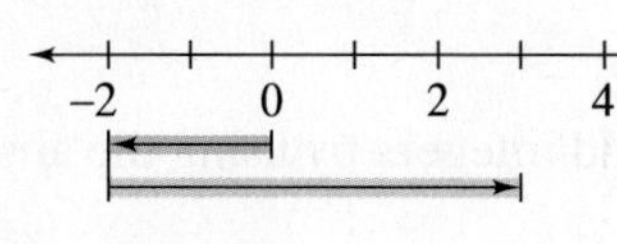

Start at 0. Move 2 units left. Then add the opposite of −5, which is 5.

$-2 - (-5) = -2 + 5 = 3$

Quick Check

3. Find −6 − 1. −7

4. Find 14 − (−7). 21

You can subtract integers to find differences between measurements.

EXAMPLE Application: Weather

10° 8° 6° 4° 2° 0° −2° −4° −6° −8° −10°

The difference is 12°F.

5 The temperature in Caribou, Maine, was 8°F at noon. By 10:00 P.M. the temperature had dropped to −4°F. Find the change in the temperatures.

8 − (−4) ← Subtract to find the difference.

8 + 4 ← Add the opposite of −4, which is 4.

12

The change in the temperatures is 12°F.

Quick Check

5. a. During the biggest drop of the Mean Streak roller coaster in Ohio, your altitude changes by −155 ft. The Texas Giant™ in Texas has a −137 ft change. You want to know how much farther you drop on the Mean Streak. Which expression can you use to solve this problem: −155 − (−137), or −137 − (−155)? −137 − (−155)

b. How much farther do you drop on the Mean Streak? 18 ft

Check Your Understanding

1. **Vocabulary** The absolute values of two numbers that are additive inverses will ___?___ be the same. **A; always**
   (A) always (B) sometimes (C) never
2. The sum of a number and −20 is 40. What is the number? **60**
3. **Reasoning** When you add a positive number and a negative number, the positive addend will ___?___ be less than the sum. **C; never**
   (A) always (B) sometimes (C) never

Find each missing number.

4. $-7 + ■ = -15$ **−8**
5. $7 - ■ = -1$ **8**
6. $-15 - ■ = 15$ **−30**

Homework Exercises

For more exercises, see Extra Skills and Word Problems.

GO for Help

For Exercises	See Examples
7–12	1
13–18	2
19–25	3–5

A **Use a number line to find each sum.**

7. $-5 + 4$ **−1**
8. $2 + (-8)$ **−6**
9. $-6 + 7$ **1**
10. $7 + 3$ **10**
11. $-2 + (-3)$ **−5**
12. $-5 + (-5)$ **−10**

Find each sum.

13. $-99 + 137$ **38**
14. $27 + (-24)$ **3**
15. $-42 + 42$ **0**
16. $-15 + 20$ **5**
17. $-28 + (-32)$ **−60**
18. $126 + (-92)$ **34**

Find each difference. You may find a number line helpful.

19. $29 - 16$ **13**
20. $-3 - (-3)$ **0**
21. $17 - (-8)$ **25**
22. $-14 - 14$ **−28**
23. $12 - (-4)$ **16**
24. $-15 - 2$ **−17**
25. In the game of billiards called 14.1, players lose points if they receive penalties. Find the difference in the scores of the winner with 50 points and the opponent with −17 points. **67 points**

B **GPS** 26. **Guided Problem Solving** On Friday, Rosa borrowed \$10 from her sister. The next day she paid back \$5. Then on Monday she borrowed \$4 more. How much did Rosa owe then? **\$9**
- How much money did Rosa owe before Monday?
- How much money did Rosa still need to repay?

27. **Temperature** (GPS) The highest temperature ever recorded in the United States was 134°F, measured at Death Valley, California. The coldest temperature, −80°F, was recorded at Prospect Creek, Alaska. What is the difference between these temperatures? **214°F**

3. Practice

Assignment Guide

Check Your Understanding

Go over Exercises 1–6 in class before assigning the Homework Exercises.

Homework Exercises

A Practice by Example	7–25
B Apply Your Skills	26–37
C Challenge	38
Test Prep and Mixed Review	39–43

Homework Quick Check

To check students' understanding of key skills and concepts, go over Exercises 8, 20, 27, 33, and 37.

Differentiated Instruction Resources

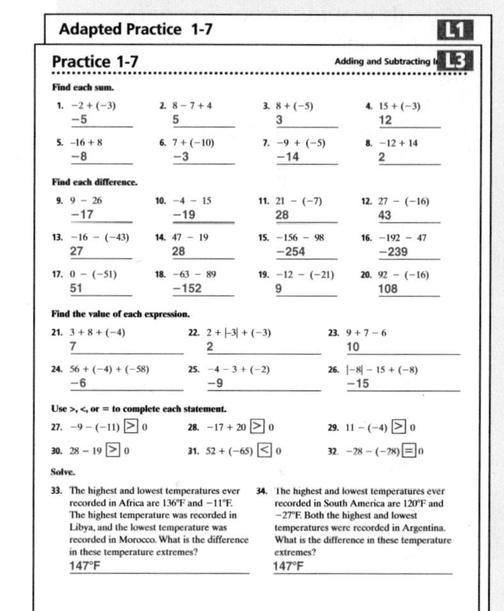

Adapted Practice 1-7 L1

Practice 1-7 Adding and Subtracting In... L3

Find each sum.

1. $-2 + (-3)$ −5
2. $8 - 7 + 4$ 5
3. $8 + (-5)$ 3
4. $15 + (-3)$ 12
5. $-16 + 8$ −8
6. $7 + (-10)$ −3
7. $-9 + (-5)$ −14
8. $-12 + 14$ 2

Find each difference.

9. $9 - 26$ −17
10. $-4 - 15$ −19
11. $21 - (-7)$ 28
12. $27 - (-16)$ 43
13. $-16 - (-43)$ 27
14. $47 - 19$ 28
15. $-156 - 98$ −254
16. $-192 - 47$ −239
17. $0 - (-51)$ 51
18. $-63 - 89$ −152
19. $-12 - (-21)$ 9
20. $92 - (-16)$ 108

Find the value of each expression.

21. $3 + 8 + (-4)$ 7
22. $2 + |-3| + (-3)$ 2
23. $9 + 7 - 6$ 10
24. $56 + (-4) + (-58)$ −6
25. $-4 - 3 + (-2)$ −9
26. $|-8| - 15 + (-8)$ −15

Use >, <, or = to complete each statement.

27. $-9 - (-11)$ [>] 0
28. $-17 + 20$ [>] 0
29. $11 - (-4)$ [>] 0
30. $28 - 19$ [>] 0
31. $52 + (-65)$ [<] 0
32. $-28 - (-28)$ [=] 0

Solve.

33. The highest and lowest temperatures ever recorded in Africa are 136°F and −11°F. The highest temperature was recorded in Libya, and the lowest temperature was recorded in Morocco. What is the difference in these temperature extremes? 147°F
34. The highest and lowest temperatures ever recorded in South America are 120°F and −27°F. Both the highest and lowest temperatures were recorded in Argentina. What is the difference in these temperature extremes? 147°F

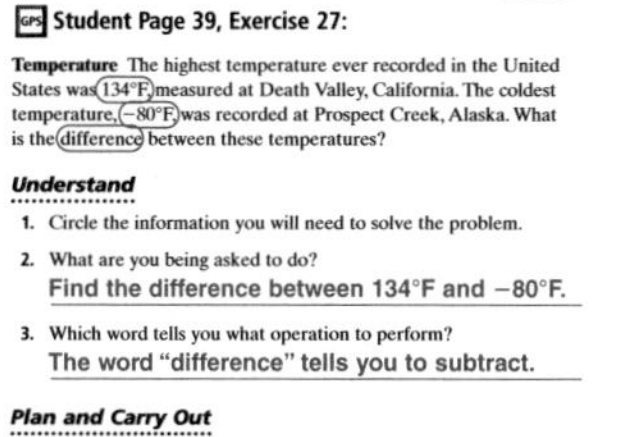

1-7 • Guided Problem Solving GPS L3

Student Page 39, Exercise 27:

Temperature The highest temperature ever recorded in the United States was 134°F, measured at Death Valley, California. The coldest temperature, −80°F, was recorded at Prospect Creek, Alaska. What is the difference between these temperatures?

Understand

1. Circle the information you will need to solve the problem.
2. What are you being asked to do?
   Find the difference between 134°F and −80°F.
3. Which word tells you what operation to perform?
   The word “difference” tells you to subtract.

Plan and Carry Out

4. Write a subtraction expression for the problem.
   $134 - (-80)$
5. Subtracting a negative number is the same as adding what type of number?
   a positive number
6. Write an addition expression that is the same as the expression you wrote in step 4.
   $134 + 80$
7. What is the difference between these temperatures?
   214°F

Check

8. What is 134°F − 214°F?
   −80°F

Solve Another Problem

9. At 6:00 A.M. the temperature was 25°F. At 9:00 P.M. the temperature was −13°F. What was the difference in the temperature?
   38°F

4. Assess & Reteach

PowerPoint

Lesson Quiz

Find each sum or difference.

1. $7 - (-4)$ 11
2. $-9 + 3$ −6
3. $-8 + (-5)$ −13
4. $-5 - 2$ −7

Alternative Assessment

Distribute counters in two colors to pairs of students. Have students model Exercises 7–12 and 19–24 using the counters. Assign one color for positive and the second color for negative. For Exercise 7, students would set out 5 negative counters and 4 positive counters. They then pair up 4 negative and 4 positive counters. The remaining 1 negative counter shows the sum of −1. Remind students that to subtract an integer, add its opposite.

GO Online
Homework Video Tutor
Visit: PHSchool.com
Web Code: are-0107

Write an addition expression for each model. Then find the sum.

28. 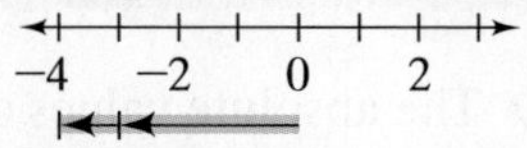 (number line from −4 to 2)

$-3 + (-1) = -4$

29.

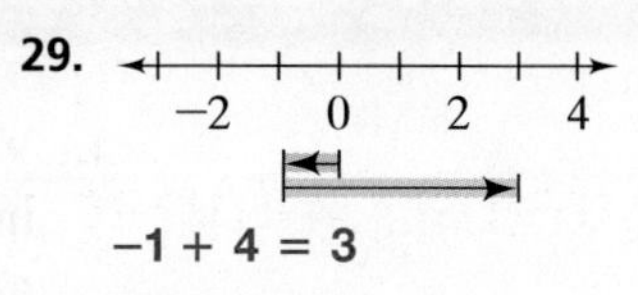

$-1 + 4 = 3$

Algebra **Find each value of *x*.**

30. $-7 + 6 = x$ −1 **31.** $x + 2 = 0$ −2 **32.** $x - 3 = -6$ −3

The continental United States has four time zones. Consider time changes as positive when going east and negative when going west. The time is given in your zone. Find the time in the indicated time zone.

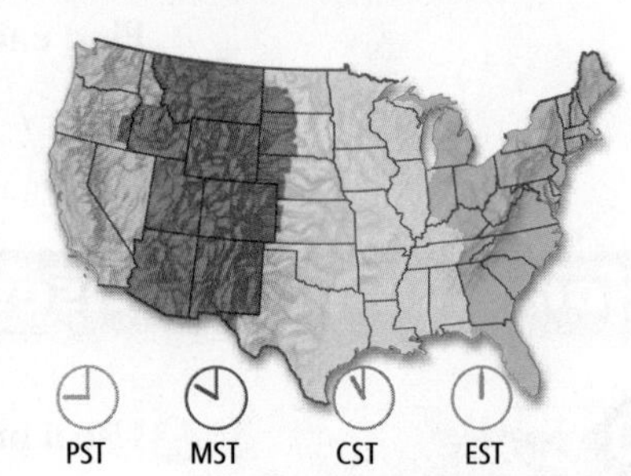

33. 6:00 A.M.; 2 time zones east
8:00 A.M.

34. 9:00 P.M.; 3 time zones west
6:00 P.M.

35. midnight; 2 time zones west
10:00 P.M.

36. 12:00 A.M.; 1 time zone east
1:00 A.M.

37. **Writing in Math** Your friend has trouble simplifying $20 - (-38)$. Write an explanation to help your friend.

37. Subtracting a number is the same as adding its opposite, so $20 - (-38) = 20 + 38 = 58$.

38. **Challenge** You earn $5.25 an hour at your job in a restaurant but pay for any food you eat. On Friday, you receive a check for 7 hours of work, minus $8.90 for food. What is the amount on your check?
$27.85

Test Prep and Mixed Review

Practice

Multiple Choice

39. Which expression is represented by the model below? C

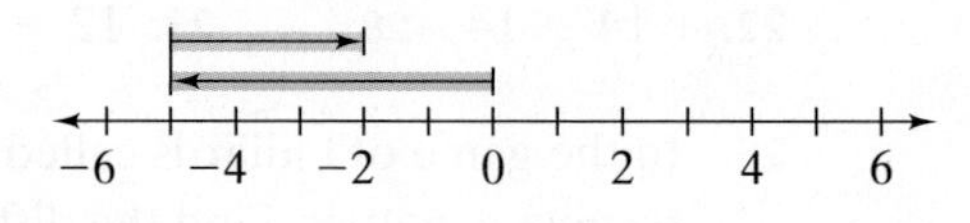

Ⓐ $-5 + 0$ Ⓒ $-5 + 3$
Ⓑ $-5 + 2$ Ⓓ $-5 + 5$

40. The Fred Hartman Bridge in Baytown, Texas, is 381 m long. The Clark Bridge in Alton, Illinois, is 1.408 km. Which method can you use to find the difference in length of the bridges in meters? H

Ⓕ Subtract 140.8 from 0.381. Ⓗ Subtract 381 from 1,408.
Ⓖ Divide 381 by 1,408. Ⓙ Multiply 1,408 by 100.

GO for Help

For Exercises	See Lesson
41–43	1-6

Compare. Use <, =, or >.

41. -4 ■ -10 > **42.** $|-3|$ ■ $|3|$ = **43.** 16 ■ $|-23|$ <

Test Prep

Resources

For additional practice with a variety of test item formats:

- Test-Taking Strategies, p. 59
- Test Prep, p. 63
- Test-Taking Strategies with Transparencies

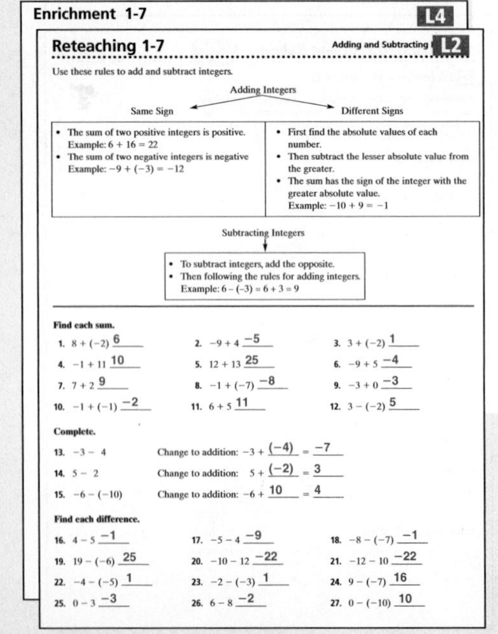

Enrichment 1-7 L4

Reteaching 1-7 Adding and Subtracting L2

Use these rules to add and subtract integers.

Adding Integers

Same Sign	Different Signs
• The sum of two positive integers is positive. Example: $6 + 16 = 22$ • The sum of two negative integers is negative Example: $-9 + (-3) = -12$	• First find the absolute values of each number. • Then subtract the lesser absolute value from the greater. • The sum has the sign of the integer with the greater absolute value. Example: $-10 + 9 = -1$

Subtracting Integers

- To subtract integers, add the opposite.
- Then following the rules for adding integers. Example: $6 - (-3) = 6 + 3 = 9$

Find each sum.

1. $8 + (-2)$ 6 2. $-9 + 4$ −5 3. $3 + (-2)$ 1
4. $-1 + 11$ 10 5. $12 + 13$ 25 6. $-9 + 5$ −4
7. $7 + 2$ 9 8. $-1 + (-7)$ −8 9. $-3 + 0$ −3
10. $-1 + (-1)$ −2 11. $6 + 5$ 11 12. $3 - (-2)$ 5

Complete.

13. $-3 - 4$ Change to addition: $-3 +$ (−4) = −7
14. $5 - 2$ Change to addition: $5 +$ (−2) = 3
15. $-6 - (-10)$ Change to addition: $-6 +$ 10 = 4

Find each difference.

16. $4 - 5$ −1 17. $-5 - 4$ −9 18. $-8 - (-7)$ −1
19. $19 - (-6)$ 25 20. $-10 - 12$ −22 21. $-12 - 10$ −22
22. $-4 - (-5)$ 1 23. $-2 - (-3)$ 1 24. $9 - (-7)$ 16
25. $0 - 3$ −3 26. $6 - 8$ −2 27. $0 - (-10)$ 10

1-8a Activity Lab

Hands On

Modeling Integer Multiplication

To remember the rules for multiplying integers, you can think of the effects that different operations would have on your bank account. The algebra tiles in the diagrams below show groups of integers added to or taken away from a bank account. Think about whether each action would make you feel positive or negative.

ACTIVITY

1. Use algebra tiles to make a "bank account" like the one at the right. Using tiles, add two groups of 3 to your account. Does adding the tiles make you feel positive or negative? Write an equation to represent this operation. **Positive; $2 \times 3 = 6$**

adding two groups of 3

Bank Account

2. Suppose you have to pay two video-rental late fees of \$3 each. This is an example of adding negative integers to your account. Would this make you feel positive or negative? Use algebra tiles to model adding two groups of -3 to your account. Use the diagram below to write an equation for the operation. **Negative; $2 \times (-3) = -6$**

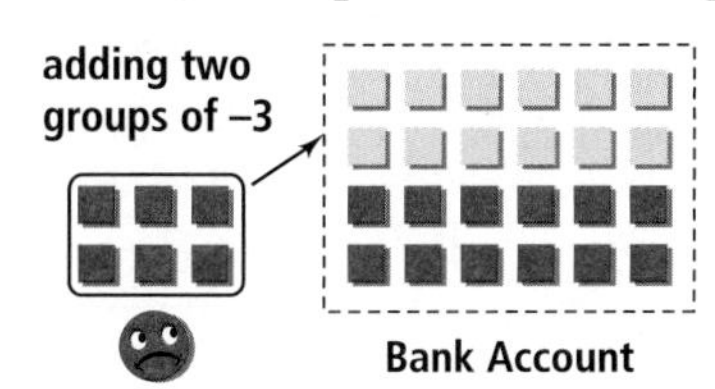

3. Take away two groups of 4 from your account. Use the diagram below to write an equation representing this operation. Describe a situation that this operation might represent. **$-3 \times 4 = -12$**

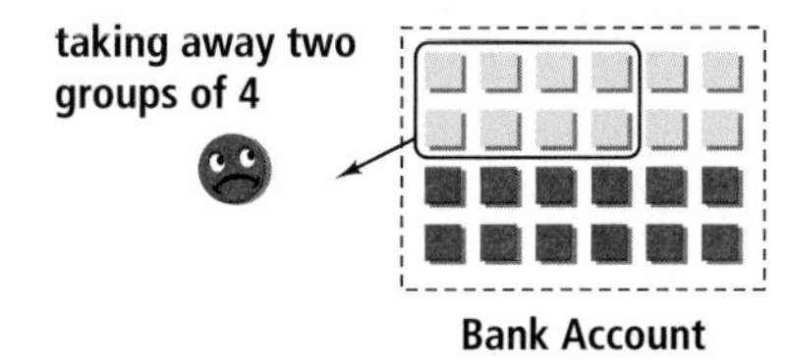

4. Suppose you have two library fines for \$4 each. You *owe* this money, so you can use the integer -4 to represent each fine. If the librarian told you that you did not have to pay the fines, how would you feel? Use the diagram at the right to write an equation representing this operation. **Positive; $-2 \times (-4) = 8$**

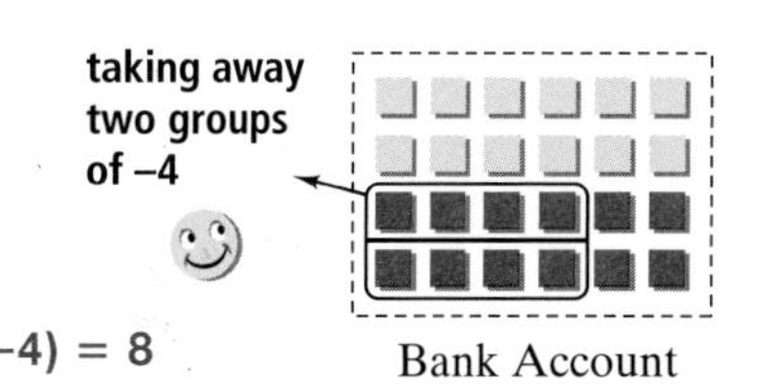

5. Use a table to summarize the rules for multiplying integers. Include all four possibilities. Describe any patterns you notice. **See margin.**

5.

Sign of First Integer	Sign of Second Integer	Sign of Product
positive	positive	positive
positive	negative	negative
negative	positive	negative
negative	negative	positive

When you multiply two integers that have the same sign, the product is positive. When you multiply two integers with different signs, the product is negative.

Modeling Integer Multiplication

Students use algebra tiles to multiply positive and negative integers.

Guided Instruction

Activity

Students should note that the accounts are "balanced" before adding positive or negative integers to the accounts. Be sure students understand that the situation in part 4 is actually a "double negative," or a positive.

Have students work in pairs to make up their own integer multiplication situations to model with algebra tiles.

Resources

- Activity Lab 1-8: Multiplication Madness
- algebra tiles
- Classroom Aid 37
- Student Manipulatives Kit

1-8

1. Plan

Objective
To multiply and divide integers and to solve problems involving integers

Examples
1 Multiplying Integers
2 Dividing Integers

Math Understandings: p. 2D

Professional Development

Math Background
You can view multiplying by a positive integer as repeated addition. So a "positive × positive" product is positive, and a "positive × negative" product is negative. Multiplying by a negative integer is the opposite of multiplying by a positive integer, so the products have opposite signs. A "negative × positive" product is negative, and a "negative × negative" product is positive.

More Math Background: p. 2D

Lesson Planning and Resources
See p. 2E for a list of the resources that support this lesson.

PowerPoint

Bell Ringer Practice
Check Skills You'll Need
Use student page, transparency, or PowerPoint. For intervention, direct students to:
Adding and Subtracting Integers
Lesson 1-7
Extra Skills and Word Problems Practice, Ch. 1

Algebra

1-8 Multiplying and Dividing Integers

Check Skills You'll Need
1. **Vocabulary Review** Two numbers that are *additive inverses* always have a sum of __?__. **zero**

Find each sum.

2. $7 + (-3)$ **4**
3. $-4 + 9$ **5**
4. $-22 + (-13)$ **−35**
5. $-17 + 17$ **0**

GO for Help

Lesson 1-7

What You'll Learn
To multiply and divide integers and to solve problems involving integers

Why Learn This?
Balloonists watch their altitude when they fly. You can multiply integers to find change in altitude.

A balloon descends at a rate of 4 ft/min for 3 min. To multiply integers, think of multiplication as repeated addition.

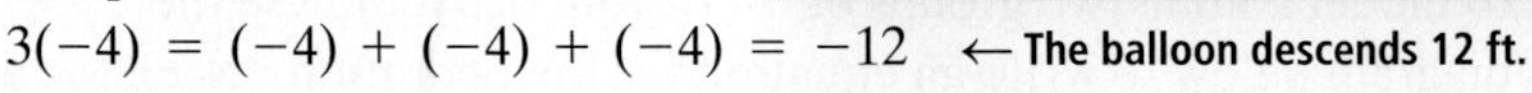
$3(-4) = (-4) + (-4) + (-4) = -12$ ← The balloon descends 12 ft.

You can use number lines to multiply integers.

3(2) means three groups of 2.

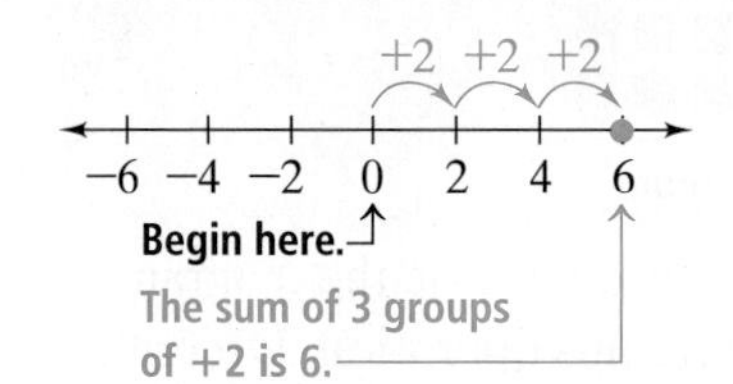

+2 +2 +2
−6 −4 −2 0 2 4 6
Begin here.
The sum of 3 groups of +2 is 6.

3(−2) means three groups of −2.

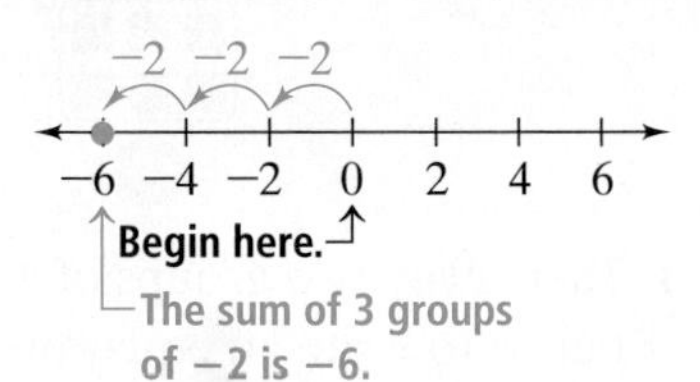

−2 −2 −2
−6 −4 −2 0 2 4 6
Begin here.
The sum of 3 groups of −2 is −6.

−3(2) is the opposite of three groups of 2.

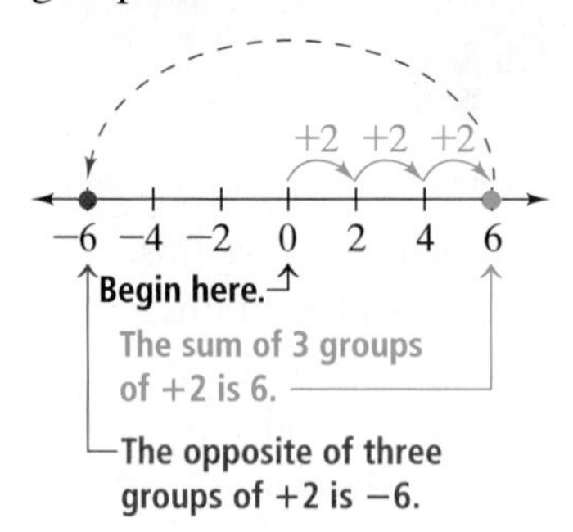

+2 +2 +2
−6 −4 −2 0 2 4 6
Begin here.
The sum of 3 groups of +2 is 6.
The opposite of three groups of +2 is −6.

−3(−2) is the opposite of three groups of −2.

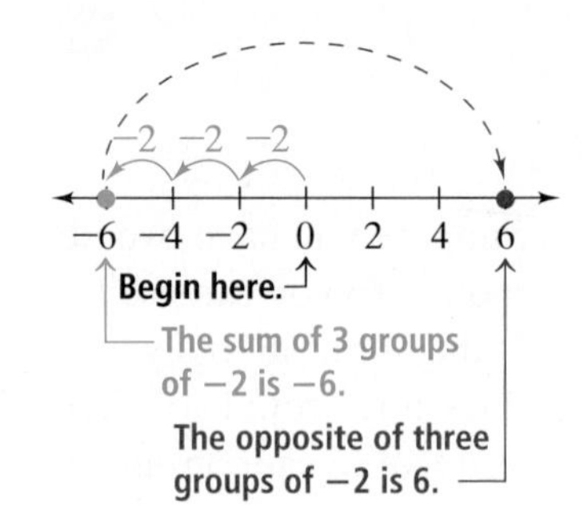

−2 −2 −2
−6 −4 −2 0 2 4 6
Begin here.
The sum of 3 groups of −2 is −6.
The opposite of three groups of −2 is 6.

This pattern suggests the rules for multiplying integers.

Differentiated Instruction Solutions for All Learners

Special Needs L1
Have students use number lines to work through Example 1, circling each group on the number line. For example, for 5(–3) they would circle a set of numbers from 0 to –3, another from –3 to –6, another from –6 to –9, and so on, until 5 sets are circled and the product is –15.
Learning style: visual

Below Level L2
Have students name all pairs of integers whose product is 15. **1, 15; 3, 5; –1, –15; –3, –5.** Then have them name all pairs of integers whose product is –15. **–1, 15; 1, –15; –3, 5; 3, –5**
learning style: verbal

KEY CONCEPTS Multiplying Integers

The product of two integers with the same sign is positive. The product of two integers with different signs is negative.

Examples $-3(-2) = 6$ $3(-2) = -6$

EXAMPLE Multiplying Integers

1 Find each product.

a. $5(3) = 15$ ← **same signs; positive product** → **b.** $-5(-3) = 15$

c. $5(-3) = -15$ ← **different signs; negative product** → **d.** $-5(3) = -15$

Quick Check

1. Simplify the expression $-4(-7)$. **28**

Since $-2(5) = -10$, you know that $-10 \div (-2) = 5$. The rules for dividing integers are similar to the rules for multiplying.

KEY CONCEPTS Dividing Integers

The quotient of two integers with the same sign is positive. The quotient of two integers with different signs is negative.

Examples $-10 \div (-2) = 5$ $10 \div (-2) = -5$

EXAMPLE Dividing Integers

2 A rock climber is at an elevation of 10,100 feet. Five hours later, she is at 7,340 feet. Use the formula below to find the climber's vertical speed.

$$\text{vertical speed} = \frac{\text{final elevation} - \text{initial elevation}}{\text{time}}$$

$$= \frac{7{,}340 - 10{,}100}{5}$$ ← **Substitute 7,340 for final elevation, 10,100 for initial elevation, and 5 for time.**

$$= \frac{-2{,}760}{5} = -552$$ ← **Simplify. The negative sign means the climber is descending.**

The climber's vertical speed is -552 feet per hour.

Quick Check

2. Find the vertical speed of a climber who goes from an elevation of 8,120 feet to an elevation of 6,548 feet in three hours. **−524 ft/h**

Advanced Learners L4

Tell whether each statement is *true* or *false*.

$|2| \cdot |3| = |2 \cdot 3|$ **true**

$|-2| \cdot |3| = |-2 \cdot 3|$ **true**

$|-2| \cdot |-3| = |-2 \cdot (-3)|$ **true**

learning style: verbal

English Language Learners ELL

Ask students to work through the Examples in pairs, discussing each. For example, 1a is 5 groups of 3, while 1b is the opposite of 5 groups of negative 3. Asking students to explain aloud will provide them with language to check integer calculations on their own.

learning style: verbal

2. Teach

Activity Lab

Use before the lesson.

Student Edition Activity Lab, Hands On 1-8a, Modeling Integer Multiplication, p. 43

All in One Teaching Resources

Activity Lab 1-8: Multiplication Madness

Guided Instruction

Example 2

Students might have difficulty with the terms *vertical speed* and *elevation.* In this problem, *vertical speed* refers to how fast the climber is rising. *Elevation* refers to the height of the mountain and climber.

Alternative Method

Have students use algebra tiles to model integer division. Let one color represent positive integers and a second color represent negative integers.

Additional Examples

1 Find each product.

a. 3(7) **21** **c.** −3(7) **−21**

b. 3(−7) **−21** **d.** −3(−7) **21**

2 You are riding your bicycle at a speed of 12 ft/s. Four seconds later, you come to a complete stop. Find the acceleration of your bicycle by using the formula:

$$\text{acceleration} = \frac{\text{final velocity} - \text{initial velocity}}{\text{time}}$$ **−3 ft/s**

All in One Teaching Resources

- Daily Notetaking Guide 1-8 L3
- Adapted Notetaking 1-8 L1

Closure

- *What are the rules for multiplying and dividing two integers?* **Sample: The product or quotient of two integers with the same sign is positive. The product or quotient of two integers with different signs is negative.**

3. Practice

Assignment Guide

Check Your Understanding
Go over Exercises 1–9 in class before assigning the Homework Exercises.

Homework Exercises

A	Practice by Example	10–28
B	Apply Your Skills	29–37
C	Challenge	38
	Test Prep and Mixed Review	39–43

Homework Quick Check
To check students' understanding of key skills and concepts, go over Exercises 16, 22, 34, 35, and 37.

Differentiated Instruction Resources

Adapted Practice 1-8 L1

Practice 1-8 — Multiplying and Dividing — L3

Complete each statement. Then write two examples to illustrate each relationship.

1. positive + positive = ? positive; Check students' answers.
2. negative · positive = ? negative; Check students' answers.
3. positive · positive = ? positive; Check students' answers.
4. negative ÷ negative = ? positive; Check students' answers.
5. negative ÷ positive = ? negative; Check students' answers.
6. positive ÷ positive = ? positive; Check students' answers.
7. positive ÷ negative = ? negative; Check students' answers.
8. negative · negative = ? positive; Check students' answers.

Estimate each product or quotient.

9. $-72 \cdot 57$ $-4{,}000$
10. $-92 \cdot (-41)$ $3{,}600$
11. $-476 \div 90$ -5
12. $-83 \cdot 52$ $-4{,}000$
13. $538 \div (-63)$ -9
14. $-803 \cdot (-106)$ $80{,}000$
15. $49 \cdot 61$ $3{,}000$
16. $479 \div (-61)$ -8

Find each product or quotient.

17. $\frac{-36}{9}$ -4
18. $\frac{-52}{-4}$ 13
19. $(-5) \cdot (-20)$ 100
20. $\frac{-63}{-9}$ 7
21. $(-15) \cdot (2)$ -30
22. $\frac{22}{-2}$ -11
23. $(13) \cdot (-6)$ -78
24. $\frac{-100}{-5}$ 20
25. $(-60) \cdot (-3)$ 180
26. $\frac{-240}{30}$ -8
27. $(43) \cdot (-8)$ -344
28. $\frac{-169}{-13}$ 13

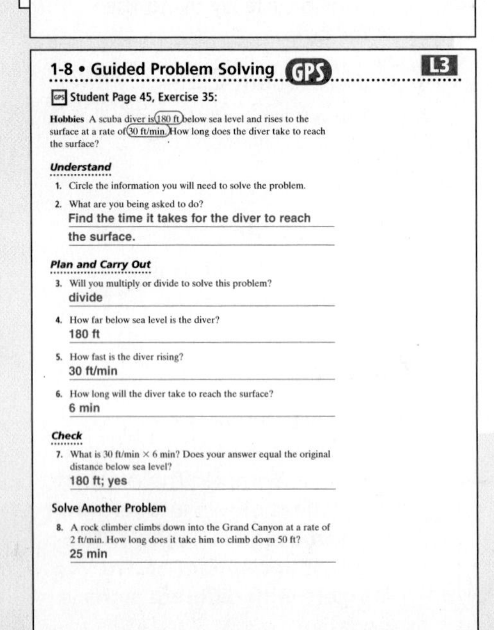

1-8 • Guided Problem Solving GPS L3

Student Page 45, Exercise 35:

Hobbies A scuba diver is 180 ft below sea level and rises to the surface at a rate of 30 ft/min. How long does the diver take to reach the surface?

Understand

1. Circle the information you will need to solve the problem.
2. What are you being asked to do? Find the time it takes for the diver to reach the surface.

Plan and Carry Out

3. Will you multiply or divide to solve this problem? divide
4. How far below sea level is the diver? 180 ft
5. How fast is the diver rising? 30 ft/min
6. How long will the diver take to reach the surface? 6 min

Check

7. What is 30 ft/min × 6 min? Does your answer equal the original distance below sea level? 180 ft; yes

Solve Another Problem

8. A rock climber climbs down into the Grand Canyon at a rate of 2 ft/min. How long does it take him to climb down 50 ft? 25 min

Check Your Understanding

1. **Number Sense** A cave explorer descends at a rate of 6 m/min. Which expression CANNOT be used to find her depth after 4 min.? **B**
   Ⓐ $-6 + (-6) + (-6) + (-6)$ Ⓒ $-6(4)$
   Ⓑ $\frac{-6}{4}$ Ⓓ $-6 - 6 - 6 - 6$
2. **Reasoning** The product of two integers is zero. What do you know about the value of at least one of the integers? Explain.

   2. At least one integer needs to be zero because of the Zero Prop. of Multiplication.

Find each missing number.

3. $-7 \times ■ = -28$ **4**
4. $-48 \div ■ = 6$ **−8**
5. $\frac{■}{-4} = -20$ **80**

Find each product or quotient.

6. $-2(-13)$ **26**
7. $22 \div (-11)$ **−2**
8. $-4(9)$ **−36**
9. $-25 \div 5$ **−5**

Homework Exercises

For more exercises, see Extra Skills and Word Problems.

GO for Help

For Exercises	See Examples
10–18	1
19–28	2

A Find each product.

10. -5×4 **−20**
11. $12(3)$ **36**
12. $6(-6)$ **−36**
13. $-7 \cdot (-3)$ **21**
14. $-21 \times (-4)$ **84**
15. $3(-33)$ **−99**
16. $-12(-17)$ **204**
17. $-35 \cdot 24$ **−840**
18. $-102(6)$ **−612**

Find each quotient.

19. $\frac{36}{12}$ **3**
20. $\frac{14}{-2}$ **−7**
21. $-42 \div 3$ **−14**
22. $-80 \div -20$ **4**
23. $-8\overline{)64}$ **−8**
24. $\frac{-27}{-9}$ **3**
25. $96 \div (-12)$ **−8**
26. $\frac{-195}{13}$ **−15**
27. $\frac{-242}{-1}$ **242**

28. **Hiking** In four hours, a hiker in a canyon goes from 892 ft to 256 ft above the canyon floor. Find the hiker's vertical speed. **−159 ft/h**

B 29. **Guided Problem Solving** A submarine takes 6 min to dive from a depth of 29 m below the water's surface to 257 m below the surface. Find the submarine's vertical speed. **−38 m/min**

$$\text{vertical speed} = \frac{\text{final depth} - \text{initial depth}}{\text{time}} = \frac{■ - (-29)}{■}$$

30. **Birds** A hawk soars at an altitude of 1,800 ft. If the hawk descends to the ground in 45 min, what is its vertical speed? **−40 ft/min**

34. If there is an even number of negative signs, the product is positive. If there is an odd number of negative signs, the product is negative.

Homework Video Tutor
Visit: PHSchool.com
Web Code: are-0108

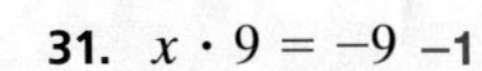

Algebra **Find each value of x.**

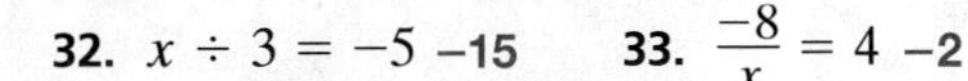

31. $x \cdot 9 = -9$ **−1** **32.** $x \div 3 = -5$ **−15** **33.** $\frac{-8}{x} = 4$ **−2**

34. **Writing in Math** Explain how you would decide whether the product of three numbers is positive or negative. **See margin.**

35. **Hobbies** A scuba diver is 180 ft below sea level and rises to the surface at a rate of 30 ft/min. How long will the diver take to reach the surface? **6 min**

36. **Open-Ended** Describe a situation that can be represented by the expression $4(-2)$. **Answers may vary. Sample: A person owes four people $2 each.**

37. In July, a sporting goods store offers a bike for $278. Over the next five months, the store reduces the price of the bike $15 each month.

a. Write an expression for the total change in price after the months of discounts. **278 − 5(15)**

b. What is the price of the bike at the end of five months? **$203**

38. **Challenge** A bank customer has $172 in a bank account. She withdraws $85 per month for the next 3 months. She also writes 4 checks for $45.75 each. How much money should she deposit to ensure that her balance is at least $25 at the end of the 3 months? **$291**

Test Prep and Mixed Review

Practice

Multiple Choice

39. Which model best represents the expression $2 \times (-4)$? **B**

Ⓐ 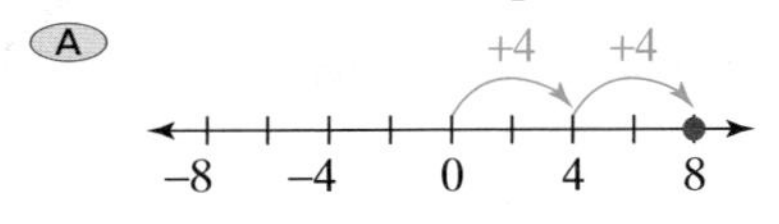

Ⓒ 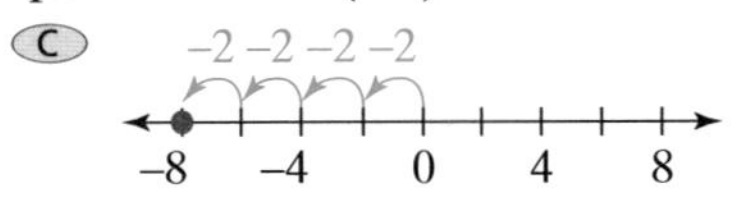

Ⓑ 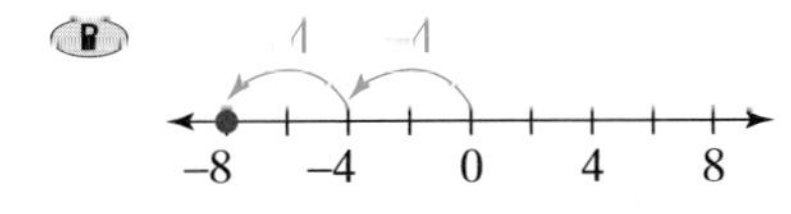

Ⓓ 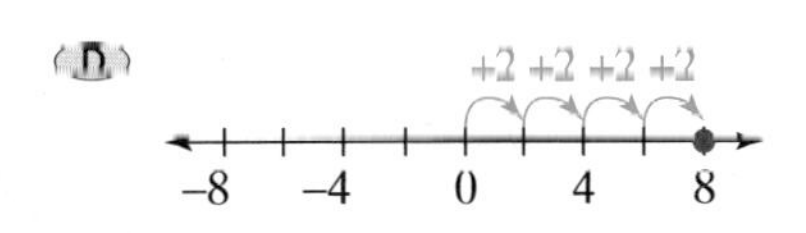

40. One day in January, five different cities had temperatures of −12°F, 5°F, −16°F, 0°F, and 73°F. Which list shows the temperatures from least to greatest? **H**

Ⓕ −16°F, 5°F, −12°F, 0°F, 73°F
Ⓖ 0°F, 5°F, −12°F, −16°F, 73°F
Ⓗ −16°F, −12°F, 0°F, 5°F, 73°F
Ⓙ 73°F, −16°F, −12°F, 0°F, 5°F

GO for Help

For Exercises	See Lesson
41–43	1-5

Write the number that makes each statement true.

41. 142 cm = ■ m **1.42** **42.** 0.67 L = ■ mL **670** **43.** ■ kg = 3,400 g **3.4**

4. Assess & Reteach

Lesson Quiz

Find each product or quotient.

1. $9 \cdot (-5)$ **−45**
2. $-63 \div (-7)$ **9**
3. $-11 \cdot (-11)$ **121**
4. $42 \div (-6)$ **−7**

Alternative Assessment

Have pairs of students use number lines to illustrate how multiplication is repeated addition. Use four related multiplication problems such as these.

$3 \cdot 4$
$3 \cdot (-4)$
$-3 \cdot 4$
$-3 \cdot (-4)$

Have students follow the models on page 44 to draw the repeated "jumps" in either the positive or negative direction. Remind students to start at zero.

Test Prep

Resources

For additional practice with a variety of test item formats:

- Test-Taking Strategies, p. 59
- Test Prep, p. 63
- Test-Taking Strategies with Transparencies

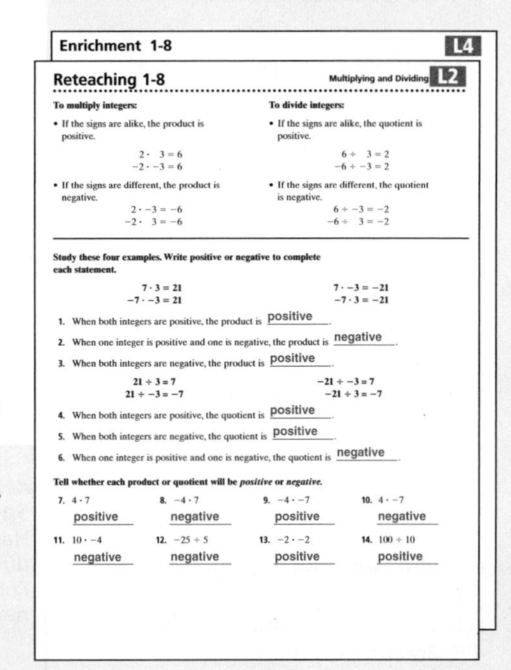

Enrichment 1-8 L4

Reteaching 1-8 Multiplying and Dividing L2

To multiply integers:

- If the signs are alike, the product is positive.

$2 \cdot 3 = 6$
$-2 \cdot -3 = 6$

- If the signs are different, the product is negative.

$2 \cdot -3 = -6$
$-2 \cdot 3 = -6$

To divide integers:

- If the signs are alike, the quotient is positive.

$6 \div 3 = 2$
$-6 \div -3 = 2$

- If the signs are different, the quotient is negative.

$6 \div -3 = -2$
$-6 \div 3 = -2$

Study these four examples. Write positive or negative to complete each statement.

$7 \cdot 3 = 21$ $7 \cdot -3 = -21$
$-7 \cdot -3 = 21$ $-7 \cdot 3 = -21$

1. When both integers are positive, the product is positive.
2. When one integer is positive and one is negative, the product is negative.
3. When both integers are negative, the product is positive.

$21 \div 3 = 7$ $-21 \div -3 = 7$
$21 \div -3 = -7$ $-21 \div 3 = -7$

4. When both integers are positive, the quotient is positive.
5. When both integers are negative, the quotient is positive.
6. When one integer is positive and one is negative, the quotient is negative.

Tell whether each product or quotient will be *positive* or *negative*.

7. $4 \cdot 7$ positive
8. $-4 \cdot 7$ negative
9. $-4 \cdot -7$ positive
10. $4 \cdot -7$ negative
11. $10 \cdot -4$ negative
12. $-25 \div 5$ negative
13. $-2 \cdot -2$ positive
14. $100 \div 10$ positive

1-9

1. Plan

Objective
To use the order of operations and the Distributive Property

Examples
1 Using the Order of Operations
2 Application: Shopping
3 The Distributive Property in Mental Math

Math Understandings: p. 2D

Math Background

Without an order of operations, an expression like $3 + 4 \cdot 5$ might have two values. One person might interpret it as the sum $3 + 4$ being multiplied by 5, with a result of 35. Another person might interpret it as the product $4 \cdot 5$ added to 3, with a result of 23. With the order of operations, the second interpretation is universally accepted as correct.

In this lesson, students learn the order of operations for expressions involving grouping symbols and the four basic operations.

More Math Background: p. 2D

Lesson Planning and Resources

See p. 2E for a list of the resources that support this lesson.

PowerPoint

Bell Ringer Practice

Check Skills You'll Need
Use student page, transparency, or PowerPoint. For intervention, direct students to:
Adding and Subtracting Decimals
Lesson 1-2
Extra Skills and Word Problems Practice, Ch. 1

1-9 Order of Operations and the Distributive Property

Check Skills You'll Need

1. **Vocabulary Review** Which property is illustrated by the statement $1.6 + 4 = 4 + 1.6$? Comm. Prop. of Add.

Use mental math to simplify.

2. $2.5 + 7.1 + 2.5$ 12.1
3. $6.4 + 6.2 + 5.6$ 18.2
4. $8.1 + 3.8 + 8.1$ 20

GO for Help Lesson 1-2

What You'll Learn

To use the order of operations and the Distributive Property

New Vocabulary order of operations, Distributive Property

Why Learn This?

Cash registers calculate the total price when a customer buys several items. You can use order of operations to be sure the total is correct.

Suppose you want to find the total price of one shirt and two hats. You can simplify $\$20 + \$5 \times 2$. If you add first, the price is \$50. If you multiply first, then the price is \$30.

To avoid confusion, mathematicians have agreed upon a particular **order of operations.**

KEY CONCEPTS **Order of Operations**

Work inside grouping symbols.
1. Multiply and divide in order from left to right.
2. Add and subtract in order from left to right.

Vocabulary Tip
Grouping symbols include parentheses, ( ), brackets, [ ], and fraction bars, $\frac{3+5}{2}$.

EXAMPLE 1 Using the Order of Operations

Find the value of each expression.

a. $30 \div 3 + 2 \cdot 6$
$10 + 12$ ← Divide and multiply.
22 ← Add.

b. $30 \div (3 + 2) \cdot 6$
$30 \div 5 \cdot 6$ ← Work inside grouping symbols.
$6 \cdot 6$ ← Divide.
36 ← Multiply.

Quick Check

1. Find the value of each expression.
a. $7(-4 + 2) - 1$ −15 b. $\frac{-40}{4} + 2 \cdot 5$ 0 c. $\frac{8 + 4}{6} - 11$ −9

Differentiated Instruction Solutions for All Learners

Special Needs L1
Have students highlight parts of an expression in different colors according to the order of operations. For example, what you do first, highlight in green; next, highlight in yellow; and last, highlight in pink.
learning style: visual

Below Level L2
Discuss this set of exercises with students.
$(6 + 2) \cdot (5 - 3)$ 16 $(6 + 2) \cdot 5 - 3$ 37
$6 + 2 \cdot (5 - 3)$ 10 $6 + 2 \cdot 5 - 3$ 13
learning style: verbal

EXAMPLE Application: Shopping

2 You want to buy two magazines and three greeting cards from the rack shown at the left. What is the total cost of the items?

Words 2 magazines plus 3 greeting cards

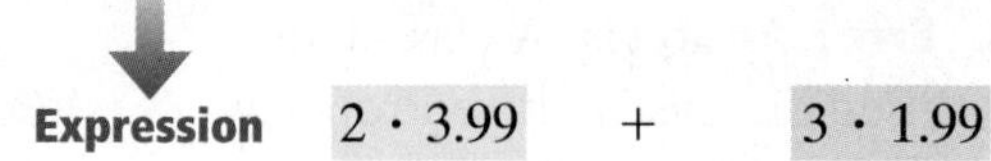

Expression $2 \cdot 3.99 \quad + \quad 3 \cdot 1.99$

$2 \cdot 3.99 + 3 \cdot 1.99 = 7.98 + 5.97$ ← **First multiply.**

$= 13.95$ ← **Then add.**

The total cost is $13.95.

Quick Check

2. What is the total cost of 5 magazines and 2 greeting cards? **$23.93**

Note that $3(5 + 4) = 3(9)$, or 27. Note also that $3(5) + 3(4) = 15 + 12$, or 27. This is an example of the Distributive Property.

KEY CONCEPTS Distributive Property

Arithmetic	Algebra
$9(4 + 5) = 9(4) + 9(5)$	$a(b + c) = a(b) + a(c)$
$5(8 - 2) = 5(8) - 5(2)$	$a(b - c) = a(b) - a(c)$

You can use the Distributive Property to multiply numbers mentally.

EXAMPLE The Distributive Property in Mental Math

3 **Mental Math** Use the Distributive Property to find 7(59).

What You Think

If I think of 59 as $(60 - 1)$, then $7(59)$ is the same as $7(60 - 1)$. I know that $7(60) - 7(1) = 420 - 7 = 413$.

Why It Works $7(59) = 7(60 - 1)$

$= 7(60) - 7(1)$ ← **Use the Distributive Property.**

$= 420 - 7$ ← **Multiply.**

$= 413$ ← **Subtract.**

Test Prep Tip

Look for opportunities for using mental math to make calculations easier.

Quick Check

3. Use the Distributive Property and mental math to find 9(14). **126**

Advanced Learners L4

Find two different ways to make a true statement by inserting +, −, ·, or ÷ into each blank:

32 □ 8 □ 4 □ 2 = 2 **Samples: $32 \div 8 \div 4 \cdot 2$; $32 - 8 \cdot 4 + 2$**

learning style: visual

English Language Learners ELL

In Example 2, have students make the translation from words to expression. For example, ask them to say *2 magazines for 3.99 each plus 3 greeting cards for 1.99 each.* This makes it easier for them to associate the terms in the expression with the words.

learning style: verbal

2. Teach

Activity Lab

Use before the lesson.

All in One Teaching Resources

Activity Lab 1-9: Order of Operations

Guided Instruction

Teaching Tip

To remember the order of operations, use mnemonic devices such as PMDAS (Parentheses, Multiplication, Division, Addition, Subtraction) or the sentence Please My Dear Aunt Sally.

Error Prevention!

Students sometimes forget to distribute the number outside the parentheses to the second number inside the parentheses. Have them write a few examples in which they draw arrows like those below:

$9(10 + 4) = 9 \cdot 10 + 9 \cdot 4$

PowerPoint

Additional Examples

1 Find the value of each expression.

a. $9 + 3 \cdot 5 - 1$ **23**

b. $9 + 3 \cdot (5 - 1)$ **21**

2 You want to buy 3 cans of soup at $.89 each and 2 boxes of crackers at $1.59 each. What is the total cost? **$5.85**

3 Use the Distributive Property to find 7(52). **364**

All in One Teaching Resources

- Daily Notetaking Guide 1-9 L3
- Adapted Notetaking 1-9 L1

Closure

- *What is the order of operations?* **Work inside grouping symbols; multiply and divide in order from left to right; add and subtract from left to right.**
- *What does the Distributive Property allow you to do?* **Sample: You can rewrite an expression like $3(10 + 2)$ as $3 \cdot 10 + 3 \cdot 2$.**

3. Practice

Assignment Guide

Check Your Understanding

Go over Exercises 1–7 in class before assigning the Homework Exercises.

Homework Exercises

A	Practice by Example	8–25
B	Apply Your Skills	26–33
C	Challenge	34
Test Prep and Mixed Review		35–41

Homework Quick Check

To check students' understanding of key skills and concepts, go over Exercises 14, 21, 30, 32, and 33.

Differentiated Instruction Resources

Adapted Practice 1-9 L1

Practice 1-9 — Order of Operations and the Distributive Property — L3

Find the value of each expression.

1. $(8 + 2) \times 9$ 90
2. $5 - 1 \div 4$ 4.75
3. $(6 + 3) \div 18$ 0.5
4. $80 - 6 \times 7$ 38
5. $4 \times 6 + 3$ 27
6. $4 \times (6 + 3)$ 36
7. $35 - 6 \times 5$ 5
8. $9 \div 3 + 6$ 9

Find the missing numbers. Then simplify.

9. $5(9 + 6) = 5(\underline{?}) + 5(\underline{?})$ 9; 6; 75
10. $4(9.7 - 8.1) = \underline{?}(9.7) - \underline{?}(8.1)$ 4; 4; 6.4
11. $\underline{?}(3.8) = 9(4) - 9(\underline{?})$ 9; 0.2; 34.2
12. $\underline{?}(17.1 + 12.6) = 6(17.1) + 6(12.6)$ 6; 178.2

Use the Distributive Property and mental math to find each product.

13. 3(6.4) 19.2
14. 5(7.1) 35.5
15. 5(8.9) 44.5
16. 4(9.2) 36.8
17. 9(11.1) 99.9
18. 7(8.9) 62.3

Copy each statement and add parentheses to make it true.

19. $6 + 6 \div 6 \times 6 + 6 = 24$ $(6 + 6) \div 6 \times (6 + 6) = 24$
20. $6 \times 6 + 6 \times 6 - 6 = 426$ $6 \times (6 + 6) \times 6 - 6 = 426$
21. $6 + 6 \div 6 \times 6 - 6 = 0$ $(6 + 6) \div 6 \times (6 - 6) = 0$
22. $6 - 6 \times 6 + 6 \div 6 = 1$ $(6 - 6) \times 6 + 6 \div 6 = 1$

23. A backyard measures 80 ft × 125 ft. A garden is planted in one corner of it. The garden measures 15 ft × 22 ft. How much of the backyard is *not* part of the garden? 9,670 ft²

1-9 • Guided Problem Solving GPS L3

GPS Student Page 49, Exercise 30:

Business A florist is buying flowers to use in centerpieces. Each centerpiece has 3 lilies. There are a total of 10 tables. Each lily costs \$.98. Use mental math to find the cost of the lilies.

Understand

1. What are you being asked to do? Determine the cost of the lilies.
2. Which method are you to use to determine the cost? mental math

Plan and Carry Out

3. How many lilies do you need in all? 30
4. Write an expression to find the total cost of the lilies. 30 × \$.98
5. The amount \$.98 can also be written as \$1.00 – \$.02. Rewrite your expression from step 4 using \$1.00 – \$.02. 30(\$1.00 – \$.02)
6. Simplify the expression using mental math. \$30 – \$.60 = \$29.40
7. How much do the lilies cost? \$29.40

Check

8. Use a calculator to determine the cost of the lilies. Is your answer from step 7 correct? \$29.40; yes

Solve Another Problem

9. Your horticulture club is planting a garden at school as a beautification project. The principal is allowing you to use an area that is 5 yd². If it costs \$8.93 to buy enough rose bulbs to plant 1 yd², how much will it cost to buy bulbs to fill 5 yd²? 5(\$9.00 – \$.07) = \$44.65

Check Your Understanding

1. **Vocabulary** The Distributive Property combines which operation with addition or subtraction? **multiplication**
2. **Reasoning** Does $6(50 + 3) = (50 + 3)6$? Explain.

   2. Yes, by the Comm. Prop. of Mult.

3. **Error Analysis** A classmate used the Distributive Property to find 11(9.2). What error did your classmate make? **11(2) should be 11(0.2).**

   $11(9.2) = 11(9) + 11(2)$
   $= 99 + 22$
   $= 121$

Use the order of operations to fill in the blanks.

4. $4 \times 5 + 7 = ■ + 7 = ■$ **20; 27**
5. $20 - 2 \cdot 8 = ■ - 16 = ■$ **20; 4**

Use mental math to find the missing numbers. Then simplify.

6. $■(4.8) = 6(5) - 6(■)$ **6; 0.2; 28.8**
7. $5(32) = 5(■) + 5(■)$ **30; 2; 160**

Homework Exercises

For more exercises, see Extra Skills and Word Problems.

GO for Help

For Exercises	See Examples
8–16	1-2
17–25	3

A **Find the value of each expression.**

8. $6 + 1 \cdot 5$ **11**
9. $-4 \div 2 + 9$ **7**
10. $5 - 8 \div 4$ **3**
11. $3 - 0 \cdot 11$ **3**
12. $18 \div 3 \cdot 2$ **12**
13. $100 - 7 \cdot 9$ **37**
14. $-12 \div 6 - (1 + 4)$ **−7**
15. $48 \div (-4 \cdot 3) + 2$ **−2**
16. **Coins** You buy some items at a store and pay with a \$5 bill. You receive two quarters, two dimes, and three pennies as change. How much money do you receive? **\$.73**

Use the Distributive Property and mental math to find each product.

17. 5(29) **145**
18. 6(3.9) **23.4**
19. $9 \cdot 2.2$ **19.8**
20. 5(42) **210**
21. $7 \cdot 2.6$ **18.2**
22. 8(87) **696**
23. $4 \cdot 10.2$ **40.8**
24. 11.6(9) **104.4**
25. 1.1(22) **24.2**

B 26. **Guided Problem Solving** The sheet of plywood at the right has a piece cut from one corner. What is the area of the plywood? **26 ft²**

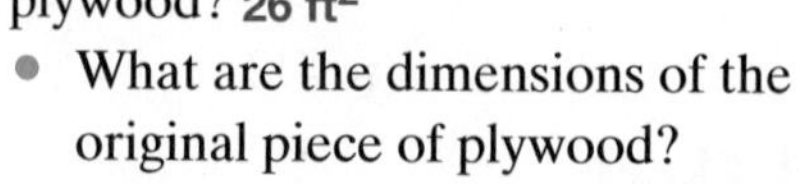

- What are the dimensions of the original piece of plywood?
- What are the dimensions of the missing piece?

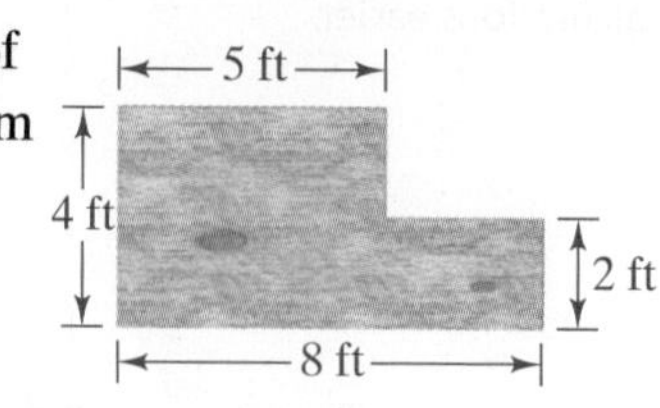

33. You could use the Dist. Prop. to rewrite 4(110.5) as 4(110 + 0.5). This gives you 440 + 2, or 442. You could also rewrite 4(110.5) as 4(111 − 0.5). This gives you 444 − 2, or 442.

Visit: PHSchool.com
Web Code: are-0109

Copy each statement. Add parentheses to make it true.

27. $4 + 4 \div 4 - 4 = -2$
$(4 + 4) \div 4 - 4 = -2$

28. $4 \cdot 4 \div 4 + 4 = 2$
$(4 \cdot 4) \div (4 + 4) = 2$

29. Architecture Use the floor plan at the right. Find the area of the floor. **1,840 ft²**

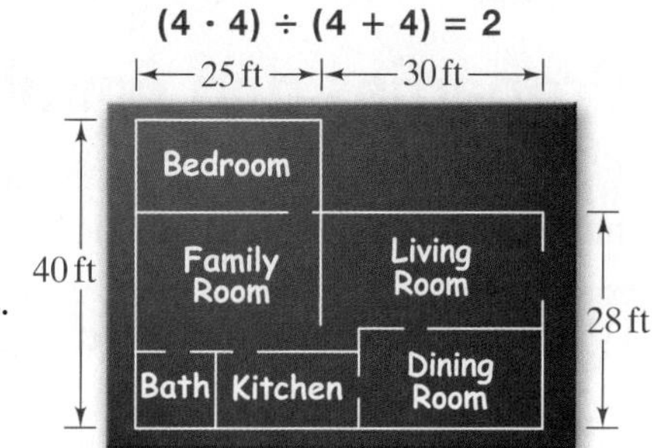

30. Business (GPS) A florist is buying flowers to use in centerpieces. Each centerpiece has 3 lilies. There are 10 tables in all. Each lily costs \$.98. Use mental math to find the cost of the lilies. **\$29.40**

31. A freight train has 62 cars and a locomotive that is 65 ft long. Each car is 50 ft long. There is 3 ft of space between the locomotive and the first car and between each pair of cars. How long is the train? **3,351 ft**

32. Geometry Write two expressions to find the total area of the figure. Then find the area. **See left.**

3 cm
5 cm 5 cm

33. Writing in Math Explain how you can use the Distributive Property in two different ways to calculate 4(110.5). **See margin.**

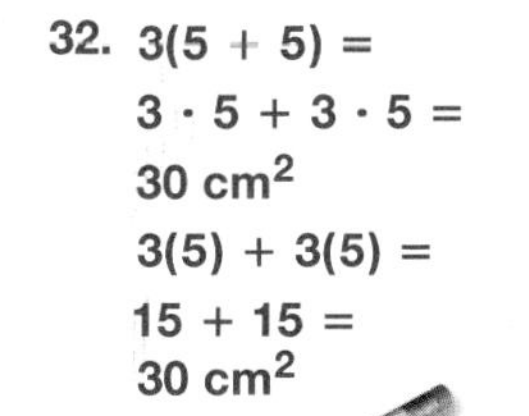

32. $3(5 + 5) =$
$3 \cdot 5 + 3 \cdot 5 =$
$30\ cm^2$
$3(5) + 3(5) =$
$15 + 15 =$
$30\ cm^2$

(C) 34. Challenge You go with five friends to an amusement park. The tickets originally cost \$36.50 each, but you receive a group discount. The total cost is \$198. What discount does each person receive? **\$3.50**

Test Prep and Mixed Review — Practice

Multiple Choice

35. What is the value of the expression $3 \times (-2) + 6 \div (-2) - 5$? **A**
(A) −14 (B) −9 (C) −5 (D) 4

36. Which expression does NOT represent the total area of the figure? **H**

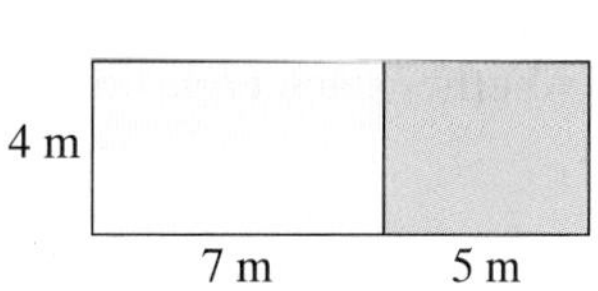

(F) $4(7 + 5)$ (H) $4(7) + 5$
(G) $28 + 20$ (J) $4(7) + 4(5)$

37. At 8 P.M., the wind-chill temperature was −9°F. One hour later, the wind-chill temperature had fallen to −29°F. Which number sentence shows this change? **A**
(A) $-29 - (-9) = -20$ (C) $-9 - (-29) = 20$
(B) $29 + (-9) = 20$ (D) $29 + (-9) = -20$

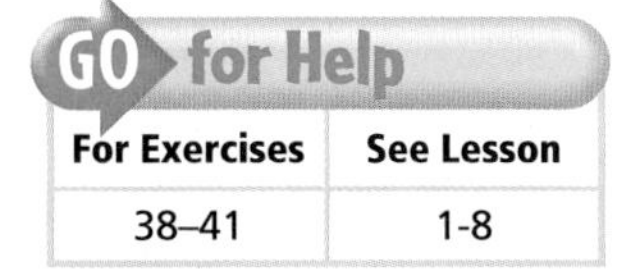

GO for Help

For Exercises	See Lesson
38–41	1-8

Find each product.

38. $-6(8)$ **−48** **39.** $12(-5)$ **−60** **40.** $-7(-9)$ **63** **41.** $11(13)$ **143**

Online lesson quiz, PHSchool.com, Web Code: ara-0109

4. Assess & Reteach

Lesson Quiz

Find the value of each expression.

1. $36 + (-2 \cdot 3) - 5$ **25**
2. $(18 - 9) \cdot 8$ **72**

Find each product mentally using the Distributive Property.

3. 3(42) **126**   4. 8(3.1) **24.8**

Alternative Assessment

Have students work in pairs. One partner writes an expression similar to those in Exercises 8–15. The other writes an expression similar to those in Exercises 17–25. Students exchange papers and simplify each other's expressions. Have partners check each other's work and work together to resolve any discrepancies. Have students alternate roles as time permits.

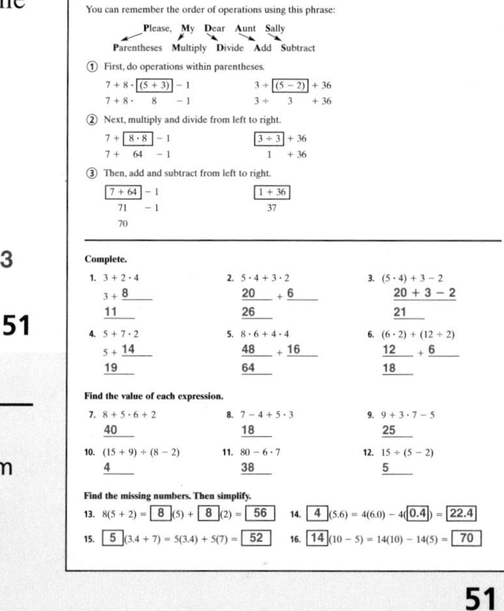

Enrichment 1-9 L4

Reteaching 1-9 Order of Operations and the Distributive L2

You can remember the order of operations using this phrase:

Please, My Dear Aunt Sally
Parentheses Multiply Divide Add Subtract

① First, do operations within parentheses.
$7 + 8 \cdot \boxed{(5 + 3)} - 1$ $3 \div \boxed{(5 - 2)} + 36$
$7 + 8 \cdot 8 - 1$ $3 \div 3 + 36$

② Next, multiply and divide from left to right.
$7 + \boxed{8 \cdot 8} - 1$ $\boxed{3 \div 3} + 36$
$7 + 64 - 1$ $1 + 36$

③ Then, add and subtract from left to right.
$\boxed{7 + 64} - 1$ $\boxed{1 + 36}$
$71 - 1$ 37
70

Complete.

1. $3 + 2 \cdot 4$; $3 +$ 8; 11
2. $5 \cdot 4 + 3 \cdot 2$; 20 + 6; 26
3. $(5 \cdot 4) + 3 - 2$; 20 + 3 − 2; 21
4. $5 + 7 \cdot 2$; $5 +$ 14; 19
5. $8 \cdot 6 + 4 \cdot 4$; 48 + 16; 64
6. $(6 \cdot 2) + (12 \div 2)$; 12 + 6; 18

Find the value of each expression.

7. $8 + 5 \cdot 6 + 2$ 40
8. $7 - 4 + 5 \cdot 3$ 18
9. $9 + 3 \cdot 7 - 5$ 25
10. $(15 + 9) \div (8 - 2)$ 4
11. $80 - 6 \cdot 7$ 38
12. $15 \div (5 - 2)$ 5

Find the missing numbers. Then simplify.

13. $8(5 + 2) = \boxed{8}(5) + \boxed{8}(2) = \boxed{56}$
14. $\boxed{4}(5.6) = 4(6.0) - 4(\boxed{0.4}) = \boxed{22.4}$
15. $\boxed{5}(3.4 + 7) = 5(3.4) + 5(7) = \boxed{52}$
16. $\boxed{14}(10 - 5) = 14(10) - 14(5) = \boxed{70}$

Test Prep

Resources
For additional practice with a variety of test item formats:
- Test-Taking Strategies, p. 59
- Test Prep, p. 63
- Test-Taking Strategies with Transparencies

Activity Lab

Properties and Equality

Students use number properties to solve algebraic equations.

Guided Instruction

Example 2
Let students add 14.75 + 3.8 and 4.75 + 13.8 to be sure the sums are the same.

Exercises
Have students work in pairs to do Exercises 1–6. For Exercise 7, have each partner make a balance-scale puzzle for his or her partner to solve.

Differentiated Instruction

Tactile Learners
Have students model balance-scale problems on an actual balance using grams, kilograms, and other metric measures.

Resources

- Classroom Aid 8
- Student Manipulatives Kit

Properties and Equality

Understanding number properties and equality will help you solve algebraic equations.

EXAMPLE Understanding Number Properties

1 Use a number property or number sense to determine whether the equation $8.86 + 12.51 + 1.23 = 1.23 + 8.86 + 12.51$ is true or false. Justify your reasoning.

$8.86 + 12.51 + 1.23 = 8.86 + 1.23 + 12.51$ ← **The order of the numbers has changed. This is an example of the Commutative Property of Addition.**

True; the Commutative Property of Addition is being used.

EXAMPLE Understanding Equality

2 Use number properties, mental math, or number sense to determine whether the scales are balanced. Justify your reasoning.

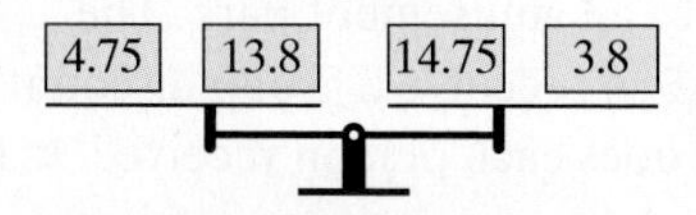

14.75 on the right is 10 more than 4.75 on the left. 13.8 on the left is 10 more than 3.8 on the right.

Each side has one block that is 10 more than a block on the other side. So the scale is balanced.

Exercises

1–6. See margin.

Determine whether each equation is true or false. Justify your reasoning.

1. $25.97 - (13 - 10) = (25.97 - 13) - 10$
2. $603 \times 9.5 = 603 \times 10 - 603 \times 0.5$
3. $530 \div 5 = 500 \div 5 + 30 \div 5$
4. $530 \div 5 = 530 \div 4 + 530 \div 1$

Determine whether each scale is balanced. Justify your reasoning.

5. 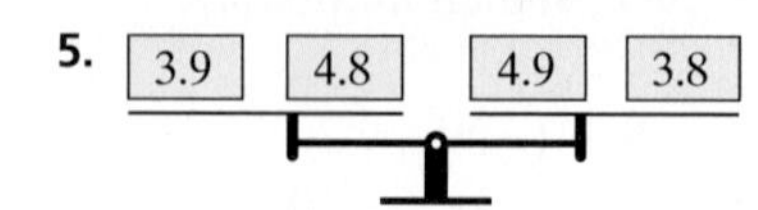

6. 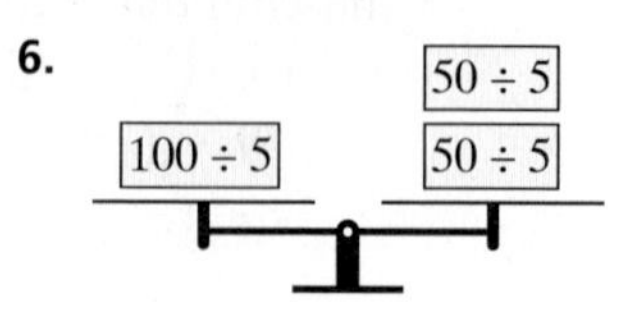

7. Make a balance-scale puzzle like those above. Write the solution. **Check students' work.**

1. False; associative prop. does not apply to subtraction.
2. true; Distributive Property
3. true; Distributive Property
4. False; the Distributive Property does not apply to division.
5. Yes; each side has one block that is 0.1 more than a block on the other side.
6. yes; Distributive Property

Algebra

1-10 Mean, Median, Mode, and Range

Check Skills You'll Need

1. **Vocabulary Review** When you simplify $2 + 6 \div 3$, which operation should you use first? division

Find the value of each expression.

2. $(7 + 19) \div 2$ 13
3. $\frac{-14 + 6}{2}$ −4
4. $-9 + 4 \cdot 2 - 1$ −2

for Help Lesson 1-9

What You'll Learn

To describe data using mean, median, mode, and range

New Vocabulary mean, outlier, median, mode, range

Why Learn This?

You can use data to model the past and predict the future.

School officials use averages to predict how many new students will enroll next year. One average is the **mean,** which is the sum of the data divided by the number of data items.

EXAMPLE Finding the Mean

1 Use the data in the graph above to find the mean number of new students per year.

$\frac{22 + 20 + 23 + 5 + 25}{5}$ ← Divide the sum by the number of items.

$\frac{95}{5} = 19$ ← Simplify.

The mean is 19 students.

Quick Check

1. Find the mean of 216, 230, 198, and 252. 224

An **outlier** is a data item that is much higher or lower than the other items in a set of data. In the data set in Example 1, 5 is an outlier.

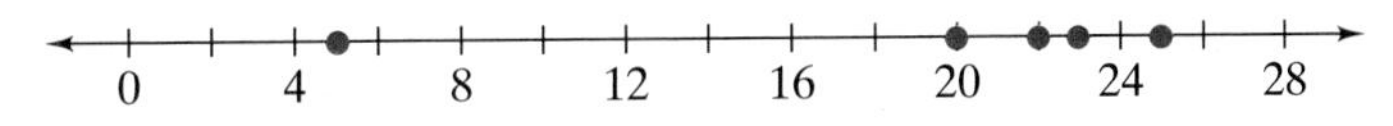

Since 5 is far less than the other values, the outlier decreases the mean. When a set of data has outliers, the mean may not be the best measure for describing the data.

GO Online

Video Tutor Help

Visit: PHSchool.com

Web Code: are-0775

1-10

1. Plan

Objective

To describe data using mean, median, mode, and range

Examples

1 Finding the Mean
2 Finding the Median
3 Finding the Mode
4 Finding Range

Math Understandings: p. 2D

Math Background

A number that describes a set of data is a *statistic.* In this lesson, students will study the *mean, median, mode,* and *range.*

The range describes the scope, or extent, of a set of data. In contrast, the mean, median, and mode describe the center of the data. For this reason, the mean, median, and mode are often called *measures of central tendency.*

More Math Background: p. 2D

Lesson Planning and Resources

See p. 2E for a list of the resources that support this lesson.

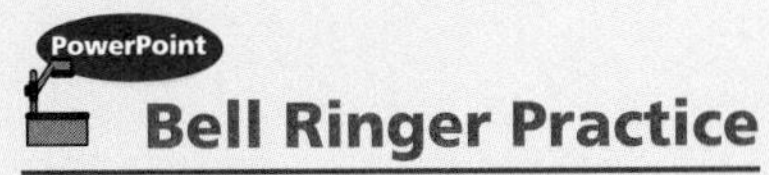

Check Skills You'll Need

Use student page, transparency, or PowerPoint. For intervention, direct students to:

Order of Operations and the Distributive Property

Lesson 1-9

Extra Skills and Word Problems Practice, Ch. 1

Differentiated Instruction Solutions for All Learners

Special Needs L1

Students will benefit from finding one measure of central tendency at a time in the practice problems that require all three. Once they find the median, they should review the meaning of mode, and find that. Then they should move on to mean.

learning style: verbal

Below Level L2

Review the process of finding the average of two numbers with exercises like these.

21, 77 49 | 16, 27 21.5

338, 604 471 | 6.15, 1.89 4.02

learning style: visual

2. Teach

Activity Lab

Use before the lesson.

All in One Teaching Resources

Activity Lab 1-10: Mean, Median, Mode, and Range

Guided Instruction

Example 1
Students might use their calculators to find the mean. If they do, remind them to use parentheses to group the sum of the data items.

Error Prevention!

Students might find a median by simply locating the middle item or items in the data set exactly as it is given to them. Stress that the process of finding a median has two steps: first rewriting the data in order from least to greatest and then identifying the middle item or items.

Example 3
Students can remember what the mode is by thinking *mode-most*.

The **median** of a data set is the middle value when the data are arranged in numerical order. The median of a set always separates the data into two groups of equal size. The median for an even number of data items is the mean of the two middle values.

Test Prep Tip
The median for an odd number of data items is the middle value.

EXAMPLE Finding the Median

2 **Multiple Choice** The table at the right shows data collected from student responses. What is the median number of times that students drank from the water fountain?

20 Responses to "How many times a day do you drink from the water fountain?"				
0	1	1	5	2
10	2	3	5	1
5	2	2	3	4
3	5	5	2	2

Ⓐ 2 Ⓑ 2.5 Ⓒ 3 Ⓓ 3.5

First write the data in order from least to greatest.

0 1 1 1 2 2 2 2 2 2 3 3 3 4 5 5 5 5 5 10

↑↑ The two middle values are 2 and 3.

$\frac{2 + 3}{2} = 2.5$ ← Find the mean of the two middle values.

The median is 2.5. The answer is B.

Quick Check

2. Find the median in the set of data: −5 −1 3 −18 −2 2. **−1.5**

The **mode** of a data set is the item that occurs with the greatest frequency. A set of data may have more than one mode. There is no mode when all the data items occur the same number of times. The mode is a useful measure for data with values that are repetitive or nonnumerical.

EXAMPLE Finding the Mode

Favorite Flowers of Ten People Surveyed	
rose	rose
pansy	peony
pansy	daisy
daisy	rose
daisy	orchid

3 Find the mode of the data at the left.

Make a table to organize the data.

Rose	Pansy	Peony	Daisy	Orchid
///	//	/	///	/

There are two modes, rose and daisy.

Quick Check

3. Find the mode(s). **See left.**
   a. 17 16 18 17 16 17
   b. 3.2 3.7 3.5 3.7 3.5 3.2
   c. pen, pencil, marker, marker, pen, pen, pen, pencil, marker

3a. 17
b. no mode
c. pen

Differentiated Instruction Solutions for All Learners

Advanced Learners L4
Find a set of six whole numbers that satisfies these conditions: The mean is 5. The mode is 4. The median is 4.5. **Samples: 4, 4, 4, 5, 6, 7; 2, 4, 4, 5, 6, 9**

learning style: visual

English Language Learners ELL
Ask students to write a data set of numbers, and find the mean, the median, the mode, and the range. Then, in their own words, have them write the meaning of each.

learning style: visual

The **range** of a data set is the difference between the greatest and the least values. The range describes the spread of the data. You find the range by subtracting the least value from the greatest value.

EXAMPLE Finding Range

4 **Climate** Temperatures at Verkhoyansk, Russia, have ranged from a low of -90°F to a high of 98°F. Find the temperature range in Verkhoyansk.

$98 - (-90) = 98 + 90$ ← **Add the opposite of -90, which is 90.**

$= 188$ ← **Simplify.**

The temperature range in Verkhoyansk is 188°F.

Quick Check

4. Record temperatures in Texas set in the 1930s were a low of -23°F and a high of 120°F. Find the temperature range. **143°F**

Check Your Understanding

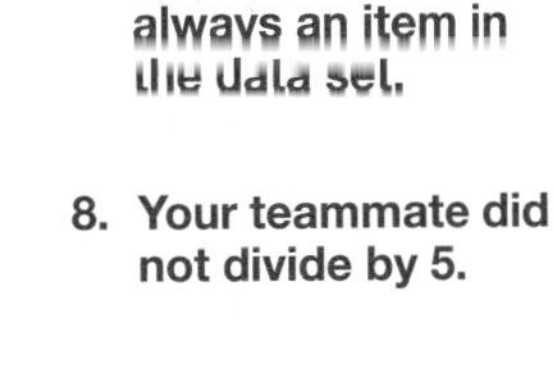

4. 5; the mode is always an item in the data set.

8. Your teammate did not divide by 5.

1. **Vocabulary** A data item that can greatly affect the mean of a set of data is called a(n) _?_. **outlier**
2. If you increase every number in a data set by 2, what will happen to the mean? **increase by 2**
3. **Mental Math** A data set consists of the integers 1 to 9. What is the median of this data set? **5**
4. A data set has a mean of 3, a median of 4, and a mode of 5. Which number *must be* in the data set—3, 4, or 5? Explain.
5. **Choose a Method** Is mean or median the better way to describe the data below? Explain your choice.

   2 4 5 5 5 6 8 10 45 **Median; the outlier 45 greatly affects the mean.**

Find the mean, median, mode, and range of each set of data.

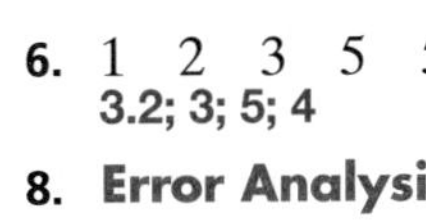

6. 1 2 3 5 5 **3.2; 3; 5; 4**
7. 3 3 3 4 4 4 **3.5; 3.5; no mode; 1**
8. **Error Analysis** Each day your tennis team practices for at least one hour and for at most five hours. Your teammate does the calculation shown. What mistake does your teammate make? **See above left.**

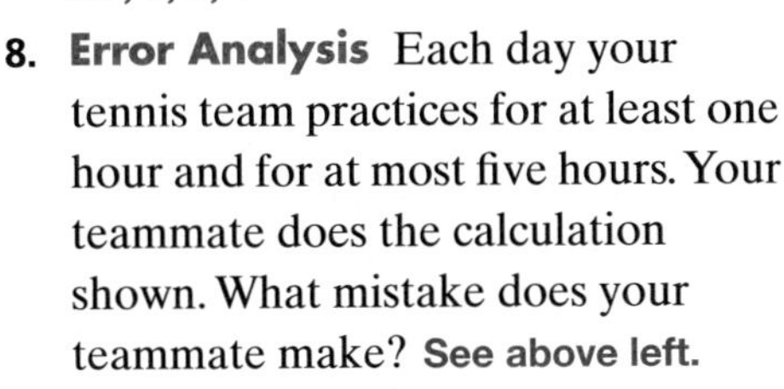

Times (hours): 1, 3, 4, 5, 3
Mean: 1 + 3 + 4 + 5 + 3
Mean = 16 hours

PowerPoint Additional Examples

1 Find the mean of 502, 477, 593, 481, 735, and 614. **567**

Use the chart below for Exercises 2 and 3.

15 Responses to "How many pets do you have?"

2	0	1	2	4
1	0	3	2	0
0	1	2	3	0

2 Find the median of the data in the chart. **1**

3 Find the mode of the data in the chart. **0**

4 Temperatures one day in January ranged from -7°F to 11°F. What was the temperature range? **18°F**

All in One Teaching Resources

- Daily Notetaking Guide 1-10 L3
- Adapted Notetaking 1-10 L1

Closure

- *What is the mean of a set of data?* the sum of the data divided by the number of data items
- *What is the median of a set of data?* the middle item in a set of data that has been written in order
- *What is the mode of a set of data?* the data item that occurs most often
- *What is the range of a set of data?* the difference between the greatest and least values

3. Practice

Assignment Guide

Check Your Understanding
Go over Exercises 1–8 in class before assigning the Homework Exercises.

Homework Exercises

A Practice by Example	9–23
B Apply Your Skills	24–32
C Challenge	33
Test Prep and Mixed Review	34–37

Homework Quick Check
To check students' understanding of key skills and concepts, go over Exercises 11, 15, 26, 31, and 32.

Differentiated Instruction Resources

Adapted Practice 1-10 L1

Practice 1-10 Mean, Median, Mode, and Range L3

The sum of the heights of all the students in a class is 1,472 in.

1. The mean height is 5 ft 4 in. How many students are in the class? (1 ft = 12 in.) 23 students
2. a. The median height is 5 ft 2 in. How many students are 5 ft 2 in. or taller? 12 students
   b. How many students are shorter than 5 ft 2 in.? 11 students

The number of pages read (to the nearest multiple of 50) by the students in history class last week are shown in the tally table at the right.

Pages	Tally
50	I
100	
150	II
200	~~IIII~~ I
250	I
300	~~IIII~~
350	III
400	IIII
450	
500	I

3. Find the mean, median, mode, and range of the data. mean: 287.5; median: 300; mode: 200
4. What is the outlier in this set of data? 50
5. Does the outlier raise or lower the mean? lower

A student hopes to have a 9-point average on his math quizzes. His quiz scores are 7, 6, 10, 8, and 9. Each quiz is worth 12 points.

6. What is his average quiz score? 8 points
7. There are two more quizzes. How many more points will be needed to give a 9-point quiz average? 23 points

Find the mean, median, mode, and range for each situation.

8. number of miles biked in one week 21, 17, 15, 18, 22, 16, 20 18.4; 18; no mode; 7
9. number of strikeouts per inning 3, 2, 0, 0, 1, 2, 3, 0, 2 1.4; 2; 2 and 0; 3

1-10 • Guided Problem Solving GPS L3

Student Page 54, Exercise 26:

Find the mean, median, and mode for the hours of practice before a concert:

2 1 0 1 5 3 4 2 0 3 1 2

Understand

1. What does this data set refer to? the number of hours of practice before a concert
2. What are you being asked to do? Find the mean, median, and mode of the data.
3. How many numbers are in the data set? 12

Plan and Carry Out

4. Find the sum of the numbers in the data set. 24
5. Find the mean by dividing the sum of the numbers by the total number of numbers. 2
6. Order the numbers in the data set in increasing order. 0 0 1 1 1 2 2 2 3 3 4 5
7. What are the two middle numbers? 2 and 2
8. Find the median of the data by finding the mean of the two middle numbers. 2
9. Find the mode by finding which number is listed most often. 1 and 2
10. How many modes are there? 2

Check

11. What do you notice about the mean, median, and mode of this data? 2 is the mean, median, and mode of the data.

Solve Another Problem

12. Millie has 3 siblings, Peggy has one sister, Larry has 5 brothers, Joey is an only child, and Marie has 6 siblings. What is the mean number of siblings for this group of people? 3 siblings

Homework Exercises

For more exercises, see Extra Skills and Word Problems.

GO for Help

For Exercises	See Examples
9–12	1
13–16	2
17–19	3
20–23	4

A **Find the mean of each set of data.**

9. 8 12 6 9 5 **8**
10. 3.4 0.53 1.3 2.9 1.47 0.24 **1.64**
11. −6 3 −2 6 7 −3 −5 8 **1**
12. −17 32 −9 0 52 12 −14 **8**

Find the median of each set of data.

13. 5 10 18 3 6 2 9 1 8 15 10 **8**
14. 23 18 67 32 54 41 70 11 56 33 41 58 **41**
15. −4 −1 −8 −5 −6 −2 7 2 0 −1 −7 2 **−1.5**
16. 2.1 −41.2 0.13 −7.1 −1.68 8.32 2.45 7.89 3.19 **2.1**

Find the mode(s).

17. 51 58 54 58 51 57 55 58 51 54 **51 and 58**
18. 1 12 2 21 22 1 13 31 32 31 12 **1, 12, and 31**
19. red, blue, white, white, red, red, red, blue, white, blue **red**

Find the range.

20. from 24 to −2 **26**
21. from −3 to 7 **10**
22. from −145 to 234 **379**
23. Video game scores vary from 6 to −6. What is the range? **12**

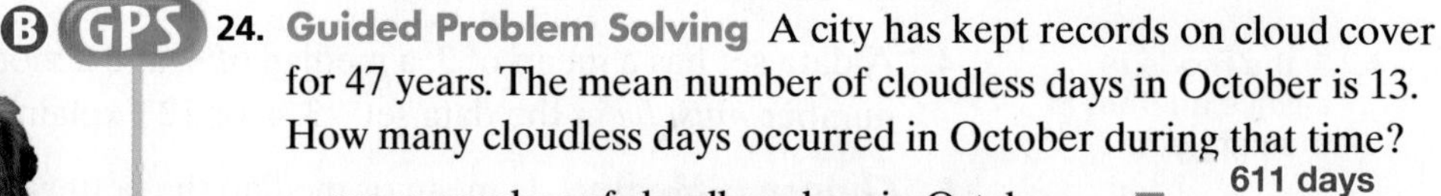

B GPS 24. **Guided Problem Solving** A city has kept records on cloud cover for 47 years. The mean number of cloudless days in October is 13. How many cloudless days occurred in October during that time? **611 days**

mean number of cloudless days in October ■
number of years of records × ■
total number of cloudless days ■

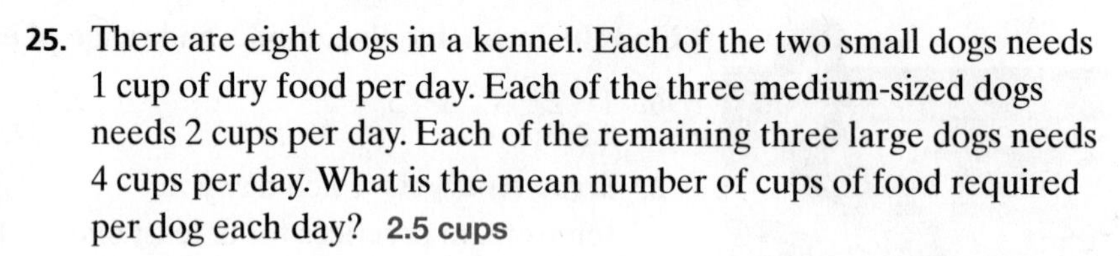

25. There are eight dogs in a kennel. Each of the two small dogs needs 1 cup of dry food per day. Each of the three medium-sized dogs needs 2 cups per day. Each of the remaining three large dogs needs 4 cups per day. What is the mean number of cups of food required per dog each day? **2.5 cups**

26. Find the mean, median, and mode for the hours of practice before a concert: 2 1 0 1 5 3 4 2 0 3 1 2. **2 h; 2 h; 1 h and 2 h**

27. **Estimation** A company's mean weekly profit is $30,021. Estimate how much profit the company will make in a year.

27. Answers may vary. Sample: $1,500,000

Visit: PHSchool.com
Web Code: are-0110

29–30. Answers may vary. Samples are given.

31. Answers may vary. Sample: Data set 2, 4, 6, 6, 12, 18 has a mean of 8, and a median and mode of 6. Data set 4, 6, 6, 6, 18 has a mean of 8, and a median and mode of 6.

28. The table at the right shows the average life expectancy for several animals.
   a. Find the mean, median, mode, and range of the data. 12.7 yr; 12 yr; 12 and 15 yr; 12 yr
   b. **Choose a Method** Which measure best describes the data? Explain. Median; 20 may be considered an outlier.

Average Life Expectancy

Animal	Years
Bison	15
Cow	15
Deer	8
Donkey	12
Elk	15
Goat	8
Horse	20
Moose	12
Pig	10
Sheep	12

SOURCE: *The World Almanac*

Make a data set for each condition.

29. mode > median
   1, 2, 3, 5, 5

30. median = mean
   5, 5, 5, 5, 5

31. **Writing in Math** Explain how it is possible for two sets of data to consist of different numbers but have the same mean, the same median, and the same mode. Give an example.

32. **Music** When you join a music club, you get six CDs for 1 cent each. You buy eight more CDs at $7.99 each. What is the mean price you pay for a CD? $4.57

33. **Challenge** According to the U.S. Department of Agriculture, the mean annual egg consumption in the United States is 258.2 eggs per person. Find the number of cartons of eggs needed in a year for an average family of four. Each carton holds one dozen eggs. 87 cartons

Test Prep and Mixed Review

Practice

Multiple Choice

34. The table shows Tammy's times in the 100-meter dash. Which number could be added to make the median and mode of the set equal? D
   A 11.6
   B 11.8
   C 12
   D 12.25

Tammy's Times in the 100-meter Dash

Trial	Time (seconds)
1	12.8
2	11.96
3	12.4
4	11.6
5	12.25
6	■

35. In a golf tournament, five players had scores of −6, +5, +1, −2, and −8. Which expression can be used to find the average score? H
   F $(6 + 5 + 1 + 2 + 8) \div 5$
   G $6 + 5 + 1 + 2 + 8 \div 5$
   H $[-6 + 5 + 1 + (-2) + (-8)] \div 5$
   J $-6 + 5 + 1 + (-2) + (-8) \div 5$

GO for Help

For Exercises	See Lesson
36–37	1-6

Order each set of numbers from least to greatest.

36. 32 35 −21 −42 29
   −42, −21, 29, 32, 35

37. 213 231 312 123 321
   123, 213, 231, 312, 321

Online lesson quiz, PHSchool.com, Web Code: ara-0110

4. Assess & Reteach

PowerPoint

Lesson Quiz

Find the mean, median, mode, and range of each set of data.

1. 25 27 21 27 20 mean, 24; median, 25; mode, 27; range, 7
2. −8 −4 2 10 −10 mean, −2; median, −4; mode, no mode; range, 20
3. 7.5 3.6 15.4 3.6 15.4 mean, 9.1; median, 7.5; mode, 3.6 and 15.4; range, 11.8

Alternative Assessment

Have students work in groups of four. Refer them to Exercises 13–16. For each set of data, have one student find the mean to the nearest hundredth, another find the median, the third find the mode, and the fourth find the range. Have students check each other's work.

Test Prep

Resources

For additional practice with a variety of test item formats:
- Test-Taking Strategies, p. 59
- Test Prep, p. 63
- Test-Taking Strategies with Transparencies

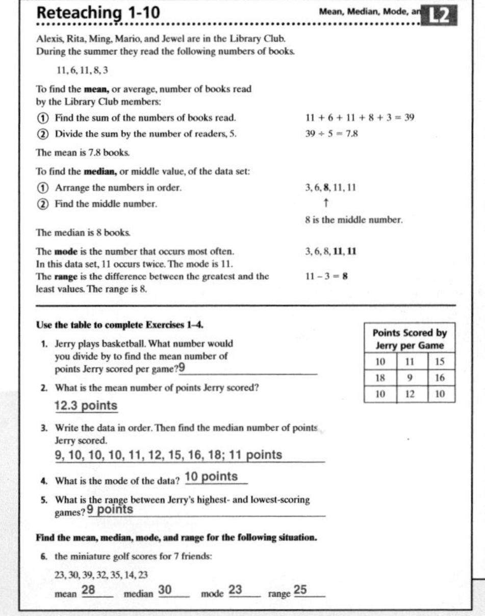

Enrichment 1-10 L4

Reteaching 1-10 Mean, Median, Mode, an[d Range] L2

Alexis, Rita, Ming, Mario, and Jewel are in the Library Club. During the summer they read the following numbers of books.

11, 6, 11, 8, 3

To find the **mean,** or average, number of books read by the Library Club members:
① Find the sum of the numbers of books read. $11 + 6 + 11 + 8 + 3 = 39$
② Divide the sum by the number of readers, 5. $39 \div 5 = 7.8$

The mean is 7.8 books.

To find the **median,** or middle value, of the data set:
① Arrange the numbers in order. 3, 6, **8**, 11, 11
② Find the middle number. ↑ 8 is the middle number.

The median is 8 books.

The **mode** is the number that occurs most often. In this data set, 11 occurs twice. The mode is 11. 3, 6, 8, **11, 11**

The **range** is the difference between the greatest and the least values. The range is 8. $11 - 3 = \mathbf{8}$

Use the table to complete Exercises 1–4.

Points Scored by Jerry per Game		
10	11	15
18	9	16
10	12	10

1. Jerry plays basketball. What number would you divide by to find the mean number of points Jerry scored per game? 9
2. What is the mean number of points Jerry scored? 12.3 points
3. Write the data in order. Then find the median number of points Jerry scored. 9, 10, 10, 10, 11, 12, 15, 16, 18; 11 points
4. What is the mode of the data? 10 points
5. What is the range between Jerry's highest- and lowest-scoring games? 9 points

Find the mean, median, mode, and range for the following situation.

6. the miniature golf scores for 7 friends:
   23, 30, 39, 32, 35, 14, 23
   mean 28 median 30 mode 23 range 25

Box-and-Whisker Plots

Students use box-and-whisker plots to show data along a number line.

Guided Instruction

Activity
In Step 2, point out that the median is the middle number so that the middle of the lower half is the lower quartile. Relate this to $\frac{1}{2}$ and $\frac{1}{4}$.

Exercises
Have students make up data for Team C and make a box-and-whisker plot. Be sure students keep data within a reasonable range for a team.

Differentiated Instruction

Visual Learners
Box-and-whisker plots are visual ways of displaying data. Have a visual learner explain what the box shows and what the whiskers show using the example on p. 58 or a box-and-whisker plot he or she has made. Encourage the use of vocabulary such as *median, upper quartile*, and *lower quartile*.

Resources

- Classroom Aid 31, 32, 33

Box-and-Whisker Plots

A **box-and-whisker plot** is a graph that summarizes a data set along a number line.

ACTIVITY

Points Scored per Player

Team A	Team B
6	5
2	9
8	7
10	9
3	13
15	11
4	13
20	15
7	14
4	10

Step 1 Arrange the data for Team A from least to greatest. Find the median. This value is called the middle quartile.

2 3 4 4 6 7 8 10 15 20

The median is 6.5.

Step 2 Find the median of the lower half of the data. This value is the lower quartile.

2 3 4 4 6

The median is 4.

Step 3 Find the median of the upper half of the data. This value is the upper quartile.

7 8 10 15 20

The median is 10.

Step 4 Identify the least and greatest values of the full data set.

least value = 2 greatest value = 20

Step 5 Draw a number line. Plot the points of the 5 values you found in Steps 1–4 above the number line.

Step 6 Draw a box using the lower and upper quartile points as the ends of the box. Then draw lines, or whiskers, from the ends of the box to the least and greatest values.

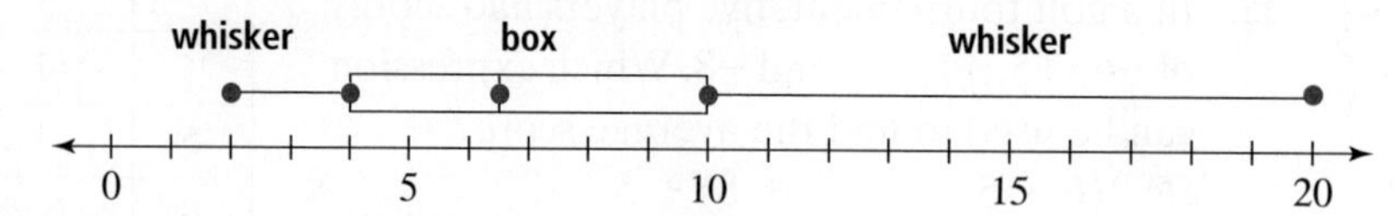

Exercises

1. Make a box-and-whisker plot using the data for Team B. **See margin.**
2. Compare your graph to the one in the activity. What do the two graphs tell you about how the players on each team score? **See right.**

2. The range is greater in the upper half of the data for Team A. Team A has low and high points, whereas Team B has a smaller range.

1.

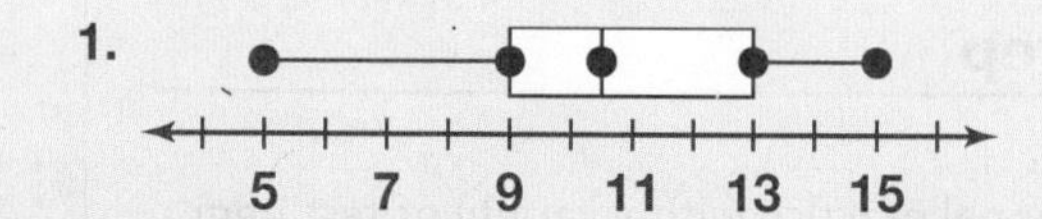

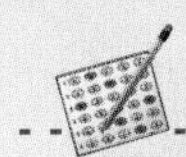

Writing Gridded Responses

Some tests call for gridded responses. You find a numerical answer. Then you write the answer at the top of the grid and fill in the corresponding bubbles below. You must use the grid correctly.

EXAMPLES

1 The mean of 0.2, 0.4, 0.6, and 0.8 is 0.5. Record this answer.

You can write the answer as 0.5 or .5. Here are the two ways to enter these answers.

	0	.	5

		.	5

2 Cindy had \$19.25 before she went shopping. She spent \$18.50 on purchases that day. How much money, in dollars, did she have left?

$$\begin{array}{r} 19.25 \\ -\ 18.50 \\ \hline 0.75 \end{array}$$

The answer is 0.75. You grid this as 0.75 or .75.

Exercises

Write the number you would grid for each answer. If you have a grid, complete it.

1. A bottle of apple juice holds 3.79 L. A bottle of orange juice holds 1.89 L. How many more liters does the bottle of apple juice hold? **1.9**
2. You are organizing a pet show. On the first morning, 40 dogs will be shown. For each dog, you will allow one minute to set up and four minutes for showing. How many minutes will the morning session last? **200**
3. A student scores 88, 93, 79, and 68 on four quizzes. What is the mean of the scores? **82**

Writing Gridded Responses

Students learn to correctly use gridded response sheets to answer test questions.

Guided Instruction

Emphasize that fully completing a gridded-response test question involves two distinct parts: (1) writing the answer in the top row, one digit or symbol to a column, and (2) filling in the circle in each column that corresponds to the digit or symbol written at the top of that column.

Teaching Tip
Show students that starting answers in any column works, as long as there is room for all digits and symbols.

Exercises
Ask: *Do any of Exercises 1–3 have two possibilities for gridding responses?* **Yes, Exercise 1: 1.9 or 1.90; Exercise 3: 82 or 82.0**

Resources

Test-Taking Strategies with Transparencies
- Transparency 2
- Practice Sheet, p. 13

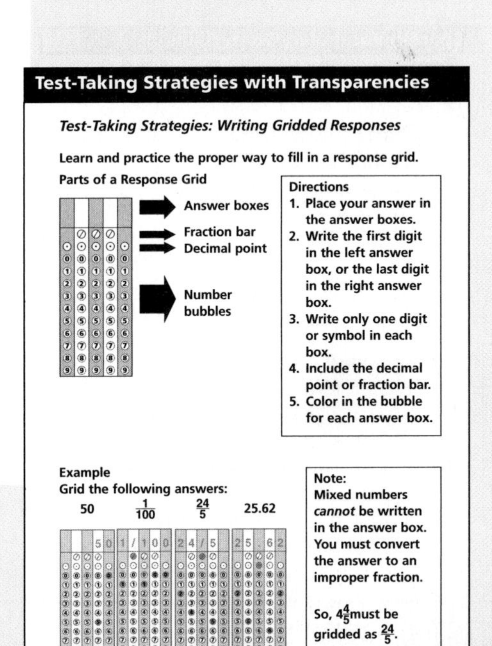
Test-Taking Strategies with Transparencies

Test-Taking Strategies: Writing Gridded Responses

Learn and practice the proper way to fill in a response grid.

Parts of a Response Grid

Directions
1. Place your answer in the answer boxes.
2. Write the first digit in the left answer box, or the last digit in the right answer box.
3. Write only one digit or symbol in each box.
4. Include the decimal point or fraction bar.
5. Color in the bubble for each answer box.

Example
Grid the following answers:
50 $\frac{1}{100}$ $\frac{24}{5}$ 25.62

Note:
Mixed numbers *cannot* be written in the answer box. You must convert the answer to an improper fraction.

So, $4\frac{4}{5}$ must be gridded as $\frac{24}{5}$.

Transparency 2

Resources

Student Edition
Extra Skills and Word Problems Practice, Ch. 1, p. 630
English/Spanish Glossary, p. 676
Formulas and Properties, p. 674
Tables, p. 670

All in One Teaching Resources
- Vocabulary and Study Skills 1F  L3

Differentiated Instruction
Spanish Vocabulary Workbook with Study Skills ELL
Interactive Textbook
- Audio Glossary

Online Vocabulary Quiz

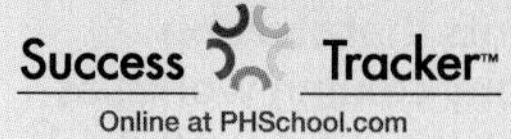

Online at PHSchool.com

Spanish Vocabulary/Study Skills ELL

Vocabulary/Study Skills L3

1F: Vocabulary Review Puzzle For use with the Chapter Review

Study Skill Vocabulary is an important part of every subject you learn. Review new words and their definitions using flashcards.

Find each of the following words in the word search. Circle the word and then cross it off the word list. Words can be displayed forwards, backwards, up, down, or diagonally.

absolute value	integers	order
median	mode	mean
opposites	outlier	range
compatible	distributive	multiplication

C	Q	E	D	E	T	A	J	P	K	A	Z	L	H	N
F	K	E	E	R	E	D	R	O	N	D	A	H	U	N
M	Z	D	W	I	T	E	P	I	E	B	B	I	P	O
R	N	O	U	R	O	L	C	C	V	E	S	E	F	I
A	O	M	B	A	P	B	M	E	I	T	O	V	X	T
N	P	R	J	N	P	I	E	P	T	I	L	A	S	A
G	F	I	R	L	O	T	T	E	U	F	U	T	E	C
E	I	N	P	M	S	A	R	T	B	E	T	K	T	I
Y	O	T	H	G	I	P	A	A	I	I	E	O	R	L
I	D	E	O	O	T	M	N	R	R	N	V	U	O	P
W	A	G	R	F	E	O	V	O	T	K	A	T	I	I
P	D	E	F	N	S	C	O	O	S	E	L	L	J	T
G	K	R	N	N	A	W	S	O	I	G	U	I	R	L
N	M	S	D	K	A	E	V	N	D	S	E	E	U	U
M	E	D	I	A	N	F	M	A	R	O	M	R	L	M

Vocabulary and Study Skills

Chapter 1 Review

Vocabulary Review

absolute value (p. 31)
additive inverses (p. 38)
Associative Property of Addition (p. 9)
Associative Property of Multiplication (p. 15)
Commutative Property of Addition (p. 9)
Commutative Property of Multiplication (p. 15)
compatible numbers (p. 5)
Distributive Property (p. 49)
Identity Property of Addition (p. 9)
Identity Property of Multiplication (p. 15)
integers (p. 31)
mean (p. 53)
median (p. 54)
mode (p. 54)
opposites (p. 31)
order of operations (p. 48)
outlier (p. 53)
range (p. 55)
Zero Property (p. 15)

Go Online PHSchool.com
For: Online vocabulary quiz
Web Code: arj-0151

Choose the correct term to complete each sentence.

1. The _?_ combines multiplication with sums and differences. **Distr. Prop.**
2. The statement $3 + (5 + 7) = (3 + 5) + 7$ demonstrates the _?_. **Assoc. Prop. of Add.**
3. The _?_ of a number is its distance from 0 on a number line. **absolute value**
4. To find the _?_, take the data item that occurs most often. **mode**
5. By the _?_, you know that $4 + 7 \cdot 3$ equals 25 and not 33. **order of operations**

Skills and Concepts

Lesson 1-1

- To estimate using rounding, front-end estimation, and compatible numbers

You can estimate decimals using rounding, front-end estimation, or **compatible numbers.** **6–9. See margin.**

Use any estimation strategy to calculate. Name the strategy you used.

6. $50.3 \div 6.9$
7. $98.52 - 46.91$
8. 6.9×8.92
9. $1.46 + 4.38$

Lessons 1-2, 1-3, 1-4

- To add and subtract decimals and to do mental math using the properties of addition
- To multiply decimals and to do mental math using the properties of multiplication
- To divide decimals and to solve problems by dividing decimals

To add or subtract decimals, align the decimal points. To multiply decimals, use the sum of the number of decimal places in the factors. To divide decimals, rewrite the problem so the divisor is a whole number.

Use the **commutative properties** to change the order in an expression. Use the **associative properties** to change the grouping.

Simplify.

10. $23.68 \div 6.4$ **3.7**
11. $0.54 + 0.027$ **0.567**
12. $4.6 - 3.87$ **0.73**
13. 2.7×6.25 **16.875**
14. **Shopping** At a grocery store, you buy hamburger weighing 1.42 lb, sausage weighing 2.16 lb, and chicken weighing 3.73 lb. How many pounds of meat do you buy? **7.31 lb**

Exercises 6–9. Answers may vary. Samples are given.

6. 7; compatible numbers
7. 52, rounding
8. 63, rounding
9. 6; front-end estimation

Lesson 1-5

- To use and convert metric units of measure

To change a measure from one unit to another, find a relationship between the two units and then multiply.

Write the number that makes each statement true.

15. 4.56 mm = ■ cm **0.456**   **16.** 14.2 L = ■ mL **14,200**   **17.** 0.34 kg = ■ g **340**

18. You have three lengths of kite string measuring 18.2 m, 927 cm, and 0.044 km. What is the total length of kite string in meters? **71.47 m**

Lesson 1-6

- To compare and order integers and to find absolute values

Opposites are two numbers that are the same distance from 0 on a number line, but in opposite directions. **Integers** are the set of positive whole numbers, their opposites, and zero. The **absolute value** of an integer is its distance from 0 on a number line.

Compare. Use <, =, or >.

19. -7 ■ 7 **<**   **20.** $|-3|$ ■ $|3|$ **=**   **21.** 9 ■ $|-4|$ **>**   **22.** $|8|$ ■ -15 **>**

Lessons 1-7, 1-8

- To add and subtract integers and to solve problems involving integers
- To multiply and divide integers and to solve problems involving integers

The sum of two positive integers is positive. The sum of two negative integers is negative. To find the sum of two integers with different signs, find the absolute value of each integer. Subtract the lesser absolute value from the greater. The sum has the sign of the integer with the greater absolute value. To subtract an integer, add its opposite.

The product or quotient of two integers with the same sign is positive. The product or quotient of two integers with different signs is negative.

Simplify.

23. $14 + (-8)$ **6**   **24.** $17 - (-12)$ **29**   **25.** $-5 \cdot 6$ **−30**   **26.** $125 \div (-5)$ **−25**

Lessons 1-9, 1-10

- To simplify numerical expressions involving order of operations
- To describe data using mean, median, mode, and range

Use the **order of operations** to simplify an expression. The **Distributive Property** combines multiplication with addition or subtraction. The **mean, median, mode,** and **range** of a set of data, along with any **outliers,** reflect the characteristics of the data.

Find the value of each expression.

27. $(7.3 + 4) \div 4 + 0.3 \cdot 2$ **3.425**   **28.** $8 - 6.2 \div 5 + 7(0.91)$ **13.13**

29. $20 + 24 \div 2 - (8 + 5)$ **19**   **30.** $8(20.3)$ **162.4**

31. Pets The data set below gives the weights, in ounces, of one-month-old hamsters. Find the mean, median, mode, and range of the data set.
4, 1, 3, 2, 2, 2, 1, 1, 2, 1, 2, 3 **2; 2; 2; 3**

Alternative Assessment **L4**

Alternative Assessment **Form C**

Chapter 1

MOVIES FOR LESS

Carla and Christian Mendoza, seventh-grade twins, love going to the movies. Their neighborhood theater was selling special summer passes for $90.00. The pass can be used to see any movies at the theater for 10 weeks. Without a pass, tickets bought before 5:00 P.M. are $4.50 each. Tickets bought at or after 5:00 P.M. are $7.00 each.

Show all of your work on a separate sheet of paper.

1. Carla and Christian want to convince their parents that they should each have a summer movie pass. Write a good presentation that they could use to convince their parents. Be sure to include some realistic data in your presentation.
2. Do you think the twins' parents should buy each twin a summer movie pass? Explain why or why not.
3. Billy and Bobbie Carter go to the movies with the Mendoza twins, but Billy and Bobbie do *not* have movie passes. Mr. and Mrs. Carter have set these guidelines:
   - Their children may see only one movie each week.
   - They may attend only early afternoon showings.
   - Each child may buy two items from the snack bar during each movie.

SNACK BAR MENU					
Popcorn	small:	$2.50	**Peanuts**	small:	$1.75
	large:	$3.50		medium:	$2.00
				large:	$2.50
Drinks	small:	$2.00			
	large:	$3.00			

Approximately how much should Mr. and Mrs. Carter budget for movie expenses for both children? Explain your answer and show your work.

Resources

- ExamView Assessment Suite CD-ROM
  - Ch. 1 Ready-Made Test
  - Make your own Ch. 1 test
- MindPoint Quiz Show CD-ROM
  - Chapter 1 Review

Differentiated Instruction

All in One Teaching Resources
- Below Level Chapter 1 Test L2
- Chapter 1 Test L3
- Chapter 1 Alternative Assessment L4

Spanish Assessment Resources ELL
- Below Level Chapter 1 Test L2
- Chapter 1 Test L3
- Chapter 1 Alternative Assessment L4

ExamView Assessment Suite CD-ROM
- Special Needs Test L1
- Special Needs Practice Bank L1

Online Chapter 1 Test at www.PHSchool.com L3

Below Level Chapter Test L2

Chapter Test L3

Chapter Test

Chapter 1

1. Use rounding to estimate the product to the nearest whole number. 110.5 × 10.2
   1110
2. Use front-end estimation to estimate the sum to the nearest whole number. 5.312 + 3.726
   9

Find each sum or difference.

3. 8.27 + 6.9 + 5 20.17
4. 19 − 12.735 6.265
5. 21.6 − 9.24 12.36
6. 0.68 + 5.681 6.361
7. 52.1 − 51.06 1.04
8. 74.5 + 0 + 17.3 91.8

Find each product or quotient.

9. $\frac{57.96}{2.8}$ 20.7
10. $\frac{408.48}{8}$ 51.06
11. 5.9 · 0.32 1.888
12. 6.73 · 0.04 0.2692
13. 3.2 ÷ 0.128 25
14. (2.6)(0.02) 0.052

Change each measurement to the given unit.

15. 24 m 5 cm to centimeters 2405 cm
16. 5 kg 240 g to kilograms 5.24 kg
17. 11 L 17 mL to liters 11.017 L

Write the metric unit that makes each statement true.

18. 12,451 dL = 1,245.1 m
19. 2.48 cm = 24.8 mm
20. In last Saturday's parade, Tom counted all of the bicycles and tricycles, for a total of 227 wheels. His count included 76 bicycles. How many tricycles were in the parade?
    25 tricycles

Chapter 1 Test

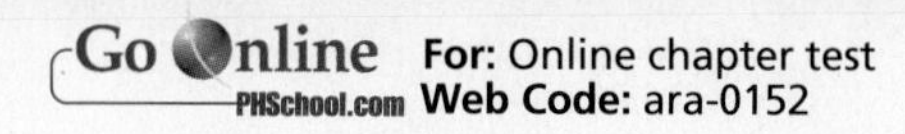

For: Online chapter test
Web Code: ara-0152

Estimate using any estimation strategy.

1. 289.76 − 52 about 240
2. 7.532 + 2.19 about 10
3. 97.6 · 3.4 about 350
4. 68.5 ÷ 7.02 about 10
5. Use front-end estimation to estimate the sum of the following grocery items to the nearest dollar: $7.99, $2.79, $4.15, $2.09. about $17

Simplify.

6. 9.53 + 3.29 12.82
7. 8 − 6.17 1.83
8. 10.5 − 9.67 0.83
9. 0.57 + 1.825 2.395
10. **Money** You have a balance of $213.15 in a savings account. You make withdrawals of $68.94 and $128.36. Find your new balance. $15.85
11. **Decorations** For a class party, the student council purchases 42 balloons at $1.85 each. Estimate the total cost of the balloons. about $78

Identify each property shown.

12. 9.5 + 6.1 + 2.3 = 9.5 + 2.3 + 6.1 Comm. Prop. of Add.
13. 7.2 × (1.6 × 3.9) = (7.2 × 1.6) × 3.9 Assoc. Prop. of Mult.
14. 5.1(7.4 − 3.1) = 5.1(7.4) − 5.1(3.1) Distr. Prop.

Find each quotient. Round to the nearest tenth.

15. 1.2 ÷ 0.3 4.0
16. 1.58 ÷ 1.1 1.4

Find the value of each expression.

17. 9.5 − 7.1 + (2.4 · 0.5) − 1.3 2.3
18. $\frac{8.25}{4} \cdot (0.6 - 0.54) + 8.3$ 8.42375

Change each measurement to the given unit.

19. 4.2 cm = ■ m 0.042
20. 5.17 kL = ■ L 5,170
21. 6 kg 14 g = ■ g 6,014
22. 2 km 7 m = ■ km 2.007

Compare. Use $<$, $=$, or $>$.

23. −12 ■ 12 $<$
24. −9 ■ −5 $<$
25. |−7| ■ −8 $>$
26. |−6| ■ 6 $=$

Mental Math **Find each product mentally using the Distributive Property.**

27. 6(10.5) 63
28. 3(98) 294

Write an expression for each model. Then find the sum.

29.

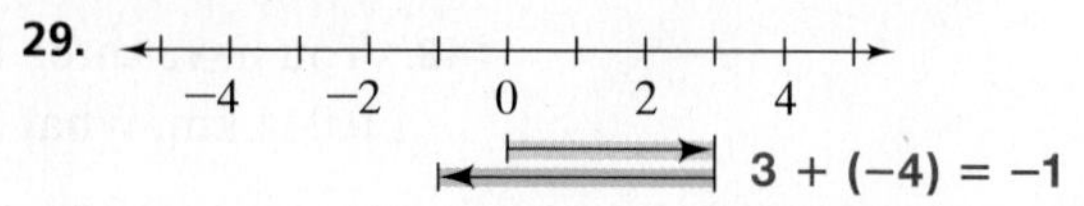

30.

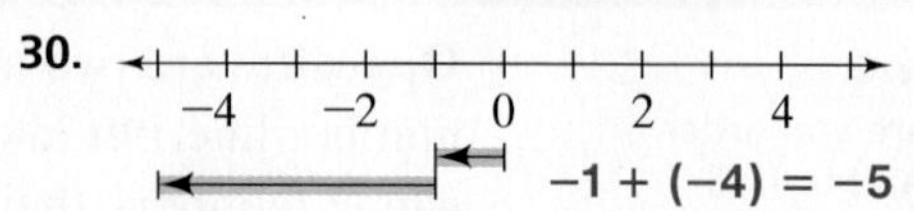

31. Write these integers in order from least to greatest: 2 5 0 −7 −3. −7, −3, 0, 2, 5

Simplify.

32. −3 + 5 2
33. −2 + (−2) −4
34. −4 − 9 −13
35. −8 · (−9) 72
36. 48 ÷ (−3) −16
37. −6(11) −66
38. **Produce** The weights of four bags of apples are 3.5 lb, 3.8 lb, 4.2 lb, and 3.5 lb. What is the median weight? 3.65 lb
39. **Fitness** Suppose you plan to ride your bicycle a total of 257.5 mi in a benefit ride. How many miles must you average each day to finish the ride in 4.5 days? about 57.2 mi
40. **Writing in Math** When you calculate a mean, the sign of the quotient always depends on the sign of the dividend. Why? See margin.
41. **Sports** On successive plays, the home football team gains 12 yd, loses 3 yd, loses 5 yd, gains 15 yd, and runs 16 yd for a touchdown. What is the average gain or loss in yards per play? 7 yd gain
42. Find the mean, median, and mode for the following junior league bowling scores: 45, 56, 134, 55, 78, 121, 38, 66, 56, 41. 69; 56; 56

40. The divisor will always be a positive number.

Test Prep Practice

Reading Comprehension

Read each passage and answer the questions that follow.

> **Big Shows** Successful films earn large amounts of money for movie studios. Five movies that have earned particularly large amounts of money in the United States, and their approximate total receipts, are: *TITANIC* (1997), $600 million; *Star Wars* (1977), $460 million; *Star Wars: The Phantom Menace* (1999), $430 million; *E.T.* (1982), $400 million; and *Jurassic Park* (1993), $360 million.

1. What are the mean total box office receipts of the five films? **C**
   - (A) $400 million
   - (B) $430 million
   - (C) $450 million
   - (D) $460 million

2. *Star Wars: The Phantom Menace* sold about $64,811,000 worth of tickets on its opening weekend. Which is the best estimate of the portion of its total sales that took place the first weekend? **G**
   - (F) 0.1
   - (G) 0.15
   - (H) 0.35
   - (J) 0.5

3. What is the median earnings of the films? **B**
   - (A) $400 million
   - (B) $430 million
   - (C) $450 million
   - (D) $460 million

4. Between which pair of films is the range in total receipts the greatest? **G**
   - (F) *Star Wars* and *Jurassic Park*
   - (G) *TITANIC* and *Star Wars*
   - (H) *Star Wars* and *E.T.*
   - (J) *E.T.* and *Jurassic Park*

> **Parity** We say that two integers have the same *parity* if they are both even or both odd. So 2 and 12 have the same parity, and 51 and 139 have the same parity. If one number is even and the other number is odd, then we say they have different or opposite parities. 2 and 51 have opposite parities.

5. Two integers have the same parity. Describe the parity of their sum. **D**
   - (A) same as the two numbers
   - (B) opposite of the two numbers
   - (C) same if the numbers are odd
   - (D) opposite if the numbers are odd

6. Two integers have the same parity. Describe the parity of their difference. **H**
   - (F) same as the two numbers
   - (G) opposite of the two numbers
   - (H) same only if the numbers are even
   - (J) same only if the numbers are odd

7. Two integers have the same parity. Describe the parity of their product. **A**
   - (A) same as the two numbers
   - (B) opposite of the two numbers
   - (C) same only if the numbers are odd
   - (D) opposite only if the numbers are odd

8. Two integers have different parity. Describe their product. **F**
   - (F) always even
   - (G) always odd
   - (H) sometimes even
   - (J) odd if the smaller number is odd

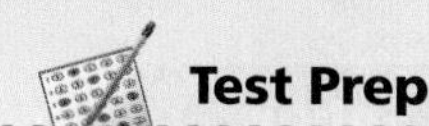

Test Prep

Resources

Test Prep Workbook

All in One Teaching Resources
- Cumulative Review L3

ExamView Assessment Suite CD-ROM
- Standardized Test Practice

Differentiated Instruction

Spanish Assessment Resources
- Spanish Cumulative Review ELL

ExamView Assessment Suite CD-ROM
- Special Needs Practice Bank L1

Spanish Cumulative Review ELL

Cumulative Review L3

Cumulative Review

Chapter 1

Multiple Choice. Choose the letter of the best answer.

1. Which of the following shows the order from least to greatest for the mean, the median, and the mode of this set of data?
   0, 1, 2, 2, 2, 3, 4, 7, 10, 11
   - (A.) mode, median, mean
   - B. median, mode, mean
   - C. mode, mean, median
   - D. mean, median, mode
2. You have saved $35.75. How many CDs can you buy if each CD costs $11.47, tax included?
   - F. 2 CDs
   - (G.) 3 CDs
   - H. 4 CDs
   - J. 5 CDs
3. Identify the property of addition shown.
   $(1.5 + 2.5) + 3 = 1.5 + (2.5 + 3)$
   - (A.) associative
   - B. distributive
   - C. commutative
   - D. identity
4. Which expression is *not* the same as $2.7 \times 3.6 \div 9$?
   - F. $2.7 \div 9 \times 3.6$
   - (G.) $3.6 \times 9 \div 2.7$
   - H. $3.6 \times 2.7 \div 9$
   - J. $3.6 \div 9 \times 2.7$
5. In the quotient $120 \div 11$, what digit is in the tenths place?
   - A. 0
   - B. 1
   - C. 8
   - (D.) 9
6. Sara saved $12.50 for five weeks. Her brother saved $13.75 for five weeks. How much more did her brother save?
   - (F.) $6.25
   - G. $62.50
   - H. $6.88
   - J. $68.75
7. Which list shows the following integers ordered from least to greatest?
   0, −1, 2, −3, −8
   - A. −1, −3, −8, 0, 2
   - (B.) −8, −3, −1, 0, 2
   - C. 0, −8, −3, −1, 2
   - D. 0, −1, 2, −3, −8
8. For a dance, the student government bought 198 carnations at $.49 each. What is the best estimate of the total amount paid for the carnations?
   - F. $60
   - G. $80
   - (H.) $100
   - J. $120
9. The temperature on Mars reaches 27°C during the day and −125°C at night. What is the average temperature?
   - (A.) −49°C
   - B. +76°C
   - C. +49°C
   - D. −76°C
10. Divide. $-10 \div (-2)$
    - F. −5
    - G. $-\frac{1}{5}$
    - H. $\frac{1}{5}$
    - (J.) 5
11. Which of the following statements is *always* true?
    I. A negative integer plus a negative integer equals a negative integer.
    II. A negative integer minus a negative integer equals a negative integer.
    - A. II only
    - B. Both I and II
    - (C.) I only
    - D. Neither I nor II

Problem Solvin

Applying Integers

Students will use the information on these two pages to answer the questions posed in Put It All Together.
Ask questions, such as:
- *What are some methods used to heat homes and buildings?*
- *What are some methods used to cool homes and buildings?*

Activating Prior Knowledge

Have students who have experienced weather conditions that are markedly different from those where you live describe them to classmates. Elicit information about how people who live in those different climates dress, eat, and act in a way that fits their environment.

Guided Instruction

Have a volunteer read the opening paragraph. Discuss students' views on what indoor temperatures feel comfortable.

Have students make a list of extreme weather conditions such as heat waves, blizzards, tornadoes, and hurricanes. Ask:
- *What are some warning signs that these weather conditions are approaching?* **Sample answer: Winds often pick up speed before the arrival of weather conditions such as tornadoes and blizzards.**
- *What are some safety precautions that can be taken ahead of these conditions arriving?* **Sample answer: preparing shelter with appropriate fortification and supplies to last through extreme weather**

Problem Solving Application

Applying Integers

Energy Field Too hot? Open a window. Too cold? Put on a sweater. Sound familiar? Heating and air-conditioning systems can let you live more comfortably in a wide range of weather conditions, but they cost money. The cost of heating or cooling a home depends on many things, including the outdoor air temperature, the indoor air temperature, and the cost of fuel.

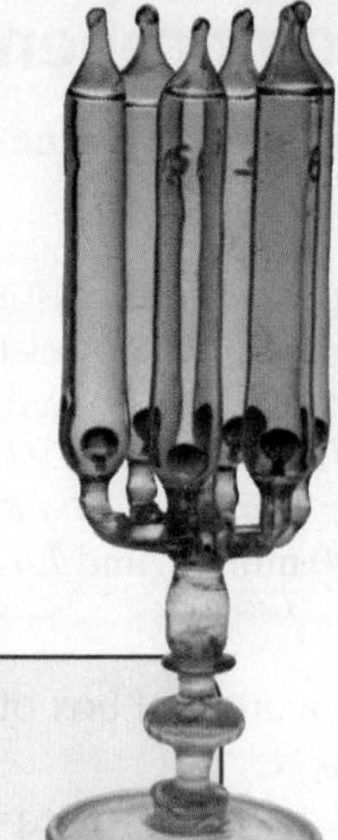

Ancient Thermometer
In this thermometer, changing temperatures cause the colored glass balls to rise and fall in the water inside the glass tubes.

Put It All Together

Data File Use the data on these two pages and on page 673 to answer these questions.

1. Use the heating cost formula to estimate the cost of heating a house to 68°F for one day when the average outside temperature is 20°F.
2. a. **Open-Ended** Choose three places from the table on page 673. Use the heating cost formula to estimate the cost of heating a house to 68°F in each place on the coldest day.
   b. Use a number line to display your answers to part (a). Include the location and the outdoor temperature.
3. You can use the equation below to calculate the cost of changing the indoor temperature from 68°F to 66°F.
   change in cost = new cost − old cost = cost at 66°F − cost at 68°F
   a. Use the low temperature for Indianapolis, Indiana. Calculate the cost of changing the temperature to 68°F.
   b. **Reasoning** Explain why your answer to part (a) is a negative number.
4. **Writing in Math** Where do you think changing the indoor temperature from 68°F to 70°F will cost the most? Explain. Check your prediction by calculating the increased cost in several locations. Explain your results.

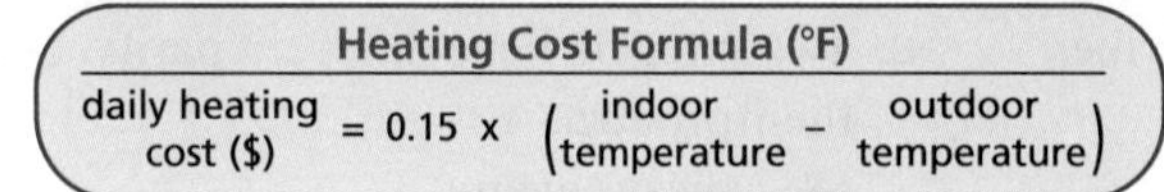

Heating Cost Formula (°F)

daily heating cost ($) = 0.15 × (indoor temperature − outdoor temperature)

Weather Around the World
Photographs from space show giant swirls of clouds around Earth. These swirls show the constant movement of gases that gives us our weather.

64

1. $0.15 \times (68 - 20) = 7.2$; the daily cost is $7.20.

2a. Answers may vary.

b. Check students' work.

3a. old cost $= 0.15 \times (68 - (-25)) = 13.95$; new cost $= 0.15 \times (66 - (-25)) = 13.65$; new cost − old cost = $13.65 - 13.95 = -0.30$; the cost is $.30 less.

b. The change in cost is $.30 less per day. The value is negative because the cost went down.

4. Answers may vary. Sample: Raising the indoor temperature 2°F gives the same increase ($.30 per day) in each location.

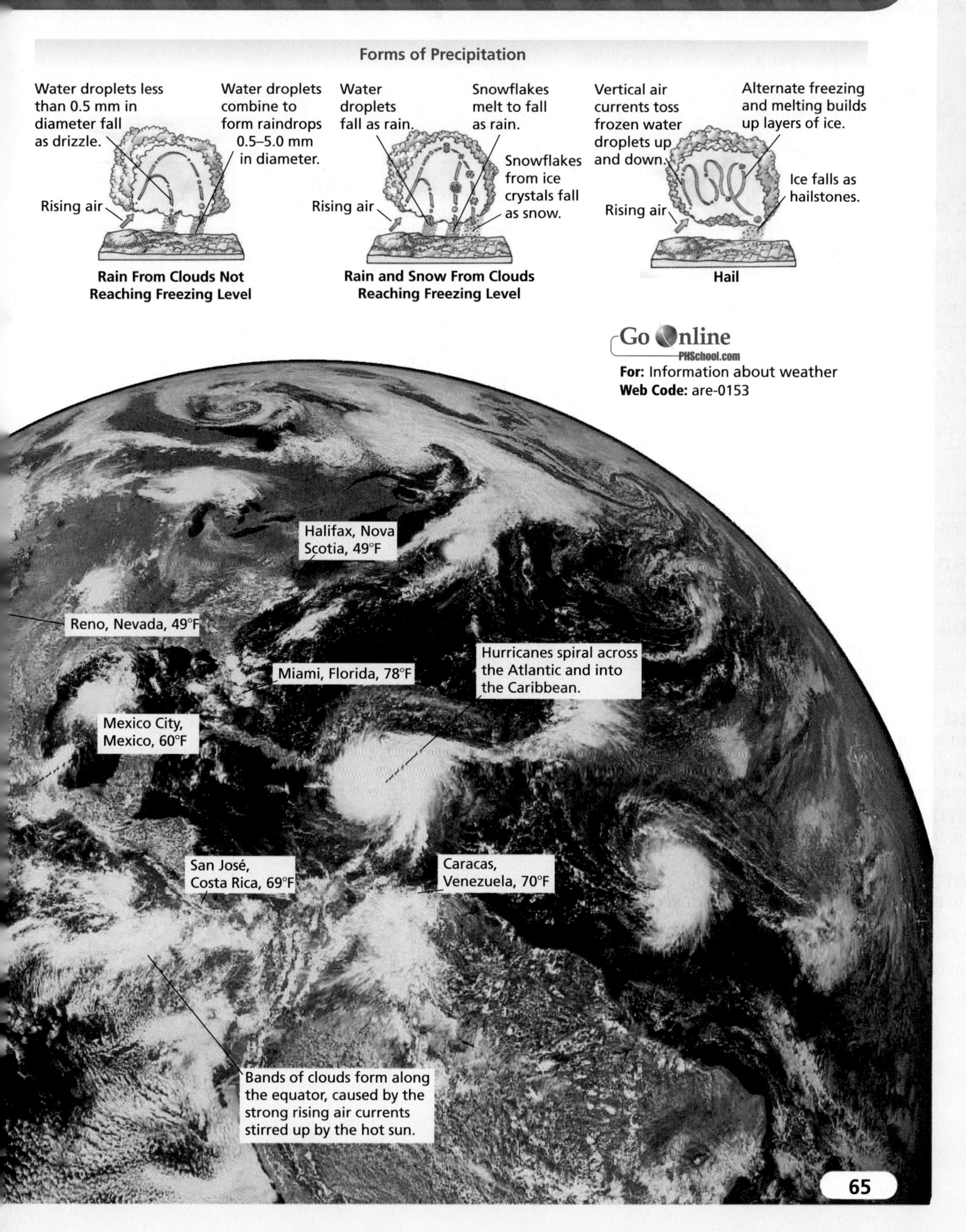

Forms of Precipitation
Water droplets less than 0.5 mm in diameter fall as drizzle.
Water droplets combine to form raindrops 0.5–5.0 mm in diameter.
Rising air
Rain From Clouds Not Reaching Freezing Level
Water droplets fall as rain.
Snowflakes melt to fall as rain.
Snowflakes from ice crystals fall as snow.
Rising air
Rain and Snow From Clouds Reaching Freezing Level
Vertical air currents toss frozen water droplets up and down.
Alternate freezing and melting builds up layers of ice.
Ice falls as hailstones.
Rising air
Hail
Halifax, Nova Scotia, 49°F
Reno, Nevada, 49°F
Miami, Florida, 78°F
Hurricanes spiral across the Atlantic and into the Caribbean.
Mexico City, Mexico, 60°F
San José, Costa Rica, 69°F
Caracas, Venezuela, 70°F
Bands of clouds form along the equator, caused by the strong rising air currents stirred up by the hot sun.
65

Go Online
PHSchool.com
For: Information about weather
Web Code: are-0153

Activity

Have students work in pairs to answer the questions.

Diversity
All temperatures are given in degrees Fahrenheit. Some students may be more familiar with degrees Celsius. Work together to come up with benchmarks they can use to help them understand the temperatures used in the questions.

Science Connection
Invite interested students to find other examples of early thermometers and to learn about how they work. Have them share what they discover with classmates. Are there any advanced thermometers on the way or in use today? Have them find out.

Geography Connection
Invite students to investigate the relationship between geography and temperature. Have them look into the effects of nearness to the equator, elevation, ocean currents, and so on.

Differentiated Instruction

Special Needs L1
Exercises 1–4 You may find it useful to work through Exercise 1 together, helping students to understand and apply the formula for heating cost.

2 Exponents, Factors, and Fractions

Chapter at a Glance

Lesson Titles, Objectives, and Features	Assessment	NCTM Standards	Local Standards
2-1 Exponents and Order of Operations • To write and simplify expressions with exponents **2-1b Activity Lab, Technology:** Using a Scientific Calculator	Lesson Quiz	1, 2, 6, 7, 8, 9, 10	
2-2a Activity Lab: Divisibility Tests **2-2 Prime Factorization** • To find multiples and factors and to use prime factorization **Guided Problem Solving:** Using LCM and GCF	Lesson Quiz Checkpoint Quiz 1	1, 3, 6, 7, 8, 9, 10	
2-3 Simplifying Fractions • To write equivalent fractions and to simplify fractions	Lesson Quiz	1, 2, 3, 6, 7, 8, 9, 10	
2-4a Activity Lab, Hands On: Comparing Fractions **2-4 Comparing and Ordering Fractions** • To compare and order fractions	Lesson Quiz	1, 2, 3, 6, 7, 8, 9, 10	
2-5 Mixed Numbers and Improper Fractions • To write mixed numbers and improper fractions	Lesson Quiz	1, 2, 3, 6, 7, 8, 9, 10	
2-6a Activity Lab: Comparing Fractions and Decimals **2-6 Fractions and Decimals** • To convert between fractions and decimals **2-6b Activity Lab, Data Analysis:** Using Tables to Compare Data	Lesson Quiz Checkpoint Quiz 2	1, 2, 6, 7, 8, 9, 10	
2-7 Rational Numbers • To compare and order rational numbers	Lesson Quiz	1, 2, 3, 6, 7, 8, 9, 10	
2-8 Scientific Notation • To write numbers in both scientific notation and standard form **Extension:** Negative Exponents	Lesson Quiz	1, 4, 6, 7, 8, 9, 10	
Problem Solving Application: Applying Fractions			

NCTM Standards 2000
1 Number and Operations
2 Algebra
3 Geometry
4 Measurement
5 Data Analysis and Probability
6 Problem Solving
7 Reasoning and Proof
8 Communication
9 Connections
10 Representation

Correlations to Standardized Tests

All content for these tests is contained in *Prentice Hall Math,* Course 2. This chart reflects coverage in this chapter only.

	2-1	2-2	2-3	2-4	2-5	2-6	2-7	2-8
Terra Nova CAT6 (Level 17)								
Number and Number Relations	✔	✔	✔	✔	✔	✔	✔	✔
Computation and Numerical Estimation	✔	✔	✔	✔	✔	✔	✔	✔
Operation Concepts	✔	✔	✔	✔	✔	✔	✔	✔
Measurement								
Geometry and Spatial Sense								
Data Analysis, Statistics, and Probability								
Patterns, Functions, Algebra	✔	✔	✔	✔	✔	✔	✔	✔
Problem Solving and Reasoning	✔	✔	✔	✔	✔	✔	✔	✔
Communication	✔	✔	✔	✔	✔	✔	✔	✔
Decimals, Fractions, Integers, and Percent	✔	✔	✔	✔	✔	✔	✔	✔
Order of Operations	✔							
Terra Nova CTBS (Level 17)								
Decimals, Fractions, Integers, Percents	✔	✔	✔	✔	✔	✔	✔	✔
Order of Operations, Numeration, Number Theory	✔	✔	✔	✔	✔	✔	✔	✔
Data Interpretation								
Pre-algebra	✔		✔	✔	✔	✔	✔	
Measurement								
Geometry								
ITBS (Level 13)								
Number Properties and Operations	✔	✔	✔	✔	✔	✔	✔	✔
Algebra	✔	✔	✔	✔	✔	✔	✔	✔
Geometry								
Measurement								
Probability and Statistics								
Estimation								
SAT10 (Int 3 Level)								
Number Sense and Operations	✔	✔	✔	✔	✔	✔	✔	✔
Patterns, Relationships, and Algebra	✔	✔	✔	✔	✔	✔	✔	✔
Data, Statistics, and Probability								
Geometry and Measurement								
NAEP								
Number Sense, Properties, and Operations	✔	✔	✔	✔	✔	✔	✔	✔
Measurement								
Geometry and Spatial Sense								
Data Analysis, Statistics, and Probability								
Algebra and Functions								

CAT6 California Achievement Test, 6th Ed.
CTBS Comprehensive Test of Basic Skills
ITBS Iowa Test of Basic Skills, Form M
SAT10 Stanford Achievement Test, 10th Ed.
NAEP National Assessment of Educational Progress 2005 Mathematics Objectives

Math Background

Skills Trace

BEFORE Chapter 2

Course 1 or Grade 6 introduced simplifying expressions with exponents.

DURING Chapter 2

Course 2 extends the Order of Operations to simplify expressions with exponents and introduces rational numbers.

AFTER Chapter 2

Throughout this course, students simplify exponential and rational expressions.

2-1 Exponents and Order of Operations

Math Understandings

- Exponentiation is a mathematical operation that indicates repeated multiplication.

An exponent shows how many times the base is used as a factor. A number that can be written using an exponent is a **power**. 81 is a power of 3 because 81 can be written as 3^4.

Example: 5^3 is $5 \cdot 5 \cdot 5$, or 125

Example: $(-2)^4$ is $(-2) \cdot (-2) \cdot (-2) \cdot (-2)$, or 16

The Order of Operations

1. Do all operations within groupings first.
2. Evaluate any term(s) with exponents.
3. Multiply and divide in order from left to right.
4. Add and subtract in order from left to right.

Example: Simplify $4^3 + (8 - 3)^2$

$4^3 + (8 - 3)^2 = 4^3 + 5^2$ ← **Do operation in parentheses.**
$= 64 + 25$ ← **Find the values of the powers.**
$= 89$ ← **Add.**

2-2 Prime Factorization

Math Understandings

- Any number is always divisible by all of its factors.
- Every composite number has a unique prime factorization.
- The number 1 is neither prime nor composite.

One whole number is **divisible** by a second whole number if the remainder is 0 when you divide the first number by the second. A **factor** is a whole number that divides another whole number with a remainder of 0. A **multiple** of a number is the product of that number and any nonzero whole number. The **least common multiple (LCM)** of two or more numbers is the least multiple that is common to all of the numbers. A **composite number** is a whole number greater than 1 that has more than two factors. A **prime number** is a whole number with exactly two factors, 1 and the number itself. The **greatest common factor (GCF)** of two or more numbers is the greatest number that is a factor of all the numbers.

2-3 2-4 Simplifying, Comparing, and Ordering Fractions

Math Understandings

- One number can be written in many different forms, all of which are equivalent.
- A fraction has a unique simplest form, which may be an improper fraction, such as $\frac{3}{2}$.

Fractions that name the same amount are equivalent fractions. Fractions obtained by multiplying (or dividing) both numerator and denominator by the same nonzero number are equivalent to each other. A fraction is written in **simplest form** when the numerator and the denominator have no common factors other than 1.

With two fractions, if the denominators are the same, the numerators tell which is greater.

Example: $\frac{5}{5} < \frac{8}{5}$ because $5 < 8$

With two fractions, if the numerators are the same, the fraction with the lesser denominator has the greater value.

Example: $\frac{3}{4} > \frac{3}{12}$ because $4 < 12$

The **least common denominator (LCD)** of two or more fractions is the least common multiple (LCM) of their denominators.

Example: Find the LCD for the fractions $\frac{2}{3}$ and $\frac{1}{4}$. The LCM of 3 and 4 is 12. So, 12 is the LCD for the denominators 3 and 4.

2-5 Mixed Numbers and Improper Fractions

Math Understandings

- A quantity greater than 1 may be written as an improper fraction or as a mixed number.
- A mixed number, such as $3\frac{1}{2}$, can be written as the equivalent sum, for example, $3 + \frac{1}{2}$.

An **improper fraction** has a numerator that is greater than or equal to its denominator.

Examples: $\frac{5}{5}$, $\frac{8}{5}$, $\frac{3}{2}$, and $\frac{25}{3}$ are all improper fractions.

A **mixed number** is the sum of a whole number and a fraction.

Examples: $1\frac{1}{3}$, $2\frac{3}{4}$, and $7\frac{2}{3}$ are mixed numbers.

2-6 2-7 Fractions, Decimals, and Rational Numbers

Math Understandings

- The number π (pi) can be represented by a nonterminating and nonrepeating decimal. Pi is an *irrational number*. It is only approximately equal to $\frac{22}{7}$ or 3.14.

A decimal that stops, or terminates, is a **terminating decimal**. If the same block of digits in a decimal repeats without end, the decimal is a **repeating decimal**. Both repeating and terminating decimals are rational numbers. A **rational number** is a number that can be written as a quotient of two integers, where the divisor is not zero.

Examples of rational numbers:
$\frac{17}{23}$, $\frac{1}{3}$, $0.\overline{6}$, $\frac{2}{9}$, $1.\overline{789}789$, -3, $-0.\overline{8}$, $-\frac{2}{3}$, -7.5079

Examples of irrational numbers:
π, 0.121221222 . . ., −530.53005300053 . . .

2-8 Scientific Notation

Math Understandings

- Scientific notation is a shorthand way to write very large and very small numbers.
- Very large and very small numbers can be compared and combined more easily when written in scientific notation.

A number in **scientific notation** is written as the product of two factors, one greater than or equal to 1 and less than 10, and the other a power of 10. For example, 3.25×10^8 is 325,000,000.

Example: 341,500,000 in scientific notation is 3.415×10^8.

Additional Professional Development Opportunities

Math Background Notes for Chapter 2: Every lesson has a Math Background in the PLAN section.

Research Overview, Mathematics Strands
Additional support for these topics and more is in the front of the Teacher's Edition.

LessonLab
LessonLab, a Pearson Education company, offers comprehensive, facilitated professional development designed to help teachers to improve student achievement. To learn more, please visit lessonlab.com.

Chapter 2 Resources

	2-1	2-2	2-3	2-4	2-5	2-6	2-7	2-8	For the Chapter
Print Resources									
L3 Practice	●	●	●	●	●	●	●	●	
L1 Adapted Practice	●	●	●	●	●	●	●	●	
L3 Guided Problem Solving	●	●	●	●	●	●	●	●	
L2 Reteaching	●	●	●	●	●	●	●	●	
L4 Enrichment	●	●	●	●	●	●	●	●	
L3 Daily Notetaking Guide	●	●	●	●	●	●	●	●	
L1 Adapted Daily Notetaking Guide	●	●	●	●	●	●	●	●	
L3 Vocabulary and Study Skills Worksheets	●		●	●		●		●	●
L3 Daily Puzzles	●	●	●	●	●	●	●	●	
L3 Activity Labs	●	●	●	●	●	●	●	●	
L3 Checkpoint Quiz		●				●			
L3 Chapter Project									●
L2 Below Level Chapter Test									●
L3 Chapter Test									●
L4 Alternative Assessment									●
L3 Cumulative Review									●
Spanish Resources ELL									
L3 Practice	●	●	●	●	●	●	●	●	
L3 Vocabulary and Study Skills Worksheets	●		●	●		●		●	●
L3 Checkpoint Quiz		●				●			
L2 Below Level Chapter Test									●
L3 Chapter Test									●
L4 Alternative Assessment									●
L3 Cumulative Review									●
Transparencies									
Check Skills You'll Need	●	●	●	●	●	●	●	●	
Additional Examples	●	●	●	●	●	●	●	●	
Problem of the Day	●	●	●	●	●	●	●	●	
Classroom Aid									
Student Edition Answers	●	●	●	●	●	●	●	●	●
Lesson Quiz	●	●	●	●	●	●	●	●	
Test-Taking Strategies									●
Technology									
Interactive Textbook Online	●	●	●	●	●	●	●	●	●
StudentExpress™ CD-ROM	●	●	●	●	●	●	●	●	●
Success Tracker™ Online Intervention	●	●	●	●	●	●	●	●	●
TeacherExpress™ CD-ROM	●	●	●	●	●	●	●	●	●
PresentationExpress™ with QuickTake Presenter CD-ROM	●	●	●	●	●	●	●	●	●
ExamView® Assessment Suite CD-ROM	●	●	●	●	●	●	●	●	●
MindPoint® Quiz Show CD-ROM									●
Prentice Hall Web Site: PHSchool.com	●	●	●	●	●	●	●	●	●

Also available: **Prentice Hall Assessment System**
- Progress Monitoring Assessments
- Skills and Concepts Review
- Test Prep Workbook

Other Resources
Algebra Readiness Tests
All-in-One Student Workbook
All-in-One Student Workbook, Adapted Version
Multilingual Handbook
Solution Key
Math Notes Study Folder
Spanish Cumulative Assessment

Where You Can Use the Lesson Resources

Here is a suggestion, following the four-step teaching plan, for how you can incorporate Differentiated Instruction Resources into your teaching.

	Instructional Resources L3	Differentiated Instruction Resources
1. Plan		
Preparation Read the Math Background in the Teacher's Edition to connect this lesson with students' previous experience. **Starting Class** **Check Skills You'll Need** Assign these exercises to review prerequisite skills. **New Vocabulary** Help students pre-read the lesson by pointing out the new terms introduced in the lesson.	**Math Background** **Math Understandings** **Transparencies & PresentationExpress™ with QuickTake Presenter CD-ROM** Check Skills You'll Need Problem of the Day **Resources** Vocabulary and Study Skills	**Spanish Support** ELL Vocabulary and Study Skills
2. Teach		
L3 Guided Instruction Use the Activity Labs to build conceptual understanding. Teach each Example. Use the Teacher's Edition side column notes for specific teaching tips, including Error Prevention notes. Use the Additional Examples found in the side column (and on transparency and PowerPoint) as an alternative presentation for the content. After each Example, assign the Quick Check exercise for that Example to get an immediate assessment of student understanding. Use the Closure activity in the Teacher's Edition to help students attain mastery of lesson content.	**Student Edition** Activity Lab **Resources** Daily Notetaking Guide Activity Lab **Transparencies & PresentationExpress™ with QuickTake Presenter CD-ROM** Additional Examples Classroom Aids **ExamView® Assessment Suite CD ROM**	**Teacher's Edition** Every lesson includes suggestions for working with students who need special attention. L1 Special Needs L2 Below Level L4 Advanced Learners ELL English Language Learners **Resources** L1 Adapted Daily Notetaking Guide **Multilingual Handbook**
3. Practice		
Assignment Guide **Check Your Understanding** Use these questions to check students' understanding before you assign homework. **Homework Exercises** Assign homework from these leveled exercises in the Assignment Guide. A Practice by Example B Apply Your Skills C Challenge Test Prep and Mixed Review **Homework Quick Check** Use these key exercises to quickly check students' homework.	**Transparencies & PresentationExpress™ with QuickTake Presenter CD-ROM** Student Answers **Resources** Practice Guided Problem Solving Vocabulary and Study Skills Activity Lab Daily Puzzles **ExamView® Assessment Suite CD-ROM**	**Spanish Support** ELL Practice ELL Vocabulary and Study Skills **Resources** L1 Adapted Practice L4 Enrichment
4. Assess & Reteach		
Lesson Quiz Assign the Lesson Quiz to assess students' mastery of the lesson content. **Checkpoint Quiz** Use the Checkpoint Quiz to assess student progress over several lessons.	**Transparencies & PresentationExpress™ with QuickTake Presenter CD-ROM** Lesson Quiz **Resources** Checkpoint Quiz	**Resources** L2 Reteaching ELL Checkpoint Quiz Success Tracker™ Online Intervention **ExamView® Assessment Suite CD-ROM**

KEY **L1** Special Needs **L2** Below Level **L3** For All Students **L4** Advanced, Gifted **ELL** English Language Learners

CHAPTER 2 Exponents, Factors, and Fractions

Exponents, Factors, and Fractions

 **Check Your Readiness**

Answers are in the back of the textbook.

For intervention, direct students to:

Adding and Subtracting Decimals
Lesson 1-2
Extra Skills and Word Problems Practice, Ch. 1

Comparing Integers
Lesson 1-6
Extra Skills and Word Problems Practice, Ch. 1

Multiplying and Dividing Integers
Lesson 1-8
Extra Skills and Word Problems Practice, Ch. 1

Order of Operations and the Distributive Property
Lesson 1-9
Extra Skills and Word Problems Practice, Ch. 1

Spanish Vocabulary/Study Skills ELL

Vocabulary/Study Skills L3

2A: Graphic Organizer — For use before Lesson 2-1

Study Skill: Develop consistent study habits. Block off the same amount of time each evening for schoolwork. Plan ahead by setting aside extra time when you have a big project or test coming up.

Write your answers.

1. What is the chapter title? Exponents, Factors, and Fractions
2. How many lessons are there in this chapter? 10
3. What is the topic of the Reading Math page? Reading a Math Lesson
4. What is the topic of the Test-Taking Strategies page? Writing Extended Responses
5. Complete the graphic organizer below as you work through the chapter.
   - In the center, write the title of the chapter.
   - When you begin a lesson, write the lesson name in a rectangle.
   - When you complete a lesson, write a skill or key concept in a circle linked to that lesson block.
   - When you complete the chapter, use this graphic organizer to help you review.

Check students' diagrams.

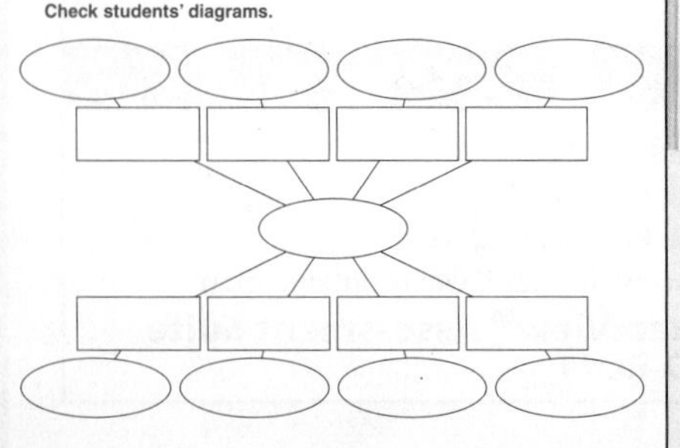

What You've Learned

- In Chapter 1, you compared and ordered integers.
- You added, subtracted, multiplied, and divided decimals and integers.
- You used the order of operations to simplify expressions involving decimals and integers.

Check Your Readiness

GO for Help

For Exercises	See Lesson
1–4	1-2
5–10	1-6
11–14	1-8
15–17	1-9

Adding and Subtracting Decimals

Find each sum or difference.

1. $2.1 + 3.4$ **5.5**
2. $6.02 - 4.597$ **1.423**
3. $7.0 - 3.11$ **3.89**
4. $671.02 + 6.427$ **677.447**

Comparing Integers

Compare. Use <, =, or >

5. $|-2|$ ■ $|-5|$ <
6. $|11|$ ■ $|-13|$ <
7. 10 ■ $|-10|$ =
8. -8 ■ $|-8|$ <
9. $|9|$ ■ $|-14|$ <
10. $|-1|$ ■ 0 >

Multiplying and Dividing Integers

Find each product or quotient.

11. $-9 \cdot 3$ **−27**
12. $-3 \cdot (-3)$ **9**
13. $27 \div (-3)$ **−9**
14. $-16 \div (-4)$ **4**

Order of Operations and the Distributive Property

Find the value of each expression.

15. $12 + 4(16 \div 4)$ **28**
16. $30 \div 3 - 4 \cdot 2$ **2**
17. $(8 + 4) \div 4 - 2$ **1**

Chapter 2 Overview

In this chapter, students extend the order of operations to include exponents. Students then simplify fractions, compare and order fractions and mixed numbers, and express fractions as decimals and decimals as fractions. They learn why fractions and decimals are rational numbers. Lastly, they express numbers in scientific notation.

Activating Prior Knowledge

In this chapter, students will build on their knowledge of basic division facts and prior work with fractions and decimals. They learn about scientific notation and develop their understanding of fraction concepts and properties. They also draw upon their understanding of the order of operations to include expressions involving exponents. Ask questions such as: *When is a fraction expressed in simplest form?* **when the only number that can divide both terms of the fraction without a remainder is 1**

What You'll Learn Next

- In this chapter, you will compare and order rational numbers.
- You will convert between fractions and decimals.
- You will use the order of operations to simplify expressions with exponents.

Key Vocabulary

- base (p. 68)
- composite number (p. 75)
- equivalent fraction (p. 82)
- exponent (p. 68)
- greatest common factor (GCF) (p. 75)
- improper fraction (p. 91)
- least common denominator (LCD) (p. 87)
- least common multiple (LCM) (p. 74)
- mixed number (p. 91)
- prime number (p. 75)
- rational number (p. 102)

Problem Solving Application On pages 116 and 117, you will work an extended activity on fractions.

2-1

1. Plan

Objective
To write and simplify expressions with exponents

Examples
1 Writing Expressions Using Exponents
2 Application: Geography
3 Simplifying Using Order of Operations

Math Understandings: p. 66C

Math Background

Raising a number to a power is a mathematical operation. Because this operation is an extension of a basic operation, namely multiplication, it is considered a *higher order* operation. The formal name of the operation is *exponentiation,* but it is also referred to as *powering.*

More Math Background: p. 66C

Lesson Planning and Resources

See p. 66E for a list of the resources that support this lesson.

Check Skills Youill Need
Use student page, transparency, or PowerPoint. For intervention, direct students to:
Order of Operations and the Distributive Property
Lesson 1-9
Extra Skills and Word Problems Practice, Ch. 1

2-1 Exponents and Order of Operations

Check Skills You'll Need

1. Vocabulary Review Using the *order of operations,* do you multiply factors before or after you add? **before**

Find the value of each expression.

2. $5 - 1 \cdot 3$ **2**

3. $(5 - 1) \cdot 3$ **12**

4. $10 \div 2 - 3 \cdot 5$ **−10**

5. $(1 + 99) \div 10 - 9$ **1**

for Help Lesson 1-9

What You'll Learn

To write and simplify expressions with exponents

New Vocabulary exponent, base, power

Why Learn This?

Googol is a large number written as the digit one followed by 100 zeros. Writing out the number takes a long time. You can use exponents to write googol and other large numbers in a easier way. For example, you can write googol as 10^{100}.

1 0 0 0 0 0 0 0 0 0 0
0 0 0 0 0 0 0 0 0 0
0 0 0 0 0 0 0 0 0 0
0 0 0 0 0 0 0 0 0 0
0 0 0 0 0 0 0 0 0 0
0 0 0 0 0 0 0 0 0 0
0 0 0 0 0 0 0 0 0 0
0 0 0 0 0 0 0 0 0 0
0 0 0 0 0 0 0 0 0 0
0 0 0 0 0 0 0 0 0 0

An **exponent** tells you how many times a number, or **base,** is used as a factor.

$$5^3 = 5 \cdot 5 \cdot 5 = 125$$

exponent → 3; base → 5; value of the expression → 125; The base is used as a factor three times.

A number that can be expressed using an exponent is called a **power.** The number 125 is a power of 5 because it can be written as 5^3.

EXAMPLE Writing Expressions Using Exponents

1 Write $3 \cdot 3 \cdot 3 \cdot 3 \cdot 3$ using an exponent.

$3 \cdot 3 \cdot 3 \cdot 3 \cdot 3 = 3^5$ ← **3 is the base. 5 is the exponent.**

Quick Check

1. Write each product using exponents.
a. $44 \cdot 44 \cdot 44 \cdot 44$ $\mathbf{44^4}$ **b.** $(-2) \cdot (-2)$ $\mathbf{(-2)^2}$

GO Online

Video Tutor Help
Visit: PHSchool.com
Web Code: are-0775

You can find the value of an expression with exponents by writing it as the product of repeated factors. You can also use a scientific calculator.

Differentiated Instruction Solutions for All Learners

Special Needs L1	**Below Level** L2
Provide students with a copy of the Homework Exercises 24–29. Before simplifying, have them underline in colored pencil or pen what they will do first.	Review multiplications involving three factors, such as the following. $2 \cdot 3 \cdot 5$ **30** $3 \cdot 3 \cdot 7$ **63**
learning style: visual	**learning style: visual**

EXAMPLE Application: Geography

2 Gibraltar is at the mouth of the Mediterranean Sea. Its area is about the same as the area of a square 1.5 mi on a side. Find Gibraltar's area.

Method 1 Since $A = s^2$, compute to find 1.5^2.

$1.5^2 = (1.5)(1.5)$ ← **Write as a product of repeated factors.**

$= 2.25$ ← **Multiply.**

Method 2 Use a geometric model. In the model, each side of the blue square has a length of 1.5. The total area in blue equals one whole square plus two half squares plus one quarter square. The total is 2.25 squares. The area of Gibraltar is about 2.25 mi^2.

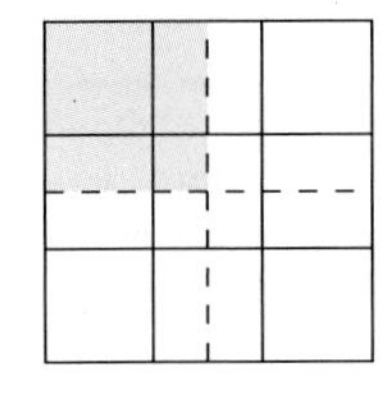

Calculator Tip

You can use the ^ or y^x key to find a power.

To find 1.5^2, use 1.5 ^ 2 = 2.25 or 1.5 y^x 2 = 2.25.

If your calculator does not have an exponent key, use 1.5 × 1.5 = 2.25.

Quick Check

2. Simplify. Use paper and pencil, a model, or a calculator.
   a. 3^5 **243** b. 10^9 **1,000,000,000** c. 3.1^2 **9.61**

The order of operations includes expressions with exponents.

KEY CONCEPTS Order of Operations

1. Do all operations within grouping symbols first.
2. Evaluate any term(s) with exponents.
3. Multiply and divide in order from left to right.
4. Add and subtract in order from left to right.

The expressions -2^4 and $(-2)^4$ are not equivalent. The expression -2^4 means the opposite, or the negative, of 2^4. The base of -2^4 is 2, not -2.

Vocabulary Tip

The phrase *Please Excuse My Dear Aunt Sally* can help you remember the order of operations. The first letter of each word in the phrase stands for *Parentheses, Exponents, Multiplication, Division, Addition, and Subtraction.*

EXAMPLE Simplifying Using Order of Operations

3 Simplify $-2^4 + (3 - 5)^4$.

$-2^4 + (3 - 5)^4 = -2^4 + (-2)^4$ ← **Do operations in parentheses.**

$= -16 + 16$ ← **Find the values of the powers.**

$= 0$ ← **Add.**

Quick Check

3. Simplify.
   a. $(-3)^3$ **−27** b. -3^3 **−27** c. $(3 + 5)^2 - 2$ **62**

2. Teach

Activity Lab

Use before the lesson.

All in One Teaching Resources

Activity Lab 2-1: Exploring Exponents

Guided Instruction

Example 3

The expression $(3 - 5)^4$ should be read as "the quantity three minus five, raised to the fourth power." If it is read as "three minus five to the fourth power," students with visual impairments might have difficulty distinguishing it from the expression $3 - 5^4$.

PowerPoint

Additional Examples

1 Write using exponents.
   a. $7 \cdot 7 \cdot 7 \cdot 7$ **7^4**
   b. $2 \cdot 2 \cdot 2 \cdot 2 \cdot 2 \cdot 2$ **2^6**

2 A seaside village has an area of 1.3^2 km^2. Find the value of 1.3^2. **1.69**

3 Simplify $2^3 \cdot (9 - 3)^2$. **288**

All in One Teaching Resources

- Daily Notetaking Guide 2-1 L3
- Adapted Notetaking 2-1 L1

Closure

- *How do you find the value of a number raised to a power?* **Perform a multiplication in which the number is used as a factor the number of times indicated by the exponent.**
- *What is the order of operations when exponents are involved?* **Do operations within grouping symbols. Evaluate terms with exponents. Multiply and divide in order from left to right. Add and subtract in order from left to right.**

Advanced Learners L4	**English Language Learners** ELL
Use four 4's and any operations to make an expression that is equal to 68. **Samples:** $4^4 \div 4 + 4$; $4 \cdot 4 \cdot 4 + 4$ learning style: visual	Make sure students understand that the word *groupings* in the order of operations frequently refers to *parentheses.* learning style: verbal

3. Practice

Assignment Guide

Check Your Understanding
Go over Exercises 1–8 in class before assigning the Homework Exercises.

Homework Exercises

A	Practice by Example	9–29
B	Apply Your Skills	30–40
C	Challenge	41
	Test Prep and Mixed Review	42–47

Homework Quick Check
To check students' understanding of key skills and concepts, go over Exercises 10, 22, 35, 36, and 39.

Differentiated Instruction Resources

Adapted Practice 2-1 L1

Practice 2-1 Exponents and Order of O... L3

Write using exponents.

1. $3 \times 3 \times 3 \times 3 \times 3$ 3^5
2. $2.7 \times 2.7 \times 2.7$ 2.7^3
3. $11.6 \times 11.6 \times 11.6 \times 11.6$ 11.6^4
4. $2 \times 2 \times 2 \times 2 \times 2 \times 2$ 2^6
5. $8.3 \times 8.3 \times 8.3 \times 8.3 \times 8.3$ 8.3^5
6. $4 \times 4 \times 4 \times 4 \times 4 \times 4 \times 4 \times 4$ 4^8

Write as the product of repeated factors. Then simplify.

7. $(0.5)^3$ $0.5 \times 0.5 \times 0.5$; 0.125
8. $(-4)^5$ $(-4) \times (-4) \times (-4) \times (-4) \times (-4)$; −1,024
9. $(2.7)^2$ 2.7×2.7; 7.29
10. 2^3 $2 \times 2 \times 2$; 8
11. $(-5)^6$ $(-5) \times (-5) \times (-5) \times (-5) \times (-5) \times (-5)$; 15,625
12. $(8.1)^3$ $8.1 \times 8.1 \times 8.1$; 531.441

Simplify. Use a calculator, paper and pencil, or mental math.

13. -4^3 −64
14. $11 + (-6^3)$ 205
15. $14 + 16^2$ 270
16. $8 + 6^4$ 1,304
17. $3^2 \cdot 5^4$ 5,625
18. $6^2 - 2^4$ 20
19. $4(0.9 + 1.3)^3$ 42.592
20. $35 - (4^2 + 5)$ 14
21. $(3^3 + 6) - 7$ 26
22. $5(0.3 \cdot 1.2)^2$ 0.648
23. $5(4 + 2)^2$ 180
24. $(8 - 6.7)^3$ 2.197
25. A cubic aquarium has edges measuring 4.3 ft each. Find the volume of the aquarium in cubic feet.
    79.507 ft^3
26. Lana is 2^3 in. taller than her little sister. How many inches taller is Lana than her sister?
    8 in.

2-1 • Guided Problem Solving GPS L3

Student Page 71, Exercise 36:

A Scanning Electron Microscope (SEM) can magnify an image to as much as 10^5 times the actual size. How many times is this?

Understand

1. What are you being asked to do?
2. What do you call the 5 in 10^5?
3. What do you call the 10 in 10^5?

Plan and Carry Out

4. The number 10^5 is what number?
5. How many zeros are in the number 10^5?
6. How many times does the SEM magnify?

Check

7. Does your answer follow the pattern of powers of 10? Explain.

Solve Another Problem

8. Lucy has a microscope that magnifies an image to as much as 10^3 times the actual size. Aaron has a microscope that magnifies an image to as much as 10^4 times the actual size. What is the difference in these two numbers?

Check Your Understanding

For: Order of Operations Activity
Use: Interactive Textbook, 2-1

1. **Vocabulary** How is an exponent different from a base? **The exponent tells you how many times the base is used as a factor.**
2. Write an expression for *the opposite of the fourth power of 2.* -2^4

Simplify each expression.

3. $-1 \cdot 5^4$ −625
4. $(-5)^4$ 625
5. $2^4 \cdot (7 - 6)^3$ 16

Science **Match each fact with the appropriate power.**

6. number of moons orbiting Earth C
7. planets in the solar system B
8. freezing point of water in degrees Fahrenheit A

A. 2^5
B. 3^2
C. 1^1

Homework Exercises

For more exercises, see Extra Skills and Word Problems.

GO for Help

For Exercises	See Examples
9–14	1
15–23	2
24–29	3

A **Write using an exponent.**

9. $2 \cdot 2 \cdot 2 \cdot 2 \cdot 2 \cdot 2$ See left.
10. $7 \cdot 7 \cdot 7 \cdot 7 \cdot 7$ 7^5
11. $6 \cdot 6 \cdot 6$ 6^3
12. $-5 \cdot -5 \cdot -5 \cdot -5$ $(-5)^4$
13. $9 \cdot 9 \cdot 9 \cdot 9 \cdot 9 \cdot 9$ 9^6
14. $12 \cdot 12 \cdot 12 \cdot 12$ 12^4

9. 2^6

Simplify. Use paper and pencil, a model, or a calculator.

15. 9^2 81
16. 10^8 100,000,000
17. 0.2^6 0.000064
18. 1.7^3 4.913
19. -3^4 −81
20. $(-3)^4$ 81
21. $(-2)^3$ −8
22. $(-4)^3$ −64
23. Each side of a sugar cube is 0.6 in. long. Find the volume of the sugar cube. 0.216 in.^{3}

Simplify using the order of operations.

24. $2^3 \cdot (6 - 3)^2$ 72
25. $(2^3 \cdot 6) - 3^2$ 39
26. $2^3 \cdot 6 - 3^2$ 39
27. $-3^2 + 2^3 \cdot 6$ 39
28. $3^2 - 2^3$ 1
29. $(2 + 1)^3 \div 3^2$ 3

B

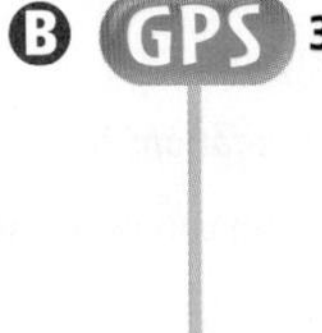

30. **Guided Problem Solving** Suppose you have a part-time job. Your boss offers to pay you \$2 the first day, with the amount to double each day. How much will you be paid for the tenth day? **\$1,024**
    - **Make a Plan** Find the amounts you will be paid on the second, third, and fourth days. Write each amount using exponents.
    - **Carry Out the Plan** On each day, you will earn ■ dollars, which can be written using exponents as $2^{■}$. Identify the pattern.
31. Write two equivalent expressions for the number of seconds in 60 hours. Use exponents in only one of the expressions. See left.

31. $60 \cdot 60 \cdot 60$; 60^3

33.

Power of 10	Value	Number of Zeros
10^1	10	1
10^2	100	2
10^3	1,000	3
10^4	10,000	4
10^5	100,000	5

Homework Video Tutor
Visit: PHSchool.com
Web Code: are-0201

32. **Algebra** Suppose a is a nonzero number. How would you write $a \cdot a \cdot a$ using an exponent? **a^3**

For Exercises 33–35, refer to the table.

Power of 10	Value	Number of Zeros
10^1	■	1
10^2	■	■
10^3	■	■
10^4	■	■
10^5	■	■

33. Copy the table. Fill in the missing values. **See margin.**
34. **Patterns** What patterns do you notice? **See left.**
35. **Reasoning** Predict the number of zeros in 10^{12}. **12**

34. The exponent is equal to the number of zeros.

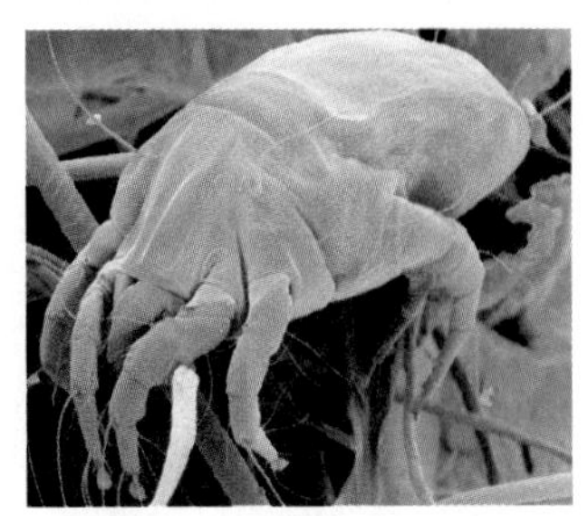

This dust mite has been magnified 1.5×10^5 times.

36. A scanning electron microscope (SEM) (GPS) can magnify an image to as much as 10^5 times the actual size. How many times is this? **100,000 times**

37. Without calculating, decide whether the value of $(-672)^2 - 192$ is positive or negative. Explain your reasoning. **Positive; $(-672)^2$ is positive and larger than 192.**

38. What exponent completes the table below? Use your answer to find $2^0, 3^0, 4^0, 5^0$, and 10^0. **0; 1, 1, 1, 1, 1**

Value	16	8	4	2	1
Power of 2	2^4	2^3	2^2	2^1	$2^{■}$

39. **Writing in Math** Write a general rule to find the value of a nonzero expression with an exponent of 0. **If $x \neq 0$, then $x^0 = 1$.**

40. a. **Geometry** How many small squares line up on one edge of the larger square? **40a–c. See left.**
b. How many small squares fit in the large square?
c. Explain why 5^2 is read as "5 squared."

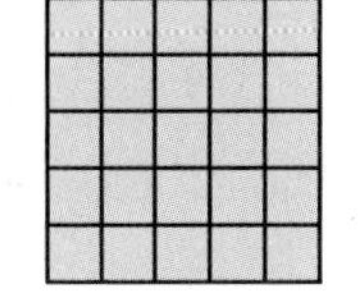

40a. 5
b. 25
c. Answers may vary. Sample: "5 squared" is the number of units of area if 5 is the length of a side.

(C) 41. **Challenge** Evaluate $g^2(f + h)$ for $f = -2$, $g = -3$, and $h = -4$. **−54**

Test Prep and Mixed Review

Practice

Multiple Choice

42. What is the value of the expression $(4 - 1)^3 - 3 \times 8 \div 6$? **D**
Ⓐ 0.5 Ⓑ 5 Ⓒ 21 Ⓓ 23

43. The table shows a plant's growth each week for four weeks. If the plant grew 4.75 cm in all, how much did it grow in week 4? **H**
Ⓕ 0.65 cm Ⓗ 1.65 cm
Ⓖ 1.35 cm Ⓙ 4.75 cm

Week	Growth (cm)
1	0.25
2	1.5
3	1.35
4	

GO for Help

For Exercises	See Lesson
44–47	1-8

Find the value of each expression.

44. $12(-2)$ **−24** 45. $-4(-10)$ **40** 46. $-8 \div (-4)$ **2** 47. $49 \div (-7)$ **−7**

Online lesson quiz, PHSchool.com, Web Code: ara-0201

4. Assess & Reteach

Lesson Quiz

Simplify.

1. 6^3 **216**
2. $2(3^2 + 12)$ **42**
3. -7^4 **−2,401**
4. $4(10 - 8)^5$ **128**

Alternative Assessment

Turn to the order of operations on page 69. For Exercises 24–29 on page 70, have students tell what operation they must perform first, second, and so on to complete each Exercise. Have them show each step of their simplification process.

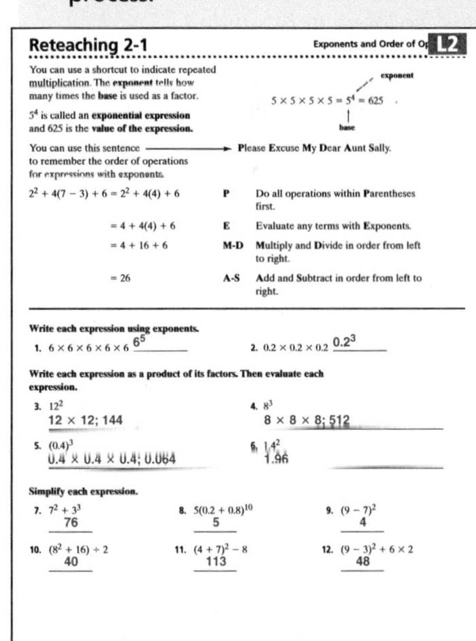

Reteaching 2-1 Exponents and Order of Op L2

You can use a shortcut to indicate repeated multiplication. The **exponent** tells how many times the **base** is used as a factor.

$5 \times 5 \times 5 \times 5 = 5^4 = 625$ (exponent; base)

5^4 is called an **exponential expression** and 625 is the **value of the expression.**

You can use this sentence to remember the order of operations for expressions with exponents. → **P**lease **E**xcuse **M**y **D**ear **A**unt **S**ally.

$2^2 + 4(7 - 3) + 6 = 2^2 + 4(4) + 6$ — **P** Do all operations within **P**arentheses first.
$= 4 + 4(4) + 6$ — **E** Evaluate any terms with **E**xponents.
$= 4 + 16 + 6$ — **M-D** **M**ultiply and **D**ivide in order from left to right.
$= 26$ — **A-S** **A**dd and **S**ubtract in order from left to right.

Write each expression using exponents.

1. $6 \times 6 \times 6 \times 6 \times 6$ **6^5**
2. $0.2 \times 0.2 \times 0.2$ **0.2^3**

Write each expression as a product of its factors. Then evaluate each expression.

3. 12^2 **12 × 12; 144**
4. 8^3 **8 × 8 × 8; 512**
5. $(0.4)^3$ **0.4 × 0.4 × 0.4; 0.064**
6. 1.4^2 **1.96**

Simplify each expression.

7. $7^2 + 3^3$ **76**
8. $5(0.2 + 0.8)^{10}$ **5**
9. $(9 - 7)^2$ **4**
10. $(8^2 + 16) \div 2$ **40**
11. $(4 + 7)^2 - 8$ **113**
12. $(9 - 3)^2 + 6 \times 2$ **48**

Enrichment 2-1 Exponents and Order of Op L4

Critical Thinking

Rewrite each of the numbers as an expression using addition, subtraction, multiplication, division, and exponents. Use only the numbers 1, 2, 3, and 4 or a combination of these numbers. Use the numbers only once each. You must use at least one operation. Don't forget the order of operations!

1–30 Sample answers given.

1	$4 - 3$	16	4^2
2	$4 - 2$	17	$4^2 + 1$
3	$4 - 1$	18	$4^2 + 3 - 1$
4	$3 + 2 - 1$	19	$4^2 + 3$
5	$3 + 2$	20	$4^2 + 3 + 1$
6	$2 + 4$	21	$4(2 + 3) + 1$
7	$4 + 2 + 1$	22	$24 - 3 + 1$
8	$4 + 1 + 3$	23	$4 \times 3 \times 2 - 1$
9	$4 + 3 + 2$	24	$4 \times 3 \times 2$
10	$4 + 1 + 3 + 2$	25	$4 \times 3 \times 2 + 1$
11	$2 \times 4 + 3$	26	$24 + 3 - 1$
12	$4^2 - (3 + 1)$	27	$(2 + 1)^3$
13	$4^2 - 3$	28	$23 + 4 + 1$
14	$4^2 - 3 + 1$	29	$4^2 + 13$
15	$4^2 - 1$	30	$13 \times 2 + 4$

Test Prep

Resources

For additional practice with a variety of test item formats:

- Test-Taking Strategies, p. 111
- Test Prep, p. 115
- Test-Taking Strategies with Transparencies

Activity Lab

Using a Scientific Calculator

Students use a scientific calculator to simplify expressions. This will help them extend their use of the order of operations in Lesson 2-1.

Guided Instruction

Explain that, to simplify expressions with multiple steps, they must enter the information into the calculator in a specific way to get the correct solution. Ask:

- *What happens after you press the first three keys in the problem in Example 1?* **The calculator displays the total.**
- *Is it necessary to insert the parentheses in Example 2? Why?* **Yes, because without parentheses the calculator first divides 9 by 2 and then subtracts.**

Exercises

Have students work on the Exercises. Review the answers and the calculator keystrokes as a class.

Differentiated Instruction

Below Level L2

Have student solve the problems in the Exercises by hand and by calculator. Help them compare the steps in their calculations with what the calculator does.

Resources

- scientific calculator

2-1b Activity Lab — Technology

Using a Scientific Calculator

Many calculators use the order of operations. To test your calculator, try to compute $3 + 5 \cdot 2$. If the answer is 13, your calculator uses the order of operations.

You can use a scientific calculator to simplify expressions that contain more than one operation.

EXAMPLE **Simplifying Expressions**

1 Find $6 + 18 \div 2$.

6 [+] 18 [÷] 2 [=] 15

The expression $6 + 18 \div 2$ simplifies to 15.

You can use a scientific calculator to simplify expressions that have grouping symbols.

EXAMPLE **Using Grouping Symbols**

2 Find $(5.5 - 9) \div 2$.

[(] 5 [.] 5 [−] 9 [)] [÷] 2 [=] −1.75

The expression simplifies to −1.75.

You can simplify expressions by inserting grouping symbols.

EXAMPLE **Inserting Parentheses**

3 Find $\frac{-7 - 5}{-6 + 3}$.

[(] [(-)] 7 [−] 5 [)] [÷] [(] [(-)] 6 [+] 3 [)] [=] 4 ← **Use the [(-)] key for negative numbers.**

The quotient is 4.

Exercises

Use a calculator to find the value of each expression.

1. $9 + 4 \cdot 2$ **17**

2. $5.6 - 9 \div 2.5$ **2**

3. $(5 - 14.9) \div 3$ **−3.3**

4. $\frac{8.2 - 16.3}{4.5}$ **−1.8**

5. $7.2 \div (4.3 - 3.7)$ **12**

6. $\frac{-9 - 6.2}{2.1 + 2.9}$ **−3.04**

2-2a Activity Lab

Divisibility Tests

One whole number is **divisible** by a second whole number if the remainder is 0 when you divide the first number by the second number. The table below shows tests to determine if a number is divisible by 2, 3, 4, 5, 8, 9, or 10.

EXAMPLES

1 Is 567 divisible by 3?

Yes, 5 + 6 + 7 = 18; 18 is divisible by 3.

2 Is 934 divisible by 4?

No it is not, since 34 is not divisible by 4.

3 Is 29,640 divisible by 8?

Yes it is, because 640 is divisible by 8.

4 Is 3,016 divisible by 9?

No, 3 + 0 + 1 + 6 = 10; 10 is not divisible by 9.

Number	Test for Divisibility
2	Number ends in 0, 2, 4, 6, or 8.
3	Sum of the digits is divisible by 3.
4	The number formed by the last two digits is divisible by 4.
5	Number ends in 0 or 5.
8	The number formed by the last three digits is divisible by 8.
9	Sum of the digits is divisible by 9.
10	Number ends in 0.

Exercises

Tell whether each number is divisible by 2, 3, 4, 5, 8, 9, or 10. Some numbers may be divisible by more than one number.

1. 324 **2, 3, 4, 9**
2. 840 **2, 3, 4, 5, 8, 10**
3. 2,724 **2, 3, 4**
4. 81,816 **2, 3, 4, 8**
5. 7,848 **2, 3, 4, 8, 9**
6. **Games** In an adventure game, you can open the door to a treasure room if you have the correct key. The number of the correct key is divisible by 3 and 4, but not 5. Which color key will open the door? **orange**

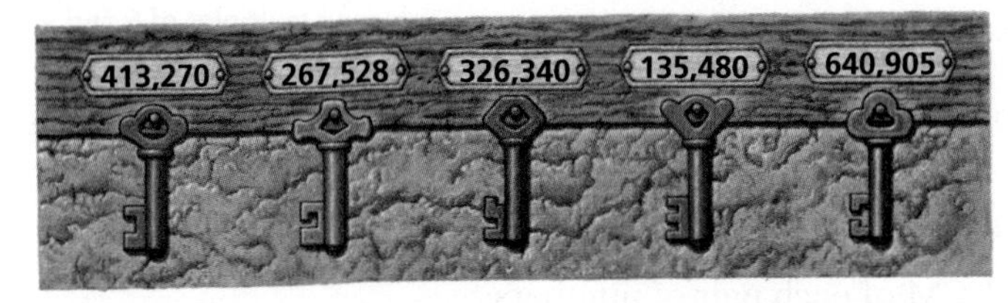

7. A *conjecture* is a prediction that suggests what can be expected to happen. Make a conjecture for the divisibility tests of 12 and 15. **Answers may vary. Sample: A number is divisible by 12 if it is divisible by 3 and 4. A number is divisible by 15 if it is divisible by 3 and 5.**

Activity Lab

Divisibility Tests

Students use rules for divisibility on a large number. This will help them work with prime factorization in Lesson 2-2.

Guided Instruction

Have students study the table. Give the example that all even numbers are divisible by 2 because there is no remainder, but odd numbers are not divisible by 2 because there will always be a remainder of 1.

Ask:

- *Are numbers divisible by 9 also divisible by 3? Why?* **Yes, 9 is divisible by 3 so numbers that divide into 9 parts can also divide into 3 parts.**
- *Are all numbers divisible by 2 also divisible by 4? Why?* **No, some even numbers like 10 are not divisible by 4.**
- *What is a good rule for divisibility by doubles? For perfect squares?* **Sample: A number divisible by an even number is also divisible by half that number. A number divisible by a perfect square is also divisible by that number's square root.**

Exercises

Have students work on the Exercises. Then have them form small groups and compare lists of numbers that each number is divisible by.

Alternative Method

Provide students with base ten manipulatives to model divisibility and help make the connection with the rules of divisibility.

Resources

- Activity Lab 2-2: Factors
- Student Manipulatives Kit

2-2

1. Plan

Objectives
To find multiples and factors and to use prime factorization

Examples
1 Finding the LCM
2 Prime Numbers and Composite Numbers
3 Writing Prime Factorization
4 Finding the GCF

Math Understandings: p. 66C

Math Background

Every integer greater than 1 has exactly one prime factorization. This idea is so important to the study of mathematics that it is sometimes called the *Fundamental Theorem of Arithmetic.* In this lesson, students learn to recognize prime numbers and to find the prime factorization of an integer.

More Math Background: p. 66C

Lesson Planning and Resources

See p. 66E for a list of the resources that support this lesson.

Bell Ringer Practice

Check Skills You'll Need
Use student page, transparency, or PowerPoint. For intervention, direct students to:
Dividing Whole Numbers
Skills Handbook, p. 657

2-2 Prime Facto

Check Skills You'll Need

1. **Vocabulary Review** A whole number is divisible by a second whole number if the remainder after division is ■. 0

Use mental math to find the quotient.

2. 48 ÷ 2 24
3. 63 ÷ 3 21
4. 72 ÷ 6 12

for Help
Skills Handbook, page 657

What You'll Learn

To find multiples and factors

New Vocabulary multiple, l mposite, prime, prime factorization

Why Learn This?

You can use factors and multi solve problems involving sche

Suppose you volunteer every third day at an animal shelter, and your friend volunteers every fourth day. You can use multiples to find the next day when you both will be at the shelter.

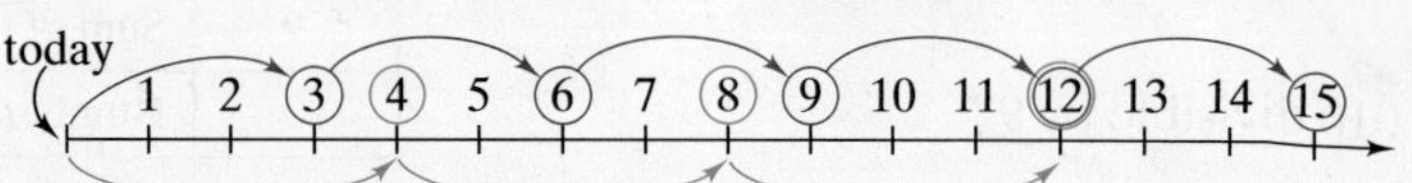

A **multiple** of a number is the product of that number and any nonzero whole number. The diagram above shows multiples of 3 and 4. The **least common multiple (LCM)** of two or more numbers is the least multiple that is common to all of the numbers.

EXAMPLE Finding the LCM

1 Scheduling A fish sandwich is on a school's lunch menu every 6 school days. Spaghetti with meat sauce is on the menu every 9 school days. If both items are on the menu today, when will both be served again?

Find the least common multiple of 6 and 9.

Multiples of 6: 6, 12, 18, 24, 30, 36, . . .
Multiples of 9: 9, 18, 27, 36, . . .
← List the first several multiples of 6 and 9.

The LCM of 6 and 9 is 18. Both will be served again in 18 school days.

Quick Check

1. Find the LCM of each pair of numbers.
   a. 4, 10 20 b. 5, 7 35 c. 12, 15 60

Differentiated Instruction Solutions for All Learners

Special Needs L1
Provide students with a multiplication grid or a table of the multiplication facts. This will help them list the multiples, and figure out the factors, or at least check once they have listed them.

learning style: visual

Below Level L2
Give students several missing-factor exercises like these.

1 • □ = 54 54 2 • □ = 54 27
3 • □ = 54 18 6 • □ = 54 9

learning style: visual

A **factor** is a whole number that divides another whole number with a remainder of 0. Any number is always divisible by all of its factors.

A **composite number** is a whole number that has more than two factors. A **prime number** is a whole number with exactly two factors, 1 and the number itself. The number 1 is neither prime nor composite.

EXAMPLE Prime Numbers and Composite Numbers

2 Determine whether each number is prime or composite.

a. 12

Look for pairs of numbers with a product of 12:

$1 \cdot 12 \quad 2 \cdot 6 \quad 3 \cdot 4$

Then list the factors in order: 1, 2, 3, 4, 6, 12.

Since 12 has factors other than 1 and itself, 12 is a composite number.

b. 13

Look for pairs of numbers with a product of 13: $1 \cdot 13$.

Since 13 has no factors other than 1 and itself, 13 is a prime number.

Quick Check

2. Is 15 prime or composite? Explain. **composite; the factors of 15 are 1, 3, 5 and 15. Since 15 has more than two factors, 15 is composite.**

Writing a composite number as the product of its prime factors shows its **prime factorization.** This product is unique except for the order of the factors. You can use a factor tree to find prime factors.

EXAMPLE Writing Prime Factorization

GO for Help For help with writing expressions involving exponents, go to Lesson 2-1, Example 1.

3 Use a factor tree to write the prime factorization of 60.

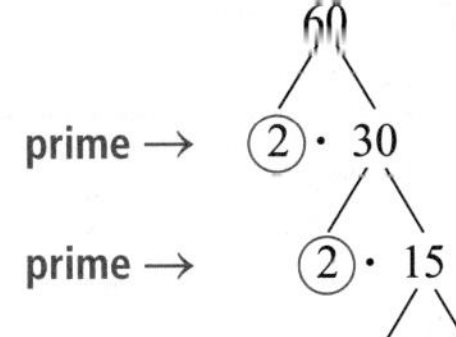

← **Write 60 as the product of any two of its factors.**

← **Write 30 as the product of two factors.**

← **Write 15 as the product of two factors.**

$60 = 2 \cdot 2 \cdot 3 \cdot 5$. Using exponents, you can write $60 = 2^2 \cdot 3 \cdot 5$.

Quick Check

3. Write the prime factorization of 72. Use exponents where possible. $72 = 2^3 \cdot 3^2$

The **greatest common factor (GCF)** of two or more numbers is the greatest number that is a factor of all the numbers.

2. Teach

Activity Lab

Use before the lesson.
Student Edition Activity Lab 2-2a, Divisibility Tests, p. 73

All in One Teaching Resources
Activity Lab 2-2: Factors

Guided Instruction

Example 1
Point out to students that they are already familiar with one special set of multiples: The *even numbers* are all the multiples of 2.

Example 2
Point out that 2 is the only prime number that is an even number. All other even numbers can be written as the product of 2 and some positive integer greater than 1.

Error Prevention!

Students might end their factor trees with one or more composite numbers. For instance, in Quick Check 3, they might write $2^3 \cdot 9$. Remind them that the purpose of the tree is finding a *prime* factorization, so all the ending numbers must be prime.

Advanced Learners L4
Find the GCF and the LCM of each set of numbers.

12, 24, 32 **4; 96**	24, 36, 48 **12; 144**
12, 15, 36 **3; 180**	9, 16, 25 **1; 3,600**

learning style: visual

English Language Learners ELL
Help students restate or rephrase the meaning of *least common multiple.* For example, for *least,* use the words *lowest* or *least value.* For *common,* use the words *the same.* An alternate definition might be: *The multiple with the lowest value that is the same for both numbers.*

learning style: verbal

Additional Examples

PowerPoint

1. Dan goes to the health club every 4 days. His sister Neesa goes there every 6 days. Dan and Neesa both went to the health club today. When will they both go there on the same day again? **12 days**

2. Tell whether each number is prime or composite.
   a. 17 **prime**
   b. 22 **composite**

3. Use a factor tree to write the prime factorization of 90.

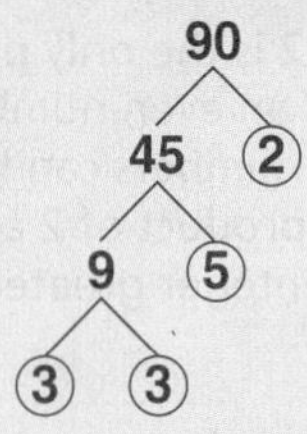

$90 = 2 \cdot 3 \cdot 3 \cdot 5$, or $2 \cdot 3^2 \cdot 5$

4. Find the GCF of 32 and 48. **16**

All in One Teaching Resources

- Daily Notetaking Guide 2-2 L3
- Adapted Notetaking 2-2 L1

Closure

- Refer students to the *New Vocabulary* list on page 74. Ask them to illustrate as many of the vocabulary words as they can using the numbers 8 and 10. **Samples: The multiples of 8 are 8,16, 24, . . . The multiples of 10 are 10, 20, 30, . . . The LCM of 8 and 10 is 40. The factors of 8 are 1, 2, 4, and 8. The factors of 10 are 1, 2, 5, and 10. Both 8 and 10 are composite numbers.**

EXAMPLE Finding the GCF

4 Find the GCF of 24 and 36.

$24 = 2 \cdot 2 \cdot 2 \cdot 3$ $36 = 2 \cdot 2 \cdot 3 \cdot 3$ ← **Write the prime factorizations.**

$\text{GCF} = 2 \cdot 2 \cdot 3 = 12$ ← **Find the product of the common factors.**

The GCF of 24 and 36 is 12.

Quick Check

- **4.** Find the GCF of 16 and 24. **8**

More Than One Way

Two different teams are marching in rows that all contain the same number of people. One team has 32 members and the other has 40. If the rows are as long as possible, how many people are in each row?

I'll write the prime factorizations of 40 and 32. Then I'll find the GCF.

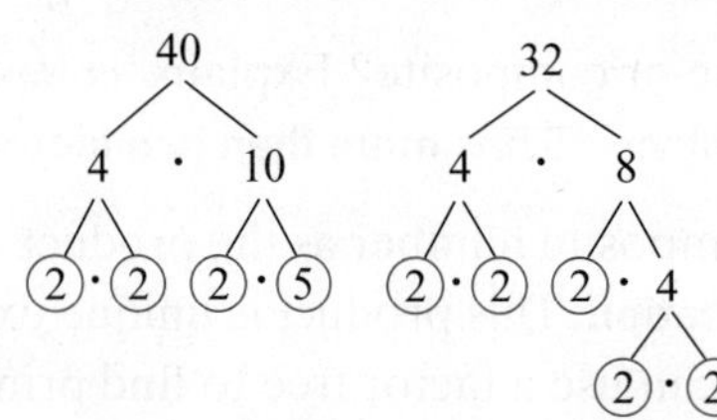

← **Write as products of two factors.**
← **Write as products of two factors.**
← **Write as a product of two factors.**

$40 = 2 \cdot 2 \cdot 2 \cdot 5$
$32 = 2 \cdot 2 \cdot 2 \cdot 2 \cdot 2$ ← **Write the prime factorizations.**

$\text{GCF} = 2 \cdot 2 \cdot 2 = 8$ ← **Multiply the common factors.**

There are 8 people in each row.

Anna's Method

First I'll list the possible sizes of rows. Then I'll choose the greatest number in both lists.

The factors of 32 are 1, 2, 4, 8, 16, and 32.
The factors of 40 are 1, 2, 4, 5, 8, 10, 20, and 40.

The greatest common factor is 8. There are 8 people in each row.

Choose a Method

Teams of 36 and 60 will march in rows of equal length. How many marchers will be in the longest row possible? **12**

Check Your Understanding

1. A factor is a whole number that divides another whole number with a remainder of 0. A multiple of a number is the product of the number and another whole number.

1. **Vocabulary** How is a factor different from a multiple? See left.
2. **Number Sense** Which numbers are factors of all even numbers? 1, 2
3. What is the only even prime number? 2

Is the number prime or composite? Explain. 4–7. See margin.

4. 29 5. 28 6. 27 7. 26

Homework Exercises

For more exercises, see Extra Skills and Word Problems.

GO for Help

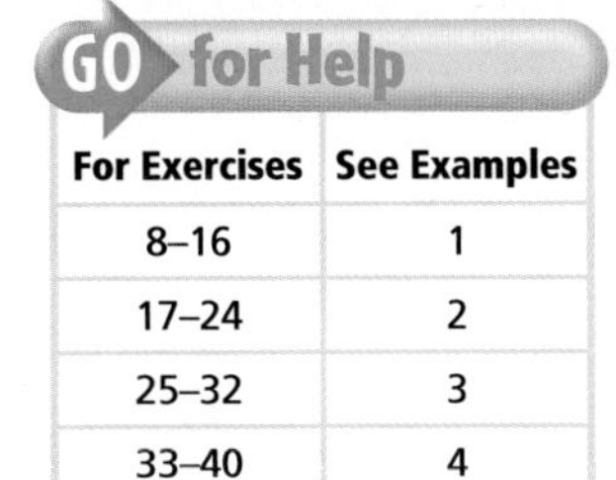

For Exercises	See Examples
8–16	1
17–24	2
25–32	3
33–40	4

A **Find the LCM of each pair of numbers.**

8. 4, 6 12 9. 9, 12 36 10. 8, 5 40 11. 2, 5 10

12. 6, 7 42 13. 5, 10 10 14. 10, 6 30 15. 24, 8 24

16. **Fitness** You have aerobics classes every 3 days and soccer practice every 7 days. Today you had both aerobics and soccer. When will you next have both activities on the same day? in 21 days

Find the factors of each number.

17. 20 1, 2, 4, 5, 10, 20 18. 23 1, 23 19. 32 1, 2, 4, 8, 16, 32 20. 62 1, 2, 31, 62

Determine whether each number is prime or composite.

21. 37 prime 22. 50 composite 23. 1 neither 24. 63 composite

Write the prime factorization of each number. Use exponents.

25. 45 $3^2 \cdot 5$ 26. 64 2^6 27. 84 $2^2 \cdot 3 \cdot 7$ 28. 111 $3 \cdot 37$

29. 52 $2^2 \cdot 13$ 30. 75 $3 \cdot 5^2$ 31. 60 $2^2 \cdot 3 \cdot 5$ 32. 132 $2^2 \cdot 3 \cdot 11$

Find the GCF of each pair of numbers.

33. 18, 32 2 34. 12, 15 3 35. 16, 80 16 36. 10, 85 5

37. 38, 76 38 38. 75, 90 15 39. 54, 80 2 40. 52, 26 26

B **GPS** 41. **Guided Problem Solving** In his art class, Raul made two rectangular mosaics with 1-cm tiles. He used 48 tiles for one mosaic and 56 tiles for the other. Both mosaics have the same length, measured in whole centimeters. What is the greatest possible length of each mosaic? 8 cm

- What are the possible lengths of each mosaic?
- What are the corresponding widths for each mosaic?

4. Prime; 1 and 29 are the only factors.
5. Composite; the factors are 1, 2, 4, 7, 14 and 28.
6. Composite; the factors are 1, 3, 9 and 27.
7. Composite; the factors are 1, 2, 13 and 26.

3. Practice

Assignment Guide

Check Your Understanding
Go over Exercises 1–7 in class before assigning the Homework Exercises.

Homework Exercises

A Practice by Example	8–40
B Apply Your Skills	41–51
C Challenge	52
Test Prep and Mixed Review	53–57

Homework Quick Check
To check students' understanding of key skills and concepts, go over Exercises 19, 35, 43, 48, and 51.

Differentiated Instruction Resources

Adapted Practice 2-2 L1

Practice 2-2 Prime Factor L3

Find the LCM of each pair of numbers.

1. 11, 5 55 2. 5, 12 60 3. 12, 7 84
4. 5, 9 45 5. 5, 18 90 6. 5, 20 20
7. 7, 10 70 8. 17, 13 221 9. 14, 8 56
10. 11, 23 253 11. 14, 5 70 12. 16, 9 144

13. Cameron is making bead necklaces. He has 90 green beads and 108 blue beads. What is the greatest number of identical necklaces he can make if he wants to use all of the beads? 18 necklaces

14. One radio station broadcasts a weather forecast every 18 minutes and another station broadcasts a commercial every 15 minutes. If the stations broadcast both a weather forecast and a commercial at noon, when is the next time that both will be broadcast at the same time? at 1:30 P.M.

Determine whether each number is prime or composite.

15. 97 prime 16. 63 composite 17. 29 prime 18. 120 composite

Write the prime factorization. Use exponents where possible.

19. 42 $2 \times 3 \times 7$ 20. 130 $2 \times 5 \times 13$
21. 78 [illegible] 22. 128 [illegible]
23. 125 5^3 24. 90 $2 \times 3^2 \times 5$
25. 92 $2^2 \times 23$ 26. 180 $2^2 \times 3^2 \times 5$

Find the GCF of each pair of numbers.

27. 45, 60 15 28. 18, 42 6 29. 32, 80 16
30. 20, 65 5 31. 24, 90 6 32. 17, 34 17
33. 14, 35 7 34. 51, 27 3 35. 42, 63 7

2-2 • Guided Problem Solving GPS L3

Student Page 78, Exercise 48:

A movie theatre is adding two rooms. One room is large enough for 125 people, and the other can seat up to 350 people. In each room, the seating is arranged in horizontal rows with the same number of seats in each row. What is the greatest number of seats that could make up each row?

Understand

1. Circle the information you will need to solve.
2. What do you need to do to answer the question? Find the greatest common factor of 350 and 125.

Plan and Carry Out

3. List the prime factors of 350. $2 \cdot 5 \cdot 5 \cdot 7$
4. List the prime factors of 125. $5 \cdot 5 \cdot 5$
5. List the factors that 350 and 125 have in common. $5 \cdot 5$
6. What is the greatest common factor of 350 and 125? 25
7. What is the largest number of seats that could make up each row? 25 seats

Check

8. What is $350 \div 25$? What is $125 \div 25$? Do these quotients have any common factors besides 1? 14; 5; no

Solve Another Problem

9. For graduation, the left side of the gymnasium can seat 228 people and the right side can seat 144 people. The principal wants the same number of chairs in each row on both sides. How many chairs does the setup committee need to put in each row? 12 chairs

4. Assess & Reteach

Lesson Quiz

Write the prime factorization. Use exponents.

1. 24 $2^3 \cdot 3$
2. 20 $2^2 \cdot 5$
3. 63 $3^2 \cdot 7$
4. 84 $2^2 \cdot 3 \cdot 7$

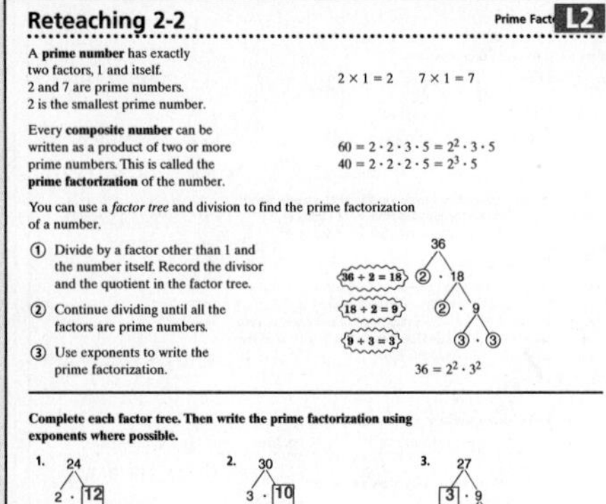

Reteaching 2-2

Prime Factorization L2

A **prime number** has exactly two factors, 1 and itself. 2 and 7 are prime numbers. 2 is the smallest prime number. $2 \times 1 = 2$ $7 \times 1 = 7$

Every **composite number** can be written as a product of two or more prime numbers. This is called the **prime factorization** of the number.

$60 = 2 \cdot 2 \cdot 3 \cdot 5 = 2^2 \cdot 3 \cdot 5$
$40 = 2 \cdot 2 \cdot 2 \cdot 5 = 2^3 \cdot 5$

You can use a *factor tree* and division to find the prime factorization of a number.

① Divide by a factor other than 1 and the number itself. Record the divisor and the quotient in the factor tree.

② Continue dividing until all the factors are prime numbers.

③ Use exponents to write the prime factorization.

$36 \div 2 = 18$ $18 \div 2 = 9$ $9 \div 3 = 3$

$36 = 2^2 \cdot 3^2$

Complete each factor tree. Then write the prime factorization using exponents where possible.

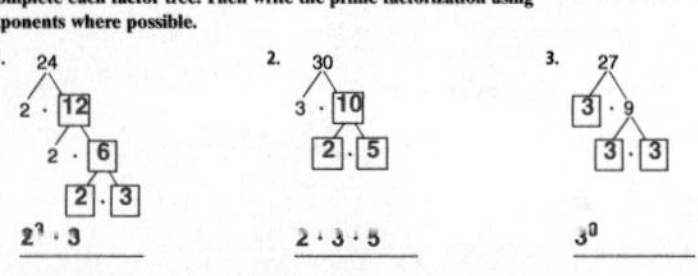

1. 24 — $2^3 \cdot 3$
2. 30 — $2 \cdot 3 \cdot 5$
3. 27 — 3^3

Write the prime factorization of each number. Use exponents where possible.

4. $20 = 2^2 \times 5$
5. $54 = 2 \times 3^3$
6. $40 = 2^3 \times 5$
7. $48 = 2^4 \times 3$
8. $56 = 2^3 \times 7$
9. $150 = 2 \times 3 \times 5^2$

Enrichment 2-2

Prime Factorization L4

Decision Making

Mr. Jones wants to arrange 24 desks into groups so that students can work together. He wants each group to be the same size.

1. What different sizes of groups can Mr. Jones use?
   2 groups of 12, 3 of 8, 4 of 6, 6 of 4, 8 of 3, 12 of 2
2. What does Mr. Jones need to consider when he moves the desks into the groups?
   Sample answer: whether or not all students can see the chalkboard; whether or not all group members can communicate; whether or not there is enough aisle space
3. The current project is a class debate. There will be four teams. In the space below, show two ways that Mr. Jones can arrange the desks so that students on each team will be sitting together. Sample answers:

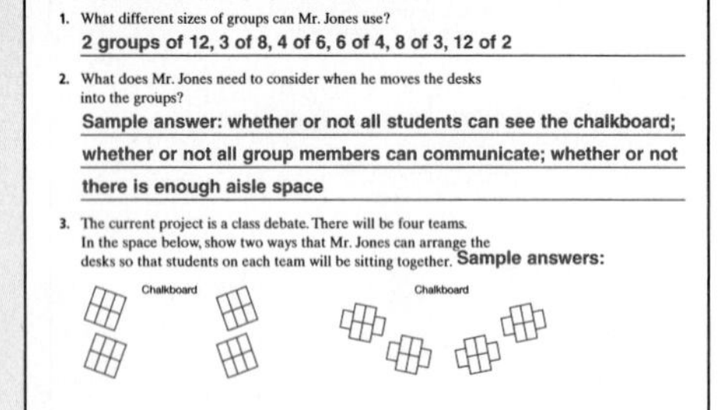

4. For the next project, each group will perform a skit showing a typical day in the life of a teenager in another country. What would be the best group sizes for this project? Explain.
   Sample answer: groups of eight, so that there will be sufficient people to research the topic and write and act in the skit
5. How can Mr. Jones arrange the desks for students to work on the skit? Use the space to the right to show the arrangement. Why is this the best seating arrangement?

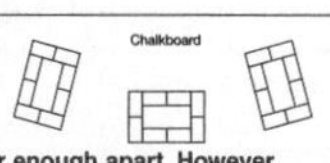

   Sample answer: It keeps the groups far enough apart. However, some students will have to move to see the chalkboard.

GO Online
Homework Video Tutor
Visit: PHSchool.com
Web Code: are-0202

49. Yes; both can be written as $2^3 \cdot 3$.

50. No; 10 is not a prime. The prime factorization of 100 is $2^2 \cdot 5^2$.

Mental Math **Find the GCF of each pair of numbers.**

42. 3, 10 1
43. 7, 12 1
44. 4, 20 4
45. 50, 1000 50

46. Find two composite numbers with a GCF of 1. Answers may vary. Sample: 21 and 22

47. **Choose a Method** You can buy juice boxes in packs of 12 or 30. Both packs contain the same number of boxes in each row. What is the greatest possible number of boxes per row? Describe your method and explain why you chose it. 6 boxes

48. A movie theater just added two rooms. One room is large enough for 125 people, and the other can seat up to 350 people. In each room, the seating is arranged in horizontal rows with the same number of seats in each row. What is the greatest number of seats that can make up each row? 25 seats

GPS

49. **Error Analysis** Two students made factor trees of the prime factors of 24. Are both correct? Explain. See left.

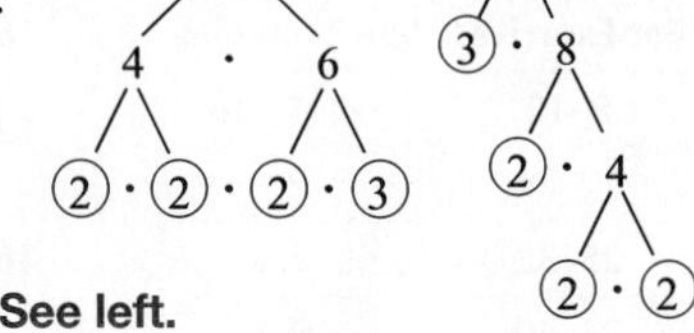

50. You can express 100 as 10^2 using exponents. Is 10^2 the same as the prime factorization of 100? Explain. See left.

51. **Writing in Math** Describe the relationships among 3, 5, and 15 using the words *factor* and *multiple*. Answers may vary. Sample: 15 is the LCM of 3 and 5, and 3 and 5 are factors of 15.

Ⓒ 52. **Challenge** Let n be any prime number. Tell whether the statement "$2n + 1$ is prime" is *sometimes, always,* or *never* true. sometimes

Test Prep and Mixed Review

Practice

Multiple Choice

53. The Stillwater High School library is arranging 108 books. Each shelf can hold as many as 10 books, and the librarian wants to place the same number of books on each shelf. What is the greatest number of books that can be placed on each shelf? B
    Ⓐ 2 Ⓑ 9 Ⓒ 10 Ⓓ 12

54. The table shows how far model cars traveled after rolling down an incline. How many centimeters farther did car 2 travel than car 4 and car 5 combined? G
    Ⓕ 16.71 Ⓗ 42.19
    Ⓖ 17.88 Ⓙ 58.9

Car	Distance (cm)
1	52.64
2	83.21
3	66.86
4	24.31
5	41.02

GO for Help

For Exercises	See Lesson
55–57	1-6

Order from least to greatest.

55. 3, −4, −5, 6
    −5, −4, 3, 6
56. −7, −10, −13
    −13, −10, −7
57. 20, −21, 21, 0
    −21, 0, 20, 21

Test Prep

Resources

For additional practice with a variety of test item formats:

- Test-Taking Strategies, p. 111
- Test Prep, p. 115
- Test-Taking Strategies with Transparencies

Alternative Assessment

Have students draw factor trees to find the prime factorization of composite numbers such as 32, 450, 396, and 384. When assisting students, ask: *What two numbers multiplied have a product of 32?* 1, 32; 2, 16; 4, 8 Make sure students arrive at a prime factorization. 2^5 Remind them to use exponents in their final answer.

Checkpoint Quiz 1

Lessons 2-1 through 2-2

Simplify each expression.

1. $8^2 + 11$ 75
2. $(-1)^4$ 1
3. $5 + (3^2 - 2)^2$ 54

Find the LCM of each pair of numbers.

4. 3, 4 12
5. 7, 10 70
6. 5, 12 60

Find the factors of each number.

7. 39 1, 3, 13, 39
8. 52 1, 2, 4, 13, 26, 52
9. 110 1, 2, 5, 10, 11, 22, 55, 110
10. 200 See margin.

Write the prime factorization of each number. Use exponents where possible.

11. 96 $2^5 \cdot 3$
12. 150 $2 \cdot 3 \cdot 5^2$
13. 225 $3^2 \cdot 5^2$
14. 333 $3^2 \cdot 37$

15. Two pieces of rope have lengths 72 ft and 96 ft. A woodcutter needs to cut the two pieces into smaller pieces all of equal lengths. What is the greatest possible length of each smaller piece? 24 ft

MATH GAMES

Factor Cards

What You'll Need

- 40 index cards, numbered from 1 to 40

How To Play

- Form two teams.
- Team A chooses a card at random.
- Team B removes the cards from the deck that are factors of Team A's card. For example, if Team A picks 10, Team B takes cards 1, 2, and 5.
- Teams switch places until all of the cards are taken. No points are given for cards that have already been chosen. The team whose cards have the highest sum wins.

10. 1, 2, 4, 5, 8, 10, 20, 25, 40, 50, 100, 200

Checkpoint Quiz

Use this Checkpoint Quiz to check students' understanding of the skills and concepts of Lessons 2-1 through 2-2.

Resources

- All-In-One Teaching Resources Checkpoint Quiz 1
- ExamView CD-ROM
- Success Tracker™ Online Intervention

Factor Cards

In this game, students practice finding the factors of numbers. One team chooses a number from 1–40 and the other team must find all the factors of that number. The game continues until all numbers from 1–40 are used.

Guided Instruction

Before they play the game, remind students of the meaning of factors. Have a volunteer read the rules. Explain that if factor-finding teams must find the factors that remain in the deck, some factors may already have been taken by the other team on a previous turn.

Resources

- 40 index cards, numbered 1–40

Using LCM and GCF

In this feature, students practice solving problems involving the least common multiple and the greatest common factor.

Guided Instruction

Remind students of the meanings of LCM and GCF. Have a volunteer read the question aloud.
Ask:

- *Why do we need to find the LCM of 6 and 8?* **Multiples of 6 will get the free popcorn, multiples of 8 will get the free rental, the LCM of 6 and 8 will be the first to get both.**
- *How would you find the next person to get both free items?* **Find the next common multiple of 6 and 8.**
- *How can you check the reasonableness of your answer?* **Check to see if 24 is divisible by 6 and 8.**

Exercises

Have students work independently on the Exercises. Circulate and help students figure out what they need to know and how they can find it. Compare answers as a class and discuss reasoning.

GPS Guided Problem Solving

Using LCM and GCF

Movie Rentals To boost sales, a video rental store offered a free bag of microwave popcorn to every sixth customer and a free movie rental to every eighth customer. What customer was the first to win both popcorn *and* a free movie rental?

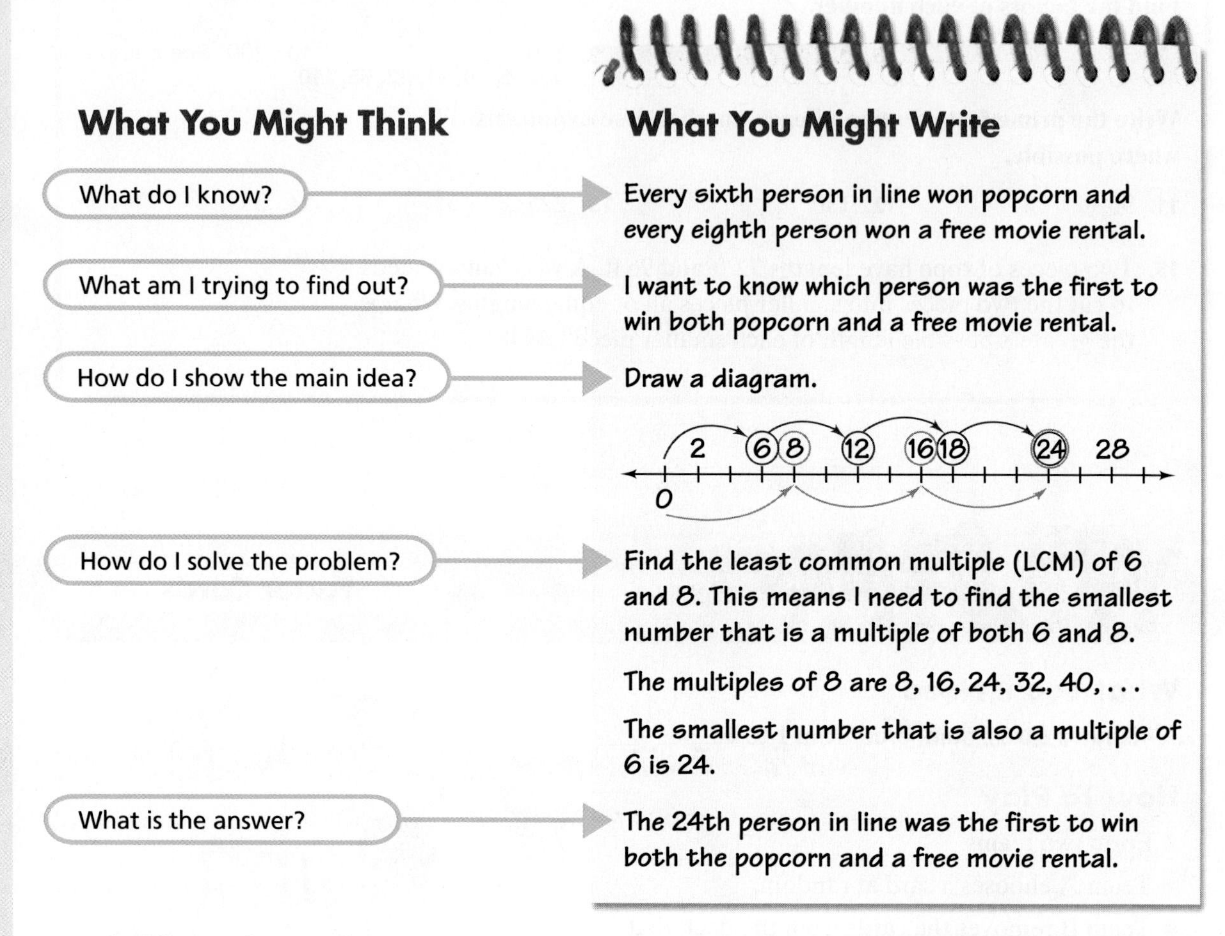

What You Might Think
What You Might Write
What do I know?
Every sixth person in line won popcorn and every eighth person won a free movie rental.
What am I trying to find out?
I want to know which person was the first to win both popcorn and a free movie rental.
How do I show the main idea?
Draw a diagram.
0
2
6
8
12
16
18
24
28
How do I solve the problem?
Find the least common multiple (LCM) of 6 and 8. This means I need to find the smallest number that is a multiple of both 6 and 8.
The multiples of 8 are 8, 16, 24, 32, 40, . . .
The smallest number that is also a multiple of 6 is 24.
What is the answer?
The 24th person in line was the first to win both the popcorn and a free movie rental.

Think It Through

1. Explain how the diagram shows multiples of 6 *and* multiples of 8. **The arrows on top point to multiples of 6, and the arrows below the line point to multiples of 8.**
2. Explain why it makes sense to look first at the multiples of the larger number when finding the LCM. **By looking at the multiples of the larger number, you don't have to check as many multiples.**
3. What is the number of the second person in line to win both prizes? The third person in line to win both prizes? **48; 72**

Exercises

For Exercises 4 and 5, answer the questions first. Then solve the problem.

4. As a promotion, a "hot wings" restaurant gave away free boxes according to the sign below.

HOT WING BOXES
6 wings
10 wings
15 wings
FREE to this customer
every 6th in line
every 10th in line
every 15th in line

Which person in line was the first to get boxes of all three sizes? **the 30th person**

a. What do you know?
b. What do you want to find out?
c. Draw a diagram that shows people winning different boxes.
d. How will finding the LCM of three numbers help you?

5. Southside Middle School held a tug-of-war contest. Seventy-eight seventh graders and seventy-two eighth graders signed up. Each grade was divided into several teams. If both grades had the same number of people on each team, what was the greatest possible number of people on a team? **6 people**

a. What do you know?
b. What do you want to find out?
c. How will finding the GCF of 72 and 78 help you?

6. Jeremiah wants to make a square pattern for an art project using tiles 8 cm by 12 cm. What is the smallest square he can design using whole tiles? **24 cm by 24 cm**

7. There are 252 beats in one passage of a musical composition. In this passage, the triangle player plays once every 12 beats. The timpani player plays once every 9 beats. How many times will they play at the same time during this passage? **7 times**

8. All the school lockers are closed at the beginning of the school day. Wayne is the first person to come into school, and he opens every tenth locker. Jake comes in after Wayne and switches every eighth locker from open to closed, or vice versa. Which locker is the first to be opened and then closed? **the 40th locker**

Differentiated Instruction

Below Level L2

If students have difficulty identifying the LCM or GCF, allow them to list multiples and factors.

2-3

1. Plan

Objective
To write equivalent fractions and to simplify fractions

Examples
1 Using Multiples
2 Using Factors
3 Simplifying by Dividing
4 Using the GCF to Simplify a Fraction

Math Understandings: p. 66C

Math Background

A *fraction* represents part of a whole. The whole could be a single object or region, as in "one fourth of a pizza." The whole might also be a group of objects, as in "half of a dozen eggs." Two fractions that represent the same part of a whole, such as $\frac{3}{4}$ and $\frac{6}{8}$, are called *equivalent fractions.*

More Math Background: p. 66C

Lesson Planning and Resources

See p. 66E for a list of the resources that support this lesson.

Bell Ringer Practice

Check Skills You'll Need
Use student page, transparency, or PowerPoint. For intervention, direct students to:
Prime Factorization
Lesson 2-2
Extra Skills and Word Problems Practice, Ch. 2

2-3 Simplifying Fractions

Check Skills You'll Need

1. **Vocabulary Review** What is special about the GCF of a pair of numbers? **See below.**

Find the GCF of each pair of numbers.

2. 6, 10 **2**
3. 12, 24 **12**
4. 45, 50 **5**
5. 17, 23 **1**

GO for Help Lesson 2-2

Check Skills You'll Need
1. **It is the greatest factor that divides both numbers.**

What You'll Learn

To write equivalent fractions and to simplify fractions

New Vocabulary equivalent fractions, simplest form

Why Learn This?

Suppose you have three identical chocolate bars. You break one bar into 4 pieces and give away 3 of them. You break the second bar into 8 pieces and give away 6. You break the third bar into 12 pieces and give away 9. The model below shows that you give away the same fraction of each bar.

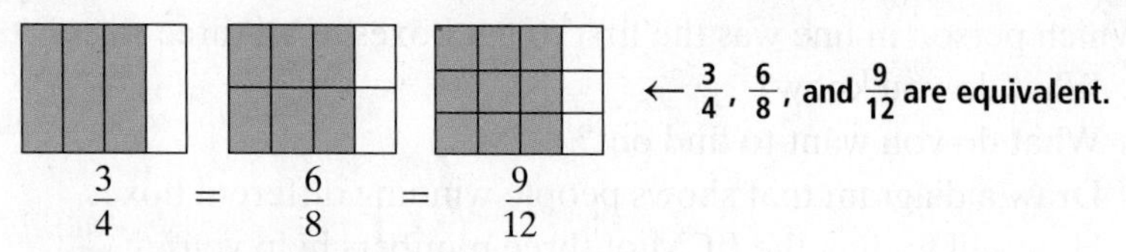

$\frac{3}{4} = \frac{6}{8} = \frac{9}{12}$

Fractions that name the same amount are **equivalent fractions.** You can write equivalent fractions by multiplying or dividing the numerator and the denominator by the same nonzero number.

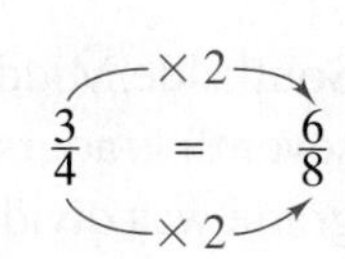

EXAMPLE Using Multiples

1 Use a table of multiples to write three fractions equivalent to $\frac{7}{8}$.

	×2	×3	×4
7	14	21	28
8	16	24	32

← Multiples in the same column form fractions equivalent to $\frac{7}{8}$.

Three fractions equivalent to $\frac{7}{8}$ are $\frac{14}{16}$, $\frac{21}{24}$, and $\frac{28}{32}$.

Quick Check

1. Use multiples to write two fractions equivalent to $\frac{4}{5}$. **Answers may vary. Sample: $\frac{8}{10}$, $\frac{12}{15}$**

A fraction is written in **simplest form** when the numerator and the denominator have no common factors other than 1. For example, $\frac{1}{3}$ and $\frac{3}{9}$ are equivalent, but only $\frac{1}{3}$ is written in simplest form.

Differentiated Instruction Solutions for All Learners

Special Needs L1	**Below Level** L2
Students can temporarily use a multiplication grid or table of multiples to find the Greatest Common Factor. This will enable them to work the process of simplifying a fraction without getting bogged down in figuring out the GCF. Eventually take the table or grid away. **learning style: visual**	Have students use fraction bars to model several sets of equivalent fractions, such as $\frac{1}{2}$, $\frac{2}{4}$, $\frac{5}{10}$. **learning style: tactile**

EXAMPLE Using Factors

2 Write three fractions equivalent to $\frac{24}{30}$.

Factors of 24: 1, 2, 3, 4, 6, 8, 12, 24

Factors of 30: 1, 2, 3, 5, 6, 10, 15, 30

List the factors of each number. Look for common factors.

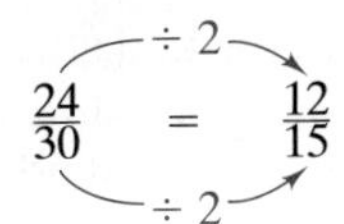

$\frac{24}{30} = \frac{12}{15}$ (÷ 2)

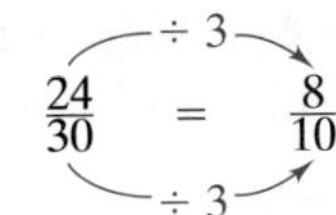

$\frac{24}{30} = \frac{8}{10}$ (÷ 3)

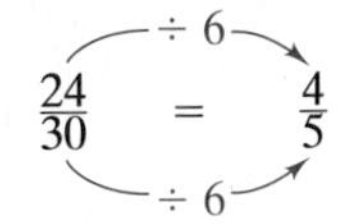

$\frac{24}{30} = \frac{4}{5}$ (÷ 6)

Three fractions equivalent to $\frac{24}{30}$ are $\frac{12}{15}$, $\frac{8}{10}$, and $\frac{4}{5}$.

Quick Check

2. Use common factors to write two fractions equivalent to $\frac{18}{30}$.
Answers may vary. Sample: $\frac{9}{15}$, $\frac{3}{5}$

EXAMPLE Simplifying by Dividing

Vocabulary Tip
Another term for simplest form is lowest terms.

3 Simplify $\frac{12}{24}$.

$\frac{12 \div 2}{24 \div 2} = \frac{6}{12}$ ← **Divide the numerator and denominator by a common factor.**

$\frac{6 \div 6}{12 \div 6} = \frac{1}{2}$ ← **If necessary, divide again by another common factor.**

In simplest form, $\frac{12}{24}$ is $\frac{1}{2}$.

Quick Check

3. Write $\frac{8}{12}$ in simplest form. $\frac{2}{3}$

EXAMPLE Using the GCF to Simplify a Fraction

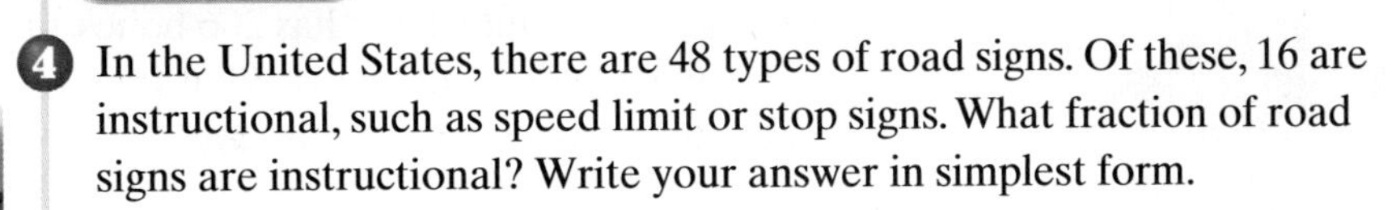

4 In the United States, there are 48 types of road signs. Of these, 16 are instructional, such as speed limit or stop signs. What fraction of road signs are instructional? Write your answer in simplest form.

$\frac{16}{48} = \frac{16 \div 16}{48 \div 16} = \frac{1}{3}$ ← **Divide both numerator and denominator by the GCF, 16.**

The fraction of road signs that are instructional is $\frac{1}{3}$.

Quick Check

4. Your class ordered 45 calculators. Of these, 18 were solar powered. What fraction of the calculators were solar powered? $\frac{2}{5}$

2. Teach

Activity Lab

Use before the lesson.

All in One Teaching Resources

Activity Lab 2-3: Fractions and Surveys

Guided Instruction

Example 1

Give students a sheet with four congruent circles that are divided respectively into 8, 16, 24, and 32 equal parts. Tell students to shade 7 of the 8 equal parts in the first circle. Then instruct them to shade the same area in each of the other three circles. Have them label each circle with the appropriate fraction equivalent to $\frac{7}{8}$. $\frac{7}{8}$, $\frac{14}{16}$, $\frac{21}{24}$, $\frac{28}{32}$

PowerPoint

Additional Examples

1. Use a table of multiples to write three fractions equivalent to $\frac{4}{5}$. **Samples: $\frac{8}{10}$, $\frac{12}{15}$, $\frac{16}{20}$**
2. Find three fractions equivalent to $\frac{16}{24}$. **Samples: $\frac{8}{12}$, $\frac{4}{6}$, $\frac{2}{3}$**
3. Write $\frac{18}{30}$ in simplest form. $\frac{3}{5}$
4. There are 28 students in Mai's homeroom. Of these, 20 students have a pet. What fraction of the students have a pet? Write your answer in simplest form. $\frac{5}{7}$

All in One Teaching Resources

- Daily Notetaking Guide 2-3 **L3**
- Adapted Notetaking 2-3 **L1**

Closure

- *What are equivalent fractions?* **fractions that name the same amount**
- *When is a fraction in simplest form?* **when the numerator and denominator have no common factors other than 1**

Advanced Learners L4

Separate these into two groups of equivalent fractions.

a. $\frac{48}{108}$ b. $\frac{54}{126}$ c. $\frac{64}{144}$ d. $\frac{72}{168}$ e. $\frac{144}{336}$

a and c; b, d, and e

learning style: visual

English Language Learners ELL

Connect the words *simplest form* to the idea of simplifying a fraction. Remind students that when they simplify a fraction they are reducing it to its simplest form.

learning style: verbal

3. Practice

Assignment Guide

Check Your Understanding

Go over Exercises 1–6 in class before assigning the Homework Exercises.

Homework Exercises

A	Practice by Example	7–28
B	Apply Your Skills	29–36
C	Challenge	37
	Test Prep and Mixed Review	38–42

Homework Quick Check

To check students' understanding of key skills and concepts, go over Exercises 17, 25, 28, 34, and 36.

Differentiated Instruction Resources

Adapted Practice 2-3 L1

Practice 2-3 Simplifying L3

Write each fraction in simplest form.

1. $\frac{8}{12}$ $\frac{2}{3}$
2. $\frac{9}{15}$ $\frac{3}{5}$
3. $\frac{16}{20}$ $\frac{4}{5}$
4. $\frac{20}{25}$ $\frac{4}{5}$
5. $\frac{15}{18}$ $\frac{5}{6}$
6. $\frac{14}{30}$ $\frac{7}{15}$
7. $\frac{11}{44}$ $\frac{1}{4}$
8. $\frac{24}{36}$ $\frac{2}{3}$

Write each fraction in simplest form. Give the GCF of the numerator and denominator.

9. $\frac{125}{200}$ $\frac{5}{8}$ GCF = 25
10. $\frac{36}{64}$ $\frac{9}{16}$ GCF = 4
11. $\frac{65}{90}$ $\frac{13}{8}$ GCF = 5
12. $\frac{45}{72}$ $\frac{5}{8}$ GCF = 9
13. $\frac{35}{85}$ $\frac{7}{17}$ GCF = 5
14. $\frac{30}{42}$ $\frac{5}{7}$ GCF = 6

Solve.

15. Emily exercised from 4:05 P.M. to 4:32 P.M. For what part of an hour did Emily exercise? Write the fraction in simplest form. $\frac{9}{20}$ h
16. Luis rode his bike after school for 48 min. For what part of an hour did he ride his bike? Write the fraction in simplest form. $\frac{4}{5}$ h
17. Philip played video games for 55 min before dinner. For what part of an hour did he play? $\frac{11}{12}$ h
18. What part of an hour is your school lunch time? Check students' answers
19. Survey 12 people to find their favorite kind of pizza from the following choices. Write the results in fraction form. Then shade the pizza shapes using different colors to indicate their choices.

Pizza Favorites

Cheese Check students' answers

Green Pepper

Olive

Mushroom

2-3 • Guided Problem Solving GPS L3

Student Page 84, Exercise 28:

Weather The city of Houston, Texas, typically has 75 clear days out of the 365 days in a year. Houston's clear days represent what fraction of a year? Write your answer in simplest form.

Understand

1. Circle the information you will need to solve.
2. What are you being asked to do? Write the number of clear days out of a year in Houston as a fraction in simplest form.

Plan and Carry Out

3. Write the fraction for the expression 75 out of 365. $\frac{75}{365}$
4. List the prime factors of 75. $5 \cdot 5 \cdot 3$
5. List the prime factors of 365. $5 \cdot 73$
6. List the factors that 75 and 365 have in common. 5
7. Divide both 75 and 365 by the common factors. 15; 73
8. Write the fraction in simplest form. $\frac{15}{73}$

Check

9. Is $365 \div 75$ the same as $73 \div 15$? Explain. yes; $365 \div 75 = (73 \times 5) \div (15 \times 5) = 73 \div 15$

Solve Another Problem

10. Gerald received a score of 66 out of 72 on his vocabulary test. Write his score as a fraction in simplest form. $\frac{11}{12}$

Check Your Understanding

1. **Vocabulary** Can two equivalent fractions both be fractions written in simplest form? Explain. See left.

   1. No; the GCF of the numerator and denominator of a fraction in simplest form is 1.

2. Are $\frac{2}{4}$ and $\frac{8}{8}$ equivalent? Explain. No; $\frac{2}{4} = \frac{1}{2}$ and $\frac{8}{8} = \frac{1}{1}$.
3. **Error Analysis** A teacher asked students to find a fraction equivalent to $\frac{5}{6}$. One answer is below. Is it correct? Explain.

   $$\frac{5}{6} = \frac{5+4}{6+4} = \frac{9}{10}$$

   No; the student added 4 instead of multiplying by 4.

For each term, write three equivalent fractions, one in simplest form.

4. $\frac{3}{9}$ $\frac{1}{3}, \frac{6}{18}, \frac{9}{27}$
5. $\frac{36}{48}$ $\frac{3}{4}, \frac{12}{16}, \frac{9}{12}$
6. $\frac{16}{36}$ $\frac{4}{9}, \frac{8}{18}, \frac{32}{72}$

4–22. Answers may vary. Samples are given.

Homework Exercises

For more exercises, see Extra Skills and Word Problems.

GO for Help

For Exercises	See Examples
7–14	1
15–22	2
23–28	3–4

A **Use multiples to write two fractions equivalent to each fraction.**

7. $\frac{5}{6}$ $\frac{10}{12}, \frac{15}{18}$
8. $\frac{3}{8}$ $\frac{6}{16}, \frac{9}{24}$
9. $\frac{2}{9}$ $\frac{4}{18}, \frac{6}{27}$
10. $\frac{7}{10}$ $\frac{14}{20}, \frac{21}{30}$
11. $\frac{4}{7}$ $\frac{8}{14}, \frac{12}{21}$
12. $\frac{3}{5}$ $\frac{6}{10}, \frac{9}{15}$
13. $\frac{6}{11}$ $\frac{12}{22}, \frac{18}{33}$
14. $\frac{1}{5}$ $\frac{2}{10}, \frac{3}{15}$

Use common factors to write two fractions equivalent to each fraction.

15. $\frac{8}{24}$ $\frac{4}{12}, \frac{1}{3}$
16. $\frac{18}{36}$ $\frac{9}{18}, \frac{1}{2}$
17. $\frac{27}{81}$ $\frac{9}{27}, \frac{1}{3}$
18. $\frac{60}{140}$ $\frac{30}{70}, \frac{3}{7}$
19. $\frac{30}{42}$ $\frac{15}{21}, \frac{5}{7}$
20. $\frac{45}{90}$ $\frac{9}{18}, \frac{1}{2}$
21. $\frac{24}{84}$ $\frac{12}{42}, \frac{2}{7}$
22. $\frac{36}{80}$ $\frac{18}{40}, \frac{9}{20}$

Write each fraction in simplest form.

23. $\frac{24}{32}$ $\frac{3}{4}$
24. $\frac{18}{27}$ $\frac{2}{3}$
25. $\frac{33}{39}$ $\frac{11}{13}$
26. $\frac{8}{18}$ $\frac{4}{9}$
27. **Biology** An adult's body has 206 bones. Of these, 106 are in the feet, ankles, wrist, and hands. What fraction of an adult's bones are in the feet, ankles, wrists, and hands? $\frac{53}{103}$
28. **Weather** (GPS) The city of Houston, Texas, typically has 90 clear days out of the 365 days in a year. Houston's clear days represent what fraction of a year? Write your answer in simplest form. $\frac{18}{73}$

B GPS

29. **Guided Problem Solving** A school offers two summer sports: lacrosse and tennis. There are 438 students in all. Of this number, 52 participate in lacrosse and 94 participate in tennis. Nobody does both. What fraction of the students participate in a sport? $\frac{1}{3}$
    - The number of students who participate in a summer sport is ■.
    - The total number of students at the school is ■.

34. Use divisibility rules to tell if the numerators and denominators are divisible by the same number(s). For the fractions $\frac{9}{16}$, $\frac{10}{24}$, and $\frac{15}{35}$, 9 and 16 have no common factors, 10 and 24 are both divisible by 2, and 15 and 35 are both divisible by 5. So $\frac{9}{16}$ is in simplest form, but $\frac{10}{24}$ and $\frac{15}{35}$ are not.

36. Answers may vary. Sample:

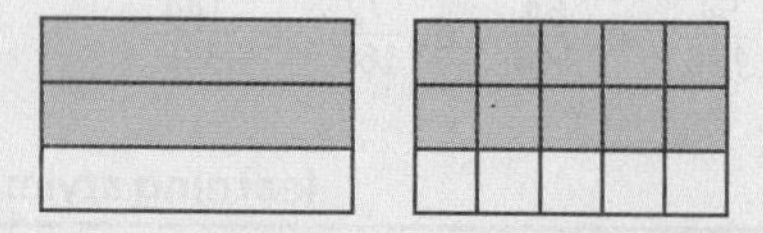

Visit: PHSchool.com
Web Code: are-0203

Write two equivalent fractions for each model.
30–32. Answers may vary. Samples are given.

30. 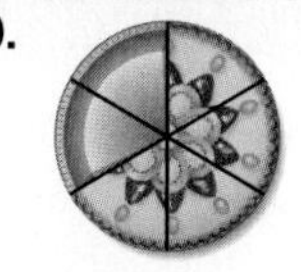
$\frac{4}{6}, \frac{2}{3}$

31. 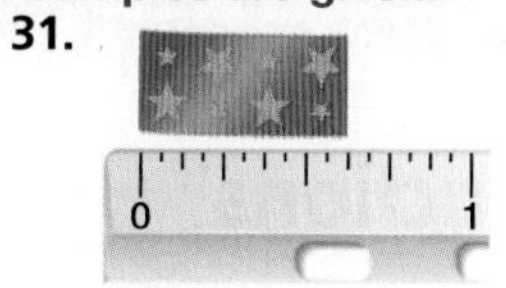

$\frac{5}{8}, \frac{10}{16}$

32. 

$\frac{1}{2}, \frac{2}{4}$

33. Which square does *not* have the same fraction shaded as the others? **D**

A. 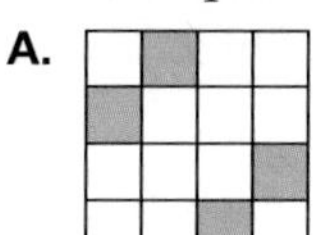
B. 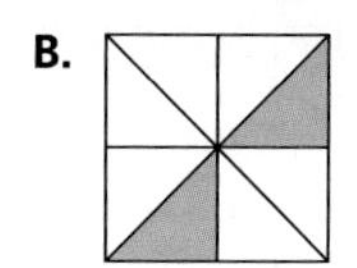
C. 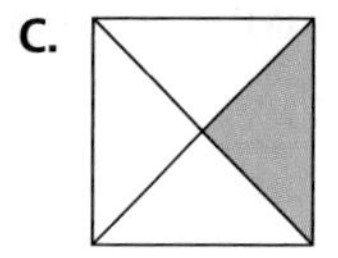
D. 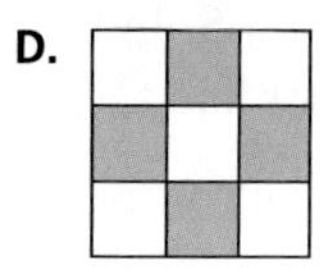

34. **Writing in Math** Are the fractions $\frac{9}{16}$, $\frac{10}{24}$, and $\frac{15}{35}$ in simplest form? Explain. See margin.

35. School $\frac{7}{24}$, Sleeping $\frac{1}{3}$, Job/Homework $\frac{1}{6}$, TV/Activity $\frac{1}{8}$, Eating $\frac{1}{12}$

35. The circle graph shows a student's daily activities. Write a fraction in simplest form to represent the time spent on each activity. See left.

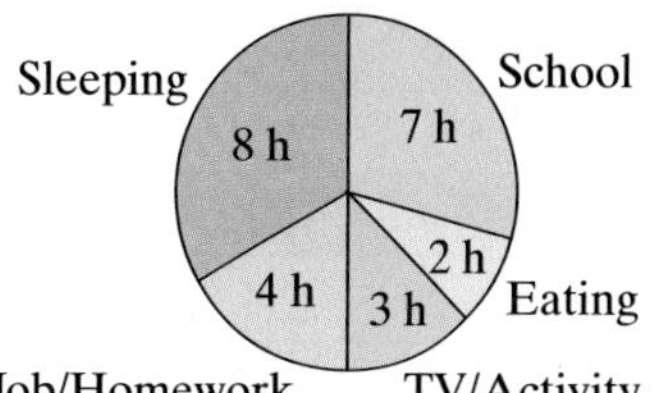

36. Draw models to show that the fractions $\frac{2}{3}$ and $\frac{10}{15}$ are equivalent. See margin.

37. **Challenge** The variables a and b represent positive integers. Name two fractions equivalent to $\frac{a}{b}$. Answers may vary. Sample: $\frac{2a}{2b}$, $\frac{3a}{3b}$

Test Prep and Mixed Review

Practice

Multiple Choice

38. If only one thousand people lived on Earth, the population would be distributed according to the table below. What fraction of the people live in Africa? **B**

Asia	Africa	Europe	South America	Australia	North America
607	132	120	57	5	79

Ⓐ $\frac{3}{25}$ Ⓑ $\frac{33}{250}$ Ⓒ $\frac{7}{50}$ Ⓓ $\frac{4}{25}$

39. Pablo spent half of his birthday money on concert tickets and one third of the remaining amount on food. After he used \$12.35 on a train ticket, he had \$8.25 left. How much money did Pablo begin with? **H**

Ⓕ \$20.60 Ⓖ \$41.20 Ⓗ \$61.80 Ⓙ \$123.60

GO for Help

For Exercises	See Lesson
40–42	1-2

Find each sum or difference.

40. $14.02 + 3.06$ 17.08
41. $25.98 - 8.89$ 17.09
42. $10.132 - 6.7$ 3.432

Online lesson quiz, PHSchool.com, Web Code: ara-0203

4. Assess & Reteach

PowerPoint

Lesson Quiz

Simplify.

1. $\frac{20}{35}$ $\frac{4}{7}$
2. $\frac{18}{42}$ $\frac{3}{7}$
3. $\frac{20}{75}$ $\frac{4}{15}$
4. $\frac{30}{48}$ $\frac{5}{8}$

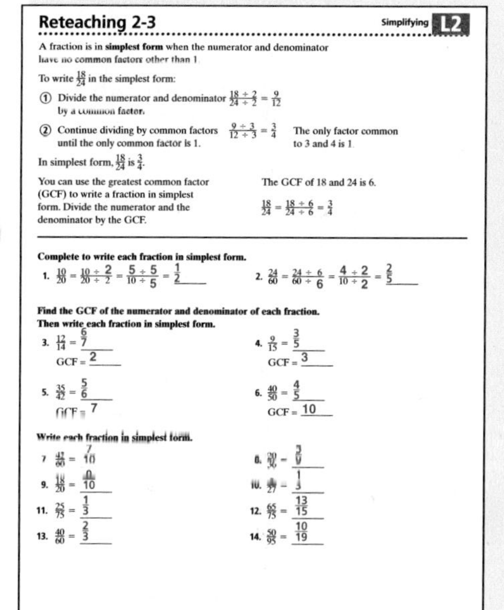

Reteaching 2-3 Simplifying L2

A fraction is in **simplest form** when the numerator and denominator have no common factors other than 1.

To write $\frac{18}{24}$ in the simplest form:

① Divide the numerator and denominator by a common factor. $\frac{18 \div 2}{24 \div 2} = \frac{9}{12}$

② Continue dividing by common factors until the only common factor is 1. $\frac{9 \div 3}{12 \div 3} = \frac{3}{4}$ The only factor common to 3 and 4 is 1.

In simplest form, $\frac{18}{24}$ is $\frac{3}{4}$.

You can use the greatest common factor (GCF) to write a fraction in simplest form. Divide the numerator and the denominator by the GCF.

The GCF of 18 and 24 is 6.

$\frac{18}{24} = \frac{18 \div 6}{24 \div 6} = \frac{3}{4}$

Complete to write each fraction in simplest form.

1. $\frac{10}{20} = \frac{10 \div 2}{20 \div 2} = \frac{5 \div 5}{10 \div 5} = \frac{1}{2}$
2. $\frac{24}{60} = \frac{24 \div 6}{60 \div 6} = \frac{4 \div 2}{10 \div 2} = \frac{2}{5}$

Find the GCF of the numerator and denominator of each fraction. Then write each fraction in simplest form.

3. $\frac{12}{14} = \frac{6}{7}$ GCF = 2
4. $\frac{9}{15} = \frac{3}{5}$ GCF = 3
5. $\frac{35}{42} = \frac{5}{6}$ GCF = 7
6. $\frac{40}{50} = \frac{4}{5}$ GCF = 10

Write each fraction in simplest form.

7. $\frac{42}{60} = \frac{7}{10}$
8. [illegible]
9. $\frac{18}{20} = \frac{9}{10}$
10. $\frac{9}{27} = \frac{1}{3}$
11. $\frac{25}{75} = \frac{1}{3}$
12. $\frac{65}{75} = \frac{13}{15}$
13. $\frac{40}{60} = \frac{2}{3}$
14. $\frac{50}{95} = \frac{10}{19}$

Enrichment 2-3 Simplifying L4

Decision Making

A grab bag at a school rummage sale costs \$15.00. Only the packers know what is inside each bag. Ferdinand and Rachel want to buy one grab bag. Ferdinand has \$10.50 and Rachel has \$6.75.

Sample answers for exercises 1–4 are shown below.

1. How much should each person contribute to the purchase? Why?
Ferdinand \$9 and Rachel \$6 so that each contributes as much as possible but still has a little left over

2. Based on how much each person contributed, what fraction of the surprises in the grab bag should each person receive? Why?
Ferdinand would get $\frac{3}{5}$ of the grab bag and Rachel would get $\frac{2}{5}$; $\frac{9}{15} = \frac{3}{5}$ and $\frac{6}{15} = \frac{2}{5}$

3. The grab bag contents are listed below.
1 wallet-size calculator, 2 tickets to a matinee, 3 school pencils, 4 pocket combs, 5 bus tokens, 2 school sports buttons
What would be a way for Ferdinand and Rachel to divide the contents so that each gets a fair share?
Assign a value to each item and divide them with Ferdinand receiving about $\frac{3}{5}$ of the value and Rachel receiving about $\frac{2}{5}$.

4. Using the way you decided, who gets which items? Explain why you chose the items for each to receive.
Ferdinand: \$12 of \$20 value—2 tickets (\$7), 2 combs (\$1), 5 tokens (\$3), and 1 button (\$1); Rachel: \$8 of \$20 value—calculator (\$5), 2 combs (\$1), 3 pencils (\$1), and 1 button (\$1)

Alternative Assessment

Provide pairs of students with a list of fractions, such as $\frac{21}{35}$, $\frac{18}{48}$, $\frac{19}{27}$, including some in simplest form. Have pairs circle the fractions already in simplest form. $\frac{19}{27}$ Then have them work together to rewrite the others in simplest form.

Test Prep

Resources

For additional practice with a variety of test item formats:

- Test-Taking Strategies, p. 111
- Test Prep, p. 115
- Test-Taking Strategies with Transparencies

Activity Lab

Comparing Fractions

Students use models to compare fractions. This will prepare them to compare and order fractions mentally in Lesson 2-4.

Guided Instruction

Explain to students that they can use a number line to compare numbers that are less than 1, just as they would use it to compare whole numbers.

Ask questions, such as:

- *Why is 10 cm a good length for the number line?* **5 is a factor of 10 so it is easy to make both fraction models for $\frac{3}{5}$ and $\frac{7}{10}$ to this scale.**
- *What is another way to determine which fraction is greater?* **Use the GCF to simplify and compare the numerators.**
- *How long would your model of $\frac{3}{4}$ be?* **7.5 cm**

Activity

Have students work in small groups on the Activity. Have each choose a representative to share their answers when you review the Activity as a class.

Alternative Method

Allow students to make models by shading graph paper. Point out that shading two squares in a ten square section makes one fifth. Have them shade three fifths of one ten square section and 7 squares of another ten square section and compare the two shaded areas.

Resources

- Activity Lab 2-4: Critical Thinking
- 10 cm strips of paper
- graph paper

2-4a Activity Lab

Hands On

Comparing Fractions

You can use a number line to compare two integers such as 3 and −5. You know 3 is greater because it lies to the right of −5. In this activity, you will use paper strips and a number line to compare two fractions.

You can compare $\frac{3}{5}$ and $\frac{7}{10}$ using two strips of paper, each 10 cm long, and a number line that is 10 cm long. On your number line, mark the points 0, $\frac{1}{2}$, and 1.

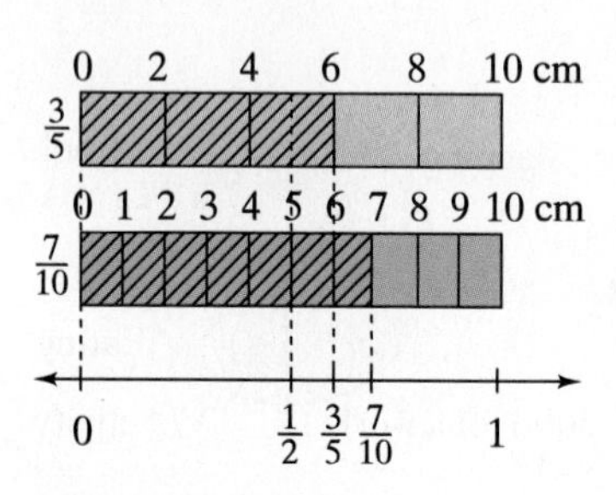

← **Divide one strip into 5 equal parts. Shade in 3 parts to represent $\frac{3}{5}$.**

← **Divide the other strip into 10 equal parts. Shade in 7 parts to represent $\frac{7}{10}$.**

← **Use the strips to mark the two fractions on a number line.**

Since $\frac{7}{10}$ is to the right of $\frac{3}{5}$ on the number line, you know that $\frac{7}{10} > \frac{3}{5}$.

ACTIVITY

Step 1 Using paper strips, compare the five pairs of fractions listed in the table at the right. For each pair, cut two strips of paper to the given length. Then use a number line and an inequality to show your comparison. **$\frac{7}{12} < \frac{3}{4}$, $\frac{1}{4} < \frac{2}{5}$, $\frac{2}{3} < \frac{6}{8}$, $\frac{5}{7} < \frac{3}{4}$, $\frac{3}{10} < \frac{2}{6}$**

Fractions	Paper Length
$\frac{3}{4}$ and $\frac{7}{12}$	12 cm
$\frac{2}{5}$ and $\frac{1}{4}$	20 cm
$\frac{6}{8}$ and $\frac{2}{3}$	24 cm
$\frac{3}{4}$ and $\frac{5}{7}$	28 cm
$\frac{3}{10}$ and $\frac{2}{6}$	30 cm

Step 2 Use 15-cm strips of paper to compare $\frac{7}{10}$ and $\frac{6}{9}$. Explain why this comparison is more difficult than the others. What would be a better length for the paper strips? Explain. **See margin.**

Step 3 Explain how the paper lengths shown in the table relate to the fractions being compared. Why do you think those lengths were chosen? **They are the LCM of the denominators, so it is easy to divide them into equal parts.**

Step 4 List several fraction pairs that could be compared using paper strips that are 32 cm long.

Step 5 How long would you make the paper strips to compare $\frac{6}{8}$, $\frac{4}{5}$, and $\frac{7}{10}$? Explain. **40 cm; 40 is divisible by 8, 5, and 10.**

Step 6 Use what you have learned to compare $\frac{7}{9}$ and $\frac{4}{5}$ without making paper strips. Explain your reasoning. **$\frac{7}{9} < \frac{4}{5}$ since $\frac{35}{45} < \frac{36}{45}$.**

4. Answers may vary. Samples are given. $\frac{3}{32}$ and $\frac{5}{8}$; $\frac{9}{32}$ and $\frac{1}{2}$; $\frac{1}{32}$ and $\frac{7}{4}$

2. This is difficult, since 15 is not the LCM of the denominators. It would be easier to use a length of 90 cm.

2-4 Comparing and Ordering Fractions

Check Skills You'll Need

1. Vocabulary Review What is the name for the smallest multiple common to two numbers? least common multiple

Find the LCM of each pair of numbers.

2. 3, 4 12
3. 4, 10 20
4. 2, 8 8
5. 9, 15 45

for Help Lesson 2-2

What You'll Learn

To compare and order fractions

New Vocabulary least common denominator (LCD)

Why Learn This?

Camera exposure times are given as fractions of a second. Knowing how to compare fractions can help you choose the correct exposure times and take better pictures.

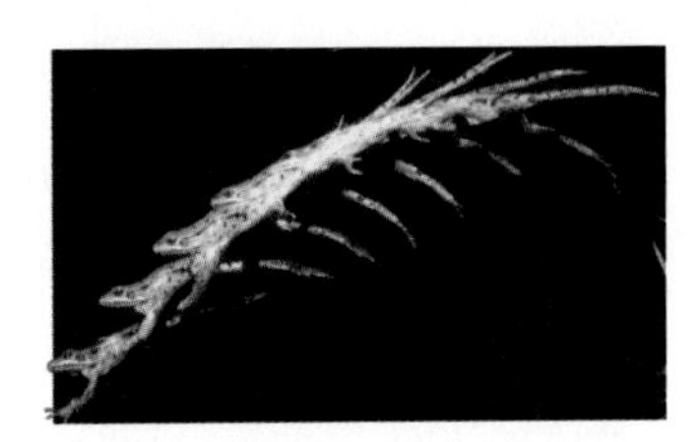

The exposure times below are listed in order of greatest to least. If the numerators of two fractions are the same, the fraction with the lesser denominator has the greater value. For example, $\frac{1}{125} > \frac{1}{500}$, because 1 divided into 125 parts is greater than 1 divided into 500 parts.

Camera Exposure Times

$\frac{1}{4}$ $\frac{1}{30}$ $\frac{1}{60}$ $\frac{1}{125}$ $\frac{1}{250}$ $\frac{1}{500}$ $\frac{1}{1000}$

Longer time—lets in more light

Shorter time—lets in less light

If the denominators are the same, the numerators show which fraction is greater. Use the "is greater than" (>) or the "is less than" (<) symbol.

You can use a number line to compare fractions. Any number to the right of any other number on a number line is the greater of the two.

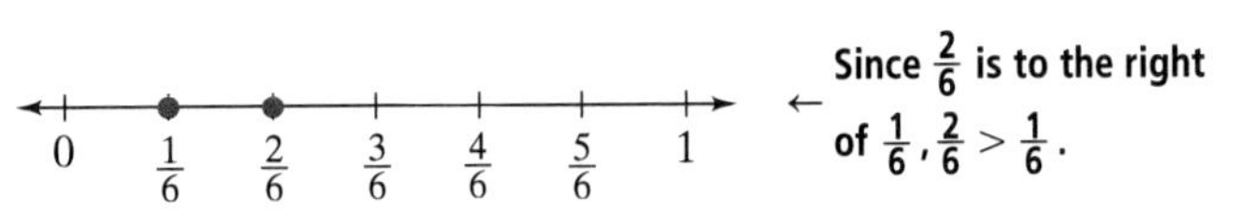

Since $\frac{2}{6}$ is to the right of $\frac{1}{6}$, $\frac{2}{6} > \frac{1}{6}$.

You can also use models to compare fractions. The fraction models show that $\frac{7}{12} > \frac{3}{8}$. To compare fractions with different denominators, rewrite each with a common denominator.

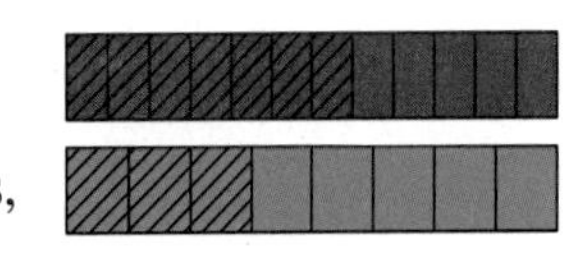

The **least common denominator (LCD)** of two or more fractions is the least common multiple (LCM) of their denominators.

GO Online

Video Tutor Help

Visit: PHSchool.com
Web Code: are-0775

Differentiated Instruction Solutions for All Learners

Special Needs L1
Provide students with paper fraction pieces, or pre-made kits. Have them compare the relative sizes of denominators such as halves, thirds, fourths, and eighths. They may need this reminder that the larger the denominator, the smaller the "piece."

learning style: tactile

Below Level L2
Review the use of a number line for comparing and ordering integers. Check that students recall the meaning of the symbols > (greater than) and < (less than).

learning style: visual

2-4

1. Plan

Objective
To compare and order fractions

Examples
1 Comparing Fractions
2 Application: Construction

Math Understanding: p. 66C

Professional Development

Math Background

There are two common methods for comparing fractions visually: If two fractions are represented by points on a number line, the fraction to the right is greater. If two fractions are represented as parts of a whole region, the fraction that represents the larger part is greater. This lesson develops a method for comparing fractions arithmetically, using common denominators.

More Math Background: p. 66C

Lesson Planning and Resources

See p. 66E for a list of the resources that support this lesson.

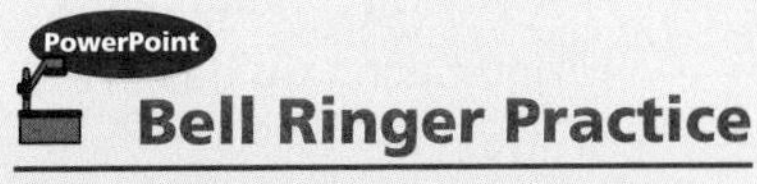

Bell Ringer Practice

Check Skills You'll Need
Use student page, transparency, or PowerPoint. For intervention, direct students to:

Prime Factorization
Lesson 2-2
Extra Skills and Word Problems Practice, Ch. 2

2. Teach

Activity Lab

Use before the lesson.
Student Edition Activity Lab Hands On 2-4a, Comparing Fractions, p. 86

All in One Teaching Resources
Activity Lab 2-4: Critical Thinking

Guided Instruction

Example 1
Students can use calculators to check the comparison.

$\frac{3}{4} \rightarrow 3 \div 4 = 0.75$

$\frac{7}{10} \rightarrow 7 \div 10 = 0.7$

Since $0.75 > 0.7$, $\frac{3}{4} > \frac{7}{10}$

Example 2
Have students first use fraction bars to order the fractions. They should keep the fraction bars in sight as you show them how to order by using common denominators.

PowerPoint

Additional Examples

1. Compare $\frac{7}{12}$ and $\frac{5}{9}$. $\frac{7}{12} > \frac{5}{9}$

2. Giselda wants to paint her room. She found $\frac{3}{5}$ gal of yellow paint, $\frac{2}{3}$ gal of blue paint, and $\frac{1}{2}$ gal of green paint.
   a. Order these amounts from least to greatest. $\frac{1}{2}$ gal $< \frac{3}{5}$ gal $< \frac{2}{3}$ gal
   b. Of which color is there the least amount? green
   c. Of which color is there the greatest amount? blue

All in One Teaching Resources
- Daily Notetaking Guide 2-4 L3
- Adapted Notetaking 2-4 L1

Closure

- Explain how to compare fractions with different denominators. Rewrite them as equivalent fractions with a common denominator, then compare the numerators.

Test Prep Tip
Draw a number line to help you visualize the order of the fractions.

EXAMPLE Comparing Fractions

1 Compare $\frac{3}{4}$ and $\frac{7}{10}$.

The denominators are 4 and 10. Their LCM is 20. So 20 is the LCD.

$\frac{3}{4} = \frac{3 \times 5}{4 \times 5} = \frac{15}{20}$

$\frac{7}{10} = \frac{7 \times 2}{10 \times 2} = \frac{14}{20}$ ← Write the equivalent fractions with a denominator of 20.

$\frac{14}{20} < \frac{15}{20}$ ← Compare the numerators.

So $\frac{7}{10} < \frac{3}{4}$.

Quick Check

1. Compare each pair of fractions. Use $<, =,$ or $>$.
   a. $\frac{3}{4}$ ■ $\frac{5}{6}$ < b. $\frac{1}{6}$ ■ $\frac{2}{9}$ < c. $\frac{4}{10}$ ■ $\frac{3}{8}$ >

You can use the LCD to order more than two fractions.

EXAMPLE Application: Construction

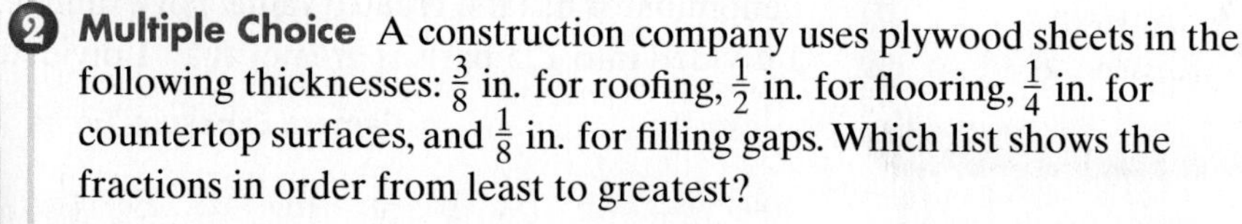

2 **Multiple Choice** A construction company uses plywood sheets in the following thicknesses: $\frac{3}{8}$ in. for roofing, $\frac{1}{2}$ in. for flooring, $\frac{1}{4}$ in. for countertop surfaces, and $\frac{1}{8}$ in. for filling gaps. Which list shows the fractions in order from least to greatest?

Ⓐ $\frac{1}{4}, \frac{3}{8}, \frac{1}{2}, \frac{1}{8}$ Ⓑ $\frac{1}{2}, \frac{1}{4}, \frac{1}{8}, \frac{3}{8}$ Ⓒ $\frac{1}{8}, \frac{1}{4}, \frac{1}{2}, \frac{3}{8}$ Ⓓ $\frac{1}{8}, \frac{1}{4}, \frac{3}{8}, \frac{1}{2}$

Order $\frac{3}{8}, \frac{1}{2}, \frac{1}{4}$, and $\frac{1}{8}$. The LCM of 8, 2, and 4 is 8. So 8 is the LCD.

Roofing: $\frac{3}{8}$

Flooring: $\frac{1}{2} = \frac{1 \times 4}{2 \times 4} = \frac{4}{8}$

Countertop: $\frac{1}{4} = \frac{1 \times 2}{4 \times 2} = \frac{2}{8}$ ← Use the LCD to write equivalent fractions.

Filling: $\frac{1}{8}$

$\frac{1}{8} < \frac{2}{8} < \frac{3}{8} < \frac{4}{8}$. So $\frac{1}{8} < \frac{1}{4} < \frac{3}{8} < \frac{1}{2}$. ← Compare the numerators.

The order is $\frac{1}{8}, \frac{1}{4}, \frac{3}{8}$, and $\frac{1}{2}$. The correct answer is choice D.

Quick Check

2. A carpenter uses four screws with diameters of $\frac{1}{4}$ in., $\frac{3}{8}$ in., $\frac{5}{16}$ in., and $\frac{5}{32}$ in. Order the diameters from least to greatest.
   $\frac{5}{32}$ in., $\frac{1}{4}$ in., $\frac{5}{16}$ in., $\frac{3}{8}$ in.

Advanced Learners L4
Have students devise a rule for comparing fractions with the same numerator. The fraction with the lesser denominator is the greater fraction.
learning style: visual

English Language Learners ELL
It is a good idea for students to read and say LCM as *least common multiple* and LCD as *least common denominator.* Otherwise, it is easy for them to lose track of the meaning of the abbreviations, especially since they have words in common.
learning style: verbal

Check Your Understanding

1. **Vocabulary** How are the least common multiple (LCM) and least common denominator (LCD) related? **The LCM of the denominators of two or more fractions is the LCD.**
2. **Reasoning** To order $\frac{4}{10}$, $\frac{3}{5}$, and $\frac{5}{25}$, it is helpful to write each fraction in simplest form first, before finding the LCD. Explain why. **See left.**

2. All of the fractions have a common denominator of 5 when in simplest form.

Find the LCD of each pair of fractions.

3. $\frac{5}{7}, \frac{2}{9}$ **63**
4. $\frac{3}{11}, \frac{2}{5}$ **55**
5. $\frac{7}{8}, \frac{1}{3}$ **24**

Compare each pair of fractions. Use <, =, or >.

6. $\frac{1}{7} \blacksquare \frac{3}{8}$ **<**
7. $\frac{3}{12} \blacksquare \frac{1}{6}$ **>**
8. $\frac{2}{10} \blacksquare \frac{3}{5}$ **<**

Homework Exercises

For more exercises, see Extra Skills and Word Problems.

GO for Help

For Exercises	See Examples
9–17	1
18–27	2

A **Compare each pair of fractions. Use <, =, or >.**

9. $\frac{5}{12} \blacksquare \frac{7}{12}$ **<**
10. $\frac{5}{6} \blacksquare \frac{3}{6}$ **>**
11. $\frac{1}{3} \blacksquare \frac{3}{4}$ **<**
12. $\frac{5}{6} \blacksquare \frac{3}{5}$ **>**
13. $\frac{3}{8} \blacksquare \frac{2}{3}$ **<**
14. $\frac{4}{7} \blacksquare \frac{4}{5}$ **<**
15. $\frac{2}{3} \blacksquare \frac{5}{8}$ **>**
16. $\frac{5}{6} \blacksquare \frac{5}{10}$ **>**
17. $\frac{1}{8} \blacksquare \frac{3}{16}$ **<**

Order from least to greatest.

18. $\frac{1}{3}, \frac{5}{6}, \frac{3}{8}$ **$\frac{1}{3}, \frac{3}{8}, \frac{5}{6}$**
19. $\frac{1}{8}, \frac{1}{6}, \frac{1}{9}$ **$\frac{1}{9}, \frac{1}{8}, \frac{1}{6}$**
20. $\frac{3}{15}, \frac{3}{10}, \frac{3}{5}$ **$\frac{3}{15}, \frac{3}{10}, \frac{3}{5}$**
21. $\frac{5}{8}, \frac{7}{9}, \frac{2}{1}$ **$\frac{5}{8}, \frac{7}{9}, \frac{2}{1}$**
22. $\frac{6}{10}, \frac{7}{12}, \frac{5}{8}$ **$\frac{7}{12}, \frac{6}{10}, \frac{5}{8}$**
23. $\frac{2}{5}, \frac{3}{20}, \frac{4}{5}$ **$\frac{3}{20}, \frac{2}{5}, \frac{4}{5}$**
24. $1, \frac{4}{6}, \frac{1}{3}$ **$\frac{1}{3}, \frac{4}{6}, 1$**
25. $\frac{10}{15}, \frac{6}{10}, \frac{1}{3}$ **$\frac{1}{3}, \frac{6}{10}, \frac{10}{15}$**
26. $\frac{1}{8}, \frac{3}{12}, \frac{4}{10}$ **$\frac{1}{8}, \frac{3}{12}, \frac{4}{10}$**

27. **Languages** At an international school, $\frac{1}{3}$ of the languages spoken are Romance languages, $\frac{2}{15}$ are Germanic, and $\frac{8}{15}$ are Balto-Slavic. Order the language categories from least to greatest. **Germanic, Romance, Balto-Slavic**

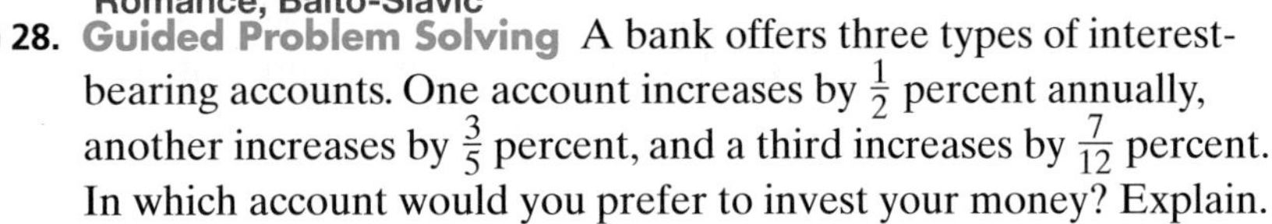

B **GPS** 28. **Guided Problem Solving** A bank offers three types of interest-bearing accounts. One account increases by $\frac{1}{2}$ percent annually, another increases by $\frac{3}{5}$ percent, and a third increases by $\frac{7}{12}$ percent. In which account would you prefer to invest your money? Explain.
- What is the least common denominator of the three fractions?
- Is it better to earn more interest or less interest? **$\frac{3}{5}$; it is the highest increase.**

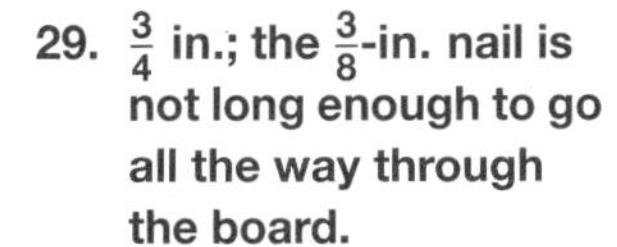

29. $\frac{3}{4}$ in.; the $\frac{3}{8}$-in. nail is not long enough to go all the way through the board.

29. **Carpentry** You want to nail a board that is $\frac{1}{2}$ in. thick onto a wall. **GPS** You can choose between nails that are $\frac{3}{8}$ in. long and $\frac{3}{4}$ in. long. Which size nail is the better choice? Explain. **See left.**

3. Practice

Assignment Guide

Check Your Understanding

Go over Exercises 1–8 in class before assigning the Homework Exercises.

Homework Exercises

A Practice by Example	9–27
B Apply Your Skills	28–39
C Challenge	40
Test Prep and Mixed Review	41–46

Homework Quick Check

To check students' understanding of key skills and concepts, go over Exercises 14, 21, 29, 36, and 39.

Differentiated Instruction Resources

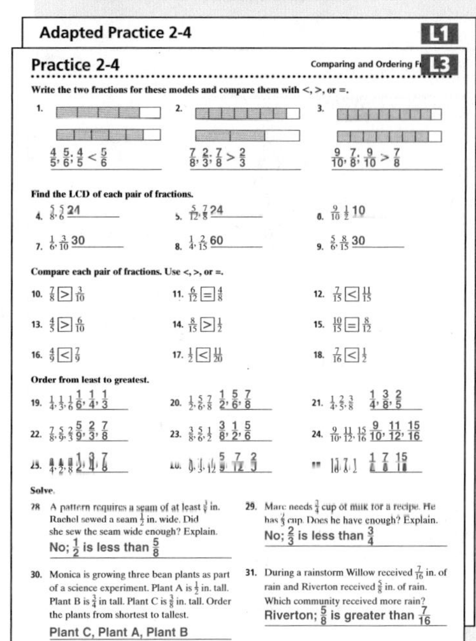

Adapted Practice 2-4 L1

Practice 2-4 Comparing and Ordering Fractions L3

Write the two fractions for these models and compare them with <, >, or =.

1. $\frac{4}{5}, \frac{5}{6}; \frac{4}{5} < \frac{5}{6}$
2. $\frac{7}{8}, \frac{2}{3}; \frac{7}{8} > \frac{2}{3}$
3. $\frac{9}{10}, \frac{7}{8}; \frac{9}{10} > \frac{7}{8}$

Find the LCD of each pair of fractions.

4. $\frac{5}{8}, \frac{5}{6}$ 24
5. $\frac{5}{12}, \frac{7}{8}$ 24
6. $\frac{9}{10}, \frac{1}{2}$ 10
7. $\frac{1}{6}, \frac{3}{10}$ 30
8. $\frac{1}{4}, \frac{2}{15}$ 60
9. $\frac{5}{6}, \frac{8}{15}$ 30

Compare each pair of fractions. Use <, >, or =.

10. $\frac{7}{8} > \frac{3}{10}$
11. $\frac{6}{12} = \frac{4}{8}$
12. $\frac{7}{15} < \frac{11}{15}$
13. $\frac{4}{5} > \frac{6}{10}$
14. $\frac{8}{15} > \frac{1}{2}$
15. $\frac{10}{15} = \frac{8}{12}$
16. $\frac{4}{9} < \frac{7}{9}$
17. $\frac{1}{2} < \frac{11}{20}$
18. $\frac{7}{16} < \frac{1}{2}$

Order from least to greatest.

19. $\frac{1}{4}, \frac{1}{3}, \frac{1}{6}$ $\frac{1}{6}, \frac{1}{4}, \frac{1}{3}$
20. $\frac{1}{2}, \frac{5}{6}, \frac{7}{8}$ $\frac{1}{2}, \frac{5}{6}, \frac{7}{8}$
21. $\frac{1}{4}, \frac{2}{5}, \frac{3}{8}$ $\frac{1}{4}, \frac{3}{8}, \frac{2}{5}$
22. $\frac{7}{8}, \frac{5}{9}, \frac{2}{3}$ $\frac{5}{9}, \frac{2}{3}, \frac{7}{8}$
23. $\frac{3}{8}, \frac{5}{6}, \frac{1}{2}$ $\frac{3}{8}, \frac{1}{2}, \frac{5}{6}$
24. $\frac{9}{10}, \frac{11}{12}, \frac{15}{16}$ $\frac{9}{10}, \frac{11}{12}, \frac{15}{16}$
25. [illegible]
26. [illegible]
27. [illegible]

Solve.

28. A pattern requires a seam of at least $\frac{5}{8}$ in. Rachel sewed a seam $\frac{1}{2}$ in. wide. Did she sew the seam wide enough? Explain. No; $\frac{1}{2}$ is less than $\frac{5}{8}$
29. Marc needs $\frac{3}{4}$ cup of milk for a recipe. He has $\frac{2}{3}$ cup. Does he have enough? Explain. No; $\frac{2}{3}$ is less than $\frac{3}{4}$
30. Monica is growing three bean plants as part of a science experiment. Plant A is $\frac{1}{2}$ in. tall. Plant B is $\frac{3}{4}$ in tall. Plant C is $\frac{3}{8}$ in. tall. Order the plants from shortest to tallest. Plant C, Plant A, Plant B
31. During a rainstorm Willow received $\frac{7}{16}$ in. of rain and Riverton received $\frac{5}{8}$ in. of rain. Which community received more rain? Riverton; $\frac{5}{8}$ is greater than $\frac{7}{16}$

2-4 • Guided Problem Solving GPS L3

GPS Student Page 89, Exercise 29:

Carpentry You want to nail a board that is $\frac{1}{2}$ in. thick onto a wall. You can choose from nails that are $\frac{3}{8}$ in. long and $\frac{3}{4}$ in. long. Which size nail is the better choice? Explain.

Understand

1. Circle the information you will need to solve.
2. What are you being asked to do?
3. In order to compare fractions what must you do?

Plan and Carry Out

4. What is the common denominator for $\frac{1}{2}, \frac{3}{8}, \frac{3}{4}$?
5. Write an equivalent fraction for $\frac{1}{2}$ and $\frac{3}{4}$ with the denominator found in Step 4.
6. Which nail is longer than $\frac{4}{8}$ in.?
7. Which size nail is the better choice, the $\frac{3}{8}$ in. nail or the $\frac{3}{4}$ in. nail?
8. Explain why you chose the nail you did in Step 8.

Check

9. What is $\frac{3}{4} - \frac{1}{2}$? What is $\frac{3}{8} - \frac{1}{2}$?

Solve Another Problem

10. Louise used the $\frac{1}{2}$ in., the $\frac{11}{16}$ in., and the $\frac{5}{8}$ in. wrench from her dad's toolbox. Now he wants her to put them back in his toolbox from smallest to largest. What order should the wrenches be in?

4. Assess & Reteach

Lesson Quiz

Order from least to greatest.

1. $\frac{3}{4}, \frac{5}{6}, \frac{1}{3}$ $\frac{1}{3} < \frac{3}{4} < \frac{5}{6}$
2. $\frac{2}{3}, \frac{5}{8}, \frac{5}{12}$ $\frac{5}{12} < \frac{5}{8} < \frac{2}{3}$
3. $\frac{7}{10}, \frac{4}{5}, \frac{1}{2}$ $\frac{1}{2} < \frac{7}{10} < \frac{4}{5}$
4. $\frac{1}{5}, \frac{6}{7}, \frac{6}{10}$ $\frac{1}{5} < \frac{6}{10} < \frac{6}{7}$

39. If two fractions have the same numerator, then the fraction with the smaller denominator is greater, because the numerator makes up a larger part of the whole.

Reteaching 2-4 — Comparing and Ordering — L2

Follow these steps to *compare* $\frac{2}{5}$ and $\frac{3}{10}$ (unlike denominators).

① Find the **least common denominator** (LCD). — The denominators are 5 and 10. The LCD of 5 and 10 is their least common multiple.

② Write the equivalent fractions using the LCD. — $\frac{2}{5} = \frac{4}{10}$ and $\frac{3}{10} = \frac{3}{10}$

③ Compare the numerators. — $4 > 3$

So, $\frac{4}{10} > \frac{3}{10}$ and $\frac{2}{5} > \frac{3}{10}$.

Follow these steps to *order* the fractions $\frac{1}{2}, \frac{3}{5}$, and $\frac{2}{3}$.

① Find the LCD. — The LCD of 2, 5, and 3 is 30.

② Write equivalent fractions using the LCD. — $\frac{1}{2} = \frac{1 \cdot 15}{2 \cdot 15} = \frac{15}{30}$; $\frac{3}{5} = \frac{3 \cdot 6}{5 \cdot 6} = \frac{18}{30}$; $\frac{2}{3} = \frac{2 \cdot 10}{3 \cdot 10} = \frac{20}{30}$

③ Order the fractions using their numerators. — $\frac{15}{30} < \frac{18}{30} < \frac{20}{30}$

So, $\frac{1}{2} < \frac{3}{5} < \frac{2}{3}$.

Find each missing number. Then compare the fractions. Use <, >, or =.

1. $\frac{2}{3}$ and $\frac{3}{5}$
   a. $\frac{2}{3} = \frac{10}{15}, \frac{3}{5} = \frac{9}{15}$
   b. $\frac{2}{3} > \frac{3}{5}$
2. $\frac{1}{2}$ and $\frac{5}{8}$
   a. $\frac{1}{2} = \frac{4}{8}, \frac{5}{8} = \frac{5}{8}$
   b. $\frac{1}{2} < \frac{5}{8}$
3. $\frac{3}{4}$ and $\frac{9}{12}$
   a. $\frac{3}{4} = \frac{9}{12}, \frac{9}{12} = \frac{9}{12}$
   b. $\frac{3}{4} = \frac{9}{12}$

Compare each pair of fractions. Use <, >, or =.

4. $\frac{3}{5} < \frac{4}{5}$
5. $\frac{3}{4} < \frac{7}{8}$
6. $\frac{8}{12} = \frac{2}{3}$
7. $\frac{1}{2} < \frac{9}{16}$
8. $\frac{2}{3} > \frac{1}{2}$
9. $\frac{5}{9} = \frac{10}{18}$
10. $\frac{6}{7} > \frac{5}{6}$
11. $\frac{3}{8} < \frac{3}{5}$

Order from least to greatest.

12. $\frac{1}{2}, \frac{4}{5}, \frac{1}{4}$ $\frac{1}{4}, \frac{1}{2}, \frac{4}{5}$
13. $\frac{2}{3}, \frac{3}{8}, \frac{1}{2}$ $\frac{3}{8}, \frac{1}{2}, \frac{2}{3}$
14. $\frac{5}{6}, \frac{7}{8}, \frac{1}{4}$ $\frac{1}{4}, \frac{5}{6}, \frac{7}{8}$
15. $\frac{1}{2}, \frac{5}{8}, \frac{5}{6}$ $\frac{1}{2}, \frac{5}{8}, \frac{5}{6}$
16. $\frac{7}{10}, \frac{2}{3}, \frac{1}{5}$ $\frac{1}{5}, \frac{2}{3}, \frac{7}{10}$
17. $\frac{2}{3}, \frac{1}{4}, \frac{11}{12}$ $\frac{1}{4}, \frac{2}{3}, \frac{11}{12}$

Enrichment 2-4 — Comparing and Ordering — L4

Critical Thinking

You can compare fractions using *cross products*.

Compare $\frac{3}{7}$ and $\frac{4}{9}$. — $\frac{3}{7} \; ? \; \frac{4}{9}$

① Write the cross products. — $27 = \frac{3}{7} \times \frac{4}{9} = 28$

② Compare the cross products. — $27 < 28$

③ Compare the fractions. — so $\frac{3}{7} < \frac{4}{9}$

Use the cross-product method to order the five fractions below from least to greatest.

$\frac{7}{9}, \quad \frac{8}{11}, \quad \frac{4}{7}, \quad \frac{13}{19}, \quad \frac{3}{4}$

1. Compare the first two fractions on the left. List them in order with the smaller fraction first. $\frac{8}{11}, \frac{7}{9}$
2. Compare the middle fraction with the smaller fraction you found in exercise 1. Is it smaller or larger? smaller
3. Starting from the left, list the first three fractions in order. $\frac{4}{7}, \frac{8}{11}, \frac{7}{9}$
4. How will you determine the position of the fraction $\frac{13}{19}$? Show your work.
   Sample answer: $\frac{13}{19} > \frac{4}{7}$; $\frac{13}{19} < \frac{8}{11}$. So $\frac{13}{19}$ falls between these two fractions.
5. List all five of the fractions in order from least to greatest.
   $\frac{4}{7}, \frac{13}{19}, \frac{8}{11}, \frac{3}{4}, \frac{7}{9}$
6. When would it be easier to use cross products to compare fractions than to find equivalent fractions with a common denominator?
   Sample answer: If two fractions have large unlike denominators, use cross products.
7. Would you ever use both methods to order a list of fractions? Explain.
   Sample answer: Yes; list common denominators for some common fractions. Cross products can be used to determine the rest.

GO Online
Homework Video Tutor
Visit: PHSchool.com
Web Code: are-0204

Compare each pair of fractions. Use <, =, or >.

30. $\frac{7}{12} \blacksquare \frac{5}{9}$ >
31. $\frac{10}{15} \blacksquare \frac{16}{24}$ =
32. $\frac{8}{16} \blacksquare \frac{15}{32}$ >
33. $\frac{22}{26} \blacksquare \frac{10}{13}$ >

Use the table at the right.

34. Do people remember more of what they say or what they do? **what they do**
35. Do people remember more of what they hear or what they say? **what they say**

Memory Facts

People remember . . .	of . . .
three fourths	what they say.
one tenth	what they hear.
nine tenths	what they do.

Write two fractions for the models and compare them. Use <, =, or >.

36. 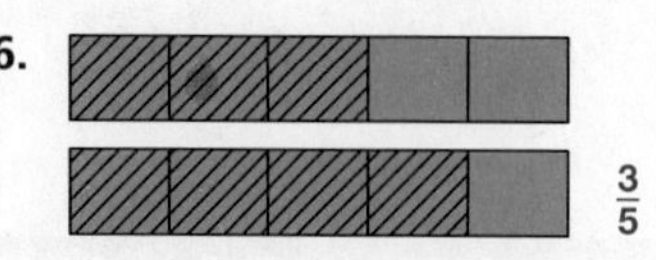 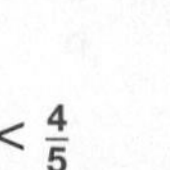 $\frac{3}{5} < \frac{4}{5}$

37. 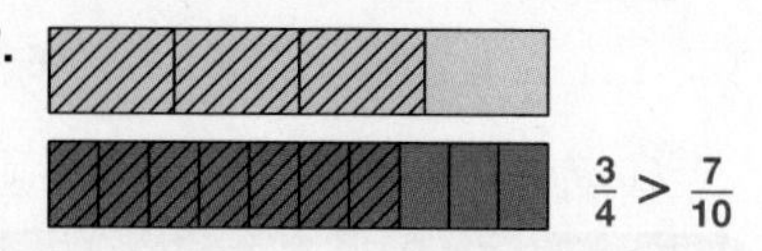 $\frac{3}{4} > \frac{7}{10}$

38. $\frac{1}{2} > \frac{1}{3}$; $\frac{1}{3} > \frac{1}{4}$; $\frac{1}{4} > \frac{1}{5}$; $\frac{1}{5} > \frac{1}{6}$

38. **Patterns** Copy the table. Compare the fractions and fill in your answers. Use <, =, or >. **See left.**

$\frac{1}{2} \blacksquare \frac{1}{3}$
$\frac{1}{3} \blacksquare \frac{1}{4}$
$\frac{1}{4} \blacksquare \frac{1}{5}$
$\frac{1}{5} \blacksquare \frac{1}{6}$

39. **Writing in Math** Describe an easy way to compare fractions that have the same numerator, such as $\frac{4}{5}$ and $\frac{4}{7}$. Explain why your method works. **See margin.**

40. **Challenge** The variable n represents a positive integer. Which is greater, $\frac{1}{n}$ or $\frac{1}{n+1}$? $\frac{1}{n}$

Test Prep and Mixed Review — Practice

Multiple Choice

41. Which fraction is found between $\frac{7}{12}$ and $\frac{5}{6}$ on a number line? **D**
    Ⓐ $\frac{13}{24}$ Ⓑ $\frac{11}{12}$ Ⓒ $\frac{5}{12}$ Ⓓ $\frac{17}{24}$

42. Eddie bought 12.9 gallons of gas that cost \$2.89 per gallon. About how much did he pay for the gas? **J**
    Ⓕ Less than \$23
    Ⓖ Between \$23 and \$30
    Ⓗ Between \$30 and \$36
    Ⓙ More than \$36

43. Which expression is represented by the model below? **B**

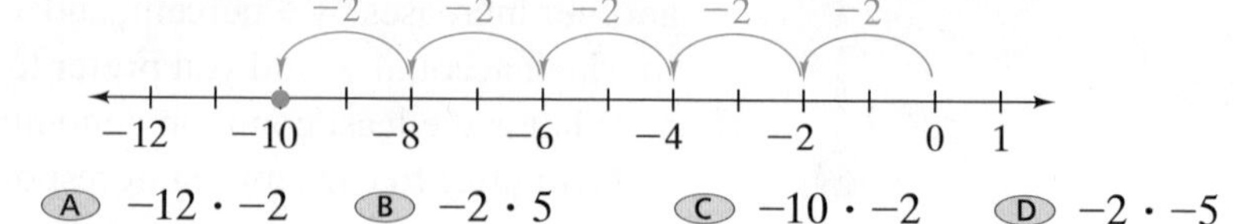

Ⓐ $-12 \cdot -2$ Ⓑ $-2 \cdot 5$ Ⓒ $-10 \cdot -2$ Ⓓ $-2 \cdot -5$

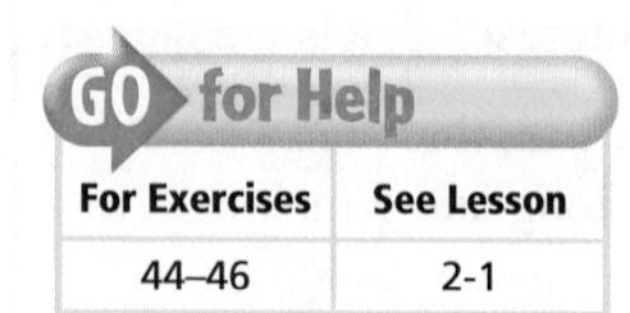

GO for Help

For Exercises	See Lesson
44–46	2-1

Simplify.

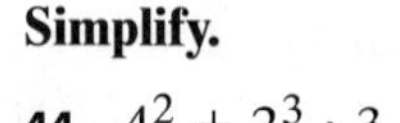

44. $4^2 + 2^3 \cdot 3$ **40**
45. $(4^2 + 2^3) \cdot 3$ **72**
46. $4^2 \cdot (2^3 + 3)^2$ **1,936**

Test Prep

Resources

For additional practice with a variety of test item formats:

- Test-Taking Strategies, p. 111
- Test Prep, p. 115
- Test-Taking Strategies with Transparencies

Alternative Assessment

Provide pairs of students with several sets of three different fractions, such as $\frac{2}{3}, \frac{3}{5}, \frac{3}{4}$. Provide fraction manipulatives such as fraction bars. Have pairs model each fraction in a set, arrange the models from least to greatest, and then record the correct order of the fractions.

2-5 Mixed Numbers and Improper Fractions

Check Skills You'll Need

1. Vocabulary Review What does it mean when a fraction is in *simplest form*? **See below.**

Write each fraction in simplest form.

2. $\frac{12}{20}$ $\frac{3}{5}$ **3.** $\frac{15}{18}$ $\frac{5}{6}$

4. $\frac{45}{60}$ $\frac{3}{4}$ **5.** $\frac{20}{48}$ $\frac{5}{12}$

for Help Lesson 2-3

Check Skills You'll Need

1. **The numerator and denominator have no common factors other than 1.**

What You'll Learn

To write mixed numbers and improper fractions

New Vocabulary improper fraction, mixed number

Why Learn This?

Suppose you are trying to share the extra pies left over from a pie eating contest. You can use improper fractions to find the number of equal-sized parts you can make from the leftovers.

The models below show that $1\frac{2}{3} = \frac{5}{3}$.

$1\frac{2}{3}$ $\frac{5}{3}$

The fraction $\frac{5}{3}$ is an improper fraction. An **improper fraction** is a fraction that has a numerator that is greater than or equal to its denominator.

The number $1\frac{2}{3}$ is a mixed number. A **mixed number** is the sum of a whole number and a fraction.

A number line can also help you understand improper fractions and mixed numbers.

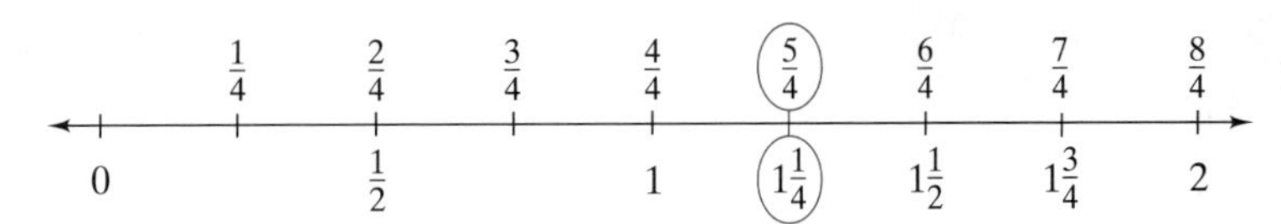

The number line shows that $1\frac{1}{4} = \frac{5}{4}$.

One way to write a mixed number as an improper fraction is to write the mixed number as a sum. Write a fraction that is equivalent to the whole number and then find the sum of the fractions.

Differentiated Instruction Solutions for All Learners

Special Needs L1	**Below Level** L2
Provide students with fraction models or kits. Alternatively, allow them to draw models to help them when rewriting numbers as improper fractions or mixed numbers.	Give students several exercises like the following. $2 = \frac{\square}{2}$ 4 $5 = \frac{\square}{3}$ 15
learning style: tactile	learning style: visual

2-5

1. Plan

Objective
To write mixed numbers and improper fractions

Examples
1 Writing an Improper Fraction
2 Writing a Mixed Number

Math Understandings: p. 66D

Math Background

A *mixed number* consists of two parts: a whole number and a fraction. It represents one or more identical whole quantities, plus some fractional part of the whole. It is possible to rewrite any mixed number as a single fraction, which is called an *improper fraction.*

More Math Background: p. 66D

Lesson Planning and Resources

See p. 66E for a list of the resources that support this lesson.

PowerPoint

Bell Ringer Practice

Check Skills You'll Need
Use student page, transparency, or PowerPoint. For intervention, direct students to:
Simplifying Fractions
Lesson 2-3
Extra Skills and Word Problems Practice, Ch. 2

2. Teach

Activity Lab

Use before the lesson.

All in One Teaching Resources

Activity Lab 2-5: Mixed Numbers and Improper Fractions

Guided Instruction

Example 1

Have students listen as you read "four and two thirds" aloud. Tell them to listen for the word "and," which signals an addition.

Error Prevention!

Using Method 2, students might multiply the numerator by the whole number, then add the denominator. For instance, they might write $3\frac{4}{5}$ as $\frac{(3 \times 4) + 5}{5}$, or $\frac{17}{5}$. Remind them that their goal is to write an improper fraction with the same denominator as the denominator in the mixed number.

PowerPoint

Additional Examples

1. Write $3\frac{2}{5}$ as an improper fraction. **$\frac{17}{5}$**

2. Karl needs $\frac{18}{4}$ cups of milk to make several loaves of bread. Express the amount of milk as a mixed number in simplest form. **$4\frac{1}{2}$ cups of milk**

All in One Teaching Resources

- Daily Notetaking Guide 2-5 **L3**
- Adapted Notetaking 2-5 **L1**

Closure

- Explain how to write a mixed number as an improper fraction. **Write the mixed number as a sum of the whole number and the fraction. Write an equivalent fraction for the whole number using the same denominator as the fraction. Find the sum of both fractions by adding numerators. Simplify the sum.**

Online active math

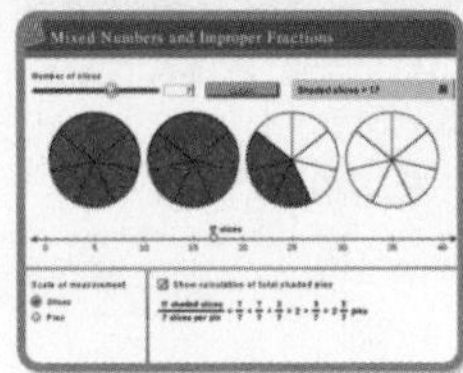

For: Fractions Activity
Use: Interactive Textbook, 2-5

EXAMPLE Writing an Improper Fraction

1 Write $4\frac{2}{3}$ as an improper fraction.

Method 1 Using addition

$4\frac{2}{3} = 4 + \frac{2}{3}$ ← **Write the mixed number as a sum.**

$= \frac{12}{3} + \frac{2}{3}$ ← **Change 4 to a fraction with the same denominator as $\frac{2}{3}$. Substitute. $4 = 4 \times \frac{3}{3} = \frac{12}{3}$**

$= \frac{12 + 2}{3} = \frac{14}{3}$ ← **Add the numerators.**

Method 2 Using multiplication

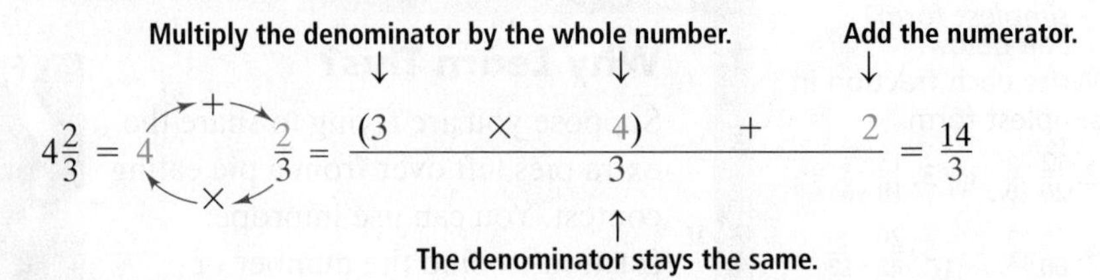

Multiply the denominator by the whole number. **Add the numerator.**

$$4\frac{2}{3} = \frac{(3 \times 4) + 2}{3} = \frac{14}{3}$$

The denominator stays the same.

Quick Check

1. **Choose a Method** Write $2\frac{5}{8}$ as an improper fraction. **$\frac{21}{8}$**

To write an improper fraction as a mixed number, divide and write the remainder as a fraction of the denominator. Then simplify the fraction.

EXAMPLE Writing a Mixed Number

2 You are planning a party and estimate that you will need 30 slices of pie for your guests. If each pie contains 8 slices, how many pies should you buy?

To find the number of pies, write $\frac{30}{8}$ as a mixed number.

$$\begin{array}{r} 3 \\ 8\overline{)30} \\ -24 \\ \hline 6 \end{array}$$

denominator → 8; 3 ← **whole number**; 6 ← **remainder**

$3\frac{6}{8} = 3\frac{3}{4}$ ← **Write the remainder as a fraction, $\frac{\text{remainder}}{\text{denominator}}$. Simplify.**

Since you cannot buy $3\frac{3}{4}$ pies, you should buy 4 pies.

Quick Check

2. A bakery sells a jumbo pie with 12 slices. If you need 30 slices, how many jumbo pies should you buy? **3**

Advanced Learners L4	**English Language Learners** ELL
Have students list ways that they use mixed numbers in their everyday activities. **Sample: measuring ingredients for a recipe** **learning style: verbal**	Allow students time to change whole numbers to fractions with given denominators. Have them work in pairs. One student names a whole number, such as 5, and a denominator, such as thirds. The other students must rename 5 as the fraction $\frac{15}{3}$ They can then switch and repeat the process. **learning style: verbal**

Check Your Understanding

1. **Vocabulary** How are mixed numbers and improper fractions related? **Both are larger than 1.**

2. Write 7 as an improper fraction. **Answers may vary. Sample: $\frac{14}{2}$**

Label each as an *improper fraction, mixed number,* or *proper fraction.*

3. $\frac{7}{11}$ **proper fraction** 4. $4\frac{1}{2}$ **mixed number** 5. $\frac{15}{6}$ **improper fraction**

Fill in the blank.

6. $7 = \frac{\blacksquare}{2}$ **14** 7. $2\frac{2}{3} = \frac{\blacksquare}{3}$ **8** 8. $\blacksquare = \frac{18}{3}$ **6**

9. **Mental Math** Without calculating, decide whether the value of $\frac{72}{12}$ is a whole number or a mixed number. Explain your reasoning.
It is a whole number because 72 is divisible by 12.

Homework Exercises

For more exercises, see Extra Skills and Word Problems.

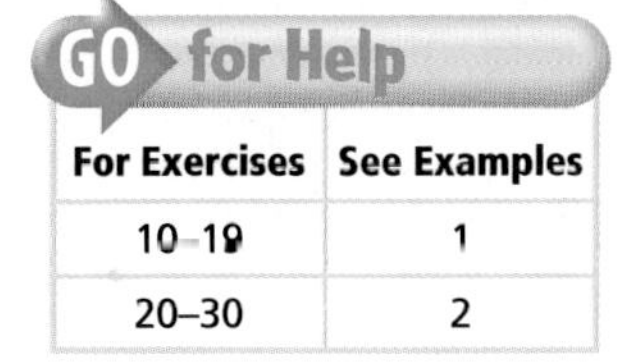

GO for Help

For Exercises	See Examples
10–19	1
20–30	2

A **Write each mixed number as an improper fraction. You may find a model helpful.**

10. $2\frac{3}{8}$ $\frac{19}{8}$ 11. $5\frac{3}{4}$ $\frac{23}{4}$ 12. $1\frac{1}{12}$ $\frac{13}{12}$ 13. $4\frac{3}{5}$ $\frac{23}{5}$ 14. $1\frac{3}{7}$ $\frac{10}{7}$

15. $4\frac{5}{8}$ $\frac{37}{8}$ 16. $3\frac{2}{5}$ $\frac{17}{5}$ 17. $2\frac{11}{12}$ $\frac{35}{12}$ 18. $5\frac{2}{3}$ $\frac{17}{3}$ 19. $9\frac{1}{4}$ $\frac{37}{4}$

Write each improper fraction as a mixed number in simplest form. You may find a model helpful.

20. $\frac{16}{3}$ $5\frac{1}{3}$ 21. $\frac{25}{3}$ $8\frac{1}{3}$ 22. $\frac{42}{4}$ $10\frac{1}{2}$ 23. $\frac{31}{12}$ $2\frac{7}{12}$ 24. $\frac{28}{6}$ $4\frac{2}{3}$

25. $\frac{49}{6}$ $8\frac{1}{6}$ 26. $\frac{40}{6}$ $6\frac{2}{3}$ 27. $\frac{45}{10}$ $4\frac{1}{2}$ 28. $\frac{48}{11}$ $4\frac{4}{11}$ 29. $\frac{15}{8}$ $1\frac{7}{8}$

30. A recipe calls for $\frac{1}{4}$ c of flour. You need $\frac{11}{4}$ c to make 11 servings. Write $\frac{11}{4}$ as a mixed number. **$2\frac{3}{4}$**

B 

31. **Guided Problem Solving** 101 faculty members attended a breakfast, and each member was served $\frac{1}{4}$ loaf of banana bread. If each loaf contains eight slices, about how many slices of bread were served? About how many loaves were baked? **about 202 slices, about 26 loaves**
    - *Draw a diagram.* If eight slices make one loaf, how many slices are in $\frac{1}{4}$ loaf?
    - 101 people were served $\frac{1}{4}$ loaf. How many whole loaves is that?

32. Marisa is sewing an outfit. She uses $2\frac{5}{6}$ yards of fabric for her blouse, 3 yards of fabric for her skirt, and $\frac{20}{3}$ yards of fabric for her jacket. Write these values in order from least to greatest. **$2\frac{5}{6}$, 3, $\frac{20}{3}$**

3. Practice

Assignment Guide

Check Your Understanding

Go over Exercises 1–9 in class before assigning the Homework Exercises.

Homework Exercises

A	Practice by Example	10–30
B	Apply Your Skills	31–41
C	Challenge	42
Test Prep and Mixed Review		43–47

Homework Quick Check

To check students' understanding of key skills and concepts, go over Exercises 12, 27, 34, 38, and 40.

Differentiated Instruction Resources

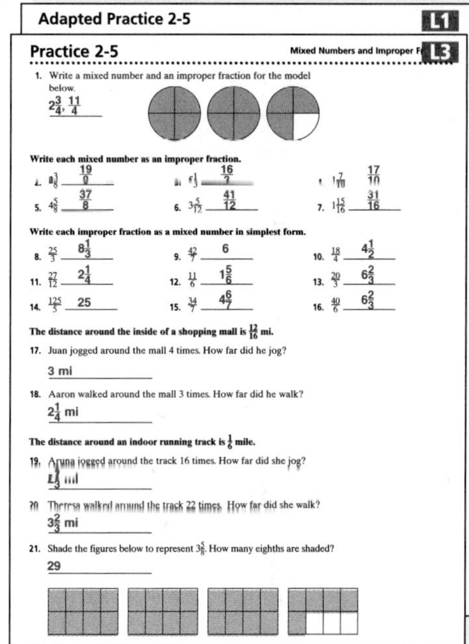

Adapted Practice 2-5 L1

Practice 2-5 Mixed Numbers and Improper F[illegible] L3

1. Write a mixed number and an improper fraction for the model below.
$2\frac{3}{4}$, $\frac{11}{4}$

Write each mixed number as an improper fraction.

2. [illegible] $\frac{19}{9}$ 3. [illegible] $\frac{16}{3}$ 4. $1\frac{7}{10}$ $\frac{17}{10}$

5. $4\frac{5}{8}$ $\frac{37}{8}$ 6. $3\frac{5}{12}$ $\frac{41}{12}$ 7. $1\frac{15}{16}$ $\frac{31}{16}$

Write each improper fraction as a mixed number in simplest form.

8. $\frac{25}{3}$ $8\frac{1}{3}$ 9. $\frac{42}{7}$ 6 10. $\frac{18}{4}$ $4\frac{1}{2}$

11. $\frac{27}{12}$ $2\frac{1}{4}$ 12. $\frac{11}{6}$ $1\frac{5}{6}$ 13. $\frac{20}{3}$ $6\frac{2}{3}$

14. $\frac{125}{5}$ 25 15. $\frac{34}{7}$ $4\frac{6}{7}$ 16. $\frac{40}{6}$ $6\frac{2}{3}$

The distance around the inside of a shopping mall is $\frac{12}{16}$ mi.

17. Juan jogged around the mall 4 times. How far did he jog?
3 mi

18. Aaron walked around the mall 3 times. How far did he walk?
$2\frac{1}{4}$ mi

The distance around an indoor running track is $\frac{1}{6}$ mile.

19. Aruna jogged around the track 16 times. How far did she jog?
[illegible] mi

20. Theresa walked around the track 22 times. How far did she walk?
$3\frac{2}{3}$ mi

21. Shade the figures below to represent $3\frac{5}{8}$. How many eighths are shaded?
29

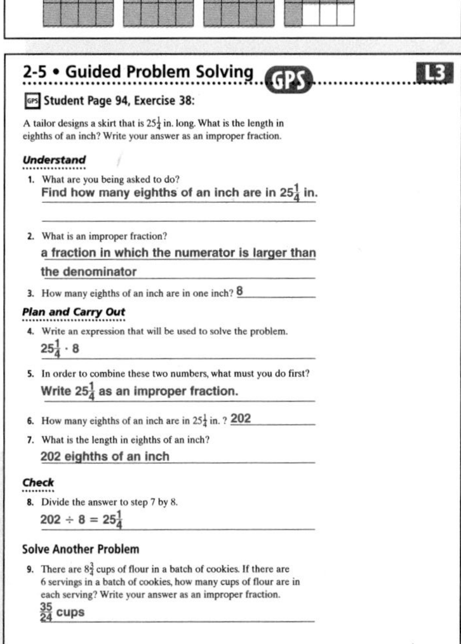

2-5 • Guided Problem Solving GPS L3

Student Page 94, Exercise 38:

A tailor designs a skirt that is $25\frac{1}{4}$ in. long. What is the length in eighths of an inch? Write your answer as an improper fraction.

Understand

1. What are you being asked to do?
Find how many eighths of an inch are in $25\frac{1}{4}$ in.

2. What is an improper fraction?
a fraction in which the numerator is larger than the denominator

3. How many eighths of an inch are in one inch? 8

Plan and Carry Out

4. Write an expression that will be used to solve the problem.
$25\frac{1}{4} \cdot 8$

5. In order to combine these two numbers, what must you do first?
Write $25\frac{1}{4}$ as an improper fraction.

6. How many eighths of an inch are in $25\frac{1}{4}$ in. ? 202

7. What is the length in eighths of an inch?
202 eighths of an inch

Check

8. Divide the answer to step 7 by 8.
$202 \div 8 = 25\frac{1}{4}$

Solve Another Problem

9. There are $8\frac{3}{4}$ cups of flour in a batch of cookies. If there are 6 servings in a batch of cookies, how many cups of flour are in each serving? Write your answer as an improper fraction.
$\frac{35}{24}$ cups

4. Assess & Reteach

Lesson Quiz

Write as an improper fraction.

1. $5\frac{3}{8}$ $\frac{43}{8}$
2. $6\frac{2}{3}$ $\frac{20}{3}$

Write as a mixed number in simplest form.

3. $\frac{34}{8}$ $4\frac{1}{4}$
4. $\frac{40}{6}$ $6\frac{2}{3}$

Reteaching 2-5 — Mixed Numbers and Improper Fractions — L2

An *improper fraction* is greater than or equal to 1. Its numerator is greater than or equal to its denominator.

Improper fractions: $\frac{6}{4}$ $\frac{8}{8}$ $\frac{10}{8}$ $\frac{7}{2}$

A *mixed number* is the sum of a whole number and a fraction.

Mixed numbers: $1\frac{2}{3}$ $5\frac{4}{9}$ $3\frac{1}{2}$

To write a mixed number as an improper fraction:

① Write the mixed number as a sum. $3\frac{1}{2} = 3 + \frac{1}{2}$

② Write both numbers as fractions. $= \frac{6}{2} + \frac{1}{2}$

③ Add the fractions. $= \frac{7}{2}$

To write an improper fraction as a mixed number:

① Divide the numerator by the denominator. $\frac{7}{2}$ (Think: 7 ÷ 2) $7 \div 2 = 3$, $7 - 6 = 1$

② Write the whole number, then the remainder over the divisor. $\frac{7}{2} = 3\frac{1}{2}$

Write each mixed number as an improper fraction.

1. $3\frac{1}{4} = \frac{13}{4}$
2. $2\frac{2}{3} = \frac{8}{3}$
3. $1\frac{3}{8} = \frac{11}{8}$
4. $5\frac{2}{7} = \frac{16}{7}$
5. $6\frac{3}{4} = \frac{27}{4}$
6. $1\frac{1}{9} = \frac{10}{9}$
7. $4\frac{1}{2} = \frac{9}{2}$
8. $3\frac{4}{5} = \frac{19}{5}$
9. $5\frac{1}{6} = \frac{31}{6}$
10. $3\frac{1}{3} = \frac{10}{3}$
11. $5\frac{7}{8} = \frac{47}{8}$
12. $4\frac{1}{8} = \frac{33}{8}$

Write each improper fraction as a mixed number in simplest form.

13. $\frac{14}{4} = 3\frac{1}{2}$
14. $\frac{12}{2} = 6$
15. $\frac{22}{5} = 4\frac{2}{5}$
16. $\frac{16}{3} = 5\frac{1}{3}$
17. $\frac{47}{8} = 5\frac{7}{8}$
18. $\frac{56}{7} = 8$
19. $\frac{17}{4} = 4\frac{1}{4}$
20. $\frac{21}{6} = 3\frac{1}{2}$
21. $\frac{13}{5} = 2\frac{3}{5}$
22. $\frac{23}{4} = 5\frac{3}{4}$
23. $\frac{13}{9} = 1\frac{4}{9}$
24. $\frac{14}{2} = 7$

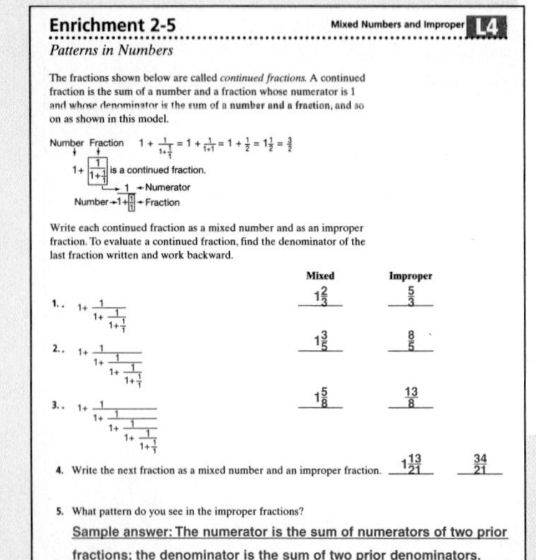

Enrichment 2-5 — Mixed Numbers and Improper Fractions — L4

Patterns in Numbers

The fractions shown below are called *continued fractions.* A continued fraction is the sum of a number and a fraction whose numerator is 1 and whose denominator is the sum of a number and a fraction, and so on as shown in this model.

Number Fraction $1 + \frac{1}{1+\frac{1}{1}} = 1 + \frac{1}{1+1} = 1 + \frac{1}{2} = 1\frac{1}{2} = \frac{3}{2}$

$1 + \frac{1}{1+\frac{1}{1}}$ is a continued fraction. (Number → 1, 1 ← Numerator, $\frac{1}{1}$ ← Fraction)

Write each continued fraction as a mixed number and as an improper fraction. To evaluate a continued fraction, find the denominator of the last fraction written and work backward.

		Mixed	Improper
1.	$1+\frac{1}{1+\frac{1}{1+\frac{1}{1}}}$	$1\frac{2}{3}$	$\frac{5}{3}$
2.	$1+\frac{1}{1+\frac{1}{1+\frac{1}{1+\frac{1}{1}}}}$	$1\frac{3}{5}$	$\frac{8}{5}$
3.	$1+\frac{1}{1+\frac{1}{1+\frac{1}{1+\frac{1}{1+\frac{1}{1}}}}}$	$1\frac{5}{8}$	$\frac{13}{8}$

4. Write the next fraction as a mixed number and an improper fraction. $1\frac{13}{21}$ $\frac{34}{21}$

5. What pattern do you see in the improper fractions?
Sample answer: The numerator is the sum of numerators of two prior fractions; the denominator is the sum of two prior denominators.

6. What would be the value of the tenth continued fraction? $\frac{233}{144}$

GO Online Homework Video Tutor
Visit: PHSchool.com
Web Code: are-0205

Algebra In Exercises 33–35, evaluate each expression for $a = 6$, $b = 3$, $c = 2$, and $d = 5$. Write your answer in simplest form.

33. $\frac{b}{a^2}$ $\frac{1}{12}$
34. $\frac{a^2 + b}{c}$ $19\frac{1}{2}$
35. $\frac{a + c}{b + d}$ 1

Write each length as a mixed number and as an improper fraction.

36. 37.

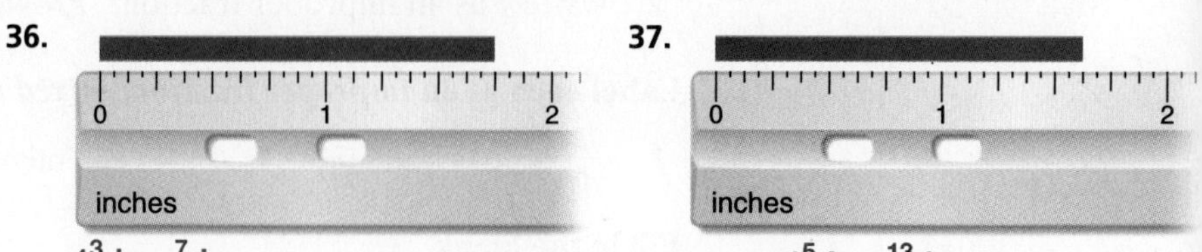

36. $1\frac{3}{4}$ in., $\frac{7}{4}$ in.
37. $1\frac{5}{8}$ in., $\frac{13}{8}$ in.

38. A tailor designs a skirt that is $25\frac{1}{4}$ in. long. What is the length in eighths of an inch? Write your answer as an improper fraction. 202 eighths; $\frac{202}{8}$ in.

39. The distance around a track is $\frac{1}{8}$ mi. A wheelchair racer completes 35 laps around the track. Write the distance he travels in miles as a mixed number. $4\frac{3}{8}$ mi

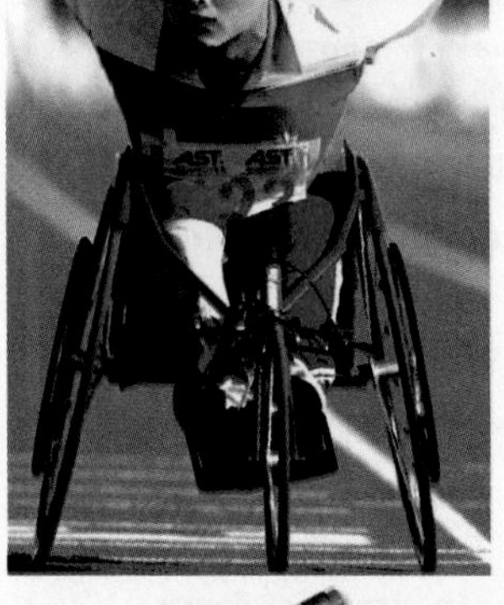

40. **Writing in Math** Which is longer, $\frac{9}{4}$ miles or $1\frac{1}{2}$ miles? Explain. See left.

40. $\frac{9}{4}$ mi; $1\frac{1}{2}$ mi $= \frac{6}{4}$ mi, which is less than $\frac{9}{4}$ mi

41. Write two mixed numbers and an improper fraction for the model.

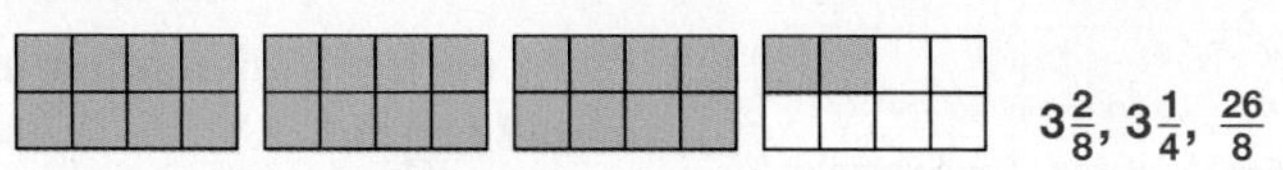

$3\frac{2}{8}$, $3\frac{1}{4}$, $\frac{26}{8}$

42. **Challenge** Using each of the digits 2, 5, and 9 exactly once, write a fraction with the greatest possible value. Then write the fraction as a mixed number. $\frac{95}{2}$; $47\frac{1}{2}$

Test Prep and Mixed Review — Practice

Multiple Choice

43. The table below shows the distances a football team ran each day last week. On which day did the team run the least? B

Football Practice

Day	Monday	Tuesday	Wednesday	Thursday
Distance Run	$\frac{7}{10}$ mile	$\frac{5}{8}$ mile	$\frac{3}{4}$ mile	$\frac{2}{3}$ mile

Ⓐ Monday Ⓑ Tuesday Ⓒ Wednesday Ⓓ Thursday

44. A 1-cup serving of mashed potatoes with whole milk and butter has 634 milligrams of sodium. How many grams of sodium are in 3 cups of mashed potatoes? G

Ⓕ 0.1902 g Ⓖ 1.902 g Ⓗ 190.2 g Ⓙ 1,902 g

GO for Help

For Exercises	See Lesson
45–47	1-7

Find each difference.

45. $-3 - 9$ -12
46. $(-10) - (-18)$ 8
47. $21 - (-7)$ 28

Test Prep

Resources

For additional practice with a variety of test item formats:

- Test-Taking Strategies, p. 111
- Test Prep, p. 115
- Test-Taking Strategies with Transparencies

Alternative Assessment

Have students work in pairs. Each student writes two improper fractions and two mixed numbers. Students exchange papers. They write each improper fraction as a mixed number in simplest form and each mixed number as an improper fraction. Have students explain to their partners the steps they took to complete each procedure.

2-6a Activity Lab

Comparing Fractions and Decimals

A baseball player's batting average is the number of hits divided by the number of times at bat. You can tell which player has the best record by comparing batting averages.

ACTIVITY

1. Copy the table below.

Name of Player	Number of Hits	Times at Bat	Batting Average	
			Fraction	Decimal
Allan	43	195	■	■
Belinda	8	127	■	■
Char	43	183	■	■
Denise	47	183	■	■
Emil	29	174	■	■
Farik	45	135	■	■

2. Write each batting average as a fraction.
3. Use a calculator to write each batting average as a decimal. Round to the nearest thousandth.

1–3.

Name of Player	Batting Average	
	Fraction	Decimal
Allan	$\frac{43}{195}$	.221
Belinda	$\frac{8}{127}$	.063
Char	$\frac{43}{183}$	.235
Denise	$\frac{47}{183}$	.257
Emil	$\frac{29}{174}$	.167
Farik	$\frac{45}{135}$	.333

Exercises

1. a. Which players had averages that are decimals with a repeating block of digits? **Emil, Farik**
   b. List the players in order, from highest to lowest batting averages. **Farik, Denise, Char, Allan, Emil, Belinda**
2. Is it easier to compare and order batting averages using fractions or decimals? Explain. **Decimals; the fractions do not have common denominators.**
3. A student thought that Belinda's average was 0.630. What error do you think she made? **Answers may vary. Sample: She used 80 instead of 8.**
4. How can you tell, without using decimals, that Denise has a higher average than Char? **They both batted the same number of times, but Denise had more hits.**
5. Can a player get 132 hits in 125 times at bat? How is your answer related to the highest possible batting average of 1.000? **No; a player cannot get more hits than times at bat.**

Activity Lab

Comparing Fractions and Decimals

Students practice ordering fractions and converting fractions to decimals. This will prepare them to compare and order fractions and decimals in Lesson 2-6.

Guided Instruction

Explain a *batting average* and that it can be expressed as a fraction or a decimal. Provide examples of record batting averages from a sports record book.
Ask:
- *Would a batting average ever be expressed as a mixed number? Why?* **No, a person cannot have more hits than number of times at bat.**
- *Batting averages are usually expressed to the thousandths place. How would you convert a fraction to make a conventional batting average?* **Convert the fraction to a decimal, round to the thousandths place or annex zeros to get to the thousandths place.**

Exercises
Have students work on the Exercises. Then have them trade papers with a partner and compare answers.

Alternative Method
Allow students to use graph paper and shade sections to model the given fractions.

Resources

- Activity Lab 2-6: Fractions and Decimals
- graph paper

2-6

1. Plan

Objective
To convert between fractions and decimals

Examples
1 Writing a Terminating Decimal
2 Writing a Repeating Decimal
3 Writing a Decimal as a Fraction
4 Ordering Fractions and Decimals
5 Application: Surveys

Math Understandings: p. 66D

Math Background

For denominators whose only prime factors are 2 or 5—denominators such as 2, 4, 5, 8, 10, 16, 20, 25, 32, 40, 50, and so on—the quotient will be a *terminating decimal.* That is, the division $a \div b$ will end, or *terminate,* with a remainder of zero. Otherwise, a block of digits in the quotient will repeat without end, resulting in a *repeating decimal.*

More Math Background: p. 66D

Lesson Planning and Resources

See p. 66E for a list of the resources that support this lesson.

PowerPoint

Bell Ringer Practice

Check Skills You'll Need
Use student page, transparency, or PowerPoint. For intervention, direct students to:
Simplifying Fractions
Lesson 2-3
Extra Skills and Word Problems Practice, Ch. 2

2-6 Fractions and Decimals

Check Skills You'll Need

1. **Vocabulary Review** How can you tell whether fractions are *equivalent fractions*? **1–5. See below.**

Write two fractions equivalent to each fraction.

2. $\frac{16}{20}$ 3. $\frac{15}{20}$
4. $\frac{6}{9}$ 5. $\frac{10}{50}$

for Help
Lesson 2-3

What You'll Learn

To convert between fractions and decimals

New Vocabulary terminating decimal, repeating decimal

Why Learn This?

When you order sandwich meat at a delicatessen, you may ask for half a pound. The scales at a deli often use decimal measures. You can convert between fractions and decimals to make sure you are receiving the correct amount.

You write a fraction as a decimal by dividing the numerator by the denominator. A decimal that stops, or terminates, is a **terminating decimal.**

Test Prep Tip
Remember to fill in the corresponding bubbles after you write your answer in the grid.

Check Skills You'll Need
1. In simplest form, they are the same.
2–5. Answers may vary. Samples are given.
2. $\frac{4}{5}, \frac{8}{10}$
3. $\frac{3}{4}, \frac{6}{8}$
4. $\frac{2}{3}, \frac{8}{12}$
5. $\frac{1}{5}, \frac{2}{10}$

EXAMPLE Writing a Terminating Decimal

1 **Gridded Response** The pull of gravity is weaker on the moon than on Earth. The fraction $\frac{4}{25}$ represents the ratio of the moon's gravity to Earth's gravity. Write this fraction as a decimal.

$\frac{4}{25}$ or $4 \div 25 = 25\overline{)4.00}$ with quotient 0.16 ← quotient

$$\begin{array}{r} 0.16 \\ 25\overline{)4.00} \\ -25 \\ 150 \\ -150 \\ 0 \end{array}$$

← The remainder is 0.

	.	1	6

The ratio of the moon's gravity to Earth's gravity as a decimal is 0.16. This is a terminating decimal because the division process stops when the remainder is 0.

Quick Check

1. The fraction of nitrogen in a chemical sample is $\frac{5}{8}$. Write the fraction as a decimal. **0.625**

Differentiated Instruction Solutions for All Learners

Special Needs L1
To help students with decimal and fraction relationships, draw a rectangular model for the fraction $\frac{2}{5}$, with 2 of the 5 bars shaded. Then have them superimpose a same-sized rectangle divided into tenths to name the decimal equivalent (4 tenths or 0.4).
learning style: tactile

Below Level L2
Review divisions in which zeros must be annexed to the dividend, such as $1.2 \div 5$.

$5\overline{)1.2} \rightarrow 5\overline{)1.20} \rightarrow 5\overline{)1.20}$ with quotient 0.24

learning style: visual

Vocabulary Tip

The symbol for a repeating decimal is a bar over the repeated digit(s), for example, $0.\overline{17}$.

If the same block of digits in a decimal repeats without end, the decimal is a **repeating decimal.** The repeating block can include one or more digits.

$5.355555555555\ldots = 5.3\overline{5}$ ← The digit 5 repeats.

$0.171717171717\ldots = 0.\overline{17}$ ← The digits 17 repeat.

EXAMPLE Writing a Repeating Decimal

2 Write $\frac{3}{11}$ as a decimal.

Method 1 Paper and Pencil

$\frac{3}{11}$ or $3 \div 11 = 11\overline{)3.00000}$, quotient 0.27272 ← The digits 27 repeat.

$$\begin{array}{r} 0.27272 \\ 11\overline{)3.00000} \\ -22 \\ \hline 80 \\ -77 \\ \hline 30 \\ -22 \\ \hline 80 \\ -77 \\ \hline 30 \end{array}$$

← There will always be a remainder of 30 or 80.

Calculator Tip

Most calculators display as many digits as possible of a repeating decimal and round off the final digit. For example, $0.\overline{27}$ might be shown as 0.272727273.

Method 2 Calculator

3 ÷ 11 = 0.27272727273

So $\frac{3}{11} = 0.\overline{27}$.

Quick Check

2. Write $\frac{5}{9}$ as a decimal. **$0.\overline{5}$**

You can write a terminating decimal as a fraction or a mixed number by writing the digits to the right of the decimal point as a fraction.

EXAMPLE Writing a Decimal as a Fraction

3 Write 1.325 as a mixed number with a fraction in simplest form.

Since $0.325 = \frac{325}{1{,}000}$, $1.325 = 1\frac{325}{1{,}000}$.

$1\frac{325}{1{,}000} = 1\frac{325 \div 25}{1{,}000 \div 25}$ ← Use the GCF to write the fraction in simplest form.

$= 1\frac{13}{40}$

Quick Check

3. Write each decimal as a mixed number in simplest form.
   a. 1.364 **$1\frac{91}{250}$** b. 2.48 **$2\frac{12}{25}$** c. 3.6 **$3\frac{3}{5}$**

2. Teach

Activity Lab

Use before the lesson.

Student Edition Activity Lab 2-6a, Comparing Fractions and Decimals, p. 95

All in One Teaching Resources

Activity Lab 2-6: Fractions and Decimals

Guided Instruction

Example 3

Some students might not recognize 25 as the GCF of 325 and 1,000. It might be easier for them to perform two successive divisions by 5.

$1\frac{325}{1{,}000} = 1\frac{325 \div 5}{1{,}000 \div 5} = 1\frac{65}{200}$

$1\frac{65}{200} = 1\frac{65 \div 5}{200 \div 5} = 1\frac{13}{40}$

Error Prevention!

In writing repeating decimals, students might place the bar over too few or too many digits. Be sure they understand that the bar must be placed over all the digits that repeat, and only those digits that repeat.

PowerPoint

Additional Examples

1. The total amount of rainfall yesterday was reported as $\frac{1}{4}$ in. Express the amount of rainfall as a decimal. **0.25 in.**
2. Write $\frac{7}{15}$ as a decimal. **$0.4\overline{6}$**
3. Write 4.105 as a fraction in simplest form. **$4\frac{21}{200}$**
4. Order from least to greatest: $3\frac{3}{4}$, 2.897, $3\frac{7}{10}$. **2.897, $3\frac{7}{10}$, $3\frac{3}{4}$**
5. In a survey of next year's seventh-grade students, 0.25 said they will come to school by bus, $\frac{5}{24}$ said they will walk, 0.375 said they will come in a car, and $\frac{1}{16}$ said they will ride their bicycles. Order the means of transportation from most used to least used. **car, bus, walk, bicycle**

Advanced Learners L4

Write each fraction as a decimal.

$\frac{1}{7}$ $0.\overline{142857}$ $\frac{1}{13}$ $0.\overline{076923}$

learning style: visual

English Language Learners ELL

Make sure students are reading the decimal numbers correctly, in tenths, hundredths, and thousandths. Discourage the use of language like *0 point 2* for two-tenths. Remind students that decimals always have denominators that are multiples of ten.

learning style: verbal

All in One Teaching Resources
- Daily Notetaking Guide 2-6 L3
- Adapted Notetaking 2-6 L1

Closure

- *What is a terminating decimal?* the decimal that results when the numerator divided by the denominator has a remainder of zero
- *What is a repeating decimal?* the decimal that results when the numerator divided by the denominator has a block of digits that repeats without end

To compare fractions and decimals, you can write the decimals as fractions or the fractions as decimals. You can decide which is easier for different numbers.

GO for Help For help with place value and decimals, go to Skills Handbook p. 658

EXAMPLE Ordering Fractions and Decimals

4 Order from greatest to least: 2, 2.55, $2\frac{6}{18}$.

$2\frac{6}{18} = 2\frac{1}{3} = 2.\overline{3}$ ← **Use a calculator to change the mixed number to a decimal.**

Use a number line to find each decimal number's relative position.

Number line from 1.0 to 2.8 (1.0, 1.2, 1.4, 1.6, 1.8, 2.0, 2.2, 2.4, 2.6, 2.8) with points marked at 2, $2.\overline{3}$, and 2.55.

So the order of the numbers from greatest to least is 2.55, $2\frac{6}{18}$, and 2.

Quick Check

4. Order from greatest to least: $1\frac{3}{8}$, $1\frac{7}{15}$, 1.862. **1.862, $1\frac{7}{15}$, $1\frac{3}{8}$**

You can order rational numbers to analyze data results.

EXAMPLE Application: Surveys

Careers Journalists gather information, analyze data, and write reports.

5 For a survey naming four animals, adults were asked to choose the animal they thought was the most endangered species. Of the adults surveyed, 0.25 chose black rhinoceros, $\frac{10}{48}$ chose tiger, $\frac{5}{12}$ chose giant panda, and 0.125 chose mako shark. List their choices in order of frequency.

First write all 4 ratios as decimals.

black rhinoceros: 0.25

tiger: $\frac{10}{48} = \frac{5}{24} = 0.208\overline{3}$

giant panda: $\frac{5}{12} = 0.41\overline{6}$

mako shark: 0.125

← **Use a calculator to change the fractions to decimals.**

Since $0.41\overline{6} > 0.25 > 0.208\overline{3} > 0.125$, the order of frequency was giant panda, black rhinoceros, tiger, and mako shark.

Quick Check

5. In a survey about pets, $\frac{2}{5}$ of students prefer cats, 0.33 prefer dogs, $\frac{3}{25}$ prefer birds, and 0.15 prefer fish. List the choices in order of preference. **cats, dogs, fish, birds**

Check Your Understanding

1. **Vocabulary** What is the difference between a terminating decimal and a repeating decimal? **A terminating decimal stops, whereas a repeating decimal has a block of digits that repeat without end.**
2. **Reasoning** Is a remainder of 0 the same as no remainder? Explain. **Yes; in each case a number is divisible by the other.**

Write each decimal as a mixed number or a fraction in simplest form.

3. 1.375 **$1\frac{3}{8}$**
4. 0.44 **$\frac{11}{25}$**
5. 3.99 **$3\frac{99}{100}$**

6. **Reasoning** Is 3.03003000300003 . . . a repeating decimal? Explain. **No; the same block of digits is not repeating.**
7. Order the following numbers from least to greatest: $1.\overline{9}$, $1\frac{1}{3}$, 2, 0.5. **0.5, $1\frac{1}{3}$, $1.\overline{9}$, 2**

Homework Exercises

For more exercises, see Extra Skills and Word Problems.

GO for Help

For Exercises	See Examples
8–15	1–2
16–23	3
24–28	4–5

A **Write each fraction as a decimal.**

8. $\frac{2}{5}$ **0.4**
9. $\frac{4}{5}$ **0.8**
10. $\frac{3}{8}$ **0.375**
11. $\frac{2}{3}$ **$0.\overline{6}$**
12. $\frac{3}{4}$ **0.75**
13. $\frac{1}{8}$ **0.125**
14. $\frac{7}{11}$ **$0.\overline{63}$**
15. $\frac{3}{16}$ **0.1875**

Write each decimal as a mixed number or a fraction in simplest form.

16. 0.125 **$\frac{1}{8}$**
17. 0.66 **$\frac{33}{50}$**
18. 2.5 **$2\frac{1}{2}$**
19. 3.75 **$3\frac{3}{4}$**
20. 0.32 **$\frac{8}{25}$**
21. 0.19 **$\frac{19}{100}$**
22. 0.8 **$\frac{4}{5}$**
23. 0.965 **$\frac{193}{200}$**

Order from greatest to least. **24–27. See left.**

24. $\frac{9}{22}$, 0.83, $\frac{7}{8}$, 0.4, $\frac{44}{44}$
25. 3.84, 3.789, 3, $3\frac{41}{50}$
26. $\frac{2}{3}$, 0.67, $\frac{5}{9}$, 0.58, $\frac{7}{12}$
27. $0.1\overline{2}$, 0.1225, $\frac{3}{25}$, $\frac{7}{125}$

24. **$\frac{44}{44}$, $\frac{7}{8}$, 0.83, $\frac{9}{22}$, 0.4**
25. **3.84, $3\frac{41}{50}$, 3.789, 3**
26. **0.67, $\frac{2}{3}$, $\frac{7}{12}$, 0.58, $\frac{5}{9}$**
27. **0.1225, $0.1\overline{2}$, $\frac{3}{25}$, $\frac{7}{125}$**

28. **Biology** DNA content in a cell is measured in picograms (pg). A sea star cell has $\frac{17}{20}$ pg of DNA, a scallop cell has $\frac{19}{20}$ pg, a red water mite cell has 0.19 pg, and a mosquito cell has 0.24 pg. Order the DNA contents from greatest to least. **$\frac{19}{20}$ pg, $\frac{17}{20}$ pg, 0.24 pg, 0.19 pg**

B 

29. **Guided Problem Solving** On an adventure trail, you biked 12 mi, walked 4 mi, ran 6 mi, and swam 2 mi. What fraction of the total distance did you bike? Write this number as a decimal. **$\frac{1}{2}$; 0.5**
    - The total distance of the adventure trail is ■ mi.
    - You biked ■ mi. As a fraction, this is ■ of the total distance.

30. **Music** To compose music on a computer, you can write notes as decimals. What decimals should you enter for a half note, a quarter note, an eighth note, and a sixteenth note? **0.5, 0.25, 0.125, 0.0625**

Online lesson quiz, PHSchool.com, Web Code: ara-0206

3. Practice

Assignment Guide

Check Your Understanding

Go over Exercises 1–7 in class before assigning the Homework Exercises.

Homework Exercises

A	Practice by Example	8–28
B	Apply Your Skills	29–40
C	Challenge	41
	Test Prep and Mixed Review	42–48

Homework Quick Check

To check students' understanding of key skills and concepts, go over Exercises 13, 21, 28, 36, and 40.

Differentiated Instruction Resources

Adapted Practice 2-6 L1

Practice 2-6 Fractions and D[ecimals] L3

Write each fraction as a decimal.

1. $\frac{3}{5}$ 0.6
2. $\frac{7}{8}$ 0.875
3. $\frac{7}{9}$ $0.\overline{7}$
4. $\frac{5}{16}$ 0.3125
5. $\frac{1}{6}$ $0.1\overline{6}$
6. $\frac{5}{8}$ 0.625
7. $\frac{1}{3}$ $0.\overline{3}$
8. $\frac{2}{3}$ $0.\overline{6}$
9. $\frac{9}{10}$ 0.9
10. $\frac{7}{11}$ $0.\overline{63}$
11. $\frac{9}{20}$ 0.45
12. $\frac{3}{4}$ 0.75
13. $\frac{4}{9}$ $0.\overline{4}$
14. $\frac{9}{11}$ $0.\overline{81}$
15. $\frac{11}{20}$ 0.55

Write each decimal as a mixed number or fraction in simplest form.

16. 0.6 $\frac{3}{5}$
17. 0.45 $\frac{9}{20}$
18. 0.62 $\frac{31}{50}$
19. 0.8 $\frac{4}{5}$
20. 0.325 $\frac{13}{40}$
21. 0.725 $\frac{29}{40}$
22. 4.75 $4\frac{3}{4}$
23. 0.33 $\frac{33}{100}$
24. 0.925 $\frac{37}{40}$
25. 3.8 $3\frac{4}{5}$
26. 4.7 $4\frac{7}{10}$
27. 0.05 $\frac{1}{20}$
28. 0.65 $\frac{13}{20}$
29. 0.855 $\frac{171}{200}$
30. 0.104 $\frac{13}{125}$
31. 0.47 $\frac{47}{100}$
32. 0.894 $\frac{447}{500}$
33. 0.276 $\frac{69}{250}$

Order from least to greatest.

34. $0.\overline{2}$, $\frac{1}{5}$, 0.02 — 0.02, $\frac{1}{5}$, $0.\overline{2}$
35. [illegible] — $1\frac{1}{10}$, 1.101, $1.\overline{1}$
36. $\frac{6}{5}$, $1\frac{5}{6}$, $1.\overline{3}$ — $\frac{6}{5}$, $1.\overline{3}$, $1\frac{5}{6}$
37. $4.\overline{3}$, $\frac{9}{2}$, $4\frac{3}{7}$ — $4.\overline{3}$, $4\frac{3}{7}$, $\frac{9}{2}$
38. A group of gymnasts were asked to name their favorite piece of equipment. 0.33 of the gymnasts chose the vault, $\frac{4}{9}$ chose the beam, and $\frac{1}{7}$ chose the uneven parallel bars. List their choices in order of preference from greatest to least. beam, vault, uneven parallel bars; $\frac{4}{9}$, 0.33, $\frac{1}{7}$

2-6 • Guided Problem Solving  L3

Student Page 99, Exercise 28:

Biology DNA content in a cell is measured in picograms (pg). A sea star cell has $\frac{17}{20}$ pg of DNA, a scallop cell has $\frac{19}{25}$ pg, a red water mite cell has 0.19 pg, and a mosquito cell has 0.024 pg. Order the DNA contents from greatest to least.

Understand

1. What are you being asked to do? Order the DNA content of the organisms.
2. To order fractions and decimals, what must you do first? Make all the numbers fractions or make all the numbers decimals.

Plan and Carry Out

3. Write the fraction $\frac{17}{20}$ as a decimal. 0.85
4. Write the fraction $\frac{19}{25}$ as a decimal. 0.76
5. Which organism has the smallest DNA content? mosquito
6. Which organism has the largest DNA content? sea star
7. Order the DNA contents from greatest to least. $\frac{17}{20}$ pg, $\frac{19}{25}$ pg, 0.19 pg, 0.024 pg

Check

8. Write 0.19 and 0.024 as fractions in simplest form. Order the DNA contents from greatest to least. Does your order check with that of step 7? $\frac{19}{100}$; $\frac{3}{125}$; $\frac{17}{20}$ pg, $\frac{19}{25}$ pg, 0.024 pg; yes

Solve Another Problem

9. A solution calls for 0.25 oz of water, $\frac{2}{3}$ oz of vinegar, 0.6 oz of carbonate, and $\frac{9}{16}$ oz of lemon juice. Order the amounts from least to greatest. 0.25 oz, $\frac{9}{16}$ oz, 0.6 oz, $\frac{2}{3}$ oz

4. Assess & Reteach

Lesson Quiz

Write each fraction as a decimal.

1. $\frac{7}{8}$ 0.875
2. $\frac{4}{9}$ $0.\overline{4}$
3. $\frac{3}{5}$ 0.6
4. $\frac{5}{16}$ 0.3125

Alternative Assessment

Have students work in pairs. Each student writes a decimal and a fraction. Students exchange papers and rewrite the decimal as a fraction and the fraction as a decimal. Then pairs work together to arrange the original decimals and fractions in order from least to greatest.

Reteaching 2-6 — Fractions and … L2

0 0.1 0.2 0.3 0.4 0.5 0.6 0.7 0.8 0.9 1.0 1.1 1.2 1.3 1.4

$\frac{1}{10}$ $\frac{1}{5}$ $\frac{1}{2}$ $\frac{3}{5}$ 1 $1\frac{1}{5}$ $1\frac{2}{5}$

To change a fraction to a decimal, divide the numerator by the denominator.

$\frac{3}{5}$ (Think: 3 ÷ 5)

$$\begin{array}{r} 0.6 \\ 5\overline{)3.0} \\ -30 \\ \hline 0 \end{array}$$

$\frac{3}{5} = 0.6$

To change a decimal to a fraction:

① Read the decimal to find the denominator. Write the decimal digits over 10, 100, or 1,000.

② 0.65 is 65 *hundredths* → $\frac{65}{100}$

Use the GCF to write the fraction in simplest form.

The GCF of 65 and 100 is 5.

$\frac{65}{100} = \frac{65 \div 5}{100 \div 5} = \frac{13}{20}$

Write each fraction as a decimal.

1. $\frac{4}{5}$ = 0.8
2. $\frac{3}{4}$ = 0.75
3. $\frac{1}{6}$ = $0.1\overline{6}$
4. $\frac{1}{4}$ = 0.25
5. $\frac{2}{3}$ = $0.\overline{6}$
6. $\frac{7}{10}$ = 0.7
7. $\frac{5}{9}$ = $0.\overline{5}$
8. $\frac{1}{5}$ = 0.2
9. $\frac{3}{8}$ = 0.375

Write each decimal as a mixed number or fraction in simplest form.

10. 0.4 = $\frac{2}{5}$
11. 0.75 = $\frac{3}{4}$
12. 1.5 = $1\frac{1}{2}$
13. 0.35 = $\frac{7}{20}$
14. 2.7 = $2\frac{7}{10}$
15. 1.8 = $1\frac{4}{5}$
16. 0.625 = $\frac{5}{8}$
17. 0.78 = $\frac{39}{50}$
18. 0.88 = $\frac{22}{25}$

Order from least to greatest.

19. $2.\overline{6}, \frac{13}{6}, 2\frac{5}{6}$ — $\frac{13}{6}, 2.\overline{6}, 2\frac{5}{6}$
20. $2.\overline{02}, 2\frac{1}{200}, 2.0202$ — $2\frac{1}{200}, 2.0202, 2.\overline{02}$
21. $\frac{5}{4}, 1\frac{4}{5}, 1.\overline{4}$ — $\frac{5}{4}, 1.\overline{4}, 1\frac{4}{5}$

Enrichment 2-6 — Fractions and … L4

Patterns in Numbers

Find a decimal for each fraction. Use your calculator to check your work.

1. $\frac{3}{8}$ 0.375
2. $\frac{5}{6}$ $0.8\overline{3}$
3. $\frac{7}{12}$ $0.58\overline{3}$
4. $\frac{4}{5}$ 0.8
5. $\frac{8}{9}$ $0.\overline{8}$
6. $\frac{3}{10}$ 0.3

7. List the denominators for the fractions that produced terminating decimals. 5, 8, 10
8. Find the prime factors of the denominators in exercise 7. 2, 5
9. List the denominators for the fractions that produced repeating decimals. 6, 9, 12
10. Find the prime factors of the denominators in exercise 9. 2, 3, 5, 11
11. What pattern do you notice? Possible answer: Denominators for terminating decimals have only 2, only 5, or 2 and 5 as prime factors.
12. Without dividing, tell whether each of the following fractions will produce a terminating or a repeating decimal.
    a. $\frac{7}{24}$ repeating
    b. $\frac{35}{64}$ terminating
    c. $\frac{17}{111}$ repeating
    d. $\frac{14}{148}$ repeating
13. Write a fraction that, as a decimal, will terminate. Use a 2-digit or a 3-digit number as the denominator. Sample answer: $\frac{102}{160}$
14. Write a fraction that, as a decimal, will repeat. Use a 2-digit or a 3-digit number as the denominator. Sample answer: $\frac{36}{101}$

GO Online
Homework Video Tutor
Visit: PHSchool.com
Web Code: are-0206

Algebra **Compare. Use <, =, or >. (*Note*: *n* is a value greater than 1.)**

31. $\frac{1}{n}$ ■ $\frac{n}{n}$ <
32. 1 ■ $\frac{n}{1}$ <
33. n ■ $\frac{1}{n}$ >
34. $\frac{n}{n^2}$ ■ $\frac{1}{n}$ =

35. **Number Sense** Examine the fractions $\frac{2}{3}$, $\frac{3}{3}$, $\frac{4}{3}$, $\frac{5}{3}$, and $\frac{6}{3}$. Explain when a denominator of 3 will result in a repeating decimal. **when the numerator is not a multiple of 3**

Greenland is the world's largest island.

36. **Geography** About 12,500 icebergs break away from Greenland each year. Of these, about 375 float into the Atlantic Ocean.
    a. What fraction of the icebergs float into the Atlantic Ocean? $\frac{3}{100}$
    b. Write your answer for part (a) as a decimal. **0.03**
    c. What fraction of the icebergs do *not* float into the Atlantic? $\frac{97}{100}$

For Exercises 37–39, use the table at the right.

37. See margin.

37. For each state, write a fraction that shows the ratio $\frac{\text{number of people under age 18}}{\text{total population}}$.
38. For most of the states, would $\frac{1}{2}$, $\frac{1}{3}$, or $\frac{1}{4}$ best describe the fraction of the population that is under age 18? $\frac{1}{4}$
39. **Calculator** Order the states from least to greatest fraction under age 18. **Fla., N.Y., Ohio, Tex., Calif.**

Population (thousands)

State	Total	Under Age 18
N.Y.	19,227	4,572
Texas	22,490	6,267
Calif.	33,893	9,596
Fla.	17,397	4,003
Ohio	11,459	2,779

SOURCE: U.S. Census Bureau. Go to **PHSchool.com** for a data update. Web Code: arg-9041

40. **Writing in Math** Describe some everyday situations in which you need to change fractions to decimals. **Answers may vary. Sample: finding the cost per ounce of cereal that costs $1.59 for $4\frac{3}{4}$ oz**

41. **Challenge** Divide 50 by these numbers: 100, 10, 1, 0.1, 0.01, 0.001. Explain how the quotient changes as the divisor approaches 0. **See left.**

41. The quotients are 0.5, 5, 50, 500, 5,000, and 50,000. As the divisor gets smaller, the quotient gets larger because divisor × quotient = dividend.

Test Prep and Mixed Review — Practice

Gridded Response

42. Bennie's teacher gave him a score of $\frac{34}{40}$ on his quiz. What decimal is equivalent to the score Bennie received? **0.85**
43. When Rahmi was 11 years old, he was 150.25 centimeters tall. For the next 4 years, he grew 6.5 centimeters a year. To the nearest centimeter, how tall was Rahmi when he was 15 years old? **176**
44. Serena has a part-time job at a supermarket. Her boss tells her to find last week's average number of smoothies sold per day. He gives her the printout of sales in the past seven days: 14, 23, 42, 22, 15, 28, and 31. Serena correctly finds the average number of smoothies sold per day. What is the number? **25**

GO for Help

For Exercises	See Lesson
45–48	1-3

Find each product.

45. $0.4 \cdot 0.9$ **0.36**
46. $1.7 \cdot 0.5$ **0.85**
47. $3.06 \cdot 0.3$ **0.918**
48. $9.013 \cdot 1.0$ **9.013**

Test Prep

Resources

For additional practice with a variety of test item formats:

- Test-Taking Strategies, p. 111
- Test Prep, p. 115
- Test-Taking Strategies with Transparencies

37. N.Y.: $\frac{4,572}{19,227}$
Tex.: $\frac{6,267}{22,490}$
Calif.: $\frac{9,596}{33,893}$
Fla.: $\frac{4,003}{17,397}$
Ohio: $\frac{2,779}{11,459}$

2-6b Activity Lab

Data Analysis

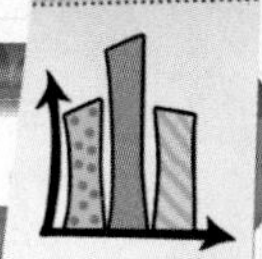

Using Tables to Compare Data

Analyzing data is easier if the data are organized. You can rearrange a table to organize and display data.

An experiment with seeds resulted in the data recorded in the table. Analyze the data and redraw the table to show which seed types sprouted most frequently.

Seed Type	A	B	C	D	E	F	G	H	I
Number Sprouted	15	5	22	17	18	21	14	18	8
Number Planted	48	20	44	35	52	63	55	35	15

1. For each seed type, find the fraction $\frac{\text{number sprouted}}{\text{number planted}}$. **See margin.**
2. Compare the fractions. Order the seed types from most frequently sprouted to least frequently sprouted. (*Hint:* Convert to decimals.) **I, H, C, D, E, F, A, G, B**
3. Redraw the table showing the seed types in order from most frequently sprouted to least frequently sprouted. **See margin.**

Checkpoint Quiz 2

Lessons 2-3 through 2-6

Simplify each fraction.

1. $\frac{18}{36}$ **$\frac{1}{2}$**
2. $\frac{42}{60}$ **$\frac{7}{10}$**
3. $\frac{35}{56}$ **$\frac{5}{8}$**

Compare. Use <, =, or >.

4. $\frac{1}{8}$ ■ $\frac{2}{100}$ **>**
5. $\frac{5}{12}$ ■ $\frac{7}{9}$ **<**
6. $\frac{12}{20}$ ■ $\frac{3}{5}$ **=**

Write each improper fraction as a mixed number and each mixed number as an improper fraction.

7. $\frac{29}{6}$ **$4\frac{5}{6}$**
8. $4\frac{1}{9}$ **$\frac{37}{9}$**
9. $\frac{82}{5}$ **$16\frac{2}{5}$**

10. Twins are born once in every 89 births. Identical twins are born 4 times in every 1,000 births. Triplets are born once in every 6,900 births. Write each birth frequency as a decimal and order the decimals from least to greatest frequency. **0.00015, 0.004, 0.0112**

1. See back of book.
3. See back of book.

Activity Lab

Using Tables to Compare Data

Students use tables to compare and order data. The data is in fraction form but to compare it, decimals are used. This will help extend the skills students learned in Lesson 2-6.

Guided Instruction

Explain that it is sometimes necessary to order data that is in the form of fractions, mixed numbers, or decimals.
Ask:
- *How can we compare the numbers in this table?* **Make a fraction for each set of data, convert each to a decimal and order to compare them.**

Differentiated Instruction

Below Level L2
Provide students with different kinds or colors of seeds to model each fraction, using a piece of string as the fraction bar. This will help them visualize the fractions.

Resources

- seeds
- string

Checkpoint Quiz

Use this Checkpoint Quiz to check students' understanding of the skills and concepts of Lessons 2-3 through 2-6.

Resources

- All-In-One Teaching Resources Checkpoint Quiz 2
- ExamView CD-ROM
- Success Tracker™ Online Intervention

2-7

1. Plan

Objective
To compare and order rational numbers

Examples
1 Comparing Rational Numbers
2 Comparing Decimals
3 Ordering Rational Numbers

Math Understandings: p. 66D

Professional Development

Math Background

A *rational number* is a number that can be expressed in the form $\frac{a}{b}$, where *a* and *b* are integers and *b* is not zero. Students learn to compare rational numbers using both fraction and decimal forms.

More Math Background: p. 66D

Lesson Planning and Resources

See p. 66E for a list of the resources that support this lesson.

PowerPoint

Bell Ringer Practice

Check Skills You'll Need
Use student page, transparency, or PowerPoint. For intervention, direct students to:
Fractions and Decimals
Lesson 2-6
Extra Skills and Word Problems Practice, Ch. 2

2-7 Rational Numbers

Check Skills You'll Need

1. **Vocabulary Review** Is 1.234 a repeating decimal or a terminating decimal? Explain. **See below.**

Write each fraction as a decimal.

2. $\frac{3}{4}$ **0.75**
3. $-\frac{7}{9}$ **$-0.\overline{7}$**
4. $1\frac{1}{3}$ **$1.\overline{3}$**
5. $\frac{12}{48}$ **0.25**

GO for Help Lesson 2-6

Check Skills You'll Need

1. **Terminating decimal; the decimal stops.**

What You'll Learn

To compare and order rational numbers

New Vocabulary rational number

Why Learn This?

Rational numbers are part of everyday life. You see them on price tags, highway signs, and charts. Rational numbers can be written in different forms. To compare rational numbers, it is easier to convert them into the same form.

A **rational number** is a number that can be written as a quotient of two integers, where the divisor is not 0. Examples are $\frac{2}{5}$, $0.\overline{3}$, -6, and $3\frac{1}{2}$.

EXAMPLE Comparing Rational Numbers

1 Compare $-\frac{1}{2}$ and $-\frac{3}{4}$.

Method 1

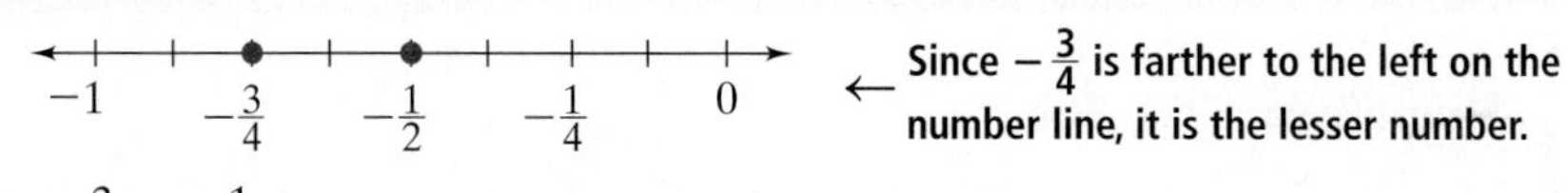

← **Since $-\frac{3}{4}$ is farther to the left on the number line, it is the lesser number.**

So $-\frac{3}{4} < -\frac{1}{2}$.

GO for Help
For help with writing equivalent fractions, go to Lesson 2-3, Example 1.

Method 2

$-\frac{1}{2} = \frac{-1}{2}$ ← **Rewrite $-\frac{1}{2}$ with −1 in the numerator.**

$= \frac{-1 \times 2}{2 \times 2}$ ← **The LCD is 4. Write an equivalent fraction.**

$= \frac{-2}{4} = -\frac{2}{4}$ ← **The fraction $-\frac{2}{4}$ is equivalent to $\frac{-2}{4}$.**

Since $-\frac{3}{4} < -\frac{2}{4}$, $-\frac{3}{4} < -\frac{1}{2}$.

Quick Check

1. Compare $-\frac{2}{3}$ and $-\frac{1}{6}$. Use <, =, or >. **$-\frac{2}{3} < -\frac{1}{6}$**

Differentiated Instruction Solutions for All Learners

Special Needs L1	**Below Level** L2
For Example 1, have students draw a number line from −1 to 1 with intervals of halves and fourths. Have them trace the distance between 0 and $-\frac{1}{2}$ and then from 0 and $-\frac{3}{4}$ with their finger. Note that the number that is the greater distance in the negative direction is smaller. **learning style: tactile**	Review the procedure for writing two fractions as equivalent fractions with a common denominator. **learning style: verbal**

EXAMPLE Comparing Decimals

2 **a.** Compare −4.4 and 4.7.

$-4.4 < 4.7$ ← **Any negative number is less than a positive number.**

b. Compare −4.4 and −4.7.

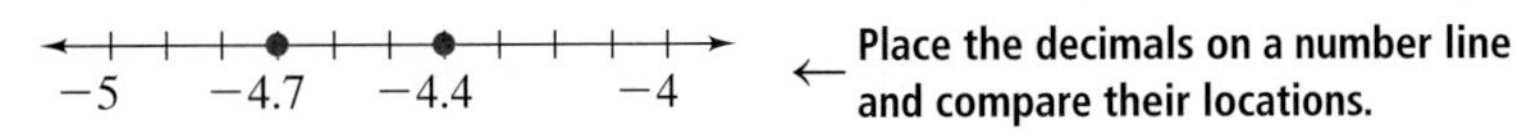

← **Place the decimals on a number line and compare their locations.**

$-4.4 > -4.7$ since −4.4 is to the right of −4.7.

Quick Check

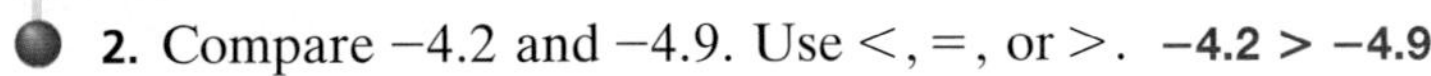

2. Compare −4.2 and −4.9. Use <, =, or >. **−4.2 > −4.9**

When you compare and order decimals and fractions, it is often helpful to write the fractions as decimals.

EXAMPLE Ordering Rational Numbers

3 **Multiple Choice** The peaks of four mountains or seamounts are located either below or above sea level as follows: $\frac{1}{4}$ mi, −0.2 mi, $-\frac{2}{9}$ mi, 1.1 mi. Which list shows the order of numbers from least to greatest?

Ⓐ $\frac{1}{4}, -0.2, -\frac{2}{9}, 1.1$ Ⓒ $-\frac{2}{9}, -0.2, \frac{1}{4}, 1.1$

Ⓑ $\frac{1}{4}, -\frac{2}{9}, 1.1, -0.2$ Ⓓ $-0.2, -\frac{2}{9}, 1.1, \frac{1}{4}$

Order these numbers from least to greatest: $\frac{1}{4}, -0.2, -\frac{2}{9}, 1.1$.

$\frac{1}{4} = 1 \div 4 = 0.250$ ← **Write as a decimal.**

$-\frac{2}{9} = -2 \div 9 = -0.22222\ldots = -0.\overline{2}$ ← **Write as a repeating decimal.**

You can use a number line to order the numbers.

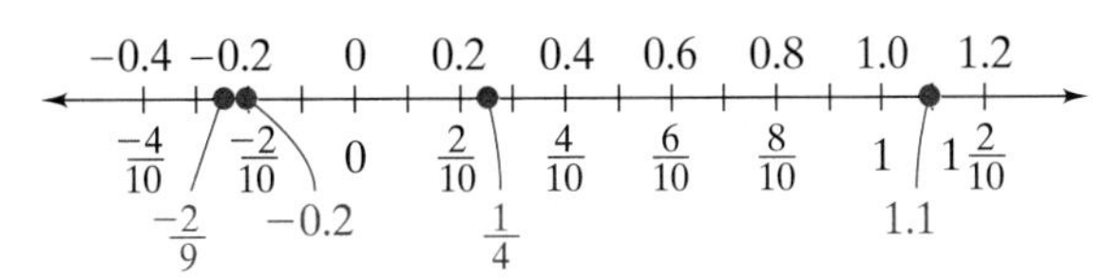

$-0.\overline{2} < -0.2 < 0.25 < 1.1$ ← **Compare the decimals.**

In order, the numbers are $-\frac{2}{9}, -0.2, \frac{1}{4}$, and 1.1. The answer is C.

Quick Check

3. The following temperatures were recorded during a science project: $12\frac{1}{2}$°C, −4°C, 6.55°C, and $-6\frac{1}{4}$°C. Order the temperatures from least to greatest. **$-6\frac{1}{4}, -4, 6.55, 12\frac{1}{2}$**

Test Prep Tip

Sometimes it is easier to order rational numbers when they are all written as decimals.

Advanced Learners L4

Find four rational numbers that are between −0.3 and $-\frac{1}{3}$. **Samples: −0.31, −0.32, −0.315, −0.325**

learning style: visual

English Language Learners ELL

For Example 3, have students draw a vertical number line, with zero as *sea level*. On the left side, have them write intervals from 2 to −2 in tenths. On the right, have them write the intervals using fractions. This will give them a sense of the relative sizes of the numbers.

learning style: visual

2. Teach

Activity Lab

Use before the lesson.

All in One Teaching Resources

Activity Lab 2-7: Rational Number Shuffle

Guided Instruction

Error Prevention!

When using common denominators to compare, students might focus so intently on rewriting the fractions that they forget whether the original numbers were positive or negative. Encourage them to use a number line to check their work.

Example 3

Divide the work into smaller tasks. Order the negative numbers first, then order the positive numbers:
$-\frac{2}{9} < -0.2$ $\frac{1}{4} < 1.1$

Negative numbers are all less than positive numbers, so the four numbers are ordered.

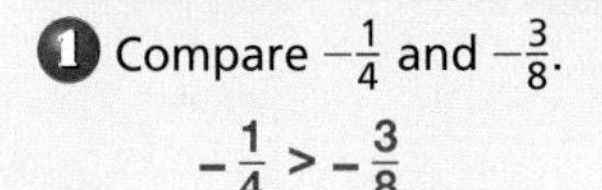

Additional Examples

1 Compare $-\frac{1}{4}$ and $-\frac{3}{8}$.

$-\frac{1}{4} > -\frac{3}{8}$

2 Compare. Use <, >, or =.

a. 8.7 ■ 8.1 **8.7 > 8.1**
b. −8.7 ■ 8.1 **−8.7 < 8.1**
c. −8.7 ■ −8.1 **−8.7 < −8.1**

3 Order these numbers from least to greatest:

$-\frac{3}{5}, 0.625, \frac{2}{3}, -0.5$

$-\frac{3}{5}, -0.5, 0.625, \frac{2}{3}$

All in One Teaching Resources

- Daily Notetaking Guide 2-7 L3
- Adapted Notetaking 2-7 L1

Closure

- *How do you compare rational numbers?* **Use a number line; express them as fractions with a common denominator and compare the numerators; or express them as decimals and compare them place-by-place.**

3. Practice

Assignment Guide

Check Your Understanding

Go over Exercises 1–6 in class before assigning the Homework Exercises.

Homework Exercises

A	Practice by Example	7–22
B	Apply Your Skills	23–30
C	Challenge	31
	Test Prep and Mixed Review	32–37

Homework Quick Check

To check students' understanding of key skills and concepts, go over Exercises 12, 21, 24, 28, and 29.

Differentiated Instruction Resources

Adapted Practice 2-7 L1

Practice 2-7 Rational L3

Compare. Use <, >, or =.

1. $-\frac{2}{9}$ [>] $-\frac{4}{9}$
2. $-\frac{1}{6}$ [>] $-\frac{2}{3}$
3. $-\frac{5}{12}$ [>] $-\frac{3}{4}$
4. -1.2 [>] -2.1
5. -0.6 [<] -0.52
6. -1.23 [>] -1.25
7. -5.3 [>] $-5.\overline{3}$
8. $-3\frac{1}{4}$ [=] -3.25
9. $-4\frac{2}{5}$ [<] -4.12

Order from least to greatest.

10. $\frac{5}{4}, 1.5, -\frac{3}{2}, -0.5$ **$-\frac{3}{2}, -0.5, \frac{5}{4}, 1.5$**
11. $\frac{1}{11}, -0.9, 0.09, \frac{1}{10}$ **$-0.9, 0.09, \frac{1}{11}, \frac{1}{10}$**
12. $0.1\overline{2}, -\frac{11}{12}, -\frac{1}{6}, -0.1$ **$-\frac{11}{12}, -\frac{1}{6}, -0.1, 0.1\overline{2}$**
13. $\frac{2}{3}, 0.6, -\frac{5}{6}, -6.6$ **$-6.6, -\frac{5}{6}, 0.6, \frac{2}{3}$**
14. $1.312, 1\frac{3}{8}, -1\frac{3}{10}, -1.33$ **$-1.33, -1\frac{3}{10}, 1.312, 1\frac{3}{8}$**
15. $1, \frac{4}{5}, -\frac{8}{9}, -1$ **$-1, -\frac{8}{9}, \frac{4}{5}, 1$**

Evaluate. Write in simplest form.

16. $\frac{y}{z}$, for $y = -6$ and $z = -20$ **$\frac{3}{10}$**
17. $\frac{2y}{-z}$, for $y = -5$ and $z = -12$ **$-\frac{5}{6}$**
18. $\frac{y+z}{2z}$, for $y = -4$ and $z = 8$ **$\frac{1}{4}$**
19. $\frac{-2y+1}{-z}$, for $y = 3$ and $z = 10$ **$\frac{1}{2}$**

Compare.

20. The temperature at 3:00 A.M. was −17.3°F. By noon the temperature was −17.8°F. At what time was it the coldest? **noon**
21. Samuel is $\frac{5}{8}$ in. taller than Jackie. Shelly is 0.7 in. taller than Jackie. Who is the tallest? **Shelly**

2-7 • Guided Problem Solving GPS L3

Student Page 105, Exercise 29:

Animals About $\frac{1}{25}$ of a toad's eggs survive to adulthood. About 0.25 of a frog's eggs and $\frac{1}{5}$ of a green turtle's eggs survive to adulthood. Which animal's eggs have the highest survival rate?

Understand

1. Circle the information you will need to solve.
2. What are you being asked to do? **Find which animal's eggs have the highest survival rate.**
3. In order to find the greatest number, what must you do first? **Change the fractions to decimals.**

Plan and Carry Out

4. Write $\frac{1}{25}$ as a decimal. **0.04**
5. Write $\frac{1}{5}$ as a decimal. **0.2**
6. Which is the largest decimal, 0.04, 0.2, or 0.25? **0.25**
7. Which animal's eggs have the highest survival rate? **frog**

Check

8. What fraction is 0.25 equal to? Is it the greatest value? **$\frac{1}{4}$; yes**

Solve Another Problem

9. In order to organize the nails in a garage, Anne and Jeff measured the nails. Anne used fractions to measure her 3 groups of nails and found that they were $\frac{3}{5}$ in., $\frac{7}{12}$ in., and $\frac{4}{9}$ in. Jeff used decimals to measure his two groups and found that they were 0.62 in., and 0.31 in. Which nail is the longest? **the 0.62 in. nail**

Check Your Understanding

1. Answers may vary. Sample: A rational number is a number that can be written as a quotient of two integers.

1. **Vocabulary** In your own words, define *rational number.*

Compare. Use <, =, or >.

2. $2\frac{1}{5}$ ■ $3\frac{1}{3}$ **<**
3. $-3\frac{1}{2}$ ■ $-3\frac{3}{4}$ **>**
4. -6.1 ■ -6 **<**

Order from least to greatest.

5. $-236, -7\frac{1}{7}, 0, \frac{41}{99}, -3.\overline{3}$ **$-236, -7\frac{1}{7}, -3.\overline{3}, 0, \frac{41}{99}$**
6. $-8, -5\frac{1}{3}, -8.22, -8\frac{1}{3}, \frac{16}{42}$ **$-8\frac{1}{3}, -8.22, -8, -5\frac{1}{3}, \frac{16}{42}$**

Homework Exercises

For more exercises, see Extra Skills and Word Problems.

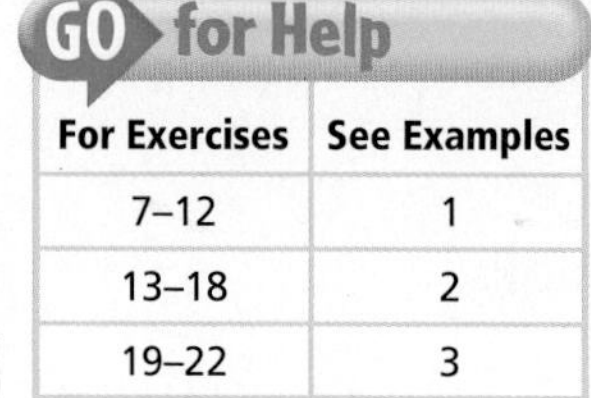

For Exercises	See Examples
7–12	1
13–18	2
19–22	3

A **Compare. Use <, =, or >.**

7. $-\frac{1}{7}$ ■ $-\frac{6}{7}$ **>**
8. $-\frac{3}{4}$ ■ -3 **>**
9. $-\frac{1}{2}$ ■ $-\frac{2}{10}$ **<**
10. $-\frac{3}{4}$ ■ -1 **>**
11. $-\frac{1}{2}$ ■ $-\frac{5}{6}$ **>**
12. $-\frac{4}{5}$ ■ $-\frac{1}{3}$ **<**
13. 5.2 ■ -8.3 **>**
14. -6.5 ■ 6.2 **<**
15. -4.9 ■ -4.3 **<**
16. 1.09 ■ -1.90 **>**
17. -1.22 ■ -6.5 **>**
18. -10.2 ■ -10.23 **>**

Order from least to greatest. **19–21. See left.**

19. $-1.0, -\frac{3}{4}, 0.25, \frac{3}{2}$
20. $-1.34, -\frac{6}{25}, \frac{7}{3}, 2.4$
21. $-1.5, \frac{1}{2}, 0.545, \frac{6}{11}$

19. $\frac{3}{2}, 0.25, -\frac{3}{4}, -1.0$
20. $\frac{7}{3}, 2.4, -\frac{6}{25}, -1.34$
21. $\frac{6}{11}, -1.5, 0.545, \frac{1}{2}$
22. $\frac{7}{6}, \frac{11}{12}, \frac{14}{24}, 1$ **$\frac{14}{24}, \frac{11}{12}, 1, \frac{7}{6}$**

B 

23. **Guided Problem Solving** You are skier A, the first skier in a skiing event with three other skiers. Compared to your time, skier B is slower, by a time of +00:28. Skier C has a time of +02:13, and skier D has a time of −01:24. Who is the fastest skier? **skier D**
    - Since all times are compared to yours, what is your time?
    - What is the order of times, from fastest to slowest?

24. The table below shows melting points of four elements. Which element has the highest melting point? **xenon**

Melting Points

Chemical Solid	Krypton	Argon	Xenon	Helium
Melting Point (°C)	−157.36	−189.35	−111.79	−272.2

SOURCE: Encyclopædia Brittanica

28. Explanations may vary. Sample: $-\frac{5}{8} > -\frac{3}{4}$. Common denominators would be easier, since 4 is a factor of 8.

Visit: PHSchool.com
Web Code: are-0207

Compare. Use <, =, or >.

25. $-5.8 \blacksquare -5\frac{9}{10}$ > **26.** $-6\frac{11}{50} \blacksquare -6.21$ < **27.** $-10.42 \blacksquare -10.4\overline{2}$ >

28. **Writing in Math** Compare $-\frac{5}{8}$ and $-\frac{3}{4}$. Is it easier to find common denominators or to write decimal equivalents? Explain. **See margin.**

29. **Animals** About $\frac{1}{25}$ of a toad's eggs survive to adulthood. About 0.25 of a frog's eggs and $\frac{1}{5}$ of a green turtle's eggs survive to adulthood. Which animal's eggs have the highest survival rate? **frog**

30. **Money** Here is part of Mr. Lostcash's checkbook register. Order his balances from greatest to least. **See below left.**

Description	Debits (−)	Credits (+)	Balance
Paycheck		122.18	122.18
Sneakers	95.00		27.18
Two outfits	68.09		−40.91
Paycheck		122.18	81.27
Insufficient funds fee	25.00		56.27
Three CDs	59.97		−3.70

30. $122.18, $81.27, $56.27, $27.18, −$3.70, −$40.91

31. **Challenge** Evaluate $\frac{m-n}{-12}$ for $m = -3$ and $n = 6$. $\frac{3}{4}$

Test Prep and Mixed Review

Practice

Multiple Choice

32. Harry's class wants to buy one slice of pie for each person in the class. The table below shows prices from a local bakery. What additional information is needed to estimate the total cost? **B**

Rainbow Bakery Pies

Pie Size	Price	Number of Pieces
[illegible]	[illegible]8	2
[illegible]	[illegible]0	3

... each pie size ... Harry's class ... ch pie

... al numbers of photos. Which of the ... ber of photos in all the albums? **J**

(H) 1,182 (J) 1,128

Check HW p. 104 #7-17 ODD, 19-22, 25-27

GO for Help

For Exercises	See
34–37	

36. $\frac{12}{56}$ $\frac{3}{14}$ **37.** $\frac{18}{48}$ $\frac{3}{8}$

Online lesson quiz, P...

4. Assess & Reteach

PowerPoint

Lesson Quiz

Compare. Use <, >, or =.

1. $-\frac{3}{4} \blacksquare -\frac{4}{7}$ <
2. $-\frac{1}{8} \blacksquare -\frac{3}{8}$ >
3. $3.4 \blacksquare -3.4$ >
4. $-8.9 \blacksquare -8.6$ <

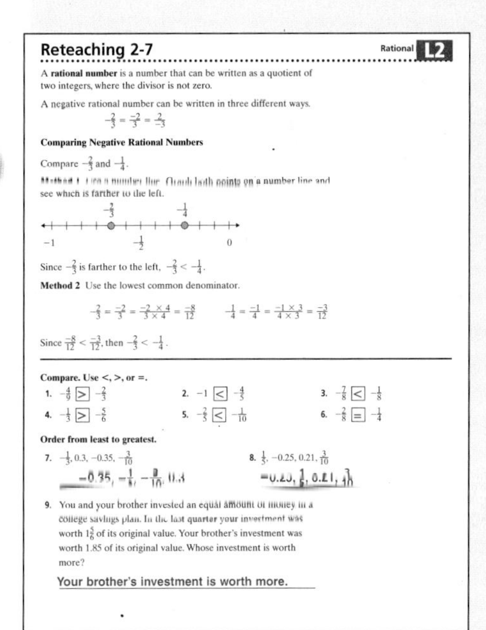
Reteaching 2-7 Rational L2

A **rational number** is a number that can be written as a quotient of two integers, where the divisor is not zero.

A negative rational number can be written in three different ways.

$-\frac{2}{3} = \frac{-2}{3} = \frac{2}{-3}$

Comparing Negative Rational Numbers

Compare $-\frac{2}{3}$ and $-\frac{1}{4}$.

Method 1 Use a number line. Graph both points on a number line and see which is farther to the left.

Since $-\frac{2}{3}$ is farther to the left, $-\frac{2}{3} < -\frac{1}{4}$.

Method 2 Use the lowest common denominator.

$-\frac{2}{3} = \frac{-2}{3} = \frac{-2 \times 4}{3 \times 4} = \frac{-8}{12}$ $-\frac{1}{4} = \frac{-1}{4} = \frac{-1 \times 3}{4 \times 3} = \frac{-3}{12}$

Since $\frac{-8}{12} < \frac{-3}{12}$, then $-\frac{2}{3} < -\frac{1}{4}$.

Compare. Use <, >, or =.

1. $-\frac{4}{9}$ > $-\frac{5}{9}$ 2. -1 < $-\frac{4}{5}$ 3. $-\frac{7}{8}$ < $-\frac{1}{8}$
4. $-\frac{1}{3}$ > $-\frac{5}{6}$ 5. $-\frac{3}{5}$ < $-\frac{1}{10}$ 6. $-\frac{2}{8}$ = $-\frac{1}{4}$

Order from least to greatest.

7. $-\frac{1}{8}$, 0.3, −0.35, $-\frac{3}{10}$ — −0.35, $-\frac{3}{10}$, $-\frac{1}{8}$, 0.3
8. $\frac{1}{5}$, −0.25, 0.21, $\frac{1}{10}$ — −0.25, $\frac{1}{10}$, $\frac{1}{5}$, 0.21

9. You and your brother invested an equal amount of money in a college savings plan. In the last quarter your investment was worth $1\frac{4}{5}$ of its original value. Your brother's investment was worth 1.85 of its original value. Whose investment is worth more?
Your brother's investment is worth more.

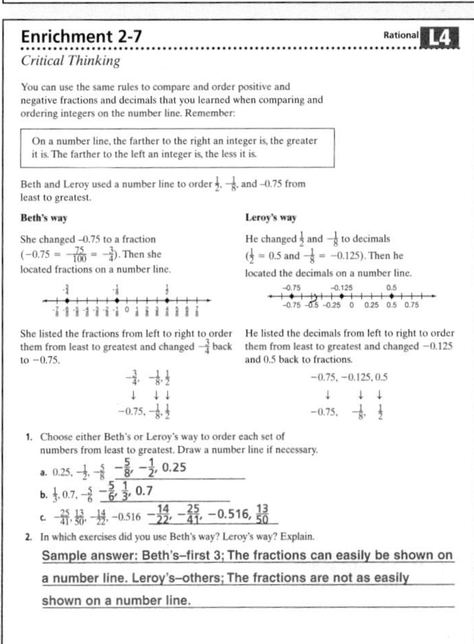
Enrichment 2-7 Rational L4

Critical Thinking

You can use the same rules to compare and order positive and negative fractions and decimals that you learned when comparing and ordering integers on the number line. Remember:

On a number line, the farther to the right an integer is, the greater it is. The farther to the left an integer is, the less it is.

Beth and Leroy used a number line to order $\frac{1}{2}$, $-\frac{1}{8}$, and −0.75 from least to greatest.

Beth's way
She changed −0.75 to a fraction ($-0.75 = -\frac{75}{100} = -\frac{3}{4}$). Then she located fractions on a number line.
She listed the fractions from left to right to order them from least to greatest and changed $-\frac{3}{4}$ back to −0.75.
−0.75, $-\frac{1}{8}$, $\frac{1}{2}$

Leroy's way
He changed $\frac{1}{2}$ and $-\frac{1}{8}$ to decimals ($\frac{1}{2} = 0.5$ and $-\frac{1}{8} = -0.125$). Then he located the decimals on a number line.
He listed the decimals from left to right to order them from least to greatest and changed −0.125 and 0.5 back to fractions.
−0.75, $-\frac{1}{8}$, $\frac{1}{2}$

1. Choose either Beth's or Leroy's way to order each set of numbers from least to greatest. Draw a number line if necessary.
a. 0.25, $-\frac{1}{2}$, $-\frac{5}{8}$ — $-\frac{5}{8}$, $-\frac{1}{2}$, 0.25
b. $\frac{1}{3}$, 0.7, $-\frac{5}{6}$ — $-\frac{5}{6}$, $\frac{1}{3}$, 0.7
c. $-\frac{25}{41}$, $\frac{13}{50}$, $-\frac{14}{22}$, −0.516 — $-\frac{14}{22}$, $-\frac{25}{41}$, −0.516, $\frac{13}{50}$

2. In which exercises did you use Beth's way? Leroy's way? Explain.
Sample answer: Beth's–first 3; The fractions can easily be shown on a number line. Leroy's–others; The fractions are not as easily shown on a number line.

Alternative Assessment

Provide pairs of students with examples such as those in Exercises 7–22. Have partners work together to use number lines to compare the rational numbers.

Test Prep

Resources

For additional practice with a variety of test item formats:

- Test-Taking Strategies, p. 111
- Test Prep, p. 115
- Test-Taking Strategies with Transparencies

2-8

1. Plan

Objective
To write numbers in both scientific notation and standard form

Examples
1 Writing in Scientific Notation
2 Writing in Standard Form

Math Understandings: p. 66D

Math Background

Scientific notation is a practical way to record the very large and very small measurements that arise in scientific endeavors. This lesson focuses on the scientific notation for numbers greater than 1, which involves positive powers of 10.

More Math Background: p. 66D

Lesson Planning and Resources

See p. 66E for a list of the resources that support this lesson.

PowerPoint

Bell Ringer Practice

Check Skills You'll Need
Use student page, transparency, or PowerPoint. For intervention, direct students to:
Prime Factorization
Lesson 2-1
Extra Skills and Word Problems Practice, Ch. 2

2-8 Scientific Notation

Check Skills You'll Need

1. **Vocabulary Review** An exponent tells you how many times a number, or base, is used as a _?_. factor

Simplify.

2. 3^3 27
3. 4^2 16
4. 10^5 100,000
5. 2^4 16

for Help Lesson 2-1

What You'll Learn

To write numbers in both scientific notation and standard form

New Vocabulary scientific notation

Why Learn This?

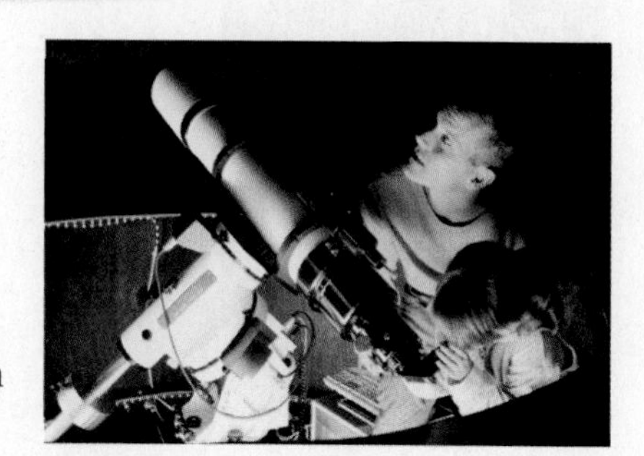

Some numbers are so large that they are difficult to write. You can use scientific notation to express large numbers, such as distances and speeds in space. Scientists use scientific notation to calculate with large numbers.

In Lesson 2-1, you learned about exponents. When the base is 10, the exponent tells you the number of zeros the number will have in standard form.

Powers of 10

Exponential Form	10^1	10^2	10^3	10^4	10^5	10^6
Number of Zeros in Standard Form	1	2	3	4	5	6
Standard Form	10	100	1,000	10,000	100,000	1,000,000

When you multiply a factor by a power of 10, you can find the product by moving the decimal point in the factor to the right. The exponent tells you how many places to move the decimal point. When you divide by a power of 10, you move the decimal point to the left.

Multiplying and Dividing by Powers of 10

Standard Form	$3.5 \div 100$	$3.5 \div 10$	3.5×10	3.5×100	$3.5 \times 1{,}000$
Exponential Form	$3.5 \div 10^2$	$3.5 \div 10^1$	3.5×10^1	3.5×10^2	3.5×10^3
Number	0.035	0.35	35	350	3,500

Scientific notation is a shorter way to write numbers using powers of 10.

Differentiated Instruction Solutions for All Learners

Special Needs L1
Show students that 10 x 10 x 10 is equal to 1,000, which is written as 10^3. Have them test this out with other powers of ten.
learning style: visual

Below Level L2
Review the process of multiplying and dividing decimals by 10, 100, and 1,000.
learning style: verbal

KEY CONCEPTS **Scientific Notation**

A number in **scientific notation** is written as the product of two factors, one greater than or equal to 1 and less than 10, and the other a power of 10.

$7{,}500{,}000{,}000{,}000 = 7.5 \times 10^{12}$

EXAMPLE Writing in Scientific Notation

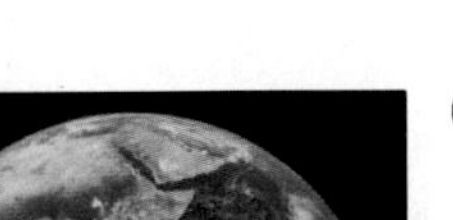

1 **Science** The moon orbits Earth at a distance of 384,000 km. Write this number in scientific notation.

3.84000. ← **Move the decimal point to get a factor greater than 1 but less than 10.**

$384{,}000 = 3.84 \times 100{,}000$ ← **Write as a product of 2 factors.**

$= 3.84 \times 10^5$ ← **Write 100,000 as a power of 10.**

The moon orbits Earth at a distance of 3.84×10^5 km.

Quick Check

1. NASA's Hubble Telescope took pictures of a supernova that is 169,000 light years away. Write this number in scientific notation. **1.69×10^5**

You can change expressions from scientific notation to standard form by simplifying the product of the two factors.

EXAMPLE Writing in Standard Form

2 **Science** The mean distance from the sun to Mars is approximately 2.3×10^8 km. Write this number in standard form.

Method 1

$2.3 \times 10^8 = 2.3 \times 100{,}000{,}000$ ← **Write as a product of 2 factors.**

$= 230{,}000{,}000$ ← **Multiply the factors.**

Method 2

$2.3 \times 10^8 = 2.30000000$ ← **The exponent is 8. Move the decimal 8 places to the right.**

$= 230{,}000{,}000$

The mean distance is approximately 230,000,000 km.

Quick Check

2. A large telescope gathers about 6.4×10^5 times the amount of light your eye receives. Write this number in standard form. **640,000**

2. Teach

Activity Lab

Use before the lesson.

All in One Teaching Resources

Activity Lab 2-8: Scientific Notation

Guided Instruction

Example 1

Students might find it helpful to quietly count the places aloud as they move the decimal point.

Error Prevention!

When changing a number from standard form to scientific notation, students might determine the exponent of 10 by counting all the digits or by counting just the zeros. Suggest that they cover the first digit with one finger and count the remaining digits.

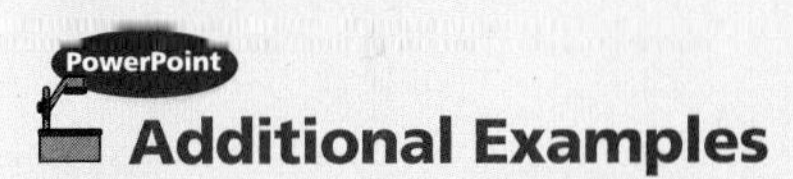

Additional Examples

1 The mean distance from Mars to the sun is about 141,750,000 mi. Write this number in scientific notation. **1.4175×10^8 mi**

2 Light from the sun reaches Earth in about 4.99012×10^2 seconds. Write this time in standard form. **499.012 s**

All in One Teaching Resources

- Daily Notetaking Guide 2-8 **L3**
- Adapted Notetaking 2-8 **L1**

Closure

- *How do you change a number from standard form to scientific notation?* **Move the decimal point to obtain a factor that is greater than or equal to 1, but less than 10. Write the second factor as a power of 10 whose exponent is the number of places that you moved the decimal point.**

Advanced Learners **L4**

Write each product in scientific notation.

$(4 \times 10^3) \times (2 \times 10^6)$ **8×10^9**

$(5 \times 10^4) \times (7 \times 10^8)$ **3.5×10^{13}**

learning style: visual

English Language Learners **ELL**

Have students work in pairs to make a table for multiplying and dividing by powers of 10 similar to the one on page 106. Have students start with standard form for $0.27 \div 100$ and go to $0.27 \times 1{,}000$. They should give exponential form and the actual number for each product.

learning style: visual

3. Practice

Assignment Guide

Check Your Understanding

Go over Exercises 1–7 in class before assigning the Homework Exercises.

Homework Exercises

A Practice by Example	8–25
B Apply Your Skills	26–33
C Challenge	34
Test Prep and Mixed Review	35–39

Homework Quick Check

To check students' understanding of key skills and concepts, go over Exercises 8, 19, 28, 29, and 31.

Differentiated Instruction Resources

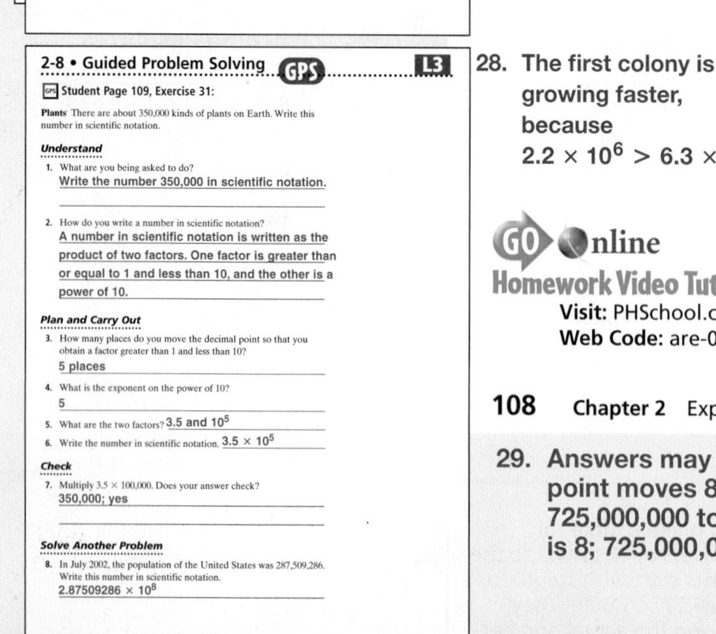

Adapted Practice 2-8 L1

Practice 2-8 Scientific L3

Write each number in scientific notation.

1. 73,000,000 7.3×10^7
2. 4,300 4.3×10^3
3. 510 5.1×10^2
4. 56,870 5.687×10^4
5. 68,900 6.89×10^4
6. 98,000,000,000 9.8×10^{10}
7. 4,890,000 4.89×10^6
8. 38 3.8×10^1
9. 120,000 1.2×10^5
10. 543,000 5.43×10^5
11. 27 2.7×10^1
12. 54,000 5.4×10^4

Write in standard form.

13. 5.7×10^6 5,700,000
14. 2.45×10^8 245,000,000
15. 4.706×10^{11} 470,600,000,000
16. 8×10^1 80
17. 7.2×10^3 7,200
18. 1.63×10^{12} 1,630,000,000,000
19. 8.03×10^{14} 803,000,000,000,000
20. 3.26×10^4 32,600
21. 5.179×10^5 517,900

Write each number in scientific notation.

22. One type of roundworm can lay 200,000 eggs each day. 2×10^5
23. The nose of a German shepherd dog has about 220 million cells that are used in picking out smells. 2.2×10^8
24. The brain contains about 100 trillion nerve connections. 1×10^{14}
25. During an average life span, the human heart will beat about 2,800,000,000 times. 2.8×10^9
26. The volume of the water behind the Grand Coulee Dam is about 10.6 million cubic yards. 1.06×10^7
27. A second has been defined as the time it takes for an atom of a particular metal to vibrate 9,192,631,770 times. 9.19263177×10^9

2-8 • Guided Problem Solving GPS L3

Student Page 109, Exercise 31:

Plants There are about 350,000 kinds of plants on Earth. Write this number in scientific notation.

Understand

1. What are you being asked to do? Write the number 350,000 in scientific notation.
2. How do you write a number in scientific notation? A number in scientific notation is written as the product of two factors. One factor is greater than or equal to 1 and less than 10, and the other is a power of 10.

Plan and Carry Out

3. How many places do you move the decimal point so that you obtain a factor greater than 1 and less than 10? 5 places
4. What is the exponent on the power of 10? 5
5. What are the two factors? 3.5 and 10^5
6. Write the number in scientific notation. 3.5×10^5

Check

7. Multiply $3.5 \times 100,000$. Does your answer check? 350,000; yes

Solve Another Problem

8. In July 2002, the population of the United States was 287,509,286. Write this number in scientific notation. 2.87509286×10^8

Check Your Understanding

1. **Vocabulary** Describe how to write a number in scientific notation. 1. It is written as the product of two factors, one greater than or equal to 1 and less than 10, and the other a power of 10.
2. **Reasoning** Is 107×10^4 written in scientific notation? Explain. No; 107 is not less than 10.

Fill in the blank.

3. $1,700 = 1.7 \times 10^{■}$ 3
4. $2,850,000 = ■ \times 10^6$ 2.85

Match the equivalent numbers.

5. 3.96×10^6 B
6. 3.96×10^7 C
7. 3.96×10^5 A

A. 396,000
B. 3,960,000
C. 39,600,000

Homework Exercises

For more exercises, see Extra Skills and Word Problems.

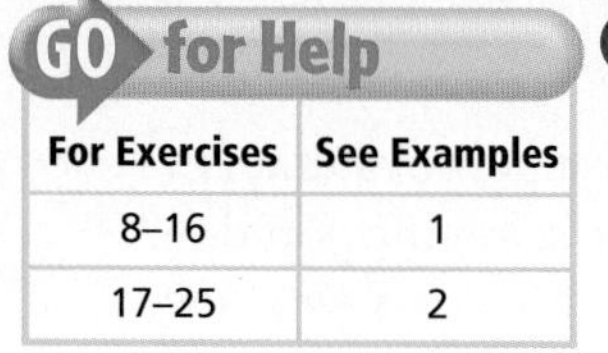

GO for Help

For Exercises	See Examples
8–16	1
17–25	2

A **Write in scientific notation.**

8. 7,500 7.5×10^3
9. 75,000,000 7.5×10^7
10. 1,250 1.25×10^3
11. 44,000 4.4×10^4
12. 149,000,000 1.49×10^8
13. 34,025 3.4025×10^4
14. 11,020 1.102×10^4
15. 120,000 1.2×10^5
16. One light year is 5,880,000,000,000 mi. Write in scientific notation. 5.88×10^{12}

Write in standard form.

17. 3.4×10^3 3,400
18. 5.9×10^2 590
19. 8.21×10^3 8,210
20. 6.678×10^2 667.8
21. 7.45×10^4 74,500
22. 9.9673×10^2 996.73
23. 5×10^{11} 500,000,000,000
24. 7.02×10^1 70.2
25. The normal red blood cell count for adult males is about 4.5×10^{12} per liter. Write this number in standard form. 4,500,000,000,000

B 

26. **Guided Problem Solving** The Folsom Dam in California holds back 319 billion gallons of water. Write this in scientific notation. 3.19×10^{11}
    - How can you rewrite 319 billion in standard form?
    - How do you write 319 billion in scientific notation?
27. **Ballooning** The first balloon to carry passengers weighed 1.6×10^3 lb. Write the number of pounds in standard form. 1,600 lb
28. A scientist observes two colonies of bacteria. The first grows at a rate of 2.2×10^6 bacteria per hour. The other grows at a rate of 6.3×10^5 bacteria per hour. Which grows faster? How do you know? See left.

28. The first colony is growing faster, because $2.2 \times 10^6 > 6.3 \times 10^5$.

GO Online
Homework Video Tutor
Visit: PHSchool.com
Web Code: are-0208

29. **Writing in Math** Explain how you would find the power of 10 needed to write 725,000,000 in scientific notation. See margin.

29. Answers may vary. Sample: The decimal point moves 8 places to the left in 725,000,000 to get 7.25, so the exponent is 8; $725,000,000 = 7.25 \times 10^8$.

30. 1.290×10^{22}, 3.303×10^{23}, 6.421×10^{23}, 4.869×10^{24}, 5.976×10^{24}, 8.686×10^{25}, 1.024×10^{26}, 5.688×10^{26}, 1.900×10^{27}

Planet	Mass (kg)
Mercury	3.303×10^{23}
Venus	4.869×10^{24}
Earth	5.976×10^{24}
Mars	6.421×10^{23}
Jupiter	1.900×10^{27}
Saturn	5.688×10^{26}
Uranus	8.686×10^{25}
Neptune	1.024×10^{26}
Pluto	1.290×10^{22}

30. Science The table at the left shows the planets in the solar system and their masses. List the masses in order from least to greatest. **See margin.**

31. Plants There are about 350,000 species of plants on Earth. Write this number in scientific notation. **3.5×10^5**

Math in the Media **Use the cartoon below for Exercises 32 and 33.**

FOX TROT **by Bill Amend**

32. How many minutes was the warning? Write in standard form. **20,000 min**

33. Convert the time to hours. Write in scientific notation. **about 3.3×10^2 h**

34. Challenge An astronomical unit (AU) is approximately 9.3×10^7 mi. Pluto is about 39 AU from the sun. Write the distance in miles using both scientific notation and standard form. **3.627×10^9, 3,627,000,000**

Test Prep and Mixed Review

Practice

Multiple Choice

35. Population density is a measure of how crowded a place is. The table below shows the population density of several states in 1990 and in 2000. Based on the information in the table, which of the following is NOT a reasonable assumption? **A**

Ⓐ Texas had the same population density in 1990 that Alabama had in 2000.

Ⓑ Alabama was more crowded in 1990 than West Virginia was in 2000.

Ⓒ In 1990, West Virginia was more crowded than Texas.

Ⓓ All of the states had a greater population density in 2000 than they did in 1990.

Population Density

State	Persons per Square Mile 1990	2000
Alabama	$79\frac{3}{5}$	87.6
Missouri	$74\frac{3}{10}$	81.2
Texas	$64\frac{9}{10}$	79.6
W. Virginia	$74\frac{1}{2}$	75.1

GO for Help

For Exercises	See Lesson
36–39	2-2

Find the LCM of each pair of numbers.

36. 7, 8 **56**
37. 6, 20 **60**
38. 11, 3 **33**
39. 9, 15 **45**

Online lesson quiz, PHSchool.com, Web Code: ara-0208

4. Assess & Reteach

PowerPoint

Lesson Quiz

Write in scientific notation.

1. 45,300 **4.53×10^4**
2. 714,900,000 **7.149×10^8**

Write in standard form.

3. 5.7×10^3 **5,700**
4. 6.034×10^6 **6,034,000**

Alternative Assessment

Provide 8 index cards to pairs of students. One student writes a decimal greater than or equal to 1 but less than 10 on each of four cards. The other writes a different power of 10 on each of the four other cards. Have students place the cards facedown in two stacks. Pairs draw a card from each stack to make a number written in scientific notation. They then work together to write the number in standard form.

Test Prep

Resources

For additional practice with a variety of test item formats:

- Test-Taking Strategies, p. 111
- Test Prep, p. 115
- Test-Taking Strategies with Transparencies

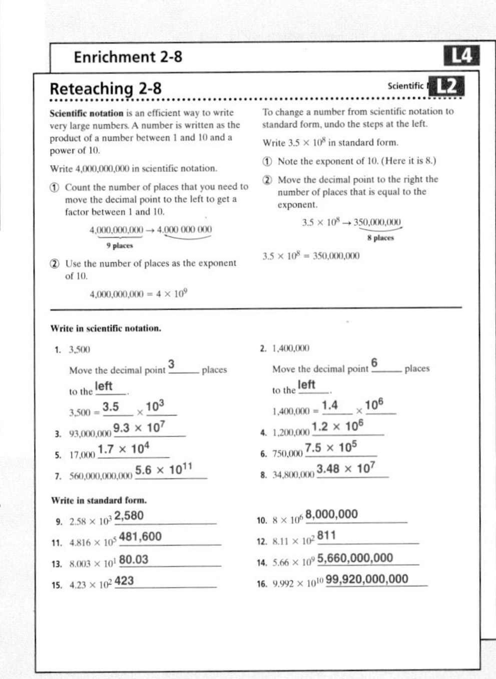
Enrichment 2-8 L4

Reteaching 2-8 Scientific L2

Scientific notation is an efficient way to write very large numbers. A number is written as the product of a number between 1 and 10 and a power of 10.

Write 4,000,000,000 in scientific notation.

① Count the number of places that you need to move the decimal point to the left to get a factor between 1 and 10. 4,000,000,000 → 4.000 000 000 (9 places)

② Use the number of places as the exponent of 10. 4,000,000,000 = 4×10^9

To change a number from scientific notation to standard form, undo the steps at the left.

Write 3.5×10^8 in standard form.

① Note the exponent of 10. (Here it is 8.)

② Move the decimal point to the right the number of places that is equal to the exponent. 3.5×10^8 → 350,000,000 (8 places)

3.5×10^8 = 350,000,000

Write in scientific notation.

1. 3,500 — Move the decimal point 3 places to the left. 3,500 = 3.5 × 10^3
2. 1,400,000 — Move the decimal point 6 places to the left. 1,400,000 = 1.4 × 10^6
3. 93,000,000 9.3×10^7
4. 1,200,000 1.2×10^6
5. 17,000 1.7×10^4
6. 750,000 7.5×10^5
7. 560,000,000,000 5.6×10^{11}
8. 34,800,000 3.48×10^7

Write in standard form.

9. 2.58×10^3 2,580
10. 8×10^6 8,000,000
11. 4.816×10^5 481,600
12. 8.11×10^2 811
13. 8.003×10^1 80.03
14. 5.66×10^9 5,660,000,000
15. 4.23×10^2 423
16. 9.992×10^{10} 99,920,000,000

Extension

Negative Exponents

This Extension shows how to use scientific notation and negative exponents to represent very small numbers. Students learn to rewrite numbers between 0 and 1 using scientific notation and, conversely, to rewrite in standard form numbers between 0 and 1 that are expressed in scientific notation.

Guided Instruction

Alternative Method
To help students grasp the meaning of negative exponents, discuss the following pattern.

$3^3 = 3 \times 3 \times 3 = 27$

$3^2 = 3 \times 3 = 9$

$3^1 = 3$

$3^0 = 1$

$3^{-1} = \frac{1}{3} = \frac{1}{3^1}$

$3^{-2} = \frac{1}{3} \times \frac{1}{3} = \frac{1}{3^2} = \frac{1}{9}$

$3^{-3} = \frac{1}{3} \times \frac{1}{3} \times \frac{1}{3} = \frac{1}{3^3} = \frac{1}{27}$

The pattern leads to: $a^{-n} = \frac{1}{a^n}$.

Exercises
Have students work on the Exercises and trade papers with a partner to correct each other's work.

Differentiated Instruction

Advanced Learners L4
Have students write several positive numbers that are less than one in standard form and in scientific notation. Have pairs trade papers and convert numbers to the appropriate form.

Extension

For Use With Lesson 2-8

Negative Exponents

To write a number between 0 and 1 in scientific notation, you use a negative exponent.

EXAMPLE Writing in Scientific Notation

1 Write 0.0084 in scientific notation.

0.008.4 ← Move the decimal point to obtain a factor greater than 1 but less than 10.

$0.0084 = 8.4 \times 0.001$ ← Write as the product of 2 factors.

$= 8.4 \times 10^{-3}$ ← The decimal point was moved 3 places to the right. Use -3 as the exponent.

In scientific notation, 0.0084 is written as 8.4×10^{-3}.

EXAMPLE Writing in Standard Form

2 Write 3.52×10^{-5} in standard form.

Method 1

$3.52 \times 10^{-5} = 3.52 \times 0.00001$ ← Write as the product of 2 factors.

$= 0.0000352$ ← Multiply the factors.

Method 2

$3.52 \times 10^{-5} = 0.00003.52$ ← The exponent of 10 is -5. Move the decimal 5 places to the left.

$= 0.0000352$

The value of 3.52×10^{-5} is 0.0000352.

Exercises

Write each number in scientific notation.

1. 0.0008 8.0×10^{-4}
2. 0.00000691 6.91×10^{-6}
3. 0.5 5.0×10^{-1}
4. 0.049562 4.9562×10^{-2}

Write each number in standard form.

5. 8.55×10^{-1} 0.855
6. 2.005×10^{-2} 0.02005
7. 6.079×10^{-5} 0.00006079
8. The width of a hair is about 3×10^{-7} meters. 0.0000003
9. A flea weighs 4.9×10^{-3} grams. 0.0049

Writing Extended Responses

An extended response question is usually worth a maximum of 4 points and has multiple parts. To get full credit, you need to answer each part and show all your work or justify your reasoning.

EXAMPLE

Without performing the division, test whether 15,534 is divisible by 3, 4, 9, and 12. Justify each response.

Below are four responses and the amount of credit each received.

4 points

Divisible by 3?
$1 + 5 + 5 + 3 + 4 = 18$
Yes, since 18 is divisible by 3.

Divisible by 4? $34 \div 4 = 8.5$
No, since 34 is not divisible by 4.

Divisible by 9? $18 \div 9 = 2$
Yes, since 18 is divisible by 9.

Divisible by 12?
No, since 15,534 is not divisible by 4.

The 4-point response shows the correct answers and justifies each one.

3 points

Divisible by 3?
$1 + 5 + 5 + 3 + 4 = 17$
No, since 17 is not divisible by 3.

Divisible by 4? $34 \div 4 = 8.5$
No, since 34 is not divisible by 4.

Divisible by 9? $17 \div 9 = 1.8$
No, since 17 is not divisible by 9.

Divisible by 12?
No, since 15,534 is not divisible by 4.

The 3-point response has a computational error, but the student completed both parts.

2 points

Divisible by 3?
$1 + 5 + 5 + 3 + 4 = 18$
Yes, since 18 is divisible by 3.

Divisible by 4? $34 \div 4 = 8$
Yes, since 34 is divisible by 4.

Divisible by 9? $18 \div 2 = 9$

Divisible by 12?
Yes, since 15,534 is divisible by 3 and 4.

The 2-point response has a computational error and is missing an answer.

1 point

Yes, 15,534 is divisible by 3.

No, 15,534 is not divisible by 4.

Yes, 15,534 is divisible by 9.

No, 15,534 is not divisible by 12.

The 1-point response shows correct answers but with no work or justification.

A 0-point response has incorrect answers and no work shown.

Writing Extended Responses

This strategy shows students what is required to get full credit on extended-response test questions.

Guided Instruction

Exercises

Review these divisibility rules, as needed: A number is divisible by 5 if it ends in 0 or 5.

A number is divisible by 8 if the number formed by the last three digits is divisible by 8.

Challenge students to come up with a rule for divisibility by 18.

A number is divisible by 18 if and only if the number is divisible by both 2 and 9 (factors of 18 with only 1 as their common factor).

Resources

Test-Taking Strategies with Transparencies
- Transparency 4
- Practice sheet, p. 14

Test-Taking Strategies: Writing Extended Responses

Test-Taking Strategies: Writing Extended Responses

1. Draw and label two rectangles of different lengths and widths, each with a perimeter of 20 units.

Scoring Guide
4 Draws 2 rectangles of different lengths and widths, with perimeter 20 units indicated by labels on sides.
3 Draws 2 identical rectangles, with perimeter 20 units, OR draws 2 rectangles of different widths and lengths, only 1 with perimeter 20 units.
2 Draws 1 or 2 rectangles, whose perimeters are not 20 units.
1 Draws 1 or 2 non-rectangles. Does not label the sides.
0 Answers inappropriately or not at all.

2. The Athletic Council is hosting a sports banquet. It costs $300 to rent a hall, plus $8 per person for food. Between 50 and 90 people will attend. What are the least and greatest amounts of money the banquet could cost? Explain.

Scoring Guide
4 Correctly computes least and greatest costs, AND explains adequately.
3 Correctly computes least and greatest costs, but explanation is inadequate.
2 Computes least cost incorrectly, OR computes greatest cost incorrectly, OR explains inadequately.
1 Computes both costs incorrectly, OR computes one cost correctly, but explains inadequately.
0 Answers inappropriately or not at all.

Transparency 4

Resources

Student Edition
Extra Skills and Word Problems Practice, Ch. 2, p. 632
English/Spanish Glossary, p. 676
Formulas and Properties, p. 674
Tables, p. 670

All in One Teaching Resources
- Vocabulary and Study Skills 2F L3

Differentiated Instruction

Spanish Vocabulary Workbook with Study Skills ELL
- Interactive Textbook
- Audio Glossary

Online Vocabulary Quiz

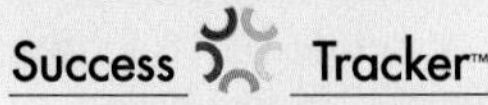

Online at PHSchool.com

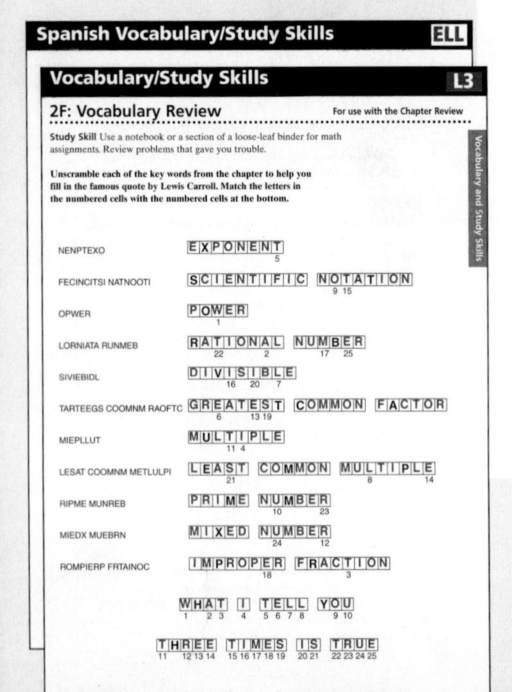

Chapter 2 Review

Vocabulary Review

- base (p. 68)
- composite number (p. 75)
- equivalent fractions (p. 82)
- exponent (p. 68)
- factor (p. 75)
- greatest common factor (GCF) (p. 75)
- improper fraction (p. 91)
- least common denominator (LCD) (p. 87)
- least common multiple (p. 74)
- mixed number (p. 91)
- multiple (p. 74)
- power (p. 68)
- prime factorization (p. 75)
- prime number (p. 75)
- rational number (p. 102)
- repeating decimal (p. 97)
- scientific notation (p. 107)
- simplest form (p. 82)
- terminating decimal (p. 96)

Go Online
PHSchool.com
For: Online vocabulary quiz
Web Code: arj-0251

Choose the correct term to complete each sentence.

1. In the expression 6^3, 3 represents the (power, exponent). **exponent**
2. A (composite, prime) number has exactly two factors, 1 and itself. **prime**
3. (Multiples, Factors) of 28 are 1, 2, 4, 7, 14, and 28. **factors**
4. At a class party, you and two friends eat $\frac{6}{8}$ of a pizza. This can also be written as the (improper fraction, equivalent fraction) $\frac{3}{4}$. **equivalent fraction**
5. The (GCF, LCM) of 24 and 36 is 12. **GCF**

Skills and Concepts

Lesson 2-1
- To write and simplify expressions with exponents

An **exponent** tells you how many times a number, or base, is used as a factor. A number expressed with an exponent is a **power**.

Simplify.

6. -2^4 **−16**
7. $(-4)^3$ **−64**
8. $5^2 + 10^2$ **125**
9. $4(5^2 - 10)$ **60**

Lesson 2-2
- To find multiples and factors and to use prime factorization

A **multiple** is the product of a number and any nonzero whole number. A **factor** is a whole number that divides another whole number with a remainder of 0. A whole number greater than 1 is **composite** if it has more than two factors and is **prime** if its only factors are 1 and itself.

Write the prime factorization. Use exponents where possible.

10. 84 **$2^2 \cdot 3 \cdot 7$**
11. 78 **$2 \cdot 3 \cdot 13$**
12. 90 **$2 \cdot 3^2 \cdot 5$**
13. 92 **$2^2 \cdot 23$**
14. 125 **5^3**
15. A grocer buys food from three suppliers. The suppliers deliver every 5 days, 6 days, and 7 days. All three came today. In how many days will they all deliver on the same day again? **210 days from now**

Lessons 2-3, 2-4

- To write equivalent fractions and to simplify fractions
- To compare and order fractions

A fraction is in **simplest form** when the numerator and denominator have no common factors other than 1. To compare and order fractions, you can use the **least common denominator (LCD),** which is the least common multiple of the fractions' denominators.

Order from least to greatest. $\frac{5}{9}, \frac{7}{12}, \frac{2}{3}$

16. $\frac{1}{4}, \frac{1}{3}, \frac{1}{6}$ $\frac{1}{6}, \frac{1}{4}, \frac{1}{3}$ **17.** $\frac{1}{4}, \frac{2}{5}, \frac{3}{8}$ $\frac{1}{4}, \frac{3}{8}, \frac{2}{5}$ **18.** $\frac{3}{8}, \frac{5}{6}, \frac{1}{2}$ $\frac{3}{8}, \frac{1}{2}, \frac{5}{6}$ **19.** $\frac{5}{9}, \frac{2}{3}, \frac{7}{12}$

20. Martha saw the same item on sale at two stores. Store A's sign read "Sale! $\frac{1}{3}$ off!" Store B's sign read "Sale! $\frac{2}{5}$ off!" Which store offered a greater discount? store B

Lesson 2-5

- To write mixed numbers and improper fractions

An **improper fraction** has a numerator that is greater than or equal to its denominator. A **mixed number** is the sum of a whole number and a fraction.

21. Tracey worked a total of 345 min. on a project. Use mixed numbers to write the time in hours. $5\frac{3}{4}$ h

Write each improper fraction as a whole or mixed number. Simplify.

22. $\frac{42}{7}$ 6 **23.** $\frac{27}{12}$ $2\frac{1}{4}$ **24.** $\frac{20}{3}$ $6\frac{2}{3}$ **25.** $\frac{125}{5}$ 25 **26.** $\frac{84}{12}$ 7

Lessons 2-6, 2-7

- To convert between fractions and decimals
- To compare and order rational numbers

To write a fraction as a decimal, you divide the numerator by the denominator. When the division ends with a remainder of 0, the quotient is a **terminating decimal.** When the same block of digits in a decimal repeats without end, the quotient is a **repeating decimal.** A **rational number** can be written as the quotient of two integers, where the denominator is not zero.

Write each fraction as a decimal.

27. $\frac{1}{3}$ $0.\overline{3}$ **28.** $\frac{5}{9}$ $0.\overline{5}$ **29.** $\frac{5}{2}$ 2.5 **30.** $\frac{16}{20}$ 0.8 **31.** $\frac{4}{50}$ 0.08

Order from least to greatest. 32–34. See margin.

32. $\frac{3}{4}, 0.\overline{3}, -\frac{7}{8}$ **33.** $2.7, -0.3, -\frac{4}{11}$ **34.** $-\frac{5}{6}, 2.2, -0.5$

Lesson 2-8

- To write numbers in both scientific notation and standard form

A number in **scientific notation** is written as a product of a factor greater than or equal to 1 but less than 10, and another factor that is a power of 10.

Write in scientific notation or in standard form. 35–38. See margin.

35. 7,123,000 **36.** 9.06×10^5 **37.** 81,900 **38.** 6.015×10^8

Alternative Assessment L4

Alternative Assessment Form C

Chapter 2

ALL IN A DAY

Everyone has 24 hours in a day, but we each use our 24 hours differently. For example, everyone spends some time asleep, but the number of hours we sleep varies from one person to another.

The table below shows the average number of hours people spend sleeping at different ages.

Age in Years	Hours Spent Sleeping
Newborn	19
6	11
12	9
25	8
40	7
48	6

Show all of your work on a separate sheet of paper.

1. Write the fraction of a day that each of the following people is likely to spend sleeping. Write each fraction in simplest form.
   a. young adult
   b. newborn
   c. first-grader
   d. seventh-grader
   e. 48-year-old man
2. Estimate the fraction of a day that you spend sleeping. Explain.
3. There are 1,440 minutes in a day. List how many minutes each day you spend doing the activities listed below. Then write a fraction that tells what part of a day you spend on each activity. Write each fraction in simplest form.
   a. eating
   b. brushing teeth
   c. doing homework
   d. talking on phone
   e. watching TV

32. $-\frac{7}{8}, 0.\overline{3}, \frac{3}{4}$

33. $-\frac{4}{11}, -0.3, 2.7$

34. $-\frac{5}{6}, -0.5, 2.2$

35. 7.123×10^6

36. 906,000

37. 8.19×10^4

38. 601,500,000

Resources

- ExamView Assessment Suite CD-ROM
  - Chapter 2 Ready-Made Test
  - Make your own Chapter 2 test
- MindPoint Quiz Show CD-ROM
  - Chapter 2 Review

Differentiated Instruction

All in One Teaching Resources
- Below Level Chapter 2 Test L2
- Chapter 2 Test L3
- Chapter 2 Alternative Assessment L4

Spanish Assessment Resources ELL
- Below Level Chapter 2 Test L2
- Chapter 2 Test L3
- Chapter 2 Alternative Assessment L4

ExamView Assessment Suite CD-ROM
- Special Needs Test L1
- Special Needs Practice Bank L1

Online Chapter 2 Test at www.PHSchool.com L3

Below Level Chapter Test L2

Chapter Test L3

Chapter Test
Chapter 2

Simplify each expression.

1. $(1 + 7)^2 \div 2$ 66
2. $(-5)^2$ 25
3. $\frac{63}{81}$ $\frac{7}{9}$
4. $\frac{35}{42}$ $\frac{5}{6}$

Order from least to greatest.

5. $\frac{7}{11}, \frac{5}{6}, \frac{2}{8}$ $\frac{2}{8}, \frac{7}{11}, \frac{5}{6}$
6. $\frac{1}{2}, -0.51, 0.\overline{5}, -\frac{1}{2}$ $-0.51, -\frac{1}{2}, \frac{1}{2}, 0.\overline{5}$

Write the prime factorization of each number.

7. 52 2 x 2 x 13
8. 60 2 x 2 x 3 x 5

Write each improper fraction as a mixed number and each mixed number as an improper fraction.

9. $3\frac{3}{10}$ $\frac{33}{10}$
10. $\frac{52}{6}$ $8\frac{4}{6}$
11. Write 1,325,000 in scientific notation. 1.325×10^6
12. Write 8.35×10^4 in standard form. 83,500
13. Is 982 divisible by 3? Explain. No; 9 + 8 + 2 = 19, and 19 is not divisible by 3
14. Write three fractions that are equivalent to $\frac{1}{3}$. Sample answer: $\frac{2}{6}, \frac{3}{9}, \frac{4}{12}$

Chapter 2 Test

Go Online PHSchool.com **For:** Online chapter test **Web Code:** ara-0252

Find the value of each expression.

1. $(3^2 - 4) + 5$ **10**
2. $5^2 - 7^2$ **−24**
3. $(6 - 9)^3$ **−27**
4. $54 + 3^2$ **63**

Write using an exponent.

5. $3 \cdot 3 \cdot 3 \cdot 3$ **3^4**
6. $11 \cdot 11 \cdot 11$ **11^3**

Write the prime factorization. Use exponents where possible.

7. 48 **$2^4 \cdot 3$**
8. 60 **$2^2 \cdot 3 \cdot 5$**
9. 121 **11^2**

List 3 factors and 3 multiples of each number. **10–13. See margin.**

10. 27
11. 36
12. 100
13. 25

Find the GCF of each pair of numbers.

14. 32, 40 **8**
15. 55, 15 **5**
16. 36, 57 **3**
17. 24, 68 **4**

18. **Reasoning** Tell whether each statement is true or false. **18a. false b. false**
    a. 2 is a composite number.
    b. Any factor of a whole number is greater than any multiple of a whole number.
    c. A number is divisible by 3 if its last digit is divisible by 3. **false**
    d. 1 is neither composite nor prime. **true**

19. Use a factor tree to write the prime factorization of 42. **See margin.**

Write two fractions equivalent to each fraction.

20. $\frac{1}{3}$
21. $-\frac{15}{24}$
22. $-\frac{4}{5}$
23. $\frac{16}{28}$

20–23. See margin.

Write each fraction in simplest form.

24. $\frac{12}{18}$ **$\frac{2}{3}$**
25. $\frac{27}{54}$ **$\frac{1}{2}$**
26. $\frac{36}{96}$ **$\frac{3}{8}$**
27. $\frac{7}{42}$ **$\frac{1}{6}$**

28. **Writing in Math** Explain how you can use prime factorization to write a fraction in simplest form. **See margin.**

29. **Modeling** Draw models to represent $\frac{3}{4}$ and $2\frac{3}{5}$. **See margin.**

30. What fraction does the shaded part of the model represent? **$\frac{3}{8}$**

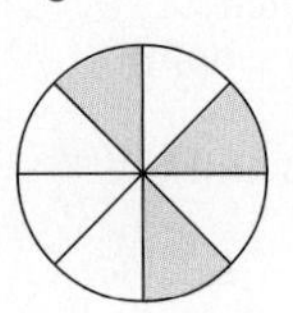

Compare. Use <, =, or >.

31. $\frac{2}{9} \blacksquare \frac{8}{9}$ **<**
32. $\frac{5}{16} \blacksquare \frac{3}{8}$ **<**
33. $\frac{7}{18} \blacksquare \frac{2}{5}$ **<**

34. **Ships** A crew finds a treasure chest with 168 gold coins and 200 silver coins. Each crew member gets the same share of gold coins and of silver coins, with none left over.
    a. What is the greatest possible number of crew members? **8 crew members**
    b. How many of each type of coin does each crew member get? **21 gold, 25 silver**

Write as an improper fraction.

35. $5\frac{2}{3}$ **$\frac{17}{3}$**
36. $4\frac{5}{6}$ **$\frac{29}{6}$**
37. $8\frac{7}{10}$ **$\frac{87}{10}$**
38. $3\frac{2}{5}$ **$\frac{17}{5}$**

Write as a whole number or a mixed number.

39. $\frac{12}{5}$ **$2\frac{2}{5}$**
40. $\frac{30}{9}$ **$3\frac{1}{3}$**
41. $\frac{48}{12}$ **4**
42. $\frac{42}{30}$ **$1\frac{2}{5}$**

Write each fraction as a decimal.

43. $\frac{2}{16}$ **0.125**
44. $\frac{6}{15}$ **0.4**
45. $\frac{5}{4}$ **1.25**
46. $\frac{8}{25}$ **0.32**

Write each decimal as a mixed number or fraction in simplest form.

47. 0.2 **$\frac{1}{5}$**
48. 1.3 **$1\frac{3}{10}$**
49. 0.35 **$\frac{7}{20}$**
50. 3.62 **$3\frac{31}{50}$**

Write using scientific notation.

51. 12,300,000 **1.23×10^7**
52. 75,462 **7.5462×10^4**

Write in standard form.

53. 2.1×10^4 **21,000**
54. 8×10^9 **8,000,000,000**

55. Order from least to greatest:
    $2.56, -2.\overline{5}, -2\frac{1}{5}, \frac{24}{10}, -2.4$
    $-2.\overline{5}, -2.4, -2\frac{1}{5}, \frac{24}{10}, 2.56$

Exercises 10–13. Answers may vary. Samples are given.

10. 1, 3, 9; 54, 81, 108
11. 1, 2, 3; 72, 108, 144
12. 1, 2, 4; 200, 300, 400
13. 1, 5, 25; 50, 75, 100

19.

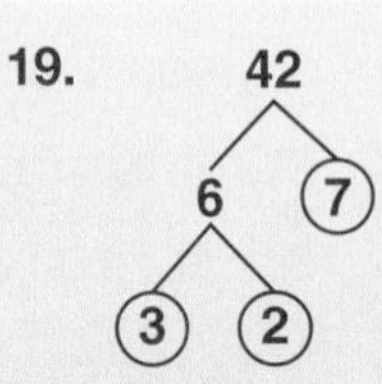

Exercises 20–23. Answers may vary. Samples are given.

20. $\frac{2}{6}, \frac{3}{9}$
21. $-\frac{5}{8}, -\frac{10}{16}$
22. $-\frac{8}{10}, -\frac{12}{15}$
23. $\frac{4}{7}, \frac{8}{14}$

28–29. See back of book.

Test Prep Practice

Reading Comprehension

Read each passage and answer the questions that follow.

Prime Construction Jackie says, "If I multiply the first two prime numbers together and add 1, I get a new prime number." Amit says, "If I multiply the first three prime numbers and add 1, I also get a prime number." Carl says, "I bet the same will happen if I multiply the first four prime numbers and add 1." "It seems to me," says Maria, "that if I multiply any two or more prime numbers, the product is never prime."

1. What prime number does Jackie get? **C**
   - A 8
   - B 6
   - C 7
   - D 11

2. What prime number does Amit get? **G**
   - F 16
   - G 31
   - H 41
   - J 43

3. What number would Carl get? **D**
   - A 107
   - B 181
   - C 210
   - D 211

4. Which numbers could NOT be used to test Maria's statement? **H**
   - F 3, 13
   - G 2, 5, 11
   - H 31, 33
   - J 17, 29

Light Reading Light travels very quickly at about 1.86×10^5 mi/s. This is fast enough that we do not notice any delay when we flip a light switch. However, light from distant objects in space does not arrive instantaneously. The sun is about 9.3×10^7 mi from Earth. The next nearest star system, Alpha Centauri, is about 2.5×10^{13} mi away.

5. About how far away would a lamp have to be for its light to take 2 s to reach our eyes? **B**
   - A 37,000 mi
   - B 370,000 mi
   - C 3,700,000 mi
   - D 37,000,000 mi

6. About how long does it take light from the sun to reach Earth? **H**
   - F 5 s
   - G 5×10 s
   - H 5×10^2 s
   - J 500 min

7. Which does NOT express the time in seconds it takes light from Alpha Centauri to reach Earth? **C**
   - A $\frac{2.5 \times 10^{13}}{186 \times 10^3}$
   - B $\frac{2.5 \times 10^{13}}{186{,}000}$
   - C $\frac{1.86}{2.5 \times 10^8}$
   - D $\frac{2.5}{1.86} \cdot \frac{10^{13}}{10^5}$

8. About how many times farther is Alpha Centauri from Earth than the sun is from Earth? **J**
   - F 4
   - G 23
   - H 2.6×10^3
   - J 270,000

Test Prep

Resources

Test Prep Workbook

All in One Teaching Resources
- Cumulative Review L3

ExamView Assessment Suite CD-ROM
- Standardized Test Practice

Differentiated Instruction

Progress Monitoring Assessments
- Benchmark Test 1 L3

Spanish Assessment Resources
- Spanish Cumulative Review ELL

ExamView Assessment Suite CD-ROM
- Special Needs Practice Bank L1

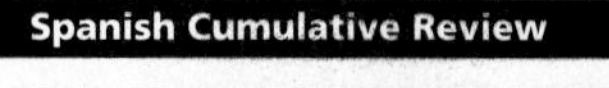

Spanish Cumulative Review ELL

Cumulative Review L3

Cumulative Review

Chapter 1–2

Multiple choice. Choose the letter of the best answer.

1. Which numbers are in order from least to greatest?
   - (A.) 1.1, 1.2, 1.25, 1.415
   - B. 0.2, 0.02, 2.2, 0.22
   - C. 0.3, 0.6, 0.51, 0.65
   - D. 3.5, 3.41, 3.02, 3.6
2. Karl's scores for five video games were 38, 57, 48, 62, and 55. What is his mean score?
   - F. 24
   - G. 38
   - (H.) 52
   - J. 55
3. Which difference is greatest?
   - A. 2.5 − 2.3
   - (B.) 5.5 − 4.15
   - C. 3.75 − 2.95
   - D. 8.05 − 7.85
4. Which number is the outlier?

   25 34 63 27 31
   - F. 34
   - G. 25
   - (H.) 63
   - J. 36
5. Ten identical coins weigh 34.5 g. How much does one coin weigh?
   - A. 0.345 g
   - (B.) 3.45 g
   - C. 345 g
   - D. not here
6. What is the mode of this data?

   0.12 0.21 0.15 0.51 0.51 0.25 0.52
   - A. 0.25
   - (B.) 0.40
   - C. 0.51
   - D. 0.52
7. Which sum is positive?
   - F. $-56 + (-4)$
   - G. $81 + (-90)$
   - (H.) $-48 + 55$
   - J. $-76 + 45$
8. Which of the following is equal to 16?
   - (A.) $(-2)^4$
   - B. -2^3
   - C. -4^2
   - D. 8^2
9. Which fraction below is not equivalent to any of the other choices?
   - F. $\frac{3}{8}$
   - G. $\frac{9}{24}$
   - (H.) $\frac{24}{72}$
   - J. $\frac{15}{40}$
10. What is the product of 2.3 and 3.45?
    - A. 0.7935
    - B. 1.15
    - C. 1.5
    - (D.) 7.935

Problem Solving

Applying Fractions

Students will use data from these two pages to answer the questions posed in Put It All Together.

Invite students to share photos they have taken.
Ask:

- *What are some of the obstacles you have encountered while taking pictures?*
- *What picture-taking techniques have you learned to overcome these obstacles?*

Discuss the differences and similarities when using a film camera versus a digital camera.

Materials

- Photos brought in by teacher or students
- Cameras brought in by teacher or students

Activating Prior Knowledge

Have students who enjoy photography share some of the challenges that all photographers face. Invite them to share photos they have taken and to explain some of the obstacles they have encountered and the picture-taking techniques they have learned or developed. Invite students to discuss the cameras they use and would like to use.

Guided Instruction

Have volunteers read aloud the opening paragraph and the information on the page.
Ask questions, such as:

- *In what situations have they varied the shutter speeds?*
- *What are zoom lenses?* **Sample answer: a camera lens that magnifies the focal image**
- *How do you use a zoom lens?* **Sample answer: The zoom lens can be turned to magnify and adjust sharpness of the focal image.**
- *What is a strobe light?*

Problem Solving Application

Applying Fractions

Photographic Memory You've probably seen pictures of athletes, animals, or cars that "freeze" the subject's motion but show all the excitement of the moment. A good photographer chooses the best shutter speed for the action. If the shutter stays open too long, the camera records too much movement, and the picture is blurry.

Cameras Then and Now
In the 1850s, leather bellows folded a camera into a protective case, making it easier to carry. Zoom lenses on digital cameras fold into the camera for protection.

Capturing Movement
For this photograph, the photographer used a very slow shutter speed and a strobe light. The shutter stayed open while the dancer moved.

Light Trail
For this photograph, the photographer used a very slow shutter speed and no flash. The shutter stayed open while traffic moved along the upper and lower levels of Interstate 5 in Seattle, Washington.

1a. 15 in.

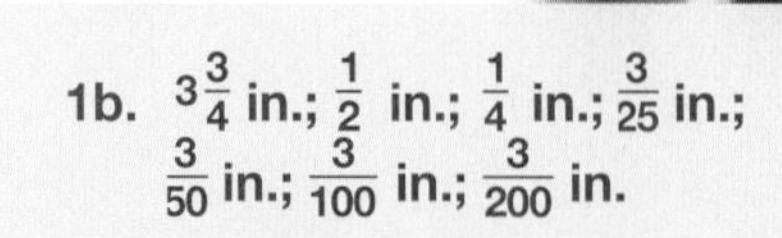

1b. $3\frac{3}{4}$ in.; $\frac{1}{2}$ in.; $\frac{1}{4}$ in.; $\frac{3}{25}$ in.; $\frac{3}{50}$ in.; $\frac{3}{100}$ in.; $\frac{3}{200}$ in.

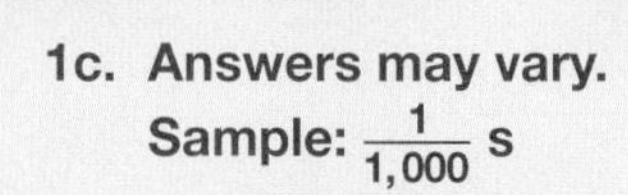

1c. Answers may vary. Sample: $\frac{1}{1{,}000}$ s

2a. 8 times; $\frac{1}{8}$

b. $\frac{1}{8}$ $\frac{1}{8} \times 10 = 1.25$; 1.25 ft.

c. 18.75 ft/s; 1,125 ft/min

Put It All Together

Data File **Use the data on these two pages and the exposure times on page 87 to answer these questions.**

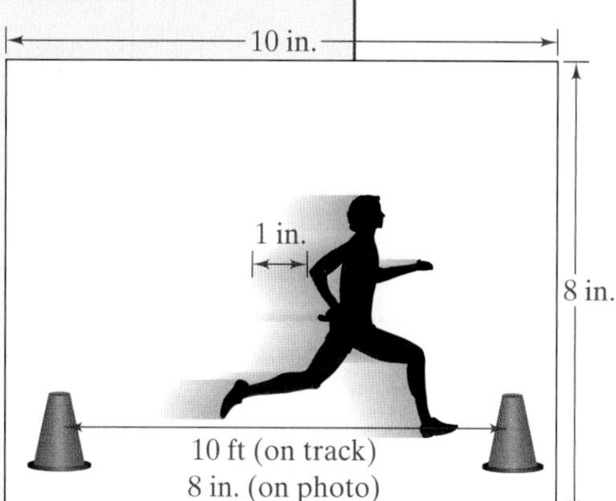

1. The image of the runner is blurry because the runner moved a visible amount during the $\frac{1}{15}$ of a second that the shutter was open.
   a. The blur for $\frac{1}{15}$ s is 1 in. long. How long would the blur be for 1 s?
   b. How long would the blur be for each exposure time on page 87?
   c. **Reasoning** At what shutter time do you think the length of the blur would be small enough that it would not show in the photo?
2. a. How many times would the blur length fit into the space between the two cones in the photo? What fraction of the distance between the cones is the blur?
   b. Use your answer to part (a) to find the distance the runner moved while the shutter was open.
   c. Maintaining the same speed, how far would the runner go in 1 s? In 1 min?

Go Online PHSchool.com
For: Information about photography
Web Code: are-0253

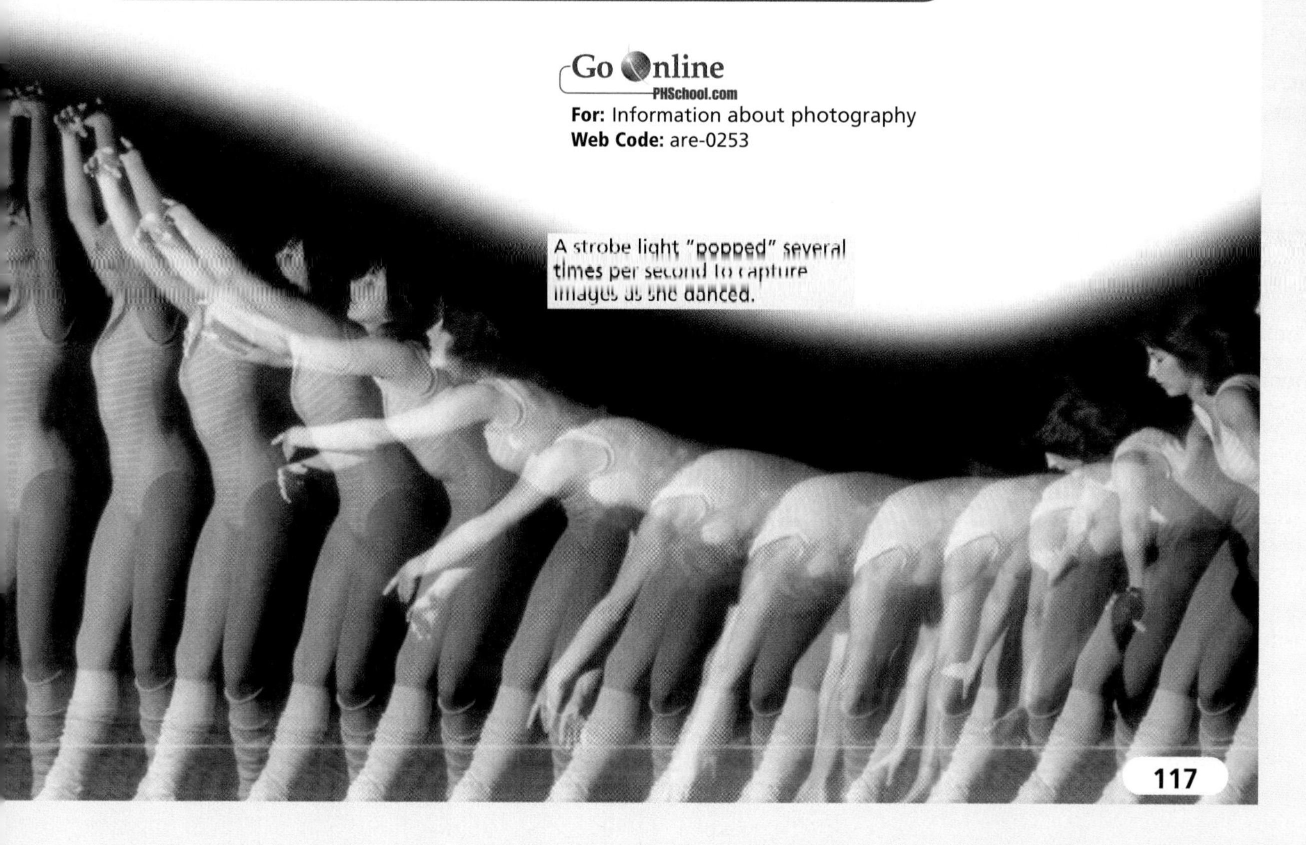
A strobe light "popped" several times per second to capture images as she danced.

Activity

Have students work in pairs to answer the questions.

Exercises 1b–1c Guide students to see that the ratio of blur lengths will be proportional to the ratio of shutter opening times. Elicit from them that although they can use an increasingly fast shutter speed, they can never entirely remove blurring because continual multiplying by $\frac{1}{2}$ will not yield a product of 0. Point out, however, that they can make the blur so slight that it will be essentially unseen.

Science Connection
Tell students that another factor in creating a sharper image is the speed of the film used. Film speed is a measure of sensitivity to light. Have interested students find out about how film speed is measured in ASA (American Standards Association) or ISO (International Standards Organisation) units. Invite them also to look into the effect of the camera's aperture, the size of the opening that lets in light, on the picture taken.

Differentiated Instruction

Special Needs L1
Exercises 1–3 As needed, review the camera exposure times on page 87. Encourage students to copy the exposure times for Exercise 1b.

3 Operations With Fractions

Chapter at a Glance

Lesson Titles, Objectives, and Features	Assessment	NCTM Standards	Local Standards
3-1 Estimating With Fractions and Mixed Numbers • To estimate sums, differences, products, and quotients of fractions **Vocabulary Builder:** High-Use Academic Words	Lesson Quiz	1, 2, 6, 7, 8, 9, 10	
3-2a Activity Lab, Hands On: Using Fraction Models **3-2 Adding and Subtracting Fractions** • To add and subtract fractions and to solve problems involving fractions	Lesson Quiz	1, 2, 6, 7, 8, 9, 10	
3-3 Adding and Subtracting Mixed Numbers • To add and subtract mixed numbers and to solve problems involving mixed numbers	Lesson Quiz Checkpoint Quiz 1	1, 2, 6, 7, 8, 9, 10	
3-4a Activity Lab, Hands On: Modeling Fraction Multiplication **3-4 Multiplying Fractions and Mixed Numbers** • To multiply fractions and mixed numbers and to solve problems by multiplying	Lesson Quiz	1, 2, 6, 7, 8, 9, 10	
3-5a Activity Lab, Hands On: Modeling Fraction Division **3-5 Dividing Fractions and Mixed Numbers** • To divide fractions and mixed numbers and to solve problems by dividing **Guided Problem Solving:** Practice Solving Problems	Lesson Quiz	1, 2, 6, 7, 8, 9, 10	
3-6 Changing Units in the Customary System • To change units of length, capacity, and weight in the customary system **3-6b Activity Lab, Algebra Thinking:** Solving Puzzles	Lesson Quiz Checkpoint Quiz 2	1, 2, 4, 6, 7, 8, 9, 10	
3-7a Activity Lab: Choosing Appropriate Units **3-7 Precision** • To find and compare the precision of measurement **3-7b Activity Lab:** Estimating in Different Systems	Lesson Quiz	1, 2, 4, 6, 7, 8, 9, 10	
Problem Solving Application: Applying Fractions			

NCTM Standards 2000
1 Number and Operations **2** Algebra **3** Geometry **4** Measurement **5** Data Analysis and Probability
6 Problem Solving **7** Reasoning and Proof **8** Communication **9** Connections **10** Representation

Correlations to Standardized Tests

All content for these tests is contained in *Prentice Hall Math,* Course 2. This chart reflects coverage in this chapter only.

	3-1	3-2	3-3	3-4	3-5	3-6	3-7
Terra Nova CAT6 (Level 17)							
Number and Number Relations	✔	✔	✔	✔	✔	✔	✔
Computation and Numerical Estimation	✔	✔	✔	✔	✔	✔	✔
Operation Concepts	✔	✔	✔	✔	✔	✔	✔
Measurement						✔	✔
Geometry and Spatial Sense							
Data Analysis, Statistics, and Probability							
Patterns, Functions, Algebra	✔	✔	✔	✔	✔	✔	✔
Problem Solving and Reasoning	✔	✔	✔	✔	✔	✔	✔
Communication	✔	✔	✔	✔	✔	✔	✔
Decimals, Fractions, Integers, and Percent	✔	✔	✔	✔	✔	✔	✔
Order of Operations							
Terra Nova CTBS (Level 17)							
Decimals, Fractions, Integers, Percents	✔	✔	✔	✔	✔	✔	✔
Order of Operations, Numeration, Number Theory	✔	✔	✔	✔	✔	✔	✔
Data Interpretation							
Pre-Algebra	✔	✔	✔	✔	✔	✔	✔
Measurement						✔	✔
Geometry							
ITBS (Level 13)							
Number Properties and Operations	✔	✔	✔	✔	✔	✔	✔
Algebra	✔	✔	✔	✔	✔	✔	✔
Geometry							
Measurement						✔	✔
Probability and Statistics							
Estimation	✔						
SAT10 (Int 3 Level)							
Number Sense and Operations	✔	✔	✔	✔	✔	✔	✔
Patterns, Relationships, and Algebra	✔	✔	✔	✔	✔	✔	✔
Data, Statistics, and Probability							
Geometry and Measurement						✔	✔
NAEP							
Number Sense, Properties, and Operations	✔	✔	✔	✔	✔		
Measurement						✔	✔
Geometry and Spatial Sense							
Data Analysis, Statistics, and Probability							
Algebra and Functions							

CAT6 California Achievement Test, 6th Ed.
SAT10 Stanford Achievement Test, 10th Ed.
CTBS Comprehensive Test of Basic Skills
NAEP National Assessment of Educational Progress 2005 Mathematics Objectives
ITBS Iowa Test of Basic Skills, Form M

Math Background

Skills Trace

BEFORE Chapter 3

Course 1 or Grade 6 introduced operations with fractions.

DURING Chapter 3

Course 2 reviews operations with fractions and extends them to solve two-step equations with fractions.

AFTER Chapter 3

Throughout this course, students perform operations with fractions and mixed numbers.

3-1 Estimating With Fractions and Mixed Numbers

Math Understandings

- You can find approximate answers by using estimation.
- You can estimate the value of fractions between 0 and 1 by comparing the size of the numerator and the denominator.
- You can estimate the value of mixed numbers by rounding to the nearest whole number.

A **benchmark** is a value that can be used as a reference point. Round to the benchmark 0 when the numerator of a fraction is very small compared to the denominator; use $\frac{1}{2}$ when the numerator is about half the denominator; and use 1 when the numerator and denominator are nearly equal.

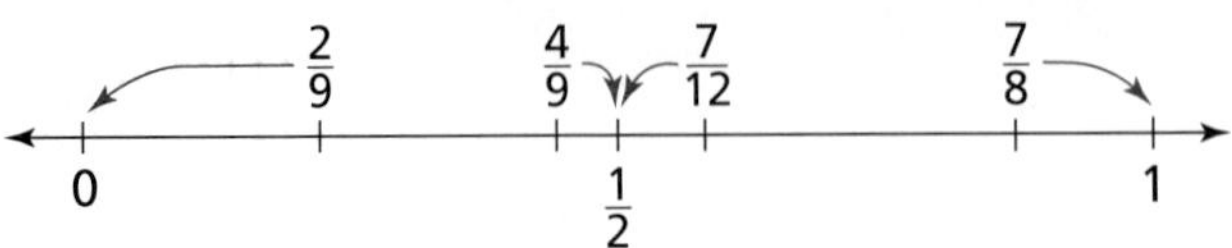

Estimate as 0 when the numerator is very small compared to the denominator.

Estimate as $\frac{1}{2}$ when the numerator is about half the denominator.

Estimate as 1 when the numerator and denominator are nearly equal.

3-2, 3-3 Adding and Subtracting Fractions and Mixed Numbers

Math Understandings

- The idea of changing something to something simpler is a powerful idea in math.
- The product of the denominators will always give a common denominator, but it may not be the Least Common Denominator.
- Any common denominator can be used, but using the Least Common Denominator may save steps.
- In order to subtract from a mixed number, it may be necessary to rename the mixed number.

Usually, to add or subtract two mixed numbers, combine the fraction parts and then combine the whole numbers. An alternative method is to write the mixed numbers as improper fractions and combine them.

3-4, 3-5 Multiplying and Dividing Fractions and Mixed Numbers

Math Understandings

- Multiplying fractions and mixed numbers does not necessarily give a product less than both factors.
- Finding half of a number is the same as multiplying that number by $\frac{1}{2}$ and as dividing that number by 2.
- A number and its reciprocal multiply to a product of 1.
- You can rewrite dividing by a number as multiplying by the reciprocal of that number.

Multiplying Fractions	
Arithmetic	**Algebra**
$\frac{3}{4} \cdot \frac{1}{2} = \frac{3 \cdot 1}{4 \cdot 2} = \frac{3}{8}$	$\frac{a}{b} \cdot \frac{c}{d} = \frac{ac}{bd}$, b and $d \neq 0$
Dividing Fractions	
Arithmetic	**Algebra**
$\frac{5}{8} \div \frac{1}{8} = \frac{5}{8} \cdot \frac{8}{1}$	$\frac{a}{b} \div \frac{c}{d} = \frac{a}{b} \cdot \frac{d}{c}$, b, c and $d \neq 0$

To multiply a fraction by a whole number, rewrite the whole number as a fraction with a denominator of 1, for example, 7 becomes $\frac{7}{1}$. To multiply mixed numbers, rewrite them as improper fractions and then multiply.

Two numbers are **reciprocals** if their product is 1. The reciprocal of a fraction is the fraction with the numerator and denominator interchanged, for example, 5 and $\frac{1}{5}$, $\frac{2}{3}$ and $\frac{3}{2}$.

3-6 Changing Units in the Customary System

Math Understandings

- To change between customary units, use multiplication or division of equivalent units.

You can use the customary unit equivalents to change from smaller units to larger units and from larger units to smaller units.

Customary Units of Measure

Type	Length	Capacity	Weight
Unit	inch (in.) foot (ft) yard (yd) mile (mi)	fluid ounce (fl oz) cup (c) pint (pt) quart (qt) gallon (gal)	ounce (oz) pound (lb) ton (t)
Equivalents	1 ft = 12 in. 1 yd = 3 ft 1 mi = 5,280 ft	1 c = 8 fl oz 1 pt = 2 c 1 qt = 2 pt 1 gal = 4 qt	1 lb = 16 oz 1 t = 2,000 lb

3-7 Precision

Math Understandings

- Every measurement is an approximation that is only as accurate as the method and the instruments used.
- The precision of a measurement tells you how finely the measurement is made.
- When you compare measurements, the more precise measurement is the one with the smaller unit of measure.

The **precision** of a measurement is its degree of exactness, which you can determine from the smallest marks on the measuring instrument used.

Example: Choose the more precise measurement.

16 cm, 15.5 cm

Tenths of a cm are smaller than the units digit, so 15.5 cm is more precise.

Additional Professional Development Opportunities

Math Background Notes for Chapter 3: Every lesson has a Math Background in the PLAN section.

Research Overview, Mathematics Strands
Additional support for these topics and more is in the front of the Teacher's Edition.

LessonLab
LessonLab, a Pearson Education company, offers comprehensive, facilitated professional development designed to help teachers to improve student achievement. To learn more, please visit lessonlab.com.

Chapter 3 Resources

	3-1	3-2	3-3	3-4	3-5	3-6	3-7	For the Chapter
Print Resources								
L3 Practice	●	●	●	●	●	●	●	
L1 Adapted Practice	●	●	●	●	●	●	●	
L3 Guided Problem Solving	●	●	●	●	●	●	●	
L2 Reteaching	●	●	●	●	●	●	●	
L4 Enrichment	●	●	●	●	●	●	●	
L3 Daily Notetaking Guide	●	●	●	●	●	●	●	
L1 Adapted Daily Notetaking Guide	●	●	●	●	●	●	●	
L3 Vocabulary and Study Skills Worksheets	●		●	●		●	●	●
L3 Daily Puzzles	●	●	●	●	●	●	●	
L3 Activity Labs	●	●	●	●	●	●	●	
L3 Checkpoint Quiz			●			●		
L3 Chapter Project								●
L2 Below Level Chapter Test								●
L3 Chapter Test								●
L4 Alternative Assessment								●
L3 Cumulative Review								●
Spanish Resources ELL								
L3 Practice	●	●	●	●	●	●	●	●
L3 Vocabulary and Study Skills Worksheets	●		●	●		●	●	●
L3 Checkpoint Quiz			●			●		
L2 Below Level Chapter Test								●
L3 Chapter Test								●
L4 Alternative Assessment								●
L3 Cumulative Review								●
Transparencies								
Check Skills You'll Need	●	●	●	●	●	●	●	
Additional Examples	●	●	●	●	●	●	●	
Problem of the Day	●	●	●	●	●	●	●	
Classroom Aid	●							
Student Edition Answers	●	●	●	●	●	●	●	●
Lesson Quiz	●	●	●	●	●	●	●	
Test-Taking Strategies								●
Technology								
Interactive Textbook Online	●	●	●	●	●	●	●	●
StudentExpress™ CD-ROM	●	●	●	●	●	●	●	●
Success Tracker™ Online Intervention	●	●	●	●	●	●	●	●
TeacherExpress™ CD-ROM	●	●	●	●	●	●	●	●
PresentationExpress™ with QuickTake Presenter CD-ROM	●	●	●	●	●	●	●	●
ExamView® Assessment Suite CD-ROM	●	●	●	●	●	●	●	●
MindPoint® Quiz Show CD-ROM								●
Prentice Hall Web Site: PHSchool.com	●	●	●	●	●	●	●	●

Also available: **Prentice Hall Assessment System**
- Progress Monitoring Assessments
- Skills and Concepts Review
- Test Prep Workbook

Other Resources
Algebra Readiness Tests
All-in-One Student Workbook
All-in-One Student Workbook, Adapted Version
Multilingual Handbook
Solution Key
Math Notes Study Folder
Spanish Cumulative Assessment

Where You Can Use the Lesson Resources

Here is a suggestion, following the four-step teaching plan, for how you can incorporate Differentiated Instruction Resources into your teaching.

	Instructional Resources L3	Differentiated Instruction Resources
1. Plan		
Preparation Read the Math Background in the Teacher's Edition to connect this lesson with students' previous experience. **Starting Class** **Check Skills You'll Need** Assign these exercises to review prerequisite skills. **New Vocabulary** Help students pre-read the lesson by pointing out the new terms introduced in the lesson.	**Math Background** **Math Understandings** **Transparencies & PresentationExpress™ with QuickTake Presenter CD-ROM** Check Skills You'll Need Problem of the Day **Resources** Vocabulary and Study Skills	**Spanish Support** ELL Vocabulary and Study Skills
2. Teach		
L3 **Guided Instruction** Use the Activity Labs to build conceptual understanding. Teach each Example. Use the Teacher's Edition side column notes for specific teaching tips, including Error Prevention notes. Use the Additional Examples found in the side column (and on transparency and PowerPoint) as an alternative presentation for the content. After each Example, assign the Quick Check exercise for that Example to get an immediate assessment of student understanding. Use the Closure activity in the Teacher's Edition to help students attain mastery of lesson content.	**Student Edition** Activity Lab **Resources** Daily Notetaking Guide Activity Lab **Transparencies & PresentationExpress™ with QuickTake Presenter CD-ROM** Additional Examples Classroom Aids **ExamView® Assessment Suite CD-ROM**	**Teacher's Edition** Every lesson includes suggestions for working with students who need special attention. L1 Special Needs L2 Below Level L4 Advanced Learners ELL English Language Learners **Resources** L1 Adapted Daily Notetaking Guide **Multilingual Handbook**
3. Practice		
Assignment Guide **Check Your Understanding** Use these questions to check students' understanding before you assign homework. **Homework Exercises** Assign homework from these leveled exercises in the Assignment Guide. A Practice by Example B Apply Your Skills C Challenge Test Prep and Mixed Review **Homework Quick Check** Use these key exercises to quickly check students' homework.	**Transparencies & PresentationExpress™ with QuickTake Presenter CD-ROM** Student Answers **Resources** Practice Guided Problem Solving Vocabulary and Study Skills Activity Lab Daily Puzzles **ExamView® Assessment Suite CD-ROM**	**Spanish Support** ELL Practice ELL Vocabulary and Study Skills **Resources** L1 Adapted Practice L4 Enrichment
4. Assess & Reteach		
Lesson Quiz Assign the Lesson Quiz to assess students' mastery of the lesson content. **Checkpoint Quiz** Use the Checkpoint Quiz to assess student progress over several lessons.	**Transparencies & PresentationExpress™ with QuickTake Presenter CD-ROM** Lesson Quiz **Resources** Checkpoint Quiz	**Resources** L2 Reteaching ELL Checkpoint Quiz Success Tracker™ Online Intervention **ExamView® Assessment Suite CD-ROM**

KEY L1 Special Needs L2 Below Level L3 For All Students L4 Advanced, Gifted ELL English Language Learners

Operations With Fractions

Answers for students are in the back of the textbook.

For intervention, direct students to:

Adding and Subtracting Integers
Lesson 1-7
Extra Skills and Word Problems Practice, Ch. 1

Multiplying and Dividing Integers
Lesson 1-8
Extra Skills and Word Problems Practice, Ch. 1

Finding the Greatest Common Factor
Lesson 2-2
Extra Skills and Word Problems Practice, Ch. 2

Simplifying Fractions
Lesson 2-3
Extra Skills and Word Problems Practice, Ch. 2

Spanish Vocabulary/Study Skills ELL

Vocabulary/Study Skills L3

3A: Graphic Organizer — For use before Lesson 3-1

Study Skill: You should fully understand the basic concepts in each chapter before moving on to more complex material. Be sure to ask questions when you are not comfortable with what you have learned.

Write your answers.

1. What is the chapter title? Operations with Fractions
2. How many lessons are there in this chapter? 9
3. What is the topic of the Math at Work page? Understanding Word Problems
4. What is the topic of the Test-Taking Strategies page? Reading-Comprehension Questions
5. Complete the graphic organizer below as you work through the chapter.
   - In the center, write the title of the chapter.
   - When you begin a lesson, write the lesson name in a rectangle.
   - When you complete a lesson, write a skill or key concept in a circle linked to that lesson block.
   - When you complete the chapter, use this graphic organizer to help you review.

Check students' diagrams.

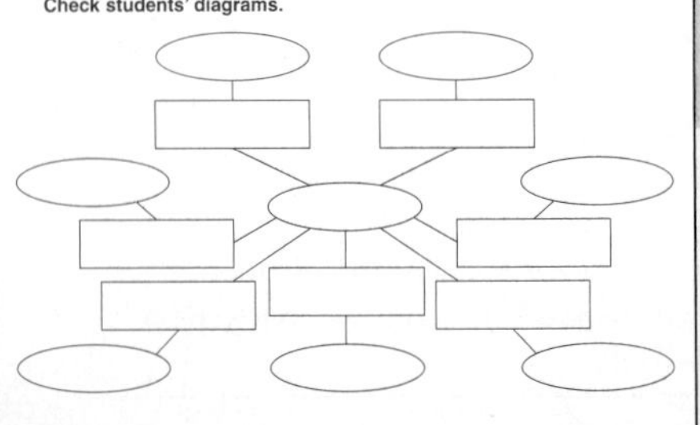

CHAPTER 3 Operations With Fractions

What You've Learned

- In Chapter 1, you estimated with decimals. You changed units in the metric system.
- In Chapter 2, you simplified fractions and wrote mixed numbers and improper fractions.

Check Your Readiness

GO for Help

For Exercises	See Lesson
1–4	1-7
5–8	1-8
9–12	2-2
13–20	2-3

Adding and Subtracting Integers

Find each sum or difference.

1. $-55 + 15$ **−40**
2. $-3 + (-21)$ **−24**
3. $58 - (-42)$ **100**
4. $-7 - (-25)$ **18**

Multiplying and Dividing Integers

Find each product or quotient.

5. $-3 \cdot 15$ **−45**
6. $-12 \cdot (-5)$ **60**
7. $39 \div (-13)$ **−3**
8. $-60 \div (-3)$ **20**

Finding the Greatest Common Factor

Use prime factorization to find the GCF of each pair of numbers.

9. 18, 27 **9**
10. 21, 42 **21**
11. 24, 36 **12**
12. 45, 36 **9**

Simplifying Fractions

Write each fraction in simplest form.

13. $\frac{12}{16}$ $\frac{3}{4}$
14. $\frac{15}{30}$ $\frac{1}{2}$
15. $\frac{42}{48}$ $\frac{7}{8}$
16. $\frac{27}{72}$ $\frac{3}{8}$
17. $\frac{19}{57}$ $\frac{1}{3}$
18. $\frac{25}{45}$ $\frac{5}{9}$
19. $\frac{34}{51}$ $\frac{2}{3}$
20. $\frac{16}{56}$ $\frac{2}{7}$

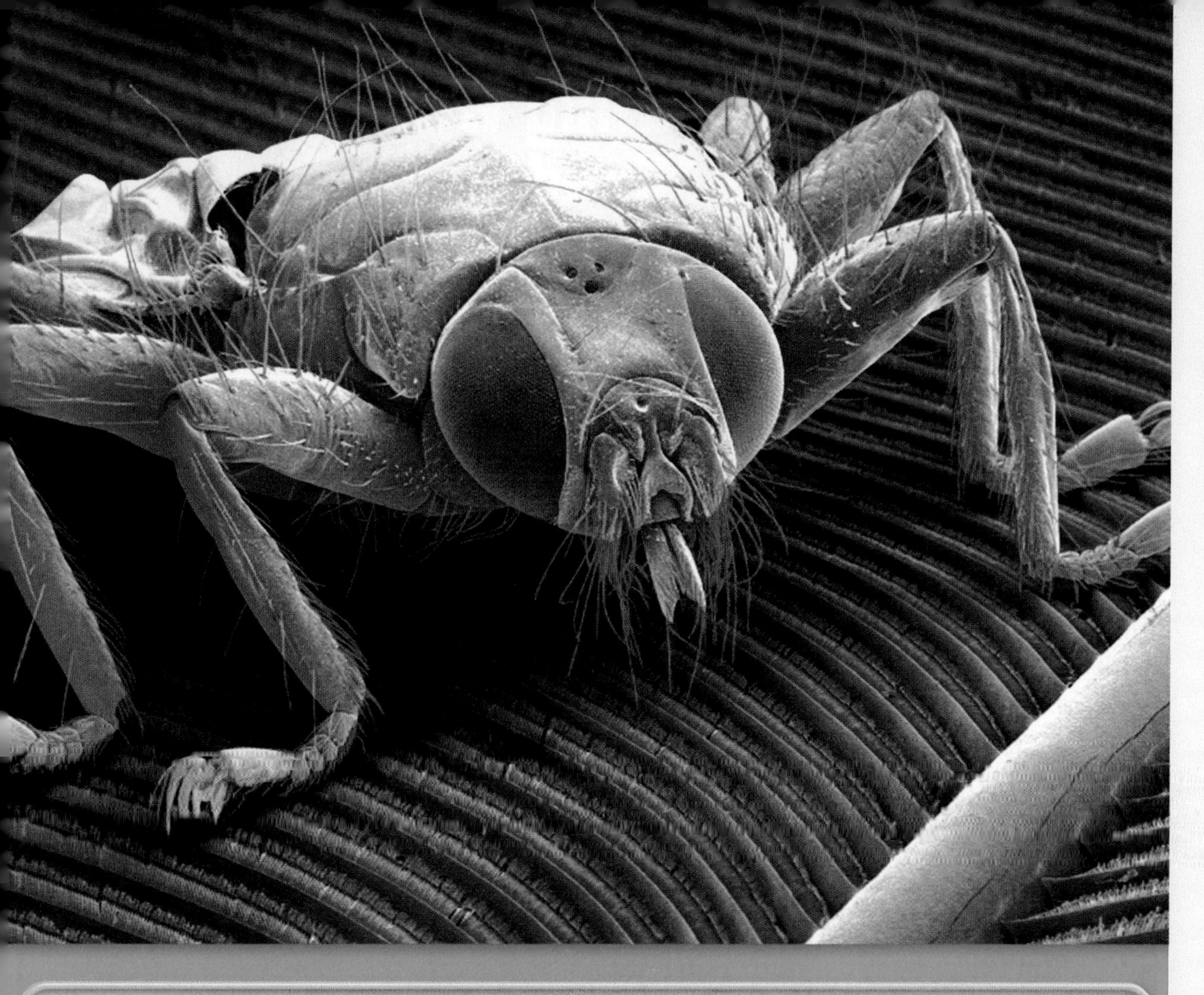

What You'll Learn Next

- In this chapter, you will estimate with fractions and mixed numbers.
- You will change units in the customary system.
- You will add, subtract, multiply, and divide fractions and mixed numbers.

Key Vocabulary

- benchmark (p. 120)
- precision (p. 154)
- reciprocal (p. 141)

Problem Solving Application On pages 164 and 165, you will work an extended activity on fractions.

Chapter 3 Overview

In this chapter, students build on fraction concepts to estimate and compute with fractions and mixed numbers. They change units within the customary measurement system. Then they learn about the concept of precision.

Activating Prior Knowledge

In this chapter, students will build on their knowledge of fraction concepts to add, subtract, multiply, and divide fractions. Ask questions such as:

- *What is the greatest common factor of two numbers?* **the greatest number that divides each number without a remainder**
- *What is the greatest common factor of 9 and 12?* **3**
- *What is the least common multiple of two numbers?* **the least number that is a multiple of both numbers**
- *What is the least common multiple of 6 and 8?* **24**

3-1

1. Plan

Objective
To estimate sums, differences, products, and quotients of fractions

Examples
1 Using Benchmarks With Fractions
2 Estimating With Mixed Numbers
3 Estimating With Compatible Numbers

Math Understandings: p. 118C

Professional Development

Math Background

A benchmark is a value that can be used as a reference point. Using benchmarks to estimate, will help students later when they evaluate the reasonableness of their answers. Students can use the benchmark values of 0, $\frac{1}{2}$, and 1 to estimate sums and differences for fractions between 0 and 1.

More Math Background: p. 118C

Lesson Planning and Resources

See p. 118E for a list of the resources that support this lesson.

PowerPoint

Bell Ringer Practice

Check Skills You'll Need
Use student page, transparency, or PowerPoint. For intervention, direct students to:
Using Estimation Strategies
Lesson 1-1
Extra Skills and Word Problems Practice, Ch. 1

3-1 Estimating With Fractions and Mixed Numbers

Check Skills You'll Need

1. **Vocabulary Review** How is estimating with *compatible numbers* different from estimating by *rounding?* **See below.**

Use compatible numbers to estimate each quotient.

2. 49.8 ÷ 6.8 **7**
3. 21.05 ÷ 5.29 **4**
4. 25.27 ÷ 2.99 **8**
5. 52.6 ÷ 8.8 **6**
6. 19.4 ÷ 10.1 **2**

GO for Help

Lesson 1-1

Check Skills You'll Need

1. **When you estimate by rounding, you round a number to the nearest unit. When you use compatible numbers, you look for numbers that are easy to compute mentally.**

What You'll Learn

To estimate sums, differences, products, and quotients of fractions

New Vocabulary benchmark

Why Learn This?

Estimation can help you find amounts that are not whole numbers.

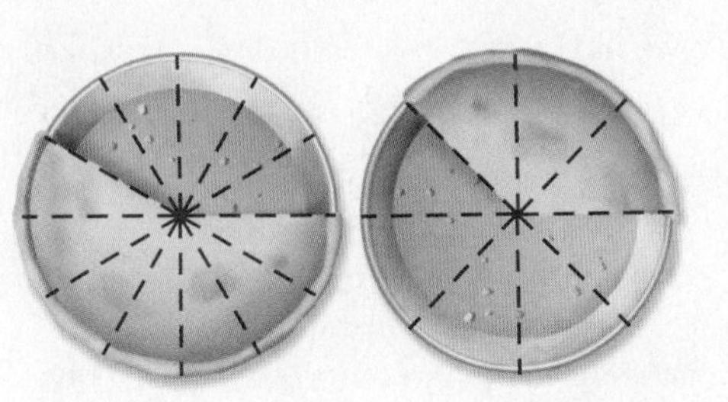

At the right, $\frac{7}{12}$ of one pie and $\frac{3}{8}$ of the other pie remain. You can estimate that a total of about one full pie remains.

A **benchmark** is a convenient number used to replace fractions less than 1. You can use the benchmarks 0, $\frac{1}{2}$, and 1 to estimate fractions.

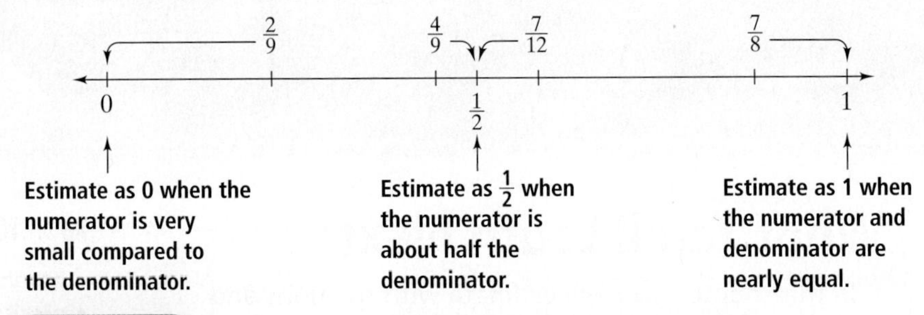

2/9
4/9
7/12
7/8
0
1/2
1
Estimate as 0 when the numerator is very small compared to the denominator.
Estimate as 1/2 when the numerator is about half the denominator.
Estimate as 1 when the numerator and denominator are nearly equal.

EXAMPLE Using Benchmarks With Fractions

1 A family bought a plot of land last year that was $\frac{4}{9}$ acre. They plan to buy an adjoining plot this year that is $\frac{7}{8}$ acre. Estimate the total amount of land the family will have.

$\frac{4}{9} + \frac{7}{8} \approx \frac{1}{2} + 1 = 1\frac{1}{2}$ ← Use benchmarks to estimate each fraction. Then add.

The family will have about $1\frac{1}{2}$ acres of land.

Quick Check

1. Use benchmarks to estimate $\frac{3}{5} - \frac{1}{8}$. **about $\frac{1}{2}$**

Differentiated Instruction Solutions for All Learners

Special Needs L1
Give students number lines and rectangles. Have them shade in about $\frac{1}{8}$ on one rectangle, about $\frac{4}{5}$ on another and about $\frac{2}{5}$ on another. Then have them place a point for each fraction on the number line closest to the benchmark number to which it corresponds.
learning style: visual

Below Level L2
Have students make lists of fractions whose benchmark values are 0, $\frac{1}{2}$, and 1.
Sample: 0: $\frac{1}{9}, \frac{3}{16}, \frac{2}{7}, \ldots$; $\frac{1}{2}$: $\frac{5}{10}, \frac{2}{5}, \frac{3}{8}, \frac{4}{9}, \ldots$; 1: $\frac{4}{5}, \frac{7}{9}, \frac{10}{13}, \frac{15}{18} \ldots$
learning style: visual

To estimate a sum, difference, or product of mixed numbers, you can first round each mixed number to the nearest whole number. Then compute.

EXAMPLES Estimating With Mixed Numbers

2 A skysurfer falls 500 feet in about $2\frac{4}{5}$ seconds. A skydiver wearing a wing suit falls 500 feet in about $4\frac{1}{6}$ seconds. Approximately how many more seconds than the skysurfer does the skydiver take to fall 500 feet?

$4\frac{1}{6} - 2\frac{4}{5}$ ← **Use subtraction.**

↓ ↓

$4 - 3 = 1$ ← **Round each mixed number. Then subtract.**

The skydiver takes about 1 more second than the skysurfer to fall 500 ft.

3 Estimate the product $2\frac{2}{5} \cdot 6\frac{1}{10}$.

$2\frac{2}{5} \cdot 6\frac{1}{10}$

↓ ↓

$2 \cdot 6 = 12$ ← **Round each mixed number. Then multiply.**

Quick Check

2. **Science** For an experiment on plant growth, you recorded the height of a plant every day. On Monday, the plant was $5\frac{1}{4}$ in. tall. A week later, it was $10\frac{7}{8}$ in. tall. About how many inches did the plant grow? **about 6 in.**
3. Use rounding to estimate each product.
   a. $3\frac{5}{6} \cdot 5\frac{1}{8}$ **about 20** b. $8\frac{1}{8} \cdot 5\frac{11}{12}$ **about 48** c. $7\frac{1}{3} \cdot 1\frac{13}{16}$ **about 14**

To estimate a quotient of mixed numbers, you can use compatible numbers. Choose numbers that are easy to divide.

EXAMPLE Estimating With Compatible Numbers

Vocabulary Tip

Compatible numbers are numbers that are easy to compute mentally.

4 Estimate the quotient $43\frac{1}{4} \div 5\frac{7}{8}$.

$43\frac{1}{4} \div 5\frac{7}{8}$

↓ ↓

$42 \div 6 = 7$ ← **Use compatible numbers. Use 42 for $43\frac{1}{4}$ and use 6 for $5\frac{7}{8}$.**

Quick Check

4. Use compatible numbers to estimate each quotient.
   a. $35\frac{3}{4} \div 5\frac{11}{12}$ **about 6** b. $22\frac{7}{8} \div 3\frac{5}{6}$ **about 6** c. $46\frac{2}{5} \div 5\frac{1}{10}$ **about 9**

2. Teach

Activity Lab

Use before the lesson.

All in One Teaching Resources

Activity Lab 3-1: Estimating With Fractions

Guided Instruction

Example 2

Ask students how long it would take them to clean their desks or lockers in hours and fractions of an hour. Use those mixed numbers to create comparison problems.

Example 4

If students are unable to recognize compatible numbers, include the intermediate step of rounding to the nearest whole number. Then ask: *What division fact is closest to this problem?* **42 ÷ 6 = 7**

Additional Examples

1. Use benchmarks to estimate $\frac{4}{7} + \frac{4}{5}$. $\frac{1}{2} + 1 = 1\frac{1}{2}$
2. Estimate $5\frac{1}{9} - 2\frac{5}{6}$. $5 - 3 = 2$
3. Estimate $6\frac{1}{8} \cdot 6\frac{5}{8}$. $6 \cdot 7 = 42$
4. Estimate $26\frac{1}{4} \div 8\frac{2}{3}$. $27 \div 9 = 3$

All in One Teaching Resources

- Daily Notetaking Guide 3-1 L3
- Adapted Notetaking 3-1 L1

Closure

- *Which fractions have a benchmark estimate of 0?* **those where the numerator is much smaller than the denominator**
- *What are the two steps in estimating a product of mixed numbers?* **Round each mixed number to the nearest whole number. Multiply.**

Advanced Learners L4

Estimate.

$2\frac{1}{2} + 3\frac{1}{5} + 1\frac{5}{20}$ 7

$2\frac{1}{2} \cdot 3\frac{1}{5} \cdot 1\frac{5}{20}$ 9

learning style: visual

English Language Learners ELL

Provide examples of fractions such as $\frac{1}{8}$, $\frac{4}{5}$, $\frac{2}{5}$. Have students decide on the benchmark number to which each is closest and answer in complete sentences. Ask: *Is $\frac{1}{8}$ closer to 0, $\frac{1}{2}$, or 1? How do you know?* **Sample: $\frac{1}{8}$ is closest to the benchmark value of 0.**

learning style: verbal

3. Practice

Assignment Guide

Check Your Understanding
Go over Exercises 1–10 in class before assigning the Homework Exercises.

Homework Exercises
A Practice by Example	11–32
B Apply Your Skills	33–43
C Challenge	44
Test Prep and Mixed Review	45–51

Homework Quick Check
To check students' understanding of key skills and concepts, go over Exercises 17, 25, 34, 37, and 43.

Differentiated Instruction Resources

Adapted Practice 3-1 L1

Practice 3-1 Estimating With Fractions and Mixed Numbers L3

Estimate each sum or difference.

1. $\frac{1}{6}+\frac{5}{8}$ $\frac{1}{2}$
2. $\frac{7}{8}-\frac{1}{16}$ 1
3. $\frac{9}{10}+\frac{7}{8}$ 2
4. $\frac{1}{10}+\frac{5}{6}$ 1
5. $\frac{4}{5}-\frac{1}{6}$ 1
6. $\frac{11}{12}-\frac{5}{16}$ $\frac{1}{2}$
7. $2\frac{1}{6}+7\frac{1}{9}$ 9
8. $4\frac{9}{10}-3\frac{5}{8}$ $1\frac{1}{2}$
9. $4\frac{7}{8}+8\frac{1}{5}$ 13
10. $14\frac{3}{4}+9\frac{7}{8}$ 25
11. $7\frac{11}{15}-6\frac{7}{16}$ $1\frac{1}{2}$
12. $3\frac{11}{15}-2\frac{9}{10}$ 1

Estimate each product or quotient.

13. $13\frac{1}{8}\div 6\frac{1}{5}$ 2
14. $5\frac{1}{6}\cdot 8\frac{4}{5}$ 45
15. $8\frac{1}{6}\div 1\frac{9}{10}$ 4
16. $27\frac{6}{7}\div 3\frac{2}{3}$ 7
17. $20\frac{4}{5}\cdot 2\frac{2}{7}$ 40
18. $9\frac{1}{3}\div 2\frac{7}{8}$ 3
19. $19\frac{4}{5}\div 4\frac{5}{8}$ 4
20. $9\frac{2}{13}\div 3\frac{1}{18}$ 3
21. $42\frac{1}{6}\div 6\frac{1}{16}$ 7
22. $15\frac{1}{20}\cdot 3\frac{1}{10}$ 45
23. $72\frac{2}{15}\div 8\frac{3}{4}$ 8
24. $3\frac{5}{6}\cdot 10\frac{1}{12}$ 40

Solve each problem.

25. Each dress for a wedding party requires $7\frac{1}{8}$ yd of material. Estimate the amount of material you would need to make 6 dresses. about 42 yd
26. A fabric store has $80\frac{3}{8}$ yd of a particular fabric. About how many pairs of curtains could be made from this fabric if each pair requires $4\frac{1}{8}$ yd of fabric? about 20 pairs
27. Adam's car can hold $16\frac{1}{10}$ gal of gasoline. About how many gallons are left if he started with a full tank and has used $11\frac{9}{10}$ gal? about 4 gal
28. Julia bought stock at $\$28\frac{1}{8}$ per share. The value of each stock increased by $\$6\frac{5}{8}$. About how much is each share of stock now worth? about $35

Estimate each answer.

29. $6\frac{2}{9}-2\frac{7}{8}$ 3
30. $\frac{1}{8}+\frac{9}{10}$ 1
31. $8\frac{2}{9}\cdot 10\frac{4}{9}$ 80
32. $6\frac{1}{4}\div 2\frac{3}{11}$ 3
33. $5\frac{1}{11}\cdot 8\frac{13}{15}$ 45
34. $\frac{21}{40}-\frac{5}{89}$ $\frac{1}{2}$
35. $\frac{81}{100}-\frac{1}{2}$ $\frac{1}{2}$
36. $11\frac{5}{9}\div 2\frac{1}{2}$ 6
37. $\frac{3}{5}+\frac{7}{8}$ $1\frac{1}{2}$

3-1 • Guided Problem Solving GPS L3

GPS Student Page 123, Exercise 43:

Writing in Math You need $9\frac{9}{16}$ lb of chicken. The store sells chicken in half-pound packages. How much chicken should you order? Explain.

Understand

1. What are you being asked to do? Determine how much chicken you should order.
2. How do you know when to round up or when to round down with a fraction? If the numerator is bigger than half of the denominator, you round to the next whole number. If the numerator is smaller than half of the denominator, you round down to the number which is exactly half the denominator.

Plan and Carry Out

3. What is the numerator of the fraction? 9
4. What is half of the denominator of the fraction? 8
5. Is the numerator bigger or smaller than half of the denominator? bigger
6. Do you round the fraction up or down? up
7. How many pounds of chicken do you need? 10 pounds
8. How many packages of chicken will you need to buy? 20 packages

Check

9. What is $9 \div 16$? Round to the nearest whole number. Does your answer make sense? 0.5625; 1; Yes, because 9 pounds + 1 pound = 10 pounds. Since the store sells chicken in half-pound packages, you will need to buy 20 packages of chicken. 10 pounds ÷ 0.5 pound = 20 packages of chicken.

Solve Another Problem

10. You are making curtains to cover the top of four windows. Each window is $15\frac{5}{8}$ in. wide. You buy material by the whole yard. How many yards should you buy? 2 yards

Check Your Understanding

1. Answers may vary. Sample: A benchmark is a convenient number used to replace a fraction. Examples are 0, $\frac{1}{2}$, and 1 for fractions between 0 and 1.

1. **Vocabulary** What is a benchmark? **See left.**
2. **Reasoning** How can you tell whether a fraction is greater than, less than, or equal to $\frac{1}{2}$? Explain. **See margin.**

Choose a benchmark for each fraction. Use 0, $\frac{1}{2}$, or 1.

3. $\frac{15}{16}$ **1**
4. $\frac{1}{10}$ **0**
5. $\frac{5}{8}$ **$\frac{1}{2}$**
6. $\frac{2}{9}$ **0**

Round each mixed number to the nearest whole number.

7. $1\frac{1}{3}$ **1**
8. $10\frac{3}{4}$ **11**
9. $6\frac{7}{12}$ **7**
10. $8\frac{2}{5}$ **8**

Homework Exercises

For more exercises, see Extra Skills and Word Problems.

GO for Help

For Exercises	See Examples
11–19	1
20–28	2–3
29–32	4

A **Use benchmarks to estimate each sum or difference.**

11. $\frac{1}{7}+\frac{3}{8}$ **about $\frac{1}{2}$**
12. $\frac{3}{5}-\frac{1}{2}$ **about 0**
13. $\frac{5}{11}-\frac{1}{5}$ **about $\frac{1}{2}$**
14. $\frac{8}{9}-\frac{5}{6}$ **about 0**
15. $\frac{1}{6}+\frac{5}{8}$ **about $\frac{1}{2}$**
16. $\frac{3}{4}+\frac{1}{5}$ **about 1**
17. $\frac{3}{4}-\frac{5}{6}$ **about 0**
18. $\frac{9}{10}+\frac{7}{16}$ **about $1\frac{1}{2}$**
19. Suppose it rains $\frac{4}{5}$ inch in the morning and $\frac{6}{7}$ inch in the afternoon. About how many inches of rain fall during the day? **about 2 in.**

Use rounding to estimate.

20. $9\frac{1}{11}-3\frac{7}{9}$ **about 5**
21. $5\frac{3}{5}+3\frac{2}{3}$ **about 10**
22. $4\frac{1}{2}-\frac{24}{25}$ **about 4**
23. $7\frac{2}{3}-2\frac{11}{12}$ **about 5**
24. $2\frac{1}{8}\cdot 3\frac{6}{7}$ **about 8**
25. $5\frac{2}{9}\cdot 4\frac{9}{10}$ **about 25**
26. $3\frac{3}{8}\cdot 5\frac{1}{6}$ **about 15**
27. $1\frac{7}{10}\cdot 8\frac{1}{12}$ **about 16**
28. Marsha bought $1\frac{3}{8}$ lb of snow peas and $3\frac{1}{10}$ lb of carrots at the market. About how many more pounds of carrots did she buy? **2 lb**

Use compatible numbers to estimate each quotient.

29. $10\frac{7}{8}\div 3\frac{1}{9}$ **about 4**
30. $36\frac{1}{3}\div 4\frac{2}{5}$ **about 9**
31. $7\frac{3}{5}\div 1\frac{1}{2}$ **about 4**
32. $8\frac{2}{12}\div 8\frac{2}{7}$ **about 1**

33. Yes; $\frac{2}{3}+\frac{3}{4}+1\frac{2}{3}=3\frac{1}{12}$, which is less than 4.

34. Low; the dividend, $30\frac{1}{7}$, becomes smaller when estimating, and the divisor, $1\frac{4}{5}$, becomes larger. When you divide by a larger number, the quotient is smaller.

B **GPS** 33. **Guided Problem Solving** You want to make three kinds of pasta salad. The recipes call for $\frac{2}{3}$ c, $\frac{3}{4}$ c, and $1\frac{2}{3}$ c of pasta. You have 4 c of pasta. Do you have enough? Explain. **See above left.**
- About how many cups of pasta do you need for each recipe?
- About how many cups of pasta do you need for all 3 recipes?

34. **Number Sense** $30\frac{1}{7}\div 1\frac{4}{5}$ is about $30 \div 2 = 15$. Is the estimate of 15 for the quotient high or low? Explain. **See above left.**

2. Compare the fraction's numerator and denominator. If the numerator is less than half the denominator, the fraction is less than $\frac{1}{2}$. If the numerator is greater than half the denominator, the fraction is greater than $\frac{1}{2}$. If the numerator is the same as half the denominator, the fraction is equal to $\frac{1}{2}$.

Homework Video Tutor
Visit: PHSchool.com
Web Code: are-0301

39. about $\frac{1}{2}$ t

Use benchmarks to estimate each sum or difference.

35. $6\frac{5}{6} + 8\frac{2}{12} + 14\frac{1}{7}$ **about 29**

36. $19\frac{9}{11} - 4\frac{5}{14} - 3\frac{7}{8}$ **about 12**

37. A teacher's school day is $8\frac{1}{4}$ h long. The teacher has six classes every day. Each class is $\frac{5}{6}$ h long. About how many hours of the school day is the teacher not in class? **2 h**

Music The table below shows the weight of the bells of Boston's Old North Church. Use the table for Exercises 38–42.

Tone	F	E	D	C	B-flat	A	G	Low F
Weight in Tons	$\frac{3}{10}$	$\frac{3}{10}$	$\frac{7}{20}$	$\frac{2}{5}$	$\frac{11}{25}$	$\frac{19}{40}$	$\frac{3}{5}$	$\frac{3}{4}$

38. Estimate the total weight of the two heaviest bells. **about $1\frac{1}{2}$ t**

39. Estimate the weight difference of the F bell and the Low F bell. **See left.**

40. Which bells weigh less than $\frac{1}{2}$ ton? **F, E, D, C, B-flat, and A**

41. Estimate the total weight of all eight bells. **about $4\frac{1}{2}$ t**

42. Estimate the average weight of the four heaviest bells. **about $\frac{1}{2}$ t**

43. Writing in Math You need $9\frac{9}{16}$ lb of chicken. The store sells chicken in half-pound quantities. How much chicken should you order? Explain. **10 lb; $9\frac{9}{16} > 9\frac{1}{2}$ so the next amount is 10 lb.**

44. Challenge Estimate the median of $9\frac{3}{5}$, $5\frac{1}{4}$, $7\frac{5}{10}$, $1\frac{7}{8}$, $6\frac{3}{4}$, and $3\frac{2}{8}$. **6**

Test Prep and Mixed Review

Practice

Multiple Choice

45. Jade needs $2\frac{2}{3}$ yards of fabric to make one tablecloth. She has $11\frac{2}{3}$ yards of fabric. What is the greatest number of tablecloths Jade can make? **B**

Ⓐ 3  Ⓑ 4  Ⓒ 5 Ⓓ 6

46. The length of a piece of fabric measures between 3 and $3\frac{1}{4}$ feet. Which length could it be? **G**

Ⓕ $\frac{72}{25}$ ft Ⓖ $\frac{25}{8}$ ft Ⓗ $\frac{18}{5}$ ft Ⓙ $\frac{26}{7}$ ft

47. Brian bought 4 burritos for \$15. He later bought another burrito for \$2.50. What was the mean cost of all the burritos? **B**

Ⓐ \$3.15 Ⓑ \$3.50 Ⓒ \$3.75 Ⓓ \$12.50

GO for Help

For Exercises	See Lesson
48–51	2-6

Write each decimal as a mixed number or fraction in simplest form.

48. 0.04 $\frac{1}{25}$ **49.** 11.125 $11\frac{1}{8}$ **50.** 0.0625 $\frac{1}{16}$ **51.** 3.408 $3\frac{51}{125}$

4. Assess & Reteach

Lesson Quiz

Estimate.

1. $\frac{1}{8} + \frac{3}{5}$ $\frac{1}{2}$
2. $4\frac{3}{4} - 1\frac{2}{3}$ 3
3. $2\frac{5}{6} \cdot 6\frac{1}{10}$ 18
4. $43\frac{7}{8} \div 10\frac{1}{3}$ 4

Alternative Assessment

Have students work in pairs. In turn, each student chooses one exercise from each group of Exercises 11–18, 20–27, and 29–32. Each student estimates, as directed, and then explains to the partner how he or she made the estimate.

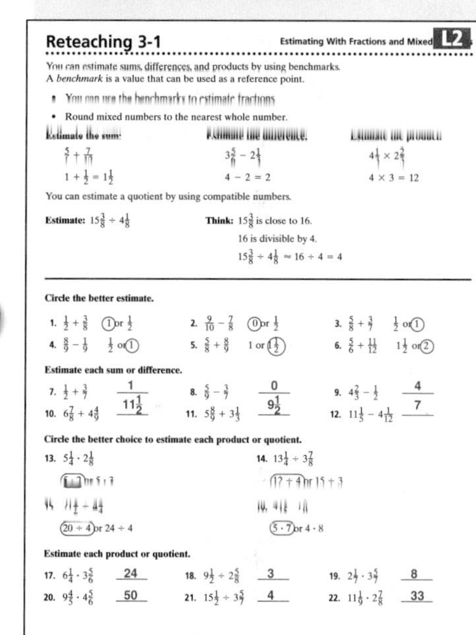

Reteaching 3-1 Estimating With Fractions and Mixed Numbers **L2**

You can estimate sums, differences, and products by using benchmarks.
A *benchmark* is a value that can be used as a reference point.

- You can use the benchmarks to estimate fractions.
- Round mixed numbers to the nearest whole number.

Estimate the sum: $\frac{5}{7} + \frac{7}{10}$; $1 + \frac{1}{2} = 1\frac{1}{2}$

Estimate the difference: $3\frac{5}{6} - 2\frac{1}{3}$; $4 - 2 = 2$

Estimate the product: $4\frac{1}{3} \times 2\frac{2}{3}$; $4 \times 3 = 12$

You can estimate a quotient by using compatible numbers.

Estimate: $15\frac{3}{8} \div 4\frac{1}{8}$

Think: $15\frac{3}{8}$ is close to 16.
16 is divisible by 4.
$15\frac{3}{8} \div 4\frac{1}{8} \approx 16 \div 4 = 4$

Circle the better estimate.

1. $\frac{1}{2} + \frac{3}{8}$ (1) or $\frac{1}{2}$
2. $\frac{9}{10} - \frac{7}{8}$ (0) or $\frac{1}{2}$
3. $\frac{5}{8} + \frac{3}{7}$ $\frac{1}{2}$ or (1)
4. $\frac{8}{9} - \frac{1}{9}$ $\frac{1}{2}$ or (1)
5. $\frac{5}{8} + \frac{8}{9}$ 1 or ($1\frac{1}{2}$)
6. $\frac{5}{6} + \frac{11}{12}$ $1\frac{1}{2}$ or (2)

Estimate each sum or difference.

7. $\frac{1}{2} + \frac{3}{7}$ 1
8. $\frac{5}{9} - \frac{3}{7}$ 0
9. $4\frac{2}{3} - \frac{1}{2}$ 4
10. $6\frac{7}{8} + 4\frac{4}{9}$ $11\frac{1}{2}$
11. $5\frac{8}{9} + 3\frac{1}{3}$ $9\frac{1}{2}$
12. $11\frac{1}{5} - 4\frac{1}{12}$ 7

Circle the better choice to estimate each product or quotient.

13. $5\frac{1}{4} \cdot 2\frac{1}{8}$
14. $13\frac{1}{4} \div 3\frac{7}{8}$ (12 ÷ 4) or 15 ÷ 3
15. (20 ÷ 4) or 24 ÷ 4
16. (5 · 7) or 4 · 8

Estimate each product or quotient.

17. $6\frac{1}{4} \cdot 3\frac{5}{6}$ 24
18. $9\frac{1}{2} \div 2\frac{5}{8}$ 3
19. $2\frac{1}{7} \cdot 3\frac{5}{7}$ 8
20. $9\frac{4}{5} \cdot 4\frac{5}{6}$ 50
21. $15\frac{1}{2} \div 3\frac{5}{7}$ 4
22. $11\frac{1}{9} \cdot 2\frac{7}{8}$ 33

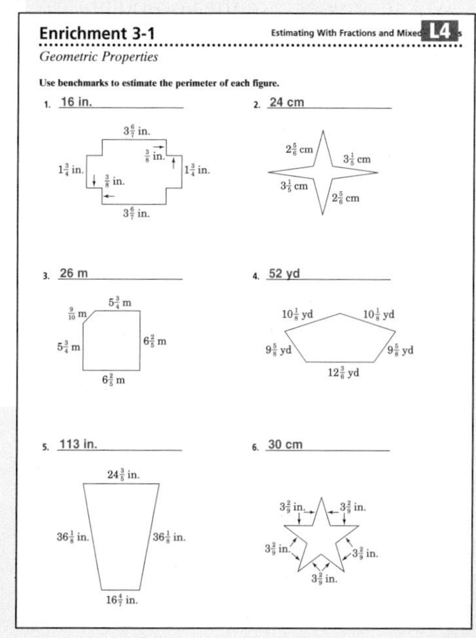

Enrichment 3-1 Estimating With Fractions and Mixed Numbers **L4**

Geometric Properties

Use benchmarks to estimate the perimeter of each figure.

1. 16 in.
2. 24 cm
3. 26 m
4. 52 yd
5. 113 in.
6. 30 cm

Test Prep

Resources

For additional practice with a variety of test item formats:

- Test-Taking Strategies, p. 159
- Test Prep, p. 163
- Test-Taking Strategies with Transparencies

Vocabulary Builder

High-Use Academic Words

Students learn words that are not necessarily math related, but that will appear frequently in their texts. Learning these words will help them understand what to do to solve problems.

Guided Instruction

Go over the words and their meanings with students. Point out that they are all verbs, or action words, that tell you to do something.
Ask:

- *How can we use* explain *in a math problem?* **Sample: Explain why we divide the numerator by the denominator to convert a fraction to a decimal.**
- *What does* find *usually mean in a math problem?* **You may need to go through several steps to obtain the correct answer.**
- *What does the* Order of Operations *do?* **It tells in which order you should do certain operations.**

Differentiated Instruction

Advanced Learners L4

Ask students to create riddles whose answers are the vocabulary words. They can trade papers with a partner and solve each other's riddles.

Vocabulary Builder

High-Use Academic Words

High-use academic words are words that you will see often in textbooks and on tests. These words are not math vocabulary terms, but knowing them will help you to succeed in mathematics.

Direction Words

Some words tell what to do in a problem. I need to understand what these words are asking so that I give the correct answer.

Word	Meaning
Explain	*To give details that make an idea easy to understand*
Find	*To obtain an answer by solving a problem*
Order	*To arrange or organize information in a sequence*

Exercises

Fill in the blank with *order, explain,* or *find*.

1. Question: What is the correct __?__ of a U.S. student's education? **order**

   Answer: elementary school, middle school, high school
2. In many middle schools, students must __?__ the classroom for each subject they take. **find**
3. On the first day of school, many teachers __?__ the rules of the class. **explain**
4. Explain how you know that the sum of $\frac{4}{5}$ and $\frac{14}{15}$ will be greater than 1. **See below.**
5. Find compatible numbers for $80\frac{5}{6}$ and $9\frac{2}{9}$ and estimate the quotient. **$81 \div 9 = 9$**
6. Order the fractions $\frac{4}{8}, \frac{2}{6}, -\frac{10}{12}, \frac{4}{4}, -\frac{1}{3}$ from least to greatest. **$-\frac{10}{12}, -\frac{1}{3}, \frac{2}{6}, \frac{4}{8}, \frac{4}{4}$**
7. **Word Knowledge** Think about the word *reasonable.* **7a–c. Check students' work.**
   a. Choose the letter for how well you know the word.
      A. I know its meaning.
      B. I've seen it, but I don't know its meaning.
      C. I don't know it.
   b. **Research** Look up and write the definition of *reasonable.*
   c. Use the word in a sentence involving mathematics.

4. Answers may vary. Sample: Their sum is greater than 1, because both $\frac{4}{5}$ and $\frac{14}{15}$ are approximately 1.

3-2a Activity Lab | Hands On

Using Fraction Models

You can use models to add and subtract fractions.

EXAMPLES Using Fraction Models

1 Find $\frac{3}{10} + \frac{1}{10}$.

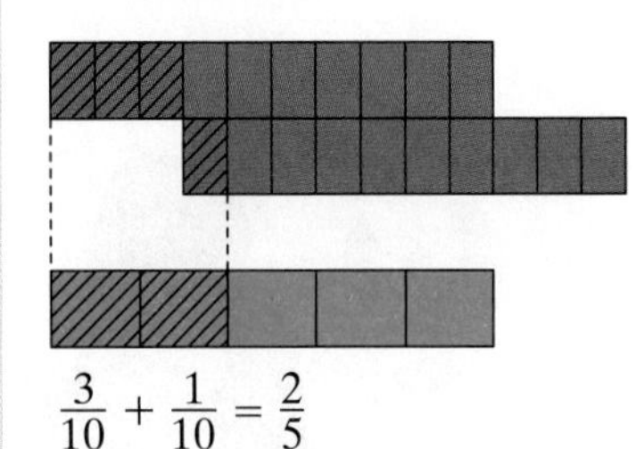

← To add, align the right side of the shaded part of the first model with the left side of the second one.

← Find a model that represents the sum of the shaded parts. If you have more than one model to choose from, use the one with the largest sections.

$\frac{3}{10} + \frac{1}{10} = \frac{2}{5}$

2 Find $\frac{3}{4} - \frac{1}{3}$.

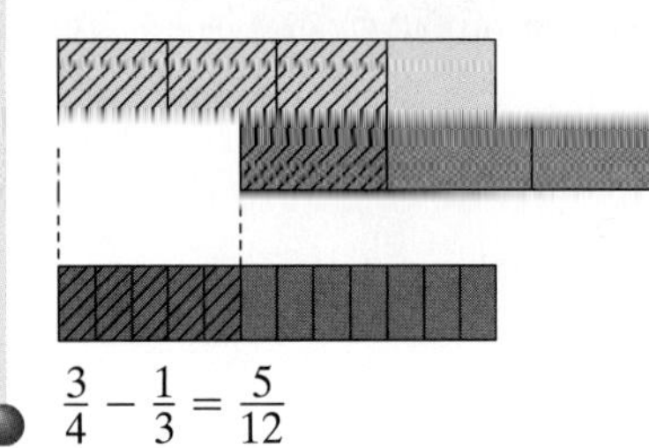

← To subtract, align the right ends of the shaded part of each model.

← Find the model that represents the difference.

$\frac{3}{4} - \frac{1}{3} = \frac{5}{12}$

Exercises

Write a number sentence for each model.

1. ? $\frac{4}{5} - \frac{2}{5} = \frac{2}{5}$

2. 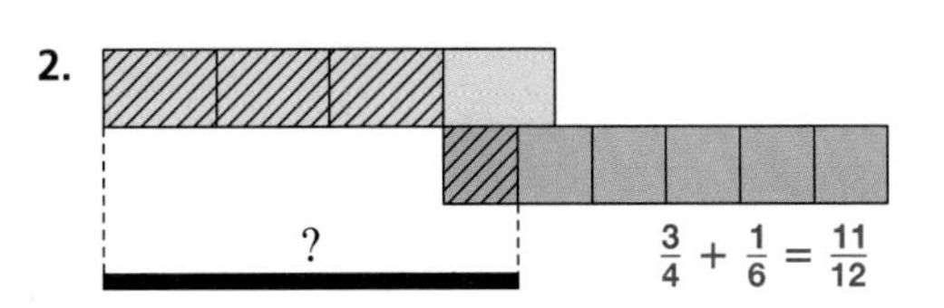

? $\frac{3}{4} + \frac{1}{6} = \frac{11}{12}$

Use models to find each sum or difference. 3–6. See margin for models.

3. $\frac{3}{8} + \frac{3}{8}$ $\frac{3}{4}$

4. $\frac{5}{6} - \frac{1}{6}$ $\frac{2}{3}$

5. $\frac{2}{5} + \frac{1}{2}$ $\frac{9}{10}$

6. $\frac{7}{10} - \frac{1}{5}$ $\frac{1}{2}$

7. You need 1 lb of sugar. You have $\frac{3}{8}$ lb and $\frac{2}{5}$ lb of sugar. Do you have enough? No; $\frac{3}{8} + \frac{2}{5} = \frac{31}{40}$, which is less than 1.

3–6. See back of book.

Activity Lab

Using Fraction Models

By using models, such as fraction bars, students can gain a concrete understanding of adding and subtracting fractions.

Guided Instruction

Provide time for students to explore modeling with their fraction bars. Circulate as students work through the examples, helping them to find and align their models.

Provide additional examples, using fractions with the same denominators such as $\frac{1}{5} + \frac{3}{5}$ $\frac{4}{5}$ and $\frac{1}{6} + \frac{3}{6}$. $\frac{2}{3}$ Discuss and model that the fractions $\frac{2}{2}$, $\frac{3}{3}$, $\frac{4}{4}$, and so on, are equivalent to 1.

Exercises
Have students work independently on the Exercises. When they have finished, have them trade papers with a partner and check each other's answers.

Alternative Method
Many kinds of fraction manipulatives are useful and readily available (for use with an overhead projector, too). Your students may have success using fraction cubes, squares, tiles, attribute blocks, or Cuisenaire® rods as alternatives.

Resources

- Activity Lab 3-2: Fraction Bar Exploration
- fraction bars
- Student Manipulatives Kit
- Classroom Aid 12–19

3-2

1. Plan

Objective
To add and subtract fractions and to solve problems involving fractions

Examples
1 Common Denominators
2 Different Denominators
3 Application: Carpentry

Math Understandings: p. 118C

Math Background

To add and subtract fractions that have a common denominator, add or subtract the numerators and simplify the result. For fractions with different denominators, the fractions must be changed to a common denominator. Then add or subtract the numerators.

More Math Background: p. 118C

Lesson Planning and Resources

See p. 118E for a list of the resources that support this lesson.

PowerPoint

Bell Ringer Practice

Check Skills You'll Need
Use student page, transparency, or PowerPoint. For intervention, direct students to:
Simplying Fractions
Lesson 2-3
Extra Skills and Word Problems Practice, Ch. 2

3-2 Adding and Subtracting Fractions

Check Skills You'll Need

1. **Vocabulary Review** Two fractions that name the same amount are called __?__. equivalent

Find each missing number.

2. $\frac{8}{16} = \frac{\blacksquare}{8}$ 4
3. $\frac{\blacksquare}{10} = \frac{4}{5}$ 8
4. $\frac{1}{6} = \frac{4}{\blacksquare}$ 24

Lesson 2-3

What You'll Learn

To add and subtract fractions and to solve problems involving fractions

Why Learn This?

Sometimes you want an exact answer when you add or subtract fractions.

When you are sewing, you should leave extra material to make a hem. You can add fractions to find how much extra material you need to make a hem of $\frac{5}{8}$ in. when you make clothing.

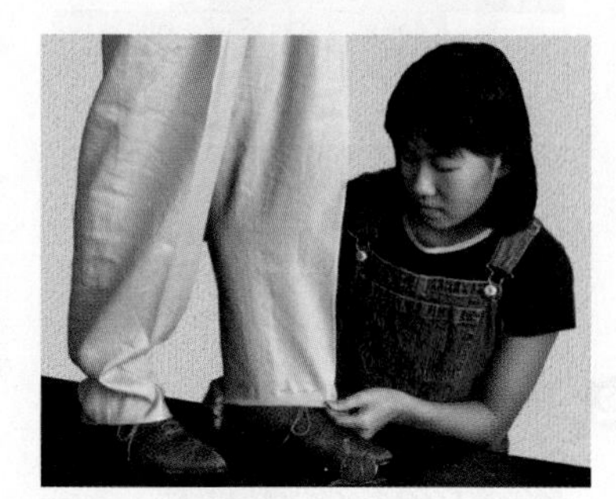

When you add or subtract fractions that have common denominators, keep the denominator the same. Add or subtract the numerators.

Video Tutor Help
Visit: PHSchool.com
Web Code: are-0775

EXAMPLE Common Denominators

1 Find $\frac{1}{8} + \frac{5}{8}$.

Estimate $\frac{1}{8} + \frac{5}{8} \approx 0 + \frac{1}{2}$, or $\frac{1}{2}$

$\frac{1}{8} + \frac{5}{8} = \frac{1+5}{8}$ ← Keep the denominator the same.

$= \frac{6}{8}$ ← Add the numerators.

$= \frac{3}{4}$ ✔ ← Simplify. The answer is close to the estimate.

Quick Check

1. Find each sum or difference.

a. $\frac{3}{5} + \frac{1}{5}$ $\frac{4}{5}$ b. $\frac{13}{16} - \frac{9}{16}$ $\frac{1}{4}$ c. $\frac{1}{4} + \frac{3}{4}$ 1

Sometimes you add or subtract fractions with different denominators. You can use models to help you find the sum or the difference.

Differentiated Instruction Solutions for All Learners

Special Needs L1
For Examples 2 and 3, some concrete or drawn representations will help students focus on what they are trying to find. You can use boards of different thickness for Example 3, and show that they are trying to find a fractional name for the difference.
learning style: visual

Below Level L2
Have students change the following fractions and the whole number 1 to fractions with the common denominator of 12: $\frac{1}{2}$, $\frac{1}{3}$, $\frac{1}{4}$, $\frac{1}{6}$, 1. $\frac{6}{12}$, $\frac{4}{12}$, $\frac{3}{12}$, $\frac{2}{12}$, $\frac{12}{12}$
learning style: visual

This model shows $\frac{1}{4} + \frac{1}{3} = \frac{7}{12}$.

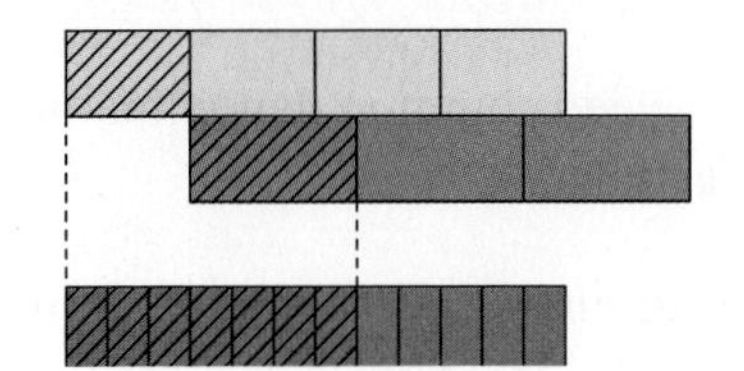

This model shows $\frac{1}{2} - \frac{1}{3} = \frac{1}{6}$.

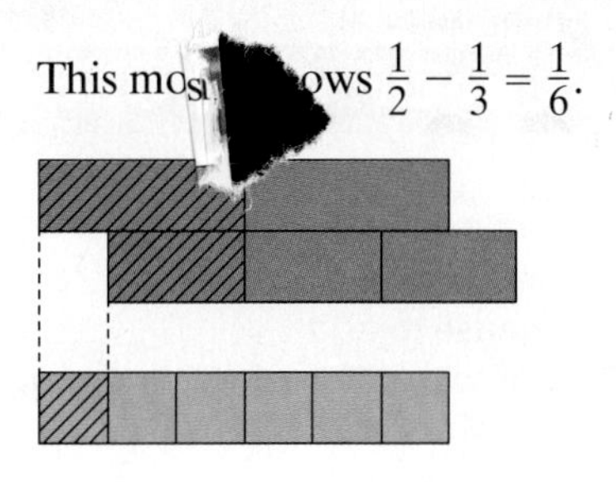

To add or subtract fractions with different denominators, first find the Least Common Denominator (LCD).

EXAMPLE Different Denominators

Test Prep Tip

The LCD of two fractions is the least common multiple of their denominators.

2 Find $\frac{4}{5} + \frac{2}{3}$.

Estimate $\frac{4}{5} + \frac{2}{3} \approx 1 + \frac{1}{2}$, or $1\frac{1}{2}$

$\frac{4}{5} = \frac{4 \cdot 3}{5 \cdot 3} = \frac{12}{15}$ ← **The LCD is 15. Write an equivalent fraction.**

$+\frac{2}{3} = \frac{2 \cdot 5}{3 \cdot 5} = +\frac{10}{15}$ ← **Write an equivalent fraction.**

$\frac{22}{15}$ ← **Add the numerators.**

$1\frac{7}{15}$ ← **Write as a mixed number. This is close to the estimate.**

Quick Check

2. a. Find $\frac{3}{4} - \frac{1}{6}$. $\frac{7}{12}$

b. Find $\frac{3}{7} + \frac{5}{14}$. $\frac{11}{14}$

EXAMPLE Application: Carpentry

3 A cabinetmaker needs a board that is $\frac{9}{16}$ in. thick. By how much must he reduce the thickness of a board that is $\frac{7}{8}$ in. thick?

Estimate $\frac{7}{8} - \frac{9}{16} \approx 1 - \frac{1}{2} = \frac{1}{2}$

$\frac{7}{8} = \frac{7 \cdot 2}{8 \cdot 2} = \frac{14}{16}$ ← **The LCD is 16. Write an equivalent fraction.**

$-\frac{9}{16} = -\frac{9}{16} = -\frac{9}{16}$ ← **Write an equivalent fraction.**

$\frac{5}{16}$ ← **Subtract the numerators.**

The cabinetmaker should reduce the thickness of the board by $\frac{5}{16}$ in.

Check for Reasonableness The answer $\frac{5}{16}$ is close to the estimate $\frac{1}{2}$.

Quick Check

3. You hiked $\frac{5}{8}$ mi and $\frac{1}{4}$ mi in the afternoon. How far did you hike? $\frac{7}{8}$ mi

Advanced Learners L4

Challenge students to write addition problems made up of three or more fractions, each with a different denominator. Together, the fractions should add to 1.

Sample: $\frac{1}{3} + \frac{5}{8} + \frac{1}{24}$

learning style: visual

English Language Learners ELL

Students may not understand the context used for the *Why Learn This?* section of the lesson. Show a hem of a garment, the stitches, and where the seam allowance is. Then add the $\frac{5}{8}$ inches to another length to show how the hem allowance works.

learning style: visual

2. Teach

Activity Lab

Use before the lesson.

Student Edition Activity Lab, Hands On 3-2a, Using Fraction Models, p. 125

All in One Teaching Resources

Activity Lab 3-2: Fraction Bar Exploration

Guided Instruction

Example 1

Encourage students to circle the common denominator as a reminder it stays the same when they add or subtract fractions.

Error Prevention!

Remind students if they change a denominator to make it a common denominator, the numerator of that fraction should change also.

Example 2

To help students find the LCD, suggest that they begin with the greater denominator. Have them recite its multiples until they reach a common multiple with the other denominator. Sample: $\frac{3}{4} + \frac{1}{6}$ "6, 12. 12 is also a multiple of 4."

PowerPoint

Additional Examples

1 Find $\frac{2}{7} + \frac{3}{7}$. $\frac{5}{7}$

2 Find $\frac{5}{9} + \frac{1}{6}$. $\frac{13}{18}$

3 A carpenter begins with a board that is $\frac{3}{4}$ in. thick. She removes $\frac{1}{16}$ in. from its thickness. How thick is the board now? $\frac{11}{16}$ in.

All in One Teaching Resources

- Daily Notetaking Guide 3-2 L3
- Adapted Notetaking 3-2 L1

Closure

- *What needs to be the same before you can add or subtract fractions?* their denominators
- *What part of the fractions do you actually add or subtract when you add or subtract fractions?* the numerators

3. Practice

Assignment Guide

Check Your Understanding
Go over Exercises 1–13 in class before assigning the Homework Exercises.

Homework Exercises

A	Practice by Example	14–34
B	Apply Your Skills	35–45
C	Challenge	46
Test Prep and Mixed Review		47–51

Homework Quick Check
To check students' understanding of key skills and concepts, go over Exercises 19, 32, 38, 40, and 42.

Differentiated Instruction Resources

Adapted Practice 3-2 L1

Practice 3-2 Adding and Subtracting [illegible]s L3

Write a number statement for each model.

1. $\frac{9}{10} - \frac{1}{2} = \frac{2}{5}$
2. $\frac{1}{3} + \frac{1}{4} = \frac{7}{12}$
3. $\frac{7}{8} - \frac{1}{4} = \frac{5}{8}$

Find each sum or difference.

4. $\frac{1}{6} + \frac{7}{8}$ $1\frac{1}{24}$
5. $\frac{9}{10} - \frac{1}{6}$ $\frac{11}{15}$
6. $\frac{1}{6} + \frac{1}{6}$ $\frac{1}{3}$
7. $\frac{1}{10} + \frac{2}{5}$ $\frac{1}{2}$
8. $\frac{5}{6} + \frac{1}{12}$ $\frac{11}{12}$
9. $\frac{2}{3} - \frac{1}{2}$ $\frac{1}{6}$
10. $\frac{7}{9} - \frac{1}{3}$ $\frac{4}{9}$
11. $\frac{3}{4} - \frac{1}{4}$ $\frac{1}{2}$
12. $\frac{1}{5} + \frac{3}{4}$ $\frac{19}{20}$
13. $\frac{1}{3} + \frac{1}{2}$ $\frac{5}{6}$
14. $\frac{1}{8} + \frac{1}{12}$ $\frac{5}{24}$
15. $\frac{7}{10} - \frac{1}{3}$ $\frac{11}{30}$

Use the table at the right for exercises 16–21. Tell which two snacks combine to make each amount.

16. $\frac{5}{6}$ c sesame sticks, dried apricots
17. $\frac{1}{2}$ c walnuts, almonds
18. $\frac{3}{4}$ c pretzels, almonds
19. $\frac{11}{12}$ c raisins, sesame sticks
20. 1 c walnuts, pretzels
21. $\frac{19}{24}$ c pretzels, dried apricots; or almonds, sesame sticks

Snack	Serving Amount
Raisins	$\frac{1}{4}$ c
Walnuts	$\frac{3}{8}$ c
Almonds	$\frac{1}{8}$ c
Sesame sticks	$\frac{2}{3}$ c
Mini pretzels	$\frac{5}{8}$ c
Dried apricots	$\frac{1}{6}$ c

Solve each equation.

22. $\frac{2}{3} + x = \frac{4}{6}$ 0
23. $s - \frac{1}{5} = \frac{2}{10}$ $\frac{2}{5}$
24. $b - \frac{4}{12} = \frac{8}{12}$ 1
25. $c + \frac{1}{6} = \frac{5}{12}$ $\frac{1}{4}$
26. $\frac{3}{8} + d = \frac{7}{8}$ $\frac{1}{2}$
27. $f - \frac{1}{10} = \frac{2}{5}$ $\frac{1}{2}$

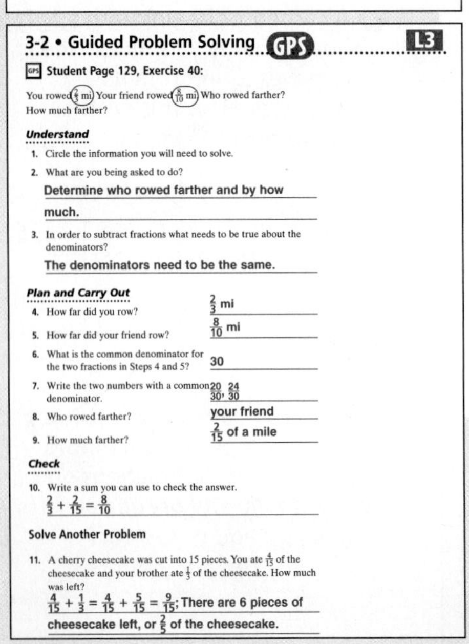
3-2 • Guided Problem Solving GPS L3

Student Page 129, Exercise 40:

You rowed $\frac{2}{3}$ mi. Your friend rowed $\frac{8}{10}$ mi. Who rowed farther? How much farther?

Understand

1. Circle the information you will need to solve.
2. What are you being asked to do?
   Determine who rowed farther and by how much.
3. In order to subtract fractions what needs to be true about the denominators?
   The denominators need to be the same.

Plan and Carry Out

4. How far did you row? $\frac{2}{3}$ mi
5. How far did your friend row? $\frac{8}{10}$ mi
6. What is the common denominator for the two fractions in Steps 4 and 5? 30
7. Write the two numbers with a common denominator. $\frac{20}{30}, \frac{24}{30}$
8. Who rowed farther? your friend
9. How much farther? $\frac{2}{15}$ of a mile

Check

10. Write a sum you can use to check the answer.
    $\frac{2}{3} + \frac{2}{15} = \frac{8}{10}$

Solve Another Problem

11. A cherry cheesecake was cut into 15 pieces. You ate $\frac{4}{15}$ of the cheesecake and your brother ate $\frac{1}{3}$ of the cheesecake. How much was left?
    $\frac{4}{15} + \frac{1}{3} = \frac{4}{15} + \frac{5}{15} = \frac{9}{15}$; There are 6 pieces of cheesecake left, or $\frac{2}{5}$ of the cheesecake.

Check Your Understanding

1. Answers may vary. Sample: It doesn't make sense to add fractions with different denominators. For example, $\frac{1}{2} + \frac{1}{3} \neq \frac{2}{5}$.

1. Why do you need common denominators to add or subtract fractions? **See left.**

Find the least common denominator of the fractions.

2. $\frac{1}{5}, \frac{3}{4}$ **20**
3. $\frac{5}{8}, \frac{1}{16}$ **16**
4. $\frac{7}{10}, \frac{1}{4}$ **20**
5. $\frac{1}{2}, \frac{2}{3}$ **6**

Find each sum or difference.

6. $\frac{1}{8} + \frac{5}{8}$ **$\frac{3}{4}$**
7. $\frac{1}{2} + \frac{1}{2}$ **1**
8. $\frac{3}{5} - \frac{2}{5}$ **$\frac{1}{5}$**
9. $\frac{5}{6} - \frac{1}{6}$ **$\frac{2}{3}$**
10. $\frac{1}{2} + \frac{5}{16}$ **$\frac{13}{16}$**
11. $\frac{1}{5} + \frac{3}{8}$ **$\frac{23}{40}$**
12. $\frac{9}{10} - \frac{2}{5}$ **$\frac{1}{2}$**
13. $\frac{3}{6} - \frac{4}{9}$ **$\frac{1}{18}$**

Homework Exercises

For more exercises, see Extra Skills and Word Problems.

GO for Help

For Exercises	See Examples
14–25	1
26–34	2–3

A **Find each sum or difference. You may find a model helpful.**

14. $\frac{1}{7} + \frac{4}{7}$ **$\frac{5}{7}$**
15. $\frac{3}{4} + \frac{3}{4}$ **$1\frac{1}{2}$**
16. $\frac{6}{8} + \frac{3}{8}$ **$1\frac{1}{8}$**
17. $\frac{1}{10} + \frac{3}{10}$ **$\frac{2}{5}$**
18. $\frac{4}{5} - \frac{1}{5}$ **$\frac{3}{5}$**
19. $\frac{7}{10} - \frac{1}{10}$ **$\frac{3}{5}$**
20. $\frac{5}{12} - \frac{1}{12}$ **$\frac{1}{3}$**
21. $\frac{7}{8} - \frac{5}{8}$ **$\frac{1}{4}$**
22. $\frac{3}{4} + \frac{1}{4}$ **1**
23. $\frac{11}{16} - \frac{1}{16}$ **$\frac{5}{8}$**
24. $\frac{8}{9} - \frac{2}{9}$ **$\frac{2}{3}$**
25. $\frac{3}{32} + \frac{1}{32}$ **$\frac{1}{8}$**

Find each sum or difference by writing equivalent fractions.

26. $\frac{7}{12} + \frac{1}{6}$ **$\frac{3}{4}$**
27. $\frac{3}{3} + \frac{5}{8}$ **$1\frac{5}{8}$**
28. $\frac{4}{5} + \frac{7}{8}$ **$1\frac{27}{40}$**
29. $\frac{1}{2} + \frac{4}{5}$ **$1\frac{3}{10}$**
30. $\frac{5}{6} - \frac{1}{3}$ **$\frac{1}{2}$**
31. $\frac{3}{5} - \frac{1}{4}$ **$\frac{7}{20}$**
32. $\frac{5}{6} - \frac{1}{2}$ **$\frac{1}{3}$**
33. $\frac{2}{3} - \frac{1}{4}$ **$\frac{5}{12}$**

34. The gas tank in your family's car was $\frac{7}{8}$ full when you left your house. When you arrived at your destination, the tank was $\frac{1}{4}$ full. What fraction of a tank of gas did you use during the trip? **$\frac{5}{8}$**

B

35. **Guided Problem Solving** One third of the students in your class got an A on a test. One fourth of them got a B. What fraction of the students got an A or a B on the test? **$\frac{7}{12}$**
    - **Make a Plan** Determine the correct operation to use. Find the LCD and write equivalent fractions.
    - **Carry Out the Plan** The correct operation is _?_. The LCD of the two fractions is ■.

GO Online
Homework Video Tutor
Visit: PHSchool.com
Web Code: are-0302

36. Suppose you are using nails $\frac{5}{6}$ in. long to nail plywood $\frac{1}{4}$ in. thick to a post. How much of the nail extends into the post? **$\frac{7}{12}$ in.**

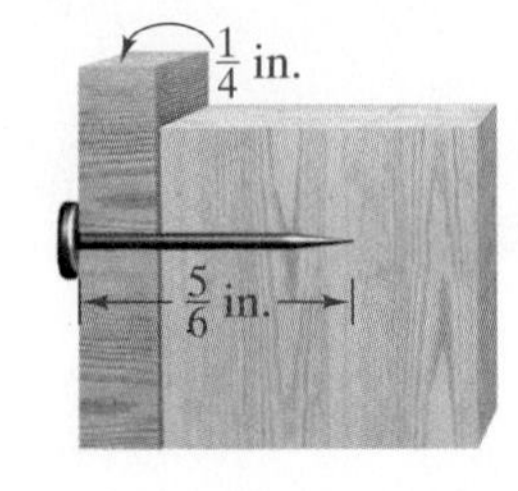

Mental Math **Will each sum be *positive, negative,* or *zero*?**

37. $-\frac{2}{3} + \frac{5}{6}$ **positive**
38. $-\frac{4}{5} + \frac{8}{10}$ **zero**
39. $-\frac{7}{8} + \frac{3}{4}$ **negative**

40. You rowed $\frac{2}{3}$ mi. Your friend rowed $\frac{8}{10}$ mi. Who rowed farther? How much farther? **your friend; $\frac{2}{15}$ mi farther**

41. What fraction, when added to $\frac{1}{6}$, results in a sum of $\frac{2}{3}$? **$\frac{1}{2}$**

42. **Error Analysis** A student added $\frac{2}{8} + \frac{3}{8}$ and got $\frac{5}{16}$. What was the student's mistake? What is the correct answer? **The student added the denominators. The correct answer is $\frac{5}{8}$.**

Use the circle graph for Exercises 43–45.

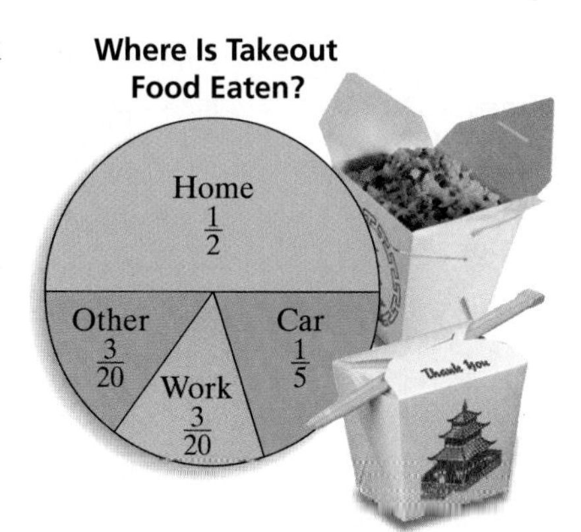

43. **Data Analysis** What fraction of takeout food is eaten at home or in a car? **$\frac{7}{10}$**

44. How much greater is the fraction of takeout food eaten at home than the fraction eaten at work? **$\frac{7}{20}$**

45. What fraction of the food is eaten at home, in a car, or at work? **$\frac{17}{20}$**

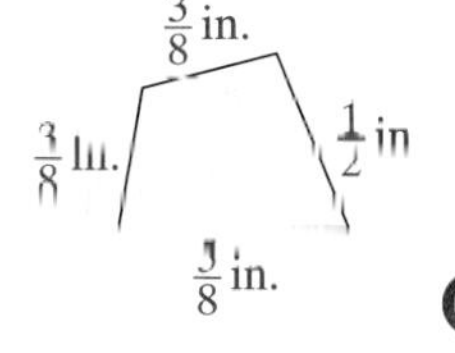

46. **Challenge** Find the perimeter of the figure at the left. **$1\frac{5}{8}$ in.**

Test Prep and Mixed Review

Practice

Multiple Choice

47. Rolf has a math test. He plans to study $\frac{5}{6}$ hour tonight and $\frac{3}{4}$ hour tomorrow. How many hours does Rolf plan to study in all? **B**

    A. $1\frac{3}{4}$ h B. $1\frac{7}{12}$ h C. $\frac{4}{5}$ h D. $\frac{9}{10}$ h

48. To make a noodle casserole, Jack uses $\frac{1}{3}$ cup of cream cheese, Myra uses $\frac{2}{5}$ cup, Donna uses $\frac{3}{4}$ cup, and Ernesto uses $\frac{1}{2}$ cup. Who uses the most cream cheese? **H**

    F. Jack G. Myra H. Donna J. Ernesto

49. Which expression can be used to find the maximum number of 0.5-inch lengths of wire that can be cut from a wire 10.2 inches long? **C**

    A. 0.5 ÷ 10.2 C. 10.2 ÷ 0.5
    B. 0.5 × 10.2 D. 10.2 × 0.5

GO for Help

For Exercises	See Lesson
50–51	2-6

Order from least to greatest.

50. $\frac{4}{3}, 0.52, -\frac{3}{4}, 1.0$ **$-\frac{3}{4}, 0.52, 1.0, \frac{4}{3}$**

51. $1.34, \frac{25}{6}, -2.4, \frac{7}{3}$ **$-2.4, 1.34, \frac{7}{3}, \frac{25}{6}$**

Online lesson quiz, **PHSchool.com**, Web Code: ara-0302

4. Assess & Reteach

Lesson Quiz

Find each sum or difference.

1. $\frac{4}{5} + \frac{3}{5}$ $1\frac{2}{5}$
2. $\frac{11}{12} - \frac{10}{12}$ $\frac{1}{12}$
3. $\frac{3}{4} + \frac{5}{6}$ $1\frac{7}{12}$
4. $\frac{7}{8} - \frac{2}{3}$ $\frac{5}{24}$

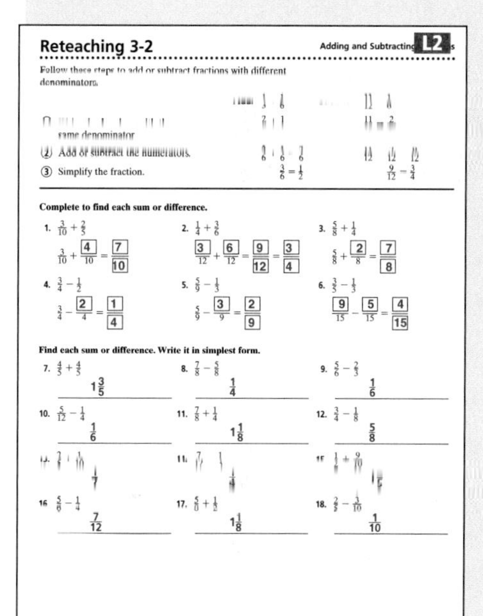

Reteaching 3-2 Adding and Subtracting Fractions **L2**

Follow these steps to add or subtract fractions with different denominators.

① ... same denominator

② Add or subtract the numerators.

③ Simplify the fraction. $\frac{3}{6} = \frac{1}{2}$ $\frac{9}{12} = \frac{3}{4}$

Complete to find each sum or difference.

1. $\frac{3}{10} + \frac{2}{5}$: $\frac{3}{10} + \frac{4}{10} = \frac{7}{10}$
2. $\frac{1}{4} + \frac{3}{6}$: $\frac{3}{12} + \frac{6}{12} = \frac{9}{12} = \frac{3}{4}$
3. $\frac{5}{8} + \frac{1}{4}$: $\frac{5}{8} + \frac{2}{8} = \frac{7}{8}$
4. $\frac{3}{4} - \frac{1}{2}$: $\frac{3}{4} - \frac{2}{4} = \frac{1}{4}$
5. $\frac{5}{9} - \frac{1}{3}$: $\frac{5}{9} - \frac{3}{9} = \frac{2}{9}$
6. $\frac{3}{5} - \frac{1}{3}$: $\frac{9}{15} - \frac{5}{15} = \frac{4}{15}$

Find each sum or difference. Write it in simplest form.

7. $\frac{4}{5} + \frac{4}{5}$ $1\frac{3}{5}$
8. $\frac{7}{8} - \frac{5}{8}$ $\frac{1}{4}$
9. $\frac{5}{6} - \frac{2}{3}$ $\frac{1}{6}$
10. $\frac{5}{12} - \frac{1}{4}$ $\frac{1}{6}$
11. $\frac{7}{8} + \frac{1}{4}$ $1\frac{1}{8}$
12. $\frac{3}{4} - \frac{1}{8}$ $\frac{5}{8}$
13. [illegible]
14. [illegible]
15. [illegible]
16. $\frac{5}{6} - \frac{1}{4}$ $\frac{7}{12}$
17. $\frac{5}{8} + \frac{1}{2}$ $1\frac{1}{8}$
18. $\frac{2}{5} - \frac{3}{10}$ $\frac{1}{10}$

Enrichment 3-2 Adding and Subtracting **L4**

Critical Thinking

Using each number only once, use the numbers 2, 4, 6, and 8 to write an expression with two proper fractions that have:

a. The largest possible sum b. The largest possible difference c. The smallest possible sum d. The smallest possible difference

1. What is a proper fraction?
   a fraction with the numerator less than the denominator
2. How many proper fractions will you write in each expression?
   2 fractions
3. How many different digits are in each pair of fractions?
   4 digits
4. List all the proper fractions you can make using the numbers 2, 4, 6, and 8.
   $\frac{2}{4}, \frac{2}{6}, \frac{2}{8}, \frac{4}{6}, \frac{4}{8}, \frac{6}{8}$
5. What are the pairs of fractions you can use when writing the expressions?
   $\frac{2}{4}, \frac{6}{8}$; $\frac{2}{6}, \frac{4}{8}$; $\frac{2}{8}, \frac{4}{6}$
6. For each pair of fractions, find the sum and the difference.
7. Write an expression for each part of the problem.
   Part a. $\frac{2}{4} + \frac{6}{8}$ Part b. $\frac{4}{6} - \frac{2}{8}$
   Part c. $\frac{2}{6} + \frac{4}{8}$ Part d. $\frac{4}{8} - \frac{2}{6}$

Fraction Pairs	Sum	Difference
$\frac{2}{4}, \frac{6}{8}$	$1\frac{1}{4}$	$\frac{1}{4}$
$\frac{2}{6}, \frac{4}{8}$	$\frac{5}{6}$	$\frac{1}{6}$
$\frac{2}{8}, \frac{4}{6}$	$\frac{11}{12}$	$\frac{5}{12}$

8. How can you tell that your answers are reasonable without calculating?
   Estimate the sums and differences. Use compensation to adjust for the values of the fractions.
9. Use the numbers 2, 3, 6, and 8 to write an expression with two proper fractions that have:
   a. The largest possible sum $\frac{2}{3} + \frac{6}{8}$
   b. The largest possible difference $\frac{3}{6} - \frac{2}{8}$
   c. The smallest possible sum $\frac{2}{6} + \frac{3}{8}$
   d. The smallest possible difference $\frac{3}{8} - \frac{2}{6}$

Alternative Assessment

Provide students with fraction models. Have them model and record the additions and subtractions for Exercises 14–33 as demonstrated in the lesson. You may help them get started by modeling one or two exercises at the overhead as students follow along at their desks.

Test Prep

Resources

For additional practice with a variety of test item formats:
- Test-Taking Strategies, p. 159
- Test Prep, p. 163
- Test-Taking Strategies with Transparencies

3-3

1. Plan

Objective
To add and subtract mixed numbers and to solve problems involving mixed numbers

Examples
1 Same Denominators
2 Different Denominators
3 Subtracting With Renaming

Math Understandings: p. 118C

Math Background

Adding or subtracting mixed numbers involves first adding or subtracting the fractions, and then adding or subtracting the whole numbers. When subtracting fractions, you may need to rename the mixed number. The steps for adding or subtracting mixed numbers are: Find a common denominator; rename the fraction, if needed, in subtraction problems; add or subtract; and simplify the resulting fraction.

More Math Background: p. 118C

Lesson Planning and Resources

See p. 118E for a list of the resources that support this lesson.

Bell Ringer Practice

Check Skills You'll Need
Use student page, transparency, or PowerPoint. For intervention, direct students to:
Comparing and Ordering Fractions
Lesson 2-4
Extra Skills and Word Problems Practice, Ch. 2

3-3 Adding and Subtracting Mixed Numbers

Check Skills You'll Need

1. **Vocabulary Review** How does the *LCD* of two fractions help you compare the fractions? **See below.**

Compare each pair of fractions. Use <, =, or >.

2. $\frac{3}{12}$ ■ $\frac{5}{12}$ <

3. $\frac{1}{6}$ ■ $\frac{1}{4}$ <

Lesson 2-4

Check Skills You'll Need

1. **When two fractions have the same denominator, you can compare them by looking at the numerators.**

What You'll Learn

To add and subtract mixed numbers and to solve problems involving mixed numbers

Why Learn This?

Adding mixed numbers can help you find the total distance you travel, such as the number of miles you run while training for a race.

To find the sum of two mixed numbers with fraction parts that have the same denominator, you can first add the fractions and then add the whole numbers.

EXAMPLE Same Denominators

1 Cross Country You are training for a race. On Monday you run $2\frac{3}{4}$ mi, and on Tuesday you run $4\frac{3}{4}$ mi. What is your total mileage?

Estimate $2\frac{3}{4} + 4\frac{3}{4} \approx 3 + 5$, or 8

$2\frac{3}{4} + 4\frac{3}{4} = 6\frac{6}{4}$ ← **Add the fractions. Add the whole numbers.**

$= 6 + 1\frac{2}{4}$ ← **Write as a sum. Write $\frac{6}{4}$ as $1\frac{2}{4}$.**

$= 6 + 1\frac{1}{2}$ ← **Simplify.**

$= 7\frac{1}{2}$ ← **Add the whole numbers.**

Your total mileage is $7\frac{1}{2}$ mi.

Check for Reasonableness The answer $7\frac{1}{2}$ is close to the estimate 8.

Quick Check

1. Find each sum or difference.
   a. $1\frac{2}{3} + 2\frac{2}{3}$ **$4\frac{1}{3}$**
   b. $1\frac{2}{5} + 3\frac{2}{5}$ **$4\frac{4}{5}$**
   c. $5\frac{1}{2} - 4\frac{1}{2}$ **1**

Differentiated Instruction Solutions for All Learners

Special Needs L1
Review how to change an improper fraction to a mixed number and vice versa. Use two clear glass jars marked off in fourths. Fill one to $\frac{4}{4}$ and ask: *How many more fourths do I need for $\frac{6}{4}$?* **two** Then fill the second jar to $\frac{2}{4}$.
learning style: visual

Below Level L2
Give students models such as shown. Have them write the related expression. **$2\frac{1}{4} + 1\frac{1}{2}$**

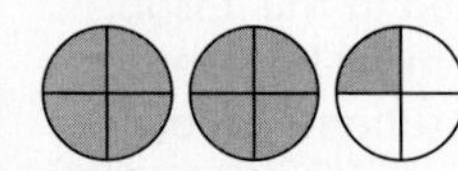
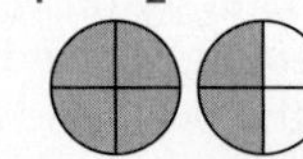

learning style: visual

Fractions in mixed numbers may have different denominators. You can use the LCD to rewrite them as fractions with common denominators.

Test Prep Tip

When you add or subtract fractions, use the least common multiple of the denominators.

EXAMPLE Different Denominators

2 Find $4\frac{1}{4} + 3\frac{2}{3}$.

Estimate $4\frac{1}{4} + 3\frac{2}{3} \approx 4 + 4$, or 8

$4\frac{1}{4} = 4\frac{3}{12}$ ← **The LCD is 12. Write an equivalent fraction.**

$+3\frac{2}{3} = +3\frac{8}{12}$ ← **Write an equivalent fraction.**

$= 7\frac{11}{12}$ ← **Add the fractions. Add the whole numbers.**

Check for Reasonableness The answer $7\frac{11}{12}$ is close to the estimate 8. The answer is reasonable.

Quick Check

2. Find each sum or difference.

a. $2\frac{3}{4} - 1\frac{3}{4}$ **1** b. $3\frac{1}{6} + 8\frac{7}{8}$ **$12\frac{1}{24}$** c. $6\frac{1}{2} - 2\frac{1}{5}$ **$4\frac{3}{10}$**

When you subtract mixed numbers, you may need to rename one of the numbers before subtracting.

EXAMPLE Subtracting With Renaming

3 **Multiple Choice** You bought $6\frac{1}{9}$ yd of ribbon and used $2\frac{2}{3}$ yd to make a pillow. How many yards of ribbon did you have left?

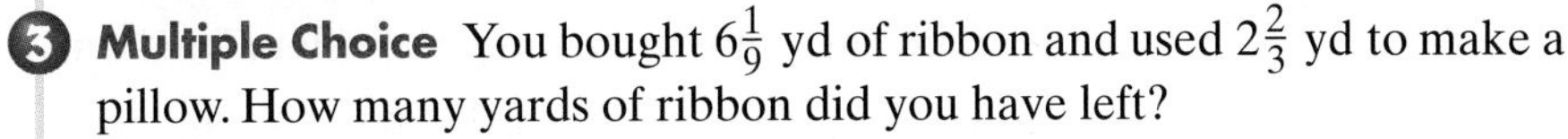

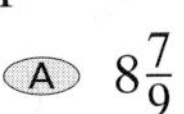

Ⓐ $8\frac{7}{9}$ Ⓑ $4\frac{5}{9}$ Ⓒ $3\frac{2}{9}$ Ⓓ $3\frac{4}{9}$

Find $6\frac{1}{9} - 2\frac{2}{3}$.

Estimate $6\frac{1}{9} - 2\frac{2}{9} \approx 6 - 3$, or 3

$6\frac{1}{9} = 5\frac{10}{9}$ ← **Rename: $6\frac{1}{9} = 5 + 1\frac{1}{9} = 5 + \frac{10}{9}$.**

$-2\frac{2}{3} = -2\frac{6}{9}$ ← **The LCD is 9. Write an equivalent fraction.**

$= 3\frac{4}{9}$ ← **Subtract.**

You have $3\frac{4}{9}$ yards of ribbon left. The correct answer is D.

Check for Reasonableness The answer $3\frac{4}{9}$ is close to the estimate 3.

Quick Check

3. Your friend bought $4\frac{1}{3}$ ft of gift-wrap and used $2\frac{5}{6}$ ft to wrap a gift. How many feet of gift-wrap did your friend have left? **$1\frac{1}{2}$ ft**

2. Teach

Activity Lab

Use before the lesson.

All in One Teaching Resources

Activity Lab 3-3: Adding and Subtracting Mixed Numbers

Guided Instruction

Example 1

Collect several empty egg cartons and use them to model addition of mixed numbers. For Quick Check 1a, cut two cartons into thirds, four egg slots to each third. Place them with whole cartons to represent $1\frac{2}{3} + 2\frac{2}{3}$. As you work through the solution on the board, invite a student to act out each step with the egg cartons.

Example 2

Have students circle the denominators to reinforce that those numbers need to be the same. Point out that when they change the fractions to the LCD, the whole numbers do not change.

PowerPoint

Additional Examples

1 Find $2\frac{3}{5} + 1\frac{4}{5}$. **$4\frac{2}{5}$**

2 Find [illegible]

3 Find $6\frac{1}{4} - 4\frac{3}{8}$. **$1\frac{7}{8}$**

All in One Teaching Resources

- Daily Notetaking Guide 3-3 **L3**
- Adapted Notetaking 3-3 **L1**

Closure

- *How do you add or subtract mixed numbers?* **Add or subtract the fractions, and then add or subtract the whole numbers.**
- *When would you need to rename a fraction in a subtraction problem?* **if the fraction being subtracted is greater than the first fraction**

Advanced Learners L4	**English Language Learners** ELL
Challenge students to write three different mixed-number subtraction problems whose answer is $3\frac{5}{6}$. **Sample: $6\frac{1}{5} - 2\frac{11}{30} = 3\frac{5}{6}$** **learning style: visual**	For Examples 1 and 2, separating the fractions from the whole numbers, and adding them separately set off by the phrases *add the whole numbers* and *add the fractions,* will help students keep track of the steps in the process. You can then recombine the sums and write as a mixed number. **learning style: verbal**

3. Practice

Assignment Guide

Check Your Understanding
Go over Exercises 1–8 in class before assigning the Homework Exercises.

Homework Exercises

A	Practice by Example	9–22
B	Apply Your Skills	23–32
C	Challenge	33
	Test Prep and Mixed Review	34–39

Homework Quick Check
To check students' understanding of key skills and concepts, go over Exercises 11, 20, 28, 29, and 30.

Differentiated Instruction Resources

Adapted Practice 3-3 L1

Practice 3-3 Adding and Subtracting Mixed ...s L3

Find each sum.

1. $5\frac{1}{3} + 3\frac{2}{3}$ 9
2. $7\frac{1}{4} + 4\frac{3}{8}$ $11\frac{5}{8}$
3. $2\frac{1}{8} + 6\frac{5}{8}$ $8\frac{3}{4}$
4. $8\frac{1}{5} + 4\frac{3}{10}$ $12\frac{1}{2}$
5. $9\frac{1}{6} + 6\frac{1}{4}$ $15\frac{5}{12}$
6. $3\frac{2}{3} + 10\frac{5}{6}$ $14\frac{1}{2}$

Find each difference.

7. $6\frac{11}{12} - 4\frac{5}{12}$ $2\frac{1}{2}$
8. $12 - 5\frac{3}{10}$ $6\frac{7}{10}$
9. $14\frac{1}{2} - 7\frac{1}{5}$ $7\frac{3}{10}$
10. $9 - 5\frac{5}{6}$ $3\frac{1}{6}$
11. $13\frac{3}{4} - 10\frac{1}{2}$ $3\frac{1}{4}$
12. $15\frac{1}{6} - 6\frac{5}{12}$ $8\frac{3}{4}$

Find each sum or difference.

13. $1\frac{1}{6} - \frac{3}{4}$ $\frac{5}{12}$
14. $4\frac{1}{2} - 2\frac{7}{8}$ $1\frac{5}{8}$
15. $9\frac{3}{4} + 7\frac{7}{8}$ $17\frac{5}{8}$
16. $5\frac{1}{6} - 4\frac{7}{12}$ $\frac{7}{12}$
17. $9\frac{8}{15} + 11\frac{5}{12}$ $20\frac{19}{20}$
18. $\frac{14}{15} - \frac{1}{2}$ $\frac{13}{30}$

Write a mixed number for each time period. Be sure each fraction is in lowest terms.

19. 8:00 A.M. to 9:20 A.M. $1\frac{1}{3}$ h
20. 9:00 A.M. to 2:45 P.M. $5\frac{3}{4}$ h
21. 11:00 A.M. to 3:55 P.M. $4\frac{11}{12}$ h
22. 8:30 A.M. to 10:40 P.M. $14\frac{1}{6}$ h

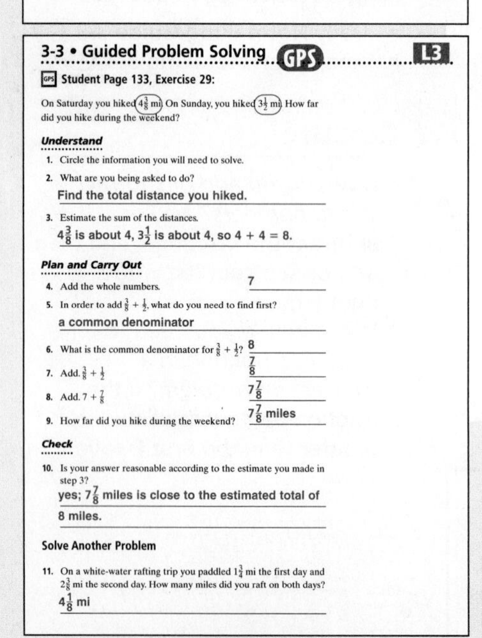

3-3 • Guided Problem Solving GPS L3

Student Page 133, Exercise 29:

On Saturday you hiked $4\frac{3}{8}$ mi. On Sunday, you hiked $3\frac{1}{2}$ mi. How far did you hike during the weekend?

Understand

1. Circle the information you will need to solve.
2. What are you being asked to do? Find the total distance you hiked.
3. Estimate the sum of the distances. $4\frac{3}{8}$ is about 4, $3\frac{1}{2}$ is about 4, so $4 + 4 = 8$.

Plan and Carry Out

4. Add the whole numbers. 7
5. In order to add $\frac{3}{8} + \frac{1}{2}$, what do you need to find first? a common denominator
6. What is the common denominator for $\frac{3}{8} + \frac{1}{2}$? 8
7. Add. $\frac{3}{8} + \frac{1}{2}$ $\frac{7}{8}$
8. Add. $7 + \frac{7}{8}$ $7\frac{7}{8}$
9. How far did you hike during the weekend? $7\frac{7}{8}$ miles

Check

10. Is your answer reasonable according to the estimate you made in step 3? yes; $7\frac{7}{8}$ miles is close to the estimated total of 8 miles.

Solve Another Problem

11. On a white-water rafting trip you paddled $1\frac{3}{4}$ mi the first day and $2\frac{3}{8}$ mi the second day. How many miles did you raft on both days? $4\frac{1}{8}$ mi

Check Your Understanding

Rewrite each mixed number so that the fraction part is a proper fraction.

1. $2\frac{7}{4}$ $3\frac{3}{4}$ 2. $3\frac{6}{5}$ $4\frac{1}{5}$ 3. $1\frac{14}{10}$ $2\frac{2}{5}$

4. **Reasoning** Do you use the skill you practiced in Exercises 1–3 when you *add* or when you *subtract* mixed numbers? Explain.

4. Add; when you add fractions, the sum may be greater than 1.

Rewrite each mixed number so that the whole-number part is less by 1.

5. $4\frac{1}{2}$ $3\frac{3}{2}$ 6. $2\frac{3}{8}$ $1\frac{11}{8}$ 7. $1\frac{9}{10}$ $\frac{19}{10}$

8. **Number Sense** Suppose you want to find the difference $6 - 3\frac{2}{5}$. How would you rewrite 6 as a mixed number? $5\frac{5}{5}$

Homework Exercises

For more exercises, see Extra Skills and Word Problems.

GO for Help

For Exercises	See Examples
9–13	1
14–17	2
18–22	3

A Find each sum.

9. $6\frac{2}{5} + 1\frac{4}{5}$ $8\frac{1}{5}$ 10. $9\frac{3}{7} + 1\frac{2}{7}$ $10\frac{5}{7}$ 11. $3\frac{3}{8} + 4\frac{5}{8}$ 8 12. $1\frac{7}{10} + 5\frac{1}{10}$ $6\frac{4}{5}$

13. To make lemonade, you use $3\frac{1}{3}$ cups of lemon concentrate and $1\frac{1}{3}$ cups of water. How many cups of lemonade do you make? $4\frac{2}{3}$ c

Find each sum by writing equivalent fractions.

14. $6\frac{2}{5} + 1\frac{4}{10}$ $7\frac{4}{5}$ 15. $9\frac{1}{2} + 9\frac{1}{3}$ $18\frac{5}{6}$ 16. $7\frac{1}{6} + 8\frac{1}{8}$ $15\frac{7}{24}$ 17. $6\frac{1}{2} + 4\frac{5}{6}$ $11\frac{1}{3}$

Find each difference.

18. $7\frac{2}{3} - 1\frac{1}{6}$ $6\frac{1}{2}$ 19. $15 - 3\frac{3}{4}$ $11\frac{1}{4}$ 20. $12\frac{1}{8} - 8\frac{3}{8}$ $3\frac{3}{4}$ 21. $14 - 5\frac{1}{5}$ $8\frac{4}{5}$

22. **Cooking** You have 4 cups of flour, and you need to use $1\frac{3}{4}$ cups for a cookie recipe. How much flour will you have left? $2\frac{1}{4}$ c

B

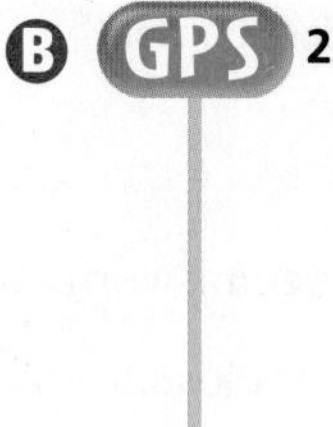

23. **Guided Problem Solving** Is a 6-quart punch bowl large enough to hold all the ingredients called for in the recipe? Explain.
    - Using mental math, what is $1\frac{1}{2} + \frac{1}{2}$?
    - What is the LCM of 3 and 4?
    - Is the sum of all the ingredients greater than or less than 6 qt?

    Yes; the total of ingredients is $5\frac{11}{12}$ qt.

$2\frac{1}{4}$ qt ginger ale • $1\frac{2}{3}$ qt lemon sherbet
$1\frac{1}{2}$ qt lemonade • $\frac{1}{2}$ qt raspberry juice

24. $21\frac{1}{8}$
25. $29\frac{3}{20}$
26. $5\frac{5}{6}$
27. $3\frac{5}{8}$

Visit: PHSchool.com
Web Code: are-0303

Find each sum or difference. 24–27. See margin.

24. $17\frac{3}{4} + 3\frac{3}{8}$ **25.** $17\frac{2}{5} + 11\frac{3}{4}$ **26.** $15\frac{1}{3} - 9\frac{1}{2}$ **27.** $6\frac{3}{8} - 2\frac{3}{4}$

28. Error Analysis A student subtracted $10\frac{1}{7} - 3\frac{5}{7}$ and got $6\frac{6}{7}$. Find the correct answer. What mistake do you think the student made? $6\frac{3}{7}$; the student borrowed $\frac{10}{7}$ instead of $\frac{7}{7}$.

29. On Saturday you hiked $4\frac{3}{8}$ mi. On Sunday, you hiked $3\frac{1}{2}$ mi. How far did you hike during the weekend? $7\frac{7}{8}$ mi

30. A bolt must go through a sign and a support that together are $2\frac{1}{8}$ in. thick. You need an additional $\frac{1}{16}$ in. for a washer and $\frac{1}{4}$ in. for a nut. How long should the bolt be? $2\frac{7}{16}$ in.

31. Geometry Figures A and B have a total area of $5\frac{3}{4}$ in.². The area of Figure A is $1\frac{1}{4}$ in.². Find the area of Figure B. $4\frac{1}{2}$ in.²

32. Calculator Use a calculator to find $5\frac{1}{2} + 4\frac{3}{4}$. $10\frac{1}{4}$

33. Challenge A newspaper has the following advertising spaces available: $2\frac{7}{8}$ c.i. (column inches), $3\frac{1}{2}$ c.i., and $4\frac{1}{4}$ c.i. What is the total number of column inches available? $10\frac{5}{8}$ c.i.

Test Prep and Mixed Review

Practice

Multiple Choice

34. How much heavier is $28\frac{1}{2}$ pounds than $15\frac{11}{16}$ pounds? B

Ⓐ $12\frac{3}{4}$ lb Ⓑ $12\frac{13}{16}$ lb Ⓒ $12\frac{7}{8}$ lb Ⓓ $13\frac{13}{16}$ lb

35. The table shows the number of runs Jason's baseball team, the Tigers, scored each game. What is the median number of runs the Tigers scored per game? J

Tigers Runs Scored

5	8	4	1	7	10	5
6	3	4	0	6	3	5

Ⓕ 4 Ⓖ 4.5 Ⓗ 4.8 Ⓙ 5

36. A submarine is located at 250 ft below sea level. At the end of the day, the submarine rises 140 ft. Which expression represents the movement of the submarine? B

Ⓐ $250 + 140$ Ⓒ $-250 + (-140)$

Ⓑ $-250 + 140$ Ⓓ $250 - (-140)$

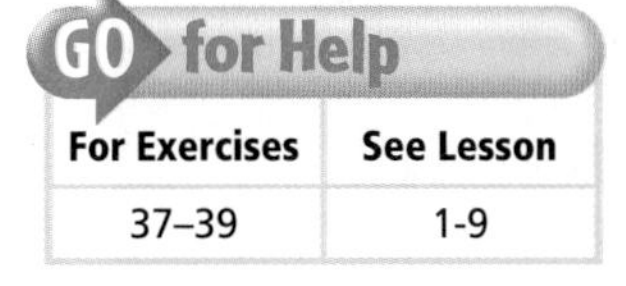

For Exercises	See Lesson
37–39	1-9

Mental Math **Find each product using the Distributive Property.**

37. 5(23) 115 **38.** 7(18) 126 **39.** 4(62) 248

4. Assess & Reteach

Lesson Quiz

Find each sum or difference.

1. $8\frac{2}{3} - 4\frac{5}{6}$ $3\frac{5}{6}$
2. $7\frac{3}{5} + 3\frac{3}{4}$ $11\frac{7}{20}$

Alternative Assessment

Have students work in pairs. Each student writes one exercise for adding mixed numbers and one exercise for subtracting mixed numbers. Partners exchange papers and find the sum and difference. Then partners discuss how they found the answers.

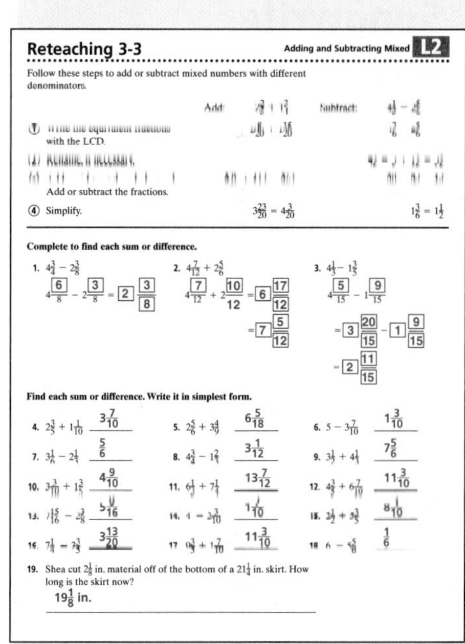

Reteaching 3-3 Adding and Subtracting Mixed L2

Follow these steps to add or subtract mixed numbers with different denominators.

Add: [illegible] Subtract: [illegible]

① Write the equivalent fractions with the LCD.

② Rename, if necessary.

③ Add or subtract the fractions.

④ Simplify. $3\frac{23}{20} = 4\frac{3}{20}$ $1\frac{3}{6} = 1\frac{1}{2}$

Complete to find each sum or difference.

1. $4\frac{3}{4} - 2\frac{3}{8}$: $4\frac{6}{8} - 2\frac{3}{8} = 2\frac{3}{8}$
2. $4\frac{7}{12} + 2\frac{5}{6}$: $4\frac{7}{12} + 2\frac{10}{12} = 6\frac{17}{12} = 7\frac{5}{12}$
3. $4\frac{1}{3} - 1\frac{3}{5}$: $4\frac{5}{15} - 1\frac{9}{15} = 3\frac{20}{15} - 1\frac{9}{15} = 2\frac{11}{15}$

Find each sum or difference. Write it in simplest form.

4. $2\frac{3}{5} + 1\frac{1}{10}$ $3\frac{7}{10}$
5. $2\frac{5}{6} + 3\frac{4}{9}$ $6\frac{5}{18}$
6. $5 - 3\frac{7}{10}$ $1\frac{3}{10}$
7. $3\frac{1}{6} - 2\frac{1}{3}$ $\frac{5}{6}$
8. $4\frac{3}{4} - 1\frac{2}{3}$ $3\frac{1}{12}$
9. [illegible] $7\frac{5}{6}$
10. $3\frac{3}{10} + 1\frac{3}{5}$ $4\frac{9}{10}$
11. $6\frac{1}{3} + 7\frac{1}{4}$ $13\frac{7}{12}$
12. $4\frac{3}{5} + 6\frac{7}{10}$ $11\frac{3}{10}$
13. $7\frac{15}{16} - 2\frac{3}{8}$ $5\frac{9}{16}$
14. $4 - 2\frac{3}{10}$ $1\frac{7}{10}$
15. $2\frac{1}{2} + 5\frac{3}{5}$ $8\frac{1}{10}$
16. $7\frac{1}{4} - 3\frac{3}{5}$ $3\frac{13}{20}$
17. $9\frac{3}{5} + 1\frac{7}{10}$ $11\frac{3}{10}$
18. [illegible] $\frac{1}{6}$
19. Shea cut $2\frac{1}{8}$ in. material off of the bottom of a $21\frac{1}{4}$ in. skirt. How long is the skirt now? $19\frac{1}{8}$ in.

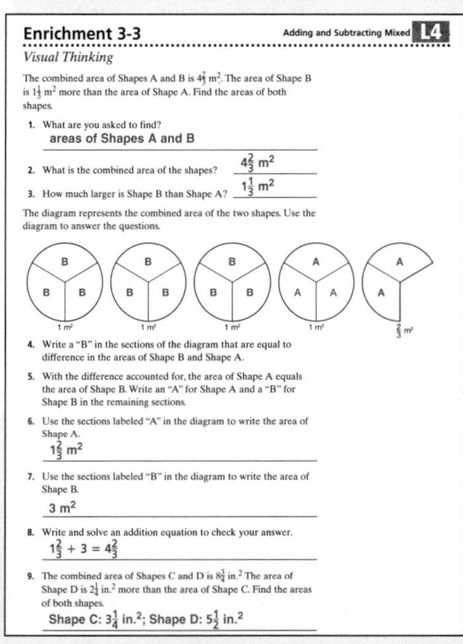

Enrichment 3-3 Adding and Subtracting Mixed L4

Visual Thinking

The combined area of Shapes A and B is $4\frac{2}{3}$ m². The area of Shape B is $1\frac{1}{3}$ m² more than the area of Shape A. Find the areas of both shapes.

1. What are you asked to find? areas of Shapes A and B
2. What is the combined area of the shapes? $4\frac{2}{3}$ m²
3. How much larger is Shape B than Shape A? $1\frac{1}{3}$ m²

The diagram represents the combined area of the two shapes. Use the diagram to answer the questions.

4. Write a "B" in the sections of the diagram that are equal to difference in the areas of Shape B and Shape A.
5. With the difference accounted for, the area of Shape A equals the area of Shape B. Write an "A" for Shape A and a "B" for Shape B in the remaining sections.
6. Use the sections labeled "A" in the diagram to write the area of Shape A. $1\frac{2}{3}$ m²
7. Use the sections labeled "B" in the diagram to write the area of Shape B. 3 m²
8. Write and solve an addition equation to check your answer. $1\frac{2}{3} + 3 = 4\frac{2}{3}$
9. The combined area of Shapes C and D is $8\frac{3}{4}$ in.² The area of Shape D is $2\frac{1}{4}$ in.² more than the area of Shape C. Find the areas of both shapes. Shape C: $3\frac{1}{4}$ in.²; Shape D: $5\frac{1}{2}$ in.²

Test Prep

Resources

For additional practice with a variety of test item formats:

- Test-Taking Strategies, p. 159
- Test Prep, p. 163
- Test-Taking Strategies with Transparencies

Use this Checkpoint Quiz to check students' understanding of the skills and concepts of Lessons 3-1 through 3-3.

Resources

-  Teaching Resources Checkpoint Quiz 1
- ExamView Assessment Suite CD-ROM
- Success Tracker™ Online Intervention

MATH AT WORK

Songwriter

Students may not be aware how math skills are important to certain attractive career possibilities. This feature explains an exciting career choice that makes math necessary every day.

Guided Instruction

Have different volunteers read each paragraph. Show a piece of sheet music. Point out quarter, half, and whole notes and how arranging notes differently makes the music sound different.
Ask questions, such as:

- *What beat patterns have you noticed in music you've heard?*
- *How do songwriters use fractions to write their music?*

Checkpoint Quiz 1

Lessons 3-1 through 3-3

The table below shows the average rainfall from July through December for a city in Florida. Use benchmarks and the table to answer Exercises 1–3.

Month	July	August	September	October	November	December
Inches	$7\frac{7}{10}$	$6\frac{4}{5}$	$6\frac{3}{5}$	$3\frac{2}{5}$	$1\frac{9}{10}$	$2\frac{1}{10}$

1. Estimate the difference between rainfall in July and in December. about $5\frac{1}{2}$ in.
2. Estimate the sum of rainfall in August and in September. about $13\frac{1}{2}$ in.
3. Estimate the average rainfall for the six months. about 5 in.

Simplify.

4. $\frac{12}{18} - \frac{9}{18}$ $\frac{1}{6}$
5. $\frac{4}{3} + \frac{2}{3}$ 2
6. $10\frac{7}{10} - 4\frac{4}{5}$ $5\frac{9}{10}$
7. $6\frac{3}{4} + 2\frac{1}{5}$ $8\frac{19}{20}$
8. $\frac{5}{9} - \frac{1}{3}$ $\frac{2}{9}$
9. $17 - 5\frac{3}{8}$ $11\frac{5}{8}$

10. **Measurement** Five years ago, Tyler's height was $43\frac{7}{8}$ in. Now his height is $65\frac{1}{2}$ in. How many inches has he grown in 5 years? $21\frac{5}{8}$ in.

MATH AT WORK

Songwriter

Songwriters need to know what types of music are "in" to compose a song that will sell. Then a songwriter must find an artist to record the song. Producers and record companies decide whether they will promote a song as a potential hit.

Songwriters use mathematics to make complex rhythms. Many hip-hop or rap beats use rhythms that follow a pattern. Songwriters write musical notes that have fractional names to create the pattern. Some common musical notes are half notes and quarter notes.

Go Online PHSchool.com For: Information on songwriters
Web Code: arb-2031

134

Modeling Fraction Multiplication

You can fold paper to model multiplication of fractions.
Use paper folding to find $\frac{1}{3}$ of $\frac{1}{4}$, or $\frac{1}{3} \cdot \frac{1}{4}$.

ACTIVITY

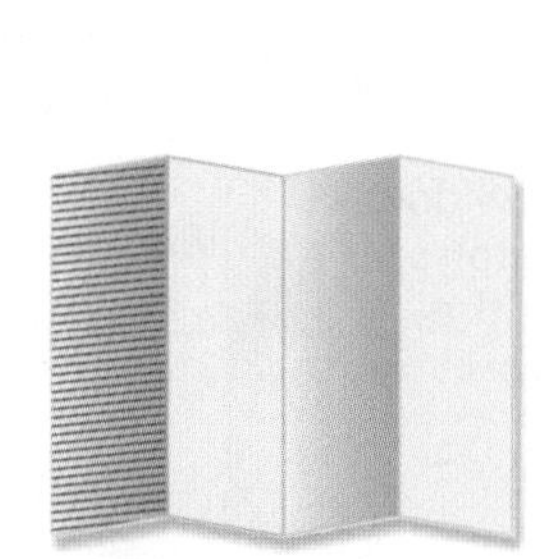

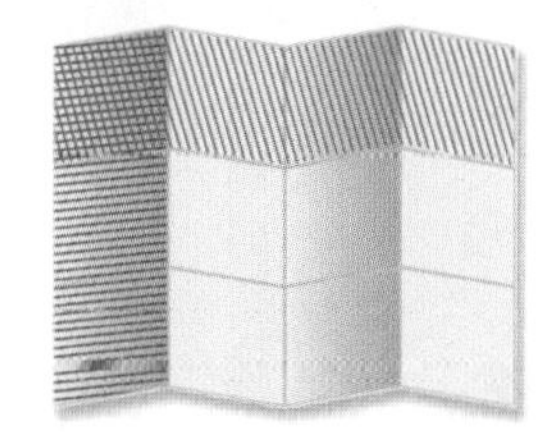

Step 1 Fold a sheet of paper into fourths as shown in the drawing at the right. Shade $\frac{1}{4}$ of it.

Step 2 Fold the same paper lengthwise into thirds as shown in the lower drawing. Shade $\frac{1}{3}$ of it.

Step 3 Count the total number of rectangles made by the folds. How many rectangles did you shade twice? What fraction of all the rectangles did you shade twice?

Step 4 Use the model to complete the equation: $\frac{1}{3} \cdot \frac{1}{4} = \blacksquare$. Explain how you found the product.

Step 5 Choose two fractions from $\frac{1}{3}$, $\frac{2}{3}$, and $\frac{3}{4}$. Use modeling to find the product of the two fractions.

Step 6 Suggest a rule for multiplying fractions.

Exercises

Complete each equation using the model shown below.

1. $\frac{1}{2} \cdot \blacksquare = \frac{1}{6}$ $\frac{1}{3}$

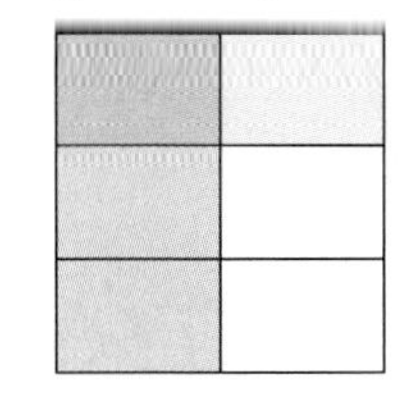

2. $\frac{2}{3} \cdot \frac{2}{3} = \blacksquare$ $\frac{4}{9}$

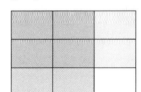

3. $\blacksquare \cdot \frac{3}{4} = \frac{3}{20}$ $\frac{1}{5}$

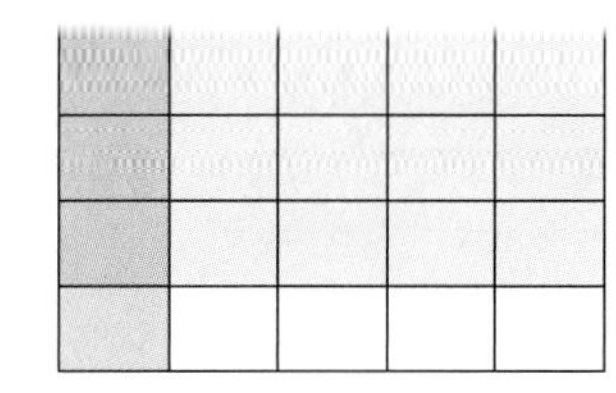

Make a model to represent each expression. Then find the product. 4–8. See margin for models.

4. $\frac{1}{3} \cdot \frac{2}{5}$ $\frac{2}{15}$
5. $\frac{5}{6} \cdot \frac{1}{2}$ $\frac{5}{12}$
6. $\frac{1}{2} \cdot \frac{1}{4}$ $\frac{1}{8}$
7. $\frac{3}{4} \cdot \frac{2}{3}$ $\frac{6}{12}$
8. $\frac{3}{8} \cdot \frac{1}{3}$ $\frac{3}{24}$

9. **Reasoning** In the exercises above, compare the product to each of its factors. Make a conjecture about the product of two numbers between 0 and 1. **The product of two numbers between 0 and 1 is less than either of the two numbers.**

4–8. See back of book.

Activity Lab

Modeling Fraction Multiplication

Students model multiplication of fractions. This will prepare them to calculate with fractions in Lesson 3-4.

Guided Instruction

Explain that you can multiply two fractions to find a certain part of a number that is less than one. Demonstrate how to fold the paper one way for one fraction and the other way for the other fraction. Make sure students are folding all vertical folds for one fraction and all horizontal folds for the other fraction. Ask:

- *How many sections did you make in total?* **12** *What does that number represent?* **greatest common denominator**
- *What does the number of vertical sections represent?* **denominator of the first fraction**
- *What does the number of shaded horizontal sections represent?* **numerator of the second fraction**

Exercises
Have students work on the Exercises. Go over the answers as a class and have them make suggestions about the formula for multiplying fractions.

Alternative Method
Have students use fraction tiles to model fraction multiplication.

Resources

- Activity Lab 3-4: Multiplying Fractions
- paper
- 2 different colored pencils

3-4

1. Plan

Objective
To multiply fractions and mixed numbers and to solve problems by multiplying

Examples
1 Multiplying Fractions
2 Multiplying by a Whole Number
3 Multiplying Mixed Numbers

Math Understandings: p. 118C

Professional Development

Math Background

Unlike addition and subtraction of fractions, a common denominator is *not* needed when you multiply fractions. Multiply the numerators and then the denominators. The product may need to be simplified before or after you multiply. When you multiply a fraction with a whole number or mixed number, first rewrite each number as an improper fraction. After you multiply, change an improper fraction to a mixed number.

More Math Background: p. 118C

Lesson Planning and Resources

See p. 118E for a list of the resources that support this lesson.

PowerPoint

Bell Ringer Practice

Check Skills You'll Need
Use student page, transparency, or PowerPoint. For intervention, direct students to:
Comparing and Ordering Fractions
Lesson 2-4
Extra Skills and Word Problems Practice, Ch. 2

3-4 Multiplying Fractions and Mixed Numbers

Check Skills You'll Need

1. **Vocabulary Review** In an *improper fraction,* how does the numerator compare to the denominator? **See below.**

Write each mixed number as an improper fraction.

2. $2\frac{3}{5}$ $\frac{13}{5}$
3. $6\frac{1}{2}$ $\frac{13}{2}$
4. $1\frac{2}{3}$ $\frac{5}{3}$
5. $4\frac{3}{4}$ $\frac{19}{4}$

 **for Help** 
Lesson 2-5

Check Skills You'll Need

1. The numerator is greater than the denominator.

What You'll Learn

To multiply fractions and mixed numbers and to solve problems by multiplying

Why Learn This?

You multiply fractions to find part of a quantity, which is useful when you cook.

A baker makes a batch of bread dough and splits it into three parts. Then she takes half of one part for one loaf. You can multiply $\frac{1}{3} \cdot \frac{1}{2}$ to find the fraction of the batch she uses for one loaf of bread.

When multiplying fractions, you do *not* need a common denominator. You multiply the numerators and multiply the denominators.

KEY CONCEPTS Multiplying Fractions

Arithmetic	Algebra
$\frac{1}{3} \cdot \frac{1}{2} = \frac{1 \cdot 1}{3 \cdot 2} = \frac{1}{6}$	$\frac{a}{b} \cdot \frac{c}{d} = \frac{ac}{bd}$, $b \neq 0$ and $d \neq 0$

EXAMPLE Multiplying Fractions

Test Prep Tip

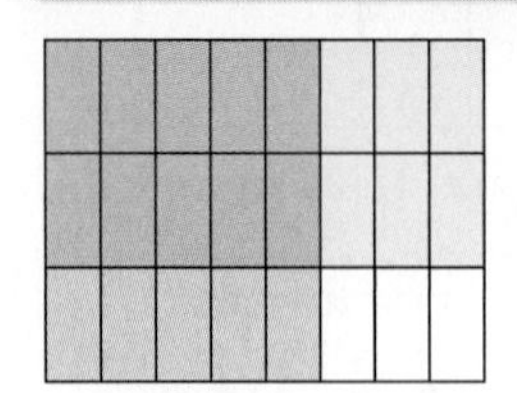

The model represents the product in Example 1.

1 Find $\frac{5}{8} \cdot \frac{2}{3}$.

$\frac{5}{8} \cdot \frac{2}{3} = \frac{5 \cdot 2}{8 \cdot 3}$ ← Multiply the numerators. Multiply the denominators.

$= \frac{10}{24}$ ← Find the two products.

$= \frac{5}{12}$ ← Simplify.

Quick Check

1. Find each product.

a. $\frac{3}{5} \cdot \frac{1}{4}$ $\frac{3}{20}$
b. $\frac{5}{6} \cdot \frac{4}{5}$ $\frac{2}{3}$
c. $\frac{2}{3} \cdot \frac{4}{5}$ $\frac{8}{15}$

Differentiated Instruction Solutions for All Learners

Special Needs L1
Help students visualize $\frac{3}{7}$ of 28 students. Ask them to look at the denominator (7), and draw rows of 7 X's until they get to 28 (4 rows of 7). Then ask them to circle columns (since 3 is the numerator). Ask: *How many X's did you circle?* 12

learning style: visual

Below Level L2
Give students several multiplication exercises involving $\frac{1}{2}$ to simplify, such as $\frac{1}{2}$ of 3 and $\frac{1}{2}$ of 7.

$\frac{1}{2}$ of $3 = \frac{1}{2} \cdot \frac{3}{1} = \frac{3}{2} = 1\frac{1}{2}$

$\frac{1}{2}$ of $7 = \frac{1}{2} \cdot \frac{7}{1} = \frac{7}{2} = 3\frac{1}{2}$

learning style: visual

To multiply a fraction by a whole number, you can write the whole number as a fraction with a denominator of 1. When a numerator and a denominator have a common factor, you can simplify before multiplying.

EXAMPLE Multiplying by a Whole Number

2 **Gridded Response** On Friday, $\frac{3}{7}$ of the 28 students in a class went to a music competition. How many students were still in class?

Find $\frac{3}{7}$ of the 28 students, or $\frac{3}{7} \cdot 28$.

$\frac{3}{7} \cdot 28 = \frac{3}{7} \cdot \frac{28}{1}$ ← **Write 28 as $\frac{28}{1}$.**

$= \frac{3}{{}_1\not{7}} \cdot \frac{\not{28}^{4}}{1}$ ← **Simplify before multiplying.**

$= \frac{12}{1}$ ← **Multiply the numerators.** ← **Multiply the denominators.**

$= 12$ ← **Simplify. There were 12 students absent.**

$28 - 12 = 16$ ← **Subtract.**

There were 16 students still in class on Friday.

		1	6

Test Prep Tip

Read the question carefully and make sure you are answering the question that is asked.

Quick Check

1. There are 168 members in an orchestra, and $\frac{3}{8}$ of them play the violin. How many members play the violin? **63 members**

To multiply mixed numbers, first write them as improper fractions, and then multiply as you do with fractions.

EXAMPLE Multiplying Mixed Numbers

3 Find $2\frac{3}{5} \cdot 4\frac{1}{2}$.

Estimate $2\frac{3}{5} \cdot 4\frac{1}{2} \approx 3 \cdot 5$, or 15

$2\frac{3}{5} \cdot 4\frac{1}{2} = \frac{13}{5} \cdot \frac{9}{2}$ ← **Write the mixed numbers as improper fractions.**

$= \frac{13 \cdot 9}{5 \cdot 2}$ ← **Multiply numerators. Multiply denominators.**

$= \frac{117}{10}$ ← **Simplify.**

$= 11\frac{7}{10}$ ← **Write as a mixed number.**

Check for Reasonableness $11\frac{7}{10}$ is close to the estimate 15.

For: Multiplying Fractions Activity
Use: Interactive Textbook, 3-4

Quick Check

3. Find each product.

a. $2\frac{1}{3} \cdot 4\frac{5}{8}$ **$10\frac{19}{24}$** **b.** $3\frac{3}{5} \cdot 1\frac{3}{10}$ **$4\frac{17}{25}$** **c.** $5\frac{3}{4} \cdot 2\frac{5}{8}$ **$15\frac{3}{32}$**

2. Teach

Activity Lab

Use before the lesson.
Student Edition Activity Lab, Hands On 3-4a, Modeling Fraction Multiplication, p. 135

All in One Teaching Resources
Activity Lab 3-4: Multiplying Fractions

Guided Instruction

Error Prevention!

Use $2\frac{1}{3} \cdot 4\frac{5}{8}$ from Quick Check 3a to demonstrate why it is necessary to change mixed numbers into improper fractions before multiplying. First model the correct method to get the answer $10\frac{19}{24}$. Next show why multiplying whole numbers and then multiplying fractions does *not* work. Multiply the whole numbers 2 and 4 to get 8. Multiply the fractions $\frac{1}{3}$ and $\frac{5}{8}$ to get $\frac{5}{24}$. Add the two parts 8 and $\frac{5}{24}$ to get $8\frac{5}{24}$. Compare this wrong answer to your correct answer to show students the method does not work.

PowerPoint

Additional Examples

1 Find $\frac{1}{9} \cdot \frac{3}{4}$. $\frac{1}{12}$

2 What is $\frac{2}{3}$ of 12? 8

3 Find $5\frac{1}{2} \cdot 3\frac{1}{3}$. $18\frac{1}{3}$

All in One Teaching Resources

- Daily Notetaking Guide 3-4 **L3**
- Adapted Notetaking 3-4 **L1**

Closure

- *How do you multiply fractions?* Sample: Multiply the numerators and multiply the denominators.
- *How can you change a whole number to a fraction?* Rewrite it with a denominator of 1.
- *What do you need to do before multiplying mixed numbers?* Change them to improper fractions.

Advanced Learners **L4**
Challenge pairs to take turns writing a multiplication problem whose answer is less than 1. Each turn must yield an answer closer to 1 than the previous answer. **Sample:** $\frac{5}{6} \cdot \frac{7}{8} = \frac{35}{48}$; $\frac{9}{10} \cdot \frac{11}{10} = \frac{99}{100}$ ...

learning style: visual

English Language Learners ELL
A distinction is made in this lesson between multiplying two fractions or multiplying a fraction and a whole number. Remind students that a whole number can be expressed as a fraction.

learning style: verbal

3. Practice

Assignment Guide

Check Your Understanding

Go over Exercises 1–6 in class before assigning the Homework Exercises.

Homework Exercises

A Practice by Example	7–31
B Apply Your Skills	32–42
C Challenge	43
Test Prep and Mixed Review	44–49

Homework Quick Check

To check students' understanding of key skills and concepts, go over Exercises 17, 28, 37, 39, and 41.

Differentiated Instruction Resources

Adapted Practice 3-4 **L1**

Practice 3-4 — Multiplying Fractions and Mixed Numbers **L3**

Find each product.

1. $\frac{5}{6} \cdot \frac{3}{5}$ $\frac{1}{2}$
2. $\frac{7}{8} \cdot \frac{4}{5}$ $\frac{7}{10}$
3. $\frac{9}{10} \cdot \frac{5}{12}$ $\frac{1}{3}$ (sic)
4. $\frac{5}{8} \cdot \frac{3}{5}$ $\frac{3}{8}$
5. $\frac{1}{6}$ of 36 6
6. $\frac{5}{9} \cdot 36$ 20
7. $\frac{3}{4} \cdot 36$ 27
8. $2 \cdot \frac{9}{10}$ $1\frac{4}{5}$
9. $8 \cdot \frac{9}{10}$ $7\frac{1}{5}$
10. $\frac{1}{3} \cdot 3\frac{1}{3}$ $1\frac{1}{9}$
11. $\frac{5}{6}$ of $1\frac{3}{5}$ $1\frac{1}{3}$
12. $\frac{1}{8}$ of $1\frac{4}{5}$ $\frac{9}{40}$
13. $3 \cdot 4\frac{1}{2}$ $13\frac{1}{2}$
14. $5 \cdot 2\frac{1}{4}$ $11\frac{1}{4}$
15. $3 \cdot 2\frac{2}{3}$ 8
16. $3\frac{2}{3} \cdot 1\frac{1}{2}$ $5\frac{1}{2}$
17. $4\frac{1}{6} \cdot 2\frac{2}{5}$ 10
18. $3\frac{1}{4} \cdot 2\frac{1}{6}$ $7\frac{1}{24}$

Solve.

19. A sheet of plywood is $\frac{5}{8}$ in. thick. How tall is a stack of 21 sheets of plywood? $13\frac{1}{8}$ in.
20. A poster measures 38 cm across. If a photocopy machine is used to make a copy that is $\frac{3}{5}$ of the original size, what is the width of the copy? $22\frac{4}{5}$ cm
21. A one-kilogram object weighs about $2\frac{1}{5}$ pounds. Find the weight, in pounds, of a computer monitor with mass $7\frac{3}{8}$ kilograms. $16\frac{9}{40}$ lb
22. The population of Sweden is about $1\frac{11}{16}$ times as great as the population of Denmark. Find the population of Sweden if the population of Denmark is about 5,190,000. about 8,760,000

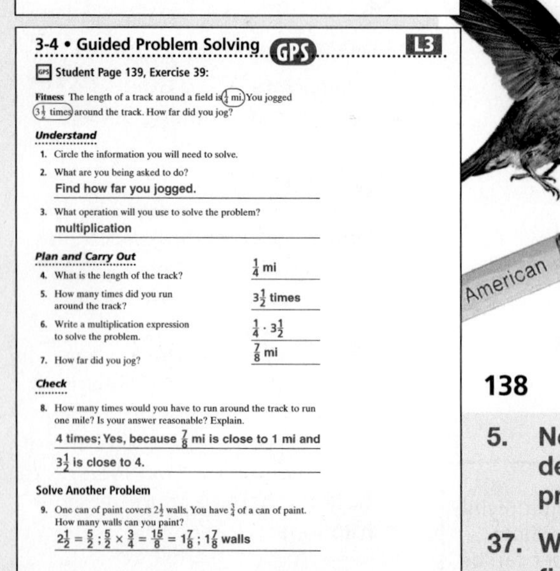

3-4 • Guided Problem Solving GPS **L3**

GPS Student Page 139, Exercise 39:

Fitness The length of a track around a field is $\frac{1}{4}$ mi. You jogged $3\frac{1}{2}$ times around the track. How far did you jog?

Understand

1. Circle the information you will need to solve.
2. What are you being asked to do? Find how far you jogged.
3. What operation will you use to solve the problem? multiplication

Plan and Carry Out

4. What is the length of the track? $\frac{1}{4}$ mi
5. How many times did you run around the track? $3\frac{1}{2}$ times
6. Write a multiplication expression to solve the problem. $\frac{1}{4} \cdot 3\frac{1}{2}$
7. How far did you jog? $\frac{7}{8}$ mi

Check

8. How many times would you have to run around the track to run one mile? Is your answer reasonable? Explain. 4 times; Yes, because $\frac{7}{8}$ mi is close to 1 mi and $3\frac{1}{2}$ is close to 4.

Solve Another Problem

9. One can of paint covers $2\frac{1}{2}$ walls. You have $\frac{3}{4}$ of a can of paint. How many walls can you paint? $2\frac{1}{2} = \frac{5}{2}; \frac{5}{2} \times \frac{3}{4} = \frac{15}{8} = 1\frac{7}{8}$; $1\frac{7}{8}$ walls

Check Your Understanding

Complete the first step in finding each product.

1. $\frac{2}{5} \cdot \frac{3}{4} = \frac{2 \cdot ■}{5 \cdot ■}$ $\frac{3}{4}$
2. $\frac{5}{6} \cdot 18 = \frac{5}{6} \cdot \frac{18}{■}$ 1
3. $2\frac{1}{4} \cdot 1\frac{2}{3} = \frac{■}{4} \cdot \frac{■}{3}$ 9 5

4. What product does the model at the left represent? $\frac{1}{4} \times \frac{5}{6}$
5. **Reasoning** When you multiply two fractions, do you need to find a common denominator first? Explain why or why not. **See margin.**
6. **Number Sense** Is $\frac{5}{7}$ of 28 *more* or *less* than 28? Explain how you know without actually multiplying.
Less; you are finding a fraction of 28.

Homework Exercises

For more exercises, see Extra Skills and Word Problems.

GO for Help

For Exercises	See Examples
7–14	1
15–23	2
24–31	3

A **Find each product. Write your result in simplest form. You may find a model helpful.**

7. $\frac{1}{2} \cdot \frac{2}{3}$ $\frac{1}{3}$
8. $\frac{1}{4} \cdot \frac{3}{5}$ $\frac{3}{20}$
9. $\frac{1}{3} \cdot \frac{5}{6}$ $\frac{5}{18}$
10. $\frac{9}{10} \cdot \frac{2}{3}$ $\frac{3}{5}$
11. $\frac{1}{6} \cdot \frac{3}{5}$ $\frac{1}{10}$
12. $\frac{1}{8} \cdot \frac{4}{5}$ $\frac{1}{10}$
13. $\frac{2}{3} \cdot \frac{2}{5}$ $\frac{4}{15}$
14. $\frac{2}{3} \cdot \frac{4}{9}$ $\frac{8}{27}$

Find each product.

15. $\frac{3}{4} \cdot 16$ 12
16. $9 \cdot \frac{1}{3}$ 3
17. $52 \cdot \frac{1}{2}$ 26
18. $\frac{2}{5} \cdot 10$ 4
19. $\frac{1}{4}$ of 12 3
20. $\frac{3}{5}$ of 15 9
21. $\frac{2}{3}$ of 18 12
22. $\frac{5}{8}$ of 64 40
23. To make a fruit punch drink, you use $\frac{1}{3}$ cup of water. You want to make 6 drinks. How many cups of water do you need? 2 c

Find each product. Write your answer as a mixed number.

24. $2\frac{2}{5} \cdot 3\frac{3}{8}$ $8\frac{1}{10}$
25. $5\frac{1}{4} \cdot 2\frac{2}{7}$ 12
26. $1\frac{3}{10} \cdot 6\frac{2}{3}$ $8\frac{2}{3}$
27. $1\frac{3}{8} \cdot 2\frac{2}{3}$ $3\frac{2}{3}$
28. $3\frac{1}{3} \cdot 1\frac{1}{4}$ $4\frac{1}{6}$
29. $2\frac{1}{2} \cdot 1\frac{3}{5}$ 4
30. $4\frac{2}{3} \cdot \frac{3}{4}$ $3\frac{1}{2}$
31. $\frac{1}{4} \cdot 3\frac{1}{5}$ $\frac{4}{5}$

B

32. **Guided Problem Solving** The largest hummingbird is the Giant Hummingbird of South America. It weighs $\frac{2}{3}$ oz. The smallest is the Bee Hummingbird of Cuba. Its weight is one tenth of the weight of a giant hummingbird. What is the total weight of both hummingbirds?
    - **Make a Plan** Find the weight of the smallest hummingbird. Then add its weight to the weight of the largest hummingbird. $\frac{11}{15}$ oz
    - **Carry Out the Plan** The weight of both hummingbirds is ■ oz.
33. **Baking** A recipe calls for $2\frac{3}{4}$ cups of flour. You want to triple the recipe and add another $\frac{1}{4}$ cup. How much flour will you need? $8\frac{1}{2}$ c

5. No; you do not need to find a common denominator, since you reduce the product anyway.

37. When multiplying, you do not need to find a common denominator.
Example: $\frac{1}{2} \cdot \frac{1}{3} = \frac{1 \cdot 1}{2 \cdot 3} = \frac{1}{6}$

Algebra **Find the value of x using mental math.**

34. $\frac{1}{4}x = \frac{1}{4}$ 1
35. $\frac{2}{3}x = 0$ 0
36. $2 + \frac{3}{5}x = 2$ 0

37. **Writing in Math** How does multiplying two fractions differ from adding two fractions? Give an example. **See margin.**

38. **Science** As a roller coaster car reaches the bottom of a slope and begins to go up the next slope, its acceleration, combined with the downward pull of gravity, can make you feel $3\frac{1}{2}$ times as heavy as you really are. This sensation is called "supergravity." How heavy would a 120-lb person experiencing supergravity feel? **420 lb**

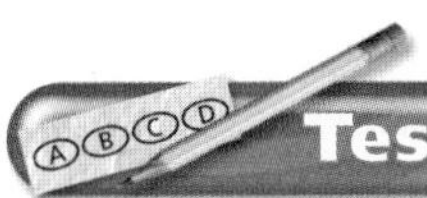

39. The length of a track around a field is $\frac{1}{4}$ mi. You jog $3\frac{1}{2}$ times around the track. How far do you jog? $\frac{7}{8}$ **mi**

40. **Error Analysis** A student multiplies two mixed numbers, $2\frac{3}{5}$ and $1\frac{1}{3}$, and finds the product to be $2\frac{3}{15}$. Explain the student's mistake. What is the correct answer? **See left.**

40. The student did not write mixed numbers as improper fractions before multiplying; $3\frac{7}{15}$.

41. Mark has 224 board games. He bought $\frac{3}{4}$ of his games on the Internet. Of the games he bought on the Internet, $\frac{1}{6}$ are card games. How many card games did he buy on the Internet? **28 games**

42. **Calculator** Use a calculator to find $10\frac{2}{5} \cdot 4\frac{1}{2}$. $46\frac{4}{5}$

43. **Challenge** A farmer built a fence around a rectangular lot. The lot's length is $2\frac{1}{2}$ mi, and its width is $\frac{4}{5}$ its length. What is the area of the lot? How long is the fence? 2 mi^2; $6\frac{3}{5}$ mi

5mi²

Test Prep and Mixed Review

Practice

Gridded Response

44. The Dancemania dancing studio sent $\frac{2}{3}$ of its dancers to a competition. If there are 36 dancers, how many dancers did the studio send to the competition? **24**

45. Damien designed a thin border for his rectangular painting. The perimeter of the painting is 22 ft. The width of the painting is 4.75 ft. What is the length of the painting, in feet? **6.25**

46. Helen received a gift certificate for her birthday. She spent $\frac{1}{3}$ of the gift certificate on lunch, and $\frac{1}{4}$ of the remaining amount on dessert. After she spent $3.50 for the tip, she had $4.00 left. What was the amount of Helen's gift certificate, in dollars? **15**

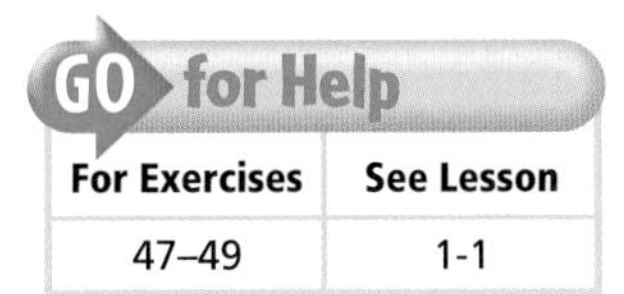

For Exercises	See Lesson
47–49	1-1

Estimate. Round to the nearest whole number before you calculate.

47. $19.5 + 56.13$ **76**
48. $34.3 - 18.9$ **15**
49. $26.7 \cdot 9.9$ **270**

4. Assess & Reteach

Lesson Quiz

Find each product.

1. $\frac{3}{5} \cdot \frac{10}{12}$ $\frac{1}{2}$
2. $\frac{5}{8}$ of 48 30
3. $4\frac{1}{2} \cdot 2\frac{2}{10}$ $9\frac{9}{10}$
4. $3\frac{1}{4} \cdot 5\frac{1}{3}$ $17\frac{1}{3}$

Alternative Assessment

Have students work in pairs. Each student writes several fractions and mixed numbers one on each small slip of paper. Partners then pair up the fractions and work together to multiply them. Have partners record their work.

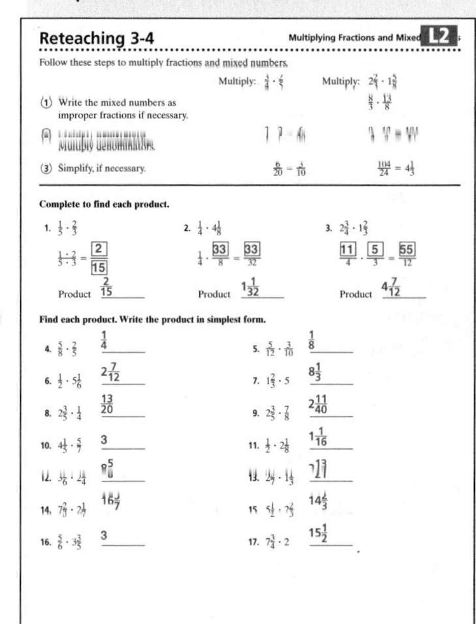

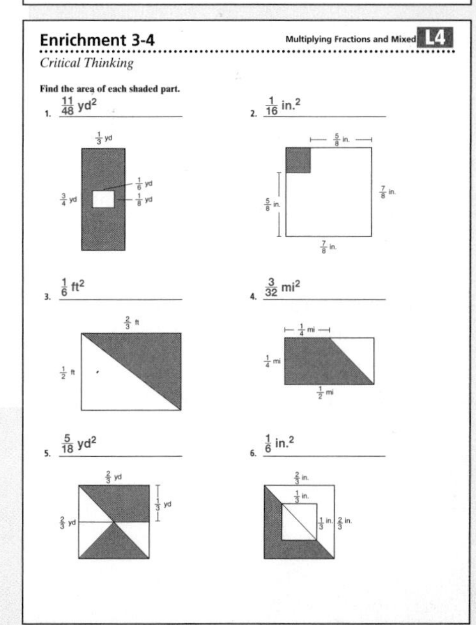

Test Prep

Resources

For additional practice with a variety of test item formats:

- Test-Taking Strategies, p. 159
- Test Prep, p. 163
- Test-Taking Strategies with Transparencies

Activity Lab

Modeling Fraction Division

Students use models to practice division of fractions. This will help them calculate with fractions in Lesson 3-5.

Guided Instruction

Explain that dividing fractions helps you figure how many pieces can be made from a mixed number amount. Have students give examples of times when this skill would be useful.

Ask:

- *Why must we divide each circle in half?* **Alfinio wants to eat $\frac{1}{2}$ pizza each meal.**
- *What is another way to say $2\frac{1}{4}$ pizzas?* **$\frac{9}{4}$ pizzas**

Exercises
Have students work independently on the Exercises. When they have finished, allow partners to share their work, discuss their answers, and how they arrived at the answers.

Alternative Method
Some students may prefer to continue to model the division problems by cutting out shapes and cutting them into pieces. Provide extra paper for these students.

Resources

- Activity Lab 3-5: Dividing Fractions
- paper
- scissors

3-5a Activity Lab — Hands On

Modeling Fraction Division

How many times does 2 go into 14? You know the answer is 7. When you divide with fractions, the answer may not be so obvious. You can use a model to help you find the answer.

EXAMPLE

Alfinio finds $2\frac{1}{4}$ pizzas left after his birthday party. If he decides to eat $\frac{1}{2}$ pizza per meal, how many meals does he have left?

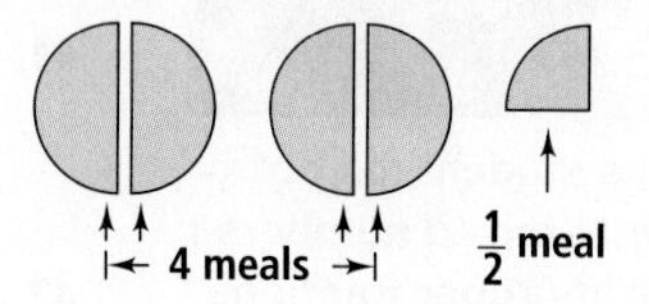

Step 1 Cut out 3 congruent circles. Then cut $\frac{1}{4}$ of one circle. You now have $2\frac{1}{4}$ circles that represent the leftover pizza.

Step 2 Cut your circles into meals for Alfinio. Remember that Alfinio wants to eat $\frac{1}{2}$ pizza per meal.

Step 3 How many $\frac{1}{2}$-pizza meals can Alfinio make from $2\frac{1}{4}$ pizzas?

4 meals — $\frac{1}{2}$ meal

Your model shows that he had $2\frac{1}{4}$ pizzas.

So there are 4 full meals plus $\frac{1}{2}$ meal left.

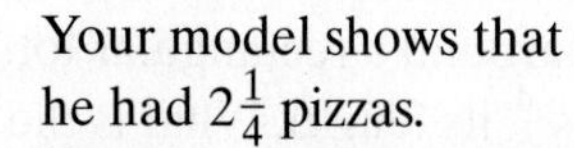
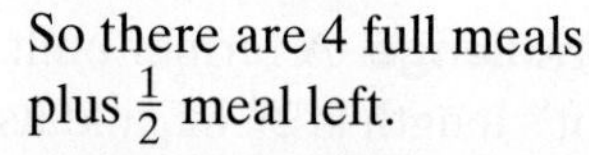
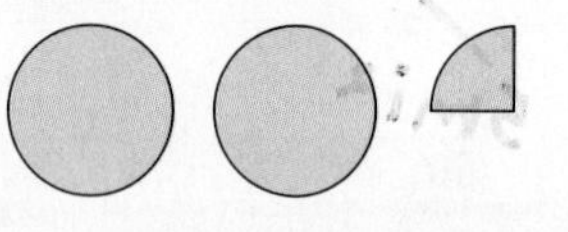
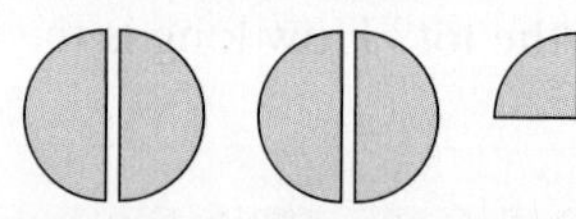

This shows that $2\frac{1}{4} \div \frac{1}{2} = 4\frac{1}{2}$.

Exercises

1. Identify the division problem shown in the model. Find the quotient. **$2\frac{1}{2} \div \frac{3}{4} = 3\frac{1}{3}$**

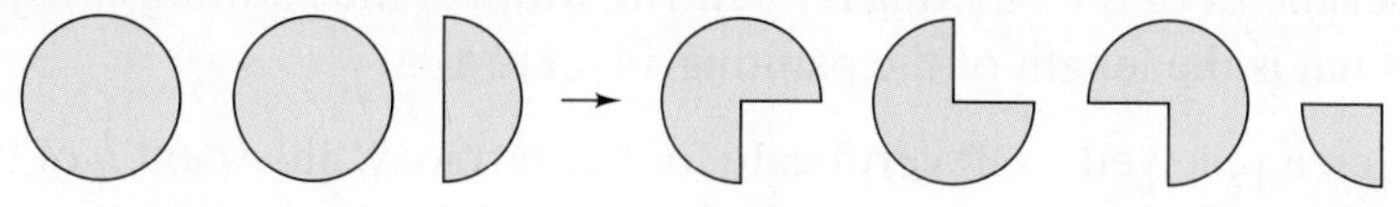

Make a model to represent each expression. Then find the quotient. **2–4. See margin for models.**

2. $1\frac{2}{3} \div \frac{1}{3}$ **5**
3. $1\frac{1}{2} \div \frac{1}{4}$ **6**
4. $3\frac{1}{2} \div \frac{3}{4}$ **$4\frac{2}{3}$**

5. a. In the example above, how does the number of meals relate to the number of pizzas? **Each pizza is two meals.**
   b. **Number Sense** Dividing by $\frac{1}{2}$ is the same as multiplying by what number? **2**

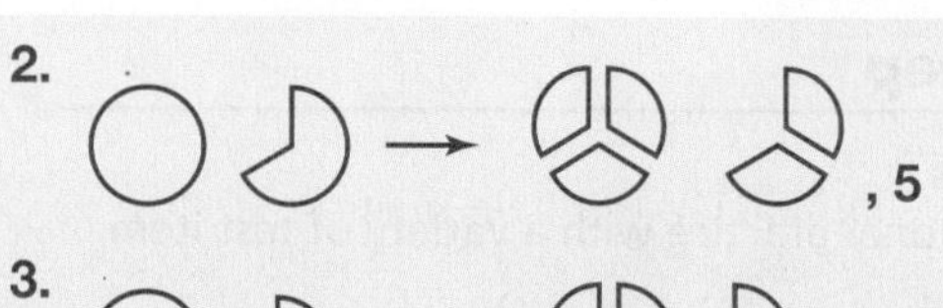

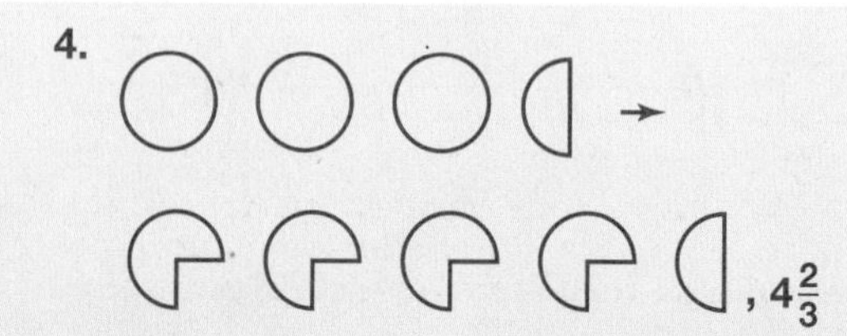

3-5 Dividing Fractions and Mixed Numbers

Check Skills You'll Need

1. **Vocabulary Review** Which operation do you use to find a *product*? multiplication

Find each product.

2. $\frac{3}{7} \cdot \frac{2}{3}$ $\frac{2}{7}$
3. $\frac{5}{9} \cdot \frac{1}{2}$ $\frac{5}{18}$
4. $50 \cdot \frac{9}{10}$ 45
5. $2\frac{2}{9} \cdot 1\frac{4}{5}$ 4

GO for Help Lesson 3-4

What You'll Learn

To divide fractions and mixed numbers and to solve problems by dividing

New Vocabulary reciprocals

Why Learn This?

When you work with measurements, you will often need to divide fractions and mixed numbers. The ruler below shows there are four $\frac{3}{4}$'s in 3 wholes. So $3 \div \frac{3}{4} = 4$. Notice that $3 \cdot \frac{4}{3}$ also equals 4. The numbers $\frac{3}{4}$ and $\frac{4}{3}$ are reciprocals. Two numbers are **reciprocals** if their product is 1. To divide by a fraction, you multiply by its reciprocal.

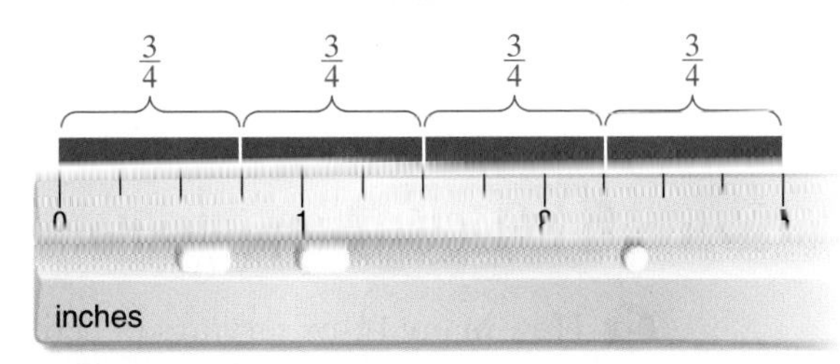

KEY CONCEPTS Dividing by Fractions

Arithmetic	Algebra
$3 \div \frac{3}{4} = 3 \cdot \frac{4}{3} = 4$	$\frac{a}{b} \div \frac{c}{d} = \frac{a}{b} \cdot \frac{d}{c}$ for b, c, and $d \neq 0$

EXAMPLE Dividing by a Fraction

1 Find $\frac{2}{3} \div \frac{5}{6}$.

$\frac{2}{3} \div \frac{5}{6} = \frac{2}{3} \cdot \frac{6}{5}$ ← Multiply by $\frac{6}{5}$, the reciprocal of $\frac{5}{6}$.

$= \frac{2 \cdot \cancel{6}^{2}}{{}_{1}\cancel{3} \cdot 5}$ ← Divide 6 and 3 by their GCF, 3.

$= \frac{4}{5}$ ← Simplify.

Video Tutor Help

Visit: PHSchool.com
Web Code: are-0775

Quick Check

1. a. Find $\frac{7}{8} \div \frac{1}{4}$. $3\frac{1}{2}$ b. Find $\frac{5}{8} \div \frac{3}{4}$. $\frac{5}{6}$ c. Find $14 \div \frac{7}{10}$. 20

3-5

1. Plan

Objective

To divide fractions and mixed numbers and to solve problems by dividing

Examples

1. Dividing by a Fraction
2. Dividing Mixed Numbers
3. Application: Meal Planning

Math Understandings: p. 118C

Math Background

Two numbers are *reciprocals* if their product is 1. For example, $\frac{2}{5}$ and $\frac{5}{2}$ are reciprocals, as are 8 and $\frac{1}{8}$. To divide by a fraction, change the fraction to its reciprocal and then multiply. When dividing mixed numbers, first change the mixed numbers to improper fractions.

More Math Background: p. 118C

Lesson Planning and Resources

See p. 118E for a list of the resources that support this lesson.

PowerPoint

Bell Ringer Practice

Check Skills You'll Need

Use student page, transparency, or PowerPoint. For intervention, direct students to:

Multiplying Fractions and Mixed Numbers
Lesson 3-4
Extra Skills and Word Problems Practice, Ch. 3

Differentiated Instruction Solutions for All Learners

Special Needs L1

Help students draw a picture for Example 2, and estimate the answer. Then ask them to work out the problem, assisting as they go through the steps. They can then compare their estimate to their calculated answer.

learning style: visual

Below Level L2

Give students fractions and have them write the reciprocal, such as these.

$\frac{2}{5}$ $\frac{5}{2}$ $\frac{3}{8}$ $\frac{8}{3}$ $\frac{1}{4}$ $\frac{4}{1}$ 6 $\frac{1}{6}$ $1\frac{1}{2}\left(=\frac{3}{2}\right)$ $\frac{2}{3}$ $1\frac{1}{3}$ $\frac{3}{4}$

$2\frac{1}{4}$ $\frac{4}{9}$ $2\frac{3}{4}$ $\frac{4}{11}$ $3\frac{5}{8}$ $\frac{8}{29}$

learning style: visual

2. Teach

Activity Lab

Use before the lesson.
Student Edition Activity Lab, Hands On 3-5a, Modeling Fraction Division, p. 140

All in One Teaching Resources
Activity Lab 3-5: Dividing Fractions

Guided Instruction

Example 1
Teach students this rhyme to help them remember how to divide fractions: *When dividing do not sigh, just invert and multiply.*

Example 2
Have students model this situation by cutting a piece of string $9\frac{1}{2}$ in. long into pieces $2\frac{3}{4}$ in. long.

Example 3
After discussing Example 3, invite students to explain alternative ways of solving the problem. For example, a student might sketch boxes for each $1\frac{1}{2}$ oz serving and keep a running total until reaching $19\frac{1}{2}$ oz.

PowerPoint
Additional Examples

1. Find $\frac{3}{10} \div \frac{2}{5}$. $\frac{3}{4}$
2. Find $4\frac{2}{3} \div 1\frac{3}{4}$. $2\frac{2}{3}$
3. You have a piece of yarn that is $12\frac{1}{2}$ ft long. You need to cut it into strips that are $1\frac{1}{4}$ ft long. How many strips will you have? 10

Online active math

For: Dividing Fractions Activity
Use: Interactive Textbook, 3-5

To divide mixed numbers, rewrite them as improper fractions.

EXAMPLE Dividing Mixed Numbers

2 Measurement You have a space of $9\frac{1}{2}$ in. on a poster for a row of photos. Each photo is $2\frac{3}{4}$ in. wide. How many photos can you fit?

To find how many photos you can fit, divide $9\frac{1}{2}$ by $2\frac{3}{4}$.

$9\frac{1}{2} \div 2\frac{3}{4} = \frac{19}{2} \div \frac{11}{4}$ ← **Write the mixed numbers as improper fractions.**

$= \frac{19}{2} \cdot \frac{4}{11}$ ← **Multiply by $\frac{4}{11}$, the reciprocal of $\frac{11}{4}$.**

$= \frac{19}{{}_1\cancel{2}} \cdot \frac{\cancel{4}^2}{11}$ ← **Divide 4 and 2 by their GCF, 2.**

$= \frac{38}{11}$ ← **Multiply.**

$= 3\frac{5}{11}$ ← **Write as a mixed number.**

You can fit 3 photos.

Quick Check

2. a. Find $5\frac{3}{4} \div 3\frac{2}{3}$. $1\frac{25}{44}$ b. Find $4\frac{1}{8} \div 5\frac{1}{2}$. $\frac{3}{4}$ c. Find $3\frac{9}{16} \div 3$. $1\frac{3}{16}$

EXAMPLE Application: Meal Planning

3 How many $1\frac{1}{2}$-oz servings of cereal are in the larger cereal box at the right?

To find how many $1\frac{1}{2}$-oz servings are in $19\frac{1}{2}$ oz, divide $19\frac{1}{2}$ by $1\frac{1}{2}$.

Estimate $19\frac{1}{2} \div 1\frac{1}{2} \approx 20 \div 2$, or 10

$19\frac{1}{2} \div 1\frac{1}{2} = \frac{39}{2} \div \frac{3}{2}$ ← **Write the mixed numbers as improper fractions.**

$= \frac{39}{2} \cdot \frac{2}{3}$ ← **Multiply by $\frac{2}{3}$, the reciprocal of $\frac{3}{2}$.**

$= \frac{{}^{13}\cancel{39} \cdot \cancel{2}^1}{{}_1\cancel{2} \cdot \cancel{3}_1}$ ← **Divide 39 and 3 by their GCF. Divide 2 by itself.**

$= \frac{13}{1} = 13$ ← **Simplify.**

There are thirteen $1\frac{1}{2}$-oz servings in the larger cereal box.

Check for Reasonableness 13 is close to 10. The answer is reasonable.

3. One can of iced tea holds 12 fl oz. A 2-liter bottle holds $67\frac{3}{5}$ fl oz. How many cans of iced tea will you need to fill a 2-liter bottle? $5\frac{19}{30}$ cans

Advanced Learners L4
Have students use different values for *a*, *b*, *c*, and *d* in this equation $\frac{a}{b} \div \frac{c}{d} = \frac{a}{b} \cdot \frac{d}{c}$ to help show why *a* may equal 0, but *b*, *c*, and *d* may not. **Division by 0 is not allowed; *a* is only a numerator.**

learning style: visual

English Language Learners ELL
To help students understand division of fractions, ask questions such as: *How many $\frac{3}{4}$-inch pieces cover 3 inches?* 4 *How many $2\frac{3}{4}$-inch sections will it take to cover $9\frac{1}{2}$ inches?* $3\frac{5}{11}$

learning style: visual

You want to put 4 lb of raisins in bags so that each bag contains $\frac{2}{3}$ lb. How many bags do you need?

More Than One Way

Kayla's Method

I can use number sense. $\frac{2}{3} + \frac{2}{3} + \frac{2}{3} = \frac{6}{3} = 2$, so three times $\frac{2}{3}$ lb is 2 lb. That means six times $\frac{2}{3}$ lb is 4 lb.

I need six bags.

Will's Method

I can divide 4 by $\frac{2}{3}$.

$4 \div \frac{2}{3} = \frac{4}{1} \div \frac{2}{3}$ ← **Write the whole number as a fraction.**

$= \frac{4}{1} \cdot \frac{3}{2}$ ← **Multiply by $\frac{3}{2}$, the reciprocal of $\frac{2}{3}$.**

$= \frac{{}^{2}\cancel{4} \cdot 3}{1 \cdot \cancel{2}_{1}}$ ← **Divide 4 and 2 by their GCF, 2.**

$= \frac{6}{1} = 6$ ← **Simplify.**

I need six bags.

Choose a Method

You want to bike 12 miles in $1\frac{1}{3}$ hours. What should your average speed be? Describe your method and explain why you chose it.
9; answers may vary. Sample: I used Kayla's method because it is quicker.

Check Your Understanding

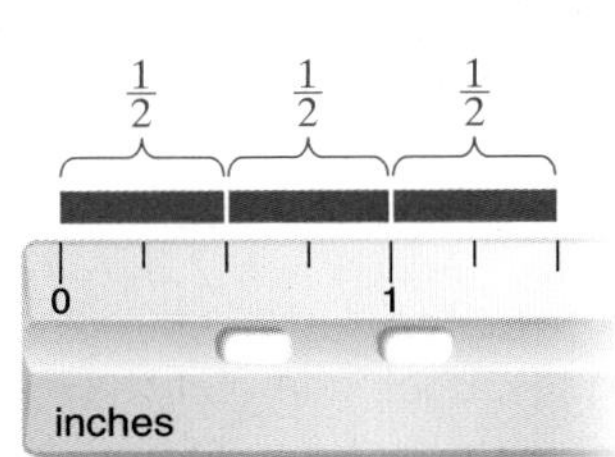

1. What quotient does the model at the left represent? **$1\frac{1}{2} \div \frac{1}{2}$**

Find the reciprocal of each number.

2. $\frac{5}{8}$ **$\frac{8}{5}$**
3. $\frac{1}{4}$ **4**
4. 9 **$\frac{1}{9}$**
5. $3\frac{1}{2}$ **$\frac{2}{7}$**

Mental Math Match each expression to the correct quotient.

6. $6 \div \frac{1}{2}$ **B**
7. $5 \div \frac{1}{5}$ **A**
8. $9 \div \frac{1}{10}$ **C**

A. 25
B. 12
C. 90

All in One Teaching Resources

- Daily Notetaking Guide 3-5 **L3**
- Adapted Notetaking 3-5 **L1**

Closure

- *How do you find the reciprocal of a fraction?* **Interchange the numerator and denominator.**
- *How do you divide by a fraction?* **Multiply by its reciprocal.**
- *What is the first step in dividing mixed numbers?* **Rewrite them as improper fractions.**

3. Practice

Assignment Guide

Check Your Understanding
Go over Exercises 1–8 in class before assigning the Homework Exercises.

Homework Exercises
A Practice by Example 9–33
B Apply Your Skills 34–47
C Challenge 48
Test Prep and Mixed Review 49–54

Homework Quick Check
To check students' understanding of key skills and concepts, go over Exercises 12, 29, 38, 44, and 45.

Differentiated Instruction Resources

Adapted Practice 3-5 L1

Practice 3-5 Dividing Fractions and Mixed Numbers L3

Find the reciprocal of each number.

1. $\frac{1}{2}$ **2**
2. $\frac{9}{16}$ $\frac{16}{9}$
3. $\frac{4}{5}$ $\frac{5}{4}$
4. $1\frac{1}{4}$ $\frac{4}{5}$
5. $2\frac{9}{10}$ $\frac{10}{29}$
6. $3\frac{1}{6}$ $\frac{6}{19}$

Find each quotient.

7. $\frac{3}{4} \div \frac{1}{4}$ **3**
8. $\frac{5}{6} \div \frac{1}{12}$ **10**
9. $\frac{1}{12} \div \frac{5}{6}$ $\frac{1}{10}$
10. $6 \div \frac{3}{4}$ **8**
11. $5 \div \frac{9}{10}$ $5\frac{5}{9}$
12. $\frac{4}{5} \div 2$ $\frac{2}{5}$
13. $\frac{7}{8} \div 3$ $\frac{7}{24}$
14. $\frac{4}{9} \div 8$ $\frac{1}{18}$
15. $1\frac{1}{2} \div \frac{2}{3}$ $2\frac{1}{4}$
16. $\frac{3}{4} \div 1\frac{1}{3}$ $\frac{9}{16}$
17. $2\frac{1}{2} \div 1\frac{1}{4}$ **2**
18. $1\frac{3}{4} \div \frac{3}{4}$ $2\frac{1}{3}$
19. $1\frac{7}{10} \div \frac{1}{2}$ $3\frac{2}{5}$
20. $4\frac{1}{2} \div 2\frac{1}{2}$ $1\frac{4}{5}$
21. $6 \div 3\frac{4}{5}$ $1\frac{11}{19}$
22. $4\frac{3}{4} \div \frac{7}{8}$ $5\frac{3}{7}$
23. $5\frac{5}{6} \div 1\frac{1}{3}$ $4\frac{3}{8}$
24. $3\frac{3}{8} \div 1\frac{1}{4}$ $2\frac{7}{10}$
25. $6\frac{1}{2} \div 1\frac{1}{2}$ $4\frac{1}{3}$
26. $2\frac{9}{10} \div 1\frac{3}{4}$ $1\frac{23}{35}$
27. $3\frac{1}{4} \div 1\frac{1}{3}$ $2\frac{7}{16}$

Solve each problem.

28. Rosa makes $2\frac{1}{2}$ c of pudding. How many $\frac{1}{3}$ c servings can she get from the pudding? **$7\frac{1}{2}$ servings**
29. One type of lightning bug glows once every $1\frac{1}{2}$ s. How many times can it glow in 1 min? **40 times**
30. Bea can run $\frac{1}{6}$ mi in 2 min. How long should it take her to run 2 mi? **24 min**
31. Joe drives 20 mi in $\frac{1}{2}$ h. How long will it take him to drive 50 mi? **$1\frac{1}{4}$ h**

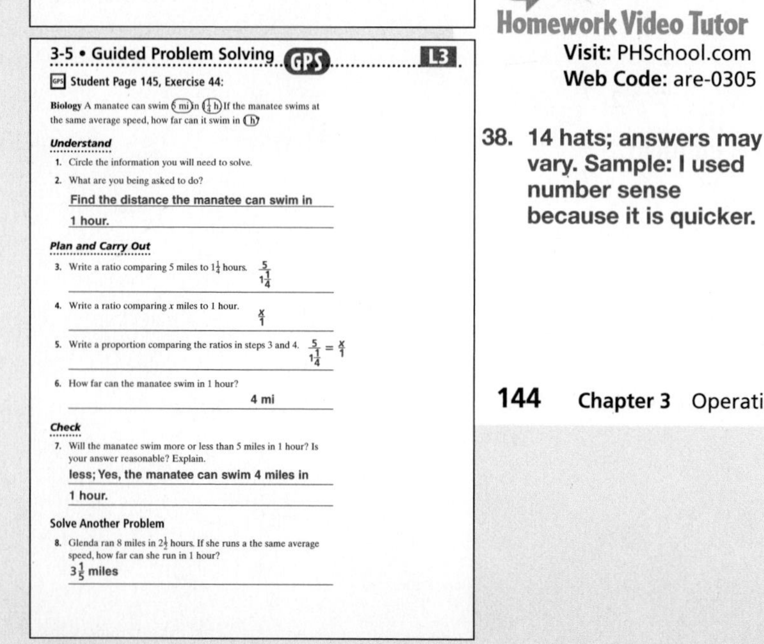
3-5 • Guided Problem Solving GPS L3

GPS Student Page 145, Exercise 44:

Biology A manatee can swim 5 mi in $1\frac{1}{4}$ h. If the manatee swims at the same average speed, how far can it swim in 1 h?

Understand

1. Circle the information you will need to solve.
2. What are you being asked to do?
Find the distance the manatee can swim in 1 hour.

Plan and Carry Out

3. Write a ratio comparing 5 miles to $1\frac{1}{4}$ hours. $\frac{5}{1\frac{1}{4}}$
4. Write a ratio comparing x miles to 1 hour. $\frac{x}{1}$
5. Write a proportion comparing the ratios in steps 3 and 4. $\frac{5}{1\frac{1}{4}} = \frac{x}{1}$
6. How far can the manatee swim in 1 hour? **4 mi**

Check

7. Will the manatee swim more or less than 5 miles in 1 hour? Is your answer reasonable? Explain.
less; Yes, the manatee can swim 4 miles in 1 hour.

Solve Another Problem

8. Glenda ran 8 miles in $2\frac{1}{2}$ hours. If she runs a the same average speed, how far can she run in 1 hour?
$3\frac{1}{5}$ miles

Homework Exercises

For more exercises, see Extra Skills and Word Problems.

GO for Help

For Exercises	See Examples
9–20	1
21–33	2–3

A **Find each quotient. You may find a model helpful.**

9. $\frac{4}{5} \div \frac{1}{10}$ **8**
10. $\frac{4}{5} \div \frac{1}{5}$ **4**
11. $\frac{5}{16} \div \frac{1}{2}$ $\frac{5}{8}$
12. $\frac{7}{10} \div \frac{1}{5}$ $3\frac{1}{2}$
13. $\frac{2}{7} \div \frac{1}{9}$ $2\frac{4}{7}$
14. $\frac{1}{9} \div \frac{5}{6}$ $\frac{2}{15}$
15. $\frac{3}{4} \div \frac{2}{3}$ $1\frac{1}{8}$
16. $\frac{1}{4} \div \frac{3}{8}$ $\frac{2}{3}$
17. $\frac{3}{8} \div \frac{1}{4}$ $1\frac{1}{2}$
18. $\frac{1}{2} \div \frac{2}{7}$ $1\frac{3}{4}$
19. $\frac{3}{5} \div \frac{2}{3}$ $\frac{9}{10}$
20. $\frac{9}{10} \div \frac{1}{5}$ $4\frac{1}{2}$

Find each quotient. Write your answer as a mixed number.

21. $1\frac{1}{2} \div 3\frac{1}{4}$ $\frac{6}{13}$
22. $2\frac{1}{6} \div 3\frac{5}{6}$ $\frac{13}{23}$
23. $5\frac{1}{4} \div 3\frac{1}{2}$ $1\frac{1}{2}$
24. $5\frac{1}{3} \div 4\frac{2}{3}$ $1\frac{1}{7}$
25. $2\frac{2}{3} \div 5\frac{1}{9}$ $\frac{12}{23}$
26. $6\frac{2}{3} \div 1\frac{1}{4}$ $5\frac{1}{3}$
27. $3\frac{3}{4} \div 4\frac{1}{2}$ $\frac{5}{6}$
28. $11\frac{1}{2} \div 3\frac{1}{2}$ $3\frac{2}{7}$
29. $12\frac{3}{4} \div 4\frac{1}{2}$ $2\frac{5}{6}$
30. $11\frac{1}{3} \div 4\frac{1}{2}$ $2\frac{14}{27}$
31. $8\frac{1}{3} \div 2\frac{1}{2}$ $3\frac{1}{3}$
32. $1\frac{5}{9} \div 2\frac{1}{3}$ $\frac{2}{3}$

33. You can jog $\frac{1}{4}$ mi in 2 min. How long will it take you to jog 2 mi? **16 min**

B GPS

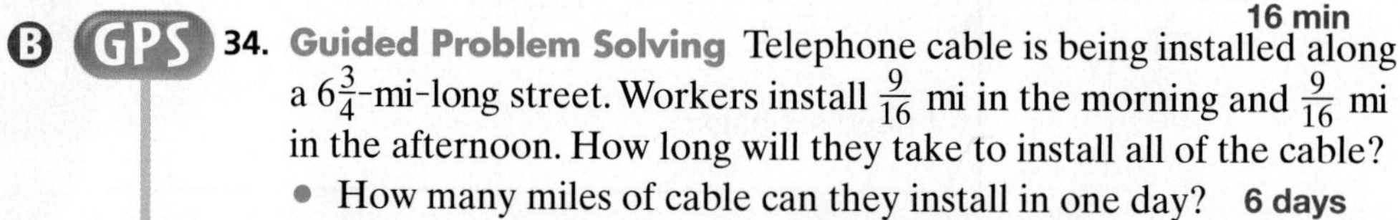
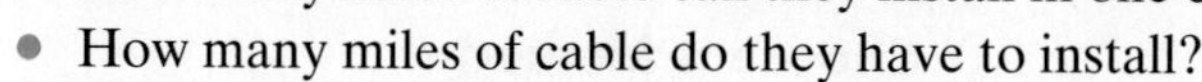

34. **Guided Problem Solving** Telephone cable is being installed along a $6\frac{3}{4}$-mi-long street. Workers install $\frac{9}{16}$ mi in the morning and $\frac{9}{16}$ mi in the afternoon. How long will they take to install all of the cable?
- How many miles of cable can they install in one day? **6 days**
- How many miles of cable do they have to install?
- What should you divide by to find how long the project will take?

35. **Error Analysis** Your friend wrote the reciprocal of $5\frac{3}{8}$ as $5\frac{8}{3}$. Explain why your friend is incorrect. What is the correct reciprocal?

35. Your friend found the reciprocal of $\frac{3}{8}$ without changing $5\frac{3}{8}$ to an improper fraction; $\frac{8}{43}$.

36. Estimate $8\frac{2}{5} \div 3\frac{1}{2}$. Compare your estimate to the exact quotient. **2; $2\frac{2}{5}$**

37. A nail weighs about $\frac{1}{10}$ oz. How many nails are in a 5-lb box? **800 nails**

38. **Choose a Method** Ranchers in the American West often wear what is called a ten-gallon hat. However, a ten-gallon hat actually holds only $\frac{3}{4}$ gallon. How many ten-gallon hats would be needed to hold 10 gallons? Describe your method and explain why you chose it. **See left.**

38. 14 hats; answers may vary. Sample: I used number sense because it is quicker.

GO Online
Homework Video Tutor
Visit: PHSchool.com
Web Code: are-0305

39. Joanne has $13\frac{1}{2}$ yd of material to make costumes. Each complete costume requires $1\frac{1}{2}$ yd for the top and $\frac{3}{4}$ yd for the bottom. How many complete costumes can she make? **6 costumes**

40. **Reasoning** Can you use the rule for multiplying by the reciprocal to divide whole numbers such as $10 \div 2$? Explain.
yes; $10 \div 2 = 10 \cdot \frac{1}{2} = 5$

Algebra **Use number sense to find the value of x.**

41. $\frac{1}{4} \div x = \frac{1}{4}$ **1**

42. $x \div \frac{33}{100} = 0$ **0**

43. $x \div \frac{2}{5} = 1$ $\frac{2}{5}$

Careers Marine biologists collect and analyze data on marine life.

44. **Biology** A manatee can swim 5 mi in $1\frac{1}{4}$ h. If the manatee swims at the same average speed, how far can it swim in 1 h? **4 mi**

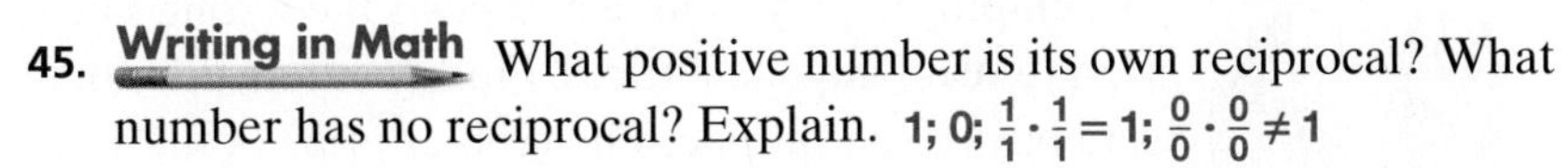

45. **Writing in Math** What positive number is its own reciprocal? What number has no reciprocal? Explain. **1; 0; $\frac{1}{1} \cdot \frac{1}{1} = 1$; $\frac{0}{0} \cdot \frac{0}{0} \neq 1$**

46. A one-serving recipe calls for $\frac{1}{3}$ c of vegetable oil for a marinade and $\frac{1}{6}$ c of oil for the sauce. You have $3\frac{1}{2}$ c of vegetable oil. How many servings can you make? **7 servings**

47. The area of a rectangle is 117 ft^2. The length of the rectangle is $9\frac{3}{4}$ ft. What is the width of the rectangle? **12 ft**

48. **Challenge** A rectangular floor measures 9 ft by 6 ft. You use $1\frac{1}{2}$-ft-square tiles to cover the floor. How many tiles will you need? **24 tiles**

Test Prep and Mixed Review — Practice

Multiple Choice

49. Which model best represents the expression $4 \div \frac{2}{3}$? **B**

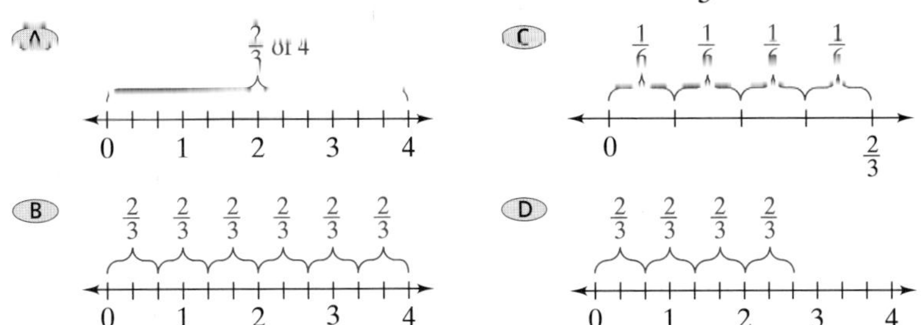

50. Eliza's grades on her daily math homework were 8, 10, 7, 9, 9, 4, 8, 2, 10, 8, 3, 7, 9, 8, 6, and 4. Which measure of the data is NOT represented by a score of 8? **F**

    F. Mean G. Median H. Mode J. Range

51. At an animal adoption center, $\frac{3}{8}$ of the animals are dogs, and $\frac{5}{16}$ are cats. How would you find the fraction of animals that are neither dogs nor cats? **D**

    A. Add the fraction of dogs and the fraction of cats.
    B. Subtract the fraction of cats from the fraction of dogs.
    C. Subtract the fraction of dogs from 1 and add the fraction of cats.
    D. Add both fractions and then subtract the result from 1.

GO for Help

For Exercises	See Lesson
52–54	2-8

Write in standard form.

52. 5.5×10^4 **55,000**

53. 2.564×10^3 **2,564**

54. 5.6302×10^1 **56.302**

4. Assess & Reteach

Lesson Quiz

Find each quotient.

1. $\frac{1}{8} \div \frac{3}{4}$ $\frac{1}{6}$
2. $\frac{2}{3} \div \frac{3}{8}$ $1\frac{7}{9}$
3. $1\frac{5}{6} \div 2\frac{2}{5}$ $\frac{55}{72}$
4. $4\frac{2}{7} \div \frac{1}{5}$ $21\frac{3}{7}$

Alternative Assessment

Make sure students have customary system rulers. Ask questions such as:

How many $\frac{1}{2}$-in. units are in $2\frac{1}{2}$ in.? 5

How many $\frac{3}{4}$-in. units are in 3 in.? 4

How many $1\frac{1}{2}$-in. units are in 9 in.? 6

Have them draw line segments of the given lengths, marking off each corresponding fractional unit. Then have students write the steps involved in performing the fractional divisions represented by each model.

Test Prep

Resources

For additional practice with a variety of test item formats:

- Test-Taking Strategies, p. 159
- Test Prep, p. 163
- Test-Taking Strategies with Transparencies

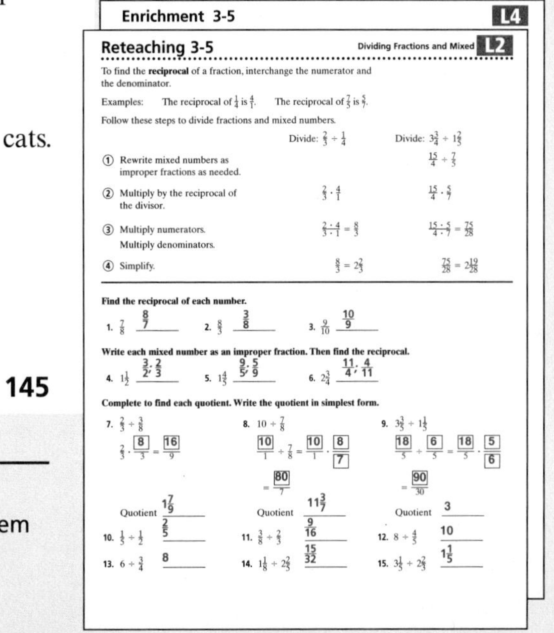

Enrichment 3-5 L4

Reteaching 3-5 Dividing Fractions and Mixed L2

To find the **reciprocal** of a fraction, interchange the numerator and the denominator.

Examples: The reciprocal of $\frac{1}{4}$ is $\frac{4}{1}$. The reciprocal of $\frac{7}{5}$ is $\frac{5}{7}$.

Follow these steps to divide fractions and mixed numbers.

	Divide: $\frac{2}{3} \div \frac{1}{4}$	Divide: $3\frac{3}{4} \div 1\frac{2}{5}$
① Rewrite mixed numbers as improper fractions as needed.		$\frac{15}{4} \div \frac{7}{5}$
② Multiply by the reciprocal of the divisor.	$\frac{2}{3} \cdot \frac{4}{1}$	$\frac{15}{4} \cdot \frac{5}{7}$
③ Multiply numerators. Multiply denominators.	$\frac{2 \cdot 4}{3 \cdot 1} = \frac{8}{3}$	$\frac{15 \cdot 5}{4 \cdot 7} = \frac{75}{28}$
④ Simplify.	$\frac{8}{3} = 2\frac{2}{3}$	$\frac{75}{28} = 2\frac{19}{28}$

Find the reciprocal of each number.

1. $\frac{7}{8}$ $\frac{8}{7}$ 2. $\frac{8}{3}$ $\frac{3}{8}$ 3. $\frac{9}{10}$ $\frac{10}{9}$

Write each mixed number as an improper fraction. Then find the reciprocal.

4. $1\frac{1}{2}$ $\frac{3}{2}, \frac{2}{3}$ 5. $1\frac{4}{5}$ $\frac{9}{5}, \frac{5}{9}$ 6. $2\frac{3}{4}$ $\frac{11}{4}, \frac{4}{11}$

Complete to find each quotient. Write the quotient in simplest form.

7. $\frac{2}{3} \div \frac{3}{8}$

   $\frac{2}{3} \cdot \frac{8}{3} = \frac{16}{9}$

   Quotient $1\frac{7}{9}$

8. $10 \div \frac{7}{8}$

   $\frac{10}{1} \div \frac{7}{8} = \frac{10}{1} \cdot \frac{8}{7}$

   $= \frac{80}{7}$

   Quotient $11\frac{3}{7}$

9. $3\frac{3}{5} \div 1\frac{1}{5}$

   $\frac{18}{5} \div \frac{6}{5} = \frac{18}{5} \cdot \frac{5}{6}$

   $= \frac{90}{30}$

   Quotient 3

10. $\frac{1}{5} \div \frac{1}{2}$ $\frac{2}{5}$ 11. $\frac{3}{8} \div \frac{2}{3}$ $\frac{9}{16}$ 12. $8 \div \frac{4}{5}$ 10

13. $6 \div \frac{3}{4}$ 8 14. $1\frac{1}{8} \div 2\frac{2}{5}$ $\frac{15}{32}$ 15. $3\frac{1}{5} \div 2\frac{2}{3}$ $1\frac{1}{5}$

Practice Solving Problems

In this feature, students practice solving problems involving multiplication with fractions and mixed numbers. This will help them see how the skills they learned in this chapter apply to everyday situations.

Guided Instruction

Explain to students that sales in stores often offer discounts of a fraction of the price of an item. Tell them they can use multiplication with fractions to determine the sale price of the item.
Ask:

- *Why is the joke sign misleading?* $\frac{3}{2}$ **of the original price is more than the original price so there is no sale.**
- *Why do we add the $21 after calculating the price Freddy paid for the suit?* **Tax is based on the sales price of the suit.**
- *What would be a good deal on the suit?* **Sample:** $\frac{1}{4}$ **or** $\frac{1}{2}$ **off the original price**

Practice Solving Problems

An employee plays a joke and posts the wrong sale sign to see if anyone notices. Freddy enters the store and buys a suit on sale. The regular price of the suit is $200. What does he pay for the suit after $21 for sales tax is added?

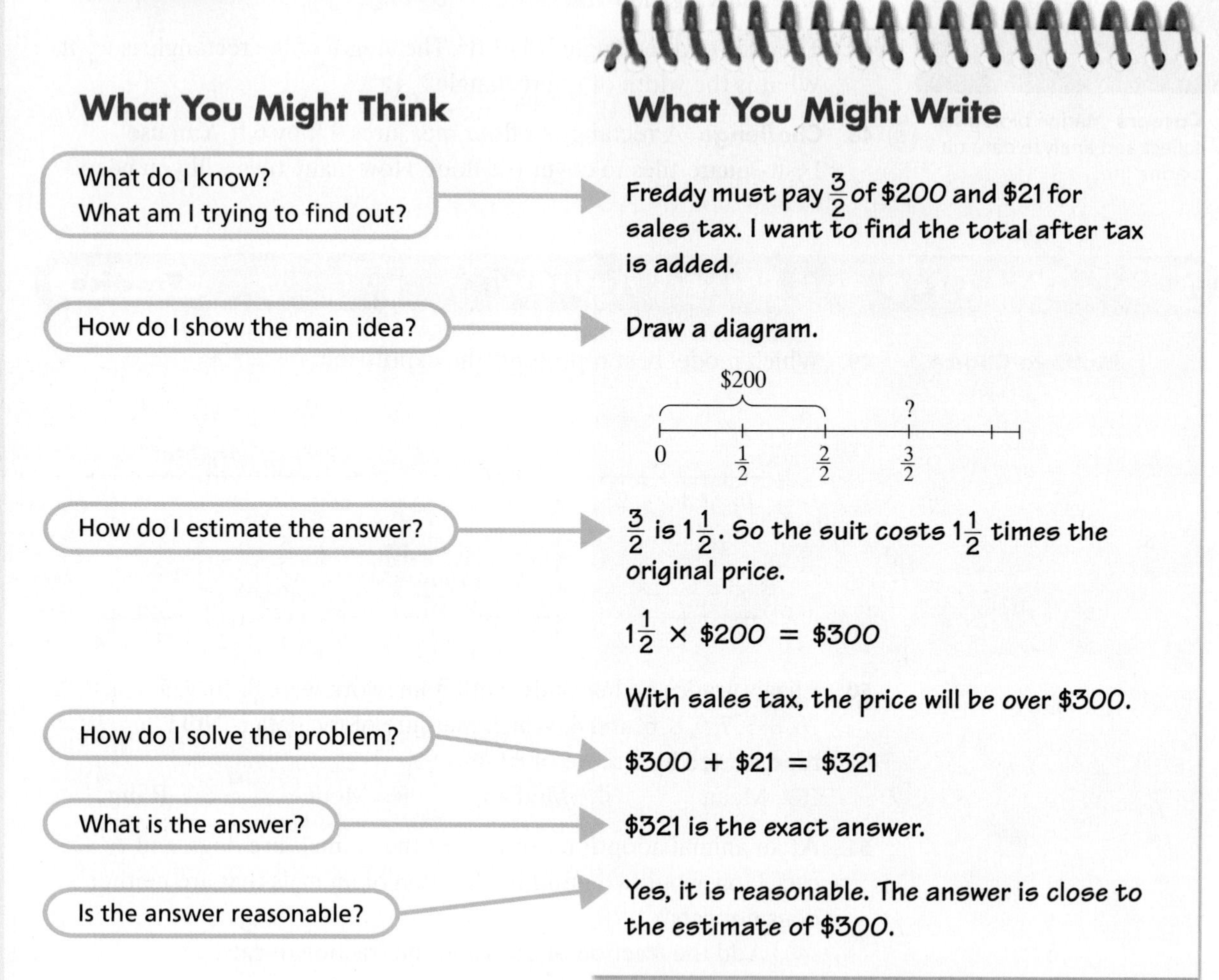

What You Might Think
What You Might Write
What do I know?
What am I trying to find out?
Freddy must pay 3/2 of $200 and $21 for sales tax. I want to find the total after tax is added.
How do I show the main idea?
Draw a diagram.
$200
?
0
1/2
2/2
3/2
How do I estimate the answer?
3/2 is 1 1/2. So the suit costs 1 1/2 times the original price.
1 1/2 × $200 = $300
With sales tax, the price will be over $300.
How do I solve the problem?
$300 + $21 = $321
What is the answer?
$321 is the exact answer.
Is the answer reasonable?
Yes, it is reasonable. The answer is close to the estimate of $300.

Think It Through

1a. The $\frac{2}{2}$ **mark represents all of the $200 before markup and tax.**

1. **a.** In the diagram, what does it mean for $200 to be equivalent to $\frac{2}{2}$?
   b. What does the $\frac{3}{2}$ mark represent? Explain. **The** $\frac{3}{2}$ **mark represents the cost including the markup.**
2. How do you know that $1\frac{1}{2} \times \$200 = \$300$?
   See margin.

2. $1\frac{1}{2} = \frac{3}{2}$, $\frac{3}{2} \times \$200 = \$300$

Exercises

For Exercises 3 and 4, answer the questions first and then solve the problem.

3. Your friend bought a shirt at the $\frac{3}{2}$-price sale. The regular price for the shirt was $60, and $4.20 for sales tax was added. If your friend paid with a $100 bill, how much change did he get? **$5.80**
   a. What do you know?
   b. What do you want to find out?
   c. Finish drawing the diagram below to help you find the answer.

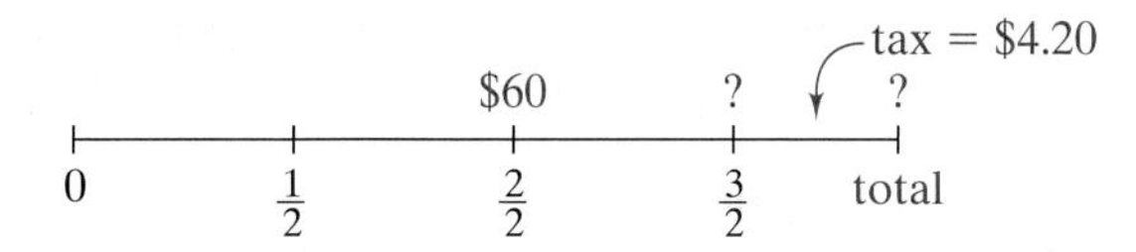

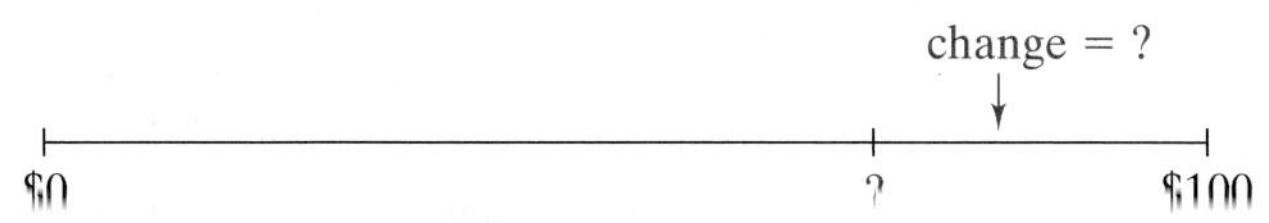

   d. Estimate the answer.

4. Harriet bought her father a tie at the $\frac{3}{2}$-price sale. The sale price was $24, not including tax. What was the regular price? **$16**
   a. What do you know?
   b. What do you want to find out?
   c. Make a drawing to show the main idea.
   d. Estimate the answer.

For Exercises 5–7, use the sale sign shown below.

SALE!	
Belts : $\frac{1}{3}$ off	Hats : $\frac{1}{4}$ off
Jewelry : $\frac{1}{2}$ off	Dresses : $\frac{1}{5}$ off

5. Mary buys a hat at the sale. The hat's regular price is $24. She will pay $1.26 for sales tax. How much change should she get from a $20 bill? **$.74**

6. Juanita buys a dress at the sale. The regular price of the dress is $45. She will pay $2.52 for sales tax. What will be the total cost? **$38.52**

7. Xavier buys a belt at the sale. The cost was $16 before tax was added. What was the regular price of the belt? **$24**

Error Prevention!

Review converting mixed numbers into improper fractions to help students understand that $\frac{3}{2}$ of the price is $1\frac{1}{2}$ times the original price.

Exercises

Have students work independently on the Exercises. Discuss the answers as a class and help students correct any mistakes they have made.

Differentiated Instruction

Advanced Learners L4

Have students create their own problems involving sales tax and sale prices. They can trade papers with a partner and solve each other's problems. Some students may want to write "joke" problems.

3-6

1. Plan

Objective
To change units of length, capacity, and weight in the customary system

Examples
1 Changing Units of Length
2 Changing Units of Capacity
3 Changing Units of Weight

Math Understandings: p. 118D

Math Background

The units of measure commonly used in the United States are known as the Customary System. Students should know how to convert between units of measure. To change from a larger unit to a smaller unit, such as pounds to ounces, multiply. To change from a smaller unit to a larger unit, such as feet to miles, divide.

More Math Background: p. 118D

Lesson Planning and Resources

See p. 118E for a list of the resources that support this lesson.

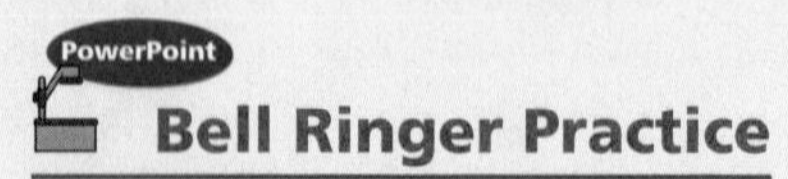

Check Skills You'll Need
Use student page, transparency, or PowerPoint. For intervention, direct students to:
Adding and Subtracting Mixed Numbers
Lesson 3-3
Extra Skills and Word Problems Practice, Ch. 3

3-6 Changing Units in the Customary System

Check Skills You'll Need

1. **Vocabulary Review** Why is $1\frac{3}{8}$ called a *mixed number?* **See below.**

Find each sum or difference.

2. $\frac{2}{3} + \frac{5}{12}$ $1\frac{1}{12}$
3. $\frac{15}{16} - \frac{5}{16}$ $\frac{5}{8}$
4. $5\frac{1}{2} + 6\frac{3}{4}$ $12\frac{1}{4}$
5. $8 - 3\frac{3}{4}$ $4\frac{1}{4}$

Lesson 3-3

Check Skills You'll Need

1. **It consists of a whole number and a fraction.**

What You'll Learn

To change units of length, capacity, and weight in the customary system

Why Learn This?

Most people in the United States use the customary system of measurement for length, capacity, and weight. Changing units allows you to compare measures and compute with them, as carpenters do.

Customary Units of Measure

Type	Length	Capacity	Weight
Unit	inch (in.) foot (ft) yard (yd) mile (mi)	fluid ounce (fl oz) cup (c) pint (pt) quart (qt) gallon (gal)	ounce (oz) pound (lb) ton (t)
Equivalents	1 ft = 12 in. 1 yd = 3 ft 1 mi = 5,280 ft	1 c = 8 fl oz 1 pt = 2 c 1 qt = 2 pt 1 gal = 4 qt	1 lb = 16 oz 1 t = 2,000 lb

Vocabulary Tip

The customary system is sometimes called the "English system," even though it is no longer used in England.

EXAMPLE Changing Units of Length

1 **Carpentry** A carpenter has a board 10 ft long. He cuts a piece 5 ft 3 in. long from the board. What is the length in feet of the remaining piece?

You need to subtract 5 ft 3 in. from 10 ft.

$5 \text{ ft } 3 \text{ in.} = 5\frac{3}{12} \text{ ft} = 5\frac{1}{4} \text{ ft}$ ← **Write 3 in. as a fraction of a foot.**

$10 - 5\frac{1}{4} = 9\frac{4}{4} - 5\frac{1}{4}$ ← **Rename 10 as $9\frac{4}{4}$.**

$= 4\frac{3}{4}$ ← **Subtract.**

The remaining piece is $4\frac{3}{4}$ ft long.

Quick Check

1. How much shorter than a board 10 ft long is a board 8 ft 5 in. long? $1\frac{7}{12}$ ft

Differentiated Instruction Solutions for All Learners

Special Needs L1
Help students see the difference between one inch and one foot by tracing those lengths with their fingers. Help them understand weight differences by holding a 1-pound weight and a 1-ounce weight. This will help them better understand why the units are unequal.
learning style: tactile

Below Level L2
Have students identify measurement abbreviations and write the conversion relationship.

in. and ft **12 inches = 1 foot**
c and fl oz **1 cup = 8 fluid ounces**
lb and oz **1 pound = 16 ounces**
learning style: verbal

To change from a smaller unit to a larger unit, you *divide*. To change from a larger unit to a smaller unit, you *multiply*.

EXAMPLE Changing Units of Capacity

2 Multiple Choice Jonah has 30 fluid ounces of juice. How many $1\frac{1}{2}$-cup servings does he have altogether?

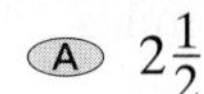

Ⓐ $2\frac{1}{2}$ Ⓑ 18 Ⓒ 20 Ⓓ $28\frac{1}{2}$

First you need to find the number of fluid ounces in $1\frac{1}{2}$ cups. Since there are 8 fluid ounces in 1 cup, you know that there are 12 fluid ounces in $1\frac{1}{2}$ cups.

Next you need to find the number of groups of 12 fluid ounces that are in 30 fluid ounces. You need to divide 30 by 12.

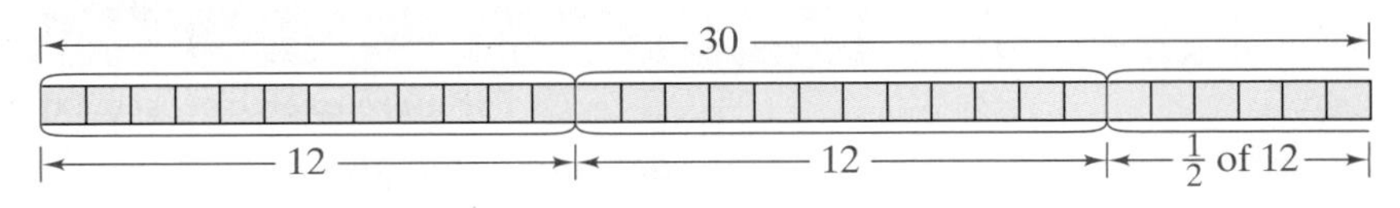

$30 \div 12 = 2.5 = 2\frac{1}{2}$ ← **Write as a mixed number.**

There are $2\frac{1}{2}$ servings in 30 fl oz of juice. The correct answer is choice A.

Test Prep Tip

Some questions require two steps to find the answer. Read carefully, and tackle one step at a time.

Quick Check

2. How many 1-cup servings are in 50 fluid ounces of juice? **$6\frac{1}{4}$ c**

EXAMPLE Changing Units of Weight

3 Cycling The lighter the frame of a mountain bike, the easier it is to ride. In the ad shown at the left, Mountain Master weighs 76 oz and Super Cycle weighs $4\frac{1}{4}$ lb. Which bike will be easier to ride?

Think of the relationship between pounds and ounces. A pound is a larger unit of measure than an ounce. There are 16 ounces in one pound.

1 lb = 16 oz (× 16)

To change $4\frac{1}{4}$ lb to ounces, multiply $4\frac{1}{4}$ by 16.

$4\frac{1}{4} \cdot 16 = \frac{17}{4} \cdot \frac{16}{1}$ ← **Write $4\frac{1}{4}$ as an improper fraction.**

$= 68$ ← **Multiply.**

The Super Cycle will be easier to ride because it is lighter.

Quick Check

3. Find the number of ounces in $4\frac{5}{8}$ lb. **74 oz**

2. Teach

Activity Lab

Use before the lesson.

All in One Teaching Resources

Activity Lab 3-6: Patterns in Measurement

Guided Instruction

Example 2

Display different-sized empty bottles and cartons to help students get a better sense of the units used to measure capacity.

Example 3

Some students prefer multiplying by a fraction of equivalent measures. For instance, to convert 40 ounces to pounds, you could set up this multiplication:

$\frac{40 \text{ oz}}{1} \cdot \frac{1 \text{ lb}}{16 \text{ oz}} = \frac{40}{16} \text{ lb} = \frac{5}{2} \text{ lb} = 2\frac{1}{2} \text{ lb}$

Multiplying by the fraction $\frac{1 \text{ lb}}{16 \text{ oz}}$ is the same as multiplying by 1 since the values in the numerator and the denominator are equal.

PowerPoint

Additional Examples

1. A carpenter cuts a 4 ft 10 in. piece from an 8 ft board. What is the length in feet of the remaining piece? **$3\frac{1}{6}$ ft**

2. How many one-cup servings are in a 30-fluid-ounce sports drink bottle? **$3\frac{3}{4}$ one-cup servings**

3. Which weighs more—a $1\frac{3}{4}$-lb book or a 24 oz catalog? **The book that weighs 28 oz**

All in One Teaching Resources

- Daily Notetaking Guide 3-6 **L3**
- Adapted Notetaking 3-6 **L1**

Closure

- *How do you change from a smaller unit to a larger unit?* **divide**
- *How do you change from a larger unit to a smaller unit?* **multiply**

Advanced Learners L4

Challenge students to find five different ways to express their height. **Sample: 5 ft 1 in.; $5\frac{1}{12}$ ft; 61 in.; $1\frac{25}{36}$ yd; about 0.0009627 mi**

learning style: visual

English Language Learners ELL

For Example 1, ask step-by-step questions. *Why is it a good idea to write 5 feet 3 inches as a mixed number?* **easier to subtract** *How do you know 3 inches is $\frac{3}{12}$ of a foot?* **There are 12 inches in one foot.** *Why did you rename 10 as $9\frac{4}{4}$?* **to subtract fourths**

learning style: verbal

3. Practice

Assignment Guide

Check Your Understanding

Go over Exercises 1–8 in class before assigning the Homework Exercises.

Homework Exercises

A	Practice by Example	9–29
B	Apply Your Skills	30–43
C	Challenge	44
Test Prep and Mixed Review		45–49

Homework Quick Check

To check students' understanding of key skills and concepts, go over Exercises 13, 21, 38, 42, and 43.

Differentiated Instruction Resources

Adapted Practice 3-6 L1

Practice 3-6 Changing Units in the Customa[ry System] L3

Tell whether you would multiply or divide to change from one unit of measure to the other.

1. tons to pounds multiply
2. pints to quarts divide
3. feet to yards divide
4. gallons to pints multiply
5. cups to quarts divide
6. pounds to ounces multiply

Change each unit of length, capacity, or weight.

7. 9 qt = $2\frac{1}{4}$ gal
8. $2\frac{1}{4}$ t = 4,500 lb
9. $3\frac{1}{2}$ yd = 126 in.
10. 4 yd = 12 ft
11. 60 c = 15 qt
12. 246 in. = $20\frac{1}{2}$ ft
13. 1,750 oz = $109\frac{3}{8}$ lb
14. 84 ft = 28 yd
15. 198 in. = $5\frac{1}{2}$ yd
16. 480 fl oz = 30 pt
17. $\frac{1}{4}$ gal = 32 fl oz
18. $\frac{1}{2}$ mi = 2,640 ft
19. $\frac{1}{10}$ mi = 6,336 in.
20. 2 lb 6 oz = $2\frac{3}{8}$ lb
21. 2 qt 8 fl oz = $2\frac{1}{4}$ qt

Solve.

22. United States farms produced 2,460,000,000 bushels of soybeans in 1994. How many quarts is this? (A bushel is 32 quarts.) 78,720,000,000 quarts
23. In 1994, Brian Berg built an 81-story "house" using playing cards. The house was $15\frac{2}{3}$ ft tall. How many inches is this? 188 in.

Choose an appropriate customary unit of measure.

24. capacity of a mug fluid ounces
25. length of a family room feet
26. distance between two capital cities miles
27. capacity of a shampoo bottle fluid ounces

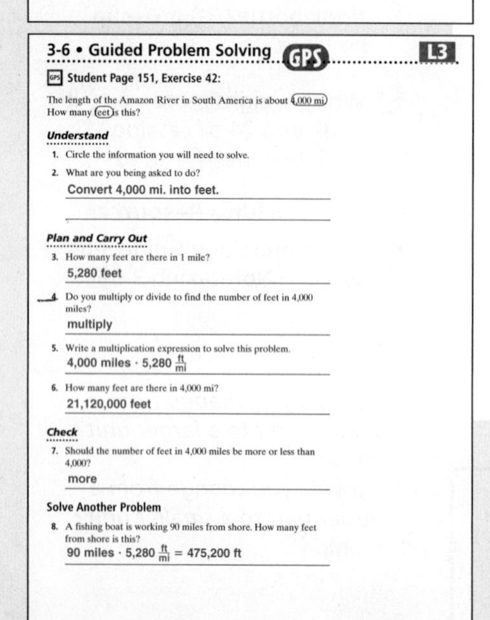

3-6 • Guided Problem Solving GPS L3

Student Page 151, Exercise 42:

The length of the Amazon River in South America is about 4,000 mi. How many feet is this?

Understand

1. Circle the information you will need to solve.
2. What are you being asked to do? Convert 4,000 mi. into feet.

Plan and Carry Out

3. How many feet are there in 1 mile? 5,280 feet
4. Do you multiply or divide to find the number of feet in 4,000 miles? multiply
5. Write a multiplication expression to solve this problem. 4,000 miles · 5,280 $\frac{ft}{mi}$
6. How many feet are there in 4,000 mi? 21,120,000 feet

Check

7. Should the number of feet in 4,000 miles be more or less than 4,000? more

Solve Another Problem

8. A fishing boat is working 90 miles from shore. How many feet from shore is this? 90 miles · 5,280 $\frac{ft}{mi}$ = 475,200 ft

Check Your Understanding

Tell whether you would *multiply* or *divide* to change from one unit of measure to the other.

1. gallons to quarts multiply
2. ounces to pounds divide
3. yards to feet multiply
4. pints to cups multiply
5. feet to miles divide
6. tons to pounds multiply
7. **Number Sense** Why do you *multiply* to change from a larger unit to a smaller unit? It takes more smaller units to equal the larger units.
8. Why do you *divide* to change from a smaller unit to a larger unit? It takes fewer larger units to equal the smaller units.

Homework Exercises

For more exercises, see Extra Skills and Word Problems.

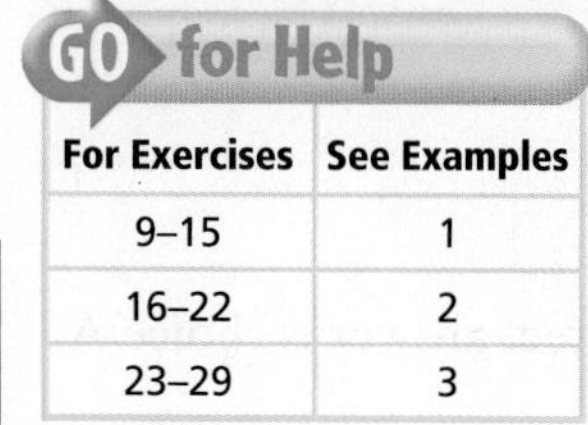

GO for Help

For Exercises	See Examples
9–15	1
16–22	2
23–29	3

A **Change each unit of length.**

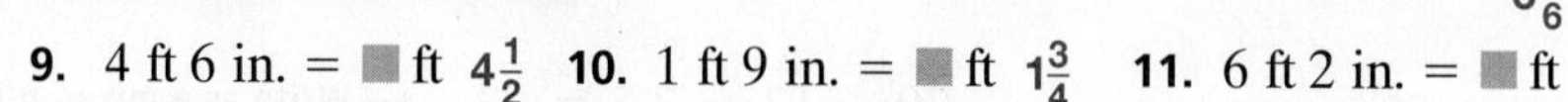

9. 4 ft 6 in. = ■ ft $4\frac{1}{2}$
10. 1 ft 9 in. = ■ ft $1\frac{3}{4}$
11. 6 ft 2 in. = ■ ft $6\frac{1}{6}$
12. 4 yd 2 ft = ■ yd $4\frac{2}{3}$
13. 1 yd 1 ft = ■ yd $1\frac{1}{3}$
14. 10 yd 3 ft = ■ yd 11
15. A carpenter cuts 2 ft 9 in. off a board that was 4 ft long. How long is the piece that is left, in feet and inches? 1 ft 3 in.

Change each unit of capacity.

16. 48 fl oz = ■ c 6
17. 6 c = ■ p 3
18. 40 qt = ■ gal 10
19. 1 c = ■ p $\frac{1}{2}$
20. 9 pt = ■ qt $4\frac{1}{2}$
21. 12 c = ■ qt 3
22. How many cups are in a 12-fl oz can of lemonade? $1\frac{1}{2}$ c

Change each unit of weight.

23. $\frac{3}{4}$ lb = ■ oz 12
24. $2\frac{1}{4}$ lb = ■ oz 36
25. $2\frac{1}{2}$ t = ■ lb 5,000
26. 32 oz = ■ lb 2
27. 80 oz = ■ lb 5
28. 1,000 lb = ■ t $\frac{1}{2}$
29. Which is lighter, 12 oz of cheese or 1 lb of cheese? 12 oz

B

30. **Guided Problem Solving** People in the United States discard an average of 75,000 t of food per day. If a garbage truck can hold 12,000 lb, how many trucks are needed to haul all of the food? 12,500 trucks
    - How many tons is 12,000 lb?
    - What operation should you use to find the number of garbage trucks needed?
31. You are hiking a 2-mi trail. A sign shows that you have hiked 1,000 ft. How many feet do you have left to hike? 9,560 ft

GO Online
Homework Video Tutor
Visit: PHSchool.com
Web Code: are-0306

Complete.

32. 500 lb = ■ t **0.25**
33. $1\frac{1}{5}$ t = ■ lb **2,400**
34. $5\frac{1}{4}$ gal = ■ qt **21**
35. 7 pt = ■ qt **$3\frac{1}{2}$**
36. 69 in. = ■ ft **$5\frac{3}{4}$**
37. 16 ft = ■ yd **$5\frac{1}{3}$**

38. Sam bought three packages of low-salt pretzels weighing 12 oz, 32 oz, and $1\frac{1}{2}$ lb. How many pounds of pretzels did Sam buy all together? **$4\frac{1}{4}$ lb**

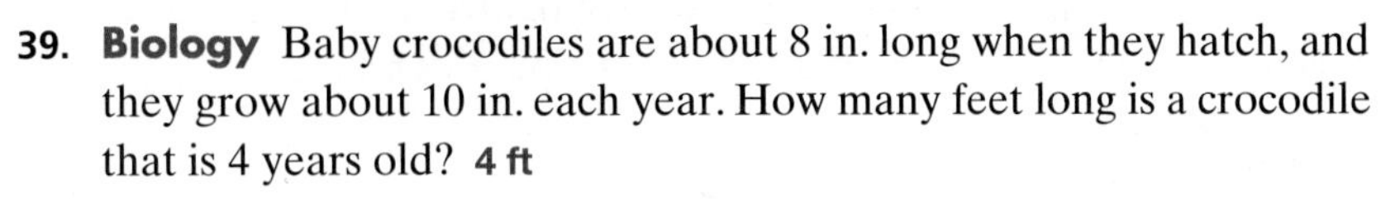

39. **Biology** Baby crocodiles are about 8 in. long when they hatch, and they grow about 10 in. each year. How many feet long is a crocodile that is 4 years old? **4 ft**

40. How are the rates "6 pounds per day" and "6 days per pound" different? When would each be useful? **Check students' work.**

41. **Error Analysis** Your friend says that a quarter-pound burger is heavier than a six-ounce burger. Is your friend correct? Explain. **no; quarter pound = 4 oz**

42. The length of the Amazon River in South America is about 4,000 mi. How many feet is this? **21,120,000 ft**

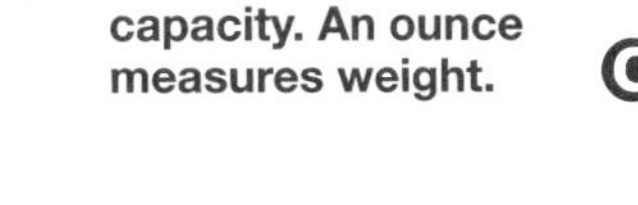

43. **Writing in Math** Explain how fluid ounces and ounces differ.

43. A fluid ounce measures capacity. An ounce measures weight.

Ⓒ 44. **Challenge** Three students recorded their heights as 64 in., $5\frac{1}{6}$ ft, and $1\frac{2}{3}$ yd. Find the average height in feet of the three students. **$5\frac{1}{6}$ ft**

Test Prep and Mixed Review

Practice

Multiple Choice

45. An assortment of cheeses is weighed on a scale. What is the best estimate of the total weight of the 8 cheeses listed at the right? **B**
   Ⓐ Less than 5 pounds
   Ⓑ Between 5 and 10 pounds
   Ⓒ Between 10 and 15 pounds
   Ⓓ More than 15 pounds

Cheese Assortment

Quantity	Weight
1	1.1 lb
2	2 lb 4 oz
5	$\frac{1}{2}$ lb

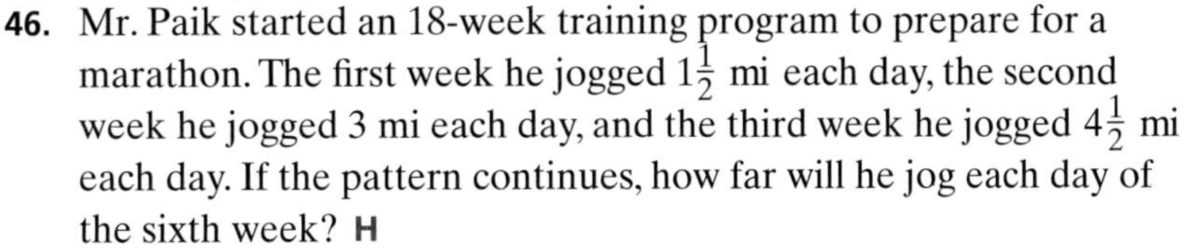

46. Mr. Paik started an 18-week training program to prepare for a marathon. The first week he jogged $1\frac{1}{2}$ mi each day, the second week he jogged 3 mi each day, and the third week he jogged $4\frac{1}{2}$ mi each day. If the pattern continues, how far will he jog each day of the sixth week? **H**
   Ⓕ 6 mi
   Ⓖ $7\frac{1}{2}$ mi
   Ⓗ 9 mi
   Ⓙ $10\frac{1}{2}$ mi

GO for Help

For Exercises	See Lesson
47–49	3-5

Find each quotient.

47. $\frac{3}{7} \div \frac{1}{7}$ **3**
48. $\frac{4}{5} \div \frac{4}{10}$ **2**
49. $\frac{3}{11} \div \frac{9}{22}$ **$\frac{2}{3}$**

4. Assess & Reteach

PowerPoint

Lesson Quiz

Complete.

1. 56 in. = ■ ft **$4\frac{2}{3}$**
2. $6\frac{1}{2}$ qt = ■ pt **13**
3. 26 fl oz = ■ c **$3\frac{1}{4}$**
4. $5\frac{1}{2}$ lb = ■ oz **88**

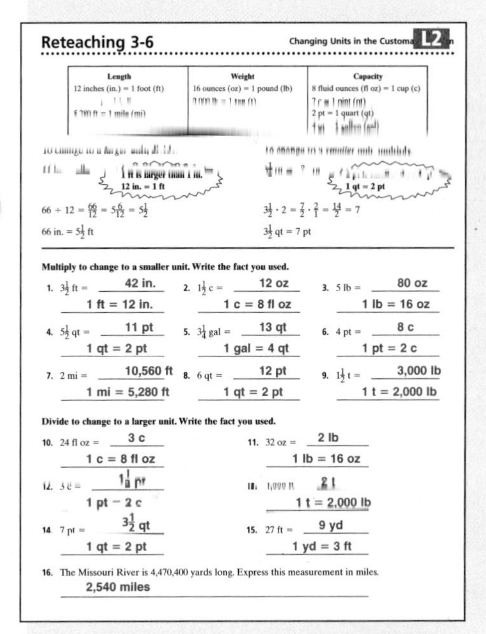

Reteaching 3-6 — Changing Units in the Customa[ry System] L2

Length	Weight	Capacity
12 inches (in.) = 1 foot (ft)	16 ounces (oz) = 1 pound (lb)	8 fluid ounces (fl oz) = 1 cup (c)
[illegible]	[illegible]	[illegible]
[illegible]		2 pt = 1 quart (qt)
		[illegible]

1 ft is larger than 1 in.
12 in. = 1 ft

$66 \div 12 = \frac{66}{12} = 5\frac{6}{12} = 5\frac{1}{2}$

66 in. = $5\frac{1}{2}$ ft

1 qt = 2 pt

$3\frac{1}{2} \cdot 2 = \frac{7}{2} \cdot \frac{2}{1} = \frac{14}{2} = 7$

$3\frac{1}{2}$ qt = 7 pt

Multiply to change to a smaller unit. Write the fact you used.

1. $3\frac{1}{2}$ ft = **42 in.** — **1 ft = 12 in.**
2. $1\frac{1}{2}$ c = **12 oz** — **1 c = 8 fl oz**
3. 5 lb = **80 oz** — **1 lb = 16 oz**
4. $5\frac{1}{2}$ qt = **11 pt** — **1 qt = 2 pt**
5. $3\frac{1}{4}$ gal = **13 qt** — **1 gal = 4 qt**
6. 4 pt = **8 c** — **1 pt = 2 c**
7. 2 mi = **10,560 ft** — **1 mi = 5,280 ft**
8. 6 qt = **12 pt** — **1 qt = 2 pt**
9. $1\frac{1}{2}$ t = **3,000 lb** — **1 t = 2,000 lb**

Divide to change to a larger unit. Write the fact you used.

10. 24 fl oz = **3 c** — **1 c = 8 fl oz**
11. 32 oz = **2 lb** — **1 lb = 16 oz**
12. 3 c = **$1\frac{1}{2}$ pt** — **1 pt = 2 c**
13. [illegible] — **1 t = 2,000 lb**
14. 7 pt = **$3\frac{1}{2}$ qt** — **1 qt = 2 pt**
15. 27 ft = **9 yd** — **1 yd = 3 ft**
16. The Missouri River is 4,470,400 yards long. Express this measurement in miles. **2,540 miles**

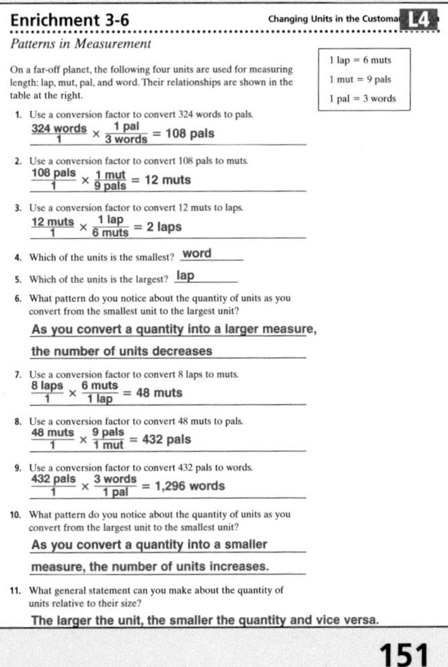

Enrichment 3-6 — Changing Units in the Customa[ry System] L4

Patterns in Measurement

On a far-off planet, the following four units are used for measuring length: lap, mut, pal, and word. Their relationships are shown in the table at the right.

1 lap = 6 muts
1 mut = 9 pals
1 pal = 3 words

1. Use a conversion factor to convert 324 words to pals. **$\frac{324 \text{ words}}{1} \times \frac{1 \text{ pal}}{3 \text{ words}} = 108$ pals**
2. Use a conversion factor to convert 108 pals to muts. **$\frac{108 \text{ pals}}{1} \times \frac{1 \text{ mut}}{9 \text{ pals}} = 12$ muts**
3. Use a conversion factor to convert 12 muts to laps. **$\frac{12 \text{ muts}}{1} \times \frac{1 \text{ lap}}{6 \text{ muts}} = 2$ laps**
4. Which of the units is the smallest? **word**
5. Which of the units is the largest? **lap**
6. What pattern do you notice about the quantity of units as you convert from the smallest unit to the largest unit? **As you convert a quantity into a larger measure, the number of units decreases**
7. Use a conversion factor to convert 8 laps to muts. **$\frac{8 \text{ laps}}{1} \times \frac{6 \text{ muts}}{1 \text{ lap}} = 48$ muts**
8. Use a conversion factor to convert 48 muts to pals. **$\frac{48 \text{ muts}}{1} \times \frac{9 \text{ pals}}{1 \text{ mut}} = 432$ pals**
9. Use a conversion factor to convert 432 pals to words. **$\frac{432 \text{ pals}}{1} \times \frac{3 \text{ words}}{1 \text{ pal}} = 1{,}296$ words**
10. What pattern do you notice about the quantity of units as you convert from the largest unit to the smallest unit? **As you convert a quantity into a smaller measure, the number of units increases.**
11. What general statement can you make about the quantity of units relative to their size? **The larger the unit, the smaller the quantity and vice versa.**

Alternative Assessment

Provide pairs of students with tape measures or yardsticks. Have them measure various classroom objects, such as the length and width of a desk, and report the measurement in inches. Then have them write each measurement in feet and, when practical, in yards.

Test Prep

Resources

For additional practice with a variety of test item formats:

- Test-Taking Strategies, p. 159
- Test Prep, p. 163
- Test-Taking Strategies with Transparencies

Activity Lab

Solving Puzzles

Students translate word problems into numerical sentences and solve them.

Guided Instruction

Explain to students that phrases like *twice as many* and *half as much* can be translated into numerical phrases. Ask volunteers to use such phrases in sentences. Then ask questions such as:

- *Why do we multiply $\frac{2}{3}$ by 12?* **Two thirds of the people are cousins and there are 12 people total.**
- *How do we express 12 as a fraction?* $\frac{12}{1}$
- *If $\frac{1}{4}$ of Sandra's cousins are boys, how many boys are there?* **2**

Differentiated Instruction

Below Level L2

Provide students with algebra tiles and help them model the Example. Divide the tiles into three groups and then count the number of tiles in two groups.

Resources

- algebra tiles
- Student Manipulatives Kit

3-6b Activity Lab — Algebra Thinking

Solving Puzzles

People often describe real-world situations using language that can be translated into numerical expressions. A class made some puzzles about the number of relatives who are coming to their family outings. Solve the puzzles below.

EXAMPLE

Sandra has a total of 12 aunts and cousins coming to her family outing. Two thirds are cousins, and half as many aunts as cousins are coming. How many aunts and how many cousins are coming to Sandra's outing?

Step 1 List what you know:

- $\frac{2}{3}$ cousins
- $\frac{1}{2}$ as many aunts as cousins
- 12 aunts and cousins all together

Step 2 Find the number of cousins:

$\frac{2}{3} \cdot 12 = \frac{24}{3}$ ← **Multiply $\frac{2}{3}$ by 12.**

$\frac{24}{3} = 8$ ← **Simplify. There are 8 cousins.**

Step 3 Find the number of aunts:

$\frac{1}{2} \cdot 8 = 4$ ← **Multiply $\frac{1}{2}$ by 8. There are 4 aunts.**

Sandra has 8 cousins and 4 aunts coming to her outing.

Exercises

1. Mark has 16 people coming to his outing. $\frac{3}{8}$ are his nieces, $\frac{1}{4}$ are his nephews, and the rest are his cousins. How many nieces, nephews, and cousins are coming to Mark's outing? **6 nieces, 4 nephews, 6 cousins**
2. Phil has 24 guests in all. Two thirds are cousins. The rest are aunts and uncles, and there are the same number of aunts as uncles. How many cousins, aunts, and uncles are coming to the outing? **16 cousins, 4 aunts, 4 uncles**
3. Create your own puzzle like the ones above. Your first clue should be the number of relatives who will come to a reunion. Then write clues using fractions, like the ones above, to describe the situation. **Check students' work.**

Checkpoint Quiz 2

Lesson 3-4 through 3-6

Find each product.

1. $\frac{1}{3} \cdot \frac{6}{7}$ $\frac{2}{7}$
2. $\frac{3}{4} \cdot 4\frac{1}{3}$ $3\frac{1}{4}$
3. $\frac{2}{5} \cdot \frac{1}{2}$ $\frac{1}{5}$
4. $3\frac{1}{4} \cdot 2\frac{7}{8}$ $9\frac{11}{32}$
5. $1\frac{1}{10} \cdot \frac{2}{3}$ $\frac{11}{15}$
6. What is the area of a courtyard that is $2\frac{2}{3}$ yd by $4\frac{1}{6}$ yd? $11\frac{1}{9}$ yd^2

Find each quotient.

7. $\frac{5}{9} \div \frac{5}{6}$ $\frac{2}{3}$
8. $12\frac{1}{2} \div 1\frac{7}{8}$ $6\frac{2}{3}$
9. $3\frac{1}{7} \div \frac{11}{7}$ 2
10. $\frac{8}{25} \div \frac{2}{5}$ $\frac{4}{5}$
11. $5\frac{1}{5} \div \frac{4}{10}$ 13
12. You have a leaking bucket with $15\frac{1}{4}$ liters of water. Water drains out of the bucket at a rate of $\frac{1}{3}$ liter per minute. In how many minutes will the bucket be empty? $45\frac{3}{4}$ min

Change each unit of length.

13. 3 ft 4 in. = ■ ft $3\frac{1}{3}$
14. 2 ft 3 in. = ■ ft $2\frac{1}{4}$
15. 8 ft 6 in. = ■ ft $8\frac{1}{2}$

3-7a Activity Lab

Choosing Appropriate Units

When you measure an object, you should measure in an appropriate unit. For example, you would measure the length of a stapler in inches, not in miles.

Number Sense **Choose the more appropriate unit of measure.**

1.	the mass of a horse	kilograms	grams	kilograms
2.	the length of a marathon road race	miles	feet	miles
3.	the capacity of a car's gas tank	ounces	gallons	gallons
4.	the length of an eyelash	meters	millimeters	millimeters
5.	the weight of a baseball	ounces	pounds	ounces

6. **Writing in Math** Choose one of your answers from Questions 1–5. Explain why you chose that unit of measure. Answers may vary. Sample: For Question 1, I chose kilograms because horses are heavy.

Checkpoint Quiz

Use this Checkpoint Quiz to check students' understanding of the skills and concepts of Lessons 3-4 through 3-6.

Resources

- All-in-One Teaching Resources Checkpoint Quiz 2
- ExamView Assessment Suite CD-ROM
- Success Tracker™ Online Intervention

Activity Lab

Choosing Appropriate Units

Students learn to choose the appropriate units to measure different objects.

Guided Instruction

Explain that objects should be measuredvcv with the unit that is appropriate to its size or length. For example, measuring a road in feet would give you a very large number, but measuring it in miles would give you a more reasonable number.

Activity

Have students work together in pairs to decide the appropriate units to measure each item. After they are finished, ask volunteers to give their answers for Exercise 6.

Differentiated Instruction

Below Level  L2

Have students point to an object in the room and name an appropriate unit of measure.

Resources

- Activity Lab 3-7: Measuring with a Ruler

3-7

1. Plan

Objective
To find and compare the precision of measurement

Examples
1 Precision in Measurement
2 Finding Precision
3 Precision and Rounding

Math Understandings: p. 118D

Math Background

Precision refers to the exactness of a measurement. For example, 98.6°F is more precise than 100°F. When comparing measurements, the more precise measurement has the smaller units.

More Math Background: p. 118D

Lesson Planning and Resources

See p. 118E for a list of the resources that support this lesson.

Bell Ringer Practice

Check Skills You'll Need
Use student page, transparency, or PowerPoint. For intervention, direct students to:
Measuring in Metric Units
Lesson 1-5
Extra Skills and Word Problems Practice, Ch. 1

3-7 Precision

Check Skills You'll Need

1. **Vocabulary Review** What are the basic metric units for *length, capacity,* and *mass*? **meter; liter; gram**

Complete.

2. 1 m = ■ cm **100**
3. 1 m = ■ km **0.001**
4. 1 cm = ■ mm **10**
5. 1 kg = ■ g **1,000**
6. 1 mL = ■ L **0.001**

for Help Lesson 1-5

What You'll Learn

To find and compare the precision of measurements

New Vocabulary precision

Why Learn This?

Every measurement has a unit. Some units are more appropriate to use than others, and some units give a more precise measurement than others.

If you measure the butterfly above to the nearest centimeter, the length is 5 cm. If you measure the butterfly to the nearest millimeter, the length is 48 mm. The measurement 48 mm is more precise than 5 cm.

The **precision** of a measurement refers to its exactness. When you compare measurements, the more precise measurement is the one that uses the smaller unit of measure.

Vocabulary Tip
Precise is the adjectival form of *precision.*

EXAMPLES Precision in Measurement

1 Choose the more precise measurement: 12 fl oz or 2 c.

A fluid ounce is a smaller unit than a cup. So the measurement 12 fl oz is more precise than 2 c.

2 Choose the more precise measurement: 3 L or 2.5 L

One tenth of a liter is a smaller unit than a liter. So the measurement 2.5 L is more precise than 3 L.

3 Choose the more precise measurement: 8 in. or 21 in.

Both measurements use the inch as a unit. So neither measurement is more precise than the other.

Quick Check

Choose the more precise measurement.

1. 2 ft, 13 in. **13 in.**
2. 12.5 g, 11 g **12.5 g**
3. $3\frac{1}{2}$ mi, $10\frac{1}{5}$ mi **$10\frac{1}{5}$ mi**

Differentiated Instruction Solutions for All Learners

Special Needs L1	Below Level L2
Have students circle more precise numbers from sets of numbers such as 245 g or 24.7 g; 43 oz or 1 qt; 1 hr 3 min or 3 min 1 s. **24.7 g; 43 oz; 3 min 1 s** **learning style: visual**	Guide students in measuring the length of their hand to the nearest inch, half-inch, quarter-inch, centimeter, and tenth-of-a-centimeter to help them better understand the concept of precision. **learning style: tactile**

There are many different measuring tools, such as rulers, scales, and thermometers. The marks on a measuring tool tell you the precision that is possible with that tool.

EXAMPLE Finding Precision

4 **Cooking** Oven thermometers measure cooking temperatures. What is the greatest precision possible with the thermometer shown at the right?

The interval from any red dot to the next black dot is 25°F. Measurements made with this thermometer are precise to the nearest 25°F.

Quick Check

4\. What is the greatest precision possible with each ruler below?

a.

0 1
inches

$\frac{1}{8}$ in.

b.

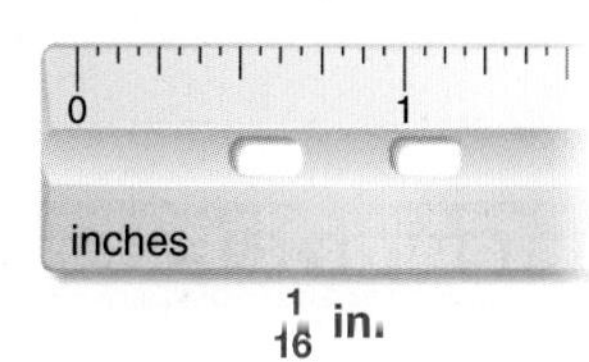

$\frac{1}{16}$ in.

A calculation is only as precise as the least precise measurement used in the calculation. When you add or subtract measurements with the same unit, round your answer to match the least precise measurement.

EXAMPLES Precision and Rounding

5 Find $8\frac{2}{5}$ mi + 5 mi. Round your answer appropriately.

$8\frac{2}{5} + 5 = 13\frac{2}{5}$

Since 5 is less precise than $8\frac{2}{5}$, round to the nearest whole number. The sum is 13 mi.

6 Find 9.97 cm − 5.9 cm. Round your answer appropriately.

$9.97 - 5.9 = 4.07$

Since 5.9 is less precise than 9.97, round to the nearest tenth. The difference is 4.1 cm.

Quick Check

Find each sum or difference. Round your answer appropriately.

5\. 11.4 g + 2.65 g **14.1 g**

6\. 45 m − 0.9 m **44 m**

Advanced Learners L4

Have students explain the difference between a rounded measurement of 1 m and 1.00 m. **Sample: 1 m is between 50 and 149 cm. 1.00 m is between 99.5 and 100.49 cm.**

learning style: verbal

English Language Learners ELL

Show students a watch with no tick marks for seconds showing. Then show them one that has the second tick marks. Point out that with one watch, you can say something took about 6 minutes. With the other, you can say it took 5 minutes and 55 seconds. That is more precise.

learning style: visual

2. Teach

Activity Lab

Use before the lesson.
Student Edition Activity Lab 3-7a, Choosing Appropriate Units, p. 153

All in One Teaching Resources

Activity Lab 3-7: Measuring with a Ruler

Guided Instruction

Example 1
Name a measurement and have students give a corresponding measurement that is more precise. For example, if you say "2 inches," a student might respond "$1\frac{7}{8}$ inches."

Error Prevention!

Suggest that students identify the less precise measurement and its units before finding the sum or difference.

Additional Examples

1 Choose the more precise measurement. 3.6 cm, 9.15 cm **9.15 cm**

2 What is the greatest precision possible with a digital clock that shows the hour and minute? **the nearest minute**

3 Find the difference: 7 mi − $2\frac{3}{4}$ mi. Round your answer to match the less precise measurement. **4 mi**

All in One Teaching Resources

- Daily Notetaking Guide 3-7 L3
- Adapted Notetaking 3-7 L1

Closure

- *What does it mean to round your answer to the less precise measurement?* **Sample: When combining two units, such as tenths and whole numbers, round the answer to the larger unit—in this case, whole numbers.**

3. Practice

Assignment Guide

Check Your Understanding
Go over Exercises 1–6 in class before assigning the Homework Exercises.

Homework Exercises

A Practice by Example	7–23
B Apply Your Skills	24–35
C Challenge	36
Test Prep and Mixed Review	37–43

Homework Quick Check
To check students' understanding of key skills and concepts, go over Exercises 11, 17, 27, 28, and 35.

Differentiated Instruction Resources

Adapted Practice 3-7 L1

Practice 3-7 L3

Underline the more precise measurement.

1. 23 oz, 20.7 oz
2. 1,830 g, 2.5 kg
3. 63.7 L, 63.70 L
4. 3.7 t, 5,610 lb
5. 58.3 cm, 4.6 m
6. 12 L, 1,735 mL
7. 3,008 pt, 0.95 pt
8. 7.3 min, 516 sec
9. 2.7 mL, 12 mL
10. 26.4 cm, 8.39 cm
11. 216 ft, 3,106 in.
12. 4.1 lb, 6.123 lb

Find each sum or difference. Round your answer to match the less precise measurement.

13. 6.35 oz + 4.2 oz **10.6 oz**
14. 83 g − 1.8 g **81 g**
15. 4.20 yd + 8.64 yd **12.84 yd**
16. 21 cm + 53.60 cm **75 cm**
17. 5.382 m + 8 m **13 m**
18. 6.4 ft + 4300 ft **4,306 ft**
19. 2.713 mL + 8.4 mL **11.1 mL**
20. 50 lb − 4.6 lb **45 lb**
21. 6.83 km + 10.3 km **17.1 km**

22. Boundary Peak in Nevada is 13,000 ft high. Guadalupe Peak in Texas is 8,749.75 ft high. How much higher than Guadalupe Peak is Boundary Peak? Round your answer to match the less precise measurement.
4,250 ft

23. You measure the area of your garden as 9 yd wide by 11 yd long. Your brother measures the garden as $27\frac{1}{2}$ ft wide by $32\frac{3}{4}$ ft long. Whose measurement is more precise? Why?
brother's measurement; since a foot is a shorter unit than a yard, the measurement in feet is more precise.

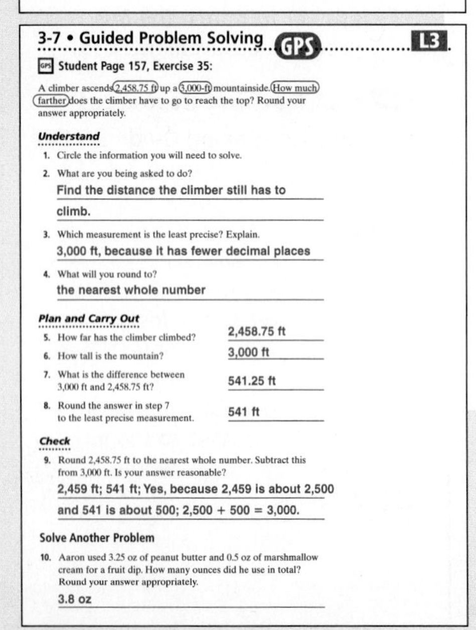

3-7 • Guided Problem Solving GPS L3

GPS **Student Page 157, Exercise 35:**

A climber ascends 2,458.75 ft up a 3,000-ft mountainside. How much farther does the climber have to go to reach the top? Round your answer appropriately.

Understand

1. Circle the information you will need to solve.
2. What are you being asked to do?
Find the distance the climber still has to climb.
3. Which measurement is the least precise? Explain.
3,000 ft, because it has fewer decimal places
4. What will you round to?
the nearest whole number

Plan and Carry Out

5. How far has the climber climbed? **2,458.75 ft**
6. How tall is the mountain? **3,000 ft**
7. What is the difference between 3,000 ft and 2,458.75 ft? **541.25 ft**
8. Round the answer in step 7 to the least precise measurement. **541 ft**

Check

9. Round 2,458.75 ft to the nearest whole number. Subtract this from 3,000 ft. Is your answer reasonable?
2,459 ft; 541 ft; Yes, because 2,459 is about 2,500 and 541 is about 500; 2,500 + 500 = 3,000.

Solve Another Problem

10. Aaron used 3.25 oz of peanut butter and 0.5 oz of marshmallow cream for a fruit dip. How many ounces did he use in total? Round your answer appropriately.
3.8 oz

Check Your Understanding

1. Measure the pencil below to the nearest inch, to the nearest $\frac{1}{2}$ inch, and to the nearest $\frac{1}{8}$ inch. **4 in., 4 in., $3\frac{6}{8}$ in.**

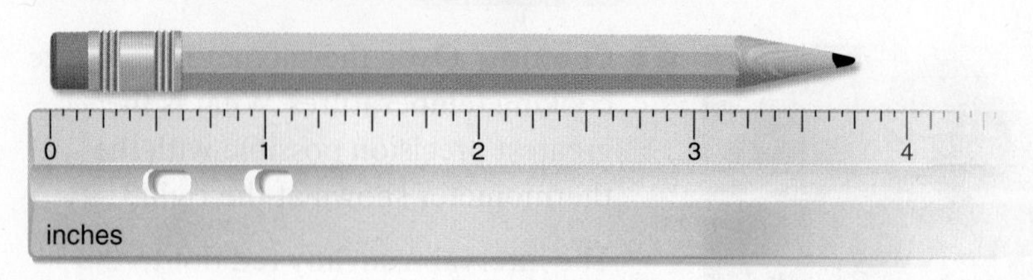

2. **Vocabulary** Which measurement in Exercise 1 is the most precise? **$3\frac{6}{8}$ in.**

3. **Number Sense** Can you use the ruler in Exercise 1 to measure the pencil to the nearest $\frac{1}{16}$ inch? Explain. **$3\frac{12}{16}$ in.**

Choose the more precise measurement.

4. 13 cm, 13 mm **13 mm**
5. 9 ft, $4\frac{1}{2}$ ft **$4\frac{1}{2}$ ft**
6. 8.5 kg, 2.25 kg **2.25 kg**

Homework Exercises

For more exercises, see Extra Skills and Word Problems.

GO for Help

For Exercises	See Examples
7–15	1–3
16–17	4
18–23	5–6

A **Choose the more precise measurement.**

7. 16 in., $11\frac{15}{16}$ in. **See left.**
8. 30 g, 2.5 kg **30 g**
9. 37 t, 56 lb **56 lb**
10. 25 qt, 38 pt **38 pt**
11. 21 L, 35 mL **35 mL**
12. $6\frac{1}{10}$ lb, 6.37 lb **6.37 lb**
13. 12 mo, 1 yr **12 mo**
14. 0.25 g, 101 mg **101 mg**
15. 12 days, 2 wk **12 days**

7. **$11\frac{15}{16}$ in.**

Find the greatest precision possible for each scale shown.

16.

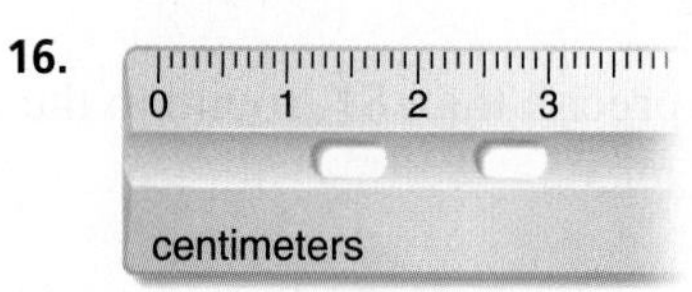

1 mm

17.

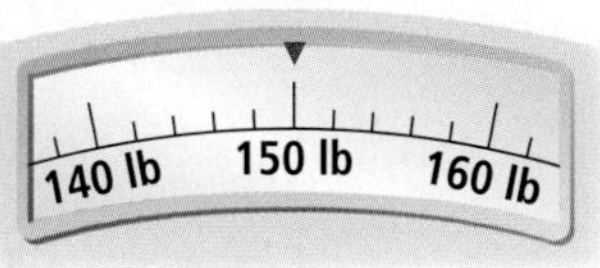

2 lb

Find each sum or difference. Round your answer appropriately.

18. $19 \text{ m} + 4\frac{9}{10} \text{ m}$ **24 m**
19. $6 \text{ ft} - 2\frac{1}{4} \text{ ft}$ **4 ft**
20. 7.4 L + 1.16 L **8.6 L**
21. $6.53 \text{ oz} + 2\frac{2}{5} \text{ oz}$ **8.9 oz**
22. 18 g − 3.8 g **14 g**
23. 6.52 in. − 5.8 in. **0.7 in.**

B

24. **Guided Problem Solving** You have 81 ft^2 of fabric. You cut a square of fabric with side length $4\frac{1}{2}$ ft. How much fabric is left? **$60\frac{3}{4}$ ft^2**
   - How many square feet of fabric did you cut out?
   - What operation tells you how much fabric is left?
   - What should you round to?

GO Online
Homework Video Tutor
Visit: PHSchool.com
Web Code: are-0307

Find the length of each segment to the greatest precision possible.

25. **26.**

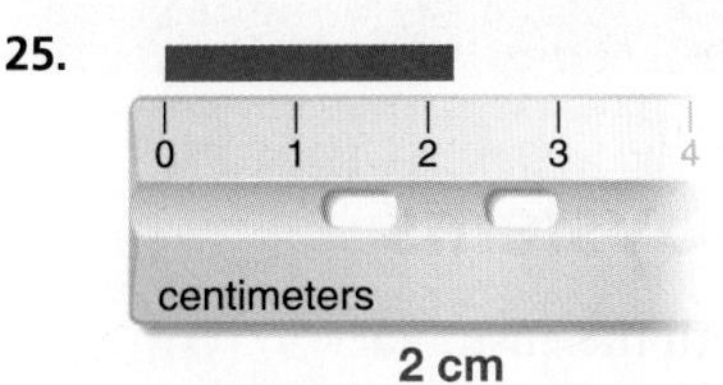

25. **2 cm**

26. **1.9 cm**

27. **Biology** The length of a chicken egg can be $2\frac{3}{16}$ in. The length of an ostrich egg can be $6\frac{3}{8}$ in. What is the difference between the lengths of the eggs? Round your answer appropriately. **about $4\frac{2}{8}$ in.**

28. **Writing in Math** Your friend says that 5.25 kg is a more precise measurement than 6.2 g because a hundredths unit is smaller than a tenths unit. Do you agree? Explain. **No; a gram is a smaller unit of measure.**

Choose the more precise unit of measure.

29. foot, meter **ft** **30.** gallon, liter **L** **31.** centimeter, inch **cm**

32. deciliter, gallon **dL** **33.** kilometer, mile **km** **34.** kilogram, pound **lb**

35. (GPS) A climber ascends 2,458.75 ft up a 3,000-ft mountainside. How much farther does the climber have to go to reach the top? Round your answer appropriately. **541 ft**

36. **Challenge** Three students measured the width of a bulletin board. The three measurements were 1 yd, 3 ft, and 36 in. Compare the measurements and the precision of the measurements. **They are all equal, but 36 in. is the most precise.**

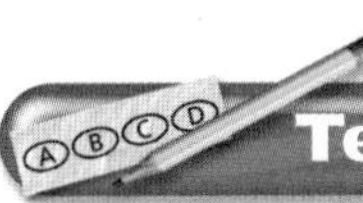

Test Prep and Mixed Review **Practice**

Multiple Choice

37. The fraction $\frac{3}{8}$ is found between which pair of fractions on a number line? **A**

Ⓐ $\frac{4}{16}$ and $\frac{13}{32}$ Ⓑ $\frac{5}{16}$ and $\frac{10}{32}$ Ⓒ $\frac{6}{16}$ and $\frac{20}{32}$ Ⓓ $\frac{7}{16}$ and $\frac{24}{32}$

38. Hector bought two bags of apples. The larger bag weighed $3\frac{3}{4}$ lb. The smaller bag weighed $\frac{5}{8}$ lb less than the larger bag. Which expression can be used to find the weight, in pounds, of the smaller bag? **H**

Ⓕ $3\frac{3}{4} + \frac{5}{8}$ Ⓖ $3\frac{3}{4} \cdot \frac{5}{8}$ Ⓗ $3\frac{3}{4} - \frac{5}{8}$ Ⓙ $3\frac{3}{4} \div \frac{5}{8}$

39. What is the value of the expression $4 + (12 \div 4)^2 - 3 \cdot 5$? **B**

Ⓐ −10 Ⓑ −2 Ⓒ 20 Ⓓ 50

GO for Help

For Exercises	See Lesson
40–43	3-1

Estimate each answer.

40. $12\frac{1}{10} - 5\frac{7}{8}$ **about 6** **41.** $2\frac{2}{9} + 4\frac{1}{8}$ **about 6** **42.** $3\frac{1}{3} \cdot 10\frac{4}{5}$ **about 33** **43.** $16\frac{1}{9} \div 3\frac{6}{7}$ **about 4**

4. Assess & Reteach

PowerPoint

Lesson Quiz

Choose the more precise measurement.

1. 12 in., $10\frac{3}{4}$ in. **$10\frac{3}{4}$ in.**
2. 7.78 kg, 7.8 kg **7.78 kg**

Find each sum or difference. Round the answer to match the less precise measurement.

3. 3.09 L + 4.7 L **7.8 L**
4. 81.2 cm − 46.08 cm **35.1 cm**

Reteaching 3-7 L2

The precision of a measurement refers to its degree of exactness. The smaller the unit of measure, the more precise the measurement. If the same unit is used in two measurements, then the measurement to the smallest decimal place is more precise.

Determine which measurement in each set is more precise.

a. [illegible]

Since the units of measure are different, the measurement with the smaller unit of measure is more precise. An inch is smaller than a yard, so 110 in. is more precise than 3 yd.

b. 45.12 cm, 45.2 cm

Since the units of measure are the same, the measurement with the smaller decimal place is more precise. Since 45.12 has the smaller decimal place, 45.12 cm is more precise than 45.2 cm.

Write the more precise measurement.

1. 1.6 mi, 8,448 ft **8,448 ft**
2. 8.9 km, 8.87 km **8.87 km**
3. 2 ft, 13 in. **13 in.**
4. 5.64 cm, 56.2 cm **5.64 cm**
5. 4.3 yd, 2 mi **4.3 yd**
6. 17.33 mm, 17 mm **17.33 mm**

Compute. Round your answer appropriately.

7. 2.3 in. + 6.31 in. **8.6 in.**
8. 17.2 cm × 5 cm **86 cm²**
9. [illegible]
10. [illegible]
11. 24 mm − 16.1 mm **8 mm**
12. 2.25 yd × 6 yd **14 yd**

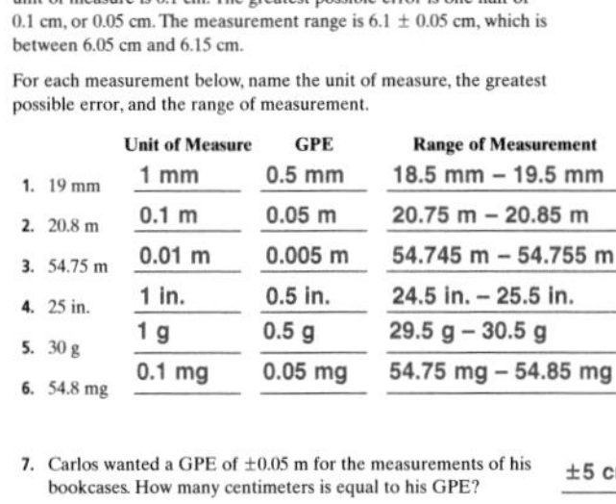

Enrichment 3-7 L4

Critical Thinking

For any measurement the **greatest possible error (GPE)** is one half the unit of measure.

For example, if the length of any object measures 5 in. to the nearest inch, the unit of measure is 1 in. The greatest possible error is one half of one inch, or $\frac{1}{2}$ in. The measurement range is $5 \pm \frac{1}{2}$, which is between $4\frac{1}{2}$ in. and $5\frac{1}{2}$ in.

If the length of an object measures 6.1 cm to the nearest 0.1 cm, the unit of measure is 0.1 cm. The greatest possible error is one half of 0.1 cm, or 0.05 cm. The measurement range is 6.1 ± 0.05 cm, which is between 6.05 cm and 6.15 cm.

For each measurement below, name the unit of measure, the greatest possible error, and the range of measurement.

	Unit of Measure	GPE	Range of Measurement
1. 19 mm	1 mm	0.5 mm	18.5 mm – 19.5 mm
2. 20.8 m	0.1 m	0.05 m	20.75 m – 20.85 m
3. 54.75 m	0.01 m	0.005 m	54.745 m – 54.755 m
4. 25 in.	1 in.	0.5 in.	24.5 in. – 25.5 in.
5. 30 g	1 g	0.5 g	29.5 g – 30.5 g
6. 54.8 mg	0.1 mg	0.05 mg	54.75 mg – 54.85 mg

7. Carlos wanted a GPE of ±0.05 m for the measurements of his bookcases. How many centimeters is equal to his GPE? **±5 cm**
8. Manuela is writing the specifications for a new product. The GPE was ±0.005 g. The lowest measure is 10.565 g. What is the largest measure? **10.575 g**
9. Pate mixed two liquids in his chemistry class. He measured each liquid to the nearest milliliter. What was his GPE? **0.5 mL**

Alternative Assessment

Have students work in pairs. One partner names something to be measured in either the customary or the metric system, for example, "in the metric system, the length of an automobile." The other student gives a precise measurement that is appropriate, such as "2.3 meters." Students then trade roles.

Test Prep

Resources

For additional practice with a variety of test item formats:

- Test-Taking Strategies, p. 159
- Test Prep, p. 163
- Test-Taking Strategies with Transparencies

Activity Lab

Estimating in Different Systems

Students practice estimating equivalent measures in the metric system and the customary system.

Guided Instruction

Have students think of examples of where and when it might be useful to approximate a measure in one measurement system when that measure is expressed in the other system. **Sample: an American traveling in Africa or a European traveling in the United States**

You may find it useful to work through how to express each of the estimates given in *reverse* order. For example, guide students to figure out how many kilometers are in a mile, or how to estimate kilograms given a weight in pounds.

Error Prevention!

As needed, review mental math procedures for multiplying and dividing by powers of 10.

Differentiated Instruction

Advanced Learners L4

Using the formula $F = \frac{9}{5}C + 32$, have students convert a few Celsius temperatures to Fahrenheit.

3-7b Activity Lab

Estimating in Different Systems

Every day you see measurements in both the customary system and the metric system. You can use estimates to help you compare measurements in the two systems.

A liter is a little more than a quart.
An inch is about 2.5 cm.
A kilometer is about 0.6 mi.
A kilogram is about 2.2 lb.

You may be more familiar with temperatures measured in Fahrenheit than in Celsius. You can use the table at the right to help you relate to Celsius temperatures.

Celsius Temperature	How It Feels
30°	hot
20°	nice
10°	cold
0°	freezing

EXAMPLE

Which measurement is longer, 15 cm or 10 in.?

One inch is about 2.5 cm.

$10 \cdot 2.5 = 25$ ← **Multiply by 10 to find the number of centimeters in 10 in.**

There are about 25 cm in 10 in. So 10 in. is longer than 15 cm.

Exercises

Which of the two measurements is greater?

1. 600 km or 200 mi **600 km**
2. 0°C or 0°F **0°C**
3. 80 L or 40 gal **40 gal**
4. 50 cm or 25 in. **25 in.**
5. 30 lb or 20 kg **20 kg**
6. 10°F or 20°C **20°C**
7. **Travel** A sign reads "Austin, 200 km." About how many miles would you have to drive to get to Austin? **about 120 mi**
8. **Clothing** A sign shows that the temperature is 27°C. Would you be more comfortable in a short-sleeved shirt or in a ski parka? **short-sleeved shirt**
9. A student tells you that he is 175 cm tall and weighs 72 kg.
   a. About how many feet tall is the student? **about 6 ft**
   b. About how many pounds does the student weigh? **about 158 lb**
10. You need to bring about a gallon of water for a hike. Bottled water is sold in liter bottles. How many liter bottles should you buy? **4 L**

Reading for Understanding

Reading-comprehension questions are based on a passage that gives you facts and information. First read the question carefully. Make sure you understand what is being asked. Then read the passage. Look for the information you need to answer the question.

EXAMPLE

Recycling Math The United States produces more than 4 pounds of trash per person each day, and recycles about one fourth of it. Canada produces about $3\frac{1}{2}$ pounds of trash per person each day, and recycles about one tenth of it. Japan produces about $2\frac{1}{2}$ of trash per person each day, and recycles about one fifth of it.

In one week, a family living in the United States produced 112 pounds of trash. About how many pounds of the trash were recycled?

What is being asked? How many pounds of trash were recycled?

Identify the information you need. The United States recycles about one fourth of the trash it produces.

Solve the problem. Pounds of recycled trash $= \frac{1}{4} \cdot 112 = 28$. About 28 pounds of the American family's trash were recycled.

Exercises

Use the passage in the example to complete Exercises 1–4.

1. a. About how many pounds of trash will a Canadian family of four produce in one week? **about 98 lb**
   b. About how many pounds of the trash will be recycled? **about 10 lb**
2. In one week, a family living in Japan produces 85 pounds of trash. About how many pounds of the trash will be recycled? **about 17 lb**
3. Suppose your family recycles one third of its trash. How much more trash does your family recycle than an average U.S. family? **$\frac{1}{12}$ more**
4. In one month, a school recycled two fifths of its paper and produced 25 pounds of paper. How many pounds were *not* recycled? **15 lb**

Reading for Understanding

This strategy provides students with pointers and practice in how to successfully approach a reading-comprehension question.

Guided Instruction

Emphasize the importance of reading the full passage carefully to identify what the question asks, what information is needed to answer the question, and what information is *not* needed. For example, students need to check whether an exact answer or an estimate is called for.

Error Prevention!

Encourage students to read the passage more than once. Encourage them to double-check that they have answered the question reasonably and not just solved the equation.

Resources

Test-Taking Strategies with Transparencies
- Transparency 5
- Practice sheet, p. 15

Test-Taking Strategies with Transparencies

Test-Taking Strategies: Reading for Understanding

Reading comprehension questions are based on a passage that gives information and facts.

To solve a problem use these steps:
- Read the directions and the passage.
- Read the questions carefully.
- Look for information that helps answer the questions.

EXAMPLE
Read the passage and answer the questions below.

Each year, more than 775,000 children and teenagers are treated in the emergency room for sports injuries. In a recent study, doctors found that 10 years ago, up to 70 percent of sports injuries in kids were acute injuries, such as a sprained ankle or a fracture. Over the past five years, however, overuse accounted for about half of all sports injuries among youngsters. Doctors are seeing a lot more elbow injuries from too much pitching and a lot more heel pain from tendinitis due to too much soccer.

This year about 25 athletes in your school will be injured. About how many of them will have injuries due to overuse?

What are you being asked?
How many of the 25 athletes will have injuries due to overuse?

What information helps you to solve the problem?
Over the past five years overuse accounted for about half of all sports injuries among youngsters.

Solve the problem: 50% of 25 students is about 13 students.

Chapter 3 Review

Resources

Student Edition
Extra Skills and Word Problem Practice, Ch. 3, p. 634
English/Spanish Glossary, p. 676
Formulas and Properties, p. 674
Tables, p. 670

All in One Teaching Resources
- Vocabulary and Study Skills 3F L3

Differentiated Instruction
Spanish Vocabulary Workbook with Study Skills ELL
Interactive Textbook
- Audio Glossary

Online Vocabulary Quiz

Online at PHSchool.com

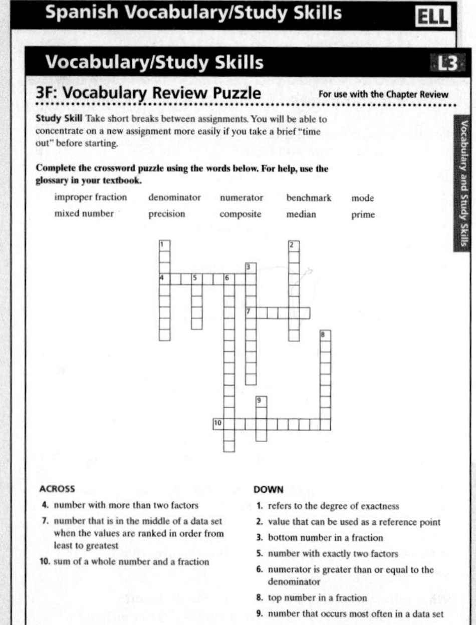
Spanish Vocabulary/Study Skills ELL

Vocabulary/Study Skills L3

3F: Vocabulary Review Puzzle — For use with the Chapter Review

Study Skill Take short breaks between assignments. You will be able to concentrate on a new assignment more easily if you take a brief "time out" before starting.

Complete the crossword puzzle using the words below. For help, use the glossary in your textbook.

improper fraction	denominator	numerator	benchmark	mode
mixed number	precision	composite	median	prime

ACROSS
4. number with more than two factors
7. number that is in the middle of a data set when the values are ranked in order from least to greatest
10. sum of a whole number and a fraction

DOWN
1. refers to the degree of exactness
2. value that can be used as a reference point
3. bottom number in a fraction
5. number with exactly two factors
6. numerator is greater than or equal to the denominator
8. top number in a fraction
9. number that occurs most often in a data set

Vocabulary and Study Skills

Vocabulary Review

benchmark (p. 120) **precision** (p. 154) **reciprocals** (p. 141)

Go Online PHSchool.com
For: Online vocabulary quiz
Web Code: arj-0351

Choose the correct vocabulary term to complete each sentence.

1. The number $\frac{7}{8}$ is the __?__ of the number $\frac{8}{7}$. **reciprocal**
2. A __?__ is a number used to replace fractions that are less than 1. **benchmark**
3. The __?__ of a measurement refers to its degree of exactness. **precision**
4. When you use a __?__ to estimate a sum, your answer has less __?__ than the exact answer. **benchmark; precision**

Skills and Concepts

Lesson 3-1
- To estimate sums, differences, products, and quotients involving fractions

You can use the **benchmarks** $0, \frac{1}{2}$, and 1 to estimate sums and difference of fractions. To estimate sums, differences, products, and quotients of mixed numbers, round to the nearest whole number or use compatible numbers.

Estimate each answer.

5. $\frac{3}{4} + \frac{1}{8}$ **about 1**
6. $\frac{5}{6} - \frac{1}{3}$ **about $\frac{1}{2}$**
7. $\frac{2}{5} + \frac{3}{8}$ **about 1**
8. $\frac{4}{9} - \frac{2}{18}$ **about $\frac{1}{2}$**

Use rounding to estimate. **9–12. See left.**

9. $4\frac{11}{12} - 2\frac{1}{10}$
10. $5\frac{2}{5} \cdot 4\frac{7}{9}$
11. $8\frac{4}{5} + 4\frac{7}{8}$
12. $24\frac{1}{6} - 13\frac{2}{3}$

9. **about 3**
10. **about 25**
11. **about 14**
12. **about 10**

13. **Maps** A map shows three hiking trails of lengths $2\frac{1}{2}$ mi, $1\frac{4}{5}$ mi, and $3\frac{3}{10}$ mi. You want to hike the entire length of each trail. Estimate the total distance. **about 8 mi**

Lessons 3-2, 3-3
- To add and subtract fractions and to solve problems involving fractions
- To add and subtract mixed numbers and to solve problems involving mixed numbers

To add or subtract fractions, you first find a common denominator and then add or subtract the numerators. To add or subtract mixed numbers, you add or subtract the fractions first, and then the whole numbers. You may need to rename a mixed number before subtracting.

Find each sum or difference.

14. $2\frac{1}{3} - \frac{3}{4}$ **$1\frac{7}{12}$**
15. $16\frac{2}{3} - 9\frac{4}{5}$ **$6\frac{13}{15}$**
16. $8\frac{1}{6} + 7\frac{3}{12}$ **$15\frac{5}{12}$**
17. $2\frac{1}{6} + 3\frac{3}{8}$ **$5\frac{13}{24}$**
18. Rich walks to school in $10\frac{1}{5}$ minutes. John walks in $21\frac{1}{4}$ minutes. How many more minutes does it take John to get to school? **$11\frac{1}{20}$ min**

Lessons 3-4, 3-5

- To multiply fractions and mixed numbers and to solve problems by multiplying
- To divide fractions and mixed numbers and to solve problems by dividing

To multiply fractions, you multiply their numerators and multiply their denominators. To divide by a fraction, you multiply by its reciprocal.

Find each product.

19. $\frac{2}{3} \cdot \frac{3}{8}$ $\frac{1}{4}$
20. $\frac{3}{5} \cdot 1\frac{1}{2}$ $\frac{9}{10}$
21. $2\frac{2}{3} \cdot 3\frac{3}{8}$ 9
22. $8\frac{5}{6} \cdot 10\frac{3}{4}$ $94\frac{23}{24}$
23. What is the area of a room that is $12\frac{2}{3}$ m long and $3\frac{1}{8}$ m wide? $39\frac{7}{12}$ m^2

Find each quotient.

24. $\frac{2}{3} \div \frac{4}{3}$ $\frac{1}{2}$
25. $5\frac{1}{4} \div \frac{7}{8}$ 6
26. $4\frac{4}{5} \div 1\frac{1}{3}$ $3\frac{3}{5}$
27. $1\frac{1}{3} \div 4\frac{4}{5}$ $\frac{5}{18}$
28. The area of Sally's garden is $17\frac{1}{2}$ square feet. If Sally's garden is $5\frac{1}{2}$ ft long, how wide is her garden? $3\frac{2}{11}$ ft

Lesson 3-6

- To change units of length, capacity, and weight in the customary system

To change from a smaller unit to a larger unit, you *divide*. To change from a larger unit to a smaller unit, you *multiply*.

Complete.

29. 42 in. = ■ ft $3\frac{1}{2}$
30. 12 fl oz = ■ c $1\frac{1}{2}$
31. 80 oz = ■ lb 5
32. 10,560 ft = ■ mi 2
33. 2 t = ■ lb 4,000
34. $3\frac{1}{2}$ qt = ■ pt 7
35. The highest annual rainfall on record for the state of Texas was about 109 in., in 1873. How many feet of rain did Texas receive that year? $9\frac{1}{12}$ ft

Lesson 3-7

- To find and compare the precision of measurements

The **precision** of a measurement refers to its degree of exactness. The smaller the units on a measuring instrument, the more precise a measurement is. When you add or subtract measurements with the same unit, round your answer to match the precision of the least precise measurement.

Choose the more precise measurement.

36. 12 c, 8 pt 12 c
37. 5.5 L, 2 L 5.5 L
38. 8.5 m, 8.75 m 8.75 m
39. 1 day, 23 h 23 h
40. 25 g, 3.5 kg 25 g
41. $11\frac{3}{16}$ in., 5 in. $11\frac{3}{16}$ in.

Find each sum or difference. Round your answer appropriately.

42. 17.3 g − 10 g 7 g
43. $5\frac{1}{3}$ yd + 8 yd 13 yd
44. 7.75 cm + 3.8 cm 11.6 cm
45. 5.25 lb + 15.75 lb
46. 8 L − 3.005 L
47. 12.175 m − 7.05 m

45. 21.00 lb
46. 5 L
47. 5.13 m

Alternative Assessment L4

Alternative Assessment Form C
Chapter 3

MARATHON TRAINING

To train for the Centerville Marathon, Keith drew a map of his favorite practice routes.

Running Times

D, C, E, A, B, Home; $2\frac{5}{6}$ h, $2\frac{1}{5}$ h, $2\frac{1}{2}$ h, $1\frac{2}{3}$ h, $1\frac{1}{6}$ h, $1\frac{1}{4}$ h, $1\frac{3}{4}$ h, $\frac{4}{5}$ h, $1\frac{1}{3}$ h

Show all your work on a separate sheet of paper.

1. Keith plans to run from Home to point *B*, then to point *A*, then back Home. How long will this route take? If he runs this route once each day from Monday to Friday, what is the total running time?
2. Yesterday Keith ran for nearly 6 hours. If he started at point *E*, name two possible routes he could have taken. Explain your thinking.
3. Keith's trainer suggested that he run from Home to point *D*, then to point *C*, then to point *B*, and then to Home. How long is the route in *hours?* How long is the route in *minutes?*
4. Design a route so that Keith runs each path once. How long will this route take?
5. If Keith runs a mile in $\frac{1}{6}$ hour, how far will he run from point *C* to point *D*? Explain your answer.

Spanish Quarter 1 Test Forms A, B ELL

Quarter 1 Test Forms D, E L2

Quarter 1 Test Forms A, B L3

Quarter 1 Test Form A
Chapters 1–3

1. Use front-end estimation to estimate the sum to the nearest whole number.
   3.59 + 44.603 + 16.24
2. At the Colorado ski meet, the French Canadian skier jumped 89.5234 meters. The American skier jumped 90.06 meters. How much farther did the American skier jump than the Canadian skier?
3. Jennifer needed material for her 4-H project. She bought 3.25 yards of material for $8.82 per yard. What was the total cost of the material? Round to the nearest cent.
4. Convert.
   25 m = ________ cm
5. Find $|-16|$.
6. Mary has a score of 7 in a game. If she loses 10 points, what is her new score?
7. A submarine descends from the surface (0 meters) to −150 meters in 3 minutes. What was the change in depth per minute?
8. Order from least to greatest.
   $\frac{1}{4}, 0.3, -\frac{7}{8}, -2$
9. Simplify. $3^3 \times (7 - 2)^2$
10. Write 7,000,000,000 in scientific notation.
11. Write a number less than 100 that is divisible by 2, 5, and 9.
12. Write the fraction $\frac{42}{49}$ in simplest form.
13. Order from least to greatest.
    $\frac{9}{10}, \frac{8}{11}, \frac{7}{9}$
14. Write $\frac{15}{7}$ as a mixed number.
15. At the local drugstore, it costs $.35 to get a negative of a picture developed. If Lauren spent $3.85 getting negatives developed, how many pictures did she get?
16. Find the GCF of this pair of numbers.
    77 and 154
17. Casey got 32 out of 40 questions correct on her social studies test. Last week she got 24 out of 30 questions correct. Which score was better? Explain.

Resources

- ExamView Assessment Suite CD-ROM
  - Chapter 3 Ready-Made Test
  - Make your own Chapter 3 test
- MindPoint Quiz Show CD-ROM
  - Chapter 3 Review

Differentiated Instruction

All in One Teaching Resources
- Below Level Chapter 3 Test L2
- Chapter 3 Test L3
- Chapter 3 Alternative Assessment L4

Spanish Assessment Resources ELL
- Below Level Chapter 3 Test L2
- Chapter 3 Test L3
- Chapter 3 Alternative Assessment L4

ExamView Assessment Suite CD-ROM
- Special Needs Test L1
- Special Needs Practice Bank L1

Online Chapter 3 Test at www.PHSchool.com L3

Below Level Chapter Test L2

Chapter Test L3

Chapter Test

Chapter 3

Estimate each answer to the nearest whole number.

1. $\frac{5}{9} + 2\frac{3}{4}$ 3
2. $4\frac{2}{5} + 5\frac{5}{9}$ 10
3. $\frac{2}{3} \cdot 119$ 80
4. $2\frac{7}{8} \cdot 4\frac{2}{3}$ 12
5. $8\frac{1}{5} - 5\frac{6}{7}$ 2
6. $23\frac{9}{10} \div 12\frac{1}{4}$ 2

Write each sum or difference as a fraction or mixed number in simplest form.

7. $\frac{2}{3} - \frac{5}{12}$ $\frac{1}{4}$
8. $\frac{3}{4} - \frac{1}{2}$ $\frac{1}{4}$
9. $2\frac{1}{3} + 4\frac{3}{5}$ $6\frac{4}{5}$
10. $7\frac{1}{4} + 5\frac{1}{2}$ $12\frac{3}{4}$
11. $\frac{1}{17} + \frac{5}{17}$ $\frac{6}{17}$
12. $\frac{7}{8} - \frac{3}{5}$ $\frac{11}{40}$
13. The price of a company's stock was $21\frac{3}{8}$ dollars a share. Then the price went up $4\frac{1}{2}$ dollars a share. What is the new price of the stock? $25\frac{7}{8}$ dollars

Find each product or quotient.

14. $\frac{3}{8} \cdot \frac{4}{9}$ $\frac{1}{6}$
15. $\frac{3}{4} \div \frac{7}{12}$ $1\frac{2}{7}$
16. $2\frac{3}{8} \cdot 4\frac{4}{5}$ $11\frac{2}{5}$
17. $2\frac{2}{7} \div 4$ $\frac{4}{7}$
18. $10\frac{5}{6} \div \frac{5}{9}$ $19\frac{1}{2}$
19. $11\frac{9}{10} \cdot \frac{2}{17}$ $1\frac{2}{5}$
20. Tiles $\frac{3}{4}$ ft long are being placed alongside a vanity top that is $4\frac{1}{2}$ ft long. How many tiles are needed? 6 tiles
21. A pattern for curtains requires $1\frac{3}{4}$ yd of fabric. How much fabric is needed to make 5 sets of these curtains? $8\frac{3}{4}$ yd

Chapter 3 Test

Go Online PHSchool.com For: Online chapter test Web Code: ara-0352

Estimate each answer.

1. $\frac{7}{8} + \frac{15}{16}$ about 2
2. $\frac{3}{5} - \frac{1}{2}$ about 0
3. $7\frac{1}{8} + 2\frac{3}{4}$ about 10
4. $8\frac{3}{8} - 5\frac{1}{3}$ about 3
5. $4\frac{5}{8} \cdot 2\frac{1}{10}$ about 10
6. $43\frac{1}{2} \div 5\frac{1}{5}$ about 9
7. **Tutoring** You tutored for $2\frac{1}{2}$ h on Monday, $2\frac{1}{6}$ h on Wednesday, and $1\frac{3}{4}$ h on Friday. Estimate the total number of hours you tutored during the week. about 7 h

Find each sum or difference.

8. $\frac{15}{16} - \frac{3}{16}$ $\frac{3}{4}$
9. $\frac{1}{4} + \frac{2}{3}$ $\frac{11}{12}$
10. $\frac{1}{2} - \frac{3}{8}$ $\frac{1}{8}$
11. $\frac{2}{3} + \frac{5}{6}$ $1\frac{1}{2}$
12. $8\frac{2}{5} + 5\frac{3}{5}$ 14
13. $1\frac{2}{3} - \frac{3}{4}$ $\frac{11}{12}$
14. $4\frac{3}{4} + 5\frac{1}{5}$ $9\frac{19}{20}$
15. $9\frac{3}{8} - 5\frac{1}{4}$ $4\frac{1}{8}$
16. At birth, a baby weighed $7\frac{3}{8}$ lb. The baby now weighs $8\frac{13}{16}$ lb. How much weight has the baby gained? $1\frac{7}{16}$ lb
17. You ride your bike $1\frac{3}{10}$ mi to school. At the end of the day, you stop at a park on the way home. The park is $\frac{2}{5}$ mi from school. How far is the park from your house? $\frac{9}{10}$ mi

Find each product or quotient.

18. $\frac{4}{5} \cdot \frac{1}{4}$ $\frac{1}{5}$
19. $\frac{4}{5} \div \frac{1}{4}$ $3\frac{1}{5}$
20. $\frac{3}{4} \cdot \frac{2}{3}$ $\frac{1}{2}$
21. $\frac{1}{4} \div \frac{4}{5}$ $\frac{5}{16}$
22. $1\frac{2}{3} \cdot 1\frac{1}{4}$ $2\frac{1}{12}$
23. $1\frac{2}{3} \div 1\frac{1}{4}$ $1\frac{1}{3}$
24. $150 \div 2\frac{2}{3}$ $56\frac{1}{4}$
25. $5\frac{3}{8} \cdot 3\frac{3}{4}$ $20\frac{5}{32}$
26. **Masonry** A brick is $1\frac{7}{8}$ in. high. Mortar that is $\frac{3}{8}$ in. thick is spread on each row of bricks. How many rows of bricks are needed to reach the top of a $7\frac{1}{2}$-ft doorway? 40 rows

Find each answer.

27. $10\frac{3}{4} + \frac{5}{8}$ $11\frac{3}{8}$
28. $6\frac{1}{2} - 4\frac{3}{5}$ $1\frac{9}{10}$
29. $10 \div \frac{1}{4}$ 40
30. $2\frac{1}{2} \cdot 12$ 30
31. $16 \div \frac{1}{3}$ 48
32. $8\frac{3}{5} + \frac{1}{7}$ $8\frac{26}{35}$
33. $2\frac{3}{10} - 1\frac{1}{2}$ $\frac{4}{5}$
34. $7\frac{3}{4} + \frac{3}{8}$ $8\frac{1}{8}$
35. **Earnings** You have a part-time job at a deli and work $12\frac{1}{2}$ hours in one week. You earn \$6.50 per hour. How much money do you earn that week? \$81.25
36. **Discounts** A store is having a $\frac{1}{3}$-off sale. You want to buy a jacket that originally cost \$60. What is the sale price of the jacket? \$40

Complete.

37. 38 in. = ■ ft $3\frac{1}{6}$
38. 60 oz = ■ lb $3\frac{3}{4}$
39. $3\frac{3}{4}$ qt = ■ c 15
40. $1\frac{2}{3}$ mi = ■ ft 8,800
41. $5\frac{1}{2}$ yd = ■ in. 198
42. 50 fl oz = ■ c $6\frac{1}{4}$
43. **Writing in Math** About how heavy should an object be before you start to measure the object in tons instead of pounds? Explain. See margin.
44. You have $1\frac{1}{2}$ lb of fish. How many 6-oz servings can you make? 4 servings
45. You jog $8\frac{1}{2}$ times around a block. The distance around the block is 770 yd. About how many miles have you jogged? (Hint: 1 mi = 1,760 yd) $3\frac{23}{32}$ mi

Choose the more precise measurement.

46. 36 min, $1\frac{1}{4}$ h 36 min
47. 1 t, 500 lb 500 lb
48. 100.5 mg, 10.67 g 100.5 mg
49. 18 months, 1 year 18 mo

43. Answers may vary. Sample: A ton is 2,000 lb, so an object should weigh at least that much.

Test Prep Practice

Multiple Choice

Read each question. Then write the letter of the correct answer on your paper.

1. Using estimation, which sum is between 21 and 22? **B**
   - (A) 13.71 + 1.5 + 8.2
   - (B) 6.75 + 9.02 + 5.838
   - (C) 5.99 + 2.69 + 15.49
   - (D) 3.772 + 12.04 + 4.009

2. Which of the following statements is true? **H**
   - (F) $-23 \geq 23$
   - (G) $14 < -21$
   - (H) $-14 < -3$
   - (J) $7 < 0$

3. Which of the following expressions is equal to 10? **D**
   - (A) $4 + 2 - 5 \div 2$
   - (B) $4 + 2 \div 2$
   - (C) $4 + 3 - 2 \cdot 2$
   - (D) (4 + [illegible] − 2) · 2

4. Karin received the following scores on her last four math tests: 87, 92, 80, 85. What was her mean test score? **F**
   - (F) 86
   - (G) 88
   - (H) 90
   - (J) 92

5. What is the value of the expression $2^3 + 3 \cdot 4$? **B**
   - (A) 18
   - (B) 20
   - (C) 28
   - (D) 44

6. How is 73,460,000,000 written in scientific notation? **J**
   - (F) $73.46 + 10^4$
   - (G) 73.46×10^9
   - (H) 7.346×10^9
   - (J) 7.346×10^{10}

7. Which of the following is NOT written in order from least to greatest? **B**
   - (A) $\frac{1}{9}, \frac{1}{8}, \frac{1}{7}$
   - (B) $\frac{2}{3}, \frac{1}{2}, \frac{3}{4}$
   - (C) $\frac{1}{2}, \frac{2}{3}, \frac{3}{4}$
   - (D) $\frac{2}{5}, \frac{1}{2}, \frac{7}{8}$

8. A student measured his height and found that it is between 5 ft and $5\frac{1}{3}$ ft. Which height could it be? **J**
   - (F) $\frac{51}{8}$ ft
   - (G) $\frac{25}{4}$ ft
   - (H) $\frac{28}{5}$ ft
   - (J) $\frac{62}{12}$ ft

9. Which unit is the most appropriate unit for the length of a pencil? **B**
   - (A) mL
   - (B) cm
   - (C) kg
   - (D) m

10. To get to your friend's house, you must travel $\frac{1}{3}$ mile down one street, $\frac{3}{4}$ mile down another street, and $1\frac{1}{2}$ mile down a third street. How many miles must you travel in all to get to your friend's house? **J**
    - (F) $1\frac{5}{9}$ mi
    - (G) $1\frac{7}{12}$ mi
    - (H) $2\frac{5}{9}$ mi
    - (J) $2\frac{7}{12}$ mi

Gridded Response

Record your answer in a grid.

11. You want to buy some helium balloons for a birthday party. A store charges $3.35 per balloon. You spend $16.75. How many balloons do you buy? **5**

Short Response

12. Is 54.5 g or 54.5 kg a more reasonable estimate for a person's weight? Explain.

12–13. See margin.

Extended Response

13. One goaltender makes 72 saves on 78 shots. Another makes 52 saves on 56 shots.
    - a. Write two fractions in simplest form to express the numbers of saves per shot.
    - b. Which goaltender has the greater number of saves per shot?
    - c. Explain how you found your answer to part (b).

Item	1	2	3	4	5	6	7	8	9	10	11	12	13
Lesson	1-1	1-6	1-9	1-10	2-1	2-8	2-4	2-5	1-5	3-3	1-4	1-5	2-6

12. [2] 54.5 kg; Answers may vary. Sample: a kg is about 2.2 lb, a g is about the weight of a paper clip
    [1] correct answer, but no explanation

13. [4] $\frac{12}{13}, \frac{13}{14}$; second goal tender; the least common denominator is 182. Rewrite $\frac{12}{13}$ as $\frac{168}{182}$ and $\frac{13}{14}$ as $\frac{169}{182}$. Comparing the numerators, $\frac{168}{182} < \frac{169}{182}$, so $\frac{12}{13} < \frac{13}{14}$.
    [3] appropriate methods, but with one computational error
    [2] correct answers, but incomplete explanation
    [1] correct answers, but no work shown

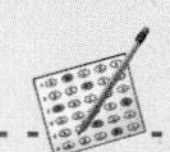

Test Prep

Resources

Test Prep Workbook

All-in-One Teaching Resources
- Cumulative Review L3

ExamView Assessment Suite CD-ROM
- Standardized Test Practice

Differentiated Instruction

Progress Monitoring Assessments
- Quarter 1 Test
  - Forms A & B L3
  - Forms D & E L2

Spanish Assessment Resources
- Spanish Cumulative Review ELL
- Spanish Quarter 1 Test
  - Forms A & B L3

Exam View Assessment Suite CD-ROM
- Special Needs Practice Bank L1

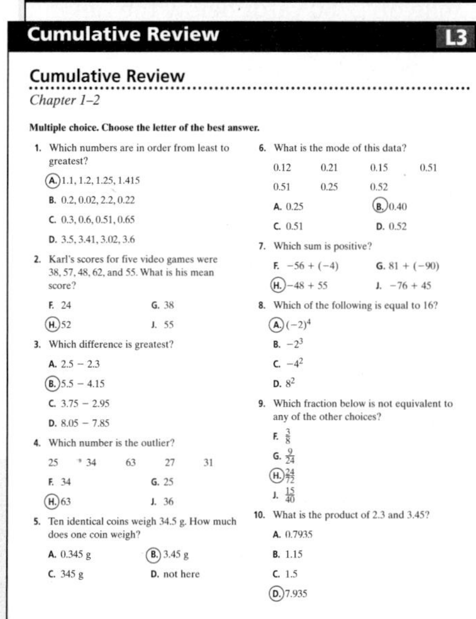

Spanish Cumulative Review ELL

Cumulative Review L3

Cumulative Review

Chapter 1–2

Multiple choice. Choose the letter of the best answer.

1. Which numbers are in order from least to greatest?
   - (A.) 1.1, 1.2, 1.25, 1.415
   - B. 0.2, 0.02, 2.2, 0.22
   - C. 0.3, 0.6, 0.51, 0.65
   - D. 3.5, 3.41, 3.02, 3.6
2. Karl's scores for five video games were 38, 57, 48, 62, and 55. What is his mean score?
   - F. 24
   - G. 38
   - (H.) 52
   - J. 55
3. Which difference is greatest?
   - A. 2.5 − 2.3
   - (B.) 5.5 − 4.15
   - C. 3.75 − 2.95
   - D. 8.05 − 7.85
4. Which number is the outlier?
   25 34 63 27 31
   - F. 34
   - G. 25
   - (H.) 63
   - J. 36
5. Ten identical coins weigh 34.5 g. How much does one coin weigh?
   - A. 0.345 g
   - (B.) 3.45 g
   - C. 345 g
   - D. not here
6. What is the mode of this data?
   0.12 0.21 0.15 0.51
   0.51 0.25 0.52
   - A. 0.25
   - (B.) 0.40
   - C. 0.51
   - D. 0.52
7. Which sum is positive?
   - F. $-56 + (-4)$
   - G. $81 + (-90)$
   - (H.) $-48 + 55$
   - J. $-76 + 45$
8. Which of the following is equal to 16?
   - (A.) $(-2)^4$
   - B. -2^3
   - C. -4^2
   - D. 8^2
9. Which fraction below is not equivalent to any of the other choices?
   - F. $\frac{3}{8}$
   - G. $\frac{9}{24}$
   - (H.) $\frac{24}{72}$
   - J. $\frac{15}{40}$
10. What is the product of 2.3 and 3.45?
    - A. 0.7935
    - B. 1.15
    - C. 1.5
    - (D.) 7.935

Applying Fractions

Students will use data from these two pages to answer the questions posed in Put It All Together.

Have students find out the length of the Grand Canyon and the Colorado River. **The Grand Canyon is about 277 miles long and the Colorado River is about 1,450 miles long.** Ask:

- *What fraction of the Colorado River runs through the Grand Canyon?* **about $\frac{1}{5}$**

Activating Prior Knowledge

Have students who have visited and hiked into the Grand Canyon or any other canyons share their experiences with classmates.

Guided Instruction

Invite students to investigate the relationship between geography and climate that explains the kinds of plants and animals that inhabit the Grand Canyon. Have them research the characteristics of these life forms that allow them to thrive in such a harsh environment.

The Colorado River runs about a mile or 1,760 yards below the rim of the Grand Canyon. Ask:

- *About how many feet is the Colorado River below the Grand Canyon?* **about 5,280 feet**
- *If you hike down 1,760 feet into the canyon, what fraction of a mile have you descended?* **about $\frac{1}{3}$ of a mile**

Problem Solving Application

Applying Fractions

Into the Earth With Integers Millions of people come to Arizona each year to stand at the edge of the Grand Canyon and look down at the Colorado River, a mile below. If you're one of them, you might decide to hike the winding trail to the canyon floor.

Collared Lizard

Adult collared lizards typically grow up to 10–13 in. long and can travel at speeds over 10 mi/h. To escape predators, they can run bipedally (on two legs).

two black bands

colorful dotted back

long tail

Put It All Together

Materials 10 index cards, markers, two boxes

What You'll Need

- Number four of the index cards -2, -1, $+1$, and $+2$. Put them in one of the boxes and label it "numerators."
- Number the remaining index cards 3, 4, 5, 6, 7, and 8. Put them in the other box and label it "denominators."

How To Play

- The goal of the game is to move from the canyon rim (0) to the canyon floor (-1).
- Draw one card from each box. Write down the fraction it represents. (For example, -2 from the numerator box and 5 from the denominator box represent $-\frac{2}{5}$, or descending $\frac{2}{5}$ mi into the canyon.) *To leave the canyon rim, you must draw a negative numerator.*
- Replace the cards and draw again. Add the fractions. Repeat until one of the players reaches -1 (or beyond). If your total rises above 0, you are back on the rim. *Start again at 0!*

1. Write an equation to show how far you go toward the bottom in your first two moves. *Use zero for each positive move before your first negative move.*
2. a. **Reasoning** How would you calculate the total distance you hiked? Explain.
   b. How far did you hike?
3. a. **Writing in Math** Choose whichever numerator and whichever denominator you want for your first move. Explain your choice.
   b. **Reasoning** Suppose you can choose for each move, but you cannot use the same denominator twice. How many moves will it take for you to get to the bottom? Explain, using a fraction sentence.

164

1. Answers may vary. Sample: $-\frac{2}{5} + \left(-\frac{1}{3}\right) = -\frac{11}{15}$

2a. Answers may vary. Sample: Take the absolute value of each fraction and add them together.

b. Check students' work.

3a. $-\frac{2}{3}$; $-\frac{2}{3}$ has the greatest absolute value

b. two moves; $-\frac{2}{3} + -\frac{2}{4} = 1\frac{1}{6}$

Riding Down

Mules are one form of transportation in the Grand Canyon.

The Pinyon Jay

Pinyon jays live in the Grand Canyon, and among the Pinyon and Juniper pines throughout the southwest.

Activity

Have students work in pairs to answer the questions.

Science Connection
Canyons are not formed entirely by the rivers that run through them, like the Colorado River at the bottom of the Grand Canyon or the Virgin River in Zion Canyon. Have students do research to learn how these wonders of nature came to be. Have them post their findings for classmates to read.

Differentiated Instruction

Special Needs L1
You may find it useful to go over the rules of the game and play a practice round or two with students. As needed, review and practice how to add and subtract fractions and integers, and how to write equations.

4 Equations and Inequalities

Chapter at a Glance

Lesson Titles, Objectives, and Features	Assessment	NCTM Standards	Local Standards
4-1a Activity Lab: Describing Patterns **4-1 Evaluating and Writing Algebraic Expressions** • To write and evaluate algebraic expressions **4-1b Activity Lab, Technology:** Using Spreadsheets	Lesson Quiz	1, 2, 6, 7, 8, 9, 10	
4-2 Using Number Sense to Solve Equations • To solve one-step equations using substitution, mental math, and estimation **4-2b Activity Lab, Algebra Thinking:** Keeping the Balance	Lesson Quiz	1, 2, 4, 6, 7, 8, 9, 10	
4-3a Activity Lab, Hands On: Modeling Equations **4-3 Solving Equations by Adding or Subtracting** • To solve equations by adding or subtracting **Vocabulary Builder:** High-Use Academic Words	Lesson Quiz	1, 2, 6, 7, 8, 9, 10	
4-4 Solving Equations by Multiplying or Dividing • To solve equations by multiplying or dividing **Guided Problem Solving:** Practice Solving Problems	Lesson Quiz Checkpoint Quiz 1	1, 2, 4, 6, 7, 8, 9, 10	
4-5 Exploring Two-Step Problems • To write and evaluate expressions with two operations and to solve two-step equations using number sense	Lesson Quiz	1, 2, 4, 6, 7, 8, 9, 10	
4-6a Activity Lab, Hands On: Modeling Two-Step Equations **4-6 Solving Two-Step Equations** • To solve two-step equations using inverse operations	Lesson Quiz	1, 2, 6, 7, 8, 9, 10	
4-7 Graphing and Writing Inequalities • To graph and write algebraic inequalities	Lesson Quiz Checkpoint Quiz 2	1, 2, 4, 6, 7, 8, 9, 10	
4-8a Activity Lab, Data Analysis: Inequalities in Bar Graphs **4-8 Solving Inequalities by Adding or Subtracting** • To solve inequalities by adding or subtracting	Lesson Quiz	1, 2, 5, 6, 7, 8, 9, 10	
4-9 Solving Inequalities by Multiplying or Dividing • To solve inequalities by multiplying or dividing	Lesson Quiz	1, 2, 4, 6, 7, 8, 9, 10	
Problem Solving Application: Applying Inequalities			

NCTM Standards 2000

1 Number and Operations **2** Algebra **3** Geometry **4** Measurement **5** Data Analysis and Probability
6 Problem Solving **7** Reasoning and Proof **8** Communication **9** Connections **10** Representation

Correlations to Standardized Tests

All content for these tests is contained in *Prentice Hall Math,* Course 2. This chart reflects coverage in this chapter only.

	4-1	4-2	4-3	4-4	4-5	4-6	4-7	4-8	4-9
Terra Nova CAT 6 (Level 17)									
Number and Number Relations	✔	✔	✔	✔	✔	✔	✔	✔	✔
Computation and Numerical Estimation	✔	✔	✔	✔	✔	✔	✔	✔	✔
Operation Concepts	✔	✔	✔	✔	✔	✔	✔	✔	✔
Measurement									
Geometry and Spatial Sense									
Data Analysis, Statistics, and Probability							✔		
Patterns, Functions, Algebra	✔	✔	✔	✔	✔	✔	✔	✔	✔
Problem Solving and Reasoning	✔	✔	✔	✔	✔	✔	✔	✔	✔
Communication	✔	✔	✔	✔	✔	✔	✔	✔	✔
Decimals, Fractions, Integers, and Percents	✔	✔	✔	✔	✔	✔	✔	✔	✔
Order of Operations									
Terra Nova CTBS (Level 17)									
Decimals, Fractions, Integers, Percents	✔	✔	✔	✔	✔	✔	✔	✔	✔
Order of Operations, Numeration, Number Theory	✔	✔	✔	✔	✔	✔	✔	✔	✔
Data Interpretation							✔		
Pre-algebra	✔	✔	✔	✔	✔	✔	✔	✔	✔
Measurement									
Geometry									
ITBS (Level 13)									
Number Properties and Operations	✔	✔	✔	✔	✔	✔	✔	✔	✔
Algebra	✔	✔	✔	✔	✔	✔	✔	✔	✔
Geometry									
Measurement									
Probability and Statistics									
Estimation									
SAT 10 (Int 3 Level)									
Number Sense and Operations	✔	✔	✔	✔	✔	✔	✔	✔	✔
Patterns, Relationships, and Algebra	✔	✔	✔	✔	✔	✔	✔	✔	✔
Data, Statistics, and Probability							✔		
Geometry and Measurement									
NAEP									
Number Sense, Properties, and Operations								✔	
Measurement									
Geometry and Spatial Sense									
Data Analysis, Statistics, and Probability									
Algebra and Functions		✔	✔	✔	✔	✔	✔	✔	✔

CAT6 California Achievement Test, 6th Ed. **CTBS** Comprehensive Test of Basic Skills **ITBS** Iowa Test of Basic Skills, Form M
SAT10 Stanford Achievement Test, 10th Ed. **NAEP** National Assessment of Educational Progress 2005 Mathematics Objectives

Math Background

Skills Trace

BEFORE Chapter 4

Course 1 or Grade 6 introduced the solving of equations and of inequalities.

DURING Chapter 4

Course 2 reviews and extends the solving equations and inequalities to include decimals, fractions, and integers.

AFTER Chapter 4

Throughout this course, students write and solve equations involving real-world problems.

4-1 Evaluating and Writing Algebraic Expressions

Math Understandings

- The value of an algebraic expression varies depending on the value of the variable.
- Within a single problem, the value of the variable usually remains the same.

A **variable** is a symbol, usually a letter, that represents one or more numbers. An **algebraic expression** is a mathematical phrase with at least one variable.

Variables	Algebraic Expressions
x, y, a, c, p, m	$3x, y - 2, a \div 5, 3p + 2, 7 - m$

4-2 Using Number Sense to Solve Equations

Math Understandings

- An equation with one or more variables is an open sentence.
- An open sentence is neither true nor false until the variable is replaced with a number value.
- Some equations are easy to solve using mental math.

An **equation** is a mathematical sentence with an equal sign. A **solution of an equation** is a value for the variable that makes the equation true. An equation with one or more variables is an **open sentence.**

Equations	Open Sentences	Solutions
$6 \cdot 8 = 48$	$6y = 42$	$y = 7$
$2x + 7 = 9$	$q - 9 = 5$	$q = 14$

4-3 4-4 Solving Equations by Adding, Subtracting, Multiplying, or Dividing

Math Understandings

- To solve an equation, you want to get the variable alone on one side of the equation by using the following properties.

Properties of Equality Arithmetic	Algebra
Addition Property of Equality	
$\frac{20}{2} = 10$, so $\frac{20}{2} + 3 = 10 + 3$	If $a = b$ then $a + c = b + c$.
Subtraction Property of Equality	
$\frac{12}{2} = 6$, so $\frac{12}{2} - 4 = 6 - 4$	If $a = b$ then $a - c = b - c$.
Division Property of Equality	
$3(2) = 6$ so, $\frac{3(2)}{2} = \frac{6}{2}$	If $a = b$ and $c \neq 0$ then $\frac{a}{c} = \frac{b}{c}$
Multiplication Property of Equality	
$\frac{12}{2} = 6$, so $\frac{12}{2} \cdot 2 = 6 \cdot 2$.	If $a = b$ then $a \cdot c = b \cdot c$.

Inverse operations are operations that undo each other. Addition and subtraction are inverse operations. Multiplication and division are inverse operations.

4-5 4-6 Exploring Two-Step Problems and Solving Two-Step Equations

Math Understandings

- To solve a two-step equation, you want to get the variable alone on one side of the equal sign using properties of equality and inverse operations.

The order for solving a two-step equation is usually the reverse of the order of operations: first undo addition or subtraction, then undo multiplication or division.

4-7 Graphing and Writing Inequalities

Math Understandings

- We can use special symbols to describe how two numbers or expressions are related to each other.

A mathematical sentence that contains one of the following symbols is an **inequality.**

$<$ is less than	$>$ is greater than
$\leq$ is less than or equal to	$\geq$ is greater than or equal to
$\neq$ is not equal to	

The **solution of an inequality** is any value that makes the inequality true. The graph of an inequality shows that an inequality can have many solutions. A closed circle indicates that a value *is* included. An open circle indicates that a value is *not* included.

Example: Graph $c \leq 3$.

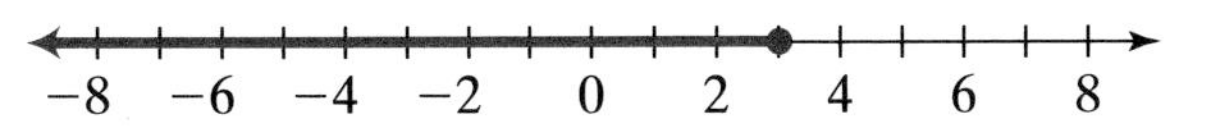

4-8 4-9 Solving Inequalities by Adding, Subtracting, Multiplying, or Dividing

Math Understandings

- The steps for solving one-step and two-step inequalities are much the same as those for solving equations.
- Adding or subtracting from both sides of an inequality by any number does not change the direction of the inequality sign.
- Multiplying or dividing both sides of an inequality by a positive number does not change the direction of the inequality sign, but multiplying or dividing by a negative number changes the direction of the inequality sign.

Properties of Inequality

Arithmetic	Algebra
Addition Property of Inequality	
Since $7 > 3$, $7 + 4 > 3 + 4$	If $a > b$, then $a + c > b + c$.
Since $1 < 3$, $1 + 4 < 3 + 4$	If $a < b$, then $a + c < b + c$.
Subtraction Property of Inequality	
Since $9 > 6$, $9 - 3 > 6 - 3$	If $a > b$, then $a - c > b - c$
Since $7 < 20$, $7 - 4 < 20 - 4$	If $a < b$, then $a - c < b - c$.
Division Property of Inequality	
$9 > 6$, so $\frac{9}{3} > \frac{6}{3}$	If $a > b$ and $c > 0$, then $\frac{a}{c} > \frac{b}{c}$.
$15 < 20$, so $\frac{15}{5} < \frac{20}{5}$.	If $a < b$ and $c > 0$, then $\frac{a}{c} < \frac{b}{c}$.
$16 > 12$, so $\frac{16}{-4} < \frac{12}{-4}$.	If $a > b$ and $c < 0$, then $\frac{a}{c} < \frac{b}{c}$.
$10 < 18$, so $\frac{10}{-2} > \frac{18}{-2}$.	If $a < b$ and $c < 0$, then $\frac{a}{c} > \frac{b}{c}$.
Multiplication Property of Inequality	
$9 > 5$, so $9 \cdot 2 > 5 \cdot 2$	If $a > b$ and $c > 0$, then $a \cdot c > b \cdot c$.
$3 < 6$, so $3 \cdot 4 < 6 \cdot 4$	If $a < b$ and $c > 0$, then $a \cdot c < b \cdot c$.
$7 > 4$, so $7 \cdot (-2) < 4 \cdot (-2)$	If $a > b$ and $c < 0$, then $a \cdot c < b \cdot c$.
$2 < 8$, so $2 \cdot (-3) > 8 \cdot (-3)$	If $a < b$ and $c < 0$, then $a \cdot c > b \cdot c$.

Additional Professional Development Opportunities

Math Background Notes for Chapter 4: Every lesson has a Math Background in the PLAN section.

Research Overview, Mathematics Strands
Additional support for these topics and more is in the front of the Teacher's Edition.

LessonLab
LessonLab, a Pearson Education company, offers comprehensive, facilitated professional development designed to help teachers to improve student achievement. To learn more, please visit lessonlab.com.

Chapter 4 Resources

	4-1	4-2	4-3	4-4	4-5	4-6	4-7	4-8	4-9	For the Chapter
Print Resources										
L3 Practice	•	•	•	•	•	•	•	•	•	
L1 Adapted Practice	•	•	•	•	•	•	•	•	•	
L3 Guided Problem Solving	•	•	•	•	•	•	•	•	•	
L2 Reteaching	•	•	•	•	•	•	•	•	•	
L4 Enrichment	•	•	•	•	•	•	•	•	•	
L3 Daily Notetaking Guide	•	•	•	•	•	•	•	•	•	
L1 Adapted Daily Notetaking Guide	•	•	•	•	•	•	•	•	•	
L3 Vocabulary and Study Skills Worksheets	•		•		•		•		•	•
L3 Daily Puzzles	•	•	•	•	•	•	•	•	•	
L3 Activity Labs	•	•	•	•	•	•	•	•	•	
L3 Checkpoint Quiz				•			•			
L3 Chapter Project										•
L2 Below Level Chapter Test										•
L3 Chapter Test										•
L4 Alternative Assessment										•
L3 Cumulative Review										•
Spanish Resources ELL										
L3 Practice	•	•	•	•	•	•	•	•	•	
L3 Vocabulary and Study Skills Worksheets	•		•		•		•		•	•
L3 Checkpoint Quiz				•			•			
L2 Below Level Chapter Test										•
L3 Chapter Test										•
L4 Alternative Assessment										•
L3 Cumulative Review										•
Transparencies										
Check Skills You'll Need	•	•	•	•	•	•	•	•	•	
Additional Examples	•	•	•	•	•	•	•	•	•	
Problem of the Day	•	•	•	•	•	•	•	•	•	
Classroom Aid	•		•	•						
Student Edition Answers	•	•	•	•	•	•	•	•	•	•
Lesson Quiz	•	•	•	•	•	•	•	•	•	
Test-Taking Strategies										•
Technology										
Interactive Textbook Online	•	•	•	•	•	•	•	•	•	•
StudentExpress™ CD-ROM	•	•	•	•	•	•	•	•	•	•
Success Tracker™ Online Intervention	•	•	•	•	•	•	•	•	•	•
TeacherExpress™ CD-ROM	•	•	•	•	•	•	•	•	•	•
PresentationExpress™ with QuickTake Presenter CD-ROM	•	•	•	•	•	•	•	•	•	•
ExamView® Assessment Suite CD-ROM	•	•	•	•	•	•	•	•	•	•
MindPoint® Quiz Show CD-ROM										•
Prentice Hall Web Site: PHSchool.com	•	•	•	•	•	•	•	•	•	•

Also available: **Prentice Hall Assessment System**
- Progress Monitoring Assessments
- Skills and Concepts Review
- Test Prep Workbook

Other Resources
Algebra Readiness Tests
All-in-One Student Workbook
All-in-One Student Workbook, Adapted Version
Multilingual Handbook
Solution Key
Math Notes Study Folder
Spanish Cumulative Assessment

Where You Can Use the Lesson Resources

Here is a suggestion, following the four-step teaching plan, for how you can incorporate Differentiated Instruction Resources into your teaching.

1. Plan

Preparation
Read the Math Background in the Teacher's Edition to connect this lesson with students' previous experience.

Starting Class
Check Skills You'll Need Assign these exercises to review prerequisite skills.

New Vocabulary Help students pre-read the lesson by pointing out the new terms introduced in the lesson.

Instructional Resources L3
Math Background

Math Understandings

Transparencies & PresentationExpress™ with QuickTake Presenter CD-ROM
Check Skills You'll Need
Problem of the Day

Resources
Vocabulary and Study Skills

Differentiated Instruction Resources
Spanish Support
ELL Vocabulary and Study Skills

2. Teach

[illegible] Guided Instruction
Use the Activity Labs to build conceptual understanding.

Teach each Example. Use the Teacher's Edition side column notes for specific teaching tips, including Error Prevention notes. Use the Additional Examples found in the side column (and on transparency and PowerPoint) as an alternative presentation for the content.

After each Example, assign the Quick Check exercise for that Example to get an immediate assessment of student understanding.

Use the Closure activity in the Teacher's Edition to help students attain mastery of lesson content.

Instructional Resources L3
Student Edition
Activity Lab

Resources
Daily Notetaking Guide
Activity Lab

Transparencies & PresentationExpress™ with QuickTake Presenter CD-ROM
Additional Examples
Classroom Aids

ExamView® Assessment Suite CD-ROM

Differentiated Instruction Resources
Teacher's Edition
Every lesson includes suggestions for working with students who need special attention.
L1 Special Needs
L2 Below Level
L4 Advanced Learners
ELL English Language Learners

Resources
L1 Adapted Daily Notetaking Guide

Multilingual Handbook

3. Practice

Assignment Guide
Check Your Understanding
Use these questions to check students' understanding before you assign homework.

Homework Exercises
Assign homework from these leveled exercises in the Assignment Guide.
- A Practice by Example
- B Apply Your Skills
- C Challenge
- Test Prep and Mixed Review

Homework Quick Check
Use these key exercises to quickly check students' homework.

Instructional Resources L3
Transparencies & PresentationExpress™ with QuickTake Presenter CD-ROM
Student Answers

Resources
Practice
Guided Problem Solving
Vocabulary and Study Skills
Activity Lab
Daily Puzzles

ExamView® Assessment Suite CD-ROM

Differentiated Instruction Resources
Spanish Support
ELL Practice
ELL Vocabulary and Study Skills

Resources
L1 Adapted Practice
L4 Enrichment

4. Assess & Reteach

Lesson Quiz Assign the Lesson Quiz to assess students' mastery of the lesson content.

Checkpoint Quiz Use the Checkpoint Quiz to assess student progress over several lessons.

Instructional Resources L3
Transparencies & PresentationExpress™ with QuickTake Presenter CD-ROM
Lesson Quiz

Resources
Checkpoint Quiz

Differentiated Instruction Resources
Resources
L2 Reteaching
ELL Checkpoint Quiz
Success Tracker™ Online Intervention

ExamView® Assessment Suite CD-ROM

KEY L1 Special Needs L2 Below Level L3 For All Students L4 Advanced, Gifted ELL English Language Learners

Equations and Inequalities

Answers are in the back of the textbook.

For intervention, direct students to:

Adding and Subtracting Decimals
Lesson 1-2
Extra Skills and Word Problems Practice, Ch. 1

Multiplying and Dividing Integers
Lesson 1-8
Extra Skills and Word Problems Practice, Ch. 1

Order of Operations
Lesson 1-9
Extra Skills and Word Problems Practice, Ch. 1

Comparing Fractions
Lesson 2-4
Extra Skills and Word Problems Practice, Ch. 2

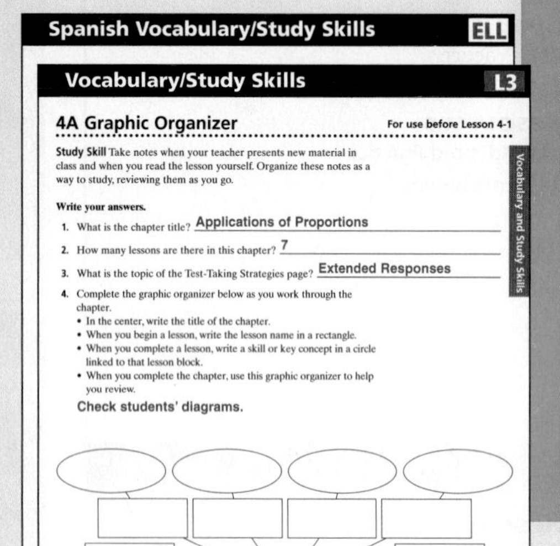
Spanish Vocabulary/Study Skills ELL

Vocabulary/Study Skills L3

4A Graphic Organizer — For use before Lesson 4-1

Study Skill Take notes when your teacher presents new material in class and when you read the lesson yourself. Organize these notes as a way to study, reviewing them as you go.

Write your answers.

1. What is the chapter title? Applications of Proportions
2. How many lessons are there in this chapter? 7
3. What is the topic of the Test-Taking Strategies page? Extended Responses
4. Complete the graphic organizer below as you work through the chapter.
   - In the center, write the title of the chapter.
   - When you begin a lesson, write the lesson name in a rectangle.
   - When you complete a lesson, write a skill or key concept in a circle linked to that lesson block.
   - When you complete the chapter, use this graphic organizer to help you review.

Check students' diagrams.

Algebra

CHAPTER 4 Equations and Inequalities

What You've Learned

- In Chapter 1, you added, subtracted, multiplied, and divided decimals and integers.
- You used number lines to model operations with integers.

Check Your Readiness

GO for Help

For Exercises	See Lesson
1–4	1-2
5–8	1-8
9–14	1-9
15–18	2-4

Adding and Subtracting Decimals

Find each sum or difference.

1. $5.304 - 0.89$ **4.414**
2. $2.35 + 1.8 + 4.45$ **8.6**
3. $2.15 - 1.36$ **0.79**
4. $3.14 + 2.67 + 9.4$ **15.21**

Multiplying and Dividing Integers

Find each product or quotient.

5. $-12 \cdot 3$ **−36**
6. $-3 \div (-3)$ **1**
7. $-6 \cdot (-7)$ **42**
8. $-54 \div 9$ **−6**

Order of Operations

Find the value of each expression.

9. $14 + 2(15 \div 5)$ **20**
10. $20 \div 2 - 3 \cdot 3$ **1**
11. $(5 + 4) \cdot 4 \div 36$ **1**
12. $(9 - 4) \div 5 \cdot 6$ **6**
13. $9 \div (18 - 15) + 4$ **7**
14. $6 + 8 \cdot 7 - 24$ **38**

Comparing Fractions

Compare each pair of fractions. Use <, =, or >.

15. $\frac{20}{24} \blacksquare \frac{28}{36}$ **>**
16. $\frac{15}{27} \blacksquare \frac{5}{9}$ **=**
17. $\frac{6}{7} \blacksquare \frac{18}{22}$ **>**
18. $\frac{8}{9} \blacksquare \frac{9}{10}$ **<**

What You'll Learn Next

- In this chapter, you will solve equations and inequalities by adding, subtracting, multiplying, or dividing.
- You will write inequalities and graph their solutions on number lines.

Key Vocabulary

- Addition Property of Equality (p. 180)
- algebraic expression (p. 169)
- Division Property of Equality (p. 186)
- equation (p. 174)
- inequality (p. 205)
- inverse operations (p. 181)
- Multiplication Property of Equality (p. 188)
- open sentence (p. 174)
- solution of an equation (p. 174)
- solution of an inequality (p. 205)
- Subtraction Property of Equality (p. 180)
- variable (p. 169)

Problem Solving Application On pages 224 and 225, you will work an extended activity on inequalities.

Chapter 4 Overview

In this chapter, students evaluate and write algebraic expressions and write and solve both one-step and two-step equations. Then they draw upon their understanding of expressions and equations to graph, write, and solve inequalities.

Activating Prior Knowledge

In this chapter, students will build on their knowledge of the order of operations and of the distributive property to write and solve equations and inequalities. They also draw upon their understanding of mental math strategies. Ask question such as:

- *How can you find the sum of 97 and 46 using mental math?* **Sample: Use compensation; add 3 to 97, subtract 3 from 46; $100 + 43 = 143$**
- *How can you use mental math to solve the following: 5 × 29?* **Sample: Use the distributive property; $5 \times 29 = (5 \times 20) + (5 \times 9) = 100 + 45 = 145$.**

Activity Lab

Describing Patterns

Students identify and describe patterns in tables. Then they continue these patterns to make predictions. They use variables in their descriptions to extend patterns beyond the numbers in the tables.

Guided Instruction

Before beginning the Activity, discuss how a plane gains altitude as it takes off before maintaining a level altitude for most of a flight.

Ask:

- *Does a plane continue its rise in altitude throughout a flight?* **No, it levels off and then decreases its altitude to land.**
- *How do the Kelvin and Celsius scales differ?* **The Kelvin temperature is always 273° higher than the Celsius temperature.**

Differentiated Instruction

Advanced Learners L4

Have students check in science books to find out about the origins of the Kelvin and Celsius scales.

4-1a Activity Lab

Describing Patterns

You can identify and describe patterns to make predictions.

ACTIVITY

During part of its flight, an airplane rises 12 feet in altitude each second. Use this information to find a pattern.

1. Look at the table at the right. Explain how to find the distance the plane rises for each amount of time. **Multiply the time by 12.**
2. Copy and complete the table at the right. **See right.**
3. At t seconds, how many feet has the plane risen? **12 times t**

Time (s)	Distance (ft)	
1	12	
5	60	
10	120	
20	■	**240**
25	■	**300**
50	■	**600**
100	■	**1,200**
200	■	**2,400**

You can use different scales to measure temperature. The Celsius and Kelvin scales have a special relationship.

4. Look at the table of temperatures at the right. Explain how to find the temperature in Kelvin for each temperature in degrees Celsius. **Add 273.**
5. Copy and complete the table. **See right.**
6. At c degrees Celsius, what is the temperature in Kelvin? **273 more than c**

Celsius (C)	Kelvin (K)	
0	273	
1	274	
5	278	
10	■	**283**
20	■	**293**
30	■	**303**
75	■	**348**
100	■	**373**

Exercises

Copy and complete each table. Describe the pattern you find.
1–3. See margin.

1.

A	B
1	8
2	9
5	12
10	■
30	■
50	■
n	■

2.

C	D
1	−4
2	−3
5	0
10	■
25	■
45	■
n	■

3.

E	F
1	40
2	80
5	200
10	■
100	■
500	■
n	■

4. **Reasoning** What does the letter n represent in each table above? How is it used to describe the pattern in the table? **See margin.**

1–4. See back of book.

4-1 Evaluating and Writing Algebraic Expressions

1. **Vocabulary Review** The set of rules for simplifying an expression is called the _?_.
   order of operations

Find the value of each expression.

2. $3 + 4 \cdot 2$ 11
3. $12 - 6 \div 3$ 10
4. $6 \cdot (5 - 7) + 2$ −10
5. $(4 + 3) \cdot 2 - 11$ 3

Lesson 1-9

What You'll Learn

To write and evaluate algebraic expressions

New Vocabulary variable, algebraic expression

Why Learn This?

You can use algebraic expressions to help you make predictions based on patterns. If you know how far you can swim in 1 minute, you can estimate how far you can swim in 5 minutes.

A **variable** is a symbol that represents one or more numbers. Variables are usually letters. An **algebraic expression** is a mathematical phrase with at least one variable.

Diagrams and algebraic expressions can represent word phrases.

Word Phrase	Diagram	Algebraic Expression
a temperature of t degrees increased by 5 degrees	t | 5	$t + 5$
five cats fewer than c cats	c / ? | 5	$c - 5$
the product of 5 and n nickels	n | n | n | n | n	$5n$
a dinner bill of d dollars divided among five friends	d / $\frac{d}{5}$ | $\frac{d}{5}$ | $\frac{d}{5}$ | $\frac{d}{5}$ | $\frac{d}{5}$	$\frac{d}{5}$

EXAMPLE Writing Algebraic Expressions

You can use diagrams to represent algebraic expressions.

1 Write an algebraic expression for each word phrase.

a. swimming m meters per minute for 3 minutes → $3m$

b. 12 heartbeats more than x heartbeats → $x + 12$

Quick Check

1. Write an algebraic expression for a price p decreased by 16. $p - 16$

4-1

1. Plan

Objective

To write and evaluate algebraic expressions

Examples

1 Writing Algebraic Expressions
2 Application: Public Service
3 Writing Word Phrases
4 Evaluating Algebraic Expressions

Math Understandings: p. 166C

Math Background

A *variable* is a placeholder for a number. A mathematical expression that contains one or more variables is called an *algebraic expression*.

When each variable in an algebraic expression is replaced by a number, the result is a numerical expression. This process is called *evaluating the algebraic expression*.

More Math Background: p. 166C

Lesson Planning and Resources

See p. 166E for a list of the resources that support this lesson.

Bell Ringer Practice

PowerPoint

Check Skills You'll Need
Use student page, transparency, or PowerPoint. For intervention, direct students to:
Order of Operations and the Distributive Property
Lesson 1-9
Extra Skills and Word Problems Practice, Ch. 1

Differentiated Instruction Solutions for All Learners

Special Needs L1
In Example 4, make sure students do not think $2p$ means $2 + 2$, since they would still get the right answer. *Ask: What if p were equal to 4, or to 5?* This lets you know they are interpreting the expression $2p$ correctly.

learning style: verbal

Below Level L2
Use several simple numerical expressions like the following to review various operation symbols.

$8 + 4$ $8 - 4$ 8×4 $8 \cdot 4$ $8(4)$

$8 \div 4$ $\frac{8}{4}$

learning style: visual

2. Teach

Activity Lab

Use before the lesson.
Student Edition Activity Lab 4-1a, Describing Patterns, p. 168

All in One Teaching Resources

Activity Lab 4-1: Solutions in Need of Equations

Guided Instruction

Example 4
Display $2(p + 7)$. Ask: *How would the answer be different if these parentheses were inserted?*
$2(2 + 7) \rightarrow 2(9) \rightarrow 18$

PowerPoint

Additional Examples

1. Write an algebraic expression for each word phrase.
   a. 6 less than *d* dollars **$d - 6$**
   b. the sum of *s* students and 9 students **$s + 9$**
   c. 12 times *b* boxes **12b**
   d. 20 books divided among *s* students **$\frac{20}{s}$**
2. The cost of a package of markers is *d* dollars. Write an algebraic expression for the total cost in dollars of 7 packages of markers. **7d**
3. Write three word phrases for 2*y*. **Sample: the product of 2 and a number *y*; 2 times a number *y*; 2 multiplied by a number *y***
4. Evaluate each expression. Use $r = 8$, $s = 1$, and $t = 3$.
   a. $6(t - 1)$ **12** b. $\frac{r}{s + t}$ **2**

All in One Teaching Resources

- Daily Notetaking Guide 4-1 L3
- Adapted Notetaking 4-1 L1

Closure

- *What is an algebraic expression?* **a mathematical phrase that uses variables, numbers, and operation symbols**
- *How do you evaluate an algebraic expression?* **Substitute a given value for each variable and simplify.**

You can use algebraic expressions to represent real-world situations.

EXAMPLE Application: Public Service

2 The Environmental Club is making posters. The materials for each poster cost \$4. Write an algebraic expression for the cost of p posters.

Words	\$4 per poster	times	the number of posters
Expression	4	$\cdot$	p

An algebraic expression for the cost of the posters is $4p$.

Quick Check

2. Nine students will hang t posters each. Write an algebraic expression for the total number of posters the students will hang. **9t**

You can translate algebraic expressions into word phrases.

EXAMPLE Writing Word Phrases

3 Write three different word phrases for $x + 2$.

A number plus two | A number increased by two | Two more than a number

Online active math

For: Expressions Activity
Use: Interactive Textbook, 4-1

Quick Check

3. Write three different word phrases for $c - 50$. **Answers may vary. Sample: a number decreased by 50, 50 less than a number, 50 subtracted from a number**

You can substitute for a variable to evaluate an algebraic expression.

EXAMPLE Evaluating Algebraic Expressions

4 Evaluate each expression. Use the values $p = 2$, $n = 3$, and $s = 5$.

a. $2p + 7$ b. $p + (n \cdot s)$

a.		b.
$2p + 7 = 2(2) + 7$	← **Substitute.** →	$p + (n \cdot s) = 2 + (3 \cdot 5)$
$= 4 + 7$	← **Multiply.** →	$= 2 + (15)$
$= 11$	← **Add.** →	$= 17$

Quick Check

4. Use the values $n = 3$, $t = 5$, and $y = 7$ to evaluate $(n + t) \cdot y$. **56**

Advanced Learners L4
Write four algebraic expressions whose value is 6 when the variable *m* is replaced by 2. **Samples:**
$m + 4$; $8 - m$; $3m$; $\frac{12}{m}$
learning style: visual

English Language Learners ELL
Alert students to translation errors. For example, *five fewer than c cats* can be misunderstood to mean 5 – *c*. Often, substituting a number for the variable and working it out clarifies the confusion.
learning style: verbal

Check Your Understanding

1. An algebraic expression differs from a numerical expression because it contains at least one variable, and the value changes.

1. **Vocabulary** A numerical expression is a mathematical phrase that uses numbers. What is the difference between an *algebraic* expression and a *numerical* expression?

Tell which operation you would use for each word phrase.

2. six goals fewer than g goals **subtraction**
3. p people increased by two **addition**

Write a word phrase for each algebraic expression. **4–7. See left.**

4. $w - 3$
5. $5 \cdot w$
6. $12 + w$
7. $\frac{w}{4}$

Exercises 4–7. Answers may vary. Samples are given:

4. three less than w
5. the product of five and w
6. twelve more than w
7. the quotient of w and four

Evaluate each expression. Use the value $p = 2$.

8. $p + 8$ **10**
9. $3 \cdot p$ **6**
10. $16 - p$ **14**
11. $\frac{12}{p}$ **6**

Homework Exercises

For more exercises, see Extra Skills and Word Problems.

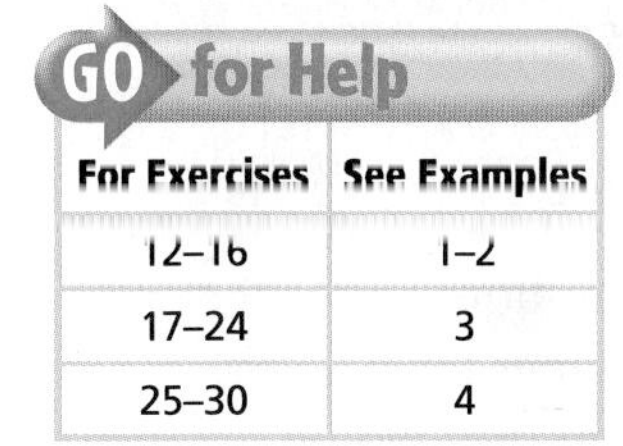

For Exercises	See Examples
12–16	1–2
17–24	3
25–30	4

A **Write an algebraic expression for each word phrase. You may find a diagram helpful.**

12. four more than s shirts **$s + 4$**
13. the quotient of p and 5 **$\frac{p}{5}$**
14. the sum of t TVs and 11 TVs **$t + 11$**
15. five times your quiz score q **$5q$**
16. Your job pays $7 per hour. Write an algebraic expression for your pay in dollars for working h hours. **$7h$**

Write a word phrase for each algebraic expression. **17–24. See margin.**

17. $d + 2$
18. $\frac{4}{n}$
19. $c - 9.1$
20. $6.5 - h$
21. $1.3 \cdot p$
22. $10 + q$
23. $\frac{w}{10}$
24. $3.5v$

Evaluate each expression. Use the values $p = 4$, $n = 6$, and $s = 2$.

25. $7n$ **42**
26. $-6.1p$ **−24.4**
27. $5 - s$ **3**
28. $\frac{n}{2}$ **3**
29. $8s - 6$ **10**
30. $1.5(p + n)$ **15**

B 

31. **Guided Problem Solving** A student mows one lawn each week day after school and two lawns on Saturday. She earns $15.75 per lawn. Write an algebraic expression for the amount of money she makes in w weeks. **$110.25w$**
   - How many lawns does she mow in one week?
   - *Draw a picture* to represent the amount of money she makes in one week.

Exercises 17–24. Answers may vary. Samples are given:

17. 2 more than a number
18. 4 divided by a number
19. $9\frac{1}{10}$ less than a number
20. a number subtracted from $6\frac{5}{10}$
21. $1\frac{3}{10}$ times a number
22. 10 more than a number
23. a number divided by 10
24. $3\frac{5}{10}$ times a number

3. Practice

Assignment Guide

Check Your Understanding

Go over Exercises 1–11 in class before assigning the Homework Exercises.

Homework Exercises

A Practice by Example	12–30
B Apply Your Skills	31–38
C Challenge	39
Test Prep and Mixed Review	40–44

Homework Quick Check

To check students' understanding of key skills and concepts, go over Exercises 13, 18, 35, 37, and 38.

Differentiated Instruction Resources

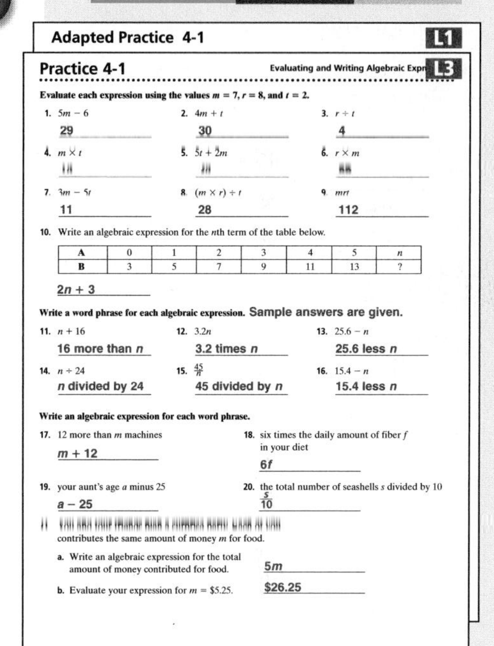

Adapted Practice 4-1 L1

Practice 4-1 Evaluating and Writing Algebraic Expressions L3

Evaluate each expression using the values $m = 7$, $r = 8$, and $t = 2$.

1. $5m - 6$ 29
2. $4m + t$ 30
3. $r \div t$ 4
4. $m \times t$ 14
5. $5t + 2m$ 24
6. $r \times m$ 56
7. $3m - 5t$ 11
8. $(m \times r) \div t$ 28
9. mrt 112
10. Write an algebraic expression for the nth term of the table below.

A	0	1	2	3	4	5	n
B	3	5	7	9	11	13	?

$2n + 3$

Write a word phrase for each algebraic expression. Sample answers are given.

11. $n + 16$ 16 more than n
12. $3.2n$ 3.2 times n
13. $25.6 - n$ 25.6 less n
14. $n \div 24$ n divided by 24
15. $\frac{45}{n}$ 45 divided by n
16. $15.4 - n$ 15.4 less n

Write an algebraic expression for each word phrase.

17. 12 more than m machines $m + 12$
18. six times the daily amount of fiber f in your diet $6f$
19. your aunt's age a minus 25 $a - 25$
20. the total number of seashells s divided by 10 $\frac{s}{10}$
21. [illegible] contributes the same amount of money m for food.
   a. Write an algebraic expression for the total amount of money contributed for food. $5m$
   b. Evaluate your expression for $m = \$5.25$. $26.25

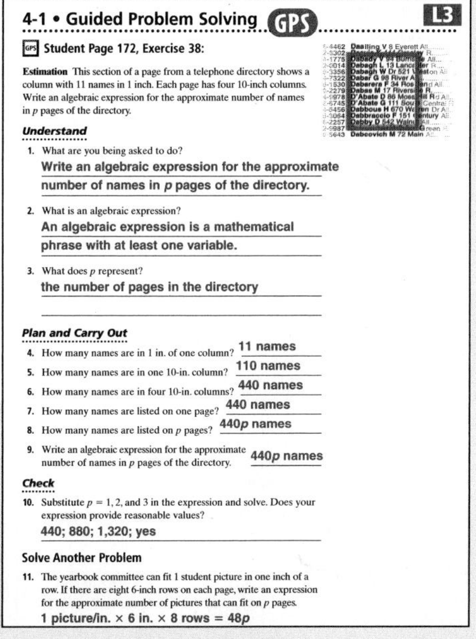

4-1 • Guided Problem Solving GPS L3

Student Page 172, Exercise 38:

Estimation This section of a page from a telephone directory shows a column with 11 names in 1 inch. Each page has four 10-inch columns. Write an algebraic expression for the approximate number of names in p pages of the directory.

Understand

1. What are you being asked to do? Write an algebraic expression for the approximate number of names in p pages of the directory.
2. What is an algebraic expression? An algebraic expression is a mathematical phrase with at least one variable.
3. What does p represent? the number of pages in the directory

Plan and Carry Out

4. How many names are in 1 in. of one column? 11 names
5. How many names are in one 10-in. column? 110 names
6. How many names are in four 10-in. columns? 440 names
7. How many names are listed on one page? 440 names
8. How many names are listed on p pages? $440p$ names
9. Write an algebraic expression for the approximate number of names in p pages of the directory. $440p$ names

Check

10. Substitute $p = 1, 2,$ and 3 in the expression and solve. Does your expression provide reasonable values? 440; 880; 1,320; yes

Solve Another Problem

11. The yearbook committee can fit 1 student picture in one inch of a row. If there are eight 6-inch rows on each page, write an expression for the approximate number of pictures that can fit on p pages. 1 picture/in. × 6 in. × 8 rows = $48p$

4. Assess & Reteach

Lesson Quiz

Evaluate when $n = 12$ and $p = 8$.

1. $2n - 9$ 15
2. $3(n + p)$ 60
3. $\frac{n}{4} + 7$ 10

Write an algebraic expression for each word phrase.

4. the number of days in w weeks $7w$
5. two less than m mice $m - 2$

Reteaching 4-1 — Evaluating and Writing Algebraic Ex — L2

To evaluate an *expression*, substitute a value for the *variable* and compute.

Evaluate $5y - 8$ for $y = 7$.

$5y - 8$
$5 \times 7 - 8$ ← **Substitute y with 7.**
$35 - 8 = 27$ ← **Compute.**

You can use key words to write a word phrase for an algebraic expression.

$a + 5$ → a plus 5 or a increased by 5

$2n$ → the product of 2 and n or 2 times n

Evaluate each expression using the values $m = 3$ and $x = 8$.

1. $4m + 9$
   Substitute m: $4 \times$ 3 $+ 9$
   Compute: 12 $+ 9 =$ 21
2. $4x - 7$
   Substitute x: $4 \times$ 8 $- 7$
   Compute: 32 $- 7 =$ 25
3. $5x + x$
   Substitute x: $5 \times$ 8 $+$ 8
   Compute: 40 $+$ 8 $=$ 48
4. $x + 2m$
   Substitute x and m: 8 $+ 2 \times$ 3
   Compute: 8 $+$ 6 $=$ 14

Evaluate each expression using the values $y = 4$, $z = 8$, and $p = 10$.

5. $3y + 6 =$ 18
6. $4z - 2 =$ 30
7. $p + 2p =$ 30
8. $3z \times z =$ 192

Write a word phrase for each algebraic expression.

9. $9 + x$ 9 plus a number
10. $6x$ 6 times a number
11. $x - 8$ 8 less than a number
12. $\frac{x}{5}$ a number divided by 5

Write an algebraic expression for each word phrase.

13. x newspapers plus 10 $x + 10$
14. 4 less than x teabags $x - 14$
15. 3 more than x envelopes $x + 3$
16. 6 times x school buses $6x$

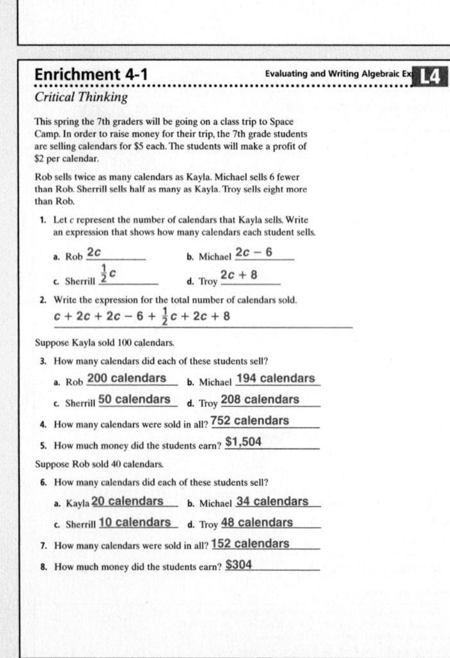

Enrichment 4-1 — Evaluating and Writing Algebraic Ex — L4

Critical Thinking

This spring the 7th graders will be going on a class trip to Space Camp. In order to raise money for their trip, the 7th grade students are selling calendars for $5 each. The students will make a profit of $2 per calendar.

Rob sells twice as many calendars as Kayla. Michael sells 6 fewer than Rob. Sherrill sells half as many as Kayla. Troy sells eight more than Rob.

1. Let c represent the number of calendars that Kayla sells. Write an expression that shows how many calendars each student sells.
   a. Rob $2c$ b. Michael $2c - 6$
   c. Sherrill $\frac{1}{2}c$ d. Troy $2c + 8$
2. Write the expression for the total number of calendars sold.
   $c + 2c + 2c - 6 + \frac{1}{2}c + 2c + 8$

Suppose Kayla sold 100 calendars.

3. How many calendars did each of these students sell?
   a. Rob 200 calendars b. Michael 194 calendars
   c. Sherrill 50 calendars d. Troy 208 calendars
4. How many calendars were sold in all? 752 calendars
5. How much money did the students earn? $1,504

Suppose Rob sold 40 calendars.

6. How many calendars did each of these students sell?
   a. Kayla 20 calendars b. Michael 34 calendars
   c. Sherrill 10 calendars d. Troy 48 calendars
7. How many calendars were sold in all? 152 calendars
8. How much money did the students earn? $304

GO Online
Homework Video Tutor
Visit: PHSchool.com
Web Code: are-0401

Write an algebraic expression for the *n*th term of each table.

32.

A	0	1	2	3	5	10	n
B	5	6	7	8	10	15	?

$n + 5$

33.

C	0	1	2	3	5	10	n
D	0	10	20	30	50	100	?

$10n$

34. Write an algebraic expression to find the number of seconds in n minutes. Evaluate the expression for $n = 20$. **$60n$; 1,200**

35. **Birds** The blue-throated hummingbird has a heart rate of about 1,260 beats per minute. Explain how you would calculate the number of times the hummingbird's heart beats in a 24-hour day. Then find the number of beats in a 24-hour day. **See left.**

36. Describe a situation that the expression $10n$ can model. **See left.**

37. **Writing in Math** You can write "twelve less than a number" as $n - 12$, but not as $12 - n$. Explain why. **See left.**

35. Answers may vary. Sample: Multiply 24(60) to find the number of minutes in 24 h. Then multiply the answer by 1,260, the number of beats per minute. The heart beats 1,814,400 times in 24 h.

36. Answers may vary. Sample: Candles cost $10 each. How much do you pay for n candles?

37. Subtraction is not commutative.

38. **Estimation** This section of a page from a telephone directory shows a column with 11 names in 1 inch. Each page has four 10-inch columns. Write an algebraic expression for the approximate number of names in p pages of the directory. **$440p$**

6-4462 **Daalling V** 8 Everett All............
2-3302 **Daavis K** 444 Greeley R..........
4-1775 **Dabady V** 94 Burnside All........
2-0014 **Dabagh L** 13 Lancaster R.......
6-3356 **Dabagh W** Dr 521 Weston All..
4-7322 **Dabar G** 98 River All.............
6-1530 **Dabarera F** 34 Roseland All....
2-2279 **Dabas M** 17 Riverside R..........
4-9978 **D'Abate D** 86 Moss Hill Rd All..
2-6745 **D'Abate G** 111 South Central R
4-5456 **Dabbous H** 670 Warren Dr All..
6-3064 **Dabbraccio F** 151 Century All..
6-2257 **Dabby D** 542 Walnut All..........
2-9987 **Dabcovich M G** 219 Green R..
6-5643 **Dabcovich M** 72 Main All.........

39. **Challenge** A student baby-sitting for $5 per hour writes the expression $5n$ to represent the money he makes for n hours. Another student writes $3n + 15$ to represent the amount in dollars she makes. What is the second student's hourly rate? What does the number 15 represent? **Answers may vary. Sample: The second student charges $15 to start and $3/h.**

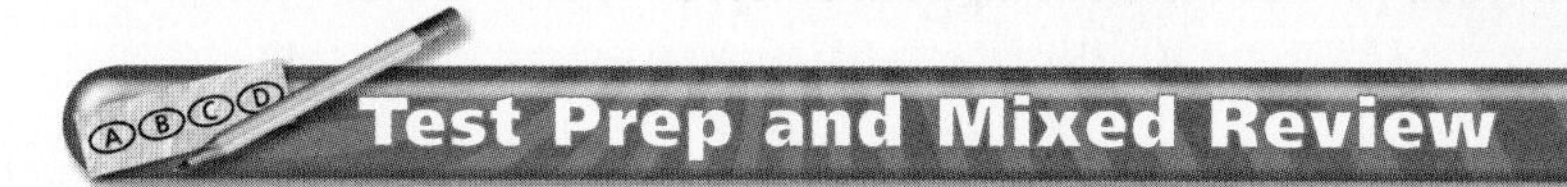

Multiple Choice

40. In ancient Greece, a measurement called a cubit equaled 18.3 inches. Which expression represents the number of inches in c cubits? **D**

Ⓐ $c - 18.3$ Ⓑ $c \div 18.3$ Ⓒ $18.3 + c$ Ⓓ $18.3c$

41. Which expression does the model represent? **F**

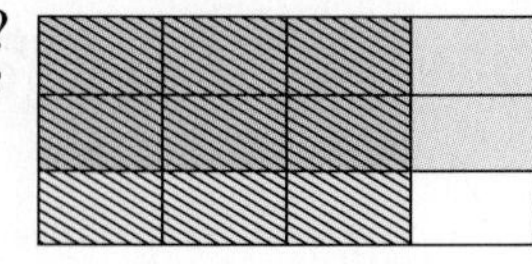

Ⓕ $\frac{2}{3} \times \frac{3}{4}$ Ⓗ $\frac{1}{2} \times \frac{2}{3}$

Ⓖ $\frac{1}{3} \times \frac{3}{4}$ Ⓙ $\frac{1}{2} \times \frac{3}{4}$

GO for Help

For Exercises	See Lesson
42–44	1-9

Find the value of each expression.

42. $7(1.2) + 7(0.5)$ **11.9**
43. $3(4 + 5) + 2(4 + 5)$ **45**
44. $5(0.25 \cdot 40)$ **50**

Test Prep

Resources
For additional practice with a variety of test item formats:
- Test-Taking Strategies, p. 219
- Test Prep, p. 223
- Test-Taking Strategies with Transparencies

Alternative Assessment

Ask each student to write a word phrase such as "3 more than s students." Have them assign a value to the variable and exchange papers with a partner. Partners then write the corresponding algebraic expression and evaluate for the given value. Partners can repeat the process as time permits.

4-1b Activity Lab

Technology

Using Spreadsheets

You can use a computer spreadsheet to keep track of the balance in a checking account. You add deposits and subtract checks. Using the formulas you supply, the spreadsheet computes values in cells.

EXAMPLE Using Spreadsheets

Use the spreadsheet. Find the balance after each entry.

	A	B	C	D
1	Date	Deposits	Checks	Balance
2				$350
3	4/29	$100		■
4	4/30		$400	■

← Use the formula "= D2 + B3". The computer finds 350 + 100 = 450.

← Use the formula "= D3 − C4". The computer finds 450 − 400 = 50.

The first balance is $450. The second balance is $50.

Exercises

Use the spreadsheet at the right.

	A	B	C	D
1	Date	Deposits	Checks	Balance
2				$250
3	11/3		$25.98	■
4	11/9		$239.40	■
5	11/10	$122.00		■
6	11/13		$54.65	■
7	11/20	$350.00		■
8	11/29		$163.80	■

1. Find the account balance after each entry. Write the formulas you used. **See margin.**
2. Which formula can you use to find the balance in cell D9, whether or not a check has been written or a deposit has been made? **D**
   - A = D8 − B9 + C9
   - B = D8 − B9 − C9
   - C = D8 + B9 + C9
   - D = D8 + B9 − C9
3. **Reasoning** Suppose the balance in cell D9 is $130.34. Was the amount of a deposit entered into cell B9, or was the amount of a check entered into cell C9? Support your answer. **See margin.**
4. **Writing in Math** Consider your answer to Exercise 2. Explain why the formula you chose works. Give examples. **See margin.**
5. **Reasoning** Suppose the balance in cell D8 is $250 and you do not know the original balance. Explain how would you calculate the original balance in cell D2. Then find the original balance.

5. Add the values in column C and subtract the values in column B; $261.83

1. $224.02(= D2 − C3)
−$15.38(= D3 − C4)
$106.62(= D4 + B5)
$51.97(= D5 − C6)
$401.97(= D6 + B7)
$238.17(= D7 − C8)

3. The amount of a check; the amount of money decreased from cell D8 to cell D9.

4. Answers may vary. Sample: The formula "D9 = D8 + B9 − C9" adds any deposit from cell B9 or subtracts any check from cell C9. For example, if $100 is in B9, then D9 = $238.17 + $100 − $0, or if $50 is in C9, then D9 = $238.17 + $0 − $50.

Activity Lab

Using Spreadsheets

Students can use spreadsheet software when repeated calculations (using the same math operations) are done over and over with different numbers. One common application of spreadsheets is to keep track of banking transactions accurately.

Guided Instruction

Error Prevention!

Be sure students understand what a cell is and how to identify a cell. Ask questions like: *What amount is in cell B3 of the spreadsheet in the Example?* **$100** *In cell C4?* **$400**

Differentiated Instruction

Visual Learners
Introduce and discuss the terminology associated with the workings of a checking account: a *deposit* is money put into a bank account; a *check* is a request for money to be withdrawn from the account; the *balance* is the amount of money remaining in an account after a deposit has been made or a check has been written. If possible, show examples of a deposit slip, a check, and a balance statement.

Resources

- any spreadsheet software program
- Classroom Aid 4

4-2

1. Plan

Objective
To solve one-step equations using substitution, mental math, and estimation

Examples
1 Solving Equations Using Substitution
2 Solving Equations Using Mental Math
3 Estimating Solutions

Math Understandings: p. 166C

Math Background

An *equation* is a mathematical sentence that contains an equal sign. This means that a statement as simple as $1 + 2 = 3$ is an equation. However, this lesson and the ones that follow focus on equations in which one side is an algebraic expression. You determine whether a number is a *solution* of such an equation by evaluating the algebraic expression when that number is substituted for the variable. If the resulting statement is true, then the number is a solution of the equation.

More Math Background: p. 166C

Lesson Planning and Resources

See p. 166E for a list of the resources that support this lesson.

Bell Ringer Practice

Check Skills You'll Need
Use student page, transparency, or PowerPoint. For intervention, direct students to:
Evaluating and Writing Algebraic Expressions
Lesson 4-1
Extra Skills and Word Problems Practice, Ch. 4

Algebra

4-2 Using Number Sense to Solve Equations

Check Skills You'll Need

1. **Vocabulary Review** A __?__ is a symbol that represents one or more numbers. **variable**

Write an algebraic expression for each word phrase.

2. four more than y **$y + 4$**
3. six less than v **$v - 6$**
4. k divided by 9 **$\frac{k}{9}$**

GO for Help Lesson 4-1

What You'll Learn

To solve one-step equations using substitution, mental math, and estimation

New Vocabulary equation, open sentence, solution of an equation

Why Learn This?

Solving equations can help you find amounts that are not easy to measure on their own. For example, if you know the weight of the bucket, you can use an equation to find the weight of the panda.

An **equation** is a mathematical sentence with an equal sign. An equation with one or more variables is an **open sentence.** The open sentence $p + 2 = 16$ is neither true nor false until p is replaced.

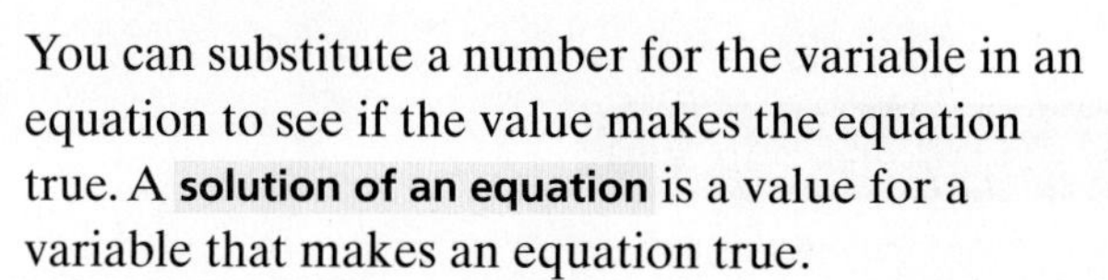

You can substitute a number for the variable in an equation to see if the value makes the equation true. A **solution of an equation** is a value for a variable that makes an equation true.

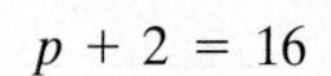

$p + 2 = 16$ $\quad$ $20 + 2 = 16$ **false** $\quad$ $14 + 2 = 16$ **true**

Since 14 makes the equation true, 14 is a solution of the equation.

EXAMPLE Solving Equations Using Substitution

1 Find the solution of $12m = 108$ from the numbers 4, 7, and 9. You can test each number by substituting for m in the equation.

$12(4) \stackrel{?}{=} 108$ $\quad$ $12(7) \stackrel{?}{=} 108$ $\quad$ $12(9) \stackrel{?}{=} 108$

$48 = 108$ ✗ **False** $\quad$ $84 = 108$ ✗ **False** $\quad$ $108 = 108$ ✓ **True**

Since the equation is true when you substitute 9 for m, the solution is 9.

Quick Check

1. Find the solution of each equation from the given numbers.
   a. $24n = 120$; 3, 5, or 11 **5**
   b. $124p = 992$; 4, 6, or 8 **8**

Differentiated Instruction Solutions for All Learners

Special Needs L1	**Below Level** L2
Ask students to draw a picture for Example 3 to help them keep track of what they know the veterinarian weighs, and what the scale shows. They can label the line drawing of a puppy w. **learning style: visual**	Give students several addition and multiplication facts. Have them write the related subtractions for the addition facts and the related divisions for the multiplication facts. **learning style: visual**

Sometimes you can solve an equation by using mental math.

EXAMPLE Solving Equations Using Mental Math

2 Use mental math to solve each equation.

a. $4y = 20$

What you think

What number times 4 equals 20?

Since $4 \cdot 5 = 20$, $y = 5$.

b. $m + 7 = 15.5$

What you think

What number plus 7 equals 15.5?

Since $7 + 8.5 = 15.5$, $m = 8.5$.

Quick Check

2. Use mental math to solve each equation.

a. $t - 3 = 7$ **10**

b. $n + 6 = -10.1$ **−16.1**

c. $\frac{h}{4} = 2.2$ **8.8**

d. $7x = -63$ **−9**

You can solve problems by using equations and estimation.

EXAMPLE Estimating Solutions

Careers A veterinarian is a health-care provider for animals.

3 **Multiple Choice** A veterinarian holds a puppy and steps on a scale. The scale reads 134.5 lb. The veterinarian weighs 125.3 lb alone. Which is the best estimate of the weight of the puppy?

Ⓐ Between 3 and 5 lb
Ⓑ Between 6 and 8 lb
Ⓒ Between 9 and 11 lb
Ⓓ Between 12 and 15 lb

Words weight plus puppy's weight equals total weight

Let w = the weight of the puppy.

Equation $125.3 + w = 134.5$

$125.3 + w = 134.5$

$125.3 \approx 125 \quad 134.5 \approx 135$ ← **Choose compatible numbers.**

$125 + w = 135$ ← **What number added to 125 is 135?**

$w = 10$ ← **Use mental math.**

The puppy weighs about 10 lb. The correct answer is choice C.

Test Prep Tip

When writing an algebraic expression, choose a letter for the variable that will remind you of what the variable represents.

Quick Check

3. A box of machine parts weighs 14.7 lb. A forklift has a maximum weight limit of 390 lb. About how many boxes of parts can the forklift carry at one time? **about 26 boxes**

2. Teach

Activity Lab

Use before the lesson.

All in One Teaching Resources

Activity Lab 4-2: Using Number Sense to Solve Equations

Guided Instruction

Example 1

Point out that the equal sign with a question mark above ($\stackrel{?}{=}$) is used immediately after a variable has been replaced by a number. It indicates that you do not yet know whether the equation is true or false.

Example 2

For part a, have students read $4y = 20$ as "4 times y equals 20." They will hear a parallel sentence structure when they read $4 \cdot 5 = 20$ as "4 times 5 equals 20."

PowerPoint

Additional Examples

1 Find the solution of $h - 18 = 54$: 3, 62, or 72? **72**

2 Use mental math to solve each equation.

a. $s - 9 = 5$ $s = 14$

b. $4z = 28$ $z = 7$

c. $g + 7 = 11$ $g = 4$

d. $\frac{c}{6} = 5$ $c = 30$

3 The weight of a packing crate is 14.65 lb. The crate and its contents together weigh 85.21 lb. Estimate the weight of the contents. **about 70 lb**

All in One Teaching Resources

- Daily Notetaking Guide 4-2 **L3**
- Adapted Notetaking 4-2 **L1**

Closure

- Explain the difference between an open sentence and a false sentence. **An open sentence is a mathematical sentence (with one or more variables) that is neither true nor false. A false sentence is false, or not true.**
- *How do you make an equation true?* **Substitute the solution for the variable in the equation.**

Advanced Learners L4	**English Language Learners** ELL
Use mental math to solve $3r - 5 = 7$. **$12 - 5 = 7$, so $3r = 12$; $3 \cdot 4 = 12$, so $r = 4$**	Ask students to read equations completely. For example, for $4y = 20$, ask them to say: *4 times a number y is equal to 20.* This will help them develop mathematical language and mathematical concepts.
learning style: visual	**learning style: verbal**

3. Practice

Assignment Guide

Check Your Understanding
Go over Exercises 1–9 in class before assigning the Homework Exercises.

Homework Exercises

A Practice by Example	10–23
B Apply Your Skills	24–36
C Challenge	37
Test Prep and Mixed Review	38–42

Homework Quick Check
To check students' understanding of key skills and concepts, go over Exercises 11, 23, 25, 34, and 35.

Differentiated Instruction Resources

Adapted Practice 4-2 L1

Practice 4-2 Using Number Sense to Solve L3

Identify a solution for each equation from the given set of numbers.

1. $30p = 900$; 3, 20, 30, or 60 — **30**
2. $\frac{h}{9} = 11$; 3, 30, 72, or 99 — **99**
3. $t + 32.4 = 62$; 29.6, 31.4, or 18.6 — **29.6**
4. $r - 17 = 40$; 23, 57 or 63 — **57**

Solve each equation using mental math.

5. $5t = 25$ — **$t = 5$**
6. $8w = 64$ — **$w = 8$**
7. $p + 5 = 12$ — **$p = 7$**
8. $a + 2 = 15$ — **$a = 13$**
9. $\frac{h}{6} = 4$ — **$h = 24$**
10. $\frac{g}{8} = 16$ — **$g = 128$**
11. $y - 11 = 28$ — **$y = 39$**
12. $d - 4 = 12$ — **$d = 16$**
13. $w - 10 = 15$ — **$w = 25$**
14. $18 - t = 14$ — **$t = 4$**
15. $21 + y = 31.64$ — **$y = 10.64$**
16. $18.43 + x = 123.4$ — **$x = 104.97$**
17. The seventh-grade class has been collecting aluminum cans for recycling. The class has collected 210 cans. Their goal is to collect 520 cans. Write an equation and estimate the number of aluminum cans needed to reach their goal.
$210 + x = 520$; about 300 cans
18. A seamstress bought some bolts of fabric at \$25.30 each. She spent a total of \$227.70. Write an equation and estimate the number of bolts of fabric that she purchased.
\$25.30x = \$227.70; about 10 bolts of fabric
19. For your party you purchased balloons for \$.79 each. You spent a total of \$11.85. Write an equation and estimate the number of balloons purchased.
\$.79x = \$11.85; about 12 balloons

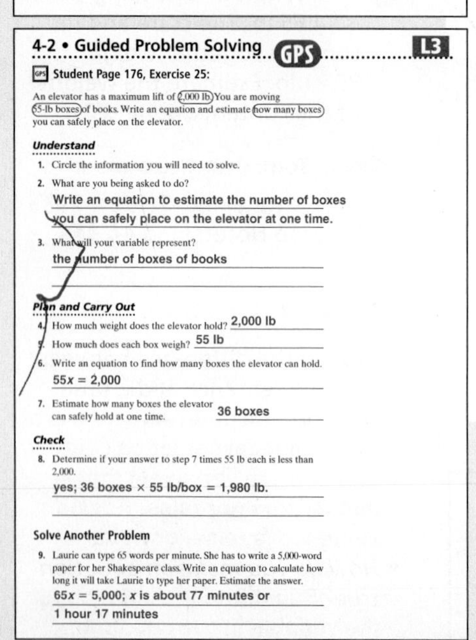
4-2 • Guided Problem Solving GPS L3

GPS Student Page 176, Exercise 25:

An elevator has a maximum lift of 2,000 lb. You are moving 55-lb boxes of books. Write an equation and estimate how many boxes you can safely place on the elevator.

Understand

1. Circle the information you will need to solve.
2. What are you being asked to do?
Write an equation to estimate the number of boxes you can safely place on the elevator at one time.
3. What will your variable represent?
the number of boxes of books

Plan and Carry Out

4. How much weight does the elevator hold? **2,000 lb**
5. How much does each box weigh? **55 lb**
6. Write an equation to find how many boxes the elevator can hold.
$55x = 2{,}000$
7. Estimate how many boxes the elevator can safely hold at one time. **36 boxes**

Check

8. Determine if your answer to step 7 times 55 lb each is less than 2,000.
yes; 36 boxes × 55 lb/box = 1,980 lb.

Solve Another Problem

9. Laurie can type 65 words per minute. She has to write a 5,000-word paper for her Shakespeare class. Write an equation to calculate how long it will take Laurie to type her paper. Estimate the answer.
$65x = 5{,}000$; x is about 77 minutes or 1 hour 17 minutes

Check Your Understanding

1. **Vocabulary** The value of a variable that makes an equation true is called the ___?___ of the equation. **solution**
2. **Number Sense** The open sentence $14k = 0$ has ___?___ solution(s).
Ⓐ no Ⓑ one Ⓒ infinitely many **B**

Tell whether each value of the variable is a solution of the equation.

3. $5x = -35$; $x = -7$ **yes**
4. $8 - y = 2$; $y = 6$ **yes**
5. $-7 + b = 4$; $b = 3$ **no**

Use mental math to solve each equation.

6. $10m = -90$ **−9**
7. $6 + w = 3$ **−3**
8. $\frac{40}{z} = 8$ **5**
9. $-5p = 20$ **−4**

Homework Exercises

For more exercises, see Extra Skills and Word Problems.

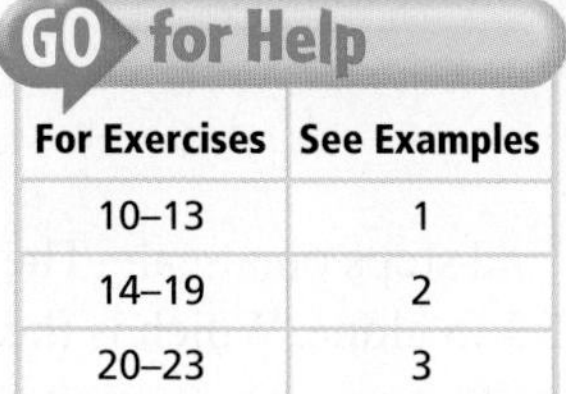
GO for Help

For Exercises	See Examples
10–13	1
14–19	2
20–23	3

A **Find the solution of each equation from the given numbers.**

10. $p + 17 = 56$; 29, 39, or 49 **39**
11. $\frac{y}{9} = 32$; 261, 279, or 288 **288**
12. $6d = -36.6$; −6.1, 0, or 6.1 **−6.1**
13. $n - 27 = 38$; 11, 51, or 65 **65**

Use mental math to solve each equation.

14. $n + 4 = 7.9$ **3.9**
15. $4d = -32$ **−8**
16. $\frac{p}{5} = 3.1$ **15.5**
17. $24 = -6w$ **−4**
18. $\frac{c}{-3} = 8$ **−24**
19. $z - 8.4 = 0$ **8.4**

Estimate the solution of each equation to the nearest whole number.

20. $3.1g = 20.9$ **about 7**
21. $h - 4.9 = 13.8$ **about 19**
22. $7.8 + n = 38.2$ **about 30**
23. A bowling ball has a mass of 5.54 kg. A bowling pin has a mass of 1.58 kg. About how many pins are equal in mass to one ball? **about 3 bowling pins**

B 

24. **Guided Problem Solving** Your class has collected 84.5 lb of canned food. The school record is 103.25 lb. Write an equation and use it to estimate the amount of canned food the class still needs to collect to match the record. **$85 + n = 103$; about 18 lb**
    - Estimate 84.5 and 103.25 by rounding.
    - Choose a variable for the amount of food the class still needs.
    - Write an equation using your variable and your estimates.
25. GPS An elevator has a maximum lift of 2,000 lb. You are moving 55-lb boxes of books. Write an equation and estimate how many boxes you can safely place on the elevator. **$55b = 2{,}000$; 36 boxes**
26. Describe a situation that can be represented by $4.5 + t = 30$.

26. Check students' work.

GO Online
Homework Video Tutor
Visit: PHSchool.com
Web Code: are-0402

Solve using mental math or estimation. If you estimate, round to the nearest whole number before you add or subtract.

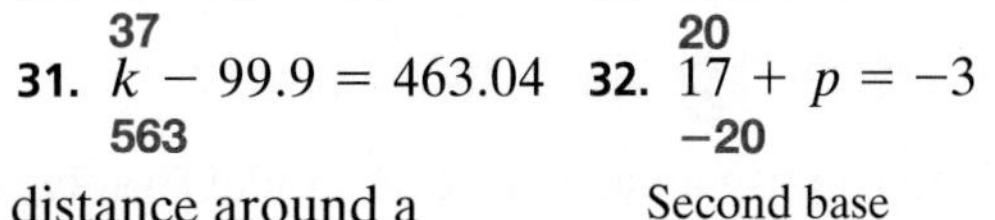

27. $p - 7.35 = 46.71$ **54**
28. $a - 20 = 17$ **37**
29. $28.71 + t = 49.43$ **20**
30. $n - 11 = 33$ **44**
31. $k - 99.9 = 463.04$ **563**
32. $17 + p = -3$ **−20**

33. **Baseball** The total distance around a baseball diamond is 360 ft. Write an equation to represent the distance from first base to second base. **$4d = 360$**

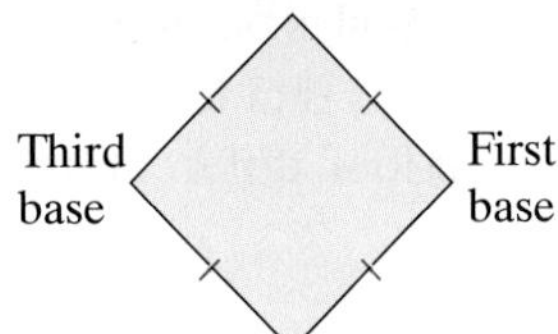

34. **Writing in Math** Equations can be true or false. Can an expression like $2a + 7$ be true or false? Explain. **See left.**

34. No; an expression has no equal sign in order to compare values.

Data Analysis Use the table below for Exercises 35 and 36.

35. The average precipitation for Houston is 1.5 times the average for Detroit. Find the average precipitation for Detroit. **31.89 in.**

36. The average precipitation for Raleigh is 4.3 times the average for San Diego. Find the average precipitation for San Diego. **10.01 in.**

Average Annual Precipitation

City	Precipitation (in.)
Detroit, Mich.	■
Houston, Tex.	47.84
Raleigh, N.C.	43.05
San Diego, Calif.	■

SOURCE: National Climatic Data Center. Go to PHSchool.com for a data update. Web Code: arg-9041

37. **Challenge** You want to buy 4 candles that cost $12.95 each. A clerk tells you that the total, including sales tax, is $67.80. Is the clerk's total reasonable? Explain. **See left.**

37. $4c = 67.80$; about $17; no; the total should be about 4($12.95), or $51.80.

Test Prep and Mixed Review

Practice

Multiple Choice

38. Jane had $22.25 before baby-sitting on Saturday. After baby-sitting, she had $48.75. Which equation can be used to find b, the amount of money Jane earned from baby-sitting? **C**
 - Ⓐ $\$22.25 - b = \$48.75$
 - Ⓑ $\$22.25 + \$48.75 = b$
 - Ⓒ $\$48.75 - b = \$22.25$
 - Ⓓ $\$48.75 + b = \$22.25$

39. The table shows the points the Cougars scored during their last 4 games. Which measure of data is represented by 64.5? **H**
 - Ⓕ mean
 - Ⓖ mode
 - Ⓗ median
 - Ⓙ range

Cougars Points Scored

Game	Points
1	72
2	60
3	47
4	69

Use a number line to find each sum.

40. $-5 + 9$ **4**
41. $-9 + 10$ **1**
42. $-2 + -8$ **−10**

GO for Help

For Exercises	See Lesson
40–42	1-7

Online lesson quiz, PHSchool.com, Web Code: ara-0402

4. Assess & Reteach

Lesson Quiz

Solve each equation. Use mental math or estimation.

1. $q - 15 = 25$ **$q = 40$**
2. $6r = 42$ **$r = 7$**
3. $17.25 + s = 21.4$ **$s \approx 4$**
4. $\frac{80}{t} = 8$ **$t = 10$**
5. $m - 39.8 = -101.03$ **$m \approx -61$**

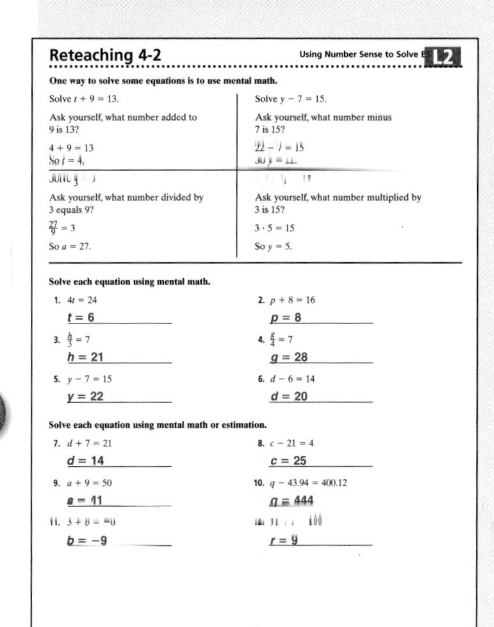

Reteaching 4-2 Using Number Sense to Solve Equations **L2**

One way to solve some equations is to use mental math.

Solve $t + 9 = 13$. Ask yourself, what number added to 9 is 13? $4 + 9 = 13$. So $t = 4$.

Solve $y - 7 = 15$. Ask yourself, what number minus 7 is 15? $22 - 7 = 15$. So $y = 22$.

Ask yourself, what number divided by 3 equals 9? $\frac{27}{3} = 9$. So $a = 27$.

Ask yourself, what number multiplied by 3 is 15? $3 \cdot 5 = 15$. So $y = 5$.

Solve each equation using mental math.

1. $4t = 24$ $t = 6$
2. $p + 8 = 16$ $p = 8$
3. $\frac{h}{3} = 7$ $h = 21$
4. $\frac{g}{4} = 7$ $g = 28$
5. $y - 7 = 15$ $y = 22$
6. $d - 6 = 14$ $d = 20$

Solve each equation using mental math or estimation.

7. $d + 7 = 21$ $d = 14$
8. $c - 21 = 4$ $c = 25$
9. $a + 9 = 50$ $a = 41$
10. $q - 43.94 = 400.12$ $q = 444$
11. $b = -9$
12. $r = 9$

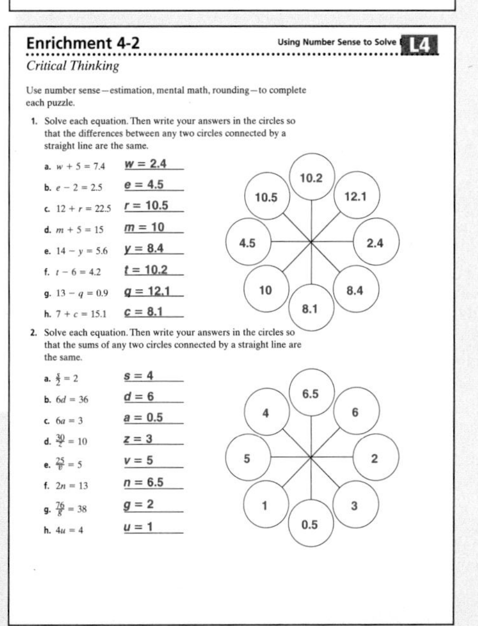

Enrichment 4-2 Using Number Sense to Solve Equations **L4**

Critical Thinking

Use number sense—estimation, mental math, rounding—to complete each puzzle.

1. Solve each equation. Then write your answers in the circles so that the differences between any two circles connected by a straight line are the same.
 a. $w + 5 = 7.4$ $w = 2.4$
 b. $e - 2 = 2.5$ $e = 4.5$
 c. $12 + r = 22.5$ $r = 10.5$
 d. $m + 5 = 15$ $m = 10$
 e. $14 - y = 5.6$ $y = 8.4$
 f. $t - 6 = 4.2$ $t = 10.2$
 g. $13 - q = 0.9$ $q = 12.1$
 h. $7 + c = 15.1$ $c = 8.1$

2. Solve each equation. Then write your answers in the circles so that the sums of any two circles connected by a straight line are the same.
 a. $\frac{s}{2} = 2$ $s = 4$
 b. $6d = 36$ $d = 6$
 c. $6a = 3$ $a = 0.5$
 d. $\frac{30}{z} = 10$ $z = 3$
 e. $\frac{25}{v} = 5$ $v = 5$
 f. $2n = 13$ $n = 6.5$
 g. $\frac{76}{g} = 38$ $g = 2$
 h. $4u = 4$ $u = 1$

Alternative Assessment

Have each student write an equation and a set of possible solutions similar to Exercises 10–13. Ask students to make sure that the solution to each equation is provided in a list of possible solutions. Students then exchange papers with a partner who tries to identify the correct solution.

Test Prep

Resources

For additional practice with a variety of test item formats:
- Test-Taking Strategies, p. 219
- Test Prep, p. 223
- Test-Taking Strategies with Transparencies

Activity Lab

Keeping the Balance

Students use pictures of a scale and weights to solve equations. First they decide which scales are balanced and explain their reasoning. Then they explore rules for balancing equations.

Guided Instruction

Before beginning the Activity Lab, show a balanced scale with different weights on each side. Ask:

- *What can be removed from each pan of the first scale so that the pans remain balanced?* the ball
- *How many cubes equal 1 rectangular prism for the second scale?* 3

Differentiated Instruction

Tactile Learners
Some students may prefer using actual balances and weights to explore balancing equations.

Resources

- balance scale
- weights

4-2b Activity Lab — Algebra Thinking

Keeping the Balance

You can use a balance-scale model to solve equations. The scales below are balanced. Use them for the activity. Objects that look identical have the same weight. Objects that look different have different weights.

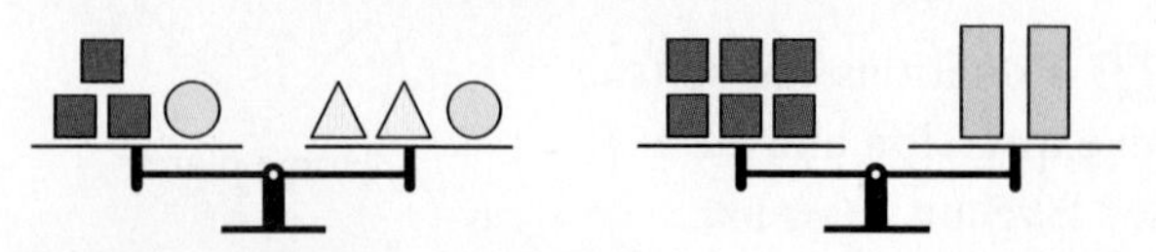

1a. Balanced; a circle was taken away from both sides.
b. Balanced; half of each side remains.
c. Balanced; a bottle was added to both sides.
d. Balanced; each side was doubled.

ACTIVITY

1. Decide whether each scale is balanced. Explain your reasoning.

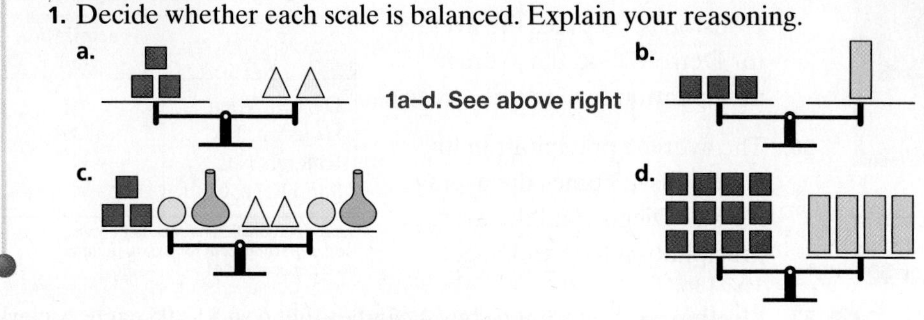

Exercises

Exercises 1–4. True. If you do the same thing to both sides of a balanced scale, it will stay balanced.

Tell whether each statement is true or false. Explain.
1–4. See right

1. A balanced scale will stay balanced if I add the same amount to each side.
2. A balanced scale will stay balanced if I remove the same amount from each side.
3. A balanced scale will stay balanced if I multiply each side by the same amount.
4. A balanced scale will stay balanced if I divide each side by the same amount.

5. Not balanced; check students' work.
6. Balanced; check students' work.
7. Balanced; check students' work.

Decide whether each scale is balanced. Explain your reasoning.
5–7. See above right.

5. 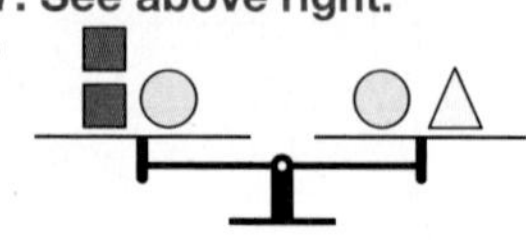

6. 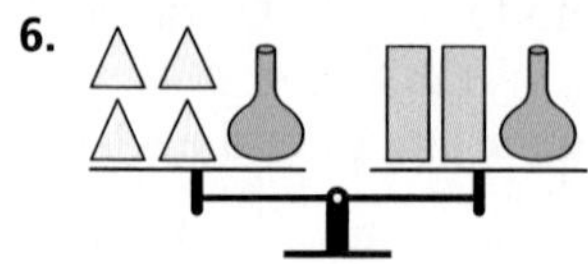

7. 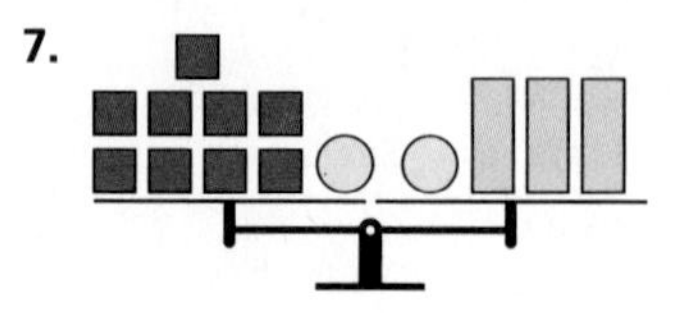

4-3a Activity Lab

Hands On

Modeling Equations

You can model and solve equations using algebra tiles.

EXAMPLE Solving Addition Equations

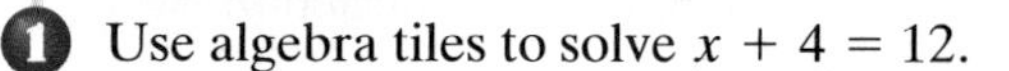

1 Use algebra tiles to solve $x + 4 = 12$.

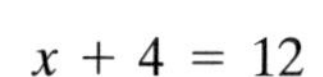

$x + 4 = 12$

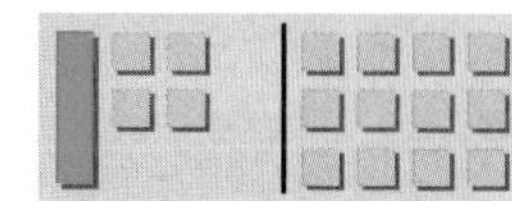

← **Model the equation. Use yellow tiles for positive integers.**

$x + 4 - 4 = 12 - 4$

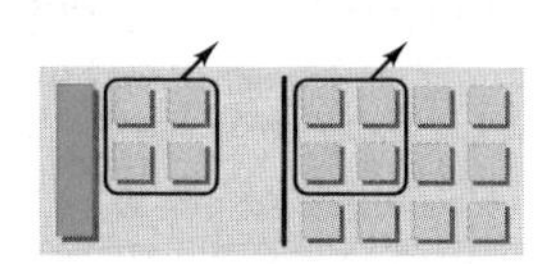

← **Remove 4 tiles from each side.**

$x = 8$

← **Simplify.**

EXAMPLE Solving Multiplication Equations

2 Use algebra tiles to solve $3x = -21$.

$3x = -21$

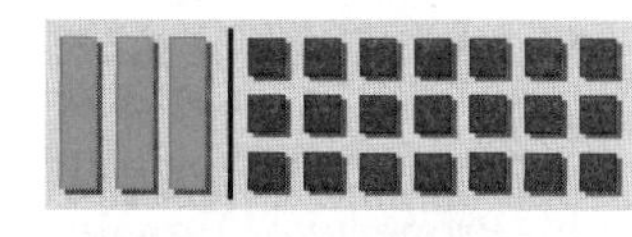

← **Model the equation. Use red tiles for negative integers.**

$\frac{3x}{3} = \frac{-21}{3}$

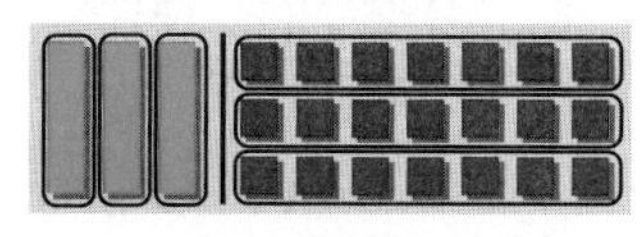

← **Divide each side into three equal groups.**

$x = -7$

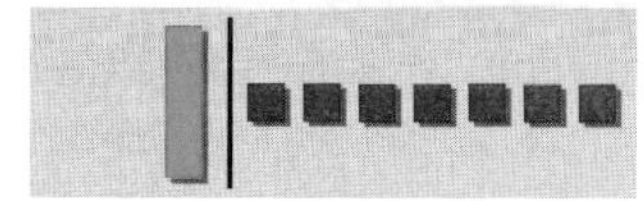

← **Simplify.**

Exercises

Use algebra tiles to solve each equation.

1. $x + 12 = 18$ **6**
2. $x + 3 = 16$ **13**
3. $x + 8 = 17$ **9**
4. $x + (-7) = 13$ **20**
5. $x + (-2) = -7$ **−5**
6. $x + (-7) = -11$ **−4**
7. $3x = 18$ **6**
8. $5x = -25$ **−5**
9. $7x = 21$ **3**
10. $2x = -18$ **−9**
11. $4x = 28$ **7**
12. $2x = 24$ **12**

Activity Lab

Modeling Equations

Students use algebra tiles to learn what an equation is and how to find its solution.

Guided Instruction

Example 2
Model the solution to the multiplication equation. Have students again follow along at their desks. Guide them to see that the equation $3x = -21$ involves multiplication. By dividing tiles into groups of equal size, they are performing the inverse operation, division.

Exercises
Have students work independently or in pairs on the Exercises. Suggest they use algebra tiles to help them with equations for which they are having difficulty.

Differentiated Instruction

Visual Learners
In Example 1, discuss with students that finding the solution to an equation means finding a value for the variable that makes the equation true. Then model the solution to the addition equation on the board or on an overhead projector. Have students follow along at their desks. Elicit from students that the equation $x + 4 = 12$ involves addition. By removing tiles they are performing the inverse operation, subtraction.

Resources

- Activity Lab 4-3: Figure It Out
- algebra tiles
- Classroom Aid 37
- Student Manipulatives Kit

4-3

1. Plan

Objective
To solve equations by adding and subtracting

Examples
1 Solving Equations by Adding
2 Solving Equations by Subtracting
3 Writing Problem Situations

Math Understandings: p. 166C

Math Background

Equations that have exactly the same solution or solutions are called *equivalent equations*. Solving a given equation is essentially a process of writing a series of equivalent equations until the variable is isolated by itself on one side of the equal sign. The equivalent equations that will achieve this goal are determined by a thoughtful application of the properties of equality.

More Math Background: p. 166C

Lesson Planning and Resources

See p. 166E for a list of the resources that support this lesson.

Bell Ringer Practice

Check Skills You'll Need
Use student page, transparency, or PowerPoint. For intervention, direct students to:
Using Number Sense to Solve Equations
Lesson 4-2
Extra Skills and Word Problems Practice, Ch. 4

Algebra

4-3 Solving Equations by Adding or Subtracting

Check Skills You'll Need

1. **Vocabulary Review** A(n) _?_ is a mathematical sentence with an equal sign. **equation**

Estimate the solution of each equation.

2. $y + 3.14 = 11.89$ **about 9**
3. $t - 4.83 = 13.12$ **about 18**
4. $14.2 + q = 38.849$ **about 25**

GO for Help Lesson 4-2

What You'll Learn

To solve equations by adding or subtracting

New Vocabulary Addition Property of Equality, Subtraction Property of Equality, inverse operations

Why Learn This?

If you can solve equations, you can use known information to find unknown information.

You can think of an equation as a balance scale. When you do something to one side of an equation, you must do the same thing to the other side of the equation to keep it balanced.

You can find the value of the unknown weight above by removing four weights from each side of the scale. The result is the scale on the right.

This illustrates the Subtraction Property of Equality. You can use this property and the Addition Property of Equality to solve equations.

GO Online

Video Tutor Help

Visit: PHSchool.com
Web Code: are-0775

KEY CONCEPTS **Properties of Equality**

Addition Property of Equality
If you add the same value to each side of an equation, the two sides remain equal.

Arithmetic	Algebra
$\frac{20}{2} = 10$, so $\frac{20}{2} + 3 = 10 + 3$.	If $a = b$, then $a + c = b + c$.

Subtraction Property of Equality
If you subtract the same value from each side of an equation, the two sides remain equal.

Arithmetic	Algebra
$\frac{12}{2} = 6$, so $\frac{12}{2} - 4 = 6 - 4$.	If $a = b$, then $a - c = b - c$.

Differentiated Instruction Solutions for All Learners

Special Needs L1
Make a double number line sketch to help students make the connection between the variables in Example 2, and the difference between the numbers. Both number lines would go from 0 – 290, but the second would stop at 245, with the difference labeled *x*.
learning style: visual

Below Level L2
Have students use a "fill-in box" to represent the solution until they become accustomed to variables.
learning style: visual

To solve an equation, you want to get the variable alone on one side of the equation. You can use **inverse operations,** operations that undo each other, to get the variable alone.

Addition and subtraction are inverse operations. You can use addition to undo subtraction.

You can represent the equation in Example 1 with this model.

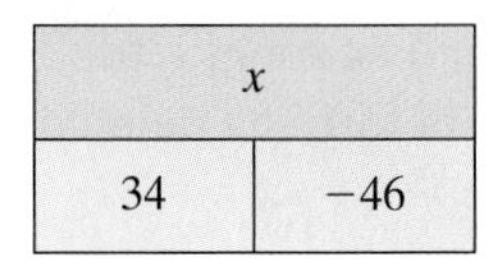

x	
34	−46

EXAMPLE Solving Equations by Adding

1 Solve $x - 34 = -46$.

$$x - 34 = -46$$

$$x - 34 + 34 = -46 + 34 \quad \leftarrow \text{Addition Property of Equality: Add 34 to each side.}$$

$$x + 0 = -12 \quad \leftarrow \text{The numbers } -34 \text{ and } 34 \text{ are additive inverses.}$$

$$x = -12 \quad \leftarrow \text{Identity Property of Addition}$$

Check $x - 34 = -46$ ← **Check the solution in the original equation.**

$-12 - 34 = -46$ ← **Substitute −12 for x.**

$-46 = -46$ ✔ ← **Subtract.**

Quick Check

1. Solve the equation $x - 104 = 64$. **168**

Just as addition undoes subtraction, subtraction undoes addition.

EXAMPLE Solving Equations by Subtracting

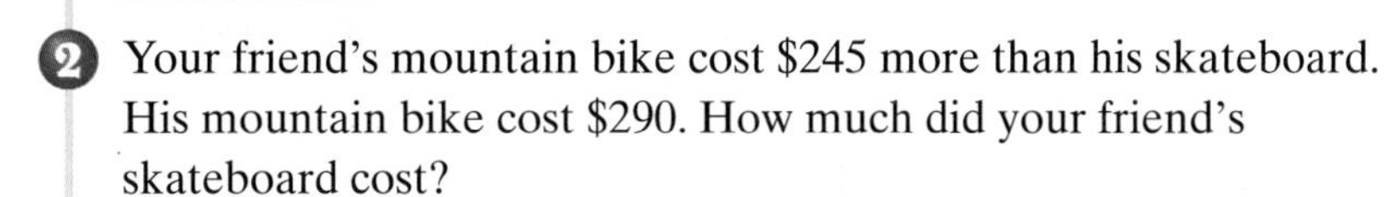

2 Your friend's mountain bike cost \$245 more than his skateboard. His mountain bike cost \$290. How much did your friend's skateboard cost?

Words cost of bike is \$245 more than cost of skateboard

Let s = the cost of the skateboard

Equation 290 = 245 + s

$$290 = 245 + s$$

$$290 - 245 = 245 - 245 + s \quad \leftarrow \text{Subtract 245 from each side.}$$

$$45 = s \quad \leftarrow \text{Simplify.}$$

The skateboard cost \$45.

Quick Check

2. A hardcover book costs \$19 more than its paperback edition. The hardcover book costs \$26.95. How much does the paperback cost? **\$7.95**

2. Teach

Activity Lab

Use before the lesson.
Student Edition Activity Lab Hands On 4-3a, Modeling Equations, p. 179

All in One Teaching Resources
Activity Lab 4-3: Figure It Out

Guided Instruction

Example 1
Some students are better able to visualize the process of adding the same number to each side when the addition is performed vertically.

$$\begin{array}{rcr} x - 34 & = & -46 \\ +\,34 & = & +\ 34 \\ \hline x & = & -12 \end{array}$$

Error Prevention!

Stress the importance of checking a proposed solution by substituting it for the variable *in the original equation.*

PowerPoint

Additional Examples

1 Solve $t - 58 = 71$. Check your solution. **$t = 129$**

2 Solve $d + 126 = 98$. Check your solution. **$d = -28$**

3 Describe a problem situation that matches the equation $p - 35 = 15$. **Sample: You spent \$35 on a sweater and had \$15 left. How much money did you have to start?**

Advanced Learners L4
Solve each equation.

$x + x = 64$ **$x = 32$**
$x - x = 64$ **no solution**

learning style: visual

English Language Learners ELL
Make the connection explicit between using inverse operations, "undoing" a subtraction or addition operation, and isolating the variable. These are often used synonymously in equations involving addition and subtraction. For example, $+3 - 3 = 0$. $x + 0$ is the same as just x by itself.

learning style: verbal

All in One Teaching Resources

- Daily Notetaking Guide 4-3 L3
- Adapted Notetaking 4-3 L1

Closure

- *What are inverse operations?* Inverse operations undo each other.
- *How can you use inverse operations to solve equations?* Use inverse operations to get the variable alone on one side of an equation. Add the same value to each side of an equation (to undo subtraction) or subtract the same value from each side of an equation (to undo addition).

EXAMPLE Writing Problem Situations

3 Describe a problem situation that matches the equation $s - 286 = 74$.

Step 1 Translate the symbols into words. The equation represents 286 subtracted from some original amount. The result of the subtraction is 74.

Step 2 Choose a context to describe the numbers and symbols.

original amount s	minus	286	is	74
↓		↓		↓
s songs	minus	286 songs	is	74 songs

Rita deleted 286 songs from her audio player. She had 74 songs left. How many songs did Rita have on her audio player before she deleted 286 songs?

You can find a solution to your equation by solving for s.

$$s - 286 = 74$$

$$s - 286 + 286 = 74 + 286 \quad \leftarrow \text{Add 286 to each side.}$$

$$s = 360 \quad \leftarrow \text{Simplify.}$$

Rita had 360 songs on her audio player.

Quick Check

3. a. Describe a problem situation that matches the equation $n + 41 = 157$. **Check students' work.**
   b. Solve the equation $n + 41 = 157$. **116**

Check Your Understanding

1. **Vocabulary** ___?___ operations are operations that undo each other. **Inverse**
2. **Error Analysis** Dylan and Jean tried to solve the equation $x + 4 = -9$. Who solved the equation correctly? Explain. **See left.**

2. Jean. Dylan's error was adding 4 to both sides instead of subtracting.

Dylan	Jean
$x + 4 = -9$	$x + 4 = -9$
$x + 4 + 4 = -9 + 4$	$x + 4 - 4 = -9 - 4$
$x = -5$	$x = -13$

Vocabulary Tip

When you get the variable alone on one side of the equation, you are "isolating the variable."

Fill in the missing numbers to solve each equation.

3. $b + 12 = 39$
   $b + 12 - \blacksquare = 39 - \blacksquare$ **12 12**

4. $y - 8 = 35$
   $y - 8 + \blacksquare = 35 + \blacksquare$ **8 8**

Homework Exercises

For more exercises, see Extra Skills and Word Problems.

GO for Help

For Exercises	See Examples
5–10	1
11–17	2
18–21	3

A **Solve each equation. Check your answer. You may find a model helpful.**

5. $x - 6 = -55$ **−49**
6. $n - 255 = -455$ **−200**
7. $-83.4 + m = 122$ **205.4**
8. $t - 32.8 = -27$ **5.8**
9. $h - 37 = -42$ **−5**
10. $q - 16 = 40$ **56**
11. $k + 17 = 29$ **12**
12. $d + 261.9 = -48$ **−309.9**
13. $x + 34 = 212$ **178**
14. $253 + c = 725$ **472**
15. $89 + y = 100$ **11**
16. $62.5 + t = -77$ **−139.5**
17. Invisible braces cost \$500 more than metal braces. Metal braces cost \$4,800. How much do invisible braces cost? **\$5,300**

Describe a problem situation that matches each equation. Then solve.
18–21. Answers may vary. Check student's work.

18. $t - 3.5 = 8$ **11.5**
19. $34 = g + 25$ **9**
20. $q + 12 = 100$ **88**
21. $63.8 = p - 17$ **80.8**

B 

22. **Guided Problem Solving** A runner's heart rate is 133 beats per minute. This is 62 beats per minute more than his resting heart rate. Write and solve an equation to find the runner's resting heart rate.
    - Choose a variable to represent the resting heart rate.
    - Write an equation that represents the information provided.
    - Check your answer. Does it fit the details of the problem?

    133 = r + 62; 71 beats/min
23. A basketball player scores 15 points in one game and p points in a second game. Her two-game total is 33 points. Write and solve an equation to find the number of points scored in the second game. **15 + p = 33; 18 points** 
24. **Biology** A student collects 12 ladybugs for a science project. This (GPS) is 9 fewer than the number of ladybugs the student collected yesterday. Write and solve an equation to find the number of ladybugs the student collected yesterday. **12 = b − 9; 21 ladybugs**

Match each equation with the graph of its solution.

25. $t + 14 = 18$ **B**
26. $x - 5 = -3$ **D**
27. $w + 4 = 2$ **A**
28. $y - 7 = -10$ **C**

A. 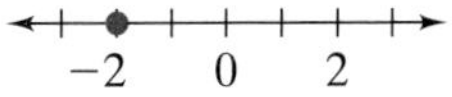 number line with labels −2, 0, 2
B. 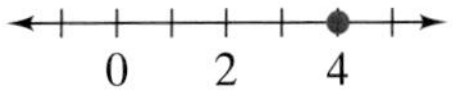 number line with labels 0, 2, 4
C. 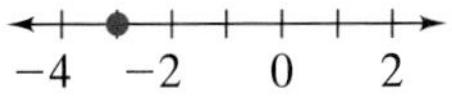 number line with labels −4, −2, 0, 2
D.  number line with labels −2, 0, 2

29. **Physics** At 20°C, the speed of sound in air is 343 m/s. This is 1,166 m/s slower than the speed of sound in water. Write and solve an equation to find the speed of sound in water. **343 = a − 1,166; 1,509 m/s**
30. **Reasoning** How can you transform the equation $5 + x = 4$ into $3 + x = 2$? Support your answer with a property of equality.

30. Subtract 2 from each side; Subtraction Property of Equality.

Online lesson quiz, PHSchool.com, Web Code: ara-0403

3. Practice

Assignment Guide

Check Your Understanding
Go over Exercises 1–4 in class before assigning the Homework Exercises.

Homework Exercises

A Practice by Example	5–21
B Apply Your Skills	22–38
C Challenge	39
Test Prep and Mixed Review	40–44

Homework Quick Check
To check students' understanding of key skills and concepts, go over Exercises 7, 18, 24, 30, and 38.

Differentiated Instruction Resources

Adapted Practice 4-3 L1

Practice 4-3 — Solving Equations by Adding or Subt… L3

Solve each equation. Check your answer.

1. $n + 2 = 5$ **3**
2. $x - 1 = -3$ **−2**
3. $7 = a + 2$ **5**
4. $p + 2 = -6$ **−8**
5. [illegible] **−5**
6. [illegible] **−4**
7. $x - (-3) = 7$ **4**
8. $a + (-6) = 5$ **11**
9. [illegible] **−10**
10. [illegible] **21**
11. [illegible] **−6**
12. [illegible] **22**

Use a calculator, paper and pencil, or mental math. Solve each equation.

13. $t + 43 = 28$ **−15**
14. $-19 = r + 6$ **−25**
15. $25 = r + 7$ **18**
16. $13 = 24 + c$ **−11**
17. $d - 19 = -46$ **−27**
18. $b + 27 = -18$ **−45**
19. $46 = f - 19$ **65**
20. $z - 74 = -19$ **55**
21. The odometer on your family car reads 20,186.7 after going 62.3 miles. Write and solve an equation to determine how many miles were on the odometer before going 62.3 miles. **m + 62.3 = 20,186.7; 20,124.4 miles**
22. Michael bought a \$25.00 gift for a friend. After he bought the gift, Michael had \$176.89. Write and solve an equation to calculate how much money Michael had before he bought the gift. **g − \$25.00 = \$176.89; \$201.89**
23. This spring it rained a total of 11.5 inches. This was 3 inches less than last spring. Write and solve an equation to find the amount of rain last season. **x − 3 = 11.5; 14.5 inches**

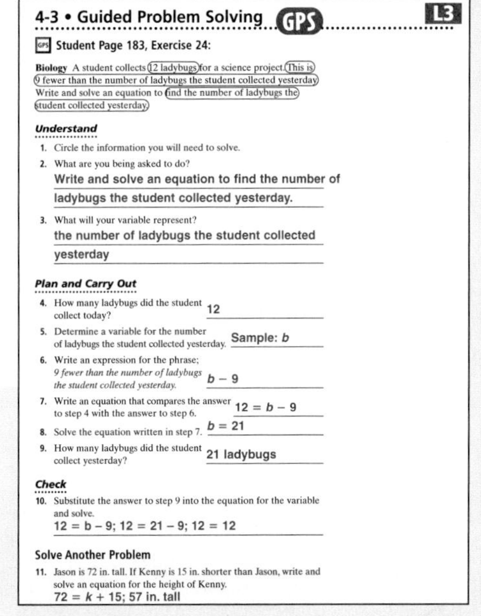

4-3 • Guided Problem Solving GPS L3

Student Page 183, Exercise 24:

Biology A student collects 12 ladybugs for a science project. This is 9 fewer than the number of ladybugs the student collected yesterday. Write and solve an equation to find the number of ladybugs the student collected yesterday.

Understand

1. Circle the information you will need to solve.
2. What are you being asked to do? **Write and solve an equation to find the number of ladybugs the student collected yesterday.**
3. What will your variable represent? **the number of ladybugs the student collected yesterday**

Plan and Carry Out

4. How many ladybugs did the student collect today? **12**
5. Determine a variable for the number of ladybugs the student collected yesterday. **Sample: b**
6. Write an expression for the phrase; *9 fewer than the number of ladybugs the student collected yesterday.* **b − 9**
7. Write an equation that compares the answer to step 4 with the answer to step 6. **12 = b − 9**
8. Solve the equation written in step 7. **b = 21**
9. How many ladybugs did the student collect yesterday? **21 ladybugs**

Check

10. Substitute the answer to step 9 into the equation for the variable and solve. **12 = b − 9; 12 = 21 − 9; 12 = 12**

Solve Another Problem

11. Jason is 72 in. tall. If Kenny is 15 in. shorter than Jason, write and solve an equation for the height of Kenny. **72 = k + 15; 57 in. tall**

4. Assess & Reteach

Lesson Quiz

Solve each equation. Check your answer.

1. $7 + a = 46$ $a = 39$
2. $b - 10 = -4$ $b = 6$
3. $9 + c = -10$ $c = -19$
4. $104 - d = 75$ $d = 29$

Reteaching 4-3

Solving Equations by Adding or S[…]g **L2**

Follow these steps to solve equations.

	Solve: $n + (-2) = 11$	Solve: $n - 6 = -36$
① Use the inverse operation on both sides of the equation.	$n + (-2) - (-2) = 11 - (-2)$	$n - 6 + 6 = -36 + 6$
② Simplify.	$n = 13$	$n = -30$
③ Check.	$n + (-2) = 11$ $13 + (-2) \stackrel{?}{=} 11$ $11 = 11$ ✔	$n - 6 = -36$ $-30 - 6 \stackrel{?}{=} -36$ $-36 = -36$ ✔

Solve each equation. Check each answer.

1. $n + 6 = 8$
   $n + 6 - 6 = 8 - \underline{6}$
   $n = \underline{2}$
2. $n - 3 = 20$
   $n - 3 + \underline{3} = 20 + 3$
   $n = \underline{23}$
3. $n - (-3) = -1$
   $n - (-3) + \underline{(-3)} = -1 + \underline{(-3)}$
   $n = \underline{-4}$
4. $-2 = n + 5$
   $-2 - \underline{5} = n + 5 - \underline{5}$
   $\underline{-7} = n$
5. $n - (-4) = -2$
   $n - (-4) + \underline{(-4)} = -2 + \underline{(-4)}$
   $n = \underline{-6}$
6. $n - 16 = -23$
   $n - 16 + \underline{16} = -23 + \underline{16}$
   $n = \underline{-7}$

Use a calculator, pencil and paper, or mental math. Solve each equation.

7. $n + 1 = 17$ **16**
8. $n - (-6) = 7$ **1**
9. $n - 8 = -12$ **−4**
10. $61 = n + 29$ **32**
11. $n + 84 = 131$ **47**
12. $-13 = n + 9$ **−22**
13. In track practice, Jesse ran a mile in 7 minutes. His mile time was $2\frac{1}{2}$ minutes faster than Michael's time. Write and solve an equation to calculate Michael's mile time.
    $m - 2\frac{1}{2} = 7$; $m = 9\frac{1}{2}$ min

Enrichment 4-3

Solving Equations by Adding or Su[…] **L4**

Decision Making

Carlos decided to join a hockey league. The annual fee is $500 and can be paid in 4 installments. He chooses to take beginner hockey lessons which cost $10 per week.

He has these choices of equipment that he can buy from each of the following stores. The more expensive equipment is usually chosen by the more experienced skaters.

Sam's Pro Shop:	skates, $395; stick, $65; helmet, $105; pads, $214; gloves, $135
Don's Sports:	skates, $120; stick, $60; helmet, $85; pads, $85; gloves, $80
Economy Sports:	skates, $65; stick, $10; helmet, $70; pads, $60; gloves, $20
Hockey Shop:	skates, $100; stick, $65; helmet, $90; pads, $80; gloves, $45

He has saved $500 and can save an additional $20 each week.

1. From which shops can Carlos buy a complete package of equipment with the money he now has?
   Don's Sports, Economy Sports, and Hockey Shop
2. From which shops can Carlos buy a complete package of equipment and have enough money left to pay the first installment for the league fee and the first week of hockey lessons? Explain.
   League installment fee and a lesson cost $135. $500 – $135 leaves $365 for equipment. Economy Sports equipment costs $225. Others cost more than $365.
3. Hockey season begins in 7 months. If Carlos chooses to buy all his equipment at Sam's Pro Shop, will he be able to afford the first installment for the league fee and the first installment for lessons? Explain.
   Yes; he will have $1,060 ($914 for equipment, $125 for fee installment, and $10 for first lesson).
4. From which shop would you advise Carlos to purchase his equipment? Explain.
   Sample answer: Economy Sports so he has money for the league fee and can begin playing immediately. He can buy better equipment when he is sure he enjoys the sport.

GO Online
Homework Video Tutor
Visit: PHSchool.com
Web Code: are-0403

Use a calculator, paper and pencil, or mental math. Solve each equation.

31. $n - 35 = 84$ **119**
32. $\frac{5}{6} = m + \frac{2}{9}$ **$\frac{11}{18}$**
33. $x + 2.5 = 1.6$ **−0.9**
34. $-\frac{1}{7} = x + 1\frac{2}{3}$ **$-1\frac{17}{21}$**
35. $\frac{3}{16} = c - \frac{7}{8}$ **$1\frac{1}{16}$**
36. $3\frac{1}{4} = 2\frac{1}{3} + r$ **$\frac{11}{12}$**

37. **Money** Use the advertisement at the right. Write and solve an equation to find the original price of the sweater.

37. $p - 8.45 = 21.50$; $29.95

38. **Writing in Math** A student is saving money for field hockey camp. Her savings are modeled by the equation $135 + d = 250$. Explain what each part of the equation represents.

38. She has saved $135. The camp costs $250. She needs d dollars.

Ⓒ 39. **Challenge** During five baseball games, your team scores 3, 4, 2, 6, and 8 runs. How many more runs must your team score to have a total of 30 runs? Write and solve an equation.

39. Let r = the number of runs needed; $3 + 4 + 2 + 6 + 8 + r = 30$; 7 runs.

Test Prep and Mixed Review

Practice

Multiple Choice

40. The model represents the equation $x - 4 = 2$. What is the value of x? **D**

Ⓐ 2 Ⓑ 4 Ⓒ 3 Ⓓ 6

Key: ⊕ = +1, ⊖ = −1

41. Which problem situation matches the equation $7.50 + x = 100$? **H**

Ⓕ Roberto bought a box of 100 baseball cards for $7.50. What is x, the price for each baseball card?
Ⓖ Will works 7.5 hours a day. He will get a pay raise when he works 100 hours. What is x, the number of days he works to get a raise?
Ⓗ Mr. Midas bought a case of fruit for $7.50. He paid the cashier with a $100 bill. What is x, the amount of change he received?
Ⓙ Jennifer walks for 7.5 minutes. What is x, the number of miles she can walk in 100 minutes?

42. The U.S. Mint made 1,303,384,000 nickels at the Denver and Philadelphia mints one year. The algebraic expression for the value in dollars of n nickels is $0.05n$. What is the value of this expression, in dollars, for the number of nickels made that year? **C**

Ⓐ 651,792,000 Ⓑ 65,179,200 Ⓒ 65,169,200 Ⓓ 6,516,920

GO for Help

For Exercises	See Lesson
43–44	2-1

Simplify each expression.

43. $3 - 4 + 5 \cdot 6 - (-4)$ **33**
44. $4 \cdot 5 - 6 + (5 - 2)^2$ **23**

Test Prep

Resources

For additional practice with a variety of test item formats:

- Test-Taking Strategies, p. 219
- Test Prep, p. 223
- Test-Taking Strategies with Transparencies

Alternative Assessment

Provide partners with algebra tiles. Have them write several simple addition and subtraction equations similar to those in Exercises 25–28. Working together, partners should use the tiles to solve each equation and record each step of their solution process.

Vocabulary Builder

High-Use Academic Words

High-use academic words are words that you will see often in textbooks and on tests. These words are not math vocabulary terms, but knowing them will help you succeed in mathematics.

Direction Words

Some words tell what to do in a problem. I need to understand what these words are asking so that I give the correct answer.

Word	Meaning
Choose	To make a selection after analyzing given information
Describe	To tell or write about something in detail
Determine	To make a decision based on investigation

Exercises

1. Choose the colors that represent the United States of America. **C**
   - (A) red, blue, green
   - (B) red, orange, yellow
   - (C) white, blue, red
   - (D) green, purple, orange
2. Describe the design of the American flag. Determine whether there are more red stripes or white stripes on the American flag.
   2. 13 red and white stripes, 50 white stars on blue background. There are more red stripes.
3. Determine whether 2 is a solution of $8 + x = 12$. **no**
4. Describe a situation that $x - 2 = 5$ can represent. **Check students' work.**
5. Choose the model that represents "3 more than a number p." **C**

(A)

p		
$\frac{p}{3}$	$\frac{p}{3}$	$\frac{p}{3}$

(B)

p	p	p

(C)

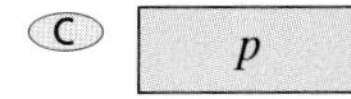

p	3

6. **Word Knowledge** Think about the word *substitute*. **6a–c. Check students' work.**
   - **a.** Choose the letter for how well you know the word.
     - **A.** I know its meaning.
     - **B.** I've seen it, but I don't know its meaning.
     - **C.** I don't know it.
   - **b.** **Research** Look up and write the definition of *substitute*.
   - **c.** Use the word in a sentence involving mathematics.

Vocabulary Builder

High-Use Academic Words

Students learn a strategy for learning words that, while not math vocabulary terms, are important for success in mathematics and on tests.

Guided Instruction

Have students look through their texts for use of the terms: *choose, describe, and determine.* Ask the following questions:

- *Where do you find the term* describe? **Sample: page 197, Exercise 31**
- *What is another way to say "Find the solution for the equation from the given numbers"?* **Sample: Determine the values for the variables that make the equation true.**
- *What direction could you write for Exercise 40 on page 184 using the term,* choose? **Sample: Choose the correct value for *x*.**

Teaching Tip
Restate directions given in the text using *choose, describe,* and *determine* as appropriate to familiarize students with these terms.

Differentiated Instruction

English Language Learners **ELL**
Encourage students to write high-use academic words in their native language as needed.

Resources

- Vocabulary and Study Skills Worksheets

4-4

1. Plan

Objective
To solve equations by multiplying or dividing

Examples
1 Solving Equations by Dividing
2 Application: Telephone Charges
3 Solving Equations by Multiplying

Math Understandings: p. 166C

Math Background

The properties of equality are sometimes summarized informally as follows: *Whatever you do to one side of an equation, you must do to the other.* While this statement is essentially correct, care must be taken to note that there is an important restriction. That is, the Division Property of Equality prohibits dividing both sides of an equation by zero. This restriction is necessary because, in the system of real numbers, division by zero is undefined.

More Math Background: p. 166C

Lesson Planning and Resources

See p. 166E for a list of the resources that support this lesson.

Bell Ringer Practice

Check Skills You'll Need
Use student page, transparency, or PowerPoint. For intervention, direct students to:

Mixed Numbers and Improper Fractions
Lesson 2-5
Extra Skills and Word Problems Practice, Ch. 2

Algebra

4-4 Solving Equations by Multiplying or Dividing

Check Skills You'll Need

1. **Vocabulary Review** When the GCF of a numerator and denominator is one, the fraction is in __?__. **simplest form**

Write each expression in simplest form.

2. $\frac{6}{24}$ $\frac{1}{4}$
3. $\frac{8(4)}{8}$ 4
4. $\frac{7(3)}{21}$ 1
5. $\frac{3a}{3}$ a

GO for Help Lesson 2-3

What You'll Learn

To solve equations by multiplying or dividing

New Vocabulary Division Property of Equality, Multiplication Property of Equality

Why Learn This?

If you know how to solve equations by multiplying or dividing, you can solve everyday problems such as sharing the cost of a meal with friends.

You and three friends go out for pizza. An extra-large pizza and four bottles of water cost \$22.68 (including tax and a \$2 coupon). How much does each person owe if you split the bill equally?

CHECK # 325 TABLE # 12

ITEMS ORDERED	AMOUNT
X-LG PIZZA	18.60
\$2 COUPON	-2.00
BOT WTR	1.25
BOT WTR	1.25
BOT WTR	1.25
BOT WTR	1.25
TAX	1.08
TOTAL	22.68

You can represent the problem with the model below. Let p represent the amount of money each person owes.

total bill	→	22.68
p p p p	→	$4p$

The model shows that you can use the equation $4p = 22.68$. To solve this equation, you can use the Division Property of Equality.

KEY CONCEPTS **Division Property of Equality**

If you divide each side of an equation by the same nonzero number, the two sides remain equal.

Arithmetic	Algebra
Since $3(2) = 6$, $\frac{3(2)}{2} = \frac{6}{2}$.	If $a = b$ and $c \neq 0$, then $\frac{a}{c} = \frac{b}{c}$.

Division is the inverse operation of multiplication. When a variable is multiplied by a number, you can use division to undo the multiplication.

Division Undoes Multiplication

$(4 \cdot 9) \div 4 = 9$ $\qquad$ $5x \div 5 = x$

Differentiated Instruction Solutions for All Learners

Special Needs L1
Have students use number sense to check the calculation for Example 2. Ask: *What would 10 minutes cost if one minute cost 39 cents?* **\$3.90** *What would 20 minutes cost?* **\$7.80** Since this is close to \$8.58, 22 minutes makes sense.
learning style: verbal

Below Level L2
Make a worksheet with several multiplication and division equations in one column and solutions in a second column. Have students match equations and solutions.
learning style: visual

Video Tutor Help

Visit: PHSchool.com
Web Code: are-0775

EXAMPLE Solving Equations by Dividing

1 Solve $4p = 22.68$.

$4p = 22.68$ ← Notice p is being *multiplied* by 4.

$\frac{4p}{4} = \frac{22.68}{4}$ ← Divide each side by 4 to get p alone.

$p = 5.67$ ← Simplify.

Check $4p = 22.68$ ← Check your solution in the original equation.

$4(5.67) \stackrel{?}{=} 22.68$ ← Replace p with 5.67.

$22.68 = 22.68$ ✔ ← The solution checks.

Quick Check

1. Solve each equation. Check your answer.
   a. $3x = -21.6$ **−7.2** b. $-12y = -108$ **9** c. $104x = 312$ **3**

You can use the Division Property of Equality and inverse operations to solve real-world applications.

EXAMPLE Application: Telephone Charges

The words *each* and *per* are frequently used in problems that involve solving an equation by multiplying or dividing.

2 **Gridded Response** Your cellular telephone bill shows that you were charged an extra $8.58 this month for going over your allotted minutes. The company charges $.39 for each extra minute. How many extra minutes did you use?

Words $0.39 times number of minutes equals extra charge

Let n = the number of extra minutes.

Equation $0.39 \cdot n = 8.58$

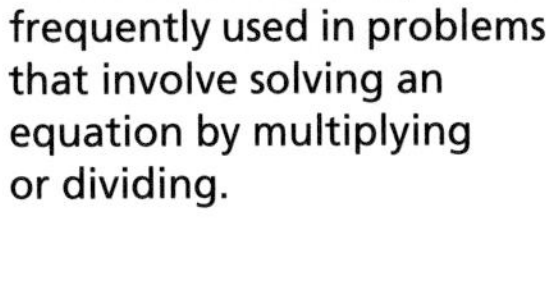

$0.39n = 8.58$

$\frac{0.39n}{0.39} = \frac{8.58}{0.39}$ ← Divide each side by 0.39.

$n = 22$ ← Simplify.

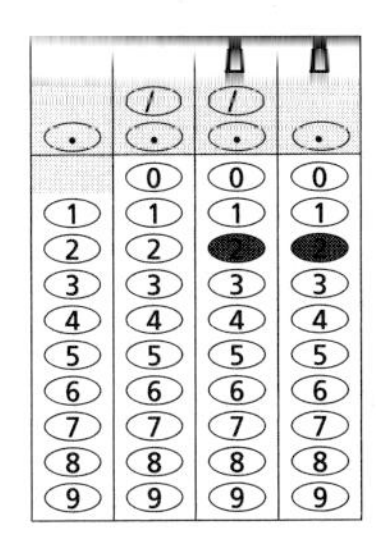

You used 22 extra minutes.

Check The charge for each extra minute is $.39. If you use 22 extra minutes, then the total charge is 22 · $.39, or $8.58. The answer checks.

Quick Check

2. Suppose you and four friends go to a baseball game. The total cost for five tickets is $110. Write and solve an equation to find the cost of one ticket. **5c = 110; $22**

2. Teach

Activity Lab

Use before the lesson.

All in One Teaching Resources
Activity Lab 4-4: Car Mileage

Guided Instruction

Example 1
Have students estimate the answer. An appropriate strategy is using compatible numbers, as follows:
Round 22.68 up to 24.
You know $4 \cdot 6 = 24$
So p is a little less than 6.

Error Prevention!

When solving equations by multiplying or dividing, students might focus on balancing the equation and neglect the sign of the solution. Have them begin their work by determining whether the solution is positive or negative and recording the sign as follows:
$r = +\blacksquare$
$n = -\blacksquare$

PowerPoint

Additional Examples

1 Solve $-3j = 44.7$. Check your answer. **$j = -14.9$**

2 The Art Club must buy 84 pieces of poster board. There are 6 pieces in a package. How many packages must the Art Club buy? **14 packages**

3 Solve $\frac{m}{-3} = 27$, **$m = -81$**

Advanced Learners L4
Write two equations with solution $c = -46$, one that is solved by multiplying and one by dividing.
Samples: $\frac{c}{-2} = 23$; $2c = -92$
learning style: visual

English Language Learners ELL
Make sure students understand what the variable represents each time you have a word problem by having them tell you. In the lesson exposition, review that the p stands for the amount of money each person owes and not *people*.
learning style: verbal

All in One Teaching Resources
- Daily Notetaking Guide 4-4 L3
- Adapted Notetaking 4-4 L1

Closure

- *Why do you use the Division Property of Equality to solve an equation with multiplication, such as $12x = 96$?* **To get the variable alone on one side of the equation, divide each side of the equation by 12 to undo the multiplication of 12x.**
- *When do you use the Multiplication Property of Equality to solve an equation?* **when a variable in an equation is divided by a number (value), as in $\frac{x}{3} = 5$**

Another property you can use to solve equations is the Multiplication Property of Equality.

KEY CONCEPTS Multiplication Property of Equality

If you multiply each side of an equation by the same number, the two sides remain equal.

Arithmetic	Algebra
$\frac{12}{2} = 6$, so $\frac{12}{2} \cdot 2 = 6 \cdot 2$.	If $a = b$, then $a \cdot c = b \cdot c$.

Multiplication is the inverse operation of division. When a variable is divided by a number, you can use multiplication to undo the division.

Multiplication Undoes Division

$\frac{3}{5} \cdot 5 = 3$ $\qquad$ $\frac{n}{3} \cdot 3 = n$

EXAMPLE Solving Equations by Multiplying

Vocabulary Tip

Read $\frac{t}{-45} = -5$ as "*t* divided by negative 45 equals negative 5."

3 Solve $\frac{t}{-45} = -5$.

$$\frac{t}{-45} = -5 \quad \leftarrow \text{Notice that } t \text{ is divided by } -45.$$

$$(-45) \cdot \left(\frac{t}{-45}\right) = (-45) \cdot (-5) \quad \leftarrow \text{Multiply each side by } -45.$$

$$t = 225 \quad \leftarrow \text{Simplify.}$$

Quick Check

3. Solve the equation $\frac{w}{26} = -15$. Check your answer. **−390**

Check Your Understanding

1. **Vocabulary** The __?__ states that if you multiply each side of an equation by the same number, the two sides remain equal. **Multiplication Property of Equality**

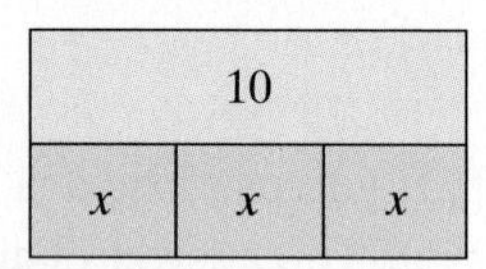

2. Write and solve the equation modeled at the left. **$10 = 3x$; $\frac{10}{3}$**

Tell which property you would use to solve each equation.

3–5. See left.

3. $8g = -25.4$
4. $\frac{w}{26} = -15$
5. $5 \cdot d = 60$

3. Division Property of Equality
4. Multiplication Property of Equality
5. Division Property of Equality

Match each equation with the correct first step of the solution.

6. $-6y = 12$ **B**
7. $12y = -6$ **C**
8. $\frac{y}{-6} = 12$ **A**

A. Multiply both sides by −6.
B. Divide both sides by −6.
C. Divide both sides by 12.

Homework Exercises

For more exercises, see Extra Skills and Word Problems.

GO for Help

For Exercises	See Examples
9–22	1–2
23–34	3

37. m represents the money the quintet needs to earn, 5 represents each member, and 50 represents the money each member receives.

38. $12x = -36$; -36 is the multiple of 12 that is closest to -38; $x \approx -3$.

A **Solve each equation. Check your answer. You may find a model helpful.**

9. $12t = 144$ **12**
10. $13e = -52$ **−4**
11. $35q = -175$ **−5**
12. $-7n = -294$ **42**
13. $0.2x = 4$ **20**
14. $-0.5r = -8$ **16**
15. $2{,}700 = -900w$ **−3**
16. $-3k = -18$ **6**
17. $8n = 112$ **14**
18. $0.4k = -40$ **−100**
19. $58 = \frac{2}{3}w$ **87**
20. $-48 = \frac{1}{2}y$ **−96**

21. **Entertainment** A local park rents paddle boats for $5.50 per hour. You have $22 to spend. For how many hours can you rent a boat? **4 h**

22. A video store charges $2.75 per day for overdue video games. You owe a late fee of $13.75. How many days overdue is the game? **5 days**

Solve each equation. Check your answer.

23. $\frac{z}{8} = -3$ **−24**
24. $\frac{n}{3} = 9$ **27**
25. $\frac{t}{8} = 12.6$ **100.8**
26. $\frac{q}{2} = -1.4$ **−2.8**
27. $\frac{k}{-6} = -5$ **30**
28. $\frac{d}{0.5} = 11$ **5.5**
29. $\frac{c}{-6} = -1$ **6**
30. $\frac{y}{-12} = -12$ **144**
31. $\frac{n}{-5} = 1.1$ **−5.5**
32. $\frac{f}{7.9} = 5$ **39.5**
33. $\frac{d}{6} = -4.25$ **−25.5**
34. $\frac{m}{-0.5} = -41$ **20.5**

B 

35. **Guided Problem Solving** Julie's car travels 27 miles per gallon of gasoline used. She recently traveled 324 miles. How many gallons of gasoline did her car use? **12 gal**
    - **Make a Plan** Choose a variable to represent the number of gallons used. Decide what operation should be used, and write an equation.
    - **Carry Out the Plan** Let ■ represent the number of gallons. The operation to use is __?__. The equation ■ can be used to solve the problem.

36. George Adrian of Indianapolis, Indiana, picked 15,830 lb of apples in 8 h. How many pounds of apples per hour is that? **about 1,979 lb of apples**

A band with five members is a quintet.

37. **Writing in Math** Explain why you can use the equation $\frac{m}{5} = 50$ to describe how much money a quintet has to earn for each member to receive $50. What does each part of the equation represent? **See above left.**

38. **Estimation** What equation would you use to estimate the solution of $12x = -38$? Explain. Estimate the solution. **See above left.**

3. Practice

Assignment Guide

Check Your Understanding

Go over Exercises 1–8 in class before assigning the Homework Exercises.

Homework Exercises

A Practice by Example	9–34
B Apply Your Skills	35–51
C Challenge	52
Test Prep and Mixed Review	53–57

Homework Quick Check

To check students' understanding of key skills and concepts, go over Exercises 15, 25, 37, 45, and 47.

Differentiated Instruction Resources

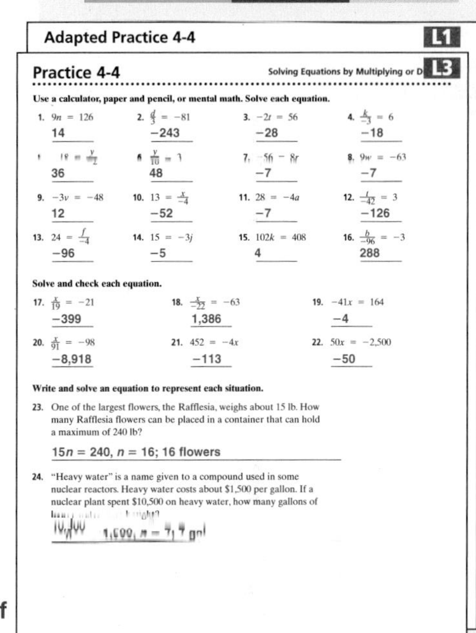

Adapted Practice 4-4 L1

Practice 4-4 Solving Equations by Multiplying or D[ividing] L3

Use a calculator, paper and pencil, or mental math. Solve each equation.

1. $9n = 126$ **14**
2. $\frac{d}{3} = -81$ **−243**
3. $-2t = 56$ **−28**
4. $\frac{k}{-3} = 6$ **−18**
5. [illegible] **36**
6. [illegible] **48**
7. $-56 = 8r$ **−7**
8. $9w = -63$ **−7**
9. $-3v = -48$ **12**
10. $13 = \frac{x}{-4}$ **−52**
11. $28 = -4a$ **−7**
12. $\frac{t}{-42} = 3$ **−126**
13. $24 = \frac{f}{-4}$ **−96**
14. $15 = -3j$ **−5**
15. $102k = 408$ **4**
16. $\frac{b}{-96} = -3$ **288**

Solve and check each equation.

17. $\frac{x}{19} = -21$ **−399**
18. $\frac{x}{-22} = -63$ **1,386**
19. $-41x = 164$ **−4**
20. $\frac{x}{91} = -98$ **−8,918**
21. $452 = -4x$ **−113**
22. $50x = -2{,}500$ **−50**

Write and solve an equation to represent each situation.

23. One of the largest flowers, the Rafflesia, weighs about 15 lb. How many Rafflesia flowers can be placed in a container that can hold a maximum of 240 lb?
$15n = 240$, $n = 16$; 16 flowers

24. "Heavy water" is a name given to a compound used in some nuclear reactors. Heavy water costs about $1,500 per gallon. If a nuclear plant spent $10,500 on heavy water, how many gallons of [illegible]
[illegible]

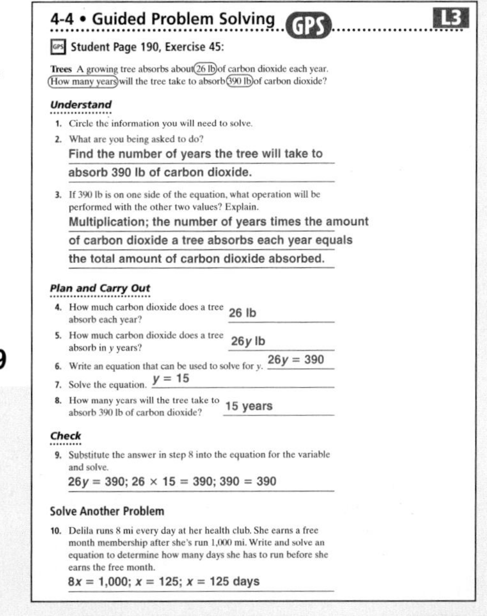

4-4 • Guided Problem Solving GPS L3

Student Page 190, Exercise 45:

Trees A growing tree absorbs about 26 lb of carbon dioxide each year. How many years will the tree take to absorb 390 lb of carbon dioxide?

Understand

1. Circle the information you will need to solve.
2. What are you being asked to do? **Find the number of years the tree will take to absorb 390 lb of carbon dioxide.**
3. If 390 lb is on one side of the equation, what operation will be performed with the other two values? Explain. **Multiplication; the number of years times the amount of carbon dioxide a tree absorbs each year equals the total amount of carbon dioxide absorbed.**

Plan and Carry Out

4. How much carbon dioxide does a tree absorb each year? **26 lb**
5. How much carbon dioxide does a tree absorb in y years? **$26y$ lb**
6. Write an equation that can be used to solve for y. **$26y = 390$**
7. Solve the equation. **$y = 15$**
8. How many years will the tree take to absorb 390 lb of carbon dioxide? **15 years**

Check

9. Substitute the answer in step 8 into the equation for the variable and solve. **$26y = 390$; $26 \times 15 = 390$; $390 = 390$**

Solve Another Problem

10. Delila runs 8 mi every day at her health club. She earns a free month membership after she's run 1,000 mi. Write and solve an equation to determine how many days she has to run before she earns the free month. **$8x = 1{,}000$; $x = 125$; $x = 125$ days**

PowerPoint

Lesson Quiz

Solve each equation.

1. $15x = 60$ **$x = 4$**
2. $\frac{y}{12} = 8$ **$y = 96$**
3. $-35 = -7b$ **$b = 5$**
4. $\frac{n}{-5} = -13$ **$n = 65$**

GO Online
Homework Video Tutor
Visit: PHSchool.com
Web Code: are-0404

Solve each equation.

39. $\frac{1}{2}m = 25$ **50**
40. $\frac{w}{-4.2} = 10.3$ **−43.26**
41. $\frac{1}{5}t = 17$ **85**
42. $\frac{k}{21.45} = 6$ **128.7**
43. $\frac{3}{4}x = 16$ **$21\frac{1}{3}$**
44. $\frac{y}{-5.22} = -3.11$ **16.2342**

45. **Trees** A growing tree absorbs about 26 lb of carbon dioxide each year. How many years will the tree take to absorb 390 lb of carbon dioxide? **15 yr**

GPS

46. The world record for playgoing is held by Dr. H. Howard Hughes of Fort Worth, Texas. He saw 6,136 plays in 31 years. How many plays did Dr. Hughes see in an average year? **about 198 plays**

47. **Reasoning** In the equation $ab = 1$, a is an integer. Explain what you know about b. **b is the reciprocal of a, and $b \neq 0$.**

48. **Error Analysis** One student's solution for the equation $\frac{n}{-6} = 12$ is $n = -2$. Explain how the student may have found this solution. Then correct the student's mistake.

48. Answers may vary. Sample: The student may have divided 12 by −6. The student should have multiplied both sides by −6 to get $n = -72$.

Open-Ended **Write a problem that can be solved using each equation.**
49–51. Check students' work.

49. $3x = 30$
50. $\frac{n}{5} = 2$
51. $2.5p = 10$

C 52. **Challenge** You are starting a savings account. You make an initial deposit of \$100. Every week after that, you deposit \$20. Your goal is to save \$1,000. Write an algebraic equation and find how many weeks you will need to reach your goal. **$100 + 20w = 1{,}000$; 45 weeks**

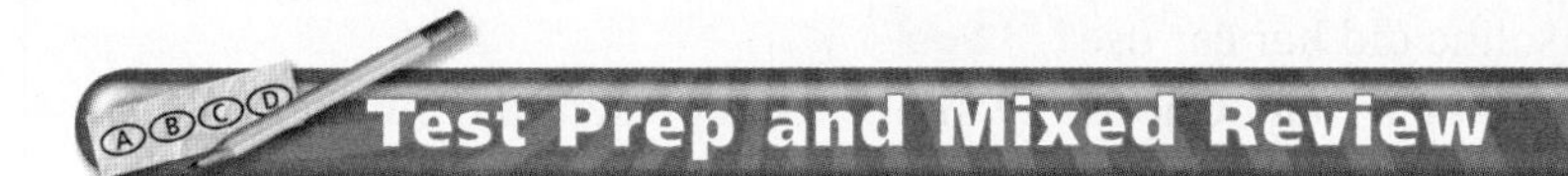

Gridded Response

53. The table at the right shows the number of cans each student collected for a school food drive. Which number could be added to the set of data in order for the median and the mode of the set to be equal? **72**

Food Drive Results

Student	Number of Cans
Abel	46
Annie	89
Cris	72
Pedro	63
Sungmee	108
Tasha	■

54. A recipe calls for 3 lb of potatoes. You have one bag of potatoes that weighs 1.25 lb, and another bag that weighs 0.65 lb. How many more pounds of potatoes do you need for the recipe? **1.1**

GO for Help

For Exercises	See Lesson
55–57	4-3

Solve each equation.

55. $p - \frac{4}{7} = \frac{1}{4}$ **$\frac{23}{28}$**
56. $\frac{1}{2} = x - \frac{3}{4}$ **$1\frac{1}{4}$**
57. $\frac{1}{2} = w + \frac{5}{12}$ **$\frac{1}{12}$**

Test Prep

Resources

For additional practice with a variety of test item formats:

- Test-Taking Strategies, p. 219
- Test Prep, p. 223
- Test-Taking Strategies with Transparencies

Alternative Assessment

Ask students to write their own word problems similar to those in Exercises 21 and 22. Then have students exchange problems with a partner. Partners should write equations for the problems, solve the equations, and check their answers.

Reteaching 4-4

Solving Equations by Multiplying o[r Dividing] L2

Follow these steps to solve equations.

	Solve: $\frac{t}{5} = -7$	Solve: $-2x = 8$
① Use the inverse operation on both sides of the equation.	$(5)\frac{t}{5} = (5)(-7)$	$\frac{-2x}{-2} = \frac{8}{-2}$
② Simplify.	$t = -35$	$x = -4$
③ Check.	$\frac{t}{5} = -7$	$-2x = 8$
	$\frac{-35}{5} \stackrel{?}{=} -7$	$-2(-4) \stackrel{?}{=} 8$
	$-7 = -7$ ✔	$8 = 8$ ✔

Solve and check each equation.

1. $-5n = 30$; $\frac{-5n}{-5} = \frac{30}{-5}$; $n =$ **−6**
2. $\frac{a}{2} = -16$; $(2)\frac{a}{2} = (2)(-16)$; $a =$ **−32**
3. $-2w = -4$; $\frac{-2w}{-2} = \frac{-4}{-2}$; $w =$ **2**
4. $8t = 32$; $\frac{8t}{8} = \frac{32}{8}$; $t =$ **4**
5. $5 = \frac{g}{6}$; $(6)(5) = (6)\frac{g}{6}$; **30** $= g$
6. $\frac{n}{-3} = -5$; $(-3)\frac{n}{-3} = (-3)(-5)$; $n =$ **15**

Use a calculator, pencil and paper, or mental math. Solve each equation.

7. $\frac{x}{4} = -1$ **−4**
8. $-5w = 125$ **−25**
9. $\frac{m}{-8} = 10$ **−80**
10. $-2 = \frac{x}{-4}$ **8**
11. $\frac{a}{-4} = 12$ **−48**
12. $-6b = 42$ **−7**
13. $-3 = \frac{c}{-8}$ **24**
14. $5 = \frac{d}{7}$ **35**
15. $2t = 38$ **19**
16. $-9 = 9q$ **−1**
17. $n + 6 = -3$ **−18**
18. $-8k = -40$ **5**

Enrichment 4-4

Solving Equations by Multiplying o[r Dividing] L4

Critical Thinking

The equation $d = r \times t$ relates rate (r), time (t) and distance (d).

1. If you are traveling at 55 miles per hour, what formula would you use to find the distance you will travel over various time intervals? **$d = 55 \times t$**
2. Sometimes you are given distances and need to know how long it would take to travel at various rates. Complete the table to find the time it would take to travel 500 miles at speeds of 40, 55, 60, 65, and 70 miles per hour. Round your times to the nearest quarter hour, if necessary.

Rate (in mi/h)	40	55	60	65	70
Time (in hours)	$12\frac{1}{2}$	9	$8\frac{1}{4}$	$7\frac{3}{4}$	$7\frac{1}{4}$
Distance (in miles)	500	500	500	500	500

3. What operation did you use to find the time? Rewrite the equation so that t stands alone on one side of the equal sign. **Division; $t = \frac{d}{r}$**
4. Rewrite the equation so that r stands alone on one side of the equal sign. Explain your reasoning. **$r = \frac{d}{t}$; Possible answer: To find the rate, divide the distance by the time.**
5. How is rewriting an equation like solving an equation? **You use inverse operations to isolate the variable.**
6. Why would you want to rewrite an equation? **Sample answer: If you needed to find many solutions using the same equation, it would be faster to have the variable isolated before doing each calculation.**
7. The equation for the perimeter of a regular polygon can be written as $P = n \times s$ where P is the perimeter, n is the number of sides and s is the length of each side. How could you rewrite the equation to find the length of a side? **$s = \frac{P}{n}$**

Checkpoint Quiz 1

Lessons 4-1 through 4-4

Write an algebraic expression for each word phrase.

1. four less than a number $x - 4$
2. three times a number $3x$
3. the quotient of 4 and a number $\frac{4}{x}$
4. nine more than a number $x + 9$

Solve each equation.

5. $g - 5 = -9.4$ -4.4
6. $y + 10.2 = 12$ 1.8
7. $-5x = 45.5$ -9.1
8. $\frac{h}{6} = 8$ 48
9. **Science** The chemical element aluminum was discovered in 1825 by Hans Christian Oersted. This was 18 years after Sir Humphry Davy discovered the element sodium. Write and solve an equation to find the year that sodium was discovered. $y + 18 = 1825; y = 1807$
10. Fingernails grow about 1.5 inches per year. How long would it take to grow nails 37 inches long? **24 years 8 months**

MATH GAMES

Evaluating Expressions

What You'll Need

- 24 index cards. Cut them in half so you have 48 smaller cards. On each card, write a different algebraic expression.
- Two number cubes

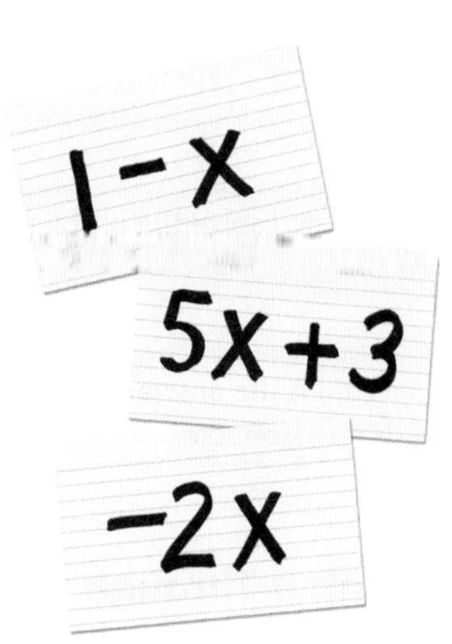

How to Play

- Deal all the cards. Each player chooses one card from his or her hand and places it face down on the table.
- A player rolls the number cubes. The sum of the numbers is the value of the variable for the first round.
- Each player turns over his or her card, evaluates the expression, and announces its value. Record the value for each player.
- Play continues until all players have rolled the number cubes.
- Find each player's total. The player with the lowest total wins.

Checkpoint Quiz

Use this Checkpoint Quiz to check students' understanding of the skills and concepts of Lessons 4-1 through 4-4.

Resources

- All-in-One Teaching Resources Checkpoint Quiz 1
- ExamView CD-ROM
- Success Tracker™ Online Intervention

MATH GAMES

Evaluating Expressions

In this game, students practice evaluating expressions. Players receive algebraic expressions written on cards. They choose an expression, roll two number cubes, find the sum, and use the sum to evaluate the expression. The player with the lowest total wins.

Guided Instruction

Before they play the game, remind students that the values found can be positive or negative. *Ask: Does a negative value decrease the total?* yes

Resources

- 24 index cards cut in half with 48 algebraic expressions
- 2 number cubes

GPS Guided Problem Solving

Practice Solving Problems

In this feature, students practice solving problems using diagrams. Then they use symbols to write equations that record their actions. They solve the equation and check the reasonableness of their answer.

Guided Instruction

Discuss with students the use of diagrams or models to show the information in a word problem. Then help them to use symbols to write an equation that describes what the situation is.

Have a volunteer read the word problem.
Ask:

- *How else could you have found the number of $\frac{1}{4}$ pounds in 4,000 lbs?* **Multiply 4,000 by 4.**
- *Why does multiplication work?* **Dividing by $\frac{1}{4}$ is the same as multiplying by 4.**

Resources

- Classroom Aid 11

Practice Solving Problems

Big Burgers The world's largest hamburger weighed 8,266 lb. If half of the weight was meat, how many regular quarter-pound hamburgers could have been made from that burger?

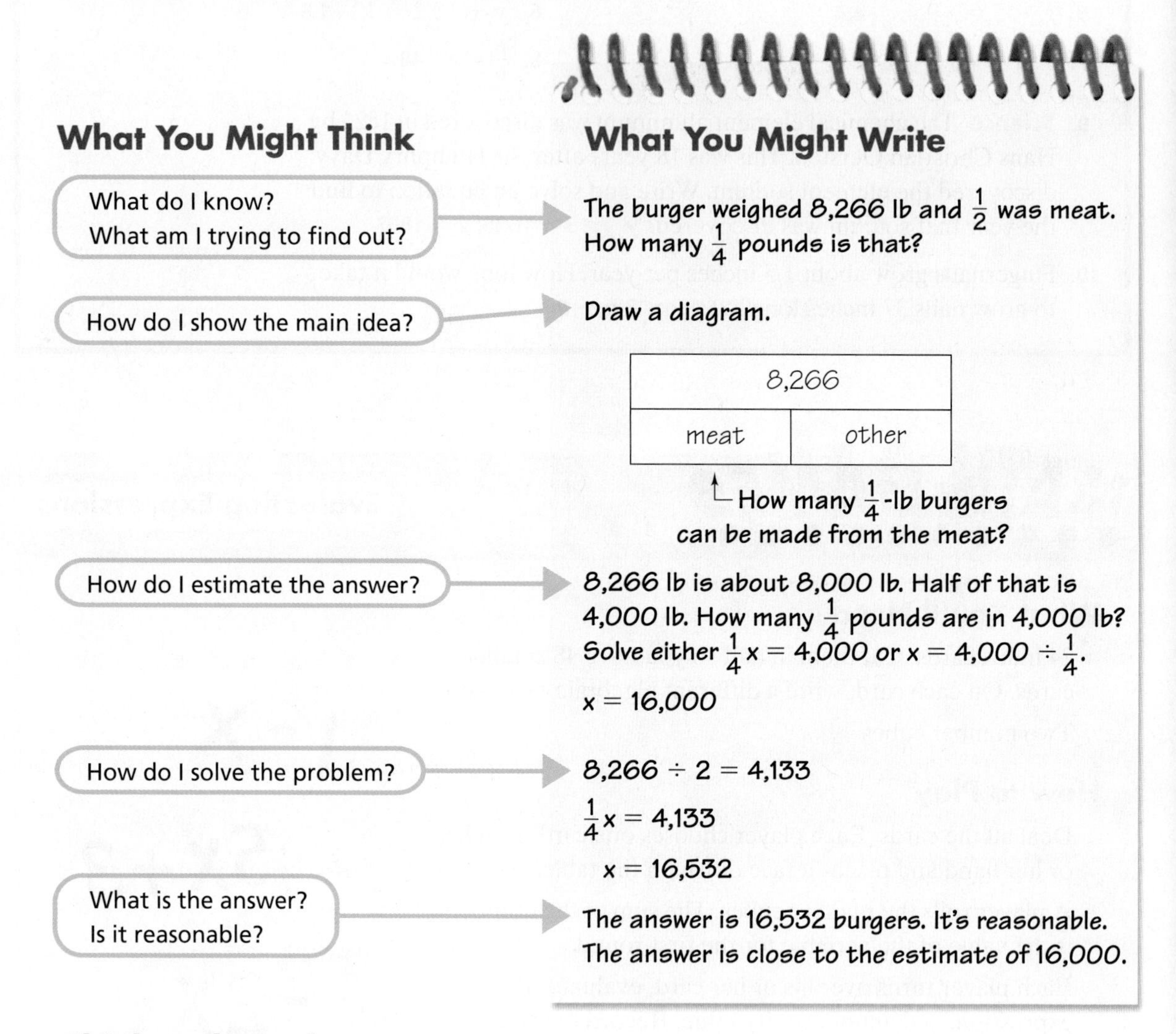

Think It Through

1. In the diagram, why do *meat* and *other* equally share the space under 8,266?
   1. Together, they are equal to 8,266 lb, and $\frac{1}{2}$ the weight is meat.
2. Explain why $\frac{1}{4}x = 4,000$ and $x = 4,000 \div \frac{1}{4}$ are equivalent.
   2. In both situations, you need to multiply 4,000 by 4.

Exercises

Solve. For Exercises 3 and 4, answer the questions first.

3. To celebrate National Hot Dog Month, a beef company made a hot dog that measured 16 feet 1 inch. A regular hot dog is $5\frac{1}{2}$ inches long. How many regular hot dogs would you need to make this hot dog?
   a. What do you know? What do you want to find out?
   b. Explain how the diagram below shows the situation. Write an equation.

16 ft 1 in.			
$5\frac{1}{2}$ in.	$5\frac{1}{2}$ in.	...	$5\frac{1}{2}$ in.

3. about 35 hot dogs

4. The world's largest glass of milk held 4,000 gallons. A gallon holds 16 cups. How many cups were in the world's largest glass of milk?
   a. Explain how the diagram below shows the situation. Write an equation.

4. 64,000 cups

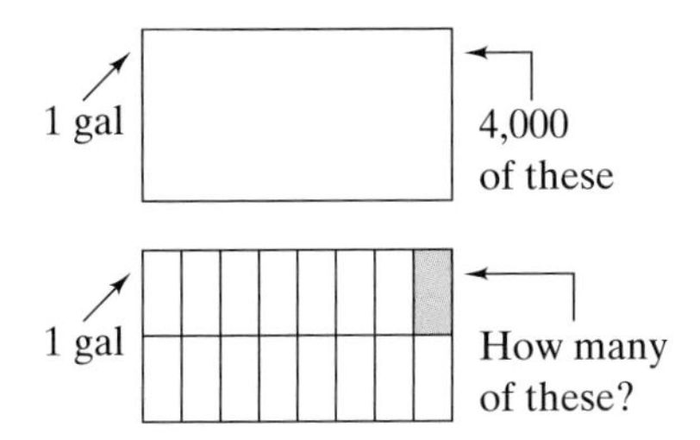

5. a. Look at the graph at the right. Estimate how many more Calories per day an American consumed in 2000 than in 1980. **600 more Calories**
   b. Suppose the increase in Calories per day was entirely from hamburgers. A quarter-pound hamburger has about 450 Calories. About how many more burgers per week did an American eat in 2000 than in 1980? **about 9 burgers**

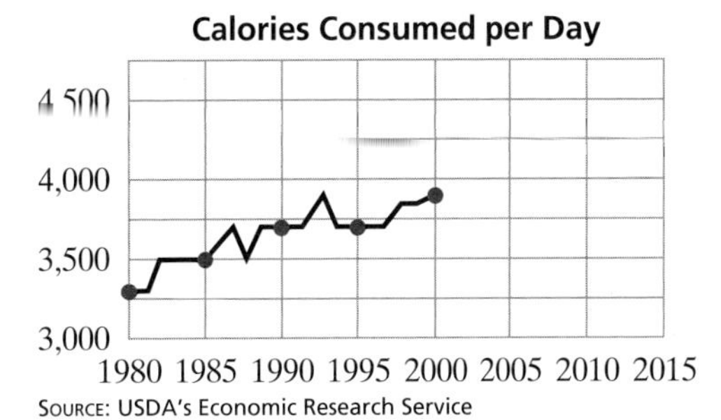

6. Assume the graph at the right keeps climbing at the same rate. Estimate the number of Calories an average American consumes each day this year. **Check students' work.**

7. A 125-pound soccer player burns about 65 Calories in 10 minutes. How long would the player need to play to burn up a snack that contains 260 Calories? **40 min**

Error Prevention!

Some students may not divide the 8,266 lb in half first. Point out that only half of the giant hamburger was meat. Ask: *How many pounds of the giant burger were not meat?* **4,133 lb**

Exercises

Have students work in pairs and answer either Exercise 3 or 4. Then have them work independently to do Exercises 5–7. They should discuss their work in groups and adjust their work based on the discussions.

Differentiated Instruction

English Language Learners ELL

Ask:

- *What are Calories and why are they important?* **The term "Calorie" refers to the amount of chemical energy found in a piece of food. When we eat, our bodies transform the food's chemical energy into energy we can use to move and stay warm.**

4-5

1. Plan

Objective
To write and evaluate expressions with two operations and to solve two-step equations using number sense.

Examples
1, 2 Writing and Evaluating Expressions
3 Using Number Sense
4 Application: Food

Math Understandings: p. 166D

Math Background

When each variable in an algebraic expression is replaced by a number, the result is a *numerical expression.* If the numerical expression involves more than one operation, its value is found by following the order of operations outlined in Lesson 1-9.

More Math Background: p. 166D

Lesson Planning and Resources

See p. 166E for a list of the resources that support this lesson.

PowerPoint

Bell Ringer Practice

Check Skills You'll Need
Use student page, transparency, or PowerPoint. For intervention, direct students to:
Evaluating and Writing Algebraic Expressions
Lesson 4-1
Extra Skills and Word Problems Practice, Ch. 4

Algebra

4-5 Exploring Two-Step Problems

Check Skills You'll Need

1. **Vocabulary Review** A(n) __?__ is a mathematical phrase with at least one variable. **algebraic expression**

Evaluate each expression. Use the value $a = -4$.

2. $a + 5$ **1**
3. $9 - a$ **13**
4. $-6a$ **24**
5. $\frac{a}{2}$ **−2**

Lesson 4-1

What You'll Learn

To write and evaluate expressions with two operations and to solve two-step equations using number sense

Why Learn This?

Suppose you are ordering roses online. Roses cost \$5 each, and shipping costs \$10. Your total cost depends on how many roses you buy. Two-step equations can help you solve everyday problems.

You can write expressions with variables using one operation. Now you will write algebraic expressions with two operations.

EXAMPLES Writing and Evaluating Expressions

1 Define a variable and write an algebraic expression for the phrase "\$10 plus \$5 times the number of roses ordered."

Let n = the number of roses ordered ← **Define the variable.**
$10 + 5 \cdot n$ ← **Write an algebraic expression.**
$10 + 5n$ ← **Rewrite $5 \cdot n$ as $5n$.**

2 Evaluate the expression for 12 roses.

$10 + 5n$
$10 + 5 \cdot 12$ ← **Evaluate the expression for 12 roses.**
$10 + 60$ ← **Multiply.**
70 ← **Simplify.**

If you order 12 roses, you will have to pay \$70.

Quick Check

1. Define a variable and write an algebraic expression for "a man is two years younger than three times his son's age." **Let s = son's age; $3s - 2$**
2. Evaluate the expression to find the man's age if his son is 13. **37**

Differentiated Instruction Solutions for All Learners

Special Needs L1
Students may need models for all the Examples in this lesson. If the models are confusing, let them come up with a different picture to represent the problems.

learning style: visual

Below Level L2
Review the order of operations by having students find the value of some simple expressions such as $3 \cdot 5 + 9$ and $8 \div 4 - 2$. **24; 0**

learning style: visual

Suppose your grandmother sends you 5 games for your birthday. Each game has the same weight. The box she mails them in weighs 8 ounces. The total weight is 48 ounces. What is the weight of one game?

You can represent this situation with the diagram below.

Let g represent the weight of a game.

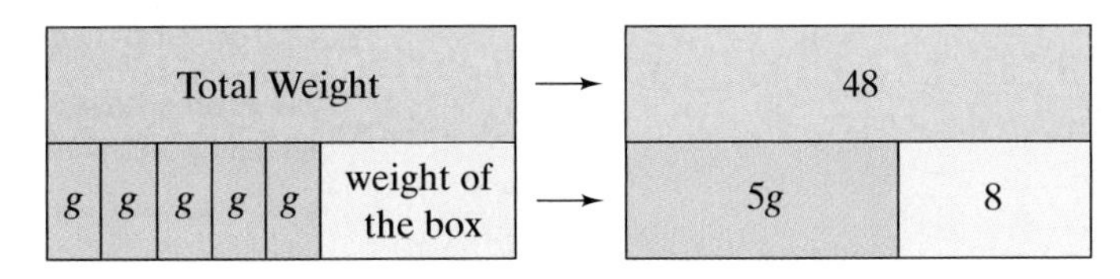

You can solve this problem using the equation $5g + 8 = 48$. Since there is more than one operation in the equation, there will be more than one step in the solution.

EXAMPLE Using Number Sense

Test Prep Tip

When solving problems, look for opportunities to use number sense and mental math.

3 Solve $5g + 8 = 48$ by using number sense.

$5g + 8 = 48$

$\blacksquare + 8 = 48$ ← **Cover 5*g*. *Think*: What number added to 8 is 48? Answer: 40**

$5g = 40$ ← **So $\blacksquare$, or 5*g*, must equal 40.**

$5 \cdot \blacksquare = 40$ ← **Now cover *g*. *Think*: What number times 5 is 40? Answer: 8**

$g = 8$ ← **So $\blacksquare$, or *g*, must equal 8.**

Check

$5g + 8 = 48$ ← **Check your solution in the original equation.**

$5(8) + 8 \stackrel{?}{=} 48$ ← **Substitute 8 for *g*.**

$40 + 8 \stackrel{?}{=} 48$ ← **Simplify.**

$48 = 48$ ✔ ← **The solution checks.**

Quick Check

3. Solve each equation using number sense.

a. $3m + 9 = 21$ **4** **b.** $8d + 5 = 45$ **5** **c.** $4y - 11 = 33$ **11**

2. Teach

Activity Lab

Use before the lesson.

All in One Teaching Resources

Activity Lab 4-5: Patterns in Data

Guided Instruction

Example 2

When evaluating two-step expressions, it is important to follow the order of operations. Ask: *When you find the value of $10 + 5 \cdot 12$, why do you first multiply $5 \cdot 12$?* **According to the order of operations, multiply and divide in order from left to right. Add and subtract in order from left to right.**

Example 3

Display the equation $x + 8 = 48$. Ask: *What number added to 8 is 48?* **40** Write $5g + 8 = 48$ directly beneath the first equation. Have students explain how the second equation is different from the first. Then proceed with the solution in the text.

PowerPoint

Additional Examples

1 2 To rent a bicycle, you pay a \$12 basic fee plus \$2 per hour. Write an expression for the total cost in dollars of a bicycle rental. Then evaluate the expression for an 8-hour bicycle rental. **$12 + 2h$, where *h* represents the number of hours; \$28**

3 Solve $9k - 4 = 14$ using number sense. **$k = 2$**

Advanced Learners L4

Have students create original problems like the one in Example 4 using costs from a restaurant or the school cafeteria.

learning style: visual

English Language Learners ELL

Ask students to work in pairs to write variables and expressions for the *Check Your Understanding* and *Homework Exercises.* Ask them to check each other's work to make sure they are not making errors going from words to symbols.

learning style: visual

Guided Instruction

Example 4
Check that students understand the problem. Ask: *What is the problem about?* **friends who are having lunch together** *How many people are going to share the cost of the chicken wings?* **three** *What does $1.50 represent?* **the cost of your lemonade** *Are the friends going to share the cost of the lemonade?* **no**

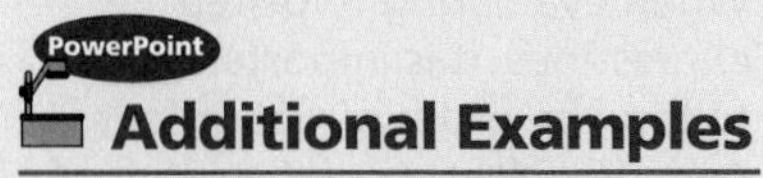

Additional Examples

4 The Healy family wants to buy a DVD player that costs $200. They have $80 saved. How much will they have to save per month for six months in order to have the whole cost saved? **$20 per month**

All in One Teaching Resources
- Daily Notetaking Guide 4-5 L3
- Adapted Notetaking 4-5 L1

Closure

- *How does solving a two-step equation differ from solving a one-step equation?* **Two-step equations combine two mathematical operations. When solving two-step equations you can usually undo addition or subtraction first, then undo multiplication or division.**

Test Prep Tip

You can represent the relationships in the problem with this model.

share of total	
lemonade cost	share of chicken cost

$5.50	
$1.50	$\frac{z}{3}$

EXAMPLE Application: Food

4 Suppose you buy a jumbo lemonade for $1.50 and divide the cost of an order of chicken wings with two friends. Your share of the total bill is $5.50. Write and solve an equation to find the cost of the chicken wings.

Words cost of lemonade plus (cost of wings ÷ 3) is $5.50

Let z = the cost of the chicken wings.

Expression 1.50 + $(z \div 3)$ = $5.50

$1.50 + \frac{z}{3} = 5.50$

$1.50 + \blacksquare = 5.50$ ← **Cover $\frac{z}{3}$. *Think*: What number added to 1.50 is 5.50? Answer: 4**

$\frac{z}{3} = 4$ ← **So $\blacksquare$, or $\frac{z}{3}$, must equal 4.**

$\frac{\blacksquare}{3} = 4$ ← **Now cover z. *Think*: What number divided by 3 is 4? Answer: 12**

$z = 12$ ← **So $\blacksquare$, or z, must equal 12.**

The cost of the chicken wings is $12.

Quick Check

4. **Basketball** During the first half of a game you scored 8 points. In the second half you made only 3-point baskets. You finished the game with 23 points. Write and solve an equation to find how many 3-point baskets you made. **Let b = number of 3-point baskets; $8 + 3b = 23$; 5 baskets.**

Check Your Understanding

1. **Vocabulary** What is the difference between a one-step expression and a two-step expression? **See below left.**

2. Write and solve the equation modeled at the left. **$16 = 4x - 12$; 7**

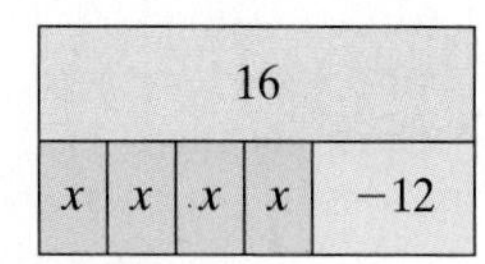

16				
x	x	x	x	-12

1. **A one-step expression uses only one operation, while a two-step expression uses two.**

Match each phrase with the correct algebraic expression.

3. 10 centimeters less than twice x, your hand length **B**
4. 10 people fewer than half x, the town's population **A**
5. 10 more than two times a number x **C**

A. $\frac{1}{2}x - 10$
B. $2x - 10$
C. $2x + 10$

Using number sense, fill in the missing number.

6. $4b + 5 = 17$
   $4b = \blacksquare$ **12**

7. $7c - 20 = 50$
   $7c = \blacksquare$ **70**

Homework Exercises

For more exercises, see Extra Skills and Word Problems.

GO for Help

For Exercises	See Examples
8–12	1
13–20	2
21–27	3–4

A **Define a variable and write an algebraic expression for each phrase.**

8. two points fewer than 3 times the number of points scored before
   Let p = number of points scored before; $3p - 2$.
9. one meter more than 6 times your height in meters
   Let h = your height; $6h + 1$.
10. seven pages fewer than half the number of pages read last week
   Let p = number of pages; $\frac{p}{2} - 7$.
11. eight pounds less than five times the weight of a chicken
   Let w = weight; $5w - 8$.
12. twice the distance in miles flown last year, plus 100 miles
   Let d = distance; $2d + 100$.

Evaluate each expression for the given value of the variable.

13. $4m - 6.5; m = 2$ 1.5
14. $5 + 2f; f = 6.1$ 17.2
15. $12 - 3b; b = 4.3$ −0.9
16. $6x + 2; x = 3$ 20
17. $5y - 5; y = 5$ 20
18. $7 + 8c; c = 7$ 63
19. $2p + 4.5; p = 5.1$ 14.7
20. $12 - 2.2s; s = 4$ 3.2

Solve each equation using number sense. You may find a model helpful.

21. $2t + 9.4 = 39.8$ 15.2
22. $4m + 12 = 52$ 10
23. $10h + 14 = 84$ 7
24. $7w + 16 = 37$ 3
25. $3y + 13.6 = 40.6$ 9
26. $5v + 19 = 24$ 1

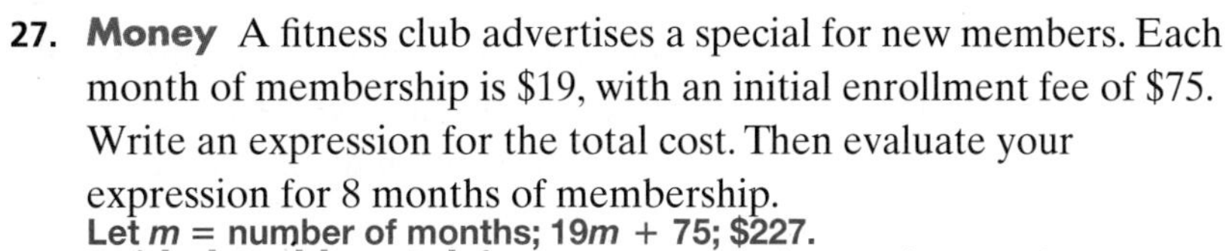

27. **Money** A fitness club advertises a special for new members. Each month of membership is $19, with an initial enrollment fee of $75. Write an expression for the total cost. Then evaluate your expression for 8 months of membership.
   Let m = number of months; $19m + 75$; $227.

B **GPS**

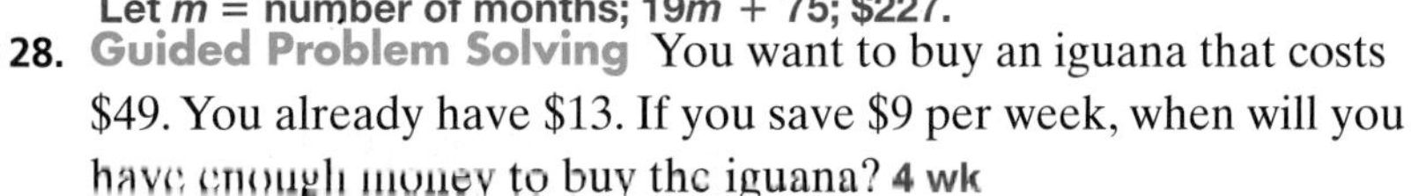

28. **Guided Problem Solving** You want to buy an iguana that costs $49. You already have $13. If you save $9 per week, when will you have enough money to buy the iguana? 4 wk
   - How much money do you already have?
   - How much more money do you need?

29. **Writing in Math** In addition to her hourly rate, an electrician charges a fee to come to your house. This can be modeled by $40h + 35 = 115$. Explain what each part of the equation represents. Then solve the equation to find the number of hours she works.
   See above left.

   29. $40 is the hourly rate, h is the number of hours the electrician works, $35 is the fee for a house call, and $115 is the total bill. The electrician works 2 h.

30. You order 3 posters advertised on the Internet. Each poster costs the same amount. The shipping charge is $5. The total cost of the posters plus the shipping charge is $41. Find the cost of one poster.
   $12
31. **Open-Ended** Describe a situation that can be modeled by the equation $\frac{b}{2} + 5 = 51$. Check students' work.

Online lesson quiz, PHSchool.com, Web Code: ara-0405

3. Practice

Assignment Guide

Check Your Understanding

Go over Exercises 1–7 in class before assigning the Homework Exercises.

Homework Exercises

A	Practice by Example	8–27
B	Apply Your Skills	28–41
C	Challenge	42
	Test Prep and Mixed Review	43–48

Homework Quick Check

To check students' understanding of key skills and concepts, go over Exercises 8, 14, 29, 39, and 41.

Differentiated Instruction Resources

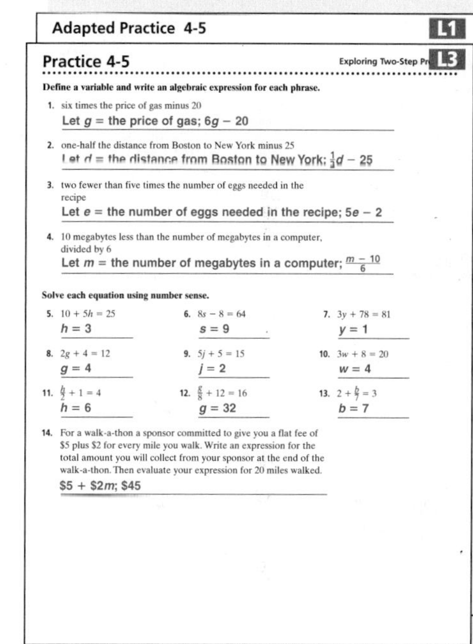

Adapted Practice 4-5 L1

Practice 4-5 Exploring Two-Step Pr L3

Define a variable and write an algebraic expression for each phrase.

1. six times the price of gas minus 20
   Let g = the price of gas; $6g - 20$
2. one-half the distance from Boston to New York minus 25
   Let d = the distance from Boston to New York; $\frac{1}{2}d - 25$
3. two fewer than five times the number of eggs needed in the recipe
   Let e = the number of eggs needed in the recipe; $5e - 2$
4. 10 megabytes less than the number of megabytes in a computer, divided by 6
   Let m = the number of megabytes in a computer; $\frac{m - 10}{6}$

Solve each equation using number sense.

5. $10 + 5h = 25$ $h = 3$
6. $8s - 8 = 64$ $s = 9$
7. $3y + 78 = 81$ $y = 1$
8. $2g + 4 = 12$ $g = 4$
9. $5j + 5 = 15$ $j = 2$
10. $3w + 8 = 20$ $w = 4$
11. $\frac{h}{2} + 1 = 4$ $h = 6$
12. $\frac{g}{8} + 12 = 16$ $g = 32$
13. $2 + \frac{b}{7} = 3$ $b = 7$
14. For a walk-a-thon a sponsor committed to give you a flat fee of $5 plus $2 for every mile you walk. Write an expression for the total amount you will collect from your sponsor at the end of the walk-a-thon. Then evaluate your expression for 20 miles walked.
   $5 + $2m; $45

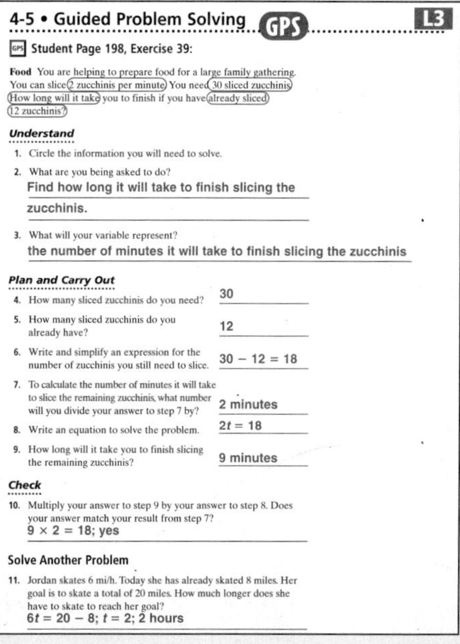

4-5 • Guided Problem Solving GPS L3

GPS Student Page 198, Exercise 39:

Food You are helping to prepare food for a large family gathering. You can slice 2 zucchinis per minute. You need 30 sliced zucchinis. How long will it take you to finish if you have already sliced 12 zucchinis?

Understand

1. Circle the information you will need to solve.
2. What are you being asked to do?
   Find how long it will take to finish slicing the zucchinis.
3. What will your variable represent?
   the number of minutes it will take to finish slicing the zucchinis

Plan and Carry Out

4. How many sliced zucchinis do you need? 30
5. How many sliced zucchinis do you already have? 12
6. Write and simplify an expression for the number of zucchinis you still need to slice. $30 - 12 = 18$
7. To calculate the number of minutes it will take to slice the remaining zucchinis, what number will you divide your answer to step 7 by? 2 minutes
8. Write an equation to solve the problem. $2t = 18$
9. How long will it take you to finish slicing the remaining zucchinis? 9 minutes

Check

10. Multiply your answer to step 9 by your answer to step 8. Does your answer match your result from step 7?
   $9 \times 2 = 18$; yes

Solve Another Problem

11. Jordan skates 6 mi/h. Today she has already skated 8 miles. Her goal is to skate a total of 20 miles. How much longer does she have to skate to reach her goal?
   $6t = 20 - 8$; $t = 2$; 2 hours

4. Assess & Reteach

Lesson Quiz

Solve each equation using number sense.

1. $5d + 3 = 53$ $d = 10$
2. $\frac{g}{6} - 4 = 2$ $g = 36$
3. $7h - 10 = 25$ $h = 5$
4. $\frac{k}{3} + 20 = 31$ $k = 33$

Reteaching 4-5 Exploring Two-Step **L2**

You can change a word expression into an algebraic expression by converting the words to variables, numbers, and operation symbols.

To write a two-step algebraic expression for *seven more than three times a number,* follow these steps.

① Define the variable. — Let n represent the number.
② Ask yourself are there any key words? — "More than" means add and "times" means multiply.
③ Write an algebraic expression. — $7 + 3 \cdot n$
④ Simplify. — $7 + 3n$

Define a variable and write an algebraic expression for each phrase.

1. 3 inches more than 4 times your height $4h + 3$
2. 4 less than 6 times the weight of a turkey $6t - 4$
3. 8 more than one-half the number of miles run last week $\frac{1}{2}m + 8$

Solve.

4. Three friends pay $4 per hour to rent a paddleboat plus $5 for snacks. Write an expression for the total cost of rental and snacks. Then evaluate the expression for 2 hours. $4h + 5$; $13
5. A lawn care service charges $10 plus $15 per hour to mow and fertilize lawns. Write an expression for the total cost of having your lawn mowed and fertilized. Then evaluate the expression for 4 hours. $10 + 15h$; $70

Solve each equation using number sense.

6. $4x - 10 = 30$ $x = 10$
7. $2n - 7 = 13$ $n = 10$
8. $\frac{s}{3} + 2 = 4$ $s = 6$

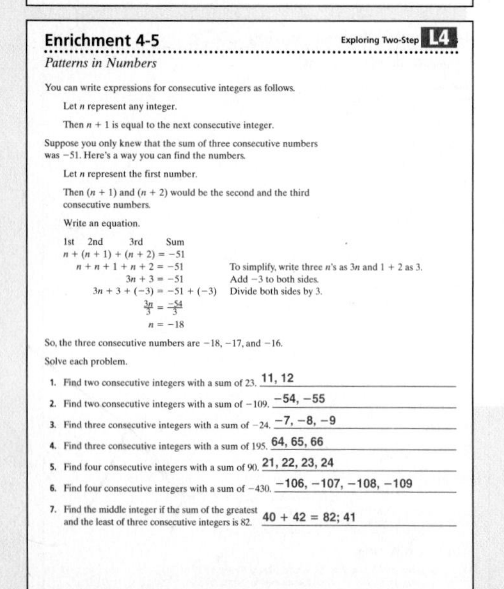

Enrichment 4-5 Exploring Two-Step **L4**

Patterns in Numbers

You can write expressions for consecutive integers as follows.

Let n represent any integer.

Then $n + 1$ is equal to the next consecutive integer.

Suppose you only knew that the sum of three consecutive numbers was -51. Here's a way you can find the numbers.

Let n represent the first number.

Then $(n + 1)$ and $(n + 2)$ would be the second and the third consecutive numbers.

Write an equation.

1st 2nd 3rd Sum

$n + (n + 1) + (n + 2) = -51$

$n + n + 1 + n + 2 = -51$ — To simplify, write three n's as $3n$ and $1 + 2$ as 3.

$3n + 3 = -51$ — Add -3 to both sides.

$3n + 3 + (-3) = -51 + (-3)$ — Divide both sides by 3.

$\frac{3n}{3} = \frac{-54}{3}$

$n = -18$

So, the three consecutive numbers are -18, -17, and -16.

Solve each problem.

1. Find two consecutive integers with a sum of 23. 11, 12
2. Find two consecutive integers with a sum of −109. −54, −55
3. Find three consecutive integers with a sum of −24. −7, −8, −9
4. Find three consecutive integers with a sum of 195. 64, 65, 66
5. Find four consecutive integers with a sum of 90. 21, 22, 23, 24
6. Find four consecutive integers with a sum of −430. −106, −107, −108, −109
7. Find the middle integer if the sum of the greatest and the least of three consecutive integers is 82. $40 + 42 = 82$; 41

GO Online
Homework Video Tutor
Visit: PHSchool.com
Web Code: are-0405

Solve each equation using number sense.

32. $5h + 3 = 18.5$ 3.1
33. $3m - 7.6 = 26.9$ 11.5
34. $8y + 17 = 65$ 6
35. $\frac{x}{3} - 3 = 12$ 45
36. $2p - 5 = 15$ 10
37. $\frac{t}{4} + 1 = 6$ 20
38. **Telephone** A cellular telephone company charges $40 per month plus a $35 activation fee. Write an expression for the total cost. Then evaluate your expression for 10 months of service. **See margin.**

39. **Food** You are helping to prepare food for a large family gathering. GPS You can slice 2 zucchinis per minute. You need 30 sliced zucchinis. How long will it take you to finish, if you have already sliced 12 zucchinis? **9 min**
40. Your family rented a car for a trip. The car rental cost $35 per day plus $.30/mile. After a one-day rental, the bill was $74. How many miles did your family drive? **130 mi**
41. **Reasoning** When you solve an equation, you must do the same operation to both sides. Explain why this is true. **See margin.**
42. Ⓒ **Challenge** You spend 5 minutes jogging as a warmup. Then you run 4 miles and cool down for 5 minutes. The total time you exercise is 54 minutes. What is your average time for running a mile? **11 min/mi**

Test Prep and Mixed Review — Practice

Multiple Choice

43. Daniel bought one dozen tennis balls priced at 3 balls for $1.99 and a tennis racquet for $49.99. What is the total amount he spent, not including tax, on tennis balls and a tennis racquet? **C**
    - Ⓐ $51.98
    - Ⓑ $55.96
    - Ⓒ $57.95
    - Ⓓ $73.87
44. Medium beverages cost $1.39 and small beverages cost $.89. Which equation can be used to find d, the total cost in dollars of 6 medium drinks? **J**
    - Ⓕ $6 = 0.89d$
    - Ⓖ $d = 0.89(6)$
    - Ⓗ $6 = 1.39d$
    - Ⓙ $d = 1.39(6)$
45. Which expression is represented by the model below? **D**

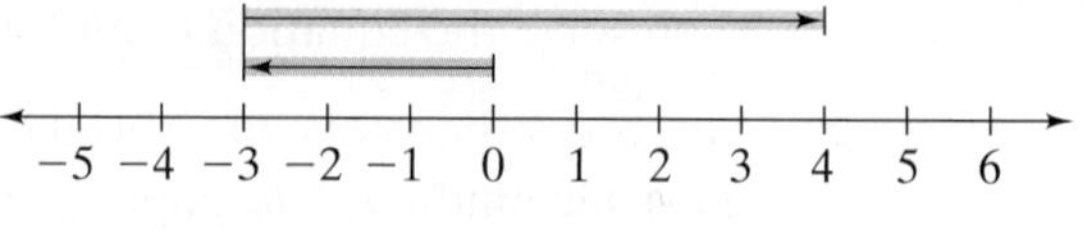

    - Ⓐ $-3 + 2$
    - Ⓑ $-3 + 4$
    - Ⓒ $-3 + 0$
    - Ⓓ $-3 + 7$

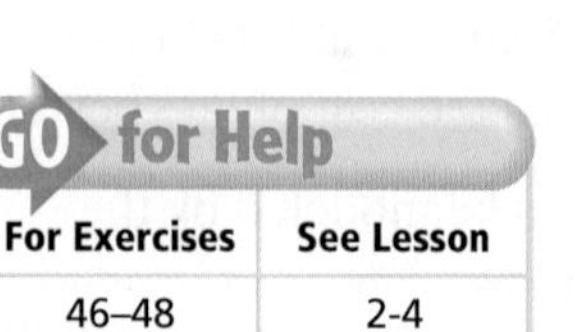

GO for Help

For Exercises	See Lesson
46–48	2-4

Compare each pair of fractions. Use <, =, or >.

46. $\frac{7}{9} < \frac{4}{5}$
47. $\frac{9}{27} = \frac{1}{3}$
48. $\frac{5}{12} > \frac{2}{6}$

Test Prep

Resources

For additional practice with a variety of test item formats:

- Test-Taking Strategies, p. 219
- Test Prep, p. 223
- Test-Taking Strategies with Transparencies

Alternative Assessment

Have students create word phrases similar to those in Exercises 8–12. Have partners work together to write an algebraic expression for each phrase. Partners should then incorporate each expression into an equation. Finally, have partners solve their equations.

Hands On

Modeling Two-Step Equations

You can use algebra tiles to solve two-step equations.

EXAMPLE Solving Two-Step Equations

Use algebra tiles to solve $2x + 1 = 7$.

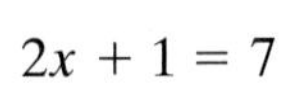

$2x + 1 = 7$

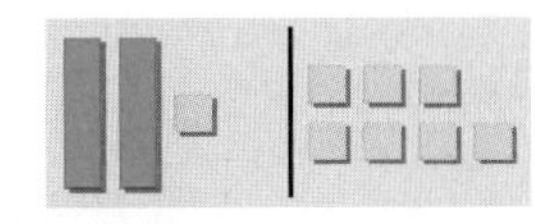

← **Model the equation. Use yellow tiles for positive integers.**

$2x + 1 - 1 = 7 - 1$
$2x = 6$

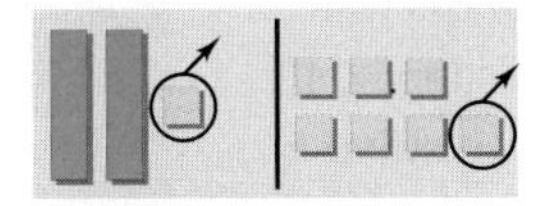

← **Remove 1 yellow tile from each side.**

$\frac{2x}{2} = \frac{6}{2}$

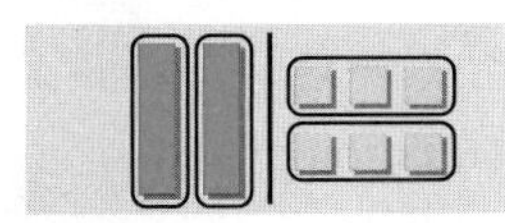

← **Divide each side into 2 equal groups.**

$x = 3$

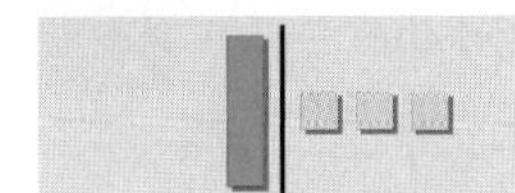

← **Simplify.**

Exercises

Write and solve the equation represented by each model.

1. 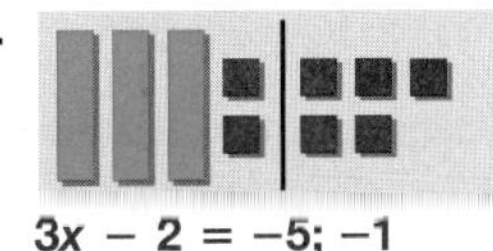

$3x - 2 = -5; -1$

2. 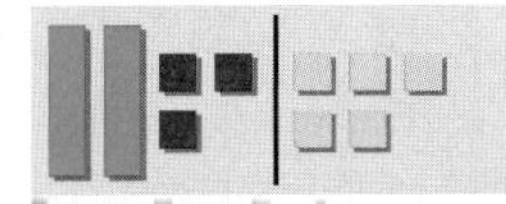

$2x - 3 = 5; 4$

3. 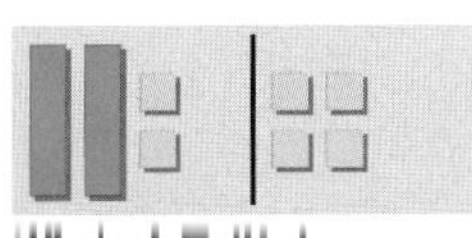

$2x + 2 = 4; 1$

Use algebra tiles to solve each equation.

4. $3x - 4 = 2$ **2**
5. $2x - 1 = 9$ **5**
6. $2x + 4 = 10$ **3**
7. $3x + 4 = 7$ **1**
8. $2x + 6 = -8$ **−7**
9. $2x - 6 = -18$ **−6**

10. At a county fair, an admission ticket costs \$5, and each ride costs \$2. You have \$13. Write an algebraic equation for the number of rides you can go on. Use algebra tiles to solve the equation. **Let x = number of rides; $5 + 2x = 13$, 4 rides**

11. **Reasoning** Suppose you have 20 green, 20 red, and 20 yellow algebra tiles. Explain how you could use the tiles to model the equation $500x - 200 = 1{,}300$. **Model an equivalent equation, $5x - 2 = 13$.**

Activity Lab

Modeling Two-Step Equations

Students use algebra tiles to model two-step equations and find their solutions.

Guided Instruction

Model the solution to the example. Have students follow along at their desks.

Ask:

Why did you subtract 1 yellow tile from each side? **so that 2x is alone on the left side**

- *What tells you that $-1 + 1 = 0$?* **the additive inverse property**
- *What two operations were used in the example?* **subtraction and division**

Alternative Method

Have students act out the two-step equation using chairs for the variables, boys for red tiles, and girls for yellow tiles.

Resources

- Activity Lab 4-6: Critical Thinking
- algebra tiles

38. Let m = number of months; $40m + 35$; \$435

41. Answers may vary. Sample: You need to keep the equation "balanced."

4-6

1. Plan

Objective
To solve two-step equations using inverse operations

Examples
1 Undoing Subtraction First
2 Undoing Addition First
3 Solving Two-Step Equations

Math Understandings: p. 166D

Math Background

The most direct approach to solving two-step equations is generally the procedure described on this page. This might be described as an "inverse order of operations." However, there are sometimes alternative procedures that are equally acceptable. For instance, in Example 2, $\frac{x}{3} + 11 = 16$, could also be solved by first multiplying both sides by 3, obtaining $x + 33 = 48$. Subtracting 33 from each side of this equation then results in the solution $x = 15$.

More Math Background: p. 166D

Lesson Planning and Resources

See p. 166E for a list of the resources that support this lesson.

PowerPoint

Bell Ringer Practice

Check Skills You'll Need
Use student page, transparency, or PowerPoint. For intervention, direct students to:
Solving Equations by Multiplying or Dividing
Lesson 4-4
Extra Skills and Word Problems Practice, Ch. 4

Algebra

4-6 Solving Two-Step Equations

Check Skills You'll Need

1. **Vocabulary Review** What property states that if $a = b$, then $a \cdot c = b \cdot c$? **See below.**

Solve each equation.

2. $4b = 24$ **6**
3. $-4d = 20$ **−5**
4. $\frac{k}{4} = -16$ **−64**
5. $\frac{h}{6} = -3$ **−18**

for Help Lesson 4-4

What You'll Learn

To solve two-step equations using inverse operations

Why Learn This?

Detectives often retrace steps to find missing information. You can use the strategy *Work Backward* to solve a two-step equation and find an unknown quantity.

You can solve a two-step equation by using inverse operations and the properties of equality to get the variable on one side of the equation.

For many equations, you can undo addition or subtraction first. Then you can multiply or divide to get the variable alone.

Check Skills You'll Need
1. Multiplication Property of Equality

EXAMPLE Undoing Subtraction First

1 Solve $5n - 18 = -33$.

$$5n - 18 = -33$$
$$5n - 18 + 18 = -33 + 18 \quad \leftarrow \text{To undo subtraction, add 18 to each side.}$$
$$5n = -15 \quad \leftarrow \text{Simplify.}$$
$$\frac{5n}{5} = \frac{-15}{5} \quad \leftarrow \text{To undo multiplication, divide each side by 5.}$$
$$n = -3 \quad \leftarrow \text{Simplify.}$$

Check $5n - 18 = -33$ ← Check your solution with the original equation.
$$5(-3) - 18 \stackrel{?}{=} -33 \quad \leftarrow \text{Substitute } -3 \text{ for } n.$$
$$-15 - 18 \stackrel{?}{=} -33 \quad \leftarrow \text{Simplify.}$$
$$-33 = -33 \checkmark \quad \leftarrow \text{The solution checks.}$$

Quick Check

1. Solve the equation $-8y - 28 = -36$. Check your answer. **1**

Differentiated Instruction Solutions for All Learners

Special Needs L1
Have students look at each Example and identify the operation they must undo first. The rules here are different from the order of operations where you multiply or divide before adding and subtracting. For many equations you undo addition or subtraction and then undo multiplication or division. **learning style: visual**

Below Level L2
Have students find the missing number in several statements like the following.
$-26 + \square = 0$ **26** $\qquad \frac{-7}{\square} = 1$ **−7**
learning style: visual

EXAMPLE Undoing Addition First

2 Solve $\frac{x}{3} + 11 = 16$.

$$\frac{x}{3} + 11 = 16$$

$$\frac{x}{3} + 11 - 11 = 16 - 11 \quad \leftarrow \text{To undo addition, subtract 11 from each side.}$$

$$\frac{x}{3} = 5 \quad \leftarrow \text{Simplify.}$$

$$3\left(\frac{x}{3}\right) = 3(5) \quad \leftarrow \text{To undo division, multiply each side by 3.}$$

$$x = 15 \quad \leftarrow \text{Simplify.}$$

Quick Check

2. Solve the equation $\frac{x}{5} + 35 = 75$. Check your answer. **200**

You can use two-step equations to solve real-world problems.

EXAMPLE Solving Two-Step Equations

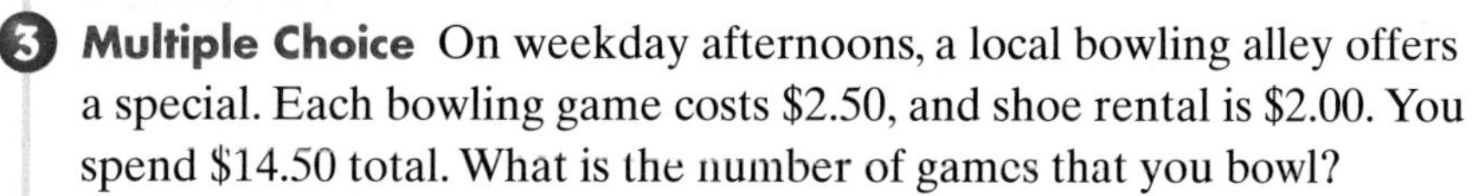

3 **Multiple Choice** On weekday afternoons, a local bowling alley offers a special. Each bowling game costs \$2.50, and shoe rental is \$2.00. You spend \$14.50 total. What is the number of games that you bowl?

 A 3  B 5  C 7  D 9

Words 2.50 times number of games plus 2.00 is 14.50

Let n = the number of games you bowl.

Equation $2.50 \cdot n + 2.00 = 14.50$

$$2.5n + 2 = 14.50$$

$$2.5n + 2 - 2 = 14.50 - 2 \quad \leftarrow \text{Subtract 2 from each side.}$$

$$2.5n = 12.50 \quad \leftarrow \text{Simplify.}$$

$$\frac{2.5n}{2.5} = \frac{12.50}{2.5} \quad \leftarrow \text{Divide each side by 2.5.}$$

$$n = 5 \quad \leftarrow \text{Simplify.}$$

You bowled 5 games. The answer is B.

Test Prep Tip

You can eliminate answer choices using number sense. Choice D can be eliminated because it is too high. If you played 9 games, your total cost would be greater than \$18.

Quick Check

3. Solomon decided to make posters for the student council election. He bought markers that cost \$.79 each and a poster board that cost \$1.25. The total cost was \$7.57. Write and solve an equation to find the number of markers that Solomon bought.
Let m = number of markers;
$0.79m + 1.25 = 7.57$; 8 markers

2. Teach

Activity Lab

Use before the lesson.
Student Edition Activity Lab Hands On 4-6a, Modeling Two-Step Equations, p. 199

All in One Teaching Resources
Activity Lab 4-6: Critical Thinking

Guided Instruction

Example 1
Remind students that they have learned to solve equations involving one operation by using the inverse operation. When an equation involves two operations, they should also use an inverse order of operations: Add or subtract first, then multiply or divide.

Example 2
Students can use an inverse order on a calculator to find the solution: start with 16, subtract 11, and multiply by 3.

Additional Examples

1 Solve $6r + 19 = 43$. $r = 4$

2 Solve $\frac{a}{5} - 4 = 10$. $a = 70$

3 At the library sale, there was a \$2.50 admission charge. Old paperbacks were on sale for \$.25 each. If you only bought paperbacks and you spent \$4.50 in all, how many paperbacks did you buy? **8 books**

Advanced Learners L4	**English Language Learners** ELL
Write a two-step equation involving multiplication and subtraction that has −5 as its solution. **Sample:** $2y - 1 = -11$	In Example 3, make sure that students understand that they do not have to rent shoes every time they play a game. Explain that it is a one-time cost per visit to the bowling alley. Show them pictures of a bowling alley in case they have never seen one.
learning style: visual	**learning style: verbal**

All in One Teaching Resources
- Daily Notetaking Guide 4-6 L3
- Adapted Notetaking 4-6 L1

Closure

- *How do you solve two-step equations?* To solve a two-step equation you must first undo addition or subtraction, then undo multiplication or division.

More Than One Way

A family expects 88 people to attend its family reunion. There will be 16 children. Picnic tables seat 8 adults per table. The children will eat on blankets. How many picnic tables does the family need?

Sarah's Method

I can use number sense. First, I know that tables are needed only for adults. There are $88 - 16$, or 72, adults. Each table holds 8 adults. Since $72 \div 8 = 9$, the family needs 9 tables.

Ryan's Method

I can write and solve an equation. Let t represent the number of tables. Then $8t$ is the number of adults.

adults + children = 88 people

$8t + 16 = 88$ ← **Write the equation.**

Each term is divisible by 8. So I will divide first.

$\frac{8t}{8} + \frac{16}{8} = \frac{88}{8}$ ← **Divide each term by 8.**

$t + 2 = 11$ ← **Simplify.**

I can use mental math to solve the equation. I know that 9 plus 2 is equal to 11. So, the family needs 9 tables.

Choose a Method

You had $25 in your savings account six weeks ago. You deposited the same amount of money each week for five weeks. Your balance is now $145. How much money did you deposit each week? Describe your method and explain why you chose it.

$24; check students' work.

Check Your Understanding

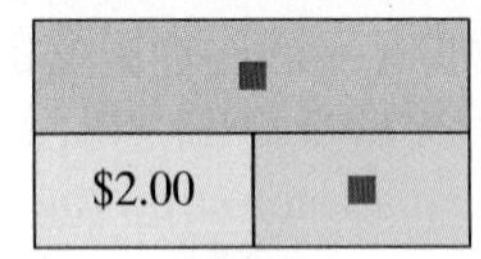

1. Fill in the missing numbers to make the diagram at the left represent the following situation: A taxi charges a flat fee of $2.00 plus $.50 for each mile. Your fare is $5.00. How many miles did you ride?
   Let m= number of miles; $2.00 + 0.50m = 5.00$; 6 mi

Solve each equation. Check your answer.

2. $5p - 2 = 18$ **4**
3. $7n - (-16) = 100$ **12**
4. $\frac{y}{4.25} + 15 = -17$ **−136**

Homework Exercises

For more exercises, see Extra Skills and Word Problems.

GO for Help

For Exercises	See Examples
5–16	1
17–29	2–3

A **Solve each equation. Check your answer. You may find a model helpful.**

5. $8r - 8 = -32$ **−3**
6. $3w - 6 = -1.5$ **1.5**
7. $4g - 4 = 28$ **8**
8. $7t - 6 = -104$ **−14**
9. $12x - 14 = -2$ **1**
10. $10m - \frac{2}{5} = 9\frac{3}{5}$ **1**
11. $0.5y - 1.1 = 4.9$ **12**
12. $-2d - 1.7 = -3.9$ **1.1**
13. $-8a - 1 = -23$ **$2\frac{3}{4}$**
14. $2h - \frac{1}{10} = \frac{5}{8}$ **$\frac{29}{80}$**
15. $6t - \frac{1}{6} = 9$ **$1\frac{19}{36}$**
16. $5q - 3.75 = 26.25$ **6**
17. $\frac{w}{5} + 3 = 6$ **15**
18. $\frac{n}{4} + 2 = 4$ **8**
19. $\frac{x}{8} + 4 = 13$ **72**
20. $\frac{a}{7} + 10 = 17$ **49**
21. $\frac{m}{-11} + 1 = -10$ **121**
22. $\frac{p}{9} + 14 = 16$ **18**
23. $\frac{c}{-7} + 3.2 = -2.2$ **37.8**
24. $\frac{v}{-3} + \frac{3}{4} = -\frac{1}{8}$ **$2\frac{5}{8}$**
25. $\frac{b}{-8} + \frac{5}{7} = \frac{11}{14}$ **$\frac{-4}{7}$**
26. $\frac{c}{2} + 7.3 = 29.3$ **44**
27. $\frac{m}{-10} + 12 = 67$ **−550**
28. $\frac{y}{-6.5} + 2 = -4$ **39**

29. Kristine bought a vase that cost \$5.99 and roses that cost \$1.25 each. The total cost was \$20.99. Write and solve an equation to find how many roses Kristine bought.
Let r = number of roses; $5.99 + 1.25r = 20.99$; 12 roses.

B **GPS** 30. **Guided Problem Solving** Renting boats on a lake costs \$22 per hour plus a flat fee of \$10 for insurance. You have \$98. Write and solve an equation to find the number of hours you can rent a boat.
Let h = number of hours; $10 + 22h = 98$; 4 h.

- insurance plus 22 times number of hours = total paid

Write and solve an equation for each situation.

31. A skating rink rents skates at \$3.95 for the first hour plus \$1.25 for each additional hour. When you returned your skates, you paid \$7.70. How many additional hours did you keep the skates?
Let h = number of additional hours; $3.95 + 1.25h = 7.70$; 3 h.

32. **Jobs** You earn \$20 per hour landscaping a yard. You pay \$1.50 in bus fare each way. How many hours must you work to earn \$117? (GPS)
Let h = number of hours; $20h - 3 = 117$; 6 h.

33. **Geometry** The sum of the measures of the angles in a triangle is 180 degrees. One angle measures 45 degrees. The measures of the other two angles are equal. What is the measure of each of the other two angles? **Let x = measure of each angle; $45 + 2x = 180$; 67.5°.**

34. **Olympics** The first modern Olympic games were held in Greece in 1896. The 2004 Olympic games in Athens had 202 participating countries. The number of countries in the 2004 Olympic games was 6 more than 14 times the number of countries in the 1896 games. How many countries participated in the 1896 Olympic games?
Let n = number of countries; $14n + 6 = 202$; 14 countries.

Online lesson quiz, PHSchool.com, Web Code: ara-0406

3. Practice

Assignment Guide

Check Your Understanding
Go over Exercises 1–4 in class before assigning the Homework Exercises.

Homework Exercises

A Practice by Example	5–29
B Apply Your Skills	30–41
C Challenge	42
Test Prep and Mixed Review	43–48

Homework Quick Check
To check students' understanding of key skills and concepts, go over Exercises 12, 22, 32, 34, and 38.

Differentiated Instruction Resources

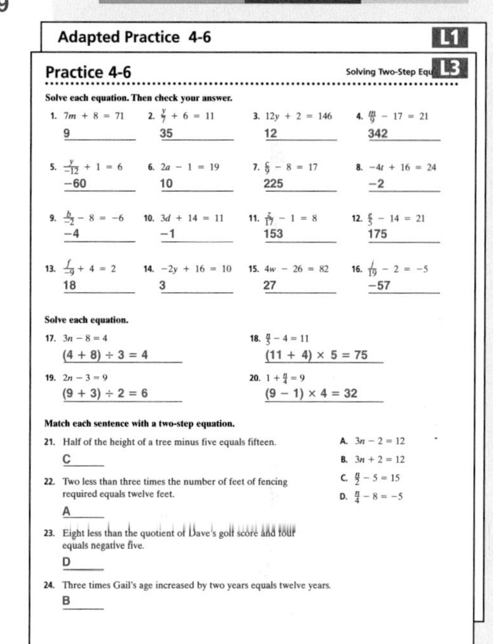

Adapted Practice 4-6 L1

Practice 4-6 Solving Two-Step Equ[ations] L3

Solve each equation. Then check your answer.

1. $7m + 8 = 71$ 9
2. $\frac{y}{7} + 6 = 11$ 35
3. $12y + 2 = 146$ 12
4. $\frac{m}{9} - 17 = 21$ 342
5. $\frac{y}{-12} + 1 = 6$ −60
6. $2a - 1 = 19$ 10
7. $\frac{c}{9} - 8 = 17$ 225
8. $-4t + 16 = 24$ −2
9. $\frac{b}{-2} - 8 = -6$ −4
10. $3d + 14 = 11$ −1
11. $\frac{z}{17} - 1 = 8$ 153
12. $\frac{c}{5} - 14 = 21$ 175
13. $\frac{f}{-9} + 4 = 2$ 18
14. $-2y + 16 = 10$ 3
15. $4w - 26 = 82$ 27
16. $\frac{j}{19} - 2 = -5$ −57

Solve each equation.

17. $3n - 8 = 4$ $(4 + 8) \div 3 = 4$
18. $\frac{n}{5} - 4 = 11$ $(11 + 4) \times 5 = 75$
19. $2n - 3 = 9$ $(9 + 3) \div 2 = 6$
20. $1 + \frac{n}{4} = 9$ $(9 - 1) \times 4 = 32$

Match each sentence with a two-step equation.

21. Half of the height of a tree minus five equals fifteen. C
22. Two less than three times the number of feet of fencing required equals twelve feet. A
23. Eight less than the quotient of Dave's golf score and four equals negative five. D
24. Three times Gail's age increased by two years equals twelve years. B

A. $3n - 2 = 12$
B. $3n + 2 = 12$
C. $\frac{n}{2} - 5 = 15$
D. $\frac{n}{4} - 8 = -5$

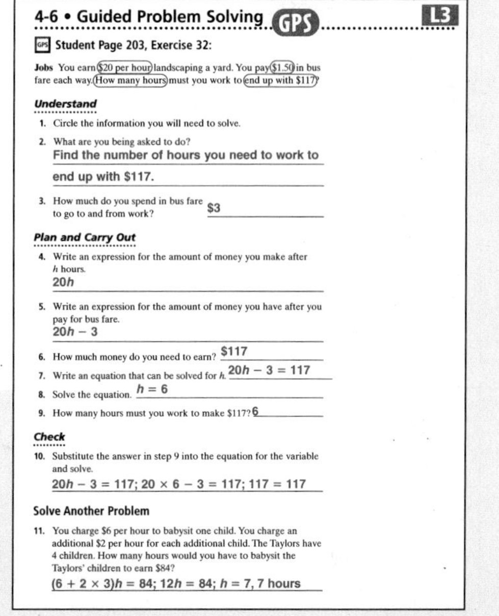

4-6 • Guided Problem Solving GPS L3

GPS Student Page 203, Exercise 32:

Jobs You earn \$20 per hour landscaping a yard. You pay \$1.50 in bus fare each way. How many hours must you work to end up with \$117?

Understand

1. Circle the information you will need to solve.
2. What are you being asked to do? Find the number of hours you need to work to end up with \$117.
3. How much do you spend in bus fare to go to and from work? \$3

Plan and Carry Out

4. Write an expression for the amount of money you make after h hours. $20h$
5. Write an expression for the amount of money you have after you pay for bus fare. $20h - 3$
6. How much money do you need to earn? \$117
7. Write an equation that can be solved for h. $20h - 3 = 117$
8. Solve the equation. $h = 6$
9. How many hours must you work to make \$117? 6

Check

10. Substitute the answer in step 9 into the equation for the variable and solve. $20h - 3 = 117$; $20 \times 6 - 3 = 117$; $117 = 117$

Solve Another Problem

11. You charge \$6 per hour to babysit one child. You charge an additional \$2 per hour for each additional child. The Taylors have 4 children. How many hours would you have to babysit the Taylors' children to earn \$84? $(6 + 2 \times 3)h = 84$; $12h = 84$; $h = 7$, 7 hours

4. Assess & Reteach

Lesson Quiz

Solve each equation.

1. $9r + 7 = 79$ **$r = 8$**
2. $\frac{s}{4} - 6 = 14$ **$s = 80$**
3. $3t - 4 = 44$ **$t = 16$**
4. $\frac{v}{5} + 6 = -1$ **$v = -35$**

38. $30 =$ the DJ's hourly wage; $x =$ hours the DJ works; $65 =$ cost of decorations; $170 =$ amount of money budgeted

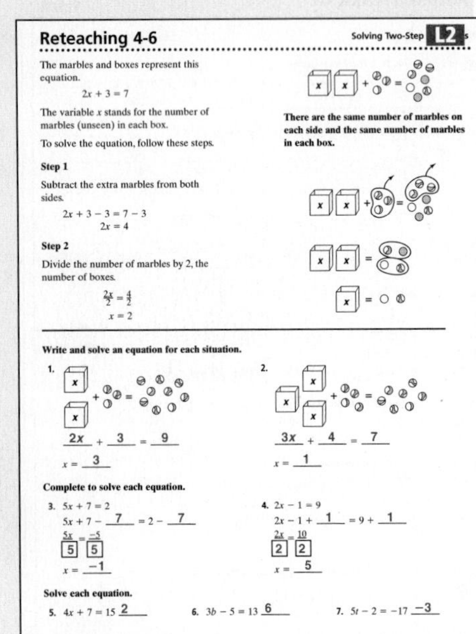

Reteaching 4-6

Solving Two-Step L2

The marbles and boxes represent this equation.

$2x + 3 = 7$

The variable x stands for the number of marbles (unseen) in each box.

To solve the equation, follow these steps.

There are the same number of marbles on each side and the same number of marbles in each box.

Step 1

Subtract the extra marbles from both sides.

$2x + 3 - 3 = 7 - 3$

$2x = 4$

Step 2

Divide the number of marbles by 2, the number of boxes.

$\frac{2x}{2} = \frac{4}{2}$

$x = 2$

Write and solve an equation for each situation.

1. $2x + 3 = 9$; $x = 3$
2. $3x + 4 = 7$; $x = 1$

Complete to solve each equation.

3. $5x + 7 = 2$
   $5x + 7 - 7 = 2 - 7$
   $\frac{5x}{5} = \frac{-5}{5}$
   $x = -1$
4. $2x - 1 = 9$
   $2x - 1 + 1 = 9 + 1$
   $\frac{2x}{2} = \frac{10}{2}$
   $x = 5$

Solve each equation.

5. $4x + 7 = 15$ 2
6. $3b - 5 = 13$ 6
7. $5t - 2 = -17$ −3

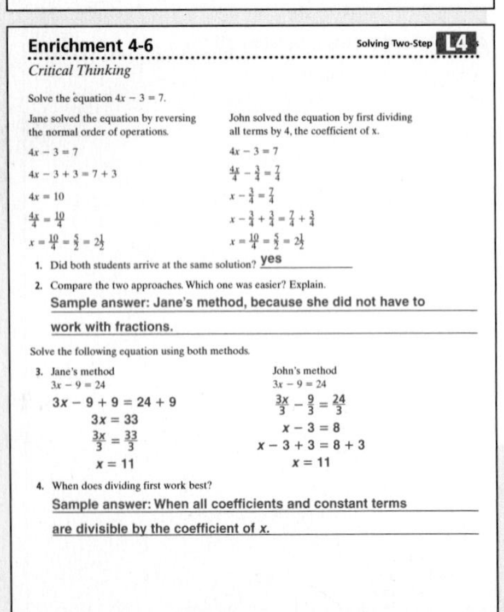

Enrichment 4-6

Solving Two-Step L4

Critical Thinking

Solve the equation $4x - 3 = 7$.

Jane solved the equation by reversing the normal order of operations.

$4x - 3 = 7$

$4x - 3 + 3 = 7 + 3$

$4x = 10$

$\frac{4x}{4} = \frac{10}{4}$

$x = \frac{10}{4} = \frac{5}{2} = 2\frac{1}{2}$

John solved the equation by first dividing all terms by 4, the coefficient of x.

$4x - 3 = 7$

$\frac{4x}{4} - \frac{3}{4} = \frac{7}{4}$

$x - \frac{3}{4} = \frac{7}{4}$

$x - \frac{3}{4} + \frac{3}{4} = \frac{7}{4} + \frac{3}{4}$

$x = \frac{10}{4} = \frac{5}{2} = 2\frac{1}{2}$

1. Did both students arrive at the same solution? yes
2. Compare the two approaches. Which one was easier? Explain.
   Sample answer: Jane's method, because she did not have to work with fractions.

Solve the following equation using both methods.

3. Jane's method
   $3x - 9 = 24$
   $3x - 9 + 9 = 24 + 9$
   $3x = 33$
   $\frac{3x}{3} = \frac{33}{3}$
   $x = 11$

   John's method
   $3x - 9 = 24$
   $\frac{3x}{3} - \frac{9}{3} = \frac{24}{3}$
   $x - 3 = 8$
   $x - 3 + 3 = 8 + 3$
   $x = 11$
4. When does dividing first work best?
   Sample answer: When all coefficients and constant terms are divisible by the coefficient of x.

GO Online
Homework Video Tutor
Visit: PHSchool.com
Web Code: are-0406

Solve each equation.

35. $\frac{x + 4}{5} = 3$ **11**
36. $\frac{t - 7}{-2} = 11$ **−15**
37. $\frac{y + \frac{1}{3}}{4} = 2\frac{1}{4}$ **$8\frac{2}{3}$**

38. **Writing in Math** Your class budgets a certain amount of money from the class treasury for a dance. Expenses will include a fixed amount for decorations, plus an hourly wage for the disc jockey. This can be represented by the equation $30x + 65 = 170$. Explain how each number in the equation relates to the problem. **See margin.**

39. **College** In college, you earn credits for courses taken. For one semester, tuition at a local college is $2,400. You have financial aid that will cover $5,160 for the semester. Refer to the information at the left. How many credits can you take? **15 credits**
40. The student council sponsored a bake sale to raise money. Kim bought a slice of cake for $1.50 and also bought six cupcakes. She spent $4.20 in all. How much did each cupcake cost? **45 cents**
41. **Open-Ended** Write two different two-step equations that both have a solution of 3. **Check students' work.**
42. **C** **Challenge** You buy 1.25 lb of apples and 2.45 lb of bananas. The total cost, after using a 75¢-off coupon, is $2.58. Apples and bananas sell for the same price per pound. Find their price per pound. **$.90/lb**

Test Prep and Mixed Review

Practice

Multiple Choice

43. The model below represents the equation $5x + 4 = 14$.

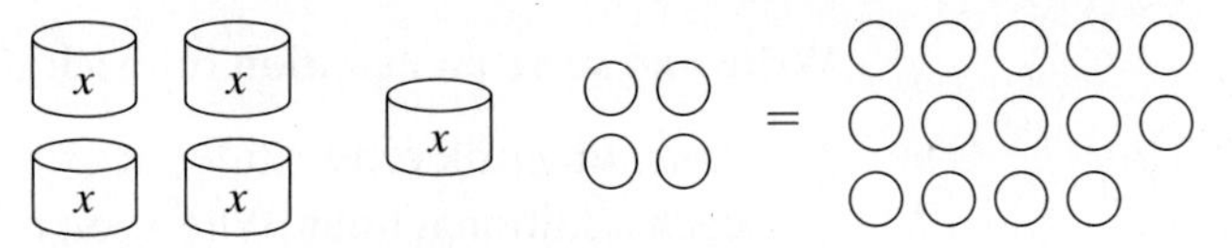

What is the value of x? **B**

Ⓐ $x = \frac{18}{5}$ Ⓑ $x = 2$ Ⓒ $x = -2$ Ⓓ $x = 5$

44. The Dead Sea is 1,345 feet below sea level. What method can be used to find the altitude of the Dead Sea in yards? **J**

Ⓕ Multiply −1,345 by 46. Ⓗ Divide −1,345 by 12.
Ⓖ Multiply −1,345 by 3. Ⓙ Divide −1,345 by 3.

45. What is the value of the expression $2(15 - 12)^2 \div 6 + 3$?

Ⓐ 12 Ⓑ 9 Ⓒ 6 Ⓓ 2 **C**

GO for Help

For Exercises	See Lesson
46–48	4-3

Describe a problem situation that matches each equation. Then solve.

46. $r + 11 = 2$
47. $h - 9 = 5.5$
48. $\frac{1}{6}q = 9$

46–48. Check students' work.

Test Prep

Resources

For additional practice with a variety of test item formats:

- Test-Taking Strategies, p. 219
- Test Prep, p. 223
- Test-Taking Strategies with Transparencies

Alternative Assessment

Have students work in pairs. Each partner chooses the two steps (adding or subtracting and multiplying or dividing) needed to solve an equation. Each partner writes a two-step equation that requires the chosen solution steps. Partners trade papers and each student solves the "made-to-order" equation. Students discuss their answers.

4-7 Graphing and Writing Inequalities

Algebra

Check Skills You'll Need

1. **Vocabulary Review** *Integers* are the set of whole numbers and their __?__. **opposites**

Compare using <, =, or >.

2. 0 ■ −2 **>**
3. 14 ■ −14 **>**
4. −4 ■ 5 **<**
5. −17 ■ −18 **>**

for Help
Lesson 1-6

What You'll Learn

To graph and write algebraic inequalities

New Vocabulary inequality, solution of an inequality

Why Learn This?

When you use an expression such as *at least* or *at most,* you are talking about an inequality. You can use inequalities to represent situations that involve minimum or maximum amounts.

YOU MUST BE AT LEAST THIS TALL (4FT) TO RIDE!

A mathematical sentence that contains $<, >, \leq, \geq$, or $\neq$ is an **inequality.** Sometimes an inequality contains a variable, as in $x \geq 2$.

A **solution of an inequality** is any value that makes the inequality true. For example, 6, 8, and 15 are solutions of $x \geq 6$ because $6 \geq 6$, $8 \geq 6$, and $15 \geq 6$.

EXAMPLE 1 Identifying Solutions of an Inequality

Vocabulary Tip

Read > as "is greater than."
Read < as "is less than."
Read ≥ as "is greater than or equal to."
Read ≤ as "is less than or equal to."

Find whether each number is a solution of $x \leq 2$; −3, 0, 2, 4.5.

Test each value by replacing the variable and evaluating the sentence.

$-3 \leq 2$ ← −3 is less than or equal to 2: true. ✔
$0 \leq 2$ ← 0 is less than or equal to 2: true. ✔
$2 \leq 2$ ← 2 is less than or equal to 2: true. ✔
$4.5 \leq 2$ ← 4.5 is less than or equal to 2: false. ✘

The numbers −3, 0, and 2 are solutions of $x \leq 2$. The number 4.5 is not a solution of $x \leq 2$.

Quick Check

1. Which numbers are solutions of the inequality $m \geq -3$; −8, −2, 1.4? **−2, 1.4**

A graph can show all the numbers in a solution. You use closed circles and open circles to show whether numbers are included in the solution.

Differentiated Instruction Solutions for All Learners

Special Needs L1
If students have difficulty graphing the solution to the inequalities, for example drawing the point or circle, ask them to work in pairs. They can provide direction while their partner graphs.

learning style: visual

Below Level L2
Read several sets of true/false exercises like these.

2 > 8 false   −2 > 8 false
2 > −8 true   −2 > −8 true

learning style: verbal

4-7

1. Plan

Objective
To graph and write algebraic inequalities

Examples
1 Identifying Solutions of an Inequality
2 Graphing Inequalities
3 Writing Inequalities
4 Application: Nutrition

Math Understandings: p. 166D

Math Background

The equations that were discussed in previous lessons had a common characteristic: Each equation had exactly one solution. In contrast, each of the inequalities presented in this lesson has infinitely many solutions. Since it is impossible to list an infinite number of solutions, a number-line graph is used as a means of picturing them.

More Math Background: p. 166D

Lesson Planning and Resources

See p. 166E for a list of the resources that support this lesson.

PowerPoint

Bell Ringer Practice

Check Skills You'll Need
Use student page, transparency, or PowerPoint. For intervention, direct students to:
Comparing and Ordering Integers
Lesson 1-6
Extra Skills and Word Problems Practice, Ch. 1

2. Teach

Activity Lab

Use before the lesson.

All in One Teaching Resources

Activity Lab 4-7: Graphing and Writing Inequalities

Guided Instruction

Example 1

Students might not understand why the statement $2 \leq 2$ is true. Tell them that $2 \leq 2$ is a short way of writing, "$2 < 2$ or $2 = 2$." A compound statement with the word *or* is true when one of the parts is true. Since $2 = 2$ is true, the entire statement is true.

Additional Examples

1 Tell whether each number is a solution of $k > -6$: $-8, -6, 0, 3, 7$
The numbers 0, 3, and 7 are solutions of $k > -6$. The numbers -8 and -6 are not.

2 Graph the solution of each inequality on a number line.

a. $r \leq 2$

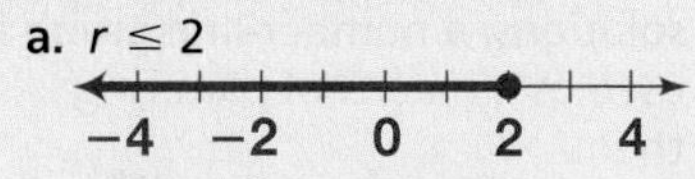

b. $m > -5$

−6 −4 −2 0 2

3 Write an inequality.

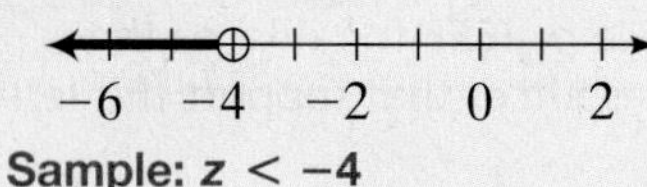

Sample: $z < -4$

4 You must be at least 18 years of age to vote. Write an inequality for this. **Let *a* stand for age in years; $a \geq 18$**

All in One Teaching Resources

- Daily Notetaking Guide 4-7 **L3**
- Adapted Notetaking 4-7 **L1**

Closure

- *What is an inequality?* **A mathematical sentence that contains $<$, $>$, $\leq$, or $\neq$.**
- Explain what a solution of an inequality is. **Any number that makes the inequality true.**

EXAMPLE **Graphing Inequalities**

2 Graph the solution of each inequality.

a. $n \geq -3$

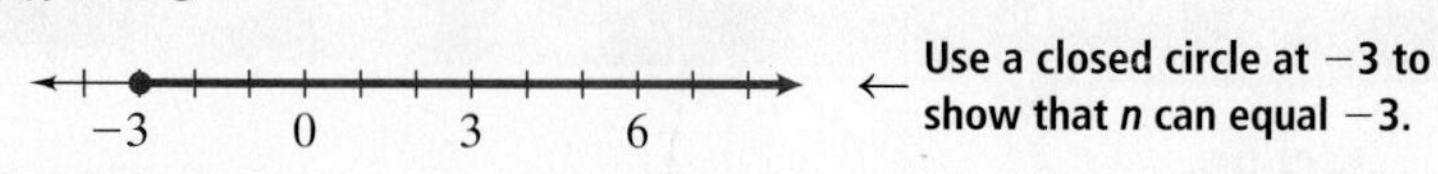

← **Use a closed circle at −3 to show that *n* can equal −3.**

b. $h < 7$

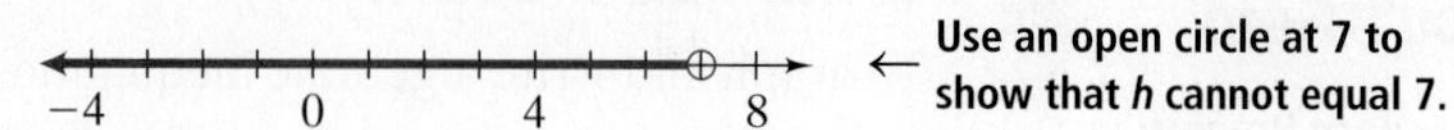

← **Use an open circle at 7 to show that *h* cannot equal 7.**

Quick Check

2. Graph the solution of the inequality $w < -3$.

−7 −6 −5 −4 −3 −2

You can write an inequality by analyzing its graph.

EXAMPLE **Writing Inequalities**

3 Write an inequality for the graph.

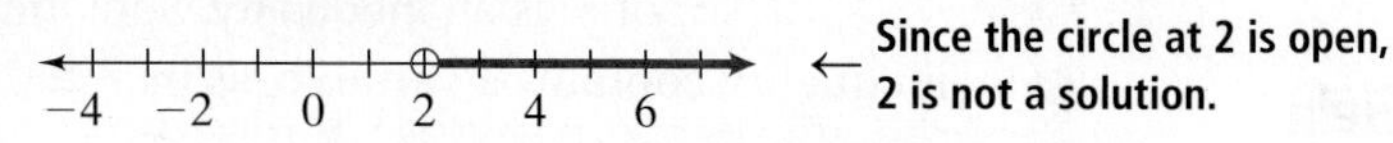

← **Since the circle at 2 is open, 2 is not a solution.**

$x > 2$ ← **Since the graph shows values greater than 2, use $>$.**

Quick Check

3. Write an inequality for the graph.

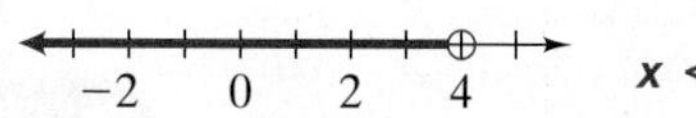

$x < 4$

You can write inequalities to describe real-world situations.

EXAMPLE **Application: Nutrition**

4 To be labeled sugar free, a food product must contain less than 0.5 g of sugar per serving. Write an inequality to describe this requirement.

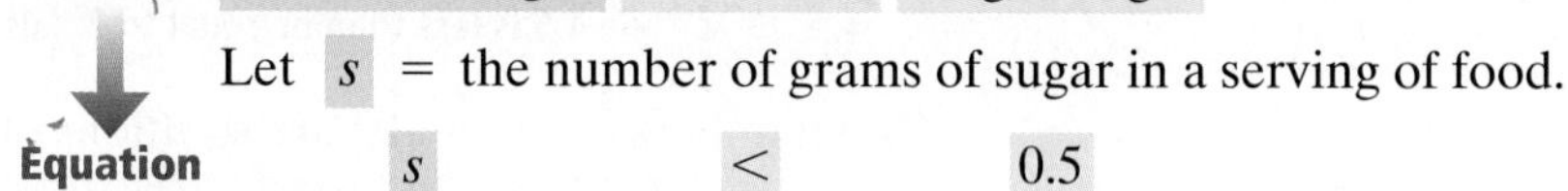

Words
amount of sugar
is less than
0.5 g of sugar
Let s = the number of grams of sugar in a serving of food.
Equation
s
<
0.5

The inequality is $s < 0.5$.

Quick Check

4. Write an inequality for "To qualify for the race, your time can be at most 62 seconds." **$t \leq 62$**

Advanced Learners **L4**
Have students write sentences about everyday experiences that can be described by inequalities. **Sample: I ride my bike at least 5 mi each week.**

learning style: visual

English Language Learners **ELL**
Have students write the symbols for *is less than* or *equal to,* and *is greater than* or *equal to* on index cards. Have them write the words for the symbols, along with some examples for each one. They can refer to these cards as needed.

learning style: visual

Check Your Understanding

1. **Vocabulary** What is the name of a mathematical sentence that contains the symbols $<$, $>$, $\leq$, $\geq$, or $\neq$? **inequality**
2. **Reasoning** Is $-4 \geq 4$? Explain. **no; $-4 < 4$**
3. Which inequality does NOT have 8 as a solution? **C**
   (A) $-3 < y$ (B) $13 \geq y$ (C) $y < 8$ (D) $y \geq 8$

Which numbers are solutions of each inequality?

4. $x \geq -5; -6, -1, 0$ **−1, 0**
5. $x \leq -1; -1, 1, 3$ **−1**
6. $x > 0; -1, 0, 1$ **1**

Homework Exercises

For more exercises, see Extra Skills and Word Problems.

GO for Help

For Exercises	See Examples
7–10	1
11–18	2
19–23	3–4

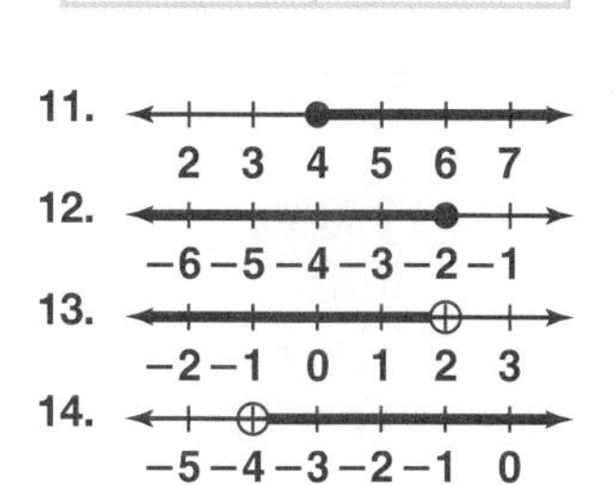

A Which numbers are solutions of each inequality?

7. $x < 1; -2, 1, 2$ **−2**
8. $x > -5; -7, -5, -1$ **−1**
9. $x \leq -9; -12, -4, 2$ **−12**
10. $x < -8; -10, -5, 0$ **−10**

Graph the solution of each inequality.
11–14. See left.

11. $x \geq 4$
12. $x \leq -2$
13. $x < 2$
14. $x > -4$

15–18. See margin.

15. $x \leq 0$
16. $h > -5$
17. $t \geq -5$
18. $p < -6$

Write an inequality for each graph.

19. **$x \leq 6$**
20. **$x < 2$**
21. **$x \geq -1$**
22. **$x \leq -1$**

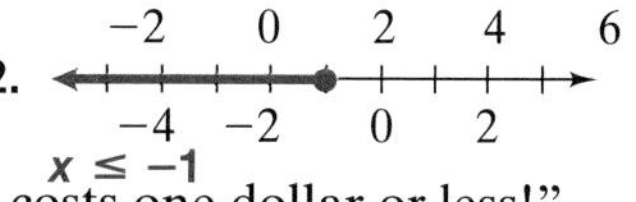

23. Write an inequality for "Every item costs one dollar or less!" **$c \leq 1.00$**

B **GPS**

24. **Guided Problem Solving** Write an inequality for "The car ride to the park will take at least 30 minutes." **$x > 30$**
    - Choose a variable: Let x represent how long the car ride will be.
    - Read for key words: Decide whether to use $<$, $>$, $\leq$, or $\geq$.
    - Write the inequality: x ■ 30.

Write an inequality for each statement. Graph the solution.
25–28. See margin.

25. To see the movie, you must be at least 17 years old. $x \geq 17$
26. The temperature is greater than 100°F.
27. A number p is not positive. $p < 0$
28. The speed limit on the highway is at most 65 mi/h.

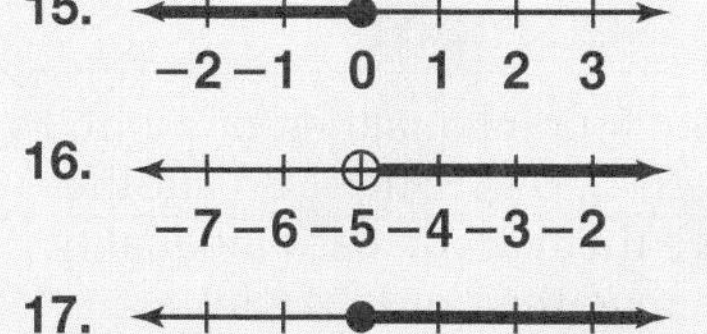

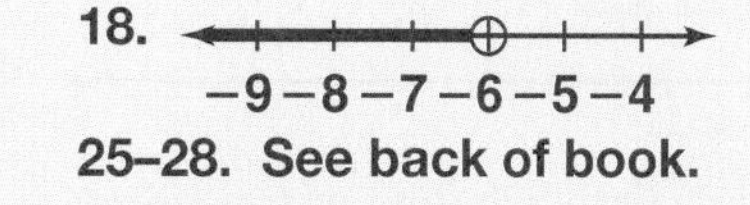

25–28. See back of book.

3. Practice

Assignment Guide

Check Your Understanding
Go over Exercises 1–6 in class before assigning the Homework Exercises.

Homework Exercises

A Practice by Example	7–23
B Apply Your Skills	24–35
C Challenge	36
Test Prep and Mixed Review	37–41

Homework Quick Check
To check students' understanding of key skills and concepts, go over Exercises 10, 19, 29, 30, and 32.

Differentiated Instruction Resources

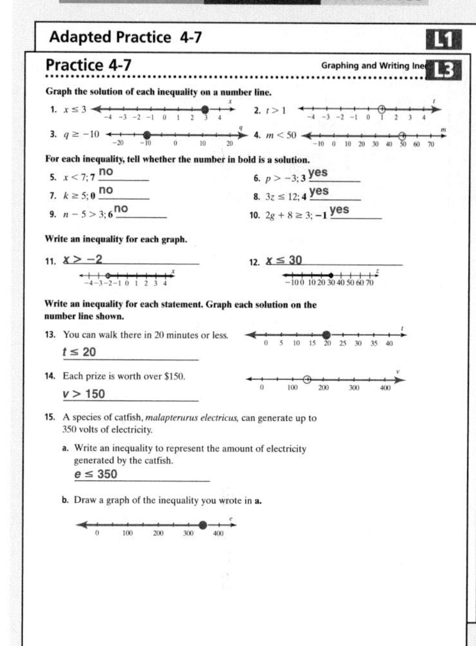

Adapted Practice 4-7 L1

Practice 4-7 Graphing and Writing Inequalities L3

Graph the solution of each inequality on a number line.

1. $x \leq 3$
2. $t > 1$
3. $q \geq -10$
4. $m < 50$

For each inequality, tell whether the number in bold is a solution.

5. $x < 7$; **7** no
6. $p > -3$; **3** yes
7. $k \geq 5$; **0** no
8. $3z \leq 12$; **4** yes
9. $n - 5 > 3$; **6** no
10. $2g + 8 \geq 3$; **−1** yes

Write an inequality for each graph.

11. $x > -2$
12. $x \leq 30$

Write an inequality for each statement. Graph each solution on the number line shown.

13. You can walk there in 20 minutes or less. $t \leq 20$
14. Each prize is worth over $150. $v > 150$
15. A species of catfish, *malapterurus electricus*, can generate up to 350 volts of electricity.
   a. Write an inequality to represent the amount of electricity generated by the catfish. $e \leq 350$
   b. Draw a graph of the inequality you wrote in **a**.

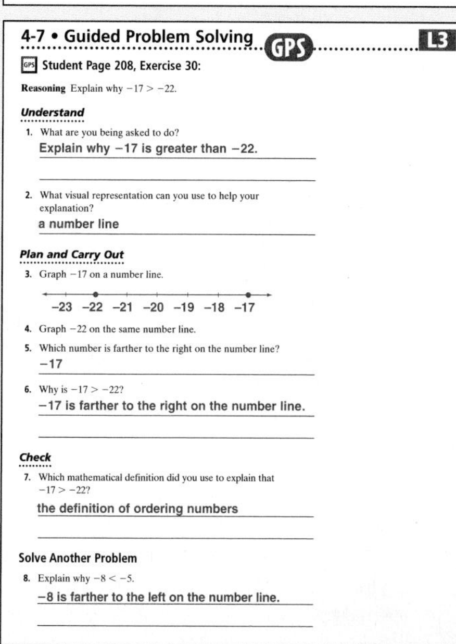

4-7 • Guided Problem Solving GPS L3

Student Page 208, Exercise 30:

Reasoning Explain why $-17 > -22$.

Understand

1. What are you being asked to do? Explain why −17 is greater than −22.
2. What visual representation can you use to help your explanation? a number line

Plan and Carry Out

3. Graph −17 on a number line.
   −23 −22 −21 −20 −19 −18 −17
4. Graph −22 on the same number line.
5. Which number is farther to the right on the number line? −17
6. Why is $-17 > -22$? −17 is farther to the right on the number line.

Check

7. Which mathematical definition did you use to explain that $-17 > -22$? the definition of ordering numbers

Solve Another Problem

8. Explain why $-8 < -5$. −8 is farther to the left on the number line.

4. Assess & Reteach

Lesson Quiz

Write an inequality for each verbal statement. Graph the solution.

1. A dog weighs less than 25 pounds. $d < 25$

   0 10 20 30

2. You must be at least 18 to vote. $v \geq 18$

   14 16 18 20 22

Reteaching 4-7 — Graphing and Writing Ine… **L2**

Two expressions separated by an inequality sign form an **inequality.** An inequality shows that the two expressions *are not* equal. Unlike the equations you have worked with, an inequality has many solutions.

The **solutions of an inequality** are the values that make the inequality true. They can be graphed on a number line. Use a closed circle (•) for ≤ and ≥ and an open circle (o) for > and <. For example:

$x > -2$ ("greater than") −5 −4 −3 −2 −1 0 1 2 3 4 5

$x < -2$ ("less than") −5 −4 −3 −2 −1 0 1 2 3 4 5

$x \geq -2$ ("greater than or equal to") −5 −4 −3 −2 −1 0 1 2 3 4 5

$x \leq -2$ ("less than or equal to") −5 −4 −3 −2 −1 0 1 2 3 4 5

Graph the inequality $x > 4$.

The inequality is read as "x is greater than 4." Since all numbers to the right of 4 are greater than 4, you can draw an arrow from 4 to the right. Since 4 is not greater than itself, use an open circle on 4.

−2 −1 0 1 2 3 4 5 6 7 8

1. **Graph the inequality $x \leq -3$.**
   a. Write the inequality in words. **x is less than or equal to −3.**
   b. Will the circle at −3 be open or closed? **closed; −3 satisfies inequality**
   c. Graph the solution. −7 −6 −5 −4 −3 −2 −1 0 1 2 3

2. **Graph the inequality $x < 3$.**
   a. Write the inequality in words. **x is less than 3.**
   b. Will the circle at 3 be open or closed? **open**
   c. Graph the solution. −4 −3 −2 −1 0 1 2 3 4 5 6

Enrichment 4-7 — Graphing and Writing In… **L4**

Exploring Inequalities

In order to create a sense of fairness in the sport, wrestling is divided into weight and age classes. The table below was used by the state of Florida for a 2001 junior wrestling tournament.

Division	Birthdate	Match Time Limit	Weight Classes
Bantam	Born 1994–1995 (State level program only)	Two 90-second periods 30-second rest between periods	40 lb, 45 lb, 50 lb, 55 lb, 60 lb 65 lb, 70 lb, 75 lb, 75+ lb
Midget	Born 1992–1993	Two 90-second periods 30-second rest between periods	50 lb, 55 lb, 60 lb, 65 lb, 70 lb 75 lb, 80 lb, 87 lb, 95 lb, 103 lb 112 lb, 120 lb, 120+ lb
Novice	Born 1990–1991	Two 2-minute periods 30-second rest between periods	60 lb, 65 lb, 70 lb, 75 lb, 80 lb 85 lb, 90 lb, 95 lb, 100 lb, 105 lb 112 lb, 120 lb, 130 lb, 140 lb, 140+ lb
Schoolboy/girl	Born 1988–1989	Two 2-minute periods 30-second rest between periods	70 lb, 75 lb, 80 lb, 85 lb, 90 lb 95 lb, 100 lb, 105 lb, 110 lb 115 lb, 120 lb, 125 lb, 130 lb 140 lb, 150 lb, 160 lb, 160+ lb

For the following questions, assume the year is 2004.

1. Write an inequality to represent the age limit of a Novice wrestler. Then write another inequality to represent the weight limit for a Novice wrestler in the 90 lb weight class.
   13 years ≤ x ≤ 14 years; 90 lb ≤ x < 95 lb
2. A wrestler is 11 years old and 68 pounds. What division is the wrestler in? What is the wrestler's match time limit?
   Midget; two 90-second periods
3. Write and graph an inequality to show the range of a wrestler's age in the Schoolboy/girl division.
   15 years ≤ x ≤ 16 years
4. You attend a wrestling meet to watch your friend wrestle. Use the following inequalities to determine what division your friend belongs. 9 years old ≤ x ≤ 10 years old; 55 lb ≤ x < 60 lb
   Bantam division

GO Online
Homework Video Tutor
Visit: PHSchool.com
Web Code: are-0407

29. Use an open circle for < or > and use a closed circle for ≤ or ≥.

30. Answers may vary. Sample: −17 is to the right of −22 on a number line.

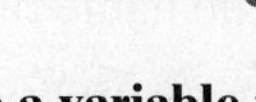

29. **Writing in Math** Explain how you know whether to draw an open or closed circle when you graph an inequality. **See left.**

30. **Reasoning** Explain why $-17 > -22$. **See left.**

GPS **Use a variable to write an inequality for each situation.**

31. $w \leq 3$

32.

$p \leq 2$

Vocabulary Tip

A *compound inequality* is a number sentence with more than one inequality symbol.

Identify an integer that is a solution for each *compound inequality*.

33. $-3 \leq x < 0$ **−3, −2, or −1**
34. $-2 < y \leq 1$ **−1, 0, or 1**
35. $-6 \leq p \leq -4$ **−6, −5, or −4**

Ⓒ 36. **Challenge** If $a \geq 9$ and $9 \geq b$, then a ■ b. Complete the statement using <, >, ≤, or ≥. **≥**

Test Prep and Mixed Review — Practice

Multiple Choice

37. The model below represents the equation $2x + 5 = 9$.

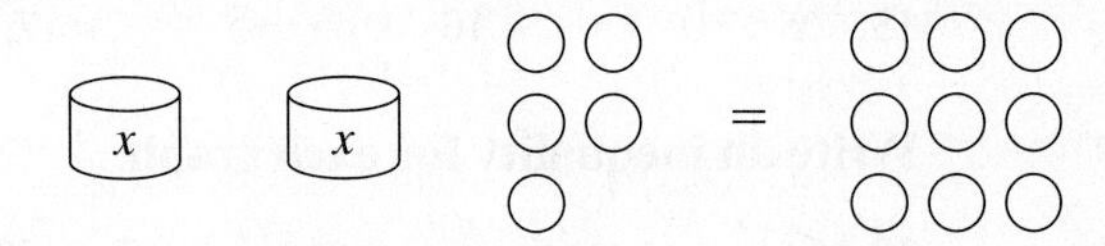

What is the first step in solving the equation? **B**

Ⓐ Add 5 to each side of the equation.
Ⓑ Subtract 5 from each side of the equation.
Ⓒ Divide each side of the equation by 5.
Ⓓ Subtract 2 from each side of the equation.

38. A recipe calls for 3 pounds of ground turkey. Carmen has one package of ground beef that weighs $1\frac{1}{2}$ pounds, one package of ground turkey that weighs $1\frac{1}{4}$ pound, and another package of ground turkey that weighs $1\frac{5}{8}$ pound. Which information is NOT necessary to find how much more ground turkey Carmen needs? **H**

Ⓕ Total amount of ground turkey needed for the recipe
Ⓖ Weight of Carmen's smaller package of ground turkey
Ⓗ Weight of Carmen's package of ground beef
Ⓙ Weight of Carmen's larger package of ground turkey

GO for Help

For Exercises	See Lesson
39–41	1-6

Order the numbers from least to greatest.

39. −2, 4, −4, 2, 7, −1 **−4, −2, −1, 2, 4, 7**
40. 3, −8, −9, 12, −6 **−9, −8, −6, 3, 12**
41. 10, 0, −5, −2, 7 **−5, −2, 0, 7, 10**

Test Prep

Resources

For additional practice with a variety of test item formats:

- Test-Taking Strategies, p. 219
- Test Prep, p. 223
- Test-Taking Strategies with Transparencies

Alternative Assessment

Ask students to write four inequalities: one with >; one with <; one with ≥, one with ≤. On another sheet of paper, have them graph each inequality. Students then exchange their graphs with a partner who writes the inequality for each graph. Students should discuss their results.

Checkpoint Quiz 2

Lessons 4-5 through 4-7

Solve each equation.

1. $3x + 4 = 19$ **5**
2. $\frac{t}{5} - 2 = 6$ **40**
3. $-2y - 5 = -9$ **2**
4. $\frac{d}{-3} + 7 = 10$ **−9**
5. $6g + 6 = -6$ **−2**
6. $\frac{f}{-1} - 8 = -3$ **−5**
7. A sweater costs $12 more than twice the cost of a skirt. The sweater costs $38. Find the cost of the skirt. **Let c = the cost of a skirt; $12 + 2c = 38$; $13.**

Graph the solution of each inequality. **8–10. See margin**

8. $x > -3$
9. $y \leq -2$
10. $d \geq -1$

Checkpoint Quiz

Use this Checkpoint Quiz to check students' understanding of the skills and concepts of Lessons 4-5 through 4-7.

Resources

- All-in-One Teaching Resources Checkpoint Quiz 2
- ExamView CD-ROM
- Success Tracker™ Online Intervention

4-8a Activity Lab

Data Analysis

Inequalities in Bar Graphs

A class conducted a survey on free time. The standard bar graph shows the average number of hours per week students spend on various activities. The floating bar graph shows the minimum and maximum number of hours students spend on each activity.

Standard Bar Graph

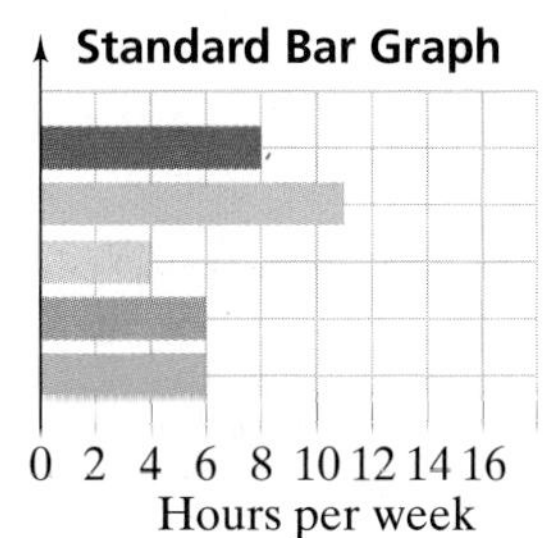

Floating Bar Graph

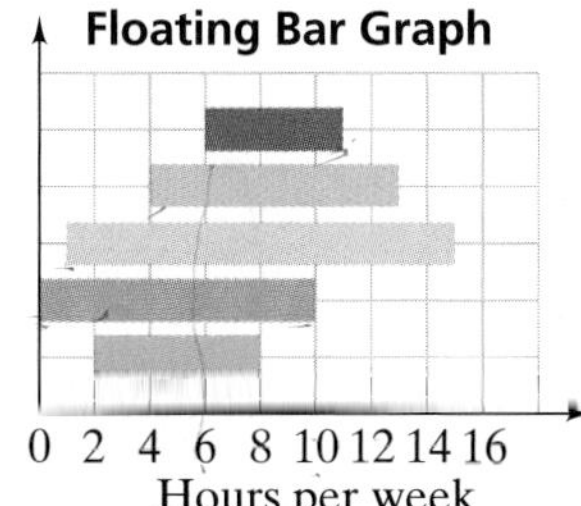

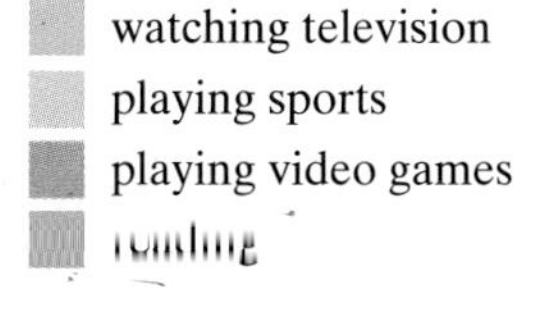

1. What are the minimum and maximum hours per week that students in the survey spend hanging out with friends? **minimum 6, maximum 10**
2. Which activity is not done by all students? Which activity shows the least variety in how students responded to the survey? Explain. **2–3. See margin.**
3. Let x represent the number of hours per week spent on an activity. Which activity in the floating bar graph is represented by the inequality $2 \leq x \leq 8$? Write inequalities to represent each activity on the floating bar graph.

Activity Lab

Inequalities in Bar Graphs

Students analyze floating bar graphs to find minimums and maximums and express these in inequalities.

Guided Instruction

Ask:

- *Which graph would you use to find out the average number of hours per week students play video games?* **standard bar graph**
- *What is the minimum and maximum number of hours per week that students play video games?* **minimum: 0, maximum: 10**

Differentiated Instruction

Advanced Learners  L4

After completing Exercise 3, have students write an inequality for another activity on the floating bar graph. **Sample: $4 \leq x \leq 12$**

Checkpoint Quiz 2

8. −5 −4 −3 −2 −1 0

9. −5 −4 −3 −2 −1 0

10. 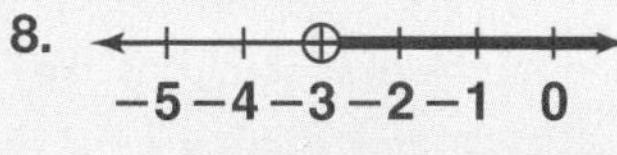 −4 −3 −2 −1 0 1

Activity Lab

2. **playing video games; hanging out with friends, because the bar is the shortest**
3. **reading; hanging out with friends: $6 \leq x \leq 11$; watching television: $4 \leq x \leq 13$; playing sports: $1 \leq x \leq 15$; playing video games: $0 \leq x \leq 10$**

4-8

1. Plan

Objective
To solve inequalities by adding or subtracting

Examples
1 Solve Inequalities by Adding
2 Solving Inequalities by Subtracting
3 Application: Transportation

Math Understandings: p. 166D

Math Background

Solving an inequality involves the same basic strategy used for solving an equation: Isolate the variable. The method is the same as that used with equations. That is, you use inverse operations to "undo" the addition or subtraction.

More Math Background: p. 166D

Lesson Planning and Resources

See p. 166E for a list of the resources that support this lesson.

Bell Ringer Practice

Check Skills You'll Need
Use student page, transparency, or PowerPoint. For intervention, direct students to:
Solving Equations by Adding or Subtracting
Lesson 4-3
Extra Skills and Word Problems Practice, Ch. 4

Algebra

4-8 Solving Inequalities by Adding or Subtracting

Check Skills You'll Need

1. **Vocabulary Review** Why are addition and subtraction called *inverse operations*? See below.

Solve each equation.

2. $c + 4 = -2$ -6
3. $x + 9 = 10$ 1
4. $b + (-2) = 7$ 9
5. $m - (-3) = 10$ 7

for Help Lesson 4-3

Check Skills You'll Need
1. because they "undo" each other

What You'll Learn

To solve inequalities by adding or subtracting

New Vocabulary Addition Property of Inequality, Subtraction Property of Inequality

Why Learn This?

Buildings and buses have limits on the number of people they can hold. You can use inequalities to find how many people can fit safely.

You can solve inequalities using properties similar to those you used solving equations.

If you add 3 to each side of the inequality $-3 < 2$, the resulting inequality, $0 < 5$, is also true.

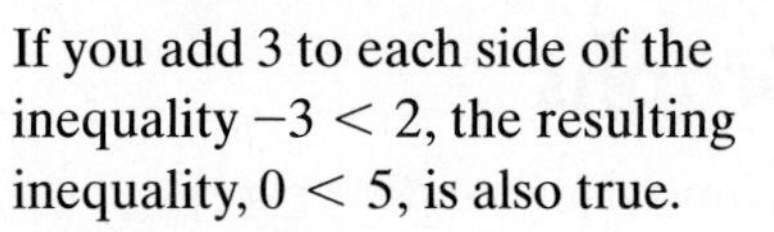

KEY CONCEPTS Addition Property of Inequality

You can add the same value to each side of an inequality.

Arithmetic	Algebra
Since $7 > 3$, $7 + 4 > 3 + 4$.	If $a > b$, then $a + c > b + c$.
Since $1 < 3$, $1 + 4 < 3 + 4$.	If $a < b$, then $a + c < b + c$.

EXAMPLE Solving Inequalities by Adding

GO for Help For help with graphing inequalities, see Lesson 4-7, Example 2.

1 Solve $n - 10 > 14$. Graph the solution.

$$n - 10 > 14$$
$$n - 10 + 10 > 14 + 10 \quad \leftarrow \text{Add 10 to each side.}$$
$$n > 24 \quad \leftarrow \text{Simplify.}$$

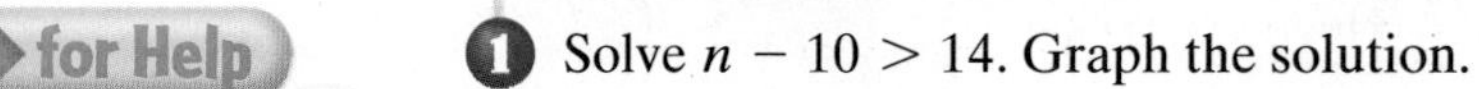

Quick Check

1. Solve $y - 3 < 4$. Graph the solution. $y < 7$;

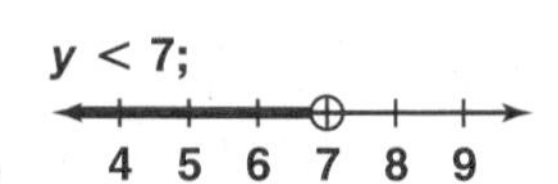

Differentiated Instruction Solutions for All Learners

Special Needs L1
Have students draw a rectangle to represent the bus, in Example 3. Have them write 19 in the rectangle. Have them use s to represent the number of students who can board the bus keeping the total to at most 76. So they should write $19 + s$ inside the bus is less than or equal to ($\leq$) 76.
learning style: visual

Below Level L2
Give students several "fill in the blanks" solutions like this.

$h - 5 < 7$
$h - 5 + \square < \square + 5$ 5; 7
$\square < \square$ h; 12

learning style: visual

To solve an inequality involving addition, use subtraction.

KEY CONCEPTS **Subtraction Property of Inequality**

You can subtract the same value from each side of an inequality.

Arithmetic	Algebra
Since $9 > 6$, $9 - 3 > 6 - 3$.	If $a > b$, then $a - c > b - c$.
Since $15 < 20$, $15 - 4 < 20 - 4$.	If $a < b$, then $a - c < b - c$.

Note: The Properties of Inequality also apply to $\leq$ and $\geq$.

EXAMPLE Solving Inequalities by Subtracting

2 Solve $y + 7 \geq 12$. Graph the solution.

$y + 7 \geq 12$

$y + 7 - 7 \geq 12 - 7$ ← **Subtract 7 from each side.**

$y \geq 5$ ← **Simplify.**

−2 0 2 4 6 8 10

Quick Check

2. Solve each inequality. Graph the solution. **See left.**
   a. $x + 9 > 5$ **b.** $y + 3 < 4$ **c.** $w + 4 \leq -5$

2 a. $x > -4$;
−6 −5 −4 −3 −2 −1
b. $y < 1$;
−2 −1 0 1 2 3
c. $w \leq -9$;
−11 −9 −7

EXAMPLE Application: Transportation

3 A school bus can safely carry as many as 76 students. If 19 students are already on the bus, how many more can board the bus?

Words	students already on bus	plus	students remaining	is at most	76
	Let s = the number of students remaining.				
Expression	19	+	s	$\leq$	76

$19 + s \leq 76$

$19 - 19 + s \leq 76 - 19$ ← **Subtract 19 from each side.**

$s \leq 57$ ← **Simplify.**

At most 57 more students can board the bus.

Quick Check

3. To get an A, you need more than 200 points on a two-part test. You score 109 on the first part. How many more points do you need?

3. Let p = number of points; $p + 109 > 200$; $p > 91$; you need more than 91 points.

Advanced Learners L4
Solve each inequality.

$x + 3 > x + 2$ **all numbers**
$x + 3 < x + 2$ **no solution**

learning style: visual

English Language Learners ELL
Have students add the terms *open circle* and *closed circle* to their index cards. Have them describe, in words and symbols, what each means.

learning style: visual

2. Teach

Activity Lab

Use before the lesson.

All in One Teaching Resources

Activity Lab 4-8: Solving Inequalities by Adding and Subtracting

Guided Instruction

Example 1
Some students better visualize the process of adding the same number to each side when the addition is performed vertically.

$$\begin{array}{rcr} n - 10 & > & 14 \\ +\,10 & = & +\,10 \\ \hline n & > & 24 \end{array}$$

Error Prevention!

To check for errors in addition or subtraction, have students use the *associated equation.* In Example 1, $n > 24$ is a reasonable solution of $n - 10 > 14$ because 24 is the solution of $n - 10 = 14$.

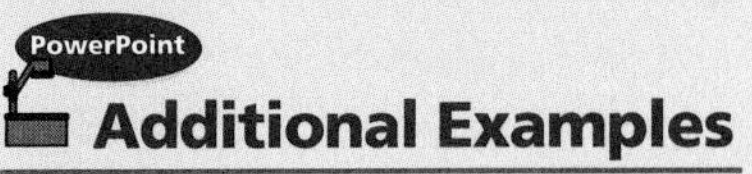

PowerPoint

Additional Examples

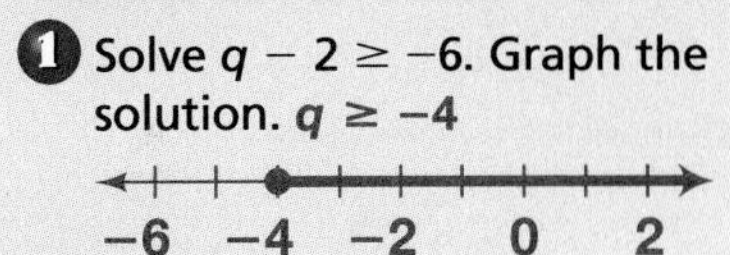

1 Solve $q - 2 \geq -6$. Graph the solution. $q \geq -4$

−6 −4 −2 0 2

2 Solve $d + 9 < 8$. Graph the solution. $d < -1$

−4 −2 0 2 4

3 The Drama Club can spend no more than $120 for costumes. They spent $79. How much more can they spend? **at most $41**

All in One Teaching Resources
- Daily Notetaking Guide 4-8 L3
- Adapted Notetaking 4-8 L1

Closure

- Explain how to solve inequalities by adding or subtracting. **Sample: Isolate the variable; If a number has been added to the variable, subtract that number from each side. If a number has been subtracted from the variable, add that number to each side.**

3. Practice

Assignment Guide

Check Your Understanding
Go over Exercises 1–6 in class before assigning the Homework Exercises.

Homework Exercises

A	Practice by Example	7–25
B	Apply Your Skills	26–38
C	Challenge	39
	Test Prep and Mixed Review	40–46

Homework Quick Check
To check students' understanding of key skills and concepts, go over Exercises 8, 17, 25, 37, and 38.

Differentiated Instruction Resources

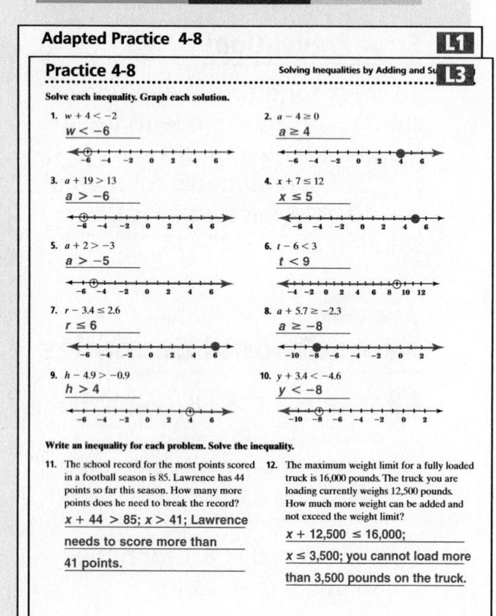
Adapted Practice 4-8 L1

Practice 4-8 Solving Inequalities by Adding and Su[btracting] L3

Solve each inequality. Graph each solution.

1. $w + 4 < -2$ — $w < -6$
2. $a - 4 \geq 0$ — $a \geq 4$
3. $a + 19 > 13$ — $a > -6$
4. $x + 7 \leq 12$ — $x \leq 5$
5. $a + 2 > -3$ — $a > -5$
6. $t - 6 < 3$ — $t < 9$
7. $r - 3.4 \leq 2.6$ — $r \leq 6$
8. $a + 5.7 \geq -2.3$ — $a \geq -8$
9. $h - 4.9 > -0.9$ — $h > 4$
10. $y + 3.4 < -4.6$ — $y < -8$

Write an inequality for each problem. Solve the inequality.

11. The school record for the most points scored in a football season is 85. Lawrence has 44 points so far this season. How many more points does he need to break the record? $x + 44 > 85$; $x > 41$; Lawrence needs to score more than 41 points.
12. The maximum weight limit for a fully loaded truck is 16,000 pounds. The truck you are loading currently weighs 12,500 pounds. How much more weight can be added and not exceed the weight limit? $x + 12{,}500 \leq 16{,}000$; $x \leq 3{,}500$; you cannot load more than 3,500 pounds on the truck.

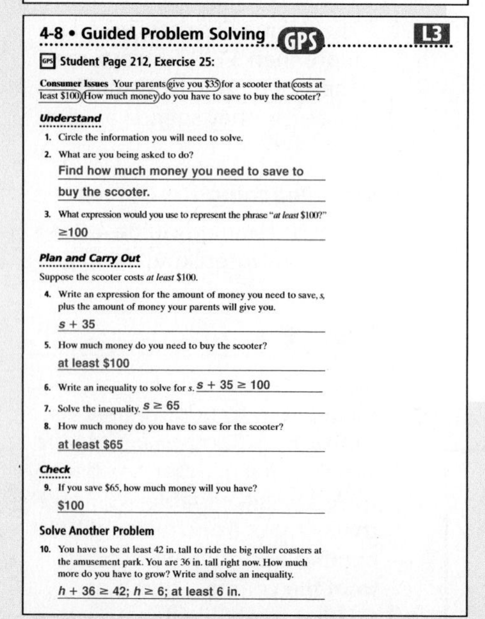
4-8 • Guided Problem Solving GPS L3

Student Page 212, Exercise 25:

Consumer Issues Your parents give you \$35 for a scooter that costs at least \$100. How much money do you have to save to buy the scooter?

Understand

1. Circle the information you will need to solve.
2. What are you being asked to do? Find how much money you need to save to buy the scooter.
3. What expression would you use to represent the phrase "*at least* \$100?" ≥ 100

Plan and Carry Out

Suppose the scooter costs *at least* \$100.

4. Write an expression for the amount of money you need to save, *s*, plus the amount of money your parents will give you. $s + 35$
5. How much money do you need to buy the scooter? at least \$100
6. Write an inequality to solve for *s*. $s + 35 \geq 100$
7. Solve the inequality. $s \geq 65$
8. How much money do you have to save for the scooter? at least \$65

Check

9. If you save \$65, how much money will you have? \$100

Solve Another Problem

10. You have to be at least 42 in. tall to ride the big roller coasters at the amusement park. You are 36 in. tall right now. How much more do you have to grow? Write and solve an inequality. $h + 36 \geq 42$; $h \geq 6$; at least 6 in.

Check Your Understanding

1. **Vocabulary** The __?__ states that you can add the same value to each side of an inequality. **Addition Property of Inequality**
2. **Reasoning** What value is a solution of $y + 7 \geq 12$ but is not a solution of $y + 7 > 12$? **5**

Match each inequality with the graph of its solution.

3. $h - 4 < 5$ **C**
4. $h + 4 \geq 5$ **D**
5. $h - 4 \leq -5$ **B**
6. $h + 4 < -5$ **A**

A. 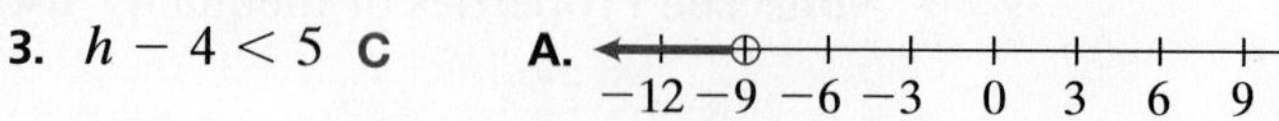
−12 −9 −6 −3 0 3 6 9

B. −3 −2 −1 0 1 2 3 4

C. −9 −6 −3 0 3 6 9 12

D. −3 −2 −1 0 1 2 3 4

Homework Exercises

For more exercises, see Extra Skills and Word Problems.

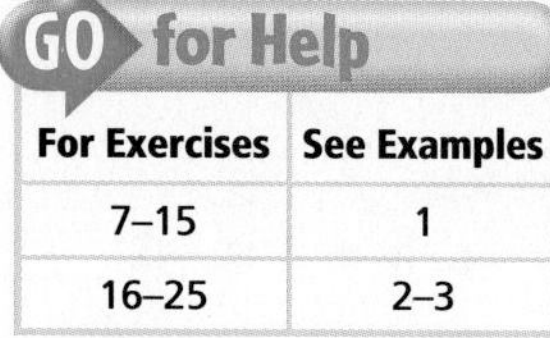
GO for Help

For Exercises	See Examples
7–15	1
16–25	2–3

A **Solve each inequality by adding. Graph the solution.**
7–9. See left.

7. $g - 2 \leq -8$
8. $m - 3 > -24$
9. $y - 5 \geq 11$

10–15. See margin.

10. $x - 7 > -11$
11. $n - 10 \leq 17$
12. $p - 9 < -9$
13. $y - 5 \geq 12$
14. $q - 2 < 4$
15. $b - 4 > -6$

7. $g \leq -6$;
−9 −8 −7 −6 −5 −4

8. $m > -21$;
−23 −21 −19

9. $y \geq 16$;
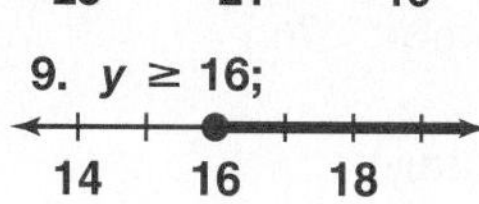
14 16 18

Solve each inequality by subtracting. Graph the solution.
16–24. See margin.

16. $h + 8 < -13$
17. $n + 3 \geq 4$
18. $r + 9 > 4$
19. $p + 10 \leq 6$
20. $b + 22 > -1$
21. $f + 5 \geq 0$
22. $m + 3 > 4$
23. $x + 10 < 11$
24. $k + 4 \leq -7$

25. **Consumer Issues** GPS Your parents give you \$35 for a scooter that costs at least \$100. How much money do you have to save to buy the scooter? **Let s = amount of money to save; $s + 35 \geq 100$; $s \geq 65$; you need to save at least \$65.**

B 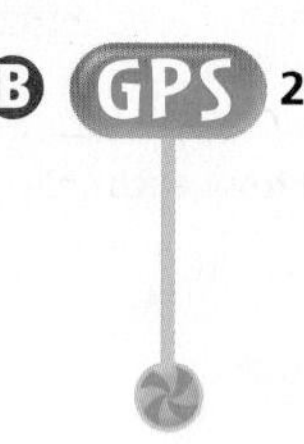

26. **Guided Problem Solving** The weight of a loaded dump truck is less than 75,000 lb. When empty, the truck weighs 32,000 lb. Write and solve an inequality to find how much the load can weigh.

words:	empty truck	plus	__?__	is less than	75,000 lb
inequality:	32,000	+	■	<	75,000

$\ell + 32{,}000 < 75{,}000$; $\ell < 43{,}000$

27. **Science** Water boils when the temperature is at least 212°F. A pot of water has a temperature of 109°F. How many degrees must the temperature rise for the water to boil? **$109 + x \geq 212$; $x \geq 103$**

10–24. See back of book.

35. No; the solution of $x + 5 \leq -2$ is $x \leq -7$, and the solution of $-2 \leq x + 5$ is $x \geq -7$.

37. $p \geq 338$; they must score at least 420 − 82, or 338, points.

GO Online
Homework Video Tutor
Visit: PHSchool.com
Web Code: are-0408

Solve each inequality by adding or subtracting.
28–30. See left.

28. $h - 9 < 1.3$
29. $w - \frac{1}{4} > -\frac{3}{4}$
30. $x + 5 \geq 98$
31. $j + 6.2 \geq 1.2$ $j \geq -5$
32. $k + 42 \geq 36$ $k \geq -6$
33. $a - 1\frac{4}{5} < 3\frac{7}{10}$ $a < 5\frac{1}{2}$

28. $h < 10.3$
29. $w > -\frac{1}{2}$
30. $x \geq 93$

34. Write an inequality for the sentence "Fifteen plus a number is greater than 10." Solve the inequality. $15 + n > 10; n > -5$

35. **Reasoning** Are the solutions of the inequalities $x + 5 \leq -2$ and $-2 \leq x + 5$ the same? Explain. See margin.

36. **Sports** To win the long jump, you need to jump a distance greater than 2.25 m. Your personal best jump is 2.1 m. Write and solve an inequality to find how much farther you need to jump to win. $d + 2.1 > 2.25; d > 0.15$

37. **Writing in Math** The basketball team needs to score at least 420 points this season in order to set a new school record. It has already scored 82 points. Four players argue about which inequality represents the number of points yet to be scored: $p \geq 338$, $338 \geq p$, $p > 338$, or $338 > p$. Which is correct? Explain. See margin.

38. A *compound inequality* is a number sentence with more than one inequality symbol. Solve the compound inequality $-3 \leq x + 4 < 9$. $-7 \leq x < 5$

C 39. **Challenge** You want to eat no more than 3,000 Calories in a day. You consume 710 Calories for breakfast and have two bowls of soup for lunch. Each bowl contains 535 Calories. How many Calories can you consume at dinner? no more than 1,220 Cal

Test Prep and Mixed Review

Practice

Gridded Response

40. A baby who weighed 7.2 pounds at birth gained about 1.4 pounds each month. How many pounds did the baby weigh when it was 6 months old? 15.6

41. Two of the driest cities in the United States are Yuma, Arizona, and Las Vegas, Nevada. Yuma averages 3.01 inches of rain each year and Las Vegas averages 4.49 inches each year. How many more inches of rain does Las Vegas receive each year than Yuma? 1.48

42. The table shows record weights of 3 types of sunfish. In decimal form, how many pounds did the redbreast sunfish weigh? Round to the nearest hundredth. 2.06

Fish	Weight (lb)
Green sunfish	$\frac{17}{8}$
Redbreast sunfish	$\frac{33}{16}$
Redear sunfish	$\frac{87}{16}$

GO for Help

For Exercises	See Lesson
43–46	2-5

Write each number as an improper fraction.

43. $3\frac{1}{5}$ $\frac{16}{5}$
44. $6\frac{3}{4}$ $\frac{27}{4}$
45. $8\frac{1}{4}$ $\frac{33}{4}$
46. $7\frac{5}{8}$ $\frac{61}{8}$

Online lesson quiz, PHSchool.com, Web Code: ara-0408

4. Assess & Reteach

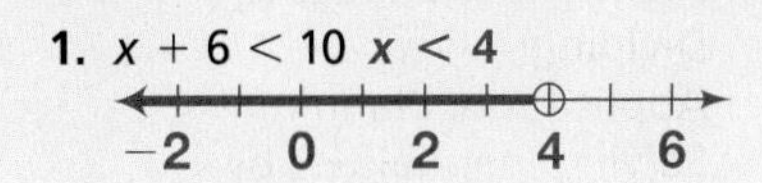

Lesson Quiz

Solve each inequality. Graph the solution.

1. $x + 6 < 10$ $x < 4$
(number line: −2, 0, 2, 4, 6)
2. $a - 9 > 14$ $a > 23$ 
(number line: 22, 24, 26, 28, 30)
3. $n + 8 \leq 25$ $n \leq 17$ 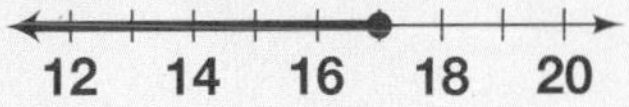
(number line: 12, 14, 16, 18, 20)
4. $p - 4 \geq 5$ $p \geq 9$
(number line: 6, 8, 10, 12, 14)

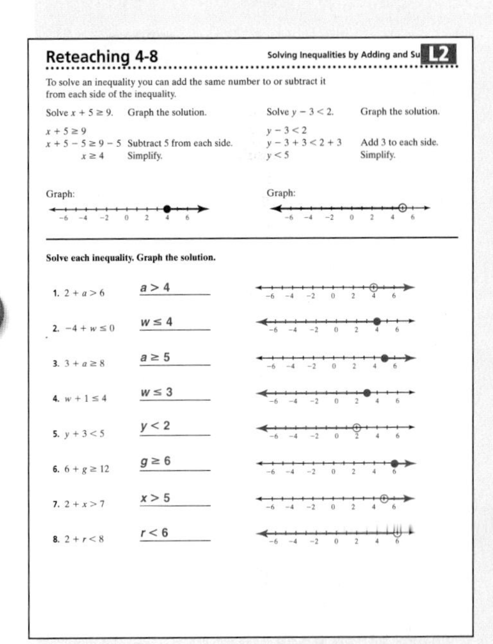

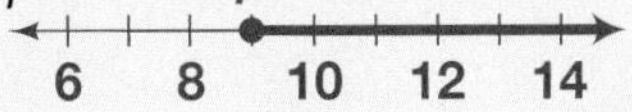

Reteaching 4-8 — Solving Inequalities by Adding and Subtracting — L2

To solve an inequality you can add the same number to or subtract it from each side of the inequality.

Solve $x + 5 \geq 9$. Graph the solution.
$x + 5 \geq 9$
$x + 5 - 5 \geq 9 - 5$ Subtract 5 from each side.
$x \geq 4$ Simplify.

Solve $y - 3 < 2$. Graph the solution.
$y - 3 < 2$
$y - 3 + 3 < 2 + 3$ Add 3 to each side.
$y < 5$ Simplify.

Solve each inequality. Graph the solution.

1. $2 + a > 6$ $a > 4$
2. $-4 + w \leq 0$ $w \leq 4$
3. $3 + a \geq 8$ $a \geq 5$
4. $w + 1 \leq 4$ $w \leq 3$
5. $y + 3 < 5$ $y < 2$
6. $6 + g \geq 12$ $g \geq 6$
7. $2 + x > 7$ $x > 5$
8. $2 + r < 8$ $r < 6$

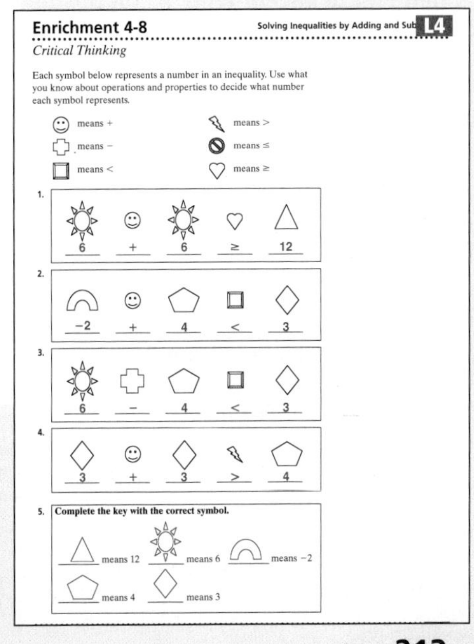

Enrichment 4-8 — Solving Inequalities by Adding and Subtracting — L4

Critical Thinking

Each symbol below represents a number in an inequality. Use what you know about operations and properties to decide what number each symbol represents.

Alternative Assessment

Ask students to write four different inequalities that all have the same solution, such as $h + 6 \leq 14$. $h \leq 8$ Two inequalities should be solved by adding and two by subtracting. Students then exchange papers with a partner who solves each inequality and graphs each solution. Have partners discuss their results.

Test Prep

Resources

For additional practice with a variety of test item formats:
- Test-Taking Strategies, p. 219
- Test Prep, p. 223
- Test-Taking Strategies with Transparencies

4-9

1. Plan

Examples

1. Solving Inequalities by Dividing
2. Application: Planning
3. Solving Inequalities by Multiplying

Math Understandings: p. 166D

Math Background

The *rules* for solving equations and inequalities refer to the properties of equality and inequality. In Lesson 4-8, the addition and subtraction rules for inequalities were essentially the same as those for equations. In this lesson, the exposition demonstrates that multiplication and division rules for inequalities differ from the rules for equations in an important way: When multiplying or dividing each side of an inequality by a negative number, the direction of the inequality must be reversed.

More Math Background: p. 166D

Lesson Planning and Resources

See p. 166E for a list of the resources that support this lesson.

Bell Ringer Practice

Check Skills You'll Need
Use student page, transparency, or PowerPoint. For intervention, direct students to:
Solving Inequalities by Adding and Subtracting
Lesson 4-8
Extra Skills and Word Problems Practice, Ch. 4

Algebra

4-9 Solving Inequalities by Multiplying or Dividing

Check Skills You'll Need

1. **Vocabulary Review** How is the *Addition Property of Equality* similar to the *Addition Property of Inequality*? **See below.**

Solve each inequality.

2. $x + 3 \leq 5$ $x \leq 2$
3. $p - 9 > -2$ $p > 7$
4. $8 \geq d + 5$ $3 \geq d$
5. $r - 2 < -8$ $r < -6$

GO for Help Lesson 4-8

Check Skills You'll Need

1. **You can add to each side of an equation or inequality without changing the relationship.**

What You'll Learn

To solve inequalities by multiplying or dividing

New Vocabulary Division Property of Inequality, Multiplication Property of Inequality

Why Learn This?

Inequalities can help you plan. You can solve inequalities to make sure you have enough ingredients when you are cooking.

Look at the pattern when you divide each side of an inequality by an integer.

$18 > 12$

$\frac{18}{6} > \frac{12}{6}$

$\frac{18}{3} > \frac{12}{3}$ ← **When the integer is positive, the direction of the inequality symbol stays the same.**

$\frac{18}{-2} < \frac{12}{-2}$

$\frac{18}{-6} < \frac{12}{-6}$ ← **When the integer is negative, the direction of the inequality symbol is reversed.**

KEY CONCEPTS Division Property of Inequality

If you divide each side of an inequality by the same positive number, the direction of the inequality symbol remains unchanged.

Arithmetic	Algebra
$9 > 6$, so $\frac{9}{3} > \frac{6}{3}$	If $a > b$, and c is positive, then $\frac{a}{c} > \frac{b}{c}$.
$15 < 20$, so $\frac{15}{5} < \frac{20}{5}$	If $a < b$, and c is positive, then $\frac{a}{c} < \frac{b}{c}$.

If you divide each side of an inequality by the same negative number, the direction of the inequality symbol is reversed.

Arithmetic	Algebra
$16 > 12$, so $\frac{16}{-4} < \frac{12}{-4}$	If $a > b$, and c is negative, then $\frac{a}{c} < \frac{b}{c}$.
$10 < 18$, so $\frac{10}{-2} > \frac{18}{-2}$	If $a < b$, and c is negative, then $\frac{a}{c} > \frac{b}{c}$.

Differentiated Instruction Solutions for All Learners

Special Needs L1
For Example 2, ask students to use a number line, and show jumps in increments of 55 until they get close to 190. Each jump represents 1 hour. It takes 3 jumps to get to 165, and you need another partial jump to get to 190, so 3 1/2 hours makes sense.

learning style: visual

Below Level L2
Have students fill in the blank for several exercises like these.

$-3t > -24$ $5w \leq -15$
t ☐ 8 < w ☐ −3 ≤

learning style: visual

EXAMPLE Solving Inequalities by Dividing

1 Solve $-3y \le -27$. Graph the solution.

$-3y \le -27$

$\frac{-3y}{-3} \ge \frac{-27}{-3}$ ← **Divide each side by −3. Reverse the direction of the symbol.**

$y \ge 9$ ← **Simplify.**

−12 −6 0 6 12 18

1a. $p > -9$;

−11 −10 −9 −8 −7 −6

b. $m \le 3$;

0 1 2 3 4 5

c. $n > -3$;

−5 −4 −3 −2 −1 0

Quick Check

1. Solve each inequality. Graph the solution. **See left.**
   a. $-4p < 36$ b. $-8m \ge -24$ c. $7n > -21$

You can solve an inequality that involves multiplication by dividing each side of the inequality by the same number.

EXAMPLE Application: Planning

2 Your class is taking a trip to a museum that is 190 miles away. The bus can travel at 55 miles per hour. At least how many hours should your class plan for the trip to the museum?

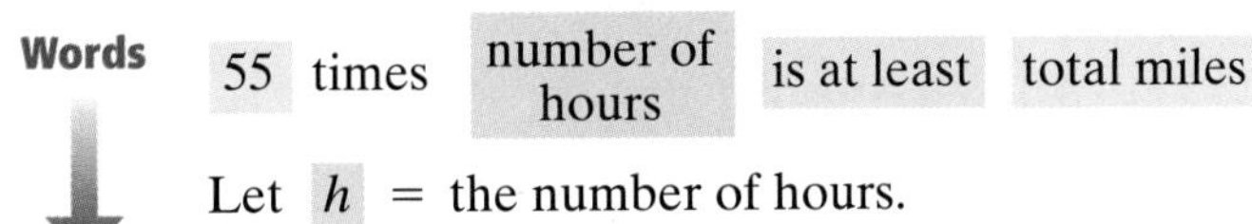

Words
55
times
number of hours
is at least
total miles
Let h = the number of hours.
Expression
55
·
h
≥
190

$55h \ge 190$

$\frac{55h}{55} \ge \frac{190}{55}$ ← **Divide each side by 55.**

$h \ge 3.4545...$ ← **Simplify.**

$h \ge 3.5$ ← **Round up to the nearest half hour.**

Your class should plan for at least 3 hours and 30 minutes.

Quick Check

2. A long-distance telephone company is offering a special rate of \$.06 per minute. Your budget for long-distance telephone calls is \$25 for the month. At most how many minutes of long distance can you use for the month with this rate? **416 min**

The properties of inequality apply to multiplication as well.

2. Teach

Guided Instruction

Example 1
Most students probably are aware that the rules for solving inequalities are very similar to the rules for solving equations. Focus their attention on the difference. Ask:

- *Why do you divide each side by −3?* **to undo the multiplication by −3**
- *Why do you reverse the direction of the inequality symbol?* **You have divided each side by a negative number.**

Error Prevention!

Watch for students who reverse the direction of the inequality whenever they see a negative number, even if the number by which they multiply or divide each side is positive.

Teaching Tip
When solving an inequality, students might find it helpful to recite the rule softly, using an abbreviated form. For example, when multiplying by a positive number, they might say, "Multiply by a positive, keep the same symbol." When dividing by a negative number, they might say, "Divide by a negative, reverse the symbol."

PowerPoint
Additional Examples

1 Solve $6a \le -18$. Graph the solution $a \le -3$

−6 −4 −2 0 2

2 A woodworker makes a profit of \$30 on each picture frame sold. How many frames must he sell to make a profit of at least \$500? **17 frames**

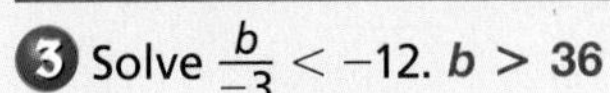

3 Solve $\frac{b}{-3} < -12$. $b > 36$

All in One **Teaching Resources**
- Daily Notetaking Guide 4-9 L3
- Adapted Notetaking 4-9 L1

Advanced Learners L4
Solve each inequality.

$2y > y$ **all positive numbers**
$2y < y$ **all negative numbers**

learning style: visual

English Language Learners ELL
Have students test out and explain why the inequality sign changes direction when you divide by a negative integer. Use $3 < 5$ as an example. If you divide each by −1, then the inequality becomes $-3 > -5$. Have them use and explain other examples to you.

learning style: verbal

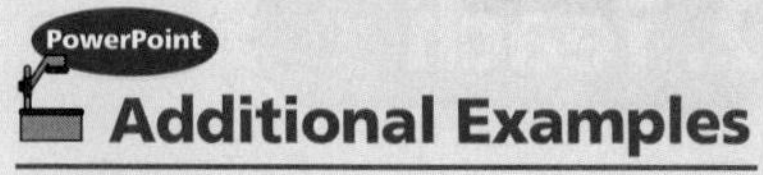

Additional Examples

❸ Solve $\frac{b}{-3} < -12$. $b > 36$

All in One Teaching Resources

- Daily Notetaking Guide 4-9 L3
- Adapted Notetaking 4-9 L1

Closure

- How do the Multiplication and Division Properties of Inequalities differ from the Multiplication and Division Properties of Equalities? **When you multiply or divide each side of an inequality by a negative number, the direction of the inequality symbol is reversed.**

KEY CONCEPTS Multiplication Property of Inequality

If you multiply each side of an inequality by the same positive number, the direction of the inequality symbol remains unchanged.

Arithmetic	Algebra
$12 > 8$, so $12 \cdot 2 > 8 \cdot 2$	If $a > b$, and c is positive, then $a \cdot c > b \cdot c$.
$3 < 6$, so $3 \cdot 4 < 6 \cdot 4$	If $a < b$, and c is positive, then $a \cdot c < b \cdot c$.

If you multiply each side of an inequality by the same negative number, the direction of the inequality symbol is reversed.

Arithmetic	Algebra
$6 > 2$, so $6(-3) < 2(-3)$	If $a > b$, and c is negative, then $a \cdot c < b \cdot c$.
$3 < 5$, so $3(-2) > 5(-2)$	If $a < b$, and c is negative, then $a \cdot c > b \cdot c$.

EXAMPLE Solving Inequalities by Multiplying

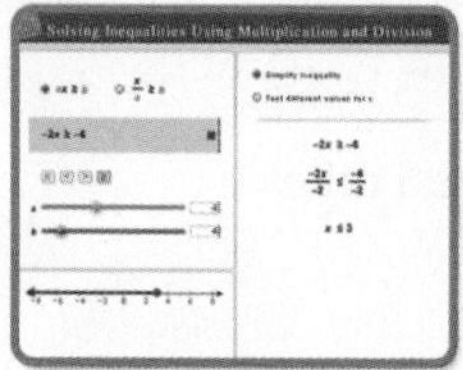

For: Inequalities Activity
Use: Interactive Textbook, 4-9

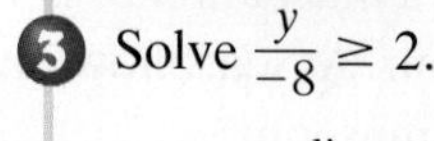

❸ Solve $\frac{y}{-8} \geq 2$.

$$\frac{y}{-8} \geq 2$$

$$-8 \cdot \frac{y}{-8} \leq -8 \cdot 2$$ ← **Multiply each side by −8. Reverse the direction of the symbol.**

$$y \leq -16$$ ← **Simplify.**

Quick Check

3. Solve $\frac{k}{-5} < -4$. Graph the solution.

$k > 20$

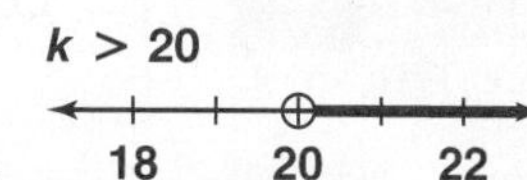

Check Your Understanding

1. **The inequality symbol is reversed.**

1. **Vocabulary** What happens when you multiply each side of an inequality by a negative number?
2. If $x > y$, which statement is NOT always true? **D**
   - Ⓐ $y < x$
   - Ⓑ $x + z > y + z$
   - Ⓒ $x - z > y - z$
   - Ⓓ $xz > yz$

Fill in the missing inequality symbol.

3. $3m > 99$
   $\frac{3m}{3}$ ■ $\frac{99}{3}$ **>**
4. $-8z \leq 80$
   $\frac{-8z}{-8}$ ■ $\frac{80}{-8}$ **≥**
5. $\frac{d}{-3} < 12$
   $\frac{d}{-3} \cdot (-3)$ ■ $12 \cdot (-3)$ **>**

Homework Exercises

For more exercises, see Extra Skills and Word Problems.

GO for Help

For Exercises	See Examples
6–18	1–2
19–30	3

6. $h < 4$;

1 2 3 4 5 6

7. $p > 12$;

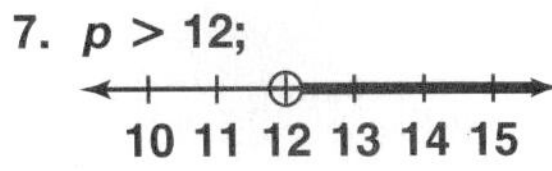

10 11 12 13 14 15

8. $n \leq -3$;

−6 −5 −4 −3 −2 −1

31. $c \cdot 70 \leq 6{,}000$; at most 85 cases

A **Solve each inequality by dividing. Graph the solution.**

6–8. See left.

6. $4h < 16$
7. $3p > 36$
8. $9n \leq -27$

9–17. See margin.

9. $6x < -48$
10. $-8b \geq -24$
11. $-5w \leq 30$
12. $-10d \geq -70$
13. $-2t > 10$
14. $5g < -35$
15. $-4.5p \leq 22.5$
16. $7y < -42.7$
17. $8.3w \geq 53.95$
18. A photo album page can hold six photographs. You have 296 photographs. How many pages do you need? at least 50 pages

19–30. See margin.

Solve each inequality by multiplying. Graph the solution.

19. $\frac{p}{5} < -3$
20. $\frac{k}{4} \geq 6$
21. $\frac{w}{7} \leq -3$
22. $\frac{y}{7} > -8$
23. $\frac{n}{-2} > -5$
24. $\frac{m}{-6} \leq 5$
25. $\frac{x}{10} < -4$
26. $\frac{c}{-5} \geq -2$
27. $\frac{x}{4} > -20$
28. $\frac{g}{1.2} \geq -7$

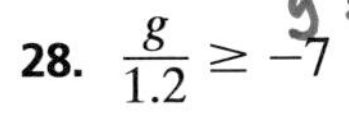

29. $\frac{p}{-8} < 2.1$
30. $\frac{f}{5} \leq -5.5$

GPS **B**

31. **Guided Problem Solving** A forklift can safely carry as much as 6,000 lb. A case of paint weighs 70 lb. At most how many cases of paint can the forklift safely carry at one time? See left.
   - Use number sense: How much do 100 cases weigh?
   - Try the strategy *Systematic Guess and Check*: How much do 90 cases weigh? 80 cases?
32. **Rides** A roller coaster can carry 36 people per run. At least how many times does the roller coaster have to run to allow 10,000 people to ride? at least 278 times
33. **Baking** A recipe for an apple pie calls for 6 apples per pie. You have 27 apples. At most how many apple pies can you make? at most 4 pies

Write an inequality for each sentence. Solve the inequality.

34. The product of −3 and a number is greater than 12. $-3x > 12$; $x < -4$
35. A number multiplied by 4 is at most −44. $4x \leq -44$; $x \leq -11$
36. A number divided by −9 is less than 10. $\frac{x}{-9} < 10$; $x > -90$
37. The quotient of a number and 5 is at least −8. $\frac{x}{5} \geq -8$; $x \geq -40$
38. The product of 2 and a number is less than −10. $2n < -10$; $n < -5$
39. **Writing in Math** Explain how solving $-5x < 25$ is different from solving $5x < 25$. See left.

39. To solve $-5x < 25$, divide each side by −5 and reverse the inequality symbol. To solve $5x < 25$, divide each side by 5 but do not reverse the symbol.

GO Online
Homework Video Tutor
Visit: PHSchool.com
Web Code: are-0409

9–17. See back of book.
19–30. See back of book.

3. Practice

Assignment Guide

Check Your Understanding
Go over Exercises 1–5 in class before assigning the Homework Exercises.

Homework Exercises

A Practice by Example	6–30
B Apply Your Skills	31–48
C Challenge	49
Test Prep and Mixed Review	50–53

Homework Quick Check
To check students' understanding of key skills and concepts, go over Exercises 13, 25, 32, 39, and 48.

Differentiated Instruction Resources

Adapted Practice 4-9 L1

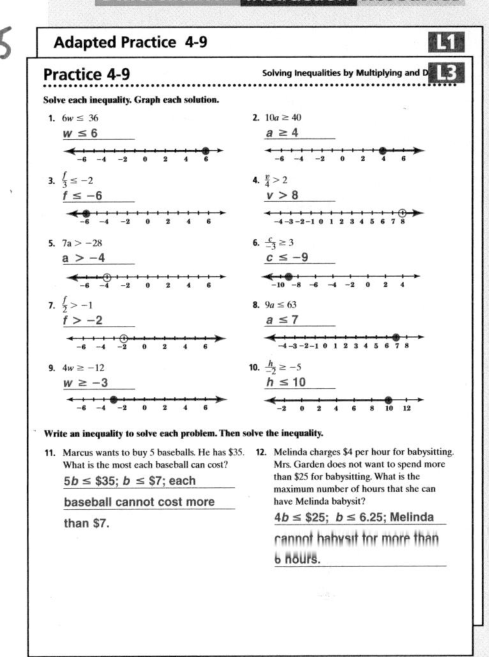

Practice 4-9 — Solving Inequalities by Multiplying and Dividing — L3

Solve each inequality. Graph each solution.

1. $6w \leq 36$ — $w \leq 6$
2. $10a \geq 40$ — $a \geq 4$
3. $\frac{f}{3} \leq -2$ — $f \leq -6$
4. $\frac{v}{4} > 2$ — $v > 8$
5. $7a > -28$ — $a > -4$
6. $\frac{c}{-3} \geq 3$ — $c \leq -9$
7. $\frac{f}{2} > -1$ — $f > -2$
8. $9a \leq 63$ — $a \leq 7$
9. $4w \geq -12$ — $w \geq -3$
10. $\frac{h}{-2} \geq -5$ — $h \leq 10$

Write an inequality to solve each problem. Then solve the inequality.

11. Marcus wants to buy 5 baseballs. He has $35. What is the most each baseball can cost? $5b \leq \$35$; $b \leq \$7$; each baseball cannot cost more than $7.
12. Melinda charges $4 per hour for babysitting. Mrs. Garden does not want to spend more than $25 for babysitting. What is the maximum number of hours that she can have Melinda babysit? $4b \leq \$25$; $b \leq 6.25$; Melinda cannot babysit for more than 6 hours.

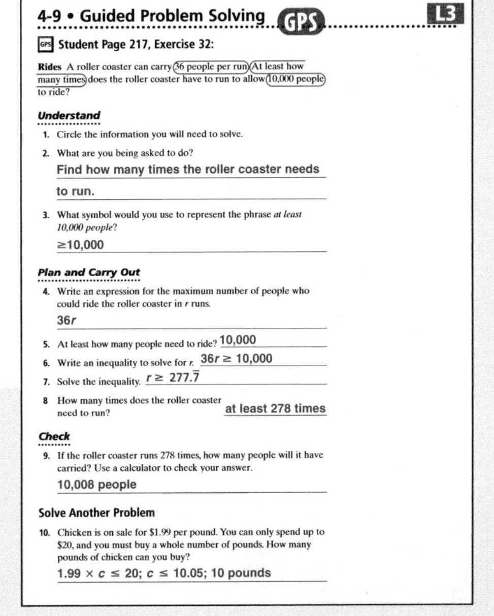

4-9 • Guided Problem Solving GPS L3

Student Page 217, Exercise 32:

Rides A roller coaster can carry 36 people per run. At least how many times does the roller coaster have to run to allow 10,000 people to ride?

Understand

1. Circle the information you will need to solve.
2. What are you being asked to do? Find how many times the roller coaster needs to run.
3. What symbol would you use to represent the phrase *at least 10,000 people*? ≥10,000

Plan and Carry Out

4. Write an expression for the maximum number of people who could ride the roller coaster in *r* runs. $36r$
5. At least how many people need to ride? 10,000
6. Write an inequality to solve for *r*. $36r \geq 10{,}000$
7. Solve the inequality. $r \geq 277.\overline{7}$
8. How many times does the roller coaster need to run? at least 278 times

Check

9. If the roller coaster runs 278 times, how many people will it have carried? Use a calculator to check your answer. 10,008 people

Solve Another Problem

10. Chicken is on sale for $1.99 per pound. You can only spend up to $20, and you must buy a whole number of pounds. How many pounds of chicken can you buy? $1.99 \times c \leq 20$; $c \leq 10.05$; 10 pounds

4. Assess & Reteach

Lesson Quiz

Solve each inequality. Graph the solution.

1. $8t < 32$ $t < 4$

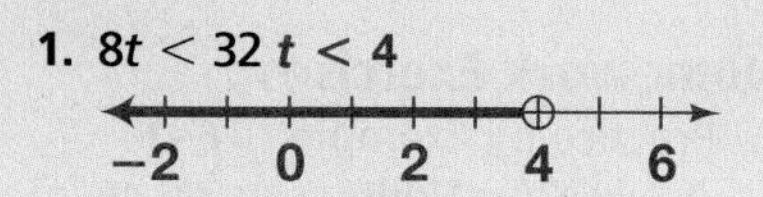

2. $\frac{s}{6} > 30$ $s > 180$

150 180 210 240

3. $-7v \geq 35$ $v \leq -5$

−8 −6 −4 −2 0

4. $\frac{x}{-3} < 7$ $x > -21$

−24 −22 −20 −18 −16

Reteaching 4-9

Solving Inequalities by Multiplying and ... L2

To solve an inequality you can multiply or divide each side by the same number. However, if the number is negative, you must also reverse the direction of the inequality sign.

Solve $-4y \geq 16$. Graph the solution.

$-4y \geq 16$

$\frac{-4y}{-4} \leq \frac{16}{-4}$ Divide each side by −4. Reverse the direction of the inequality symbol.

$y \leq -4$ Simplify.

Graph:

−6 −4 −2 0 2 4 6

Solve $\frac{w}{3} > 2$. Graph the solution.

$\frac{w}{3} > 2$

$(3)\frac{w}{3} > 2(3)$ Multiply each side by 3.

$w > 6$ Simplify.

Graph:

−6 −4 −2 0 2 4 6

Solve each inequality. Graph the solution.

1. $2a > 10$ $a > 5$
2. $-4w < 16$ $w > -4$
3. $\frac{r}{2} \geq -2$ $r \geq -4$
4. $\frac{a}{3} < 1$ $a < 3$
5. $6g < 6$ $g < 1$
6. $-3x \geq -6$ $x \leq 2$
7. $\frac{m}{-2} > 0$ $m < 0$

Enrichment 4-9

Solving Inequalities by Multiplying and ... L4

Critical Thinking

Many companies throughout the United States and Europe have been rating their customer management performance through a benchmark program. This program measures the average performance of customer satisfaction in various different categories.

The table to the right shows the average scores a computer company has received so far this year in their Information and Technology department. How does the company need to score in *managing customer information* to achieve an overall average score greater than 40?

Information and Technology	Average Score
Acquiring customer information	49
Managing customer information	?
Current system functions	32
Developing new systems	52

Source: http://www.qci.co.uk/Home_page.asp

1. Write an inequality to represent the overall score that the company wants to achieve. **average score > 40**
2. What operations do you use to find an average? **addition and division**
3. Let x represent the score of *managing customer information*. Write an inequality that represents the desired average of the Information and Technology department. $\frac{x + 49 + 32 + 52}{4} > 40$
4. Solve the inequality in step 3. $x > 27$
5. Based on the other scores in the table, do you think the company can achieve their goal? Explain why or why not. **Sample answer: Yes; all of the other averages listed in the table are well above 27. So it does not appear that the company should have trouble achieving their goal.**
6. To be awarded a national certificate of achievement, the computer company must have an overall average in their Information and Technology Department of no less than 51.
   a. Write and solve an inequality to find what the company must score in *managing customer information* to achieve the certificate. $\frac{x + 49 + 32 + 52}{4} \geq 51$; $x \geq 71$
   b. Based on the other scores in the table, do you think the company can achieve the certificate? Explain why or why not. **Sample answer: No; the averages listed in the table are well below 71. It does not appear that the company will achieve the certificate.**

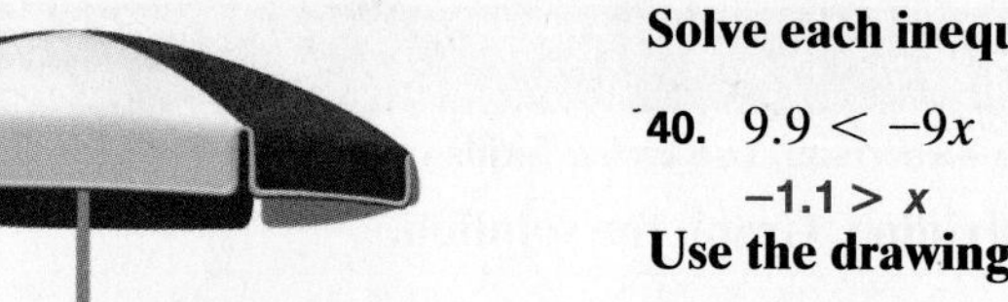

Solve each inequality.

40. $9.9 < -9x$ **$-1.1 > x$**
41. $-25.1 > \frac{t}{-2}$ **$50.2 < t$**
42. $-5.6 \leq -8p$ **$0.7 \geq p$**

Use the drawing at the left for Exercises 43–45. You have $15.

43. At most how many hot dogs can you buy? **at most 3 hot dogs**
44. At most how many bags of peanuts can you buy? **at most 12 bags of peanuts**
45. You buy two hot dogs. How many bags of peanuts can you buy? How much money do you have left? **4 bags of peanuts; $.50**
46. A 1-ton truck can haul 2,000 lb. A refrigerator weighs 302 lb. How many refrigerators can the truck carry? **at most 6 refrigerators**
47. **Error Analysis** A student solves the inequality $5n > -25$. He says the solution is $n < -5$. Explain the student's error. **See left.**
48. **Reasoning** Solve and graph $-18 \geq -2y$ and $-2y < -18$. Are the solutions the same? Explain. **See back of book.**
49. Ⓒ **Challenge** Ten more than −3 times a number is greater than 19. Write and solve an inequality. Graph the solution. **See back of book.**

47. The student should not have reversed the inequality symbol.

Test Prep and Mixed Review

Practice

Multiple Choice

50. Which problem situation matches the equation below? **A**

$$2x + 3 = 21$$

Ⓐ Janice is 3 years more than twice as old as Rashon. Janice is 21 years old. How old is Rashon?
Ⓑ Felicity has scored 2 points more than 3 times as many points as Kendra. Kendra has scored 21 points. How many points has Felicity scored?
Ⓒ Drew paid $2 more than Adam for tickets to a play. Richard's tickets cost him 3 times as much as Drew's tickets. Richard paid $21. How much did Adam pay for his tickets?
Ⓓ Nate worked 2 hours more than 3 times as many hours as Chad. Nate worked 21 hours. How many hours did Chad work?

51. Arnold had a collection of baseball cards that he divided evenly among 5 friends. Each friend received 17 baseball cards. Which equation can be used to find y, the number of baseball cards Arnold had? **G**

Ⓕ $5y = 17$ Ⓗ $17y = 5$
Ⓖ $\frac{y}{5} = 17$ Ⓙ $\frac{y}{5} - 5 = 17$

GO for Help

For Exercises	See Lesson
52–53	1-10

Find the mean, median, and mode for each set of data.

52. 4 6 3 7 4 7 4 4 **4.875; 4; 4**
53. 2.2 6.4 5 2.2 2.2 5 **$3.8\overline{3}$; 3.6; 2.2**

Test Prep

Resources

For additional practice with a variety of test item formats:

- Test-Taking Strategies, p. 219
- Test Prep, p. 223
- Test-Taking Strategies with Transparencies

Alternative Assessment

Ask students to write four different inequalities that all have the same solution. Two inequalities should be solved by multiplying. The two other inequalities should be solved by dividing. Students then trade papers with a partner who solves each inequality and graphs the solution. Have students discuss their results.

Writing Short Responses

Short-response questions are usually worth a maximum of 2 points. To get full credit, you need to give the correct answer, including appropriate units. You may also need to show your work or justify your reasoning.

The cost for using a phone card is 45¢ per call plus 5¢ per minute. A recent call cost $2.05. Write and solve an equation to find the length of the call.

To get full credit you must use a variable to set up an equation, solve the equation, and find the length of the call. Below is a scoring guide that shows the number of points awarded for different answers.

Scoring

[2] The equation and the solution are correct. The call took 32 minutes.

[1] There is no equation, but there is a method to show that the call took 32 minutes, OR There is an equation and a solution, both of which may contain minor errors.

[0] There is no response, or the response is completely incorrect.

Three responses are shown below with the points each one received.

2 points	1 point	0 points
Let m represent the number of minutes. $205 = 45 + 5m$ $160 = 5m$ $32 = m$ The call took 32 minutes.	$\frac{2.05 - 0.45}{0.05}$ 32 minutes	30 minutes

Exercises

Use the scoring guide above to answer each question. 1–2. See margin.

1. Explain why each response above received the indicated points.
2. Write a 2-point response that includes the equation $0.05n + 0.45 = 2.05$.

1. Answers may vary. Sample: The 2-point answer defines and uses a variable, solves an equation, and finds the length of the call. The 1-point answer finds the length of the call, but there is no equation or defined variable. The 0-point answer shows an incorrect answer and no work.

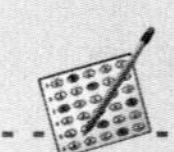

Writing Short Responses

This strategy provides students with a rubric and an example showing how to get full credit for answers to short-response questions.

Guided Instruction

Teaching Tip

Guide students to use estimation and common number sense to make sure their answers make sense. In this case, 32 minutes is a reasonable answer, whereas, 0.032 minutes, 0.32 minutes, 320, or 3,200 minutes are not reasonable times for a call costing $2.05.

Error Prevention!

Some students may not notice that the equation $205 = 45 + 5m$ uses whole numbers to make the computations easier.

2. Answers may vary. Sample: Let n = minutes.

$$0.05n + 0.45 = 2.05$$
$$0.05n + 0.45 - 0.45 = 2.05 - 0.45$$
$$0.05n = 1.6$$
$$\frac{0.05n}{0.05} = \frac{1.6}{0.05}$$
$$n = 32$$

The call took 32 min.

Test-Taking Strategies with Transparencies

Test-Taking Strategies: Writing Short Responses

Estimate 98.57 × 206. Write your estimate and explain in writing how you got it.

Scoring Guide

2 Explains method, with answer that matches method.

1 Gives estimate with no explanation, OR gives explanation with no answer, OR shows computation, OR rounds first, but not enough to make computation easy.

0 Computes exact answer, then rounds, OR gives incorrect response.

Answer earning 2 points	Answer earning 1 point
Round: 98.57 → 100 206 → 200 Estimate = 20,000 First round, then multiply the rounded numbers.	Estimate: 98.57 206 206 × 99 1854 1854 20394

Answer earning 0 points

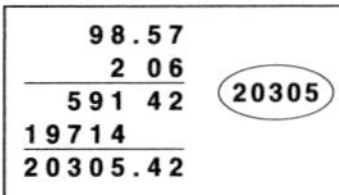

98.57 × 206 = 59142, 19714, 20305.42 → 20305

Transparency 3

Chapter 4 Review

Resources

Student Edition
Extra Skills and Word Problems Practice, Ch. 4, p. 630
English/Spanish Glossary, p. 676
Formulas and Properties, p. 674
Tables, p. 670

All in One Teaching Resources
- Vocabulary and Study Skills 4F 

Differentiated Instruction
Spanish Vocabulary Workbook with Study Skills ELL
Interactive Textbook
- Audio Glossary

Online Vocabulary Quiz

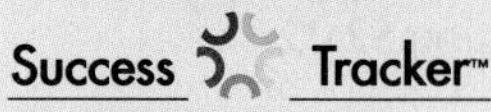

Online at PHSchool.com

Spanish Vocabulary/Study Skills ELL

Vocabulary/Study Skills L3

4F: Vocabulary For use with the Chapter Review

Study Skill: Review notes that you have taken in class as soon as possible to clarify any points you missed and to refresh your memory.

Circle the word that best completes the sentence.

1. A (variable, expression) is a letter that stands for a number.
2. An (expression, equation) is a mathematical statement with an equal sign.
3. A (solution, sentence) is a value for a variable that makes an equation true.
4. To solve an equation, use (inverse, variable) operations.
5. A mathematical statement that contains < or > is called an (equation, inequality).
6. The statement 4 + (9 + 3) = (4 + 9) + 3 is an example of the (Commutative Property of Addition, Associative Property of Addition).
7. The (opposite, absolute value) of 15 is −15.
8. The (mean, median) is the middle number in a data set when the data is arranged from least to greatest.
9. You can use the (commutative, identity) property to change the order in an expression.
10. The statement $a + 0 = a$ is an example of the (Identity Property of Zero, Identity Property of Multiplication).
11. The (absolute value, opposite) of a number is its distance from 0 on a number line.
12. (Whole, Integers) are the set of whole numbers, their opposites, and zero.
13. Two numbers whose sum is 0 are (additive, opposite) inverses.
14. A(n) (outlier, range) is a data value that is much greater or less than the other values in the data set.
15. Using the (distributive property, order of operations), you can calculate that 12 + 5 · 2 equals 22.

Vocabulary and Study Skills

Chapter 4 Review

Vocabulary Review

Addition Property of Equality (p. 180)
Addition Property of Inequality (p. 210)
algebraic expression (p. 169)
equation (p. 174)
Division Property of Equality (p. 186)
Division Property of Inequality (p. 214)
inequality (p. 205)
inverse operations (p. 181)
Multiplication Property of Equality (p. 188)
Multiplication Property of Inequality (p. 216)
open sentence (p. 174)
solution of an equation (p. 174)
solution of an inequality (p. 205)
Subtraction Property of Equality (p. 180)
Subtraction Property of Inequality (p. 211)
variable (p. 169)

Choose the correct term to complete each sentence.

1. A letter that represents a number is called a(n) (equation, variable). variable
2. If you multiply both sides of an (equation, inequality) by the same negative number, the direction of the inequality symbol is reversed. inequality
3. A mathematical sentence with an equal sign is called a(n) (equation, inverse operation). equation
4. A mathematical phrase with at least one variable in it is a(n) (algebraic expression, solution of an equation). algebraic expression
5. A(n) (open sentence, solution of an inequality) is a value that makes an inequality true. solution of an inequality

Go Online PHSchool.com
For: Online vocabulary quiz
Web Code: arj-0451

Skills and Concepts

Lesson 4-1
- To write and evaluate algebraic expressions

A **variable** is a letter that stands for a number. An **algebraic expression** is a mathematical phrase that uses variables, numbers, and operation symbols. To evaluate an expression, substitute a given value for each variable and then simplify.

Evaluate each expression for $n = 3$, $p = 5$, and $w = 2$.

6. $3n - 2w$ 5
7. $\frac{4n}{w}$ 6
8. $p + 4w$ 13
9. $7w - 2p$ 4

Lesson 4-2
- To solve one-step equations using substitution, mental math, and estimation

An **equation** is a mathematical sentence with an equal sign. An equation with one or more variables is an **open sentence.** A **solution of an equation** is a value for a variable that makes an equation true.

Use number sense to solve each equation.

10. $p + 5 = -2$ −7
11. $m - 12 = 8$ 20
12. $7t = 28$ 4
13. $\frac{w}{8} = 9$ 72

Lessons 4-3, 4-4

- To solve equations by adding or subtracting
- To solve equations by mulitplying or dividing

Whatever you do to one side of the equation, you must do the same thing to the other side of the equation. To solve a one-step equation, use **inverse operations.**

Use inverse operations to solve each equation.

14. $y + 14 = 38$ 24 **15.** $p - 12 = 72$ 84 **16.** $\frac{m}{11} = 9$ 99 **17.** $-7b = 84$ −12

18. $x - 8 = 44$ 52 **19.** $12h = 60$ 5 **20.** $k - 14 = 29$ 43 **21.** $\frac{n}{6} = -9$ −54

22. A local band had a concert and made $1,824 from tickets sales. If each ticket cost $16, how many tickets did the band sell? **114 tickets**

Lessons 4-5, 4-6

- To write and evaluate expressions with two operations and to solve two-step equations using number sense
- To solve two-step equations using inverse operations

You can also use inverse operations to solve two-step equations.

Solve each equation.

23. $4d + 7 = 11$ 1 **24.** $2m - 21 = 3$ 12 **25.** $-5y + 8 = 23$ −3

26. Savings You save $35 each week. You now have $140. You plan to save enough money for a cruise that costs $1,050. Write and solve an equation to find the number of weeks it will take to save for the cruise. **140 + 35w = 1,050; 26 wk**

Lesson 4-7

- To graph and write inequalities

A mathematical sentence that contains $<$, $>$, $\leq$, $\geq$, or $\neq$ is called an **inequality.** A **solution of an inequality** is any number that makes an inequality true. When graphing, use an open circle for $>$ and $<$ and use a closed circle for $\geq$ and $\leq$.

27–28. See margin.

Write an inequality for each statement. Then graph the inequality.

27. The ticket is at most $10. **28.** The race is less than 5 miles.

Lessons 4-8, 4-9

- To solve inequalities by adding or subtracting
- To solve inequalities by multiplying or dividing

You solve inequalities just as you solve equations, but with one exception. When you multiply or divide by a negative number, you must reverse the direction of the inequality symbol.

Solve each inequality. Then graph the inequality. 29–33. See margin.

29. $h + 7 < -15$ **30.** $\frac{p}{5} \geq -3$ **31.** $g - 14 > 3$ **32.** $-4m \leq 28$

33. A farmer needs no less than 20 lb of seeds to plant a crop. Write an inequality for this situation. Then graph the inequality.

27–33. See back of book.

Alternative Assessment L4

Alternative Assessment Form C

Chapter 4

ALGEBRA AT THE AMUSEMENT PARK

You just walked into the Amazing Amusement Park and are ready to buy tickets for the rides and games. The park sells tickets based on the "class" of a ride or a game. The chart below lists the names of the rides and games in the park in three classes: A, B, and C. One Class A ticket costs $1.50, one Class B ticket costs $2.00, and one Class C ticket costs $3.00.

Class A	Class B	Class C
Spider Swing	Bump Cars	Roller Coaster
Wiggly Worm	MicroCoaster	Ferris Wheel
Creepy Crawler	Water Roller	Wacky Wheel
Bongo Bears	Balloon Pop	Haunted House
Dancing Does	Basket Throw	Magic Mouse
Hoppity Hip	Jungle Gym	On the Wet Side

Show all of your work on a separate sheet of paper.

1. You asked your friend Marina how much she was going to spend on tickets. She decided to play a game with you. On a piece of paper, she wrote this algebraic expression:

$$2A + 3B + 2C$$

She said, "A is the cost of a Class A ticket, B is the cost of a Class B ticket, and C is the cost of a Class C ticket." How much was Marina going to spend altogether? Show your work.

2. You may buy a total of six tickets. Which six activities would you choose? List each activity and the class of ticket it requires. (Although there may be some rides or games listed that you have never heard of, you should be able to imagine what each is like based on its name.)

3. a. Marina asked you how much you were going to spend. Playing Marina's game, write the algebraic expression that shows how much you would have to spend on the six rides and games you chose in Question 2 above.

   b. How much will all these activities cost you?

4. Max asked how much both you and Marina were going to spend on tickets. Marina continued to play her game. She wrote an expression with three terms that represented the amount the two of you were going to spend together. What expression might she have written? Be sure to write the expression in simplest terms.

Resources

- ExamView Assessment Suite CD-ROM
  - Chapter 4 Ready-Made Test
  - Make your own Chapter 4 test
- MindPoint Quiz Show CD-ROM
  - Chapter 4 Review

Differentiated Instruction

All in One Teaching Resources

- Below Level Chapter 4 Test L2
- Chapter 4 Test L3
- Chapter 4 Alternative Assessment L4

Spanish Assessment Resources ELL

- Below Level Chapter 4 Test L2
- Chapter 4 Test L3
- Chapter 4 Alternative Assessment L4

ExamView Assessment Suite CD-ROM

- Special Needs Test L1
- Special Needs Practice Bank L1

Online Chapter 4 Test at www.PHSchool.com L3

Below Level Chapter Test L2

Chapter Test L3

Chapter Test Form A

Chapter 4

Evaluate each expression using the values $w = 4$, $a = 2$, and $h = 5$.

1. $w + a$ 6
2. $h + 2w$ 13
3. $w \div a$ 2
4. $6w + 2a$ 28
5. $3(h + w)$ 27
6. $5w + 2a$ 24

Solve each equation.

7. $x + 8 = 13$ $x = 5$
8. $m - 9 = 21$ $m = 30$
9. $5a = 55$ $a = 11$
10. $9y = 81$ $y = 9$
11. $6 + n = 20$ $n = 14$
12. $\frac{y}{4} = 8$ $y = 32$
13. $\frac{y}{-3} + 2 = -13$ $y = 45$
14. $16 + 2n = 10$ $n = -3$

Write an equation for each problem. Solve the equation.

15. The math club made a total of $1,155 from selling 210 T-shirts. How much did they sell each T-shirt for? $210p = 1{,}155$; $5.50
16. A youth baseball league recently completed registration. The league was able to form 15 teams of 11 players each. How many players registered? $p \div 11 = 15$; 165 players

Chapter 4 Test

For: Online chapter test
Web Code: ara-0452

Evaluate each expression for $n = 3$, $t = -2$, and $y = 4$.

1. $3n + 2t$ 5
2. $5y - 4n$ 8
3. $2(n + 3y)$ 30
4. $ny - 6$ 6
5. $\frac{2ny}{t}$ −12
6. $-2t + 6n$ 22

Solve each equation.

7. $x + 7 = 12$ 5
8. $m - \frac{1}{3} = \frac{1}{6}$ $\frac{1}{2}$
9. $13 + d = 44$ 31
10. $p - 1.8 = 6.2$ 8
11. $5n = 45.5$ 9.1
12. $\frac{h}{7} = 8$ 56
13. $-\frac{1}{3}t = 24$ −72
14. $\frac{k}{-4} = -12$ 48
15. $14 + 3n = 8$ −2
16. $9h - 21 = 24$ 5
17. $\frac{w}{5} - 10 = -4$ 30
18. $14 + \frac{y}{8} = 10$ −32

Define a variable. Then write and solve an equation for each problem. 19–24. See margin.

19. **Masonry** A mason is laying a brick foundation 72 in. wide. Each brick is 6 in. wide. How many bricks will the mason need across the width of the foundation?
20. **Groceries** You buy 12 apples. You also buy a box of cereal that costs $3.35. The bill is $8.75. How much does each apple cost? $0.45
21. The music boosters sell 322 music buttons and raise $483 for the music department. How much does each button cost?
22. Your family drives from Austin, Texas, to Tampa, Florida. The trip is about 1,145 mi and lasts four days. How many miles must your family drive each day? 286.25 miles
23. **Sports** A youth soccer league recently held registration. The number of players was divided into 13 teams of 12 players. How many players registered?
24. Six friends split the cost of a party. Each person also spends $65 for a hotel room. Each person spends $160 on the party and a room. What is the cost of the party?

25. **Open-Ended** Write a problem you can represent with $2k - 10 = 6$. Solve the equation. Show your work. 8; check students' work.

Define a variable and write an inequality for each statement.

26. The game's duration is at most 3 hours. Let d = game's duration; $d \le 3$.
27. To rent a car, you must be at least 25 years old. Let a = age; $a \ge 25$.
28. The truck can haul more than 20,000 lb. Let h = weight hauled; $h > 20{,}000$.
29. There are fewer than 10 tickets available for the concert tonight. Let t = number of available tickets; $t < 10$.

Solve each inequality. Graph your solution. 30–39. See margin.

30. $n + 12 \ge 15$
31. $y - 14 > 10$
32. $m - 8 \le -17$
33. $w + 22 < 14$
34. $12x < -48$
35. $10h \ge 90$
36. $\frac{v}{-8} > 6$
37. $\frac{p}{11} \le -6$
38. $-7k \le -84$
39. $\frac{x}{-15} < -4$

Define a variable. Then write and solve an inequality for each problem. 40–43. See margin.

40. **Transportation** A ferry can safely transport at most 220 people. There are already 143 people aboard. How many more people can the ferry take aboard?
41. **Money** A sports drink costs $1.49 per bottle. At most how many bottles can you buy if you have $12?
42. **Shopping** Alex and his mother spent at least $130 while shopping for new clothes. Alex spent $52. How much money did his mother spend?
43. **Writing in Math** Describe the similarities and differences between solving inequalities and equations. Include examples.

19–24. See back of book.
30–43. See back of book.

Multiple Choice

Read each question. Then write the letter of the correct answer on your paper.

1. Which pair of numbers has a product that is greater than its sum? **A**
   (A) $-2, -5$ (B) $-3, 3$ (C) $4, -2$ (D) $0, 1$

2. What are the prime factors of 136? **G**
   (F) 2, 2, 34 (G) 2, 2, 2, 17 (H) 2, 2, 3, 17 (J) 2, 3, 3, 15

3. What is the mean of the temperatures shown below? **B**
   92° 87° 79° 85° 92°
   (A) 85° (B) 87° (C) 90° (D) 92°

4. You buy a package of socks for \$4.89, a T-shirt for \$7.75, and a pair of shorts for \$14.95. Which is the best estimate of the amount of change you will get back if you give the cashier \$30? **G**
   (F) about \$1 (G) about \$2 (H) about \$3 (J) about \$4

5. What is the solution of $5y + 11 = 56$? **B**
   (A) 8 (B) 9 (C) 13 (D) 10

6. Which list is in order from least to greatest? **G**
   (F) $\frac{1}{3}, \frac{3}{4}, \frac{1}{5}$ (G) $5\frac{1}{8}, 5\frac{1}{4}, 5\frac{2}{3}$ (H) $\frac{13}{12}, \frac{4}{5}, \frac{3}{7}$ (J) $\frac{8}{9}, \frac{2}{5}, \frac{1}{2}$

7. The average person drinks about 2.5 quarts of water each day. How many pints is this? **D**
   (A) 2 pt (B) 3 pt (C) 4 pt (D) 5 pt

8. Which variable expression can be described by the word phrase "r increased by 2"? **H**
   (F) $2r$ (G) $r - 2$ (H) $r + 2$ (J) $r \cdot 2$

9. Which decimal is closest to $\frac{11}{16}$? **A**
   (A) 0.687 (B) 0.69 (C) 0.68 (D) 0.7

10. Which is the best estimate of $6\frac{3}{4} \cdot 3\frac{1}{5}$? **G**
    (F) 18 (G) 21 (H) 24 (J) 27

11. Your friend divided 0.56 by 0.7 and got 8 for an answer. This answer is not correct. Which answer is correct? **C**
    (A) 0.08 (B) 0.008 (C) 0.8 (D) 80

12. Which graph shows the solution of $d + 10 \leq 19$? **J**

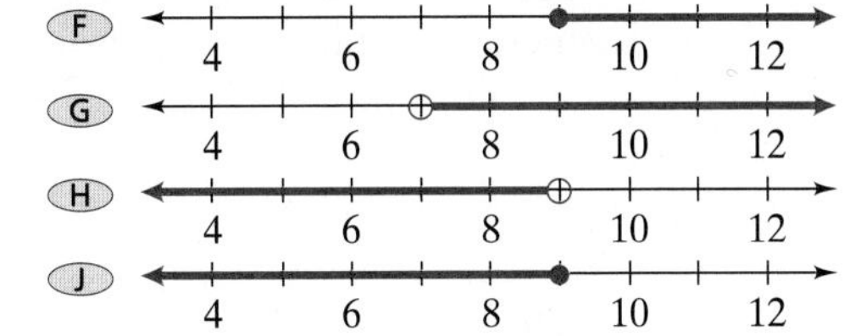

Gridded Response

Record your answer in a grid.

13. During the summer, you work 27 hours per week. Each week, you earn \$168.75. How many dollars do you earn per hour? **6.25**

Short Response 14–15. See margin.

14. Define a variable and write an inequality to model the word sentence "Today's temperature will be at least 55°F."

15. You and four friends are planning a surprise birthday party. Each of you contributes the same amount of money m for the food.
    a. Write a variable expression for the total amount of money contributed for food.
    b. Evaluate your expression for $m = \$7.75$.

Extended Response 16a–b. See margin.

16. Eli buys 2 vinyl records each month. He started with 18 vinyl records.
    a. Write an expression to model the number of records Eli has after x months.
    b. How many records will Eli have after 22 months? Justify your answer.

Item	1	2	3	4	5	6	7	8	9	10	11	12	13	14	15	16
Lesson	1-8	2-2	1-10	1-1	4-5	2-4	3-6	4-1	2-6	3-1	1-4	4-7	1-4	4-7	4-1	4-5

14. [2] Let t = temperature; $t \geq 55$.
[1] does not define variable OR does not write inequality

15. [2] $5m$; $5(\$7.75) = \$38.75$
[1] appropriate variable expression with a computational error OR incorrect expression but evaluated correctly

Test Prep

Resources

Test Prep Workbook

All in One Teaching Resources
- Cumulative Review L3

ExamView Assessment Suite CD-ROM
- Standardized Test Practice

Differentiated Instruction

Progress Monitoring Assessments
- Benchmark Test 2 L3

Spanish Assessment Resources
- Spanish Cumulative Review ELL

ExamView Assessment Suite CD-ROM
- Special Needs Practice Bank L1

16. [4] $18 + 2x$; $18 + 2(22) = 62$; 62 records
[3] appropriate methods, but with one computational error
[2] incorrect expression evaluated correctly OR correct expression evaluated incorrectly
[1] expression with no work OR answer with no work

Spanish Cumulative Review ELL

Cumulative Review L3

Cumulative Review

Chapter 1–2

Multiple choice. Choose the letter of the best answer.

1. Which numbers are in order from least to greatest?
   (A.) 1.1, 1.2, 1.25, 1.415
   B. 0.2, 0.02, 2.2, 0.22
   C. 0.3, 0.6, 0.51, 0.65
   D. 3.5, 3.41, 3.02, 3.6
2. Karl's scores for five video games were 38, 57, 48, 62, and 55. What is his mean score?
   F. 24 G. 38 (H.) 52 J. 55
3. Which difference is greatest?
   A. $2.5 - 2.3$
   (B.) $5.5 - 4.15$
   C. $3.75 - 2.95$
   D. $8.05 - 7.85$
4. Which number is the outlier?
   25 34 63 27 31
   F. 34 G. 25 (H.) 63 J. 36
5. Ten identical coins weigh 34.5 g. How much does one coin weigh?
   A. 0.345 g (B.) 3.45 g C. 345 g D. not here
6. What is the mode of this data?
   0.12 0.21 0.15 0.51 0.51 0.25 0.52
   A. 0.25 (B.) 0.40 C. 0.51 D. 0.52
7. Which sum is positive?
   F. $-56 + (-4)$ G. $81 + (-90)$ (H.) $-48 + 55$ J. $-76 + 45$
8. Which of the following is equal to 16?
   (A.) $(-2)^4$
   B. -2^3
   C. -4^2
   D. 8^2
9. Which fraction below is not equivalent to any of the other choices?
   F. $\frac{3}{8}$
   G. $\frac{9}{24}$
   (H.) $\frac{24}{72}$
   J. $\frac{15}{40}$
10. What is the product of 2.3 and 3.45?
    A. 0.7935
    B. 1.15
    C. 1.5
    (D.) 7.935

Problem Solving

Applying Inequalities

Students will use data from these two pages to answer the questions posed in Put It All Together.

Have students let *b* stand for the number of dog breeds Ask:

- *What would the inequality for the statement, "More than 200 different breeds of dogs exist." look like?* **$b > 200$**
- *Which numbers are possible solutions for the inequality: −215, 199, 215, 380?* **only 215 and 380 are possible solutions because they are greater than 200**

Activating Prior Knowledge

Have students who have or have had dogs as pets share their experiences feeding and caring for them. Have them discuss what they have noticed about the different personalities and characteristics of the breeds of dogs they have known.

Guided Instruction

Have a volunteer read aloud the opening paragraph. Discuss students' views on working dogs and other working animals. Ask them to brainstorm a list of the kinds of jobs dogs do.
Ask:

- *Which jobs are dogs better suited to do than people? Explain.* **Answers will vary. Sample: seeing-eye guide, wilderness rescue search.**
- *Why are dogs better suited to do these jobs?* **Answers will vary. Sample: seeing-eye guides can be a constant companion to visually impaired people, St. Bernards, used for search and rescue missions, have a keen sense of smell, hearing, and sight.**

Problem Solving Application

Applying Inequalities

It's a Dog's Life Dogs come in all shapes and sizes. Different breeds have different personalities and sometimes special skills as well. Some make good guard dogs, some hunt or race, and some can pull heavy sleds. Some dogs are very friendly and make great pets. Dogs generally require a lot of attention — exercise, grooming, and, especially, feeding.

Different breeds
More than 200 different breeds of dogs exist. Most scientists believe that dogs are descendants of the wolf.

A German Shepherd's nose is usually black.

Dalmatians are born white, then develop faint smudges that become their distinctive markings.

1-2. Answers may vary. Samples are given based on a 20-lb Boston Terrier.

1a. 20-lb

b. $1\frac{3}{4}$ cups

c. about 4.6 meals

d. about $291.51

e. It costs more to feed two 20-lb dogs than it costs to feed one 40-lb dog.

2a. about $11.07

b. Check students' work.

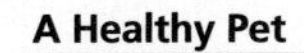

A Healthy Pet

A healthy diet for a dog includes protein, carbohydrates, fats, fiber, water, vitamins, and minerals. In general, a puppy needs about 100 Cal/lb (Calories per pound of body weight) daily, an adult dog needs about 60 Cal/lb, and a senior dog needs about 25 Cal/lb.

Put It All Together

Data File Use the information on these two pages to answer these questions.

GROCERY
$3.69

Dog Food

NET WT 4 LB (1.81 kg)

Adult Dog Size	Daily Feeding
(pounds)	Dry (cups)
3–12 lb	1/2 to 1 1/4
13–20 lb	1 1/4 to 1 3/4
21–35 lb	1 3/4 to 2 2/3
36–50 lb	2 2/3 to 3 1/2
51–75 lb	3 1/2 to 4 3/4
76–100 lb	4 3/4 to 5 3/4
Over 100 lb	5 3/4 + 2/3 c for each 10 lb body weight over 100 lb

1. **Research** Use your own dog, or find information about a specific dog.
   a. How much does the dog weigh?
   b. Use the dog food label. Find the amount of food the dog will eat in a day.
   c. One cup of dog food weighs about 8 oz. About how many meals will one bag contain?
   d. About how much will it cost to feed the dog for a year?
   e. **Reasoning** Which would cost more to feed for a year: two dogs this size or one dog that is twice as large?
2. A friend decides to start a dog-care business for people who travel.
   a. Use the same dog as for Question 1. How much would it cost to feed the dog for two weeks?
   b.  **Writing in Math** Suppose the dog's owners are going on vacation for two weeks. Your friend decides to charge $50/wk, which includes food, grooming, and walking the dog twice a day. Are the dog's owners likely to hire your friend? Why or why not?

Go Online

PHSchool.com

For: Information about dogs
Web Code: are-0453

Activity

Have students work in pairs to answer the questions.

Science Connection
Discuss what a "Calorie" is. **the amount of heat needed to raise the temperature of one kilogram of water one degree Celsius** Then have students find out how the recommended daily Calorie intake of puppies and full-grown dogs compares to their own daily Calorie intake and that of adults.

Exercise 2a Have students share their strategies for answering the question. Invite interested students to visit or contact a dog-sitting service to find out what they charge and what services they provide dog owners.

Differentiated Instruction

Special Needs L1
You may find it useful to review some of the measures mentioned on the page, such as cups, pounds, ounces, and so on. Provide examples of each. In addition, help students to read the data in the table, as needed.

5 Ratios, Rates, and Proportions

Chapter at a Glance

Lesson Titles, Objectives, and Features	Assessment	NCTM Standards	Local Standards
5-1 Ratios • To write ratios and use them to compare quantities	Lesson Quiz	1, 2, 6, 7, 8, 9, 10	
5-2 Unit Rates and Proportional Reasoning • To find unit rates and unit costs using proportional reasoning **Extension:** Using Conversion Factors	Lesson Quiz Checkpoint Quiz 1	1, 2, 6, 7, 8, 9, 10	
5-3 Proportions • To test whether ratios form a proportion by using equivalent ratios and cross products **5-3b Activity Lab, Algebra Thinking:** Interpreting Rates Visually	Lesson Quiz	1, 2, 6, 7, 8, 9, 10	
5-4a Activity Lab, Data Collection: Using Proportions With Data **5-4 Solving Proportions** • To solve proportions using unit rates, mental math, and cross products **Guided Problem Solving:** Proportions and Equations	Lesson Quiz	1, 2, 6, 7, 8, 9, 10	
5-5a Activity Lab, Hands On: Exploring Similar Figures **5-5 Using Similar Figures** • To use proportions to find missing lengths in similar figures **5-5b Activity Lab, Technology:** Drawing Similar Figures **Vocabulary Builder:** Learning Vocabulary	Lesson Quiz Checkpoint Quiz 2	1, 2, 3, 4, 6, 7, 8, 9, 10	
5-6a Activity Lab, Hands On: Scale Drawings and Models **5-6 Maps and Scale Drawings** • To use proportions to solve problems involving scale **5-6b Activity Lab:** Plan a Trip	Lesson Quiz	1, 2, 3, 4, 6, 7, 8, 9, 10	
Problem Solving Application: Applying Ratios			

NCTM Standards 2000
1 Number and Operations 2 Algebra 3 Geometry 4 Measurement 5 Data Analysis and Probability
6 Problem Solving 7 Reasoning and Proof 8 Communication 9 Connections 10 Representation

Correlations to Standardized Tests

All content for these tests is contained in *Prentice Hall Math,* Course 2. This chart reflects coverage in this chapter only.

	5-1	5-2	5-3	5-4	5-5	5-6
Terra Nova CAT6 (Level 17)						
Number and Number Relations	✔	✔	✔	✔	✔	✔
Computation and Numerical Estimation	✔	✔	✔	✔	✔	✔
Operation Concepts	✔	✔	✔	✔	✔	✔
Measurement					✔	✔
Geometry and Spatial Sense					✔	✔
Data Analysis, Statistics, and Probability						
Patterns, Functions, Algebra	✔	✔	✔	✔	✔	✔
Problem Solving and Reasoning	✔	✔	✔	✔	✔	✔
Communication	✔	✔	✔	✔	✔	✔
Decimals, Fractions, Integers, and Percent	✔	✔	✔	✔	✔	✔
Order of Operations						
Terra Nova CTBS (Level 17)						
Decimals, Fractions, Integers, Percents	✔	✔	✔	✔	✔	✔
Order of Operations, Numeration, Number Theory	✔	✔	✔	✔	✔	✔
Data Interpretation						
Pre-algebra	✔	✔	✔	✔	✔	✔
Measurement					✔	✔
Geometry					✔	✔
ITBS (Level 13)						
Number Properties and Operations	✔	✔	✔	✔	✔	✔
Algebra	✔	✔	✔	✔	✔	✔
Geometry					✔	✔
Measurement					✔	✔
Probability and Statistics						
Estimation						
SAT10 (Int 3 Level)						
Number Sense and Operations	✔	✔	✔	✔	✔	✔
Patterns, Relationships, and Algebra	✔	✔	✔	✔	✔	✔
Data, Statistics, and Probability						
Geometry and Measurement					✔	✔
NAEP						
Number Sense, Properties, and Operations		✔			✔	✔
Measurement					✔	✔
Geometry and Spatial Sense						
Data Analysis, Statistics, and Probability						
Algebra and Functions						

CAT6 California Achievement Test, 6th Ed. **CTBS** Comprehensive Test of Basic Skills **ITBS** Iowa Test of Basic Skills, Form M
SAT10 Stanford Achievement Test, 10th Ed. **NAEP** National Assessment of Educational Progress 2005 Mathematics Objectives

Math Background

Skills Trace

BEFORE Chapter 5

Course 1 or Grade 6 introduced ratios, rates, and proportions.

DURING Chapter 5

Course 2 reviews these ideas and then extends their use to various applications.

AFTER Chapter 5

These ideas are used throughout the remainder of this book with percent, geometry, measurement, and algebra.

5-1 5-2 Ratios, Unit Rates and Proportional Reasoning

Math Understandings

- A ratio is a multiplicative relationship between quantities.
- All ratios can be written in fraction form $\left(\frac{a}{b}\right)$.
- All fractions are ratios but not all ratios are fractions.
- Equivalent ratios can be generated using multiplication or division, just as with equivalent fractions.
- Rates are a special type of ratio.
- Unit rates and unit prices are special kinds of ratios.

A **ratio** is a pair of numbers used to show a comparison between like or unlike quantities, written x to y, $\frac{x}{y}$, $x \div y$, or $x : y (y \neq 0)$. The numbers x and y are called the **terms** of the ratio.

Two ratios are **equivalent ratios** if their corresponding fractions are equivalent or if the quotients of the respective terms are the same.

Different Types of Ratios

There are 36 seats inside a restaurant and 24 seats on the terrace.

Part to part: The ratio of indoor seats to outdoor seats is $36 : 24$ or $\frac{36}{24}$.

Part to whole: The ratio of indoor seats to the total number of seats is $36 : 60$ or $\frac{36}{60}$.

Whole to part: The ratio of the total number of seats to the number of seats indoors is $60 : 36$ or $\frac{60}{36}$.

A **rate** is a ratio comparing quantities involving different units. Twenty miles per (one) hour is an example of **unit rate** you are familiar with. It is a unit rate because the first quantity, 20 miles, is compared to one (1) of the second quantity. A **unit price** is also a unit rate, for example, \$1.75 for 1 pound.

Fractions and Ratios

Students first interpret a fraction as a comparison of a part of a whole. Yet all of the ratios above were written as $\frac{x}{y}$. Remember that while all ratios can be written in $\frac{x}{y}$ form (fraction form), all ratios do not behave like fractions. This is why we use "terms" rather than numerator and denominator when working with ratios.

Example: Suppose a student played softball on two days. On each day, she got 1 hit for 2 times at bat $\left(\frac{1}{2}\right)$, so for the two days, she got 2 hits for 4 times at bat or $\frac{2}{4}$. So, for the two days, it is tempting to write $\frac{1}{2} + \frac{1}{2} = \frac{2}{4}$ but this is not correct; if we were adding fractions, we know that $\frac{1}{2} + \frac{1}{2} = \frac{2}{2} = 1$.

So, an important point in developing an understanding of ratios is that all part to whole comparisons (what we typically think of as a fraction) represent only one kind of ratio, but there are other kinds of ratios.

Models and Ratios

Models are very helpful for understanding unit rate and unit price. If 8 oz cost \$2.48, the model below shows how we can think about the unit price.

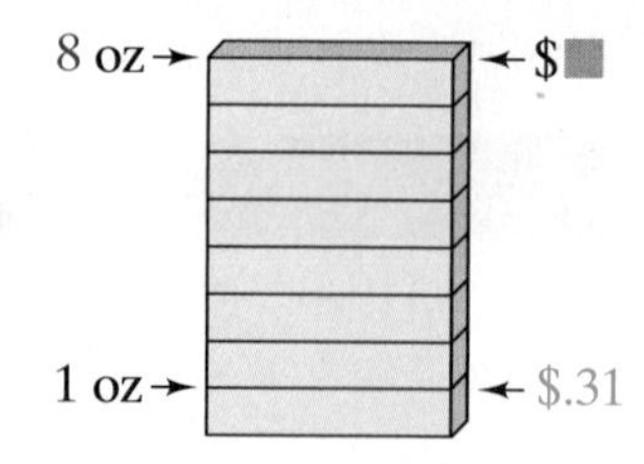

The model shows why a ratio is a multiplicative relationship. One ounce costs \$.31 and 8 oz cost 8 times as much.

5-3 5-4 Proportions and Solving Proportions

Math Understandings

- Equal ratios behave like equivalent fractions.
- There are different related methods for solving proportions.
- If $\frac{a}{b} = \frac{c}{d}$ (where b and d are not 0), then we can use the Multiplication Property of Equality to show cross products, ad and bc, and that $ad = bc$.
- Unit rates and unit prices are helpful for making comparisons.

A **proportion** is an equation stating that two ratios are equal. Quantities **vary proportionally** because as their corresponding values increase or decrease, the ratio of the two quantities is always equivalent.

Example: One ticket for a play costs $12. The ratio $\frac{1}{12}$ shows this comparison. The table below shows three other ratios that are equivalent to $\frac{1}{12}$.

Number of tickets	1	2	3	4
Total cost	$12	$24	$36	$48

Any pair of these equal ratios can be pulled out and written as an equation, called a proportion.

$\frac{2}{24} = \frac{4}{48}$

Also notice that the quotient of the total cost to the number of tickets is 12 for all of the equivalent ratios.

Methods for Solving Proportions

There are three methods commonly used for solving proportions:

- finding the unit amount
- finding the common multiplier
- using cross products

The development on p. 239 shows why the cross product method works. Be sure to connect the other two methods to the cross product method. In Example 1 on p. 244, notice that $\frac{\$2.34}{6}$ gives the unit amount and $\frac{8}{6}$ gives the common multiplier.

$\frac{\$2.34}{6}$ oranges $= \frac{n}{8}$ oranges

$\$2.34 \times 8 = 6 \times n$ using cross products

$\$2.34 \times \frac{8}{6} = n$ dividing both sides by 6

5-5 5-6 Using Similar Figures, Maps and Scale Drawings

Math Understandings

- Similar figures, maps, and scale drawings have corresponding quantities that vary proportionally.

Maps and scale drawings are special types of similar figures. For all similar figures, corresponding lengths are proportional.

Proportions used for similar figures can be written in different ways as long as corresponding parts of the figures are corresponding terms in the proportion.

Example: Proportions for this situation (found on p. 253) can be set up in either of these equivalent ways.

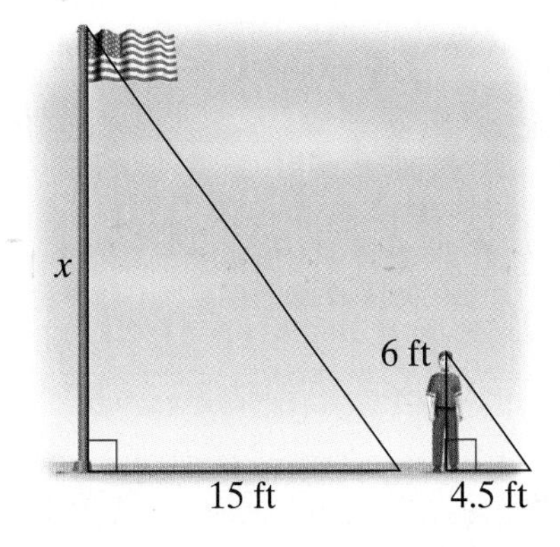

$$\frac{\text{height of pole}}{\text{shadow length of pole}} = \frac{\text{height of boy}}{\text{shadow length of boy}}$$

$$\frac{\text{height of pole}}{\text{height of boy}} = \frac{\text{shadow length of pole}}{\text{shadow length of boy}}$$

Additional Professional Development Opportunities

Math Background Notes for Chapter 5: Every lesson has a Math Backgound in the PLAN section.

Research Overview, Mathematics Strands
Additional support for these topics and more is in the front of the Teacher's Edition.

LessonLab
LessonLab, a Pearson Education company, offers comprehensive, facilitated professional development designed to help teachers to improve student achievement. To learn more, please visit lessonlab.com.

Chapter 5 Resources

	5-1	5-2	5-3	5-4	5-5	5-6	For the Chapter
Print Resources							
L3 Practice	●	●	●	●	●	●	
L1 Adapted Practice	●	●	●	●	●	●	
L3 Guided Problem Solving	●	●	●	●	●	●	
L2 Reteaching	●	●	●	●	●	●	
L4 Enrichment	●	●	●	●	●	●	
L3 Daily Notetaking Guide	●	●	●	●	●	●	
L1 Adapted Daily Notetaking Guide	●	●	●	●	●	●	
L3 Vocabulary and Study Skills Worksheets	●		●	●	●	●	●
L3 Daily Puzzles	●	●	●	●	●	●	
L3 Activity Labs			●	●	●	●	
L3 Checkpoint Quiz		●			●		
L3 Chapter Project							●
L2 Below Level Chapter Test							●
L3 Chapter Test							●
L4 Alternative Assessment							●
L3 Cumulative Review							●
Spanish Resources ELL							
L3 Practice	●	●	●	●	●	●	
L3 Vocabulary and Study Skills Worksheets	●		●	●	●	●	●
L3 Checkpoint Quiz		●			●		
L2 Below Level Chapter Test							●
L3 Chapter Test							●
L4 Alternative Assessment							●
L3 Cumulative Review							●
Transparencies							
Check Skills You'll Need	●	●	●	●	●	●	
Additional Examples	●	●	●	●	●	●	
Problem of the Day	●	●	●	●	●	●	
Classroom Aid							
Student Edition Answers	●	●	●	●	●	●	●
Lesson Quiz	●	●	●	●	●	●	
Test-Taking Strategies							●
Technology							
Interactive Textbook Online	●	●	●	●	●	●	●
StudentExpress™ CD-ROM	●	●	●	●	●	●	●
Success Tracker™ Online Intervention	●	●	●	●	●	●	●
ExamView® Assessment Suite CD-ROM	●	●	●	●	●	●	●
TeacherExpress™ CD-ROM	●	●	●	●	●	●	●
PresentationExpress™ with QuickTake Presenter CD-ROM	●	●	●	●	●	●	●
MindPoint Quiz® Show CD-ROM							●
Prentice Hall Web Site: PHSchool.com	●	●	●	●	●	●	●

Also available:

Prentice Hall Assessment System
- Progress Monitoring Assessments
- Skills and Concepts Review
- Test Prep Workbook

Other Resources
Algebra Readiness Tests
All-in-One Student Workbook
All-in-One Student Workbook, Adapted Version
Multilingual Handbook
Solution Key
Math Notes Study Folder
Spanish Cumulative Assessment

Where You Can Use the Lesson Resources

Here is a suggestion, following the four-step teaching plan, for how you can incorporate Differentiated Instruction Resources into your teaching.

	Instructional Resources L3	Differentiated Instruction Resources
1. Plan		
Preparation Read the Math Background in the Teacher's Edition to connect this lesson with students' previous experience. **Starting Class** **Check Skills You'll Need** Assign these exercises to review prerequisite skills. **New Vocabulary** Help students pre-read the lesson by pointing out the new terms introduced in the lesson.	**Math Background** **Math Understandings** **Transparencies & PresentationExpress™ with QuickTake Presenter CD-ROM** Check Skills You'll Need Problem of the Day **Resources** Vocabulary and Study Skills	**Spanish Support** ELL Vocabulary and Study Skills
2. Teach		
L3 **Guided Instruction** Use the Activity Labs to build conceptual understanding. Teach each Example. Use the Teacher's Edition side column notes for specific teaching tips, including Error Prevention notes. Use the Additional Examples found in the side column (and on transparency and PowerPoint) as an alternative presentation for the content. After each Example, assign the Quick Check exercise for that Example to get an immediate assessment of student understanding. Use the Closure activity in the Teacher's Edition to help students attain mastery of lesson content.	**Student Edition** Activity Lab **Resources** Daily Notetaking Guide Activity Lab **Transparencies & PresentationExpress™ with QuickTake Presenter CD-ROM** Additional Examples Classroom Aids **ExamView® Assessment Suite CD-ROM**	**Teacher's Edition** Every lesson includes suggestions for working with students who need special attention. L1 Special Needs L2 Below Level L4 Advanced Learners ELL English Language Learners **Resources** L1 Adapted Daily Notetaking Guide **Multilingual Handbook**
3. Practice		
Assignment Guide **Check Your Understanding** Use these questions to check students' understanding before you assign homework. **Homework Exercises** Assign homework from these leveled exercises in the Assignment Guide. A Practice by Example B Apply Your Skills C Challenge Test Prep and Mixed Review **Homework Quick Check** Use these key exercises to quickly check students' homework.	**Transparencies & PresentationExpress™ with QuickTake Presenter CD-ROM** Student Answers **Resources** Practice Guided Problem Solving Vocabulary and Study Skills Activity Lab Daily Puzzles **ExamView® Assessment Suite CD-ROM**	**Spanish Support** ELL Practice ELL Vocabulary and Study Skills **Resources** L1 Adapted Practice L4 Enrichment
4. Assess & Reteach		
Lesson Quiz Assign the Lesson Quiz to assess students' mastery of the lesson content. **Checkpoint Quiz** Use the Checkpoint Quiz to assess student progress over several lessons.	**Transparencies & PresentationExpress™ with QuickTake Presenter CD-ROM** Lesson Quiz **Resources** Checkpoint Quiz	**Resources** L2 Reteaching ELL Checkpoint Quiz Success Tracker™ Online Intervention **ExamView® Assessment Suite CD-ROM**

KEY L1 Special Needs L2 Below Level L3 For All Students L4 Advanced, Gifted ELL English Language Learners

Ratios, Rates, and Proportions

Answers are in the back of the textbook.

For intervention, direct students to:

Multiplying and Dividing Decimals
Lesson 1-4
Extra Skills and Word Problems Practice, Ch. 1

Solving Equations by Multiplying or Dividing
Lesson 4-4
Extra Skills and Word Problems Practice, Ch. 4

Comparing and Ordering Fractions
Lesson 2-4
Extra Skills and Word Problems Practice, Ch. 2

Simplifying Fractions
Lesson 2-3
Extra Skills and Word Problems Practice, Ch. 2

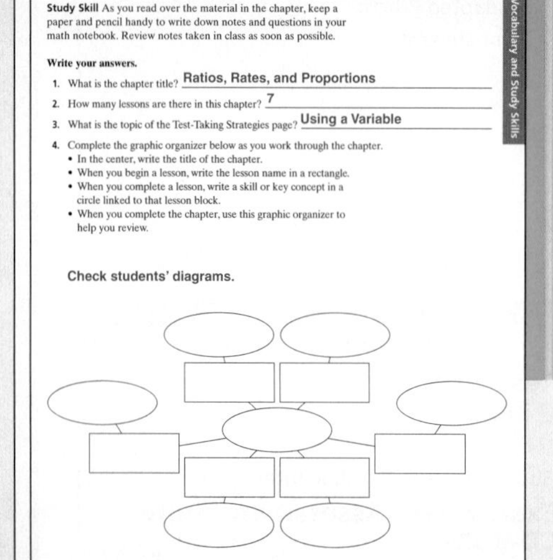
Spanish Vocabulary/Study Skills ELL

Vocabulary/Study Skills L3

5A: Graphic Organizer — For use before Lesson 5-1

Study Skill As you read over the material in the chapter, keep a paper and pencil handy to write down notes and questions in your math notebook. Review notes taken in class as soon as possible.

Write your answers.

1. What is the chapter title? Ratios, Rates, and Proportions
2. How many lessons are there in this chapter? 7
3. What is the topic of the Test-Taking Strategies page? Using a Variable
4. Complete the graphic organizer below as you work through the chapter.
   - In the center, write the title of the chapter.
   - When you begin a lesson, write the lesson name in a rectangle.
   - When you complete a lesson, write a skill or key concept in a circle linked to that lesson block.
   - When you complete the chapter, use this graphic organizer to help you review.

Check students' diagrams.

CHAPTER 5 Ratios, Rates, and Proportions

What You've Learned

- In Chapter 2, you compared and ordered rational numbers.
- You converted between fractions and decimals.
- In Chapter 3, you used addition, subtraction, multiplication, and division to solve problems involving fractions.

Check Your Readiness

GO for Help

For Exercises	See Lesson
1–6	1-4
7–10	4-4
11–14	2-4
15–19	2-3

Multiplying and Dividing Decimals

Find each product or quotient.

1. $(3.6)(4)$ **14.4**
2. $(12.74)(9)$ **114.66**
3. $(15.9)(3)$ **47.7**
4. $8.96 \div 8$ **1.12**
5. $10.4 \div 4$ **2.6**
6. $52.6 \div 5$ **10.52**

Solving Equations by Multiplying or Dividing

Algebra **Solve.**

7. $4n = -32$ **−8**
8. $\frac{a}{6} = 10$ **60**
9. $18 = -9z$ **−2**
10. $-15 = \frac{p}{-5}$ **75**

Comparing and Ordering Fractions

Compare. Use <, =, or >.

11. $\frac{8}{9} \blacksquare \frac{3}{4}$ **>**
12. $\frac{7}{12} \blacksquare \frac{4}{5}$ **<**
13. $\frac{6}{3} \blacksquare \frac{24}{12}$ **=**
14. $\frac{1}{6} \blacksquare \frac{1}{7}$ **>**

Simplifying Fractions

Write each fraction in simplest form.

15. $\frac{12}{16}$ **$\frac{3}{4}$**
16. $\frac{15}{30}$ **$\frac{1}{2}$**
17. $\frac{27}{72}$ **$\frac{3}{8}$**
18. $\frac{19}{57}$ **$\frac{1}{3}$**
19. $\frac{16}{56}$ **$\frac{2}{7}$**

Chapter 5 Overview

In this chapter, students learn to find equivalent ratios and to see if ratios can form a proportion. They use unit rates to solve proportions involving variables, and they use proportions to solve problems involving scale. They solve proportions to find missing lengths in similar figures. They solve problems using proportions and equations.

Activating Prior Knowledge

In this chapter, students will build on their knowledge of fractions by writing ratios and unit rates, solving proportions, and finding values in similar figures, maps, and scale models. Ask questions such as:

- *What are equivalent fractions?* **fractions that are equal**
- *How do you find equivalent fractions?* **Multiply or divide the numerator and the denominator by the same nonzero number.**
- *How do you simplify fractions?* **Divide the numerator and the denominator by their GCF (greatest common factor).**

What You'll Learn Next

- In this chapter, you will write ratios and unit rates.
- You will write and solve proportions.
- You will use rates and proportions to solve problems involving similar figures, maps, and scale models.

Key Vocabulary

- cross products (p. 239)
- equivalent ratios (p. 229)
- indirect measurement (p. 253)
- polygon (p. 252)
- proportion (p. 238)
- rate (p. 232)
- ratio (p. 228)
- scale (p. 259)
- scale drawing (p. 259)
- similar polygons (p. 252)
- unit cost (p. 233)
- unit rate (p. 232)

Problem Solving Application On pages 270 and 271, you will work an extended activity on ratios.

5-1

1. Plan

Objective
To write ratios and use them to compare quantities

Examples
1 Writing Ratios
2 Writing Equivalent Ratios
3 Comparing Ratios

Math Understandings: p. 226C

Math Background

A ratio is a comparison of two quantities by *division.* Using a ratio to compare 2 lb and 6 lb, you can express the result in one of these forms.

2 lb to 6 lb 2 lb : 6 lb $\frac{2\text{ lb}}{6\text{ lb}}$

All three forms are acceptable, but in this textbook ratios are most often written in the familiar form of a fraction. Students can then connect their knowledge of fraction concepts to the task of writing equivalent ratios and writing ratios in simplest form.

More Math Background: p. 226C

Lesson Planning and Resources

See p. 226E for a list of the resources that support this lesson.

Bell Ringer Practice

Check Skills You'll Need
Use student page, transparency, or PowerPoint. For intervention, direct students to:
Simplifying Fractions
Lesson 2-3
Extra Skills and Word Problems Practice, Ch. 2

5-1 Ratios

Check Skills You'll Need

1. **Vocabulary Review** What are *equivalent fractions?* **See below.**

Write each fraction in simplest form.

2. $\frac{2}{4}$ $\frac{1}{2}$ 3. $\frac{21}{27}$ $\frac{7}{9}$
4. $\frac{36}{63}$ $\frac{4}{7}$ 5. $\frac{8}{24}$ $\frac{1}{3}$

GO for Help Lesson 2-3

Check Skills You'll Need
1. **Equivalent fractions are fractions that name the same amount.**

What You'll Learn

To write ratios and use them to compare quantities

New Vocabulary ratio, equivalent ratios

Why Learn This?

The keys on a music keyboard have a repeating pattern of five black keys and seven white keys. You can use ratios to describe patterns.

The ratio of black keys to white keys in the pattern is 5 to 7. What is the ratio of black keys to all keys in the pattern?

KEY CONCEPTS Ratio

A **ratio** is a comparison of two quantities by division. You can write a ratio in three ways.

Arithmetic	Algebra
5 to 7 5 : 7 $\frac{5}{7}$	a to b $a : b$ $\frac{a}{b}$ where $b \neq 0$

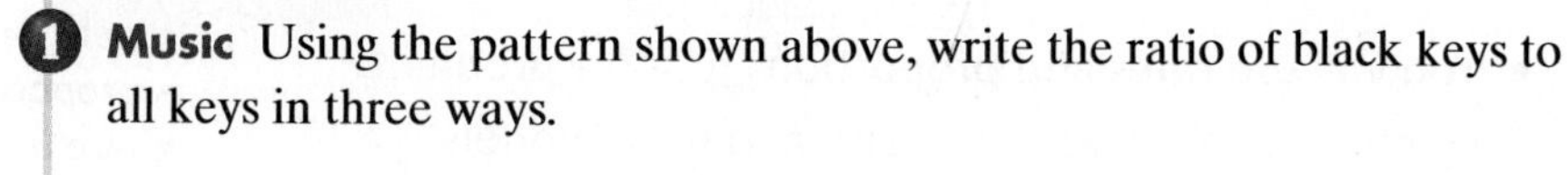

EXAMPLE Writing Ratios

1 **Music** Using the pattern shown above, write the ratio of black keys to all keys in three ways.

black keys → 5 to 12 ← all keys
black keys → 5 : 12 ← all keys
$\frac{5}{12}$ ← black keys / ← all keys

Go Online
Video Tutor Help
Visit: PHSchool.com
Web Code: are-0775

Quick Check

1. Write each ratio in three ways. Use the pattern of keys shown above.
   a. white keys to all keys **7 to 12, 7 : 12, $\frac{7}{12}$**
   b. white keys to black keys **7 to 5; 7 : 5; $\frac{7}{5}$**

Differentiated Instruction Solutions for All Learners

Special Needs L1	**Below Level** L2
Provide students with index cards that show different ways to write ratios. Make sure they understand that while ratios can be written like fractions, fractions are not written like ratios. **learning style: visual**	Review equivalent fractions. Remind students that you may multiply or divide both numerator and denominator by the same nonzero number. **learning style: verbal**

Two ratios that name the same number are **equivalent ratios.** In Chapter 2, you learned to write equivalent fractions. You can find equivalent ratios by writing a ratio as a fraction and finding an equivalent fraction.

EXAMPLES Writing Equivalent Ratios

2 Find a ratio equivalent to $\frac{4}{5}$.

$\frac{4 \times 2}{5 \times 2} = \frac{8}{10}$ ← **Multiply the numerator and denominator by 2.**

GO for Help
For help converting customary units, go to Lesson 3-6.

3 Write the ratio 2 yd to 20 ft as a fraction in simplest form.

$\frac{2\text{ yd}}{20\text{ ft}} = \frac{2 \times 3\text{ ft}}{20\text{ ft}}$ ← **There are 3 ft in each yard.**

$= \frac{6\text{ ft}}{20\text{ ft}}$ ← **Multiply.**

$= \frac{6\text{ ft} \div 2}{20\text{ ft} \div 2}$ ← **Divide by the GCF, 2.**

$= \frac{3}{10}$ ← **Simplify.**

Quick Check

2. Find a ratio equivalent to $\frac{7}{9}$. **Answers may vary. Samples: $\frac{14}{18}$, $\frac{35}{45}$**
3. Write the ratio 3 gal to 10 qt as a fraction in simplest form. **$\frac{6}{5}$**

You can use decimals to express and compare ratios.

EXAMPLE Comparing Ratios

Online active math
For: Ratios Activity
Use: Interactive Textbook, 5-1

4 **Social Studies** An official U.S. flag has a length-to-width ratio of 19 : 10. The largest U.S. flag measures 505 ft by 255 ft. Is this an official U.S. flag?

official flag ↓ $\frac{19}{10}$ ← **length** → / ← **width** → **largest flag** ↓ $\frac{505}{255}$

$\frac{19}{10} = 1.9$ ← **Write as a decimal. Round if necessary.** → $\frac{505}{255} \approx 1.98$

Since 1.98 is not equal to 1.9, the largest flag is *not* an official U.S. flag.

Quick Check

4. Tell whether the ratios are *equivalent* or *not equivalent.*
   a. 7 : 3, 128 : 54 **not equivalent**
   b. $\frac{180}{240}$, $\frac{25}{34}$ **not equivalent**
   c. 6.1 to 7, 30.5 to 35 **equivalent**

Advanced Learners L4
Challenge students to write a ratio in simplest form when a decimal is involved.
Example:

$\frac{2.4}{16} = \frac{2.4 \times 10}{16 \times 10} = \frac{24}{160} = \frac{24 \div 8}{160 \div 8} = \frac{3}{20}$

learning style: visual

English Language Learners ELL
For Example 1, show students a picture of a piano, with the number of keys. Say: *The number of black keys compared to all keys is a part to whole comparison. The number of black keys to white keys is a part to part comparison. Both are ratios.*

learning style: visual

2. Teach

Activity Lab

Use before the lesson.

All in One Teaching Resources
Activity Lab 5-1: Golden Ratios

Guided Instruction

Example 1
Stress the importance of the order of the numbers when writing a ratio. Ask: *Why would it not be correct to express this ratio as 12 to 5?* **The ratio 12 to 5 is the ratio of all keys to black keys.**

Example 2
Suggest that students use a colored pencil to write the number by which they are multiplying or dividing the numerator and denominator. This will help check that both multipliers or both divisors are the same.

PowerPoint

Additional Examples

1 There are 7 red stripes and 6 white stripes on the flag of the United States. Write the ratio of red stripes to white stripes in three ways. **7 to 6; 7 : 6; $\frac{7}{6}$**

2 Find two ratios equivalent to $\frac{14}{1}$. **Answers may vary. Samples: $\frac{7}{2}$ and $\frac{28}{8}$**

3 The ratio of girls to boys enrolled at King Middle School is 15 : 16. There are 195 girls and 208 boys in Grade 8. Is the ratio of girls to boys in Grade 8 the same as the ratio of girls to boys in the entire school? **yes**

All in One Teaching Resources
- Daily Notetaking Guide 5-1 L3
- Adapted Notetaking 5-1 L1

Closure

- *What is a ratio?* **a comparison of two quantities by division**
- *What are three different ways to write the ratio "a to b"?* **a to b; a : b; $\frac{a}{b}$**

3. Practice

Assignment Guide

Check Your Understanding

Go over Exercises 1–11 in class before assigning the Homework Exercises.

Homework Exercises

A Practice by Example	12–25
B Apply Your Skills	26–31
C Challenge	32
Test Prep and Mixed Review	33–38

Homework Quick Check

To check students' understanding of key skills and concepts, go over Exercises 15, 23, 28, 29, and 31.

Differentiated Instruction Resources

Adapted Practice 5-1 L1

Practice 5-1 L3

Write a ratio for each situation in three ways.

1. Ten years ago in Louisiana, schools averaged 182 pupils for every 10 teachers. 182 to 10; 182 : 10; $\frac{182}{10}$
2. Between 1899 and 1900, 284 out of 1,000 people in the United States were 5–17 years old. 284 to 1,000; 284 : 1,000; $\frac{284}{1,000}$

Use the chart below for Exercises 3–4.

Three seventh-grade classes were asked whether they wanted chicken or pasta served at their awards banquet.

Room Number	Chicken	Pasta
201	10	12
202	8	17
203	16	10

3. In room 201, what is the ratio of students who prefer chicken to students who prefer pasta? 10 : 12; or 5 : 6
4. Combine the totals for all three rooms. What is the ratio of the number of students who prefer pasta to the number of students who prefer chicken? 39 : 34

Write each ratio as a fraction in simplest form.

5. 12 to 18 $\frac{2}{3}$ 6. 81 : 27 $\frac{3}{1}$ 7. $\frac{6}{28}$ $\frac{3}{14}$

Tell whether the ratios are *equivalent* or *not equivalent*.

8. 12 : 24, 50 : 100 Yes, they are equivalent.
9. $\frac{22}{1}$, $\frac{1}{22}$ No, they are not equivalent.
10. 2 to 3, 24 to 36 Yes, they are equivalent.
11. A bag contains green, yellow, and orange marbles. The ratio of green marbles to yellow marbles is 2 : 5. The ratio of yellow marbles to orange marbles is 3 : 4. What is the ratio of green marbles to orange marbles? 3 : 1

5-1 • Guided Problem Solving GPS L3

GPS Student Page 231, Exercise 27:

Cooking To make pancakes, you need 2 cups of water for every 3 cups of flour. Write an equivalent ratio to find how much water you will need with 9 cups of flour.

Understand

1. Circle the information you will need to solve.
2. What are you being asked to do? Write a ratio for the number of cups of water to the number of cups of flour.
3. Why will a ratio help you to solve the problem? A ratio will help determine how much water you need to use with the 9 cups of flour.

Plan and Carry Out

4. What is the ratio of the cups of water to the cups of flour? $\frac{2}{3}$ or 2 : 3
5. How many cups of flour are you using? 9 cups
6. Write an equivalent ratio to use 9 cups of flour. $\frac{2}{3} = \frac{6}{9}$
7. How many cups of water are needed for 9 cups of flour? 6 cups

Check

8. Why is the number of cups of water triple the number of cups needed for 3 cups of flour? Since 9 cups is three times 3 cups, the number of cups of water needed is also tripled.

Solve Another Problem

9. Rebecca is laying tile in her bathroom. She needs 4 black tiles for every 16 white tiles. How many black tiles are needed if she uses 128 white tiles? $\frac{4}{16} = \frac{x}{128}$; 32 black tiles

Check Your Understanding

1. Equivalent ratios name the same amount, as do equivalent fractions.
2. No; ratios can compare any two quantities. For example, a ratio may compare two parts.

1. **Vocabulary** How are equivalent ratios like equivalent fractions?
2. **Number Sense** Do all ratios compare a part to a whole? Explain. See left.

Find an equivalent ratio for each ratio. Answers may vary. Samples are given.

3. $\frac{1}{8}$ 2 to 16
4. 2 to 7 $\frac{4}{14}$
5. 10 : 9 20 to 18

Write each ratio as a fraction in simplest form.

6. 2 gal to 14 qt $\frac{4}{7}$
7. 34 in. to 4 ft $\frac{17}{24}$
8. $\frac{4 \text{ min}}{90 \text{ s}}$ $\frac{8}{3}$

Tell whether the ratios are *equivalent* or *not equivalent*.

9. $\frac{12}{24}, \frac{50}{100}$ equivalent
10. 1 to 3, 2 to 9 not equivalent
11. 2 : 3, 24 : 36 equivalent

Homework Exercises

For more exercises, see Extra Skills and Word Problems.

GO for Help

For Exercises	See Examples
12–13	1
14–16	2
17–22	3
23–25	4

12. 5 to 2, 5 : 2, $\frac{5}{2}$
13. 21 to 25, 21 : 25, $\frac{21}{25}$

A **Write a ratio in three ways, comparing the first quantity to the second.**

12. A week has five school days and two weekend days. See left.
13. About 21 out of 25 Texans live in an urban area. See left.

Find an equivalent ratio for each ratio. 14–16. Answers may vary. Samples are given.

14. $\frac{14}{28}$ $\frac{1}{2}$
15. 6 to 7 12 to 14
16. 4 : 5 8 : 10

Write each ratio as a fraction in simplest form.

17. $\frac{4 \text{ ft}}{8 \text{ ft}}$ $\frac{1}{2}$
18. 10 s : 1 min $\frac{1}{6}$
19. $\frac{30 \text{ mL}}{2 \text{ L}}$ $\frac{3}{200}$
20. 12 oz : 3 lb $\frac{1}{4}$
21. 2 ft to 30 in. $\frac{4}{5}$
22. $\frac{1 \text{ m}}{300 \text{ cm}}$ $\frac{1}{3}$

Tell whether the ratios are *equivalent* or *not equivalent*.

23. $\frac{18}{24}, \frac{3}{4}$ equivalent
24. 6 : 7, 30 : 36 not equivalent
25. 16 to 3, 27 to 5 not equivalent

B GPS 26. **Guided Problem Solving** The students in Room 101 and Room 104 have one class together. Write the ratio of girls to boys for the combined class. $\frac{21}{36}$ or $\frac{7}{12}$

	Room 101	Room 104
Girls	12	9
Boys	16	20

- **Make a Plan** First find the total numbers of girls and boys. Then find the ratio of girls to boys for the combined class.

27. $\frac{2}{3} = \frac{6}{9}$; you need 6 c of water.

27. **Cooking** To make pancakes, you need 2 cups of water for every GPS 3 cups of flour. Write an equivalent ratio to find how much water you will need with 9 cups of flour.

28. With the new students, the ratio of girls to boys is 16 : 11 ≈ 1.45. Without the new students, the ratio of girls to boys was 15 : 10 = 1.5.
29. Answers may vary. Sample: if the GCF of the numerator and denominator is 1

31a. 8 : 4, 7.5 : 3, 3.5 : 1

b. 10 qt antifreeze, 5 qt water

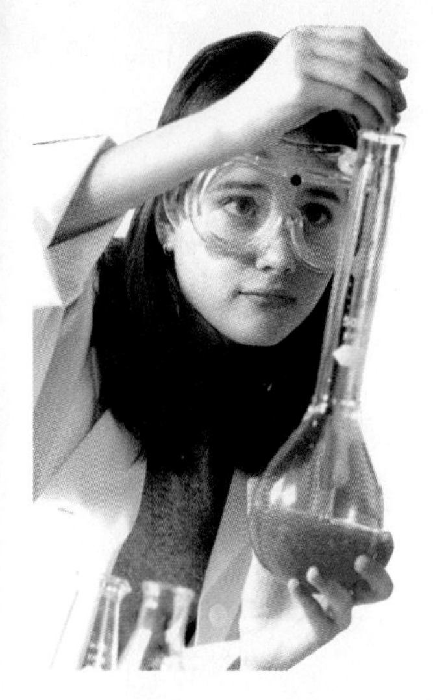

Online
Homework Video Tutor
Visit: PHSchool.com
Web Code: are-0501

28. **Error Analysis** Your math class includes 15 girls and 10 boys. Two new students, a girl and a boy, join the class. Your friend says the ratio of girls to boys is the same as before. Explain your friend's error. **See margin.**

29. **Writing in Math** How can you tell when a ratio is in simplest form? **See margin.**

30. **Chemistry** A chemical formula shows the ratio of atoms in a substance. The formula for carbon dioxide, CO_2, tells you that there is 1 atom of carbon (C) for every 2 atoms of oxygen (O). Write the ratio of hydrogen (H) atoms to oxygen atoms in water, H_2O. $\frac{2}{1}$

31. Antifreeze protects a car's radiator from freezing. In extremely cold weather, you must mix at least 2 parts antifreeze with every 1 part water.

Mixing Antifreeze

Antifreeze (qt)	Water (qt)
8	4
7.5	3
12	8
3.5	1
9	18

a. List all of the ratios in the table that provide the necessary protection.

b. **Reasoning** How much antifreeze and how much water should you use to protect a 15-qt radiator?

31a–b. See margin.

C 32. **Challenge** A bag contains colored marbles. The ratio of red marbles to blue marbles is 1 : 4. The ratio of blue marbles to yellow marbles is 2 : 5. What is the ratio of red marbles to yellow marbles? **1 : 10**

Test Prep and Mixed Review

Practice

Multiple Choice

33. Maria tossed a coin 20 times and got 12 heads. What is the first step to find the ratio of the number of tails to the total number of tosses? **B**

(A) Divide 12 by 20.
(B) Subtract 12 from 20.
(C) Multiply 12 by 20.
(D) Add 12 to 20.

34. The model represents the equation $3x - 2 = -8$. What is the value of x? **J**

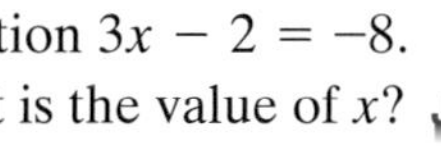

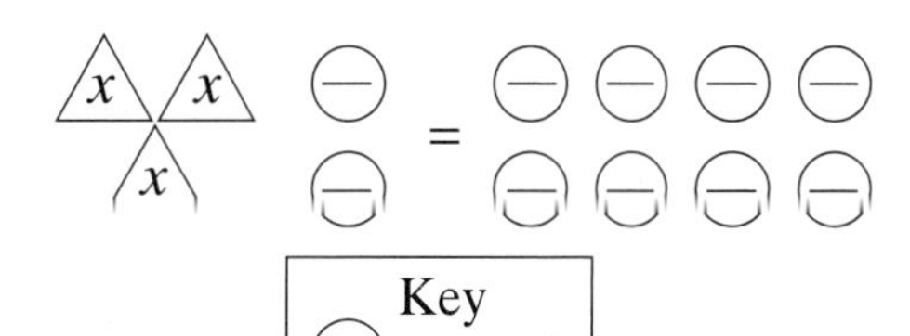

Key: ⊖ = −1

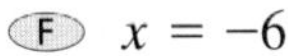

(F) $x = -6$
(G) $x = -3$
(H) $x = -\frac{10}{3}$
(J) $x = -2$

35. Emily ran $4\frac{1}{8}$ miles at track practice one day and $3\frac{3}{4}$ miles the next day. How many miles did she run in those two days? **C**

(A) $7\frac{1}{2}$ miles (B) $7\frac{3}{4}$ miles (C) $7\frac{7}{8}$ miles (D) $8\frac{1}{8}$ miles

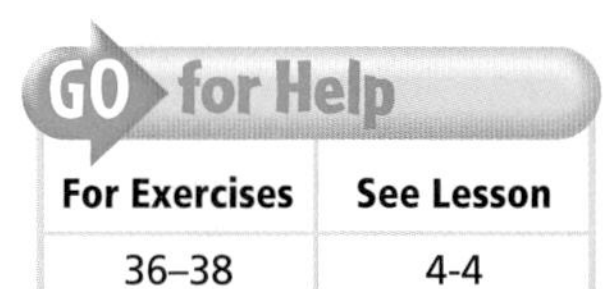

GO for Help

For Exercises	See Lesson
36–38	4-4

Algebra **Solve each equation.**

36. $72 = 8k$ **9**
37. $\frac{y}{3} = 15$ **45**
38. $-5 = \frac{q}{7}$ **−35**

Online lesson quiz, PHSchool.com, Web Code: ara-0501

4. Assess & Reteach

PowerPoint

Lesson Quiz

1. Write the ratio $\frac{9}{7}$ in two other ways. **9 to 7; 9 : 7**
2. Write two other ratios equivalent to $\frac{8}{14}$. **Samples: $\frac{4}{7}$, $\frac{16}{28}$**
3. Compare the ratios $\frac{18}{9}$ and $\frac{16}{4}$. Are they equivalent? **no**

Alternative Assessment

Have students write ratios to compare various quantities within the classroom. For instance, they might write the ratio comparing the number of doors to the number of windows, or the ratio comparing the width of the door to the height of the door. Remind students that all their ratios should be written in simplest form.

Test Prep

Resources

For additional practice with a variety of test item formats:

- Test-Taking Strategies, p. 265
- Test Prep, p. 269
- Test-Taking Strategies with Transparencies

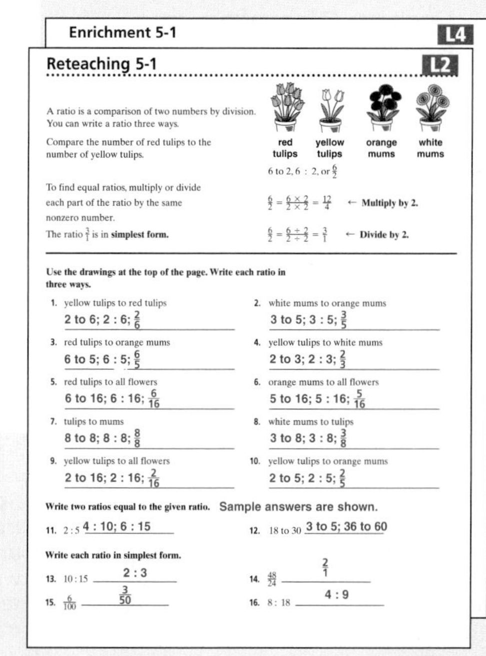

Enrichment 5-1 L4

Reteaching 5-1 L2

A ratio is a comparison of two numbers by division. You can write a ratio three ways.

Compare the number of red tulips to the number of yellow tulips.

red tulips, yellow tulips, orange mums, white mums

6 to 2, 6 : 2, or $\frac{6}{2}$

To find equal ratios, multiply or divide each part of the ratio by the same nonzero number.

$\frac{6}{2} = \frac{6 \times 2}{2 \times 2} = \frac{12}{4}$ ← **Multiply by 2.**

The ratio $\frac{3}{1}$ is in **simplest form.**

$\frac{6}{2} = \frac{6 \div 2}{2 \div 2} = \frac{3}{1}$ ← **Divide by 2.**

Use the drawings at the top of the page. Write each ratio in three ways.

1. yellow tulips to red tulips — 2 to 6; 2 : 6; $\frac{2}{6}$
2. white mums to orange mums — 3 to 5; 3 : 5; $\frac{3}{5}$
3. red tulips to orange mums — 6 to 5; 6 : 5; $\frac{6}{5}$
4. yellow tulips to white mums — 2 to 3; 2 : 3; $\frac{2}{3}$
5. red tulips to all flowers — 6 to 16; 6 : 16; $\frac{6}{16}$
6. orange mums to all flowers — 5 to 16; 5 : 16; $\frac{5}{16}$
7. tulips to mums — 8 to 8; 8 : 8; $\frac{8}{8}$
8. white mums to tulips — 3 to 8; 3 : 8; $\frac{3}{8}$
9. yellow tulips to all flowers — 2 to 16; 2 : 16; $\frac{2}{16}$
10. yellow tulips to orange mums — 2 to 5; 2 : 5; $\frac{2}{5}$

Write two ratios equal to the given ratio. Sample answers are shown.

11. 2 : 5 — 4 : 10; 6 : 15
12. 18 to 30 — 3 to 5; 36 to 60

Write each ratio in simplest form.

13. 10 : 15 — 2 : 3
14. $\frac{48}{24}$ — $\frac{2}{1}$
15. $\frac{6}{100}$ — $\frac{3}{50}$
16. 8 : 18 — 4 : 9

5-2

1. Plan

Objective
To find unit rates and unit costs using proportional reasoning

Examples
1 Finding a Unit Rate
2 Using Unit Cost to Find Total Cost
3 Using Unit Cost to Compare

Math Understandings: p. 226C

Math Background

The ratios in Lesson 5-1 involved quantities of the same type. That is, a number of students was compared to a number of students.

In this lesson, ratios compare different types of quantities. For instance, grams of fat are compared to servings. These ratios are called *rates.* A rate calculated for one unit of a given quantity is called a *unit rate.*

More Math Background: p. 226C

Lesson Planning and Resources

See p. 226E for a list of the resources that support this lesson.

Bell Ringer Practice

Check Skills You'll Need
Use student page, transparency, or PowerPoint. For intervention, direct students to:
Ratios
Lesson 5-1
Extra Skills and Word Problems Practice, Ch. 5

5-2 Unit Rates and Proportional Reasoning

Check Skills You'll Need

1. **Vocabulary Review** A *ratio* is a comparison of two quantities by __?__. **See below.**

Write each ratio in simplest form.

2. $\frac{15}{25}$ **$\frac{3}{5}$**
3. $\frac{21}{7}$ **$\frac{3}{1}$**
4. $\frac{22}{16}$ **$\frac{11}{8}$**
5. $\frac{4}{36}$ **$\frac{1}{9}$**

GO for Help Lesson 5-1

Check Skills You'll Need
1. **division**

What You'll Learn

To find unit rates and unit costs using proportional reasoning

New Vocabulary rate, unit rate, unit cost

Why Learn This?

You make decisions about the foods you eat every day. Looking at rates such as grams of fat per serving can help you stay healthy.

A **rate** is a ratio that compares two quantities measured in different units. There are 15 grams of fat in 5 servings of canned soup. The rate of grams of fat per serving is $\frac{\text{15 grams of fat}}{\text{5 servings}}$.

The rate for one unit of a given quantity is the **unit rate.** To find a unit rate, divide the first quantity by the second quantity. For a rate of $\frac{\text{15 grams of fat}}{\text{5 servings}}$, the unit rate is 3 grams of fat per serving.

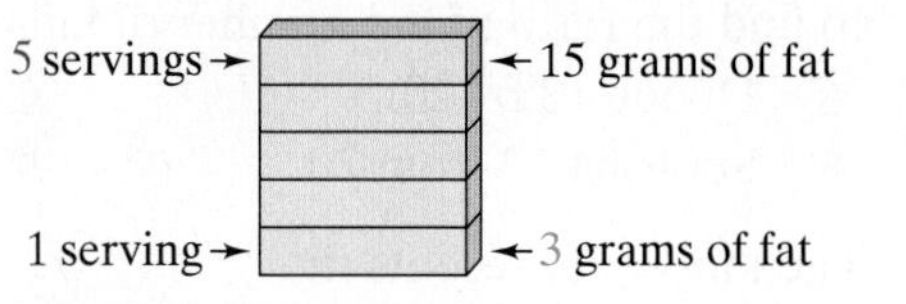

The model shows that

$$\text{total fat} \div \frac{\text{number of}}{\text{servings}} = \frac{\text{fat per}}{\text{serving}}$$

EXAMPLE Finding a Unit Rate

1 A package of cheddar cheese contains 15 servings and has a total of 147 grams of fat. Find the unit rate of grams of fat per serving.

$$\begin{matrix}\text{grams} \rightarrow \\ \text{servings} \rightarrow\end{matrix} \frac{147}{15} = 9.8$$ ← **Divide the first quantity by the second quantity.**

The unit rate is $\frac{\text{9.8 grams}}{\text{1 serving}}$, or 9.8 grams of fat per serving.

Quick Check

1. Find the unit rate for 210 heartbeats in 3 minutes.
70 heartbeats per min

Differentiated Instruction Solutions for All Learners

Special Needs L1
Help students understand that unit rates are understood to have a denominator of 1. Have them write a (1) in parentheses in colored ink. When they see a rate like miles/hour, have them rewrite it to read miles/(1) hour.
learning style: visual

Below Level L2
Check that students recall relationships between units of time, such as 1 h = 60 min. Also review the rules for rounding decimals.
learning style: verbal

A unit rate that gives the cost per unit is a **unit cost.** Suppose a box of cereal weighs 8 oz and has a unit cost of \$.31 per ounce.

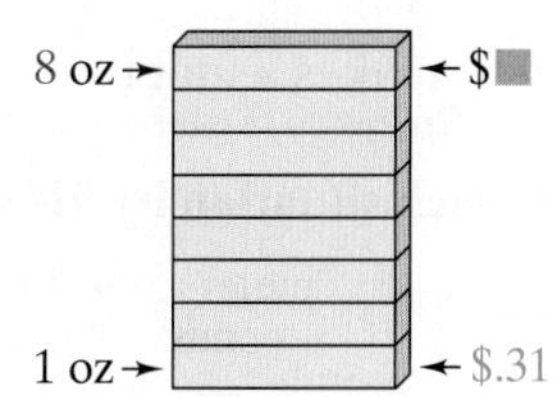

The model shows that

unit cost · number of ounces = total cost

EXAMPLE Using Unit Cost to Find Total Cost

2 Food Use the information at the right to find the cost of the box of cereal.

Estimate \$.31 · 8 ≈ \$.30 · 8, or \$2.40

\$.31 · 8 = \$2.48 ← **unit cost · number of units = total cost**

Check for Reasonableness The total cost of \$2.48 is close to the estimate of \$2.40. So \$2.48 is reasonable.

Quick Check

2. Dog food costs \$.35/lb. How much does a 20-lb bag cost? **\$7.00**

To find the unit cost, divide the total cost of the item by the number of units in the item.

EXAMPLE Using Unit Cost to Compare

Test Prep Tip

The "better buy" is the item that has the lower unit cost.

3 Smart Shopping Two sizes of shampoo bottles are shown. Which size is the better buy? Round to the nearest cent.

shampoo² 16 oz \$6.19 — shampoo² 13.5 oz \$3.99

Divide to find the unit cost of each size.

cost → / size → $\frac{\$3.99}{13.5 \text{ fl oz}} \approx \$.30/\text{fl oz}$

cost → / size → $\frac{\$6.19}{16 \text{ fl oz}} \approx \$.39/\text{fl oz}$

Since \$.30 < \$.39, the 13.5-fl-oz bottle is the better buy.

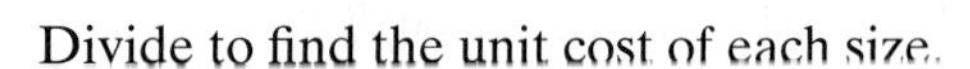

Quick Check

3. Which bottle of apple juice is the better buy: 48 fl oz for \$3.05 or 64 fl oz for \$3.59? **\$.064/fl oz, \$.056/fl oz; the 64-fl-oz bottle is the better buy.**

Advanced Learners L4

Suppose you could bicycle from Earth to the moon. The distance is about 239,000 mi, and your speed is 12 mi/h. How long would it take to get to the moon? ≈ **830 days**

learning style: verbal

English Language Learners ELL

As students are learning English, they might equate better with bigger, both because of examples they have seen and the similarity between the words. Make sure they understand the *better* buy has the *smallest* price per unit.

learning style: verbal

2. Teach

Activity Lab

Use before the lesson.

All in One Teaching Resources

Activity Lab 5-2: Unit Rates

Guided Instruction

Example 3

Have students use calculators to find unit prices. Focus attention on the concept of unit prices. Stress the importance of writing the $\frac{\text{price}}{\text{size}}$ ratio, to identify which order they should enter the numbers into the calculator.

Error Prevention!

Remind students to assess their work for reasonableness using general guidelines.

$\text{speed} = \frac{\text{distance}}{\text{time}}$

$\text{unit price} = \frac{\text{total price}}{\text{size}}$

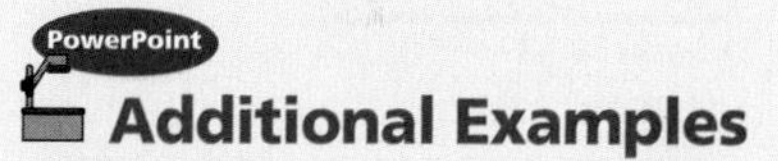

Additional Examples

1 Find the unit rate: earn \$33 for 4 hours of work **\$8.25/h**

2 Use the unit cost to find the total cost: 7 yd of ribbon at \$.39 per yard **\$2.73**

3 Find each unit cost. Then determine the better buy.
3 lb of potatoes for \$.89
5 lb of potatoes for \$1.59
≈ \$.30/lb; ≈ \$.32/lb; The better buy is 3 lb for \$.89.

All in One Teaching Resources

- Daily Notetaking Guide 5-2 L3
- Adapted Notetaking 5-2 L1

Closure

- *What is the relationship between a rate and a ratio?* **A rate is a ratio that compares two different types of quantities.**
- *What is a unit rate?* **a rate that compares a quantity to one unit of a second quantity**
- *What is a unit price?* **a unit rate that identifies the cost of one unit of a given item**

3. Practice

Assignment Guide

Check Your Understanding
Go over Exercises 1–5 in class before assigning the Homework Exercises.

Homework Exercises
A Practice by Example 6–18
B Apply Your Skills 19–28
C Challenge 29
Test Prep and Mixed Review 30–35

Homework Quick Check
To check students' understanding of key skills and concepts, go over Exercises 12, 17, 25, 27, and 28.

Differentiated Instruction Resources

Adapted Practice 5-2 L1

Practice 5-2 L3 Unit Rates and Proportional R...

Write the unit rate for each situation.

1. travel 250 mi in 5 h — 50 mi/h
2. earn $75.20 in 8 h — $9.40/h
3. read 80 pages in 2 h — 40 pages/h
4. type 8,580 words in 2 h 45 min — 52 words/min or 3,120 words/h
5. manufacture 2,488 parts in 8 h — 311 parts/h
6. 50 copies of a book on 2 shelves — 25 books/shelf

Find each unit price. Then determine the better buy.

7. paper: 100 sheets for $.99; 500 sheets for $4.29 — $.0099/sheet; $.00858/sheet; 500 sheets
8. peanuts: 1 lb for $1.29; 12 oz for $.95 — $1.29/lb; $1.267/lb; 12 oz
9. crackers: 15 oz for $1.79; 12 oz for $1.49 — $.1193/oz; $.1242/oz; 15 oz
10. apples: 3 lb for $1.89; 5 lb for $2.49 — $.63/lb; $.498/lb; 5 lb
11. mechanical pencils: 4 for $1.25; 25 for $5.69 — $.3125/pencil; $.2276/pencil; 25 pencils
12. bagels: 4 for $.89; 6 for $1.39 — $.2225/bagel; $.2317/bagel; 4 bagels
13. a. Yolanda and Yoko ran in a 100-yd dash. When Yolanda crossed the finish line, Yoko was 10 yd behind her. The girls then repeated the race, with Yolanda starting 10 yd behind the starting line. If each girl ran at the same rate as before, who won the race? By how many yards? — Yolanda; 1 yd
    b. Assuming the girls run at the same rate as before, how far behind the starting line should Yolanda be in order for the two to finish in a tie? — $11\frac{1}{9}$ yd, or 11 yd 4 in.

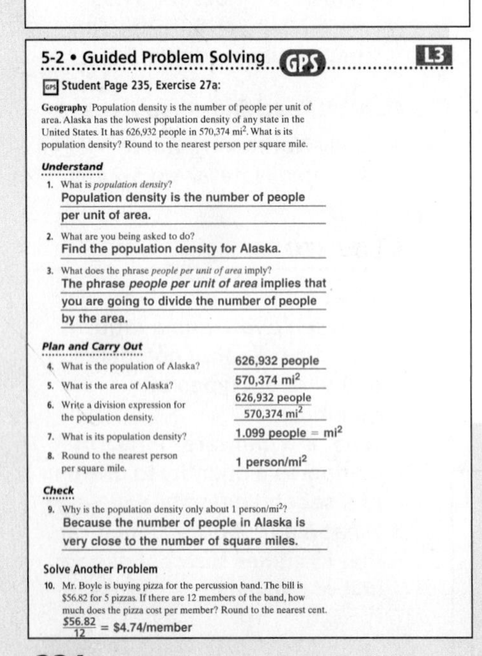

5-2 • Guided Problem Solving GPS L3

Student Page 235, Exercise 27a:

Geography Population density is the number of people per unit of area. Alaska has the lowest population density of any state in the United States. It has 626,932 people in 570,374 mi². What is its population density? Round to the nearest person per square mile.

Understand

1. What is *population density*? — Population density is the number of people per unit of area.
2. What are you being asked to do? — Find the population density for Alaska.
3. What does the phrase *people per unit of area* imply? — The phrase *people per unit of area* implies that you are going to divide the number of people by the area.

Plan and Carry Out

4. What is the population of Alaska? — 626,932 people
5. What is the area of Alaska? — 570,374 mi²
6. Write a division expression for the population density. — $\frac{626{,}932 \text{ people}}{570{,}374 \text{ mi}^2}$
7. What is its population density? — 1.099 people = mi²
8. Round to the nearest person per square mile. — 1 person/mi²

Check

9. Why is the population density only about 1 person/mi²? — Because the number of people in Alaska is very close to the number of square miles.

Solve Another Problem

10. Mr. Boyle is buying pizza for the percussion band. The bill is $56.82 for 5 pizzas. If there are 12 members of the band, how much does the pizza cost per member? Round to the nearest cent. — $\frac{\$56.82}{12}$ = $4.74/member

Check Your Understanding

1. **Vocabulary** What makes a unit rate a unit cost? **Answers may vary. Sample: It gives the cost of one item.**

Find the unit rate for each situation by filling in the blanks.

2. skating 1,000 m in 200 s: $\frac{\text{meters}}{\text{seconds}} \to \frac{1{,}000}{\blacksquare}$ = 5 meters per second **200**
3. earning $147 in 21 h: $\frac{\text{dollars}}{\text{hours}} \to \frac{\blacksquare}{21}$ = ■ dollars per hour **147; 7**
4. spending $89 in 5 h: $\frac{\text{dollars}}{\text{hours}} \to \frac{89}{\blacksquare}$ = ■ dollars per hour **5; 17.80**
5. Use the model at the right to find the total amount of fat in 4 servings. **8 g**

4 servings → ← ■ grams
1 serving → ← 2 grams

Homework Exercises

For more exercises, see Extra Skills and Word Problems.

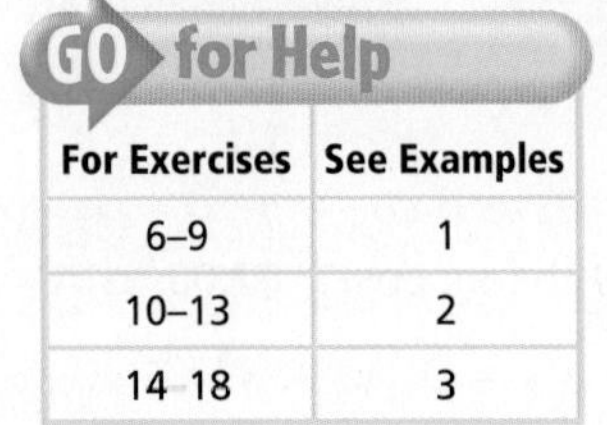

GO for Help

For Exercises	See Examples
6–9	1
10–13	2
14–18	3

A **Find the unit rate for each situation. Round to the nearest hundredth, if necessary.**

6. traveling 1,200 mi in 4 h **300 mi/h**
7. scoring 96 points in 6 games **16 points/game**
8. reading 53 pages in 2 h **26.5 pages/h**
9. 592 students in 17 classrooms **34.82 students/classroom**

Find the total cost using each unit cost.

10. 5 ft at $3 per foot **$15**
11. 15 yd² at $2 per square yard **$30**
12. 10 gal at $2.40 per gallon **$24**
13. 26 oz at $.15 per ounce **$3.90**

Test Prep Tip

Drawing a model can help you find unit rates.

Find each unit cost.

14. $12 for 4 yd² **$3/yd²**
15. $3.45 for 3.7 oz **$.93/oz**
16. $9 for 5 L **$1.80/L**

Which size is the better buy? **17–18. See margin.**

17. detergent: 32 fl oz for $1.99; 50 fl oz for $2.49
18. crackers: 12 oz for $2.69; 16 oz for $3.19

B

19. **Guided Problem Solving** A school has 945 students and 35 teachers. If the numbers of teachers and students both increase by 5, does the unit rate remain the same? Explain. **See margin.**
    - Find unit rates. $\frac{\text{students}}{\text{teachers}} \to \frac{\blacksquare}{\blacksquare}$ = ■ students per teacher
20. **Biking** You bike 18.25 km in 1 h 45 min. What is the unit rate in kilometers per minute? In meters per minute? **See left.**
21. **Crafts** The costs for three different types of ribbon are $.79 for 1 yd, $1.95 for 3 yd, and $2.94 for 6 yd. Which is the best buy?

20. 0.1738 km/min; 173.8 m/min

21. $2.94 for 6 yd

17. $.06/fl oz, $.05/fl oz; the 50-fl-oz detergent is the better buy.
18. $.22/oz, $.20/oz; the 16-oz crackers are the better buy.
19. No; the new student-to-teacher ratio would be 950 students : 40 teachers, or 23.75 students : 1 teacher.

GO Online
Homework Video Tutor
Visit: PHSchool.com
Web Code: are-0502

28. A rate is a ratio that compares two quantities measured in different units. A unit rate is the rate for one unit of something.

Find each unit cost.

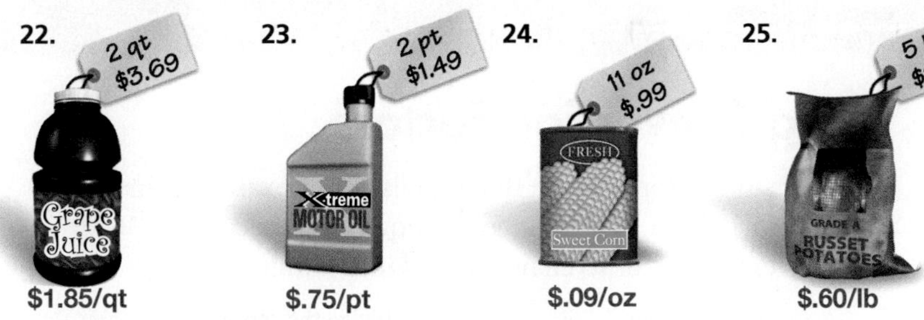

22. $1.85/qt **23.** $.75/pt **24.** $.09/oz **25.** $.60/lb

26. The world record for the women's 3,000-m steeplechase is 9 min 1.59 s. Find the runner's speed in meters per second. Round your answer to the nearest hundredth. **5.54 m/s**

27. Geography Population density is the number of people per unit of area. **1 person/mi^2**
GPS
a. Alaska has the lowest population density of any state in the United States. It has 626,932 people in 570,374 mi^2. What is its population density? Round to the nearest person per square mile.
b. Reasoning New Jersey has 1,134.5 people/mi^2. Can you conclude that 1,134.5 people live in every square mile in New Jersey? Explain. **No; answers may vary. Sample: Some regions of the state are more densely populated than others.**

28. Writing in Math Explain the difference between a rate and a unit rate. **See above left.**

C 29. Challenge A human heart beats an average of 2,956,575,000 times in 75 years. About how many times does a heart beat in one year? In one day? In one minute?
about 39,421,000 times; about 108,000 times; about 75 times

Test Prep and Mixed Review — Practice

Multiple Choice

30. A grocery store sells apple juice in these sizes: 64 ounces for $2.48, 128 ounces for $4.48, and 48 ounces for $1.92. Which size has the lowest unit cost? **C**
(A) 48-oz only (C) 128-oz only
(B) 64-oz only (D) 128-oz and 64-oz

31. There are 12 teams in a soccer league. Each team must play every other team once. How many games will be played in all? **G**
(F) 60 games (G) 66 games (H) 72 games (J) 132 games

32. Aaron received the following scores on different tests: 98, 79, 85, 92, 88. Which measure of the data is represented by 19 points? **B**
(A) median (B) range (C) mean (D) mode

GO for Help

For Exercises	See Lesson
33–35	2-1

Use paper and pencil, mental math, or a calculator to simplify.

33. $3^2 + 4 \cdot 5$ **29** **34.** $(6 - 3)^3 - 1$ **26** **35.** $2 \cdot 4 - 5^2$ **–17**

Online lesson quiz, **PHSchool.com,** **Web Code:** ara-0502

4. Assess & Reteach

PowerPoint

Lesson Quiz

1. It took Anthony 5 h 45 min to complete a 25-mi walkathon. What was his average speed in miles per hour? Round your answer to the nearest hundredth. **≈ 4.35 mi/h**
2. A 20-lb bag of dog food costs $21.50. A 30-lb bag of the same dog food costs $32.90. Find each unit cost. Then determine the better buy.
≈ $1.08/lb;
≈ $1.10/lb; 20 lb for $21.50

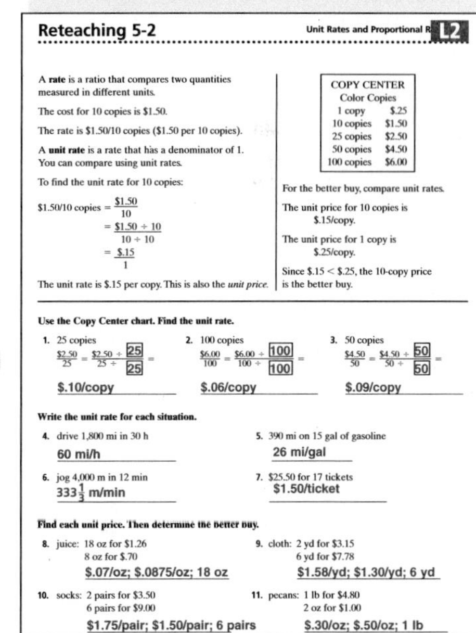

Reteaching 5-2 Unit Rates and Proportional R... L2

A **rate** is a ratio that compares two quantities measured in different units.

The cost for 10 copies is $1.50.

The rate is $1.50/10 copies ($1.50 per 10 copies).

A **unit rate** is a rate that has a denominator of 1. You can compare using unit rates.

To find the unit rate for 10 copies:

$1.50/10 copies $= \frac{\$1.50}{10} = \frac{\$1.50 \div 10}{10 \div 10} = \frac{\$.15}{1}$

The unit rate is $.15 per copy. This is also the *unit price.*

COPY CENTER Color Copies	
1 copy	$.25
10 copies	$1.50
25 copies	$2.50
50 copies	$4.50
100 copies	$6.00

For the better buy, compare unit rates.

The unit price for 10 copies is $.15/copy.

The unit price for 1 copy is $.25/copy.

Since $.15 < $.25, the 10-copy price is the better buy.

Use the Copy Center chart. Find the unit rate.

1. 25 copies $\frac{\$2.50}{25} = \frac{\$2.50 \div 25}{25 \div 25} =$ **$.10/copy**
2. 100 copies $\frac{\$6.00}{100} = \frac{\$6.00 \div 100}{100 \div 100} =$ **$.06/copy**
3. 50 copies $\frac{\$4.50}{50} = \frac{\$4.50 \div 50}{50 \div 50} =$ **$.09/copy**

Write the unit rate for each situation.

4. drive 1,800 mi in 30 h **60 mi/h**
5. 390 mi on 15 gal of gasoline **26 mi/gal**
6. jog 4,000 m in 12 min **$333\frac{1}{3}$ m/min**
7. $25.50 for 17 tickets **$1.50/ticket**

Find each unit price. Then determine the better buy.

8. juice: 18 oz for $1.26, 8 oz for $.70 **$.07/oz; $.0875/oz; 18 oz**
9. cloth: 2 yd for $3.15, 6 yd for $7.78 **$1.58/yd; $1.30/yd; 6 yd**
10. socks: 2 pairs for $3.50, 6 pairs for $9.00 **$1.75/pair; $1.50/pair; 6 pairs**
11. pecans: 1 lb for $4.80, 2 oz for $1.00 **$.30/oz; $.50/oz; 1 lb**

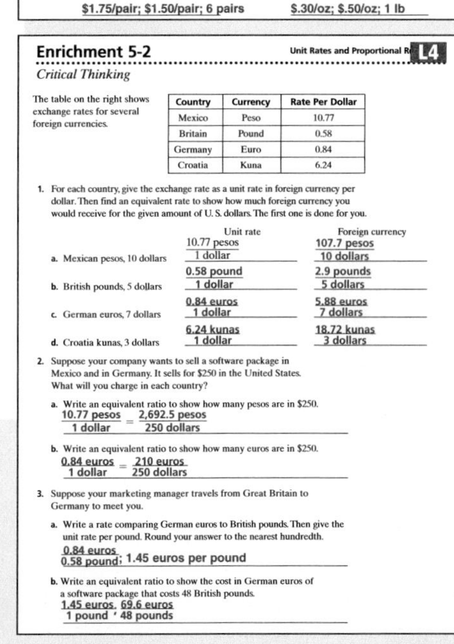

Enrichment 5-2 Unit Rates and Proportional R... L4

Critical Thinking

The table on the right shows exchange rates for several foreign currencies.

Country	Currency	Rate Per Dollar
Mexico	Peso	10.77
Britain	Pound	0.58
Germany	Euro	0.84
Croatia	Kuna	6.24

1. For each country, give the exchange rate as a unit rate in foreign currency per dollar. Then find an equivalent rate to show how much foreign currency you would receive for the given amount of U. S. dollars. The first one is done for you.

	Unit rate	Foreign currency
a. Mexican pesos, 10 dollars	10.77 pesos / 1 dollar	**107.7 pesos / 10 dollars**
b. British pounds, 5 dollars	**0.58 pound / 1 dollar**	**2.9 pounds / 5 dollars**
c. German euros, 7 dollars	**0.84 euros / 1 dollar**	**5.88 euros / 7 dollars**
d. Croatia kunas, 3 dollars	**6.24 kunas / 1 dollar**	**18.72 kunas / 3 dollars**

2. Suppose your company wants to sell a software package in Mexico and in Germany. It sells for $250 in the United States. What will you charge in each country?
a. Write an equivalent ratio to show how many pesos are in $250.
10.77 pesos / 1 dollar = 2,692.5 pesos / 250 dollars
b. Write an equivalent ratio to show how many euros are in $250.
0.84 euros / 1 dollar = 210 euros / 250 dollars

3. Suppose your marketing manager travels from Great Britain to Germany to meet you.
a. Write a rate comparing German euros to British pounds. Then give the unit rate per pound. Round your answer to the nearest hundredth.
0.84 euros / 0.58 pound; 1.45 euros per pound
b. Write an equivalent ratio to show the cost in German euros of a software package that costs 48 British pounds.
1.45 euros / 1 pound , 69.6 euros / 48 pounds

Alternative Assessment

Ask each student to write the steps needed to find the better buy, given two costs. Have students trade steps with a partner and use the partner's steps to do Exercises 17 and 18. Afterwards, partners should discuss their steps and revise them accordingly.

Test Prep

Resources
For additional practice with a variety of test item formats:
- Test-Taking Strategies, p. 265
- Test Prep, p. 269
- Test-Taking Strategies with Transparencies

Extension

Using Conversion Factors

This extension of Lesson 5-2 teaches students how to generate formulas involving unit conversions. Students learn to use ratios as a means of converting a measurement from one system of measurement to another.

Guided Instruction

Ask students what measures they often see in metric units and why they might want to convert to customary units. Then ask what measures they see in customary units that they might want to convert to metric.

Example 1

Ask: *Is an inch or a centimeter longer?* inch *How do you know?* There are 2.54 cm in an inch. *Should there be more or fewer centimeters in 15 inches?* more

Science Connection

The metric system measures mass, not weight, in grams and kilograms. In order to make conversions, the language "weighs as much as" is used.

Using Conversion Factors

A conversion factor is a rate that equals 1. For example, since 60 min = 1 h, both $\frac{60\text{ min}}{1\text{ h}}$ and $\frac{1\text{ h}}{60\text{ min}}$ equal 1. You can use $\frac{60\text{ min}}{1\text{ h}}$ as a conversion factor to change hours into minutes.

$$7\text{ h} = \frac{7\text{ h}}{1} \cdot \frac{60\text{ min}}{1\text{ h}} = 420\text{ min}$$ ← **Divide the common unit, hours (h). The result is in minutes.**

The table shows some common conversion factors for converting between the metric system and the customary system.

Conversion Factors

Length
1 in. = 2.54 cm
1 km ≈ 0.62 mi
Capacity
1 L ≈ 1.06 qt
Weight and Mass
1 oz ≈ 28 g
1 kg ≈ 2.2 lb

EXAMPLE 1 Converting to the Metric System

Convert 15 inches to centimeters.

$15\text{ in.} = \frac{15\text{ in.}}{1} \cdot \frac{2.54\text{ cm}}{1\text{ in.}}$ ← **Use $\frac{2.54\text{ cm}}{1\text{ in.}}$ since conversion is to centimeters. Divide the common units.**

$= (15)(2.54)\text{ cm}$ ← **Simplify.**

$= 38.1\text{ cm}$ ← **Multiply.**

EXAMPLE 2 Converting to the Customary System

The mass of a western diamondback rattlesnake is about 6.7 kg. How many pounds does the snake weigh? Round to the nearest tenth.

$6.7\text{ kg} = \frac{6.7\text{ kg}}{1} \cdot \frac{2.2\text{ lb}}{1\text{ kg}}$ ← **Use $\frac{2.2\text{ lb}}{1\text{ kg}}$ since conversion is to pounds.**

$= (6.7)(2.2)\text{ lb}$ ← **Simplify.**

$= 14.74\text{ lb}$ ← **Multiply.**

$\approx 14.7\text{ lb}$ ← **Round to the nearest tenth.**

Exercises

Write a conversion factor you can use to convert each measure.

1. kilometers to miles $\frac{0.62\text{ mi}}{1\text{ km}}$
2. liters to quarts $\frac{1.06\text{ qt}}{1\text{ L}}$
3. ounces to grams $\frac{28\text{ g}}{1\text{ oz}}$

Use a conversion factor to convert each measure. Round to the nearest tenth.

4. 22 in. ≈ ■ cm 55.9
5. 26.4 lb ≈ ■ kg 12
6. 20.5 oz ≈ ■ g 574
7. 500 g ≈ ■ oz 17.9
8. 5 km ≈ ■ mi 3.1
9. 20 L ≈ ■ qt 21.2

Checkpoint Quiz 1

Lessons 5-1 through 5-2

1. Write the ratio 7 : 52 in two other ways. **7 to 52, $\frac{7}{52}$**

Write each ratio in simplest form.

2. $\frac{4}{6}$ **$\frac{2}{3}$**

3. $\frac{16}{48}$ **$\frac{1}{3}$**

4. 24 to 14 **12 to 7**

5. 18 : 27 **2 : 3**

Write a unit rate for each situation.

6. typing 126 words in 3 min **42 words/min**

7. scoring 45 points in 5 games **9 points/game**

Find each unit cost. Which is the better buy?

8. 3 for $.79, 4 for $.99 **$.2633, $.2475; the second item is the better buy.**

9. 5 for $39, 7 for $46 **$7.80, $6.57; the second item is the better buy.**

10. The last time you bought pizza, 3 pizzas were just enough for 7 people. At that rate, how many pizzas should you buy for a party for 35 people? **15 pizzas**

MATH AT WORK

Automotive Engineer

Automotive engineers design, develop, and test all kinds of vehicles. They also test and evaluate a design's cost, reliability, and safety.

Engineers have many opportunities to use math skills. They use problem-solving skills to find out why cars break down.

Computer-aided design systems help automotive engineers plan the cars of the future. Computer simulations allow them to test for quality and to predict how their designs will work in the real world.

Go Online PHSchool.com **For:** Information on automotive engineers **Web Code:** arb-2031

Checkpoint Quiz

Use this Checkpoint Quiz to check students' understanding of the skills and concepts of Lessons 5-1 through 5-2.

Resources

- All-in-One Teaching Resources Checkpoint Quiz 1
- ExamView CD-ROM
- Success Tracker™ Online Intervention

MATH AT WORK

Automotive Engineer

Students may not be aware of some unusual jobs where they can use their math skills. The application of math to this career should generate enthusiasm.

Guided Instruction

Have different students read each paragraph. After students finish reading the selection, discuss designing cars for the future.

Then ask questions, such as:

- *What types of skills does an automotive engineer need?*
- *How does an automotive engineer use math to design models?*

5-3

1. Plan

Objective
To test whether ratios form a proportion by using equivalent ratios and cross products

Examples
1 Writing Ratios in Simplest Form
2 Using Cross Products

Math Understandings: p. 226D

Professional Development

Math Background

A ratio is an expression that shows a comparison of two numbers by division. That is, a ratio is a "mathematical phrase." In contrast, a proportion is a "mathematical sentence." It is an equation that declares the equality of two ratios. You can determine if two ratios form a proportion by looking for equivalent fractions or by using *cross products.*

More Math Background: p. 226D

Lesson Planning and Resources

See p. 226E for a list of the resources that support this lesson.

Bell Ringer Practice

Check Skills You'll Need
Use student page, transparency, or PowerPoint. For intervention, direct students to:
Comparing and Ordering Fractions
Lesson 2-4
Extra Skills and Word Problems Practice, Ch. 2

Algebra

5-3 Proportions

Check Skills You'll Need

1. **Vocabulary Review** The *LCD* of two fractions is the __?__ of the fractions' denominators. **1–5. See below.**

Compare. Use <, =, or >.

2. $\frac{3}{4}$ ■ $\frac{5}{8}$ 3. $\frac{3}{7}$ ■ $\frac{12}{28}$

4. $\frac{2}{3}$ ■ $\frac{11}{15}$ 5. $\frac{1}{6}$ ■ $\frac{1}{12}$

GO for Help Lesson 2-4

Check Skills You'll Need
1. LCM
2. >
3. =
4. <
5. >

GO Online

Video Tutor Help
Visit: PHSchool.com
Web Code: are-0775

What You'll Learn

To test whether ratios form a proportion by using equivalent ratios and cross products

New Vocabulary proportion, cross products

Why Learn This?

Pollsters conduct surveys. They ask different groups of people the same questions. You can use proportions to compare the answers from the different groups.

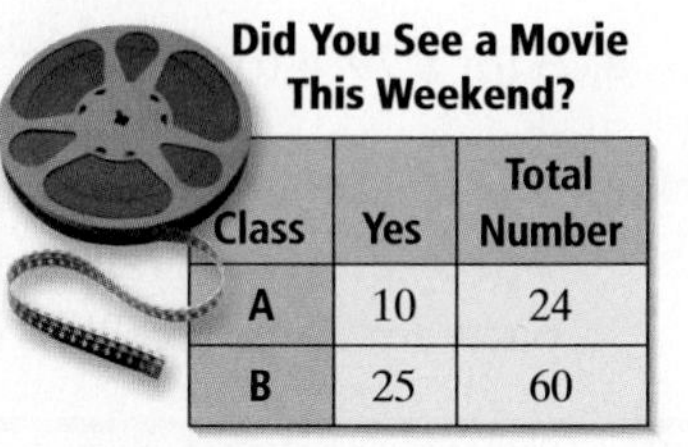

Did You See a Movie This Weekend?

Class	Yes	Total Number
A	10	24
B	25	60

A **proportion** is an equation stating that two ratios are equal. One method of testing whether ratios form a proportion is to write both ratios in simplest form. Then see if they are equal.

EXAMPLE Writing Ratios in Simplest Form

1 Surveys Refer to the table above. For each class, write the ratio of the number of students who saw a movie to the total number of students. Do the ratios form a proportion?

Class A: $\frac{10}{24} = \frac{10 \div 2}{24 \div 2} = \frac{5}{12}$ ← **Divide 10 and 24 by their GCF, which is 2.**

Class B: $\frac{25}{60} = \frac{25 \div 5}{60 \div 5} = \frac{5}{12}$ ← **Divide 25 and 60 by their GCF, which is 5.**

Since both ratios are equal to $\frac{5}{12}$, the ratios are proportional.

Quick Check

1. Do $\frac{10}{12}$ and $\frac{40}{56}$ form a proportion? **no; $\frac{5}{6} \neq \frac{5}{7}$**

KEY CONCEPTS Proportion

Ratios that are equal form a proportion.

Arithmetic	Algebra
$\frac{6}{8} = \frac{9}{12}$	$\frac{a}{b} = \frac{c}{d}, b \neq 0, d \neq 0$

Differentiated Instruction Solutions for All Learners

Special Needs L1
To visualize proportions, show students a one inch length and a one foot length. Ask: *If we wanted to measure the distance across your desk, we would need more of one of these units than the other. Which one? Why?* **The shorter one. It takes up less distance so you need more.**
learning style: visual

Below Level L2
Give students a ratio such as $\frac{12}{18}$ and have them write as many ratios equal to it as they can in two minutes. Discuss their results.

$\frac{12}{18} = \frac{6}{9} = \frac{4}{6} = \frac{2}{3} = \frac{8}{12} = \frac{16}{24} = \frac{24}{36} \ldots$

learning style: visual

You can use the properties of equality to discover another way to determine whether ratios form a proportion.

$\frac{6}{8} = \frac{9}{12}$ ← Use the ratios from the beginning of the lesson.

$\frac{6}{8}\left(\frac{8}{1} \cdot \frac{12}{1}\right) = \frac{9}{12}\left(\frac{8}{1} \cdot \frac{12}{1}\right)$ ← Use the Multiplication Property of Equality. Multiply each side by both denominators.

$\frac{6}{\cancel{8}_1}\left(\frac{\cancel{8}^1}{1} \cdot \frac{12}{1}\right) = \frac{9}{\cancel{12}_1}\left(\frac{8}{1} \cdot \frac{\cancel{12}^1}{1}\right)$ ← Divide numerators and denominators by their GCF.

$6 \cdot 12 = 9 \cdot 8$

The products $6 \cdot 12$ and $9 \cdot 8$ are called cross products. For two ratios, the **cross products** are the two products found by multiplying the denominator of each ratio by the numerator of the other ratio.

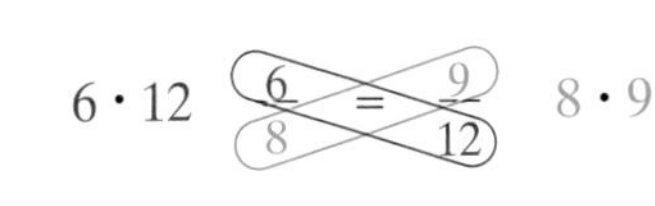

KEY CONCEPTS **Cross-Products Property**

If two ratios form a proportion, the cross products are equal. If two ratios have equal cross products, they form a proportion.

Arithmetic	Algebra
$\frac{6}{8} = \frac{9}{12}$	$\frac{a}{b} = \frac{c}{d}$
$6 \cdot 12 = 8 \cdot 9$	$ad = bc$, where $b \neq 0$ and $d \neq 0$

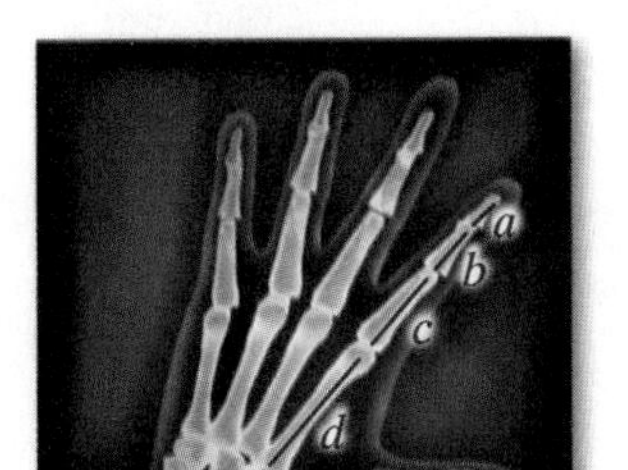

The lengths of the bones in your hands form a proportion.

SOURCE: *Fascinating Fibonaccis*

You can use the Cross-Products Property to determine whether ratios form a proportion.

EXAMPLE Using Cross Products

2 Do the ratios in each pair form a proportion?

a. $\frac{5}{9}, \frac{30}{54}$

$\frac{5}{9} \stackrel{?}{=} \frac{30}{54}$ ← Test each pair of ratios.

$5 \cdot 54 \stackrel{?}{=} 9 \cdot 30$ ← Write cross products.

$270 = 270$ ← Simplify.

Yes, $\frac{5}{9}$ and $\frac{30}{54}$ form a proportion.

b. $\frac{7}{8}, \frac{55}{65}$

$\frac{7}{8} \stackrel{?}{=} \frac{55}{65}$

$7 \cdot 65 \stackrel{?}{=} 8 \cdot 55$

$455 \neq 440$

No, $\frac{7}{8}$ and $\frac{55}{65}$ do *not* form a proportion.

Quick Check

2. Determine whether the ratios form a proportion.

a. $\frac{3}{8}, \frac{6}{16}$ yes; $48 = 48$ **b.** $\frac{6}{9}, \frac{4}{6}$ yes; $36 = 36$ **c.** $\frac{4}{8}, \frac{5}{9}$ no; $36 \neq 40$

2. Teach

Activity Lab

Use before the lesson.

All in One Teaching Resources

Activity Lab 5-3: Math Reasoning

Guided Instruction

Example 1

Students might recall that another method for determining whether two ratios are equivalent is to write each as a decimal. If students use calculators to do this, suggest that they store the decimal for the first ratio in the calculator's memory in order to have it available for comparison.

Example 2

Reinforce the concept of cross products by pointing out that the ovals connecting the pairs of numbers form a large "X," which is a symbol for multiplication.

PowerPoint

Additional Examples

Determine if the ratios can form a proportion.

1 $\frac{42}{56}, \frac{56}{64}$ no

2 **a.** $\frac{2}{3}, \frac{18}{24}$ no **b.** $\frac{5}{4}, \frac{30}{24}$ yes

3 **a.** $\frac{4}{10}, \frac{6}{15}$ yes **b.** $\frac{8}{6}, \frac{9}{7}$ no

All in One Teaching Resources

- Daily Notetaking Guide 5-3 L3
- Adapted Notetaking 5-3 L1

Closure

- Define proportion. **An equation stating that two ratios are equal.**
- Summarize the three methods for determining whether a pair of ratios can form a proportion. **Write ratios in simplest form and check for equality, find a common multiplier between numerators and denominators, and find cross products and check for equality.**

Advanced Learners L4

Have students write four proportions for which $4 \cdot 20$ and $5 \cdot 16$ can be the cross products.

$\frac{4}{5} = \frac{16}{20}$; $\frac{5}{4} = \frac{20}{16}$; $\frac{4}{16} = \frac{5}{20}$; $\frac{16}{4} = \frac{20}{5}$

learning style: visual

English Language Learners ELL

Remind students that when they convert units, this rule usually applies: To convert from larger to smaller units, multiply. To convert from smaller to larger, divide. Have them talk about why this is true, and use several examples to verify it.

learning style: verbal

3. Practice

Assignment Guide

Check Your Understanding

Go over Exercises 1–9 in class before assigning the Homework Exercises.

Homework Exercises

A Practice by Example	10–27
B Apply Your Skills	28–38
C Challenge	39
Test Prep and Mixed Review	40–45

Homework Quick Check

To check students' understanding of key skills and concepts, go over Exercises 14, 22, 29, 34, and 38.

Differentiated Instruction Resources

Adapted Practice 5-3 L1

Practice 5-3 L3

Determine if the ratios in each pair are proportional.

1. $\frac{12}{16}, \frac{30}{40}$ yes
2. $\frac{8}{12}, \frac{15}{21}$ no
3. $\frac{27}{21}, \frac{81}{56}$ no
4. $\frac{45}{24}, \frac{75}{40}$ yes
5. $\frac{5}{9}, \frac{80}{117}$ no
6. $\frac{15}{25}, \frac{75}{125}$ yes
7. $\frac{2}{14}, \frac{20}{35}$ no
8. $\frac{9}{6}, \frac{21}{14}$ yes
9. $\frac{24}{15}, \frac{16}{10}$ yes
10. $\frac{3}{4}, \frac{8}{10}$ no
11. $\frac{20}{4}, \frac{17}{3}$ no
12. $\frac{25}{6}, \frac{9}{8}$ no

Decide if each pair of ratios is proportional.

13. $\frac{14}{10} \stackrel{?}{=} \frac{9}{7}$ not proportional
14. $\frac{18}{8} \stackrel{?}{=} \frac{36}{16}$ proportional
15. $\frac{6}{10} \stackrel{?}{=} \frac{15}{25}$ proportional
16. $\frac{7}{16} \stackrel{?}{=} \frac{4}{9}$ not proportional
17. $\frac{6}{4} \stackrel{?}{=} \frac{12}{8}$ proportional
18. $\frac{19}{3} \stackrel{?}{=} \frac{114}{8}$ not proportional
19. $\frac{5}{14} \stackrel{?}{=} \frac{6}{15}$ not proportional
20. $\frac{6}{27} \stackrel{?}{=} \frac{8}{36}$ proportional
21. $\frac{27}{15} \stackrel{?}{=} \frac{45}{25}$ proportional
22. $\frac{3}{18} \stackrel{?}{=} \frac{4}{20}$ not proportional
23. $\frac{5}{2} \stackrel{?}{=} \frac{15}{6}$ proportional
24. $\frac{20}{15} \stackrel{?}{=} \frac{4}{3}$ proportional

Solve.

25. During the breaststroke competitions of the 1992 Olympics, Nelson Diebel swam 100 meters in 62 seconds, and Mike Bowerman swam 200 meters in 130 seconds. Are the rates proportional?
no

26. During a vacation, the Vasquez family traveled 174 miles in 3 hours on Monday, and 290 miles in 5 hours on Tuesday. Are the rates proportional?
yes

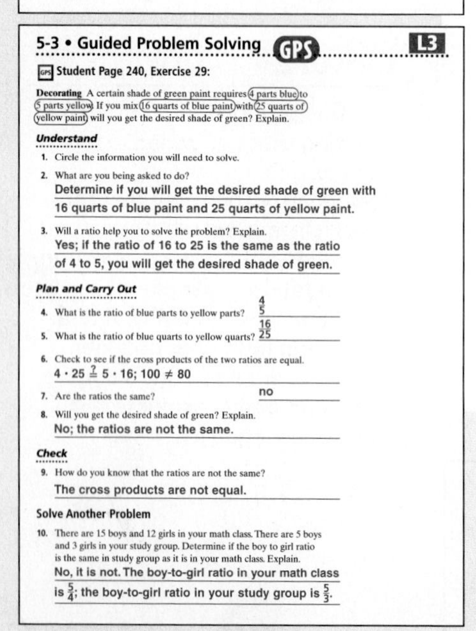

5-3 • Guided Problem Solving GPS L3

GPS Student Page 240, Exercise 29:

Decorating A certain shade of green paint requires 4 parts blue to 5 parts yellow. If you mix 16 quarts of blue paint with 25 quarts of yellow paint, will you get the desired shade of green? Explain.

Understand

1. Circle the information you will need to solve.
2. What are you being asked to do?
Determine if you will get the desired shade of green with 16 quarts of blue paint and 25 quarts of yellow paint.
3. Will a ratio help you to solve the problem? Explain.
Yes; if the ratio of 16 to 25 is the same as the ratio of 4 to 5, you will get the desired shade of green.

Plan and Carry Out

4. What is the ratio of blue parts to yellow parts? $\frac{4}{5}$
5. What is the ratio of blue quarts to yellow quarts? $\frac{16}{25}$
6. Check to see if the cross products of the two ratios are equal.
$4 \cdot 25 \stackrel{?}{=} 5 \cdot 16$; $100 \neq 80$
7. Are the ratios the same? no
8. Will you get the desired shade of green? Explain.
No; the ratios are not the same.

Check

9. How do you know that the ratios are not the same?
The cross products are not equal.

Solve Another Problem

10. There are 15 boys and 12 girls in your math class. There are 5 boys and 3 girls in your study group. Determine if the boy to girl ratio is the same in study group as it is in your math class. Explain.
No, it is not. The boy-to-girl ratio in your math class is $\frac{5}{4}$; the boy-to-girl ratio in your study group is $\frac{5}{3}$.

Check Your Understanding

2. If you multiply both the numerator and denominator of the first fraction by 3, you get the second fraction. $\frac{3}{5} \times \frac{3}{3} = \frac{9}{15}$

9. No; you need to multiply the numerator of one by the denominator of the other.
$\frac{3}{4} \stackrel{?}{=} \frac{12}{16}$
$3 \cdot 16 \stackrel{?}{=} 4 \cdot 12$
$48 = 48$
They do form a proportion.

1. **Vocabulary** A proportion states that two __?__ are equal. **ratios**

2. **Number Sense** Without writing $\frac{3}{5}$ and $\frac{9}{15}$ in simplest form or using the Cross-Products Property, how can you tell whether the ratios form a proportion? **See left.**

Fill in the blank so that each pair of ratios forms a proportion.

3. $\frac{1}{2}, \frac{4}{\blacksquare}$ **8**
4. $\frac{3}{3}, \frac{9}{\blacksquare}$ **9**
5. $\frac{3}{4}, \frac{\blacksquare}{12}$ **9**
6. $\frac{4}{7}, \frac{8}{\blacksquare}$ **14**
7. $\frac{2}{3}, \frac{\blacksquare}{18}$ **12**
8. $\frac{\blacksquare}{5}, \frac{6}{10}$ **3**

9. **Error Analysis** A student used the Cross-Products Property to determine that $\frac{3}{4}$ and $\frac{12}{16}$ do not form a proportion. His work is shown at the right. Is he correct? Explain. **See left.**

$\frac{3}{4} \stackrel{?}{=} \frac{12}{16}$
$3 \cdot 12 \stackrel{?}{=} 4 \cdot 16$
$36 \neq 64$

Homework Exercises

For more exercises, see Extra Skills and Word Problems.

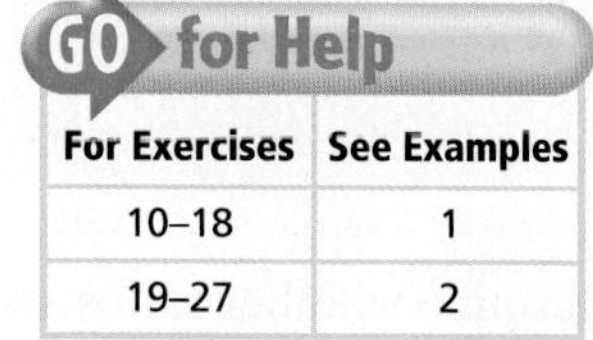

For Exercises	See Examples
10–18	1
19–27	2

A Determine whether the ratios can form a proportion.

10. $\frac{1}{2}, \frac{14}{28}$ **yes**
11. $\frac{6}{8}, \frac{4}{3}$ **no**
12. $\frac{8}{18}, \frac{20}{45}$ **yes**
13. $\frac{21}{24}, \frac{56}{64}$ **yes**
14. $\frac{15}{45}, \frac{3}{15}$ **no**
15. $\frac{45}{9}, \frac{10}{2}$ **yes**
16. $\frac{19}{76}, \frac{5}{20}$ **yes**
17. $\frac{17}{34}, \frac{2}{3}$ **no**
18. $\frac{40}{12}, \frac{160}{3}$ **no**
19. $\frac{6}{10}, \frac{9}{15}$ **yes**
20. $\frac{4}{5}, \frac{10}{13}$ **no**
21. $\frac{7}{8}, \frac{15}{18}$ **no**
22. $\frac{6}{14}, \frac{3}{7}$ **yes**
23. $\frac{7}{22}, \frac{28}{77}$ **no**
24. $\frac{12}{15}, \frac{20}{25}$ **yes**
25. $\frac{6}{10}, \frac{24}{42}$ **no**
26. $\frac{5}{9}, \frac{15}{27}$ **yes**
27. $\frac{3}{10}, \frac{15}{25}$ **no**

28. No; $\frac{5}{2}$ and $\frac{34}{12}$ are not proportional because $5 \times 12 \neq 2 \times 34$.

29. No; $\frac{4}{5}$ and $\frac{16}{25}$ are not proportional because $4 \times 25 \neq 5 \times 16$.

B 

28. **Guided Problem Solving** Your boat engine needs 5 fl oz of oil mixed with every 2 gal of gas. A gas container has 12 gal of gas mixed with 34 fl oz of oil. Is this the correct mixture? Explain.
   - **Make a Plan** Write the ratio of oil to gas for the boat and for the gas container. Determine whether the ratios form a proportion.
   - **Carry Out the Plan** The ratios of oil to gas are $\frac{\blacksquare \text{ fl oz}}{2 \text{ gal}}$ and $\frac{\blacksquare \text{ fl oz}}{12 \text{ gal}}$. **See above left.**

29. **Decorating** A certain shade of green paint requires 4 parts blue to 5 parts yellow. If you mix 16 quarts of blue paint with 25 quarts of yellow paint, will you get the desired shade of green? Explain. **See left.**

34. Answers may vary. Sample: Use cross products to find that $ab = ab + b^2$, which is only true if $b = 0$, which makes the ratios undefined.

35. Yes; the weight ratios $\frac{174}{29}$ and $\frac{102}{17}$ are proportional because $174 \cdot 17 = 102 \cdot 29$.

36. Yes; $\frac{b}{h} = \frac{20}{15} = \frac{12}{9}$ is a proportion.

37. yes; $\frac{26}{1} = \frac{104}{4}$

38. No, the ratios $\frac{18}{6}$ and $\frac{12}{3}$ are not proportional because $18 \times 3 \neq 6 \times 12$.

39. $\frac{4n}{3}$ and $\frac{12n}{9}$ will always form a proportion because $36n = 36n$ for all values of n.

Visit: PHSchool.com
Web Code: are-0503

Do the ratios in each pair form a proportion?

30. $\frac{56}{2}, \frac{110}{3}$ no 31. $\frac{18}{12}, \frac{4.8}{3.6}$ no 32. $\frac{20}{1.5}, \frac{60}{4.5}$ yes 33. $\frac{3.5}{35}, \frac{2.04}{204}$ no

34. **Writing in Math** Explain why $\frac{a}{b}$ and $\frac{a+b}{b}$ can *never* form a proportion. **See margin.**

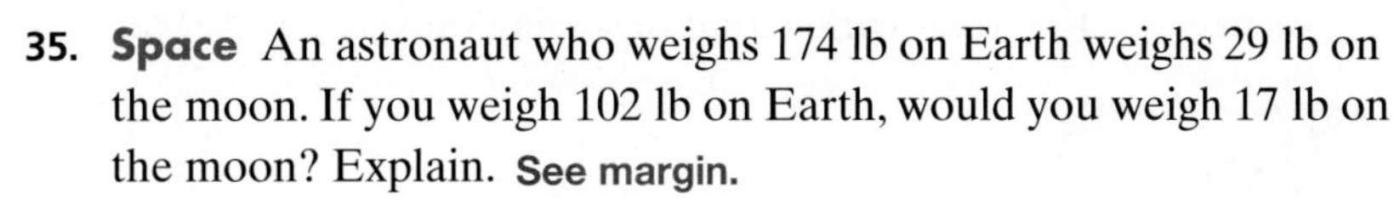

35. **Space** An astronaut who weighs 174 lb on Earth weighs 29 lb on the moon. If you weigh 102 lb on Earth, would you weigh 17 lb on the moon? Explain. **See margin.**

36. **Geometry** Is the ratio of b to h the same in both triangles? Explain your reasoning. **See margin.**

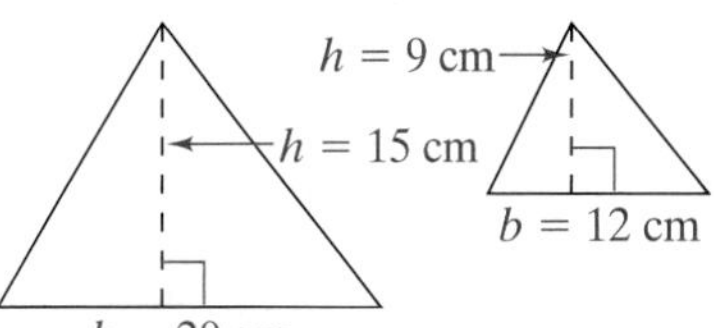

37. An elephant's heart rate was measured at 26 beats in 1 minute and 104 beats in 4 minutes. Are the rates proportional? **See margin.**

38. **Physical Science** Eighteen-karat gold contains 18 parts gold and 6 parts other metals. A ring contains 12 parts gold and 3 parts other metals. Is the ring eighteen-karat gold? Explain. **See margin.**

Careers Astronauts pilot spacecraft or work on science projects in space.

C 39. **Challenge** Determine whether $\frac{4n}{3}$ and $\frac{12n}{9}$ *always, sometimes,* or *never* form a proportion. Explain. **See margin.**

Test Prep and Mixed Review

Practice

Multiple Choice

40. Which ratio does NOT form a proportion with $\frac{5}{8}$? C

Ⓐ $\frac{20}{32}$ Ⓑ $\frac{100}{160}$ Ⓒ $\frac{45}{56}$ Ⓓ $\frac{10}{16}$

41. Which expression does the model below represent? H

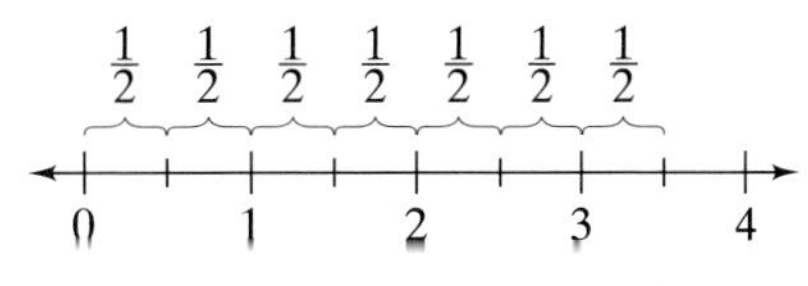

Ⓕ $3\frac{1}{2} - \frac{1}{2}$ Ⓖ $\frac{1}{2} \div 3\frac{1}{2}$ Ⓗ $3\frac{1}{2} \div \frac{1}{2}$ Ⓙ $\frac{1}{2} \times 3\frac{1}{2}$

42. Carrie buys T-shirts for \$6.25 each. After paying to have her school mascot printed on the shirts, she sells each one for \$9.50. Carrie plans to sell 12 T-shirts at a basketball game. What missing information is needed to find how much profit per shirt she will make? A

Ⓐ Cost to print the mascot Ⓒ Sizes of the T-shirts
Ⓑ Number of fans at the game Ⓓ Price of each T-shirt

GO for Help

For Exercises	See Lesson
43–45	4-3

Algebra **Solve each equation.**

43. $y - 37 = 68$ 105 44. $m + 59 = -348$ −407 45. $b + 175 = 102$ −73

Online lesson quiz, PHSchool.com, Web Code: ara-0503

4. Assess & Reteach

Lesson Quiz

Determine if the ratios can form a proportion.

1. $\frac{35}{56}, \frac{45}{63}$ no 2. $\frac{9}{13}, \frac{18}{26}$ yes
3. $\frac{12}{9}, \frac{16}{12}$ yes 4. $\frac{30}{16}, \frac{45}{12}$ no

Alternative Assessment

Give students the following ratios.
$\frac{2}{3}$ $\frac{10}{16}$ $\frac{48}{56}$ $\frac{24}{36}$ $\frac{6}{4}$ $\frac{15}{24}$ $\frac{36}{42}$ $\frac{26}{36}$
Have them use the ratios to form as many proportions as possible.
$\frac{2}{3} = \frac{24}{36}$; $\frac{10}{16} = \frac{15}{24}$; $\frac{48}{56} = \frac{36}{42}$

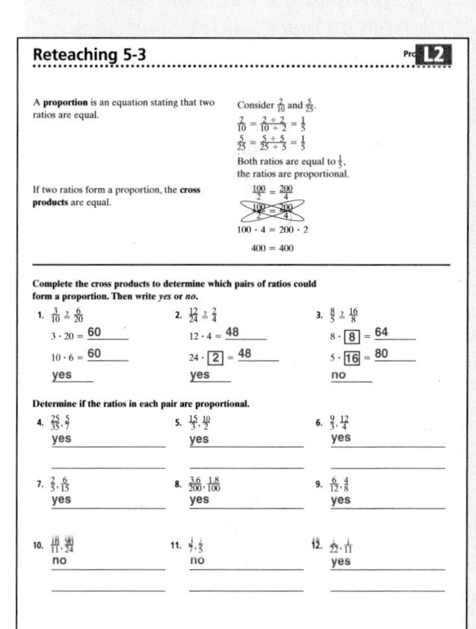

Reteaching 5-3 L2

A **proportion** is an equation stating that two ratios are equal.

Consider $\frac{2}{10}$ and $\frac{5}{25}$.
$\frac{2}{10} = \frac{2 \div 2}{10 \div 2} = \frac{1}{5}$
$\frac{5}{25} = \frac{5 \div 5}{25 \div 5} = \frac{1}{5}$
Both ratios are equal to $\frac{1}{5}$, the ratios are proportional.

If two ratios form a proportion, the **cross products** are equal.

$\frac{100}{2} = \frac{200}{4}$
$100 \cdot 4 = 200 \cdot 2$
$400 = 400$

Complete the cross products to determine which pairs of ratios could form a proportion. Then write *yes* or *no*.

1. $\frac{3}{10} \stackrel{?}{=} \frac{6}{20}$
$3 \cdot 20 =$ 60
$10 \cdot 6 =$ 60
yes

2. $\frac{12}{24} \stackrel{?}{=} \frac{2}{4}$
$12 \cdot 4 =$ 48
$24 \cdot$ 2 $=$ 48
yes

3. $\frac{8}{5} \stackrel{?}{=} \frac{16}{8}$
$8 \cdot$ 8 $=$ 64
$5 \cdot$ 16 $=$ 80
no

Determine if the ratios in each pair are proportional.

4. $\frac{25}{35}, \frac{5}{7}$ yes
5. $\frac{15}{3}, \frac{10}{2}$ yes
6. $\frac{9}{3}, \frac{12}{4}$ yes
7. $\frac{2}{5}, \frac{6}{15}$ yes
8. $\frac{3.6}{200}, \frac{1.8}{100}$ yes
9. $\frac{6}{12}, \frac{4}{8}$ yes
10. [illegible] no
11. [illegible] no
12. [illegible] yes

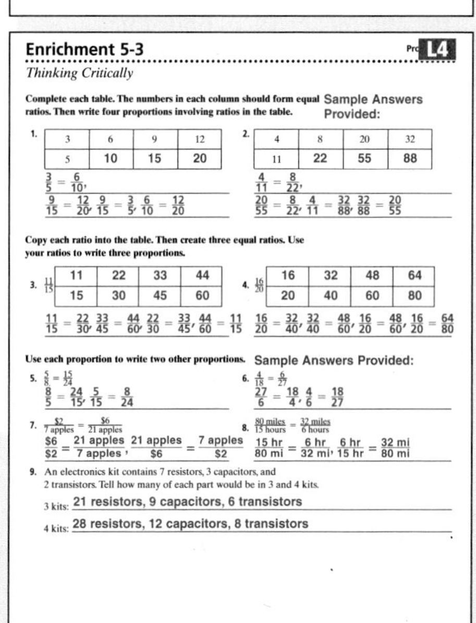

Enrichment 5-3 L4

Thinking Critically

Complete each table. The numbers in each column should form equal ratios. Then write four proportions involving ratios in the table. Sample Answers Provided:

1.

3	6	9	12
5	10	15	20

$\frac{3}{5} = \frac{6}{10}$, $\frac{9}{15} = \frac{12}{20}$, $\frac{9}{15} = \frac{3}{5}$, $\frac{6}{10} = \frac{12}{20}$

2.

4	8	20	32
11	22	55	88

$\frac{4}{11} = \frac{8}{22}$, $\frac{20}{55} = \frac{8}{22}$, $\frac{4}{11} = \frac{32}{88}$, $\frac{32}{88} = \frac{20}{55}$

Copy each ratio into the table. Then create three equal ratios. Use your ratios to write three proportions.

3. $\frac{11}{15}$

11	22	33	44
15	30	45	60

$\frac{11}{15} = \frac{22}{30}$, $\frac{33}{45} = \frac{44}{60}$, $\frac{22}{30} = \frac{33}{45}$, $\frac{44}{60} = \frac{11}{15}$

4. $\frac{16}{20}$

16	32	48	64
20	40	60	80

$\frac{16}{20} = \frac{32}{40}$, $\frac{32}{40} = \frac{48}{60}$, $\frac{16}{20} = \frac{48}{60}$, $\frac{16}{20} = \frac{64}{80}$

Use each proportion to write two other proportions. Sample Answers Provided:

5. $\frac{5}{8} = \frac{15}{24}$
$\frac{8}{5} = \frac{24}{15}$, $\frac{5}{15} = \frac{8}{24}$

6. $\frac{4}{18} = \frac{6}{27}$
$\frac{27}{6} = \frac{18}{4}$, $\frac{4}{6} = \frac{18}{27}$

7. $\frac{\$2}{7 \text{ apples}} = \frac{\$6}{21 \text{ apples}}$
$\frac{\$6}{\$2} = \frac{21 \text{ apples}}{7 \text{ apples}}$, $\frac{21 \text{ apples}}{\$6} = \frac{7 \text{ apples}}{\$2}$

8. $\frac{80 \text{ miles}}{15 \text{ hours}} = \frac{32 \text{ miles}}{6 \text{ hours}}$
$\frac{15 \text{ hr}}{80 \text{ mi}} = \frac{6 \text{ hr}}{32 \text{ mi}}$, $\frac{6 \text{ hr}}{15 \text{ hr}} = \frac{32 \text{ mi}}{80 \text{ mi}}$

9. An electronics kit contains 7 resistors, 3 capacitors, and 2 transistors. Tell how many of each part would be in 3 and 4 kits.
3 kits: 21 resistors, 9 capacitors, 6 transistors
4 kits: 28 resistors, 12 capacitors, 8 transistors

Test Prep

Resources

For additional practice with a variety of test item formats:

- Test-Taking Strategies, p. 265
- Test Prep, p. 269
- Test-Taking Strategies with Transparencies

Activity Lab

Interpreting Rates Visually

Students use graphs and tables to visually display information about rates. This will help them extend the concepts they learned in Lesson 5-3.

Guided Instruction

Before beginning the Activity, ask students about jobs they have done in their homes or for neighbors and how they were paid. Ask questions, such as:

- *If you were paid a certain amount of money for each dog you walked, what is the unit rate you were paid?* **dollars per dog**
- *How would you label a graph of your pay rate?* **one axis labeled dollars, one axis labeled dogs**

Exercises

Have students work independently on the exercises. Discuss answers as a class. Have different volunteers read each question and their answer and explain how they arrived at their answer.

Differentiated Instruction

Advanced Learners L4

Have students graph another babysitter who charges a $5 travel fee and makes $9 an hour for 4 hours, but must pay $20 for a broken window.

Resources

- graph paper
- Student Manipulatives Kit

Interpreting Rates Visually

You can describe real-world situations using rates.

ACTIVITY

Manny mowed lawns to earn spending money. He charged $10 per hour. Ilene earned money by baby-sitting at $8 per hour, plus a $5 travel fee. Who earned the most for a 3-hour job?

1. Copy and complete the table below to show how much each person earns. **See margin.**

Earnings

	1 hour	2 hours	3 hours	4 hours	5 hours
Manny	■	■	■	■	■
Ilene	■	■	■	■	■

2. Graph the data above on a coordinate system like the one below. Make Manny's line solid and make Ilene's line dashed. **See margin.**

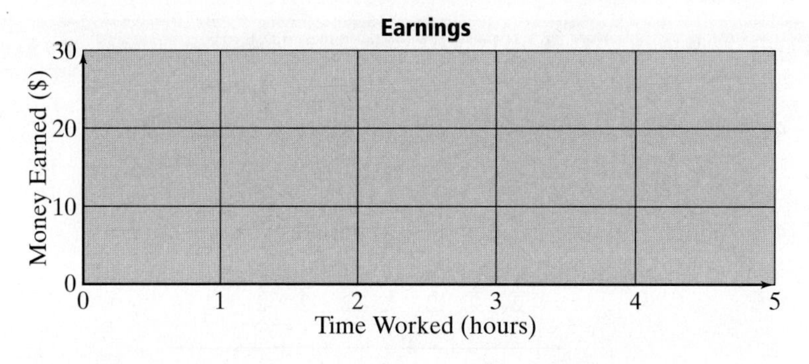

3. Why does the line for Manny's earnings start at $0 for zero hours worked? Why does the line for Ilene's earnings start at $5 for zero hours worked? **Manny does not charge a travel fee. Ilene charges a $5 travel fee.**
4. An expression for Manny's earnings is 10x$, where x is the number of hours worked. Write an expression for Ilene's earnings. **$5 + 8x$**
5. Estimate the time at which the lines cross. What does this point represent? **2.5 hrs. It means they are earning the same amount for the same number of hours worked.**
6. According to the graph, who earned more for a 3-hour job? **Manny earned more.**
7. Estimate each person's earnings for a $1\frac{1}{2}$-hour job. Estimate each person's earnings for a $2\frac{1}{2}$-hour job. **Manny: $15, Ilene $17; Manny: $25, Ilene $25**

1–2. See back of book.

5-4a Activity Lab

Data Collection

Using Proportions With Data

You can use proportions to estimate the number of times your heart beats in one minute, which is called your heart rate.

ACTIVITY

Copy and complete the table below. Use your data from Steps 1–4.

My Heart Rate Data

Ratio: $\frac{\text{counted beats}}{\text{10 seconds}}$	■
Resting Heart Rate	■
Minimum Target Value	■
Maximum Target Value	■
Exercising Heart Rate	■

Step 1 Count the number of times your heart beats in 10 seconds. Write this value as a ratio.

Step 2 Let x = your heart rate. Use your data to write a proportion.

$$\frac{\text{counted beats}}{\text{10 seconds}} = \frac{x \text{ beats}}{\text{60 seconds}}$$

Use mental math to find x. This is your resting heart rate.

Step 3 When you exercise, your heart rate should fall within a target zone. Calculate the values for your target zone.

Minimum value = 0.6(220 − your age)
Maximum value = 0.8(220 − your age)

Step 4 Jog in place for one minute. Repeat Steps 1 and 2 to estimate your exercising heart rate.

Exercises

1. Compare your exercising rate with your target zone. **Answers may vary. Sample: My exercising heart rate is less than my minimum target rate.**
2. Explain why counting your heartbeats for 10 seconds, rather than for a full minute, gives you a more accurate estimate of your exercising heart rate. **Your heart rate will decrease rapidly when you stop exercising.**
3. Express your resting heart rate as a unit rate and estimate the number of times your heart beats in 24 hours. **Check students' work.**

Activity Lab

Using Proportions With Data

Students use a sample of data to arrive at a proportional unit rate. This will prepare them to solve proportions in Lesson 5-4.

Guided Instruction

Before beginning the Activity, ask students to name different kinds of exercise and their unit rates (*e.g.* jogging in miles per hour). Ask:

- *Why would you monitor your heart rate when you exercise?* to make sure your heart is beating at a healthy rate
- *What other body activities could you measure with a unit rate?* breaths per minute, eye blinks per minute

Exercises
Have students work independently on the Exercises. Then have them trade papers with a partner and compare answers. Students with errors should correct their papers.

Alternative Method
Some students may prefer to represent data visually to arrive at an answer. Have them create a table for number of heartbeats at ten second intervals and help them find a unit rate.

Resources

- Activity Lab 5-4: Decision Making
- Student Manipulatives Kit

5-4

1. Plan

Objective
To solve proportions using unit rates, mental math, and cross products

Examples
1 Using Unit Rates
2 Solving Using Mental Math
3 Solving Using Cross Products

Math Understandings: p. 226D

Math Background

In a proportion, such as $\frac{3}{4} = \frac{18}{24}$, each of the four *terms* is a number, and the proportion is a true statement. In many situations one of the terms of a proportion is unknown. The unknown term is represented by a variable, as in $\frac{r}{3} = \frac{36}{27}$. This statement is neither true nor false. Solve the proportion by finding the value of the variable that makes the statement true. Since $\frac{4}{3} = \frac{36}{27}$, the solution of $\frac{r}{3} = \frac{36}{27}$ is 4.

More Math Background: p. 226D

Lesson Planning and Resources

See p. 226E for a list of the resources that support this lesson.

Bell Ringer Practice

Check Skills You'll Need
Use student page, transparency, or PowerPoint. For intervention, direct students to:
Unit Rates and Proportional Reasoning
Lesson 5-2
Extra Skills and Word Problems Practice, Ch. 5

Algebra

5-4 Solving Proportions

Check Skills You'll Need

1. **Vocabulary Review** When is a ratio a *unit rate*?
1–5. See below.
Write the unit rate for each situation.
2. 192 km in 24 d
3. 248 mi in 4 h
4. 50 push-ups in 2 min
5. 180 words in 3 min

GO for Help Lesson 5-2

Check Skills You'll Need
1. **A ratio is a unit rate when you are finding the rate for one unit of something.**
2. **8 km/d**
3. **62 mi/h**
4. **25 push-ups/min**
5. **60 words/min**

What You'll Learn

To solve proportions using unit rates, mental math, and cross products

Why Learn This?

You know the price of six oranges, but you want to buy eight. You can solve a proportion to find the total cost of the quantity that you want to buy.

You can use unit rates to solve a proportion. First find the unit rate. Then multiply to solve the problem.

EXAMPLE Using Unit Rates

1 Shopping Use the information above to find the cost in dollars of 8 oranges.

Solve the proportion $\frac{2.34 \text{ dollars}}{6 \text{ oranges}} = \frac{x \text{ dollars}}{8 \text{ oranges}}$.

Step 1 Find the unit price.

$\frac{2.34 \text{ dollars}}{6 \text{ oranges}}$

$\$2.34 \div 6 \text{ oranges}$ ← **Divide to find the unit price.**

$.39/orange

Step 2 You know the cost of one orange. Multiply to find the cost of 8 oranges.

$\$.39 \cdot 8 = \$3.12$ ← **Multiply the unit rate by the number of oranges.**

The cost of 8 oranges is $3.12.

Quick Check

1. a. Postcards cost $2.45 for 5 cards. How much will 13 cards cost? **$6.37**
   b. Swimming goggles cost $84.36 for 12. At this rate, how much will new goggles for 17 members of a swim team cost? **$119.51**

Differentiated Instruction Solutions for All Learners

Special Needs L1
To help students with their proportional reasoning, have them use fraction kits or strips to show how $\frac{1}{4}$ and $\frac{9}{12}$ are equal and therefore proportional. Repeat with $\frac{4}{6}$ and $\frac{10}{15}$.
learning style: tactile

Below Level L2
Review the process of solving simple multiplication equations such as $3x = 12$ and $54 = -9y$. $x = 4$, $y = -6$
learning style: visual

You can use mental math to solve some proportions. When a proportion involves a variable, you solve the proportion by finding the value of the variable.

EXAMPLE Solving Using Mental Math

2 **Algebra** Solve each proportion using mental math.

a. $\frac{z}{12} = \frac{21}{36}$

$\frac{z}{12} = \frac{21}{36}$ (×3 top and bottom) ← Since 12 × 3 = 36, the common multiplier is 3.

$z = 7$ ← Use mental math to find what number times 3 equals 21.

b. $\frac{8}{10} = \frac{n}{40}$

$\frac{8}{10} = \frac{n}{40}$ (×4 top and bottom) ← Since 10 × 4 = 40, 8 × 4 = *n*.

$n = 32$ ← Use mental math.

Quick Check

2. Solve each proportion using mental math.

a. $\frac{3}{8} = \frac{b}{24}$ 9 **b.** $\frac{m}{5} = \frac{16}{40}$ 2 **c.** $\frac{15}{30} = \frac{5}{p}$ 10

Many proportions cannot easily be solved with mental math. In these situations, you can use cross products to solve a proportion.

EXAMPLE Solving Using Cross Products

Test Prep Tip

Before you use cross products to solve a proportion, check whether you can use mental math.

3 **Gridded Response** Solve $\frac{25}{38} = \frac{15}{x}$ using cross products.

$\frac{25}{38} = \frac{15}{x}$

$25x = 38(15)$ ← Write the cross products.

$25x = 570$ ← Simplify.

$\frac{25x}{25} = \frac{570}{25}$ ← Divide each side by 25.

$x = 22.8$ ← Simplify.

2 2 . 8

Quick Check

3. Solve each proportion using cross products.

a. $\frac{12}{15} = \frac{x}{21}$ 16.8 **b.** $\frac{16}{30} = \frac{d}{51}$ 27.2 **c.** $\frac{20}{35} = \frac{110}{m}$ 192.5

2. Teach

Activity Lab

Use before the lesson.
Student Edition Activity Lab, Data Collection 5-4a, Using Proportions With Data, p. 243

All in One Teaching Resources
Activity Lab 5-4: Decision Making

Guided Instruction

Example 2
Have students describe each part orally, using a parallel structure to state the relationship between the denominators and the relationship between the numerators. Have them say for part a: 12 times 3 equals 36.
z times 3 equals 21.
z must be what number? 7

The method of solving proportions using a common multiplier is sometimes called the "factor of change" method. You can scale a ratio up or down to an equivalent ratio by multiplying its terms by a common multiplier, just as you can generate equivalent fractions by multiplying the numerator and denominator of a fraction by a common multiplier.

Error Prevention!

When solving a proportion, students might multiply all the numbers. Stress that the process of solving a proportion involves a series of equations. It is important to write an *equation* involving the cross products. After writing the correct cross-product equation, highlight the number next to the variable. Focus attention on this number and use it as a divisor, not a multiplier.

Advanced Learners L4	**English Language Learners** ELL
Challenge students to solve this proportion. $\frac{32}{m} = \frac{m}{2}$ $m = 8$ or $m = -8$	Make sure to clearly distinguish between a common *multiple* and a common *multiplier.* The similarity in the terms can cause confusion.
learning style: visual	learning style: verbal

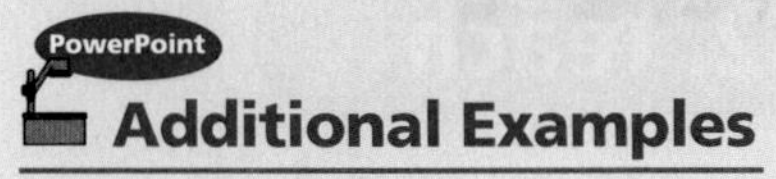

Additional Examples

1. The cost of 4 light bulbs is \$3. At this rate, what is the cost of 10 light bulbs? **\$7.50**
2. Solve each proportion using mental math.
   a. $\frac{5}{c} = \frac{30}{42}$ **$c = 7$**
   b. $\frac{9}{4} = \frac{72}{t}$ **$t = 32$**
3. Solve $\frac{6}{8} = \frac{9}{a}$ using cross products. **$a = 12$**

All in One Teaching Resources

- Daily Notetaking Guide 5-4 L3
- Adapted Notetaking 5-4 L1

Closure

Give students the following equations.

$\frac{4}{5} = \frac{20}{25}$ $\frac{3}{8} = \frac{18}{w}$

- Ask: *Why is each equation a proportion?* Each is a statement that shows two ratios are equivalent. *How is the second proportion different from the first?* One number in the proportion is unknown and is represented by a variable.
- Discuss the three methods for solving proportions that were presented in Examples 1 through 3. Tell students to choose the method they think most appropriate and use it to solve $\frac{3}{8} = \frac{18}{w}$. **$w = 48$**

More Than One Way

Nature An oyster bed covers 36 m². Your class studies 4 m² of the oyster bed. In those 4 m² you count 96 oysters. Predict the number of oysters in the entire bed.

Carlos's Method

I will let x represent the number of oysters in the 36-m² bed.

oysters → $\frac{96}{4}$ ← area = $\frac{x}{36}$ ← oysters / area		← Write a proportion.
$96(36) = 4x$		← Write the cross products.
$3{,}456 = 4x$		← Simplify.
$\frac{3{,}456}{4} = \frac{4x}{4}$		← Divide each side by 4.
$864 = x$		← Simplify.

There are about 864 oysters in the oyster bed.

Brianna's Method

Since $9 \cdot 4\text{ m}^2 = 36\text{ m}^2$, the entire bed is 9 times as large as the portion studied. So I know there should be about $9 \cdot 96$, or 864, oysters in the oyster bed.

Choose a Method

You buy a bag of 400 marbles. In a handful of 20 marbles, you find 8 red marbles. About how many red marbles are in the bag? Explain why you chose the method you used. **Check students' methods: about 160 red marbles.**

Check Your Understanding

1. **Writing in Math** How does a unit rate help you solve a proportion?
2. **Number Sense** Does the proportion $\frac{3}{7} = \frac{x}{21}$ have the same solution as $\frac{7}{3} = \frac{21}{x}$? Explain. **Yes; both give the cross products of $7x = 3 \cdot 21$.**

1. Answers may vary. Sample: Finding a unit rate gives you a denominator of 1, so you only need to multiply to solve the proportion. $\frac{5}{1} = \frac{x}{8}$, $x = 40$

Solve each proportion using mental math.

3. $\frac{2}{5} = \frac{m}{10}$ (× 2) $m = ■$ **4**
4. $\frac{1}{6} = \frac{4}{y}$ (× 4) $y = ■$ **24**
5. $\frac{7}{3} = \frac{28}{b}$ (× 4) $b = ■$ **12**

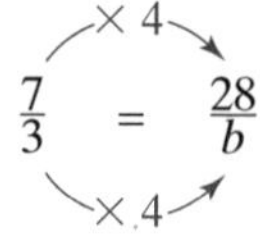

Homework Exercises

For more exercises, see Extra Skills and Word Problems.

GO for Help

For Exercises	See Examples
6–10	1
11–16	2
17–25	3

A **Solve each problem by finding a unit rate and multiplying.**

6. If 5 goldfish cost \$6.45, what is the cost of 8 goldfish? **\$10.32**
7. If 12 roses cost \$18.96, what is the cost of 5 roses? **\$7.90**
8. If 3 onions weigh 0.75 lb, how much do 10 onions weigh? **2.5 lb**
9. If 13 key chains cost \$38.35, what is the cost of 20 key chains? **\$59.00**
10. At a telethon, a volunteer can take 48 calls over a 4-hour shift. At this rate, how many calls can 12 volunteers take in a 4-hour shift? **576 calls**

Solve each proportion using mental math.

11. $\frac{2}{7} = \frac{x}{21}$ **6**
12. $\frac{18}{32} = \frac{m}{16}$ **9**
13. $\frac{c}{10} = \frac{36}{60}$ **6**
14. $\frac{c}{35} = \frac{4}{7}$ **20**
15. $\frac{16}{38} = \frac{b}{19}$ **8**
16. $\frac{9}{w} = \frac{36}{20}$ **5**

Solve each proportion using cross products.

17. $\frac{8}{12} = \frac{y}{30}$ **20**
18. $\frac{15}{33} = \frac{m}{22}$ **10**
19. $\frac{c}{28} = \frac{49}{16}$ **85.75**
20. $\frac{y}{18} = \frac{21}{63}$ **6**
21. $\frac{14}{34} = \frac{x}{51}$ **21**
22. $\frac{9}{30} = \frac{p}{16}$ **4.8**
23. $\frac{20}{w} = \frac{12}{3}$ **5**
24. $\frac{27}{20} = \frac{36}{v}$ **$26\frac{2}{3}$**
25. $\frac{19}{r} = \frac{152}{4}$ **0.5**

29. Yes; when you simplify a ratio, it is still equivalent to the original ratio.

B **GPS** 26. **Guided Problem Solving** You received \$57.04 for working 8 h. At that rate, how much would you receive for working 11 h? **\$78.43**

hours → $\frac{8}{\blacksquare} = \frac{\blacksquare}{x}$ ← pay
← You would receive *x* dollars for ■ hours.

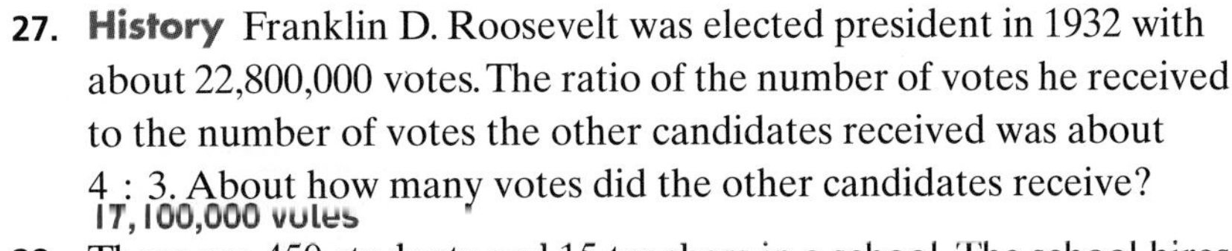

President Franklin D. Roosevelt

27. **History** Franklin D. Roosevelt was elected president in 1932 with about 22,800,000 votes. The ratio of the number of votes he received to the number of votes the other candidates received was about 4 : 3. About how many votes did the other candidates receive? **17,100,000 votes**
28. There are 450 students and 15 teachers in a school. The school hires 2 new teachers. To keep the student-to-teacher ratio the same, how many students in all should attend the school? **510 students** (GPS)
29. **Reasoning** Brian solved the proportion $\frac{18}{45} = \frac{a}{20}$ at the right. His first step was to simplify the ratio $\frac{18}{45}$. Is his answer correct? Explain. **See above left.**

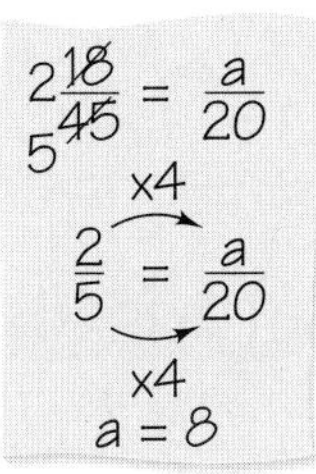

30. A jet takes $5\frac{3}{4}$ h to fly 2,475 mi from New York City to Los Angeles. About how many hours will a jet flying at the same average rate take to fly 5,452 mi from Los Angeles to Tokyo? **about $12\frac{1}{2}$ h**

Online lesson quiz, PHSchool.com, Web Code: ara-0504

3. Practice

Assignment Guide

Check Your Understanding
Go over Exercises 1–5 in class before assigning the Homework Exercises.

Homework Exercises

A Practice by Example	6–25
B Apply Your Skills	26–37
C Challenge	38
Test Prep and Mixed Review	39–44

Homework Quick Check
To check students' understanding of key skills and concepts, go over Exercises 8, 21, 28, 36, and 37.

Differentiated Instruction Resources

Adapted Practice 5-4 L1

Practice 5-4 Solving Prop L3

Use mental math to solve for each value of *n*.

1. $\frac{n}{14} = \frac{20}{35}$ **8**
2. $\frac{9}{6} = \frac{21}{n}$ **14**
3. $\frac{24}{n} = \frac{16}{10}$ **15**
4. $\frac{3}{4} = \frac{n}{10}$ **7.5**

Solve each proportion using cross products.

5. $\frac{k}{8} = \frac{14}{4}$ $k =$ **28**
6. $\frac{u}{3} = \frac{10}{5}$ $u =$ **6**
7. $\frac{14}{6} = \frac{d}{15}$ $d =$ **35**
8. $\frac{5}{1} = \frac{m}{4}$ $m =$ **20**
9. $\frac{36}{32} = \frac{n}{8}$ $n =$ **9**
10. $\frac{5}{30} = \frac{1}{x}$ $x =$ **6**
11. $\frac{t}{4} = \frac{5}{10}$ $t =$ **2**
12. $\frac{9}{2} = \frac{v}{4}$ $v =$ **18**

Solve.

13. A contractor estimates it will cost \$2,400 to build a deck to a customer's specifications. How much would it cost to build five similar decks? **\$12,000**
14. A recipe requires 3 c of flour to make 27 dinner rolls. How much flour is needed to make 9 rolls? **1c**

Solve using a calculator, paper and pencil, or mental math.

15. Mandy runs 4 km in 18 min. She plans to run in a 15 km race. How long will it take her to complete the race? **67.5 min**
16. Ken's new car can go 26 miles per gallon of gasoline. The car's gasoline tank holds 14 gal. How far will he be able to go on a full tank? **364 mi**
17. Eleanor can complete two skirts in 15 days. How long will it take her to complete eight skirts? **60 days**
18. Three eggs are required to make two dozen muffins. How many eggs are needed to make 12 dozen muffins? **18 eggs**

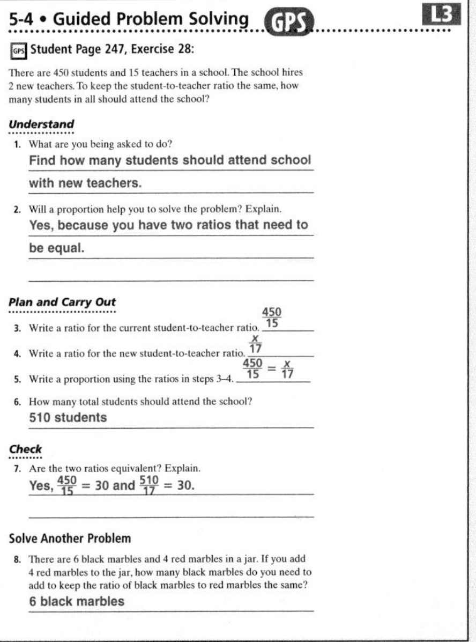

5-4 • Guided Problem Solving GPS L3

GPS **Student Page 247, Exercise 28:**

There are 450 students and 15 teachers in a school. The school hires 2 new teachers. To keep the student-to-teacher ratio the same, how many students in all should attend the school?

Understand

1. What are you being asked to do? **Find how many students should attend school with new teachers.**
2. Will a proportion help you to solve the problem? Explain. **Yes, because you have two ratios that need to be equal.**

Plan and Carry Out

3. Write a ratio for the current student-to-teacher ratio. $\frac{450}{15}$
4. Write a ratio for the new student-to-teacher ratio. $\frac{x}{17}$
5. Write a proportion using the ratios in steps 3–4. $\frac{450}{15} = \frac{x}{17}$
6. How many total students should attend the school? **510 students**

Check

7. Are the two ratios equivalent? Explain. **Yes, $\frac{450}{15} = 30$ and $\frac{510}{17} = 30$.**

Solve Another Problem

8. There are 6 black marbles and 4 red marbles in a jar. If you add 4 red marbles to the jar, how many black marbles do you need to add to keep the ratio of black marbles to red marbles the same? **6 black marbles**

4. Assess & Reteach

Lesson Quiz

1. The cost of 5 CDs is \$42. At this rate, what is the cost of 7 CDs? **\$58.80**

Solve each proposition.

2. $\frac{72}{45} = \frac{8}{n}$ **$n = 5$**
3. $\frac{b}{18} = \frac{9}{5}$ **$b = 32.4$**

Alternative Assessment

Have each student in a pair write two equivalent ratios and use them to form a proportion. Each student should then erase one of the four numbers and replace it with a variable. Now have students exchange papers and solve each other's proportions. Partners should discuss their results.

Enrichment 5-4 L4

Reteaching 5-4 Solving Pro L2

Solving a proportion means finding a missing part of the proportion. You can use unit rates to solve a proportion. First find the unit rate. Then multiply to solve the proportion.

Shawn filled 8 bags of leaves in 2 hours. At this rate, how many bags would he fill in 6 hours?

① Find a unit rate for the number of bags per hour. Divide by the denominator.

$\frac{8 \text{ bags}}{2 \text{ hours}} = \frac{8 \text{ bags} \div 2}{2 \text{ hours} \div 2} = \frac{4 \text{ bags}}{1 \text{ hour}}$ The unit rate is 4 bags per hour.

② Multiply the unit rate by 6 to find the number of bags he will fill in 6 hours.

Unit rate ↓ 4 × Number of hours ↓ 6 = Total ↓ 24

At this rate, Shawn can fill 24 bags in 6 hours.

If two ratios form a proportion, the **cross products** are equal.

Solve. $\frac{5}{15} = \frac{n}{3}$

① Write the cross products. $5 \cdot 3 = 15 \cdot n$

② Simplify. $15 = 15n$

③ Solve the equation. $n = 1$

Solve.

1. The bookstore advertises 5 notebooks for \$7.75. At this rate, how much will 7 notebooks cost? **\$10.85**
2. Leroy can lay 144 bricks in 3 hours. At this rate, how many bricks can he lay in 7 hours? **336**

Solve each proportion using cross products.

3. $\frac{4}{24} = \frac{n}{6}$ $4 \cdot$ **6** $= 24 \cdot$ **n**; $n =$ **1**
4. $\frac{30}{5} = \frac{6}{n}$ **30** $=$ **$30n$**; $n =$ **1**
5. $\frac{n}{6} = \frac{27}{9}$ **162** $=$ **$9n$**; $n =$ **18**

Solve each proportion.

6. $\frac{4}{10} = \frac{n}{15}$ **$n = 6$**
7. $\frac{4}{200} = \frac{n}{100}$ **$n = 2$**
8. $\frac{6}{n} = \frac{5}{10}$ **$n = 12$**
9. $\frac{32}{22} = \frac{96}{n}$ **$n = 66$**
10. $\frac{6}{3} = \frac{n}{5}$ **$n = 10$**
11. $\frac{2}{n} = \frac{4}{10}$ **$n = 5$**

GO Online
Homework Video Tutor
Visit: PHSchool.com
Web Code: are-0504

Solve each proportion using cross products. Round to the nearest tenth, if necessary.

31. $\frac{1.7}{2.5} = \frac{3.4}{d}$ **5**
32. $\frac{y}{9.3} = \frac{12.6}{5.4}$ **21.7**
33. $\frac{33.1}{x} = \frac{6.2}{1.3}$ **6.9**
34. $\frac{16.9}{13.5} = \frac{t}{7.4}$ **9.3**

35. **Error Analysis** A videocassette recorder uses 2 m of tape in 3 min when set on extended play. To determine how many minutes a tape that is 240 m long can record on extended play, one student wrote the proportion $\frac{2}{3} = \frac{n}{240}$. Explain why this proportion is incorrect. Then write a correct proportion. **240 should be the numerator. $\frac{2}{3} = \frac{240}{n}$**

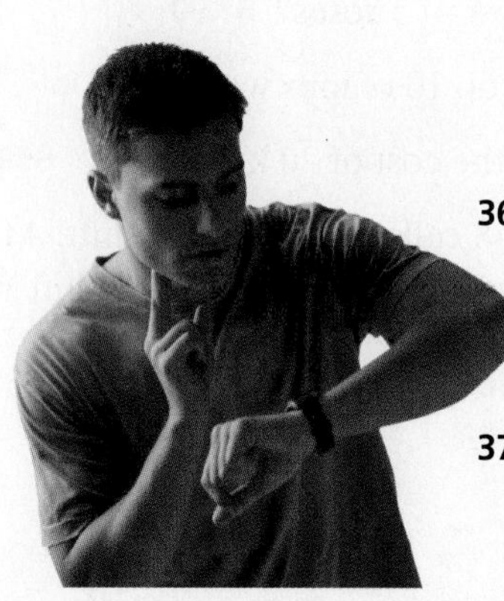

36. **Health** Your heart rate is the number of heartbeats per minute.
    a. What is your heart rate if you count 18 beats in 15 seconds?
    b. **Choose a Method** How many beats do you count in 15 seconds if your heart rate is 96 beats/min? Explain the method you chose. **36a–b. See left.**

36a. 72 beats/min

b. 24; answers may vary.

37. **Writing in Math** You estimate you will take 75 min to bike 15 mi to a state park. After 30 min, you have traveled 5 mi. Are you on schedule? Explain. **See margin.**

Ⓒ 38. **Challenge** A recipe for fruit salad serves 4 people. It calls for $2\frac{1}{2}$ oranges and 16 grapes. You want to serve 11 people. How many oranges and how many grapes will you need? **7 oranges, 44 grapes**

Test Prep and Mixed Review

Practice

Gridded Response

39. An antelope ran 237 ft in 3 seconds. If the antelope continued to run at the same rate, how many seconds would it take him to run 790 ft? **10**

40. The table below shows the leaders in punt returns for the National Football Conference during one season.

Punt Return Leaders

Name	Number of Returns	Yards
Brian Westbrook	20	306
Allen Rossum	39	545
Reggie Swinton	23	318
R. W. McQuarters	37	452

What was the unit rate of yards per return for Reggie Swinton? Round to the nearest hundredth. **13.83**

41. A restaurant bill for four people is \$37.80. How much money, in dollars, should each person contribute to share the cost evenly? **9.45**

GO for Help

For Exercises	See Lesson
42–44	1-6

Order the numbers from least to greatest.

42. 16, −12, 10, −3 **−12, −3, 10, 16**
43. −6, −3, 8, −2, 1 **−6, −3, −2, 1, 8**
44. 5, 0, −1, 2, −5 **−5, −1, 0, 2, 5**

Test Prep

Resources

For additional practice with a variety of test item formats:

- Test-Taking Strategies, p. 265
- Test Prep, p. 269
- Test-Taking Strategies with Transparencies

37. No; answers may vary. Sample: 15 mi in 75 min gives a unit rate of 0.2 mi/min. 5 mi in 30 min is behind schedule with a unit rate of 0.17 mi/min.

Proportions and Equations

Often you can translate a proportion problem into an equation. You can then use the equation to solve similar problems.

Muscles For each 5 pounds of body weight, about 2 pounds is muscle. How much of a 125-pound student is muscle?

What You Might Think

What You Might Write

What do I know? → Each 5 lb of weight includes 2 lb of muscle.

What am I trying to find out? → How many pounds of muscle are in 125 lb of body weight?

How do I solve the problem? → I can write an equation to find the muscle weight for any student weight. A proportion that represents the situation is

$\frac{x}{125} = \frac{2}{5}$.

To solve for x, I can multiply each side by 125.

$x = \frac{2}{5}(125)$

125 represents the student's weight. So if a student's weight is w, then the equation $x = \frac{2}{5}w$ gives the amount of muscle for any weight w.

$x = \frac{2}{5}(125)$

$= 250 \div 5$

$= 50$

What is the answer? → The 125-lb student has about 50 lb of muscle.

Think It Through When you find the cross products, you get $2w = 5x$. To solve for x, you divide by 5, which results in $x = \frac{2}{5}w$.

1. Show how solving the equation $\frac{2}{5} = \frac{x}{w}$ for x gives the equation $x = \frac{2}{5}w$.
2. Explain why you can use the equation $x = \frac{2}{5}w$ as a shortcut for finding the amount of muscle for any student.
If you substitute the weight of the student, you only need to multiply.

Proportions and Equations

In this feature, students write a proportion for a situation and translate the proportion into an equation. Then they use the equation to solve the problem. Students apply the problem solving steps to this process.

Guided Instruction

Discuss with students situations in which they can make proportions, then use equations to solve real-world problems.

Have a volunteer read the problem. Ask:

- *What do you know from the problem that will let you set up a proportion?* for each 5 pounds of body weight about 2 pounds is muscle
- *What ratio can you set up?* 2 to 5 or $\frac{2}{5}$
- *What are you trying to find out?* how many pounds of muscle a 125-pound student has
- What proportion can you set up? $\frac{2}{5} = \frac{x}{125}$

Error Prevention!

Some students may have difficulty understanding the equation $x = \frac{2}{5}w$ where w is a student's weight. You may want to make a table with several weights and have students find the amount of muscle for each weight.

Exercises
Have students work through Exercises 3 and 4 in pairs. For Exercises 5–7, you may wish to ask questions similar to the **a** and **b** questions in Exercises 3 and 4.

Differentiated Instruction

English Language Learners ELL

You may need to explain some of the terms in Exercises 6 and 7, for example *concession, equator,* and *rotation*. You may want to draw a picture for *equator* and show *rotation* with arrows.

Exercises

Solve each problem. For Exercises 3 and 4, answer the questions first.

3. During one season, a basketball player made 353 of 765 free throws. At that rate, how many free throws would the player make in another 600 attempts? **about 277 free throws**
   a. What do you know? What do you want to find out?
   b. Can you write an equation to predict the number of free throws x the player will make for any number of shots s?

4. Jess wants to sell fruit punch on a hot day. The punch is 2 parts grape juice to 3 parts apple juice. Jess has 5 quarts of apple juice. How many quarts of grape juice does she need? **$3\frac{1}{3}$ qt**
   a. What do you know? What do you want to find out?
   b. Can you write an equation to find the number of quarts of grape juice g needed for a quarts of apple juice?

5. Jess makes the graph below so that she does not have to keep calculating amounts of juice. Use the graph to find how much grape juice is needed for 15 quarts of apple juice. Then find how much apple juice is needed for 12 quarts of grape juice. **10 qt; 18 qt**

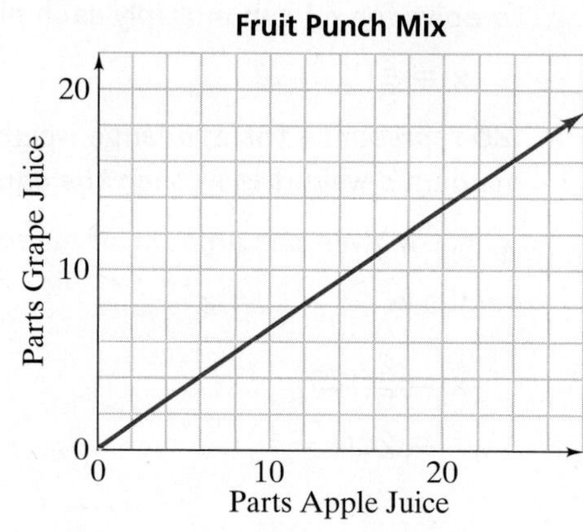

6. A concession stand needs 3 hot dogs for every 7 people who attend a football game. If 1,400 fans are expected to attend the game, how many hot dogs are needed? **600 hot dogs**

7. Earth is about 25,000 mi around at the equator. If you are sitting on the equator, about how fast are you traveling? (*Hint:* Earth makes one rotation in 24 hours.) **about 1,042 mi/h**

5-5a Activity Lab

Hands On

Exploring Similar Figures

In everyday language, two items are *similar* if they are the same in some, but not necessarily all, ways. In math, similarity has a related meaning. Complete the activity to find out what makes two figures mathematically similar.

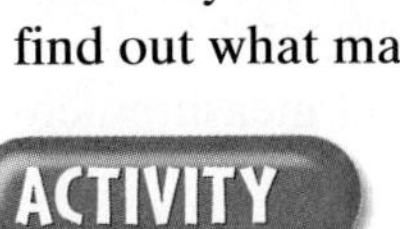

ACTIVITY

Figures 1 and 2 below are similar. Figures 3 and 4 are similar, too. You can test whether two figures are similar by measuring their angles and side lengths.

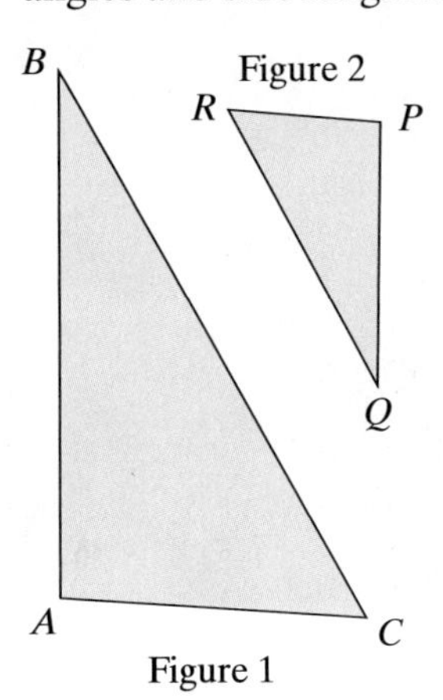

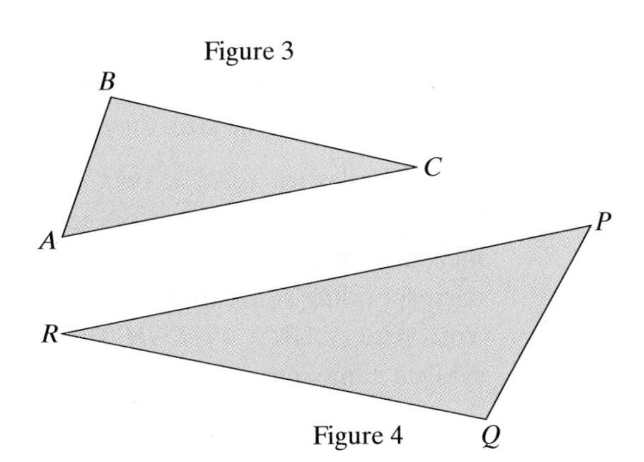

1. Copy the table at the right. Use a protractor to measure each angle of Figure 1 and Figure 2 above. Fill in the missing information in the table. What do you notice about the angle measures in each row of your table? **See margin.**

Fig. 1	Angle Measure	Fig. 2	Angle Measure
$\angle A$	■	$\angle P$	■
$\angle B$	■	$\angle Q$	■
$\angle C$	■	$\angle R$	■

2. Copy the table at the right. Use a centimeter ruler to measure each side of Figure 1 and Figure 2 to the nearest tenth of a centimeter. Fill in the missing information in the table. What pattern do you notice in the length of the sides in each row of the table? **See margin.**

Fig. 1	Side Length (cm)	Fig. 2	Side Length (cm)
AB	■	PQ	■
BC	■	QR	■
CA	■	RP	■

3. Repeat Steps 1 and 2 for Figures 3 and 4. **See margin.**
4. Make a conjecture about the angle measures and side lengths of similar figures. **See margin.**

1–4. See back of book.

Activity Lab

Exploring Similar Figures

Students learn a mathematical definition for similar figures and learn what makes figures similar. This will help them work with similar figures in Lesson 5-5.

Guided Instruction

Before beginning the Activity, ask students what the word *similar* means. Ask them to name pairs of things that are similar and tell why they are similar. Then ask questions, such as:

- *What makes two shapes similar?* **same number of sides, same number of angles, same shape**
- *Are all triangles similar?* **No, corresponding angles must have the same measure and lengths of corresponding sides must form equal ratios.**

Activity

Have students work in pairs on the Activity. Circulate and provide assistance where needed with measuring angles and lengths. Make sure students have the numbers in their proportions in the proper positions.

Alternative Method

Students may prefer to work with paper models of the triangles. Provide paper triangle cutouts that are to scale with the triangles on the page.

Resources

- Activity Lab 5-5: Indirect Measurement
- protractor
- centimeter ruler
- Student Manipulatives Kit

5-5

1. Plan

Objective
To use proportions to find missing lengths in similar figures

Examples
1 Finding a Missing Measure
2 Application: Indirect Measurement

Math Understandings: p. 226D

Math Background

Similar figures have the same shape but not necessarily the same size. For two *polygons* to be similar, corresponding angles must have the same measure, and the lengths of corresponding sides must be proportional. When it is known that two polygons are similar, it may be possible to find an unknown length by writing and solving a proportion.

More Math Background: p. 226D

Lesson Planning and Resources

See p. 226E for a list of the resources that support this lesson.

PowerPoint

Bell Ringer Practice

Check Skills You'll Need
Use student page, transparency, or PowerPoint. For intervention, direct students to:
Solving Proportions
Lesson 5-4
Extra Skills and Word Problems Practice, Ch. 5

5-5 Using Similar Figures

Check Skills You'll Need

1. **Vocabulary Review** An equation stating that two ratios are equal is a __?__.
1–5. See below.

Solve each proportion.

2. $\frac{x}{4} = \frac{12}{8}$
3. $\frac{4}{b} = \frac{16}{48}$
4. $\frac{84}{12} = \frac{g}{6}$
5. $\frac{6}{3} = \frac{19}{m}$

Lesson 5-4

What You'll Learn

To use proportions to find missing lengths in similar figures

New Vocabulary polygon, similar polygons, indirect measurement

Why Learn This?

The heights of objects such as totem poles may be difficult to measure directly. You can measure indirectly by using figures that have the same shape. When two figures have the same shape, but not necessarily the same size, they are similar.

In the similar triangles below, corresponding angles have the same measure. Since $\frac{40}{60} = \frac{50}{75} = \frac{34}{51}$, the corresponding sides are proportional. You write $\triangle ABC \sim \triangle FGH$. The symbol $\sim$ means "is similar to."

Test Prep Tip
When solving problems involving similar figures, make sure you are working with corresponding angles and sides.

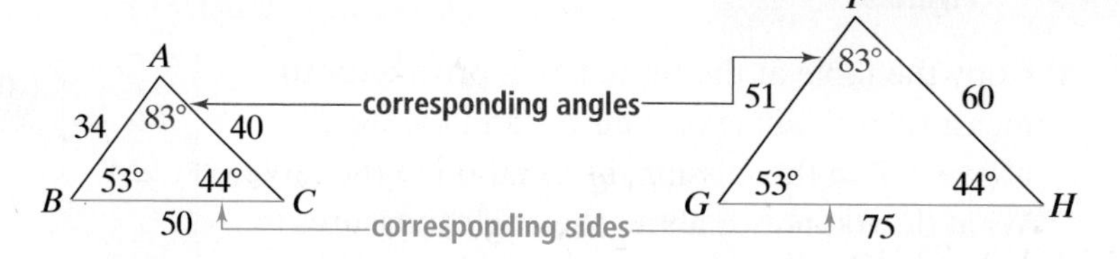

Check Skills You'll Need
1. proportion
2. 6
3. 12
4. 42
5. 9.5

A **polygon** is a closed plane figure formed by three or more line segments that do not cross.

KEY CONCEPTS Similar Polygons

Two polygons are **similar polygons** if
- corresponding angles have the same measure, and
- the lengths of the corresponding sides form equivalent ratios.

You can use proportions to find missing side lengths in similar polygons.

Differentiated Instruction Solutions for All Learners

Special Needs L1
If available, take students out to see a flagpole and its shadow, or a tree and its shadow. Let them stand next to the taller structure, and show them how you might actually determine the height of the flagpole by using corresponding sides.

learning style: visual

Below Level L2
In Example 1, students may better visualize the proportion by aligning the triangle names vertically.

$\triangle C A T$
$\triangle D O G$ $\rightarrow \frac{CA}{DO} = \frac{AT}{OG}$

learning style: visual

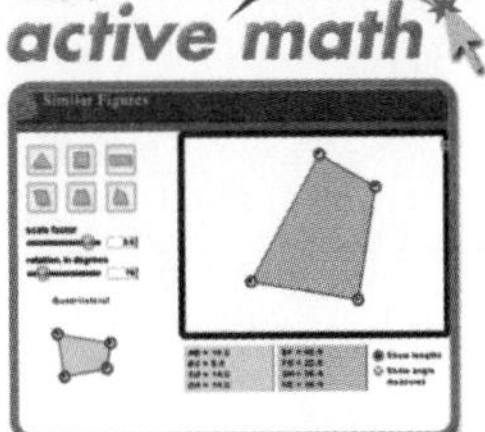

For: Similarity Activity
Use: Interactive Textbook, 5-5

EXAMPLE Finding a Missing Measure

1 **Algebra** $\triangle ACT$ and $\triangle ODG$ are similar. Find the value of x.

$\frac{AC}{OD} = \frac{AT}{OG}$ ← **Write a proportion.**

$\frac{x}{50} = \frac{24}{30}$ ← **Substitute.**

$\frac{x}{50} = \frac{4}{5}$ ← **Write $\frac{24}{30}$ in simplest form.**

$\frac{x}{50} = \frac{4}{5}$ (×10) ← **Find the common multiplier.**

$x = 40$ ← **Use mental math.**

Quick Check

1. The trapezoids are similar. Find x. 20

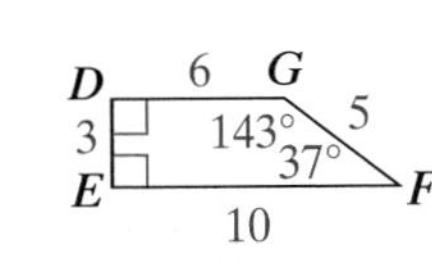

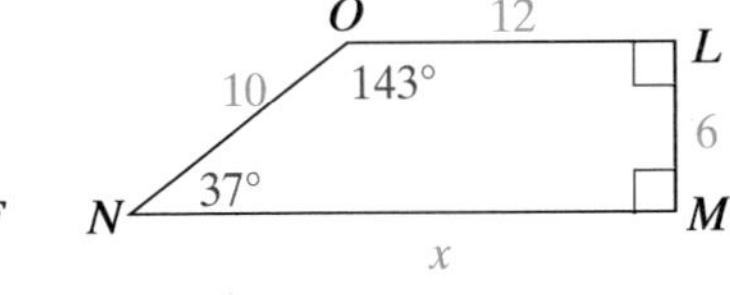

You can use **indirect measurement** to measure distances that are difficult to measure directly. You do this by using proportions and similar figures.

EXAMPLE Application: Indirect Measurement

2 **Multiple Choice** A 6-ft-tall person standing near a flagpole casts a shadow 4.5 ft long. The flagpole casts a shadow 15 ft long. What is the height of the flagpole?

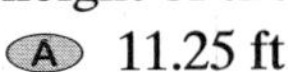

Ⓐ 11.25 ft Ⓑ 18 ft Ⓒ 20 ft Ⓓ 360 ft

Draw a picture and let x represent the height of the flagpole.

$\frac{x}{6} = \frac{15}{4.5}$ ← **Write a proportion.**

$4.5x = 6 \cdot 15$ ← **Write the cross products.**

$\frac{4.5x}{4.5} = \frac{6 \cdot 15}{4.5}$ ← **Divide each side by 4.5.**

$x = 20$ ← **Simplify.**

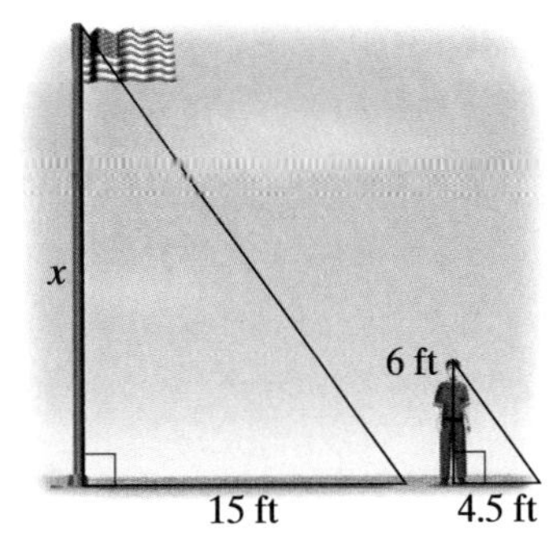

The height of the flagpole is 20 ft. The answer is C.

Test Prep Tip

Drawing a picture can help you see what quantities you know and what quantities you are looking for.

Quick Check

2. A 6-ft person has a shadow 5 ft long. A nearby tree has a shadow 30 ft long. What is the height of the tree? 36 ft

2. Teach

Activity Lab

Use before the lesson.
Student Edition Activity Lab, Hands On 5-5a, Exploring Similar Figures, p. 251

All in One Teaching Resources
Activity Lab 5-5: Indirect Measurement

Guided Instruction

Example 2
Some students might perceive the heights and lengths as a maze, and have difficulty writing a proportion. Suggest that they first write the proportion using words.

$\frac{\text{height of flagpole}}{\text{height of person}} = \frac{\text{shadow of flagpole}}{\text{shadow of person}}$

Substitute the numbers and variable for the words they represent.

PowerPoint
Additional Examples

1 $\triangle ABC$ is similar to $\triangle DEF$. Find the value of c. 12

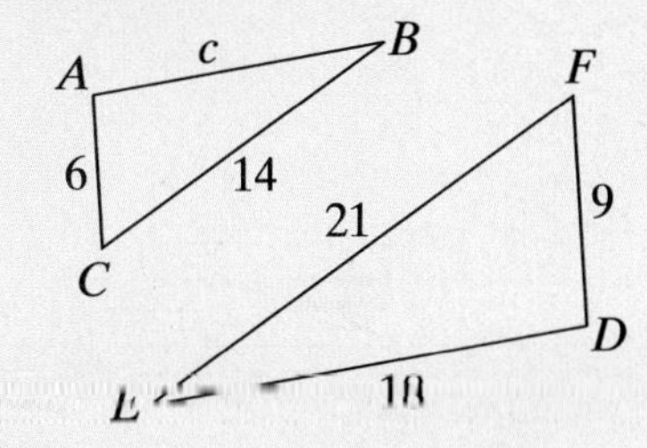

2 A 5-ft person near a tree has a shadow 12 ft long. The tree has a shadow 42 ft long. What is the height of the tree? $17\frac{1}{2}$ ft

All in One Teaching Resources
- Daily Notetaking Guide 5-5 L3
- Adapted Notetaking 5-5 L1

Closure

- *The angle measures of a triangle are 40°, 60°, and 80°. What are the angle measures of a similar triangle?* 40°, 60°, and 80°
- *The side lengths of a triangle are 40 cm, 60 cm, and 80 cm. The smallest side length of a similar triangle is 120 cm. What are the lengths of the other two sides?* 180 cm, 240 cm

Advanced Learners L4
Have students decide whether each statement is *true* or *false*.

All squares are similar. true
All rectangles are similar. false

learning style: visual

English Language Learners ELL
Make sure students can identify which sides and angles are corresponding. They can label or name each side, and make an organized list of which sides correspond.

learning style: verbal

3. Practice

Assignment Guide

Check Your Understanding

Go over Exercises 1–5 in class before assigning the Homework Exercises.

Homework Exercises

A Practice by Example	6–10
B Apply Your Skills	11–18
C Challenge	19
Test Prep and Mixed Review	20–25

Homework Quick Check

To check students' understanding of key skills and concepts, go over Exercises 7, 10, 13, 17, and 18.

Differentiated Instruction Resources

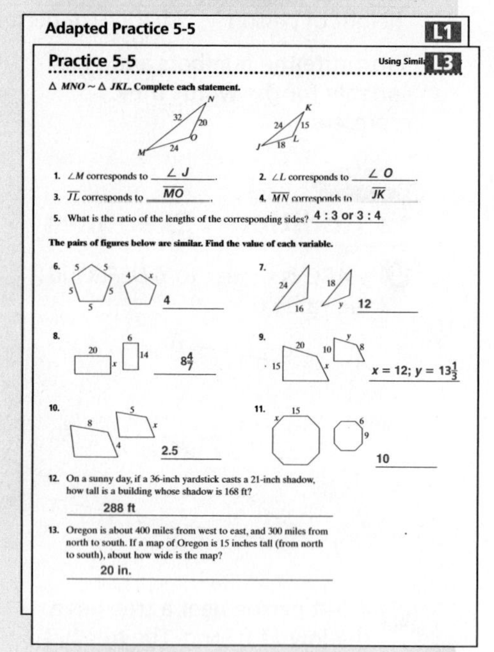

Adapted Practice 5-5 L1

Practice 5-5 Using Simil... L3

$\triangle MNO \sim \triangle JKL$. Complete each statement.

1. $\angle M$ corresponds to $\angle J$.
2. $\angle L$ corresponds to $\angle O$.
3. $\overline{JL}$ corresponds to $\overline{MO}$.
4. $\overline{MN}$ corresponds to $\overline{JK}$
5. What is the ratio of the lengths of the corresponding sides? 4 : 3 or 3 : 4

The pairs of figures below are similar. Find the value of each variable.

6. 4
7. 12
8. $8\frac{4}{7}$
9. $x = 12$; $y = 13\frac{1}{3}$
10. 2.5
11. 10
12. On a sunny day, if a 36-inch yardstick casts a 21-inch shadow, how tall is a building whose shadow is 168 ft? 288 ft
13. Oregon is about 400 miles from west to east, and 300 miles from north to south. If a map of Oregon is 15 inches tall (from north to south), about how wide is the map? 20 in.

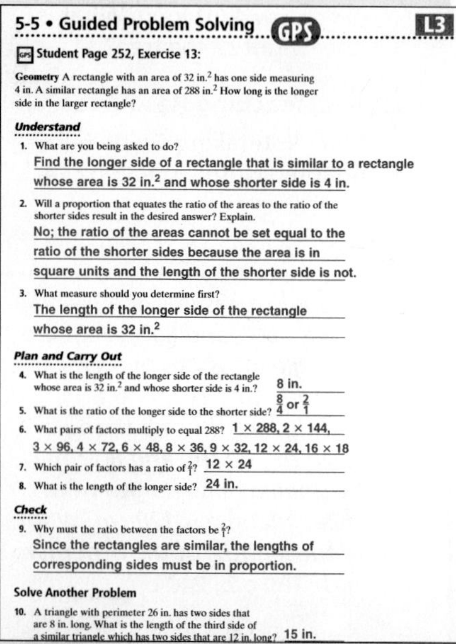

5-5 • Guided Problem Solving GPS L3

Student Page 252, Exercise 13:

Geometry A rectangle with an area of 32 in.^{2} has one side measuring 4 in. A similar rectangle has an area of 288 in.^{2} How long is the longer side in the larger rectangle?

Understand

1. What are you being asked to do? Find the longer side of a rectangle that is similar to a rectangle whose area is 32 in.^{2} and whose shorter side is 4 in.
2. Will a proportion that equates the ratio of the areas to the ratio of the shorter sides result in the desired answer? Explain. No; the ratio of the areas cannot be set equal to the ratio of the shorter sides because the area is in square units and the length of the shorter side is not.
3. What measure should you determine first? The length of the longer side of the rectangle whose area is 32 in.^{2}

Plan and Carry Out

4. What is the length of the longer side of the rectangle whose area is 32 in.^{2} and whose shorter side is 4 in.? 8 in.
5. What is the ratio of the longer side to the shorter side? $\frac{8}{4}$ or $\frac{2}{1}$
6. What pairs of factors multiply to equal 288? 1×288, 2×144, 3×96, 4×72, 6×48, 8×36, 9×32, 12×24, 16×18
7. Which pair of factors has a ratio of $\frac{2}{1}$? 12×24
8. What is the length of the longer side? 24 in.

Check

9. Why must the ratio between the factors be $\frac{2}{1}$? Since the rectangles are similar, the lengths of corresponding sides must be in proportion.

Solve Another Problem

10. A triangle with perimeter 26 in. has two sides that are 8 in. long. What is the length of the third side of a similar triangle which has two sides that are 12 in. long? 15 in.

Check Your Understanding

1. Two polygons are similar if corresponding angles have the same measure and the lengths of the corresponding sides form equivalent ratios.

4. No; the lengths of corresponding sides do not form equivalent ratios.

1. **Vocabulary** What must be true about the corresponding angles and the corresponding sides for two polygons to be similar?

$\triangle ABC$ is similar to $\triangle RST$. Complete each statement.

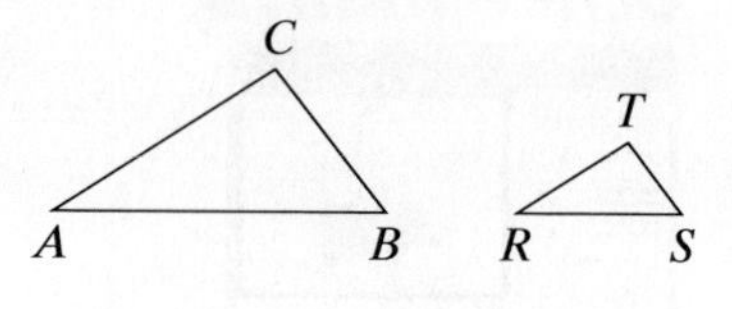

2. $\angle B$ corresponds to __?__. $\angle S$
3. $\overline{RS}$ corresponds to __?__. $\overline{AB}$
4. **Geometry** Are the rectangles similar? Explain. See left.

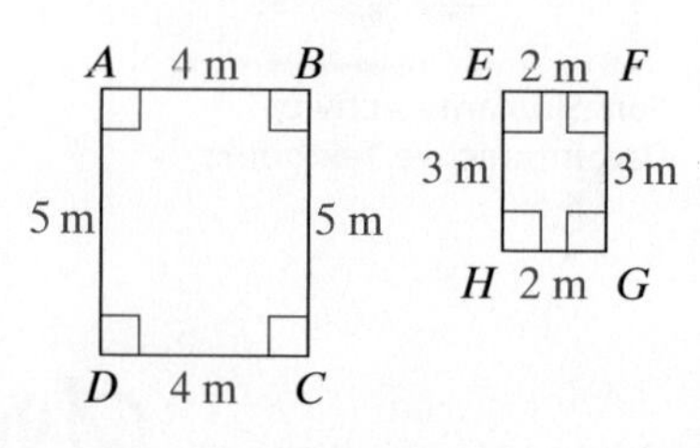

5. **Open-Ended** Think of a distance that is difficult to measure directly. How would you find it indirectly? Check students' work.

Homework Exercises

For more exercises, see Extra Skills and Word Problems.

GO for Help

For Exercises	See Examples
6–9	1
10	2

A **Algebra** **$\triangle ABC$ is similar to $\triangle PQR$. Find each measure.**

6. length of $\overline{AB}$ 12 ft
7. length of $\overline{RP}$ 11.2 ft
8. measure of $\angle A$ 92°
9. measure of $\angle Q$ 53°

10. A woman is 5 ft tall and her shadow is 4 ft long. A nearby tree has a shadow 30 ft long. How tall is the tree? 37.5 ft

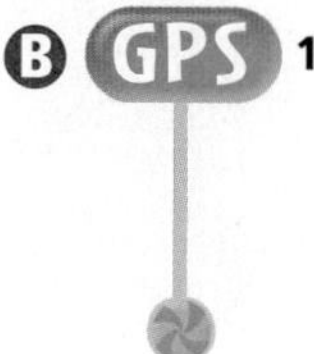

B **GPS** 11. **Guided Problem Solving** An image is 16 in. by 20 in. You want to make a copy that is similar. Its longer side will be 38 in. The copy costs \$.60 per square inch. Estimate the copy's total cost. about \$700

- What is the approximate length of the shorter side of the copy?
- What is the approximate area of the copy?

12. **Social Studies** You want to enlarge a copy of the flag of the Philippines that is 4 in. by 8 in. The two flags will be similar. How long should you make the shorter side if the long side is 6 ft? 3 ft

13. **Geometry** A rectangle with an area of 32 in.^{2} has one side measuring 4 in. A similar rectangle has an area of 288 in.^{2}. How long is the longer side in the larger rectangle? 24 in.

GO Online
Homework Video Tutor
Visit: PHSchool.com
Web Code: are-0505

Each pair of figures below is similar. Find the value of each variable.

14.

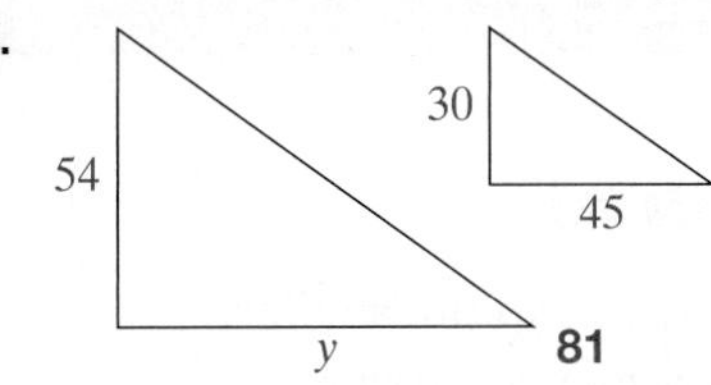

81

15.

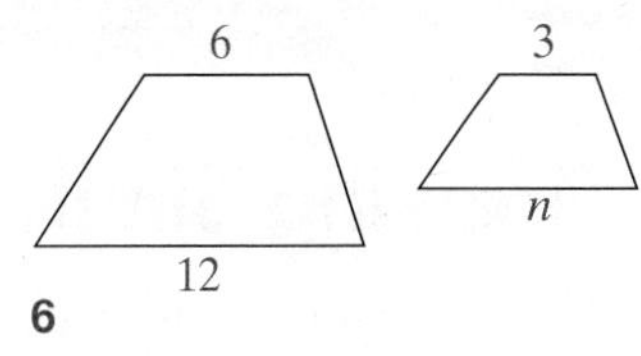

6

16. A burro is standing near a cactus. The burro is 59 in. tall. His shadow is 4 ft long. The shadow of the cactus is 7 ft long. Estimate the height of the cactus. **about 9 ft**

17. **Surveying** Surveyors know that $\triangle PQR$ and $\triangle STR$ are similar. They cannot measure the distance d across the lake directly. Find the distance across the lake. **2.88 km**

Q 2.32 km S R 2.16 km Elk Lake d T 1.74 km P

18. **Writing in Math** Give three examples of real-world objects that are similar. Explain why they are similar. **Check students' work.**

Ⓒ 19. **Challenge** The ratio of the corresponding sides of two similar triangles is 4 : 9. The sides of the smaller triangle are 10 cm, 16 cm, and 18 cm. Find the perimeter of the larger triangle. **99 cm**

Test Prep and Mixed Review

Practice

Multiple Choice

20. In the figure, $\triangle LMN \sim \triangle TVU$. What is the length of $\overline{UT}$? **A**

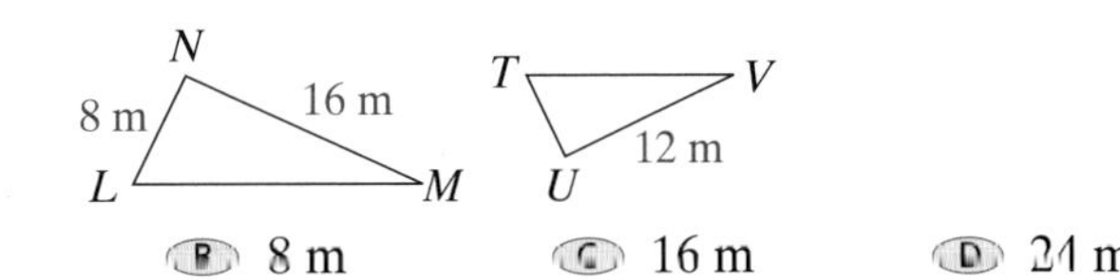

Ⓐ 6 m Ⓑ 8 m Ⓒ 16 m Ⓓ 24 m

21. The cost of two pounds of pears is \$1.78. How much will five pounds of pears cost? **H**

Ⓕ \$0.71 Ⓖ \$2.67 Ⓗ \$4.45 Ⓙ \$8.90

22. Which algebraic expression represents five inches less than twice last year's rainfall f in inches? **B**

Ⓐ $5 - 2f$ Ⓑ $2f - 5$ Ⓒ $2 - 5f$ Ⓓ $5f - 2$

GO for Help

For Exercises	See Lesson
23–25	4-6

Algebra **Solve each equation.**

23. $3x + 2 = 17$ **5**

24. $\frac{x}{5} + 5 = 21$ **80**

25. $2a - 4 = 8$ **6**

4. Assess & Reteach

Lesson Quiz

In the figure below, $\triangle JKL \sim \triangle XYZ$.

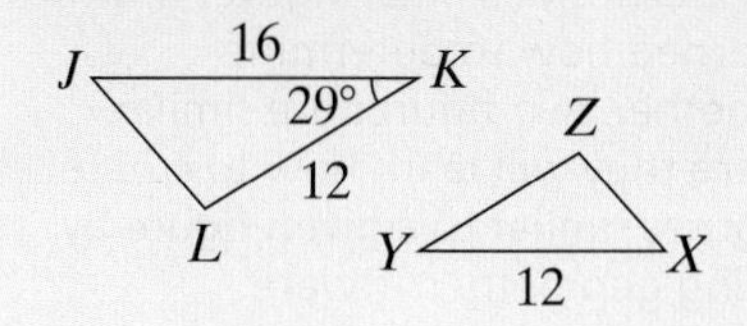

1. Which angle besides $\angle K$ has measure 29°? **$\angle Y$**
2. Find YZ. **9**
3. A 4-ft person standing near a telephone pole has a shadow 3 ft long. At the same time, the telephone pole has a shadow 18 ft long. What is the height of the telephone pole? **24 ft**

Alternative Assessment

Have students work in pairs. Each partner should draw two figures that appear to be similar and two figures that appear not to be similar. Then have the partners explain to each other why the pairs of figures they drew are similar or not similar.

Test Prep

Resources

For additional practice with a variety of test item formats:

- Test-Taking Strategies, p. 265
- Test Prep, p. 269
- Test-Taking Strategies with Transparencies

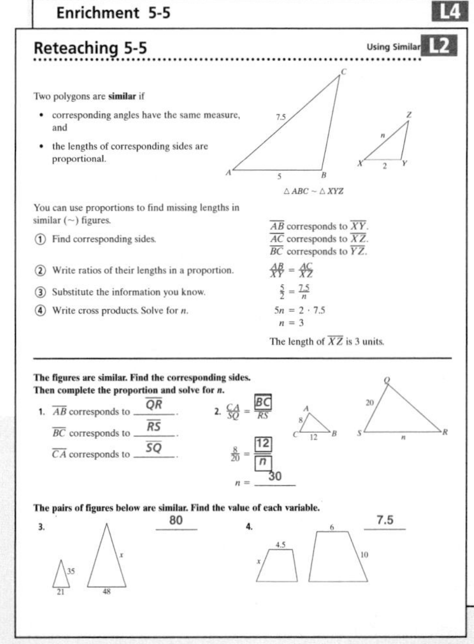

Enrichment 5-5 L4

Reteaching 5-5 Using Similar L2

Two polygons are **similar** if

- corresponding angles have the same measure, and
- the lengths of corresponding sides are proportional.

$\triangle ABC \sim \triangle XYZ$

You can use proportions to find missing lengths in similar (~) figures.

① Find corresponding sides. — $\overline{AB}$ corresponds to $\overline{XY}$. $\overline{AC}$ corresponds to $\overline{XZ}$. $\overline{BC}$ corresponds to $\overline{YZ}$.

② Write ratios of their lengths in a proportion. — $\frac{AB}{XY} = \frac{AC}{XZ}$

③ Substitute the information you know. — $\frac{5}{2} = \frac{7.5}{n}$

④ Write cross products. Solve for n. — $5n = 2 \cdot 7.5$, $n = 3$

The length of $\overline{XZ}$ is 3 units.

The figures are similar. Find the corresponding sides. Then complete the proportion and solve for n.

1. $\overline{AB}$ corresponds to $\overline{QR}$. $\overline{BC}$ corresponds to $\overline{RS}$. $\overline{CA}$ corresponds to $\overline{SQ}$.

2. $\frac{CA}{SQ} = \frac{BC}{RS}$; $\frac{8}{20} = \frac{12}{n}$; $n = 30$

The pairs of figures below are similar. Find the value of each variable.

3. 80

4. 7.5

Activity Lab

Drawing Similar Figures

In Lesson 5-5, students studied the properties of similar figures and learned how to determine whether two figures are similar. Here they will learn how to create figures similar to a given figure by using geometry software.

Guided Instruction

Error Prevention!

Some students might have difficulty using the software. Have students do the Activity with a partner. Pair students who are less proficient on the computer with those who are comfortable using the geometry software.

Teaching Tip
Depending on the size of $\triangle ABC$ and the position of point D, the sides of the similar triangle might extend beyond the bounds of the viewing window. If this occurs, suggest that students select the entire drawing, click on any part of it, and drag it until it is entirely visible. If the drawing is too large to be displayed on the screen in its entirety, suggest that students open a new window and begin with a smaller $\triangle ABC$.

Exercises
For Exercise 3, note that many geometry software programs do not display arrowheads at the ends of lines. Watch for students who believe they have made an error because they do not see the arrowheads.

Resources

- computer
- geometry software program
- Students Manipulatives Kit

5-5b Activity Lab — Technology

Drawing Similar Figures

Using a computer is an excellent way to explore similar figures. For this activity you need geometry software.

ACTIVITY

Follow these steps to draw two similar triangles like those at the right.

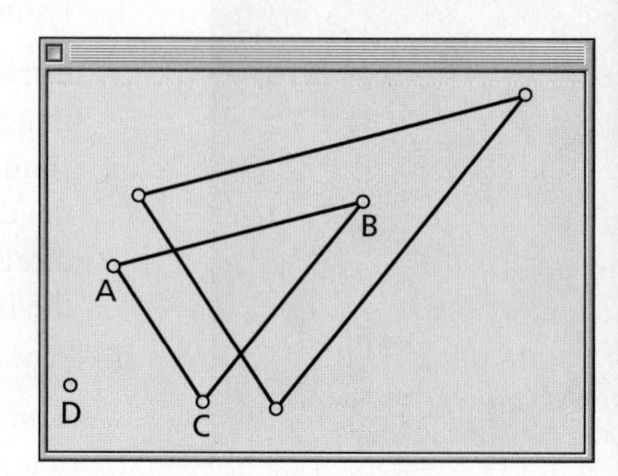

Step 1 Draw a triangle and label it ABC.

Step 2 Draw a point D not on the triangle.

Step 3 Draw a triangle similar to $\triangle ABC$ using the dilation command. To use this command, you will need to enter a scale factor and name one point as the center of the dilation. Use 2 as a scale factor and name point D as the center.

Exercises

1. Change the shape of $\triangle ABC$ by dragging point A, B, or C. What happens to the larger triangle each time? **The larger triangle is always similar to $\triangle ABC$.**
2. Drag point D to different locations. Describe what happens in each case below.
   a. D is inside $\triangle ABC$. **The larger triangle is outside $\triangle ABC$.**
   b. D is on $\overline{AB}$. **One side of the larger triangle contains $\overline{AB}$.**
   c. D is on top of point C. **The larger triangle has $\angle C$ as one of its angles.**
3. Use the software to draw $\overleftrightarrow{AD}$ (line AD), $\overleftrightarrow{BD}$, and $\overleftrightarrow{CD}$.
   a. What do you notice about the lines and the larger triangle? **Each line goes through a vertex of the larger triangle.**
   b. Why do you think that point D is called the center of the dilation? **Answers may vary. Sample: The lines that connect corresponding vertices meet at D.**
4. Draw a third triangle similar to $\triangle ABC$. This time use 0.5 as the scale factor. Keep D as the center. What do you notice about the new triangle? **The new triangle is smaller than $\triangle ABC$.**
5. a. **Open-Ended** Draw more triangles similar to $\triangle ABC$ by choosing other scale factors. Again keep D as the center. **Check students' work.**
   b. **Writing in Math** Explain how your choice of scale factor affects the final figure. **See margin.**
   c. Use what you have learned to write a definition for scale factor. **See margin.**

5b. Answers may vary. Sample: A scale factor greater than 1 creates a triangle larger than $\triangle ABC$; a scale factor less than 1 creates a triangle smaller than $\triangle ABC$.

c. Answers may vary. Sample: The scale factor is the ratio of the sides of the created triangle to the corresponding sides of $\triangle ABC$.

Checkpoint Quiz 2

Lessons 5-3 through 5-5

Determine whether the ratios form a proportion.

1. $\frac{5}{8}, \frac{12}{20}$ no
2. $\frac{4}{10}, \frac{2}{5}$ yes
3. $\frac{24}{15}, \frac{4}{3}$ no
4. $\frac{8}{3}, \frac{48}{18}$ yes
5. The last time your family bought hamburgers, 5 hamburgers cost \$5.25. At this rate, what is the cost of 8 hamburgers? **\$8.40**
6. 7 movie tickets cost \$57.75. What is the cost of 2 movie tickets? **\$16.50**

$\triangle ABC$ is similar to $\triangle WXY$. Find each measure.

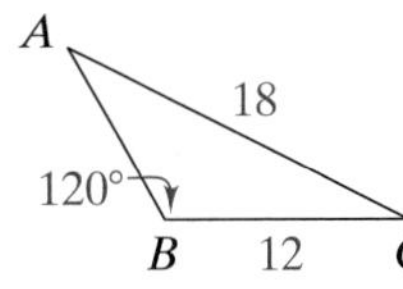

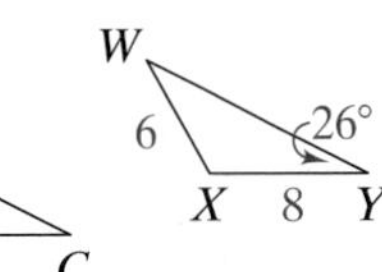

7. $\angle X$ **120°**
8. $\angle C$ **26°**
9. length of $\overline{AB}$ **9**
10. length of $\overline{WY}$ **12**

Vocabulary Builder

Learning Vocabulary

You can make your own dictionary of new math vocabulary terms.

EXAMPLE

Write entries for your dictionary for the terms *ratio* and *rate.*

Term	Definition	Example
Ratio	Comparison of two different quantities by division	$\frac{\text{20 people}}{\text{5 people}}$
Rate	Ratio that compares two quantities measured in different units	$\frac{\text{20 people}}{\text{5 cars}}$

← **Make a table with 3 columns. Label the columns "Term," "Definition," and "Example."**

← **Write the vocabulary term and its definition. Then give an example of the term.**

Exercises

Write an entry for your dictionary for each term. 1–3. Check students' work.

1. proportion
2. similar figures
3. cross products

Checkpoint Quiz

Use this Checkpoint Quiz to check students' understanding of the skills and concepts of Lessons 5-3 through 5-5.

Resources

- All-in-One Teaching Resources Checkpoint Quiz 2
- ExamView CD-ROM
- Success Tracker™ Online Intervention

Vocabulary Builder

Learning Vocabulary

Students create their own dictionaries with examples to learn math vocabulary words.

Guided Instruction

Ask students how they use a dictionary to learn new words. Point out the definitions in the glossary at the back of this book. Ask:

- *How does the example help you to understand the word's meaning?* **It shows how the word is used, demonstrates its meaning.**
- *Why is a dictionary a useful tool?* **You can use it at any time to help you understand how a word is being used.**

Exercises

Have students share their dictionary entries with a partner. Invite pairs to discuss ways to improve each other's entries.

Scale Drawings and Models

Students use drawings and measurements to make accurate scale models. This will help them understand the concept of maps and scale drawings in Lesson 5-6.

Guided Instruction

Before beginning the Activity, discuss models that students have seen or built themselves. Ask:

- *Why is it important for models to be to scale?* **People use models to learn information about the actual item the model represents. To have accurate information about that item, the model must have the same proportions.**
- *Who would use a model in their work? Why?* **Sample: an architect to plan a building he or she is designing; a doctor to study a body part she or he is going to work with**

Exercises

Have students work independently on the Exercises. Have them compare models and discuss why they are the same or why they are different.

Alternative Method

Some students may prefer to use three-dimensional objects, such as blocks, to build a model. Provide these materials to help students understand the concept of models.

Resources

- Activity Lab 5-6: Exploring Scale
- grid paper
- Student Manipulatives Kit

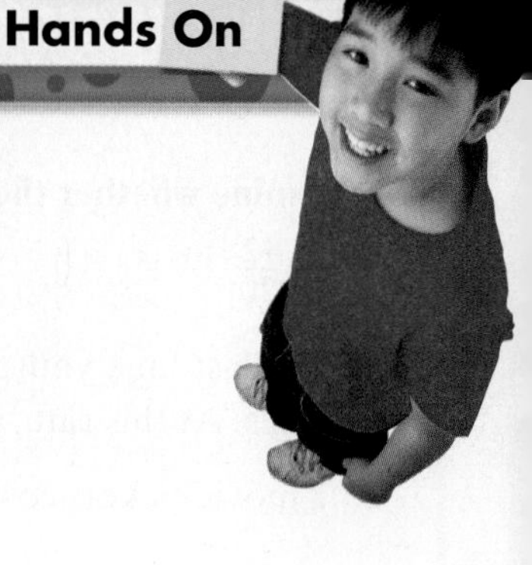

Scale Drawings and Models

The sketches below show the measurements of Jackie's bedroom. You can use these sketches to make a model of Jackie's bedroom.

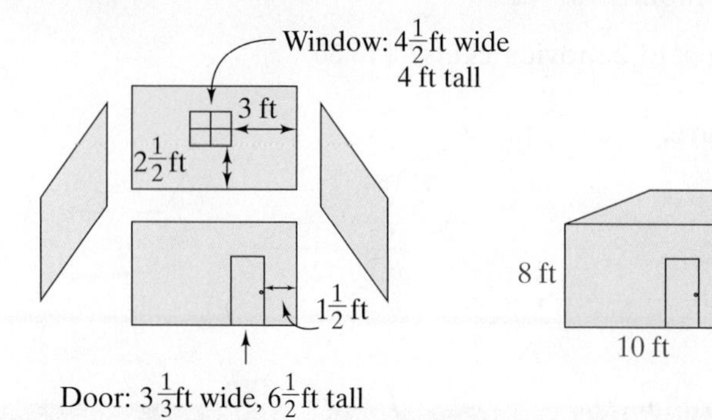

ACTIVITY

Step 1 Make a drawing on graph paper of the floor and walls of Jackie's room as shown. Let one unit on the graph paper represent one foot. Include the windows and doors.

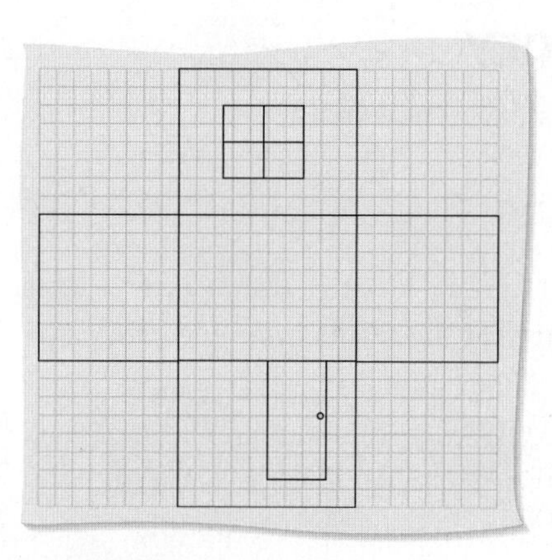

Step 2 Cut around your drawing and tape up the walls to make your model.

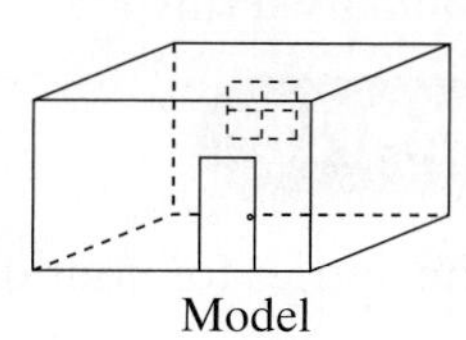

Model

Exercises

1–4. Check students' work.

1. Jackie's bedroom bureau is shown at the right. Use the same steps as above to make a drawing and a model of the bureau.

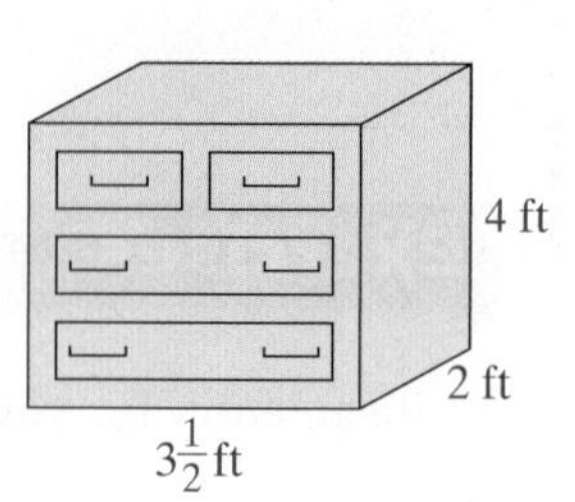

2. Measure the heights and lengths of the walls of your classroom. Use graph paper to make an accurate drawing of the floor and walls.
3. Cut around your drawing and make a model of your classroom.
4. Explain how you can use ratios and proportions to find the dimensions of your model.

5-6 Maps and Scale Drawings

Check Skills You'll Need

1. **Vocabulary Review** What are *cross products?* See back of book.

Solve each proportion.

2. $\frac{2}{3} = \frac{x}{12}$ 8
3. $\frac{9}{5} = \frac{27}{y}$ 15
4. $\frac{5}{m} = \frac{25}{5}$ 1
5. $\frac{t}{21} = \frac{6}{18}$ 7

GO for Help Lesson 5-4

What You'll Learn

To use proportions to solve problems involving scale

New Vocabulary scale drawing, scale

Why Learn This?

When you know how scales work, you can see them in everything from maps to giant sculptures.

A **scale drawing** is an enlarged or reduced drawing of an object that is similar to the actual object.

A **scale** is the ratio that compares a length in a drawing or model to the corresponding length in the actual object. If a 15-foot boat is 1 inch long on a drawing, you can write the scale of the drawing in these three ways.

1 in. : 15 ft (↑ drawing, ↑ actual) $\quad \frac{1 \text{ in.}}{15 \text{ ft}} \quad$ 1 in. = 15 ft

Vocabulary Tip

On a map, the equal sign in 1 in. = 15 ft does not mean that the two quantities are equal, as it would in an equation.

EXAMPLE Using a Scale Drawing

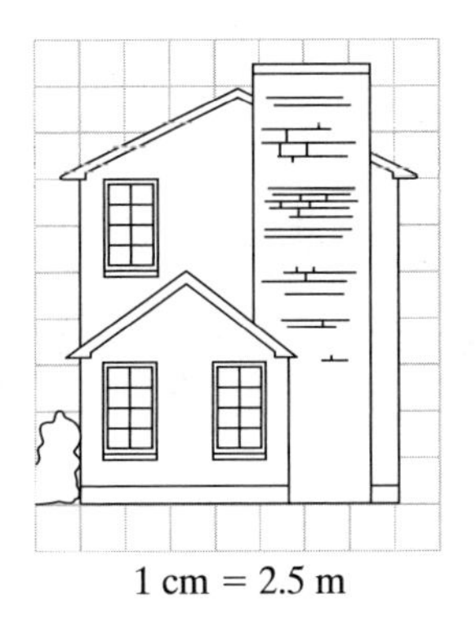

1 cm = 2.5 m

1 (Algebra) The length of the side of a house is 3 cm on a scale drawing. What is the actual length of the side of the house?

You can write the scale of the drawing as $\frac{1 \text{ cm}}{2.5 \text{ m}}$. Then write a proportion. Let n represent the actual length of the house.

drawing (cm) → / actual (m) → $\frac{1}{2.5} = \frac{3}{n}$ ← drawing (cm) / ← actual (m)

$1n = 2.5(3)$ ← Write the cross products.

$n = 7.5$ ← Simplify.

The actual length is 7.5 m.

Quick Check

1. The chimney of the house is 4 cm tall on the drawing. How tall is the chimney of the actual house? 10 m

Differentiated Instruction Solutions for All Learners

Special Needs L1

Provide students with experiences using map scales. Let them measure the actual distance between two points on a map, then compare that distance to the scale and find the actual distance.

learning style: visual

Below Level L2

Give students blank templates for organizing their work.

drawing → ■/■ = ■/■ ← drawing
actual → ← actual

learning style: visual

5-6

1. Plan

Objective

To use proportions to solve problems involving scale

Examples

1 Using a Scale Drawing
2 Application: Geography
3 Finding the Scale
4 Application: Models

Math Understandings: p. 226D

Math Background

A *scale drawing* is a two-dimensional representation of a real object. The ratio $\frac{\text{drawing length}}{\text{corresponding actual length}}$ is the same throughout the drawing. This ratio is called the *scale* of the drawing. An artist uses the scale and actual lengths to create the drawing; the reader uses the scale and drawing lengths to interpret the drawing. Both creating and interpreting the drawing require proportional reasoning.

More Math Background: p. 226D

Lesson Planning and Resources

See p. 226E for a list of the resources that support this lesson.

PowerPoint

Bell Ringer Practice

Check Skills You'll Need

Use student page, transparency, or PowerPoint. For intervention, direct students to:

Using Similar Figures
Lesson 5-5
Extra Skills and Word Problems Practice, Ch. 5

2. Teach

Activity Lab

Use before the lesson.
Student Edition Activity Lab, Hands On 5-6a, Scale Drawings and Models, p. 258

All in One Teaching Resources
Activity Lab 5-6: Exploring Scale

Guided Instruction

Example 2

Students will need metric rulers. The map distance between Charlotte and Raleigh is about 2.8 cm. The map distance between Raleigh and Winston-Salem is about 1.9 cm.

Error Prevention!

In Quick Check, students might record the map distance in millimeters and use this number in their proportions. For instance, if they record the distance from Charlotte to Raleigh as 28 mm, they might write $\frac{1}{75} = \frac{28}{n}$. Suggest that they include the units in their proportions and highlight the units. This should focus their attention on the need to convert the measure.

$$\frac{1\text{ cm}}{75\text{ km}} = \frac{28\text{ mm}}{n\text{ km}} \rightarrow \frac{1\text{ cm}}{75\text{ km}} = \frac{2.8\text{ cm}}{n\text{ km}}$$

EXAMPLE Application: Geography

2 Find the actual distance from Charlotte to Winston-Salem.

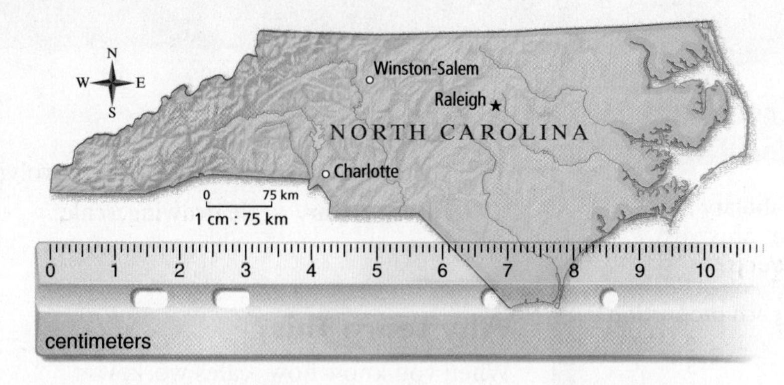

Step 1 Use a centimeter ruler to find the map distance from Charlotte to Winston-Salem. The map distance is about 1.6 cm.

Step 2 Use a proportion to find the actual distance. Let n represent the actual distance.

Test Prep Tip
When writing a proportion, make sure each ratio compares the same types of quantities.

$$\text{map (cm)} \rightarrow \frac{1}{75} = \frac{1.6}{n} \begin{matrix}\leftarrow \text{map (cm)} \\ \leftarrow \text{actual (km)}\end{matrix} \quad \text{actual (km)} \rightarrow \qquad \leftarrow \textbf{Write a proportion.}$$

$$1n = 75(1.6) \qquad \leftarrow \textbf{Write the cross products.}$$

$$n = 120 \qquad \leftarrow \textbf{Simplify.}$$

The actual distance from Charlotte to Winston-Salem is about 120 km.

Quick Check

2. Find the actual distance from Charlotte to Raleigh.

You can use the GCF to find the scale of a drawing or a model.

EXAMPLE Finding the Scale

3 **Models** Refer to the model boxcar shown at the right. The actual length of a boxcar is 609 in. What is the scale of the model?

$$\frac{7}{609} = \frac{7 \div 7}{609 \div 7} = \frac{1}{87} \qquad \leftarrow \textbf{Write the ratio in simplest form.}$$

(scale length → 7; actual length → 609)

The scale is 1 in. : 87 in.

Quick Check

3. The length of a room in an architectural drawing is 10 in. Its actual length is 160 in. What is the scale of the drawing?

Advanced Learners L4	**English Language Learners** ELL
The scale of a map is 1 : 250,000. What is the actual distance in miles for each inch of map distance? **about 3.9 miles**	Have students read scales on atlases, maps, and globes. Ask them to talk about what they notice about the map scales. *How are the map scales different? How are they similar? Why is a map scale useful? Could a map be drawn with the real distance?* **no**
learning style: visual	**learning style: verbal**

EXAMPLE Application: Models

4 **Multiple Choice** You want to make a scale model of a sailboat that is 51 ft long and 48 ft tall. You plan to make the model 17 in. long. Which equation can you use to find x, the height of the model?

Ⓐ $\frac{48}{51} = \frac{17}{x}$ Ⓑ $\frac{17}{51} = \frac{x}{48}$ Ⓒ $\frac{48}{17} = \frac{x}{51}$ Ⓓ $\frac{x}{17} = \frac{51}{48}$

$$\frac{\text{model (in.)} \to 17}{\text{actual (ft)} \to 51} = \frac{\blacksquare \leftarrow \text{model (in.)}}{\blacksquare \leftarrow \text{actual (ft)}}$$ ← Write a proportion.

$$\frac{17}{51} = \frac{x}{48}$$ ← Fill in the information you know. Use x for the information you do not know.

The correct answer is choice B.

You can solve the proportion to find the height of the model. You can simplify $\frac{17}{51}$ to make calculating easier.

$$\frac{\overset{1}{\cancel{17}}}{\underset{3}{\cancel{51}}} = \frac{x}{48}$$ ← Simplify $\frac{17}{51}$ by dividing by the GCF, 17.

$$\frac{1}{3} = \frac{x}{48}$$

$$\frac{1}{3} \overset{\times 16}{=} \frac{x}{48}$$ ← Find the common multiplier, 16.

$$x = 16$$ ← Use mental math.

The height of the model is 16 inches.

GO for Help

For help with simplifying ratios, go to Lesson 5-3, Example 1.

Quick Check

4. If the sailboat is 15 ft wide, how wide should the model be? 5 in.

Check Your Understanding

1. **Vocabulary** A scale is a __?__ that compares a length in a drawing to the corresponding __?__ in the actual object. ratio; length

2. **Reasoning** The scale of a drawing is 5 cm : 1 mm. Is the scale drawing larger or smaller than the actual figure? Explain. See left.

2. Larger; since a scale is a ratio of $\frac{\text{drawing}}{\text{actual}}$, then in $\frac{5 \text{ cm}}{1 \text{ mm}}$, the drawing is larger than the actual figure.

3. **Error Analysis** A student wants to write the scale of a statue of President Kennedy, who was 6 ft tall. The statue is 8 ft tall. The student writes the scale as 6 ft : 8 ft. Is the student correct? Explain. No; the scale should be written as model : actual, or 8 ft : 6 ft.

The scale of a map is 1 inch : 5 miles. Find each distance.

4. A road is 3 in. long on the map. Find the actual length of the road. 15 mi

5. A lake is 35 miles long. Find the length of the lake on the map. 7 in.

PowerPoint

Additional Examples

1. The scale of a drawing is 1 in. : 6 ft. Find the actual length for a drawing length of 4.5 in. 27 ft

2. On the map of North Carolina, the map distance from Asheville to Raleigh would be about 4.4 cm. Find the actual distance from Asheville to Raleigh. about 330 km

3. The actual length of the wheelbase of a mountain bike is 260 cm. The length of the wheelbase in a scale drawing is 4 cm. Find the scale of the drawing. 1 cm : 65 cm

4. An architect's model of a house is 44 in. high. The actual house is 20 ft high and 45 ft wide. What should the width of the architect's model be? 99 in.

All in One Teaching Resources

- Daily Notetaking Guide 5-6 L3
- Adapted Notetaking 5-6 L1

Closure

- Explain how maps and scale drawings are related to ratios and proportions. Sample: The scale of a map or drawing is a ratio that compares each length in the map or drawing to a corresponding actual length. You use a proportion involving the scale and a known map or drawing length to find an unknown actual length.

3. Practice

Assignment Guide

Check Your Understanding

Go over Exercises 1–5 in class before assigning the Homework Exercises.

Homework Exercises

A	Practice by Example	6–17
B	Apply Your Skills	18–25
C	Challenge	26
Test Prep and Mixed Review		27–30

Homework Quick Check

To check student's understanding of key skills and concepts, go over Exercises 9, 15, 21, 24, and 25.

Differentiated Instruction Resources

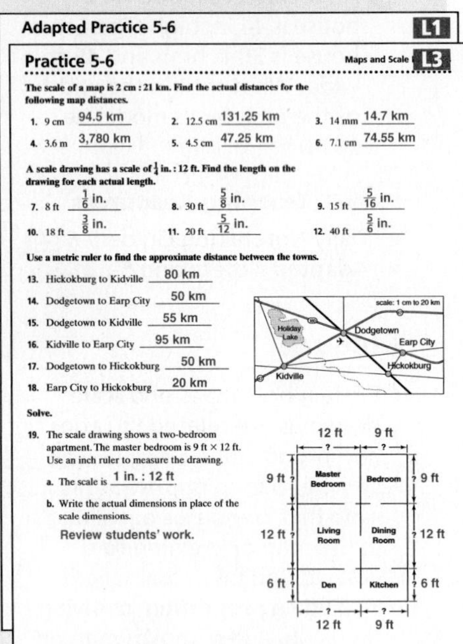

Adapted Practice 5-6 L1

Practice 5-6 Maps and Scale L3

The scale of a map is 2 cm : 21 km. Find the actual distances for the following map distances.

1. 9 cm 94.5 km
2. 12.5 cm 131.25 km
3. 14 mm 14.7 km
4. 3.6 m 3,780 km
5. 4.5 cm 47.25 km
6. 7.1 cm 74.55 km

A scale drawing has a scale of $\frac{1}{4}$ in. : 12 ft. Find the length on the drawing for each actual length.

7. 8 ft $\frac{1}{6}$ in.
8. 30 ft $\frac{5}{8}$ in.
9. 15 ft $\frac{5}{16}$ in.
10. 18 ft $\frac{3}{8}$ in.
11. 20 ft $\frac{5}{12}$ in.
12. 40 ft $\frac{5}{6}$ in.

Use a metric ruler to find the approximate distance between the towns.

13. Hickokburg to Kidville 80 km
14. Dodgetown to Earp City 50 km
15. Dodgetown to Kidville 55 km
16. Kidville to Earp City 95 km
17. Dodgetown to Hickokburg 50 km
18. Earp City to Hickokburg 20 km

Solve.

19. The scale drawing shows a two-bedroom apartment. The master bedroom is 9 ft × 12 ft. Use an inch ruler to measure the drawing.
   a. The scale is 1 in. : 12 ft.
   b. Write the actual dimensions in place of the scale dimensions.
   Review students' work.

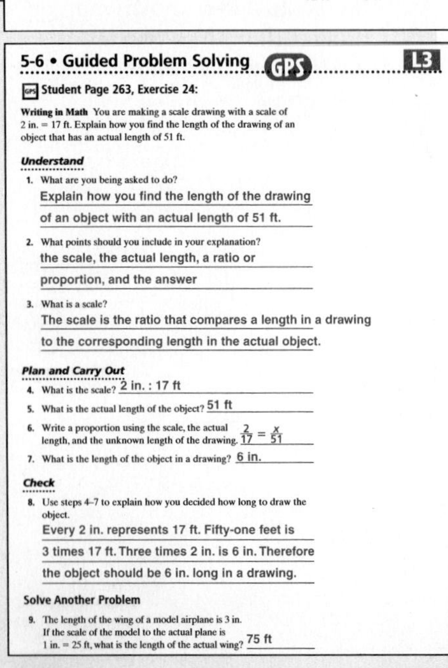

5-6 • Guided Problem Solving GPS L3

Student Page 263, Exercise 24:

Writing in Math You are making a scale drawing with a scale of 2 in. = 17 ft. Explain how you find the length of the drawing of an object that has an actual length of 51 ft.

Understand

1. What are you being asked to do?
   Explain how you find the length of the drawing of an object with an actual length of 51 ft.
2. What points should you include in your explanation?
   the scale, the actual length, a ratio or proportion, and the answer
3. What is a scale?
   The scale is the ratio that compares a length in a drawing to the corresponding length in the actual object.

Plan and Carry Out

4. What is the scale? 2 in. : 17 ft
5. What is the actual length of the object? 51 ft
6. Write a proportion using the scale, the actual length, and the unknown length of the drawing. $\frac{2}{17} = \frac{x}{51}$
7. What is the length of the object in a drawing? 6 in.

Check

8. Use steps 4–7 to explain how you decided how long to draw the object.
   Every 2 in. represents 17 ft. Fifty-one feet is 3 times 17 ft. Three times 2 in. is 6 in. Therefore the object should be 6 in. long in a drawing.

Solve Another Problem

9. The length of the wing of a model airplane is 3 in. If the scale of the model to the actual plane is 1 in. = 25 ft, what is the length of the actual wing? 75 ft

Homework Exercises

For more exercises, see Extra Skills and Word Problems.

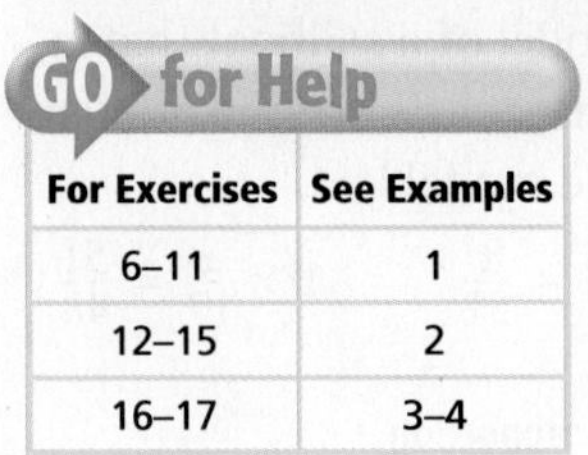

GO for Help

For Exercises	See Examples
6–11	1
12–15	2
16–17	3–4

A (Algebra) **A scale drawing has a scale of 1 in. : 11 ft. Find the actual length for each drawing length.**

6. 21 in. **231 ft**
7. 15 in. **165 ft**
8. 6 in. **66 ft**
9. 45 in. **495 ft**
10. 13.5 in. **148.5 ft**
11. 1.5 in. **16.5 ft**

Geography Find the actual distance between each pair of cities. Use a ruler to measure. Round to the nearest mile.

Hartford
CONNECTICUT
Norwich
Danbury
New Haven
N
W
E
S
0 22 mi
1 cm : 22 mi

12. Hartford and Danbury **about 47 mi**
13. Norwich and Hartford **about 38 mi**
14. New Haven and Norwich **about 47 mi**
15. New Haven and Danbury **about 28 mi**

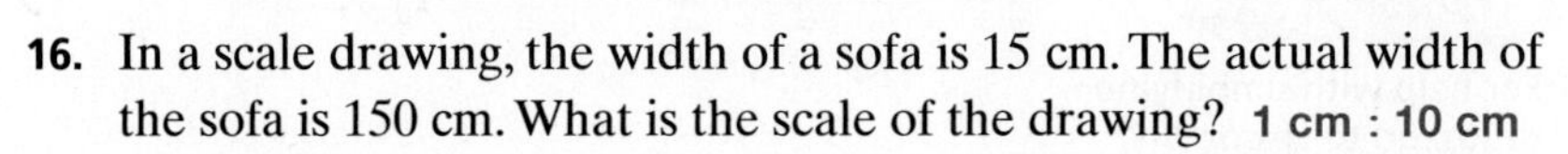

16. In a scale drawing, the width of a sofa is 15 cm. The actual width of the sofa is 150 cm. What is the scale of the drawing? **1 cm : 10 cm**
17. A certain car is about 100 in. long and 60 in. wide. You plan to make a model of this car that is 9 in. wide. How long will your model be? **15 in.**

B 

18. **Guided Problem Solving** The height of a building on a blueprint is 10 in. Its actual height is 150 ft. What is the scale of the drawing?
   - What ratio can you use to find the scale? **1 in. : 15 ft**
   - How can finding the GCF help you to simplify your answer?
19. The scale of a model is 0.5 in. : 6 ft. Find the length of the model for an actual length of 204 ft. **17 in.**
20. **Geography** Dallas is $6\frac{3}{4}$ in. from Houston on a map of Texas with a scale of $\frac{3 \text{ in.}}{100 \text{ mi}}$. What is the actual distance from Dallas to Houston? **225 mi**
21. **Architecture** The blueprint below is a scale drawing of an apartment. The scale is $\frac{1}{4}$ in. : 4 ft. Sketch a copy of the floor plan. Write the actual dimensions in place of the scale dimensions.

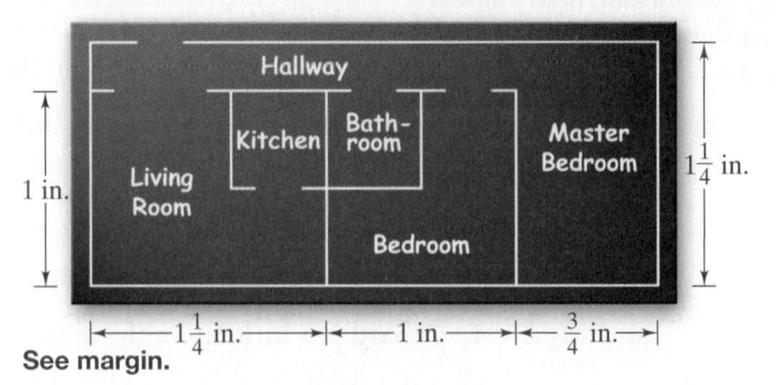

See margin.

GO Online
Homework Video Tutor
Visit: PHSchool.com
Web Code: are-0506

21.

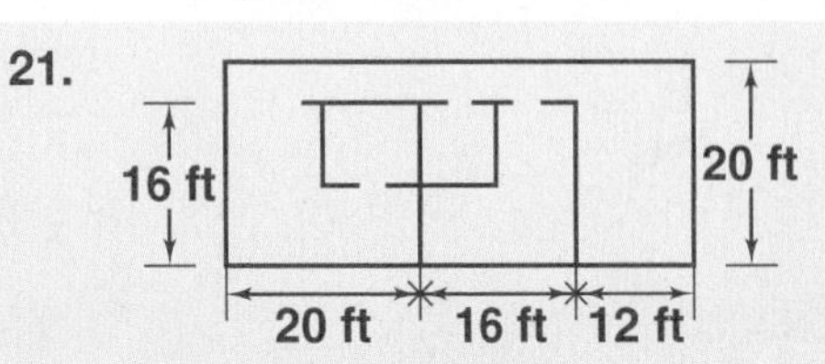

24. Answers may vary. Sample: Set up the proportion $\frac{2 \text{ in.}}{17 \text{ ft}} = \frac{x}{51 \text{ ft}}$. Solve the proportion by mental math or cross products. $x = 6$, so the drawing will have a length of 6 in.

Use a centimeter ruler to measure the length of the segment shown in each figure below. Find the scale of each drawing. Answers may vary. Samples are given.

22. Peach Aphid

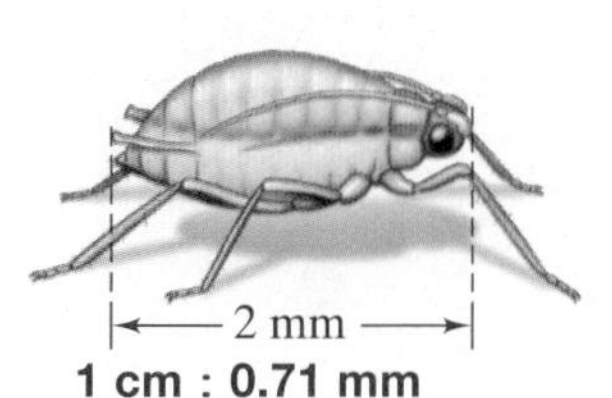

1 cm : 0.71 mm

23. Killer Whale

1 cm : 1.9 m

24. **Writing in Math** You are making a scale drawing with a scale of 2 in. = 17 ft. Explain how you find the length of the drawing of an object that has an actual length of 51 ft. **See margin.**

GPS

25. **Special Effects** A special-effects artist has made a scale model of a dragon for a movie. In the movie, the dragon will appear to be 16 ft tall. The model is 4 in. tall.
    a. What scale has the artist used? **1 in. : 4 ft**
    b. The same scale is used for a model of a baby dragon, which will appear to be 2 ft tall. What is the height of the model? **0.5 in.**

C 26. **Challenge** A building is drawn with a scale of 1 in. : 3 ft. The height of the drawing is 1 ft 2 in. After a design change, the scale is modified to be 1 in. : 4 ft. What is the height of the new drawing? **$10\frac{1}{2}$ in.**

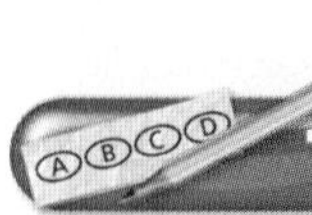

Test Prep and Mixed Review

Practice

Multiple Choice

27. Doug drew a map with a scale of 1 inch : 5 miles. What distance on Doug's map should represent 4.5 miles? **B**
    A 0.45 in.   B 0.9 in.   C 0.95 in.   D 4.5 in.

28. In the figure at the right, $ABCD \sim STUV$. Which of the following statements is NOT true? **J**
    F $\overline{UV}$ corresponds to $\overline{CD}$.
    G $\overline{AD}$ corresponds to $\overline{SV}$.
    H $\angle A$ corresponds to $\angle S$.
    J $\angle D$ corresponds to $\angle U$.

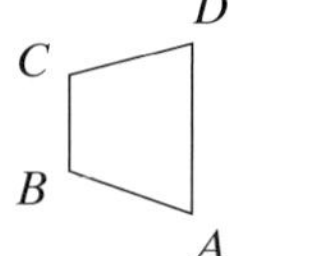

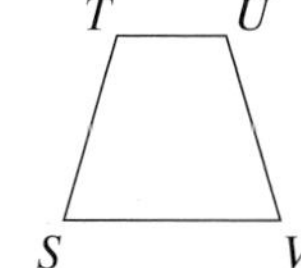

29. A cycling route is 56 miles long. There is a water station every $3\frac{3}{4}$ mi. Which equation represents the total number of water stations w? **B**
    A $56w = 3\frac{3}{4}$
    B $3\frac{3}{4}w = 56$
    C $3\frac{3}{4} - w = 56$
    D $w + 3\frac{3}{4} = 56$

GO for Help

For Exercise	See Lesson
30	5-5

30. Find the values of x and y in the similar triangles at the right. **$x = 40$, $y = 32$**

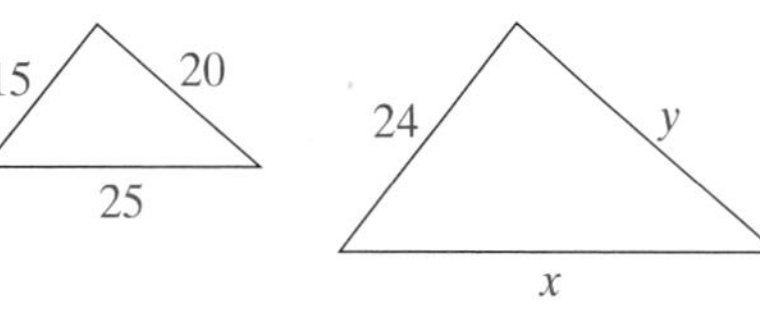

Alternative Assessment

Display a map of your state and assign students to groups of four. Ask students to find and explain the scale on the map. Then have them find the distance between two cities.

Test Prep

Resources

For additional practice with a variety of test item formats:

- Test-Taking Strategies, p. 265
- Test Prep, p. 269
- Test-Taking Strategies with Transparencies

4. Assess & Reteach

1. The scale of a map is 1 cm : 25 km. Find the actual distance for a map distance of 6.2 cm. **155 km**
2. The actual length of a machine part is 40 in. The length of the machine part in a scale drawing is 5 in. Find the scale of the drawing. **1 in. : 8 in.**
3. A scale drawing of a playground is 4.7 cm wide and 6.2 cm long. The playground is 15 m wide. How long is it? **19.8 m**

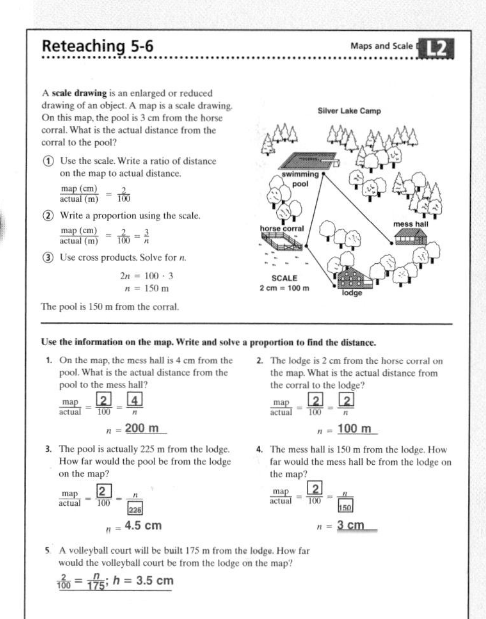

Reteaching 5-6 Maps and Scale L2

A **scale drawing** is an enlarged or reduced drawing of an object. A map is a scale drawing. On this map, the pool is 3 cm from the horse corral. What is the actual distance from the corral to the pool?

① Use the scale. Write a ratio of distance on the map to actual distance.

$\frac{\text{map (cm)}}{\text{actual (m)}} = \frac{2}{100}$

② Write a proportion using the scale.

$\frac{\text{map (cm)}}{\text{actual (m)}} = \frac{2}{100} = \frac{3}{n}$

③ Use cross products. Solve for n.

$2n = 100 \cdot 3$
$n = 150$ m

The pool is 150 m from the corral.

Silver Lake Camp — swimming pool, horse corral, mess hall, lodge. SCALE 2 cm = 100 m

Use the information on the map. Write and solve a proportion to find the distance.

1. On the map, the mess hall is 4 cm from the pool. What is the actual distance from the pool to the mess hall?
   $\frac{\text{map}}{\text{actual}} = \frac{2}{100} = \frac{4}{n}$
   $n =$ **200 m**
2. The lodge is 2 cm from the horse corral on the map. What is the actual distance from the corral to the lodge?
   $\frac{\text{map}}{\text{actual}} = \frac{2}{100} = \frac{2}{n}$
   $n =$ **100 m**
3. The pool is actually 225 m from the lodge. How far would the pool be from the lodge on the map?
   $\frac{\text{map}}{\text{actual}} = \frac{2}{100} = \frac{n}{225}$
   $n =$ **4.5 cm**
4. The mess hall is 150 m from the lodge. How far would the mess hall be from the lodge on the map?
   $\frac{\text{map}}{\text{actual}} = \frac{2}{100} = \frac{n}{150}$
   $n =$ **3 cm**
5. A volleyball court will be built 175 m from the lodge. How far would the volleyball court be from the lodge on the map?
   $\frac{2}{100} = \frac{n}{175}$; $h = 3.5$ cm

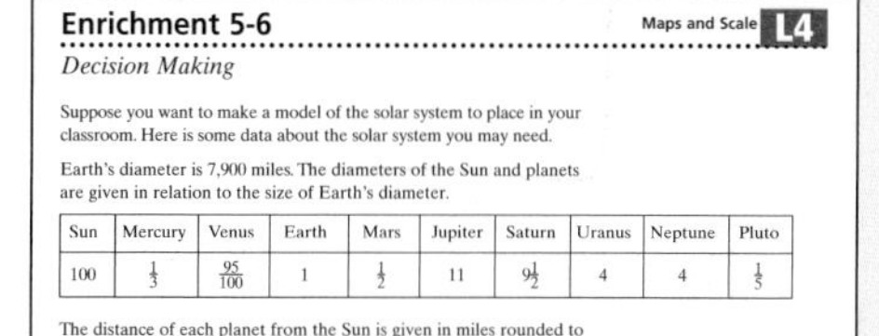

Enrichment 5-6 Maps and Scale L4

Decision Making

Suppose you want to make a model of the solar system to place in your classroom. Here is some data about the solar system you may need.

Earth's diameter is 7,900 miles. The diameters of the Sun and planets are given in relation to the size of Earth's diameter.

Sun	Mercury	Venus	Earth	Mars	Jupiter	Saturn	Uranus	Neptune	Pluto
100	$\frac{1}{3}$	$\frac{95}{100}$	1	$\frac{1}{2}$	11	$9\frac{1}{2}$	4	4	$\frac{1}{5}$

The distance of each planet from the Sun is given in miles rounded to nearest million.

Sun	Mercury	Venus	Earth	Mars	Jupiter	Saturn	Uranus	Neptune	Pluto
0	36	67	93	142	484	887	1,782	2,794	3,666

1. You decide to use a baseball (about 2.9 inches in diameter) to represent Earth.
   a. To the nearest whole number, what will 1 inch on the scale represent if you use the baseball to represent Earth?
   2,724 miles
   b. Write the scale distance of the diameter for the Sun.
   The diameter of the Sun is 290 in. (2.9 × 100).
   c. About how many inches from the Sun must you place Earth? Is this reasonable? Explain.
   Sample answer: about 34,140 in. No, you can't place the model in the classroom since 34,140 in. is approximately equal to 948 yd, over 9 football fields in length.
2. Choose other objects to represent Earth. Can you devise a model that is true to scale that can be contained in your classroom? Explain.
   Sample answer: No, even if Earth is reduced to a speck of dust, the solar system is a circle with a 6-ft diameter. The sun and planets would be nearly invisible to the naked eye.

Activity Lab

Plan a Trip

Students use the scale of a map to find the lengths of different routes between two cities, and then to calculate the amount of time to drive each route.

Guided Instruction

Before assigning the Exercises, be sure students understand the map scale: 1 in. : 85 mi.

Ask questions such as:

- *How many miles does 1 inch represent?* **85 mi**
- *How many miles do 3 inches represent?* **255 mi**
- *Why can't you measure the shortest straight line from Indianapolis to Cleveland to plan your trip?* **Sample: You have to drive on the roads, and there is no road shown that goes straight from Indianapolis to Cleveland.**

Activity

Have students work in pairs. For Exercise 1, one partner can make the measurements and the other can record the measurements.

Exercises 2–4 The partners should work independently and then compare results.

Exercise 5 can be used as a short-term project that can be displayed in the classroom.

Differentiated Instruction

Advanced Learners L4

Have students calculate the map distance and actual distance of a round trip from Indianapolis to Columbus to Cleveland and back to Indianapolis through Toledo and Fort Wayne.

Resources

- inch ruler
- map of the United States
- Student Manipulatives Kit

5-6b Activity Lab

Plan a Trip

Located on the shore of Lake Erie, the 150,000-square-foot Rock and Roll Hall of Fame and Museum is a landmark for the city of Cleveland, Ohio.

ACTIVITY

Answers may vary. Samples are given.

Use the map above to plan a trip from Indianapolis to Cleveland.

1. Measure the distances on Routes 69 and 90 from Indianapolis to Fort Wayne, Fort Wayne to Toledo, and Toledo to Cleveland. Add them to get the total map distance. **3.75 in.**
2. Locate the scale on the map. Use the scale to convert the map distance into the actual distance for your trip. **320 mi**
3. Use an average speed of 50 mi/h. Find the time it will take to drive the total actual distance. If you plan to drive no more than 4 hours per day, how many days will your trip take? **6.3 h; 2 days**
4. Suppose you want to make the same trip, but you want to visit a friend in Columbus. Plan your trip from Indianapolis to Cleveland to pass through Columbus. Compare the two trips. **The trip will be a little shorter.**
5. **Research** Now suppose that you are planning a cross-country road trip from Trenton, New Jersey, to San Francisco, California. Locate a map of the United States and plan your trip. Decide which cities you would like to visit along the way. How long will your trip take you? **Check students' work.**

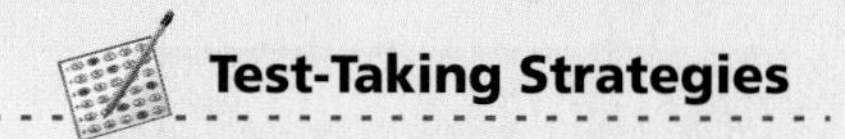

Using a Variable

You can solve many problems by using a variable to represent an unknown quantity.

EXAMPLE

Purple is a mixture of the primary colors red and blue. A certain shade of purple paint requires 6 parts red paint to 7 parts blue paint. If you have 16 quarts of red paint, how many quarts of blue paint do you need to make the desired shade of purple? How many quart cans of blue paint do you need to buy?

The problem is asking for the amount of blue paint needed to make a shade of purple. You can write and solve a proportion.

Let b represent the number of quarts of blue paint you need.

red paint → / blue paint → $\frac{6}{7} = \frac{16}{b}$ ← red paint / ← blue paint ← **Write a proportion.**

$6b = 112$ ← **Write the cross products and simplify.**

$\frac{6b}{6} = \frac{112}{6}$ ← **Divide each side by 6.**

$b = 18\frac{2}{3}$ ← **Simplify.**

You need $18\frac{2}{3}$ quarts of blue paint. You should buy 19 cans.

Exercises

Use a variable to write and solve an equation.

1. To serve 16 people, 96 pieces of fruit are needed. How many pieces of fruit are needed to serve 22 people? **C**
   (A) $18\frac{2}{3}$ (B) 11 (C) 132 (D) 144

2. A paper distributor is shipping mailing tubes to an art gallery. The distributor has previously shipped 560 tubes in 16 boxes. At this rate, how many boxes would the distributor need to ship 112 tubes? **F**
   (F) 4 (G) 5 (H) 12 (J) 35

3. A class of 37 students is making origami boxes. Each box requires 8 pieces of origami paper. The class will split into groups of at most 3 students per group. If each group is going to make one box, what is the least number of pieces of paper needed by the class? **C**
   (A) 24 (B) 96 (C) 104 (D) 296

Using a Variable

Students learn to use a variable as the placeholder for an unknown quantity in a problem. They then write and solve an equation involving that variable. The solution of the equation leads to the solution of the problem.

Guided Instruction

After students read the problem, write $\frac{\text{red}}{\text{blue}} = \frac{\text{red}}{\text{blue}}$ on the chalkboard or on a transparency. As students watch, erase the words one at a time, replacing each with the corresponding number until $\frac{6}{7} = \frac{16}{\text{blue}}$ is displayed. Lead students to see that the remaining word *blue* represents the unknown number of quart-cans of blue paint needed to make the shade of purple. Now replace *blue* with the variable *b*. Proceed to solve the proportion and answer the question.

Resources

Test-Taking Strategies with Transparencies
- Transparency 11
- Practice sheet, p. 17

Test-Taking Strategies with Transparencies

Test-Taking Strategies: Using a Variable

You can solve many problems by using a variable to represent an unknown quantity.

Let the variable be the quantity that you are looking for.

Examples
Let x = the number of hours worked.
Let t = the amount of time spent.
Let A = the area of the backyard.

Then use the variable to write an equation.

You attend a cheerleading clinic with your team every Saturday morning. The clinic costs $6.50 per person. It costs $78 for your entire team to attend the clinic. How many cheerleaders are on your team?

Let c = the number of cheerleaders.

Write and solve an equation to find the number of cheerleaders.

$6.50c = 78$
$\frac{6.50}{6.50}c = \frac{78}{6.50}$ There are 12 cheerleaders on your team.
$c = 12$

Define a variable. Then, write and solve an equation for each problem.

1. An illustrator charges $75 per picture. Find the total number of pictures in a children's book if the illustrator earned $3,000.
2. There are 12 songs on a CD. The CD lasts 42 minutes. About how long is the average song on the CD?

Chapter 5 Review

Resources

Student Edition
Extra Skills and Word Problems Practice, Ch. 5, p. 638
English/Spanish Glossary, p. 676
Formulas and Properties, p. 674
Tables, p. 670

All in One Teaching Resources
- Vocabulary and Study Skills 5F L3

Differentiated Instruction
Spanish Vocabulary Workbook with Study Skills ELL
Interactive Textbook
- Audio Glossary

Online Vocabulary Quiz

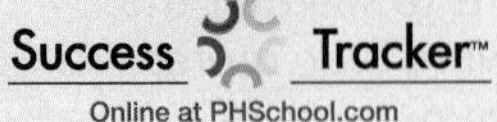

Spanish Vocabulary/Study Skills ELL

Vocabulary/Study Skills L3

5F: Vocabulary Review Puzzle For use with the Chapter Review

Study Skill Use a special notebook or section of a loose-leaf binder for math.

Complete the crossword puzzle. For help, use the Glossary in your textbook.

Here are the words you will use to complete this crossword puzzle:

equation, factor, figures, fraction, inequality, mixed number, prime, proportion, ratio, scale drawing

ACROSS
5. enlarged or reduced drawing of an object
7. equation stating two ratios are equal
8. a statement of two equal expressions
10. a whole number that divides another whole number evenly

DOWN
1. Similar ______ have the same shape but not necessarily the same size.
2. a statement that two expressions are not equal
3. a number made up of a nonzero whole number and a fraction
4. a number with only two factors, one and itself
6. a number in the form $\frac{a}{b}$
9. a comparison of two numbers by division

Chapter 5 Review

Vocabulary Review

cross products (p. 239)
equivalent ratios (p. 229)
indirect measurement (p. 253)
polygon (p. 252)
proportion (p. 238)
rate (p. 232)
ratio (p. 228)
scale (p. 259)
scale drawing (p. 259)
similar polygons (p. 252)
unit cost (p. 233)
unit rate (p. 232)

Go Online PHSchool.com
For: Online vocabulary quiz
Web Code: arj-0551

Choose the correct term to complete each sentence.

1. A street map is an example of (a scale drawing, cross products). a scale drawing
2. Knowing a(n) (indirect measurement, unit cost) is helpful for getting the best buy when you shop. unit cost
3. A speed limit of 40 mi/h is an example of a (rate, proportion). rate
4. A ratio that compares a length in a drawing to the actual length of an object is a (scale, proportion). scale
5. When two polygons have corresponding angles with equal measures and corresponding sides with proportional lengths, the polygons are (cross products, similar). similar

Skills and Concepts

Lesson 5-1
- To write ratios and use them to compare quantities

A **ratio** is a comparison of two quantities by division. You can write the same ratio in three ways. To find equal ratios, multiply or divide the numerator and denominator by the same nonzero number.

Write each ratio in simplest form.

6. $\frac{9}{30}$ $\frac{3}{10}$ 7. $\frac{64}{20}$ $\frac{16}{5}$ 8. 99 : 33 3 : 1 9. 75 : 20 15 : 4 10. $\frac{45}{180}$ $\frac{1}{4}$

11. **Sports** During a recent Olympics, the United States won 97 medals, including 39 gold medals. Write the ratio of gold medals to total medals in three ways. $\frac{39}{97}$, 39 to 97, 39 : 97

Lesson 5-2
- To find unit rates and unit costs using proportional reasoning

A **rate** is a ratio that compares two quantities measured in different units. A **unit rate** has a denominator of 1. Find a **unit cost** by dividing the price of an item by the size of the item.

Write the unit rate for each situation.

12. 282 passengers in 47 cars 6 passengers/car
13. 600 Calories in 8 servings 75 Cal/serving
14. 414 students in 18 classrooms 23 students/classroom
15. $136 for 34 kg $4/kg

Alternative Assessment L4

Alternative Assessment — Form C

Chapter 5

SUPER COOPER ATHLETES

Each year thousands of student athletes take the Cooper test. Its purpose is to measure an athlete's physical condition. Test results can be used to plan an exercise program.

The Cooper test is very simple. It measures how far a person can run in 12 min. An athlete who is in good physical condition can run farther than an athlete whose physical condition is not as good. Since it is easier to measure laps than to measure miles, athletes usually take the test on a school track. One lap equals $\frac{1}{4}$ mi on a standard track.

This chart shows how to use the Cooper test results to determine an individual's physical condition.

Condition	Number of Laps Run in 12 min	Miles
Excellent	7–8	1.75–2.00
Good	6–7	1.50–1.75
Fair	5–6	1.25–1.50
Poor	4–5	1.00–1.25

Solve.

1. Martina ran $6\frac{1}{2}$ laps in 12 min. Approximately how many miles did she run? According to the Cooper test results, what is Martina's physical condition?
2. Jeb is getting in shape to play baseball this summer. He can't run on the school track. However, there is a measured mile on a road near his house. He can run that mile in 10 min. What kind of shape is Jeb in? Explain your thinking.

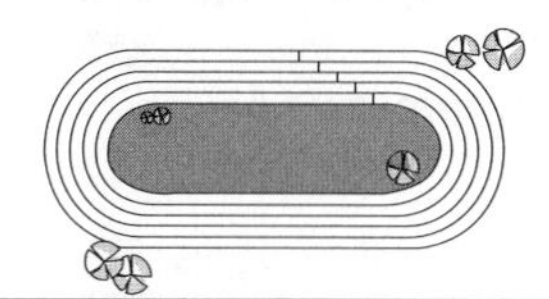

Lesson 5-3
- To test whether ratios form a proportion by using equivalent ratios and cross products

16. $.28/oz, $.31/oz; the 10-oz size is the better buy.

16. Shopping A 10-oz box of cereal costs $2.79. A 13-oz box of the same brand of cereal costs $3.99. Find the unit cost for each item and determine which is the better buy. **See left.**

17. A carpenter renovating a house is sanding the dining room floor. She sands 300 ft^2 of wood floor in 1 h 40 min. What is the unit rate in square feet per minute? **3 ft^2 per minute**

Lesson 5-4
- To solve proportions using unit rates, mental math, and cross products

22. $\frac{5}{3} = \frac{x}{250{,}000}$; 417,000 board feet

A proportion is an equation stating that two ratios are equal. If two ratios form a proportion, the **cross products** are equal.

Solve each proportion.

18. $\frac{3}{7} = \frac{n}{28}$ **12** **19.** $\frac{3}{5} = \frac{15}{x}$ **25** **20.** $\frac{a}{18} = \frac{12}{72}$ **3** **21.** $\frac{32}{c} = \frac{4}{17}$ **136**

22. Wood A local lumberyard sells a total of 250,000 board feet of hardwood each year. The ratio of softwood sold to hardwood sold is 5 : 3. Write and solve a proportion to find the amount of softwood sold each year. Round your answer to the nearest thousand. **See left.**

23. The ratio of the width of a rectangle to its length is 5 : 8. What is the width in feet if the length is 12 ft? **7.5 ft**

Lesson 5-5
- To use proportions to find missing lengths in similar figures

Two polygons are **similar polygons** if corresponding angles have equal measures and the lengths of corresponding sides form equivalent ratios.

24. A fire hydrant is 30 in. tall and casts a shadow 8 in. long. How tall is a nearby tree that casts a shadow 4 ft long? **15 ft** (or 180in.)

Each pair of figures is similar. Find each missing value.

25.

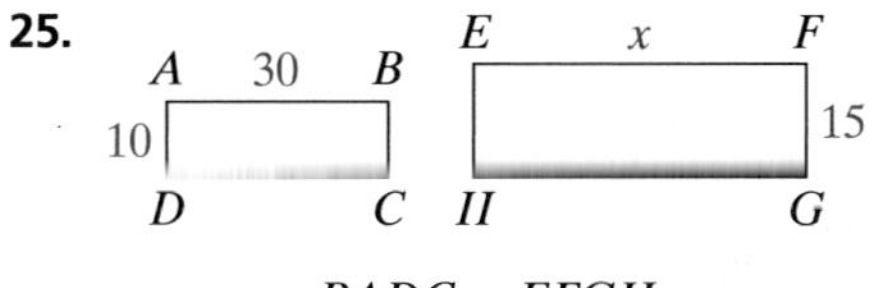

$BADC \sim EFGH$

45

26.

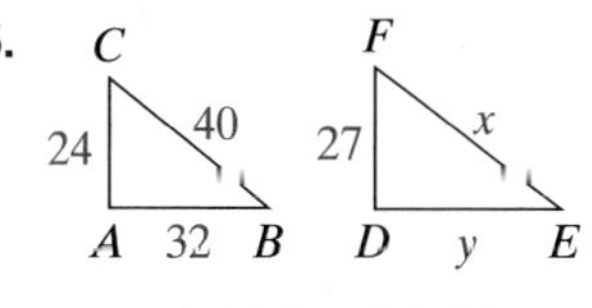

$\triangle ABC \sim \triangle DEF$

$x = 45$, $y = 36$

Lesson 5-6
- To use proportions to solve problems involving scale

A **scale drawing** is an enlarged or reduced drawing of an object that is similar to the actual object. A **scale** is a ratio that compares a length in a drawing to the corresponding length in an actual object.

27. Maps The scale on a map is 1 in. : 525 mi. The map distance from Chicago to Tokyo is 12 in. Find the actual distance between the cities. **6,300 mi**

28. Architecture A drawing's scale is 0.5 in. : 10 ft. A room is 15 ft long. How long is the room on the drawing? **0.75 in.**

Chapter 5 Test

Resources

- ExamView Assessment Suite CD-ROM
  - Ch. 5 Ready-Made Test
  - Make your own Ch. 5 test
- MindPoint Quiz Show CD-ROM
  - Ch. 5 Review

Differentiated Instruction

All in One Teaching Resources
- Below Level Chapter 5 Test L2
- Chapter 5 Test L3
- Chapter 5 Alternative Assessment L4

Spanish Assessment Resources ELL
- Below Level Chapter 5 Test L2
- Chapter 5 Test L3
- Chapter 5 Alternative Assessment L4

ExamView Assessment Suite CD-ROM
- Special Needs Test L1
- Special Needs Practice Bank L1

Online Chapter 5 Test at www.PHSchool.com L3

Below Level Chapter Test L2

Chapter Test L3

Chapter Test Form A

Chapter 5

1. There are 14 boys and 16 girls in a class. Write the ratio of boys to girls in simplest form.
   7 : 8

Determine whether the ratios in each pair are proportional.

2. 1 : 10; 2 : 5 no
3. 4 to 7; 28 to 49 yes
4. $\frac{6}{19}$; $\frac{2}{5}$ no
5. Which is a better buy, 6 apples for $1.74 or 5 apples for $1.20? Why?
   5 apples for $1.20; the unit rate, $.24 < $.29.

Write the unit rate for each situation.

6. bike 21 km in 50 min 0.42 km/min
7. earn $18 for 1 hr 15 min work $.24/min
8. sew 9 curtain panels in 15 days 0.6 curtain panels per day
9. Wyatt spent $\frac{1}{4}$ of his money on a train set and $\frac{1}{2}$ on tickets to a play. He now has $18. How much money did Wyatt start with? Explain why you chose the method you did.
   $72; sample answer: Work backwards.

Write each ratio in two other ways.

10. $\frac{5}{9}$ 5 to 9; 5 : 9
11. 8 to 3 8 : 3; $\frac{8}{3}$
12. 12 : 15 12 to 15; $\frac{12}{15}$

Solve each proportion.

13. $\frac{n}{6} = \frac{25}{30}$ $n = 5$
14. $\frac{6.1}{x} = \frac{2}{3}$ $x = 9.15$
15. $\frac{100}{25} = \frac{t}{4}$ $t = 16$
16. Alonso can jog 2 miles in 16 minutes. Is it reasonable to assume that it will take him 24 minutes to jog 3 miles? Explain.
    Yes; sample answer: The ratios form a proportion; 24 min/3 mi = 16 min/2 mi.

Chapter 5 Test

Go Online PHSchool.com For: Online chapter test Web Code: ara-0552

1. Write a ratio for the following information in three ways: In the United States, 98 million out of 99.6 million homes have at least one television. **98 : 99.6, $\frac{98}{99.6}$, 98 to 99.6**

Write each ratio in two other ways. 2–7. See margin.

2. $\frac{9}{7}$
3. 48 : 100
4. 33 to 9
5. 46 : 50
6. 19 to 91
7. 6 : 7

8. **Gas Mileage** Engineers test four cars to find their fuel efficiency. Use the information in the table below. Which car gets the most miles per gallon? **Car C**

Car	Miles	Gallons Used
A	225	14
B	312	15
C	315	10
D	452	16

Find the unit rate for each situation.

9. running 2.3 km in 7 min **0.33 km/min**
10. earning $36.00 in 4 h **$9.00/h**
11. **Writing in Math** You need to buy 10 lb of rice. A 2-lb bag costs $1.29. A 10-lb bag costs $6.99. You want to buy a 10-lb bag. Your friend thinks buying five 2-lb bags is a better deal because the unit rate is lower. Is your friend correct? Explain. **See margin.**

Solve each proportion.

12. $\frac{6}{5} = \frac{n}{7}$ **8.4**
13. $\frac{3}{7} = \frac{8}{x}$ **$18\frac{2}{3}$**
14. $\frac{k}{4} = \frac{9}{32}$ **$1\frac{1}{8}$**
15. $\frac{3.5}{d} = \frac{14}{15}$ **3.75**
16. $\frac{80}{35} = \frac{w}{7}$ **16**
17. $\frac{y}{18} = \frac{2.4}{15}$ **2.88**
18. The ratio of teachers to students in a middle school is 2 to 25. There are 350 students in the school. Find the number of teachers. **28 teachers**

19. **Maps** A map with a scale of 1 in. : 175 mi shows two cities 5 in. apart. How many miles apart are the cities? **875 mi**

Write each ratio in simplest form.

20. $\frac{7}{49}$ **$\frac{1}{7}$**
21. $\frac{12}{4}$ **$\frac{3}{1}$**
22. $\frac{8}{2}$ **$\frac{4}{1}$**
23. $\frac{14}{18}$ **$\frac{7}{9}$**
24. $\frac{68}{100}$ **$\frac{17}{25}$**
25. $\frac{15}{95}$ **$\frac{3}{19}$**

26. In the figure below, $\triangle JKL \sim \triangle PQR$. Find x and y. **$x = 12$, $y = 14$**

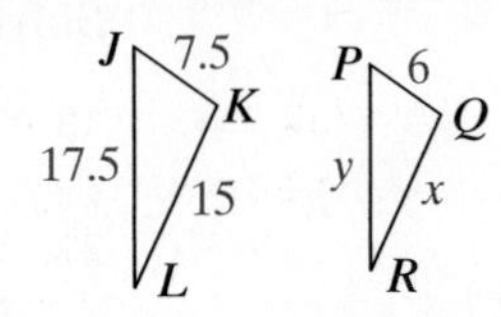

27. **Utilities** Last month, your electric bill was $25.32 for 450 kilowatt-hours of electricity. At that rate, what would be the bill for 240 kilowatt-hours? **$13.50**
28. A person who is 60 in. tall casts a shadow that is 15 in. long. How tall is a nearby tree that casts a shadow that is 40 in. long? **160 in.**

Use mental math to solve each proportion.

29. $\frac{14}{36} = \frac{7}{x}$ **18**
30. $\frac{40}{16} = \frac{x}{2}$ **5**
31. $\frac{x}{1} = \frac{24}{2}$ **12**
32. $\frac{4}{9} = \frac{12}{x}$ **27**

33. Suppose you are making a scale drawing of a giraffe that is 5.5 m tall. The drawing is 7 cm tall. Find the scale of the drawing. **See margin.**
34. **Ballooning** A hot-air balloon 2,100 ft above the ground can descend at the rate of 1.5 ft/s. The balloon is scheduled to land at 3:30 P.M. When should the balloonist start descending? **about 3:07 P.M.**
35. **Science** A 380-cubic-centimeter sample of titanium has a mass of 1,170 g. Find the mass of a titanium sample that has a volume of 532 cubic centimeters. **1,638 g**

2. 9 : 7, 9 to 7
3. $\frac{48}{100}$, 48 to 100
4. 33 : 9, $\frac{33}{9}$
5. $\frac{46}{50}$, 46 to 50
6. 19 : 91, $\frac{19}{91}$
7. $\frac{6}{7}$, 6 to 7
11. Yes; the 2-lb bag for $.65/lb is a better buy than the 10-lb bag for $.70/lb.
33. Answers may vary. Sample: 1 cm : 78.6 cm

Reading Comprehension

Read each passage and answer the questions that follow.

Europe Goes Euro Until recently, you needed different types of money (francs, marks, punts, etc.) to travel through Europe. Now all you need to get by in most European countries is the new European currency, the euro. One U.S. dollar buys about 0.81 euros. One French franc was worth about 0.15 euros, and an Irish punt was worth about 1.27 euros.

1. About how many euros were 20 French francs worth? **B**
   - Ⓐ 0.15 euros
   - Ⓑ 3.00 euros
   - Ⓒ 20.00 euros
   - Ⓓ 150.00 euros

2. Hsio came home from a trip with currency worth 20 U.S. dollars. She had 10 euros and the rest in U.S. currency. About how many dollars did she have? **J**
   - Ⓕ \$12.35
   - Ⓖ \$11.90
   - Ⓗ \$8.10
   - Ⓙ \$7.65

3. About how many U.S. dollars would you get if you exchanged 15 euros? **B**
   - Ⓐ \$12.15
   - Ⓑ \$18.52
   - Ⓒ \$22.50
   - Ⓓ \$27.28

4. Which expression shows how many French francs you could get for one Irish punt? **J**
   - Ⓕ 0.15×1.27
   - Ⓖ $\frac{0.15}{1.27}$
   - Ⓗ $0.15 + 1.27$
   - Ⓙ $\frac{1.27}{0.15}$

More Money Notes In the United States, paper money is all the same size: 2.61 inches wide by 6.14 inches long and 0.0043 inch thick. But did you know that American banknotes used to be bigger? Until 1929, they were 3.125 inches wide by 7.4218 inches long. By the way, it costs about 4.2 cents to produce one paper note.

5. What is the current ratio of length to width for U.S. paper money? **B**
   - Ⓐ 0.43 : 1
   - Ⓑ 2.35 : 1
   - Ⓒ 2.61 : 1
   - Ⓓ 6.14 : 1

6. Suppose you make a stack of one thousand \$100 bills. How tall is the stack? **H**
   - Ⓕ 0.043 in.
   - Ⓖ 0.43 in.
   - Ⓗ 4.3 in.
   - Ⓙ 43 in.

7. What is the best approximation of the ratio of the current length of U.S. notes to the length before 1929? **C**
   - Ⓐ 1 : 2
   - Ⓑ 3 : 4
   - Ⓒ 5 : 6
   - Ⓓ 9 : 10

8. The government prints about 12 billion paper notes each year. What is the best estimate of how much this costs? **J**
   - Ⓕ \$.5 million
   - Ⓖ \$5 million
   - Ⓗ \$50 million
   - Ⓙ \$500 million

Test Prep

Resources

Test Prep Workbook

All in One Teaching Resources
- Cumulative Review **L3**

ExamView Assessment Suite CD-ROM
- Standardized Test Practice

Differentiated Instruction

Spanish Assessment Resources
- Spanish Cumulative Review **ELL**

ExamView Assessment Suite CD-ROM
- Special Needs Practice Bank **L1**

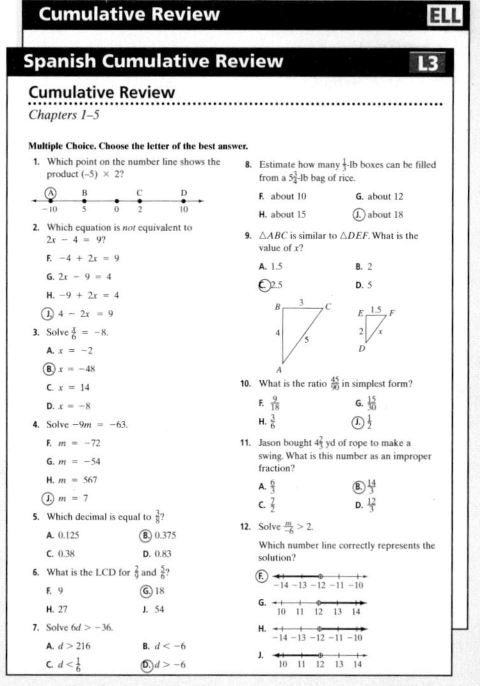

Cumulative Review ELL

Spanish Cumulative Review L3

Cumulative Review

Chapters 1–5

Multiple Choice. Choose the letter of the best answer.

1. Which point on the number line shows the product $(-5) \times 2$?
   (A) A: −10; B: 5; 0; C: 2; D: 10
2. Which equation is *not* equivalent to $2x - 4 = 9$?
   F. $-4 + 2x = 9$
   G. $2x - 9 = 4$
   H. $-9 + 2x = 4$
   (J.) $4 - 2x = 9$
3. Solve $\frac{x}{6} = -8$.
   A. $x = -2$
   (B.) $x = -48$
   C. $x = 14$
   D. $x = -8$
4. Solve $-9m = -63$.
   F. $m = -72$
   G. $m = -54$
   H. $m = 567$
   (J.) $m = 7$
5. Which decimal is equal to $\frac{3}{8}$?
   A. 0.125 (B.) 0.375
   C. 0.38 D. 0.83
6. What is the LCD for $\frac{2}{9}$ and $\frac{5}{6}$?
   F. 9 (G.) 18
   H. 27 J. 54
7. Solve $6d > -36$.
   A. $d > 216$ B. $d < -6$
   C. $d < \frac{1}{6}$ (D.) $d > -6$
8. Estimate how many $\frac{1}{3}$-lb boxes can be filled from a $5\frac{3}{4}$-lb bag of rice.
   F. about 10 G. about 12
   H. about 15 (J.) about 18
9. $\triangle ABC$ is similar to $\triangle DEF$. What is the value of x?
   A. 1.5 B. 2
   (C.) 2.5 D. 5
   B 3 C; 4; 5; A — E 1.5 F; 2; x; D
10. What is the ratio $\frac{45}{90}$ in simplest form?
    F. $\frac{9}{18}$ G. $\frac{15}{30}$
    H. $\frac{3}{6}$ (J.) $\frac{1}{2}$
11. Jason bought $4\frac{2}{3}$ yd of rope to make a swing. What is this number as an improper fraction?
    A. $\frac{6}{3}$ (B.) $\frac{14}{3}$
    C. $\frac{7}{2}$ D. $\frac{12}{3}$
12. Solve $\frac{m}{-6} > 2$.
    Which number line correctly represents the solution?
    (F.) −14 −13 −12 −11 −10
    G. 10 11 12 13 14
    H. −14 −13 −12 −11 −10
    J. 10 11 12 13 14

Applying Ratios

Students will use the information on these two pages to answer the questions posed in Put It All Together.

Have students think big. A blueprint of a school floor plan is scaled at $\frac{1}{4}$ in. : 4 ft. The music room perimeter measure is $\frac{1}{2}$ the perimeter of the communications/media room. Ask questions such as:

- *If the blueprint shows the music room is $1\frac{1}{2}$ in. × $2\frac{1}{4}$ in., what are the actual dimensions of the music room?* **24 ft × 36 ft**
- *What are the dimensions of the communications/media room?* **48 ft × 72 ft**

Activating Prior Knowledge

Have students who have seen movies about people who suddenly shrink share their knowledge with the class. Students might also be familiar with special-effects models used in popular movies, such as those in stories about the *Titanic.*

Guided Instruction

Have a volunteer read the introductory paragraph. Ask questions such as:

- *What are some careers that make use of models?* **Sample: car makers, architects, home decorators, stage designers**
- *What does "using a consistent scale" mean?* **Sample: keeping everything in proportion to each other**
- *Why are scaled models used instead of actual dimensions?* **Answers will vary. Check students' work.**
- *What are some books where the characters find themselves in a larger or smaller scale to the setting around them?* ***Alice's Adventures in Wonderland* by Lewis Carroll, *Gulliver's Travels* by Jonathan Swift**

Applying Ratios

A Matter of Scale Some movies are about people suddenly shrinking in size. They find themselves in environments both familiar and unfamiliar. A piece of furniture becomes a cliff to climb. Insects become beasts to fear. Although special-effects technicians create many of the visuals on computers, model makers provide scale models for some objects. Using a consistent scale for the models keeps the illusion believable.

Put It All Together

1. You have been sent on a mission to learn as much as you can about a giant and her home.
   a. **Research** Use the clues in the "Giant's World" caption. Research the sizes of things mentioned in the caption. Then find the scale of the giant's world to your world.
   b. Use your scale to find the dimensions of at least three items from the giant's kitchen. Sketch the items and label their dimensions.
   c. Find the height of the giant.
2. Your next trip has brought you to a home inhabited by tiny people.
   a. **Research** Use the clues in the "Miniature World" caption and your research to find the scale of the miniature room to your world.
   b. Use your scale to find the dimensions of at least three items in the miniature living room. Sketch these items and label their dimensions.
   c. Find the height of a person who lives in the tiny house.
3. **Reasoning** Suppose someone in your class is 5 ft 5 in. tall, and another person is 4 ft 10 in. tall. Will these two people get the same answers in Exercises 1 and 2? Explain.

Giant's World

When standing on your toes, you can just barely reach the top of a cereal box.

When stood on end, a teaspoon reaches to the bottom of your rib cage.

1a. **Answers may vary. Sample: height of cereal box : height of reach = 12 in. : 72 in. Scale factor is 1 : 6.**

b. **Answers may vary. Sample: Chair is 35 in. high; giant chair is 17 ft 6 in. high. Table is 30 in. high; giant table is 15 ft high. Desk is 52 in. long; giant desk is 26 ft long.**

c. **Answers may vary. Sample: Person is 6 ft tall; giant is 36 ft tall.**

2a. **Answers may vary. Sample: diameter of bicycle wheel : length of index finger = 28 in. : 3 in. Scale factor is 28 : 3.**

b. **Answers may vary. Sample: Coffee table is 18 in. high; miniature**

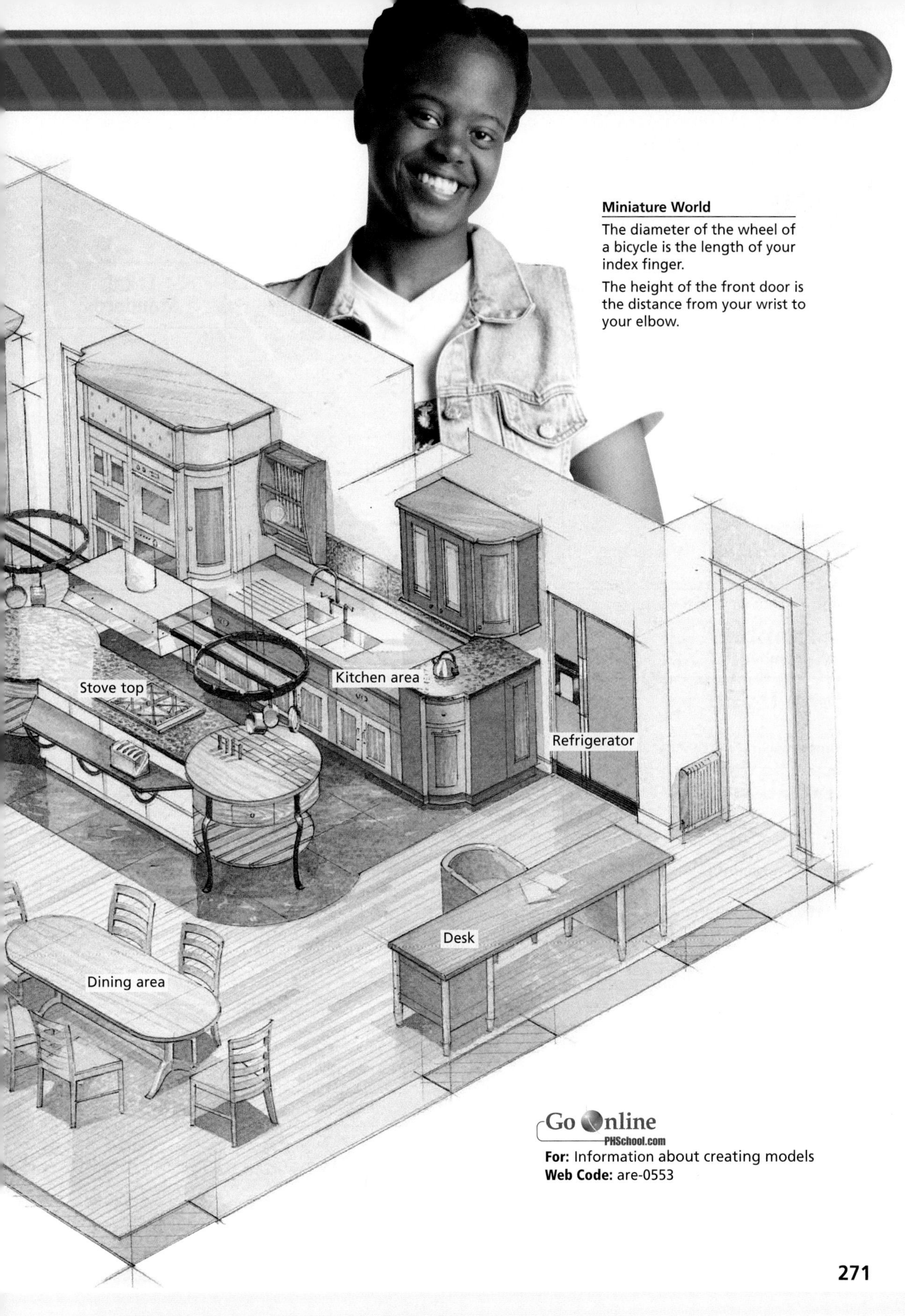
Miniature World
The diameter of the wheel of a bicycle is the length of your index finger.
The height of the front door is the distance from your wrist to your elbow.
Stove top
Kitchen area
Refrigerator
Desk
Dining area

Go Online
PHSchool.com
For: Information about creating models
Web Code: are-0553

Activity

Have students work in pairs to answer the questions.

Exercise 1a Call attention to the fact that each student must determine scale by analyzing the clues in the Giant's World. Students can use either the customary system of measurement or the metric system. Remind them, though, to use a *consistent scale*.

Exercise 2a Call attention to the fact that each student must now determine his or her own scale by analyzing the clues in the Miniature World.

Exercise 3 Have students answer the question independently and then share their responses with the class.

Differentiated Instruction

Special Needs L1

Since students must determine the size of the giant's world to their own world, you might have students work in pairs and use the size of one of the partners as the basis for the scale.

coffee table is 1.3 in. high. Sofa is 32 in. high; miniature sofa is 2.3 in. high. Clock is 12 in. high; miniature clock is 0.9 in. high.

c. Answers may vary. Sample: Person is 6 ft tall; miniature person is 5.1 in. tall.

3. No. The scale factors will be different.

6 Percents

Chapter at a Glance

Lesson Titles, Objectives, and Features	Assessment	NCTM Standards	Local Standards
6-1 Understanding Percents • To model percents and to write percents using equivalent ratios	Lesson Quiz	1, 2, 6, 7, 8, 9, 10	
6-2a Activity Lab, Hands On: Rational Number Cubes **6-2 Percents, Fractions, and Decimals** • To convert between fractions, decimals, and percents	Lesson Quiz	1, 2, 6, 7, 8, 9, 10	
6-3 Percents Greater Than 100% or Less Than 1% • To convert between fractions, decimals, and percents greater than 100% or less that 1%	Lesson Quiz Checkpoint Quiz 1	1, 2, 6, 7, 8, 9, 10	
6-4a Activity Lab, Data Analysis: Using Percent Data in a Graph **6-4 Finding a Percent of a Number** • To find and estimate the percent of a number	Lesson Quiz	1, 2, 5, 6, 7, 8, 9, 10	
6-5 Solving Percent Problems Using Proportions • To use proportions to solve problems involving percent	Lesson Quiz	1, 2, 6, 7, 8, 9, 10	
6-6 Solving Percent Problems Using Equations • To use equations to solve problems involving percent **Guided Problem Solving:** Solving Percent Problems	Lesson Quiz	1, 2, 6, 7, 8, 9, 10	
6-7 Applications of Percent • To find and estimate solutions to application problems involving percent **6-7b Activity Lab, Algebra Thinking:** Percent Equations	Lesson Quiz Checkpoint Quiz 2	1, 2, 6, 7, 8, 9, 10	
6-8a Activity Lab, Data Analysis: Exploring Percent of Change **6-8 Finding Percent of Change** • To find percents of increase and percents of decrease	Lesson Quiz	1, 2, 5, 6, 7, 8, 9, 10	
Problem Solving Application: Applying Percents			

NCTM Standards 2000

1 Number and Operations **2** Algebra **3** Geometry **4** Measurement **5** Data Analysis and Probability
6 Problem Solving **7** Reasoning and Proof **8** Communication **9** Connections **10** Representation

Correlations to Standardized Tests

All content for these tests is contained in *Prentice Hall Math,* Course 2. This chart reflects coverage in this chapter only.

	6-1	6-2	6-3	6-4	6-5	6-6	6-7	6-8
Terra Nova CAT6 (Level 17)								
Number and Number Relations	✔	✔	✔	✔	✔	✔	✔	✔
Computation and Numerical Estimation	✔	✔	✔	✔	✔	✔	✔	✔
Operation Concepts	✔	✔	✔	✔	✔	✔	✔	✔
Measurement								
Geometry and Spatial Sense								
Data Analysis, Statistics, and Probability				✔				✔
Patterns, Functions, Algebra	✔	✔	✔	✔	✔	✔	✔	✔
Problem Solving and Reasoning	✔	✔	✔	✔	✔	✔	✔	✔
Communication	✔	✔	✔	✔	✔	✔	✔	✔
Decimals, Fractions, Integers, and Percent	✔	✔	✔	✔	✔	✔	✔	✔
Order of Operations								
Terra Nova CTBS (Level 17)								
Decimals, Fractions, Integers, Percents	✔	✔	✔	✔	✔	✔	✔	✔
Order of Operations, Numeration, Number Theory	✔	✔	✔	✔	✔	✔	✔	✔
Data Interpretation				✔				✔
Pre-algebra	✔	✔	✔	✔	✔	✔	✔	✔
Measurement								
Geometry								
ITBS (Level 13)								
Number Properties and Operations	✔	✔	✔	✔	✔	✔	✔	✔
Algebra	✔	✔	✔	✔	✔	✔	✔	✔
Geometry								
Measurement								
Probability and Statistics								
Estimation			✔		✔		✔	
SAT10 (Int 3 Level)								
Number Sense and Operations	✔	✔	✔	✔	✔	✔	✔	✔
Patterns, Relationships, and Algebra	✔	✔	✔	✔	✔	✔	✔	✔
Data, Statistics, and Probability				✔				✔
Geometry and Measurement								
NAEP								
Number Sense, Properties, and Operations	✔		✔	✔	✔		✔	✔
Measurement								
Geometry and Spatial Sense								
Data Analysis, Statistics, and Probability								
Algebra and Functions						✔		

CAT6 California Achievement Test, 6th Ed.
SAT10 Stanford Achievement Test, 10th Ed.
CTBS Comprehensive Test of Basic Skills
NAEP National Assessment of Educational Progress 2005 Mathematics Objectives
ITBS Iowa Test of Basic Skills, Form M

Math Background

Skills Trace

BEFORE Chapter 6

Course 1 or Grade 6 modeled percents and solved basic percent problems.

DURING Chapter 6

Course 2 reviews and extends the solving of percent problems to percent of change and real-world applications such as commissions.

AFTER Chapter 6

Throughout this course, students solve real-world applications that involve percents.

6-1 Understanding Percents

Math Understandings

- A percent expresses parts per 100. The percent symbol, %, means "per 100" or "/100."
- A percent is a relative, rather than an absolute amount. For example, 25% of a greater number represents a different value from 25% of a lesser number.

A **percent** is a ratio that compares a number to 100. For example, you can express 3 out of 4 as the equivalent ratio 75 out of 100, or $\frac{75}{100}$, which is 75%.

6-2 Percents, Fractions, and Decimals

Math Understandings

- You can represent a percent in different but related ways, such as a ratio, a fraction, and a decimal.
- You can rewrite a decimal that names hundredths directly as the same percent.

Example: You say 0.75 as "seventy-five hundredths," which you can write as the fraction $\frac{75}{100}$ or 75%.

Converting Percents, Decimals, and Fractions	
To rewrite this:	**Do this:**
a decimal as a percent	multiply by 100 OR move the decimal point two places to the right
a percent as a decimal	divide by 100 OR move the decimal point two places to the left
a fraction as a percent	rewrite with a denominator of 100 OR convert to a decimal and write as a percent
a percent as a fraction	rewrite with a denominator of 100 and simplify

6-3 Percents Greater Than 100% or Less Than 1%

Math Understandings

- A fraction or a decimal that is less than $\frac{1}{100}$ or 0.01 can be expressed as a percent less than 1%.
- A number that is greater than 1 can be expressed as a percent greater than 100%.

If the number compared to 100 is less than 1, the percent is less than 1%. If the number compared to 100 is greater than 100, the percent is greater than 100%. A mixed number such as $4\frac{1}{2}$ represents a percent greater than 100%. A fraction less than $\frac{1}{100}$ represents a percent less than 1%.

6-4 Finding a Percent of a Number

Math Understandings

- You can find a given percent of a number by rewriting the percent as either a decimal or a fraction and then multiplying.
- You can use compatible numbers (numbers easy to calculate with mental math) to estimate a percent.

Knowing the fraction equivalents of common percents can make it easier to find percents of a number using mental math.

Percent	10%	25%	33.33%	50%	66.67%	75%
Fraction	$\frac{1}{10}$	$\frac{1}{4}$	$\frac{1}{3}$	$\frac{1}{2}$	$\frac{2}{3}$	$\frac{3}{4}$

6-5 6-6 Solving Percent Problems Using Proportions and Equations

Math Understandings

- Percent problems involve a part, a percent, and a whole.
- If you know two of the three quantities, you can find the missing one by using either a proportion or an equation.
- You can write proportions in different equivalent ways.

In a problem, you can identify the percent by the % sign, and the whole usually follows the word *of.* The model below shows the relationship among the part, whole, and percent.

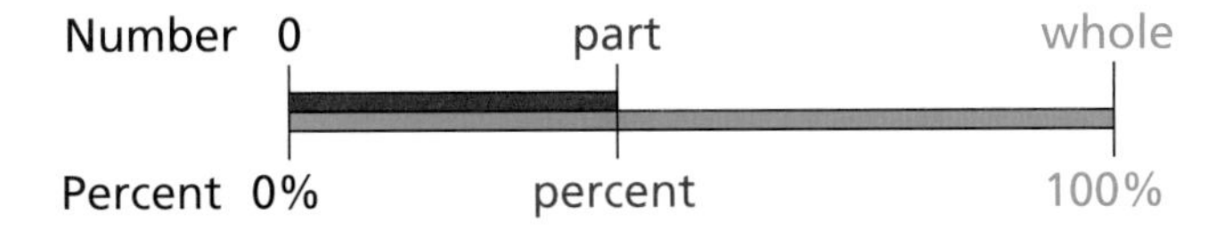

You can write and solve the proportion $\frac{\text{part}}{\text{whole}} = \frac{\text{percent}}{100}$ to find the missing number, percent, or whole. To find the missing number, percent, or whole, you can also write and solve the equation (whole) $\times$ (percent) $=$ part.

6-7 Applications of Percent

Math Understandings

- You often use percent in daily transactions such as paying sales tax, estimating the amount of the tip on a bill, and figuring commission.

In many towns and states, you pay a sales tax that is a percent of the purchase price. A **tip** is a percent of a bill that you give to the person providing a service. Some sales jobs pay a **commission**, a percent of the amount you sell.

Example: A lunch costs \$7.95. The sales tax is 5% and you want to leave a 20% tip. What is the total you pay for lunch?
Total = Cost + Tax + Tip = \$7.95 + \$.40 + \$1.59 = \$9.94

6-8 Finding Percent of Change

Math Understandings

- A percent of change may be an increase or a decrease.
- The percent of change is a ratio, expressed as a percent. It compares the amount of change to the original amount. The amount of change is the difference between the original amount and the amount that results after the change.

The **percent of change** is the percent a quantity increases or decreases from its original amount. The difference between the selling price and the store's original cost of an item is called the **markup**. So a common percent of increase is the ratio of the amount of markup to the original cost of an item. The difference between the original price and the sale price of an item is called a **discount**.

Additional Professional Development Opportunities

Math Background Notes for Chapter 6: Every lesson has a Math Background in the PLAN section.

Research Overview, Mathematics Strands
Additional support for these topics and more is in the front of the Teacher's Edition.

LessonLab
LessonLab, a Pearson Education company, offers comprehensive, facilitated professional development designed to help teachers to improve student achievement. To learn more, please visit lessonlab.com.

Chapter 6 Resources

	6-1	6-2	6-3	6-4	6-5	6-6	6-7	6-8	For the Chapter
Print Resources									
L3 Practice	●	●	●	●	●	●	●	●	
L1 Adapted Practice	●	●	●	●	●	●	●	●	
L3 Guided Problem Solving	●	●	●	●	●	●	●	●	
L2 Reteaching	●	●	●	●	●	●	●	●	
L4 Enrichment	●	●	●	●	●	●	●	●	
L3 Daily Notetaking Guide	●	●	●	●	●	●	●	●	
L1 Adapted Daily Notetaking Guide	●	●	●	●	●	●	●	●	
L3 Vocabulary and Study Skills Worksheet	●		●		●		●	●	●
L3 Daily Puzzles	●	●	●	●	●	●	●	●	
L3 Activity Labs	●	●	●	●	●	●	●	●	
L3 Checkpoint Quiz			●				●		
L3 Chapter Project									●
L2 Below Level Chapter Test									●
L3 Chapter Test									●
L4 Alternative Assessment									●
L3 Cumulative Review									●
Spanish Resources ELL									
L3 Practice	●	●	●	●	●	●	●	●	
L3 Vocabulary and Study Skills Worksheets	●		●		●		●	●	●
L3 Checkpoint Quiz			●				●		
L2 Below Level Chapter Test									●
L3 Chapter Test									●
L4 Alternative Assessment									●
L3 Cumulative Review									●
Transparencies									
Check Skills You'll Need	●	●	●	●	●	●	●	●	
Additional Examples	●	●	●	●	●	●	●	●	
Problem of the Day	●	●	●	●	●	●	●	●	
Classroom Aid									
Student Edition Answers	●	●	●	●	●	●	●	●	●
Lesson Quiz	●	●	●	●	●	●	●	●	
Test-Taking Strategies									●
Technology									
Interactive Textbook Online	●	●	●	●	●	●	●	●	●
StudentExpress™ CD-ROM	●	●	●	●	●	●	●	●	●
Success Tracker™ Online Intervention	●	●	●	●	●	●	●	●	●
TeacherExpress™ CD-ROM	●	●	●	●	●	●	●	●	●
PresentationExpress™ with QuickTake Presenter CD-ROM	●	●	●	●	●	●	●	●	●
ExamView® Assessment Suite CD-ROM	●	●	●	●	●	●	●	●	●
MindPoint® Quiz Show CD-ROM									●
Prentice Hall Web Site: PHSchool.com	●	●	●	●	●	●	●	●	●

Also available:

Prentice Hall Assessment System
- Progress Monitoring Assessments
- Skills and Concepts Review
- Test Prep Workbook

Other Resources
Algebra Readiness Tests
All-in-One Student Workbook
All-in-One Student Workbook, Adapted Version
Multilingual Handbook
Solution Key
Math Notes Study Folder
Spanish Cumulative Assessment

Where You Can Use the Lesson Resources

Here is a suggestion, following the four-step teaching plan, for how you can incorporate Differentiated Instruction Resources into your teaching.

	Instructional Resources L3	Differentiated Instruction Resources
1. Plan		
Preparation Read the Math Background in the Teacher's Edition to connect this lesson with students' previous experience. **Starting Class** **Check Skills You'll Need** Assign these exercises to review prerequisite skills. **New Vocabulary** Help students pre-read the lesson by pointing out the new terms introduced in the lesson.	**Math Background** **Math Understandings** **Transparencies & PresentationExpress™ with QuickTake Presenter CD-ROM** Check Skills You'll Need Problem of the Day **Resources** Vocabulary and Study Skills	**Spanish Support** ELL Vocabulary and Study Skills
2. Teach		
L3 **Guided Instruction** Use the Activity Labs to build conceptual understanding. Teach each Example. Use the Teacher's Edition side column notes for specific teaching tips, including Error Prevention notes. Use the Additional Examples found in the side column (and on transparency and PowerPoint) as an alternative presentation for the content. After each Example, assign the Quick Check exercise for that Example to get an immediate assessment of student understanding. Use the Closure activity in the Teacher's Edition to help students attain mastery of lesson content.	**Student Edition** Activity Lab **Resources** Daily Notetaking Guide Activity Lab **Transparencies & PresentationExpress™ with QuickTake Presenter CD-ROM** Additional Examples Classroom Aids **ExamView® Assessment Suite CD-ROM**	**Teacher's Edition** Every lesson includes suggestions for working with students who need special attention. L1 Special Needs L2 Below Level L4 Advanced Learners ELL English Language Learners **Resources** L1 Adapted Daily Notetaking Guide **Multilingual Handbook**
3. Practice		
Assignment Guide **Check Your Understanding** Use these questions to check students' understanding before you assign homework. **Homework Exercises** Assign homework from these leveled exercises in the Assignment Guide. A Practice by Example B Apply Your Skills C Challenge Test Prep and Mixed Review **Homework Quick Check** Use these key exercises to quickly check students' homework.	**Transparencies & PresentationExpress™ with QuickTake Presenter CD-ROM** Student Answers **Resources** Practice Guided Problem Solving Vocabulary and Study Skills Activity Lab Daily Puzzles **ExamView® Assessment Suite CD-ROM**	**Spanish Support** ELL Practice ELL Vocabulary and Study Skills **Resources** L1 Adapted Practice L4 Enrichment
4. Assess & Reteach		
Lesson Quiz Assign the Lesson Quiz to assess students' mastery of the lesson content. **Checkpoint Quiz** Use the Checkpoint Quiz to assess student progress over several lessons.	**Transparencies & PresentationExpress™ with QuickTake Presenter CD-ROM** Lesson Quiz **Resources** Checkpoint Quiz	**Resources** L2 Reteaching ELL Checkpoint Quiz Success Tracker™ Online Intervention **ExamView® Assessment Suite CD-ROM**

KEY L1 Special Needs L2 Below Level L3 For All Students L4 Advanced, Gifted ELL English Language Learners

Percents

Answers are in the back of the textbook.

For intervention, direct students to:

Using Estimation Strategies
Lesson 1-1
Extra Skills and Word Problems Practice, Ch. 1

Fractions and Decimals
Lesson 2-6
Extra Skills and Word Problems Practice, Ch. 2

Solving Equations by Multiplying or Dividing
Lesson 4-4
Extra Skills and Word Problems Practice, Ch. 4

Using Proportional Reasoning
Lesson 5-4
Extra Skills and Word Problems Practice, Ch. 5

Spanish Vocabulary/Study Skills ELL

Vocabulary/Study Skills L3

6A: Graphic Organizer For use before Lesson 6-1

Study Skill As you read over the material in the chapter, keep a paper and pencil handy to write down notes and questions that you have.

Write your answers.

1. What is the chapter title? Percents
2. How many lessons are there in this chapter? 9
3. What is the topic of the Test-Taking Strategies page? Working Backward
4. Complete the graphic organizer below as you work through the chapter.
   - In the center, write the title of the chapter.
   - When you begin a lesson, write the lesson name in a rectangle.
   - When you complete a lesson, write a skill or key concept in a circle linked to that lesson block.
   - When you complete the chapter, use this graphic organizer to help you review.

Check students' diagrams.

Vocabulary and Study Skills

CHAPTER 6

Percents

What You've Learned

- In Chapter 2, you compared, ordered, and converted between fractions and decimals.
- In Chapter 4, you solved equations.
- In Chapter 5, you solved proportions.

Check Your Readiness

GO for Help

For Exercises	See Lesson
1–4	1-1
5–9	2-6
10–13	4-4
14–16	5-4

Using Estimation Strategies

Use compatible numbers to estimate each quotient.

1. $74.89 \div 14.7$ **5**

2. $1{,}409 \div 102.4$ **14**

3. $495.89 \div 99.3$ **5**

4. $1{,}913 \div 188$ **10**

Fractions and Decimals

Write each decimal as a fraction in simplest form.

5. 0.85 $\frac{17}{20}$

6. 0.4 $\frac{2}{5}$

7. 0.68 $\frac{17}{25}$

8. 1.25 $\frac{5}{4}$

9. 0.01 $\frac{1}{100}$

Solving Equations by Multiplying or Dividing

Algebra **Solve each equation.**

10. $0.8t = 24$ **30**

11. $0.35w = 280$ **800**

12. $\frac{n}{0.6} = 14$ **8.4**

13. $\frac{z}{0.25} = 12$ **3**

Using Proportional Reasoning

Solve each proportion.

14. $\frac{4}{5} = \frac{n}{100}$ **80**

15. $\frac{x}{8} = \frac{27}{100}$ **2.16**

16. $\frac{6}{a} = \frac{3}{100}$ **200**

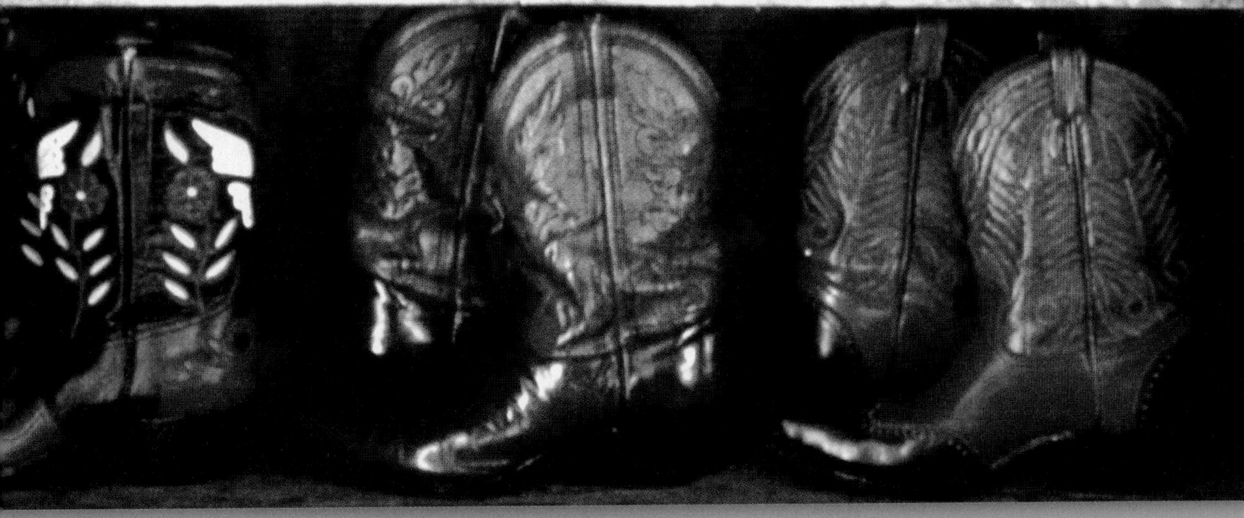

What You'll Learn Next

- In this chapter, you will compare, order, and convert between fractions, decimals, and percents.
- You will solve percent problems using equations and proportions.

 **Key Vocabulary**

- commission (p. 305)
- discount (p. 311)
- markup (p. 311)
- percent (p. 274)
- percent of change (p. 310)

 **Problem Solving Application** On pages 320 and 321, you will work an extended activity on percents.

Chapter 6 Overview

In this chapter, students work with percents, including percents less than 1 or greater than 100. They begin by comparing percents to fractions and decimals. Then, using real-world applications, they solve problems involving percents, including finding a percent of a number and finding percent of change.

Activating Prior Knowledge

In this chapter, students will build on their knowledge of proportions and proportional reasoning to solve problems involving applications of percents. They also draw upon their understanding of the relationship between decimals and fractions to express both as percents. Ask:

- *How can you write a decimal as a fraction?* **Sample: Write a fraction in tenths, hundredths, thousandths, and so on, and then express that fraction in simplest form.**
- *What is the missing term in the proportion:* $\frac{n}{12} = \frac{5}{6}$? **$n = 10$**

6-1

1. Plan

Objective

To model percents and to write percents using equivalent ratios

Examples

1, 2 Using Models With Percents
3 Finding Percents Using Models
4 Using Equivalent Ratios

Math Understandings: p. 272C

Math Background

While a ratio compares any two numbers, such as $\frac{4}{5}$, a percent is a ratio that compares a number to 100, such as $\frac{80}{100}$ or 80%. Any ratio can be expressed as a percent by writing an equal ratio with a denominator of 100.

$\frac{4}{5} = \frac{4 \cdot 20}{5 \cdot 20} = \frac{80}{100} = 80\%$

More Math Background: p. 272C

Lesson Planning and Resources

See p. 272E for a list of the resources that support this lesson.

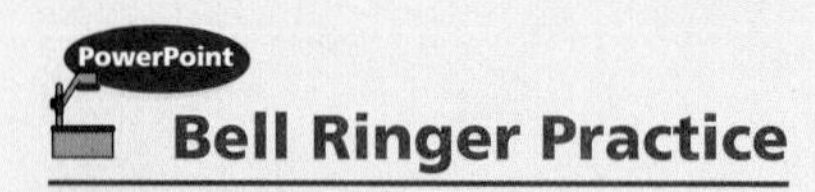

Bell Ringer Practice

Check Skills You'll Need
Use student page, transparency, or PowerPoint. For intervention, direct students to:
Ratios
Lesson 5-1
Extra Skills and Word Problems Practice, Ch. 5

6-1 Understanding Percents

Check Skills You'll Need

1. **Vocabulary Review** What are *equivalent ratios?* **1–5. See below.**

Find two equivalent ratios for each ratio.

2. $\frac{2}{5}$ **$\frac{4}{10}, \frac{6}{15}$**
3. $\frac{13}{50}$ **$\frac{26}{100}, \frac{39}{150}$**
4. $\frac{3}{25}$ **$\frac{6}{50}, \frac{9}{75}$**
5. $\frac{1}{10}$ **$\frac{2}{20}, \frac{3}{30}$**

Lesson 5-1

What You'll Learn

To model percents and to write percents using equivalent ratios

New Vocabulary percent

Why Learn This?

Percents make ratios easier to understand and compare. You can use a percent to represent the floor space needed for each piece of furniture in a room.

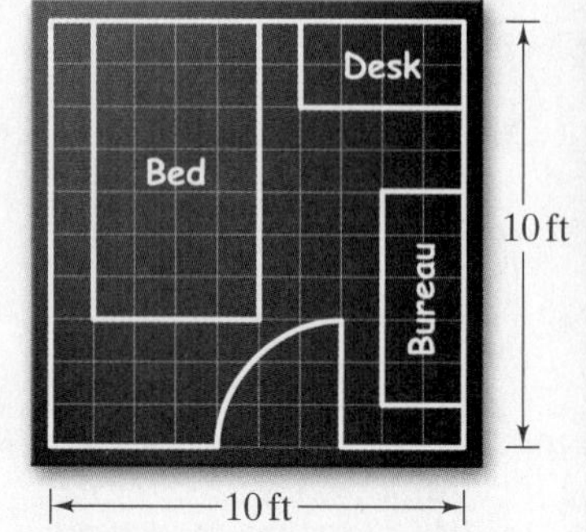

A **percent** is a ratio that compares a number to 100. You can use a model to find a percent. You can model a percent using a 10 × 10 grid.

Check Skills You'll Need

1. **Equivalent ratios have the same value.**
2–5. **Answers may vary. Samples are given.**

Vocabulary Tip

Percent means "per hundred." The root *cent* appears in words such as century, centimeter, and centipede.

EXAMPLES Using Models With Percents

1 The floor plan above shows a bedroom with an area of 100 ft^2. Find the percent of floor space needed for each piece of furniture.

Count the number of grid spaces for each piece. Write each number as a ratio to the total number of grid spaces, 100. Then write it as a percent.

Bed $\frac{28}{100} = 28\%$ **Bureau** $\frac{10}{100} = 10\%$ **Desk** $\frac{8}{100} = 8\%$

2 Model 25% on a 10 × 10 grid.

Shade 25 of the 100 grid spaces. →

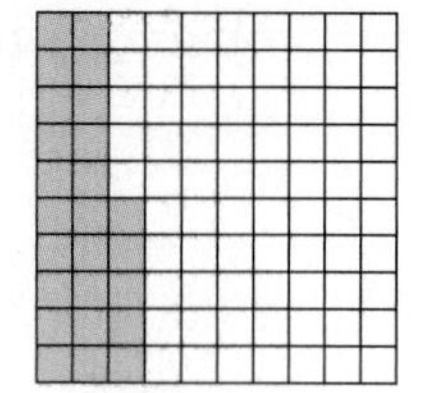

2.

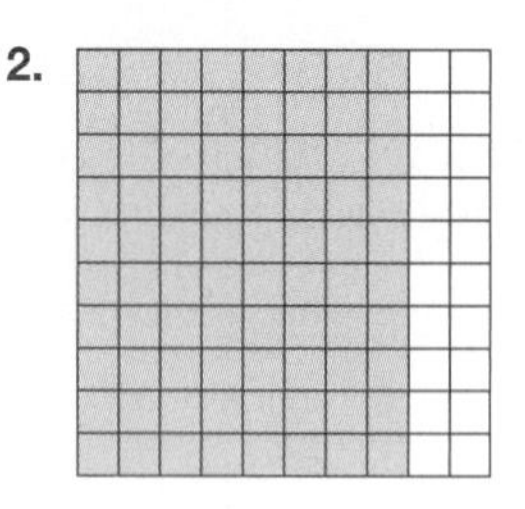

Quick Check

1. Write a ratio and a percent to represent the unused floor space. **$\frac{54}{100}$; 54%**
2. Model 80% on a 10 × 10 grid. **See left.**

Differentiated Instruction Solutions for All Learners

Special Needs L1
For Quick Check 3, have students overlay a hundred grid on the grids shown. Then they can see, for example, that a ratio of $\frac{3}{4}$ is equal to $\frac{75}{100}$ or 75%.

learning style: visual

Below Level L2
The following numbers are denominators in ratios. By what number do you need to multiply each denominator to find a percent?

5 **20** 20 **5** 25 **4** 4 **25**

learning style: verbal

For help with finding equivalent fractions, go to Lesson 2-3, Example 1.

The factors of 100 are 1, 2, 4, 5, 10, 20, 25, 50, and 100. Any ratio written as a fraction with a denominator that is a factor of 100 is easy to write as a percent. You can use common multiples to write an equivalent ratio that has a denominator of 100. When you have a fraction with a denominator of 100, look at the numerator to find the percent.

EXAMPLE Finding Percents Using Models

3 What percent does each shaded area represent?

a.

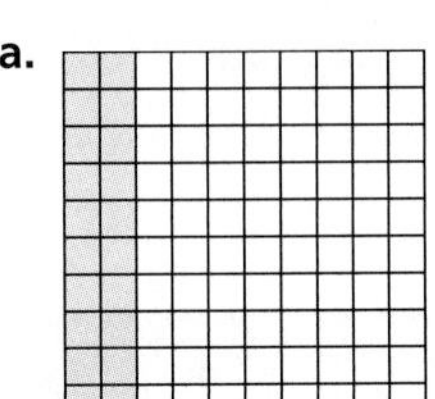

$\frac{20}{100} = 20\%$

b.

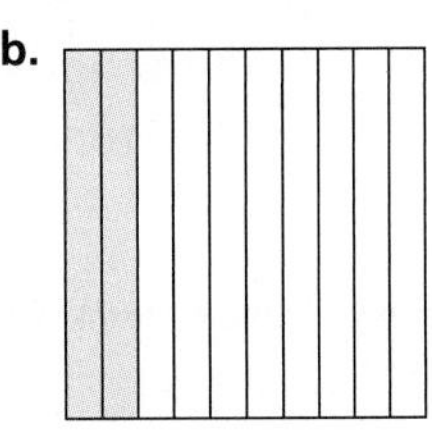

$\frac{2}{10} = \frac{20}{100} = 20\%$

c.

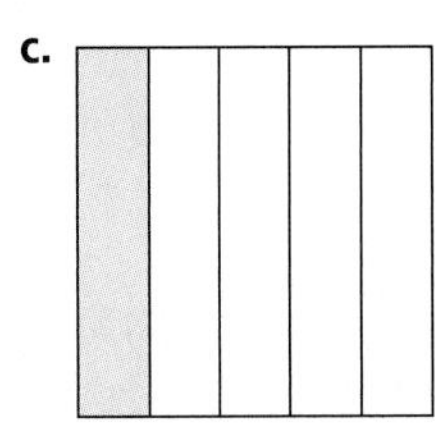

$\frac{1}{5} = \frac{20}{100} = 20\%$

Quick Check

3. Write a ratio and a percent for each shaded area.

a.

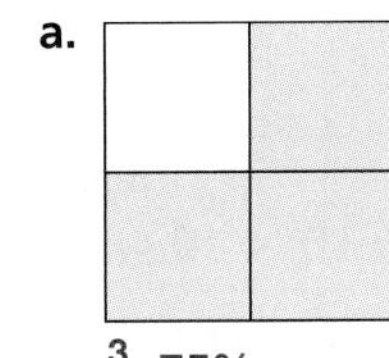

$\frac{3}{4}$; 75%

b.

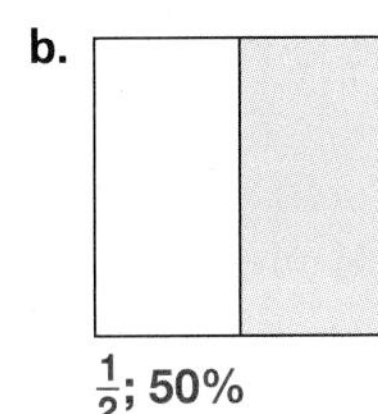

$\frac{1}{2}$; 50%

c.

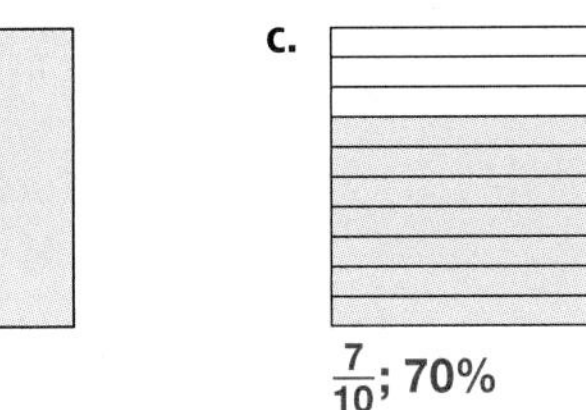

$\frac{7}{10}$; 70%

EXAMPLE Using Equivalent Ratios

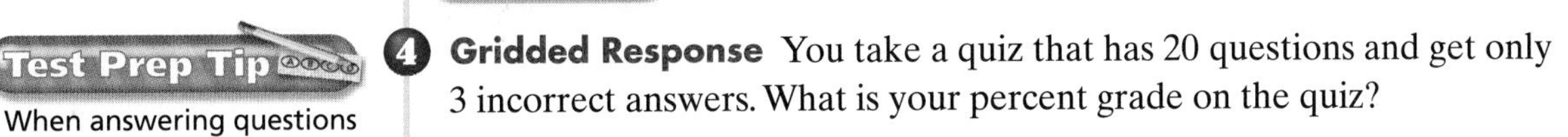

Test Prep Tip

When answering questions in gridded format, enter only the numerical portion of your answer in the grid.

4 **Gridded Response** You take a quiz that has 20 questions and get only 3 incorrect answers. What is your percent grade on the quiz?

First find the number of correct answers. Then write a ratio.

$\frac{17}{20}$ ← number of correct answers / ← total number of answers

$\frac{17 \cdot 5}{20 \cdot 5}$ ← Since 20 · 5 is 100, multiply numerator and denominator by 5.

$\frac{85}{100}$ ← Simplify.

85% ← Write as a percent.

Your percent grade is 85%. Fill in 85 on your answer grid.

Quick Check

4. A tennis team played a total of 25 games and won 20 of them. What percent of the games did the team win? 80%

2. Teach

Activity Lab

Use before the lesson.

All in One Teaching Resources

Activity Lab 6-1: Fractions, Decimals, Percents

Guided Instruction

Example 3

Have students practice repeating the factor pairs of 100. **2, 50; 4, 25; 5, 20; 10,10**

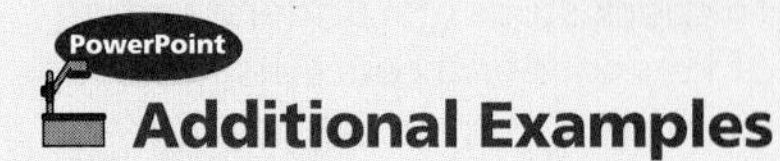

Additional Examples

1 In a floor plan of a 10 × 10 room, a chair takes up 4 spaces. Write the ratio and percent of the chair space to the total floor space. $\frac{4}{100}$; 4%

2 Model 82% in a 10 × 10 grid.

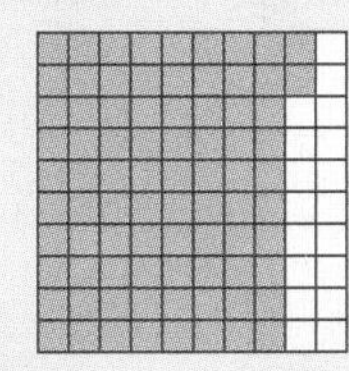

3 What percent does the shaded area represent? 60%

4 Write each ratio as a percent.

a. $\frac{3}{10}$ 30%   b. $\frac{4}{5}$ 80%

All in One Teaching Resources

- Daily Notetaking Guide 6-1 L3
- Adapted Notetaking 6-1 L1

Closure

- *What is a percent?* a ratio that compares a number to 100
- *How do you write a ratio with any denominator as a percent?* Write as an equal ratio with a denominator of 100. The numerator is the percent.

Advanced Learners L4

Your collection of books has five types: $\frac{1}{5}$ are mysteries, $\frac{8}{50}$ are biographies, $\frac{2}{10}$ are joke books, and $\frac{10}{100}$ are books about animals. What percent are science fiction, the last type? 34%

learning style: verbal

English Language Learners ELL

Some students have a difficult time understanding that $\frac{17}{20}$ is = 85%, even when they do the calculations correctly. Use alternate ways to explain that no matter what the denominator is, the ratio is equal to a part out of a whole with a denominator of 100.

learning style: verbal

3. Practice

Assignment Guide

Check Your Understanding
Go over Exercises 1–6 in class before assigning the Homework Exercises.

Homework Exercises

A	Practice by Example	7–18
B	Apply Your Skills	19–33
C	Challenge	34
	Test Prep and Mixed Review	35–40

Homework Quick Check
To check students' understanding of key skills and concepts, go over Exercises 8, 13, 26, 30, and 32.

Differentiated Instruction Resources

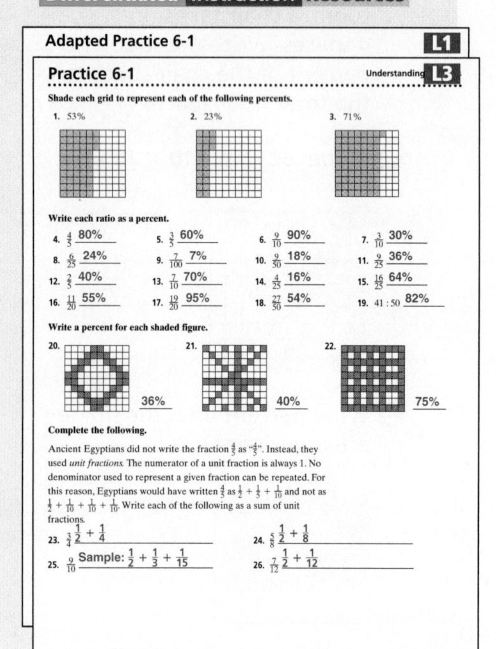

Adapted Practice 6-1 L1

Practice 6-1 Understanding L3

Shade each grid to represent each of the following percents.

1. 53% 2. 23% 3. 71%

Write each ratio as a percent.

4. $\frac{4}{5}$ 80% 5. $\frac{3}{5}$ 60% 6. $\frac{9}{10}$ 90% 7. $\frac{3}{10}$ 30%
8. $\frac{6}{25}$ 24% 9. $\frac{7}{100}$ 7% 10. $\frac{9}{50}$ 18% 11. $\frac{9}{25}$ 36%
12. $\frac{2}{5}$ 40% 13. $\frac{7}{10}$ 70% 14. $\frac{4}{25}$ 16% 15. $\frac{16}{25}$ 64%
16. $\frac{11}{20}$ 55% 17. $\frac{19}{20}$ 95% 18. $\frac{27}{50}$ 54% 19. 41 : 50 82%

Write a percent for each shaded figure.

20. 36% 21. 40% 22. 75%

Complete the following.

Ancient Egyptians did not write the fraction $\frac{4}{5}$ as "$\frac{4}{5}$". Instead, they used *unit fractions.* The numerator of a unit fraction is always 1. No denominator used to represent a given fraction can be repeated. For this reason, Egyptians would have written $\frac{4}{5}$ as $\frac{1}{2}+\frac{1}{5}+\frac{1}{10}$ and not as $\frac{1}{2}+\frac{1}{10}+\frac{1}{10}+\frac{1}{10}$. Write each of the following as a sum of unit fractions.

23. $\frac{3}{4}$ $\frac{1}{2}+\frac{1}{4}$ 24. $\frac{5}{8}$ $\frac{1}{2}+\frac{1}{8}$
25. $\frac{9}{10}$ Sample: $\frac{1}{2}+\frac{1}{3}+\frac{1}{15}$ 26. $\frac{7}{12}$ $\frac{1}{2}+\frac{1}{12}$

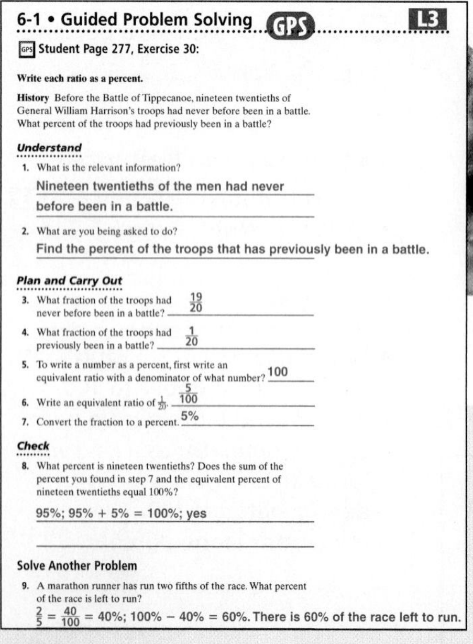

6-1 • Guided Problem Solving GPS L3

Student Page 277, Exercise 30:

Write each ratio as a percent.

History Before the Battle of Tippecanoe, nineteen twentieths of General William Harrison's troops had never before been in a battle. What percent of the troops had previously been in a battle?

Understand

1. What is the relevant information? Nineteen twentieths of the men had never before been in a battle.
2. What are you being asked to do? Find the percent of the troops that has previously been in a battle.

Plan and Carry Out

3. What fraction of the troops had never before been in a battle? $\frac{19}{20}$
4. What fraction of the troops had previously been in a battle? $\frac{1}{20}$
5. To write a number as a percent, first write an equivalent ratio with a denominator of what number? 100
6. Write an equivalent ratio of $\frac{1}{20}$. $\frac{5}{100}$
7. Convert the fraction to a percent. 5%

Check

8. What percent is nineteen twentieths? Does the sum of the percent you found in step 7 and the equivalent percent of nineteen twentieths equal 100%? 95%; 95% + 5% = 100%; yes

Solve Another Problem

9. A marathon runner has run two fifths of the race. What percent of the race is left to run? $\frac{2}{5}=\frac{40}{100}$ = 40%; 100% − 40% = 60%. There is 60% of the race left to run.

Check Your Understanding

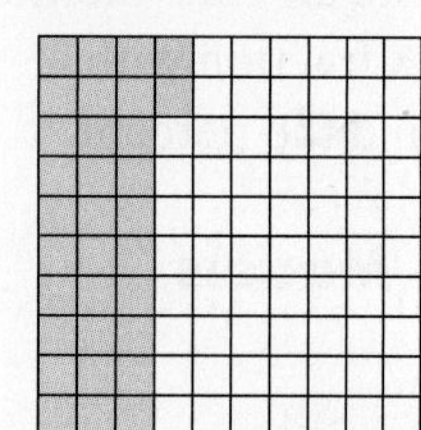

1. Write a ratio and a percent for the shaded area in the diagram. $\frac{32}{100}$; 32%
2. Model 15% on a 10 × 10 grid. **See margin.**

Write each ratio as a percent.

3. $\frac{67}{100}$ **67%** 4. $\frac{4}{5}$ **80%** 5. $\frac{9}{10}$ **90%**

6. **Reasoning** How can you use percents to compare two ratios with different denominators? **Answers may vary. Sample: Write each ratio as a percent and compare the two percents.**

Homework Exercises

For more exercises, see Extra Skills and Word Problems.

GO for Help

For Exercises	See Examples
7–9	1
10–14	2
15–18	3–4

A **Write a ratio and a percent for each shaded figure.**

7. 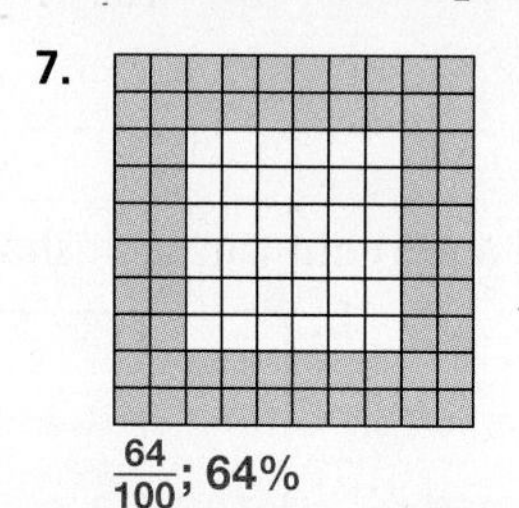 $\frac{64}{100}$; 64%

8. 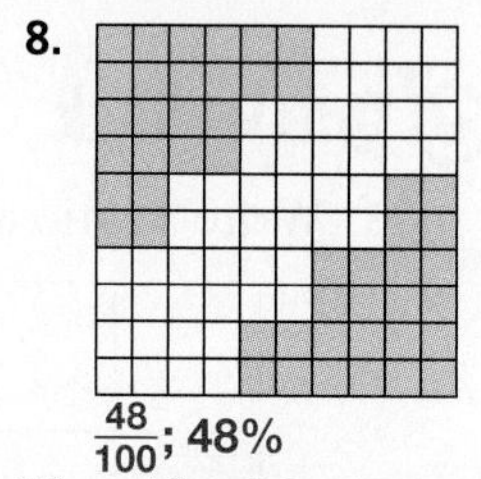 $\frac{48}{100}$; 48%

9. $\frac{60}{100}$; 60%

Model each percent on a 10 × 10 grid. **10–14. See margin.**

10. 35% 11. 78% 12. 10% 13. 8% 14. 90.5%

Write a ratio and a percent for each shaded area.

15. 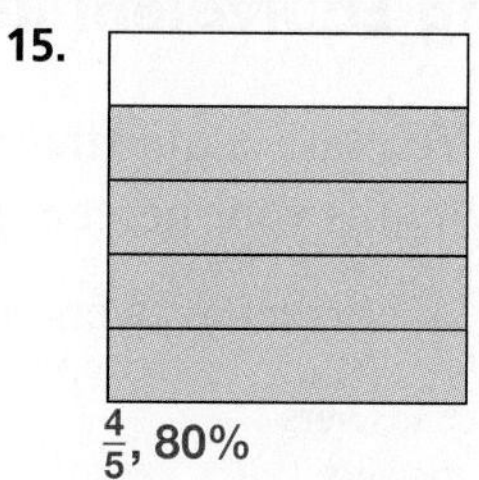 $\frac{4}{5}$, 80%

16. 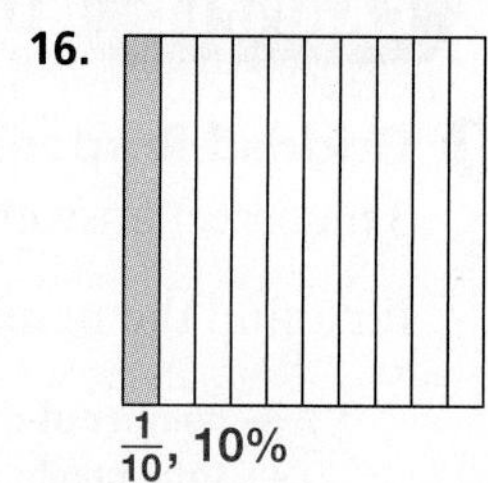 $\frac{1}{10}$, 10%

17. 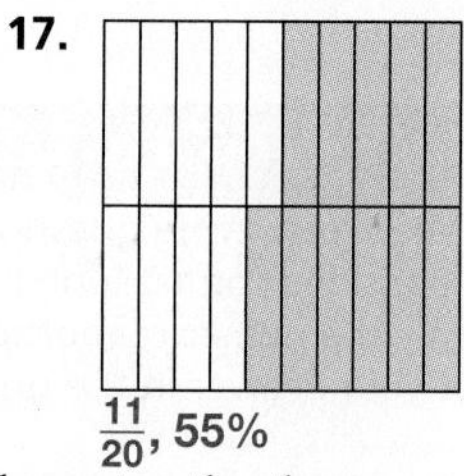 $\frac{11}{20}$, 55%

18. **Geography** The area of Argentina is about three tenths the area of the United States. Write this ratio as a percent. **30%**

B **GPS** 19. **Guided Problem Solving** In a litter of 10 puppies, exactly six puppies are black. What percent of the puppies are not black? **40%**
- How many puppies are not black?
- How many puppies are in the litter?

20. **Government** An amendment to the U.S. Constitution must be ratified by at least three fourths of the states to become law. Write this ratio as a percent. **75%**

2. 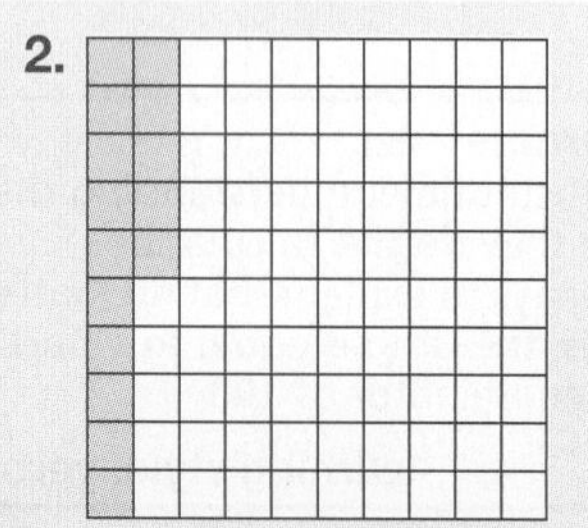

10–14. See back of book.

Go Online
Homework Video Tutor
Visit: PHSchool.com
Web Code: are-0601

Write each ratio as a percent.

21. $\frac{3}{5}$ 60%   22. $\frac{1}{2}$ 50%   23. $\frac{21}{25}$ 84%   24. $\frac{9}{50}$ 18%   25. $\frac{11}{20}$ 55%

26. **Writing in Math** Explain how to model 25% two different ways. See left.

26. Answers may vary. Sample: Shade 25 squares on a 10 × 10 grid or shade 1 section of a square divided into 4 equal areas.

Find what percent of a dollar each set of coins makes.

27. 2 quarters and 2 dimes 70%
28. 3 quarters, 1 dime, 3 pennies 88%
29. Your class has 12 boys and 13 girls. What percent of the students in your class are girls? 52%
30. **History** Before the Battle of Tippecanoe, nineteen twentieths of General William Harrison's troops had never before been in a battle. What percent of the troops had previously been in a battle? 5%

Estimate the percent of each figure that is shaded.

31.
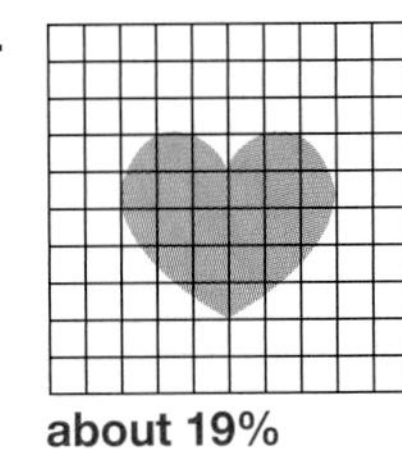
about 19%

32.
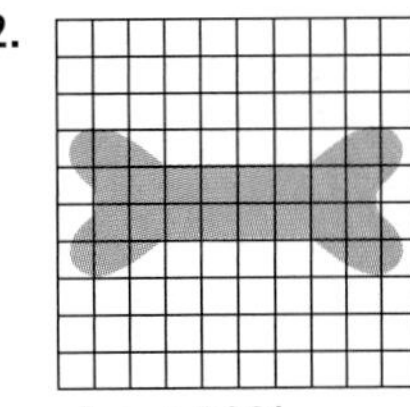
about 22%

33.
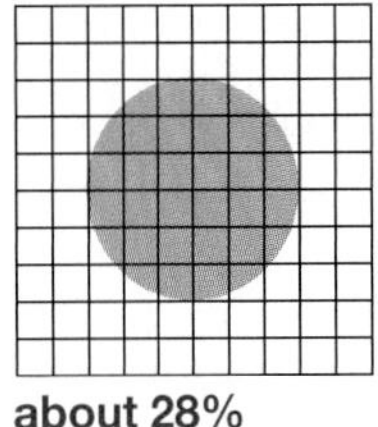
about 28%

34. **Challenge** You shade four squares in a grid. How many squares are there if the shaded portion represents 20% of the grid? 20 squares

Test Prep and Mixed Review

Practice

Gridded Response

35. The table shows the ratios of people who prefer four popular yogurt flavors. What is the percent of people surveyed who prefer blueberry? 12

Most Popular Yogurt Flavors

Personal Favorite	Ratio
Strawberry	$\frac{2}{5}$
Blueberry	$\frac{3}{25}$
Vanilla	$\frac{3}{50}$
Peach	$\frac{3}{100}$

36. Five students in Ms. Power's class ran for charity. The distances they ran were as follows: 5.8 mi, $4\frac{1}{2}$ mi, 2.4 mi, $3\frac{9}{10}$ mi, and 7 mi. What was the distance, in miles, the students ran altogether? 23.6

37. Arthur scored 87 on each of his first three tests. He scored 93 and 95 on his next two tests. What was the mean score of all his tests? 89.8

Go for Help

For Exercises	See Lesson
38–40	4-7

Which numbers are solutions of each inequality?

38. $x > -3$; $-4, 0, 3$ — 0, 3
39. $x \leq -2$; $0, -2, -3$ — −2, −3
40. $x < 5$; $5, 3, -1$ — 3, −1

Alternative Assessment

Provide student pairs with hundredths-decimal grids. Each partner shades different parts of five grids. Partners exchange squares and write a ratio and a percent for the shaded part of each square.

Test Prep

Resources

For additional practice with a variety of test item formats:

- Test-Taking Strategies, p. 315
- Test Prep, p. 319
- Test-Taking Strategies with Transparencies

4. Assess & Reteach

Lesson Quiz

1. Write a ratio and a percent for the shaded figure. $\frac{30}{100}$; 30%

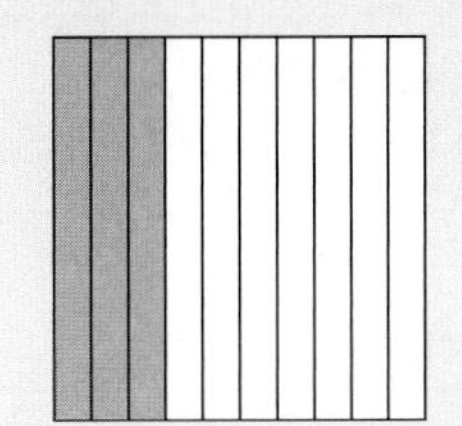

Write each ratio as a percent.

2. $\frac{1}{10}$ 10%   3. $\frac{11}{50}$ 22%

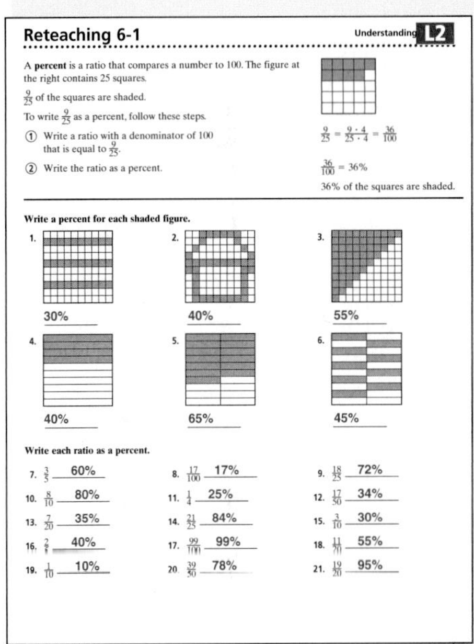

Reteaching 6-1 Understanding L2

A **percent** is a ratio that compares a number to 100. The figure at the right contains 25 squares.

$\frac{9}{25}$ of the squares are shaded.

To write $\frac{9}{25}$ as a percent, follow these steps.

① Write a ratio with a denominator of 100 that is equal to $\frac{9}{25}$. $\frac{9}{25} = \frac{9 \cdot 4}{25 \cdot 4} = \frac{36}{100}$

② Write the ratio as a percent. $\frac{36}{100} = 36\%$

36% of the squares are shaded.

Write a percent for each shaded figure.

1. 30%   2. 40%   3. 55%
4. 40%   5. 65%   6. 45%

Write each ratio as a percent.

7. $\frac{3}{5}$ 60%   8. $\frac{17}{100}$ 17%   9. $\frac{18}{25}$ 72%
10. $\frac{8}{10}$ 80%   11. $\frac{1}{4}$ 25%   12. $\frac{17}{50}$ 34%
13. $\frac{7}{20}$ 35%   14. $\frac{21}{25}$ 84%   15. $\frac{3}{10}$ 30%
16. $\frac{2}{5}$ 40%   17. $\frac{99}{100}$ 99%   18. $\frac{11}{20}$ 55%
19. $\frac{1}{10}$ 10%   20. $\frac{39}{50}$ 78%   21. $\frac{19}{20}$ 95%

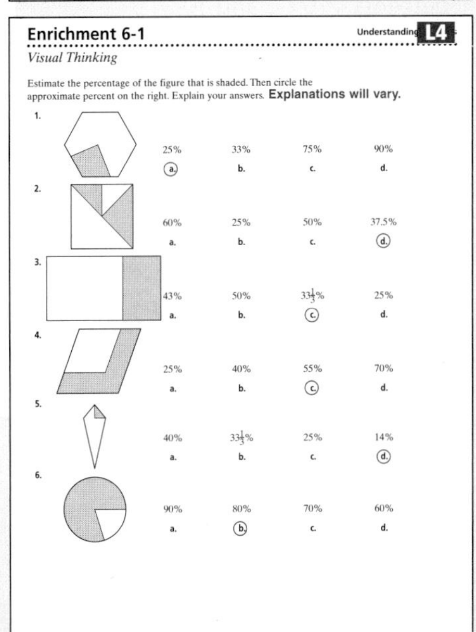

Enrichment 6-1 Understanding L4

Visual Thinking

Estimate the percentage of the figure that is shaded. Then circle the approximate percent on the right. Explain your answers. Explanations will vary.

1. a. 25% (circled) b. 33% c. 75% d. 90%
2. a. 60% b. 25% c. 50% d. 37.5% (circled)
3. a. 43% b. 50% c. $33\frac{1}{3}$% (circled) d. 25%
4. a. 25% b. 40% c. 55% (circled) d. 70%
5. a. 40% b. $33\frac{1}{3}$% c. 25% d. 14% (circled)
6. a. 90% b. 80% (circled) c. 70% d. 60%

Activity Lab

Rational Number Cubes

Students order rational numbers using greater than, less than, and equal symbols. This will help them become familiar with how numbers can be equivalent fractions, decimals, and percents.

Guided Instruction

Explain the steps of the Activity. Point out that the numbers on each cube are equivalent.
Ask:

- *If you rolled 3 cubes and got $\frac{1}{8}$, =, and 0.75, could you use that roll?* **no** *Why?* **$\frac{1}{8}$ does not equal 0.75**
- *How many squares of the grid should you shade to equal 12.5%?* **twelve full squares and one half square**

Activity

Have students work in pairs. They should toss the cubes until all the numbers and symbols fit in the chain. They can take turns tossing the cubes and recording the results and work together to decide where each number or symbol should be placed.

Alternative Method
Some students may have an easier time with simpler fractions. Provide these students with cubes reflecting decimals, fractions, and percents of 0, $\frac{1}{2}$, and 1.

Teaching Tip
Have students study the table to help them convert mentally between fractions, decimals, and percents.

Resources

- Activity Lab 6-2: Critical Thinking
- number cube patterns
- scissors
- tape

6-2a Activity Lab

Rational Number Cubes

You can write equivalent rational numbers in several different ways. The table below shows common rational numbers in equivalent forms. The fraction, decimal, and percent in each column are equivalent.

Fraction	$\frac{1}{8}$	$\frac{1}{4}$	$\frac{1}{3}$	$\frac{1}{2}$	$\frac{2}{3}$	$\frac{3}{4}$
Decimal	0.125	0.25	$0.33\overline{3}$	0.5	$0.66\overline{6}$	0.75
Percent	12.5%	25%	$33\frac{1}{3}$%	50%	$66\frac{2}{3}$%	75%

Check students' work.

Step 1 A net of a cube is shown at the right. Make six copies of the net.

Step 2 Use the first column from the table above. Fill in the squares of the first net. Write $\frac{1}{8}$ in the first empty square. Write 0.125 and 12.5% in the other two empty squares. Then shade the 10×10 grid to model this rational number.

Step 3 Repeat Step 2 with the other nets and the other columns of the table.

Step 4 Cut out each net. Fold each net along its dotted lines. Tape the edges to form a cube.

Step 5 Work in pairs. Toss the 12 cubes. Use the top views of the cubes. Arrange as many as you can into a chain that reads correctly from left to right. (You can use the inequality symbol as either "less than" or "greater than.")

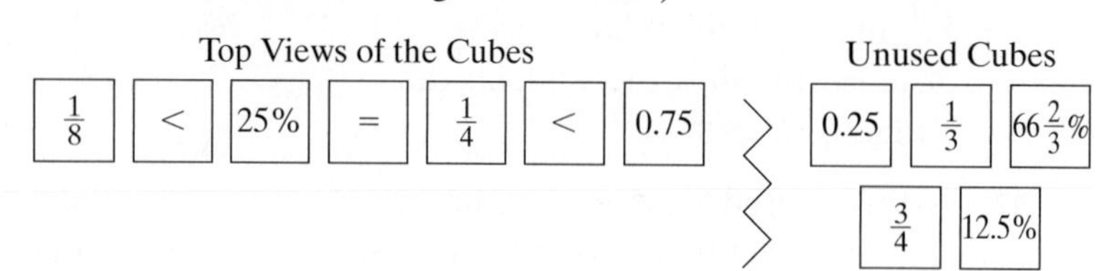

Step 6 Pick up the unused cubes and toss them again. Insert these cubes into the chain wherever they fit. Repeat until you can successfully place all 12 cubes in the chain.

6-2 Percents, Fractions, and Decimals

Check Skills You'll Need

1. Vocabulary Review What is a *repeating decimal?*
1–2. See below.

Write each fraction as a decimal.

2. $\frac{5}{16}$
3. $\frac{11}{40}$ **0.275**
4. $\frac{4}{9}$ **$0.\overline{4}$**
5. $\frac{2}{15}$ **$0.1\overline{3}$**

GO for Help Lesson 2-6

Check Skills You'll Need

1. **A repeating decimal is a decimal that repeats without end.**
2. **0.3125**

What You'll Learn

To convert between fractions, decimals, and percents

Why Learn This?

The food labels below all use a different form of $\frac{1}{2}$. Any rational number can be written as a fraction, a decimal, or a percent.

50% of the minimum daily amount

Same taste
½ the sugar

This product has 0.5g of fat.

KEY CONCEPTS Fractions, Decimals, and Percents

The model at the right shows 21 out of 100 squares shaded. You can write the shaded part of the model as a fraction, a decimal, or a percent.

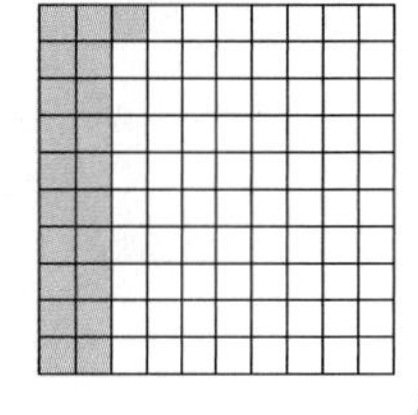

Fraction	Decimal	Percent
$\frac{21}{100}$	0.21	21%

Test Prep Tip

When you multiply a decimal by 100, the decimal point moves 2 places to the right.

To write a decimal as a percent, you can multiply the decimal by 100.

EXAMPLE Writing Decimals as Percents

1 Write 0.759 as a percent.

$$0.759 = \frac{759}{1{,}000} \quad \leftarrow \text{Write as a fraction.}$$

$$= \frac{75.9}{100} \quad \leftarrow \text{Write an equivalent fraction with 100 in the denominator.}$$

$$= 75.9\% \quad \leftarrow \text{Write as a percent.}$$

Quick Check

1. Write 0.607 as a percent. **60.7%**

6-2

1. Plan

Objective
To convert between fractions, decimals, and percents

Examples
1 Writing Decimals as Percents
2 Writing Percents as Decimals
3 Writing Fractions as Percents
4 Writing Percents as Fractions
5 Ordering Rational Numbers

Math Understandings: p. 272C

Math Background

You can convert between decimals, fractions, and percent by using the definition of percent as a quantity over 100. For instance, 0.5 is $\frac{5}{10}$, which is the same as $\frac{50}{100}$ or 50%. A decimal can easily be changed to a percent by multiplying by 100, or by moving the decimal point two places to the right and using the % symbol.

More Math Background: p. 272C

Lesson Planning and Resources

See p. 272E for a list of the resources that support this lesson.

PowerPoint

Bell Ringer Practice

Check Skills You'll Need
Use student page, transparency, or PowerPoint. For intervention, direct students to:
Fractions and Decimals
Lesson 2-6
Extra Skills and Word Problems Practice, Ch. 2

Differentiated Instruction Solutions for All Learners

Special Needs L1
Make sure students understand that moving the decimal point two places to the right is the same as multiplying by 100. Underline the two 0's in 100, and underline the two decimal places so they see the connection.

learning style: visual

Below Level L2
Have students complete a table such as the one below.

Fraction	$\frac{1}{10}$	$\frac{1}{5}$	$\frac{1}{4}$	$\frac{3}{4}$
Decimal	0.1	0.2	0.25	0.75
Percent	10%	20%	25%	75%

learning style: visual

2. Teach

Activity Lab

Use before the lesson.
Student Edition Activity Lab, Hands On 6-2a, Rational Number Cubes, p. 278

All in One Teaching Resources
Activity Lab 6-2: Critical Thinking

Guided Instruction

Error Prevention!

In Quick Check 2, make sure students do not keep the percent sign when they convert percents to decimals.

Example 4
If students have difficulty finding the GCF (Greatest Common Factor), have them factor the numerator and use those factors to rewrite 100 as shown below:

$\frac{15}{100} = \frac{5 \cdot 3}{5 \cdot 20} = \frac{3}{20}$

Example 5
To help students remember which way to move the decimal point, tell them this story. When the percent symbol (%) is written to the right of a number, the decimal point moves two places to be closer to the symbol. When the symbol is removed, the decimal point "goes back home" two places to the left.

Online active math

For: Rational Number Activity
Use: Interactive Textbook, 6-2

To write a percent as a decimal, you can divide it by 100, or move the decimal point two places to the left.

EXAMPLE Writing Percents as Decimals

2 Write 47.5% as a decimal.

$47.5\% = \frac{47.5}{100}$ ← **Write the percent as a fraction.**

$= 0.475$ ← **Divide.**

Quick Check

2. Write each percent as a decimal.
a. 35% **0.35** **b.** 12.5% **0.125** **c.** 7.8% **0.078**

When the denominator of a fraction is a factor of 100, you can easily use equivalent ratios to convert the fraction to a percent. For fractions with other denominators, you can use a calculator to convert the fraction into a decimal, and then rewrite the decimal as a percent.

EXAMPLE Writing Fractions as Percents

3 **Nutrition** In a slice of cheese pizza, 45 Calories are from fat. The total number of Calories in each slice is 158. About what percent of the Calories are *not* from fat? Round to the nearest tenth of a percent.

Step 1 Find the number of Calories that are not from fat.

$158 - 45 = 113$

Step 2 Estimate.

$\frac{113}{158} \approx \frac{120}{160}$, which is $\frac{3}{4}$, or 75%.

Step 3 Write the ratio.

$\frac{113}{158}$ ← **Calories from fat** / ← **total Calories**

$113 \div 158 = 0.71518987$ ← **Use a calculator.**

$= 71.518987\%$ ← **Write as a percent.**

$\approx 71.5\%$ ← **Round to the nearest tenth of a percent.**

About 71.5% of the Calories are not from fat.

Check for Reasonableness Since 71.5% is close to the estimate 75%, the answer is reasonable.

Quick Check

3. Write $\frac{21}{40}$ as a percent. Round to the nearest tenth of a percent. **52.5%**

Advanced Learners L4
You go to a restaurant with \$8.73. You want to have enough money to give your waitress a 15% tip for your lunch. What is the most money you can spend on the meal? **\$7.59**

learning style: verbal

English Language Learners ELL
Ask: *What happens to the decimal point when you multiply or divide a decimal by 100?* **Answers may vary.** Have students work with a partner to explain and test their ideas. They will be less prone to errors if they understand the movement of the decimal point.

learning style: verbal

You can write a percent as a fraction. First write the percent as a fraction with a denominator of 100. Then simplify the fraction.

EXAMPLE Writing Percents as Fractions

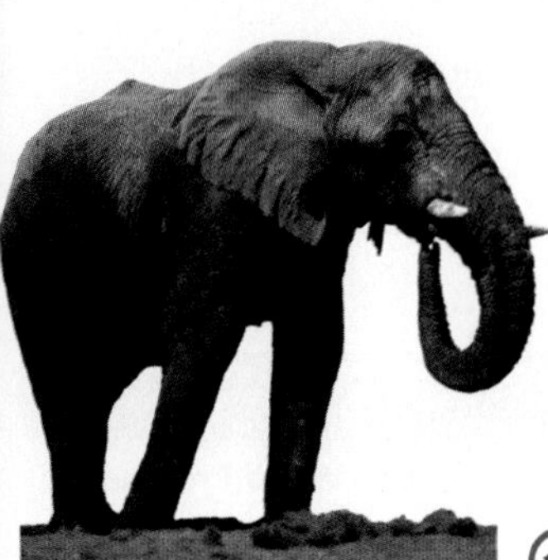

4 Science Behavioral scientists observed an elephant that slept about 12.5% of each day. What fraction of each day did the elephant sleep?

$12.5\% = \frac{12.5}{100}$ ← **Write 12.5% as a fraction with a denominator of 100.**

$= \frac{12.5 \times 10}{100 \times 10}$ ← **Multiply the numerator and denominator by 10.**

$= \frac{125 \div 125}{1{,}000 \div 125}$ ← **Divide both numerator and denominator by the GCF, 125.**

$= \frac{1}{8}$ ← **Simplify the fraction.**

The elephant slept about $\frac{1}{8}$ of each day.

Quick Check

4. An elephant eats about 6% of its body weight in vegetation each day. Write this as a fraction in simplest form. $\frac{3}{50}$

To compare rational numbers in different forms, you can write all the numbers in the same form. Then graph each number on a number line.

EXAMPLE Ordering Rational Numbers

5 Order 0.52, 37%, 0.19, and $\frac{1}{4}$ from least to greatest.

Write all the numbers as decimals. Then graph them.

0.52 ← **This number is already in decimal form.**

$37\% = 0.37$ ← **Move the decimal point two places to the left.**

0.19 ← **This number is already in decimal form.**

$\frac{1}{4} = 0.25$ ← **Divide the numerator by the denominator.**

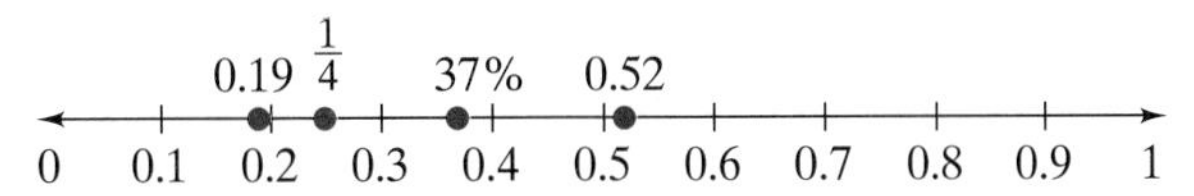

From least to greatest, the numbers are 0.19, $\frac{1}{4}$, 37%, and 0.52.

Quick Check

5. Order from least to greatest.

a. $\frac{3}{10}$, 0.74, 29%, $\frac{11}{25}$
29%, $\frac{3}{10}$, $\frac{11}{25}$, 0.74

b. 15%, $\frac{7}{20}$, 0.08, 50%
0.08, 15%, $\frac{7}{20}$, 50%

GO for Help

For help converting a fraction to a decimal, go to Lesson 2-6.

PowerPoint Additional Examples

1 Write each decimal as a percent.
a. 0.06 **6%** **b.** 0.523 **52.3%**
c. 0.5 **50%** **d.** 0.95 **95%**

2 Write each percent as a decimal.
a. 4% **0.04** **b.** 34.3% **0.343**
c. 16% **0.16** **d.** 6.4% **0.064**

3 Write each fraction as a percent. Round to the nearest tenth of a percent.
a. $\frac{3}{10}$ **30%** **b.** $\frac{37}{49}$ **75.5%**
c. $\frac{6}{30}$ **20%** **d.** $\frac{14}{65}$ **21.5%**

4 Write each percent as a fraction in simplest form.
a. 18% $\frac{9}{50}$ **b.** 12% $\frac{3}{25}$
c. 45% $\frac{9}{20}$

5 Order from least to greatest.
a. $\frac{3}{5}$, $\frac{2}{10}$, 0.645, 13%
13%, $\frac{2}{10}$, $\frac{3}{5}$, 0.645
b. 40%, $\frac{10}{20}$, 18%, 0.082
0.082, 18%, 40%, $\frac{10}{20}$

All in One Teaching Resources
- Daily Notetaking Guide 6-2 L3
- Adapted Notetaking 6-2 L1

Closure

- *How do you write a percent as a fraction?* **Write the percent as a fraction with denominator of 100 and simplify.**
- *How do you write a decimal as a percent?* **Move the decimal point two places to the right.**
- *How do you write a percent as a decimal?* **Divide by 100, or move the decimal point two places to the left.**

3. Practice

Assignment Guide

Check Your Understanding
Go over Exercises 1–4 in class before assigning the Homework Exercises.

Homework Exercises

A Practice by Example	5–29
B Apply Your Skills	30–35
C Challenge	36
Test Prep and Mixed Review	37–41

Homework Quick Check
To check students' understanding of key skills and concepts, go over Exercises 12, 21, 31, 34, and 35.

Differentiated Instruction Resources

Adapted Practice 6-2 L1

Practice 6-2 Percents, Fractions, and ... L3

Write each percent as a fraction in simplest form and as a decimal.

1. 65% $\frac{13}{20}$; 0.65 2. 37.5% $\frac{3}{8}$; 0.375 3. 80% $\frac{4}{5}$; 0.8 4. 25% $\frac{1}{4}$; 0.25
5. 18% $\frac{9}{50}$; 0.18 6. 46% $\frac{23}{50}$; 0.46 7. 87% $\frac{87}{100}$; 0.87 8. 8% $\frac{2}{25}$; 0.08
9. 43% $\frac{43}{100}$; 0.43 10. 55% $\frac{11}{20}$; 0.55 11. 94% $\frac{47}{50}$; 0.94 12. 36% $\frac{9}{25}$; 0.36

Write each number as a percent. Round to the nearest tenth of a percent where necessary.

13. $\frac{8}{15}$ 53.3% 14. $\frac{7}{50}$ 14% 15. 0.56 56%
16. 0.0413 4.1% 17. $\frac{3}{8}$ 37.5% 18. $\frac{7}{12}$ 58.3%
19. 0.387 38.7% 20. 0.283 28.3% 21. $\frac{2}{9}$ 22.2%

Write each number as a percent. Place the number into the puzzle without using the percent sign or decimal point.

22.

Across: 1. 0.134 3. $\frac{53}{100}$ 5. 0.565 7. $1\frac{7}{50}$ 9. 0.456 10. 0.63 11. $\frac{11}{200}$ 13. 0.58 14. $\frac{191}{200}$ 16. 0.605

Down: 2. 0.346 4. 0.324 5. $\frac{1}{2}$ 6. 0.515 8. $\frac{33}{200}$ 9. 0.4385 10. $\frac{659}{1,000}$ 12. $\frac{1,087}{20,000}$ 15. $\frac{14}{25}$

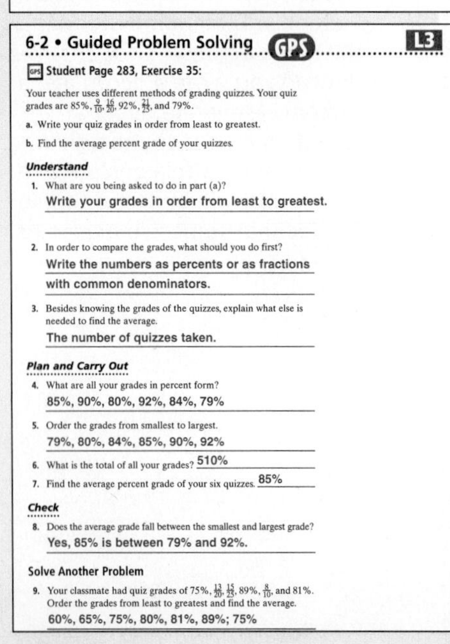

6-2 • Guided Problem Solving GPS L3

GPS Student Page 283, Exercise 35:

Your teacher uses different methods of grading quizzes. Your quiz grades are 85%, $\frac{4}{5}$, $\frac{18}{20}$, 92%, $\frac{21}{25}$, and 79%.

a. Write your quiz grades in order from least to greatest.
b. Find the average percent grade of your quizzes.

Understand

1. What are you being asked to do in part (a)? Write your grades in order from least to greatest.
2. In order to compare the grades, what should you do first? Write the numbers as percents or as fractions with common denominators.
3. Besides knowing the grades of the quizzes, explain what else is needed to find the average. The number of quizzes taken.

Plan and Carry Out

4. What are all your grades in percent form? 85%, 90%, 80%, 92%, 84%, 79%
5. Order the grades from smallest to largest. 79%, 80%, 84%, 85%, 90%, 92%
6. What is the total of all your grades? 510%
7. Find the average percent grade of your six quizzes. 85%

Check

8. Does the average grade fall between the smallest and largest grade? Yes, 85% is between 79% and 92%.

Solve Another Problem

9. Your classmate had quiz grades of 75%, $\frac{13}{20}$, $\frac{15}{25}$, 89%, $\frac{8}{10}$, and 81%. Order the grades from least to greatest and find the average. 60%, 65%, 75%, 80%, 81%, 89%; 75%

Check Your Understanding

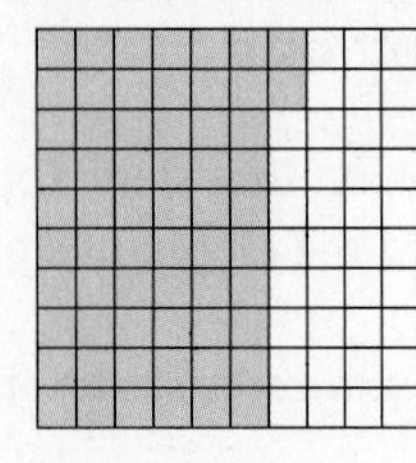

1. Write the shaded part of the model at the left as a percent, a fraction, and a decimal. **62%, $\frac{62}{100}$, 0.62**
2. In each set, find the number that does *not* equal the other two.
   a. 9.9%, $\frac{99}{100}$, 0.99 **9.9%** b. 64%, $\frac{16}{50}$, 0.64 **$\frac{16}{50}$** c. 12%, $\frac{3}{25}$, 1.2 **1.2**
3. **Mental Math** Order 0.54, 55%, and $\frac{1}{2}$ from least to greatest. **See left.**
4. **Reasoning** When you write a percent as a decimal, why do you move the decimal point 2 units to the left? **You are dividing by 100.**

3. $\frac{1}{2}$, 0.54, 55%

Homework Exercises

For more exercises, see Extra Skills and Word Problems.

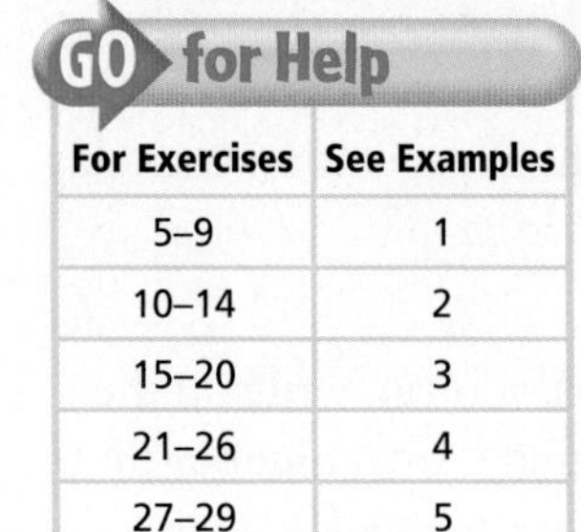

GO for Help

For Exercises	See Examples
5–9	1
10–14	2
15–20	3
21–26	4
27–29	5

A **Write each decimal as a percent.**

5. 0.57 **57%** 6. 0.375 **37.5%** 7. 0.09 **9%** 8. 0.155 **15.5%** 9. 0.6 **60%**

Write each percent as a decimal.

10. 32% **0.32** 11. 88% **0.88** 12. 19.1% **0.191** 13. 3% **0.03** 14. 1.25% **0.0125**

Write each fraction as a percent to the nearest tenth of a percent.

15. $\frac{45}{50}$ **90%** 16. $\frac{7}{8}$ **87.5%** 17. $\frac{1}{12}$ **8.3%** 18. $\frac{5}{6}$ **83.3%** 19. $\frac{3}{11}$ **27.3%**

20. Out of 49 fish, 31 are goldfish. About what percent are goldfish? **about 63%**

Write each percent as a fraction in simplest form.

21. 15% **$\frac{3}{20}$** 22. 6% **$\frac{3}{50}$** 23. 20% **$\frac{1}{5}$** 24. 37.5% **$\frac{3}{8}$** 25. 17% **$\frac{17}{100}$**

26. **Computers** A computer screen shows the print on a page at 78% of its actual size. Write 78% as a fraction in simplest form. **$\frac{39}{50}$**

Order from least to greatest.

27. $\frac{1}{2}$, 12%, 0.25 **12%, 0.25, $\frac{1}{2}$** 28. 68%, 0.37, $\frac{3}{10}$ **$\frac{3}{10}$, 0.37, 68%** 29. 0.81, $\frac{4}{5}$, 90% **$\frac{4}{5}$, 0.81, 90%**

B GPS 30. **Guided Problem Solving** In your class, 9 of the 26 students are in the chorus. What percent of your class is in the chorus? Round to the nearest tenth of a percent. **34.6%**
- What ratio can help to find the percent of students in the chorus?
- Should you multiply or divide by 100 to find the percent?

31. **Writing in Math** Does 0.4 equal 0.4%? Explain. **See left.**

31. No, 0.4% is equal to 0.004 as a decimal.

32. Compare 0.32 and 3.2%. Use <, =, or >. **0.32 > 3.2%**

33. See back of book.

Homework Video Tutor
Visit: PHSchool.com
Web Code: are-0602

33. **Math in the Media** Use the cartoon below to complete the table.

Topping	With Olives	Plain	With Onions and Green Peppers
Percent of the Pizza	■	■	■
Number of Slices	■	■	■

See margin.

Macaroni & Cheese

Nutrition	
Fat	33 g
Carbohydrates	40 g
Protein	17 g

Spaghetti & Meat Sauce

Nutrition	
Fat	12 g
Carbohydrates	39 g
Protein	19 g

34. **Nutrition** The tables at the left give data on two different foods. A gram of fat has 9 Calories. A gram of carbohydrates and a gram of protein both have 4 Calories. What percent of the Calories in each food are from carbohydrates? Round to the nearest percent. **macaroni and cheese, 30%; spaghetti and meat sauce, 46%**

35. Your teacher uses different methods of grading quizzes. Your quiz grades are 85%, $\frac{9}{10}$, $\frac{16}{20}$, 92%, $\frac{21}{25}$, and 79%. **79%, $\frac{16}{20}$, $\frac{21}{25}$, 85%, $\frac{9}{10}$, 92%**
   GPS
   a. Write your quiz grades in order from least to greatest.
   b. Find the average percent grade of your quizzes. **85%**

36. **Challenge** Write 3.75% as a fraction in simplest form. **$\frac{3}{80}$**
   C

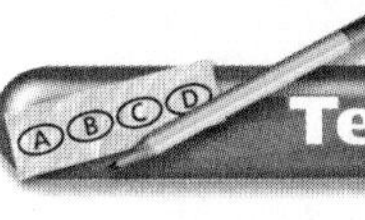

Test Prep and Mixed Review

Practice

Multiple Choice

37. Nathan runs m miles each weekday and $3\frac{1}{4}$ times farther on Saturday. He runs $6\frac{1}{2}$ miles on Saturday. Which equation can be used to find the number of miles Nathan runs each weekday? **A**

   (A) $3\frac{1}{4}m = 6\frac{1}{2}$ (C) $m + 3\frac{1}{4} = 6\frac{1}{2}$

   (B) $6\frac{1}{2}m = 3\frac{1}{4}$ (D) $m \div 3\frac{1}{4} = 6\frac{1}{2}$

38. The prices of 3 different bags of onions are given in the table below. Which size bag has the lowest price per pound? **H**

   (F) The 5-lb bag only
   (G) The 5-lb bag and the 15-lb bag
   (H) The 10-lb bag only
   (J) The 15-lb bag only

Bag (lb)	Price
5	\$3.49
10	\$6.70
15	\$10.33

GO for Help

For Exercises	See Lesson
39–41	4-9

Solve each inequality.

39. $6.3 \geq -7x$ **$-0.9 \leq x$**

40. $-12 < \frac{m}{2}$ **$-24 < m$**

41. $-10.2 \leq -0.2y$ **$51 \geq y$**

Online lesson quiz, PHSchool.com, Web Code: ara-0602

4. Assess & Reteach

Lesson Quiz

Write in order from least to greatest.

1. 30%, 0.24, $\frac{2}{5}$ **0.24, 30%, $\frac{2}{5}$**
2. 0.9, $\frac{7}{8}$, 9% **9%, $\frac{7}{8}$, 0.9**
3. $\frac{1}{4}$, 40%, 0.35 **$\frac{1}{4}$, 0.35, 40%**
4. $\frac{7}{10}$, 0.77, 7% **7%, $\frac{7}{10}$, 0.77**

Alternative Assessment

Each student in a pair writes one decimal, one fraction, and one percent. Partners exchange papers and write the partner's decimals and fractions as percents. Students write percents as decimals and each other's fractions in simplest form. Then partners work together to arrange the six original fractions, decimals, and percents in order from least to greatest.

Test Prep

Resources

For additional practice with a variety of test item formats:

- Test-Taking Strategies, p. 315
- Test Prep, p. 319
- Test-Taking Strategies with Transparencies

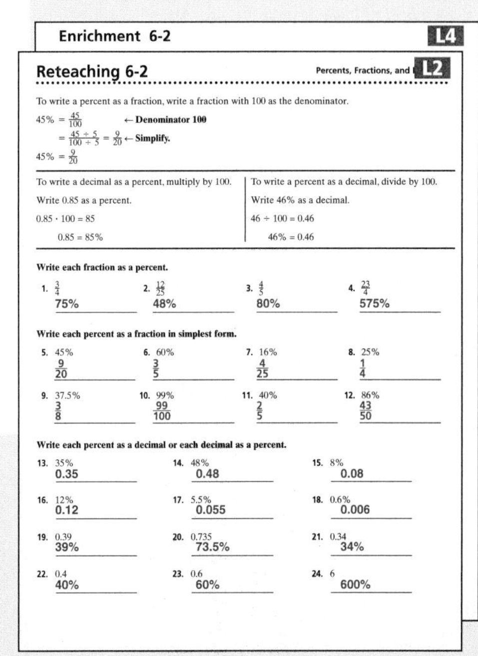

Enrichment 6-2 L4

Reteaching 6-2 Percents, Fractions, and L2

To write a percent as a fraction, write a fraction with 100 as the denominator.

$45\% = \frac{45}{100}$ ← Denominator 100

$= \frac{45 \div 5}{100 \div 5} = \frac{9}{20}$ ← Simplify.

$45\% = \frac{9}{20}$

To write a decimal as a percent, multiply by 100.
Write 0.85 as a percent.
$0.85 \cdot 100 = 85$
$0.85 = 85\%$

To write a percent as a decimal, divide by 100.
Write 46% as a decimal.
$46 \div 100 = 0.46$
$46\% = 0.46$

Write each fraction as a percent.

1. $\frac{3}{4}$ 75%
2. $\frac{12}{25}$ 48%
3. $\frac{4}{5}$ 80%
4. $\frac{23}{4}$ 575%

Write each percent as a fraction in simplest form.

5. 45% $\frac{9}{20}$
6. 60% $\frac{3}{5}$
7. 16% $\frac{4}{25}$
8. 25% $\frac{1}{4}$
9. 37.5% $\frac{3}{8}$
10. 99% $\frac{99}{100}$
11. 40% $\frac{2}{5}$
12. 86% $\frac{43}{50}$

Write each percent as a decimal or each decimal as a percent.

13. 35% 0.35
14. 48% 0.48
15. 8% 0.08
16. 12% 0.12
17. 5.5% 0.055
18. 0.6% 0.006
19. 0.39 39%
20. 0.735 73.5%
21. 0.34 34%
22. 0.4 40%
23. 0.6 60%
24. 6 600%

6-3

1. Plan

Objective
To convert between fractions, decimals, and percents greater than 100% or less than 1%

Examples
1, 2 Rewriting Percents
3 Writing Mixed Numbers as Percents
4 Application: Government

Math Understandings: p. 272C

Math Background

A percent can be greater than 100% or less than 1%. Percents greater than 100% convert to mixed numbers. Percents less than 1% convert to fractions less than $\frac{1}{100}$.

$150\% \rightarrow \frac{150}{100} \rightarrow 1\frac{1}{2}$

$0.15\% \rightarrow \frac{15}{10{,}000} \rightarrow \frac{3}{2{,}000}$

To convert a mixed number to a percent, first write it as a decimal.

More Math Background: p. 272C

Lesson Planning and Resources

See p. 272E for a list of the resources that support this lesson.

PowerPoint

Bell Ringer Practice

Check Skills You'll Need
Use student page, transparency, or PowerPoint. For intervention, direct students to:
Understanding Percents
Lesson 6-1
Extra Skills and Word Problems Practice, Ch. 6

6-3 Percents Greater Than 100% or Less Than 1%

Check Skills You'll Need

1. **Vocabulary Review** A *percent* is a ratio that compares a number to _?_. 100

Write each ratio as a percent.

2. $\frac{1}{100}$ 1% 3. $\frac{49}{50}$ 98%

4. $\frac{19}{20}$ 95% 5. $\frac{2}{25}$ 8%

GO for Help Lesson 6-1

What You'll Learn

To convert between fractions, decimals, and percents greater than 100% or less than 1%

Why Learn This?

You can get 110% of one day's Recommended Dietary Allowance of vitamin C by eating one half of a grapefruit.

In a percent, if the number compared to 100 is greater than 100, the percent is greater than 100%. If the number compared to 100 is less than 1, the percent is less than 1%.

You can rewrite these percents as decimals or fractions.

EXAMPLES Rewriting Percents

1 Write 110% as a decimal and as a fraction.

$110\% = 1.10$ ← **Move the decimal point two places to the left.**

$110\% = \frac{110}{100} = \frac{11}{10} = 1\frac{1}{10}$ ← **Use the definition of percent. Simplify the fraction.**

110% equals 1.10 in decimal form and $1\frac{1}{10}$ in fraction form.

2 Write 0.7% as a decimal and as a fraction in simplest form.

$0.7\% = 0.007$ ← **Move the decimal point two places to the left.**

$0.7\% = \frac{0.7}{100}$ ← **Use the definition of percent.**

$= \frac{0.7 \cdot 10}{100 \cdot 10} = \frac{7}{1{,}000}$ ← **Multiply numerator and denominator by 10 to get a whole number numerator. Simplify.**

0.7% equals 0.007 in decimal form and $\frac{7}{1{,}000}$ in fraction form.

Quick Check

1. Write 125% as a decimal and as a fraction. **1.25; $\frac{5}{4}$ or $1\frac{1}{4}$**
2. Write 0.35% as a decimal and as a fraction in simplest form. **0.0035; $\frac{7}{2{,}000}$**

Differentiated Instruction Solutions for All Learners

Special Needs L1
Give students a 100-grid and ask them to color in 120% of the whole grid. As they run out of room, ask them whether they need another grid. Show them that 120% is one whole and 20% of another whole.

learning style: visual

Below Level L2
To change percents less than 1 into fractions, review place values with students. "Count off" the decimal places (*tenths, hundredths, thousandths*) and write the correct number as the denominator.

learning style: verbal

A mixed number represents a percent greater than 100%.

EXAMPLE Writing Mixed Numbers as Percents

3 **Entertainment** A movie ticket costs $1\frac{7}{8}$ times as much as renting a video. Write this mixed number as a percent.

$1\frac{7}{8} = 1$ + 7 ÷ 8 = 1.875 ← Use a calculator.

$= 1.87.5 = 187.5\%$ ← Move the decimal point two places to the right.

A movie ticket costs 187.5% of the cost of renting a video.

Quick Check

3. You plan to run $2\frac{4}{5}$ times the distance you ran yesterday. Write this number as a percent. **280%**

A proper fraction represents a percent less than 100%.

EXAMPLE Application: Government

Test Prep Tip

A fraction less than $\frac{1}{100}$ represents a percent less than 1%. You can use percents less than 1% to describe small numbers.

4 **Multiple Choice** West Virginia has 3 members in the U.S. House of Representatives. There are 432 other representatives who are not from West Virginia. What percent of the representatives are from West Virginia?

Ⓐ 70% Ⓑ 7% Ⓒ 0.7% Ⓓ 0.007%

The total number of representatives is 432 + 3, or 435.

Estimate Round 435 to 400. 4 is close to 3. Then $\frac{4}{400} = 0.01$, or 1%.

$\frac{\text{West Virginia representatives}}{\text{total number of representatives}} = \frac{3}{435}$ ← Write the fraction.

$= 0.0068965517$ ← Use a calculator.

$\approx 0.7\%$ ← Write as a percent and round.

About 0.7% of the representatives are from West Virginia. The correct answer is choice C.

Check for Reasonableness Since 0.7% is close to the estimate 1%, the answer is reasonable.

Quick Check

4. Idaho has 2 members in the U.S. House of Representatives. What percent of the representatives are from Idaho? Round to the nearest hundredth of a percent. **0.46%**

Advanced Learners L4

Change each to a decimal. Place in order from least to greatest. If necessary, round to the nearest hundredth.

$\frac{7}{400}$ 0.0175 0.08% 0.0008 $2\frac{3}{5}$ 2.60 6.2% 0.062

0.0008, 0.0175, 0.062, 2.60

learning style: visual

English Language Learners ELL

To clarify the explanations for percents greater or less than 100%, make sure that students understand that 100% can be thought of as one whole. Percents greater or less than 100% are greater or less than one whole.

learning style: verbal

2. Teach

Activity Lab

Use before the lesson.

All in One Teaching Resources

Activity Lab 6-3: Percents Greater than 100 or Less than 1

Guided Instruction

Example 1

Provide students with two 10 × 10 grids to model percents greater than 100. One grid will become completely shaded and the part greater than 100 will be shaded in the second grid.

Example 2

Remind students that they can write zeros to the right of decimals without changing the value of the decimals. Encourage students to insert these zeros before moving the decimal point.

PowerPoint

Additional Examples

Write each percent as a decimal and a fraction in simplest form.

1. 140% **1.40; $1\frac{2}{5}$**
2. 0.75% **0.0075; $\frac{3}{400}$**

Write each fraction or decimal as a percent. Round to the nearest hundredth of a percent.

3. a. $4\frac{2}{3}$ **466.67%**
   b. 2.557 **255.7%**
4. a. $\frac{1}{325}$ **0.31%** b. $\frac{6}{875}$ **0.69%**

All in One Teaching Resources

- Daily Notetaking Guide 6-3 L3
- Adapted Notetaking 6-3 L1

Closure

- *How do you write a percent greater than 100 as a mixed number?* **Sample: Rewrite the percent as a fraction with denominator of 100, simplify, and write the improper fraction as a mixed number.**
- *How do you write a percent greater than 100 or less than 1 as a decimal?* **Move the decimal point two places to the left.**

3. Practice

Assignment Guide

Check Your Understanding
Go over Exercises 1–7 in class before assigning the Homework Exercises.

Homework Exercises
A	Practice by Example	8–32
B	Apply Your Skills	33–49
C	Challenge	50
Test Prep and Mixed Review		51–56

Homework Quick Check
To check students' understanding of key skills and concepts, go over Exercises 15, 28, 40, 43, and 46.

Differentiated Instruction Resources

Adapted Practice 6-3 L1

Practice 6-3 — Percents Greater Than 100% or Less Than 1% L3

Classify each of the following as: (A) less than 1%, (B) greater than 100%, or (C) between 1% and 100%.

1. $\frac{1}{2}$ C 2. $\frac{4}{3}$ B 3. $\frac{2}{300}$ A 4. $\frac{3}{10}$ C
5. 1.03 B 6. 0.009 A 7. 0.635 C 8. 0.0053 A

Use >, <, or = to compare the numbers in each pair.

9. $\frac{1}{4}$ [>] 20% 10. $\frac{1}{2}$% [<] 50 11. 0.008 [<] 8%
12. 150% [>] $\frac{5}{4}$ 13. 3 [=] 300% 14. $\frac{7}{250}$ [>] 0.3%

Write each fraction as a percent. Round to the nearest tenth of a percent if necessary.

15. $\frac{7}{5}$ 140% 16. $\frac{137}{100}$ 137% 17. $\frac{0.8}{100}$ 0.8%
18. $\frac{21}{4}$ 525% 19. $\frac{17}{10}$ 170% 20. $\frac{65}{40}$ 106.3%
21. $\frac{37}{20}$ 185% 22. $\frac{7}{500}$ 1.4% 23. $\frac{9}{8}$ 112.5%

Write each decimal as a percent.

24. 0.003 0.3% 25. 1.8 180% 26. 0.0025 0.25%
27. 5.3 530% 28. 0.0041 0.41% 29. 0.083 8.3%
30. 0.0009 0.09% 31. 0.83 83% 32. 20 2,000%

Write each percent as a decimal and as a fraction in simplest form.

33. 175% 1.75; $1\frac{3}{4}$ 34. 120% 1.2; $1\frac{1}{5}$ 35. $\frac{2}{5}$% 0.004, $\frac{1}{250}$
36. $\frac{5}{8}$% 0.00625, $\frac{1}{160}$ 37. 750% 7.5; $7\frac{1}{2}$ 38. $8\frac{1}{4}$% 0.0825, $\frac{33}{400}$

39. In 1990, the population of Kansas was 2,477,574, which included 21,965 Native Americans. What percent of the people living in Kansas were Native Americans?
about 0.89%

40. The mass of Earth is $\frac{1}{318}$ of the mass of Jupiter. What percent is this?
about 0.31%

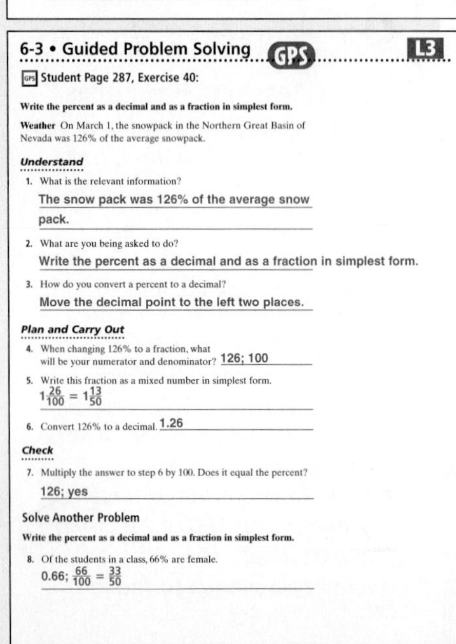

6-3 • Guided Problem Solving GPS L3

Student Page 287, Exercise 40:

Write the percent as a decimal and as a fraction in simplest form.

Weather On March 1, the snowpack in the Northern Great Basin of Nevada was 126% of the average snowpack.

Understand

1. What is the relevant information?
The snow pack was 126% of the average snow pack.
2. What are you being asked to do?
Write the percent as a decimal and as a fraction in simplest form.
3. How do you convert a percent to a decimal?
Move the decimal point to the left two places.

Plan and Carry Out

4. When changing 126% to a fraction, what will be your numerator and denominator? 126; 100
5. Write this fraction as a mixed number in simplest form.
$1\frac{26}{100} = 1\frac{13}{50}$
6. Convert 126% to a decimal. 1.26

Check

7. Multiply the answer to step 6 by 100. Does it equal the percent?
126; yes

Solve Another Problem

Write the percent as a decimal and as a fraction in simplest form.

8. Of the students in a class, 66% are female.
0.66; $\frac{66}{100} = \frac{33}{50}$

Check Your Understanding

Write each percent as a decimal and as a fraction in simplest form.

1. 150% **1.5; $\frac{3}{2}$ or $1\frac{1}{2}$**
2. 0.2% **0.002; $\frac{1}{500}$**

Write each decimal or fraction as a percent.

3. 8.25 **825%**
4. $\frac{1}{160}$ **0.625%**
5. **Mental Math** Write 400% as a whole number. **4**
6. **Number Sense** Compare the numerator and denominator of a fraction that is greater than 100%. **The numerator is greater than the denominator.**
7. How can you decide whether a decimal is less than 1%? More than 100%? Give an example for each case. **See left.**

7. Answers may vary. Sample: A decimal that is less than 1% has zeros in the tenths and hundredths place, such as 0.009. A decimal that is greater than 100% has a number other than zero before the decimal point, such as 1.01.

Homework Exercises

For more exercises, see Extra Skills and Word Problems.

GO for Help

For Exercises	See Examples
8–13	1
14–19	2
20–32	3–4

A **Write each percent as a decimal and as a fraction in simplest form.** **8–15. See left.**

8. 180% 9. 130% 10. 175% 11. 345%

12. 240% 13. 452% 14. 0.1% 15. 0.75%
16. 0.09% **0.0009; $\frac{9}{10,000}$** 17. 0.16% **0.0016; $\frac{1}{625}$** 18. 0.5% **0.005; $\frac{1}{200}$** 19. 0.05% **0.0005; $\frac{1}{2,000}$**

8. 1.8; $1\frac{4}{5}$
9. 1.3; $1\frac{3}{10}$
10. 1.75; $1\frac{3}{4}$
11. 3.45; $3\frac{9}{20}$
12. 2.4; $2\frac{2}{5}$
13. 4.52; $4\frac{13}{25}$
14. 0.001; $\frac{1}{1,000}$
15. 0.0075; $\frac{3}{400}$

Write each number as a percent to the nearest hundredth of a percent.

20. $4\frac{3}{4}$ **475%** 21. $1\frac{3}{5}$ **160%** 22. $1\frac{1}{100}$ **101%** 23. $2\frac{29}{50}$ **258%**
24. $3\frac{7}{20}$ **335%** 25. $2\frac{3}{8}$ **237.5%** 26. $\frac{5}{684}$ **0.73%** 27. $\frac{2}{329}$ **0.61%**
28. $\frac{7}{1,000}$ **0.7%** 29. $\frac{1}{400}$ **0.25%** 30. $\frac{3}{500}$ **0.6%** 31. $\frac{7}{998}$ **0.70%**

32. **Social Studies** About 4 of every 804 people in the world are citizens of Canada. Write this number as a percent. Round to the nearest hundredth of a percent. **0.50%**

B

33. **Guided Problem Solving** Interlibrary loans of reference books have increased to $1\frac{9}{10}$ of what they were 15 years ago. If a library loaned 100 books 15 years ago, how many would it loan today? **190 books**
   - What is $1\frac{9}{10}$, written as a decimal?
   - How do you write a decimal as a percent?

34. **Geography** The world's total land area is about 57.9 million square miles. Luxembourg has an area of 999 square miles. What percent of the world's total land area does Luxembourg occupy? **0.001725%**

41–44. See back of book.

Visit: PHSchool.com
Web Code: are-0603

Write each mixed number as a percent.

35. $2\frac{7}{10}$ **270%** **36.** $5\frac{6}{100}$ **506%** **37.** $10\frac{1}{100}$ **1,001%** **38.** $10\frac{3}{20}$ **1,015%**

Write each percent as a decimal and as a fraction in simplest form.

39. Jewelry sales in December were 166% of sales in November. **1.66; $1\frac{33}{50}$**

40. **Weather** On March 1, the snowpack in the Northern Great Basin of Nevada was 126% of the average snowpack. **1.26; $1\frac{13}{50}$**

GPS

Model each percent using one or more 10 × 10 grids. **41–44. See margin.**

41. 175% **42.** 120% **43.** 200% **44.** 0.5%

45. **Science** The number of known living species is about 1.7 million. About 4,500 species are mammals. What percent of known living species are mammals? Round to the nearest hundredth of a percent. **0.26%**

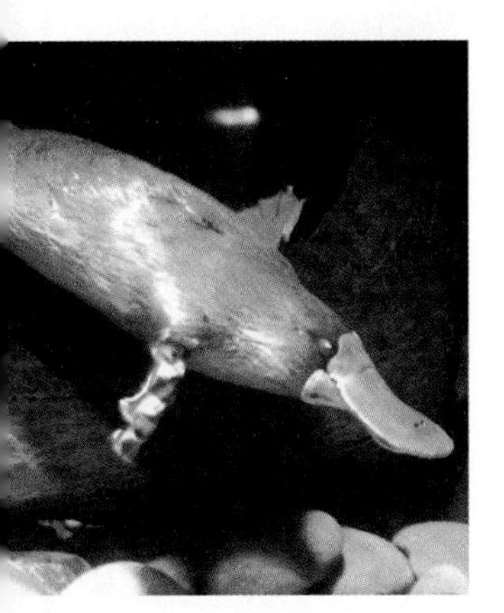

The male platypus has a poisonous spur to use against attackers.

46. **Writing in Math** What does it mean to reach 120% of a savings goal? **Answers may vary. Sample: Your actual savings were 1.2 times your goal.**

Decide whether each percent is reasonable. Explain why or why not.

47. Buttermilk is 105% milk fat. **No; it cannot be more than 100% fat.**

48. Rainfall in Oregon this year is reported to be 160% of the average. **Yes; the rainfall can be 1.6 times the average.**

49. A scientific study concluded that 0.5% of the seeds will not grow. **Yes; it is reasonable that $\frac{1}{2}$ of 1% of the seeds will not grow.**

Ⓒ **50.** **Challenge** Recently, a near-mint copy of the Baltimore Orioles 1966 Yearbook was auctioned for $15. The yearbook cost $.50 in 1966. Write a ratio of the auction price to the original price. Find the percent. **$\frac{15}{0.5}$, or $\frac{150}{5}$, or 30; 3,000%**

Test Prep and Mixed Review

Practice

Multiple Choice

51. The arctic and antarctic icecaps and glaciers make up about 2.3% of the world's water. Which fraction equals 2.3%? **C**

Ⓐ $\frac{23}{100}$ Ⓑ $\frac{1}{5}$ Ⓒ $\frac{23}{1,000}$ Ⓓ $\frac{23}{10,000}$

52. Which expression is represented by the model below? **F**

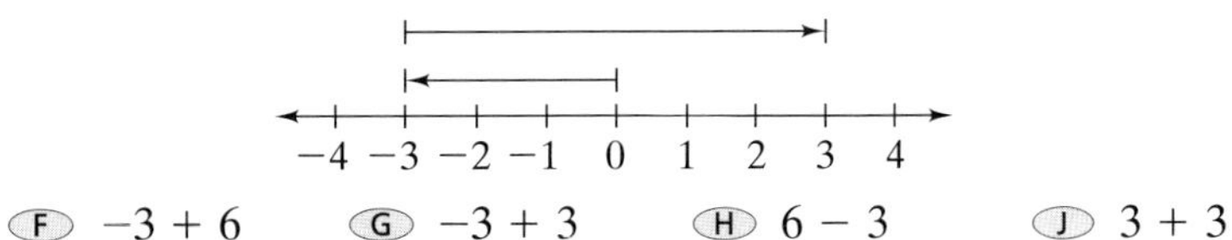

Ⓕ $-3 + 6$ Ⓖ $-3 + 3$ Ⓗ $6 - 3$ Ⓙ $3 + 3$

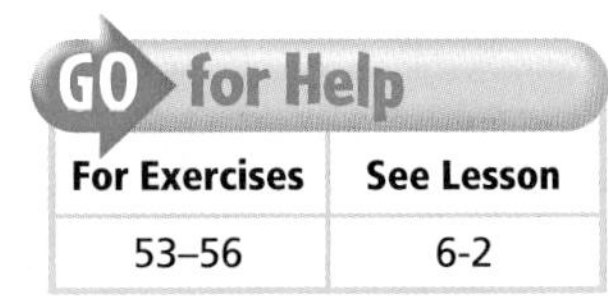

For Exercises	See Lesson
53–56	6-2

Write each percent as a fraction in simplest form.

53. 28% **$\frac{7}{25}$** **54.** 37.5% **$\frac{3}{8}$** **55.** 80% **$\frac{4}{5}$** **56.** 74% **$\frac{37}{50}$**

Alternative Assessment

Each student in a pair writes two percents greater than 100 and two percents less than 1. Partners exchange papers and write each percent as a decimal and as a fraction. Then partners work together to arrange the eight original percents in order from least to greatest.

Test Prep

Resources

For additional practice with a variety of test item formats:

- Test-Taking Strategies, p. 315
- Test Prep, p. 319
- Test-Taking Strategies with Transparencies

4. Assess & Reteach

Lesson Quiz

Write each as a decimal and a fraction or mixed number in simplest form.

1. 112% **1.12; $1\frac{3}{25}$**
2. 0.8% **0.008; $\frac{1}{125}$**

Write each as a percent.

3. $3\frac{1}{2}$ **350%**
4. 0.0045 **0.45%**

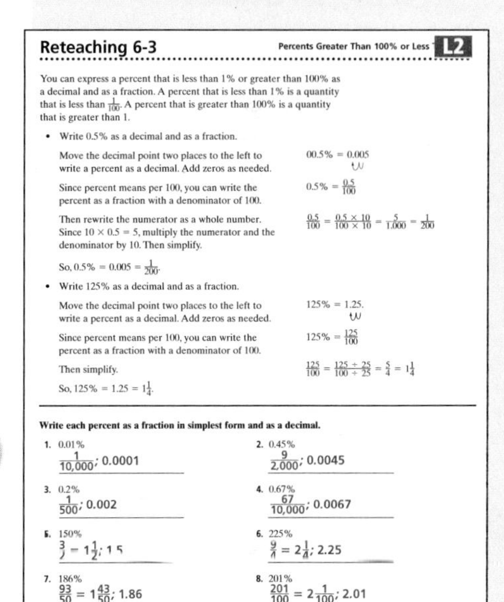

Reteaching 6-3 Percents Greater Than 100% or Less L2

You can express a percent that is less than 1% or greater than 100% as a decimal and as a fraction. A percent that is less than 1% is a quantity that is less than $\frac{1}{100}$. A percent that is greater than 100% is a quantity that is greater than 1.

- Write 0.5% as a decimal and as a fraction.

  Move the decimal point two places to the left to write a percent as a decimal. Add zeros as needed. $00.5\% = 0.005$

  Since percent means per 100, you can write the percent as a fraction with a denominator of 100. $0.5\% = \frac{0.5}{100}$

  Then rewrite the numerator as a whole number. Since $10 \times 0.5 = 5$, multiply the numerator and the denominator by 10. Then simplify. $\frac{0.5}{100} = \frac{0.5 \times 10}{100 \times 10} = \frac{5}{1,000} = \frac{1}{200}$

  So, $0.5\% = 0.005 = \frac{1}{200}$.

- Write 125% as a decimal and as a fraction.

  Move the decimal point two places to the left to write a percent as a decimal. Add zeros as needed. $125\% = 1.25.$

  Since percent means per 100, you can write the percent as a fraction with a denominator of 100. $125\% = \frac{125}{100}$

  Then simplify. $\frac{125}{100} = \frac{125 \div 25}{100 \div 25} = \frac{5}{4} = 1\frac{1}{4}$

  So, $125\% = 1.25 = 1\frac{1}{4}$.

Write each percent as a fraction in simplest form and as a decimal.

1. 0.01% $\frac{1}{10,000}$; 0.0001
2. 0.45% $\frac{9}{2,000}$; 0.0045
3. 0.2% $\frac{1}{500}$; 0.002
4. 0.67% $\frac{67}{10,000}$; 0.0067
5. 150% $\frac{3}{2} = 1\frac{1}{2}$; 1.5
6. 225% $\frac{9}{4} = 2\frac{1}{4}$; 2.25
7. 186% $\frac{93}{50} = 1\frac{43}{50}$; 1.86
8. 201% $\frac{201}{100} = 2\frac{1}{100}$; 2.01

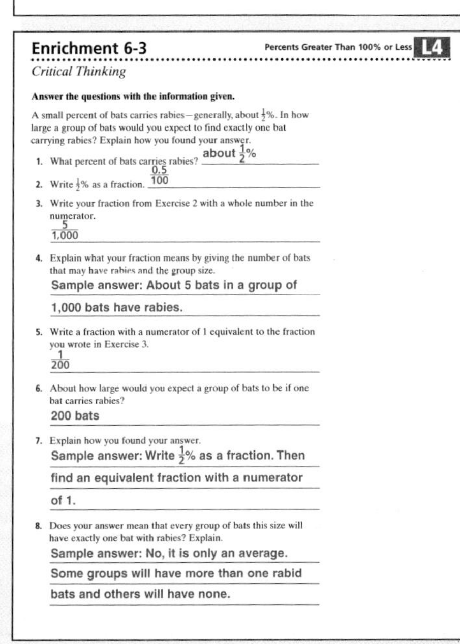

Enrichment 6-3 Percents Greater Than 100% or Less L4

Critical Thinking

Answer the questions with the information given.

A small percent of bats carries rabies—generally, about $\frac{1}{2}$%. In how large a group of bats would you expect to find exactly one bat carrying rabies? Explain how you found your answer.

1. What percent of bats carries rabies? about $\frac{1}{2}$%
2. Write $\frac{1}{2}$% as a fraction. $\frac{0.5}{100}$
3. Write your fraction from Exercise 2 with a whole number in the numerator. $\frac{5}{1,000}$
4. Explain what your fraction means by giving the number of bats that may have rabies and the group size. Sample answer: About 5 bats in a group of 1,000 bats have rabies.
5. Write a fraction with a numerator of 1 equivalent to the fraction you wrote in Exercise 3. $\frac{1}{200}$
6. About how large would you expect a group of bats to be if one bat carries rabies? 200 bats
7. Explain how you found your answer. Sample answer: Write $\frac{1}{2}$% as a fraction. Then find an equivalent fraction with a numerator of 1.
8. Does your answer mean that every group of bats this size will have exactly one bat with rabies? Explain. Sample answer: No, it is only an average. Some groups will have more than one rabid bats and others will have none.

Checkpoint Quiz

Use this Checkpoint Quiz to check students' understanding of the skills and concepts of Lessons 6-1 through 6-3.

Resources

- All-in-One Teaching Resources Checkpoint Quiz 1
- ExamView CD-ROM
- Success Tracker™ Online Intervention

Order, Please!

In this game, students practice comparing and ordering fractions, decimals, and percents. Each player receives a number and the players must order themselves from least to greatest.

Guided Instruction

Before they play the game, remind students that they may need to convert their numbers to different forms in order to figure out the order.

Ask:

- *Is a mixed number greater than or less than 100%?* **greater than**
- *Is a decimal that is less than 1 greater than or less than 100%?* **less than**

Resources

- construction paper

Checkpoint Quiz 1

Lessons 6-1 through 6-3

Write each percent as a decimal and as a fraction in simplest form.

1. 45% **0.45; $\frac{9}{20}$**
2. 135% **1.35; $1\frac{7}{20}$**
3. 0.98% **0.0098; $\frac{49}{5{,}000}$**
4. Write $\frac{14}{25}$ as a percent. **56%**
5. Order 0.245, $\frac{1}{6}$, 20%, and $\frac{1}{4}$ from least to greatest. **$\frac{1}{6}$, 20%, 0.245, $\frac{1}{4}$**
6. Write a ratio and a percent for the shaded area below. **$\frac{16}{25}$; 64%**

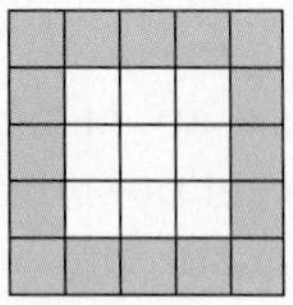

7.

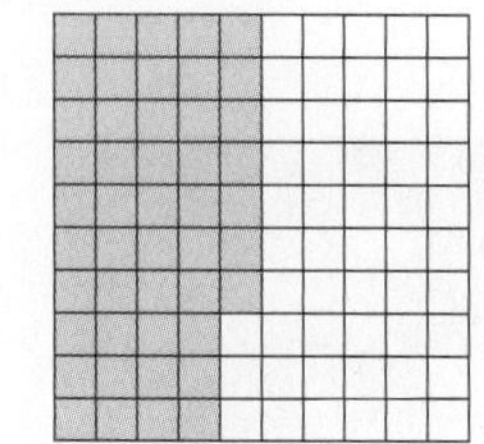

7. Model 47% on a 10 × 10 grid.
8. One hour is what percent of one week? Round to the nearest tenth of a percent. **about 0.6%**
9. A club has 100 members. Five of the members are officers. Each officer gets six other members to help them decorate for a club party. What percent of the club members help decorate? **35%**
10. You walked to school on 135 days out of 180 days. What percent of the days did you walk? Round to the nearest tenth of a percent. **75%**

MATH GAMES — Order, Please!

What You'll Need

- 30 pieces of construction paper, each with a fraction, decimal, or a percent written on it. Include mixed numbers or the equivalent decimals and percents.

How To Play

- Select two teams of five players. Each player receives one piece of construction paper.
- When play begins, team members must order their numbers from the least to the greatest number.
- The first team to order their numbers correctly is the winner.

Data Analysis

Using Percent Data in a Graph

You can use percent data in a graph to interpret and understand the information that is shown.

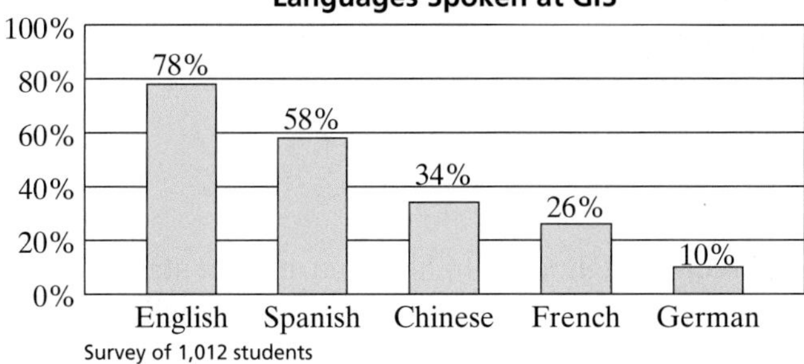

The bar graph at the left shows which languages are spoken at Gould International School.

ACTIVITY

Work with a partner to analyze the graph and answer the following questions.

1. What language do most of the students speak? **English**
2. How many students were surveyed? **1,012 students**
3. Copy and complete the table at the right. Estimate the number of students who speak each language. **See margin.**
4. What language do about $\frac{1}{3}$ of the students speak? **Chinese**
5. What language do about $\frac{1}{4}$ of the students speak? **French**
6. Why do you think the sum of the percents is greater than 100%? **Answers may vary. Sample: Some students speak more than one language, so they were counted more than once.**

Language	Percent	Number of Students
English	78%	■
Spanish	■	■
Chinese	■	■
French	■	■
German	10%	~101

Exercises

Based on the graph above, tell whether each of the following statements is reasonable. Explain.

1. More than half of the students at Gould speak English or Spanish. **Yes; both 78% and 58% are greater than 50%.**
2. Everyone who speaks French also speaks German. **No; there are no data about which students speak both French and German.**
3. Exactly 20 students speak English but not Spanish. **No; there are more than 200 more students who speak English than Spanish.**
4. There could be a student at Gould International School who speaks all five languages shown in the bar graph. **Yes; the total is more than 100%, so some students speak more than one language.**

3. See back of book.

Activity Lab

Using Percent Data in a Graph

Students analyze data in a graph that is in the form of percents. This will prepare them to compare percents in Lesson 6-4.

Guided Instruction

Before beginning the Activity, review the information in the graph. You may wish to draw the graph on the board. Ask:

- *What does having the longest bar mean for that language?* **Most students speak that language.**
- *Do the percents on the bars add to more than 100%?* **yes**
- *Why is that?* **Some students speak more than one language.**
- *What important pieces of information does the graph tell us?* **number of students at the school, percent of total students that speak each language**

Activity

Have students work together to answer the questions. Circulate and check students' work. Discuss the Exercises as a class explaining why the statements are reasonable or unreasonable.

Alternative Method
Students may be more comfortable with another kind of graph. Provide them with a circle graph, pictogram, or other visual data display.

Resources

- Activity Lab 6-4: Decision Making

6-4

1. Plan

Objective
To find and estimate the percent of a number

Examples
1 Finding Percent of a Number
2 Using Mental Math
3 Estimating a Percent

Math Understandings: p. 272C

Math Background

To find a percent of a number, simply multiply the number by the decimal or fraction equivalent of the percent. In word problems, it is helpful to know that the word *of* indicates multiplication.

More Math Background: p. 272C

Lesson Planning and Resources

See p. 272E for a list of the resources that support this lesson.

Check Skills You'll Need
Use student page, transparency, or PowerPoint. For intervention, direct students to:
Order of Operations and the Distributive Property
Lesson 1-9
Extra Skills and Word Problems Practice, Ch. 1

6-4 Finding a Percent of a Number

Check Skills You'll Need

1. **Vocabulary Review** The *Distributive Property* combines the operation of _?_ with addition or subtraction. **multiplication**

Find each product using the Distributive Property and mental math.

2. 80 · 15 **1,200**
3. 75 · 18 **1,350**
4. 78 · 0.9 **70.2**
5. 27 · 1.1 **29.7**

Lesson 1-9

What You'll Learn

To find and estimate the percent of a number

Why Learn This?

You can use a percent of a number to help you analyze statistics, such as how many students in each grade participate in school activities.

Suppose your choir has 44 students, and 25% of the students are in eighth grade. Then 25% of 44 is the number of eighth-graders in the choir.

To find 25% of 44, you can write 25% as a decimal or as a fraction, and then multiply by 44.

EXAMPLE Finding Percent of a Number

1 Find 25% of 44.

For help in writing a percent as a decimal, go to Lesson 6-2, Example 2.

Method 1 Write the percent as a decimal.

$25\% = 0.25$ ← **Change 25% to an equivalent form.** →

$0.25 \cdot 44 = 11$ ← **Multiply.** →

25% of 44 is 11.

Method 2 Write the percent as a fraction.

$25\% = \frac{1}{4}$

$\frac{1}{4} \cdot 44 = 11$

Quick Check

1. Find 75% of 140. **105**

Differentiated Instruction Solutions for All Learners

Special Needs L1
In Example 3, check for students who might write $\frac{3}{2}$ or $\frac{2}{3}$ for 32%. To help them, provide a list of commonly used benchmark fraction, percent, and decimal equivalencies.

learning style: visual

Below Level L2
Have students practice basic percents for mental math.

10% of 65 **6.5**	50% of 22 **11**
10% of 349 **34.9**	50% of 640 **320**
10% of 784 **78.4**	50% of 584 **292**

learning style: visual

You can use mental math with some percents.

- 100% of a number is the number itself. 100% of 190 is 190.
- 50% of a number is $\frac{1}{2}$ of the number. 50% of 190 is 95.
- 10% of a number is 0.1 of the number. 10% of 190 is 19.
- 1% of a number is 0.01 of the number. 1% of 190 is 1.9.

EXAMPLE Using Mental Math

2 Find 11% of 840.

What You Think

11% = 10% + 1%.

10% of 840 is 0.1 · 840, or 84.

1% of 840 is 0.01 · 840, or 8.4.

So 84 + 8.4 = 92.4.

Why It Works

$$\begin{aligned} 11\% \text{ of } 840 &= 0.11 \cdot 840 && \leftarrow \text{Write 11\% as a decimal.} \\ &= (0.10 + 0.01) \cdot 840 && \leftarrow \text{Substitute } 0.10 + 0.01 \text{ for } 0.11. \\ &= (0.10 \cdot 840) + (0.01 \cdot 840) && \leftarrow \text{Use the Distributive Property.} \\ &= 84 + 8.4 && \leftarrow \text{Multiply.} \\ &= 92.4 && \leftarrow \text{Add.} \end{aligned}$$

Quick Check

2. Use mental math to find 40% of 2,400. **960**

For help with estimating using compatible numbers, go to Lesson 1-1, Example 3.

You can use compatible numbers to estimate a percent.

EXAMPLE Estimating a Percent

3 **Elections** The candidate you voted for received 32% of the votes in an election. If 912 votes were counted, about how many votes did your candidate receive?

$$\begin{aligned} &32\% \cdot 912 && \leftarrow \text{Write an expression.} \\ &\downarrow \quad\ \ \downarrow \\ &\tfrac{1}{3} \cdot 900 = 300 && \leftarrow \text{Use compatible numbers such as } \tfrac{1}{3} \text{ and } 900. \end{aligned}$$

Your candidate received about 300 votes.

Quick Check

3. Estimate each answer.
   a. 24% of 238 **about 60**
   b. 19% of 473 **about 100**
   c. 82% of 747 **about 600**

Advanced Learners L4

Which expression will *not* help you find the price of a $60 item on sale for 25% off? **D**

A 60 − 0.25(60) B 0.75(60)
C 60(1 − 0.25) D 60 − 0.75(60)

learning style: visual

English Language Learners ELL

For Example 1, ask students to solve the problem in different ways. Ask them what fraction 25% is equal to, then ask them to mentally calculate what $\frac{1}{4}$ of 44 is. Finally, have them draw it.

learning style: visual

2. Teach

Activity Lab

Use before the lesson.
Student Edition Activity Lab, Data Analysis 6-4a, Using Percent Data in a Graph, p. 289

All in One Teaching Resources
Activity Lab 6-4: Decision Making

Guided Instruction

Example 3
Remind students that compatible numbers are numbers that are easy to divide. For example, the numbers 20 and 4 are compatible but 20 and 6 are not.

PowerPoint

Additional Examples

1 Find each answer.
a. 15% of 40 **6**
b. 80% of 460 **368**

2 Find each answer using mental math.
a. 49% of 300 **147**
b. 11% of 720 **79.2**

3 Estimate each answer.
a. 76% of 405 **300**
b. 12% of 5,575 **560**
c. 48% of 925 **450**

All in One Teaching Resources
- Daily Notetaking Guide 6-4 L3
- Adapted Notetaking 6-4 L1

Closure

- *How can you find a percent of a number?* **Sample: Change the percent into either a fraction or a decimal. Then multiply and simplify the product.**
- *How can you use mental math to find a percent of a number?* **Sample: Use the Distributive Property.**
- *How can you estimate a percent of a number?* **Sample: Use compatible numbers.**

3. Practice

Assignment Guide

Check Your Understanding

Go over Exercises 1–9 in class before assigning the Homework Exercises.

Homework Exercises

A	Practice by Example	10–34
B	Apply Your Skills	35–44
C	Challenge	45
Test Prep and Mixed Review		46–50

Homework Quick Check

To check students' understanding of key skills and concepts, go over Exercises 24, 31, 41, 43, and 44.

Differentiated Instruction Resources

Adapted Practice 6-4 L1

Practice 6-4 Finding a Percent of L3

Find each answer.

1. 20% of 560 — 112
2. 42% of 200 — 84
3. 9% of 50 — 4.5
4. 40% of 70 — 28
5. 25% of 80 — 20
6. 50% of 80 — 40
7. 40% of 200 — 80
8. 5% of 80 — 4
9. 75% of 200 — 150

Find each answer using mental math.

10. 14% of 120 — 16.8
11. 30% of 180 — 54
12. 62.5% of 24 — 15
13. 34% of 50 — 17
14. 25% of 240 — 60
15. 85.5% of 23 — 19.665
16. 120% of 56 — 67.2
17. 80% of 90 — 72
18. 42% of 120 — 50.4

Solve.

19. A farmer grew a watermelon that weighed 20 lb. From his experience with growing watermelons, he estimated that 95% of the watermelon's weight is water.
    a. How much of the watermelon is water? 19 lb
    b. How much of the watermelon is not water? 1 lb
    c. The watermelon was shipped off to market, where it sat until it had dehydrated (lost water). If the watermelon is still 90% water, what percent of it is not water? 10%
    d. The solid part of the watermelon still weighs the same. What was the weight of the watermelon at this point? 18.1 lb
20. A bicycle goes on sale at 75% of its original price of $160. What is its sale price? $120

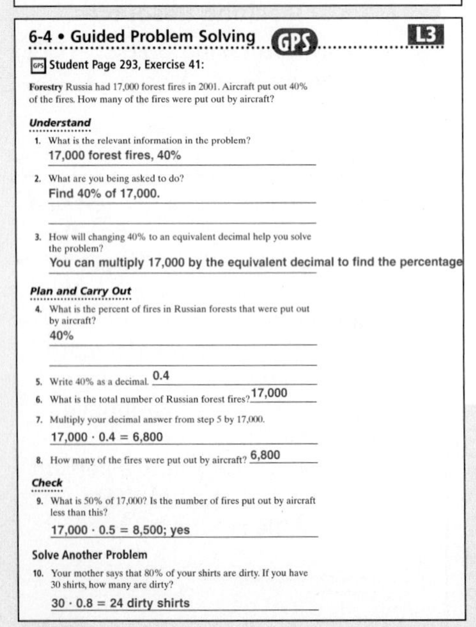
6-4 • Guided Problem Solving GPS L3

GPS Student Page 293, Exercise 41:

Forestry Russia had 17,000 forest fires in 2001. Aircraft put out 40% of the fires. How many of the fires were put out by aircraft?

Understand

1. What is the relevant information in the problem? 17,000 forest fires, 40%
2. What are you being asked to do? Find 40% of 17,000.
3. How will changing 40% to an equivalent decimal help you solve the problem? You can multiply 17,000 by the equivalent decimal to find the percentage

Plan and Carry Out

4. What is the percent of fires in Russian forests that were put out by aircraft? 40%
5. Write 40% as a decimal. 0.4
6. What is the total number of Russian forest fires? 17,000
7. Multiply your decimal answer from step 5 by 17,000. $17{,}000 \cdot 0.4 = 6{,}800$
8. How many of the fires were put out by aircraft? 6,800

Check

9. What is 50% of 17,000? Is the number of fires put out by aircraft less than this? $17{,}000 \cdot 0.5 = 8{,}500$; yes

Solve Another Problem

10. Your mother says that 80% of your shirts are dirty. If you have 30 shirts, how many are dirty? $30 \cdot 0.8 = 24$ dirty shirts

Check Your Understanding

Use mental math, paper and pencil, or a calculator to find each answer.

1. 58% of 50 — 29
2. 8% of 400 — 32
3. 48% of 121 — 58.08
4. 10% of 70 — 7

Estimation **Write an expression and estimate each answer.**

5. 19% of 63 — $\frac{1}{5} \cdot 60 = 12$
6. 73% of 80 — $\frac{3}{4} \cdot 80 = 60$
7. 15% of 39 — $\frac{3}{20} \cdot 40 = 6$
8. What is 12% of 100? What is n% of 100? 12; n
9. **Reasoning** When would you use a decimal to find a percent of a number? When would you use a fraction instead of a decimal? See left.

9. Answers may vary. Sample: You would probably use a fraction when the percent can be written as a fraction that is compatible with the other number. You would use a decimal for all other cases.

Homework Exercises

For more exercises, see Extra Skills and Word Problems.

GO for Help

For Exercises	See Examples
10–18	1
19–27	2
28–34	3

A **Find each answer.**

10. 6% of 90 — 5.4
11. 20% of 80 — 16
12. 27% of 120 — 32.4
13. 12% of 230 — 27.6
14. 75% of 240 — 180
15. 15% of 45 — 6.75
16. 3% of 12 — 0.36
17. 150% of 17 — 25.5
18. 7% of 300 — 21

Mental Math **Find each answer using mental math.**

19. 11% of 520 — 57.2
20. 9% of 780 — 70.2
21. 50% of 948 — 474
22. 40% of 216 — 86.4
23. 51% of 840 — 428.4
24. 100% of 194 — 194
25. 60% of 520 — 312
26. 49% of 150 — 73.5
27. 90% of 345 — 310.5

Estimate each answer.

28. 27% of 162 — about 40
29. 33% of 88 — about 30
30. 53% of 721 — about 360
31. 19% of 399 — about 80
32. 98% of 65 — about 64
33. 73% of 522 — about 400
34. There are 75 students at tryouts for the basketball team. Of this number, about 65% are in the seventh grade. About how many seventh-grade students are trying out for the team? about 49 seventh-graders

B

35. **Guided Problem Solving** Of 90 coins in a piggy bank, 20% are quarters. What is the least possible amount of money in the bank? $5.22
    - How many coins in the bank are quarters?
    - What is the value of the coins that are quarters?
    - What is the least possible value of the coins that are not quarters?
36. **Sales** The regular price of a calculator is $15.98. Today you can buy it on sale for 70% of the regular price. Estimate the sale price. about $11.00

43. Yes; $0.03 \times 96 = 2.88$ and $0.96 \times 3 = 2.88$.

44a. Ceramics: $23.\overline{3}$%; Typing: $11.\overline{3}$%; Cooking: $45.\overline{3}$%; Others: 20%

b. Ceramics: about 583 students; Typing: about 283 students; Cooking: about 1,133 students; Others: about 500 students

Homework Video Tutor
Visit: PHSchool.com
Web Code: are-0604

First estimate. Then check your estimate by finding the answer.

37. 66% of 243 **160.38** **38.** 48% of 658 **315.84** **39.** 13% of 326 **42.38**

40. A nurse earning an annual salary of $39,235 gets a 4% raise. What is the amount of the raise? **$1,569.40**

41. Forestry Russia had 17,000 forest fires in 2001. Aircraft put out GPS 40% of the fires. How many of the fires were put out by aircraft? **6,800 forest fires**

42. You take a test with 25 questions on it. Your grade on the test is 84%. How many questions do you get right? **21 questions**

43. Writing in Math Is 3% of 96 the same value as 96% of 3? Explain. **See margin.**

44. Data Analysis At a high school, 150 students are surveyed about their electives.

a. Find the percent of the students surveyed taking each elective. **44a–b. See margin.**

b. **Estimation** The school population is about 2,500. Estimate the number of students taking each elective.

What Elective Do You Take?

Elective	Number of Students
Ceramics	35
Typing	17
Cooking	68
Others	30

C 45. Challenge The number of students in this year's class is 110% of the number in last year's class. If last year's class had 260 students, how many more students are in this year's class? **26 students**

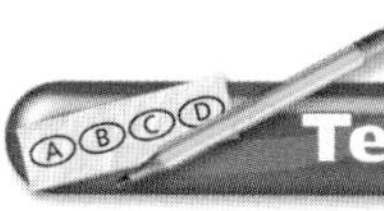

Test Prep and Mixed Review

Practice

Multiple Choice

46. At a sale, everything is 70% of the original price. Marc bought a jacket originally priced at $58.90 and a pair of jeans originally priced at $29.95. Which is the best estimate of the total cost? **B**

Ⓐ $30 Ⓑ $60 Ⓒ $90 Ⓓ $120

47. Which model best represents the expression $\frac{1}{2} \times \frac{2}{5}$? **H**

Ⓕ 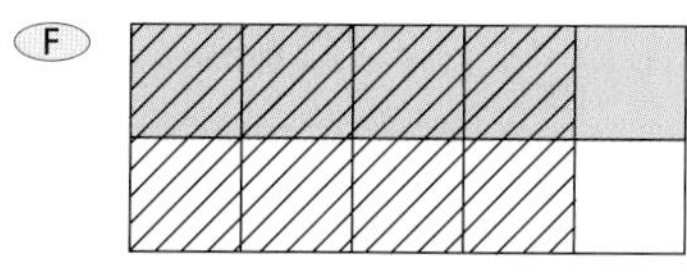

Ⓗ 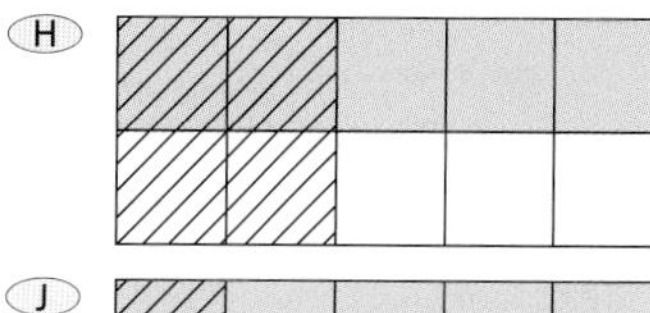

Ⓖ 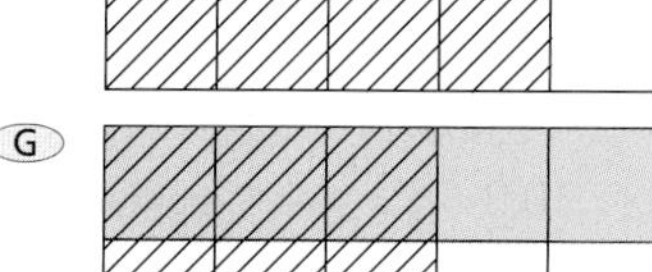

Ⓙ 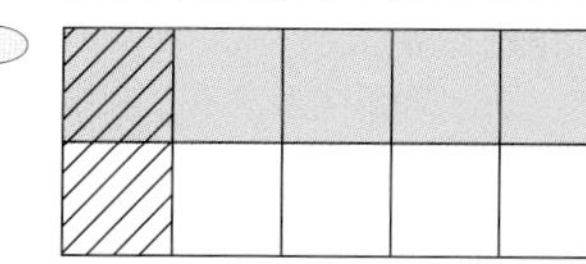

GO for Help

For Exercises	See Lesson
48–50	5-4

Algebra **Solve each proportion.**

48. $\frac{13}{39} = \frac{n}{60}$ **20** **49.** $\frac{7}{15} = \frac{28}{m}$ **60** **50.** $\frac{21}{x} = \frac{5}{8}$ **33.6**

4. Assess & Reteach

PowerPoint

Lesson Quiz

Find the percent of each number.

1. 40% of 90 **36**
2. 120% of 45 **54**
3. 3.5% of 700 **24.5**
4. 90% of 650 **585**

Alternative Assessment

Have students work in pairs on Exercises 10–18 as you randomly write the answers on the board. Have pairs check their work by matching their answers to those on the board. Direct students to show their work.

Reteaching 6-4 Finding a Percent of a **L2**

	Find 12% of 50.	Find 150% of 90.
① Write the percent as a decimal.	0.12	1.5
② Multiply.	0.12 · 50 = 6	1.5 · 90 = 135
	12% of 50 is 6.	150% of 90 is 135.

Complete to find each answer.

1. 15% of 80
   15% = **0.15**
   0.15 · 80 = **12**
2. 4% of 70
   4% = **0.04**
   0.04 · 70 = **2.8**
3. 70% of 20
   70% = **0.7**
   0.7 · 20 = **14**

Find each answer.

4. 10% of 80 **8**
5. 20% of 80 **16**
6. 50% of 80 **40**
7. 9% of 70 **6.3**
8. 2% of 66 **1.32**
9. 28% of 50 **14**
10. 16% of 35 **5.6**
11. 94% of 22 **20.68**
12. 33% of 50 **16.5**
13. 120% of 30 **36**
14. 110% of 70 **77**
15. 160% of 200 **320**
16. 145% of 78 **113.1**
17. 187% of 40 **74.8**
18. 164% of 350 **574**

Solve.

19. Pablo's weekly salary is $105. Each week he saves 60% of his salary. How much does he save each week? **$63**
20. The sixth-grade class is selling magazine subscriptions to raise money for charity. They will give 55% of the money they raise to the homeless. If they raise $2,670, how much do they give to the homeless? **$1,468.50**

Enrichment 6-4 Finding a Percent of a **L4**

Patterns in Data

You can graph some percents and use the graph to estimate other values. Use these interest rates and principal amounts to find the annual amount of interest due on each loan. Then graph your results.

1. You borrow money and pay 15% interest. Solve to find the interest due for each loan amount.
   a. 15% of 100 **15**
   b. 15% of 80 **12**
   c. 15% of 40 **6**
   d. 15% of 60 **9**
2. Graph each of your solutions from exercise 1. Connect the points. Label the line 15%.

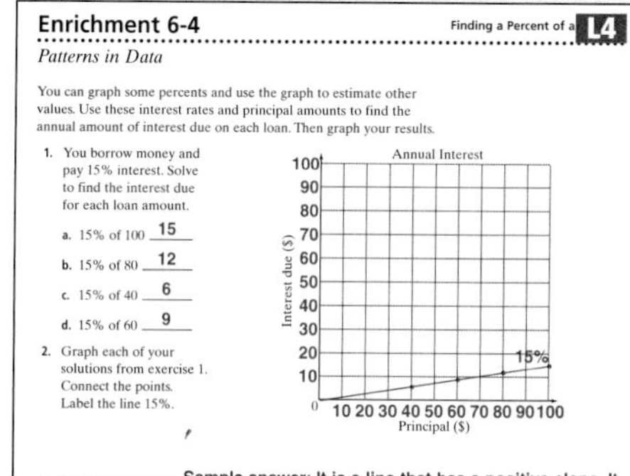

3. Describe the graph. **Sample answer: It is a line that has a positive slope. It does not have a steep slope.**
4. How can you use the graph to estimate 15% of 65.2? **Sample answer: Find the point on the graph that is above 65.2. Interest is the corresponding value on the vertical axis. The interest is about $10.00.**
5. Why might it be helpful to graph this information? **Sample answer: To help you decide if you can afford the interest on a specific loan.**
6. Choose another interest rate and find the interest for these principal amounts. Then graph your solutions. **Check students' answers.**
   a. ______ of 100 ______ b. ______ of 80 ______
   c. ______ of 40 ______ d. ______ of 60 ______
7. Use the graph for the interest rate you chose to predict the annual interest due for a loan of $25. **Check students' answers.**

Test Prep

Resources

For additional practice with a variety of test item formats:

- Test-Taking Strategies, p. 315
- Test Prep, p. 319
- Test-Taking Strategies with Transparencies

6-5

1. Plan

Objective
To use proportions to solve problems involving percent

Examples
1 Finding a Percent
2, 3 Finding a Part and the Whole

Math Understandings: p. 272D

Math Background

A proportion is an equation stating that two ratios are equal. Because the ratios part to whole and percent to 100 are equal, they form the proportion $\frac{\text{part}}{\text{whole}} = \frac{\text{percent}}{100}$. Proportions can be solved using cross products, which are numerators that result if you multiply denominators to find a common denominator. For $\frac{3}{4} = \frac{9}{12}$, the cross products are $3 \cdot 12 = 4 \cdot 9$, or $36 = 36$. And 36 is the numerator for each fraction when the common denominator is $4 \cdot 12$, or 48.

More Math Background: p. 272D

Lesson Planning and Resources

See p. 272E for a list of the resources that support this lesson.

Bell Ringer Practice

Check Skills You'll Need
Use student page, transparency, or PowerPoint. For intervention, direct students to:
Solving Proportions
Lesson 5-4
Extra Skills and Word Problems Practice, Ch. 5

Algebra

6-5 Solving Percent Problems Using Proportions

Check Skills You'll Need

1. **Vocabulary Review** A *proportion* is an equation stating that two __?__ are equal. **ratios**

Solve each proportion.

2. $\frac{n}{32} = \frac{1}{4}$ **8**
3. $\frac{6}{n} = \frac{2}{5}$ **15**
4. $\frac{7}{8} = \frac{n}{100}$ **87.5**

for Help Lesson 5-4

What You'll Learn

To use proportions to solve problems involving percent

Why Learn This?

Survey and poll results are often reported using percents.

In a survey of 2,000 people in the United States, 204 said they are left-handed. You can use this information to find the percent of people who are left-handed.

You can use a model to help find this percent.

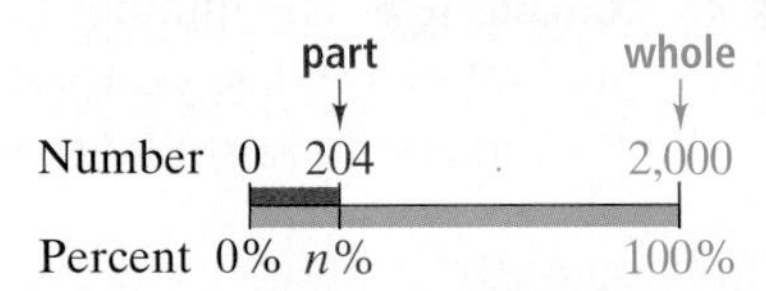

$\frac{204}{2{,}000} = \frac{n}{100}$ ← **The part, 204, corresponds to *n* in the model.** ← **The whole, 2,000, corresponds to 100.**

Online active math

For: Percents Activity
Use: Interactive Textbook, 6-5

EXAMPLE Finding a Percent

1 What percent of 2,000 is 204?

Using the model above, you can write and solve a proportion.

$\frac{204}{2{,}000} = \frac{n}{100}$ ← **Write a proportion.**

$2000n = 204(100)$ ← **Write cross products.**

$\frac{2{,}000n}{2{,}000} = \frac{204(100)}{2{,}000}$ ← **Divide each side by 2,000.**

$n = 10.2$ ← **Simplify.**

204 is 10.2% of 2,000.

Quick Check

1. What percent of 92 is 23? **25%**

Differentiated Instruction Solutions for All Learners

Special Needs L1
Help students draw pictures for each of the examples in this lesson, using hundreds grids. Then help them transition to the percent model.

learning style: visual

Below Level L2
Have students use the proportion $\frac{\text{is}}{\text{of}} = \frac{\text{percent}}{100}$ to solve percent problems such as these.

50% of what number is 75? **150**
What percent of 90 is 18? **20%**

learning style: verbal

EXAMPLES Finding a Part and the Whole

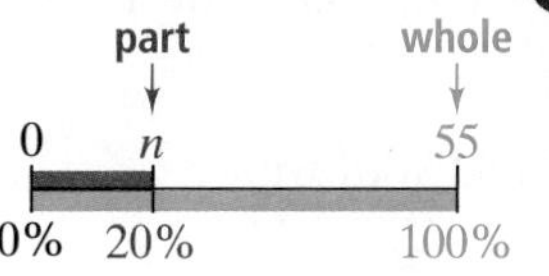

2 20% of 55 is what number?

$\frac{n}{55} = \frac{20}{100}$ ← Write a proportion.

$\frac{n}{55} = \frac{1}{5}$ ← Simplify the fraction.

$\frac{n}{55} = \frac{1}{5}$ (×11) ← Use the common multiplier, 11.

$n = 11$ ← Simplify.

11 is 20% of 55.

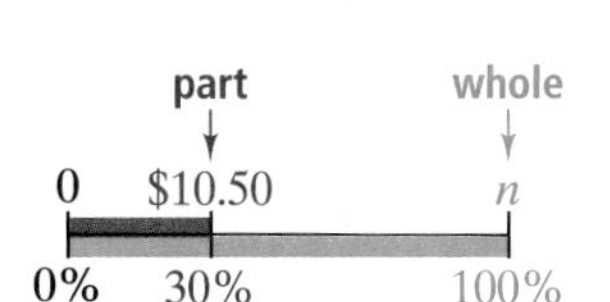

3 **Budgets** Suppose your entertainment budget is 30% of your weekly wages from a job. You plan to spend \$10.50 on a movie night. How much will you need to earn at your job in order to stay within your budget?

$\frac{10.50}{n} = \frac{30}{100}$ ← Write a proportion.

$30n = 10.50(100)$ ← Write cross products.

$\frac{30n}{30} = \frac{10.50(100)}{30}$ ← Divide.

$n = 35$ ← Simplify.

You need to earn \$35 to stay within your budget.

Quick Check

2. 85% of 20 is what number? **17**
3. Your math teacher assigns 25 problems for homework. You have done 60% of them. How many problems have you done? **15 problems**

KEY CONCEPTS Percents and Proportions

Finding a Percent	Finding a Part	Finding a Whole
What percent of 25 is 5?	What is 20% of 25?	20% of what is 5?
0, 5, 25 / 0%, n%, 100%	0, n, 25 / 0%, 20%, 100%	0, 5, n / 0%, 20%, 100%
$\frac{5}{25} = \frac{n}{100}$	$\frac{n}{25} = \frac{20}{100}$	$\frac{5}{n} = \frac{20}{100}$
$n = 20$	$n = 5$	$n = 25$
20% of 25 is 5.	5 is 20% of 25.	20% of 25 is 5.

2. Teach

Activity Lab

Use before the lesson.

All in One Teaching Resources

Activity Lab 6-5: Using Proportions to Find Percents

Guided Instruction

Example 1

Review the concept of cross products by circling the terms of the diagonals, making two ovals. Point out that the ovals connecting the pairs of numbers form a large "X," which is a symbol for multiplication.

Example 3

If students are having trouble writing the proportion for Quick Check 3, have them draw models. Remind them that the whole line segment represents the whole amount that they are looking for.

PowerPoint

Additional Examples

1 What percent of 150 is 45? **30%**

2 35% of 90 is what number? **31.5**

3 117 is 45% of what number? **260**

All in One Teaching Resources

- Daily Notetaking Guide 6-5 L3
- Adapted Notetaking 6-5 L1

Closure

- *What are the three basic types of percent problems?* **finding a part, finding a percent, and finding a whole**
- *How do you use a proportion to solve each type of percent problem?* **Sample: For each type of problem, you use the proportion $\frac{\text{part}}{\text{whole}} = \frac{\text{percent}}{100}$, substitute for two pieces, and solve for the missing piece using cross products.**

Advanced Learners L4

Rob's score on a math test was 88% of Cathy's score. Cathy's score was 97% of Bill's score. Bill's score was 92% of Lynn's score. Lynn's score was 94 out of 100. *What was Rob's score?* **74**

learning style: verbal

English Language Learners ELL

Take some time talking about the scales in the Key Concepts section. Ask students questions like: *Why is the 25 in the same location as the 100%? Why does n change location in the proportions?* Make sure they understand part and whole to prevent errors in setting up proportions.

learning style: visual

3. Practice

Assignment Guide

Check Your Understanding
Go over Exercises 1–7 in class before assigning the Homework Exercises.

Homework Exercises

A Practice by Example	8–24
B Apply Your Skills	25–35
C Challenge	36
Test Prep and Mixed Review	37–43

Homework Quick Check
To check students' understanding of key skills and concepts, go over Exercises 10, 18, 33, 34, and 35.

Differentiated Instruction Resources

Adapted Practice 6-5 L1

Practice 6-5 — Solving Percent Problems Using Pr... L3

Use a proportion to solve.

1. 48 is 60% of what number? 80
2. What is 175% of 85? 148.75
3. What percent of 90 is 50? 55.6%
4. 76 is 80% of what number? 95
5. What is 50% of 42.88? 21.44
6. 96 is 160% of what number? 60
7. What percent of 24 is 72? 300%
8. What is 85% of 120? 102
9. What is 80% of 12? 9.6
10. 56 is 75% of what number? $74\frac{2}{3}$

Solve.

11. The sale price of a bicycle is $120. This is 75% of the original price. Find the original price. $160
12. The attendance at a family reunion was 160 people. This was 125% of last year's attendance. How many people attended the reunion last year? 128 people
13. A company has 875 employees. On "Half-Price Wednesday," 64% of the employees eat lunch at the company cafeteria. How many employees eat lunch at the cafeteria on Wednesdays? 560 employees
14. There are 1,295 students attending a small university. There are 714 women enrolled. What percentage of students are women? about 55%

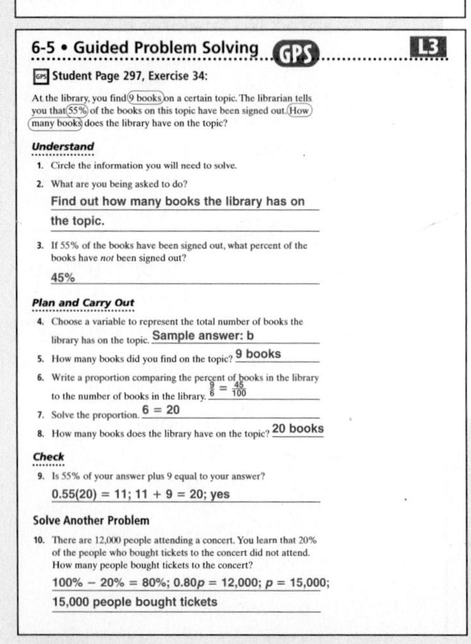

6-5 • Guided Problem Solving GPS L3

Student Page 297, Exercise 34:

At the library, you find 9 books on a certain topic. The librarian tells you that 55% of the books on this topic have been signed out. How many books does the library have on the topic?

Understand

1. Circle the information you will need to solve.
2. What are you being asked to do? Find out how many books the library has on the topic.
3. If 55% of the books have been signed out, what percent of the books have *not* been signed out? 45%

Plan and Carry Out

4. Choose a variable to represent the total number of books the library has on the topic. Sample answer: b
5. How many books did you find on the topic? 9 books
6. Write a proportion comparing the percent of books in the library to the number of books in the library. $\frac{9}{b} = \frac{45}{100}$
7. Solve the proportion. b = 20
8. How many books does the library have on the topic? 20 books

Check

9. Is 55% of your answer plus 9 equal to your answer? 0.55(20) = 11; 11 + 9 = 20; yes

Solve Another Problem

10. There are 12,000 people attending a concert. You learn that 20% of the people who bought tickets to the concert did not attend. How many people bought tickets to the concert? 100% − 20% = 80%; 0.80p = 12,000; p = 15,000; 15,000 people bought tickets

Check Your Understanding

Tell whether the answer to the question is a *percent,* a *part,* or a *whole.* Then answer the question.

1. What percent of 200 is 50? **percent; 25%**
2. 12 is 80% of what? **whole; 15**
3. 50% of what number is 8? **whole; 16**
4. What is 30% of 90? **part; 27**

Match each question with the proportion you could use to answer it.

5. What is 40% of 15? **C**
6. 15 is what percent of 40? **A**
7. 40% of what number is 15? **B**

A. $\frac{15}{40} = \frac{n}{100}$

B. $\frac{15}{n} = \frac{40}{100}$

C. $\frac{n}{15} = \frac{40}{100}$

Homework Exercises

For more exercises, see Extra Skills and Word Problems.

GO for Help

For Exercises	See Examples
8–13	1
14–19	2
20–24	3

A **Use a proportion to find the percent.**

8. 24 is what percent of 32? **75%**
9. What percent of 230 is 23? **10%**
10. What percent of 25 is 23? **92%**
11. 8 is what percent of 400? **2%**
12. What percent of 600 is 84? **14%**
13. 21 is what percent of 168? **12.5%**

Use a proportion to find the part.

14. What is 4% of 350? **14**
15. 1% of 500 is what number? **5**
16. What number is 62% of 50? **31**
17. 15% of 15 is what number? **2.25**
18. 40% of 25 is what number? **10**
19. What is 37.5% of 8? **3**

Test Prep Tip
You can use a model to help you solve percent problems.

Use a proportion to find the whole.

20. 36 is 72% of what number? **50**
21. 80% of what number is 15? **18.75**
22. 21 is 84% of what number? **25**
23. 28 is 35% of what number? **80**
24. A sweater is on sale for $33. This is 75% of the original price. Find the original price. **$44**

B 

25. **Guided Problem Solving** In a market, 44 of the 80 types of vegetables are grown locally. What percent of the vegetables are grown locally? **55%**
    - Identify the part. Identify the whole.
    - Complete and solve the proportion: $\frac{\blacksquare}{\blacksquare} = \frac{\blacksquare}{100}$.
26. A school holds classes from 8:00 A.M. to 2:00 P.M. For what percent of a 24-hour day does this school hold classes? **25%**

Online
Homework Video Tutor
Visit: PHSchool.com
Web Code: are-0605

Write a proportion for each model. Solve for n. 27–30. See margin.

27. 0, 90, n; 0%, 40%, 100%

28. 0, n, 80; 0%, 70%, 100%

29. 0, 54, 144; 0%, n%, 100%

30. 0, 139.1, n; 0%, 65%, 100%

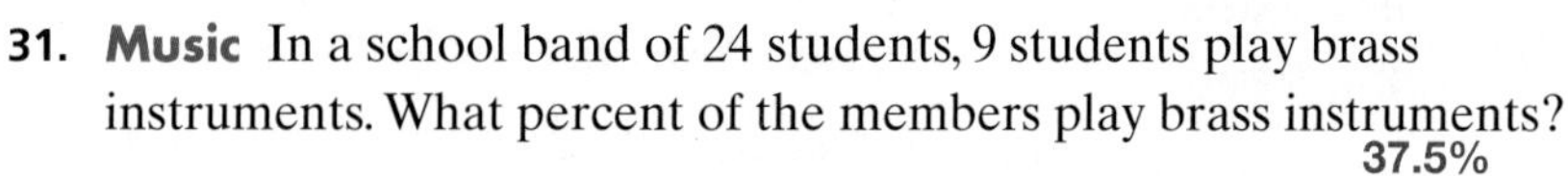

31. **Music** In a school band of 24 students, 9 students play brass instruments. What percent of the members play brass instruments? 37.5%

32. You purchase a telescope in a state with a 5% sales tax. You pay \$14.85 in tax. Estimate the price of the telescope. about \$300

33. **Open-Ended** Write a percent problem that compares the number of boys to the number of girls in your class. Check students' work.

34. GPS At the library, you find 9 books on a certain topic. The librarian tells you that 55% of the books on this topic have been signed out. How many books does the library have on the topic? 20 books

35. **Writing in Math** A proportion that models a percent problem has four numbers. One of the numbers is always the same. Explain why. See left.

35. The number 100 always appears as the denominator of one of the ratios, since percent means "out of 100."

36. **Challenge** A car dealer advertises "All cars 19% off sticker price!" A buyer pays \$15,930.95 for a car. Estimate the sticker price. about \$20,000

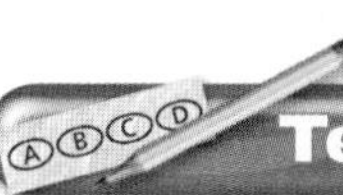

Test Prep and Mixed Review

Practice

Multiple Choice

37. Out of 45 students, 29 go on a field trip. Which best represents the percent of the students who do NOT go on the trip? B
    - (A) 16%
    - (B) 36%
    - (C) 64%
    - (D) 84%

38. In 2001, 56.5% of households in the United States had a computer. Which expression provides the best estimate for the number of households with computers in a survey of 621 households in 2001? H
    - (F) 60% of 650
    - (G) 50% of 600
    - (H) 60% of 600
    - (J) 50% of 650

39. The model below represents $4n + 6 = 18$. What is the value of n? C

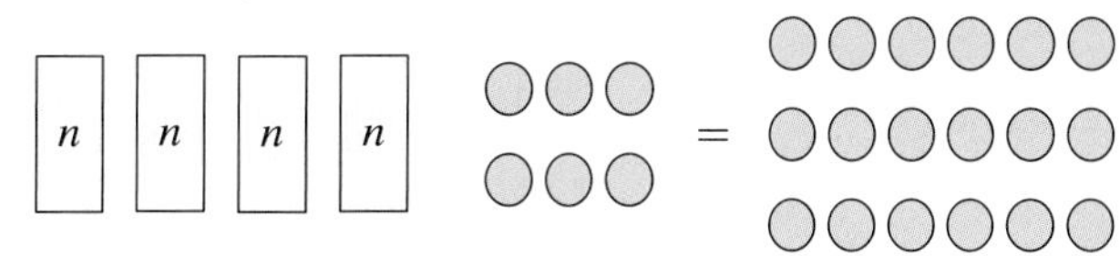

- (A) $n = 24$
- (B) $n = 6$
- (C) $n = 3$
- (D) $n = -3$

GO for Help

For Exercises	See Lesson
40–43	2-3

Write each fraction in simplest form.

40. $\frac{8}{10}$ $\frac{4}{5}$

41. $\frac{4}{12}$ $\frac{1}{3}$

42. $\frac{5}{100}$ $\frac{1}{20}$

43. $\frac{16}{24}$ $\frac{2}{3}$

Online lesson quiz, PHSchool.com, Web Code: ara-0605

27. $\frac{90}{n} = \frac{40}{100}$; 225

28. $\frac{n}{80} = \frac{70}{100}$; 56

29. $\frac{54}{144} = \frac{n}{100}$; 37.5

30. $\frac{139.1}{n} = \frac{65}{100}$; 214

4. Assess & Reteach

Lesson Quiz

Use a proportion to solve.

1. What percent of 240 is 60? 25%
2. 90 is what percent of 120? 75%
3. 75% of what number is 66? 88
4. 24 is 96% of what number? 25

Alternative Assessment

Have students work in pairs. Partners take turns drawing models for Exercises 8–23. Then they work together to complete the exercises.

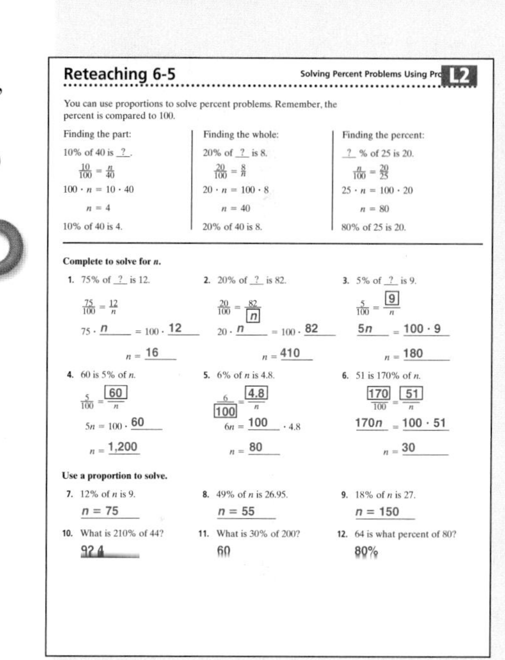

Reteaching 6-5 Solving Percent Problems Using Pro L2

You can use proportions to solve percent problems. Remember, the percent is compared to 100.

Finding the part:	Finding the whole:	Finding the percent:
10% of 40 is _?_.	20% of _?_ is 8.	_?_ % of 25 is 20.
$\frac{10}{100} = \frac{n}{40}$	$\frac{20}{100} = \frac{8}{n}$	$\frac{n}{100} = \frac{20}{25}$
$100 \cdot n = 10 \cdot 40$	$20 \cdot n = 100 \cdot 8$	$25 \cdot n = 100 \cdot 20$
$n = 4$	$n = 40$	$n = 80$
10% of 40 is 4.	20% of 40 is 8.	80% of 25 is 20.

Complete to solve for n.

1. 75% of _?_ is 12.
   $\frac{75}{100} = \frac{12}{n}$
   $75 \cdot$ n $= 100 \cdot$ 12
   $n =$ 16

2. 20% of _?_ is 82.
   $\frac{20}{100} = \frac{82}{\boxed{n}}$
   $20 \cdot$ n $= 100 \cdot$ 82
   $n =$ 410

3. 5% of _?_ is 9.
   $\frac{5}{100} = \frac{\boxed{9}}{n}$
   5n = 100 · 9
   $n =$ 180

4. 60 is 5% of n.
   $\frac{5}{100} = \frac{\boxed{60}}{n}$
   $5n = 100 \cdot$ 60
   $n =$ 1,200

5. 6% of n is 4.8.
   $\frac{6}{\boxed{100}} = \frac{\boxed{4.8}}{n}$
   $6n =$ 100 $\cdot 4.8$
   $n =$ 80

6. 51 is 170% of n.
   $\frac{\boxed{170}}{100} = \frac{\boxed{51}}{n}$
   170n = 100 · 51
   $n =$ 30

Use a proportion to solve.

7. 12% of n is 9. $n = 75$
8. 49% of n is 26.95. $n = 55$
9. 18% of n is 27. $n = 150$
10. What is 210% of 44? 92.4
11. What is 30% of 200? 60
12. 64 is what percent of 80? 80%

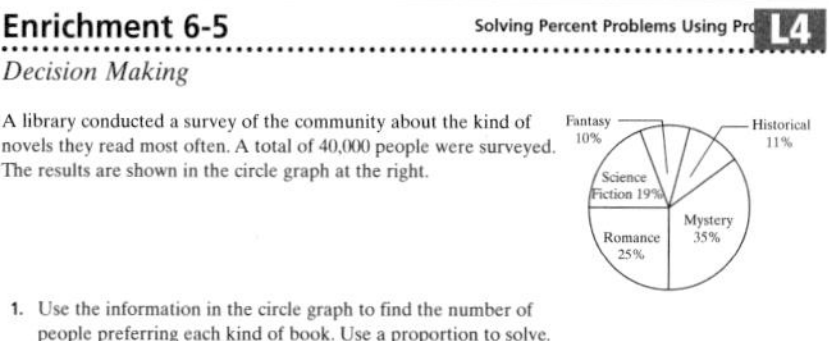

Enrichment 6-5 Solving Percent Problems Using Pro L4

Decision Making

A library conducted a survey of the community about the kind of novels they read most often. A total of 40,000 people were surveyed. The results are shown in the circle graph at the right.

1. Use the information in the circle graph to find the number of people preferring each kind of book. Use a proportion to solve.

Kind of Novel	Percent of Community	Number of People in Community	Budget
Fantasy	10%	4,000	\$25,000
Historical	11%	4,400	\$27,500
Science Fiction	19%	7,600	\$47,500
Romance	25%	10,000	\$62,500
Mystery	35%	14,000	\$87,500

Sample answers: Exercises 2, 3, and 4.

2. Why is a circle graph a good way to show percent information?
   Since the circle represents 100% of the whole, it gives a visual idea of how the data relates to the whole.

3. You can spend \$250,000 on new library books. Write the amount you would spend for each kind of book in the table. Explain.
   Would spend the same percentage as the kind of books preferred.

4. A patron gave the library \$50,000 to expand the children's fiction section. How would you decide which kinds of books to buy? Would you use the survey results above?
   Would use children's books recommended by various national groups. Survey results probably are not valid for children, since adults are more likely to be surveyed.

Test Prep

Resources

For additional practice with a variety of test item formats:

- Test-Taking Strategies, p. 315
- Test Prep, p. 319
- Test-Taking Strategies with Transparencies

1. Plan

Objective
To use equations to solve problems involving percent

Examples
1 Finding a Whole
2 Finding a Part
3 Finding a Percent

Math Understandings: p. 272D

Math Background

You can solve many percent problems by directly translating the words into an equation. In these equations, two quantities are known and the third unknown quantity is represented by a variable.

More Math Background: p. 272D

Lesson Planning and Resources

See p. 272E for a list of the resources that support this lesson.

PowerPoint

Bell Ringer Practice

Check Skills You'll Need
Use student page, transparency, or PowerPoint. For intervention, direct students to:
Solving Equations by Multiplying or Dividing
Lesson 4-4
Extra Skills and Word Problems Practice, Ch. 4

Algebra

6-6 Solving Percent Problems Using Equations

Check Skills You'll Need

1. **Vocabulary Review** State the *Division Property of Equality.* **See below.**

Solve each equation.

2. $3n = 51$ **17**
3. $\frac{x}{4} = 12$ **48**

for Help Lesson 4-4

What You'll Learn

To use equations to solve problems involving percent

Why Learn This?

Suppose a ski resort reports that 60% of its trails are open. If you know how many trails are open, you can solve an equation to find the total number of trails in the park.

You can translate percent problems into equations to find parts, wholes, or percents.

GO Online

Video Tutor Help
Visit: PHSchool.com
Web Code: are-0775

Check Skills You'll Need

1. If both sides of an equation are divided by the same nonzero number, the results are equal.

EXAMPLE Finding a Whole

1 **Multiple Choice** A ski resort in New Hampshire begins the season with 60% of its trails open. There are 27 trails open. How many trails does the ski resort have?

Ⓐ 5 Ⓑ 16 Ⓒ 22 Ⓓ 45

Words 60% of the number of trails is 27

Let x = the number of trails at this ski resort.

Equation $0.60 \cdot x = 27$

$0.60x = 27$ ← **Write the equation.**

$\frac{0.60x}{0.60} = \frac{27}{0.60}$ ← **Divide each side by 0.60.**

$x = 45$ ← **Simplify.**

The ski resort has 45 trails. The correct answer is choice D.

Quick Check

1. A plane flies with 54% of its seats empty. If 81 seats are empty, what is the total number of seats on the plane? **150 seats**

Differentiated Instruction Solutions for All Learners

Special Needs L1
For each of the examples, ask students to predict whether the product will be less than or greater than the given number in the problem. For example: *Will the product be greater than 27 in Example 1, greater than 377 in Example 2? Why?*
learning style: verbal

Below Level L2
Use a single set of values to underscore how to use the equations, such as these.
What is 20% of 65? **13**
20% of what number is 13? **65**
What percent of 65 is 13? **20%**
learning style: verbal

You can use an equation to find a whole, a part, or a percent.

EXAMPLE Finding a Part

Test Prep Tip

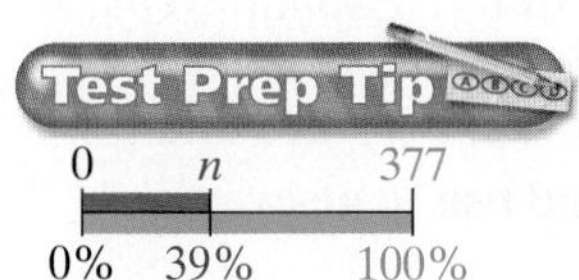

Remember that you can draw a model to help you. This model shows that n is 39% of 377.

2 What number is 39% of 377?

Words A number is 39% of 377

Let n = the number.

Equation $n = 0.39 \cdot 377$

$n = 0.39 \cdot 377 = 147.03$ ← **Simplify.**

Quick Check

2. 27% of 60 is what number? **16.2**

EXAMPLE Finding a Percent

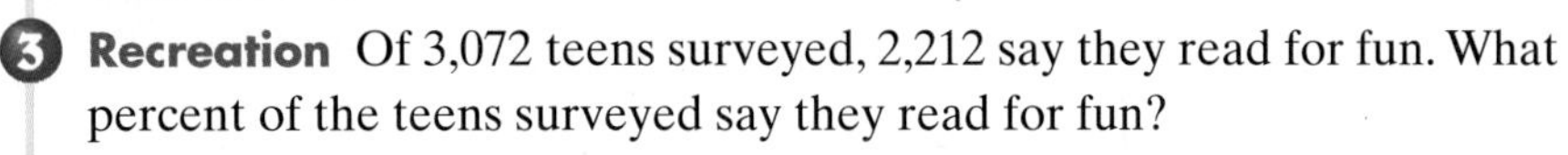

3 **Recreation** Of 3,072 teens surveyed, 2,212 say they read for fun. What percent of the teens surveyed say they read for fun?

Estimate About 2,000 of 3,000 teens read for fun.

$\frac{2,000}{3,000} = \frac{2}{3} \approx 0.67 = 67\%$

$3,072p = 2,212$ ← **Write an equation. Let p = the percent of teens who read for fun.**

$\frac{3,072p}{3,072} = \frac{2,212}{3,072}$ ← **Divide each side by 3,072.**

$p \approx 0.7200520833$ ← **Use a calculator.**

$p \approx 72\%$ ← **Write the decimal as a percent.**

About 72% of the teens surveyed say they read for fun.

Check for Reasonableness 72% is close to the estimate 67%.

Quick Check

3. It rained 75 days last year. About what percent of the year was rainy? **about 20.5%**

KEY CONCEPTS **Percents and Equations**

Finding a Percent	Finding a Part	Finding a Whole
What percent of 25 is 5?	What is 20% of 25?	20% of what is 5?
$n \cdot 25 = 5$	$n = 0.2 \cdot 25$	$0.2 \cdot n = 5$
$n = 0.2$	$n = 5$	$n = 25$
5 is 20% of 25.	5 is 20% of 25.	20% of 25 is 5.

2. Teach

Activity Lab

Use before the lesson.

All in One Teaching Resources

Activity Lab 6-6: Patterns in Data

Guided Instruction

Example 2

Write the "general form" equation on the board in words. Box the parts shown. As a student reads Quick Check 2, have another student write the equation below the sentence as shown.

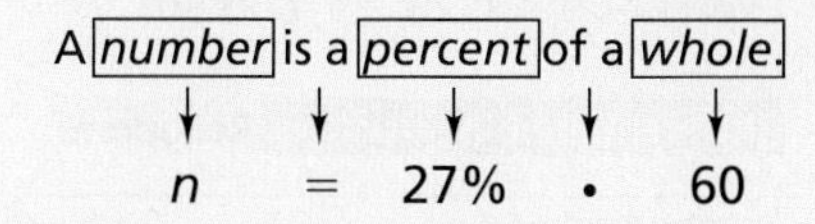

A number is a percent of a whole.
n = 27% • 60

Error Prevention!

Remind students that a percent needs to be changed to a decimal before they do any calculations.

PowerPoint

Additional Examples

1 Use an equation to solve: 18% of what number is 81? **$81 = 18\% \cdot n$; 450**

2 Use an equation to solve: 32% of 40 is what number? **$n = 32\% \cdot 40$; 12.8**

3 Use an equation to solve: What percent of 200 is 45? **$45 = n\% \cdot 200$; 22.5%**

All in One Teaching Resources

- Daily Notetaking Guide 6-6 **L3**
- Adapted Notetaking 6-6 **L1**

Closure

- *How can you solve percent problems with equations?* **Translate the words into an equation that relates the percent, the part, and the whole. Solve for the unknown quantity.**

Advanced Learners L4

Cheryl bought stock for $20 and sold it for $40, saying her stock went up 100%. Mark said she sold it for 200% of its value. Who is right? Explain. **both; $20 gain is 100%; 40 is 200% of 20**

learning style: verbal

English Language Learners ELL

Spend some time translating percent word problems to equations before asking the students to calculate. Students need practice analyzing the word problems and rewording them, as in Example 1, so that they then lead more naturally to an equation.

learning style: verbal

3. Practice

Assignment Guide

Check Your Understanding

Go over Exercises 1–6 in class before assigning the Homework Exercises.

Homework Exercises

A	Practice by Example	7–20
B	Apply Your Skills	21–31
C	Challenge	32
Test Prep and Mixed Review		33–39

Homework Quick Check

To check students' understanding of key skills and concepts, go over Exercises 9, 17, 28, 29, and 30.

Differentiated Instruction Resources

Adapted Practice 6-6 L1

Practice 6-6 Solving Percent Problems Using E... L3

Write and solve an equation. Round answers to the nearest tenth.

1. What percent of 64 is 48? **75%**
2. 16% of 130 is what number? **20.8**
3. 25% of what number is 24? **96**
4. What percent of 18 is 12? **66.7%**
5. 48% of 83 is what number? **39.8**
6. 40% of what number is 136? **340**
7. What percent of 530 is 107? **20.2%**
8. 74% of 643 is what number? **475.8**
9. 62% of what number is 84? **135.5**
10. What percent of 84 is 50? **59.5%**
11. 37% of 245 is what number? **90.7**
12. 12% of what number is 105? **875**

Solve.

13. A cafe offers senior citizens a 15% discount off its regular price of $8.95 for the dinner buffet.
    a. What percent of the regular price is the price for senior citizens? **85%**
    b. What is the price for senior citizens? **$7.61**
14. In 1990, 12.5% of the people in Oregon did not have health insurance. If the population of Oregon was 2,880,000, how many people were uninsured? **360,000 people**

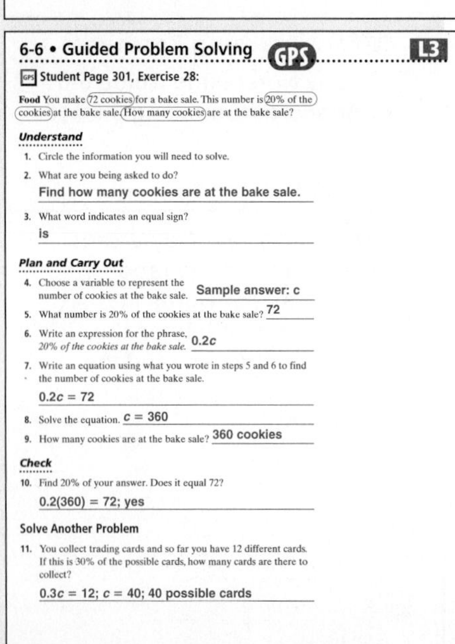

6-6 • Guided Problem Solving GPS L3

GPS Student Page 301, Exercise 28:

Food You make 72 cookies for a bake sale. This number is 20% of the cookies at the bake sale. How many cookies are at the bake sale?

Understand

1. Circle the information you will need to solve.
2. What are you being asked to do? **Find how many cookies are at the bake sale.**
3. What word indicates an equal sign? **is**

Plan and Carry Out

4. Choose a variable to represent the number of cookies at the bake sale. **Sample answer: c**
5. What number is 20% of the cookies at the bake sale? **72**
6. Write an expression for the phrase, *20% of the cookies at the bake sale.* **$0.2c$**
7. Write an equation using what you wrote in steps 5 and 6 to find the number of cookies at the bake sale. **$0.2c = 72$**
8. Solve the equation. **$c = 360$**
9. How many cookies are at the bake sale? **360 cookies**

Check

10. Find 20% of your answer. Does it equal 72? **0.2(360) = 72; yes**

Solve Another Problem

11. You collect trading cards and so far you have 12 different cards. If this is 30% of the possible cards, how many cards are there to collect? **$0.3c = 12$; $c = 40$; 40 possible cards**

Check Your Understanding

1. Do the following questions mean the same thing? Explain. *What is 20% of 40? 20 is what percent of 40? 20 is 40% of what number?* **See left.**

1. No; "20% of 40" asks for a part, "20 is what percent of 40" asks for a percent, and "20 is 40% of what number" asks for a whole.

Match each question with the equation you could use to answer it.

2. What is 16% of 200? **B**
3. 16 is what percent of 200? **C**
4. 16% of what number is 200? **A**

A. $0.16n = 200$
B. $n = 0.16(200)$
C. $16 = 200n$

Write an equation for each question. Then answer the question.

5. What percent of 625 is 500? **$625p = 500$; 80%**
6. What number is 5% of 520? **$n = 0.05 \cdot 520$; 26**

Homework Exercises

For more exercises, see Extra Skills and Word Problems.

GO for Help

For Exercises	See Examples
7–11	1
12–15	2
16–20	3

A Use an equation to find the whole.

7. 96% of what number is 24? **$0.96x = 24$; 25**
8. 40% of what number is 30? **$0.40x = 30$; 75**
9. 50.4 is 36% of what number? **$50.4 = 0.36x$; 140**
10. 12.8 is 32% of what number? **$12.8 = 0.32x$; 40**
11. You answered 22 questions correctly and scored 88% on a test. How many questions were on the test in all? **25 questions**

Use an equation to find a part.

12. 18% of 90 is what number? **$0.18 \cdot 90 = x$; 16.2**
13. What number is 41% of 800? **$x = 0.41 \cdot 800$; 328**
14. What number is 56% of 48? **$x = 0.56 \cdot 48$; 26.88**
15. 70% of 279 is what number? **$0.70 \cdot 279 = x$; 195.3**

Use an equation to find the percent.

16. What percent of 496 is 124? **$496x = 124$; 25%**
17. 18 is what percent of 48? **$18 = 48x$; 37.5%**
18. 39 is what percent of 260? **$39 = 260x$; 15%**
19. What percent of 620 is 372? **$620x = 372$; 60%**
20. A sports team has won 21 out of the 40 games it has played. About what percent of the games has the team won? **about 52.5**

B

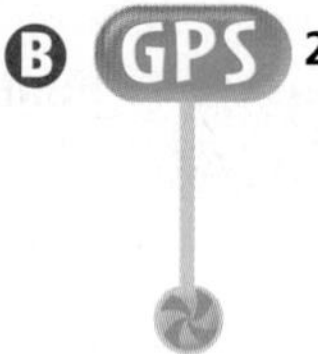

21. **Guided Problem Solving** Suppose 24% of a 1,500-Calorie diet is from protein. How many Calories are *not* from protein? **1,140 Calories**
    - **Make a Plan** Find the number of Calories that are from protein. Subtract that number from the daily total number of Calories.
    - **Carry Out the Plan** 24% of 1,500 is ■. Subtract ■ from 1,500.
22. A water tank containing 496 gallons is 62% full. How many more gallons are needed to fill the tank? **304 gallons**

30. If 25% of a number is 45, the number is more than 45, because 45 is a part of the whole number. If 150% of a number is 45, the number is less than 45, because 150% is more than the whole number.

Visit: PHSchool.com
Web Code: are-0606

Use the table at the right. Find the percent of days in a 365-day year that are school days in each country. Round to the nearest percent.

Length of School Year

Country	Days
China	251
Israel	215
Russia	210
Scotland	200
United States	180

SOURCE: *The Top 10 of Everything*

23. China **69%** **24.** Israel **59%** **25.** Russia **58%**

26. Scotland **55%** **27.** United States **49%**

28. Food You make 72 cookies for a bake sale. GPS This is 20% of the cookies at the bake sale. How many cookies are at the bake sale? **360 cookies**

29. The attendance at the school play on Friday was 95% of the attendance on Saturday night. If 203 people attended on Saturday night, estimate how many attended on Friday night. **about 190 people**

30. **Writing in Math**  If 25% of a number is 45, is the number greater than or less than 45? If 150% of a number is 45, is the number greater than or less than 45? Explain how you can tell. **See margin.**

31. Of the 60 members of a choir, 30% sing alto and 45% sing soprano. How many members of the choir sing alto or soprano? **45 members**

C **32. Challenge** You plant 40 pots with seedlings. Eight of the pots contain tomato plants. What percent of your seedlings are *not* tomato plants? **80%**

Test Prep and Mixed Review — Practice

Multiple Choice

33. About 12% of an iceberg's mass is above water. If the mass above water is 9,000,000 kg, what is the mass of the entire iceberg? **C**

Ⓐ 108,000 kg Ⓒ 75,000,000 kg
Ⓑ 1,080,000 kg Ⓓ 120,000,000 kg

34. On average, a group of 25 college students contains 14 females. Which equation can be used to find x, the percent of females in a typical group of college students? **F**

Ⓕ $\frac{x}{100} = \frac{14}{25}$ Ⓖ $\frac{x}{25} = \frac{14}{100}$ Ⓗ $\frac{x}{25} = \frac{14}{39}$ Ⓙ $\frac{x}{14} = \frac{25}{100}$

35. The model at the right represents which expression? **C**

Ⓐ $\frac{1}{5} \times \frac{3}{5}$ Ⓒ $\frac{1}{2} \times \frac{3}{5}$
Ⓑ $\frac{1}{3} \times \frac{3}{5}$ Ⓓ $\frac{2}{3} \times \frac{7}{10}$

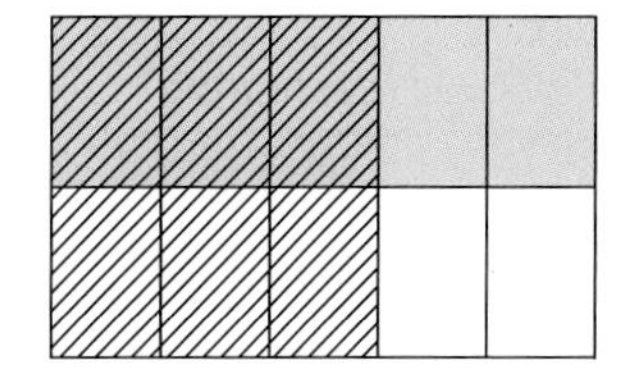

GO for Help

For Exercises	See Lesson
36–39	5-3

Determine whether the ratios in each pair can form a proportion.

36. $\frac{4}{12}, \frac{140}{360}$ **no** **37.** $\frac{7}{9}, \frac{35}{45}$ **yes** **38.** $\frac{12}{28}, \frac{3}{7}$ **yes** **39.** $\frac{45}{60}, \frac{3}{4}$ **yes**

Online lesson quiz, PHSchool.com, Web Code: ara-0606

4. Assess & Reteach

PowerPoint

Lesson Quiz

Write and solve an equation.

1. 6% of 38 is what number?
   $n = 6\% \cdot 38$; 2.28
2. 40% of what number is 24?
   $24 = 40\% \cdot n$; 60
3. What percent of 52 is 13?
   $13 = n\% \cdot 52$; 25%
4. 96% of what number is 240?
   $240 = 96\% \cdot n$; 250

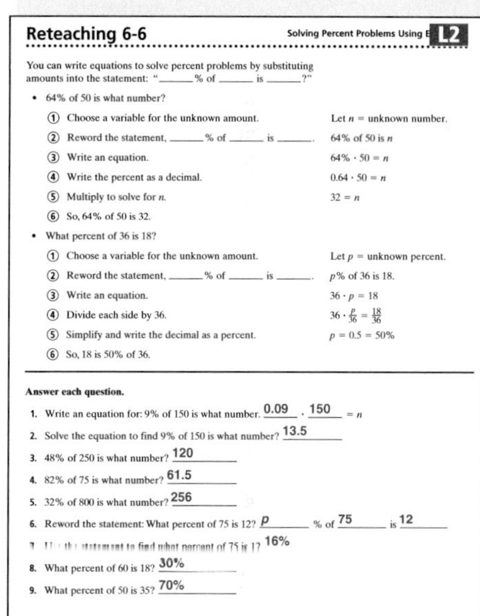

Reteaching 6-6 Solving Percent Problems Using E… L2

You can write equations to solve percent problems by substituting amounts into the statement: "____% of ____ is ____?"

- 64% of 50 is what number?
  1. Choose a variable for the unknown amount. — Let n = unknown number.
  2. Reword the statement, ____% of ____ is ____. — 64% of 50 is n
  3. Write an equation. — $64\% \cdot 50 = n$
  4. Write the percent as a decimal. — $0.64 \cdot 50 = n$
  5. Multiply to solve for n. — $32 = n$
  6. So, 64% of 50 is 32.
- What percent of 36 is 18?
  1. Choose a variable for the unknown amount. — Let p = unknown percent.
  2. Reword the statement, ____% of ____ is ____. — p% of 36 is 18.
  3. Write an equation. — $36 \cdot p = 18$
  4. Divide each side by 36. — $36 \cdot \frac{p}{36} = \frac{18}{36}$
  5. Simplify and write the decimal as a percent. — $p = 0.5 = 50\%$
  6. So, 18 is 50% of 36.

Answer each question.

1. Write an equation for: 9% of 150 is what number. 0.09 · 150 = n
2. Solve the equation to find 9% of 150 is what number? 13.5
3. 48% of 250 is what number? 120
4. 82% of 75 is what number? 61.5
5. 32% of 800 is what number? 256
6. Reword the statement: What percent of 75 is 12? p % of 75 is 12
7. [illegible] 16%
8. What percent of 60 is 18? 30%
9. What percent of 50 is 35? 70%

Test Prep

Resources

For additional practice with a variety of test item formats:

- Test-Taking Strategies, p. 315
- Test Prep, p. 319
- Test-Taking Strategies with Transparencies

Alternative Assessment

Students work in pairs and take turns posing questions such as those in Exercises 7–20. The partner writes an equation to solve. Together they solve the equation.

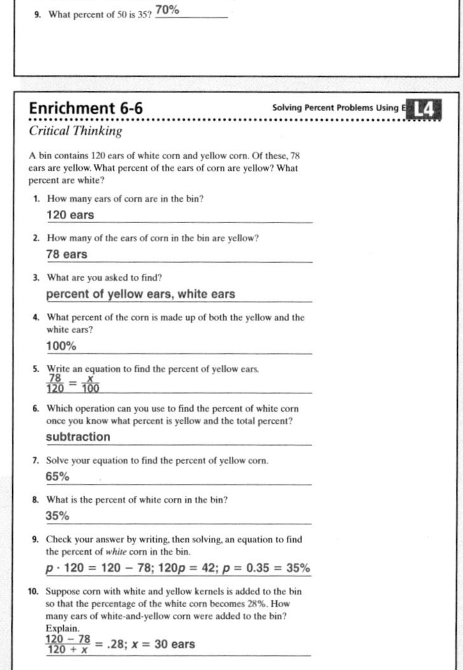

Enrichment 6-6 Solving Percent Problems Using E… L4

Critical Thinking

A bin contains 120 ears of white corn and yellow corn. Of these, 78 ears are yellow. What percent of the ears of corn are yellow? What percent are white?

1. How many ears of corn are in the bin?
   120 ears
2. How many of the ears of corn in the bin are yellow?
   78 ears
3. What are you asked to find?
   percent of yellow ears, white ears
4. What percent of the corn is made up of both the yellow and the white ears?
   100%
5. Write an equation to find the percent of yellow ears.
   $\frac{78}{120} = \frac{x}{100}$
6. Which operation can you use to find the percent of white corn once you know what percent is yellow and the total percent?
   subtraction
7. Solve your equation to find the percent of yellow corn.
   65%
8. What is the percent of white corn in the bin?
   35%
9. Check your answer by writing, then solving, an equation to find the percent of *white* corn in the bin.
   $p \cdot 120 = 120 - 78$; $120p = 42$; $p = 0.35 = 35\%$
10. Suppose corn with white and yellow kernels is added to the bin so that the percentage of the white corn becomes 28%. How many ears of white-and-yellow corn were added to the bin? Explain.
    $\frac{120 - 78}{120 + x} = .28$; $x = 30$ ears

GPS Guided Problem Solving

Solving Percent Problems

In this feature, students practice solving application problems involving percents. They estimate using benchmark fractions and then multiply to find parts of a whole.

Guided Instruction

Explain to students that percents can be applied to real-life situations, such as survey data. Have a volunteer read the problem aloud. Ask:

- *What information do you need to solve the problem?* **number of total children with braces, percent of mothers who said children "felt cool," number of children each mother has**
- *How would the answer be different if the mothers had two children each and both said they "felt cool" with braces?* **The 406 children would double to 812 and the answer would be 69% of 812 or double 280.**
- *How do you know that your answer is reasonable?* **280 is close to the estimate we got by rounding the percent and the total number of children.**

Error Prevention!

Some students may be tempted to divide the total by the percent. Have them set up an equation and remind them that they must cross multiply to find the answer.

GPS Guided Problem Solving

Solving Percent Problems

Braces are cool! A recent survey of 406 mothers by the American Association of Orthodontists reported that 69% felt that wearing braces makes their children feel cool. If each mother had only one child with braces, how many children would that be?

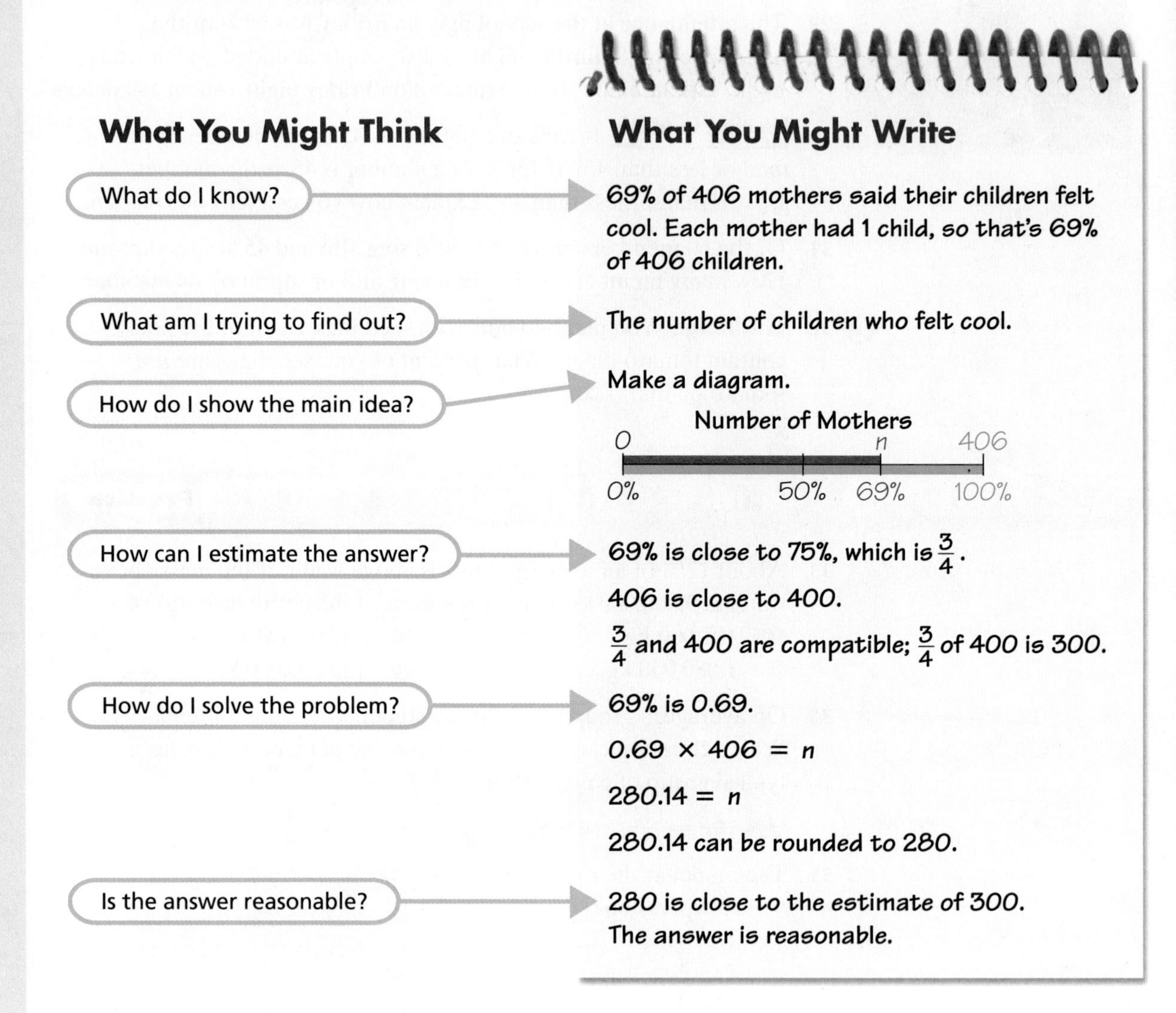

What You Might Think
What You Might Write
What do I know?
69% of 406 mothers said their children felt cool. Each mother had 1 child, so that's 69% of 406 children.
What am I trying to find out?
The number of children who felt cool.
How do I show the main idea?
Make a diagram.
Number of Mothers
0
n
406
0%
50%
69%
100%
How can I estimate the answer?
69% is close to 75%, which is $\frac{3}{4}$.
406 is close to 400.
$\frac{3}{4}$ and 400 are compatible; $\frac{3}{4}$ of 400 is 300.
How do I solve the problem?
69% is 0.69.
$0.69 \times 406 = n$
$280.14 = n$
280.14 can be rounded to 280.
Is the answer reasonable?
280 is close to the estimate of 300.
The answer is reasonable.

Think It Through

1. What percent of 406 mothers did not say their children "felt cool?" How many children is that? **31%; about 126 children**
2. How does the diagram above help show the main idea? **See margin.**

2. **Answers may vary. Sample: It compares the number of mothers who said their children felt cool to the total number of mothers.**

Exercises

For Exercises 3 and 4, answer the questions first, and then solve the problem.

3. A solo guitarist received a royalty payment of 9% based on the sales of her CD. If she received a check for $5,238, what were the total sales of her CD? **$58,200**
   a. What do you know?
   b. What do you want to find out?
   c. How can the diagram below help you write an equation?

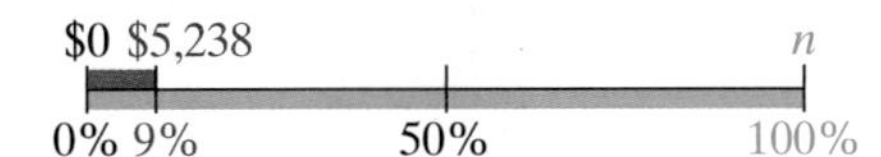

4. The human body is about 67% water. A student weighs 130 pounds. How much of his weight is water? **about 87.1 lb**
   a. What do you know?
   b. What do you want to find out?
   c. What diagram would help you show the main idea?

5. Teenagers were asked how many hours per week they worked. The data are shown below. Assume they were paid $5.50 per hour. About how much per week would the largest category of students make? The second-largest? **between $49.50 and $66.00; between $33.00 and $49.50**

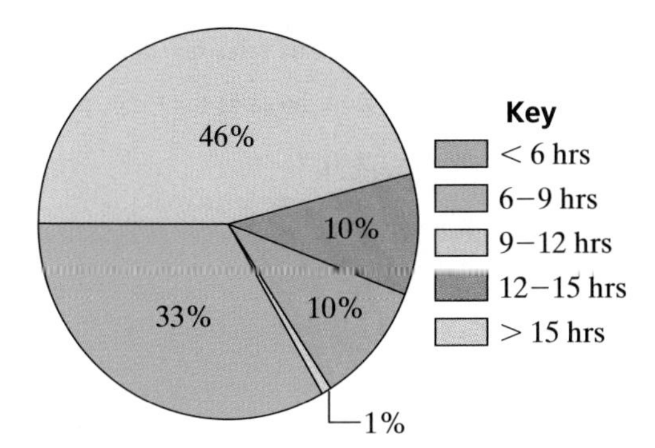

6. When water freezes, its volume increases by 9%. If you freeze one gallon of water to make ice for a party, how many cubic inches of ice will you have? (*Hint*: 1 gallon = 231 in.^{3}) **about 252 in.^{3}**

7. Research says that humans learn through listening 11% of the time and through observing 83% of the time. In a 50-minute math class, how much of the time would you be learning by listening? By observing? **5.5 minutes; 41.5 minutes**

Exercises
Have students work on the Exercises independently. Encourage them to identify the important information before solving the problem. Allow students to trade papers and compare answers when they have finished the Exercises.

Differentiated Instruction

Below Level 

Some students may need to model the problems to help solve them. Help these students to draw graphs or tables to display the data before they work on the Exercises.

6-7

1. Plan

Objective
To find and estimate solutions to application problems involving percent

Examples
1 Finding Sales Tax
2 Estimating a Tip
3 Finding a Commission

Math Understandings: p. 272D

Math Background

Sales tax can be calculated using the word equation *sales tax* = *tax%* · *purchase price*. Similarly, a tip is a percent of a bill. You can easily calculate tips of 15% by finding 10% of the bill (move the decimal 1 place to the left) and adding $\frac{1}{2}$ of this result. Commission is a percent of sales earned by a salesperson as part of a salary.

More Math Background: p. 272D

Lesson Planning and Resources

See p. 272E for a list of the resources that support this lesson.

Bell Ringer Practice

Check Skills You'll Need
Use student page, transparency, or PowerPoint. For intervention, direct students to:
Percent, Fractions, and Decimals
Lesson 6-2
Extra Skills and Word Problems Practice, Ch. 6

Algebra

6-7 Applications of Percent

Check Skills You'll Need

1. **Vocabulary Review** To write a *percent* as a decimal, move the decimal point two places to the __?__. **left**

Write each percent as a decimal.

2. 6.5% **0.065**
3. 4.25% **0.0425**
4. 15% **0.15**
5. 20% **0.2**

Lesson 6-2

What You'll Learn

To find and estimate solutions to application problems involving percent

New Vocabulary commission

Why Learn This?

You use percents to calculate taxes, tips, and commissions.

In many states, you must pay a sales tax on items you buy. The sales tax is a percent of the purchase price. A tax percent is also called a tax rate.

To find sales tax, you can use the formula *sales tax* = *tax rate* · *purchase price*.

EXAMPLE Finding Sales Tax

1 **Shopping** The price of a bicycle you plan to buy is \$159.99. The sales tax rate is 6%. How much will you pay for the bicycle?

$0.06 \cdot 159.99 \approx 9.60$ ← Find the sales tax. Round to the nearest cent.

$159.99 + 9.60 = 169.59$ ← Add the sales tax to the purchase price.

You will pay \$169.59 for the bicycle.

Quick Check

1. Find the total cost for a purchase of \$185 if the sales tax rate is 5.5%. **\$195.18**

A tip is a percent of a bill that you give to someone who provides a service. You can use estimation and mental math to find a 15% tip.

Step 1 Round the bill to the nearest dollar.

Step 2 Find 10% of the bill by moving the decimal point one place to the left.

Step 3 Find 5% of the bill by taking one half of the result of Step 2.

Step 4 Add the amounts of Step 2 and Step 3 together to find 15%.

Differentiated Instruction Solutions for All Learners

Special Needs L1
For Example 1, draw a number line that sets \$159.99 equal to 100%. Show students that they are finding the cost of the bike (100%) plus the tax, so their answer must be greater than the cost of the bike.

learning style: visual

Below Level L2
Have students practice manipulating decimal points by calculating 10% tips. Ask: *How would you calculate a 10% tip?* **Sample: Move the decimal point one place to the left and round to hundredths.**

learning style: verbal

EXAMPLE Estimating a Tip

2 Your family takes a taxi to the train station. The taxi fare is $17.85. Estimate a 15% tip to give the driver.

$17.85 \approx 18$ ← **Round to the nearest dollar.**

$0.1 \cdot 18 = 1.8$ ← **Find 10% of the bill.**

$\frac{1}{2} \cdot 1.8 = 0.9$ ← **Find 5% of the bill. 5% is $\frac{1}{2}$ of the 10% amount.**

$1.8 + 0.9 = 2.7$ ← **Add the 10% and 5% amounts to get 15%.**

For a $17.85 taxi fare, a 15% tip is about $2.70.

Quick Check

2. Estimate a 15% tip for each amount.
   a. $58.20 **about $8.70** b. $61.80 **about $9.30** c. $49.75 **about $7.50**

Some sales jobs pay you a **commission**, a percent of the amount of your sales. To find a commission, use *commission* = *commission rate* · *sales*.

EXAMPLES Finding a Commission

3 Find the commission on a $500 sale with a commission rate of 12.5%.

$0.125 \cdot 500 = 62.5$ ← **Write 12.5% as 0.125 and multiply.**

The commission on the sale is $62.50.

Video Tutor Help

Visit: PHSchool.com
Web Code: are-0775

4 A sales agent earns a weekly salary of $650, plus a commission of 4% on all sales. His sales this week are $1,250. How much does he earn?

Words total earnings = salary + commission

Let t = total earnings.

Equation $t = 650 + 0.04 \cdot 1{,}250$

$t = 650 + 0.04 \cdot 1{,}250$ ← **Write the equation.**

$= 650 + 50$ ← **Multiply.**

$= 700$ ← **Simplify.**

The sales agent earns $700 this week.

Quick Check

3. Find the commission on a $3,200 sale with a commission rate of 6%.
4. Suppose you earn a weekly salary of $800 plus a commission of 3.5% on all sales. Find your earnings for a week with total sales of $1,400.

3. **$192**
4. **$849**

Advanced Learners L4
Ben's weekly salary is $850. He also earns a 6% commission on sales. Last week, his total salary with commission was $959.80. What were Ben's sales?
$1,830

learning style: verbal

English Language Learners ELL
Act out examples of someone giving or getting a tip, paying sales tax, or getting a commission on a sale. These words and concepts can be difficult to grasp. Students need experience with the language and the concept before they can do the calculation.

learning style: visual

2. Teach

Activity Lab

Use before the lesson.

All in One Teaching Resources
Activity Lab 6-7: Commission

Guided Instruction

Example 2
Explain that a tip is money given to someone, such as a waitperson or taxi driver, for doing a good job. Emphasize that a tax is a required amount, while a tip is voluntary and is generally determined by the person giving the tip.

PowerPoint
Additional Examples

1. A video game costs $34.98. The sales tax rate is 5.5%. Find the total cost. **$36.90**
2. Use estimation to calculate a 15% tip for each amount.
   a. $34.50 **$5.25**
   b. $11.84 **$1.80**
3. Find the commission on a $450 sale when the commission rate is 7%. **$31.50**
4. Find the total earnings for a salesperson with a salary of $550 plus a 4% commission on sales of $1,485. **$609.40**

All in One Teaching Resources
- Daily Notetaking Guide 6-7 L3
- Adapted Notetaking 6-7 L1

Closure

- *What is the word equation you use to find sales tax?* **sales tax = percent of tax · purchase price**
- *What is the word equation you use to find commissions?* **commission = percent of sales · sales**
- *What is the word equation for the total cost of a meal and tip?* **total = cost of food + (percent of tip · cost of food)**

3. Practice

Assignment Guide

Check Your Understanding

Go over Exercises 1–7 in class before assigning the Homework Exercises.

Homework Exercises

A	Practice by Example	8–19
B	Apply Your Skills	20–27
C	Challenge	28
Test Prep and Mixed Review		29–33

Homework Quick Check

To check student's understanding of key skills and concepts, go over Exercises 9, 14, 22, 24, and 26.

Differentiated Instruction Resources

Adapted Practice 6-7 L1

Practice 6-7 Applications c L3

Find the total cost.

1. $17.50 with a 7% sales tax **$18.73**
2. $21.95 with a 4.25% sales tax **$22.88**
3. $52.25 with an 8% sales tax **$56.43**
4. $206.88 with a 5.75% sales tax **$218.78**
5. The price of a pair of shoes is $85.99 before sales tax. The sales tax is 7.5%. Find the total cost of the shoes. **$92.44**

Estimate a 15% tip for each amount.

6. $12.68 **Sample: $1.95**
7. $18.25 **Sample: $2.70**
8. $15.00 **Sample: $2.25**

Find each commission.

9. 2% on $1,500 in sales **$30**
10. 8% on $80,000 in sales **$6,400**
11. 5% on $600 in sales **$30**
12. 12% on $3,200 in sales **$384**

Find the total earnings when given the salary, commission rate, and sales.

13. $1,000 plus 6% on sales of $2,000 **$1,120**
14. $500 plus 10% on sales of $1,400 **$640**
15. $850 plus 8% on sales of $8,000 **$1,490**
16. $1,200 plus 4.5% on sales of $6,500 **$1,492.50**
17. To recover a large chair in your home, you purchase $9\frac{1}{2}$ yards of upholstery fabric at $11.00 per yard. If there is a 7% sales tax, what is the total cost of the fabric? **$111.82**
18. You and your sister rake and pick up leaves from your grandmother's yard. She says she'll pay you $35.00 for the job. Upon completion of your work, she also decides to add a separate 20% tip for each of you. You and your sister each keep your individual tips, and you decide to split your other earnings in half. How much will each of you earn from the job? **You and your sister each earn $24.50.**

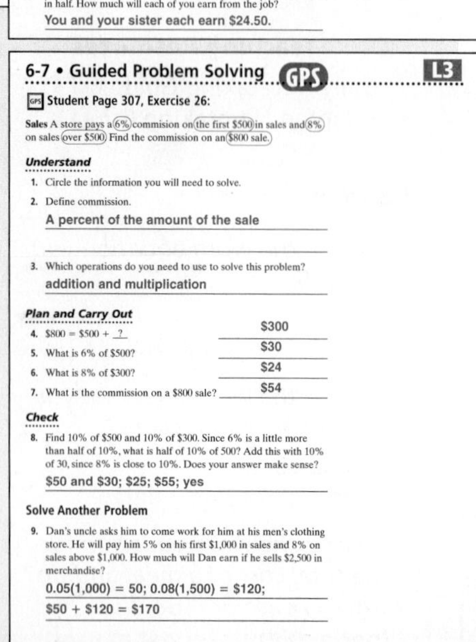

6-7 • Guided Problem Solving GPS L3

Student Page 307, Exercise 26:

Sales A store pays a 6% commision on the first $500 in sales and 8% on sales over $500. Find the commission on an $800 sale.

Understand

1. Circle the information you will need to solve.
2. Define commission. **A percent of the amount of the sale**
3. Which operations do you need to use to solve this problem? **addition and multiplication**

Plan and Carry Out

4. $800 = $500 + ? **$300**
5. What is 6% of $500? **$30**
6. What is 8% of $300? **$24**
7. What is the commission on a $800 sale? **$54**

Check

8. Find 10% of $500 and 10% of $300. Since 6% is a little more than half of 10%, what is half of 10% of 500? Add this with 10% of 30, since 8% is close to 10%. Does your answer make sense? **$50 and $30; $25; $55; yes**

Solve Another Problem

9. Dan's uncle asks him to come work for him at his men's clothing store. He will pay him 5% on his first $1,000 in sales and 8% on sales above $1,000. How much will Dan earn if he sells $2,500 in merchandise? **0.05(1,000) = 50; 0.08(1,500) = $120; $50 + $120 = $170**

Check Your Understanding

1. **Vocabulary** What does it mean to earn an 8% commission? **You earn 8% of the amount you sell.**
2. **Number Sense** Is 4% sales tax on a $250 item *greater than, less than,* or *equal to* 4% commission on a $250 sale? **equal to**
3. **Mental Math** Calculate a 15% tip on a restaurant bill of $24. **$3.60**

Find the sales tax for each item. The tax rate is 6%.

4. a CD priced at $12.99 **$.78**
5. a $450 TV **$27**

Find each commission, given the sale.

6. 5% on a $900 sale **$45**
7. 2% on a $35.50 sale **$.71**

Homework Exercises

For more exercises, see Extra Skills and Word Problems.

GO for Help

For Exercises	See Examples
8–10	1
11–14	2
15–19	3–4

A **Find the total cost.**

8. $35.99 with a 5% sales tax **$37.79**
9. $72.75 with a 6% sales tax **$77.12**
10. The price of a coat is $114 before sales tax. The sales tax is 7%. Find the total cost of the coat. **$121.98**

Estimate a 15% tip for each amount.

11. $68.50 **about $10.35**
12. $30.80 **about $4.65**
13. $9.89 **about $1.50**
14. $27.59 **about $4.20**

Find each commission, given the sale and the commission rate.

15. $800, 12% **$96**
16. $2,500, 8% **$200**
17. $2,000, 7.5% **$150**
18. $600, 4.5% **$27**
19. Suppose your boss owes you $800, plus a commission of 2.5% on a sale of $1,000. How much does your boss owe you? **$825**

B 

20. **Guided Problem Solving** Your lunch bill is $19.75. A 5% sales tax will be added, and you want to give a tip of about 20% of $19.75. Estimate how much you will pay for lunch. **about $25**
    - To what number should you round the bill?
    - About how much tip should you give?
21. Your neighbor pays $40 to have her lawn mowed and always adds a 15% tip. You and your friend decide to mow the lawn together and split the earnings evenly. How much will each of you make? **$23**
22. **Writing in Math** Explain how you can use estimation and mental math to calculate a 20% tip. **See margin.**

GO Online
Homework Video Tutor
Visit: PHSchool.com
Web Code: are-0607

22. Answers may vary. Sample: Find 10% of the bill by moving the decimal point one place to the left. Then double this number.

23. A purchase costs \$25.79 with a tax of \$1.29. Find the sales tax rate. **5%**

24. **Art** For a craft project, you select the four packages of modeling clay and the set of tools shown at the right. If there is a 6% sales tax, what is the total cost? **\$13.73**

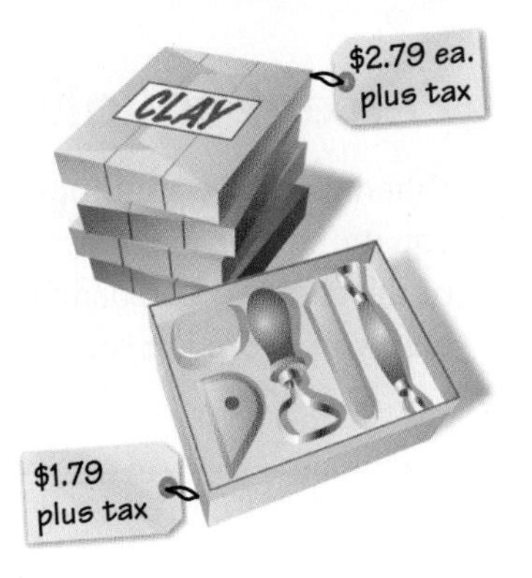

25. A real estate agent earns a weekly salary of \$200. This week, the agent sold a home for \$120,000 and was paid a 5.5% commission. Find the agent's earnings for the week. **\$6,800**

26. **Sales** A store pays a 6% commission on the first \$500 in sales and GPS 8% on sales over \$500. Find the commission on an \$800 sale. **\$54**

27. **Open-Ended** If your employer gave you a choice between earning a fixed salary and earning a commission based on your sales, how would you choose? Explain your reasoning and give an example. **Check students' work.**

C 28. **Challenge** Find the commission rate if the total earnings are \$970, including a salary of \$350 and a commission on sales of \$12,400. **5%**

Test Prep and Mixed Review

Practice

Multiple Choice

29. Laundry workers can expect a tip between 15% and 20%. Which is closest to the amount Diane should offer in order to give the minimum tip for a laundry service of \$11.50? **D**
    - (A) \$4.00
    - (B) \$2.50
    - (C) \$2.00
    - (D) \$1.75

30. A surveyor drew this diagram to find the distance across a small lake. If $\triangle RSV$ is similar to $\triangle UST$, what is the width w? **H**
    - (F) 44 m
    - (H) 41.25 m
    - (G) 21.8 m
    - (J) 11.7 m

U, w, T, 30 m, S, V, 16 m, 22 m, R

31. Raul bought 3 posters for \$2.59 each and 2 posters for \$1.98 each. He paid 59 cents tax. What other information is necessary to find Raul's correct change? **A**
    - (A) The amount of money Raul gave the cashier
    - (B) Whether the posters were on sale
    - (C) The total amount of money Raul spent
    - (D) How Raul got to the store

GO for Help

For Exercises	See Lesson
32–33	5-2

Write each unit rate.

32. 408 mi on 12 gal of gasoline **34 mi/gal**

33. \$16.45 for 7 lb of fish **\$2.35/lb**

4. Assess & Reteach

Lesson Quiz

1. Find the payment for a \$75 purchase with a sales tax rate of 5%. **\$78.75**
2. Find a 15% tip for a \$27.60 bill. **\$4.14**
3. Find the commission on \$3,580 in sales when commission is paid at 6%. **\$214.80**

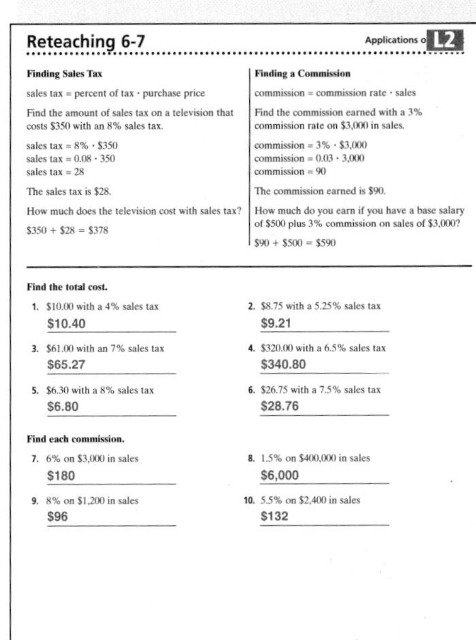

Reteaching 6-7 Applications o L2

Finding Sales Tax

sales tax = percent of tax · purchase price

Find the amount of sales tax on a television that costs \$350 with an 8% sales tax.

sales tax = 8% · \$350
sales tax = 0.08 · 350
sales tax = 28

The sales tax is \$28.

How much does the television cost with sales tax?

\$350 + \$28 = \$378

Finding a Commission

commission = commission rate · sales

Find the commission earned with a 3% commission rate on \$3,000 in sales.

commission = 3% · \$3,000
commission = 0.03 · 3,000
commission = 90

The commission earned is \$90.

How much do you earn if you have a base salary of \$500 plus 3% commission on sales of \$3,000?

\$90 + \$500 = \$590

Find the total cost.

1. \$10.00 with a 4% sales tax **\$10.40**
2. \$8.75 with a 5.25% sales tax **\$9.21**
3. \$61.00 with an 7% sales tax **\$65.27**
4. \$320.00 with a 6.5% sales tax **\$340.80**
5. \$6.30 with a 8% sales tax **\$6.80**
6. \$26.75 with a 7.5% sales tax **\$28.76**

Find each commission.

7. 6% on \$3,000 in sales **\$180**
8. 1.5% on \$400,000 in sales **\$6,000**
9. 8% on \$1,200 in sales **\$96**
10. 5.5% on \$2,400 in sales **\$132**

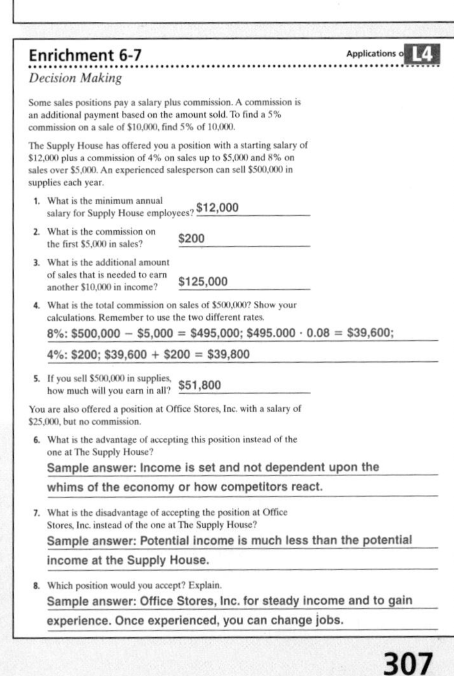

Enrichment 6-7 Applications o L4

Decision Making

Some sales positions pay a salary plus commission. A commission is an additional payment based on the amount sold. To find a 5% commission on a sale of \$10,000, find 5% of 10,000.

The Supply House has offered you a position with a starting salary of \$12,000 plus a commission of 4% on sales up to \$5,000 and 8% on sales over \$5,000. An experienced salesperson can sell \$500,000 in supplies each year.

1. What is the minimum annual salary for Supply House employees? **\$12,000**
2. What is the commission on the first \$5,000 in sales? **\$200**
3. What is the additional amount of sales that is needed to earn another \$10,000 in income? **\$125,000**
4. What is the total commission on sales of \$500,000? Show your calculations. Remember to use the two different rates.
   8%: \$500,000 − \$5,000 = \$495,000; \$495.000 · 0.08 = \$39,600;
   4%: \$200; \$39,600 + \$200 = \$39,800
5. If you sell \$500,000 in supplies, how much will you earn in all? **\$51,800**

You are also offered a position at Office Stores, Inc. with a salary of \$25,000, but no commission.

6. What is the advantage of accepting this position instead of the one at The Supply House?
   Sample answer: Income is set and not dependent upon the whims of the economy or how competitors react.
7. What is the disadvantage of accepting the position at Office Stores, Inc. instead of the one at The Supply House?
   Sample answer: Potential income is much less than the potential income at the Supply House.
8. Which position would you accept? Explain.
   Sample answer: Office Stores, Inc. for steady income and to gain experience. Once experienced, you can change jobs.

Alternative Assessment

Each student in a pair writes three problems: one for finding total cost including sales tax; one for finding a tip; and one for finding a commission. Partners exchange papers and solve each other's problems. Partners check each other's work and discuss their results.

Test Prep

Resources

For additional practice with a variety of test item formats:

- Test-Taking Strategies, p. 315
- Test Prep, p. 319
- Test-Taking Strategies with Transparencies

Activity Lab

Percent Equations

Students are given percent equations and answer multiple-choice questions about them. They eliminate answer choices by telling whether they are true or false. This will extend the percent skills they learned in Lesson 6-7.

Guided Instruction

Review the Example with students. Explain that only one choice can be correct and they must find the correct answer by process of elimination. Ask questions, such as:

- *How can you use estimation to help solve this problem?* **49% is about 50% and $\frac{1}{2}$ of x is 30. D is the only correct choice because 60 > 30. So x must be about 60.**
- *How would you solve the equation to prove that D is true?* **Convert 49% to 0.49. Divide 30 by the 0.49. $30 \div 0.49 \approx 61.2$, $61.2 > 30$**

Differentiated Instruction

Below Level L2

Some students may have difficulty because of the number of possible answers. Provide these students with one true option and one false option for each question.

6-7b Activity Lab — Algebra Thinking $x=?$

Percent Equations

When you have a multiple-choice question with an equation using percent, you can quickly eliminate some of the choices. Use benchmark numbers for percents to help you. Here are some benchmarks:

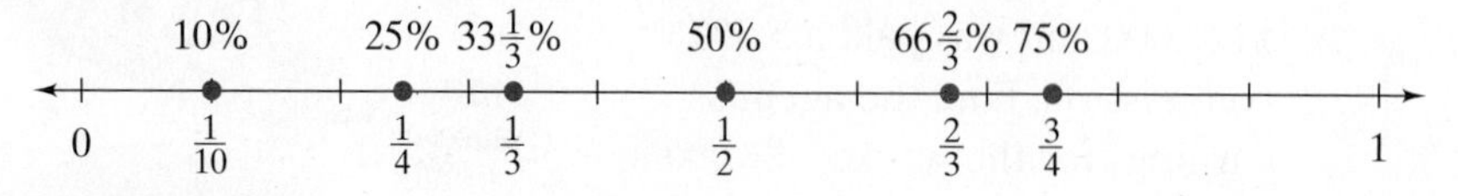

EXAMPLE

Which statement is true for $49\% \times x = 30$? Explain.

Ⓐ $x \approx 30$ Ⓑ x is negative. Ⓒ $x < 60$ Ⓓ $x > 30$

Choice A is not true. ← **49% is about $\frac{1}{2}$ and $\frac{1}{2}$ of 30 is 15. So $x \approx 30$ isn't close.**

Choice B is not true. ← **That would make the left side a negative number.**

Choice C is not true. ← **49% is a little less than $\frac{1}{2}$, and $\frac{1}{2}$ of 60 is 30.**

Choice D is true. ← ***x* must be big enough that half of it is 30. *x* must be about twice 30.**

The correct answer is choice D.

Exercises

Reasoning Which choice is true for the equation? Use benchmark numbers to explain. Check each answer choice. Do not compute.

1. $31\% \times x = 23.1$ **B; $x < 100$ is true because $\frac{1}{3}$ of 100 is more than 23.1.**
   Ⓐ $x < 23.1$ Ⓑ $x < 100$ Ⓒ $x \approx 90$ Ⓓ $x \approx \frac{2}{3}$
2. $90 \times x\% = 8.9$ **F; $x \approx 10$ is true because 8.9 is about $\frac{1}{10}$ of 90.**
   Ⓕ $x \approx 10$ Ⓖ $x > 30$ Ⓗ $x \approx 50$ Ⓙ $x > 10$
3. $251 \div 500 = x\%$ **B; $x \approx 50$ is true because 251 is about $\frac{1}{2}$ of 500.**
   Ⓐ $x < 0$ Ⓑ $x \approx 50$ Ⓒ $x \approx 2$ Ⓓ $x < 45$

Write three statements about the variable for each equation. Make some statements true and some false. Circle the ones that are true. **4–6. Check students' work.**

4. $26\% \times x = 101$
5. $500 \div 1007 = x\%$
6. $20 \times x\% = 10.5$

Checkpoint Quiz 2

Lessons 6-4 through 6-7

Find each answer using a proportion.

1. 35 is what percent of 60? **58.$\bar{3}$%**
2. 14.4 is 90% of what number? **16**
3. What percent of 75 is 63? **84%**

Find each answer using an equation.

4. What percent of 120 is 54? **45%**
5. What is 72% of 95? **68.4**
6. What is 120% of 185? **222**

Find the price of each item.

7. originally $299; 5% markup **$313.95**
8. originally $97; 15% discount **$82.45**
9. price $32.79; 8.25% tax **$35.50**
10. **Sales** Find the commission for a rate of 5.5% and $1,400 in sales. **$77**

6-8a Activity Lab

Data Analysis

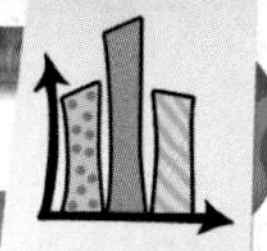

Exploring Percent of Change

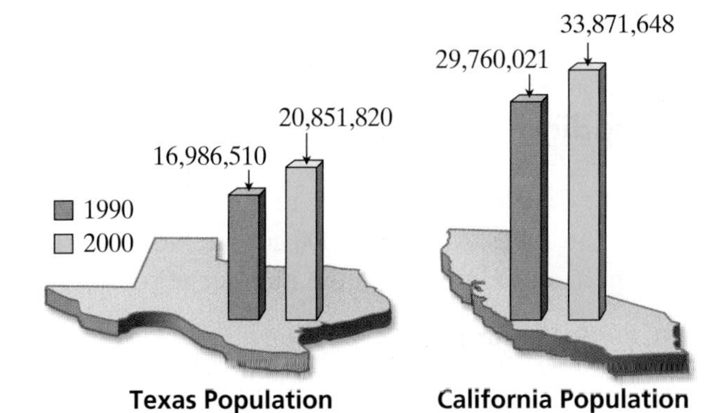

Texas Population

California Population

Use the graph above to complete the following exercises.

1. Find the population change from 1990 to 2000 for each state. **Texas: 4,111,**
2. Which state had the greater change in population? **California**
3. For each state, write the ratio $\frac{\text{change in population}}{\text{1990 population}}$. Write each as a percent. **Texas:** **Califo**
4. Which state had the greater population change in terms of percen
5. **Reasoning** Why are your answers to 2 and 4 above different? **See margin.**

-make roster for HR
-LP (download pic)

5. **Answers may vary. Sample: California had a greater increase in population, but because it started with a greater population, it had a lesser percent of change.**

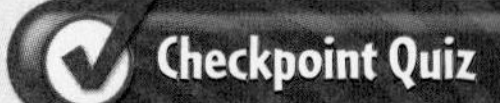

Checkpoint Quiz

Use this Checkpoint Quiz to check students' understanding of the skills and concepts of Lessons 6-4 through 6-7.

Resources

- All-in-One Teaching Resources Checkpoint Quiz 2
- ExamView CD-ROM
- Success Tracker™ Online Intervention

Activity Lab

Exploring Percent of Change

Students compare data to find a percent of change over time. This will familiarize them with the concept that they will work on in Lesson 6-8.

Guided Instruction

Define percent of change for students. Explain that percent of change is found by using a proportion.

Exercises
Work on the exercises as a class. Ask volunteers to provide each answer and discuss why that answer is correct or incorrect.

Differentiated Instruction

Advanced Learners L4
Have students create a graph for 2000 to 2010, using the same percent of change rates.

Resources

- Activity Lab 6-8: Percent Change

6-8

1. Plan

Objective
To find percents of increase and percents of decimals

Examples
1 Finding a Percent of Increase
2 Finding a Percent of Markup
3 Finding a Percent of Discount

Math Understandings: p. 272D

Math Background

The *percent of change* is the percent a quantity increases or decreases from its original amount. To find the percent of change, you first find the amount of change and divide it by the original amount. A *markup* is the amount of increase in the price of an item. A *discount* is the amount of decrease in an item. So, a *percent of markup* is the percent increase in price and a *percent of discount* is the percent decrease in price.

More Math Background: p. 272D

Lesson Planning and Resources

See p. 272E for a list of the resources that support this lesson.

Check Skills You'll Need
Use student page, transparency, or PowerPoint. For intervention, direct students to:
Solving Proportions
Lesson 5-4
Extra Skills and Word Problems Practice, Ch. 5

Algebra

6-8 Finding Percent of Change

Check Skills You'll Need

1. **Vocabulary Review** What are the *cross products* in the proportion $\frac{5}{16} = \frac{n}{100}$? **500 and 16n**

Solve each proportion.

2. $\frac{2}{20} = \frac{n}{100}$ **10**
3. $\frac{1}{8} = \frac{n}{100}$ **12.5**
4. $\frac{6.5}{13} = \frac{n}{100}$ **50**

GO for Help Lesson 5-4

What You'll Learn

To find percents of increase and percents of decrease

New Vocabulary percent of change, markup, discount

Why Learn This?

You can use percent of change to describe how much an amount increases or decreases over time. For example, every 10 years, the number of U.S. representatives for a state may change, based on the change in the state's population.

A **percent of change** is the percent a quantity increases or decreases from its original amount. Use a proportion to find a percent of change.

$$\frac{\text{amount of change}}{\text{original amount}} = \frac{\text{percent of change}}{100}$$

EXAMPLE Finding a Percent of Increase

1 Government North Carolina had 12 seats in the U.S. House of Representatives in the 1990s. After the 2000 census, North Carolina had 13 seats. Find the percent of increase in the number of representatives.

$13 - 12 = 1$ ← **Find the amount of change.**

$\frac{1}{12} = \frac{n}{100}$ ← **Write a proportion. Let n = percent of change.**

$100 \cdot \frac{1}{12} = \frac{n}{100} \cdot 100$ ← **Multiply each side by 100.**

$\frac{100}{12} = n$ ← **Simplify.**

$8.3 \approx n$ ← **Divide.**

The number of North Carolina representatives increased by about 8%.

Quick Check

1. In 2000, California went from 52 to 53 representatives. Find the percent of increase in the number of representatives. **1.9%**

Differentiated Instruction Solutions for All Learners

Special Needs L1	**Below Level** L2
Help students draw double number lines for Examples 1–3, with one of them showing percents 0–150%. Highlight the distance that corresponds to the percent students are trying to find.	Remind students that the amount of increase and the percent of increase are not the same. Ask: *A store buys an item for \$4 and sells it for \$6. Is \$2 the percent of increase or the markup?* **the markup**
learning style: visual	learning style: verbal

To make a profit, stores charge more for items than they pay for them. The difference between the selling price and the store's cost of an item is called the **markup.** The percent of markup is a percent of increase.

$$\frac{\text{amount of markup}}{\text{original cost}} = \frac{\text{percent of markup}}{100}$$

EXAMPLE Finding a Percent of Markup

2 An electronics store orders sets of walkie-talkies for \$14.85 each. The store sells each set for \$19.90. What is the percent of markup?

$19.90 - 14.85 = 5.05$ ← **Find the amount of markup.**

$\frac{5.05}{14.85} = \frac{n}{100}$ ← **Write a proportion. Let *n* be the percent of markup.**

$14.85n = 5.05(100)$ ← **Write cross products.**

$\frac{14.85n}{14.85} = \frac{5.05(100)}{14.85}$ ← **Divide each side by 14.85.**

$n \approx 34$ ← **Simplify.**

The percent of markup is about 34%.

Quick Check

2. Find the percent of markup for a \$17.95 headset marked up to \$35.79. 99.4%

The difference between the original price and the sale price of an item is called a **discount.** The percent of discount is a percent of decrease.

$$\frac{\text{amount of discount}}{\text{original cost}} = \frac{\text{percent of discount}}{100}$$

EXAMPLE Finding a Percent of Discount

Before you find the percent of change, decide whether the change is an increase or a decrease.

3 **Music** During a clearance sale, a keyboard that normally sells for \$49.99 is discounted to \$34.99. What is the percent of discount?

$49.99 - 34.99 = 15.00$ ← **Find the amount of discount.**

$\frac{15}{49.99} = \frac{n}{100}$ ← **Write a proportion. Let *n* be the percent of discount.**

$49.99n = 15(100)$ ← **Write cross products.**

$\frac{49.99n}{49.99} = \frac{15(100)}{49.99}$ ← **Divide each side by 49.99.**

$n \approx 30$ ← **Simplify.**

The percent of discount for the keyboard is about 30%.

Quick Check

3. Find the percent of discount of a \$24.95 novel on sale for \$14.97. 40%

2. Teach

Activity Lab

Use before the lesson.
Student Edition Activity Lab, Data Analysis 6-8a, Exploring Percent of Change, p. 309

All in One Teaching Resources
Activity Lab 6-8: Percent Change

Guided Instruction

Example 1
Be sure students understand the difference between the *amount of change* and the *percent of change*. Amount of change is found by subtraction and refers to the number of seats (or individuals). The percent of change is found by solving the proportion and refers to the percent of increase in the number of seats, which relates a part to a whole (the number of new seats to the original number of seats).

Example 3
Help students understand that the same steps are used to find percent of markup and percent of discount. You may want them to write the following proportions to help illustrate this point.

$$\frac{\text{amount of change}}{\text{original amount}} = \frac{\text{percent of change}}{100}$$

$$\frac{\text{amount of markup}}{\text{original cost}} = \frac{\text{percent of markup}}{100}$$

$$\frac{\text{amount of discount}}{\text{original cost}} = \frac{\text{percent of discount}}{100}$$

PowerPoint

Additional Examples

1. Last year, a school had 632 students. This year the school has 670 students. Find the percent of increase in the number of students. 6%

2. Find the percent markup for a car that a dealer buys for \$10,590 and sells for \$13,775. 30%

3. Find the percent of discount for a \$74.99 tent that is discounted to \$48.75. 35%

Advanced Learners L4	**English Language Learners** ELL
A storeowner buys three dozen T-shirts for \$185.75. He wants to sell the T-shirts at a 45% profit. What is the price of each shirt if sales tax is 5.5%? \$7.89 learning style: verbal	Orally, connect the words *increase* to *markup*, and *decrease* to *discount*. This will reinforce students' understanding of those words. learning style: verbal

All in One Teaching Resources
- Daily Notetaking Guide 6-8 L3
- Adapted Notetaking 6-8 L1

Closure

- *What is the percent of change?* **the percent a quantity increases or decreases from its original amount**
- *What is the proportion you use to find percent increase?* $\frac{\text{amount of change}}{\text{original amount}} = \frac{\text{percent of change}}{100}$
- *What is the word equation you use to find markup?* **markup = selling price − original price**

More Than One Way

Answers may vary. Sample: Chris's method is simpler because you can use mental math to find that you have to pay 80% of the original price, and then you need only one step to calculate (0.80)($27.50) = $22.

More Than One Way

A jacket goes on sale with a discount of 40% off the original price. The original price of the jacket is $42.95. What is the sale price of the jacket?

Anna's Method

I can find the amount of the discount by multiplying $42.95 by 40%. Then I will subtract the discount from the original price.

$42.95 \cdot 0.40 = 17.18$ ← **Find the amount of the discount.**

$42.95 - 17.18 = 25.77$ ← **Subtract the discount from the original price.**

The sale price of the jacket is $25.77.

Chris's Method

The jacket is discounted by 40%, so I will pay 60% of the original price. I can multiply the original price of $42.95 by the percent I need to pay.

$42.95 \cdot 0.60 = 25.77$ ← **Find the discounted price.**

The sale price of the jacket is $25.77.

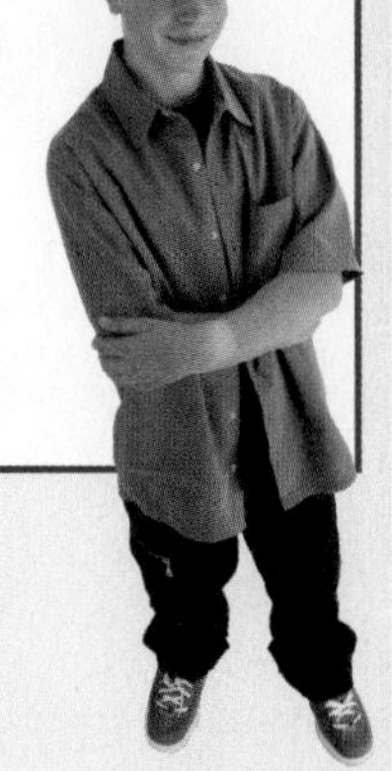

Choose a Method

You get a discount of 20% on a $27.50 ticket. How much will your ticket cost? Describe your method and explain why it is appropriate.

Check Your Understanding

1. **Answers may vary. Sample: They both involve the difference between the original price and the selling price. Percent of markup is a percent of increase and percent of discount is a percent of decrease.**

1. **Vocabulary** How are percent of markup and percent of discount similar? How are they different? **See left.**
2. **Number Sense** Is it possible for a markup to be 200%? Give an example and explain. **Yes; for example, an item's cost could be $10, and it could be sold for $30, which is a markup price of $20 or 200%.**

Write the proportion to find percent of change.

3. $35 to $50 **$\frac{15}{35}$, 43% increase**
4. 98 to 72 **$\frac{26}{98}$, 27% decrease**
5. 748 to 374 **$\frac{374}{748}$, 50% decrease**

Matching **Match each situation with the correct percent of change.**

6. Boots first priced at $110 go on sale for $88. **B**
7. A radio costs a store $88 but sells for $110. **A**

A. 25% markup
B. 20% discount

31. **Find the difference in the number of students for last year and this year, divide that by last year's number, and express the result as a percent. If this year's number is greater than last year's number, the change is an increase. If this year's number is less, the change is a decrease.**

Homework Exercises

For more exercises, see Extra Skills and Word Problems.

GO for Help

For Exercises	See Examples
8–16	1
17–21	2
22–26	3

A **Find each percent of increase. Round to the nearest percent.**

8. 60 to 75 **25%**
9. 88 to 99 **13%**
10. 135 to 200 **48%**
11. 12 to 18 **50%**
12. 2 to 7 **250%**
13. 12 to 63 **425%**
14. 120 to 240 **100%**
15. 15 to 35 **133%**

16. **Business** A worker earning \$5.15/h receives a raise. She now earns \$6/h. Find the percent of increase in her hourly rate of pay. **17%**

Find each percent of markup. Round to the nearest percent.

17. \$22 marked up to \$33 **50%**
18. \$15 marked up to \$60 **300%**
19. \$13.50 marked up to \$25 **85%**
20. \$40 marked up to \$59.75 **49%**

21. **Clothing** Find the percent of markup for a shirt that a store buys for \$3.25 and sells for \$7.50. **131%**

Find each percent of discount. Round to the nearest percent.

22. \$70 discounted to \$63 **10%**
23. \$9 discounted to \$4 **56%**
24. \$10 discounted to \$7 **30%**
25. \$480 discounted to \$300 **38%**

26. **Crafts** A package of poster board usually sells for \$8.40. This week the package is on sale for \$6.30. What is the percent of discount? **25%**

B 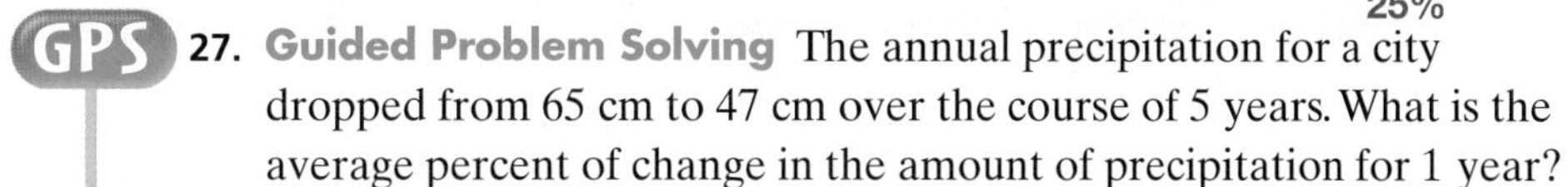

27. **Guided Problem Solving** The annual precipitation for a city dropped from 65 cm to 47 cm over the course of 5 years. What is the average percent of change in the amount of precipitation for 1 year?
- What was the amount of change in the precipitation over 5 years?
- What was the percent of change over 5 years? **5.5% decrease**

28. A scientist earning an annual salary of \$49,839 gets a 4% raise. Estimate the new annual salary for this scientist. **about \$52,000**

29. **Sports** A football player gained 1,200 yd last season and 900 yd this season. Find the percent of change. State whether the change is an increase or a decrease. **25% decrease**

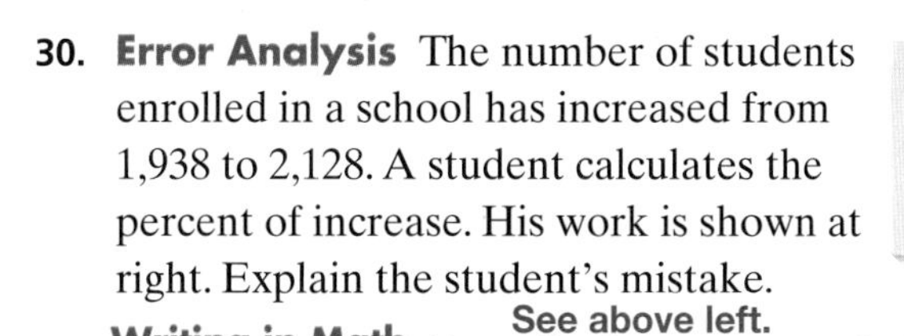

30. **Error Analysis** The number of students enrolled in a school has increased from 1,938 to 2,128. A student calculates the percent of increase. His work is shown at right. Explain the student's mistake. **See above left.**

$2{,}128 - 1{,}938 = 190$
$190 \div 2{,}128 \approx 0.089$
$0.089 = 8.9\%$

31. **Writing in Math** Describe how you can find the percent of change in the number of students in your school from last year to this year. **See margin.**

30. The student should have divided by 1,938 instead of 2,128.

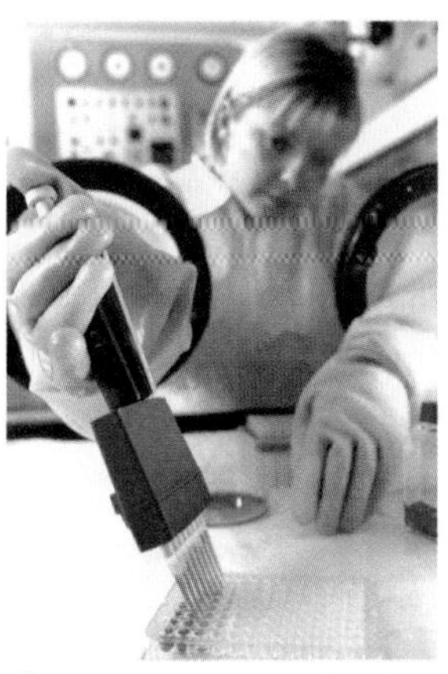

Careers Scientists develop vaccines and treatments.

3. Practice

Assignment Guide

Check Your Understanding

Go over Exercises 1–7 in class before assigning the Homework Exercises.

Homework Exercises

A Practice by Example	8–26
B Apply Your Skills	27–35
C Challenge	36
Test Prep and Mixed Review	37–41

Homework Quick Check

To check students' understanding of key skills and concepts, go over Exercises 14, 22, 29, 31, and 34.

Differentiated Instruction Resources

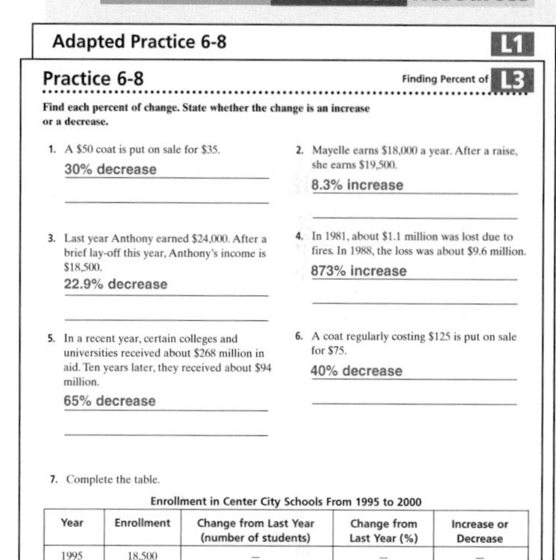

Adapted Practice 6-8 L1

Practice 6-8 Finding Percent of L3

Find each percent of change. State whether the change is an increase or a decrease.

1. A \$50 coat is put on sale for \$35. **30% decrease**
2. Mayelle earns \$18,000 a year. After a raise, she earns \$19,500. **8.3% increase**
3. Last year Anthony earned \$24,000. After a brief lay-off this year, Anthony's income is \$18,500. **22.9% decrease**
4. In 1981, about \$1.1 million was lost due to fires. In 1988, the loss was about \$9.6 million. **873% increase**
5. In a recent year, certain colleges and universities received about \$268 million in aid. Ten years later, they received about \$94 million. **65% decrease**
6. A coat regularly costing \$125 is put on sale for \$75. **40% decrease**
7. Complete the table.

Enrollment in Center City Schools From 1995 to 2000

Year	Enrollment	Change from Last Year (number of students)	Change from Last Year (%)	Increase or Decrease
1995	18,500	–	–	–
1996	19,300	800	4%	increase
1997	19,700	400	2%	increase
1998	19,500	200	1%	decrease
1999	19,870	370	2%	increase
[illegible]	[illegible]	[illegible]	3%	decrease

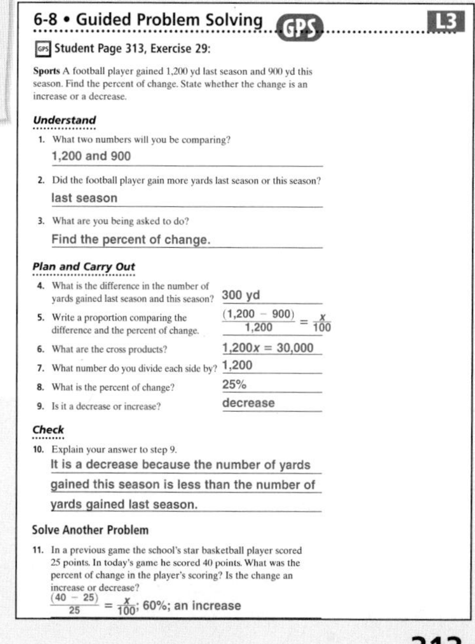

6-8 • Guided Problem Solving GPS L3

Student Page 313, Exercise 29:

Sports A football player gained 1,200 yd last season and 900 yd this season. Find the percent of change. State whether the change is an increase or a decrease.

Understand

1. What two numbers will you be comparing? **1,200 and 900**
2. Did the football player gain more yards last season or this season? **last season**
3. What are you being asked to do? **Find the percent of change.**

Plan and Carry Out

4. What is the difference in the number of yards gained last season and this season? **300 yd**
5. Write a proportion comparing the difference and the percent of change. $\frac{(1{,}200 - 900)}{1{,}200} = \frac{x}{100}$
6. What are the cross products? $1{,}200x = 30{,}000$
7. What number do you divide each side by? **1,200**
8. What is the percent of change? **25%**
9. Is it a decrease or increase? **decrease**

Check

10. Explain your answer to step 9. **It is a decrease because the number of yards gained this season is less than the number of yards gained last season.**

Solve Another Problem

11. In a previous game the school's star basketball player scored 25 points. In today's game he scored 40 points. What was the percent of change in the player's scoring? Is the change an increase or decrease? $\frac{(40 - 25)}{25} = \frac{x}{100}$; **60%; an increase**

4. Assess & Reteach

Lesson Quiz

Find each percent of change and round to the nearest tenth. State whether it is an increase or decrease.

1. 10 to 25 **150% increase**
2. 95 to 76 **20% decrease**
3. 36 to 30 **16.7% decrease**
4. 70 to 82 **17.1% increase**

35. See back of book.

Reteaching 6-8 — Finding Percent of Change — L2

Percent of change is the percent something increases or decreases from its original amount.

	Find the percent of increase from 12 to 18.	Find the percent of decrease from 20 to 12.
① Subtract to find the amount of change.	$18 - 12 = 6$	$20 - 12 = 8$
② Write a proportion. $\frac{\text{change}}{\text{original}} = \frac{\text{percent}}{100}$	$\frac{6}{12} = \frac{n}{100}$	$\frac{8}{20} = \frac{n}{100}$
	$6 \cdot 100 = 12n$	$8 \cdot 100 = 20n$
③ Solve for n.	$n = 50$	$n = 40$
	The percent of increase is 50%.	The percent of decrease is 40%.

State whether the change is an *increase* or *decrease*. Complete to find the percent of change.

1. 40 to 60
   $60 - 40 =$ 20
   $\frac{20}{40} = \frac{n}{100}$
   20 $\cdot 100 = 40n$
   $n =$ 50
   increase of 50%
2. 15 to 9
   $15 - 9 =$ 6
   $\frac{6}{15} = \frac{n}{100}$
   6 $\cdot 100 = 15n$
   $n =$ 40
   decrease of 40%
3. 0.4 to 0.9
   $0.9 - 0.4 =$ 0.5
   $\frac{0.5}{0.4} = \frac{n}{100}$
   0.5 · 100 $= 0.4n$
   $n =$ 125
   increase of 125%

Find the percent of *increase*.

4. 16 to 40 — 150%
5. 20 to 22 — 10%
6. 9 to 18 — 100%
7. 28 to 35 — 25%
8. 80 to 112 — 40%
9. 150 to 165 — 10%

Find the percent of *decrease*.

10. 20 to 15 — 25%
11. 100 to 57 — 43%
12. 52 to 26 — 50%
13. 140 to 126 — 10%
14. 75 to 72 — 4%
15. 1000 to 990 — 1%

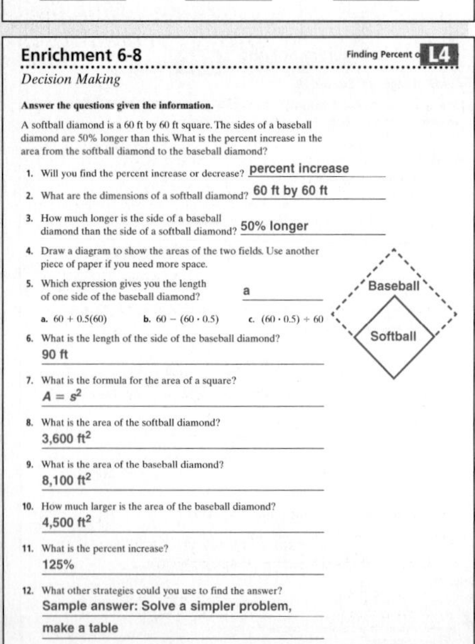

Enrichment 6-8 — Finding Percent of Change — L4

Decision Making

Answer the questions given the information.

A softball diamond is a 60 ft by 60 ft square. The sides of a baseball diamond are 50% longer than this. What is the percent increase in the area from the softball diamond to the baseball diamond?

1. Will you find the percent increase or decrease? percent increase
2. What are the dimensions of a softball diamond? 60 ft by 60 ft
3. How much longer is the side of a baseball diamond than the side of a softball diamond? 50% longer
4. Draw a diagram to show the areas of the two fields. Use another piece of paper if you need more space.
5. Which expression gives you the length of one side of the baseball diamond? a
   a. $60 + 0.5(60)$ b. $60 - (60 \cdot 0.5)$ c. $(60 \cdot 0.5) + 60$
6. What is the length of the side of the baseball diamond? 90 ft
7. What is the formula for the area of a square? $A = s^2$
8. What is the area of the softball diamond? 3,600 ft²
9. What is the area of the baseball diamond? 8,100 ft²
10. How much larger is the area of the baseball diamond? 4,500 ft²
11. What is the percent increase? 125%
12. What other strategies could you use to find the answer? Sample answer: Solve a simpler problem, make a table

GO Online
Homework Video Tutor
Visit: PHSchool.com
Web Code: are-0608

Find the price of each item.

32. originally \$35.75; 65% markup **\$58.99**
33. originally \$82; 35% discount **\$53.30**
34. **Choose a Method** A TV goes on sale with a discount of 28%. The original price of the TV is \$942. What is the sale price of the TV? **\$678.24**
35. **Business** A toy store opened five years ago. The owner uses a computer to track sales. She uses a program that prints @@@ in some cells instead of numbers. Copy and complete the spreadsheet. **See margin.**

	A	B	C	D
1	Year	Sales (\$)	Change From Last Year (\$)	Change From Last Year (%)
2	1	200,000	(not open last year)	(not open last year)
3	2	240,000	40,000	@@@
4	3	300,000	@@@	@@@
5	4	330,000	@@@	@@@

Ⓒ 36. **Challenge** A storeowner buys a case of 144 pens for \$28.80. Tax and shipping cost an additional \$8.64. He sells the pens for \$.59 each. What is the markup per pen? What is the percent of markup? **\$.33; 126.9%**

Test Prep and Mixed Review — Practice

Multiple Choice

37. Which of the following represents the least percent of change? **A**
    Ⓐ A child grew from 40 inches to 46 inches in one year.
    Ⓑ Internet service costs increased from \$21 to \$25 per month.
    Ⓒ An after-school program enrollment was 72 and is now 84.
    Ⓓ A child's weekly allowance is changed from \$5 to \$6.
38. Mr. Chun earns a salary of \$150 a week plus 8% commission on all sales. How much will he earn if his sales in one week are \$2,990? **G**
    Ⓕ \$3,002 Ⓖ \$389.20 Ⓗ \$251.20 Ⓙ \$239.20
39. Julie took a taxi from school to her home. The taxi rate started at \$2.00 and then \$0.50 was added for every $\frac{1}{4}$ mile traveled. What information is needed to find the cost of the taxi ride? **B**
    Ⓐ Number of minutes she rode in the taxi
    Ⓑ Number of miles Julie's home is from the school
    Ⓒ Number of gallons of gasoline used for the trip
    Ⓓ Average speed of the taxi

GO for Help

For Exercises	See Lesson
40–41	6-7

Algebra **Find each payment.**

40. \$218 with a 6.25% sales tax **\$231.63**
41. \$451 with a 4.5% sales tax **\$471.30**

Test Prep

Resources

For additional practice with a variety of test item formats:

- Test-Taking Strategies, p. 315
- Test Prep, p. 319
- Test-Taking Strategies with Transparencies

Alternative Assessment

Provide pairs of students with 10 index cards. Each partner writes a different number on each of five of the cards. Partners mix the cards, make two piles, and designate one as the original number. Partners work together to find the percent of increase or decrease for each pair of numbers.

Working Backward

A useful problem solving strategy for answering multiple-choice questions is to *Work Backward*. Check to see which choice results in a correct answer by substituting the answers into the problem.

EXAMPLE

In a pile of dimes and quarters, there are twice as many dimes as quarters. The total value of the coins is $9.45. How many quarters are in the pile?

Ⓐ 11 Ⓑ 18 Ⓒ 21 Ⓓ 24

Check each answer to see whether it works.

Choice A	11 quarters: 2.75	22 dimes: 2.20	2.75 + 2.20 = 4.95 ✗
Choice B	18 quarters: 4.50	36 dimes: 3.60	4.50 + 3.60 = 8.10 ✗
Choice C	21 quarters: 5.25	42 dimes: 4.20	5.25 + 4.20 = 9.45 ✓
Choice D	24 quarters: 6.00	48 dimes: 4.80	6.00 + 4.80 = 10.80 ✗

The correct answer is choice C.

Exercises

Solve each problem by working backward.

1. What is the greatest number of movie tickets you can buy if you have $33.48 and each movie ticket costs $6.75? B
   Ⓐ 3 Ⓑ 4 Ⓒ 5 Ⓓ 6
2. Your grades on four math tests are 97, 88, 79, and 92. What grade do you need on the fifth test to have a mean of 90? H
   Ⓕ 90 Ⓖ 92 Ⓗ 94 Ⓙ 96
3. If you start with a number, add 5, and then multiply by 7, the result is 133. What is the number? B
   Ⓐ 12 Ⓑ 14 Ⓒ 15 Ⓓ 21
4. For your birthday, you receive $48 and a $15 gift certificate to a department store. The store is having a sale that takes 40% off the price of all items. What is the total value of the merchandise you can buy and still have $7.50 left for lunch? H
   Ⓕ $70.50 Ⓖ $88.20 Ⓗ $92.50 Ⓙ $100

Working Backward

In general, students can use the Working Backward strategy effectively when they know a solution but need to find out information that led to that solution. This feature guides students to work backwards to solve multiple-choice questions.

Guided Instruction

Guide students to stop checking answer choices once their substitution gives the desired result.

Assign the Exercises for independent work. Elicit from students that the answer choices in each are too close to leave to educated guesses or estimates.

Resources

Test-Taking Strategies with Transparencies
- Transparency 7
- Practice sheet, p. 18

Test-Taking Strategies with Transparencies

Test-Taking Strategies: Working Backward

Sometimes it is easier to start with an end result and work backward.

Example What is the prime factorization of 48?

A. 6×2^3 B. $2^4 \times 3$ C. 4×12 D. 2^5

Try each choice:

Since 6 is not a prime number, choice A is wrong.

Since 4 is not a prime number, choice C is wrong.

Compare choice B and choice D. Is 48 divisible by 3? Yes. So, choice D is wrong.

The answer is $2^4 \times 3$, or choice B.

Work backward to find the answer. Explain your reasoning.

1. Which number is composite, less than 50, and has a tens digit greater than its ones digit?
   A. 51 B. 13 C. 28 D. 32
2. Carlos has $\frac{2}{3}$ of a bushel of apples, weighing 26 pounds. How much does a bushel of apples weigh?
   F. 20 pounds G. 50 pounds
   H. 39 pounds J. 33 pounds

Resources

Student Edition
Extra Skills and Word Problems Practice, Ch. 6, p. 640
English/Spanish Glossary, p. 676
Formulas and Properties, p. 674
Tables, p. 670

All in One Teaching Resources
Vocabulary and Study Skills 6F — L3

Differentiated Instruction

Spanish Vocabulary Workbook with Study Skills — ELL
Interactive Textbook
- Audio Glossary

Online Vocabulary Quiz

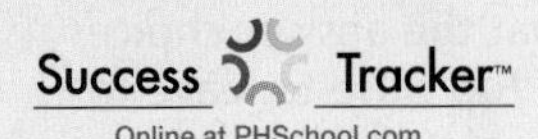

Online at PHSchool.com

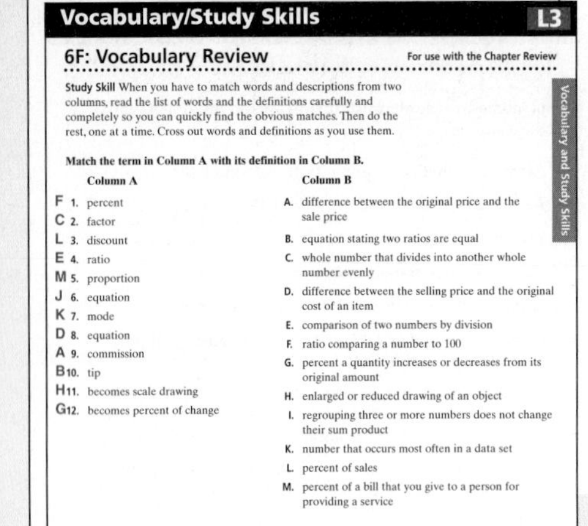
Spanish Vocabulary/Study Skills — ELL

Vocabulary/Study Skills — L3

6F: Vocabulary Review — For use with the Chapter Review

Study Skill When you have to match words and descriptions from two columns, read the list of words and the definitions carefully and completely so you can quickly find the obvious matches. Then do the rest, one at a time. Cross out words and definitions as you use them.

Match the term in Column A with its definition in Column B.

Column A

F 1. percent
C 2. factor
L 3. discount
E 4. ratio
M 5. proportion
J 6. equation
K 7. mode
D 8. equation
A 9. commission
B 10. tip
H 11. becomes scale drawing
G 12. becomes percent of change

Column B

A. difference between the original price and the sale price
B. equation stating two ratios are equal
C. whole number that divides into another whole number evenly
D. difference between the selling price and the original cost of an item
E. comparison of two numbers by division
F. ratio comparing a number to 100
G. percent a quantity increases or decreases from its original amount
H. enlarged or reduced drawing of an object
I. regrouping three or more numbers does not change their sum product
K. number that occurs most often in a data set
L. percent of sales
M. percent of a bill that you give to a person for providing a service

Vocabulary and Study Skills

Chapter 6 Review

Vocabulary Review

commission (p. 305)
discount (p. 311)
markup (p. 311)
percent (p. 274)
percent of change (p. 310)

Choose the vocabulary term from the column on the right that completes the sentence.

1. The difference between the selling price and a store's cost is the ___?___. **C**
2. A ___?___ can be an increase or a decrease. **E**
3. A ___?___ is a ratio that compares a number to 100. **D**
4. The difference between the original price and the sale price is the ___?___. **B**
5. A ___?___ is a percent of the sales made by a salesperson. **A**

A. commission
B. discount
C. markup
D. percent
E. percent of change

Go Online PHSchool.com
For: Online vocabulary quiz
Web Code: arj-0651

Skills and Concepts

Lessons 6-1, 6-2

- To model percents and to write percents using equivalent ratios
- To convert between fractions, decimals, and percents

A **percent** is a ratio that compares a number to 100.

To write a decimal as a percent, multiply the decimal by 100, or move the decimal point two places to the right. To write a percent as a decimal, divide by 100, or move the decimal point two places to the left.

To write a fraction as a percent, first convert the fraction into a decimal. To write a percent as a fraction, write the percent with a denominator of 100 and simplify.

Write each percent as a decimal and as a fraction in simplest form.

6. 65% **0.65; $\frac{13}{20}$**
7. 2% **0.02; $\frac{1}{50}$**
8. 1.8% **0.018; $\frac{9}{500}$**
9. $62\frac{1}{2}\%$ **0.625; $\frac{5}{8}$**
10. Write $\frac{3}{8}$ as a percent. **37.5%**
11. Write 0.16 as a percent. **16%**

Lessons 6-3, 6-4

- To convert between fractions, decimals, and percents greater than 100% or less than 1%
- To find and estimate the percent of a number

A mixed number represents a percent greater than 100%. A proper fraction represents a percent less than 100%. To find a percent of a whole, write the percent as a decimal or fraction and then multiply.

Find each answer.

12. Find 83% of 54. **44.82**
13. What is 4% of 16? **0.64**
14. Find 135% of 72. **97.2**

Lessons 6-5, 6-6

- To use proportions to solve problems involving percent
- To use equations to solve problems involving percent

Percent problems are solved by using a proportion or an equation.

Use a proportion or an equation to solve.

15. What percent of 40 is 28? **70%**

16. 38 is 80% of what number? **47.5**

17. What is 60% of 420? **252**

18. 80% of 15 is what number? **12**

19. 0 54 n / 0% 75% 100% **72**

20. 0 36 180 / 0% n% 100% **20%**

21. Technology The price of a new version of a computer game is 120% of the price of the original version. The original version cost $48. What is the cost of the new version? **$57.60**

Lesson 6-7

- To find and estimate solutions to application problems involving percent

A tip is a percent of a bill that you give to the person providing a service. A **commission** is a percent of a sale.

22. You go to a restaurant with four other people. The total for the food is $43.85. You need to add 5% for tax and 15% for tip. If you decide to split the bill evently, estimate how much you will pay. **about $10.52**

23. Diamonds Find the commission on a diamond that is sold for $6,700 when the commission paid is 4%. **$268**

Find each commission, given the sale and the commission rate.

24. $700, 9% **$63**

25. $3,600, 6% **$216**

26. $5,000, 5.5% **$275**

27. Insurance An insurance company pays its agents 40% commission on the first year's premium and 5% on the second year's premium for life insurance policies. If the premiums are $500 per year, what is the total commission that will be paid during the two years? **$225**

Lesson 6-8

- To find percents of increase and percents of decrease

A **percent of change** is the percent a quantity increases or decreases from its original amount. Use the proportion $\frac{\text{amount of change}}{\text{original amount}} = \frac{\text{percent change}}{100}$. **Markup** is an example of a percent of increase. **Discount** is an example of a percent of decrease.

Find each percent of change. Tell whether it is an increase or a decrease.

28. $90 to $75 **16.7% decrease**

29. 3.5 ft to 4.2 ft **20% increase**

30. 120 lb to 138 lb **15% increase**

31. 300 cm to 420 cm **40% increase**

32. 80.5 g to 22.5 g **72% decrease**

33. 108 kg to 90 kg **16.7% decrease**

34. Shopping The sale price of a game is $24.95. Its original price was $36.00. Find the percent of change. Round to the nearest percent.

34. 31% decrease

Alternative Assessment L4

Alternative Assessment Form C

Chapter 6

SAVING FOR COLLEGE

Michelle has to make a presentation to parents, advocating savings accounts to help pay for their children's college. She knows that interest rates change all the time. Michelle created this table to give parents an idea of how much different amounts of money would be worth after a year at different simple interest rates.

1. Copy and complete the table.

	1%	2%	3%	4%	5%	6%	7%	8%	9%
$1,000	$1,010	$1,020	$1,030						
$2,000	$2,020	$2,040							
$3,000	$3,030								
$4,000									
$5,000									
$6,000									
$7,000									
$8,000									
$9,000									

2. Suppose Michelle had written the interest rates in the table as decimals rather than percents. What decimals would appear?
3. How can you use the table to find the amount after one year if $1,000 is invested at $1\frac{1}{2}$%? If $5,000 is invested at $8\frac{1}{2}$%?
4. How can you use the table to find the amount after one year if $1,500 is invested at 1%? If $6,500 is invested at 6%?
5. How can you use the table to find the amount after one year if $1,000 is invested at 10%? If $5,000 is invested at 12%?
6. Use the information in the table to write a general formula for calculating the amount saved (S) after 1 year in terms of the interest rate (R) and the initial amount invested (I).

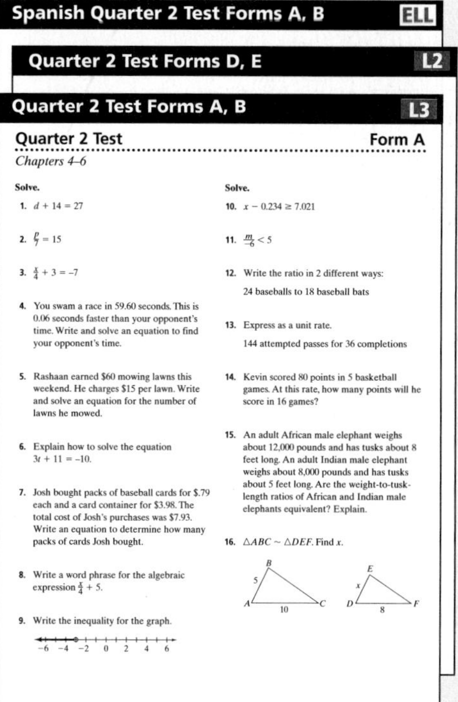

Spanish Quarter 2 Test Forms A, B ELL

Quarter 2 Test Forms D, E L2

Quarter 2 Test Forms A, B L3

Quarter 2 Test Form A

Chapters 4–6

Solve.

1. $d + 14 = 27$
2. $\frac{p}{7} = 15$
3. $\frac{x}{4} + 3 = -7$
4. You swam a race in 59.60 seconds. This is 0.06 seconds faster than your opponent's time. Write and solve an equation to find your opponent's time.
5. Rashaan earned $60 mowing lawns this weekend. He charges $15 per lawn. Write and solve an equation for the number of lawns he mowed.
6. Explain how to solve the equation $3t + 11 = -10$.
7. Josh bought packs of baseball cards for $.79 each and a card container for $3.98. The total cost of Josh's purchases was $7.93. Write an equation to determine how many packs of cards Josh bought.
8. Write a word phrase for the algebraic expression $\frac{x}{4} + 5$.
9. Write the inequality for the graph.

Solve.

10. $x - 0.234 \geq 7.021$
11. $\frac{m}{-6} < 5$
12. Write the ratio in 2 different ways:

    24 baseballs to 18 baseball bats
13. Express as a unit rate.

    144 attempted passes for 36 completions
14. Kevin scored 80 points in 5 basketball games. At this rate, how many points will he score in 16 games?
15. An adult African male elephant weighs about 12,000 pounds and has tusks about 8 feet long. An adult Indian male elephant weighs about 8,000 pounds and has tusks about 5 feet long. Are the weight-to-tusk-length ratios of African and Indian male elephants equivalent? Explain.
16. $\triangle ABC \sim \triangle DEF$. Find x.

Resources

- ExamView Assessment Suite CD-ROM
  - Chapter 6 Ready-Made Test
  - Make your own Chapter 6 test
- MindPoint Quiz Show CD-ROM
  - Chapter 6 Review

Differentiated Instruction

All in One Teaching Resources
- Below Level Chapter 6 Test L2
- Chapter 6 Test L3
- Chapter 6 Alternative Assessment L4

Spanish Assessment Resources ELL
- Below Level Chapter 6 Test L2
- Chapter 6 Test L3
- Chapter 6 Alternative Assessment L4

ExamView Assessment Suite CD-ROM
- Special Needs Test L1
- Special Needs Practice Bank L1

Online Chapter 6 Test at www.PHSchool.com L3

Below Level Chapter Test L2

Chapter Test L3

Chapter Test — Form A
Chapter 6

Write each decimal as a percent and write each percent as a decimal.
1. 4% 0.04 2. 0.6 60% 3. 225% 2.25
4. 0.0032 0.32% 5. 0.72% 0.0072 6. 5.2 520%

Write each fraction as a percent and each percent as a fraction.
7. 85% $\frac{17}{20}$ 8. $\frac{3}{5}$ 60% 9. 6% $\frac{3}{50}$
10. $\frac{3}{8}$ 37.5% 11. 350% $3\frac{1}{2}$ 12. $\frac{8}{5}$ 160%

Model each percent on these 10 × 10 grids.
13. 60% 14. 15%

Write a proportion and solve.
15. What percent of 80 is 42? 52.5%
16. 80% of 70 is what number? 56

Write an equation for each question. Then solve the equation.
17. 63 is 25% of what number? $63 = 0.25x$; $x = 252$
18. 20% of what number is 80? $0.2x = 80$; $x = 400$
19. What percent of 72 is 12? $\frac{x}{100} \cdot 72 = 12$; $x = 16\frac{2}{3}\%$
20. 12% of 600 is what number? $0.12 \cdot 600 = x$; $x = 72$

Chapter 6 Test

Go Online PHSchool.com — For: Online chapter test — Web Code: ara-0652

Write each decimal as a percent and write each percent as a decimal.

1. 5% **0.05**
2. 0.3 **30%**
3. 125% **1.25**
4. 0.0045 **0.45%**
5. 0.39% **0.0039**
6. 3.4 **340%**

Write each fraction as a percent. Write each percent as a fraction.

7. 35% $\frac{7}{20}$
8. 125% $\frac{5}{4}$
9. 2% $\frac{1}{50}$
10. $\frac{7}{8}$ **87.5%**
11. $\frac{3}{4}$ **75%**
12. $\frac{6}{5}$ **120%**
13. According to the U.S. Census Bureau, 0.98% of females in the United States in 1990 were named Barbara. Express this percent as a fraction. $\frac{49}{5{,}000}$

Model each percent on a 10 × 10 grid.

14–16. See margin.

14. 34%
15. 285%
16. $12\frac{1}{2}\%$
17. You work 20 hours per week at a grocery store during the summer. Sixty percent of your job is restocking the shelves. How many hours per week do you spend restocking the shelves? **12 h**
18. Draw a model and write a proportion to find the answer to "25 percent of what number is 30?" **See margin.**

Write an equation for each question. Then solve the equation.

19. What percent of 82 is 10.25? $82x = 10.25$; 12.5%
20. 108% of 47 is what number? $(1.08)(47) = x$; 50.76
21. 99 is 72% of what number? $99 = 0.72x$; 137.5
22. 12 is what percent of 1,920? $12 = 1{,}920x$; 0.625%
23. What is 62% of 128? $x = (0.62)(128)$; 79.36
24. 168% of what number is 714? $1.68x = 714$; 425
25. In a grade of 250 students, there are 6 sets of twins. What percent of the students in this grade have a twin? **4.8%**

Write a proportion for each model. Solve for *n*.

26. 0, 50, 80; 0%, n%, 100% $\frac{50}{80} = \frac{n}{100}$; 62.5
27. 0, 60, n; 0%, 80%, 100% $\frac{60}{n} = \frac{80}{100}$; 75
28. **Shopping** You buy a sweater for $18.75, which is 25% off the original price. What was the original price? **$25**
29. To prepare for competitions, your swimming coach required you to swim 8 lengths in the pool. You swam 10 lengths. What percent of the required practice did you swim? **125%**

Find each percent of change. Round to the nearest tenth of a percent. State whether the change is an increase or a decrease.

30. 4.15 to 4.55 **9.6% increase**
31. 379 to 302 **20.3% decrease**
32. 72 to 102 **41.7% increase**
33. **Jobs** According to the U.S. Department of Labor, total employment is expected to increase from 146 million in 2000 to 168 million in 2010. Find the percent of increase. **15.1% increase**
34. **Restaurants** You order items from a menu that total $7.85. Your bill comes to $8.30, including tax. What is the percent of the tax? Round to the nearest tenth of a percent. **5.7%**
35. A salesperson receives a salary of $300 per week and a 6% commission on all sales. How much does this salesperson earn in a week with $2,540 in sales? **$452.40**
36. A bicycle store pays $29.62 for a helmet. The store sells the helmet for $39.99. Find the percent of markup. **35% markup**
37. **Writing in Math** How do you determine whether you are finding a percent of increase or a percent of decrease between two values? Explain. **See margin.**

14–16. See back of book.

18.
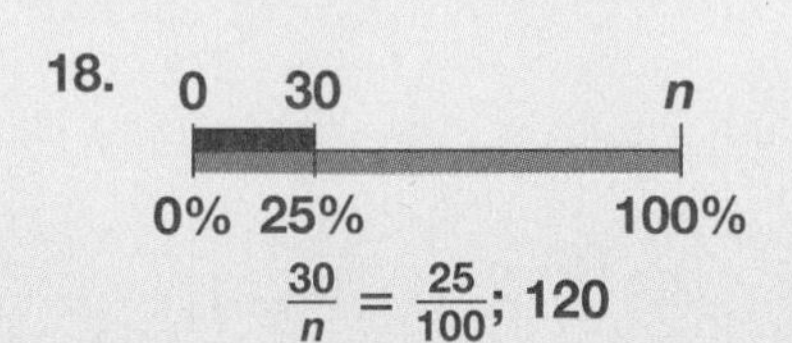

$\frac{30}{n} = \frac{25}{100}$; 120

37. If the original value is less than the new value, it is an increase. If the original value is greater than the new value, it is a decrease.

Test Prep

Multiple Choice

Read each question. Then write the letter of the correct answer on your paper.

1. Which number is closest to 35% of 1,291? **B**
   (A) 400 (B) 450 (C) 500 (D) 550

2. Which equation is NOT equivalent to $2x - 3 = 5$? **G**
   (F) $2x = 8$ (H) $2x - 4 = 4$
   (G) $4x - 3 = 10$ (J) $x - 1.5 = 2.5$

3. Which expression equals $3 \times 3 \times 3 \times 3$? **A**
   (A) 3^4 (B) 4^3 (C) 4×3 (D) 3^3

4. In which set of numbers is 9 a factor of all the numbers? **G**
   (F) 36, 18, 21 (H) 98, 81, 450
   (G) 108, 252, 45 (J) 120, 180, 267

5. Which point shows the product $\left(1\frac{7}{8}\right)\left(2\frac{1}{5}\right)$? **C**

   A B C D
   0 1 2 3 4 5

6. Which fraction is closest in value to 0.46? **G**
   (F) $\frac{19}{50}$ (G) $\frac{22}{50}$ (H) $\frac{25}{50}$ (J) $\frac{28}{50}$

7. You buy a sandwich for \$3.45, a salad for \$2.25, and a drink for \$.89. How much change do you receive from a \$10 bill? **C**
   (A) \$16.59 (B) \$4.30 (C) \$3.41 (D) \$2.59

8. Which statement is NOT true? **J**
   (F) $\frac{12}{16} = \frac{9}{12}$ (H) $\frac{12}{9} = \frac{16}{12}$
   (G) $\frac{12 + 16}{16} = \frac{9 + 12}{12}$ (J) $\frac{12 + 1}{16} = \frac{9 + 1}{12}$

9. What is $\frac{5}{8}$ written as a percent? **C**
   (A) 625% (B) 160% (C) $62\frac{1}{2}\%$ (D) 16%

10. What is the order of the numbers 0.361×10^7, 4.22×10^7, and 13.5×10^6 from least to greatest? **H**
    (F) $13.5 \times 10^6, 0.361 \times 10^7, 4.22 \times 10^7$
    (G) $4.22 \times 10^7, 13.5 \times 10^6, 0.361 \times 10^7$
    (H) $0.361 \times 10^7, 13.5 \times 10^6, 4.22 \times 10^7$
    (J) $13.5 \times 10^6, 4.22 \times 10^7, 0.361 \times 10^7$

11. What is the value of $\frac{2m}{m + 2n}$ when $m = -4$ and $n = 3$? **B**
    (A) -8 (B) -4 (C) 0 (D) 4

12. Which is the best estimate of $92.56 \cdot 37.1$? **G**
    (F) 2,700 (G) 3,600 (H) 4,000 (J) 4,500

13. Which expression has the greatest value? **A**
    (A) $32 - (-12)$ (C) $-32 - (-12)$
    (B) $32 - |-12|$ (D) $|-32 - (-12)|$

14. The mean of six numbers is 9. Five of the numbers are 4, 7, 9, 10, and 11. What is the sixth number? **J**
    (F) 6 (G) 9 (H) 12 (J) 13

Gridded Response

Record your answer in a grid.

15. A map's scale is 1 in. : 15 mi. Two towns are 3.5 in. apart on the map. How many miles apart are the two towns? **52.5**

16. A video store charges \$.75 per day for overdue videos. Your friend has a video that was due on Sunday. She returns it on the following Friday. How much does she owe? **3.75**

Short Response **17–19. See margin.**

17. Eighteen students in a class of 25 students plan to go on a hiking trip. What percent of the students plan to go on the trip? Show your work.

18. A blue shark swims about 2.26 mi in 10 min. What is the speed of the shark in miles per minute and in miles per hour?

Extended Response

19. A movie theater charges \$9 for admission and \$4.50 for a bucket of popcorn. Write an expression for the total cost for a group of friends to see a movie and split one bucket of popcorn. Then evaluate your expression for five friends.

Item	1	2	3	4	5	6	7	8	9	10	11	12	13	14	15	16	17	18	19
Lesson	6-4	4-6	2-1	2-2	3-4	2-6	1-2	5-3	6-2	2-8	4-1	1-1	1-6	1-10	5-6	1-3	6-5	5-2	4-1

17. [2] $\frac{18}{25} = 0.72 = 72\%$
72% plan to go.
[1] minor error OR answer only

18. [2] $\frac{2.26 \text{ mi}}{10 \text{ min}} \cdot \frac{6}{6} = \frac{13.56 \text{ mi}}{60 \text{ min}}$
The speed of the shark is 13.56 miles per hour.
[1] minor error OR answer only

19. [4] Let n be the number of friends and T be the total cost. Then $T = 9n + 4.50$. If $n = 5$, then $T = 9(5) + 4.50 = 45 + 4.50 = 49.50$. For 5 friends, the total cost would be \$49.50.
[3] minor error
[2] incomplete explanation but correct answer
[1] answer only

Resources

Test Prep Workbook

All in One Teaching Resources
- Cumulative Review L3

ExamView Assessment Suite CD-ROM
- Standardized Test Practice

Differentiated Instruction

Progress Monitoring Assessments
- Benchmark Test 3 L3
- Quarter 2 Test
  - Forms A & B L3
  - Forms D & E L2
- Mid-Course Test
  - Forms A & B L3
  - Forms D & E L2

Spanish Assessment Resources
- Spanish Cumulative Review ELL
- Spanish Quarter 2 Test
  - Forms A & B L3
- Spanish Mid-Course Test
  - Forms A & B L3

ExamView Assessment Suite CD-ROM
- Special Needs Practice Bank L1

Spanish Cumulative Review ELL

Cumulative Review L3

Cumulative Review

Chapter 1–6

Multiple choice. Circle the letter of the best answer.

1. Which number is 1.92509 rounded to the nearest ten-thousandth?
   (A) 1.9251 B. 1.9250 C. 1.93000 D. 1.90000
2. Find the value of $(3.1 + 9.8) + 0.9$.
   F. 138 (G) 13.8 H. 12.9 J. 12
3. Find the product. 2.4×0.06
   A. 0.0144 (B) 0.144 C. 1.44 D. 14.4
4. Find the quotient. $0.45 \div 9$
   F. 5 G. 0.5 (H) 0.05 J. 0.005
5. Which variable expression could you use to find the number of pencils in b boxes if there are 12 pencils in a box?
   A. $b + 12$ B. $b - 12$ C. $\frac{b}{12}$ (D) $12b$
6. Which of the following is in order from least to greatest?
   (F) $-9, -5, 0, |-2|$
   G. $|-5|, -4, 2, 6.7$
   H. $1.5, 0.9, -4, -6$
   J. $-1, 0, -0.3, 7$
7. Evaluate $-3 - (-9) + 4$.
   A. -10 B. -8 (C) 10 D. 8
8. Solve the equation $-9x = -72$.
   F. -9 G. -8 (H) 8 J. 9
9. Solve $\frac{x}{5} + 8 = 3$.
   A. 1 B. 0 C. -1 (D) -25
10. Find the LCM of 16 and 20.
    F. 20 G. 40 H. 60 (J) 80
11. Order these fractions from least to greatest. $\frac{2}{5}, \frac{1}{3}$, and $\frac{3}{8}$
    A. $\frac{3}{8}, \frac{2}{5}, \frac{1}{3}$ (B) $\frac{1}{3}, \frac{3}{8}, \frac{2}{5}$ C. $\frac{3}{8}, \frac{1}{3}, \frac{2}{5}$ D. $\frac{2}{5}, \frac{3}{8}, \frac{1}{3}$
12. Evaluate $2^4 \times 3^2$.
    (F) 144 G. 96 H. 72 J. 48
13. Find the GCF of 48 and 60.
    A. 20 B. 40 C. 60 (D) 12
14. Write $\frac{8}{15}$ as a decimal.
    F. 1.875 G. 0.5 H. 0.53 (J) $0.5\overline{3}$
15. Find the sum. $\frac{5}{6} + \frac{1}{3}$
    A. $\frac{4}{9}$ B. $\frac{2}{3}$ (C) $1\frac{1}{6}$ D. $\frac{6}{7}$
16. Find the difference. $8\frac{1}{4} - 5\frac{3}{5}$
    F. $3\frac{13}{20}$ G. $3\frac{2}{9}$ (H) $2\frac{13}{20}$ J. $2\frac{2}{9}$
17. Solve $x - \frac{5}{6} = 2\frac{1}{3}$.
    A. $1\frac{1}{2}$ B. $2\frac{2}{3}$ C. 3 (D) $3\frac{1}{6}$

Applying Percents

Students will use data from these two pages to answer the questions posed in Put It All Together.

Have students examine the photo of the fern and river delta on page 321. Tell them that these are not the only shapes in nature that repeat a design at smaller levels. Ask:

- *What are some features of land or water that, like a river delta, repeat a design at smaller and smaller levels?* **Sample: mountains, coastlines**
- *Can you think of some man-made objects or systems that are designed similarly?* **Sample: roadways, cable networks, and water systems**

Discuss how the flow of traffic, information, or water in man-made systems echoes the flow of water in nature.

Materials

- Photos, maps or diagrams of roadway systems, cable networks, or water systems

Activating Prior Knowledge

Have students describe the human circulatory system, starting with the heart and major arteries, continuing to capillaries in the skin, and returning back to the heart via veins of increasing size. Diagram the system on the board. Ask students to compare the circulatory system to a river, roadway, or cable network.

Guided Instruction

Have a volunteer read aloud the opening paragraphs. Invite students who have seen the Mississippi Delta to describe what it looks like at ground level. Ask:

- *What does the Mississippi River look like when it enters New Orleans?*
- *What does it look like as it flows out of New Orleans and into the Gulf of Mexico?*

Problem Solving Application

Applying Percents

Fractal Facts A fractal is a design that repeats itself at smaller and smaller levels. Fractals give us beautiful, intricate pictures of things like ferns and rivers. They also provide a practical way to increase surface area. For example, the circulatory system branches from arteries into smaller and smaller blood vessels called capillaries. Because there are so many of them, capillaries have a much greater surface area than arteries and can absorb nutrients more effectively.

The activity models how the length of a "blood vessel" increases as it branches out into smaller capillaries.

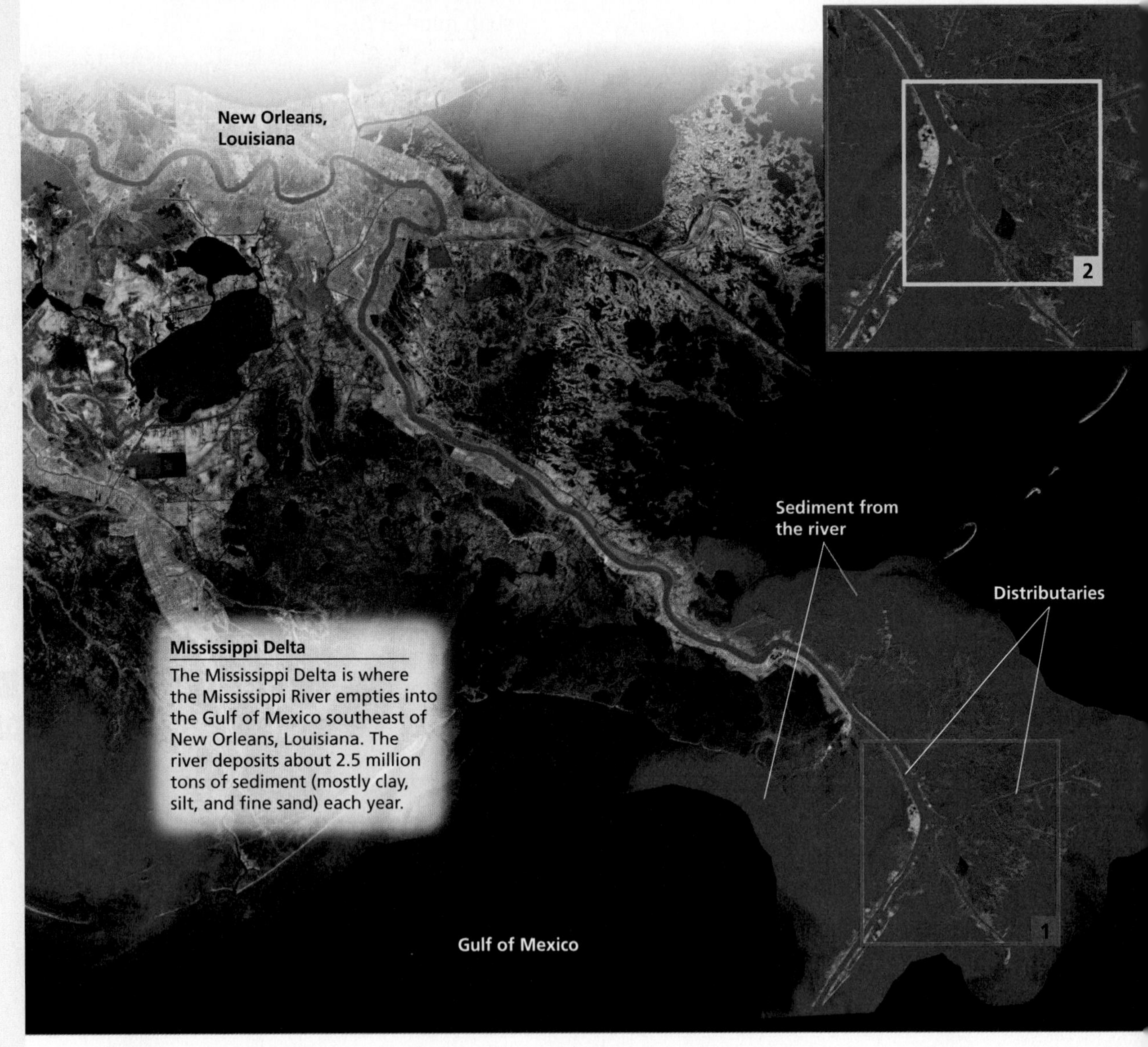

Mississippi Delta

The Mississippi Delta is where the Mississippi River empties into the Gulf of Mexico southeast of New Orleans, Louisiana. The river deposits about 2.5 million tons of sediment (mostly clay, silt, and fine sand) each year.

1a-b. and 2a. Answers may vary. Samples are given.

1a. 27 cm
b. 36 cm; 9 cm
c. 33.3%

2a. 48 cm; 12 cm
b. 33.3%

3. The length increases by 33.3% after each step. The next length would be 64 cm.

4. 77.7%

Fractals in Nature

Each fern leaf (called a "frond") has many small fronds along its main vein. Each of the small fronds also has many even smaller fronds.

Main vein

Fern frond

Smaller fronds

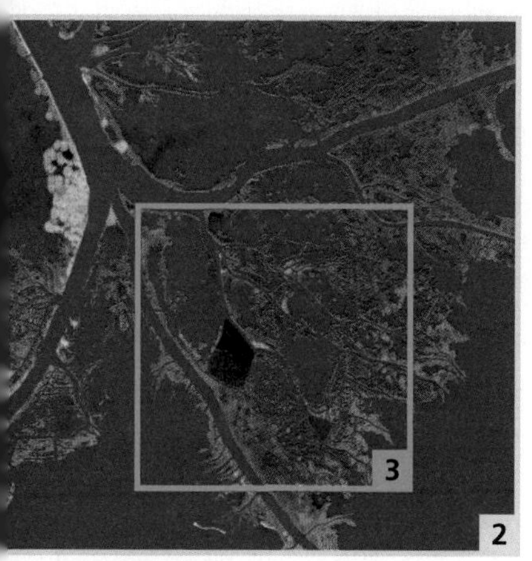

Fractal Structure

At the delta, the Mississippi River splits into distributaries, or branches. Many of the distributaries split into smaller branches. The smallest branches of the river look the same as the larger ones.

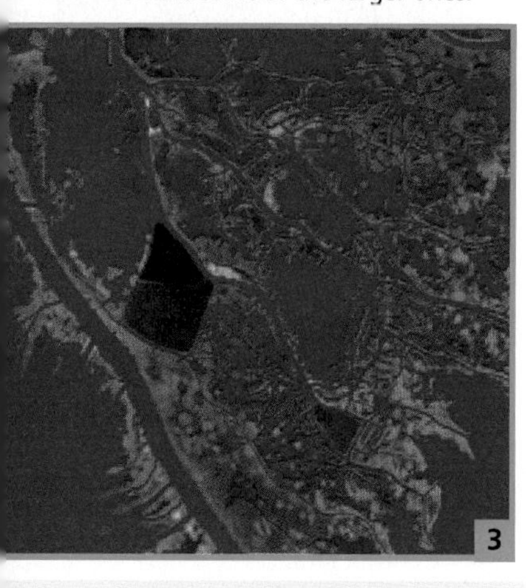

Put It All Together

Materials ruler, scissors, tape

1. Cut four thin strips of paper of equal lengths. Choose a length that is easy to divide into thirds. Use one strip to model a simple blood vessel. Mark it to show three equal segments.

a. How long is your blood vessel?

b. Fold an unmarked strip into thirds and tape it to form a triangle. Attach this triangle to the center segment of your blood vessel. Measure the total length of the paper blood vessel after you add the triangle. How much did the length increase?

length

c. Use your answers to parts (a) and (b). Find the percent of increase.

2. Use the last two strips to make four new triangles with sides that are $\frac{1}{9}$ the length of the marked strip. Attach each triangle to the center of each of the the four segments of your blood vessel.

a. How long is the blood vessel after you add the four smaller triangles? How much did the length increase?

b. What is the percent increase?

3. Patterns Describe the pattern as a percent increase from one step to the next. Predict the total length if you were to repeat the pattern one more time.

4. Find the percent increase in length of the blood vessel from Exercise 1, part (b) to Exercise 3.

PHSchool.com

For: Information about fractals
Web Code: are-0653

Activity

Have students work in pairs to answer the questions.

Exercise 4 The percent increase is about 77.8%. To provide the basis for later study of compound interest (Chapter 9), help students understand that they cannot simply add the percents to find the percent increase.

Science Connection
Invite interested students to investigate how scientists use fractals to help them understand natural phenomena. For instance, students can learn about how they model soil erosion and seismic patterns.

Differentiated Instruction

Special Needs L1

Circulate as pairs cut, tape, and measure to make their self-similar blood vessels. As needed, help students to find the percent of increase.

7 Geometry

Chapter at a Glance

Lesson Titles, Objectives, and Features	Assessment	NCTM Standards	Local Standards
7-1 Lines and Planes • To identify segments, rays, and lines **Vocabulary Builder:** High-Use Academic Words	Lesson Quiz	3, 6, 9, 10	
7-2a Activity Lab, Hands On: Measuring Angles **7-2 Identifying and Classifying Angles** • To classify angles and to work with pairs of angles	Lesson Quiz	3, 6, 9, 10	
7-3a Activity Lab, Hands On: Sides and Angles of a Triangle **7-3 Triangles** • To classify triangles and to find the angle measures of triangles	Lesson Quiz	3, 6, 9, 10	
7-4 Quadrilaterals and Other Polygons • To classify polygons and special quadrilaterals	Lesson Quiz Checkpoint Quiz 1	3, 6, 9, 10	
7-5 Congruent Figures • To identify congruent figures and to use them to find missing measures	Lesson Quiz	3, 6, 9, 10	
7-6 Circles • To identify parts of a circle	Lesson Quiz	3, 6, 9, 10	
7-7 Circle Graphs • To analyze and construct circle graphs **7-7b Activity Lab, Data Collection:** Making a Circle Graph **Guided Problem Solving:** Percents and Circle Graphs	Lesson Quiz Checkpoint 2	3, 5, 6, 9, 10	
7-8 Constructions • To construct congruent segments and perpendicular bisectors	Lesson Quiz	3, 6, 9, 10	
Problem Solving Application: Applying Geometry			

NCTM Standards 2000
1 Number and Operations **2** Algebra **3** Geometry **4** Measurement **5** Data Analysis and Probability
6 Problem Solving **7** Reasoning and Proof **8** Communication **9** Connections **10** Representation

Correlations to Standardized Tests

All content for these tests is contained in *Prentice Hall Math,* Course 2. This chart reflects coverage in this chapter only.

	7-1	7-2	7-3	7-4	7-5	7-6	7-7	7-8
Terra Nova CAT6 (Level 17)								
Number and Number Relations								
Computation and Numerical Estimation								
Operation Concepts								
Measurement		✔						
Geometry and Spatial Sense	✔	✔	✔	✔	✔	✔	✔	✔
Data Analysis, Statistics, and Probability								
Patterns, Functions, Algebra								
Problem Solving and Reasoning	✔	✔	✔	✔	✔	✔	✔	✔
Communication								
Decimals, Fractions, Integers, and Percents							✔	
Order of Operations								
Terra Nova CTBS (Level 17)								
Decimals, Fractions, Integers, Percents								
Order of Operations, Numeration, and Number Theory								
Data Interpretation							✔	
Pre-Algebra								
Measurement		✔						
Geometry	✔	✔	✔	✔	✔	✔	✔	✔
ITBS (Level 13)								
Number Properties and Operations								
Algebra								
Geometry	✔	✔	✔	✔	✔	✔	✔	✔
Measurement		✔						
Probability and Statistics							✔	
Estimation								
SAT10 (Int 3 Level)								
Number Sense and Operations								
Patterns, Relationships, and Algebra								
Data, Statistics, and Probability							✔	
Geometry and Measurement	✔	✔	✔	✔	✔	✔	✔	✔
NAEP								
Number Sense, Properties, and Operations								
Measurement		✔						
Geometry and Spatial Sense	✔							
Data Analysis, Statistics, and Probability							✔	
Algebra and Functions								

CAT6 California Achievement Test, 6th Ed.
CTBS Comprehensive Test of Basic Skills
ITBS Iowa Test of Basic Skills, Form M
SAT10 Stanford Achievement Test, 10th Ed.
NAEP National Assessment of Educational Progress 2005 Mathematics Objectives

Math Background

Skills Trace

BEFORE Chapter 7

Course 1 or Grade 6 introduced the basic figures of geometry such as squares, rectangles, and circles.

DURING Chapter 7

Course 2 reviews these ideas and then extends their use to various applications.

AFTER Chapter 7

Throughout this course, students apply geometric skills, concepts, and vocabulary.

7-1 Lines and Planes

Math Understandings

- *Line, point,* and *plane* are the basic geometric concepts used to define all other geometric terms.
- Points, lines, and planes are ideas that do not physically exist.

A **plane** is a flat surface that extends indefinitely in all directions. **Intersecting lines** have exactly one point in common. **Parallel lines** are lines in the same plane that never intersect. **Skew lines** lie in different planes and are neither parallel nor intersecting.

7-2 Identifying and Classifying Angles

Math Understandings

- The measure of an angle describes the opening between the two sides, or rays, that form the angle. It does not depend on the lengths of the sides.
- You usually measure angles in terms of a circle, using the unit of a degree, or $\frac{1}{360}$, of a complete rotation about a point.

An **angle** is formed by two rays with a common endpoint called the **vertex.** You can classify angles by their measures.

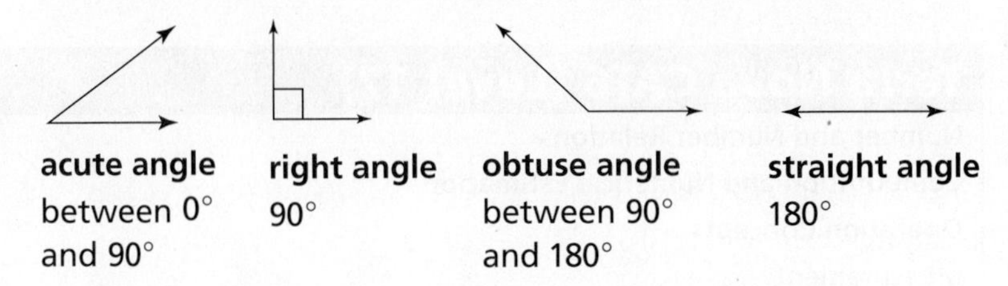

acute angle	**right angle**	**obtuse angle**	**straight angle**
between 0° and 90°	90°	between 90° and 180°	180°

If the sum of the measures of two angles is 90°, the angles are **complementary** angles. If the sum of two angles is 180°, the angles are **supplementary. Adjacent angles** share a vertex and a side but have no interior points in common. **Vertical angles** are formed by two intersecting lines and are opposite each other. Vertical angles have equal measures. Angles with equal measures are **congruent angles.**

7-3 Triangles

Math Understandings

- You can classify a triangle by the number of congruent sides it has or by its angle measures.

Angle Sum of a Triangle
The sum of the measures of the angles of a triangle is 180°.

Congruent sides have the same length. A **scalene triangle** has no congruent sides. An **isosceles triangle** has at least two congruent sides. An **equilateral triangle** has three congruent sides. An equilateral triangle is always an isosceles triangle.

A **right triangle** has one right angle. An **acute triangle** has three acute angles. An **obtuse triangle** has one obtuse angle.

7-4 Quadrilaterals and Other Polygons

Math Understandings

- You can classify polygons by the relationships among sides and angles.

A **polygon** is a closed plane figure with sides formed by three or more line segments. The sides meet only at endpoints.

Polygon Names	Number of Sides
triangle	3
quadrilateral	4
pentagon	5
hexagon	6
octagon	8
decagon	10

A **parallelogram** is a quadrilateral with both pairs of opposite sides parallel. There are three special types of parallelograms: a **rectangle** has four right angles; a **rhombus** has four congruent sides; a **square** has four right angles and four congruent sides.

7-5 Congruent Figures

Math Understandings

- Corresponding parts (sides and angles) of congruent polygons are congruent.

Congruent polygons are polygons with the same size and shape. Two congruent figures have exactly the same size and shape, and you can move them so that they will coincide exactly with each other.

7-6 7-7 Circles and Circle Graphs

Math Understandings

- In every circle, the radius is half the diameter and the diameter is twice the radius.
- A circle graph represents one whole. Percents can be used to describe parts of the whole.

A **circle** is the set of points in a plane that are all the same distance from a given point, called the *center*. A **radius** is a segment that connects the center of a circle to the circle. A **chord** is a segment that has both endpoints on the circle. A **diameter** is a segment that passes through the center of a circle and has both endpoints on the circle. A **semicircle** is an arc that is half of the circumference of a circle. A **central angle** is an angle with its vertex at the center of the circle.

7-8 Constructions

Math Understandings

- You can draw many geometric constructions using only an unmarked straightedge and a compass.
- You can use a compass and a straightedge to construct a segment congruent to a given segment.
- Any point on the perpendicular bisector of a line segment is equidistant from the endpoints of the segment.

The **midpoint** of a segment is the point that divides the segment into two segments of equal length. A **segment bisector** is a line, segment, or ray that goes through the midpoint of a segment. **Perpendicular lines** intersect to form right angles. A segment bisector that is perpendicular to a segment is the **perpendicular bisector** of the segment.

A **compass** is a geometric tool used to draw circles and arcs. An **arc** is part of a circle.

Additional Professional Development Opportunities

Math Background Notes for Chapter 7: Every lesson has a Math Backgound in the PLAN section.

Research Overview, Mathematics Strands
Additional support for these topics and more is in the front of the Teacher's Edition.

LessonLab
LessonLab, a Pearson Education company, offers comprehensive, facilitated professional development designed to help teachers to improve student achievement. To learn more, please visit lessonlab.com.

Chapter 7 Resources

	7-1	7-2	7-3	7-4	7-5	7-6	7-7	7-8	For the Chapter
Print Resources									
L3 Practice	●	●	●	●	●	●	●	●	
L1 Adapted Practice	●	●	●	●	●	●	●	●	
L3 Guided Problem Solving	●	●	●	●	●	●	●	●	
L2 Reteaching	●	●	●	●	●	●	●	●	
L4 Enrichment	●	●	●	●	●	●	●	●	
L3 Daily Notetaking Guide	●	●	●	●	●	●	●	●	
L1 Adapted Daily Notetaking Guide	●	●	●	●	●	●	●	●	
L3 Vocabulary and Study Skills Worksheets	●		●	●		●		●	●
L3 Daily Puzzles	●	●	●	●	●	●	●	●	
L3 Activity Labs	●	●	●	●	●	●	●	●	
L3 Checkpoint Quiz				●			●		
L3 Chapter Project									●
L2 Below Level Chapter Test									●
L3 Chapter Test									●
L4 Alternative Assessment									●
L3 Cumulative Review									●
Spanish Resources ELL									
L3 Practice	●	●	●	●	●	●	●	●	●
L3 Vocabulary and Study Skills Worksheet	●		●	●		●		●	●
L3 Checkpoint Quiz				●			●		
L2 Below Level Chapter Test									●
L3 Chapter Test									●
L4 Alternative Assessment									●
L3 Cumulative Review									●
Transparencies									
Check Skills You'll Need	●	●	●	●	●	●	●	●	
Additional Examples	●	●	●	●	●	●	●	●	
Problem of the Day	●	●	●	●	●	●	●	●	
Classroom Aid							●		
Student Edition Answers	●	●	●	●	●	●	●	●	●
Lesson Quiz	●	●	●	●	●	●	●	●	
Test-Taking Strategies									●
Technology									
Interactive Textbook Online	●	●	●	●	●	●	●	●	●
StudentExpress™ CD-ROM	●	●	●	●	●	●	●	●	●
Success Tracker™ Online Intervention	●	●	●	●	●	●	●	●	●
TeacherExpress™ CD-ROM	●	●	●	●	●	●	●	●	●
PresentationExpress™ with QuickTake Presenter CD-ROM	●	●	●	●	●	●	●	●	●
ExamView® Assessment Suite CD-ROM	●	●	●	●	●	●	●	●	●
MindPoint® Quiz Show CD-ROM									●
Prentice Hall Web Site: PHSchool.com	●	●	●	●	●	●	●	●	●

Also available: **Prentice Hall Assessment System**

- Progress Monitoring Assessments
- Skills and Concepts Review
- Test Prep Workbook

Other Resources

Algebra Readiness Tests
All-in-One Student Workbook
All-in-One Student Workbook, Adapted Version
Multilingual Handbook
Solution Key
Math Notes Study Folder
Spanish Cumulative Assessment

Where You Can Use the Lesson Resources

Here is a suggestion, following the four-step teaching plan, for how you can incorporate Differentiated Instruction Resources into your teaching.

	Instructional Resources L3	Differentiated Instruction Resources
1. Plan		
Preparation Read the Math Background in the Teacher's Edition to connect this lesson with students' previous experience. **Starting Class** **Check Skills You'll Need** Assign these exercises to review prerequisite skills. **New Vocabulary** Help students pre-read the lesson by pointing out the new terms introduced in the lesson.	**Math Background** **Math Understandings** **Transparencies & PresentationExpress™ with QuickTake Presenter CD-ROM** Check Skills You'll Need Problem of the Day **Resources** Vocabulary and Study Skills	**Spanish Support** ELL Vocabulary and Study Skills
2. Teach		
L3 **Guided Instruction** Use the Activity Labs to build conceptual understanding. Teach each Example. Use the Teacher's Edition side column notes for specific teaching tips, including Error Prevention notes. Use the Additional Examples found in the side column (and on transparency and PowerPoint) as an alternative presentation for the content. After each Example, assign the Quick Check exercise for that Example to get an immediate assessment of student understanding. Use the Closure activity in the Teacher's Edition to help students attain mastery of lesson content.	**Student Edition** Activity Lab **Resources** Daily Notetaking Guide Activity Lab **Transparencies & PresentationExpress™ with QuickTake Presenter CD-ROM** Additional Examples Classroom Aids **ExamView® Assessment Suite CD-ROM**	**Teacher's Edition** Every lesson includes suggestions for working with students who need special attention. L1 Special Needs L2 Below Level L4 Advanced Learners ELL English Language Learners **Resources** L1 Adapted Daily Notetaking Guide **Multilingual Handbook**
3. Practice		
Assignment Guide **Check Your Understanding** Use these questions to check students' understanding before you assign homework. **Homework Exercises** Assign homework from these leveled exercises in the Assignment Guide. A Practice by Example B Apply Your Skills C Challenge Test Prep and Mixed Review **Homework Quick Check** Use these key exercises to quickly check students' homework.	**Transparencies & PresentationExpress™ with QuickTake Presenter CD-ROM** Student Answers **Resources** Practice Guided Problem Solving Vocabulary and Study Skills Activity Lab Daily Puzzles **ExamView® Assessment Suite CD-ROM**	**Spanish Support** ELL Practice ELL Vocabulary and Study Skills **Resources** L1 Adapted Practice L4 Enrichment
4. Assess & Reteach		
Lesson Quiz Assign the Lesson Quiz to assess students' mastery of the lesson content. **Checkpoint Quiz** Use the Checkpoint Quiz to assess student progress over several lessons.	**Transparencies & PresentationExpress™ with QuickTake Presenter CD-ROM** Lesson Quiz **Resources** Checkpoint Quiz	**Resources** L2 Reteaching ELL Checkpoint Quiz Success Tracker™ Online Intervention **ExamView® Assessment Suite CD-ROM**

KEY L1 Special Needs L2 Below Level L3 For All Students L4 Advanced, Gifted ELL English Language Learners

Geometry

Answers are in the back of the textbook.

For intervention, direct students to:

Comparing Integers
Lesson 1-6
Extra Skills and Word Problems Practice, Ch. 1

Solving One-Step Equations
Lesson 4-3
Extra Skills and Word Problems Practice, Ch. 4

Solving Proportions
Lesson 5-4
Extra Skills and Word Problems Practice, Ch. 5

Finding a Percent of a Number
Lesson 6-4
Extra Skills and Word Problems Practice, Ch. 6

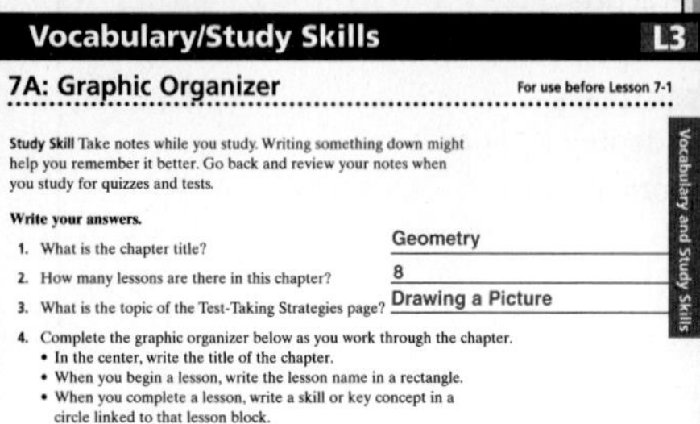

Spanish Vocabulary/Study Skills ELL

Vocabulary/Study Skills L3

7A: Graphic Organizer — For use before Lesson 7-1

Study Skill Take notes while you study. Writing something down might help you remember it better. Go back and review your notes when you study for quizzes and tests.

Write your answers.

1. What is the chapter title? Geometry
2. How many lessons are there in this chapter? 8
3. What is the topic of the Test-Taking Strategies page? Drawing a Picture
4. Complete the graphic organizer below as you work through the chapter.
   - In the center, write the title of the chapter.
   - When you begin a lesson, write the lesson name in a rectangle.
   - When you complete a lesson, write a skill or key concept in a circle linked to that lesson block.
   - When you complete the chapter, use this graphic organizer to help you review.

   Check students' diagrams.

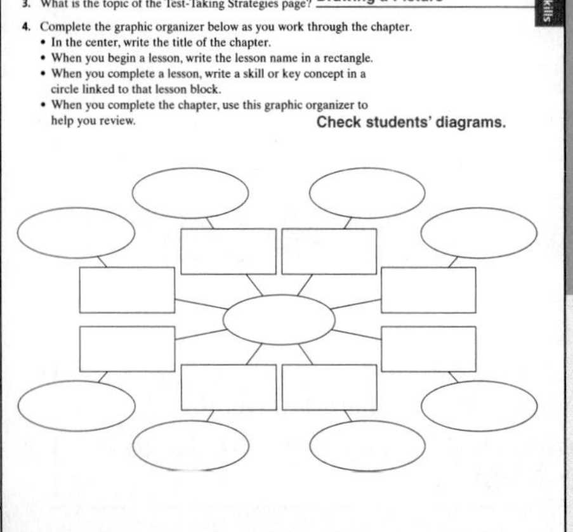

CHAPTER 7 Geometry

What You've Learned

- In Chapter 5, you identified similar figures and used proportions to find missing lengths.
- In Chapter 6, you converted between fractions, decimals, and percents.

Check Your Readiness

GO for Help

For Exercises	See Lesson
1–4	1-6
5–10	4-3
11–14	5-4
15–18	6-4

Comparing Integers

Compare using <, =, or >.

1. 83 ■ 90 <
2. 120 ■ 99 >
3. 0 ■ −47 >
4. −21 ■ −11 <

Solving One-Step Equations

Solve each equation.

5. $d + 17 = 19$ 2
6. $m - 12 = 3$ 15
7. $j - 5 = 7$ 12
8. $m - 15 = 90$ 105
9. $58 + n = 63$ 5
10. $y + 86 = 180$ 94

Solving Proportions

Solve each proportion.

11. $\frac{5}{16} = \frac{25}{w}$ 80
12. $\frac{n}{12} = \frac{20}{15}$ 16
13. $\frac{18}{k} = \frac{6}{37}$ 111
14. $\frac{23}{12} = \frac{x}{24}$ 46

Finding a Percentage of a Number

Find each answer.

15. 30% of 360 108
16. 24% of 360 86.4
17. 4.5% of 360 16.2
18. 18% of 360 64.8

Chapter 7 Overview

In this chapter, students work with plane geometry concepts. They learn about the properties of lines and angles. Then they classify angles according to their degree measures. They work with triangles, quadrilaterals, and other polygons, including congruent figures. Next, they investigate circles and circle graphs. They conclude the chapter by focusing on constructions.

Activating Prior Knowledge

In this chapter, students will build on their knowledge of angles such as adjacent angles, supplementary angles, and complementary angles. They learn to classify angles by their degree measures and their relationship to other angles. They draw upon their knowledge of triangles to classify and name them by the measures of their angles and lengths of their sides.

Ask questions such as:

- *What are some names for different types of triangles?* **Sample: an equilateral triangle, an isosceles triangle, a scalene triangle**
- *Describe an angle.* **Sample: the opening formed by two lines that intersect**

What You'll Learn Next

- In this chapter, you will classify angles, triangles, and quadrilaterals.
- You will identify congruent figures and find missing measures.
- You will analyze and construct circle graphs.

Problem Solving Application On pages 370 and 371, you will work an extended activity on geometry.

Key Vocabulary

- acute triangle (p. 337)
- circle graph (p. 354)
- complementary (p. 331)
- equilateral triangle (p. 336)
- isosceles triangle (p. 336)
- obtuse triangle (p. 337)
- parallel lines (p. 325)
- parallelogram (p. 341)
- rectangle (p. 341)
- regular polygon (p. 340)
- rhombus (p. 341)
- right triangle (p. 337)
- scalene triangle (p. 336)
- square (p. 341)
- supplementary (p. 331)
- trapezoid (p. 341)

7-1

1. Plan

Objective
To identify segments, rays, and lines

Examples
1 Naming Segments, Rays, and Lines
2 Intersecting, Parallel, and Skew

Math Understandings: p. 322C

Math Background

Points are the basis of all geometry. Points have no height, width, or length so they are called zero-dimensional. A *point* indicates an exact position. A *line* is a series of points that extends in opposite directions without end, making a line one-dimensional. A *ray* is part of a line with one endpoint and extends forever in one direction. A *line segment* is a part of a line with two endpoints. A *plane* is a two-dimensional figure that extends indefinitely.

More Math Background: p. 322C

Lesson Planning and Resources

See p. 322E for a list of the resources that support this lesson.

Bell Ringer Practice

Check Skills You'll Need
Use student page, transparency, or PowerPoint. For intervention, direct students to:
Graphing and Writing Inequalities
Lesson 4-7
Extra Skills and Word Problems Practice, Ch. 4

7-1 Lines and Planes

Check Skills You'll Need

1. **Vocabulary Review** List the five *inequality* symbols. $<, >, \leq, \geq, \neq$

Graph the solution of each inequality.

2. $x \leq -3$ 3. $x < 1$

4. $x \leq 2$ 5. $x \geq 5$

6. $x > 1$ 7. $x < 4$

2–7. See back of book.

GO for Help Lesson 4-7

What You'll Learn

To identify segments, rays, and lines

New Vocabulary point, line, ray, segment, plane, intersecting lines, parallel lines, skew lines

Why Learn This?

City maps show parallel and intersecting streets. A statement such as "W. 4th St. runs parallel to W. 3rd St." can help you give and understand directions.

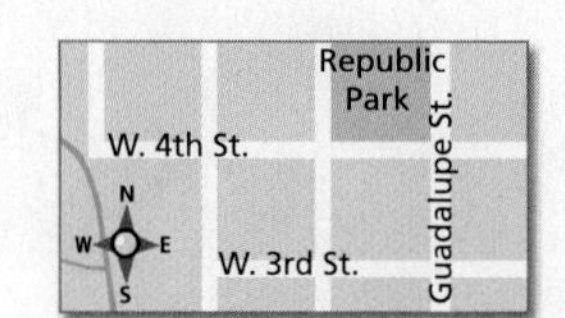

A **point** is a location. A point has no size. You name a point by a capital letter.

point A point B

A **line** is a series of points that extend in opposite directions without end. You name a line by any two points on the line or by a lowercase letter.

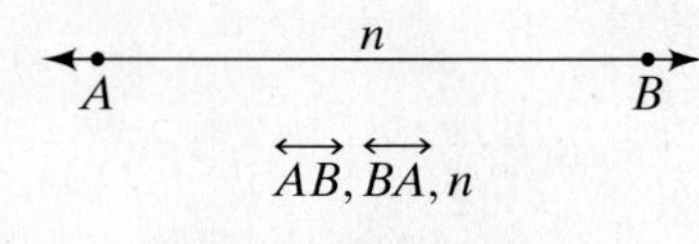

$\overleftrightarrow{AB}, \overleftrightarrow{BA}, n$

A **ray** is part of a line with one endpoint and all the points of the line on one side of the endpoint. You name a ray using two points, starting with the endpoint.

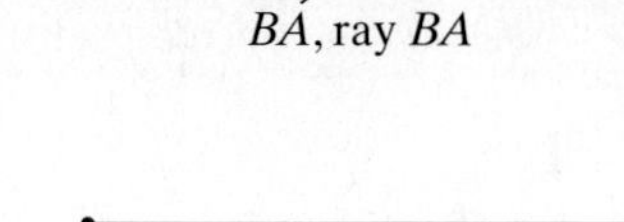

$\overrightarrow{BA}$, ray BA

A **segment** is part of a line with two endpoints and all points in between. You name a segment by its endpoints.

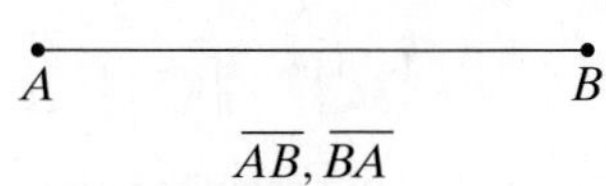

$\overline{AB}, \overline{BA}$

EXAMPLE Naming Segments, Rays, and Lines

1 Use the points in each diagram to name the figure shown.

a. X Y — $\overline{XY}$

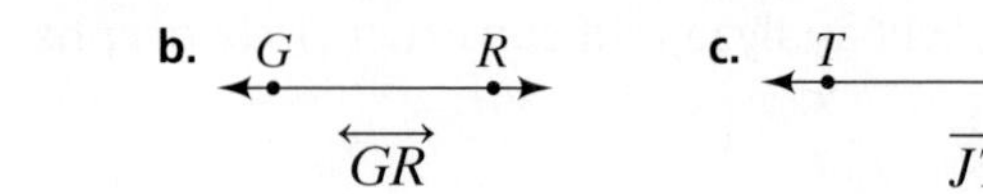

b. $\overleftrightarrow{GR}$

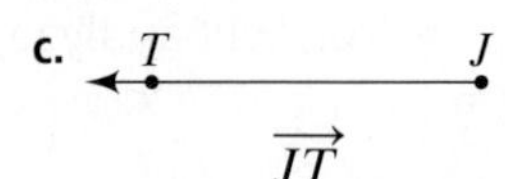

c. $\overrightarrow{JT}$

Quick Check

1. Use the points in each diagram to name the figure shown.

a. P D $\overrightarrow{PD}$ **b.** R S $\overline{RS}$ **c.**

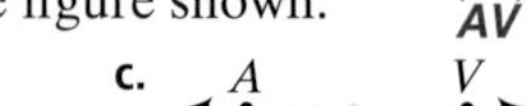

$\overleftrightarrow{AV}$

Differentiated Instruction Solutions for All Learners

Special Needs L1	**Below Level** L2
Students often have a difficult time visualizing skew lines, since they are neither parallel nor intersecting. Take opportunities to show them skew lines around the classroom. Have them trace with their finger a pair of skew lines on their desks. **learning style: tactile**	Have students model parallel, intersecting, and skew lines as shown in Example 2 with toothpicks or pipe cleaners. **learning style: tactile**

A **plane** is a flat surface that extends indefinitely in all directions and has no thickness. There are two planes in the diagram below.

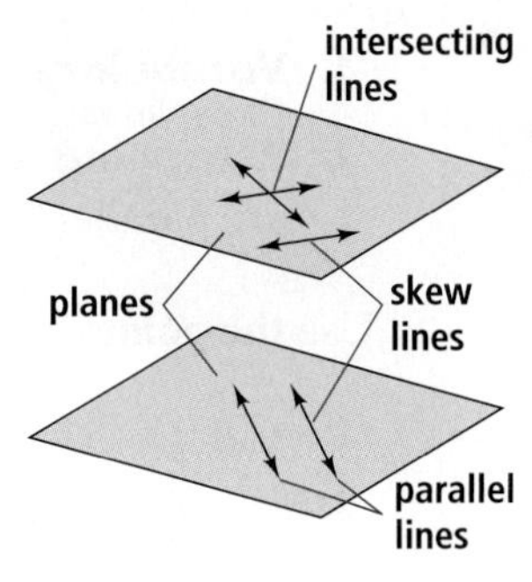

Intersecting lines lie in the same plane. **Intersecting lines** have exactly one point in common. **Parallel lines** are lines in the same plane that never intersect. Parallel segments and rays lie in parallel lines.

Skew lines lie in different planes. They are neither parallel nor intersecting.

Vocabulary Tip

The word *skew* comes from a root word meaning "to avoid."

EXAMPLE Intersecting, Parallel, and Skew

2 **Architecture** Use the information in the photograph and diagram below to name a segment with the given description.

a. a segment parallel to $\overline{AB}$

$\overline{GH}$ lies in the same plane as $\overline{AB}$. $\overline{GH}$ does not intersect $\overline{AB}$.

$\overline{GH}$ is parallel to $\overline{AB}$.

b. a segment skew to $\overline{AB}$

$\overline{EF}$ lies in a different plane. $\overline{EF}$ is not parallel to $\overline{AB}$ and does not intersect $\overline{AB}$.

$\overline{EF}$ is skew to $\overline{AB}$.

Quick Check

2. Name the segments in the diagram that fit each description.
   a. parallel to $\overline{BC}$ **b.** intersect $\overline{BH}$ **c.** skew to $\overline{AG}$

2a. $\overline{EF}, \overline{HD}$
b. $\overline{EB}, \overline{CB}, \overline{AB}, \overline{GH}, \overline{DH}$
c. $\overline{FC}, \overline{EF}, \overline{BC}, \overline{HD}$

2. Teach

Activity Lab

Use before the lesson.

All in One Teaching Resources

Activity Lab 7-1: Lines and Planes

Guided Instruction

Example 1

- *Which of the figures could be named differently?* **$\overline{XY}$ can be $\overline{YX}$ and $\overleftrightarrow{GR}$ can be $\overleftrightarrow{RG}$.**
- *Why can ray JT not be named differently?* **The name of the ray begins with the endpoint.**

PowerPoint

Additional Examples

1 Use the points in each diagram to name the figure.

a. R S **$\overline{RS}$**

b. J K **$\overleftrightarrow{JK}$**

c. A B **$\overrightarrow{BA}$**

2 Name all the segments that have each characteristic.

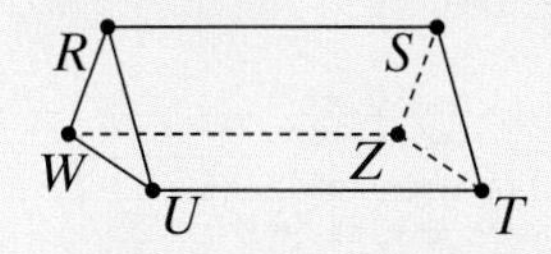

a. parallel to $\overline{UT}$ **$\overline{RS}, \overline{WZ}$**

b. intersecting $\overline{RU}$ **$\overline{WU}, \overline{TU}, \overline{WR}, \overline{RS}$**

c. skew to $\overline{RS}$ **$\overline{WU}, \overline{ZT}$**

All in One Teaching Resources

- Daily Notetaking Guide 7-1 L3
- Adapted Notetaking 7-1 L1

Closure

- Describe a line, a line segment, and a ray. **A line is a series of points that extends in two directions indefinitely. A line segment is part of a line with two endpoints. A ray is part of a line with one endpoint.**

Advanced Learners L4	**English Language Learners** ELL
How many ways can you name a line that has 2 points labeled? **2** 3 points labeled? **6** 4 points labeled? **12**	Ask students to write the terms used in this lesson, along with pictorial representations on an index card. Give them opportunities to "test" each other in pairs. One partner can read the word, the other can draw.
learning style: verbal	**learning style: visual**

3. Practice

Assignment Guide

Check Your Understanding
Go over Exercises 1–8 in class before assigning the Homework Exercises.

Homework Exercises

A	Practice by Example	9–17
B	Apply Your Skills	18–30
C	Challenge	31
	Test Prep and Mixed Review	32–36

Homework Quick Check
To check students' understanding of key skills and concepts, go over Exercises 11, 16, 23, 27, and 30.

Differentiated Instruction Resources

Adapted Practice 7-1 L1

Practice 7-1 Lines a... L3

Describe the lines or line segments as *parallel* or *intersecting*.

1. the rows on a spreadsheet parallel
2. the marks left by a skidding car parallel
3. sidewalks on opposite sides of a street parallel
4. the cut sides of a wedge of apple pie intersecting
5. the wires suspended between telephone poles parallel
6. the hands of a clock at 7:00 intersecting
7. the trunks of grown trees in a forest parallel

Use the diagram below for exercises 8–12.

8. ame a pair of parallel lines. $\overleftrightarrow{CD}$ and $\overleftrightarrow{EF}$
9. ame a segment. Sample answer: $\overline{HG}$
10. ame three points. Sample answer: A, G, B
11. ame two rays. Sample answer: $\overrightarrow{AB}$, $\overrightarrow{CD}$
12. ame a pair of intersecting lines. Sample answer: $\overleftrightarrow{AB}$, $\overleftrightarrow{CD}$

Use a straightedge to draw each figure.

13. a line parallel to $\overleftrightarrow{UV}$
14. a line intersecting $\overleftrightarrow{XY}$

7-1 • Guided Problem Solving GPS L3

Student Page 327, Exercise 23:

Are the rungs on a stepladder parallel, intersecting, or skew?

Understand

1. What is a stepladder? raw a sketch of one. A stepladder is a device that helps you reach things that are high off the ground. Check students' drawings.
2. What are the rungs of a ladder? Circle the rungs on your sketch. They are the steps that you climb. (Check students' drawings for circled rungs.)
3. What are you being asked to do? Determine if the rungs are parallel, intersecting, or skew.

Plan and Carry Out

4. o the rungs have any points in common? no
5. Are the rungs in the same plane? yes
6. Are the rungs intersecting? no
7. Are the rungs skew? no
8. Are the rungs parallel? yes

Check

9. efine *parallel*. oes this describe the rungs? Parallel lines are lines in the same plane that do not intersect; yes

Solve Another Problem

10. Are the lines on your palm parallel, intersecting, or skew? Answers will vary.

Check Your Understanding

2. No. The lines could be either skew or parallel. If they were in the same plane, then they would be parallel.

1. **Vocabulary** __?__ are lines that are neither parallel nor intersecting. Skew lines
2. **Reasoning** Two lines do not intersect. Can you conclude that they are parallel? Explain. See left.

Use the points in each diagram to name the figure shown.

3. L C $\overline{LC}$
4. M R $\overleftrightarrow{MR}$
5. K E $\overleftrightarrow{KE}$
6. O D $\overrightarrow{OD}$

Use the diagram for Exercises 7 and 8.

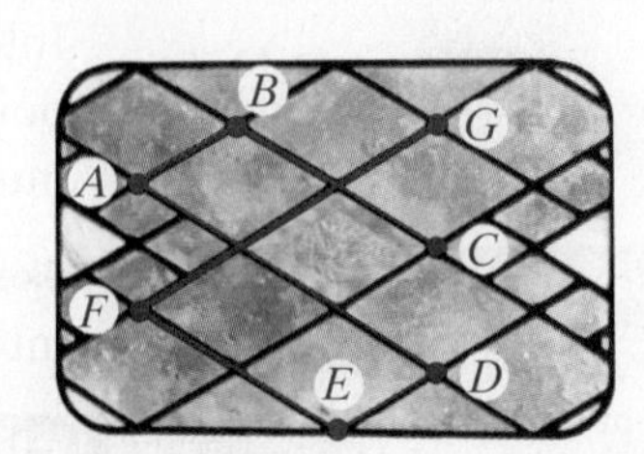

7. Name all the segments parallel to $\overline{AD}$. $\overline{BC}$, $\overline{FE}$
8. Name all the segments intersecting $\overline{FG}$. $\overline{BC}$, $\overline{AD}$, $\overline{FE}$

Homework Exercises

For more exercises, see Extra Skills and Word Problems.

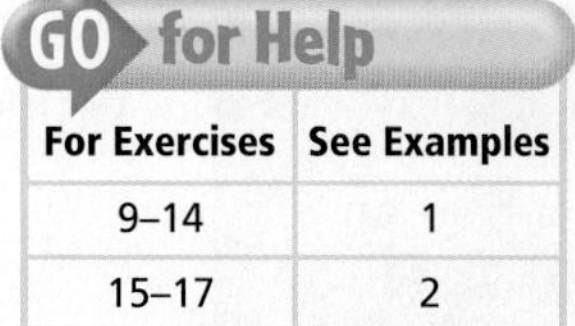

GO for Help

For Exercises	See Examples
9–14	1
15–17	2

A **Use the points in each diagram to name the figure shown.** 9–14. See left.

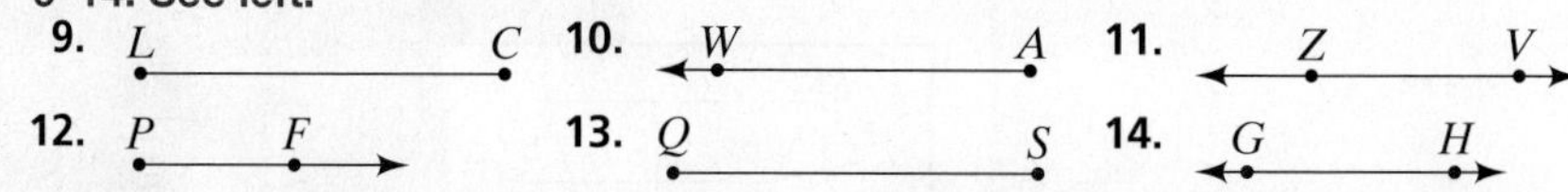

9. $\overline{LC}$
10. $\overrightarrow{AW}$
11. $\overleftarrow{ZV}$
12. $\overrightarrow{PF}$
13. $\overline{QS}$
14. $\overleftrightarrow{GH}$

Buildings Use the diagram at the right for Exercises 15–17.

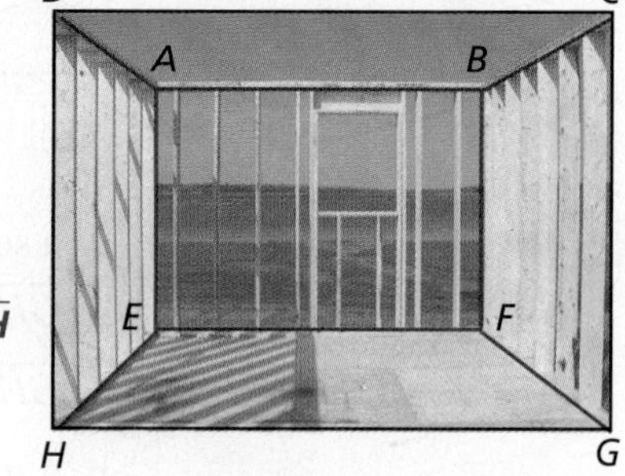

15. Name all the segments skew to $\overline{BC}$. $\overline{AE}$, $\overline{EF}$, $\overline{DH}$, $\overline{GH}$
16. Name all the segments intersecting $\overline{AD}$. $\overline{AB}$, $\overline{AE}$, $\overline{CD}$, $\overline{DH}$
17. Name all the segments parallel to $\overline{EH}$. $\overline{AD}$, $\overline{FG}$

B 

18. **Guided Problem Solving** Describe a route from A to B that includes Main St. and a street parallel to Main St.
    - What street is parallel to Main St.?
    - What street(s) connect Main St. and Lee Ave.? 18–19. See margin.

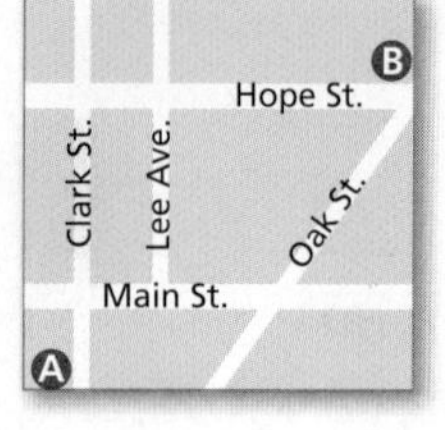

19. Use the map shown in Exercise 18. Describe a route from A to B that includes Oak St. and a street that intersects Oak St.

Exercises 18–19. Answers may vary. Samples are given.

18. From point A, go along Clark St to Main Street. Turn right on Main Street. Turn left on Lee Ave. Then turn right on Hope Street. Continue on Hope Street to point B at the corner of Hope and Oak Streets.

19. From point A, go along Clark Street to Main Street. Turn right on Main Street. Then turn left on Oak Street. Continue on Oak Street to point B at the corner of Hope and Oak Streets.

Homework Video Tutor
Visit: PHSchool.com
Web Code: are-0701

Tell whether each figure below contains ray AB.

20. A —— B
no

21. ← A —— B →
yes

22. ← A —— B
no

23. Are the rungs on a ladder parallel, intersecting, or skew?
GPS parallel

Draw each figure. 24–28. See margin.

24. $\overleftrightarrow{AD}$ 25. $\overline{QW}$ 26. $\overrightarrow{PR}$

27. $\overline{FG}$ intersecting $\overleftrightarrow{TU}$ 28. $\overrightarrow{TB}$ and $\overrightarrow{TA}$ on the same line

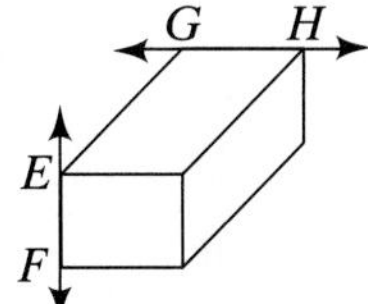

29. **Error Analysis** Jim says that in the box shown at the left $\overleftrightarrow{EF}$ and $\overleftrightarrow{GH}$ are parallel, since they do not intersect. Why is Jim incorrect? See left.

30. **Writing in Math** Describe examples of parallel, intersecting, and skew lines in your classroom. See left.

29. $\overleftrightarrow{EF}$ and $\overleftrightarrow{GH}$ do not lie in the same plane.

30. Answers may vary. Samples are given. Parallel: top and bottom edges of a doorway; Intersecting: the lines in the addition symbol; Skew: the top of a doorway and the side of a desk

C 31. **Challenge** Name each segment, ray, and line in the figure below.

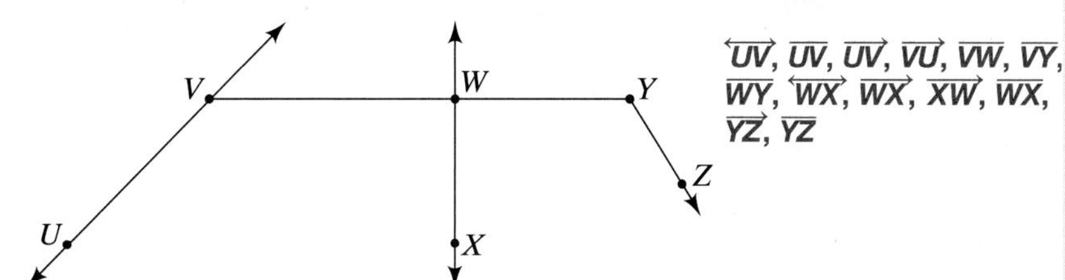

$\overleftrightarrow{UV}$, $\overline{UV}$, $\overrightarrow{UV}$, $\overrightarrow{VU}$, $\overline{VW}$, $\overline{VY}$, $\overline{WY}$, $\overleftrightarrow{WX}$, $\overrightarrow{WX}$, $\overrightarrow{XW}$, $\overline{WX}$, $\overrightarrow{YZ}$, $\overline{YZ}$

Test Prep and Mixed Review

Practice

Multiple Choice

32. An advertisement offers you 0.3% off your next purchase. What fraction of the original price does this represent? C

Ⓐ $\frac{3}{10}$ Ⓑ $\frac{3}{100}$ Ⓒ $\frac{3}{1{,}000}$ Ⓓ $\frac{3}{10{,}000}$

33. Which expression does the model represent? F

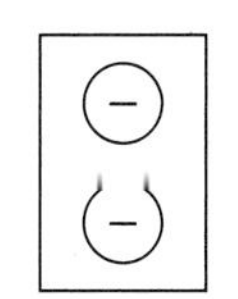
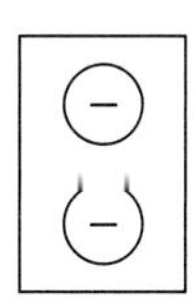
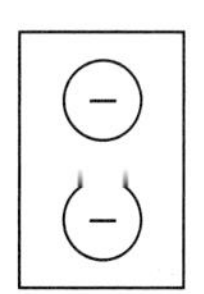
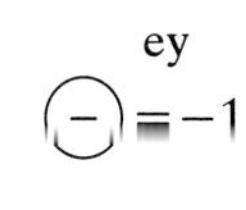

Ⓕ 3×-2 Ⓖ 2×-3 Ⓗ 2×-2 Ⓙ 3×3

34. Which formula shows z, the number of milliliters that result when you add x milliliters to y liters? D

Ⓐ $z = 100x + y$ Ⓒ $z = 100y + x$
Ⓑ $z = 1{,}000x + y$ Ⓓ $z = 1{,}000y + x$

GO for Help

For Exercises	See Lesson
35–36	6-8

Find the percent of change. Is the change an increase or a decrease?

35. original 5.75 new 6.25
about 9% increase

36. original 380 ft new 320 ft
about 16% decrease

Online lesson quiz, PHSchool.com, Web Code: ara-0701

24–28. See back of book.

4. Assess & Reteach

Lesson Quiz

Use the diagram for questions 1–3.

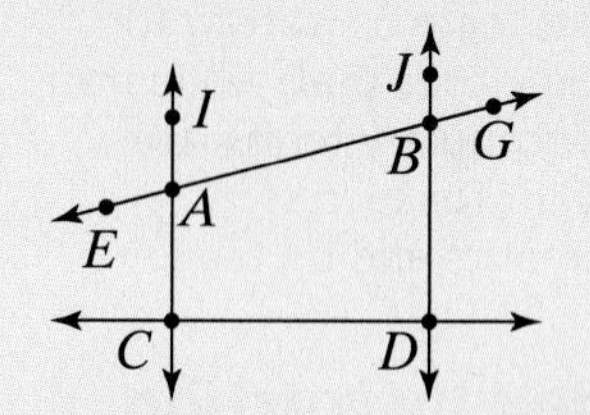

1. parallel lines Sample: $\overleftrightarrow{AC}$ and $\overleftrightarrow{BD}$
2. all the rays that have B as an endpoint $\overrightarrow{BJ}$, $\overrightarrow{BG}$, $\overrightarrow{BD}$, $\overrightarrow{BA}$ or $\overrightarrow{BE}$
3. a segment on $\overleftrightarrow{AC}$ $\overline{AC}$ or $\overline{CA}$, $\overline{AI}$ or $\overline{IA}$, $\overline{CI}$ or $\overline{IC}$

Alternative Assessment

Have students work in pairs. Each student takes a turn drawing and labeling one of the figures discussed in this lesson. His or her partner names the figure verbally and with symbolic notation.

Test Prep

Resources
For additional practice with a variety of test item formats:
- Test-Taking Strategies, p. 365
- Test Prep, p. 369
- Test-Taking Strategies with Transparencies

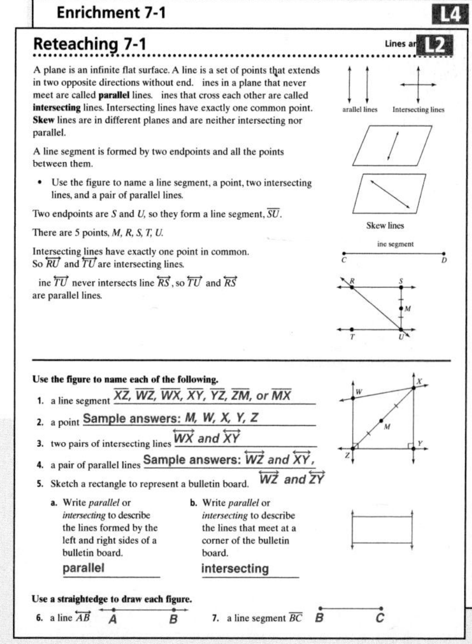

Enrichment 7-1 L4

Reteaching 7-1 Lines ar L2

A plane is an infinite flat surface. A line is a set of points that extends in two opposite directions without end. ines in a plane that never meet are called **parallel** lines. ines that cross each other are called **intersecting** lines. Intersecting lines have exactly one common point. **Skew** lines are in different planes and are neither intersecting nor parallel.

arallel lines Intersecting lines

Skew lines

A line segment is formed by two endpoints and all the points between them.

ine segment

- Use the figure to name a line segment, a point, two intersecting lines, and a pair of parallel lines.

Two endpoints are S and U, so they form a line segment, $\overline{SU}$.

There are 5 points, M, R, S, T, U.

Intersecting lines have exactly one point in common. So $\overleftrightarrow{RU}$ and $\overleftrightarrow{TU}$ are intersecting lines.

ine $\overleftrightarrow{TU}$ never intersects line $\overleftrightarrow{RS}$, so $\overleftrightarrow{TU}$ and $\overleftrightarrow{RS}$ are parallel lines.

Use the figure to name each of the following.

1. a line segment $\overline{XZ}$, $\overline{WZ}$, $\overline{WX}$, $\overline{XY}$, $\overline{YZ}$, $\overline{ZM}$, or $\overline{MX}$
2. a point Sample answers: M, W, X, Y, Z
3. two pairs of intersecting lines $\overleftrightarrow{WX}$ and $\overleftrightarrow{XY}$
4. a pair of parallel lines Sample answers: $\overleftrightarrow{WZ}$ and $\overleftrightarrow{XY}$, $\overleftrightarrow{WZ}$ and $\overleftrightarrow{ZY}$
5. Sketch a rectangle to represent a bulletin board.
   a. Write *parallel* or *intersecting* to describe the lines formed by the left and right sides of a bulletin board. parallel
   b. Write *parallel* or *intersecting* to describe the lines that meet at a corner of the bulletin board. intersecting

Use a straightedge to draw each figure.

6. a line $\overleftrightarrow{AB}$ A B
7. a line segment $\overline{BC}$ B C

High-Use Academic Words

Students learn a strategy for learning words that, while not math vocabulary terms, are important for success in mathematics and on tests.

Guided Instruction

Have students look through their texts for use of the terms: *classify, sketch,* and *identify.* Ask the following questions:

- *Where do you find the term* classify? **Sample: page 13, above Exercises 6–8**
- *What is another way to say "Which of the triangles are congruent?"* **Sample: Identify the congruent triangles.**
- *If you sketched a quadrilateral, what would you sketch?* **Sample: a closed plane figure with sides formed by four line segments**

Teaching Tip
Restate directions given in the text using *classify, sketch,* and *identify* as appropriate to familiarize students with these terms.

Differentiated Instruction

English Language Learners **ELL**
Encourage students to write high-use academic words in their native language as needed.

Resources

- Vocabulary and Study Skills Worksheets

High-Use Academic Words

High-use academic words are words that you will see often in textbooks and on tests. These words are not math vocabulary terms, but knowing them will help you succeed in mathematics.

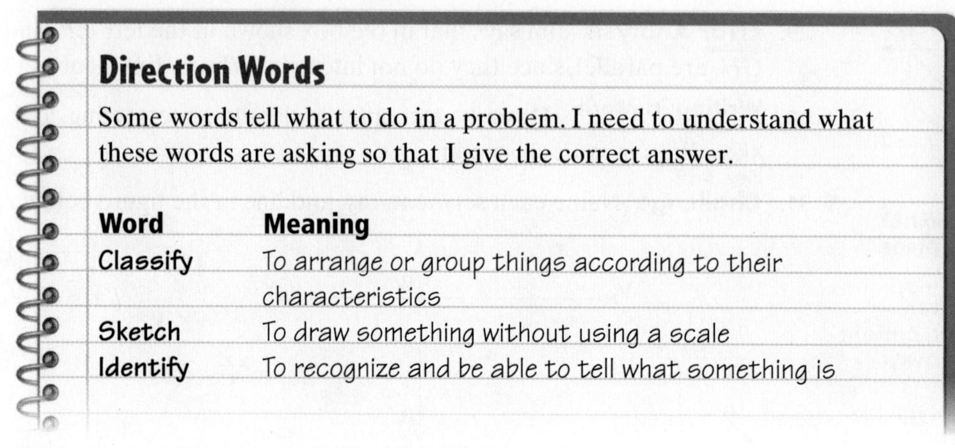

Direction Words

Some words tell what to do in a problem. I need to understand what these words are asking so that I give the correct answer.

Word	Meaning
Classify	To arrange or group things according to their characteristics
Sketch	To draw something without using a scale
Identify	To recognize and be able to tell what something is

Exercises

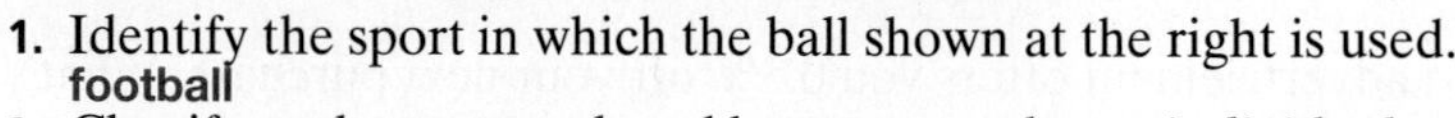

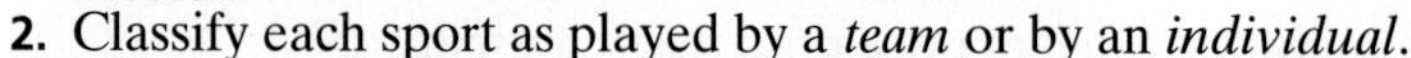

1. Identify the sport in which the ball shown at the right is used. **football**
2. Classify each sport as played by a *team* or by an *individual.*
   a. basketball **team** b. golf **individual** c. volleyball **team** d. billiards **individual**
3. There are six pairs of lines in the diagram at the right. Classify each pair of lines as *intersecting* or *parallel*. **See margin.**
4. Sketch a floor plan of your school's cafeteria. **Check students' work.**
5. Sketch a line segment and a ray that share endpoint T. **Check students' work.**
6. Identify the ray shown at the right. **$\overrightarrow{FG}$**

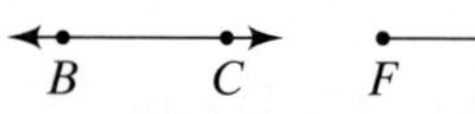

7. **Word Knowledge** Think about the word *justify*. **7a–c. Check students' work.**
   a. Choose the letter for how well you know the word.
      A. I know its meaning.
      B. I've seen it, but I don't know its meaning.
      C. I don't know it.
   b. **Research** Look up and write the definition of *justify*.
   c. Use the word in a sentence involving mathematics.

3. $\overleftrightarrow{FG}$ and $\overleftrightarrow{RS}$, parallel
$\overleftrightarrow{FG}$ and $\overleftrightarrow{BC}$, intersect
$\overleftrightarrow{FG}$ and $\overleftrightarrow{KL}$, intersect
$\overleftrightarrow{RS}$ and $\overleftrightarrow{BC}$, intersect
$\overleftrightarrow{RS}$ and $\overleftrightarrow{KL}$, intersect
$\overleftrightarrow{BC}$ and $\overleftrightarrow{KL}$, parallel

7-2a Activity Lab

Hands On

Measuring Angles

An angle ∠ has two sides and a vertex. Angles are measured in degrees°. You can estimate the measure of an angle before measuring.

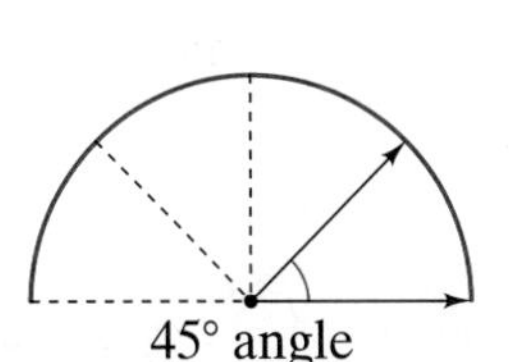
45° angle (half of a right angle)

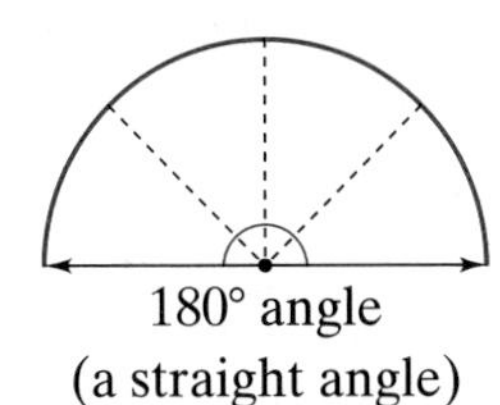
180° angle (a straight angle)

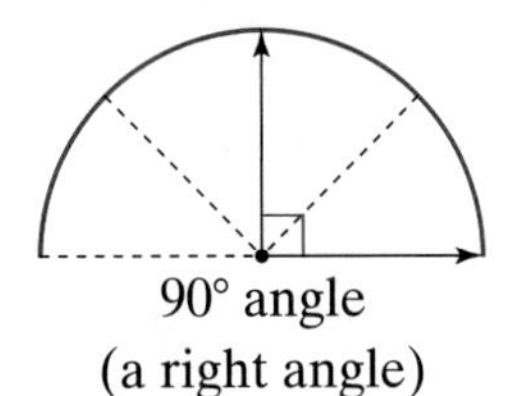
90° angle (a right angle)

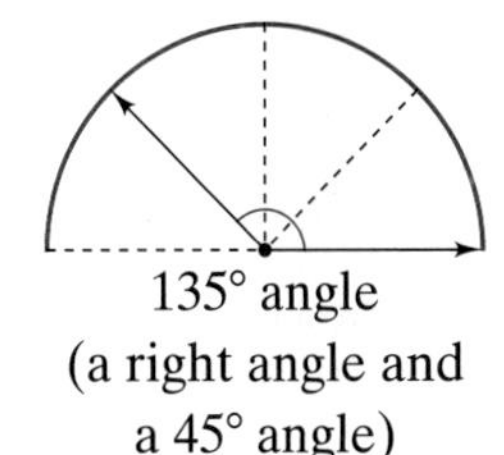
135° angle (a right angle and a 45° angle)

EXAMPLE Measuring Angles

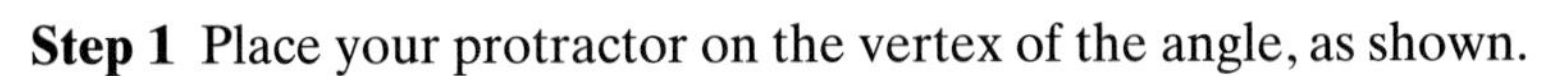

What is the measure of $\angle X$ at the right?

Estimate $\angle X$ is larger than a 90° angle and smaller than a 135° angle. You can estimate that the measure of $\angle X$ is between 90° and 135°.

Step 1 Place your protractor on the vertex of the angle, as shown.

Step 2 Make sure that one side of the angle passes through zero on one of the protractor's scales.

Step 3 Read the same scale where it intersects the second side of the angle.

The measure of $\angle X$ is 110°. You can write this as $m\angle X = 110°$.

Check for Reasonableness The measure of $\angle X$ is between 90° and 135°. The protractor measure of 110° is reasonable.

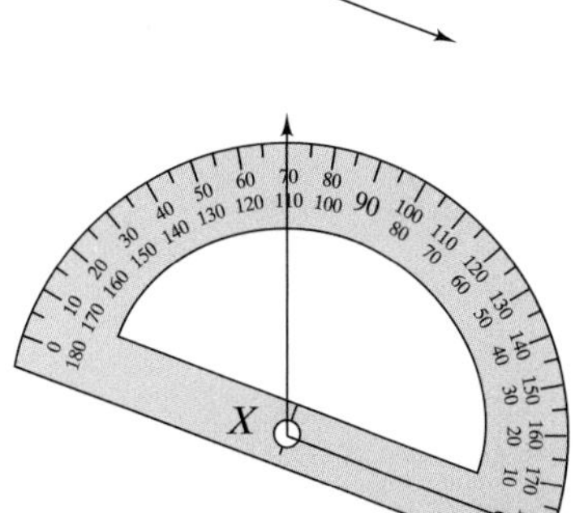

Exercises

Estimate each angle. Then use your protractor to measure each angle.

1. 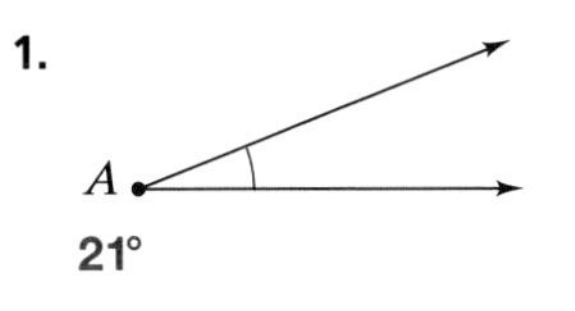
21°

2. 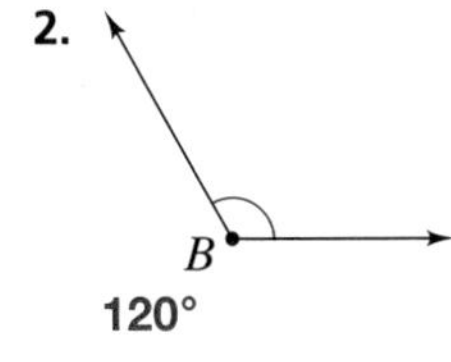
120°

3. 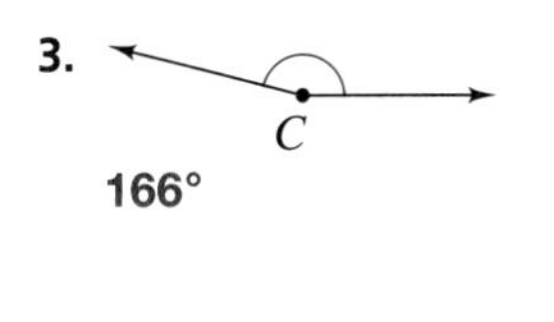
166°

4. 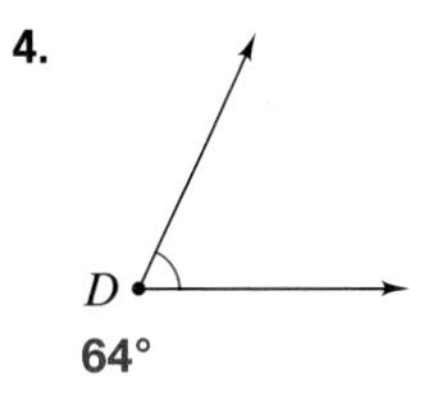
64°

Activity Lab

Measuring Angles

Guided Instruction

Ask: *Which scale on the protractor are you using to measure the angle, the upper scale or lower scale?* **lower**

Activity

Have students work in pairs to do Exercises 1–4. One student estimates the angle for Exercises 1 and 2; the other measures it. Students reverse roles for Exercises 3 and 4.

Resources

- Activity Lab 7-2: Angles and Parallel Lines
- protractor

7-2

1. Plan

Objective
To classify angles and to work with pairs of angles

Examples
1 Identifying Angles
2 Finding Complements and Supplements
3 Finding Angle Measures

Math Understandings: p. 322C

Professional Development

Math Background

An *angle* is formed when two lines, line segments, or rays meet at a common point called the *vertex*. An angle formed by rays $\overrightarrow{BA}$ and $\overrightarrow{BC}$ can be named: $\angle ABC$, $\angle CBA$, or $\angle B$ for short. The *measure* of an angle is indicated by the lowercase *m*, as in $m\angle B$, in degrees.

When two lines intersect, four pairs of adjacent angles and two pairs of vertical angles are formed. *Adjacent angles* share a side and are supplementary (add to 180°). *Vertical angles* are opposite each other and are congruent (have equal measures).

More Math Background: p. 322C

Lesson Planning and Resources

See p. 2E for a list of the resources that support this lesson.

PowerPoint

Bell Ringer Practice

Check Skills You'll Need
Use student page, transparency, or PowerPoint. For intervention, direct students to:
Using Number Sense to Solve Equations
Lesson 4-2
Extra Skills and Word Problems Practice, Ch. 4

7-2 Identifying and Classifying Angles

Check Skills You'll Need

1. **Vocabulary Review** What do you call a mathematical sentence with an equal sign? equation

Use mental math to solve each equation.

2. $\ell + 20 = 90$ 70
3. $n + 40 = 180$ 140
4. $p + 16 = 90$ 74
5. $s + 22 = 180$ 158

GO for Help Lesson 4-2

What You'll Learn

To classify angles and to work with pairs of angles

New Vocabulary angle, vertex, acute angle, right angle, obtuse angle, straight angle, complemetary, supplementary, adjacent angles, vertical angles, congruent angles

Why Learn This?

Architects think about angles in the structures they design. If you can measure angles, you can predict how different geometric figures can fit together.

The design of a geodesic dome requires triangles, because triangles have exactly three angles. They make the dome stable.

An **angle** ($\angle$) is a figure formed by two rays with a common endpoint. You can call the angle below $\angle DCE$, $\angle ECD$, $\angle C$, or $\angle 1$.

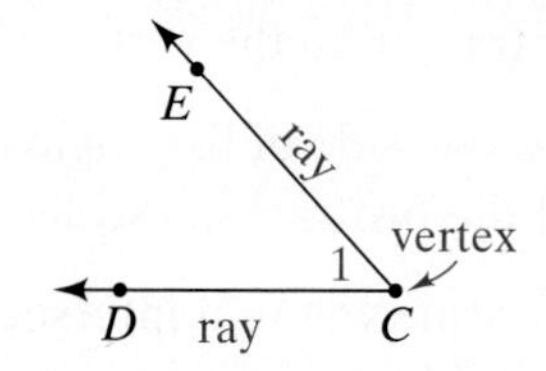

A **vertex** is the point of intersection of two sides of an angle or figure. The plural of *vertex* is *vertices*.

You can classify angles by their measures.

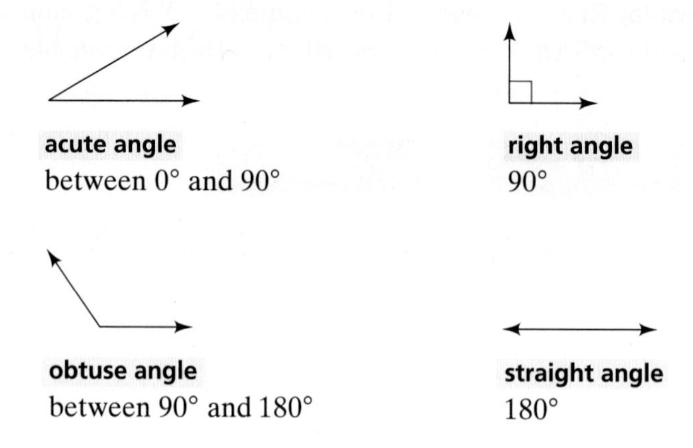

Vocabulary Tip
The ⌝ symbol indicates a right angle. Write "the measure of $\angle C$" as $m\angle C$.

Differentiated Instruction Solutions for All Learners

Special Needs L1
It may be hard for some students to see the angle in the straight line. Redraw the angles, and erase everything but the vertical line along the bottom. Explain that the angles form this line which is equal to 180°.
learning style: visual

Below Level L2
Have students fold three pieces of cardstock in half and turn them on edge. Students can adjust the fold to model acute, right, and obtuse angles.
learning style: tactile

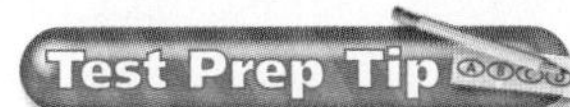

To estimate an angle measure, compare the angle to the corner of a page. If the angle is larger than the corner, its measure will be greater than 90°.

EXAMPLE Identifying Angles

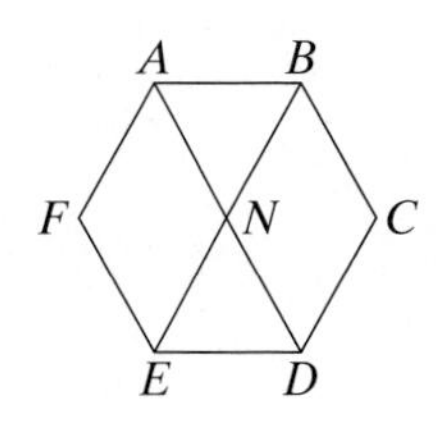

1 **Architecture** Part of a geodesic dome is shown at the right. Identify all the acute angles.

$\angle NEF$, $\angle FAN$, $\angle NAB$, $\angle ABN$, $\angle BNA$, $\angle NBC$, $\angle CDN$, $\angle NDE$, $\angle DEN$, and $\angle END$ are acute.

Quick Check

1. Classify $\angle AFE$ as *acute, right, obtuse,* or *straight.* **obtuse**

If the sum of the measures of two angles is 90°, the angles are **complementary.** If the sum is 180°, the angles are **supplementary.**

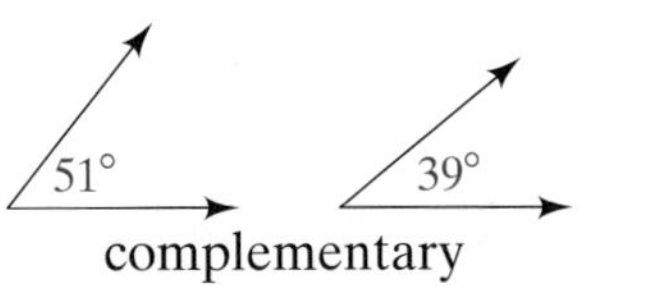

complementary

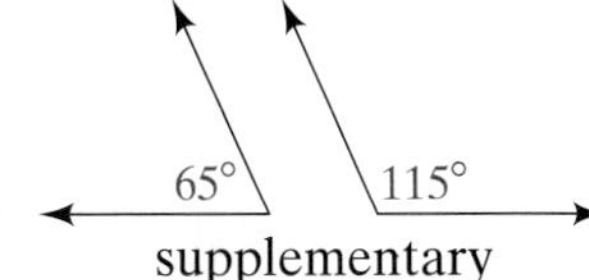

supplementary

EXAMPLE Finding Complements and Supplements

GO Online

Video Tutor Help

Visit: PHSchool.com
Web Code: are-0775

2 **Multiple Choice** If $\angle A$ and $\angle B$ are supplementary and the measure of $\angle A$ is 37°, what is the measure of $\angle B$?

Ⓐ 43° Ⓑ 53° Ⓒ 143° Ⓓ 153°

Write an equation. Let x = the measure of $\angle B$.

$x + 37° = 180°$ ← **The angles are supplementary.**

$x + 37° - 37° = 180° - 37°$ ← **Subtract 37° from each side.**

$x = 143°$ ← **Simplify.**

The measure of $\angle B$ is 143°. The answer is C.

Quick Check

2. Find the measure of the complement of $\angle A$ in Example 2. **53°**

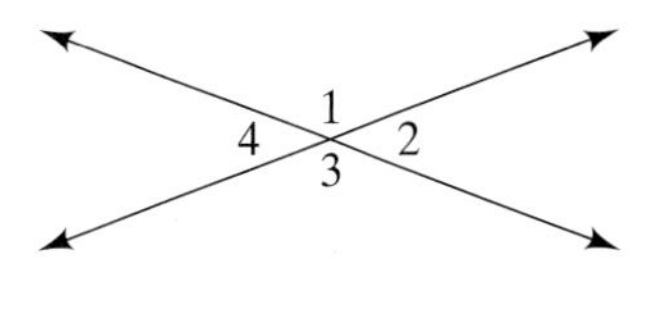

Adjacent angles share a vertex and a side but have no interior points in common. Angles 1 and 2 are adjacent angles. Adjacent angles formed by two intersecting lines are supplementary.

Angles 1 and 3 above are vertical angles. **Vertical angles** are formed by two intersecting lines and are opposite each other. Vertical angles have equal measures. Angles with equal measures are **congruent angles.**

2. Teach

Activity Lab

Use before the lesson.
Student Edition Activity Lab, Hands On 7-2a Measuring Angles, p. 329

All in One Teaching Resources

Activity Lab 7-2: Angles and Parallel Lines

Guided Instruction

Example 1

Ask: *Put the four types of angles in order from greatest measure to least measure.* **straight, obtuse, right, acute**

Error Prevention!

Students may confuse the terms *supplementary* and *complementary.* Invite them to suggest mnemonic devices to remember how they differ. For example, *complementary* is before *supplementary* alphabetically, and 90 is before 180 numerically.

Science Connection

An astrolabe measures the angles formed by a star in the sky and the horizon. Early sailors used the astrolabe to navigate the seas.

PowerPoint Additional Examples

1 Identify the acute angles, obtuse angles, and straight angles in the figure.

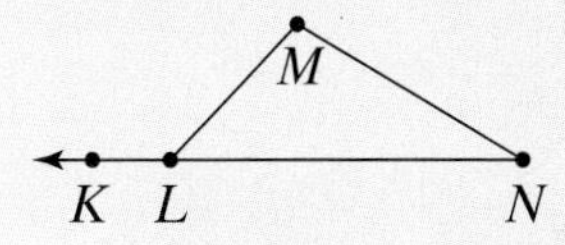

acute: $\angle MLN$, $\angle MNL$
obtuse: $\angle KLM$, $\angle LMN$
straight: $\angle KLN$

2 The measure of $\angle Q$ is 49°. Find the measure of its complement. **41°**

Advanced Learners L4	**English Language Learners** ELL
Can an angle be larger than 180°? Explain. **Sample: Yes, the exterior angle for any angle less than 180° is greater than 180°.**	Help students use language to connect terms to one another. For example, ask them to state the relationship between vertical angles and congruent angles. Vertical angles are congruent, but not all congruent angles are vertical.
learning style: visual	**learning style: verbal**

Additional Examples

3 Find the measures of $\angle 1$, $\angle 2$, and $\angle 4$, if $m\angle 3 = 32°$. **$m\angle 1 = 32°$, $m\angle 2 = 148°$, and $m\angle 4 = 148°$**

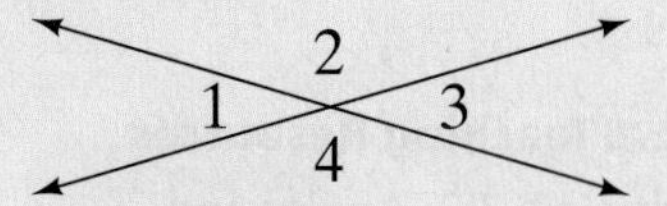

All in One Teaching Resources

- Daily Notetaking Guide 7-2 L3
- Adapted Notetaking 7-2 L1

Closure

- *What are acute, obtuse, right, and straight angles?* **Acute angles are greater than 0° and less than 90°; obtuse angles are greater than 90° and less than 180°; right angles are equal to 90°, and straight angles are equal to 180° or a straight line.**
- *What are complementary and supplementary angles?* **Complementary angles are two angles that add to 90°. Supplementary angles are two angles that add to 180°.**
- *What are vertical and adjacent angles?* **When two lines intersect, vertical angles are opposite each other; adjacent angles share a side and a vertex and are supplementary.**

More Than One Way

155°; methods and explanations may vary.

EXAMPLE Finding Angle Measures

3 (Algebra) Find the measures of $\angle 1$, $\angle 2$, and $\angle 3$, for $m\angle 4 = 128°$.

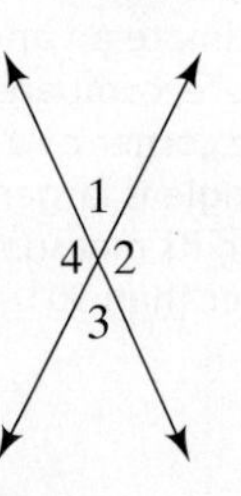

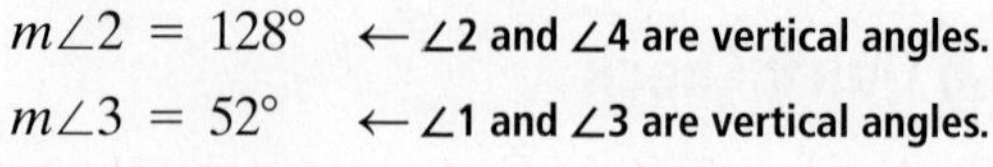

$m\angle 1 = 180° - 128°$ ← **$\angle 1$ and $\angle 4$ are supplementary.**

$= 52°$

$m\angle 2 = 128°$ ← **$\angle 2$ and $\angle 4$ are vertical angles.**

$m\angle 3 = 52°$ ← **$\angle 1$ and $\angle 3$ are vertical angles.**

Quick Check

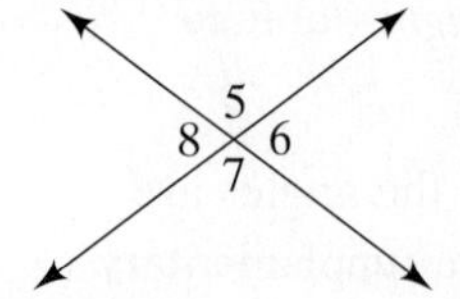

3. In the diagram at the left, $m\angle 8 = 72°$. Find the measures of $\angle 5$, $\angle 6$, and $\angle 7$. **108°; 72°; 108°**

More Than One Way

If $m\angle 1 = 140°$ and $m\angle 2 = 40°$, what is $m\angle 3$?

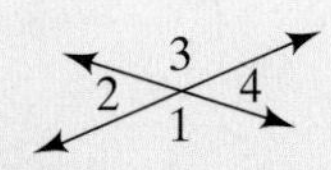

Carlos's Method

$\angle 3$ and $\angle 1$ are across from each other, so they are vertical angles. Since vertical angles have the same measure, $m\angle 3 = m\angle 1$.

So $m\angle 3 = 140°$.

Brianna's Method

$\angle 2$ and $\angle 3$ together form a straight angle, so their measures add up to 180°. This means that they are supplementary angles.

$40° + m\angle 3 = 180°$

$40° - 40° + m\angle 3 = 180° - 40°$ ← **Subtract 40° from each side.**

$m\angle 3 = 140°$ ← **Simplify.**

So $m\angle 3$ is 140°.

Choose a Method

In the figure at the right, $m\angle BEC = 25°$ and $m\angle CED = 155°$. Find $m\angle AEB$. Explain why you chose the method you used.

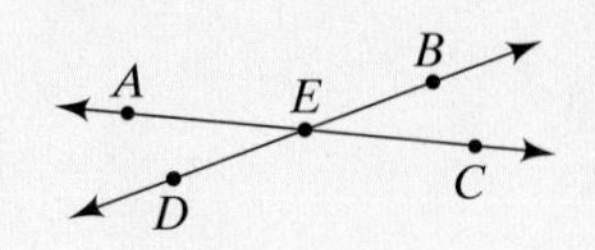

Check Your Understanding

1. Vertical angles lie opposite each other, while adjacent angles lie next to each other.

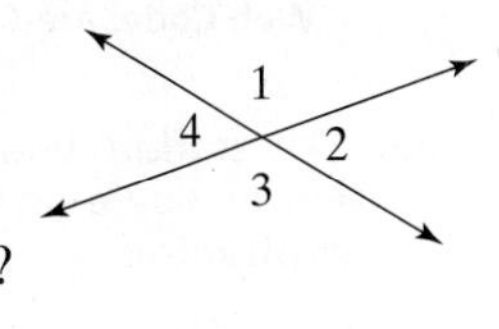

1. **Vocabulary** How are vertical angles and adjacent angles different? **See left.**
2. **Reasoning** What is the sum of the measures of the four angles formed by intersecting lines? **360°**

Classify each angle as *acute, right, obtuse,* or *straight.* Then find the measures of the complement and the supplement of each angle.

3. $m\angle A = 45°$ **acute, 45°, 135°**
4. $m\angle B = 105°$ **obtuse, none, 75°**
5. $m\angle C = 75°$ **acute, 15°, 105°**

Homework Exercises

For more exercises, see Extra Skills and Word Problems.

GO for Help

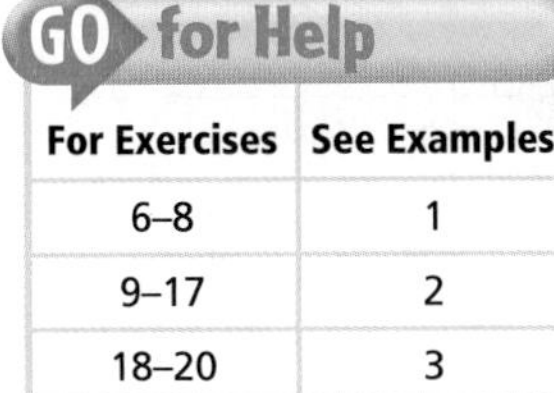

For Exercises	See Examples
6–8	1
9–17	2
18–20	3

A **Classify each angle as *acute, right, obtuse,* or *straight.***

6. 7.

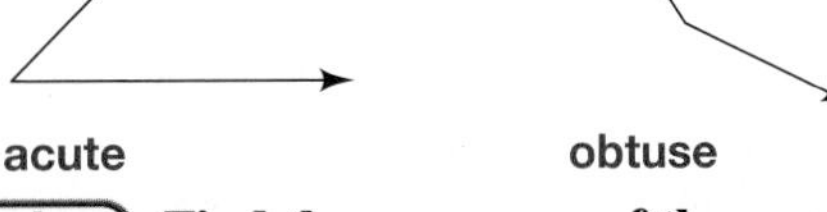

8.

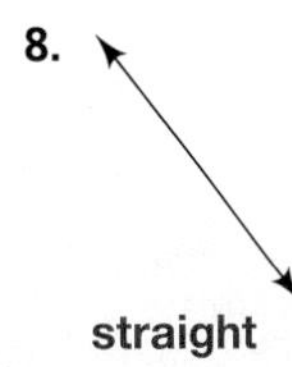

6. acute 7. obtuse 8. straight

Algebra **Find the measures of the complement and the supplement of each angle.** 9–17. See left.

9. $m\angle G = 79°$
10. $m\angle H = 67°$
11. $m\angle J = 12°$
12. $m\angle A = 23.5°$
13. $m\angle B = 37.6°$
14. $m\angle C = 47.9°$
15. $m\angle D = 56.4°$
16. $m\angle E = 75.1°$
17. $m\angle F = 82.2°$

9. 11°; 101°
10. 23°; 113°
11. 78°; 168°
12. 66.5°; 156.5°
13. 52.4°; 142.4°
14. 42.1°; 132.1°
15. 33.6°; 123.6°
16. 14.9°; 104.9°
17. 7.8°; 97.8°

Algebra **In the diagram at the right, $m\angle 2 = 123°$. Find the measure of each angle.**

18. $m\angle 1$ **57°**
19. $m\angle 3$ **57°**
20. $m\angle 4$ **123°**

B 21. **Guided Problem Solving** Engineers designed a metal support that forms a 65° angle with a dam. Find the measure of the angle's supplement. **115°**

$$x + 65° = \blacksquare$$
$$x + 65° - \blacksquare = 180° - \blacksquare$$
$$x = \blacksquare$$

Online lesson quiz, PHSchool.com, Web Code: ara-0702

3. Practice

Assignment Guide

Check Your Understanding
Go over Exercises 1–5 in class before assigning the Homework Exercises.

Homework Exercises

A Practice by Example	6–20
B Apply Your Skills	21–31
C Challenge	32
Test Prep and Mixed Review	33–35

Homework Quick Check
To check students' understanding of key skills and concepts, go over Exercises 7, 13, 25, 26, and 28.

Differentiated Instruction Resources

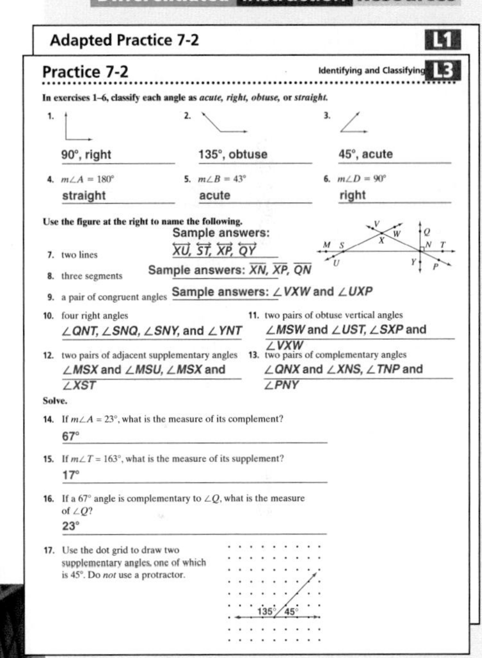

Adapted Practice 7-2 L1

Practice 7-2 Identifying and Classifying L3

In exercises 1–6, classify each angle as *acute, right, obtuse,* or *straight.*

1. 90°, right
2. 135°, obtuse
3. 45°, acute
4. $m\angle A = 180°$ straight
5. $m\angle B = 43°$ acute
6. $m\angle D = 90°$ right

Use the figure at the right to name the following.

7. two lines Sample answers: $\overleftrightarrow{XU}, \overleftrightarrow{ST}, \overleftrightarrow{XP}, \overleftrightarrow{QY}$
8. three segments Sample answers: $\overline{XN}, \overline{XP}, \overline{QN}$
9. a pair of congruent angles Sample answers: $\angle VXW$ and $\angle UXP$
10. four right angles $\angle QNT, \angle SNQ, \angle SNY,$ and $\angle YNT$
11. two pairs of obtuse vertical angles $\angle MSW$ and $\angle UST$, $\angle SXP$ and $\angle VXW$
12. two pairs of adjacent supplementary angles $\angle MSX$ and $\angle MSU$, $\angle MSX$ and $\angle XST$
13. two pairs of complementary angles $\angle QNX$ and $\angle XNS$, $\angle TNP$ and $\angle PNY$

Solve.

14. If $m\angle A = 23°$, what is the measure of its complement? 67°
15. If $m\angle T = 163°$, what is the measure of its supplement? 17°
16. If a 67° angle is complementary to $\angle Q$, what is the measure of $\angle Q$? 23°
17. Use the dot grid to draw two supplementary angles, one of which is 45°. Do *not* use a protractor. 135° 45°

7-2 • Guided Problem Solving GPS L3

Student Page 334, Exercise 26:

Writing in Math Can an angle ever have the same measure as its complement? Explain.

Understand

1. What are you being asked to do?
   Determine if an angle can ever have the same measure as its complement.
2. What do you have to do to explain your answer?
   If there is an angle that has the same measure as its complement, show that the sum of the angles is 90°. If there is not an angle that has the same measure as its complement, explain why.

Plan and Carry Out

3. What is the definition of complementary angles?
   two angles whose sum measures 90°
4. If an angle and its complement have the same measure, explain the relationship between the angle and 90°.
   The angle is half of 90°.
5. Determine the measure of the angle. 45°
6. Can an angle ever have the same measure as its complement? yes

Check

7. Explain your answer.
   Sample answer: The complement of a 45° angle is a 45° angle and the sum of 45° and 45° is 90°.

Solve Another Problem

8. Can an angle ever have the same measure as its supplement? Explain.
   Yes; 90° + 90° = 180° which is the definition of supplementary angles.

4. Assess & Reteach

Lesson Quiz

Use the diagram for questions 1–3.

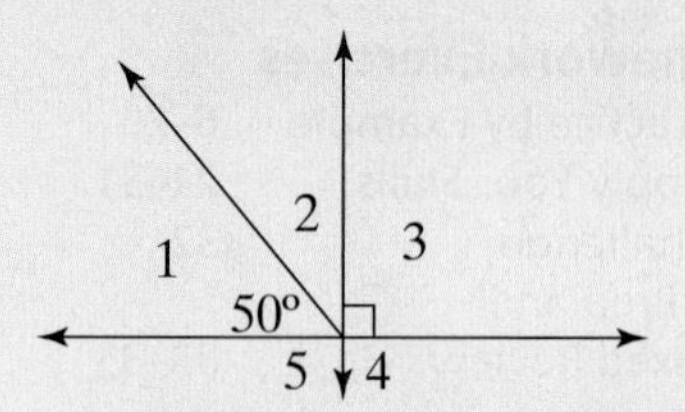

1. Find the measures of ∠2 and ∠3. **m∠2 = 40° m∠3 = 90°**
2. Name a pair of adjacent, complementary angles. **∠1 and ∠2**
3. Name a pair of supplementary angles. **∠3 and ∠4 or ∠4 and ∠5**

Alternative Assessment

Each student in a pair uses a protractor to draw three adjacent angles whose sum is greater than 180° and less than 360°. Students exchange papers and find the measure of each other's angles. Then they classify each angle as acute, obtuse, right, or straight.

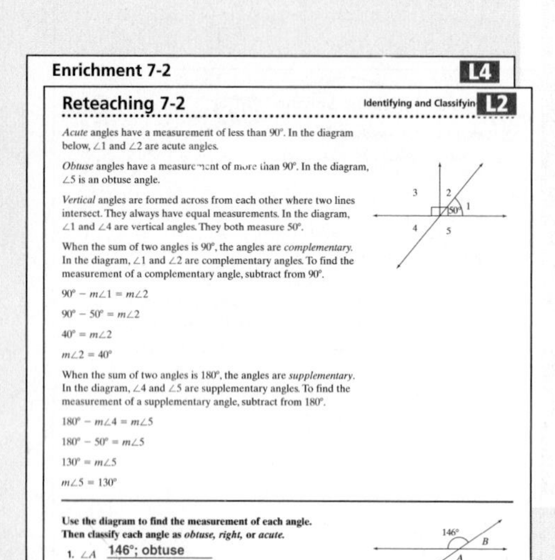

Enrichment 7-2 L4

Reteaching 7-2 Identifying and Classifyin[g] L2

Acute angles have a measurement of less than 90°. In the diagram below, ∠1 and ∠2 are acute angles.

Obtuse angles have a measurement of more than 90°. In the diagram, ∠5 is an obtuse angle.

Vertical angles are formed across from each other where two lines intersect. They always have equal measurements. In the diagram, ∠1 and ∠4 are vertical angles. They both measure 50°.

When the sum of two angles is 90°, the angles are *complementary*. In the diagram, ∠1 and ∠2 are complementary angles. To find the measurement of a complementary angle, subtract from 90°.

90° − m∠1 = m∠2
90° − 50° = m∠2
40° = m∠2
m∠2 = 40°

When the sum of two angles is 180°, the angles are *supplementary*. In the diagram, ∠4 and ∠5 are supplementary angles. To find the measurement of a supplementary angle, subtract from 180°.

180° − m∠4 = m∠5
180° − 50° = m∠5
130° = m∠5
m∠5 = 130°

Use the diagram to find the measurement of each angle. Then classify each angle as *obtuse, right,* or *acute.*

1. ∠A 146°; obtuse
2. ∠B 34°; acute
3. ∠C 50°; acute
4. ∠D 90°; right

GO Online
Homework Video Tutor
Visit: PHSchool.com
Web Code: are-0702

25. The student used the wrong scale on the protractor.

Find the measures of the supplement and the complement of each angle.

22. m∠Q = 48° **132° and 42°**
23. m∠R = 20.2° **159.8° and 69.8°**
24. m∠S = 77.7° **102.3° and 12.3°**

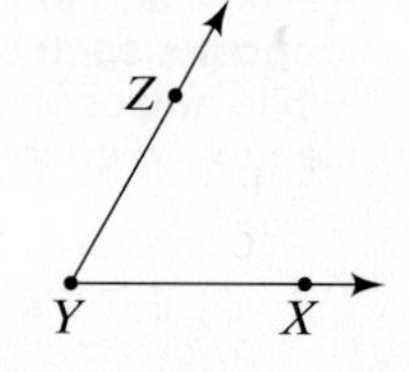

25. **Error Analysis** A student measured ∠XYZ and said that m∠XYZ = 120°. Explain the student's error. **See left.**
26. **Writing in Math** Can an angle ever have the same measure as its complement? Explain. **See margin.**
27. **Physical Therapy** Physical therapists use goniometers to measure the amount of motion a person has in a joint, such as an elbow or a knee. Estimate the measure of the angle shown by the goniometer in the photo at the left. **about 65°**

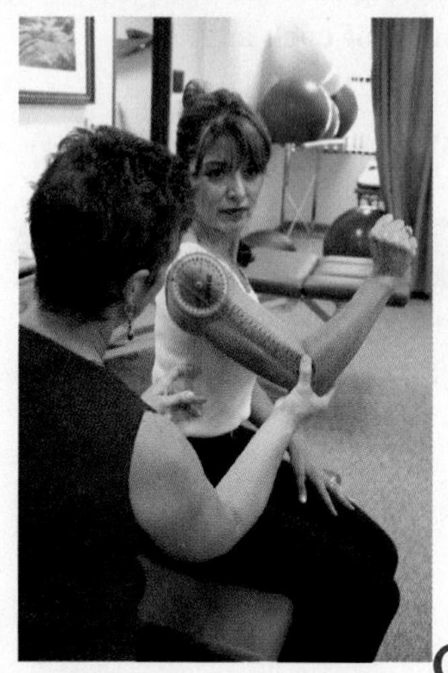

Goniometer

Use the figure to name the following.
28–29. See margin.

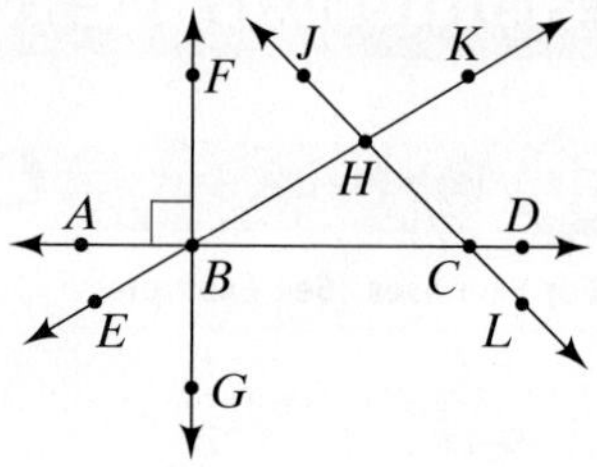

28. two pairs of adjacent supplementary angles
29. two pairs of obtuse vertical angles
30. two pairs of complementary angles **∠ABE and ∠EBG, ∠FBH and ∠HBC**

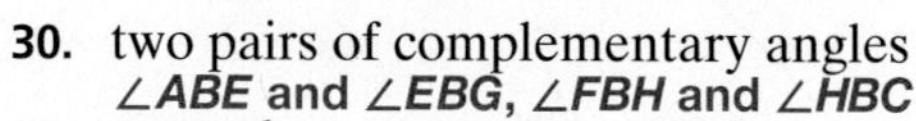

31. an angle congruent to ∠DCL **∠HCB**
32. **Challenge** You know that ∠B is the complement of ∠A, that m∠B = 51°, and that m∠A = (3x − 12)°. Find x. **See margin.**

Test Prep and Mixed Review

Practice

Multiple Choice

33. In the diagram at the right, which pair of angles form complementary angles? **C**
    - (A) ∠AFB and ∠AFC
    - (B) ∠BFC and ∠CFD
    - (C) ∠AFB and ∠CFD
    - (D) ∠AFB and ∠EFD

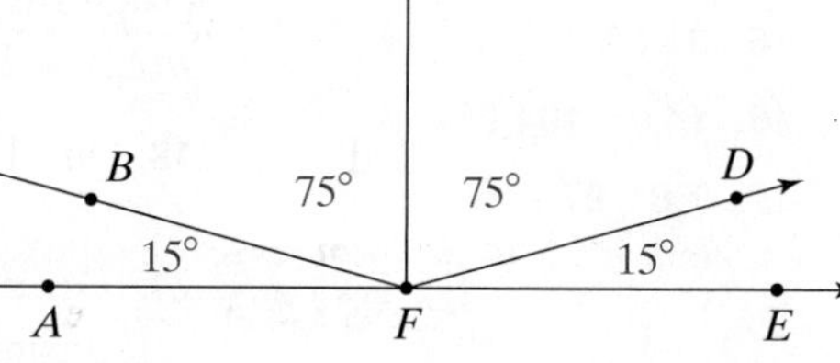

34. The five countries in northern Africa that border the Mediterranean Sea are shown in the table, along with the lowest elevation in each country. Which country contains the lowest point? **H**
    - (F) Algeria
    - (H) Egypt
    - (G) Morocco
    - (J) Tunisia

Country	Elevation
Algeria	–40 m
Egypt	–133 m
Libya	–47 m
Morocco	–55 m
Tunisia	–17 m

GO for Help

For Exercise	See Lesson
35	5-6

35. For a scale of 1 cm : 12 km, find the actual length represented by the length 1.7 cm in a drawing. **20.4 km**

Test Prep

Resources

For additional practice with a variety of test item formats:

- Test-Taking Strategies, p. 365
- Test Prep, p. 369
- Test-Taking Strategies with Transparencies

26. Yes, a 45° angle has the same measure as its complement.
28. Answers may vary. Sample: ∠BHJ and ∠JHK, ∠JHK and ∠KHC
29. Answers may vary. Sample: ∠JHK and ∠BHC, ∠HCD and ∠BCL
32. See back of book.

7-3a Activity Lab

Hands On

Sides and Angles of a Triangle

You need three sides and three angles to form a triangle. Can you make a triangle with *any* three sides? The activity below will help you understand the relationship between the side lengths of a triangle and its angles.

Check students' work.

Step 1 Cut five straws to the following lengths: 2 in., 4 in., 5 in., 8 in., and 10.5 in.

Step 2 Using any three pieces, try to form a triangle. When you find three lengths that can form a triangle, trace the perimeter of the triangle on a sheet of paper. Label the triangles $\triangle A$, $\triangle B$, $\triangle C$, and so on.

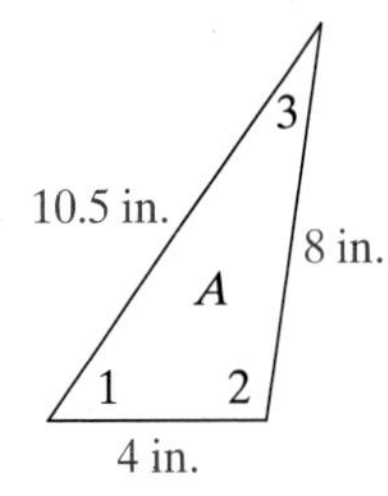

Step 3 Using all possible combinations of three pieces, repeat Step 2 and make as many triangles as you can.

Step 4 Make a list of the side lengths of all the triangles that you made. Record your results in a table like the one below.

Triangle	Shortest Side	Middle Side	Longest Side
A	4 in.	8 in.	10.5 in.
B	■	■	■
C	■	■	■

Step 5 Analyze your table. For each triangle, make a conjecture about the sum of any two side lengths compared to the length of the third side.

Step 6 Now use a protractor to measure the three angles in each triangle. Record your results in a table like the one below.

Triangle	$m\angle 1$	$m\angle 2$	$m\angle 3$	Total
A	■	■	■	■
B	■	■	■	■
C	■	■	■	■

Step 7 Analyze your table. Make a conjecture about the sum of the angles in each triangle.

Activity Lab

Sides and Angles of a Triangle

Students use pipe cleaners to discover relationships among the sides and angles of triangles.

Guided Instruction

Before beginning the Activity, have students decide if they think a triangle can be made with any three sides. Tell them they will answer this question as they do the Activity. They will also learn what the sum of the angles of any triangle is.

You may want students to write their opinions before doing the Activity and revise them afterwards.

Differentiated Instruction

Tactile Learners
If students have trouble seeing that the sum of the measures of any triangle is 180°, have them cut out a paper triangle, number the angles of the triangle and cut them off. Then they place the three angles side by side to form non-overlapping adjacent angles. The three angles will form a straight angle, which will measure 180°. So the sum of the angles of the triangle is 180°.

Visual Learners
Using pipe cleaners of different colors will help students to differentiate lengths more easily.

Resources

- Activity Lab 7-3: Triangles
- pipe cleaners
- protractor

7-3

1. Plan

Objective
To classify triangles and to find the angle measures of triangles

Examples
1 Classifying Triangles by Sides
2 Classifying Triangles by Angles
3 Finding an Angle Measure

Math Understandings: p. 322C

Math Background

Triangles can be classified by the number of their *congruent sides. Scalene triangles* have no congruent sides. *Isosceles triangles* have two or more congruent sides. *Equilateral triangles* have three congruent sides and are therefore also isosceles triangles.

Triangles can also be classified by their angle measures. A *right triangle* has one right angle. An *acute triangle* has three acute angles. An *obtuse triangle* has one obtuse angle.

More Math Background: p. 322C

Lesson Planning and Resources

See p. 322E for a list of the resources that support this lesson.

Check Skills You'll Need
Use student page, transparency, or PowerPoint. For intervention, direct students to:
Identifying and Classifying Angles
Lesson 7-2
Extra Skills and Word Problems Practice, Ch. 7

7-3 Triangles

Check Skills You'll Need

1. Vocabulary Review How does the measure of an *acute angle* compare to 90°? **See below.**

Classify each angle as *acute, right, obtuse,* or *straight.*

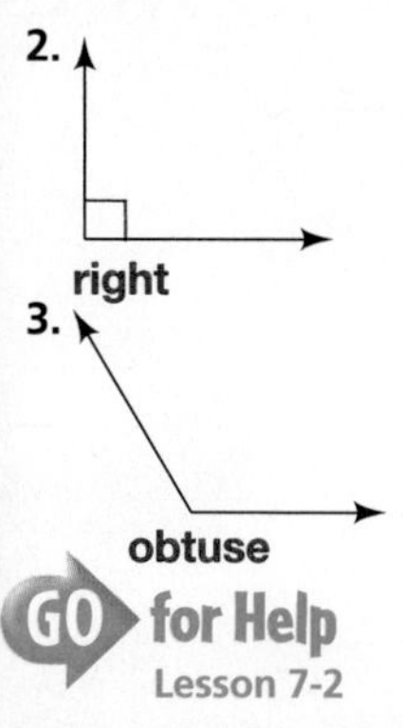

GO for Help Lesson 7-2

Check Skills You'll Need

1. an angle with a measure greater than 0° and less than 90°

Test Prep Tip
You can eliminate choice B because it describes the angles of a triangle, not the sides.

What You'll Learn

To classify triangles and to find the angle measures of triangles

New Vocabulary congruent sides, scalene triangle, isosceles triangle, equilateral triangle, right triangle, acute triangle, obtuse triangle

Why Learn This?

When you can classify triangles, you will recognize them in art and architecture.

Origami is the art of paper folding. At the right is an origami crane. The folds form a variety of triangles.

Congruent sides have the same length. You can classify a triangle by the number of congruent sides it has.

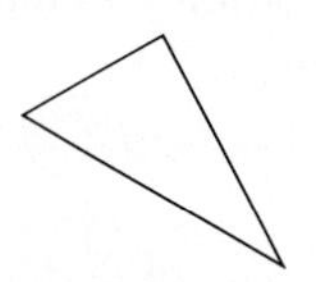

scalene triangle
no congruent sides

isosceles triangle
at least two congruent sides

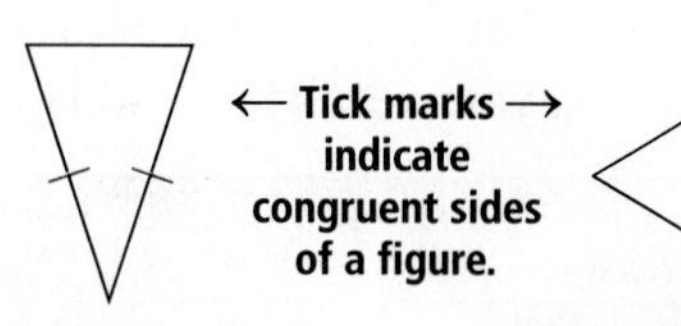

equilateral triangle
three congruent sides

EXAMPLE **Classifying Triangles by Sides**

1 **Multiple Choice** Which of the following best describes $\triangle ABC$ based on its sides?

Ⓐ Scalene Ⓒ Isosceles
Ⓑ Right Ⓓ Equilateral

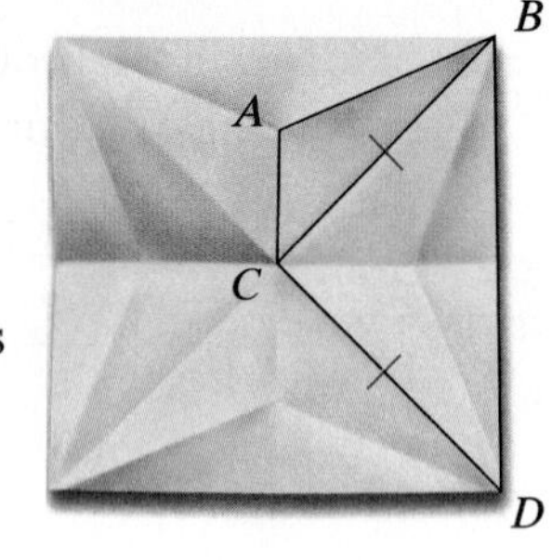

$\triangle ABC$ has no congruent sides. Therefore, it is a scalene triangle. The correct answer is A.

Quick Check

1. Classify $\triangle BCD$ by its sides. **isosceles**

You can also classify a triangle by its angle measures.

Differentiated Instruction Solutions for All Learners

Special Needs L1
Have students highlight the angles when classifying triangles by their angle measures. Have them highlight the sides when classifying triangles by sides. This will help them keep track of their triangles.
learning style: visual

Below Level L2
Have students cut out a variety of triangles. Then have them use a ruler and protractor to classify their triangles by side and angle measures.
learning style: tactile

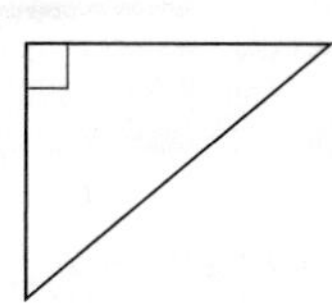

right triangle
one right angle

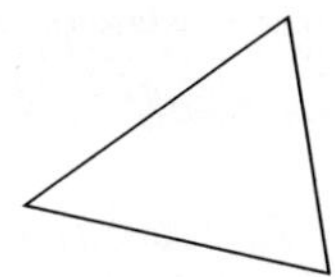

acute triangle
three acute angles

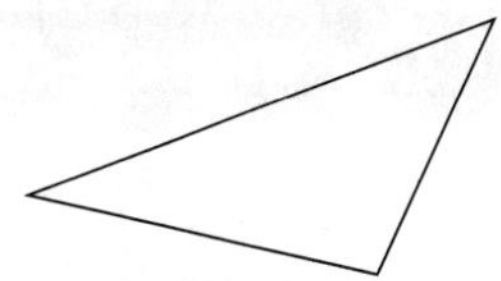

obtuse triangle
one obtuse angle

EXAMPLE Classifying Triangles by Angles

Online active math

For: Triangle Activity
Use: Interactive Textbook, 7-3

2 Classify the triangle shown at the right by its angle measures.

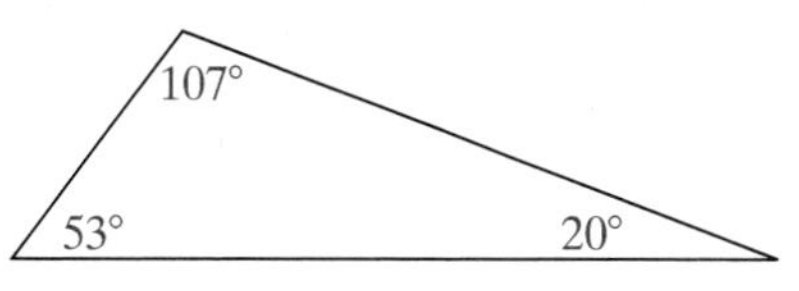

The triangle has one obtuse angle, so it is an obtuse triangle.

Quick Check

2. Classify each triangle by its angle measures.

a.

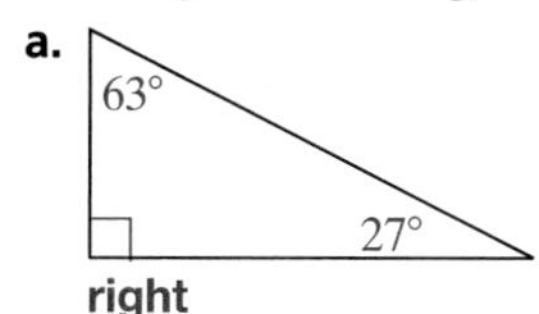

right

b.

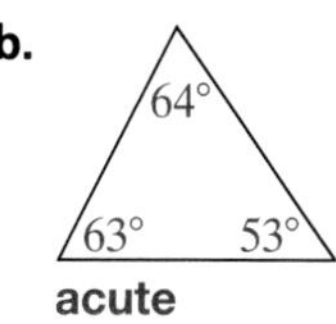

acute

There is a very important property of the angles of a triangle: the sum of the measures of the angles of every triangle is the same.

KEY CONCEPTS **Angle Sum of a Triangle**

The sum of the measures of the angles of any triangle is 180°.

Suppose you know the measures of two of the angles of a triangle. You can write and solve an equation to find the third angle measure.

EXAMPLE Finding an Angle Measure

GO for Help

For help solving equations, go to Lesson 4-3, Example 2.

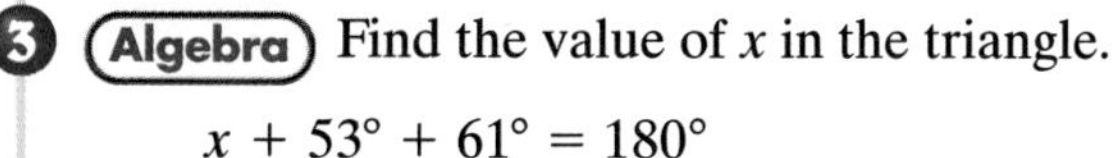

3 (Algebra) Find the value of x in the triangle.

$$x + 53° + 61° = 180°$$
$$x + 114° = 180°$$
$$x + 114° - 114° = 180° - 114°$$
$$x = 66°$$

Quick Check

3. Find the value of x in the triangle.
22°

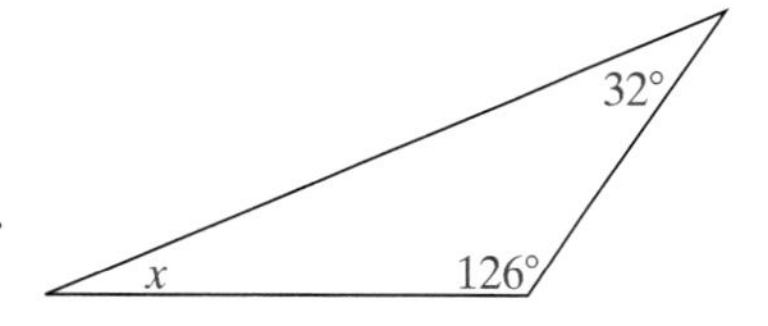

Advanced Learners L4
Is it possible for a triangle to have two right angles? Two obtuse angles? Justify your response. No. The sum of the angles in a triangle is always 180°. The sum of any two angles must be < 180°.

learning style: visual

English Language Learners ELL
Have students draw six triangles on an index card. Three should be classified by their angles (right, obtuse, acute), and three by their sides (scalene, isosceles, equilateral).

learning style: visual

2. Teach

Activity Lab

Use before the lesson.
Student Edition Activity Lab, Hands On 7-3a, Sides and Angles of a Triangle, p. 335

All in One Teaching Resources
Activity Lab 7-3: Triangles

Guided Instruction

Example 2
One right angle classifies a triangle as a right triangle; one obtuse angle classifies it as an obtuse triangle; but three acute angles are necessary for it to be classified as an acute triangle.

PowerPoint

Additional Examples

1 Classify each triangle by its sides.

a. **b.**

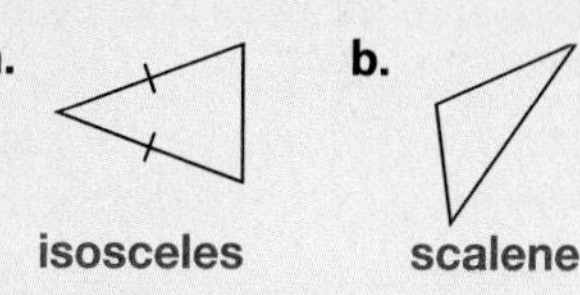

isosceles scalene

2 Classify each by its angle measures.

a. **b.**

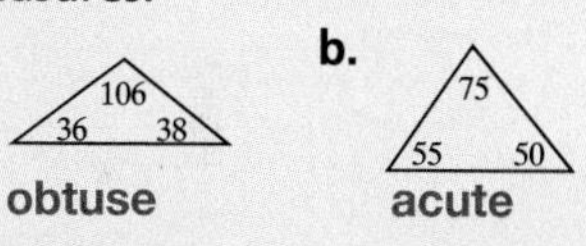

obtuse acute

3 Find the value of x below. 46°

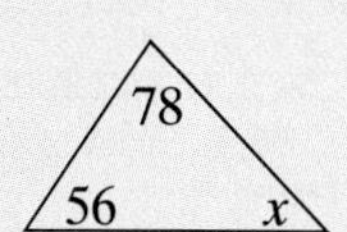

All in One Teaching Resources
- Daily Notetaking Guide 7-3 L3
- Adapted Notetaking 7-3 L1

Closure

- *How are triangles classified as scalene, isosceles, and equilateral?* See back of book.
- *How are triangles classified as acute, obtuse, and right?* See back of book.
- *How can you find the measure of a missing angle in a triangle?* See back of book.

3. Practice

Assignment Guide

Check Your Understanding

Go over Exercises 1–5 in class before assigning the Homework Exercises.

Homework Exercises

A	Practice by Example	6–14
B	Apply Your Skills	15–22
C	Challenge	23
	Test Prep and Mixed Review	24–30

Homework Quick Check

To check students' understanding of key skills and concepts, go over Exercises 8, 13, 19, 21, and 22.

Differentiated Instruction Resources

Adapted Practice 7-3 L1

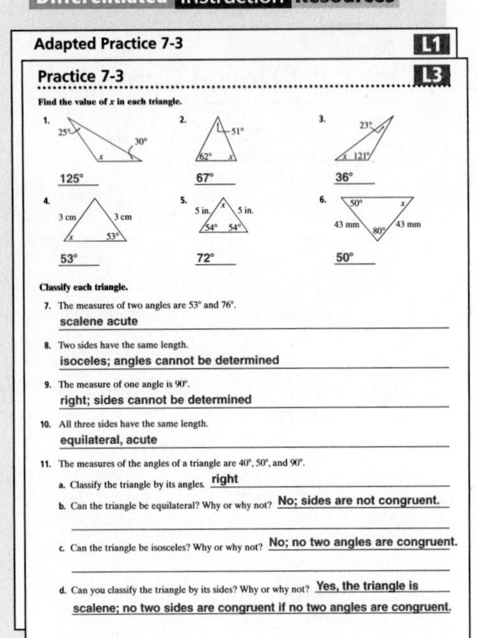
Practice 7-3 L3

Find the value of x in each triangle.

1. 125°
2. 67°
3. 36°
4. 53°
5. 72°
6. 50°

Classify each triangle.

7. The measures of two angles are 53° and 76°. scalene acute
8. Two sides have the same length. isoceles; angles cannot be determined
9. The measure of one angle is 90°. right; sides cannot be determined
10. All three sides have the same length. equilateral, acute
11. The measures of the angles of a triangle are 40°, 50°, and 90°.
    a. Classify the triangle by its angles. right
    b. Can the triangle be equilateral? Why or why not? No; sides are not congruent.
    c. Can the triangle be isosceles? Why or why not? No; no two angles are congruent.
    d. Can you classify the triangle by its sides? Why or why not? Yes, the triangle is scalene; no two sides are congruent if no two angles are congruent.

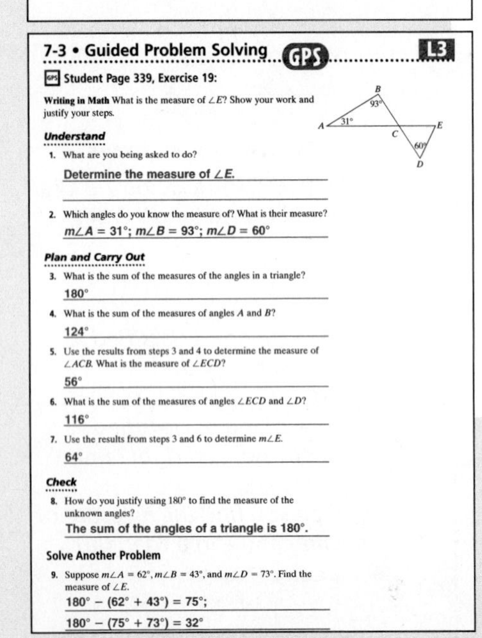
7-3 • Guided Problem Solving GPS L3

Student Page 339, Exercise 19:

Writing in Math What is the measure of ∠E? Show your work and justify your steps.

Understand

1. What are you being asked to do? Determine the measure of ∠E.
2. Which angles do you know the measure of? What is their measure? m∠A = 31°; m∠B = 93°; m∠D = 60°

Plan and Carry Out

3. What is the sum of the measures of the angles in a triangle? 180°
4. What is the sum of the measures of angles A and B? 124°
5. Use the results from steps 3 and 4 to determine the measure of ∠ACB. What is the measure of ∠ECD? 56°
6. What is the sum of the measures of angles ∠ECD and ∠D? 116°
7. Use the results from steps 3 and 6 to determine m∠E. 64°

Check

8. How do you justify using 180° to find the measure of the unknown angles? The sum of the angles of a triangle is 180°.

Solve Another Problem

9. Suppose m∠A = 62°, m∠B = 43°, and m∠D = 73°. Find the measure of ∠E. 180° − (62° + 43°) = 75°; 180° − (75° + 73°) = 32°

Check Your Understanding

1. **Vocabulary** A scalene triangle has __?__ congruent sides. **no**

2. **Open-Ended** Draw an isosceles right triangle. Label the angle measures and use tick marks to indicate congruent sides.

2. Check students' work. An example is given.

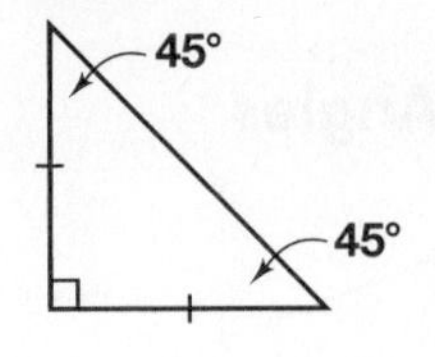

Classify each triangle by its angle measures and by its sides.

3. 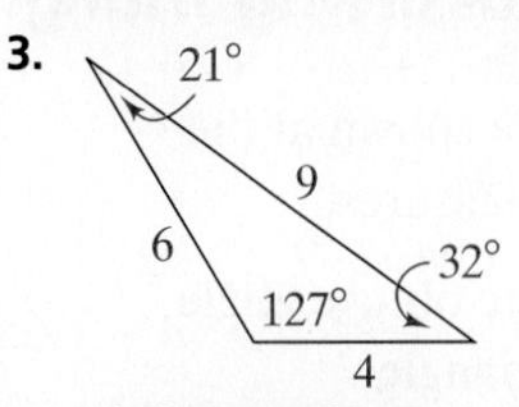

obtuse scalene

4. 45°, 7, 5, 45°, 5
right isosceles

5. 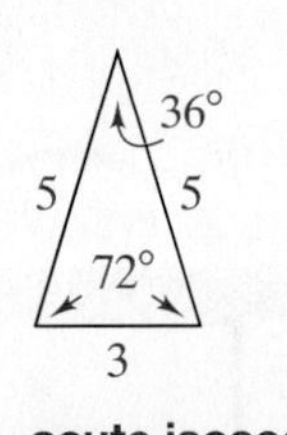

acute isosceles

Homework Exercises

For more exercises, see Extra Skills and Word Problems.

GO for Help

For Exercises	See Examples
6–8	1
9–11	2
12–14	3

A **Classify each triangle by its sides.**

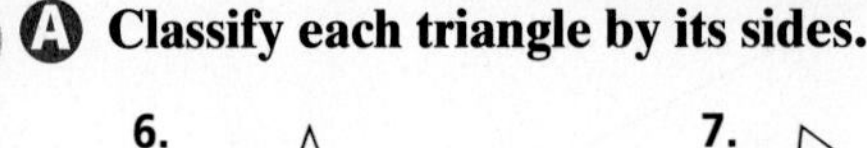

6. 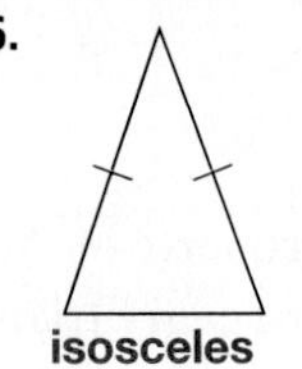
isosceles

7. 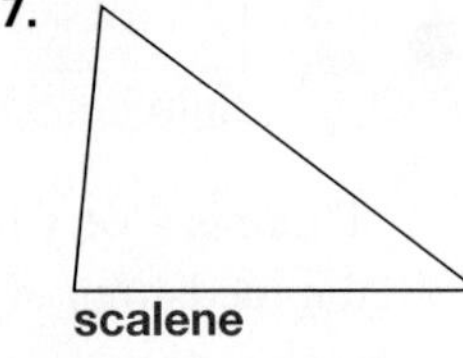
scalene

8. 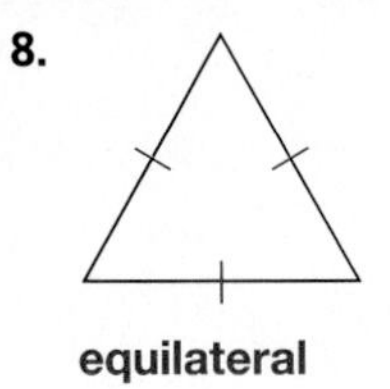
equilateral

Classify each triangle by its angle measures.

9. 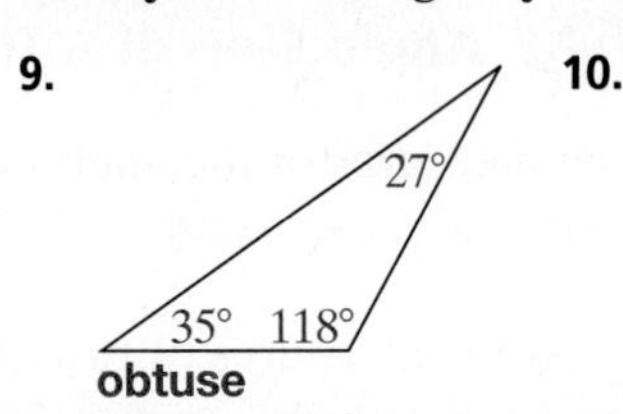

obtuse

10. acute

11. 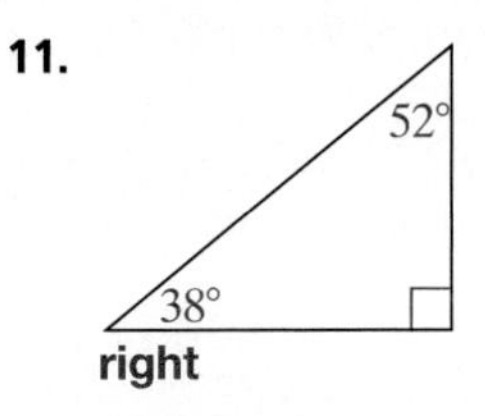

right

Algebra **Find the value of x in each triangle.**

12. 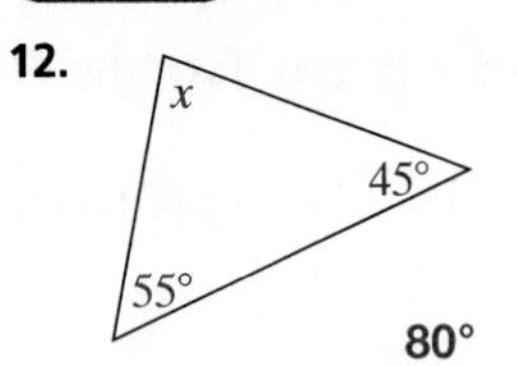

80°

13. x, 42°, 74°
64°

14. 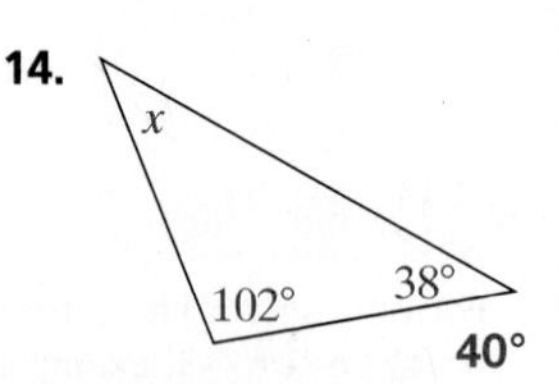

40°

B **GPS** 15. **Guided Problem Solving** The angles of a triangle measure $x°$, $2x°$, and $3x°$. Classify the triangle by its angles. **See left.**

- Use the strategy *Systematic Guess and Check.*
- Continue until the sum of the measures is 180°.
- Classify the triangle.

$x°$	$2x°$	$3x°$	Sum of Angle Measures	Result
10	20	30	60	Too low
■	■	■	■	■

15. right triangle

22b. Complementary; the sum of the angles of a triangle is 180°, and if one of the angles is 90°, the other two must add up to 90°.

23. An equilateral triangle is always isosceles because it has 3 congruent sides. An isosceles triangle is not always equilateral because it has at least 2 congruent sides.

GO Online
Homework Video Tutor
Visit: PHSchool.com
Web Code: are-0703

Algebra **Suppose the sides of a triangle have each of the given measures. Classify each triangle by its sides.**

16. j, j, j **equilateral**
17. $3a, 3a, 5a$ **isosceles**
18. $3w, 4w, 6w$ **scalene**

19. **Writing in Math** What is the measure of $\angle E$? Show your work and justify your steps. **See left.**

19. 64°; 180° − (93° + 31°) = 56°; 56° + 60° = 116°; 180° − 116° = 64°

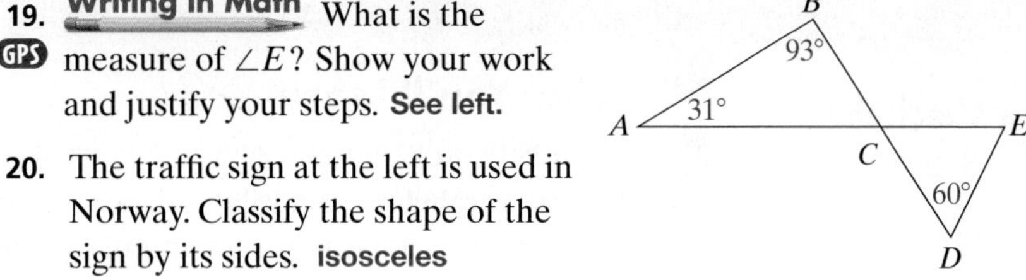
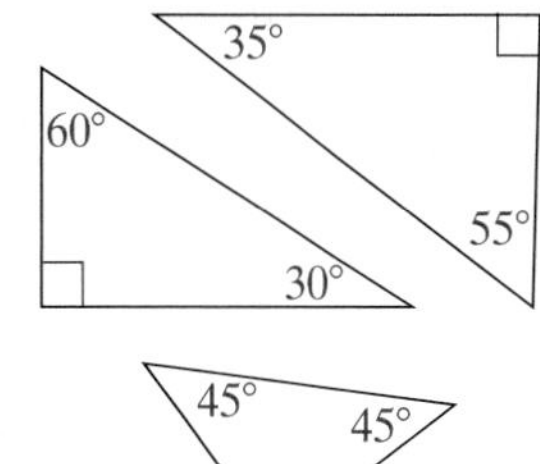
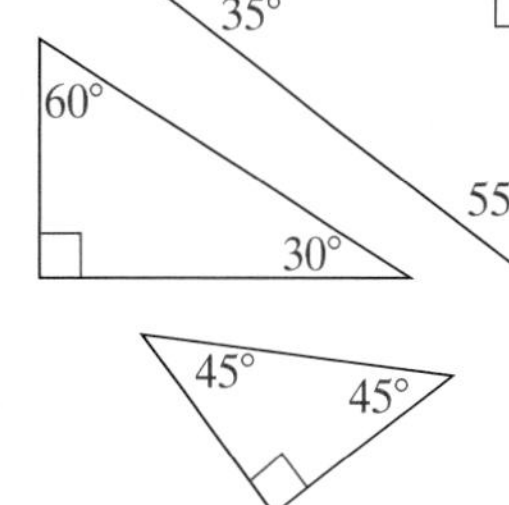

20. The traffic sign at the left is used in Norway. Classify the shape of the sign by its sides. **isosceles**

21. **Algebra** A triangle has two angles that both measure 68°. What is the measure of the third angle? **44°**

22. The triangles shown are right triangles.
    a. What is the sum of the measures of the two acute angles in each triangle? **90°**
    b. **Reasoning** What is the relationship between the two acute angles in any right triangle? Explain. **See margin.**

23. **Challenge** Are all equilateral triangles isosceles? Are all isosceles triangles equilateral? Explain. **See margin.**

Test Prep and Mixed Review

Practice

Multiple Choice

24. The side view of a ramp has the shape of a triangle. Which of the following best describes the triangle with the given measures? **B**
    - (A) Right isosceles triangle
    - (B) Right scalene triangle
    - (C) Acute isosceles triangle
    - (D) Acute scalene triangle

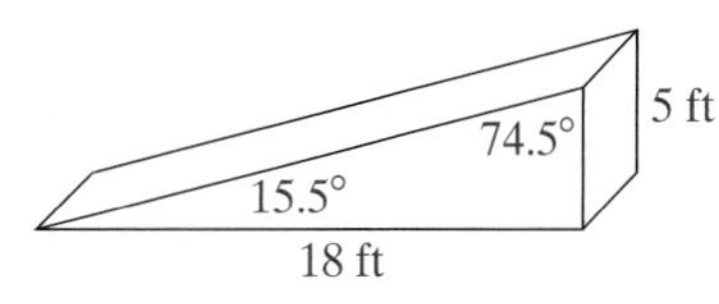

25. On a scale drawing, one side of a box is 2 inches long. The actual length of the box is 6 feet. What is the scale of the drawing? **G**
    (F) 6 in. : 2 ft (G) 1 in. : 3 ft (H) 3 in. : 1 ft (J) 3 ft : 2 in.

26. Three seventh-grade classes go on a camping trip together. The classes have 21, 20, and 19 students. If four students sleep in each tent, how many tents are needed? **C**
    (A) 10 (B) 12 (C) 15 (D) 20

GO for Help

For Exercises	See Lesson
27–30	6-3

Write each percent as a decimal and as a fraction in simplest form.

27. 116% **1.16; $\frac{29}{25}$**
28. 137% **1.37; $\frac{137}{100}$**
29. 155% **1.55; $\frac{31}{20}$**
30. 175% **1.75; $\frac{7}{4}$**

Online lesson quiz, PHSchool.com, Web Code: ara-0703

PowerPoint

Lesson Quiz

Classify each triangle by its sides and by its angle measures. Then find the measure of the missing angle.

1. **isosceles; acute; 30°**

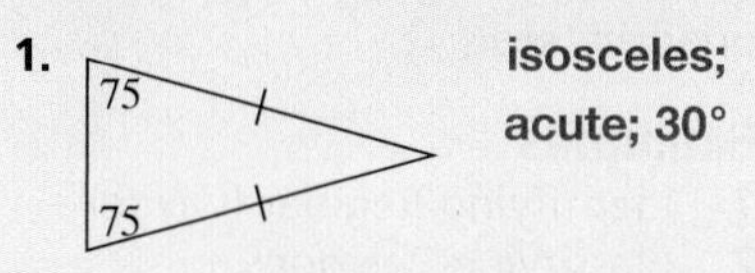

2.

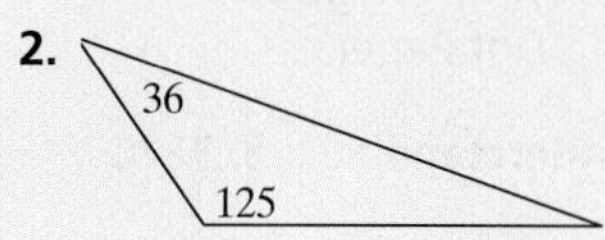

scalene; obtuse; 19°

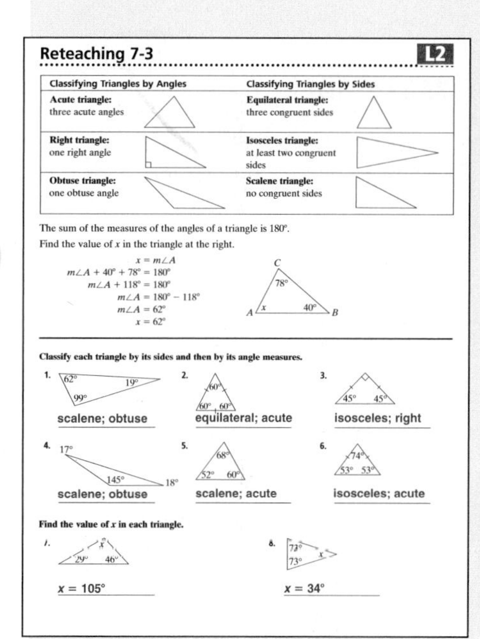

Reteaching 7-3 L2

Classifying Triangles by Angles		Classifying Triangles by Sides	
Acute triangle: three acute angles		**Equilateral triangle:** three congruent sides	
Right triangle: one right angle		**Isosceles triangle:** at least two congruent sides	
Obtuse triangle: one obtuse angle		**Scalene triangle:** no congruent sides	

The sum of the measures of the angles of a triangle is 180°.
Find the value of x in the triangle at the right.

$x = m\angle A$
$m\angle A + 40° + 78° = 180°$
$m\angle A + 118° = 180°$
$m\angle A = 180° - 118°$
$m\angle A = 62°$
$x = 62°$

Classify each triangle by its sides and then by its angle measures.

1. scalene; obtuse
2. equilateral; acute
3. isosceles; right
4. scalene; obtuse
5. scalene; acute
6. isosceles; acute

Find the value of x in each triangle.

7. x = 105°
8. x = 34°

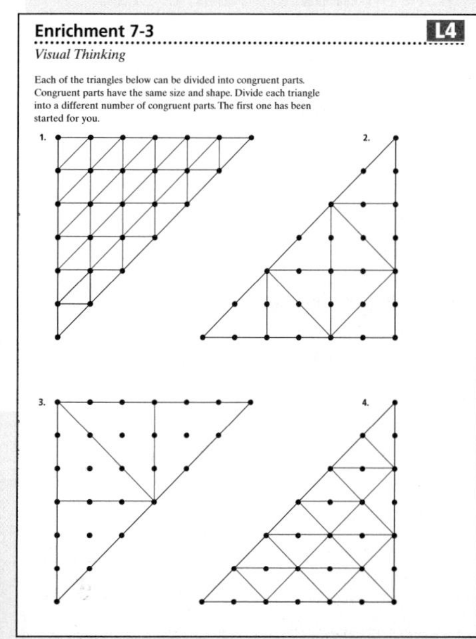

Enrichment 7-3 L4

Visual Thinking

Each of the triangles below can be divided into congruent parts. Congruent parts have the same size and shape. Divide each triangle into a different number of congruent parts. The first one has been started for you.

Alternative Assessment

Ask each student to draw five different triangles and then exchange papers with a partner. Partners classify each triangle by its sides and by the measures of its angles. To extend the activity, have students measure and label each angle.

Test Prep

Resources

For additional practice with a variety of test item formats:
- Test-Taking Strategies, p. 365
- Test Prep, p. 369
- Test-Taking Strategies with Transparencies

7-4

1. Plan

Objective
To classify polygons and special quadrilaterals

Examples
1 Identifying Regular Polygons
2 Classifying Polygons
3 Using Dot Paper

Math Understandings: p. 322C

Math Background

A *polygon* is a closed, two-dimensional figure classified by its number of sides.

The only regular quadrilateral is a *square*. A *rhombus* has four congruent sides, and opposite parallel sides, so it is a parallelogram. A quadrilateral with parallel opposite sides can also be a *parallelogram*. A *trapezoid* is not a parallelogram because only one pair of sides is parallel. A parallelogram with four right angles is a *rectangle*. So a square is also a rectangle.

More Math Background: p. 322C

Lesson Planning and Resources

See p. 322E for a list of the resources that support this lesson.

Bell Ringer Practice

Check Skills You'll Need
Use student page, transparency, or PowerPoint. For intervention, direct students to:
Triangles
Lesson 7-3
Extra Skills and Word Problems Practice, Ch. 7

7-4 Quadrilaterals and Other Polygons

Check Skills You'll Need

1. **Vocabulary Review** What kind of triangle will always have three congruent sides? **1–4. See below.**

Classify each triangle by its sides.

2.
3.

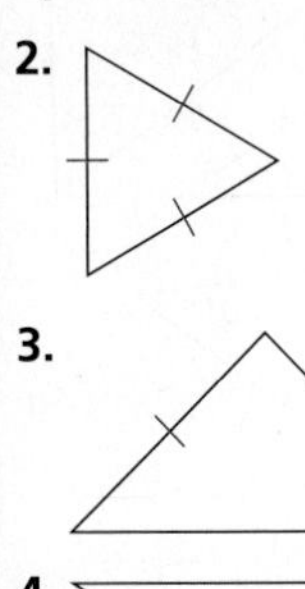

4.

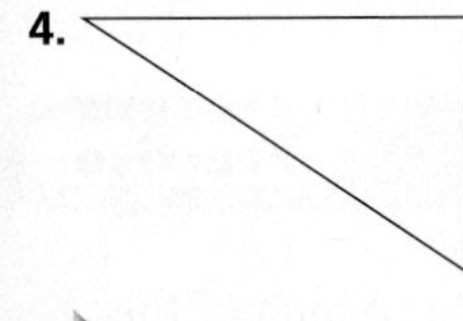

GO for Help Lesson 7-3

Check Skills You'll Need

1. an equilateral triangle
2. equilateral
3. isosceles
4. scalene

What You'll Learn

To classify polygons and special quadrilaterals

New Vocabulary quadrilateral, pentagon, hexagon, octagon, decagon, regular polygon, irregular polygon, trapezoid, parallelogram, rectangle, rhombus, square

Why Learn This?

Artists and designers use quadrilaterals and other polygons in their work because these shapes are pleasing to the eye. The artist Piet Mondrian is known for painting rectangles.

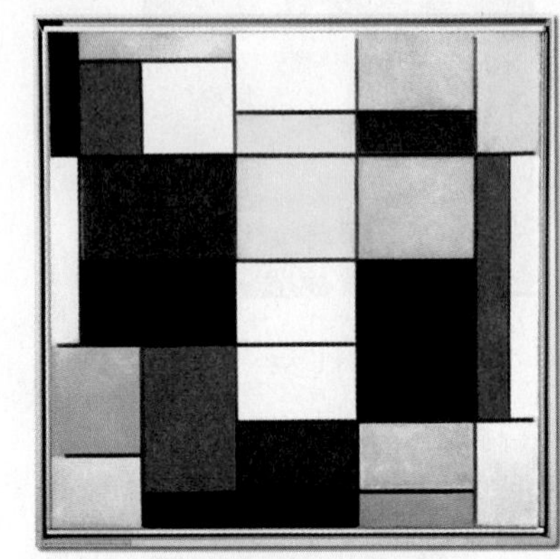

Composition A. 1920. Oil on canvas 35 1/2 x 35 7/8 inches. © 2006 Mondrian/Holtzman Trust c/o HCR International, Warrenton, VA

A polygon is a closed plane figure with sides formed by three or more line segments. The sides meet only at their endpoints.

You classify polygons by their number of sides.

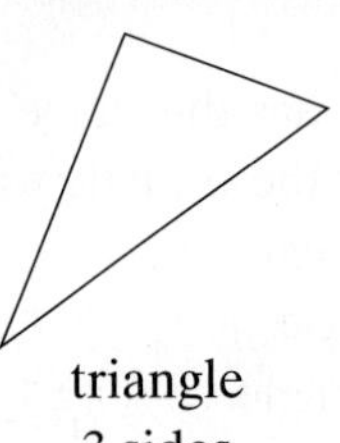

triangle
3 sides

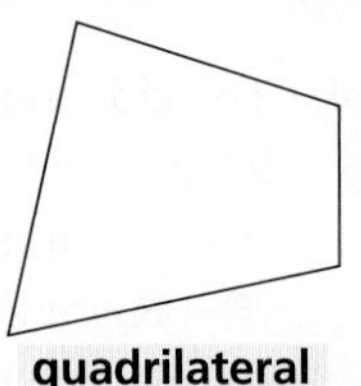

quadrilateral
4 sides

pentagon
5 sides

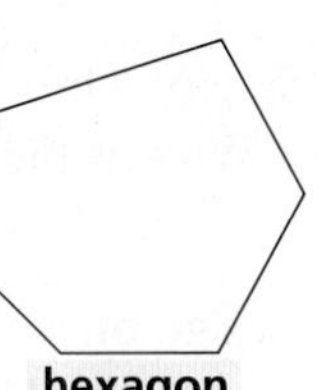

hexagon
6 sides

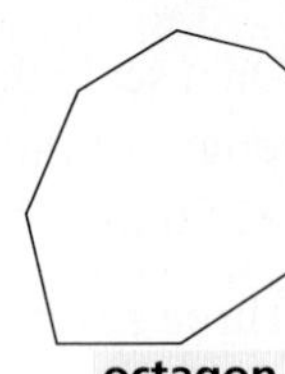

octagon
8 sides

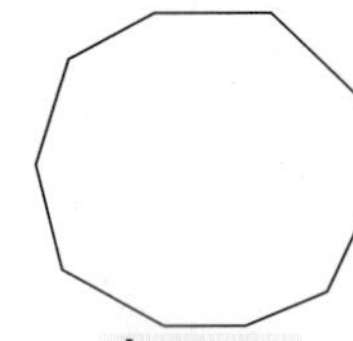

decagon
10 sides

A **regular polygon** is a polygon with all sides congruent and all angles congruent. An **irregular polygon** is a polygon with sides that are not all congruent or angles that are not all congruent.

Differentiated Instruction Solutions for All Learners

Special Needs L1	**Below Level** L2
Some students my have difficulty drawing the quadrilaterals in Example 3. Have them trace the drawn figure with their fingers, counting each side as they trace. Then, pair them up with a student who can draw. **learning style: tactile**	Have students cut various polygons from index cards and label them on one side. Have students mix up the cards and use them like flashcards to practice the naming of polygons. **learning style: tactile**

EXAMPLE Identifying Regular Polygons

Vocabulary Tip

Regular means "consistent" or "the same."

1 Identify each polygon and classify it as *regular* or *irregular*.

a.

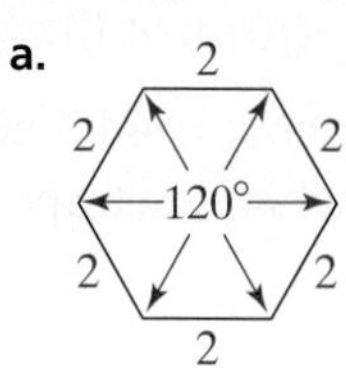

The figure has 6 sides. All sides are congruent. The hexagon is regular.

b.

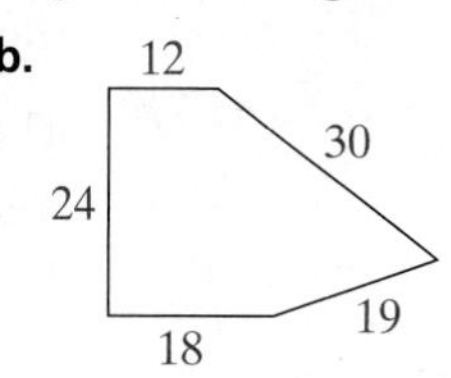

The figure has 5 sides. Not all sides are congruent. The pentagon is irregular.

Quick Check

1. Identify each polygon and classify it as *regular* or *irregular*.

a.

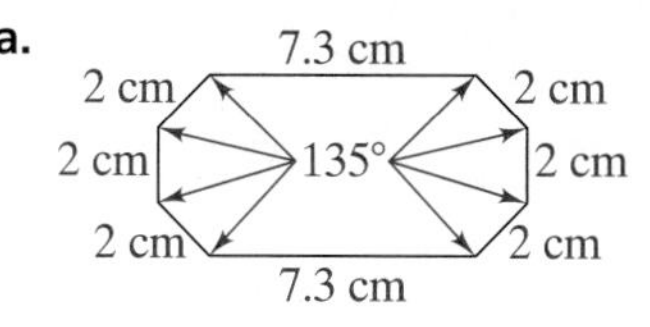

b.

c. **Reasoning** How would you find the perimeter of a regular polygon? Explain. **Multiply the side length by the number of sides the polygon has.**

1a. **The octagon is irregular because the sides are not all congruent.**

b. **The triangle is regular because all sides and all angles are congruent.**

Some quadrilaterals have special names.

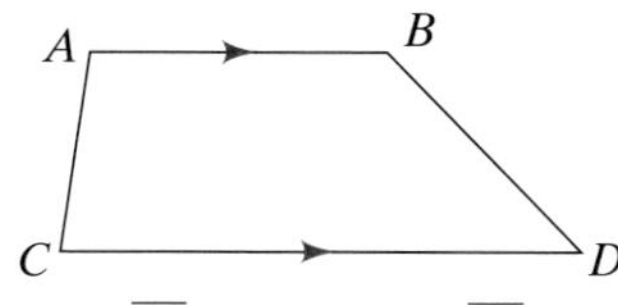

$\overline{AB}$ is parallel to $\overline{CD}$.

A **trapezoid** is a quadrilateral with exactly one pair of parallel sides. The arrows indicate parallel sides.

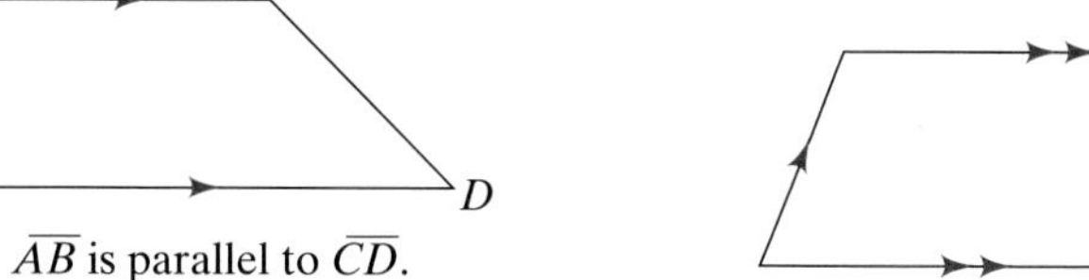

A **parallelogram** is a quadrilateral with both pairs of opposite sides parallel.

GO for Help

For help with parallel lines, go to Lesson 7-1, Example 2.

There are three special types of parallelograms.

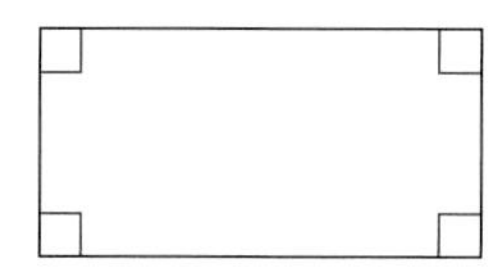

rectangle
four right angles

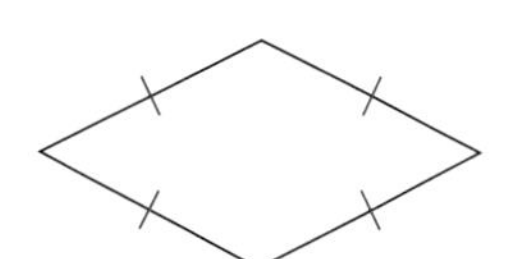

rhombus
four congruent sides

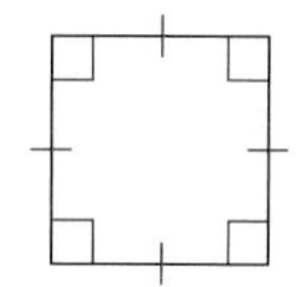

square
four right angles and four congruent sides

2. Teach

Activity Lab

Use before the lesson.

All in One Teaching Resources

Activity Lab 7-4: Shapes and Angles

Guided Instruction

Error Prevention!
Some students may have difficulty drawing the figures. Provide them with a set of pattern blocks to trace. You may want students to measure the sides and angles of their tracings to reinforce the relationships between sides and angles of a given figure.

PowerPoint

Additional Examples

1 Identify the polygon and classify it as *regular* or *irregular.*

a.

hexagon; irregular

b.

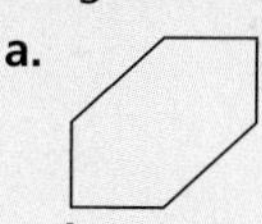

triangle; regular

2 Identify the polygons in the design. **The outside polygon is a decagon. Inside are octagons, quadrilaterals, triangles and a hexagon.**

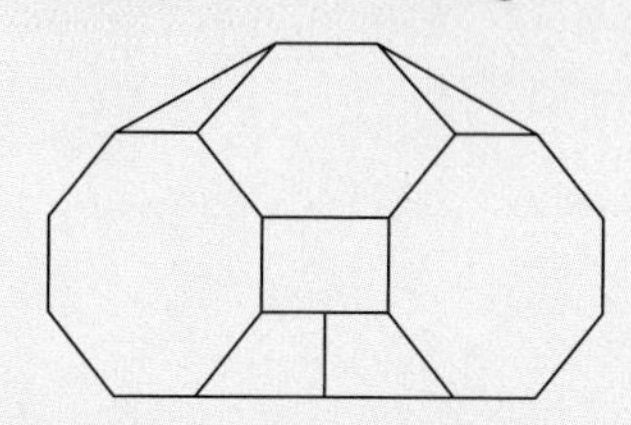

3 Draw a regular parallelogram on graph or dot paper.

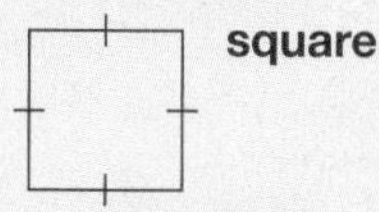

square

Advanced Learners L4	**English Language Learners** ELL
What are the different ways you can classify a square? **A square is a rectangle, a rhombus, a quadrilateral, a parallelogram, and a regular polygon.**	Have students draw a classification chart for quadrilaterals. It should show rectangles, rhombi, and squares as subsets of parallelograms.
learning style: visual	**learning style: visual**

All in One Teaching Resources
- Daily Notetaking Guide 7-4 L3
- Adapted Notetaking 7-4 L1

Closure

- *What is a regular polygon?* a closed plane figure with all sides and all angles congruent
- *How do you classify a polygon by its number of sides?* Sample: A triangle has 3 sides; a quadrilateral has 4 sides; a pentagon has 5 sides; a hexagon has 6 sides; etc.
- *What are the five special quadrilaterals?* trapezoid, parallelogram, rectangle, rhombus, square

3. Check student's work. An example is given.

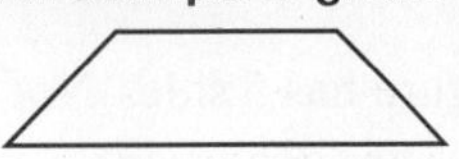

EXAMPLE Classifying Polygons

2 **Architecture** Use the best names to identify the polygons in the window shown at the right.

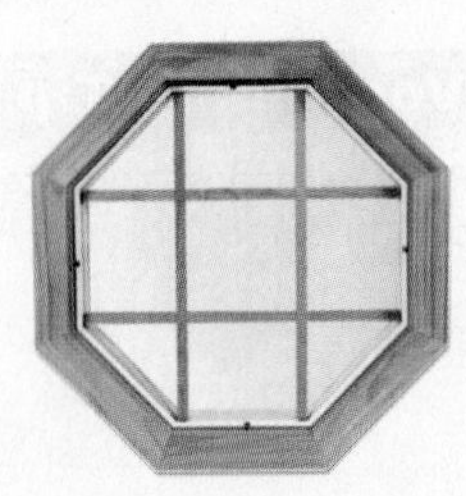

The outside frame is an octagon. Inside the frame there are triangles, rectangles, trapezoids, and squares.

Quick Check

2. Use the best names to identify the polygons in each pattern.

a. 

hexagons, triangles

b. 

squares, triangles, rhombuses

EXAMPLE Using Dot Paper

Online active math

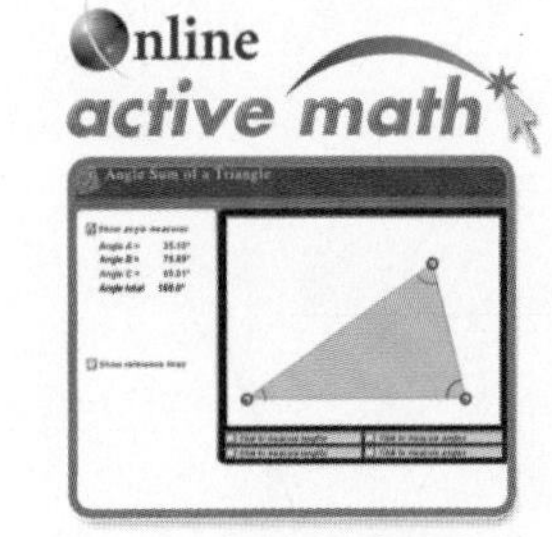

For: Quadrilateral Activity
Use: Interactive Textbook, 7-4

3 Draw each of the following figures on dot paper.

a. a parallelogram that is not a rectangle or a rhombus

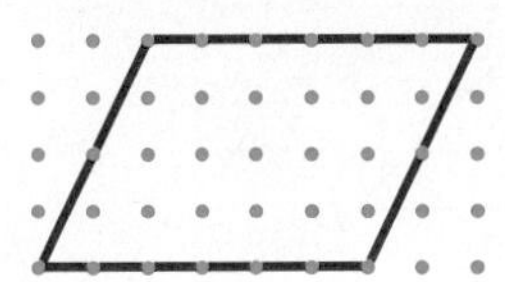

b. a rhombus that is not a square

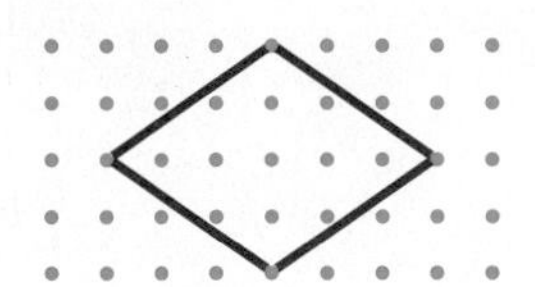

Quick Check

3. Draw a trapezoid with a pair of congruent opposite sides.
See above left.

Check Your Understanding

1. A trapezoid is a quadrilateral with only one pair of parallel sides. A parallelogram is a quadrilateral with two pairs of parallel sides.

1. **Vocabulary** How do parallelograms and trapezoids differ? Explain.

2. **Reasoning** Can you draw a square that is not a rhombus? Explain.
2–5. See margin.

List all the names that apply to each figure.

3. 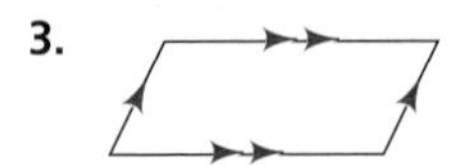

4. 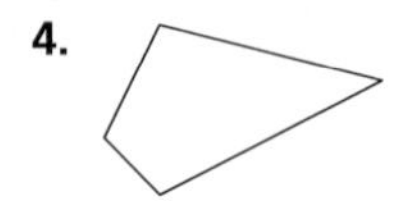

5. 

2. No. All squares have four congruent sides. Every square is a rhombus.
3. quadrilateral, parallelogram
4. quadrilateral
5. quadrilateral, parallelogram, rhombus, rectangle, square

Homework Exercises

For more exercises, see Extra Skills and Word Problems.

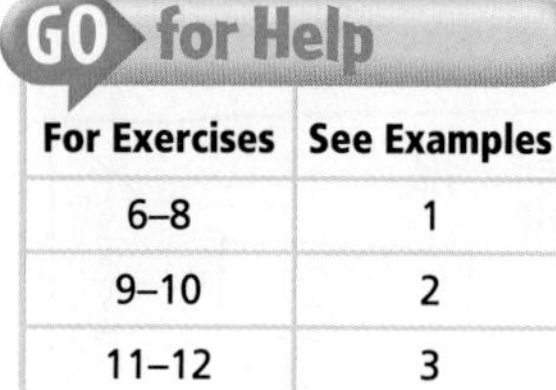

GO for Help

For Exercises	See Examples
6–8	1
9–10	2
11–12	3

A **Identify each polygon and classify it as *regular* or *irregular*. Explain.**

6.

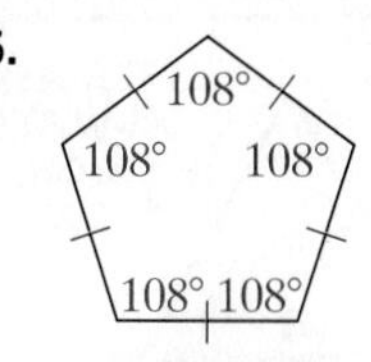

7. 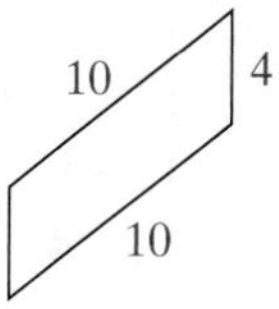 (parallelogram with sides 10, 4, 4, 10)

8.

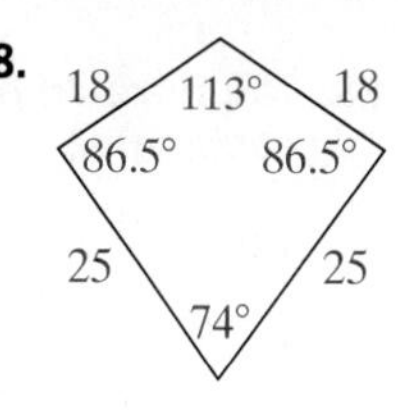

6–8. See left.

6. The pentagon is regular because all sides and angles are congruent.

7. The parallelogram is irregular because the sides are not all congruent.

8. The quadrilateral is irregular because the sides are not all congruent.

Use the best names to identify the polygons in each pattern.

9.

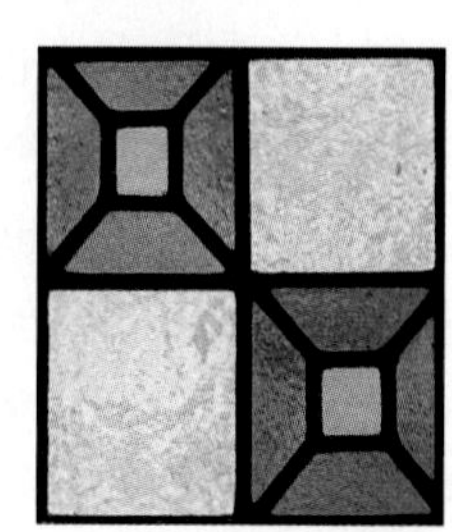

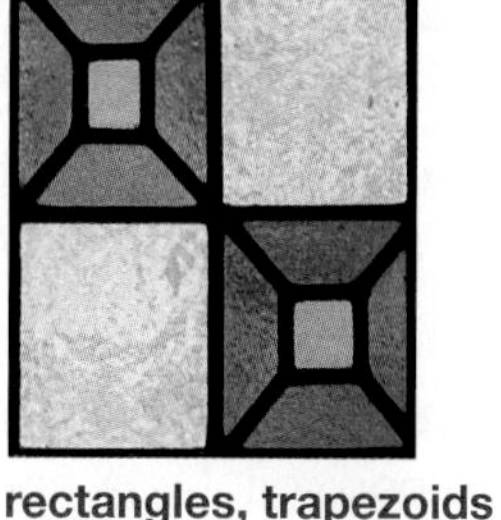

rectangles, trapezoids

10.

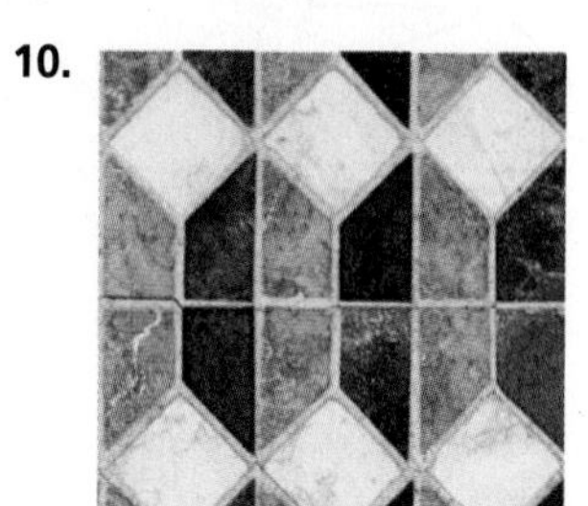

squares, trapezoids, triangles

Use dot paper or graph paper to draw each quadrilateral.

11. a trapezoid with vertical sides parallel 11–12. See margin.

12. two squares, the second of which has twice the perimeter of the first

B

13. **Guided Problem Solving** Find the length of a side of a regular decagon that has a perimeter of 22 ft. 2.2 ft

- How many sides does a decagon have?
- Can you write a formula for the perimeter of a regular decagon?

Judging by appearance, classify each quadrilateral in the photos below. Then name the parallel sides.

14.

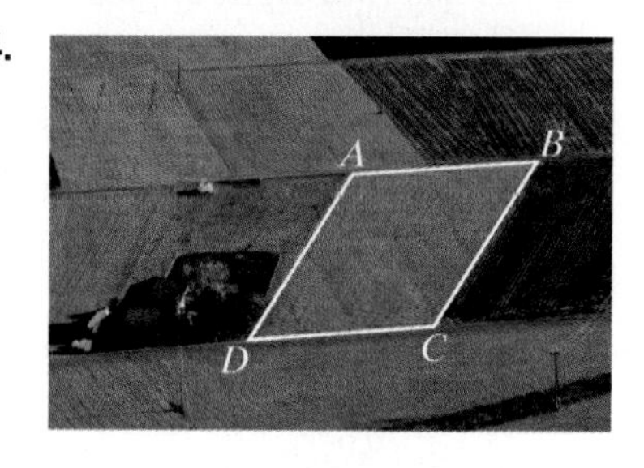

15.

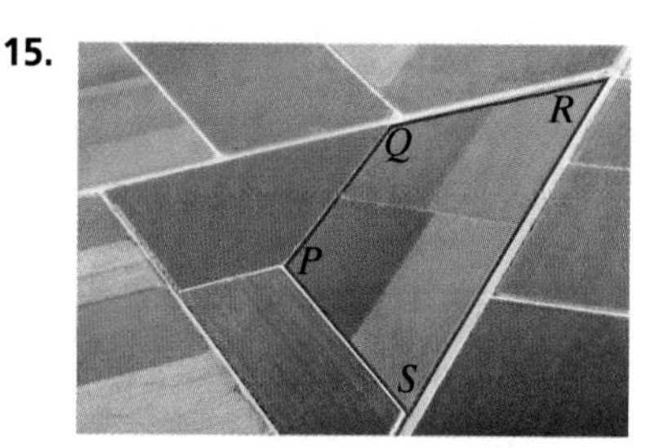

14. Parallelogram; $\overline{AB}$ is parallel to $\overline{DC}$, and $\overline{AD}$ is parallel to $\overline{BC}$.

15. Trapezoid; $\overline{PQ}$ is parallel to $\overline{RS}$.

List all side lengths and angle measures you can find for each polygon.

16. rhombus $WXYZ$, $WX = 4$ cm
$XY = YZ = ZW = 4$ cm

17. rectangle $JKLN$, $KL = 5$ in.

18. parallelogram $ABCD$, $AB = 6$ cm, $m\angle A = 115°$
$DC = 6$ cm; $m\angle C = 115°$ $m\angle B = m\angle D = 65°$

17. $JN = 5$ in.; $m\angle J = m\angle K = m\angle L = m\angle N = 90°$

3. Practice

Assignment Guide

Check Your Understanding
Go over Exercises 1–5 in class before assigning the Homework Exercises.

Homework Exercises

A Practice by Example	6–12
B Apply Your Skills	13–24
C Challenge	25
Test Prep and Mixed Review	26–32

Homework Quick Check
To check students' understanding of key skills and concepts, go over Exercises 9, 11, 22, 23, and 24.

Differentiated Instruction Resources

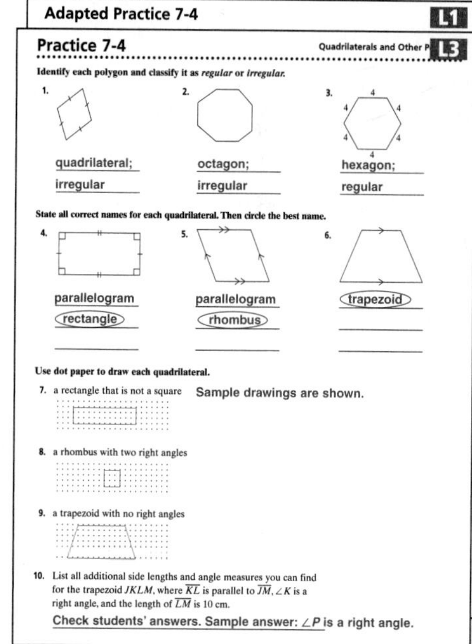

Adapted Practice 7-4 L1

Practice 7-4 Quadrilaterals and Other Polygons L3

Identify each polygon and classify it as *regular* or *irregular*.

1. quadrilateral; irregular
2. octagon; irregular
3. hexagon; regular

State all correct names for each quadrilateral. Then circle the best name.

4. parallelogram, rectangle
5. parallelogram, rhombus
6. trapezoid

Use dot paper to draw each quadrilateral.

7. a rectangle that is not a square Sample drawings are shown.
8. a rhombus with two right angles
9. a trapezoid with no right angles
10. List all additional side lengths and angle measures you can find for the trapezoid $JKLM$, where $\overline{KL}$ is parallel to $\overline{JM}$, $\angle K$ is a right angle, and the length of $\overline{LM}$ is 10 cm.
Check students' answers. Sample answer: $\angle P$ is a right angle.

7-4 • Guided Problem Solving GPS L3

Student Page 344, Exercise 22:

Writing in Math Can a quadrilateral be both a rhombus and a rectangle? Explain.

Understand

1. What are you being asked to do?
Determine if a quadrilateral can be both a rhombus and a rectangle.
2. In order to answer this question, what definitions do you need to know?
quadrilateral, rhombus, and rectangle

Plan and Carry Out

3. What is a quadrilateral?
A quadrilateral is a polygon that has 4 sides.
4. What is a rhombus?
A rhombus is a parallelogram with 4 congruent sides.
5. What is a rectangle?
A rectangle is a parallelogram with 4 right angles.
6. Can a quadrilateral be both a rhombus and a rectangle?
yes

Check

7. Explain your answer to step 6 by giving an example of a shape that is both a rhombus and a rectangle.
A square is a quadrilateral because it has four sides. It is a rhombus because it has four congruent sides. It is a rectangle because it has four right angles.

Solve Another Problem

8. If all squares are types of rectangles, are all rhombuses types of rectangles? Explain.
No, rectangles are rhombuses with four right angles.

For Exercises 11 and 12, check students' work. Examples are given.

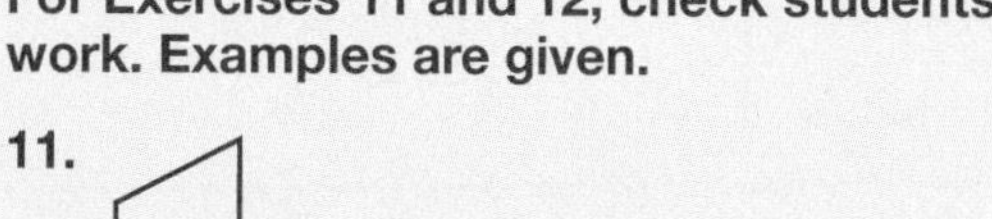

11. (trapezoid drawing)

12.

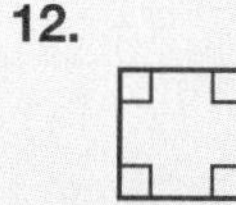

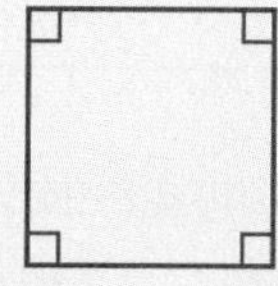

4. Assess & Reteach

PowerPoint

Lesson Quiz

Identify each polygon. Then classify it as *regular* or *irregular.*

1. rectangle; irregular

2. hexagon; regular

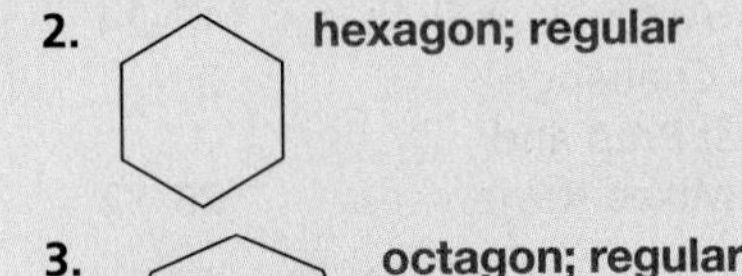

3. octagon; regular

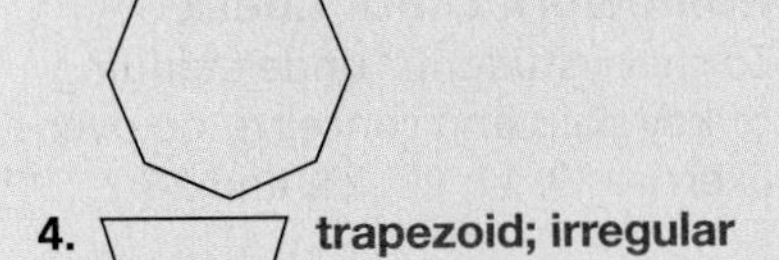

4. trapezoid; irregular

Alternative Assessment

Provide pairs of students with graph paper or dot paper. One partner names one of the polygons in this lesson while the other draws and labels the polygon. Have partners alternate roles.

19.

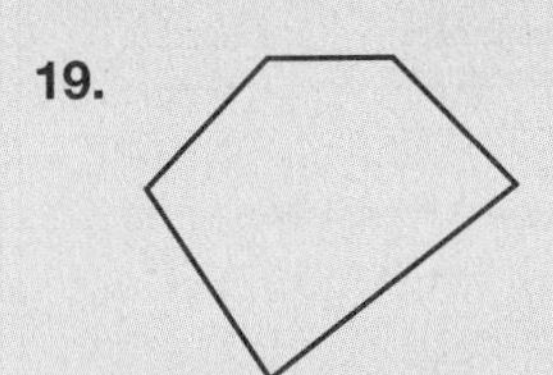

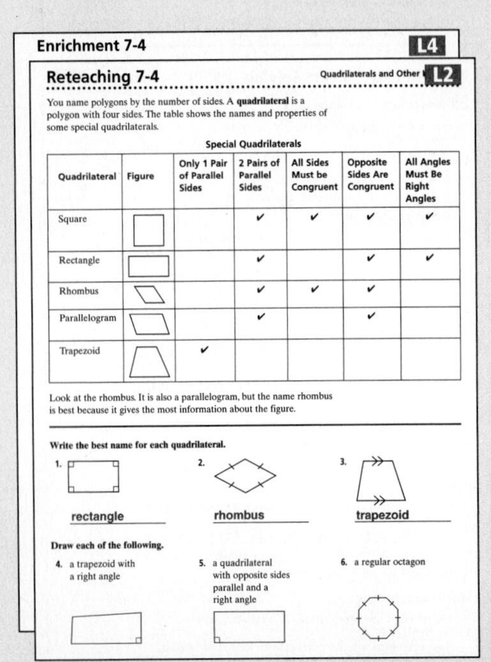

Enrichment 7-4 L4

Reteaching 7-4 — Quadrilaterals and Other Polygons L2

You name polygons by the number of sides. A **quadrilateral** is a polygon with four sides. The table shows the names and properties of some special quadrilaterals.

Special Quadrilaterals

Quadrilateral	Figure	Only 1 Pair of Parallel Sides	2 Pairs of Parallel Sides	All Sides Must be Congruent	Opposite Sides Are Congruent	All Angles Must Be Right Angles
Square			✔	✔	✔	✔
Rectangle			✔		✔	✔
Rhombus			✔	✔	✔	
Parallelogram			✔		✔	
Trapezoid		✔				

Look at the rhombus. It is also a parallelogram, but the name rhombus is best because it gives the most information about the figure.

Write the best name for each quadrilateral.

1. rectangle
2. rhombus
3. trapezoid

Draw each of the following.

4. a trapezoid with a right angle
5. a quadrilateral with opposite sides parallel and a right angle
6. a regular octagon

GO Online
Homework Video Tutor
Visit: PHSchool.com
Web Code: are-0704

23. No; it would be a rectangle.

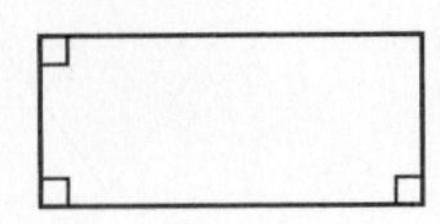

24a.

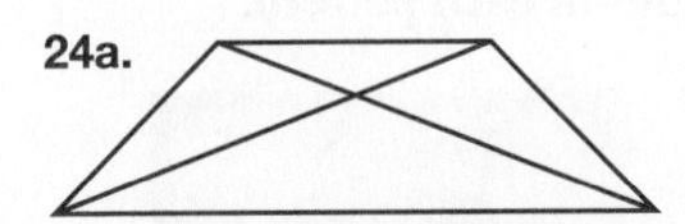

b. The diagonals have the same length.

25.

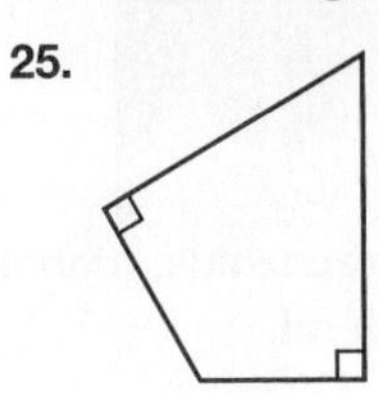

Sketch each polygon. 19–21. See margin.

19. pentagon
20. octagon
21. regular quadrilateral

Math in the Media Use the cartoon for Exercises 22 and 23.

LET'S ASK SQUARE... HE ALWAYS HAS THE RIGHT ANGLE ON THINGS.
THAVES 6-13

22. **Writing in Math** Can a quadrilateral be both a rhombus and a rectangle? Explain. **Yes; a square is both a rhombus and a rectangle.**

23. **Reasoning** Can a trapezoid have three right angles? Draw a diagram to support your answer. **See left.**

24. a. Use a ruler to draw a trapezoid with opposite sides congruent.
    b. A diagonal joins two vertices that are not endpoints of the same side of a polygon. Draw and measure the diagonals of the figure you drew in part (a). What do you notice? **See left.**

25. **Challenge** Draw a quadrilateral with exactly one pair of congruent opposite angles. **See left.**

Test Prep and Mixed Review

Practice

Multiple Choice

26. Which statement is always true about a rhombus? **B**
    - (A) It has four congruent angles.
    - (B) It has four congruent sides.
    - (C) It has four right angles.
    - (D) It has exactly one pair of parallel sides.

27. A punch recipe calls for 6 cups of pineapple juice. Alicia has 1 quart of pineapple juice. How much more does she need? **F**
    - (F) $\frac{1}{2}$ qt
    - (G) $1\frac{1}{2}$ qt
    - (H) 3 c
    - (J) 2 pt

28. An art collector bought five paintings for a total of \$3,600. Then he bought another painting for \$1,200. What was the mean cost for all of the paintings? **B**
    - (A) \$720
    - (B) \$800
    - (C) \$960
    - (D) \$4,800

GO for Help

For Exercises	See Lesson
29–32	2-6

Write each fraction as a decimal.

29. $\frac{5}{3}$ $1.\overline{6}$
30. $\frac{7}{16}$ 0.4375
31. $\frac{15}{18}$ $0.8\overline{3}$
32. $\frac{10}{6}$ $1.\overline{6}$

Test Prep

Resources
For additional practice with a variety of test item formats:
- Test-Taking Strategies, p. 365
- Test Prep, p. 369
- Test-Taking Strategies with Transparencies

20.

21.

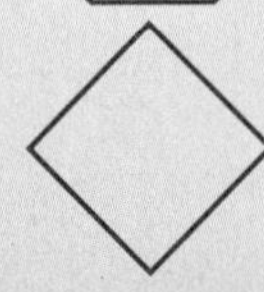

Checkpoint Quiz 1

Lessons 7-1 through 7-4

Use the points in each diagram to name the figure shown.

1. X Z **$\overrightarrow{XZ}$**
2. N M **$\overleftrightarrow{NM}$**
3. K V **$\overline{KV}$**

Find the measures of the complement and the supplement of each angle.

4. $m\angle T = 12°$ **78°; 168°**
5. $m\angle R = 47°$ **43°; 133°**
6. $m\angle U = 65°$ **25°; 115°**

7. A triangle has angles that measure 63° and 47°. What is the measure of the third angle? **70°**

Classify the triangles in the figure at the right.

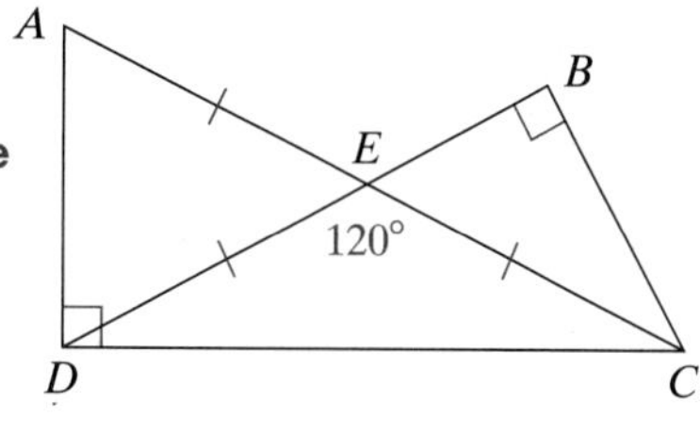

8. $\triangle ADE$ **equilateral**
9. $\triangle DEC$ **obtuse isosceles**
10. $\triangle ACD$ **right scalene**

11. In the figure below, all six segments are congruent. Identify the three polygons that are formed by the segments and classify them as *regular* or *irregular*.

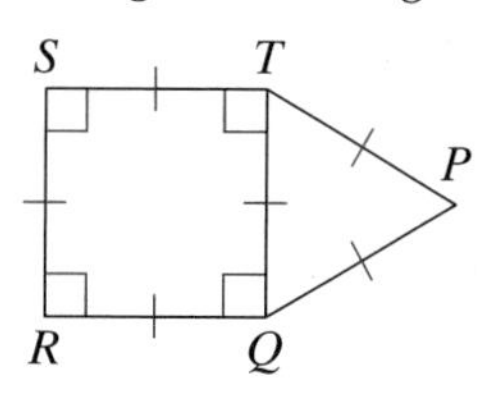

Equilateral $\triangle PQT$ is regular, square $QRST$ is regular, and pentagon $PQRST$ is irregular.

Math at Work — Architect

Do you have an eye for design? If so, then architecture could be a career for you. Architects use creativity, math, science, and art to plan buildings that are beautiful, functional, safe, and economical.

Architects use geometry to understand spatial relationships. They use ratios and percents to plan scale drawings and to build scale models, too.

Architects also must be able to manage projects, supervise people, and communicate complex ideas.

For: Information about architects
Web Code: arb-2031

Checkpoint Quiz

Use this Checkpoint Quiz to check students' understanding of the skills and concepts of Lessons 7-1 through 7-4.

Resources

- All-in-One Teaching Resources Checkpoint Quiz 1
- ExamView CD-ROM
- Success Tracker™ Online Intervention

Math at Work

Architect

Relate the drawings an architect does to the geometry covered in this chapter. The drawings and models architects make rely on many geometric principles.

Guided Instruction

After students read the paragraphs, have them discuss any famous or local architects they know about. Then ask:

- *What do architects design besides buildings?* **Sample: bridges, roadways, tunnels**
- *Name a building or other work for which you admire the design. Explain why.* **Answers will vary.**
- *Name geometric elements in the building or other work.* **Sample: triangles, parallel lines, rectangles**

7-5

1. Plan

Objective
To identify congruent figures and to use them to find missing measures

Examples
1 Identifying Congruent Figures
2 Application: Manufacturing
3 Working With Congruent Figures

Math Understandings: p. 322D

Math Background

Two congruent polygons have *corresponding parts*: each angle and each side in the first polygon is congruent to an angle and a side in the second polygon. The symbol ≅ means "is congruent to." To name congruent figures, write vertices in corresponding order: $\triangle ABC \cong \triangle XYZ$. Tick marks can indicate corresponding parts.

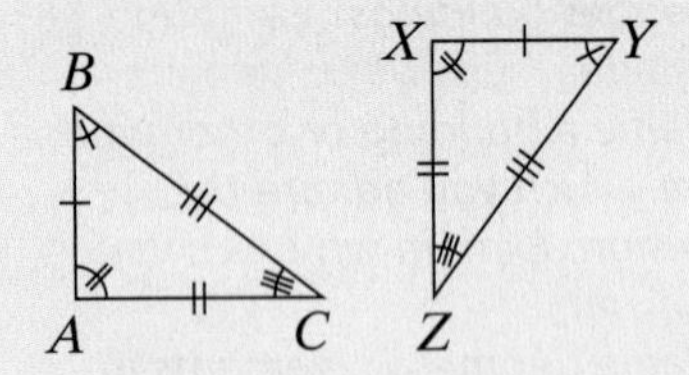

More Math Background: p. 322D

Lesson Planning and Resources

See p. 322E for a list of the resources that support this lesson.

Check Skills You'll Need
Use student page, transparency, or PowerPoint. For intervention, direct students to:
Using Similar Figures
Lesson 5-5
Extra Skills and Word Problems Practice, Ch. 5

7-5 Congruent Figures

Check Skills You'll Need

1. **Vocabulary Review** What makes two polygons *similar*? See below.

$\triangle ABC \sim \triangle PQR$. Find each measure.

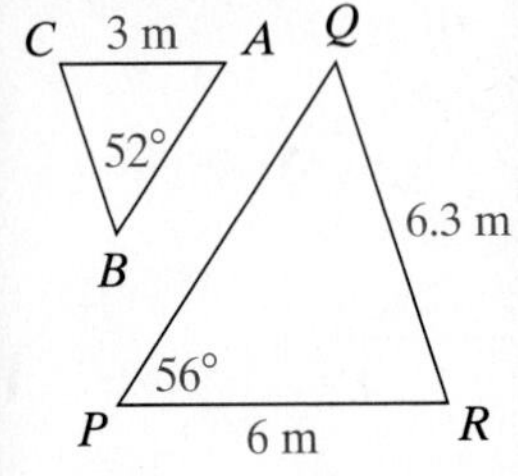

2. length of $\overline{BC}$ 3.15 m
3. $m\angle A$ 56°
4. $m\angle Q$ 52°

What You'll Learn

To identify congruent figures and to use them to find missing measures

New Vocabulary congruent polygons

Why Learn This?

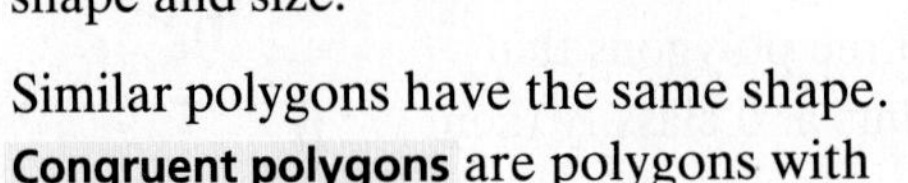

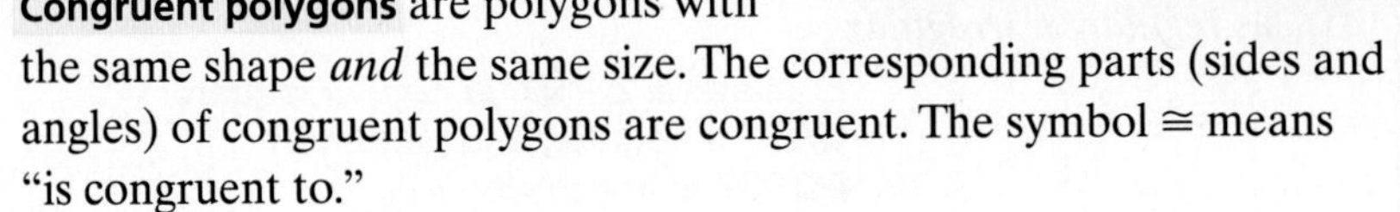

A manufacturer that makes large quantities of the same item must be sure that the items are all the same shape and size.

Similar polygons have the same shape. **Congruent polygons** are polygons with the same shape *and* the same size. The corresponding parts (sides and angles) of congruent polygons are congruent. The symbol ≅ means "is congruent to."

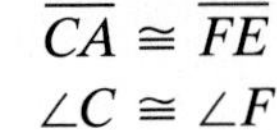

$\overline{AB} \cong \overline{ED}$ $\quad \overline{BC} \cong \overline{DF}$ $\quad \overline{CA} \cong \overline{FE}$

$\angle A \cong \angle E$ $\quad \angle B \cong \angle D$ $\quad \angle C \cong \angle F$

$\triangle ABC \cong \triangle EDF$

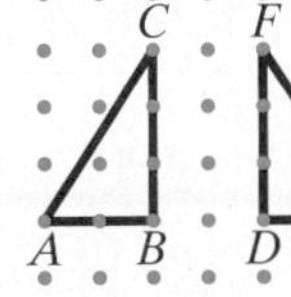

Write the vertices of congruent triangles in corresponding order.

Check Skills You'll Need

1. Two polygons are similar if corresponding angles are congruent and corresponding side lengths have the same proportion.

EXAMPLE Identifying Congruent Figures

1 Are the figures *congruent* or *not congruent?* Explain.

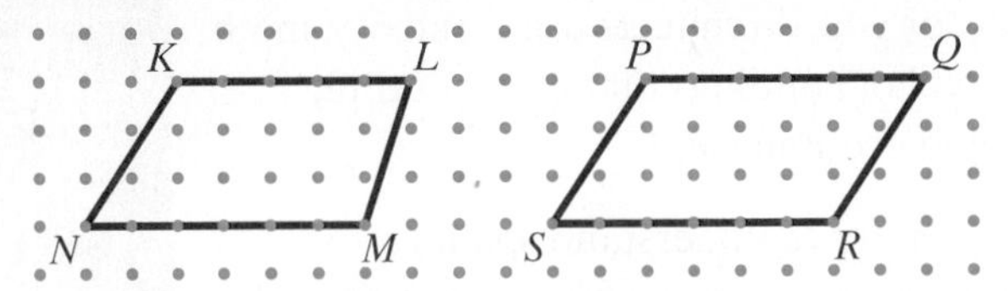

$\overline{KN} \cong \overline{PS}$ and $\overline{NM} \cong \overline{SR}$, but $\overline{LM}$ is not congruent to $\overline{QR}$.

The figures are not congruent.

1. Congruent; the sides and angles all have the same measure.

Quick Check

1. Are the figures *congruent* or *not congruent*? Explain. See left.

Differentiated Instruction Solutions for All Learners

Special Needs L1	**Below Level** L2
Have students cut out congruent figures drawn on grid or dot paper. Ask them to place one on top of another to show that they have equal angle measures and equal side lengths. learning style: visual	Have students take turns describing a polygon while the rest of the group draws it on graph paper. Have students compare their drawings to see which ones are congruent. learning style: visual

EXAMPLE Application: Manufacturing

2 Assembly-line workers compare manufactured parts to a sample part to see if they are congruent. Is $\triangle UVW$ congruent to the sample triangle, $\triangle RST$?

$\overline{UV} \cong \overline{RS}, \overline{VW} \cong \overline{ST}, \overline{WU} \cong \overline{TR},$
$\angle U \cong \angle R, \angle V \cong \angle S,$ and $\angle W \cong \angle T$.

$\triangle UVW \cong \triangle RST$

Quick Check

2. Is the right figure congruent to the sample figure? Explain.

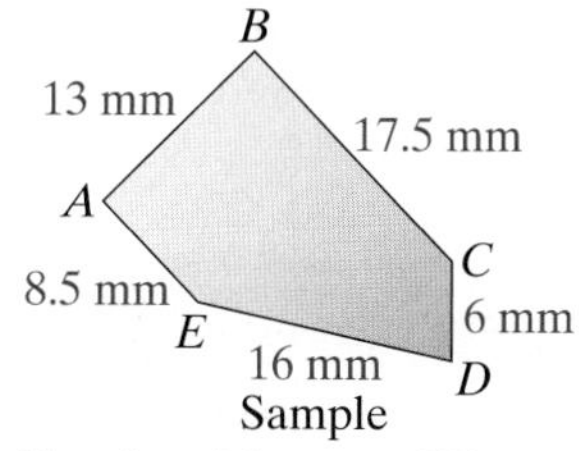

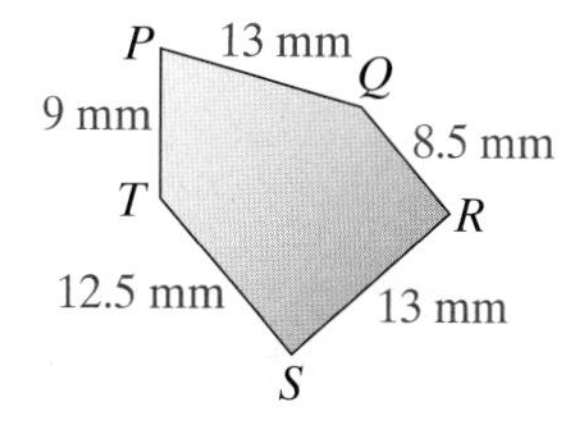

No; the sides are different lengths.

If you know that figures are congruent, you can use information about one figure to find information about the other.

EXAMPLE Working With Congruent Figures

If two angles in one triangle are congruent to two angles in another triangle, then both triangles' third angles must also be congruent.

3 The triangles at the right are congruent.

a. Write six congruences for the corresponding parts of the triangles.

$\angle X \cong \angle T \quad \angle Y \cong \angle S \quad \angle Z \cong \angle R$
$\overline{XY} \cong \overline{TS} \quad \overline{ZY} \cong \overline{RS} \quad \overline{ZX} \cong \overline{RT}$

b. Find ZY and $m\angle X$.

$ZY = 3$ cm $\leftarrow \overline{ZY} \cong \overline{RS}$, so $ZY = RS$

$m\angle X = 47°$ $\leftarrow \angle X \cong \angle T$, so $m\angle X = m\angle T$

Quick Check

3. The quadrilaterals are congruent. **See left.**
   a. Write the congruences for the corresponding parts.
   b. Find AD and $m\angle G$.

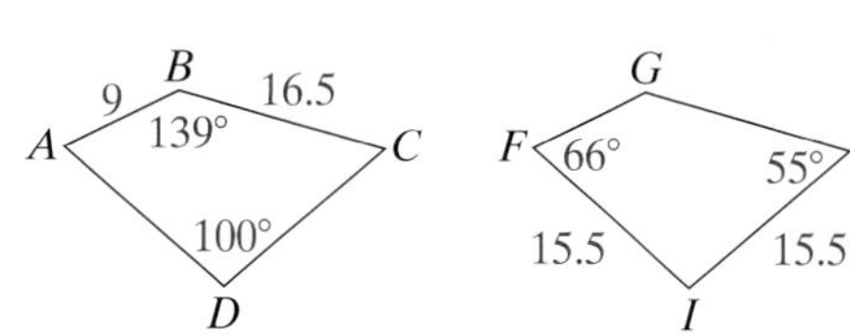

3a. $\overline{AB} \cong \overline{FG}; \overline{BC} \cong \overline{GH}; \overline{CD} \cong \overline{HI}; \overline{DA} \cong \overline{IF}; \angle A \cong \angle F; \angle B \cong \angle G; \angle C \cong \angle H; \angle D \cong \angle I$

b. 15.5; 139°

Advanced Learners L4
Have students predict how many ways a square can be divided into two congruent figures. **There are an infinite number of ways.**

learning style: visual

English Language Learners ELL
Have students label each of the following symbols: =, ?, ≥, ≤, ≈, ≅, ≠. Have them practice reading the symbols to each other.

learning style: verbal

2. Teach

Activity Lab

Use before the lesson.

All in One Teaching Resources

Activity Lab 7-5: Congruent Figures

Guided Instruction

Example 3
Have students repeat aloud, "triangle *XYZ* is congruent to triangle *TSR*." This helps students identify corresponding angles.

PowerPoint

Additional Examples

1 Are the figures *congruent*? Explain. **See back of book.**

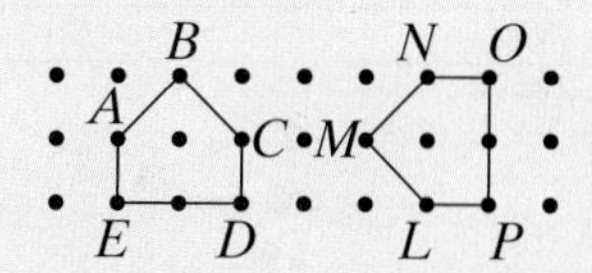

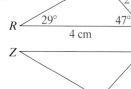

2 Are the figures congruent? **yes**

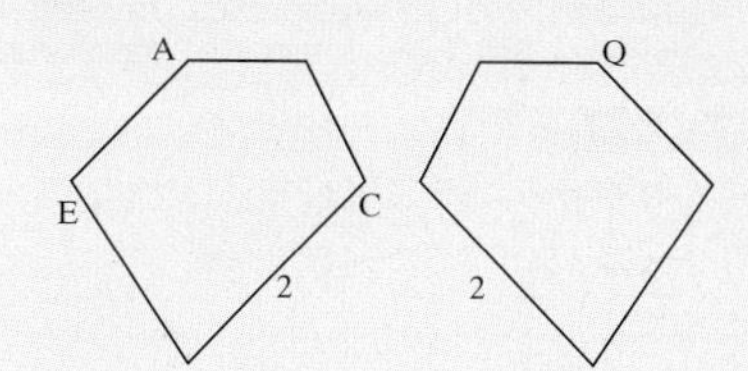

3 The triangles are congruent.

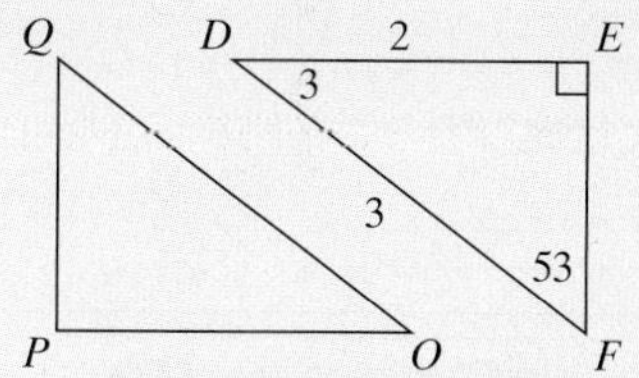

a. Write six congruences for the corresponding parts. **See back of book.**

b. Find *OP* and $m\angle Q$ **See back of book.**

All in One Teaching Resources

- Daily Notetaking Guide 7-5 L3
- Adapted Notetaking 7-5 L1

Closure

- *What are congruent polygons?* **polygons with the same size and shape**
- *How do you identify congruent polygons?* **Compare that corresponding angle and side measures are the same.**

3. Practice

Assignment Guide

Check Your Understanding
Go over Exercises 1–4 in class before assigning the Homework Exercises.

Homework Exercises

A Practice by Example	5–8
B Apply Your Skills	9–21
C Challenge	22
Test Prep and Mixed Review	23–26

Homework Quick Check
To check students' understanding of key skills and concepts, go over Exercises 5, 7, 13, 20, and 21.

Differentiated Instruction Resources

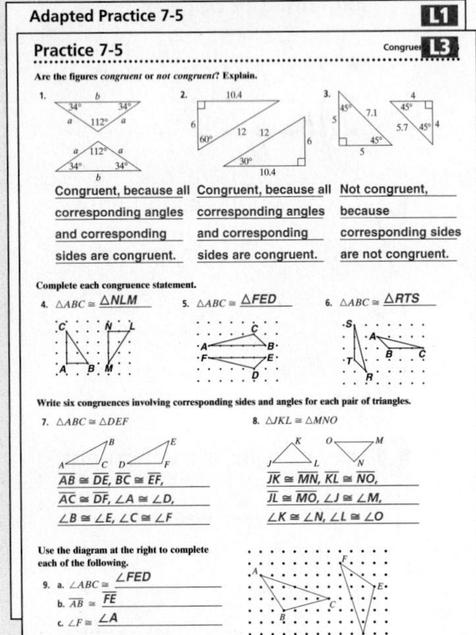

Adapted Practice 7-5 L1

Practice 7-5 Congruent L3

Are the figures *congruent* or *not congruent*? Explain.

1. Congruent, because all corresponding angles and corresponding sides are congruent.
2. Congruent, because all corresponding angles and corresponding sides are congruent.
3. Not congruent, because corresponding sides are not congruent.

Complete each congruence statement.

4. $\triangle ABC \cong$ $\triangle NLM$
5. $\triangle ABC \cong$ $\triangle FED$
6. $\triangle ABC \cong$ $\triangle RTS$

Write six congruences involving corresponding sides and angles for each pair of triangles.

7. $\triangle ABC \cong \triangle DEF$
$\overline{AB} \cong \overline{DE}$, $\overline{BC} \cong \overline{EF}$, $\overline{AC} \cong \overline{DF}$, $\angle A \cong \angle D$, $\angle B \cong \angle E$, $\angle C \cong \angle F$

8. $\triangle JKL \cong \triangle MNO$
$\overline{JK} \cong \overline{MN}$, $\overline{KL} \cong \overline{NO}$, $\overline{JL} \cong \overline{MO}$, $\angle J \cong \angle M$, $\angle K \cong \angle N$, $\angle L \cong \angle O$

Use the diagram at the right to complete each of the following.

9. a. $\angle ABC \cong$ $\angle FED$
b. $\overline{AB} \cong$ $\overline{FE}$
c. $\angle F \cong$ $\angle A$

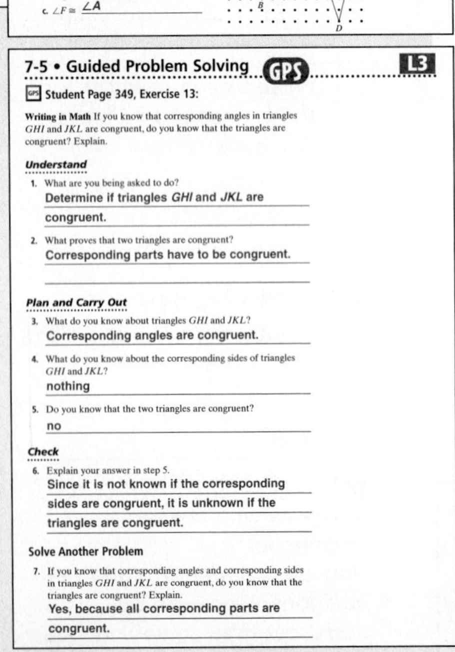

7-5 • Guided Problem Solving GPS L3

Student Page 349, Exercise 13:

Writing in Math If you know that corresponding angles in triangles *GHI* and *JKL* are congruent, do you know that the triangles are congruent? Explain.

Understand

1. What are you being asked to do?
Determine if triangles *GHI* and *JKL* are congruent.
2. What proves that two triangles are congruent?
Corresponding parts have to be congruent.

Plan and Carry Out

3. What do you know about triangles *GHI* and *JKL*?
Corresponding angles are congruent.
4. What do you know about the corresponding sides of triangles *GHI* and *JKL*?
nothing
5. Do you know that the two triangles are congruent?
no

Check

6. Explain your answer in step 5.
Since it is not known if the corresponding sides are congruent, it is unknown if the triangles are congruent.

Solve Another Problem

7. If you know that corresponding angles and corresponding sides in triangles *GHI* and *JKL* are congruent, do you know that the triangles are congruent? Explain.
Yes, because all corresponding parts are congruent.

Check Your Understanding

1. Yes. Similar polygons have congruent angles and sides in proportion. If the proportion is 1 to 1, then the polygons are not just similar, but congruent.
3. Congruent; all corresponding sides and angles are congruent.
4. Not congruent; not all corresponding sides and angles are congruent.

1. **Vocabulary** Is it possible for two similar polygons to be congruent polygons? Explain.
2. **Reasoning** If you know that two polygons are congruent, can you conclude that they are similar? Explain.
Yes. The ratio is 1 : 1, and the angles are congruent.

Are the figures *congruent* or *not congruent*? Explain.

3. 4.

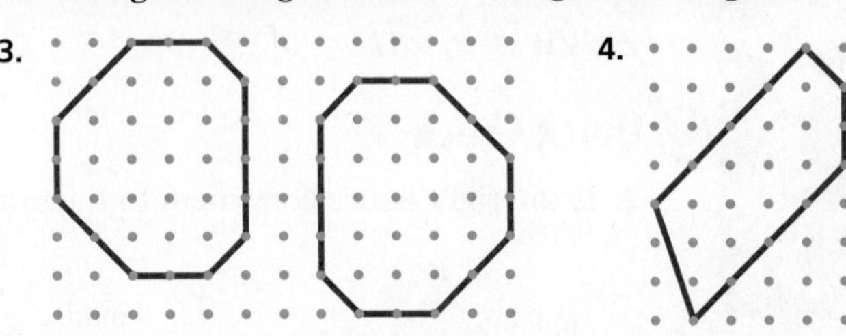

Homework Exercises

For more exercises, see Extra Skills and Word Problems.

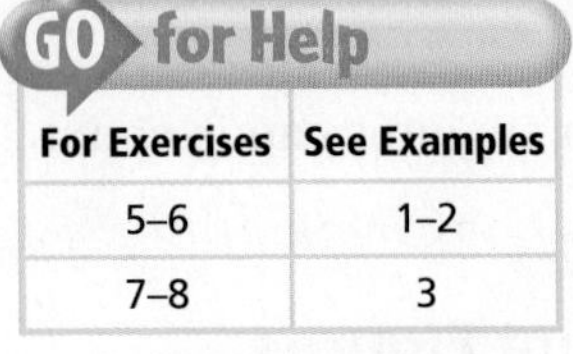

GO for Help

For Exercises	See Examples
5–6	1–2
7–8	3

5. Not congruent; not all corresponding sides and angles are congruent.

A **Are the figures *congruent* or *not congruent*? Explain.**

5. 6.

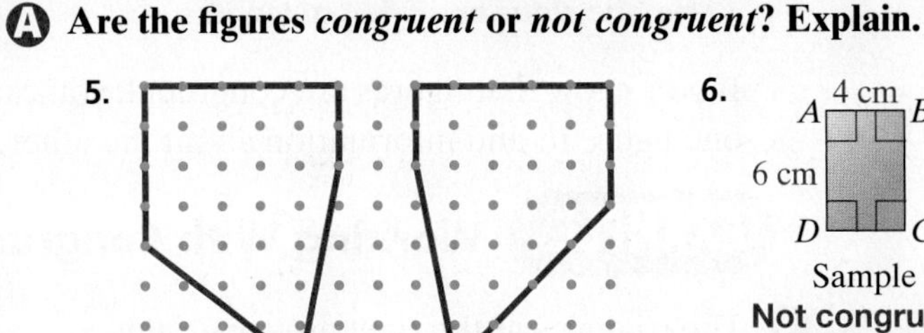

See left.

Not congruent; not all corresponding sides and angles are congruent.

Each pair of figures is congruent. Write six congruences for the corresponding parts of the figures. Then find the missing side lengths and angle measures. 7–8. See margin.

7.

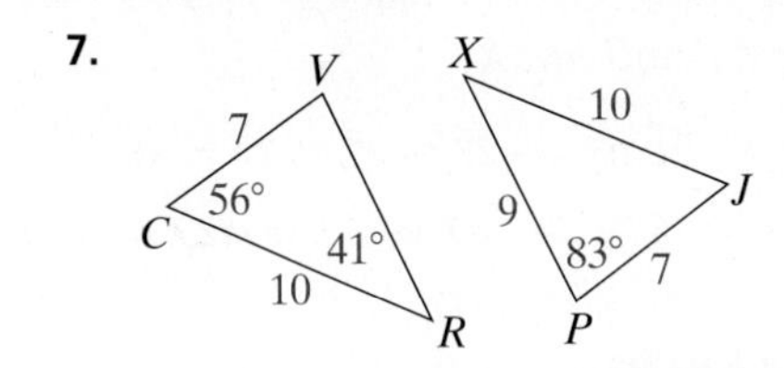

8. E 8 D, 11, 12, F 108°, 135°, C, 9, A 8 B; K 8 J, 135°, L, 9, 117°, I, 6, G 8 H

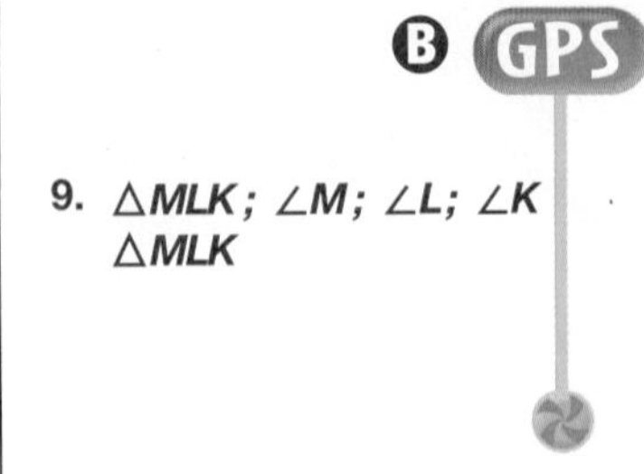

9. $\triangle MLK$; $\angle M$; $\angle L$; $\angle K$
$\triangle MLK$

B GPS 9. **Guided Problem Solving** Complete the congruence statement: $\triangle RST \cong$ ■.

- **Make a Plan** List pairs of congruent corresponding angles to find corresponding vertices.
- **Carry Out the Plan**
$\angle R \cong$ ■, $\angle S \cong$ ■, $\angle T \cong$ ■; $\triangle RST \cong$ ■ See left.

R 100°, 3, S, T 27°, K, 53°, M 100°, 3, L

7–8. See back of book.

Visit: PHSchool.com
Web Code: are-0705

Complete each congruence statement.

10. $\triangle ABC \cong$ ■ $\triangle FDE$

11. $\triangle ABC \cong$ ■ $\triangle SRT$

12. $\triangle ABC \cong$ ■ $\triangle ZXY$

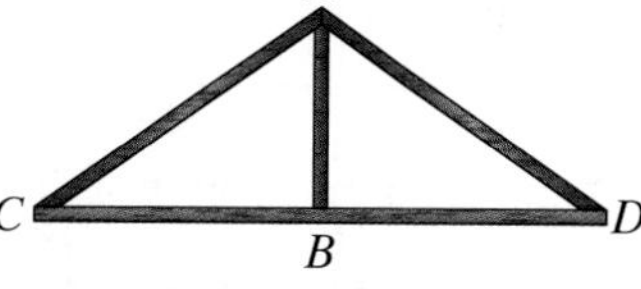

13. **Writing in Math** If you know that corresponding angles in triangles *GHI* and *JKL* are congruent, do you know that the triangles are congruent? Explain. **No; you must also know that the sides are congruent.**

Architecture **Complete each congruence statement. $\triangle ABC \cong \triangle ABD$**

14. $\overline{AC} \cong$ ■ $\overline{AD}$

15. $\overline{BC} \cong$ ■ $\overline{BD}$

16. $\overline{AB} \cong$ ■ $\overline{AB}$

17. $\angle D \cong$ ■ $\angle C$

18. $\angle CAB \cong$ ■ $\angle DAB$

19. $\angle ABC \cong$ ■ $\angle ABD$

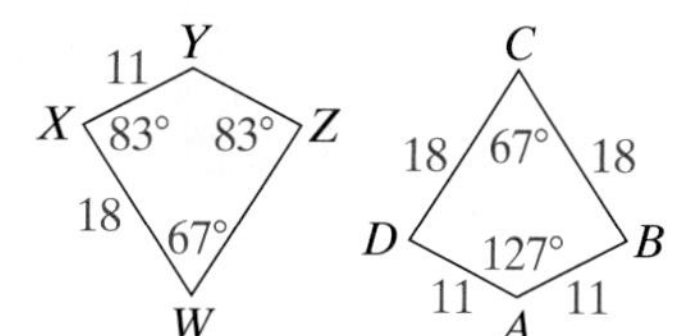

20. **Open-Ended** Draw a pair of congruent triangles. Label the vertices. Then list the pairs of congruent sides and congruent angles. **Answers will vary. Check students' work.**

21. The pair of figures is congruent. Write six congruences for the corresponding parts of the figures. Then find the missing side lengths and angle measures.

22. **Challenge** Does the statement $\triangle JKL \cong \triangle MNO$ say the same thing as the statement $\triangle JKL \cong \triangle NOM$? Explain.

21. $\overline{XY} \cong \overline{BA}$; $\overline{YZ} \cong \overline{AD}$; $\overline{ZW} \cong \overline{DC}$; $\overline{WX} \cong \overline{CB}$; $\angle Y \cong \angle A$; $\angle Z \cong \angle D$; $\angle W \cong \angle C$; $\angle X \cong \angle B$; $YZ = 11$; $ZW = 18$; $m\angle Y = 127°$; $m\angle D = 83°$; $m\angle B = 83°$

22. No; the order of letters is different, so different angles are congruent.

Test Prep and Mixed Review

Practice

Multiple Choice

23. $\triangle QRS$ is similar to $\triangle UTS$. Which of the following is NOT true about $\triangle QRS$ and $\triangle UTS$? **D**
   - (A) $\overline{QS}$ corresponds to $\overline{US}$.
   - (B) $\overline{RQ}$ corresponds to $\overline{TU}$.
   - (C) $\angle S$ corresponds to $\angle S$.
   - (D) $\angle Q$ corresponds to $\angle T$.

24. The measure of $\angle DHE$ is 65°. Which angle is supplementary to $\angle DHE$? **J**
   - (F) $\angle CHD$
   - (G) $\angle CHB$
   - (H) $\angle DHB$
   - (J) $\angle DHA$

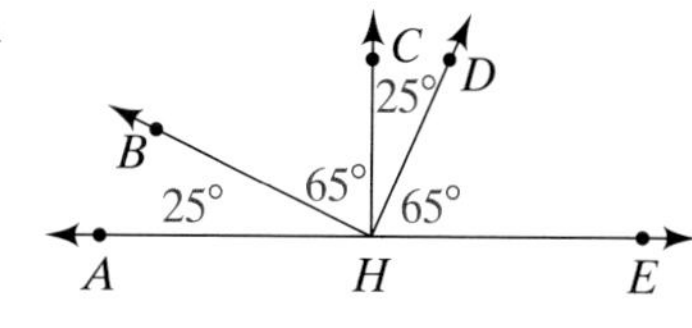

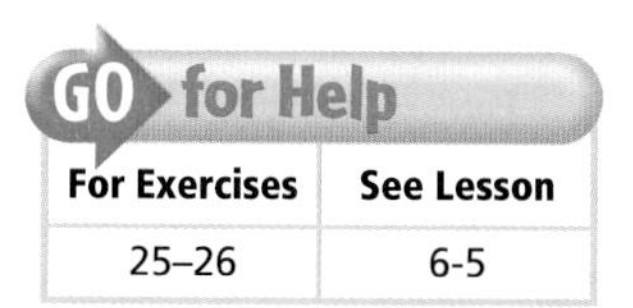

For Exercises	See Lesson
25–26	6-5

Use a proportion to find the percent.

25. 12 is what percent of 48? **25%**

26. 2 is what percent of 100? **2%**

Online lesson quiz, PHSchool.com, Web Code: ara-0705

4. Assess & Reteach

Lesson Quiz

$\triangle ABC \cong \triangle QRP$.

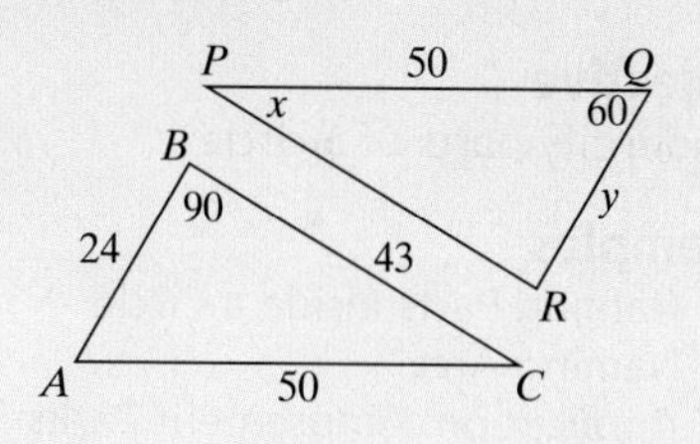

1. Write the congruences.

   $\overline{AB} \cong \overline{QR}$; $\overline{BC} \cong \overline{RP}$; $\overline{CA} \cong \overline{PQ}$; $\angle A \cong \angle Q$; $\angle B \cong \angle R$; $\angle C \cong \angle P$

2. Find the values of x and y.

   $x = 30°$; $y = 24$ units

Alternative Assessment

Have students work in pairs with geoboards. One student makes a figure and challenges the partner to make one congruent and one non-congruent figure. Partners must agree which figure is congruent and which is non-congruent. Partners trade roles as time permits.

Test Prep

Resources

For additional practice with a variety of test item formats:
- Test-Taking Strategies, p. 365
- Test Prep, p. 369
- Test-Taking Strategies with Transparencies

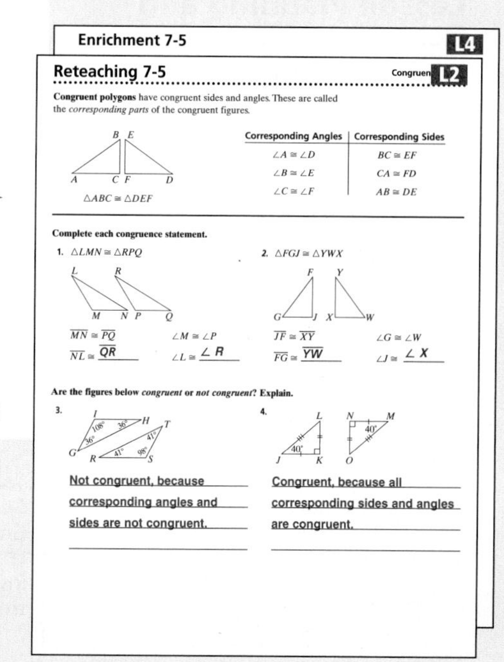
Enrichment 7-5 L4

Reteaching 7-5 L2

Congruent polygons have congruent sides and angles. These are called the *corresponding parts* of the congruent figures.

$\triangle ABC \cong \triangle DEF$

Corresponding Angles	Corresponding Sides
$\angle A \cong \angle D$	$BC \cong EF$
$\angle B \cong \angle E$	$CA \cong FD$
$\angle C \cong \angle F$	$AB \cong DE$

Complete each congruence statement.

1. $\triangle LMN \cong \triangle RPQ$
   $\overline{MN} \cong \overline{PQ}$ $\angle M \cong \angle P$
   $\overline{NL} \cong \overline{QR}$ $\angle L \cong \angle R$

2. $\triangle FGJ \cong \triangle YWX$
   $\overline{JF} \cong \overline{XY}$ $\angle G \cong \angle W$
   $\overline{FG} \cong \overline{YW}$ $\angle J \cong \angle X$

Are the figures below *congruent* or *not congruent*? Explain.

3. Not congruent, because corresponding angles and sides are not congruent.

4. Congruent, because all corresponding sides and angles are congruent.

7-6

1. Plan

Objective
To identify parts of a circle

Examples
1 Naming Parts Inside a Circle
2 Naming Arcs
3 Application: Amusement Parks

Math Understandings: p. 322D

Math Background

A *circle* is formed by a set of points that are all equidistant from the center point. If the chord passes through the center of the circle, then it is a *diameter*. Two radii drawn on a circle form a *central angle*. The central angle forms two arcs, one inside the angle, and the other outside the angle. A diameter forms two *semicircles*, one on each side of the *diameter*.

More Math Background: p. 322D

Lesson Planning and Resources

See p. 322E for a list of the resources that support this lesson.

Check Skills You'll Need
Use student page, transparency, or PowerPoint. For intervention, direct students to:
Lines and Planes
Lesson 7-1
Extra Skills and Word Problems Practice, Ch. 7

7-6 Circles

Check Skills You'll Need

1. **Vocabulary Review** How many endpoints does a *segment* have? 2

Use the points in each diagram to name the figure shown.

2. F R
3. Z J
4. N K

2–4. See below.

GO for Help Lesson 7-1

Check Skills You'll Need
2. $\overline{FR}$
3. $\overrightarrow{ZJ}$
4. $\overleftrightarrow{NK}$

What You'll Learn

To identify parts of a circle

New Vocabulary circle, radius, diameter, central angle, chord, arc, semicircle

Why Learn This?

In a pizza, each slice is part of a whole pie. In geometry, a central angle is part of a full circle. Just as you can use slices to measure a portion of a pizza, you can use central angles to measure a portion of a circle.

A **circle** is the set of points in a plane that are all the same distance from a given point, called the center. You name a circle by its center. Circle O is shown at the right.

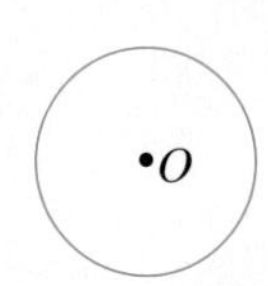

A **radius** is a segment that connects the center of a circle to the circle.

$\overline{OB}$ is a radius of circle O.

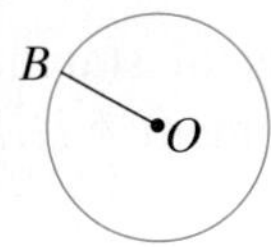

A **diameter** is a segment that passes through the center of a circle and has both endpoints on the circle.

$\overline{AC}$ is a diameter of circle O.

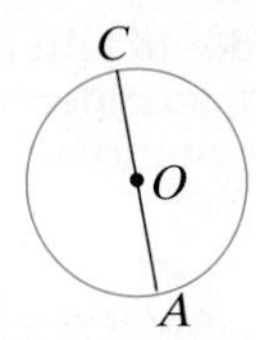

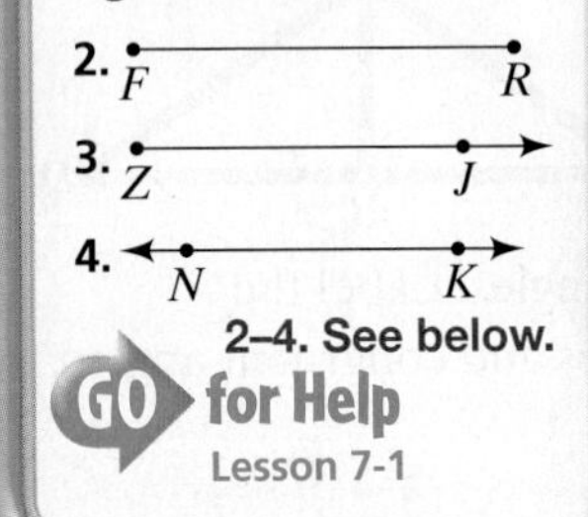

Any two hands of a clock form a central angle on the circular face of the clock.

A **central angle** is an angle with its vertex at the center of a circle.

$\angle AOB$ is a central angle of circle O.

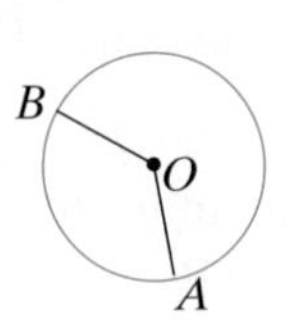

A **chord** is a segment that has both endpoints on the circle.

$\overline{AD}$ is a chord of circle O.

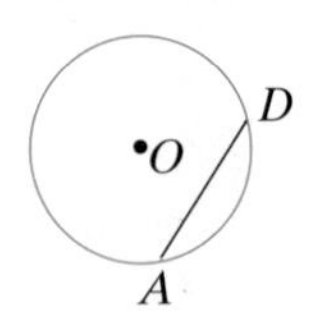

Notice that a central angle is formed by two radii. Two radii lying on the same line form a diameter of a circle. So a radius is half a diameter.

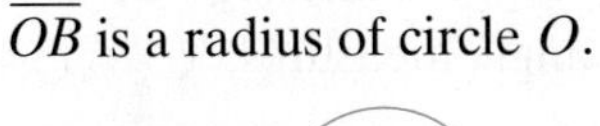

Differentiated Instruction Solutions for All Learners

Special Needs L1
Give students a picture of a circle with all the parts labeled. Include circumference and area by having the inside of the circle shaded. They can use this to help them work through the exercises.
learning style: visual

Below Level L2
Have students draw circles that include a radius, a diameter, and a chord with endpoints labeled. Students should then name every segment and arc.
learning style: visual

You can use points on a circle as well as the circle's center to name parts of the circle.

EXAMPLE 1 Naming Parts Inside a Circle

1 Name all the radii, diameters, and chords shown for circle *O*.

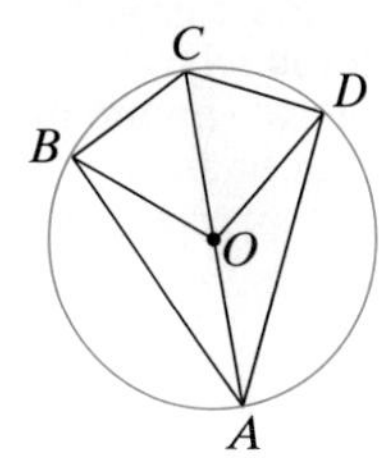

radii: $\overline{OA}$, $\overline{OB}$, $\overline{OC}$, and $\overline{OD}$

diameter: $\overline{AC}$

chords: $\overline{AB}$, $\overline{BC}$, $\overline{CD}$, $\overline{DA}$, and $\overline{AC}$

Vocabulary Tip

The plural of *radius* is *radii* (RAY dee eye).

Quick Check

1. Name all the central angles shown in circle O. **$\angle AOB$, $\angle AOC$, $\angle AOD$, $\angle BOC$, $\angle BOD$, $\angle COD$**

An **arc** is part of a circle. A **semicircle** is half of a circle. In circle P, $\overset{\frown}{ST}$ and $\overset{\frown}{TW}$ are arcs less than the length of a semicircle. $\overset{\frown}{STW}$ is a semicircle. You use three letters to name an arc that is a semicircle or longer.

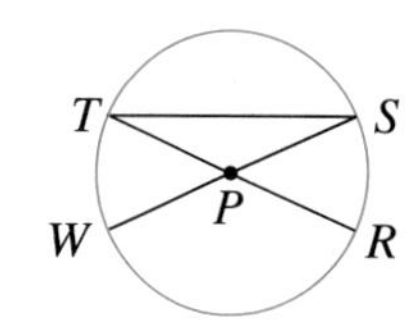

EXAMPLE 2 Naming Arcs

2 Name three of the arcs in circle O.

Three arcs are $\overset{\frown}{XZ}$, $\overset{\frown}{XY}$, and $\overset{\frown}{ZXY}$.

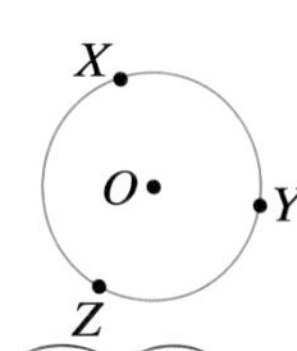

Quick Check

2. Name three other arcs in circle O. **$\overset{\frown}{ZY}$, $\overset{\frown}{ZYX}$, $\overset{\frown}{YZX}$**

You can use arcs to describe real-world situations involving circles.

EXAMPLE 3 Application: Amusement Parks

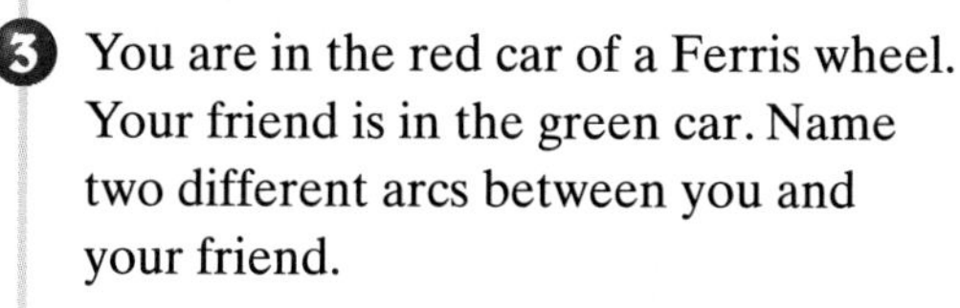

3 You are in the red car of a Ferris wheel. Your friend is in the green car. Name two different arcs between you and your friend.

The shorter arc is $\overset{\frown}{DE}$.

The longer arc is $\overset{\frown}{DCE}$.

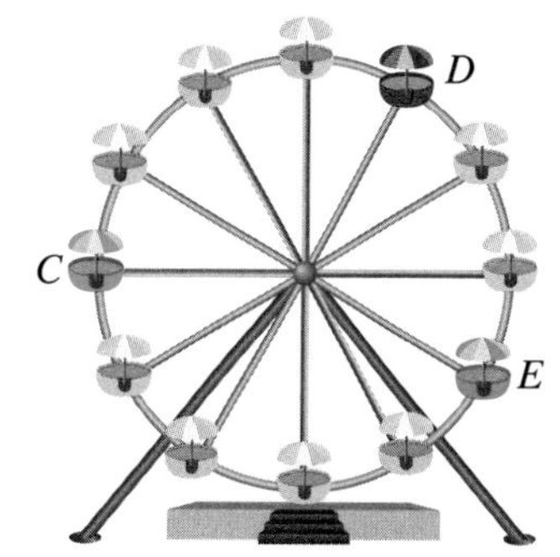

Quick Check

3. Name two different arcs from the blue car to the green car. **$\overset{\frown}{CDE}$, $\overset{\frown}{CE}$**

2. Teach

Activity Lab

Use before the lesson.

All in One Teaching Resources
Activity Lab 7-6: Circles

Guided Instruction

Error Prevention!

Watch for students who confuse the terms *chord* and *diameter.*

Additional Examples

Use circle *F* to answer 1–3.

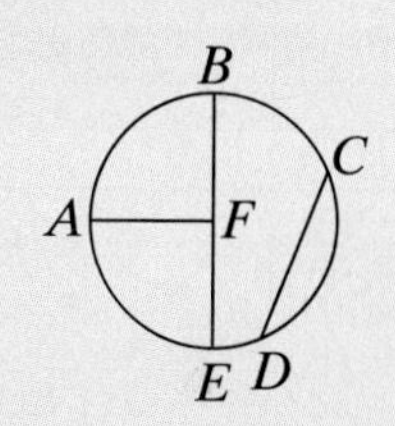

1 Name all the radii, diameters, central angles, and chords shown in circle *F*. **radii: $\overline{FB}$, $\overline{FE}$, $\overline{FA}$; diameter: $\overline{EB}$; chords: $\overline{EB}$, $\overline{CD}$; central angles: $\angle BFA$, $\angle BFE$, $\angle EFA$**

2 Name three of the arcs in circle *F*. **Sample: $\overset{\frown}{AB}$, $\overset{\frown}{AE}$, and $\overset{\frown}{BCE}$**

3 Name two different arcs between points *C* and *D*. **Sample: $\overset{\frown}{CD}$, and $\overset{\frown}{CBD}$**

All in One Teaching Resources

- Daily Notetaking Guide 7-6 **L3**
- Adapted Notetaking 7-6 **L1**

Closure

- *What is a circle?* **a set of points in a plane that are all the same distance from the center**
- *What is the difference between a central angle and an arc?* **A central angle is an angle formed by two radii. The part of the circle cut by the angle is an arc.**

Advanced Learners L4

If two circles have congruent central angles, are the circles congruent? Explain. **No. The size of the angle is not related to the length of the radius or diameter.**

learning style: visual

English Language Learners ELL

Ask students to tell you what is similar and different between a chord, a radius, and a diameter. Ask them to use the words *endpoints* and *segment* in their explanations.

learning style: verbal

3. Practice

Assignment Guide

Check Your Understanding

Go over Exercises 1–7 in class before assigning the Homework Exercises.

Homework Exercises

A Practice by Example	8–16
B Apply Your Skills	17–30
C Challenge	31
Test Prep and Mixed Review	32–37

Homework Quick Check

To check students' understanding of key skills and concepts, go over Exercises 10, 15, 25, 26, and 29.

Differentiated Instruction Resources

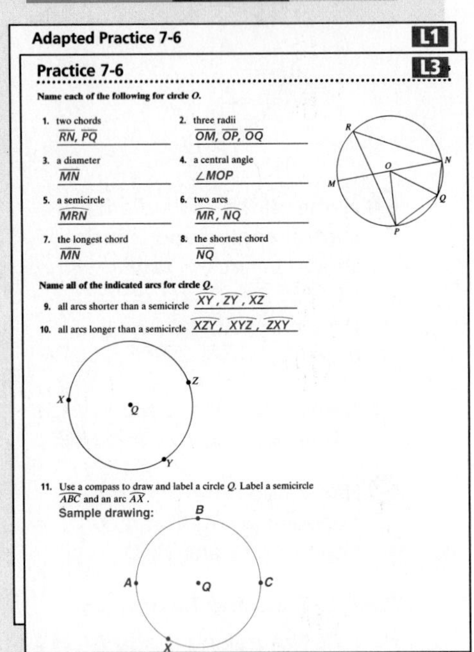

Adapted Practice 7-6 L1

Practice 7-6 L3

Name each of the following for circle O.

1. two chords $\overline{RN}, \overline{PQ}$
2. three radii $\overline{OM}, \overline{OP}, \overline{OQ}$
3. a diameter $\overline{MN}$
4. a central angle $\angle MOP$
5. a semicircle $\widehat{MRN}$
6. two arcs $\widehat{MR}, \widehat{NQ}$
7. the longest chord $\overline{MN}$
8. the shortest chord $\overline{NQ}$

Name all of the indicated arcs for circle Q.

9. all arcs shorter than a semicircle $\widehat{XY}, \widehat{ZY}, \widehat{XZ}$
10. all arcs longer than a semicircle $\widehat{XZY}, \widehat{XYZ}, \widehat{ZXY}$

11. Use a compass to draw and label a circle Q. Label a semicircle $\widehat{ABC}$ and an arc $\widehat{AX}$.
Sample drawing:

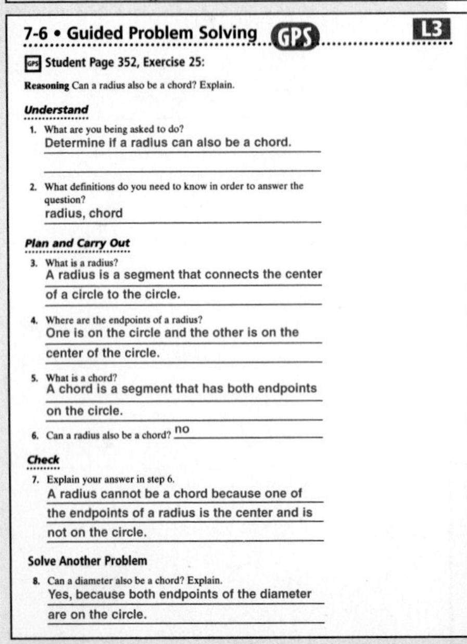

7-6 • Guided Problem Solving GPS L3

Student Page 352, Exercise 25:

Reasoning Can a radius also be a chord? Explain.

Understand

1. What are you being asked to do?
Determine if a radius can also be a chord.
2. What definitions do you need to know in order to answer the question?
radius, chord

Plan and Carry Out

3. What is a radius?
A radius is a segment that connects the center of a circle to the circle.
4. Where are the endpoints of a radius?
One is on the circle and the other is on the center of the circle.
5. What is a chord?
A chord is a segment that has both endpoints on the circle.
6. Can a radius also be a chord? no

Check

7. Explain your answer in step 6.
A radius cannot be a chord because one of the endpoints of a radius is the center and is not on the circle.

Solve Another Problem

8. Can a diameter also be a chord? Explain.
Yes, because both endpoints of the diameter are on the circle.

Check Your Understanding

1. A radius is a line from the center of a circle to a point on the circle. A diameter is a line between two points on a circle that passes through the center of the circle.

1. **Vocabulary** Explain how a radius is different from a diameter.
2. **Reasoning** Must a diameter also be a chord? Explain. **See margin.**

Name each of the following for circle M.

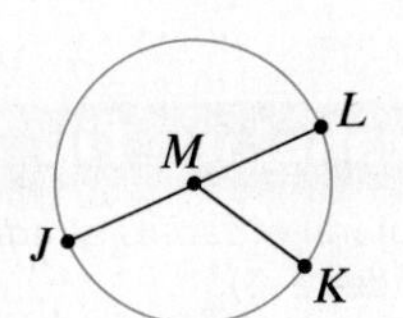

3. center M
4. radii $\overline{ML}, \overline{MK}, \overline{MJ}$
5. chords $\overline{JL}$
6. diameter $\overline{JL}$
7. central angles $\angle JMK, \angle KML, \angle JML$

Homework Exercises

For more exercises, see Extra Skills and Word Problems.

GO for Help

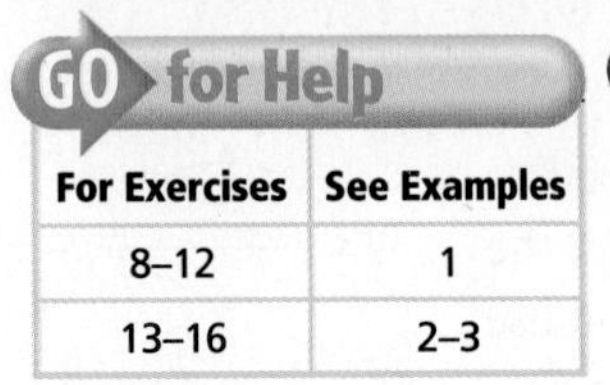

For Exercises	See Examples
8–12	1
13–16	2–3

8. O
9. $\overline{OC}, \overline{OK}, \overline{OD}$
10. $\overline{CK}, \overline{KD}, \overline{DE}, \overline{CE}, \overline{CD}$
11. $\overline{CD}$
12. $\angle COK, \angle KOD, \angle COD$
13. $\widehat{TR}, \widehat{RS}, \widehat{ST}$
14. $\widehat{TRS}, \widehat{RST}, \widehat{STR}$
15. $\widehat{CD}, \widehat{DB}, \widehat{CB}$
16. $\widehat{DBC}, \widehat{DCB}, \widehat{BDC}$

A **Name each of the following for circle O.**
8–12. See left.

8. center
9. radii
10. chords
11. diameter
12. central angles

Name the following arcs for circle Q.
13–14. See left.

13. all arcs shorter than a semicircle
14. all arcs longer than a semicircle

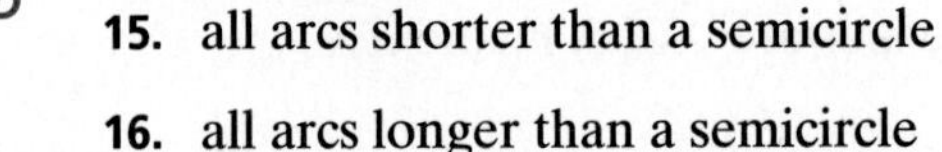

Name the following arcs for circle A.
15–16. See left.

15. all arcs shorter than a semicircle
16. all arcs longer than a semicircle

B GPS

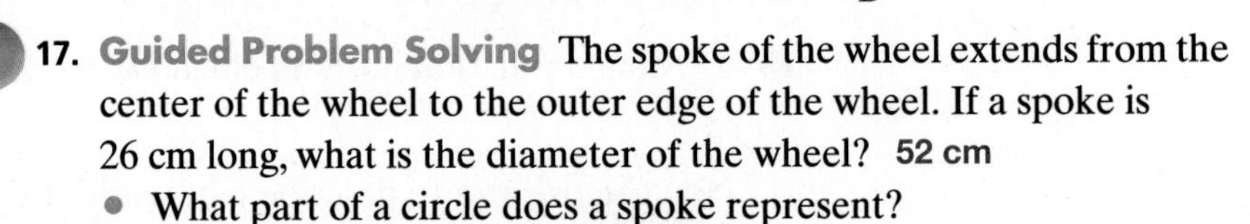

17. **Guided Problem Solving** The spoke of the wheel extends from the center of the wheel to the outer edge of the wheel. If a spoke is 26 cm long, what is the diameter of the wheel? **52 cm**
   - What part of a circle does a spoke represent?

Name each of the following for circle D.
18–23. See margin.

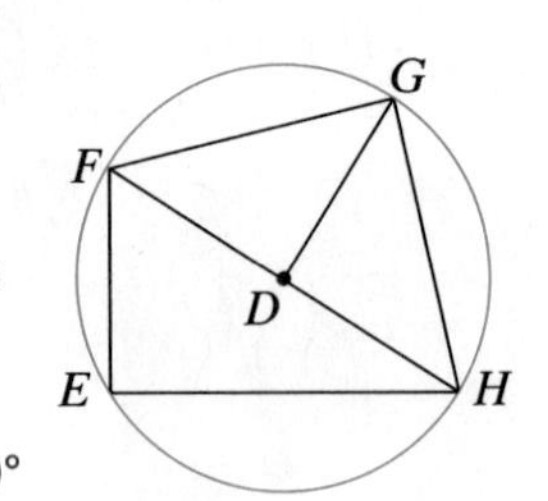

18. two chords
19. two central angles
20. a diameter
21. an isosceles triangle
22. five arcs
23. the longest chord
24. If $m\angle FDG = 90°$, find $m\angle HDG$. **90°**
25. **Reasoning** Can a radius also be a chord? Explain. GPS
No; only one endpoint is on the circle.
26. a. **Open-Ended** Draw a design that includes a quadrilateral with vertices on a circle. **a–b. Check students' work.**
b.  **Writing in Math** Describe your design so someone can draw it without looking at your drawing.

GO Online
Homework Video Tutor
Visit: PHSchool.com
Web Code: are-0706

2. Yes. The diameter is the longest chord in a circle. It has both endpoints on the circle.

Exercises 18–22. Answers may vary. Samples are given.

18. $\overline{FG}, \overline{EF}$
19. $\angle FDG, \angle GDH$
20. $\overline{FH}$
21. $\triangle FDG$
22. $\widehat{FG}, \widehat{FGH}, \widehat{GH}, \widehat{HE}, \widehat{GHE}$
23. $\overline{FH}$

29–30. See back of book.

Careers Fabric designers use geometric shapes, including circles, to create beautiful patterns on fabrics.

Find each length for radius r and diameter d.

27. $d = 45.2$ cm, $r = ■$ **22.6 cm**
28. $r = 2.9$ mm, $d = ■$ **5.8 mm**

Draw several diagrams that fit each description. Then make a conjecture. **29–30. See margin.**

29. A quadrilateral has vertices that are the endpoints of two diameters.
30. A triangle has all its vertices on a circle. One of its sides is a diameter of the circle.

Ⓒ 31. **Challenge** You are running around a circular track. The distance around the track is 200 m. Points A, B, C, D, and E are spaced evenly around the track. If you run along $\widehat{ABD}$, $\widehat{DEB}$, and $\widehat{BCA}$, how far have you run? **400 m**

Test Prep and Mixed Review

Practice

Multiple Choice

32. Jorge drew the figure shown at the right, with O at the center of the circle. What kind of triangle is $\triangle COB$? **D**
    - Ⓐ Acute isosceles triangle
    - Ⓑ Right isosceles triangle
    - Ⓒ Obtuse scalene triangle
    - Ⓓ Obtuse isosceles triangle

33. Tickets to a sporting event that cost \$25 last year cost \$32 now. Which equation can be used to find n, the percent of increase in the ticket price? **J**
    - Ⓕ $\frac{25}{32} = \frac{n}{100}$
    - Ⓖ $\frac{32}{25} = \frac{n}{100}$
    - Ⓗ $\frac{7}{32} = \frac{n}{100}$
    - Ⓙ $\frac{7}{25} = \frac{n}{100}$

34. Kiera bought a new computer on sale for \$50 less than the original selling price. What other information is necessary to find the percent of the discount? **B**
    - Ⓐ The store where the computer was purchased
    - Ⓑ The original price of the computer
    - Ⓒ The amount of tax charged
    - Ⓓ The reason for buying a new computer

GO for Help

For Exercises	See Lesson
35–37	7-2

Classify each angle as *acute, right, obtuse,* or *straight*.

35.
36.
37.

35. acute
36. obtuse
37. right

4. Assess & Reteach

Lesson Quiz

Use circle O for questions 1–3.

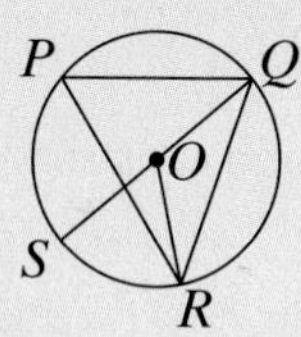

1. Name the radii, diameter, and chords. **radii: $\overline{OQ}$, $\overline{OR}$, $\overline{OS}$; diameter: $\overline{QS}$; chords: $\overline{PQ}$, $\overline{PR}$, $\overline{QR}$, $\overline{SQ}$**
2. Name two central angles. **Sample: $\angle SOR$, $\angle QOR$**
3. Name three arcs. **Sample: $\widehat{PQ}$, $\widehat{QRS}$, $\widehat{SP}$**

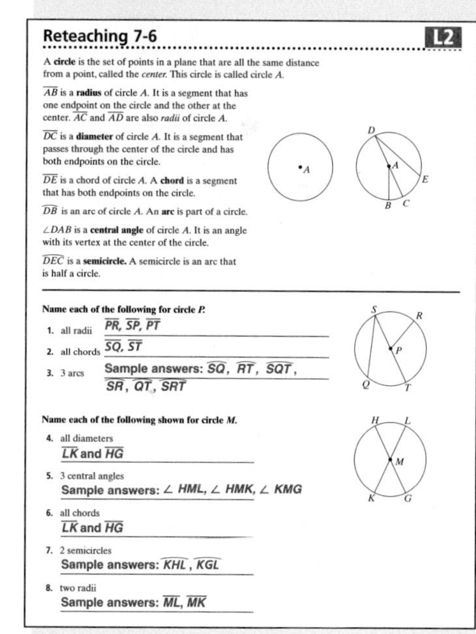

Reteaching 7-6 L2

A **circle** is the set of points in a plane that are all the same distance from a point, called the *center*. This circle is called circle *A*.

$\overline{AB}$ is a **radius** of circle *A*. It is a segment that has one endpoint on the circle and the other at the center. $\overline{AC}$ and $\overline{AD}$ are also *radii* of circle *A*.

$\overline{DC}$ is a **diameter** of circle *A*. It is a segment that passes through the center of the circle and has both endpoints on the circle.

$\overline{DE}$ is a chord of circle *A*. A **chord** is a segment that has both endpoints on the circle.

$\widehat{DB}$ is an arc of circle *A*. An **arc** is part of a circle.

$\angle DAB$ is a **central angle** of circle *A*. It is an angle with its vertex at the center of the circle.

$\widehat{DEC}$ is a **semicircle**. A semicircle is an arc that is half a circle.

Name each of the following for circle *P*.

1. all radii $\overline{PR}$, $\overline{SP}$, $\overline{PT}$
2. all chords $\overline{SQ}$, $\overline{ST}$
3. 3 arcs Sample answers: $\widehat{SQ}$, $\widehat{RT}$, $\widehat{SQT}$, $\widehat{SR}$, $\widehat{QT}$, $\widehat{SRT}$

Name each of the following shown for circle *M*.

4. all diameters $\overline{LK}$ and $\overline{HG}$
5. 3 central angles Sample answers: $\angle HML$, $\angle HMK$, $\angle KMG$
6. all chords $\overline{LK}$ and $\overline{HG}$
7. 2 semicircles Sample answers: $\widehat{KHL}$, $\widehat{KGL}$
8. two radii Sample answers: $\overline{ML}$, $\overline{MK}$

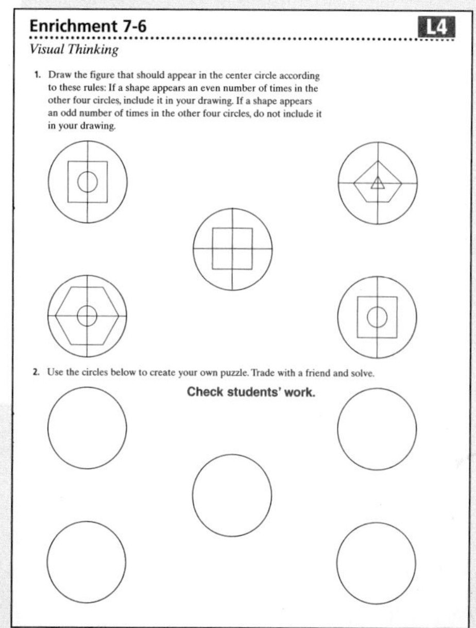

Enrichment 7-6 L4

Visual Thinking

1. Draw the figure that should appear in the center circle according to these rules: If a shape appears an even number of times in the other four circles, include it in your drawing. If a shape appears an odd number of times in the other four circles, do not include it in your drawing.

2. Use the circles below to create your own puzzle. Trade with a friend and solve. Check students' work.

Alternative Assessment

Each student draws a circle and labels several of its parts. Partners exchange circles and identify all diameters, radii, chords, central angles, and arcs.

Test Prep

Resources

For additional practice with a variety of test item formats:
- Test-Taking Strategies, p. 365
- Test Prep, p. 369
- Test-Taking Strategies with Transparencies

7-7

1. Plan

Objective
To analyze and construct circle graphs

Examples
1 Analyzing a Circle Graph
2 Constructing a Circle Graph

Math Understandings: p. 322D

Professional Development

Math Background

Circle graphs provide a quick way to visualize the size of parts of a whole. To make a circle graph you need to find the central angle for each part or section by solving the proportion $\frac{\text{part}}{\text{whole}} = \frac{\text{angle}}{360°}$.

More Math Background: p. 322D

Lesson Planning and Resources

See p. 322E for a list of the resources that support this lesson.

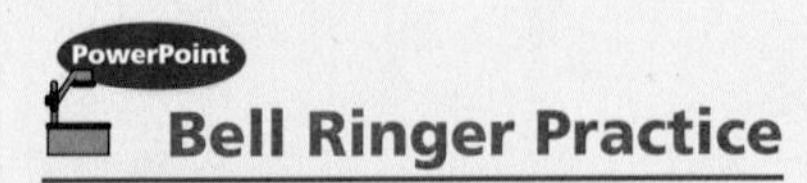

PowerPoint

Bell Ringer Practice

Check Skills You'll Need
Use student page, transparency, or PowerPoint. For intervention, direct students to:
Finding a Percent of a Number
Lesson 6-4
Extra Skills and Word Problems Practice, Ch. 6

7-7 Circle Graphs

Check Skills You'll Need

1. **Vocabulary Review** How is finding the percent of a number like multiplying decimals? **See below.**

Find each answer.

2. 25% of 360 **90**
3. 60% of 360 **216**
4. 72% of 360 **259.2**

for Help
Lesson 6-4

Check Skills You'll Need

1. **Every percent can be written as a decimal by moving the decimal point two places to the left. Thus 25% becomes 0.25. So taking the percent of a number is the same as multiplying by a decimal.**

What You'll Learn

To analyze and construct circle graphs

New Vocabulary circle graph

Why Learn This?

You can use a circle graph like the one below to display survey or research results. You can see at a glance how the parts compare to one another and to the whole amount.

Music That People Buy

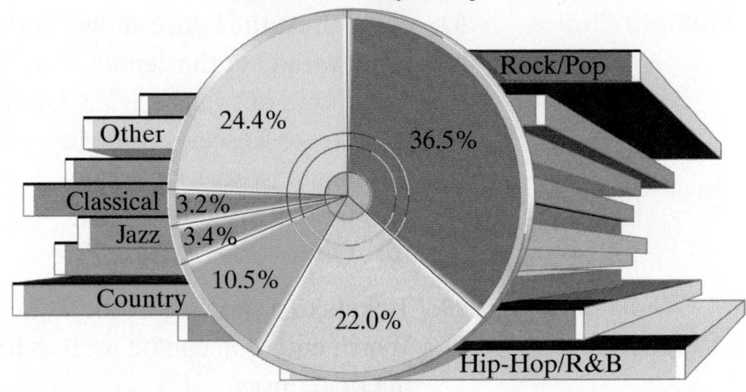

SOURCE: Recording Industry Association of America

A **circle graph** is a graph of data in which a circle represents the whole. Each wedge, or sector, is part of the whole. The total must be 100%. You can use a circle graph to see how the whole breaks down into parts.

GO Online
Video Tutor Help
Visit: PHSchool.com
Web Code: are-0775

EXAMPLE Analyzing a Circle Graph

1 **Gridded Response** In a recent year, consumers spent \$13.7 billion on music recordings. Use the circle graph above to find how many dollars, to the nearest tenth of a billion, were spent on country music.

Find 10.5% of \$13.7 billion.

$0.105 \cdot \$13.7 \text{ billion} \approx \$1.4 \text{ billion}$

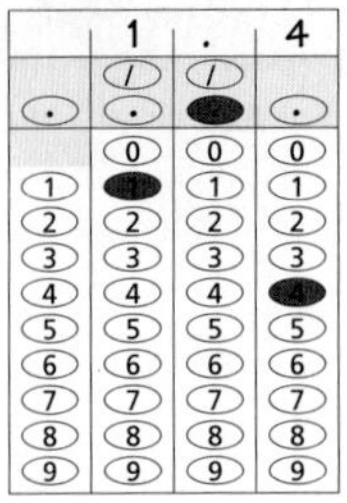

(Gridded answer: 1 . 4)

Quick Check

1. Approximately how much money was spent on jazz? **about 470 million**

Differentiated Instruction Solutions for All Learners

Special Needs L1	**Below Level** L2
Some students may have difficulty drawing a circle graph. If so, pair them up with students who can do the drawing while they check the sections against the data in the table. **learning style: visual**	Use circle graph paper for Example 2. Shade the angle measures on the circle graph based on the scale of the graph paper rather than using compass measurements. **learning style: visual**

A circle graph is divided into sectors. Each sector is determined by a central angle. The sum of the measures of the central angles is 360°.

EXAMPLE Constructing a Circle Graph

2 Science Use the information in the table below to make a circle graph.

NASA Space Shuttle Expenditures

Category	Cost (millions of dollars)
Orbiter	698.8
Propulsion	1,053.1
Operations	738.8
Upgrades	488.8

SOURCE: *Statistical Abstract of the United States*
Go to **PHSchool.com** for a data update.
Web Code: arg-9041

Step 1 Add to find the total space shuttle expenditures.
698.8 + 1,053.1 + 738.8 + 488.8 = 2,979.5

Step 2 For each central angle, set up a proportion to find the angle measure. Use a calculator to solve. Round to the nearest tenth.

$\frac{698.8}{2,979.5} = \frac{a}{360}$ $a \approx 84.4°$ $\frac{1,053.1}{2,979.5} = \frac{b}{360}$ $b \approx 127.2°$

$\frac{738.8}{2,979.5} = \frac{c}{360}$ $c \approx 89.3°$ $\frac{488.8}{2,979.5} = \frac{d}{360}$ $d \approx 59.1°$

Step 3 Draw a circle. Draw the central angles using the measures found in Step 2. Label each section. Include a title and a key.

NASA Space Shuttle Expenditures (millions of dollars)

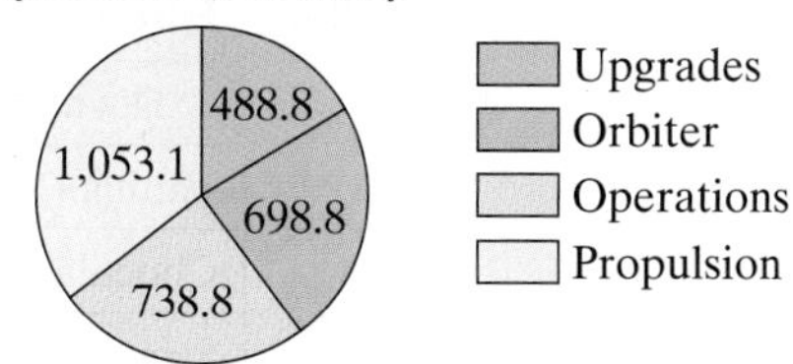

Quick Check

2. a. Find the measure of the central angle that you would draw to represent summer. **144°**

b. Use the information in the table at the right to make a circle graph. **See back of book.**

Favorite Season

Season	Percent
Summer	40%
Spring	11%
Winter	4%
Fall	45%

Test Prep Tip

To find a percent of a number, change the percent to a decimal and then multiply.

2. Teach

Activity Lab

Use before the lesson.

All in One Teaching Resources
Activity Lab 7-7: Circle Graphs

Guided Instruction

Technology
Students can use a spreadsheet program to draw circle graphs.

PowerPoint

Additional Examples

1 Use the circle graph below.

November Weather (30 days)

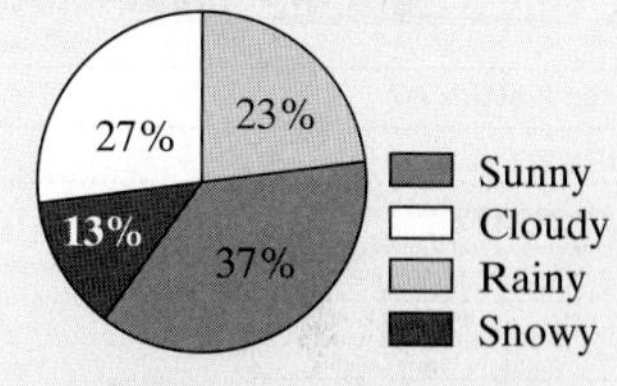

a. What percent of the days was snowy? **13%**

b. How many days were cloudy? **8 days**

2 Use the information in the table to make a circle graph.

Size of U.S. Households (2000)

No. of People in Household	Number (in thousands)
1	26,724
2	34,666
3	17,152
4	15,309
5 or more	10,854

See back of book.

All in One Teaching Resources
- Daily Notetaking Guide 7-7 **L3**
- Adapted Notetaking 7-7 **L1**

Closure

- *How do you make a circle graph?* **See back of book.**

Advanced Learners **L4**
Have students make and conduct a survey, such as the number of siblings or pets of classmates. Students then construct a circle graph to illustrate their results.

learning style: visual

English Language Learners **ELL**
Students will have to deal with words that may be unfamiliar to them. For example, words such as *typically, approximately, display,* are used to analyze circle graphs. Make sure they understand what the questions are asking.

learning style: verbal

3. Practice

Assignment Guide

Check Your Understanding
Go over Exercises 1–6 in class before assigning the Homework Exercises.

Homework Exercises
A Practice by Example 7–11
B Apply Your Skills 12–18
C Challenge 19
Test Prep and Mixed Review 20–26

Homework Quick Check
To check students' understanding of key skills and concepts, go over Exercises 8, 11, 14, 17, and 18.

Differentiated Instruction Resources

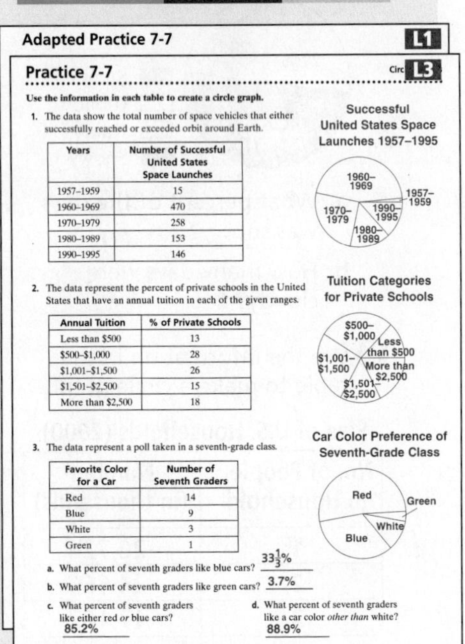

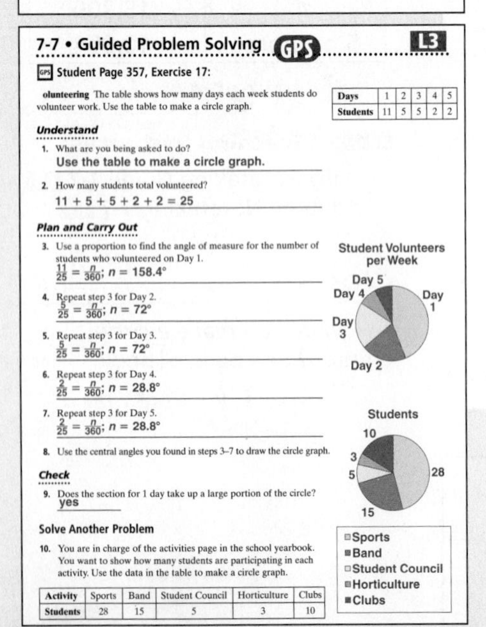

Check Your Understanding

1. **Vocabulary** In a circle graph, the sum of the sectors must equal ■. **the whole circle, 360°, 100%**
2. **Number Sense** A circle graph has five sectors. Two sectors are equal to 25% each. The third sector is equal to 35%. Which CANNOT be the value of either of the remaining two sectors? **A**
   Ⓐ 22% Ⓑ 11% Ⓒ 10% Ⓓ 1%

Find the measure of the central angle that you would draw to represent each percent in a circle graph.

3. 25% **90°**
4. 50% **180°**
5. 10% **36°**
6. 12.5% **45°**

Homework Exercises

For more exercises, see Extra Skills and Word Problems.

GO for Help

For Exercises	See Examples
7–9	1
10–11	2

A **Use the circle graph for Exercises 7– .**

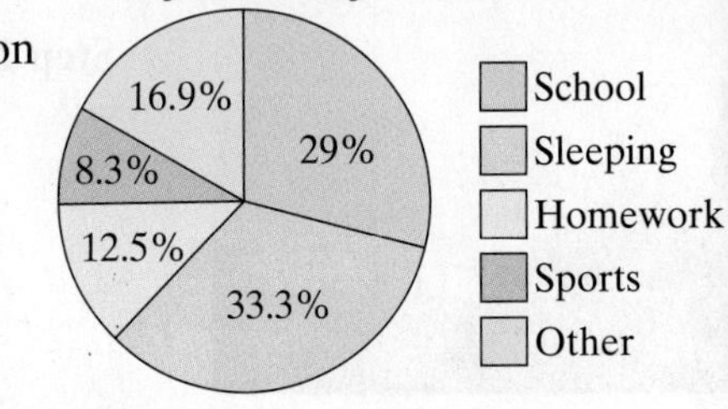

7. What takes up the largest portion of Royston's day? **sleeping**
8. What percent of the day does Royston typically spend doing homework? **12.5%**
9. How many hours a day does Royston spend sleeping? **about 8 hours**

Use the information in each table to make a circle graph.

10. **Frozen Yogurt Sales**

Flavor	Scoops
Vanilla	84
Chocolate	107
Strawberry	43

11. **Movie Rentals**

Type	Number
Action	7
Comedy	9
Other	5

10–11. See margin.

B **GPS** 12. **Guided Problem Solving** Use the data to draw a circle graph. **See margin.**

Transportation Mode	Walk	Bicycle	Bus	Car
Number of Students	252	135	432	81

- **Make a Plan** First find the total number of students. Then set up a proportion for each mode to find its central angle measure.
- **Carry Out the Plan** The total number of students is ■. Use the proportions below.

$\frac{252}{■} = \frac{w}{360}$ $\frac{135}{■} = \frac{b}{360}$ $\frac{432}{■} = \frac{u}{■}$ $\frac{■}{■} = \frac{c}{■}$

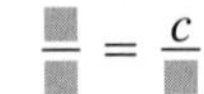

10–12. See back of book.

Online
Homework Video Tutor
Visit: PHSchool.com
Web Code: are-0707

14. No, "other" and Europe both indicate many different countries.

Use the circle graph for Exercises 13–16.

U.S. Foreign Travel (millions of people per year)

13, 17.7, 11.6, 16

Mexico, Europe, Canada, Other

SOURCE: Statistical Abstract of the United States

13. Which country or region is most visited by United States travelers? **Mexico**
14. **Reasoning** Can you tell which country or region is least visited by United States travelers? Explain.
15. Approximately how many people travel to Europe from the United States each year? **11.6 million people**
16. **Writing in Math** Why use Europe rather than list every country individually? **The angles would be too small to display.**

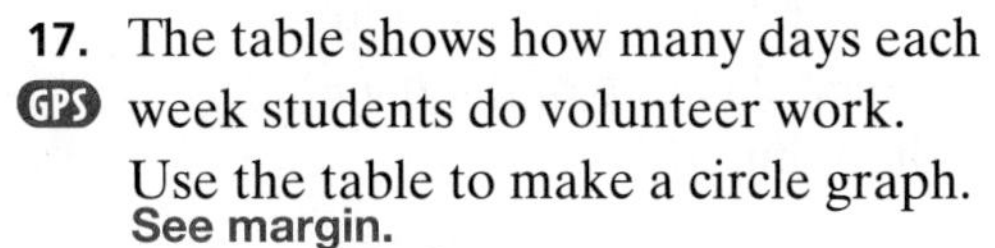

17. The table shows how many days each week students do volunteer work. Use the table to make a circle graph. **See margin.**

Days	1	2	3	4	5
Students	11	5	5	2	2

18. **Open-Ended** Describe a situation for which a circle graph is an appropriate display to represent the data. Describe a situation for which a circle graph is *not* an appropriate display to represent the data. **See margin.**
19. **Challenge** A circle graph shows a survey of the favorite ice cream flavors of all the students at Park School. Vanilla is the favorite flavor of 28 students. The central angle measure for vanilla is 72° in the circle graph. How many students are at Park School? **140 students**

Test Prep and Mixed Review

Practice

Gridded Response

20. There are 150 animals in Sal's Pet Store. According to the circle graph, how many of the animals are dogs? **45**

Sal's Pet Store

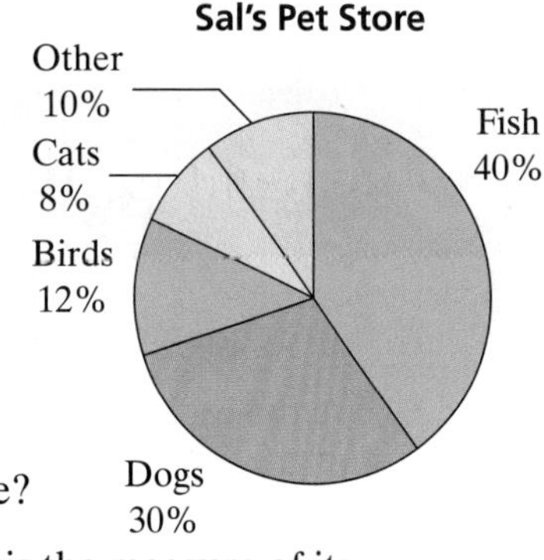

21. A carpenter renovating a house is sanding the dining room floor. She sands 300 ft^2 of wood floor in 1 hour and 40 minutes. What is the unit rate in square feet per minute? **3**
22. The measure of $\angle ABC$ is 128°. What is the measure of its supplement in degrees? **52**

GO for Help

For Exercises	See Lesson
23–26	3-3

Find each sum or difference.

23. $\frac{2}{3} + \frac{3}{8}$ $1\frac{1}{24}$
24. $4\frac{1}{6} + 6\frac{2}{9}$ $10\frac{7}{18}$
25. $\frac{7}{8} - \frac{1}{4}$ $\frac{5}{8}$
26. $14\frac{5}{8} - 6\frac{5}{12}$ $8\frac{5}{24}$

4. Assess & Reteach

Lesson Quiz

The circle graph below shows how a class of 48 students spends free time after lunch.

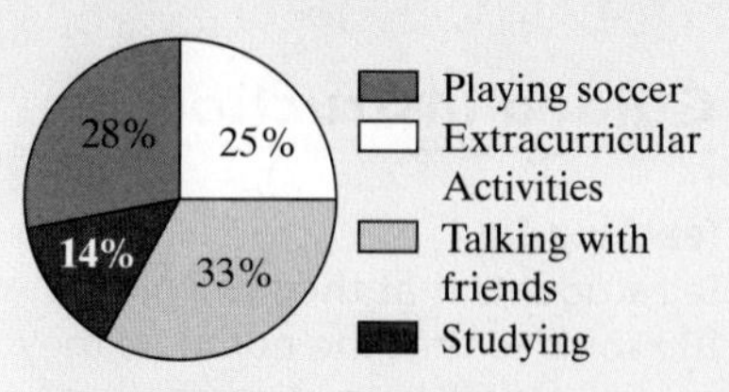

1. What percent study or do extracurricular activities? **39%**
2. How many students spend time on extracurricular activities? **12**
3. Find the measure of the central angle that you would draw to represent 15% in a circle graph. **54°**

17–18. See back of book.

Alternative Assessment

Have students work in pairs to complete Exercises 10 and 11. First they list the steps they need to take before they can make the graph, such as: find the total, find the percent each item is of the total, and find the measure of the central angle to the nearest whole degree. Then each pair makes its graphs.

Test Prep

Resources

For additional practice with a variety of test item formats:

- Test-Taking Strategies, p. 365
- Test Prep, p. 369
- Test-Taking Strategies with Transparencies

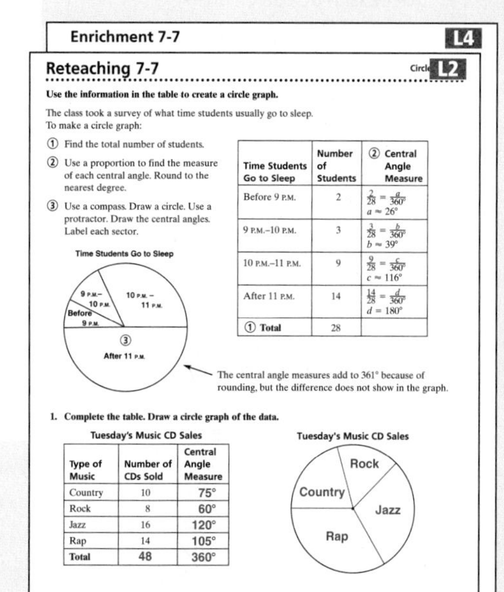

Enrichment 7-7 L4

Reteaching 7-7 L2

Use the information in the table to create a circle graph.

The class took a survey of what time students usually go to sleep. To make a circle graph:

① Find the total number of students.
② Use a proportion to find the measure of each central angle. Round to the nearest degree.
③ Use a compass. Draw a circle. Use a protractor. Draw the central angles. Label each sector.

Time Students Go to Sleep	Number of Students	② Central Angle Measure
Before 9 P.M.	2	$\frac{2}{28} = \frac{a}{360°}$ $a \approx 26°$
9 P.M.–10 P.M.	3	$\frac{3}{28} = \frac{b}{360°}$ $b \approx 39°$
10 P.M.–11 P.M.	9	$\frac{9}{28} = \frac{c}{360°}$ $c \approx 116°$
After 11 P.M.	14	$\frac{14}{28} = \frac{d}{360°}$ $d = 180°$
① Total	28	

Time Students Go to Sleep

The central angle measures add to 361° because of rounding, but the difference does not show in the graph.

1. Complete the table. Draw a circle graph of the data.

Tuesday's Music CD Sales

Type of Music	Number of CDs Sold	Central Angle Measure
Country	10	75°
Rock	8	60°
Jazz	16	120°
Rap	14	105°
Total	48	360°

Tuesday's Music CD Sales

Activity Lab

Making a Circle Graph

Students will use a spinner to generate data that they will then present in a circle graph.

Guided Instruction

Teaching Tip
Tell students that they will make 100 spins so that the numbers they work with will be easy to convert to decimals or percents.

Resources

- spinner
- paper and writing instruments
- Classroom Aid 27

Checkpoint Quiz

Use this Checkpoint Quiz to check students' understanding of the skills and concepts of Lessons 7-5 through 7-7.

Resources

- All-in-One Teaching Resources Checkpoint Quiz 1
- ExamView Assessment Suite CD-ROM
- Success Tracker Online Intervention

Activity Lab

1. The *tallies* column should total 100. The *fraction* column should total 1. The *angle* column should total 360°.
3. Answers may vary. Sample: The results are very close. The circle graph might be even closer if I used 1,000 spins. The fact that each spin is random means the experimental probability might be different from the theoretical probability.

7-7b Activity Lab

Data Collection

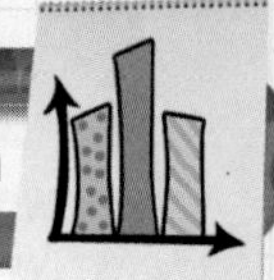

Making a Circle Graph

You can use a spinner to generate data. You can use circle graphs to present the data.

ACTIVITY

Step 1 Copy the spinner at the right. Use a protractor to measure and copy the central angles of the spinner.

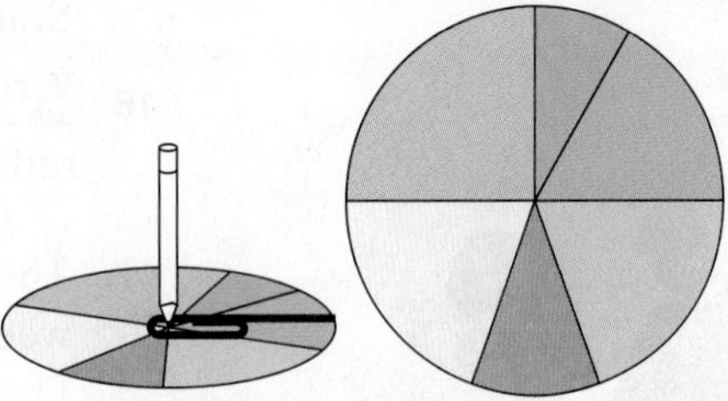

Step 2 Straighten half a paper clip to use as the pointer. Use a pencil point to keep the loop of the paper clip at the center of the spinner.

Step 3 Make a table like the one at the right. Spin the paper clip. Use your frequency column to record the results of 100 spins.

Category	Tallies	Fraction	Angle
Red	■	■	■
Blue	■	■	■
Green	■	■	■
Yellow	■	■	■
Gray	■	■	■
Orange	■	■	■

1. Fill in the fraction of tallies for each color and the number of degrees in that fraction of a circle. What should be the sum of each column? **See margin.**
2. Make a circle graph for your data. Use a protractor to draw the central angles for your graph. Label each sector of your graph. **Check students' work.**
3. Compare your circle graph to the spinner that you used. Are they exactly the same? Explain. **See margin.**

Checkpoint Quiz 2

Lessons 7-5 through 7-7

1. If two squares have the same area, are they congruent? Explain. **See margin.**
2. Use the information in the table to make a circle graph. **See margin.**

Town Middle School

Grade	6	7	8
Number of Students	75	100	125

Name each of the following for circle *P*.

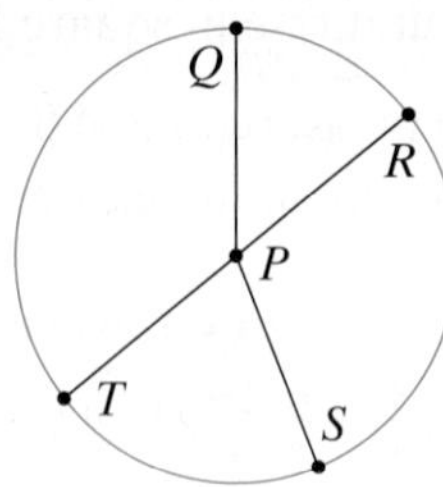

3. two radii **$\overline{QP}$, $\overline{RP}$, $\overline{SP}$, and $\overline{TP}$**
4. a diameter **$\overline{TR}$**
5. two central angles **$\angle RPS$, $\angle SPT$, $\angle TPQ$, and $\angle QPR$**

Checkpoint Quiz 2

1. Yes. The square root of the area is the side length. Since both squares have the same area, they have the same side length. Being squares, they both have four right angles. Congruent sides and congruent angles means the squares are congruent.
2. See back of book.

Percents and Circle Graphs

The circle graph shows people's milk preferences. If your school cafeteria ordered pints of milk for 450 milk-drinking students, how many pints of each type should the cafeteria order?

School Milk Preferences

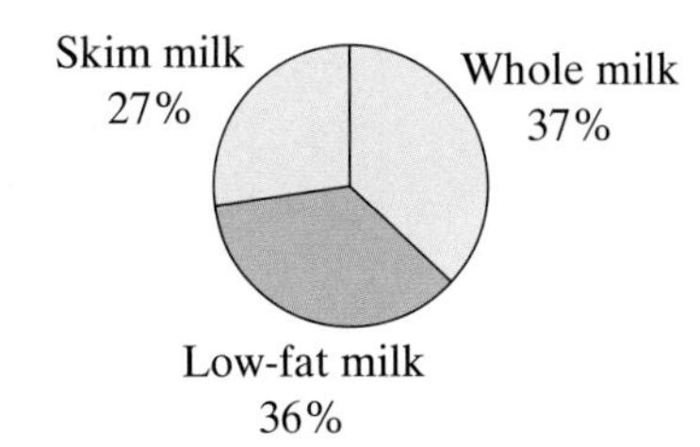

What You Might Think

What do I know?
What do I want to find out?

How do I solve the problem?

What is the answer?

Is it reasonable?

What You Might Write

The graph shows whole, low-fat, and skim milk preferences. I want to find each percent for 450 students.

Whole milk: $37\% \cdot 450 = 166.5$
Low-fat milk: $36\% \cdot 450 = 162$
Skim milk: $27\% \cdot 450 = 121.5$

Round: 166.5 rounds to 170.
162 rounds to 160.
121.5 rounds to 120.

Order 170 pints of whole milk, 160 pints of low-fat milk, and 120 pints of skim milk.

The numbers add to 450, with skim milk a smaller number than the other two. The answer is reasonable.

Think It Through

1. Do you think these data will change in ten years? In what way? Explain your reasoning. **See margin.**

1. **Answers may vary. Sample: Yes; students are more diet conscious, and preferences for low-fat and skim milk would be increasing.**

GPS Guided Problem Solving

Percent and Circle Graphs

In this feature, students learn to solve problems using percents and circle graphs. The circle graph shows milk preferences in percents using 450 as a base.

Guided Instruction

Discuss with students the percent equation: percent × base = part. Have them identify or find each percent for the various milk types.

Resources

- Classroom Aid 27, 28

Error Prevention!

Be sure students understand that the base remains the same although the percents and parts change.

Exercises

Work through Exercise 1 as a class. Then assign students to work with partners to complete Exercises 2–4.

For Exercise 4, have partners work together to choose five categories. If they disagree on the categories, help them broaden the categories, for example, from soccer and band to after school activities.

Differentiated Instruction

English Language Learners ELL

For Exercise 3, be sure students understand what blood typing is. Ask if any students know their own blood type.

Exercises

Solve the problems. For Exercises 2 and 3, answer the questions first.

2. The circle graph shows different age groups in the population of the United States as reported in the 2000 U.S. Census. If the population of the United States is about 281 million, approximately how many people are in your age group? **about 31.2 million**

Age Groups

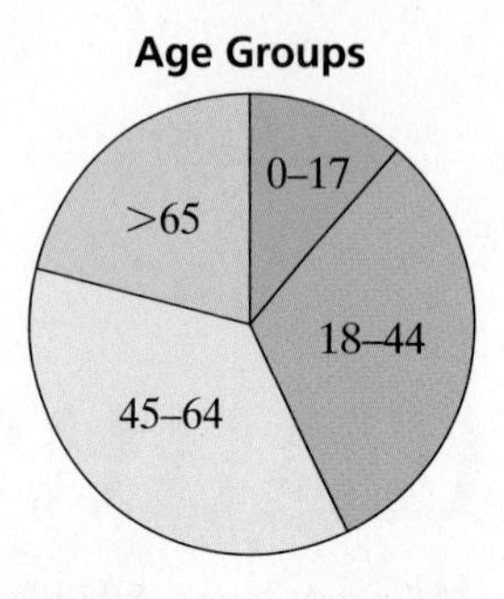

a. What do you know? What do you want to find out?

b. How can you estimate the percent of people in your age group?

3. A study collected percents of male and female teachers across the nation. The results are shown in the table below. If you have five classes each year, how many male teachers can you expect during your three middle school years? How many female teachers can you expect? **4 male teachers; 11 female teachers**

Teachers

Female	Male
72%	28%

a. What do you know? What do you want to find out?

b. Can a model help you find the answers?

4. Blood typing is important in transfusions and organ transplants. Use the data in the table to make a circle graph showing the percents of blood types in the United States. Use a calculator to find the number of U. S. citizens with each blood type. Use 281 million as the population of the United States. **See back of book.**

U.S. Blood Type Percentages

Type	Positive	Negative
O	38%	7%
A	34%	6%
B	9%	2%
AB	3%	1%

5. Make a circle graph of how you spend time during a typical day in school. Limit your graph to five categories. **Check students' work.**

7-8 Constructions

Check Skills You'll Need

1. **Vocabulary Review** What do two *intersecting lines* share? **a common point**

Use the points in each diagram to name the figure shown.

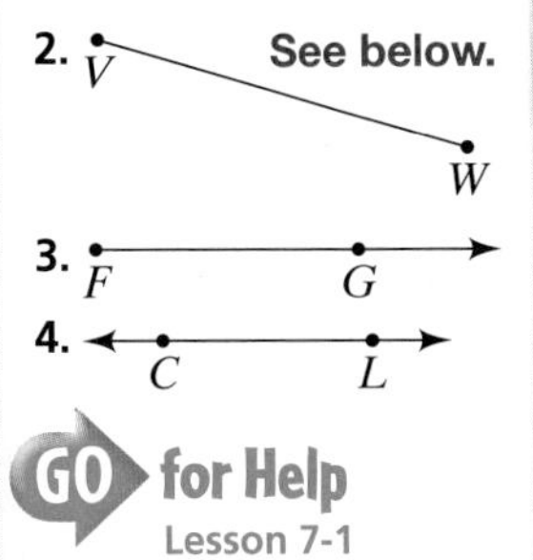

GO for Help Lesson 7-1

Check Skills You'll Need
2. $\overline{VW}$
3. $\overrightarrow{FG}$
4. $\overleftrightarrow{CL}$

What You'll Learn

To construct congruent segments and perpendicular bisectors

New Vocabulary compass, midpoint, segment bisector, perpendicular lines, perpendicular bisector

Why Learn This?

Architects use congruent segments and perpendicular lines when they design buildings. Congruent segments and perpendicular lines can give a design a sense of balance.

You can use a geometric tool called a **compass** to draw a circle, or a part of a circle, called an arc.

You can use a compass and a straightedge (an unmarked ruler) to construct a congruent segment for a given segment.

EXAMPLE 1 Constructing a Congruent Segment

Construct segment $\overline{CD}$ congruent to $\overline{AB}$.

A B

Step 1 Draw a ray with endpoint C.

C

Step 2 Open the compass to the length of $\overline{AB}$.

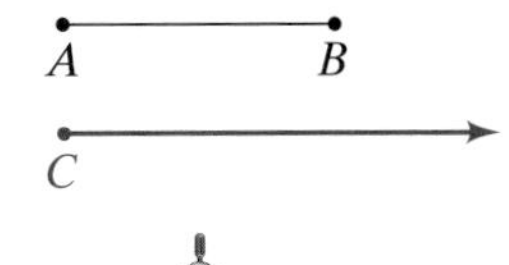

Step 3 Keep the compass open to the same width. Put the compass point on point C. Draw an arc that intersects the ray. Label the point of intersection D.

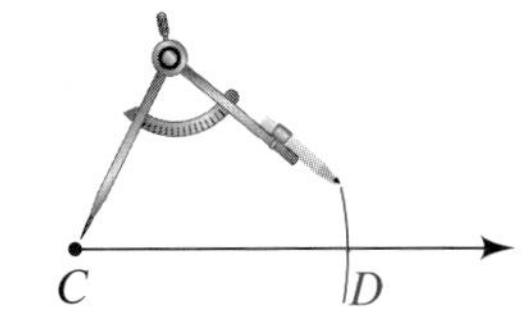

$\overline{CD}$ is congruent to $\overline{AB}$.

Quick Check

1. Draw a segment $\overline{TR}$ 25 mm long. Construct $\overline{SV}$ congruent to $\overline{TR}$. See left.

1.
T 25 mm R

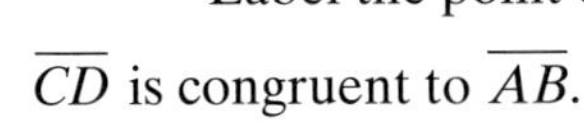

7-8

1. Plan

Objective
To construct congruent segments and perpendicular bisectors

Examples
1 Constructing a Congruent Segment
2 Constructing a Perpendicular Bisector

Math Understandings: p. 322D

Professional Development

Math Background

You can use a *compass* to draw circles and parts of circles, called *arcs*. Some basic constructions are constructing congruent segments and bisecting a segment. A *segment bisector* is a line, or segment, or ray that divides a segment at its midpoint into two congruent segments. An infinite number of segment bisectors can divide one segment into two equal segments. However, it has only one *perpendicular bisector* that intersects the segment at a right angle.

More Math Background: p. 322D

Lesson Planning and Resources

See p. 322E for a list of the resources that support this lesson.

PowerPoint

Bell Ringer Practice

Check Skills You'll Need
Use student page, transparency, or PowerPoint. For intervention, direct students to:

Lines and Planes
Lesson 7-1
Extra Skills and Word Problems Practice, Ch. 7

Differentiated Instruction Solutions for All Learners

Special Needs L1	**Below Level** L2
Some students may have difficulty handling the compass to complete the constructions. Pair them up with a student who can draw while they read the directions out loud. They can trace the drawing after it is constructed.	Use a ruler to measure the segments in Example 2 and find the midpoint arithmetically. Students should verify that the midpoint by construction matches the measured one.
learning style: tactile	learning style: tactile

2. Teach

Activity Lab

Use before the lesson.

All in One Teaching Resources

Activity Lab 7-8: Constructions

Guided Instruction

Example 1

Have students practice drawing circles with a compass. Show students that the tip must remain anchored while drawing an arc. Make sure students draw without putting too much pressure on the compass because this can change the size of the opening.

Technology Tip

Geometry software programs allow students to perform compass and straightedge constructions.

PowerPoint

Additional Examples

1 Construct a segment $\overline{RS}$ that is congruent to $\overline{PQ}$.

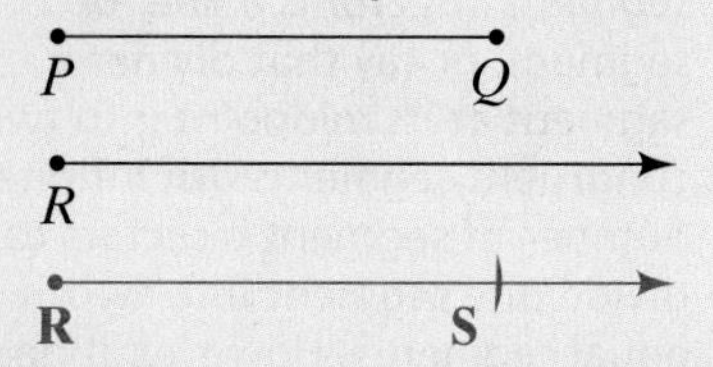

2 Construct the perpendicular bisector of $\overline{PQ}$.

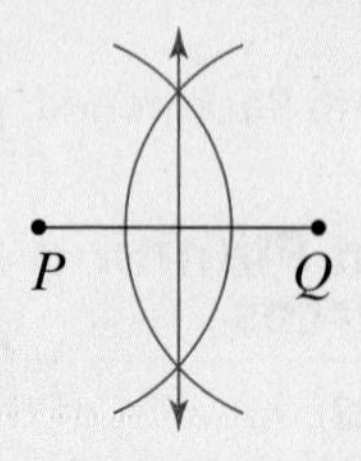

All in One Teaching Resources

- Daily Notetaking Guide 7-8 L3
- Adapted Notetaking 7-8 L1

Closure

- *How are a segment bisector and midpoint of a segment related?* **The bisector cuts the segment into two segments of equal length by intersecting the segment at the midpoint.**

The **midpoint** of a segment is the point that divides the segment into two segments of equal length.

A **segment bisector** is a line, segment, or ray that goes through the midpoint of a segment.

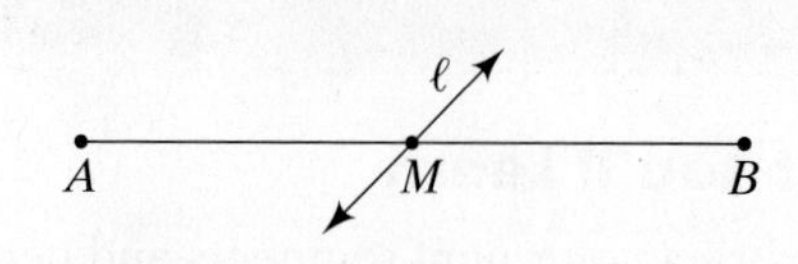

$\overline{AM}$ is congruent to $\overline{MB}$.

M is the midpoint of $\overline{AB}$.

Line ℓ is a segment bisector of $\overline{AB}$.

Perpendicular lines are lines that intersect to form right angles.

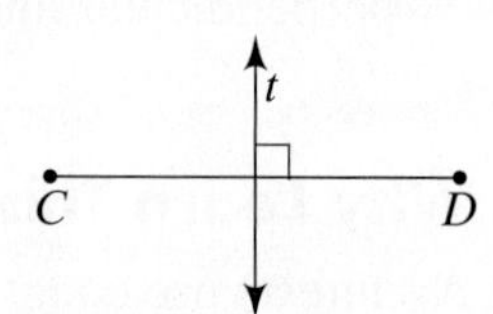

A segment bisector that is perpendicular to a segment is the **perpendicular bisector** of the segment. You can use a compass and a straightedge to construct the perpendicular bisector of a given segment.

EXAMPLE Constructing a Perpendicular Bisector

2 Construct the perpendicular bisector of $\overline{AB}$.

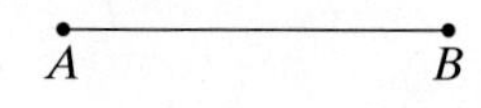

Step 1 Set the compass to more than half the length of $\overline{AB}$. Put the tip of the compass at A and draw an arc intersecting $\overline{AB}$.

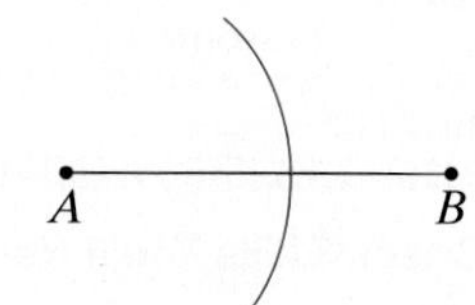

Step 2 Keeping the compass set at the same width, put the tip at B and draw another arc intersecting $\overline{AB}$. Points C and D are where the arcs intersect.

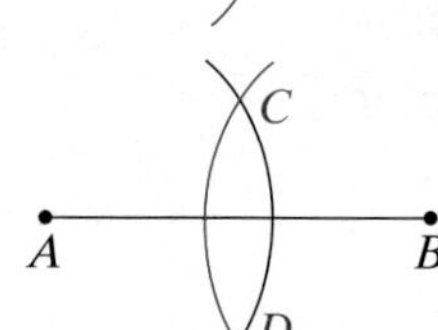

Step 3 Draw $\overleftrightarrow{CD}$. The intersection of $\overline{AB}$ and $\overleftrightarrow{CD}$ is point M. $\overleftrightarrow{CD}$ is the perpendicular bisector of $\overline{AB}$. Point M is the midpoint of $\overline{AB}$.

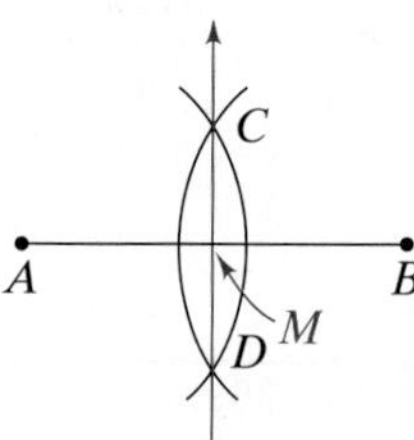

This 1482 edition of Euclid's *Elements of Geometry* was the first math book produced on a printing press.

SOURCE: Wellesley College Library

Quick Check

2. Draw a segment 3 in. long. Label the segment $\overline{XY}$. Construct the perpendicular bisector of $\overline{XY}$.

2.

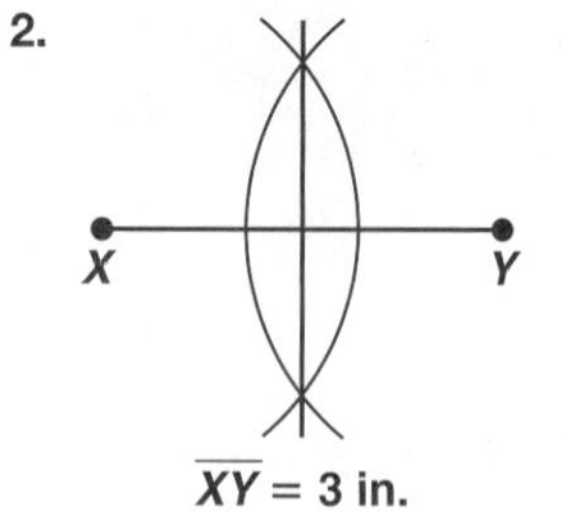

$\overline{XY}$ = 3 in.

Advanced Learners L4	**English Language Learners** ELL
How many segment bisectors of $\overline{AB}$ are there? **an infinite number** *How many perpendicular segment bisectors of $\overline{AB}$ are there?* **one**	Have students pair up and take turns, with one partner reading the data and one drawing. Then switch roles. Both students practice speaking about, and listening to, the terms used in the constructions.
learning style: visual	**learning style: verbal**

Check Your Understanding

1. Answers may vary. Since Q is the midpoint of $\overline{PR}$, $\overline{PQ}$ and $\overline{QR}$ are congruent.

Vocabulary Tip

AB means the length of $\overline{AB}$.

1. **Vocabulary** Q is the midpoint of $\overline{PR}$. Write a statement about two congruent segments having endpoint Q.
2. **Reasoning** Can you construct a different perpendicular bisector of $\overline{AB}$ in Example 2? Explain. **No. The perpendicular bisector is unique in the plane.**

Point B is the midpoint of $\overline{AC}$. Complete each statement.

3. $AB = 4$ in., $AC = \blacksquare$ **8 in.**
4. $AC = 9$ m, $AB = \blacksquare$ $4\frac{1}{2}$ **m**
5. $AC = 7$ m, $BC = \blacksquare$ $3\frac{1}{2}$ **m**
6. $BC = 5$ ft, $AB = \blacksquare$ **5 ft**

Homework Exercises

For more exercises, see Extra Skills and Word Problems.

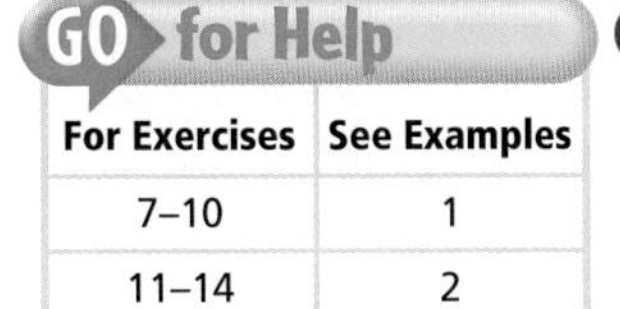

GO for Help

For Exercises	See Examples
7–10	1
11–14	2

7–14. Check students' work.

A **Copy each segment. Then construct a congruent segment.**

7. A ———— B
8. S ———— T

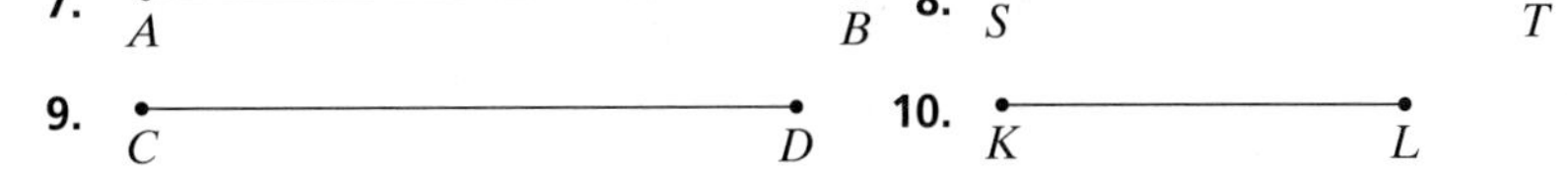

9. C ———— D
10. K ———— L

Draw each segment. Then construct its perpendicular bisector.

11. a segment 4 in. long
12. a segment 10 cm long
13. a segment 5 in. long
14. a segment 13.5 cm long

B 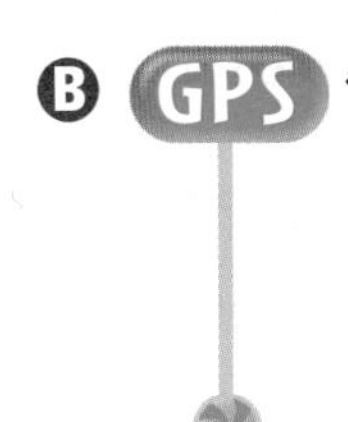

15. **Guided Problem Solving** Draw $\overline{CD}$ at least 3 in. long. Construct and label a segment one fourth as long as $\overline{CD}$. **See margin.**
    - **Make a Plan** One half of one half is one fourth. So bisecting half of a line segment will give you one fourth of a line segment.
    - **Carry Out the Plan** Bisect $\overline{CD}$ and label the midpoint E. Then bisect $\overline{CE}$. That segment will be one fourth as long as $\overline{CD}$.
16. Draw $\overline{MN}$ about 4 in. long. Then construct $\overline{JK}$ two and one half times as long as $\overline{MN}$. **See margin.**
17. Draw $\overline{AB}$ about 5 in. long. Then construct $\overline{XY}$ so that the two line segments are perpendicular bisectors of each other. **See margin.**
18. **Patterns** Draw two large triangles, one acute and one obtuse. Construct the perpendicular bisector of each side. Make a conjecture about the perpendicular bisectors of the sides of any triangle. **See margin.**
19. **Writing in Math** How many midpoints does a segment have? How many segment bisectors does it have? How many perpendicular bisectors does it have? Explain. **See left.**

19. 1, infinite, 1; a segment can only have one exact midpoint or perpendicular bisector. A segment can be bisected by an infinite number of lines.

Online lesson quiz, PHSchool.com, Web Code: ara-0708

15–18. See back of book.

3. Practice

Assignment Guide

Check Your Understanding

Go over Exercises 1–6 in class before assigning the Homework Exercises.

Homework Exercises

A	Practice by Example	7–14
B	Apply Your Skills	15–27
C	Challenge	28
	Test Prep and Mixed Review	29–34

Homework Quick Check

To check students' understanding of key skills and concepts, go over Exercises 8, 12, 16, 19, and 26.

Differentiated Instruction Resources

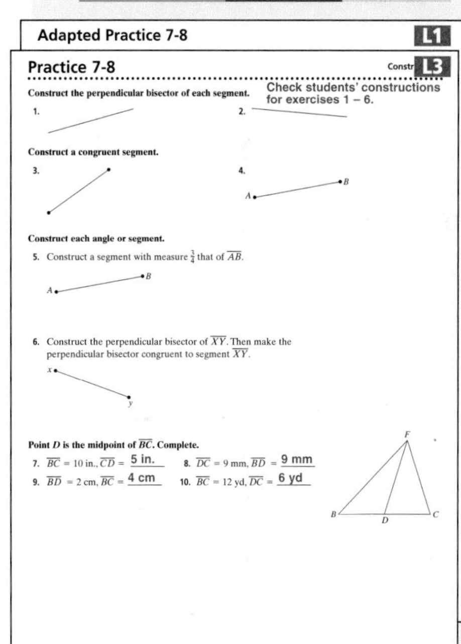

Adapted Practice 7-8 L1

Practice 7-8 L3

Construct the perpendicular bisector of each segment. Check students' constructions for exercises 1 – 6.

Construct a congruent segment.

Construct each angle or segment.

5. Construct a segment with measure $\frac{3}{4}$ that of $\overline{AB}$.

6. Construct the perpendicular bisector of $\overline{XY}$. Then make the perpendicular bisector congruent to segment $\overline{XY}$.

Point D is the midpoint of $\overline{BC}$. Complete.

7. $BC = 10$ in., $CD =$ 5 in.
8. $DC = 9$ mm, $BD =$ 9 mm
9. $BD = 2$ cm, $BC =$ 4 cm
10. $BC = 12$ yd, $DC =$ 6 yd

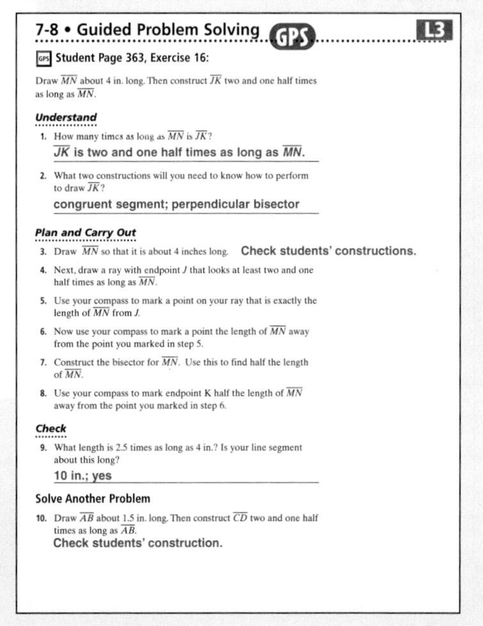

7-8 • Guided Problem Solving GPS L3

Student Page 363, Exercise 16:

Draw $\overline{MN}$ about 4 in. long. Then construct $\overline{JK}$ two and one half times as long as $\overline{MN}$.

Understand

1. How many times as long as $\overline{MN}$ is $\overline{JK}$? $\overline{JK}$ is two and one half times as long as $\overline{MN}$.
2. What two constructions will you need to know how to perform to draw $\overline{JK}$? congruent segment; perpendicular bisector

Plan and Carry Out

3. Draw $\overline{MN}$ so that it is about 4 inches long. Check students' constructions.
4. Next, draw a ray with endpoint J that looks at least two and one half times as long as $\overline{MN}$.
5. Use your compass to mark a point on your ray that is exactly the length of $\overline{MN}$ from J.
6. Now use your compass to mark a point the length of $\overline{MN}$ away from the point you marked in step 5.
7. Construct the bisector for $\overline{MN}$. Use this to find half the length of $\overline{MN}$.
8. Use your compass to mark endpoint K half the length of $\overline{MN}$ away from the point you marked in step 6.

Check

9. What length is 2.5 times as long as 4 in.? Is your line segment about this long? 10 in.; yes

Solve Another Problem

10. Draw $\overline{AB}$ about 1.5 in. long. Then construct $\overline{CD}$ two and one half times as long as $\overline{AB}$. Check students' construction.

4. Assess & Reteach

PowerPoint

Lesson Quiz

1. Construct a segment congruent to this segment.

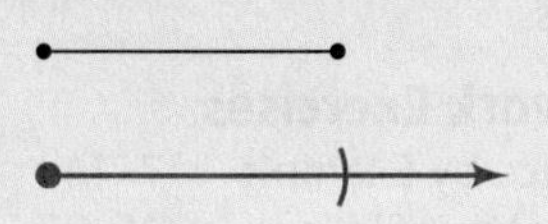

2. Draw a segment 9.5 cm long. Label the segment $\overline{CD}$. Then construct the perpendicular bisector of $\overline{CD}$.

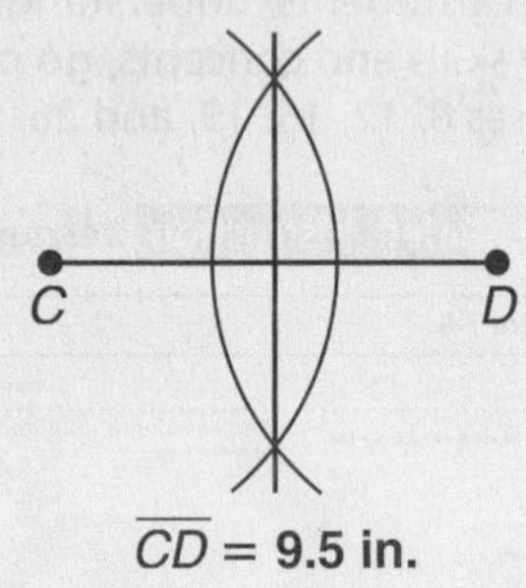

$\overline{CD}$ = 9.5 in.

31. $\frac{5}{7}$, 5 : 7
32. 6 to 13, $\frac{6}{13}$
33. 9 to 4, 9 : 4
34. 16 to 25, 16 : 25

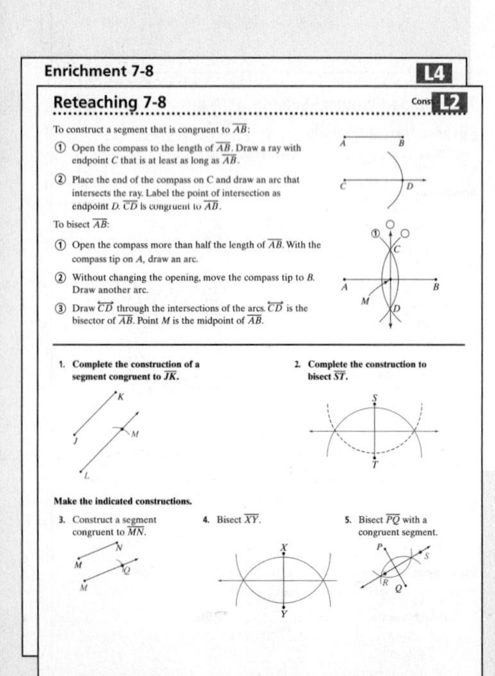

Enrichment 7-8 L4

Reteaching 7-8 Cons L2

To construct a segment that is congruent to $\overline{AB}$:

① Open the compass to the length of $\overline{AB}$. Draw a ray with endpoint C that is at least as long as $\overline{AB}$.

② Place the end of the compass on C and draw an arc that intersects the ray. Label the point of intersection as endpoint D. $\overline{CD}$ is congruent to $\overline{AB}$.

To bisect $\overline{AB}$:

① Open the compass more than half the length of $\overline{AB}$. With the compass tip on A, draw an arc.

② Without changing the opening, move the compass tip to B. Draw another arc.

③ Draw $\overleftrightarrow{CD}$ through the intersections of the arcs. $\overleftrightarrow{CD}$ is the bisector of $\overline{AB}$. Point M is the midpoint of $\overline{AB}$.

1. **Complete the construction of a segment congruent to $\overline{JK}$.**

2. **Complete the construction to bisect $\overline{ST}$.**

Make the indicated constructions.

3. Construct a segment congruent to $\overline{MN}$.

4. Bisect $\overline{XY}$.

5. Bisect $\overline{PQ}$ with a congruent segment.

GO Online
Homework Video Tutor
Visit: PHSchool.com
Web Code: are-0708

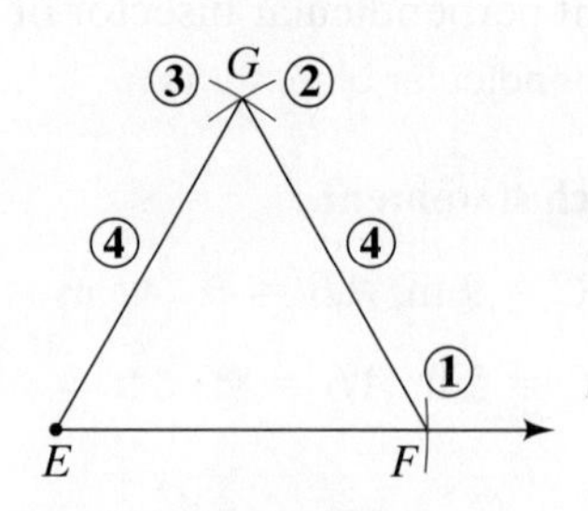

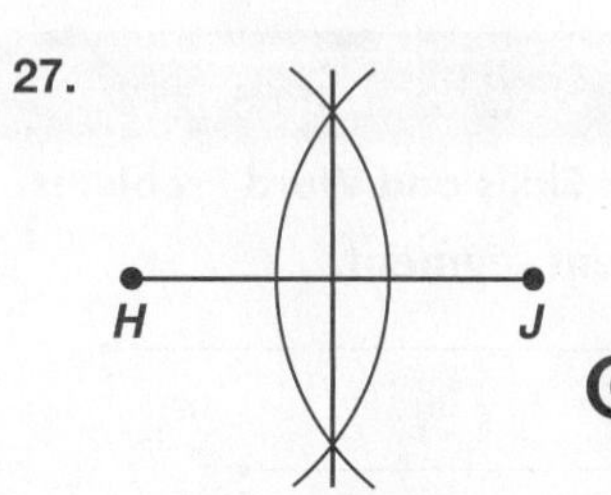

Point B is the midpoint of AC. Complete each statement.

20. $AB = 2.25$ in., $AC = ■$ **4.5 in.**
21. $AC = 8.4$ cm, $AB = ■$ **4.2 cm**
22. $BC = 1.7$ ft, $AB = ■$ **1.7 ft**
23. $AB = 17$ mm, $AC = ■$ **34 mm**
24. $AC = 3$ in., $BC = ■$ **1.5 in.**
25. $BC = 75$ cm, $AC = ■$ **150 cm**

26. Follow the steps to construct $\triangle EFG$ with all sides congruent to $\overline{XY}$.

Step 1 Draw a segment $\overline{XY}$. Use a compass to construct $\overline{EF} \cong \overline{XY}$.

Step 2 Using the same compass width, place the compass tip on point F and draw an arc above $\overline{EF}$.

Step 3 Using the same compass width, place the compass tip on point E and make another arc above $\overline{EF}$, intersecting the first arc. Label the intersection G.

Step 4 Draw $\overline{EG}$ and $\overline{FG}$ to form $\triangle EFG$. **Check students' work.**

27. Use the construction techniques you learned in this lesson to draw a 90° angle without using a protractor. **See left.**

Ⓒ 28. **Challenge** A is the midpoint of $\overline{XY}$. Y is the midpoint of $\overline{XZ}$. Z is the midpoint of $\overline{AB}$. $\overline{XA}$ is 2 cm long. How long is $\overline{XB}$? **14 cm**

Test Prep and Mixed Review

Practice

Multiple Choice

29. Don bought items that cost \$2.59, \$3.48, \$1.75, \$0.63, and \$0.98. He used a coupon worth \$0.80. The tax totaled \$0.73. If he gave the clerk a \$10 bill, how much change did he receive? **A**

Ⓐ \$0.64 Ⓑ \$0.96 Ⓒ \$1.37 Ⓓ \$9.36

30. Which problem situation matches the equation $0.80x = 400$? **J**

Ⓕ In a recent election, 80% of the people voted for a new tax to repair streets. Four hundred people voted. What is x, the number who voted for the tax?

Ⓖ Eighty out of 400 of the items at one manufacturing plant were found to have flaws. What is x, the percent of items that contained flaws?

Ⓗ The Drake family has driven 80 miles. They need to travel 400 miles to reach their destination. What is x, the percent of their trip completed?

Ⓙ In a school survey, 400 students said they would like more variety in the cafeteria. This was 80% of those surveyed. What is x, the total number of students who were surveyed?

GO for Help

For Exercises	See Lesson
31–34	5-1

Write each ratio in two other ways. **31–34. See margin.**

31. 5 to 7
32. 6 : 13
33. $\frac{9}{4}$
34. $\frac{16}{25}$

Test Prep

Resources

For additional practice with a variety of test item formats:

- Test-Taking Strategies, p. 365
- Test Prep, p. 369
- Test-Taking Strategies with Transparencies

Alternative Assessment

Each student in a pair draws a line segment. Partners exchange drawings. Each student constructs a segment congruent to the partner's segment and then constructs the perpendicular bisector of the segment. Students can share their work and their methods with the class.

Drawing a Picture

Sometimes a picture is not supplied with a problem. Then you can draw a picture to help you solve the problem. Make sure your picture is large enough to allow you to label all the parts.

EXAMPLE

$\triangle CRT \cong \triangle POV$. The measure of $\angle C$ is 41°, and the measure of $\angle T$ is 104°. What is the measure of $\angle O$?

Draw and label triangles CRT and POV. Label angles C and T. Label the corresponding angles in $\triangle POV$.

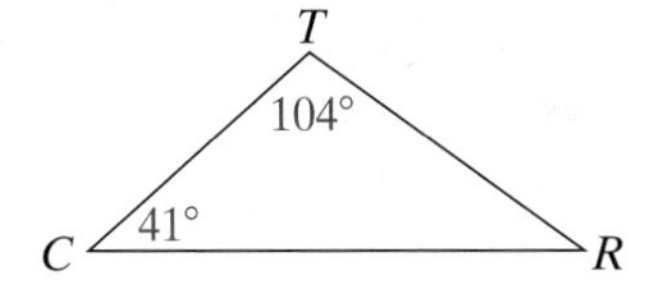

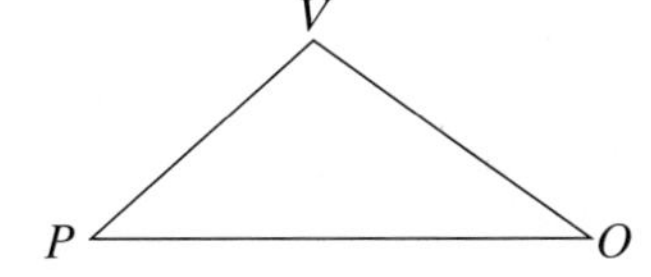

Write an equation to find the measure of $\angle O$.

Let $x = m\angle O = m\angle R$.

$$x + 41° + 104° = 180°$$
$$x + 145° = 180°$$
$$x + 145° - 145° = 180° - 145°$$
$$x = 35°$$

The measure of $\angle O$ is 35°.

Exercises

Draw a picture to solve each problem.

1. $\overleftrightarrow{QR}$ intersects $\overleftrightarrow{ST}$ at point U. The measure of $\angle QUT$ is 123°, and the measure of $\angle TUR$ is 57°. What is the measure of $\angle SUR$? **C**
   (A) 57° (B) 90° (C) 123° (D) 133°

2. A right triangle has one angle that measures 16°. Which of the following could be the measures of the other two angles? **G**
   (F) 16°, 148° (G) 74°, 90° (H) 84°, 90° (J) 82°, 82°

3. Rectangle $ABCD$ shares $\overline{BC}$ with equilateral triangle BCE. The length of $\overline{CD}$ is 4 cm, and the perimeter of $ABCD$ is 14 cm. What is the perimeter of $\triangle BCE$? **C**
   (A) 3 cm (B) 6 cm (C) 9 cm (D) 14 cm

Drawing a Picture

This feature presents the effective test-taking strategy of drawing a picture to help solve a problem.

Guided Instruction

Discuss with students that drawing a picture or sketch can help them more clearly understand the relationships presented in words.

Error Prevention!

Stress that the picture they draw accurately represents the data provided in the text. Guide students to reread the text after drawing the picture to make sure it expresses the identical information.

Resources

Test-Taking Strategies with Transparencies
- Transparency 9
- Practice sheet, p. 19

Test-Taking Strategies with Transparencies

Test-Taking Strategies: Drawing a Picture

A diagram can help you see how to solve a problem.

Example A square quilt has 6 in. by 6 in. fabric squares. The outside perimeter is 144 in. How many squares are on the quilt perimeter?

A. 144 B. 36 C. 24 D. 20

Each side must be $144 \div 4 = 36$ in.

Let one unit on graph paper represent 6 in.

There must be $36 \div 6 = 6$ squares on each side.

Draw a diagram, and count the perimeter squares.

The answer is 20, or choice D.

Use a diagram to find the answer. Explain your reasoning.

1. Tess wants to fence in her garden, which is 5 ft long and 4 ft wide. She will put a post at each corner and at every foot. How many fence posts will she need?
   A. 14 B. 20 C. 18 D. 19

2. What is the area of the shaded square in the figure?
   F. 50 cm^2
   G. 100 cm^2
   H. 200 cm^2
   J. 400 cm^2

0 cm
0 cm

Resources

Student Edition
Extra Skills and Word Problems Practice, Ch. 7, p. 642
English/Spanish Glossary, p. 675
Formulas and Properties, p. 673
Tables, p. 670

All in One Teaching Resources
- Vocabulary and Study Skills 7F L3

Differentiated Instruction

Spanish Vocabulary Workbook with Study Skills ELL
Interactive Textbook
- Audio Glossary

Online Vocabulary Quiz

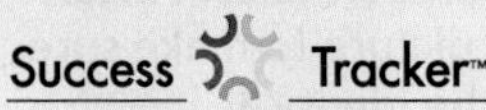

Online at PHSchool.com

Spanish Vocabulary/Study Skills ELL

Vocabulary/Study Skills L3

7F: Vocabulary Review For use with the Chapter Review

Study Skill Write assignments down; do not rely only on your memory.

Below is a list of clues grouped by the number of letters in the answer. Identify the word each clue represents, and fit each word into the puzzle grid.

3 letters
- part of a circle

5 letters
- angle that measures between 0° and 90°
- formed by two rays with a common endpoint
- segment that has both endpoints on the circle
- flat surface that extends indefinitely in all directions

6 letters
- point of intersection of two sides on an angle or figure
- set of all points in a plane that are the same distance from a given point
- angle that measures between 90° and 180°

7 letters
- tool used to draw circles and arcs
- polygon with 10 sides
- type of triangle with no congruent sides
- polygon with all sides and angles congruent

8 letters
- point that divides a segment into two segments of equal length
- polygon with 5 sides

9 letters
- sides that have the same length
- type of triangle with at least two sides congruent
- parallelogram with four right angles
- quadrilateral with exacly one pair of parallel sides

11 letters
- triangle with three congruent sides

12 letters
- lines that have exactly one point in common

13 letters
- two angles whose sum is 180°
- two angles whose sum is 90°

Chapter 7 Review

Vocabulary Review

acute angle (p. 330)
acute triangle (p. 337)
adjacent angles (p. 331)
angle (p. 330)
arc (p. 351)
central angle (p. 350)
chord (p. 350)
circle (p. 350)
circle graph (p. 354)
compass (p. 361)
complementary (p. 331)
congruent angles (p. 331)
congruent polygons (p. 346)
congruent sides (p. 336)
decagon (p. 340)
diameter (p. 350)
equilateral triangle (p. 336)
hexagon (p. 340)
intersecting lines (p. 325)
irregular polygon (p. 340)
isosceles triangle (p. 336)
line (p. 324)
midpoint (p. 362)
obtuse angle (p. 330)
obtuse triangle (p. 337)
octagon (p. 340)
parallel lines (p. 325)
parallelogram (p. 341)
pentagon (p. 340)
perpendicular bisector (p. 362)
perpendicular lines (p. 362)
plane (p. 325)
point (p. 324)
quadrilateral (p. 340)
radius (p. 350)
ray (p. 324)
rectangle (p. 341)
regular polygon (p. 340)
rhombus (p. 341)
right angle (p. 330)
right triangle (p. 337)
scalene triangle (p. 336)
segment (p. 324)
segment bisector (p. 362)
semicircle (p. 351)
skew lines (p. 325)
square (p. 341)
straight angle (p. 330)
supplementary (p. 331)
trapezoid (p. 341)
vertex (p. 330)
vertical angles (p. 331)

Go Online PHSchool.com
For: Online vocabulary quiz
Web Code: arj-0751

Choose the correct term to complete each sentence.

1. (Parallel, Skew) lines lie in the same plane. **parallel**
2. A (decagon, pentagon) is a polygon with five sides. **pentagon**
3. Angles whose sum is 180° are (complementary, supplementary). **supplementary**
4. A(n) (isosceles, scalene) triangle has no congruent sides. **scalene**
5. An (acute, obtuse) angle measures less than 90°. **acute**

Skills and Concepts

Lessons 7-1, 7-2

- To identify segments, rays, and lines
- To classify angles and to work with pairs of angles

A **plane** is a flat surface that extends indefinitely in all directions and has no thickness. **Parallel lines** are lines that lie in the same plane and have no points in common. **Skew lines** do not lie in the same plane.

An **angle** is formed by two rays with a common endpoint. The sum of two **complementary** angles is 90°. The sum of two **supplementary** angles is 180°.

Find the complement and the supplement of each angle.

6. $m\angle A = 55°$ **35°, 125°**
7. $m\angle B = 27°$ **63°, 153°**
8. $m\angle C = 87°$ **3°, 93°**
9. $m\angle D = 12°$ **78°, 168°**

Lessons 7-3, 7-4

- To classify triangles and to find the angle measures of triangles
- To classify polygons and special quadrilaterals

You can classify a triangle by the measure of its sides and angles. The sum of the measures of the angles of any triangle is 180°. A **polygon** is classified by the number of sides it has. A **regular polygon** has congruent sides and congruent angles.

Find the value of x. Then classify each triangle by its sides and angles.

10. 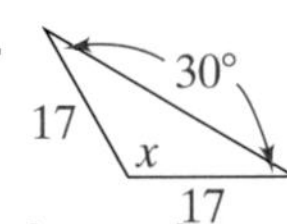

120°; isosceles; obtuse

11. 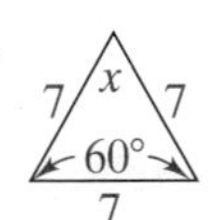

60°; equilateral; acute

12. 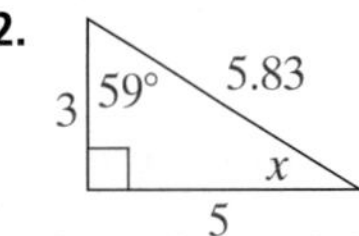

31°; scalene; right

Identify each polygon and classify it as *regular* or *irregular.*

13. **pentagon; regular**

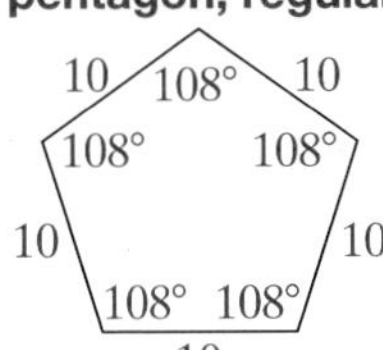

14. **square, regular**

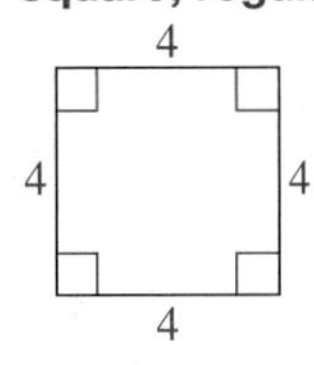

15. **octagon, irregular**

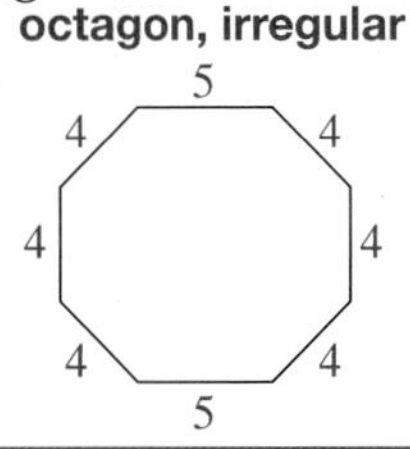

16. $m\angle M = 30°$, $m\angle C = m\angle P = 46°$, $m\angle B = 104°$, $AC = 22$, $BC = 12$, $MN = 17$

Lesson 7-5

- To identify congruent figures and to use them to find missing measures

Congruent polygons have the same size and shape. **Corresponding parts** of congruent polygons are congruent.

16. In the diagram, $\triangle ABC \cong \triangle MNP$. Find the missing side lengths and angle measures for the figures. **See left.**

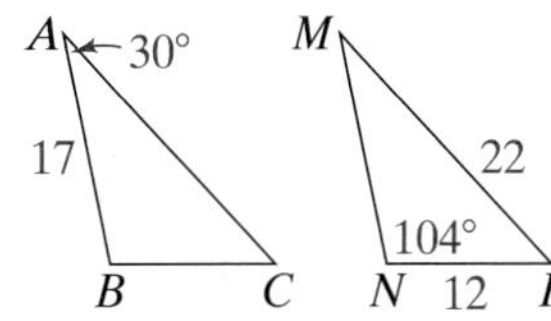

Lessons 7-6, 7-7

- To identify parts of a circle
- To analyze and construct circle graphs

A **circle** is the set of points in a plane that are all the same distance from the center. A circle can have **radii, diameters, central angles, chords,** and **arcs. Circle graphs** present data as percents or fractions of a total.

Name each of the following for circle T.

17. radii **$\overline{TW}, \overline{TY}, \overline{TK}, \overline{TM}$**

18. diameters **$\overline{MY}$**

19. center **T**

20. chords **$\overline{WY}$; $\overline{KY}$** $\overline{MY}$

21. arcs longer than a semicircle **$\overset{\frown}{WYM}$, $\overset{\frown}{YKW}$, $\overset{\frown}{KMY}$, $\overset{\frown}{MWK}$, $\overset{\frown}{WYK}$**

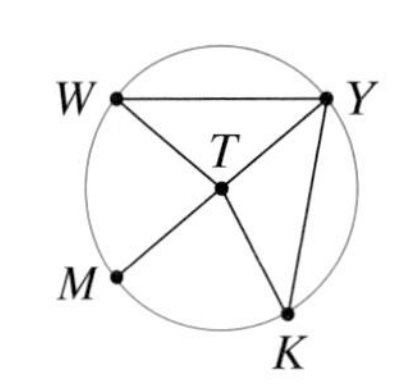

Lesson 7-8

- To construct congruent segments and perpendicular bisectors

You can use a **compass** to construct congruent segments and **perpendicular bisectors.** **22–23. See margin.**

22. Draw a segment. Construct a segment congruent to it.

23. Draw a segment. Construct its perpendicular bisector.

22. A B C D

23.

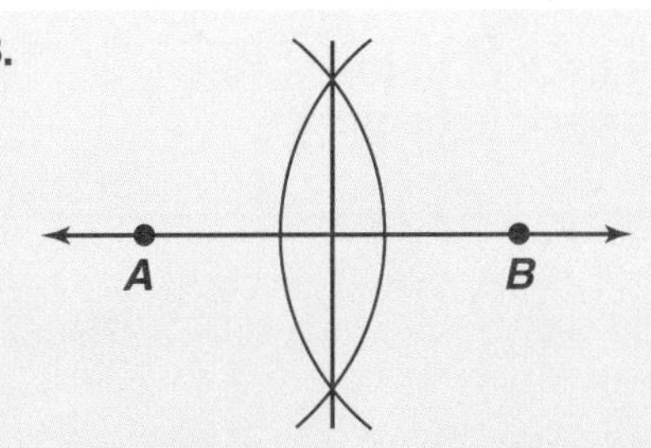

Alternative Assessment L4

Alternative Assessment

Chapter 7

POL ONS TO WEAR

An art class is making T-shirt designs. The teacher has given the following guidelines:

- Create your design by repeating *one regular polygon.*
- The polygons that make up your design must completely cover the design area; there may not be any space between the polygons. In other words, the sides of each polygon must meet exactly with the sides of other polygons.
- Do not worry about the edges of the patterns because fabrics will be cut where necessary.

Show all of your work on a separate sheet of paper.

1. Tamika drew the design shown below. On dot paper, make a *different* design, using the same type of polygon that Tamika used.

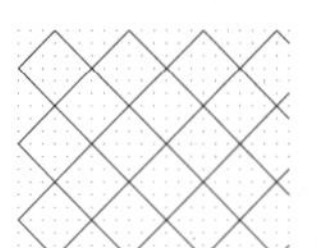

2. Travis decided to pattern his design after Tamika's by adding vertical lines to her pattern, as shown below. Did Travis's design meet the teacher's requirements? Explain.

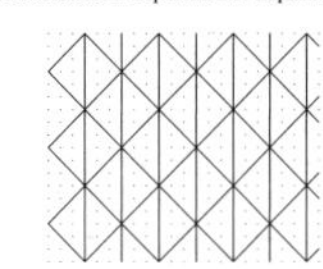

Resources

- ExamView Assessment Suite CD-ROM
  - Ch. 7 Ready-Made Test
  - Make your own Ch. 7 test
- MindPoint Quiz Show CD-ROM
  - Chapter 7 Review

Differentiated Instruction

All-in-One Teaching Resources
- Below Level Chapter 7 Test L2
- Chapter 7 Test L3
- Chapter 7 Alternative Assessment L4

Spanish Assessment Resources ELL
- Below Level Chapter 7 Test L2
- Chapter 7 Test L3
- Chapter 7 Alternative Assessment L4

ExamView Assessment Suite CD-ROM
- Special Needs Test L1
- Special Needs Practice Bank L1

Online Chapter 7 Test at www.PHSchool.com L3

Below Level Chapter Test L2

Chapter Test L3

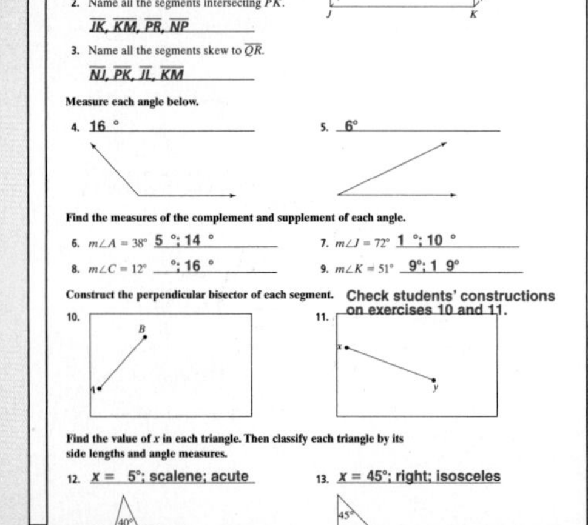
Chapter Test Form A

Chapter 7

Use the diagram to the right for excercises 1–3.

1. Name all the segments parallel to $\overline{NP}$. $\overline{JK}$, $\overline{QR}$, $\overline{LM}$
2. Name all the segments intersecting $\overline{PK}$. $\overline{JK}$, $\overline{KM}$, $\overline{PR}$, $\overline{NP}$
3. Name all the segments skew to $\overline{QR}$. $\overline{NJ}$, $\overline{PK}$, $\overline{JL}$, $\overline{KM}$

Measure each angle below.

4. 16 °
5. 6°

Find the measures of the complement and supplement of each angle.

6. $m\angle A = 38°$ 5 °; 14 °
7. $m\angle J = 72°$ 1 °; 10 °
8. $m\angle C = 12°$ °; 16 °
9. $m\angle K = 51°$ 9°; 1 9°

Construct the perpendicular bisector of each segment. Check students' constructions on exercises 10 and 11.

10.

11.

Find the value of x in each triangle. Then classify each triangle by its side lengths and angle measures.

12. $x =$ 5°; scalene; acute
13. $x =$ 45°; right; isosceles

Chapter 7 Test

Go Online PHSchool.com For: Online chapter test Web Code: ara-0752

Use the diagram for Exercises 1–3.

1. Name all the segments parallel to $\overline{AB}$. **$\overline{MH}$, $\overline{GC}$, $\overline{FD}$**
2. Name all the segments intersecting $\overline{FG}$. **$\overline{MF}$, $\overline{FD}$, $\overline{AG}$, $\overline{GC}$**
3. Name all the segments skew to $\overline{DH}$. **$\overline{GC}$, $\overline{AB}$, $\overline{MA}$, $\overline{FG}$**

Use a protractor to measure each angle.

4. **40°**
5. **100°**

Find the measures of the complement and the supplement of each angle.

6. $m\angle H = 45°$ **45°, 135°**
7. $m\angle R = 7°$ **83°, 173°**
8. $m\angle K = 89°$ **1°, 91°**
9. $m\angle P = 25°$ **65°, 155°**
10. What is the supplement of a 102° angle? **a 78° angle**
11. Does a 98° angle have a complement? Explain. **No, it would have to be a −8° angle.**
12. Draw a segment. Construct its perpendicular bisector. **See margin.**
13. Draw a segment $\overline{AB}$. Construct a congruent segment $\overline{CD}$. **See margin.**

Find the value of x in each triangle. Classify each triangle by its side lengths and its angle measures.

14. 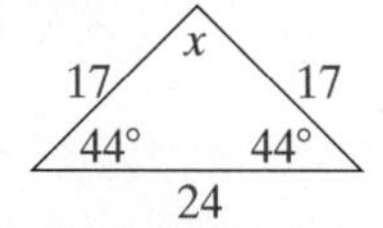

92°; isosceles; obtuse

15. **150°; scalene; obtuse**

Identify each polygon.

16. 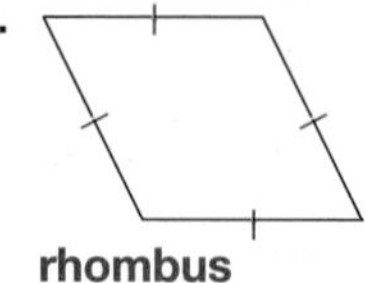
rhombus

17. 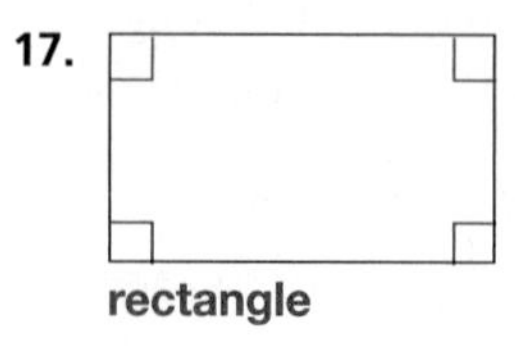
rectangle

18. What is the name of a polygon with one pair of parallel sides? **trapezoid**

In the diagram below, $\triangle ABC \cong \triangle MNP$. Complete each statement.

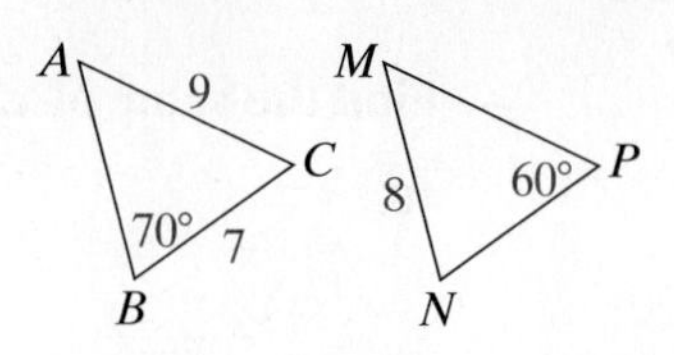

19. $\overline{AC} \cong$ ■ **$\overline{MP}$**
20. $\angle P \cong$ ■ **$\angle C$**
21. ■ $\cong \overline{PN}$ **$\overline{CB}$**
22. $\angle B \cong$ ■ **$\angle N$**
23. ■ $\cong \angle M$ **$\angle A$**
24. $AB =$ ■ **8**
25. $m\angle N =$ ■ **70°**
26. $MP =$ ■ **9**

Name each of the following for circle W.

27. three radii **$\overline{WA}$, $\overline{WT}$, $\overline{WM}$**
28. two chords **Any two of $\overline{AC}$, $\overline{TM}$, $\overline{AM}$**
29. two central angles **$\angle AWT$, $\angle TWM$**
30. one diameter **$\overline{AM}$**
31. three arcs shorter than half the circle **any three of $\overset{\frown}{AC}$, $\overset{\frown}{CM}$, $\overset{\frown}{MT}$, $\overset{\frown}{TA}$**
32. **Class Trip** Students earned the following amounts of money to pay the transportation costs of a class trip. Make a circle graph for the data. **See margin.**

Fundraiser	Money
Car wash	$150
Paper drive	$75
Book sale	$225
Food stand	$378

33. **Writing in Math** Briefly explain the differences and similarities among a rectangle, a rhombus, and a square. **See margin.**

12–13. See back of book.
32–33. See back of book.

Reading Comprehension

Read each passage and answer the questions that follow.

Energetic Math The amount of energy that Americans use each year varies greatly from state to state. People in Alaska use about 1,143.7 million BTU (British thermal units) per person. People in Hawaii use about 200.9 million BTU per person. People in Texas, Ohio, Vermont, and New York use about 587.8 million BTU, 370.2 million BTU, 283.8 million BTU, and 225.5 million BTU per person, respectively.

1. Energy consumption in Hawaii is, on average, about what percent of the energy consumption in Alaska? A

   (A) 18% (B) 21% (C) 25% (D) 550%

2. Energy consumption in Texas is, on average, about what percent of the consumption in Vermont? J

   (F) 50% (H) 150%
   (G) 100% (J) 200%

3. People moving from New York to Ohio might expect their energy consumption to increase by about what percent? C

   (A) 40% (C) 64%
   (B) 61% (D) 164%

4. For the states mentioned in the passage, what is the median energy consumption per person in millions of BTU? G

   (F) 320 (G) 327 (H) 389 (J) 469

The Value of Education People with more education generally earn more money. On average, college graduates make about $21 per hour. If you begin college, but don't finish, you can expect an average of about $13 per hour. High school graduates with no college earn about $11 per hour, and those who don't finish high school average about $8 per hour.

5. For the average earnings of high school dropouts compared to earnings of college graduates, which expression could you use to find the percent of increase? A

   (A) $\frac{(21-8)}{8}$ (C) $\frac{(21-11)}{11}$
   (B) $\frac{(13-11)}{11}$ (D) $\frac{(21-11)}{21}$

6. Which is the best estimate of how much the average college graduate earns in a year? (Use 8 hours per day, 5 days per week.) H

   (F) $8,400 (H) $42,000
   (G) $21,000 (J) $68,000

7. A person who drops out of high school will earn, on average, about what percent of the earnings of someone who completes high school but does not go to college? C

   (A) 40% (C) 73%
   (B) 62% (D) 138%

8. A job pays $546 for 40 hours of work in one week. What percent is this of the average wage for a person who started college but didn't finish? G

   (F) 95% (H) 155%
   (G) 105% (J) 215%

Resources

Test Prep Workbook

All-in-One Teaching Resources
- Cumulative Review L3

ExamView Assessment Suite CD-ROM
- Standardized Test Practice

Differentiated Instruction

Spanish Assessment Resources
- Spanish Cumulative Review ELL

ExamView Assessment Suite CD-ROM
- Special Needs Practice Bank L1

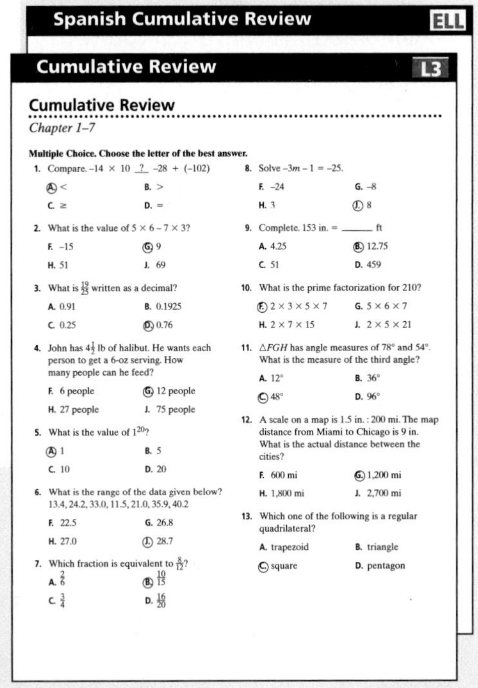

Spanish Cumulative Review ELL

Cumulative Review L3

Cumulative Review

Chapter 1–7

Multiple Choice. Choose the letter of the best answer.

1. Compare. -14×10 __?__ $-28 + (-102)$
   (A) < B. > C. ≥ D. =
2. What is the value of $5 \times 6 - 7 \times 3$?
   F. −15 (G) 9 H. 51 J. 69
3. What is $\frac{19}{25}$ written as a decimal?
   A. 0.91 B. 0.1925 C. 0.25 (D) 0.76
4. John has $4\frac{1}{2}$ lb of halibut. He wants each person to get a 6-oz serving. How many people can he feed?
   F. 6 people (G) 12 people H. 27 people J. 75 people
5. What is the value of 1^{20}?
   (A) 1 B. 5 C. 10 D. 20
6. What is the range of the data given below? 13.4, 24.2, 33.0, 11.5, 21.0, 35.9, 40.2
   F. 22.5 G. 26.8 H. 27.0 (J) 28.7
7. Which fraction is equivalent to $\frac{8}{12}$?
   A. $\frac{2}{6}$ (B) $\frac{10}{15}$ C. $\frac{3}{4}$ D. $\frac{16}{20}$
8. Solve $-3m - 1 = -25$.
   F. −24 G. −8 H. 3 (J) 8
9. Complete. 153 in. = ______ ft
   A. 4.25 (B) 12.75 C. 51 D. 459
10. What is the prime factorization for 210?
    (F) $2 \times 3 \times 5 \times 7$ G. $5 \times 6 \times 7$ H. $2 \times 7 \times 15$ J. $2 \times 5 \times 21$
11. $\triangle FGH$ has angle measures of 78° and 54°. What is the measure of the third angle?
    A. 12° B. 36° (C) 48° D. 96°
12. A scale on a map is 1.5 in. : 200 mi. The map distance from Miami to Chicago is 9 in. What is the actual distance between the cities?
    F. 600 mi (G) 1,200 mi H. 1,800 mi J. 2,700 mi
13. Which one of the following is a regular quadrilateral?
    A. trapezoid B. triangle (C) square D. pentagon

Problem Solving

Applying Geometry

Students will use data from these two pages to answer the questions posed in Put It All Together.

Invite students who have played miniature golf to describe the game. Ask:

- *What equipment do you need to play the game?* **a golf club, a golf ball, a score card**
- *How do you play?* **Answers will vary. Starting with the ball on the tee, you use a club to hit the ball toward the hole. Wherever the ball stops is where you take your next shot. You keep taking shots until you hit the ball into the hole.**
- *How do you keep score?* **A course usually has 18 different holes. For each one, you record how many shots it takes you to get the ball into the hole. At the end of the course, you add up your total for all the holes. The person with the lowest score wins.**

Discuss the differences and similarities between golf and miniature golf.

Activating Prior Knowledge

Have students who have played memorable miniature golf courses describe what about the courses made them so much fun to play. Ask them to describe what they like most and least about miniature golf.

Guided Instruction

Have a volunteer read the opening paragraph. Ask students to describe some of the more unusual miniature golf course obstacles they have encountered. Ask:

- *What kind of obstacles do miniature golf courses have?*
- *What's the most unusual obstacle you've encountered?*

Problem Solving Application

Applying Geometry

Golf Course Math A good miniature golf course should be challenging and creative. The best courses have some clever twist to delight even the most experienced players. For example, consider a course where you tee off right next to the hole and have to go all the way around, avoiding obstacles, to finish where you started.

Put It All Together

Materials ruler, protractor

1. Design a hole for a miniature golf course so that the tee (start) and the hole (finish) are right next to each other, and you end where you began. Shape the hole and arrange obstacles so it takes 5 strokes to play.
2. On your hole diagram, use a ruler to draw the path the ball might travel from the tee around the obstacles and back to the hole. Mark the starting (and ending) point A and label the others B, C, D, and E. What is the name of the polygon you drew?

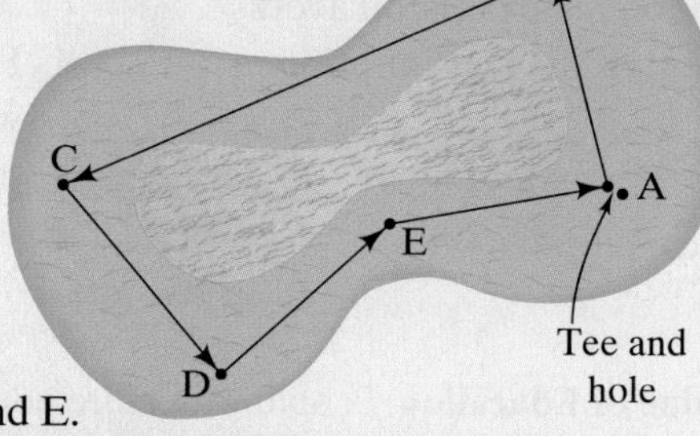

3. Use a protractor to measure each of the internal angles of the course. Find the sum of the internal angles of your polygon.
4. Draw a different five-stroke path that the ball could follow. Find the sum of the internal angles. How does your answer compare with the first total? How does it compare with the totals that other students in your class are getting for their courses?
5. a. Draw two lines from one of the vertices to the two other non-adjacent vertices (for example, by connecting B to D and B to E). Make sure your lines stay inside the polygon. How many triangles did you make?
   b. Recall that the sum of the measures of the interior angles of one triangle is 180°. Calculate the sum of the measures of the interior angles of your polygon.

Sand Trap

Hole

Miniature Golfing
There are about 150 professional miniature golfers in the United States.

1. **Check students' work.**
2. **Check students' work; pentagon**
3. **Check students' work; 540°**
4. **Check students' work; 540°, both are the same; same as classmates**

5a. **three**

b. **$3 \cdot 180° = 540°$**

Water hazard
Miniature Golf in America
Americans play about 11,550,000 rounds of miniature golf each year at an average cost of $4.25 per round.
Green
Putting green
Go Online
PHSchool.com
For: Information about miniature golf
Web Code: are-0753
371

Activity

Have students work in pairs to answer the questions and design their miniature golf holes.

Diversity
Miniature golf courses are often designed with themes in mind. Have a class discussion of the themes they have seen used. Then invite pairs to "design" their own course based upon a theme of their choice. They need not lay out the whole course, but they should describe its theme-related features. Encourage each student to use imagination and a sense of humor.

Art Connection
Invite students to construct a scale model of a miniature golf hole. Suggest that, if possible, students should visit a local course to get ideas about course lay-outs and materials used.

Differentiated Instruction

Special Needs L1
As needed, help students to correctly use a protractor and to understand any unfamiliar terms used in the activity.

8 Measurement

Chapter at a Glance

Lesson Titles, Objectives, and Features	Assessment	NCTM Standards	Local Standards
8-1 Estimating Perimeter and Area • To estimate length, perimeter, and area.	Lesson Quiz	3, 4, 6, 7, 8, 9, 10	
8-2a Activity Lab, Hands On: Generating Formulas for Area **8-2 Area of a Parallelogram** • To find the area and perimeter of a parallelogram	Lesson Quiz	3, 4, 6, 7, 8, 9, 10	
8-3 Perimeter and Area of a Triangle • To find the perimeter and area of a triangle	Lesson Quiz	3, 4, 6, 7, 8, 9, 10	
8-4 Areas of Other Figures • To find the area of a trapezoid and the areas of irregular figures	Lesson Quiz Checkpoint Quiz 1	3, 4, 6, 7, 8, 9, 10	
8-5a Activity Lab, Hands On: Modeling a Circle **8-5 Circumference and Area of a Circle** • To find the circumference and area of a circle **Guided Problem Solving:** Areas of Irregular Figures	Lesson Quiz	3, 4, 6, 7, 8, 9, 10	
8-6 Square Roots and Irrational Numbers • To find and estimate square roots and to classify numbers as rational or irrational	Lesson Quiz	3, 4, 6, 7, 8, 9, 10	
8-7a Activity Lab, Hands On: Exploring Right Triangles **8-7 The Pythagorean Theorem** • To use the Pythagorean Theorem to solve real-world problems	Lesson Quiz	2, 3, 4, 6, 7, 8, 9, 10	
8-8a Activity Lab, Hands On: Three Views of an Object **8-8 Three-Dimensional Figures** • To classify and draw three-dimensional figures	Lesson Quiz	3, 4, 6, 7, 8, 9, 10	
8-9 Surface Areas of Prisms and Cylinders • To find the surface areas of prisms and cylinders using nets **Extension:** Patterns in Three-Dimensional Figures	Lesson Quiz Checkpoint Quiz 2	3, 4, 6, 7, 8, 9, 10	
8-10 Volumes of Prisms and Cylinders • To find the volumes of prisms and cylinders **8-10b Activity Lab, Hands On:** Generating Formulas for Volume	Lesson Quiz	3, 4, 6, 7, 8, 9, 10	
Problem Solving Applications: Applying Volume			

NCTM Standards 2000
1 Number and Operations **2** Algebra **3** Geometry **4** Measurement **5** Data Analysis and Probability
6 Problem Solving **7** Reasoning and Proof **8** Communication **9** Connections **10** Representation

Correlations to Standardized Tests

All content for these tests is contained in *Prentice Hall Math,* Course 2. This chart reflects coverage in this chapter only.

	8-1	8-2	8-3	8-4	8-5	8-6	8-7	8-8	8-9	8-10
Terra Nova CAT6 (Level 17)										
Number and Number Relations										
Computation and Numerical Estimation										
Operation Concepts										
Measurement	✔	✔	✔	✔	✔	✔	✔	✔	✔	✔
Geometry and Spatial Sense	✔	✔	✔	✔	✔	✔	✔	✔	✔	✔
Data Analysis, Statistics, and Probability										
Patterns, Functions, Algebra										
Problem Solving and Reasoning	✔	✔	✔	✔	✔	✔	✔	✔	✔	✔
Communication	✔	✔	✔	✔	✔	✔	✔	✔	✔	✔
Decimals, Fractions, Integers, and Percent	✔	✔	✔	✔	✔	✔	✔	✔	✔	✔
Order of Operations										
Terra Nova CTBS (Level 17)										
Decimals, Fractions, Integers, Percents										
Order of Operations, Numeration, Number Theory										
Data Interpretation										
Pre-algebra							✔			
Measurement	✔	✔	✔	✔	✔	✔	✔	✔	✔	✔
Geometry	✔	✔	✔	✔	✔	✔	✔	✔	✔	✔
ITBS (Level 13)										
Number Properties and Operations										
Algebra										
Geometry	✔	✔	✔	✔	✔	✔	✔	✔	✔	✔
Measurement	✔	✔	✔	✔	✔	✔	✔	✔	✔	✔
Probability and Statistics										
Estimation	✔									
SAT10 (Int 3 Level)										
Number Sense and Operations										
Patterns, Relationships, and Algebra										
Data, Statistics, and Probability										
Geometry and Measurement	✔	✔	✔	✔	✔	✔	✔	✔	✔	✔
NAEP										
Number Sense, Properties, and Operations										
Measurement	✔	✔	✔	✔	✔				✔	✔
Geometry and Spatial Sense							✔	✔		
Data Analysis, Statistics, and Probability										
Algebra and Functions										

CAT6 California Achievement Test, 6th Ed.
SAT10 Stanford Achievement Test, 10th Ed.
CTBS Comprehensive Test of Basic Skills
NAEP National Assessment of Educational Progress 2005 Mathematics Objectives
ITBS Iowa Test of Basic Skills, Form M

Math Background

Skills Trace

BEFORE Chapter 8

Course 1 introduced area and volume for two and three-dimensional figures.

DURING Chapter 8

Course 2 reviews and extends the study of area of two-dimensional figures and surface area and volume for three-dimensional figures.

AFTER Chapter 8

Throughout this course, students use area and volume to solve real-world problems.

8-1 Estimating Perimeter and Area

Math Understandings

- You can estimate the perimeter of a figure in units and its area in square units.

8-2 Area of Parallelogram
8-3 Perimeter and Area of a Triangle
8-4 Areas of Other Figures

Math Understandings

- You can find the area of an irregular figure by separating it into familiar figures, finding the area of each smaller figure, and adding the areas together.
- Any side of a parallelogram or triangle can be its base.

Area of a Parallelogram
The area of a parallelogram is equal to the product of any base b and the corresponding height h.

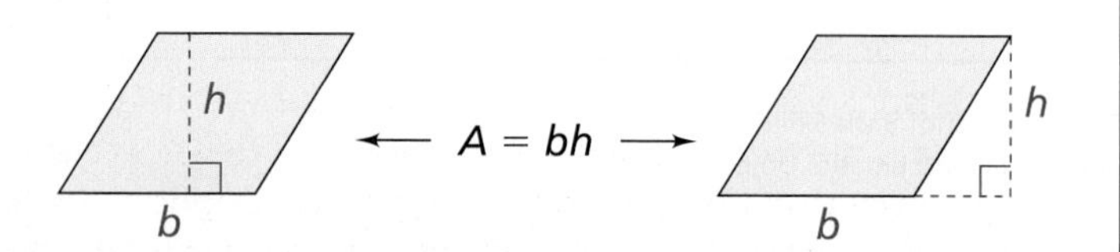

The **height of a parallelogram** is the perpendicular distance from one **base of a parallelogram** to the other. Any side of a triangle can be considered the **base of a triangle**. The **height of a triangle** is the length of the perpendicular segment from a vertex to the base opposite the vertex.

Area of a Triangle	Area of a Trapezoid
The area of a triangle is equal to half the product of any base and the corresponding height.	The area of a trapezoid is one half the product of the height and the sum of the lengths of the bases.
$A = \frac{1}{2}bh$	$A = \frac{1}{2}h(b_1 + b_2)$

The two parallel sides of a trapezoid are the **bases**, with lengths b_1 and b_2. The **height** h is the length of a perpendicular segment connecting the bases.

8-5 Circumference and Area of a Circle

Math Understandings

- Although π does not equal $\frac{22}{7}$ or 3.14, you can use these numbers as good approximations for the value of π.
- For every circle, the ratio of the circumference to the diameter is the same.
- The circumference of a circle is about three times its diameter or six times its radius.

Circumference is the distance around a circle. *Pi* (π) is a nonterminating, nonrepeating decimal (irrational number), the ratio of the circumference to its diameter.

Circumference of a Circle	Area of a Circle
The circumference of a circle is π times the diameter.	The area of a circle is π times the square of the radius r.
$C = \pi d = 2\pi r$	$A = \pi r^2$

8-6 Square Roots and Irrational Numbers
8-7 The Pythagorean Theorem

Math Understandings

- Mathematicians have agreed that the symbol $\sqrt{\ }$ indicates the nonnegative square root of a number. So, while $x^2 = 9$ has two roots, $+3$ and -3, the symbol $\sqrt{9}$ has only one meaning: $+3$.

A number that is the square of an integer is a **perfect square**. The inverse of squaring a number is finding a **square root**. If a positive integer is not a perfect square, its square root is *irrational*. An **irrational number** is a number that cannot be written as a ratio of two integers. As decimals, irrational numbers neither terminate nor repeat.
In a right triangle, the two shortest sides are **legs**. The side opposite the right angle is the **hypotenuse**.

Pythagorean Theorem	
The Pythagorean Theorem states that, in any right triangle, the sum of the squares of the lengths of the legs is equal to the square of the length of the hypotenuse.	$a^2 + b^2 = c^2$

8-8 Three-Dimensional Figures
8-9 Surface Areas and Volumes of
8-10 Rectangular Prisms and Cylinders

Math Understandings

- Prisms and pyramids have faces that are flat surfaces. Cones, cylinders, and spheres have at least one surface that is curved.
- One way to find the surface area of a three-dimensional solid is to find the area of its net.
- You can find the volume of a rectangular prism and a cylinder by using the same formula, $V = B \cdot h$, where B is the area of the base and h is height.

A **three-dimensional figure**, or solid, is a figure that does not lie in a plane. A flat surface shaped like a polygon is called a **face**. Each segment formed by the intersection of two faces is an **edge**. A **prism** is a three-dimensional figure with two parallel and congruent polygonal faces, called **bases**. The other faces are rectangles. The **height** of a prism is the length of a perpendicular segment that joins the bases. A **cube** is a rectangular prism with square faces. A **cylinder** has two congruent parallel **bases** that are circles. A cylinder's **height** is the length of a perpendicular segment that joins the bases.

A **pyramid** is a three-dimensional figure with triangular faces that meet at one point, a **vertex**, and a base that is a polygon. A **cone** has one circular **base** and one **vertex**. A **sphere** is the set of all points in space that are the same distance from a **center** point.

A **net** is a two-dimensional pattern that you can fold to form a three-dimensional figure. The **surface area** of a prism is the sum of the areas of its faces. The **volume** of a three-dimensional figure is the number of cubic units needed to fill the space.

Volume of a Rectangular Prism	Volume of a Cylinder
V = area of base • height $= Bh$ $= lwh$	V = area of base • height $= Bh$ $= \pi r^2 h$

Additional Professional Development Opportunities

Math Background Notes for Chapter 8: Every lesson has a Math Background in the PLAN section.

Research Overview, Mathematics Strands
Additional support for these topics and more is in the front of the Teacher's Edition.

LessonLab
LessonLab, a Pearson Education company, offers comprehensive, facilitated professional development designed to help teachers to improve student achievement. To learn more, please visit lessonlab.com.

Chapter 8 Resources

	8-1	8-2	8-3	8-4	8-5	8-6	8-7	8-8	8-9	8-10	For the Chapter
Print Resources											
L3 Practice	●	●	●	●	●	●	●	●	●	●	
L1 Adapted Practice	●	●	●	●	●	●	●	●	●	●	
L3 Guided Problem Solving	●	●	●	●	●	●	●	●	●	●	
L2 Reteaching	●	●	●	●	●	●	●	●	●	●	
L4 Enrichment	●	●	●	●	●	●	●	●	●	●	
L3 Daily Notetaking Guide	●	●	●	●	●	●	●	●	●	●	
L1 Adapted Daily Notetaking Guide	●	●	●	●	●	●	●	●	●	●	
L3 Vocabulary and Study Skills Worksheets	●		●			●		●		●	●
L3 Daily Puzzles	●	●	●	●	●	●	●	●	●	●	
L3 Activity Labs	●	●	●	●	●	●	●	●	●	●	
L3 Checkpoint Quiz				●					●		
L3 Chapter Project											●
L2 Below Level Chapter Test											●
L3 Chapter Test											●
L4 Alternative Assessment											●
L3 Cumulative Review											●
Spanish Resources ELL											
L3 Practice	●	●	●	●	●	●	●	●	●	●	●
L3 Vocabulary and Study Skills Worksheets	●		●			●		●		●	●
L3 Checkpoint Quiz				●					●		
L2 Below Level Chapter Test											●
L3 Chapter Test											●
L4 Alternative Assessment											●
L3 Cumulative Review											●
Transparencies											
Check Skills You'll Need	●	●	●	●	●	●	●	●	●	●	
Additional Examples	●	●	●	●	●	●	●	●	●	●	
Problem of the Day	●	●	●	●	●	●	●	●	●	●	
Classroom Aid							●	●			
Student Edition Answers	●	●	●	●	●	●	●	●	●	●	●
Lesson Quiz	●	●	●	●	●	●	●	●	●	●	
Test-Taking Strategies											●
Technology											
Interactive Textbook Online	●	●	●	●	●	●	●	●	●	●	●
StudentExpress™ CD-ROM	●	●	●	●	●	●	●	●	●	●	●
Success Tracker™ Online Intervention	●	●	●	●	●	●	●	●	●	●	●
TeacherExpress™ CD-ROM	●	●	●	●	●	●	●	●	●	●	●
PresentationExpress™ with QuickTake Presenter CD-ROM	●	●	●	●	●	●	●	●	●	●	●
ExamView® Assessment Suite CD-ROM	●	●	●	●	●	●	●	●	●	●	●
MindPoint® Quiz Show CD-ROM											●
Prentice Hall Web Site: PHSchool.com	●	●	●	●	●	●	●	●	●	●	●

Also available: **Prentice Hall Assessment System**
- Progress Monitoring Assessments
- Skills and Concepts Review
- Test Prep Workbook

Other Resources
Algebra Readiness Tests
All-in-One Student Workbook
All-in-One Student Workbook, Adapted Version
Multilingual Handbook
Solution Key
Math Notes Study Folder
Spanish Cumulative Assessment

Where You Can Use the Lesson Resources

Here is a suggestion, following the four-step teaching plan, for how you can incorporate Differentiated Instruction resources into your teaching.

	Instructional Resources L3	**Differentiated Instruction Resources**
1. Plan		
Preparation Read the Math Background in the Teacher's Edition to connect this lesson with students' previous experience. **Starting Class** **Check Skills You'll Need** Assign these exercises to review prerequisite skills. **New Vocabulary** Help students pre-read the lesson by pointing out the new terms introduced in the lesson.	**Math Background** **Math Understandings** **Transparencies & PresentationExpress™ with QuickTake Presenter CD-ROM** Check Skills You'll Need Problem of the Day **Resources** Vocabulary and Study Skills	**Spanish Support** ELL Vocabulary and Study Skills
2. Teach		
L3 **Guided Instruction** Use the Activity Labs to build conceptual understanding. Teach each Example. Use the Teacher's Edition side column notes for specific teaching tips, including Error Prevention notes. Use the Additional Examples found in the side column (and on transparency and PowerPoint) as an alternative presentation for the content. After each Example, assign the Quick Check exercise for that Example to get an immediate assessment of student understanding. Use the Closure activity in the Teacher's Edition to help students attain mastery of lesson content.	**Student Edition** Activity Lab **Resources** Daily Notetaking Guide Activity Lab **Transparencies & PresentationExpress™ with QuickTake Presenter CD-ROM** Additional Examples Classroom Aids **ExamView® Assessment Suite CD-ROM**	**Teacher's Edition** Every lesson includes suggestions for working with students who need special attention. L1 Special Needs L2 Below Level L4 Advanced Learners ELL English Language Learners **Resources** L1 Adapted Daily Notetaking Guide **Multilingual Handbook**
3. Practice		
Assignment Guide **Check Your Understanding** Use these questions to check students' understanding before you assign homework. **Homework Exercises** Assign homework from these leveled exercises in the Assignment Guide. **A** Practice by Example **B** Apply Your Skills **C** Challenge Test Prep and Mixed Review **Homework Quick Check** Use these key exercises to quickly check students' homework.	**Transparencies & PresentationExpress™ with QuickTake Presenter CD-ROM** Student Answers **Resources** Practice Guided Problem Solving Vocabulary and Study Skills Activity Lab Daily Puzzles **ExamView® Assessment Suite CD-ROM**	**Spanish Support** ELL Practice ELL Vocabulary and Study Skills **Resources** L1 Adapted Practice L4 Enrichment
4. Assess & Reteach		
Lesson Quiz Assign the Lesson Quiz to assess students' mastery of the lesson content. **Checkpoint Quiz** Use the Checkpoint Quiz to assess student progress over several lessons.	**Transparencies & PresentationExpress™ with QuickTake Presenter CD-ROM** Lesson Quiz **Resources** Checkpoint Quiz	**Resources** L2 Reteaching ELL Checkpoint Quiz Success Tracker™ Online Intervention **ExamView® Assessment Suite CD-ROM**

KEY L1 Special Needs L2 Below Level L3 For All Students L4 Advanced, Gifted ELL English Language Learners

Geometry and Measurement

Answers are in the back of the textbook.

For intervention, direct students to:

Multiplying Decimals
Lesson 1-3
Extra Skills and Word Problems Practice, Ch. 1

Order of Operations
Lesson 2-1
Extra Skills and Word Problems Practice, Ch. 2

Changing Units in the Customary System
Lesson 3-6
Extra Skills and Word Problems Practice, Ch. 3

Finding the Measures of Angles in Right Triangles
Lesson 7-3
Extra Skills and Word Problems Practice, Ch. 7

Spanish Vocabulary/Study Skills ELL

Vocabulary/Study Skills L3

8A: Graphic Organizer — For use before Lesson 8-1

Study Skill Take a few minutes to relax before and after studying. Your mind will absorb and retain more information if you alternate studying with brief rest intervals.

Write your answers.

1. What is the chapter title? Measurement
2. How many lessons are there in this chapter? 10
3. What is the topic of the Test-Taking Strategies page? Measuring to Solve
4. Complete the graphic organizer below as you work through the chapter.
   - In the center, write the title of the chapter.
   - When you begin a lesson, write the lesson name in a rectangle.
   - When you complete a lesson, write a skill or key concept in a circle linked to that lesson block.
   - When you complete the chapter, use this graphic organizer to help you review.

Check students' diagrams.

CHAPTER 8 Measurement

What You've Learned

- In Chapter 5, you identified similar figures.
- In Chapter 7, you classified triangles, quadrilaterals, and other polygons.
- You identified the parts of a circle.

Check Your Readiness

GO for Help

For Exercises	See Lesson
1–2	1-3
3–5	2-1
6–9	3-6
10–12	7-3

Multiplying Decimals

Find each product.

1. $0.25 \cdot 3.14 \cdot 4$ **3.14** **2.** $3 \cdot 20.5 \cdot 2$ **123**

Order of Operations

Simplify.

3. $3^3 \cdot (8 - 6)^2$ **108** **4.** $(2^3 \cdot 5) - 6^2$ **4** **5.** $6^2 \cdot 2 + 5^2$ **97**

Changing Units in the Customary System

Complete.

6. 3 c = ■ fl oz **24** **7.** 12 ft = ■ in. **144** **8.** 48 oz = ■ lb **3** **9.** 96 in. = ■ ft **8**

Finding the Measures of Angles in Triangles

Algebra **Find the value of *x* in each triangle.**

10. **58°**

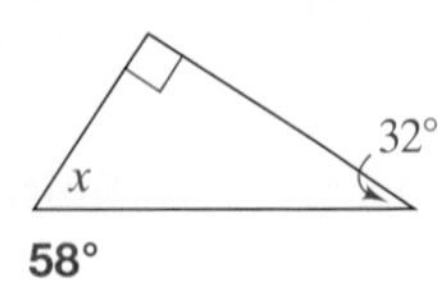

11. **63°**

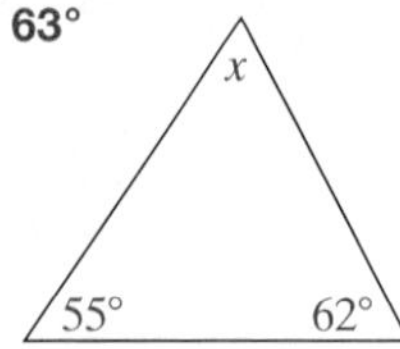

12. **50°**

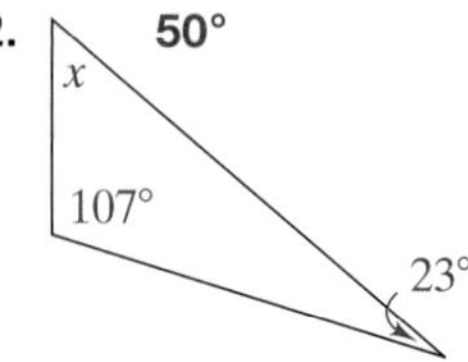

Chapter 8 Overview

In this chapter, students build on their knowledge of the geometric concepts from the previous chapter to estimate and find areas of triangles, parallelograms, and other polygons. They also find the circumference and area of circles, as well as the surface area and volume of rectangular prisms and cylinders. In addition, students learn about squares and square roots in preparation for using the Pythagorean Theorem to find lengths of sides of right triangles.

Activating Prior Knowledge

In this chapter, students will build on their knowledge of geometric concepts to compute areas and volumes of geometric figures. They use their understanding of the properties and characteristics of triangles when they apply the Pythagorean Theorem to solve problems.

Ask questions such as:

- *How many right angles are in a right triangle?* **one**
- *How many acute angles are in a right triangle?* **two**
- *What is the sum of the measures of the angles of any triangle?* **180°**

What You'll Learn Next

- In this chapter, you will find the areas of polygons, including triangles, parallelograms, and trapezoids.
- You will find the circumferences and areas of circles.
- You will classify three-dimensional figures and find their surface areas and volumes.

Problem Solving Application On pages 432 and 433, you will work an extended activity on volume.

Key Vocabulary

- area (p. 375)
- circumference (p. 394)
- cone (p. 411)
- cylinder (p. 410)
- hypotenuse (p. 405)
- irrational number (p. 401)
- net (p. 414)
- perimeter (p. 375)
- prism (p. 410)
- pyramid (p. 410)
- Pythagorean Theorem (p. 405)
- sphere (p. 411)
- square root (p. 400)
- surface area (p. 415)
- three-dimensional figure (p. 410)
- volume (p. 421)

8-1

1. Plan

Objective
To estimate length, perimeter, and area

Examples
1 Choosing Reasonable Estimates
2 Estimating Perimeter
3 Estimating Area
4 Using Area and Perimeter

Math Understandings: p. 372C

Math Background
Estimation is a fundamental number sense skill. A meaningful estimate for measurements, such as length and area, must include a unit label.

More Math Background: p. 372C

Lesson Planning and Resources
See p. 372E for a list of the resources that support this lesson.

PowerPoint

Bell Ringer Practice

Check Skills You'll Need
Use student page, transparency, or PowerPoint. For intervention, direct students to:
Changing Units in the Customary System
Lesson 3-6
Extra Skills and Word Problems Practice, Ch. 3

8-1 Estimating Perimeter and Area

Check Skills You'll Need

1. **Vocabulary Review** Name four units of length in the *customary system*. **inch, foot, yard, mile**

Complete.

2. 6 ft = ■ in. **72**
3. 48 in. = ■ ft **4**
4. 17 ft = ■ in. **204**

GO for Help Lesson 3-6

What You'll Learn
To estimate length, perimeter, and area

New Vocabulary area, perimeter

Why Learn This?

In some situations, an estimate of length or area will be enough to solve a problem. Painters estimate the area of a wall to make sure they buy enough paint.

A measurement must include a unit of measure to make sense. When you estimate, you can use familiar objects whose lengths you know.

length of a paper clip ≈ 1 inch

length of a textbook ≈ 1 foot

length of a baseball bat ≈ 1 yard

EXAMPLE Choosing Reasonable Estimates

Choose a reasonable estimate. Explain your choice.

a. the length of a new pencil: 9 in. or 9 ft

A new pencil is about 9 paper clips long. So 9 in. is reasonable.

b. the height of a flagpole: 10 in. or 10 yd

A flagpole is many baseball-bat lengths high. So 10 yd is reasonable.

Quick Check

1. Which is a reasonable estimate for the distance between Boston, Massachusetts, and Washington, D.C., 400 ft or 400 mi? Explain. **400 mi; driving distances are usually measured in miles.**

Differentiated Instruction Solutions for All Learners

Special Needs L1	**Below Level** L2
In Example 1, ask students to line up some paper clips next to their pencils to see about how many it takes to cover the length of their pencils. Ask them to hold 1 paper clip against a ruler. **learning style: tactile**	Review with students customary and metric measures (p. 148). Encourage students to use these tables for the lesson. **learning style: visual**

The **perimeter** of a figure is the total distance around the figure.

EXAMPLE Estimating Perimeter

GO for Help

The formula for perimeter P of a rectangle is
$P = 2\ell + 2w$ or
$P = 2(\ell + w)$.

For a list of formulas, go to p. 674.

2 Estimate the perimeter of the rectangle.

Estimate length and width.

The length is about 7 units. The width is about 5 units. →

Use the formula for perimeter of a rectangle.

$2(7 + 5) = 24$ ← **Substitute for ℓ and w.**

The perimeter is about 24 units long.

Quick Check

2. Estimate the perimeter of the rectangle.
Answers may vary. Sample: about 24 units

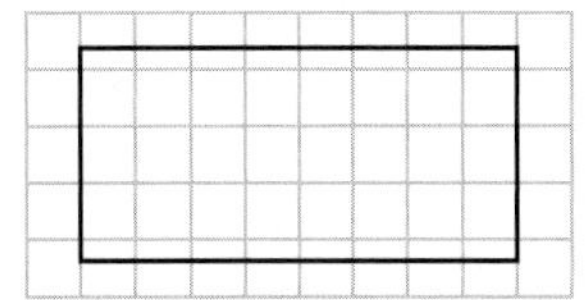

The **area** of a figure is the number of square units a figure encloses.

EXAMPLE Estimating Area

3 **Geography** Estimate the area of Lake Superior. Each square represents 900 mi^2.

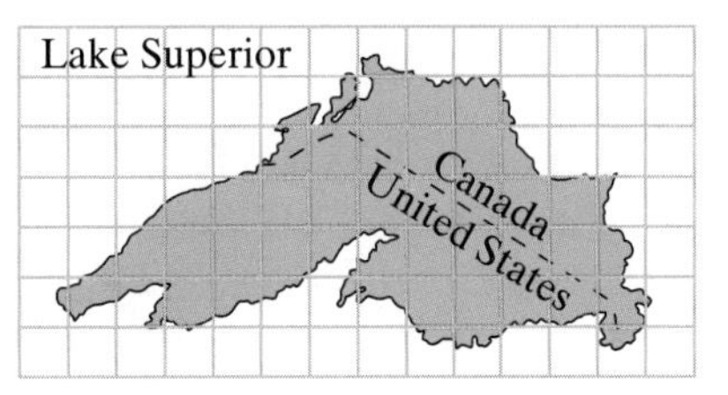

Count the number of squares filled or almost filled. Then count the number of squares that are about half filled.

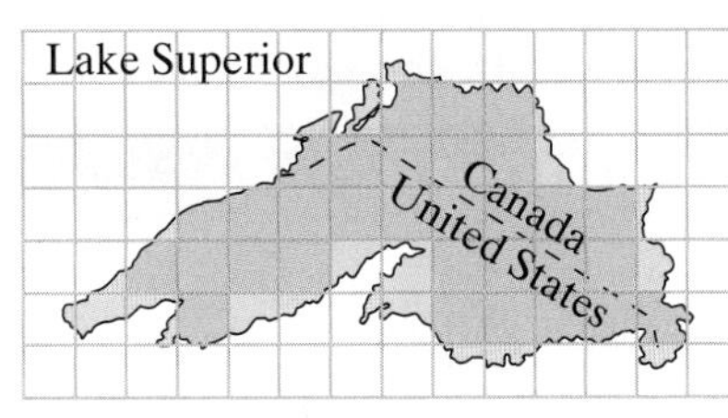

3	1
5	1
9	0
8	3
6	3
31	8

Add the filled squares and the half-filled squares. The total is $31 + \frac{1}{2}(8)$, or 35. Since each square represents 900 mi^2, multiply 35 by 900 mi^2.

The area is about 31,500 mi^2.

Quick Check

3. Estimate the area of the shaded region. Each square represents 4 yd^2.
Answers may vary. Sample: about 92 yd^2

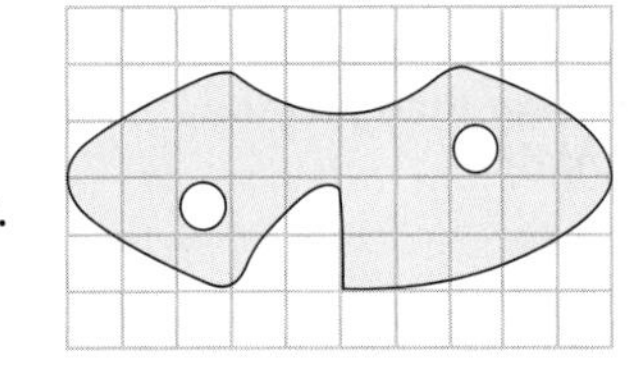

2. Teach

Activity Lab

Use before the lesson.

All in One Teaching Resources

Activity Lab 8-1: Estimating Area and Perimeter

Guided Instruction

Example 3
Use plain language to help students understand the notation for square units: ft^2 (square feet), yd^2 (square yards), and so on.

Additional Examples

1 Choose a reasonable estimate for the perimeter of a piece of paper: 40 in. or 40 ft. **40 in.**

2 Each square represents 1 in^2. Estimate the perimeter of the rectangle.

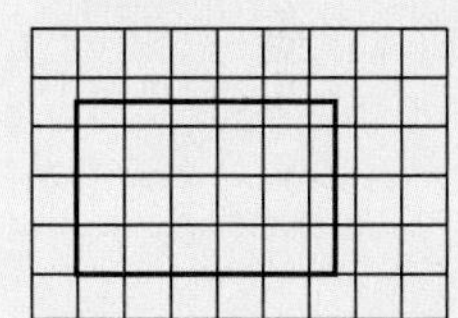

about 18 inches

3 Each square represents 50 yd^2. Estimate the area of the shaded figure. **about 250 yd^2**

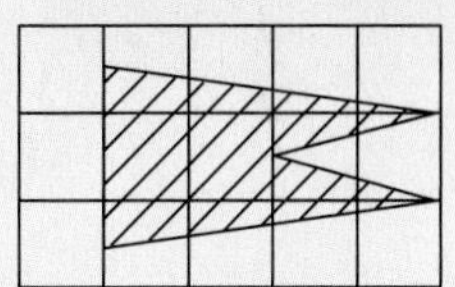

4 Each square represents 4 ft^2. Estimate the area and perimeter of the figure.

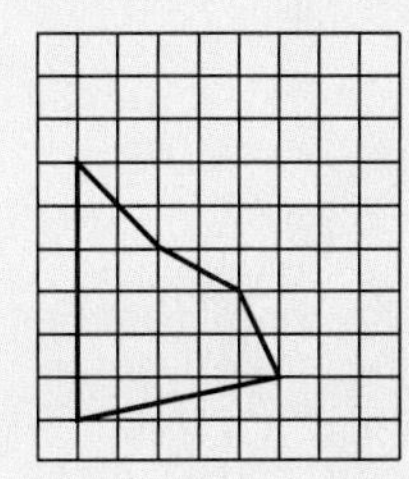

perimeter about 32 ft, area about 64 ft^2

Advanced Learners L4
Have students draw three different triangles on graph paper, each having an area of 18 units. **Check students' drawings.**

learning style: visual

English Language Learners ELL
For Example 1, ask students to give alternate reasons for why the estimates are reasonable. For example, ask them to show how long 9 feet might be, and to compare that length to their pencils.

learning style: visual

All in One Teaching Resources
- Daily Notetaking Guide 8-1 L3
- Adapted Notetaking 8-1 L1

Closure

- *How can you estimate an appropriate unit of length?* Sample: Identify possible units and choose the best one.
- *How can you estimate the area of an irregular figure?* Count the number of squares that the figure occupies and multiply by the number of square units that each square represents.

EXAMPLE Using Area and Perimeter

4 **Multiple Choice** Carl wants to paint a landscape. He will choose one of the four canvases shown below.

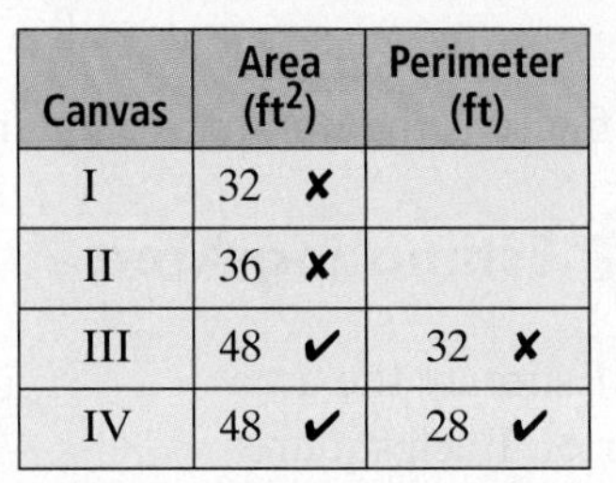

Each square on the grid represents 4 square feet. Which canvas has an area of about 48 ft^2 and a perimeter of about 28 ft?

Ⓐ Canvas I Ⓒ Canvas III
Ⓑ Canvas II Ⓓ Canvas IV

You can estimate each area by counting the number of filled squares. You can estimate each perimeter by finding the sum of the side lengths. Since each square represents 4 ft^2, the side length of each square is 2 ft.

Test Prep Tip
Making a table can help you decide which answer choices to eliminate.

Canvas	Area (ft^2)	Perimeter (ft)
I	32 ✗	
II	36 ✗	
III	48 ✔	32 ✗
IV	48 ✔	28 ✔

Since Canvases I and II do not have an area of about 48 ft^2, you can eliminate them.

Canvas III has an area of about 48 ft^2, but does not have a perimeter of about 28 ft.

Canvas IV has an area of about 48 ft^2 and has a perimeter of about 28 ft.

Canvas IV has the correct area and perimeter, so the answer is D.

Quick Check

4. Each square represents 25 square feet. Estimate the area and perimeter of the figure.
   Answers may vary. Sample: area: about 300 ft^2; perimeter: about 100 ft

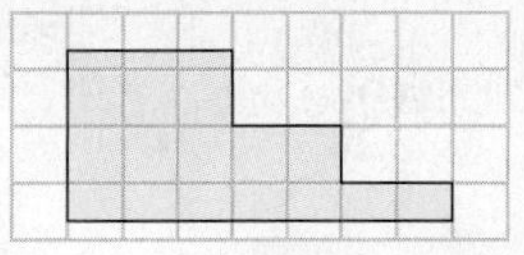

Check Your Understanding

1. **Vocabulary** Area is the number of _?_ units a figure encloses.
2. **Reasoning** Is it possible for the length of a rectangle to be greater than its perimeter? Explain.

Match each item with a reasonable estimate of its length.

3. length of a bowling lane C
4. height of a coffee mug B
5. length of a car A

A. 16 ft
B. 4 in.
C. 20 yd

1. square
2. No; the perimeter is twice the rectangle's length plus twice its width, so the perimeter must be greater than just the length.

Homework Exercises

For more exercises, see Extra Skills and Word Problems.

GO for Help

For Exercises	See Examples
6–9	1
10–11	2
12–13	3
14–15	4

8. 10 ft; it is deeper than the height of a person.

A **Choose a reasonable estimate. Explain your choice.**

6. length of a spoon: 4 in. or 4 ft **4 in.; a spoon is about as long as 4 paperclips.**
7. width of your hand: 6 in. or 6 ft **6 in.; it is less than a foot.**
8. depth of an in-ground swimming pool: 10 in. or 10 ft **See left.**
9. length of a mouse's tail: 2 in. or 2 yd **2 in.; the tail is less than a foot.**

Estimate the perimeter of each figure. The length of one side of each square represents 1 ft. **10–15. Answers may vary. Samples are given.**

10.

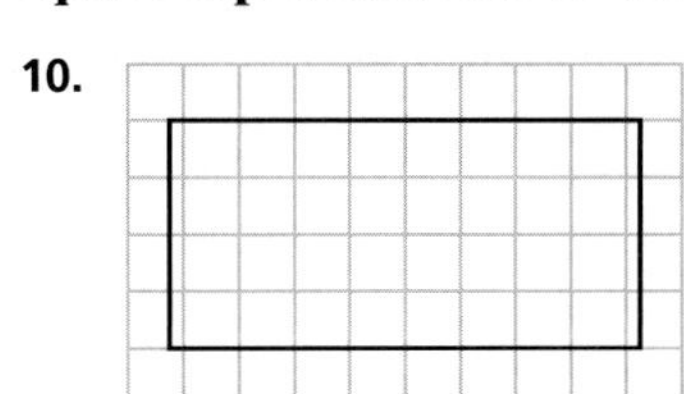

about 24 ft

11.

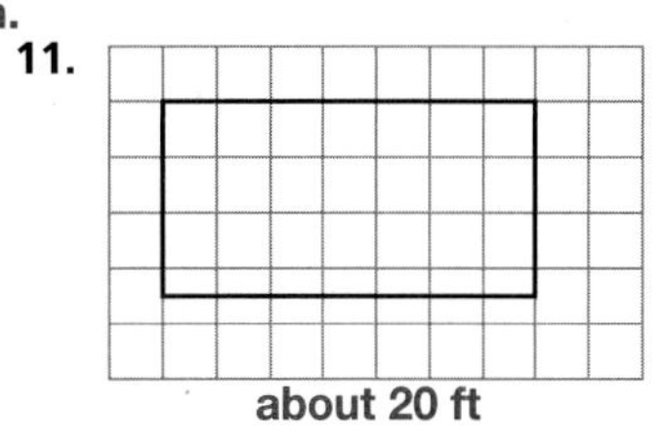

about 20 ft

Estimate the area of each shaded region. Each square represents 25 mi^2.

12.

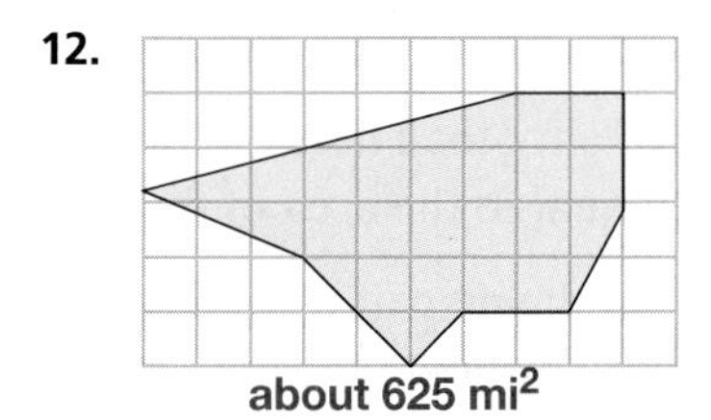

about 625 mi^2

13.

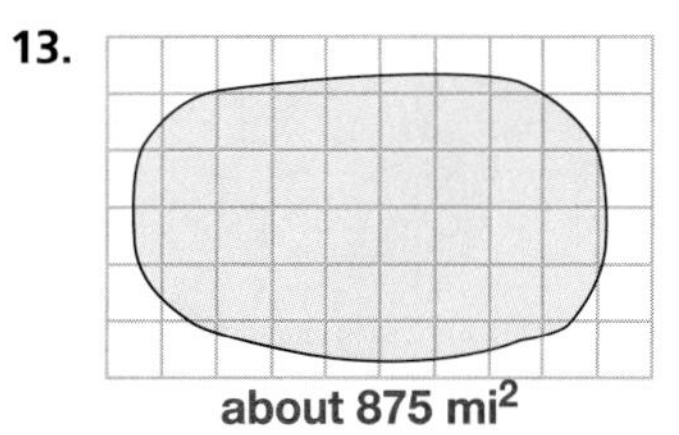

about 875 mi^2

Estimate the area and perimeter. Each square represents 9 yd^2.

14. area: about 135 yd^2; perimeter: about 75 yd

15. area: about 198 yd^2; perimeter: about 81 yd

14.

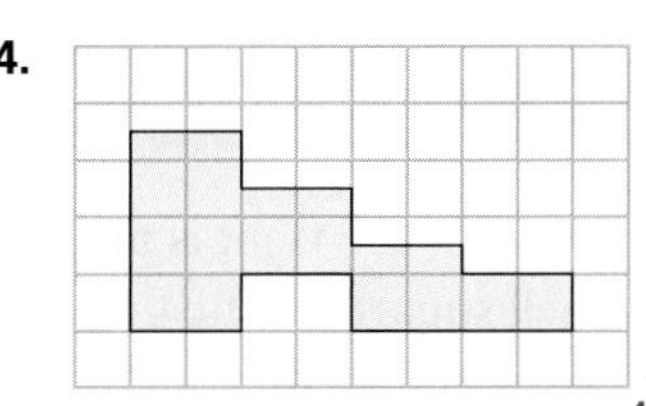

15.

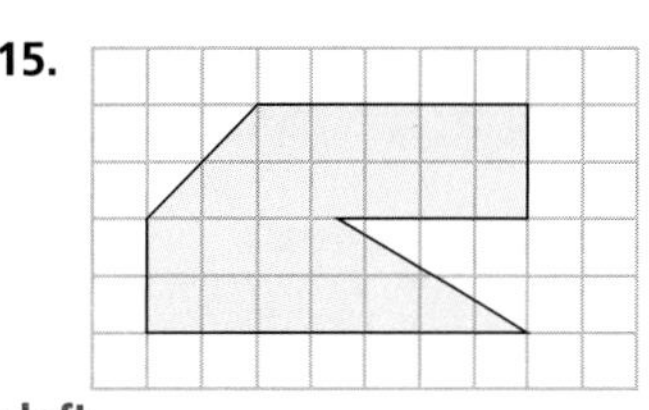

14–15. See left.

B **GPS**

16. **Guided Problem Solving** In the diagram of the fish pond at the right, each square represents 1 ft^2. Estimate the area of the pond. **See left.**
    - The number of full squares in the pond = ■.
    - The number of half squares in the pond = ■.
    - The area of the pond ≅ ■ + $\frac{1}{2}$ ■.

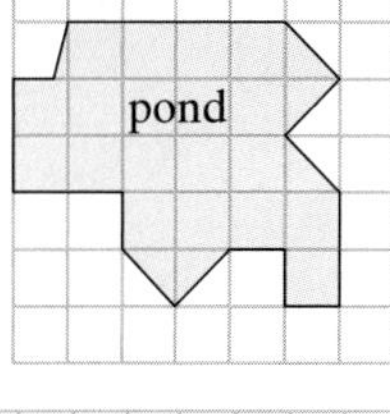

16. Answers may vary. Sample: about $21\frac{1}{2}$ ft^2

17. Katy painted the window frame at the right. She did not paint the glass inside. If each square represents 1 ft^2, what is the area of the glass? **about 8 ft^2**

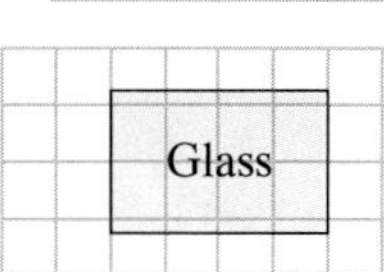

Online lesson quiz, PHSchool.com, Web Code: ara-0801

3. Practice

Assignment Guide

Check Your Understanding

Go over Exercises 1–5 in class before assigning the Homework Exercises.

Homework Exercises

A Practice by Example	6–15
B Apply Your Skills	16–27
C Challenge	28
Test Prep and Mixed Review	29–34

Homework Quick Check

To check students' understanding of key skills and concepts, go over Exercises 11, 14, 17, 26, and 27.

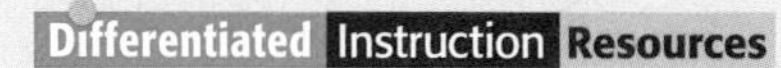

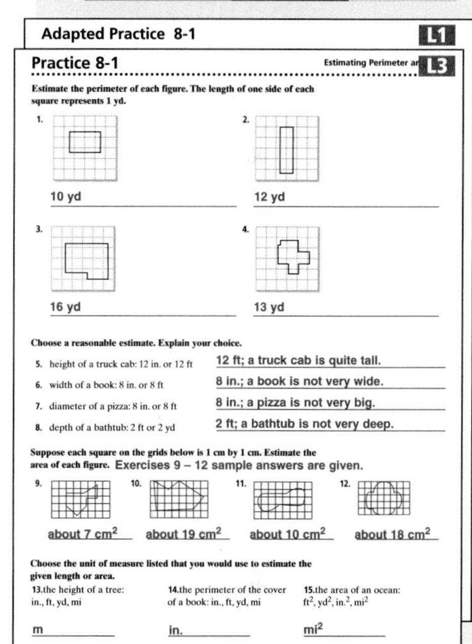

Adapted Practice 8-1 L1

Practice 8-1 Estimating Perimeter ar L3

Estimate the perimeter of each figure. The length of one side of each square represents 1 yd.

1. 10 yd
2. 12 yd
3. 16 yd
4. 13 yd

Choose a reasonable estimate. Explain your choice.

5. height of a truck cab: 12 in. or 12 ft **12 ft; a truck cab is quite tall.**
6. width of a book: 8 in. or 8 ft **8 in.; a book is not very wide.**
7. diameter of a pizza: 8 in. or 8 ft **8 in.; a pizza is not very big.**
8. depth of a bathtub: 2 ft or 2 yd **2 ft; a bathtub is not very deep.**

Suppose each square on the grids below is 1 cm by 1 cm. Estimate the area of each figure. **Exercises 9 – 12 sample answers are given.**

9. about 7 cm^2
10. about 19 cm^2
11. about 10 cm^2
12. about 18 cm^2

Choose the unit of measure listed that you would use to estimate the given length or area.

13. the height of a tree: in., ft, yd, mi **m**
14. the perimeter of the cover of a book: in., ft, yd, mi **in.**
15. the area of an ocean: ft^2, yd^2, $in.^2$, mi^2 **mi^2**

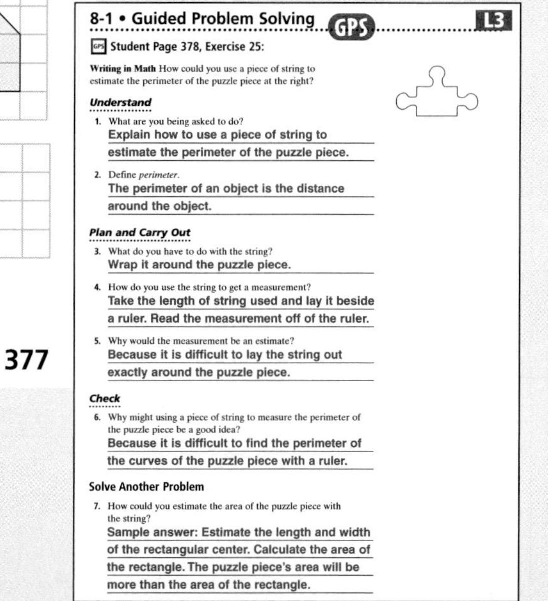

8-1 • Guided Problem Solving GPS L3

Student Page 378, Exercise 25:

Writing in Math How could you use a piece of string to estimate the perimeter of the puzzle piece at the right?

Understand

1. What are you being asked to do?
   Explain how to use a piece of string to estimate the perimeter of the puzzle piece.
2. Define *perimeter*.
   The perimeter of an object is the distance around the object.

Plan and Carry Out

3. What do you have to do with the string?
   Wrap it around the puzzle piece.
4. How do you use the string to get a measurement?
   Take the length of string used and lay it beside a ruler. Read the measurement off of the ruler.
5. Why would the measurement be an estimate?
   Because it is difficult to lay the string out exactly around the puzzle piece.

Check

6. Why might using a piece of string to measure the perimeter of the puzzle piece be a good idea?
   Because it is difficult to find the perimeter of the curves of the puzzle piece with a ruler.

Solve Another Problem

7. How could you estimate the area of the puzzle piece with the string?
   Sample answer: Estimate the length and width of the rectangular center. Calculate the area of the rectangle. The puzzle piece's area will be more than the area of the rectangle.

4. Assess & Reteach

Lesson Quiz

Choose a reasonable estimate for the length of each object.

1. bed: 5 ft or 5 in. **5 ft**
2. soccer field: 100 in. or 100 yd **100 yd**
3. Estimate the perimeter. Each square represents 4 ft^2. **35 ft**

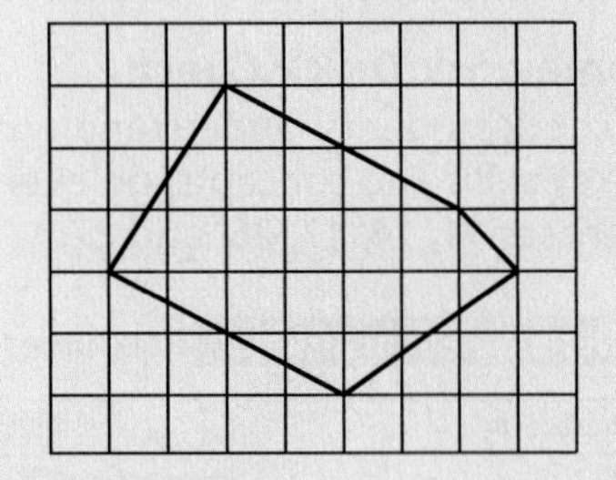

4. Estimate the area. Each square represents 5 cm^2. **about 75 cm^2**

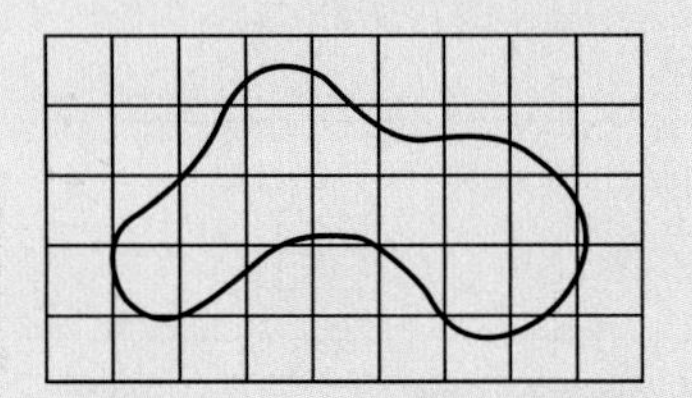

31. 150% increase
32. 75% decrease
33. 44.4% decrease
34. 46.7% increase

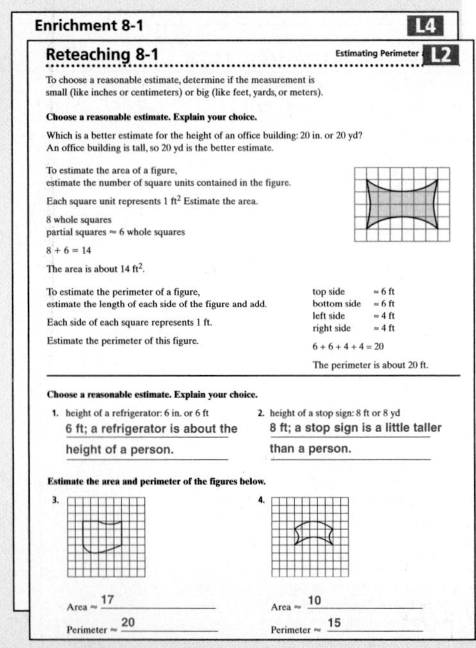

Enrichment 8-1 L4

Reteaching 8-1 — Estimating Perimeter L2

To choose a reasonable estimate, determine if the measurement is small (like inches or centimeters) or big (like feet, yards, or meters).

Choose a reasonable estimate. Explain your choice.

Which is a better estimate for the height of an office building: 20 in. or 20 yd?
An office building is tall, so 20 yd is the better estimate.

To estimate the area of a figure,
estimate the number of square units contained in the figure.

Each square unit represents 1 ft^2 Estimate the area.

8 whole squares
partial squares ≈ 6 whole squares

$8 + 6 = 14$

The area is about 14 ft^2.

To estimate the perimeter of a figure,
estimate the length of each side of the figure and add.

Each side of each square represents 1 ft.

Estimate the perimeter of this figure.

top side ≈ 6 ft
bottom side ≈ 6 ft
left side ≈ 4 ft
right side ≈ 4 ft

$6 + 6 + 4 + 4 = 20$

The perimeter is about 20 ft.

Choose a reasonable estimate. Explain your choice.

1. height of a refrigerator: 6 in. or 6 ft
   6 ft; a refrigerator is about the height of a person.
2. height of a stop sign: 8 ft or 8 yd
   8 ft; a stop sign is a little taller than a person.

Estimate the area and perimeter of the figures below.

3. Area ≈ 17
   Perimeter ≈ 20
4. Area ≈ 10
   Perimeter ≈ 15

Homework Video Tutor
Visit: PHSchool.com
Web Code: are-0801

23. Answers may vary. Sample: about 500 yd^2
24. Answers may vary. Sample: 2 ft × 5 ft
25. Lay the string around the edge of the puzzle piece. Mark the length and then measure the string.
26.

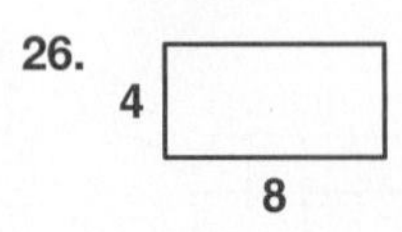

27a. about 7,840,000 ft^2

Estimate each length in inches.

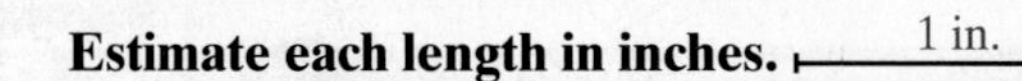

18. about 2 in.
19. about $\frac{1}{2}$ in.
20. about $1\frac{1}{2}$ in.
21. about 1 in.
22. **Estimation** Measure the length of your shoe. Use your shoe length to estimate the length of your desk. **Check students' work.**
23. A diagram of a golf fairway is shown at the right. Each square represents 20 yd^2. Estimate the area. **See left.**

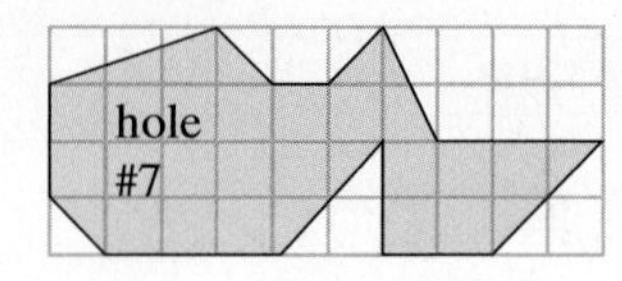

24. A rectangle has a perimeter of 14 ft. Write whole-number dimensions for another rectangle with the same perimeter. **See left.**
25. **Writing in Math** How could you use a piece of string to estimate the perimeter of the puzzle piece at the right? **See left.**

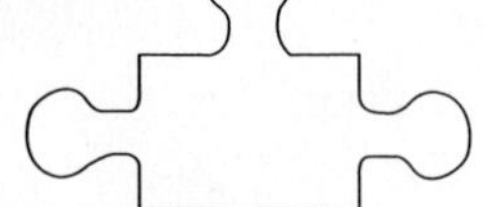

26. **Open Ended** Draw a rectangle on graph paper with a perimeter of 24 units and an area of 32 units^2. **See left.**
27. An acre equals 43,560 ft^2. A theme park covers about 180 acres.
    a. Estimate the number of square feet in the theme park. **See left.**
    b. The area of a football field is 57,600 ft^2. About how many football fields are equal to the area of the theme park? **about 136**
28. **Challenge** In the map at the left, each square represents 5,575 mi^2. Use the map to estimate the area of Florida. **about 61,325 mi^2**

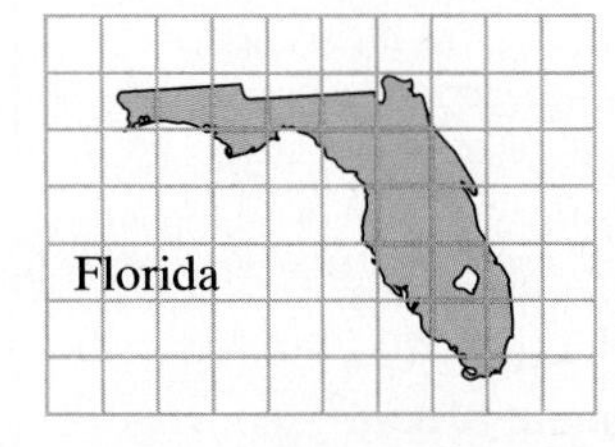

Test Prep and Mixed Review — Practice

Multiple Choice

29. The flag of Nepal at the right is not rectangular in shape. If each square on the grid represents 4 square inches, what is the approximate area of the flag shown? **D**

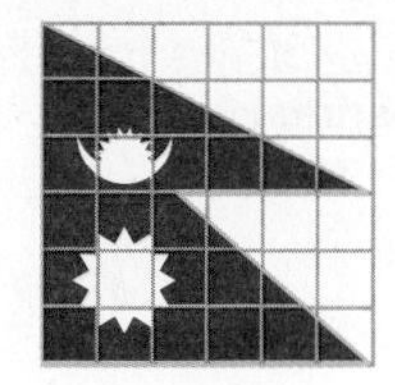

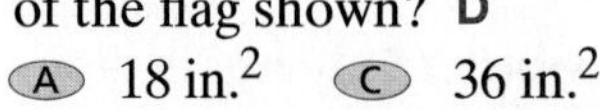

Ⓐ 18 in.^{2} Ⓒ 36 in.^{2}
Ⓑ 44 in.^{2} Ⓓ 84 in.^{2}

30. Stewart Middle School has 243 seventh-grade students. About 32% of them attended the high school football game. About how many seventh-grade students attended the football game? **G**
    Ⓕ Fewer than 75 Ⓗ Between 85 and 100
    Ⓖ Between 75 and 85 Ⓙ More than 100

GO for Help

For Exercises	See Lesson
31–34	6-8

Find each percent of increase or decrease. Round to the nearest tenth. **31–34. See margin.**

31. 20 to 50
32. 32 to 8
33. 99 to 55
34. 75 to 110

Test Prep

Resources

For additional practice with a variety of test item formats:
- Test-Taking Strategies, p. 427
- Test Prep, p. 431
- Test-Taking Strategies with Transparencies

Alternative Assessment

Each student in a pair draws on graph paper a shaded picture, such as a pond, leaf, or geometric figure. Each designates the measurement that a square unit represents. Then partners estimate the area of each other's figures. Pairs then exchange drawings with other pairs to check their estimates.

Hands On

Generating Formulas for Area

You can generate the area of a figure by separating or combining the areas of two figures you know.

ACTIVITY

1. Using graph paper, draw a parallelogram like the one at the right. When you draw the perpendicular segment, what two polygons are formed? **trapezoid, triangle**
2. Cut out the parallelogram and then cut along the perpendicular segment. **Check students' work.**

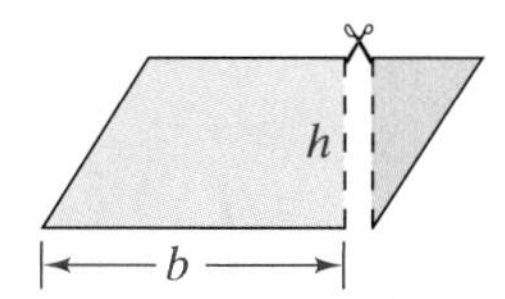

3. Rearrange the pieces to form a rectangle.
   a. What is the area of the rectangle?
   b. What was the area of the parallelogram?

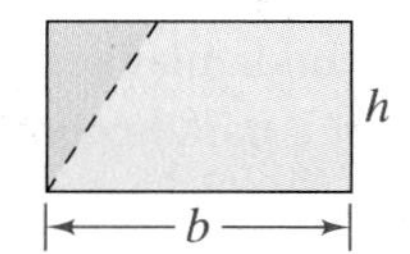

The area of the rectangle and the area of the parallelogram should be the same.

4. How do b and h relate to the length and width of the rectangle? Write a formula for the area of a parallelogram. **The rectangle has length b and height h; $A = bh$.**

ACTIVITY

5–6. Check students' work.

5. Fold a piece of graph paper in half. On one side, draw a right triangle like the one at the right.

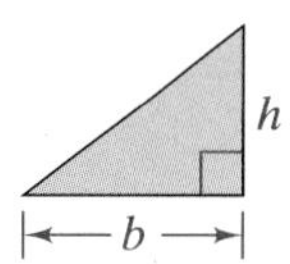

6. Cut out the triangle, cutting through both layers of the folded paper. You now have two congruent triangles.

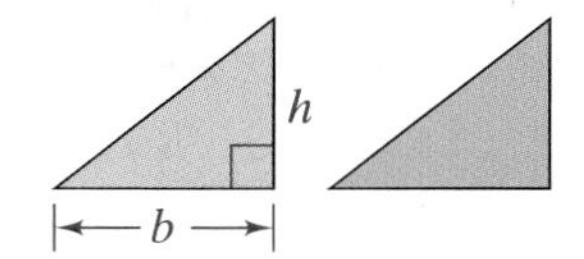

7. Arrange the pieces to form a rectangle.
   a. What is the area of the rectangle?
   b. What is the area of one triangle?

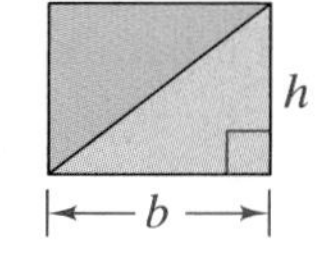

The area of the triangle should be half the area of the rectangle.

8. How do b and h relate to the length and width of the rectangle? Write a formula for the area of a triangle. **The rectangle has length b and height h; $A = \frac{1}{2}bh$.**

Activity Lab

Generating Formulas for Area

Students learn to find the area of an unfamiliar figure by rearranging its pieces into familiar figures. This will help them understand the formula for area used in Lesson 8-2.

Guided Instruction

Before beginning the Activity, explain to students that any parallelogram can be separated into two pieces and those pieces will form a rectangle. Demonstrate separating a triangle from a parallelogram and attaching it to the other side to form a rectangle.
Ask:

- *What dimension is the same height as the height of the rectangle?* **the height of the triangle**

Activity
Distribute graph paper and work through the Activity as a class. Have volunteers answer the questions in steps 3, 4, 7, and 8. Make sure each student has enough time to complete and understand each step before proceeding to the next step.

Alternative Method
Some students may need to count the squares to verify that the height and base of the parallelogram is the same as that of the rectangle.

Resources

- Activity Lab 8-2: Critical Thinking
- graph paper
- scissors

8-2

1. Plan

Objective
To find the area and perimeter of a parallelogram

Examples
1 Finding the Area of a Parallelogram
2 Relating Perimeter and Area

Math Understandings: p. 372C

Math Background

A *parallelogram* is a four-sided figure whose opposite sides are parallel. The formula for the area of a parallelogram is $A = b \cdot h$, where b is the base length and h is the height, or perpendicular distance, from one base to the other. Note that any side of a parallelogram can be considered the base.

More Math Background: p. 372C

Lesson Planning and Resources

See p. 372E for a list of the resources that support this lesson.

PowerPoint

Bell Ringer Practice

Check Skills You'll Need
Use student page, transparency, or PowerPoint. For intervention, direct students to:
Multiplying Decimals
Lesson 1-3
Extra Skills and Word Problems Practice, Ch. 1

8-2 Area of a Parallelogram

Check Skills You'll Need

1. **Vocabulary Review** State the *Commutative Property of Multiplication.* **See below.**

Find each product.

2. 100 × 4.5 **450**
3. 2.34 × 12 **28.08**
4. 10.2 × 5.6 **57.12**
5. 0.6 × 3.4 **2.04**

Check Skills You'll Need
1. **Changing the order of the factors does not change the product.**

Vocabulary Tip
Perpendicular segments intersect to form right angles.

What You'll Learn

To find the area and perimeter of a parallelogram

New Vocabulary height of a parallelogram, base of a parallelogram

Why Learn This?

The floor plan of the building at the right is in the shape of a parallelogram. You can calculate the area of the parallelogram to determine how much office space is available on a given floor of the building.

The **height of a parallelogram** is the length of a perpendicular segment connecting one **base of a parallelogram** to the other.

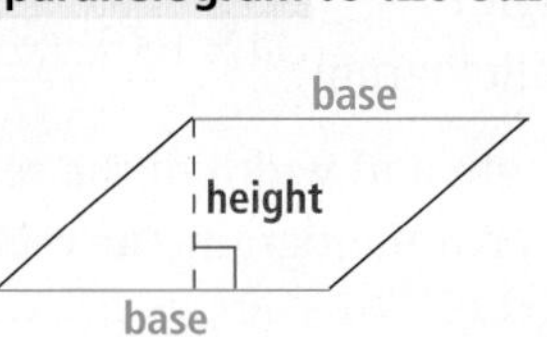

The diagram below relates the formula for the area of a rectangle to the formula for the area of a parallelogram.

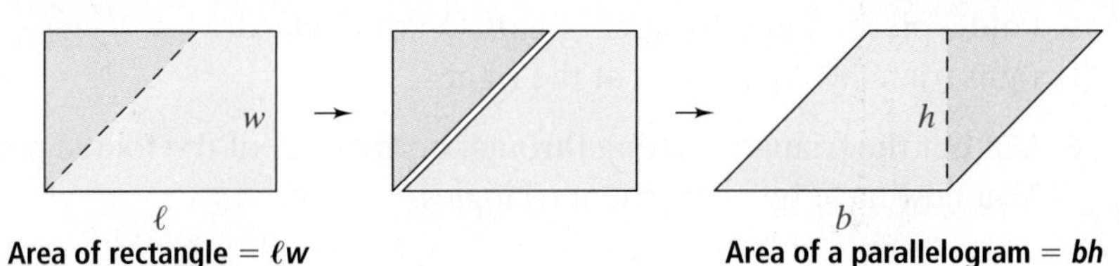

Video Tutor Help
Visit: PHSchool.com
Web Code: are-0775

KEY CONCEPTS **Area of a Parallelogram**

The area of a parallelogram is equal to the product of any base b and the corresponding height h.

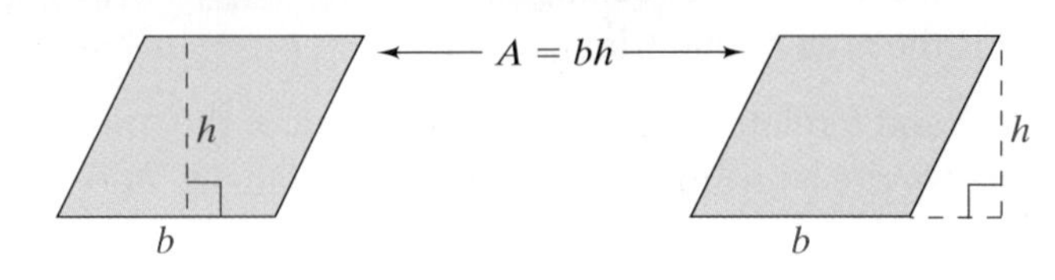

Differentiated Instruction Solutions for All Learners

Special Needs L1
Show students that when you mark off the height on both sides of a parallelogram, the resulting interior shape is a rectangle. This will help them understand why bh works for both shapes, although we customarily use lw for the area of rectangles.

learning style: visual

Below Level L2
Review with students the meaning of *parallel* and *perpendicular.* Have them identify a vertex in a parallelogram.

learning style: verbal

EXAMPLE Finding the Area of a Parallelogram

1 Find the area of the parallelogram.

$A = bh$ ← **Use the area formula.**
$= (9)(15)$ ← **Substitute.**
$= 135$ ← **Simplify.**

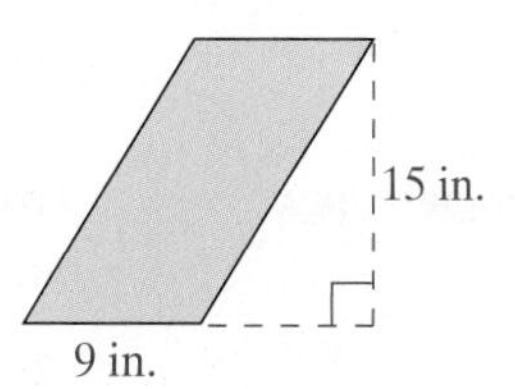

The area is 135 in.^{2}.

Quick Check

1. Find the area of the parallelogram.
90 cm^2

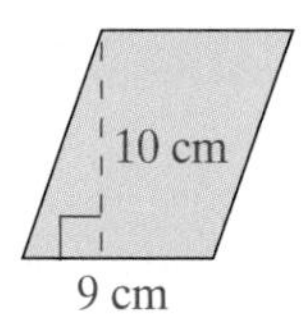

You can also use lengths of the sides of a rectangle to find perimeter.

EXAMPLE Relating Perimeter and Area

2 **Multiple Choice** Melinda wants to plant a rectangular garden and put a fence around it. She has 34 ft of fencing and she wants her garden to be as big as possible. Which dimensions should she use?

(A) Length of 9 ft and width of 8 ft
(B) Length of 10 ft and width of 7 ft
(C) Length of 12 ft and width of 6 ft
(D) Length of 14 ft and width of 5 ft

Since all answer choices give the length ℓ and width w, you can calculate both the perimeter $2\ell + 2w$ and the area $\ell \times w$.

Perimeter	**Area**	
$2(9) + 2(8) = 34$ ✔	$9 \times 8 = 72$	← **Perimeter is correct; find the area.**
$2(10) + 2(7) = 34$ ✔	$10 \times 7 = 70$	← **Perimeter is correct; the area in choice A is greater.**
$2(12) + 2(6) = 36$ ✘		← **Perimeter is greater than 34 ft.**
$2(14) + 2(5) = 38$ ✘		← **Perimeter is greater than 34 ft.**

The rectangle with a length of 9 ft and a width of 8 ft will have the correct perimeter and the greatest area. The answer is A.

Test Prep Tip

If the first part in an answer choice is incorrect, do not bother to calculate the second part.

Quick Check

2. What is the perimeter of the rectangle?
22 cm

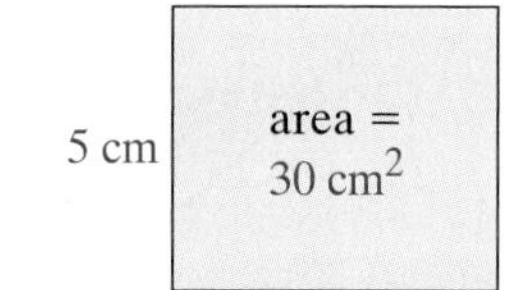

2. Teach

Activity Lab

Use before the lesson.
Student Edition Activity Lab, Hands On 8-2a, Generating Formulas for Area, p. 379

All in One Teaching Resources
Activity Lab 8-2: Critical Thinking

Guided Instruction

Error Prevention!

Some students confuse *base* and *height.* Have them cut two strips of paper and tape them together to make an upside-down *T.* Have them label the horizontal strip *base* and the vertical strip *height.* They can tape their "reference charts" in their notebooks.

Example 1
Remind students that the units for area are square units. You can illustrate this by including the units in the calculation.
9 in. × 15 in. = 135 in.^{2} because in. × in. = in.^{2}

PowerPoint

Additional Examples

1 Find the area of the parallelogram. **216 ft^2**

2 A backyard pool has a perimeter of 90 ft and an area of 500 ft^2. What are the pool's length and width? **25 ft, 20 ft**

All in One Teaching Resources
- Daily Notetaking Guide 8-2 **L3**
- Adapted Notetaking 8-2 **L1**

Closure

- *How do you find the area of a parallelogram?* **Multiply any base times the corresponding height (perpendicular distance from one base to the other).**

Advanced Learners **L4**
Three identical parallelograms are placed side by side. A diagonal connects opposite corners and forms two triangles with an area of 21 ft^2 each. What is the area of each parallelogram? **14 ft^2**

learning style: visual

English Language Learners **ELL**
For Example 2, provide graph paper and have students draw shapes according to the dimensions in each of the choices. Ask them to compare their pictures to their calculations.

learning style: visual

3. Practice

Assignment Guide

Check Your Understanding

Go over Exercises 1–5 in class before assigning the Homework Exercises.

Homework Exercises

A	Practice by Example	6–16
B	Apply Your Skills	17–24
C	Challenge	25
Test Prep and Mixed Review		26–33

Homework Quick Check

To check students' understanding of key skills and concepts, go over Exercises 6, 13, 22, 23, and 24.

Differentiated Instruction Resources

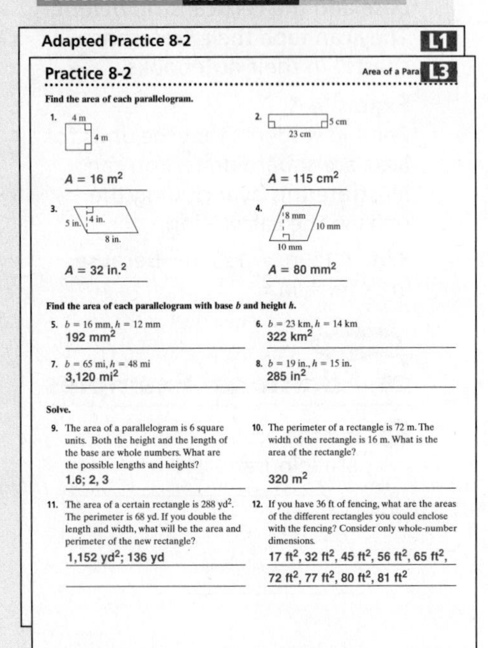

Adapted Practice 8-2 L1

Practice 8-2 — Area of a Para[llelogram] L3

Find the area of each parallelogram.

1. 4 m, 4 m — $A = 16$ m²
2. 23 cm, 5 cm — $A = 115$ cm²
3. 5 in., 4 in., 8 in. — $A = 32$ in.²
4. 8 mm, 10 mm, 10 mm — $A = 80$ mm²

Find the area of each parallelogram with base b and height h.

5. $b = 16$ mm, $h = 12$ mm — 192 mm²
6. $b = 23$ km, $h = 14$ km — 322 km²
7. $b = 65$ mi, $h = 48$ mi — 3,120 mi²
8. $b = 19$ in., $h = 15$ in. — 285 in²

Solve.

9. The area of a parallelogram is 6 square units. Both the height and the length of the base are whole numbers. What are the possible lengths and heights? 1.6; 2, 3
10. The perimeter of a rectangle is 72 m. The width of the rectangle is 16 m. What is the area of the rectangle? 320 m²
11. The area of a certain rectangle is 288 yd². The perimeter is 68 yd. If you double the length and width, what will be the area and perimeter of the new rectangle? 1,152 yd²; 136 yd
12. If you have 36 ft of fencing, what are the areas of the different rectangles you could enclose with the fencing? Consider only whole-number dimensions. 17 ft², 32 ft², 45 ft², 56 ft², 65 ft², 72 ft², 77 ft², 80 ft², 81 ft²

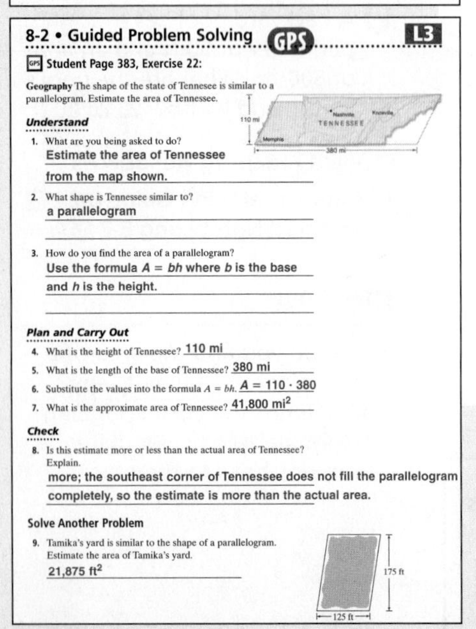

8-2 • Guided Problem Solving GPS L3

Student Page 383, Exercise 22:

Geography The shape of the state of Tennesee is similar to a parallelogram. Estimate the area of Tennessee.

Understand

1. What are you being asked to do? Estimate the area of Tennessee from the map shown.
2. What shape is Tennessee similar to? a parallelogram
3. How do you find the area of a parallelogram? Use the formula $A = bh$ where b is the base and h is the height.

Plan and Carry Out

4. What is the height of Tennessee? 110 mi
5. What is the length of the base of Tennessee? 380 mi
6. Substitute the values into the formula $A = bh$. $A = 110 \cdot 380$
7. What is the approximate area of Tennessee? 41,800 mi²

Check

8. Is this estimate more or less than the actual area of Tennessee? Explain. more; the southeast corner of Tennessee does not fill the parallelogram completely, so the estimate is more than the actual area.

Solve Another Problem

9. Tamika's yard is similar to the shape of a parallelogram. Estimate the area of Tamika's yard. 21,875 ft²

Check Your Understanding

1. **Vocabulary** What kind of angle is formed by perpendicular lines? right

Two parallelograms have a base and a height that are equal. Tell whether each statement is true or false. Explain your answer.

2. The two parallelograms must be congruent.
3. The areas of the two parallelograms are equal.

2. False; angles may differ.
3. True; $A = bh$, so if the bases are equal and the heights are equal, the areas will be equal.

Use the parallelogram at the right. Fill in the blank.

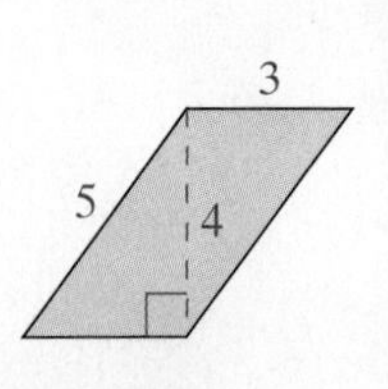

4. The formula for the area is $A = bh = (3)(\blacksquare)$. 4
5. The formula for the perimeter is $P = 2(\blacksquare) + 2(5)$. 3

Homework Exercises

For more exercises, see Extra Skills and Word Problems.

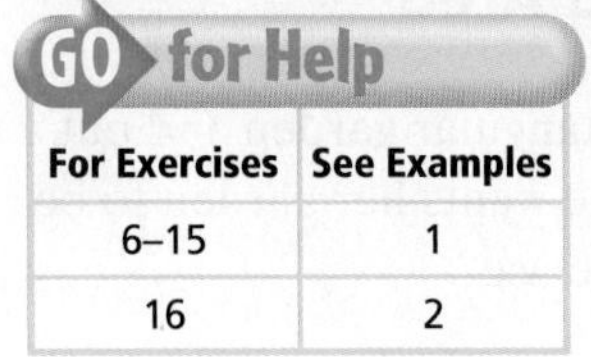

GO for Help

For Exercises	See Examples
6–15	1
16	2

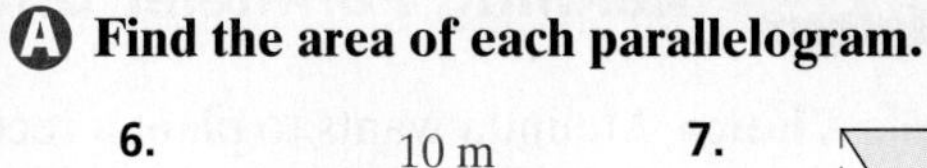

A Find the area of each parallelogram.

6. 10 m, 6 m — 60 m²
7. 5 m, 5 m — 25 m²

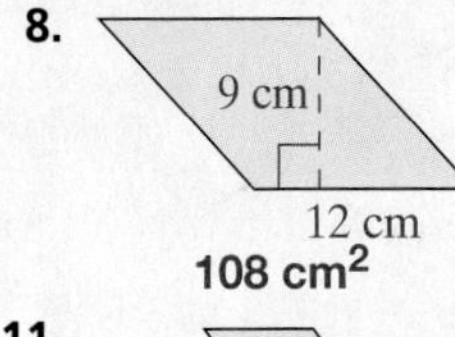

8. 108 cm²

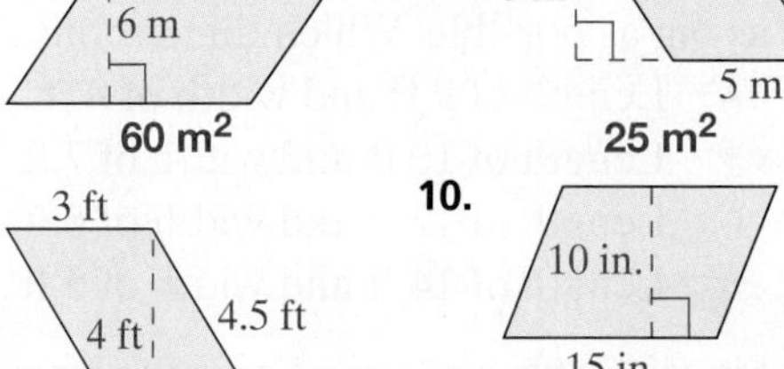

9. 12 ft²
10. 150 in.²

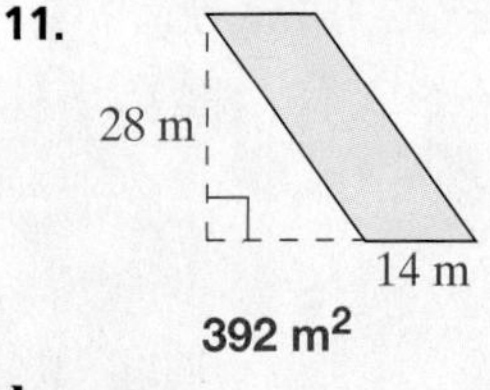

11. 392 m²

Find each area for base b and height h of a parallelogram.

12. $b = 14$ in. $h = 6$ in. 84 in.²
13. $b = 25$ mi $h = 25$ mi 625 mi²
14. $h = 40$ cm $b = 0.5$ cm 20 cm²
15. $h = 1{,}000$ m $b = 20$ m 20,000 m²

16. A rectangular fish pond is 21 ft² in area. If the owner can surround the pond with a 20-foot fence, what are the dimensions of the pond? 3 ft by 7 ft

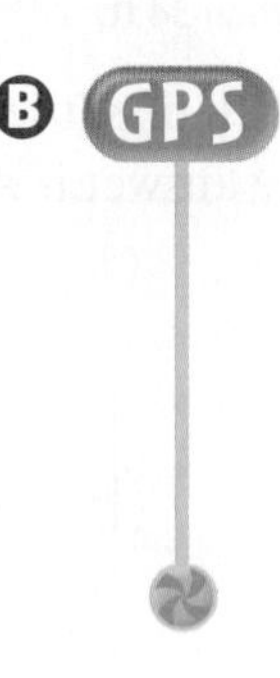

B GPS

17. **Guided Problem Solving** The diagram shows a park bounded by streets. The park is in the shape of a parallelogram. Each square is 10 yards on a side. What is the area of the park? 3,600 yd²
   - What are the base and the height of the parallelogram?
   - What formula should you use?

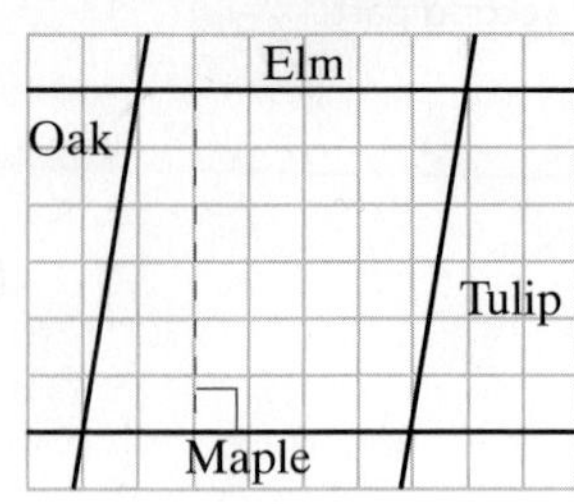

29. 2; 2
30. 1.35; $1\frac{7}{20}$
31. 1.52; $1\frac{13}{25}$
32. 0.0003; $\frac{3}{10{,}000}$
33. 0.0045; $\frac{9}{2{,}000}$

GO Online
Homework Video Tutor
Visit: PHSchool.com
Web Code: are-0802

Find the missing measures for each rectangle.

18. $\ell = 14$ in.; $w = ■$; $A = ■$; $P = 34$ in. — **3 in.; 42 in.^{2}**

19. $\ell = ■$; $w = 4.2$ m; $A = 37.8$ m^2; $P = ■$ — **9 m; 26.4 m**

20. $\ell = 7$ ft; $w = ■$; $A = 18.2$ ft^2; $P = ■$ — **2.6 ft; 19.2 ft**

21. $\ell = ■$; $w = 2$ cm; $A = ■$; $P = 25$ cm — **10.5 cm; 21 cm^2**

22. Geography The shape of the state of Tennessee is similar to a parallelogram. Estimate the area of Tennessee. **41,800 mi^2**

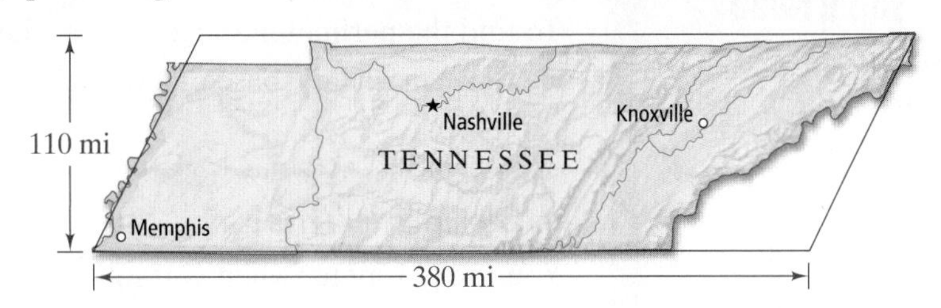

4 m; 2 m; 4 m; 1.6 m; 2 m

23. Reasoning The rectangle and the parallelogram at the left have the same perimeter. How do you know that the area of the rectangle is greater than the area of the parallelogram? **See left.**

23. They have the same bases, but the height of the parallelogram must be less than the height of the rectangle because a leg is shorter than the hypotenuse.

24. Writing in Math A rectangular lot is 70.2 m long and 59.8 m wide. Is 42,000 m^2 a reasonable estimate for the area of the lot? Explain. **See left.**

24. No; the approximate area is 60 m × 70 m or 4,200 m^2.

C 25. Challenge Find the area and perimeter of the figure at the right.
area = 161 in.^{2}
perimeter = 62 in.

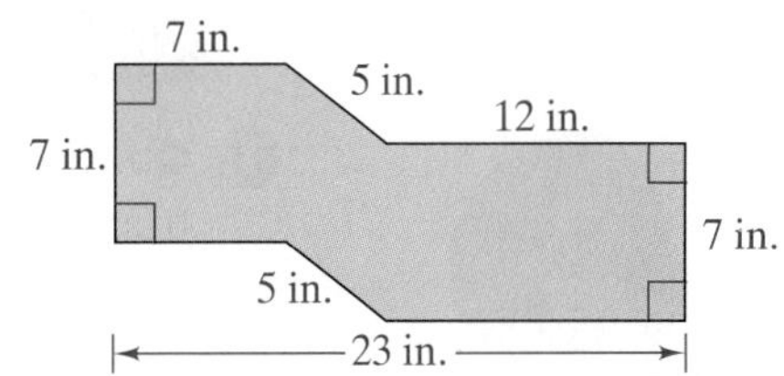

Test Prep and Mixed Review

Practice

Multiple Choice

26. A playground has the shape of a parallelogram. If the base is 30 feet, and the corresponding height is 25 feet, what is its area? **D**
(A) 55 ft^2 (B) 187.5 ft^2 (C) 375 ft^2 (D) 750 ft^2

27. Which of the following expressions CANNOT be used to find the perimeter of a regular hexagon with sides of length h? **H**
(F) $6h$ (H) $3h \times 3h$
(G) $2h + 2h + 2h$ (J) $6 \times h$

28. In Solomon's stamp collection, $\frac{4}{25}$ of the stamps are international. What percent of Solomon's stamps are international? **B**
(A) 4% (B) 16% (C) 21% (D) 29%

GO for Help

For Exercises	See Lesson
29–33	6-3

Write each percent as a decimal and as a fraction in simplest form.
29–33. See margin.

29. 200% **30.** 135% **31.** 152% **32.** 0.03% **33.** 0.45%

Online lesson quiz, PHSchool.com, Web Code: ara-0802

4. Assess & Reteach

PowerPoint
Lesson Quiz

Find the area and perimeter of the figure.

1.
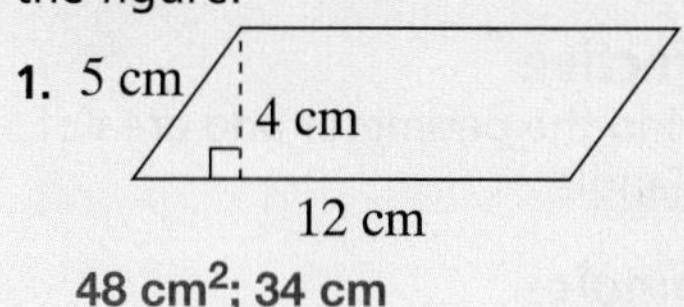

48 cm^2; 34 cm

2. Find the area for a parallelogram with $b = 7$ ft and $h = 4$ ft.
area = 28 ft^2

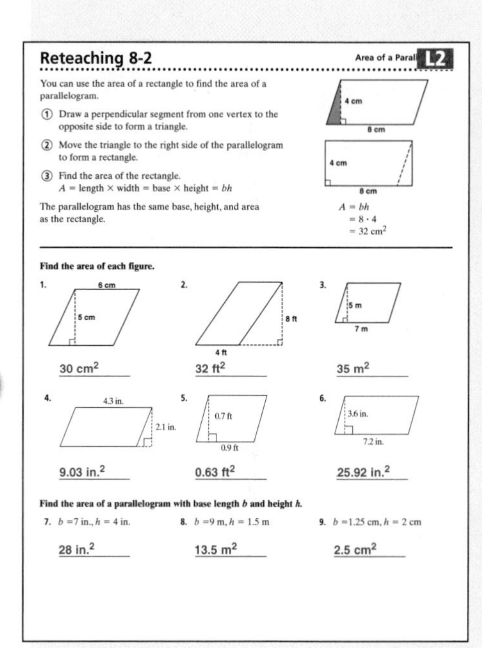
Reteaching 8-2 Area of a Parallelogram L2

You can use the area of a rectangle to find the area of a parallelogram.
1. Draw a perpendicular segment from one vertex to the opposite side to form a triangle.
2. Move the triangle to the right side of the parallelogram to form a rectangle.
3. Find the area of the rectangle. A = length × width = base × height = bh

The parallelogram has the same base, height, and area as the rectangle.

$A = bh = 8 \cdot 4 = 32$ cm^2

Find the area of each figure.

1. 30 cm^2 2. 32 ft^2 3. 35 m^2
4. 9.03 in.^{2} 5. 0.63 ft^2 6. 25.92 in.^{2}

Find the area of a parallelogram with base length b and height h.

7. $b = 7$ in., $h = 4$ in. 28 in.^{2}
8. $b = 9$ m, $h = 1.5$ m 13.5 m^2
9. $b = 1.25$ cm, $h = 2$ cm 2.5 cm^2

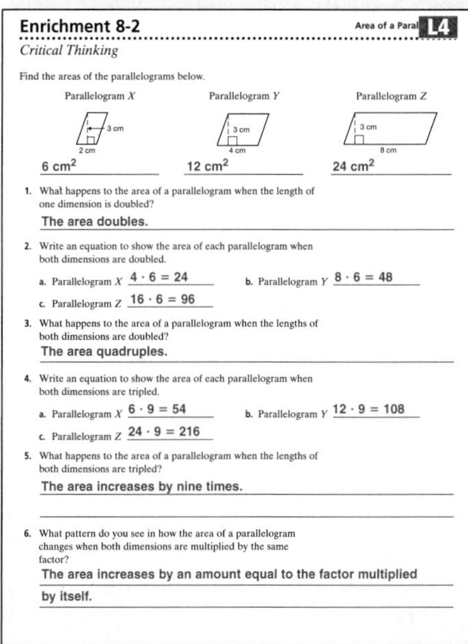
Enrichment 8-2 Area of a Parallelogram L4
Critical Thinking

Find the areas of the parallelograms below.

Parallelogram X: 6 cm^2 Parallelogram Y: 12 cm^2 Parallelogram Z: 24 cm^2

1. What happens to the area of a parallelogram when the length of one dimension is doubled?
The area doubles.
2. Write an equation to show the area of each parallelogram when both dimensions are doubled.
a. Parallelogram X $4 \cdot 6 = 24$ b. Parallelogram Y $8 \cdot 6 = 48$
c. Parallelogram Z $16 \cdot 6 = 96$
3. What happens to the area of a parallelogram when the lengths of both dimensions are doubled?
The area quadruples.
4. Write an equation to show the area of each parallelogram when both dimensions are tripled.
a. Parallelogram X $6 \cdot 9 = 54$ b. Parallelogram Y $12 \cdot 9 = 108$
c. Parallelogram Z $24 \cdot 9 = 216$
5. What happens to the area of a parallelogram when the lengths of both dimensions are tripled?
The area increases by nine times.
6. What pattern do you see in how the area of a parallelogram changes when both dimensions are multiplied by the same factor?
The area increases by an amount equal to the factor multiplied by itself.

Alternative Assessment

Each student in a pair draws a parallelogram and labels the base and height. Partners exchange drawings and find the area of each other's figure. If necessary, direct students to find the area to the nearest tenth. Remind students to report area in square units.

Test Prep

Resources
For additional practice with a variety of test item formats:
- Test-Taking Strategies, p. 427
- Test Prep, p. 431
- Test-Taking Strategies with Transparencies

8-3

1. Plan

Objective
To find the perimeter and area of a triangle

Examples
1 Finding the Perimeter of a Triangle
2 Finding the Area of a Triangle

Math Understandings: p. 372C

Math Background

The perimeter of a triangle is found by adding the lengths of all three sides. The area of a triangle is always one half the area of a parallelogram with the same base and height. To find the formula for the area of a triangle, multiply the formula for the area of a parallelogram, bh, by $\frac{1}{2}$. The area of a triangle is $\frac{1}{2}bh$.

More Math Background: p. 372C

Lesson Planning and Resources

See p. 372E for a list of the resources that support this lesson.

Bell Ringer Practice

Check Skills You'll Need
Use student page, transparency, or PowerPoint. For intervention, direct students to:

Multiplying Fractions and Mixed Numbers
Lesson 3-4
Extra Skills and Word Problems Practice, Ch. 3

8-3 Perimeter and Area of a Triangle

Check Skills You'll Need

1. **Vocabulary Review** What is the *reciprocal* of 2? $\frac{1}{2}$

Find each product.

2. $\frac{1}{4} \cdot 12$ 3
3. $20 \cdot \frac{1}{10}$ 2
4. $15 \cdot \frac{1}{3}$ 5
5. $\frac{1}{2} \cdot 9$ 4.5

Lesson 3-4

What You'll Learn

To find the perimeter and area of a triangle

New Vocabulary base of a triangle, height of a triangle

Why Learn This?

You use perimeter when you solve problems involving borders. You can find perimeter to see if you have enough material to sew a border around the triangular quilt piece at the right.

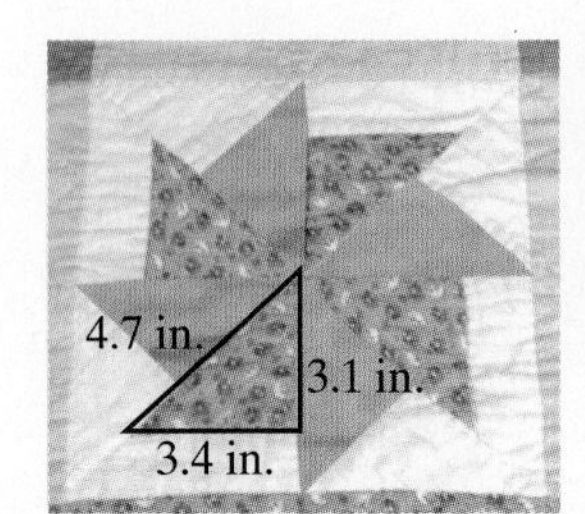

To find the perimeter of a triangle, you can add the side lengths.

EXAMPLE Finding the Perimeter of a Triangle

1 **Quilting** How much material do you need to sew a border around the triangular piece on the quilt above?

Estimate $3.4 + 3.1 + 4.7 \approx 3 + 3 + 5 = 11$

$P = 3.4 + 3.1 + 4.7$ ← **Find the perimeter.**

$= 11.2$ ← **Simplify.**

You need 11.2 in. of material to border the triangular piece.

Check for Reasonableness 11.2 is close to 11. The answer is reasonable.

Quick Check

1. **Mental Math** How much fabric do you need to border a triangular quilt piece whose sides are 6 cm, 8 cm, and 10 cm long? **24 cm**

Any side of a triangle can be considered the **base of a triangle.** The **height of a triangle** is the length of the perpendicular segment from a vertex to the base opposite the vertex or to an extension of the base.

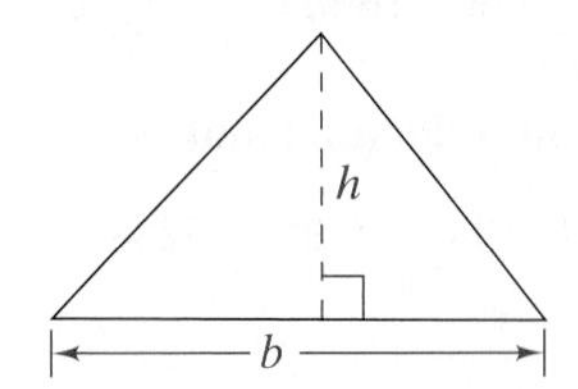

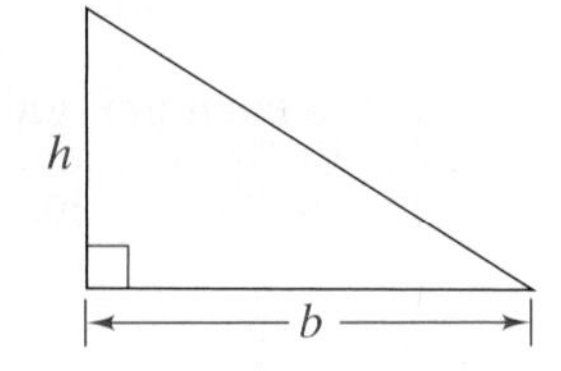

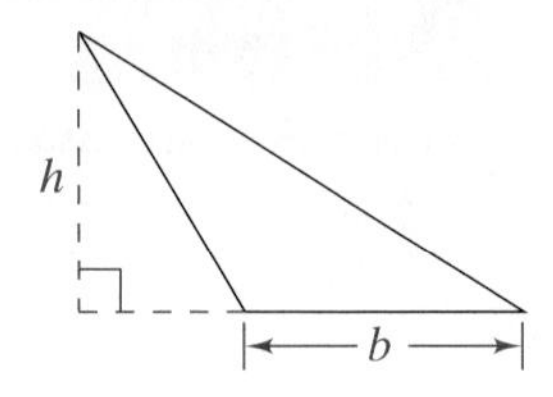

Differentiated Instruction Solutions for All Learners

Special Needs L1
To help students see the relationship between parallelograms and triangles, give them a drawing of a parallelogram. Ask them to cut off the triangles.

learning style: visual

Below Level L2
Provide students with several triangle cutouts, so they can trace them, label the base and height, and trace the other half of a parallelogram with the same base and height.

learning style: tactile

The formula for the area of a triangle follows from the formula for the area of a parallelogram.

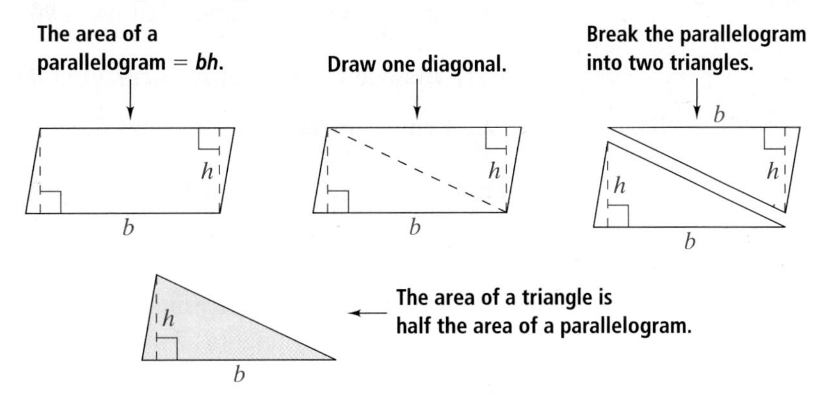

KEY CONCEPTS Area of a Triangle

The area of a triangle is equal to half the product of any base b and the corresponding height h.

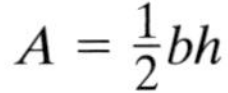
$A = \frac{1}{2}bh$

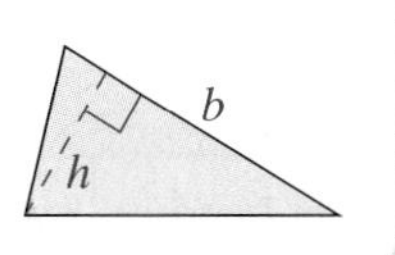

When you find the area of a triangle, remember that you only need the length of the base and the perpendicular height of the triangle.

EXAMPLE Finding the Area of a Triangle

2 Find the area of each triangle.

a.

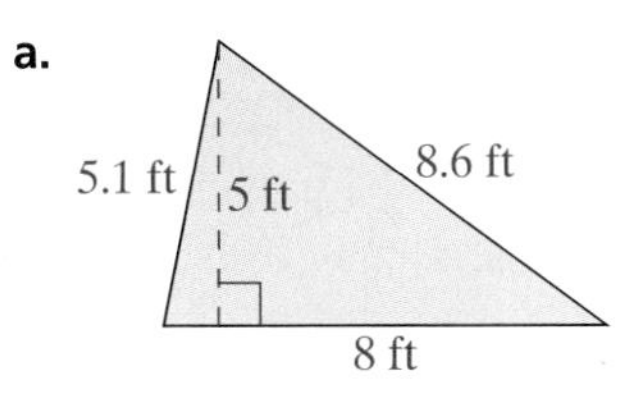

b.

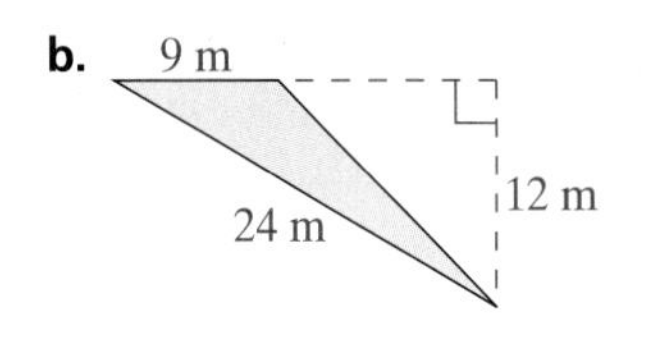

a.		b.
$A = \frac{1}{2}bh$	← Use the area formula. →	$A = \frac{1}{2}bh$
$= \frac{1}{2}(8)(5)$	← Substitute. →	$= \frac{1}{2}(9)(12)$
$= 20$	← Simplify. →	$= 54$
The area is 20 ft^2.		The area is 54 m^2.

Vocabulary Tip

Areas are always measured in *square units*.

Quick Check

2. Find the area of each triangle.

a.

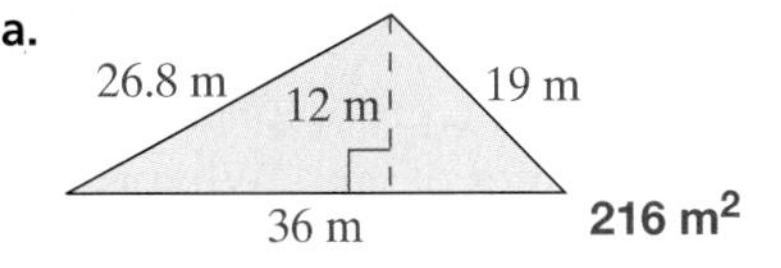

216 m^2

b. **48 cm^2**

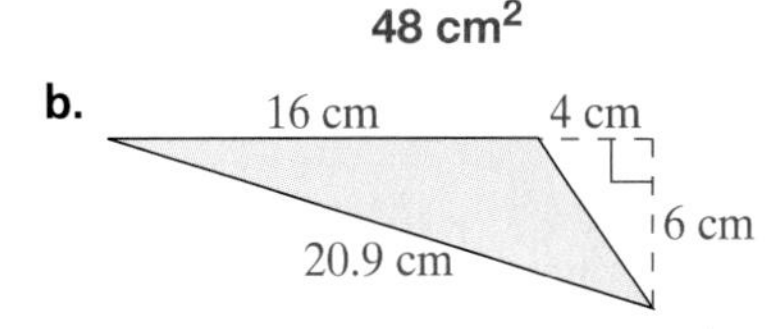

2. Teach

Activity Lab

Use before the lesson.

All in One Teaching Resources

Activity Lab 8-3: Perimeter and Area of Triangles

Guided Instruction

Example 2

Help make the connection that the area of each triangle is one half the area of the related parallelogram. Ask:

- *How many triangles form a parallelogram?* **2**
- *Write an equation to represent this. Use t for the area of the triangle and p for the area of the parallelogram.* **$2t = p$**
- *Solve for t.* **$t = \frac{p}{2}$ or $\frac{1}{2}p$**

PowerPoint Additional Examples

1 Find the perimeter of the triangle. **31 ft**

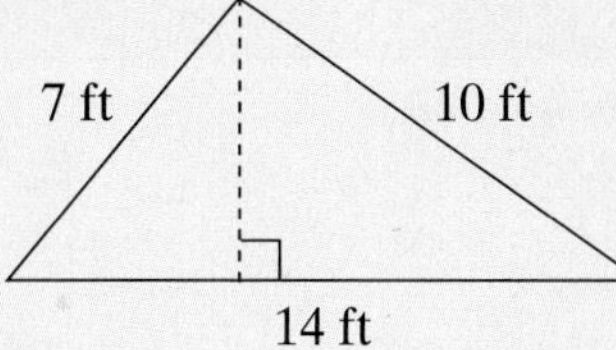

2 A school is creating a triangular playground. Find the area of the playground if the base is 80 yd and the height is 40 yd. **1,600 yd^2**

All in One Teaching Resources

- Daily Notetaking Guide 8-2 **L3**
- Adapted Notetaking 8-2 **L1**

Closure

- *How is the formula for the area of a triangle related to the formula for an area of a parallelogram?* **Sample: Both formulas use the base and height of the figure to calculate area. Both have measurements expressed in square units. The formula for the area of a triangle is $A = \frac{1}{2}bh$ or the area is one half that of a parallelogram.**

Advanced Learners L4

Jake has a garden shaped like a parallelogram. Its base is 6 m and its height is 4.5 m. If he plants carrots in half of his garden, what is the area of the carrot patch? **13.5 m^2**

learning style: visual

English Language Learners ELL

Ask students to *restate*, in their own words, what the base and height of a triangle are. Ask them to draw a triangle and label the base and height of it.

learning style: visual

3. Practice

Assignment Guide

Check Your Understanding

Go over Exercises 1–7 in class before assigning the Homework Exercises.

Homework Exercises

A Practice by Example	8–17
B Apply Your Skills	18–27
C Challenge	28
Test Prep and Mixed Review	29–33

Homework Quick Check

To check students' understanding of key skills and concepts, go over Exercises 10, 15, 24, 25, and 26.

Differentiated Instruction Resources

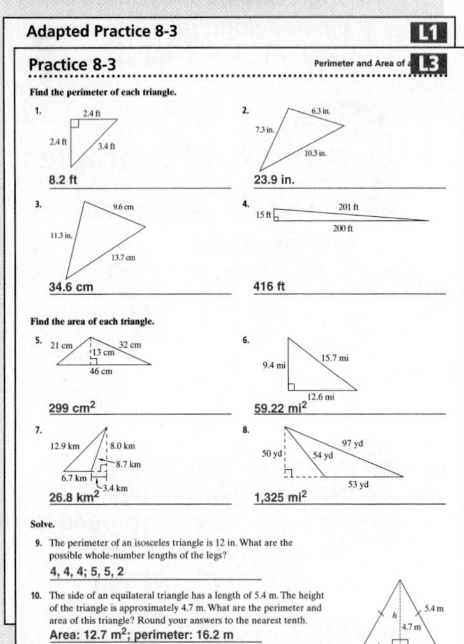

Adapted Practice 8-3 L1

Practice 8-3 — Perimeter and Area of a... L3

Find the perimeter of each triangle.

1. 2.4 ft, 2.4 ft, 3.4 ft — 8.2 ft
2. 6.3 in., 7.3 in., 10.3 in. — 23.9 in.
3. 9.6 cm, 11.3 in., 13.7 cm — 34.6 cm
4. 15 ft, 201 ft, 200 ft — 416 ft

Find the area of each triangle.

5. 21 cm, 32 cm, 13 cm, 46 cm — 299 cm²
6. 9.4 mi, 15.7 mi, 12.6 mi — 59.22 mi²
7. 12.9 km, 8.0 km, 8.7 km, 6.7 km, 3.4 km — 26.8 km²
8. 50 yd, 97 yd, 54 yd, 53 yd — 1,325 mi²

Solve.

9. The perimeter of an isosceles triangle is 12 in. What are the possible whole-number lengths of the legs?
4, 4, 4; 5, 5, 2
10. The side of an equilateral triangle has a length of 5.4 m. The height of the triangle is approximately 4.7 m. What are the perimeter and area of this triangle? Round your answers to the nearest tenth.
Area: 12.7 m²; perimeter: 16.2 m

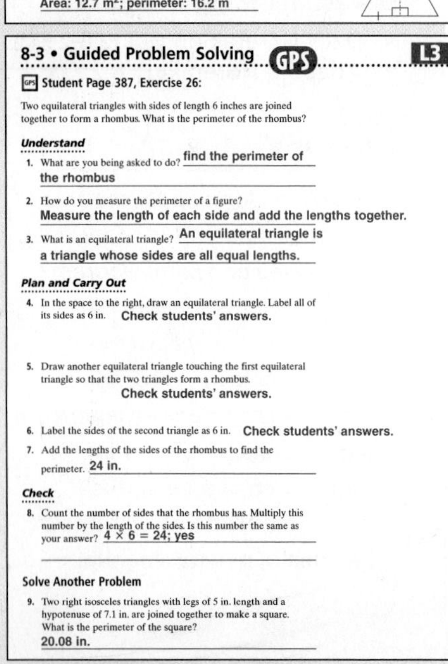

8-3 • Guided Problem Solving GPS L3

Student Page 387, Exercise 26:

Two equilateral triangles with sides of length 6 inches are joined together to form a rhombus. What is the perimeter of the rhombus?

Understand

1. What are you being asked to do? find the perimeter of the rhombus
2. How do you measure the perimeter of a figure? Measure the length of each side and add the lengths together.
3. What is an equilateral triangle? An equilateral triangle is a triangle whose sides are all equal lengths.

Plan and Carry Out

4. In the space to the right, draw an equilateral triangle. Label all of its sides as 6 in. Check students' answers.
5. Draw another equilateral triangle touching the first equilateral triangle so that the two triangles form a rhombus. Check students' answers.
6. Label the sides of the second triangle as 6 in. Check students' answers.
7. Add the lengths of the sides of the rhombus to find the perimeter. 24 in.

Check

8. Count the number of sides that the rhombus has. Multiply this number by the length of the sides. Is this number the same as your answer? 4 × 6 = 24; yes

Solve Another Problem

9. Two right isosceles triangles with legs of 5 in. length and a hypotenuse of 7.1 in. are joined together to make a square. What is the perimeter of the square?
20.08 in.

Check Your Understanding

1. **Vocabulary** A triangle that has a 90° angle is a(n) __?__ triangle. **right**

Each triangle's perimeter is 15 cm. Find the length of the missing side.

2. 4 cm, 5 cm, ■ **6 cm**
3. 2 cm, 7 cm, ■ **6 cm**
4. 1 cm, 7 cm, ■ **7 cm**

Find the area of each triangle.

5. $b = 4$ cm, $h = 5$ cm **10 cm²**
6. $b = 2$ in., $h = 7$ in. **7 in.²**
7. A carpenter has blueprints for a wooden triangular patio. The base is 5 m and the height is 7 m. What is the area of the patio? **17.5 m²**

Homework Exercises

For more exercises, see Extra Skills and Word Problems.

GO for Help

For Exercises	See Examples
8–11	1
12–17	2

A **Find the perimeter of each triangle.**

8. 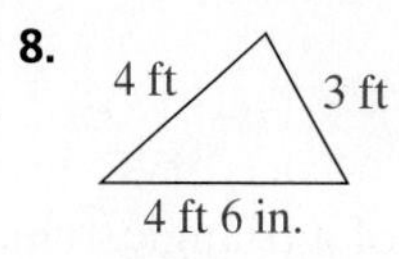

11 ft 6 in.

9. 5 ft, 4 ft, 3.5 ft **12.5 ft**

10. 5.5 cm, 5 cm, 2 cm **12.5 cm**

11. You bend a drinking straw into the shape of an equilateral triangle. Each side is 4.5 cm long. How long is the straw? **13.5 cm**

Find the area of each triangle.

12. 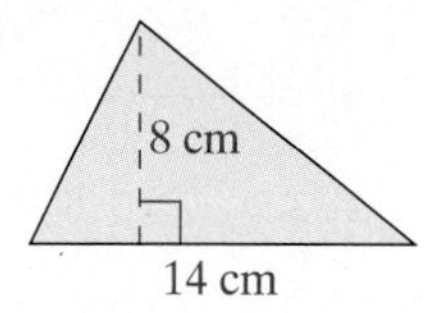

56 cm²

13. 60 yd, 48 yd **1,440 yd²**

14. 30 m, 33 m, 18 m **270 m²**

15. 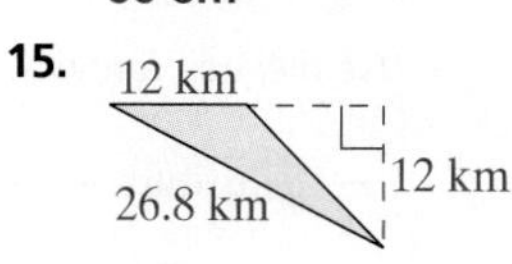

72 km²

16. 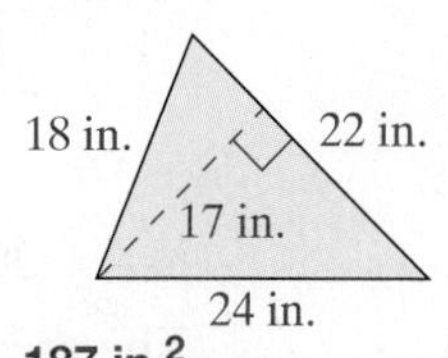

187 in.²

17. 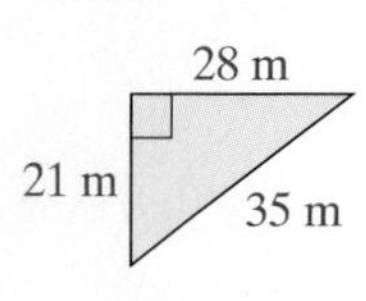

294 m²

B GPS 18. **Guided Problem Solving** An equilateral triangle's perimeter is 27 ft. The height of the triangle is 7.8 ft. What is the triangle's area?
- You can *Draw a Picture* to solve this problem. Sketch and label the triangle. Find the perimeter and then use the area formula.

35.1 ft²

19. A conservation group plans to buy a triangular plot of land. What is the area of the plot of land in the diagram? **See left.**

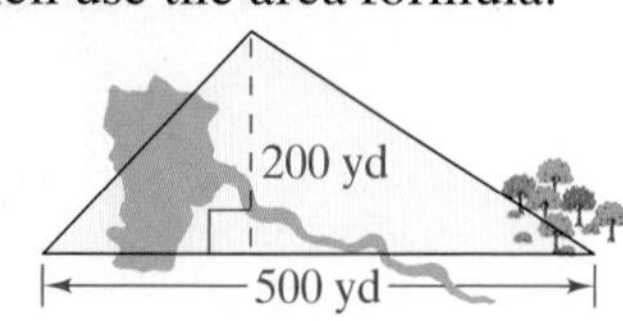

19. **50,000 yd²**

24. The area is doubled. If $b = 3$ and $h = 10$, then the area of the triangle is $A = \frac{1}{2}(3)(10) = 15$. If you double the base, then the area is $A = \frac{1}{2}(6)(10) = 30$.

27. 3 ft; if $b = 4$ and $h =$ the corresponding height, then the area of the triangle is $\frac{1}{2}(4)h$, or $2h$. Since the area is 6 ft², $2h = 6$. So $h = 3$.

Visit: PHSchool.com
Web Code: are-0803

Find the area for base b and height h of each triangle.

20. $b = 4.2$ in. $h = 6.3$ in. **13.23 in.^{2}**
21. $b = 12$ m $h = 17$ m **102 m^2**
22. $h = 6.2$ ft $b = 2.5$ ft **7.75 ft^2**
23. $h = 100$ km $b = 200$ km **10,000 km^2**

24. **Writing in Math** The base of a triangle is doubled and the height remains the same. Explain how the area changes. Use examples. **See margin.**

25. A rescue helicopter receives a distress call from a ship at sea. The diagram at the right displays the search pattern the helicopter will use. Each pass from a central point forms an equilateral triangle. What is the area of one of the triangular regions? **27.68 km^2**

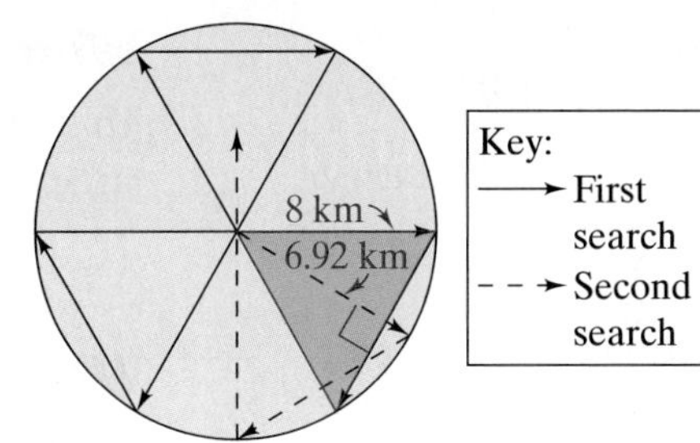

Key:
→ First search
- - → Second search

Careers Rescue swimmer is one of the jobs offered in the coast guard.

26. Two equilateral triangles with sides of length 6 inches are joined together to form a rhombus. What is the perimeter of the rhombus? **24 in.**

27. **Reasoning** One base of a triangle has a length of 6 ft and a corresponding height of 2 ft. This means that the area of the triangle is $\frac{1}{2}(6 \text{ ft} \cdot 2 \text{ ft}) = 6 \text{ ft}^2$. Another base of the same triangle has a length of 4 ft. What is its corresponding height? Explain. **See margin.**

28. **Challenge** The area of an isosceles right triangle is 121 ft^2. What is the approximate length of each of the two equal sides? **15.6 ft**

Test Prep and Mixed Review

Practice

Multiple Choice

29. Which two of the figures shown have the same area? **C**

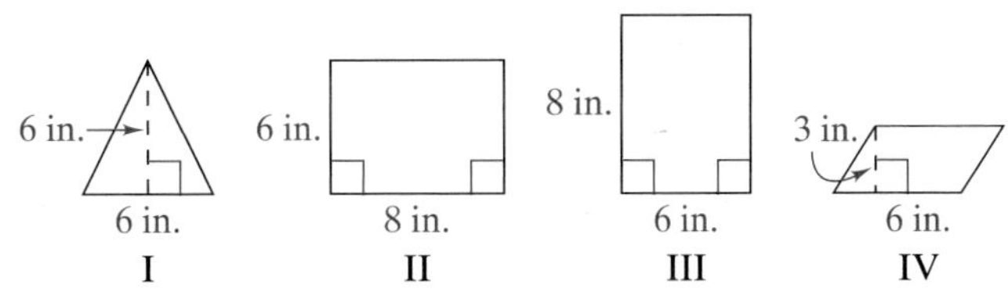

Ⓐ Figures I and II
Ⓑ Figures I and III
Ⓒ Figures I and IV
Ⓓ Figures II and IV

30. A square has a perimeter of x feet. What is its area in terms of x? **F**

Ⓕ $\frac{x^2}{16}$ Ⓖ $4x^2$ Ⓗ $\frac{x^2}{4}$ Ⓙ $\frac{x}{16}$

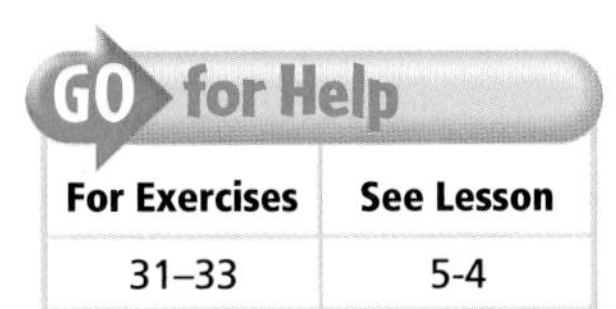

For Exercises	See Lesson
31–33	5-4

Algebra **Solve each proportion using mental math.**

31. $\frac{m}{35} = \frac{4}{5}$ **28**
32. $\frac{55}{99} = \frac{5}{x}$ **9**
33. $\frac{9}{p} = \frac{180}{200}$ **10**

Online lesson quiz, PHSchool.com, Web Code: ara-0803

Alternative Assessment

Each student in a pair draws a triangle and labels the base and lengths of each leg. Partners exchange drawings and find the perimeter and area of each other's figure. If necessary, direct students to find the area to the nearest tenth. Remind students to report area in square units.

Test Prep

Resources

For additional practice with a variety of test item formats:
- Test-Taking Strategies, p. 427
- Test Prep, p. 431
- Test-Taking Strategies with Transparencies

4. Assess & Reteach

Lesson Quiz

Use the figure below.

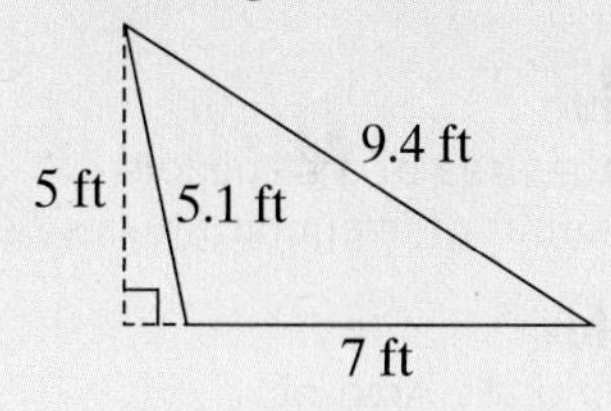

1. Find the perimeter. **21.5 ft**
2. Find the area. **17.5 ft^2**

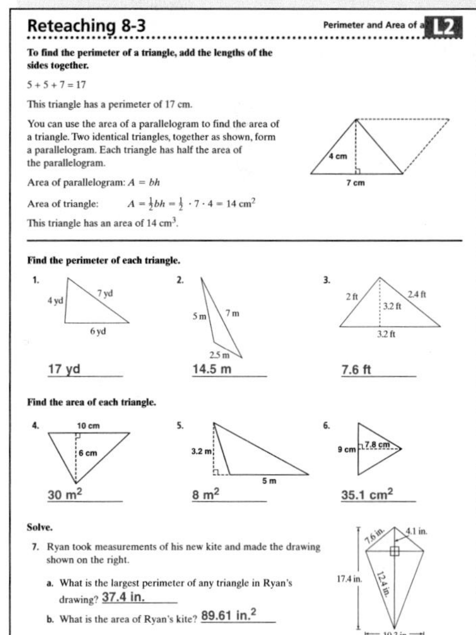

Reteaching 8-3 Perimeter and Area of a Triangle **L2**

To find the perimeter of a triangle, add the lengths of the sides together.

$5 + 5 + 7 = 17$

This triangle has a perimeter of 17 cm.

You can use the area of a parallelogram to find the area of a triangle. Two identical triangles, together as shown, form a parallelogram. Each triangle has half the area of the parallelogram.

4 cm
7 cm

Area of parallelogram: $A = bh$

Area of triangle: $A = \frac{1}{2}bh = \frac{1}{2} \cdot 7 \cdot 4 = 14 \text{ cm}^2$

This triangle has an area of 14 cm^3.

Find the perimeter of each triangle.

1. 4 yd, 7 yd, 6 yd — **17 yd**
2. 5 m, 7 m, 2.5 m — **14.5 m**
3. 2 ft, 2.4 ft, 3.2 ft, 3.2 ft — **7.6 ft**

Find the area of each triangle.

4. 10 cm, 6 cm — **30 m^2**
5. 3.2 m, 5 m — **8 m^2**
6. 9 cm, 7.8 cm — **35.1 cm^2**

Solve.

7. Ryan took measurements of his new kite and made the drawing shown on the right.
   a. What is the largest perimeter of any triangle in Ryan's drawing? **37.4 in.**
   b. What is the area of Ryan's kite? **89.61 in.^{2}**

7.6 in., 4.1 in., 17.4 in., 12.4 in., 10.3 in.

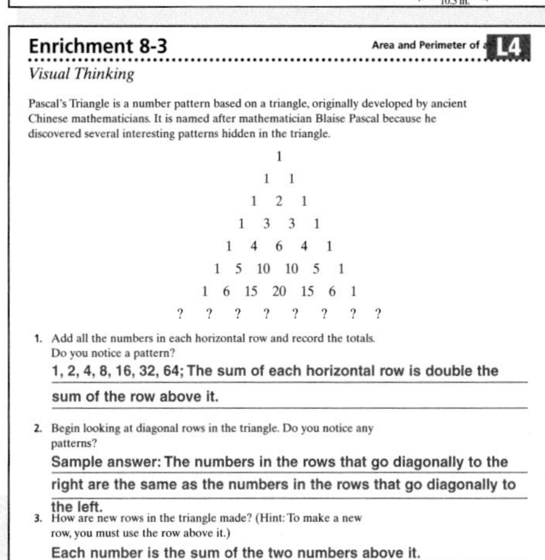

Enrichment 8-3 Area and Perimeter of a Triangle **L4**

Visual Thinking

Pascal's Triangle is a number pattern based on a triangle, originally developed by ancient Chinese mathematicians. It is named after mathematician Blaise Pascal because he discovered several interesting patterns hidden in the triangle.

1
1 1
1 2 1
1 3 3 1
1 4 6 4 1
1 5 10 10 5 1
1 6 15 20 15 6 1
? ? ? ? ? ? ? ?

1. Add all the numbers in each horizontal row and record the totals. Do you notice a pattern?
   1, 2, 4, 8, 16, 32, 64; The sum of each horizontal row is double the sum of the row above it.
2. Begin looking at diagonal rows in the triangle. Do you notice any patterns?
   Sample answer: The numbers in the rows that go diagonally to the right are the same as the numbers in the rows that go diagonally to the left.
3. How are new rows in the triangle made? (Hint: To make a new row, you must use the row above it.)
   Each number is the sum of the two numbers above it.
4. Write the values for the next row in the triangle.
   1, 7, 21, 35, 35, 21, 7, 1

8-4

1. Plan

Objective
To find the area of a trapezoid and the areas of irregular figures

Examples
1 Finding the Area of a Trapezoid
2 Application: Geography

Math Understandings: p. 372C

Math Background

A *trapezoid* is a four-sided figure with one pair of parallel sides that form the bases. The height of a trapezoid is the perpendicular distance between bases. Take two identical trapezoids, rotate one, and put their sides together to form a parallelogram. Since the formula for the area of a parallelogram is $A = b \times h$, the formula for the area of a trapezoid is $\frac{1}{2}$ of the height multiplied by the sum of the bases: $A = \frac{1}{2}h(b_1 + b_2)$, read as "area equals one half height times the quantity of base one plus base two."

More Math Background: p. 372C

Lesson Planning and Resources

See p. 372E for a list of the resources that support this lesson.

Bell Ringer Practice

Check Skills You'll Need
Use student page, transparency, or PowerPoint. For intervention, direct students to:
Area of a Parallelogram
Lesson 8-2
Extra Skills and Word Problems Practice, Ch. 8

8-4 Areas of Other Figures

Check Skills You'll Need

1. **Vocabulary Review** What is the *base of a parallelogram?* **See below.**

Find the area of each figure.

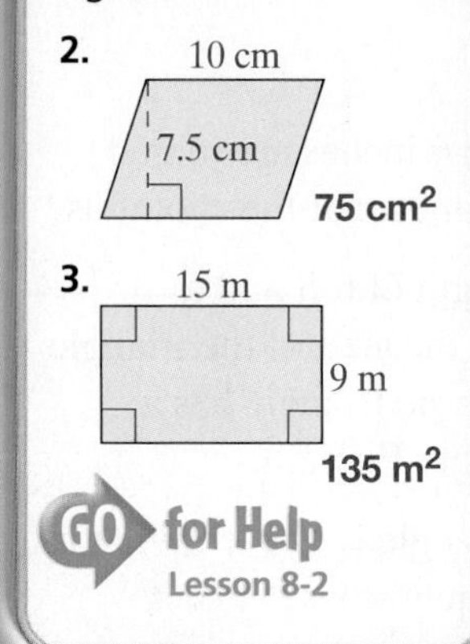

GO for Help Lesson 8-2

What You'll Learn

To find the area of a trapezoid and the areas of irregular figures

New Vocabulary base of a trapezoid, height of a trapezoid

Why Learn This?

If you know how to find the area of simple figures, you can find the area of an irregular figure, such as the area of a backyard deck.

The formula for the area of a trapezoid follows from the formula for the area of a parallelogram.

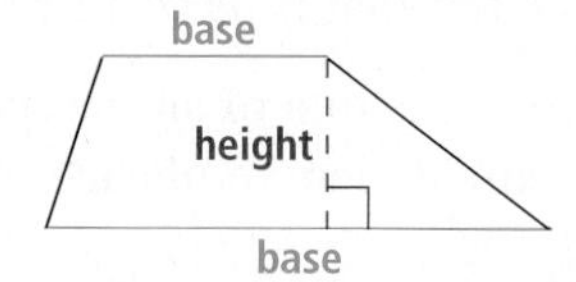

The two parallel sides of a trapezoid are the **bases of a trapezoid**, with lengths b_1 and b_2. The **height of a trapezoid** h is the length of a perpendicular segment connecting the bases.

If you put two identical trapezoids together, you get a parallelogram. The area of the parallelogram is $(b_1 + b_2)h$. The area of one trapezoid equals $\frac{1}{2}(b_1 + b_2)h$.

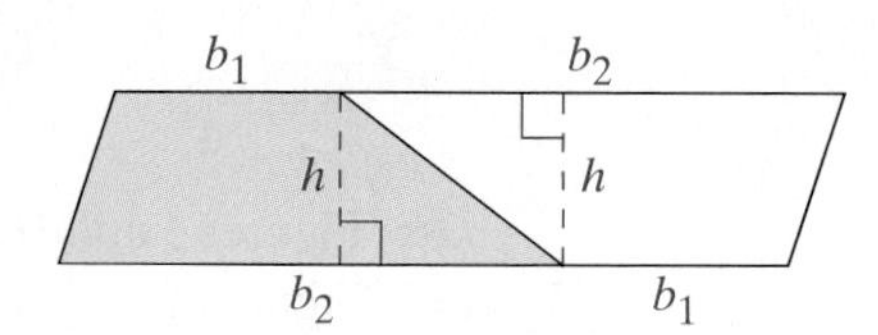

Vocabulary Tip
A *trapezoid* is a quadrilateral that has exactly one pair of parallel sides. Read b_1 as "base 1" and b_2 as "base 2."

Check Skills You'll Need
1. It is one of a pair of opposite sides.

KEY CONCEPTS **Area of a Trapezoid**

The area of a trapezoid is one half the product of the height and the sum of the lengths of the bases.

$$A = \frac{1}{2}h(b_1 + b_2)$$

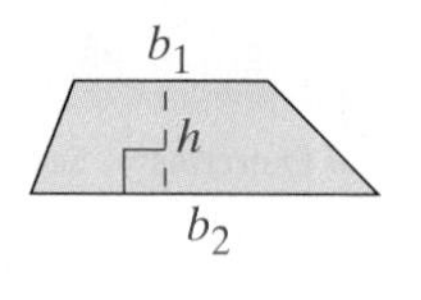

Differentiated Instruction Solutions for All Learners

Special Needs L1
Ask students to draw a triangle, parallelogram, and trapezoid, one on each of three index cards. Ask them to write the formula for the area of each next to the drawing. Suggest that students refer to the cards as they work through the chapter.
learning style: visual

Below Level L2
Review with students the Order of Operations and multiplying a number by $\frac{1}{2}$.
learning style: verbal

EXAMPLE Finding the Area of a Trapezoid

1 Find the area of the trapezoid shown at the right.

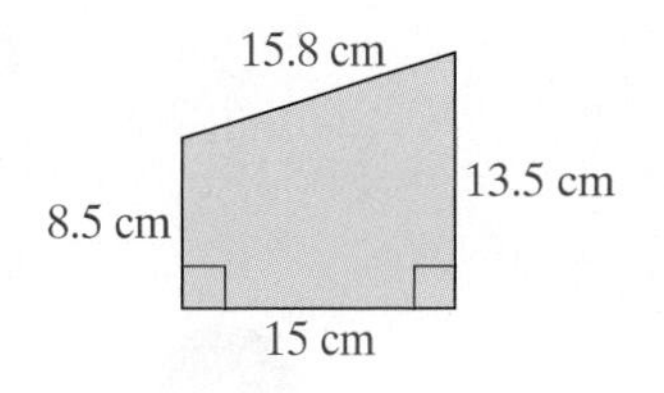

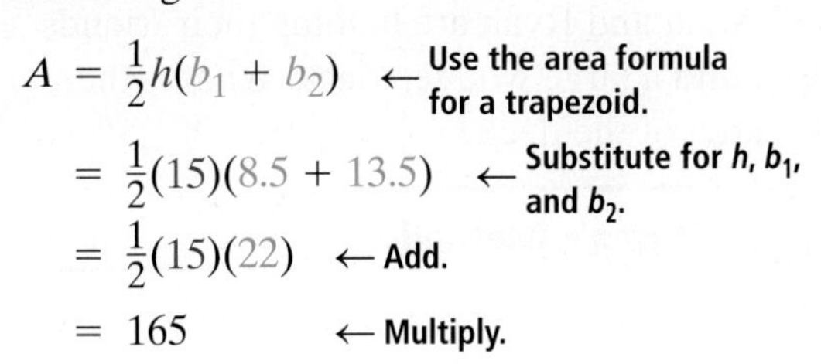

$A = \frac{1}{2}h(b_1 + b_2)$ ← **Use the area formula for a trapezoid.**

$= \frac{1}{2}(15)(8.5 + 13.5)$ ← **Substitute for h, b_1, and b_2.**

$= \frac{1}{2}(15)(22)$ ← **Add.**

$= 165$ ← **Multiply.**

The area of the trapezoid is 165 cm^2.

Quick Check

1. Find the area of each trapezoid.

a.

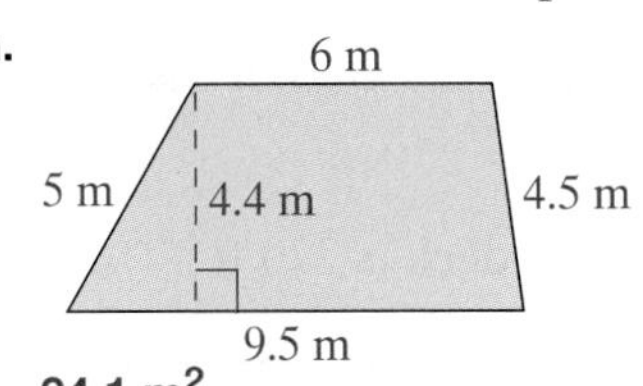

34.1 m^2

b.

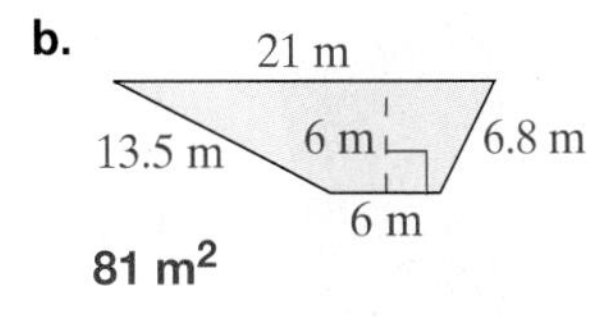

81 m^2

You can estimate the area of states shaped like trapezoids.

EXAMPLE Application: Geography

2 Estimate the area of Arkansas by finding the area of the trapezoid shown.

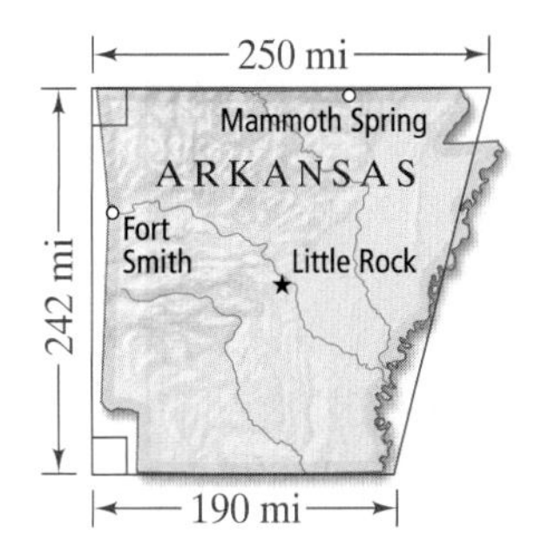

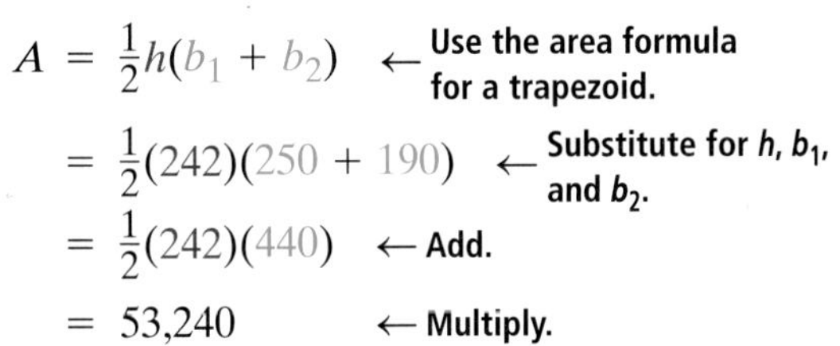

$A = \frac{1}{2}h(b_1 + b_2)$ ← **Use the area formula for a trapezoid.**

$= \frac{1}{2}(242)(250 + 190)$ ← **Substitute for h, b_1, and b_2.**

$= \frac{1}{2}(242)(440)$ ← **Add.**

$= 53{,}240$ ← **Multiply.**

The area of Arkansas is about 53,240 mi^2.

At Crater of Diamonds State Park in Arkansas, visitors can search for and keep diamonds and other gems.

Quick Check

2. Estimate the area of the figure at the right by finding the area of the trapezoid. **8.75 $in.^2$**

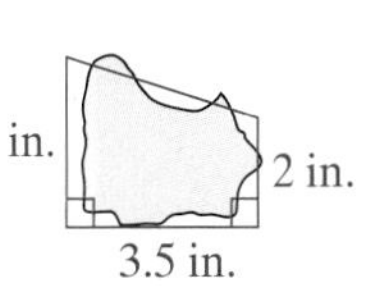

Activity Lab

Use before the lesson.

All in One Teaching Resources

Activity Lab 8-4: Extend Your Thinking About Areas

Guided Instruction

Error Prevention!

In Example 1, students may think that the bases are always the top and bottom sides. Remind them that the bases need to be parallel. If students are used to considering the bases as the top and bottom, have them simply rotate their books.

Additional Examples

1 Find the area of each trapezoid. **48 cm^2**

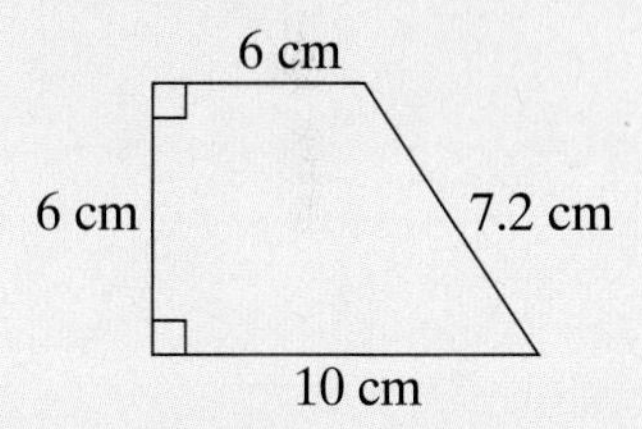

2 Estimate the area of the piece of metal by using the trapezoid shown. **126 cm^2**

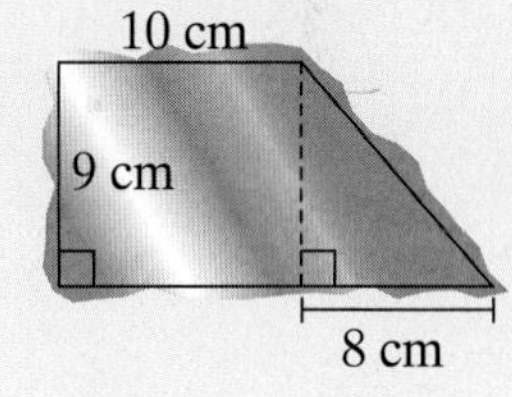

Advanced Learners L4
A trapezoid has an area of 96 m^2, a height of 12 m, and one base of 6 m. What is the measure of the other base? **10 m**

learning style: visual

English Language Learners ELL
Ask students to read the formulas given for each of the shapes they have learned. Make sure they use words for A, $\frac{1}{2}$, b, h, l and w. Ask them to show you, on drawings, where each part is on particular figures.

learning style: verbal

All in One Teaching Resources
- Daily Notetaking Guide 8-4 L3
- Adapted Notetaking 8-4 L1

Closure

- *What is a trapezoid?* a four-sided figure in which exactly two sides are parallel
- *How can you calculate the area of a trapezoid?* $A = \frac{1}{2}h(b_1 + b_2)$

You can find the area of any figure by separating it into familiar figures.

More Than One Way

Anna and Ryan are helping their friends build a large wooden deck. What is the area of the deck?

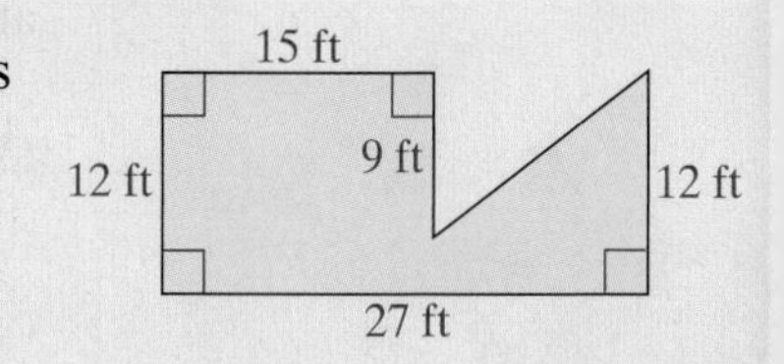

Anna's Method

I'll subtract the area of the triangle from the area of the rectangle.

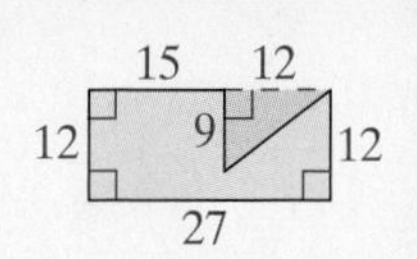

Area of the rectangle:

$A = bh$
$= (27)(12) = 324$

Area of the triangle:

$A = \frac{1}{2}bh$
$= \frac{1}{2}(12)(9) = 54$

Now I'll subtract the area of the triangle from the area of the rectangle.

$A = 324 - 54 = 270$

The area of the deck is 270 ft^2.

Ryan's Method

I'll add the areas of the rectangle and the trapezoid.

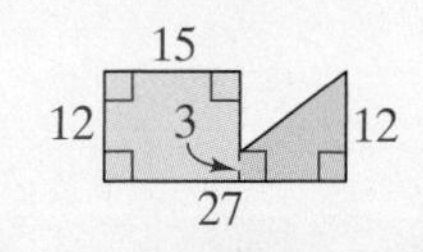

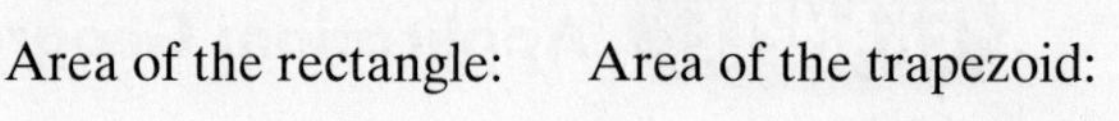

Area of the rectangle: Area of the trapezoid:

$+ b_2)$

$3 + 12)$

5 m

p. 391-392
2-6, 10-11, 12, 18
12) 15m²
18.) 3300 cm² = A
260 cm = p

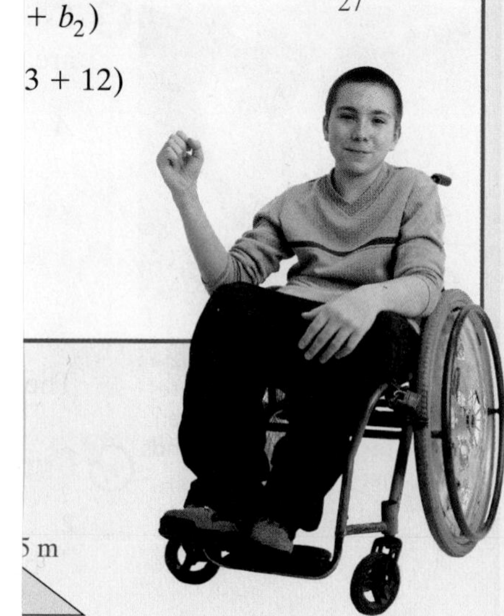

Check Your Understanding

1. **Vocabulary** The perpendicular distance between the two parallel sides of a trapezoid is called the __?__ of the trapezoid. **height**

Identify the bases b_1 and b_2 and height h of each trapezoid below.

2. 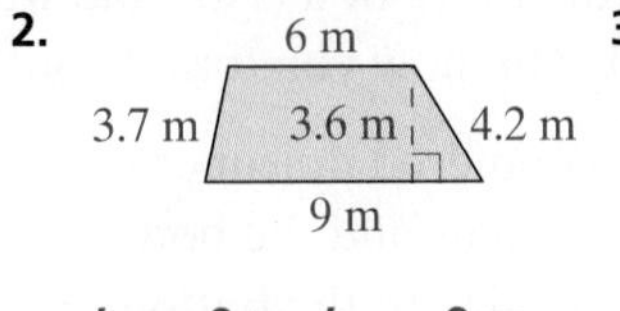

$b_1 = 6$ m, $b_2 = 9$ m, $h = 3.6$ m

3. 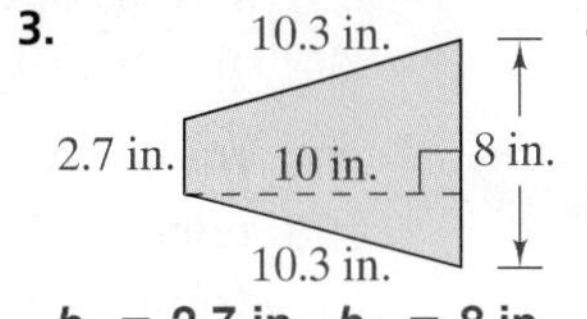

$b_1 = 2.7$ in., $b_2 = 8$ in., $h = 10$ in.

4. 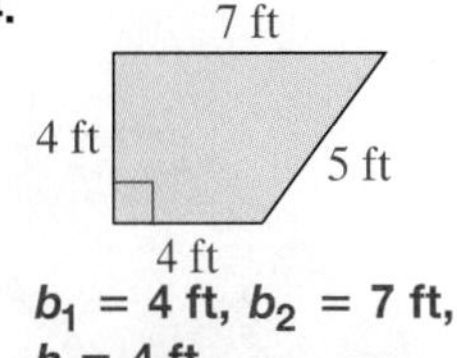

$b_1 = 4$ ft, $b_2 = 7$ ft, $h = 4$ ft

Homework Exercises

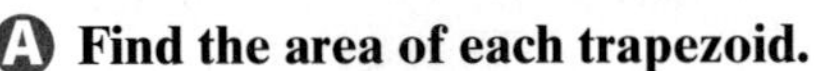

For more exercises, see Extra Skills and Word Problems.

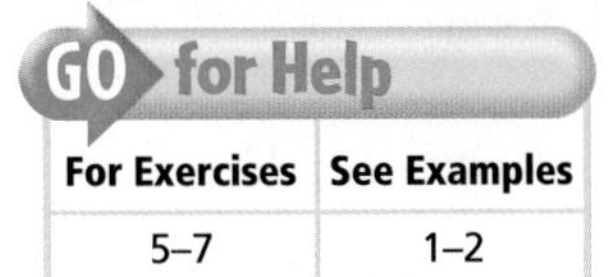

GO for Help

For Exercises	See Examples
5–7	1–2

A **Find the area of each trapezoid.**

5. 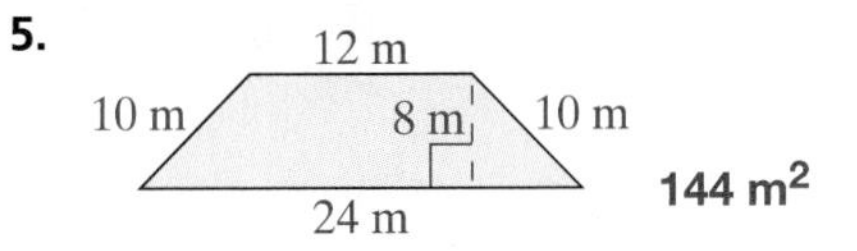

144 m²

6. 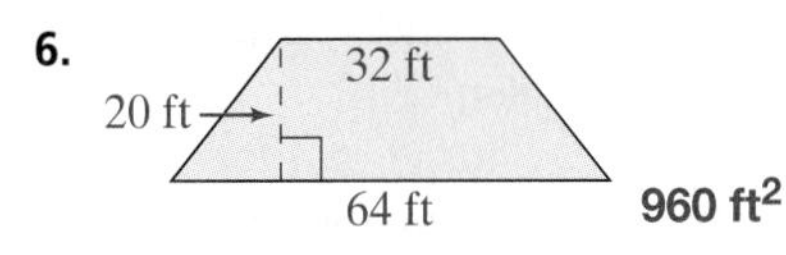

960 ft²

7. **Engineering** When the Erie Canal opened in 1825, it was hailed as an engineering marvel. Find the area of the trapezoidal cross section of the Erie Canal at the right. **136 ft²**

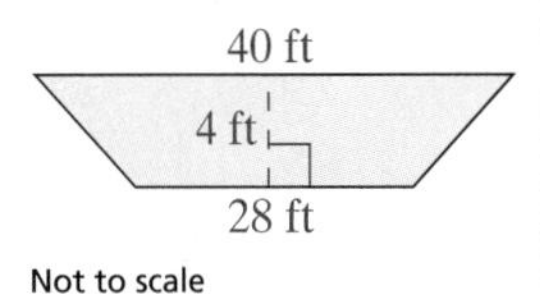

Not to scale

B 

8. **Guided Problem Solving** Estimate the area of Nevada by finding the area of the trapezoid shown at the right. **110,622 mi²**
   - Which measurements in the diagram will you use for the bases and the height?
   - How will you use the bases and height to calculate the area?

9. **Choose a Method** You plan to replace the carpeting in the room shown at the left. What is the area of the room? **198 ft²**

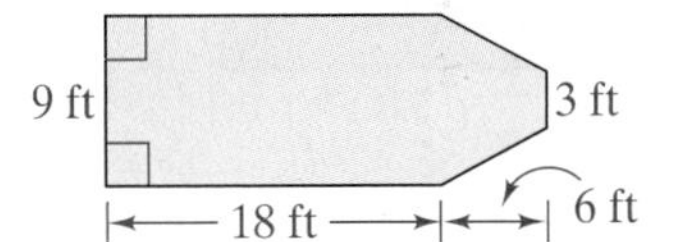

Use familiar figures to find the area of each irregular figure.

10. 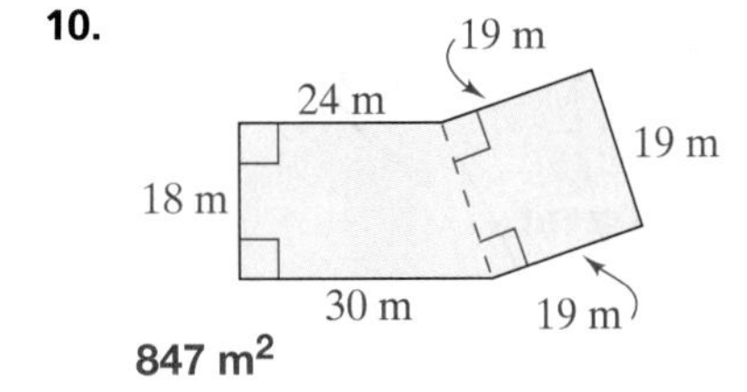

847 m²

11. 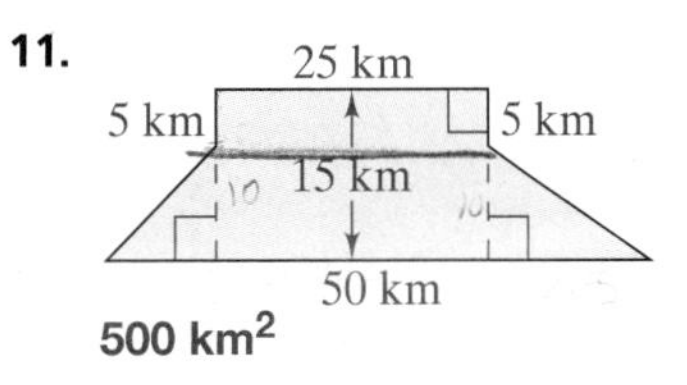

500 km²

3. Practice

Assignment Guide

Check Your Understanding

Go over Exercises 1–4 in class before assigning the Homework Exercises.

Homework Exercises

A	Practice by Example	5–7
B	Apply Your Skills	8–19
C	Challenge	20
	Test Prep and Mixed Review	21–24

Homework Quick Check

To check students' understanding of key skills and concepts, go over Exercises 5, 7, 9, 16, 17, and 18.

Differentiated Instruction Resources

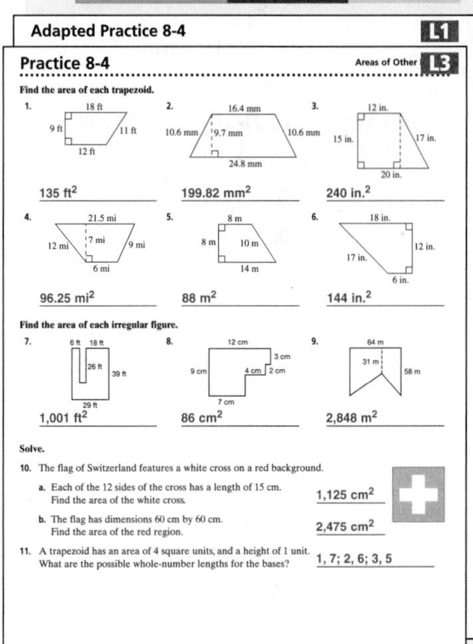

Adapted Practice 8-4 L1

Practice 8-4 Areas of Other Figures L3

Find the area of each trapezoid.

1. 135 ft²
2. 199.82 mm²
3. 240 in.²
4. 96.25 mi²
5. 88 m²
6. 144 in.²

Find the area of each irregular figure.

7. 1,001 ft²
8. 86 cm²
9. 2,848 m²

Solve.

10. The flag of Switzerland features a white cross on a red background.
    a. Each of the 12 sides of the cross has a length of 15 cm. Find the area of the white cross. 1,125 cm²
    b. The flag has dimensions 60 cm by 60 cm. Find the area of the red region. 2,475 cm²
11. A trapezoid has an area of 4 square units, and a height of 1 unit. What are the possible whole-number lengths for the bases? 1, 7; 2, 6; 3, 5

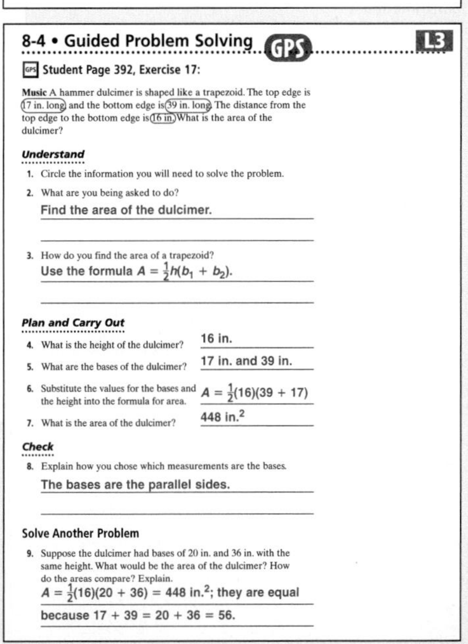

8-4 • Guided Problem Solving GPS L3

Student Page 392, Exercise 17:

Music A hammer dulcimer is shaped like a trapezoid. The top edge is 17 in. long and the bottom edge is 39 in. long. The distance from the top edge to the bottom edge is 16 in. What is the area of the dulcimer?

Understand

1. Circle the information you will need to solve the problem.
2. What are you being asked to do? Find the area of the dulcimer.
3. How do you find the area of a trapezoid? Use the formula $A = \frac{1}{2}h(b_1 + b_2)$.

Plan and Carry Out

4. What is the height of the dulcimer? 16 in.
5. What are the bases of the dulcimer? 17 in. and 39 in.
6. Substitute the values for the bases and the height into the formula for area. $A = \frac{1}{2}(16)(39 + 17)$
7. What is the area of the dulcimer? 448 in.²

Check

8. Explain how you chose which measurements are the bases. The bases are the parallel sides.

Solve Another Problem

9. Suppose the dulcimer had bases of 20 in. and 36 in. with the same height. What would be the area of the dulcimer? How do the areas compare? Explain. $A = \frac{1}{2}(16)(20 + 36) = 448$ in.²; they are equal because 17 + 39 = 20 + 36 = 56.

4. Assess & Reteach

Lesson Quiz

Find the area of each figure.

1. 148.5 cm^2

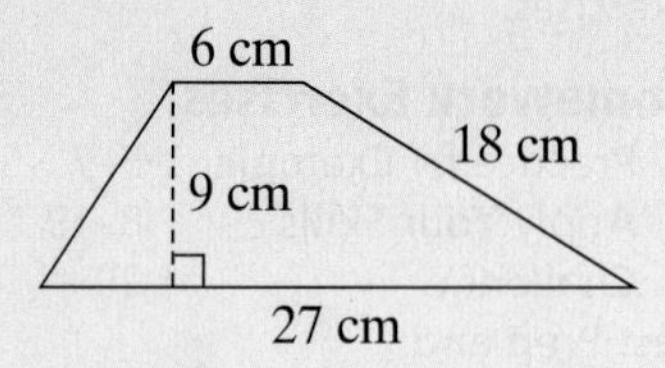

2. 240 mm^2

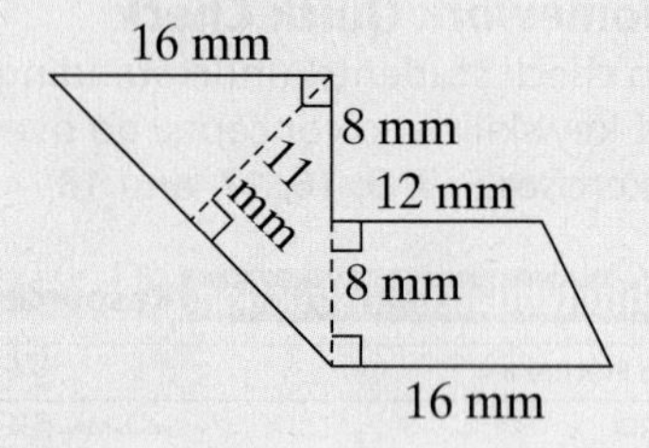

Alternative Assessment

Have students work in pairs. Each student draws and labels the bases and height of two trapezoids. Partners exchange drawings and find the areas to the nearest tenth. Then each student draws an irregular figure similar to those in Exercises 18 and 19. Again partners exchange drawings and find the areas.

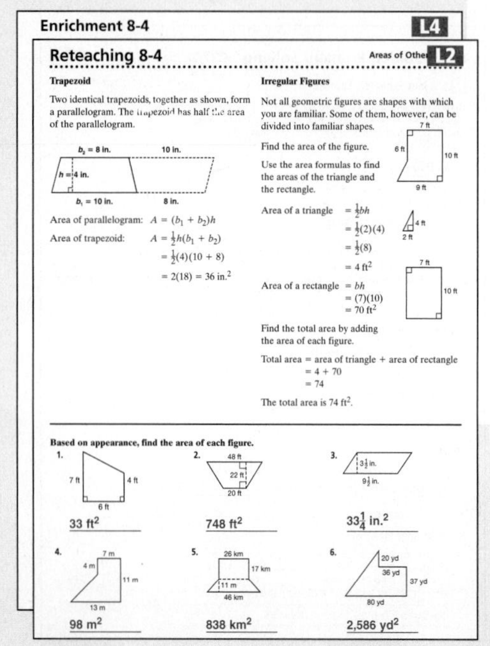

Enrichment 8-4 L4

Reteaching 8-4 Areas of Other Figures L2

Trapezoid

Two identical trapezoids, together as shown, form a parallelogram. The trapezoid has half the area of the parallelogram.

Area of parallelogram: $A = (b_1 + b_2)h$

Area of trapezoid: $A = \frac{1}{2}h(b_1 + b_2)$
$= \frac{1}{2}(4)(10 + 8)$
$= 2(18) = 36 \text{ in.}^2$

Irregular Figures

Not all geometric figures are shapes with which you are familiar. Some of them, however, can be divided into familiar shapes.

Find the area of the figure.

Use the area formulas to find the areas of the triangle and the rectangle.

Area of a triangle $= \frac{1}{2}bh$
$= \frac{1}{2}(2)(4)$
$= \frac{1}{2}(8)$
$= 4 \text{ ft}^2$

Area of a rectangle $= bh$
$= (7)(10)$
$= 70 \text{ ft}^2$

Find the total area by adding the area of each figure.

Total area = area of triangle + area of rectangle
$= 4 + 70$
$= 74$

The total area is 74 ft^2.

Based on appearance, find the area of each figure.

1. 33 ft^2
2. 748 ft^2
3. $33\frac{3}{4} \text{ in.}^2$
4. 98 m^2
5. 838 km^2
6. $2{,}586 \text{ yd}^2$

GO Online
Homework Video Tutor
Visit: PHSchool.com
Web Code: are-0804

15. $5\frac{3}{4} \text{ ft}^2$

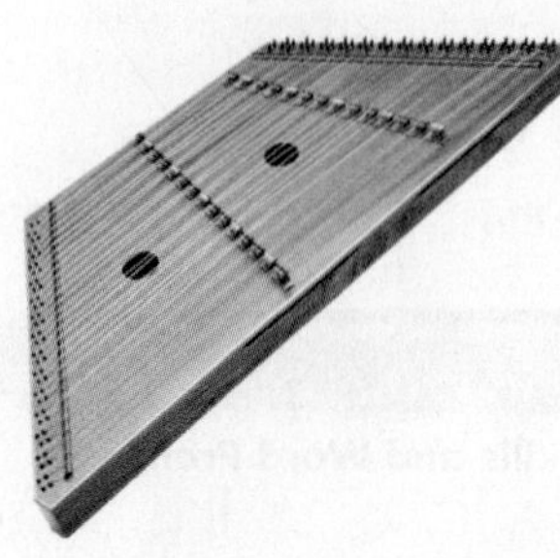

16. 1 and 5; 2 and 4; 3 and 3
By solving
$3 = \frac{1}{2} \cdot 1(b_1 + b_2)$ for $b_1 + b_2$, you find the sum of the numbers must be 6.

Find the area of each trapezoid, given the bases b_1 and b_2 and height h.

12. $b_1 = 3$ m
$b_2 = 7$ m
$h = 3$ m
15 m^2

13. $b_1 = 11$ in.
$b_2 = 16$ in.
$h = 9$ in.
121.5 in.^2

14. $b_1 = 5.6$ cm
$b_2 = 8.5$ cm
$h = 6$ cm
42.3 cm^2

15. $b_1 = 3\frac{1}{2}$ ft
$b_2 = 2\frac{1}{4}$ ft
$h = 2$ ft
See left.

16. **Writing in Math** Find the whole-number possibilities for the lengths of the bases of a trapezoid with a height of 1 m and an area of 3 m^2. Explain how you found your answer. See left.

17. **Music** A hammered dulcimer is shaped like a trapezoid. The top edge is 17 in. long, and the bottom edge is 39 in. long. The distance from the top edge to the bottom edge is 16 in. What is the area of the dulcimer? 448 in.^2

Use familiar figures to find the area and perimeter of each figure.

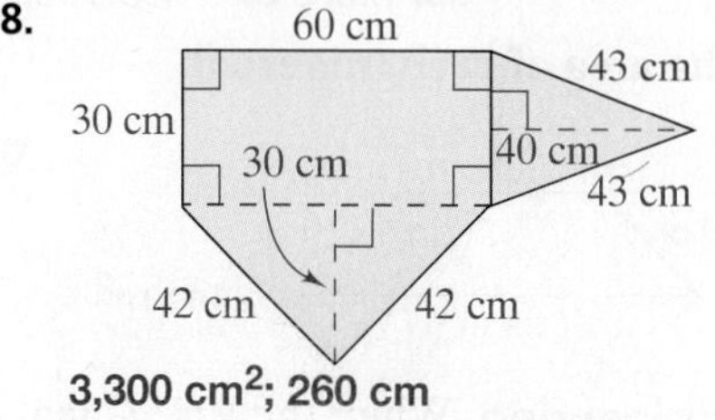

18. $3{,}300 \text{ cm}^2$; 260 cm

19. 104 m^2; 45.6 m

20. **Challenge** A trapezoid has an area of 184 in.^2. The height is 8 in. and the length of one base is 16 in. Write and solve an equation to find the length of the other base. $184 = \frac{1}{2}(8)(16 + b_2)$; 30 in.

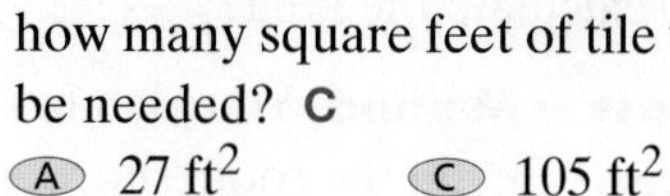

Test Prep and Mixed Review

Practice

Multiple Choice

21. The Hernandez family is purchasing tile for the kitchen shown in the diagram. If tile is not needed for the island area, how many square feet of tile will be needed? C

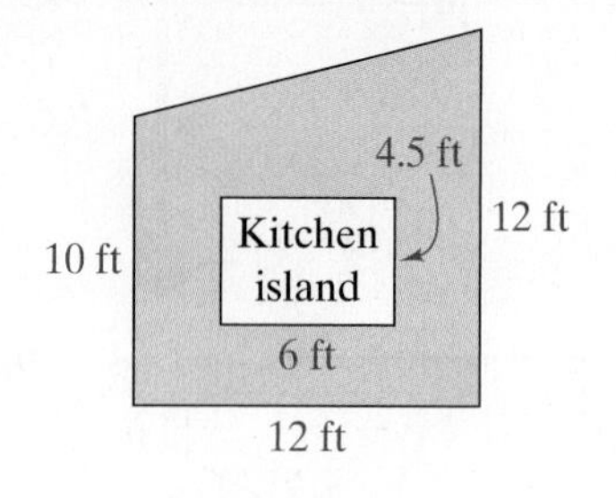

Ⓐ 27 ft^2 Ⓒ 105 ft^2
Ⓑ 93 ft^2 Ⓓ 120 ft^2

22. A carpenter finds that the measure of an angle formed between a vaulted ceiling and one wall is 67°. How many degrees is the supplement of this angle? H

Ⓕ 23° Ⓖ 103° Ⓗ 113° Ⓙ 157°

GO for Help

For Exercises	See Lesson
23–24	6-7

Find each payment.

23. \$453 with a 6% sales tax \$480.18

24. \$49.95 with a 5.5% sales tax \$52.70

Test Prep

Resources

For additional practice with a variety of test item formats:

- Test-Taking Strategies, p. 427
- Test Prep, p. 431
- Test-Taking Strategies with Transparencies

Checkpoint Quiz 1

Lessons 8-1 through 8-4

Each square below represents 20 km². Estimate the area of each shaded region.

1. 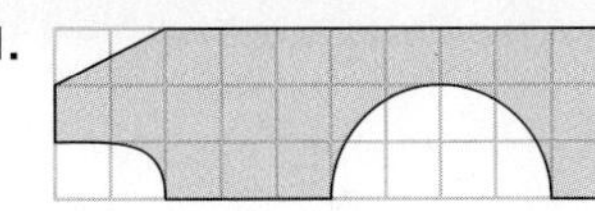
   420 km²

2. 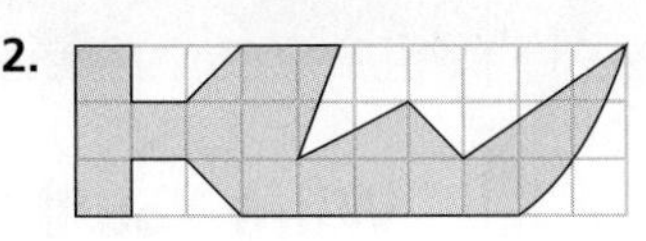
   360 km²

Find the area of each figure. Where necessary, use familiar figures.

3. 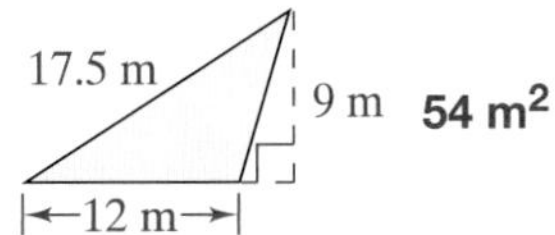

   54 m²

4. 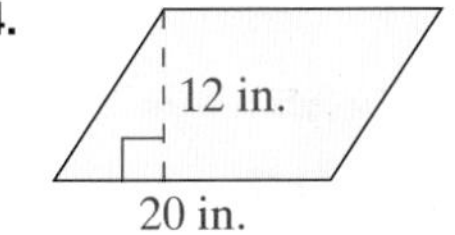

   240 in.²

5. 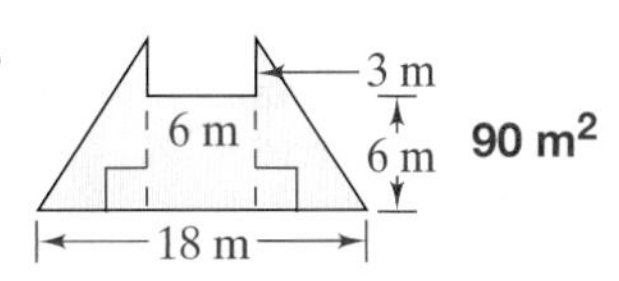

   90 m²

8-5a Activity Lab

Hands On

Modeling a Circle

When a regular polygon has many sides, it can be a model for a circle.

ACTIVITY

1. Form a chain of drinking straws by stapling them e...
2. Make a regular polygon using a chain of about 15 straws. Use another chain of straws to measure the widest distance across the polygon.
3. Record the number of straws you used in the table at the right. Measure the length of each straw to the nearest centimeter, and calculate the distance around and across each polygon.

Number of Straws		Distance (cm)		Around ÷ Across (cm)
Around	Across	Around	Across	
■	■	■	■	■
■	■	■	■	■

4. Repeat steps 2 and 3 using 20 straws and 30 straws.
   5–6. See margin.
5. Calculate and record the ratio of "distance around ÷ distance across" for each polygon. What pattern do you notice?
6. Suppose a regular polygon of 100 sides has a distance across of 100 cm. What is the distance around the polygon? Explain.

5. The ratio of distance around ÷ distance across is about 3 for each polygon.

6. About 300 cm; the distance around a regular polygon is about 3 times the distance across.

Checkpoint Quiz

Use this Checkpoint Quiz to check students' understanding of the skills and concepts of Lessons 8-1 through 8-4.

Resources

- All-in-One Teaching Resources Checkpoint Quiz 1
- ExamView Assessment Suite CD-ROM
- Success Tracker™ Online Intervention

Activity Lab

Modeling a Circle

Use a model of a polygon with many sides to help to understand the dimensions of a circle and prepare to calculate the circumference and area of a circle in Lesson 8-5.

Guided Instruction

Ask questions such as:

- *Why is it difficult to measure a circle with a ruler?* **It does not have straight sides.**
- *Why is a model with many sides better than a model with fewer sides?* **It more closely approximates a circle.**

Activity
Have students in pairs take turns measuring and recording results, then together answer questions.

Differentiated Instruction

Below Level L2
Some students may have difficulty with the straw chain. Provide string for the chain of segments.

Resources

- Activity Lab 8-5: Areas of Circles
- straws, stapler, string

8-5

1. Plan

Objective
To find the circumference and area of a circle

Examples
1 Finding the Circumference of a Circle
2 Finding the Area of a Circle

Math Understandings: p. 372C

Math Background

Similar to perimeter the *circumference* is the distance around a circle. It is calculated by using the ratio $\pi = \frac{C}{d}$, (C is circumference, d is diameter, and π is pi). Pi is a constant for all circles and is approximated by $\frac{22}{7}$ or 3.14. So the circumference of a circle is $C = \pi d$. Because the radius of a circle is half its diameter, the circumference can also be expressed as $C = 2\pi r$. The *area* of a circle is $A = \pi r^2$ and is measured in square units.

More Math Background: p. 372C

Lesson Planning and Resources

See p. 372E for a list of the resources that support this lesson.

Bell Ringer Practice

Check Skills You'll Need
Use student page, transparency, or PowerPoint. For intervention, direct students to:
Circles
Lesson 7-6
Extra Skills and Word Problems Practice, Ch. 7

8-5 Circumference and Area of a Circle

Check Skills You'll Need

1. **Vocabulary Review** What is the name of the segment that connects a circle to its center? **radius**

Name each segment for circle O.

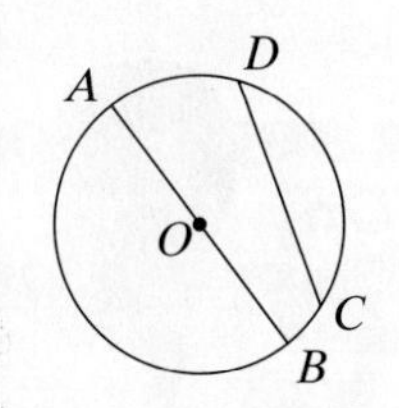

2. radius $\overline{OA}$ or $\overline{OB}$
3. chord $\overline{DC}$ or $\overline{AB}$
4. diameter $\overline{AB}$

Lesson 7-6

What You'll Learn

To find the circumference and area of a circle

New Vocabulary circumference, pi

Why Learn This?

If you know how to find the circumference of a circle, you can find how far you must travel to move all the way around the circle.

In the picture below, a Sacagawea dollar rolls along a surface. The distance the dollar rolls is the same as the distance around the edge of the dollar. This distance is the coin's circumference. **Circumference** is the distance around a circle.

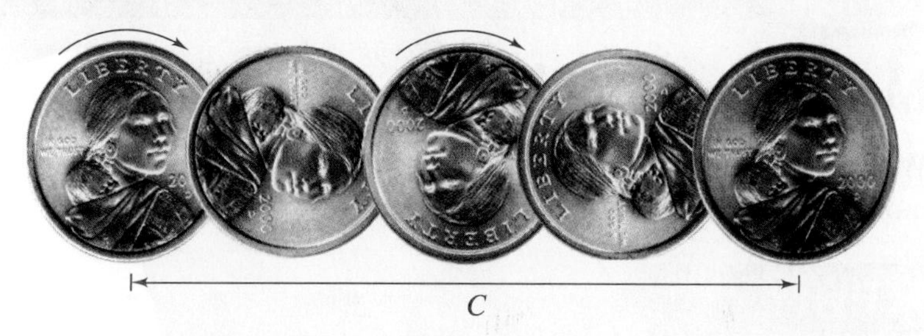

Pi is the ratio of a circle's circumference C to its diameter d. Use the symbol π for this ratio. So, $\pi = \frac{C}{d}$. The formula for the circumference comes from this ratio.

KEY CONCEPTS **Circumference of a Circle**

The circumference of a circle is π times the diameter d.

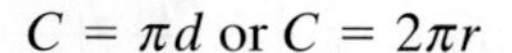

$C = \pi d$ or $C = 2\pi r$

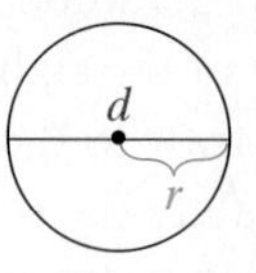

Pi is a nonterminating and nonrepeating decimal. Both $\frac{22}{7}$ and 3.14 are good approximations for π. Many calculators have a key for π and display it to nine decimal places. Your results will vary slightly, depending on which value for π you use.

Differentiated Instruction Solutions for All Learners

Special Needs L1
Ask students to identify which part of the circle has the greatest and least length: radius, diameter, or circumference. Ask them to prove their ideas using a picture of a circle and a piece of string.

learning style: tactile

Below Level L2
Review with students the meaning of *radius, diameter,* and *center of a circle.* Relate circumference of a circle to perimeter of a rectangle.

learning style: verbal

EXAMPLE Finding the Circumference of a Circle

1 **a.** Find the circumference of the circle using 3.14 for π.

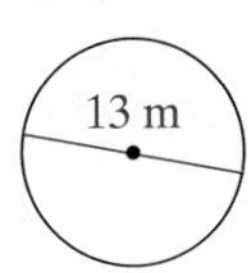

$C = \pi d$ ← **Use the formula for a circumference.**

$= 3.14(13)$ ← **Substitute.**

$= 40.82$

The circumference is about 40.8 m.

b. Find the circumference of the circle using a calculator's π key.

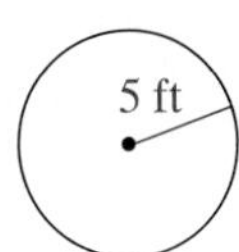

$C = 2\pi r$ ← **Use the formula for a circumference.**

$= 2\pi(5)$ ← **Substitute.**

Use a calculator. → ≈ 31.41592654

The circumference is about 31.4 ft.

Quick Check

1. Find the circumference of the circle at the right. Round to the nearest tenth. **28.3 m**

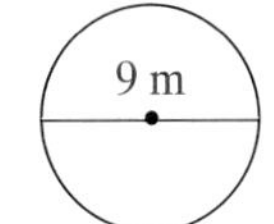

KEY CONCEPTS Area of a Circle

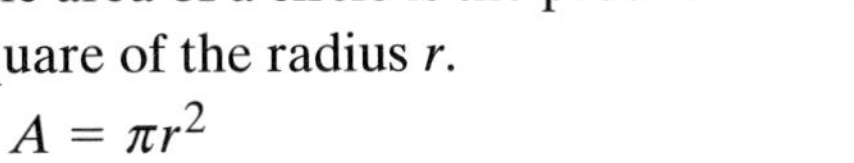

The area of a circle is the product of π and the square of the radius r.

$A = \pi r^2$

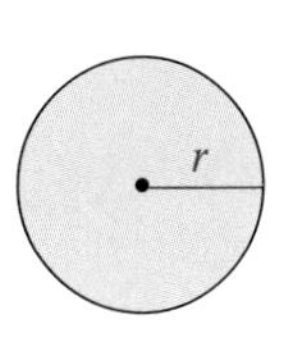

EXAMPLE Finding the Area of a Circle

2 A standard circus ring has a diameter of 13 m. What is the area of the ring? Round to the nearest tenth.

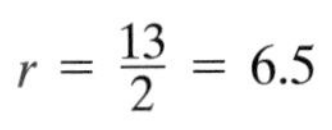

$r = \frac{13}{2} = 6.5$ ← **The radius is half of the diameter.**

$A = \pi r^2$ ← **Use the formula for the area of a circle.**

$= \pi(6.5)^2$ ← **Substitute 6.5 for the radius.**

$= 132.73228$ ← **Use a calculator.**

≈ 132.7 ← **Round to the nearest tenth.**

The area of a standard circus ring is about 132.7 m^2.

Quick Check

2. Find the area of the circle. Round to the nearest square unit. **452 m^2**

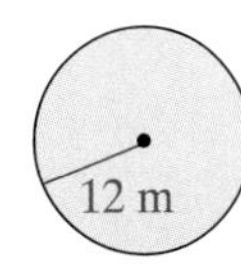

2. Teach

Activity Lab

Use before the lesson.
Student Edition Activity Lab, Hands On 8-5a, Modeling a Circle, p. 393

All in One Teaching Resources
Activity Lab 8-5: Areas of Circles

Guided Instruction

Error Prevention!

Some students may multiply π times the radius and then square the entire product. Have students rewrite the equation as $A = \pi \times r \times r$ as a reminder that only r should be squared. Students can substitute the value for the radius into this equation.

PowerPoint Additional Examples

1 Find the circumference.

a. **56.5 yd**

b. **125.7 cm**

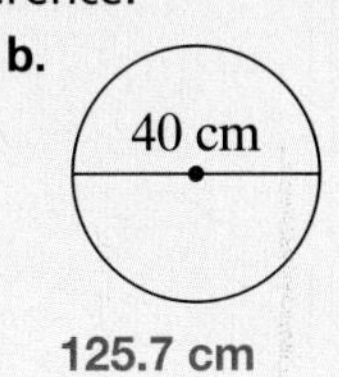

2 Find the area of the circle to the nearest unit. **227 m^2**

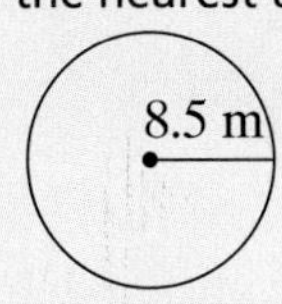

All in One Teaching Resources
- Daily Notetaking Guide 8-5 L3
- Adapted Notetaking 8-5 L1

Closure

- *What is the formula for the circumference of a circle?* **$C = 2\pi r$, where r is the radius of the circle and π is 3.14; or $C = \pi d$ where d is the diameter**
- *How do you find the area of a circle given its radius?* **Use the formula $A = \pi r^2$ (r is the radius).**

Advanced Learners L4

Solve for the radius of a circle given the area. $r = \sqrt{\frac{A}{\pi}}$ Use your equation to find the diameter of a circle with an area of 128.5 m^2. **about 12.8 m**

learning style: visual

English Language Learners ELL

Have students draw a circle. Ask them to draw an arrow around it, and label it *circumference*. Then have them shade the inside, and label that *area*. Make sure they identify the diameter and the radius. They can write the formulas for area and circumference next to the circle.

learning style: visual

3. Practice

Assignment Guide

Check Your Understanding

Go over Exercises 1–5 in class before assigning the Homework Exercises.

Homework Exercises

A	Practice by Example	6–18
B	Apply Your Skills	19–27
C	Challenge	28
	Test Prep and Mixed Review	29–33

Homework Quick Check

To check students understanding of key skills and concepts, go over Exercises 7, 13, 22, 25, and 27.

Differentiated Instruction **Resources**

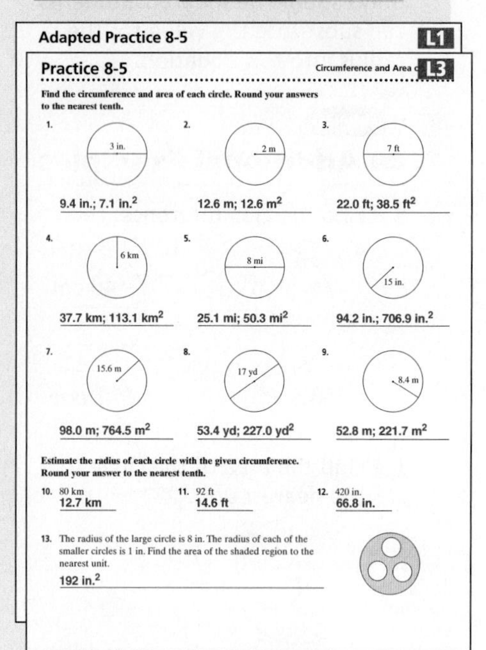

Adapted Practice 8-5 L1

Practice 8-5 Circumference and Area of Circles L3

Find the circumference and area of each circle. Round your answers to the nearest tenth.

1. 3 in. — 9.4 in.; 7.1 in.^{2}
2. 2 m — 12.6 m; 12.6 m^2
3. 7 ft — 22.0 ft; 38.5 ft^2
4. 6 km — 37.7 km; 113.1 km^2
5. 8 mi — 25.1 mi; 50.3 mi^2
6. 15 in. — 94.2 in.; 706.9 in.^{2}
7. 15.6 m — 98.0 m; 764.5 m^2
8. 17 yd — 53.4 yd; 227.0 yd^2
9. 8.4 m — 52.8 m; 221.7 m^2

Estimate the radius of each circle with the given circumference. Round your answer to the nearest tenth.

10. 80 km — 12.7 km
11. 92 ft — 14.6 ft
12. 420 in. — 66.8 in.

13. The radius of the large circle is 8 in. The radius of each of the smaller circles is 1 in. Find the area of the shaded region to the nearest unit. 192 in.^{2}

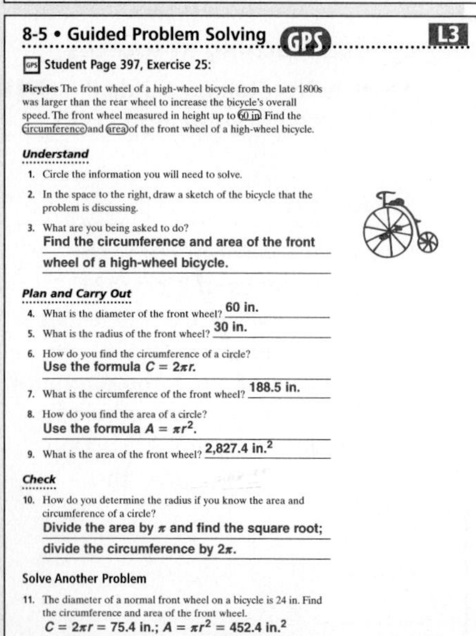

8-5 • Guided Problem Solving GPS L3

Student Page 397, Exercise 25:

Bicycles The front wheel of a high-wheel bicycle from the late 1800s was larger than the rear wheel to increase the bicycle's overall speed. The front wheel measured in height up to 60 in. Find the circumference and area of the front wheel of a high-wheel bicycle.

Understand

1. Circle the information you will need to solve.
2. In the space to the right, draw a sketch of the bicycle that the problem is discussing.
3. What are you being asked to do? Find the circumference and area of the front wheel of a high-wheel bicycle.

Plan and Carry Out

4. What is the diameter of the front wheel? 60 in.
5. What is the radius of the front wheel? 30 in.
6. How do you find the circumference of a circle? Use the formula $C = 2\pi r$.
7. What is the circumference of the front wheel? 188.5 in.
8. How do you find the area of a circle? Use the formula $A = \pi r^2$.
9. What is the area of the front wheel? 2,827.4 in.^{2}

Check

10. How do you determine the radius if you know the area and circumference of a circle? Divide the area by π and find the square root; divide the circumference by 2π.

Solve Another Problem

11. The diameter of a normal front wheel on a bicycle is 24 in. Find the circumference and area of the front wheel. $C = 2\pi r = 75.4$ in.; $A = \pi r^2 = 452.4$ in.^{2}

Check Your Understanding

1. **Number Sense** Is it possible to write out the exact value of pi as a decimal? Explain. **No; π is nonrepeating and nonterminating.**

Identify the radius, the diameter, and the circumference.

2. $1, 2\pi, 2$ — **1; 2; 2π**
3. $4\pi, 4, 2$ — **2; 4; 4π**
4. $7, 3.5, 22$ — **3.5; 7; 22**
5. $\frac{2}{\pi}, \frac{1}{\pi}, 2$ — **$\frac{1}{\pi}$; $\frac{2}{\pi}$; 2**

Homework Exercises

For more exercises, see Extra Skills and Word Problems.

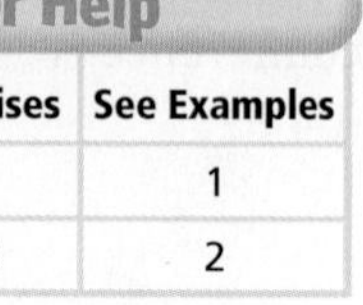

GO for Help

For Exercises	See Examples
6–11	1
12–18	2

Find the circumference of each circle. Round to the nearest tenth.

A 6. 50 cm — **157.1 cm**

7. 17 mm — **53.4 mm**

8. 27 m — **84.8 m**

9. 40 in. — **251.3 in.**

10. 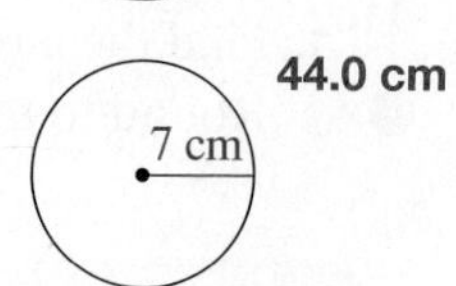

44.0 cm

11. 8 mi — **50.3 mi**

Find the area of each circle. Round to the nearest square unit.

12. 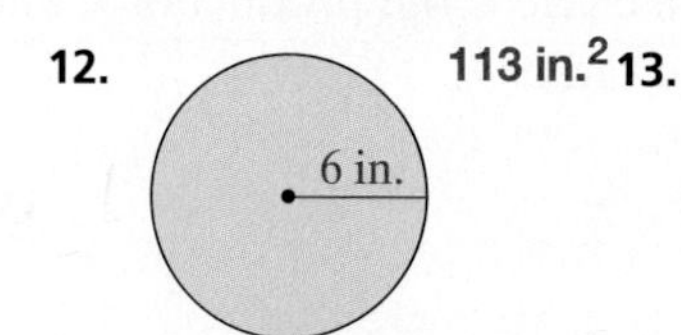

113 in.^{2}

13. 10 m — **314 m^2**

14. 25 cm — **1,963 cm^2**

15. 30 ft — **707 ft^2**

16. 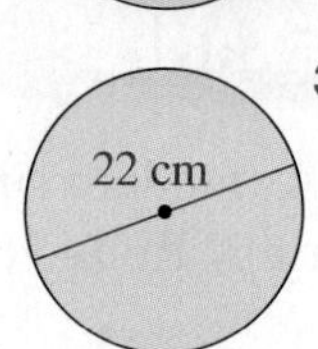

380 cm^2

17. 15 km — **177 km^2**

18. **Social Studies** The circular bases of the traditional tepees of the Sioux and Cheyenne tribes have a diameter of about 15 ft. What is the area of the base? Round to the nearest square unit. **177 ft^2**

B 

19. **Guided Problem Solving** A Ferris wheel has a diameter of 135 m. How far does a rider travel in one full revolution of the wheel? Round to the nearest unit. **424 m**
   - What is the diameter of the circle?
   - What is the formula for circumference, using diameter?

20. In a circle with radius 5 cm, how long can the longest chord be? **10 cm**

Visit: PHSchool.com
Web Code: are-0805

Use $\pi \approx \frac{22}{7}$ to estimate the circumference and area for each circle. Where necessary, round to the nearest tenth.

21. $r = 14$ m
88 m; 616 m²

22. $r = \frac{7}{10}$ cm
4.4 cm; 1.54 cm²

23. $d = 22$ in.
69.1 in.; 380.3 in.²

24. $d = 12$ ft
37.7 ft; 113.1 ft²

25. **Bicycles** The front wheel of a high-wheel bicycle from the late 1800s was larger than the rear wheel to increase the bicycle's overall speed. The front wheel measured in height up to 60 in. Find the circumference and area of the front wheel of a high-wheel bicycle. about 188 in.; about 2,827 in.²

26. about 707 m²

26. **Archaeology** The large stones of Stonehenge are arranged in a circle about 30 m in diameter. Find the area of the circle. See left.

27. **Writing in Math** Use the π key to calculate the area of the circle at the right to the nearest hundredth. Which is the better estimate, 98 m² or 99 m²? Explain. 99 m²; 98.52 rounds to 99.

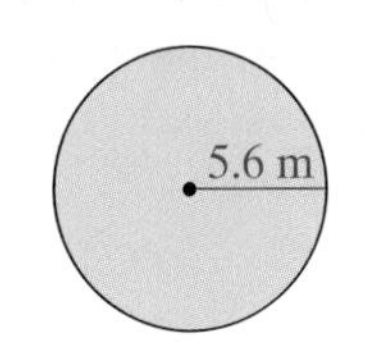

28. **Challenge** The diagram shows a fountain at the center of a circular park. The radius of the circle is 30 ft. The circular region is divided into six equal parts. What is the length of the arc in the shaded region? Round to the nearest tenth. 31.4 ft

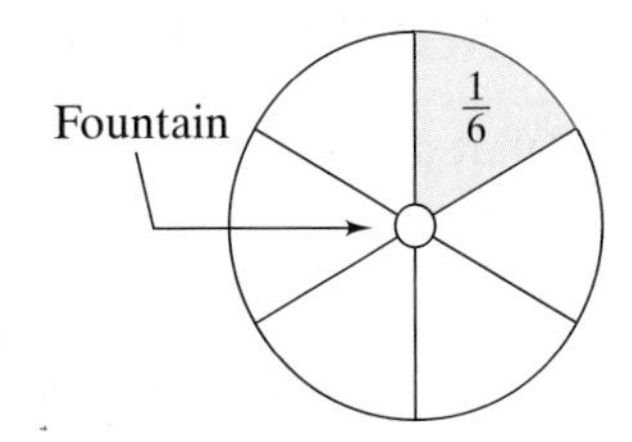

Test Prep and Mixed Review

Practice

Multiple Choice

29. Use a centimeter ruler to measure the radius of the button. What is the area of the button to the nearest square centimeter? C

A 3 cm²
B 22 cm²
C 7 cm²
D 89 cm²

30. A homebuilder wants to use the logo shown on a sign. Which of the following expressions can be used to find the perimeter of the logo? G

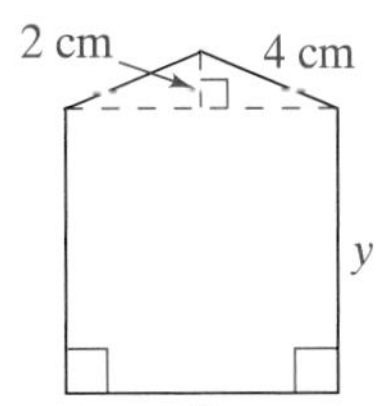

F $2 + 4 + x + y$
G $4 + 4 + x + 2y$
H $4 + 4 + 2x + 2y$
J $2 + 4 + 2x + 4y$

GO for Help

For Exercises	See Lesson
31–33	7-5

$\triangle CAT \cong \triangle DOG$. Complete each congruence statement.

31. $\angle A \cong$ ■ $\angle O$

32. $\angle G \cong$ ■ $\angle T$

33. $\overline{CT} \cong$ ■ $\overline{DG}$

Online lesson quiz, PHSchool.com, Web Code: ara-0805

4. Assess & Reteach

PowerPoint

Lesson Quiz

Use 3.14 for π. Find the area and the circumference of each circle to the nearest tenth.

1. $r = 10$ m $A = 314$ m²; $C = 62.8$ m
2. $d = 14$ ft $A = 153.9$ ft²; $C = 44.0$ ft

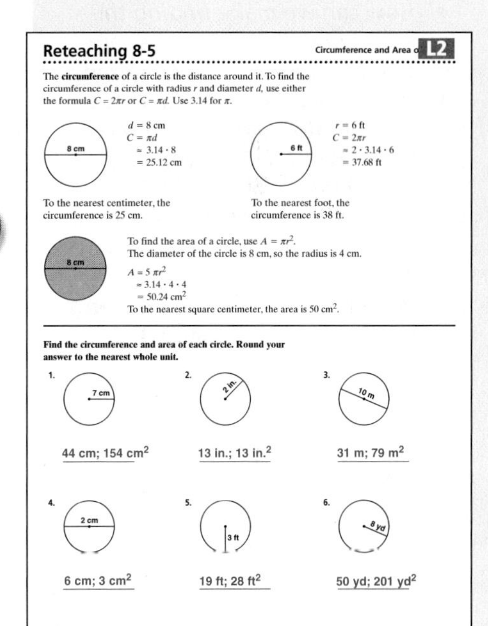

Reteaching 8-5 Circumference and Area of ... L2

The **circumference** of a circle is the distance around it. To find the circumference of a circle with radius r and diameter d, use either the formula $C = 2\pi r$ or $C = \pi d$. Use 3.14 for π.

$d = 8$ cm
$C = \pi d$
$\approx 3.14 \cdot 8$
$= 25.12$ cm

To the nearest centimeter, the circumference is 25 cm.

$r = 6$ ft
$C = 2\pi r$
$\approx 2 \cdot 3.14 \cdot 6$
$= 37.68$ ft

To the nearest foot, the circumference is 38 ft.

To find the area of a circle, use $A = \pi r^2$.
The diameter of the circle is 8 cm, so the radius is 4 cm.

$A = 5\ \pi r^2$
$\approx 3.14 \cdot 4 \cdot 4$
$= 50.24$ cm²

To the nearest square centimeter, the area is 50 cm².

Find the circumference and area of each circle. Round your answer to the nearest whole unit.

1. 7 cm — 44 cm; 154 cm²
2. 2 in. — 13 in.; 13 in.²
3. 10 m — 31 m; 79 m²
4. 2 cm — 6 cm; 3 cm²
5. 3 ft — 19 ft; 28 ft²
6. 8 yd — 50 yd; 201 yd²

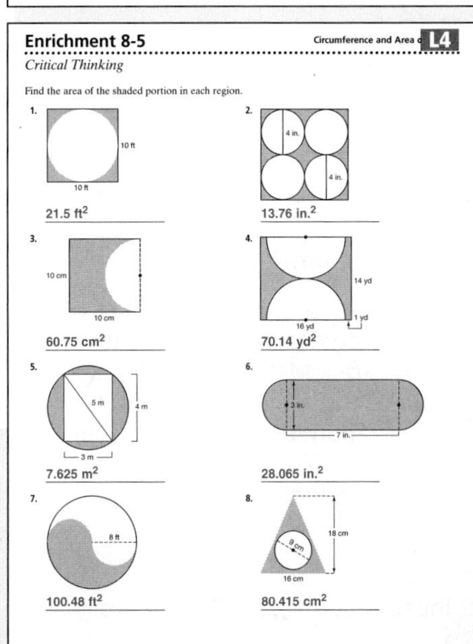

Enrichment 8-5 Circumference and Area of ... L4

Critical Thinking

Find the area of the shaded portion in each region.

1. 21.5 ft²
2. 13.76 in.²
3. 60.75 cm²
4. 70.14 yd²
5. 7.625 m²
6. 28.065 in.²
7. 100.48 ft²
8. 80.415 cm²

Alternative Assessment

Have students work in pairs. Each student draws four circles, labeling either the radius or diameter in each circle. Partners exchange circles and find the circumference and the area of each circle. Partners must agree on the results.

Test Prep

Resources

For additional practice with a variety of test item formats:

- Test-Taking Strategies, p. 427
- Test Prep, p. 431
- Test-Taking Strategies with Transparencies

GPS Guided Problem Solving

Areas of Irregular Figures

In this feature, students work with problems involving area of irregular figures.

Guided Instruction

Discuss with students why it is helpful to know how to calculate the areas of irregular figures. Explain that most objects in everyday life are not in the shape of regular polygons.

Have a volunteer read the problem aloud.
Ask:

- *How can we make finding the area simpler?* **Divide the shape into familiar geometric figures.**
- *Are there other ways to divide the art into figures? What are they?* **Yes. Answers will vary.**
- *Why multiply by 1.10?* **Erin wants to buy 10% more tiles than the area to be covered. 10% is 0.10 as a decimal. (1 + 0.10) times the total area will give the total amount of tiles to buy.**

Error Prevention!

Some students may multiply by the wrong decimal to find 110% of the area. Make sure students understand that they will find the total amount of tile by adding the amount of tile for the art and 10% more.

Exercises

Have students work independently on the Exercises.

GPS Guided Problem Solving

Areas of Irregular Figures

You can find the area of an irregular-shaped figure by combining basic shapes such as rectangles or triangles, or removing them from an existing figure.

Mosaic Erin is creating a mosaic and needs to buy tiles for her artwork. She wants to buy 10% more than the area to be covered. How many square inches of tile does she need to purchase?

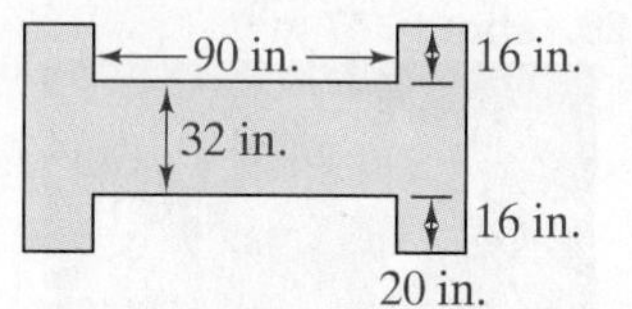

What You Might Think	What You Might Write
What do I know? What am I trying to find out?	I know the dimensions of the art work. I need to find the area, plus 10%.
How do I solve the problem?	I'll subtract the area of the two smaller rectangles from the area of the whole rectangle. Then I'll add 10%.
What do I need to calculate?	The area of each smaller rectangle is 90 in. × 16 in., or 1,440 in.^{2} The area of the larger rectangle is 130 in. × 64 in., or 8,320 in.^{2}. The area of the artwork is 8,320 in.^{2} − 2(1,440 in.^{2}), or 5,440 in.^{2}. Then I need to add 10%. 5,440 in.^{2} × 1.10 = 5,984 in.^{2}
What is the answer?	Erin needs to buy 5,984 in.^{2} of tile.

90
16
32
16
20

Think It Through

1. Why does multiplying by 1.1 add 10% to the area? **1.1 is the same as the sum of one whole and 0.1 (10%).**
2. Is there a way to find the total area by adding instead of subtracting? Explain. **Yes; make three rectangles and add the areas.**

Exercises

Solve the problems. For Exercises 3 and 4, answer the questions first.

3. You plan to replace the carpeting in the room shown below. What is the area of the room? **575 ft^2**

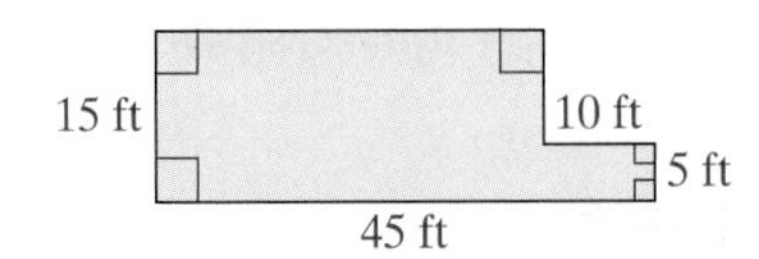

   a. What do you know? What do you need to find out?
   b. Can you find the answer in two different ways? Explain.

4. Sam wants to paint the wall below. What is the area of the wall to the nearest tenth? **76.9 ft^2**

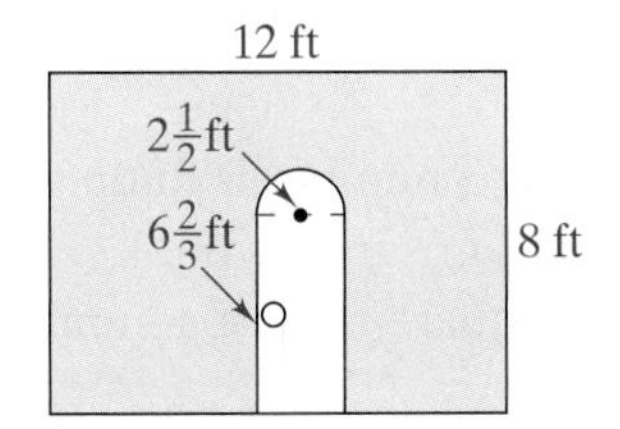

   a. What shapes make up the door? **rectangle and semicircle**

5. Data from the 2000 U.S. Census for four states are in the table below. Order the states from smallest population per square mile, to greatest. Explain what this means. **Alaska, Texas, Florida, and New Jersey; New Jersey is the most densely populated.**

State	Population	Area (mi^2)
Alaska	626,932	663,267
Florida	15,982,378	65,755
New Jersey	8,414,350	8,721
Texas	20,851,820	268,581

6. Wilma agreed to cut the grass on the infield of the school track for \$0.35 /yd^2 on her riding lawnmower. How much would she make each time she cut the grass? **about \$40.65**

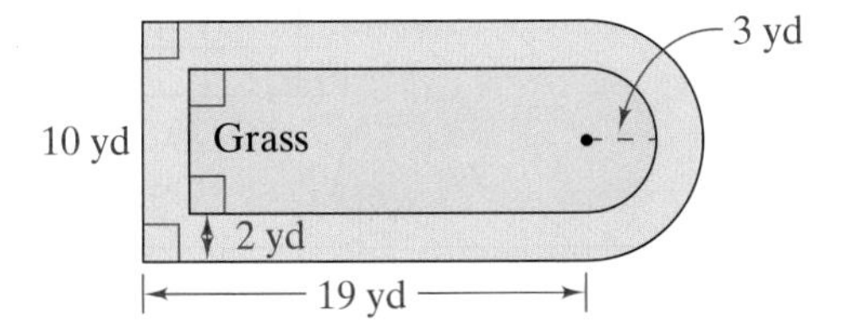

Students may compare papers when they have finished and discuss how they arrived at their answers.

Differentiated Instruction

Below Level L2

Provide students with paper models of the shapes shown in the problems. Allow them to cut the models into pieces to help them understand how to divide shapes into geometric figures whose area they know how to find.

8-6

1. Plan

Objective
To find and estimate square roots and to classify numbers as rational or irrational

Examples
1 Finding Square Roots of Perfect Squares
2 Estimating Square Roots
3 Classifying Numbers

Math Understandings: p. 372D

Math Background

The number 25 is called a *perfect square* because it is the square of 5, or 5×5. The inverse operation of squaring (multiplying a number by itself) is taking a *square root,* finding the number that is multiplied by itself to make the square. So, the square root of 25 is 5, or $\sqrt{25} = 5$. The symbol $\sqrt{\ }$ indicates a positive square root. Numbers that are not perfect squares, such as $\sqrt{21}$, are classified as *irrational numbers* because they cannot be stated in the form $\frac{a}{b}$ where a is an integer and b is any nonzero integer.

More Math Background: p. 372D

Lesson Planning and Resources

See p. 372E for a list of the resources that support this lesson.

Bell Ringer Practice

Check Skills You'll Need
Use student page, transparency, or PowerPoint. For intervention, direct students to:
Exponents and Order of Operations
Lesson 2-1
Extra Skills and Word Problems Practice, Ch. 2

8-6 Square Roots and Irrational Numbers

Check Skills You'll Need

1. Vocabulary Review How do you find the *square* of a number? **See below.**

Simplify.

2. 8^2 **64** **3.** 12^2 **144**

4. 2^2 **4** **5.** 7^2 **49**

GO for Help Lesson 2-1

Check Skills You'll Need
1. **Multiply the number by itself.**

What You'll Learn

To find and estimate square roots and to classify numbers as rational or irrational

New Vocabulary perfect square, square root, irrational number

Why Learn This?

If you know the area of a square, you can find its square root to find the side lengths. You can use square roots when you install flooring.

A number that is the square of an integer is a **perfect square.** For example, the square of 2 is 4, so 4 is a perfect square.

The inverse of squaring a number is finding a **square root.** The symbol $\sqrt{\ }$ in this book indicates the positive square root of a number.

Test Prep Tip
You can use the 7-by-7 grid below to represent 7^2 or $\sqrt{49} = 7$.

EXAMPLE Finding Square Roots of Perfect Squares

1 Mental Math Simplify $\sqrt{49}$.

Since $7^2 = 49$, $\sqrt{49} = 7$.

Quick Check

1. Simplify.

a. $\sqrt{64}$ **8** **b.** $\sqrt{81}$ **9** **c.** $\sqrt{225}$ **15**

If a number is not a perfect square, you can estimate its square root.

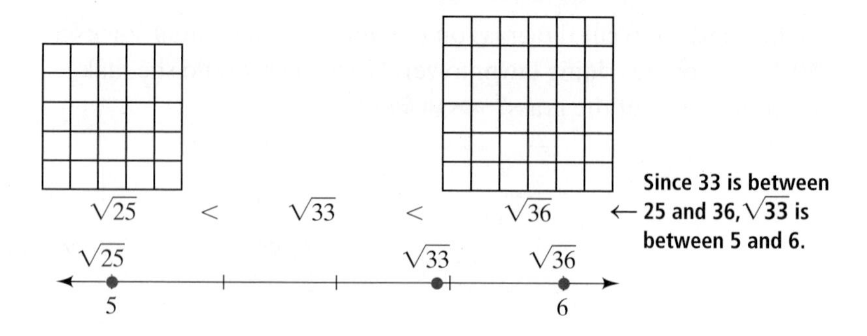
$\sqrt{25}$ < $\sqrt{33}$ < $\sqrt{36}$
Since 33 is between 25 and 36, $\sqrt{33}$ is between 5 and 6.
$\sqrt{25}$
$\sqrt{33}$
$\sqrt{36}$
5
6

Differentiated Instruction Solutions for All Learners

Special Needs L1
Given 10, 36, 56, 64, and 100, have students use tiles, equations, or pictures to show whether each number is a perfect square. **36, 64, 100** For each perfect square, they should find the square root. **6, 8, 10**

learning style: tactile

Below Level L2
Revisit pictures that represent numbers that are perfect squares. For example, provide a list that shows the numbers 1, 4, 9, 16, 25, through 100, along with the squares drawn (1x1, 2x2, and so on.) Ask students to identify the square root for each.

learning style: visual

EXAMPLE Estimating Square Roots

2 **Tiles** Juanita bought 40 tiles on sale. They measured 1 ft^2 each. What is the largest square bathroom floor she can tile?

$A = s^2$ ← **Use the formula for the area of a square.**

$40 = s^2$ ← **Substitute 40 for the area.**

$\sqrt{40} = \sqrt{s^2}$ ← **Take the square root of each side.**

$\sqrt{40} = s$ ← **Simplify.**

$\sqrt{36} < \sqrt{40} < \sqrt{49}$ ← **Find the perfect squares close to 40.**

$6 < \sqrt{40} < 7$ ← **Simplify.**

$\sqrt{40}$ is between 6 and 7. Since 40 is closer to 36 than to 49, $\sqrt{40}$ is about 6. So the largest floor she can tile is about 6 ft × 6 ft.

Vocabulary Tip

Read $\sqrt{40}$ as "the square root of 40."

Quick Check

2. Suppose Juanita bought twice the number of tiles above. Estimate the dimensions of the largest square floor she can tile.
about 9 ft × 9 ft

An **irrational number** is a number that cannot be written as a ratio of two integers. As decimals, irrational numbers neither terminate nor repeat. The diagram below summarizes these relationships.

GO for Help

For help with terminating decimals and repeating decimals, go to Lesson 2-6, Examples 1 and 2.

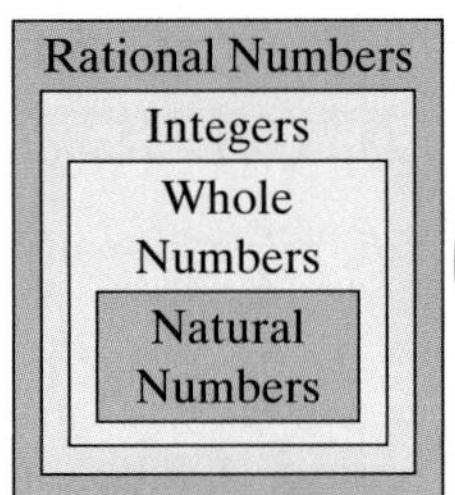

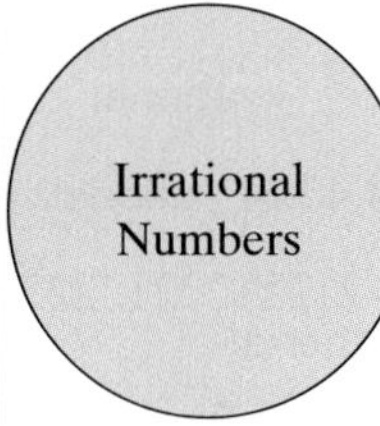

If a positive integer is not a perfect square, its square root is irrational.

Rational → $\sqrt{4}, \sqrt{9}, \sqrt{16}$

Irrational → $\sqrt{2}, \sqrt{3}, \sqrt{27}$

EXAMPLE Classifying Numbers

3 Identify each number as *rational* or *irrational*.

a. $\sqrt{14}$ irrational ← **14 is not a perfect square.**

b. 0.323223222 . . . irrational ← **The decimal neither terminates nor repeats.**

c. −0.98 rational ← **It is a terminating decimal.**

Quick Check

3. Identify each number as *rational* or *irrational*.

a. $\sqrt{2}$ **irrational** **b.** $\sqrt{81}$ **rational** **c.** $0.\overline{6}$ **rational** **d.** $1\frac{2}{7}$ **rational**

2. Teach

Activity Lab

Use before the lesson.

All in One Teaching Resources

Activity Lab 8-6: Patterns in Numbers

Guided Instruction

Example 1

Give each student a piece of graph paper. Demonstrate $\sqrt{49}$ by having students draw a 7 by 7 square of exactly 49 units. Have them verify that the number of enclosed square units is 49. Then ask: *What is the length of each side of the square?* **7 units**

Example 3

Help students understand that a rational number can be written as a ratio $\frac{a}{b}$ where a is an integer (whole numbers and their opposites) and b is a nonzero integer.

PowerPoint

Additional Examples

1 Simplify $\sqrt{81}$. **9**

2 Estimate the value of $\sqrt{60}$. **about 8**

3 Identify each number as rational or irrational.

a. $\sqrt{121}$ **rational**

b. $\sqrt{30}$ **irrational**

c. −0.5167 **rational**

d. 29.2992999 . . . **irrational**

All in One Teaching Resources

- Daily Notetaking Guide 8-6 **L3**
- Adapted Notetaking 8-6 **L1**

Closure

- *What is an irrational number? Give an example.* **a number that cannot be written as a ratio $\frac{a}{b}$ where a is an integer and b is a nonzero integer; as a decimal it neither terminates nor repeats; $\sqrt{5}$**
- *What is the square root of a number? Give an example.* **a number that when multiplied by itself equals the given number; $\sqrt{25} = 5$ because $5^2 = 25$**

Advanced Learners **L4**

The area of a square garden is 170 ft^2. Have students estimate the perimeter for the garden. **about 52 ft**

learning style: visual

English Language Learners **ELL**

Ask students to work in pairs on Check Your Understanding exercises 1 and 2. One partner works on number 1, and the other on 2. Then ask them to explain their responses to each other in words and pictures.

learning style: verbal

3. Practice

Assignment Guide

Check Your Understanding

Go over Exercises 1–5 in class before assigning the Homework Exercises.

Homework Exercises

A	Practice by Example	6–29
B	Apply Your Skills	30–45
C	Challenge	46
	Test Prep and Mixed Review	47–52

Homework Quick Check

To check students' understanding of key skills and concepts, go over Exercises 13, 19, 32, 33, and 35.

Differentiated Instruction Resources

Adapted Practice 8-6 L1

Practice 8-6 — Square Roots and Irrational Numbers L3

Simplify each square root.

1. $\sqrt{64}$ 8 2. $\sqrt{81}$ 9 3. $\sqrt{100}$ 10
4. $\sqrt{121}$ 11 5. $\sqrt{1}$ 1 6. $\sqrt{36}$ 6
7. $\sqrt{25}$ 5 8. $\sqrt{16}$ 4 9. $\sqrt{256}$ 16
10. $\sqrt{196}$ 14 11. $\sqrt{49}$ 7 12. $\sqrt{225}$ 15

Identify each number as rational or irrational.

13. 0.363636 . . . rational 14. $\sqrt{10}$ irrational 15. $-\frac{1}{9}$ rational

For each number, write all the sets to which it belongs. Choose from rational number, irrational number, whole number, and integer.

16. $\frac{3}{8}$ rational 17. $\sqrt{49}$ rational, integer, whole 18. $\sqrt{98}$ irrational

Find the length of the side of a square with the given area.

19. 64 km^2 8 km 20. 81 m^2 9 m 21. 121 ft^2 11 ft
22. 225 $in.^2$ 15 in. 23. 196 yd^2 14 yd 24. 169 cm^2 13 cm

Solve.

25. The square of a certain number is the same as three times the number. What is the number? 0 or 3
26. The area of a square lawn is 196 yd^2. What is the perimeter of the lawn? 56 yd

Find two consecutive whole numbers that each number is between.

27. $\sqrt{80}$ 8, 9 28. $\sqrt{56}$ 7, 8 29. $\sqrt{130}$ 11, 12
30. $\sqrt{70}$ 8, 9 31. $\sqrt{190}$ 13, 14 32. $\sqrt{204}$ 14, 15

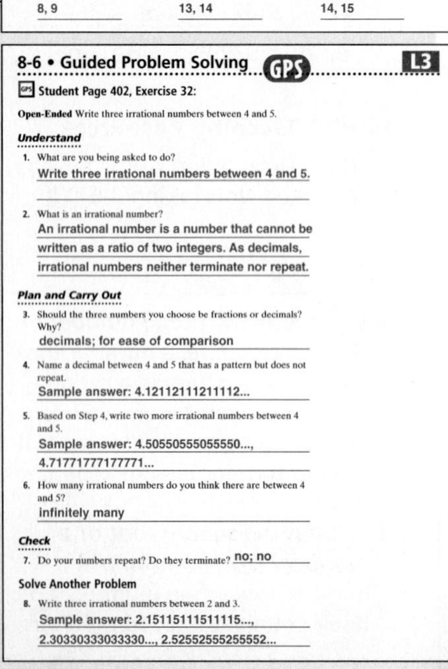

8-6 • Guided Problem Solving GPS L3

GPS **Student Page 402, Exercise 32:**

Open-Ended Write three irrational numbers between 4 and 5.

Understand

1. What are you being asked to do? Write three irrational numbers between 4 and 5.
2. What is an irrational number? An irrational number is a number that cannot be written as a ratio of two integers. As decimals, irrational numbers neither terminate nor repeat.

Plan and Carry Out

3. Should the three numbers you choose be fractions or decimals? Why? decimals; for ease of comparison
4. Name a decimal between 4 and 5 that has a pattern but does not repeat. Sample answer: 4.12112111211112...
5. Based on Step 4, write two more irrational numbers between 4 and 5. Sample answer: 4.50550555055550..., 4.71771777177771...
6. How many irrational numbers do you think there are between 4 and 5? infinitely many

Check

7. Do your numbers repeat? Do they terminate? no; no

Solve Another Problem

8. Write three irrational numbers between 2 and 3. Sample answer: 2.15115111511115..., 2.30330333033330..., 2.52552555255552...

Check Your Understanding

1. **Vocabulary** The square root of the __?__ of a square is the same as the length of one side of the square. **area**
2. **Number Sense** If you double the length of each side of a square, what happens to the area of the square? **It quadruples.**

Determine whether or not each decimal has a repeating pattern.

3. 2.3423423423 . . . **yes**
4. 0.1234567891011 . . . **no**
5. 0.17893624775 . . . **no**

Homework Exercises

For more exercises, see Extra Skills and Word Problems.

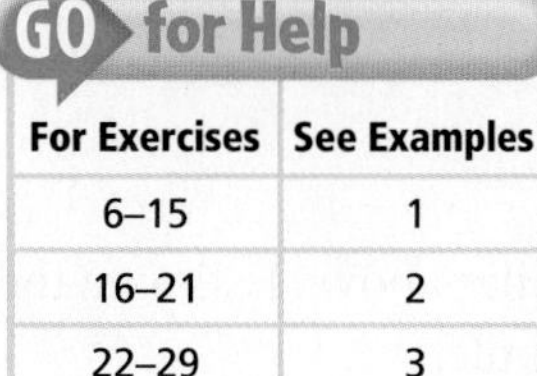

GO for Help

For Exercises	See Examples
6–15	1
16–21	2
22–29	3

A **Simplify each square root.**

6. $\sqrt{16}$ **4**
7. $\sqrt{100}$ **10**
8. $\sqrt{36}$ **6**
9. $\sqrt{25}$ **5**
10. $\sqrt{81}$ **9**
11. $\sqrt{169}$ **13**
12. $\sqrt{144}$ **12**
13. $\sqrt{121}$ **11**
14. $\sqrt{1}$ **1**
15. $\sqrt{9}$ **3**

Estimate the value of each square root.

16. $\sqrt{18}$ **about 4**
17. $\sqrt{5}$ **about 2**
18. $\sqrt{41}$ **about 6**
19. $\sqrt{54}$ **about 7**
20. $\sqrt{75}$ **about 9**
21. **Industrial Arts** A square carpet covers an area of 64 ft^2. What is the length of each side of the carpet? **8 ft**

Identify each number as *rational* or *irrational*.

22. $\sqrt{99}$ **irrational**
23. $\sqrt{41}$ **irrational**
24. $\sqrt{49}$ **rational**
25. -0.4744 **rational**
26. $-\frac{3}{2}$ **rational**
27. $-0.666666\ldots$ **rational**
28. 0.12122122212222 . . . **irrational**
29. $0.\overline{35}$ **rational**

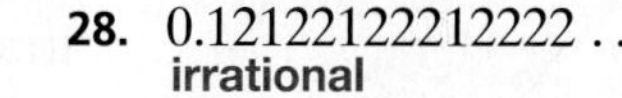

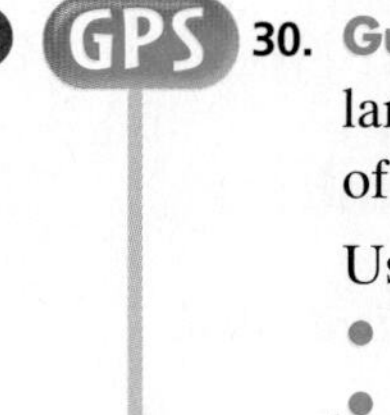

B GPS 30. **Guided Problem Solving** A fence surrounds a square plot of land. The area of the land is 324 yd^2. Find the length of each side of the fence. **18 yd**

Use the strategy *Systematic Guess and Check.*

- $10 \times 10 =$ ■ The initial guess is (correct, high, low).
- $20 \times 20 =$ ■ This guess is (correct, high, low).
- ■ × ■ = 324 This guess is correct.

31. The area of a square cover for a whirlpool bath is 144 ft^2. What is the length of each side of the cover? **12 ft**
32. **Open-Ended** Write three irrational numbers between 4 and 5. **See left.**
33. GPS **Writing in Math** What are the square roots of $\frac{1}{4}$ and $\frac{4}{9}$? Write a method of finding the square root of a fraction. **See left.**

32. Answers may vary. Sample: 4.112123 . . .; 4.535335 . . .; 4.91911 . . .

33. $\frac{1}{2}$; $\frac{2}{3}$; take the square root of both the numerator and denominator.

41–44. See back of book.

Homework Video Tutor
Visit: PHSchool.com
Web Code: are-0806

34. rational
35. rational
36. rational, whole, integer, natural
37. rational
38. rational

For each number, write all the sets to which it belongs. Choose from *irrational, rational, whole,* or *natural numbers,* or *integers.*

34. $\frac{3}{5}$ **35.** $0.\overline{23}$ **36.** $\sqrt{36}$ **37.** 4.5 **38.** $\frac{22}{7}$

34–38. See left.

Math in the Media **Use the cartoon below for Exercises 39 and 40.**

FOXTROT *by Bill Amend.*

39. Does the cartoon suggest that π is *rational* or *irrational*? **irrational**

40. What rational numbers can you use to approximate π?
Answers may vary. Sample: 3.14; $\frac{22}{7}$

Draw and label a square to model each area. 41–44. See margin.

41. 64 km^2 **42.** 81 m^2 **43.** 121 ft^2 **44.** 4 mi^2

Press 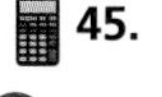 7 2 3
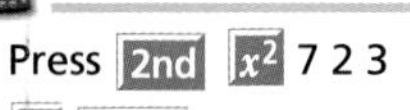
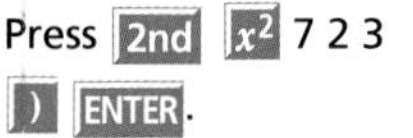

45. Use a calculator to estimate $\sqrt{723}$ to the nearest tenth. **about 26.9**

46. Challenge When an object falls, the distance that it falls is given by the formula $d = 16t^2$, where d is the distance in feet and t is the time in seconds. A stone falls from a bridge 1,600 ft above the water. How long does it take for the stone to reach the water? **10 s**

Test Prep and Mixed Review

Practice

Multiple Choice

47. The model represents $\sqrt{36} = 6$. Which arrangement can be used to represent $\sqrt{225}$? **C**
Ⓐ 2 rows of 15 small squares
Ⓑ 2 rows of 25 small squares
Ⓒ 15 rows of 15 small squares
Ⓓ 25 rows of 25 small squares

48. A standard basketball rim has a circumference of about 56.6 inches. Which expression can be used to find the diameter of the rim? **H**
Ⓕ $\frac{56.6}{2\pi}$ Ⓗ $\frac{56.6}{\pi}$
Ⓖ $56.6 \times \pi$ Ⓙ $56.6 \times 2\pi$

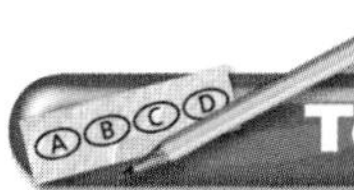

For Exercises	See Lesson
49–52	6-4

Find each answer.

49. 5% of 40 **2** **50.** 60% of 90 **54** **51.** 75% of 15 **11.25** **52.** 30% of 120 **36**

Online lesson quiz, **PHSchool.com, Web Code:** ara-0806

4. Assess & Reteach

PowerPoint

Lesson Quiz

Estimate each square root.

1. $\sqrt{6}$ about 2
2. $\sqrt{22}$ about 5

Identify each number as rational or irrational.

3. 0.625 rational
4. $\sqrt{150}$ irrational

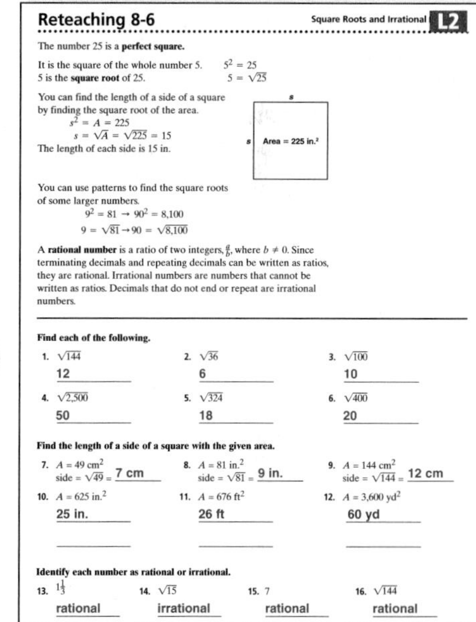

Reteaching 8-6 Square Roots and Irrational **L2**

The number 25 is a **perfect square.**

It is the square of the whole number 5. $5^2 = 25$
5 is the **square root** of 25. $5 = \sqrt{25}$

You can find the length of a side of a square by finding the square root of the area.
$s^2 = A = 225$
$s = \sqrt{A} = \sqrt{225} = 15$
The length of each side is 15 in.

Area = 225 in.²

You can use patterns to find the square roots of some larger numbers.
$9^2 = 81 \rightarrow 90^2 = 8{,}100$
$9 = \sqrt{81} \rightarrow 90 = \sqrt{8{,}100}$

A **rational number** is a ratio of two integers, $\frac{a}{b}$, where $b \neq 0$. Since terminating decimals and repeating decimals can be written as ratios, they are rational. Irrational numbers are numbers that cannot be written as ratios. Decimals that do not end or repeat are irrational numbers.

Find each of the following.

1. $\sqrt{144}$ 12
2. $\sqrt{36}$ 6
3. $\sqrt{100}$ 10
4. $\sqrt{2{,}500}$ 50
5. $\sqrt{324}$ 18
6. $\sqrt{400}$ 20

Find the length of a side of a square with the given area.

7. $A = 49$ cm², side = $\sqrt{49}$ = 7 cm
8. $A = 81$ in.², side = $\sqrt{81}$ = 9 in.
9. $A = 144$ cm², side = $\sqrt{144}$ = 12 cm
10. $A = 625$ in.² 25 in.
11. $A = 676$ ft² 26 ft
12. $A = 3{,}600$ yd² 60 yd

Identify each number as rational or irrational.

13. $1\frac{1}{3}$ rational
14. $\sqrt{15}$ irrational
15. 7 rational
16. $\sqrt{144}$ rational

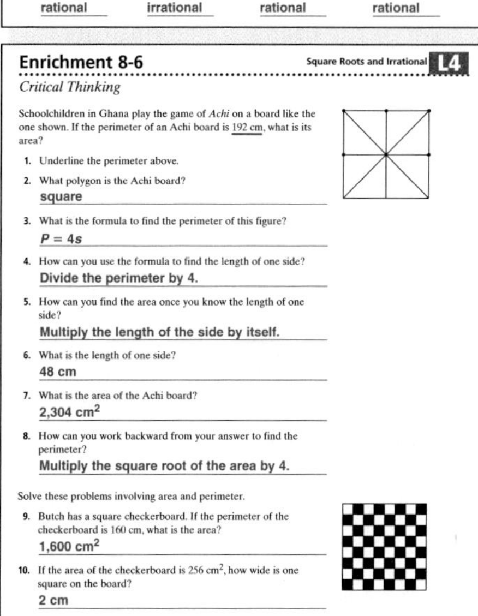

Enrichment 8-6 Square Roots and Irrational **L4**

Critical Thinking

Schoolchildren in Ghana play the game of *Achi* on a board like the one shown. If the perimeter of an Achi board is 192 cm, what is its area?

1. Underline the perimeter above.
2. What polygon is the Achi board?
square
3. What is the formula to find the perimeter of this figure?
$P = 4s$
4. How can you use the formula to find the length of one side?
Divide the perimeter by 4.
5. How can you find the area once you know the length of one side?
Multiply the length of the side by itself.
6. What is the length of one side?
48 cm
7. What is the area of the Achi board?
2,304 cm²
8. How can you work backward from your answer to find the perimeter?
Multiply the square root of the area by 4.

Solve these problems involving area and perimeter.

9. Butch has a square checkerboard. If the perimeter of the checkerboard is 160 cm, what is the area?
1,600 cm²
10. If the area of the checkerboard is 256 cm², how wide is one square on the board?
2 cm

Alternative Assessment

Students work with a partner to find and list the perfect squares from 1 to 225 and their square roots. You may wish to have them use calculators. Students can then refer to their lists to complete Exercises 6–15 and 22–29.

Test Prep

Resources
For additional practice with a variety of test item formats:
- Test-Taking Strategies, p. 427
- Test Prep, p. 431
- Test-Taking Strategies with Transparencies

Activity Lab

Exploring Right Triangles

In this Activity, students explore the relationship between the lengths of the sides of a right triangle by investigating the areas of squares formed from the legs and the hypotenuse of the triangle. Working through this Activity prepares students for exploring and using the Pythagorean Theorem in Lesson 8-7.

Guided Instruction

Before beginning the Activity, discuss the features of triangles with students. You may wish to review the meanings of *leg* and *hypotenuse* at this time.

Ask:

- *What characteristics do all triangles share?* **three sides, three angles whose sum is 180°**
- *How are right triangles different from other triangles?* **A right triangle must contain one 90° angle.**

Exercises

Have students work in pairs on the Exercises. Circulate to check students' work.

Alternative Method

You can have some students use a compass and straightedge to construct the squares from the sides of the triangle. Demonstrate or ask a volunteer to demonstrate this procedure.

Resources

- Activity Lab 8-7: Patterns in Geometry
- centimeter graph paper
- straightedge
- compass
- scissors
- Classroom Aid 2

8-7a Activity Lab — Hands On

Exploring Right Triangles

Complete this activity to discover a special relationship among the side lengths of right triangles.

ACTIVITY

Step 1 Use centimeter graph paper to draw a right triangle with perpendicular sides that are 3 cm and 4 cm long.

Step 2 Draw squares that have the horizontal and vertical sides of the right triangle as sides.

Step 3 Use another piece of the graph paper to make a square on the side opposite the right angle, as shown at the right.

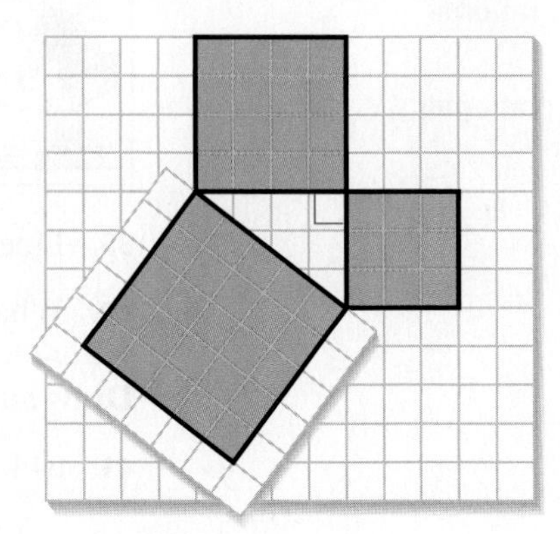

Exercises

1. What is the length of the side of the square opposite the right angle? **5 cm**
2. What is the area of each of the three squares? **9 cm^2; 16 cm^2; 25 cm^2**
3. Draw a right triangle on graph paper with perpendicular sides that are 8 cm and 15 cm long. Repeat Steps 2 and 3.
   a. What is the length of each side of the square made on the side opposite the right angle? **17 cm**
   b. What is the area of each of the three squares? **64 cm^2; 225 cm^2; 289 cm^2**
4. Construct three squares on the sides of the triangle at the right. What are the areas of the three squares? **36 cm^2; 64 cm^2; 100 cm^2**

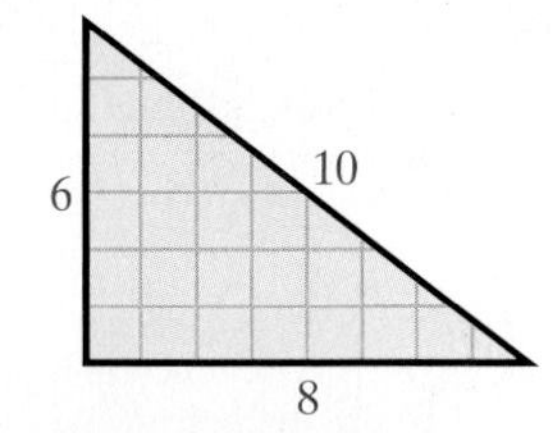

5. a. **Patterns** What seems to be the relationship of the areas of the smaller two squares and the third square made on the sides of a right triangle? **See margin.**
   b. Write an equation for each triangle that compares the areas of its squares. **$3^2 + 4^2 = 5^2$; $8^2 + 15^2 = 17^2$; $6^2 + 8^2 = 10^2$**
6. A rectangular park is shown at the right. The park is 630 yd long and 430 yd wide. How long is the path connecting opposite corners of the park? Round to the nearest tenth. **762.8 yd**

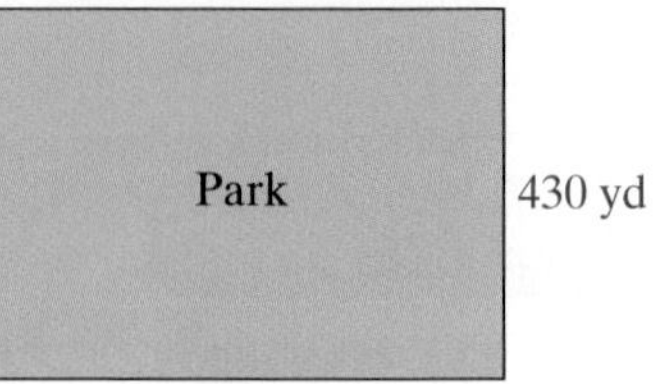

7. **Writing in Math** A triangle has side lengths 6 cm, 10 cm, and 12 cm. Explain why this triangle is not a right triangle. **See margin.**

5a. The sum of the areas of the two smaller squares is equal to the area of the larger square.

7. It is not a right triangle, because the sum of the squares of the shorter sides does not equal the square of the longest side. $36 + 100 \neq 144$

8-7 The Pythagorean Theorem

Check Skills You'll Need

1. **Vocabulary Review** A *perfect square* is the square of what kind of number? an integer

Simplify each square root.

2. $\sqrt{4}$ 2
3. $\sqrt{16}$ 4
4. $\sqrt{36}$ 6
5. $\sqrt{49}$ 7

GO for Help Lesson 8-6

What You'll Learn

To use the Pythagorean Theorem to solve real-world problems

New Vocabulary legs, hypotenuse, Pythagorean Theorem

Why Learn This?

The lengths of the sides of a right triangle are related. When you understand the Pythagorean Theorem, you can use right triangles to find unknown distances.

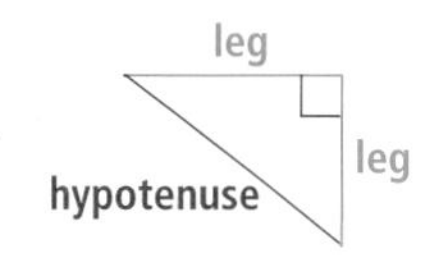

In a right triangle, the two shortest sides are the **legs**. The side opposite the right angle is the **hypotenuse**.

The **Pythagorean Theorem** states that in any right triangle, the sum of the squares of the lengths of the legs equals the square of the length of the hypotenuse.

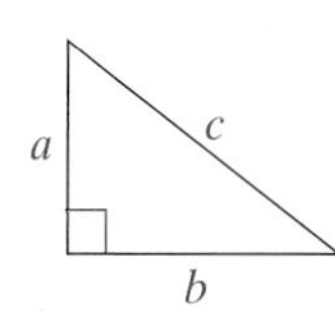

$$a^2 + b^2 = c^2$$

You can find the length of a hypotenuse with the Pythagorean Theorem.

EXAMPLE 1 Finding the Length of a Hypotenuse

The catcher throws a ball from home plate to second base. How far does the catcher throw?

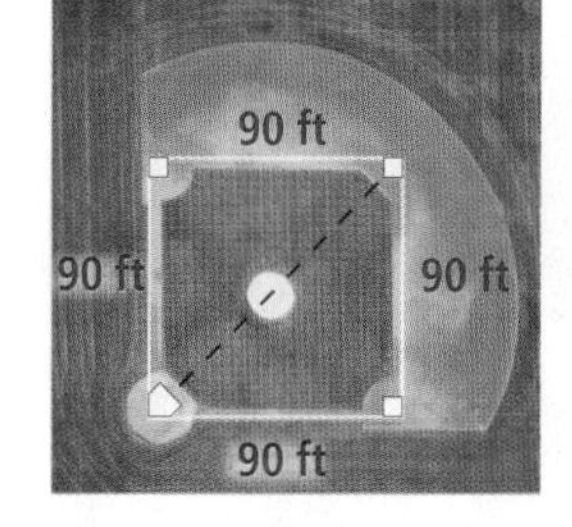

$c^2 = a^2 + b^2$ ← Pythagorean Theorem

$c^2 = 90^2 + 90^2$ ← Substitute.

$c^2 = 8{,}100 + 8{,}100$ ← Simplify.

$c^2 = 16{,}200$

$\sqrt{c^2} = \sqrt{16{,}200} \approx 127.3$ ← Take the square root of each side.

To reach second base from home plate, the catcher throws about 127 ft.

Online active math

For: Triangle Activity
Use: Interactive Textbook, 8-7

Quick Check

1. Find the length of the hypotenuse in the triangle at the right. 17 in.

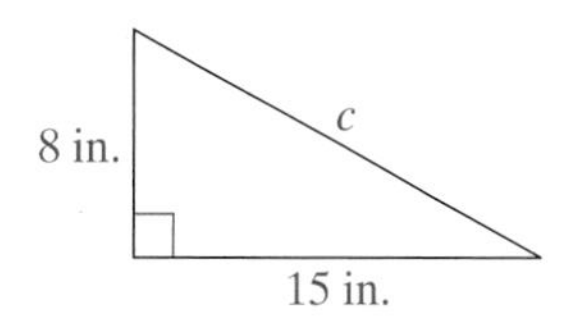

Differentiated Instruction Solutions for All Learners

Special Needs L1	**Below Level** L2
Give students a picture of a right triangle with side lengths that make it easy to compute square roots. For example, a 3, 4, 5 unit triangle whose hypotenuse is 5 units can help them practice the calculation. learning style: visual	Review with students the characteristics of triangles. Elicit the description of a right triangle. Help students identify the legs and the hypotenuse. learning style: visual

8-7

1. Plan

Objective
To use the Pythagorean Theorem to solve real-world problems

Examples
1 Finding the Length of a Hypotenuse
2 Finding the Length of a Leg
3 Application: Recreation

Math Understandings: p. 372D

Professional Development

Math Background

In a right triangle, the two short sides, or legs, form the 90° angle. The longest side, the hypotenuse, is opposite the 90° (and largest) angle. If you know the length of any two sides, you can calculate the "missing" side length by using the Pythagorean Theorem: $a^2 + b^2 = c^2$, where a and b are lengths of the legs and c is the length of the hypotenuse.

More Math Background: p. 372D

Lesson Planning and Resources

See p. 372E for a list of the resources that support this lesson.

PowerPoint

Bell Ringer Practice

Check Skills You'll Need
Use student page, transparency, or PowerPoint. For intervention, direct students to:

Square Roots and Irrational Numbers
Lesson 8-6
Extra Skills and Word Problems Practice, Ch. 8

2. Teach

Activity Lab

Use before the lesson.
Student Edition Activity Lab, Hands On 8-7a, Exploring Right Triangles, p. 404

All in One Teaching Resources
Activity Lab 8-7: Patterns in Geometry

Guided Instruction

Example 1
Point out that you can substitute the lengths of the legs into the theorem in either order.

Error Prevention!
In Quick Check 3, students might substitute 60 for the hypotenuse. Point out that this is the height of the tower.

Additional Examples

1. Find the length of the hypotenuse. **29 in.**

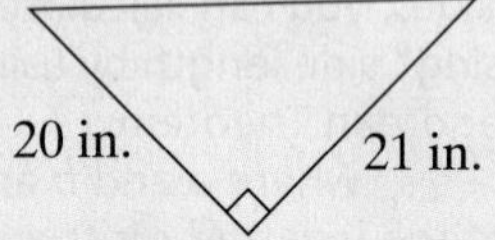

2. Find the missing leg of the triangle. **20 ft**

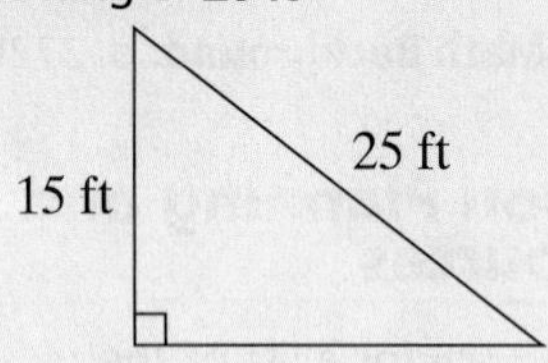

3. A ladder, placed 4 ft from a wall, touches the wall 11.3 ft above the ground. What is the approximate length of the ladder? **about 12 ft**

All in One Teaching Resources
- Daily Notetaking Guide 8-7 L3
- Adapted Notetaking 8-7 L1

Closure

- *How can you find the length of a leg of a right triangle if one leg and the hypotenuse are known?* **Substitute the length of one leg and the hypotenuse in $a^2 + b^2 = c^2$; solve for the missing leg, either *a* or *b*.**

You can use the Pythagorean Theorem to find the length of a leg of a right triangle.

EXAMPLE Finding the Length of a Leg

2 Find the length of the missing leg of the triangle.

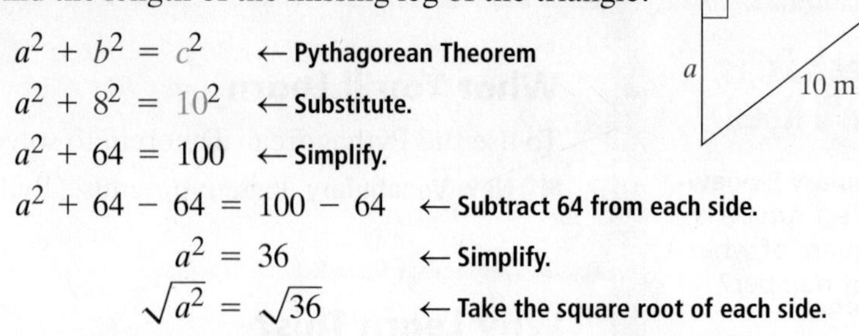

$a^2 + b^2 = c^2$ ← **Pythagorean Theorem**

$a^2 + 8^2 = 10^2$ ← **Substitute.**

$a^2 + 64 = 100$ ← **Simplify.**

$a^2 + 64 - 64 = 100 - 64$ ← **Subtract 64 from each side.**

$a^2 = 36$ ← **Simplify.**

$\sqrt{a^2} = \sqrt{36}$ ← **Take the square root of each side.**

$a = 6$

The length of the leg is 6 m.

Quick Check

2. Find the missing length in the triangle below. **24 mi**

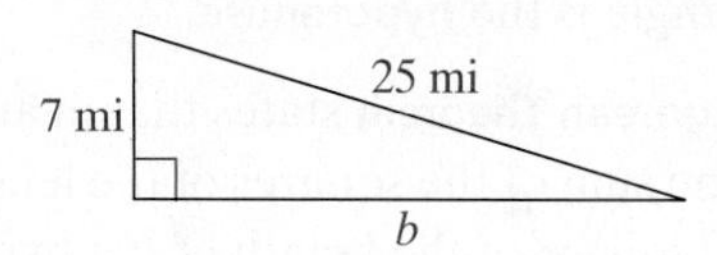

EXAMPLE Application: Recreation

3 A water slide starts 6 m above the water and extends 11 m horizontally. What is the length of the slide to the nearest tenth of a meter?

A sketch shows that the length of the slide is the length of the hypotenuse.

$a^2 + b^2 = c^2$ ← **Use the Pythagorean Theorem.**

$6^2 + 11^2 = c^2$ ← **Substitute.**

$157 = c^2$ ← **Simplify.**

$\sqrt{157} = \sqrt{c^2}$ ← **Take the square root of each side.**

$c \approx 12.529964$

The slide is about 12.5 m.

Some water slides are straight, while others have curves.

Quick Check

3. A support wire is attached to the top of a 60-m tower. It meets the ground 25 m from the base of the tower. How long is the wire? **65 m**

Advanced Learners L4
A right triangle has an area of 241 cm². The length of one of its legs is 26 cm. What is the length of the hypotenuse? **about 32 cm**

learning style: visual

English Language Learners ELL
For most of the practice problems and homework, where triangles are described, ask students to draw the triangles before computing. Then ask them to identify the hypotenuse, and label the lengths they know.

learning style: visual

Check Your Understanding

1. hypotenuse
2. No; you are taking the square root of the squares of the numbers, so the length of t will increase by less than 2.

1. **Vocabulary** The side opposite the right angle in a triangle is called the __?__.

2. **Number Sense** If the longer leg of the triangle at the right is increased by 2, will the length of t increase by 2? Explain.

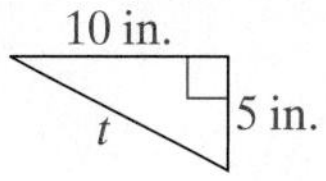

The lengths of two legs of a right triangle are given. Find the length of the hypotenuse to the nearest tenth.

3. 8 m and 11 m **13.6 m**
4. 4 in. and 3 in. **5 in.**
5. 12 ft and 20 ft **23.3 ft**

Homework Exercises

For more exercises, see Extra Skills and Word Problems.

GO for Help

For Exercises	See Examples
6–8	1
9–15	2–3

A **Find the length of the hypotenuse of each triangle. Round to the nearest tenth of a unit, if necessary.**

6. 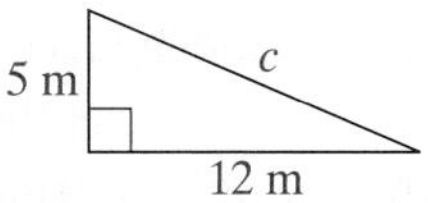

13 m

7. 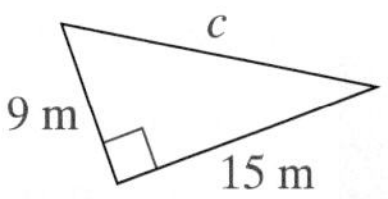

17.5 m

8. 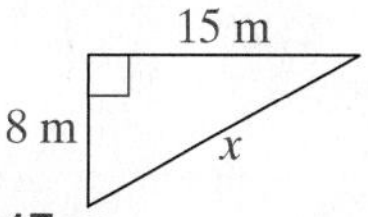

17 m

Find each missing length. Round to the nearest tenth, if necessary.

9. 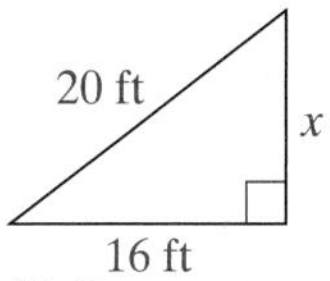

12 ft

10. 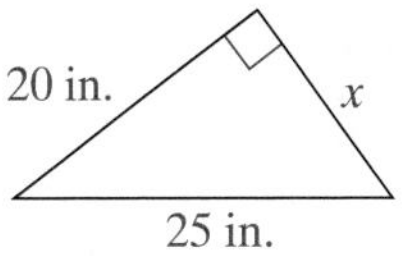

15 in.

11. 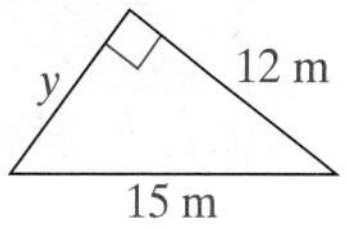

9 m

12. 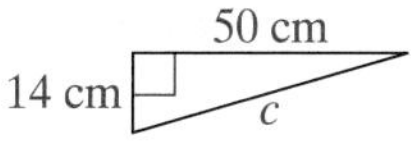

51.9 cm

13. 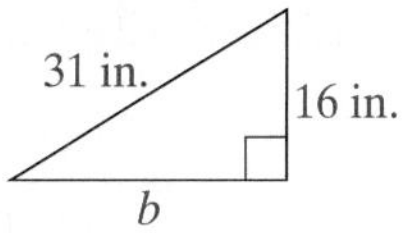

26.6 in.

14. 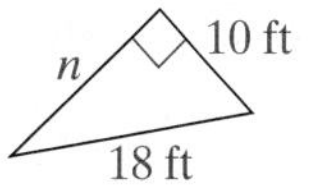

15.0 ft

15. A tennis court is 78 ft long and 27 ft wide. To the nearest foot, what is the length of the diagonal of a tennis court? **83 ft**

B **GPS** 16. **Guided Problem Solving** A rectangular park is 600 m long and 300 m wide. You walk diagonally across the park from corner to corner. How far do you walk, to the nearest meter? **671 m**
   - You can use the strategy *Draw a Diagram*. Label the diagram. Then use the Pythagorean Theorem.

17. A ladder is 6 m long. How much farther up a wall does the ladder reach when the base of the ladder is 2 m from the wall than when it is 3 m from the wall? Round to the nearest tenth of a meter. **0.5 m**

Online lesson quiz, **PHSchool.com, Web Code:** ara-0807

3. Practice

Assignment Guide

Check Your Understanding

Go over Exercises 1–5 in class before assigning the Homework Exercises.

Homework Exercises

A	Practice by Example	6–15
B	Apply Your Skills	16–24
C	Challenge	25
	Test Prep and Mixed Review	26–29

Homework Quick Check

To check students' understanding of key skills and concepts, go over Exercises 6, 13, 22, 23, and 24.

Differentiated Instruction Resources

Adapted Practice 8-7 L1

Practice 8-7 The Pythagorean T L3

The lengths of two sides of a right triangle are given. Find the length of the third side to the nearest tenth of a unit.

1. legs: 5 ft and 12 ft **13.0 ft**
2. legs: 13 cm and 9 cm **15.8 cm**
3. leg: 7 m; hypotenuse: 14 m **12.1 m**

Find each missing length. Round to the nearest tenth of a unit, if necessary.

4. 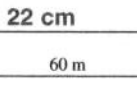

x = 22 cm

5. 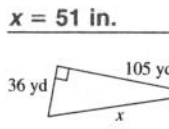

x = 51 in.

6. 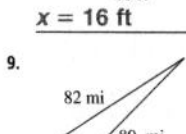

x = 16 ft

7. 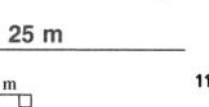

x = 25 m

8. 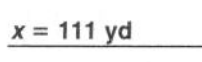

x = 111 yd

9. 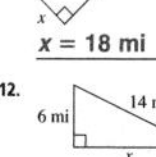

x = 18 mi

10. 

x = 23.0 m

11. 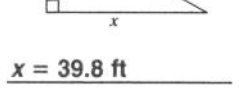

x = 39.8 ft

12. 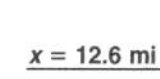

x = 12.6 mi

Solve.

13. A playground is 50 yd by 50 yd. Amy walked across the playground from one corner to the opposite corner. How far did she walk?
70.7 yd

14. A 70-ft ladder is mounted 10 ft above the ground on a fire truck. The bottom of the ladder is 40 ft from the wall of a building. The top of the ladder is touching the building. How high off the ground is the top of the ladder?
67.4 ft

8-7 • Guided Problem Solving GPS L3

Student Page 408, Exercise 21:

Camping A large tent has an adjustable center pole. A rope 26 ft long connects the top of the pole to a peg 24 ft from the bottom of the pole. What is the height of the pole? Round to the nearest hundredth if necessary.

Understand

1. Circle the information you will need to solve.
2. What are you being asked to do?
Find the height of the pole.
3. The right triangle shown models the tent pole and rope. Label it with the correct values.

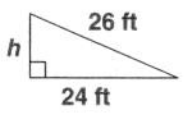

Plan and Carry Out

4. Write the formula for the Pythagorean Theorem. **$a^2 + b^2 = c^2$**
5. What variable does 26 ft represent in the Pythagorean Theorem? **c**
6. What variable does 24 ft represent in the Pythagorean Theorem? **a or b**
7. Substitute these values into the Pythagorean Theorem formula and find the height of the pole. **10 ft**

Check

8. Substitute your answer from step 7 and 24 ft into the Pythagorean Theorem formula for a and b. Do you get a true statement?
yes; $10^2 + 24^2 = 26^2$

Solve Another Problem

9. You're building a right-triangular brace for a basketball hoop. You have the piece for the hypotenuse and for one of the legs. The hypotenuse measures 10 in. long, and the leg measures 6 in. long. How long must the other leg of the brace be?
$10^2 - 6^2 = 64 = 8^2$; 8 in.

4. Assess & Reteach

Lesson Quiz

Find the length of the missing side of each right triangle to the nearest tenth.

1. legs of 4 cm and 6 cm **7.2 cm**
2. leg of 9 m, hypotenuse of 15 m **12 m**
3. leg of 36 ft, hypotenuse of 39 ft **15 ft**
4. legs of 28 cm and 45 cm **53 cm**

28. $n = 0.05 \cdot 225$; 11.25

29. $n = 0.60 \cdot 40$; 24

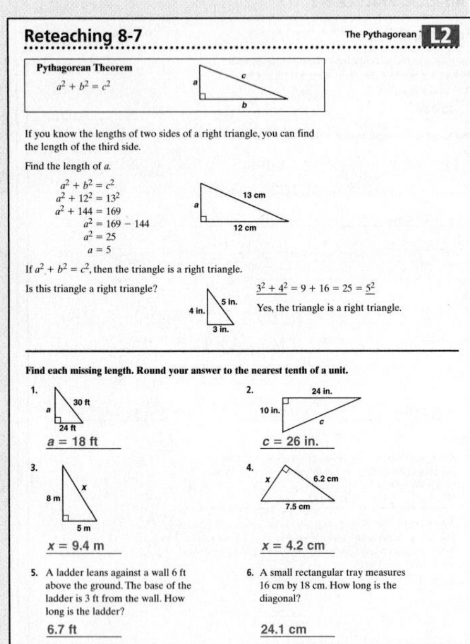

Reteaching 8-7 — The Pythagorean — L2

Pythagorean Theorem

$a^2 + b^2 = c^2$

If you know the lengths of two sides of a right triangle, you can find the length of the third side.

Find the length of a.

$$a^2 + b^2 = c^2$$
$$a^2 + 12^2 = 13^2$$
$$a^2 + 144 = 169$$
$$a^2 = 169 - 144$$
$$a^2 = 25$$
$$a = 5$$

If $a^2 + b^2 = c^2$, then the triangle is a right triangle.

Is this triangle a right triangle? (4 in., 3 in., 5 in.)

$3^2 + 4^2 = 9 + 16 = 25 = 5^2$

Yes, the triangle is a right triangle.

Find each missing length. Round your answer to the nearest tenth of a unit.

1. (30 ft, 24 ft, a) $a = 18$ ft
2. (24 in., 10 in., c) $c = 26$ in.
3. (8 m, 5 m, x) $x = 9.4$ m
4. (x, 6.2 cm, 7.5 cm) $x = 4.2$ cm
5. A ladder leans against a wall 6 ft above the ground. The base of the ladder is 3 ft from the wall. How long is the ladder? 6.7 ft
6. A small rectangular tray measures 16 cm by 18 cm. How long is the diagonal? 24.1 cm

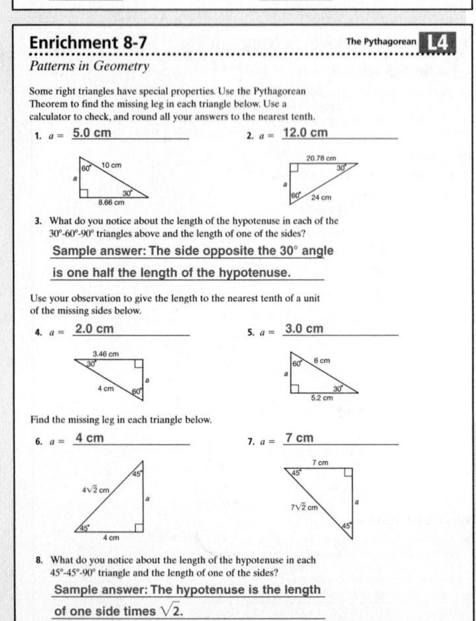

Enrichment 8-7 — The Pythagorean — L4

Patterns in Geometry

Some right triangles have special properties. Use the Pythagorean Theorem to find the missing leg in each triangle below. Use a calculator to check, and round all your answers to the nearest tenth.

1. $a =$ 5.0 cm (10 cm, 8.66 cm, 60°, 30°)
2. $a =$ 12.0 cm (20.78 cm, 24 cm, 30°, 60°)
3. What do you notice about the length of the hypotenuse in each of the 30°-60°-90° triangles above and the length of one of the sides? Sample answer: The side opposite the 30° angle is one half the length of the hypotenuse.

Use your observation to give the length to the nearest tenth of a unit of the missing sides below.

4. $a =$ 2.0 cm (3.46 cm, 4 cm, 30°, 60°)
5. $a =$ 3.0 cm (6 cm, 5.2 cm, 60°, 30°)

Find the missing leg in each triangle below.

6. $a =$ 4 cm ($4\sqrt{2}$ cm, 4 cm, 45°, 45°)
7. $a =$ 7 cm (7 cm, $7\sqrt{2}$ cm, 45°, 45°)
8. What do you notice about the length of the hypotenuse in each 45°-45°-90° triangle and the length of one of the sides? Sample answer: The hypotenuse is the length of one side times $\sqrt{2}$.

Homework Video Tutor
Visit: PHSchool.com
Web Code: are-0807

24. Check to see if the sum of the squares of the two shortest sides equals the square of the longest side.
$10^2 + 24^2 \stackrel{?}{=} 26^2$
$100 + 576 \stackrel{?}{=} 676$
$676 = 676$
It is a right triangle.

The lengths of two sides of a right triangle are given. Find the length of the third side to the nearest tenth.

18. legs 8 yd and 11 yd **13.6 yd**

19. leg 18 m and hypotenuse 28 m **21.4 m**

20. Find the perimeter and area of the triangle at the right. Round to the nearest tenth. **33.7 in.; 48.3 in.²**

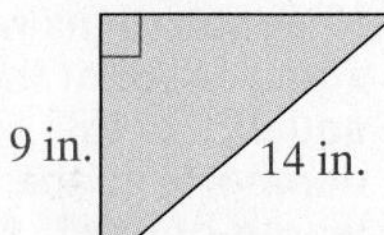

21. Camping A large tent has an adjustable center pole. A rope 26 ft long connects the top of the pole to a peg 24 ft from the bottom of the pole. What is the height of the pole? Round to the nearest hundredth if necessary. **10 ft**

22. The rectangular section of fencing at the right is reinforced with wood nailed across the diagonal of the rectangle. What is the length of the diagonal? **10 ft**

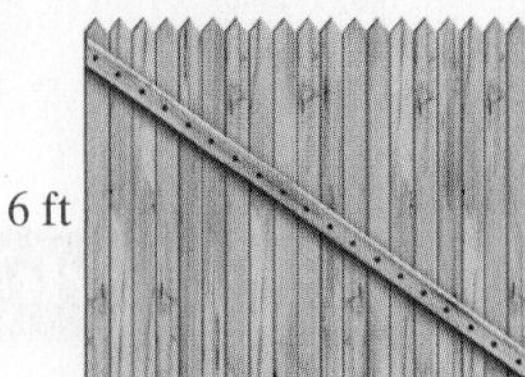

23. Navigation A dock is located 24 km directly east of a lighthouse. A sailboat is directly north of the lighthouse. The sailboat is 25 km from the dock. How far away from the lighthouse is the sailboat? **7 km**

24. Writing in Math A triangle has side lengths measuring 10 m, 24 m, and 26 m. Explain how you use the Pythagorean Theorem to determine whether or not the triangle is a right triangle. **See left.**

25. Challenge Find x. **3**

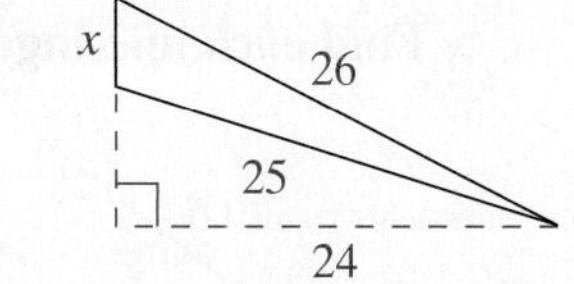

Multiple Choice

26. The hypotenuse of a right triangle is 12 cm long. Another side of the triangle is 8 cm long. Which equation can be used to find n, the length of the third side of the triangle? **D**

- Ⓐ $n = 12^2 + 8^2$
- Ⓑ $n = \sqrt{12^2 + 8^2}$
- Ⓒ $n = 12^2 - 8^2$
- Ⓓ $n = \sqrt{12^2 - 8^2}$

27. Computers are used to design and improve spacesuits. An astronaut's arm is 32 inches long. What is the arm length in the computer image, if the computer image is $\frac{1}{8}$ of the actual length? **G**

- Ⓕ $\frac{1}{8}$ in.
- Ⓖ 4 in.
- Ⓗ 32 in.
- Ⓙ 256 in.

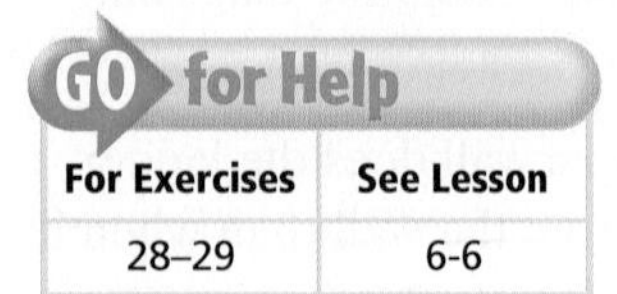

For Exercises	See Lesson
28–29	6-6

Algebra **Write and solve an equation to find the part of a whole.** **28–29. See margin.**

28. What number is 5% of 225? 11.25

29. What number is 60% of 40? 24

Test Prep

Resources

For additional practice with a variety of test item formats:

- Test-Taking Strategies, p. 427
- Test Prep, p. 431
- Test-Taking Strategies with Transparencies

Alternative Assessment

Have each student in a pair draw a triangle and label two sides. Partners exchange triangles and solve for the missing side length. Partners check each other's results and repeat the activity as time permits.

8-8a Activity Lab

Hands On

Three Views of an Object

Three-dimensional objects can be drawn to show length, width, and height. You can make drawings that show the *top view*, *front view*, and the *right side view*.

EXAMPLE Drawing Three Views

Draw the top, front, and right side views of the figure at the right.

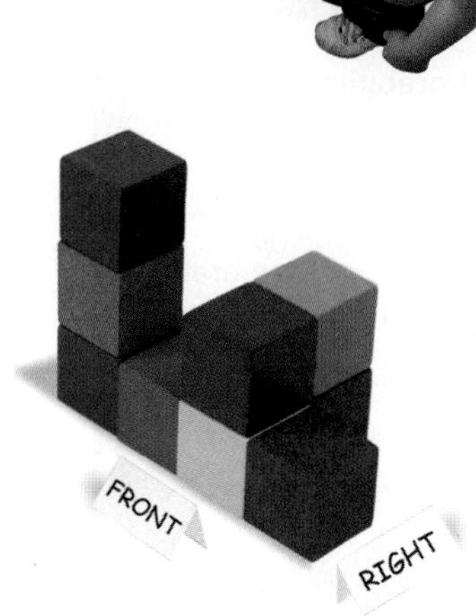

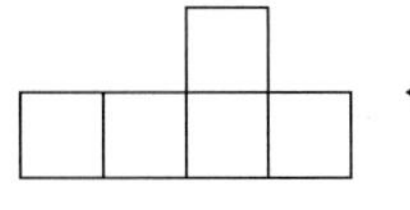

← **Draw the top view as if you are looking down on the blocks.**

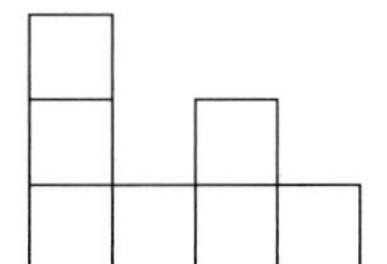

← **Draw the front view as if you are in front of the blocks.**

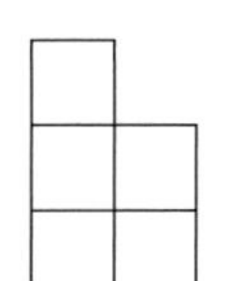

← **Draw the right side view as if you are on the right side of the blocks.**

Exercises

Draw the top, front, and right side views of each figure. 1–5. See margin.

1.

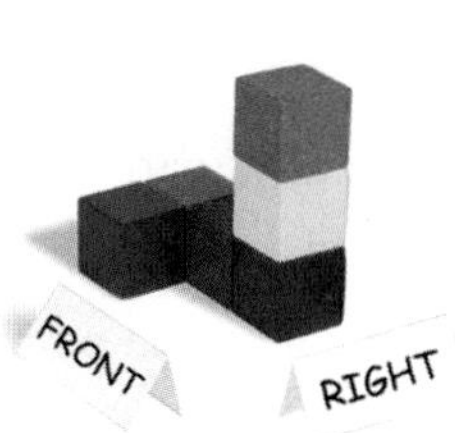

2.

3.

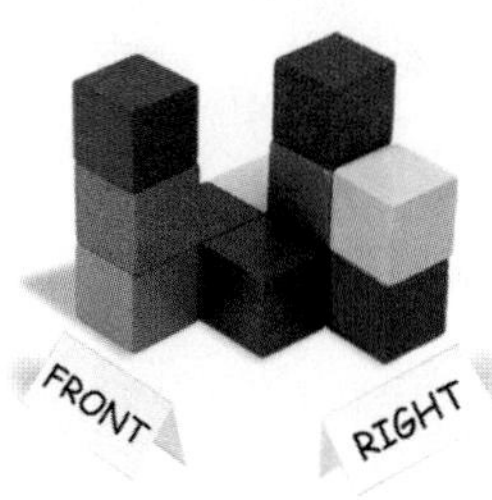

Use the given views to draw a three-dimensional figure.

4.

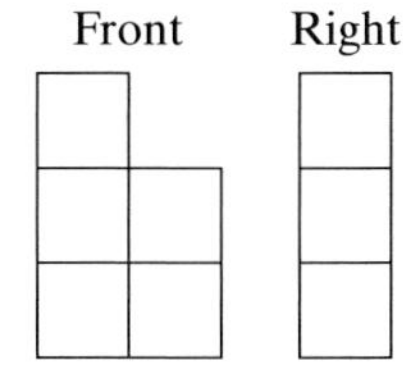

5.

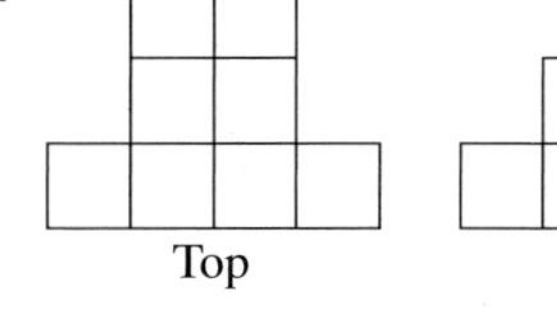

Front

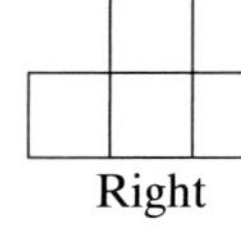

1–5. See back of book.

Activity Lab

Three Views of an Object

The ability to visualize and analyze different views of three-dimensional objects is a valuable life skill. Architects and construction workers are among those who regularly make use of it in a work setting. This Activity Lab prepares students for Lesson 8-8 and addresses the skill of viewing objects from the top, front, and sides.

Guided Instruction

Before beginning the Activity, ask students to describe classroom objects that look different when viewed from different positions. Encourage students to draw one of these objects from different views.

Ask questions such as:

- *Why does each view look different?* **because you can see different parts of the object from each view.**
- *How can each view help you construct a model of the object?* **Each view shows what one side of the model should look like.**

Exercises

Have students work in pairs and compare their drawings for each exercise.

Alternative Method

Invite students to use cubes to construct the figures in the Exercises before they make their drawings. Encourage them to build different figures for others to draw from the top, sides, and front.

Resources

- Activity Lab 8-8: Three Dimensional Figures
- cubes
- graph paper
- Classroom Aid 2

8-8

1. Plan

Objective
To classify and draw three-dimensional figures

Examples
1 Naming Figures
2 Drawing Three-Dimensional Figures

Math Understandings: p. 372D

Math Background

Because a parallelogram has only height and width, it is called *two-dimensional*. Figures with three dimensions do not lie in a plane. A flat surface on a three-dimensional figure is called a *face*. Prisms, cubes, and pyramids are composed of only flat faces. Cones, cylinders, and spheres are all three-dimensional objects that include curved surfaces.

More Math Background: p. 372D

Lesson Planning and Resources

See p. 372E for a list of the resources that support this lesson.

Bell Ringer Practice

Check Skills You'll Need
Use student page, transparency, or PowerPoint. For intervention, direct students to:
Quadrilaterals and Other Polygons
Lesson 7-4
Extra Skills and Word Problems Practice, Ch. 7

8-8 Three-Dimensional Figures

Check Skills You'll Need

1. Vocabulary Review
A __?__ is a polygon with all sides and angles congruent. **regular polygon**

Use dot paper to draw each figure. **2–4. See back of book.**

2. rhombus

3. trapezoid

4. rectangle

Lesson 7-4

What You'll Learn

To classify and draw three-dimensional figures

New Vocabulary three-dimensional figure, face, edge, bases, prism, height, cube, cylinder, pyramid, vertex, cone, sphere, center

Why Learn This?

You already know about some three-dimensional figures. You see them in many ordinary objects around you. If you know how to classify three-dimensional figures, you can describe the shapes of the objects you see.

A **three-dimensional figure,** or solid, is a figure that does not lie in a plane. A flat surface of a solid shaped like a polygon is called a **face.** Each segment formed by the intersection of two faces is an **edge.**

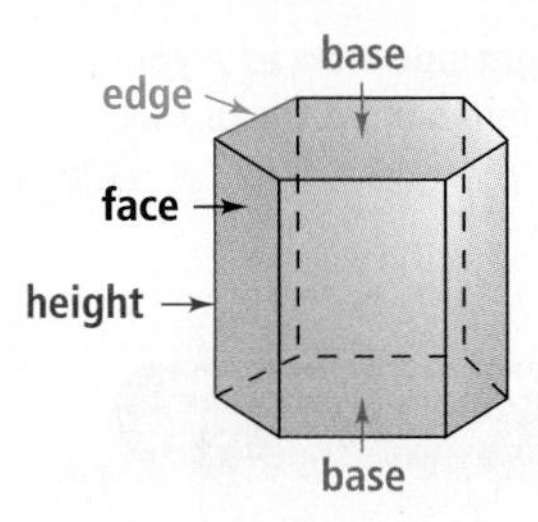

A glass prism can refract, or bend, light.

A **prism** is a three-dimensional figure with two parallel and congruent polygonal faces, called **bases.** The other faces are rectangles. The **height** of a prism is the length of a perpendicular segment that joins the bases. A prism is named for the shape of its bases.

A **cube** is a rectangular prism with faces that are all squares.

A **cylinder** has two congruent parallel **bases** that are circles. The **height** of a cylinder is the length of a perpendicular segment that joins the bases.

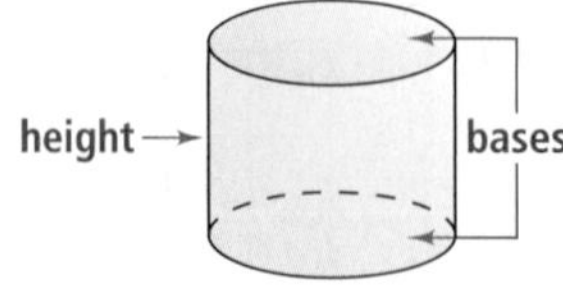
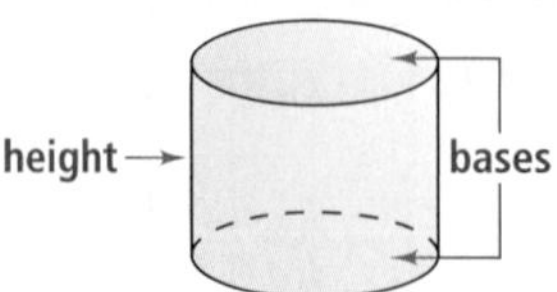

A **pyramid** has triangular faces that meet at one point, a **vertex**, and a **base** that is a polygon. A pyramid is named for the shape of its base.

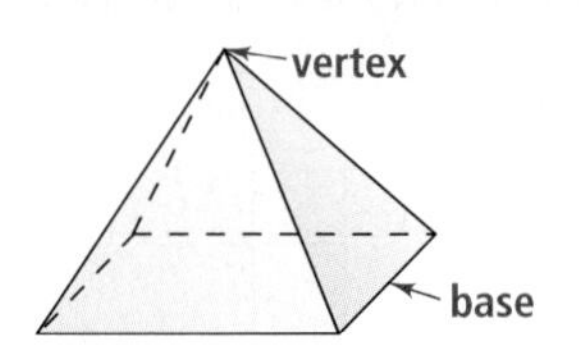

Differentiated Instruction Solutions for All Learners

Special Needs L1
Drawing the figures may be too difficult for some students. If so, provide them nets so that they can form the solids. If that is also difficult, pair them up with students who can do the drawing. They can then check the drawing against the definition.

learning style: tactile

Below Level L2
Students can work in groups to create charts. Charts should have 4 columns: *two-dimensional figures, prisms, pyramids, other three-dimensional objects.* Have them draw at least three figures in each column.

learning style: visual

A **cone** has one circular **base** and one **vertex.**

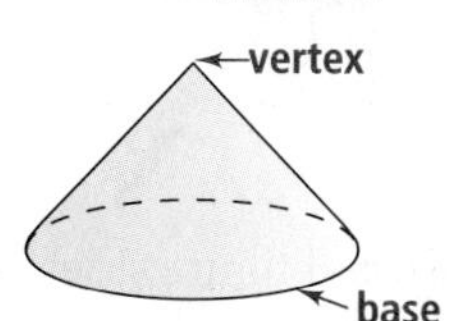

A **sphere** is the set of all points in space that are the same distance from a **center** point.

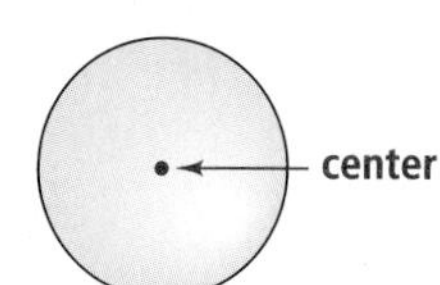

EXAMPLE Naming Figures

1 **Architecture** Look at the architectural blocks. Name Figure 3.

Figure 3 has two parallel, congruent bases that are circles.

Figure 3 is a cylinder.

Quick Check

1. Name Figures 1 and 2. **triangular prism; cone**

You can use graph paper to draw three-dimensional figures.

EXAMPLE Drawing Three-Dimensional Figures

Vocabulary Tip

Notice the word hexagon inside *hexagonal.* A hexagon has six sides, so a *hexagonal* prism is a prism with a six-sided base.

2.

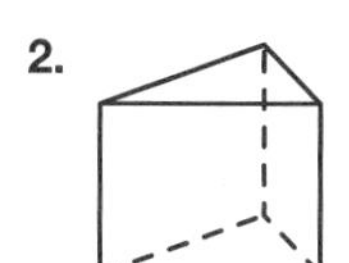

2 Draw a hexagonal prism.

Step 1 Draw a hexagon.

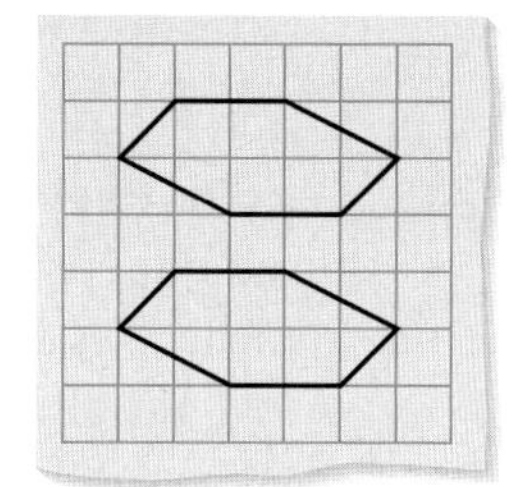

Step 2 Draw a second hexagon congruent to the first.

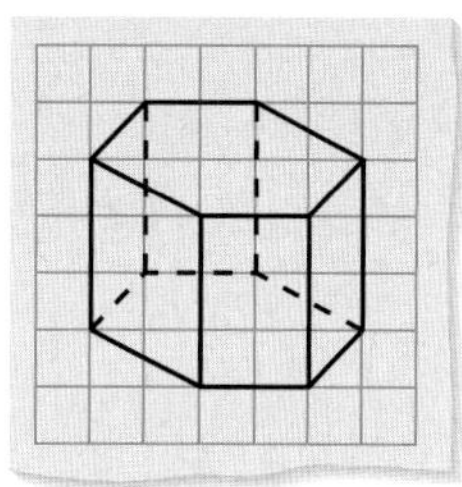

Step 3 Connect the vertices. Use dashed lines for hidden edges.

Quick Check

2. Use graph paper to draw a triangular prism. **See left.**

2. Teach

Activity Lab

Use before the lesson.
Student Edition Activity Lab, Hands On 8-8a, Three Views of an Object, p. 409

All in One Teaching Resources
Activity Lab 8-8: Three Dimensional Figures

Guided Instruction

Error Prevention!

Students may confuse *prism* and *pyramid.* Have students ask themselves: *Can the figure stand on both of its bases?* If the answer is yes, it's a prism.

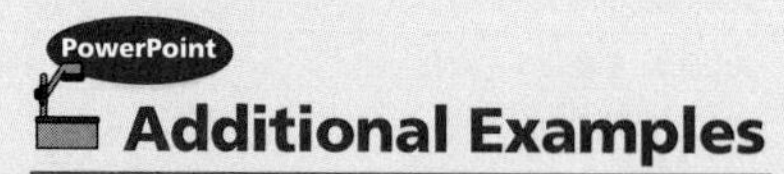

Additional Examples

1 Name the geometric figure. **sphere**

2 Draw a pentagonal prism.

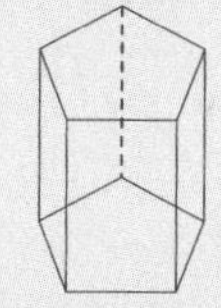

All in One Teaching Resources
- Daily Notetaking Guide 8-8 L3
- Adapted Notetaking 8-8 L1

Closure

- *What three-dimensional figures are named for their bases?* **prisms and pyramids**
- *What three-dimensional figures have only flat faces?* **cube, prism, pyramid**
- *What three-dimensional figures have curved surfaces?* **sphere, cone, cylinder**

Advanced Learners L4
True or false? Explain. *Every rectangular prism is a cube.* **False; a cube must have six square faces.** *Some pyramids are prisms.* **False; pyramids have only one base and prisms have two.**

learning style: verbal

English Language Learners ELL
Divide the students into three groups and distribute index cards. One group copies the definitions of prisms. One group draws the shapes. The third group writes the names of the shapes. Ask students to correctly find the name, picture, and definition of each shape.

learning style: visual

3. Practice

Assignment Guide

Check Your Understanding
Go over Exercises 1–5 in class before assigning the Homework Exercises.

Homework Exercises

A	Practice by Example	6–14
B	Apply Your Skills	15–24
C	Challenge	25
	Test Prep and Mixed Review	26–29

Homework Quick Check
To check students' understanding of key skills and concepts, go over Exercises 8, 14, 16, 23, and 24.

Differentiated Instruction Resources

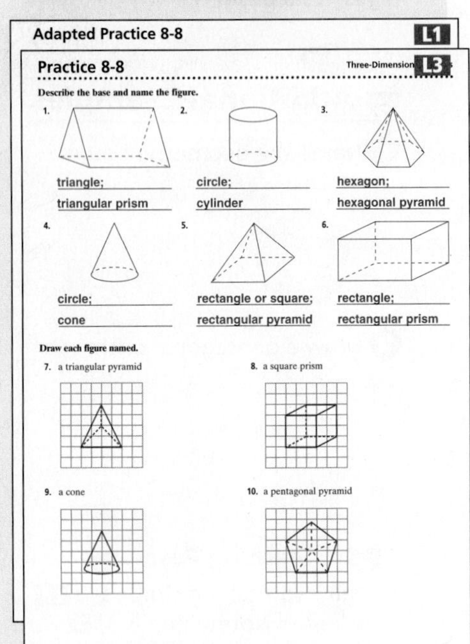
Adapted Practice 8-8 L1

Practice 8-8 Three-Dimensional Figures L3

Describe the base and name the figure.

1. triangle; triangular prism
2. circle; cylinder
3. hexagon; hexagonal pyramid
4. circle; cone
5. rectangle or square; rectangular pyramid
6. rectangle; rectangular prism

Draw each figure named.

7. a triangular pyramid
8. a square prism
9. a cone
10. a pentagonal pyramid

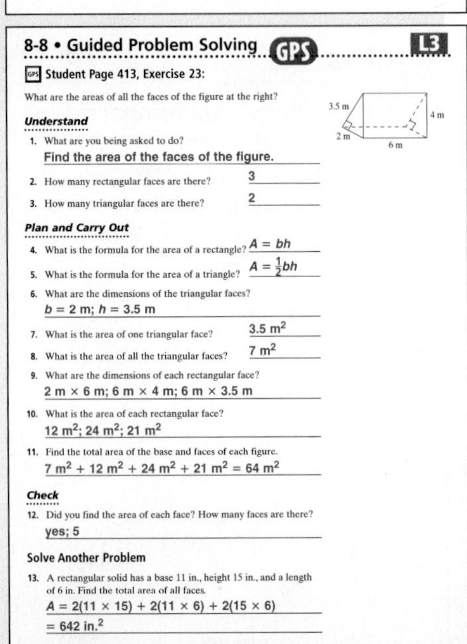
8-8 • Guided Problem Solving GPS L3

GPS Student Page 413, Exercise 23:

What are the areas of all the faces of the figure at the right?

Understand

1. What are you being asked to do? Find the area of the faces of the figure.
2. How many rectangular faces are there? 3
3. How many triangular faces are there? 2

Plan and Carry Out

4. What is the formula for the area of a rectangle? $A = bh$
5. What is the formula for the area of a triangle? $A = \frac{1}{2}bh$
6. What are the dimensions of the triangular faces? $b = 2$ m; $h = 3.5$ m
7. What is the area of one triangular face? 3.5 m^2
8. What is the area of all the triangular faces? 7 m^2
9. What are the dimensions of each rectangular face? 2 m × 6 m; 6 m × 4 m; 6 m × 3.5 m
10. What is the area of each rectangular face? 12 m^2; 24 m^2; 21 m^2
11. Find the total area of the base and faces of each figure. 7 m^2 + 12 m^2 + 24 m^2 + 21 m^2 = 64 m^2

Check

12. Did you find the area of each face? How many faces are there? yes; 5

Solve Another Problem

13. A rectangular solid has a base 11 in., height 15 in., and a length of 6 in. Find the total area of all faces.
$A = 2(11 \times 15) + 2(11 \times 6) + 2(15 \times 6)$
$= 642$ in.^{2}

Check Your Understanding

1. **Vocabulary** A _?_ has two congruent parallel bases that are circles. **cylinder**

2. Which three-dimensional figure does NOT have a base? **D**
 (A) cone (B) prism (C) pyramid (D) sphere

Describe each base and name each prism.

3. 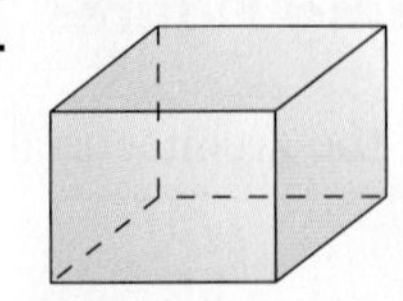
rectangle; rectangular prism

4. 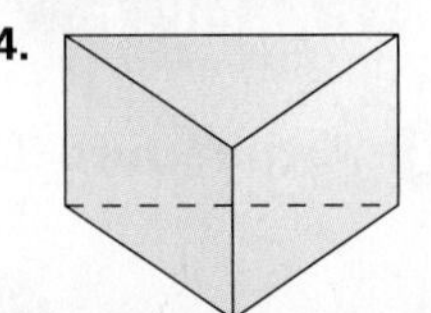
triangle; triangular prism

5. 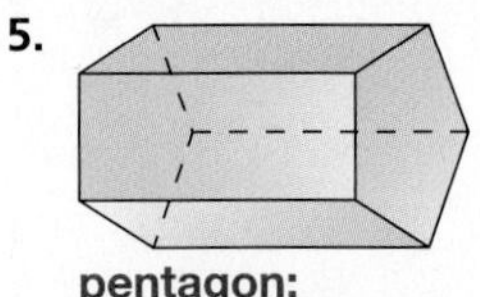
pentagon; pentagonal prism

Homework Exercises

For more exercises, see Extra Skills and Word Problems.

GO for Help

For Exercises	See Examples
6–11	1
12–14	2

A **Name each figure.** **6–11. See left.**

6. 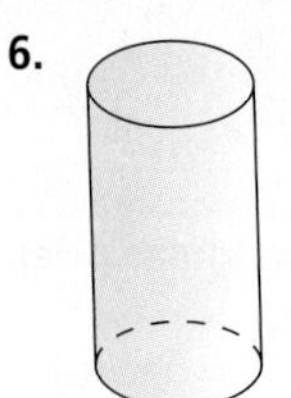
7. 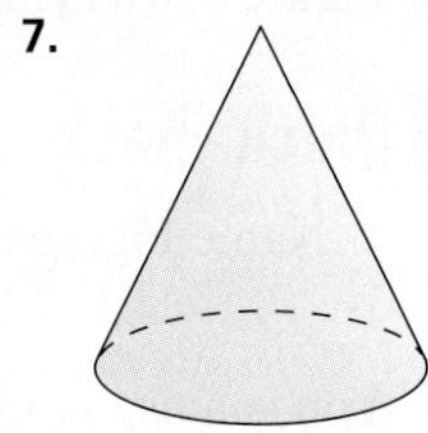
8. 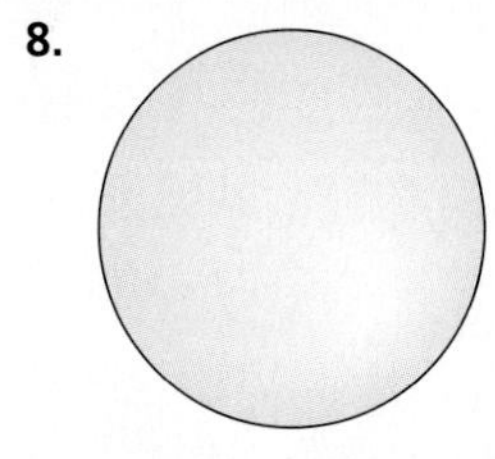
9. 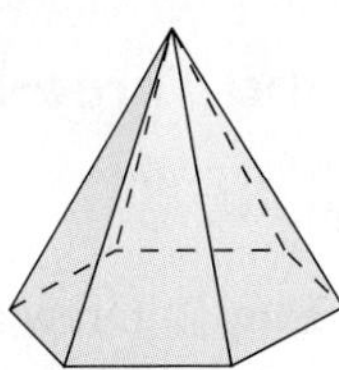
10. 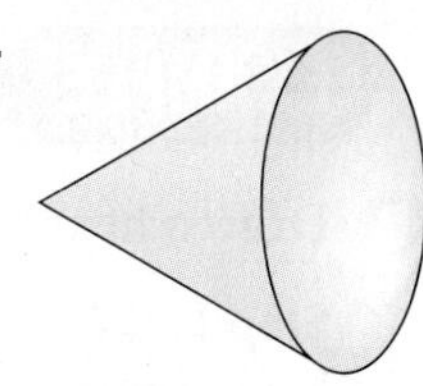
11. 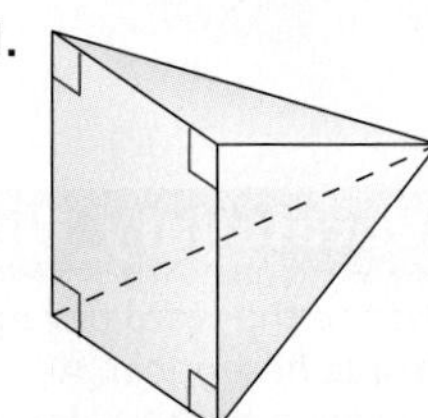

6. cylinder
7. cone
8. sphere
9. hexagonal pyramid
10. cone
11. rectangular pyramid

Use graph paper to draw each figure. **12–14. See margin.**

12. cylinder 13. pentagonal prism 14. square pyramid

B 

15. **Guided Problem Solving** Refer to the three-dimensional figure shown at the right. Which two solids make up the figure? **See left.**
 - What polygon is the base of the lower portion?
 - What polygon is the base of the upper portion?

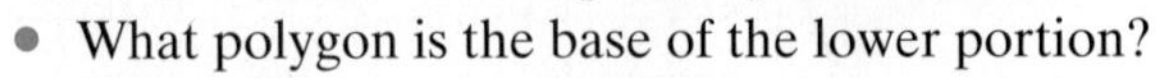

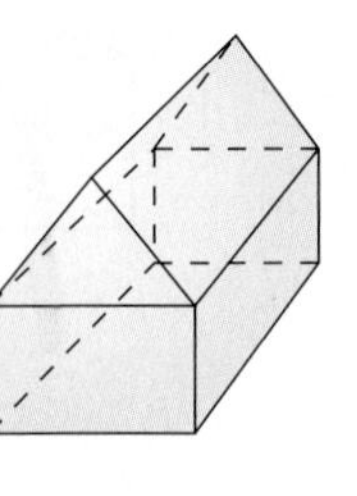

15. rectangular prism; triangular prism

16. A solid has a rectangle for its base and four faces that are triangles. What is the solid? **rectangular pyramid**

12–14. See back of book.

GO Online
Homework Video Tutor
Visit: PHSchool.com
Web Code: are-0808

Name the three-dimensional figure in each photograph.

17. 
rectangular prism

18. 
sphere

19. 
cone

Use the pentagonal pyramid at the right.

20. Name four edges that intersect $\overline{AB}$. $\overline{MA}$, $\overline{BP}$, $\overline{KA}$, $\overline{KB}$

21. Name any edges that are parallel to $\overline{AB}$. none

22. Name the five edges that are *not* parallel to $\overline{AB}$ and do *not* intersect $\overline{AB}$.
$\overline{MD}$, $\overline{DP}$, $\overline{KP}$, $\overline{KM}$, $\overline{KD}$

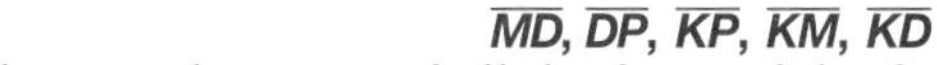

23. What are the areas of all the faces of the figure at the left? 64 m^2

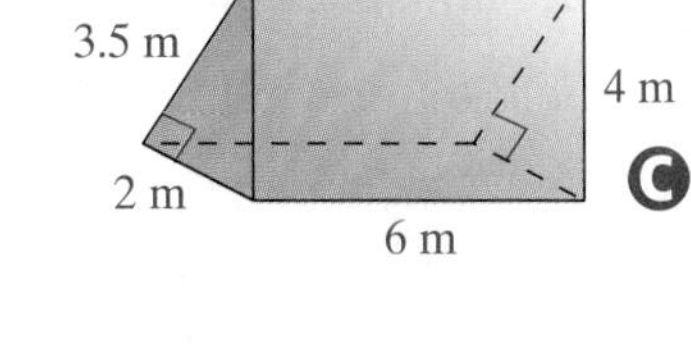

24. **Writing in Math** Are the edges of a cube congruent? Explain. See left.

24. Yes; the edges are all part of square faces of the same size.

25. **Challenge** Identify the number of faces, edges, and vertices a hexagonal pyramid has. 7 faces; 12 edges; 7 vertices

Multiple Choice

26. Which model represents 5^2? C

Ⓐ 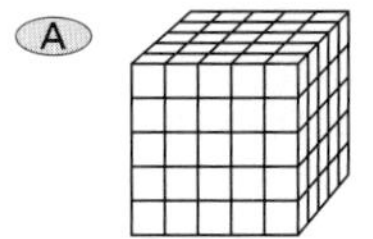
Ⓑ 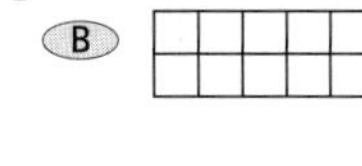
Ⓒ 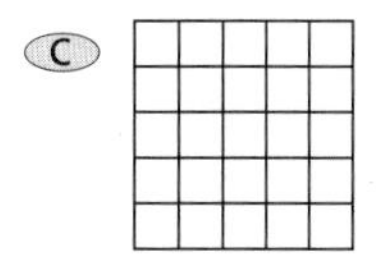
Ⓓ 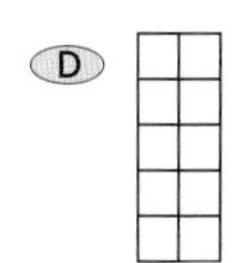

27. What is the measure of each angle of an equilateral triangle? H
Ⓕ 180°, 90°, and 90°, because the sum of the angles in a triangle is 360° and two angles are congruent
Ⓖ 45°, 45°, and 90°, because the sum of the angles in a triangle is 180° and two angles are congruent
Ⓗ 60°, 60°, and 60°, because the sum of the angles in a triangle is 180° and all three angles are congruent
Ⓙ 45°, 45°, and 45°, because the sum of the angles in a triangle is 135° and all three angles are congruent

28. Which of the following has two bases that are regular polygons? D
Ⓐ Pyramid Ⓑ Cylinder Ⓒ Cone Ⓓ Prism

29. The hypotenuse of a right triangle is 61 m long. One leg is 60 m long. What is the length of the third side? 11 m

GO for Help

For Exercise	See Lesson
29	8-7

4. Assess & Reteach

Lesson Quiz

Name each figure.

1. cylinder

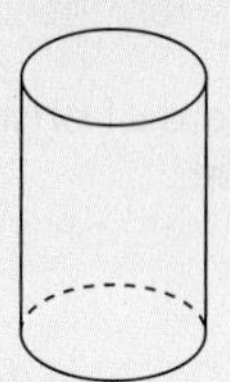

2. hexagonal prism

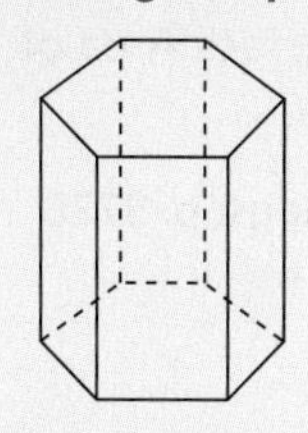

3. Draw a square pyramid.
Sample:

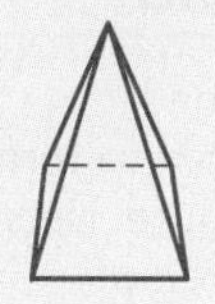

Alternative Assessment

Each student in a pair names one of the three-dimensional figures in this lesson. The partner draws the figure on graph paper or makes a model.

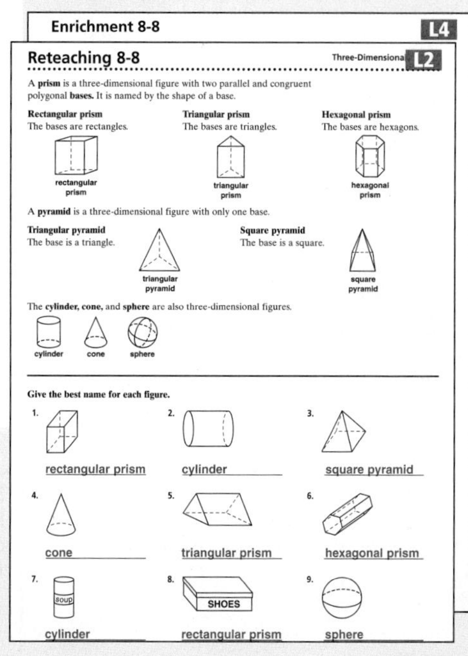

Enrichment 8-8 L4

Reteaching 8-8 Three-Dimensional L2

A **prism** is a three-dimensional figure with two parallel and congruent polygonal **bases.** It is named by the shape of a base.

Rectangular prism The bases are rectangles. (rectangular prism)

Triangular prism The bases are triangles. (triangular prism)

Hexagonal prism The bases are hexagons. (hexagonal prism)

A **pyramid** is a three-dimensional figure with only one base.

Triangular pyramid The base is a triangle. (triangular pyramid)

Square pyramid The base is a square. (square pyramid)

The **cylinder, cone,** and **sphere** are also three-dimensional figures.

cylinder cone sphere

Give the best name for each figure.

1. rectangular prism
2. cylinder
3. square pyramid
4. cone
5. triangular prism
6. hexagonal prism
7. SOUP cylinder
8. SHOES rectangular prism
9. sphere

Test Prep

Resources
For additional practice with a variety of test item formats:
- Test-Taking Strategies, p. 427
- Test Prep, p. 431
- Test-Taking Strategies with Transparencies

8-9

1. Plan

Objective
To find the surface areas of prisms and cylinders using nets

Examples
1 Drawing a Net
2 Finding the Surface Area of a Prism
3 Finding the Surface Area of a Cylinder

Math Understandings: p. 372D

Professional Development

Math Background

A *net* is a two-dimensional pattern that you can fold to form a three-dimensional figure. Nets illustrate the *surface area* of three-dimensional figures. Surface area is the two-dimensional "surface" of a three-dimensional figure. Surface area is measured in square units like the area of any two-dimensional figure. The surface area of three-dimensional figures can be found by finding the area of each face and finding the sum of all these areas. So the surface area of a figure's net is also the surface area of the figure.

More Math Background: p. 372D

Lesson Planning and Resources

See p. 372E for a list of the resources that support this lesson.

Bell Ringer Practice

Check Skills You'll Need
Use student page, transparency, or PowerPoint. For intervention, direct students to:

Perimeter and Area of a Triangle
Lesson 8-3
Extra Skills and Word Problems Practice, Ch. 8

8-9 Surface Areas of Prisms and Cylinders

Check Skills You'll Need

1. **Vocabulary Review** What is the *height of a triangle*? **See below.**

Find the area of each triangle.

2. **48 m²**

3. **6 ft²**

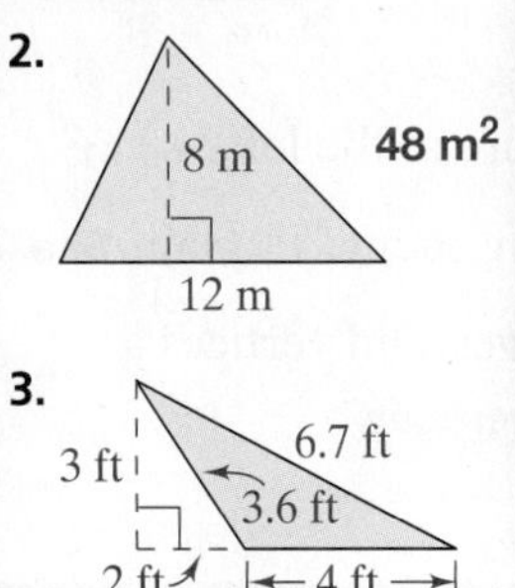

GO for Help Lesson 8-3

What You'll Learn

To find the surface areas of prisms and cylinders using nets

New Vocabulary net, surface area

Why Learn This?

When you wrap a birthday gift or cover a textbook, you are working with surface area. Surface area tells you how much material you need to cover something. You can use a net to solve surface area problems.

A **net** is a two-dimensional pattern that you can fold to form a three-dimensional figure. You can use nets to design boxes.

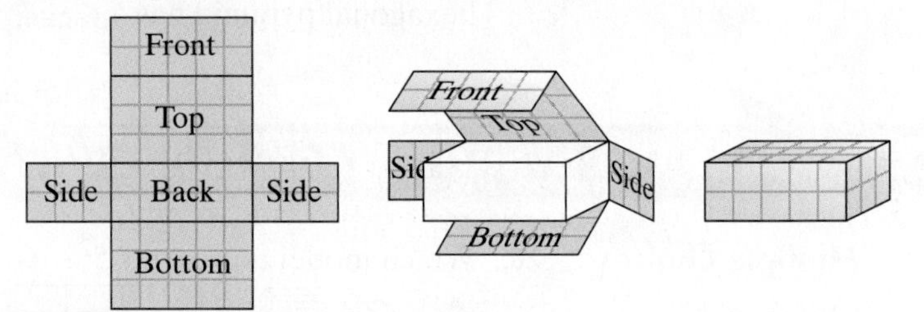

You can draw many different nets for a three-dimensional figure.

Check Skills You'll Need

1. **The height of a triangle is the length of the perpendicular segment from a vertex to the base opposite the vertex or to an extension of the base.**

EXAMPLE 1 Drawing a Net

Draw a net for the triangular prism at the right.

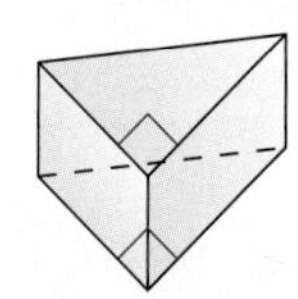

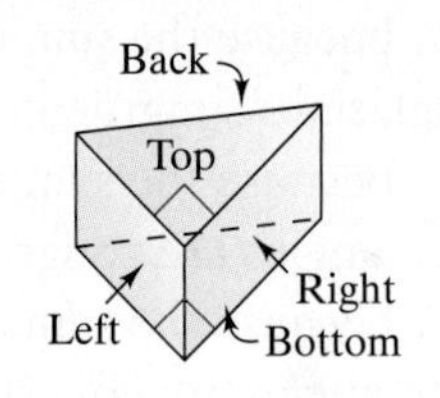

Begin by labeling the bases and faces.

First draw one base. Then draw one face that connects both bases. Next, draw the other base. Draw the remaining faces.

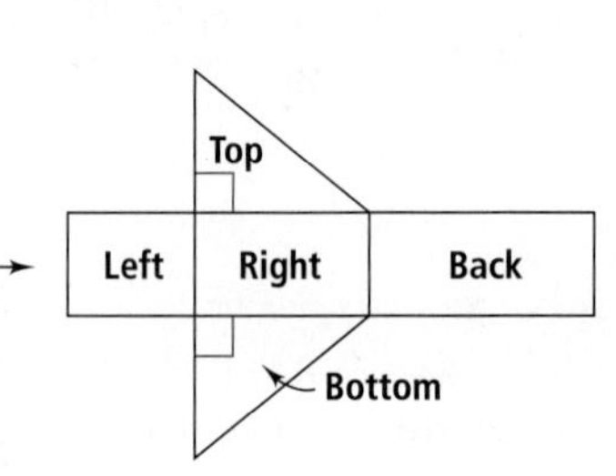

Quick Check
See back of book.

1. Draw a different net for the right triangular prism in Example 1.

Differentiated Instruction Solutions for All Learners

Special Needs L1	**Below Level** L2
Have students label the shape of each face of the different solids. Then ask them to write the formula for the area of each face. **learning style: visual**	Display a box and several possible nets for the box. Ask students which net matches the box. Students can trace each possible net, fold along the lines, and see if the net will fit on the box. Continue with other shapes. **learning style: visual**

The **surface area** of a prism is the sum of the areas of its faces. You measure surface area of a prism in square units. You can find the surface area by finding the area of its net.

EXAMPLE Finding the Surface Area of a Prism

GO for Help
For help with finding the area of a triangle, go to Lesson 8-3, Example 2.

2 Find the surface area of the triangular prism.

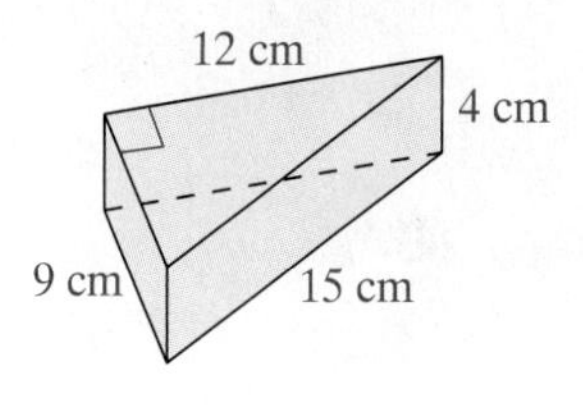

First draw a net for the prism.

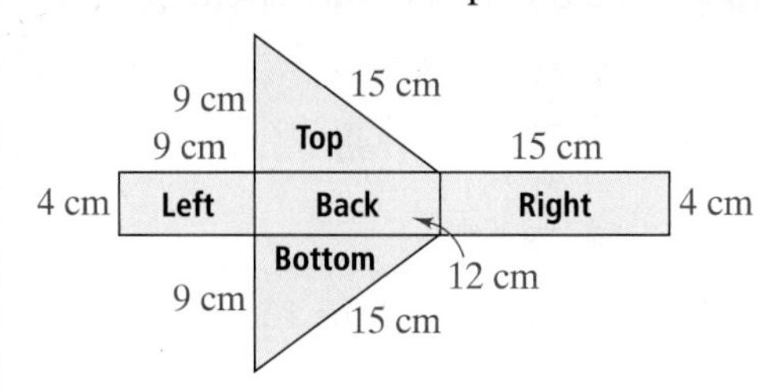

Then find the total area of the five faces.

left side		back		right side		top		bottom
4(9)	+	4(12)	+	4(15)	+	$\frac{1}{2}(12)(9)$	+	$\frac{1}{2}(12)(9) = 252$

The surface area of the triangular prism is 252 cm^2.

Quick Check

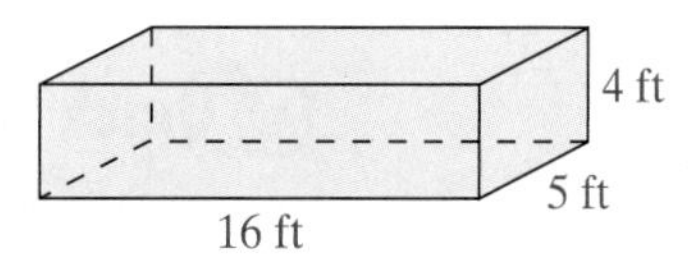

2. Find the surface area of the rectangular prism. **328 ft^2**

If you cut a label from a can, you will see that the label is a rectangle. The height of the rectangle is about the height of the can. The base length of the rectangle is the circumference of the can.

Similarly, if you cut up a cylinder, you get a rectangle and two circles.

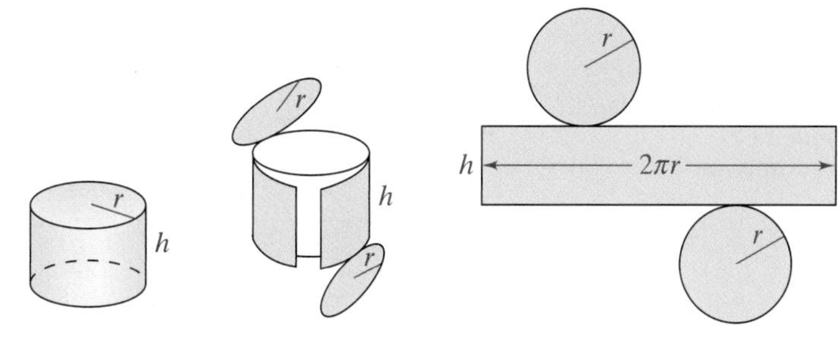

You can use a net of a cylinder to find its surface area.

2. Teach

Activity Lab

Use before the lesson.

All in One Teaching Resources
Activity Lab 8-9: Surface Area

Guided Instruction

Example 1
Some students have difficulty visualizing a three-dimensional figure given a net. Use common objects, such as a tissue box, that can be unfolded to make nets.

Error Prevention!

Review the formulas for finding the area of a rectangle ($A = \ell w$ or $A = bh$) and the area of a triangle ($A = \frac{1}{2}bh$).

PowerPoint Additional Examples

1 Draw a net for the square prism.

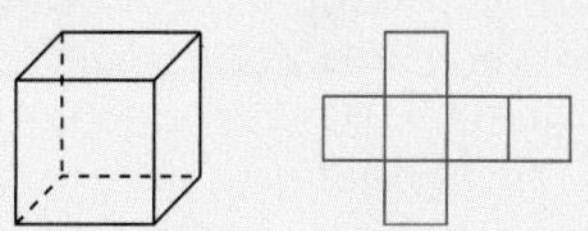

2 Find the surface area of the prism. **840 cm^2**

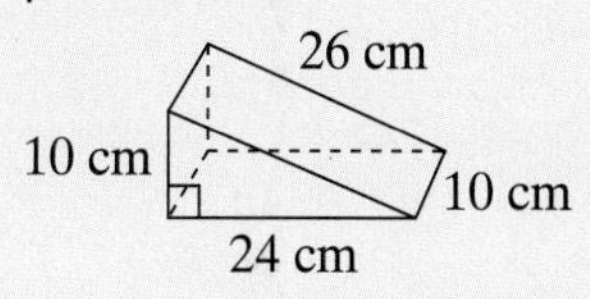

3 Find the surface area of the cylinder. Round to the nearest tenth. **94.2 cm^2**

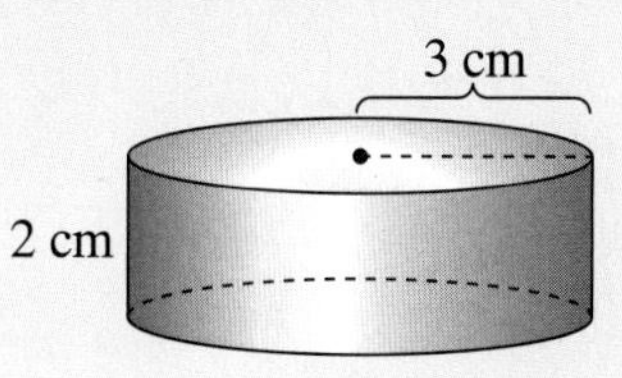

Advanced Learners L4
Explain why the shape of a net for a cylinder can or cannot vary. **Sample: The circles and the rectangle can vary in size, but the shape of the net is always two circles and a rectangle.**

learning style: verbal

English Language Learners ELL
Students often have a difficult time with the term *surface area*, since it appears that the area of a polygon is also surface area. Use solids, cereal boxes, or desks, and ask them to touch all the surfaces. Remind them that the faces are polygons.

learning style: tactile

All in One Teaching Resources

- Daily Notetaking Guide 8-9 L3
- Adapted Notetaking 8-9 L1

Closure

- *What is a net?* a two-dimensional pattern that you can fold to form a three-dimensional figure
- *How do you find the surface area of a prism?* Find the area of each face and add the values together.
- *How do you find the surface area of a cylinder?* Find the area of one circle and the area of the rectangle. Add the areas of 2 circles and the rectangle.

EXAMPLE Finding the Surface Area of a Cylinder

3 **Crafts** You plan to make a birthday present for your friend. The first step is to cover a coffee can with construction paper. How much construction paper do you need?

Step 1 Draw a net.

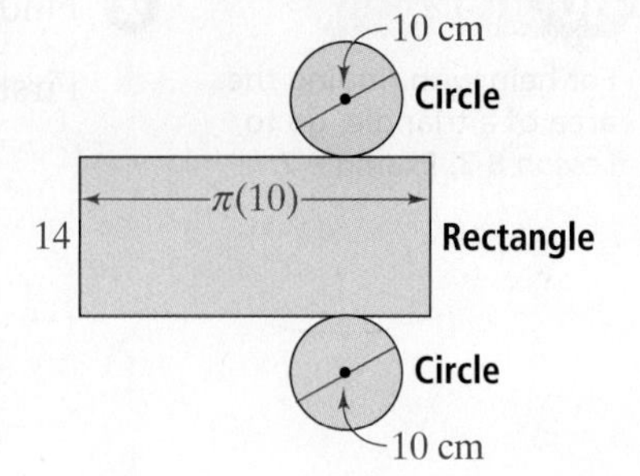

Step 2 Find the area of one circle.

$$A = \pi r^2$$
$$= \pi(5)^2$$
$$= \pi(25)$$
$$\approx 78.54$$

Step 3 Find the area of the rectangle.

$$(\pi d)h = \pi(10)(14)$$
$$= 140\pi$$
$$\approx 439.82$$

Step 4 Add the areas of the two circles and the rectangle.

Surface area $= 78.54 + 78.54 + 439.82 = 596.9$

The amount of construction paper needed is about 597 cm^2.

Quick Check

3. What is the surface area of the cylinder at the right? Round to the nearest tenth. **3,455.8 m^2**

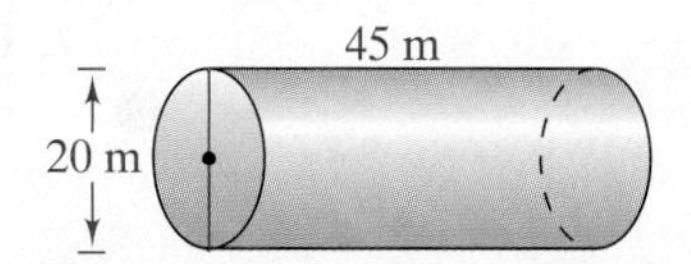

Check Your Understanding

1. **Vocabulary** The ___?___ of a prism is the sum of the areas of its faces. **surface area**
2. **Reasoning** What kind of polygon is included in both a net of a triangular prism and a net of a rectangular prism? **See left.**

2. rectangle; the net of a triangular prism consists of 3 rectangles and 2 triangles. The net of a rectangular prism consists of 6 rectangles.

Identify the figure formed by each net. Then find its surface area.

3.

36 in.
27 in.

cylinder; about 5,089.4 in.^{2}

4. 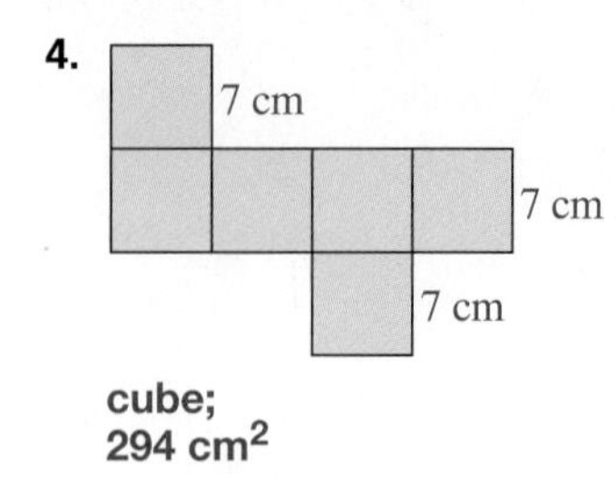

cube; 294 cm^2

Homework Exercises

For more exercises, see Extra Skills and Word Problems.

GO for Help

For Exercises	See Examples
5–8	1
9–11	2
12–14	3

A **Draw a net for each three-dimensional figure.** 5–8. See margin.

5.

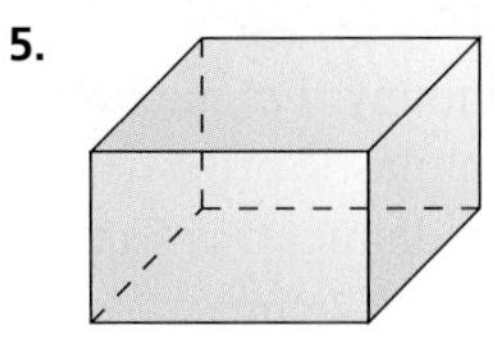

6.

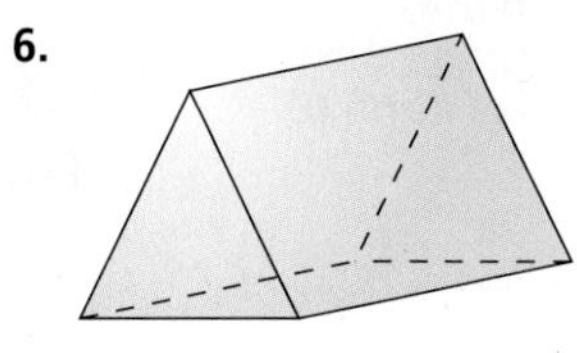

7.

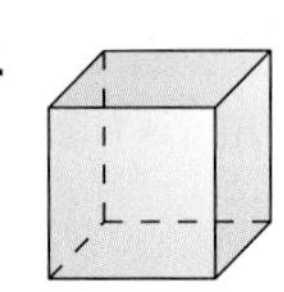

8.

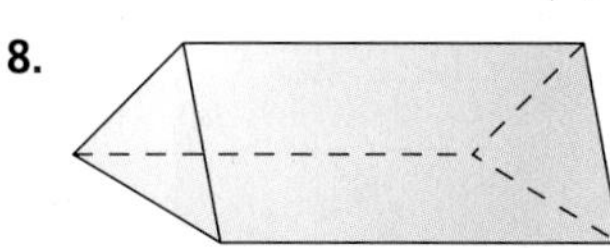

Find the surface area of each prism.

9. 9 m, 6 m, 10 m, 8 m — 264 m^2

10.

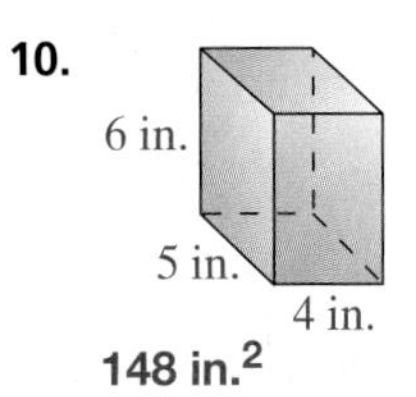

148 in.^2

11.

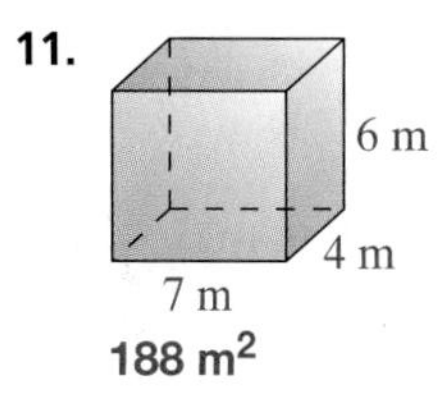

188 m^2

Vocabulary Tip

Cylindrical means "in the shape of a cylinder."

Find the surface area of each cylinder. Round to the nearest tenth.

12.

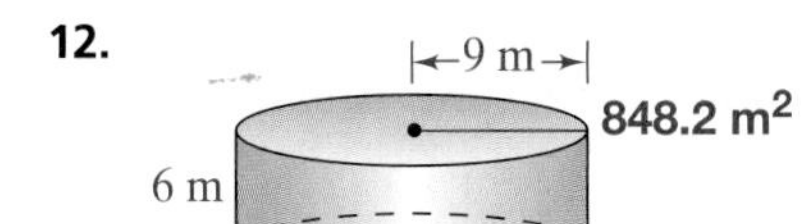

848.2 m^2

13.

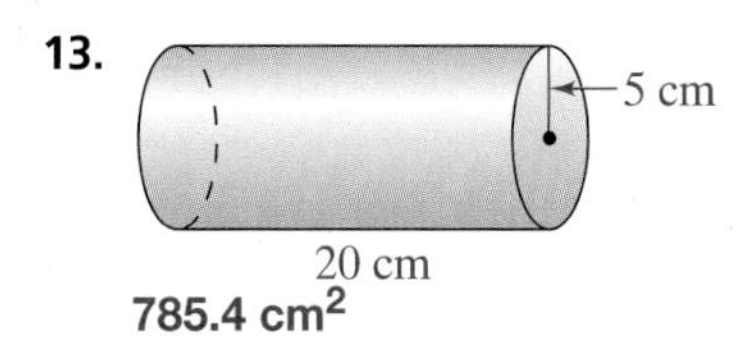

785.4 cm^2

14. The diameter of the base of a cylindrical can is 4 in. The height of the can is 6.5 in. Find the can's surface area to the nearest tenth. 106.8 in.^2

B **GPS**

15. **Guided Problem Solving** The tent at the right is similar to a triangular prism. Calculate the surface area of the tent to find the amount of fabric needed to make the tent. 136 ft^2

- A triangular prism has ■ faces.
- Find and add the areas of the faces: surface area = ■ + ■ + ■ + ■ . . .

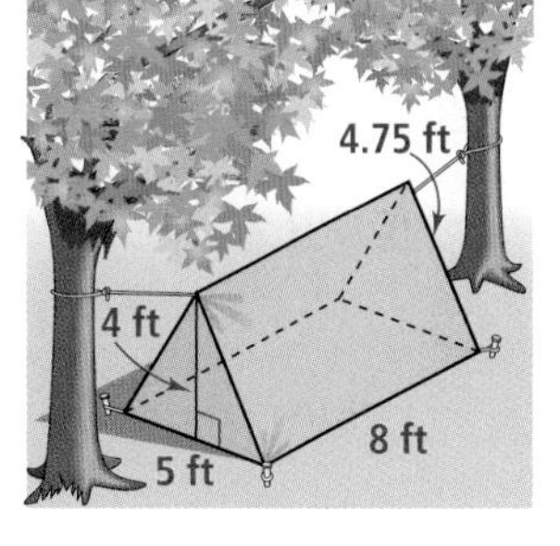

16. Some cans are cut from a large sheet of metal. Find the amount of metal needed to make a can similar to the one at the right. Round to the nearest tenth. 282.7 cm^2

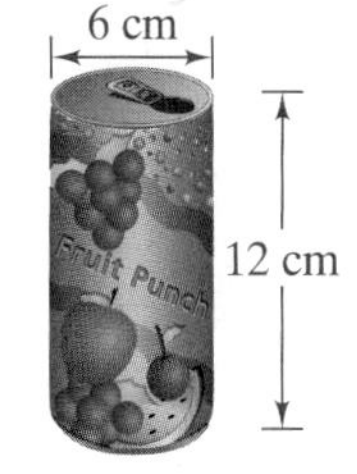

17. Calculate the surface area of a rectangular prism with a height of 4 in., a width of 16 in., and a length of 10 in. 528 in.^2

Online lesson quiz, PHSchool.com, Web Code: ara-0809

5–8. See back of book.

3. Practice

Assignment Guide

Check Your Understanding

Go over Exercises 1–4 in class before assigning the Homework Exercises.

Homework Exercises

A Practice by Example	5–14
B Apply Your Skills	15–26
C Challenge	27
Test Prep and Mixed Review	28–32

Homework Quick Check

To check students' understanding of key skills and concepts, go over Exercises 9, 12, 16, 22, and 25.

Differentiated Instruction Resources

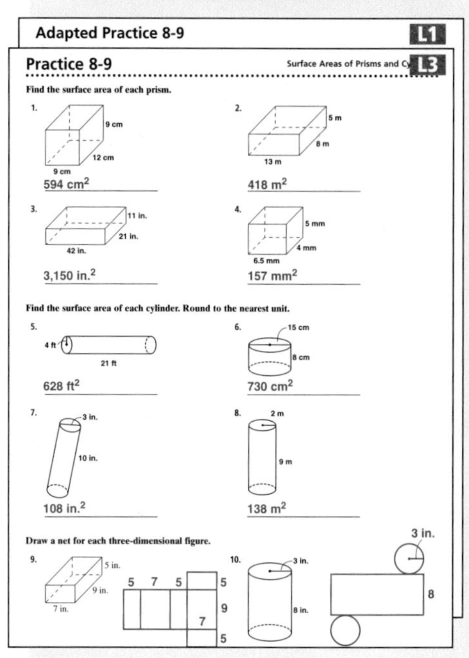

Adapted Practice 8-9 L1

Practice 8-9 Surface Areas of Prisms and Cylinders L3

Find the surface area of each prism.

1. 594 cm²
2. 418 m²
3. 3,150 in.²
4. 157 mm²

Find the surface area of each cylinder. Round to the nearest unit.

5. 628 ft²
6. 730 cm²
7. 108 in.²
8. 138 m²

Draw a net for each three-dimensional figure.

9.
10.

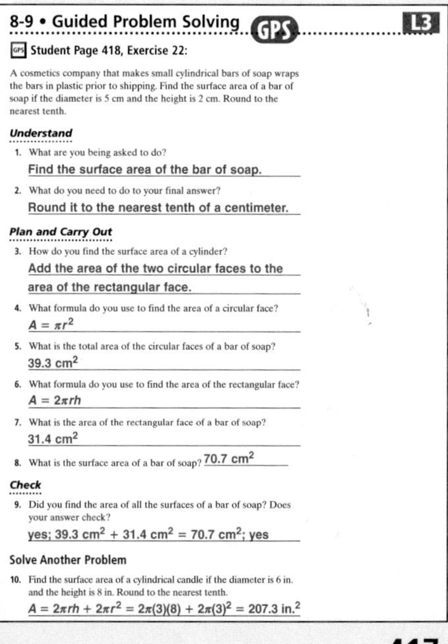

8-9 • Guided Problem Solving GPS L3

Student Page 418, Exercise 22:

A cosmetics company that makes small cylindrical bars of soap wraps the bars in plastic prior to shipping. Find the surface area of a bar of soap if the diameter is 5 cm and the height is 2 cm. Round to the nearest tenth.

Understand

1. What are you being asked to do? Find the surface area of the bar of soap.
2. What do you need to do to your final answer? Round it to the nearest tenth of a centimeter.

Plan and Carry Out

3. How do you find the surface area of a cylinder? Add the area of the two circular faces to the area of the rectangular face.
4. What formula do you use to find the area of a circular face? $A = \pi r^2$
5. What is the total area of the circular faces of a bar of soap? 39.3 cm^2
6. What formula do you use to find the area of the rectangular face? $A = 2\pi rh$
7. What is the area of the rectangular face of a bar of soap? 31.4 cm^2
8. What is the surface area of a bar of soap? 70.7 cm^2

Check

9. Did you find the area of all the surfaces of a bar of soap? Does your answer check? yes; $39.3 \text{ cm}^2 + 31.4 \text{ cm}^2 = 70.7 \text{ cm}^2$; yes

Solve Another Problem

10. Find the surface area of a cylindrical candle if the diameter is 6 in. and the height is 8 in. Round to the nearest tenth. $A = 2\pi rh + 2\pi r^2 = 2\pi(3)(8) + 2\pi(3)^2 = 207.3 \text{ in.}^2$

PowerPoint

Lesson Quiz

1. Draw a net for a triangular prism. **Sample:**

Find the surface area of each. Round to the nearest tenth.

2. rectangular prism: 12 cm by 8 cm by 9 cm. **552 cm^2**
3. cylinder: radius 6 cm, height 10 cm **603.2 cm^2**

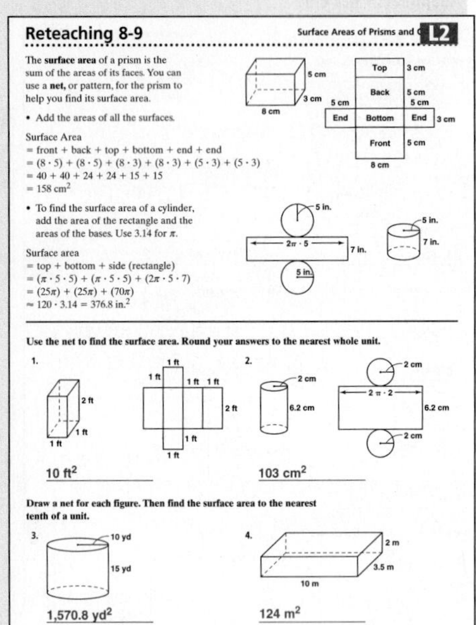

Reteaching 8-9 — Surface Areas of Prisms and C… L2

The **surface area** of a prism is the sum of the areas of its faces. You can use a **net,** or pattern, for the prism to help you find its surface area.

- Add the areas of all the surfaces.

Surface Area
= front + back + top + bottom + end + end
$= (8 \cdot 5) + (8 \cdot 5) + (8 \cdot 3) + (8 \cdot 3) + (5 \cdot 3) + (5 \cdot 3)$
$= 40 + 40 + 24 + 24 + 15 + 15$
$= 158\ cm^2$

- To find the surface area of a cylinder, add the area of the rectangle and the areas of the bases. Use 3.14 for π.

Surface area
= top + bottom + side (rectangle)
$= (\pi \cdot 5 \cdot 5) + (\pi \cdot 5 \cdot 5) + (2\pi \cdot 5 \cdot 7)$
$= (25\pi) + (25\pi) + (70\pi)$
$\approx 120 \cdot 3.14 = 376.8\ in.^2$

Use the net to find the surface area. Round your answers to the nearest whole unit.

1. **10 ft^2**
2. **103 cm^2**

Draw a net for each figure. Then find the surface area to the nearest tenth of a unit.

3. **1,570.8 yd^2**
4. **124 m^2**

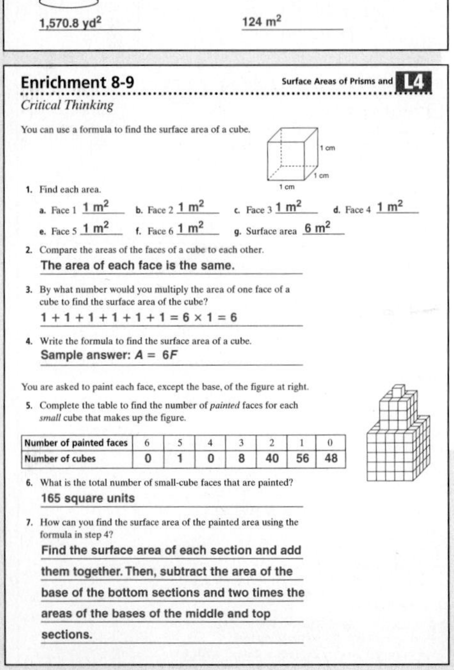

Enrichment 8-9 — Surface Areas of Prisms and … L4

Critical Thinking

You can use a formula to find the surface area of a cube.

1. Find each area.
   a. Face 1 **1 m^2** b. Face 2 **1 m^2** c. Face 3 **1 m^2** d. Face 4 **1 m^2**
   e. Face 5 **1 m^2** f. Face 6 **1 m^2** g. Surface area **6 m^2**
2. Compare the areas of the faces of a cube to each other.
   The area of each face is the same.
3. By what number would you multiply the area of one face of a cube to find the surface area of the cube?
   $1 + 1 + 1 + 1 + 1 + 1 = 6 \times 1 = 6$
4. Write the formula to find the surface area of a cube.
   Sample answer: $A = 6F$

You are asked to paint each face, except the base, of the figure at right.

5. Complete the table to find the number of *painted* faces for each *small* cube that makes up the figure.

Number of painted faces	6	5	4	3	2	1	0
Number of cubes	0	1	0	8	40	56	48

6. What is the total number of small-cube faces that are painted?
   165 square units
7. How can you find the surface area of the painted area using the formula in step 4?
   Find the surface area of each section and add them together. Then, subtract the area of the base of the bottom sections and two times the areas of the bases of the middle and top sections.

Visit: PHSchool.com
Web Code: are-0809

24. You cannot interchange the values of r and h.

Find the surface area of each cylinder given the base radius and height of the cylinder. Round to the nearest square unit.

18. $r = 3$ cm, $h = 10$ cm **245 cm^2**
19. $r = 7$ ft, $h = 25$ ft **1,407 ft^2**
20. $r = 12$ m, $h = 16$ m **2,111 m^2**
21. $r = 10$ in., $h = 3$ ft **2,890 $in.^2$**

22. GPS A cosmetics company that makes small cylindrical bars of soap wraps the bars in plastic prior to shipping. Find the surface area of a bar of soap if the diameter is 5 cm and the height is 2 cm. Round to the nearest tenth. **70.7 cm^2**

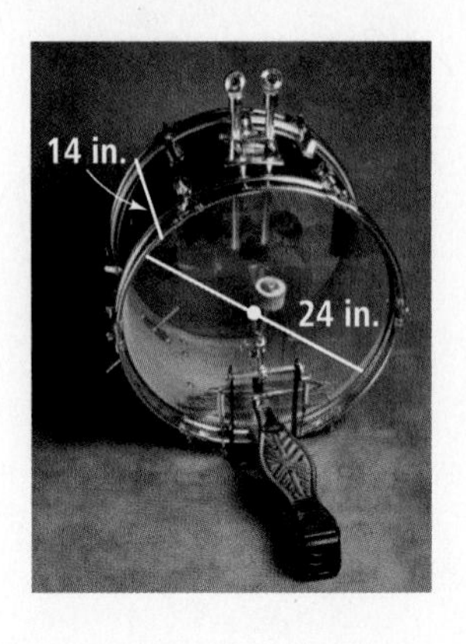

23. Suppose you wish to make a cylindrical case that will exactly fit the bass drum at the left. What is the surface area of the case to the nearest tenth? **1960.4 $in.^2$**

24. **Error Analysis** A student says the two cylinders at the right have the same surface area. Explain the student's error. **See above left.**

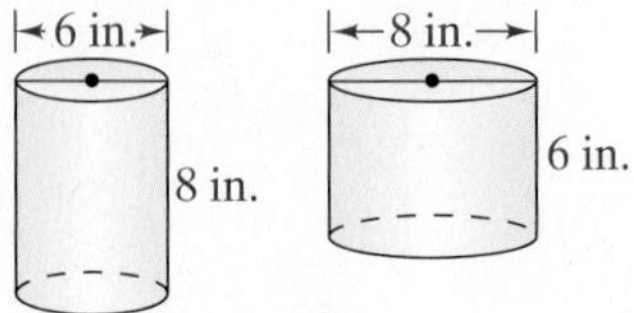

25. **Writing in Math** Which has a greater effect on the surface area of a cylinder—doubling the radius or doubling the height? Explain. **See margin.**

26. **Open-Ended** Draw a net for a prism that has a surface area of 72 cm^2. **See margin.**

27. (C) **Challenge** Find the surface area of the figure at the right. Round to the nearest tenth. **990 ft^2**

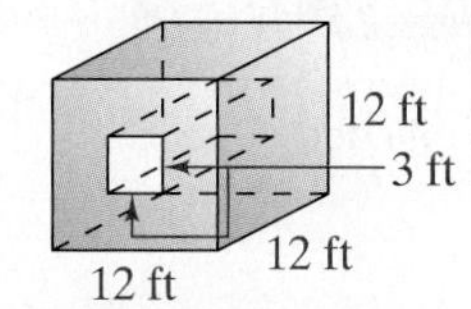

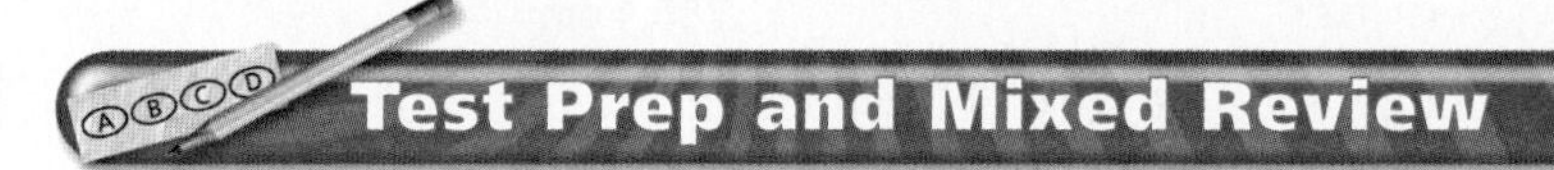

Multiple Choice

28. Which solid can be formed from the net shown at the right? **A**
   - (A) Triangular prism
   - (B) Rectangular prism
   - (C) Triangular pyramid
   - (D) Rectangular pyramid

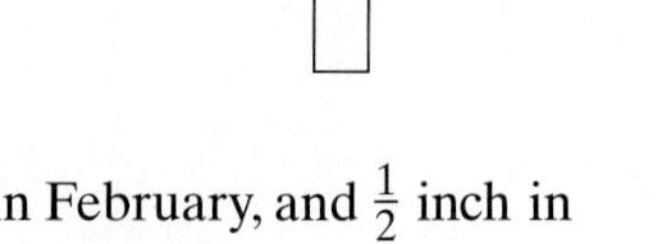

29. Chip grows $\frac{3}{4}$ inch in January, $\frac{5}{8}$ inch in February, and $\frac{1}{2}$ inch in March. If the pattern continues, how much will he grow in May? **G**
   - (F) $\frac{1}{8}$ in.
   - (G) $\frac{1}{4}$ in.
   - (H) $\frac{3}{8}$ in.
   - (J) $\frac{1}{2}$ in.

GO for Help

For Exercises	See Lesson
30–32	8-6

Identify each number as *rational* or *irrational*.

30. 2.22222 . . . **rational**
31. $\sqrt{625}$ **rational**
32. $\sqrt{18}$ **irrational**

Test Prep

Resources

For additional practice with a variety of test item formats:
- Test-Taking Strategies, p. 427
- Test Prep, p. 431
- Test-Taking Strategies with Transparencies

Alternative Assessment

Provide pairs of students with graph paper and exercises similar to Exercises 9–13. Partners check each otherÕs work and must agree on the results

25–26. See back of book.

Patterns in 3-Dimensional Figures

You can explore number patterns using unit cubes.

Suppose you use unit cubes to make a larger cube with two unit cubes on an edge. You paint the outside of the larger cube. You need eight unit cubes to form the larger cube. Each cube has three sides painted.

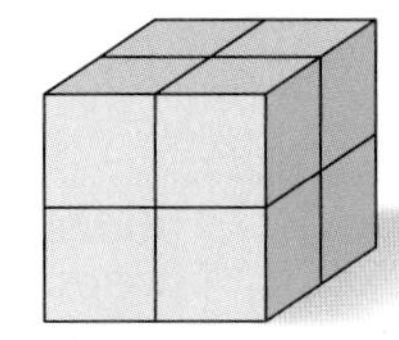

Exercises

Use the table for Exercises 1–3.

Number of Unit Cubes on an Edge	Total Number of Unit Cubes	Total Number Expressed as a Power	Number of Unit Cubes With Given Number of Sides Painted			
			0	1	2	3
2	8	2^3	0	0	0	8
3	■	■	■	■	■	■
4	■	■	■	■	■	■
5	■	■	■	■	■	■
6	■	■	■	■	■	■
7	■	■	■	■	■	■

2a. $(n-2)^3$; $6(n-2)^2$; $12(n-2)$; 8

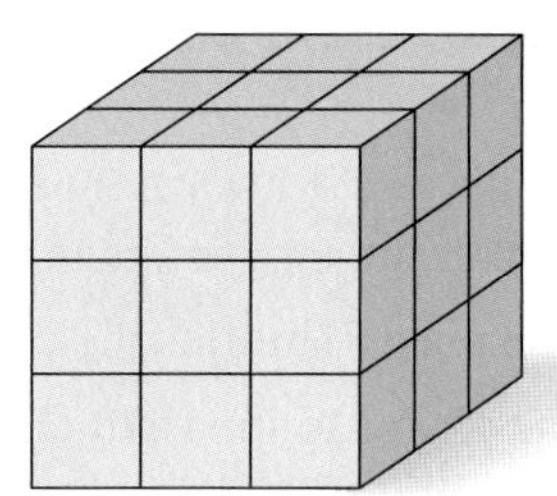

1. Copy and complete the table above. Use the figure at the right to help you fill in the row for 3 unit cubes on an edge. **See back of book.**
2. a. **Patterns** Describe the number pattern you see in each of the last four columns of your table. **See above right.**
   b. Use the number patterns and extend the table for 8 number cubes on an edge. **8, 512, 8^3, 216, 216, 72, 8**
3. a. **Number Sense** What is the total number of unit cubes in a cube with 10 unit cubes on an edge? **1,000**
   b. If there are 15 unit cubes on an edge, how many unit cubes will have no side painted? One side painted? Two sides painted? **2,197; 1,014; 156**
   c. **Reasoning** If 144 unit cubes have two sides painted, how many unit cubes are on one edge of the cube? **14**

Extension

Patterns in 3-Dimensional Figures

In this Extension of Lesson 8-9, students use unit cubes to explore number patterns in three-dimensional figures.

Guided Instruction

Review the meanings of the terms *edge* and *face* with students. Show examples of each. Have students construct the model in the example with unit cubes. Ask students what they observe about the cube. Ask questions such as:

- *Why is one face of the large cube made of 4 small cube faces?* **The face of a cube is a square. The edge of the cube is 2 units. 2 squared is 4.**
- *What is the total surface area of the large cube?* **24 square units**

Exercises

Have students work with partners. Guide pairs to construct models of the 27-cube larger cube to fill in the table and complete Exercise 2. Encourage tactile and visual learners to make models using the small cubes to complete the exercises.

Teaching Tip

Have students connect their models to formulas for the volume and surface area of rectangular prisms.

Differentiated Instruction

Advanced Learners L4

Challenge your advanced students to explore with models to see how *removing* one or more unit cubes affects the surface area of the large figures. Have them share their findings.

Resources

- unit cubes
- Student Manipulatives Kit

Checkpoint Quiz

Use this Checkpoint Quiz to check students' understanding of the skills and concepts of Lessons 8-5 through 8-9.

Resources

- All-in-One Teaching Resources Checkpoint Quiz 2
- ExamView Assessment Suite CD-ROM
- Success Tracker™ Online Intervention

Square Root Bingo

This game will help students associate perfect squares with their square roots.

Guided Instruction

Students play in pairs.

Have them read the rules before beginning. Review the definition of *perfect square* and *square root.* Ask for examples of perfect squares.

Differentiated Instruction

English Language Learners ELL
Review the terms *diagonally, horizontally,* and *vertically.* Ask students to use their hands to show their meanings.

Resources

- 20 index cards with a perfect square of a number from 1–20 under the square root symbol on each

Checkpoint Quiz 2

Lessons 8-5 through 8-9

1. Find two consecutive whole numbers that $\sqrt{77}$ falls between. **$8 < \sqrt{77} < 9$**
2. Find the length of the sides of a square with an area of 324 cm^2. **18 cm**

Find each missing length.

3. 15 cm, x, 36 cm **39 cm**

4. 5 yd, 3 yd, y **4 yd**

5. 15 m, z, 17 m **8 m**

Name each figure.

6. **sphere**

7. 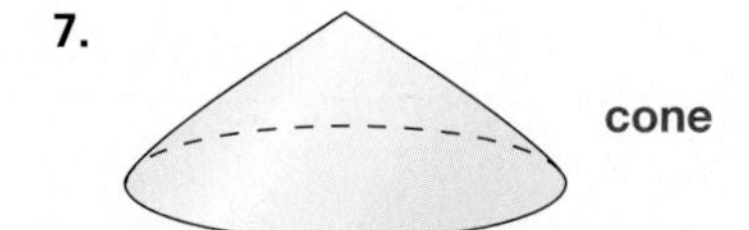 **cone**

8. 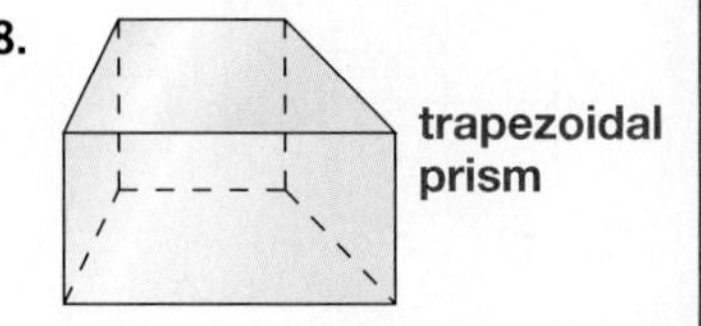 **trapezoidal prism**

Find the surface area of each figure. Round to the nearest tenth.

9. 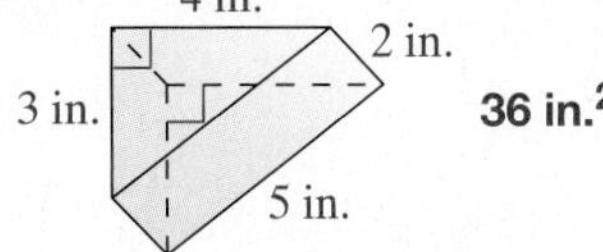

36 $in.^2$

10. 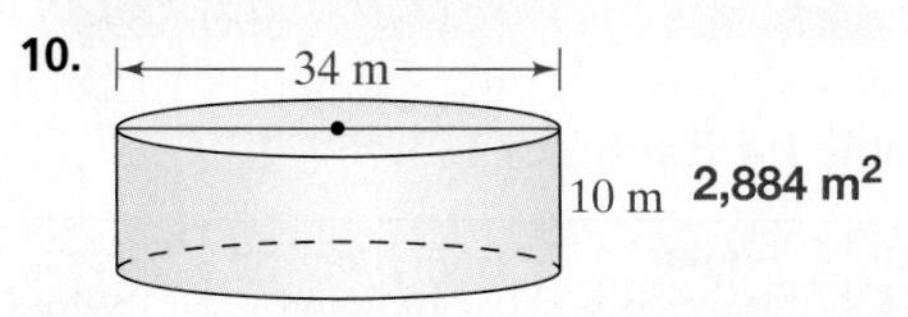

2,884 m^2

MATH GAMES — Square Root Bingo

What You'll Need

- 20 index cards. On each card write one of the first 20 perfect squares (not including 0) under the square root symbol.

How To Play

- All players draw a 16-square playing board. In each square, they write a number from 1–20 without repeats.
- One player shuffles the cards and places them face down.
- The player to the right chooses the top card and places it face up. Players with the matching square root on their board make a mark on the appropriate square.
- The player to the right chooses the next card.
- The winner is the first player who has four marks diagonally, horizontally, or vertically.

420

8-10 Volumes of Prisms and Cylinders

Check Skills You'll Need

1. Vocabulary Review How is π related to the circumference and the diameter of a circle? $\pi = \frac{C}{d}$

Find the area of each circle. Round to the nearest square unit.

2. 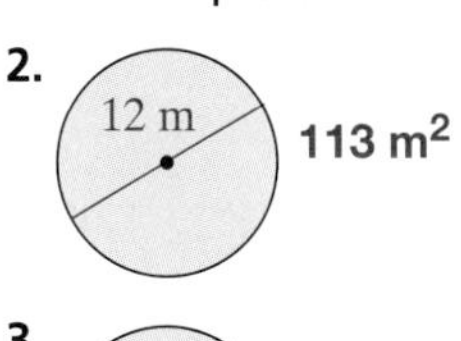

113 m^2

3. 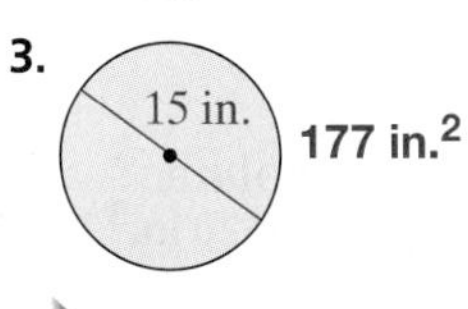

177 in.^2

Lesson 8-5

What You'll Learn

To find the volumes of prisms and cylinders

New Vocabulary volume, cubic unit

Why Learn This?

Many storage silos are shaped like cylinders. You can use volume formulas to find the amount of storage space inside a figure like a silo.

The **volume** of a three-dimensional figure is the number of cubic units needed to fill the space inside the figure.

A **cubic unit** is a cube with edges one unit long. A cubic centimeter is a cube with edges one centimeter long.

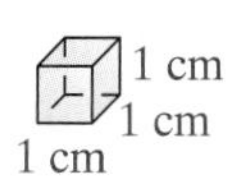

Consider filling the rectangular prism below with cubic centimeters.

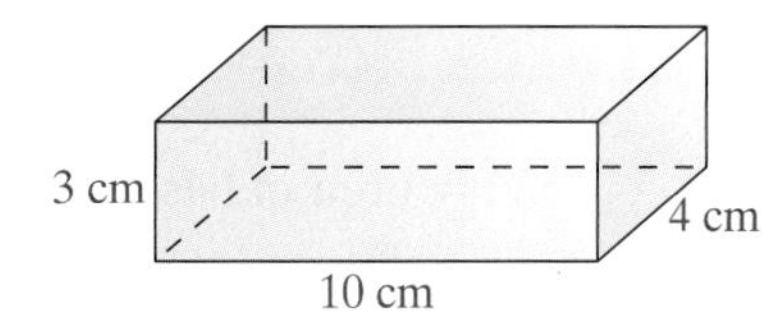

The bottom layer of the prism contains $10 \cdot 4$, or 40, cubes.

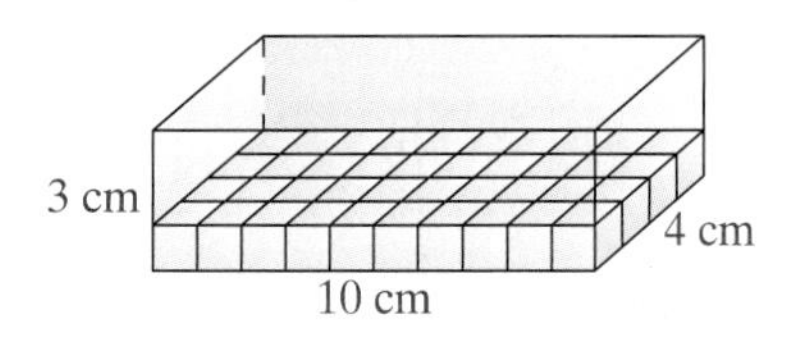

Three layers of 40 cubes fit in the prism. $3 \cdot 40 = 120$

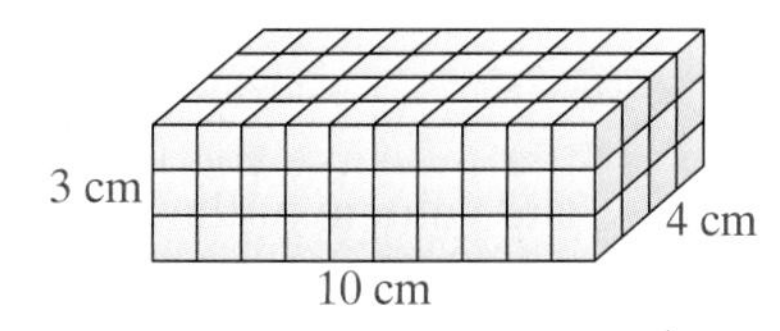

The volume of the prism is 120 cm^3.

GO Online

Video Tutor Help

Visit: PHSchool.com
Web Code: are-0775

Differentiated Instruction Solutions for All Learners

Special Needs L1

Have students hold a cube that is formed by smaller linking cubes, or a box with graph paper for sides. Ask them to undo the cube figure (or take out the cubes) and put it back together. Then ask the students how many little cubes formed the solid. That is the volume of the larger cube.

learning style: tactile

Below Level L2

Help students differentiate the concepts of perimeter (distance around), area (surface covered), and volume (space filled).

learning style: verbal

8-10

1. Plan

Objective

To find the volumes of prisms and cylinders

Examples

1 Finding Volume of a Rectangular Prism
2 Finding the Volume of a Triangular Prism
3 Finding the Volume of a Cylinder

Math Understandings: p. 372D

Math Background

When you pick up a drinking glass, you are touching its surface (area), which is measured in square units. The capacity or volume of water contained in the glass is measured in cubic units. To calculate the volume of prisms and cylinders, find the area of the base and multiply it by the height. The formula for volume is $V = Bh$ where B is the area of the base and h is the height.

More Math Background: p. 372D

Lesson Planning and Resources

See p. 372E for a list of the resources that support this lesson.

PowerPoint

Bell Ringer Practice

Check Skills You'll Need

Use student page, transparency, or PowerPoint. For intervention, direct students to:

Circumference and Area of a Circle
Lesson 8-5
Extra Skills and Word Problems Practice, Ch. 8

2. Teach

Activity Lab

Use before the lesson.

All in One Teaching Resources

Activity Lab 8-10: Patterns and Prisms

Guided Instruction

Example 1

Some students may think of volume as only an amount of liquid and may have trouble thinking of volume in terms of cubic units. Tell them that liquids like juice assume the shape of the container that holds it and could easily be poured into vessels of other shapes, like a juice carton.

Example 3

Point out that volume is expressed in cubic units, or units^3. Ask: Why do you think the volume is 452 in^3 instead of 452 in^2? **Sample: Volume tells how many cubes a container can hold.**

Error Prevention!

Students may multiply π and r and then square the entire product before multiplying by height. Have students write $V = \pi \cdot r \cdot r \cdot h$. Then have them substitute the values into the formula and multiply.

Teaching Tips

- Have students use circular chips to build cylinders and triangular chips to build triangular prisms. Discuss the fact that the total volume is the volume of one chip multiplied by how many chips high the cylinder or triangular prism is.
- Have students use models to show why the formula for the volume of a prism is similar to the formula for the volume of a cylinder.

The previous calculation of volume suggests the following formula.

KEY CONCEPTS **Volume of a Rectangular Prism**

V = area of base · height
$= Bh$
$= \ell wh$

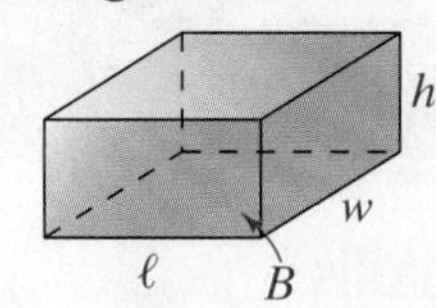

EXAMPLE Finding Volume of a Rectangular Prism

1 **Gridded Response** Mr. Cho is building a craft box like the one shown at the right. What is the volume of the craft box in cubic inches?

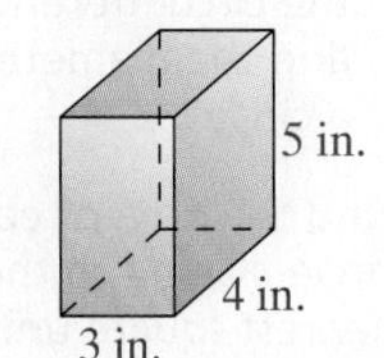

$V = \ell wh$ ← **Use the formula.**
$= (3)(4)(5)$ ← **Substitute.**
$= 60$ ← **Multiply.**

The volume of the craft box is 60 cubic inches.

		6	0

1. If the height of the prism above is doubled, what is the volume? **120 in.^{3}**

The volume formulas for rectangular and triangular prisms are similar.

When finding volume, remember to calculate the area of the base first.

KEY CONCEPTS **Volume of a Triangular Prism**

V = area of base · height
$= Bh$

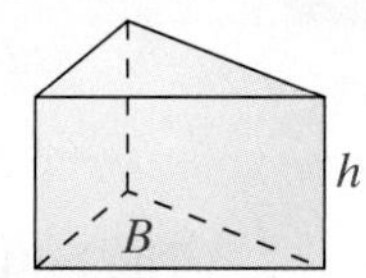

EXAMPLE Finding the Volume of a Triangular Prism

2 Find the volume of the triangular prism.

6 cm
3 cm
4 cm

$V = Bh$ ← **Use the formula.**
$= (6)(6)$ ← **Substitute:** $B = \frac{1}{2} \times 3 \times 4 = 6$.
$= 36$ ← **Multiply.**

The volume of the triangular prism is 36 cm^3.

2. If the height of the prism above is doubled, what is the volume? **72 cm^3**

Advanced Learners L4	**English Language Learners** ELL
A rectangular prism has two square faces and four rectangular faces. Its longest edge is 9 cm, and its surface area is 230 cm^2. What is its volume? **225 cm^3** **learning style: visual**	Tell students that although volume is measured in cubic units, containers are rarely filled with cubic units. They are filled with various products or items, such as cereal, or popcorn. Remind them that a cubic unit is a standard unit of measure. **learning style: visual**

The volume formula for a cylinder is also similar to the volume formula for a prism.

KEY CONCEPTS **Volume of a Cylinder**

V = area of base · height
$= Bh$
$= \pi r^2 h$

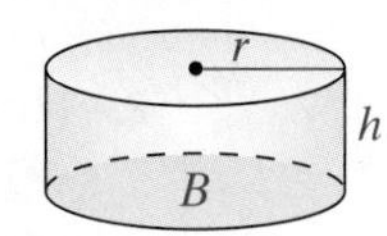

EXAMPLE Finding the Volume of a Cylinder

3 **Painting** Estimate the volume of the cylindrical paint can. Then find the volume to the nearest cubic unit.

Estimate

$V = \pi r^2 h$ ← Use the formula.
$\approx (3)(4)^2(9)$ ← Use 3 to estimate π.
$\approx (50)(9)$ ← Use 50 to estimate 48 (3 · 16).
≈ 450

The estimated volume is 450 in.^{3}.

Calculate

$V = \pi r^2 h$ ← Use the formula.
$\approx (\pi)(4)^2(9)$ ← Substitute.
$\approx$ 452.38934 ← Use a calculator.
≈ 452 ← Round to the nearest whole number.

The calculated volume is about 452 in.^{3}.

Check for Reasonableness The calculated volume is close to the estimated volume, so the answer is reasonable.

Quick Check

3. a. Estimate the volume of the cylinder. Then find the volume to the nearest cubic centimeter. **about 6,000 cm^3; 6,107 cm^3**

24 cm
18 cm

b. Reasoning Suppose you estimate using 20 for 4^2 instead of 16 in Example 3. Will your estimate be reasonable? **The estimate will be a little large.**

Online active math

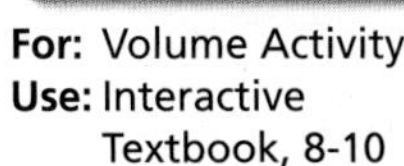

For: Volume Activity
Use: Interactive Textbook, 8-10

PowerPoint

Additional Examples

1 Find the volume of the rectangular prism. **960 cm^3**

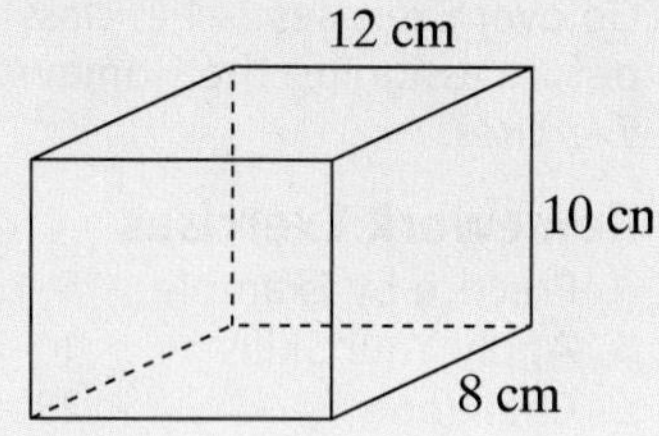

2 Find the volume of the triangular prism. **120 cm^3**

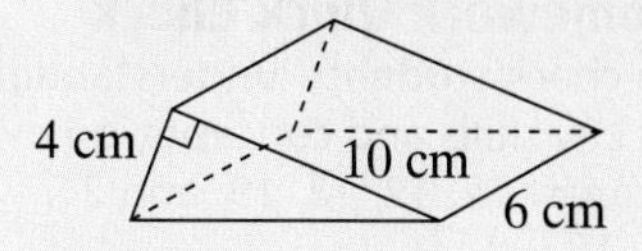

3 Find the volume of the cylinder. Round to the nearest cubic centimeter. **113 cm^3**

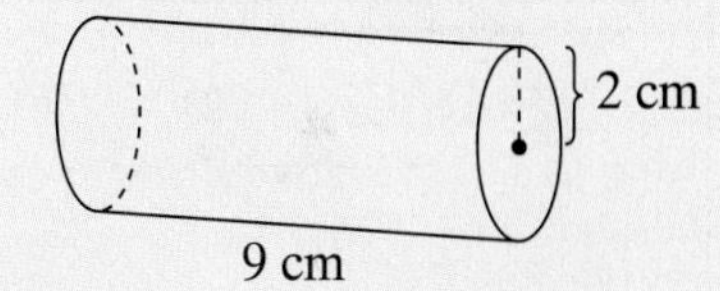

All in One Teaching Resources

- Daily Notetaking Guide 8-10 L3
- Adapted Notetaking 8-10 L1

Closure

- *What is volume for a three-dimensional figure?* **the number of cubic units needed to fill the space inside the figure**
- *How do you find the volume of a rectangular prism?* **Use $V = \ell wh$ and write the units^3.**
- *How do you find the volume of a cylinder?* **Use the formula $V = \pi r^2 h$, where r is the radius and h is the height of the cylinder; write the units^3.**

3. Practice

Assignment Guide

Check Your Understanding

Go over Exercises 1–4 in class before assigning the Homework Exercises.

Homework Exercises

A	Practice by Example	5–13
B	Apply Your Skills	14–22
C	Challenge	23
	Test Prep and Mixed Review	24–30

Homework Quick Check

To check students' understanding of key skills and concepts, go over Exercises 7, 12, 18, 19, and 21.

Differentiated Instruction Resources

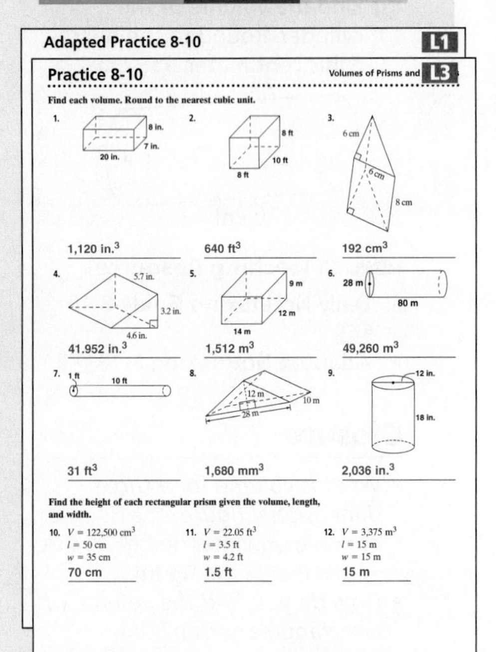

Adapted Practice 8-10 L1

Practice 8-10 Volumes of Prisms and Cylinders L3

Find each volume. Round to the nearest cubic unit.

1. 1,120 in.^{3} 2. 640 ft^3 3. 192 cm^3
4. 41.952 in.^{3} 5. 1,512 m^3 6. 49,260 m^3
7. 31 ft^3 8. 1,680 mm^3 9. 2,036 in.^{3}

Find the height of each rectangular prism given the volume, length, and width.

10. $V = 122{,}500$ cm^3, $l = 50$ cm, $w = 35$ cm — 70 cm
11. $V = 22.05$ ft^3, $l = 3.5$ ft, $w = 4.2$ ft — 1.5 ft
12. $V = 3{,}375$ m^3, $l = 15$ m, $w = 15$ m — 15 m

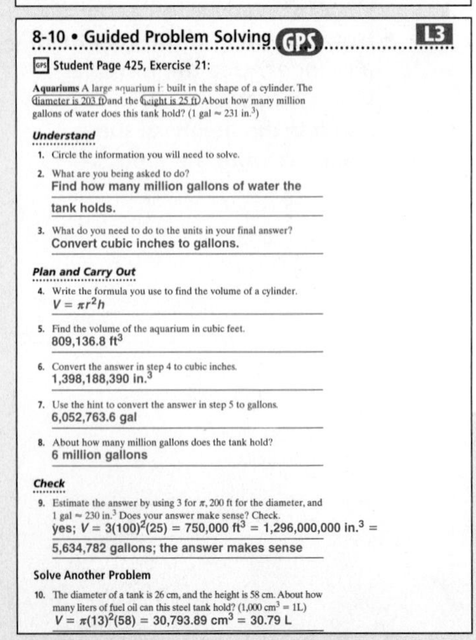

8-10 • Guided Problem Solving GPS L3

Student Page 425, Exercise 21:

Aquariums A large aquarium is built in the shape of a cylinder. The diameter is 203 ft and the height is 25 ft. About how many million gallons of water does this tank hold? (1 gal ≈ 231 in.^{3})

Understand

1. Circle the information you will need to solve.
2. What are you being asked to do? Find how many million gallons of water the tank holds.
3. What do you need to do to the units in your final answer? Convert cubic inches to gallons.

Plan and Carry Out

4. Write the formula you use to find the volume of a cylinder. $V = \pi r^2 h$
5. Find the volume of the aquarium in cubic feet. 809,136.8 ft^3
6. Convert the answer in step 4 to cubic inches. 1,398,188,390 in.^{3}
7. Use the hint to convert the answer in step 5 to gallons. 6,052,763.6 gal
8. About how many million gallons does the tank hold? 6 million gallons

Check

9. Estimate the answer by using 3 for π, 200 ft for the diameter, and 1 gal ≈ 230 in.^{3} Does your answer make sense? Check. yes; $V = 3(100)^2(25) = 750{,}000$ ft^3 = 1,296,000,000 in.^{3} = 5,634,782 gallons; the answer makes sense

Solve Another Problem

10. The diameter of a tank is 26 cm, and the height is 58 cm. About how many liters of fuel oil can this steel tank hold? (1,000 cm^3 = 1L) $V = \pi(13)^2(58) = 30{,}793.89$ cm^3 = 30.79 L

Check Your Understanding

1. **Vocabulary** The number of cubic units needed to fill the space inside a three-dimensional figure is called the ___?___. volume

2. How does the volume of a cylinder change if the height is doubled? B
   - (A) It does not change.
   - (B) It doubles.
   - (C) It quadruples.
   - (D) It halves.

Find the volume of each figure, given the following dimensions.

3. triangular prism $B = 20$ in.^{2}; $h = 3$ in. 60 in.^{3}

4. cylinder $r = 5$ cm; $h = 7$ cm 549.8 cm^3 549.5 cm^3

Homework Exercises

For more exercises, see Extra Skills and Word Problems.

GO for Help

For Exercises	See Examples
5–7	1
8–10	2
11–13	3

A Find the volume of each rectangular prism.

5. 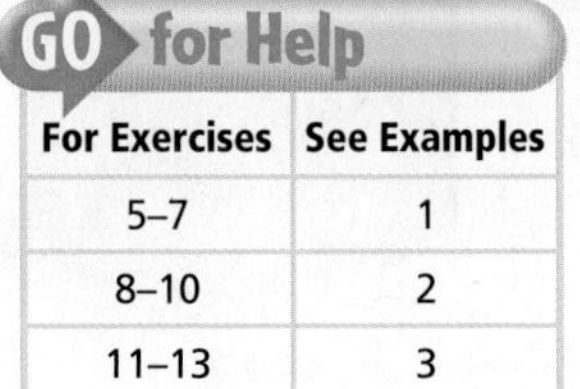

166.375 in.^{3}

6. 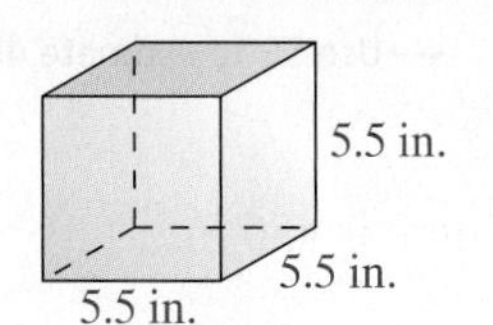

24 cm^3

7. 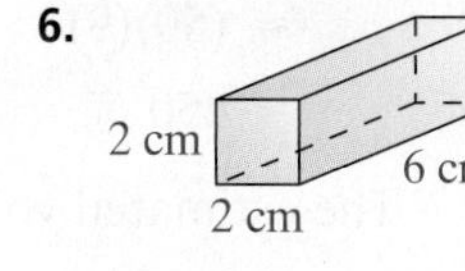

421.875 ft^3

Find the volume of each triangular prism.

8. 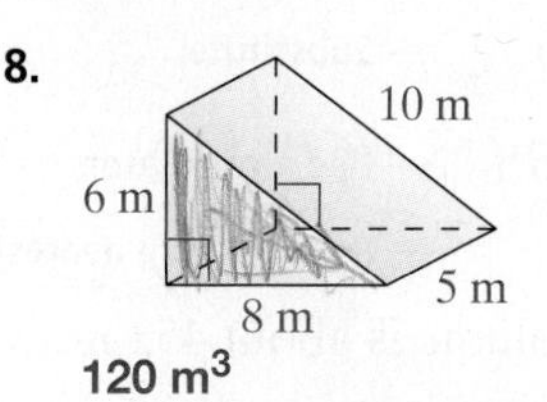

120 m^3

9. 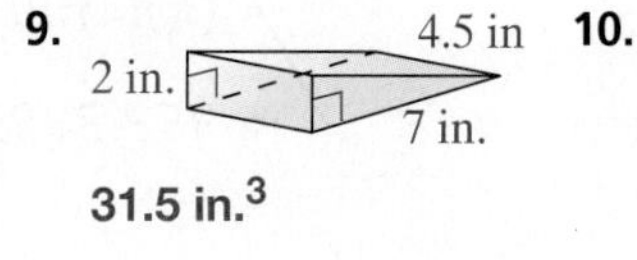

31.5 in.^{3}

10. 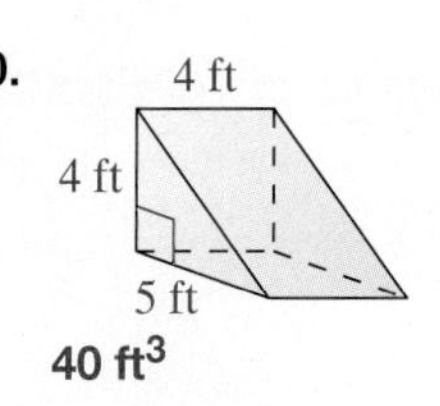

40 ft^3

Packaging **Estimate the volume of each cylinder. Then find the volume to the nearest cubic unit.**

11. 

about 84 in.^{3}; 88 in.^{3}

12. about 15,000 cm^3; 17,521 cm^3

13. 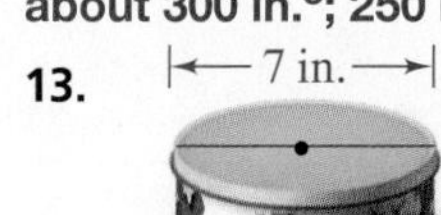

about 300 in.^{3}; 250 in.^{3}

B GPS 14. **Guided Problem Solving** What is the volume of the object shown at the right? 1,759 mm^3
- Identify the two familiar solids that make up the object. What are they?
- What is the volume of each familiar solid?

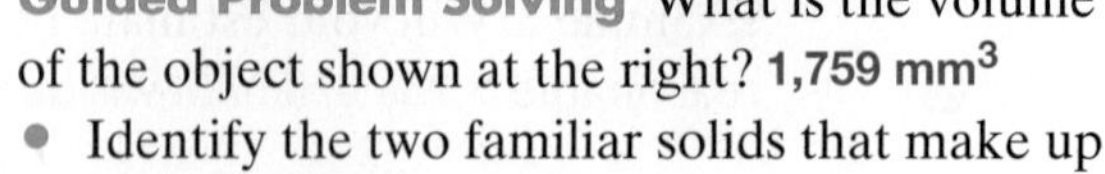
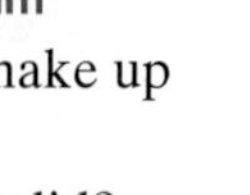
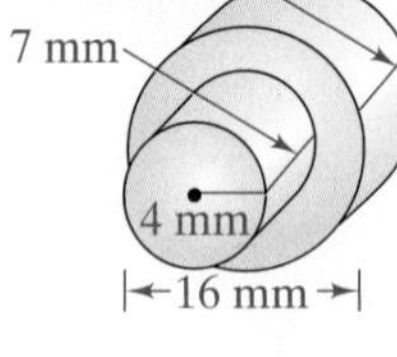

19. Since the volume of a cylinder is $V = \pi r^2 h$, you can substitute the values for V and h.
$385 = 3.14 \cdot 10r^2$
$385 = 31.4r^2$
$12.26 = r^2$
$3.5 \approx r$
The radius is about 3.5 in.

23. Yes; the height of the can is less than the height of the case. Since $3(2.5) < 7.6$, 3 cans will fit along the width of the case. Since $4(2.5) < 11$, 4 cans will fit along the length of the case. Then the case can hold 3×4, or 12, cans.

Online
Homework Video Tutor
Visit: PHSchool.com
Web Code: are-0810

Find the height of each rectangular prism.

15. $V = 455 \text{ cm}^3$
$\ell = 10$ cm
$w = 7$ cm
6.5 cm

16. $V = 525 \text{ m}^3$
$\ell = 7.5$ m
$w = 3.5$ m
20 m

17. $V = 5{,}832 \text{ in.}^3$
$\ell = 18$ in.
$w = 18$ in.
18 in.

18. The cylinder has a volume of 130.7 cm^3. The cube has a volume of 166.4 cm^3. The cylinder is missing the "corners."

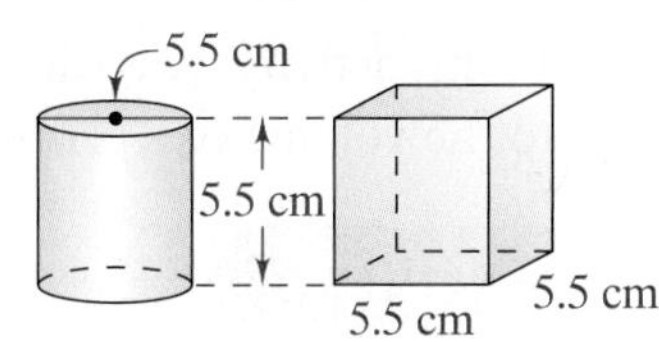

18. **Reasoning** Compare the volumes of the figures. Why are their volumes different? **See left.**

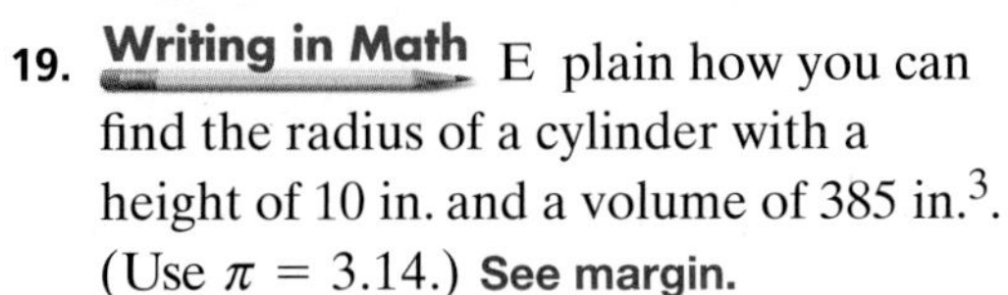

19. **Writing in Math** E plain how you can find the radius of a cylinder with a height of 10 in. and a volume of 385 in.^{3}. (Use $\pi = 3.14$.) **See margin.**

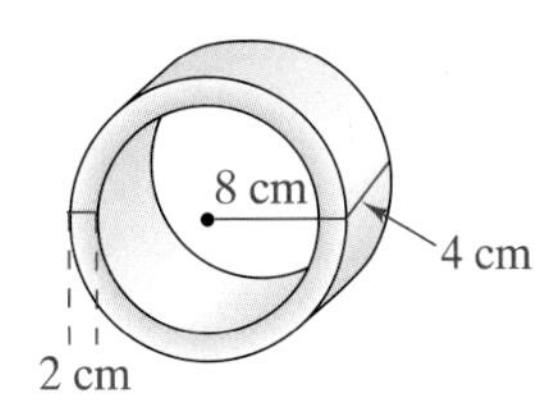

20. Find the volume of the figure at the left to the nearest cubic centimeter. **352 cm^3**

21. **Aquariums** A large aquarium is built in the shape of a cylinder. GPS The diameter is 203 ft and the height is 25 ft. About how many million gallons of water does this tank hold? (1 gal $\approx$ 231 in.^{3}) **about 6 million gal**

22. A rectangular prism has a length of 3.1 m, a width of 2.2 m, and a height of 5.6 m. Find the volume to the nearest cubic meter. **38 m^3**

C 23. **Challenge** A soft-drink can has a height of about 4.8 in. and a diameter of about 2.5 in. uppose you need to put 12 cans in a rectangular case about 5 in. tall, 11 in. long, and 7.6 in. wide. Will the cans fit in the case? E plain. **See margin.**

Test Prep and Mixed Review

Practice

Gridded Response

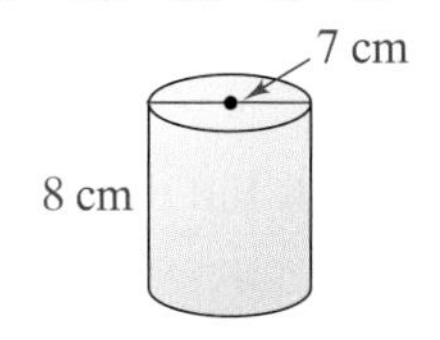

24. rs. anosian has a can of concentrated orange juice like the one at the right. What is the volume of the can in cubic centimeters? **308**

25. ina created a project on water conservation. he found that you use 1.6 gallons of water each time you flush a toilet. At this rate, how many gallons of water would you use flushing 7 times? **11.2**

26. Ethan collected five rocks that weighed a total of 43.4 kg. Two of the rocks were identical, and each weighed 13.3 kg. If two of the other rocks weighed 4.7 kg and 5.3 kg, what was the weight of the fifth rock in kilograms? **6.8**

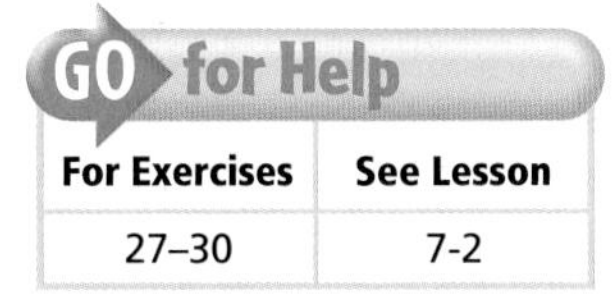

For Exercises	See Lesson
27–30	7-2

Find the measures of the complement and supplement of each angle.

27. $m\angle A = 40°$ **50°; 140°**
28. $m\angle B = 65°$ **25°; 115°**
29. $m\angle C = 37°$ **53°; 143°**
30. $m\angle D = 5°$ **85°; 175°**

4. Assess & Reteach

PowerPoint
Lesson Quiz

Find the volume of each figure to the nearest cubic unit.

1. **2,304 ft^3**

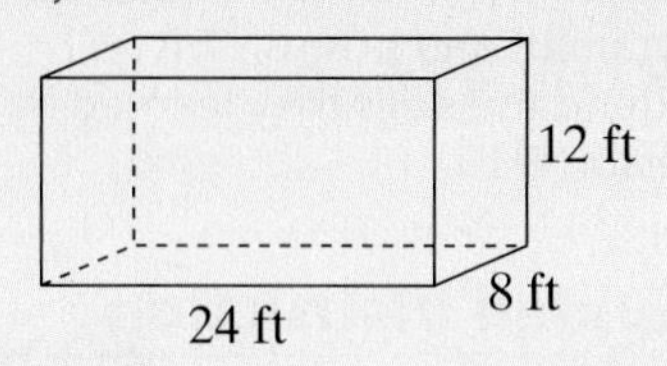

2. **90 ft^3**

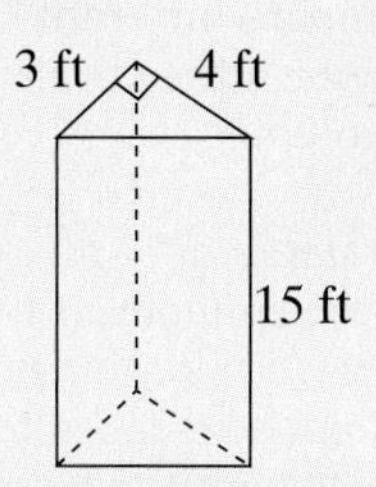

3. **8,743 cm^3**

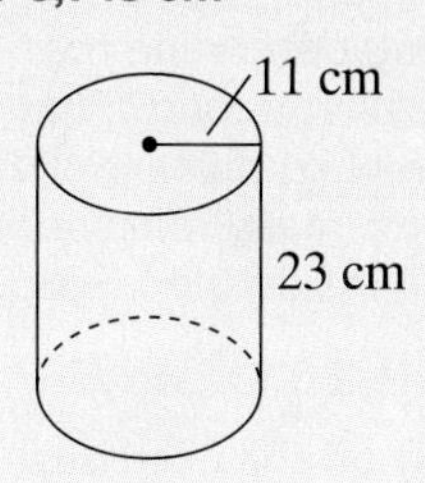

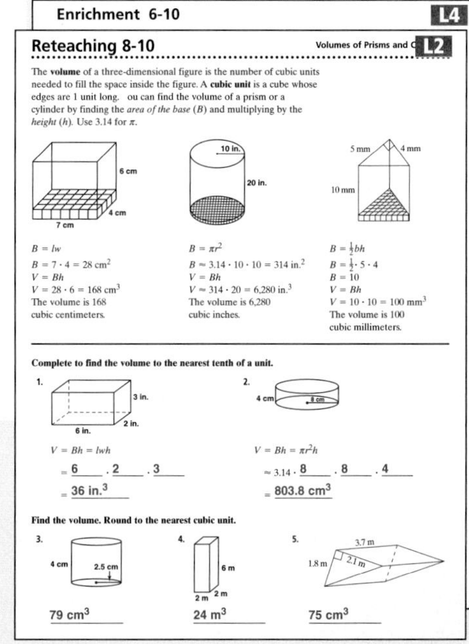

Enrichment 6-10 L4

Reteaching 8-10 Volumes of Prisms and C L2

The **volume** of a three-dimensional figure is the number of cubic units needed to fill the space inside the figure. A **cubic unit** is a cube whose edges are 1 unit long. ou can find the volume of a prism or a cylinder by finding the *area of the base* (*B*) and multiplying by the *height* (*h*). Use 3.14 for π.

$B = lw$
$B = 7 \cdot 4 = 28$ cm^2
$V = Bh$
$V = 28 \cdot 6 = 168$ cm^3
The volume is 168 cubic centimeters.

$B = \pi r^2$
$B = 3.14 \cdot 10 \cdot 10 = 314$ in.^{2}
$V = Bh$
$V \approx 314 \cdot 20 = 6{,}280$ in.^{3}
The volume is 6,280 cubic inches.

$B = \frac{1}{2}bh$
$B = \frac{1}{2} \cdot 5 \cdot 4$
$B = 10$
$V = Bh$
$V = 10 \cdot 10 = 100$ mm^3
The volume is 100 cubic millimeters.

Complete to find the volume to the nearest tenth of a unit.

1. $V = Bh = lwh$
= 6 . 2 . 3
= 36 in.^{3}

2. $V = Bh = \pi r^2 h$
≈ 3.14 · 8 . 8 . 4
≈ 803.8 cm^3

Find the volume. Round to the nearest cubic unit.

3. 79 cm^3
4. 24 m^3
5. 75 cm^3

Alternative Assessment

Have students work in pairs and take turns completing Exercises 5–13. Suggest that they write each formula at the top of their papers for quick reference. Allow pairs to discuss their methods as they work. You may wish to have them use calculators to aid them in their computations.

Test Prep

Resources
For additional practice with a variety of test item formats:
- Test-Taking Strategies, p. 427
- Test Prep, p. 431
- Test-Taking Strategies with Transparencies

Activity Lab

Generating Formulas For Volume

Students will relate volumes of pyramids and prisms. This will extend the skills they learned in Lesson 8-10.

Guided Instruction

Before beginning the Activity, ask students to compare different solids and tell which they think has a greater volume.
Ask:

- *Name an example of comparing the volumes of two solids.* **Sample: comparing how many small cubes it takes to fill a large rectangular box or prism**
- *How can you determine the volume of a large solid?* **Find the volume of a smaller solid and find how many times that solid will fit inside the larger solid.**

Have students work independently on the Exercises. Then conduct a class discussion about how the volumes of different solids are related.

Differentiated Instruction

Below Level L2
Provide paper models of the nets of the cone and cylinder in Exercise 2. Allow students to construct the solids and fill them with rice as in the Activity to reinforce the connection between the solids and the formula.

Resources

- posterboard
- scissors
- tape
- rice
- paper

8-10b Activity Lab Hands On

Generating Formulas for Volume

In this activity, you will relate the volume of a pyramid to the volume of a prism.

ACTIVITY

Step 1 Using poster board, draw and cut out four congruent isosceles triangles like the one below.

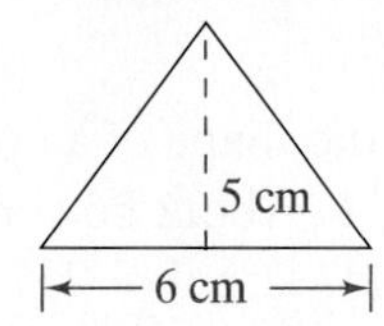

Step 2 Tape the edges of the four triangles to form a pyramid without a base. What is the area of the missing base?

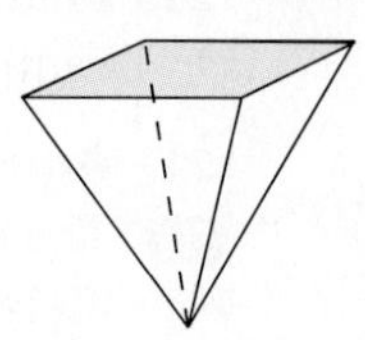

Step 3 Using poster board, draw and cut out four congruent rectangles and one square like the one below.

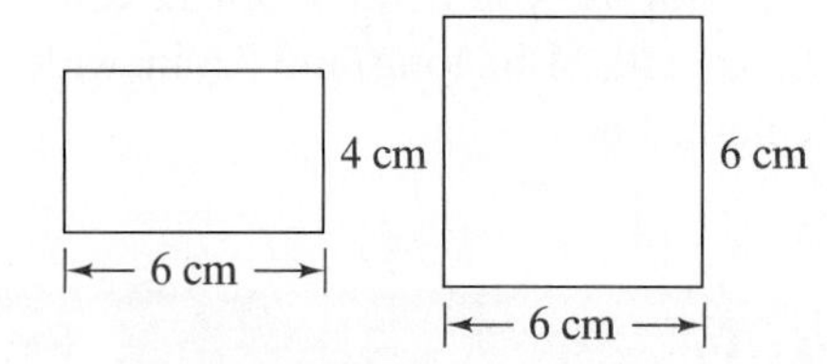

Step 4 Tape the edges of the polygons to form a prism without a base. Compare the areas of the missing base of the prism and the missing base of the pyramid.

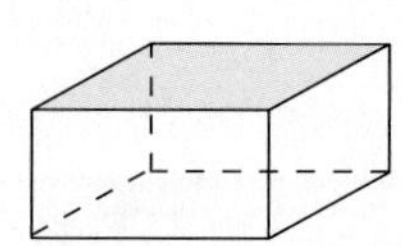

Step 5 Place the pyramid and the prism side by side. What do you notice about their heights?

Step 6 Fill the pyramid with rice. Pour the rice from the pyramid into the prism. Repeat until the prism is full.

Exercises

1. a. How many pyramids full of rice did you need to fill the prism? **3 pyramids**
   b. How does the volume of the pyramid compare to the volume of the prism? **See margin.**
   c. Make a conjecture about the formula for volume of a pyramid. $V = \frac{1}{3}Bh$

2. **Reasoning** To fill a cylinder with base area B and height h, you need 3 cones as shown at the right. Make a conjecture about the formula for volume of a cone. **See margin.**

+
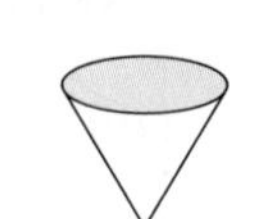
+
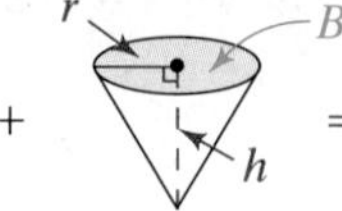

=
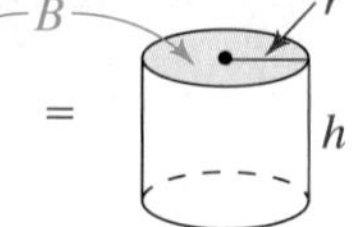

1b. The volume of the pyramid is one third the volume of the prism.

2. $V = \frac{1}{3}Bh$ **or** $V = \frac{1}{3}\pi r^2 h$

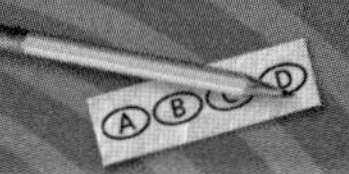

Test-Taking Strategies

Measuring to Solve

Some test questions ask you to measure with a centimeter ruler before solving a problem.

EXAMPLE

The bottom of a bottle is circular, as shown at the right. Measure the radius of the bottle in centimeters.

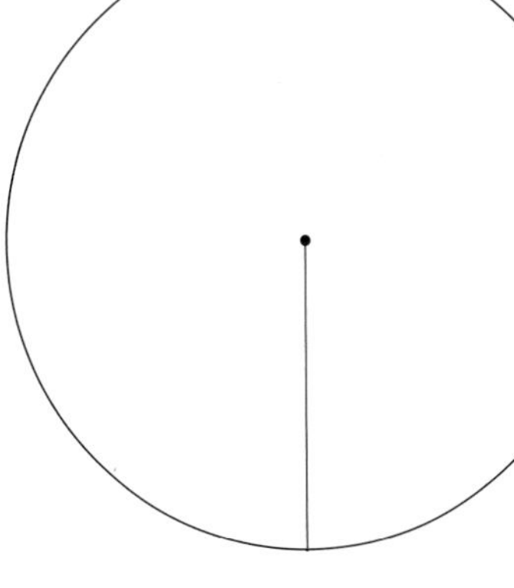

Which of the following is closest to the circumference of the bottom of the bottle?

(A) 6 cm (C) 15 cm
(B) 9 cm (D) 19 cm

Use a centimeter ruler to measure the radius of the bottle. Label the radius. To find the circumference of the bottle, use the formula for the circumference of a circle.

$C = 2\pi r$, or πd
$C = 2\pi r = 2 \times \pi \times 3 \approx 18.84955\ldots$

The circumference of the bottle is about 19 cm. The answer is D.

Exercises

1. A vase has a circular base, as shown at the right. Measure the radius of the base in centimeters. Which of the following is the closest to the area of the circular base? **C**

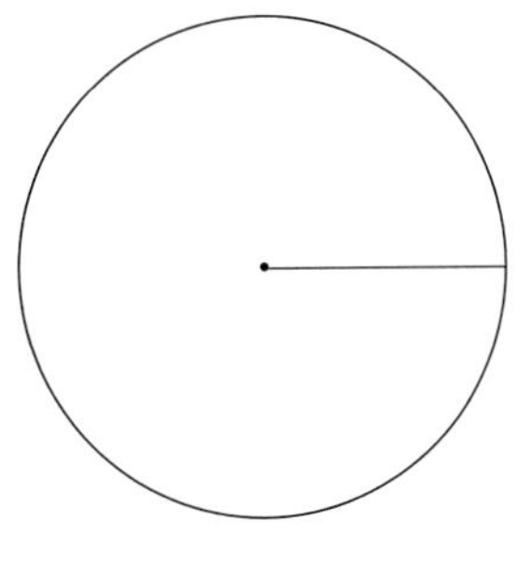

(A) 2 cm^2
(B) 6 cm^2
(C) 12 cm^2
(D) 24 cm^2

2. Celine bought earrings and made a gift box for them that is 4 cm high. The base of the box is shown at the right. Measure the dimensions of the base in centimeters. Which best represents the volume of the box? **J**

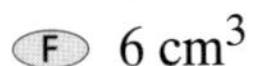

(F) 6 cm^3 (H) 12 cm^3
(G) 8 cm^3 (J) 32 cm^3

Test-Taking Strategies

Measuring to Solve

On some tests, students are asked to measure objects in order to determine an answer. This feature provides practice in measuring test items and using the measurements to answer questions.

Guided Instruction

Explain to students that they may not always be able to measure the dimension in the question, but they can use their ruler and knowledge of formulas to find the answer. Measuring a drawing in a question may also give them information that they need for a formula to answer the question.

Exercises
You may want students to write formulas they need to solve Exercises 1 and 2 before they do the problems.

Resources

Test-Taking Strategies with Transparencies
- Transparency 13
- Practice sheet, p. 20

Test-Taking Strategies with Transparencies

Test-Taking Strategies: Measuring to Solve

Some questions ask you to measure with a protractor or ruler to solve a problem.

Example What is the area of the circle below, to the nearest square centimeter?

A. 9 cm^2 B. 19 cm^2
C. 28 cm^2 D. 113 cm^2

Use a centimeter ruler to measure the radius of the circle. To find the area of the circle, use the formula for the area of a circle.

$A = \pi r^2$
$A = \pi(3)^2 = \pi \times 9 \approx 28.27$
The area of the circle is about 28 cm^2. The answer is choice C.

Example Find the measure of angle *S* in the polygon below.

F. 83° G. 87° H. 93° J. 97°

The side of the angle falls between the 85° and 90° marks on the same scale that side SW crosses at its zero point. The measure of the angle is about 87°. The answer is choice G.

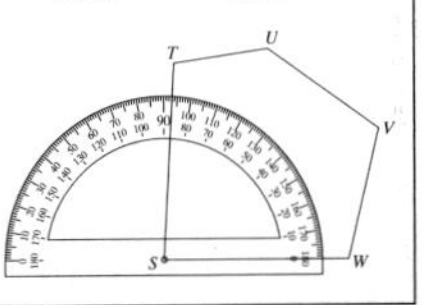

Resources

Student Edition
Extra Skills and Word Problems Practice, Ch. 8, p. 644
English/Spanish Glossary, p. 676
Formulas and Properties, p. 674
Tables, p. 670

All in One Teaching Resources
- Vocabulary and Study Skills 8F — L3

Differentiated Instruction

Spanish Vocabulary Workbook with Study Skills — ELL
Interactive Textbook
- Audio Glossary

Online Vocabulary Quiz

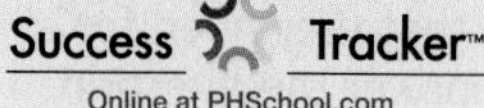

Online at PHSchool.com

Spanish Vocabulary/Study Skills — ELL

Vocabulary/Study Skills — L3

8F: Vocabulary Review — For use with the Chapter Review

Vocabulary and Study Skills

Study Skill Participating in class discussions will help you remember new material. Do not be afraid to express your thoughts when your teacher asks for questions, answers, or discussion.

Circle the word that best completes the sentence.

1. The longest side of a right triangle is the (*leg*, *hypotenuse*).
2. (*Parallel*, *Perpendicular*) lines lie in the same plane and do not intersect.
3. A (*solution*, *statement*) is a value of a variable that makes an equation true.
4. Figures that are the same size and shape are (*similar*, *congruent*).
5. (*Complementary*, *Supplementary*) angles are two angles whose sum is 90°.
6. A (*circle*, *sphere*) is the set of all points in space that are the same distance from a center point.
7. The perimeter of a circle is the (*circumference*, *circumcenter*).
8. The (*area*, *volume*) of a figure is the number of square units it encloses.
9. A(n) (*isosceles*, *scalene*) triangle has no congruent sides.
10. A (*rhombus*, *square*) is a parallelogram with four right angles and four congruent sides.
11. A number that is the square of an integer is a (*perfect square*, *square root*).
12. A (*pyramid*, *prism*) is a three-dimensional figure with triangular faces that meet at one point.
13. A speed limit of 65 mi/h is an example of a (*ratio*, *rate*).
14. A (*cone*, *cylinder*) has two congruent parallel bases that are circles.

Chapter 8 Review

Vocabulary Review

area (p. 375)
base(s) (pp. 380, 384, 388, 410, 411)
center of a sphere (p. 411)
circumference (p. 394)
cone (p. 411)
cube (p. 410)
cubic unit (p. 421)
cylinder (p. 410)
edge (p. 410)
face (p. 410)
height (pp. 380, 384, 388, 410)
hypotenuse (p. 405)
irrational number (p. 401)
legs (p. 405)
net (p. 414)
perfect square (p. 400)
perimeter (p. 375)
pi (p. 394)
prism (p. 410)
pyramid (p. 410)
Pythagorean Theorem (p. 405)
sphere (p. 411)
square root (p. 400)
surface area (p. 415)
three-dimensional figure (p. 410)
vertex (pp. 410, 411)
volume (p. 421)

Go Online PHSchool.com
For: Online Vocabulary Quiz
Web Code: arj-0851

Choose the correct term to complete each sentence.

1. A(n) (edge, vertex) is the intersection of two faces. **edge**
2. The longest side of a right triangle is the (hypotenuse, leg). **hypotenuse**
3. A (prism, pyramid) has two parallel and congruent bases. **prism**
4. The perimeter of a circle is the (area, circumference) of the circle. **circumference**
5. A (cone, cylinder) has one circular base and one vertex. **cone**

Skills and Concepts

Lesson 8-1
- To estimate length, perimeter, and area

The **area** of a figure is the number of square units it encloses.

6. Estimate the area of the shaded region. Each square represents 20 in.^{2}. **about 400 in.^{2}**

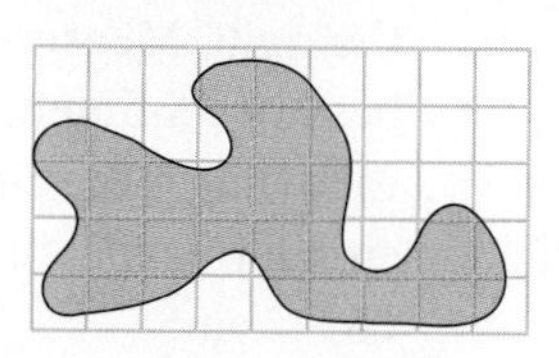

7. Choose a reasonable estimate for the width of a book—7 in. or 7 ft. Explain. **7 in.; the width is less than a foot.**

Lessons 8-2, 8-3, 8-4
- To find the area and perimeter of a parallelogram
- To find the perimeter and area of a triangle
- To find the area of a trapezoid and the areas of irregular figures

To find the area of an irregular figure, first separate it into familiar figures and find the area of each piece. Then add the areas.

parallelogram	triangle	trapezoid
$A = bh$	$A = \frac{1}{2}bh$	$A = \frac{1}{2}h(b_1 + b_2)$

Use familiar figures to find the area of each figure. **8–10. See left.**

8. 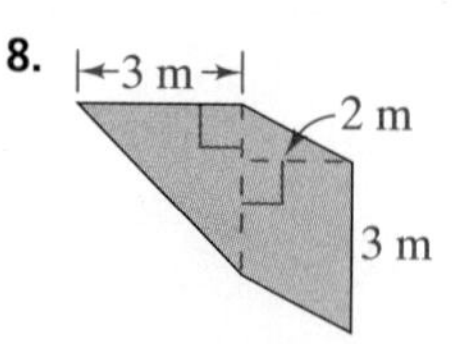

9. 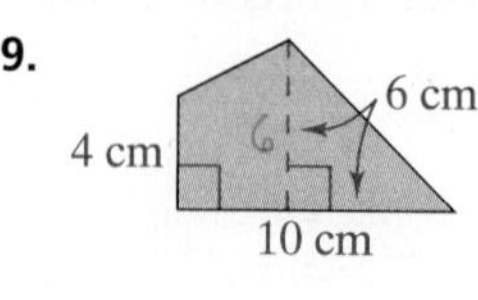

10. 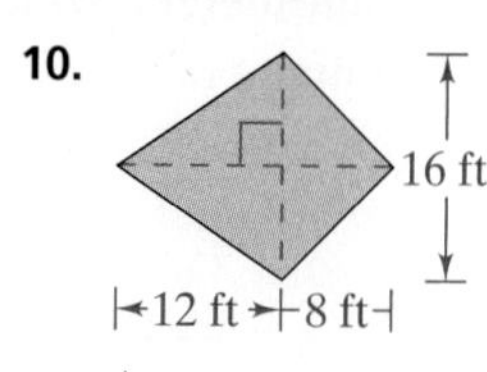

8. **10.5 m^2**
9. **38 cm^2**
10. **160 ft^2**

Lesson 8-5

- To find the circumference and area of a circle

To find the **circumference** of a circle, use the formula $C = \pi d = 2\pi r$. To find the area of a circle, use the formula $A = \pi r^2$.

Find the circumference and area of each circle.

11.

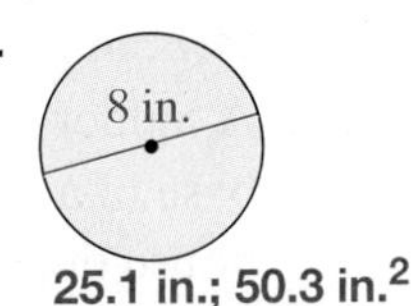

25.1 in.; 50.3 in.^{2}

12.

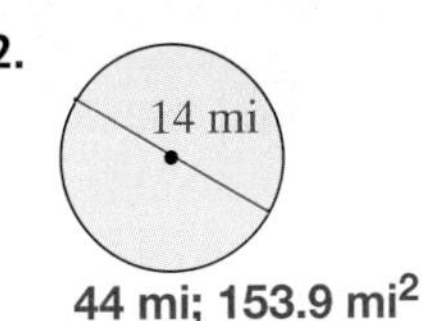

44 mi; 153.9 mi^2

13.

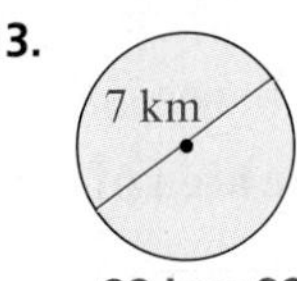

22 km; 38.5 km^2

Lessons 8-6, 8-7

- To find and estimate square roots and to classify numbers as rational or irrational
- To use the Pythagorean Theorem to solve real-world problems

A **perfect square** is a square of an integer. The opposite of squaring a number is finding its **square root.** Many square roots are **irrational numbers,** or numbers that cannot be written as the ratio of two integers.

The **Pythagorean Theorem,** $a^2 + b^2 = c^2$, relates the lengths of the **legs** of a right triangle to the length of its **hypotenuse.**

14. Art A square piece of glass in a picture frame covers an area of 36 in.^{2}. What is the length of each side of the glass? **6 in.**

15. A pipeline is placed diagonally across a rectangular field that is 25 yd by 30 yd. How long is the pipeline, to the nearest yard? **39 yd**

Lesson 8-8

- To classify and draw three-dimensional figures

Some **three-dimensional figures** have only flat surfaces. **Prisms** and **pyramids** are named for the shape of their bases. **Cones** and pyramids have one **vertex.**

Name each figure.

16.

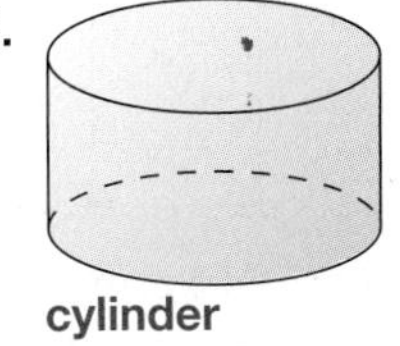

cylinder

17.

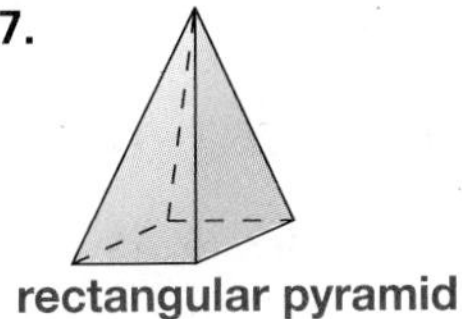

rectangular pyramid

18.

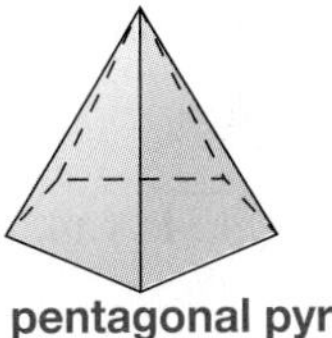

pentagonal pyramid

Lessons 8-9, 8-10

- To find the surface areas of prisms and cylinders using nets
- To find the volumes of prisms and cylinders

To find the **surface area** of a prism or cylinder, draw a **net** and find the area of the net. To find the **volume** of a prism, use the formula $V = Bh$. To find the volume of a cylinder, use the formula $V = \pi r^2 h$.

Find the surface area and volume for each figure.

19.

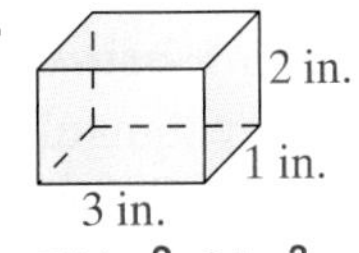

22 in.^{2}; 6 in.^{3}

20.

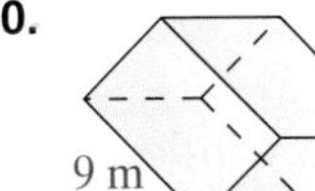

288 m^2; 324 m^3

21.

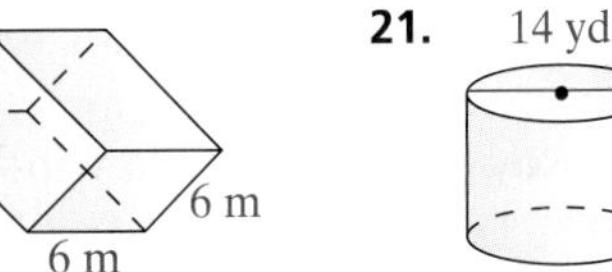

747.7 yd^2; 1,539.4 yd^3

747.3 yd^2; 1538.6 yd^3

Alternative Assessment L4

Alternative Assessment

Chapter 8

ZIGZAG SAILING

Dwayne wants to sail his boat across a circular lake that has a diameter of 1,000 yards. He has one problem: The wind is blowing directly toward him. To solve the problem, Dwayne learned to zigzag, or "tack," his boat at a 45° angle to the wind, as shown in this picture.

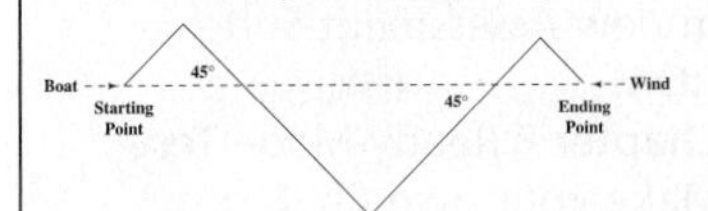

Show all of your work on a separate sheet of paper.

1. Make a scale drawing of a circular lake with a 1,000-yd diameter on grid paper. Then draw a path across the lake, using the zigzag method Dwayne was taught for tacking. Start on one side of the lake and go to the opposite side. Label the scale for your lake and the approximate length of each segment of your path. Use a ruler, a compass, and a protractor.

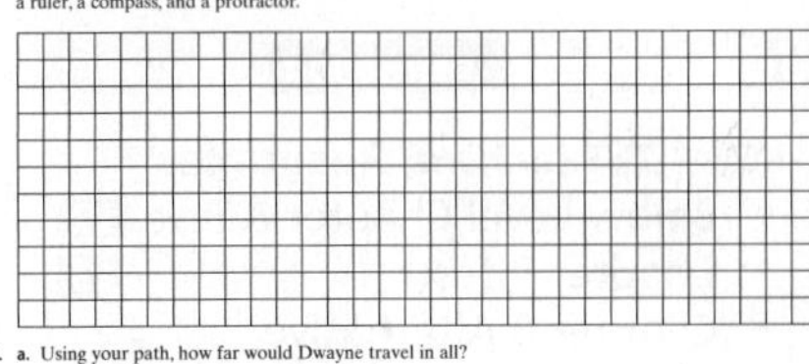

2. a. Using your path, how far would Dwayne travel in all?
   b. With 45° tacking, are there other paths shorter than this path?
   c. With 45° tacking, are there longer paths?
   d. Draw several more paths to confirm your decisions. What can you conclude? Explain.

Chapter 8 Test

Resources

- ExamView Assessment Suite CD-ROM
  - Chapter 8 Ready-Made Test
  - Make your own Ch. 8 test
- MindPoint Quiz Show CD-ROM
  - Chapter 8 Review

Differentiated Instruction

All in One Teaching Resources
- Below Level Chapter 8 Test L2
- Chapter 8 Test L3
- Chapter 8 Alternative Assessment L4

Spanish Assessment Resources ELL
- Below Level Chapter 8 Test L2
- Chapter 8 Test L3
- Chapter 8 Alternative Assessment L4

ExamView Assessment Suite CD-ROM
- Special Needs Test L1
- Special Needs Practice Bank L1

Online Chapter 8 Test at www.PHSchool.com L3

Below Level Chapter Test L2

Chapter Test L3

Chapter Test — Form A

Chapter 8

Estimate the area of each shaded region. Each square represents 25 in.^{2}

1. $1{,}200$ in.^{2}
2. 900 in.^{2}

Find the area of each figure.

3. 216 ft^2 (12 ft, 36 ft)
4. 60 in.^{2} (8 in., 10 in., 4 in.)
5. 96 m^2 (8 m, 12 m)

Find the circumference and area of each circle. Round to the nearest tenth.

6. 25.1 cm; 50.7 cm^2 (4 cm)
7. 37.7 cm; 113.1 cm^2 (6 cm)

Simplify each square root.

8. $\sqrt{225}$ 15
9. $\sqrt{16}$ 4
10. $\sqrt{324}$ 18
11. $\sqrt{400}$ 20
12. $\sqrt{289}$ 17
13. $\sqrt{36}$ 6
14. A square indoor play space has an area of 121 ft^2. What is the perimeter of the play space? 44 ft

Identify each number as *rational* or *irrational*.

15. $\sqrt{64}$ rational
16. $2\frac{2}{5}$ rational
17. $-1.\overline{2}$ rational
18. $\sqrt{50}$ irrational

Chapter 8 Test

Go Online PHSchool.com — For: Online chapter test — Web Code: ara-0852

Estimate the area of each shaded region. Each square represents 50 in.^{2}.

1. about 800 in.^{2}

2. 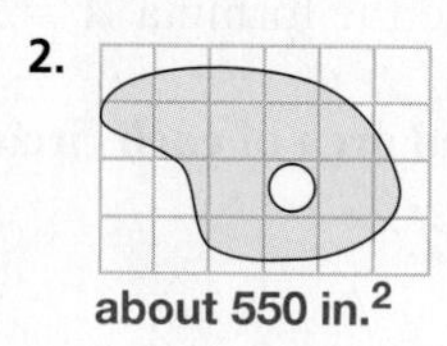 about 550 in.^{2}

Find the area of each figure.

3. 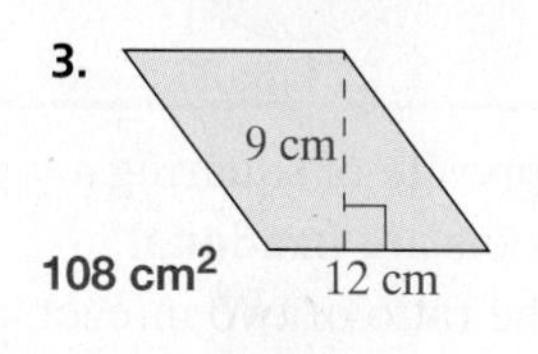

108 cm^2

4. 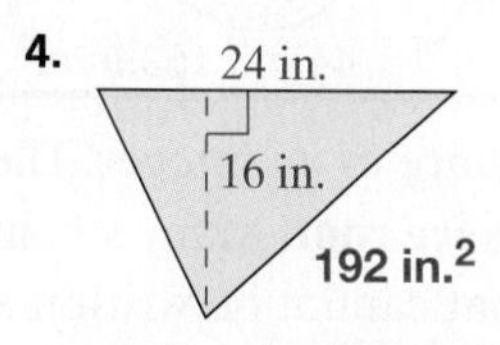

192 in.^{2}

5. 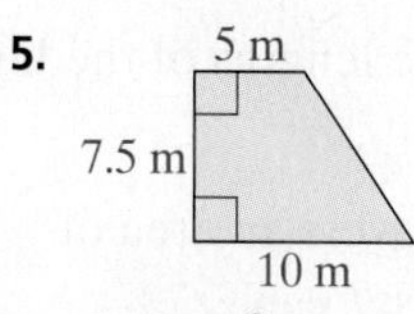

56.25 m^2

6. 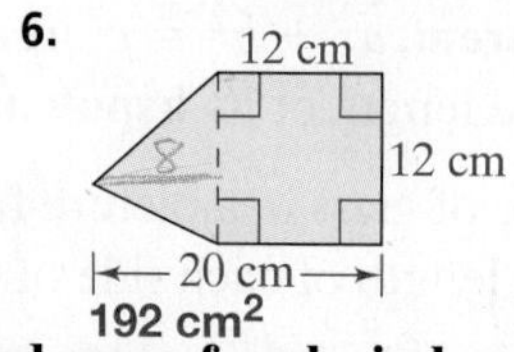

192 cm^2

Find the circumference and area of each circle. Round to the nearest tenth.

7. 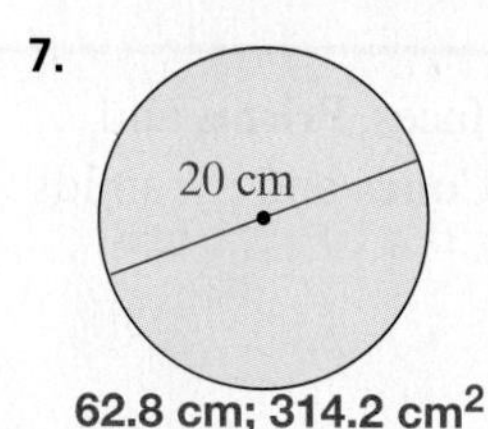

62.8 cm; 314.2 cm^2

8. 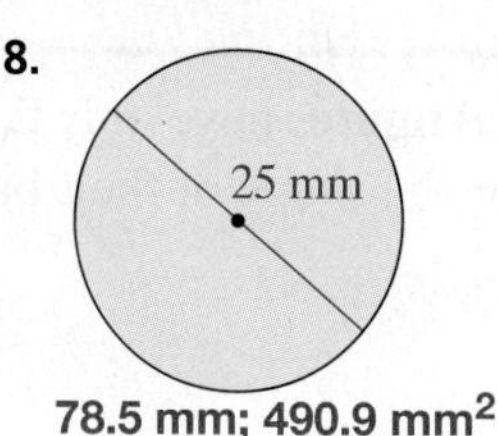

78.5 mm; 490.9 mm^2

Simplify each square root.

9. $\sqrt{9}$ 3
10. $\sqrt{25}$ 5
11. $\sqrt{49}$ 7
12. $\sqrt{100}$ 10
13. $\sqrt{121}$ 11
14. $\sqrt{1}$ 1
15. $\sqrt{64}$ 8
16. $\sqrt{81}$ 9
17. A square plot of land has an area of 100 m^2. What is the perimeter of the plot? 40 m

Find two consecutive whole numbers that each number falls between. Then estimate the number's value. 18–21. See margin.

18. $\sqrt{55}$
19. $\sqrt{63}$
20. $\sqrt{8}$
21. $\sqrt{45}$

22. **Construction** The area of a window is 18 ft^2. The length of the window is two times the width of the window. What are the dimensions of the window? 3 ft × 6 ft

Find each missing length to the nearest tenth.

23. 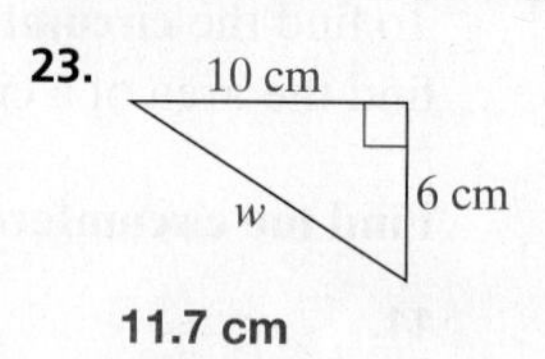

11.7 cm

24. 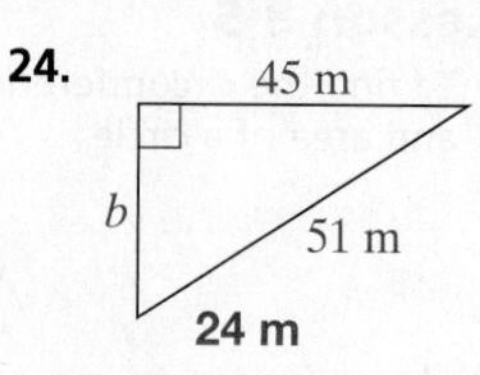

24 m

25. A ladder 26 ft long is placed 10 ft from the base of a house. How high up the side of the house does the ladder reach? 24 ft

26. A support cable connects the top of a 30-m pole to an anchor 20 m from the base of the pole. How long is the support cable, to the nearest tenth of a meter? 36.1 m

Identify each number as *rational* or *irrational*.

27. $\sqrt{30}$ irrational
28. $3.\overline{7}$ rational
29. $\frac{22}{7}$ rational
30. π irrational

Find the surface area to the nearest whole unit.

31. 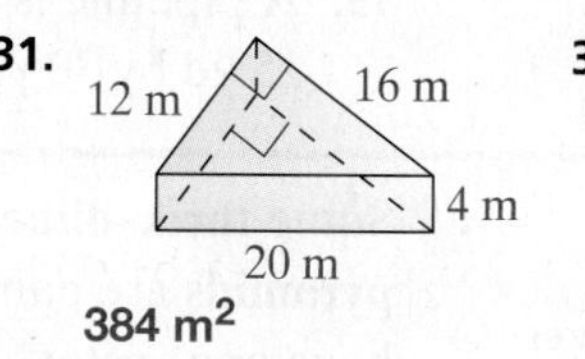

384 m^2

32. 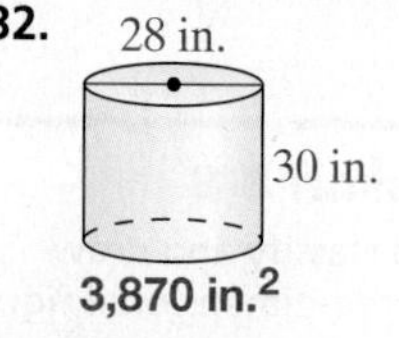

3,870 in.^{2}

Find the volume to the nearest whole unit.

33. 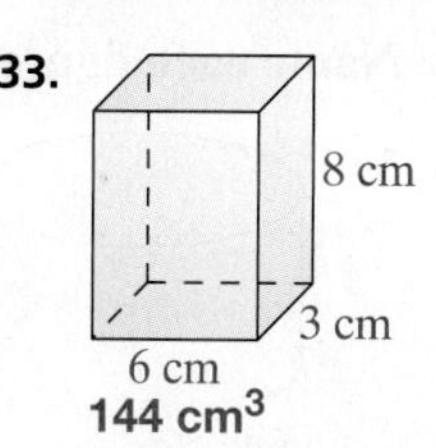

144 cm^3

34. 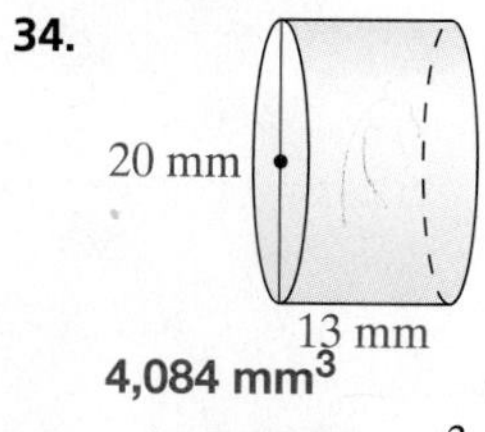

4,084 mm^3

35. A triangular prism has a volume of 96 m^3. The area of the base is 16 m^2. What is the height of the prism? 6 m

36. **Writing in Math** Explain how you would show that a triangle with side lengths 7 in., 24 in., and 25 in. is a right triangle. See margin.

37. **Measurement** The volume of a rectangular prism is 2,058 cm^3. The length of the prism is three times the width. The height is 14 cm. Find the other dimensions. length 21 cm, width 7 cm

18. $7 < \sqrt{55} < 8$; about 7
19. $7 < \sqrt{63} < 8$; about 8
20. $2 < \sqrt{8} < 3$; about 3
21. $6 < \sqrt{45} < 7$; about 7

36. Use the Pythagorean Theorem.
$7^2 + 24^2 = 25^2$
$49 + 576 = 625$
$625 = 625$

Multiple Choice

Read each question. Then write the letter of the correct answer on your paper.

1. Which number has the greatest value? **A**
   (A) 2 (B) 3^3 (C) 5^2 (D) 20^1

2. Rectangle $ABCD$ has dimensions 3 in. × 4 in. What are the area and perimeter of $ABCD$? **G**
   (F) $A = 12$ in.^{2}, $P = 12$ in.
   (G) $A = 12$ in.^{2}, $P = 14$ in.
   (H) $A = 6$ in.^{2}, $P = 12$ in.
   (J) $A = 12$ in.^{2}, $P = 7$ in.

3. The rectangles are similar. Which proportion could NOT be used to find the value of x? **B**

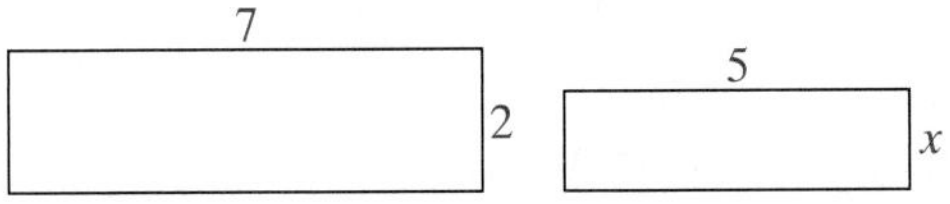

   (A) $\frac{7}{5} = \frac{2}{x}$ (C) $\frac{x}{5} = \frac{2}{7}$
   (B) $\frac{x}{2} = \frac{7}{5}$ (D) $\frac{2}{7} = \frac{x}{5}$

4. Which of the following could NOT be the length of the sides of a right triangle? **H**
   (F) 8, 15, 17 (H) 15, 35, 40
   (G) 10, 24, 26 (J) 12, 16, 20

5. Which expression could you use to find the area of the cylinder's base? **B**
   (A) $2 \cdot \pi \cdot 5$
   (B) $\pi \cdot 2.5 \cdot 2.5$
   (C) $\pi \cdot 5 \cdot 5$
   (D) $2 \cdot \pi \cdot 2.5 \cdot 6$

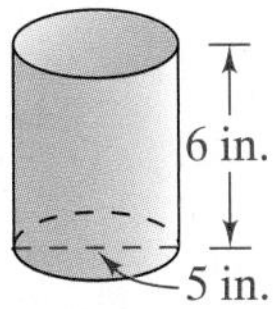

6. Which figure has the greatest volume? **J**

(F)
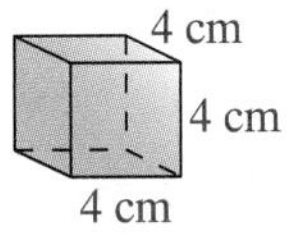

(H)
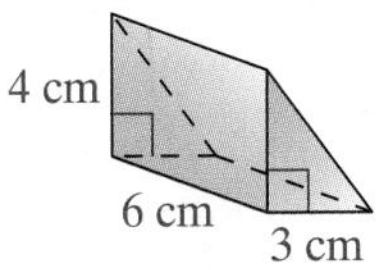

(G)
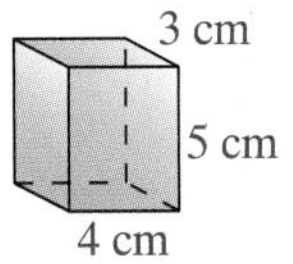

(J)
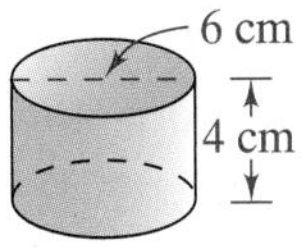

7. Which equation can you use to represent the following? Five more than half of the people (p) on the bus are students (s). **A**
   (A) $\frac{p}{2} + 5 = s$ (C) $\frac{p}{2} - 5 = s$
   (B) $(p - 5) \div 2 = s$ (D) $(p + 5) \div 2 = s$

8. What is the value of x to the nearest tenth? **G**
   (F) 17.0 (H) 8.5
   (G) 12.7 (J) 7

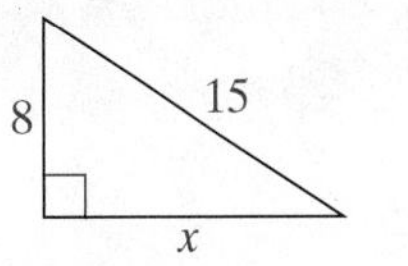

9. Which expression does NOT equal 12? **C**
   (A) $\sqrt{144}$ (C) $\sqrt{4} + \sqrt{64}$
   (B) $\sqrt{36} + \sqrt{36}$ (D) $\sqrt{81} + \sqrt{9}$

10. What number is 75% of 150? **H**
    (F) 11.25 (H) 112.5
    (G) 20 (J) 11,250

Gridded Response

Record your answer in a grid.

11. A triangle has a height of 8 ft and a base of 15 ft. Find the triangle's area in square feet. **60**

12. **Sewing** You sew 34 squares for a quilt. This is 5% of the squares used in the quilt. How many squares are there to sew in all? **680**

Short Response **13–14. See margin.**

13. Define a variable and write an inequality to model "To qualify for the long-jump finals, I need to jump at least 14 ft."

14. The area of a circular rug is 113.04 ft^2. What is the diameter of the rug? Use $\pi = 3.14$. Show your work.

Extended Response

15. The ratio of the corresponding sides of two similar triangles is 3 : 5. The sides of the smaller triangle are 9 m, 12 m, and 18 m. Find the perimeter of the larger triangle. Show your work. **See margin.**

Item	1	2	3	4	5	6	7	8	9	10	11	12	13	14	15
Lesson	2-1	8-2	5-5	8-7	8-9	8-10	2-5	8-7	8-6	6-4	8-3	6-5	4-7	8-5	5-5

13. [2] Let d = distance jumped. $d \geq 14$
[1] variable defined but incorrect inequality

14. [2] $A = \pi r^2$
$113.04 = 3.14r^2$
$36 = r^2$
$6 = r$
The diameter of the rug is 12 ft.
[1] correct procedure with one minor computational error OR correct answer without work shown

Test Prep

Resources

Test Prep Workbook

All in One Teaching Resources
- Cumulative Review **L3**

ExamView Assessment Suite CD-ROM
- Standardized Test Practice

Differentiated Instruction

Progress Monitoring Assessments
- Benchmark Test 4 **L3**

Spanish Assessment Resources
- Spanish Cumulative Review **ELL**

ExamView Assessment Suite CD-ROM
- Special Needs Practice Bank **L1**

15. [4] $\frac{3}{5} = \frac{9}{x}$, so $x = 15$
$\frac{3}{5} = \frac{12}{x}$, so $x = 20$
$\frac{3}{5} = \frac{18}{x}$, so $x = 30$
The perimeter of the larger triangle is $15 + 20 + 30 = 65$ m
[3] correct procedure with one minor computational error
[2] sides are correct with wrong perimeter
[1] correct answer with no work shown

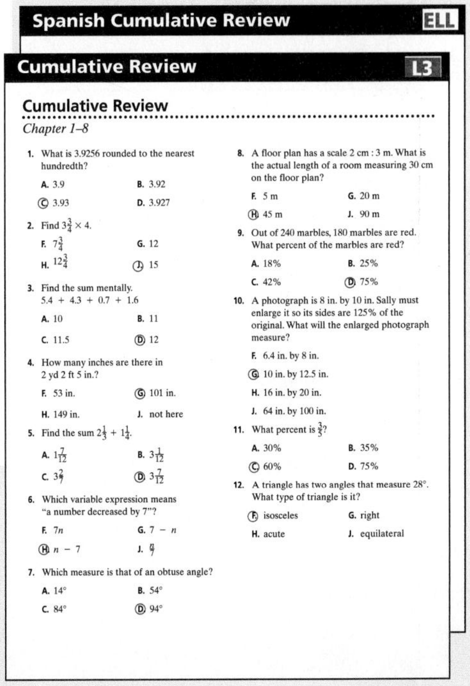
Spanish Cumulative Review ELL

Cumulative Review L3

Cumulative Review

Chapter 1–8

1. What is 3.9256 rounded to the nearest hundredth?
   A. 3.9 B. 3.92 (C) 3.93 D. 3.927
2. Find $3\frac{3}{4} \times 4$.
   F. $7\frac{3}{4}$ G. 12 H. $12\frac{3}{4}$ (J) 15
3. Find the sum mentally. $5.4 + 4.3 + 0.7 + 1.6$
   A. 10 B. 11 C. 11.5 (D) 12
4. How many inches are there in 2 yd 2 ft 5 in.?
   F. 53 in. (G) 101 in. H. 149 in. J. not here
5. Find the sum $2\frac{1}{3} + 1\frac{1}{4}$.
   A. $1\frac{7}{12}$ B. $3\frac{1}{12}$ C. $3\frac{2}{7}$ (D) $3\frac{7}{12}$
6. Which variable expression means "a number decreased by 7"?
   F. $7n$ G. $7 - n$ (H) $n - 7$ J. $\frac{n}{7}$
7. Which measure is that of an obtuse angle?
   A. 14° B. 54° C. 84° (D) 94°
8. A floor plan has a scale 2 cm : 3 m. What is the actual length of a room measuring 30 cm on the floor plan?
   F. 5 m G. 20 m (H) 45 m J. 90 m
9. Out of 240 marbles, 180 marbles are red. What percent of the marbles are red?
   A. 18% B. 25% C. 42% (D) 75%
10. A photograph is 8 in. by 10 in. Sally must enlarge it so its sides are 125% of the original. What will the enlarged photograph measure?
    F. 6.4 in. by 8 in. (G) 10 in. by 12.5 in. H. 16 in. by 20 in. J. 64 in. by 100 in.
11. What percent is $\frac{3}{5}$?
    A. 30% B. 35% (C) 60% D. 75%
12. A triangle has two angles that measure 28°. What type of triangle is it?
    (F) isosceles G. right H. acute J. equilateral

Problem Solving

Applying Volume

Students will use data from these two pages to answer the questions posed in Put It All Together.

Invite students to share any experiences they have had with older music storage devices such as compact discs (CDs), cassette tapes, and vinyl records. Ask:

- *How are CDs and vinyl records played?*
- *How is the tape in a cassette tape stored?*
- *How is a cassette tape played?*

Discuss how music is stored on CDs, cassette tapes, and vinyl records.

Materials

- Compact discs
- Cassette tapes
- Vinyl records

Activating Prior Knowledge

Have students brainstorm a list of the many forms in which music comes—from vinyl records to CDs, and from phonographs to stereo systems to tape players. Have them discuss how they themselves most often listen to music, and how their listening habits compare with those described by the data about typical teens in the Musical Interlude on page 433.

Guided Instruction

Have a volunteer read the opening paragraph about the shapes of records and CDs. Then invite them to discuss the information about CDs and portable music on the next page. Ask questions such as:

- *How do the shapes of a CD and a vinyl record compare?* **They are both discs.**
- *How do their sizes compare?* **The vinyl record is much larger than the CD.**
- *How many minutes of music does each one hold?* **Answers will vary.**

Problem Solving Application

Applying Volume

Musical Shapes Today's music comes in many forms. You're probably most familiar with compact discs (CDs) and cassette tapes. Maybe you've also seen older vinyl records or the newer mini-discs. These forms of music look and play differently, but you may have noticed that the recorded areas all have the same geometric shape.

Put It All Together

Materials centimeter ruler, cassette tape, CDs

1. Wind the tape in a cassette completely around one of the spools, making a cylinder.

a. Measure the diameter of the cylinder of tape and the height of the tape. Because the tape is inside a plastic cover, you may have to approximate these measurements.

b. Find the volume of the cylinder.

c. Notice that the center of the cylinder is a spool. Measure the diameter of the spool. Find its volume.

d. Subtract the volume of the spool from the total volume. What is the volume of the magnetic tape?

Cassette tape

Diameter of tape

Diameter of spool

2. a. Measure the radius of a CD.

b. Find the height of the CD. (*Hint:* Stack several CDs, measure their combined height, and divide by the number of discs.)

c. Find the volume of the CD.

d. Measure the radius from the center of the CD to the beginning of the music area. (See the photo at the right.) This section contains no music. Find its volume.

e. Subtract your answer to part (d) from your answer to part (c) to find the volume used for music.

CD

Radius from center to music

Radius of CD

3. a. How many minutes of music are on the tape? How many minutes are stored in each cubic centimeter of volume?

b. How many minutes of music are on the CD? How many minutes are stored in each cubic centimeter of volume?

c. **Writing in Math** Which format stores music more efficiently, a cassette or a CD? Explain.

Answers may vary. Samples are given using a 90-min cassette tape and a 60-min CD.

1a. diameter of tape ≈ 4.8 cm; height of tape ≈ 0.4 cm
b. overall volume ≈ 7.2 cm^3
c. diameter of spool ≈ 2.2 cm; volume of spool ≈ 1.5 cm^3
d. volume of tape ≈ 5.7 cm^3

2a. CD radius ≈ 6 cm
b. height of 6 CDs ≈ 0.5 cm; height of 1 CD ≈ 0.12 cm
c. total volume ≈ 13.6 cm^3

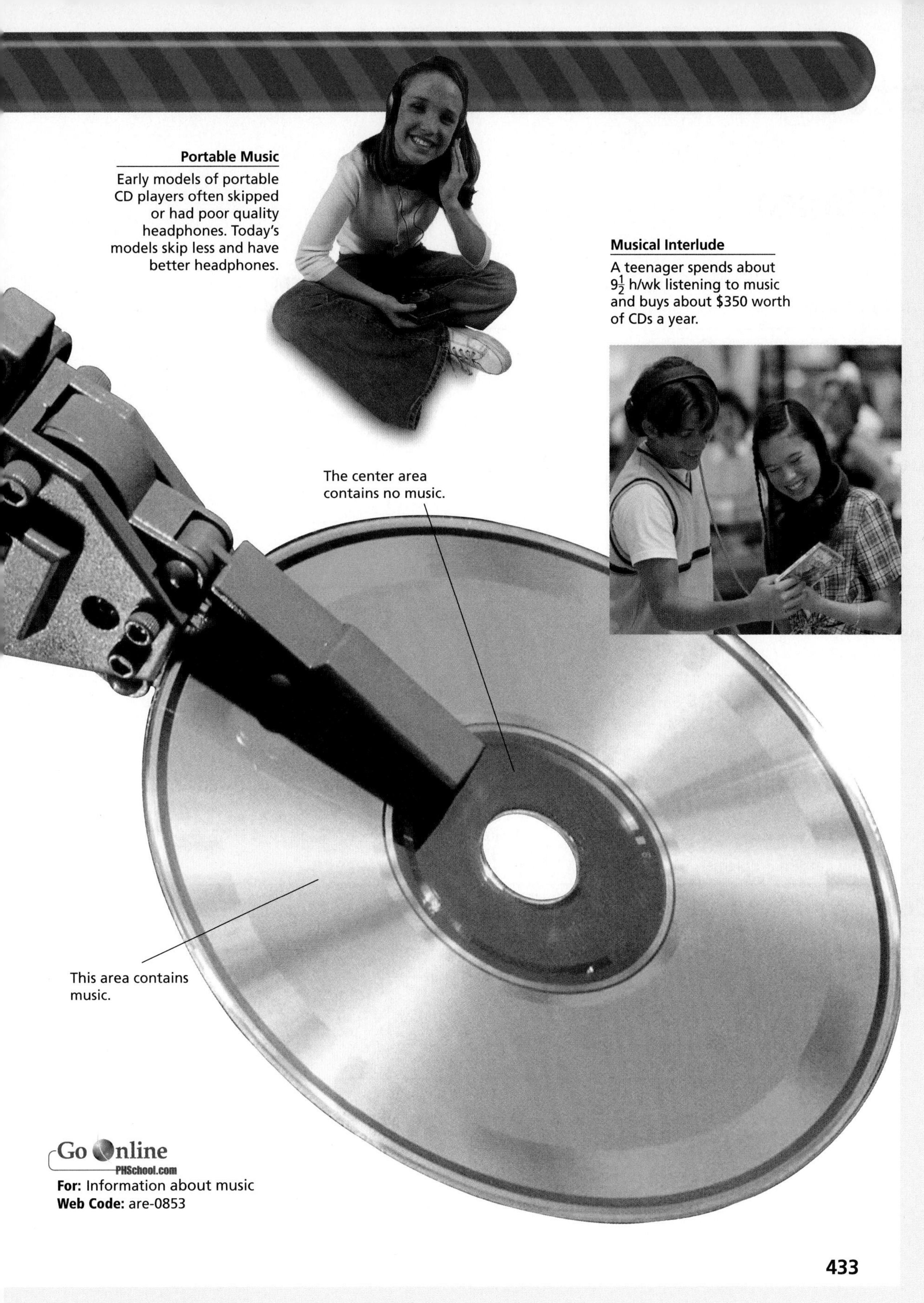

Portable Music
Early models of portable CD players often skipped or had poor quality headphones. Today's models skip less and have better headphones.
Musical Interlude
A teenager spends about $9\frac{1}{2}$ h/wk listening to music and buys about $350 worth of CDs a year.
The center area contains no music.
This area contains music.

Go Online
PHSchool.com
For: Information about music
Web Code: are-0853

Activity

Have students work in pairs to answer the questions. Guide them to record data as they measure and accumulate it.

Exercises 1–2 Elicit from students that when they compute volume, they measure it in cubic units; in this case, cubic centimeters, which are abbreviated cm^3.

Science Connection
Invite interested students to find out how CD players and CDs work. Once they understand this, have them share their knowledge with classmates.

Differentiated Instruction

Special Needs L1
As needed, help students to compute volume and to measure and estimate lengths in centimeters.

d. center-to-lead-in volume $\approx 2.0\ cm^3$
e. program area volume $\approx 11.6\ cm^3$

3a. 90 min; 90 min $\div$ $7.2\ cm^3 \approx 13\ min/cm^3$
b. 60 min; 60 min $\div$ $11.6\ cm^3 \approx 5\ min/cm^3$
c. Check students' reasoning.

9 Patterns and Rules

Chapter at a Glance

Lesson Titles, Objectives, and Features	Assessment	NCTM Standards	Local Standards
9-1a Activity Lab, Data Analysis: Choosing Scales and Intervals **9-1 Patterns and Graphs** • To graph data and to use graphs to make predictions	Lesson Quiz	1, 2, 3, 6, 7, 8, 9, 10	
9-2a Activity Lab, Hands On: Finding Patterns **9-2 Number Sequences** • To describe the patterns in arithmetic and geometric sequences and use the patterns to find terms	Lesson Quiz	1, 2, 3, 6, 7, 8, 9, 10	
9-3 Patterns and Tables • To use tables to represent and describe patterns	Lesson Quiz Checkpoint Quiz 1	1, 2, 6, 7, 8, 9, 10	
9-4a Activity Lab: Generating Formulas From a Table **9-4 Function Rules** • To write and evaluate functions	Lesson Quiz	1, 2, 6, 7, 8, 9, 10	
9-5 Using Tables, Rules, and Graphs • To find solutions to application problems using tables, rules, and graphs **9-5b Activity Lab, Technology:** Three Views of a Function	Lesson Quiz	1, 2, 5, 6, 7, 8, 9, 10	
9-6 Interpreting Graphs • To describe and sketch graphs that represent real-world situations **Guided Problem Solving:** Solving Pattern Problems	Lesson Quiz Checkpoint Quiz 2	1, 2, 5, 6, 7, 8, 9, 10	
9-7 Simple and Compound Interest • To find simple interest and compound interest	Lesson Quiz	1, 2, 6, 7, 8, 9, 10	
9-8 Transforming Formulas • To solve for a variable **9-8b Activity Lab, Algebra Thinking:** More About Formulas	Lesson Quiz	1, 2, 3, 4, 5, 6, 7, 8, 9, 10	
Problem Solving Application: Applying Graphs			

NCTM Standards 2000
1 Number and Operations **2** Algebra **3** Geometry **4** Measurement **5** Data Analysis and Probability
6 Problem Solving **7** Reasoning and Proof **8** Communication **9** Connections **10** Representation

Correlations to Standardized Tests

All content for these tests is contained in *Prentice Hall Math,* Course 2. This chart reflects coverage in this chapter only.

	9-1	9-2	9-3	9-4	9-5	9-6	9-7	9-8
Terra Nova CAT6 (Level 17)								
Number and Number Relations	✔	✔	✔	✔	✔	✔	✔	✔
Computation and Numerical Estimation	✔	✔	✔	✔	✔	✔	✔	✔
Operation Concepts	✔	✔	✔	✔	✔	✔	✔	✔
Measurement								
Geometry and Spatial Sense	✔	✔						✔
Data Analysis, Statistics, and Probability								✔
Patterns, Functions, Algebra	✔	✔	✔	✔	✔	✔	✔	✔
Problem Solving and Reasoning	✔	✔	✔	✔	✔	✔	✔	✔
Communication	✔	✔	✔	✔	✔	✔	✔	✔
Decimals, Fractions, Integers, and Percent	✔	✔	✔	✔	✔	✔	✔	✔
Order of Operations								
Terra Nova CTBS (Level 17)								
Decimals, Fractions, Integers, Percents	✔	✔	✔	✔	✔	✔	✔	✔
Order of Operations, Numeration, Number Theory	✔	✔	✔	✔	✔	✔	✔	✔
Data Interpretation				✔				✔
Pre-algebra	✔	✔	✔	✔	✔	✔	✔	✔
Measurement								
Geometry	✔	✔						✔
ITBS (Level 13)								
Number Properties and Operations	✔	✔	✔	✔	✔	✔	✔	✔
Algebra	✔	✔	✔	✔	✔	✔	✔	✔
Geometry	✔	✔						✔
Measurement								
Probability and Statistics								
Estimation								
SAT10 (Int 3 Level)								
Number Sense and Operations	✔	✔	✔	✔	✔	✔	✔	✔
Patterns, Relationships, and Algebra	✔	✔	✔	✔	✔	✔	✔	✔
Data, Statistics, and Probability								
Geometry and Measurement	✔	✔						✔
NAEP								
Number Sense, Properties, and Operations	✔		✔	✔	✔		✔	✔
Measurement								
Geometry and Spatial Sense								
Data Analysis, Statistics, and Probability								
Algebra and Functions						✔		

CAT6 California Achievement Test, 6th Ed. **CTBS** Comprehensive Test of Basic Skills **ITBS** Iowa Test of Basic Skills, Form M
SAT10 Stanford Achievement Test, 10th Ed. **NAEP** National Assessment of Educational Progress 2005 Mathematics Objectives

Math Background

Skills Trace

BEFORE Chapter 9

Course 1 introduced using patterns with tables and graphs of functions.

DURING Chapter 9

Course 2 reviews and extends patterns and number sequences using tables, graphs, formulas, and function rules.

AFTER Chapter 9

Throughout this course, students apply number patterns to solve real-world problems.

9-1 Patterns and Graphs

Math Understandings

- You can visually represent the relationship between two or more sets of numbers by drawing a graph.
- The scales on each axis can have different increments and can represent different measures. However, the increments on each axis must be of equal size.
- You can use a graph to make estimates between or beyond data points.

You can use a graph to make a prediction by extending the graph by finding the corresponding value on the appropriate axis.

Example: Graph the data in the table and predict from the graph what the height of the plant will be after 5 weeks.
2.5 in.

Plant Growth	
Age (weeks)	Height (in.)
1	0.5
2	1.0
3	1.5

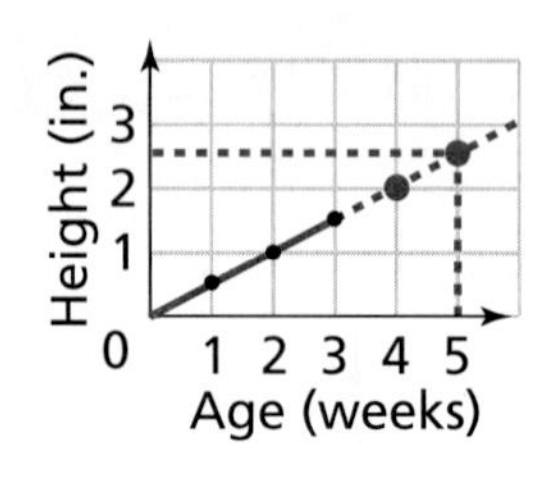

9-2 Number Sequences

Math Understandings

- You can write a sequence if you know the pattern it follows and the first term.
- In an arithmetic sequence, there is a common difference between successive terms; in a geometric sequence, there is a common ratio between successive terms.
- A conjecture assumes that a given pattern will continue (inductive reasoning).

A **sequence** is a set of numbers that follow a pattern. You can find each term of an **arithmetic sequence** by adding a fixed number to the previous term. In a **geometric sequence,** you find each term by multiplying the previous term by a fixed number. A **conjecture** is a prediction that suggests what you expect will happen.

9-3 Patterns and Tables
9-4 Function Rules

Math Understandings

- You can represent the relationship between two quantities using a table or a rule.
- When the value of one quantity, y, depends on one value of another quantity, x, then y is a function of x.

You can pair the term numbers in a sequence with the values in the sequence. The rule for the relationship in the table below is multiply the term number by 3, or $3n$.

Term number	1	2	3	4
Value	3	6	9	12

A **function** is a relationship that assigns exactly one output value for each input value.

Example: Is the relationship in each table a function?

input	1	2	3	4
output	0	2	0	2

This is a function because each input has exactly one output. Note that two inputs can have the same output.

input	1	1	3	4
output	0	2	0	2

This is NOT a function because one input is associated with two outputs.

9-5 Using Tables, Rules, and Graphs
9-6 Interpreting Graphs

Math Understandings

- You can represent the relationship between two quantities using a table, a rule, or a graph.
- The graph of a function shows the relationship between inputs and outputs.

Example: Write and graph the function $y = 8x$ that relates the total amount earned, y, to the number of $8 shirts sold, x.

Make a table of values.

Number of Shirts Sold	1	10	20	30
Total Earned ($)	8	80	160	240

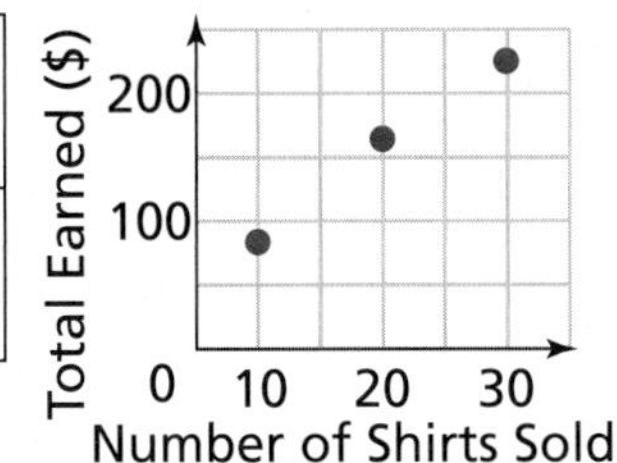

9-7 Simple and Compound Interest

Math Understandings

- Interest is money that you pay to use someone else's money, or money that you earn when someone else uses your money.

Simple Interest Formula	Compound Interest Formula
$I = prt$	$B = p(1 + r)^t$
I is interest, p is principal, r is annual interest rate, and t is time in years.	B is balance, p is principal, r is annual interest rate, and t is time in years.

The original amount you deposit or borrow is the **principal.** Interest calculated on the principal is **simple interest.** In accounts where money is not withdrawn, compound interest is paid on the original principal and on any interest that has been left in the account. The **balance** of an account is the principal plus the interest earned.

9-8 Transforming Formulas

Math Understandings

- When you solve an equation or formula for a variable, you use inverse operations and the properties of equality.

A **formula** is a rule that shows the relationship between two or more quantities. You can use the properties of equality to transform a formula and solve for a variable. When you transform a formula, you rewrite it in an equivalent form so that the variable for which you are solving is isolated on one side of the equation.

Example: Solve the formula $F = \frac{9}{5}C + 32$ for C.

$$F = \frac{9}{5}C + 32$$

$$F - 32 = \frac{9}{5}C$$ ← Subtract 32 from each side.

$$\frac{5}{9}(F - 32) = \frac{5}{9} \cdot \frac{9}{5}C$$ ← Multiply each side by $\frac{5}{9}$, the reciprocal of $\frac{9}{5}$.

$$\frac{5}{9}(F - 32) = C$$

Additional Professional Development Opportunities

Math Background Notes for Chapter 9: Every lesson has a Math Background in the PLAN section.

Research Overview, Mathematics Strands
Additional support for these topics and more is in the front of the Teacher's Edition.

LessonLab
LessonLab, a Pearson Education company, offers comprehensive, facilitated professional development designed to help teachers to improve student achievement. To learn more, please visit lessonlab.com.

Chapter 9 Resources

	9-1	9-2	9-3	9-4	9-5	9-6	9-7	9-8	For the Chapter
Print Resources									
L3 Practice	●	●	●	●	●	●	●	●	
L1 Adapted Practice	●	●	●	●	●	●	●	●	
L3 Guided Problem Solving	●	●	●	●	●	●	●	●	
L2 Reteaching	●	●	●	●	●	●	●	●	
L4 Enrichment	●	●	●	●	●	●	●	●	
L3 Daily Notetaking Guide	●	●	●	●	●	●	●	●	
L1 Adapted Daily Notetaking Guide	●	●	●	●	●	●	●	●	
L3 Vocabulary and Study Skills Worksheet	●		●	●		●		●	●
L3 Daily Puzzles	●	●	●	●	●	●	●	●	
L3 Activity Labs	●	●	●	●	●	●	●	●	
L3 Checkpoint Quiz			●			●			
L3 Chapter Project									●
L2 Below Level Chapter Test									●
L3 Chapter Test									●
L4 Alternative Assessment									●
L3 Cumulative Review									●
Spanish Resources ELL									
L3 Practice	●	●	●	●	●	●	●	●	●
L3 Vocabulary and Study Skills Worksheets	●		●	●		●		●	●
L3 Checkpoint Quiz			●			●			
L2 Below Level Chapter Test									●
L3 Chapter Test									●
L4 Alternative Assessment									●
L3 Cumulative Review									●
Transparencies									
Check Skills You'll Need	●	●	●	●	●	●	●	●	
Additional Examples	●	●	●	●	●	●	●	●	
Problem of the Day	●	●	●	●	●	●	●	●	
Classroom Aid					●				
Student Edition Answers	●	●	●	●	●	●	●	●	●
Lesson Quiz	●	●	●	●	●	●	●	●	
Test-Taking Strategies									●
Technology									
Interactive Textbook Online	●	●	●	●	●	●	●	●	●
StudentExpress™ CD-ROM	●	●	●	●	●	●	●	●	●
Success Tracker™ Online Intervention	●	●	●	●	●	●	●	●	●
TeacherExpress™ CD-ROM	●	●	●	●	●	●	●	●	●
PresentationExpress™ with QuickTake Presenter CD-ROM	●	●	●	●	●	●	●	●	●
ExamView® Assessment Suite CD-ROM	●	●	●	●	●	●	●	●	●
MindPoint® Quiz Show CD-ROM									●
Prentice Hall Web Site: PHSchool.com	●	●	●	●	●	●	●	●	●

Also available:

Prentice Hall Assessment System
- Progress Monitoring Assessments
- Skills and Concepts Review
- Test Prep Workbook

Other Resources
Algebra Readiness Tests
All-in-One Student Workbook
All-in-One Student Workbook, Adapted Version
Multilingual Handbook
Solution Key
Math Notes Study Folder
Spanish Cumulative Assessment

Where You Can Use the Lesson Resources

Here is a suggestion, following the four-step teaching plan, for how you can incorporate Differentiated Instruction Resources into your teaching.

	Instructional Resources L3	Differentiated Instruction Resources
1. Plan		
Preparation Read the Math Background in the Teacher's Edition to connect this lesson with students' previous experience. **Starting Class** **Check Skills You'll Need** Assign these exercises to review prerequisite skills. **New Vocabulary** Help students pre-read the lesson by pointing out the new terms introduced in the lesson.	**Math Background** **Math Understandings** **Transparencies & PresentationExpress™ with QuickTake Presenter CD-ROM** Check Skills You'll Need Problem of the Day **Resources** Vocabulary and Study Skills	**Spanish Support** ELL Vocabulary and Study Skills
2. Teach		
L3 Guided Instruction Use the Activity Labs to build conceptual understanding. Teach each Example. Use the Teacher's Edition side column notes for specific teaching tips, including Error Prevention notes. Use the Additional Examples found in the side column (and on transparency and PowerPoint) as an alternative presentation for the content. After each Example, assign the Quick Check exercise for that Example to get an immediate assessment of student understanding. Use the Closure activity in the Teacher's Edition to help students attain mastery of lesson content.	**Student Edition** Activity Lab **Resources** Daily Notetaking Guide Activity Lab **Transparencies & PresentationExpress™ with QuickTake Presenter CD-ROM** Additional Examples Classroom Aids **ExamView® Assessment Suite CD-ROM**	**Teacher's Edition** Every lesson includes suggestions for working with students who need special attention. L1 Special Needs L2 Below Level L4 Advanced Learners ELL English Language Learners **Resources** L1 Adapted Daily Notetaking Guide **Multilingual Handbook**
3. Practice		
Assignment Guide **Check Your Understanding** Use these questions to check students' understanding before you assign homework. **Homework Exercises** Assign homework from these leveled exercises in the Assignment Guide. **A** Practice by Example **B** Apply Your Skills **C** Challenge Test Prep and Mixed Review **Homework Quick Check** Use these key exercises to quickly check students' homework.	**Transparencies & PresentationExpress™ with QuickTake Presenter CD-ROM** Student Answers **Resources** Practice Guided Problem Solving Vocabulary and Study Skills Activity Lab Daily Puzzles **ExamView® Assessment Suite CD-ROM**	**Spanish Support** ELL Practice ELL Vocabulary and Study Skills **Resources** L1 Adapted Practice L4 Enrichment
4. Assess & Reteach		
Lesson Quiz Assign the Lesson Quiz to assess students' mastery of the lesson content. **Checkpoint Quiz** Use the Checkpoint Quiz to assess student progress over several lessons.	**Transparencies & PresentationExpress™ with QuickTake Presenter CD-ROM** Lesson Quiz **Resources** Checkpoint Quiz	**Resources** L2 Reteaching ELL Checkpoint Quiz Success Tracker™ Online Intervention **ExamView® Assessment Suite CD-ROM**

KEY **L1** Special Needs **L2** Below Level **L3** For All Students **L4** Advanced, Gifted **ELL** English Language Learners

CHAPTER 9

Patterns and Rules

Answers are in the back of the textbook.

For intervention, direct students to:

Exponents and Order of Operations
Lesson 2-1
Extra Skills and Word Problems Practice, Ch. 2

Evaluating Algebraic Expressions
Lesson 4-1
Extra Skills and Word Problems Practice, Ch. 4

Solving Two-Step Equations
Lesson 4-6
Extra Skills and Word Problems Practice, Ch. 4

Percents, Fractions, and Decimals
Lesson 6-2
Extra Skills and Word Problems Practice, Ch. 6

Spanish Vocabulary/Study Skills ELL

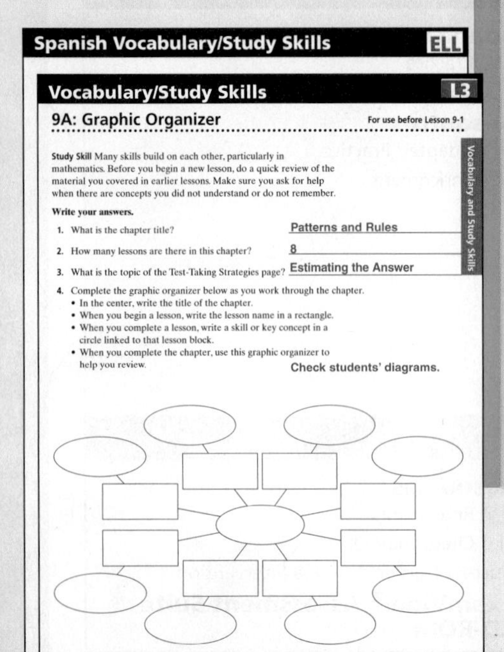
Vocabulary/Study Skills L3

9A: Graphic Organizer — For use before Lesson 9-1

Study Skill Many skills build on each other, particularly in mathematics. Before you begin a new lesson, do a quick review of the material you covered in earlier lessons. Make sure you ask for help when there are concepts you did not understand or do not remember.

Write your answers.

1. What is the chapter title? Patterns and Rules
2. How many lessons are there in this chapter? 8
3. What is the topic of the Test-Taking Strategies page? Estimating the Answer
4. Complete the graphic organizer below as you work through the chapter.
   - In the center, write the title of the chapter.
   - When you begin a lesson, write the lesson name in a rectangle.
   - When you complete a lesson, write a skill or key concept in a circle linked to that lesson block.
   - When you complete the chapter, use this graphic organizer to help you review.

Check students' diagrams.

CHAPTER 9 Patterns and Rules

What You've Learned

- In Chapter 4, you wrote algebraic expressions and equations to represent patterns and real-world situations.
- In Chapter 6, you solved application problems involving percents.

Check Your Readiness

GO for Help

For Exercises	See Lesson
1–4	2-1
5–8	4-1
9–14	4-6
15–19	6-2

Exponents and Order of Operations

Simplify.

1. $2^3 \cdot 2 - 4^2$ 0
2. $2^3 \cdot (2 - 4)^2$ 32
3. $(3 - 2)^2 - 2^2$ −3
4. $4^3 + 4 \div 4$ 65

Evaluating Algebraic Expressions

Evaluate each expression using $r = 4$, $s = -2$, and $t = 5.1$.

5. $3r - t$ 6.9
6. rst −40.8
7. $8s^2 + rt$ 52.4
8. $1.5(1 + s)^r$ 1.5

Solving Two-Step Equations

Solve each equation.

9. $3x - 1 = 14$ 5
10. $10 + 3n = 25$ 5
11. $4(b - 3) = 7$ $4\frac{3}{4}$
12. $\frac{2}{3}n - 10 = 14$ 36
13. $\frac{x}{7} = 49$ 343
14. $1.5 + \frac{4}{5}a = 21$ $24\frac{3}{8}$

Percents, Fractions, and Decimals

Write each percent as a decimal.

15. 4% 0.04
16. 12% 0.12
17. 3.58% 0.0358
18. 4.05% 0.0405
19. 10.3% 0.103

Chapter 9 Overview

In this chapter, students study patterns, sequences, and functions as they interpret and use tables, rules, graphs, and formulas. They draw upon their knowledge of percents and decimals to find simple and compound interest.

Activating Prior Knowledge

In this chapter, to find and use patterns to solve problems, students will use their number sense and build on their knowledge of equations, expressions, decimals, fractions, and percents. Ask questions such as:

- *How do you find 20% of a number?* **Multiply the number by 0.2 or $\frac{1}{5}$.**
- *What is the value of the expression* $y + 3x + 2$ *when* $y = 4$ and $x = 12$? **42**

What You'll Learn Next

- In this chapter, you will describe arithmetic and geometric sequences.
- You will represent patterns using tables, rules, and graphs.
- You will find simple and compound interest.

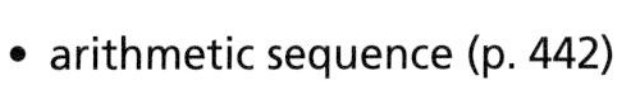

Key Vocabulary

- arithmetic sequence (p. 442)
- balance (p. 469)
- compound interest (p. 469)
- conjecture (p. 443)
- formula (p. 472)
- function (p. 452)
- geometric sequence (p. 442)
- principal (p. 468)
- sequence (p. 442)
- simple interest (p. 468)

Problem Solving Application On pages 482 and 483, you will work an extended activity on graphs.

Activity Lab

Choosing Scales and Intervals

Students learn to choose appropriate scales and intervals to accurately display data on a graph. This will prepare them to work with graphs in Lesson 9-1.

Guided Instruction

Display a graph and use it to demonstrate the meanings of *scale* and *interval.* Point out that a graph has two scales, horizontal and vertical. Ask questions such as:

- *Why is it important to find the right scale and interval to display data accurately?* **A graph with the right scale and intervals shows all the relevant data at a level of detail that makes it easy to see the relationship between sets of data.**
- *Why do we divide to find the scale?* **to find the number of intervals and how far apart they should be**

Exercises
Have students work independently on the Exercises. Have students form small groups and compare graphs when they have finished.

Alternative Method
Some students may find it easier to make the calculations using a calculator. Make sure they are available to students who want them.

Resources

- Activity Lab 9-1: Patterns in Data
- graph paper
- calculator

9-1a Activity Lab — Data Analysis

Choosing Scales and Intervals

To display data on a graph, you need to choose scales and intervals. A graph includes two *scales*, the horizontal axis and the vertical axis. An *interval* is the difference between the values on a scale.

EXAMPLE

Graph the data in the table at the right.

Interest on Savings

Amount Saved ($)	Interest Earned ($)
200	8
350	14
500	20
750	30
900	36

Step 1 Choose the scales and intervals.
Use the horizontal scale for the amount saved. Use the vertical scale for the interest earned. Start both scales at 0. Graphs that have from 6 to 10 intervals are easy to read. Since the greatest amount saved is $900, the horizontal scale needs to range from $0 to at least $900. Divide 900 by a factor from 6 to 10. Choose 9, since 900 is divisible by 9.

$900 \div 9 = 100$ ← **Divide the greatest amount by a compatible number.**

Use 9 intervals of $100 for the horizontal scale.
Since the greatest interest earned is $36, the vertical scale needs to range from $0 to at least $36. Divide 36 by a factor from 6 to 10.

$36 \div 9 = 4$ ← **Divide the greatest amount by a compatible number.**

Use 9 intervals of $4 for the vertical scale.

Step 2 Use points to represent the data.
The red dashes show how to plot the point representing interest earned of $14 for an amount saved of $350.

Interest on Savings

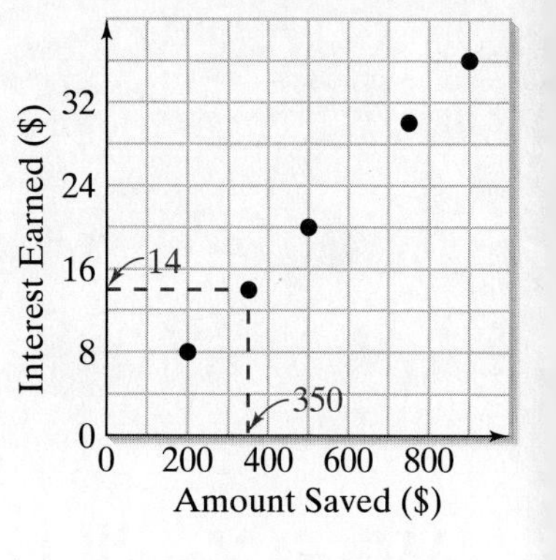

Exercises

1. a. **Number Sense** Use the table at the right. What interval can you use for time? For distance? **Answers may vary. Sample: 8 intervals of 10; 6 intervals of 10**
   b. Graph the data. **See margin.**

Travel Speed

Time (min)	Distance (mi)
0	0
20	15
40	30
60	45
80	60

2. a. Graph the data in the Example using a vertical interval of $5.
   b. **Reasoning** Which interval is easier to use, $4 or $5? Explain.
   2a. See margin.
   b. An interval of $4; more of the numbers are divisible by 4 than by 5.

1b. See back of book.
2a. See back of book.

9-1 Patterns and Graphs

Check Skills You'll Need

1. Vocabulary Review Is a *repeating decimal* a rational number? Explain. **1–5. See back of book.**

Graph and label each point.

2. 7 **3.** $3\frac{1}{2}$

4. 2.3 **5.** 0.8

GO for Help Lesson 2-7

What You'll Learn

To graph data and to use graphs to make predictions

Why Learn This?

You may have heard that "a picture is worth a thousand words." Graphs can help you see patterns in data. The table at the right shows the conversion of yards to inches. A graph of the same data can be easier to understand.

Yards and Inches

Number of Yards	Number of Inches
1	36
2	72
3	108
4	144
5	180

EXAMPLE Graphing Data

1 Graph the data in the table above.

The pattern in the first column of data suggests a horizontal interval of 1.

Graphs that have from 6 to 10 intervals are easy to read. The greatest value in the second column is 180. Divide 180 by a factor from 6 to 10. Choose 9, since 180 is divisible by 9.

$180 \div 9 = 20$ ← **Divide the greatest amount by a compatible number.**

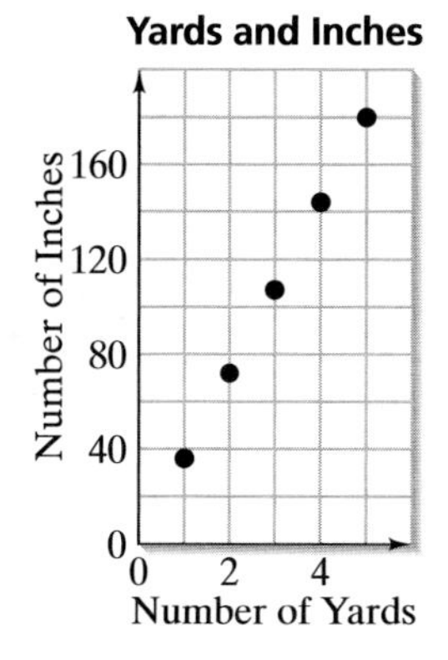

Use 9 intervals of 20 for the vertical scale. Use points to represent the data.

Quick Check

1. Graph the data in the table below. **See back of book.**

Yogurt Costs

Amount of Yogurt (c)	Price ($)
50	26
100	49
150	72
200	95

Differentiated Instruction Solutions for All Learners

Special Needs L1
If students have a difficult time drawing graphs, have them work with a partner. The partner can graph the data while the student makes decisions about interval sizes and checks the graph against the data table.

learning style: visual

Below Level L2
What intervals would you use to graph the following data?

Number of CDs	1	3	7
Cost ($)	4.99	14.97	34.93

Sample: 1 for *x*-axis, $5 for *y*-axis

learning style: visual

9-1

1. Plan

Objective
To graph data and to use graphs to make predictions

Examples
1 Graphing Data
2 Estimating on a Graph
3 Making a Prediction

Math Understandings: p. 434C

Math Background

The key to making a useful graph is choosing appropriate *scales* and *intervals.* The scale is the units used for an axis. The interval is the distance between values on a scale. For instance, a scale for an axis that represents distance might be feet or miles, and its interval might be 5 feet, 20 feet, or 1 mile.

More Math Background: p. 434C

Lesson Planning and Resources

See p. 434E for a list of the resources that support this lesson.

PowerPoint

Bell Ringer Practice

Check Skills You'll Need
Use student page, transparency, or PowerPoint. For intervention, direct students to:
Rational Numbers
Lesson 2-7
Extra Skills and Word Problems Practice, Ch. 2

2. Teach

Activity Lab

Use before the lesson.
Student Edition Activity Lab, Data Analysis 9-1a, Choosing Scales and Intervals, p. 436

All in One Teaching Resources
Activity Lab 9-1: Patterns in Data

Guided Instruction

Example 3
Students might extend graphs in the wrong direction. Remind them that data, such as temperature, can be extended into quadrants with negative numbers. Other types of data, such as height and length, cannot have negative values.

PowerPoint

Additional Examples

1. Graph the data in the table. **See back of book.**

No. of Boxes	20	40	60	80	100
Cost ($)	15	25	35	40	50

Use the graph for Exercises 2–3.

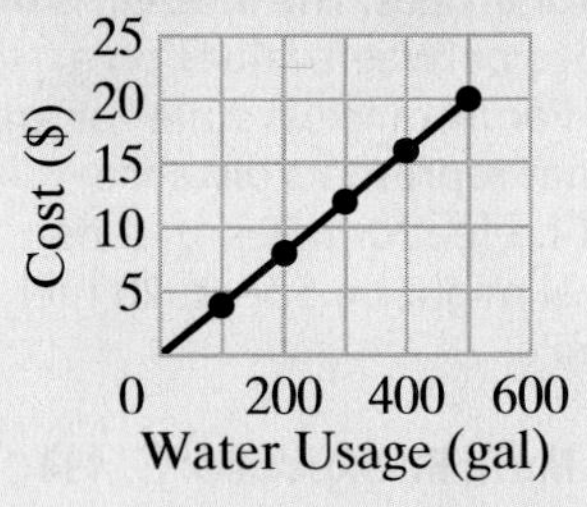

2. A water company charges a set fee for each gallon of water. How much do 250 gallons of water cost? **about $10**

3. Estimate the cost for 600 gallons of water. **about $24**

All in One Teaching Resources
- Daily Notetaking Guide 9-1 L3
- Adapted Notetaking 9-1 L1

Closure

- Explain how to choose an appropriate interval for a scale. **Sample: Subtract the smallest data value from the largest; divide by the number of intervals desired and round to a convenient number.**

You can use a graph to make estimates between data points.

EXAMPLE Estimating on a Graph

2 **Multiple Choice** The graph shows the cost of renting a personal watercraft. Which is the best estimate for the cost of a $3\frac{1}{2}$-hour rental?

Ⓐ about $200 Ⓑ about $225 Ⓒ about $250 Ⓓ about $350

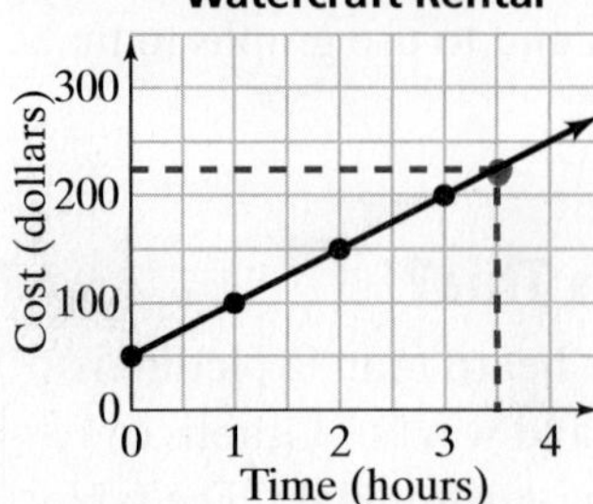

Draw lines to locate the value on the vertical axis that corresponds to $3\frac{1}{2}$ on the horizontal axis.

The cost is greater than $200, but less than $250. Estimate the cost as $225.

The cost of a $3\frac{1}{2}$-hour rental is about $225. The correct answer is B.

Quick Check

2. Use the graph in Example 2 to estimate the cost of a $1\frac{1}{2}$-hour rental. **about $125**

You can use a graph to make a prediction. Extend the graph and find a corresponding value on the appropriate axis.

EXAMPLE Making a Prediction

Test Prep Tip
Make sure the lines you use to locate a point are parallel to the scales of the graph.

3 The graph shows the relationship between Celsius and Fahrenheit temperatures. Estimate the Celsius temperature for 160°F.

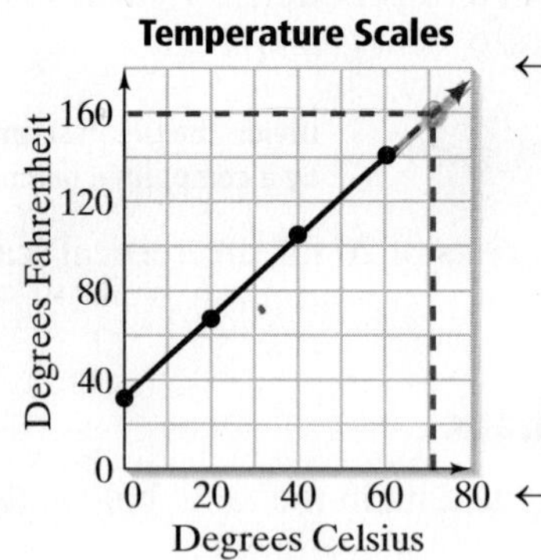

← Extend the graph beyond 160°F.

← For 160°F, the Celsius temperature is slightly more than 70°. Estimate the answer.

A temperature of 160°F is about 71°C.

Quick Check

3. Estimate the Fahrenheit temperature for 80°C. **about 175°F**

Advanced Learners L4	**English Language Learners** ELL
Given $x = -1$, 2, and 4 when $y = -6$, 6, and 14, what equation describes the pattern? $y = 4x - 2$ **learning style: verbal**	Help students make sense of the choices made in Example 1. Ask questions like, *Why choose an interval of one along the horizontal axis?* **Data suggests it.** *What would have happened if we had chosen 6 or 10 as a factor to divide 180 by?* **different interval sizes** **learning style: verbal**

Check Your Understanding

Shoveling Snow

Hours Worked	Wages ($)
2	18
4	36
5	45
6	54

1. **Vocabulary** How can a graph help you make a prediction? **A graph shows how the data are changing.**
2. Use the data in the table at the left. If you were to graph wages on the vertical axis, what scale and interval would you use? **Answers may vary. Sample: $0–$54; 6 intervals of $9**
3. If you were to graph hours worked on the horizontal axis, what scale and interval would you use? **Answers may vary. Sample: 0–8; 4 intervals of 2**

Choose a reasonable scale and interval to graph each set of data.
4–5. Answers may vary. Samples are given.

4. 70, 35, 55, 10, 43, 25, 80 **0–80; 8 intervals of 10**
5. 4,700; 2,000; 3,400; 1,650; 2,800 **0–5,000; 10 intervals of 500**

Homework Exercises

For more exercises, see Extra Skills and Word Problems.

GO for Help

For Exercises	See Examples
6–7	1
8–9	2
10–12	3

A **Graph the data in each table.** **6–7. See margin.**

6. **Plant Growth**

Age (yr)	Height (cm)
5	90
7	95
9	102
11	110

7. **Used Dirt Bike Prices**

Age (yr)	Price ($)
2	43
4	37
6	30

Estimate using your graphs from Exercises 6 and 7.

8. the height of an 8-year-old plant **about 100 cm**
9. the age of a bike that is sold for $40 **about 3 yr**

Electricity Charges

Cost ($): 0, 2, 4, 6
Electricity (kWh): 0, 25, 50, 75

Estimate using the graph at the left.

10. the cost of 85 kilowatt-hours of electricity **about $7**
11. the cost of 100 kilowatt-hours of electricity **about $8**
12. the number of kilowatt-hours that cost $6.50 **about 81 h**
13. **Writing in Math** Describe what a graph looks like when both sets of values increase. **Answers may vary. Sample: Each point is to the right and above the previous point.**

B **GPS**

14. **Guided Problem Solving** Graph the data in the table. Then estimate the missing value. **See margin.**
    - **Make a Plan** Draw a graph. Extend the graph to locate the missing value.
    - **Carry Out the Plan** For a value of 6 on the horizontal scale, the value on the vertical scale is ■.

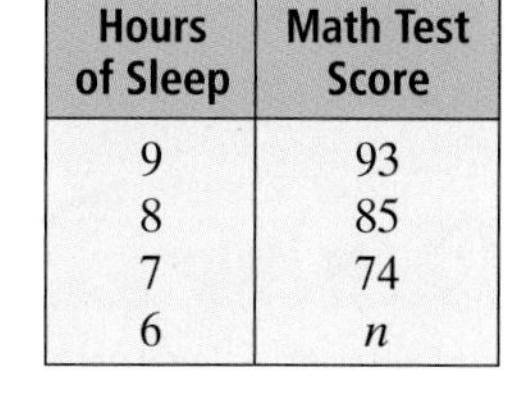

Hours of Sleep	Math Test Score
9	93
8	85
7	74
6	n

6.

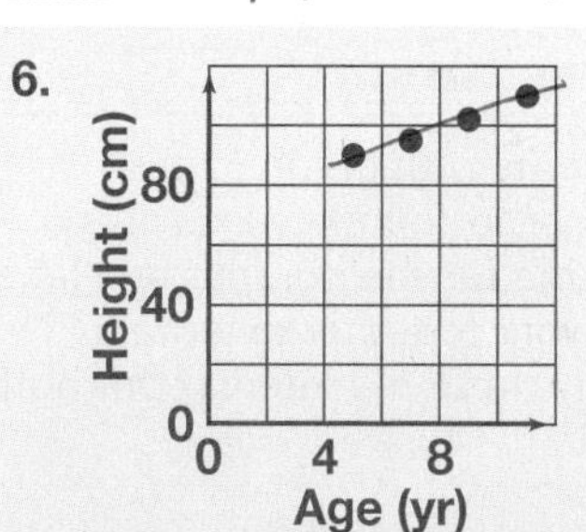

7.

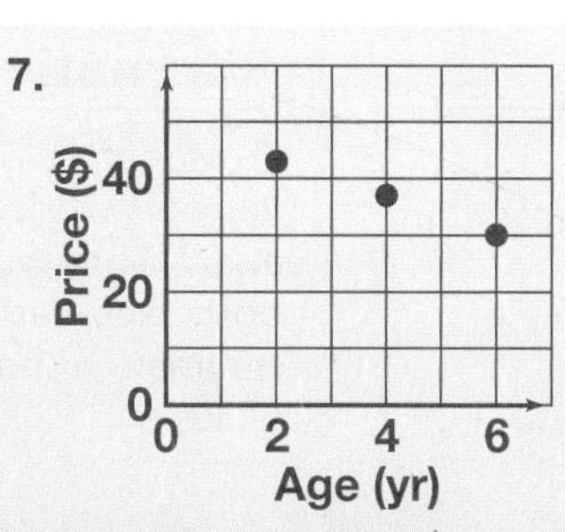

14. See back of book.

3. Practice

Assignment Guide

Check Your Understanding
Go over Exercises 1–5 in class before assigning the Homework Exercises.

Homework Exercises

A	Practice by Example	6–13
B	Apply Your Skills	14–18
C	Challenge	19
	Test Prep and Mixed Review	20–25

Homework Quick Check
To check students' understanding of key skills and concepts, go over Exercises 6, 8, 13, 17, and 18.

Differentiated Instruction Resources

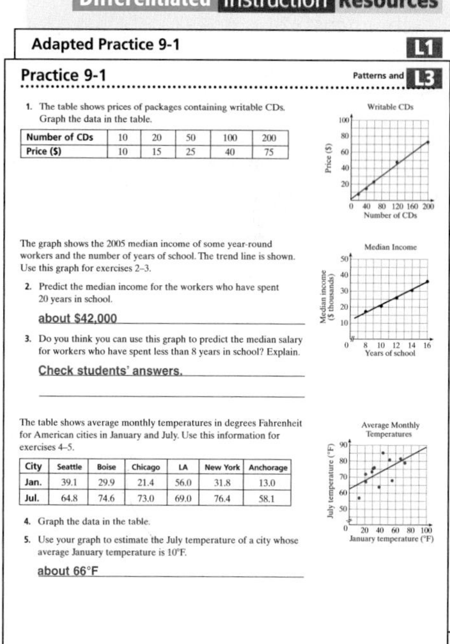

Adapted Practice 9-1 L1

Practice 9-1 Patterns and Graphs L3

1. The table shows prices of packages containing writable CDs. Graph the data in the table.

Number of CDs	10	20	50	100	200
Price ($)	10	15	25	40	75

The graph shows the 2005 median income of some year-round workers and the number of years of school. The trend line is shown. Use this graph for exercises 2–3.

2. Predict the median income for the workers who have spent 20 years in school. about $42,000
3. Do you think you can use this graph to predict the median salary for workers who have spent less than 8 years in school? Explain. Check students' answers.

The table shows average monthly temperatures in degrees Fahrenheit for American cities in January and July. Use this information for exercises 4–5.

City	Seattle	Boise	Chicago	LA	New York	Anchorage
Jan.	39.1	29.9	21.4	56.0	31.8	13.0
Jul.	64.8	74.6	73.0	69.0	76.4	58.1

4. Graph the data in the table.
5. Use your graph to estimate the July temperature of a city whose average January temperature is 10°F. about 66°F

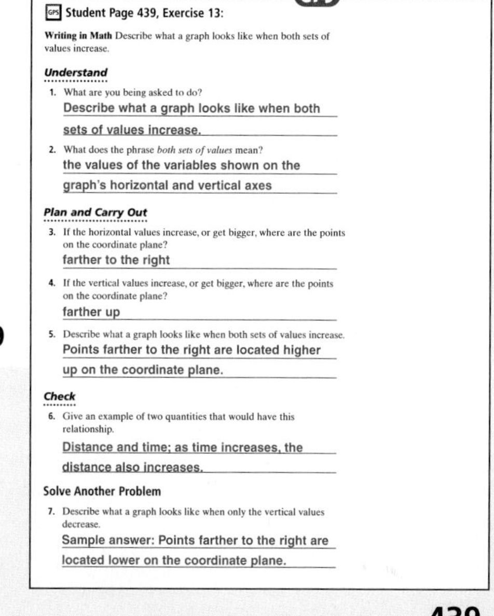

9-1 • Guided Problem Solving GPS L3

Student Page 439, Exercise 13:

Writing in Math Describe what a graph looks like when both sets of values increase.

Understand

1. What are you being asked to do? Describe what a graph looks like when both sets of values increase.
2. What does the phrase *both sets of values* mean? the values of the variables shown on the graph's horizontal and vertical axes

Plan and Carry Out

3. If the horizontal values increase, or get bigger, where are the points on the coordinate plane? farther to the right
4. If the vertical values increase, or get bigger, where are the points on the coordinate plane? farther up
5. Describe what a graph looks like when both sets of values increase. Points farther to the right are located higher up on the coordinate plane.

Check

6. Give an example of two quantities that would have this relationship. Distance and time; as time increases, the distance also increases.

Solve Another Problem

7. Describe what a graph looks like when only the vertical values decrease. Sample answer: Points farther to the right are located lower on the coordinate plane.

4. Assess & Reteach

PowerPoint

Lesson Quiz

1. Graph the data in the table.

Chain (ft)	15	20	30	40
Cost ($)	36	48	72	96

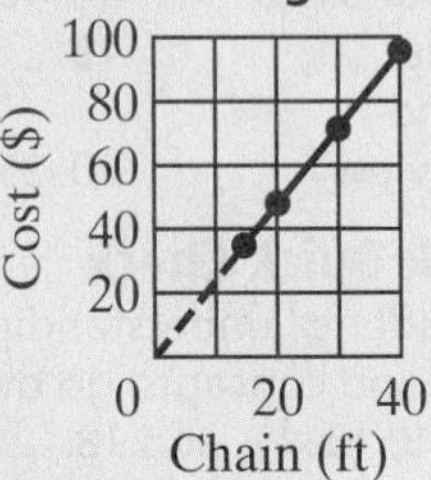

2. About how much do 25 ft of chain cost? **about $60**

3. About how much do 5 ft of chain cost? **about $12**

15–16. See back of book.
17a. See back of book.
18–19b. See back of book.

Enrichment 9-1 L4

Reteaching 9-1 Patterns and ... L2

Graphs can help you visualize the relationship between data. A graph includes two scales, one on the horizontal axis and one on the vertical axis. An interval is the distance between the values on a scale.

Graph the data in the table.

Step 1 Choose the scales and intervals.

Graph the data in the first column on the horizontal axis. Graph the data in the second column on the vertical axis.

Choose the interval for the scale on the horizontal axis. The greatest number of hours worked is 9. If each interval is 1 hour, then the number of intervals is 9 ÷ 1 = 9.

Choose the interval for the scale on the vertical axis. The greatest amount of money earned is $74.25. If each interval is $5, then the number of intervals is 74.25 ÷ 5 = 14.85, or 15 intervals.

Step 2 Draw the graph and plot the data. Estimate the position of data points that fall between intervals.

Hours Worked	Money Earned
5	$41.25
6	$49.50
7	$57.75
8	$66.00
9	$74.25

Graph the data in each table.

1.

Hours	Miles Run
2	10
2.8	14
3.2	16
3.6	18

2.

Gallons of Gas	Miles Driven
0.5	13
1.5	39
2.5	65
3	78

GO Online
Homework Video Tutor
Visit: PHSchool.com
Web Code: are-0901

Graph the data. Use the graph to estimate the missing value.
15–16. See margin.

15.

Time (h)	Temp. (°C)
1	12
2	15
5	24
8	n

16.

Gallons	Quarts
2	8
3	n
5	20
6	24

17. a. **Geometry** Graph the perimeters of squares with side lengths of 1, 2, 3, 4, and 5 in. Use the graph for parts (b) and (c). **See margin.**
    b. Estimate the side length of a square with perimeter 9.6 in. **2.4 in.**
    c. Estimate the perimeter of a square with side length 3.5 in. **14 in.**
    d. **Calculator** Test your estimates with a calculator. Were your estimates correct? **Check students' work.**

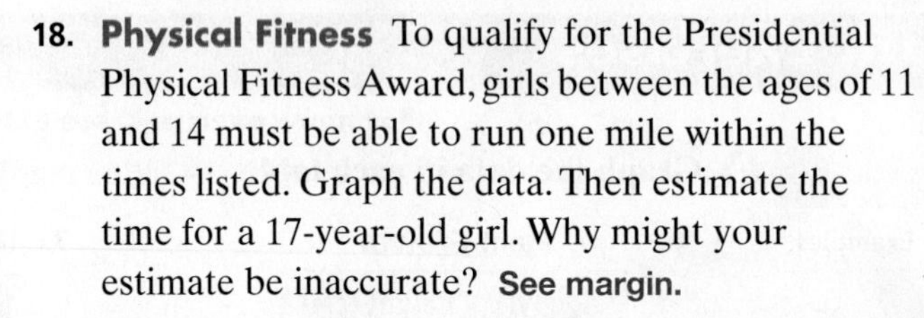

The Presidential Fitness Challenge consists of five fitness tests, including pull-ups.

18. **Physical Fitness** To qualify for the Presidential Physical Fitness Award, girls between the ages of 11 and 14 must be able to run one mile within the times listed. Graph the data. Then estimate the time for a 17-year-old girl. Why might your estimate be inaccurate? **See margin.**

Age (yr)	Time (min)
11	9.02
12	8.23
13	8.13
14	7.59

C 19. **Challenge** Suppose a neighbor will pay you $10 per week to wash windows. Another neighbor will give you $40 plus $7 per week.
    a. Make two tables, one for each neighbor, showing the amount you receive from each neighbor for 1, 2, 3, 4, and 5 weeks of work.
    b. Graph both sets of data on the same axes. **19a–b. See margin.**
    c. How much will you receive from each neighbor after 10 weeks? Which job would you prefer? **first neighbor: $100; second neighbor: $110; working for the second neighbor**

Test Prep and Mixed Review

Practice

Multiple Choice

20. Which of the following relationships is best represented by the data in the graph at the right? **C**
    Ⓐ Conversion of quarts to pints
    Ⓑ Conversion of gallons to cups
    Ⓒ Conversion of cups to fluid ounces
    Ⓓ Conversion of gallons to fluid ounces

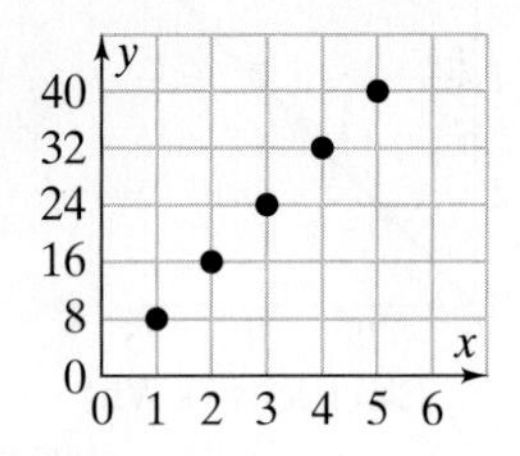

21. A small box measures 3 inches on each side. A larger box holds 125 of these small boxes. What is the volume of the larger box? **F**
    Ⓕ 3,375 in.^{3} Ⓖ 1,125 in.^{3} Ⓗ 375 in.^{3} Ⓙ 26 in.^{3}

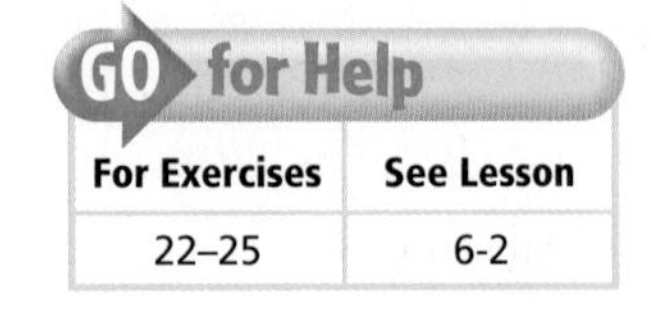

GO for Help

For Exercises	See Lesson
22–25	6-2

Write each percent as a decimal.

22. 38% **0.38** 23. 5% **0.05** 24. 150% **1.50** 25. 6.2% **0.062**

Test Prep

Resources
For additional practice with a variety of test item formats:
- Test-Taking Strategies, p. 477
- Test Prep, p. 481
- Test-Taking Strategies with Transparencies

Alternative Assessment

Provide pairs of students with graph paper. Partners work together to complete Exercises 6 and 7. They write a title for their graphs and label both axes. Partners work together to write a summary of how they chose the intervals for both scales.

9-2a Activity Lab

Hands On

Finding Patterns

ACTIVITY

Use pattern blocks or draw diagrams.

1. Make the next two figures in the pattern below.

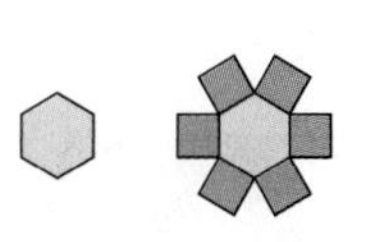 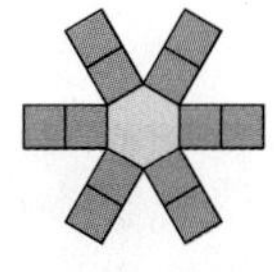

Figure 1 Figure 2 Figure 3

1.

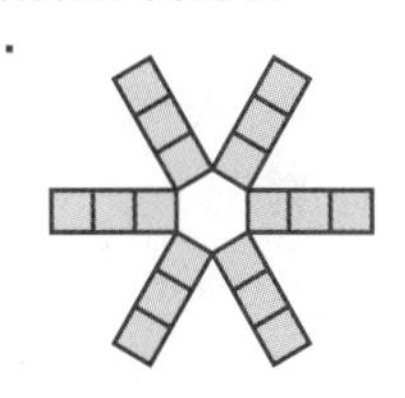 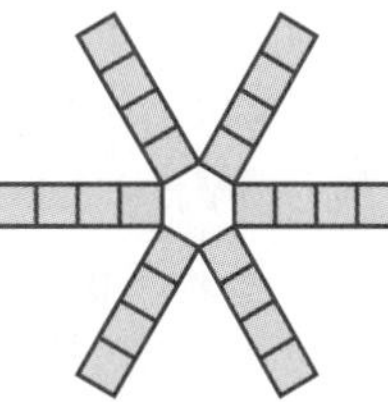

2. How many blocks do you add to each figure to make the next figure in the pattern? **6 blocks**
3. Copy and complete the table. **See margin.**
4. Describe any patterns you notice in your table.
   Answers may vary. Sample: Each figure has 6 more blocks than the previous figure.

Figure	1	2	3	4	5	6	7	8
Number of Blocks	1	■	■	■	■	■	■	■

Exercises

Use the pattern of cubes pictured below for Exercises 1–4.

1. a. **Number Sense** How many cubes will be in the fourth prism? **64**
   b. Build the fourth prism using colored cubes. **Check students' work.**
2. a. **Number Sense** How many cubes will be in the fifth prism? **125**
   b. Build the fifth prism using colored cubes. **Check students' work.**
3. Copy and complete the table at the right. **See margin.**
4. **Algebra** Write a formula that relates the prism number p to the number of blocks n. $n = p^3$

Prism	1	2	3	4	5
Number of Blocks	1	8	27	■	■

Activity Lab

Finding Patterns

Students use pattern blocks or drawings to study patterns and draw conclusions. This Activity will help them recognize patterns in number sequences in Lesson 9-2.

Guided Instruction

Before beginning the Activity, discuss patterns they have seen in everyday life. Display a quilt or other item containing a repeating pattern and have students identify the pattern. Ask questions, such as:

- *What is the pattern in the Activity?* **one central block plus six new blocks for each new figure**
- *What expression would you write to show that pattern?* **$1 + 6x$ or $6x + 1$**
- *What does the variable stand for?* **the number of times the pattern repeats**
- *How many blocks would there be in Figure 10?* **61**

Exercises

Have students work independently on the Exercises. When they have finished, have them compare answers with a partner to see if their patterns are the same.

Alternative Method

Some students may be able to see the pattern from the numbers in the table. Have them look at the numbers in the bottom row of the table and ask if they can see the pattern that way.

Resources

- Activity Lab 9-2: Patterns in Geometry
- pattern blocks

Activity
3.

Figure	1	2	3	4	5	6	7	8
Number of Blocks	1	7	13	19	25	31	37	43

Exercises
3.

Prism	1	2	3	4	5
Number of Blocks	1	8	27	64	125

9-2

1. Plan

Objective
To describe the patterns in arithmetic and geometric sequences and use the patterns to find terms

Examples
1 Describing an Arithmetic Sequence
2 Describing Geometric Sequence
3 Geometry

Math Understandings: p. 434C

Math Background

A *sequence* is a set of numbers that follows a pattern. In an *arithmetic sequence*, each term is derived from the one before it by consistently adding the same amount. In a *geometric sequence*, each term is derived from the one before it by multiplying the term by the same number. A sequence can be arithmetic, geometric, both, or neither.

More Math Background: p. 434C

Lesson Planning and Resources

See p. 434E for a list of the resources that support this lesson.

PowerPoint

Bell Ringer Practice

Check Skills You'll Need
Use student page, transparency, or PowerPoint. For intervention, direct students to:
Adding and Subtracting Integers
Lesson 1-7
Extra Skills and Word Problems Practice, Ch. 1

Algebra

9-2 Number Sequences

Check Skills You'll Need

1. Vocabulary Review The *additive inverse* of −8 is ■. 8

Add.

2. −3 + 3 0

3. −3 + 2 −1

4. −3 + 1 −2

5. −3 + 0 −3

Lesson 1-7

What You'll Learn

To describe the patterns in arithmetic and geometric sequences and use the patterns to find terms

New Vocabulary sequence, arithmetic sequence, geometric sequence, conjecture

Why Learn This?

Some patterns are found in nature. If you can recognize a pattern in a list of numbers, you can make predictions about how the list will continue.

A **sequence** is a set of numbers that follow a pattern. Each number in a sequence is called a *term*. The set of numbers 1, 3, 5, 7, 9, . . . has a pattern. If you add 2 to any number, you get the next number in the set.

In an **arithmetic sequence,** you find each term by adding a fixed number (called the common difference) to the previous term.

EXAMPLE Describing an Arithmetic Sequence

Vocabulary Tip
You pronounce *arithmetic sequence* as "ar ith MET ik SEE kwuns."

1 Describe the pattern in the sequence below. Then find the next three terms in the sequence.

Position	1	2	3	4
Value of Term	12	7	2	−3

+(−5) +(−5) +(−5) ← **Find the common difference.**

Start with 12 and add −5 repeatedly.

The next three terms are −8, −13, and −18.

Quick Check

1. Describe the pattern in the sequence. Find the next 3 terms.

Position	1	2	3	4	5
Value of Term	44	35	26	17	8

Start with 44 and add −9 repeatedly; −1, −10, −19.

In a **geometric sequence,** you find each term by multiplying the previous term by a fixed number (called the common ratio). In the sequence 2, 7, 24.5, 85.75, . . . you multiply each term by 3.5 to get the next term.

Differentiated Instruction Solutions for All Learners

Special Needs L1
Have students draw number lines and trace the common difference with their finger, between one term and the next for some of the arithmetic sequences. They can understand the idea of adding positive or negative numbers to get the next term.
learning style: tactile

Below Level L2
Find the next three terms in each sequence and give the rule.
2, 5, 8, 11 **14, 17, 20; Rule: +3**
1, 10, 100 **1,000, 10,000, 100,000; Rule: ×10**
10, 9, 8, 7 **6, 5, 4; Rule: +(−1)**
learning style: verbal

EXAMPLE Describing Geometric Sequences

2 Describe the pattern in 27, 9, 3, 1, . . . Find the next three terms.

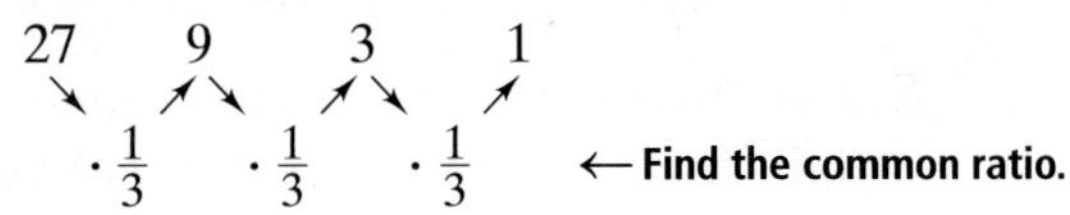

← **Find the common ratio.**

Start with 27 and multiply by $\frac{1}{3}$ repeatedly.

$1 \cdot \frac{1}{3} = \frac{1}{3}$ $\quad \frac{1}{3} \cdot \frac{1}{3} = \frac{1}{9}$ $\quad \frac{1}{9} \cdot \frac{1}{3} = \frac{1}{27}$ ← **Find the next three terms.**

The next three terms are $\frac{1}{3}$, $\frac{1}{9}$, and $\frac{1}{27}$.

Test Prep Tip

Dividing is the same as multiplying by a reciprocal.

Quick Check

2. Describe the pattern in 1,000; 100; 10; . . . Find the next 3 terms.
   Start with 1,000 and multiply by $\frac{1}{10}$ repeatedly; 1, $\frac{1}{10}$, $\frac{1}{100}$.

A sequence is neither arithmetic nor geometric if there is no common difference or common ratio.

A **conjecture** is a prediction that suggests what you expect to happen. When you describe a pattern in a sequence, you are using *inductive reasoning*. Check your results whenever possible.

EXAMPLE Geometry

GO for Help

For help with ratios, go to Lesson 5-1, Example 1.

3 Describe the pattern to find the number of circles in each figure. Is the resulting sequence *arithmetic, geometric, both,* or *neither*?

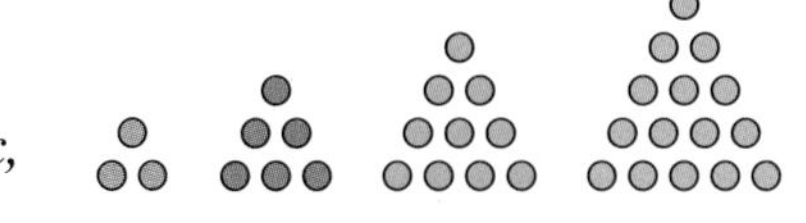

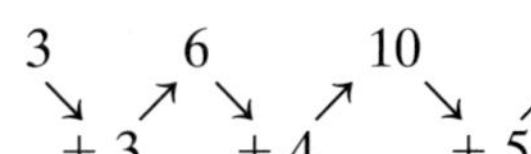

← **number of circles in each figure**

← **Look for a common difference or a common ratio.**

Start with 3. Add consecutive integers. First add 3, then add 4, and so on.

The sequence is neither arithmetic nor geometric. ← **conjecture**

Check Is there a common ratio?

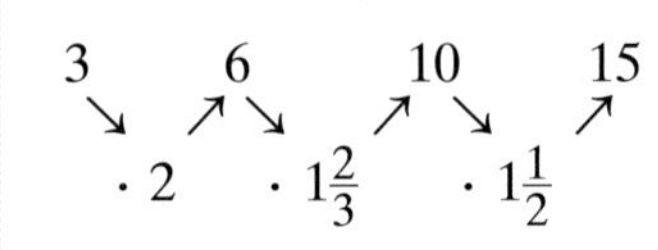

← **You cannot multiply by or add the same number to each term to find the next term. The sequence is neither arithmetic nor geometric. The conjecture is correct.**

Quick Check

3. Identify each sequence as *arithmetic, geometric, both,* or *neither*.
   a. 1, 2, 6, 24, . . . **neither**
   b. 2, 3, 6, 11, . . . **neither**
   c. 10, 9, 8, 7, . . . **arithmetic**

2. Teach

Activity Lab

Use before the lesson.
Student Edition Activity Lab, Hands On 9-2a, Finding Patterns, p. 441

All in One Teaching Resources
Activity Lab 9-2: Patterns in Geometry

Guided Instruction

Example 1
If your school has a long hallway with square tiles, use a section of hall that is at least 55 tiles long. Have students start on one end and place counters on tiles 44, 35, 26, 17, ... Relate the tiles to a number line to help them follow the pattern into negative numbers.

PowerPoint

Additional Examples

1 Write a rule for the sequence 100, 93, 86, 79 . . . . Find the next three terms. **Start with 100 and add −7 repeatedly. 72, 65, and 58**

2 Write a rule for the sequence 2, 6, 18, 54 . . . . Find the next three terms. **Start with 2 and multiply by 3 repeatedly. 162, 486, and 1,458**

3 Identify each sequence as *arithmetic, geometric, both,* or *neither.*
   a. 2, 8, 32, 128, . . . **geometric**
   b. 15, 16, 18, 21, . . . **neither**
   c. 40, 37, 34, . . . **arithmetic**

All in One Teaching Resources
- Daily Notetaking Guide 9-2 L3
- Adapted Notetaking 9-2 L1

Closure

- *What makes a sequence an arithmetic sequence?* **Each term is found by adding the same number to the previous term.**
- *What makes a geometric sequence?* **Each term is found by multiplying the same number by the previous term.**

Advanced Learners L4
Have students write a sequence that is both geometric and arithmetic, such as 8, 8, 8, 8, ... Have them give the two rules for each sequence. **arithmetic: +0; geometric: ×1**

learning style: visual

English Language Learners ELL
Differentiate between the terms in arithmetic and geometric sequences, and the terms in algebraic expressions. Students have learned that terms have variables in expressions, but that is not always the case in these sequences.

learning style: verbal

3. Practice

Assignment Guide

Check Your Understanding
Go over Exercises 1–5 in class before assigning the Homework Exercises.

Homework Exercises
A Practice by Example	6–19
B Apply Your Skills	20–30
C Challenge	31
Test Prep and Mixed Review	32–37

Homework Quick Check
To check students' understanding of key skills and concepts, go over Exercises 10, 18, 21, 26, and 28.

Differentiated Instruction Resources

Adapted Practice 9-2 L1

Practice 9-2 Number S L3

Identify each sequence as *arithmetic, geometric,* or *neither.* Write a rule for each sequence.

1. 2, 6, 18, 54, ... geometric; start with 2 and multiply by 3 repeatedly
2. 5, −10, 20, −40, ... geometric; start with 5 and multiply by −2 repeatedly
3. 3, 5, 7, 9, ... arithmetic; start with 3 and add 2 repeatedly
4. 5, 6, 8, 11, 15, ... neither
5. 1, 2, 6, 24, ... neither
6. 17, 16, 15, 14, ... arithmetic; start with 17 and add −1 repeatedly
7. 50, −50, 50, −50, ... geometric; start with 50 and multiply by −1 repeatedly
8. 1, 2, 4, 5, 10, 11, 22, ... neither

Find the next three terms in each sequence.

9. 15, −14, 13, −12, ... 11, −10, 9
10. 243, 81, 27, ... 9, 3, 1
11. 5, 12, 26, ... 54, 110, 222
12. 2, 5, 9, 14, ... 20, 27, 35

Write the first five terms in the sequence described by the rule. Identify the sequence as *arithmetic, geometric,* or *neither.*

13. Start with 2 and multiply by −3 repeatedly. 2, −6, 18, −54, 162; geometric
14. Start with 27 and add −9 repeatedly. 27, 18, 9, 0, −9; arithmetic
15. Start with 18 and multiply by 0.1, then by 0.2, then by 0.3, and so on. 18, 1.8, 0.36, 0.108, 0.0432; neither

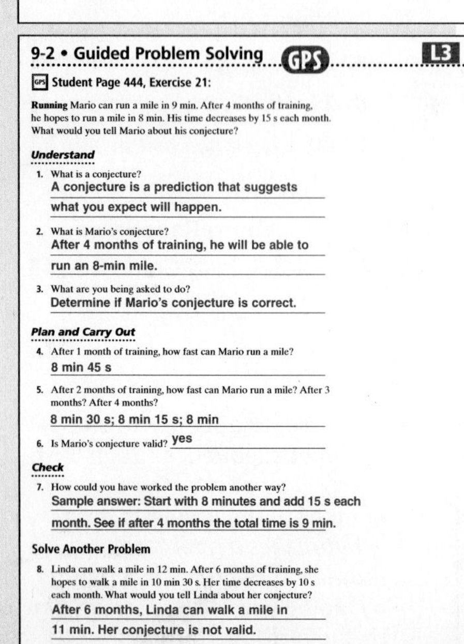

9-2 • Guided Problem Solving GPS L3

Student Page 444, Exercise 21:

Running Mario can run a mile in 9 min. After 4 months of training, he hopes to run a mile in 8 min. His time decreases by 15 s each month. What would you tell Mario about his conjecture?

Understand

1. What is a conjecture? A conjecture is a prediction that suggests what you expect will happen.
2. What is Mario's conjecture? After 4 months of training, he will be able to run an 8-min mile.
3. What are you being asked to do? Determine if Mario's conjecture is correct.

Plan and Carry Out

4. After 1 month of training, how fast can Mario run a mile? 8 min 45 s
5. After 2 months of training, how fast can Mario run a mile? After 3 months? After 4 months? 8 min 30 s; 8 min 15 s; 8 min
6. Is Mario's conjecture valid? yes

Check

7. How could you have worked the problem another way? Sample answer: Start with 8 minutes and add 15 s each month. See if after 4 months the total time is 9 min.

Solve Another Problem

8. Linda can walk a mile in 12 min. After 6 months of training, she hopes to walk a mile in 10 min 30 s. Her time decreases by 10 s each month. What would you tell Linda about her conjecture? After 6 months, Linda can walk a mile in 11 min. Her conjecture is not valid.

Check Your Understanding

1. Yes; dividing by −4 is the same as multiplying by $-\frac{1}{4}$.

1. **Vocabulary** A pattern for a sequence is described as *start with 12 and divide by −4 repeatedly*. Is this a geometric sequence? Explain.
2. Can two different arithmetic sequences have the same common difference? Explain. **Yes; they can begin with different numbers.**
3. What is the common difference in the sequence 35, 29, 23, 17, . . . ? **−6**

Find the missing term in each sequence.

4. 6, 13, 20, ■, 34, 41 **27**
5. 100, 50, 25, ■, 6.125 **12.5**

Homework Exercises

For more exercises, see Extra Skills and Word Problems.

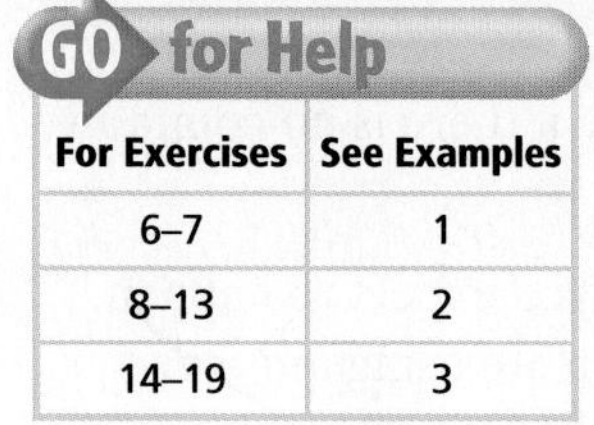

GO for Help

For Exercises	See Examples
6–7	1
8–13	2
14–19	3

A **Describe the pattern for each sequence. Then find the next three terms.** 6–13. See margin.

6.

Position	1	2	3	4
Value of Term	−8	−1	6	13

7.

Position	1	2	3	4
Value of Term	25	21	17	13

8. 1, 2, 4, 8, . . .
9. 2, −6, 18, −54, . . .
10. 600, −300, 150, . . .
11. $\frac{1}{2}, \frac{1}{4}, \frac{1}{8}, \frac{1}{16}, \ldots$
12. −2, 4, −8, 16, . . .
13. $\frac{1}{4}, \frac{1}{12}, \frac{1}{36}, \frac{1}{108}, \ldots$

Identify each sequence as *arithmetic, geometric, both,* or *neither.*

14. 2, 5, 10, 17, 26, . . . **neither**
15. 1, 4, 9, 16, 25, . . . **neither**
16. 7, 14, 28, 56, . . . **geometric**
17. −2, −2, −2, . . . **both**
18. 300, 60, 12, 2.4, . . . **geometric**
19. 84, 63, 42, 21, . . . **arithmetic**

B 

20. **Guided Problem Solving** A female bee has two biological parents—a female and a male. A male bee has only one biological parent, a female. The numbers of ancestors form a number sequence. Is the sequence *arithmetic, geometric, both,* or *neither?* **neither**

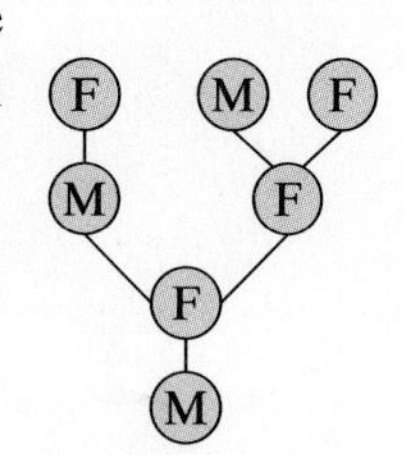

- **Make a Plan** Make a family tree of a male bee's ancestors. Show five generations. Find the number of ancestors in each generation.

21. Answers may vary. Sample: He may be correct, but he cannot keep decreasing his time by 15 s indefinitely.

21. **Running** Mario can run a mile in 9 min. After 4 months of training, he hopes to run a mile in 8 min. His time decreases by 15 s each month. What would you tell Mario about his conjecture? **See left.**
22. A sequence starts with 7. The common difference is *D*. Write expressions for the next three terms in the sequence. **7 + D, 7 + D + D, 7 + D + D + D**

6–13. See back of book.

Homework Video Tutor
Visit: PHSchool.com
Web Code: are-0902

27a. 17; 16

b. Blue tiles: start with 1, add 0, 4, 0, 4 and so on; yellow tiles: start with 0, add 4, 0, 4, 0 and so on.

The spiral of a chambered nautilus shell follows the Fibonacci sequence.

28. The sequence is both arithmetic and geometric. You can add 0 repeatedly or multiply by 1 repeatedly to get any term.

29. $123{,}456 \times 9 = 1{,}111{,}104$

Calculator **Make a conjecture about the next term in each sequence. Test your conjecture with a calculator.**

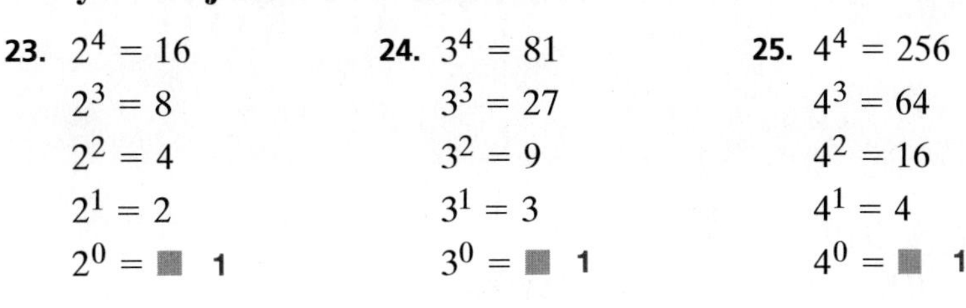

23.	24.	25.
$2^4 = 16$	$3^4 = 81$	$4^4 = 256$
$2^3 = 8$	$3^3 = 27$	$4^3 = 64$
$2^2 = 4$	$3^2 = 9$	$4^2 = 16$
$2^1 = 2$	$3^1 = 3$	$4^1 = 4$
$2^0 = \blacksquare$ 1	$3^0 = \blacksquare$ 1	$4^0 = \blacksquare$ 1

26. The Fibonacci sequence 1, 1, 2, 3, 5, 8, . . . occurs in nature. Find the ninth and tenth terms in the Fibonacci sequence. Is the Fibonacci sequence *arithmetic, geometric, both,* or *neither?* **34, 55; neither**

27. a. How many blue tiles will be in the ninth figure of the pattern? How many yellow tiles will there be?
b. Describe the pattern. **27a–b. See left.**

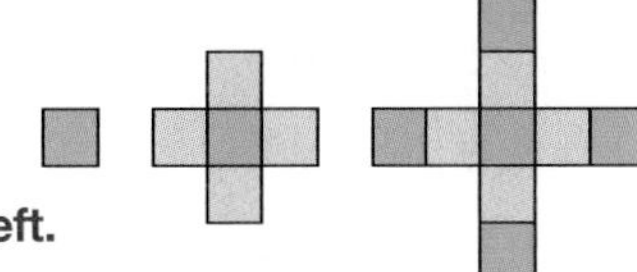

28. **Writing in Math** Every term in a sequence is 1. Is the sequence *arithmetic, geometric, both,* or *neither*? Explain. **See left.**

29. **Patterns** Look at the pattern below. Make a conjecture about the next term in the sequence. Test your conjecture with a calculator.
$123 \times 9 = 1{,}107$ $\quad 1{,}234 \times 9 = 11{,}106$ $\quad 12{,}345 \times 9 = 111{,}105$
See left.

30. A sequence starts with 2. The common ratio is R. Write expressions for the next three terms in the sequence. **$2R$, $2R^2$, $2R^3$**

Ⓒ 31. **Challenge** Make a conjecture about the next term in the sequence, and then find it. **−5**

Position	1	2	3	4	5	6
Value of Term	1	4	−1	6	−3	8

Test Prep and Mixed Review

Practice

Multiple Choice

32. Which sequence follows the rule $3n + 2$, where n is the position of a term in the sequence? **D**
Ⓐ 3, 9, 27, 81, . . .
Ⓑ 5, 7, 9, 11, . . .
Ⓒ 3, 6, 9, 12, . . .
Ⓓ 5, 8, 11, 14, . . .

33. Which expression does the model at the right represent? **F**
Ⓕ 9^2
Ⓖ 9^3
Ⓗ 2^3
Ⓙ 2^9

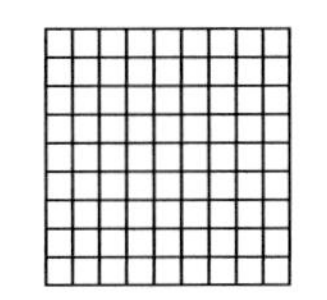

GO for Help

For Exercises	See Lesson
34–37	5-6

The scale on a map is 2 in. : 50 mi. Find the actual distance for each map distance. Round your answer to the nearest tenth of a mile.

34. 3 in. **75 mi** 35. $\frac{1}{2}$ in. **12.5 mi** 36. $1\frac{3}{4}$ in. **43.8 mi** 37. 5 in. **125 mi**

Alternative Assessment

Each student in a pair writes one arithmetic sequence, one geometric sequence, and one sequence that is neither arithmetic nor geometric. Partners exchange papers, write the rules for each other's sequences, and then extend the sequences by three terms.

Test Prep

Resources
For additional practice with a variety of test item formats:
- Test-Taking Strategies, p. 477
- Test Prep, p. 481
- Test-Taking Strategies with Transparencies

4. Assess & Reteach

Lesson Quiz

Identify each sequence as *arithmetic, geometric,* or *neither.* Give the next two terms in each sequence.

1. 2, 6, 18, 54, . . . **geometric; 162, 486**
2. 5, 6, 8, 11, . . . **neither; 15, 20**
3. 42, 37, 32, 27, . . . **arithmetic; 22, 17**
4. 1, 1, 2, 3, 5, 8, . . . **neither; 13, 21**

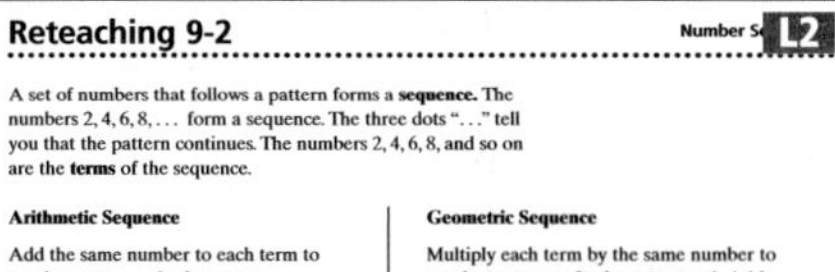

Reteaching 9-2 Number Se L2

A set of numbers that follows a pattern forms a **sequence.** The numbers 2, 4, 6, 8, . . . form a sequence. The three dots ". . ." tell you that the pattern continues. The numbers 2, 4, 6, 8, and so on are the **terms** of the sequence.

Arithmetic Sequence

Add the same number to each term to get the next term. In the sequence 2, 4, 6, 8, . . . , you add 2 to each term to get the next term.

Write a rule to describe this sequence, and find the next three terms.
5, 10, 15, 20, . . .

5 (+5) 10 (+5) 15 (+5) 20

Start with 5 and add 5 repeatedly.

To find the next three terms, add 5.

$20 + 5 = 25$
$25 + 5 = 30$
$30 + 5 = 35$

The next three terms are 25, 30, and 35.

Geometric Sequence

Multiply each term by the same number to get the next term. In the sequence 1, 4, 16, 64, . . . , you multiply each term by 4 to get the next term.

Write a rule to describe this sequence, and find the next three terms.
2, 4, 8, 16, . . .

2 (×2) 4 (×2) 8 (×2) 16

Start with 2 and multiply by 2 repeatedly.

To find the next three terms, multiply by 2.

$16 \times 2 = 32$
$32 \times 2 = 64$
$64 \times 2 = 128$

The next three terms are 32, 64, and 128.

Write a rule for each arithmetic sequence. Then find the next three terms.

1. 4, 7, 10, 13, . . . Start with 4 and add 3 repeatedly. 16, 19, 22
2. 2, 4, 6, 8, . . . Start with 2 and add 2 repeatedly. 10, 12, 14
3. 20, 35, 50, . . . Start with 20 and add 15 repeatedly. 65, 80, 95

Write a rule for each geometric sequence. Then find the next three terms.

4. 5, 25, 125, 625, . . . Start with 5 and multiply by 5 repeatedly. 3,125; 15,625; 78,125
5. 7, 49, 343, 2,401, . . . Start with 7 and multiply by 7 repeatedly. 16,807; 117,649; 823,543
6. 0.3, 0.9, 2.7, 8.1, . . . Start with 0.3 and multiply by 3 repeatedly. 24.3, 72.9, 218.7

Enrichment 9-2 Number L4

Patterns in Data

In 1202, Leonardo Fibonacci, an Italian mathematician, described a mathematical sequence that is named after him, the *Fibonacci Sequence.*

The first ten numbers in the sequence are given below.

1 1 2 3 5 8 13 21 34 55

1. How are the first two terms related to the third term?
The third term is equal to the sum of the first two terms.
2. How are the second and third terms related to the fourth term?
The fourth term is equal to the sum of the second and third terms.
3. How are the third and fourth terms related to the fifth term?
The fifth term is equal to the sum of the third and fourth terms.
4. What general rule can you make for finding each successive term?
Add the last two known terms to find the next term.

Additional patterns can be found in the *Fibonacci Sequence.*

5. Complete the table for the given three consecutive terms of the sequence. Then choose two additional sets of three consecutive terms of the sequence and complete the table for them. Sample answers are:

Consecutive terms	Product of first and third terms	Square of second term
1, 2, 3	3	4
2, 3, 5	10	9
3, 5, 8	24	25
5, 8, 13	65	64

6. What pattern do you notice in the numbers in the second and third columns for each set of consecutive terms? Explain.
They have a difference of one.
7. Is the smaller number always in the same column? Explain.
No, it alternates between columns.

9-3

1. Plan

Objective
To use tables to represent and describe patterns

Examples
1 Representing a Pattern
2 Finding the Value of a Term
3 Using a Table With a Sequence

Math Understandings: p. 434C

Math Background

In Lesson 9-1, graphs were drawn to illustrate patterns in data. Number patterns can also be represented in tables. You can find values for unknown quantities in a table by applying the rule that describes the relationship between values in the table.

More Math Background: p. 434C

Lesson Planning and Resources

See p. 434E for a list of the resources that support this lesson.

Bell Ringer Practice

Check Skills You'll Need
Use student page, transparency, or PowerPoint. For intervention, direct students to:
Solving Equations by Adding or Subtracting
Lesson 4-3
Extra Skills and Word Problems Practice, Ch. 4

Algebra

9-3 Patterns and Tables

Check Skills You'll Need

1. **Vocabulary Review** The *inverse operation* of subtraction is __?__. addition

Solve each equation.

2. $x + 4 = -6$ −10
3. $7 + t = 11$ 4
4. $y - 5 = -13$ −8
5. $a - (-3) = 17$ 14

GO for Help Lesson 4-3

What You'll Learn

To use tables to represent and describe patterns

Why Learn This?

At a market, it is common to display costs for different items. Sometimes making a table is the easiest way to organize data. Often you can make a table as your first step in solving a problem.

You can use a table to represent a pattern.

EXAMPLE Representing a Pattern

1 **Groceries** The table below shows the costs for different quantities of fish. Find the cost of 20 lb of fish.

Pounds of Fish	Price ($)	
1	6.50	$= 1 \times 6.5$
2	13.00	$= 2 \times 6.5$
3	19.50	$= 3 \times 6.5$
4	26.00	$= 4 \times 6.5$

← The values in the second column are 6.5 times the values in the first column.

To find the cost of 20 lb of fish, multiply 20 by 6.5.

$20 \cdot 6.5 = 130$

The cost of 20 lb of fish is $130.

Check for Reasonableness 20 lb is 10 times 2 lb. The cost for 2 lb is $13. $130 is 10 times $13, so the answer is reasonable.

Quick Check

Amount of Gas (gal)	Miles Driven
1	18.1
2	36.2
3	54.3
4	■
5	■

1. The table at the left shows the number of miles a car can travel using different amounts of gasoline. Copy and complete the table. Find the distance the car can travel using 15 gallons of gasoline. See back of book.

Differentiated Instruction Solutions for All Learners

Special Needs L1
If students have a difficult time copying the tables in the Homework Exercises, ask a partner to copy them. Alternatively, provide a copy of the page with the Homework Exercises.

learning style: visual

Below Level L2
The first term in a sequence is 4. The rule is n + 1. *Find the next 3 terms.* **5, 6, 7** *What is the 100th term?* **103**

learning style: verbal

Given a rule, you can find the value of a term using the position.

EXAMPLE Finding the Value of a Term

Test Prep Tip

You can eliminate choices as you find each value. Since the first term in the sequence is 2, you can eliminate choices C and D.

2 **Multiple Choice** Which sequence follows the rule $-3n + 5$, where n represents the position of a term in a sequence?

(A) 2, −1, −4, −7 . . .
(B) 2, 1, 4, 7 . . .
(C) −3, −6, −9, −12 . . .
(D) 8, 11, 14, 17 . . .

Position	$-3n + 5$	Value of Term
1	$-3(1) + 5$	2
2	$-3(2) + 5$	−1
3	$-3(3) + 5$	−4
4	$-3(4) + 5$	−7

← Substitute 1, 2, 3, and 4 for n.

The correct answer is choice A.

Quick Check

2. Use the rule $2n + 3$, where n represents the position of a term in a sequence. Find the first four terms in the sequence. **5, 7, 9, 11**

A table can help you write a variable expression to describe a sequence.

EXAMPLE Using a Table With a Sequence

Vocabulary Tip

Using a pattern to find any term in a sequence is often referred to as "finding the nth term."

3 Write an expression to describe the sequence 8, 16, 24, 32, . . . Then find the 10th term in the sequence.

8	16	24	32	. . .
Position 1	Position 2	Position 3	Position 4	and so on

Make a table that pairs the position of each term with its value.

Position	1	2	3	4	⋯	10
	↓·8	↓·8	↓·8	↓·8	↓·8	↓·8
Value of Term Sequence	8	16	24	32	⋯	■

The relationship is *Multiply the term number by 8.*

Let n = the term number. You can write the expression $n \cdot 8$, or $8n$.

$n \cdot 8 = 10 \cdot 8 = 80$ ← Substitute 10 for n to find the 10th term.

Quick Check

3. Write an expression to describe the sequence −8, −7, −6, −5, . . . Find the 10th term in the sequence. **$n - 9$; 1**

2. Teach

Activity Lab

Use before the lesson.

All in One Teaching Resources

Activity Lab 9-3: Patterns in Numbers

Guided Instruction

Example 1

Some students may benefit from "reasoning aloud" as they look at the table. Have them look for the common difference.

Example 2

For Quick Check 2, have students make a table to find the first four terms in the sequence. Then discuss how to check the value of the term.

PowerPoint

Additional Examples

1 Copy and complete this table.

Number of horses	1	3	5	9
Number of horse shoes	4	12	■ 20	■ 36

2 Find the values of the variables in the table below. **$s = 5$, $t = 84$**

N	2	s	6	12
P	14	35	42	t

3 Write an expression for the rule for 2, 4, 6, 8, 10 . . . . Find the hundredth term in the sequence. **$2n$; 200**

All in One Teaching Resources

- Daily Notetaking Guide 9-3 L3
- Adapted Notetaking 9-3 L1

Closure

- Explain how to find values of variables in a table. **Sample: First, figure out the relationship between the quantities in a row by trying out rules for each row. Then use that relationship to find the values of variables.**

Advanced Learners L4

Write a rule that shows the relationship between x and y. **$y = \frac{1}{2}x - 2$**

x	−2	0	4	10
y	−3	−2	0	3

learning style: visual

English Language Learners ELL

Have students work in pairs to come up with words to describe the patterns. One partner can write the words while the other translates to an algebraic rule, and then they can switch roles. This translation is important for developing mathematical language.

learning style: verbal

3. Practice

Assignment Guide

Check Your Understanding

Go over Exercises 1–4 in class before assigning the Homework Exercises.

Homework Exercises

A Practice by Example	5–18
B Apply Your Skills	19–26
C Challenge	27
Test Prep and Mixed Review	28–34

Homework Quick Check

To check students' understanding of key skills and concepts, go over Exercises 5, 12, 20, 24, and 26.

Differentiated Instruction Resources

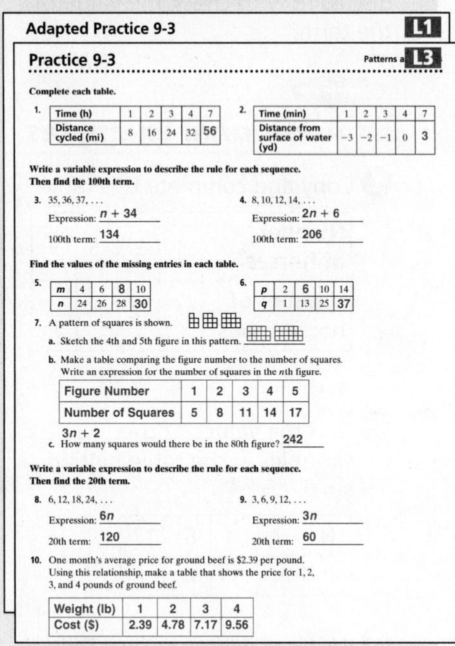

Adapted Practice 9-3 L1

Practice 9-3 Patterns a... L3

Complete each table.

1.

Time (h)	1	2	3	4	7
Distance cycled (mi)	8	16	24	32	56

2.

Time (min)	1	2	3	4	7
Distance from surface of water (yd)	−3	−2	−1	0	3

Write a variable expression to describe the rule for each sequence. Then find the 100th term.

3. 35, 36, 37, . . . Expression: $n + 34$ 100th term: 134

4. 8, 10, 12, 14, . . . Expression: $2n + 6$ 100th term: 206

Find the values of the missing entries in each table.

5.

m	4	6	8	10
n	24	26	28	30

6.

p	2	6	10	14
q	1	13	25	37

7. A pattern of squares is shown.

a. Sketch the 4th and 5th figure in this pattern.

b. Make a table comparing the figure number to the number of squares. Write an expression for the number of squares in the nth figure.

Figure Number	1	2	3	4	5
Number of Squares	5	8	11	14	17

$3n + 2$

c. How many squares would there be in the 80th figure? 242

Write a variable expression to describe the rule for each sequence. Then find the 20th term.

8. 6, 12, 18, 24, . . . Expression: $6n$ 20th term: 120

9. 3, 6, 9, 12, . . . Expression: $3n$ 20th term: 60

10. One month's average price for ground beef is $2.39 per pound. Using this relationship, make a table that shows the price for 1, 2, 3, and 4 pounds of ground beef.

Weight (lb)	1	2	3	4
Cost ($)	2.39	4.78	7.17	9.56

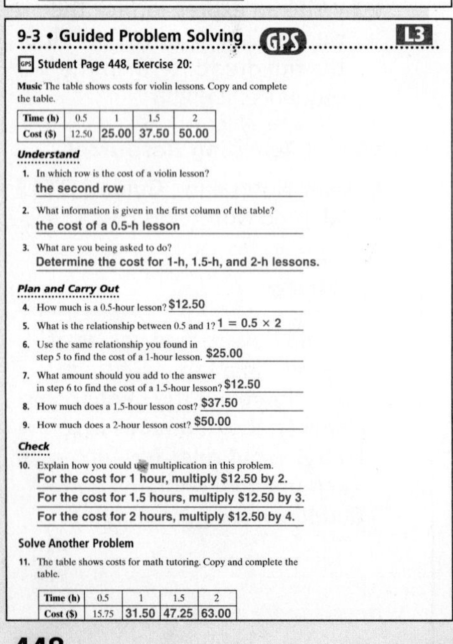

9-3 • Guided Problem Solving GPS L3

Student Page 448, Exercise 20:

Music The table shows costs for violin lessons. Copy and complete the table.

Time (h)	0.5	1	1.5	2
Cost ($)	12.50	25.00	37.50	50.00

Understand

1. In which row is the cost of a violin lesson? the second row
2. What information is given in the first column of the table? the cost of a 0.5-h lesson
3. What are you being asked to do? Determine the cost for 1-h, 1.5-h, and 2-h lessons.

Plan and Carry Out

4. How much is a 0.5-hour lesson? $12.50
5. What is the relationship between 0.5 and 1? $1 = 0.5 \times 2$
6. Use the same relationship you found in step 5 to find the cost of a 1-hour lesson. $25.00
7. What amount should you add to the answer in step 6 to find the cost of a 1.5-hour lesson? $12.50
8. How much does a 1.5-hour lesson cost? $37.50
9. How much does a 2-hour lesson cost? $50.00

Check

10. Explain how you could use multiplication in this problem. For the cost for 1 hour, multiply $12.50 by 2. For the cost for 1.5 hours, multiply $12.50 by 3. For the cost for 2 hours, multiply $12.50 by 4.

Solve Another Problem

11. The table shows costs for math tutoring. Copy and complete the table.

Time (h)	0.5	1	1.5	2
Cost ($)	15.75	31.50	47.25	63.00

Check Your Understanding

1. **Vocabulary** The expression $5n$ describes a sequence. When $n = 10$, the term position is ■ and the term value is ■. **10; 50**

2. **Mental Math** A sequence follows the rule $100n$, where n is the position of a term. Find the 10th term. **1,000**

Find the next three numbers in each sequence.

3. 4, 12, 36, 108, ■, ■, ■ **324; 972; 2,916**

4. 4, 12, 20, 28, ■, ■, ■ **36, 44, 52**

Homework Exercises

For more exercises, see Extra Skills and Word Problems.

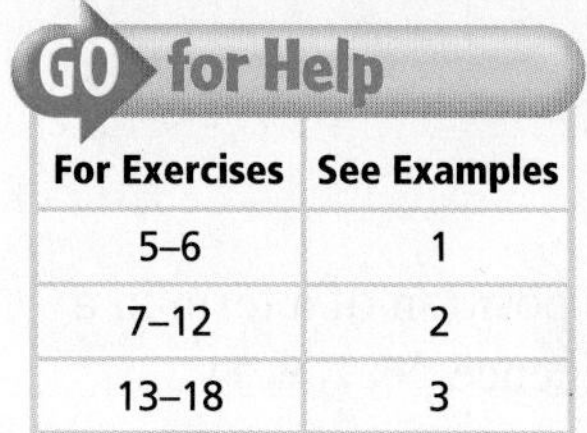

GO for Help

For Exercises	See Examples
5–6	1
7–12	2
13–18	3

A **Copy and complete each table.** **5–6. See margin.**

5.

Cans of Soup	Number of Servings
3	9
4	12
5	15
6	■
7	■

6.

Dozens of Beads	Cost ($)
1	0.48
2	0.96
3	1.44
4	■
5	■

For Exercises 7–12, *n* represents the position of a term in a sequence. Find the first four terms in each sequence.

7. $n + 3$ **4, 5, 6, 7**

8. $4n - 2$ **2, 6, 10, 14**

9. $5n$ **5, 10, 15, 20**

10. $6n + 1$ **7, 13, 19, 25**

11. $n \div 4$ **$\frac{1}{4}, \frac{1}{2}, \frac{3}{4}, 1$**

12. $n^2 - 3$ **−2, 1, 6, 13**

Write an expression to describe each sequence. Then find the 10th term.

13. 11, 22, 33, 44, . . . **$11n$; 110**

14. −19, −18, −17, −16, . . . **$n - 20$; −10**

15. $\frac{1}{2}, 1, 1\frac{1}{2}, 2, \ldots$ **$\frac{n}{2}$; 5**

16. −3, −6, −9, −12, . . . **$-3n$; −30**

17. −18, −36, −54, −72, . . . **$-18n$; −180**

18. 100, 200, 300, 400, . . . **$100n$; 1,000**

B **GPS** 19. **Guided Problem Solving** Suppose the average price for regular unleaded gasoline is $2.20 per gallon. Make a table that shows the price for 5, 10, 15, and 25 gallons of regular gasoline. **See left.**

- *Make a Table* that shows the price for 1, 2, and 3 gallons.
- What is the pattern in the table?

19.

Gallons of Gas	Price ($)
5	11
10	22
15	33
25	55

20. **Music** The table shows costs for violin lessons. Copy and complete the table. **See margin.**

Time (h)	0.5	1	1.5	2
Cost ($)	12.50	■	■	■

5.

Cans of Soup	Number of Servings
3	9
4	12
5	15
6	18
7	21

6.

Beads (doz)	Cost ($)
1	0.48
2	0.96
3	1.44
4	1.92
5	2.40

20–27a–b. See back of book.

Online Homework Video Tutor
Visit: PHSchool.com
Web Code: are-0903

Copy and complete each table. 21–22. See margin.

21.

Miles	Time (h)
10	0.4
20	0.8
30	1.2
40	■
50	■

22.

Change in a Parking Meter ($)	Time Allowed to Park (h)
0.25	0.5
0.50	1
0.75	■
1.25	■

The freezing point of water is 0°C, or 273 K.

23. **Temperature** The relationship between Kelvin (K) and Celsius (C) temperatures is $K = 273 + C$. Make a table of Kelvin temperatures for Celsius temperatures of 0°, 20°, 40°, 80°, and 120°. See margin.

24. a. Copy and complete the table.
b. **Writing in Math** Explain how to find y when you know x.
24a–b. See margin.

x	0	1	2	3	4	■
y	−1	2	5	8	■	14

25. **Geometry** Use the relationship between side length and area to make a table for the areas of squares with side lengths of 2, 3, 5, 8, 10, and 12 in. See margin.

26. Make a table showing the number of blue and yellow squares in each group. How many blue squares will be in group 10? See margin.

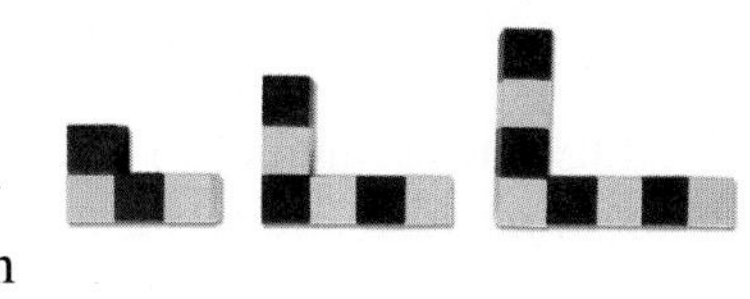

Ⓒ 27. **Challenge** Use the relationship $y = \frac{3}{2}x$.
a. Make a table that shows the values of y for $x = -2, -4, 2$, and 4.
b. Find the value of x when $y = 0$. 27a–b. See margin.

Test Prep and Mixed Review — Practice

Gridded Response

28. Isabel wants to cover her rectangular dining table with mosaic tiles that are 1 inch on each side. The table measures 3 feet by 4.5 feet. How many tiles does she need to cover the table? 1,944

29. Chencha has a part-time job after school. She earns $110.50 for 17 hours of work. How many dollars per hour does she earn? 6.50

30. Carlos plans to read every day for 6 days. He reads 9 pages on the first day, 18 pages on the second day, and 27 pages on the third day. If the pattern continues, how many pages will Carlos read on the sixth day? 54

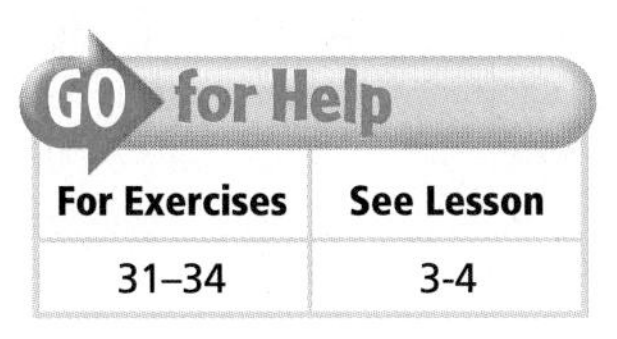
GO for Help

For Exercises	See Lesson
31–34	3-4

Find each product.

31. $\frac{1}{4}$ of 28 7
32. $\frac{3}{5} \cdot 25$ 15
33. $\frac{2}{7} \cdot \frac{21}{50}$ $\frac{3}{25}$
34. $4\frac{2}{3} \cdot 4\frac{1}{2}$ 21

4. Assess & Reteach

Lesson Quiz

Write a variable expression to describe the rule for each sequence. Then find the hundredth term.

1. 3, 6, 9, 12, . . . $3n$; 300
2. 14, 15, 16, 17, . . . $n + 13$; 113
3. 21, 42, 63, 84, . . . $21n$; 2,100
4. −10, −9, −8, −7, . . . $n - 11$; 89

Alternative Method
Have students solve Exercise 26 using the strategy *Work a Simpler Problem.*

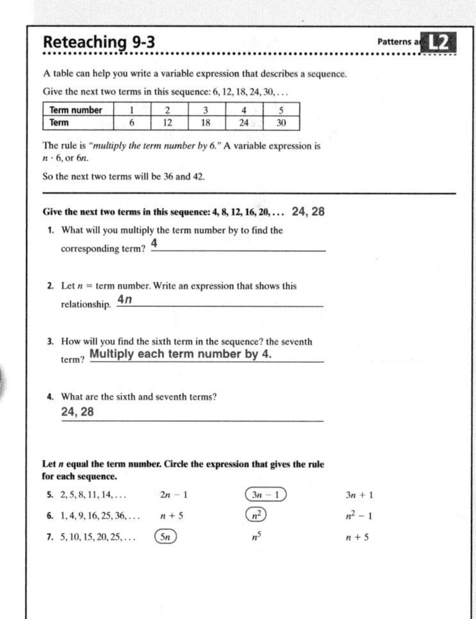

Reteaching 9-3 Patterns a[nd Tables] L2

A table can help you write a variable expression that describes a sequence.

Give the next two terms in this sequence: 6, 12, 18, 24, 30, . . .

Term number	1	2	3	4	5
Term	6	12	18	24	30

The rule is "*multiply the term number by 6.*" A variable expression is $n \cdot 6$, or $6n$.

So the next two terms will be 36 and 42.

Give the next two terms in this sequence: 4, 8, 12, 16, 20, . . . 24, 28

1. What will you multiply the term number by to find the corresponding term? 4
2. Let n = term number. Write an expression that shows this relationship. $4n$
3. How will you find the sixth term in the sequence? the seventh term? Multiply each term number by 4.
4. What are the sixth and seventh terms? 24, 28

Let n equal the term number. Circle the expression that gives the rule for each sequence.

5. 2, 5, 8, 11, 14, . . . $2n - 1$ ($3n - 1$) $3n + 1$
6. 1, 4, 9, 16, 25, 36, . . . $n + 5$ (n^2) $n^2 - 1$
7. 5, 10, 15, 20, 25, . . . ($5n$) n^5 $n + 5$

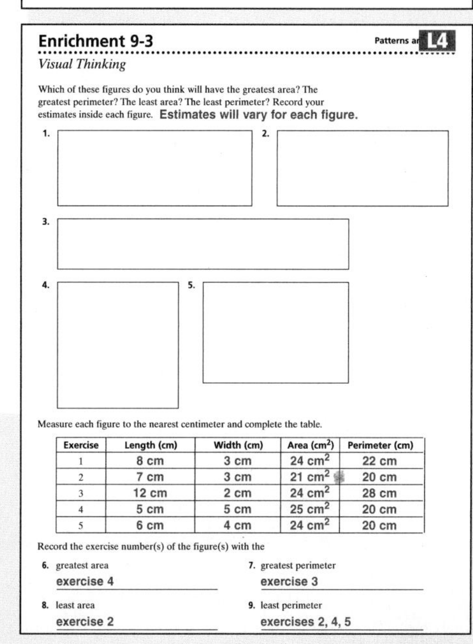

Enrichment 9-3 Patterns a[nd Tables] L4

Visual Thinking

Which of these figures do you think will have the greatest area? The greatest perimeter? The least area? The least perimeter? Record your estimates inside each figure. Estimates will vary for each figure.

1. 2. 3. 4. 5.

Measure each figure to the nearest centimeter and complete the table.

Exercise	Length (cm)	Width (cm)	Area (cm²)	Perimeter (cm)
1	8 cm	3 cm	24 cm²	22 cm
2	7 cm	3 cm	21 cm²	20 cm
3	12 cm	2 cm	24 cm²	28 cm
4	5 cm	5 cm	25 cm²	20 cm
5	6 cm	4 cm	24 cm²	20 cm

Record the exercise number(s) of the figure(s) with the

6. greatest area exercise 4
7. greatest perimeter exercise 3
8. least area exercise 2
9. least perimeter exercises 2, 4, 5

Alternative Assessment

Provide pairs of students with exercises similar to Exercises 13–18. Partners work together to make a table for each exercise and describe the rule. Then students use the expression to find the 20th term, the 50th term, and the 100th term.

Test Prep

Resources
For additional practice with a variety of test item formats:
- Test-Taking Strategies, p. 477
- Test Prep, p. 481
- Test-Taking Strategies with Transparencies

Checkpoint Quiz

Use this Checkpoint Quiz to check students' understanding of the skills and concepts of Lessons 9-1 through 9-3.

Resources

- All-in-One Teaching Resources Checkpoint Quiz 1
- ExamView CD-ROM
- Success Tracker™ Online Intervention

MATH AT WORK

Artists

This feature describes how math is used in a career field that may interest many students. It should generate enthusiasm about math among students.

Guided Instruction

Have volunteers read each paragraph. Discuss with students where they may have seen math in art. If possible, display a few pieces of art for examples. Ask questions, such as:

- *How would an artist use math to mix his or her materials?*
- *What other math skills might an artist use?*
- *Have you seen art with patterns in it? Did you like it? Why or why not?*

Checkpoint Quiz 1

Lessons 9-1 through 9-3

Use the table below for Exercises 1–3.

Tablespoons	Teaspoons
1	3
3	9
4	12
6	18

1. 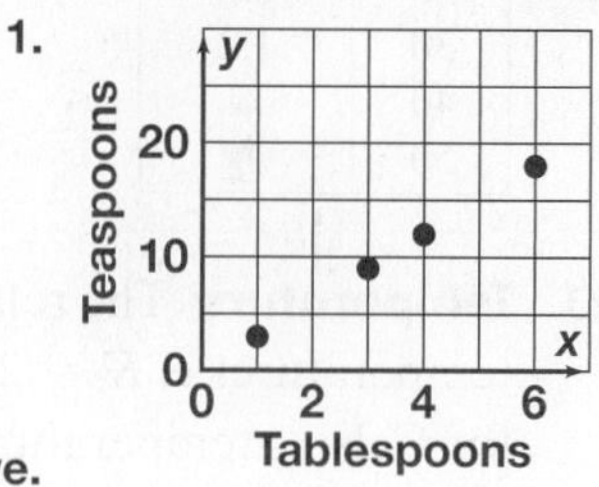

1. Graph the data in the table. **See above.**
2. Use the graph to estimate the number of teaspoons in 10 tablespoons. **30 teaspoons**
3. Use the graph to estimate the number of tablespoons in 20 teaspoons. **7 tablespoons**

Describe the pattern for each sequence. Then find the next three terms.

4. 7, 14, 21, 28, . . . **Start with 7 and add 7 repeatedly; 35, 42, 49.**
5. 250, 220, 190, 160, . . . **Start with 250 and add −30 repeatedly; 130, 100, 70.**
6. 2, 5, 11, 23, . . . **Start with 2, then add 3, 6, 12, 24 and so on; 47, 95, 191.**
7. −4, 12, −36, 108, . . . **Start with −4 and multiply by −3 repeatedly; −324, 972, −2,916.**
8. Identify each sequence in Exercises 4–7 as *arithmetic, geometric, both,* or *neither.* **arithmetic; arithmetic; neither; geometric**
9. Write an expression to describe the sequence 50, 100, 150, 200, . . . Then find the 10th term. **50*n*; 500**

MATH AT WORK

Artists

Artists use a variety of materials to make images, such as oils, watercolors, plaster, or clay. Recently, many artists have begun to use computers. Graphic artists work for businesses. Fine artists display their works in galleries or museums.

Artists use mathematics when they mix different materials, sell their work, or estimate costs of their materials. Artists also create designs with patterns.

Go Online PHSchool.com **For:** more information about artists. **Web Code:** arb-2031

9-4a Activity Lab

Generating Formulas From a Table

You can write a formula for the area of a square that is missing its corners.

ACTIVITY

The corners of each square shown below are shaded.

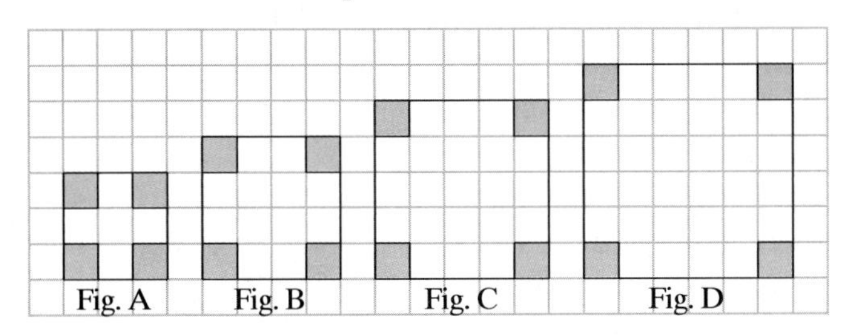

1. Copy and complete the table at the right to find the unshaded area of each square. **See margin.**
2. For each square, how is the area of the entire square related to the length of a side? $A = s^2$
3. How many corners are shaded in each square? **4 corners**
4. What pattern do you notice about the unshaded area of each square? **The area of the unshaded area is 4 less than the area of the entire square.**

Figure	A	B	C	D
Side Length	3			
Area of Square	9			
Unshaded Area	5			

Exercises

Find the area of a square that is missing its corners for each side length.

1. 7 units **45 units^2**
2. 8 units **60 units^2**
3. 10 units **96 units^2**
4. Suppose a square has a side length of n units and all four corners are shaded. Write a formula for the unshaded area of the square. $n^2 - 4$

Use your formula from Exercise 4 to find the unshaded area of each square with the given side length.

5. 100 units **9,996 units^2**
6. 500 units **249,996 units^2**
7. 1,000 units **999,996 units^2**

Activity Lab

Generating Formulas From a Table

Students use a table to generate formulas for area. This will prepare them to work with formulas in Lesson 9-4.

Guided Instruction

Before beginning the Activity, review the formula for the area of a square, $A = s^2$. Explain that you can use that formula with a table to find the area of parts of a square.

Ask:

- *How can you find the area of the unshaded area using the formula for area?* **Find the area of the square, subtract 4 units.**
- *If you shaded 4 squares in each corner, how could you find the area of the unshaded area?* **Find the total area, subtract 16.**

Exercises

Have students work independently on the Exercises. When they have finished, discuss the answers as a class and come to a conclusion about the area of the unshaded areas.

Alternative Method

Some students may prefer to use algebra tiles. Use different colors for the shaded and unshaded areas and physically remove the "shaded" tiles to help students see the connection

Resources

- Activity Lab 9-4: Function Rules
- algebra tiles
- Student Manipulatives Kit

1.

Figure	A	B	C	D
Side Length	3	4	5	6
Area of Square	9	16	25	36
Unshaded Area	5	12	21	32

9-4

1. Plan

Objective
To write and evaluate functions

Examples
1 Writing a Function Rule
2 Using Tables to Analyze Functions
3 Evaluating Functions

Math Understandings: p. 434C

Math Background

A function is a relationship in which an output value (y) depends on an input value (x) in such a way that only one output is possible for each input. This means that two different input values can have the same output value. But two different output values cannot have the same input value.

More Math Background: p. 434C

Lesson Planning and Resources

See p. 434E for a list of the resources that support this lesson.

Bell Ringer Practice

Check Skills You'll Need
Use student page, transparency, or PowerPoint. For intervention, direct students to:
Evaluating and Writing Algebraic Expressions
Lesson 4-1
Extra Skills and Word Problems Practice, Ch. 4

Algebra

9-4 Function Rules

Check Skills You'll Need

1. **Vocabulary Review** Why is 5 + 2 not an *algebraic expression*? **See below.**

Evaluate $-4x + 1$ for each value of x.

2. -2 **9**
3. 0 **1**
4. $\frac{1}{4}$ **0**
5. $-\frac{1}{4}$ **2**

GO for Help Lesson 4-1

Check Skills You'll Need

1. **The expression has no variable.**

What You'll Learn

To write and evaluate functions

New Vocabulary function

Why Learn This?

The distance you travel in a car depends on the driving time. When one quantity depends on another, you say that one is a *function* of the other. So distance is a function of time. You can use functions to help you make predictions.

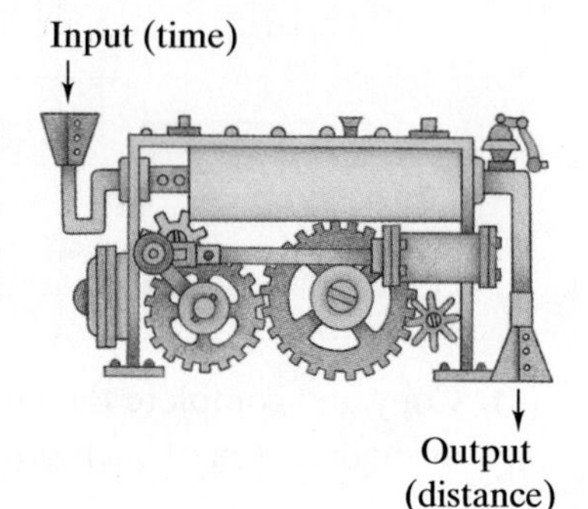

In the diagram at the right, an input goes through the "function machine" to produce an output.

A **function** is a relationship that assigns exactly one output value for each input value.

EXAMPLE Writing a Function Rule

1 **Cars** You are traveling in a car at an average speed of 55 mi/h. Write a function rule that describes the relationship between the time and the distance you travel.

You can *make a table* to solve this problem.

Input: time (h)	1	2	3	4
Output: distance (mi)	55	110	165	220

distance in miles = 55 · time in hours ← **Write the function rule in words.**

$d = 55t$ ← **Use variables *d* and *t* for distance and time.**

Quick Check

1. Write a function rule for the relationship between the time and the distance you travel at an average speed of 62 mi/h. $d = 62t$

Differentiated Instruction Solutions for All Learners

Special Needs L1
Provide students with counters or paper tiles to represent the numbers in the function tables. One color should represent the input, while the other represents the output. For example, 2, 3, and 4 red tiles and 4, 5, and 6 blue tiles in a table can represent the function $y = x + 2$.
learning style: tactile

Below Level L2
Use each function rule to find y when x = − 2, 0, 2, and 4.
$y = x + 2$ **0, 2, 4, 6**
$y = 2x$ **−4, 0, 4, 8**
learning style: visual

The variables x and y are often used to represent input and output. ou can describe the relationship between the values in the table in three ways.

Input x	Output y
1	4
2	5
3	6
4	7

Each output is 3 greater than the input.

$$\text{output} = \text{input} + 3$$
$$y = x + 3$$

EXAMPLE Using Tables to Analyze Functions

2 Write a rule for the function represented by each table.

a.

x	y
0	0
1	−4
2	−8
3	−12

When $x = 0$, $y = 0$. Each y equals −4 times x.

The function rule is $y = -4x + 0$, or $y = -4x$.

b.

x	y
0	−3
1	−1
2	1
3	3

When $x = 0$, $y = -3$. Each y equals 2 times x, plus −3.

The function rule is $y = 2x + (-3)$, or $y = 2x - 3$.

Quick Check

2. Write a rule for the function represented by the table below. $y = 4x + 1$

x	0	1	2	3
y	1	5	9	13

Given a function rule, you can evaluate the function for any input value.

EXAMPLE Evaluating Functions

3 Use the function $y = -3x + 5$. Find y for $x = 0, 1, 2$, and 3. Then make a table for the function.

$y = -3(0) + 5 = 5$ ← **Substitute 0, 1, 2, and 3 for x.**
$y = -3(1) + 5 = 2$ **List the values in a table.** →
$y = -3(2) + 5 = -1$
$y = -3(3) + 5 = -4$

x	$y = -3x + 5$
0	5
1	2
2	−1
3	−4

Quick Check

3. Use the function $y = 2x - 4$. Find y for $x = 0, 1, 2$, and 3. Then make a table for the function. **See left.**

3.

x	y
0	−4
1	−2
2	0
3	2

2. Teach

Activity Lab

Use before the lesson.
Student Edition Activity Lab 9-4a, Generating Formulas From a Table, p. 451

All in One Teaching Resources
Activity Lab 9-4: Function Rules

Guided Instruction

Example 2
Make sure students understand that the function rule must work for every row in the table. Have students test the rule on each row to help them understand this fact.

PowerPoint

Additional Examples

1 Suppose you can bicycle at an average speed of 10 mi/h. Write a function rule for the relationship between the time and distance you travel. $d = 10t$

2 Write a rule for the function represented by the table.

	0	1	2	3	4
	−2	2	6	10	14

$y = 4x - 2$

3 Use the function $y = 2x - 3$. Find y when $x = -1, 0, 2$, and 3. Then make a table for the function.

	−1	0	2	3
	−5	−3	1	3

All in One Teaching Resources
- Daily Notetaking Guide 9-4 **L3**
- Adapted Notetaking 9-4 **L1**

Closure

- *What is a function?* **a relationship that assigns one output value for each input value**
- *How do you evaluate a function?* **Substitute a given input value for x and simplify.**

Advanced Learners L4
Have students make a table and evaluate the function $y = (x + 1)^2 + 2x$ for 0, positive and negative integers, and positive and negative fractions. **Check students' work.**

learning style: visual

English Language Learners ELL
Explicitly state to students that when x and y are used as variables in function tables and equations, x is customarily the input, and y is the output.

learning style: verbal

3. Practice

Assignment Guide

Check Your Understanding
Go over Exercises 1–4 in class before assigning the Homework Exercises.

Homework Exercises

A	Practice by Example	5–19
B	Apply Your Skills	20–27
C	Challenge	28
	Test Prep and Mixed Review	29–32

Homework Quick Check
To check students' understanding of key skills and concepts, go over Exercises 9, 16, 21, 22, and 25.

Differentiated Instruction Resources

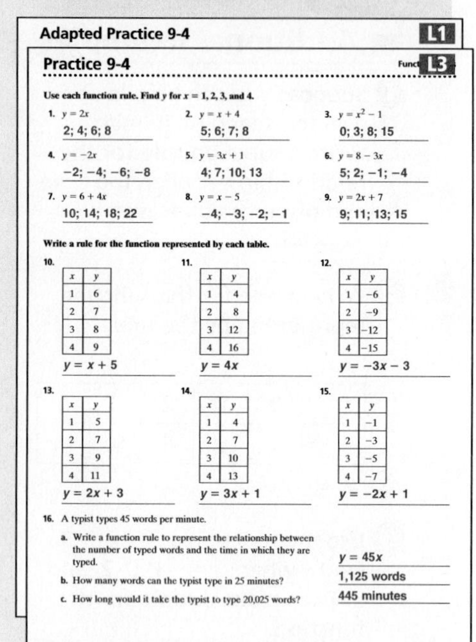

Adapted Practice 9-4 L1

Practice 9-4 Func... L3

Use each function rule. Find y for x = 1, 2, 3, and 4.

1. $y = 2x$ 2; 4; 6; 8
2. $y = x + 4$ 5; 6; 7; 8
3. $y = x^2 - 1$ 0; 3; 8; 15
4. $y = -2x$ −2; −4; −6; −8
5. $y = 3x + 1$ 4; 7; 10; 13
6. $y = 8 - 3x$ 5; 2; −1; −4
7. $y = 6 + 4x$ 10; 14; 18; 22
8. $y = x - 5$ −4; −3; −2; −1
9. $y = 2x + 7$ 9; 11; 13; 15

Write a rule for the function represented by each table.

10.

x	y
1	6
2	7
3	8
4	9

$y = x + 5$

11.

x	y
1	4
2	8
3	12
4	16

$y = 4x$

12.

x	y
1	−6
2	−9
3	−12
4	−15

$y = -3x - 3$

13.

x	y
1	5
2	7
3	9
4	11

$y = 2x + 3$

14.

x	y
1	4
2	7
3	10
4	13

$y = 3x + 1$

15.

x	y
1	−1
2	−3
3	−5
4	−7

$y = -2x + 1$

16. A typist types 45 words per minute.
   a. Write a function rule to represent the relationship between the number of typed words and the time in which they are typed. $y = 45x$
   b. How many words can the typist type in 25 minutes? 1,125 words
   c. How long would it take the typist to type 20,025 words? 445 minutes

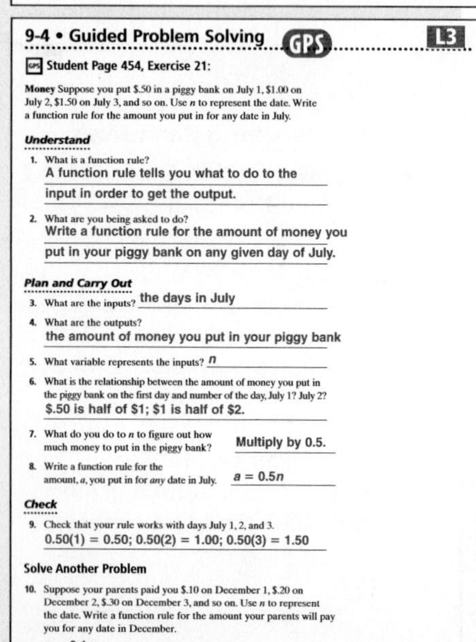

9-4 • Guided Problem Solving GPS L3

Student Page 454, Exercise 21:

Money Suppose you put \$.50 in a piggy bank on July 1, \$1.00 on July 2, \$1.50 on July 3, and so on. Use n to represent the date. Write a function rule for the amount you put in for any date in July.

Understand

1. What is a function rule? A function rule tells you what to do to the input in order to get the output.
2. What are you being asked to do? Write a function rule for the amount of money you put in your piggy bank on any given day of July.

Plan and Carry Out

3. What are the inputs? the days in July
4. What are the outputs? the amount of money you put in your piggy bank
5. What variable represents the inputs? n
6. What is the relationship between the amount of money you put in the piggy bank on the first day and number of the day, July 1? July 2? \$.50 is half of \$1; \$1 is half of \$2.
7. What do you do to n to figure out how much money to put in the piggy bank? Multiply by 0.5.
8. Write a function rule for the amount, a, you put in for *any* date in July. $a = 0.5n$

Check

9. Check that your rule works with days July 1, 2, and 3. $0.50(1) = 0.50; 0.50(2) = 1.00; 0.50(3) = 1.50$

Solve Another Problem

10. Suppose your parents paid you \$.10 on December 1, \$.20 on December 2, \$.30 on December 3, and so on. Use n to represent the date. Write a function rule for the amount your parents will pay you for any date in December. $a = 0.1n$

Check Your Understanding

1. **Vocabulary** Complete with *at least one* or *exactly one*: Every function pairs __?__ output value with each input value. **exactly one**

x	y
0	0
1	5
2	10
3	15

Use the function table at the left for Exercises 2–4.

2. What number does the function pair with 3? **15**
3. Describe the pattern in the table using words. **Each output is 5 times the input.**
4. Write a rule for the function represented by the table. **$y = 5x$**

Homework Exercises

For more exercises, see Extra Skills and Word Problems.

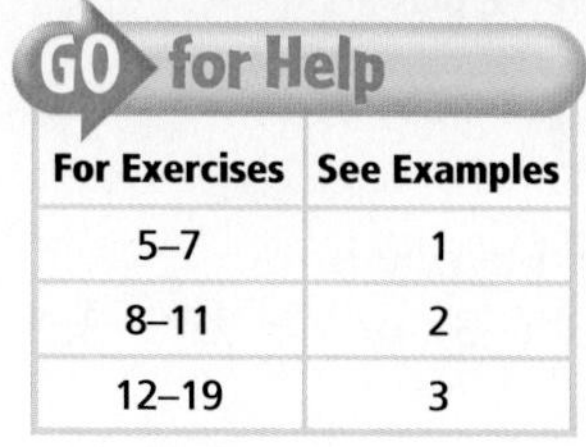

GO for Help

For Exercises	See Examples
5–7	1
8–11	2
12–19	3

A **Write a function rule for each relationship.**

5. the time t and distance d you travel at an average speed of 30 mi/h **$d = 30t$**
6. the number n of words you type and the time t it takes, if you type at a rate of 32 words/min **$n = 32t$**
7. the amount c of energy you burn and the time t you spend exercising, if you burn Calories at a rate of 12 Cal/min **$c = 12t$**

Write a rule for the function represented by each table.

8.

x	y
0	4
1	5
2	6
3	7

$y = x + 4$

9.

x	y
0	5
1	8
2	11
3	14

$y = 3x + 5$

10.

x	y
0	1
1	−8
2	−17
3	−26

$y = -9x + 1$

11.

x	y
0	0
1	−8
2	−16
3	−24

$y = -8x$

Use each function rule. Find y for $x = 0, 1, 2,$ and 3. Then make a table for the function. **12–19. See margin.**

12. $y = x + 2$
13. $y = 9 - x$
14. $y = 4x$
15. $y = x \div 2$
16. $y = -3x$
17. $y = 2x + 1$
18. $y = 4x - 2$
19. $y = x^2 + 1$

B 

20. **Guided Problem Solving** Write a function rule for the number n of inches in f feet. Then find the number of inches in 7 feet. **$n = 12f$; 84 in.**
   - How many inches are in 1 foot?
   - How many inches are in f feet?

21. **Money** Suppose you put \$.50 in a piggy bank on July 1, \$1.00 on July 2, \$1.50 on July 3, and so on. Use n to represent the date. Write a function rule for the amount you put in for any date in July. **$y = (0.50)n$**

12–19. See back of book.

26.

cm³	L
3	0.003
300	0.3
3,000	3

$L = \frac{c}{1{,}000}$

Visit: PHSchool.com
Web Code: are-0904

Write a rule for the function represented by each table.

22.

Laundry Loads	Cost ($)
1	2.75
2	5.50
3	8.25
4	11.00

$c = 2.75\ell$

23.

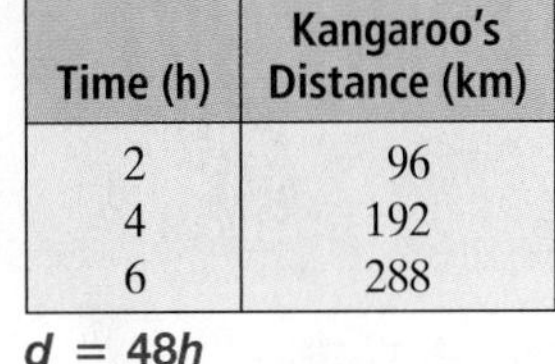

Time (h)	Kangaroo's Distance (km)
2	96
4	192
6	288

$d = 48h$

24. Reading A student can read 150 words in one minute.

a. Write a function rule to represent the relationship between the number of words and the time in which they are read. $w = 150m$

b. How many words can the student read in 8 minutes? **1,200 words**

c. How long would it take the student to read 2,850 words? **19 minutes**

25. Writing in Math Use the function table at the right. Describe the patterns you notice and explain how to find the function rule. **See left.**

x	y
0	0
1	$\frac{1}{2}$
2	1
3	$1\frac{1}{2}$

26. One cm^3 is equal to 1 mL. Use this relationship to make a table showing the number of liters in 3 cm^3, 300 cm^3, and 3,000 cm^3. Then write a function rule. **See margin.**

27. Use the rule $y = 2x^2 - 4$. Evaluate the function for $x = -0.25, -\frac{1}{2}$, and $1\frac{1}{4}$. **See left.**

C 28. Challenge Use the function machine at the right to make a table for integer inputs from −5 to 5. Which two input values result in an output of 22? **−5, 4**

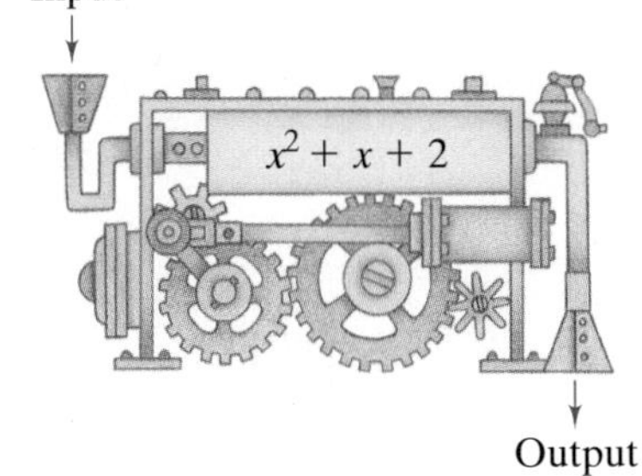

25. Answers may vary. Sample: The y-values increase by $\frac{1}{2}$ when the x-values increase by 1.

27. −3.875, −3.5, −0.875

Test Prep and Mixed Review

Practice

Multiple Choice

29. Which description shows the relationship between a term and n, its position in the sequence? **D**

Position	1	2	3	4	n
Value of Term	6.50	13	19.50	26	■

Ⓐ Add 6.50 to n.
Ⓑ Add 26 to n.
Ⓒ Multiply n by 26.
Ⓓ Multiply n by 6.50.

30. Which of the following *never* names a quadrilateral with four congruent sides and four congruent angles? **F**

Ⓕ Trapezoid Ⓖ Rectangle Ⓗ Rhombus Ⓙ Square

GO for Help

For Exercises	See Lesson
31–32	8-1

Choose a reasonable estimate. Explain your choice.

31. width of a nail: 1 mm or 1 m
1 mm; it is less than 1 m.

32. length of a car: 4 m or 4 km
4 m; it is less than 1 km.

Online lesson quiz, PHSchool.com, Web Code: ara-0904

4. Assess & Reteach

PowerPoint

Lesson Quiz

1. Make a table for the function $y = 3x - 2$. Find y for $x = 0, 1, 2,$ and 3.

	0	1	2	3
	−2	1	4	7

2. Write a rule for the function represented in the table.

	0	1	2	3
	4	6	8	10

$y = 2x + 4$

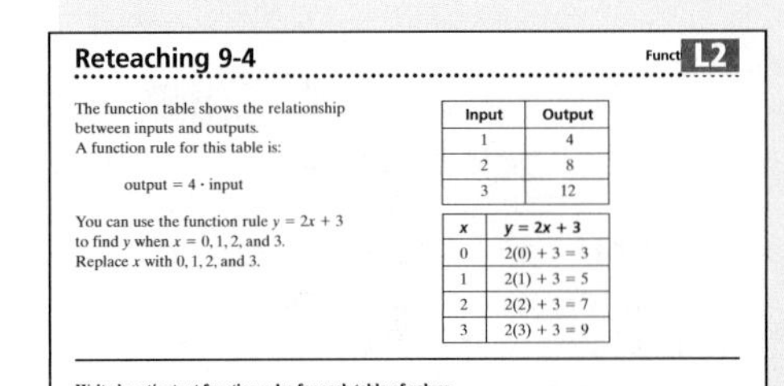

Reteaching 9-4 L2

The function table shows the relationship between inputs and outputs.
A function rule for this table is:

output = 4 · input

Input	Output
1	4
2	8
3	12

You can use the function rule $y = 2x + 3$ to find y when $x = 0, 1, 2,$ and 3.
Replace x with 0, 1, 2, and 3.

x	$y = 2x + 3$
0	2(0) + 3 = 3
1	2(1) + 3 = 5
2	2(2) + 3 = 7
3	2(3) + 3 = 9

Write input/output function rules for each table of values.

1.

Input	Output
3	6
4	8
5	10
6	12

output = 2 · input

2.

Input	Output
1	45
2	90
3	135
4	180

output = 45 · input

Make a table for the function represented by each rule. Find y when $x = 0, 1, 2,$ and 3.

3. $y = 10x$

x	y
0	0
1	10
2	20
3	30

4. $y = x - 4$

x	y
0	−4
1	−3
2	−2
3	−1

5. $y = 3x - 1$

x	y
0	−1
1	2
2	5
3	8

6. A printer can print 9 black-and-white pages per minute.
a. Write a function rule to represent the relationship between the number of black-and-white printed pages and the time it takes to print them. $y = 9x$
b. How many black-and-white pages can be printed in 15 minutes? 135 pages
c. How long would it take to print a 75-page black-and-white report? $8\frac{1}{3}$ minutes

Enrichment 9-4 L4

Decision Making

Many banks have different types of checking accounts available. The type of account you open depends upon how many checks you write each month and how much money you plan to keep in your account.

Here are the charges from one bank:

Basic Account — \$5 per month, plus \$0.10 per check
Student Account — \$0.20 per check for the first 10 checks each month, \$1 per check for each check over 10
\$1,000 Minimum Balance Account — \$0.10 per check

1. Make a graph that shows the monthly fees related to the number of checks written for each type of account.

Checking Account Fees
Fees (dollars): 5, 10, 15, 20
Number of Checks Written per Month: 0, 5, 10, 15, 20, 25
Key Basic —— Student · · · · \$1,000 ——

2. Suppose you are a student who writes about 5 checks per month.
a. Which account will you open? student account
b. About how much will your monthly fees be? about \$1

3. Suppose you are a student who writes 20 checks per month.
a. Which account will you open? Sample answer: basic account
b. How much will your monthly fees be? Sample answer: \$7.00
c. Suppose you have \$1,000 earning 4.8% interest each year. Should you open the \$1,000 minimum balance checking account instead of the basic account? Explain. If so, how much money will you save?
Sample answer: Yes, savings in fees by switching from basic to minimum balance are \$5.00. You lose \$4.00/month interest you were earning, bringing the actual savings per month to \$1.00.

Alternative Assessment

Each student in a pair writes a function rule expressed in terms of x and y. Partners exchange papers and make function tables for $x = 0, 1, 2, 3, 5,$ and 10.

Test Prep

Resources
For additional practice with a variety of test item formats:
- Test-Taking Strategies, p. 477
- Test Prep, p. 481
- Test-Taking Strategies with Transparencies

9-5

1. Plan

Objective
To find solutions to application problems using tables, rules, and graphs

Examples
1 Graphing Using a Table
2 Application: Plants

Math Understandings: p. 434D

Math Background

Functions can be represented by equations, word descriptions, tables, and graphs. When graphing functions, the input or independent variable is plotted on the horizontal or *x*-axis.

More Math Background: p. 434D

Lesson Planning and Resources

See p. 434E for a list of the resources that support this lesson.

Check Skills You'll Need
Use student page, transparency, or PowerPoint. For intervention, direct students to:
Patterns and Tables
Lesson 9-3
Extra Skills and Word Problems Practice, Ch. 9

9-5 Using Tables, Rules, and Graphs

Check Skills You'll Need

1. **Vocabulary Review** What is a *sequence*? See below.

Write an expression to describe each sequence.

2. 5, 10, 15, 20, . . . $5n$
3. −7, −5, −3, −1, . . . $2n - 9$
4. 101, 202, 303, . . . $101n$

GO for Help Lesson 9-3

What You'll Learn

To find solutions to application problems using tables, rules, and graphs

Why Learn This?

If you know your height in inches, you can find your height in feet. You can show the relationship between units such as inches and feet using a table, a rule, or a graph.

A graph can show the relationship between inputs and outputs.

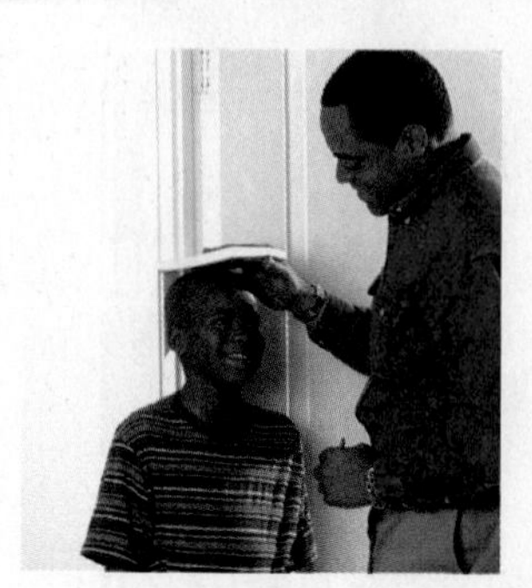

EXAMPLE Graphing Using a Table

Number of Inches	Number of Feet
12	1
24	2
36	3
48	4

1 The table at the left shows the relationship between the number of inches (input) and the number of feet (output). The rule is $f = \frac{n}{12}$, where f represents the number of feet, and n represents the number of inches. Graph the relationship represented by the table.

Check Skills You'll Need

1. a set of numbers that follow a pattern

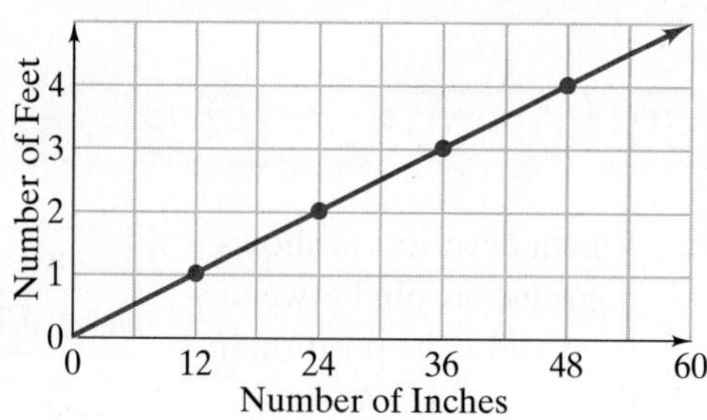

Graph inches on the horizontal axis and feet on the vertical axis.

Draw a line through the points.

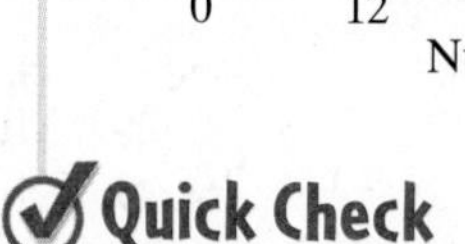

1. Graph the function represented by the table below.

Input *x*	Output *y*
0	3
1	5
2	7
3	9

1.

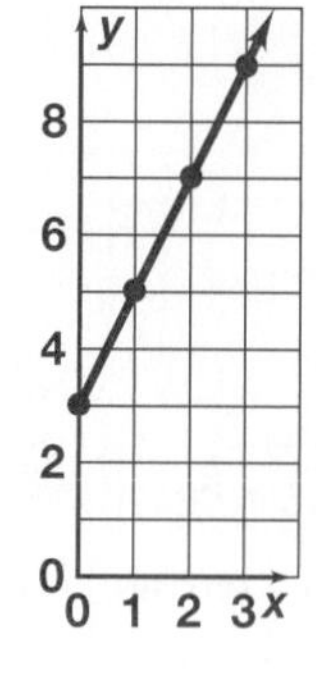

Differentiated Instruction Solutions for All Learners

Special Needs L1
If students have difficulty drawing the graphs, pair them up with students who can draw the graphs. The pairs can share the task of making a table of values, and coming up with word rules for the relationships.

learning style: visual

Below Level L2
Write the ordered pairs that you would graph for the function in the table. (−1, 2), (0, 3), (2, 5), (4, 7)

	−1	0	2	4
	2	3	5	7

learning style: visual

EXAMPLE Application: Plants

Careers Botanists study the life and growth of plants.

2 A plant grows 1.38 cm for each hour of sunlight it receives. Write and graph a rule to find the growth of the plant when it receives 4 h of sunlight.

Step 1 Write a rule.

Words growth equals 1.38 times hours of sunlight

Let g = growth in centimeters.

Let s = hours of sunlight.

Equation $g = 1.38 \cdot s$

Step 2 Make a table of values.

Hours of Sunlight (s)	$1.38s$	Centimeters of Growth (g)
1	1.38(1)	1.38
2	1.38(2)	2.76
3	1.38(3)	4.14
4	1.38(4)	5.52

Step 3 Make a graph.

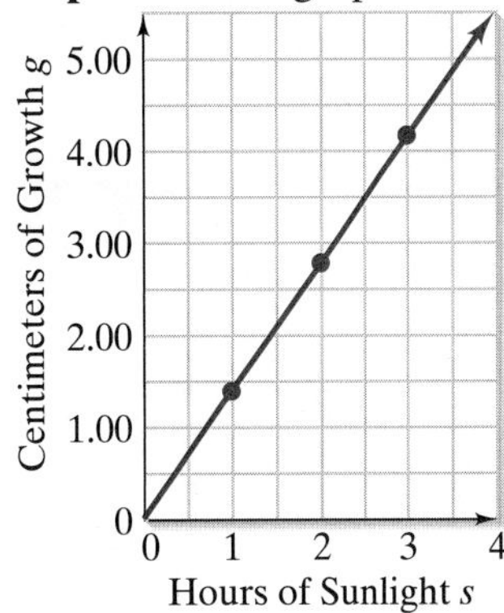

The plant grows 5.52 cm when it receives 4 h of sunlight.

Quick Check

2. A bus travels 60 miles per hour. Write and graph a rule to find the number of miles the bus can travel in 4.5 hours. **See back of book.**

Check Your Understanding

1. **Vocabulary** What is a table of values? **a table showing the relationship between the input and output of a function**
2. **Reasoning** Does every relationship have an output of 0 for an input value of 0? Explain. **No; any function that involves adding or subtracting from the input will not give you 0 as the output.**
3. Graph the relationship represented by the table. **See left.**

Input x	0	1	2	3
Output y	5	8	11	14

3.

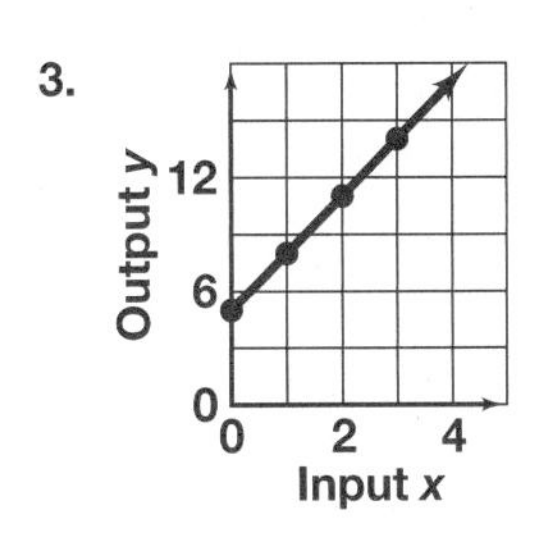

2. Teach

Activity Lab

Use before the lesson.

All in One Teaching Resources

Activity Lab 9-5: Tables and Graphs

Guided Instruction

Error Prevention!

Students may confuse the x-axis and the y-axis. Make sure students are graphing all the values for x on the horizontal axis and all the values for y on the vertical axis.

Example 2

Ask students to evaluate the effectiveness of the rule, table, and graph to communicate the growth of the plant.

Additional Examples

1 Graph the function represented by the table.

Input	0	1	2	3
Output	4	6	8	10

See back of book.

2 A plant grows 3 in. in one week. The second week, its height is 5 in., and the third week its height is 7 in. Write and evaluate a function rule to find the height of a plant in week 4. Then graph the function.
See back of book.

All in One Teaching Resources

- Daily Notetaking Guide 9-5 L3
- Adapted Notetaking 9-5 L1

Closure

- *What are some ways you can represent functions?* **Sample: You can represent functions with equations, tables, graphs, and word descriptions.**

Advanced Learners L4

Have students write original functions and make tables based on their functions. Students can graph their functions. Have them use both positive and negative values for x.

learning style: visual

English Language Learners ELL

The most difficult part of this lesson is writing relationships in words that translate easily to tables and functions. Have students work in pairs to develop rules in words.

learning style: verbal

3. Practice

Assignment Guide

Check Your Understanding

Go over Exercises 1–3 in class before assigning the Homework Exercises.

Homework Exercises

A	Practice by Example	4–7
B	Apply Your Skills	8–20
C	Challenge	21
Test Prep and Mixed Review		22–26

Homework Quick Check

To check students' understanding of key skills and concepts, go over Exercises 4, 6, 9, 13, and 20.

Differentiated Instruction Resources

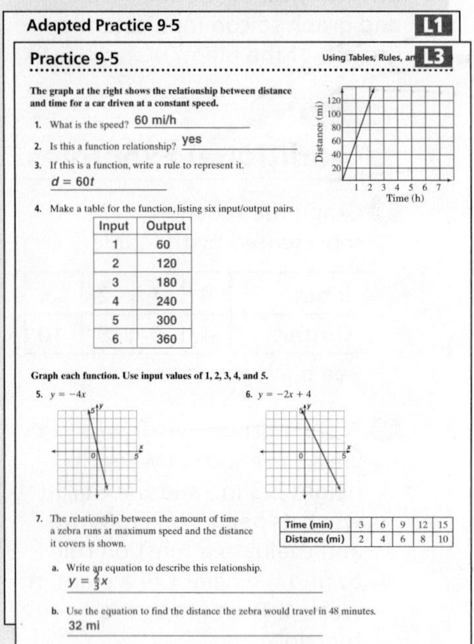

Adapted Practice 9-5 L1

Practice 9-5 Using Tables, Rules, and Graphs L3

The graph at the right shows the relationship between distance and time for a car driven at a constant speed.

1. What is the speed? 60 mi/h
2. Is this a function relationship? yes
3. If this is a function, write a rule to represent it. $d = 60t$
4. Make a table for the function, listing six input/output pairs.

Input	Output
1	60
2	120
3	180
4	240
5	300
6	360

Graph each function. Use input values of 1, 2, 3, 4, and 5.

5. $y = -4x$
6. $y = -2x + 4$

7. The relationship between the amount of time a zebra runs at maximum speed and the distance it covers is shown.

Time (min)	3	6	9	12	15
Distance (mi)	2	4	6	8	10

a. Write an equation to describe this relationship. $y = \frac{2}{3}x$

b. Use the equation to find the distance the zebra would travel in 48 minutes. 32 mi

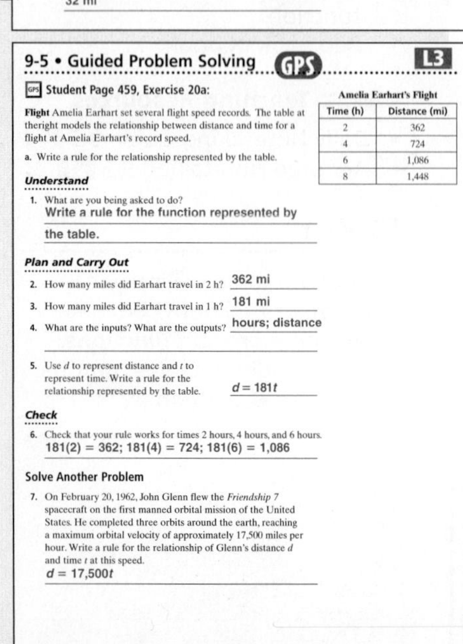

9-5 • Guided Problem Solving GPS L3

Student Page 459, Exercise 20a:

Flight Amelia Earhart set several flight speed records. The table at the right models the relationship between distance and time for a flight at Amelia Earhart's record speed.

a. Write a rule for the relationship represented by the table.

Amelia Earhart's Flight

Time (h)	Distance (mi)
2	362
4	724
6	1,086
8	1,448

Understand

1. What are you being asked to do? Write a rule for the function represented by the table.

Plan and Carry Out

2. How many miles did Earhart travel in 2 h? 362 mi
3. How many miles did Earhart travel in 1 h? 181 mi
4. What are the inputs? What are the outputs? hours; distance
5. Use d to represent distance and t to represent time. Write a rule for the relationship represented by the table. $d = 181t$

Check

6. Check that your rule works for times 2 hours, 4 hours, and 6 hours. $181(2) = 362$; $181(4) = 724$; $181(6) = 1{,}086$

Solve Another Problem

7. On February 20, 1962, John Glenn flew the *Friendship 7* spacecraft on the first manned orbital mission of the United States. He completed three orbits around the earth, reaching a maximum orbital velocity of approximately 17,500 miles per hour. Write a rule for the relationship of Glenn's distance d and time t at this speed. $d = 17{,}500t$

Homework Exercises

For more exercises, see Extra Skills and Word Problems.

GO for Help

For Exercises	See Examples
4–5	1
6–7	2

A **Graph the relationship represented by each table.** 4–5. See margin.

4.

Side Length of Square s	0	1	2	3	4
Perimeter of Square P	0	4	8	12	16

5.

Input x	Output y
0	1
1	4
2	7
3	10

For Exercises 6 and 7, write and graph a function to find the output for an input of 10 h. 6–7. See margin.

6. **Employment** Total earnings S depends on the number t of hours worked. You earn \$6/h.

7. **Air Travel** Total distance depends on the number t of hours traveled. Airplane speed averages 320 mi/h.

B 

8. **Guided Problem Solving** It costs \$120 per year to feed a cat. Write and evaluate a rule to find the cost of feeding a cat that lives to the maximum life span of 28 years. $c = 120y$; \$3,360
   - **Make a Plan** Write a rule for the cost c of feeding a cat during its lifetime y.
   - **Carry Out the Plan** Substitute ■ for y to find the cost of feeding a cat that lives to the maximum life span.

9. The number of Calories c that are burned by walking depends on t, the number of hours spent walking. If you burn 300 Cal/h, how many Calories do you burn in 2.5 hours of walking? 750 Calories

Match each graph with a rule.

10. 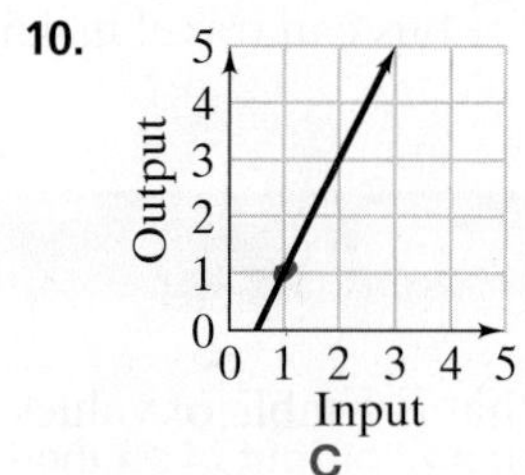

C

11. 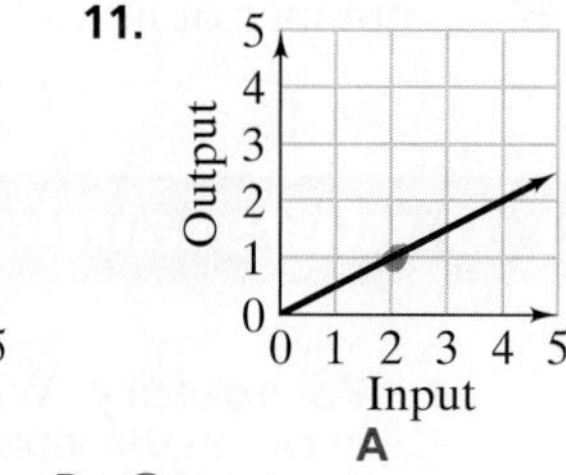

A

12. 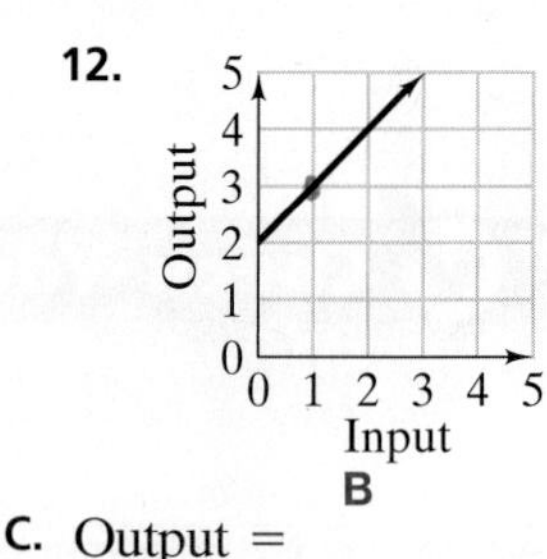

B

A. Output $= \frac{1}{2} \cdot$ Input

B. Output $=$ Input $+ 2$

C. Output $= 2 \cdot$ Input $- 1$

13. **Writing in Math** For each situation, which would best represent the relationship—a table, a rule, or a graph? Explain your choice.
   a. You only have a few values or you do not know the rule.
   b. You have many input and output values. 13a–c. See left.
   c. You want to see the relationship between the values.

GO Online Homework Video Tutor
Visit: PHSchool.com
Web Code: are-0905

13a–c. Answers may vary. Samples are given.

13a. A table; it can be drawn with only a few values and without a rule.

b. A rule; the rule can represent infinitely many values.

c. A graph; it visually represents the relationship between values.

4–7. See back of book.
14a–18. See back of book.
20a–21. See back of book.

14. a. Make a function table for the graph.
b. Write a rule for the function.
14a–b. See margin.

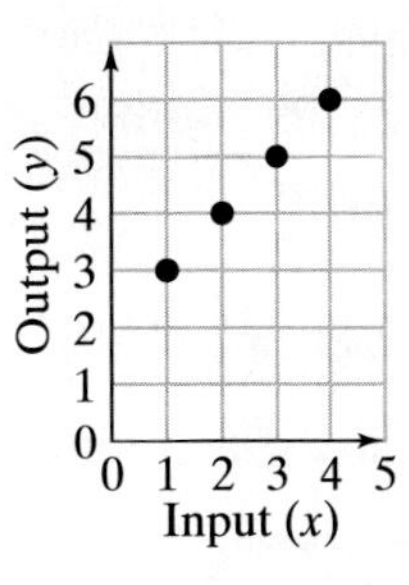

Graph each rule. Use input values of 1, 2, 3, 4, and 5. **15–18. See margin.**

15. $y = 5x$
16. $y = 2x + 1$
17. $y = x \div 2$
18. $y = x - 3$

19. **Open-Ended** Choose one rule from Exercises 15–18. Describe a real-world situation that the rule could represent. **Check students' work.**

In 1928, Amelia Earhart became the first woman to fly across the Atlantic Ocean.

20. **Flight** Amelia Earhart set several flight speed records. The table at the right models the relationship between distance and time for a flight at Amelia Earhart's record speed. **20a–d. See margin.**

Amelia Earhart's Flight

Time (h)	Distance (mi)
2	362
4	724
6	1,086
8	1,448

a. Write a rule for the relationship represented by the table.
b. Find the average speed. Justify your answer.
c. Estimate the number of hours it would take to fly 1,890 mi.
d. Graph the rule.

21. **Challenge** The area of an equilateral triangle depends on the side length s. The rule is $A = \frac{\sqrt{3}}{4}s^2$. Evaluate the rule for several values and make a table. Use 0.433 as an approximation of $\frac{\sqrt{3}}{4}$. Then graph the rule. Describe the shape of your graph. **See margin.**

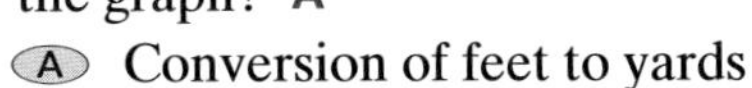

Test Prep and Mixed Review

Practice

Multiple Choice

22. Which of the following relationships is best represented by the data in the graph? **A**

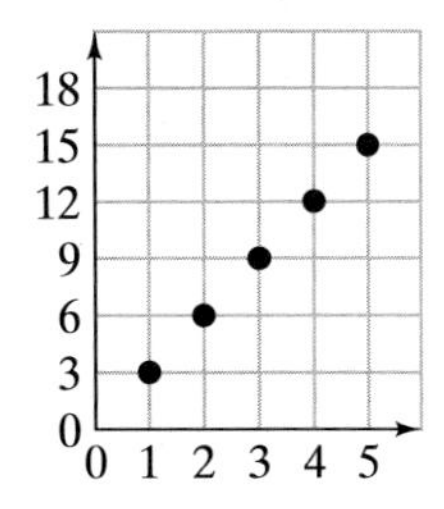

A. Conversion of feet to yards
B. Conversion of feet to inches
C. Conversion of miles to feet
D. Conversion of inches to yards

23. The circular base of a dome has a diameter of 60 ft. Which expression can be used to find the area of the base? **H**

F. $2 \cdot 30 \cdot \pi$
G. $2 \cdot 60 \cdot \pi$
H. $30 \cdot 30 \cdot \pi$
J. $60 \cdot 60 \cdot \pi$

Find the area of a triangle with the given base and height.

24. 5 m, 2 m **5 m²**
25. 10 ft, 3 ft **15 ft²**
26. 18 cm, 6 cm **54 cm²**

GO for Help

For Exercises	See Lesson
24–26	8-3

4. Assess & Reteach

Lesson Quiz

A carpenter uses 6 nails for each shelf.

1. Write and evaluate a function to find the number of nails for 14 shelves. **$n = 6s$; 84 nails**
2. Graph the function using input values of 1, 2, 3, and 4.

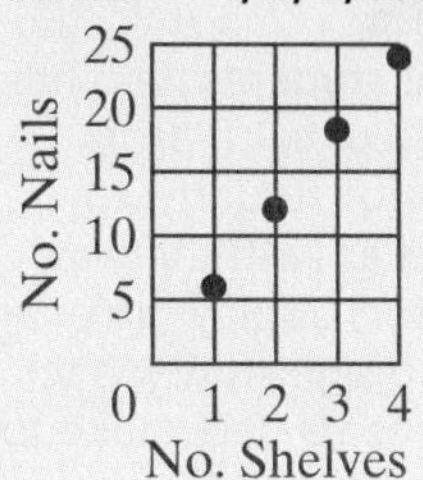

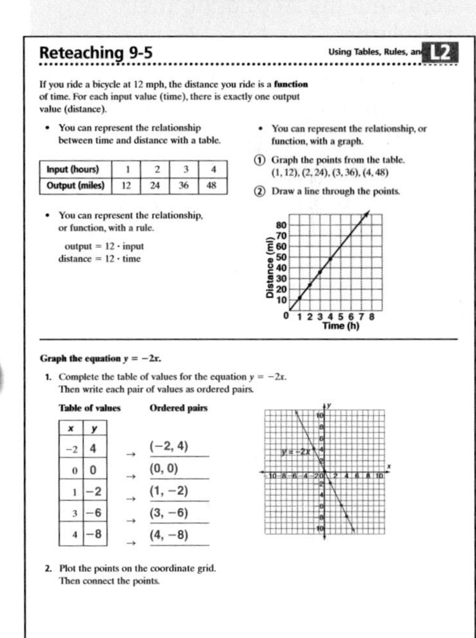

Reteaching 9-5 Using Tables, Rules, and Graphs **L2**

If you ride a bicycle at 12 mph, the distance you ride is a **function** of time. For each input value (time), there is exactly one output value (distance).

- You can represent the relationship between time and distance with a table.

Input (hours)	1	2	3	4
Output (miles)	12	24	36	48

- You can represent the relationship, or function, with a rule.

output = 12 · input
distance = 12 · time

- You can represent the relationship, or function, with a graph.

① Graph the points from the table. (1, 12), (2, 24), (3, 36), (4, 48)
② Draw a line through the points.

Distance (mi)
Time (h)

Graph the equation $y = -2x$.

1. Complete the table of values for the equation $y = -2x$. Then write each pair of values as ordered pairs.

Table of values / **Ordered pairs**

x	y	Ordered pair
−2	4	(−2, 4)
0	0	(0, 0)
1	−2	(1, −2)
3	−6	(3, −6)
4	−8	(4, −8)

2. Plot the points on the coordinate grid. Then connect the points.

$y = -2x$

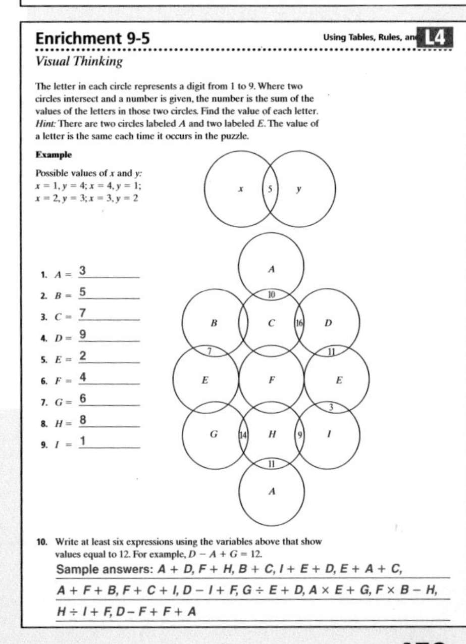

Enrichment 9-5 Using Tables, Rules, and Graphs **L4**

Visual Thinking

The letter in each circle represents a digit from 1 to 9. Where two circles intersect and a number is given, the number is the sum of the values of the letters in those two circles. Find the value of each letter. *Hint:* There are two circles labeled *A* and two labeled *E*. The value of a letter is the same each time it occurs in the puzzle.

Example

Possible values of x and y:
$x = 1, y = 4; x = 4, y = 1;$
$x = 2, y = 3; x = 3, y = 2$

1. $A =$ 3
2. $B =$ 5
3. $C =$ 7
4. $D =$ 9
5. $E =$ 2
6. $F =$ 4
7. $G =$ 6
8. $H =$ 8
9. $I =$ 1

10. Write at least six expressions using the variables above that show values equal to 12. For example, $D - A + G = 12$.
Sample answers: $A + D$, $F + H$, $B + C$, $I + E + D$, $E + A + C$, $A + F + B$, $F + C + I$, $D - I + F$, $G \div E + D$, $A \times E + G$, $F \times B - H$, $H \div I + F$, $D - F + F + A$

Alternative Assessment

Provide graph paper to pairs of students. Partners work together to complete Exercises 6–7. Before they draw each graph, have them discuss appropriate axes labels, scales, and intervals. Have pairs compare their graphs with classmates.

Test Prep

Resources

For additional practice with a variety of test item formats:

- Test-Taking Strategies, p. 477
- Test Prep, p. 481
- Test-Taking Strategies with Transparencies

Activity Lab

Three Views of a Function

In this Activity, students use a graphing calculator to view a graph and a table of values for a function.

Guided Instruction

Technology Tip
Group or pair students who have the same kind of calculator. Provide time for students to become familiar with the way their calculators display tables and graphs. Encourage students to share adjustments they make to complete the steps using their calculator.

Teaching Tip
Work through Steps 1–6 with the class. If you have an overhead calculator, demonstrate each step and the corresponding display. Otherwise, circulate, or have a volunteer show the students what the calculator should display after each step. Make sure students understand what a "range" for a table of values is.

Exercises
Have students work in pairs. Challenge them to predict what the graphs of each of the functions will look like and to explain their reasoning. Invite partners to give each other another function to graph and make a table of values.

Differentiated Instruction

Advanced Learners L4
Have students write their own functions. They can trade functions with a partner and create the three views of their partner's function.

Resources

- graphing calculator
- Classroom Aid 36

9-5b Activity Lab Technology

Three Views of a Function

You can use a graphing calculator to graph a function.

EXAMPLE

Graph $y = 9 - x$ and make a table of values.

Step 1 Press WINDOW to set the range.

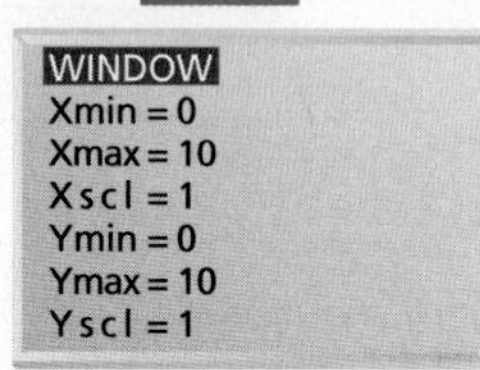

Step 2 Press Y= to enter the function.

Plot1 Plot2 Plot3
\Y1 = 9−X
\Y2 =
\Y3 =
\Y4 =
\Y5 =
\Y6 =

Step 3 Press GRAPH to view the graph.

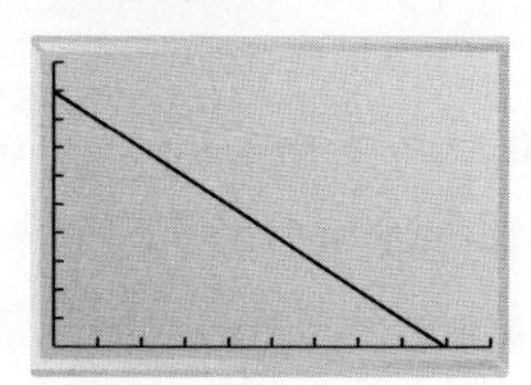

Step 4 Use the TblSet feature. Set TblStart = 0 and ΔTbl = 1.

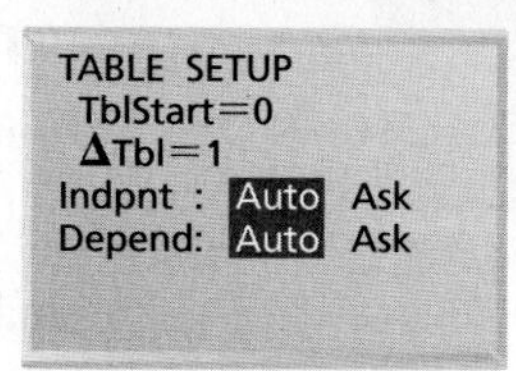

Step 5 Use the TABLE feature to make a table of values.

X	Y1
0	9
1	8
2	7
3	6
4	5
5	4
6	3

X=0

Step 6 Sketch the graph. Copy the table.

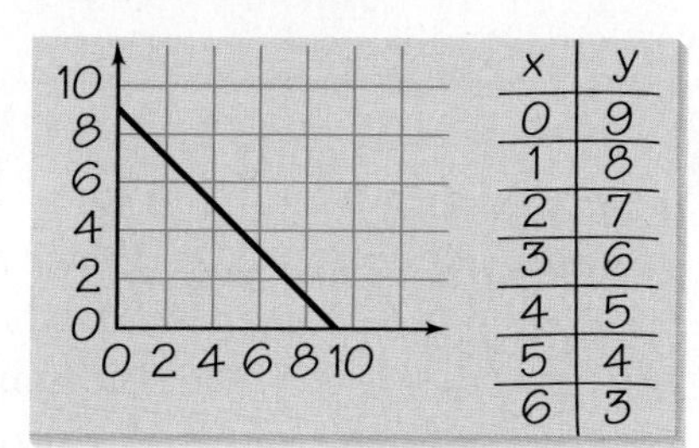

Exercises

Use a graphing calculator. Graph each function and make a table of values. Sketch the graph and copy the table of values. 1–7. See margin.

1. $y = 2x$
2. $y = x - 3$
3. $y = 13 - 2x$
4. $y = x + 1$
5. $y = 3x - 4$
6. $y = 0.5x + 6$
7. **Reasoning** What values in the WINDOW feature would you use to view the graph of $y = 100x$? Explain.

1–7. See back of book.

Algebra

9-6 Interpreting Graphs

Check Skills You'll Need

1. Vocabulary Review Give an example of an *input*. **Check students' work.**

Find the distance for each time. Use $d = rt$ with $r = 40$ mi/h.

2. 2.5 h **100 mi** **3.** 3.5 h **140 mi**

4. $\frac{1}{4}$ h **10 mi** **5.** $\frac{1}{2}$ h **20 mi**

GO for Help Lesson 9-5

What You'll Learn

To describe and sketch graphs that represent real-world situations

Why Learn This?

You can see the history of an event by looking at a graph. You can use a graph like the one at the right to show your distance from home when you take a trip or run an errand.

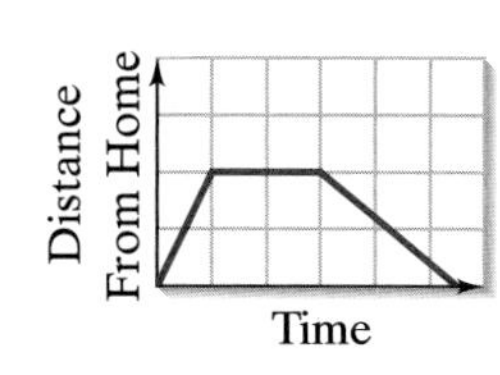

When you graph a relationship, you can see how one quantity changes compared to another.

EXAMPLE Describing a Graph

1 Shopping The graph above relates time and your distance from home. What can you tell about the trip from the steepness of the lines?

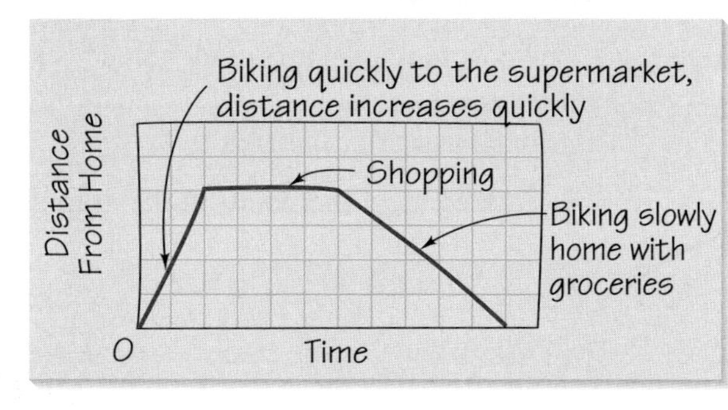

A steeper line on the graph shows faster speed. A horizontal line represents a period of no change in distance from home.

Quick Check

1. You live 6 blocks from school. The graph at the right shows your walk home on a sunny day. Describe what the graph shows. **See left.**

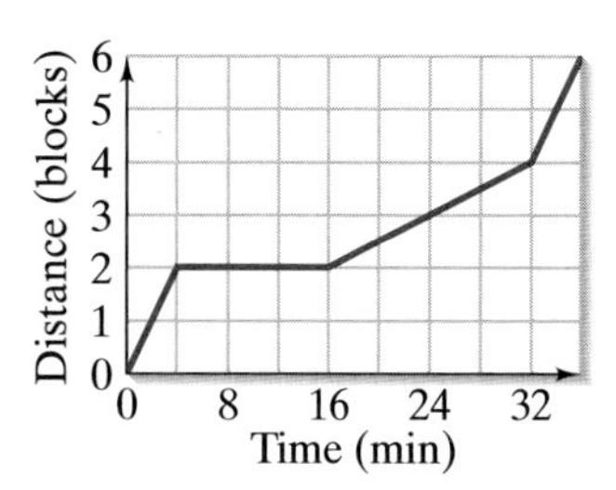

1. **You walk two blocks at a fast pace and then stop for 12 min. Then you walk 2 blocks at a slower pace and 2 blocks at the faster pace.**

Differentiated Instruction Solutions for All Learners

Special Needs L1
Students who have difficulty sketching graphs can work with a partner. The partner can draw the graphs, while the student labels the sections and axes.

learning style: visual

Below Level L2
Should each situation have a graph of points or a line?

no. of visitors each hour **points**
puppy's weight by month **line**
no. of students in each grade **points**

learning style: verbal

9-6

1. Plan

Objective
To describe and sketch graphs that represent real-world situations

Examples
1 Describing a Graph
2 Sketching a Graph
3 Graphing Data

Math Understandings: p. 434D

Math Background

A graph can provide a visual description of a situation. Graphs show how one quantity changes in relation to another quantity. A horizontal line shows no change in the variable on the vertical axis.

More Math Background: p. 434D

Lesson Planning and Resources

See p. 434E for a list of the resources that support this lesson.

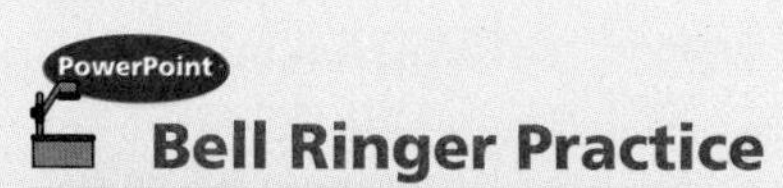

Bell Ringer Practice

Check Skills You'll Need
Use student page, transparency, or PowerPoint. For intervention, direct students to:
Using Tables, Rules, and Graphs
Lesson 9-5
Extra Skills and Word Problems Practice, Ch. 9

2. Teach

Activity Lab

Use before the lesson.

All in One Teaching Resources

Activity Lab 9-6: Critical Thinking and Graphs

Guided Instruction

Example 2

Make sure students understand the connection between the slope of the line and the speed of Ciara's travel. The slope for the drive is steeper than the slope for the walk because driving is faster than walking. The slope for time at school is horizontal because her distance did not change. She stayed in the same place.

PowerPoint

Additional Examples

1 Nancy took a roundtrip walk from home. Describe what the graph shows. **See back of book.**

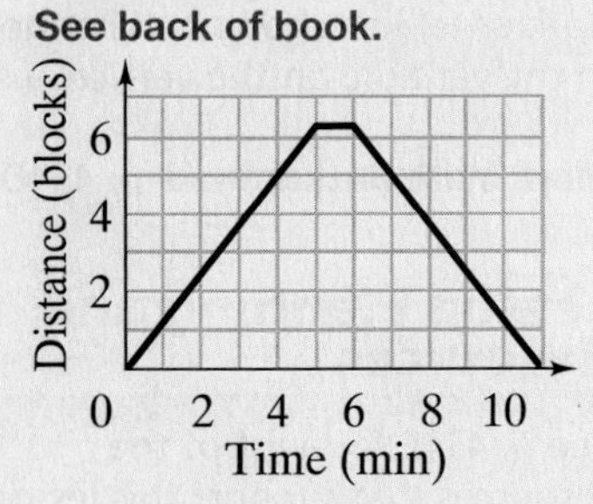

2 Suppose that after 5 minutes, Nancy ran home. Draw the new graph. **See back of book.**

3 Make a graph of this data.

No. Mugs	1	2	3	4
Cost ($)	6	10	15	18

See back of book.

All in One Teaching Resources

- Daily Notetaking Guide 9-6 L3
- Adapted Notetaking 9-6 L1

Closure

- *What can you learn about a situation from looking at a graph?* **Sample: You can learn how one quantity changes relative to another.**

You can sketch a graph to describe a real-world situation.

EXAMPLE Sketching a Graph

2 **Transportation** Ciara's mother drove her part of the way to school. Ciara waited for a friend and walked the rest of the way to school. She took a bus home. Sketch a graph to show the distance Ciara traveled compared to time.

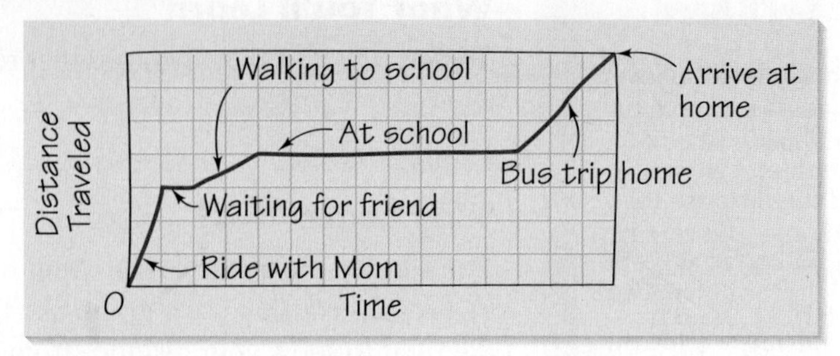

Quick Check

2. Sketch a graph of the situation in Example 2 using *Distance from Home* instead of *Distance Traveled* for the vertical axis. **See back of book.**

When you draw a graph, you may need to consider what is reasonable.

EXAMPLE Graphing Data

3 The cost to play games at an arcade is given in the table. Make a graph of the cost to play 1, 2, 3, and 4 games at the arcade.

Number of Games	Cost ($)
1	2.00
2	3.50
3	5.00
4	6.50

Do not connect the points in the graph.

Check for Reasonableness Each cost is for playing an entire game. Since you cannot pay for part of a game, connecting the points would not be meaningful.

Online active math

For: Graphing Activity
Use: Interactive Textbook, 9-6

Quick Check

3. The table shows the number of cans in the cafeteria juice machine over time. Graph the data. **See left.**

Time	Number of Cans
8 A.M.	30
9 A.M.	20
10 A.M.	19
11 A.M.	19

Advanced Learners L4

Sketch a graph for this situation: Turtle and Rabbit raced. Rabbit was faster, but slept for 30 minutes. The race ended in a tie. **Check students' work.**

learning style: visual

English Language Learners ELL

Have students generate situations, in words, in which they would connect the points on a graph. Have them generate situations in which the points should not be connected. Make sure they understand the difference between graphs of continuous and discrete data.

learning style: verbal

Check Your Understanding

Match each situation with the appropriate graph.

1. height of a person from birth to age 20 C
2. air temperature in a 24-hour period starting at midnight A
3. distance raced with a fall over a hurdle B

A.
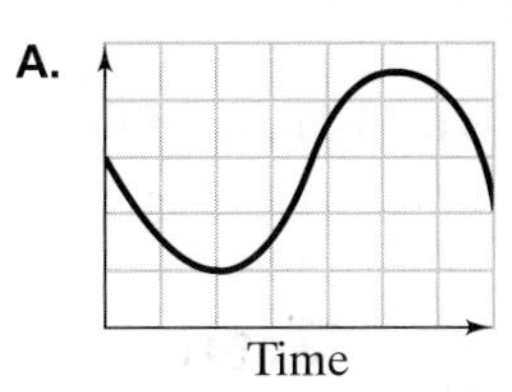

B.
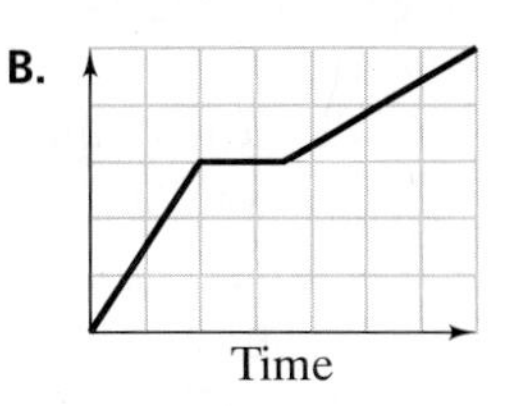

C.
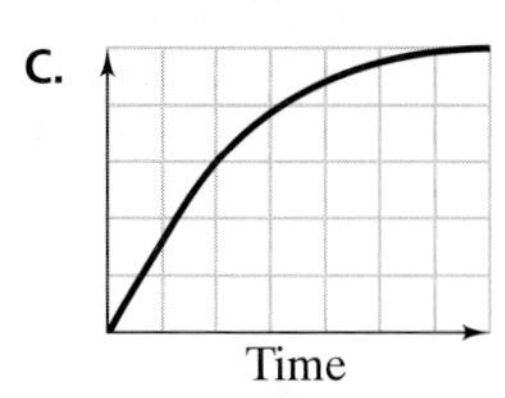

Homework Exercises

For more exercises, see Extra Skills and Word Problems.

GO for Help

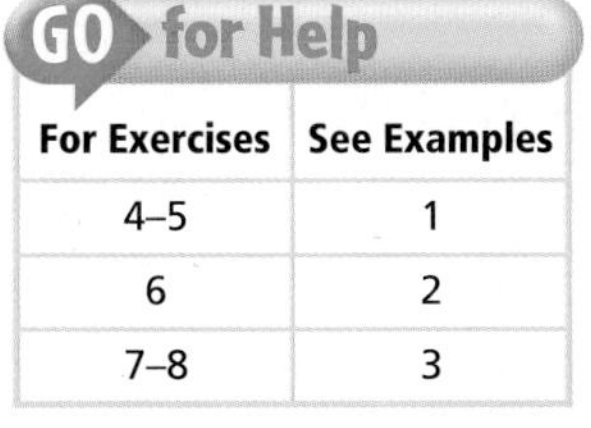

For Exercises	See Examples
4–5	1
6	2
7–8	3

A **Describe what each graph shows.** 4–5. See left.

4. **Walking Home From School**

Distance (blocks) vs. Time (min)

5. **Distance From Home**

Distance (mi) vs. Time (h)

4. You walk at a steady pace, traveling 6 blocks in 16 min.
5. You travel at a steady pace for 2 h, stop for $\frac{1}{2}$ h, and then travel at the same steady pace for 2 h.

6. You ride your bike slowly up a steep hill and then quickly down the other side. Sketch a graph for the situation. Label each section and each axis. Show your speed on the bicycle on the vertical axis. See margin.

Graph the data. Should you connect the points on each graph? Explain. 7–8. See margin.

7. **Lemonade Sales**

Cups Sold	Income ($)
1	0.75
2	1.50
3	2.25
4	3.00
5	3.75

8. **Miles From Home**

Time (h)	Miles
1	60
2	85
3	120
4	180

B

9. **Guided Problem Solving** You pay 5 cents a day for overdue library books. Make a graph of the fines for 1–5 days. See margin.
   - How can a table help you plot the points on the graph?
   - Should you connect the points on your graph?

6–9. See back of book.

3. Practice

Assignment Guide

Check Your Understanding

Go over Exercises 1–3 in class before assigning the Homework Exercises.

Homework Exercises

A	Practice by Example	4–8
B	Apply Your Skills	9–14
C	Challenge	15
	Test Prep and Mixed Review	16–21

Homework Quick Check

To check students' understanding of key skills and concepts, go over Exercises 5, 8, 10, 11, and 13.

Differentiated Instruction Resources

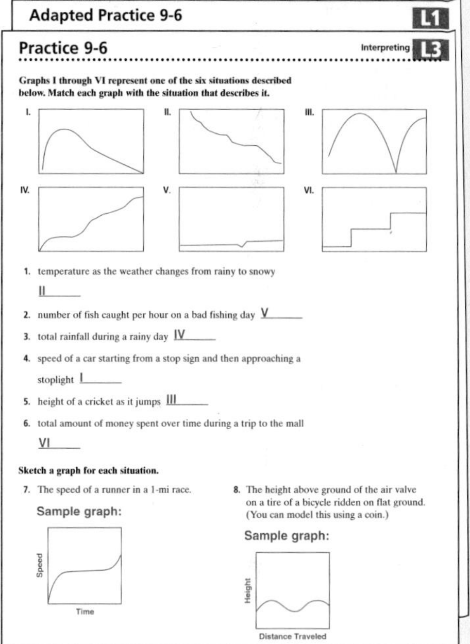

Adapted Practice 9-6 L1

Practice 9-6 Interpreting L3

Graphs I through VI represent one of the six situations described below. Match each graph with the situation that describes it.

1. temperature as the weather changes from rainy to snowy II
2. number of fish caught per hour on a bad fishing day V
3. total rainfall during a rainy day IV
4. speed of a car starting from a stop sign and then approaching a stoplight I
5. height of a cricket as it jumps III
6. total amount of money spent over time during a trip to the mall VI

Sketch a graph for each situation.

7. The speed of a runner in a 1-mi race. Sample graph:
8. The height above ground of the air valve on a tire of a bicycle ridden on flat ground. (You can model this using a coin.) Sample graph:

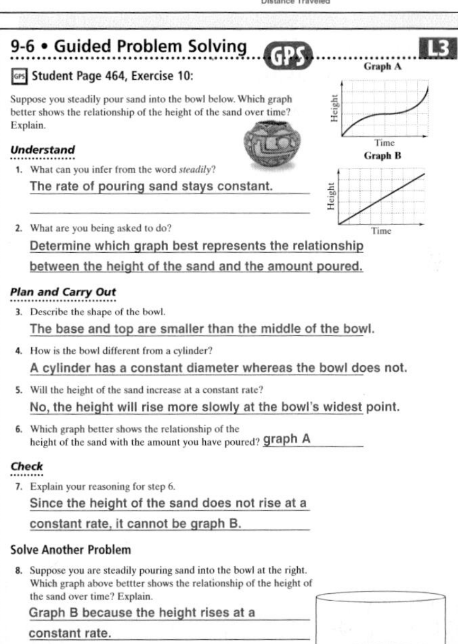

9-6 • Guided Problem Solving GPS L3

Student Page 464, Exercise 10:

Suppose you steadily pour sand into the bowl below. Which graph better shows the relationship of the height of the sand over time? Explain.

Graph A

Graph B

Understand

1. What can you infer from the word *steadily*? The rate of pouring sand stays constant.
2. What are you being asked to do? Determine which graph best represents the relationship between the height of the sand and the amount poured.

Plan and Carry Out

3. Describe the shape of the bowl. The base and top are smaller than the middle of the bowl.
4. How is the bowl different from a cylinder? A cylinder has a constant diameter whereas the bowl does not.
5. Will the height of the sand increase at a constant rate? No, the height will rise more slowly at the bowl's widest point.
6. Which graph better shows the relationship of the height of the sand with the amount you have poured? graph A

Check

7. Explain your reasoning for step 6. Since the height of the sand does not rise at a constant rate, it cannot be graph B.

Solve Another Problem

8. Suppose you are steadily pouring sand into the bowl at the right. Which graph above bettter shows the relationship of the height of the sand over time? Explain. Graph B because the height rises at a constant rate.

4. Assess & Reteach

PowerPoint

Lesson Quiz

1. Describe what the graph shows.

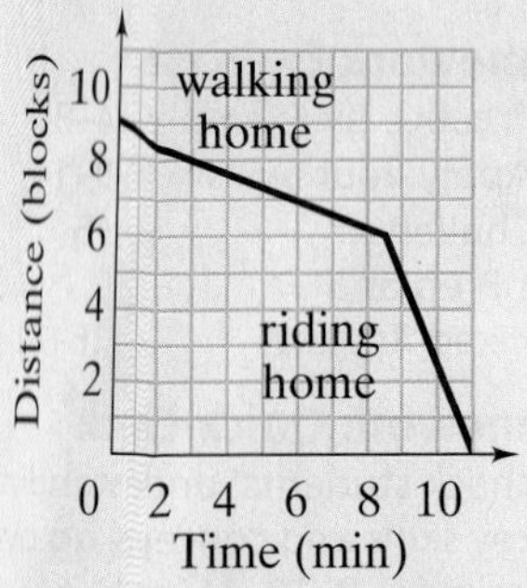

Sample: Someone walked for 3 blocks then rode for 6 blocks.

2. Graph the data in the table.

No. Copies	1	2	3	4
Cost (¢)	8	16	24	32

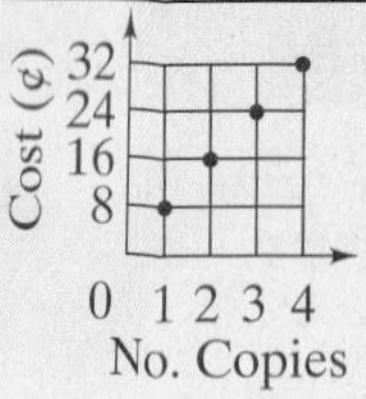

10–11c. See back of book.

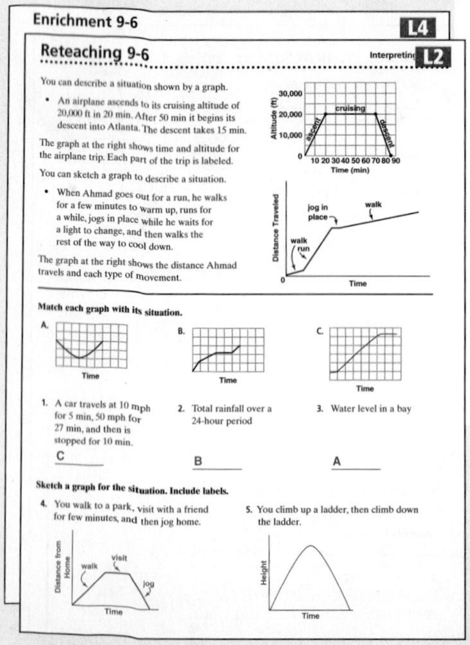

Enrichment 9-6 L4

Reteaching 9-6 Interpreting L2

You can describe a situation shown by a graph.

- An airplane ascends to its cruising altitude of 20,000 ft in 20 min. After 50 min it begins its descent into Atlanta. The descent takes 15 min.

The graph at the right shows time and altitude for the airplane trip. Each part of the trip is labeled.

You can sketch a graph to describe a situation.

- When Ahmad goes out for a run, he walks for a few minutes to warm up, runs for a while, jogs in place while he waits for a light to change, and then walks the rest of the way to cool down.

The graph at the right shows the distance Ahmad travels and each type of movement.

Match each graph with its situation.

1. A car travels at 10 mph for 5 min, 50 mph for 27 min, and then is stopped for 10 min. C
2. Total rainfall over a 24-hour period B
3. Water level in a bay A

Sketch a graph for the situation. Include labels.

4. You walk to a park, visit with a friend for few minutes, and then jog home.
5. You climb up a ladder, then climb down the ladder.

10. Suppose you steadily pour sand into the bowl at the left. Which graph below better shows the relationship of the height of the sand over time? Explain. **See margin.**

GPS

(A.)

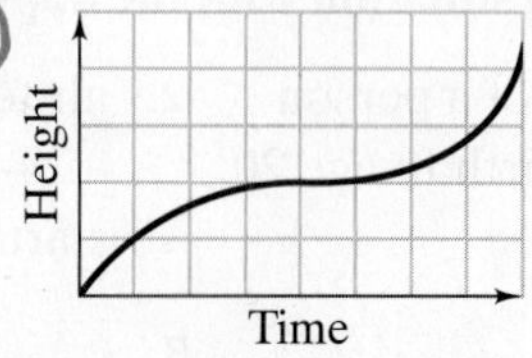

B.

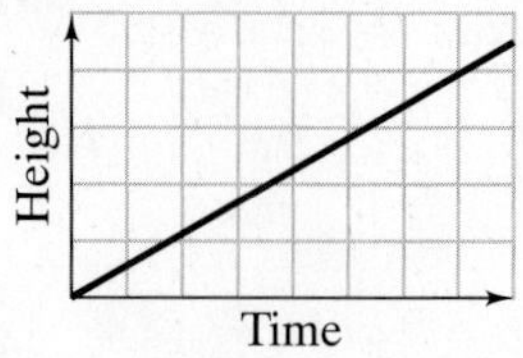

11. The graph shows a 90-m race. One student starts 5 s after the other.
   a. Describe what the graph shows.
   b. Who wins the race? Josh
   c. **Writing in Math** If the lines were parallel, what would the graph tell you about who wins the race? **11a–c. See margin.**

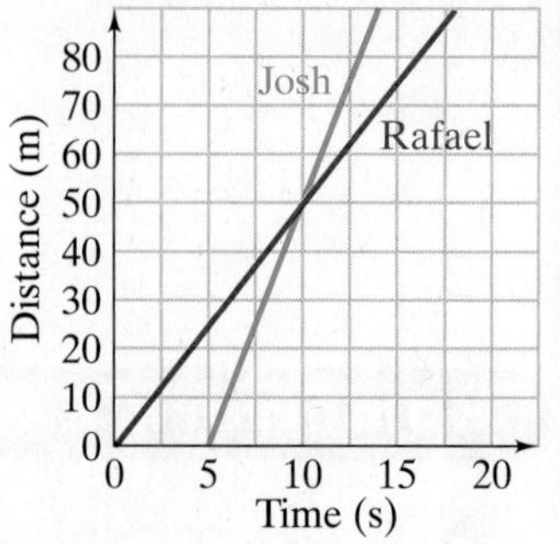

GO Online
Homework Video Tutor
Visit: PHSchool.com
Web Code: are-0906

Estimation The graph below shows what happens when a ball is thrown in the air.

12. When does the ball hit the ground? **about 2.75 s**
13. Why are there two times when the ball's height is 20 ft? What are they? **See left.**
14. When the time is 0, the height of the ball is *not* 0. Explain. **The ball starts at a height of about 5 ft.**
15. **Challenge** Describe what might have happened to make the data in the graph. **Answers may vary. Sample: A person threw a ball into the air. It left the person's hand 5 feet above the ground and landed on the ground about 2.75 seconds later.**

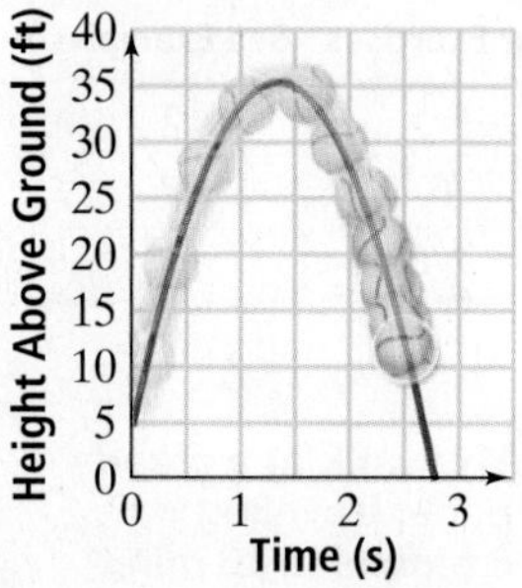

13. The ball is 20 ft high on the way up and on the way down. It reaches that height after about 0.4 s and about 2.4 s.

Test Prep and Mixed Review — Practice

Multiple Choice

16. Lee is starting his own business by selling personalized photo frames to his class for $19 each. It costs Lee $14 for materials to make one frame. Which equation can be used to find p, the amount of profit he makes by selling 25 frames? **B**
   (A) $p = (25 \times 19) - 14$
   (B) $p = 25(19 - 14)$
   (C) $p = 25 \times 19 \times 14$
   (D) $p = 14(25 - 19)$

17. Which problem situation is NOT modeled by $y = 4x$? **H**
   (F) Cost of a call at $0.04/min
   (G) Janet's pay after working 4 h
   (H) Perimeter of a trapezoid
   (J) Number of pens in 4 packs

GO for Help

For Exercises	See Lesson
18–21	2-4

Compare. Use <, =, or >.

18. $\frac{7}{9} \blacksquare \frac{3}{4}$ **>**
19. $\frac{5}{14} \blacksquare \frac{2}{6}$ **>**
20. $\frac{7}{30} \blacksquare \frac{1}{3}$ **<**
21. $\frac{8}{24} \blacksquare \frac{2}{6}$ **=**

Test Prep

Resources

For additional practice with a variety of test item formats:

- Test-Taking Strategies, p. 477
- Test Prep, p. 481
- Test-Taking Strategies with Transparencies

Alternative Assessment

Provide students with graph paper. Partners separately list their activities and the amount of time each takes from the time they leave home in the morning to the time they reach their first class. Students exchange lists and sketch a graph to represent each other's situation.

Checkpoint Quiz 2

Lessons 9-4 through 9-6

Write a rule for each table.

1.

x	0	1	2	3
y	0	5	10	15

$y = 5x$

2.

x	0	1	2	3
y	−3	0	3	6

$y = 3x - 3$

3.

x	0	1	2	3
y	9	10	11	12

$y = x + 9$

4. Graph $y = 2x - 4$. See margin.

For Exercises 5–7, use the graph at the right.

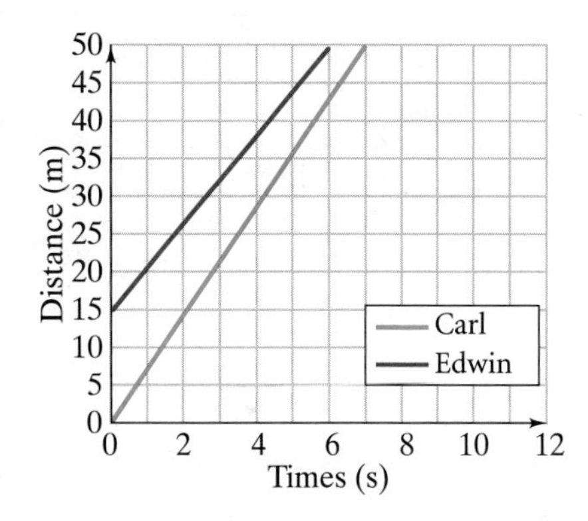

5. The graph shows a 50-m race. In this race Edwin had a 15-m head start over Carl. Who won the race? Edwin
6. By how many seconds did the winner win the race? 1 s
7. You walk for 3 h, eat lunch for 1 h, bike for 1 h, and then do homework for 2 h. Sketch a graph that describes your speed. See margin.

MATH GAMES

Lines in Space

What You'll Need

- 2 players and 2 coordinate grids similar to the one at the right

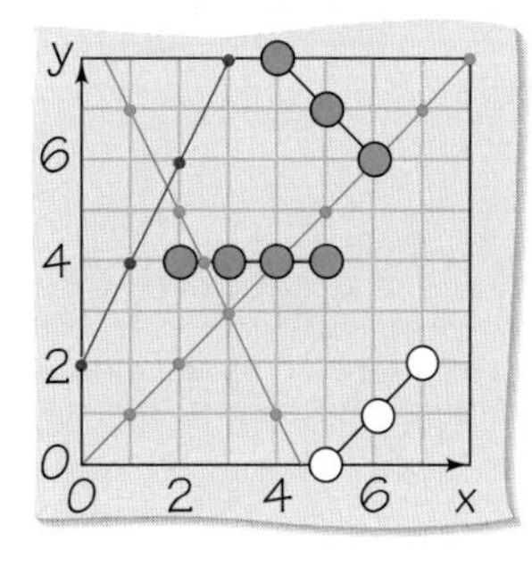

How To Play

- Without showing each other, Players A and B put three spaceships on their own grids. The lengths of the spaceships are four points, three points, and three points.
- Ships occupy adjacent grid points either left and right, up and down, or diagonally. Players choose whole-number points. They take turns trying to locate each other's ships by calling out function rules.
- If a function rule hits one or more sections of a spaceship, the player must announce a hit. See the table at the right.
- After 7 guesses each, the player who has hit the most sections of the other player's spaceships wins.

Rule	Result
$y = 2x + 2$?	Miss!
$y = -2x + 9$?	1 Hit!
$y = x$	2 Hits!

Checkpoint Quiz

Use this Checkpoint Quiz to check students' understanding of the skills and concepts of Lessons 9-4 through 9-6.

Resources

- All-in-One Teaching Resources Checkpoint Quiz 2
- ExamView CD-ROM
- Success Tracker™ Online Intervention

Lines in Space

In this game, students practice using functions to see where they land on a graph. Players mark their graphs with three lines and then try to guess functions that might go through their opponent's lines.

Guided Instruction

Have a volunteer read the rules. Explain to students that they may want to sketch their guesses on their own grid to make it easier to track their guesses. Provide colored pencils for this purpose.

Resources

- coordinate grids
- colored pencils

4.

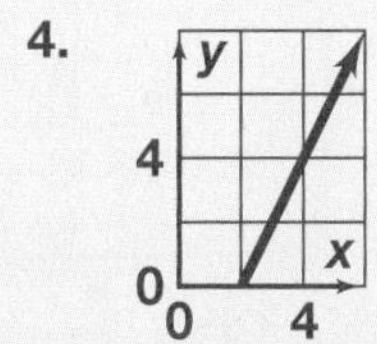

7.

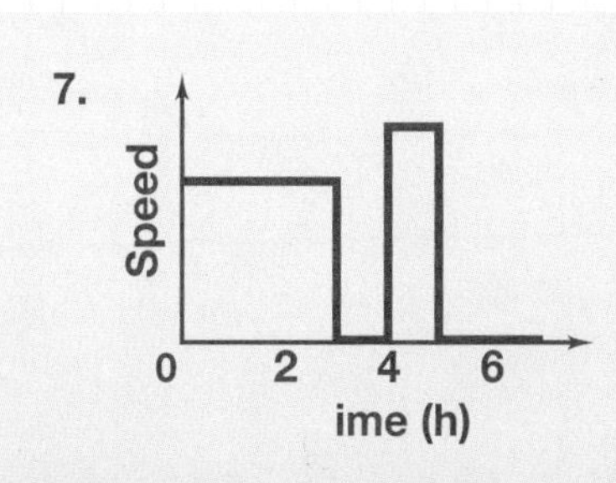

Solving Pattern Problems

In this feature, students practice solving problems involving patterns. They identify patterns, find the next terms in the pattern, and generate formulas based on the patterns.

Guided Instruction

Review with students the ways of finding patterns they have learned throughout the chapter. Have a volunteer read the problem aloud.

Ask:

- *What is the pattern of the dominoes?* Each row has one more domino than the row preceding it.
- *What is the function for the pattern of dominoes?* $y = x + 1$
- *What kind of number sequence is this?* arithmetic

GPS Guided Problem Solving

Solving Pattern Problems

Alondra lines up dominoes to see how many she can knock over by touching only one. She lines them up as shown below. If Alondra makes 20 rows, how many dominoes can she knock over?

What You Might Think	What You Might Write
What do I know? What do I want to find out?	I can see one domino in the first row, two in the second, three in the third, and so on. I want to know how many are in 20 rows.
How can I show the main idea?	I see a pattern. I see these numbers of dominoes: 1, 2, 3, . . . 20. So, add 1 + 2 + 3 + . . . + 20.
How do I solve the problem?	One way to add the numbers is to make "pairs of 20s," such as 1 + 19 and 2 + 18. I get nine 20s, plus 10 and 20. Since $9 \cdot 20 + 10 + 20 = 210$, the answer is 210 dominoes.
How can I check my answer?	I can add the numbers from 1 to 20 in order, using a calculator. I still get 210.

Think It Through

1. Write the numbers from 1 to 20. Show that you have nine pairs that add up to 20, plus 10 and 20. **Answers may vary. Sample: 1 + 19, 2 + 18, 3 + 17, 4 + 16, 5 + 15, 6 + 14, 7 + 13, 8 + 12, 9 + 11, 10, 20**
2. If Alondra had made 30 rows, how many dominoes did she use? **465 dominoes**
3. **Number Sense** If Alondra had 300 dominoes, how many rows could she make? Explain. **24 rows; since 210 dominoes make 20 rows, you can add row by row until you reach 300.**

Exercises

Solve each problem. For Exercises 4 and 5, answer the questions first.

4. William stacked cans of soup as shown below. What is the total number of cans would he need to make the stack 7 rows high? **140 cans**

a. What do you know? What do you want to find out?

b. Is there a pattern? If so, how can you find the sum?

5. One of the largest passenger jets, the Airbus A380, first flew in 2005. The dimensions of the A380 and the Wright Flyer are shown in the table below. Assume that Wright Flyers can be stacked on top of each other and that they always face the same direction inside a hangar. How many Wright Flyers could fit in a hangar built to house the A380? **480 or 528; answers may vary depending on how planes are stacked.**

Plane Dimensions

Plane	Length (ft)	Width (ft)	Height (ft)
Airbus A380	239	262	79
Wright Flyer	21	40	9

a. A sketch of a box representing the dimensions of the Airbus A380 appears at the right. How can drawing a box that represents the Wright Flyer help you find the answer?

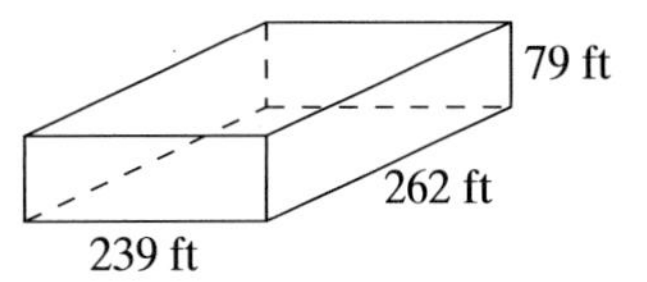

6. A lawnmower engine turns at about 2,500 rpm (revolutions per minute). How many times does the blade turn around every second? How many times does the blade turn around in half an hour? **$41\frac{2}{3}$ revolutions per second; 75,000 revolutions per half hour**

7. A sheet of notebook paper is about 0.0025 inch thick. It is physically impossible to fold a piece of paper in half 13 times. If you could, how thick would the stack of folded paper be? If you could fold a piece of paper 25 times, how thick would the stack be? **20.48 in.; 83,886.08 in. or about 1.3 mi**

Error Prevention!

Remind students of the difference between an arithmetic sequence and a geometric sequence. In an arithmetic sequence, they must add the *same* number to each term to get the next term. In a geometric sequence, they must multiply each term by the *same* number to get the next term.

Exercises

Have students work independently on the Exercises. When they have finished, have them form small groups to compare and discuss their answers. Have them elect a representative to report their results and call on different groups to answer each question.

Teaching Tip

For Exercise 4, have students work a simpler problem first. Have them find the total number of cans needed to make a stack 1, 2, 3, and 4 rows high.

Differentiated Instruction

Advanced Learners L4

Have students write their own similar problems and trade papers with a partner to solve.

9-7

1. Plan

Objective
To find simple interest and compound interest

Examples
1 Finding Simple Interest
2 Graphing Simple Interest
3 Finding Compound Interest

Math Understandings: p. 434D

Math Background

Interest is both charged to those who borrow money and paid to those who allow others to use their money. Unlike simple interest, compound interest calculates interest on both the principal and the interest paid. Simple and compound interest can both be calculated with formulas.

More Math Background: p. 434D

Lesson Planning and Resources

See p. 434E for a list of the resources that support this lesson.

Bell Ringer Practice

Check Skills You'll Need
Use student page, transparency, or PowerPoint. For intervention, direct students to:
Percents, Fractions, and Decimals
Lesson 6-2
Extra Skills and Word Problems Practice, Ch. 6

9-7 Simple and Compound Interest

Check Skills You'll Need

1. **Vocabulary Review** What is a *percent*? See below.

Change each percent to a decimal.

2. 4% 0.04
3. 9% 0.09
4. 2.0% 0.020
5. 6.5% 0.065

GO for Help Lesson 6-2

Check Skills Youíll Need

1. A percent is a ratio that compares a number to 100.

What You'll Learn

To find simple interest and compound interest

New Vocabulary principal, simple interest, compound interest, balance

Why Learn This?

Money may not grow on trees, but it can grow in a bank. When you deposit money, you earn money called interest. When you borrow money, you pay interest on your loan.

The original amount you deposit or borrow is the **principal.** Interest earned only on the principal is **simple interest.**

You can use a formula to calculate simple interest.

KEY CONCEPTS Simple Interest Formula

$$I = prt$$

I is the interest earned, p is the principal, r is the interest rate per year, and t is the time in years.

EXAMPLE Finding Simple Interest

1 Gridded Response You borrow \$300 for 5 years at an annual interest rate of 4%. What is the simple interest you pay in dollars?

$I = prt$ ← **Write the formula.**

$I = (300)(0.04)(5) = 60$ ← **Substitute. Use 0.04 for 4%.**

The interest is \$60.

		6	0

Quick Check

1. Find the simple interest you pay on a \$220 loan at a 5% annual interest rate for 4 years. **\$44.00**

GO Online

Video Tutor Help
Visit: PHSchool.com
Web Code: are-0775

Differentiated Instruction Solutions for All Learners

Special Needs L1
When students use the simple and compound interest formulas, have them label the numbers that correspond to the time, the principal, interest rate, and balance. You can also encourage them to color code the numbers in their formulas.

learning style: visual

Below Level L2
Remind students to change percents to hundredths. 4%, for instance, is 0.04. *Find the simple interest on \$400 borrowed for 3 years at a 5% annual interest rate.* \$60

learning style: visual

A graph can show the increase in interest earned over time.

EXAMPLE Graphing Simple Interest

For: Interest Activity
Use: Interactive Textbook, 9-7

2 ou have $500 in an account that earns an annual rate of 5.1%. At the end of each year, you withdraw the interest you have earned. Graph the total interest you earn after 1, 2, 3, and 4 years.

Step 1 Make a table.

Time (yr)	Interest ($)
1	25.50
2	51.00
3	76.50
4	102.00

Step 2 Draw a graph.

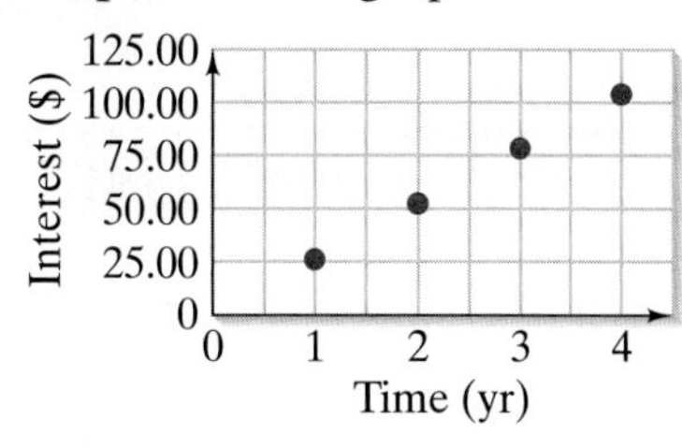

Quick Check

2. Graph the simple interest earned on $950 at an annual rate of 4.2%. **See left.**

2. 

Compound interest is interest that is paid on the original principal and on any interest that has been left in the account. The **balance** of an account is the principal plus the interest earned.

KEY CONCEPTS Compound Interest Formula

$$B = p(1 + r)^t$$

B is the balance, p is the principal, r is the annual interest rate, and t is the time in years.

EXAMPLE Finding Compound Interest

GO for Help
For help using the order of operations with exponents, go to Lesson 2-1, Example 3.

3 **Banking** ou deposit $5,000 in a bank account that pays 3.75% compound interest. What is your balance after 9 years?

$B = p(1 + r)^t$ ← **Write the formula.**
$= 5{,}000(1 + 0.0375)^9$ ← **Substitute. Use 0.0375 for 3.75%.**
$\approx 5{,}000(1.392813439)$ ← **Use a calculator to simplify the power.**
$= 6{,}964.07$ ← **Round to the nearest cent.**

The balance after 9 years is $6,964.07.

Quick Check

3. ou deposit $3,000 in a bank account that pays 4.25% compound interest. What is your balance after 12 years? **$4,943.49**

2. Teach

Activity Lab

Use before the lesson.

All in One Teaching Resources
Activity Lab 9-7: Investigating Interest

Guided Instruction

Example 1
Some students may be suspicious of banks. Discuss the benefits of saving and borrowing money in financial institutions.

Example 3
Students may make errors applying the order of operations.

PowerPoint

Additional Examples

1. Find the simple interest on $500 invested at a 3% annual interest rate for 4 years. **$60**

2. Graph the simple interest you earn on $500 at an annual rate of 3% over 4 years.

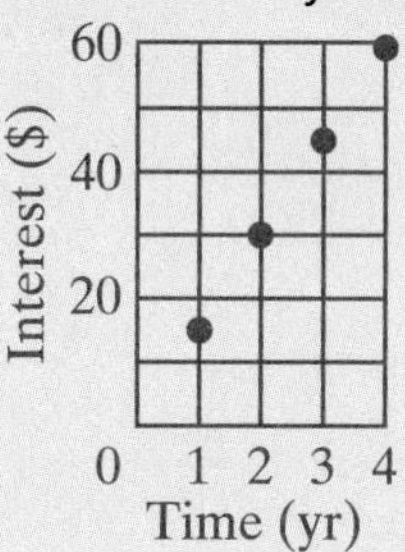

3. You deposit $500 in a bank account that pays 3% interest compounded annually. What is your balance after 4 years? **$562.75**

All in One Teaching Resources
- Daily Notetaking Guide 9-7 L3
- Adapted Notetaking 9-7 L1

Closure

- *What is the difference between simple and compound interest?* **Sample: Simple interest is calculated on the principal. Compound interest is paid on the original principal and any interest left in the account.**

Advanced Learners L4
You deposit money in an account that pays 4.25% interest compounded annually. After 3 years, the balance is $7,081.22. *What was your initial deposit?* **$6,250**

learning style: visual

English Language Learners ELL
Help students connect the term *compound interest* to their understanding of compound words. *Compound words* are composed of more than one word, put together. Compound interest is interest based on more than one total.

learning style: verbal

3. Practice

Assignment Guide

Check Your Understanding

Go over Exercises 1–4 in class before assigning the Homework Exercises.

Homework Exercises

A	Practice by Example	5–19
B	Apply Your Skills	20–29
C	Challenge	30
	Test Prep and Mixed Review	31–35

Homework Quick Check

To check students' understanding of key skills and concepts, go over Exercises 7, 13, 21, 23, and 29.

Differentiated Instruction Resources

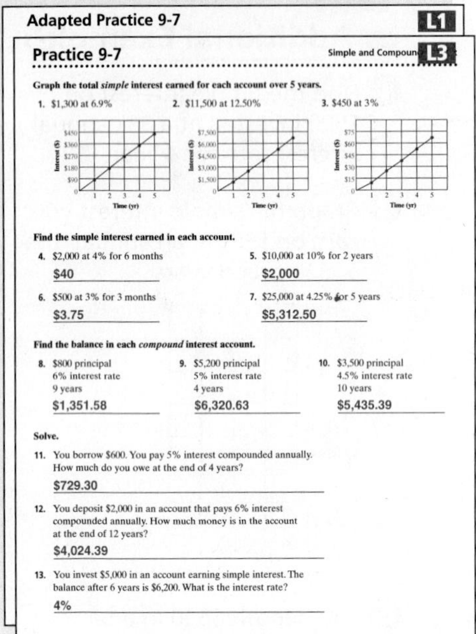

Adapted Practice 9-7 L1

Practice 9-7 Simple and Compoun[d] L3

Graph the total *simple* interest earned for each account over 5 years.

1. \$1,300 at 6.9%
2. \$11,500 at 12.50%
3. \$450 at 3%

Find the simple interest earned in each account.

4. \$2,000 at 4% for 6 months **\$40**
5. \$10,000 at 10% for 2 years **\$2,000**
6. \$500 at 3% for 3 months **\$3.75**
7. \$25,000 at 4.25% for 5 years **\$5,312.50**

Find the balance in each *compound* interest account.

8. \$800 principal, 6% interest rate, 9 years **\$1,351.58**
9. \$5,200 principal, 5% interest rate, 4 years **\$6,320.63**
10. \$3,500 principal, 4.5% interest rate, 10 years **\$5,435.39**

Solve.

11. You borrow \$600. You pay 5% interest compounded annually. How much do you owe at the end of 4 years? **\$729.30**
12. You deposit \$2,000 in an account that pays 6% interest compounded annually. How much money is in the account at the end of 12 years? **\$4,024.39**
13. You invest \$5,000 in an account earning simple interest. The balance after 6 years is \$6,200. What is the interest rate? **4%**

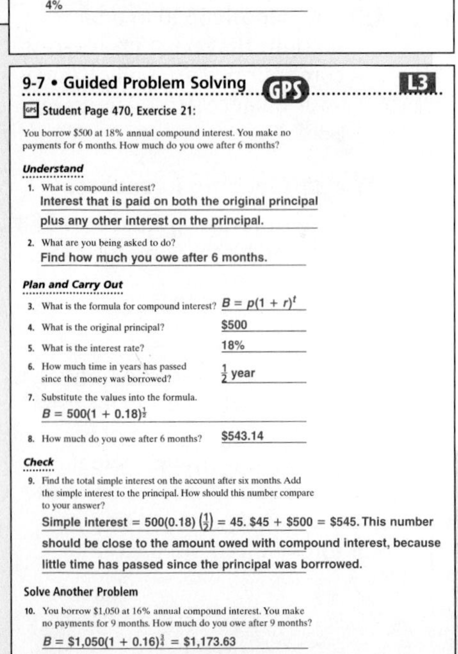

9-7 • Guided Problem Solving GPS L3

GPS Student Page 470, Exercise 21:

You borrow \$500 at 18% annual compound interest. You make no payments for 6 months. How much do you owe after 6 months?

Understand

1. What is compound interest? **Interest that is paid on both the original principal plus any other interest on the principal.**
2. What are you being asked to do? **Find how much you owe after 6 months.**

Plan and Carry Out

3. What is the formula for compound interest? $B = p(1 + r)^t$
4. What is the original principal? **\$500**
5. What is the interest rate? **18%**
6. How much time in years has passed since the money was borrowed? $\frac{1}{2}$ **year**
7. Substitute the values into the formula. $B = 500(1 + 0.18)^{\frac{1}{2}}$
8. How much do you owe after 6 months? **\$543.14**

Check

9. Find the total simple interest on the account after six months. Add the simple interest to the principal. How should this number compare to your answer? **Simple interest** $= 500(0.18)\left(\frac{1}{2}\right) = 45$. **\$45 + \$500 = \$545. This number should be close to the amount owed with compound interest, because little time has passed since the principal was borrowed.**

Solve Another Problem

10. You borrow \$1,050 at 16% annual compound interest. You make no payments for 9 months. How much do you owe after 9 months? $B = \$1,050(1 + 0.16)^{\frac{3}{4}} = \$1,173.63$

Check Your Understanding

1. **Simple interest is calculated only on the principal. Compound interest is calculated on the principal and any interest left in the account.**

1. **Vocabulary** How do simple interest and compound interest differ?
2. Find the simple interest earned on \$2000 at 10% for 6 years. **\$1,200**
3. Find the compound interest earned on \$2000 at 10% for 6 years. **\$1,543.12**
4. True or False: Doubling the principal will double the balance. **True**

Homework Exercises

For more exercises, see Extra Skills and Word Problems.

GO for Help

For Exercises	See Examples
5–8	1
9–14	2
15–19	3

A **Find the simple interest on a \$340 loan at each rate.**

5. 7% annual interest, 3 years **\$71.40**
6. 12% annual interest, 5 years **\$204.00**
7. 15% annual interest, 1 year **\$51.00**
8. 4.6% annual interest, 6 years **\$93.84**

Graph the total simple interest earned for each amount over 4 years.
9–14. See margin.

9. \$500 at 4.5%
10. \$1,200 at 6.5%
11. \$375 at 5.75%
12. \$200 at 5.0%
13. \$2,000 at 10%
14. \$2,000 at 0.5%

Find the balance in each compound interest account.

15. \$1,400 after 3 years at 5.5% **\$1,643.94**
16. \$1,800 after 11 years at 6.0% **\$3,416.94**
17. \$900 after 10 years at 4.62% **\$1,413.81**
18. \$2,500 after 50 years at 2.2% **\$7,421.38**
19. You deposit \$1,000 in a certificate of deposit that pays 5.9% compound interest. What is your balance after 3 years? **\$1,187.65**

B **GPS** 20. **Guided Problem Solving** You have \$7,500 in a college savings account that earns 4.25% compound interest. What will the account balance be at the end of 12 years? **\$12,358.74**
- What is 4.25% expressed as a decimal?
- What value do you substitute for each variable in the formula $B = p(1 + r)^t$?

21. (GPS) You borrow \$500 at 18% annual compound interest. You make no payments for 6 months. How much do you owe after 6 months? **\$543.14**

22. Suppose you invest \$2,000 for 5 years at 4% compounded annually. Which would increase your balance in 5 years the most? **A**
    - (A) Doubling the starting amount from \$2,000 to \$4,000
    - (B) Doubling the interest rate to 8% annual interest
    - (C) Doubling the time from 5 years to 10 years

23. **Writing in Math** Would you prefer \$2,000 at 6% compound interest for 5 years or \$2,000 at 5% compound interest for 6 years? Explain. **See margin.**

9–14. See back of book.
23. Answers may vary. Sample: I prefer \$2,000 at 5% compound interest for 6 years because I will eventually make more money.
34–35. See back of book.

Visit: PHSchool.com
Web Code: are-0907

The spreadsheet shows calculations using the compound interest formula. State which column corresponds to each variable.

	A	B	C	D	E
1	Year	Balance at Start of Year	Rate	Interest	Balance at End of Year
2	1st	$3,000.00	0.04	$120.00	$3,120.00
3	2nd	$3,120.00	0.04	$124.80	$3,244.80
4	3rd	$3,244.80	0.04	$129.79	$3,374.59

24. p B **25.** r C **26.** t A **27.** B E

28. Show how to calculate the amount in E4 in the spreadsheet above. E4 = B4 + D4

29. Calculator You invest $4,000 at 3% compound interest. What is the balance after 3 years? $4,370.91

30. Challenge You invest $2,000 in a simple interest account. The balance after 8 years is $2,720. What is the interest rate? 4.5%

Test Prep and Mixed Review

Practice

Multiple Choice

31. When a principal p has a compound interest rate r for t years, the balance B is given by $B = p(1 + r)^t$. Which expression represents the balance for $200 invested for 6 years at a 5% compound interest rate? C

- A $200(1.06)^5$
- B $200(0.06)(5)$
- C $200(1.05)^6$
- D $200(1.05)(6)$

32. Which situation is best represented by the graph of babies born in Texas? G

- F About 85 babies were born every hour.
- G About 85 babies were born every 2 hours.
- H About 85 babies were born every 3 hours.
- J About 85 babies were born every 4 hours.

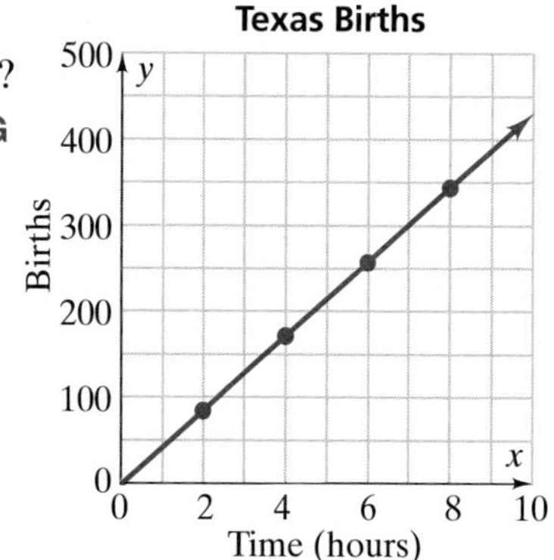

33. $\angle A$ and $\angle B$ are supplementary. The measure of $\angle A$ is 24°. What is the measure of $\angle B$? C

- A 24°
- B 66°
- C 156°
- D 336°

Use graph paper to draw each figure. 34–35. See margin.

34. triangular prism **35.** pentagonal pyramid

GO for Help

For Exercises	See Lesson
34–35	8-8

Online lesson quiz, PHSchool.com, Web Code: ara-0907

4. Assess & Reteach

Lesson Quiz

Krista opened a savings account with $100.

1. Find the simple interest earned at a 6% annual rate for 5 years. $30
2. Find the balance at 6% interest compounded annually for 5 years. $133.82

Reteaching 9-7

Simple and Compound L2

When you deposit money in a bank, the bank pays interest. **Simple interest** is interest paid only on the amount you deposited, called the **principal. Compound interest** is paid on the original principal and on any interest that has been left in the account.

Simple Interest

To find simple interest, use this formula.

Interest = principal · rate · time in years

$I = p \cdot r \cdot t$

Find the simple interest on $1,800 invested at 5% annual interest for 3 years.

$I = p \cdot r \cdot t$
$= 1,800 \cdot 0.05 \cdot 3$ ← **Use 0.05 for 5%.**
$= 270$

The interest is $270. (The balance will be $1,800 + $270, or $2,070.)

Compound Interest

To find compound interest, use this formula.

Balance = principal · (1 + rate)$^{time\ in\ years}$

$B = p \cdot (1 + r)^t$

You put $1,800 in the bank. The interest rate is 5% compounded annually. How much will be in the account after 3 years?

$B = p \cdot (1 + r)^t$
$= 1,800(1 + 0.05)^3$ ← **Use 0.05 for 5%.**
$= 1,800 \cdot (1.05)^3$
$\approx 2,083.73$

The balance is $2,083.73.

Find the simple interest earned by each account.

1. $800 principal, 4% interest rate, 5 years
   $I = p \cdot r \cdot t$
   = 800 · 0.04 · 5
   = $160
2. $800 principal, 3% interest, 4 years
   $96
3. $1,900 principal, 4.5% interest, 20 years
   $1,710
4. $20,000 principal, 3.5% interest, 15 years
   $10,500

Find the balance of each account earning compound interest.

5. $600 principal, 6% interest rate, 3 years
   $B = p(1 + r)^t$
   = 600 (1 + 0.06)^3
   = $714.61
6. $9,000 principal, 5% interest rate, 4 years
   $B = p(1 + r)^t$
   = 9,000 (1 + 0.05)^4
   = $10,939.56

Enrichment 9-7

Simple and Compound L4

Critical Thinking

Solve each problem by writing an equation.

1. You have $3,200 to invest. You decide to put some of your money in a money market fund that makes 8% simple interest and the rest in a bond that makes 5% interest. How much did you invest in each account if after 1 year you earned $220 in interest?
   $2,000 at 8% and $1,200 at 5%
2. You have 2 options for saving $12,000 in a bank. You can put it in a savings account that earns 6% simple interest for 1 year or in another account that earns 2% compound interest for 3 years. Which account earns you the most interest?
   the 2% compound interest account
3. Your parents invested $5,000 in one account for 6 years and $3,000 in another account for 4 years. If the $5,000 account earned $1,080 more in simple interest and both accounts have the same interest rate, what is the interest rate?
   6%
4. Your grandparents invest $4,200 in an account for 3 years that earns 2.5% compound interest. Your aunt earns the same amount of interest as your grandparents, but invests $3,500 in an account for 2 years earning compound interest. What is the interest rate on your aunt's account?
   4.5%
5. Your uncle has $2,500 to invest at 3.2% compound interest, but he can only invest it for 6 months. Write an equation using the formula for compound interest to show the balance.
   $B = 2,500(1 + 0.032)^{0.5}$ or $B = 2,500(1 + 0.032)^{\frac{1}{2}}$
6. If an exponent of 2 means you "square" the base, what does an exponent of the multiplicative inverse of 2, or $\frac{1}{2}$, mean?
   to find the square root of the base
7. Simplify the equation you wrote in exercise 5 to find the balance.
   $2,539.69
8. Suppose you deposit $12,000 in a bank account that pays 2.75% interest compounded annually. What is your balance after 6 months?
   $12,163.88

Alternative Assessment

Provide pairs of students with index cards. One student writes a percent less than 10% on each card while the partner writes different whole-dollar amounts on each card. Students place the cards facedown in separate stacks. They choose a card from each stack and use the percents as rates to find the simple interest on the dollar amount.

Test Prep

Resources

For additional practice with a variety of test item formats:

- Test-Taking Strategies, p. 477
- Test Prep, p. 481
- Test-Taking Strategies with Transparencies

9-8

1. Plan

Objective
To solve for a variable

Examples
1 Transforming a Formula
2 Application: Savings

Math Understandings: p. 434D

Math Background

A *formula* is a rule that shows the relationship between two or more quantities. The formula for the area of a rectangle, for instance, is $A = \ell \cdot w$, where A is area, ℓ is length, and w is width. But what if the area and length are known and the width is unknown? The formula has to be transformed. First, solve for the unknown value by using the properties of equality: $w = \frac{A}{\ell}$. Then substitute the values you know and simplify.

More Math Background: p. 434D

Lesson Planning and Resources

See p. 434E for a list of the resources that support this lesson.

Bell Ringer Practice

Check Skills You'll Need
Use student page, transparency, or PowerPoint. For intervention, direct students to:
Solving Equations by Multiplying or Dividing
Lesson 4-4
Extra Skills and Word Problems Practice, Ch. 4

Algebra

9-8 Transforming Formulas

Check Skills You'll Need

1. **Vocabulary Review** What is the *Division Property of Equality?* See below.

Solve each equation.

2. $5a = 9$ $\frac{9}{5}$
3. $12 = 4.5t$ $2\frac{2}{3}$
4. $\frac{p}{4} = -8$ -32

GO for Help Lesson 4-4

Check Skills You'll Need

1. **If you divide each side of an equation by the same nonzero number, the two sides remain equal.**

Test Prep Tip
To solve a formula for a variable, you have to get the variable alone on one side of the equation.

What You'll Learn

To solve for a variable

New Vocabulary formula

Why Learn This?

Albert Einstein discovered the relationship between mass, energy, and the speed of light. The formula $E = mc^2$ shows this relationship.

A **formula** is a rule that shows the relationship between two or more quantities.

You can use the properties of equality to transform a formula and solve for a variable.

EXAMPLE Transforming a Formula

1 The formula for the perimeter of a rectangle is $P = 2\ell + 2w$. Solve the formula for ℓ.

$P = 2\ell + 2w$ ← **Write the formula.**

$P - 2w = 2\ell$ ← **Use the Subtraction Property of Equality.**

$\frac{P - 2w}{2} = \frac{2\ell}{2}$ ← **Use the Division Property of Equality.**

$\frac{P - 2w}{2} = \ell$ ← **Simplify.**

Quick Check

1. Solve each equation for x.
   a. $y = 2x - 4$
   b. $y = x + 3$
   c. $4y = 2x + 10$

a. $x = \frac{y + 4}{2}$ b. $x = y - 3$ c. $x = 2y - 5$

You can transform formulas to solve real-world problems. First solve for the desired variable. Then substitute the values you know.

Differentiated Instruction Solutions for All Learners

Special Needs L1	**Below Level** L2
Have students draw a rectangle and assign length measures to the sides. Have them label *l* and *w*, and substitute their numbers in the transformed formula in Example 1. This will help them see how the formulas work. **learning style: visual**	Give students practice transforming simple equations. *Solve each equation for* x. $y = x + 3$, $x = y - 3$; $y = 4x$, $x = \frac{y}{4}$; $y = 3x + 2$, $x = \frac{y - 2}{3}$ **learning style: visual**

Video Tutor Help
Visit: PHSchool.com
Web Code: are-0775

EXAMPLE Application: Savings

2 Your bank account has an annual interest rate of 5.7%. How much should you invest in order to earn $100 in interest each year?

$\frac{I}{rt} = p$ ← **Use the simple interest formula $I = prt$ and solve for p.**

$\frac{100}{(0.057)(1)} = p$ ← **Substitute 100 for I, 0.057 for r, and 1 for t.**

$1{,}754.39 \approx p$ ← **Simplify.**

You should invest about $1,755.

Quick Check

2. Find the interest rate that yields $120 interest each year on $2,000.
6%

More Than One Way

Your first four test scores were 85, 98, 79, and 92. To get an average score of at least 90, what minimum score do you need on the fifth test?

Will's Method

I'll use a variable for each score and let z be the missing score.

Formula for mean → $90 = \frac{v + w + x + y + z}{5}$

Mult. Prop. of Equality → $5(90) = v + w + x + y + z$

$5(90) - (v + w + x + y) = z$ ← **Subtraction Prop. of Equality**

$5(90) - (85 + 98 + 79 + 92) = z = 96$ ← **Substitute and simplify.**

I need a minimum score of 96 on the fifth test.

Sarah's Method

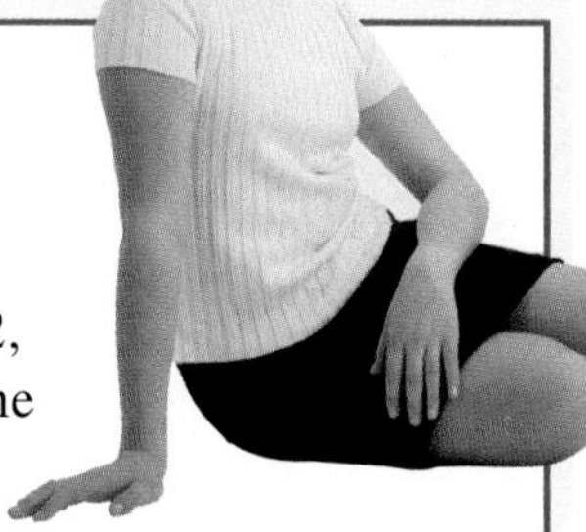

I'll work backward. The sum of the 5 scores divided by the number of scores is 90. Since $5 \cdot 90 = 450$, the sum is 450.

The sum of the first scores is $85 + 98 + 79 + 92$, or 354. So, for a total of 450, I need a score on the fifth test of $450 - 354$, or 96.

Choose a Method

Your scores on three tests are 95, 89, and 75. You can replace your lowest score with the mean of that score and a retest. For an average score of 90, what is the minimum you must score on the retest?
The minimum score must be 97.

2. Teach

Activity Lab

Use before the lesson.

All in One Teaching Resources
Activity Lab 9-8: Critical Thinking with Formulas

Guided Instruction

Example 1
Students may benefit from a quick review of inverse operations and solving two-step equations (see Lessons 4-3, 4-4, and 4-6). Remind students to begin with undoing addition or subtraction, and then undoing multiplication or division.

Example 2
Help students transform the equation $I = prt$ by undoing each variable with its own step. For instance, first divide each side of the equation by r and then divide each side by t.

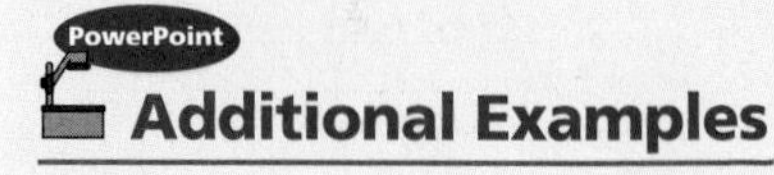

Additional Examples

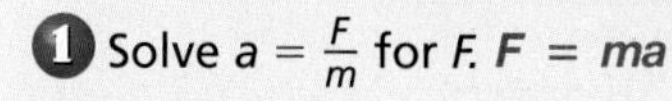

1 Solve $a = \frac{F}{m}$ for F. $F = ma$

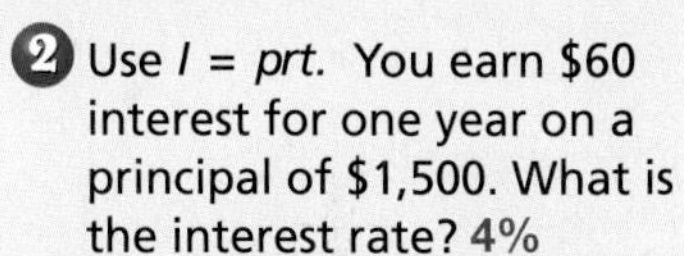

2 Use $I = prt$. You earn $60 interest for one year on a principal of $1,500. What is the interest rate? 4%

All in One Teaching Resources
- Daily Notetaking Guide 9-8 L3
- Adapted Notetaking 9-8 L1

Closure

- *How can you transform a formula to solve a problem?* **Sample: Use the properties of equality to solve the formula for the unknown variable. Then substitute the values you know and simplify.**

Advanced Learners L4
The formula for Celsius temperature given Fahrenheit temperature is $C = \frac{5}{9}(F - 32)$. *Solve the formula for* F. $F = \frac{9}{5}C + 32$

learning style: verbal

English Language Learners ELL
The More Than One Way problem may be difficult linguistically. Review what an *average* or *mean* is, and have students discuss in pairs to compare Will's and Sarah's methods for solving the problem.

learning style: verbal

3. Practice

Assignment Guide

Check Your Understanding
Go over Exercises 1–6 in class before assigning the Homework Exercises.

Homework Exercises

A Practice by Example	7–18
B Apply Your Skills	19–28
C Challenge	29
Test Prep and Mixed Review	30–34

Homework Quick Check
To check students' understanding of key skills and concepts, go over Exercises 12, 17, 23, 27, and 28.

Differentiated Instruction Resources

Adapted Practice 9-8 L1

Practice 9-8 Transforming L3

Solve each formula for the indicated variable.

1. $d = rt$, for r $r = \frac{d}{t}$
2. $P = 4s$, for s $s = \frac{P}{4}$
3. $K = C + 273$, for C $C = K - 273$
4. $S = 180(n - 2)$, for n $n = \frac{S + 360}{180}$
5. $m = \frac{a + b + c}{3}$, for a $a = 3m - b - c$
6. $P = 2b + 2h$, for b $b = \frac{P}{2} - h$
7. $V = \frac{1}{3}Bh$, for B $B = \frac{3V}{h}$
8. $A = 2(\ell w + wh + \ell h)$, for ℓ given, $w = 5$, $h = 3$, and $A = 158$
   $\ell = 8$
9. $C = \frac{5}{9}(F - 32)$, for F, given $C = 25$
   $F = 77$
10. $F = ma$, for m, given $a = 9.8$ and $F = 117.6$
    $m = 12$

Solve.

11. In 1989, Dutch ice skater Dries van Wijhe skated 200 km at an average speed of 35.27 km/hr. How long was he skating?
    5.7 h
12. A roofer calculates his bid price using the formula $P = 1.85s + 4.2f$, where s is the area of the roof in square feet and f is the length of the fascia in feet. Find the area of the roof with 190 feet of fascia and a price of $4,148.
    1,811 sq. ft

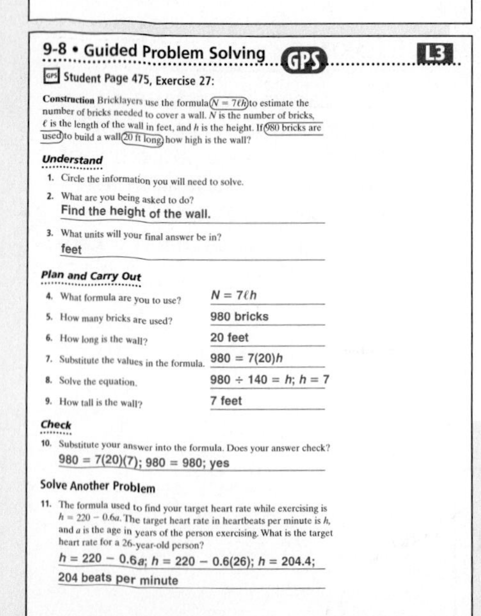

9-8 • Guided Problem Solving GPS L3

Student Page 475, Exercise 27:

Construction Bricklayers use the formula $N = 7\ell h$ to estimate the number of bricks needed to cover a wall. N is the number of bricks, ℓ is the length of the wall in feet, and h is the height. If 980 bricks are used to build a wall 20 ft long, how high is the wall?

Understand

1. Circle the information you will need to solve.
2. What are you being asked to do?
   Find the height of the wall.
3. What units will your final answer be in?
   feet

Plan and Carry Out

4. What formula are you to use? $N = 7\ell h$
5. How many bricks are used? 980 bricks
6. How long is the wall? 20 feet
7. Substitute the values in the formula. $980 = 7(20)h$
8. Solve the equation. $980 \div 140 = h$; $h = 7$
9. How tall is the wall? 7 feet

Check

10. Substitute your answer into the formula. Does your answer check?
    $980 = 7(20)(7)$; $980 = 980$; yes

Solve Another Problem

11. The formula used to find your target heart rate while exercising is $h = 220 - 0.6a$. The target heart rate in heartbeats per minute is h, and a is the age in years of the person exercising. What is the target heart rate for a 26-year-old person?
    $h = 220 - 0.6a$; $h = 220 - 0.6(26)$; $h = 204.4$; 204 beats per minute

Check Your Understanding

1. No; it cannot be used to show the relationship between two quantities.
2. Yes; it can be used to show the relationship between t and u.

Reasoning **Could the equation be a formula? Explain.** 1–2. See left.

1. $3 + 4 = 7$
2. $3t + 4u = 7$
3. $3 + 4 = 7v$

3. No; it cannot be used to show the relationship between two quantities.

Which operation must you use to solve each equation for x?

4. $1 + x = 13$ subtraction
5. $x \div 4 = 52$ multiplication
6. $7x = 35$ division

Homework Exercises

For more exercises, see Extra Skills and Word Problems.

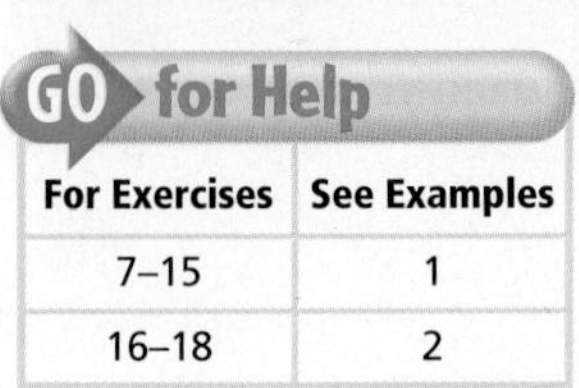

GO for Help

For Exercises	See Examples
7–15	1
16–18	2

A **Solve each equation for the variable in red.**

7. $x = yz$ $y = \frac{x}{z}$
8. $t = \frac{u + v}{2}$ $u = 2t - v$
9. $p = 3r - 5$ $r = \frac{p + 5}{3}$
10. $P = 4s$ $s = \frac{P}{4}$
11. $q = \frac{p}{r}$ $p = qr$
12. $p = s - c$ $s = p + c$
13. $A = \frac{1}{2}bh$ $h = \frac{2A}{b}$
14. $h = \frac{k}{j}$ $k = hj$
15. $I = prt$ $t = \frac{I}{pr}$

16. How long would it take to earn $6,000 in interest on a principal of $9,000 at an annual simple interest rate of 4.1%? a little longer than 16.26 years
17. You earn $1,400 simple interest on a principal of $12,500 in 4 years. What is the interest rate on your account? 2.8%
18. Suppose you borrow money for a year at a simple interest rate of 7.2%. You pay $86.40 in interest. How much have you borrowed? $1,200

B 

19. **Guided Problem Solving** A real estate agent sells a house and earns 7% commission, or $8,400. Find the selling price of the house. $120,000
    - What is a formula for the agent's commission? Let c represent the commission. Let s represent the selling price.
    - For which variable should you solve?
20. **Geometry** Find the radius of the circle at the right. about 4 cm

21. The formula for converting F degrees Fahrenheit to C degrees Celsius is $C = \frac{5}{9}(F - 32)$. Solve the equation for F to generate the formula for converting C degrees Celsius into F degrees Fahrenheit. $F = \frac{9}{5}C + 32$

Homework Video Tutor
Visit: PHSchool.com
Web Code: are-0908

22. **Choose a Method** During a 3-day premiere, a theater must sell an average of 200 tickets per night to make a profit. For the first two nights, 206 and 185 tickets were sold. How many tickets must be sold for the third night for the theater to make a profit? 209 tickets
23. **Writing in Math** Solve the formula $V = \ell wh$ for w. Then write and solve a problem involving the transformed version of the formula. $w = \frac{V}{\ell \cdot h}$; check students' work.

Solve each equation for the variable in red.

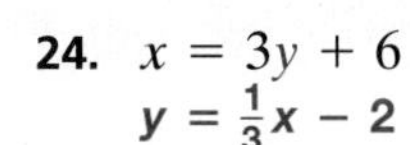

24. $x = 3y + 6$ **$y = \frac{1}{3}x - 2$**
25. $t = \frac{1}{2}r$ **$r = 2t$**
26. $w = 3n + 5m$ **$n = \frac{w - 5m}{3}$**

27. **Construction** Bricklayers use the formula $N = 7\ell h$ to estimate the number of bricks needed to cover a wall. N is the number of bricks, ℓ is the length of the wall in feet, and h is the height. If 980 bricks are used to build a wall 20 feet long, how high is the wall? **7 ft**

28. Weekly pay w is given by the formula $w = rh + 1.5rv$, where r is the regular hourly wage for a 40-h week, h is the number of regular hours you work, and v is the number of overtime hours you work beyond 40. Suppose your regular wage is \$9/h.
    a. If you work 45 hours in one week, what do you earn? **\$427.50**
    b. For weekly pay of \$468, how many overtime hours do you work? **8 h**

Ⓒ 29. **Challenge** The surface area S of a cube with side e is $S = 6e^2$. Find the side length of a cube with a surface area of 150 cm^2. **5 cm**

Test Prep and Mixed Review

Practice

Multiple Choice

30. You want to paint a wall of the playhouse but not the door in the wall. How many square feet of wall do you need to paint? **B**

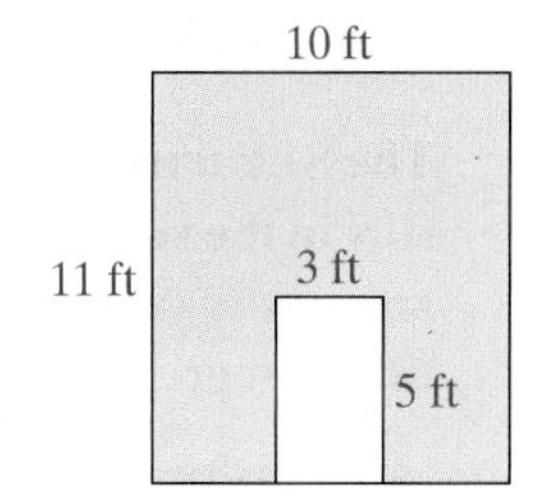

    Ⓐ 15 ft^2
    Ⓑ 95 ft^2
    Ⓒ 110 ft^2
    Ⓓ 125 ft^2

31. Ezra can swim 50 meters in 30 seconds. About how many seconds will it take him to swim a 200-meter race? **H**
    Ⓕ 60 sec Ⓖ 80 sec Ⓗ 120 sec Ⓙ 150 sec

32. Which description shows the relationship between a term and n, its position in the sequence? **A**

Term Number	1	2	3	4	5	n
Value of Term	12	24	36	48	60	■

    Ⓐ Multiply n by 12.
    Ⓑ Divide n by 12.
    Ⓒ Add 12 to n.
    Ⓓ Subtract 12 from n.

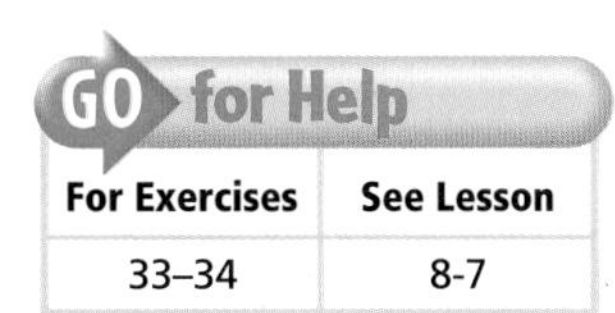

For Exercises	See Lesson
33–34	8-7

Find the missing side length. Round your answer to the nearest tenth.

33. 8 m, c, 12 m **14.4 m**

34. 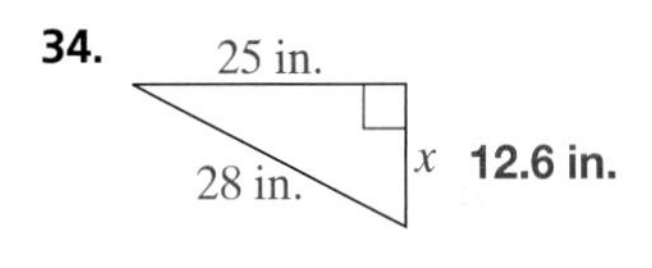

12.6 in.

Online lesson quiz, PHSchool.com, Web Code: ara-0908

4. Assess & Reteach

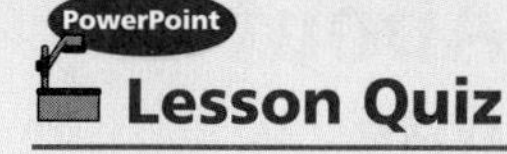

Lesson Quiz

1. Solve $P = 2RT$ for T. $T = \frac{P}{2R}$
2. Use $I = prt$. You earn \$50 interest for one year with an interest rate of 2.5%. What is the principal? **\$2,000**

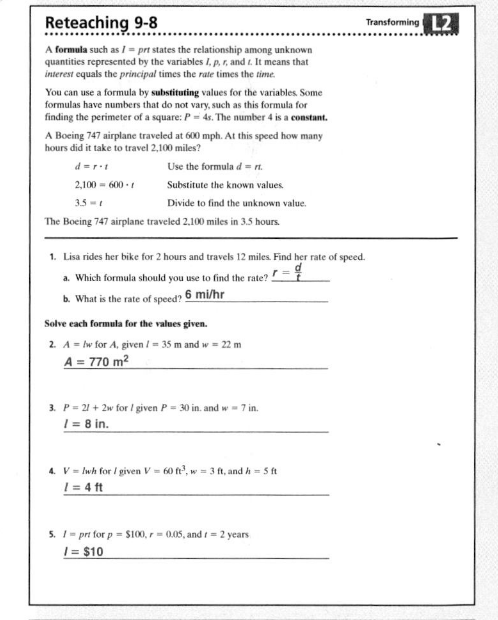

Reteaching 9-8 Transforming L2

A **formula** such as $I = prt$ states the relationship among unknown quantities represented by the variables I, p, r, and t. It means that *interest* equals the *principal* times the *rate* times the *time*.

You can use a formula by **substituting** values for the variables. Some formulas have numbers that do not vary, such as this formula for finding the perimeter of a square: $P = 4s$. The number 4 is a **constant.**

A Boeing 747 airplane traveled at 600 mph. At this speed how many hours did it take to travel 2,100 miles?

$d = r \cdot t$ Use the formula $d = rt$.
$2{,}100 = 600 \cdot t$ Substitute the known values.
$3.5 = t$ Divide to find the unknown value.

The Boeing 747 airplane traveled 2,100 miles in 3.5 hours.

1. Lisa rides her bike for 2 hours and travels 12 miles. Find her rate of speed.
   a. Which formula should you use to find the rate? $r = \frac{d}{t}$
   b. What is the rate of speed? 6 mi/hr

Solve each formula for the values given.

2. $A = lw$ for A, given $l = 35$ m and $w = 22$ m
   $A = 770\ m^2$
3. $P = 2l + 2w$ for l given $P = 30$ in. and $w = 7$ in.
   $l = 8$ in.
4. $V = lwh$ for l given $V = 60\ ft^3$, $w = 3$ ft, and $h = 5$ ft
   $l = 4$ ft
5. $I = prt$ for $p = \$100$, $r = 0.05$, and $t = 2$ years
   $I = \$10$

Enrichment 9-8 Transforming L4

Critical Thinking

The equation $d = r \times t$ relates rate (r), time (t), and distance (d).

1. If you are traveling at 55 mph, what formula would you use to find the distance you will travel over various lengths of time?
   $d = 55 \times t$
2. Sometimes you are given distances and need to know how long it would take to travel at various rates. Complete the table to find the time it would take to travel 500 miles at speeds of 40, 55, 60, 65, and 70 mph. Round your times to the nearest quarter hour, if necessary.

Rate (mph)	40	55	60	65	70
Time (hours)	$12\frac{1}{2}$	9	$8\frac{1}{4}$	$7\frac{3}{4}$	$7\frac{1}{4}$
Distance (miles)	500	500	500	500	500

3. What operation did you use to find the time? Rewrite the equation so that t stands alone on one side of the equal sign.
   Division: $t = \frac{d}{r}$
4. Rewrite the equation so that r stands alone on one side of the equal sign. Explain your reasoning.
   $r = \frac{d}{t}$; Sample answer: To find the rate, divide the distance by the time.
5. How is rewriting an equation like solving an equation?
   Sample answer: You use inverse operations to isolate the variable.
6. Why would you want to rewrite an equation?
   Sample answer: to find many solutions using the same equation; it would be faster to isolate the variable before doing each calculation.
7. The equation to find the perimeter of a regular polygon can be written as $P = n \times s$ where P is the perimeter, n is the number of sides, and s is the length of each side. How could you rewrite the equation to find the length of a side?
   $s = \frac{P}{n}$

Alternative Assessment

Each student in a pair writes a word problem modeled on Example 2 in this lesson. Partners exchange papers and write a function rule to solve each other's problem. Partners explain their function rules to each other.

Test Prep

Resources

For additional practice with a variety of test item formats:

- Test-Taking Strategies, p. 477
- Test Prep, p. 481
- Test-Taking Strategies with Transparencies

More About Formulas

Students practice solving problems involving formulas as they apply to everyday life. This will extend the skills they have learned throughout the chapter.

Guided Instruction

Before beginning the Activity, discuss formulas that apply to students' lives. Ask volunteers for examples, such as miles per hour and dollars per hour.
Ask questions such as:

- *What is an easy way to translate the formula to solve for* k? **Multiply by the reciprocal.**
- *What is another way to display the data in the graph?* **Make a table.**
- *Does it make sense to predict for k past 100%? Why?* **No, an item cannot be made of more than 100% gold.**

Activity

Have students work together as a class on the Activity. Display the pertinent data on the board or on a transparency and have volunteers solve each problem on the display.

Differentiated Instruction

Advanced Learners L4
Have students write everyday problems for the class to solve.

Resources

- chalkboard or overhead projector
- chalk or transparencies/markers

More About Formulas

There are some formulas that are not just interesting mathematically, but are also part of your life. Consider the formulas below.

ACTIVITY

Use the graph at the right for Exercises 1–3.

1. Gold is measured in karats. The formula $P = \frac{25k}{6}$ gives you the percent of gold in k karats. Graph the percents of gold for 10, 12, 14, and 18 karats. Use the graph started at the right. **See margin.**

Gold

Percent Gold P: 0, 20, 40, 60, 80, 100
Karats k: 0, 4, 8, 12, 16, 20, 24

2. Extend the graph until you reach 100% gold. Which karat value corresponds to 100% gold? **24 karat**

3. Transform the formula $P = \frac{25k}{6}$ by solving for k. Which karat value corresponds to 100% gold? $k = \frac{6P}{25}$; **24 karat**

The SPF number of sunscreen s tells you how long you can stay in the sun safely. The formula $t = sn$ represents the amount of time t it takes skin to burn with sunscreen. The amount of time n represents how long it takes skin to burn without sunscreen.

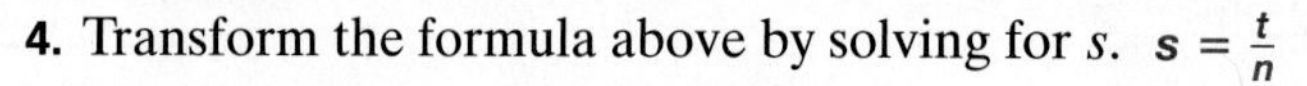

4. Transform the formula above by solving for s. $s = \frac{t}{n}$

5. A summer pool party might last 3 hours. Use the transformed formula to find what SPF number you should apply for $n = 15$ min. **SPF 12**

Data Analysis **Peter Brancazio developed the formula $v = 126 - \frac{5}{4}h$. This formula determines the vertical leap v needed for people of different heights h to slam-dunk a basketball. Both v and h are in inches.** **6–9. See margin.**

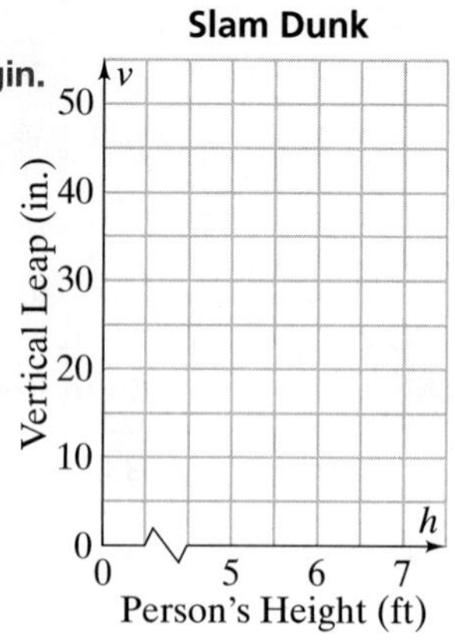

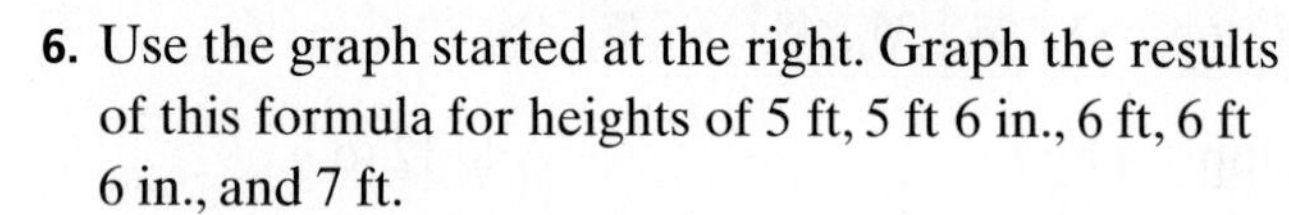

6. Use the graph started at the right. Graph the results of this formula for heights of 5 ft, 5 ft 6 in., 6 ft, 6 ft 6 in., and 7 ft.

7. What do you think 126 represents in the formula? Most basketball hoops are about 10 ft high.

8. **Reasoning** Explain why the formula uses $\frac{5}{4}$ of height instead of exact height.

9. Transform the formula above to solve for h.

1. See back of book.
6–9. See back of book.

Estimating the Answer

Using estimation can help you find an answer, check an answer, or eliminate one or more answer choices.

EXAMPLES

1. A store is having a 30%-off sale on all of its cross-training sneakers. What is the sale price on a pair of sneakers that regularly costs $84.99?
   (A) $25.50 (B) $51.99 (C) $59.49 (D) $79.99

You can estimate by changing $84.99 to a number that is easy to multiply in your head, such as $90. A 30% discount will result in a sale price that is 70% of the regular price.

$s = 0.7c$ ← **Write a function rule for sale price. Let s = sale price. Let c = regular cost.**

$\approx 0.7(90)$ ← **Substitute the estimated value.**

$= 63$ ← **Use mental math.**

The sale price will be a little less than $63. The answer is choice C.

2. The formula for converting Celsius temperatures to Fahrenheit temperatures is $F = \frac{9}{5}C + 32$. Sterling silver melts at approximately 893°C. What is the approximate Fahrenheit temperature?
   (F) 998°F (G) 1,422°F (H) 1,639°F (J) 1,995°F

$F = \frac{9}{5}(893) + 32$ ← **Substitute into the formula.**

$\approx 2(900) + 32$ ← **Estimate. $\frac{9}{5} \approx \frac{10}{5}$, or 2, and $893 \approx 900$.**

$= 1,832$ ← **This is an overestimate, since $900 > 893$ and $2 > \frac{9}{5}$.**

According to your estimate, choices F and G are too low. Since 1,832 is an overestimate, you can eliminate choice J. The correct answer is choice H.

Exercises

1. A salon is offering a 20% discount on all haircuts. What is the discount price of a cut that regularly costs $23.50? **C**
   (A) $12.50 (B) $14.75 (C) $18.80 (D) $22.00

2. The melting point of pure gold is 1,945°F. What is the approximate melting point in Celsius using the formula $C = \frac{5}{9}(F - 32)$? **F**
   (F) 1,063°C (G) 1,159°C (H) 1,205°C (J) 1,495°C

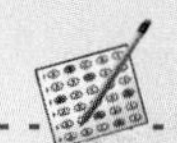

Estimating the Answer

This feature presents the valuable test-taking strategy of using estimation to find answers, eliminate answers, or check answers.

Guided Instruction

Present another way to estimate 30% of $84.99: *Think:* 30% is a little less than one third; $84 is a little less than $90. Using the estimating strategy of compatible numbers, find one third of 90, or 30. So, the answer must be about $90 − $30, or $60. Or point out how to use percent number sense to rule out choices A and D—the former is far too low and the latter, far too high.

Teaching Tip

Emphasize that eliminating answer choices is a powerful test-taking strategy.

Resources

Test-Taking Strategies with Transparencies
- Transparency 10
- Practice Sheet, p. 21

Test-Taking Strategies with Transparencies

Test-Taking Strategies: Estimating the Answer

Sometimes you can estimate to find the answer.

Example Find the sum: 0.75 + 8.23 + 5.5

A. 15.53 B. 14.48 C. 9.53 D. 21.23

Estimate: Round to the nearest whole number.

0.75 + 8.23 + 5.5

1 + 8 + 6 = 15

Both A and B are near 15, so round to the nearest tenth.

0.8 + 8.2 + 5.5 must be less than 15.

The answer is 14.48, or choice B.

Estimate to find the answer. Explain your reasoning.

1. The area of a square with side 2.7 cm is
   A. 5.4 cm^2. B. 7.29 cm^2. C. 54 cm^2. D. 72.9 cm^2.

2. Reese went grocery shopping to buy spaghetti sauce, spaghetti noodles, and a loaf of french bread. These items cost $1.59, $1.79, and $1.89. About how much should Reese's grocery bill be?
   F. less than $5 G. between $5 and $6
   H. between $6 and $7 J. more than $7

Chapter 9 Review

Resources

Student Edition
Extra Skills and Word Problems Practice, Ch. 9, p. 646
English/Spanish Glossary, p. 676
Formulas and Properties, p. 674
Tables, p. 670

All in One Teaching Resources
- Vocabulary and Study Skills 9F L3

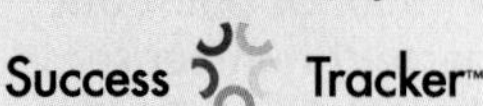

Spanish Vocabulary Workbook with Study Skills ELL
Interactive Textbook
- Audio Glossary

Online Vocabulary Quiz

Success Tracker™
Online at PHSchool.com

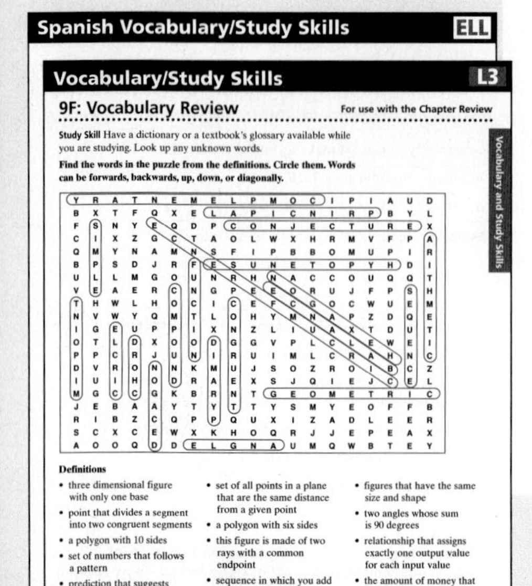

Chapter 9 Review

Vocabulary Review

arithmetic sequence (p. 442)
balance (p. 469)
compound interest (p. 469)
conjecture (p. 443)
formula (p. 472)
function (p. 452)
geometric sequence (p. 442)
principal (p. 468)
sequence (p. 442)
simple interest (p. 468)

Go Online PHSchool.com
For: Vocabulary Quiz
Web Code: arj-0951

Choose the correct term to complete each sentence.

1. A sequence is (arithmetic, geometric) if each term is found by adding the same number to the previous term. **arithmetic**
2. A (formula, function) has only 1 output value for each input value. **function**
3. (Balance, Principal) is an amount deposited or borrowed. **principal**
4. Interest paid on an original deposit and on any interest that has been left in an account is (compound, simple) interest. **compound**
5. A (conjecture, formula) is a prediction. **conjecture**

Skills and Concepts

Lesson 9-1
- To graph data and to use graphs to make predictions

Graphs can help you visualize the relationship between data. Graphs have horizontal and vertical scales. Each scale is divided into intervals.

Graph the data in each table. **6–7. See margin.**

6.

Servings	1	2	3
Calories	280	560	840

7.

Time (days)	2	4	6
Pay ($)	15	30	45

8. Use your graph from Exercise 7 to estimate the pay for 11 days. **$82.50**

Lessons 9-2, 9-3
- To describe the patterns in arithmetic and geometric sequences and use the patterns to find terms
- To use tables to represent and describe patterns

A **sequence** is a set of numbers that follow a pattern. Find each term of an **arithmetic sequence** by adding a common difference to each term. Find each term of a **geometric sequence** by multiplying each term by a common ratio. Use a table to show patterns and find unknown quantities.

Identify each sequence as *arithmetic, geometric, both,* or *neither.*

9. $2, 10, 18, 26, \dots$ **arithmetic**
10. $48, 4, \frac{1}{3}, \dots$ **geometric**
11. $0, 1, 4, 13, 40, \dots$ **neither**
12. Write a variable expression to describe the sequence $-6, -12, -18, -24, \dots$ Then find the 10th term. **$-6n$; -60**

6.
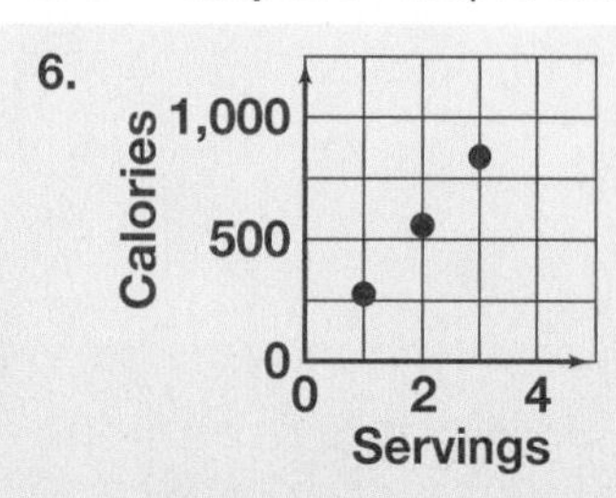

7.
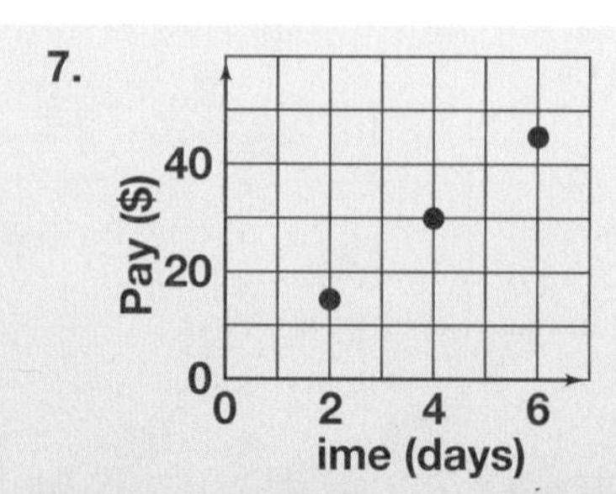

Alternative Assessment L4

Alternative Assessment

Chapter 9

HOW OLD IS THAT TREE?

Do you know a candidate for the National Register of Big Trees? The Register is a list of the largest reported specimen of each tree species in the United States. The list below shows some of these champion trees.

Tree and Location	Height	Spread
Pecan Mer Rouge, Louisiana	160 ft	95 ft
American Beech Jamaica Plain, Massachusetts	144 ft	96 ft
Red Maple Macomb County, Michigan	136 ft	95 ft
Tulip Tree Macon County, North Carolina	135 ft	55 ft
California Plane Tree Santa Barbara, California	116 ft	158 ft

Show all of your work on a separate sheet of paper.

- A tree's height is the distance from the highest point on its crown to the spot where the soil meets the trunk.
- A tree's spread is the width of the area covered by its branches.

The diagram at the right illustrates height and spread.

Height

Spread

1. Based on the data above, do you think there is a pattern related to the height and spread of these champion trees? Explain.
2. Suppose you took one kind of tree, like a maple or an oak, and measured ten examples of that kind of tree. Do you think that you would find a pattern related to height and spread? Explain.
3. The National Register requires information about a tree trunk's circumference. It is always measured at a point 4.5 ft above the ground. You can use this information to measure the tree's age. A typical tree grows about 1 in. in circumference each year. For example, if a tree has a circumference of 2 ft, the tree is about 24 years old. In metric units, a tree's circumference increases about 2.5 cm each year.
   a. Write a function table that shows the relationship between the age of a tree and its circumference, in inches.
   b. Write a function table that shows the relationship between the age of a tree and its circumference, in centimeters.

Lessons 9-4, 9-5

- To write and evaluate functions
- To find solutions to application problems using tables, rules, and graphs

A **function** is a relationship that assigns one output value for each input value. Write a function rule by looking for patterns in a table. A graph can show the relationship between inputs and outputs.

Use the graph for Exercises 13–14 and the table for Exercises 15–16.

x	y
0	2
1	4
2	6
3	8

13. Make a table for the graph. **See margin.**
14. Write a rule for the graph. $y = \frac{1}{2}x + 3$
15. Write a rule for the relationship represented by the table. $y = 2x + 2$
16. Graph the rule represented by the table. **See margin.**
17. Evaluate $y = -2x + 5$ for x-values -1, 0, 1, 2, and 3. **7, 5, 3, 1, –1**

Lesson 9-6

- To describe and sketch graphs that represent real-world situations

A graph shows how one quantity changes relative to another. In a real-world context, you need to consider what is reasonable.

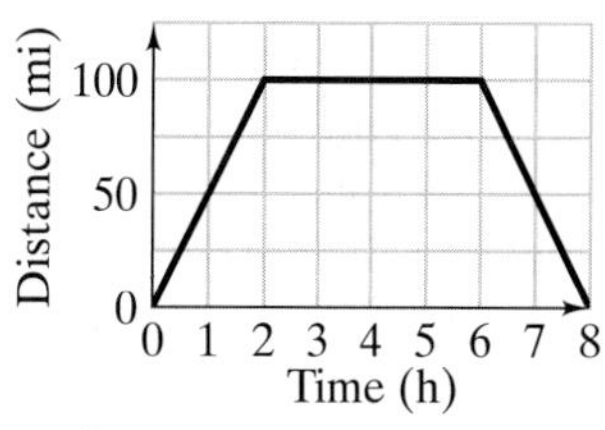

18. Describe a situation that the graph at the right might represent. **See margin.**
19. You walk at a rate of 3 mi/h for 3 h. You rest for 1 h and then walk 3 mi/h for an hour. Sketch a graph that shows the distance you travel over time. **See margin.**

Lesson 9-7

- To find simple interest and compound interest

Use the formula $I = prt$ to find **simple interest.** Use the formula $B = p(1 + r)^t$ to find the account **balance** with **compound interest.**

20. You deposit \$1,500 in an account that earns 6% simple interest. How much interest do you earn in five years? **\$450.00**
21. You deposit \$2,500 in an account that pays 5.7% interest compounded annually. What is the balance after five years? **\$3,298.49**

Lesson 9-8

- To solve for a variable

A **formula** is a rule that shows the relationship between quantities. Use properties of equality to solve for any variable in a formula.

Solve each formula for x.

22. $z = 3x + y$ $\quad x = \frac{z - y}{3}$
23. $k = -4xyz$ $\quad x = -\frac{k}{4yz}$
24. $\frac{1}{9}x - 4 = \frac{z}{3}$ $\quad x = 9\left(\frac{z}{3} + 4\right)$, or $x = 3z + 36$
25. You borrow \$200 at a 3% simple interest rate. About how much interest will you owe in 18 months? **\$9.00**

13. See back of book.
16. See back of book.
18. Answers may vary. Sample: You travel for 2 h at a constant rate of 50 mi/h. You stop for 4 h, and then return to your starting place at a constant rate of 50 mi/h.

19.

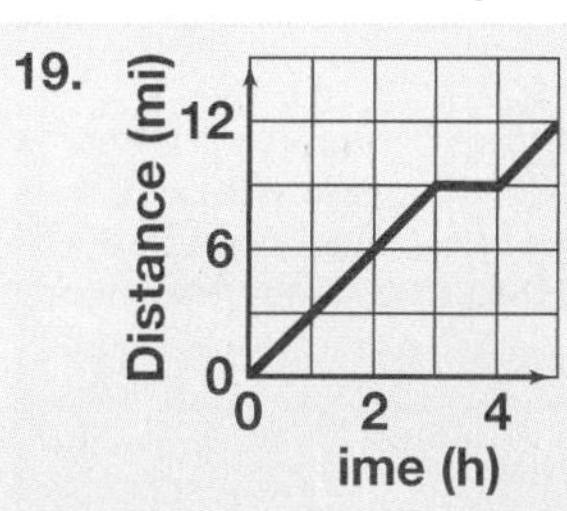

Spanish Quarter 3 Test Forms A & B ELL

Quarter 3 Test Forms D & E L2

Quarter 3 Test Forms A & B L3

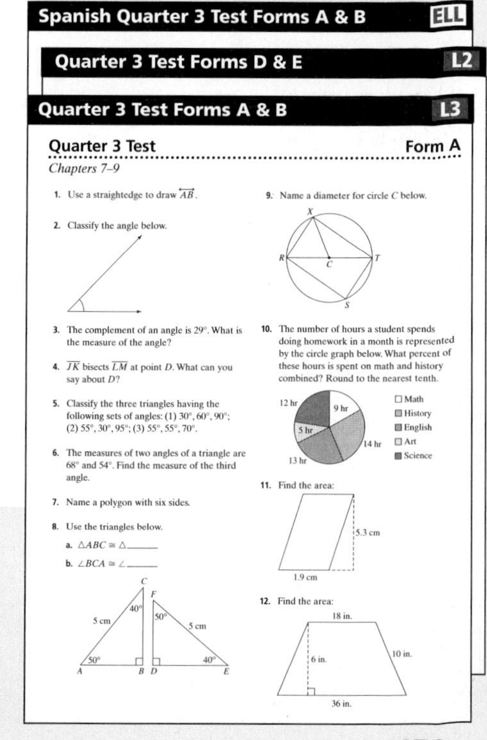
Quarter 3 Test Form A

Chapters 7–9

1. Use a straightedge to draw $\overrightarrow{AB}$.
2. Classify the angle below.
3. The complement of an angle is 29°. What is the measure of the angle?
4. $\overline{JK}$ bisects $\overline{LM}$ at point D. What can you say about D?
5. Classify the three triangles having the following sets of angles: (1) 30°, 60°, 90°; (2) 55°, 30°, 95°; (3) 55°, 55°, 70°.
6. The measures of two angles of a triangle are 68° and 54°. Find the measure of the third angle.
7. Name a polygon with six sides.
8. Use the triangles below.
   a. $\triangle ABC \cong \triangle$ ____
   b. $\angle BCA \cong \angle$ ____
9. Name a diameter for circle C below.
10. The number of hours a student spends doing homework in a month is represented by the circle graph below. What percent of these hours is spent on math and history combined? Round to the nearest tenth.
11. Find the area:
12. Find the area:

Resources

- ExamView Assessment Suite CD-ROM
  - Ch. 9 Ready-Made Test
  - Make your own Ch. 9 test
- MindPoint Quiz Show CD-ROM
  - Chapter 9 Review

Differentiated Instruction

All in One Teaching Resources
- Below Level Chapter 9 Test L2
- Chapter 9 Test L3
- Chapter 9 Alternative Assessment L4

Spanish Assessment Resources ELL
- Below Level Chapter 9 Test L2
- Chapter 9 Test L3
- Chapter 9 Alternative Assessment L4

ExamView Assessment Suite CD-ROM
- Special Needs Test L1
- Special Needs Practice Bank L1

Online Chapter 9 Test at www.PHSchool.com L3

Below Level Chapter Test L2

Chapter Test L3

Chapter Test — Form A

Chapter 9

Write a rule for each sequence. Find the next three terms.

1. 8, 14, 20, 26, ... Start with 8 and add 6 repeatedly; 32, 38, 44
2. 96, 84, 72, 60, ... Start with 96 and subtract 12 repeatedly; 48, 36, 24
3. 1, 4, 9, 16, ... Start with 1 and square the number of the term; 25, 36, 49
4. 300, 150, 75, 37.5, ... Start with 300 and divide by 2 repeatedly; 18.75, 9.375, 4.6875
5. 248, 246, 242, 236, ... Start with 248 and subtract the next even number each time; 228, 218, 206
6. 6, 12, 24, 48, ... Start with 6 and multiply by 2; 96, 192, 384

Identify each sequence in Exercises 1–6 as either *arithmetic, geometric,* or *neither.*

7. Exercise 1: arithmetic 8. Exercise 2: arithmetic 9. Exercise 3: neither
10. Exercise 4: geometric 11. Exercise 5: neither 12. Exercise 6: geometric

Suppose you get an allowance starting at \$0.05 a week. Each week thereafter, the amount you get doubles.

13. What kind of sequence is this? geometric
14. During which week will your allowance be over \$10? the ninth week

Evaluate each function for $x = 0, 1,$ and 2.

15. $y = x^2 - 10$ −10, −9, −6 16. $y = 3x + 2$ 2, 5, 8 17. $y = x - 3$ −3, −2, −1

Solve each formula for *n*.

18. $2n = 4xp - 6$ $n = 2xp - 3$ 19. $t = 100(n + 3)$ $n = \frac{t}{100} - 3$ 20. $8a + 2n = 6b$ $n = 3b - 4a$

Chapter 9 Test

Go Online PHSchool.com For: Online chapter test Web Code: ara-0952

Describe the pattern in each sequence. Find the next three terms. 1–6. See margin.

1. 1, 3, 9, 27, ...
2. 4, 9, 14, 19, ...
3. 3, 4, 6, 9, ...
4. 10, 8, 6, 4, ...
5. −23, −19, −15, ...
6. 6, 3, 1.5, 0.75, ...
7. Identify each sequence in Exercises 1–6 as *arithmetic, geometric, both,* or *neither.* See margin.
8. a. Graph the data. See margin.

Picture Framing

Photo Width (in.)	Framed Width (in.)
5	4.17
8	6.67
12	10

b. Estimate the framed width for a photo that is 9.5 in. wide. about 8 in.

c. Estimate the framed width for a photo 18 in. wide. about 15 in.

Make a function table for each function. 9–10. See margin.

9. the cost of 1 to 5 books at \$2.95 each
10. the perimeters of squares with sides of 5, 6, 7, 8, and 9 in.

Write a rule for each table. 11–13. See margin.

11.

x	y
0	−2
1	−7
2	−12
3	−17

12.

x	y
0	1
1	3
2	5
3	7

13.

x	y
0	0
1	3
2	6
3	9

14. Graph the rules in Exercises 12 and 13. See margin.

Evaluate for $x = -2$, 0, and 5.

15. $y = x - 5$ −7, −5, 0
16. $y = 9 + x$ 7, 9, 14
17. $y = 2x + 1$ −3, 1, 11
18. $y = x^2 - 1$ 3, −1, 24

19. Which is not an output for $y = 2x^2 - 5$? D
A. −3 B. 45 C. 27 D. −8

Use the graph at the right.

20. Make a table for the graph. See margin.

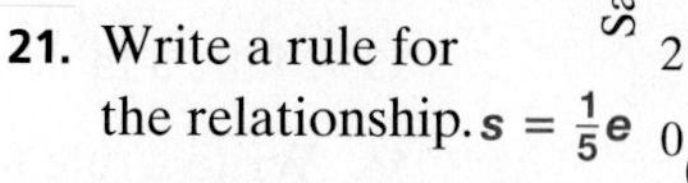

21. Write a rule for the relationship. $s = \frac{1}{5}e$
22. Predict the amount saved when \$100 is earned. \$20
23. **Sports** You dribble a basketball five times, pause briefly, and then shoot it into the basket. Sketch a graph that describes the ball's height as a function of time. See margin.
24. Describe what the graph below shows. See margin.

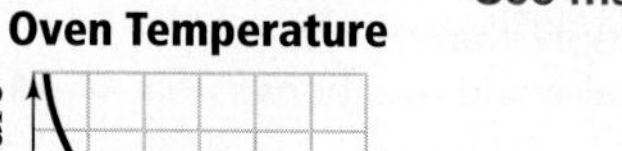

Oven Temperature

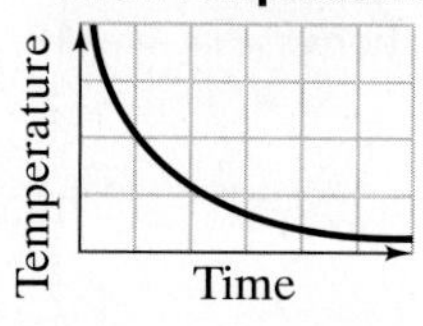

25. Suppose you borrow \$500 from a bank that charges 14.5% compound interest. What do you owe after 4 years? \$859.39

Solve each formula for *n*.

26. $3n - p = 6m$ 26. $n = \frac{6m + p}{3}$, or $n = 2m + \frac{1}{3}p$
27. $PV = nRT$ $n = \frac{PV}{RT}$
28. $s = (n - 2)180$ $n = \frac{s}{180} + 2$
29. $(m + 1)n = b$ $n = \frac{b}{m + 1}$
30. **Home Repairs** A plumber charges customers using the formula $C = 30t + 65$, where *C* is the amount he charges, and *t* is the number of hours he works. How many hours does he work when he charges \$185? 4 h
31. **Science** Density is found using the formula $D = \frac{m}{V}$, where *m* is mass in grams (g), and *V* is volume in cubic centimeters (cm^3). What is the volume of a pearl with a density of 2.72 g/cm^3 and a mass of 1.768 g? 0.65 cm^3

1–8a. See back of book.
9–14. See back of book.
20. See back of book.
23–24. See back of book.

Test Prep Practice

Reading Comprehension

Read each passage and answer the questions that follow.

Grade A To calculate grades for report cards, Ms. Sammler uses students' three test scores during the semester. She also gives credit for class participation. First, she finds the mean test score for each student, which she calls T. She then adds in the P factor—zero, three, or five points for class participation. She adds those items to get G, the grade.

1. Hari's test scores are 80, 85, and 90. He never contributes in class, so he gets zero for participation. What will his grade G be? **B**
   A 83 B 85 C 88 D 90

2. Ms. Sammler writes on the board and says, "Here is a mathematical equation that describes my system." What does she write? **G**
   F $G = \frac{1}{3}T + P$ H $3G = T + P$
   G $G = T + P$ J $T = G + P$

3. Jennifer knows her test average is 88, so she is pleasantly surprised when she gets a 93 on her report card. What is her P factor? **C**
   A 0 B 3 C 5 D 6

4. Jaime gets five points for class participation and receives an 85 on his report card. Which set could NOT have been his test scores? **J**
   F 60, 80, 100 H 79, 80, 81
   G 70, 70, 100 J 80, 81, 83

Archaeology Archaeologists find the age of materials like bone and wood using carbon-14 (C-14) dating. A tiny fraction (about one out of a trillion) of carbon atoms are radioactive C-14 that decays over time. Scientists measure the amount of C-14 left in an object to calculate its age. C-14 has a half-life of 5,700 years. This means that half of it remains after 5,700 years. In another 5,700 years, half of the remaining half will remain, and so on.

5. A 5,700-year-old bone has 10^{13} C-14 atoms. How much C-14 did it have originally? **B**
   A 10^{13} atoms C 10^{14} atoms
   B 2×10^{13} atoms D 10^{26} atoms

6. A wood fragment is about 11,000 years old. About what part of its C-14 has decayed? **H**
   F 0.25 G 0.50 H 0.75 J 1.0

7. How old would an object be if only $\frac{1}{8}$ of its original carbon-14 atoms remained? **B**
   A $2 \times 5{,}700$ years C $4 \times 5{,}700$ years
   B $3 \times 5{,}700$ years D $8 \times 5{,}700$ years

8. Radioactive potassium-40 (K-40) is found naturally in the human body. Its half-life is 1.3 billion years. After 1.3 billion years, how would the remaining percent of K-40 compare to the remaining percent of C-14? **F**
   F The percent of K-40 would be greater.
   H The percent of C-14 would be greater.
   G The same percent of each would remain.
   J There would be more K-40 atoms than C-14 atoms.

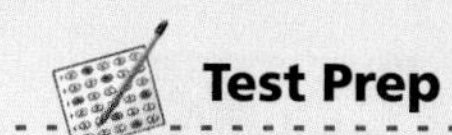

Test Prep

Resources

Test Prep Workbook

All in One Teaching Resources
- Cumulative Review L3

ExamView Assessment Suite CD-ROM
- Standardized Test Practice

Differentiated Instruction

Progress Monitoring Assessments
- Quarter 3 Test
  - Forms A & B L3
  - Forms D & E L2

Spanish Assessment Resources
- Spanish Cumulative Review ELL
- Spanish Quarter 3 Test
  - Forms A & B L3

ExamView Assessment Suite CD-ROM
- Special Needs Practice Bank L1

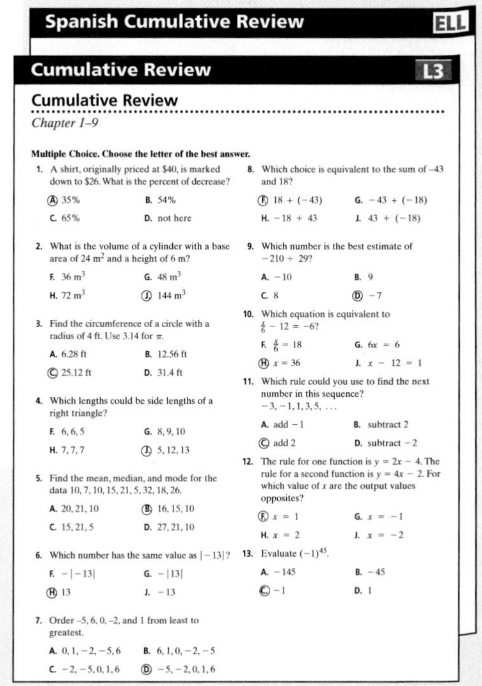

Spanish Cumulative Review ELL

Cumulative Review L3

Cumulative Review

Chapter 1–9

Multiple Choice. Choose the letter of the best answer.

1. A shirt, originally priced at \$40, is marked down to \$26. What is the percent of decrease?
   (A) 35% B. 54% C. 65% D. not here
2. What is the volume of a cylinder with a base area of 24 m^2 and a height of 6 m?
   F. 36 m^3 G. 48 m^3 H. 72 m^3 (J) 144 m^3
3. Find the circumference of a circle with a radius of 4 ft. Use 3.14 for π.
   A. 6.28 ft B. 12.56 ft (C) 25.12 ft D. 31.4 ft
4. Which lengths could be side lengths of a right triangle?
   F. 6, 6, 5 G. 8, 9, 10 H. 7, 7, 7 (J) 5, 12, 13
5. Find the mean, median, and mode for the data 10, 7, 10, 15, 21, 5, 32, 18, 26.
   A. 20, 21, 10 (B) 16, 15, 10 C. 15, 21, 5 D. 27, 21, 10
6. Which number has the same value as $|-13|$?
   F. $-|-13|$ G. $-|13|$ (H) 13 J. -13
7. Order –5, 6, 0, –2, and 1 from least to greatest.
   A. 0, 1, −2, −5, 6 B. 6, 1, 0, −2, −5 C. −2, −5, 0, 1, 6 (D) −5, −2, 0, 1, 6
8. Which choice is equivalent to the sum of –43 and 18?
   (F) $18 + (-43)$ G. $-43 + (-18)$ H. $-18 + 43$ J. $43 + (-18)$
9. Which number is the best estimate of $-210 \div 29$?
   A. −10 B. 9 C. 8 (D) −7
10. Which equation is equivalent to $\frac{x}{6} - 12 = -6$?
    F. $\frac{x}{6} = 18$ G. $6x = 6$ (H) $x = 36$ J. $x - 12 = 1$
11. Which rule could you use to find the next number in this sequence? −3, −1, 1, 3, 5, …
    A. add −1 B. subtract 2 (C) add 2 D. subtract −2
12. The rule for one function is $y = 2x - 4$. The rule for a second function is $y = 4x - 2$. For which value of x are the output values opposites?
    (F) $x = 1$ G. $x = -1$ H. $x = 2$ J. $x = -2$
13. Evaluate $(-1)^{45}$.
    A. −145 B. −45 (C) −1 D. 1

Problem Solvin

Applying Graphs

Students will use data from these two pages to answer the questions posed in Put It All Together.

Make sure all students are familiar with the saying that one dog year is like seven human years. Invite students to share stories about dogs they have had as pets. Ask questions such as:

- *If the puppy was two months old, to what human age does that correspond?* **14 months**
- *To what human age does a 1-year-old dog correspond?* **7 years**
- *a 3-year-old dog?* **21 years**
- *a 12-year-old dog?* **84 years**

Discuss how these comparisons between humans and dogs are only rough estimates based on differences in life expectancies.

Materials

- Photographs of dogs at different ages: puppy to old age
- Photographs of people at different ages: toddler, adolescent, adult, and senior

Activating Prior Knowledge

Have students call upon their experiences working with animals or as pet owners to share their knowledge of how long birds, cats, fish, horses, hamsters, gerbils, and turtles live.

Guided Instruction

Have volunteers read the paragraphs and "call-outs." Ask students to show with their hands how big a dragonfly with a 27-inch wingspan would have been.

Then focus them on the enlarged photo of the dragonfly. Ask:

- *To what part of the dragonfly's body are its wings attached?* **thorax**
- *How many wings does the dragonfly have?* **4**
- *How many legs does the dragonfly have?* **6**

Problem Solving Application

Applying Graphs

Through the Ages Different animals have different life expectancies, which means they live different lengths of time. Reptiles, such as turtles, can live for more than a hundred years. Some insects, such as dragonflies, may live only a year or two.

You may know the saying that one dog year is like seven human years. That's because, on average, people live about seven times as long as dogs. But how do other animals compare? How old is a five-year-old cat or horse in human years? You can use graphs to make these comparisons.

The Long and the Short of It

Dragonflies can live as long as 6 or 7 years or as short as 6 or 7 months.

Ancient Insects

Archaeologists have found fossils of dragonflies over 200 million years old. Those dragonflies had wingspans up to 27 in. across. Today, the largest dragonflies have wingspans about 5 or 6 in. across.

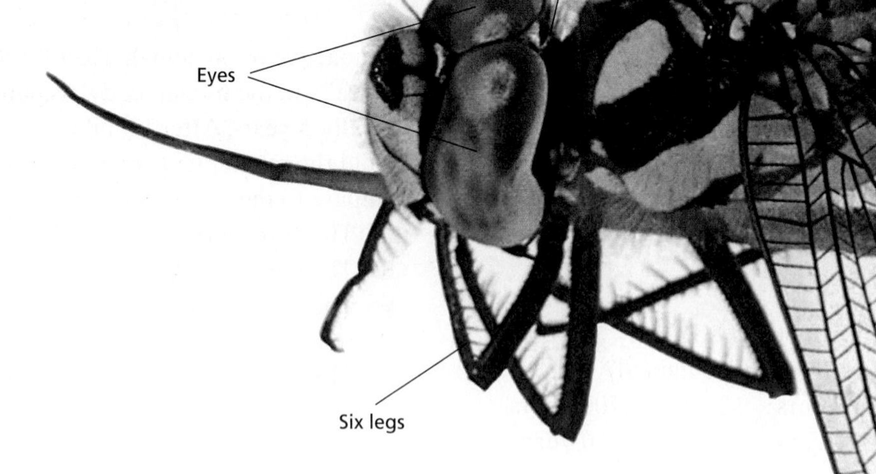

1a-b.

Animal-Human Age

Human Age (years): 0, 10, 20, 30, 40, 50, 60, 70, 80, 90, 100

Animal Age (years): 0, 4, 8, 12, 16

c. about 56 years old

d. Answers may vary. Sample: $13 \div 7 \approx 2$ years old.

Put It All Together

Data File Use the information on these two pages and on page 673 to make a graph comparing animal and human life expectancies.

1. **a.** Start your graph by labeling the x-axis "Animal Age" and labeling the y-axis "Human Age." Use a scale up to 40 on the x-axis and 100 on the y-axis.
   b. Graph the point $(5, 35)$ to show that 5 dog years are equivalent to 35 human years. Draw a line through the origin and this point.
   c. Use the line you graphed to find the "human age" of an 8-year-old dog.
   d. Use the line you graphed to find your age in "dog years."
2. Use your graph from Question 1. Use 100 years as a human's maximum life span and choose at least three animals. Plot the points that compare each animal's maximum life span to a human's maximum life span. Use the points and the origin to draw lines for each animal.
3. Pick one of the animals from your graph. Compare the animal's life span to a human's life span. At what age would the animal be likely to start kindergarten? At what age would it graduate from high school? Mark those points on your graph.
4. Suppose you get a newborn kitten when you are 32 years old. How old will you be when you and the cat are the same age in human years?
5. **Reasoning** Many animals mature more quickly than people. They learn to walk a few hours after birth, and they are able to care for themselves in less than a year. Of the animals in your graph, which animal matures most rapidly compared to people? How can you tell by looking at the graph?

Go Online PHSchool.com
For: Information about dragonflies
Web Code: are-0953

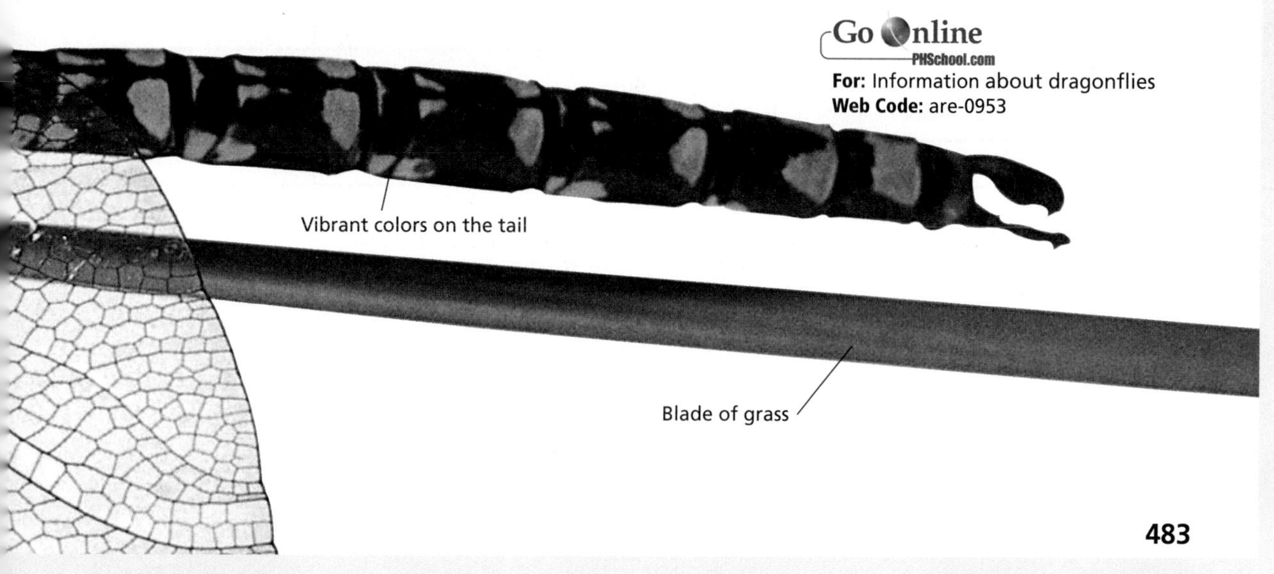

2. Check students' graph.
3. Answers may vary. Samples are in table.

Animal	Kindergarten	Graduate H.S.
Human	5	18
Dragonfly	0.005	0.0018
Rat	0.165	0.594
Salmon	0.65	2.34
Spoonbill	0.5	1.8
Rabbit	0.65	2.34
Cat	1.4	5.04
Dog(Spaniel)	1.0	3.6

4. just over 44 years old
5. Check students' answer; the animal with the steepest line graph matures most rapidly compared to people.

Activity

Help students work in pairs to make the graphs and answer the questions.

Exercise 2 Have students predict how the slopes of the lines of the graphs of the animals they pick will compare with that of the line representing the dog's age. Discourage students from selecting *Dragonfly* or *Rat* because of their very short life spans.

Exercise 3 Remind students to start at the y-axis and move horizontally to find the correct points on the lines. Invite students to pick other age benchmarks to use to compare animal life spans to human life spans.

Exercise 4 Students can solve this problem by drawing a double line graph, making a table, or using *try, check, and revise.* Start by having students calculate the number of "cat years" in one human year.

Science Connection
Dragonflies have been around for 200 million years. Invite interested students to research how old other animals are.

Differentiated Instruction

Special Needs L1
Review how to make and label a line graph. Help students understand the different intervals on the two axes.

10 Graphing in the Coordinate Plane

Chapter at a Glance

Lesson Titles, Objectives, and Features	Assessment	NCTM Standards	Local Standards
10-1 Graphing Points in Four Quadrants • To name and graph points on a coordinate plane **10-1b Activity Lab:** Geometry in the Coordinate Plane	Lesson Quiz	1, 2, 3, 6, 7, 8, 9, 10	
10-2 Graphing Linear Equations • To find solutions of linear equations and to graph linear equations **10-2b Activity Lab, Data Analysis:** Representing Data **Guided Problem Solving:** Practice Solving Problems	Lesson Quiz	1, 2, 3, 6, 7, 8, 9, 10	
10-3 Finding the Slope of a Line • To find the slope of a line and use it to solve problems **10-3b Activity Lab, Technology:** Exploring Slope	Lesson Quiz Checkpoint Quiz 1	1, 2, 3, 6, 7, 8, 9, 10	
10-4 Graphing NonLinear Relationships • To graph nonlinear relationships **Vocabulary Builder:** High-Use Academic Words	Lesson Quiz	1, 2, 3, 6, 7, 8, 9, 10	
10-5a Activity Lab: Slides, Flips, and Turns **10-5 Translations** • To graph and write rules for translations	Lesson Quiz	1, 2, 3, 6, 7, 8, 9, 10	
10-6 Line Symmetry and Reflections • To identify lines of symmetry and to graph reflections **Extension:** Exploring Tessellations	Lesson Quiz Checkpoint Quiz 2	1, 2, 3, 6, 7, 8, 9, 10	
10-7 Rotational Symmetry and Rotations • To identify rotational symmetry and to rotate a figure about a point	Lesson Quiz	1, 2, 3, 6, 7, 8, 9, 10	
Problem Solving Application: Applying Coordinates			

NCTM Standards 2000
1 Number and Operations **2** Algebra **3** Geometry **4** Measurement **5** Data Analysis and Probability
6 Problem Solving **7** Reasoning and Proof **8** Communication **9** Connections **10** Representation

Correlations to Standardized Tests

All content for these tests is contained in *Prentice Hall Math,* Course 2. This chart reflects coverage in this chapter only.

	10-1	10-2	10-3	10-4	10-5	10-6	10-7
Terra Nova CAT6 (Level 17)							
Number and Number Relations							
Computation and Numerical Estimation							
Operation Concepts							
Measurement							
Geometry and Spatial Sense	✔	✔	✔	✔	✔	✔	✔
Data Analysis, Statistics, and Probability							
Patterns, Functions, Algebra	✔	✔	✔	✔	✔	✔	✔
Problem Solving and Reasoning	✔	✔	✔	✔	✔	✔	✔
Communication	✔	✔	✔	✔	✔	✔	✔
Decimals, Fractions, Integers, and Percent							
Order of Operations							
Terra Nova CTBS (Level 17)							
Decimals, Fractions, Integers, Percents							
Order of Operations, Numeration, Number Theory							
Data Interpretation							
Pre-algebra	✔	✔	✔	✔	✔	✔	✔
Measurement							
Geometry	✔	✔	✔	✔	✔	✔	✔
ITBS (Level 13)							
Number Properties and Operations							
Algebra	✔	✔	✔	✔	✔	✔	✔
Geometry	✔	✔	✔	✔	✔	✔	✔
Measurement							
Probability and Statistics							
Estimation							
SAT10 (Int 3 Level)							
Number Sense and Operations							
Patterns, Relationships, and Algebra	✔	✔	✔	✔	✔	✔	✔
Data, Statistics, and Probability							
Geometry and Measurement	✔	✔	✔	✔	✔	✔	✔
NAEP							
Number Sense, Properties, and Operations							
Measurement							
Geometry and Spatial Sense					✔	✔	✔
Data Analysis, Statistics, and Probability							
Algebra and Functions	✔	✔	✔	✔			

CAT6 California Achievement Test, 6th Ed. **CTBS** Comprehensive Test of Basic Skills **ITBS** Iowa Test of Basic Skills, Form M
SAT10 Stanford Achievement Test, 10th Ed. **NAEP** National Assessment of Educational Progress 2005 Mathematics Objectives

Math Background

Skills Trace

BEFORE Chapter 10

Course 1 introduced graphing in the coordinate plane and basic reflections.

DURING Chapter 10

Course 2 reviews and extends graphing in the coordinate plane with linear and nonlinear relationships, slope, and transformations.

AFTER Chapter 10

Throughout this course, students represent data in graphs, tables, and equations.

10-1 Graphing Points in Four Quadrants

Math Understandings

- You can name any point on a Cartesian coordinate plane by an ordered pair, and you can graph any ordered pair of real numbers as a point on the plane.
- In a coordinate plane, the interval for the negative half of each axis is the same as the corresponding positive half. However, the *x*-axis and the *y*-axis may use different intervals.

A **coordinate plane** is a grid formed by a **horizontal** number line called the ***x*-axis** and a **vertical** number line called the ***y*-axis.** An **ordered pair** gives the coordinates of the location of a point. The ordered pair (0, 0) indicates the **origin** *0*, where the axes intersect. The first number in an ordered pair is the ***x*-coordinate,** which tells the number of horizontal units a point is from 0. The second number is the ***y*-coordinate,** which tells the number of vertical units a point is from 0.

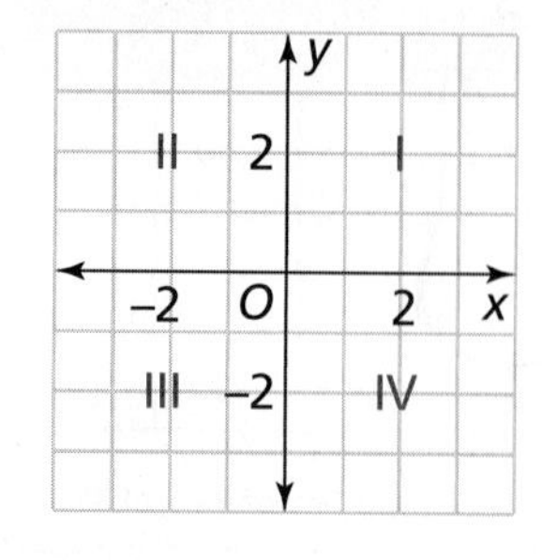

Lines that are parallel to the *x*-axis are **horizontal lines**. Lines that are parallel to the *y*-axis are **vertical lines**. The *x*-axis and *y*-axis divide the coordinate plane into four **quadrants.**

10-2 Graphing Linear Equations

Math Understandings

- Every point on the graph of a linear equation represents a solution, and every solution to the equation represents a point on the graph.
- Although you usually graph points where both values are integers, a linear equation can have infinitely many solutions that are not integers and could be used to graph the equation.

Any ordered pair that makes an equation true is a solution of the equation. The **graph of an equation** is the graph of all the points with coordinates that are solutions of the equation. An equation is a **linear equation** when the graph of its solutions is a line.

10-3 Finding the Slope of a Line

Math Understandings

- The graph of a linear equation has a constant slope.
- To determine the slope, use any pair of points on a line.
- When you move from left to right along a line, if it goes upward, the slope is positive; if it goes downward, the slope is negative.

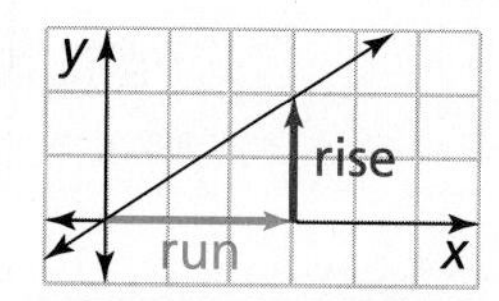

$\text{slope} = \frac{\text{rise}}{\text{run}}$

Slope is a ratio that describes the steepness of a line. The slope is the ratio between the vertical change **(rise)** and the horizontal change **(run)**. A line parallel to the *y*-axis is infinitely steep and has no slope. A line parallel to the *x*-axis is level and has a slope of zero.

10-4 Graphing Nonlinear Relationships

Math Understandings

- A graph that is not a straight line may be represented by a nonlinear equation.

A **nonlinear equation** is an equation with a graph that is not a straight line.

10-5 Translations

Math Understandings

- Translations, reflections, and rotations are called rigid motion transformations because they affect only position and leave shape and size unchanged.

A **transformation** is a change of the position, shape, or size of a figure. Three types of transformations that change position only are slides (translations), flips (reflections), and turns (rotations).

A **translation** is a transformation that moves every point of a figure the same distance and in the same direction. The figure you get after a translation is the **image** of the original. You use prime notation (A', said as "A prime") to identify an image point. ABC has been translated 2 units right and 4 units down. The rule for this translation is $(x, y) \rightarrow (x + 2, y - 4)$.

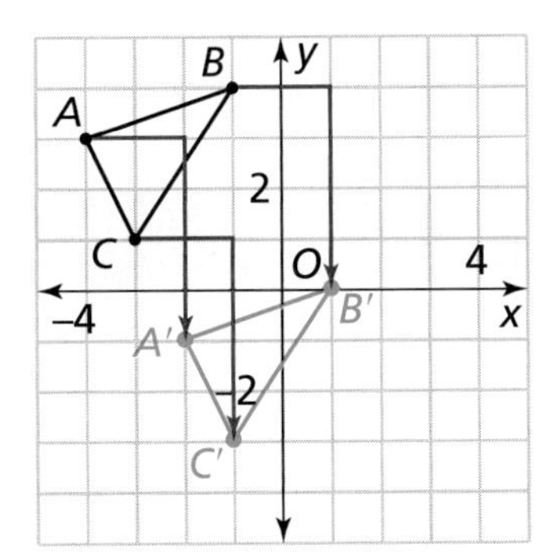

10-6 Line Symmetry and Reflections

Math Understandings

- To reflect a figure, you need to know the line of reflection.

A figure has **line symmetry** if a line can be drawn through the figure so that one side is a mirror image of the other. A **line of symmetry,** such as the dashed lines in this figure, divides a figure into mirror images.

If you fold along a line of symmetry, one side of the figure fits exactly onto the other side. As shown in this figure, some figures have more than one line of symmetry.

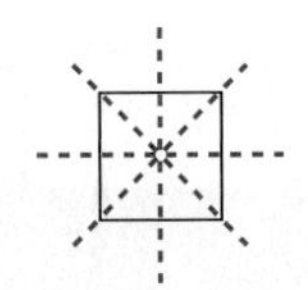

A **reflection** is a transformation that flips a figure over a **line of reflection**. When a figure is reflected, the image is congruent to the original.

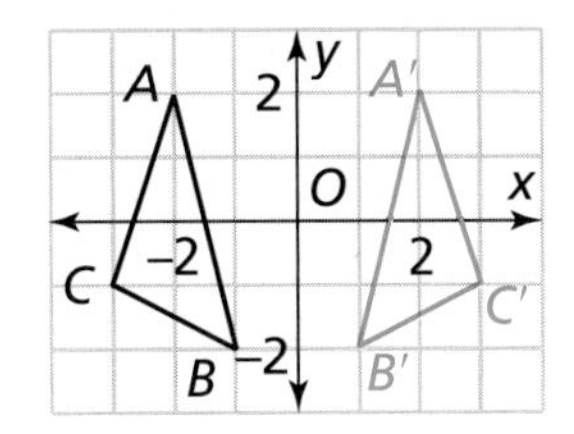

10-7 Rotational Symmetry and Rotations

Math Understandings

- To rotate a figure, you need to know the center of rotation and the angle of rotation.

A **rotation** is a transformation that turns a figure about a fixed point called the **center of rotation**. In this text, all rotations are counterclockwise. A figure has **rotational symmetry** if it can be rotated 180° or less and match the original figure. When a figure has rotational symmetry, the angle measure that it rotates is the **angle of rotation**.

Example: Does this figure have rotational symmetry? Yes. Its angle of rotation is 360° ÷ 2, or 180°.

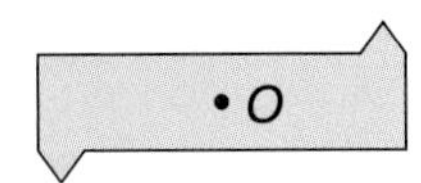

Additional Professional Development Opportunities

Math Background Notes for Chapter 10: Every lesson has a Math Background in the PLAN section.

Research Overview, Mathematics Strands
Additional support for these topics and more is in the front of the Teacher's Edition.

LessonLab
LessonLab, a Pearson Education company, offers comprehensive, facilitated professional development designed to help teachers to improve student achievement. To learn more, please visit lessonlab.com.

Chapter 10 Resources

	10-1	10-2	10-3	10-4	10-5	10-6	10-7	For the Chapter
Print Resources								
L3 Practice	•	•	•	•	•	•	•	
L1 Adapted Practice	•	•	•	•	•	•	•	
L3 Guided Problem Solving	•	•	•	•	•	•	•	
L2 Reteaching	•	•	•	•	•	•	•	
L4 Enrichment	•	•	•	•	•	•	•	
L3 Daily Notetaking Guide	•	•	•	•	•	•	•	
L1 Adapted Daily Notetaking Guide	•	•	•	•	•	•	•	
L3 Vocabulary and Study Skills Worksheets	•		•	•	•		•	•
L3 Daily Puzzles	•	•	•	•	•	•	•	
L3 Activity Labs	•	•	•	•	•	•	•	
L3 Checkpoint Quiz			•			•		
L3 Chapter Project								•
L2 Below Level Chapter Test								•
L3 Chapter Test								•
L4 Alternative Assessment								•
L3 Cumulative Review								•
Spanish Resources ELL								
L3 Practice	•	•	•	•	•	•	•	•
L3 Vocabulary and Study Skills Worksheets	•		•	•	•		•	•
L3 Checkpoint Quiz			•			•		
L2 Below Level Chapter Test								•
L3 Chapter Test								•
L4 Alternative Assessment								•
L3 Cumulative Review								•
Transparencies								
Check Skills You'll Need	•	•	•	•	•	•	•	
Additional Examples	•	•	•	•	•	•	•	
Problem of the Day	•	•	•	•	•	•	•	
Classroom Aid	•		•					
Student Edition Answers	•	•	•	•	•	•	•	•
Lesson Quiz	•	•	•	•	•	•	•	
Test-Taking Strategies								•
Technology								
Interactive Textbook Online	•	•	•	•	•	•	•	•
StudentExpress™ CD-ROM	•	•	•	•	•	•	•	•
Success Tracker™ Online Intervention	•	•	•	•	•	•	•	•
TeacherExpress™ CD-ROM	•	•	•	•	•	•	•	•
PresentationExpress™ with QuickTake Presenter CD-ROM	•	•	•	•	•	•	•	•
ExamView® Assessment Suite CD-ROM	•	•	•	•	•	•	•	•
MindPoint® Quiz Show CD-ROM								•
Prentice Hall Web Site: PHSchool.com	•	•	•	•	•	•	•	•

Also available: **Prentice Hall Assessment System**
- Progress Monitoring Assessments
- Skills and Concepts Review
- Test Prep Workbook

Other Resources
Algebra Readiness Tests
All-in-One Student Workbook
All-in-One Student Workbook, Adapted Version
Multilingual Handbook

Solution Key
Math Notes Study Folder
Spanish Cumulative Assessment

Where You Can Use the Lesson Resources

Here is a suggestion, following the four-step teaching plan, for how you can incorporate Differentiated Instruction Resources into your teaching.

	Instructional Resources L3	Differentiated Instruction Resources
1. Plan		
Preparation Read the Math Background in the Teacher's Edition to connect this lesson with students' previous experience. **Starting Class** **Check Skills You'll Need** Assign these exercises to review prerequisite skills. **New Vocabulary** Help students pre-read the lesson by pointing out the new terms introduced in the lesson.	**Math Background** **Math Understandings** **Transparencies & PresentationExpress™ with QuickTake Presenter CD-ROM** Check Skills You'll Need Problem of the Day **Resources** Vocabulary and Study Skills	**Spanish Support** ELL Vocabulary and Study Skills
2. Teach		
L3 Guided Instruction Use the Activity Labs to build conceptual understanding. Teach each Example. Use the Teacher's Edition side column notes for specific teaching tips, including Error Prevention notes. Use the Additional Examples found in the side column (and on transparency and PowerPoint) as an alternative presentation for the content. After each Example, assign the Quick Check exercise for that Example to get an immediate assessment of student understanding. Use the Closure activity in the Teacher's Edition to help students attain mastery of lesson content.	**Student Edition** Activity Lab **Resources** Daily Notetaking Guide Activity Lab **Transparencies & PresentationExpress™ with QuickTake Presenter CD-ROM** Additional Examples Classroom Aids **ExamView® Assessment Suite CD-ROM**	**Teacher's Edition** Every lesson includes suggestions for working with students who need special attention. L1 Special Needs L2 Below Level L4 Advanced Learners ELL English Language Learners **Resources** L1 Adapted Daily Notetaking Guide **Multilingual Handbook**
3. Practice		
Assignment Guide **Check Your Understanding** Use these questions to check students' understanding before you assign homework. **Homework Exercises** Assign homework from these leveled exercises in the Assignment Guide. A Practice by Example B Apply Your Skills C Challenge Test Prep and Mixed Review **Homework Quick Check** Use these key exercises to quickly check students' homework.	**Transparencies & PresentationExpress™ with QuickTake Presenter CD-ROM** Student Answers **Resources** Practice Guided Problem Solving Vocabulary and Study Skills Activity Lab Daily Puzzles **ExamView® Assessment Suite CD-ROM**	**Spanish Support** ELL Practice ELL Vocabulary and Study Skills **Resources** L1 Adapted Practice L4 Enrichment
4. Assess & Reteach		
Lesson Quiz Assign the Lesson Quiz to assess students' mastery of the lesson content. **Checkpoint Quiz** Use the Checkpoint Quiz to assess student progress over several lessons.	**Transparencies & PresentationExpress™ with QuickTake Presenter CD-ROM** Lesson Quiz **Resources** Checkpoint Quiz	**Resources** L2 Reteaching ELL Checkpoint Quiz Success Tracker™ Online Intervention **ExamView® Assessment Suite CD-ROM**

KEY **L1** Special Needs **L2** Below Level **L3** For All Students **L4** Advanced, Gifted **ELL** English Language Learners

Graphing in the Coordinate Plane

Answers are in the back of the textbook.

For intervention, direct students to:

Graphing Integers
Lesson 1-6
Extra Skills and Word Problems Practice, Ch. 1

Evaluating Algebraic Expressions
Lesson 4-1
Extra Skills and Word Problems Practice, Ch. 4

Using Exponents
Lesson 2-1
Extra Skills and Word Problems Practice, Ch. 2

Classifying Angles
Lesson 7-2
Extra Skills and Word Problems Practice, Ch. 7

Spanish Vocabulary/Study Skills ELL

Vocabulary/Study Skills L3

10A: Graphic Organizer For use before Lesson 10-1

Study Skill As you learn new skills, practice them regularly. Each time you work a problem, it should seem easier than the time before. Keep a list of problems that you want to spend extra time practicing.

Write your answers.

1. What is the chapter title? Graphing in the Coordinate Plane
2. How many lessons are there in this chapter? 7
3. What is the topic of the Test-Taking Strategies page? Answering the Question Asked
4. Complete the graphic organizer below as you work through the chapter.
   - In the center, write the title of the chapter.
   - When you begin a lesson, write the lesson name in a rectangle.
   - When you complete a lesson, write a skill or key concept in a circle linked to that lesson block.
   - When you complete the chapter, use this graphic organizer to help you review. Check students' diagrams.

Vocabulary and Study Skills

Algebra

CHAPTER 10 Graphing in the Coordinate Plane

What You've Learned

- In Chapter 4, you graphed inequalities on a number line.
- In Chapter 9, you represented patterns using tables, rules, and graphs.

GO for Help

For Exercises	See Lesson
1–4	1-6
5–8	4-1
9–12	2-1
13–18	7-2

Graphing Integers

Algebra **Graph each integer and its opposite.**

1. 7 1–4. See margin.
2. -5
3. -3
4. 6

Evaluating Algebraic Expressions

Evaluate each expression using the values $a = 5$, $c = 2$, and $g = 7$.

5. $9g$ 63
6. $-2a$ −10
7. $8c - 10$ 6
8. $2a - 5c$ 0

Using Exponents

Simplify each expression.

9. 5^3 125
10. $(-2)^6$ 64
11. 11^3 1,331
12. $(-10)^4$ 10,000

Classifying Angles

Classify each angle as *acute, right, obtuse,* or *straight.*

13. $m\angle A = 43^\circ$ acute
14. $m\angle B = 90^\circ$ right
15. $m\angle C = 148^\circ$ obtuse
16. $m\angle D = 180^\circ$ straight
17. $m\angle E = 167^\circ$ obtuse
18. $m\angle F = 79^\circ$ acute

1. 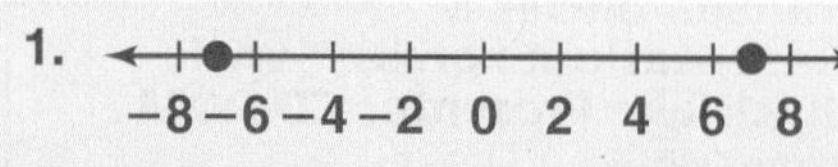

2. 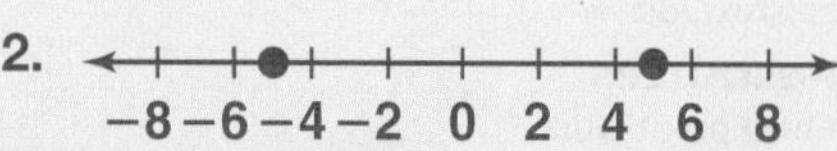

3. 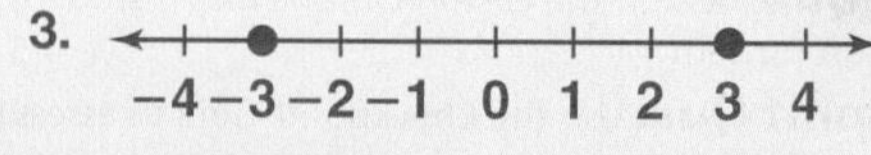

4. 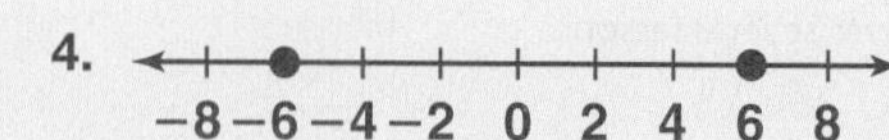

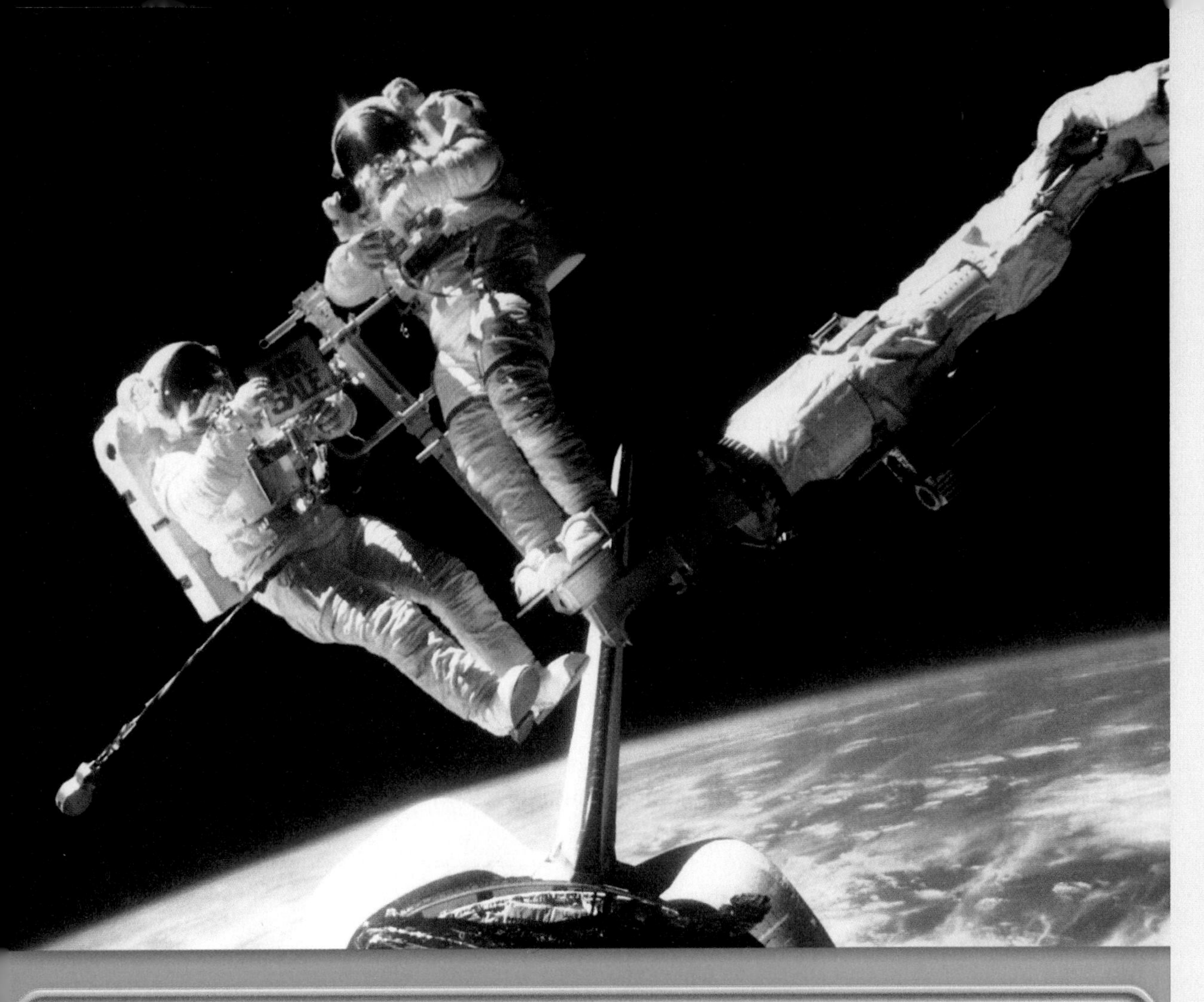

Chapter 10 Overview

In this chapter, students study graphing in the coordinate plane. They explore linear relationships and the concept of slope as they graph points and lines in all four quadrants. They also investigate nonlinear relationships, translations, reflections, rotations, and symmetry.

Activating Prior Knowledge

In this chapter, students will build on their knowledge of coordinates and ordered pairs, graphing, and using patterns and tables. They will graph and explore relationships of points on the coordinate plane.
Ask questions such as:

- *How can you describe the line formed by connecting points (3,1), (3,2), and (3,3)?* **vertical line; line parallel to *y*-axis**
- *Using the formula* d = rt, *what related formula can you write to solve for* r? $r = \frac{d}{t}$
- *What is the next point in this sequence: (2,3), (3,4), (4,5)?* **(5,6)**

What You'll Learn Next

- In this chapter, you will graph and name points in the coordinate plane.
- You will graph linear and nonlinear relationships.
- You will graph transformations and identify symmetry.

Key Vocabulary

- coordinate plane (p. 486)
- image (p. 510)
- linear equation (p. 492)
- line symmetry (p. 514)
- nonlinear equation (p. 504)
- ordered pair (p. 486)
- reflection (p. 515)
- rotation (p. 519)
- rotational symmetry (p. 519)
- transformation (p. 510)
- translation (p. 510)
- *x*-axis (p. 486)
- *x*-coordinate (p. 486)
- *y*-axis (p. 486)
- *y*-coordinate (p. 486)

Problem Solving Application On pages 528 and 529, you will work an extended activity on coordinates.

10-1

1. Plan

Objective
To name and graph points on a coordinate plane

Examples
1 Naming Coordinates
2 Graphing Points
3 Graphing Polygons

Math Understandings: p. 484C

Math Background

Maps allow users to find the locations of places. A coordinate plane, such as a map, allows users to identify the locations of points. The plane is formed by a horizontal number line (*x*-axis) and a vertical number line (*y*-axis). A point on the plane is described by coordinates, which give the position of the point in relation to the origin. The coordinates are identified by an ordered pair. The first number of the pair describes a point's location to the left or right of the origin. The second number describes a point's location up or down.

More Math Background: p. 484C

Lesson Planning and Resources

See p. 484E for a list of the resources that support this lesson.

Bell Ringer Practice

Check Skills You'll Need
Use student page, transparency, or PowerPoint. For intervention, direct students to:
Comparing and Ordering Integers
Lesson 1-6
Extra Skills and Word Problems Practice, Ch. 1

Algebra

10-1 Graphing Points in Four Quadrants

Check Skills You'll Need

1. **Vocabulary Review** How can you use a number line to show that 4 and −4 are *opposites*? **1–7. See back of book.**

Graph each number and its opposite on a number line.

2. 5	3. −3
4. 1	5. −10
6. 12	7. 8

GO for Help
Lesson 1-6

What You'll Learn

To name and graph points on a coordinate plane

New Vocabulary coordinate plane, *x*-axis, *y*-axis, ordered pair, quadrants, origin, *x*-coordinate, *y*-coordinate

Why Learn This?

Maps use coordinates to help you locate streets and buildings. You can use coordinates to describe the location of a point on a grid.

A **coordinate plane** is a grid formed by a horizontal number line called the ***x*-axis** and a vertical number line called the ***y*-axis.**

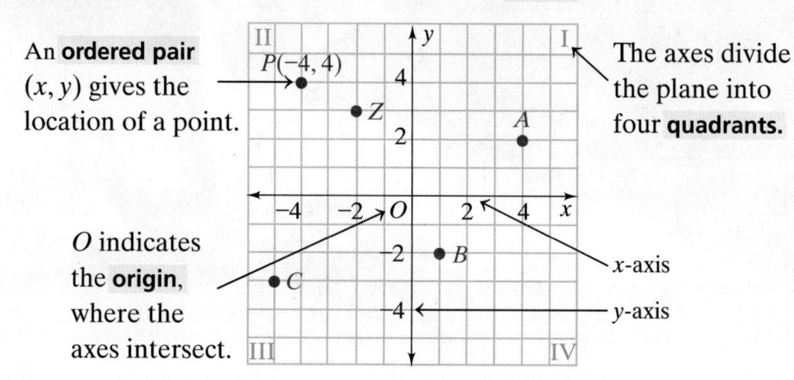

An ordered pair (x, y) gives the location of a point.
O indicates the origin, where the axes intersect.
The axes divide the plane into four quadrants.
x-axis
y-axis
II
I
III
IV
P(−4, 4)
Z
A
B
C

The first number of an ordered pair is the ***x*-coordinate.** It tells the number of horizontal units a point is from O. The second number is the ***y*-coordinate.** It tells the number of vertical units a point is from O.

Test Prep Tip
You can remember that the *x*-coordinate is first in an ordered pair because *x* comes before *y* in the alphabet.

EXAMPLE 1 Naming Coordinates

Multiple Choice Name the coordinates of point Z in the graph above.

Ⓐ (2, 3) Ⓑ (−2, −3) Ⓒ (−2, 3) Ⓓ (2, −3)

Point Z is 2 units to the left of the y-axis, so the x-coordinate is −2.
Point Z is 3 units up from the x-axis, so the y-coordinate is 3.

The coordinates of point Z are (−2, 3). The correct answer is C.

Quick Check

1. Name the coordinates of A, B, and C in the graph above. **(4, 2), (1, −2), (−5, −3)**

Differentiated Instruction Solutions for All Learners

Special Needs L1
For Example 1, have students run their finger along the *x*-axis. Have them run their finger along the *y*-axis. Have students point to quadrant 1, 2, 3, and 4 when you name them.
learning style: tactile

Below Level L2
Draw a coordinate plane on the board. Call out ordered pairs for students to point out and graph. Turn the activity into a game in which students gain points for plotting ordered pairs correctly.
learning style: tactile

You can use an ordered pair to graph a point in a coordinate plane.

EXAMPLE Graphing Points

2 Graph point $A(3, -5)$ in a coordinate plane. In which quadrant does the point lie?

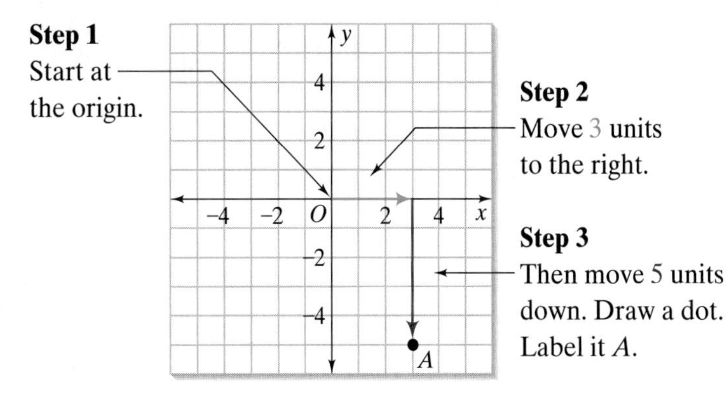

Step 1
Start at the origin.
Step 2
Move 3 units to the right.
Step 3
Then move 5 units down. Draw a dot. Label it A.

The point $A(3, -5)$ lies in quadrant IV.

Quick Check

2. Graph point $R(-3, 5)$. In which quadrant does the point lie? **See left.**

2. Quadrant II

When you graph a polygon in a coordinate plane, first graph each vertex. Then draw line segments to connect adjacent vertices.

EXAMPLE Graphing Polygons

3 **Archaeology** Archaeologists record the location of objects they find by making a grid. Suppose you use graph paper to represent a rectangular dig that measures 9 ft by 6 ft. Draw a rectangle in a coordinate plane. Use (0, 0) as one vertex and label all vertices.

Mark (0, 0) as one vertex.

From the origin, count 9 units right and mark a vertex at (9, 0).

From the origin, count 6 units up and mark a vertex at (0, 6).

Mark the fourth vertex at (9, 6).

Draw the sides of the rectangle.

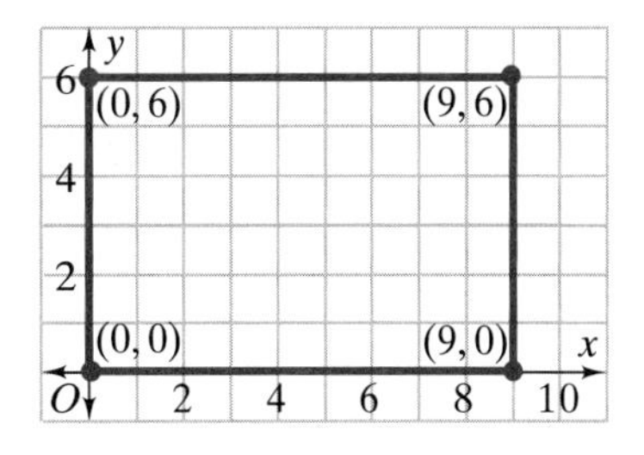

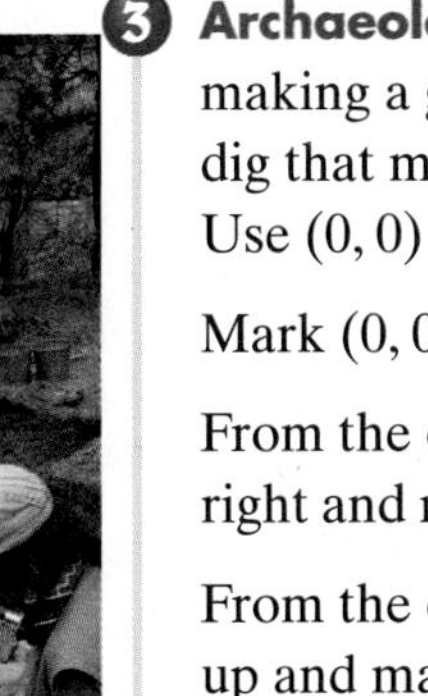

Careers Archaeologists excavate artifacts to study how people lived in past cultures.

Quick Check

3. In a coordinate plane, draw a different rectangle for the dig described above. Use (0, 0) as one vertex and label all vertices. **See back of book.**

2. Teach

Activity Lab

Use before the lesson.

All in One Teaching Resources

Activity Lab 10-1: A Classroom Coordinate Plane

Guided Instruction

Example 1
Trace the path from the origin to the point. Begin by tracing along the x-axis to line up with the point. Then slide up or down to reach the point.

PowerPoint

Additional Examples

1 Write the coordinates of A and B.
$A(-2, 2)$, $B(-3, -2)$

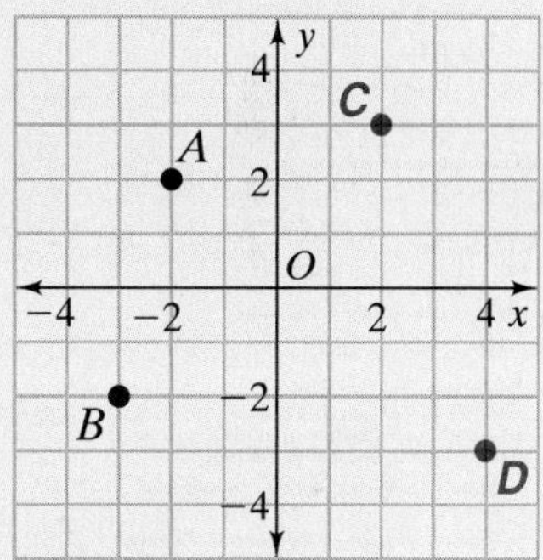

2 Graph each point on a coordinate plane. **See above.**

a. $C(2, 3)$ **b.** $D(4, -3)$

3 Graph the points and connect to make a rectangle:
$(2, 6), (-4, 6), (2, -3), (-4, -3)$.
See back of the book.

All in One Teaching Resources

- Daily Notetaking Guide 10-1 L3
- Adapted Notetaking 10-1 L1

Closure

- *What is a coordinate plane?* **Sample: a grid formed by a horizontal number line (x-axis) and a vertical number line (y-axis).**
- *What is an ordered pair and what do the two numbers represent?* **An ordered pair gives the location of a point. The first number is the location to the left or right of the origin. The second number is the point up or down.**

Advanced Learners L4
For each ordered pair, give two more: one that would be on a vertical line and one on a horizontal line.
Samples:
$(-5, 4)$ vert: $(-5, 2)$, horiz: $(4, 4)$
$(a, 2)$ vert: $(a, -3)$, horiz: $(6, 2)$
learning style: visual

English Language Learners ELL
Use terms that are synonymous with vertical and horizontal to describe the x-axis and y-axis. For example, explain: *The x-axis is a number line that goes west and east or left and right. This means it is horizontal. The y-axis goes up and down or north and south. This means it is vertical.* **learning style: verbal**

3. Practice

Assignment Guide

Check Your Understanding
Go over Exercises 1–6 in class before assigning the Homework Exercises.

Homework Exercises

A	Practice by Example	7–26
B	Apply Your Skills	27–36
C	Challenge	37
	Test Prep and Mixed Review	38–40

Homework Quick Check
To check students' understanding of key skills and concepts, go over Exercises 13, 20, 33, 34, and 35.

Differentiated Instruction Resources

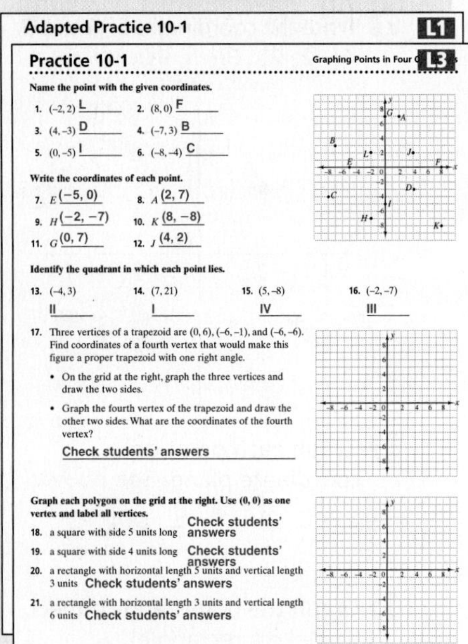

Adapted Practice 10-1 L1

Practice 10-1 Graphing Points in Four Quadrants L3

Name the point with the given coordinates.

1. (−2, 2) L
2. (8, 0) F
3. (4, −3) D
4. (−7, 3) B
5. (0, −5) I
6. (−8, −4) C

Write the coordinates of each point.

7. *E* (−5, 0)
8. *A* (2, 7)
9. *H* (−2, −7)
10. *K* (8, −8)
11. *G* (0, 7)
12. *J* (4, 2)

Identify the quadrant in which each point lies.

13. (−4, 3) II
14. (7, 21) I
15. (5, −8) IV
16. (−2, −7) III

17. Three vertices of a trapezoid are (0, 6), (−6, −1), and (−6, −6). Find coordinates of a fourth vertex that would make this figure a proper trapezoid with one right angle.
    - On the grid at the right, graph the three vertices and draw the two sides.
    - Graph the fourth vertex of the trapezoid and draw the other two sides. What are the coordinates of the fourth vertex?

    Check students' answers

Graph each polygon on the grid at the right. Use (0, 0) as one vertex and label all vertices.

18. a square with side 5 units long Check students' answers
19. a square with side 4 units long Check students' answers
20. a rectangle with horizontal length 5 units and vertical length 3 units Check students' answers
21. a rectangle with horizontal length 3 units and vertical length 6 units Check students' answers

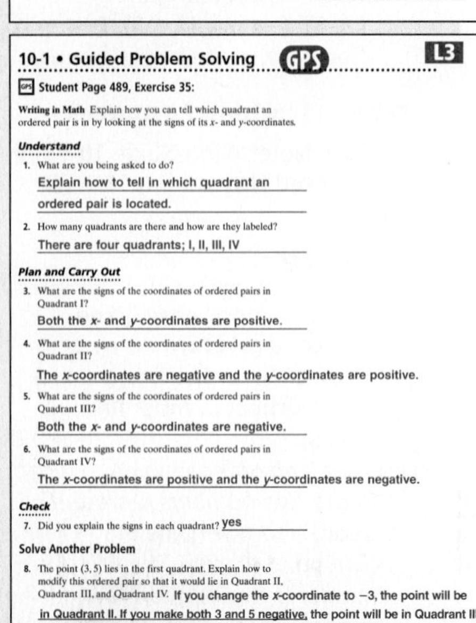

10-1 • Guided Problem Solving GPS L3

Student Page 489, Exercise 35:

Writing in Math Explain how you can tell which quadrant an ordered pair is in by looking at the signs of its *x*- and *y*-coordinates.

Understand

1. What are you being asked to do?
   Explain how to tell in which quadrant an ordered pair is located.
2. How many quadrants are there and how are they labeled?
   There are four quadrants; I, II, III, IV

Plan and Carry Out

3. What are the signs of the coordinates of ordered pairs in Quadrant I?
   Both the *x*- and *y*-coordinates are positive.
4. What are the signs of the coordinates of ordered pairs in Quadrant II?
   The *x*-coordinates are negative and the *y*-coordinates are positive.
5. What are the signs of the coordinates of ordered pairs in Quadrant III?
   Both the *x*- and *y*-coordinates are negative.
6. What are the signs of the coordinates of ordered pairs in Quadrant IV?
   The *x*-coordinates are positive and the *y*-coordinates are negative.

Check

7. Did you explain the signs in each quadrant? yes

Solve Another Problem

8. The point (3, 5) lies in the first quadrant. Explain how to modify this ordered pair so that it would lie in Quadrant II, Quadrant III, and Quadrant IV. If you change the *x*-coordinate to −3, the point will be in Quadrant II. If you make both 3 and 5 negative, the point will be in Quadrant III. If you change the *y*-coordinate to −5, the point will be in Quadrant IV.

Check Your Understanding

1. **Vocabulary** Name the coordinates of the origin. **(0, 0)**

Match each point with its coordinates.

2. *J* **B**
3. *K* **A**
4. *L* **C**
5. *M* **D**

A. (3, 4)
B. (−3, 4)
C. (3, −4)
D. (−3, −4)

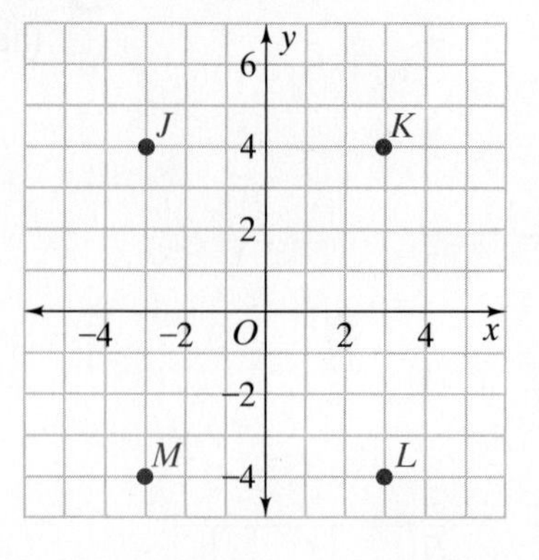

6. Which axis in a coordinate plane is vertical? Which axis is horizontal? ***y*-axis; *x*-axis**

Homework Exercises

For more exercises, see Extra Skills and Word Problems.

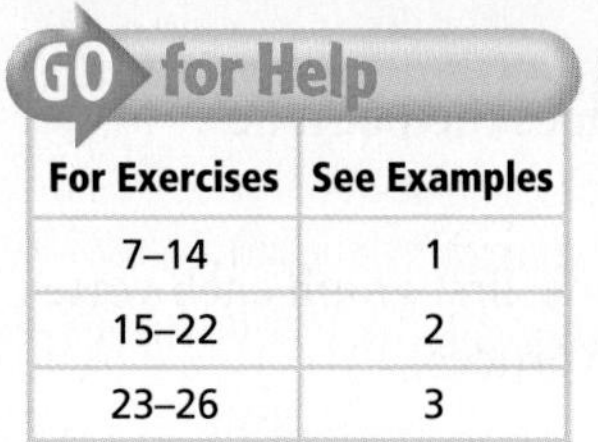

GO for Help

For Exercises	See Examples
7–14	1
15–22	2
23–26	3

A **Name the coordinates of each point.**

7. *G* **(−2, 2)**
8. *H* **(2, −3)**
9. *J* **(3, 0)**
10. *K* **(4, 4)**
11. *L* **(−3, −4)**
12. *M* **(−4, 1)**
13. *N* **(1, 3)**
14. *P* **(0, −2)**

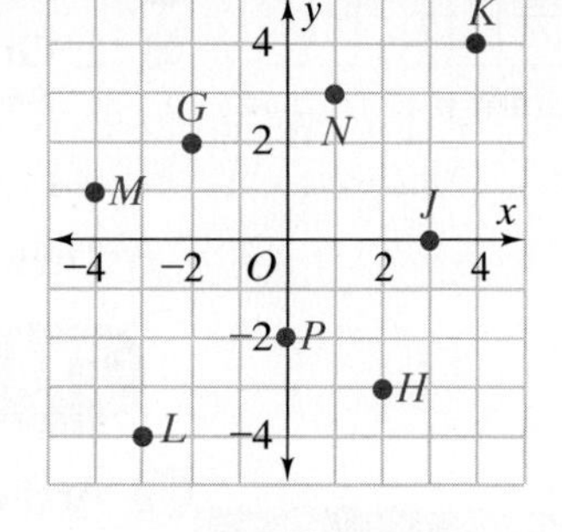

Graph each point on the same coordinate plane. Name the quadrant in which each point lies. See margin for graph.

15. $Q(1, -4)$ **IV**
16. $R(-5, 3)$ **II**
17. $S(-3, -2)$ **III**
18. $T(-6, 2)$ **II**
19. $U(5, -3)$ **IV**
20. $V(2, 6)$ **I**
21. $W(6, 2)$ **I**
22. $Z(-4, -4)$ **III**

Graph each polygon. Use (0, 0) as one vertex. Label all vertices.
23–26. See margin.

23. a square with side 3 units long
24. a square with side 6 units long
25. a rectangle with horizontal length 4 units and vertical length 2 units
26. a rectangle with horizontal length 2 units and vertical length 4 units

B

27. **Guided Problem Solving** Three vertices of a square are (−2, −4), (3, −4), and (3, 1). What are the coordinates of the fourth vertex?
    - *Draw a Picture* by graphing the three vertices and drawing the two sides of the square.
    - Graph the fourth vertex of the square. What are its coordinates? **(−2, 1)**

15–22. See back of book.
23–26. See back of book.
31–33. See back of book.
35–36. See back of book.

GO Online
Homework Video Tutor
Visit: PHSchool.com
Web Code: are-1001

Without graphing, name the quadrant in which each point (x, y) lies.

28. $x < 0$ and $y < 0$ III
29. $x > 0$ and $y < 0$ IV
30. $x < 0$ and $y > 0$ II

31. Geometry Graph a triangle with vertices (2, 5), (8, 3), and (2, 1). Classify the triangle by its angles and by its sides. **See margin.**

32. Science A bee's wings move at a rate of about 150 beats per second. The expression $150x$ gives the number of beats in x seconds. Graph the ordered pairs $(x, 150x)$ for $x = 1$, $x = 2$, and $x = 3$. **See margin.**

33. Reasoning List the coordinates of three points on the red line. If the x-coordinate of a point on the line is 37, what is the y-coordinate? Explain. **See margin.**

34. A scale drawing of a rectangular board shows three vertices at $(-3, -2)$, $(-3, 2)$, and $(3, 2)$. Each unit represents 2 ft. Find the board's dimensions. **12 ft by 8 ft**

GPS **35. Writing in Math** Explain how you can tell which quadrant an ordered pair is in by looking at the signs of its x- and y-coordinates. **See margin.**

36. Open-Ended Graph a parallelogram in a coordinate plane so that each vertex is in a different quadrant. Label all vertices. **See margin.**

37. Answers may vary. Sample: $(-6, -2) \rightarrow (-6, -4) \rightarrow (-2, -4) \rightarrow (-2, 0) \rightarrow (-6, 0) \rightarrow (-6, 6) \rightarrow (-3, 6) \rightarrow (-3, 4)$

C **37. Challenge** A robot arm must move the black peg in the diagram at the right onto the white square. The peg must be moved around—not over—the red walls. List the coordinates of the vertices of a path the robot arm might follow to move the peg. **See left.**

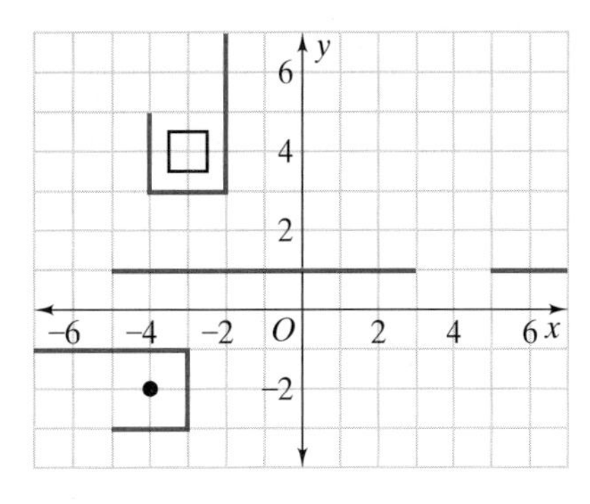

Test Prep and Mixed Review — Practice

Multiple Choice

38. Which point has the coordinates $(-3, 5)$ in the graph at the right? **C**

Ⓐ Point L Ⓒ Point N
Ⓑ Point M Ⓓ Point P

39. Which three-dimensional figure has only squares as faces? **H**

Ⓕ Pyramid Ⓗ Cube
Ⓖ Cone Ⓙ Cylinder

GO for Help

For Exercise	See Lesson
40	8-5

40. The radius of a circle is 11 cm long. Find the area of the circle to the nearest tenth. **380.1 cm²**

Online lesson quiz, PHSchool.com, Web Code: ara-1001

4. Assess & Reteach

Lesson Quiz

Write the coordinates of each point.

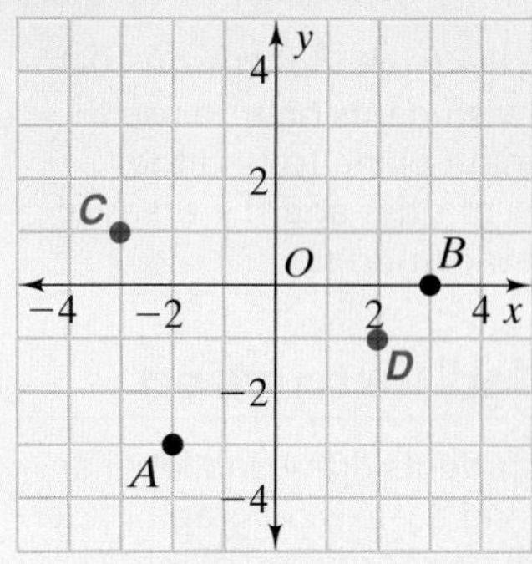

1. $A(-2, -3)$ **2.** $B(3, 0)$

Graph each point on the same coordinate plane. **See above.**

3. $C(-3, 1)$ **4.** $D(2, -1)$

Name the coordinates necessary to complete the figure named.

5. Square with sides of 4 units, one vertex at (6, 6) **Sample: (2, 2), (6, 2), (2, 6)**

6. Rectangle 3 units by 5 units, one vertex at (−5, 3) **Sample: (−5, 6), (0, 3), (0, 6)**

Alternative Assessment

Partners take turns. One partner gives the coordinates for a point, and the other graphs it. The partner graphing the point verbalizes how the point is being placed, such as, "right four, down two" for (4, −2).

Test Prep

Resources

For additional practice with a variety of test item formats:

- Test-Taking Strategies, p. 523
- Test Prep, p. 527
- Test-Taking Strategies with Transparencies

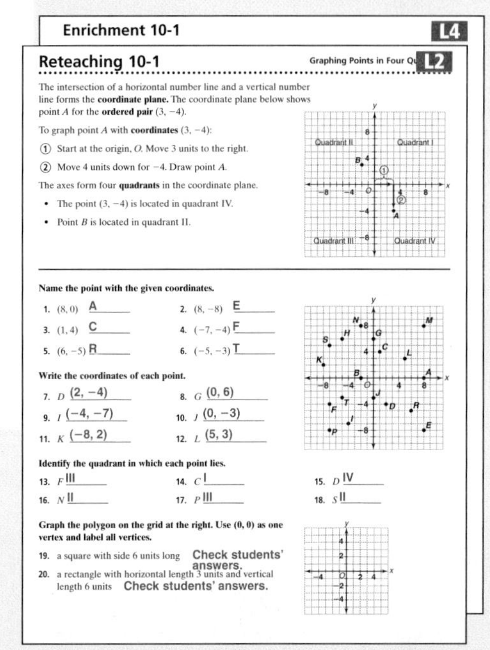

Enrichment 10-1 L4

Reteaching 10-1 Graphing Points in Four Qu L2

The intersection of a horizontal number line and a vertical number line forms the **coordinate plane.** The coordinate plane below shows point A for the **ordered pair** (3, −4).

To graph point A with **coordinates** (3, −4):

① Start at the origin, O. Move 3 units to the right.

② Move 4 units down for −4. Draw point A.

The axes form four **quadrants** in the coordinate plane.

- The point (3, −4) is located in quadrant IV.
- Point B is located in quadrant II.

Name the point with the given coordinates.

1. (8, 0) A 2. (8, −8) E
3. (1, 4) C 4. (−7, −4) F
5. (6, −5) R 6. (−5, −3) T

Write the coordinates of each point.

7. D (2, −4) 8. G (0, 6)
9. I (−4, −7) 10. J (0, −3)
11. K (−8, 2) 12. L (5, 3)

Identify the quadrant in which each point lies.

13. F III 14. C I 15. D IV
16. N II 17. P III 18. S II

Graph the polygon on the grid at the right. Use (0, 0) as one vertex and label all vertices.

19. a square with side 6 units long **Check students' answers.**
20. a rectangle with horizontal length 3 units and vertical length 6 units **Check students' answers.**

Activity Lab

Geometry in the Coordinate Plane

This Activity extends Lesson 10-1 to show students how to use a coordinate plane to find the lengths of sides and the areas of geometric figures.

Guided Instruction

Have students draw and label an *x*-axis and a *y*-axis on graph paper to create a coordinate plane. They should plot and then connect the vertices to form $\triangle LMN$, as you do the same on an overhead transparency. Review, as needed, the meanings of *base* and *height* of a triangle. Discuss how to identify the height of any triangle.

Exercises
Have students work independently on the Exercises. When they are finished have them compare answers with a partner to check that their answers are the same.

Differentiated Instruction

Advanced Learners L4
Have students use the Pythagorean Theorem to find the length of the hypotenuse of $\triangle LMN$.

Resources

- graph paper
- straightedge
- Classroom Aid 2, 3

10-1b Activity Lab

Geometry in the Coordinate Plane

Graphing a figure in a coordinate plane can help you find the figure's perimeter or area.

EXAMPLES Finding Perimeter and Area

1 A map has a coordinate plane printed over it. Town Hall *T* is at the origin. Julio rides his bike from home *H* to school *S*. After school, Julio rides to the library *L* and then to his friend's house *F*. Finally, he rides back home. Each unit in the graph represents 1 kilometer. Use the graph to find the number of kilometers Julio rides.

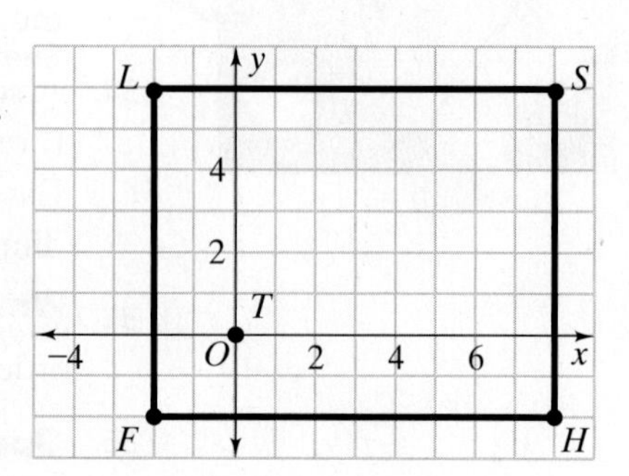

The perimeter of the rectangle represents the distance Julio rides. Since $8 + 10 + 8 + 10$ is 36, Julio rides 36 km.

2 Connect the points $L(-3, -2)$, $M(-1, 3)$, and $N(4, -2)$ to form $\triangle LMN$. Find the area of the triangle using the formula $A = \frac{1}{2}bh$.

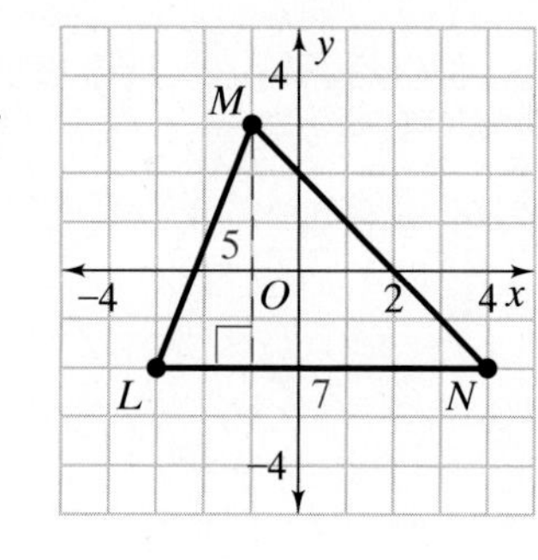

The base of the triangle is 7 units. The distance from point *M* to the base is 5 units, so the height of the triangle is 5 units.

$$A = \frac{1}{2}bh$$
$$= \frac{1}{2} \cdot 7 \cdot 5 \quad \leftarrow \textbf{Substitute.}$$
$$= 17.5 \quad \leftarrow \textbf{Simplify.}$$

The area of the triangle is 17.5 units^2.

Exercises

1. Yvonne walks from her house (0, 0) to her aunt's house (4, 0). She then walks to a park (4, 2). From the park, she walks to the store (0, 2) and then returns home. Graph and connect the points. If each unit in the graph represents 1 kilometer, how many kilometers does Yvonne walk in all?
2. Connect the points $A(4, -4)$, $B(4, 4)$, and $C(-2, 3)$ to form $\triangle ABC$. Find the area of the triangle. **See margin.**
3. Use graphing to find the area of a pentagon with vertices at $(-3, 4)$, $(-3, -9)$, $(6, -4)$, $(8, 0)$ and $(6, 9)$. **See margin.**

1. (0, 2), (4, 2), (0, 0), (4, 0) **12 km**

2–3. See back of book.

Algebra

10-2 Graphing Linear Equations

Check Skills You'll Need

1. **Vocabulary Review** What does an *equation* have that an *expression* does not have? an equal sign

Evaluate each expression for $p = 2$.

2. $p + 7$ 9
3. $3p$ 6
4. $p - 1$ 1
5. $12 - p$ 10

Lesson 4-1

What You'll Learn

To find solutions of linear equations and to graph linear equations

New Vocabulary graph of an equation, linear equation

Why Learn This?

Sometimes you can use an equation to describe a relationship between two quantities. Graphing the equation can make the relationship easier to see.

To make a swing, you need enough rope to reach the branch plus 5 feet to tie the rope. You can use $y = x + 5$ to describe the relationship between the number of feet x the tire hangs below the branch and the number of feet y of rope you need.

A solution of a linear equation is any ordered pair (x, y) that makes the equation true. To find a solution, choose a value of x and substitute it into the equation. Then find the corresponding value of y.

EXAMPLE 1 Finding Solutions

Find three solutions of $y = x + 5$. Organize your solutions in a table.

x	$x + 5$	y	Solution (x, y)	Interpretation
6	$6 + 5$	11	(6, 11)	If the tire hangs 6 ft below the branch, you need 11 ft of rope.
7	$7 + 5$	12	(7, 12)	If the tire hangs 7 ft below the branch, you need 12 ft of rope.
10	$10 + 5$	15	(10, 15)	If the tire hangs 10 ft below the branch, you need 15 ft of rope.

Quick Check

1a–c. See left.

1. Find three solutions of each equation. Use $x = -2$, $x = 0$, and $x = 2$.
   a. $y = x + 8$ b. $y = x - 1$ c. $y = -2x$

1a. (−2, 6), (0, 8), (2, 10)
b. (−2, −3), (0, −1), (2, 1)
c. (−2, 4), (0, 0), (2, −4)

Differentiated Instruction Solutions for All Learners

Special Needs L1
Some students may have a difficult time graphing ordered pairs. Have them work with a partner who can graph the points as they give the coordinates. They can also check and test the solutions.

learning style: visual

Below Level L2
Tell whether each ordered pair is a solution of each equation.
$y = 3x$ (0, 3) no (2, 6) yes (3, 3) no
$y = x - 2$ (4, 2) yes (0, 2) no (0, −2) yes

learning style: verbal

10-2

1. Plan

Objective
To find solutions of linear equations and to graph linear equations

Examples
1 Finding Solutions
2 Graphing to Test Solutions
3 Graphing a Linear Equation

Math Understandings: p. 484C

Math Background

In an equation such as $y = x + 4$ any ordered pair (x, y) that makes the equation true is a *solution* of the equation. Such ordered pairs can be graphed on a coordinate plane. So the *graph of an equation* is the graph of all the points with coordinates that are solutions of that equation. If the graph of solutions is a line, the equation is a *linear equation.* Linear equations have an infinite number of possible solutions, comparable to lines that are composed of an infinite number of points.

More Math Background: p. 484C

Lesson Planning and Resources

See p. 484E for a list of the resources that support this lesson.

Bell Ringer Practice

Check Skills You'll Need
Use student page, transparency, or PowerPoint. For intervention, direct students to:
Evaluating and Writing Algebraic Expressions
Lesson 4-1
Extra Skills and Word Problems Practice, Ch. 4

2. Teach

Activity Lab

Use before the lesson.

All in One Teaching Resources

Activity Lab 10-2: Graphing Linear Equations

Guided Instruction

Example 2
Graphing calculators allow for graphing equations expressed in the $y =$ format.

PowerPoint

Ad[...]

1 Find t[...] $y = x$[...] (5, 0)[...]

2 Deter[...] ordere[...] $y = x$[...]

a. (−2[...]

3 Graph[...] $y = x$[...]

solve algebra EQ (inverse op.)
combining like terms
proportions
probability
surface area (apply geometry formulas)
triangle - Pythag
mean, median, mode, range
read graph

All in One Teaching Resources

- Daily Notetaking Guide 10-2 L3
- Adapted Notetaking 10-2 L1

Closure

- *How can you find solutions for equations with two variables?* **Sample: Choose values for x and make a table. Substitute for x to find the corresponding y-values. Write the ordered pairs.**
- *How can you tell from a graph whether an equation is linear?* **The graph of a linear equation is a straight line.**

Vocabulary Tip

Notice the word *line* in the word *linear.* A linear equation has a graph that is a line.

Go Online

Video Tutor Help
Visit: PHSchool.com
Web Code: are-0775

The **graph of an equation** is the graph of all the points with coordinates that are solutions of the equation. An equation is a **linear equation** when the graph of its solutions lies on a line.

EXAMPLE Graphing to Test Solutions

2 Use the solutions from Example 1 to graph the equation $y = x + 5$. Use the graph to test whether (9, 13) is a solution to the equation.

Step 1 Plot the three ordered-pair solutions.

Step 2 Draw a line through the points.

Step 3 Test (9, 13) by plotting the point in the same coordinate plane. Look to see if the point lies on the line of the graph of the equation.

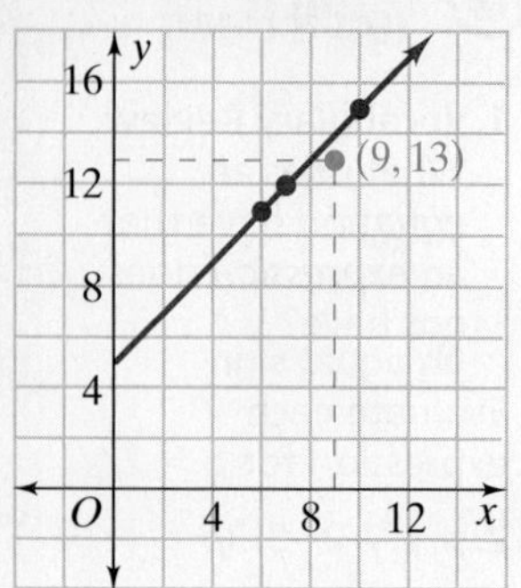

Since (9, 13) is not on the line, (9, 13) is not a solution of $y = x + 5$. For the tire to hang 9 ft below the branch, you need 14 ft of rope.

Quick Check

2. Tell whether (7, 12) is a solution of $y = 3x - 1$. **no**

Linear equations may have negative values in their solutions. You can graph linear equations using positive values, negative values, and zero.

EXAMPLE Graphing a Linear Equation

3 Graph the linear equation $y = 2x + 1$.

Step 1 Make a table of solutions. Use zero as well as positive and negative values for x.

x	$y = 2x + 1$	y	(x, y)
−1	$y = 2(-1) + 1$	−1	(−1, −1)
0	$y = 2(0) + 1$	1	(0, 1)
1	$y = 2(1) + 1$	3	(1, 3)
2	$y = 2(2) + 1$	5	(2, 5)

Step 2 Graph the points. Draw a line through the points.

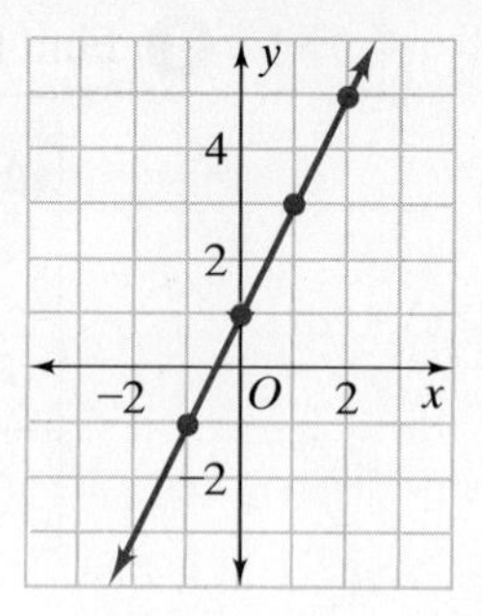

Quick Check

3. Graph each linear equation. **3a–c. See back of book.**

a. $y = x + 4$ b. $y = \frac{1}{2}x$ c. $y = -x$

Differentiated Instruction Solutions for All Learners

Advanced Learners L4
Write equations for lines that pass through the following quadrants. **Samples:**
Quadrants I and II $y = 1$
Quadrants I and III $y = \frac{3}{2}x$
learning style: visual

English Language Learners ELL
Help students make a connection between the term *ordered pair,* and the order in which they find the point named by two coordinates. Let them know the first number is always the x-coordinate, and the second is always the y-coordinate. Order matters.
learning style: verbal

Check Your Understanding

1. **Vocabulary** What does the graph of the solution of a linear equation look like? **a line**

x	x − 7	y	(x, y)
0	■	−7	(0, −7)
−3	■	−10	(−3, −10)
10	■	3	(10, 3)

Use the equation $y = x - 7$ for Exercises 2–5.

2. Copy and complete the table at the left to find three solutions of the equation. **See margin.**
3. Graph the equation. **See margin.**
4. Is $(-1, -6)$ a solution of the equation? **no**
5. How does a graph of the equation show that $(5, -2)$ is a solution? **The point (5, −2) lies on the line.**

Homework Exercises

For more exercises, see Extra Skills and Word Problems.

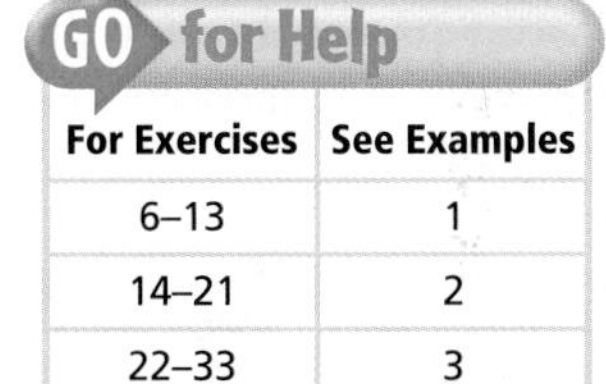

GO for Help

For Exercises	See Examples
6–13	1
14–21	2
22–33	3

A **Find three solutions of each equation.** **6–13. See margin.**

6. $y = x - 2$ 7. $y = x + 9$ 8. $y = x$ 9. $y = -x + 4$

10. $y = 5x$ 11. $y = -8x$ 12. $y = 3x + 1$ 13. $y = 4x - 5$

Tell whether each ordered pair is a solution of $y = x + 12$.

14. $(-12, 24)$ **no** 15. $(12, 24)$ **yes** 16. $(0, -12)$ **no** 17. $(-12, 0)$ **yes**

18. $(7, 19)$ **yes** 19. $(24, 12)$ **no** 20. $(6, 15)$ **no** 21. $(9, 21)$ **yes**

Graph each linear equation. **22–33. See margin.**

22. $y = x - 1$ 23. $y = x - 3$ 24. $y = 3x$ 25. $y = 5 + x$

26. $y = x + 4$ 27. $y = -5x$ 28. $y = 2 - x$ 29. $y = 4 - x$

30. $y = -x - 5$ 31. $y = \frac{1}{3}x$ 32. $y = 2x - 1$ 33. $y = \frac{1}{2}x + 4$

B 

34. **Guided Problem Solving** A shipping company charges \$10 for delivery and \$5 per pound shipped. The linear equation $y = 5x + 10$ models the cost of shipping an object that weighs x pounds. Graph the equation to find the cost to ship a vase that weighs 4 pounds. **See margin.**
   - Make a table of solutions using the ordered pair (x, y).
   - Graph the equation.
   - Which point on your graph shows the cost for a 4-lb vase?

35. **Pets** You are building a fence around a square pen for your pig. You want to use 46 ft of fencing, which accounts for a 2-ft opening. Graph the equation $P = 4s - 2$. Use your graph to find the length of each side of the pen. **See margin.**

Online lesson quiz, PHSchool.com, Web Code: ara-1002

2–3. See back of book.
6–13. See back of book.
22–35. See back of book.

3. Practice

Assignment Guide

Check Your Understanding
Go over Exercises 1–5 in class before assigning the Homework Exercises.

Homework Exercises

A Practice by Example	6–33
B Apply Your Skills	34–42
C Challenge	43
Test Prep and Mixed Review	44–45

Homework Quick Check
To check students' understanding of key skills and concepts, go over Exercises 15, 28, 39, 41, and 42.

Differentiated Instruction Resources

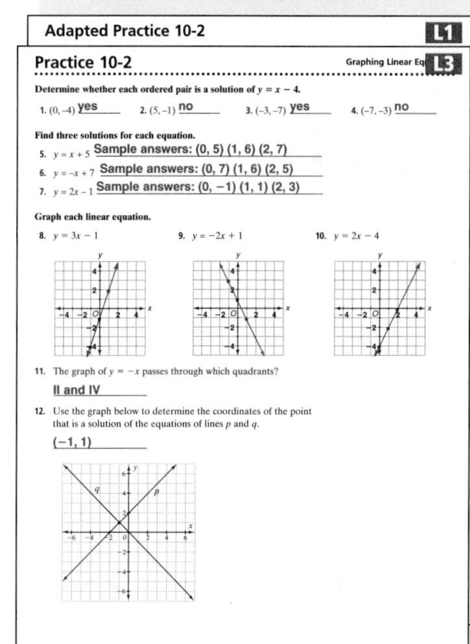

Adapted Practice 10-2 L1

Practice 10-2 Graphing Linear Eq L3

Determine whether each ordered pair is a solution of $y = x - 4$.

1. (0, −4) yes 2. (5, −1) no 3. (−3, −7) yes 4. (−7, −3) no

Find three solutions for each equation.

5. $y = x + 5$ Sample answers: (0, 5) (1, 6) (2, 7)
6. $y = -x + 7$ Sample answers: (0, 7) (1, 6) (2, 5)
7. $y = 2x - 1$ Sample answers: (0, −1) (1, 1) (2, 3)

Graph each linear equation.

8. $y = 3x - 1$ 9. $y = -2x + 1$ 10. $y = 2x - 4$

11. The graph of $y = -x$ passes through which quadrants? II and IV

12. Use the graph below to determine the coordinates of the point that is a solution of the equations of lines p and q. (−1, 1)

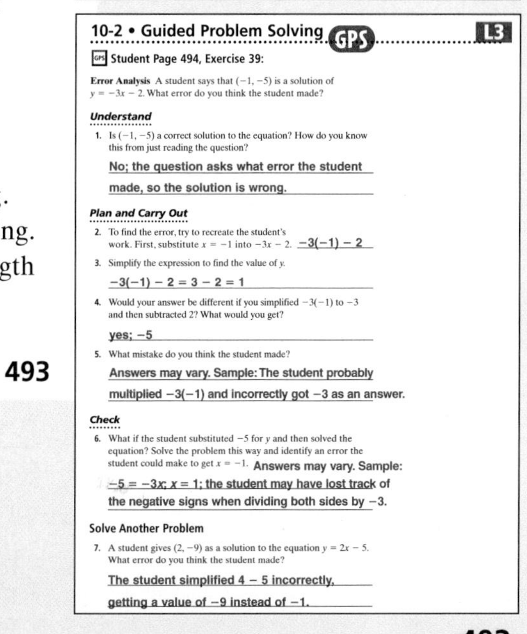

10-2 • Guided Problem Solving GPS L3

Student Page 494, Exercise 39:

Error Analysis A student says that (−1, −5) is a solution of $y = -3x - 2$. What error do you think the student made?

Understand

1. Is (−1, −5) a correct solution to the equation? How do you know this from just reading the question? No; the question asks what error the student made, so the solution is wrong.

Plan and Carry Out

2. To find the error, try to recreate the student's work. First, substitute $x = -1$ into $-3x - 2$. $-3(-1) - 2$
3. Simplify the expression to find the value of y. $-3(-1) - 2 = 3 - 2 = 1$
4. Would your answer be different if you simplified $-3(-1)$ to -3 and then subtracted 2? What would you get? yes; −5
5. What mistake do you think the student made? Answers may vary. Sample: The student probably multiplied $-3(-1)$ and incorrectly got -3 as an answer.

Check

6. What if the student substituted −5 for y and then solved the equation? Solve the problem this way and identify an error the student could make to get $x = -1$. Answers may vary. Sample: $-5 = -3x$; $x = 1$; the student may have lost track of the negative signs when dividing both sides by −3.

Solve Another Problem

7. A student gives (2, −9) as a solution to the equation $y = 2x - 5$. What error do you think the student made? The student simplified 4 − 5 incorrectly, getting a value of −9 instead of −1.

4. Assess & Reteach

PowerPoint

Lesson Quiz

1. Is (−3, 5) a solution for $y = x - 7$? **no**
2. Find three solutions of $y = -2x$ **Sample: (−1, 2), (0, 0), (1, −2)**
3. Graph the equation $y = x + 2$. Is it a linear equation? **yes**

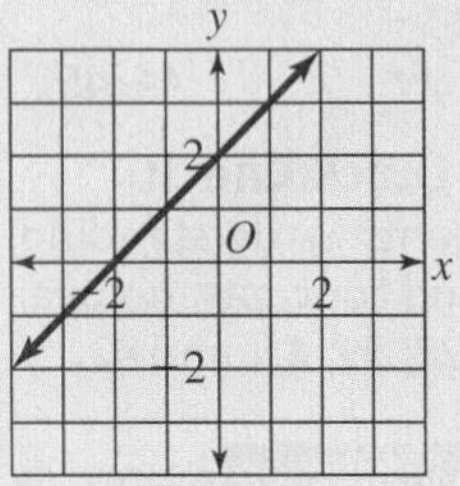

Reteaching 10-2

Graphing Linear E… L2

The **solutions** of $y = x + 3$ are the (x, y) pairs that make the equation true.

The solutions can be listed in a table.

x	$x + 3$	y	(x, y)
0	0 + 3	3	(0, 3)
1	1 + 3	4	(1, 4)
−2	−2 + 3	1	(−2, 1)

If all the solutions lie on a line, the equation is a **linear equation** and the line is its **graph**.

$y = x + 3$ is a linear equation.

The solutions can be graphed in the coordinate plane, as shown.

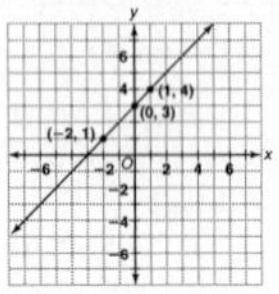

Complete each table.

1. $y = x - 4$

x	$x - 4$	y	(x, y)
2	2 − 4	−2	(2, −2)
4	4 − 4	0	(4, 0)
6	6 − 4	2	(6, 2)

2. $y = 3x$

x	$3x$	y	(x, y)
−1	3(− 1)	−3	(−1, −3)
0	3(0)	0	(0, 0)
3	3(3)	9	(3, 9)

3. $y = -x + 1$

x	$-x + 1$	y	(x, y)
0	0 + 1	1	(0, 1)
2	−2 + 1	−1	(2, −1)
−3	−(−3) + 1	4	(−3, 4)

Graph each linear equation.

4. $y = x - 5$

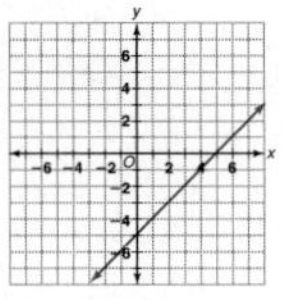

5. $y = 3x - 4$

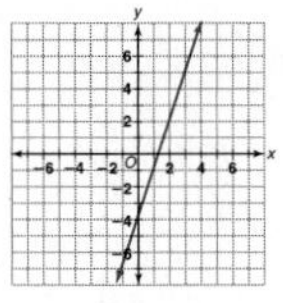

Enrichment 10-2

Graphing Linear E… L4

Critical Thinking

When an insect is active, its heart can beat at a rate of 140 beats per minute. When it is inactive and cold, its heart rate can slow to an average rate of 24 beats per day.

1. How many heartbeats per hour are there when an insect is active? **8,400 beats/h**
2. How many heartbeats per hour are there when an insect is inactive? **1 beat/h**
3. Use b for beats and m for minutes. Write the equation that represents the
   a. active heartbeat per minute. **$b = 140$ m**
   b. inactive heartbeat per minute. **$b = \frac{m}{60}$**
4. Graph the equations on the same coordinate plane.

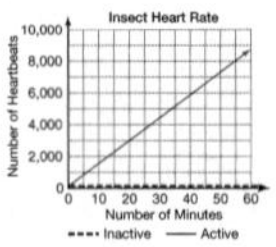

   a. Describe the similarities and differences of each graph. **one is steep; other is almost horizontal**
   b. How can you tell, by looking, which graph represents the lesser rate? **The graph with the smaller slope represents the lesser rate.**
   c. Do you have to graph the equations to tell which rate is the lesser one? Explain. **No, the equations show which is a lesser rate.**
   d. For an active insect, how many heartbeats would you expect in 50 min? in 2 h? **about 7,000; about 16,800**
5. An adult's heart can beat at a rate of 4,800 times per hour. A newborn baby's heart can beat at a rate of 140 times per minute. Write an equation to represent each of these as $\frac{\text{beats}}{\text{hour}}$. Which represents the greater rate? **$r = 4{,}800\ \frac{beats}{h}$; $r = 8400\ \frac{beats}{h}$; new baby's heart rate**

GO Online
Homework Video Tutor
Visit: PHSchool.com
Web Code: are-1002

On which of the following lines does each point lie? A point may lie on more than one line.

36. (0, 0) **none**
37. (3, 9) **I, III**
38. (−2, −1) **III**

I. $y = x + 6$ II. $y = x - 6$ III. $y = 2x + 3$

39. **Error Analysis** A student says that (−1, −5) is a solution of $y = -3x - 2$. What error do you think the student made? **The student thought the product of −3 and −1 was −3; it is 3.**

GPS

40. Honey bees produce about 50 lb of honey per hive each year. This situation can be represented by $y = 50x$, where x is the number of hives and y is the amount of honey in pounds. Make a table. Can 800 lb of honey be produced in one year with 16 hives? **See margin.**
41. **Writing in Math** Why is it a good idea to plot at least three points when you graph a linear equation? **Answers may vary. Sample: Plotting three points will help you check whether the graph is correct.**
42. **Measurement** Graph the equations $y = 12x$ and $y = \frac{1}{12}x$ on the same coordinate plane. Which of these equations models the conversion of x inches to y feet? **See margin.**
43. Ⓒ **Challenge** Tell whether the graph of $y = x - 5$ passes through the second quadrant. Explain how you know. **No; for $x < 0$, the y-values are negative.**

Test Prep and Mixed Review

Practice

Multiple Choice

44. The data in the table at the left show the relationship between x, the diameter of a circle, and y, the radius of a circle. Which graph best represents the data? **D**

Diameter, x (cm)	Radius, y (cm)
0	0
2	1
4	2
6	3

Ⓐ
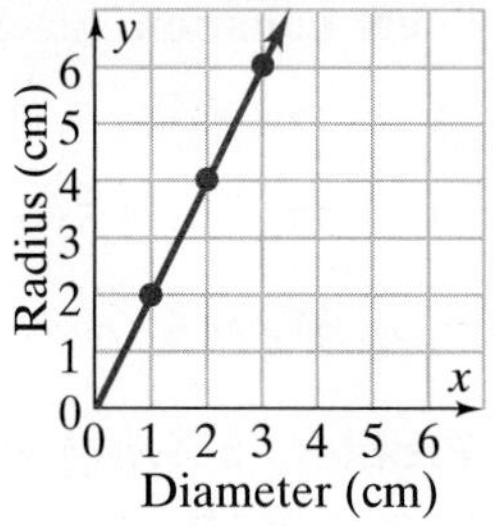

Ⓒ
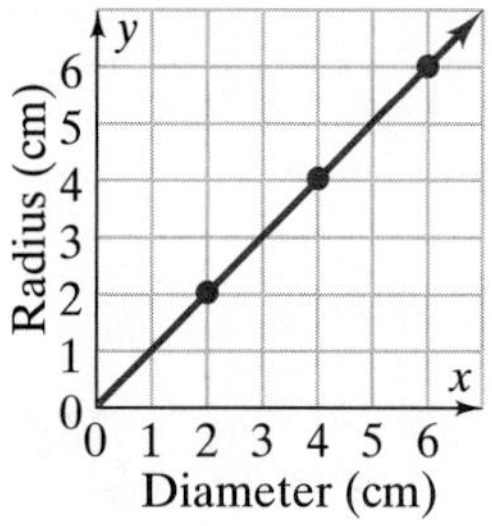

Ⓑ
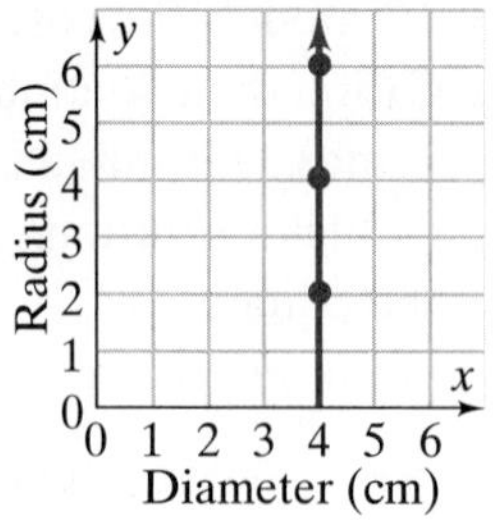

Ⓓ
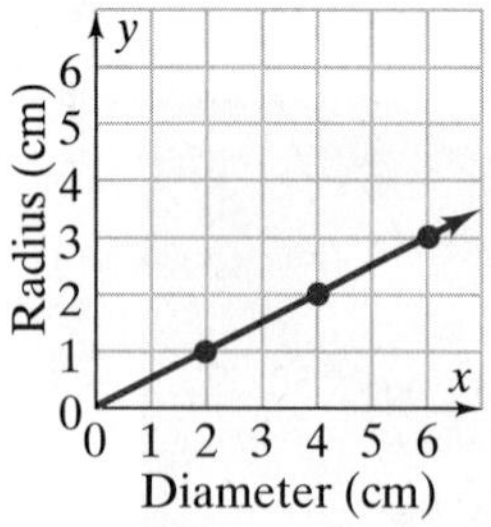

GO for Help

For Exercise	See Lesson
45	9-7

45. You have \$480 in an account that pays 8% compound interest annually. What is your balance after 3 years? **\$604.66**

Test Prep

Resources

For additional practice with a variety of test item formats:

- Test-Taking Strategies, p. 523
- Test Prep, p. 527
- Test-Taking Strategies with Transparencies

Alternative Assessment

Provide pairs of students with blank coordinate grids. Have partners work together to make a table of values and graph the linear equations in Exercises 22–33. Ask pairs to compare their graphs with other groups and resolve any discrepancies.

40. See back of book.
42. See back of book.

10-2b Activity Lab

Data Analysis

Representing Data

A table is an organized way to represent data. Sometimes it is helpful to visualize the data by creating a graph.

EXAMPLE

The data in the table show the relationship between the diameter d and the circumference C of a circle.

Which graph best represents the data in the table at the right?

Diameter d (cm)	2	4	6	8
Circumference C (cm)	6.3	12.6	18.8	25.1

Ⓐ
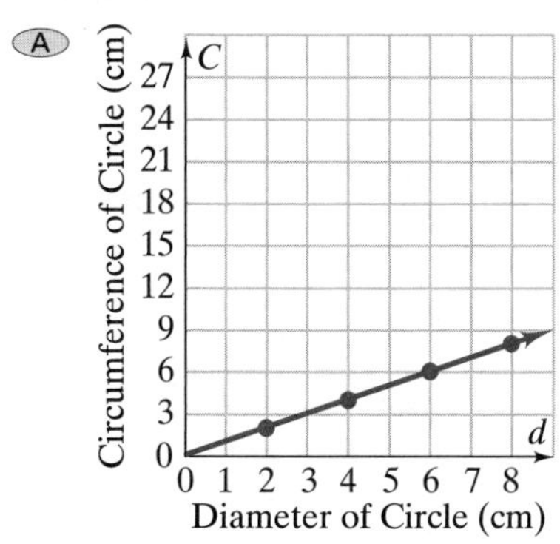

Ⓒ
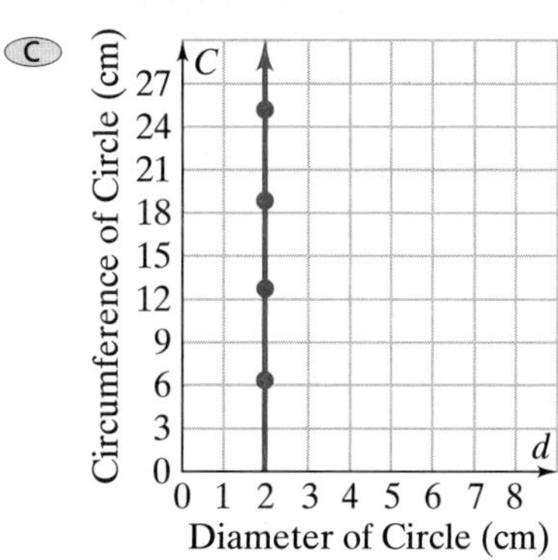

Ⓑ
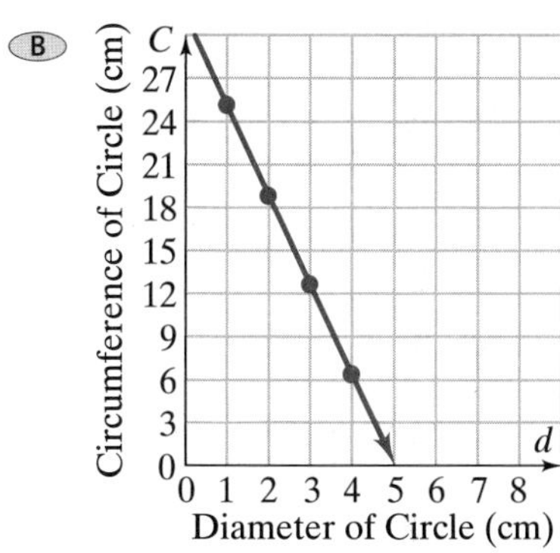

Ⓓ
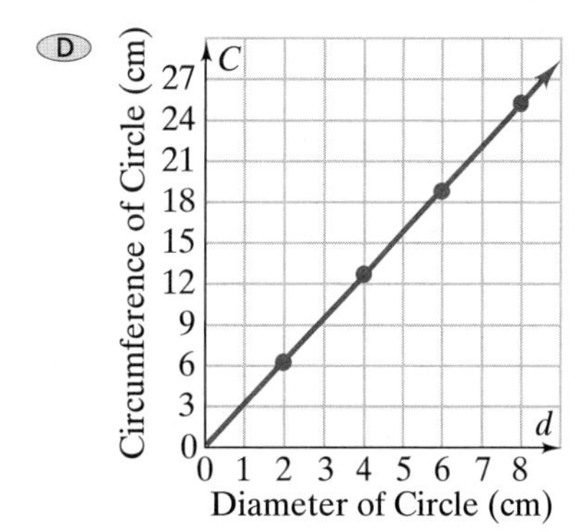

According to the data in the table and what you know to be true about circles, the circumference increases as the diameter increases.

Choice B shows the circumference *decreasing* as the diameter increases. Eliminate choice B.

Choice C shows a vertical line, which means the diameter is constant. A circle with one diameter cannot have several different circumferences. Eliminate choice C.

The first ordered pair is (2, 6.3). In the remaining graphs, this point appears only in choice D. The correct answer is choice D.

Activity Lab

Representing Data

Students extend their graphing skills from Lesson 10-2 by using graphs to represent data from tables.

Guided Instruction

Explain to students that there is often a relationship between data in a table and that sometimes graphing that data makes the relationship easier to see.

Ask questions, such as:

- *How do we know the line in Choice B shows an inverse relationship?* **as one set of coordinates increases, the other decreases**
- *What does the vertical line in Choice C mean?* **There is no change in the x-coordinate as the y-coordinate moves up or down.**
- *What general conclusion can you draw from these graphs?* **You can tell the relationship between two sets of data based on the direction and angle of the line.**

Differentiated Instruction

Below Level L2

Allow students to create their own graph of the data in the table to help them connect the relationship between two sets of data and the appearance of the graph.

Practice Solving Problems

In this feature, students graph information to help them solve problems. They write an equation and then solve the equation for different values in order to graph the equation.

Guided Instruction

Explain to students that they can find additional information by graphing information from a problem. Tell them they can use information from a problem to create an equation that will allow them to graph.

Ask:

- *Why do we add $3 to the amount Mandy earns per hour?* **She gets a $3-dollar transportation fee per job, plus her hourly rate.**
- *Why should we solve the equation for* $h = 1$ *and* $h = 5$*?* **to find out how much money, m, Mandy makes for a 1-hour job and a 5-hour job**
- *What would the equation look like if Mandy charged $5 for transportation and $4.50 per hour? How much would she earn for a 4-hour babysitting job?* **$m = 4.5h + 5$; $23**
- *What are the steps required to solve the problem?* **find the important information, write an equation, solve the equation for different values, graph the equation**

Practice Solving Problems

Jobs Mandy earns $6 per hour babysitting, plus $3 for transportation. How much will she earn for a 1-hour job and a 5-hour job this weekend?

What You Might Think	What You Might Write
What do I know? What am I trying to find out?	Mandy gets $3 to start and earns $6 per hour. I want to find out how much she will earn for a 1-hour job and for a 5-hour job.
How do I show the main idea?	I will graph $m = 6h + 3$ to show how much money m she earns after h hours of work.
What does the graph look like?	Solve for $h = 1$ and $h = 5$. If $h = 1$, then $m = 6(1) + 3$, or 9. If $h = 5$, then $m = 6(5) + 3$, or 33. Graph (1, 9) and (5, 33). I can connect the points with a solid line because Mandy can get paid for working part of an hour.
What is the answer?	Mandy will earn $9 for the 1-hour job and $33 for the 5-hour job.

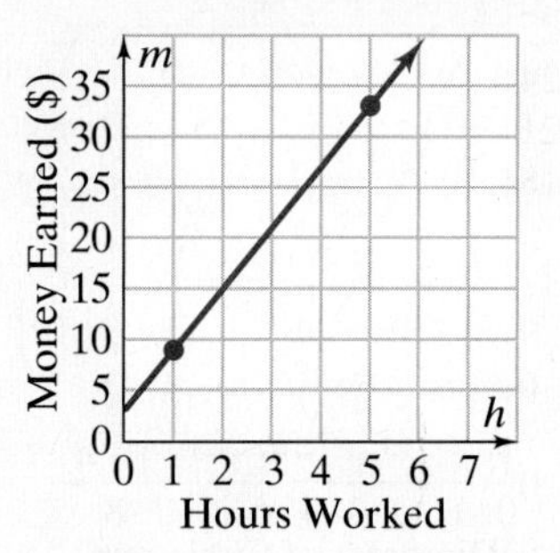

Think It Through

1. How does the equation $m = 6h + 3$ represent the amount of money m Mandy earns for h hours of work? Explain. **The $3 transportation fee does not change, but Mandy earns $6 for every hour h she works.**
2. How much does Mandy earn for 2 hours, $3\frac{1}{2}$ hours, 4 hours, and $6\frac{1}{4}$ hours of work? **$15; $24; $27; $40.50**

Exercises

Solve each problem. For Exercises 3 and 4, answer the questions first.

3. Fingernails grow an average of 1.5 in. per year. Suppose your nails are $\frac{1}{2}$ in. long on your tenth birthday. If you do not cut your nails, how long will they be on your sixteenth birthday? **$9\frac{1}{2}$ in.**

a. What do you know? What do you want to find out?

b. What equation can you write to find the length of your fingernails at age 10 and age 16?

c. How does graphing the equation give you additional information?

3b. Let f = length of your fingernails. Let y = number of years. $f = 1.5y + 0.5$

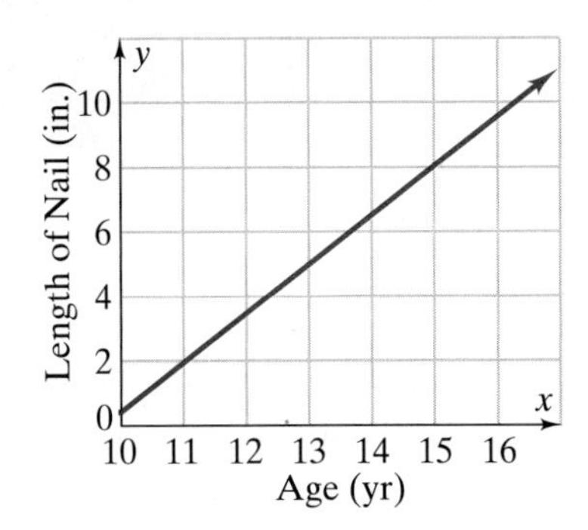

4. Romesh Sharma of India set a record for the longest fingernails. He had five fingernails on his left hand that had a total length of 33 ft. If his nails grew $\frac{1}{8}$ in. each month, about how long did it take to grow those nails? Assume that he started when he was born and never cut his fingernails. **almost 53 years**

a. If the five nails were about the same length, about how long was each nail? **about $6\frac{1}{2}$ ft**

b. How does the diagram below help you decide what to do?

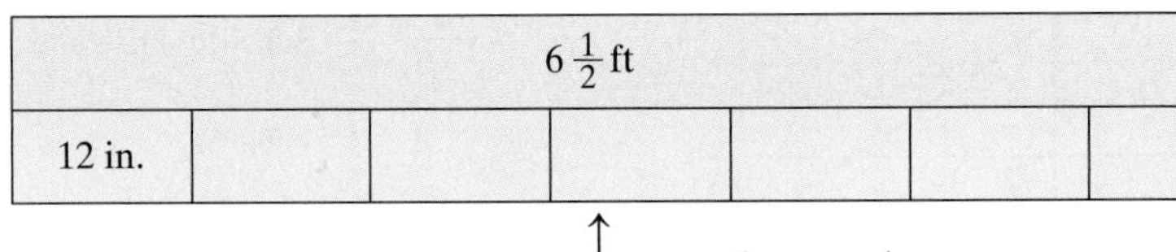

How many inches are in $6\frac{1}{2}$ feet?

5. In 1982, Larry Walters tied 42 helium balloons to an aluminum lawn chair. He floated above Los Angeles International Airport in his chair for 45 min. If Larry and all his equipment weighed 168 lb, how much did each balloon lift on average? **4 lb**

Error Prevention!

Students may have trouble graphing variables other than *x* and *y*. Allow them to substitute these variables for other variables in the problem.

Exercises

Have students work independently on the Exercises. Circulate to check students' work and help them remember the necessary problem solving steps.

Differentiated Instruction

Advanced Learners L4

Provide students with coordinates to test to decide whether or not they fit in the equation.

10-3

1. Plan

Objective
To find the slope of a line and use it to solve problems

Examples
1 Finding Slope
2 Application: Avalanches
3 Drawing Lines on a Graph

Math Understandings: p. 484C

Professional Development

Math Background

A ratio expresses the relationship between two values. For a line in the coordinate plane, the steepness of the line is expressed by a ratio. The ratio of the line's rise (the vertical change in a line) to its run (its horizontal change) is called *slope.* The slope for any given line is constant. To find the slope of a line, divide the line's rise by its run. Given a value for slope, you can draw a line with that slope through a given point.

More Math Background: p. 484C

Lesson Planning and Resources

See p. 484E for a list of the resources that support this lesson.

PowerPoint

Bell Ringer Practice

Check Skills You'll Need
Use student page, transparency, or PowerPoint. For intervention, direct students to:
Simplifying Fractions
Lesson 2-3
Extra Skills and Word Problems Practice, Ch. 2

Algebra

10-3 Finding the Slope of a Line

Check Skills You'll Need

1. **Vocabulary Review** What does it mean to say that the fraction $\frac{5}{9}$ is written in *simplest form*? **See below.**

Write each fraction in simplest form.

2. $\frac{6}{8}$ $\frac{3}{4}$
3. $\frac{8}{-12}$ $-\frac{2}{3}$
4. $\frac{-4}{-12}$ $\frac{1}{3}$
5. $\frac{-15}{3}$ -5

Lesson 2-3

Check Skills You'll Need
1. The numerator and denominator have no common factors other than 1.

What You'll Learn

To find the slope of a line and use it to solve problems

New Vocabulary slope, rise, run

Why Learn This?

Elephants will not climb a ramp with more than a 33-degree slope. When ramps are constructed, they must meet safety standards that ensure they are not too steep. You can understand what factors affect steepness when you understand slope.

Slope is a ratio that describes the steepness of a line. Slope compares the vertical change in a line, called the **rise,** to the horizontal change, called the **run.** For any two points on a line, the ratio of rise to run is the same.

$$\text{slope} = \frac{\text{rise}}{\text{run}}$$

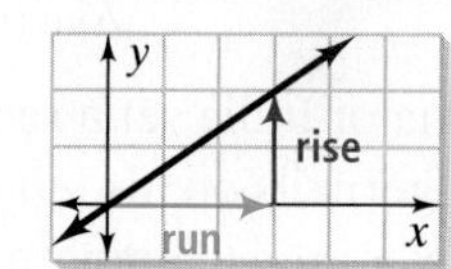

Online active math

For: Slope Activity
Use: Interactive Textbook, 10-5

EXAMPLE Finding Slope

1 Find the slope of the line.

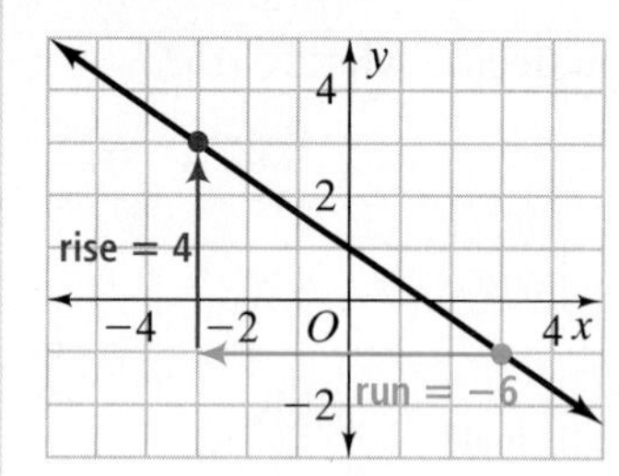

$$\begin{aligned}\text{slope} &= \frac{\text{rise}}{\text{run}} \\ &= \frac{4}{-6} \quad \leftarrow \textbf{Substitute rise and run.} \\ &= -\frac{2}{3} \quad \leftarrow \textbf{Simplify.}\end{aligned}$$

The slope of the line is $-\frac{2}{3}$.

Quick Check

1. Find the slope of the line at the right. $\frac{1}{2}$

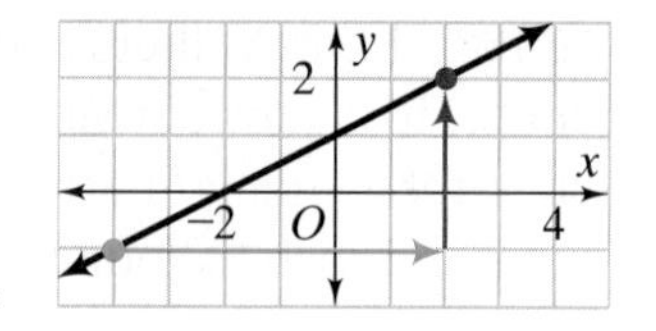

Differentiated Instruction Solutions for All Learners

Special Needs L1
Assist students, as needed, to draw and connect points. If they have difficulty, give them a straightedge, and pair them up with a student who can help plot the points and hold the straightedge as they connect the points.
learning style: visual

Below Level L2
Review the rules for subtraction of positive and negative integers. Have students find the rise and run with simple pairs of coordinates, such as (0, 2) and (4, 6). **rise: 4, run: 4**
learning style: visual

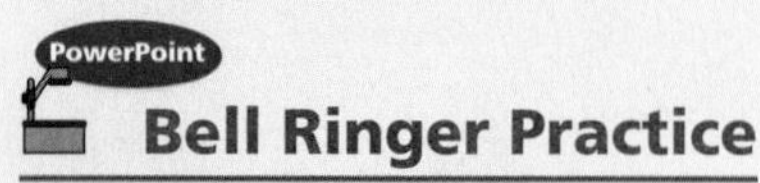

The slope of a line can be positive, negative, zero, or undefined. A line that goes upward from left to right has positive slope. A line that goes downward has negative slope. A horizontal line has a slope of 0. The slope of a vertical line is undefined because you cannot divide by 0.

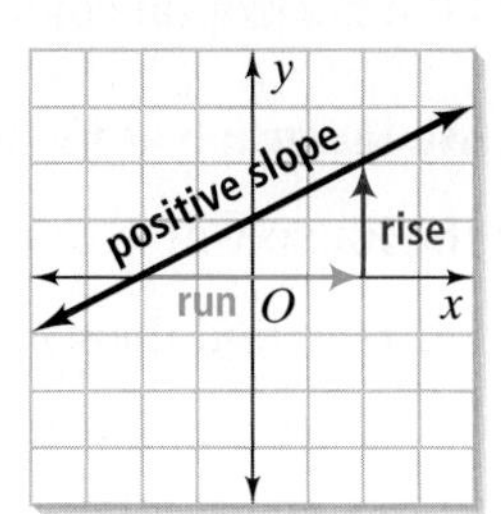

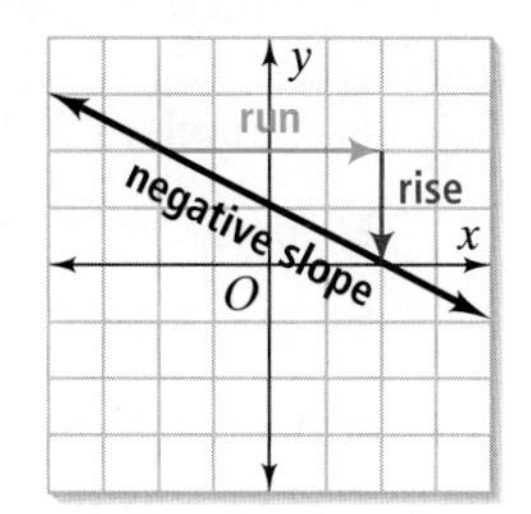

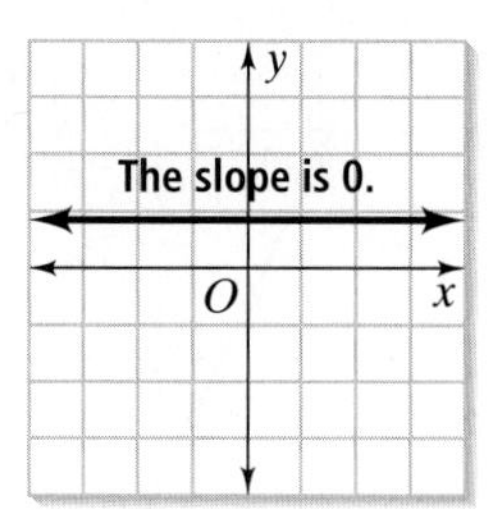

EXAMPLE Application: Avalanches

2 Avalanches are likely to occur on trails where the absolute value of the slope is between 0.5 and 1. Find the slope of the trail and determine whether an avalanche is likely.

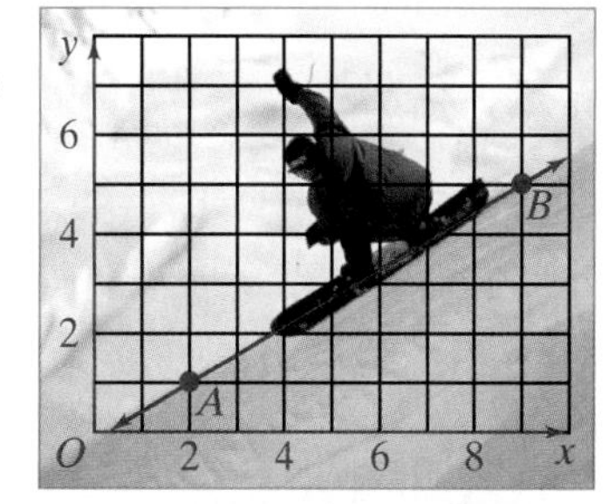

From point A to point B, the rise is 4 and the run is 7.

$$\text{slope} = \frac{\text{rise}}{\text{run}} = \frac{4}{7} \approx 0.57$$

The slope is about 0.57, so an avalanche is likely.

Vocabulary Tip

Snowboarders and skiers often use the word *slope* to refer to the side of a mountain.

Quick Check

2. Slope is used to find the pitch, or steepness, of a roof. Roof A has a pitch of 3 to 12, which means it rises 3 in. for every 12 in. of run. What is the slope of Roof A? $\frac{1}{4}$

3.

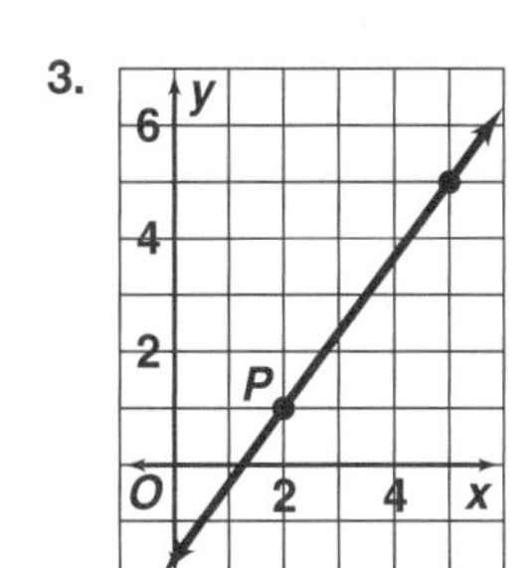

You can use slope to graph a line if you know a point on the line.

EXAMPLE Drawing Lines on a Graph

3 Draw a line through the origin with a slope of $-\frac{3}{2}$.

Step 1 Graph a point at (0, 0).

Step 2 Move 3 units down and 2 units to the right. Graph a second point.

Step 3 Connect the points to draw a line.

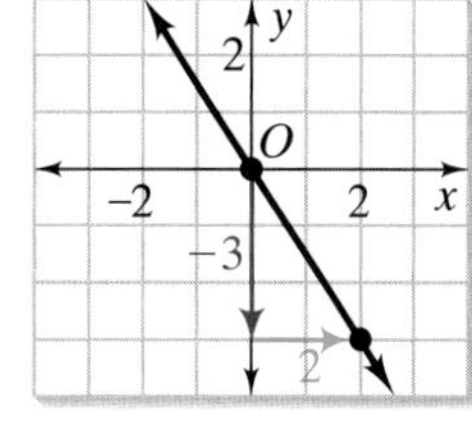

Test Prep Tip

You can also move 2 units to the left and 3 units up.

Quick Check

3. Draw a line through $P(2, 1)$ with a slope of $\frac{4}{3}$. **See above left.**

Advanced Learners L4

Challenge students to think of relationships that would yield lines with positive slopes (increasing speed and distance), and negative slopes (value depreciation).

learning style: visual

English Language Learners ELL

Have students use index cards to draw and label examples of a positive, negative, zero, and undefined slope. They might also draw and label examples for the terms *rise* and *run*.

learning style: visual

2. Teach

Activity Lab

Use before the lesson.

All in One Teaching Resources

Activity Lab 10-3: Finding the Slope of a Line

Guided Instruction

Example 3

Students may forget to write the negative sign for a line with a negative slope. Remind them that if a line goes downward from left to right, its slope is negative. If a line goes upward from left to right, its slope is positive. Draw both a positive slope line and a negative slope line on the board. Write "+" and "−" to indicate each lines' slope.

PowerPoint

Additional Examples

1 Find the slope of the line. **−2**

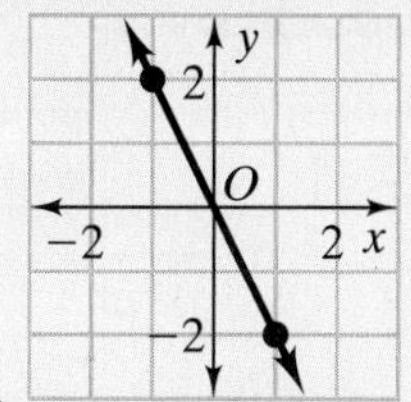

2 Find the slope of the roof. $\frac{4}{3}$

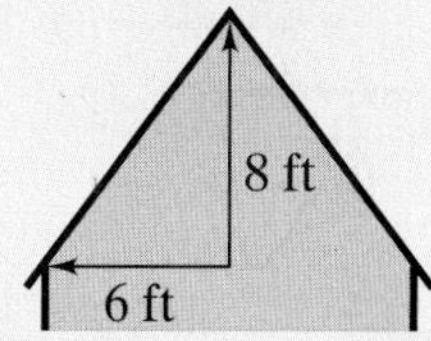

3 Draw a line through the origin with a slope of $\frac{-5}{4}$. **See back of book for graph.**

All in One Teaching Resources

- Daily Notetaking Guide 10-3 L3
- Adapted Notetaking 10-3 L1

Closure

- *What is the slope of a line?* **Sample: Slope is the ratio of the vertical change to horizontal change, or rise over run.**
- *What is the rise of a line?* **The rise is its vertical change.**
- *What is the run of a line?* **The run is its horizontal change.**

3. Practice

Assignment Guide

Check Your Understanding

Go over Exercises 1–4 in class before assigning the Homework Exercises.

Homework Exercises

A	Practice by Example	5–11
B	Apply Your Skills	12–21
C	Challenge	22
	Test Prep and Mixed Review	23–25

Homework Quick Check

To check students' understanding of key skills and concepts, go over Exercises 6, 10, 19, 20, and 21.

Differentiated Instruction Resources

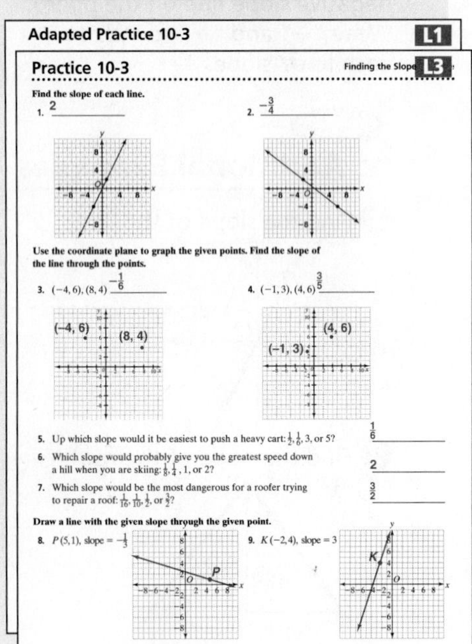
Adapted Practice 10-3 L1

Practice 10-3 Finding the Slope L3

Find the slope of each line.

1. 2
2. $-\frac{3}{4}$

Use the coordinate plane to graph the given points. Find the slope of the line through the points.

3. (−4, 6), (8, 4) $-\frac{1}{6}$
4. (−1, 3), (4, 6) $\frac{3}{5}$

5. Up which slope would it be easiest to push a heavy cart: $\frac{1}{3}$, $\frac{1}{6}$, 3, or 5? $\frac{1}{6}$
6. Which slope would probably give you the greatest speed down a hill when you are skiing: $\frac{1}{5}$, $\frac{1}{4}$, 1, or 2? 2
7. Which slope would be the most dangerous for a roofer trying to repair a roof: $\frac{1}{10}$, $\frac{1}{10}$, $\frac{1}{3}$, or $\frac{3}{2}$? $\frac{3}{2}$

Draw a line with the given slope through the given point.

8. $P(5, 1)$, slope = $-\frac{1}{3}$
9. $K(-2, 4)$, slope = 3

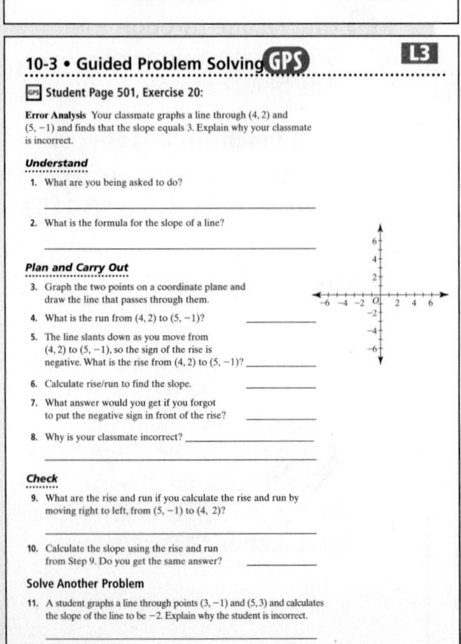
10-3 • Guided Problem Solving GPS L3

Student Page 501, Exercise 20:

Error Analysis Your classmate graphs a line through (4, 2) and (5, −1) and finds that the slope equals 3. Explain why your classmate is incorrect.

Understand

1. What are you being asked to do?
2. What is the formula for the slope of a line?

Plan and Carry Out

3. Graph the two points on a coordinate plane and draw the line that passes through them.
4. What is the run from (4, 2) to (5, −1)?
5. The line slants down as you move from (4, 2) to (5, −1), so the sign of the rise is negative. What is the rise from (4, 2) to (5, −1)?
6. Calculate rise/run to find the slope.
7. What answer would you get if you forgot to put the negative sign in front of the rise?
8. Why is your classmate incorrect?

Check

9. What are the rise and run if you calculate the rise and run by moving right to left, from (5, −1) to (4, 2)?
10. Calculate the slope using the rise and run from Step 9. Do you get the same answer?

Solve Another Problem

11. A student graphs a line through points (3, −1) and (5, 3) and calculates the slope of the line to be −2. Explain why the student is incorrect.

Check Your Understanding

1. **Vocabulary** The vertical change between two points on a line is called _?_, and the horizontal change is called _?_. **rise; run**

For Exercises 2–4, use the graph at the left.

2. Is the slope of the line positive or negative? **negative**
3. Use points A and D to find the slope of the line. **−2**
4. **Reasoning** Use points B and C to find the slope of the line. Does using different points affect the slope? Explain. **−2; the slope is the same because the ratio is always the same.**

Homework Exercises

For more exercises, see Extra Skills and Word Problems.

GO for Help

For Exercises	See Examples
5–7	1
8–9	2
10–11	3

A **Find the slope of each line.** 5. −1 6. $-\frac{1}{3}$ 7. 2

5. 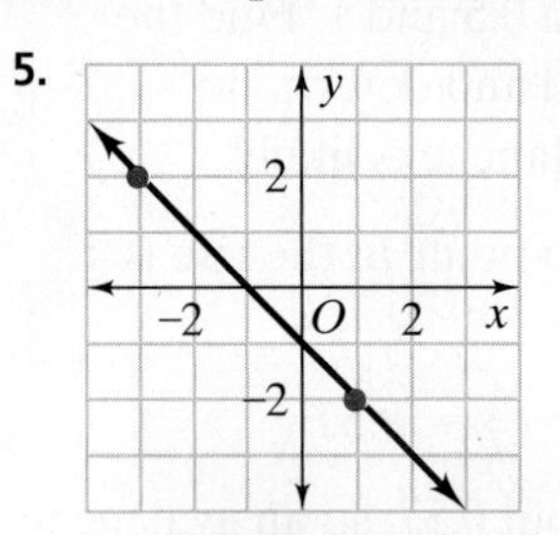

6. 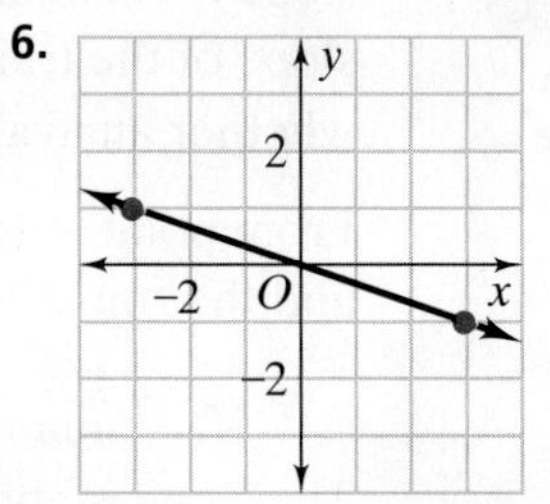

7. 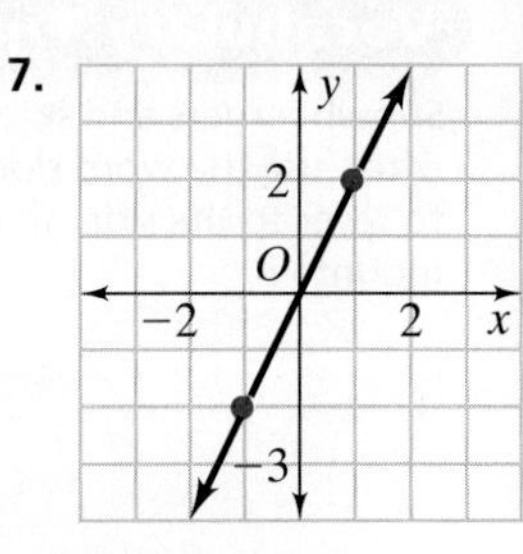

8. 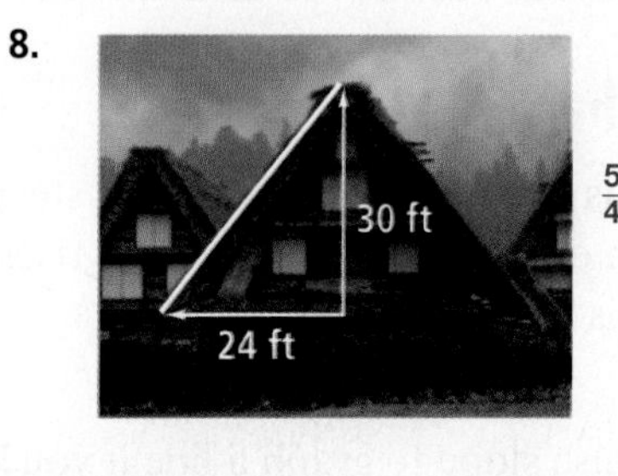

$\frac{5}{4}$

9. 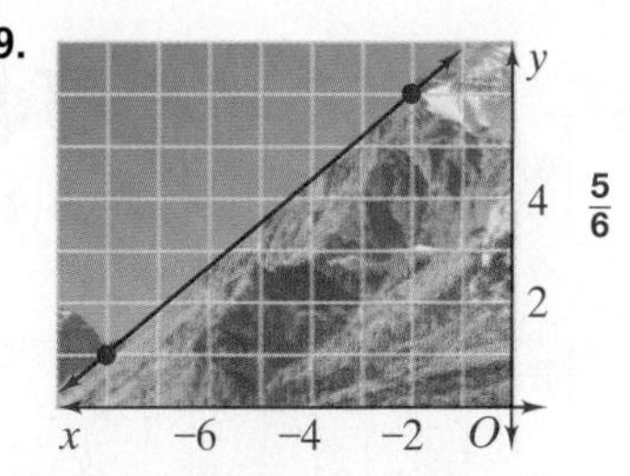

$\frac{5}{6}$

Draw a line with the given slope through the given point.
10–11. See left.

10. $P(0, 0)$, slope $= -3$
11. $R(6, 6)$, slope $= \frac{5}{2}$

10. 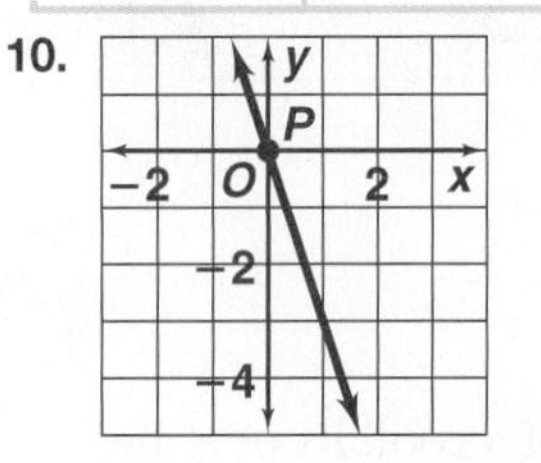

11. 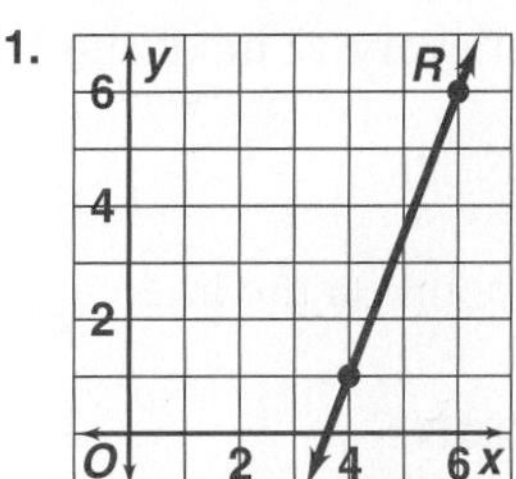

B **GPS** 12. **Guided Problem Solving** Some skiers prefer steep trails. Which trails would you recommend to those skiers? Explain. **See left.**

- What is the slope of each trail?
- Which trails are the steepest?

Ski Trail	Total Rise	Total Run
Alpine	1,800 ft	7,200 ft
Diamond	1,840 ft	3,680 ft
Bear	1,900 ft	4,750 ft
Donner	750 ft	3,000 ft

12. Diamond and Bear; their slopes are $\frac{1}{2}$ and $\frac{2}{5}$, which are steeper than the other trails, whose slopes are $\frac{1}{4}$.

13. **Reasoning** Draw a horizontal line through (1, 3). Use two points on the line to explain why the slope of the line equals 0. **See margin.**

13. See back of book.
17–18. See back of book.
25. See back of book.

GO Online
Homework Video Tutor
Visit: PHSchool.com
Web Code: are-1003

Graph the given points. Find the slope of the line through the points.

14. (4, 8), (5, 10) **2**   15. (4, −1), (−4, 1) $-\frac{1}{4}$ 16. (2, 7), (3, −1) **−8**

17. Roof A has a rise of 5 and a run of 3. Roof B has a rise of 3 and a run of 5. Which roof is steeper? Explain. **See margin.**

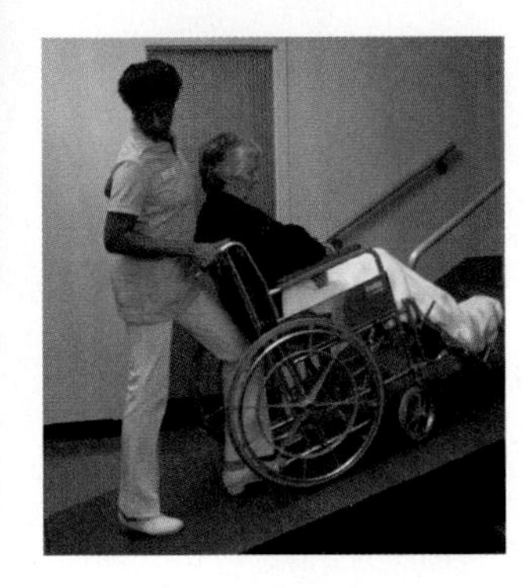

18. **Ramps** Guidelines for a wheelchair ramp allow a maximum of 1 in. of rise for every 12 in. of run. The ramp at the left runs 6 ft 8 in. and rises 2 ft 9 in. Does the ramp meet the guidelines? Explain. **See margin.**

19. **Cars** The graph shows the value of a car for the first seven years of ownership.
   a. What was the value of the car when it was new? **$24,000**
   b. What is the slope of the graph? **−2**
   c. What does the slope tell you about the relationship between the age of the car and its value? **Every year, the value of the car decreases $2,000.**

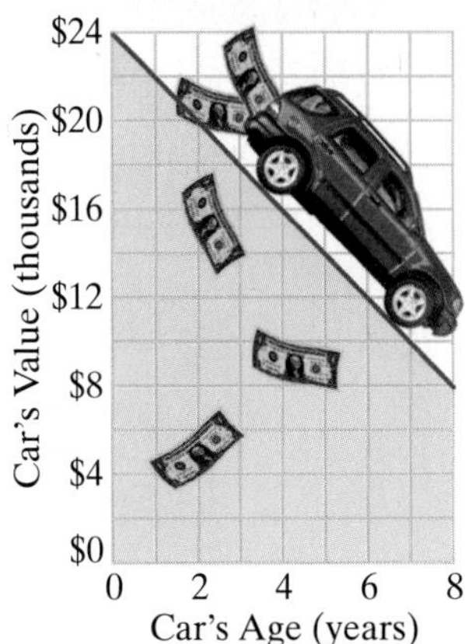

20. **Error Analysis** Your classmate graphs a line through (4, 2) and (5, −1) and finds that the slope equals 3. Explain why your classmate is incorrect.

20. The line slants down from left to right, so the slope should be negative 3.

21. **Writing in Math** Explain why it is more difficult to run up a hill with a slope of $\frac{1}{2}$ than a hill with a slope of $\frac{1}{6}$.

21. It would be more difficult, because a slope of $\frac{1}{2}$ is steeper than a slope of $\frac{1}{6}$.

Ⓒ 22. **Challenge** On the same coordinate plane, graph line r through points (0, 4) and (3, −3) and line s through points (1, 7) and (4, 0). Find the slopes of r and s. What are lines with equal slopes called?
$-\frac{7}{3}$; $-\frac{7}{3}$; parallel lines

Test Prep and Mixed Review

Practice

Multiple Choice

23. Which line in the graph at the right contains the ordered pair (0, −4)? **C**
   Ⓐ Line j Ⓒ Line l
   Ⓑ Line k Ⓓ Line m

24. The lowest point in a city is 8 ft below sea level, or −8 ft. A building at that point has 5 stories, and each story is 15 ft high. What is the elevation of the top of the building? **H**
   Ⓕ 12 ft   Ⓖ 20 ft   Ⓗ 67 ft   Ⓙ 75 ft

25. △ABC is an isosceles triangle. △XYZ has angle measures 45°, 70°, and 65°. Are the two triangles congruent? Explain. **See margin.**

GO for Help

For Exercise	See Lesson
25	7-5

4. Assess & Reteach

Lesson Quiz

1. Find the slope of the line. **2**

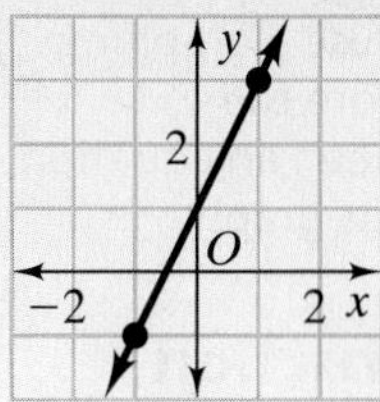

2. (1, 3) and (2, 6) **3**
3. (−3, 4) and (3, −5) $-\frac{3}{2}$

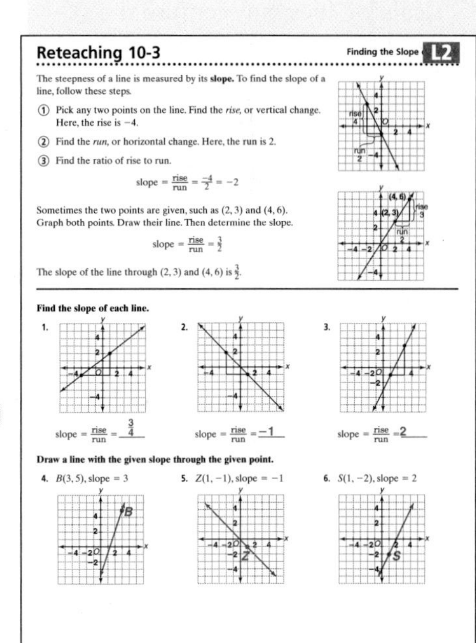

Reteaching 10-3 — Finding the Slope **L2**

The steepness of a line is measured by its **slope.** To find the slope of a line, follow these steps.

① Pick any two points on the line. Find the *rise*, or vertical change. Here, the rise is −4.

② Find the *run*, or horizontal change. Here, the run is 2.

③ Find the ratio of rise to run.

$$\text{slope} = \frac{\text{rise}}{\text{run}} = \frac{-4}{2} = -2$$

Sometimes the two points are given, such as (2, 3) and (4, 6). Graph both points. Draw their line. Then determine the slope.

$$\text{slope} = \frac{\text{rise}}{\text{run}} = \frac{3}{2}$$

The slope of the line through (2, 3) and (4, 6) is $\frac{3}{2}$.

Find the slope of each line.

1. slope = $\frac{\text{rise}}{\text{run}}$ = $\frac{3}{4}$
2. slope = $\frac{\text{rise}}{\text{run}}$ = −1
3. slope = $\frac{\text{rise}}{\text{run}}$ = 2

Draw a line with the given slope through the given point.

4. $B(3, 5)$, slope = 3
5. $Z(1, -1)$, slope = −1
6. $S(1, -2)$, slope = 2

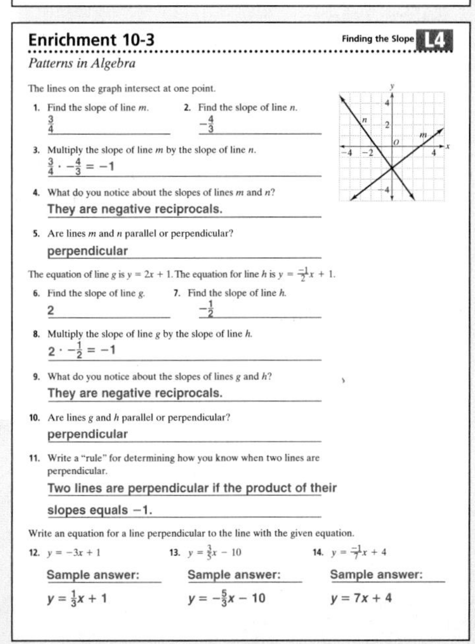

Enrichment 10-3 — Finding the Slope **L4**

Patterns in Algebra

The lines on the graph intersect at one point.

1. Find the slope of line m. $\frac{3}{4}$
2. Find the slope of line n. $-\frac{4}{3}$
3. Multiply the slope of line m by the slope of line n. $\frac{3}{4} \cdot -\frac{4}{3} = -1$
4. What do you notice about the slopes of lines m and n? They are negative reciprocals.
5. Are lines m and n parallel or perpendicular? perpendicular

The equation of line g is $y = 2x + 1$. The equation for line h is $y = \frac{-1}{2}x + 1$.

6. Find the slope of line g. 2
7. Find the slope of line h. $-\frac{1}{2}$
8. Multiply the slope of line g by the slope of line h. $2 \cdot -\frac{1}{2} = -1$
9. What do you notice about the slopes of lines g and h? They are negative reciprocals.
10. Are lines g and h parallel or perpendicular? perpendicular
11. Write a "rule" for determining how you know when two lines are perpendicular. Two lines are perpendicular if the product of their slopes equals −1.

Write an equation for a line perpendicular to the line with the given equation.

12. $y = -3x + 1$ — Sample answer: $y = \frac{1}{3}x + 1$
13. $y = \frac{3}{5}x - 10$ — Sample answer: $y = -\frac{5}{3}x - 10$
14. $y = \frac{-1}{7}x + 4$ — Sample answer: $y = 7x + 4$

Alternative Assessment

Have students work in pairs to complete Exercises 5–9. Partners work together to copy each line on graph paper. Students then use red pencil to draw and label the run and blue pencil to draw and label the rise. They then express the slope as a fraction in simplest form.

Test Prep

Resources

For additional practice with a variety of test item formats:
- Test-Taking Strategies, p. 523
- Test Prep, p. 527
- Test-Taking Strategies with Transparencies

Activity Lab

Exploring Slope

In Lesson 10-3, students found and used the slope of a line. In this activity, they use a graphing calculator to explore the relationship between an equation and the slope of its graph.

Guided Instruction

Revisit the meaning of *linear equation* and *slope*, and *rise* and *run*. Review the steps for finding the slope of a line.

Error Prevention!

Exercises 1–4
When students use a graphing calculator, they may confuse the steps involved in finding slope. To help, pair students and guide them to use the five steps presented for the Example. In addition, you can have students write the keystrokes they will use to find the slope in each exercise.
Exercises 6–9
Have students share their methods for finding the slopes of the linear equations. Ask: *What are the features all of these forms of equations have in common?* **Sample: Each has the $y = mx + b$ form; each equation provides the slope and y-intercept of its graph.**

Differentiated Instruction

Below Level L2
Students may have trouble using the graphing calculator. Have students complete each step by hand and by calculator before going onto the next step.

Resources

- graphing calculator
- Classroom Aid 36

10-3a Activity Lab — Technology

Exploring Slope

You can use a graphing calculator to explore the relationship between an equation and the slope of its graph.

EXAMPLE

Use a graphing calculator to find the slope of $y = 2x + 1$.

Step 1 Use Y= to enter the equation.

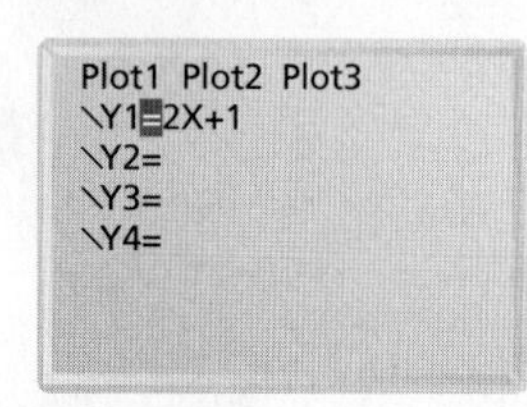

Step 2 Press ZOOM 0 ENTER for the integer mode.

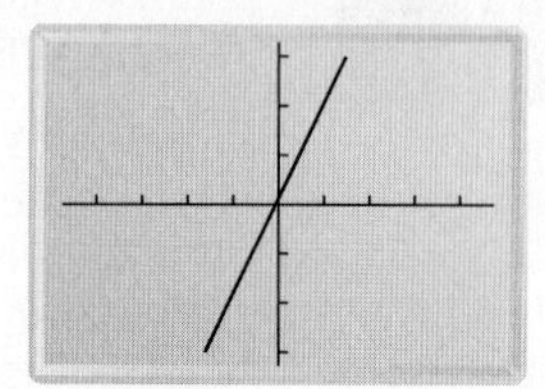

Step 3 Press TRACE. The x-coordinate is 0, and the y-coordinate is 1.

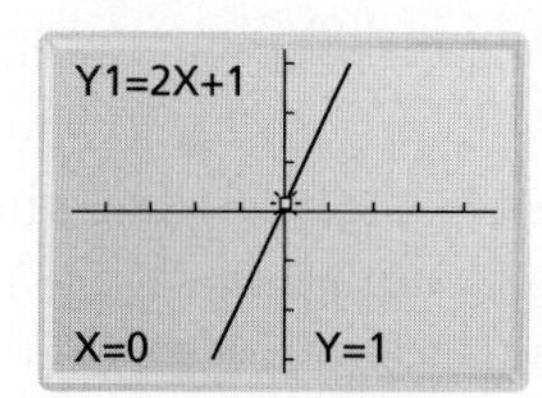

Step 4 Use ▶ to move the cursor to the right 1 unit.

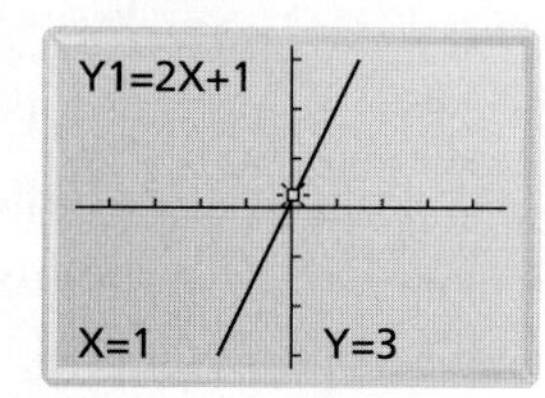

Step 5 Repeat step 4 and examine how the coordinates change. As the x-coordinate increases 1 unit, the y-coordinate increases 2 units. Another way to say this is that for every 1 unit of run, there are 2 units of rise.
slope $= \frac{\text{rise}}{\text{run}} = \frac{2}{1}$, or 2.

Exercises

Use a graphing calculator to find the slope of each equation.

1. $y = 2x - 3$ **2**
2. $y = x + 3$ **1**
3. $y = \frac{1}{3}x + 2$ **$\frac{1}{3}$**
4. $y = -3x$ **−3**
5. **Reasoning** Look at each slope you found. How is the slope of a line represented in each equation? **It is the coefficient of x.**

Without graphing, find the slope of each line.

6. $y = 8x + 3$ **8**
7. $y = -x - 1$ **−1**
8. $y = \frac{1}{2}x + 9$ **$\frac{1}{2}$**
9. $y = \frac{5}{2}x - 3$ **$\frac{5}{2}$**

Checkpoint Quiz 1

Lessons 10-1 through 10-3

Graph each point in the same coordinate plane. Identify the quadrant in which the point lies. See margin for graph.

1. $K(-2, -5)$ III **2.** $Q(-3, 4)$ II **3.** $D(2, 1)$ I

Graph each linear equation. 4–6. See margin.

4. $y = x + 7$ **5.** $y = -3x + 2$ **6.** $y = x - 3$

7. A dragonfly can reach a speed of 50 km/h. The dragonfly's speed can be represented using $d = 50t$, where d represents the distance in kilometers and t represents the time in hours. Graph this equation. See margin.

8. Find the slope of the line on the coordinate plane at the right. $-\frac{3}{2}$

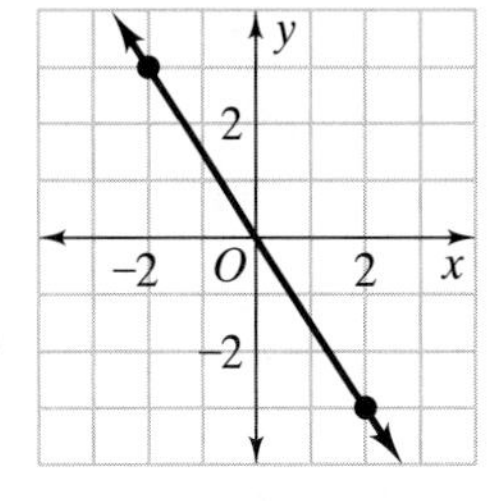

Draw a line with the given slope through the given point. 9–10. See margin.

9. $P(-2, 0)$, slope $= 3$ **10.** $A(0, -3)$, slope $= \frac{1}{2}$

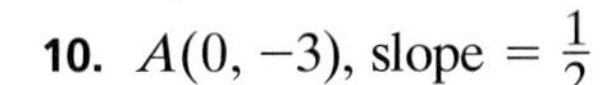

MATH GAMES

Hide and Seek

What You'll Need

- 2 players, two sheets of graph paper

How To Play

- Using graph paper, draw the first quadrant of the coordinate plane. Label the x- and y-axes from 0 to 10.
- Draw a triangle or rectangle on your graph. The ordered pairs for each vertex must be whole numbers.
- You and your opponent take turns trying to locate each vertex of the other's shape by calling out ordered pairs that are whole numbers.
- If your opponent's guess falls inside your figure, you say "inside." If it falls outside your figure, you say "outside." If it is on the border, you say "border." If the guess falls on a vertex, you must say "bingo."
- The first player to name all the vertices of the opponent's figure wins.

1–7. See back of book.
9–10. See back of book.

Use this Checkpoint Quiz to check students' understanding of the skills and concepts of Lessons 10-1 through 10-3.

Resources

- All-in-One Teaching Resources Checkpoint Quiz 1
- ExamView CD-ROM
- Success Tracker™ Online Intervention

Hide and Seek

In this game, students become more familiar with the locations of ordered pairs on a graph. Players draw a figure on a graph, then take turns trying to guess the coordinates of the vertices of the other player's figure.

Guided Instruction

Have a volunteer read the rules of the game. Explain that it may help students keep track of their guesses by recording them on a graph and marking them inside, outside, border, or "bingo."

Resources

- graph paper

10-4

1. Plan

Objective
To graph nonlinear relationships

Examples
1 Graphing a Nonlinear Equation
2 Graphing Absolute Value Equations

Math Understandings: p. 484C

Math Background

The graph of every linear equation is a straight line. The graph of every *nonlinear* equation is not a straight line, although it can be a combination of straight lines as well as curves. This lesson introduces two nonlinear equations: a parabola, or U-shaped curve; and an absolute-value equation, which is V-shaped.

More Math Background: p. 484C

Lesson Planning and Resources

See p. 484E for a list of the resources that support this lesson.

PowerPoint

Bell Ringer Practice

Check Skills You'll Need
Use student page, transparency, or PowerPoint. For intervention, direct students to:
Graphing Linear Equations
Lesson 10-2
Extra Skills and Word Problems Practice, Ch. 10

Algebra

10-4 Graphing Nonlinear Relationships

Check Skills You'll Need

1. **Vocabulary Review** What is a *linear equation*? **1–4. See back of book.**

Find three solutions of each equation.

2. $y = x + 12$
3. $y = x - 20$
4. $y = 15x$

Lesson 10-2

What You'll Learn

To graph nonlinear relationships

New Vocabulary nonlinear equation

Why Learn This?

When you kick a ball in the air, the path the ball follows is a curve called a parabola. A parabola is a sign of a relationship that is not linear.

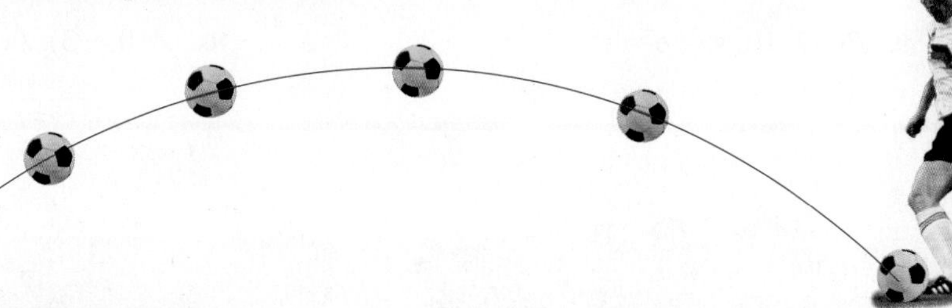

A **nonlinear equation** is an equation whose graph is not a line.

EXAMPLE Graphing a Nonlinear Equation

Test Prep Tip
When substituting a value for x in $-x^2$, make sure you square the value *before* finding its opposite.

1 Graph $y = -x^2$ using integer values of x from -3 to 3.

Step 1 Make a table of solutions.

x	$-x^2$	y	(x, y)
-3	$-(-3)^2$	-9	$(-3, -9)$
-2	$-(-2)^2$	-4	$(-2, -4)$
-1	$-(-1)^2$	-1	$(-1, -1)$
0	$-(0)^2$	0	$(0, 0)$
1	$-(1)^2$	-1	$(1, -1)$
2	$-(2)^2$	-4	$(2, -4)$
3	$-(3)^2$	-9	$(3, -9)$

Step 2 Use each solution (x, y) to graph a point. Draw a curve through the points.

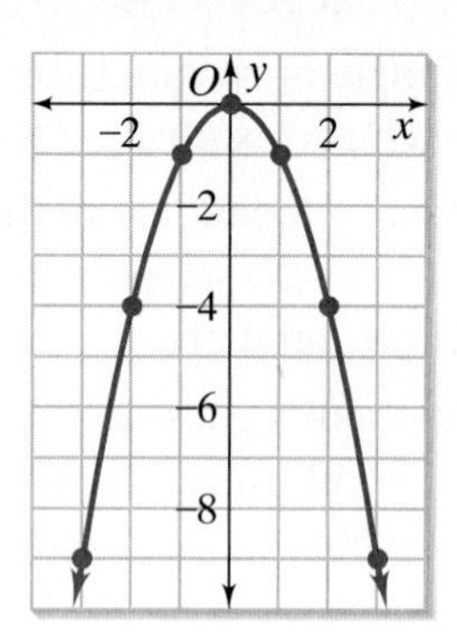

Quick Check

1. Graph $y = 2x^2$ using integer values of x from -3 to 3. **See back of book.**

Differentiated Instruction Solutions for All Learners

Special Needs L1
For some exercises, students who have difficulty graphing can make a table of solutions. Pair them up with a student who can graph, then use the table of solutions to graph.

learning style: visual

Below Level L2
Make a table for each equation for integer values of x from -3 to 3.

$y = 2x$ −6, −4, −2, 0, 2, 4, 6
$y = x - 2$ −5, −4, −3, −2, −1, 0, 1

learning style: visual

Video Tutor Help

Visit: PHSchool.com
Web Code: are-0775

2.

x	−2	−1	0	1	2
y	4	2	0	2	4

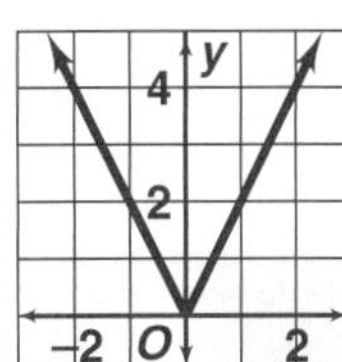

Answers may vary. Sample: No; by substituting (−3, 10) into the equation, you can determine that (−3, 10) is not a solution of $y = x^2 + 2$.

EXAMPLE Graphing Absolute Value Equations

2 Graph $y = |x|$ using integer values of x from −2 to 2.

x	$\lvert x\rvert$	y	(x, y)
−2	$\lvert -2\rvert$	2	(−2, 2)
−1	$\lvert -1\rvert$	1	(−1, 1)
0	$\lvert 0\rvert$	0	(0, 0)
1	$\lvert 1\rvert$	1	(1, 1)
2	$\lvert 2\rvert$	2	(2, 2)

← Make a table of values.

Graph the ordered pairs. Then connect the points. →

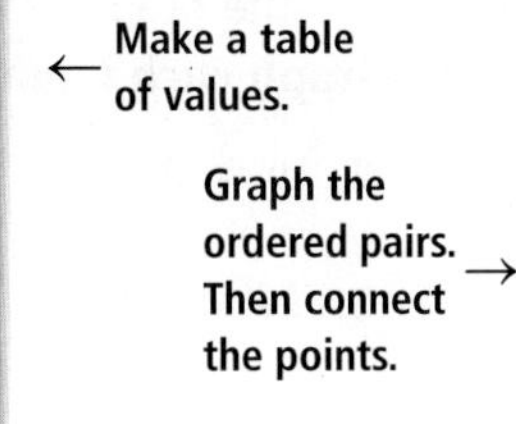

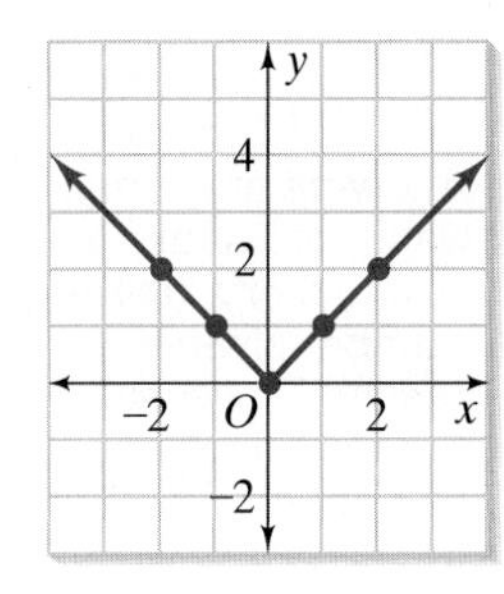

Quick Check

2. Graph $y = 2|x|$ using integer values of x from −2 to 2. **See left.**

More Than One Way

The equation $y = x^3$ shows the nonlinear relationship between edge length x and volume y of a cube. Determine whether (3, 12) is a solution of $y = x^3$.

Kayla's Method

I will use my graphing calculator. I press Y= and enter $y = x^3$. Then I select the integer mode under ZOOM. I can use the TRACE feature to get the screen at the right.

At $x = 3$, $y = 27$, so (3, 12) is not a solution.

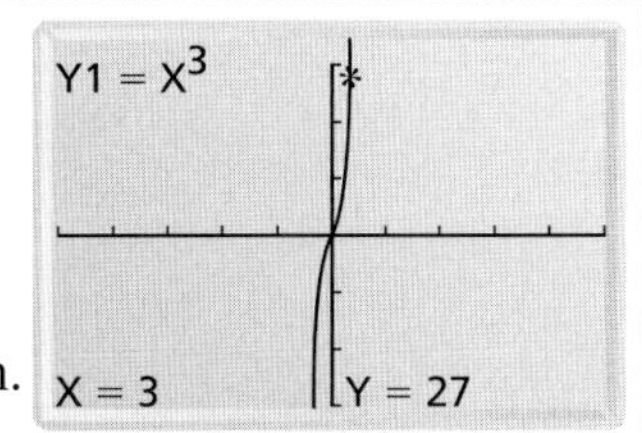

Carlos's Method

I will substitute 3 for x and 12 for y in the equation.

$y = x^3$

$12 \stackrel{?}{=} 3^3$ ← Substitute.

$12 \stackrel{?}{=} 27$ ← Simplify.

$12 \neq 27$

The equation is not true, so (3, 12) is not a solution.

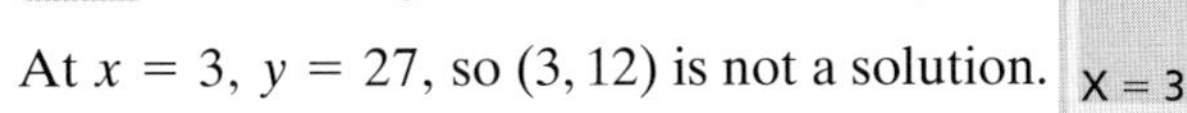

Choose a Method

Determine whether (−3, 10) is a solution of $y = x^2 + 2$. Describe your method and explain why you chose it.

2. Teach

Activity Lab

Use before the lesson.

All in One Teaching Resources

Activity Lab 10-4: Critical Thinking

Guided Instruction

Example 1

For Quick Check 1, remind students to square the value of x *before* multiplying by 2.

Draw a table on the board and record the values of x and y as students supply the y-values.

Example 2

Have students use their arms to model basic shapes of functions. As you read equations aloud, students can hold their arms in *U* shapes or *V* shapes to indicate the shape of the function. For instance: $y = 3x^2$ **U-shaped**
$y = x^2 + 3$ **U-shaped**
$y = |x| + 3$ **V-shaped**

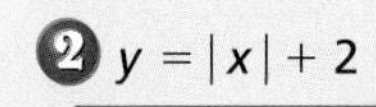

Additional Examples

Graph each equation using integer values of x from −3 to 3.

1 $y = x^2 - 2$

2 $y = |x| + 2$

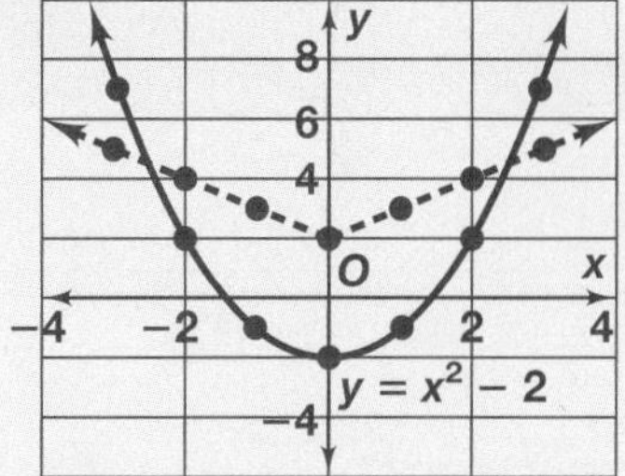

All in One Teaching Resources

- Daily Notetaking Guide 10-4 L3
- Adapted Notetaking 10-4 L1

Closure

- Describe the graph of a parabola. **a U-shaped curve**
- Describe the graph of an absolute-value equation. **a V-shaped curve**

Advanced Learners L4

Write an equation whose solutions include each set of ordered pairs.

(−4, 2), (−3, 1), (−1, 1), (3, 5) $y = |x + 2|$

(−3, 25), (−1, 9), (0, 4), (2, 0) $y = (x - 2)^2$

learning style: visual

English Language Learners ELL

Write "linear" and "nonlinear" on the board. Write "line" and "not a line" under each, respectively. Show students that the prefix *non* often refers to the word *not*.

learning style: verbal

3. Practice

Assignment Guide

Check Your Understanding

Go over Exercises 1–7 in class before assigning the Homework Exercises.

Homework Exercises

A	Practice by Example	8–22
B	Apply Your Skills	23–31
C	Challenge	32
	Test Prep and Mixed Review	33–35

Homework Quick Check

To check students' understanding of key skills and concepts, go over Exercises 13, 21, 24, 25, and 30.

Differentiated Instruction Resources

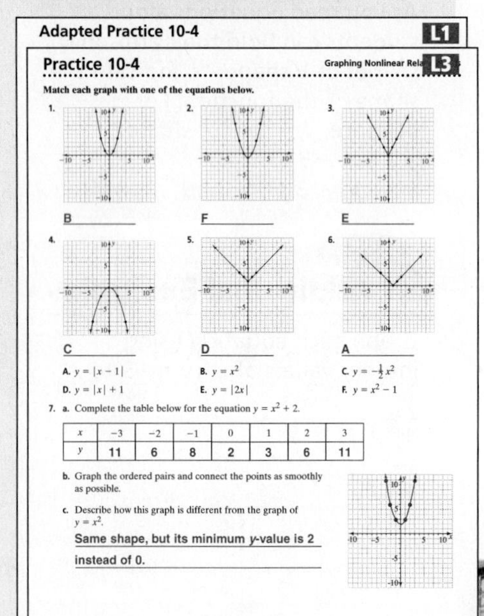

Adapted Practice 10-4 L1

Practice 10-4 Graphing Nonlinear Rela[tionships] L3

Match each graph with one of the equations below.

1. B
2. F
3. E
4. C
5. D
6. A

A. $y = |x - 1|$ B. $y = x^2$ C. $y = -\frac{1}{2}x^2$

D. $y = |x| + 1$ E. $y = |2x|$ F. $y = x^2 - 1$

7. a. Complete the table below for the equation $y = x^2 + 2$.

x	−3	−2	−1	0	1	2	3
y	11	6	8	2	3	6	11

b. Graph the ordered pairs and connect the points as smoothly as possible.

c. Describe how this graph is different from the graph of $y = x^2$.

Same shape, but its minimum y-value is 2 instead of 0.

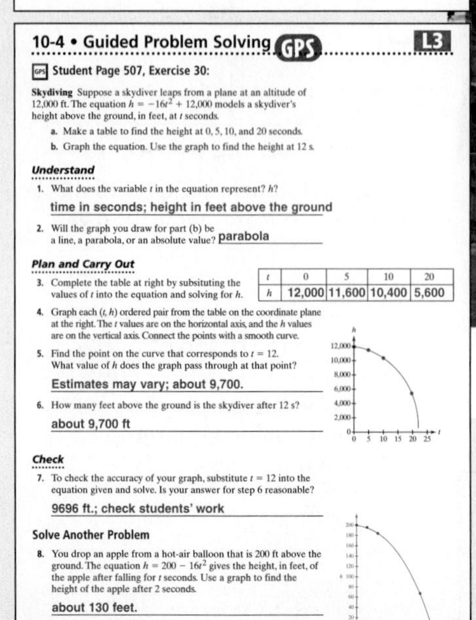

10-4 • Guided Problem Solving GPS L3

Student Page 507, Exercise 30:

Skydiving Suppose a skydiver leaps from a plane at an altitude of 12,000 ft. The equation $h = -16t^2 + 12{,}000$ models a skydiver's height above the ground, in feet, at t seconds.

a. Make a table to find the height at 0, 5, 10, and 20 seconds.

b. Graph the equation. Use the graph to find the height at 12 s.

Understand

1. What does the variable t in the equation represent? h?

time in seconds; height in feet above the ground

2. Will the graph you draw for part (b) be a line, a parabola, or an absolute value? parabola

Plan and Carry Out

3. Complete the table at right by subsituting the values of t into the equation and solving for h.

t	0	5	10	20
h	12,000	11,600	10,400	5,600

4. Graph each (t, h) ordered pair from the table on the coordinate plane at the right. The t values are on the horizontal axis, and the h values are on the vertical axis. Connect the points with a smooth curve.

5. Find the point on the curve that corresponds to $t = 12$. What value of h does the graph pass through at that point?

Estimates may vary; about 9,700.

6. How many feet above the ground is the skydiver after 12 s?

about 9,700 ft

Check

7. To check the accuracy of your graph, substitute $t = 12$ into the equation given and solve. Is your answer for step 6 reasonable?

9696 ft.; check students' work

Solve Another Problem

8. You drop an apple from a hot-air balloon that is 200 ft above the ground. The equation $h = 200 - 16t^2$ gives the height, in feet, of the apple after falling for t seconds. Use a graph to find the height of the apple after 2 seconds.

about 130 feet.

Check Your Understanding

1. **Vocabulary** How can you tell whether an equation is linear or nonlinear? **If the solutions of an equation form a line, it is linear. If the graph of the solutions is not a line, it is nonlinear.**

Graph each equation using integer values of x from −3 to 3. 2–4. See margin.

2. $y = x^2$
3. $y = |x|$
4. $y = -x^3$

5. **Number Sense** Why does the graph of the equation $y = x^2$ not fall in Quadrant III or Quadrant IV? **See left.**

5. Answers may vary. Sample: Squaring a positive or a negative number always results in a positive number. So for all values of x, x^2 is positive, and y is always positive.

6. Describe the shape of the graph of the absolute value equation $y = |x|$. **It is V-shaped and opens upward.**

7. **Mental Math** Is $P(1, -3)$ a solution of the equation $y = -x^3$? **no**

Homework Exercises

For more exercises, see Extra Skills and Word Problems.

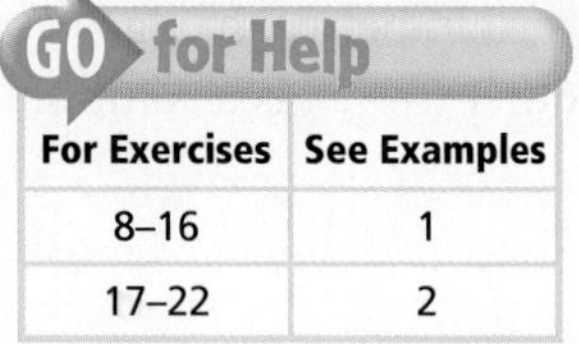

GO for Help

For Exercises	See Examples
8–16	1
17–22	2

A **Graph each equation using integer values of x from −3 to 3.** 8–16. See margin.

8. $y = 4x^2$
9. $y = -3x^2$
10. $y = \frac{1}{2}x^2$
11. $y = x^2 - 2$
12. $y = x^2 + 2$
13. $y = -x^2 + 3$
14. $y = -x^2 - 4$
15. $y = (x + 1)^2$
16. $y = (x - 1)^2$

Graph each equation using integer values of x from −3 to 3. 17–22. See margin.

17. $y = 2|x|$
18. $y = \frac{1}{3}|x|$
19. $y = -|x|$
20. $y = -|x| + 2$
21. $y = |x - 1|$
22. $y = |x + 1|$

B GPS 23. **Guided Problem Solving** According to legend, Galileo dropped two different objects from the Tower of Pisa to prove that they would fall at the same rate. The equation $d = 16t^2$ can be used to find the distance d, in feet, that the objects fall in t seconds. Graph the equation showing how far the objects fall between 0 and 5 seconds. **See margin.**

- Make a table of solutions. What values of t will you use?
- Graph the points. What is the greatest value of d in your graph?

24. Graph the equation $A = \pi r^2$ to show the relationship between radius r and area A of a circle. Use 3.14 as an approximation for π. **See margin.**

25. **Writing in Math** Explain why the graph of $y = -|x|$ has no points in Quadrant I or Quadrant II. **See margin.**

26. Use integer values of −3 to 3 to graph $y = 3|x| + 5$. **See margin.**

2–4. See back of book.
8–26. See back of book.
30a–30b. See back of book.
32. See back of book.

Go Online

Homework Video Tutor
Visit: PHSchool.com
Web Code: are-1004

Match each graph with an equation.

27. 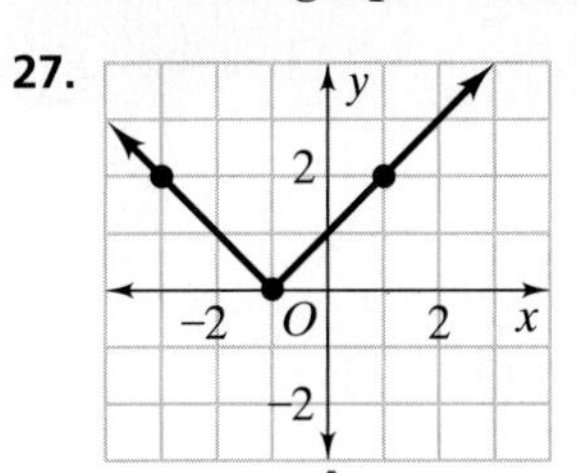
A

28. 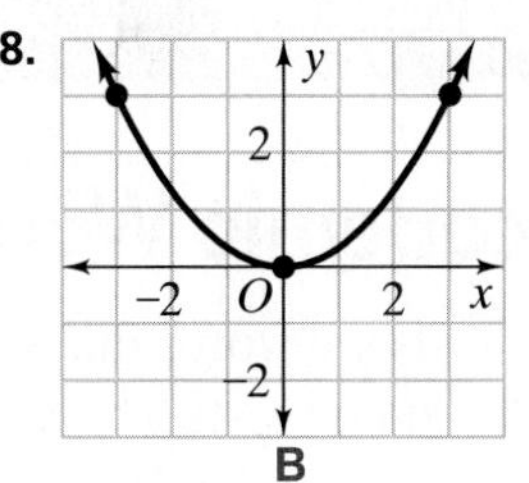
B

29. 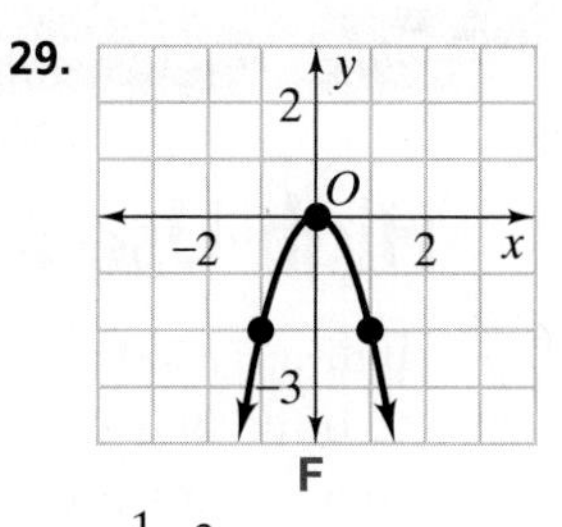
F

A. $y = |x + 1|$ B. $y = \frac{1}{3}x^2$ C. $y = \frac{1}{2}x^2$
D. $y = |x| - 1$ E. $y = 2x^2$ F. $y = -2x^2$

30. **Skydiving** Suppose a skydiver leaps from a plane at an altitude of 12,000 ft. The equation $h = -16t^2 + 12{,}000$ models the skydiver's height above the ground, in feet, at t seconds. **30a–b. See margin.**
  a. Make a table to find the height at 0, 5, 10, and 20 seconds.
  b. Graph the equation. Use the graph to find the height at 12 s.

31. **Choose a Method** The equation $A = s^2$ shows the nonlinear relationship between side length s and area A of a square. Determine whether (6.5, 42.25) is a solution of $A = s^2$. **yes; $6.5^2 = 42.25$**

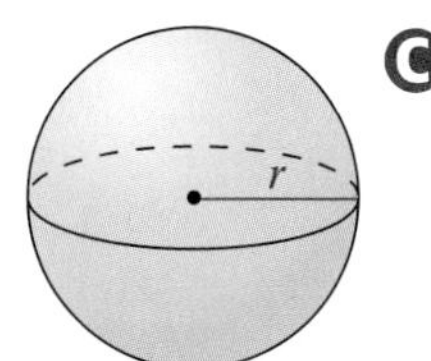

32. **Challenge** The equation $S = 4\pi r^2$ gives the surface area S of a sphere with radius r. The equation $V = \frac{4}{3}\pi r^3$ gives the volume V of a sphere with radius r. For each equation, make a table of values using integer values of r from 0 to 5. Use 3 as an approximation for π. For what value of r will the graphs of the equations intersect? **See margin.**

Test Prep and Mixed Review — Practice

Multiple Choice

33. Which of the following relationships is represented in the graph? **A**

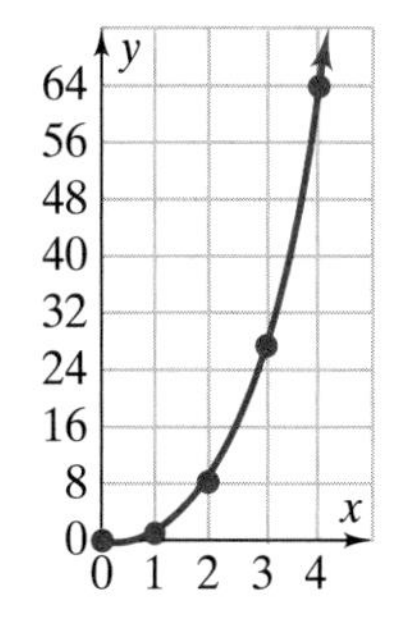

  A. The relationship between the edge length and the volume of a cube
  B. The relationship between the side length and the surface area of a cube
  C. The relationship between the side length and the area of a square
  D. The relationship between the radius and the area of a circle

34. On Monday, $\frac{4}{7}$ of the students at school bought a hot lunch. About what percent of the students did NOT buy a hot lunch? **G**
  F. 37% G. 43% H. 57% J. 62%

35. Find the height of a rectangular prism with a volume of 357 cm^3, a length of 6 cm, and a width of 8.5 cm. **7 cm**

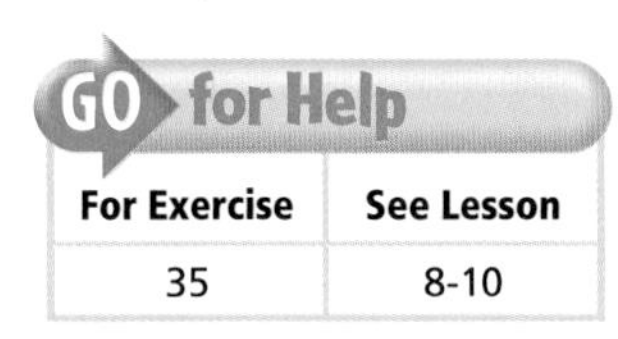

GO for Help

For Exercise	See Lesson
35	8-10

Online lesson quiz, PHSchool.com, Web Code: ara-1004

4. Assess & Reteach

Lesson Quiz

Make a table of values. Then graph each equation.

1. $y = x^2 + 1$ 2. $y = 5|x|$

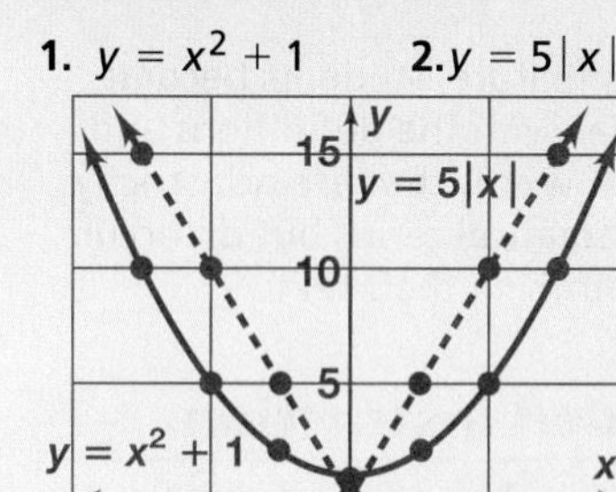

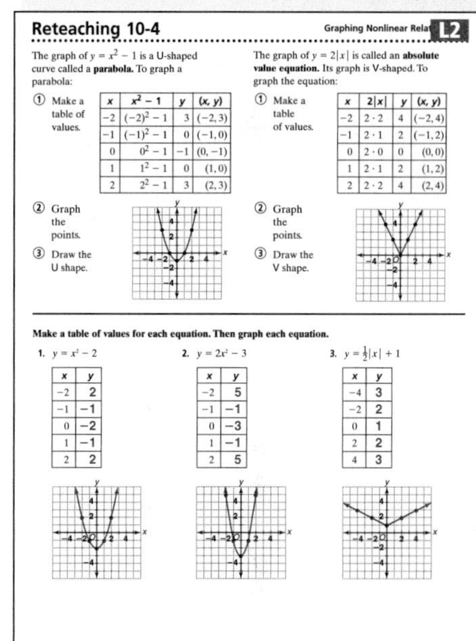

Reteaching 10-4 Graphing Nonlinear Rela... L2

The graph of $y = x^2 - 1$ is a U-shaped curve called a **parabola.** To graph a parabola:

① Make a table of values.

x	$x^2 - 1$	y	(x, y)
−2	$(-2)^2 - 1$	3	(−2, 3)
−1	$(-1)^2 - 1$	0	(−1, 0)
0	$0^2 - 1$	−1	(0, −1)
1	$1^2 - 1$	0	(1, 0)
2	$2^2 - 1$	3	(2, 3)

② Graph the points.
③ Draw the U shape.

The graph of $y = 2|x|$ is called an **absolute value equation.** Its graph is V-shaped. To graph the equation:

① Make a table of values.

x	$2\|x\|$	y	(x, y)
−2	2 · 2	4	(−2, 4)
−1	2 · 1	2	(−1, 2)
0	2 · 0	0	(0, 0)
1	2 · 1	2	(1, 2)
2	2 · 2	4	(2, 4)

② Graph the points.
③ Draw the V shape.

Make a table of values for each equation. Then graph each equation.

1. $y = x^2 - 2$

x	y
−2	2
−1	−1
0	−2
1	−1
2	2

2. $y = 2x^2 - 3$

x	y
−2	5
−1	−1
0	−3
1	−1
2	5

3. $y = \frac{1}{2}|x| + 1$

x	y
−4	3
−2	2
0	1
2	2
4	3

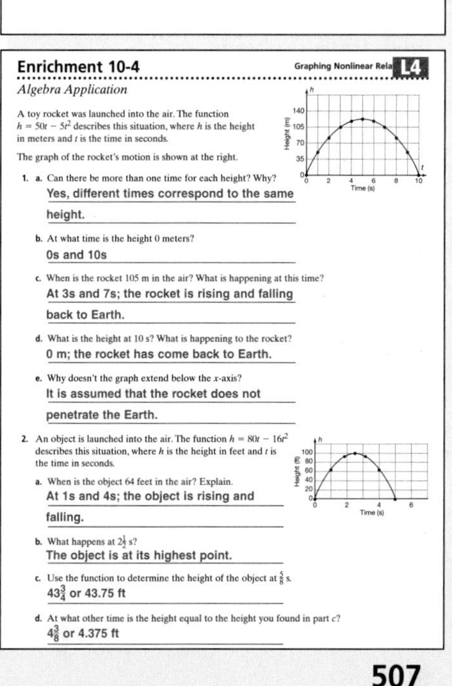

Enrichment 10-4 Graphing Nonlinear Rela... L4

Algebra Application

A toy rocket was launched into the air. The function $h = 50t - 5t^2$ describes this situation, where h is the height in meters and t is the time in seconds.

The graph of the rocket's motion is shown at the right.

1. a. Can there be more than one time for each height? Why? **Yes, different times correspond to the same height.**
  b. At what time is the height 0 meters? **0s and 10s**
  c. When is the rocket 105 m in the air? What is happening at this time? **At 3s and 7s; the rocket is rising and falling back to Earth.**
  d. What is the height at 10 s? What is happening to the rocket? **0 m; the rocket has come back to Earth.**
  e. Why doesn't the graph extend below the x-axis? **It is assumed that the rocket does not penetrate the Earth.**

2. An object is launched into the air. The function $h = 80t - 16t^2$ describes this situation, where h is the height in feet and t is the time in seconds.
  a. When is the object 64 feet in the air? Explain. **At 1s and 4s; the object is rising and falling.**
  b. What happens at $2\frac{1}{2}$ s? **The object is at its highest point.**
  c. Use the function to determine the height of the object at $\frac{2}{5}$ s. **$43\frac{3}{4}$ or 43.75 ft**
  d. At what other time is the height equal to the height you found in part c? **$4\frac{3}{5}$ or 4.375 ft**

Alternative Assessment

Have pairs of students work together to complete Exercises 8–19. Have partners alternate roles of making a table and graphing the equation.

Test Prep

Resources
For additional practice with a variety of test item formats:
- Test-Taking Strategies, p. 523
- Test Prep, p. 527
- Test-Taking Strategies with Transparencies

Vocabulary Builder

High-Use Academic Words

In this feature, students become familiar with the definitions and uses of words that are not strictly mathematical terms but do occur frequently in their texts.

Guided Instruction

Explain to students that some words have multiple meanings, some of which apply to math and others that do not. Have volunteers read each word and its definition aloud. Then ask questions, such as:

- *What is another way to say "Make a diagram using the following coordinates."?* **Draw a diagram using the following coordinates.**
- *How do we use* predict *when graphing lines?* **We can use the slope to *predict* where other points will fall on the line.**
- *How can you use* compare *to describe two lines on a graph?* **You can *compare* the slopes to see which is greater.**

Differentiated Instruction

Advanced Learners L4

Have students create their own math problems using the vocabulary words. Then have them trade problems with a partner and solve.

Vocabulary Builder

High-Use Academic Words

High-use academic words are words that you will see often in textbooks and on tests. These words are not math vocabulary terms, but knowing them will help you to succeed in mathematics.

Direction Words

Some words tell what to do in a problem. I need to understand what these words are asking so that I give the correct answer.

Word	Meaning
Represent	To replace with other words or symbols
Predict	To say in advance the outcomes or effects
Compare	To say how two things are similar or different

Exercises

1–3. Check students' work.

1. Draw a picture to represent the weather outside today.
2. Predict the weather tomorrow.
3. Compare the weather today with the weather six months ago.
4. A puppy weighed about 24 lb. She then gained about 12 lb per month, for x months. Which equation represents her growth? **B**

   Ⓐ $y = 24x + 12$ Ⓑ $y = 12x + 24$ Ⓒ $y = 24x - 12$ Ⓓ $y = 12x - 24$

5. Use the graph at the right. Predict the value of y when x is 10. **800**

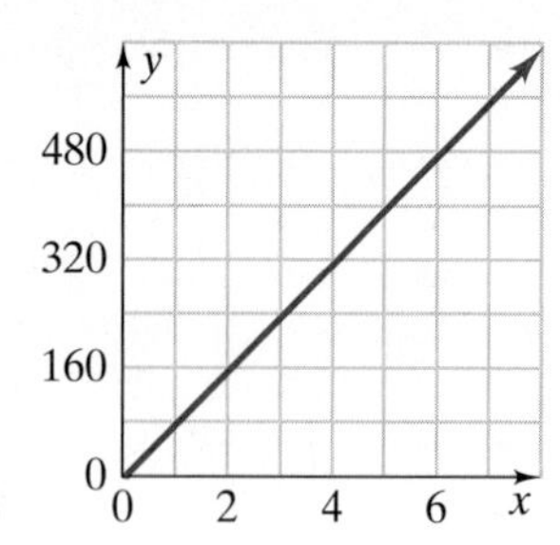

6. Compare the slope of the line that goes through (0, 0) and (5, 2) with the slope of the line that goes through (1, 1) and (2, 5). Which slope is greater? **See above right.**

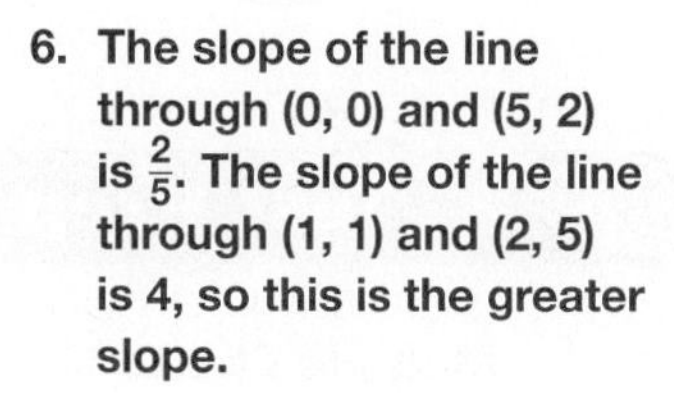

6. The slope of the line through (0, 0) and (5, 2) is $\frac{2}{5}$. The slope of the line through (1, 1) and (2, 5) is 4, so this is the greater slope.

7. **Word Knowledge** Think about the word *conclude*.
   a. Choose the letter for how well you know the word. **7a–c. Check students' work.**
      A. I know its meaning.
      B. I've seen it, but I don't know its meaning.
      C. I don't know it.
   b. **Research** Look up and write the definition of *conclude*.
   c. Use the word *conclude* in a sentence involving mathematics.

Slides, Flips, and Turns

To change the position of a geometric figure, you can use slides, flips, and turns.

A *slide* moves a figure so that every point moves the same direction and the same distance.

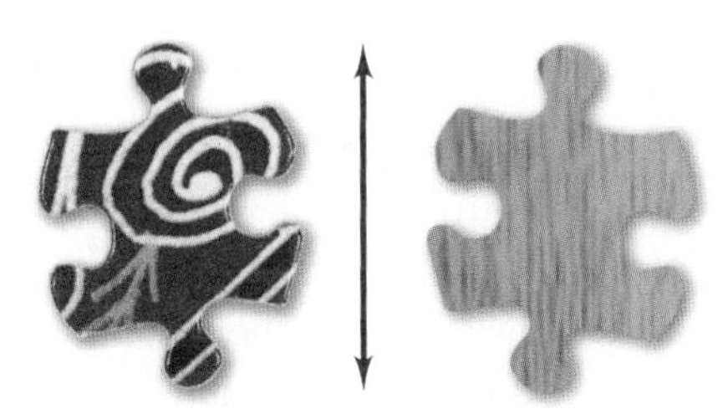

A *flip* reflects a figure over a line.

A *turn* rotates a figure around a point.

4. Answers may vary. Sample: By flipping a figure twice, you have moved it to a new position, which is a slide.

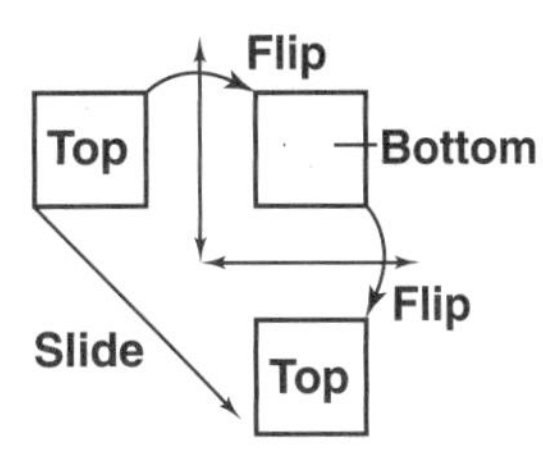

Exercises

Describe each transformation as a *slide, flip,* or *turn.*

1.

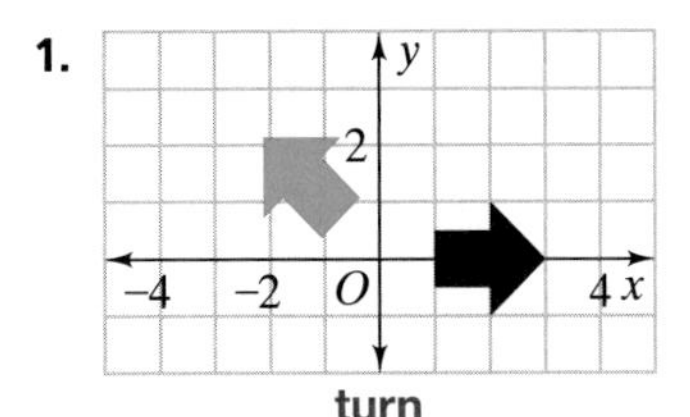

turn

2.

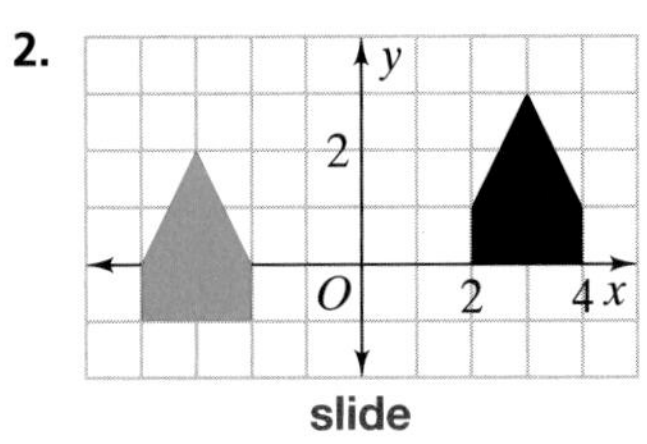

slide

3.

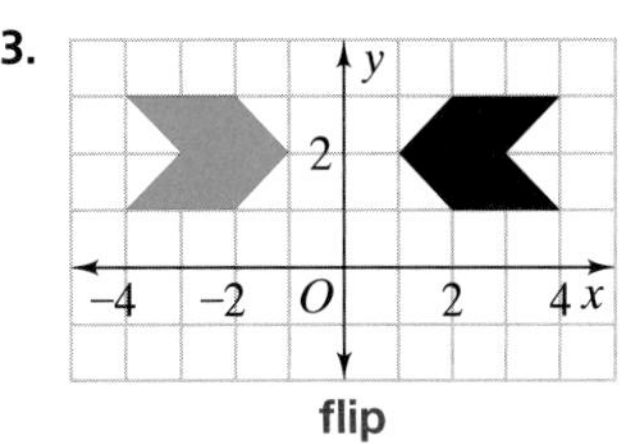

flip

4. **Writing in Math** Describe how two flips can have the same effect as one slide. Include a drawing to illustrate your example. **See above.**

5. **Open-Ended** Give an example that involves a slide, a flip, or a turn. **Check students' work.**

Activity Lab

Slides, Flips, and Turns

Students investigate the transformation of figures. They begin here to explore translations (slides), reflections (flips), and rotations (turns).

Guided Instruction

Explain that figures can change position—move, or *transform*. To identify the three kinds of rigid motion in the transformations presented, draw two congruent irregular figures on the board in different positions (perhaps rotated). Label the vertices (corners) of each. Have students identify the corresponding vertices.

Guide students to see the distinction between *slides, flips,* and *turns* as you discuss the transformations. Provide examples on the board. Ask: *Are transformed figures congruent with the original figures? Explain.* **Yes; the figures change in position and orientation only and not in their size and shape.**

Exercises

Have students work on the Exercises. When they have finished, they may discuss their answers with a partner and correct their work, if necessary.

Alternative Method

Have student pairs work on the Exercises. Provide pattern blocks to explore transformations. Encourage them to use these manipulatives to model the movements of the figures.

Resources

- Activity Lab 10-5: Translations
- pattern blocks

10-5

1. Plan

Objective
To graph and write rules for translations

Examples
1 Translating a Point
2 Translating a Figure
3 Writing a Rule for a Translation

Math Understandings: p. 484D

Professional Development

Math Background

A *translation* moves each point in a figure the same distance and direction. The *image* of the original figure is the figure you get after a translation. The image does not change in its size or shape—it is a copy of the original figure that has simply moved. Use *prime notation* to identify an image point. Read A' as "A prime." A translation that moves each point in a figure 1 unit right and 2 units up can be expressed as $(x, y) \rightarrow (x + 1, y + 2)$. So, point $B(3, -5)$ becomes the image point $B'(3 + 1, -5 + 2)$, or $B'(4, -3)$.

More Math Background: p. 484D

Lesson Planning and Resources

See p. 484E for a list of the resources that support this lesson.

Bell Ringer Practice

Check Skills You'll Need
Use student page, transparency, or PowerPoint. For intervention, direct students to:
Graphing Points in Four Quadrants
Lesson 10-1
Extra Skills and Word Problems Practice, Ch. 10

10-5 Translations

Check Skills You'll Need

1. Vocabulary Review Name the *x-coordinate* and the *y-coordinate* of point $A(2, -5)$.
1–5. See back of book.
Graph each point on the same coordinate plane.

2. $E(3, 1)$
3. $R(1, 0)$
4. $G(-3, -1)$
5. $S(2, -2)$

Lesson 10-1

What You'll Learn

To graph and write rules for translations

New Vocabulary transformation, translation, image, prime notation

Why Learn This?

Patterns like the one at the right use translations. You can describe translations mathematically by graphing in a coordinate plane.

A **transformation** is a change in the position, shape, or size of a figure. Three types of transformations that change only the position are slides, flips, and turns. A slide is also known as a translation. A **translation** is a transformation that moves each point of a figure the same distance and in the same direction.

The result of a transformation is the **image** of the original. **Prime notation** is the way to name an image point. A' is read as "A prime."

EXAMPLE Translating a Point

1 Multiple Choice Translate point $F(4, 1)$ left 3 units and up 2 units. What are the coordinates of the image F'?

Ⓐ $(1, -1)$ Ⓑ $(1, 3)$ Ⓒ $(7, -1)$ Ⓓ $(7, 3)$

Test Prep Tip
Graphing helps you visualize a translation.

Locate point F at $(4, 1)$.

From point F, move 3 units left and 2 units up. Graph the image point F'.

The coordinates of F' are $(1, 3)$. The answer is B.

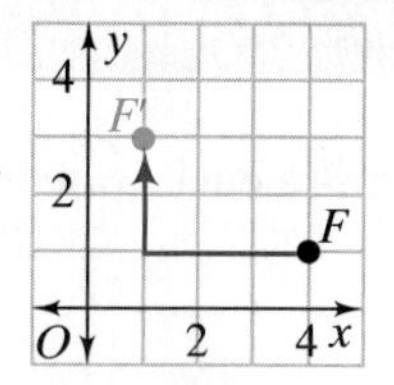

Quick Check

1. Translate point $G(-4, 1)$ right 1 unit and down 4 units. What are the coordinates of the image G'? **(−3, −3)**

To show a translation, you can use arrow notation. For the translation of F to F' in Example 1, you write $F(4, 1) \rightarrow F'(1, 3)$.

Differentiated Instruction Solutions for All Learners

Special Needs L1
Pair students who have difficulty drawing on the grid with students who have facility. One can draw while the other writes the arrow notation to show the translation.
learning style: visual

Below Level L2
Have students draw horizontal and vertical arrows along their graphs to show the horizontal and vertical translations for each point.
learning style: visual

Vocabulary Tip

A vertex is the point of intersection of two sides of a figure.

To translate a geometric figure, first translate each vertex of the figure. Then connect the image points. When a geometric figure is translated, the image is congruent to the original figure.

EXAMPLE Translating a Figure

2 The vertices of $\triangle ABC$ are $A(-4, 3)$, $B(-1, 4)$, and $C(-3, 1)$. Translate $\triangle ABC$ right 2 units and down 4 units. Use arrow notation to describe the translation.

Graph and label vertices A, B, and C. Draw $\triangle ABC$.

From each vertex, move right 2 units and down 4 units, and then draw an image point.

Label A', B', and C'. Draw $\triangle A'B'C'$.

Use arrow notation:
$A(-4, 3), B(-1, 4), C(-3, 1) \rightarrow A'(-2, -1), B'(1, 0), C'(-1, -3)$

Quick Check

2. Graph $\triangle ABC$ from Example 2. Translate it left 3 units and up 1 unit. Use arrow notation to describe the translation. **See back of book.**

You can also use arrow notation to write a rule for a translation.

EXAMPLE Writing a Rule for a Translation

3 **Animation** Computer animators use translations to move objects to new positions on the computer screen. Write a rule for the translation of spaceship A that a computer could apply to the other spaceships.

The horizontal change from A to A' is 6 units right, so $x \rightarrow x + 6$.

The vertical change from A to A' is 3 units down, so $y \rightarrow y - 3$.

The rule for the translation is
$(x, y) \rightarrow (x + 6, y - 3)$.

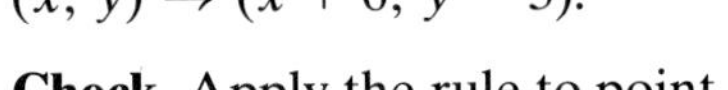

Check Apply the rule to point B.
$(x, y) \rightarrow (x + 6, y - 3)$

$B(-4, 1) \rightarrow B'(-4 + 6, 1 - 3) = B'(2, -2)$ ✔ The answer checks.

Quick Check

3. Write a rule for the translation of spaceship A to $(4, 3)$.
$(x, y) \rightarrow (x + 6, y + 1)$

2. Teach

Activity Lab

Use before the lesson.
Student Edition Activity Lab 10-5a, Slides, Flips, and Turns, p. 509

All in One Teaching Resources
Activity Lab 10-5: Translations

Guided Instruction

Example 2
Have students shade each vertex a different color. Have them use the same color to trace the path of translation.

PowerPoint

Additional Examples

1. Translate point $A(-3, -1)$ left 2 units and up 5 units. What are the coordinates of image A'? **(−5, 4)**

2. The vertices of $\triangle ABC$ are $A(-4, -3)$, $B(-3, 1)$, $C(-1, -2)$. Graph $\triangle ABC$ and translate it right 5 and down 2. Use arrow notation to show the translation. **See back of book.**

3. Write a rule for the translation.
$(x, y) \rightarrow (x + 3, y - 2)$

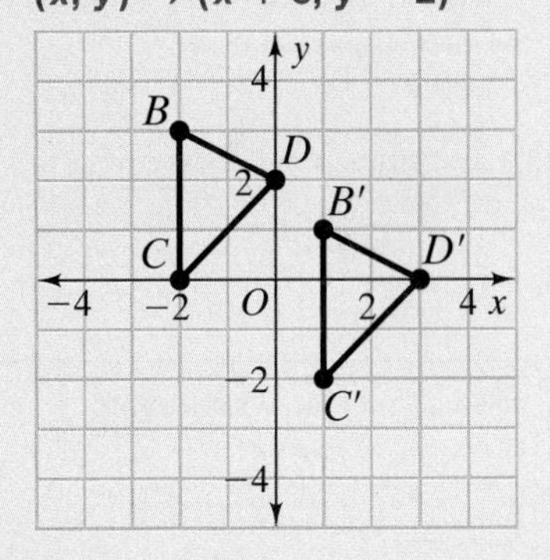

All in One Teaching Resources
- Daily Notetaking Guide 10-5 L3
- Adapted Notetaking 10-5 L1

Closure

- *What is a transformation?* **a change of the position, shape, or size of a figure**
- *What is a translation?* **a transformation that moves each point in a figure the same distance and same direction**

Advanced Learners L4
Have students find how the translation rule for moving an image back to the original relates to the original rule. **The rules are opposites.**

learning style: visual

English Language Learners ELL
Help students make a connection between the translations in this lesson to translations of language. The translated images only change location, not shape. Translated words change language, not meaning.

learning style: verbal

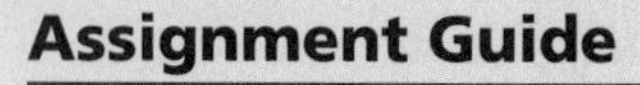

3. Practice

Assignment Guide

Check Your Understanding
Go over Exercises 1–4 in class before assigning the Homework Exercises.

Homework Exercises

A	Practice by Example	5–16
B	Apply Your Skills	17–26
C	Challenge	27
	Test Prep and Mixed Review	28–33

Homework Quick Check
To check students' understanding of key skills and concepts, go over Exercises 12, 16, 18, 22, and 26.

Differentiated Instruction Resources

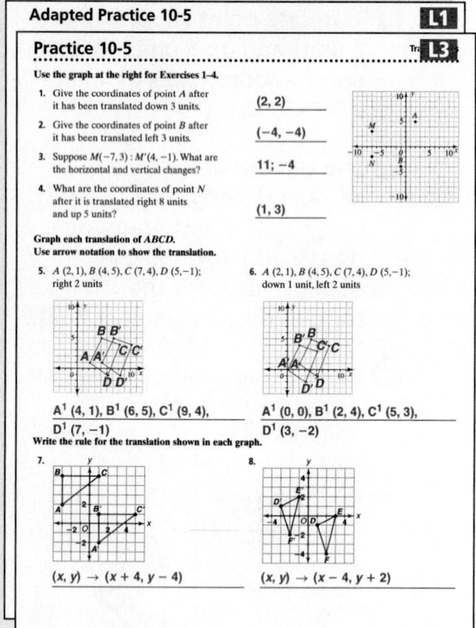

Adapted Practice 10-5 L1

Practice 10-5 L3

Use the graph at the right for Exercises 1–4.

1. Give the coordinates of point A after it has been translated down 3 units. (2, 2)
2. Give the coordinates of point B after it has been translated left 3 units. (−4, −4)
3. Suppose $M(-7, 3)$: $M'(4, -1)$. What are the horizontal and vertical changes? 11; −4
4. What are the coordinates of point N after it is translated right 8 units and up 5 units? (1, 3)

Graph each translation of $ABCD$. Use arrow notation to show the translation.

5. $A(2, 1), B(4, 5), C(7, 4), D(5, -1)$; right 2 units
   $A'(4, 1), B'(6, 5), C'(9, 4), D'(7, -1)$
6. $A(2, 1), B(4, 5), C(7, 4), D(5, -1)$; down 1 unit, left 2 units
   $A'(0, 0), B'(2, 4), C'(5, 3), D'(3, -2)$

Write the rule for the translation shown in each graph.

7. $(x, y) \to (x + 4, y - 4)$
8. $(x, y) \to (x - 4, y + 2)$

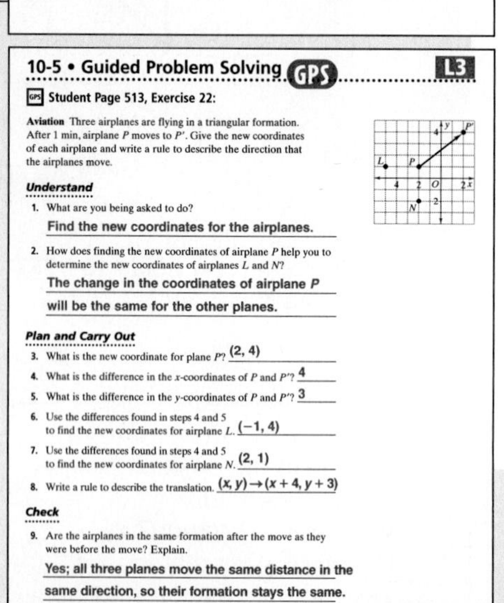

10-5 • Guided Problem Solving GPS L3

Student Page 513, Exercise 22:

Aviation Three airplanes are flying in a triangular formation. After 1 min, airplane P moves to P'. Give the new coordinates of each airplane and write a rule to describe the direction that the airplanes move.

Understand

1. What are you being asked to do? Find the new coordinates for the airplanes.
2. How does finding the new coordinates of airplane P help you to determine the new coordinates of airplanes L and N? The change in the coordinates of airplane P will be the same for the other planes.

Plan and Carry Out

3. What is the new coordinate for plane P? (2, 4)
4. What is the difference in the x-coordinates of P and P'? 4
5. What is the difference in the y-coordinates of P and P'? 3
6. Use the differences found in steps 4 and 5 to find the new coordinates for airplane L. (−1, 4)
7. Use the differences found in steps 4 and 5 to find the new coordinates for airplane N. (2, 1)
8. Write a rule to describe the translation. $(x, y) \to (x + 4, y + 3)$

Check

9. Are the airplanes in the same formation after the move as they were before the move? Explain. Yes; all three planes move the same distance in the same direction, so their formation stays the same.

Solve Another Problem

10. Three airplanes are flying in a triangular formation. If airplane P moves to P', find the new coordinates of the other two planes and write a rule to describe the move. $R'(1, 0)$, $S'(4, 0)$; $(x, y) \to (x + 2, y - 3)$

Check Your Understanding

1. **Vocabulary** A translation is a type of __?__. transformation
2. **Mental Math** Translate the origin (0, 0) left 3 units and up 2 units. What are the coordinates of the image? (−3, 2)

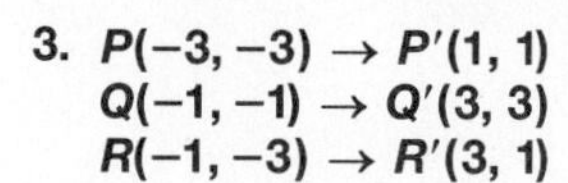

3. $P(-3, -3) \to P'(1, 1)$
   $Q(-1, -1) \to Q'(3, 3)$
   $R(-1, -3) \to R'(3, 1)$

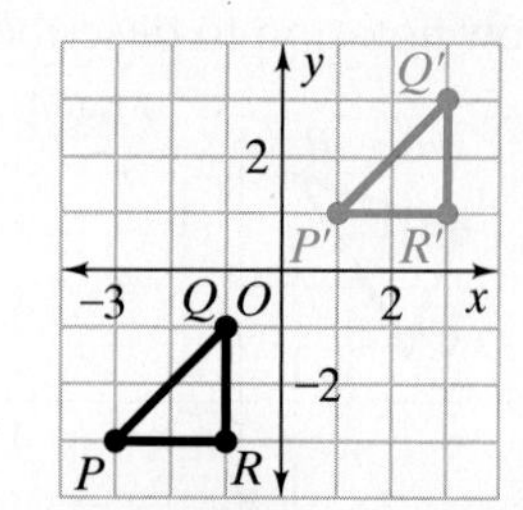

For Exercises 3–4, use the graph at the left.

3. Use arrow notation to show the translation. See left.
   $P(■, ■), Q(■, ■), R(■, ■) \to P'(■, ■), Q'(■, ■), R'(■, ■)$
4. Use words to describe the translation. Each point moves __?__ 4 units and __?__ 4 units. right; up

Homework Exercises

For more exercises, see Extra Skills and Word Problems.

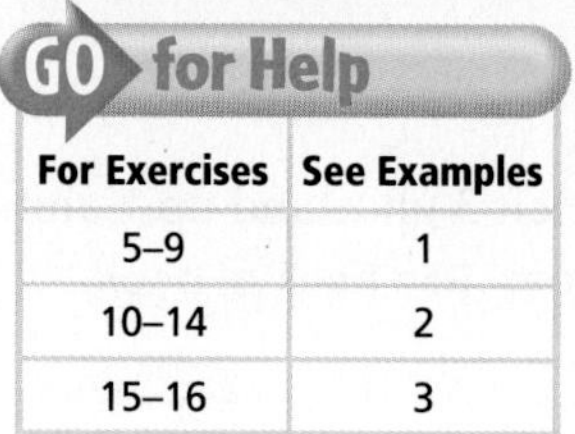

GO for Help

For Exercises	See Examples
5–9	1
10–14	2
15–16	3

A **Translate each point left 2 units and down 5 units. Write the coordinates of the image point.**

5. (3, 3) (1, −2)
6. (0, 0) (−2, −5)
7. (−3, 2) (−5, −3)
8. (−6, −1) (−8, −6)
9. (6, −1) (4, −6)

Graph each translation of $\triangle ABC$. Use arrow notation to show the translation. 10–14. See margin.

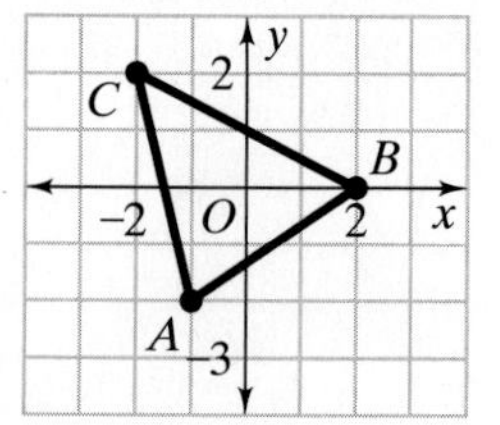

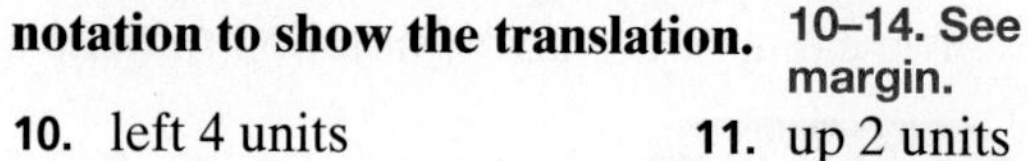

10. left 4 units
11. up 2 units
12. right 6 units, up 1 unit
13. down 2 units
14. left 3 units, up 3 units

Write a rule for the translation shown in each graph.

15. 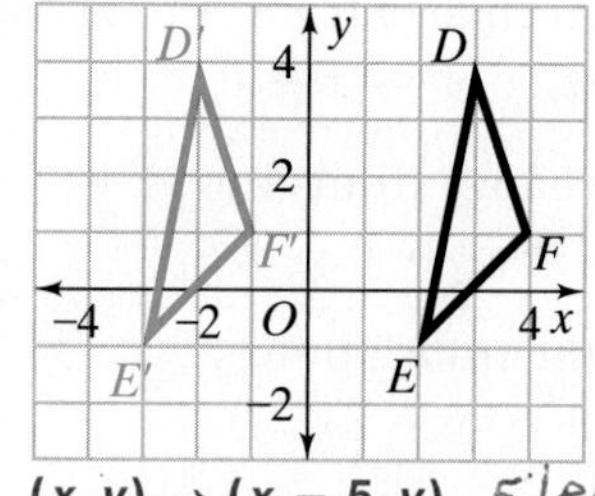

$(x, y) \to (x - 5, y)$ 5 left

16. 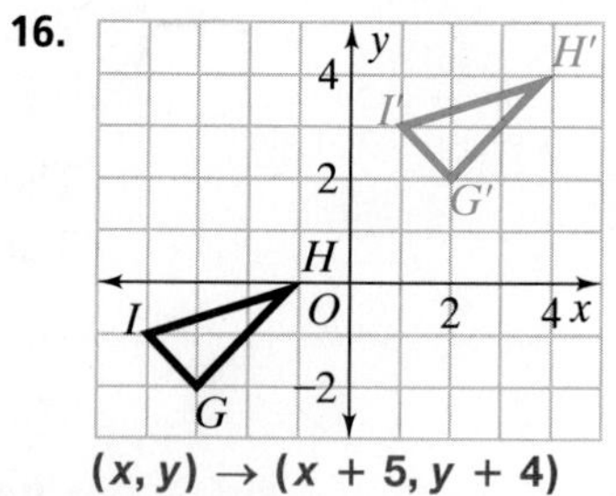

$(x, y) \to (x + 5, y + 4)$

17. $(x, y) \to (x + 2, y - 4)$;
    $(x, y) \to (x + 3, y)$

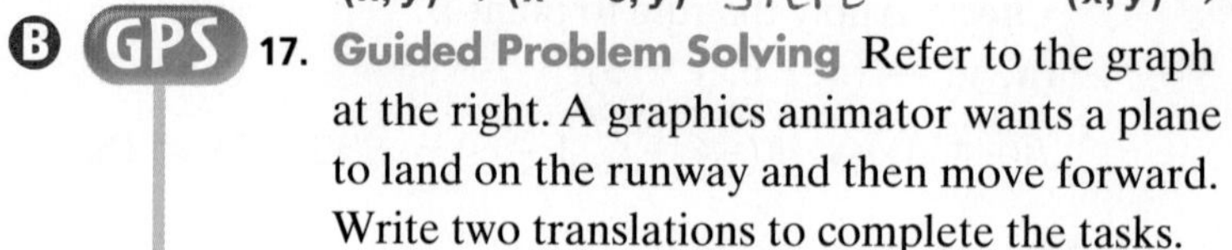

B GPS 17. **Guided Problem Solving** Refer to the graph at the right. A graphics animator wants a plane to land on the runway and then move forward. Write two translations to complete the tasks.

- What are the plane's coordinates? See left.
- What are the runway's coordinates?
- What translation will move the plane horizontally?

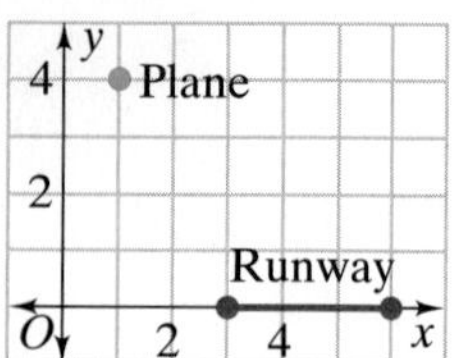

10–14. See back of book.
18–22. See back of book.
26. See back of book.

Visit: PHSchool.com
Web Code: are-1005

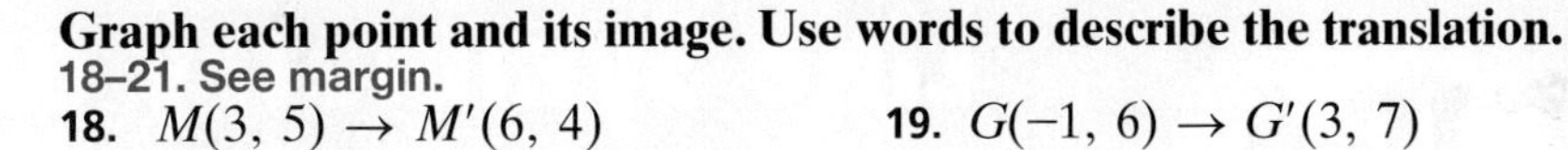

Graph each point and its image. Use words to describe the translation.
18–21. See margin.

18. $M(3, 5) \rightarrow M'(6, 4)$
19. $G(-1, 6) \rightarrow G'(3, 7)$
20. $H(-2, -5) \rightarrow H'(-1, -3)$
21. $J(4, 0) \rightarrow J'(3, 4)$

22. **Aviation** Three airplanes are flying in a triangular formation. After 1 min, airplane P moves to P'. Give the new coordinates of each airplane and write a rule to describe the direction that the airplanes move. See margin.

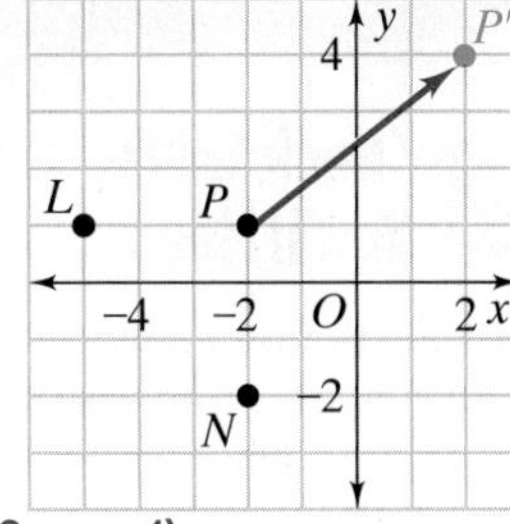

Write a rule for the translation described.

23. right 3 units and down 1 unit $(x, y) \rightarrow (x + 3, y - 1)$
24. right 4 units and up 1 unit $(x, y) \rightarrow (x + 4, y + 1)$
25. left 1 unit and up 4 units $(x, y) \rightarrow (x - 1, y + 4)$

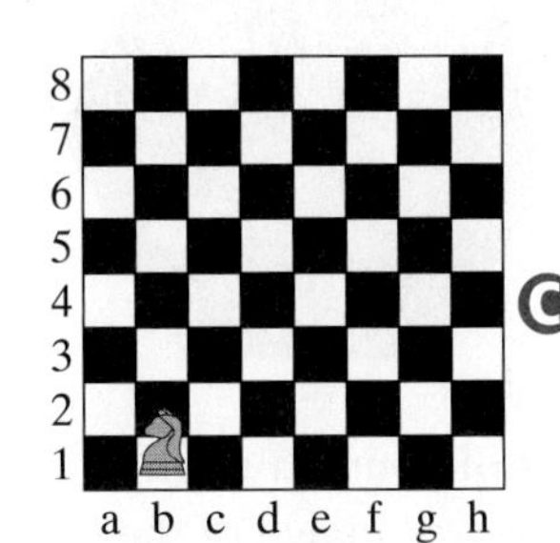

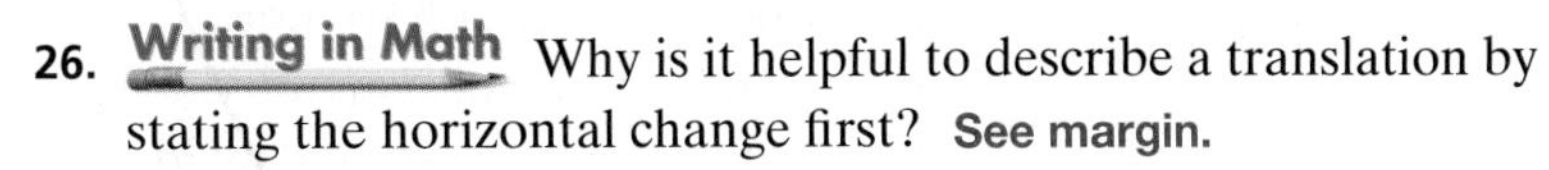

26. **Writing in Math** Why is it helpful to describe a translation by stating the horizontal change first? See margin.

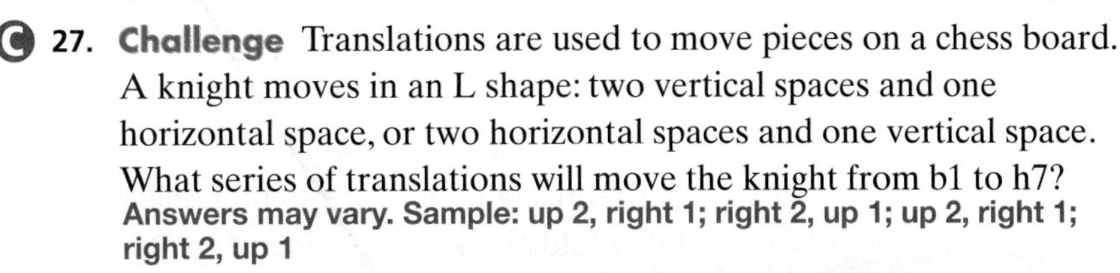

27. **Challenge** Translations are used to move pieces on a chess board. A knight moves in an L shape: two vertical spaces and one horizontal space, or two horizontal spaces and one vertical space. What series of translations will move the knight from b1 to h7? Answers may vary. Sample: up 2, right 1; right 2, up 1; up 2, right 1; right 2, up 1

Test Prep and Mixed Review

Practice

Multiple Choice

28. If the quadrilateral shown at the right is translated 3 units to the left and 4 units down, what will be the new coordinates of point A? C
    - A) $(-2, 0)$
    - B) $(-5, 4)$
    - C) $(-8, 0)$
    - D) $(-8, 8)$

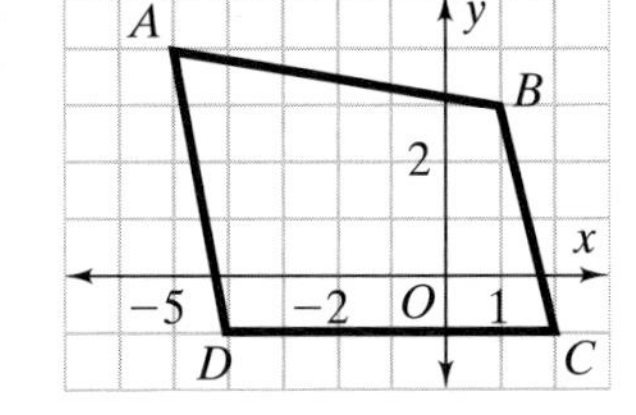

29. A container of juice holds 64 ounces. Which equation shows the remaining amount of juice j after you drink d ounces? G
    - F) $\frac{64}{d} = j$
    - G) $64 - d = j$
    - H) $64 + d = j$
    - J) $64d = j$

30. Jaleel spent \$39.95 on a pair of shoes. The sales tax was 8.5%. About how much did Jaleel pay for the shoes, including tax? B
    - A) Less than \$42
    - B) Between \$42 and \$45
    - C) Between \$45 and \$48
    - D) More than \$48

GO for Help

For Exercises	See Lesson
31–33	10-2

Algebra **Find three solutions of each equation.**

31. $y = 4x - 2$

x	y
0	−2
1	2
2	6

32. $y = -3x - 6$
(−1, −3); (0, −6); (1, −9)

33. $y = 5x + 4$
(−1, −1); (0, 4); (1, 9)

Online lesson quiz, PHSchool.

4. Assess & Reteach

Lesson Quiz

1. Graph $K(-1, 4)$ and translate it right 2 units and down 3 units. Write a rule to describe the translation.

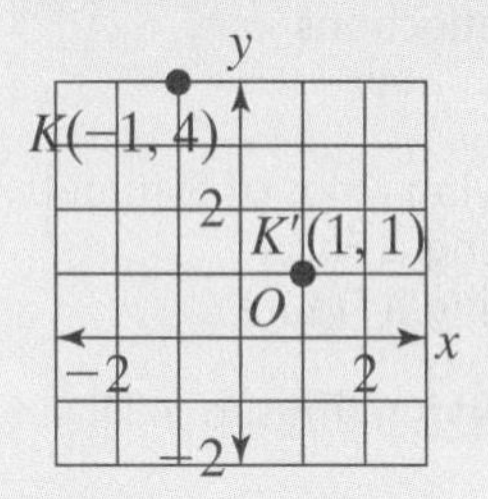

$K'(1, 1)$; $(x, y) \rightarrow (x + 2, y - 3)$

2. Translate point $M(3, -4)$ left 5 units and down 2 units. What are the coordinates of image M'? $M'(-2, -6)$
3. A figure is translated to the right 3 units and up 6 units. Write the general rule to show how to find the image. $(x, y) \rightarrow (x + 3, y + 6)$

Alternative Assessment

Each student in a pair dra... gives the coordinates of t... vertices.

...t Prep

...urces

...dditional practice with a variety of test item ...ats:

- ...t-Taking Strategies, p. 523
- Test Prep, p. 527
- Test-Taking Strategies with Transparencies

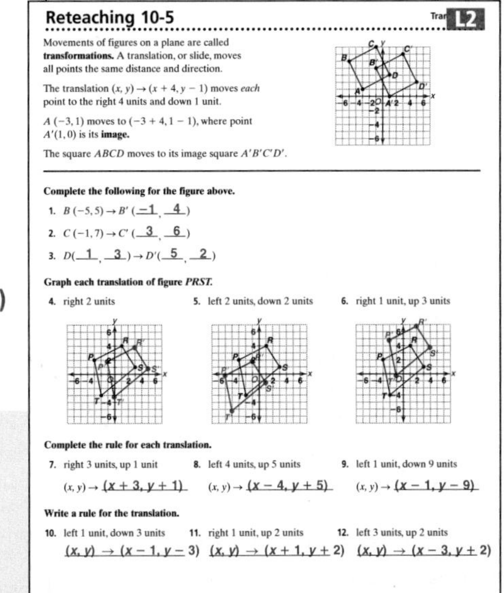

Enrichment 10-5 L4

Reteaching 10-5 L2

Movements of figures on a plane are called **transformations.** A translation, or slide, moves all points the same distance and direction.

The translation $(x, y) \rightarrow (x + 4, y - 1)$ moves *each* point to the right 4 units and down 1 unit.

$A(-3, 1)$ moves to $(-3 + 4, 1 - 1)$, where point $A'(1, 0)$ is its **image.**

The square $ABCD$ moves to its image square $A'B'C'D'$.

Complete the following for the figure above.

1. $B(-5, 5) \rightarrow B'(\underline{-1}, \underline{4})$
2. $C(-1, 7) \rightarrow C'(\underline{3}, \underline{6})$
3. $D(\underline{1}, \underline{3}) \rightarrow D'(\underline{5}, \underline{2})$

Graph each translation of figure PRST.

4. right 2 units
5. left 2 units, down 2 units
6. right 1 unit, up 3 units

Complete the rule for each translation.

7. right 3 units, up 1 unit $(x, y) \rightarrow \underline{(x + 3, y + 1)}$
8. left 4 units, up 5 units $(x, y) \rightarrow \underline{(x - 4, y + 5)}$
9. left 1 unit, down 9 units $(x, y) \rightarrow \underline{(x - 1, y - 9)}$

Write a rule for the translation.

10. left 1 unit, down 3 units $\underline{(x, y) \rightarrow (x - 1, y - 3)}$
11. right 1 unit, up 2 units $\underline{(x, y) \rightarrow (x + 1, y + 2)}$
12. left 3 units, up 2 units $\underline{(x, y) \rightarrow (x - 3, y + 2)}$

10-6

1. Plan

Objective
To identify lines of symmetry and to graph reflections

Examples
1 Identifying Lines of Symmetry
2 Reflecting a Point
3 Reflecting a Figure

Math Understandings: p. 484D

Math Background

A figure has *line symmetry* when one or more sides of the figure are mirror images of the other. A *line of symmetry* divides a figure into two mirror images. A *reflection* is a transformation of an entire figure that is flipped over a line called the *line of reflection.*

More Math Background: p. 484D

Lesson Planning and Resources

See p. 484E for a list of the resources that support this lesson.

Bell Ringer Practice

Check Skills You'll Need
Use student page, transparency, or PowerPoint. For intervention, direct students to:
Translations
Lesson 10-5
Extra Skills and Word Problems Practice, Ch. 10

10-6 Line Symmetry and Reflections

Check Skills You'll Need

1. **Vocabulary Review** When you translate a point, what do you call the new point? **image**

Translate point $P(1, 3)$ as described. Write the coordinates of P'.

2. 4 units left **(−3, 3)**
3. 2 units up **(1, 5)**
4. 4 units left and 2 units up **(−3, 5)**

What You'll Learn

To identify lines of symmetry and to graph reflections

New Vocabulary line symmetry, line of symmetry, reflection, line of reflection

Why Learn This?

Symmetry is often seen in nature. A snowflake forms when water vapor freezes. Most snowflakes have line symmetry and rotational symmetry. The red line drawn on the snowflake divides it in half, and the two halves are mirror images of each other.

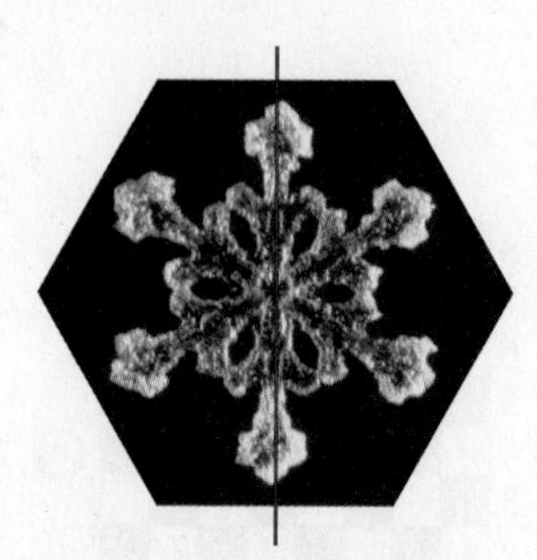

A figure has **line symmetry** if a line, called a **line of symmetry,** can be drawn through the figure so that one side is a mirror image of the other. If you fold along the red line of symmetry drawn on the snowflake, the right side of the snowflake fits exactly onto the left side.

EXAMPLE Identifying Lines of Symmetry

1 **Gridded Response** How many lines of symmetry does the flower have?

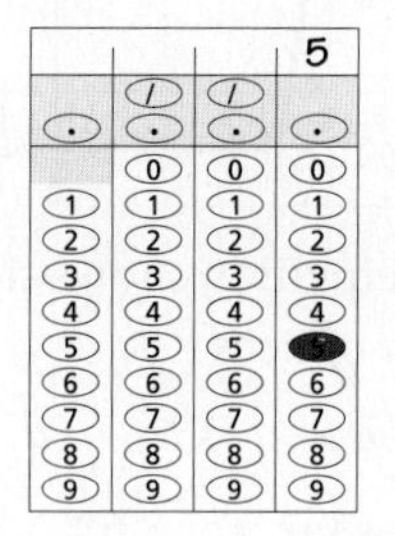

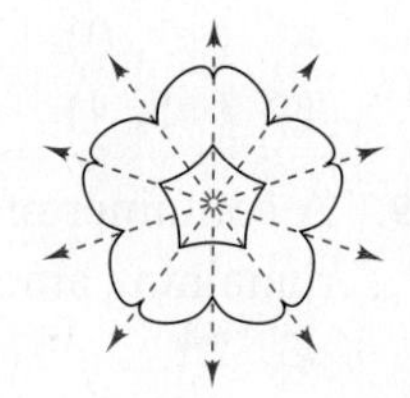

The flower has 5 lines of symmetry, so grid the answer 5 as shown.

Quick Check

1. **Art** Sketch the mask and draw the line(s) of symmetry. Does the mask have line symmetry?
See back of book.

Differentiated Instruction Solutions for All Learners

Special Needs L1
Pair students who have difficulty drawing a reflection with a partner who can draw. After their partner draws, have them identify and check the axis of the drawn reflected image and its coordinates.

learning style: visual

Below Level L2
Have students reflect a letter such as Z or R in a mirror. Elicit the fact that the image is reversed.

learning style: tactile

A **reflection** is a transformation that flips a figure over a line called a **line of reflection.**

EXAMPLE Reflecting a Point

2 Graph the point $A(3, -2)$ and its reflection over the indicated axis. Write the coordinates of the image.

a. y-axis

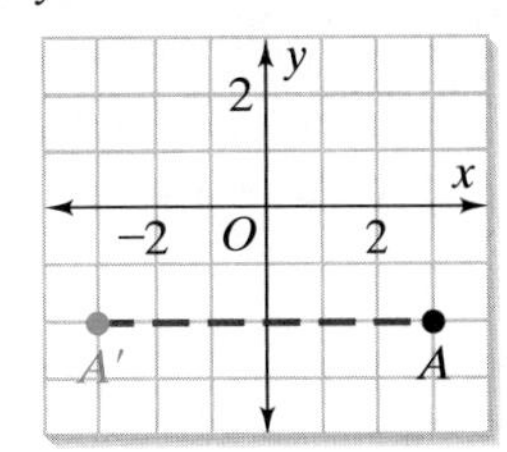

A is 3 units to the right of the y-axis, so A' is 3 units to the left of the y-axis.

A' has coordinates $(-3, -2)$.

b. x-axis

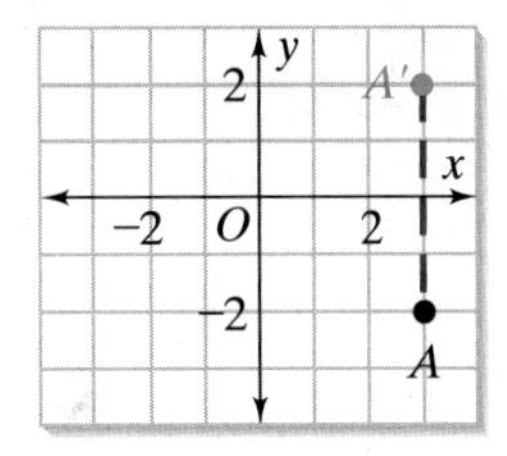

A is 2 units below the x-axis, so A' is 2 units above the x-axis.

A' has coordinates $(3, 2)$.

Test Prep Tip

When a point is reflected over the x-axis, the y-coordinate changes. When a point is reflected over the y-axis, the x-coordinate changes.

Quick Check

2. Graph the point $P(-4, 1)$ and its reflection over the indicated axis. Write the coordinates of the image. **2a–b. See back of book.**

a. y-axis **b.** x-axis

When a figure is reflected, the image is congruent to the original figure.

EXAMPLE Reflecting a Figure

3 Draw the image of $\triangle ABC$ reflected over the y-axis. Use arrow notation to describe the original triangle and its image.

A is 1 unit to the left of the y-axis, so A' is 1 unit to the right of the y-axis.

B is 4 units to the left of the y-axis, so B' is 4 units to the right of the y-axis.

C is 2 units to the left of the y-axis, so C' is 2 units to the right of the y-axis.

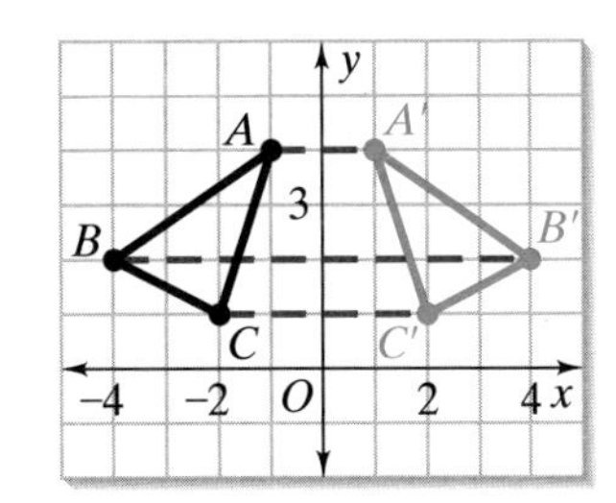

Draw $\triangle A'B'C'$ and use arrow notation:

$A(-1, 4), B(-4, 2), C(-2, 1) \rightarrow A'(1, 4), B'(4, 2), C'(2, 1)$

Quick Check

3. Graph $\triangle ABC$ and its reflection over the x-axis. Use arrow notation to describe the original triangle and its image. **See back of book.**

2. Teach

Activity Lab

Use before the lesson.

All in One Teaching Resources

Activity Lab 10-6: Patterns in Geometry

Guided Instruction

Example 1

Have students trace figures on paper and fold or use mirrors to test for symmetry.

Additional Examples

1 Draw a capital letter A and draw the line(s) of symmetry. If there are no lines of symmetry, write *none*.

2 Graph $K(3, -4)$ and its reflection over the indicated axis. Write the coordinate of the reflected point.

a. x-axis
(3, 4)

b. y-axis
(−3, −4)

See back of the book for graph.

3 $\triangle ABC$ has coordinates $A(-3, 1)$, $B(-1, 4)$, $C(2, 1)$. Graph $\triangle ABC$ and its reflection over the x-axis. Use arrow notation to describe $\triangle ABC$ and its reflection. **$A(-3, 1)$, $B(-1, 4)$, $C(2, 1) \rightarrow A'(-3, -1)$, $B'(-1, -4)$, $C'(2, -1)$ See back of the book for graph.**

All in One Teaching Resources

- Daily Notetaking Guide 10-6 **L3**
- Adapted Notetaking 10-6 **L1**

Closure

- *When does a figure have line symmetry?* **when one or more sides are mirror images of each other**
- *What is a reflection?* **a transformation that flips a figure over a line of reflection**

Advanced Learners L4

Have students draw figures with no lines of symmetry. Then have them reflect their figures so they have 1, 2, 3, and 4 or more lines of symmetry.

learning style: visual

English Language Learners ELL

Show students what it means to be a mirror image. Hold a mirror in the center of a flat object and show the students the mirror image. Connect the center of the object with the line of symmetry, and call the image a reflection.

learning style: visual

3. Practice

Assignment Guide

Check Your Understanding

Go over Exercises 1–5 in class before assigning the Homework Exercises.

Homework Exercises

A	Practice by Example	6–18
B	Apply Your Skills	19–28
C	Challenge	29
Test Prep and Mixed Review		30–35

Homework Quick Check

To check students' understanding of key skills and concepts, go over Exercises 9, 16, 21, 25, and 27.

Differentiated Instruction Resources

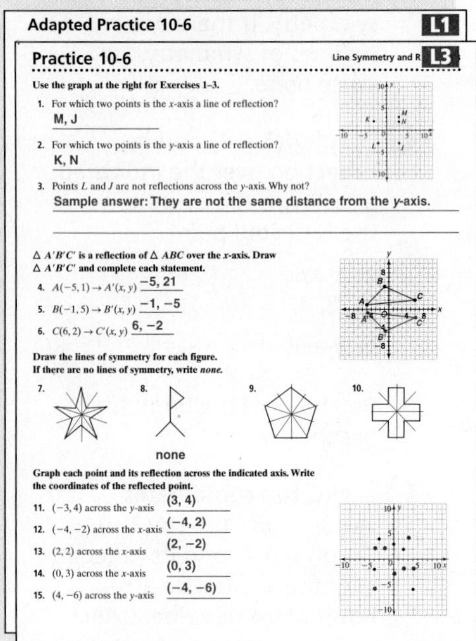

Adapted Practice 10-6 L1

Practice 10-6 Line Symmetry and R... L3

Use the graph at the right for Exercises 1–3.

1. For which two points is the *x*-axis a line of reflection? M, J
2. For which two points is the *y*-axis a line of reflection? K, N
3. Points *L* and *J* are not reflections across the *y*-axis. Why not? Sample answer: They are not the same distance from the *y*-axis.

△ A′B′C′ is a reflection of △ ABC over the *x*-axis. Draw △ A′B′C′ and complete each statement.

4. $A(-5, 1) \rightarrow A'(x, y)$ −5, 21
5. $B(-1, 5) \rightarrow B'(x, y)$ −1, −5
6. $C(6, 2) \rightarrow C'(x, y)$ 6, −2

Draw the lines of symmetry for each figure. If there are no lines of symmetry, write *none*.

7. 8. 9. 10.

none

Graph each point and its reflection across the indicated axis. Write the coordinates of the reflected point.

11. (−3, 4) across the *y*-axis (3, 4)
12. (−4, −2) across the *x*-axis (−4, 2)
13. (2, 2) across the *x*-axis (2, −2)
14. (0, 3) across the *x*-axis (0, 3)
15. (4, −6) across the *y*-axis (−4, −6)

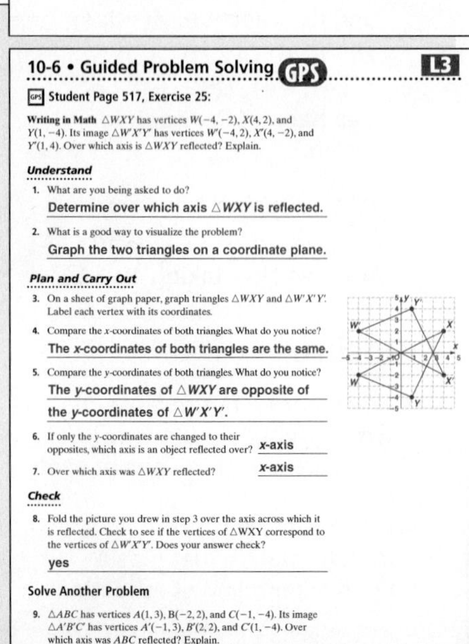

10-6 • Guided Problem Solving GPS L3

Student Page 517, Exercise 25:

Writing in Math △*WXY* has vertices *W*(−4, −2), *X*(4, 2), and *Y*(1, −4). Its image △*W′X′Y′* has vertices *W′*(−4, 2), *X′*(4, −2), and *Y′*(1, 4). Over which axis is △*WXY* reflected? Explain.

Understand

1. What are you being asked to do? Determine over which axis △*WXY* is reflected.
2. What is a good way to visualize the problem? Graph the two triangles on a coordinate plane.

Plan and Carry Out

3. On a sheet of graph paper, graph triangles △*WXY* and △*W′X′Y′*. Label each vertex with its coordinates.
4. Compare the *x*-coordinates of both triangles. What do you notice? The *x*-coordinates of both triangles are the same.
5. Compare the *y*-coordinates of both triangles. What do you notice? The *y*-coordinates of △*WXY* are opposite of the *y*-coordinates of △*W′X′Y′*.
6. If only the *y*-coordinates are changed to their opposites, which axis is an object reflected over? *x*-axis
7. Over which axis was △*WXY* reflected? *x*-axis

Check

8. Fold the picture you drew in step 3 over the axis across which it is reflected. Check to see if the vertices of △*WXY* correspond to the vertices of △*W′X′Y′*. Does your answer check? yes

Solve Another Problem

9. △*ABC* has vertices *A*(1, 3), *B*(−2, 2), and *C*(1, −4). Its image △*A′B′C′* has vertices *A′*(−1, 3), *B′*(2, 2), and *C′*(1, −4). Over which axis was *ABC* reflected? Explain. *y*-axis; the *x*-coordinates were changed to their opposites.

Check Your Understanding

1. A line of reflection produces a mirror image of the figure, and a line of symmetry divides a figure into mirror images.

3. $A(-1, 1)$, $B(-3, 4)$, $C(-5, -2) \rightarrow A'(1, 1)$, $B'(3, 4)$, $C'(5, -2)$

1. **Vocabulary** How is a line of reflection like a line of symmetry?
2. How many lines of symmetry does an equilateral triangle have? **three**

Use the graph at the right.

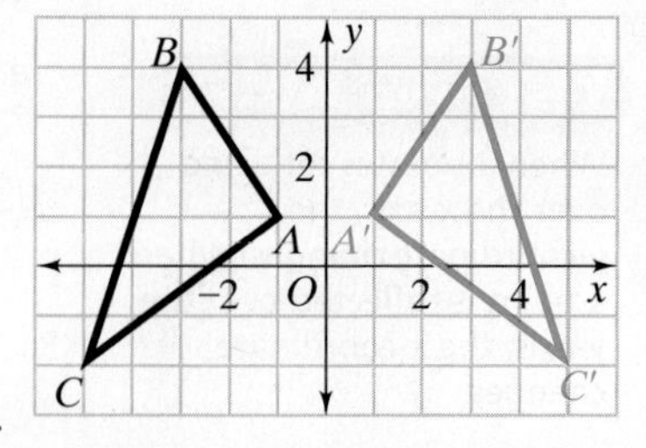

3. Use arrow notation to show the reflection. **See left.**

   $A(■, ■)$, $B(■, ■)$, $C(■, ■) \rightarrow$
   $A'(■, ■)$, $B'(■, ■)$, $C'(■, ■)$

4. △*ABC* is reflected over the __?__-axis. ***y*-axis**
5. Compare the coordinate pair of each vertex of △*ABC* and its image. **The *x*-coordinates are opposites; the *y*-coordinates are the same.**

Homework Exercises

For more exercises, see Extra Skills and Word Problems.

GO for Help

For Exercises	See Examples
6–9	1
10–15	2
16–18	3

A **Trace each figure and draw the line(s) of symmetry. If there are no lines of symmetry, write *none*.** 6–9 See margin.

6. 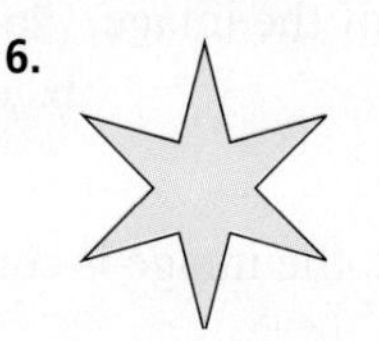
7. 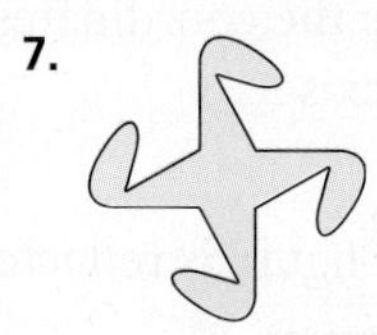
8. 
9. 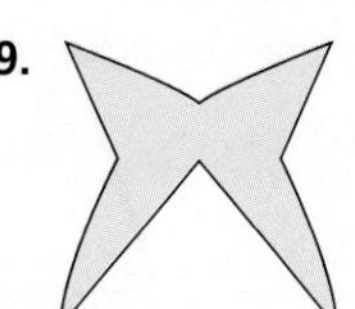

Graph each point and its reflection over the indicated axis. Write the coordinates of the image. 10. (4, −2) 11. (1, 5) 12. (3, −2)

10. $D(4, 2)$, *x*-axis
11. $F(-1, 5)$, *y*-axis
12. $G(-3, -2)$, *y*-axis
13. $H(2, -6)$, *x*-axis (2, 6)
14. $J(0, 3)$, *x*-axis (0, −3)
15. $K(-4, 0)$, *y*-axis (4, 0)

The vertices of a triangle are given. Graph the triangle, its reflection over the *x*-axis, and its reflection over the *y*-axis. 16–17. See margin.

16. $P(1, 6)$, $Q(6, 2)$, $R(2, 0)$
17. $S(-5, 1)$, $T(-3, 5)$, $V(-3, 1)$

18. The vertices of rectangle *ABCD* are $A(-3, 3)$, $B(-1, 3)$, $C(-1, -1)$, and $D(-3, -1)$. Use arrow notation to describe the original rectangle and its reflection over the *y*-axis. **See margin.**

B GPS

19. **Guided Problem Solving** A figure is a parallelogram but not a rectangle. It has two lines of symmetry. What type of figure is it?
    - *Act It Out* by drawing and cutting out several parallelograms.
    - Check for symmetry by folding. **rhombus**

6–9. See back of book.
16–18. See back of book.

Visit: PHSchool.com
Web Code: are-1006

Mental Math Without graphing, name the coordinates of the point's image after it is reflected over the x-axis and over the y-axis.

20. $(6, -1)$ **(6, 1), (−6, −1)**
21. $(-3, -4)$ **(−3, 4), (3, −4)**
22. $(-5, 8)$ **(−5, −8), (5, 8)**
23. $(7, 2)$ **(7, −2), (−7, 2)**

24. **Geometry** How many lines of symmetry does an isosceles right triangle have? **1**

25. **Writing in Math** $\triangle WXY$ has vertices $W(-4, -2)$, $X(4, 2)$, and $Y(1, -4)$. Its image $\triangle W'X'Y'$ has vertices $W'(-4, 2)$, $X'(4, -2)$, and $Y'(1, 4)$. Over which axis is $\triangle WXY$ reflected? Explain. **Over the x-axis; the y-coordinates are opposite.**

26. **Natural Science** Give an example (other than a butterfly) of line symmetry seen in a plant or animal. Describe the line symmetry. **Check students' work.**

Use words and arrow notation to describe the transformation. **27–28. See left.**

27. 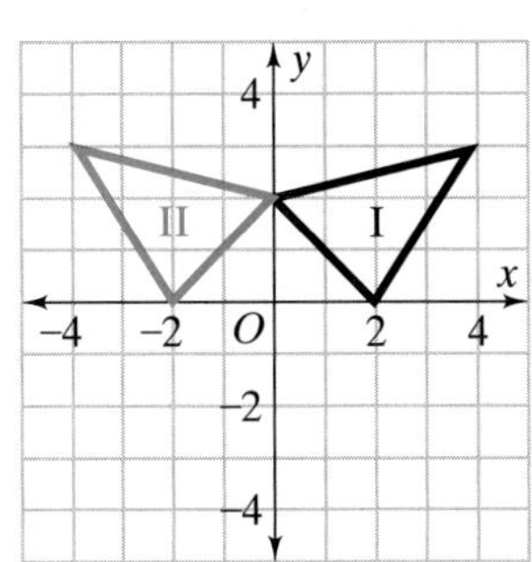

28. 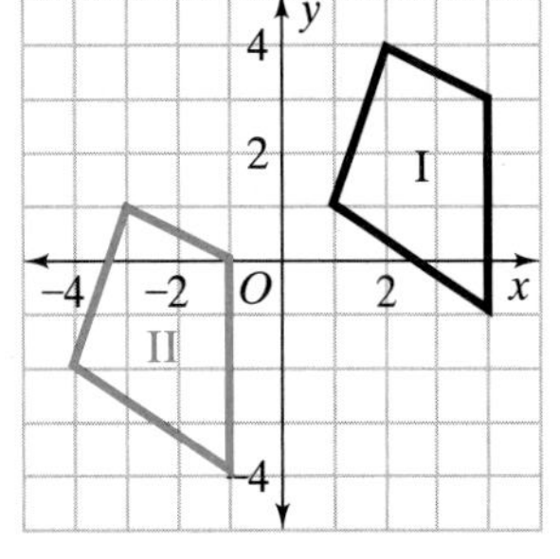

27. **reflection over the y-axis; $(x, y) \rightarrow (-x, y)$**

28. **translation 5 units left and 3 units down; $(x, y) \rightarrow (x - 5, y - 3)$**

C 29. **Challenge** $L(0, 0)$, $M(0, 5)$, and $N(4, 0)$ are vertices of $\triangle LMN$. Graph $\triangle LMN$ in a coordinate plane. Reflect $\triangle LMN$ over the x-axis to form a larger triangle. Reflect the larger triangle over the y-axis to form a quadrilateral. What type of quadrilateral is formed? **See margin.**

Test Prep and Mixed Review

Practice

Gridded Response

30. $\triangle BCD$ has vertices $B(4, 3)$, $C(6, 3)$, and $D(1, 4)$. What is the x-coordinate of B' after $\triangle BCD$ is reflected over the x-axis? **4**

31. A farmer has eight cows. Two are 3 years old, two are 5 years old, one is 6 years old, one is 7 years old, and one is 9 years old. The mean age of the farmer's eight cows is 5 years. What is the age in years of the remaining cow? **2**

32. If $\angle B$ and $\angle C$ are complementary, and the measure of $\angle B$ is 27°, what is the measure of $\angle C$ in degrees? **63**

GO for Help

For Exercises	See Lesson
33–35	9-3

Algebra **Find the first four terms in each sequence.** **33–35. See margin.**

33. $3n + 2$
34. $-n + 4$
35. $-2n - 1$

Online lesson quiz, PHSchool.com, Web Code: ara-1006

29. See back of book.
33–35. See back of book.

4. Assess & Reteach

PowerPoint

Lesson Quiz

1. Draw the lines of symmetry of a regular hexagon.

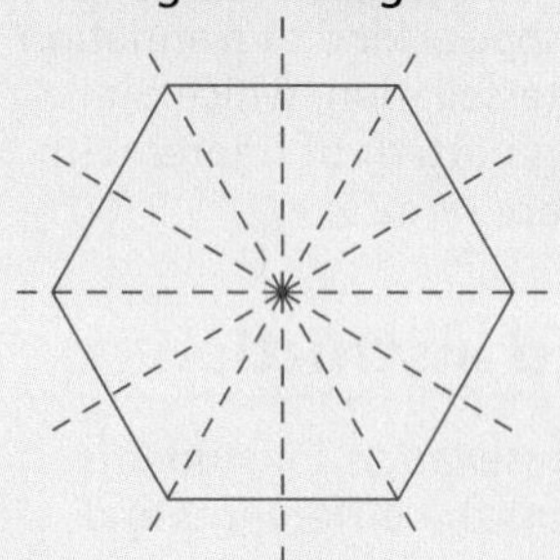

2. $\triangle ABC$ has vertices at $A(1, 4)$, $B(4, 3)$, and $C(3, 1)$. Find the vertices of its reflection over the y-axis.
   $A'(-1, 4)$, $B'(-4, 3)$, $C'(-3, 1)$

Alternative Assessment

Without marking the lines of symmetry, each student in a pair draws four figures: one with no lines of symmetry, one with one line of symmetry, one with two lines of symmetry, and one with more than two lines of symmetry. Partners exchange drawings and mark the lines of symmetry on each other's figures.

Test Prep

Resources

For additional practice with a variety of test item formats:

- Test-Taking Strategies, p. 523
- Test Prep, p. 527
- Test-Taking Strategies with Transparencies

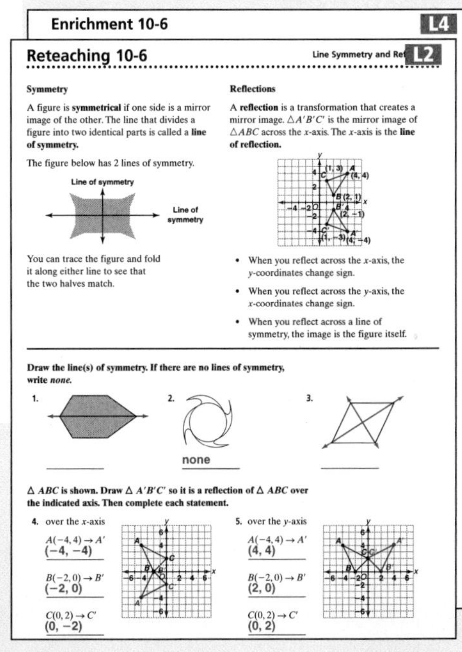

Enrichment 10-6 L4

Reteaching 10-6 Line Symmetry and Ref L2

Symmetry

A figure is **symmetrical** if one side is a mirror image of the other. The line that divides a figure into two identical parts is called a **line of symmetry.**

The figure below has 2 lines of symmetry.

Line of symmetry / Line of symmetry

You can trace the figure and fold it along either line to see that the two halves match.

Reflections

A **reflection** is a transformation that creates a mirror image. $\triangle A'B'C'$ is the mirror image of $\triangle ABC$ across the x-axis. The x-axis is the **line of reflection.**

- When you reflect across the x-axis, the y-coordinates change sign.
- When you reflect across the y-axis, the x-coordinates change sign.
- When you reflect across a line of symmetry, the image is the figure itself.

Draw the line(s) of symmetry. If there are no lines of symmetry, write *none*.

1. ______
2. none
3. ______

$\triangle ABC$ is shown. Draw $\triangle A'B'C'$ so it is a reflection of $\triangle ABC$ over the indicated axis. Then complete each statement.

4. over the x-axis
   $A(-4, 4) \rightarrow A'$ (−4, −4)
   $B(-2, 0) \rightarrow B'$ (−2, 0)
   $C(0, 2) \rightarrow C'$ (0, −2)

5. over the y-axis
   $A(-4, 4) \rightarrow A'$ (4, 4)
   $B(-2, 0) \rightarrow B'$ (2, 0)
   $C(0, 2) \rightarrow C'$ (0, 2)

Extension

Exploring Tessellations

In this feature, students learn about a special kind of translation called a tessellation, which is a repeating pattern of figures with no overlaps or gaps.

Guided Instruction

Define *tessellation* for students and discuss the differences and similarities between tessellations and translations. Point out that all tessellations are translations, but not all translations are tessellations. If you have it available, display some famous tessellations such as the artwork of M. C. Escher.

Activity

Work on the activity as a class. Have each student draw his or her own figure as you display one on the board or on a transparency.

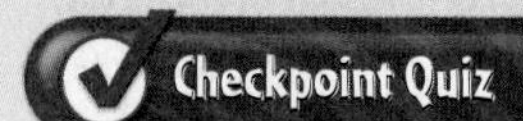

Use this Checkpoint Quiz to check students' understanding of the skills and concepts of Lessons 10-4 through 10-6.

Resources

- All-in-One Teaching Resources Checkpoint Quiz 2
- ExamView CD-ROM
- Success Tracker™ Online Intervention

Extension — For Use With Lesson 10-6

Exploring Tessellations

A *tessellation* is a repeating pattern of figures that has no gaps or overlaps. Tessellations are made using transformations. An example of a tessellation is shown at the right.

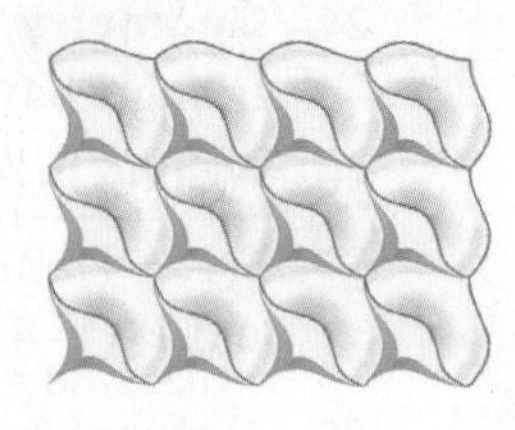

Use a square piece of cardboard.
1–5. Check students' work.

1. Draw a curve from one vertex of the square to a neighboring vertex, as shown at the right.

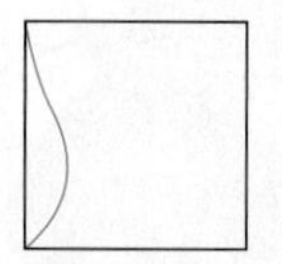

2. Cut along the curve you drew. Translate the cutout piece to the opposite side of the square, and tape it down.
3. Draw the same curve on the bottom and repeat the process.
4. Trace around your figure on a piece of paper. Carefully translate the figure to the right so that the edges touch.
5. After you have covered a row, translate your figure downward to start a new row. Continue tracing until you have covered the paper.

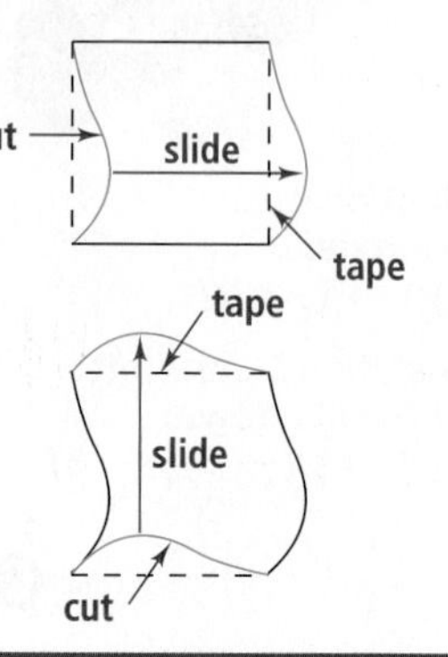

6. **Reasoning** Can you make a tessellation with your figure using reflections? Explain. **No; any pattern made by reflecting the figure will have gaps.**

Checkpoint Quiz 2

Lessons 10-4 through 10-6

Graph each equation. **1–4. See margin.**

1. $y = 3x^2$
2. $y = |x| - 2$
3. $y = 2x^2 + 3$
4. $y = x - 1$

5. Is $(3, -5)$ a solution of $y = x^2 - 4$? Explain. **No; $3^2 - 4 \neq -5$.**

Translate each point right 3 units and up 7 units. Write the new coordinates.

6. $(-6, 2)$ **(−3, 9)**
7. $(9, -1)$ **(12, 6)**
8. $(0, -7)$ **(3, 0)**
9. $(-2, -3)$ **(1, 4)**

10. Which of the following digits have lines of symmetry: 2, 3, 6, 8? Draw the lines of symmetry. **See margin.**

1.
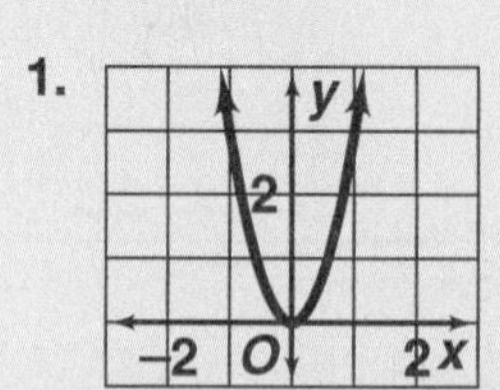

2.
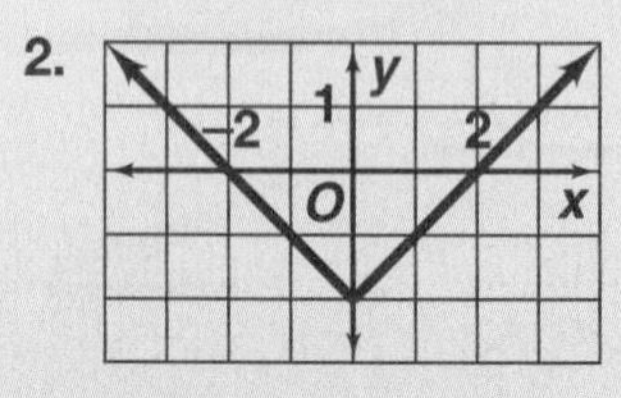

3.
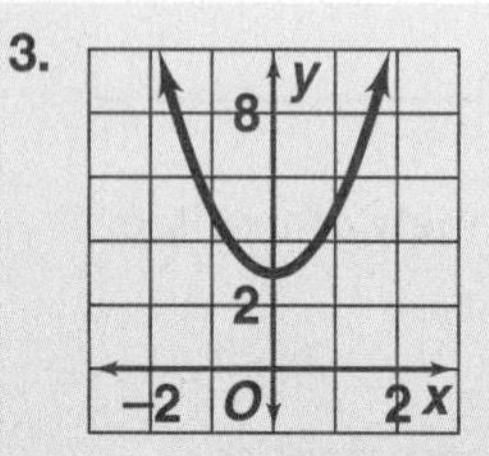

4.
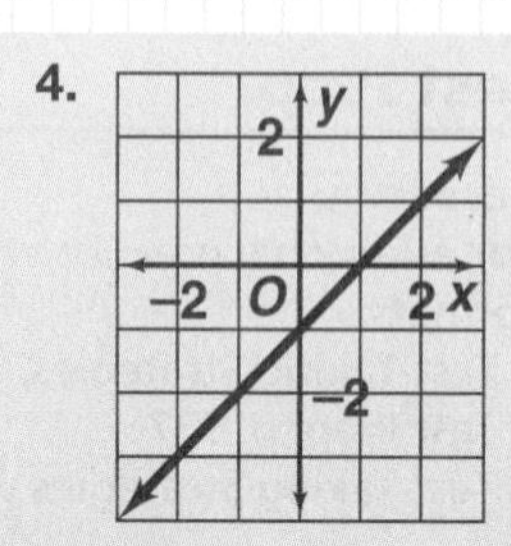

10. See back of book.

10-7 Rotational Symmetry and Rotations

10-7

1. Plan

Check Skills You'll Need

1. Vocabulary Review What is the measure of any *right angle*? 90°

Draw an angle for each measure. Then classify each angle as *acute, right, obtuse*, or *straight*. **2–5. See back of book.**

2. 60° **3.** 120°

4. 90° **5.** 180°

GO for Help Lesson 7-2

What You'll Learn

To identify rotational symmetry and to rotate a figure about a point

New Vocabulary rotation, center of rotation, rotational symmetry, angle of rotation

Why Learn This?

Some objects rotate, which means to move in a circular manner. When you ride a unicycle, the wheel rotates, and you move!

A **rotation** is a transformation that turns a figure about a fixed point called the **center of rotation.** You describe a rotation by its angle measure and its direction.

The direction of every rotation in this book is counterclockwise unless noted as clockwise. If a figure can be rotated 180° or less and match the original figure, it has **rotational symmetry.**

Vocabulary Tip

Counterclockwise means the opposite of the direction in which a clock's hands move.

EXAMPLE Identifying Rotational Symmetry

1 **Windmills** Do the sails of the windmill have rotational symmetry? Use points O, A, A', A'', and A''' to explain.

Yes, the sails have rotational symmetry. The center of rotation is O.

If A is rotated 90°, the image is A'.
If A is rotated 180°, the image is A''.
If A is rotated 270°, the image is A'''.
If A is rotated 360°, it returns to its original position.

Quick Check

1. Does the figure have rotational symmetry? Explain. **1a–c. See left.**

a. 

b. 

c. 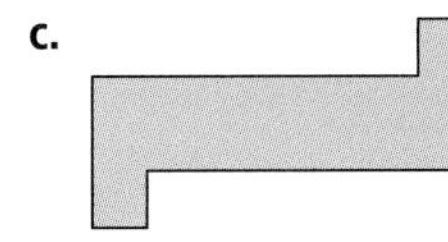

1a. Yes; it looks the same as you move each point to another point.
b. No; it does not match the original figure after any rotation of 180° or less.
c. Yes; it looks the same as the original when it is rotated 180°.

Objective

To identify rotational symmetry and to rotate a figure about a point

Examples

1 Identifying Rotational Symmetry
2 Finding an Angle of Rotation
3 Rotating a Figure

Math Understandings: p. 484D

Professional Development

Math Background

A *rotation* is a transformation that turns a figure about a fixed point called a *center of rotation*. If a figure can be rotated 180° or less and fit exactly on top of the original figure, then that figure has *rotational symmetry*. The angle of rotation is measured in a counterclockwise direction.

More Math Background: p. 484D

Lesson Planning and Resources

See p. 484E for a list of the resources that support this lesson.

PowerPoint

Bell Ringer Practice

Check Skills You'll Need
Use student page, transparency, or PowerPoint. For intervention, direct students to:
Identifying and Classifying Angles
Lesson 7-2
Extra Skills and Word Problems Practice, Ch. 7

Differentiated Instruction Solutions for All Learners

Special Needs L1
Pair students who have difficulty graphing the images with a partner who can draw them. Provide tracing paper to the student who has difficulty. Have them trace the original and rotated image to check the accuracy of the transformation.
learning style: visual

Below Level L2
Review angle measures and the meaning of counterclockwise. Then have students rotate a point such as (3, 4) about the origin for 90°, 180°, and 270°.
learning style: visual

2. Teach

Activity Lab

Use before the lesson.

All in One Teaching Resources

Activity Lab 10-7: Rotational Symmetry and Rotations

Guided Instruction

Example 1
Discuss the meaning of *rotate*. Show things that rotate (fan blades, hands on a clock) and elicit suggestions from students.

Example 3
Have students trace the figure on their own papers. They can then rotate their figures by holding their pencil tips on point *O*.

PowerPoint

Additional Examples

1. Do the letters "H" and "N" have rotational symmetry? Explain. **Yes, both letters look the same when rotated 180°.**

2. Does ☼ have rotational symmetry? If it does, find the angle of rotation. **yes, 45°**

3. Choose a nonsymmetrical letter of the alphabet, such as P, and use it to illustrate a rotation of 90°, 180°, and 270° about its center.

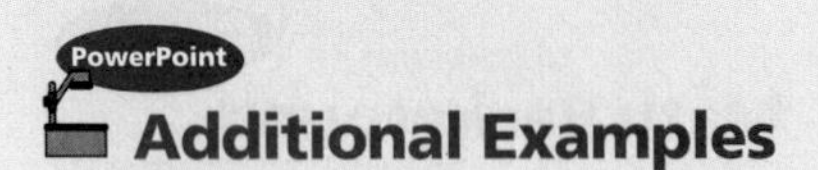

90° 180° 270°

All in One Teaching Resources
- Daily Notetaking Guide 10-7 L3
- Adapted Notetaking 10-7 L1

Closure

- *What is a rotation?* **A transformation that turns a figure around a fixed point is called a rotation.**
- Define *rotational symmetry.* **A figure with rotational symmetry can be rotated 180° or less and match the original figure.**

The number of degrees a figure rotates is the **angle of rotation**. When a figure has rotational symmetry, the angle of rotation is the angle measure the figure must rotate to match the original figure.

EXAMPLE Finding an Angle of Rotation

2 The wheel below has rotational symmetry. Find the angle of rotation.

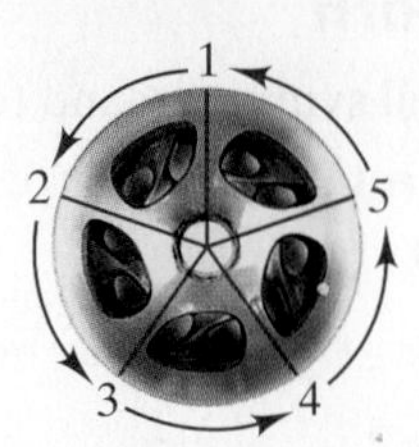

The wheel matches itself in 5 positions. The angle of rotation is $360° \div 5$, or $72°$.

Quick Check

2. Find the angle of rotation of the flower at the left. **60°**

You can rotate a figure in a coordinate plane. Use the center of rotation and an angle.

EXAMPLE Rotating a Figure

3 Graph rectangle $ABCD$ with vertices $A(0, 0)$, $B(0, 2)$, $C(4, 2)$, and $D(4, 0)$. Rotate the rectangle as described. Write the coordinates of the vertices of the image.

a. 180° about point A

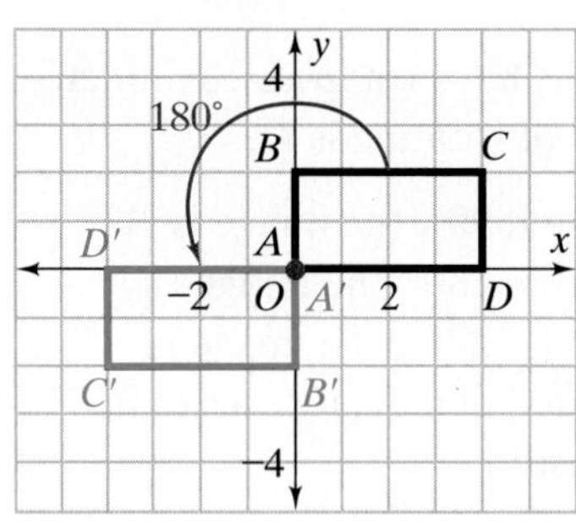

$A'(0, 0)$, $B'(0, -2)$, $C'(-4, -2)$, and $D'(-4, 0)$

b. 90° about its center, $(2, 1)$.

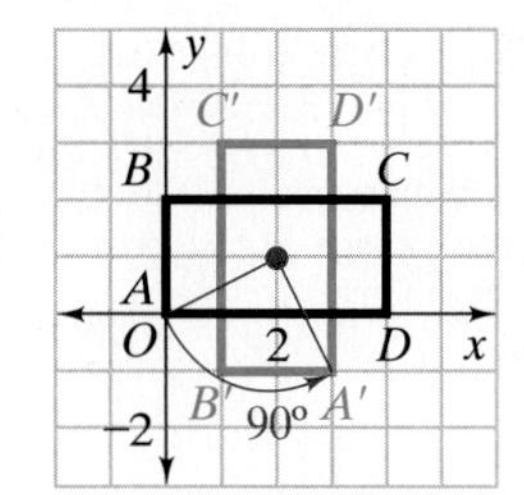

$A'(3, -1)$, $B'(1, -1)$, $C'(1, 3)$, and $D'(3, 3)$

Online active math

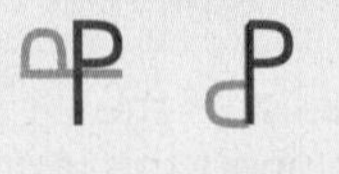

For: Transformations Activity
Use: Interactive Textbook, 10-7

Quick Check

See back of book.

3. Graph $\triangle TRG$ with vertices $T(0, 0)$, $R(3, 3)$, and $G(5, 1)$. Rotate $\triangle TRG$ 180° about T. Write the coordinates of T', R', and G'.

Advanced Learners L4
What letters of the alphabet have *both* line symmetry and rotational symmetry? **H, I, O, X**

learning style: visual

English Language Learners ELL
Provide students with a paper square. Have them place the tip of their pencil in the center, and turn it 90 degrees. Ask: *Does it still look like a square?* **Yes.** Do it again several times. Let them know the square has rotational symmetry, and the pencil point was on the center of rotation.

learning style: tactile

Check Your Understanding

1. **Vocabulary** Name two types of symmetry.
   line symmetry and rotational symmetry
2. Name three types of transformations.
   translation, reflection, and rotation
3. Does the figure at the right have rotational symmetry? If so, find the angle of rotation.
   yes; 120°

Estimation **Estimate the angle measure of the rotation of $\triangle RSO$.**

4. 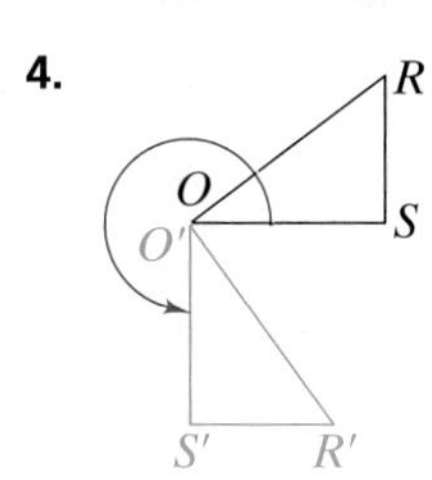

270°

5. 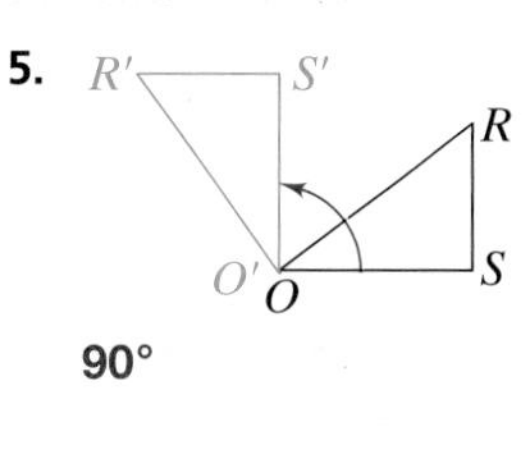

90°

6. 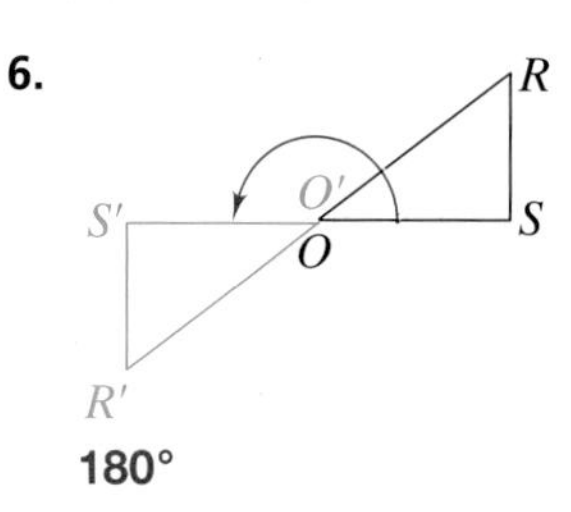

180°

Homework Exercises

For more exercises, see Extra Skills and Word Problems.

GO for Help

For Exercises	See Examples
7–9	1
10–12	2
13–16	3

A **Does the figure have rotational symmetry? Explain.** 7–9. See left.

7. 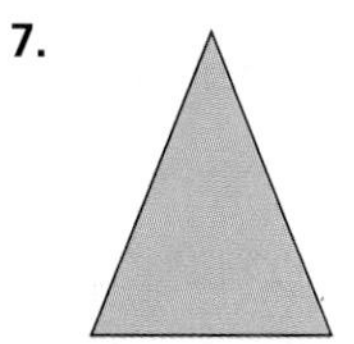

8. 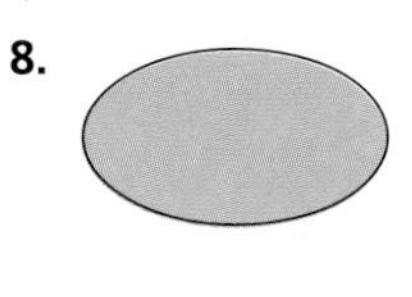

9. 

7. No; it does not match the original figure after any rotation of 180° or less.
8. Yes; when the figure is rotated 180°, it looks the same.
9. Yes; the figure looks the same after being rotated 90°, 180°, or 270°.

Each figure has rotational symmetry. Find the angle of rotation.

10. 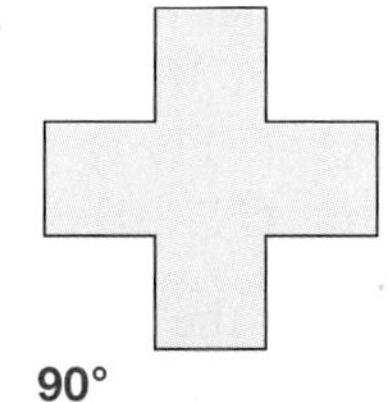
90°

11. 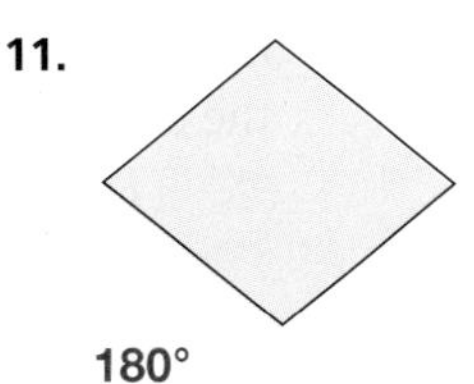
180°

12. 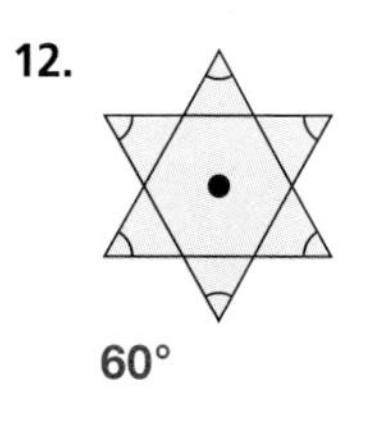
60°

Graph rectangle *PQRS* with vertices $P(0, 0)$, $Q(2, 0)$, $R(2, 6)$, and $S(0, 6)$. Rotate the rectangle as described. Write the new coordinates.
13–16. See margin.

13. 180° about point P
14. 90° about its center, $(1, 3)$
15. 180° about its center, $(1, 3)$
16. 270° about its center, $(1, 3)$

B **GPS** 17. **Guided Problem Solving** What is the clockwise angle of rotation of the hour hand on a clock as it moves from 4:00 to 5:00? 30°
- To which number does the hour hand point at 4:00? At 5:00?
- How many degrees does the hour hand rotate in a full circle?

13–16. See back of book.

3. Practice

Assignment Guide

Check Your Understanding

Go over Exercises 1–6 in class before assigning the Homework Exercises.

Homework Exercises

A	Practice by Example	7–16
B	Apply Your Skills	17–26
C	Challenge	27
	Test Prep and Mixed Review	28–33

Homework Quick Check

To check students' understanding of key skills and concepts, go over Exercises 11, 15, 23, 25, and 26.

Differentiated Instruction Resources

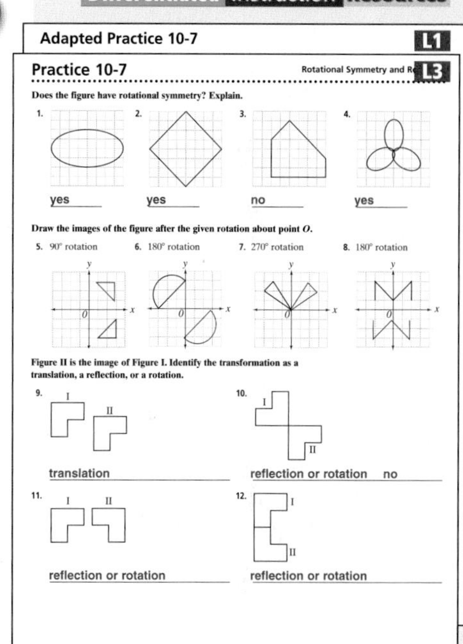

Adapted Practice 10-7 L1

Practice 10-7 Rotational Symmetry and R... L3

Does the figure have rotational symmetry? Explain.

1. yes
2. yes
3. no
4. yes

Draw the images of the figure after the given rotation about point O.

5. 90° rotation
6. 180° rotation
7. 270° rotation
8. 180° rotation

Figure II is the image of Figure I. Identify the transformation as a translation, a reflection, or a rotation.

9. translation
10. reflection or rotation no
11. reflection or rotation
12. reflection or rotation

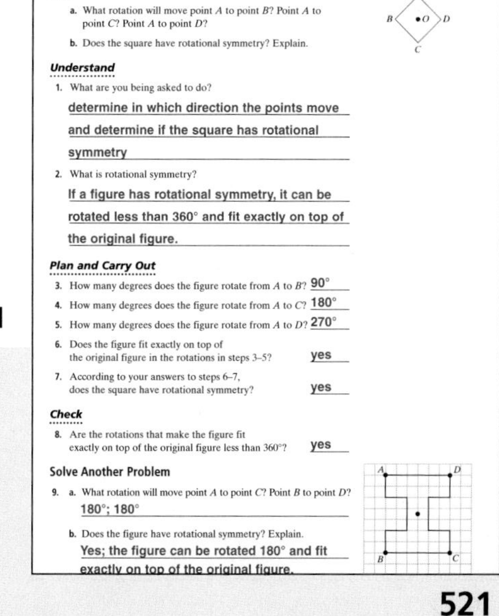

10-7 • Guided Problem Solving GPS L3

Student Page 522, Exercise 23:

a. What rotation will move point A to point B? Point A to point C? Point A to point D?
b. Does the square have rotational symmetry? Explain.

Understand

1. What are you being asked to do?
   determine in which direction the points move and determine if the square has rotational symmetry
2. What is rotational symmetry?
   If a figure has rotational symmetry, it can be rotated less than 360° and fit exactly on top of the original figure.

Plan and Carry Out

3. How many degrees does the figure rotate from A to B? 90°
4. How many degrees does the figure rotate from A to C? 180°
5. How many degrees does the figure rotate from A to D? 270°
6. Does the figure fit exactly on top of the original figure in the rotations in steps 3–5? yes
7. According to your answers to steps 6–7, does the square have rotational symmetry? yes

Check

8. Are the rotations that make the figure fit exactly on top of the original figure less than 360°? yes

Solve Another Problem

9. a. What rotation will move point A to point C? Point B to point D?
   180°; 180°
   b. Does the figure have rotational symmetry? Explain.
   Yes; the figure can be rotated 180° and fit exactly on top of the original figure.

4. Assess & Reteach

PowerPoint

Lesson Quiz

1. Which of the following math symbols have rotational symmetry?

   a. > b. ÷ c. %

   b and c

2. Graph $\triangle XYZ$ with vertices of $X(0, 0)$, $Y(3, 4)$, and $Z(4, 0)$. Then draw the three images formed by rotating the triangle 90°, 180°, and 270° about point X.

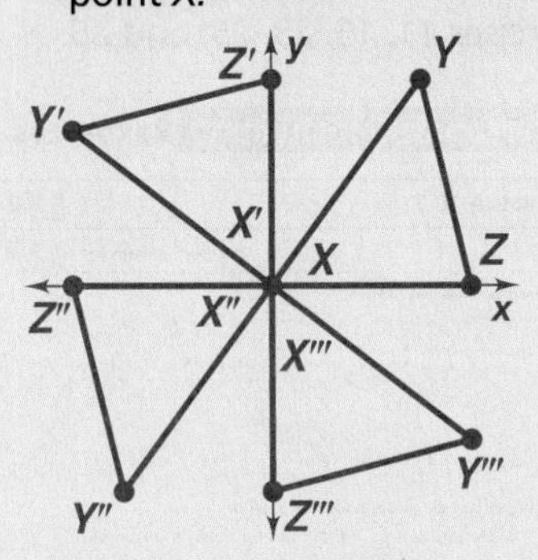

Alternative Assessment

Each student in a pair draws and cuts out two or three geometric figures. Partners exchange figures and determine whether they have rotational symmetry. On each figure that has rotational symmetry, students write the angle of rotation.

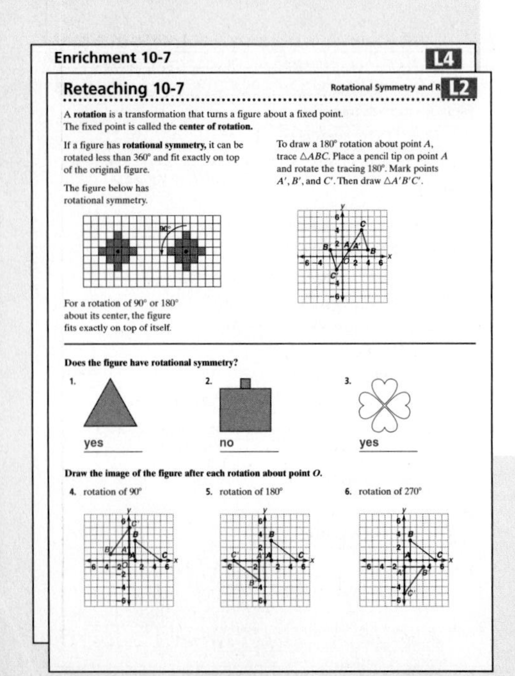

Enrichment 10-7 L4

Reteaching 10-7 — Rotational Symmetry and R... L2

A **rotation** is a transformation that turns a figure about a fixed point. The fixed point is called the **center of rotation.**

If a figure has **rotational symmetry,** it can be rotated less than 360° and fit exactly on top of the original figure.

The figure below has rotational symmetry.

For a rotation of 90° or 180° about its center, the figure fits exactly on top of itself.

To draw a 180° rotation about point A, trace $\triangle ABC$. Place a pencil tip on point A and rotate the tracing 180°. Mark points A', B', and C'. Then draw $\triangle A'B'C'$.

Does the figure have rotational symmetry?

1. yes
2. no
3. yes

Draw the image of the figure after each rotation about point O.

4. rotation of 90°
5. rotation of 180°
6. rotation of 270°

GO Online
Homework Video Tutor
Visit: PHSchool.com
Web Code: are-1007

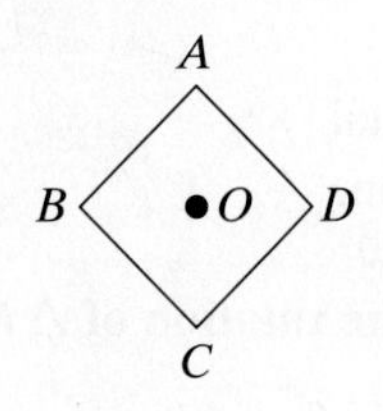

23 a. 90°; 180°; 270°

b. Yes; the image stays the same after rotation.

Mental Math **A triangle lies entirely in Quadrant I. In which quadrant will the triangle lie after each rotation about (0, 0)?**

18. 90° **II** **19.** 180° **III** **20.** 270° **IV** **21.** 360° **I**

22. Open-Ended Draw a capital letter of the alphabet that has rotational symmetry. Mark a point at the center of rotation. **Check students' work.**

23. GPS a. What rotation will move point A to point B? Point A to point C? Point A to point D?

b. Does the square have rotational symmetry? Explain. **23a–b. See left.**

24. Clocks The second hand of a clock moves clockwise. It makes a full revolution once every minute. What is its clockwise angle of rotation after 20 seconds? After 45 seconds? **120°; 270°**

25. Writing in Math Describe an object in your classroom that has rotational symmetry. Explain how it shows rotational symmetry. **Check students' work.**

26. The arrow notation given shows how the vertices of a triangle are moved. Name the type of transformation and describe it with words. $A(2, 4) \rightarrow A'(4, -4)$, $B(4, 4) \rightarrow B'(2, -4)$, and $C(3, 0) \rightarrow C'(3, 0)$ **Rotation; it has been rotated 180° around point C.**

27. Challenge Graph the point $E(-6, -2)$ and its three images after 90°, 180°, and 270° rotations about the origin. Connect the four points. What type of quadrilateral is formed? **See margin.**

Test Prep and Mixed Review

Practice

Multiple Choice

28. If $\overline{MN}$ is translated 5 units to the right and 2 units up, what will be the coordinates of point M? **A**

Ⓐ (1, 1) Ⓑ (7, 7) Ⓒ (−11, 1) Ⓓ (11, −1)

29. Which expression does the model best represent? **F**

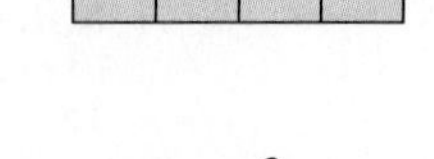

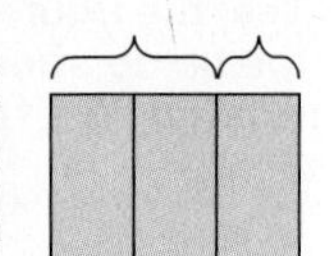

Ⓕ $1\frac{3}{4} \div \frac{1}{2}$ Ⓖ $1\frac{3}{4} \times \frac{1}{2}$ Ⓗ $1.5 \div \frac{1}{2}$ Ⓙ $7 \times 3\frac{1}{2}$

30. What is the value of the expression $(4 + 2)^2 \times 2 + 9 \div 3$? **D**

Ⓐ 7 Ⓑ 27 Ⓒ 42 Ⓓ 75

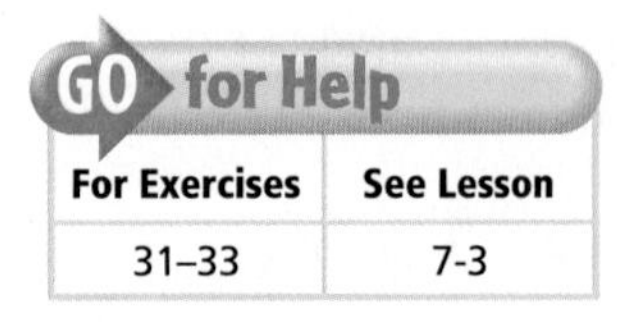

GO for Help

For Exercises	See Lesson
31–33	7-3

Algebra **Suppose the sides of a triangle have the given measures. Classify each triangle by its sides.**

31. $4s, 4s, 4s$ **equilateral** **32.** $2.2y, 1.5y, 2.2y$ **isoceles** **33.** $3k, 4k, 5k$ **right**

Test Prep

Resources

For additional practice with a variety of test item formats:

- Test-Taking Strategies, p. 523
- Test Prep, p. 527
- Test-Taking Strategies with Transparencies

27. See back of book.

Test-Taking Strategies

Answering the Question Asked

When answering a question, be sure to answer the question that is asked. Read the question carefully and identify the information you need to find. Eliminate answer choices that are not related to the question that is asked.

EXAMPLE

Point M is translated 3 units to the right and 2 units down.

What are the coordinates of M'?

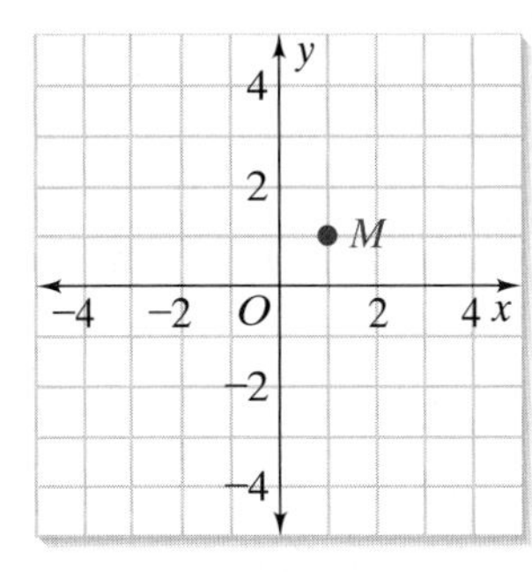

(A) (4, 3)　　(C) (4, −1)
(B) (−2, −1)　　(D) (3, −2)

Choice A translates point M 3 units to the right and 2 units *up*. Choice B translates M 3 units to the *left* and 2 units down. Choice C translates M 3 units to the right and 2 units down. Choice D translates M *2 units* to the right and *3 units* down. The correct answer is C.

Exercises

1. Which line contains the ordered pair (−1, −3)? **D**
   (A) Line a
   (B) Line b
   (C) Line c
   (D) Line d

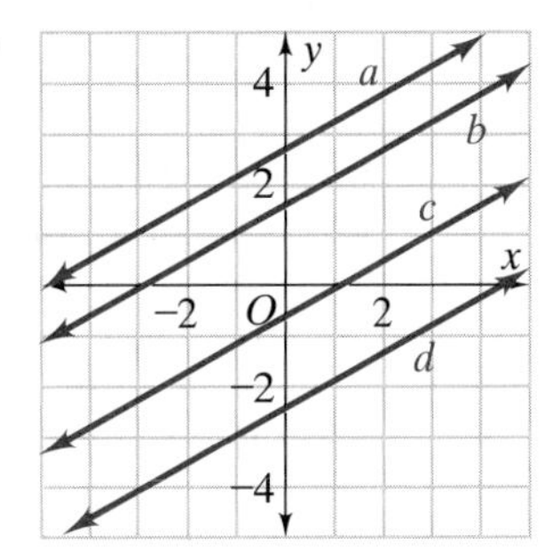

Answering the Question Asked

In this feature, students learn to read the question carefully. They discover that they can eliminate answer choices that are not related to the question that was asked.

Guided Instruction

Explain to students that often answer choices contain information that is close to what is needed. In order to correctly answer the question, they must first carefully read and determine what is being asked. Ask:

- *What are you being asked to find?* **the coordinates of M'**
- *How can you find those coordinates?* **Go 3 units right and 2 units down.**
- *How do you know which are the incorrect choices?* **They translate M either in an incorrect direction or an incorrect distance.**
- *What would be another way to find the answer?* **graph M'**

Resources

Test-Taking Strategies with Transparencies
- Transparency 8
- Practice sheet, p. 22

Test-Taking Strategies with Transparencies

Test-Taking Strategies: Answering the Question Asked

Incorrect choices may answer related questions.

Example Midori owes \$20 on a restaurant bill, and wants to tip the server 15%. How much should Midori pay altogether?

A. \$3　B. \$15　C. \$20　D. \$23

Calculate the tip: 0.15 × 20 = \$3

Choice A is \$3, but this is how much Midori should leave for a tip, not how much to pay altogether.

Calculate the total bill: 20 + 3 = 23

The answer is \$23, or choice D.

Answer the question asked. Explain your reasoning.

1. Clarisse earns \$10 per hour and works about 25 hours per week. How much could she earn in a year?
   A. \$250　B. \$1,000　C. \$3,000　D. \$13,000

2. Find the area of the triangle.

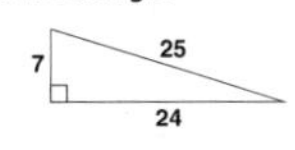

F. 84 units^2　G. 56 units^2　H. 186 units^2　J. 4,200 units^2

Chapter 10 Review

Chapter 10 Review

Resources

Student Edition
Extra Skills and Word Problems Practice, Ch. 10, p. 648
English/Spanish Glossary, p. 675
Formulas and Properties, p. 673
Tables, p. 670

All in One Teaching Resources
- Vocabulary and Study Skills 10F 

Differentiated Instruction

Spanish Vocabulary Workbook with Study Skills ELL
Interactive Textbook
- Audio Glossary

Online Vocabulary Quiz

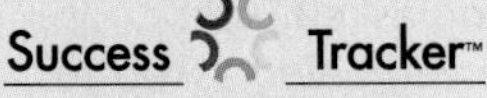
Online at PHSchool.com

Spanish Vocabulary/Study Skills ELL

Vocabulary/Study Skills L3

10F: Vocabulary Review For use with the Chapter Review

Study Skill Use a highlighter to mark material that is important in your notes or on teacher handouts. When you review for tests, pay special attention to this highlighted information.

Match the term in Column A with its definition in Column B.

	Column A	Column B
E	1. transformation	A. transformation that creates symmetry
B	2. decagon	B. polygon with 10 sides
A	3. reflection	C. segment that has both endpoints on the circle
C	4. chord	D. quadrilateral with four congruent sides
D	5. rhombus	E. change of position, shape, or size of a figure
G	6. slope	F. quadrilateral with exactly one pair of parallel sides
F	7. trapezoid	G. ratio that describes the steepness of a line

Match the term in Column A with its definition in Column B.

	Column A	Column B
L	8. complementary	H. two angles whose sum is 180 degrees
N	9. rotation	I. identifies the location of a point
M	10. hypotenuse	J. one side of a figure is the mirror image of the other side
K	11. translation	K. transformation that moves points the same distance and in the same direction
J	12. symmetry	L. two angles whose sum is 90 degrees
I	13. ordered pair	M. the longest side of a right triangle
H	14. supplementary	N. transformation that turns a figure about a fixed point
O	15. radius	O. segment with one endpoint at the center and the other endpoint on the circle

Vocabulary and Study Skills

Vocabulary Review

angle of rotation (p. 520)
center of rotation (p. 519)
coordinate plane (p. 486)
graph of an equation (p. 492)
image (p. 510)
linear equation (p. 492)
line of reflection (p. 515)
line of symmetry (p. 514)
line symmetry (p. 514)
nonlinear equation (p. 504)
ordered pair (p. 486)
origin (p. 486)
prime notation (p. 510)
quadrants (p. 486)
reflection (p. 515)
rise (p. 498)
rotation (p. 519)
rotational symmetry (p. 519)
run (p. 498)
slope (p. 498)
transformation (p. 510)
translation (p. 510)
***x*-axis** (p. 486)
***x*-coordinate** (p. 486)
***y*-axis** (p. 486)
***y*-coordinate** (p. 486)

Choose the correct term to complete each sentence.

1. A flip over a line is a (translation, reflection). **reflection**
2. A rotation turns a figure about a fixed point called the (center of rotation, angle of rotation). **center of rotation**
3. (Rotation, Slope) compares the vertical change, called the rise, to the horizontal change, called the run. **slope**
4. The second number in a(n) (coordinate plane, ordered pair) is the y-coordinate. **ordered pair**
5. If a graph of the solutions of an equation is a line, then the equation is a (linear equation, nonlinear equation). **linear equation**

Go Online
PHSchool.com
For: Vocabulary Quiz
Web Code: arj-1051

Skills and Concepts

Lessons 10-1, 10-2

- To name and graph points on a coordinate plane
- To find solutions of linear equations and to graph linear equations

An **ordered pair** (x, y) gives the coordinates of a point. Any ordered pair that makes an equation true is a solution of the equation. The **graph of an equation** is the graph of coordinates that are solutions of the equation.

Graph each point on the same coordinate plane. Name the quadrant in which each point lies. See margin for graph.

6. $A(1, -5)$ **IV**
7. $B(-3, -4)$ **III**
8. $C(-2, 3)$ **II**

Find three solutions of each equation. **9–12. Answers may vary. Samples are given.**

9. $y = x + 3$
   (0, 3), (−2, 1), (5, 8)
10. $y = x - 5$
   (0, −5), (5, 0), (−1, −6)
11. $y = 2x + 1$
   (0, 1), (2, 5), (−3, −5)
12. $y = -x - 2$
   (0, −2), (4, −6), (−3, 1)

6–8. See back of book.
13–22. See back of book.

Lessons 10-3, 10-4

- To find the slope of a line and use it to solve problems
- To graph nonlinear relationships

Slope is a ratio that describes the steepness of a line.

$$\text{slope} = \frac{\text{rise}}{\text{run}}$$

A **nonlinear equation** is an equation with a graph that is not a straight line.

Draw a line with the given slope through the given point. 13–14. See margin.

13. $B(2, 4)$, slope $= 2$ **14.** $R(-1, 2)$, slope $= \frac{1}{3}$

Make a table of values for each equation. Use integer values of x from -3 to 3. Then graph the equation. 15–18. See margin.

15. $y = x^2 - 1$ **16.** $y = |x| + 1$ **17.** $y = 2x^2 - 2$ **18.** $y = 4|x|$

Lessons 10-5, 10-6

- To graph and write rules for translations
- To identify lines of symmetry and to graph reflections

A **transformation** is the change of the position, shape, or size of a figure. A **translation** is a transformation that moves every point of a figure the same distance and in the same direction.

A figure has **line symmetry** when one side of the figure is a mirror image of the other side. A **reflection** is a transformation that flips a figure over a line.

Graph each transformation of $\triangle ABC$. Use arrow notation to show the translation. 19–22. See margin.

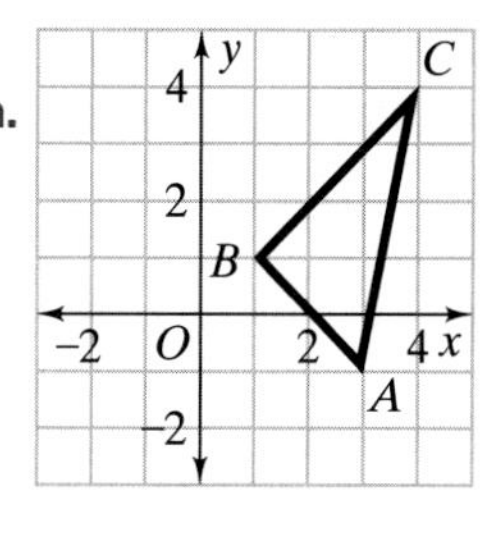

19. down 2 units **20.** left 3 units

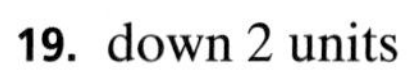

21. right 3 units and down 5 units

22. Graph $\triangle ABC$ from Exercises 19–21. Then graph its reflection over the x-axis. Use arrow notation to describe the transformation.

Lesson 10-7

- To identify rotational symmetry and to rotate a figure about a point

A **rotation** is a transformation that turns a figure about a fixed point.

Does each figure have rotational symmetry? If it does, find the angle of rotation.

23. yes; 90°

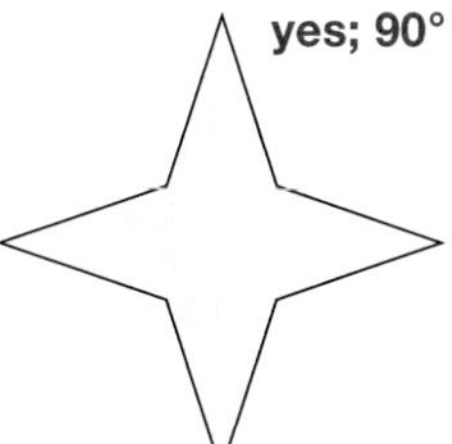

24. yes; 180°

Alternative Assessment L4

Alternative Assessment Form C

Chapter 10

HIT THE SLOPES

Did you ever consider the fun that you can have on a mountain? Going up mountains is fun for hikers. Going down mountains is fun for skiers.

Show all of your work on a separate sheet of paper.

1. The steeper a mountain is, the harder it is to climb up or down. Look at these two mountain profiles and decide which one looks harder to climb. Explain.

Mountain A Mountain B

2. Ski resorts mark their trails with symbols that show the levels of difficulty of the trails. Circles represent the easiest trails, squares represent somewhat harder trails, and diamonds represent the most difficult. The map below shows several trails at Mount Ski. Choose one circle trail, one square trail, and one diamond trail. Circle the trails you choose and give them each an appropriate name.

3. Copy the chart below. Draw the profile of each of your three trails in the chart. (A profile of a ski trail shows how steep it is.) A sample profile of the beginner trail appears on the chart.

Profile Chart		
Type of trail	**Name**	**Profile**
Beginner	Easy as Pie	

Chapter 10 Test

Resources

- ExamView Assessment Suite CD-ROM
  - Ch. 10 Ready-Made Test
  - Make your own Ch. 10 test
- MindPoint Quiz Show CD-ROM
  - Chapter 10 Review

Differentiated Instruction

All in One Teaching Resources
- Below Level Chapter 10 Test L2
- Chapter 10 Test L3
- Chapter 10 Alternative Assessment L4

Spanish Assessment Resources ELL
- Below Level Chapter 10 Test L2
- Chapter 10 Test L3
- Chapter 10 Alternative Assessment L4

ExamView Assessment Suite CD-ROM
- Special Needs Test L1
- Special Needs Practice Bank L1

Online Chapter 10 Test at www.PHSchool.com L3

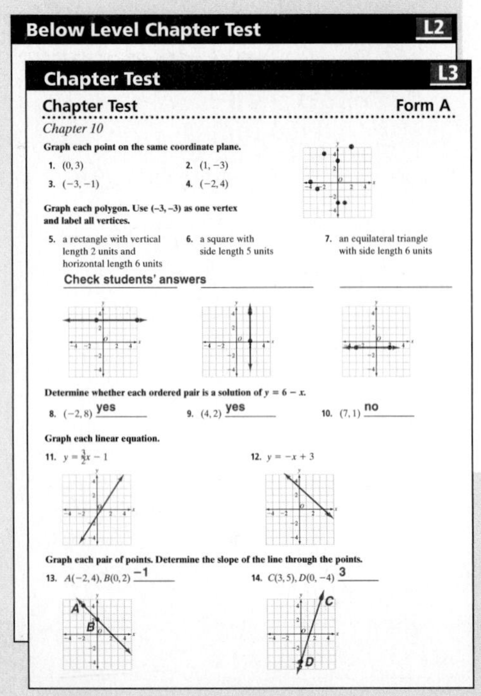
Below Level Chapter Test L2

Chapter Test L3

Chapter Test Form A

Chapter 10

Graph each point on the same coordinate plane.

1. (0, 3) 2. (1, −3)
3. (−3, −1) 4. (−2, 4)

Graph each polygon. Use (−3, −3) as one vertex and label all vertices.

5. a rectangle with vertical length 2 units and horizontal length 6 units 6. a square with side length 5 units 7. an equilateral triangle with side length 6 units

Check students' answers

Determine whether each ordered pair is a solution of $y = 6 - x$.

8. (−2, 8) yes 9. (4, 2) yes 10. (7, 1) no

Graph each linear equation.

11. $y = \frac{3}{2}x - 1$ 12. $y = -x + 3$

Graph each pair of points. Determine the slope of the line through the points.

13. $A(-2, 4), B(0, 2)$ −1 14. $C(3, 5), D(0, -4)$ 3

Graph each point on the same coordinate plane. 1–4. See margin.

1. $A(1, 4)$
2. $B(-2, -1)$
3. $C(3, -2)$
4. $D(-3, 2)$

Graph each polygon. Use (0, 0) as one vertex and label all vertices. 5–6. See margin.

5. a square with side 4 units long
6. a rectangle with horizontal length 3 units and vertical length 5 units.

Determine whether each ordered pair is a solution of $y = -2x + 5$.

7. $(3, -5)$ no
8. $(2.5, 0)$ yes
9. $(4, -3)$ yes
10. $(0, 5)$ yes

Graph each linear equation. 11–14. See margin.

11. $y = x - 3$
12. $y = 3x + 1$
13. $y = -x + 2$
14. $y = 2x - 4$

Graph each pair of points. Determine the slope of the line through the points. 15. $-\frac{1}{5}$ 16. −3

15. $E(7, 1), F(-3, 3)$
16. $G(-2, 6), H(0, 0)$
17. $L(-4, 0), M(0, 2)$ $\frac{1}{2}$
18. $S(8, 5), T(1, -1)$ $\frac{6}{7}$

Draw a line with the given slope through the given point. 19–20. See margin.

19. $P(-2, -1)$, slope $\frac{1}{4}$
20. $R(2, 4)$, slope $-\frac{2}{3}$

Graph each equation for integer values of x from −3 to 3. 21–22. See margin.

21. $y = x^2 - 2$
22. $y = 2|x| + 1$

23. **Advertising** To advertise in the classified section of a local paper costs \$2 plus \$.25 for each word. Make a table to graph the equation $y = 0.25x + 2$, where x is the number of words and y is the total cost. See margin.

24. Draw the images of the triangle after rotations of 90°, 180°, and 270° about O. See margin.

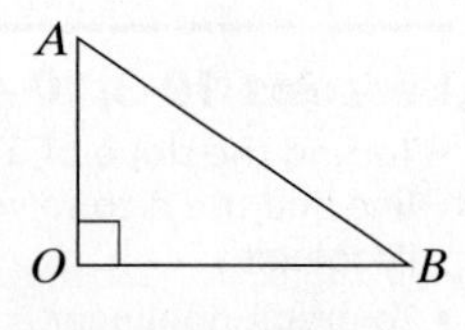

The graph of $\triangle ABC$ has vertices at $A(1, 3)$, $B(5, 8)$, and $C(7, 1)$. Graph each image of $\triangle ABC$ for each transformation described. Use arrow notation to show the transformation. 25–27. See margin.

25. translated right 1 unit and down 3 units
26. translated left 3 units and up 2 units
27. reflected over the y-axis

Write a rule for the translation shown in each graph.

28.

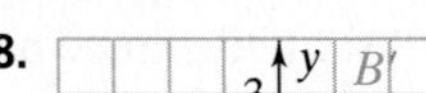
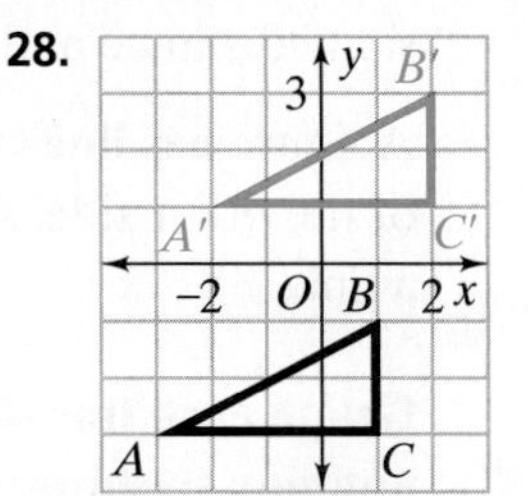

$(x, y) \rightarrow (x + 1, y + 4)$

29. $(x, y) \rightarrow (x - 3, y - 2)$

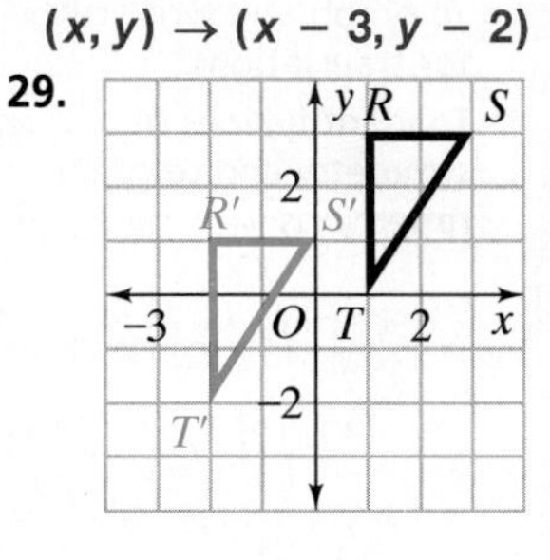

Use the two figures below for Exercises 30 and 31.

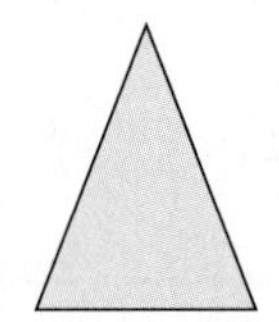

30. Trace each figure and draw the line(s) of symmetry. If there are no lines of symmetry, write *none*. See margin.
31. Does each figure have rotational symmetry? If so, what is the angle of rotation? yes, 90°; no
32. **Writing in Math** Give an example of a rotation, a reflection, and a translation you might see in the real world. Check students' work.

1–6. See back of book.
11–14. See back of book.
19–27. See back of book.
30. See back of book.

Multiple Choice

Read each question. Then write the letter of the correct answer on your paper.

1. Which rule best describes the function in the table? **D**

x	y
0	0
1	1
2	4
3	9

   Ⓐ $y = x + 2$
   Ⓑ $y = x$
   Ⓒ $y = 2x$
   Ⓓ $y = x^2$

2. A bus company charges $10.50 per ticket for a trip between two cities. Each trip costs the company $200. Which equation describes the profit the company makes on each trip? **G**
   Ⓕ $P = 10.5x + 200$
   Ⓖ $P = 10.5x - 200$
   Ⓗ $P = 200x + 10.5$
   Ⓙ $P = 10.5 + x - 200$

3. A muffin recipe calls for $2\frac{1}{4}$ c of flour and makes 12 muffins. How many muffins can you make with 6 c of flour? **C**
   Ⓐ 24 Ⓑ 30 Ⓒ 32 Ⓓ 45

4. What is the ones digit of 7^{23}? **F**
   Ⓕ 3 Ⓖ 5 Ⓗ 7 Ⓙ 9

5. For which linear equation is $(-3, 0.5)$ NOT a solution? **A**
   Ⓐ $x - 2y = 4$ Ⓒ $x = -6y$
   Ⓑ $4y = 3x + 11$ Ⓓ $x + 6y = 0$

6. A bag of 15 lemons costs $2.30. What is the approximate unit price of a lemon? **H**
   Ⓕ $.075 Ⓗ $.15
   Ⓖ $.13 Ⓙ $.30

7. Which translation moves $\triangle ABC$ to $\triangle A'B'C'$? **C**
   Ⓐ $(x, y) \rightarrow (x - 3, y)$
   Ⓑ $(x, y) \rightarrow (x + 3, y)$
   Ⓒ $(x, y) \rightarrow (x, y - 3)$
   Ⓓ $(x, y) \rightarrow (x, y + 3)$

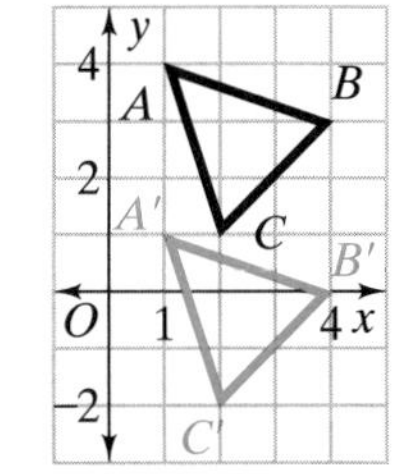

8. Order the numbers 0.361×10^7, 4.22×10^7, and 13.5×10^6 from least to greatest. **H**
   Ⓕ $13.5 \times 10^6, 0.361 \times 10^7, 4.22 \times 10^7$
   Ⓖ $4.22 \times 10^7, 13.5 \times 10^6, 0.361 \times 10^7$
   Ⓗ $0.361 \times 10^7, 13.5 \times 10^6, 4.22 \times 10^7$
   Ⓙ $13.5 \times 10^6, 4.22 \times 10^7, 0.361 \times 10^7$

9. Of 26 letters in the alphabet, 5 are vowels. What percent are vowels? **A**
   Ⓐ about 19% Ⓒ about 30%
   Ⓑ about 21% Ⓓ about 33%

10. Which expression has the greatest value? **H**
   Ⓕ $\frac{3}{4}(8)$ Ⓖ $2 \cdot 3.1$ Ⓗ 23 Ⓙ $\frac{2}{3} \cdot \frac{6}{5}$

11. A map has a scale of 2 in. : $\frac{1}{2}$ mi. Find the actual distance for a map distance of $5\frac{1}{2}$ in. **D**
   Ⓐ $13\frac{3}{4}$ mi Ⓒ $5\frac{1}{2}$ mi
   Ⓑ 11 mi Ⓓ $1\frac{3}{8}$ mi

Gridded Response

Record your answer in a grid.

12. What is the next term in this pattern?
   500, 250, 125, . . . **62.5**

13. Find the circumference in centimeters of a circle with a radius of 5 cm. Use 3.14 for π. **31.4**

Short Response 14–16. See margin.

14. A right triangle has legs of 12 ft and 16 ft. Find the longest side of the triangle.

15. A rectangular yard has a perimeter of 96 ft. To support a fence, 12 posts will be placed at equal intervals. How far apart are the posts?

Extended Response

16. You paid $385 for car repairs. The garage charged $125 for parts and $65 per hour for labor. Write and solve an equation to find the number of hours the mechanic worked on your car. Show your work.

Item	1	2	3	4	5	6	7	8	9	10	11	12	13	14	15	16
Lesson	9-4	4-6	5-4	9-2	10-2	5-2	10-5	2-8	6-6	3-4	5-6	9-2	8-5	8-7	8-2	4-6

14.[2] $a^2 + b^2 = c^2$
$12^2 + 16^2 = c^2$
$144 + 256 = c^2$
$400 = c^2$
$c = 20$ ft

[1] correct answer, without work shown

Test Prep

Resources

Test Prep Workbook

All in One **Teaching Resources**
- Cumulative Review L3

ExamView Assessment Suite CD-ROM
- Standardized Test Practice

Differentiated Instruction

Progress Monitoring Assessments
- Benchmark Test 5 L3

Spanish Assessment Resources
- Spanish Cumulative Review ELL

ExamView Assessment Suite CD-ROM
- Special Needs Practice Bank L1

15.[2] 96 ft ÷ 12 posts = 8 ft per post

[1] correct answer, without work shown

16[4] Let h = the number of hours worked

$65h + 125 = 385$
$-125 \quad -125$
$\frac{65h}{65} = \frac{260}{65}$
$h = 4$ h

Spanish Cumulative Review ELL

Cumulative Review L3

Cumulative Review

Chapters 1–10

Multiple Choice. Choose the letter of the best answer.

1. What is the next term in this pattern? 500, 250, 125, . . .
   A. 100 B. 75 Ⓒ 62.5 D. 50
2. What is 1.47×10^8 written in standard notation?
   F. 14,700,000 Ⓖ 147,000,000 H. 1,470,000,000 J. 14,700,000,000
3. Use the formula $B = p(1 + r)^t$ to find the amount of money in your savings account after ten years if you deposited $100 at an interest rate of 5% compounded annually.
   A. $111.01 B. $112.74 C. $150.00 Ⓓ $162.89
4. Find the circumference of a circle with a diameter of 10 cm. Use 3.14 for π.
   F. 15.7 cm Ⓖ 31.4 cm H. 78.5 cm J. 314 cm
5. Which of the situations below could this graph describe?

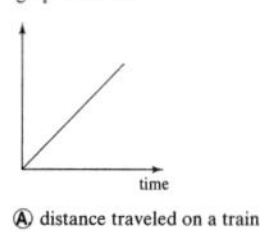

   Ⓐ distance traveled on a train
   B. height of a baseball after a pitch
   C. weight of a person on a diet
   D. amount of money after paying bills
6. What is the approximate unit price of a lemon if a bag of 15 lemons costs $2?
   F. $0.075 Ⓖ $0.13 H. $0.15 J. $0.30
7. A map of a city has a scale of 2 in. : $\frac{1}{2}$ mi. The distance on the map from the school to your house is $5\frac{1}{2}$ in. What is the actual distance from the school to your house?
   A. $13\frac{3}{4}$ mi B. 11 mi C. $5\frac{1}{2}$ mi Ⓓ $1\frac{3}{8}$ mi
8. How far apart are the 12 posts used to fence in a region with a perimeter of 96 ft if they are at equal intervals?
   F. 12 ft G. 10 ft Ⓗ 8 ft J. 6 ft
9. Which proportion will help you find 36% of 90?
   A. $\frac{90}{60} = \frac{n}{100}$
   Ⓑ $\frac{n}{90} = \frac{36}{100}$
   C. $\frac{90}{100} = \frac{36}{n}$
   D. $\frac{100}{n} = \frac{36}{90}$
10. The sale price of a TV is $279. This is 80% of the original price. What was the original price?
   F. $55.80 G. $286.74 Ⓗ $348.75 J. $400
11. In which quadrant is the point $A(3, -1)$ located?
   A. I B. II C. III Ⓓ IV

Problem Solving

Applying Coordinates

Students will use data from these two pages to answer the questions posed in the Activity.

Invite students to imagine an intersection on a multi-lane highway where a variety of cars, trucks, and motorcycles are all stopped. Ask:

- *When the traffic light turns green, do the vehicles that were stopped all begin to move at the same time?* **No**
- *Which vehicles tend to move away from the light faster?* **Sample: small cars, sports cars, motorcycles** *more slowly?* **Sample: trucks and older cars**

Discuss how these differences in acceleration are caused by differences in engines and vehicle weight.

Materials

- Photographs or diagrams of the engines of a motorcycle, car, small truck, and large truck

Activating Prior Knowledge

Not all engines are alike. Ask students to share what they know about working with engines and how to make them work efficiently and powerfully. Invite students to talk about the engines automobiles, trucks, motorcycles, and racing cars have. Discuss the similarities and differences.

Guided Instruction

Have volunteers read the opening paragraph and captions. If a copy of *Star Wars Episode 1* is available show students the pod-racing scene. Ask:

- *In a race, when would a pod with Engine 3 pass a pod using Engine 2?* **Between Checkpoints C and D**
- *If the race were extended, based on your graph, can you predict what time each engine would reach 5,000 m?* **Engine 1: 625 s; Engine 2: 1,000 s; Engine 3: 430 s; Engine 4: 1,300 s**

Problem Solving Application

Applying Coordinates

On Your Mark! Do you remember the great pod-racing scene in *Star Wars Episode I*? The racers built their own Pod racers, so each one looked and flew differently. Some Pod racers got off to a fast start, but couldn't maintain their speed. Others started out slowly, then sped up. To build a winning Pod racer, you'd want to know the length of the race, so you could choose the best engine.

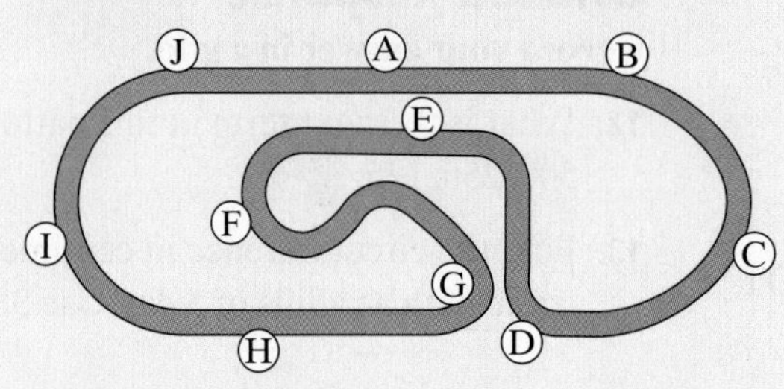

The ring rotates for stability, keeping the pod upright.

Anakin Skywalker's control pod with cockpit computer

Anakin's fuel atomizer and distribution system make his engines perform better than some larger engines.

Anakin's Podracer

Each of the two Radon-Ulzer 620C racing engines, modified by Anakin Skywalker, is 7 m long. The estimated top speed of the Pod racer is 947 km/h.

Put It All Together

Suppose you are designing a pod for the big race. You have four engines to choose from. Each performs differently. The table shows test results for each engine recorded at ten checkpoints around the track.

J A B E F I C G H D

1. Make a graph, labeling the x-axis from 0 to 900 s, and the y-axis from 0 to 4,000 m. Plot the time–distance points for each of the four engines. (*Hint:* Use a different color for each engine.)
2. **a.** Connect the points for each engine.
   b. How are the graphs for the engines similar? How are they different?
3. **a.** Which engine starts a pod at the fastest speed?
   b. Which engine starts a pod at the slowest speed?
   c. Which two engines move the pod at a constant speed?
   d. Which engines speed up or slow down during the race?
   e. Which engine makes the pod go fastest at the end of 4,000 m?
4. **Reasoning** Suppose the big race is 2,000 m long. Which engine would you choose to complete the course the fastest? Would your answer change if the race were 3,000 m? Explain.

528

1-2a. See back of book.
2b. They are not straight lines.
3a. engine 4
b. engine 3
c. engine 1 and engine 2
d. engine 3 and engine 4
e. engine 3
4. engine 4; yes, the pod covers 3,000 meters in the least time with engine 3

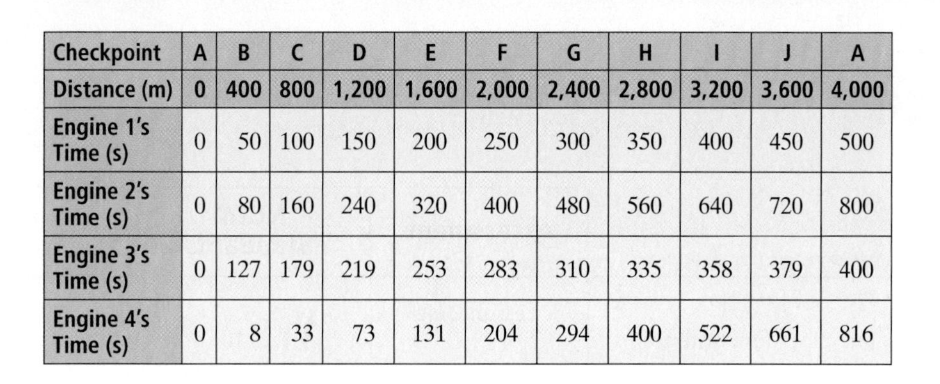

Checkpoint	A	B	C	D	E	F	G	H	I	J	A
Distance (m)	0	400	800	1,200	1,600	2,000	2,400	2,800	3,200	3,600	4,000
Engine 1's Time (s)	0	50	100	150	200	250	300	350	400	450	500
Engine 2's Time (s)	0	80	160	240	320	400	480	560	640	720	800
Engine 3's Time (s)	0	127	179	219	253	283	310	335	358	379	400
Engine 4's Time (s)	0	8	33	73	131	204	294	400	522	661	816

Go Online PHSchool.com
For: Information about flight
Web Code: are-1053

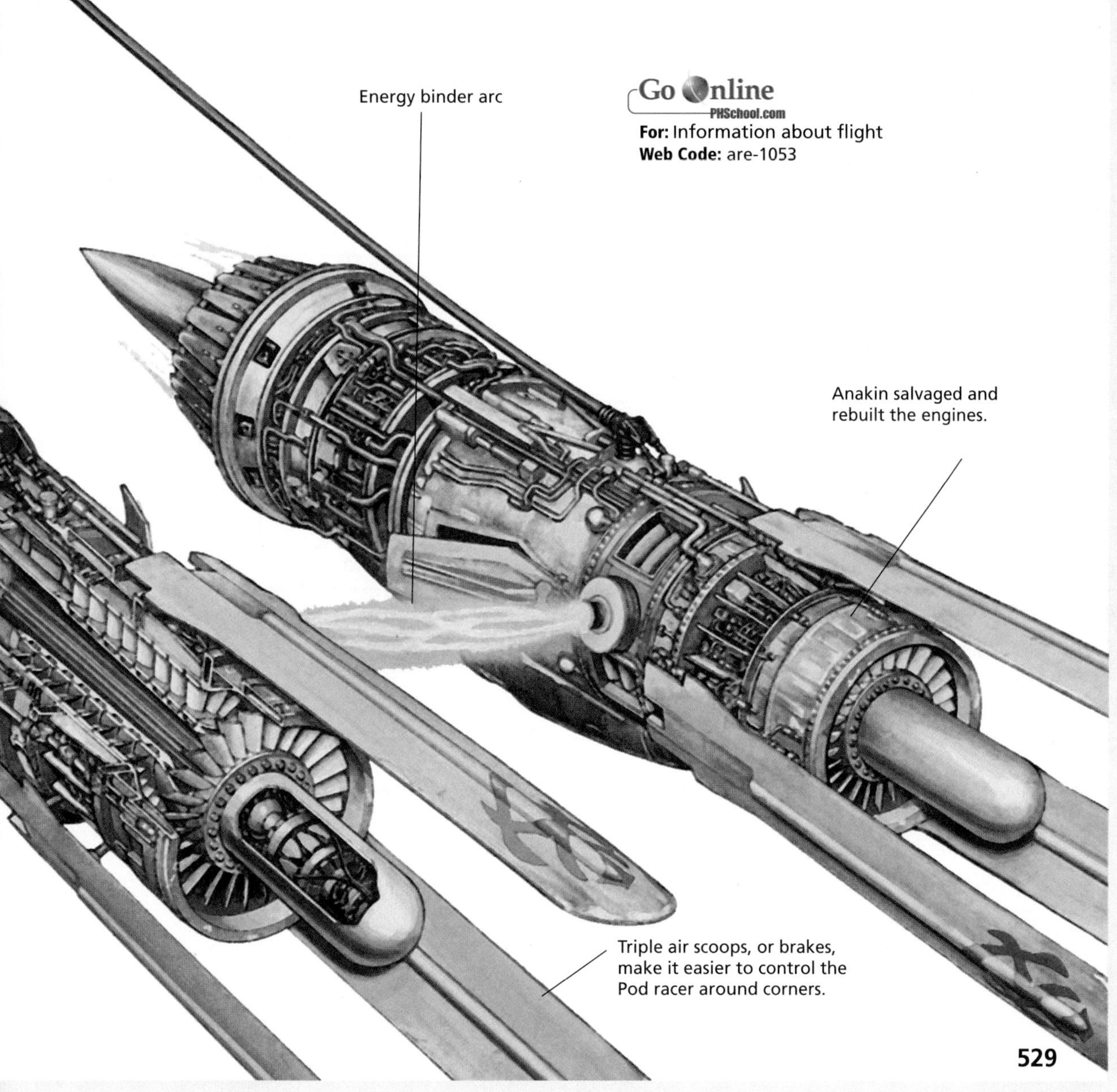
Energy binder arc
Anakin salvaged and rebuilt the engines.
Triple air scoops, or brakes, make it easier to control the Pod racer around corners.

Activity

Have students work in pairs to study the table and answer the questions.

Exercise 1 Have students examine the information in the table. Ask them to tell, just by reading the data, which of the graphs of engine speeds will be straight lines. **the graphs for engines 1 and 2**

Exercise 3 Ask: *Which point on the course is the starting point? How do you know?* **Point A; the times and distances for all engines are 0.**

Science Connection
How does a gasoline engine work? Invite interested students to find out and share what they learn in a poster or display. You may wish to invite an auto mechanic to visit the class and answer students' questions about car engines, emissions, and car care.

Differentiated Instruction

Special Needs L1
Help pairs, as needed, to read and understand the table. Point out, for instance, that the distances are given in meters (m), not miles (mi), and that "s" means "seconds."

529

11 Displaying and Analyzing Data

Chapter at a Glance

Lesson Titles, Objectives, and Features	Assessment	NCTM Standards	Local Standards
11-1 Reporting Frequency • To represent data using frequency tables, line plots, and histograms **11-1b Activity Lab:** Venn Diagrams	Lesson Quiz	1, 5, 6, 7, 8, 9, 10	
11-2 Spreadsheets and Data Displays • To interpret spreadsheets, double bar graphs, and double line graphs **11-2b Activity Lab, Technology:** Graphing Using Spreadsheets	Lesson Quiz	1, 2, 5, 6, 7, 8, 9, 10	
11-3 Stem-and-Leaf Plots • To represent and interpret data using stem-and-leaf plots **11-3b Activity Lab, Data Analysis:** Choosing the Best Display	Lesson Quiz Checkpoint Quiz 1	1, 2, 5, 6, 7, 8, 9, 10	
11-4a Activity Lab, Data Collection: Writing Survey Questions **11-4 Random Samples and Surveys** • To identify a random sample and to write a survey question	Lesson Quiz	5, 6, 7, 8, 9, 10	
11-5 Estimating Population Size • To estimate population size using proportions **Guided Problem Solving:** Describing Data **11-5b Activity Lab, Data Analysis:** Graphing Population Data	Lesson Quiz	1, 2, 4, 5, 6, 7, 8, 9, 10	
11-6 Using Data to Persuade • To identify misleading graphs and statistics	Lesson Quiz Checkpoint Quiz 2	1, 5, 6, 7, 8, 9, 10	
11-7a Activity Lab, Data Collection: Two-Variable Data Collection **11-7 Exploring Scatter Plots** • To draw and interpret scatter plots	Lesson Quiz	1, 2, 5, 6, 7, 8, 9, 10	
Problem Solving Application: Applying Data Analysis			

NCTM Standards 2000
1 Number and Operations **2** Algebra **3** Geometry **4** Measurement **5** Data Analysis and Probability
6 Problem Solving **7** Reasoning and Proof **8** Communication **9** Connections **10** Representation

Correlations to Standardized Tests

All content for these tests is contained in *Prentice Hall Math,* Course 2. This chart reflects coverage in this chapter only.

	11-1	11-2	11-3	11-4	11-5	11-6	11-7
Terra Nova CAT6 (Level 17)							
Number and Number Relations	✔	✔	✔	✔	✔	✔	✔
Computation and Numerical Estimation	✔	✔	✔	✔	✔	✔	✔
Operation Concepts	✔	✔	✔	✔	✔	✔	✔
Measurement							
Geometry and Spatial Sense							
Data Analysis, Statistics, and Probability	✔	✔	✔	✔	✔	✔	✔
Patterns, Functions, Algebra	✔	✔	✔	✔	✔	✔	✔
Problem Solving and Reasoning	✔	✔	✔	✔	✔	✔	✔
Communication	✔	✔	✔	✔	✔	✔	✔
Decimals, Fractions, Integers, and Percent	✔	✔	✔	✔	✔	✔	✔
Order of Operations							
Terra Nova CTBS (Level 17)							
Decimals, Fractions, Integers, Percents	✔	✔	✔	✔	✔	✔	✔
Order of Operations, Numeration, Number Theory	✔	✔	✔	✔	✔	✔	✔
Data Interpretation	✔	✔	✔	✔	✔	✔	✔
Pre-algebra		✔	✔		✔		✔
Measurement							
Geometry							
ITBS (Level 13)							
Number Properties and Operations	✔	✔	✔	✔	✔	✔	✔
Algebra	✔	✔	✔	✔	✔	✔	✔
Geometry							
Measurement							
Probability and Statistics	✔	✔	✔	✔	✔	✔	✔
Estimation							
SAT10 (Int 3 Level)							
Number Sense and Operations	✔	✔	✔	✔	✔	✔	✔
Patterns, Relationships, and Algebra	✔	✔	✔	✔	✔	✔	✔
Data, Statistics, and Probability	✔	✔	✔	✔	✔	✔	✔
Geometry and Measurement							
NAEP							
Number Sense, Properties, and Operations					✔		
Measurement							
Geometry and Spatial Sense							
Data Analysis, Statistics, and Probability	✔	✔	✔	✔		✔	✔
Algebra and Functions							

CAT6 California Achievement Test, 6th Ed. **CTBS** Comprehensive Test of Basic Skills **ITBS** Iowa Test of Basic Skills, Form M
SAT10 Stanford Achievement Test, 10th Ed. **NAEP** National Assessment of Educational Progress 2005 Mathematics Objectives

Math Background

Skills Trace

BEFORE Chapter 11

Course 1 introduced graphical displays of data as well as misleading graphs.

DURING Chapter 11

Course 2 reviews and extends graphical displays to include histograms and scatter plots, and it introduces random samples and surveys.

AFTER Chapter 11

Throughout this course, students make and interpret graphical displays of data.

11-1 Reporting Frequency

Math Understandings

- A frequency table shows how often each data value occurs.
- A line plot visually represents a frequency table on a number line and is best used for a small number of data values.
- A histogram can help you make a clear visual comparison of frequency by grouping many data points into bars of equal width.

A **frequency table** is a table that lists each item in a data set with the number of times the item occurs. A **line plot** is a graph that shows the shape of a data set by stacking Xs above each data value on a number line. A **histogram** is a bar graph with no spaces between the bars. The height of each bar shows the frequency of data within that interval. The intervals of a histogram are of equal size and do not overlap.

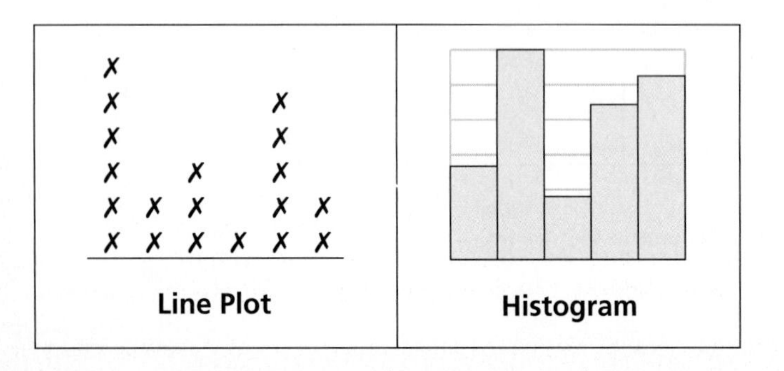

11-2 Spreadsheets and Data Displays

Math Understandings

- A spreadsheet is an electronic workspace that organizes information in cells that can be used in calculations and graphs.
- You can use a double bar graph or a double line graph to compare two sets of data.

A **spreadsheet** is a tool for organizing and analyzing data. Spreadsheets are arranged in numbered rows and lettered columns. A **cell** is a box where a row and column meet. For example, B6 represents the cell at the intersection of column B and row 6. A bar graph uses vertical or horizontal bars to display numerical information. A **double bar graph** uses bars to compare two sets of data. The **legend,** or key, identifies the data that are compared. A line graph uses a series of line segments to show changes in data over time. A **double line graph** compares changes in two sets of data over time.

11-3 Stem-and-Leaf Plots

Math Understandings

- A stem-and-leaf plot can quickly show the distribution of a data set by retaining each data value.

A **stem-and-leaf plot** is a graph that uses the digits of each number to show the shape of the data set. Each data value is broken into a "stem" (digit or digits on the left) and a "leaf" (digit or digits on the right). A back-to-back stem-and-leaf plot compares two sets of data.

11-4 Random Samples and Surveys

Math Understandings

- In most situations, it is extremely difficult and costly to survey every member of a population. You can select a random sample to accurately represent the entire population.
- A survey question should not influence responses by making one answer appear more attractive.

A **population** is a group of objects or people. Pollsters select a **sample,** or part of the population. In a **random sample,** each member of population has an equal chance of being selected for the sample. A **biased question** is a question that makes an unjustified assumption or makes one answer appear better than another.

Example: Is the following question biased or fair? Explain.

Do you prefer bright lighting or gentle lighting?

This question may be biased because "bright" and "gentle" are not clearly defined and may influence responses.

11-5 Estimating Population Size

Math Understandings

- The validity of the capture/recapture method depends on how well both the first and second samples accurately represent the population.

The capture/recapture method is frequently used to estimate animals populations. Researchers collect and mark a sample of animals and release them back in the population. Then, they again capture a sample and count the number of marked animals. Finally, they use a proportion to estimate the animal population.

$$\frac{\text{number of marked animals counted}}{\text{total number of animals counted}} = \frac{\text{number of animals marked}}{\text{estimate of animal population}}$$

11-6 Using Data to Persuade

Math Understandings

- Choices you make about the labels, scale intervals, and style of a graph can mislead the viewer of the information.

To evaluate a graph, check to see whether both scales start at zero, or whether the axes have breaks. Examine the intervals to see if they have equal intervals.

11-7 Exploring Scatter Plots

Math Understandings

- A scatter plot may show a relationship that is not obvious from looking at the numerical data.

A **scatter plot** is a graph that relates two sets of data. When a scatter plot shows a **positive trend,** one set of values increases as the other set tends to increase. When a scatter plot shows a **negative trend,** one set of values increases as the other tends to decrease. When the points in a scatter plot do not cluster along a trend line, the points show no relationship and **no trend.**

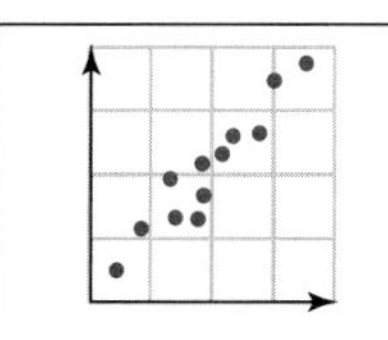

Positive trend

As one set of values increases, the other set tends to increase.

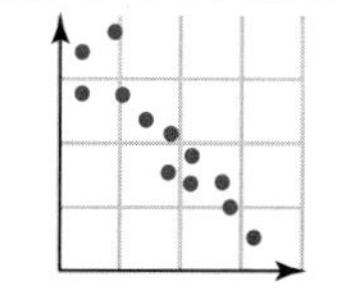

Negative trend

As one set of values increases, the other set tends to decrease.

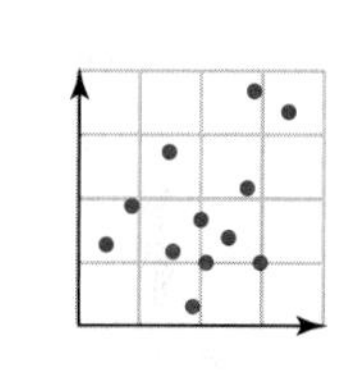

No trend

The points show no relationship.

Additional Professional Development Opportunities

Math Background Notes for Chapter 11: Every lesson has a Math Background in the PLAN section.

Research Overview, Mathematics Strands
Additional support for these topics and more is in the front of the Teacher's Edition.

LessonLab
LessonLab, a Pearson Education company, offers comprehensive, facilitated professional development designed to help teachers to improve student achievement. To learn more, please visit lessonlab.com.

Chapter 11 Resources

	11-1	11-2	11-3	11-4	11-5	11-6	11-7	For the Chapter
Print Resources								
L3 Practice	●	●	●	●	●	●	●	
L1 Adapted Practice	●	●	●	●	●	●	●	
L3 Guided Problem Solving	●	●	●	●	●	●	●	
L2 Reteaching	●	●	●	●	●	●	●	
L4 Enrichment	●	●	●	●	●	●	●	
L3 Daily Notetaking Guide	●	●	●	●	●	●	●	
L1 Adapted Daily Notetaking Guide	●	●	●	●	●	●	●	
L3 Vocabulary and Study Skills Worksheets	●		●	●	●		●	●
L3 Daily Puzzles	●	●	●	●	●	●	●	
L3 Activity Labs	●	●	●	●	●	●	●	
L3 Checkpoint Quiz			●			●		
L3 Chapter Project								●
L2 Below Level Chapter Test								●
L3 Chapter Test								●
L4 Alternative Assessment								●
L3 Cumulative Review								●
Spanish Resources ELL								
L3 Practice	●	●	●	●	●	●	●	
L3 Vocabulary and Study Skills Worksheets	●		●	●	●		●	●
L3 Checkpoint Quiz			●			●		
L2 Below Level Chapter Test								●
L3 Chapter Test								●
L4 Alternative Assessment								●
L3 Cumulative Review								●
Transparencies								
Check Skills You'll Need	●	●	●	●	●	●	●	
Additional Examples	●	●	●	●	●	●	●	
Problem of the Day	●	●	●	●	●	●	●	
Classroom Aid								
Student Edition Answers	●	●	●	●	●	●	●	●
Lesson Quiz	●	●	●	●	●	●	●	
Test-Taking Strategies								●
Technology								
Interactive Textbook Online	●	●	●	●	●	●	●	●
StudentExpress™ CD-ROM	●	●	●	●	●	●	●	●
Success Tracker™ Online Intervention	●	●	●	●	●	●	●	●
TeacherExpress™ CD-ROM	●	●	●	●	●	●	●	●
PresentationExpress™ with QuickTake Presenter CD-ROM	●	●	●	●	●	●	●	●
ExamView® Assessment Suite CD-ROM	●	●	●	●	●	●	●	●
MindPoint® Quiz Show CD-ROM								●
Prentice Hall Web Site: PHSchool.com	●	●	●	●	●	●	●	●

Also available: **Prentice Hall Assessment System**
- Progress Monitoring Assessments
- Skills and Concepts Review
- Test Prep Workbook

Other Resources
Algebra Readiness Tests
All-in-One Student Workbook
All-in-One Student Workbook, Adapted Version
Multilingual Handbook
Solution Key
Math Notes Study Folder
Spanish Cumulative Assessment

Where You Can Use the Lesson Resources

Here is a suggestion, following the four-step teaching plan, for how you can incorporate Differentiated Instruction resources into your teaching.

	Instructional Resources L3	Differentiated Instruction Resources
1. Plan		
Preparation Read the Math Background in the Teacher's Edition to connect this lesson with students' previous experience. **Starting Class** **Check Skills You'll Need** Assign these exercises to review prerequisite skills. **New Vocabulary** Help students pre-read the lesson by pointing out the new terms introduced in the lesson.	**Math Background** **Math Understandings** **Transparencies & PresentationExpress™ with QuickTake Presenter CD-ROM** Check Skills You'll Need Problem of the Day **Resources** Vocabulary and Study Skills	**Spanish Support** ELL Vocabulary and Study Skills
2. Teach		
L3 Guided Instruction Use the Activity Labs to build conceptual understanding. Teach each Example. Use the Teacher's Edition side column notes for specific teaching tips, including Error Prevention notes. Use the Additional Examples found in the side column (and on transparency and PowerPoint) as an alternative presentation for the content. After each Example, assign the Quick Check exercise for that Example to get an immediate assessment of student understanding. Use the Closure activity in the Teacher's Edition to help students attain mastery of lesson content.	**Student Edition** Activity Lab **Resources** Daily Notetaking Guide Activity Lab **Transparencies & PresentationExpress™ with QuickTake Presenter CD-ROM** Additional Examples Classroom Aids **ExamView® Assessment Suite CD-ROM**	**Teacher's Edition** Every lesson includes suggestions for working with students who need special attention. L1 Special Needs L2 Below Level L4 Advanced Learners ELL English Language Learners **Resources** L1 Adapted Daily Notetaking Guide **Multilingual Handbook**
3. Practice		
Assignment Guide **Check Your Understanding** Use these questions to check students' understanding before you assign homework. **Homework Exercises** Assign homework from these leveled exercises in the Assignment Guide. **A** Practice by Example **B** Apply Your Skills **C** Challenge Test Prep and Mixed Review **Homework Quick Check** Use these key exercises to quickly check students' homework.	**Transparencies & PresentationExpress™ with QuickTake Presenter CD-ROM** Student Answers **Resources** Practice Guided Problem Solving Vocabulary and Study Skills Activity Lab Daily Puzzles **ExamView® Assessment Suite CD-ROM**	**Spanish Support** ELL Practice ELL Vocabulary and Study Skills **Resources** L1 Adapted Practice L4 Enrichment
4. Assess & Reteach		
Lesson Quiz Assign the Lesson Quiz to assess students' mastery of the lesson content. **Checkpoint Quiz** Use the Checkpoint Quiz to assess student progress over several lessons.	**Transparencies & PresentationExpress™ with QuickTake Presenter CD-ROM** Lesson Quiz **Resources** Checkpoint Quiz	**Resources** L2 Reteaching ELL Checkpoint Quiz Success Tracker™ Online Intervention **ExamView® Assessment Suite CD-ROM**

KEY **L1** Special Needs **L2** Below Level **L3** For All Students **L4** Advanced, Gifted **ELL** English Language Learners

Displaying and Analyzing Data

Answers are in the back of the textbook.

For intervention, direct students to:

Comparing Numbers
Lesson 1-6
Extra Skills and Word Problems Practice, Ch. 1

Finding the Median
Lesson 1-10
Extra Skills and Word Problems Practice, Ch. 1

Solving Proportions
Lesson 5-4
Extra Skills and Word Problems Practice, Ch. 5

Graphing on the Coordinate Plane
Lesson 10-1
Extra Skills and Word Problems Practice, Ch. 10

Spanish Vocabulary/Study Skills ELL

Vocabulary/Study Skills L3

11A: Graphic Organizer For use before Lesson 11-1

Study Skill As your teacher presents new material in the chapter, keep a paper and pencil handy to write down notes and questions. If you miss class, borrow a classmate's notes so you will not fall behind.

Write your answers.

1. What is the chapter title? Displaying and Analyzing Data
2. How many lessons are there in this chapter? 7
3. What is the topic of the Test-Taking Strategies page? Interpreting Data
4. Complete the graphic organizer below as you work through the chapter.
   - In the center, write the title of the chapter.
   - When you begin a lesson, write the lesson name in a rectangle.
   - When you complete a lesson, write a skill or key concept in a circle linked to that lesson block.
   - When you complete the chapter, use this graphic organizer to help you review. Check students' diagrams.

Vocabulary and Study Skills

CHAPTER 11 Displaying and Analyzing Data

What You've Learned

- In Chapter 1, you described data using mean, median, mode, and range.
- In Chapter 7, you interpreted circle graphs and used them to represent data.
- In Chapter 10, you graphed points in the coordinate plane.

Check Your Readiness

GO for Help

For Exercises	See Lesson
1–2	1-6
3–6	1-10
7–10	5-4
11–14	10-1

Comparing Numbers

Order the numbers from least to greatest.

1. 32, −31, 34, −30, 13, 33 **−31, −30, 13, 32, 33, 34**
2. 11.1, 10.9, 11.3, 11.5, 10.2 **10.2, 10.9, 11.1, 11.3, 11.5**

Finding the Median

Find the median of each set of data.

3. 15, 9, 16, 12, 8, 10, 13 **12**
4. 27, 35, 24, 56, 29, 37 **32**
5. 55, 69, 112, 67, 32, 123, 45 **67**
6. 8.9, 8.5, 7.6, 8.4, 9.1, 8.5 **8.5**

Solving Proportions

Algebra **Solve each proportion.**

7. $\frac{3}{4} = \frac{a}{24}$ **18**
8. $\frac{2}{b} = \frac{3}{21}$ **14**
9. $\frac{n}{52} = \frac{17}{13}$ **68**
10. $\frac{12}{5} = \frac{a}{45}$ **108**

Graphing on the Coordinate Plane

Graph each point on the same coordinate plane. Name the quadrant in which each point lies or the axis on which the point lies. 11–14. See margin.

11. $(-4, 7)$
12. $(0, -5)$
13. $(6, 3)$
14. $(2, 0)$

11–14.

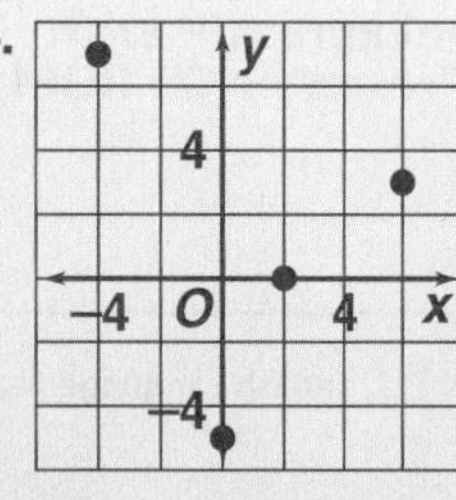

11. II
12. *y*-axis
13. I
14. *x*-axis

Chapter 11 Overview

In this chapter, students build on their knowledge of graphing and statistics as they investigate different ways of obtaining and displaying data. They work with frequency tables and data displays like spreadsheets, histograms, stem-and-leaf plots, and scatter plots. They learn about sampling, surveys, and how to use data to make predictions.

Activating Prior Knowledge

In this chapter, to make and use different kinds of data displays, students will build on their knowledge of graphs, ordered pairs, and measures of central tendency. They will draw upon their understanding of ratios and proportions to estimate populations. Ask questions such as:

- *What is the median of a set of numbers?* **The middle value when the data values are arranged in numerical order.**
- *What is the mode of this set of data: 2, 4, 4, 8, 9, 12?* **4**
- *A model airplane is built using the scale 1 in. = 12 ft. If the wingspan of the model is 6 inches, what is the wingspan of the plane?* **72 ft**

What You'll Learn Next

- In this chapter, you will represent data using line plots, line graphs, bar graphs, stem-and-leaf plots, scatter plots, and Venn diagrams.
- You will interpret double bar graphs and double line graphs.
- You will identify misleading graphs and statistics.

Problem Solving Application On pages 576 and 577, you will work an extended activity on data analysis.

Key Vocabulary

- biased question (p. 551)
- cell (p. 538)
- double bar graph (p. 539)
- double line graph (p. 539)
- frequency table (p. 532)
- histogram (p. 533)
- legend (p. 539)
- line plot (p. 533)
- negative trend (p. 568)
- no trend (p. 568)
- population (p. 550)
- positive trend (p. 568)
- random sample (p. 550)
- sample (p. 550)
- scatter plot (p. 567)
- spreadsheet (p. 538)
- stem-and-leaf plot (p. 544)

11-1

1. Plan

Objective
To represent data using frequency tables, line plots, and histograms

Examples
1 Making a Frequency Table
2 Making a Line Plot
3 Making a Histogram

Math Understandings: p. 530C

Math Background

Frequency refers to how often something happens. A *frequency table* organizes each item in a data set by how often it occurs. A *line plot* is a quick way to visualize the frequency distribution of a data set using Xs on a number line.

A *histogram* is a bar graph that uses equal intervals to illustrate the frequency distribution of a data set. No spaces are used between consecutive bars because the intervals are continuous (and non-overlapping).

More Math Background: p. 530C

Lesson Planning and Resources

See p. 530E for a list of the resources that support this lesson.

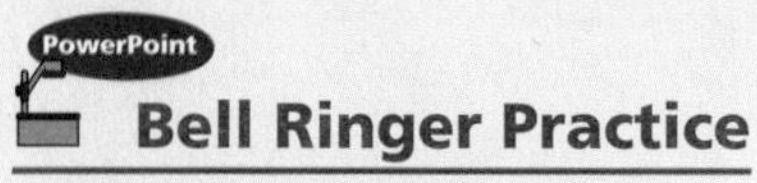

Bell Ringer Practice

Check Skills You'll Need
Use student page, transparency, or PowerPoint. For intervention, direct students to:
Comparing and Ordering Integers
Lesson 1-6
Extra Skills and Word Problems Practice, Ch. 1

11-1 Reporting Frequency

Check Skills You'll Need

1. **Vocabulary Review** What is an *integer?* **1–3. See below.**

Order the numbers from least to greatest.

2. 23, 45, 61, 87, 91, 16, 22, 52
3. −41, 42, −43, 45, 43, −47

GO for Help Lesson 1-6

Check Skills You'll Need

1. An integer is a positive or negative whole number, or 0.
2. 16, 22, 23, 45, 52, 61, 87, 91
3. −47, −43, −41, 42, 43, 45

What You'll Learn

To represent data using frequency tables, line plots, and histograms

New Vocabulary frequency table, line plot, histogram

Why Learn This?

When you count the number of times a particular event or item occurs, you measure frequency. You can use frequency to analyze events such as extreme weather events.

A **frequency table** is a table that lists each item in a data set and the number of times each item occurs.

EXAMPLE Making a Frequency Table

1 **Hurricanes** The data below show the number of hurricanes in the Atlantic Ocean each year for a period of 30 years. Make a frequency table of the data.

7 4 9 8 8 10 3 9 11 3 4 4 4 8 7
5 3 4 7 5 2 2 7 9 5 5 5 6 6 4

Atlantic Ocean Hurricanes

Number	2	3	4	5	6	7	8	9	10	11
Tally	//	///	~~////~~ /	~~////~~	//	////	///	///	/	/
Frequency	2	3	6	5	2	4	3	3	1	1

Quick Check

1. The data below show the number of U.S. Representatives for 22 states. Make a frequency table of the data. **See back of book.**

4 3 1 2 1 5 4 1 7 2 8 1 3 8 5 9 3 5 7 3 1 9

Differentiated Instruction Solutions for All Learners

Special Needs L1
Help students keep track of the Xs they draw in line plots, and keep track of the data in their frequency tables. Provide them with graph paper so that they can use the grid squares for orientation.
learning style: visual

Below Level L2
Have students make a histogram using birth dates of students in the class as the data and using the months as the intervals.
learning style: visual

A **line plot** is a graph that shows the shape of a data set by stacking ✗'s above each data value on a number line.

EXAMPLE Making a Line Plot

2 Make a line plot of the data in Example 1.

Step 1 Draw a number line from the least to the greatest value (from 2 to 11).

Step 2 Write an ✗ above each value for each time the value occurs in the data.

Atlantic Ocean Hurricanes

```
      X
      X  X
      X  X     X
   X  X  X     X  X  X
X  X  X  X  X  X  X  X
X  X  X  X  X  X  X  X  X  X
2  3  4  5  6  7  8  9  10 11
```

Quick Check

2. Make a line plot of the number of students in math classes: 24 27 21 25 25 28 22 23 25 25 28 22 23 25 22 24 25 28 27 22.
See back of book.

Vocabulary Tip
The *histo* in histogram is short for *history.*

A **histogram** is a bar graph with no spaces between the bars. The height of each bar shows the frequency of data within that interval. The intervals of a histogram are of equal size and do not overlap.

EXAMPLE Making a Histogram

3 Make a histogram of the data in Example 1.

Make a frequency table. Use the equal-sized intervals 2–3, 4–5, 6–7, 8–9, and 10–11. Then make a histogram.

Atlantic Ocean Hurricanes

Number	Frequency
2–3	𝍸
4–5	𝍸 𝍸 /
6–7	𝍸 /
8–9	𝍸 /
10–11	//

Atlantic Ocean Hurricanes

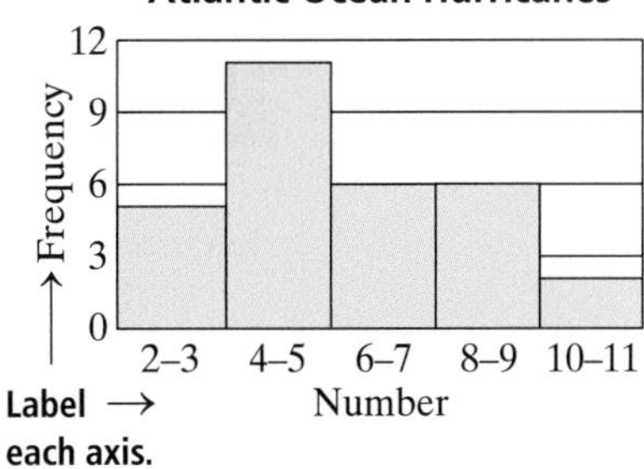

Quick Check

3. Make a histogram of the ages of employees at a retail store: 28 20 44 72 65 40 59 29 22 36 28 61 30 27 33 55 48 24 28 32.
See back of book.

2. Teach

Activity Lab

Use before the lesson.

All in One Teaching Resources
Activity Lab 11-1: Reporting Frequency

Guided Instruction

Example 2
Have students examine the line plot. Ask:
- *Which number of hurricanes occurs most often?* 4
- *Which number of hurricanes occurs least often?* 10 or 11

Error Prevention!

Help students make a clear distinction between histograms and bar graphs. A histogram has no space between bars because the intervals are continuous. The height of a histogram's bars show frequency instead of amounts.

PowerPoint

Additional Examples

1 The numbers below represent the ages of the students in a karate class. Make a frequency table of the data.
11 5 9 13 8 9 9 11 10 8 6 7 12 11 13 12 7 6 11 12 10 8

Age	5	6	7	8	9
Tally	I	II	II	III	III
Freq.	1	2	2	3	3

Age	10	11	12	13
Tally	II	IIII	III	II
Freq.	2	4	3	2

2 Make a line plot of the data in Exercise 1.

Ages of Students in a Karate Class

```
                  X
         X  X     X  X
      X  X  X  X  X  X  X
   X  X  X  X  X  X  X  X  X
5  6  7  8  9  10 11 12 13
```

Advanced Learners L4
What happens to a histogram as the number of intervals is reduced? **the histogram is narrow and high** *as the number of intervals is increased?* **the histogram is wider and flatter**

learning style: visual

English Language Learners ELL
Have students describe the parts of a histogram and a line plot. Have them work in pairs to come up with a list of similarities and differences between the two displays of data.

learning style: verbal

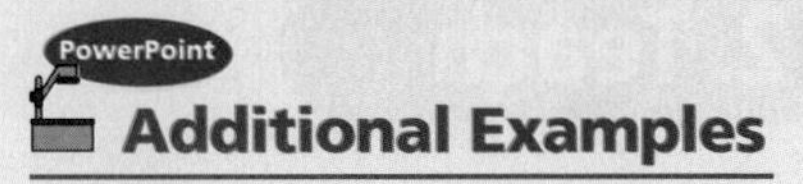

Additional Examples

3 Make a histogram of the data below.

Hours Spent on the Internet

Hours	Tally	Hours	Tally
1	I	11	III
2		12	I
3	II	13	II
4	IIII	14	
5	III	15	I
6	III	16	II
7	IIII	17	III
8	I	18	IIII
9		19	II
10	IIII	20	I

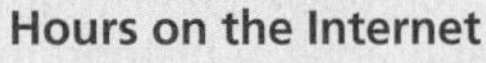

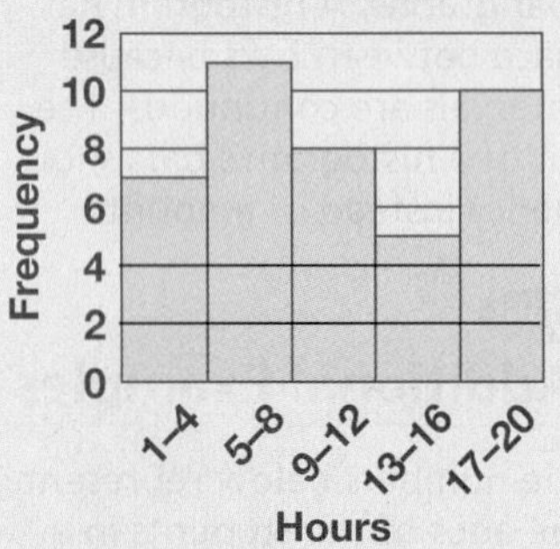

All in One Teaching Resources

- Daily Notetaking Guide 11-1 L3
- Adapted Notetaking 11-1 L1

Closure

- *What is a frequency table?* a table that lists each item in a data set with the number of times the item occurs
- *What do histograms and line plots illustrate?* the shape of a data set

Voting-Age Population Registered to Vote (percent)

State	Percent
Calif.	54
Fla.	63
Ga.	62
Ind.	67
N.C.	69
Ore.	75
Tex.	61
Va.	64

More Than One Way

Use the table at the left to make a data display.

Anna's Method

I can make a line plot. I'll use six intervals of 4.

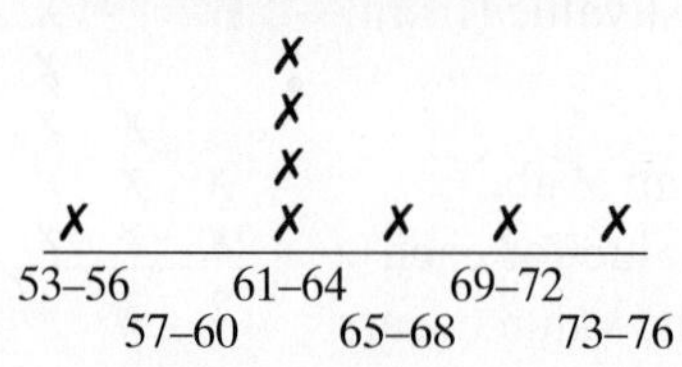

Ryan's Method

I can make a histogram. I'll use five intervals of 5.

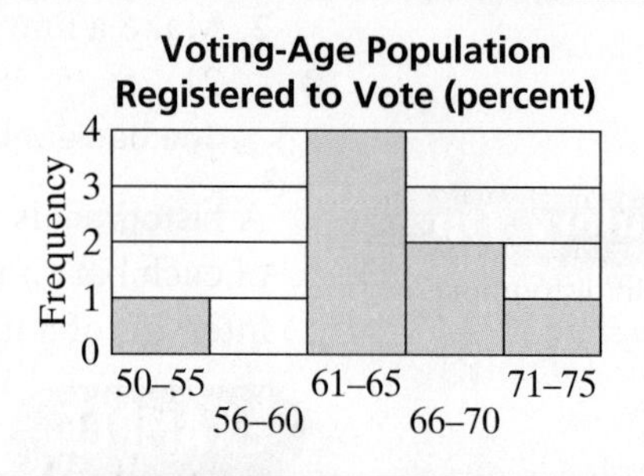

Choose a Method

Display the data in the table. Explain your choice of data display.

Number of Hours Worked per Week

Number	35	36	37	38	39	40	41	42	43	44
Tally	𝍸	III	𝍸 II	𝍸 II	𝍸 I	𝍸 𝍸	III	𝍸		I

See margin.

Check Your Understanding

Use the data in the frequency table below. 1–2. See margin.

Number	15	16	17	18	19	20	21	22	23	24
Tally	𝍸 I	I	III		II	III	I	IIII	𝍸	II

1. Make a line plot of the data.
2. Make a histogram of the data.

More Than One Way. See back of book.

1.

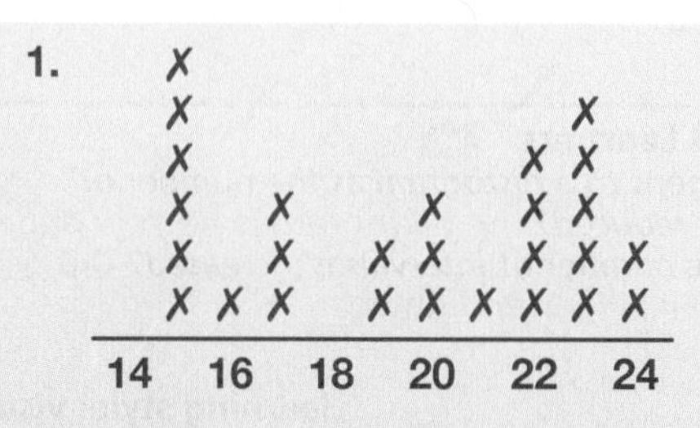

Homework Exercises

For more exercises, see Extra Skills and Word Problems.

GO for Help

For Exercises	See Examples
3–5	1
6–8	2
9–10	3

A **Make a frequency table of the data.** 3–5. See margin.

3. tickets sold: 45 48 51 53 50 46 46 50 51 48 46 45 50 49 46
4. number of TVs: 1 3 2 2 1 4 1 2 2 1 3 1 3 3 2 2 3 1
5. student ages:
   13 12 14 12 11 12 13 14 13 13 14 11 12 12 13 11 11

Make a line plot of the data. 6–8. See margin.

6. number of plants sold per person:
   5 10 11 8 7 11 9 8 6 7 12 10 10 9 8 7 6
7. miles from home to shopping center:
   2 4 10 5 4 6 7 9 5 5 3 1 10 8 6 4 3
8. blocks walked from home to school:
   2 8 10 1 2 9 8 7 6 8 4 3 8 9 1 3 10 12 6 4 8

Make a histogram of the data. 9–10. See margin.

9. **How Many Amusement Parks Did You Visit Last Year?**

Number of Parks	0–2	3–5	6–8	9–11
Frequency	10	4	3	1

10. **How Many Hours Do You Sleep Each Night?**

Number of Hours	5–6	7–8	9–10	11–12
Frequency	4	12	9	1

B GPS 11. **Guided Problem Solving** At the right are the times at which 16 people get up each morning. Display the data to show the most common half-hour interval.

5:30	6:45	5:45	6:15
6:25	6:20	7:15	7:45
8:00	7:00	8:00	7:30
6:00	7:10	7:50	6:10

- What type of display will you use?
- Use the interval 5:30–5:59 first. What are the other intervals? See margin.

Use the graph at the right.

Number of Books Purchased

```
X
X  X
X  X  X
X  X  X           X
X  X  X  X  X  X
1  2  3  4  5  6
```

12. **Books** The line plot shows the number of books each bookstore customer bought. How many customers bought more than three books? 4 customers
13. How many customers bought an even number of books? 8 customers

2–11. See back of book.

3. Practice

Assignment Guide

Check Your Understanding
Go over Exercises 1–2 in class before assigning the Homework Exercises.

Homework Exercises

A Practice by Example	3–10
B Apply Your Skills	11–17
C Challenge	18
Test Prep and Mixed Review	19–24

Homework Quick Check
To check students' understanding of key skills and concepts, go over Exercises 6, 10, 12, 15, and 16.

Differentiated Instruction Resources

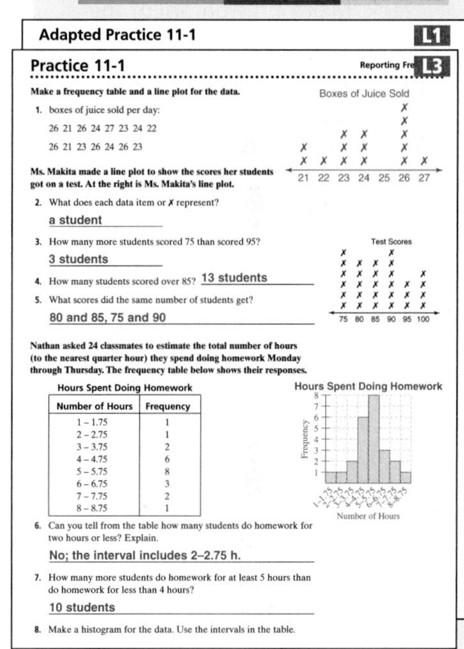

Adapted Practice 11-1 L1

Practice 11-1 Reporting Fre... L3

Make a frequency table and a line plot for the data.

1. boxes of juice sold per day:
   26 21 26 24 27 23 24 22
   26 21 23 26 24 26 23

Boxes of Juice Sold

Ms. Makita made a line plot to show the scores her students got on a test. At the right is Ms. Makita's line plot.

2. What does each data item or ✗ represent? a student
3. How many more students scored 75 than scored 95? 3 students
4. How many students scored over 85? 13 students
5. What scores did the same number of students get? 80 and 85, 75 and 90

Test Scores

Nathan asked 24 classmates to estimate the total number of hours (to the nearest quarter hour) they spend doing homework Monday through Thursday. The frequency table below shows their responses.

Hours Spent Doing Homework

Number of Hours	Frequency
1 – 1.75	1
2 – 2.75	1
3 – 3.75	2
4 – 4.75	6
5 – 5.75	8
6 – 6.75	3
7 – 7.75	2
8 – 8.75	1

Hours Spent Doing Homework

6. Can you tell from the table how many students do homework for two hours or less? Explain. No; the interval includes 2–2.75 h.
7. How many more students do homework for at least 5 hours than do homework for less than 4 hours? 10 students
8. Make a histogram for the data. Use the intervals in the table.

11-1 • Guided Problem Solving GPS L3

GPS Student Page 535, Exercise 12:

Books The line plot shows the number of books each bookstore customer bought. How many customers bought more than three books?

Number of Books Purchased

Understand

1. In the line plot, what do the numbers at the bottom represent? the number of books purchased
2. In the line plot, what do the X's represent? The number of customers who bought a certain number of books.

Plan and Carry Out

3. To find the number of customers who bought more than three books, in which columns do you have to look? columns 4, 5, and 6
4. How many customers bought 4 books? 1 customer
5. How many customers bought 5 books? 1 customer
6. How many customers bought 6 books? 2 customers
7. How many customers bought more than 3 books? 4 customers

Check

8. Why do you not include the customers who bought 3 books? The expression *more than 3* does not include 3.

Solve Another Problem

Number of CDs Purchased

9. The line plot shows how many CDs each customer in a music store bought. How many customers bought two or fewer CDs? 2 + 4 + 2 = 8 customers

4. Assess & Reteach

Lesson Quiz

PowerPoint

1. Make a frequency table and a line plot for the number of students in each class:

   25 24 25 23 24 23 24 24 23

Students	23	24	25
Class	3	4	2

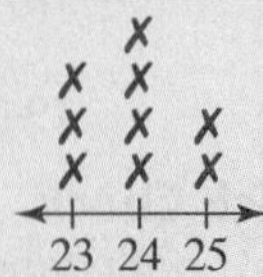

2. The data table below shows how many hours of TV different people watch each week. Make a histogram of the data.

Hours	0–5	6–10	11–15
Frequency	3	5	10

Hours	16–20	21–25
Frequency	24	15

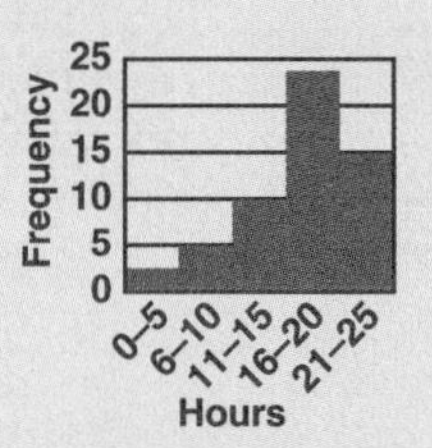

17–18. See back of book.
22–24. See back of book.

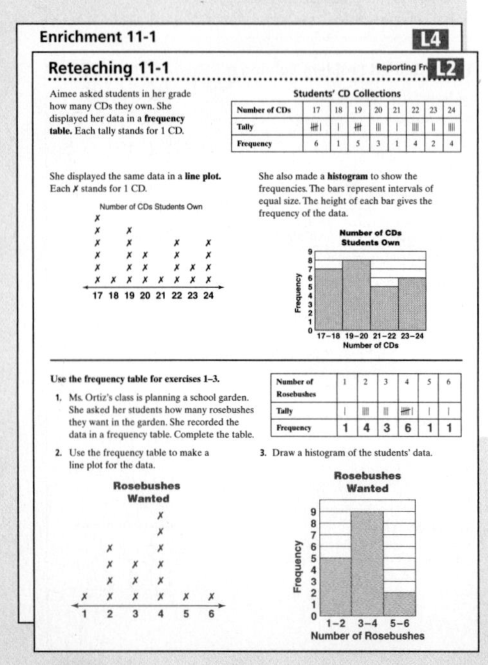

Enrichment 11-1 L4

Reteaching 11-1 Reporting Fr L2

Aimee asked students in her grade how many CDs they own. She displayed her data in a **frequency table.** Each tally stands for 1 CD.

Students' CD Collections

Number of CDs	17	18	19	20	21	22	23	24
Tally	𝍸 I	I	𝍸	III	I	IIII	II	IIII
Frequency	6	1	5	3	1	4	2	4

She displayed the same data in a **line plot.** Each ✗ stands for 1 CD.

She also made a **histogram** to show the frequencies. The bars represent intervals of equal size. The height of each bar gives the frequency of the data.

Use the frequency table for exercises 1–3.

1. Ms. Ortiz's class is planning a school garden. She asked her students how many rosebushes they want in the garden. She recorded the data in a frequency table. Complete the table.

Number of Rosebushes	1	2	3	4	5	6
Tally	I	IIII	III	𝍸 I	I	I
Frequency	1	4	3	6	1	1

2. Use the frequency table to make a line plot for the data.

3. Draw a histogram of the students' data.

GO Online
Homework Video Tutor
Visit: PHSchool.com
Web Code: are-1101

14. about 12 people
15. Find the frequency of each interval and add them all together.
16. No; about 9 people saw either 6 or 7 movies.

Use the histogram for Exercises 14–16.

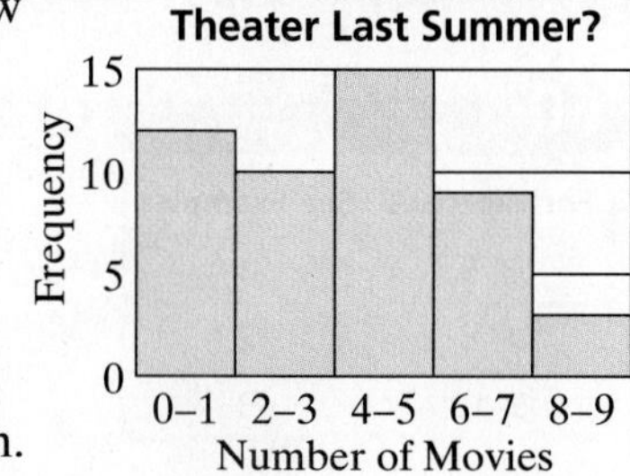

14. **Movies** About how many people saw fewer than two movies? **See left.**
15. **Writing in Math** How can you find the number of people who answered the survey? Explain. **See left.**
16. **Reasoning** Can you tell how many people saw exactly 7 movies? Explain. **See left.**

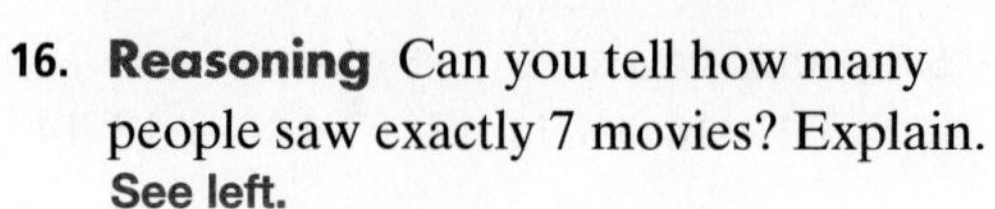

17. **Choose a Method** Display the data below, which give 26 responses to the question, "How many siblings do you have?"
    1 3 4 2 2 1 3 1 0 2 0 1 3 4 2 1 0 1 2 0 3 4 2 5 2 6. **See margin.**
18. **Challenge** The table shows home prices. Draw two histograms for the data, one with 4 intervals and one with 8. **See margin.**

\$129,000	\$132,000	\$121,000	\$115,000
\$138,000	\$152,000	\$147,000	\$136,000
\$137,000	\$148,000	\$175,000	\$127,000
\$192,000	\$133,000	\$167,000	\$154,000

Test Prep and Mixed Review

Practice

Multiple Choice

19. The line plot shows the number of children of each of 12 recent U.S. presidents. Which set of data does the line plot show? **B**

   Ⓐ 1 6 2 2 4 4 2 2 3 2 2 6
   Ⓑ 1 6 2 2 6 4 2 3 1 2 4 4
   Ⓒ 1 6 2 2 2 4 2 3 3 2 4 4
   Ⓓ 1 5 6 2 2 4 2 3 2 2 4 4

20. A square park measures 45 m on each side. The sidewalk around the park's edge encloses a square area of grass that is 42 m on each side. What is the area of the sidewalk? **F**
   Ⓕ 261 m^2 Ⓖ 1,764 m^2 Ⓗ 1,890 m^2 Ⓙ 2,025 m^2

21. Which expression can be used to find the perimeter of a rectangular trampoline with a length of 16 feet and width w? **B**
   Ⓐ $16 + 2w$ Ⓑ $32 + 2w$ Ⓒ $16w$ Ⓓ $32 + w$

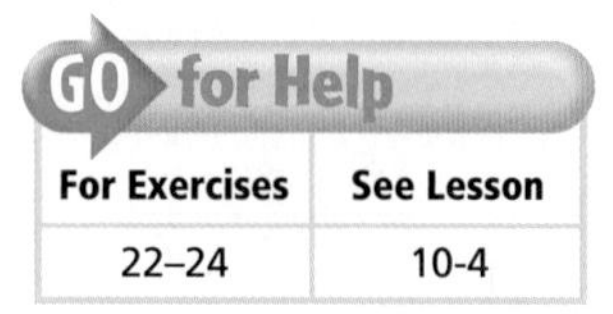
GO for Help

For Exercises	See Lesson
22–24	10-4

Algebra **Make a table of values for each equation. Use integer values of x from -3 to 3. Then graph each parabola.** 22–24. **See margin.**

22. $y = x^2 - 1$
23. $y = 3x^2$
24. $y = -3x^2$

Test Prep

Resources
For additional practice with a variety of test item formats:
- Test-Taking Strategies, p. 571
- Test Prep, p. 575
- Test-Taking Strategies with Transparencies

Alternative Assessment

Students work in pairs to collect or make up data of their choosing, such as how many classmates participate in various sports. Together the partners make a table of the data and a histogram. Students can share their findings with the class.

11-1b Activity Lab

Venn Diagrams

A Venn diagram shows relationships between sets of items. Each set is represented separately. Items that belong to both sets are represented by the intersection.

EXAMPLE Using a Venn Diagram

Geography There are 22 states that are all or partly in the eastern time zone, and 15 states that are all or partly in the central time zone. This includes the 5 states that are in both the eastern and central time zones. How many states are in at least one of the two time zones?

Draw a Venn diagram.

There are $17 + 5 + 10$, or 32 states in either or both the eastern and central time zones.

Eastern Time Zone States $22 - 5 = 17$ | 5 | Central Time Zone States $15 - 5 = 10$

Exercises

Forty students have pets. Thirty-two students have cats or dogs or both, but no birds. Eight students have birds. One student has all three kinds of pets.

cats 15 | ■ | dogs 14 | ■ | 0 | ■ | 4 birds

1. Copy and complete the Venn diagram. **See above right.**
2. How many students have only dogs and birds? Only dogs and cats? **3 students; 3 students**
3. **Reasoning** A new student who has only fish joins the class. How many of the existing sets would this new set overlap? **none**
4. **Geography** There are 10 states that are completely in the central time zone, and 6 states that are completely in the mountain time zone. There are 21 states that are completely in one of the time zones, or in both. Use a Venn diagram to find how many states are in both time zones.
5. **Data Collection** Conduct a survey of your class. Ask students whether they have brothers, sisters, both, or neither. Use a Venn diagram to display the results. **Check students' work.**

1.

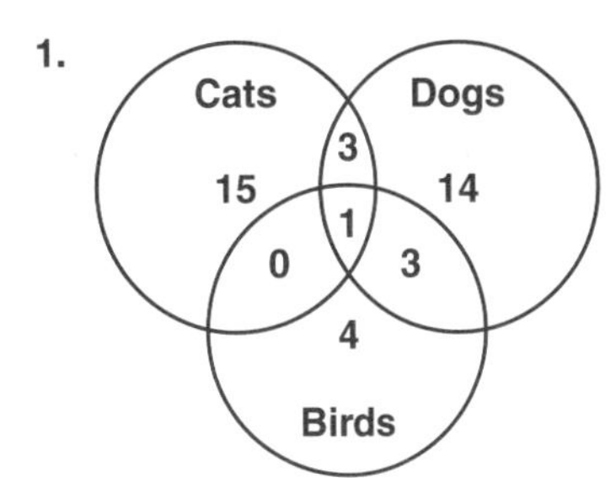

4.

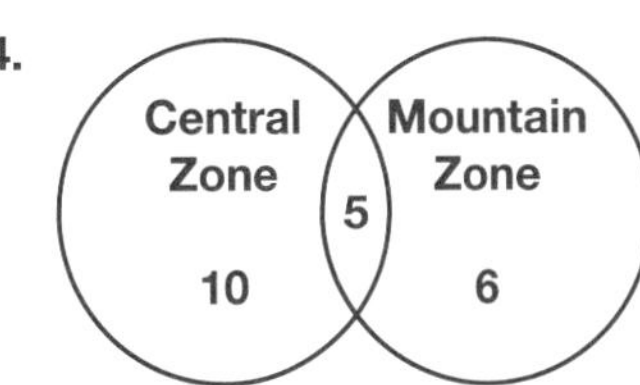

Activity Lab

Venn Diagrams

Students learn to use Venn diagrams to represent the relationship between sets of data. This will extend the skills they used in Lesson 11-1.

Guided Instruction

To work with intersecting sets, write two lists of numbers on the board: the whole numbers 1–11 and the even numbers 2–12. Ask:

- *Which numbers are common to both lists?* **2, 4, 6, 8, 10**
- *Which are in the first list only?* **1, 3, 5, 7, 9, 11**
- *Which are in the second list only?* **12**

Example

Discuss the information the Venn diagram illustrates.

Ask:

- *What do the overlapping circles represent?* **states in both time zones**
- *How can you find the number of states only in the eastern time zone or only in the central time zone?* **Subtract from each total the 5 states in both time zones (the intersection of the sets).**

English Language Learners ELL

Review the distinction between eastern and central time zones. Discuss the time zone you are in and identify the time zones of familiar cities and states.

Exercises

Pair or group students with varying degrees of proficiency in logical reasoning. Have them work together helping one another to use Venn Diagrams to organize the data logically.

Resources

- circle manipulatives

11-2

1. Plan

Objective
To interpret spreadsheets, double bar graphs, and double line graphs

Examples
1 Using a Spreadsheet
2 Using a Double Bar Graph
3 Predicting With a Double Line Graph

Math Understandings: p. 530C

Professional Development

Math Background

A *spreadsheet* is an electronic worksheet that organizes data in columns and rows. Spreadsheets can perform calculations and graph data using different types of graphs. Line graphs show trends over time. A *double line graph* compares two lines on the same set of axes. A bar graph allows you to show the relative size of different categories. A *double bar graph* compares the relative sizes of two sets of data. When more than one set of data is graphed, *a legend* identifies the different data sets, or series.

More Math Background: p. 530C

Lesson Planning and Resources

See p. 530E for a list of the resources that support this lesson.

Check Skills You'll Need
Use student page, transparency, or PowerPoint. For intervention, direct students to:
Patterns and Graphs
Lesson 9-1
Extra Skills and Word Problems Practice, Ch. 9

11-2 Spreadsheets and Data Displays

Check Skills You'll Need

1. **Vocabulary Review** A(n) __?__ is the difference between values on a scale. **interval**

Graph the data.

2.

Age (yr)	Height (in.)
1	29
2	33
3	36
4	39

See back of book.

for Help Lesson 9-1

What You'll Learn

To interpret spreadsheets, double bar graphs, and double line graphs

New Vocabulary spreadsheet, cell, double bar graph, legend, double line graph

Why Learn This?

You can use spreadsheets to draw graphs and compare data, such as the number of households with VCRs or DVD players.

A **spreadsheet** is a tool for organizing and analyzing data. Spreadsheets are arranged in lettered columns and numbered rows.

A **cell** is a box where a column and a row of a spreadsheet meet. You use a letter and a number to identify each cell.

EXAMPLE Using a Spreadsheet

1 Electronics The spreadsheet below shows the number of U.S. households with VCRs and the number with DVD players.

U.S. Households (millions)

	A	B	C
1	Year	VCRs	DVDs
2	1999	89	2
3	2000	92	6
4	2001	95	15
5	2002	96	28
6	2003	96	42
7	2004	97	62

a. What value is in cell C3? What does the value represent?

Column C and row 3 meet at cell C3. The value is 6. The value represents the number of U.S. households, in millions, with DVD players in 2000.

b. How many U.S. households had VCRs in 2004?

Cell B7 shows that 97 million U.S. households had VCRs in 2004.

Quick Check

1. **a.** What value is in cell B4? What does the value represent? **95; 95 million households with VCRs in 2001**
   b. Which cell shows the number of DVD players in 2003? **C6**

Differentiated Instruction Solutions for All Learners

Special Needs L1
Show students how to run their fingers across the columns and rows of a spreadsheet to meet at a cell. Make a connection between this intersection, and locating points on a coordinate grid.
learning style: tactile

Below Level L2
Review simple bar and line graphs. Help students correlate the height of a bar with a number from the vertical axis. Help students find vertical axis values for specific points on a line graph.
learning style: visual

A **double bar graph** is a graph that uses bars to compare two sets of data. The **legend,** or key, identifies the data that are compared.

EXAMPLE Using a Double Bar Graph

Technology Tip

Using a spreadsheet, you can choose the data you want to graph. Then you can select the type of graph for the data.

2 In which year did the number of households with DVD players first exceed 50 million?

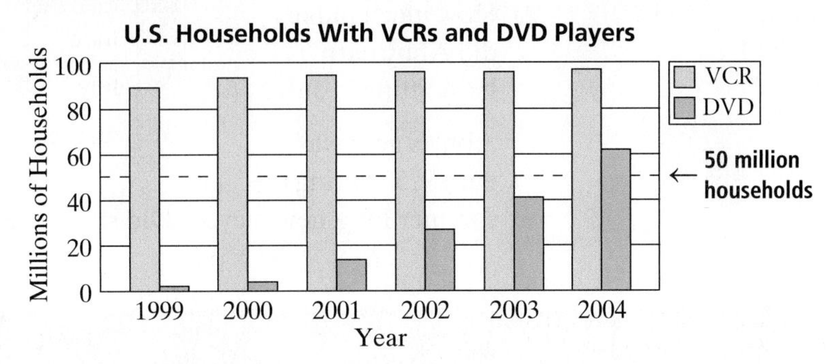

U.S. households with DVD players first exceeded 50 million in 2004.

Quick Check

2. In which year did the number of U.S. households with DVD players first exceed 10 million? **2001**

A **double line graph** is a graph that compares changes in two sets of data over time.

EXAMPLE Predicting With a Double Line Graph

3 **Estimation** Use the graph to estimate the year in which the number of households with DVD players equals the number of households with VCRs.

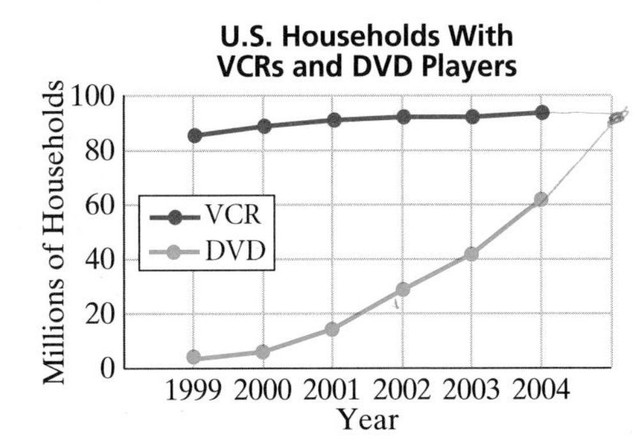

The two lines appear to intersect in 2005.

You can estimate that the number of households with DVD players equals the number with VCRs in 2005.

Quick Check

3. Copy and extend the graph. Estimate the number of DVD players in U.S. households in 2006. **Answers may vary. Sample: about 130 million**

2. Teach

Activity Lab

Use before the lesson.

All in One Teaching Resources

Activity Lab 11-2: Spreadsheets and Data Displays

Guided Instruction

Example 1

Have students trace down column C and across row 3 with their index fingers so that they intersect at C3.

Example 3

Students may confuse the use of bar graphs and line graphs. Bar graphs are used to compare amounts, while line graphs show changes over time. A double bar graph compares two sets of data. A double line graph compares the trends of two sets of data.

PowerPoint

Additional Examples

1 The spreadsheet below shows the number of graduates in dentistry and medicine.

a. Find the value in cell B5. **43**

b. In what year were the number of graduates from both schools the same? **1996**

	A	B	C
1	Year	Dentistry	Medicine
2	1996	48	48
3	1997	36	51
4	1998	40	56
5	1999	43	42
6	2000	65	38
7	2001	61	46

Advanced Learners L4

Explain when a triple bar graph might be used. **when comparing three sets of data**

Have students construct a triple bar graph for a topic of their choice.

learning style: visual

English Language Learners ELL

Make sure that students understand the difference between columns and rows. Show them pictures of columns in front of homes, and point out that columns are vertical. Show them rows of chairs or tables, and point out that rows are horizontal.

learning style: visual

PowerPoint

Additional Examples

2. When did more people first obtain news from the radio than the newspaper? **1990**

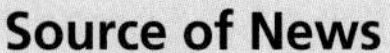

Source of News

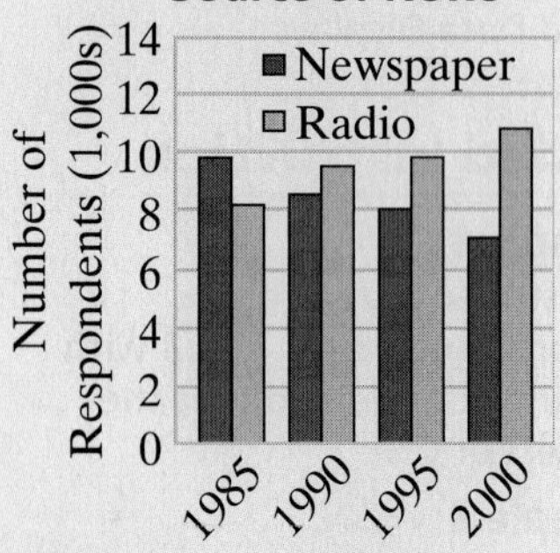

3. During what period of time did whole milk consumption decrease the most? **between 1980 and 1985**

Milk Consumption

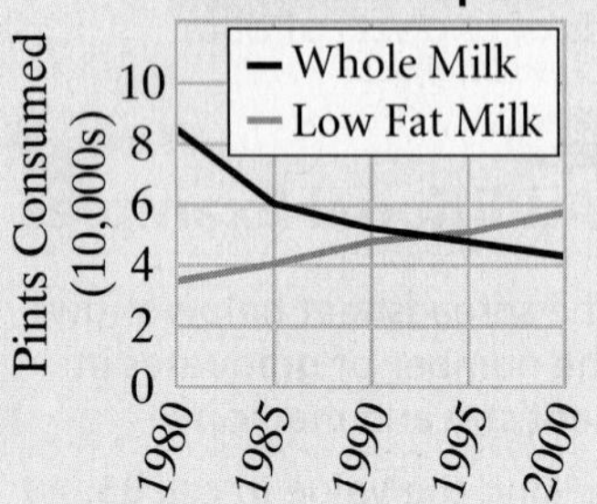

All in One Teaching Resources

- Daily Notetaking Guide 11-2 L3
- Adapted Notetaking 11-2 L1

Closure

- *What is a spreadsheet?* **an electronic tool for organizing and analyzing data**
- *How does a legend help you interpret a double bar graph or a double line graph?* **A legend identifies each set of data.**

Check Your Understanding

Use the data in the spreadsheet.

Abby's Summer Earnings ($)

	A	B	C
1	Month	Babysitting	Lawn Mowing
2	June	36	50
3	July	62	40
4	August	55	40

1. How did Abby earn $50 in June? **lawn mowing**
2. How much money did Abby earn babysitting in July? **$62**
3. Abby's goal was to earn $250 during the summer for a new bicycle. Did she reach her goal? **yes**

Homework Exercises

For more exercises, see Extra Skills and Word Problems.

GO for Help

For Exercises	See Examples
4–12	1
13–17	2
18–21	3

A **Give the content or value of each cell.**

Nutrition in One Brand's Dog and Cat Foods (Calories per 100 Calories)

	A	B	C	D
1	Animal	Fat	Carbohydrates	Protein
2	Dog	37	37	26
3	Cat	46	26	28

4. A2 **dog**
5. D2 **26**
6. B3 **46**
7. C2 **37**

Which cell contains the given word or value?

8. Protein **D1**
9. Animal **A1**
10. 28 **D3**
11. 46 **B3**
12. Cat **A3**

Libraries **Use the double bar graph below for Exercises 13–17.**

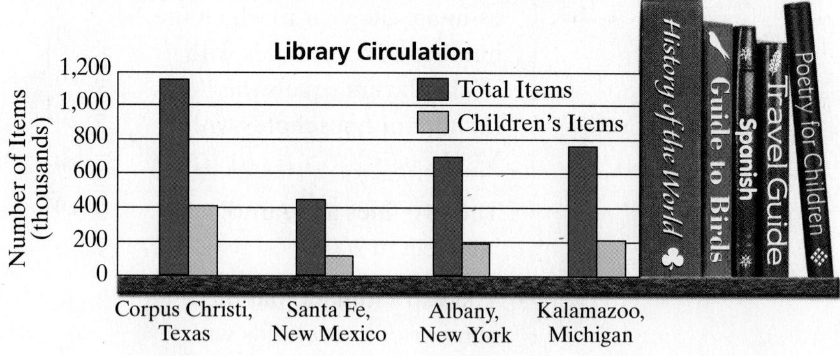

13. Which library circulated the most children's items? **Corpus Christi, Texas**
14. Which library had the greatest total circulation? **Corpus Christi, Texas**
15. Which libraries circulated fewer than 600,000 children's items? **all of them**
16. Which libraries circulated more than 500,000 total items? **16–17. See left.**
17. **Estimation** Which two libraries had about the same circulation?

16. **Corpus Christi, Texas; Albany, New York; and Kalamazoo, Michigan**

17. **Albany, New York, and Kalamazoo, Michigan**

Go Online
Homework Video Tutor
Visit: PHSchool.com
Web Code: are-1102

Industry **Use the double line graph below for Exercises 18–21.**

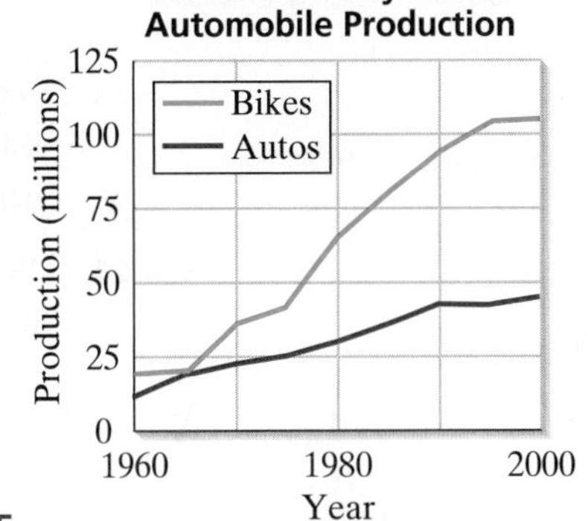

18. About how many bikes were produced in 1960? **about 20 million**

19. In which years did automobile production exceed 23 million? **1975–2000**

20. Reasoning The two lines begin to separate widely after 1967. Why might this be? **See left.**

20. Answers may vary. Sample: Because of gas shortages in the 1970s, more people may have started riding bikes.

21. Estimation About what year was bicycle production closest to automobile production? **1965**

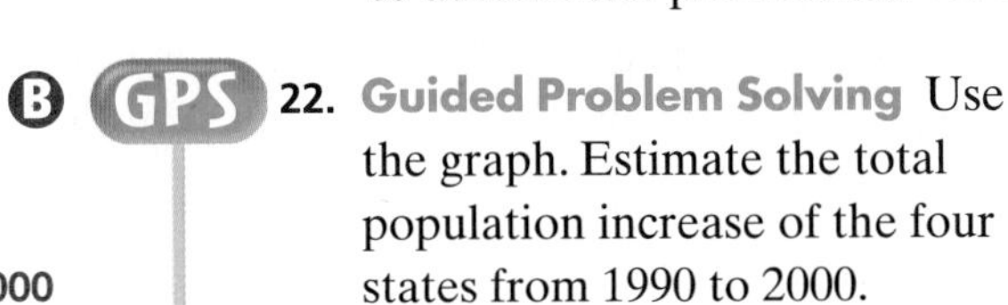

B **GPS** **22. Guided Problem Solving** Use the graph. Estimate the total population increase of the four states from 1990 to 2000.

- The 1990 total population of the four states was ■.
- The 2000 total population of the four states was ■.
- Find the population increase.

22. about 2,300,000

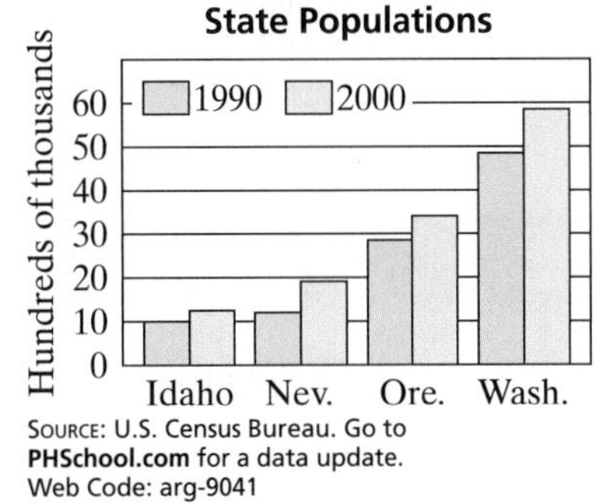

SOURCE: U.S. Census Bureau. Go to **PHSchool.com** for a data update.
Web Code: arg-9041

23. Writing in Math Describe how you would display measurements of a pet's growth over several years. What measurements would you use? How would you display them? **See margin.**

24. Multiple Choice Which graph best represents the data at the left? **A**

Sacajawea Middle School Enrollment

Grade	Girls	Boys
Grade 6	37	48
Grade 7	50	44
Grade 8	45	40

Ⓐ
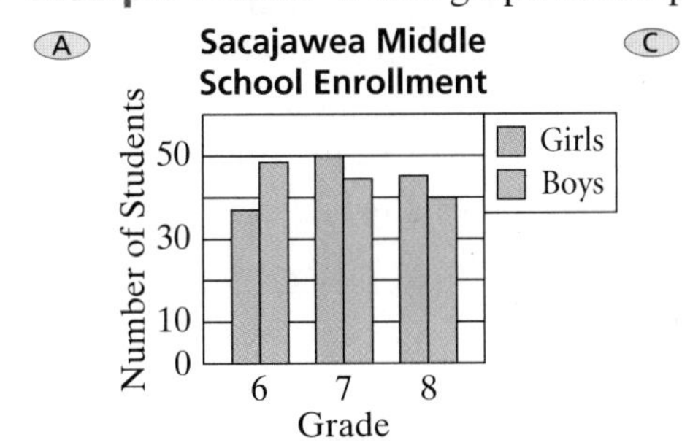

Ⓑ
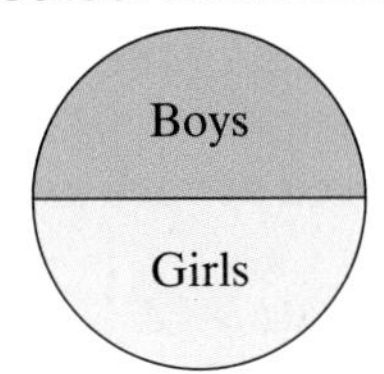

Ⓒ
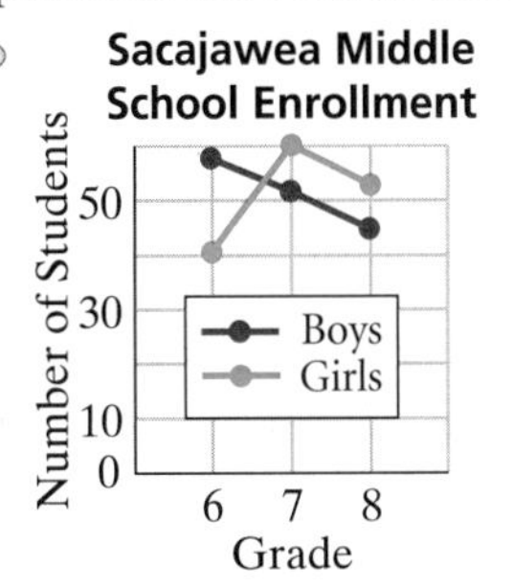

Ⓓ
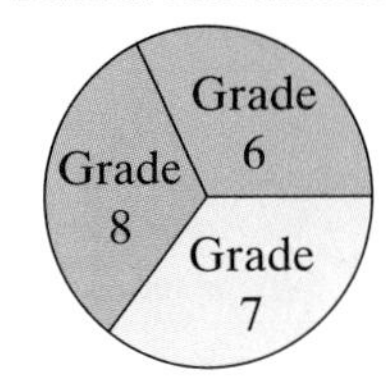

Online lesson quiz, PHSchool.com, Web Code: ara-1102

23. Answers may vary. Sample: I would use the measurement of my pet every six months and display the data on a line graph.

3. Practice

Assignment Guide

Check Your Understanding

Go over Exercises 1–3 in class before assigning the Homework Exercises.

Homework Exercises

A	Practice by Example	4–21
B	Apply Your Skills	22–26
C	Challenge	27
	Test Prep and Mixed Review	28–32

Homework Quick Check

To check students' understanding of key skills and concepts, go over Exercises 15, 21, 23, 24, and 25.

Differentiated Instruction Resources

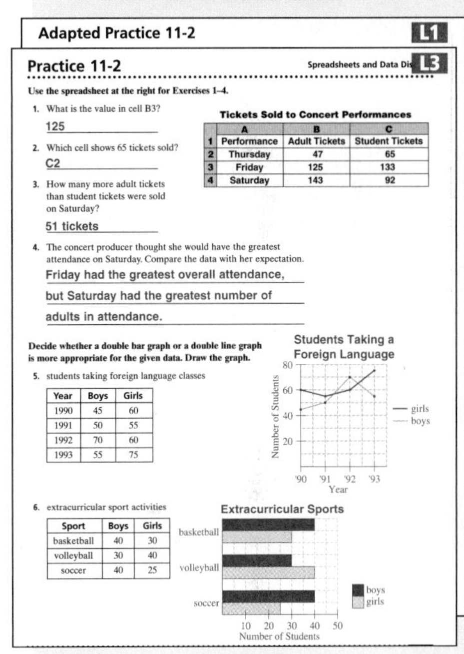
Adapted Practice 11-2 L1

Practice 11-2 Spreadsheets and Data Displays L3

Use the spreadsheet at the right for Exercises 1–4.

1. What is the value in cell B3? 125
2. Which cell shows 65 tickets sold? C2
3. How many more adult tickets than student tickets were sold on Saturday? 51 tickets
4. The concert producer thought she would have the greatest attendance on Saturday. Compare the data with her expectation. Friday had the greatest overall attendance, but Saturday had the greatest number of adults in attendance.

Tickets Sold to Concert Performances

	A	B	C
1	Performance	Adult Tickets	Student Tickets
2	Thursday	47	65
3	Friday	125	133
4	Saturday	143	92

Decide whether a double bar graph or a double line graph is more appropriate for the given data. Draw the graph.

5. students taking foreign language classes

Year	Boys	Girls
1990	45	60
1991	50	55
1992	70	60
1993	55	75

Students Taking a Foreign Language

6. extracurricular sport activities

Sport	Boys	Girls
basketball	40	30
volleyball	30	40
soccer	40	25

Extracurricular Sports

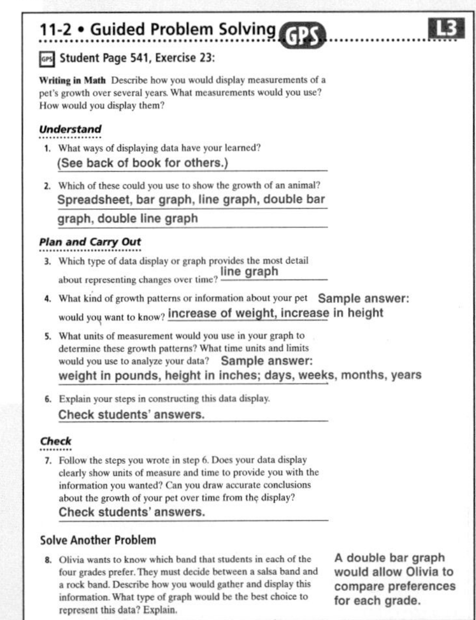
11-2 • Guided Problem Solving L3

Student Page 541, Exercise 23:

Writing in Math Describe how you would display measurements of a pet's growth over several years. What measurements would you use? How would you display them?

Understand

1. What ways of displaying data have you learned? (See back of book for others.)
2. Which of these could you use to show the growth of an animal? Spreadsheet, bar graph, line graph, double bar graph, double line graph

Plan and Carry Out

3. Which type of data display or graph provides the most detail about representing changes over time? line graph
4. What kind of growth patterns or information about your pet would you want to know? Sample answer: increase of weight, increase in height
5. What units of measurement would you use in your graph to determine these growth patterns? What time units and limits would you use to analyze your data? Sample answer: weight in pounds, height in inches; days, weeks, months, years
6. Explain your steps in constructing this data display. Check students' answers.

Check

7. Follow the steps you wrote in step 6. Does your data display clearly show units of measure and time to provide you with the information you wanted? Can you draw accurate conclusions about the growth of your pet over time from the display? Check students' answers.

Solve Another Problem

8. Olivia wants to know which band that students in each of the four grades prefer. They must decide between a salsa band and a rock band. Describe how you would gather and display this information. What type of graph would be the best choice to represent this data? Explain. A double bar graph would allow Olivia to compare preferences for each grade.

4. Assess & Reteach

Lesson Quiz

	A	B	C
1	Date	Key West	Juneau
2	9/1	71	70
3	9/2	69	69
4	9/3	70	69
5	9/4	70	71
6	9/5	71	72

1. What value is in cell B4? **70**
2. In which cell is the value 72? **C6**
3. Make a double bar graph to display the data contained in the spreadsheet.

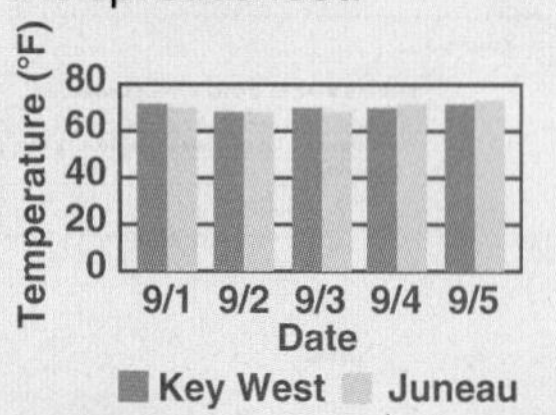

27a–b. See back of book.

30. 216 in.^{2}

31. 1,332 m^2

32. 1,536 ft^2

Enrichment 11-2 L4

Reteaching 11-2 Spreadsheets and Data L2

A **spreadsheet** is one way to organize data.

Columns are labeled A, B, C, and so on. Rows are numbered 1, 2, 3, and so on. The box where column B and row 3 meet is called **cell** B3. The *value* in cell B3 is 10.

Weekly Butter and Margarine Sales

	A	B	C
1	Day	Butter	Margarine
2	Monday	9	7
3	Tuesday	10	9
4	Wednesday	7	6
5	Thursday	9	6
6	Friday	10	8
7	Saturday	11	9

- Spreadsheet column A gives the labels for the horizontal axis.
- Spreadsheet column B gives the heights for one set of bars and one set of points.
- Spreadsheet column C gives the heights for another set of bars and another set of points.

You can use the data from this spreadsheet to make a **double bar graph.** A double bar graph compares two sets of data. The **legend,** or key, tells what kinds of data the graph is comparing.

Weekly Butter and Margarine Sales

Butter Margarine

Number of Packages: 0 2 4 6 8 10 12

Mon Tues Wed Thurs Fri Sat

Day

You can compare changes over time of two sets of data with a double line graph.

Weekly Butter and Margarine Sales

Butter

Margarine

Sales: 0 2 4 6 8 10 12

Mon Tues Wed Thurs Fri Sat

Day

Use the data in the spreadsheet for exercises 1–4.

1. What is the value in cell B4? cell C2?
   78; 39
2. In which cell is the year 2000? 1997?
   A5; A2
3. Make a double bar graph from the spreadsheet. Include a legend.
   See Answers pages.
4. Make a double line graph from the spreadsheet. Include a legend.
   See Answers pages.

Percents of Families Who Prefer Frozen and Fresh Vegetables

	A	B	C
1	Year	Frozen	Fresh
2	1997	61	39
3	1998	70	30
4	1999	78	22
5	2000	84	16

25. Double line graph; line graphs are best for showing change over time.

25. **Reasoning** You want to show grade averages in English and math every month throughout the year. Should you use a double bar or a double line graph? Explain.

26. **Data Collection** Find the number of students in several grades of your school last year. Find the number of students in those same grades this year. Record your data in a spreadsheet. **Check students' work.**

GO for Help

For help finding the intersection of two lines, go to Lesson 7-1, Example 2.

Ⓒ 27. **Challenge** Make a double line graph of the data in the spreadsheet. Data points are (x_1, y_1) and (x_2, y_2).

a. From the graph, find the missing y-values in the spreadsheet.

b. **Estimation** Estimate the coordinates of the point of intersection of the lines.

27a–b. See margin.

	A	B	C	D
1	x_1	y_1	x_2	y_2
2	0	1	0	4
3	1	?	1	5
4	2	?	2	?
5	3	10	3	?
6	4	13	4	8
7	5	16	5	9
8	6	19	6	10

Test Prep and Mixed Review

Practice

Multiple Choice

28. Which statement is NOT supported by the graph? **D**

Computer Use by Country

2001 2004

Millions of People: 0 100 200

United States Japan Germany United Kingdom

Ⓐ Computer use increased in Japan from 2001 to 2004.

Ⓑ The United States had about 6 times as many computer users as the United Kingdom in 2001.

Ⓒ There were more U.S. computer users in 2004 than in all the other listed countries combined.

Ⓓ Germany had twice as many computer users as Japan in 2001.

29. In quadrilateral $FGHJ$, the measures of $\angle F$, $\angle G$, and $\angle H$ are 95°, 105°, and 45°. What is the measure of $\angle J$? Justify your reasoning. **F**

Ⓕ 115°; the sum of the angle measures of a quadrilateral is 360°.

Ⓖ 125°; the sum of the angle measures of a quadrilateral is 360°.

Ⓗ 135°; $\angle J$ is supplementary to $\angle H$.

Ⓙ 175°; the sum of the angle measures of a quadrilateral is 420°.

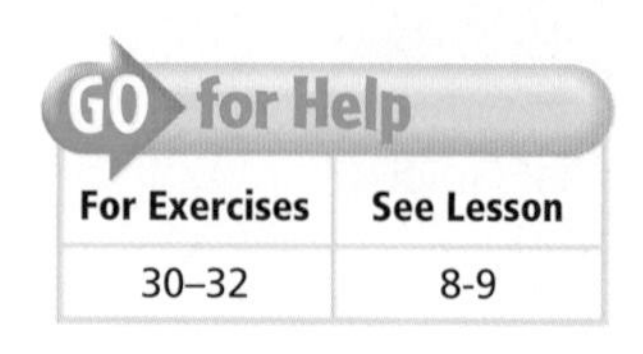

For Exercises	See Lesson
30–32	8-9

Find the surface area of each prism. **30–32. See margin.**

30. 31. 32.

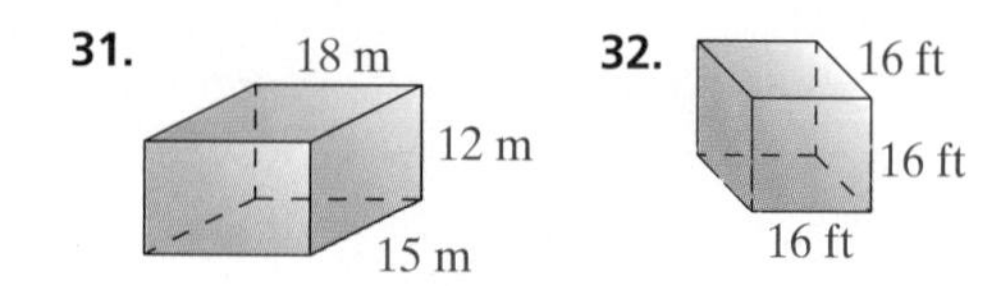

Test Prep

Resources

For additional practice with a variety of test item formats:

- Test-Taking Strategies, p. 571
- Test Prep, p. 575
- Test-Taking Strategies with Transparencies

Alternative Assessment

Have students work in pairs. Each student writes a sample spreadsheet using real-world data. Partners exchange spreadsheets and make either a double bar or double line graph using the data in their partner's spreadsheets. Students check each other's graphs and share their results with the class.

11-2b Activity Lab

Technology

Graphing Using Spreadsheets

You can use spreadsheet software to graph data.

ACTIVITY

1. Enter the data in the table below into a spreadsheet. Use three column headings: Year, Men's Winning Time, and Women's Winning Time.

Olympic Winning Times in 400 Meters Freestyle (min)

Year	1980	1984	1988	1992	1996	2000	2004
Men	3.51	3.51	3.47	3.45	3.48	3.41	3.43
Women	4.09	4.07	4.04	4.07	4.07	4.06	4.05

1–3. Check students' work.

2. Use the graphing capability of the software. Show the data as a double bar graph. Label the horizontal scale "Year" and label the vertical scale "Men's and Women's Winning Times." Title the graph "Olympic Winning Times in 400 Meters Freestyle."
3. **Data Analysis** Use the graph to compare the winning times for men and women.
4. Make a double line graph. Label the scales as you did in Exercise 2. **Check students' work.**
5. **Data Analysis** How is the double line graph similar to the double bar graph in Exercise 2? **It compares the same information.**
6. Which display, a double bar graph or a double line graph, do you prefer to use for this set of data? Explain. **See margin.**

Use the bar graph displayed at the right.

7. Use spreadsheet software to make the graph. **Check students' work.**
8. Use the software to convert the graph to a double line graph. **Check students' work.**
9. Which display, a double bar graph or a double line graph, do you prefer for this set of data? Explain. **See margin.**
10. **Algebra** Suppose 380 boys were surveyed. About how many boys chose tomatoes as their favorite burger topping? **about 228 boys**

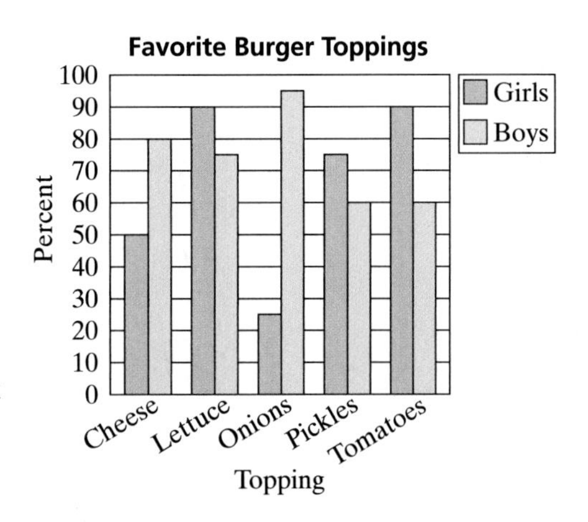

6. **A double line graph better shows the differences between the winning times of men and of women for each year.**

9. **A double bar graph better shows the difference in topping preferences between girls and boys.**

Activity Lab

Graphing Using Spreadsheets

Students use spreadsheets to graph data. This will extend the work they did with spreadsheets in Lesson 11-2.

Guided Instruction

Review with students how spreadsheets display data. Conduct steps 1–6 of the Activity as a class and ask questions, such as:

- *How would it change the appearance of the bar graph if you labeled the axes the opposite way?* **The bars would be horizontal instead of vertical.**
- *What does each line on the line graph represent?* **One line shows data for men and the other lines shows data for women.**
- *How would both graphs show a third column for children?* **as a third bar or line**

Exercises

Have students work independently on the Exercises. Circulate and check students' work. Keep in mind, students may be unfamiliar with the software program.

Differentiated Instruction

Advanced Learners L4

Have students create their own set of data and represent it in all three forms, spreadsheet, double bar graph, and double line graph.

Resources

- spreadsheet software with graphing capability

11-3

1. Plan

Objective
To represent and interpret data using stem-and-leaf plots

Examples
1. Making a Stem-and-Leaf Plot
2. Analyzing a Stem-and-Leaf Plot
3. Application: U.S. Presidential Elections

Math Understandings: p. 530C

Math Background

A stem-and-leaf plot separates each data value into a stem and a leaf. The leaf lists the values as often as they appear in the data. A *double stem-and-leaf plot* compares two sets of data in one diagram. Stem-and-leaf plots are much like line plots because each data value is represented in the plot.

More Math Background: p. 530C

Lesson Planning and Resources

See p. 530E for a list of the resources that support this lesson.

Bell Ringer Practice

Check Skills You'll Need
Use student page, transparency, or PowerPoint. For intervention, direct students to:

Mean, Median, Mode, and Range
Lesson 1-10
Extra Skills and Word Problems Practice, Ch. 1

11-3 Stem-and-Leaf Plots

Check Skills You'll Need

1. **Vocabulary Review** What is a *median*? **1–4. See below.**

Find the median of each set of data.

2. 24, 42, 51, 25, 63
3. 5.8, 6.9, 7.4, 3.9, 6.4
4. 110, 120, 130, 125

Check Skills You'll Need

1. the middle value of a set of data
2. 42
3. 6.4
4. 122.5

What You'll Learn

To represent and interpret data using stem-and-leaf plots

New Vocabulary stem-and-leaf plot

Why Learn This?

You can use a stem-and-leaf plot to compare measurements such as the different heights of teammates.

A **stem-and-leaf plot** is a graph that uses the digits of each number to show the data distribution. Each data value is broken into a "stem" (digit or digits on the left) and a "leaf" (digit or digits on the right).

EXAMPLE Making a Stem-and-Leaf Plot

1 Sports The list gives the height in inches of each player on the San Antonio Spurs basketball team during a season. Make a stem-and-leaf plot of the data: 77 74 81 83 78 84 79 82 75 82 79 79.

Step 1 Write the stems. All the data values are in the 70s and 80s, so use 7 to represent 70 and 8 to represent 80. Draw a vertical line to the right of the stems.

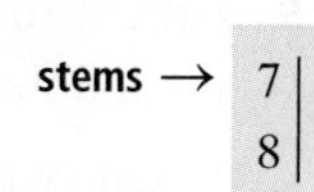

Step 2 Write the leaves. For these data, the leaves are the values in the ones place.

7	7 4 8 9 5 9 9	← leaves
8	1 3 4 2 2	

Step 3 Make the stem-and-leaf plot with the leaves in order from least to greatest. Add a key to explain the leaves. Add a title.

San Antonio Spurs Heights of Players

7	4 5 7 8 9 9 9
8	1 2 2 3 4

Key → **Key:** 7 | 4 means 74 in.

Quick Check

1. Make a stem-and-leaf plot of the wind speeds (in miles per hour) recorded during a storm: 9, 14, 30, 16, 18, 25, 29, 25, 38, 34, 33. **See back of book.**

Differentiated Instruction Solutions for All Learners

Special Needs L1
Provide a copy of Example 1 to the students. Have them circle the digits that will represent the "stem" in the stem-and-leaf plot, and underline the digits that will be the "leaves."
learning style: visual

Below Level L2
Give students slips of paper with each player's height on them. Have them cut the slips in the appropriate place, discard all but one of the first halves and align the second halves appropriately to make a stem-and-leaf plot.
learning style: tactile

You can use a stem-and-leaf plot to compare two sets of data. From a stem, the leaves increase in value outward in each direction.

EXAMPLE Analyzing a Stem-and-Leaf Plot

2 **Weather** Use the stem-and-leaf plot below. How many times did each city receive 3.5 in. of precipitation in a month?

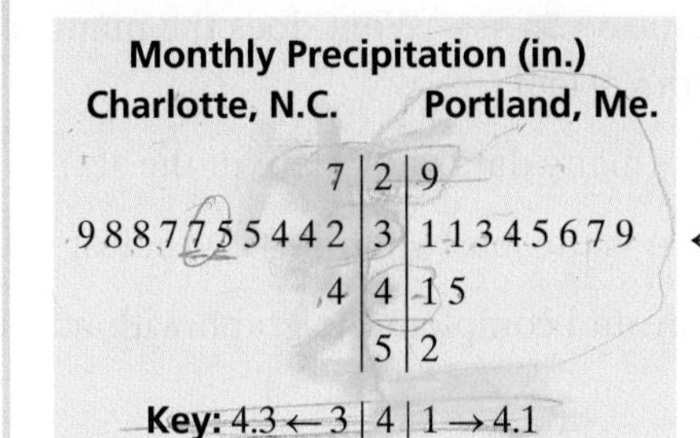

Monthly Precipitation (in.)

Charlotte, N.C.		Portland, Me.
7	2	9
9 8 8 7 7 5 5 4 4 2	3	1 1 3 4 5 6 7 9
4	4	1 5
	5	2

Key: 4.3 ← 3 | 4 | 1 → 4.1

← **Find the values with a stem of 3 and leaves of 5.**

Charlotte twice received 3.5 in. of precipitation in a month. Portland received 3.5 in. of precipitation in a month only once.

2. **Charlote, N.C.; Charlotte had a median of 3.6 in., while Portland had a median of 3.55 in.**

Quick Check

2. Which city had a higher median monthly precipitation? Explain. **See left.**

EXAMPLE Application: U.S. Presidential Elections

3 **Multiple Choice** The plot below shows the number of electoral votes for president each state (plus the District of Columbia) has.

Number of Electoral Votes by State

Stem	Leaves
0	3 3 3 3 3 3 3 3 4 4 4 4 4 5 5 5 5 5 6 6 6 7 7 7 7 8 8 9 9 9
1	0 0 0 0 1 1 1 1 2 3 5 5 5 7
2	0 1 1 7
3	1 4
4	
5	5

Key: 0 | 3 means 3 votes

Test Prep Tip
List the data from the least to the greatest value. Then find the mean, median, mode, or range.

Which statement is best supported by the stem-and-leaf plot?

(A) The mean is 2.5 votes.
(B) The median is 8 votes.
(C) The mode is 8 votes.
(D) The range is 55 votes.

The middle value is 8, so the median is 8. The correct answer is choice B.

Quick Check

3. Find the correct values for mean, mode, and range in Example 3. **10.5; 3; 52**

2. Teach

Activity Lab

Use before the lesson.

All in One Teaching Resources

Activity Lab 11-3: Stem-and-Leaf Plots

Guided Instruction

Example 2

Point out that the double stem-and-leaf plot uses one set of stems for both sets of leaves.

PowerPoint

Additional Examples

1 Use the table below. Draw a stem-and-leaf plot of the data.

High Temperatures (°F)

Week 1	81	73	67	81	77	79	73
Week 2	80	74	61	66	70	67	73

6 | 1 6 7 7
7 | 0 3 3 3 4 7 9
8 | 0 1 1

6 | 1 means 61

2 Compare the two sets of data in the stem-and-leaf plot.

Waiting for the Doctor

Monday		Friday
2 1	0	5 6
0	1	1 1 4
6 1	2	3 4 6 6
3	3	5 5
9	4	7 8 8
1 min ← 1	0	5 → 5 min

a. Which day had the longest wait time? **Monday**

b. How long was the longest wait? **49 min**

3 Does the data above support this statement? The average wait time for the doctor is about 24 minutes. **yes**

All in One Teaching Resources

- Daily Notetaking Guide 11-3 L3
- Adapted Notetaking 11-3 L1

Closure

- *What is a stem-and-leaf plot?* **a graph that uses the digits of each number to show the shape of the data**

Advanced Learners L4
Create a stem-and-leaf plot for the prices: $1.56, $4.22, $3.78, $2.80, $4.50, $7.25, $6.10, $3.40, $5.31, $3.60 **See back of book.**

learning style: visual

English Language Learners ELL
Ensure that students construct their stem-and-leaf plots with the understanding that the leaves will be the ones digit in the whole number examples, but for the decimal examples, the ones digits will be the stems. Ask: *What would the stem be for a data point of 112?* **(11)**

learning style: verbal

3. Practice

Assignment Guide

Check Your Understanding
Go over Exercises 1–5 in class before assigning the Homework Exercises.

Homework Exercises

A Practice by Example	6–13
B Apply Your Skills	14–16
C Challenge	17
Test Prep and Mixed Review	18–20

Homework Quick Check
To check students' understanding of key skills and concepts, go over Exercises 7, 9, 12, 15, and 16.

Differentiated Instruction Resources

Adapted Practice 11-3 L1

Practice 11-3 Stem-and-L L3

The stem-and-leaf plot at the right shows the number of baskets scored by one of ten intramural teams last season. Use it for Exercises 1–4.

5	2 6 9
6	0 4 6
7	1 5
8	4 8

8|4 means 84

1. How many data items are there? 10
2. What is the least measurement given? 52
3. What is the greatest measurement given? 88
4. In how many games did the team score less than 70 baskets? 6 games
5. Draw a stem-and-leaf plot for the set of data.
   science test scores: 83 73 78 60 85 92 95 85 99 68

6	0 8
7	3 8
8	3 5 5
9	2 5 9

6|0 means 60.

Use the stem-and-leaf plot from Exercise 5 to answer Exercises 6–8.

6. Find the mode of the test scores. 85
7. How many data items are in the set? 10
8. Find the mean of the test scores. 81.8
9. You have a spreadsheet showing how many DVDs people bought in 1998, 2000, and 2002. The spreadsheet also shows how many videocassettes were bought in 1998, 2000, and 2002. Which is the most effective data display choice: a double line graph or a stem-and-leaf plot? Explain your answer.
   The most effective data display choice is the double line graph because it shows that the number of DVDs bought is consistently greater than the number of video cassettes bought.

11-3 • Guided Problem Solving GPS L3

GPS Student Page 546, Exercise 8:

Height How many males are 65 in. tall?

Student Height (in.)

Female		Male
7 4 3 1 0 0	5	6 7
8 5 4 1 0	6	2 3 5 5 6 7 9
0	7	1 2 3 4 6

Key: 61 ← 1 | 6 | 3 → 63

Understand

1. What data does the stem-and-leaf plot above display? The heights of the female and male students
2. What is the question asking you to find? How many males are 65 in. tall?

Plan and Carry Out

3. Circle the part of the stem-and-leaf plot that pertains to males.
4. According to the key, what place value do the numbers on the right have? ones place
5. According to the key, what place value do the numbers between the stems have? tens place
6. Underline the row that has a 6 in the tens place.
7. Circle the fives in the row you underlined in step 6.
8. Count the number of fives you circled in the male half of the stem-and-leaf plot. How many males are 65 in. tall? 2

Check

9. Write out the heights of all of the males in the stem-and-leaf plot. Count the number of males who are 65 in. tall.
   56 in.; 57 in.; 62 in.; 63 in.; 65 in.; 65 in.; 66 in.; 67 in.; 69 in.; 71 in.; 72 in.; 73 in.; 74 in.; 76 in.; two

Solve Another Problem

10. In the stem-and-leaf plot at the right, how many female students scored 91 on their test? 4

Test Scores

Female		Male
0	5	
7	6	1 3
9 4	7	6 8
9 6 1 1	8	2 8 8 9 9
3 1 1 1 1 0	9	2 2 9

Key: 79 ← 9 | 7 | 3 → 73

Check Your Understanding

1. **Vocabulary** In a stem-and-leaf plot, what is the difference between a stem and a leaf? **The stem is the digit or digits on the left and the leaf is the digit or digits on the right.**

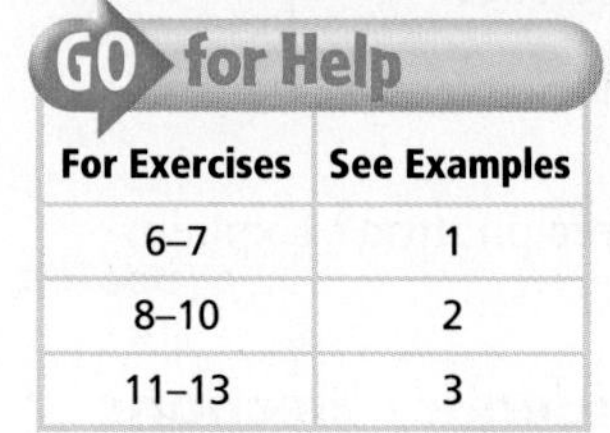

4	3 6 7
5	1 2
6	1 7
7	1 8
8	2 6 8

Key: 8 | 2 means 82

Use the stem-and-leaf plot at the left.

2. **Number Sense** What does the number 6 represent when it is a stem? A leaf? **60; 6**
3. How many data values are in the set? **12**
4. What is the least value? The greatest value? **43; 88**
5. Copy and complete the graph with new data items 60 and 72. **See margin.**

Homework Exercises

For more exercises, see Extra Skills and Word Problems.

GO for Help

For Exercises	See Examples
6–7	1
8–10	2
11–13	3

A Draw a stem-and-leaf plot for each set of data. 6-7. See margin.

6. sales of twelve companies (millions of dollars):
   1.3 1.4 2.3 1.4 2.4 2.5 3.9 1.4 1.3 2.5 3.6 1.4
7. high temperatures (°F) in a desert:
   99 113 112 98 100 103 101 111 104 108 109 112 113 118

Height Use the stem-and-leaf plot at the right to answer each question.

Student Height (in.)

Female		Male
7 4 3 1 0 0	5	6 7
8 5 4 1 0	6	2 3 5 5 6 7 9
0	7	1 2 3 4 6

Key: 61 ← 1 | 6 | 3 → 63

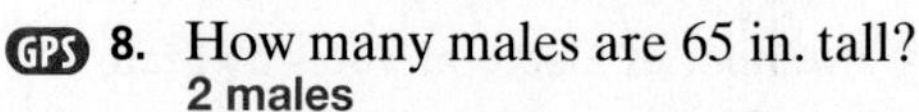

GPS 8. How many males are 65 in. tall? **2 males**

9. What is the height of the shortest male? The shortest female? **56 in.; 50 in.**
10. What is the tallest female's height? **70 in.**

Is the statement supported by the graph above? Explain.

11. The median height for females is 57 in. **No; the median height for females is 58.5 in.**
12. The mode of all heights is 65 in. **Yes; 65 in. occurs the largest number of times.**
13. The range of heights for males is 20 in. **Yes; 76 − 56 = 20.**

GPS B

14. **Guided Problem Solving** Find the mean, median, and range of the data in the graph.
    - What is the place value of the leaves?
    - What is the order of the data from least to greatest?
    - How many data items are there?

    18, 18.4, 3.8

Distances Walked by Fundraisers (km)

16	1 1 2 3 5 5
17	0 2 2
18	4 5 8 9
19	3 6 7 9 9 9

Key: 19 | 3 means 19.3

5–7. See back of book.

SOURCE: National Association of Counties

15. **Multiple Choice** Which graph best represents the data? **C**

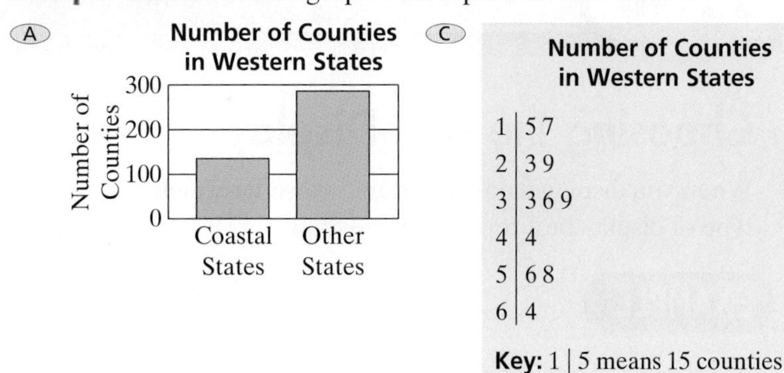

Ⓒ **Number of Counties in Western States**

1	5 7
2	3 9
3	3 6 9
4	4
5	6 8
6	4

Key: 1 | 5 means 15 counties

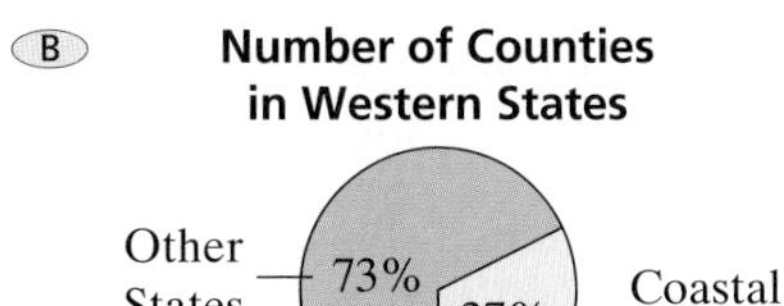

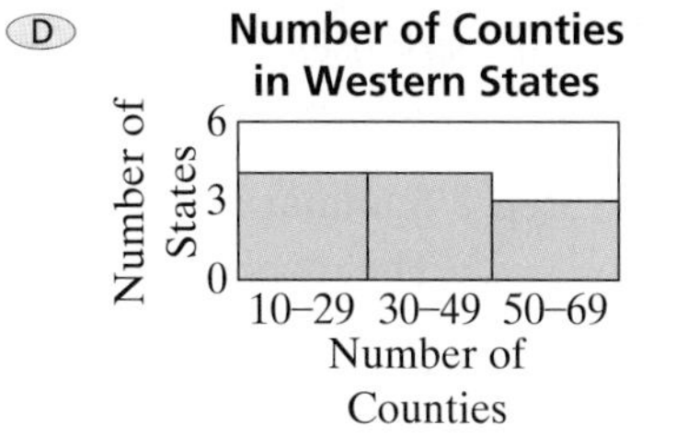

Homework Video Tutor
Visit: PHSchool.com
Web Code: are-1103

16. a. **Data Collection** Measure the width of the hands of at least 10 people. Use metric units. Make a stem-and-leaf plot of the data. **16a–b. Check students' work.**
    b. **Writing in Math** Write two true statements about the data.

17. **Challenge** The median of a data set is 48. Find the value of the missing data item: 23 34 42 ■ 62 67. **54**

Test Prep and Mixed Review

Practice

Multiple Choice

18. The amount of iron in high-protein foods is shown. Which statement is NOT supported by the data? **C**
    Ⓐ There are three foods with 2.1 mg of iron or more.
    Ⓑ The modes are both 0.7 and 2.6 mg.
    Ⓒ The mean and median are the same.
    Ⓓ The median is 1.45 mg.

Iron in Three Ounces of High-Protein Foods (mg)

0	7 7 9
1	1 4 5 6
2	1 6 6

Key: 0 | 7 means 0.7

19. A store is having a 30%-off sale. Hunter opens a store charge card and gets an additional 20% off his first purchase. What is the final cost of an item that was priced originally at $25? **H**
    Ⓕ $10.00 Ⓖ $12.50 Ⓗ $14.00 Ⓙ $17.50

GO for Help

For Exercise	See Lesson
20	10-7

20. A figure lies entirely in Quadrant I of a coordinate plane. In which quadrant will the figure lie after a 270° rotation about (0, 0)? **IV**

PowerPoint

Lesson Quiz

1. Make a stem-and-leaf plot for this set of data.
   points scored by winning teams in the first fifteen Super Bowls:
   35 33 16 23 16 24 14 24 16 21 32 27 35 31 27

1	4 6 6 6
2	1 3 4 4 7 7
3	1 2 3 5 5

2. Using the stem-and-leaf plot, find the mean, median, and mode scores. **mean 24.9, median 24, mode 16**

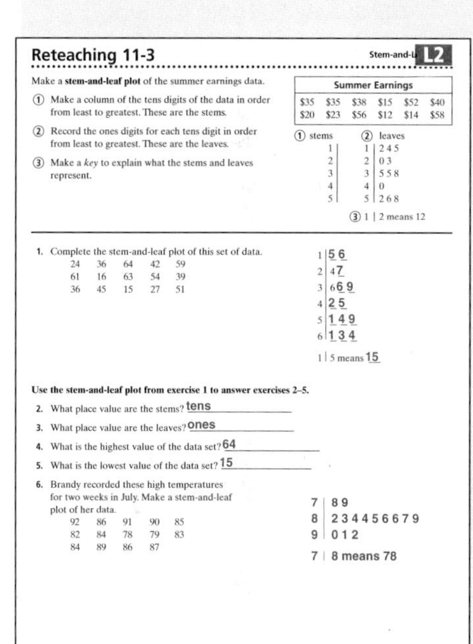
Reteaching 11-3 Stem-and-L L2

Make a **stem-and-leaf plot** of the summer earnings data.

① Make a column of the tens digits of the data in order from least to greatest. These are the stems.

② Record the ones digits for each tens digit in order from least to greatest. These are the leaves.

③ Make a *key* to explain what the stems and leaves represent.

Summer Earnings					
$35	$35	$38	$15	$52	$40
$20	$23	$56	$12	$14	$58

① stems: 1, 2, 3, 4, 5

② leaves:

1	2 4 5
2	0 3
3	5 5 8
4	0
5	2 6 8

③ 1 | 2 means 12

1. Complete the stem-and-leaf plot of this set of data.
   24 36 64 42 59
   61 16 63 54 39
   36 45 15 27 51

1	5 6
2	4 7
3	6 6 9
4	2 5
5	1 4 9
6	1 3 4

1 | 5 means 15

Use the stem-and-leaf plot from exercise 1 to answer exercises 2–5.

2. What place value are the stems? tens
3. What place value are the leaves? ones
4. What is the highest value of the data set? 64
5. What is the lowest value of the data set? 15
6. Brandy recorded these high temperatures for two weeks in July. Make a stem-and-leaf plot of her data.
   92 86 91 90 85
   82 84 78 79 83
   84 89 86 87

7	8 9
8	2 3 4 4 5 6 6 7 9
9	0 1 2

7 | 8 means 78

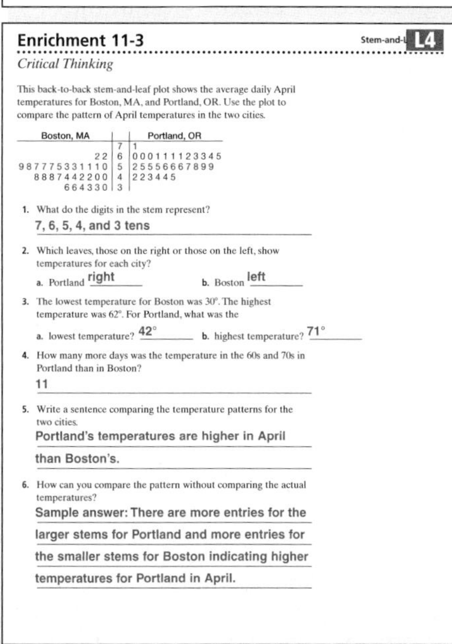
Enrichment 11-3 Stem-and-L L4

Critical Thinking

This back-to-back stem-and-leaf plot shows the average daily April temperatures for Boston, MA, and Portland, OR. Use the plot to compare the pattern of April temperatures in the two cities.

Boston, MA		Portland, OR
	7	1
2 2	6	0 0 0 1 1 1 1 2 3 3 4 5
9 8 7 7 7 5 3 3 1 1 1 0	5	2 5 5 5 6 6 6 7 8 9 9
8 8 8 7 4 4 2 2 0 0	4	2 2 3 4 4 5
6 6 4 3 3 0	3	

1. What do the digits in the stem represent?
   7, 6, 5, 4, and 3 tens
2. Which leaves, those on the right or those on the left, show temperatures for each city?
   a. Portland right b. Boston left
3. The lowest temperature for Boston was 30°. The highest temperature was 62°. For Portland, what was the
   a. lowest temperature? 42° b. highest temperature? 71°
4. How many more days was the temperature in the 60s and 70s in Portland than in Boston?
   11
5. Write a sentence comparing the temperature patterns for the two cities.
   Portland's temperatures are higher in April than Boston's.
6. How can you compare the pattern without comparing the actual temperatures?
   Sample answer: There are more entries for the larger stems for Portland and more entries for the smaller stems for Boston indicating higher temperatures for Portland in April.

Alternative Assessment

Have students work in pairs to find data in almanacs, textbooks, or encyclopedias that can be displayed in a stem-and-leaf plot. Have pairs share their displays with the class.

Test Prep

Resources

For additional practice with a variety of test item formats:
- Test-Taking Strategies, p. 571
- Test Prep, p. 575
- Test-Taking Strategies with Transparencies

Activity Lab

Choosing the Best Display

Students learn to choose the most appropriate method of displaying sets of data among all the methods they have learned.

Guided Instruction

Review all the methods of data display that students have learned so far: line plot, Venn diagram, stem-and-leaf plot, double line graph, circle graph, double bar graph. As you go over the Example with the class, ask:

- *Why should a Venn diagram be eliminated for the results of the survey?* **People can only choose one use for the park so there will never be any overlapping responses.**
- *What situation could you represent with a double line graph?* **Sample: people who bike in the park and people who bring dogs to the park over a span of time**
- *If you used a circle graph to represent the data, what would it look like?* **The circle would be split into four sections, each proportional to a number of responses.**

Exercises

Review the Exercises as a class, having volunteers give each answer and the reasoning for their choice. Have students evaluate the effectiveness of the different representations to communicate the data.

Differentiated Instruction

Advanced Learners L4

Have students create a set of data for one of the situations in the Exercises and represent the data with the appropriate display.

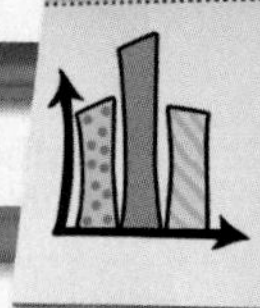

Choosing the Best Display

When you display data, you should consider which type of display best represents the data.

EXAMPLE

A city council conducts a survey to decide how to develop a new public park. The council asks people to choose one of four park uses, as shown in the table. Choose the best display to represent the data. Explain your choice.

City Park Use Survey Results

Use	18 and Under	Over 18
Tennis	72	86
Basketball	114	95
Skate Park	173	57
Garden	48	139

Step 1 Summarize the purpose of the data display.
The display must compare data for two age groups.

Step 2 Narrow down your options.
There are too many responses to use a line plot. Each person chose one use for the park. Since there is no overlap, you can eliminate a Venn diagram. You are not showing the distribution of data over a range, so you can eliminate a stem-and-leaf plot. There is no change over time, so you can eliminate a line graph.

Step 3 Choose one of the remaining displays.
Both a circle graph and a double bar graph can compare parts to wholes. In this situation, the table compares two sets of data across the same four categories, so the best option is a double bar graph.

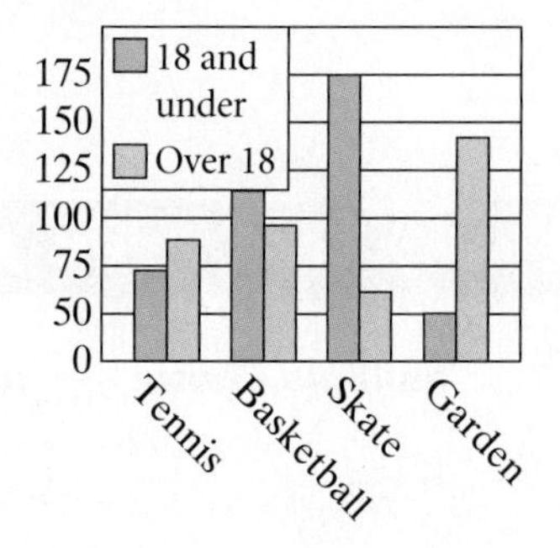

Exercises

Reasoning Match each data set at the left with the best type of data display at the right. Explain your choices. 1–6. See margin.

1. the number of students who hold part-time jobs, tutor, play sports, or do a combination of all three activities
2. favorite pizza topping, by percent of students in a class
3. the popularity of different bikes among boys and among girls
4. changes in desktop and laptop computer prices over time
5. the height in centimeters of each student in a class
6. the number of letters in the first names of students in your class

A. line plot
B. double line graph
C. double bar graph
D. stem-and-leaf plot
E. circle graph
F. Venn diagram

1. F; there is an overlap in the categories.
2. E; the data show parts of a whole.
3. C; the data compare categories of two groups.
4. B; a line graph shows change in data over time.
5. D; stem-and-leaf plots best show distribution of data.
6. A; a line plot compares frequency of the data.

Checkpoint Quiz 1

Lessons 11-1 through 11-3

1. Make a frequency table and a line plot of the data below.
   5 7 8 3 5 4 6 7 8 9 1 2 5 4 2 1 3 **See margin.**

2. Graph the data below. Explain why you chose the graph you drew.
   See margin.

Art Show Attendance

Day	Sun.	Mon.	Tue.	Wed.	Thur.	Fri.	Sat.
Number of Adults	54	29	22	28	12	15	49
Number of Children	32	21	16	20	8	10	36

Give the content or value (millions of dollars) of each cell in the spreadsheet.

3. B3 **$6,527 million**
4. A1 **Industry**
5. A3 **Health**
6. B1 **Revenue (millions)**

Tell which cell has the given word or value in the spreadsheet.

7. 6,527 **B3**
8. Health **A3**
9. 983 **B2**

U.S. Service Industries Revenue 1998–2003

	A	B
1	Industry	Revenue (millions)
2	Trucking	$983
3	Health	$6,527
4	Broadcasting	$2,734

10. Make a stem-and-leaf plot of these 18-hole golf scores.
    93 120 112 89 87 124 117 95 117 121 113 95
    See margin.

11-4a Activity Lab

Data Collection

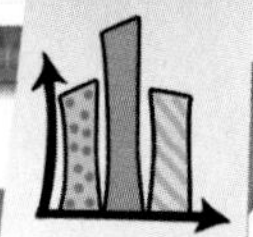

Writing Survey Questions

1–5. Check students' work.

1. Choose a topic about which you would like people's opinions. Write five survey questions that do not influence the answer.
2. You want to survey people who will give you all kinds of responses. What group of people should you survey?
3. How many people should you survey?
4. **Data Collection** Conduct the survey. Display the results of your survey in a graph.
5. Consider the results of your survey. Do you think your questions influenced the answers? If so, how would you change the questions?

1–2. See back of book.
10. See back of book.

Checkpoint Quiz

Use this Checkpoint Quiz to check students' understanding of the skills and concepts of Lessons 11-1 through 11-3.

Resources

- All-in-One Teaching Resources Checkpoint Quiz 1
- ExamView CD-ROM
- Success Tracker™ Online Intervention

Writing Survey Questions

Students learn to tailor surveys to gain meaningful results. This will prepare them to work with random samples and surveys in Lesson 11-4.

Guided Instruction

Explain to students that surveys must be written in a certain way to get information. Ask:

- *How can you limit the size of your graph?* **Use a multiple-choice question rather than an open-ended question.**
- *Why is it important to use neutral questions?* **so you do not influence the answer**

Activity

Have students work in pairs on the Activity. You may wish to extend it to a homework assignment so that students have a chance to survey people outside their class.

Differentiated Instruction

Below Level L2

See limits for the number of responses and number of people to survey to make conducting the survey simpler.

Resources

- Activity Lab 11-4: Decision Making

11-4

1. Plan

Objective
To identify a random sample and to write a survey question

Examples
1 Identifying a Random Sample
2 Identifying Biased Questions

Math Understandings: p. 530D

Professional Development

Math Background

Surveys are used to gather information about certain topics. A survey focuses on a certain group, or *population*. Since it is often difficult to survey each member of the population, a *random sample* is chosen. A random sample gives each member of the population an equal chance of being selected. If the sample is asked a *biased question*, or if the sample is not selected at random, the results of a survey may be misleading.

More Math Background: p. 530D

Lesson Planning and Resources

See p. 530E for a list of the resources that support this lesson.

Bell Ringer Practice

Check Skills You'll Need
Use student page, transparency, or PowerPoint. For intervention, direct students to:
Understanding Percents
Lesson 6-1
Extra Skills and Word Problems Practice, Ch. 6

11-4 Random Samples and Surveys

Check Skills You'll Need

1. **Vocabulary Review** A _?_ is a ratio that compares a number to 100. **percent**

Write each ratio as a percent.

2. 4 out of 5 **80%**
3. 10 out of 40 **25%**
4. 14 out of 200 **7%**

for Help
Lesson 6-1

What You'll Learn

To identify a random sample and to write a survey question

New Vocabulary population, sample, random sample, biased question

Why Learn This?

You can use a survey to gather information from a group of people. Pollsters use surveys to understand group preferences.

A **population** is a group of objects or people. The population of an election is all the people who vote in that election. It is not practical to ask all the voters how they expect to vote. Pollsters select a **sample,** or a part of the population. A sample is called a **random sample** when each member of a population has the same chance of being selected.

EXAMPLE Identifying a Random Sample

1 You survey customers at a mall. You want to know which stores they shop at the most. Which sample is more likely to be random? Explain.

a. You survey shoppers in a computer store.

Customers that shop in a particular store may not represent all the shoppers in the entire mall. This sample is not random.

b. You walk around the mall and survey shoppers.

By walking around, you give everyone in the mall the same chance to be surveyed. This sample is more likely to be random.

Quick Check

1. You survey a store's customers. You ask why they chose the store. Which sample is more likely to be random? Explain.
   a. You survey 20 people at the entrance from 5:00 P.M. to 8:00 P.M.
   b. You survey 20 people at the entrance throughout the day.

1a–b. See left.

1a. Answers may vary. Sample: Less likely to be random; you may get only people shopping after work.
b. More likely to be random; you won't just get people shopping after work.

Differentiated Instruction Solutions for All Learners

Special Needs L1
Ask student to work in pairs. Have them generate lists of adjectives that might make questions biased or unfair in the Homework Exercises. For example, they might list words such as *unhealthy* and *nutritious,* or *harsh* and *inspiring.*
learning style: verbal

Below Level L2
Have students make a list of words and phrases that may indicate a biased question, such as *all* or *never*. Students should also include double negatives.
learning style: verbal

Vocabulary Tip

Bias means "slant." A biased question slants the answers in one direction.

When you conduct a survey, ask questions that do not influence the answer. A **biased question** is a question that makes an unjustified assumption or makes some answers appear better than others.

EXAMPLE Identifying Biased Questions

2 **Music** Is each question *biased* or *fair?* Explain.

a. "Do you think that soothing classical music is more pleasing than the loud, obnoxious pop music that teenagers listen to?"

This question is biased against pop music. It implies that all pop music is loud and that only teenagers listen to it. The adjectives "soothing" and "obnoxious" may also influence responses.

b. "Which do you think is the most common age group of people who like pop music?"

This question is fair. It does not assume that listeners of pop music fall into only one age group.

c. "Do you prefer classical music or pop music?"

This question is fair. It does not make any assumptions about classical music, pop music, or people.

Quick Check

2. Is each question *biased* or *fair?* Explain. **2a–b. See left.**

a. Do you prefer greasy meat or healthy vegetables on your pizza?

b. Which pizza topping do you like best?

2a. Biased; the question implies that meat is greasy and vegetables are healthy.

b. Fair; the question makes no assumptions about pizza toppings.

Check Your Understanding

Vocabulary **Match each statement with the appropriate term.**

1. a group of objects or people **C**

2. makes some answers appear better **A**

3. gives members of a group the same chance to be selected **B**

A. biased question
B. random sample
C. population

You want to determine the favorite spectator sport of seventh-graders at your school. You ask the first 20 seventh-graders who arrive at a soccer game, "Is soccer your favorite sport to watch?"

4. What was the population of your survey? What was the sample?

5. The survey (was, was not) random. **was not**

6. You used a (biased, fair) question. **biased**

4. seventh-graders; seventh-graders attending a soccer game

Advanced Learners L4

Differences exist within every population. How do you account for differences that do not matter in a survey? **randomly select the sample for differences that affect responses**

learning style: verbal

English Language Learners ELL

Help students understand the relationship between a *pollster,* a *poll,* and a *survey.* Make sure they understand that the word *biased* is used as the opposite of *fair.*

learning style: verbal

2. Teach

Activity Lab

Use before the lesson.
Student Edition Activity Lab, Data Collection 11-4a, Writing Survey Questions, p. 549

All in One Teaching Resources

Activity Lab 11-4: Decision Making

Guided Instruction

Example 2

Point out to students that Example 2c is a fair way to ask the question in Example 2a. Ask volunteers for other examples of biased questions and a fair way to ask them.

PowerPoint

Additional Examples

1 Suppose you survey students in your school about their snacking habits. Would you get a random sample if you questioned different English classes? Explain. **probably yes; accept all reasonable answers.**

2 Is each question *biased* or *fair*? Explain.

a. Which is a brighter color, pink or green? **fair; the question presents the choices equally**

b. Is an electric pink shirt brighter than a green shirt? **biased; it makes pink sound brighter, influencing response**

All in One Teaching Resources

- Daily Notetaking Guide 11-4 L3
- Adapted Notetaking 11-4 L1

Closure

- *How is a random sample taken?* **by making sure that each member of a population has an equal chance of being surveyed**
- *How do you write a fair survey question?* **Write a question that does not influence the answer by making one answer appear preferable.**

3. Practice

Assignment Guide

Check Your Understanding
Go over Exercises 1–6 in class before assigning the Homework Exercises.

Homework Exercises

A	Practice by Example	7–15
B	Apply Your Skills	16–21
C	Challenge	22
	Test Prep and Mixed Review	23–27

Homework Quick Check
To check students' understanding of key skills and concepts, go over Exercises 9, 13, 17, 19, and 20.

Differentiated Instruction Resources

Adapted Practice 11-4 L1

Practice 11-4 Random Samples and Surveys L3

You want to survey students in your school about their exercise habits. Tell whether the situations described in exercises 1 and 2 are likely to give a random sample of the population. Explain.

1. You select every tenth student on an alphabetical list of the students in your school. You survey the selected students in their first-period classes.
   random sample; the selected students represent the population.
2. At lunchtime you stand by a vending machine. You survey every student who buys something from the vending machine.
   Not a random sample; students that use the vending machine may not represent all types of students.

Is each question *biased* or *fair*? Rewrite biased questions as fair questions.

3. Do you think bike helmets should be mandatory for all bike riders?
   Fair
4. Do you prefer the natural beauty of hardwood floors in your home?
   Biased; do you prefer hardwood floors in your home?
5. Do you exercise regularly?
   Fair
6. Do you eat at least the recommended number of servings of fruits and vegetables to ensure a healthy and long life?
   Biased; how many servings of fruits and vegetables do you eat?
7. Do you prefer the look and feel of thick lush carpeting in your living room?
   Biased; do you prefer thick carpeting?
8. Do you take a daily multiple vitamin to supplement your diet?
   Fair
9. Do you read the newspaper to be informed about world events?
   Fair
10. Do you feel that the TV news is a sensational portrayal of life's problems?
    Biased; does TV news portray life accurately?

11-4 • Guided Problem Solving GPS L3

GPS Student Page 553, Exercise 19:

Parks Suppose you are gathering information about visitors to Yosemite National Park. You survey every tenth person entering the park. Would you get a random sample of visitors? Explain.

Understand

1. What is a random sample?
   In a random sample, each member of a population has an equal chance of being selected.
2. What are you being asked to do?
   Determine if this sampling is random and explain the answer.

Plan and Carry Out

3. What is the population you are surveying?
   people who visit Yosemite National Park
4. Does every person in the population have an equal chance of being surveyed? yes
5. Is this a random sample? Why or why not? yes; Each person has an equal chance of being surveyed.

Check

6. How else could you randomly survey the people at Yosemite National Park?
   Sample answer: Survey visitors as they leave the park.

Solve Another Problem

7. You want to survey the people at the local pool about the food served in the snack shack. You decide to walk around the kiddy pool and survey parents. Is this a random sample? Why or why not?
   No, this is not a random sample. People who are not parents might eat at the snack shack.

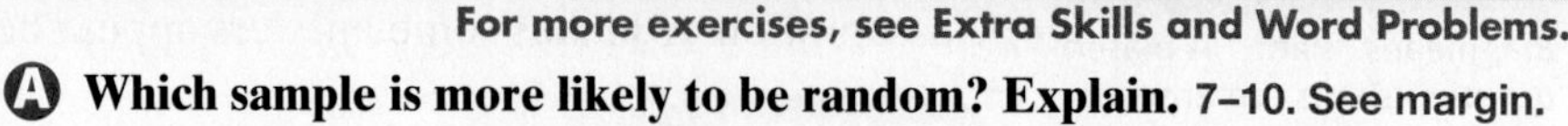

For more exercises, see Extra Skills and Word Problems.

GO for Help

For Exercises	See Examples
7–10	1
11–15	2

A **Which sample is more likely to be random? Explain.** 7–10. See margin.

7. You want to survey teens about their snacking habits.
   a. You ask people at a party to name their favorite snack.
   b. You ask several teens entering a grocery store.
8. You want to know the most popular book among all the students at your school.
   a. You ask students from different grades at your school.
   b. You ask a group of your friends.
9. You want to know which baseball team is regarded as the best.
   a. You ask everyone seated in your section of the ballpark.
   b. You ask several visitors at a tourist attraction.
10. You want to survey seventh-grade students about computer use.
    a. You ask seventh-graders leaving the cafeteria after lunch.
    b. You ask seventh-graders entering a library on Friday night.

Is each question *biased* or *fair*? Explain.

11. Do you prefer to exercise or to watch television?
    Fair; the question makes no assumptions.
12. Do you prefer rock music or jazz?
    Fair; the question makes no assumptions.
13. Do you prefer harsh rock music or inspiring jazz?
    Biased; the question uses the terms *harsh* and *inspiring*.
14. Do you prefer unhealthy snacks or nutritious snacks?
    Biased; the question uses the terms *nutritious* and *unhealthy*.
15. What type of snack do you prefer?
    Fair; the question makes no assumptions.

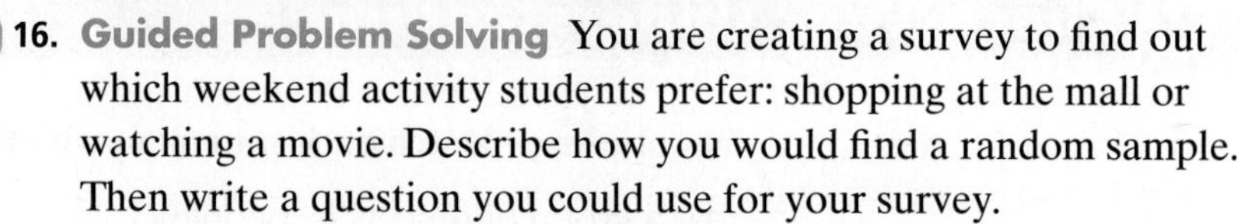

B GPS 16. **Guided Problem Solving** You are creating a survey to find out which weekend activity students prefer: shopping at the mall or watching a movie. Describe how you would find a random sample. Then write a question you could use for your survey.
- What is the population you will need to survey?
- Is your question fair or biased? Check students' work.

17. **Writing in Math** Which of the two surveys below would you use to determine attitudes about carnivals? Explain. See margin.

Survey	Question	Yes	No	Don't Know
A	Do you like going to noisy, overpriced carnivals?	53%	46%	1%
B	Do you like going to carnivals?	72%	27%	1%

18. Suppose you study the eating habits of college students. What question would you ask someone to determine whether he or she is a member of the population you want to study? See left.

18. Are you a college student?

Visit: PHSchool.com
Web Code: are-1104

7–10. See back of book.
17. See back of book.
19–20. See back of book.
25–27. See back of book.

Careers A park ranger needs a college degree related to park management, natural history, forestry, or outdoor recreation.

19. **Parks** Suppose you are gathering information about visitors to Yosemite National Park. You survey every tenth person entering the park. Would you get a random sample of visitors? Explain. **See margin.**

Clothes A clothing company surveys women ages 18 to 35 to decide the price of a suit. Is each method a random sample? Explain.

20. Select names at random from a national telephone directory. Call these people and survey the person that answers. **See margin.**

21. Select names from a national telephone directory. Call these people and survey any woman 18 to 35 who answers. **Yes; you will survey people in your population without any bias.**

22. **Challenge** A newspaper surveys 3 out of every 100 people of voting age in a community. The community has 38,592 people of voting age. How many people are surveyed? **about 1,158 people**

Test Prep and Mixed Review — Practice

Multiple Choice

23. A class is surveyed about computers and services. Which of the following gives the most detailed information about the results? **C**

A. **Home Computer Use**

Activity	Number of Students
Internet	6
Printer	4
Both	4

B. **Home Computer Use**

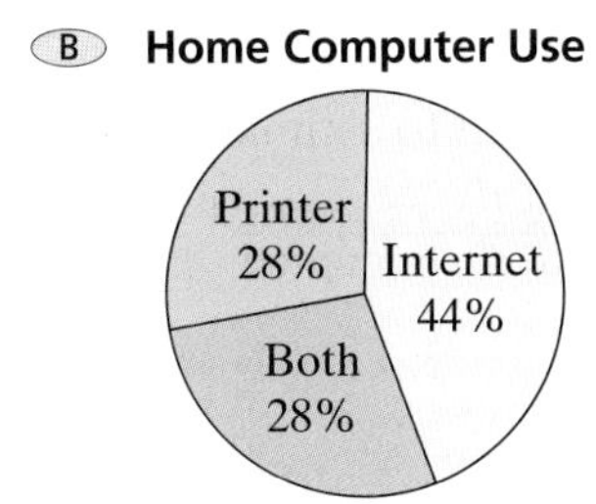

C. **Home Computer Use**

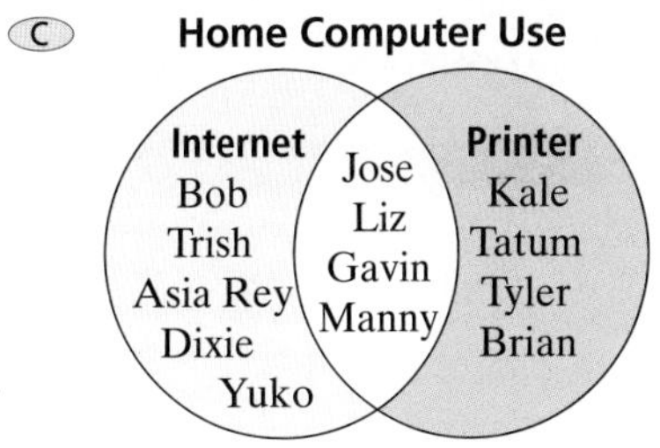

D. **Home Computer Use**

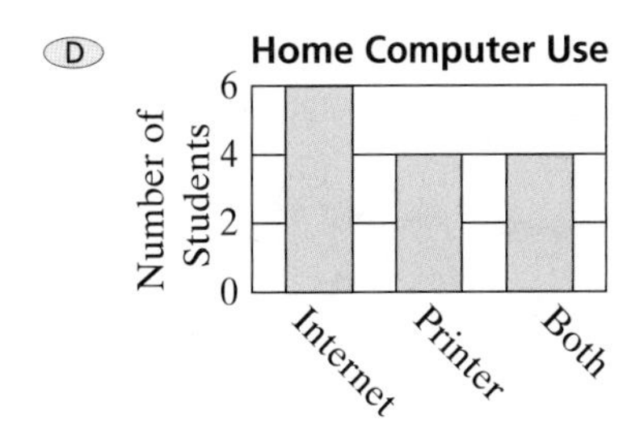

24. Ann, Betty, Cat, and Diane are married to Liem, Martin, Nate, and Pedro, but not in that order. Martin is Cat's brother. Cat is not married to Pedro. Ann and Nate are married. Diane's husband is an only child. Who is married to Betty? **G**

   F. Liem G. Martin H. Nate J. Pedro

GO for Help

For Exercises	See Lesson
25–27	9-2

Write a rule for each sequence. Then find the next three terms. **25–27. See margin.**

25. 3, 8, 13, 18, . . . 26. $-2, 1, 4, 7, \dots$ 27. $27, 16, 5, -6, \dots$

4. Assess & Reteach

PowerPoint Lesson Quiz

1. You want to survey people about their favorite exercise. Which sample below is more likely to be random? Explain.
   A. You ask people on a jogging track in a park to name their favorite exercise.
   B. You ask people on their lunch hour at a downtown intersection to name their favorite exercise. **B; Sample: You are more likely to get a variety of answers by interviewing people away from an exercise site.**
2. Explain why this question is either *biased* or *fair.* Do you prefer to play games on your computer or do boring homework? **Biased. Sample: Homework is said to be boring, and games are assumed to be fun.**

Alternative Assessment

Have each student in a pair write a survey question. The question can be biased or fair. Partners exchange questions, rewrite them to be the opposite, and explain their changes. Pairs continue the activity as time permits.

Test Prep

Resources

For additional practice with a variety of test item formats:

- Test-Taking Strategies, p. 571
- Test Prep, p. 575
- Test-Taking Strategies with Transparencies

Enrichment 11-4 L4

Reteaching 11-4 Random Samples and L2

Carlos is curious about sports that students in his school like best. He cannot interview every student in the school. But he could interview a sample of the school **population.**

Carlos wants a **random sample.** A sample is random if everyone has an equal chance of being selected. How will Carlos get a random sample? He considers two possibilities:

- He can interview 30 students at a soccer game.
- He can interview 5 students in each of 6 class changes.

Carlos realizes that students at a soccer game probably like soccer better than other sports. That would not be a random sample. He decides on the interviews during class changes.

What question will he ask? He considers two possibilities:

- "Which sport do you prefer, football, soccer, baseball, or tennis?"
- "Which do you enjoy most, the slow sport of baseball or one of the more exciting sports like football, soccer, or tennis?"

The second question is **biased.** It makes one answer seem better than another. Carlos decides to ask the first question.

1. You want to find how many people in your community are vegetarian. Where would be the best place to take a survey? **a shopping mall**

Is each question biased or fair?

2. Will you vote for the young, inexperienced candidate, Mr. Soong, or the experienced candidate, Ms. Lopez? **biased**
3. Will you vote for Mr. Soong or Ms. Lopez? **fair**

You plan to survey people to see what percent own their home and what percent rent. Tell whether the following will give a random sample. Explain.

4. You interview people outside a pool supply store in the suburbs. **No; you are more likely to interview homeowners.**
5. You interview people in the street near an apartment complex. **No; you are more likely to interview renters.**
6. You mail a survey to every 20th person in the telephone book. **Yes; you can't tell if people own or rent.**

11-5

1. Plan

Objective
To estimate population size using proportions

Example
1 Using the Capture/Recapture Method

Math Understandings: p. 530D

Math Background

See p. 530D for content support.

Lesson Planning and Resources

See p. 530E for a list of the resources that support this lesson.

Bell Ringer Practice

Check Skills You'll Need
Use student page, transparency, or PowerPoint. For intervention, direct students to:
Using Similar Figures
Lesson 5-5
Extra Skills and Word Problems Practice, Ch. 5

2. Teach

Activity Lab

Use before the lesson.

All in One Teaching Resources
Activity Lab 11-5: Using Random Samples

Guided Instruction

Error Prevention!

Students may place the data in the proportion incorrectly. Both ratios contain marked deer in the numerators. The left ratio is *counted;* the right is *actual*. The left ratio should be lesser.

11-5 Estimating Population Size

Check Skills You'll Need

1. **Vocabulary Review** A *proportion* is an equation stating that two _?_ are equal. **ratios**

Solve each proportion.

2. $\frac{2}{3} = \frac{a}{15}$ **10**
3. $\frac{n}{36} = \frac{11}{9}$ **44**
4. $\frac{42}{63} = \frac{6}{k}$ **9**

Lesson 5-4

What You'll Learn

To estimate population size using proportions

Why Learn This?

Researchers use the *capture/recapture method* to estimate animal population size. They collect, mark, and release animals. Then they capture another group of animals. The number of marked animals in the second group indicates the population size.

The following proportion is used to estimate a deer population.

$$\frac{\text{number of marked deer counted}}{\text{total number of deer counted}} = \frac{\text{total number of marked deer}}{\text{estimate of deer population}}$$

Online active math

For: Population Activity
Use: Interactive Textbook, 11-5

EXAMPLE Using the Capture/Recapture Method

1 **Gridded Response** Researchers count 48 marked deer and a total of 638 deer on a flight over an area. They know there are 105 marked deer. Write a proportion to estimate the deer population in the area.

$$\frac{\text{number of marked deer counted}}{\text{total number of deer counted}} = \frac{\text{total number of marked deer}}{\text{estimate of deer population}}$$

$\frac{48}{638} = \frac{105}{x}$ ← **Write a proportion.**

$48x = 105 \cdot 638$ ← **Write the cross products.**

$48x = 66{,}990$ ← **Multiply.**

$\frac{48x}{48} = \frac{66{,}990}{48}$ ← **Divide each side by 48.**

$x \approx 1{,}396$ ← **Round to the nearest integer.**

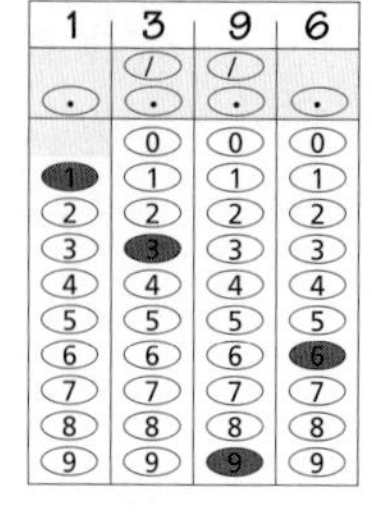

There are about 1,396 deer.

Quick Check

1. Suppose the researchers in Example 1 count 638 deer, but only 35 marked deer. Estimate the total deer population in the area. **about 1,914 deer**

Differentiated Instruction Solutions for All Learners

Special Needs L1
Point out that in the Example in this lesson, since the tagged or marked animals are a subset of the total populations, the proportions will always have a greater number on the bottom, and a lesser number on top. This will help students set up the proportions accurately.
learning style: visual

Below Level L2
Have students use four different colors to represent each part of the proportion in words. They then color the numbers appropriately as they solve each proportion.
learning style: visual

Check Your Understanding

1. **Vocabulary** What is the capture/recapture method of estimating an animal population? **See back of book.**

Use a proportion to estimate each animal population. Exercises 2 and 3 have been started for you. **2–3. See left.**

Test Prep Tip
You can check your answer to a proportion by verifying that the cross products are equal.

2. total trout counted: 2,985
   tagged trout counted: 452
   total tagged trout: 1,956
   $\frac{\text{tagged trout counted}}{\text{total trout counted}} = \frac{\text{total tagged trout}}{x}$

3. total bass counted: 3,102
   tagged bass counted: 198
   total tagged bass: 872
   $\frac{\text{tagged bass counted}}{\text{total bass counted}} = \frac{\text{total tagged bass}}{x}$

2. about 12,917 trout
3. about 13,661 bass

4. total rabbits counted: 5,804
   marked rabbits counted: 3,214
   total marked rabbits: 5,398
   about 9,748 rabbits

5. total black bears counted: 218
   marked black bears counted: 25
   total marked black bears: 35
   about 305 bears

Homework Exercises

For more exercises, see Extra Skills and Word Problems.

GO for Help

For Exercises	See Examples
6–14	1

6. about 1,823 deer
7. about 2,010 deer
8. about 2,273 deer
9. about 2,048 deer
10. about 2,110 deer
11. about 2,535 deer
12. about 1,822 deer
13. about 2,081 deer
14. about 2,345 deer

A **Estimate the total deer population for each year in the table.**

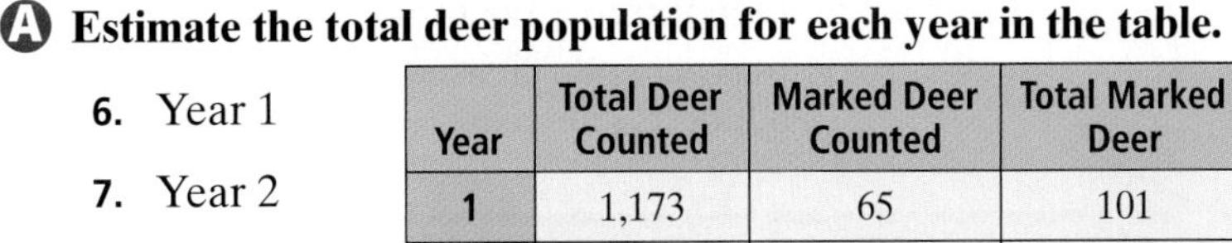

6. Year 1
7. Year 2
8. Year 3
9. Year 4
10. Year 5
11. Year 6
12. Year 7
13. Year 8
14. Year 9

Year	Total Deer Counted	Marked Deer Counted	Total Marked Deer
1	1,173	65	101
2	1,017	42	83
3	1,212	32	60
4	1,707	30	36
5	1,612	68	89
6	1,590	37	59
7	1,417	42	54
8	1,608	85	110
9	1,469	52	83

B **GPS**

15. **Guided Problem Solving** In a study, a fish and game department worker catches, tags, and frees 124 catfish in a lake. A few weeks later, he catches and frees 140 catfish. Thirty-five have tags. Estimate the number of catfish in the lake. **about 496 catfish**
    - number of marked catfish counted = ■
    - total number of catfish counted = ■
    - total number of marked catfish = ■

16. The population increased by about 522.

16. **Data Analysis** Use your answers to Exercises 6–14 above. Describe how the deer population changed over time. **See left.**

PowerPoint

Additional Examples

1. There are 30 marked seals. In a survey, you count 58 seals, 16 of which were marked. About how many seals are in the area? **about 109 seals**

All in One Teaching Resources
- Daily Notetaking Guide 11-5 L3
- Adapted Notetaking 11-5 L1

Closure

- *How do you estimate populations with the capture/recapture method?* **See back of book.**

3. Practice

Assignment Guide

Check Your Understanding
Go over Exercises 1–5 in class before assigning the Homework Exercises.

Homework Exercises

A Practice by Example	6–14
B Apply Your Skills	15–20
C Challenge	21
Test Prep and Mixed Review	22–29

Homework Quick Check
To check student's understanding of key skills and concepts, go over Exercises 8, 11, 18, 19, and 20.

4. Assess & Reteach

PowerPoint

Lesson Quiz

Use a proportion to estimate.

1. total bison counted: 350 tagged bison counted: 45 total tagged bison: 100 **about 778 bison**
2. total elk counted: 675; marked elk counted: 123; total marked elk: 465 **about 2,552 elk**

Advanced Learners L4
Quality control relates to capture/recapture. If all randomly selected widgets are acceptable, what can be deduced about the total "population" of widgets? **They are acceptable.**

learning style: verbal

English Language Learners ELL
Make sure students understand the numbers in the proportions. For example, be explicit that the 638 total deer counted includes the 48 marked deer. They can then understand that the estimate of the deer population includes the total number of marked deer.

learning style: verbal

Differentiated Instruction Resources

Adapted Practice 11-5 — L1

Practice 11-5 — Estimating Popul... — L3

Workers at a state park caught, tagged, and set free the species shown at the right. Later that same year, the workers caught the number of animals shown in the table below and counted the tagged animals. Use a proportion to estimate the park population of each species.

Tagged Animals	
Bears	12
Squirrels	50
Raccoons	23
Rabbits	42
Trout	46
Skunks	21

	Caught	Counted Tagged	Estimated Population
1. Bears	30	9	40
2. Squirrels	1,102	28	1,968
3. Raccoons	412	10	948
4. Rabbits	210	2	4,410
5. Trout	318	25	585
6. Skunks	45	6	158

A park ranger tags 100 animals. Use a proportion to estimate the total population for each sample.

7. 23 out of 100 animals are tagged **435 animals**
8. 12 out of 75 animals are tagged **625 animals**
9. 8 out of 116 animals are tagged **1,450 animals**
10. 5 out of 63 animals are tagged **1,260 animals**
11. 4 out of 83 animals are tagged **2,075 animals**
12. 3 out of 121 animals are tagged **4,033 animals**
13. 83 out of 125 animals are tagged **151 animals**
14. 7 out of 165 animals are tagged **2,357 animals**

Use a proportion to estimate each animal population.

15. Total ducks counted: 1,100
Marked ducks counted: 257
Total marked ducks: 960
4,109 ducks

16. Total alligators counted: 310
Marked alligators counted: 16
Total marked alligators: 90
1,744 alligators

11-5 • Guided Problem Solving GPS — L3

Student Page 556, Exercise 19:

Sharks A biologist is studying the shark populaton off the Florida coast. He captures, tags, and sets free 38 sharks. A week later, 8 out of 25 sharks captured have tags. He uses the proportion $\frac{25}{8} = \frac{38}{x}$ to estimate that the population is about 12.

a. **Error Analysis** Find the error in the biologist's proportion.

b. Estimate the shark population.

Understand

1. What are you being asked to do? **Find the error; make a correct estimate.**

Plan and Carry Out

2. If x represents all the sharks off the coast of Florida, write the ratio of the sharks the biologist originally tagged to the number of all the sharks off the coast of Florida. $\frac{38}{x}$
3. How many sharks did the biologist capture the second time? How many were tagged? **25 sharks; 8 sharks**
4. Write a ratio comparing the sharks the biologist found tagged and the number that were captured the second time. $\frac{8}{25}$
5. Set the ratios from steps 2 and 4 equal to form the correct proportion. $\frac{8}{25} = \frac{38}{x}$
6. What is wrong with the biologist's proportion? **The 8 and 25 are reversed.**
7. Solve this proportion to find the correct estimate. **119 sharks**

Check

8. Explain why the biologist should have known the estimate was wrong. **He captured 38 to begin with, so he knew there had to be more than 12 in the population.**

Solve Another Problem

9. A ranger traps, tags, and releases 32 jackrabbits. Later she captures 12 jackrabbits, of which 4 are tagged. The ranger estimates that there are 96 jackrabbits in the area. Write a proportion and check the ranger's estimate. Is she correct? $\frac{32}{x} = \frac{4}{12}$; **96 jack rabbits; yes**

Enrichment 11-5 — L4

Reteaching 11-5 — Estimating Popul... — L2

Researchers tagged 100 fish in a pond and then released them back into the pond. Later they captured 60 fish and found that 3 were tagged. Estimate the number of fish in the pond.

① Write a proportion of tagged fish to total fish. $\frac{\text{tagged fish (pond)}}{\text{total fish (pond)}} = \frac{\text{tagged fish (sample)}}{\text{total fish (sample)}}$, $\frac{100}{n} = \frac{3}{60}$

② Write cross products. $3n = 6{,}000$

③ Solve. $n = 2{,}000$

There are about 2,000 fish in the pond.

Complete to estimate the number of deer in the woods.

1. One year researchers tagged 80 deer. They later captured 15 deer and found 5 were tagged. Estimate the number of deer in the woods.
$\frac{80}{n} = \frac{5}{15}$
$5n = 15 \cdot$ **80**
$n =$ **240**
There are about **240** deer.

2. Two years later, researchers tagged 45 deer. They later captured 20 and found 3 were tagged. Estimate the number of deer in the woods then.
$\frac{45}{n} = \frac{3}{20}$
$3n =$ **20 · 45**
$n =$ **300**
There are about **300** deer.

Use a proportion to estimate each animal population.

3. In another project, researchers caught and tagged 85 sea lions in a bay. Later they caught and released 50 sea lions. Of those, 9 had tags. Estimate the sea lion population in the bay. **about 472 sea lions**

4. Other researchers caught and tagged 5 spotted owls. Later they caught 7 owls. Of those, 4 were tagged. Estimate the number of spotted owls in that forest. **about 9 owls**

5. An ecology class helped researchers determine the rabbit population in a nature preserve. One weekend, the students captured, tagged, and set free 32 rabbits. A month later, they captured 27 rabbits, including 16 with tags. Estimate the number of rabbits in the nature preserve. **about 54 rabbits**

GO Online
Homework Video Tutor
Visit: PHSchool.com
Web Code: are-1105

17. **63 alligators**
18. **It may underestimate the population.**

20. **It would make the estimate appear too large.**

Use the report for Exercises 17 and 18.

Alligator Population	
Number caught, tagged, and set free	[illegible]
Number recaptured	105
Number recaptured with tags	50
Estimated total population	132

17. A biologist spilled juice on the report. Find the number of alligators that were caught, tagged, and set free. **See left.**

18. **Writing in Math** Explain how counting a high percent of the marked alligators affects your alligator estimate. **See left.**

19. **Sharks** (GPS) A biologist is studying the shark population off the Florida coast. He captures, tags, and sets free 38 sharks. A week later, 8 out of 25 sharks captured have tags. He uses the proportion $\frac{25}{8} = \frac{38}{x}$ to estimate that the population is about 12.
a. **Error Analysis** Find the error in the biologist's proportion.
b. Estimate the shark population. **19a–b. See margin.**

20. **Reasoning** A class helps determine the squirrel population in a park. Students capture, tag, and free squirrels. A few squirrels lose their tags. How will this affect the population estimate? Explain. **See left.**

(C) 21. **Challenge** In a capture/recapture program, 30% of the animals recaptured have tags. Suppose 70 animals were originally captured, tagged, and released. Estimate the population. **about 233 animals**

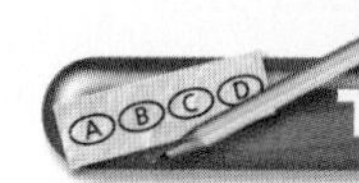

Test Prep and Mixed Review — Practice

Gridded Response

22. The table at the right shows the results of a biologist who captures, tags, and sets free lake trout. About what is the trout population in the lake? **250**

Trout Population

Number Tagged	150
Number Recaptured	100
Number Recaptured with Tags	60

23. Alexandro needs 300 cubic inches of clay to make a sculpture. The clay comes in blocks that are 5 inches wide, 3 inches tall, and 4 inches deep. How many blocks does Alexandro need? **5**

24. Partners A and B split their profits in a ratio of 2 : 3. If Partner A makes a profit of $2,700, how much profit, in dollars, does Partner B make? Assume Partner B makes more than Partner A. **4,050**

GO for Help

For Exercises	See Lesson
25–29	10-1

Algebra **Graph each point on the same coordinate plane.**

25. $(3, 4)$ 26. $(0, 0)$ 27. $(-5, -2)$ 28. $(-1, -2)$ 29. $(4, -3)$
25–29. See margin.

Test Prep

Resources

For additional practice with a variety of test item formats:

- Test-Taking Strategies, p. 571
- Test Prep, p. 575
- Test-Taking Strategies with Transparencies

Alternative Assessment

Have students work in pairs. Each student writes a problem similar to Exercises 2–5. Partners exchange papers and use their partner's data to estimate each population. Partners explain their reasoning to each other.

11-5b Activity Lab

Data Analysis

Graphing Population Data

You can represent population data in different ways. The table below shows data about U.S. households.

U.S. Households

Year	Number of Households (millions)	People per Household
1950	42.9	3.38
1960	53.0	3.29
1970	63.4	3.11
1980	80.3	2.75
1990	91.9	2.63
2000	105.5	2.59

You can also represent the data above using a double line graph.

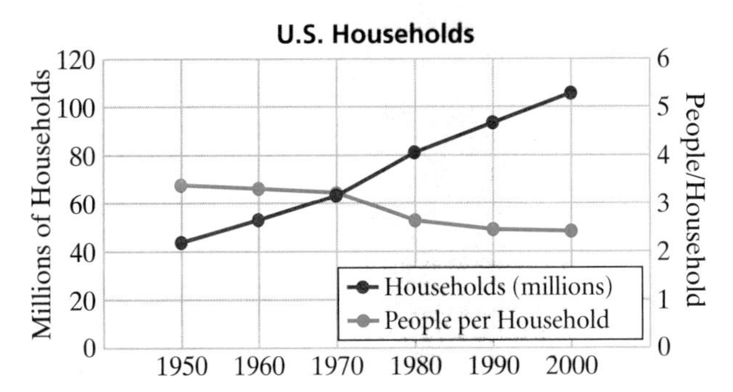

ACTIVITY

Use the data in the table above for Exercises 1–3.

1. Make a line graph. Label the horizontal axis "Millions of Households." Label the vertical axis "People per Household." **See margin.**
2. Describe the relationship that the line graph in Exercise 1 shows between the number of people per household and the number of households. **As the number of households increases, the average number of people per household decreases.**
3. Use unit rates. Estimate the U. S. population for each year listed. **See above.**
4. Extend your graph in Exercise 1 to 2010. Estimate the number of households and the number of people per household in 2010. **about 118 million; about 2.5**
5. Use unit rates to estimate the U.S. population in 2010. **about 295 million**

3. 1950: 145 million
1960: 174.4 million
1970: 197.2 million
1980: 220.8 million
1990: 241.697 million
2000: 273.2 million

4–5. Answers may vary. Samples are given.

19a–b. See back of book.
25–29. See back of book.

1. See back of book.

Activity Lab

Graphing Population Data

Students graph population data. This extends both their skills with population data and their skills with data display.

Guided Instruction

Explain to students that they can make a graph to show population data. Tell them to make the graph to scale to make the graph a reasonable size.

Ask:

- *What does each line in the double line graph represent?* **number of households, number of people in a household**
- *Why is a line graph the best choice to represent this data?* **line graphs are useful to show change over time, as in this situation**
- *What would change about the graph if you reversed the axis labels?* **the slope of the line**

Exercises

Have students work independently. Have them form small groups to compare their graphs and the answers to the questions.

Differentiated Instruction

Below Level L2

Rephrase the question to show the population of a small town for students who might have trouble visualizing millions of households.

GPS Guided Problem Solving

Describing Data

In this feature, students interpret data to answer questions. They study and analyze data to find missing numbers and interpret whether statements about the data are valid.

Guided Instruction

Explain to students that data can often be confusing or misleading. They often need to analyze it in order to understand it. Have a volunteer read the problem aloud.

Ask:

- *Why do you multiply the number of students that gave a score by the score?* **to find the total score for this plan**
- *Why is the median or mode more appropriate to represent how the students feel?* **More students objected to the change than supported the change. 4.5 and 4 are closer to 0, which represents students who oppose the change.**
- *What makes this survey misleading?* **Students who are for the change give a higher score than students who are against the change. If only a few students are for the change, the plan still gets a high score.**
- *What would be a fair way to conduct this survey and represent the data?* **Sample: Ask students if they are for or against the change, make a double bar graph.**

Exercises

Have students work independently on the Exercises. When they have finished, have them trade papers with a partner and check each other's work.

GPS Guided Problem Solving

Describing Data

Poll A middle-school class was polled on whether the school should change its mascot. A 0 rating meant the student was strongly against the change. A 10 rating meant the student was strongly in favor of the change. The line plot below shows the data. Whitney computed the mean. Was the mean a good descriptor of how the students felt?

Mascot Survey Result

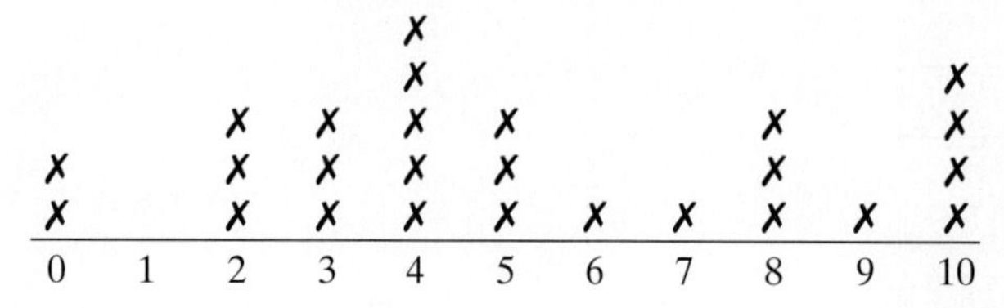

What You Might Think

What do I know? What do I want to find out?

What is the mean?

Does the mean represent the students' feelings?

What You Might Write

I know the students' ratings. I need to find the mean, which is the sum of the ratings divided by 26. I need to find out whether the mean represents how the students felt.

$\text{total} = 3 \times 2 + 3 \times 3 + 5 \times 4 + 3 \times 5 + 6 + 7 + 3 \times 8 + 9 + 4 \times 10$

$= 136$

$\text{mean} = 136 \div 26$, or about 5.2

I do not think so. Thirteen students were against the change. Ten students were for the change. The median of 4.5 or mode of 4 would better show how students felt.

Think It Through

1. **Reasoning** How did Whitney know to divide by 26 to find the mean? Explain. **1. There were 26 X's on the line, which represent 26 student responses.**
2. Explain how to find the median of 4.5 and the mode of 4. **See margin.**

2. Since there are 26 pieces of data, the median is the average of the 13th and 14th pieces of ordered data, which were 4 and 5. The mode was 4, since it was the most commonly reported piece of data.

Exercises

Solve each problem. For Exercises 3 and 4, answer the questions first.

3. Jax has a weekly test in math. His scores on the last five tests were 78, 92, 86, 94, and 95. What score does he need on his next exam to have an average of 90? **95**
   a. What do you know? What do you want to find out?
   b. How does a diagram like the one below help you understand what to do?

$600 \times 90\% = 540$					
78	92	86	94	95	■

4. Suppose a fox and a lizard compete in a 300-ft race. The fox runs at a rate of 30 ft/s. The lizard runs at a rate of 10 ft/s and starts 15 seconds before the fox. Who wins the race? Explain.
   a. What do you know? What do you want to find out?
   b. How does a table like the one below help you understand what to do?

4. the fox; the lizard finishes the race in 30 s. After a 15 s head start, the lizard still runs for 15 s. The fox runs the entire race in 10 s, so the fox wins.

Animal	Distance (ft)	Rate (ft/s)	Time (s)
Fox	300	30	■
Lizard	300	10	■

5. A tsunami is a big wave. A tsunami that struck land in 1883 was 120 ft high. An office building is about 12 ft high per floor. How many floors high was this wave? **10 floors**

6. Parents were asked in a survey, "Should the school require uniforms?" A 0 rating meant the parent was strongly against uniforms. A 5 rating meant the parent was strongly for uniforms. The frequency table below shows the survey results. Which measure—the mean, the median, or the mode—would best describe the data? Explain.

6. Answers may vary. Sample: The mean of 2.03 and median of 2 best describe the data because they show there are parents who feel strongly both ways.

Parent Survey Results

Rating	0	1	2	3	4	5
Tally	𝍸 𝍸 ///	/	///	///	//	𝍸 //

Differentiated Instruction

Advanced Learners L4

Have students conduct a poll of the class, as described in the example, and represent the results as they see fit. Remind them about unbiased questions and random samples.

11-6

1. Plan

Objective
To identify misleading graphs and statistics

Examples
1 Redrawing Misleading Graphs
2 Misleading Intervals
3 Misleading Use of Data Measures
4 Application: Advertising

Math Understandings: p. 530D

Professional Development

Math Background

Graphs can either intentionally or unintentionally be misleading. For example, some graphs are drawn with a vertical scale that skips over values instead of starting at zero. Others use uneven intervals that can exaggerate perceived differences and mislead a reader.

More Math Background: p. 530D

Lesson Planning and Resources

See p. 530E for a list of the resources that support this lesson.

PowerPoint

Bell Ringer Practice

Check Skills You'll Need
Use student page, transparency, or PowerPoint. For intervention, direct students to:
Mean, Median, Mode, and Range
Lesson 1-10
Extra Skills and Word Problems Practice, Ch. 1

2a–b. See back of book.

11-6 Using Data to Persuade

Check Skills You'll Need

1. **Vocabulary Review** An *outlier* mostly affects the _?_ of a data set. **mean**
2. Find the mean, median, and mode. 122, 106, 113, 116, 120, 123, 119, 117, 123, 111 **117; 118; 123**

GO for Help Lesson 1-10

What You'll Learn

To identify misleading graphs and statistics

Why Learn This?

A graph can be a powerful way to present data. How you draw a graph can affect the impression you give about the data.

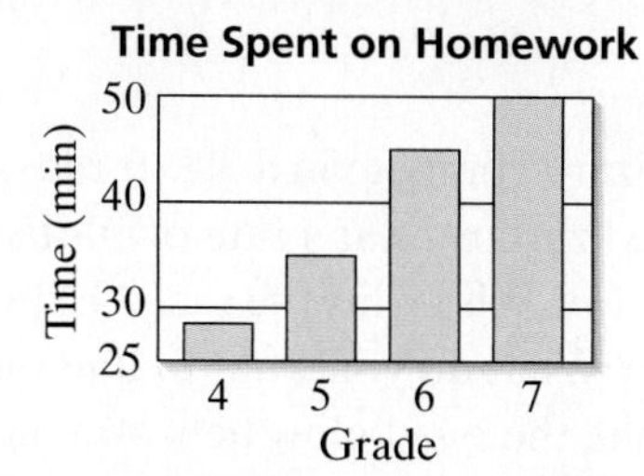

The graph above is misleading. It leads you to think that fourth-grade students spend almost no time on homework.

EXAMPLE Redrawing Misleading Graphs

1 Redraw the graph above so it is not misleading.

Method 1 Start the vertical scale at 0.

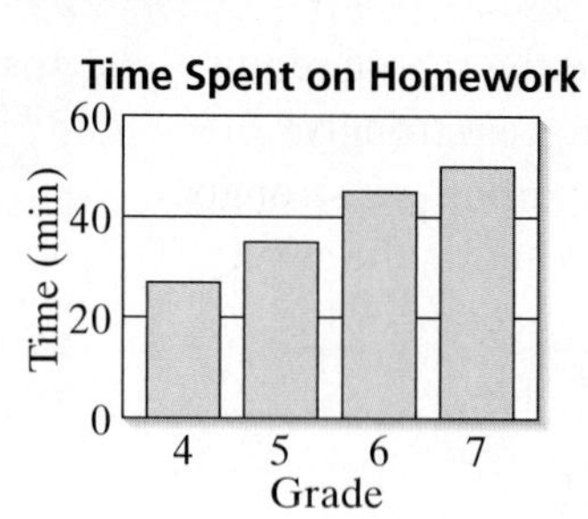

Method 2 Use a break in the vertical scale.

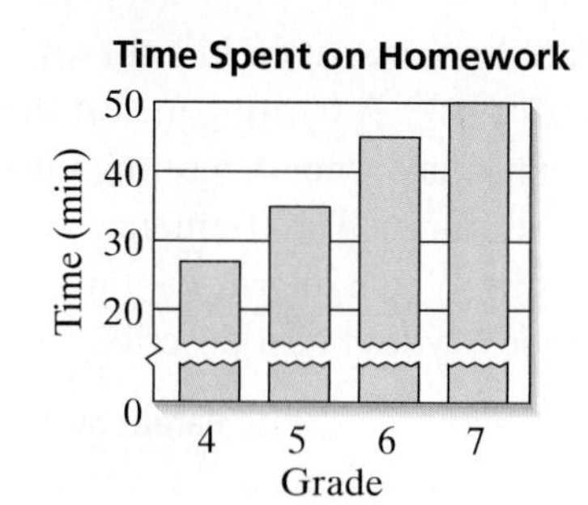

Quick Check

1. When is a break in a vertical scale especially useful? **when space is limited or when you want to focus on a narrow range of data**

Differentiated Instruction Solutions for All Learners

Special Needs L1
Some students might have a difficult time drawing or redrawing graphs. If so, have them work in pairs and make the decisions about intervals and breaks in the data, while their partner draws the graphs.
learning style: visual

Below Level L2
Have students brainstorm a list of ways data are used to mislead on television, such as "discounts up to 50% and more." Students can compare lists and refine them as needed.
learning style: verbal

Graphs can also mislead if the vertical or horizontal axes have unequal or very large intervals.

EXAMPLE Misleading Intervals

2 **Profit** Marketers use the graph below to show investors that annual profits increase steadily. Explain why the graph is misleading.

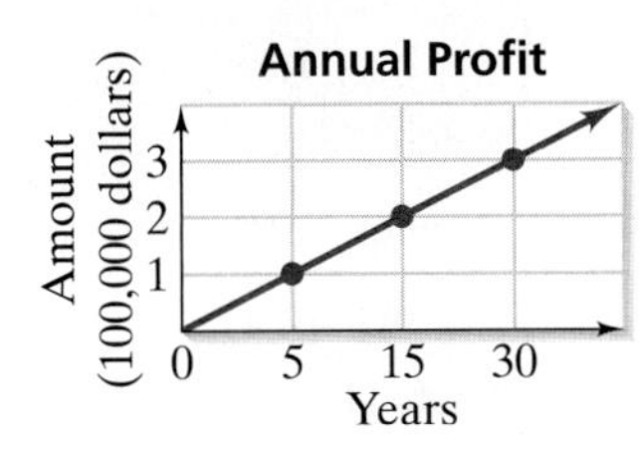

The intervals of the horizontal axis are not equal. Annual profits increase $100,000 over the first 5 years. Profits then increase only $100,000 in 10 years. Annual profits increase at a slower rate.

Quick Check **2a–b. See back of book.**

2. a. Redraw the graph in Example 2. Use equal intervals on both axes.
   b. Redraw the graph at the right so it does not mislead.

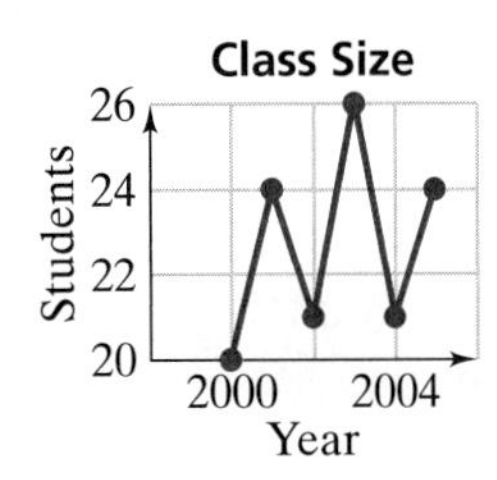

Use of the mean, median, or mode of a data set can inform or mislead.

EXAMPLE Misleading Use of Data Measures

3 **Multiple Choice** You survey students in your class to find the number of pets per student. The results are shown. You want to convince your parents to let you have four pets. Which measure of the data should you use?

How Many Pets Do You Have?

0, 0, 0, 1, 1, 1, 1, 1, 2, 2, 3, 4, 7, 15, 27

Ⓐ Mean Ⓒ Mode
Ⓑ Median Ⓓ Range

The median and the mode are both 1 pet. Although these measures represent the typical number of pets, they are too low to convince your parents to let you have 4 pets. The range, 27 pets, is too large and only shows how much the number of pets varies. The mean is 4 pets. You can use this measure to influence your parents. The correct answer is A.

Calculate each data measure. Then think about what kind of information or impression each measure provides.

Quick Check

3. Which two values in the data above would you consider outliers? **15 and 27**

2. Teach

Activity Lab

Use before the lesson.

All in One Teaching Resources

Activity Lab 11-6: Using Data to Persuade

Guided Instruction

Example 4

Have students find the mean, median, and mode to determine how the data can be displayed more accurately.

PowerPoint

Additional Examples

1 The graph shows the weekly earnings of four brothers.

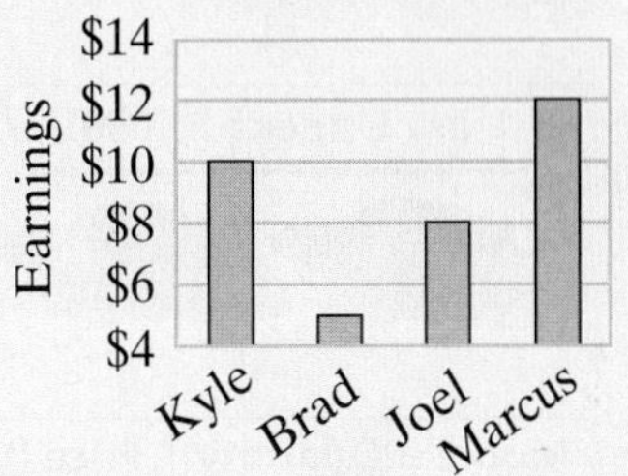

a. How is this graph misleading? **The graph starts at 4 instead of 0. It makes the lower values look lower than they are, and the difference between the high and low numbers is exaggerated.**

b. Redraw the graph so it is not misleading. **Check students' graphs.**

2 A television station presented this graph to its advertisers to show how viewership has increased. Explain how the graph is misleading.

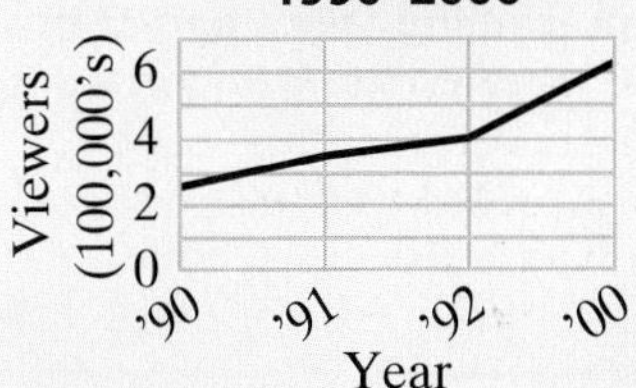

See back of book.

Advanced Learners L4
Have students write sample scenarios and data sets that mislead by presenting only the mean, median, or mode as in Examples 3 and 4.
learning style: visual

English Language Learners ELL
Have students work in pairs to talk about the words *misleading, impression,* and *influence.* Have them explain how each of these words relates to the kind of data displays people use in different situations.
learning style: verbal

PowerPoint

Additional Examples

3 A salary survey of a television station is shown in the table. A reporter wants a raise, so he tells his boss, "The mean salary here is $55,000." How is this misleading?

Position	No.	Salary
Owner	1	$250,000
Anchor	4	$70,000
Reporter	6	$37,000
Producer	2	$35,000
Camera Operator	4	$29,000

See back of book.

4 The sign below is advertising for an auto dealership. Below are the data for the cars at the dealership. Explain whether or not the ad is misleading.

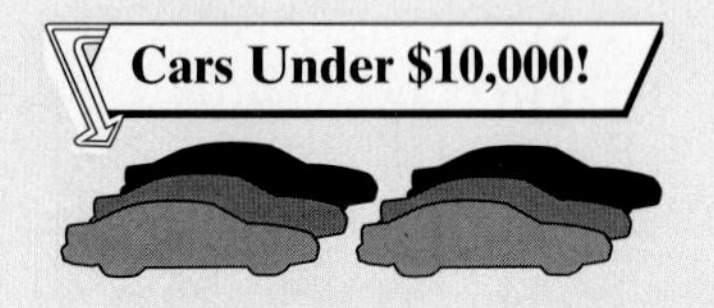

Model	Quantity	Price ($)
Pico	2	9,999
Toro	7	13,999
Enviro	9	15,999
Urban	4	22,999

See back of book.

All in One Teaching Resources

- Daily Notetaking Guide 11-6 L3
- Adapted Notetaking 11-6 L1

Closure

- *How can graphs be misleading?* **Sample: The graph can start at a value other than zero or have uneven or broken intervals.**
- *How can statistics be misleading?* **Sample: by presenting only a part of the data or omitting important facts**

You can mislead by showing part of the data or by leaving out facts.

HARRY'S
HAMBURGER LAND
"Home of the Under-500-Calorie Burger"

EXAMPLE Application: Advertising

4 Below are data about Harry's hamburgers. Is the advertisement at the left for Harry's Hamburger Land misleading?

Item	Calories	Fat (g)	Average Number Sold Daily
Kiddie Burger	480	35	80
Hamburger Plus	575	42	68
Health Burger	580	40	65
Golden Fries Hamburger	660	57	43
Burger Deluxe	700	55	75

Only one hamburger has less than 500 Calories. The ad is misleading.

Quick Check

4. How could you change the poster to better reflect the data? **Answers may vary. Sample: "Home of the under-500-Cal Kiddie Burger."**

Check Your Understanding

Use the graph at the right.

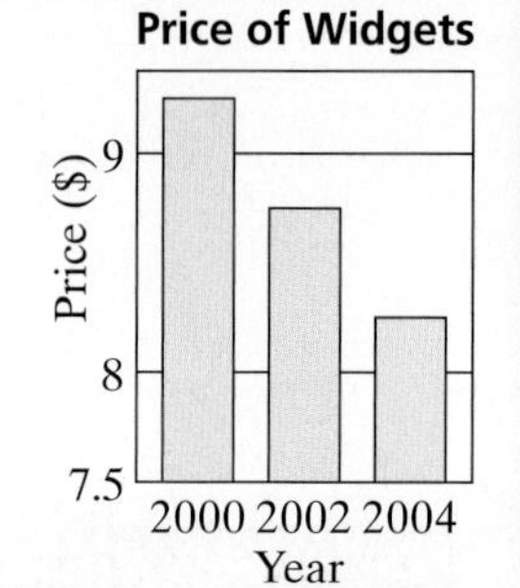

1. Does the vertical scale start at 0? **no**
2. Are equal intervals used on both the horizontal and vertical axes? **no**
3. **Number Sense** The graph makes the price change seem (greater, less) than it really is. **greater**
4. Redraw the graph so that it does not mislead. **See margin.**

Homework Exercises

For more exercises, see Extra Skills and Word Problems.

GO for Help

For Exercises	See Examples
5–7	1–2
8–11	3–4

A 5. **Nutrition** Farnaz asks students at her school, "What's your favorite fruit?" She draws the graph below. Why is it misleading? **See margin.**

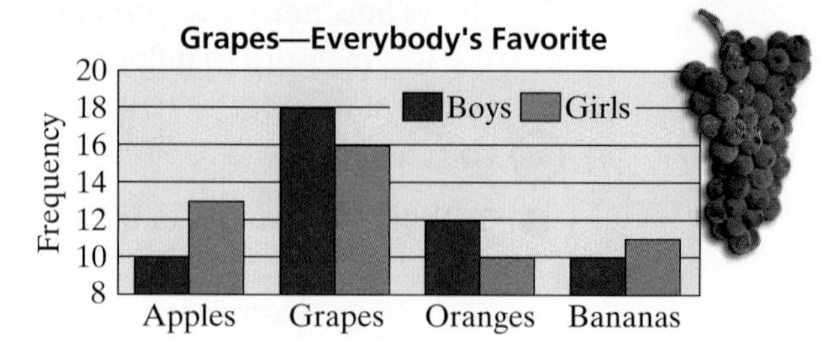

4–7. See back of book.
12. See back of book.

6. The graph gives the impression that vehicle production in 2000 was about 5 times higher than in 1980. The vertical axis has a break in the scale.

7. The graph gives the impression that the largest group of students received the highest scores. There are uneven intervals on the *x*-axis.

For help with calculating the mean, median, or mode of a data set, go to Lesson 1-10, Examples 1–3.

8a. Mean; it is the highest.

b. Mode; it is the lowest.

10. Median; the $189 bike costs only $10 more than the median.

11. No; you have to subtract expenses of $15. You actually raised $113 for charity.

For Exercises 6 and 7, tell what impression the graph gives and tell how the graph makes this impression. Then use the data to draw a graph that does not mislead. 6–7. See margin for graphs.

6.

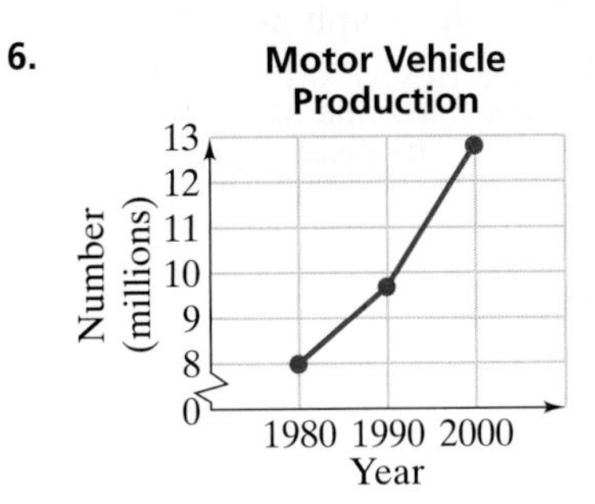

7.

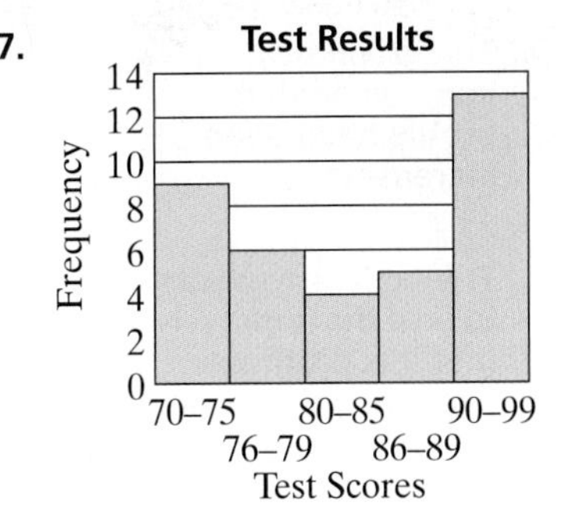

8. You score 93, 83, 76, 92, and 76 on five science exams.
   a. You want to show your parents how well you are doing in class. Should you use the mean, median, or mode? Explain. See left.
   b. Your teacher wants to encourage you to work harder. Which measure should your teacher use? Explain. See left.

9. **Bowling** The table shows the scores of two students in a bowling match.
   a. Bill says he won the match because he has a higher average. Is he correct? yes
   b. Kisha says she won the match. How can she justify this statement? She won two of three strings.

Bowling Scores

Game	1	2	3
Kisha	81	60	93
Bill	78	95	91

10. **Reasoning** You shop for bikes and find 7 types priced at $119, $139, $149, $179, $189, $199, and $209. You want to buy the $189 bike. Should you use the mean, median, or mode to convince your parents that this is a reasonable price for a bike? Explain.

11. **Number Sense** To raise money for charity, your class holds a car wash. You pay $15 for washing materials and collect $128 in donations. Can you say you raised $128 for charity? Explain.

B GPS 12. **Guided Problem Solving** Use the data at the right to make two line graphs. The first graph should show great change in the winning times and the second graph should show little change. See margin.
- **Make a Plan** Use a break for the vertical scale of the first graph. Use increments of 0.01 to 0.05 s. Start the vertical scale of the next graph at 0. Use increments of 0.5 to 2 s.
- **Check the Answer** The first graph should suggest a greater variation in winning times than the second graph shows.

Men's 100-Meter Winning Times

Year	Time (s)
1988	9.92
1992	9.96
1996	9.84
2000	9.87
2004	9.85

SOURCE: *Sports Almanac*

13. **Open-Ended** Find a graph in a newspaper or magazine. Is the graph trying to mislead you? Explain. Check students' work.

3. Practice

Assignment Guide

Check Your Understanding

Go over Exercises 1–4 in class before assigning the Homework Exercises.

Homework Exercises

A	Practice by Example	5–11
B	Apply Your Skills	12–19
C	Challenge	20
	Test Prep and Mixed Review	21–25

Homework Quick Check

To check students' understanding of key skills and concepts, go over Exercises 6, 9, 16, 17, and 19.

Differentiated Instruction Resources

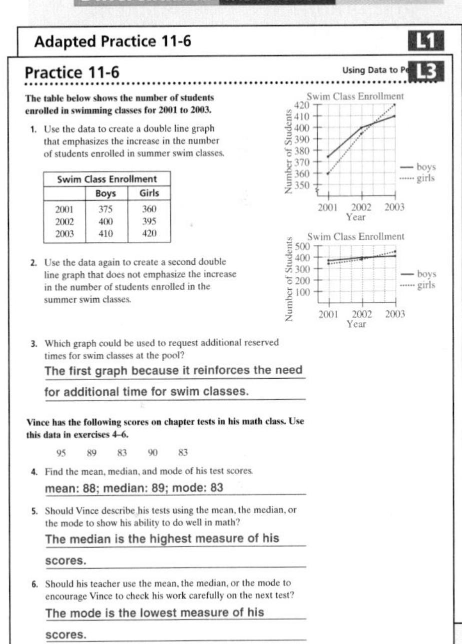
Adapted Practice 11-6 L1

Practice 11-6 — Using Data to Persuade L3

The table below shows the number of students enrolled in swimming classes for 2001 to 2003.

1. Use the data to create a double line graph that emphasizes the increase in the number of students enrolled in summer swim classes.

Swim Class Enrollment

	Boys	Girls
2001	375	360
2002	400	395
2003	410	420

2. Use the data again to create a second double line graph that does not emphasize the increase in the number of students enrolled in the summer swim classes.

3. Which graph could be used to request additional reserved times for swim classes at the pool? The first graph because it reinforces the need for additional time for swim classes.

Vince has the following scores on chapter tests in his math class. Use this data in exercises 4–6.

95 89 83 90 83

4. Find the mean, median, and mode of his test scores. mean: 88; median: 89; mode: 83

5. Should Vince describe his tests using the mean, the median, or the mode to show his ability to do well in math? The median is the highest measure of his scores.

6. Should his teacher use the mean, the median, or the mode to encourage Vince to check his work carefully on the next test? The mode is the lowest measure of his scores.

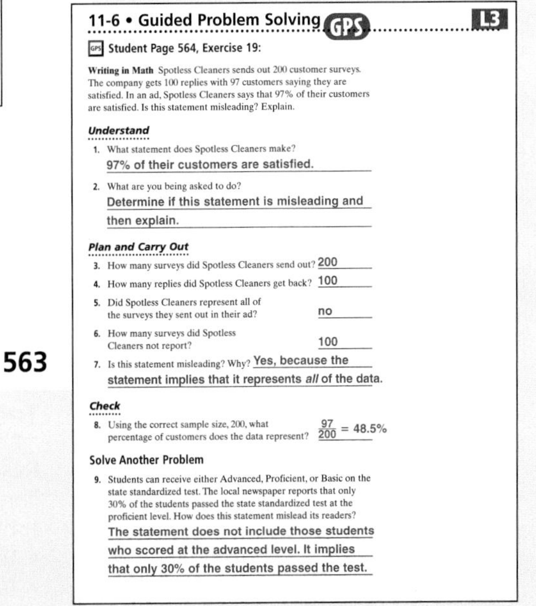
11-6 • Guided Problem Solving GPS L3

Student Page 564, Exercise 19:

Writing in Math Spotless Cleaners sends out 200 customer surveys. The company gets 100 replies with 97 customers saying they are satisfied. In an ad, Spotless Cleaners says that 97% of their customers are satisfied. Is this statement misleading? Explain.

Understand

1. What statement does Spotless Cleaners make? 97% of their customers are satisfied.
2. What are you being asked to do? Determine if this statement is misleading and then explain.

Plan and Carry Out

3. How many surveys did Spotless Cleaners send out? 200
4. How many replies did Spotless Cleaners get back? 100
5. Did Spotless Cleaners represent all of the surveys they sent out in their ad? no
6. How many surveys did Spotless Cleaners not report? 100
7. Is this statement misleading? Why? Yes, because the statement implies that it represents *all* of the data.

Check

8. Using the correct sample size, 200, what percentage of customers does the data represent? $\frac{97}{200}$ = 48.5%

Solve Another Problem

9. Students can receive either Advanced, Proficient, or Basic on the state standardized test. The local newspaper reports that only 30% of the students passed the state standardized test at the proficient level. How does this statement mislead its readers? The statement does not include those students who scored at the advanced level. It implies that only 30% of the students passed the test.

4. Assess & Reteach

Lesson Quiz

Study the bar graph below.

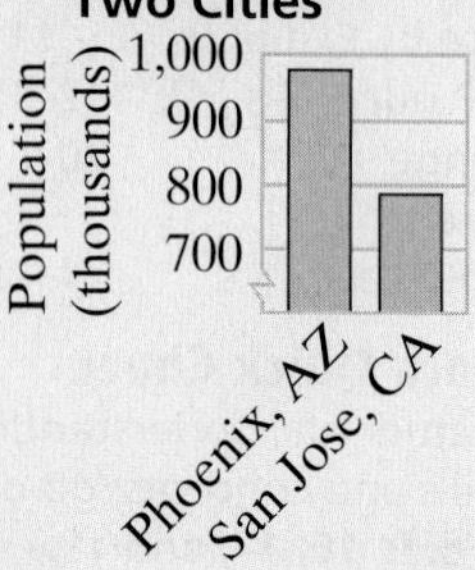

1. What impression is given by this graph? **The population of Phoenix is double that of San Jose.**
2. How would you change the graph to better reflect the data? **Change the vertical scale by eliminating the break in the scale.**
3. Redraw the graph to accurately reflect the data. **Check students' graphs.**

16. See back of book.
18. See back of book.

Enrichment 11-6 L4

Reteaching 11-6 Using Data to P[…] L2

There are 3 ways that graphs can be drawn to be misleading.

1. The interval on the vertical axis may not start at zero.
2. There may be a break in the graph.
3. The intervals on the horizontal or vertical axis may be unequal.

Average Annual Rainfall (three graphs: Avg. Annual Rainfall (in.) vs. Cities — Albany, Atlanta, Chicago, Houston, St. Louis, Seattle)

Mean, median, and mode can also be used to mislead. Consider a set of data where most of the numbers are in a certain range. There are a few numbers that are either way above or way below the range. The mean is not a good measure of the data in this case.

For each graph, do the following:
(a) Tell what the graph shows. (b) What can you say about the graph?

1. Rhythm & Blues Performers (Number of #1 Singles vs. Performer — James Brown, Aretha Franklin, Marvin Gaye, Temptations, Stevie Wonder)

   a. It appears that Aretha Franklin had many more #1 singles than anyone else.

   b. The vertical axis does not start at zero.

2. Civilian Staff in the Military (Civilian Staff (thousands) vs. Year — '85, '87, '89, '91, '93)

   a. It appears that there has been a great amount of change in the civilian staff.

   b. There is a break in the vertical axis.

GO Online
Homework Video Tutor
Visit: PHSchool.com
Web Code: are-1106

Business Use the graph below for Exercises 14–18.

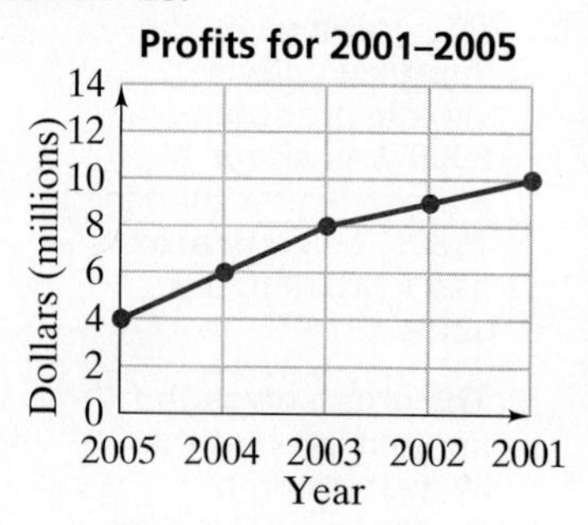

14. What does the graph suggest? **See left.**
15. Why does the graph give the impression described in Exercise 14? **It reverses the scale on the *x*-axis.**
16. Use the data to draw a graph that does not mislead. **See margin.**
17. **Reasoning** Why would a company draw the misleading graph? Explain.
18. You want to show that the company is successful. Should you use the mean, median, or mode to suggest the greatest profit? Explain. **See margin.**
19. **Writing in Math** Spotless Cleaners sends out 200 customer surveys. GPS The company gets 100 replies with 97 customers saying they are satisfied. In an ad, Spotless Cleaners says that 97% of its customers are satisfied. Is this statement misleading? Explain.
20. **Challenge** Research the cost of U.S. postage stamps for the past six years. Use your data to draw a graph that gives the impression of a large increase in stamp cost over time. **Check students' work.**

14. The graph gives the impression that profits have been increasing.

17. The company wants to give the impression that it is profitable even though it's not.

19. Yes; the 100 customers who did not respond may not be satisfied.

Test Prep and Mixed Review

Practice

Multiple Choice

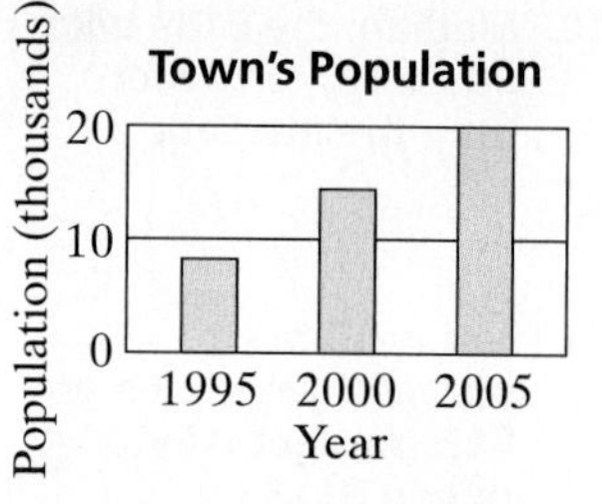

21. A town's population is shown. If the trend continues, which is the best prediction of the population in 2010? **B**
    - (A) Fewer than 22,000 people
    - (B) Between 22,000 and 29,000 people
    - (C) Between 29,000 and 36,000 people
    - (D) More than 36,000 people
22. Which of the following can have a single base that is a hexagon? **J**
    (F) Prism (G) Cone (H) Cylinder (J) Pyramid
23. What is the value of the expression $18 \div (9 - 6)^2 + 2 \times 6$? **A**
    (A) 14 (B) 24 (C) 30 (D) 108

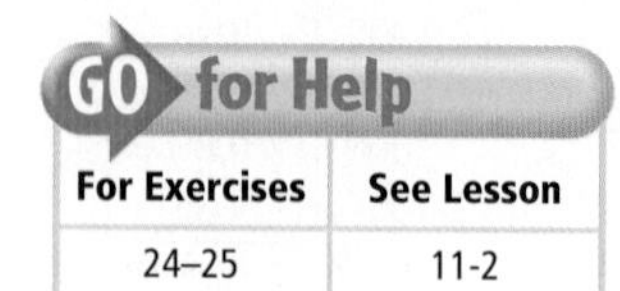

GO for Help

For Exercises	See Lesson
24–25	11-2

Estimation Use the graph.

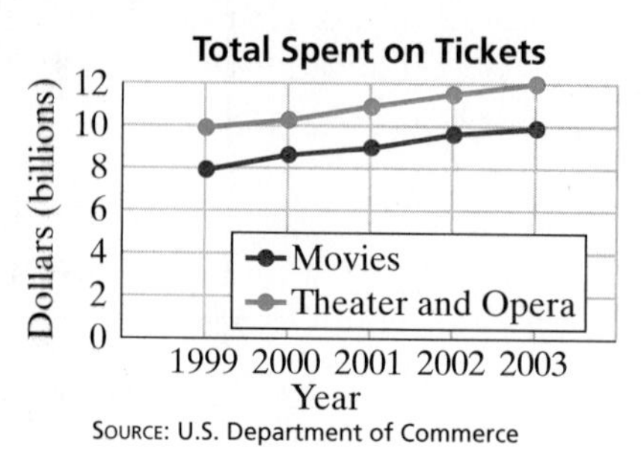

SOURCE: U.S. Department of Commerce

24. Estimate the total amount spent on theater and opera from 1999 to 2003. **about 54 billion**
25. Estimate the total amount spent on movie tickets from 1999 to 2003. **about 45 billion**

Test Prep

Resources

For additional practice with a variety of test item formats:

- Test-Taking Strategies, p. 571
- Test Prep, p. 575
- Test-Taking Strategies with Transparencies

Alternative Assessment

Have each student in a pair write a data set that has a very different mean, median, and mode. Partners exchange data sets and try to find which statistic best represents the data and which statistic is especially misleading. Students check each other's results.

Checkpoint Quiz 2

Lessons 11-4 through 11-6

1. Use the data at the right. Make a graph that would persuade filmmakers to make more movies in Georgia. **See margin.**

2. The data below show the class sizes at a school. Would you use the mean, the median, or the mode to convince the principal that there is a high number of students per class? Explain.
   27 29 34 24 29 19 30 19 25 22 27 19
   Median; it is a higher value than mean or mode.

3. **Open-Ended** Choose a survey topic and write a biased survey question and a fair survey question. **Check students' work.**

4. Which is the best way to survey a random sample of students from your school about their favorite radio station? Explain.
   (A) Survey 5 students in each first-period class.
   (B) Survey 12 students in the band.
   (C) Call 25 friends.
   (D) Survey each student in your math class.
   A; it is a more diverse sample.

Movies Filmed in Georgia

Year	Number
1997	10
1998	5
1999	4
2000	8
2001	4
2002	4
2003	6
2004	8

SOURCE: 2005 Georgia Film Sourcebook

Use a proportion to estimate each animal population.

5. total wild horses counted: 1,583
   marked wild horses counted: 496
   total marked wild horses: 1,213
   about 3,871 horses

6. total turtles counted: 51
   marked turtles counted: 32
   total marked turtles: 108
   about 172 turtles

MATH AT WORK

Pollsters

Pollsters interview people to find out their opinions and their preferences about specific topics. They must understand the topic of interest well in order to ask the right questions. If pollsters ask poor-quality questions, they will not get accurate responses.

Pollsters use math to analyze the data they collect. They find the mean, median, mode, and range of the data. Their analysis gives us an overall view of the data. They often graph data to display their findings.

For: more information about pollsters
Web Code: arb-2031

1. See back of book.

Use this Checkpoint Quiz to check students' understanding of the skills and concepts of Lessons 11-4 through 11-6.

Resources

- All-in-One Teaching Resources Checkpoint Quiz 2
- ExamView CD-ROM
- Success Tracker™ Online Intervention

Pollsters

Students may not understand how the skills they have learned in this chapter can be applied to a career field. This feature should generate interest in applying math skills to real-world situations.

Guided Instruction

Define *pollster* for students. Discuss polls they have taken in class during the chapter and other polls they may have taken outside of class. Have volunteers read each paragraph. Ask questions, such as:

- *What kind of jobs might require data from a pollster?*
- *What skills that you have learned in this chapter are useful to a pollster?*
- *Do you think being a pollster would be an interesting job for you? Why or why not?*

Activity Lab

Two-Variable Data Collection

Students study two sets of data in terms of how they affect each other. This will prepare them to study trends and scatter plots in Lesson 11-7.

Guided Instruction

Explain to students that sometimes two sets of data will affect each other. For example, the Number of Peanut Butter Sandwiches Eaten affects the Amount of Peanut Butter Remaining in the Jar.

Draw graphs to illustrate the changes in the wolf and moose populations. Ask:

- *What is the relationship between the populations?* **As the wolf population increases, the moose population decreases.**
- *What is a possible explanation for this relationship?* **Sample: Wolves hunt the moose so there are fewer moose while the wolves thrive.**

Exercises

Have students work in pairs on Exercises 1-4. When students have finished, have volunteers share their answers and discuss Exercise 4 as a class.
Repeat with Exercises 5-7.

Alternative Method

Allow students to make other kinds of graphs with which they are comfortable. For example, using a double bar graph in the Activity will display the same relationship, but may be easier for students to understand.

Resources

- Activity Lab 11-7: Exploring Scatter Plots

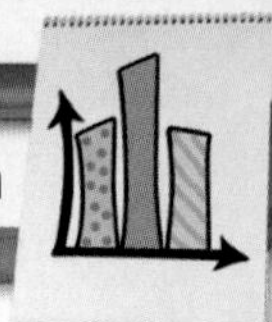

Two-Variable Data Collection

Isle Royale, the largest island in Lake Superior, is a favorable location for studying wolf and moose populations. The table shows wolf and moose population estimates during a 20-year period.

Moose and Wolf Populations on Isle Royale, Michigan

Year	Wolf	Moose	Year	Wolf	Moose
1985	22	1,115	1995	16	2,422
1986	20	1,192	1996	22	1,163
1987	16	1,268	1997	24	500
1988	12	1,335	1998	14	699
1989	12	1,397	1999	25	750
1990	15	1,216	2000	29	850
1991	12	1,313	2001	19	900
1992	12	1,590	2002	17	1,100
1993	13	1,879	2003	19	900
1994	17	1,770	2004	29	750

SOURCE: National Park Service

1. **The wolf population tended to decrease from 1985 to 1993 and tended to increase after 1993.**
2. **The moose population tended to increase from 1985 to 1995. After dropping in 1997 and 1998, it began to increase again.**

ACTIVITY

1. Describe the changes in the wolf population over the years. **See above right.**
2. Describe the changes in the moose population over the years. **See above right.**
3. Make a graph that shows the wolf population on the horizontal scale and the moose population on the vertical scale. Graph the ordered pairs (wolf, moose) for each year. **See margin.**
4. **Data Analysis** Do you see a relationship between the two populations? Explain. **As the wolf population increases, the moose population decreases.**

Exercises

5–7. Check students' work.

5. **Data Collection** Collect data on the length of your classmates' feet and the length of their forearms. Record the data in a table.
6. Display the data in an appropriate graph.
7. Do you see a relationship between the two lengths? Explain.

3. See back of book.

11-7 Exploring Scatter Plots

Check Skills You'll Need

1. Vocabulary Review Is the *x-axis* vertical or horizontal? **horizontal**

Graph each point on the same coordinate plane.

2. (4, 0) **3.** (2, 5)

4. (0, 3) **5.** (1, 4)

6. (2, 3) **7.** (0, 0)

2–7. See back of book.

GO for Help Lesson 10-1

What You'll Learn

To draw and interpret scatter plots

New Vocabulary scatter plot, positive trend, negative trend, no trend

Why Learn This?

You can graph data, such as the data shown below, as points in a coordinate plane. The graph may show an important pattern.

Book Bag Weights

Number of Books	3	3	4	4	5	6	6	7	7	8
Weight (lb)	6	8	6.5	9	10	7.5	12	9.5	11	12

A **scatter plot** is a graph that relates two sets of data. To make a scatter plot, graph the two sets of data as ordered pairs.

EXAMPLE Making Scatter Plots

1 Weights Graph the data in the table above in a scatter plot.

Each column in the table represents a point on the scatter plot.

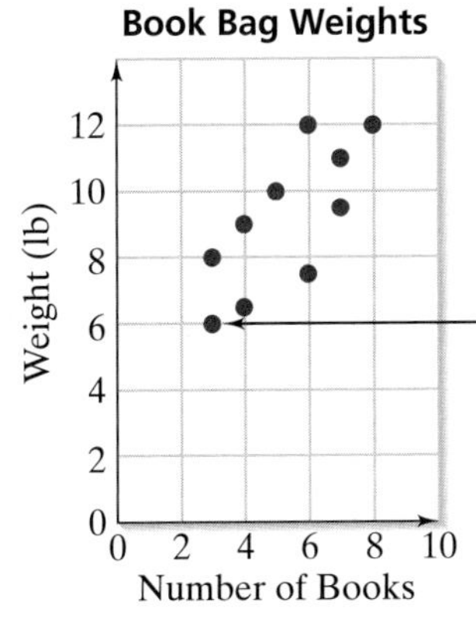

This point is for the book bag that holds 3 books and weighs 6 lb.

GO Online

Video Tutor Help

Visit: PHSchool.com
Web Code: are-0775

Quick Check

1. Graph the data in the table below in a scatter plot.

Height (in.)	58	64.5	67.5	65.5	63.5	64	71	62.5	69
Arm Span (in.)	57.5	64	68.5	66	62.5	66	72	63	70

1.

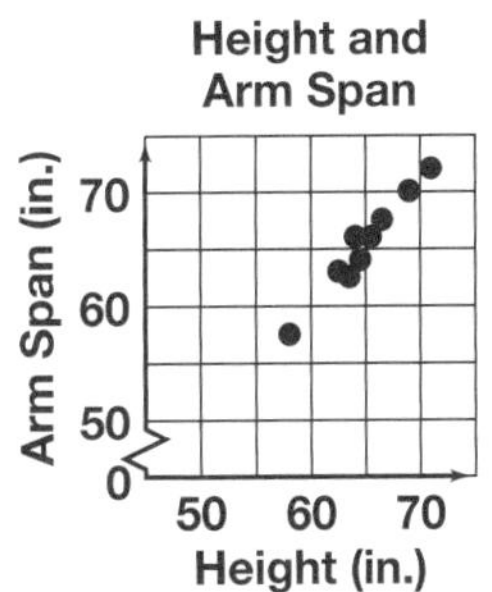

11-7

1. Plan

Objective

To draw and interpret scatter plots

Examples

1 Making Scatter Plots
2 Describing Trends in Scatter Plots

Math Understandings: p. 530D

Math Background

A *scatter plot* is a set of points that shows a relationship between two sets of data by plotting them on a coordinate grid. Sometimes a scatter plot forms a pattern called a *trend*. *No trend* indicates that no relationship exists between the data. Note that the trend is independent of which variable is placed along the *x*-axis or *y*-axis. Because scatter plots are not functions, there can be multiple data points for any horizontal axis value.

More Math Background: p. 530D

Lesson Planning and Resources

See p. 530E for a list of the resources that support this lesson.

PowerPoint

Bell Ringer Practice

Check Skills You'll Need

Use student page, transparency, or PowerPoint. For intervention, direct students to:

Graphing Points in Four Quadrants
Lesson 10-1
Extra Skills and Word Problems Practice, Ch. 10

Differentiated Instruction Solutions for All Learners

Special Needs L1

Ask students to use their arms and hands to show you what a positive trend would look like on a scatter plot, emulating the general direction of the points. Repeat for a negative trend.

learning style: tactile

Below Level L2

Review with students positive-sloped lines and negative-sloped lines. Relate these lines to positive trends and negative trends.

learning style: verbal

2. Teach

Activity Lab

Use before the lesson.
Student Edition Activity Lab, Data Collection 11-7a, Two-Variable Data Collection, p. 566

All in One Teaching Resources
Activity Lab 11-7: Exploring Scatter Plots

Guided Instruction

Example 1
In Quick Check 1, watch for students who switch the *x*-coordinates and *y*-coordinates. Write the ordered pair as a guide.

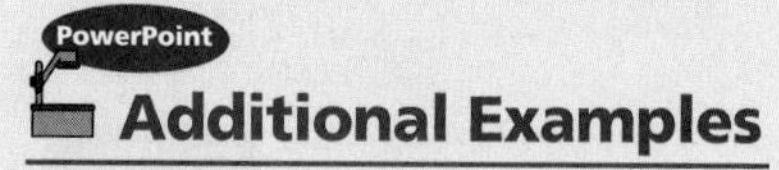

Additional Examples

1 Graph in a scatter plot the data from the table.

Height (in)	Test Scores
58	90
60	95
63	90
64	70
65	100
65	90
66	60
67	90
68	80
68	75
70	85
70	100
71	80
72	70

See back of book.

2 Describe the trend in the scatter plot for Exercise 1. no trend

All in One Teaching Resources
- Daily Notetaking Guide 11-7 L3
- Adapted Notetaking 11-7 L1

Closure

- *What is a scatter plot?* See back of book.
- *Identify a positive, negative, and no trend.* See back of book.

The scatter plot in Example 1 shows a relationship, or *trend*. As the number of books increases, the weight of the bag generally increases.

You can examine a scatter plot to see what kind of trend is shown.

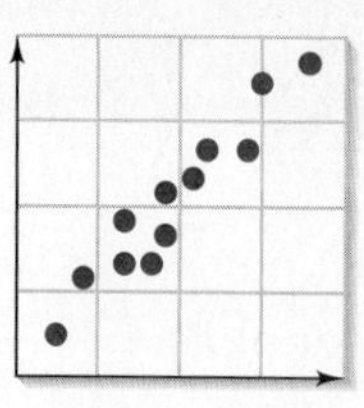

Positive trend
As one set of values increases, the other set tends to increase.

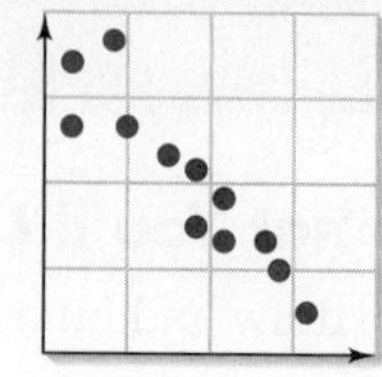

Negative trend
As one set of values increases, the other set tends to decrease.

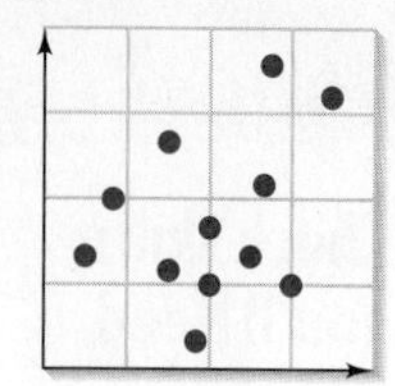

No trend
There is no apparent relationship among the data.

EXAMPLE Describing Trends in Scatter Plots

2 Trees Describe the trend in the scatter plot.

As the age of a tree increases, the diameter of the tree tends to increase.

The scatter plot shows a positive trend.

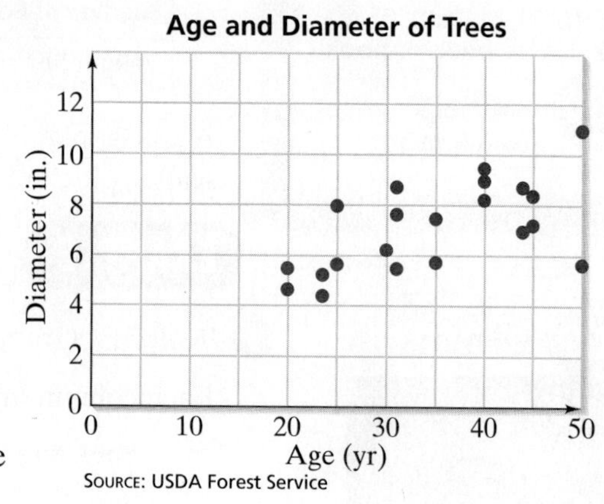

Quick Check

2. Suppose you draw a line that follows the trend shown in the scatter plot. Is the slope of the line positive or negative? **positive**

Check Your Understanding

1. **Vocabulary** How does a scatter plot relate two sets of data?

Use the scatter plot in Example 2 for Exercises 2–5.

2. What does the point at (25, 8) represent? **See left.**
3. A tree that is 40 years old has a diameter of about 9 in. Is this data point shown on the scatter plot? **yes**
4. **Estimation** A tree is 28 years old. Estimate its diameter. **about 7 in.**
5. **Estimation** A tree has a diameter of 10 in. Estimate its age. **about 47 years**

1. It illustrates what happens to one set of data when the other increases.
2. A tree that is 25 years old has a diameter of 8 inches.

Advanced Learners L4	**English Language Learners** ELL
Have students find the definition of *lines of best fit*. Students can fit lines to several scatter plots. Students can then speculate how the lines can help them make predictions. learning style: visual	Have students work in pairs to generate examples of relationships that would show up as positive, negative, or no trends in a scatter plot. learning style: visual

Homework Exercises

For more exercises, see Extra Skills and Word Problems.

GO for Help

For Exercises	See Examples
6–8	1
9–14	2

A **Graph each set of data in a scatter plot.** 6–8. See margin.

6.

Temperature (°F)	Weight of Clothing (lb)
60	5.5
58	5.2
50	6.2
42	6.8
36	7.8
32	7.4
30	8.4
26	9.9
22	10.9
20	12

7.

Hours Studying	Test Grade
0.5	68
0.75	70
1	82
1	78
1.25	78
1.25	86
1.25	94
1.5	82
1.75	90
2	88

8.

Price of CD ($)	14	13	7	10	18	12	23	17	12	19	15
Number of Songs	15	10	11	12	14	19	20	9	7	10	11

Describe the trend in each scatter plot.

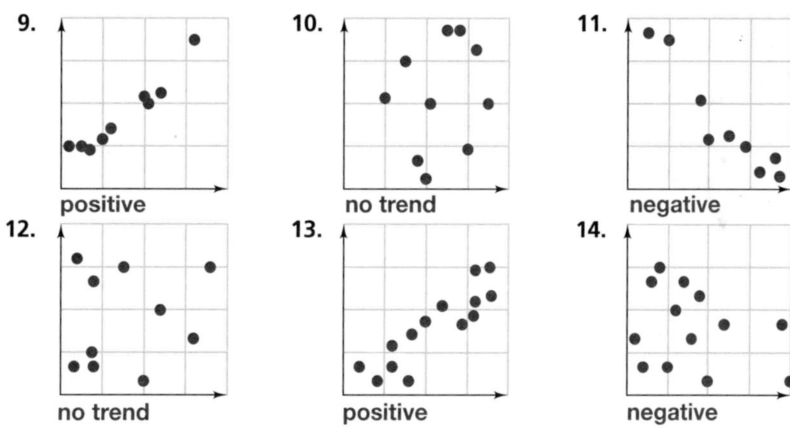

B GPS 15. **Guided Problem Solving** Use the data below. Predict the electrical use in a month that has an average temperature of 60°F. about 190 kWh

Monthly Electrical Use

Average Temperature (°F)	77	72	68	45	39	50	35	30	47
Electricity Use (kWh)	170	143	168	236	260	196	244	309	266

- **Make a Plan** Make a scatter plot with °F on the horizontal axis. Draw a line that is close to most of the data points. Use the line to estimate the electricity use at 60°F.
- **Carry Out the Plan** The scatter plot shows a _?_ trend. The ordered pair (60, ■) represents the electricity use at 60°F.

6–8. See back of book.

3. Practice

Assignment Guide

Check Your Understanding

Go over Exercises 1–5 in class before assigning the Homework Exercises.

Homework Exercises

A Practice by Example	6–14
B Apply Your Skills	15–18
C Challenge	19
Test Prep and Mixed Review	20–23

Homework Quick Check

To check students' understanding of key skills and concepts, go over Exercises 7, 11, 16, 17, and 18.

Differentiated Instruction Resources

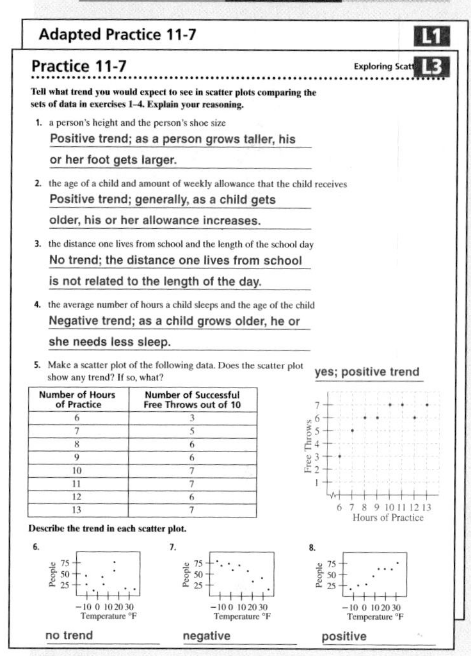

Adapted Practice 11-7 L1

Practice 11-7 Exploring Scatt L3

Tell what trend you would expect to see in scatter plots comparing the sets of data in exercises 1–4. Explain your reasoning.

1. a person's height and the person's shoe size
Positive trend; as a person grows taller, his or her foot gets larger.
2. the age of a child and amount of weekly allowance that the child receives
Positive trend; generally, as a child gets older, his or her allowance increases.
3. the distance one lives from school and the length of the school day
No trend; the distance one lives from school is not related to the length of the day.
4. the average number of hours a child sleeps and the age of the child
Negative trend; as a child grows older, he or she needs less sleep.
5. Make a scatter plot of the following data. Does the scatter plot show any trend? If so, what? yes; positive trend

Number of Hours of Practice	Number of Successful Free Throws out of 10
6	3
7	5
8	6
9	6
10	7
11	7
12	6
13	7

Describe the trend in each scatter plot.

6. no trend
7. negative
8. positive

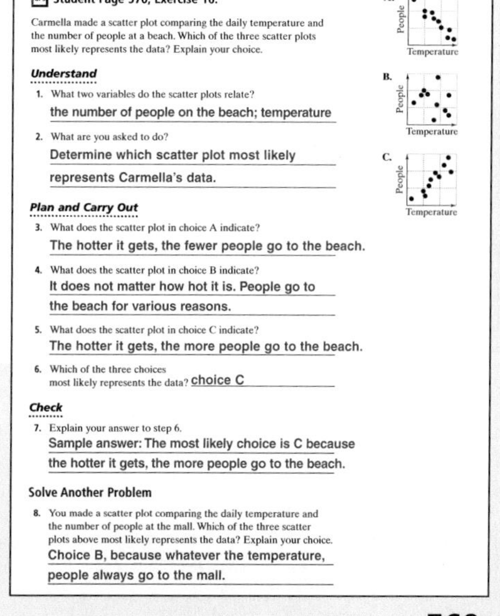

11-7 • Guided Problem Solving GPS L3

Student Page 570, Exercise 16:

Carmella made a scatter plot comparing the daily temperature and the number of people at a beach. Which of the three scatter plots most likely represents the data? Explain your choice.

Understand

1. What two variables do the scatter plots relate?
the number of people on the beach; temperature
2. What are you asked to do?
Determine which scatter plot most likely represents Carmella's data.

Plan and Carry Out

3. What does the scatter plot in choice A indicate?
The hotter it gets, the fewer people go to the beach.
4. What does the scatter plot in choice B indicate?
It does not matter how hot it is. People go to the beach for various reasons.
5. What does the scatter plot in choice C indicate?
The hotter it gets, the more people go to the beach.
6. Which of the three choices most likely represents the data? choice C

Check

7. Explain your answer to step 6.
Sample answer: The most likely choice is C because the hotter it gets, the more people go to the beach.

Solve Another Problem

8. You made a scatter plot comparing the daily temperature and the number of people at the mall. Which of the three scatter plots above most likely represents the data? Explain your choice.
Choice B, because whatever the temperature, people always go to the mall.

PowerPoint

Lesson Quiz

Time	Temperature
7 A.M.	13°F
8 A.M.	18°F
9 A.M.	25°F
10 A.M.	32°F

1. Make a scatter plot of the data in the table and describe the trend. **a positive trend**

Temperature

Temp. (°F) 40 30 20 10

7 8 9 10

Time (A.M.)

19. **See back of book.**

Enrichment 11-7 L4

Reteaching 11-7

Exploring Scat L2

Gilbert is investigating the relationship between the number of credit cards a person has and the amount of credit card debt.

First, he made a table from his data.

Credit Cards and Credit Card Debt

Number of Cards	Amount of Debt
1	$0
1	$1,000
1	$5,000
2	$3,000
2	$5,000
3	$10,000
3	$5,000
3	$8,000
4	$10,000
5	$19,000

Then he plotted the data in a scatter plot.

Credit Card Debt

Amount of Debt: $20,000, 15,000, 10,000, 5,000, 0

Credit Cards: 1 2 3 4 5

Gilbert's scatter plot shows a **positive trend** in the data. That means as the number of credit cards goes up, so does the amount of debt. As one value goes up, so does the other.

In a **negative trend,** one value goes up while the other goes down.

Complete the scatter plot for the data.

1. Dana surveyed her friends about how much TV they watch and their average test scores. Her results are shown below.

Test Scores and TV

TV Hours Per Day	Average Test Score	TV Hours Per Day	Average Test Score
1	98	3	79
1	86	3	73
2	90	3	75
2	82	4	62
2	85	5	68

Test Scores and TV

Test Scores: 100, 90, 80, 70, 60, 50, 0

Hours of TV: 1 2 3 4 5

2. Is the trend in the data negative or positive? Explain.
Negative; as one value goes up, the other goes down.

3. Describe the relationship Dana likely found between test scores and TV time.
The more TV students watch, the lower their test scores.

GO Online
Homework Video Tutor
Visit: PHSchool.com
Web Code: are-1107

16. **C; as the temperature rises, more people will go to the beach.**

17. **Positive; the values are both increasing.**

18. **As ticket cost increased, admissions would also; no; as the values in one data set increase, so do values in the other.**

16. (GPS) Carmella made a scatter plot comparing the daily temperature and the number of people at a beach. Which of the three scatter plots below most likely represents the data? Explain your choice.

A.
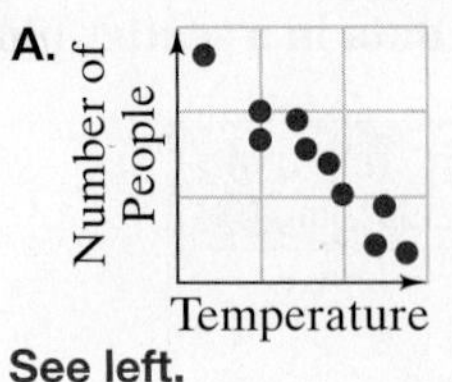

B.
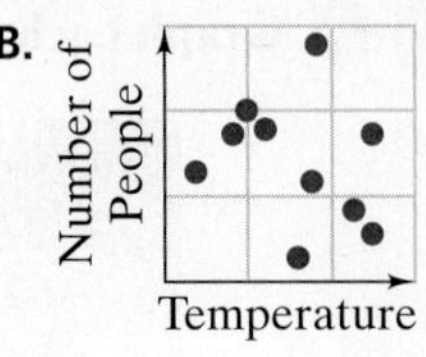

C.
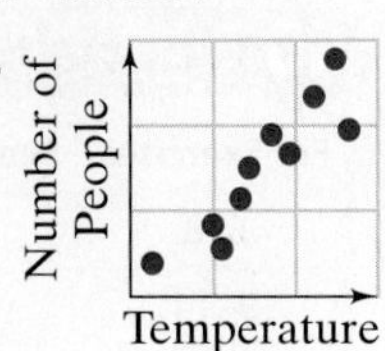

See left.

Make a scatter plot using the data in the table. Use the average ticket cost and the number of admissions.

Movie Attendance

Year	Average Ticket Cost (dollars)	Admissions (millions)
1997	4.59	1,388
1998	4.69	1,481
1999	5.08	1,465
2000	5.39	1,421
2001	5.66	1,487
2002	5.81	1,689
2003	6.03	1,574
2004	6.21	1,536

SOURCE: Motion Picture Association of America

17. **Data Analysis** What kind of trend do you see? Explain.

18. **Writing in Math** How would the scatter plot change if you switched the two axes? Would the trend change? Explain.

Ⓒ 19. **Challenge** Make a scatter plot with "Year" as the horizontal scale and "Admissions" as the vertical scale. Predict the number of admissions in 2010. **See margin.**

Test Prep and Mixed Review

Practice

Multiple Choice

20. The survey results of 23 students are shown. Which statement is supported by the graph? **B**

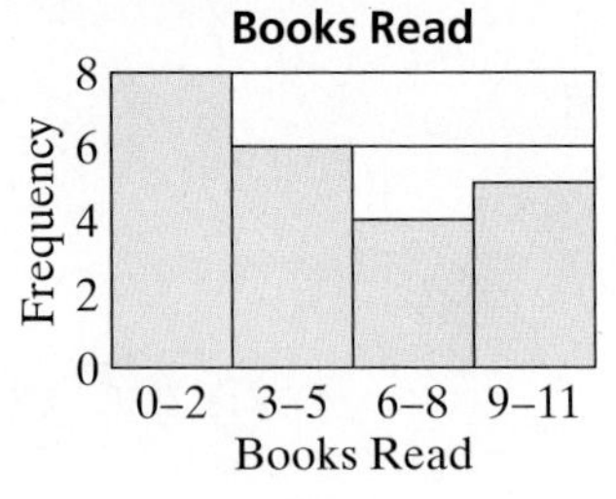

Ⓐ Most students read 6–11 books.
Ⓑ Most students read 0–5 books.
Ⓒ The median was about 7 books.
Ⓓ The mean was about 6 books.

21. Which sequence follows the rule $3n + 1$, where n represents the position of a term in the sequence? **F**

Ⓕ 4, 7, 10, 13, 16, . . .
Ⓖ 4, 8, 13, 16, 20, . . .
Ⓗ 1, 7, 10, 13, 19, . . .
Ⓙ 4, 6, 7, 8, 9, . . .

22. Which statement is always true about a right isosceles triangle? **D**

Ⓐ It has 3 congruent sides.
Ⓑ It has no congruent sides.
Ⓒ It has three 60° angles.
Ⓓ It has two acute angles.

23. There are 35 marked seals in a region. Biologists count 36 seals, of which 8 are marked. About how many seals are in the region? **158 seals**

GO for Help

For Exercise	See Lesson
23	11-5

Test Prep

Resources
For additional practice with a variety of test item formats:
- Test-Taking Strategies, p. 571
- Test Prep, p. 575
- Test-Taking Strategies with Transparencies

Alternative Assessment

Have students work in pairs. Each student writes a set of data for a scatter plot. Partners exchange papers, make a scatter plot using their partner's data, and describe any trends. Students check each other's scatter plot and descriptions.

Test-Taking Strategies

Interpreting Data

Before you answer a question that involves data, make sure you understand the information displayed in the graph. Then try to relate each of the answer choices to the data.

EXAMPLE

The stem-and-leaf plot at the right shows the different costs of hair dryers. What is the mode hair dryer price?

Ⓐ $4 Ⓑ $20 Ⓒ $24 Ⓓ $25

The stems are tens digits. So the hair dryer prices range from $15 to $35. Choice A is the mode of the leaves, not of the data. Choice B is the range of the data. Choice D is the median. The correct answer is C.

Hair Dryer Prices

Stem	Leaves
1	5 7
2	0 4 4 6
3	0 1 3 5

Key: 1 | 7 means $17.

Exercises

Use the double bar graph.

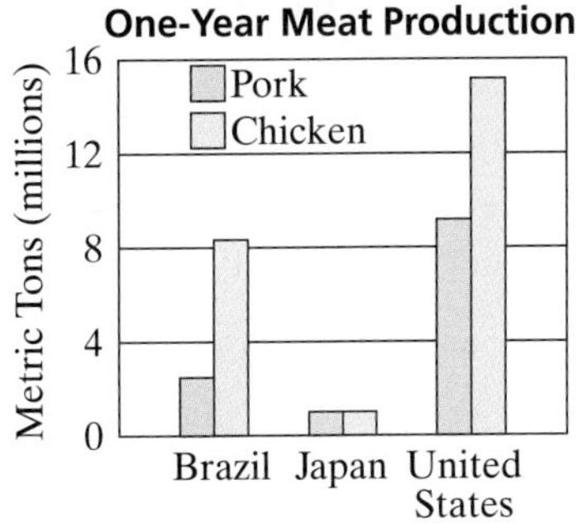

SOURCE: U.S. Department of Agriculture

1. What is the approximate range of chicken production? **A**
   Ⓐ 14,000,000 metric tons
   Ⓑ 11,000,000 metric tons
   Ⓒ 7,000,000 metric tons
   Ⓓ 4,000,000 metric tons

2. Which statement is NOT supported by the graph? **H**
   Ⓕ The mean of pork produced is about 4,000,000 metric tons.
   Ⓖ The mean of chicken produced is about 8,000,000 metric tons.
   Ⓗ All countries produced more chicken than pork.
   Ⓙ Brazil produced more pork than Japan.

3. What is the approximate median of pork production? **B**
   Ⓐ 1,200,000 metric tons
   Ⓑ 2,600,000 metric tons
   Ⓒ 6,700,000 metric tons
   Ⓓ 9,300,000 metric tons

Interpreting Data

Students learn to use and interpret data to answer questions. They learn to analyze the data that is presented and then relate it to the answer choices.

Guided Instruction

Explain to students that data can often help them choose the correct answer to a question. Remind them about misleading ways to present data and how they can redraw a graph to be more accurate.

Example

Look at the stem-and-leaf plot and make sure students understand what data it shows. Then ask:

- *How is the mode represented in the stem-and-leaf plot?* **The number that repeats most often in the data is the mode.**

Resources

Test-Taking Strategies with Transparencies

- Transparency 12
- Practice Sheet, p. 23

Test-Taking Strategies with Transparencies

Test-Taking Strategies: Interpreting Data

Before you answer a question that involves interpreting data from a table, graph, or plot, make sure that you understand the information displayed.

Example The bar graph below shows the favorite fruit choices of a class of students. Which statement is best supported by the graph?

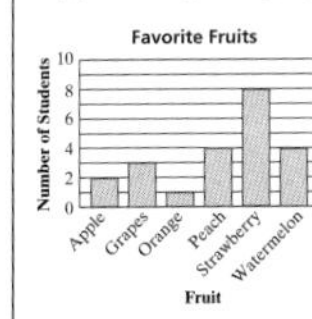

A. More students chose grapes than chose apple and orange combined.

B. The fewest students chose apple.

C. The same number of students chose peach as chose watermelon.

D. More than half the students chose strawberry.

- 3 students chose grapes, 2 students chose apple, and 1 chose orange. Since $3 \not> 2 + 1$, eliminate choice A.
- 2 students chose apple. This is not the fewest, since only 1 student chose orange. Eliminate choice B.
- The total number of students choosing was $2 + 3 + 1 + 4 + 8 + 4 = 22$. 8 is not more than half of 22, so eliminate choice D.
- 4 students chose peach, and 4 students chose watermelon. This is the same number, so the correct answer is choice C.

Resources

Student Edition
Extra Skills and Word Problems Practice, Ch. 11, p. 650
English/Spanish Glossary, p. 675
Formulas and Properties, p. 673
Tables, p. 670

All-in-One Teaching Resources
Vocabulary and Study Skills 11F

Differentiated Instruction
Spanish Vocabulary Workbook with Study Skills **ELL**
Interactive Textbook
- Audio Glossary

Online Vocabulary Quiz

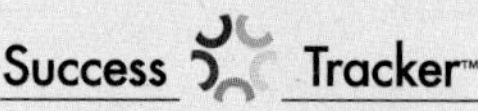

Online at PHSchool.com

Spanish Vocabulary/Study Skills ELL

Vocabulary/Study Skills L3

11F: Vocabulary Review For use with the Chapter Review

Study Skill Take notes while you study. Use a highlighter to emphasize important material in your notes.

Circle the term that correctly completes each sentence.

1. A flip that creates symmetry is a (*translation*, *reflection*).
2. The first number in an ordered pair is the (x, y) coordinate.
3. Lines in a coordinate plane that are parallel to the y-axis are (*horizontal*, *vertical*).
4. A (*line plot*, *frequency table*) uses a number line with "x" marks to represent each data item.
5. You can use a (*bar*, *circle*) graph to easily compare amounts.
6. The (*mean*, *median*) is the middle number in a data set when the values are written in order from least to greatest.
7. The (*mode*, *range*) of a data set is the difference between the greatest and least data values.
8. A (*line graph*, *scatter plot*) can be used to investigate the relationship between two sets of data.
9. A sequence is (*arithmetic*, *geometric*) if each term is found by adding the same number to the previous term.
10. (*Principal*, *Interest*) is the amount of money borrowed or deposited.
11. The (*area*, *surface area*) of a prism is the sum of the areas of the faces.
12. (*Circumference*, *Area*) is the distance around a circle.
13. The opposite of squaring a number is finding its (*square root*, *perfect square*).
14. The (*slope*, *bisector*) of a line segment is a line, segment, or ray that goes through the midpoint of the segment.

Vocabulary and Study Skills

Chapter 11 Review

Vocabulary Review

biased question (p. 551)
cell (p. 538)
double bar graph (p. 539)
double line graph (p. 539)
frequency table (p. 532)
histogram (p. 533)
legend (p. 539)
line plot (p. 533)
negative trend (p. 568)
no trend (p. 568)
population (p. 550)
positive trend (p. 568)
random sample (p. 550)
sample (p. 550)
scatter plot (p. 567)
spreadsheet (p. 538)
stem-and-leaf plot (p. 544)

Go Online PHSchool.com
For: Online vocabulary quiz
Web Code: arj-1151

Choose the correct term to complete each sentence.

1. A (line, stem-and-leaf) plot separates the digits of the data. **stem-and-leaf**
2. A (negative, positive) trend involves a set of values that increases as another set of values decreases. **negative**
3. A (cell, legend) identifies the data being compared. **legend**
4. A (line plot, scatter plot) shows data by stacking ✗s above values. **line plot**
5. A (population, random sample) is a whole group. **population**

Skills and Concepts

Lesson 11-1
- To represent data using frequency tables, line plots, and histograms

A **frequency table** lists data items with the number of times each item occurs. A **line plot** shows data by stacking ✗'s above data values on a number line. A **histogram** is a bar graph with no spaces between bars.

6. Make a frequency table and a line plot for the number of hours of TV watched per person per week:
5 7 9 5 3 6 8 6 5 7 6 8 7 7 6 5 4 4 5 6. **See margin.**
7. Use the table below to make a histogram. **See margin.**

How Many Pencils or Pens Are in Your Backpack?

Number of Pencils or Pens	0–4	5–9	10–14	15–19
Frequency	6	13	7	4

Airplanes The line plot at the right shows responses to a survey.

How Many Times Have You Flown in an Airplane?

✗		✗			
✗	✗	✗			
✗	✗	✗	✗		
✗	✗	✗	✗		✗
1	2	3	4	5	6

8. What do the numbers in the line plot represent? **8. the number of times a person has flown**
9. How many people answered the survey? **14 people**

6–7. See back of book.
13. See back of book.
17. The y-axis interval starts at 93, so it appears that the student who studied 6 h did twice as well as the student who studied 4 h.

Lessons 11-2, 11-3

- To interpret spreadsheets, double bar graphs, and double line graphs
- To represent and interpret data using stem-and-leaf plots

A **double bar graph** compares two sets of data. A **double line graph** compares changes over time of two sets of data. A **stem-and-leaf plot** uses the digits of each number to show the shape of the data.

Education **Use the double bar graph for Exercises 10–12.**

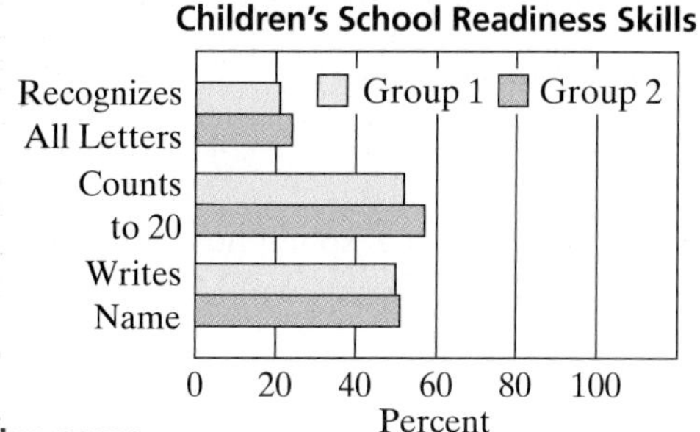

10. In which skill did Group 2 excel the most? **counts to 20**
11. In which skill did the two groups differ the most? **counts to 20**
12. In which skill did the two groups differ the least? **writes name**
13. Draw a stem-and-leaf plot for the temperatures below. **See margin.**

    48 54 45 60 50 70 66 69 40 61 50 60 58
    47 40 27 23 60 47 40 29 16 55 36 19 27

Lessons 11-4, 11-5

- To identify a random sample and to write a survey question
- To estimate population size using proportions

A **biased question** makes some answers appear better than others. You can use the capture/recapture method to estimate **population** size.

Tell whether each question is *biased* or *fair.* Explain.

14. What is your favorite activity after school?
    Fair; the question makes no assumptions about the activity.
15. Do you like the calm, soothing ocean?
    Biased; the question assumes the ocean is calm and soothing.
16. **Biology** Researchers know that there are 53 marked wolves in an area. On a flight over the area, they count 18 marked wolves and a total of 125 wolves. Estimate the total wolf population.
    about 368 wolves

Lessons 11-6, 11-7

- To identify misleading graphs and statistics
- To draw and interpret scatter plots

Graphs can mislead if they use unequal intervals or improper breaks on an axis. A **scatter plot** relates two sets of data, showing whether the data have a **positive trend,** a **negative trend,** or **no trend.**

17. How is the graph at the right misleading?
    See margin.

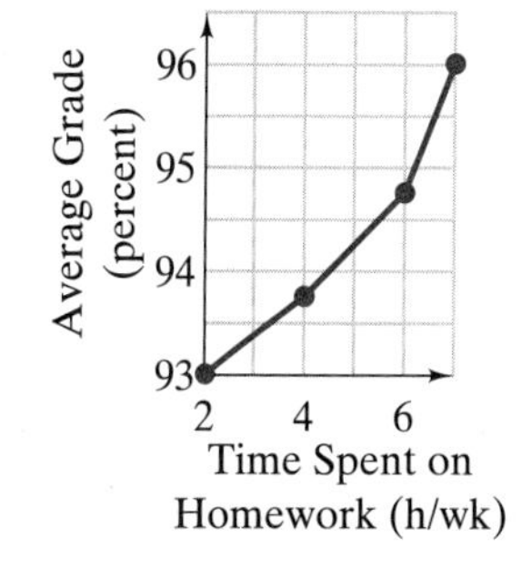

Describe the trend in each scatter plot.

18. 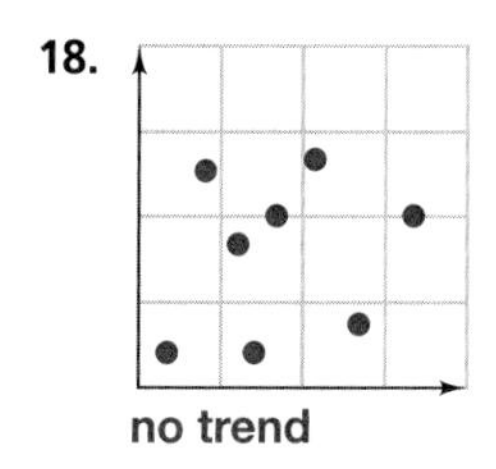

    no trend

19. 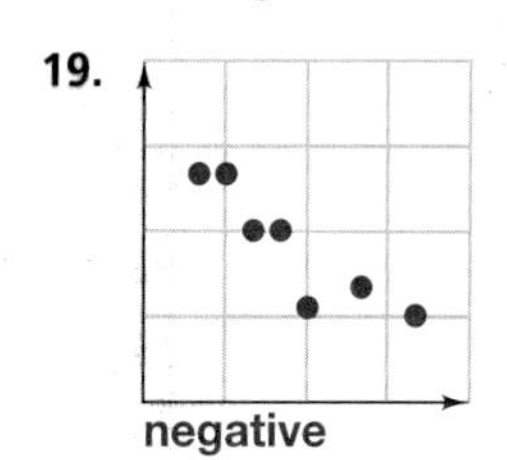

    negative

Alternative Assessment L4

Alternative Assessment Form C

Chapter 11

AT YOUR SERVICE

Jill owns an ice cream parlor. She needs your help in making a schedule to show which employees will work on which Saturdays. First she studied her cash register receipts from last Saturday. Then she made the following chart to show the number of customers she had each hour.

CUSTOMERS AT JILL'S ICE CREAM PARLOR		
From	Until	Number of Customers
11:00 A.M.	12:00 Noon	63
12:00 Noon	1:00 P.M.	120
1:00 P.M.	2:00 P.M.	102
2:00 P.M.	3:00 P.M.	86
3:00 P.M.	4:00 P.M.	50
4:00 P.M.	5:00 P.M.	29
5:00 P.M.	6:00 P.M.	56
6:00 P.M.	7:00 P.M.	87
7:00 P.M.	8:00 P.M.	143
8:00 P.M.	9:00 P.M.	119
9:00 P.M.	10:00 P.M.	86
10:00 P.M.	11:00 P.M.	58

Show all of your work on a separate sheet of paper.

1. Make a graph to illustrate the data above.
2. When planning each Saturday's schedule, Jill likes to follow certain guidelines. Using the guidelines listed below, make up a time schedule for this Saturday, based on the number of customers Jill had last Saturday.
   - No person works more than 4 hours in a day.
   - No person works fewer than 2 hours in a day.
   - In addition to the servers, there is always 1 person working the cash register each hour.
   - There is always 1 server working for every 30 customers in the store. Jill always rounds up the number of servers needed to the nearest whole number. So, if there are 31 customers in an hour, 2 servers should be working, and so on.

Resources

- ExamView Assessment Suite CD-ROM
  - Chapter 11 Ready-Made Test
  - Make your own Chapter 11 test
- MindPoint Quiz Show CD-ROM
  - Chapter 11 Review

Differentiated Instruction

All in One Teaching Resources
- Below Level Chapter 11 Test L2
- Chapter 11 Test L3
- Chapter 11 Alternative Assessment L4

Spanish Assessment Resources ELL
- Below Level Chapter 11 Test L2
- Chapter 11 Test L3
- Chapter 11 Alternative Assessment L4

ExamView Assessment Suite CD-ROM
- Special Needs Test L1
- Special Needs Practice Bank L1

Online Chapter 11 Test at www.PHSchool.com L3

Below Level Chapter Test L2

Chapter Test L3

Chapter Test Form A

Chapter 11

Fifteen students were asked, "How many hours of TV did you watch last night?" Use the responses below to make each data display.

1 0 1 4 6 3.5 2.5 2 1 1.5 2.5 3.5 3.5 4 5.5

1. frequency table

Number of Hours	Tally	Frequency
0	|	1
1	|||	3
1.5	|	1
2	|	1
2.5	||	2
3.5	|||	3
4	||	2
5.5	|	1
6	|	1

2. line plot

3. histogram

4. stem-and-leaf plot

Key: 6 | 0 means 6.0

Stem	Leaves
0	0
1	0 0 0 5
2	0 5 5
3	5 5 5
4	0 0
5	5
6	0

5. The following are your scores on science tests.
86 80 91 86 95 100 Sample answers given.

a. Which is the most impressive representation of how well you are doing in class: the mean, the mode, or the median of your scores? Explain.
The mean, 89.66, is higher than the median, 88.5, or the mode, 86.

b. Which is the most accurate? Explain.
The mean of 89.66 averages the actual scores.

Chapter 11 Test

Go Online PHSchool.com For: Online chapter test Web Code: ara-1152

A pollster asks 20 people how many hours they sleep each night. Use the data below to draw each data display in Exercises 1–5.

8 7.5 9 7 8 7.5 6 6.5 9.5 7.5
8 7.5 8 7 8 6.5 8 8.5 8.5 7.5

1. frequency table
2. line plot
3. histogram
4. stem-and-leaf plot
5. **Reasoning** Explain how the stem-and-leaf plot in Exercise 4 would change if the person who said 9 had said 12.

1–5. See margin.

Use the spreadsheet below for Exercises 6–10.

	A	B	C	D
1	Student	Test 1	Test 2	Quiz
2	Alice	86	85	8
3	Xavier	91	89	9
4	Timotheo	79	84	8

6. What is the content of cell A3? **Xavier**
7. What is the content of cell C2? **85**
8. In which cell is the word *Quiz?* **D1**
9. In which cell is the value 79? **B4**
10. In which cell is the word *Student?* **A1**

Is each question *biased* or *fair?* Explain. 11–14. See margin.

11. Do you prefer watching comedies or dramas?
12. Do you like watching violent sports or informative documentaries on TV?
13. Do you like sunny summers or dark winters?
14. Which season do you like best?
15. **Writing in Math** Explain how you can get a random sample of the people who use a town library. See margin.

Sales **Use the double line graph below.**

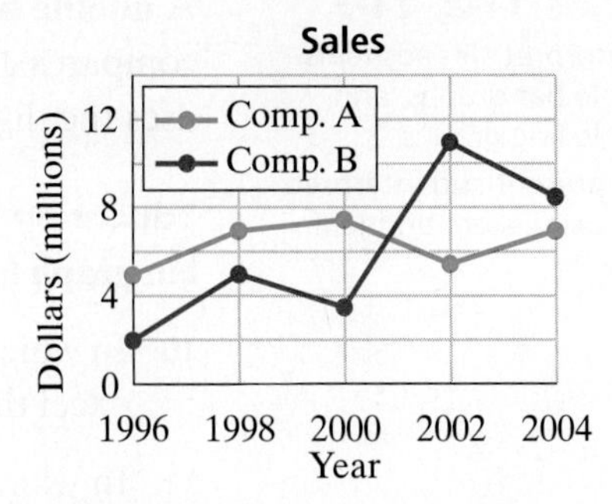

16. Estimate the year in which the sales were equal. **2001**
17. In which year were sales lowest for each company? **1996**
18. Redraw the graph to emphasize the highest sales for each company. See margin.
19. **Wages** Five students work at a store. Their hourly earnings are \$8, \$7.50, \$12, \$8.50, and \$8. One student who earns \$8 wants a raise. Should he use the mean, the median, or the mode to convince the boss? **mean**

Use a proportion to estimate each population.

20. total counted: 102 **about 150**
    tagged counted: 38
    total tagged: 56
21. total counted: 958 **about 1,391**
    tagged counted: 210
    total tagged: 305
22. a. Make a scatter plot of the data below.

Students' Ages and Heights

Age	12	13	12	14	14	13
Height (in.)	57	60	56	63	65	61

b. Describe the trend in the scatter plot.
22a–b. See margin.

1–5. See back of book.
11–15. See back of book.
18. See back of book.
22a–b. See back of book.

Reading Comprehension

Read each passage and answer the questions that follow.

Baker Coordinates Mr. Baker, a math teacher, experimented with points on a coordinate plane. He invented an operation that he named after himself. To "bake" the point (x, y), make a new point with an x-coordinate that is the square of the original x-coordinate and with a y-coordinate that is the original y-coordinate.

1. If you bake the point (x, y), what ordered pair describes the coordinates of the new point? **B**
   - (A) (x, y)
   - (B) (x^2, y)
   - (C) (x, y^2)
   - (D) (x^2, y^2)

2. If you bake a point in the second quadrant, in which quadrant is the new point located? **F**
   (F) I (G) II (H) III (J) IV

3. If you bake a point in the third quadrant, in which quadrant is the new point located? **D**
   (A) I (B) II (C) III (D) IV

4. If a point not on an axis has been baked, in which quadrants could the new point be located? **H**
   - (F) either the first or second
   - (G) either the first or third
   - (H) either the first or fourth
   - (J) either the second or third

Capital Geometry In 1790, George Washington hired Pierre L'Enfant to design a capital city. L'Enfant created an orderly grid for Washington, D.C. Streets that run north and south are numbered, and streets running east and west have letter names. Numbers and letters begin at the Capitol and run in both directions, so there are two 10th Streets and two K Streets. When you give an address, you also have to state its quadrant. The quadrants correspond to the four quadrants on a coordinate plane, but people call them by their compass directions: NE, SE, NW, and SW.

5. Ford's Theatre is located at 511 10th St. NW. In what quadrant is this? **B**
   (A) I (B) II (C) III (D) IV

6. Which word describes the location of the Capitol on the street grid of Washington, D.C.? **G**
   - (F) intercept
   - (G) origin
   - (H) perimeter
   - (J) vertex

7. How many points are there where a 4th Street intersects an I Street? **C**
   (A) one (B) two (C) four (D) eight

8. A race is run along K Street. It starts at the intersection of 6th Street in NW, and goes to the point where K Street meets 6th Street in NE. What distance does the race cover? **H**
   - (F) 4 blocks
   - (G) 8 blocks
   - (H) 12 blocks
   - (J) 24 blocks

Resources

Test Prep Workbook

All in One Teaching Resources
- Cumulative Review L3

ExamView Assessment Suite CD-ROM
- Standardized Test Practice

Differentiated Instruction

Spanish Assessment Resources
- Spanish Cumulative Review ELL

ExamView Assessment Suite CD-ROM
- Special Needs Practice Bank L1

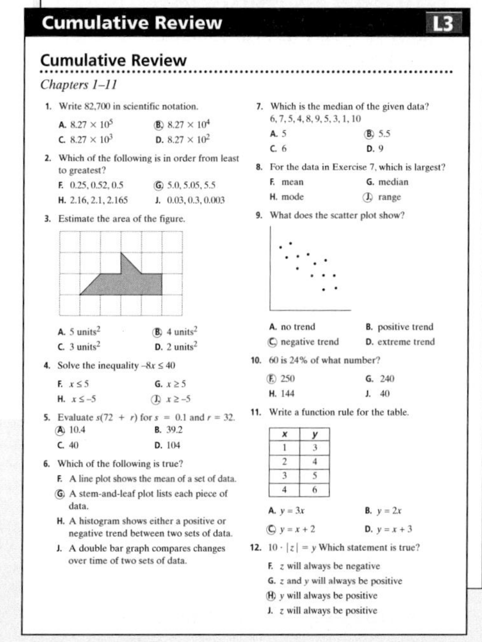

Spanish Cumulative Review ELL

Cumulative Review L3

Cumulative Review

Chapters 1–11

1. Write 82,700 in scientific notation.
   A. 8.27×10^5 (B.) 8.27×10^4 C. 8.27×10^3 D. 8.27×10^2
2. Which of the following is in order from least to greatest?
   F. 0.25, 0.52, 0.5 (G.) 5.0, 5.05, 5.5 H. 2.16, 2.1, 2.165 J. 0.03, 0.3, 0.003
3. Estimate the area of the figure.
   A. 5 units² (B.) 4 units² C. 3 units² D. 2 units²
4. Solve the inequality $-8x \leq 40$
   F. $x \leq 5$ G. $x \geq 5$ H. $x \leq -5$ (J.) $x \geq -5$
5. Evaluate $s(72 + r)$ for $s = 0.1$ and $r = 32$.
   (A.) 10.4 B. 39.2 C. 40 D. 104
6. Which of the following is true?
   F. A line plot shows the mean of a set of data.
   (G.) A stem-and-leaf plot lists each piece of data.
   H. A histogram shows either a positive or negative trend between two sets of data.
   J. A double bar graph compares changes over time of two sets of data.
7. Which is the median of the given data? 6, 7, 5, 4, 8, 9, 5, 3, 1, 10
   A. 5 (B.) 5.5 C. 6 D. 9
8. For the data in Exercise 7, which is largest?
   F. mean G. median H. mode (J.) range
9. What does the scatter plot show?
   A. no trend B. positive trend (C.) negative trend D. extreme trend
10. 60 is 24% of what number?
   (F.) 250 G. 240 H. 144 J. 40
11. Write a function rule for the table.

x	y
1	3
2	4
3	5
4	6

   A. $y = 3x$ B. $y = 2x$ (C.) $y = x + 2$ D. $y = x + 3$
12. $10 \cdot |z| = y$ Which statement is true?
   F. z will always be negative
   G. z and y will always be positive
   (H.) y will always be positive
   J. z will always be positive

Problem Solving

Applying Data Analysis

Students will use data from these two pages to answer the questions posed in Put It All Together.

Remind students that a way to change the appearance of a graph is to shorten or compress one or both of the scales. Graph the same simple set of data twice. Show a dramatic rise in one graph and a mild rise in the other. Show how the changes affect the appearance of the data.

Materials
- graph paper

Activating Prior Knowledge

Have students draw on their experiences to discuss trends in products that are popular in their age group. Ask:
- *Can you name a style or brand of a tech item, such as a cell phone, that was really popular among students a few years ago but is less popular now?* **Answers will vary.**
- *Why did people stop buying it?* **Sample: People got tired of that style, brand, or model; A better one came out; A new style was introduced.**

Guided Instruction

In the activity, students manipulate the horizontal or vertical scale of a graph to create a particular impression of a set of data. Plot the data for all three bicycle companies on a single line graph. Then plot it again on a graph where the scale for sales is doubled or tripled along the vertical axis. Ask:
- *According to the first graph, which company's sales are the best?* **Deals on Wheels** *the worst?* **Better Bikes**
- *According to the second graph, which company's sales are the best?* **Deals on Wheels** *the worst?* **Better Bikes**

Compare the similarities and differences of the two graphs.

DK Problem Solving Application

Applying Data Analysis

Bicycle Business Line graphs can communicate information more quickly than data tables because plots show how the data points are related. But line graphs can also be misleading. They can make a downward slide seem slight or an upward spike seem like a significant trend. You have to look critically at line graphs to see whether you're being informed . . . or misinformed.

Put It All Together

Materials graph paper

1. The table shows sales data for three bicycle companies. Pick one of the bicycle companies and activities on page 577. In each case, put the year on the horizontal axis and the sales on the vertical axis.
2. **Writing in Math** Compare your two line graphs. Which gives the more accurate view of the company's actual performance? Explain.
3. **Research** Find a graph in a newspaper. Is the graph a fair representation of the data? How could you change the graph to give a different impression?

Annual Sales for Three Bicycle Manufacturers (thousands of dollars)

Year	Better Bikes	Deals on Wheels	Super Cycles
1996	$2,520	$2,920	$3,210
1997	$2,369	$3,008	$3,082
1998	$2,298	$2,978	$2,958
1999	$2,160	$2,978	$3,106
2000	$2,073	$3,007	$2,920
2001	$1,991	$3,158	$2,832
2002	$1,871	$3,316	$2,889
2003	$1,778	$3,448	$3,062

Answers may vary. Samples are given for Exercises 1–2.

1. **See margin p. 577.**
2. **The first graph gives a slightly more accurate impression of the company's sales performance. Sales have declined steadily by about 30% from 1996 to 2003. The second graph gives a more dramatic impression of the same decrease because the vertical scale is smaller than the vertical scale of the first graph.**
3. **Check students' work.**

BETTER BIKES
Sales have gone down since 1996.
A. Draw an optimistic line graph for the next stockholders' meeting.
B. Draw another line graph for the president of the company. Make the drop in sales look very serious.
Deals on Wheels
Sales have generally gone up over the last eight years.
A. Draw a line graph that will help convince the bank to give the company a big loan.
B. Draw a line graph showing employees why they can't have big raises this year.
Super Cycles
Although sales have gone up and down over the last eight years, sales rose from 2001 to 2003.
A. Draw a line graph that makes Super Cycles look incredibly successful by the year 2010. Use a dashed line to extend the graph.
B. Draw a line graph that a possible buyer of the company might use.
Go Online
PHSchool.com
For: Information about bicycles
Web Code: are-1153

Activity

Have a volunteer read aloud the opening paragraph about data analysis and the bicycle business. Then discuss with students why anyone would wish to manipulate the data in a graph to create a particular appearance or impression. Then have students work in a pairs to answer the questions and make the graphs.

Exercise 1 Discuss that each money amount in the table is $\frac{1}{1,000}$ of the actual amount. Each is presented in this form for convenience of use. Ask: *What were the sales for Super Cycles in 1997?* **$3,082,000**

Differentiated Instruction

Special Needs L1

As needed, review the distinction between *horizontal* and *vertical*. Discuss the meaning of the term "impression." Help students to read the table. Then help pairs to choose sensible intervals and to set up the scales for their graphs.

1.

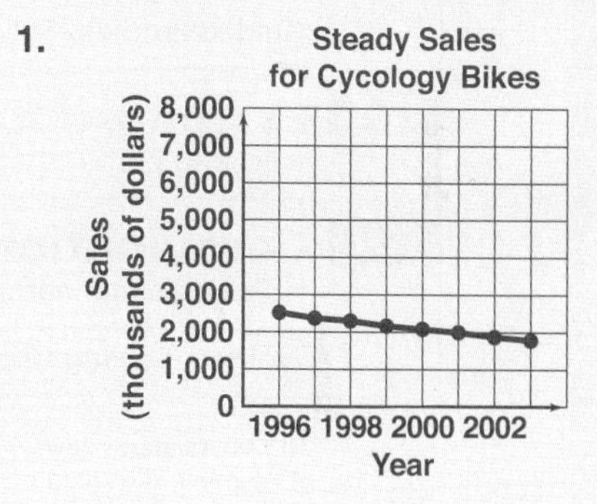

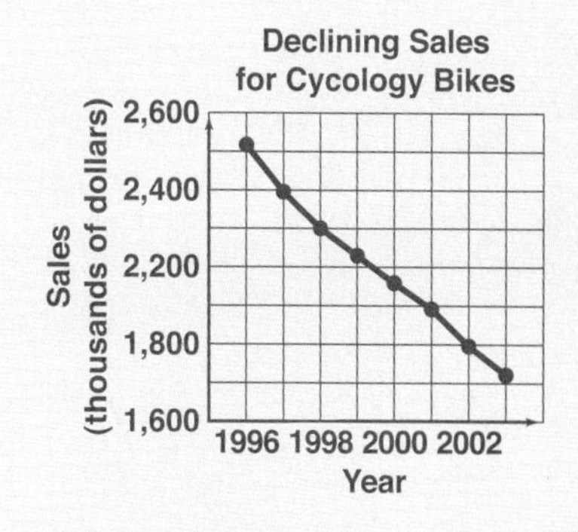

12 Using Probability

Chapter at a Glance

Lesson Titles, Objectives, and Features	Assessment	NCTM Standards	Local Standards
12-1 Probability • To find the probability and complement of an event **Extension:** Odds	Lesson Quiz	1, 2, 5, 6, 7, 8, 9, 10	
12-2a Activity Lab, Hands On: Exploring Probability **12-2 Experimental Probability** • To find the experimental probability and to use simulations **12-2b Activity Lab, Technology:** Random Numbers	Lesson Quiz	1, 2, 5, 6, 7, 8, 9, 10	
12-3 Sample Spaces • To make and use sample spaces and to use the counting principle **12-3b Activity Lab, Data Analysis:** Using Data to Predict	Lesson Quiz Checkpoint Quiz 1	1, 2, 3, 5, 6, 7, 8, 9, 10	
12-4a Activity Lab: Exploring Multiple Events **12-4 Compound Events** • To find the probability of independent and dependent events **Vocabulary Builder:** High-Use Academic Words **Guided Problem Solving:** Practice With Probability	Lesson Quiz	1, 2, 5, 6, 7, 8, 9, 10	
12-5 Permutations • To find permutations	Lesson Quiz	1, 2, 5, 6, 7, 8, 9, 10	
12-6 Combinations • To find combinations	Lesson Quiz Checkpoint Quiz 2	1, 2, 5, 6, 7, 8, 9, 10	
Problem Solving Application: Applying Probability			

NCTM Standards 2000
1 Number and Operations **2** Algebra **3** Geometry **4** Measurement **5** Data Analysis and Probability
6 Problem Solving **7** Reasoning and Proof **8** Communication **9** Connections **10** Representation

Correlations to Standardized Tests

All content for these tests is contained in *Prentice Hall Math,* Course 2. This chart reflects coverage in this chapter only.

	12-1	12-2	12-3	12-4	12-5	12-6
Terra Nova CAT6 (Level 17)						
Number and Number Relations	✔	✔	✔	✔	✔	✔
Computation and Numerical Estimation	✔	✔	✔	✔	✔	✔
Operation Concepts	✔	✔	✔	✔	✔	✔
Measurement						
Geometry and Spatial Sense						
Data Analysis, Statistics, and Probability	✔	✔	✔	✔	✔	✔
Patterns, Functions, Algebra	✔	✔	✔	✔	✔	✔
Problem Solving and Reasoning	✔	✔	✔	✔	✔	✔
Communication	✔	✔	✔	✔	✔	✔
Decimals, Fractions, Integers, and Percent	✔	✔	✔	✔	✔	✔
Order of Operations						
Terra Nova CTBS (Level 17)						
Decimals, Fractions, Integers, Percents	✔	✔	✔	✔	✔	✔
Order of Operations, Numeration, Number Theory	✔	✔	✔	✔	✔	✔
Data Interpretation	✔	✔	✔	✔	✔	✔
Pre-algebra	✔	✔	✔	✔	✔	✔
Measurement						
Geometry						
ITBS (Level 13)						
Number Properties and Operations	✔	✔	✔	✔	✔	✔
Algebra	✔	✔	✔	✔	✔	✔
Geometry						
Measurement						
Probability and Statistics	✔	✔	✔	✔	✔	✔
Estimation						
SAT10 (Int 3 Level)						
Number Sense and Operations	✔	✔	✔	✔	✔	✔
Patterns, Relationships, and Algebra	✔	✔	✔	✔	✔	✔
Data, Statistics, and Probability	✔	✔	✔	✔	✔	✔
Geometry and Measurement						
NAEP						
Number Sense, Properties, and Operations						
Measurement						
Geometry and Spatial Sense						
Data Analysis, Statistics, and Probability	✔	✔	✔	✔	✔	✔
Algebra and Functions						

CAT6 California Achievement Test, 6th Ed.
CTBS Comprehensive Test of Basic Skills
ITBS Iowa Test of Basic Skills, Form M
SAT10 Stanford Achievement Test, 10th Ed.
NAEP National Assessment of Educational Progress 2005 Mathematics Objectives

Math Background

Professional Development

Skills Trace

BEFORE Chapter 12

Course 1 introduced probability, including simulation and independent events.

DURING Chapter 12

Course 2 reviews and extends the study of probability to include dependent events, compound events, and combinations.

AFTER Chapter 12

Throughout this course, students build a foundation for probability by using ratios, proportions, percents, fractions, and decimals.

12-1 Probability

Math Understandings

- In mathematics, probability is expressed as a number that estimates how often an event will occur.
- Probabilities range from 0 (impossible) to 1 (certain).

An **outcome** is the result of an action. An **event** is an outcome or group of outcomes. If all the outcomes are equally likely, you can use a formula to find the theoretical probability.

Theoretical Probability
theoretical probability = $P(\text{event}) = \dfrac{\text{number of favorable outcomes}}{\text{total number of possible outcomes}}$

The probability of an impossible event is 0. The probability of a certain event is 1.

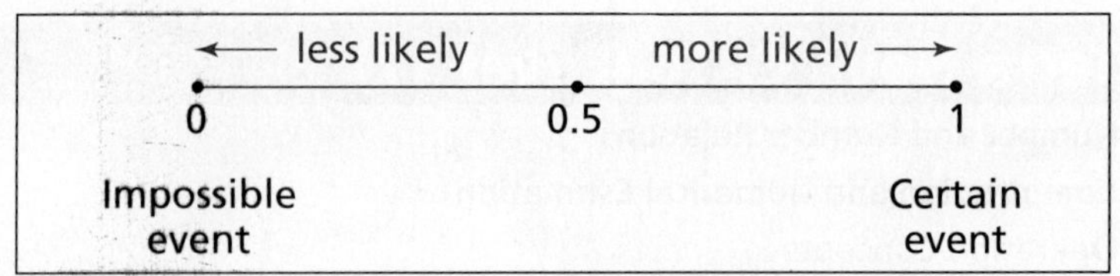

The **complement** of an event is the collection of outcomes not contained in the event.

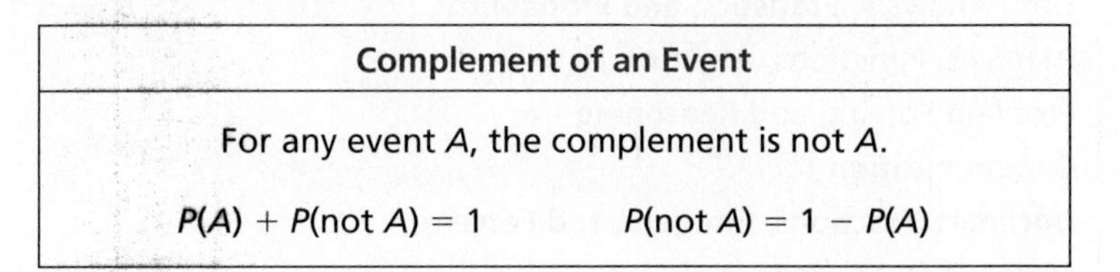

Complement of an Event
For any event A, the complement is not A.
$P(A) + P(\text{not } A) = 1$ $P(\text{not } A) = 1 - P(A)$

12-2 Experimental Probability

Math Understandings

- Experimental probability is based on the results of an actual experiment; theoretical probability is based on the assumption that certain outcomes are equally likely.
- As the number of events increases, the ratio of the experimental results approaches the ratio that is the theoretical probability (the Law of Large Numbers).
- It is a fallacy that, when tossing a fair coin, you are more likely to get heads after tossing a series of tails. The probability remains one out of two.

Probability based on experimental data or observations is called **experimental probability.** Some events are too difficult or time-consuming to perform. You can simulate, or model, problems to find experimental probabilities.

Experimental Probability
$P(\text{event}) = \dfrac{\text{number of favorable outcomes}}{\text{total number of possible outcomes}}$

12-3 Sample Spaces

Math Understandings

- To find the probability of an event, you need to know the total number of outcomes, or sample space.

The collection of all possible outcomes in an experiment is the **sample space.** You can use a tree diagram or the counting principle to find the total number of outcomes for an event.

The Counting Principle
Suppose there are m ways of making one choice and n ways of making a second choice. Then there are $m \times n$ ways to make the first choice followed by the second choice.

Example: If you can choose a shirt from 5 sizes and 7 colors, then you can choose 5×7, or 35 different shirts.

12-4 Compound Events

Math Understandings

- Two events are independent if one event does not affect the sample space of the other event.
- Two events are dependent if one event affects the sample space of the other event.

A **compound event** consists of two or more events. These events either depend on or do not depend on each other. Two events are **independent events** if the occurrence of one event does not affect the probability of the occurrence of the other. Two events are **dependent events** if the occurrence of one event affects the probability of the occurrence of the other event.

Example: If you draw a letter from a set of alphabet cards and, without replacing, draw a second letter, the probability of drawing a Q followed by a U is $\frac{1}{26} \times \frac{1}{25}$ or $\frac{1}{650}$.

12-5 Permutations

Math Understandings

- Order makes a difference in permutations.

A **permutation** is an arrangement of objects in a particular order. The product of all positive integers less than or equal to a number is a **factorial**. You write 5 factorial as 5! The number of permutations of n items, using all n in each arrangement, is n! (read as "n factorial").

Example: Five students present reports to a class in $5! = 5 \times 4 \times 3 \times 2 \times 1$, or 120 different ways.

The number of permutations of n items, using r of them at a time, is n! divided by $(n - r)!$

Example: Three out of five students present reports to a class in $\frac{5!}{2!} = 5 \times 4 \times 3$, or 60 different ways.

12-6 Combinations

Math Understandings

- Order does not matter in combinations.

A **combination** is a grouping of objects in which the order of the objects does not matter. The number of combinations of n items, using r of them at a time, is the number of permutations of n items taken r at a time divided by r!

Example: The combination of 6 objects taken 4 at a time is 15.

$$\frac{6!}{(6-4)!} \div 4! = 15$$

Additional Professional Development Opportunities

Math Background Notes for Chapter 12: Every lesson has a Math Background in the PLAN section.

Research Overview, Mathematics Strands
Additional support for these topics and more is in the front of the Teacher's Edition.

LessonLab
LessonLab, a Pearson Education company, offers comprehensive, facilitated professional development designed to help teachers to improve student achievement. To learn more, please visit lessonlab.com.

Chapter 12 Resources

	12-1	12-2	12-3	12-4	12-5	12-6	For the Chapter
Print Resources							
L3 Practice	●	●	●	●	●	●	
L1 Adapted Practice	●	●	●	●	●	●	
L3 Guided Problem Solving	●	●	●	●	●	●	
L2 Reteaching	●	●	●	●	●	●	
L4 Enrichment	●	●	●	●	●	●	
L3 Daily Notetaking Guide	●	●	●	●	●	●	
L1 Adapted Daily Notetaking Guide	●	●	●	●	●	●	
L3 Vocabulary and Study Skills Worksheets	●		●	●	●	●	●
L3 Daily Puzzles	●	●	●	●	●	●	
L3 Activity Labs	●	●	●	●	●	●	
L3 Checkpoint Quiz			●			●	
L3 Chapter Project							●
L2 Below Level Chapter Test							●
L3 Chapter Test							●
L4 Alternative Assessment							●
L3 Cumulative Review							●
Spanish Resources ELL							
L3 Practice	●	●	●	●	●	●	
L3 Vocabulary and Study Skills Worksheets	●		●	●	●	●	●
L3 Checkpoint Quiz			●			●	
L2 Below Level Chapter Test							●
L3 Chapter Test							●
L4 Alternative Assessment							●
L3 Cumulative Review							●
Transparencies							
Check Skills You'll Need	●	●	●	●	●	●	
Additional Examples	●	●	●	●	●	●	
Problem of the Day	●	●	●	●	●	●	
Classroom Aid							
Student Edition Answers	●	●	●	●	●	●	●
Lesson Quiz	●	●	●	●	●	●	
Test-Taking Strategies							●
Technology							
Interactive Textbook Online	●	●	●	●	●	●	●
StudentExpress™ CD-ROM	●	●	●	●	●	●	●
Success Tracker™ Online Intervention	●	●	●	●	●	●	●
TeacherExpress™ CD-ROM	●	●	●	●	●	●	●
PresentationExpress™ with QuickTake Presenter CD-ROM	●	●	●	●	●	●	●
ExamView® Assessment Suite CD-ROM	●	●	●	●	●	●	●
MindPoint® Quiz Show CD-ROM							●
Prentice Hall Web Site: PHSchool.com	●	●	●	●	●	●	●

Also available:

Prentice Hall Assessment System
- Progress Monitoring Assessments
- Skills and Concepts Review
- Test Prep Workbook

Other Resources
Algebra Readiness Tests
All-in-One Student Workbook
All-in-One Student Workbook, Adapted Version
Multilingual Handbook
Solution Key
Math Notes Study Folder
Spanish Cumulative Assessment

Where You Can Use the Lesson Resources

Here is a suggestion, following the four-step teaching plan, for how you can incorporate Differentiated Instruction Resources into your teaching.

	Instructional Resources L3	Differentiated Instruction Resources
1. Plan		
Preparation Read the Math Background in the Teacher's Edition to connect this lesson with students' previous experience. **Starting Class** **Check Skills You'll Need** Assign these exercises to review prerequisite skills. **New Vocabulary** Help students pre-read the lesson by pointing out the new terms introduced in the lesson.	**Math Background** **Math Understandings** **Transparencies & PresentationExpress™ with QuickTake Presenter CD-ROM** Check Skills You'll Need Problem of the Day **Resources** Vocabulary and Study Skills	**Spanish Support** ELL Vocabulary and Study Skills
2. Teach		
L3 **Guided Instruction** Use the Activity Labs to build conceptual understanding. Teach each Example. Use the Teacher's Edition side column notes for specific teaching tips, including Error Prevention notes. Use the Additional Examples found in the side column (and on transparency and PowerPoint) as an alternative presentation for the content. After each Example, assign the Quick Check exercise for that Example to get an immediate assessment of student understanding. Use the Closure activity in the Teacher's Edition to help students attain mastery of lesson content.	**Student Edition** Activity Lab **Resources** Daily Notetaking Guide Activity Lab **Transparencies & PresentationExpress™ with QuickTake Presenter CD-ROM** Additional Examples Classroom Aids **ExamView® Assessment Suite CD-ROM**	**Teacher's Edition** Every lesson includes suggestions for working with students who need special attention. L1 Special Needs L2 Below Level L4 Advanced Learners ELL English Language Learners **Resources** L1 Adapted Daily Notetaking Guide **Multilingual Handbook**
3. Practice		
Assignment Guide **Check Your Understanding** Use these questions to check students' understanding before you assign homework. **Homework Exercises** Assign homework from these leveled exercises in the Assignment Guide. A Practice by Example B Apply Your Skills C Challenge Test Prep and Mixed Review **Homework Quick Check** Use these key exercises to quickly check students' homework.	**Transparencies & PresentationExpress™ with QuickTake Presenter CD-ROM** Student Answers **Resources** Practice Guided Problem Solving Vocabulary and Study Skills Activity Lab Daily Puzzles **ExamView® Assessment Suite CD-ROM**	**Spanish Support** ELL Practice ELL Vocabulary and Study Skills **Resources** L1 Adapted Practice L4 Enrichment
4. Assess & Reteach		
Lesson Quiz Assign the Lesson Quiz to assess students' mastery of the lesson content. **Checkpoint Quiz** Use the Checkpoint Quiz to assess student progress over several lessons.	**Transparencies & PresentationExpress™ with QuickTake Presenter CD-ROM** Lesson Quiz **Resources** Checkpoint Quiz	**Resources** L2 Reteaching ELL Checkpoint Quiz Success Tracker™ Online Intervention **ExamView® Assessment Suite CD-ROM**

KEY L1 Special Needs  L2 Below Level L3 For All Students L4 Advanced, Gifted ELL English Language Learners

Using Probability

Answers are in the back of the textbook.

For intervention, direct students to:

Writing Ratios
Lesson 5-1
Extra Skills and Word Problems Practice, Ch. 5

Using Proportional Reasoning
Lesson 5-4
Extra Skills and Word Problems Practice, Ch. 5

Percents, Fractions, and Decimals
Lesson 6-2
Extra Skills and Word Problems Practice, Ch. 6

Spanish Vocabulary/Study Skills ELL

Vocabulary/Study Skills L3

12A: Graphic Organizer — For use before Lesson 12-1

Study Skill Try to read new lessons before your teacher presents them in class. Important information is sometimes printed in **boldface** type or highlighted inside a box or with color. Pay special attention to this information.

Write your answers.

1. What is the chapter title? Using Probability
2. How many lessons are there in this chapter? 7
3. What is the topic of the Test-Taking Strategies page? Eliminating Answers
4. Complete the graphic organizer below as you work through the chapter.
   - In the center, write the title of the chapter.
   - When you begin a lesson, write the lesson name in a rectangle.
   - When you complete a lesson, write a skill or key concept in a circle linked to that lesson block.
   - When you complete the chapter, use this graphic organizer to help you review. Check students' diagrams.

Vocabulary and Study Skills

CHAPTER 12 Using Probability

What You've Learned

- In Chapter 3, you simplified fractions.
- In Chapter 6, you converted between fractions, decimals, and percents.
- In Chapter 11, you made line plots, line graphs, bar graphs, stem-and-leaf plots, and scatter plots to represent data.

Check Your Readiness

GO for Help

For Exercises	See Lesson
1–6	5-1
7–12	5-4
13–22	6-2

Writing Ratios

Write each ratio in simplest form.

1. $\frac{9}{24}$ $\frac{3}{8}$
2. $\frac{20}{54}$ $\frac{10}{27}$
3. $\frac{15}{65}$ $\frac{3}{13}$
4. $\frac{16}{22}$ $\frac{8}{11}$
5. $\frac{21}{84}$ $\frac{1}{4}$
6. $\frac{18}{42}$ $\frac{3}{7}$

Using Proportional Reasoning

Algebra **Solve each proportion.**

7. $\frac{3}{10} = \frac{x}{30}$ 9
8. $\frac{n}{14} = \frac{25}{8}$ 43.75
9. $\frac{22}{c} = \frac{66}{15}$ 5
10. $\frac{16}{35} = \frac{20}{y}$ 43.75
11. $\frac{a}{24} = \frac{24}{9}$ 64
12. $\frac{19}{38} = \frac{f}{21}$ 10.5

Percents, Fractions, and Decimals

Write each decimal as a percent.

13. 0.46 46%
14. 0.265 26.5%
15. 0.07 7%
16. 0.256 25.6%
17. 0.82 82%

Write each fraction as a percent. When necessary, round to the nearest tenth of a percent.

18. $\frac{4}{5}$ 80%
19. $\frac{5}{11}$ 45.5% or 45%
20. $\frac{8}{14}$ 57.1%
21. $\frac{12}{30}$ 40%
22. $\frac{15}{32}$ 46.9%

Chapter 12 Overview

In this chapter, students work with both theoretical and experimental probability. They find the probability of both simple and compound events. They conclude the chapter by learning about permutations and combinations.

Activating Prior Knowledge

In this chapter, as they work with probability concepts, students will build on their knowledge of fractions, decimals, percents, ratios, and proportions. Ask questions such as:

- *How would you express the ratio $\frac{10}{16}$ in simplest form?* $\frac{5}{8}$
- *What is the value of x in the proportion $\frac{6}{x} = \frac{24}{8}$?* $x = 2$
- *What percent is equivalent to the fraction $\frac{7}{20}$?* 35%

What You'll Learn Next

- In this chapter, you will find the probabilities of independent and dependent events.
- You will make sample spaces to represent all the possible outcomes in a probability experiment.
- You will find permutations and combinations.

Key Vocabulary

- combination (p. 610)
- complement (p. 581)
- compound event (p. 598)
- counting principle (p. 592)
- dependent events (p. 599)
- event (p. 580)
- experimental probability (p. 586)
- factorial (p. 606)
- independent events (p. 598)
- outcome (p. 580)
- permutation (p. 606)
- sample space (p. 591)
- theoretical probability (p. 580)

Problem Solving Application On pages 622 and 623, you will work an extended activity on probability.

12-1

1. Plan

Objective
To find the probability and the complement of an event

Examples
1 Finding Probability
2, 4 Finding Probability From 0 to 1

Math Understandings: p. 578C

Math Background

In the language of mathematical probability, the *theoretical probability* of an event, usually written as a fraction, is $P(\text{event}) = \frac{\text{number of favorable outcomes}}{\text{total number of possible outcomes}}$.
The probability of any event can be represented by a rational number from 0 to 1, so $0 \le P(\text{event}) \le 1$ where 0 is $P(\text{impossible outcome})$ and 1 is $P(\text{certain outcome})$. The *complement* of an event, A, is the collection of unfavorable outcomes. This means that $P(A) + P(\text{not A}) = 1$.

More Math Background: p. 578C

Lesson Planning and Resources

See p. 578E for a list of the resources that support this lesson.

Bell Ringer Practice

Check Skills You'll Need
Use student page, transparency, or PowerPoint. For intervention, direct students to:
Percents, Fractions, and Decimals
Lesson 6-2
Extra Skills and Word Problems Practice, Ch. 6

12-1 Probability

Check Skills You'll Need

1. **Vocabulary Review** In what three forms can you write a *rational number*?

1–5. See below.
Write each fraction as a decimal and as a percent.

2. $\frac{31}{50}$
3. $\frac{19}{20}$
4. $\frac{11}{40}$
5. $\frac{11}{10}$

for Help Lesson 6-2

Check Skills You'll Need
1. fraction, decimal, percent
2. 0.62; 62%
3. 0.95; 95%
4. 0.275; 27.5%
5. 1.1; 110%

What You'll Learn

To find the probability and the complement of an event

New Vocabulary outcome, event, theoretical probability, complement

Why Learn This?

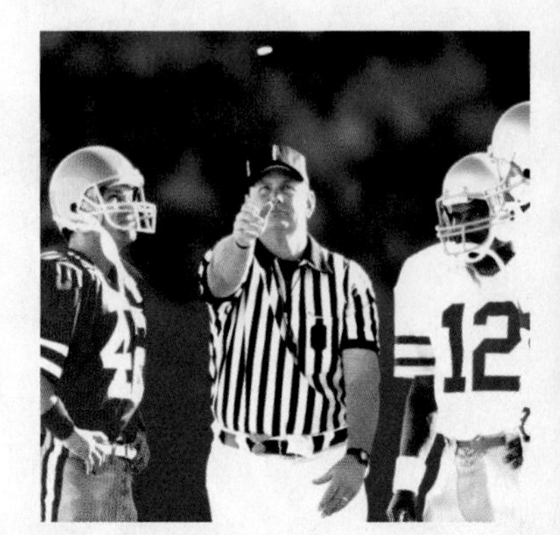

In sports, a coin toss often determines which team gets the ball first.

An **outcome** is the result of an action. For example, getting tails is a possible outcome of flipping a coin. An **event** is a collection of possible outcomes. If all the outcomes are equally likely, you can use a formula to find the theoretical probability.

KEY CONCEPTS **Theoretical Probability**

theoretical probability $= P(\text{event}) = \frac{\text{number of favorable outcomes}}{\text{total number of possible outcomes}}$

You can express probability as a fraction, a decimal, or a percent.

EXAMPLE **Finding Probability**

1 You select a letter at random from the letters shown. Find the probability of selecting a vowel. Express the probability as a fraction, a decimal, and a percent.

The event *vowel* has 2 outcomes, A and E, out of 5 possible outcomes.

$P(\text{vowel}) = \frac{2}{5}$ ← number of favorable outcomes / ← total number of possible outcomes

$= \frac{2}{5}$, 0.4, or 40% ← Write as a fraction, decimal, and percent.

Vocabulary Tip
You read $P(\text{vowel})$ as "the probability of a vowel."

Quick Check

1. Find $P(\text{consonant})$ as a fraction for the letters in Example 1. $\frac{3}{5}$

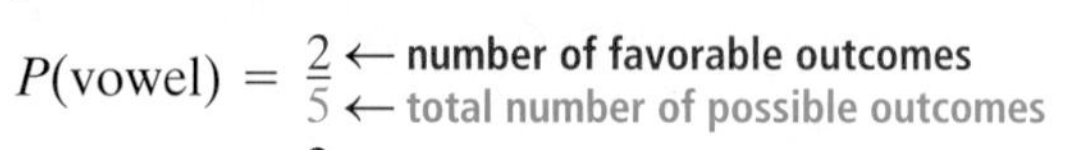

Differentiated Instruction Solutions for All Learners

Special Needs L1
For Quick Check 2, have students actually roll a number cube several times. Have them list the numbers they get, and think about the numbers they might get on any given roll.
learning style: tactile

Below Level L2
Have students convert these decimals to fractions: 0.25, 0.50, 0.75, 0.90. $\frac{1}{4}, \frac{1}{2}, \frac{3}{4}, \frac{9}{10}$
learning style: visual

Video Tutor Help

Visit: PHSchool.com
Web Code: are-0775

All probabilities range from 0 to 1. The probability of rolling a 7 on a number cube is 0, so that is an *impossible* event. The probability of rolling a positive integer less than 7 is 1, so that is a *certain* event.

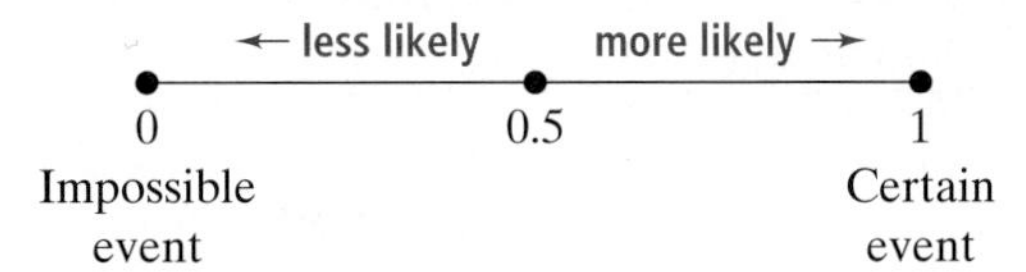

The **complement** of an event is the collection of outcomes not contained in the event. The sum of the probabilities of an event and its complement is 1. So $P(\text{event}) + P(\text{not event}) = 1$.

EXAMPLES Finding Probabilities From 0 to 1

2 **Clothes** The picture shows the jeans in Juanita's closet. She selects a pair of jeans with her eyes shut. Find $P(\text{dark color})$.

There are 8 possible outcomes. Since there are 3 black pairs and 2 blue pairs, the event *dark color* has 5 favorable outcomes.

$P(\text{dark color}) = \frac{5}{8}$ ← number of favorable outcomes ← total number of possible outcomes

3 Refer to Juanita's closet. Find $P(\text{red})$.

The event *red* has no favorable outcome.

$P(\text{red}) = \frac{0}{8}$, or 0 ← number of favorable outcomes ← total number of possible outcomes

4 Refer to Juanita's closet. Find $P(\text{not dark color})$.

$P(\text{dark color}) + P(\text{not dark color}) = 1$ ← The sum of probabilities of an event and its complement is 1.

$\frac{5}{8} + P(\text{not dark color}) = 1$ ← Substitute $\frac{5}{8}$ for P(dark color).

$\frac{5}{8} - \frac{5}{8} + P(\text{not dark color}) = 1 - \frac{5}{8}$ ← Subtract $\frac{5}{8}$ from each side.

$P(\text{not dark color}) = \frac{3}{8}$ ← Simplify.

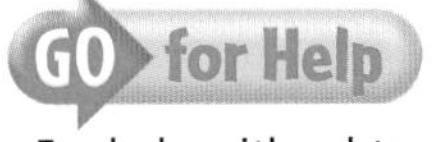

For help with subtracting fractions, go to Lesson 3-2, Example 3.

Quick Check

You roll a number cube once. Find each probability.

2. $P(\text{multiple of } 3)$ $\frac{1}{3}$ **3.** $P(\text{not multiple of } 2)$ $\frac{1}{2}$ **4.** $P(9)$ 0

2. Teach

Activity Lab

Use before the lesson.

All in One Teaching Resources

Activity Lab 12-1: Critical Thinking

Guided Instruction

Example 1

To learn the vocabulary of probability, have students define aloud the event **(selecting a letter at random from those shown)**, the favorable outcome **(vowel)**, and the unfavorable outcome **(consonant)**.

Error Prevention!

Ask students to say "the probability of a vowel" for P(vowel) rather than using shortcuts such as "pea-vowel."

1 Find the probability of selecting a vowel from the letters F, G, H, I. Express the probability as a fraction, decimal, and a percent. **$\frac{1}{4}$, 0.25, 25%**

2 Jacques has 1 blue shirt, 5 white shirts, 3 green shirts, and 2 brown shirts. He selects a shirt from his closet with his eyes shut. Find each probability.

a. P(white shirt) $\frac{5}{11}$

b. P(not green shirt) $\frac{8}{11}$

c. P(blue shirt) $\frac{1}{11}$

All in One Teaching Resources

- Daily Notetaking Guide 12-1 L3
- Adapted Notetaking 12-1 L1

Closure

- *How do you find the theoretical probability of an event?* **Divide the number of favorable outcomes by the total number of possible outcomes.**
- *How do you find the probability of the complement of an event?* **Subtract the probability of the event from 1.**

Advanced Learners L4	**English Language Learners** ELL
Have students write a problem similar to Example 2, showing examples of no favorable outcomes and the sum of the probabilities of an event and its complement. **learning style: verbal**	The notion of complements is conceptually and linguistically complex. Help students differentiate between the word *complement* here and the word *complement* in complementary angles. **learning style: verbal**

3. Practice

Assignment Guide

Check Your Understanding
Go over Exercises 1–6 in class before assigning the Homework Exercises.

Homework Exercises

A	Practice by Example	7–21
B	Apply Your Skills	22–32
C	Challenge	33
	Test Prep and Mixed Review	34–39

Homework Quick Check
To check students' understanding of key skills and concepts, go over Exercises 11, 18, 23, 29, and 32.

Differentiated Instruction Resources

Adapted Practice 12-1 L1

Practice 12-1 L3

You spin a spinner numbered 1 through 10. Each outcome is equally likely. Find the probabilities below as a fraction, decimal, and percent.

1. $P(9)$ $\frac{1}{10}$; 0.1; 10%
2. $P(\text{even})$ $\frac{1}{2}$ 0.5; 50%
3. $P(\text{number greater than } 0)$ 1; 1.0; 100%
4. $P(\text{multiple of } 4)$ $\frac{1}{5}$; 0.2; 20%

There are eight blue marbles, nine orange marbles, and six yellow marbles in a bag. You draw one marble at random. Find each probability.

5. $P(\text{blue marble})$ $\frac{8}{23}$ or 34.8%
6. $P(\text{yellow marble})$ $\frac{6}{23}$ or 26.1%
7. What marble could you add or remove so that the probability of drawing a blue marble is $\frac{1}{3}$? Add a marble that is not blue.

A box contains 12 slips of paper as shown. Each slip of paper is equally likely to be drawn. Find each probability.

red	blue	yellow	blue
yellow	red	blue	red
red	red	red	yellow

8. $P(\text{red})$ $\frac{1}{2}$
9. $P(\text{blue})$ $\frac{1}{4}$
10. $P(\text{yellow})$ $\frac{1}{4}$
11. $P(\text{red or blue})$ $\frac{3}{4}$
12. $P(\text{red or yellow})$ $\frac{3}{4}$
13. $P(\text{blue or yellow})$ $\frac{1}{2}$
14. $P(\text{not red})$ $\frac{1}{2}$
15. $P(\text{not blue})$ $\frac{3}{4}$
16. $P(\text{not yellow})$ $\frac{3}{4}$

You select a letter randomly from a bag containing the letters S, P, I, N, N, E, and R. Find the odds in favor of each outcome.

17. selecting an N $\frac{2}{5}$
18. selecting an S $\frac{1}{6}$

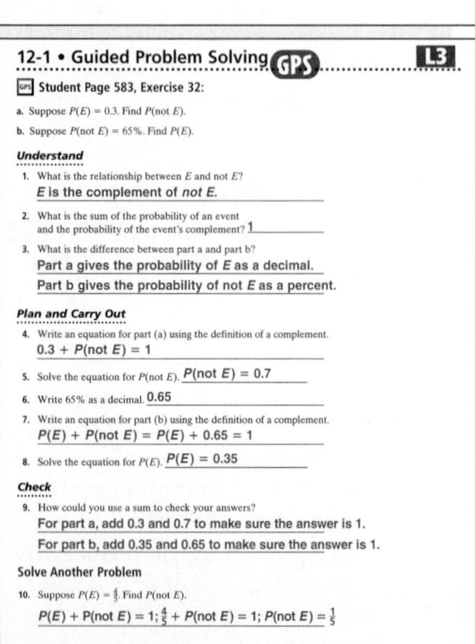

12-1 • Guided Problem Solving GPS L3

GPS **Student Page 583, Exercise 32:**

a. Suppose $P(E) = 0.3$. Find $P(\text{not } E)$.

b. Suppose $P(\text{not } E) = 65\%$. Find $P(E)$.

Understand

1. What is the relationship between E and not E? E is the complement of *not E*.
2. What is the sum of the probability of an event and the probability of the event's complement? 1
3. What is the difference between part a and part b? Part a gives the probability of E as a decimal. Part b gives the probability of not E as a percent.

Plan and Carry Out

4. Write an equation for part (a) using the definition of a complement. $0.3 + P(\text{not } E) = 1$
5. Solve the equation for $P(\text{not } E)$. $P(\text{not } E) = 0.7$
6. Write 65% as a decimal. 0.65
7. Write an equation for part (b) using the definition of a complement. $P(E) + P(\text{not } E) = P(E) + 0.65 = 1$
8. Solve the equation for $P(E)$. $P(E) = 0.35$

Check

9. How could you use a sum to check your answers? For part a, add 0.3 and 0.7 to make sure the answer is 1. For part b, add 0.35 and 0.65 to make sure the answer is 1.

Solve Another Problem

10. Suppose $P(E) = \frac{4}{5}$. Find $P(\text{not } E)$. $P(E) + P(\text{not } E) = 1; \frac{4}{5} + P(\text{not } E) = 1; P(\text{not } E) = \frac{1}{5}$

Check Your Understanding

1. Answers may vary. Sample: the result or group of results of an action

1. **Vocabulary** Define *event* without using the word *outcome*.
2. $P(\text{A}) = \frac{1}{3}$. Write an expression for $P(\text{not A})$. $\frac{2}{3}$

You select a marble from those shown. Match each event with its probability.

3. $P(\text{red})$ B
4. $P(\text{yellow})$ C
5. $P(\text{blue})$ A
6. $P(\text{red or blue})$ D

A. $\frac{5}{7}$

B. $\frac{2}{7}$

C. 0

D. 1

Homework Exercises

For more exercises, see Extra Skills and Word Problems.

GO for Help

For Exercises	See Examples
7–12	1
13–21	2

A **You mix the letters A, C, Q, U, A, I, N, T, A, N, C, and E thoroughly. Without looking, you select one letter. Find the probability of each event as a fraction, a decimal, and a percent.**

7. $P(T)$ $\frac{1}{12}$; $0.08\overline{3}$; about 8.3%
8. $P(A)$ $\frac{1}{4}$; 0.25; 25%
9. $P(\text{vowel})$ $\frac{1}{2}$; 0.5; 50%
10. $P(\text{consonant})$ $\frac{1}{2}$; 0.5; 50%
11. $P(N)$ $\frac{1}{6}$; $0.1\overline{6}$; about 16.7%
12. $P(Q \text{ or } C)$

12. $\frac{1}{4}$; 0.25; 25%

You spin the spinner once. Find each probability.

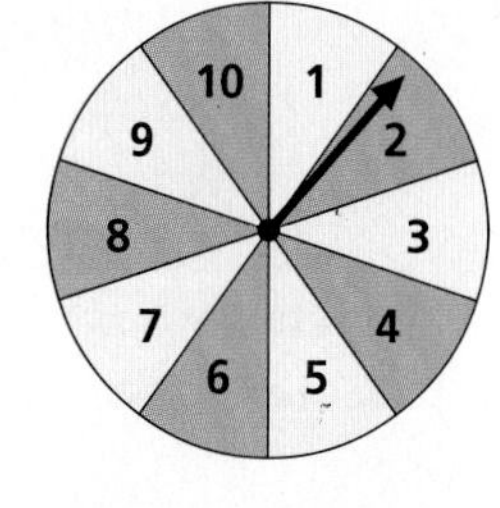

13. $P(12)$ 0
14. $P(2 \text{ or } 4)$ $\frac{2}{10}$ or $\frac{1}{5}$
15. $P(\text{multiple of } 3)$ $\frac{3}{10}$
16. $P(\text{even})$ $\frac{5}{10}$ or $\frac{1}{2}$
17. $P(\text{not } 1)$ $\frac{9}{10}$
18. $P(\text{not a factor of } 10)$
19. $P(\text{less than } 11)$ $\frac{10}{10}$ or 1
20. $P(\text{not divisible by } 3)$ $\frac{7}{10}$

18. $\frac{6}{10}$ or $\frac{3}{5}$

21. **Science** Six out of the 111 elements are noble gases. You write the names of all the elements on cards and select a card at random. What is the probability of *not* picking a noble gas? $\frac{105}{111}$

B 

22. **Guided Problem Solving** The table shows data about a group of people's hair colors. You select a person at random from the group. What is $P(\text{not black hair})$? $\frac{122}{219}$
    - How many people are in the group?
    - How many people do *not* have black hair?

Hair Color

Color	Number
Blond	58
Brown	64
Black	97

23. **Writing in Math** Describe a real-life situation where the probability of an event is 1. Then describe the complement of that situation. See margin.

GO Online
Homework Video Tutor
Visit: PHSchool.com
Web Code: are-1201

23. Answers may vary. Sample: The probability that a female student will be chosen from an all-female school is 1. The complement is 0; it is impossible for a male student to be chosen.

You spin the spinner once. Find each probability.

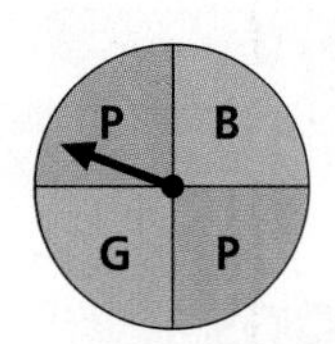

24. $P(\text{not green})$ $\frac{3}{4}$
25. $P(\text{purple or blue})$ $\frac{3}{4}$
26. $P(\text{white})$ 0
27. $P(\text{not purple})$ $\frac{2}{4}$ or $\frac{1}{2}$

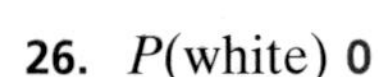

Government **The U.S. House of Representatives has 435 members. Each member's name is put into a hat and one name is chosen at random. Find each probability as a decimal to the nearest hundredth.**

28. $P(\text{Florida})$ 0.06
29. $P(\text{Texas})$ 0.07
30. $P(\text{not Illinois})$ 0.96
31. $P(\text{Pennsylvania})$ 0.04

U.S. House of Representatives

State	Number	State	Number
Florida	25	Illinois	19
Pennsylvania	19	Texas	32

SOURCE: U.S. Census Bureau. Go to **PHSchool.com** for a data update. Web Code arg-9041

32. a. Suppose $P(E) = 0.3$. Find $P(\text{not } E)$. 0.7
GPS b. Suppose $P(\text{not } E) = 65\%$. Find $P(E)$. 35%

C 33. **Challenge** A bag contains an unknown number of marbles. You know that $P(\text{red}) = \frac{1}{4}$ and $P(\text{green}) = \frac{1}{4}$. What can you conclude about how many marbles are in the bag? **Any positive integer divisible by 4 can be the number of marbles, since $\frac{1}{4}$ of that number must be a positive integer.**

Test Prep and Mixed Review

Practice

Multiple Choice

34. The model represents the equation $3x + 5 = 14$. What is the value of x? **C**

Ⓐ $x = \frac{11}{5}$ Ⓒ $x = 3$
Ⓑ $x = 6$ Ⓓ $x = \frac{19}{3}$

35. Suppose 2 out of every 25 people in your state are 10 to 14 years old. Which equation can be used to find x, the percent of people in your state who are 10 to 14 years old? **J**

Ⓕ $\frac{x}{100} = \frac{25}{2}$ Ⓖ $\frac{100}{x} = \frac{10}{14}$ Ⓗ $\frac{100}{25} = \frac{2}{x}$ Ⓙ $\frac{x}{100} = \frac{2}{25}$

36. Karl buys a pair of jeans that regularly costs \$38. They are on sale for 25% off. Karl also buys a shirt that costs \$23. The sales tax is 4.5%. What other information is necessary to find Karl's correct change? **D**

Ⓐ The sale price of the jeans
Ⓑ The total cost of the purchase
Ⓒ The amount he paid for the sales tax
Ⓓ The amount he gave the cashier

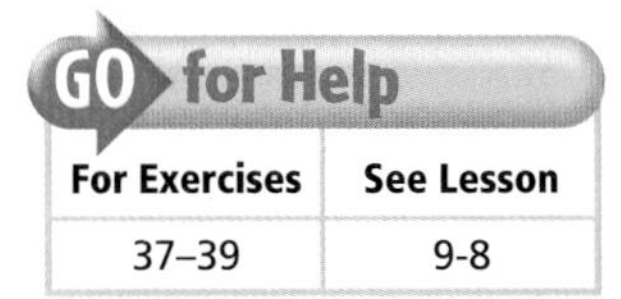

For Exercises	See Lesson
37–39	9-8

Algebra **Solve each equation for the variable in red.**

37. $a = bc$ $\frac{a}{b}$
38. $n = \frac{g + w}{4}$ $4n - w$
39. $q = 2t + 8$ $\frac{q - 8}{2}$

Online lesson quiz, **PHSchool.com**, **Web Code:** ara-1201

4. Assess & Reteach

Lesson Quiz

In a stack are several number cards: three 1s, four 2s, three 3s, two 4s, two 6s, and six 7s. You pick a card at random.

1. Write $P(3)$ as a fraction, a decimal, and a percent.
$\frac{3}{20}$, 0.15, 15%
2. Write $P(\text{not } 7)$ as a fraction, a decimal, and a percent.
$\frac{7}{10}$, 0.7, 70%

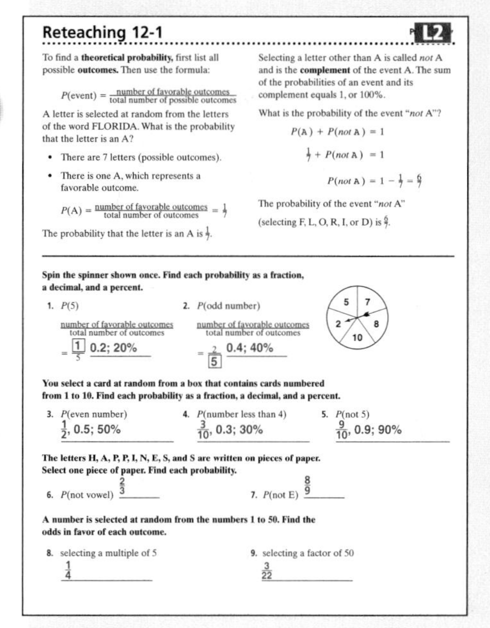

Reteaching 12-1

L2

To find a **theoretical probability,** first list all possible **outcomes.** Then use the formula:

$P(\text{event}) = \frac{\text{number of favorable outcomes}}{\text{total number of possible outcomes}}$

A letter is selected at random from the letters of the word FLORIDA. What is the probability that the letter is an A?

- There are 7 letters (possible outcomes).
- There is one A, which represents a favorable outcome.

$P(\text{A}) = \frac{\text{number of favorable outcomes}}{\text{total number of outcomes}} = \frac{1}{7}$

The probability that the letter is an A is $\frac{1}{7}$.

Selecting a letter other than A is called *not* A and is the **complement** of the event A. The sum of the probabilities of an event and its complement equals 1, or 100%.

What is the probability of the event "*not* A"?

$P(\text{A}) + P(\text{not A}) = 1$

$\frac{1}{7} + P(\text{not A}) = 1$

$P(\text{not A}) = 1 - \frac{1}{7} = \frac{6}{7}$

The probability of the event "*not* A" (selecting F, L, O, R, I, or D) is $\frac{6}{7}$.

Spin the spinner shown once. Find each probability as a fraction, a decimal, and a percent.

1. $P(5)$
$\frac{\text{number of favorable outcomes}}{\text{total number of outcomes}} = \frac{1}{5}$ 0.2; 20%
2. $P(\text{odd number})$
$\frac{\text{number of favorable outcomes}}{\text{total number of outcomes}} = \frac{2}{5}$ 0.4; 40%

You select a card at random from a box that contains cards numbered from 1 to 10. Find each probability as a fraction, a decimal, and a percent.

3. $P(\text{even number})$ $\frac{1}{2}$, 0.5; 50%
4. $P(\text{number less than 4})$ $\frac{3}{10}$, 0.3; 30%
5. $P(\text{not 5})$ $\frac{9}{10}$, 0.9; 90%

The letters H, A, P, P, I, N, E, S, and S are written on pieces of paper. Select one piece of paper. Find each probability.

6. $P(\text{not vowel})$ $\frac{2}{3}$
7. $P(\text{not E})$ $\frac{8}{9}$

A number is selected at random from the numbers 1 to 50. Find the odds in favor of each outcome.

8. selecting a multiple of 5 $\frac{1}{4}$
9. selecting a factor of 50 $\frac{3}{22}$

Enrichment 12-1

L4

Critical Thinking

Odds The probability ratio compares favorable outcomes to total outcomes. When outcomes are equally likely, you can write ratios, called *odds,* that compare favorable outcomes to unfavorable outcomes.

Odds in favor of an event = the ratio of the number of favorable outcomes to the number of unfavorable outcomes

Odds against an event = the ratio of the number of unfavorable outcomes to the number of favorable outcomes

Suppose a bag contains ten slips with classroom math challenge tasks you can do for extra credit. The tasks include problems related to probability, geometry, and algebra. The number of tasks related to probability is twice the number of tasks related to algebra. There is at least one task devoted to each type of problem.

1. If the odds of choosing an algebra task are 1 : 9, how many slips for each task will there be?
2 probability tasks, 1 algebra task, 7 geometry tasks
2. If the odds of choosing an algebra task are 1 : 4, how many slips for each task will there be?
4 probability tasks, 2 algebra tasks, 4 geometry tasks
3. If the odds of choosing a probability task are 3 : 2, how many slips for each task will there be?
6 probability tasks, 3 algebra tasks, 1 geometry task

Suppose a bag contains 15 slips, each listing one of these sports: basketball, baseball, football, volleyball. There are three times as many slips for basketball as baseball and twice as many slips for football as for volleyball. There is at least one slip for each sport.

4. What are the odds of drawing a slip that shows each of the following sports?
basketball 9 : 6 or 3 : 2 baseball 3 : 12 or 1 : 4
football 2 : 13 volleyball 1 : 14
5. Which sport is most likely to be picked in a random drawing? Explain.
Basketball, since 9 chances out of 15 is more likely to happen than 3 out of 15, 2 out of 15, or 1 out of 15.

Alternative Assessment

Each student in a pair writes a probability problem for a standard number cube, such as "find $P(2 \text{ or } 4)$." Partners trade problems and find the probability as a decimal, fraction, and percent. Partners then find the complement. Partners discuss and agree on their results.

Test Prep

Resources

For additional practice with a variety of test item formats:

- Test-Taking Strategies, p. 615
- Test Prep, p. 619
- Test-Taking Strategies with Transparencies

Extension

Odds

Students learn about odds and how to write them. This will extend the work they did with probability in Lesson 12-1.

Guided Instruction

Define *odds* for students. Explain that this is a common way of stating a probability ratio. Ask volunteers for examples of odds in everyday life. Make sure students understand the difference between odds and probability. Ask questions such as:

- *Can odds be even?* **yes** *How?* **If there are an equal number of favorable and unfavorable events, the odds are even. It is equally likely that either will happen.**
- *How are the odds in favor and the odds against the same event related?* **They are reciprocals.**
- *If you are given the odds in favor of an event, how can you find the total number of possible outcomes?* **Add both terms. For example, if the odds are 2 : 3, the total number of possible outcomes is 5.**

Exercises

Have students work independently on the Exercises. When they have finished, have them compare answers with a partner and discuss any differences.

Differentiated Instruction

Advanced Learners L4

Have students create their own odds problems, trade them with a partner, and solve their partner's problems.

Odds

The probability ratio compares favorable outcomes to all possible outcomes. When outcomes are equally likely, you can write ratios, called *odds*, that compare favorable outcomes to unfavorable outcomes.

Odds in favor of an event = the ratio of the number of favorable outcomes to the number of unfavorable outcomes

Odds against an event = the ratio of the number of unfavorable outcomes to the number of favorable outcomes

EXAMPLE **Finding Odds**

Coins Five quarters are shown below. Find the odds that a quarter you select at random shows at least one musical instrument.

odds in favor = 2 to 3 or 2 : 3 ← **Two have an instrument. Three do not.**

The odds that a quarter shows at least one musical instrument are 2 to 3 in favor.

Exercises

1. Refer to the example above. Find the odds that a quarter selected at random shows a race car. **1 : 4**

You roll a number cube once. Find the odds in favor of each outcome.

2. rolling a 5 **1 : 5**
3. rolling a multiple of 3 **2 : 4**
4. rolling an odd number **3 : 3**

5. You spin a spinner with equal sections lettered A–Z once. The spinner lands on the first letter of your name. Calculate the odds for and against this event. **1 : 25; 25 : 1**

12-2a Activity Lab

Hands On

Exploring Probability

If you toss one coin 100 times, you might expect to get heads and tails about the same number of times. What happens when you toss *two* coins 100 times?

EXAMPLE

Toss two coins 10 times. Record the results. Use the results to determine which event is most likely: two heads (HH), two tails (TT), or one head and one tail (HT).

The table below shows one set of results.

Toss	1	2	3	4	5	6	7	8	9	10
Result	HH	HT	HT	TT	HH	HT	TT	HT	HH	HT

In all, there are three outcomes of HH, two outcomes of TT, and five outcomes of HT. One head and one tail is the most likely event.

Exercises

1. Conduct an experiment in which you use two coins, such as a penny and a nickel. Place them in a small paper cup. Cover the top, and shake the cup before each coin toss. Toss both coins 100 times. Make a table to record each result. **Check students' work.**
2. Are your results similar to other students' results? Explain. **Check students' work.**
3. **Reasoning** Are the three outcomes all equally likely? Or is one outcome more likely to occur than the others? Explain. **See right.**

Three students play the game at the right.

4. Is a game with these rules fair? Explain why or why not. **See right.**
5. **Open-Ended** How might you change the rules to make the game fair? **See right.**
6. Conduct an experiment to test your new game. Do you still think your game is fair? Explain. **Check students' work.**

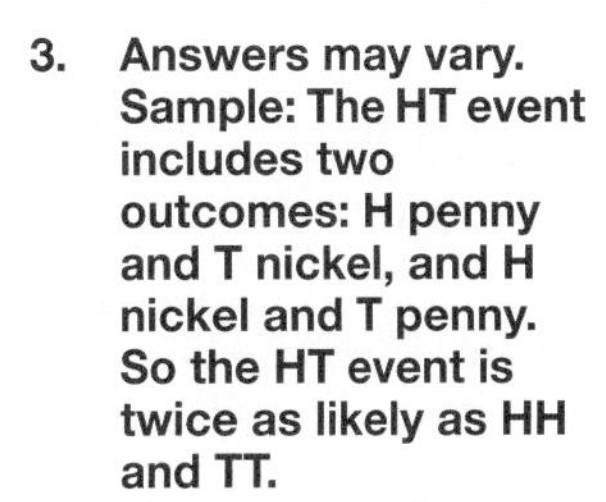

GAME RULES

1. Player A receives 1 point if two heads (HH) are tossed.
2. Player B receives 1 point if two tails (TT) are tossed.
3. Player C receives 1 point if one head and one tail (HT) are tossed.

3. **Answers may vary. Sample: The HT event includes two outcomes: H penny and T nickel, and H nickel and T penny. So the HT event is twice as likely as HH and TT.**
4. **Not fair; player C will score about twice as many points as player A or player B.**
5. **Answers may vary. Sample: players A and B get two points for HH or TT.**

Activity Lab

Exploring Probability

This Activity lays the foundation for Lesson 12-2 on experimental probability. Students investigate the idea of doing an experiment to observe the frequency of events occurring. They explore the concept of equally likely outcomes.

Guided Instruction

Example
Review and discuss the theoretical probability of an event, $P(E)$. Then discuss the meaning of *equally likely events.* Explain that the events occur with the same relative frequency. Challenge students to give examples.

Special Needs L1
Point out that the outcomes—two tails, a head and a tail, and two heads—are not equally likely. $P(\text{two heads})$ is $\frac{1}{4}$, as is $P(\text{two tails})$. But the probability of getting a head and a tail is $\frac{1}{2}$.

Exercises
Have one partner shake and toss, while the other records results.
Exercises 4–6 Elicit from students that for a game to be fair, each player must have an equal chance of success according to its rules.

Alternative Method
Students may prefer to draw from a bag containing equal numbers of two different colored marbles.

Resources

- Activity Lab 12-2: Experimental and Theoretical Probability
- coins
- paper cup
- equal numbers of two different colors of marbles
- small bag

12-2

1. Plan

Objective
To find experimental probability and to use simulations

Examples
1 Finding Experimental Probability
2 Application: Manufacturing
3 Simulating an Event

Math Understandings: p. 578C

Math Background

Experimental probability is used to monitor the quality of manufacturing processes and products. These processes have small but acceptable variations. Probability is used to monitor and correct variations beyond acceptable measures. Such "quality control" methods rely on finding experimental probability by sampling and simulations. *Simulations* are models used to calculate probabilities for experimental situations that are too difficult or costly to perform.

More Math Background: p. 578C

Lesson Planning and Resources

See p. 578E for a list of the resources that support this lesson.

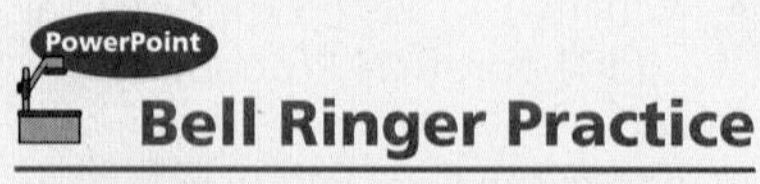

Bell Ringer Practice

Check Skills You'll Need
Use student page, transparency, or PowerPoint. For intervention, direct students to:
Probability
Lesson 12-1
Extra Skills and Word Problems Practice, Ch. 12

12-2 Experimental Probability

Check Skills You'll Need

1. **Vocabulary Review** Explain the difference between an *event* and an *outcome*. **See below.**

You roll a number cube once. Find each probability.

2. $P(4)$ $\frac{1}{6}$
3. $P(\text{multiple of 2})$ $\frac{1}{2}$
4. $P(8)$ 0

Lesson 12-1

Check Skills You'll Need

1. **Answers may vary. Sample: An outcome is a single result, but an event can be a group of results.**

Video Tutor Help
Visit: PHSchool.com
Web Code: are-0775

What You'll Learn

To find experimental probability and to use simulations

New Vocabulary experimental probability

Why Learn This?

Manufacturers collect data on the quality of their products. They use experimental probability to determine how many defective items they can expect to produce.

Probability based on experimental data or observations is called **experimental probability.**

KEY CONCEPTS Experimental Probability

$$P(\text{event}) = \frac{\text{number of times an event occurs}}{\text{total number of trials}}$$

EXAMPLE Finding Experimental Probability

1 You attempt 16 free throws in a basketball game. Your results are shown. What is the experimental probability of making a free throw?

Results of Free Throw Attempts

0 = miss				1 = make			
0	0	1	1	1	0	1	0
0	1	0	1	1	0	0	1

$P(\text{free throw}) = \frac{8}{16}$ ← number of throws made / ← total number of attempted free throws

$= \frac{1}{2}$ ← Simplify.

The experimental probability of making a free throw is $\frac{1}{2}$.

Quick Check

1. In 60 coin tosses, 25 are tails. Find the experimental probability. $\frac{5}{12}$

Differentiated Instruction Solutions for All Learners

Special Needs L1
For Example 3, provide students with coins, and let them act out the simulation by tossing the coins 20 times. Let them try this several times. Have them record their results.
learning style: tactile

Below Level L2
Have students write these fractions in simplest form: $\frac{4}{6}, \frac{10}{15}, \frac{8}{16}, \frac{7}{21}, \frac{2}{3}, \frac{2}{3}, \frac{1}{2}, \frac{1}{3}$
learning style: visual

EXAMPLE Application: Manufacturing

Test Prep Tip

You can write and solve a proportion using a probability ratio to make a prediction.

2 Multiple Choice A bicycle company checks a random sample of bikes. The results are shown. If the trend continues, which is the best prediction of the number of defective bikes in a batch of 1,300?

Quality Control Results

Defective Bikes	Bikes Checked
12	400

Ⓐ 430 bikes Ⓑ 390 bikes Ⓒ 43 bikes Ⓓ 39 bikes

The experimental probability that a bike is defective is $\frac{12}{400}$, or $\frac{3}{100}$.

Let x represent the predicted number of defective bikes.

defective bikes → $\frac{3}{100}$ = $\frac{x}{1,300}$ ← defective bikes
bikes checked → ← bikes checked   ← **Write a proportion.**

$3(1,300) = 100x$ ← **Write the cross products.**

$3,900 = 100x$ ← **Simplify.**

$\frac{3,900}{100} = \frac{100x}{100}$ ← **Divide each side by 100.**

$39 = x$ ← **Simplify.**

You can predict that 39 bikes are defective. The correct answer is D.

Quick Check

2. Predict the number of defective bikes in a batch of 3,500. **105 bikes**

You can simulate, or model, events to find experimental probabilities.

EXAMPLE Simulating an Event

3 Find the experimental probability that 2 of 3 children in a family are girls. Assume that girls and boys are equally likely.

Simulate the problem by tossing three coins. Let "heads" represent a girl and "tails" represent a boy. A sample of 20 coin tosses is shown.

T T H	T T T	(H T H)	H T T	(H T H)
T T H	(H H T)	H T T	T H T	H H H
(H H T)	T T H	(T H H)	(H T H)	T H T
T H T	T H T	T H T	H H H	H H H

$P(\text{two girls}) = \frac{6}{20}$, or $\frac{3}{10}$ ← number of times *two heads* occur
← total number of tosses

The experimental probability that 2 of 3 children are girls is $\frac{3}{10}$.

Quick Check

3. What is the experimental probability that 3 children are all boys? $\frac{1}{20}$

2. Teach

Activity Lab

Use before the lesson.
Student Edition Activity Lab, Hands On 12-2a, Exploring Probability, p. 585

All in One Teaching Resources

Activity Lab 12-2: Experimental and Theoretical Probability

Guided Instruction

Example 3
Explain that a simulation uses an artificial situation to imitate or model a situation in the world that cannot be controlled or observed as easily as the model. Simulating an event is a way of acting it out.

PowerPoint

Additional Examples

1. A player makes 7 free throws out of 12 attempts. Based on this, what is the experimental probability of this player making a free throw? $\frac{7}{12}$

2. A manufacturer of computer parts checks 100 parts each day. On Monday, 2 of the checked parts are defective.
   a. What is the experimental probability that a part is defective? $\frac{1}{50}$
   b. Predict the probable number of defective parts in Monday's total production of 1,250 parts. **25 parts**

3. It is equally likely that a puppy will be born male or female. Use a simulation to find the experimental probability that, in a litter of four puppies, all four will be male. **See back of book.**

All in One Teaching Resources

- Daily Notetaking Guide 12-2 L3
- Adapted Notetaking 12-2 L1

Closure

- *What is experimental probability?* **See back of book.**
- *What are the steps in using a simulation to find experimental probability?* **See back of book.**

Advanced Learners L4
Have students find the experimental probability of getting a total of 8 when you roll two standard number cubes. $\frac{5}{36}$

learning style: visual

English Language Learners ELL
Students often mistakenly think that probability indicates what will always happen. Explain that probability indicates what is very likely to happen and how often, under certain circumstances, but it is not a guarantee.

learning style: verbal

3. Practice

Assignment Guide

Check Your Understanding

Go over Exercises 1–5 in class before assigning the Homework Exercises.

Homework Exercises

A	Practice by Example	6–12
B	Apply Your Skills	13–22
C	Challenge	23
	Test Prep and Mixed Review	24–29

Homework Quick Check

To check students' understanding of key skills and concepts, go over Exercises 7, 10, 14, 15, and 22.

Differentiated Instruction Resources

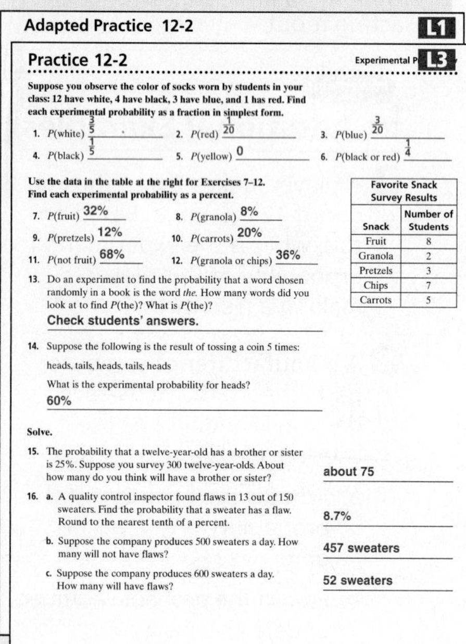

Adapted Practice 12-2 L1

Practice 12-2 Experimental P[robability] L3

Suppose you observe the color of socks worn by students in your class: 12 have white, 4 have black, 3 have blue, and 1 has red. Find each experimental probability as a fraction in simplest form.

1. P(white) $\frac{3}{5}$
2. P(red) $\frac{1}{20}$
3. P(blue) $\frac{3}{20}$
4. P(black) $\frac{1}{5}$
5. P(yellow) 0
6. P(black or red) $\frac{1}{4}$

Use the data in the table at the right for Exercises 7–12. Find each experimental probability as a percent.

Favorite Snack Survey Results	
Snack	Number of Students
Fruit	8
Granola	2
Pretzels	3
Chips	7
Carrots	5

7. P(fruit) 32%
8. P(granola) 8%
9. P(pretzels) 12%
10. P(carrots) 20%
11. P(not fruit) 68%
12. P(granola or chips) 36%
13. Do an experiment to find the probability that a word chosen randomly in a book is the word *the*. How many words did you look at to find P(the)? What is P(the)?
    Check students' answers.
14. Suppose the following is the result of tossing a coin 5 times:
    heads, tails, heads, tails, heads
    What is the experimental probability for heads?
    60%

Solve.

15. The probability that a twelve-year-old has a brother or sister is 25%. Suppose you survey 300 twelve-year-olds. About how many do you think will have a brother or sister? **about 75**
16. a. A quality control inspector found flaws in 13 out of 150 sweaters. Find the probability that a sweater has a flaw. Round to the nearest tenth of a percent. **8.7%**
    b. Suppose the company produces 500 sweaters a day. How many will not have flaws? **457 sweaters**
    c. Suppose the company produces 600 sweaters a day. How many will have flaws? **52 sweaters**

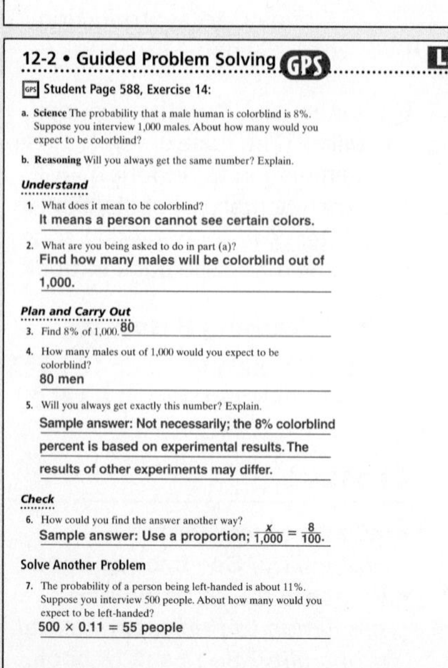

12-2 • Guided Problem Solving GPS L3

GPS Student Page 588, Exercise 14:

a. **Science** The probability that a male human is colorblind is 8%. Suppose you interview 1,000 males. About how many would you expect to be colorblind?

b. **Reasoning** Will you always get the same number? Explain.

Understand

1. What does it mean to be colorblind?
   It means a person cannot see certain colors.
2. What are you being asked to do in part (a)?
   Find how many males will be colorblind out of 1,000.

Plan and Carry Out

3. Find 8% of 1,000. **80**
4. How many males out of 1,000 would you expect to be colorblind?
   80 men
5. Will you always get exactly this number? Explain.
   Sample answer: Not necessarily; the 8% colorblind percent is based on experimental results. The results of other experiments may differ.

Check

6. How could you find the answer another way?
   Sample answer: Use a proportion; $\frac{x}{1,000} = \frac{8}{100}$.

Solve Another Problem

7. The probability of a person being left-handed is about 11%. Suppose you interview 500 people. About how many would you expect to be left-handed?
   500 × 0.11 = 55 people

Check Your Understanding

1. **Vocabulary** What is the difference between theoretical probability and experimental probability? Explain. **See margin.**

You toss a coin 40 times and get 18 tails. Find each experimental probability.

2. $P(\text{heads}) = \frac{22}{\blacksquare}$ **40**
3. $P(\text{tails}) = \frac{\blacksquare}{40}$ **18**
4. **Mental Math** In a bird sanctuary, the experimental probability that any bird you see is a robin is about $\frac{1}{8}$. Suppose this trend continues. There are 48 birds. Predict the number of robins you see. **6 robins**
5. You want to find the probability that three out of five babies are boys. You decide to toss coins to simulate the problem. How many coins would you use? Explain. **5 coins; one for each baby**

Homework Exercises

For more exercises, see Extra Skills and Word Problems.

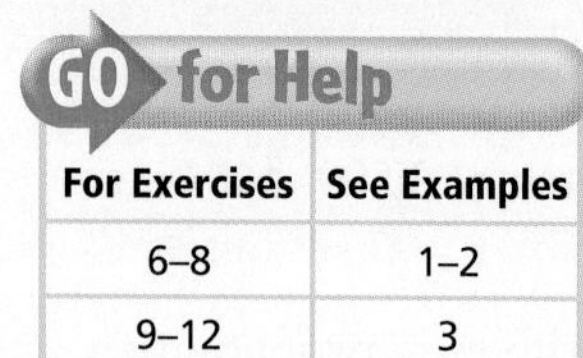

GO for Help

For Exercises	See Examples
6–8	1–2
9–12	3

A **Find each experimental probability.**

6. tosses: 80; tails: 40; $P(\text{tails}) = \underline{\ ?\ }$ $\frac{1}{2}$
7. tosses: 250; heads: 180; $P(\text{heads}) = \underline{\ ?\ }$ $\frac{18}{25}$
8. **Manufacturing** The quality-control engineer of Top Notch Tool Company finds flaws in 8 of 60 wrenches examined. Predict the number of flawed wrenches in a batch of 2,400. **320 wrenches**

Baseball **A baseball team averages one win to every one loss. Use a simulation to find each experimental probability for three games.**

9. P(three wins)
10. P(1 win and 2 losses)
11. P(2 wins and 1 loss)
12. P(three losses)

9–12. Check students' work.

B 

13. **Guided Problem Solving** During hockey practice, Yuri blocked 19 out of 30 shots and Gene blocked 17 out of 24 shots. For the first game, the coach wants to choose the goalie with the greater probability of blocking a shot. Which player should he choose? **Gene**
    - **Make a Plan** Find the experimental probability that each player will block the shot.
14. a. **Science** The probability that a male human is colorblind is 8%. Suppose you interview 1,000 males. About how many would you expect to be colorblind? **80 males**
    GPS
    b. **Reasoning** Will you always get the same number? Explain. **See left.**

14b. Not necessarily; the 8% colorblind percent is based on experimental results. The results of other experiments may differ.

1. Theoretical probability is computed by the formula $P(\text{event}) = \frac{\text{number of favorable outcomes}}{\text{total number of possible outcomes}}$. Experimental probability is based on experimental data or observation.

22. Answers may vary. Sample: Toss six coins, record data, and repeat several times. Calculate the experimental probability of two coins being heads.

GO Online
Homework Video Tutor
Visit: PHSchool.com
Web Code: are-1202

15. A company checks washers at four plants and records the number of defective washers. Find each experimental probability.

Plant	Number of Washers	Number Defective	P(Defective)
1	2,940	588	■ $\frac{1}{5}$
2	1,860	93	■ $\frac{1}{20}$
3	640	26	■ $\frac{13}{320}$
4	3,048	54	■ $\frac{9}{508}$

Data Analysis **Use the data shown. Find each experimental probability.**

Students' Birthdays

Su	M	Tu	W	Th	F	Sa
3 X	6 X	5 X	2 X	3 X	2 X	4 X

16. $P(\text{Sunday})$ $\frac{3}{25}$
17. $P(\text{Monday})$ $\frac{6}{25}$
18. $P(\text{Tuesday})$ $\frac{1}{5}$
19. $P(\text{Friday})$ $\frac{2}{25}$
20. $P(\text{weekday})$ $\frac{18}{25}$
21. $P(\text{weekend})$ $\frac{7}{25}$

22. **Writing in Math** Describe a possible simulation to solve the following problem. You guess on six true-or-false questions. What is the probability that you guess exactly two answers correctly? See margin.

Ⓒ 23. **Challenge** On any day, a company has x torn posters in stock. On Monday, the total number of torn posters is 252. Express $P(\text{torn})$ on Monday in terms of x. If $P(\text{torn}) = \frac{1}{42}$, what is x? $\frac{x}{252}$; 6

Test Prep and Mixed Review

Practice

Multiple Choice

24. The picture shows the number of colored shirts sold this week at Joan's Clothing Shop. If the trend continues, which is the best prediction of the number of orange shirts sold in one year? C

Ⓐ 20 Ⓑ 130 Ⓒ 260 Ⓓ 1,362

25. The scale on a map of Texas is 1 in. : 31 mi. Which distance on the map should represent the 420 mi from Amarillo to El Paso? G

Ⓕ 1.25 in. Ⓖ 13.5 in. Ⓗ 135 in. Ⓙ 13,824 in.

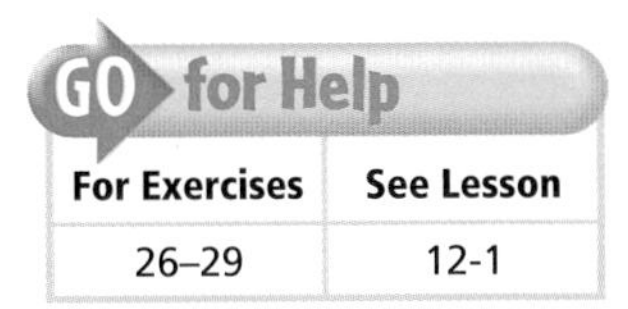

GO for Help

For Exercises	See Lesson
26–29	12-1

You spin the spinner once. Find each probability.

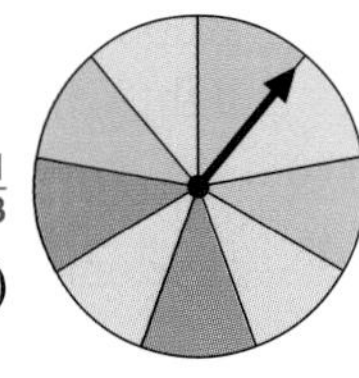

26. $P(\text{purple})$ $\frac{2}{9}$
27. $P(\text{blue})$ $\frac{3}{9}$ or $\frac{1}{3}$
28. $P(\text{blue or yellow})$ $\frac{7}{9}$
29. $P(\text{not yellow})$ $\frac{5}{9}$

Online lesson quiz, PHSchool.com, Web Code: ara-1202

4. Assess & Reteach

Lesson Quiz

PowerPoint

A manufacturer makes computer circuit boards. A random check of 5,000 circuit boards shows that 25 are defective.

1. What is the experimental probability that a circuit board is defective? $\frac{1}{200}$
2. Predict the number of defective circuit boards per month if the company manufactures 41,000 circuit boards in July. 205 defective circuit boards

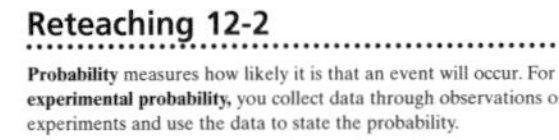

Reteaching 12-2 Experimental Pr **L2**

Probability measures how likely it is that an event will occur. For an **experimental probability,** you collect data through observations or experiments and use the data to state the probability.

The jar contains red, green, and blue chips. You shake the jar, draw a chip, note its color, and then put it back. You do this 20 times with these results: 7 blue chips, 5 red chips, and 8 green chips. The experimental probability of drawing a green chip is

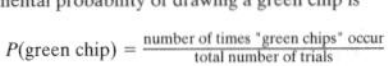

$P(\text{green chip}) = \frac{\text{number of times "green chips" occur}}{\text{total number of trials}}$

$P(\text{green chip}) = \frac{8}{20} = \frac{2}{5} = 0.4 = 40\%$

The probability of drawing a green chip is $\frac{2}{5}$, or 0.4, or 40%.

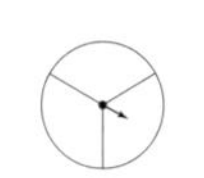

Sometimes a model, or simulation, is used to represent a situation. Then, the simulaton is used to find the experimental probability. For example, spinning this spinner can simulate the probability that 1 of 3 people is chosen for president of the student body.

Use the 20 draws above to complete each exercise.

1. What is the experimental probability of drawing a red chip? Write the probability as a fraction.
$P(\text{red chip}) = \frac{5}{20} = \frac{1}{4}$
2. What is the experimental probability of drawing a blue chip? Write the probability as a percent.
$P(\text{blue chip}) = \frac{7}{20} =$ 35%

Suppose you have a bag with 30 chips: 12 red, 8 white, and 10 blue. You shake the jar, draw a chip, note its color, and then put it back. You do this 30 times with these results: 10 blue chips, 12 red chips, and 8 white chips. Write each probability as fraction in simplest form.

3. $P(\text{red})$ $\frac{2}{5}$
4. $P(\text{white})$ $\frac{4}{15}$
5. $P(\text{blue})$ $\frac{1}{3}$

Describe a probability simulation for each situation.

6. You guess the answers on a true/false test with 20 questions.
Sample answer: Flip a coin 20 times.
7. One student out of 6 is randomly chosen to be the homeroom representative.
Sample answer: Spin a spinner divided into six equal sections.

Enrichment 12-2 Experimental Pr **L4**

Applications of Probability

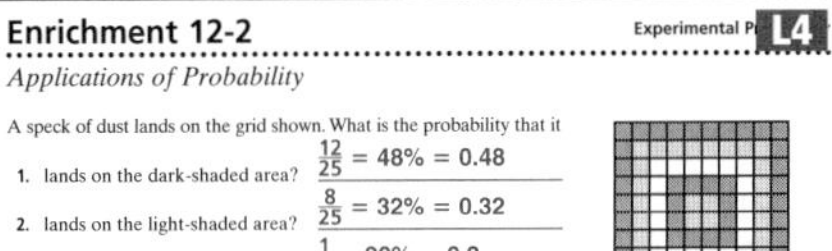

A speck of dust lands on the grid shown. What is the probability that it

1. lands on the dark-shaded area? $\frac{12}{25} = 48\% = 0.48$
2. lands on the light-shaded area? $\frac{8}{25} = 32\% = 0.32$
3. lands on the unshaded area? $\frac{1}{5} = 20\% = 0.2$
4. lands on the 4 center squares? $\frac{1}{25} = 4\% = 0.04$
5. does not land on the unshaded area?
$\frac{4}{5} = 80\% = 0.80$
6. does not land on the dark-shaded area?
$\frac{13}{25} = 52\% = 0.52$

In a coin toss game, you earn points for landing on the shaded figures. Assume coins land randomly in the large square. What is the probability that a coin

7. lands on the trapezoid?
$\frac{9}{125} = 7.2\% = 0.072$
8. lands on the circle?
$\approx 6.75\% = 0.0675$
9. lands on the L-shape?
$\frac{7}{100} = 7\% = 0.07$
10. does not land on the trapezoid?
$92.8\% = 0.928$
11. does not land on the circle or the L-shape?
$86.25\% = 0.8625$
12. lands in the unshaded area?
$79.05\% = 0.7905$
13. lands in the square?
100%

Alternative Assessment

Have pairs of students survey twenty classmates or friends to find the month of each person's birthday. Pairs record the data in a line plot. They then find the experimental probability of a birthday occurring in each month.

Test Prep

Resources

For additional practice with a variety of test item formats:

- Test-Taking Strategies, p. 615
- Test Prep, p. 619
- Test-Taking Strategies with Transparencies

Activity Lab

Random Numbers

Students worked with simulations in Lesson 12-2. This Activity introduces them to random number generation using spreadsheet software as a useful tool for simulating events.

Guided Instruction

Technology Tip
Elicit what the advantages are of using computer simulation to do a probability experiment. Guide students to appreciate that spreadsheets can generate random numbers rapidly and can keep track of the results.

Exercises
Have students work in pairs. Then pairs can discuss their results with the class. For the Writing in Math activity, have students ask classmates for their predictions before presenting those results.

Differentiated Instruction

Advanced Learners L4
Have students generate their own spreadsheet of random numbers and write problems. Have them trade papers with a partner and solve each other's problems.

Resources

- spreadsheet software

12-2b Activity Lab — Technology

Random Numbers

You can use a random number table to simulate some problems. To generate a random number table in a spreadsheet, follow these steps.

Step 1 Highlight the group of cells to use for your table.

Step 2 Select the Format Cells menu.

Step 3 Choose the category Custom and enter 0000. Click OK.

Step 4 Use the formula RAND()*10,000. This will make a group of 4 digits in each cell of a spreadsheet. (*Note:* Each time you generate a random number table, you will get a different group of digits.)

	A	B	C	D
1	2260	1927	7807	0912
2	8879	6235	5897	8068
3	8121	4646	8368	1613
4	0821	8911	3022	0307
5	9393	5403	4930	4898

EXAMPLE

A rare lily bulb has a 50% chance of growing. You plant four bulbs. What is the experimental probability that all four will grow?

2260	1927	7807	0912
8879	6235	5897	8068
8121	4646	8368	1613
0821	8911	3022	0307
9393	5403	4930	4898

← **Any group of 4 even digits represents *all four grow.***

Use the random number table. Let even digits represent *grows,* and let odd digits represent *does not grow.* Then a 4-digit number of all even digits represents the event *all four grow*. Of 20 groups, 3 consist entirely of even digits. So the experimental probability of four bulbs growing is $\frac{3}{20}$, or 15%.

Exercises

Use the random number table above or generate your own.

1. Suppose there is a 30% probability of being stopped by a red light at each of four stoplights. What is the experimental probability of being stopped by at least two red lights? Let 0, 1, and 2 represent red lights. $\frac{9}{20}$
2. **Reasoning** Suppose there is a 60% chance of a red light at each stoplight. How many digits would you use to represent getting a red light? Explain. **6 digits; 6 out of 10 digits represent 60%.**
3. **Writing in Math** Write a probability problem you can solve using a random number table. Solve your problem. **Check students' work.**

12-3 Sample Spaces

Check Skills You'll Need

1. **Vocabulary Review** What diagram can you use to find the *prime factors* of a number? **factor tree**

Write the prime factorization of each number.

2. 15 **3 · 5** 3. 14 **2 · 7**

4. 26 **2 · 13** 5. 55 **5 · 11**

GO for Help Lesson 2-2

What You'll Learn

To make and use sample spaces and to use the counting principle

New Vocabulary sample space, counting principle

Why Learn This?

When you are at a salad bar, you can choose from different vegetables, fruits, and dressings. You may want to know all the possible combinations of ingredients you can use.

The collection of all possible outcomes in an experiment is the **sample space.** You can use the sample space to find the probability of an event.

EXAMPLE Finding a Sample Space

Test Prep Tip

When you construct a sample space for two events, pair each possible outcome for the first event with each possible outcome for the second event.

1 **a.** Make a table to find the sample space for rolling two number cubes colored red and blue. Write the outcomes as ordered pairs.

	1	2	3	4	5	6
1	(1, 1)	(2, 1)	(3, 1)	(4, 1)	(5, 1)	(6, 1)
2	(1, 2)	(2, 2)	(3, 2)	(4, 2)	(5, 2)	(6, 2)
3	(1, 3)	(2, 3)	(3, 3)	(4, 3)	(5, 3)	(6, 3)
4	(1, 4)	(2, 4)	(3, 4)	(4, 4)	(5, 4)	(6, 4)
5	(1, 5)	(2, 5)	(3, 5)	(4, 5)	(5, 5)	(6, 5)
6	(1, 6)	(2, 6)	(3, 6)	(4, 6)	(5, 6)	(6, 6)

← There are 36 possible outcomes.

b. Find the probability of rolling at least one 3.

There are 11 outcomes with at least one 3. There are 36 possible outcomes. So the probability of rolling at least one 3 is $\frac{11}{36}$.

Quick Check

1. Give the sample space for tossing two coins. Find the probability of getting two heads. **See left.**

1.

	H	T
H	HH	HT
T	TH	TT

; $\frac{1}{4}$

12-3

1. Plan

Objective

To make and use sample spaces and to use the counting principle

Examples

1. Finding a Sample Space
2. Using a Tree Diagram
3. Using the Counting Principle

Math Understandings: p. 578D

Math Background

The counting principle states that when *m* choices can be made for the first place and *n* choices can be made for the second place, then the total number of choices is the product, *mn*. The *sample space* of an event is all possible outcomes that can be shown in a table, tree diagram, and set of ordered pairs.

More Math Background: p. 578D

Lesson Planning and Resources

See p. 578E for a list of the resources that support this lesson.

Bell Ringer Practice

Check Skills You'll Need

Use student page, transparency, or PowerPoint. For intervention, direct students to:

Prime Factorization

Lesson 2-2

Extra Skills and Word Problems Practice, Ch. 2

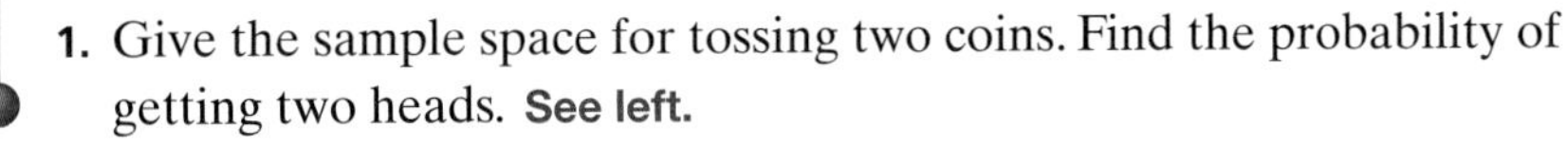

Differentiated Instruction Solutions for All Learners

Special Needs L1

Students may have a difficult time drawing tree diagrams. Provide them with rulers, or graph paper to help them. If they still have trouble, have them work with a partner who can draw the tree diagram while they label the outcomes from each branch.

learning style: visual

Below Level L2

Review the sample space for rolling a standard number cube and the probability of various outcomes, such as rolling an even number.

$P(\text{even}) = \frac{3}{6}$ or $\frac{1}{2}$

learning style: verbal

2. Teach

Activity Lab

Use before the lesson.

All in One Teaching Resources
Activity Lab 12-3: Decision Making

Guided Instruction

Example 2
Ask a volunteer to explain the differences between a kayak, a canoe, and a rowboat.

Example 3
Explain that since the Commutative Property states that $5 \times 6 = 6 \times 5$, you could choose the meat first, followed by the bread, with the same results.

PowerPoint

Additional Examples

1. **a.** Make a table to show the sample space for tossing two coins. Write the outcomes as ordered pairs. {(H, H), (H, T), (T, H), (T, T)}
   b. Find P(T, T), the probability of tossing two tails. $\frac{1}{4}$

2. Suppose you can go west or northwest by train, bus, or car.
   a. Draw a tree diagram to show the sample space. {(w, t), (w, b), (w, c), (nw, t), (nw, b), (nw, c)}
   b. What is the probability of a random selection that results in a bus trip west? $\frac{1}{6}$

3. How many kinds of coin purses are available if the purses come in small or large sizes and colors red, blue, yellow, and black? 8

You can also show a sample space by using a tree diagram. Each branch of the tree represents one choice.

EXAMPLE Using a Tree Diagram

2 **River Travel** Suppose you are going to travel on a river. You have two choices of boats—a kayak or a rowboat. You can go upstream on three smaller streams, to the north, northwest, and northeast.

a. What is the sample space for your journey?

Make a tree diagram for the possible outcomes.

Boat	Stream	Outcome
Kayak	North	Kayak, North
	Northwest	Kayak, Northwest
	Northeast	Kayak, Northeast
Rowboat	North	Rowboat, North
	Northwest	Rowboat, Northwest
	Northeast	Rowboat, Northeast

← There are six possible outcomes.

b. Suppose you select a trip at random. What is the probability of selecting a kayak and going directly north?

There is one favorable outcome (kayak, north) out of six possible outcomes. The probability is $\frac{1}{6}$.

Quick Check

2. **a.** Suppose a canoe is added as another choice of boats in Example 2. Draw a tree diagram to show the sample space. **See back of book.**
   b. Find the probability of selecting a canoe at random for the trip. $\frac{1}{3}$

In Example 2 above, there are 2 choices of boats and 3 choices of direction. There are 2×3, or 6, total possible choices. This suggests a simple way to find the number of outcomes—using the **counting principle.**

KEY CONCEPTS The Counting Principle

Suppose there are m ways of making one choice and n ways of making a second choice. Then there are $m \times n$ ways to make the first choice followed by the second choice.

Example
If you can choose a shirt in 5 sizes and 7 colors, then you can choose among 5×7, or 35, shirts.

Advanced Learners L4
Have students find the number of sandwich choices if there are 4 kinds of bread, 5 kinds of meat, and 3 kinds of cheese. **60 choices**

learning style: visual

English Language Learners ELL
For Example 1, make sure to point out that the ordered pairs are formed by using the intersection of the outcomes for each pair of number cubes. Students may have a difficult time understanding why 1, 3 and 3, 1 are different outcomes. Show them using number cubes.

learning style: visual

EXAMPLE Using the Counting Principle

3 Gridded Response How many different sandwiches can you order when you choose one bread and one meat from the menu?

THE
DELI COUNTER
SANDWICHES
FRESH BREADS
Rye
Wheat
White
Pita
Wrap
DELI MEATS
Roast Beef
Turkey
Ham
Pastrami
Salami
Liverwurst

Use the counting principle.

Bread		Meat		
number of choices	×	number of choices		
5	×	6	=	30

There are 30 different sandwiches available.

3 0

Quick Check

3. A manager at the Deli Counter decides to add chicken to the list of meat choices. How many different sandwiches are now available? **35 sandwiches**

Check Your Understanding

1. **Vocabulary** What is a sample space? **the collection of all possible outcomes in an experiment**
2. Complete the tree diagram for tossing a coin three times. **See margin.**

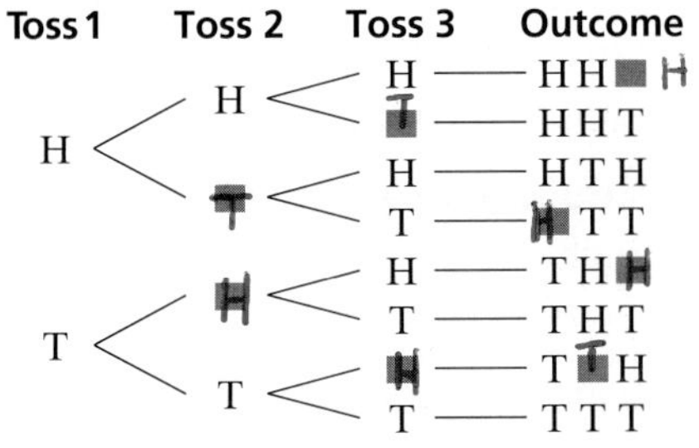

Use your completed diagram from Exercise 2 to find each probability.

3. $P(\text{HHH}) = \frac{■}{8}$ **1**
4. $P(\text{TTT}) = \frac{1}{■}$ **8**
5. $P(\text{at least one H}) = \frac{■}{8}$ **7**
6. $P(\text{exactly 2 T's}) = \frac{■}{8}$ **3**
7. If you toss 4 coins, how many possible outcomes are there? **16**
8. Find the number of different couches that you can make using 16 different fabrics and 8 patterns. **128 couches**
9. **Multiple Choice** An architect has 3 different widths he can use for a rectangular building design. He also has 4 different lengths to use for the design. How many different designs are possible? **D**

   (A) 3 (B) 4 (C) 7 (D) 12

2. See back of book.

All in One Teaching Resources

- Daily Notetaking Guide 12-3 L3
- Adapted Notetaking 12-3 L1

Closure

- *What is the sample space?* **all possible outcomes of an event**
- *What is the counting principle?* **The number of ways to make *m* choices followed by *n* choices is $m \times n$.**

3. Practice

Assignment Guide

Check Your Understanding
Go over Exercises 1–9 in class before assigning the Homework Exercises.

Homework Exercises

A Practice by Example	10–16
B Apply Your Skills	17–27
C Challenge	28
Test Prep and Mixed Review	29–32

Homework Quick Check
To check students' understanding of key skills and concepts, go over Exercises 13, 16, 23, 24, and 27.

Differentiated Instruction Resources

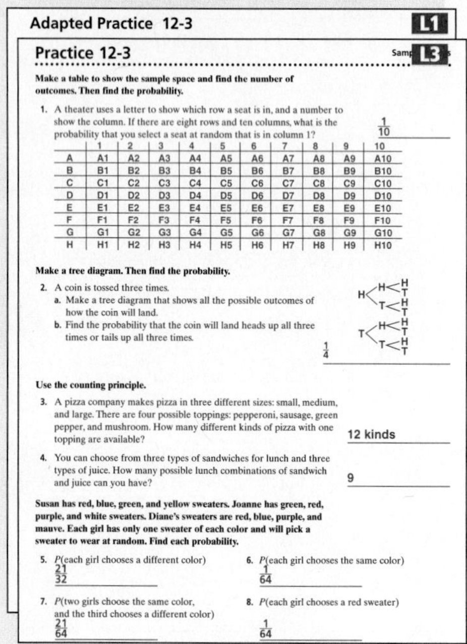

Adapted Practice 12-3 L1

Practice 12-3 L3

Make a table to show the sample space and find the number of outcomes. Then find the probability.

1. A theater uses a letter to show which row a seat is in, and a number to show the column. If there are eight rows and ten columns, what is the probability that you select a seat at random that is in column 1? $\frac{1}{10}$

	1	2	3	4	5	6	7	8	9	10
A	A1	A2	A3	A4	A5	A6	A7	A8	A9	A10
B	B1	B2	B3	B4	B5	B6	B7	B8	B9	B10
C	C1	C2	C3	C4	C5	C6	C7	C8	C9	C10
D	D1	D2	D3	D4	D5	D6	D7	D8	D9	D10
E	E1	E2	E3	E4	E5	E6	E7	E8	E9	E10
F	F1	F2	F3	F4	F5	F6	F7	F8	F9	F10
G	G1	G2	G3	G4	G5	G6	G7	G8	G9	G10
H	H1	H2	H3	H4	H5	H6	H7	H8	H9	H10

Make a tree diagram. Then find the probability.

2. A coin is tossed three times.
   a. Make a tree diagram that shows all the possible outcomes of how the coin will land.
   b. Find the probability that the coin will land heads up all three times or tails up all three times. $\frac{1}{4}$

Use the counting principle.

3. A pizza company makes pizza in three different sizes: small, medium, and large. There are four possible toppings: pepperoni, sausage, green pepper, and mushroom. How many different kinds of pizza with one topping are available? 12 kinds
4. You can choose from three types of sandwiches for lunch and three types of juice. How many possible lunch combinations of sandwich and juice can you have? 9

Susan has red, blue, green, and yellow sweaters. Joanne has green, red, purple, and white sweaters. Diane's sweaters are red, blue, purple, and mauve. Each girl has only one sweater of each color and will pick a sweater to wear at random. Find each probability.

5. P(each girl chooses a different color) $\frac{21}{32}$
6. P(each girl chooses the same color) $\frac{1}{64}$
7. P(two girls choose the same color, and the third chooses a different color) $\frac{21}{64}$
8. P(each girl chooses a red sweater) $\frac{1}{64}$

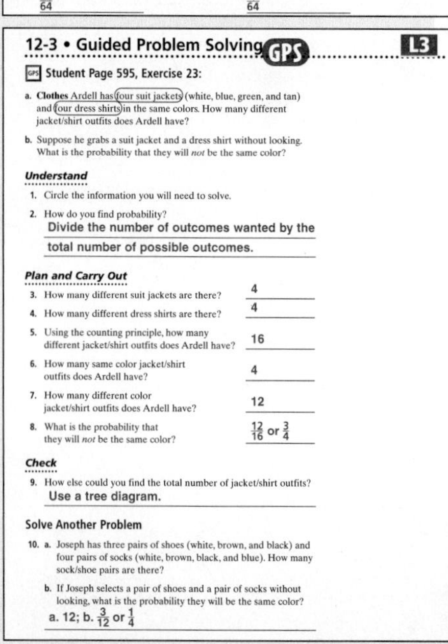

12-3 • Guided Problem Solving GPS L3

Student Page 595, Exercise 23:

a. Clothes Ardell has four suit jackets (white, blue, green, and tan) and four dress shirts in the same colors. How many different jacket/shirt outfits does Ardell have?

b. Suppose he grabs a suit jacket and a dress shirt without looking. What is the probability that they will *not* be the same color?

Understand

1. Circle the information you will need to solve.
2. How do you find probability? Divide the number of outcomes wanted by the total number of possible outcomes.

Plan and Carry Out

3. How many different suit jackets are there? 4
4. How many different dress shirts are there? 4
5. Using the counting principle, how many different jacket/shirt outfits does Ardell have? 16
6. How many same color jacket/shirt outfits does Ardell have? 4
7. How many different color jacket/shirt outfits does Ardell have? 12
8. What is the probability that they will *not* be the same color? $\frac{12}{16}$ or $\frac{3}{4}$

Check

9. How else could you find the total number of jacket/shirt outfits? Use a tree diagram.

Solve Another Problem

10. a. Joseph has three pairs of shoes (white, brown, and black) and four pairs of socks (white, brown, black, and blue). How many sock/shoe pairs are there?
    b. If Joseph selects a pair of shoes and a pair of socks without looking, what is the probability they will be the same color?
    a. 12; b. $\frac{3}{12}$ or $\frac{1}{4}$

Homework Exercises

For more exercises, see Extra Skills and Word Problems.

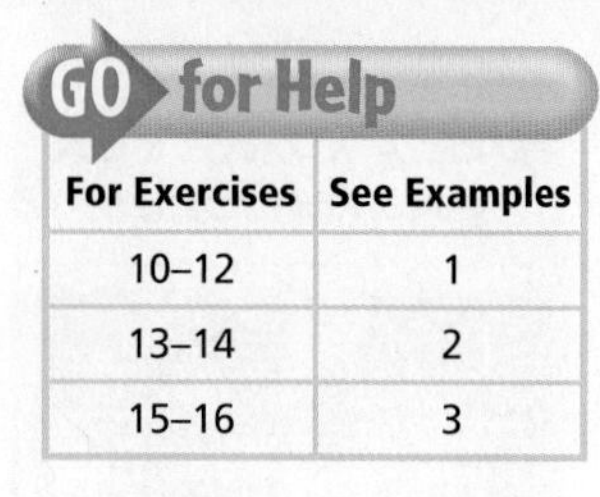

GO for Help

For Exercises	See Examples
10–12	1
13–14	2
15–16	3

10. H T ; $\frac{1}{2}$

	H	T
H	HH	HT
T	TH	TT

12. 1 2 3 4 ; $\frac{1}{8}$

	1	2	3	4
H	H1	H2	H3	H4
T	T1	T2	T3	T4

13. 1st Spin 2nd Spin ; $\frac{1}{4}$
red — red, blue
blue — red, blue

A Make a table to show the sample space for each situation and find the number of outcomes. Then find the probability.

10. You toss two coins. What is the probability of getting one tail and one head? **See left.**
11. You roll a number cube once. What is the probability of rolling a number less than 4? **1 2 3 4 5 6; $\frac{3}{6}$ or $\frac{1}{2}$**
12. You toss a coin and spin a spinner. The spinner has four equal sections that are numbered from 1 to 4. Find the probability of getting tails and spinning a 4. **See left.**

Make a tree diagram. Then find the probability of each event.

13. A spinner is half red and half blue. If you spin the spinner twice, what is the probability that you will get red both times? **See left.**
14. You choose at random from the letters A, B, C, and D, and you roll a number cube once. What are the chances you get A and 5? **See margin.**

Use the counting principle.

15. **Cooking** You make a recipe with herbs and spices for a party. You have four herbs—basil, bay leaves, chives, and dill. You also have three spices—paprika, pepper, and garlic powder. How many different recipes with one herb and one spice can you make? **12 recipes**
16. **Education** A school has four art teachers, three music teachers, and eight history teachers. In how many ways can a student be assigned an art teacher, a music teacher, and a history teacher? **96 ways**

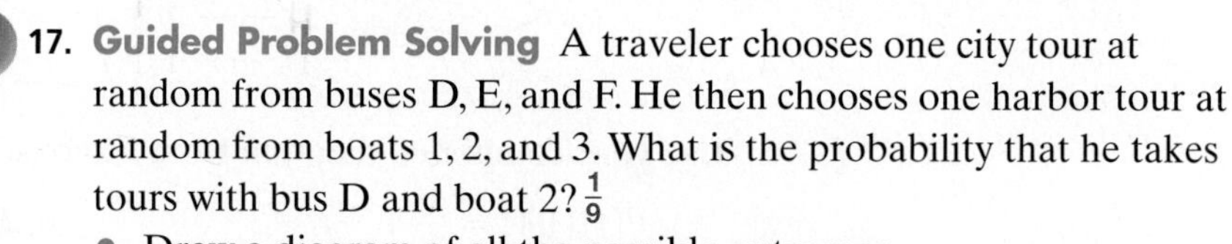

B GPS 17. **Guided Problem Solving** A traveler chooses one city tour at random from buses D, E, and F. He then chooses one harbor tour at random from boats 1, 2, and 3. What is the probability that he takes tours with bus D and boat 2? **$\frac{1}{9}$**
- Draw a diagram of all the possible outcomes.
- How many outcomes include tours with bus D and boat 2?

A spinner has four equal sections numbered 1 through 4. You spin it twice. Use the sample space below to find each probability.

18. $P(1, 2)$ **$\frac{1}{16}$**
19. $P(1, \text{odd})$ **$\frac{2}{16}$ or $\frac{1}{8}$**
20. $P(\text{even, odd})$ **$\frac{4}{16}$ or $\frac{1}{4}$**

First Spin \ Second Spin	1	2	3	4
1	(1, 1)	(1, 2)	(1, 3)	(1, 4)
2	(2, 1)	(2, 2)	(2, 3)	(2, 4)
3	(3, 1)	(3, 2)	(3, 3)	(3, 4)
4	(4, 1)	(4, 2)	(4, 3)	(4, 4)

GO Online
Homework Video Tutor
Visit: PHSchool.com
Web Code: are-1203

14. See back of book.

Find the number of outcomes for each situation.

21. Pick one of 7 boys and one of 12 girls. **84**

22. Toss five coins once each. **32**

23. a. **Clothes** Ardell has four suit jackets (white, blue, green, and tan) and four dress shirts in the same colors. How many different jacket-and-shirt outfits does Ardell have? **16 outfits**
    b. Suppose he grabs a suit jacket and a dress shirt without looking. What is the probability that they will *not* be the same color? $\frac{12}{16}$ **or** $\frac{3}{4}$

24. **Writing in Math** Explain how to use the counting principle to find the number of outcomes in a sample space. **See left.**

Use the menu for Exercises 25–28.

25. List all the possible drink orders. **See margin.**

26. You order lemonade and popcorn. Draw a tree diagram to show the sample space. **See margin.**

27. **Reasoning** A manager uses the counting principle to find P(small popcorn, medium lemonade) = $\frac{1}{24}$. Do you agree? Explain. **See margin.**

28. **Challenge** Find the probability that you randomly select the same size popcorn and drink. $\frac{6}{24}$ **or** $\frac{1}{4}$

Careers Tailors fit designer clothes for important events.

24. **Multiply the number of choices for each part of the sample space.**

Test Prep and Mixed Review

Practice

Gridded Response

29. The data in the stem-and-leaf plot at the right show the lengths of principal rivers in Africa. What is the median length, in miles, of the rivers? **1,100**

African River Lengths (hundreds of miles)

Stem	Leaf
0	6 7 7
1	0 0 0 1 2 3 7
2	6 7
4	1

Key: 2 | 6 means 2,600 miles

30. Suppose $\frac{21}{25}$ of your classmates can attend your birthday party. What is this fraction expressed as a decimal? **0.84**

31. Trista has purple, yellow, and red wrapping paper. She can use blue or white ribbon. She has four shapes of gift tags. In how many different ways can she choose one wrapping paper, one ribbon, and one gift tag? **24**

32. A hockey player makes 3 goals out of 9 shots in a game. What is the experimental probability that he does not make a goal? $\frac{6}{9}$ **or** $\frac{2}{3}$

GO for Help

For Exercise	See Lesson
32	12-2

4. Assess & Reteach

PowerPoint

Lesson Quiz

You pick a card at random from cards labeled 1, 2, or 3 and then spin a spinner divided into thirds, each of which is labeled with a number 1, 2, or 3.

1. Make a table showing the sample space. **{(1, 1), (1, 2), (1, 3), (2, 1), (2, 2), (2, 3), (3, 1), (3, 2), (3, 3)}**
2. How many possible outcomes are there? **9**
3. What is the probability of two 3s? $\frac{1}{9}$
4. What is the probability of picking a card and spinning the same number? $\frac{1}{3}$

25. **small fruit punch, small lemonade, medium fruit punch, medium lemonade, large fruit punch, large lemonade, jumbo fruit punch, jumbo lemonade**
26. **See back of book.**
27. **Yes; there are 24 possible orders, so the probability for each order is $\frac{1}{24}$.**

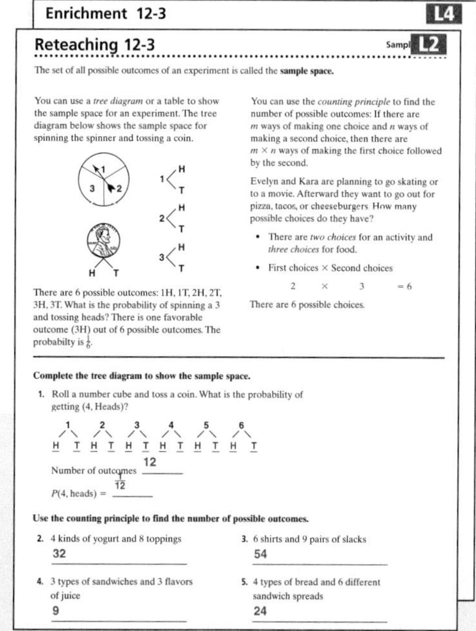

Enrichment 12-3 L4

Reteaching 12-3 Sample L2

The set of all possible outcomes of an experiment is called the **sample space.**

You can use a *tree diagram* or a table to show the sample space for an experiment. The tree diagram below shows the sample space for spinning the spinner and tossing a coin.

There are 6 possible outcomes: 1H, 1T, 2H, 2T, 3H, 3T. What is the probability of spinning a 3 and tossing heads? There is one favorable outcome (3H) out of 6 possible outcomes. The probabilty is $\frac{1}{6}$.

You can use the *counting principle* to find the number of possible outcomes: If there are m ways of making one choice and n ways of making a second choice, then there are $m \times n$ ways of making the first choice followed by the second.

Evelyn and Kara are planning to go skating or to a movie. Afterward they want to go out for pizza, tacos, or cheeseburgers. How many possible choices do they have?

- There are *two choices* for an activity and *three choices* for food.
- First choices × Second choices

  2 × 3 = 6

There are 6 possible choices.

Complete the tree diagram to show the sample space.

1. Roll a number cube and toss a coin. What is the probability of getting (4, Heads)?

   1: H T; 2: H T; 3: H T; 4: H T; 5: H T; 6: H T

   Number of outcomes 12

   P(4, heads) = $\frac{1}{12}$

Use the counting principle to find the number of possible outcomes.

2. 4 kinds of yogurt and 8 toppings **32**
3. 6 shirts and 9 pairs of slacks **54**
4. 3 types of sandwiches and 3 flavors of juice **9**
5. 4 types of bread and 6 different sandwich spreads **24**

Alternative Assessment

Each student in a pair writes a problem that can be solved by either making a tree diagram or using the counting principle. Partners exchange papers and solve each other's problem. Have students share their problems and solutions with the class.

Test Prep

Resources

For additional practice with a variety of test item formats:
- Test-Taking Strategies, p. 615
- Test Prep, p. 619
- Test-Taking Strategies with Transparencies

Activity Lab

Using Data to Predict

Students use data from tables to predict outcomes. This will extend their work with sample spaces from Lesson 12-3.

Guided Instruction

Explain to students that they study the data in a table to find a trend and make predictions about the data. Have a volunteer read the problem aloud. Then ask questions, such as:

- *What is the probability of Katie mowing the lawn?* **$\frac{2}{5}$**
- *What are the odds in favor of Tim mowing the lawn?* **9 to 6**
- *What conclusion can you draw about how often Tim will mow the lawn compared to how often Katie will mow the lawn?* **Tim will mow the lawn $1\frac{1}{2}$ times as often as Katie will.**

Activity

Have students work in pairs on the Activity. When they have completed the Activity, review students' conclusions as a class.

Differentiated Instruction

Advanced Learners L4

Have students estimate how many times each Tim and Katie will mow the lawn in 75 trials, 105 trials, and 135 trials. Ask them what trend they notice.

Using Data to Predict

Katie and Tim play the following game to decide who should mow the lawn each week.

- Katie or Tim places three black marbles and three white marbles into a bag.
- Tim pulls out two marbles.
- If the marbles match, Katie mows the lawn. Otherwise, Tim mows the lawn.

The table below shows the results of the first 15 trials.

Marble Selection Results

Event	Number of Times
Marbles matched	6
No match	9

ACTIVITY

1. **Estimation** Use the data from the table above. Assume the data trend continues. For Katie and Tim, estimate the number of times that each one mows the lawn in 45 trials. **Katie: 18; Tim: 27**
2. **Data Collection** Simulate the game with three red cubes and three yellow cubes. Put the cubes in a bag. Select two cubes at random. Record whether or not they match. Return the cubes to the bag, and repeat until you have recorded 45 trials. **Check students' work.**
3. **Reasoning** Describe how the results of your experiment compare to your predictions. Is the game fair? Explain. **Check students' work.**
4. Tim thinks the probability that the marbles match equals the probability that they do not match. Use diagrams and what you have learned to show why the probabilities are not equal. Find the probability that the two marbles match and the probability that they do not match. **See margin.**
5. Add a fourth cube of each color into the bag. Find the probability of selecting two cubes that match. Predict the number of matches in 49 trials. **$\frac{3}{7}$; 21 matches**
6. **Data Analysis** Test your prediction. Record the results of 49 trials. Compare your results to your prediction. Is this game more fair or less fair than the original game? Explain. **Check students' work.**

4. See back of book.

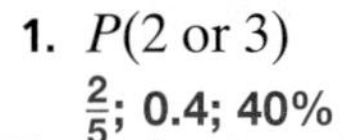

Checkpoint Quiz 1

Lessons 12-1 through 12-3

You spin the spinner at the right once. Write each probability as a fraction, a decimal, and a percent.

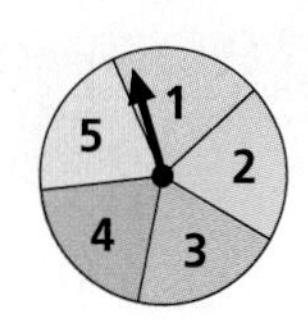

1. P(2 or 3) $\frac{2}{5}$; 0.4; 40%
2. P(even) $\frac{2}{5}$; 0.4; 40%
3. P(not 4) $\frac{4}{5}$; 0.8; 80%
4. P(not even) $\frac{3}{5}$; 0.6; 60%

You spin the spinner twice.

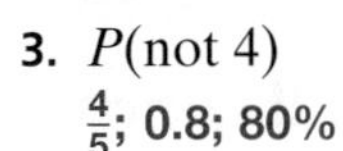

5. Give the sample space.
6. Find P(green, then green)
7. Find P(purple, then blue) $\frac{1}{16}$

5. PP, PB, PR, PG
BB, BR, BG, BP
GG, GP, GB, GR
RR, RB, RP, RG

6. $\frac{1}{16}$

8. **Forestry** The table shows a sample of the number of spruce trees counted in a forest area. Find the experimental probability of selecting a Serbian spruce. $\frac{20}{119}$

Spruce Trees

Tree Type	Number
Norway spruce	32
Serbian spruce	20
Colorado spruce	67

9. A true-or-false quiz has five questions. Use a simulation or a sample space to find the probability of guessing at random and getting exactly three correct answers. $\frac{5}{16}$

Use this Checkpoint Quiz to check students' understanding of the skills and concepts of Lessons 12-1 through 12-3.

Resources

- All-in-One Teaching Resources Checkpoint Quiz 1
- ExamView CD-ROM
- Success Tracker™ Online Intervention

12-4a Activity Lab

Exploring Multiple Events

You want to make a necklace using two colors of beads. You decide which colors to use by selecting from the beads at the right. You select the first bead at random. You put the bead back and make another selection at random.

1. Give the sample space for the colors of the two beads. Write the outcomes as ordered pairs. **See margin.**
2. What is the probability of selecting a red bead first? $\frac{4}{16}$ or $\frac{1}{4}$
3. What is the probability of selecting red beads twice? $\frac{1}{16}$
4. Suppose you now select a red bead and do not replace it. How many beads of each color are left? **1 red, 2 green, 2 blue, 2 yellow**
5. **Reasoning** Does the probability of selecting a red bead second depend on whether you replace the first red bead? Explain. **See margin.**

Exploring Multiple Events

Students experiment with the probability of multiple events. This will prepare them to calculate the probability of compound events in Lesson 12-4.

Guided Instruction

Explain to students that they can find the probability of more than one event. Introduce the Activity and work through it as a class. Elicit answers to each question from volunteers.

Differentiated Instruction

Below Level L2
Provide actual beads or colored counters to help visual and tactile learners connect to the problem.

Resources

- Activity Lab 12-4: Compound Events
- colored beads or counters

1. RR, RB, RG, RY, BR, BB, BG, BY, GR, GB, GG, GY, YR, YB, YG, YY

5. Yes; if you replace the first red bead, then P(red for second draw) = $\frac{2}{8}$, or $\frac{1}{4}$, but if you do not replace the first red bead, then P(red for second draw) = $\frac{1}{7}$.

12-4

1. Plan

Objective
To find the probability of independent and dependent events

Examples
1 Probability of Independent Events
2, 3 Probability of Dependent Events

Math Understandings: p. 578D

Professional Development

Math Background

A *compound event* consists of two or more events. If the probability of the second event is not affected by the first event, the two events are *independent events.* The probability of independent events, $P(A \text{ and } B)$, is the product of the probabilities for each separate event, $P(A) \times P(B)$

Two events are *dependent events* if the occurrence of the first event changes the probability of the second event. The probability of dependent events is also the product of the separate probabilities.

More Math Background: p. 578D

Lesson Planning and Resources

See p. 578E for a list of the resources that support this lesson.

PowerPoint

Bell Ringer Practice

Check Skills You'll Need
Use student page, transparency, or PowerPoint. For intervention, direct students to:
Multiplying Fractions and Mixed Numbers
Lesson 3-4
Extra Skills and Word Problems Practice, Ch. 3

12-4 Compound Events

Check Skills You'll Need

1. **Vocabulary Review** Describe multiplying fractions using the terms *denominator* and *numerator.* **1–5. See below.**

Find each product.

2. $\frac{3}{4} \cdot \frac{3}{4}$
3. $\frac{3}{5} \cdot \frac{2}{5}$
4. $\frac{1}{5} \cdot \frac{1}{4}$
5. $\frac{3}{7} \cdot \frac{2}{7}$

for Help Lesson 3-4

Check Skills You'll Need

1. **To multiply fractions, multiply the numerators and multiply the denominators.**
2. $\frac{9}{16}$
3. $\frac{6}{25}$
4. $\frac{1}{20}$
5. $\frac{6}{49}$

What You'll Learn

To find the probability of independent and dependent events

New Vocabulary compound event, independent events, dependent events

Why Learn This?

You can find the probability of more than one event, such as winning a game twice.

A **compound event** consists of two or more events. Two events are **independent events** if the occurrence of one event does not affect the probability of the occurrence of the other.

KEY CONCEPTS **Probability of Independent Events**

If A and B are independent events, then $P(A, \text{then } B) = P(A) \times P(B)$.

EXAMPLE **Probability of Independent Events**

1 **Multiple Choice** You and a friend play a game twice. What is the probability that you win both games? Assume $P(\text{win})$ is $\frac{1}{2}$.

Ⓐ $\frac{1}{2}$ Ⓑ $\frac{4}{9}$ Ⓒ $\frac{1}{4}$ Ⓓ $\frac{1}{8}$

$P(\text{win, then win}) = P(\text{win}) \times P(\text{win})$ ← **Winning is the first and second event.**

$= \frac{1}{2} \times \frac{1}{2}$ ← **Substitute $\frac{1}{2}$ for $P(\text{win})$.**

$= \frac{1}{4}$ ← **Multiply.**

The probability of winning both games is $\frac{1}{4}$. The correct answer is C.

Quick Check

1. Find $P(\text{win, then lose})$. $\frac{1}{4}$

Differentiated Instruction **Solutions for All Learners**

Special Needs L1
To model the idea of *with* and *without replacement,* place some counters or marbles, in two colors, in a bag. Have students do experiments in which they take some marbles out and do not replace them. Then repeat the experiment, taking some out and replacing them. They should record their data.
learning style: tactile

Below Level L2
Review multiplication of fractions. Using a calculator will help students multiply fractions with greater denominators, as in $\frac{1}{26} \cdot \frac{1}{26} \cdot \frac{1}{676}$
learning style: visual

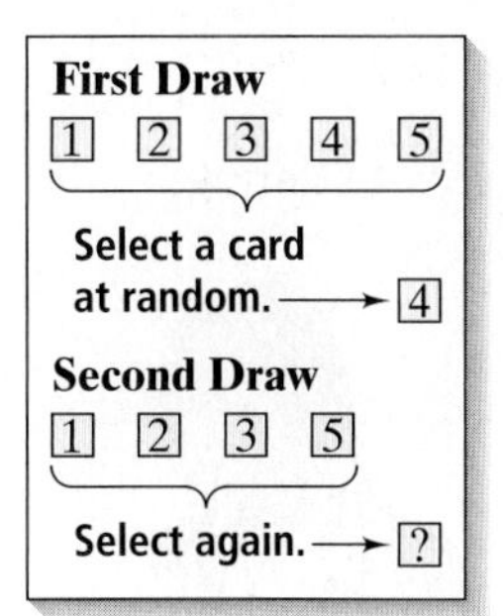

Suppose you play a game with cards numbered 1–5. You draw two cards at random. You draw the first card and do not replace it. The probability in the second draw depends on the result of the first draw.

Two events are **dependent events** if the occurrence of one event affects the probability of the occurrence of the other event.

KEY CONCEPTS **Probability of Dependent Events**

If event B depends on event A, then
$P(A, \text{then } B) = P(A) \times P(B \text{ after } A)$.

EXAMPLES Probability of Dependent Events

2 You select a card at random from those below. The card has the letter M. Without replacing the M card, you select a second card. Find the probability that you select a card with the letter A after you select M.

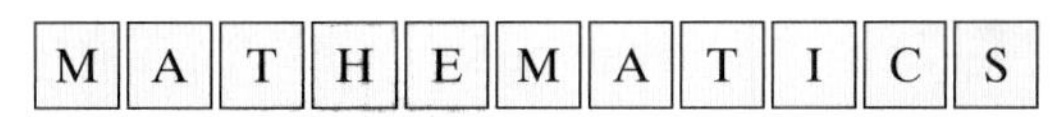

There are 10 cards remaining after you select an M card.

$P(A) = \frac{2}{10}$ ← number of cards with the letter A / ← number of cards remaining

$= \frac{1}{5}$ ← Simplify.

The probability of selecting an A for the second card is $\frac{1}{5}$.

3 You select a card from a bucket that contains 26 cards lettered A–Z without looking. Without replacing the first card, you select a second one. Find the probability of choosing C and then M.

The events are dependent. After the first selection, 25 letters remain.

$P(\text{C, then M}) = P(\text{C}) \times P(\text{M after C})$ ← Use the formula for dependent events.

$= \frac{1}{26} \times \frac{1}{25}$ ← Substitute.

$= \frac{1}{650}$ ← Multiply.

The probability of choosing C and then M is $\frac{1}{650}$.

For: Probability Activity
Use: Interactive Textbook, 12-4

Quick Check

2. Use the cards in Example 2. You select a T card at random. Without replacing the T card, you select a second card. Find $P(S)$. $\frac{1}{10}$
3. Suppose another 26 cards lettered A–Z are put in the bucket in Example 3. Find $P(\text{J, then J})$. $\frac{1}{1{,}326}$

2. Teach

Activity Lab

Use before the lesson.
Student Edition Activity Lab 12-4a, Exploring Multiple Events, p. 597

All in One Teaching Resources
Activity Lab 12-4: Compound Events

Guided Instruction

Error Prevention!

Note that dependent events must have a change in sample space.

Example 3
Ask: *What words in the problem make it clear that these two selections are dependent events?* **"without" and "replacing"**

PowerPoint

Additional Examples

1 A box contains the same number of green marbles, orange marbles, and blue marbles. You draw one marble, replace it, and draw a second marble. What is the probability that both marbles you draw are blue? $\frac{1}{9}$

2 You select a card at random from those having A, E, I, O, U, P, C. The card has the letter E. Without replacing the E card, you select a second card. Find the probability of selecting a card that does not have a vowel. $\frac{1}{3}$

3 A bag contains 3 red marbles, 4 white marbles, and 1 blue marble. You draw one marble. Without replacing it, you draw a second marble. What is the probability that the two marbles you draw are red, then white? $\frac{3}{14}$

Differentiated Instruction Solutions for All Learners

Advanced Learners L4	**English Language Learners** ELL
In Example 3, if you replaced the first card before drawing the second, what would the probability of winning be? $\frac{1}{676}$	Relate the phrase *compound* event to compound words, which are composed of more than one word. Likewise, relate *independent events* to independent people, who do not depend on others to do their tasks.
learning style: visual	learning style: verbal

All in One Teaching Resources
- Daily Notetaking Guide 12-4 L3
- Adapted Notetaking 12-4 L4

Closure

- *How do you find the probability of a compound event?* **Find the product of the probability of each event.**
- *Explain the difference between independent and dependent events.* **Sample: For independent events, the occurrence of one event does not affect the probability of the other event; for dependent events, one event does affect the probability of the other event.**

More Than One Way

You toss a coin three times. What is the probability of getting three heads?

Brianna's Method

Each toss of a coin is an independent event. The probability of getting heads for one coin toss is $\frac{1}{2}$. I can multiply the probabilities of the three coin tosses.

$P(\text{three heads}) = \frac{1}{2} \times \frac{1}{2} \times \frac{1}{2} = \frac{1}{8}$

The probability of three heads is $\frac{1}{8}$.

Chris's Method

I can make a tree diagram for the coin tosses. A favorable outcome is one with 3 heads.

The tree diagram shows 1 favorable outcome out of 8 possible outcomes. The probability of three heads is $\frac{1}{8}$.

Toss 1	Toss 2	Toss 3	Outcome
H	H	H	HHH
		T	HHT
	T	H	HTH
		T	HTT
T	H	H	THH
		T	THT
	T	H	TTH
		T	TTT

Choose a Method

You toss a coin four times. What is the probability of getting tails all four times? Describe your method and explain why you chose it.

$\frac{1}{16}$; answers may vary. Sample: It is quicker to find the product of probabilities.

Check Your Understanding

1. **Vocabulary** How do independent and dependent events differ? **See margin.**
2. **Multiple Choice** Two independent events A and B both have a probability of $\frac{1}{3}$. Which expression represents $P(\text{A, then B})$? **B**

   Ⓐ $\frac{1}{3} + \frac{1}{3}$ Ⓑ $\frac{1}{3} \times \frac{1}{3}$ Ⓒ $\frac{1}{3} + \frac{1}{2}$ Ⓓ $\frac{1}{3} \times \frac{1}{2}$

Are the two events *independent* or *dependent?*

3. You toss a nickel. Then you toss a dime. **independent**
4. You select a card. Then you select again without replacement.

4. dependent

1. Two events are independent if the occurrence of one does not affect the probability of the other occurring; two events are dependent if the occurrence of one does affect the probability of the other occurring.

Homework Exercises

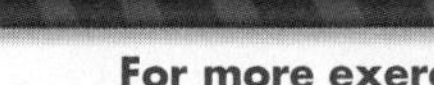

For more exercises, see Extra Skills and Word Problems.

GO for Help

For Exercises	See Examples
5–10	1
11–19	2
20–22	3

A You roll a number cube twice. Find each probability.

5. $P(1, \text{then } 2)$ $\frac{1}{36}$
6. $P(3, \text{then even})$ $\frac{3}{36}$ or $\frac{1}{12}$
7. $P(\text{less than } 4, \text{then } 1)$ $\frac{3}{36}$ or $\frac{1}{12}$
8. $P(\text{odd, then even})$ $\frac{9}{36}$ or $\frac{1}{4}$
9. $P(\text{divisible by } 2, \text{then } 5)$ $\frac{3}{36}$ or $\frac{1}{12}$
10. $P(\text{greater than } 2, \text{then odd})$ $\frac{12}{36}$ or $\frac{1}{3}$

An arrangement of 8 students is shown below. The numbers of all the students are in a basket. The teacher selects a number and replaces it. Then the teacher selects a second number. Find each probability.

Row	Student			
A	1	2	3	4
B	5	6	7	8

11. $P(\text{student } 1, \text{then student } 8)$ $\frac{1}{64}$
12. $P(\text{student in row A, then student in row B})$
13. $P(\text{student in row A, then student } 6, 7, \text{or } 8)$ $\frac{12}{64}$ or $\frac{3}{16}$

12. $\frac{16}{64}$ or $\frac{1}{4}$

You select the letter A from the group. Without replacing the A, you select a second letter. Find each probability.

14. $P(Z)$ $\frac{1}{7}$
15. $P(\text{vowel})$ $\frac{2}{7}$
16. $P(\text{red})$ $\frac{2}{7}$
17. $P(\text{blue})$ $\frac{5}{7}$
18. $P(\text{consonant})$ $\frac{5}{7}$
19. $P(\text{not K})$ $\frac{6}{7}$

A box contains 20 cards numbered 1–20. You select a card. Without replacing the first card, you select a second card. Find each probability.

20. $P(1, \text{then } 20)$ $\frac{1}{380}$
21. $P(3, \text{then even})$ $\frac{10}{380}$ or $\frac{1}{38}$
22. $P(\text{even, then } 7)$ $\frac{10}{380}$ or $\frac{1}{38}$

B GPS

23. **Guided Problem Solving** Five girls and seven boys want to be the two broadcasters for a school show. To be fair, a teacher puts their names in a hat and selects two. Find $P(\text{girl, then boy})$. $\frac{35}{132}$
    - **Make a Plan** The selections of the two names are (dependent, independent) events. Find the probability of selecting a girl first. Then find the probability of selecting a boy after selecting a girl.
    - **Carry Out the Plan** $P(\text{girl first}) = \frac{5}{■}$; $P(\text{boy after girl}) = \frac{7}{■}$

Two coins are dropped at random into the boxes in the diagram below.

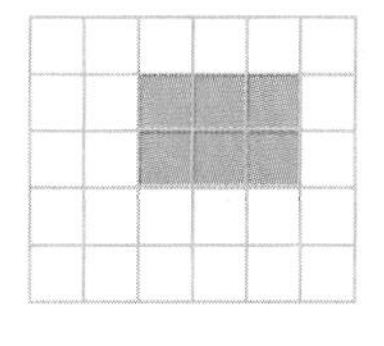

24. What is the theoretical probability that both coins fall into a shaded box? $\frac{1}{25}$
25. In 50 trials, both coins land in a shaded box twice. What is the experimental probability that both coins fall into a shaded box? $\frac{1}{25}$

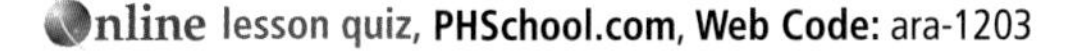

3. Practice

Assignment Guide

Check Your Understanding

Go over Exercises 1–4 in class before assigning the Homework Exercises.

Homework Exercises

A Practice by Example	5–22
B Apply Your Skills	23–30
C Challenge	31
Test Prep and Mixed Review	32–35

Homework Quick Check

To check students' understanding of key skills and concepts, go over Exercises 7, 16, 26, 29, and 30.

Differentiated Instruction Resources

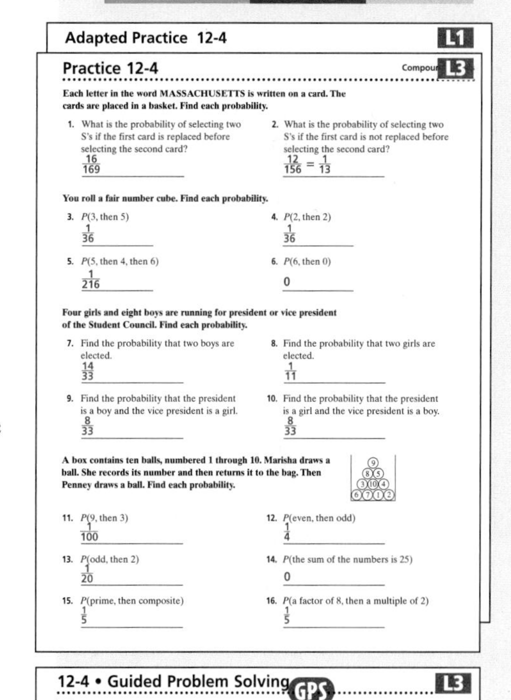

Adapted Practice 12-4 L1

Practice 12-4 Compoun[d Events] L3

Each letter in the word MASSACHUSETTS is written on a card. The cards are placed in a basket. Find each probability.

1. What is the probability of selecting two S's if the first card is replaced before selecting the second card? $\frac{16}{169}$
2. What is the probability of selecting two S's if the first card is not replaced before selecting the second card? $\frac{12}{156} = \frac{1}{13}$

You roll a fair number cube. Find each probability.

3. $P(3, \text{then } 5)$ $\frac{1}{36}$
4. $P(2, \text{then } 2)$ $\frac{1}{36}$
5. $P(5, \text{then } 4, \text{then } 6)$ $\frac{1}{216}$
6. $P(6, \text{then } 0)$ 0

Four girls and eight boys are running for president or vice president of the Student Council. Find each probability.

7. Find the probability that two boys are elected. $\frac{14}{33}$
8. Find the probability that two girls are elected. $\frac{1}{11}$
9. Find the probability that the president is a boy and the vice president is a girl. $\frac{8}{33}$
10. Find the probability that the president is a girl and the vice president is a boy. $\frac{8}{33}$

A box contains ten balls, numbered 1 through 10. Marisha draws a ball. She records its number and then returns it to the bag. Then Penney draws a ball. Find each probability.

11. $P(9, \text{then } 3)$ $\frac{1}{100}$
12. $P(\text{even, then odd})$ $\frac{1}{4}$
13. $P(\text{odd, then } 2)$ $\frac{1}{20}$
14. $P(\text{the sum of the numbers is } 25)$ 0
15. $P(\text{prime, then composite})$ $\frac{1}{5}$
16. $P(\text{a factor of } 8, \text{then a multiple of } 2)$ $\frac{1}{5}$

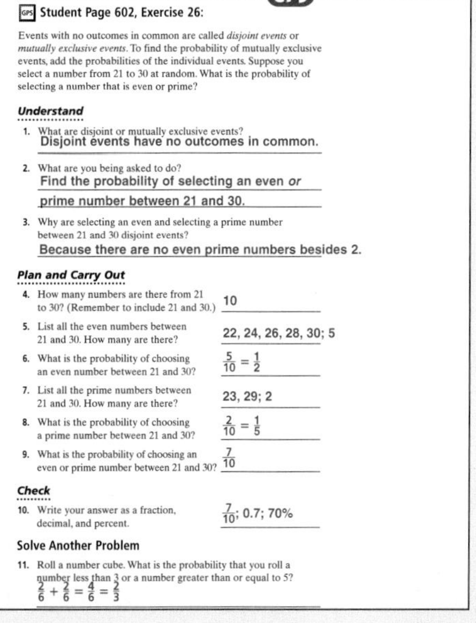

12-4 • Guided Problem Solving GPS L3

GPS Student Page 602, Exercise 26:

Events with no outcomes in common are called *disjoint events* or *mutually exclusive events*. To find the probability of mutually exclusive events, add the probabilities of the individual events. Suppose you select a number from 21 to 30 at random. What is the probability of selecting a number that is even or prime?

Understand

1. What are disjoint or mutually exclusive events? Disjoint events have no outcomes in common.
2. What are you being asked to do? Find the probability of selecting an even *or* prime number between 21 and 30.
3. Why are selecting an even and selecting a prime number between 21 and 30 disjoint events? Because there are no even prime numbers besides 2.

Plan and Carry Out

4. How many numbers are there from 21 to 30? (Remember to include 21 and 30.) 10
5. List all the even numbers between 21 and 30. How many are there? 22, 24, 26, 28, 30; 5
6. What is the probability of choosing an even number between 21 and 30? $\frac{5}{10} = \frac{1}{2}$
7. List all the prime numbers between 21 and 30. How many are there? 23, 29; 2
8. What is the probability of choosing a prime number between 21 and 30? $\frac{2}{10} = \frac{1}{5}$
9. What is the probability of choosing an even or prime number between 21 and 30? $\frac{7}{10}$

Check

10. Write your answer as a fraction, decimal, and percent. $\frac{7}{10}$; 0.7; 70%

Solve Another Problem

11. Roll a number cube. What is the probability that you roll a number less than 3 or a number greater than or equal to 5? $\frac{2}{6} + \frac{2}{6} = \frac{4}{6} = \frac{2}{3}$

PowerPoint

Lesson Quiz

Find each probability.

1. You roll a number cube twice. What is P(2, then even)? $\frac{1}{12}$
2. You draw a card at random from a stack of ten cards, each labeled with a number from 1 through 10. Then you draw a second card. What is P(5, then 3)? $\frac{1}{90}$

29–30. See back of book.

35. See back of book.

Reteaching 12-4

Compoun L2

If you toss a coin and roll a number cube, the events are **independent.** The outcome of one event does not affect the outcome of the second event.

Find the probability of tossing a heads (H) and rolling an even number (E).

Find P(H and E). H and E are independent.

① Find P(H):

$P(\text{H}) = \frac{1 \text{ heads}}{2 \text{ sides}} = \frac{1}{2}$

② Find P(E):

$P(\text{E}) = \frac{3 \text{ evens}}{6 \text{ faces}} = \frac{1}{2}$

③ $P(\text{H and E}) = P(\text{H}) \times P(\text{E}) = \frac{1}{2} \times \frac{1}{2} = \frac{1}{4}$

If the outcome of the first event affects the outcome of the second event, the events are **dependent.**

A bag contains 3 blue and 3 red marbles. Draw a marble, then draw a second marble without replacing the first marble. Find the probability of drawing 2 blue marbles.

① Find P(blue).

$P(\text{blue}) = \frac{3 \text{ blue}}{6 \text{ marbles}} = \frac{1}{2}$

② Find P(blue after blue).

$P(\text{blue after blue}) = \frac{2 \text{ blue}}{5 \text{ marbles}} = \frac{2}{5}$

③ Find P(blue, then blue)

P(blue, then blue)
$= P(\text{blue}) \times P(\text{blue after blue})$
$= \frac{1}{2} \times \frac{2}{5} = \frac{1}{5}$

In Exercises 1–3, you draw a marble at random from the bag of marbles shown. Then, you replace it and draw again. Find each probability.

1. P(blue, then red) $\frac{6}{25}$
2. P(2 reds) $\frac{9}{25}$
3. P(2 blues) $\frac{4}{25}$

Next, you draw two marbles randomly *without* replacing the first marble. Find each probability.

4. P(blue, then red) $\frac{4}{15}$
5. P(2 reds) $\frac{1}{3}$
6. P(2 blues) $\frac{2}{15}$

You draw two letters randomly from a box containing the letters M, I, S, S, O, U, R, and I.

7. Suppose you do not replace the first letter before drawing the second. What is P(M, then I)? $\frac{1}{28}$
8. Suppose you replace the first letter before drawing the second. What is P(M, then I)? $\frac{1}{32}$

Enrichment 12-4

Compoun L4

Critical Thinking

A spinning top has four sides with one letter on each side: A, B, C, D. When a child spins the top, any one of the four letters is equally likely to come up.

1. How many sides does the top have? **4 sides**
2. Will the result of the first spin change the possible outcome for the second spin? **no**
3. How many possible outcomes are there for one spin? **4 outcomes**
4. How many ways can you spin a B? **1 way**
5. What is the probability of spinning a B on the
   a. first spin? $\frac{1}{4}$ b. second spin? $\frac{1}{4}$
6. How can you find the probability of two events? **Multiply the probabilities.**
7. Are the spins of the top dependent or independent events? **independent**
8. What is the probability of spinning 2 Bs in a row? $\frac{1}{4} \times \frac{1}{4} = \frac{1}{16}$
9. How did you decide whether or not the events were dependent or independent? **The probability of the 2nd outcome does *not* depend on the result of the 1st outcome.**
10. What is the probability of spinning an A, then spinning a B, on the top? $\frac{1}{16}$

Visit: PHSchool.com
Web Code: are-1204

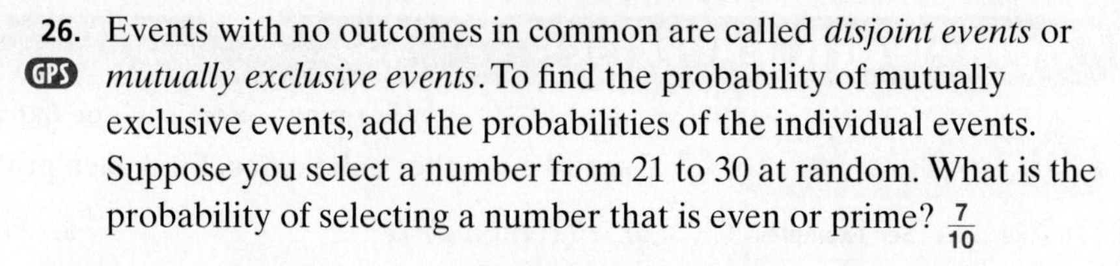

26. Events with no outcomes in common are called *disjoint events* or *mutually exclusive events*. To find the probability of mutually exclusive events, add the probabilities of the individual events. Suppose you select a number from 21 to 30 at random. What is the probability of selecting a number that is even or prime? $\frac{7}{10}$

Choose a Method A bag contains 3 blue marbles, 4 red marbles, and 2 white marbles. Three times you draw a marble and return it. Find each probability.

27. P(red, then white, then blue) $\frac{24}{729}$ or $\frac{8}{243}$
28. P(all white) $\frac{8}{729}$
29. **Reasoning** Events are complementary if they cover all possibilities with no overlap. Are the following sets of events *complementary, mutually exclusive,* or *neither?* Explain. **29a–b. See margin.**
    a. A traffic light shows red, yellow, or green or is broken.
    b. A student receives an A, B, or C on a test.
30. **Writing in Math** When you select marbles without replacing them, are the events *independent* or *dependent?* Explain. **See margin.**
31. **Challenge** You have two spinners with colors on them. The probability of spinning green on both spinners is $\frac{5}{21}$. The probability of spinning green on the first one alone is $\frac{1}{3}$. What is the probability of spinning green on the second spinner alone? $\frac{5}{7}$

Test Prep and Mixed Review

Practice

Multiple Choice

32. Dominica rolls a number cube once and spins the spinner at the right once. Which choice shows all the possible outcomes when she rolls an odd number and spins an odd number? **C**

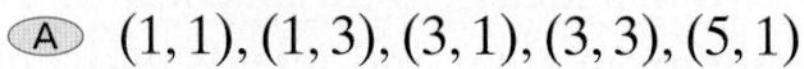

    A (1, 1), (1, 3), (3, 1), (3, 3), (5, 1)
    B (1, 1), (2, 2), (3, 3), (4, 3), (5, 3), (6, 3)
    C (1, 1), (1, 3), (3, 1), (3, 3), (5, 1), (5, 3)
    D (1, 1), (3, 3), (5, 5)

33. Which two angles are NOT complementary? **J**
    F 1°, 89° G 33°, 57° H 22°, 68° J 42°, 56°
34. William worked $2\frac{3}{4}$ hours on Saturday and $3\frac{3}{4}$ hours on Sunday. On Monday, he worked half as many hours as he did on Saturday and Sunday combined. How many hours did he work Monday? **C**
    A $1\frac{3}{8}$ hours B $1\frac{7}{8}$ hours C $3\frac{1}{4}$ hours D $6\frac{1}{2}$ hours

GO for Help

For Exercise	See Lesson
35	12-3

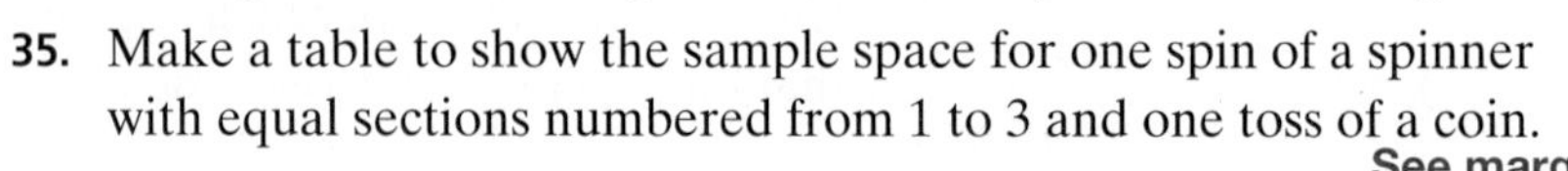

35. Make a table to show the sample space for one spin of a spinner with equal sections numbered from 1 to 3 and one toss of a coin. **See margin.**

Test Prep

Resources

For additional practice with a variety of test item formats:
- Test-Taking Strategies, p. 615
- Test Prep, p. 619
- Test-Taking Strategies with Transparencies

Alternative Assessment

Have students work in small groups. Provide groups with coins, index cards, spinners, and/or number cubes. Groups write one problem dealing with dependent events and one problem dealing with independent events. Then groups challenge other groups to solve their problems.

Vocabulary Builder

High-Use Academic Words

High-use academic words are words that you will see often in textbooks and on tests. These words are not math vocabulary terms, but knowing them will help you succeed in mathematics.

Direction Words

Some words tell what to do in a problem. I need to understand what these words are asking so that I give the correct answer.

Word	Meaning
Analyze	To examine in detail to determine relationships
List	To present information in some order or to give examples
Persuade	To cause someone to do or believe something, especially by reasoning

Exercises

Match each situation with the correct word.

A. analyze
B. list
C. persuade

1. You do a survey in which you write the types of cereal people prefer and the number of times people choose each type of cereal. **B**
2. You convince the manager of a diner to serve a certain type of cereal based on a survey. **C**
3. You determine the cereal preferences of people based on a survey. **A**
4. List all the possible outcomes of flipping a coin and rolling a number cube. Then find the probability that an outcome is heads and even. **See margin.**
5. Analyze the data at the right. How would you use the data to persuade a disc jockey to play hip-hop music? Explain. **See margin.**
6. **Word Knowledge** Think about the word *outcome*.
   a. Choose the letter for how well you know the word.
      A. I know its meaning.
      B. I've seen it, but I don't know its meaning.
      C. I don't know it.
   6a–c. Check students' work.
   b. **Research** Look up and write the definition of *outcome*.
   c. Use the word in a sentence involving mathematics.

Music Preference Survey

Type of Music	Frequency
Hip-hop	卌 卌 卌
House	///
Country	////
Rock	卌 ////

4. H1, H2, H3, H4, H5, H6, T1, T2, T3, T4, T5, T6; $P(\text{heads and even}) = \frac{3}{12}$, or $\frac{1}{4}$

5. When asked about music preference, there is the greatest probability that the response is hip-hop.

Vocabulary Builder

High-Use Academic Words

In this feature, students are introduced to words that are not strictly mathematical but appear frequently in their texts.

Guided Instruction

Explain that the vocabulary words in this feature have multiple meanings and some can apply to math problems. Have volunteers read each word and its definition aloud. Ask:

- *Which word is the closest synonym for study?* **analyze**
- *How would you use list as it relates to probability?* **Sample: List all the potential outcomes to form a sample space.**
- *How could you use persuade in a math problem?* **Sample: The experimental probability of a coin flip landing heads is $\frac{93}{100}$. Would this persuade you to predict heads or tails on the next flip?**

Exercises

Have students work independently, review the answers as a class, and discuss any differences.

Differentiated Instruction

Advanced Learners L4

Have students create their own fill-in-the-blank questions for the vocabulary words. Have them trade papers with a partner and answer their partner's questions.

Practice With Probability

In this feature, students practice solving problems involving probability. They find the probability of independent and dependent events.

Guided Instruction

Explain to students that they can use probability to make predictions in everyday life. Ask volunteers for examples of using probability in their lives. Then have a volunteer read the problem aloud.
Ask:

- *What kind of probability is this question asking you to find?* **dependent probability**
- *Why do we multiply the three probabilities?* **Dependent probability is the product of the probability of the first event and all subsequent events.**
- *How would you find the probability of Anna winning first place and Cole winning second place?* **Multiply $\frac{60}{300}$ by $\frac{51}{240}$.**
- *How would you express the probability of Anna winning first place as a percent?* **20%** *The probability of Anna winning first place and Cole winning second place?* **4.3%**

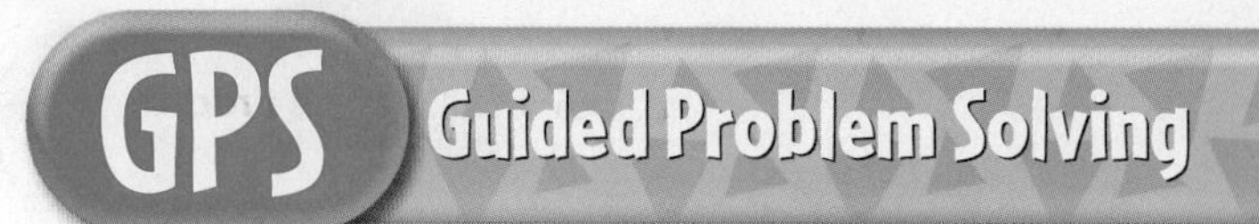

Practice With Probability

Members of a math club solve problems worth 1 to 4 points. Then the club has a drawing for three prizes based on the number of points members have earned.

Each student's name is put in a hat once for every point he or she earns. A student can win only one prize. Based on the table below, what are the chances that Anna wins first prize, Cole wins second prize, and Dillon wins third prize?

Points Earned

Name	Frequency	Name	Frequency
Anna	𝍸 𝍸 𝍸 𝍸 𝍸 𝍸 𝍸 𝍸 𝍸 𝍸 𝍸 𝍸	Cole	𝍸 𝍸 𝍸 𝍸 𝍸 𝍸 𝍸 𝍸 𝍸 𝍸 𝍷
Dillon	𝍸 𝍸 𝍸 𝍸 𝍸 𝍸 𝍸 𝍸 𝍸 𝍸 𝍷𝍷𝍷	Raja	𝍸 𝍸 𝍸 𝍸 𝍸 𝍸 𝍸 𝍸 𝍸 𝍷𝍷
Bailey	𝍸 𝍸 𝍸 𝍸 𝍸 𝍸 𝍸 𝍷	Rosa	𝍸 𝍸 𝍸 𝍸 𝍸 𝍸 𝍸 𝍸 𝍸 𝍸 𝍷𝍷𝍷

What You Might Think

What do I know? What do I want to find out?

How can I solve the problem?

What is the answer?

What You Might Write

I know each person's number of points. I want to know the chances that Anna wins first, Cole wins second, and Dillon wins third.

Anna's chances are $\frac{60}{300}$. After Anna wins first prize, she cannot win again. So Cole's chances are $\frac{51}{240}$. Then Dillon's chances are $\frac{53}{189}$.

The chances of the three winning in that order are $\frac{60}{300} \times \frac{51}{240} \times \frac{53}{189}$. This is about 0.01, or 1%.

Think It Through

1. Why are Anna's chances $\frac{60}{300}$? Why are Cole's chances $\frac{51}{240}$ if Anna wins first? Why are Dillon's chances $\frac{53}{189}$ if Cole wins second? **See margin.**
2. How can you use a calculator to verify $\frac{60}{300} \times \frac{51}{240} \times \frac{53}{189} \approx 0.01$? **Type in (60 × 51 × 53) ÷ (300 × 240 × 189).**

1. Anna's chances to win are $\frac{60}{300}$ because she has earned 60 of the total 300 points; Cole's chances to win are $\frac{51}{240}$ because he has earned 51 out of the remaining 240 points (after Anna's are removed); Dillon's chances to win are $\frac{53}{189}$ because he has earned 53 of the remaining 189 points (after Anna's and Cole's are removed).

Exercises

Solve each problem. For Exercise 3, answer the questions first.

3. Fran's chances of making free throws in basketball are 80% based on her performance this season. If the trend continues, what are the chances she will make both of her next two free throws? **64%**
   a. What do you know? What do you want to find out?
   b. How can you solve the problem? Explain.

4. You play the card game In Between with the following rules.
   - Use 30 cards numbered 1–30.
   - You turn over two cards at random.
   - You win if the next card turned over is between the two cards. **$\frac{6}{28}$ or $\frac{3}{14}$**

   For the cards displayed at the right, what is the probability you will win?

5. Janice agrees to work for her neighbor for a year. In return, her neighbor will pay her $480 and give her a car. Janice has to stop working after 9 months. Since she does not work the full year, she gets only $60 and the car. How much is the car worth? **$1,200**

6. George participates in a 3-mile walk for hunger. He averages 6 miles per hour for the first mile, 5 miles per hour for the next mile, and 4 miles per hour for the last mile. How long does it take him to walk the 3 miles? **37 min**

7. Students study the advantages and disadvantages of year-round schools (YRS). They then express their opinions in a poll. A "0" means the student is strongly against YRS. A "10" means the student is strongly for YRS.

   YRS Survey Results

   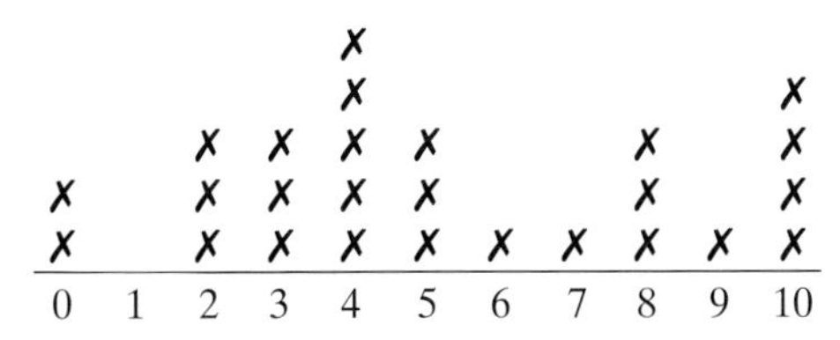

   The line plot shows the data collected. Find the mean, median, and mode of the data. Which measure best represents how the students feel about YRS? Explain.

7. About 5.23; 4.5; 4; answers may vary. Sample: the median, because half the students are on either side of that response

Exercises

Have students work independently on the Exercises. When they have finished, they can form small groups to compare answers and elect a representative to report their results to the class.

Differentiated Instruction

Below Level 

Provide students with problems involving multiple independent events, such as repeatedly spinning a spinner to get a certain result, to allow them to practice simpler problems.

12-5

1. Plan

Objective
To find permutations

Examples
1 Finding Permutations
2 Finding Permutations Using Factorials
3 Application: Olympics

Math Understandings: p. 578D

Math Background

The simplest permutation that students have used is the ordered pair that name the coordinates of a point on a plane. In this case, the other possible permutations, (y, x), (y, y), and (x, x) are not generally used. The letters in MATH have, for example, $4 \cdot 3 \cdot 2 \cdot 1$, or 24, permutations. The total number of permutations of n objects is $n!$ while the number of permutations of n objects taken r at a time is $\frac{n!}{(n-r)!}$

More Math Background: p. 578D

Lesson Planning and Resources

See p. 578E for a list of the resources that support this lesson.

PowerPoint

Bell Ringer Practice

Check Skills You'll Need
Use student page, transparency, or PowerPoint. For intervention, direct students to:
Multiplying and Dividing Integers
Lesson 1-8
Extra Skills and Word Problems Practice, Ch. 1

12-5 Permutations

Check Skills You'll Need

1. **Vocabulary Review** What is a *product*? **See below.**

Find each product.

2. $10 \cdot 9$ **90**
3. $20 \cdot 19$ **380**
4. $8 \cdot 7 \cdot 6$ **336**
5. $10 \cdot 9 \cdot 8$ **720**
6. $5 \cdot 4 \cdot 3$ **60**

Lesson 1-8

What You'll Learn

To find permutations

New Vocabulary permutation, factorial

Why Learn This?

Sometimes the order of the outcomes is important, such as the order of letters in words.

A **permutation** is an arrangement of items in a particular order. Suppose you arrange the four letters O, P, S, and T in two ways. The permutation STOP is different from the permutation POTS because the order of the letters is different.

Online active math

For: Permutations Activity
Use: Interactive Textbook, 12-5

EXAMPLE **Finding Permutations**

1 Find the number of permutations of the letters L, I, K, and E. The first letter can be any of the four letters. You have three choices for the second, two for the third, and one for the fourth letter.

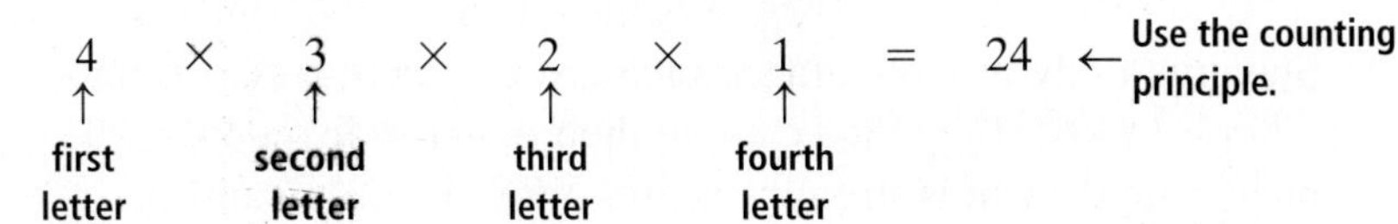

There are 24 different permutations.

Quick Check

1. Find the number of permutations of the letters H, A, N, D, L, and E. **720**

Check Skills You'll Need

1. the result of multiplication

There are $4 \times 3 \times 2 \times 1$ permutations of the letters C, A, R, and E. The product of all positive integers less than or equal to a number is the **factorial** for that number. You write 4 factorial as 4!.

$$4! = 4 \times 3 \times 2 \times 1$$

Differentiated Instruction Solutions for All Learners

Special Needs L1	**Below Level** L2
Provide students with square paper tiles. Have them write the letters listed in Example 1, and arrange them, recording their arrangement each time. This will let them see the counting principle in action. **learning style: visual**	Allow students to use calculators to find factorials. Many calculators have a factorial key. **learning style: tactile**

EXAMPLE Finding Permutations Using Factorials

Calculator Tip

To evaluate 5!, press 5 MATH. Scroll right to option PRB. Select option 5—"!". Press ENTER.

2 **Television** A TV station has five shows to broadcast on a weeknight. How many different arrangements of the shows can they make?

$$5! = 5 \times 4 \times 3 \times 2 \times 1 = 120$$

5! ↑ ways to arrange shows; 5 ↑ first show; 4 ↑ second show; 3 ↑ third show; 2 ↑ fourth show; 1 ↑ fifth show

The station can make 120 different arrangements of the shows.

Quick Check

2. Write the number of permutations for the letters G, R, A, V, I, E, and S in factorial form. Then multiply. **7! = 5,040**

The sample space below shows the two-letter permutations of the 4 letters in STOP. There are 4 possible choices for the first letter and 3 possible choices for the second letter. The number of permutations is $4 \times 3 = 12$.

ST	OP	TP	TS	PO	PT
TO	PS	OS	OT	SP	SO

EXAMPLE Application: Olympics

3 Suppose a men's Olympic ice hockey tournament has 12 teams. Find the number of different ways that teams can win the gold, silver, and bronze medals.

There are 12 possible teams that can win the gold medal. After that, there are 11 teams that can win the silver medal. Finally, there are 10 teams that can win the bronze medal.

$$12 \times 11 \times 10 = 1{,}320$$ ← Use the counting principle.

12 ↑ gold medal; 11 ↑ silver medal; 10 ↑ bronze medal

There are 1,320 different ways that teams can win the three medals.

Quick Check

3. a. Women's Olympic ice hockey tournaments have eight teams. Find the number of different ways that teams can win the gold, silver, and bronze medals. **336**

b. **Reasoning** In Example 3, the number is not found by finding 12!. Explain why. **See left.**

3b. The example selects only 3 teams from the 12 choices, so the answer is 12 × 11 × 10, not 12!.

2. Teach

Activity Lab

Use before the lesson.

All in One Teaching Resources

Activity Lab 12-5: Exploring Order

Guided Instruction

Example 1

Have students model the Example by making four cards or pieces of paper with the letters on them.

Example 3

Discuss why the number decreases by one each time.

PowerPoint

Additional Examples

1 Find the number of permutations of the letters T, I, G, E, and R. **120**

2 How many different ways can you arrange a nickel, a dime, a penny, and a quarter in a row? **24**

3 Find the number of permutations for the blue, red, and green ribbons for 12 horses at a show. **1,320**

All in One Teaching Resources

- Daily Notetaking Guide 12-5 L3
- Adapted Notetaking 12-5 L1

Closure

- *What is a permutation?* **an arrangement of objects in a particular order**
- Explain how to use factorial form to find the permutations of the letters in F, R, O, and G. **4! = 4 · 3 · 2 · 1 = 24**

Advanced Learners L4

Have students explore with examples whether $a! + b!$ has the same value as $(a + b)!$. **The values are not always the same.**

learning style: visual

English Language Learners ELL

Have students explain when they use factorials to find permutations and when they do not. In Example 3, there are not 12 medal choices, there are only 3.

learning style: verbal

3. Practice

Assignment Guide

Check Your Understanding
Go over Exercises 1–6 in class before assigning the Homework Exercises.

Homework Exercises
A Practice by Example 7–25
B Apply Your Skills 26–39
C Challenge 40
Test Prep and Mixed Review 41–45

Homework Quick Check
To check students' understanding of key skills and concepts, go over Exercises 9, 16, 33, 34, and 39.

Differentiated Instruction Resources

Adapted Practice 12-5 L1

Practice 12-5 L3

Find the number of permutations of each group of letters.

1. C, H, A, I, R 120
2. L, I, G, H, T, S 720
3. C, O, M, P, U, T, E, R 40,320

Write the number of permutations in factorial form. Then simplify.

4. S, P, A, C, E $5! = 120$
5. P, L, A, N $4! = 24$
6. S, A, M, P, L, E $6! = 720$

Find the number of three-letter permutations of the letters.

7. A, P, Q, M 24
8. L, S, U, V, R 60
9. M, B, T, O, D, K 120

Find the value of each factorial expression.

10. 9! 362,880
11. 7! 5,040
12. 6! 720

Solve.

13. Suppose that first-, second-, and third-place winners of a contest are to be selected from eight students who entered. In how many ways can the winners be chosen? 336
14. Antonio has nine different sweat shirts that he can wear for his job doing yardwork. He has three pairs of jeans and two pairs of sweat pants. How many different outfits can Antonio wear for the yardwork? 45
15. Ramona has a combination lock for her bicycle. She knows the numbers are 20, 41, and 6, but she can't remember the order. How many different arrangements are possible? 6 arrangements
16. Travis is planting 5 rosebushes along a fence. Each rosebush has a different flower color: red, yellow, pink, peach, and white. If he wants to plant 3 rosebushes in between white and yellow rosebushes, in how many ways can he plant the 5 rosebushes? 12 ways

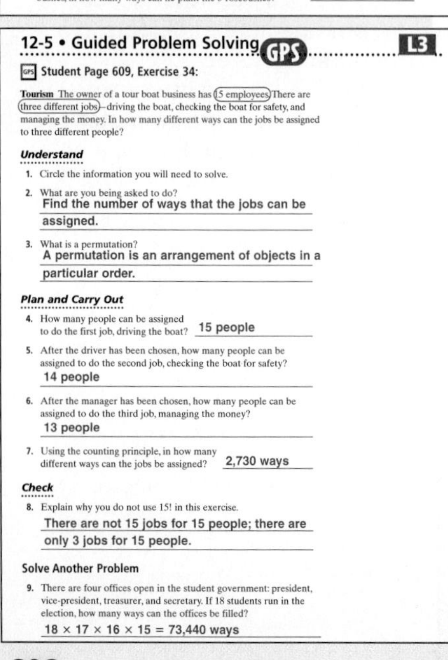

12-5 • Guided Problem Solving GPS L3

GPS Student Page 609, Exercise 34:

Tourism The owner of a tour boat business has 5 employees. There are three different jobs—driving the boat, checking the boat for safety, and managing the money. In how many different ways can the jobs be assigned to three different people?

Understand

1. Circle the information you will need to solve.
2. What are you being asked to do? Find the number of ways that the jobs can be assigned.
3. What is a permutation? A permutation is an arrangement of objects in a particular order.

Plan and Carry Out

4. How many people can be assigned to do the first job, driving the boat? 15 people
5. After the driver has been chosen, how many people can be assigned to do the second job, checking the boat for safety? 14 people
6. After the manager has been chosen, how many people can be assigned to do the third job, managing the money? 13 people
7. Using the counting principle, in how many different ways can the jobs be assigned? 2,730 ways

Check

8. Explain why you do not use 15! in this exercise. There are not 15 jobs for 15 people; there are only 3 jobs for 15 people.

Solve Another Problem

9. There are four offices open in the student government: president, vice-president, treasurer, and secretary. If 18 students run in the election, how many ways can the offices be filled? $18 \times 17 \times 16 \times 15 = 73{,}440$ ways

Check Your Understanding

1. **Vocabulary** What is a permutation? **an arrangement of items in a particular order**
2. **Multiple Choice** How many permutations are there of the words *star*, *square*, and *triangle?* **C**
   (A) 1 (B) 3 (C) 6 (D) 9

Match each set of digits to the correct number of permutations.

3. even digits 2 through 8 **B**
4. odd digits 1 through 9 **A**
5. all digits 1 through 9 **C**

A. 120
B. 24
C. 362,880

6. **Reasoning** Does $3! + 2! = 5!$? Justify your answer. **No, $3! = 6$ and $2! = 2$, so $3! + 2! = 8$, but $5! = 120$.**

Homework Exercises

For more exercises, see Extra Skills and Word Problems.

GO for Help

For Exercises	See Examples
7–12	1
13–19	2
20–25	3

14. $5! = 120$

A Find the number of permutations of each group of letters.

7. W, O, R, L, D **120**
8. H, U, M, A, N **120**
9. T, O, Y **6**
10. P, I, C, K, L, E **720**
11. L, U, N, C, H, E, S **5,040**
12. M, A, R, S **24**

Write the number of permutations in factorial form. Then simplify.

13. C, A, T **$3! = 6$**
14. R, A, T, E, S
15. P, A, C, K **$4! = 24$**
16. D, E, P, A, R, T **$6! = 720$**
17. I, N, C, L, U, D, E **$7! = 5{,}040$**
18. L, U, C, K, Y **$5! = 120$**
19. **Planning** You plan to shop, call a friend, study, and exercise, all in one day. How many arrangements of activities can you plan? **24 arrangements**

Find the number of two-letter permutations of the letters.

20. R, E, P, S **12**
21. Q, I, E, R, T, Y, U **42**
22. G, D, X, Z, C **20**
23. A, E, I, O, U, Y **30**
24. M, A, P, L, E **20**
25. L, A, P **6**

B 26. **Guided Problem Solving** Two sisters and a brother line up for movie tickets. Find the probability that they line up boy-girl-girl. $\frac{1}{3}$
- **Make a Plan** Make a sample space. Remember, each sister counts separately, so use B, G, and g for the three siblings. Use the sample space to find the probability.
- **Carry Out the Plan** Write a ratio to find probability.

Find the value of each factorial expression.

27. 4! **24**
28. 6! **720**
29. 7! **5,040**
30. 8! **40,320**
31. 10! **3,628,800**

33. 6 MATH ! = or 6 × 5 × 4 × 3 × 2 × 1 =

39a. BS, GS, RS, BD, GD, RD, BP, GP, RP

b. $3 \times 3 = 9$ possibilities

44. $\frac{64}{256}$ or $\frac{1}{4}$

45. $\frac{8}{256}$ or $\frac{1}{32}$

Homework Video Tutor
Visit: PHSchool.com
Web Code: are-1205

32. The password to access a computer consists of 3 lower-case letters. You do not have the password. What is the greatest number of passwords you can try before you have access? **17,576 passwords**

33. **Writing in Math** Describe two different ways you can use a calculator to find the value of 6!. **See margin.**

34. **Tourism** The owner of a tour boat business has 15 employees. There are three different jobs—driving the boat, checking the boat for safety, and managing the money. In how many different ways can the jobs be assigned to three different people? **2,730 ways**

Find the value of each expression.

35. $5! \div 2!$ **60**
36. $4! \div 3!$ **4**
37. $6! \div 4!$ **30**
38. $7! \div 3!$ **840**

39. **Soccer** The team names and uniform colors for a soccer league are shown.
   a. Show the sample space of the different name-color possibilities.
   b. Use the counting principle to support your answer to part (a). **39a–b. See margin.**

Team Name	Uniform Color
Scorers	Blue
Defenders	Green
Passers	Red

C 40. **Challenge** Write an algebraic expression to show how many two-letter permutations are possible with $(n - 2)$ different letters. **$(n - 2)(n - 3)$**

Test Prep and Mixed Review — Practice

Multiple Choice

41. You can play one sport and join one music group. The sports are football, soccer, and tennis. The groups play rock or hip-hop. Which list shows all possible arrangements of a sport and a music group? **A**
   A (football, rock), (football, hip-hop), (soccer, rock), (soccer, hip-hop), (tennis, rock), (tennis, hip-hop)
   B (football, rock), (football, soccer), (football, tennis), (soccer, rock), (tennis, hip-hop)
   C (football, rock), (soccer, rock), (tennis, rock)
   D (football, rock), (football, hip-hop), (soccer, tennis), (tennis, rock), (tennis, hip-hop)

42. A Fahrenheit temperature is 32° more than $\frac{9}{5}$ of a Celsius temperature C. Which algebraic expression represents the Fahrenheit temperature for a given Celsius temperature? **G**
   F $32 + \frac{9}{5}$ G $\frac{9}{5}C + 32$ H $(32)\frac{9}{5} + C$ J $\frac{9}{5}C - 32$

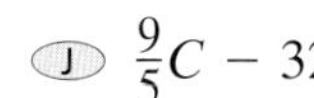

GO for Help

For Exercises	See Lesson
43–45	12-4

You have a set of 16 cards numbered 1–16. You select a card and put it back into the set. Then you select another card. Find each probability. **44–45. See margin.**

43. $P(2, \text{then } 6)$ $\frac{1}{256}$
44. $P(\text{even, then odd})$
45. $P(16, \text{then odd})$

4. Assess & Reteach

Lesson Quiz

1. Find the number of permutations for the letters A, B, C, and D. **24**
2. Find the value of 5!. **120**
3. Write the number of permutations in factorial form of the letters U, N, I, T, E, and D. **6!**
4. Find the number of three-letter permutations of the letters E, N, O, U, G, and H. **120**

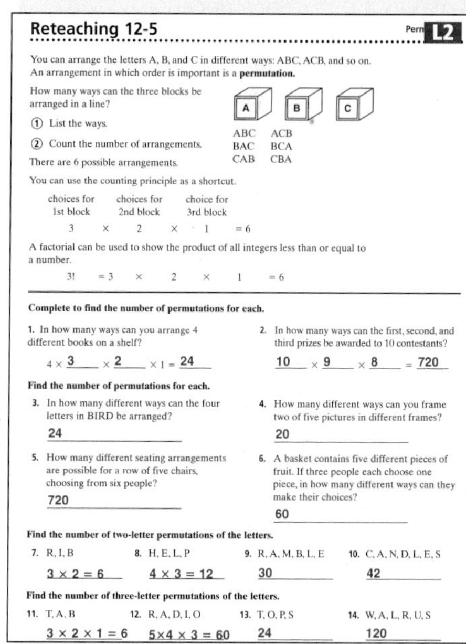

Reteaching 12-5 L2

You can arrange the letters A, B, and C in different ways: ABC, ACB, and so on. An arrangement in which order is important is a **permutation.**

How many ways can the three blocks be arranged in a line?

① List the ways.
② Count the number of arrangements.

ABC ACB
BAC BCA
CAB CBA

There are 6 possible arrangements.

You can use the counting principle as a shortcut.

choices for 1st block		choices for 2nd block		choice for 3rd block	
3	×	2	×	1	= 6

A factorial can be used to show the product of all integers less than or equal to a number.

3! = 3 × 2 × 1 = 6

Complete to find the number of permutations for each.

1. In how many ways can you arrange 4 different books on a shelf?
   4 × **3** × **2** × 1 = **24**
2. In how many ways can the first, second, and third prizes be awarded to 10 contestants?
   10 × **9** × **8** = **720**

Find the number of permutations for each.

3. In how many different ways can the four letters in BIRD be arranged? **24**
4. How many different ways can you frame two of five pictures in different frames? **20**
5. How many different seating arrangements are possible for a row of five chairs, choosing from six people? **720**
6. A basket contains five different pieces of fruit. If three people each choose one piece, in how many different ways can they make their choices? **60**

Find the number of two-letter permutations of the letters.

7. R, I, B **3 × 2 = 6**
8. H, E, L, P **4 × 3 = 12**
9. R, A, M, B, L, E **30**
10. C, A, N, D, L, E, S **42**

Find the number of three-letter permutations of the letters.

11. T, A, B **3 × 2 × 1 = 6**
12. R, A, D, I, O **5×4 × 3 = 60**
13. T, O, P, S **24**
14. W, A, L, R, U, S **120**

Enrichment 12-5 L4

Critical Thinking

How many ways could you select ③ possible components for a sound system from a tape deck, CD player, laser disk, equalizer, and surround-sound stereo?

1. Circle the number of components that will be in the sound system.
2. Underline the possible components to include in the system.
3. Are you asked for how the system can be assembled or the number of ways it can be assembled? **number of ways**
4. Write each in factorial notation.
   a. number of ways to select and arrange all components **5!**
   b. number of ways to arrange the selected components **3!**
   c. number of ways to arrange the components that are *not* selected **2!**
5. Write an expression that shows the number of ways to assemble the sound system. $\frac{5!}{3! \times 2!}$
6. How many ways could you select 3 components for the sound system? **10 ways**
7. Check your answer by making a list of the possible sound-system components. Use T, C, L, E, and S to represent the components. **TCL; TCE; TCS; TLE; TLS; TES; CLE; CLS; CES; LES**
8. How many ways could you select 4 possible combinations for a sound system? **5 ways**
9. How many ways could you select 2 possible combinations for a sound system? **10 ways**

Alternative Assessment

Have students work in pairs or small groups. Each student writes a word with four to eight letters in which no letters are repeated. Students exchange words with members of their group and find the number of permutations of the letters.

Test Prep

Resources
For additional practice with a variety of test item formats:
- Test-Taking Strategies, p. 615
- Test Prep, p. 619
- Test-Taking Strategies with Transparencies

12-6

1. Plan

Objective
To find combinations

Examples
1 Application: Clothing
2 Application: Careers

Math Understandings: p. 578D

Math Background

In the permutations of the letters in BLACK, for example, the order makes a difference because BLAKC is not the same as BLKAC. A *combination* is the grouping of objects in which the order of the objects does not matter; so BLAKC is considered the same combination as BLKAC. You can find the number of combinations by finding the number of permutations and eliminating duplicates.

More Math Background: p. 578D

Lesson Planning and Resources

See p. 578E for a list of the resources that support this lesson.

Bell Ringer Practice

Check Skills You'll Need
Use student page, transparency, or PowerPoint. For intervention, direct students to:
Permutations
Lesson 12-5
Extra Skills and Word Problems Practice, Ch. 12

12-6 Combinations

Check Skills You'll Need

1. **Vocabulary Review** Explain why CAT and ACT are not the same *permutations.* **See below.**

Find the number of permutations of each group of letters.

2. G, I, R, L **24**
3. H, I, K, E, R, S **720**

Lesson 12-5

What You'll Learn

To find combinations

New Vocabulary combination

Why Learn This?

Sometimes the order of objects does not matter. You may only care about the combination of objects.

Whether you pack your hat before your scarf, or your scarf before your hat, you have still packed the two items. A **combination** is a grouping of objects in which the order of the objects does not matter.

Check Skills You'll Need

1. **The letters are not in the same order.**

Vocabulary Tip

Combination comes from a Latin word that means "together."

EXAMPLE Application: Clothing

1 The colors of four scarves are listed. You decide to pack two scarves. How many different combinations of two scarves are possible?

Color	Letter
Blue	b
Yellow	y
Green	g
Red	r

Step 1 Let letters represent the scarves. Make a list of all the possible permutations.

(b, y) (y, b) (g, b) (r, b) (b, g) (y, g)
(g, y) (r, y) (b, r) (y, r) (g, r) (r, g)

Step 2 Cross out any group containing the same letters as another group.

(b, y) ~~(y, b)~~ (g, b) (r, b) ~~(b, g)~~ (y, g)
~~(g, y)~~ (r, y) ~~(b, r)~~ ~~(y, r)~~ (g, r) ~~(r, g)~~

Six different combinations of two colors are possible.

Quick Check

1. How many different combinations of two seashells can you make from three seashells? **3**

Differentiated Instruction Solutions for All Learners

Special Needs L1
Provide students with a copy of Example 1. Have them color-code the pairs of letters, matching the color of their marker or crayon to the letter (blue, yellow, green, and red.) This will let them visualize why certain pairs of letters are crossed out.

learning style: visual

Below Level L2
Review how to simplify fractions, such as

$$\frac{120}{6} = \frac{120 \div 6}{6 \div 6} = \frac{20}{1} = 20.$$

learning style: visual

For help with permutations, go to Lesson 12-5, Example 1.

In Example 1, the total number of permutations for four scarves taken two at a time is 4×3. The number of permutations of the smaller group, two scarves, is 2×1. You can find the number of combinations by dividing the total number of permutations by the number of permutations for the smaller group.

$$\text{combinations} = \frac{\text{total number of permutations}}{\text{number of permutations of smaller group}} = \frac{4 \times 3}{2 \times 1} = 6$$

EXAMPLE Application: Careers

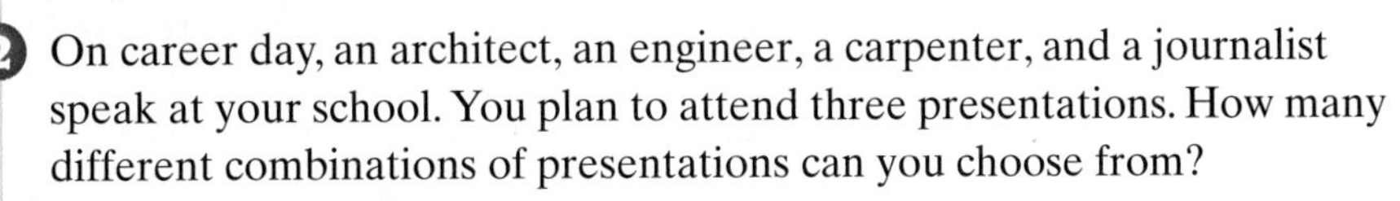

2 On career day, an architect, an engineer, a carpenter, and a journalist speak at your school. You plan to attend three presentations. How many different combinations of presentations can you choose from?

Step 1 Find the total number of permutations.

$$\underset{\text{first choice}}{4} \times \underset{\text{second choice}}{3} \times \underset{\text{third choice}}{2} = 24 \text{ permutations} \quad \leftarrow \textbf{Use the counting principle.}$$

Step 2 Find the number of permutations of the smaller group.

$$3 \times 2 \times 1 = 6 \text{ permutations} \quad \leftarrow \textbf{Use the counting principle.}$$

Step 3 Find the number of combinations.

$$\frac{\text{total number of permutations}}{\text{number of permutations of smaller group}} = \frac{24}{6} \quad \leftarrow \textbf{Divide.}$$

$$= 4 \quad \leftarrow \textbf{Simplify.}$$

You can attend 4 combinations of presentations.

Quick Check

2. If you go to two presentations, how many different combinations of presentations can you choose from? **6**

Check Your Understanding

1. **Vocabulary** Why is order not important in finding combinations? **In a combination, only the grouping of the items matters.**

Would you solve Exercises 2–4 using a *combination* or a *permutation*?

2. choosing three pieces of fruit from seven pieces **combination**
3. giving first and second place awards **permutation**
4. choosing five books from a list of ten books **combination**
5. Write an expression for each combination in Exercises 2–4.

5. $\frac{7 \times 6 \times 5}{3 \times 2 \times 1}$, $\frac{10 \times 9 \times 8 \times 7 \times 6}{5 \times 4 \times 3 \times 2 \times 1}$

2. Teach

Activity Lab

Use before the lesson.

All in One Teaching Resources

Activity Lab 12-6: Arranging Items

Guided Instruction

Example 1

Verify students' understanding of "number of permutations of the smaller group" by having them work through the pattern for Quick Check 1, three shells taken two at a time.

PowerPoint

Additional Examples

1 You have one pen in each of these colors: red, green, blue, and purple. You lend three to a friend. How many combinations of colors are possible in the pens you lend? **4**

2 The county fair has 10 rides. You have time for 3 of them. How many different combinations of rides are available to you? **120**

All in One Teaching Resources

- Daily Notetaking Guide 12-6 **L3**
- Adapted Notetaking 12-6 **L1**

Closure

- *How do you find the combinations of four shirts taken two at a time?* **Sample: Make an organized list and cross out duplicates; or divide the total number of permutations, 4 × 3 or 12, by the number of permutations of the smaller group, 2 × 1 or 2, which is $\frac{12}{2}$ or 6.**

Advanced Learners L4

How many basketball teams of 5 players each can you select from 12 players? **792**

learning style: visual

English Language Learners ELL

In Example 1, point out that the pairs of letters being crossed out are duplicates of pairs of letters that are not crossed out. The pairs of letters crossed out, which correspond to colors, are the same letters but in different order than the ones remaining.

learning style: verbal

3. Practice

Assignment Guide

Check Your Understanding
Go over Exercises 1–5 in class before assigning the Homework Exercises.

Homework Exercises
A	Practice by Example	6–16
B	Apply Your Skills	17–30
C	Challenge	31
	Test Prep and Mixed Review	32–36

Homework Quick Check
To check students' understanding of key skills and concepts, go over Exercises 8, 15, 22, 24, and 27.

Differentiated Instruction Resources

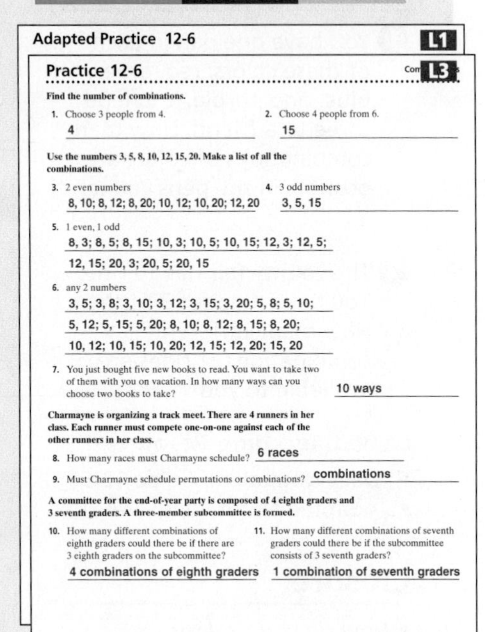

Adapted Practice 12-6 L1

Practice 12-6 L3

Find the number of combinations.

1. Choose 3 people from 4. 4
2. Choose 4 people from 6. 15

Use the numbers 3, 5, 8, 10, 12, 15, 20. Make a list of all the combinations.

3. 2 even numbers 8, 10; 8, 12; 8, 20; 10, 12; 10, 20; 12, 20
4. 3 odd numbers 3, 5, 15
5. 1 even, 1 odd 8, 3; 8, 5; 8, 15; 10, 3; 10, 5; 10, 15; 12, 3; 12, 5; 12, 15; 20, 3; 20, 5; 20, 15
6. any 2 numbers 3, 5; 3, 8; 3, 10; 3, 12; 3, 15; 3, 20; 5, 8; 5, 10; 5, 12; 5, 15; 5, 20; 8, 10; 8, 12; 8, 15; 8, 20; 10, 12; 10, 15; 10, 20; 12, 15; 12, 20; 15, 20
7. You just bought five new books to read. You want to take two of them with you on vacation. In how many ways can you choose two books to take? 10 ways

Charmayne is organizing a track meet. There are 4 runners in her class. Each runner must compete one-on-one against each of the other runners in her class.

8. How many races must Charmayne schedule? 6 races
9. Must Charmayne schedule permutations or combinations? combinations

A committee for the end-of-year party is composed of 4 eighth graders and 3 seventh graders. A three-member subcommittee is formed.

10. How many different combinations of eighth graders could there be if there are 3 eighth graders on the subcommittee? 4 combinations of eighth graders
11. How many different combinations of seventh graders could there be if the subcommittee consists of 3 seventh graders? 1 combination of seventh graders

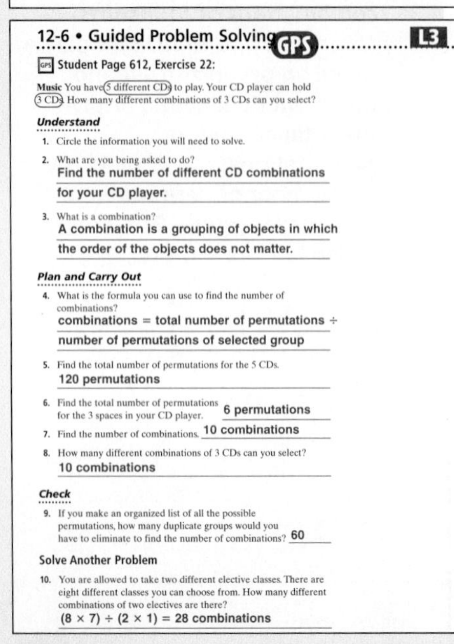

12-6 • Guided Problem Solving GPS L3

Student Page 612, Exercise 22:

Music You have 5 different CDs to play. Your CD player can hold 3 CDs. How many different combinations of 3 CDs can you select?

Understand

1. Circle the information you will need to solve.
2. What are you being asked to do? Find the number of different CD combinations for your CD player.
3. What is a combination? A combination is a grouping of objects in which the order of the objects does not matter.

Plan and Carry Out

4. What is the formula you can use to find the number of combinations? combinations = total number of permutations ÷ number of permutations of selected group
5. Find the total number of permutations for the 5 CDs. 120 permutations
6. Find the total number of permutations for the 3 spaces in your CD player. 6 permutations
7. Find the number of combinations. 10 combinations
8. How many different combinations of 3 CDs can you select? 10 combinations

Check

9. If you make an organized list of all the possible permutations, how many duplicate groups would you have to eliminate to find the number of combinations? 60

Solve Another Problem

10. You are allowed to take two different elective classes. There are eight different classes you can choose from. How many different combinations of two electives are there? (8 × 7) ÷ (2 × 1) = 28 combinations

Homework Exercises

For more exercises, see Extra Skills and Word Problems.

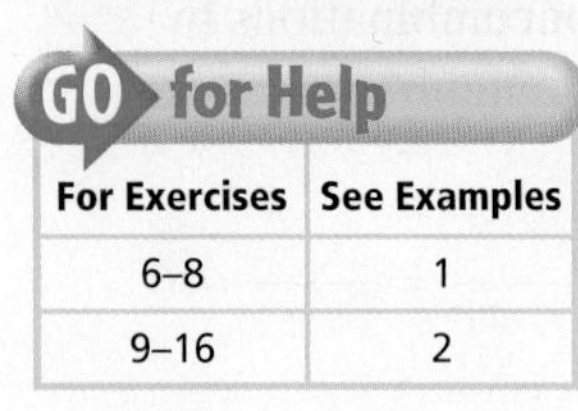

GO for Help

For Exercises	See Examples
6–8	1
9–16	2

A For Exercises 6 and 7, use the table to make an organized list.

Swimmer	Letter
Noah	N
Olivia	O
Kevin	K
Chloe	C

6. **Swimming** The four swimmers listed at the right are trying out for the swim team. Two will make the team. How many different combinations of two swimmers are possible? 6

7. If 3 swimmers will make the team, how many combinations of 3 swimmers are possible? 4

8. Five campers want to go on a hiking trip. A camp counselor will choose three of them. How many different combinations are possible? 10

Find the number of combinations.

9. Choose two people from three. 3
10. Choose three people from five. 10
11. Choose two people from six. 15
12. Choose four people from six. 15

Sports Use the poster for Exercises 13–16. Find the number of combinations for each situation.

13. Participate in 2 track events. 6
14. Participate in 2 games. 1
15. Participate in 2 field events. 3
16. Participate in 3 track events. 4

B GPS 17. **Guided Problem Solving** You have six pizza toppings. How many different 3-topping pizzas can you make? 20 pizzas
- total permutations: 6 × ■ × ■
- permutations of smaller group: ■ × 2 × 1

GO Online
Homework Video Tutor
Visit: PHSchool.com
Web Code: are-1206

Evaluate each expression.

18. $\frac{4!}{3!}$ 4
19. $\frac{5!}{2!}$ 60
20. $\frac{6!}{4!}$ 30
21. $\frac{7!}{3!}$ 840

22. **Music** You have 5 different CDs to play. Your CD player can hold 3 CDs. How many different combinations of 3 CDs can you select? 10 combinations

23. Twelve students organize a trip. Two of them are assigned to collect money. In how many ways can these two students be chosen? 66 ways

24. **Reasoning** To open a combination lock, you must dial the numbers in the right order. Explain why "permutation lock" might be more appropriate than "combination lock" as a name for the lock shown at the left. **Answers may vary. Sample: If the lock's combination is 5-10-15, you cannot open the lock with the combination 10-15-5.**

Determine whether each situation involves a combination or a permutation. Then answer the question.

25. You select three books from a bookshelf that holds eight books. How many different sets of books can you choose? **combination; 56 sets**
26. Four students stand beside one another for a photograph. How many different orders are possible? **permutation; 24 orders**
27. **Writing in Math** Use your own words and an example to explain the difference between a permutation and a combination. **Check students' work.**

Calculator Tip

To find ${}_4C_2$, press 4 MATH. Scroll right to option PRB. Select option 4—*nCr*. Press 2 ENTER.

You can use a graphing calculator to find the number of combinations. Use ${}_nC_r$ where *n* is the total number of items, C is combinations, and *r* is the number of items in a grouping. Evaluate each combination.

28. ${}_7C_3$ **35**
29. ${}_6C_3$ **20**
30. ${}_8C_5$ **56**

Ⓒ 31. **Challenge** You want to mix two of the paint colors below. Use the number of possible combinations to find *P*(blue and green). $\frac{1}{6}$

Test Prep and Mixed Review — Practice

Multiple Choice

32. Julia sold magazine subscriptions for a fundraiser. The numbers of subscriptions she sold in the last seven days were 5, 2, 3, 2, 7, 1, and 4. Which measure of data is represented by 2 subscriptions? **B**

Ⓐ Mean Ⓑ Mode Ⓒ Median Ⓓ Range

33. The top, side, and front views of an object are shown at the right. Which solid matches the views? **G**

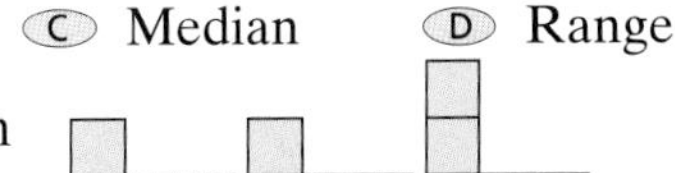

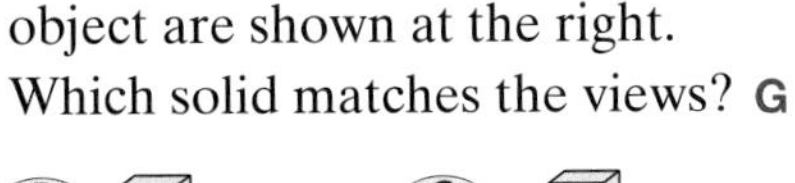
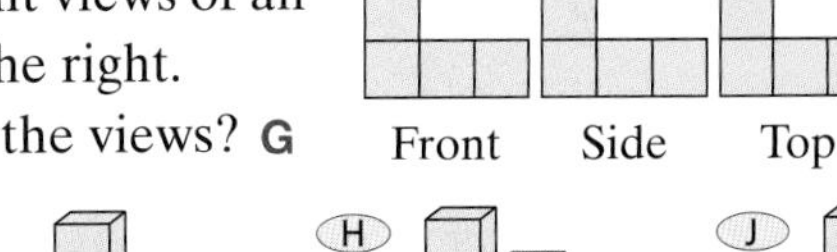

Ⓕ Ⓖ Ⓗ Ⓙ

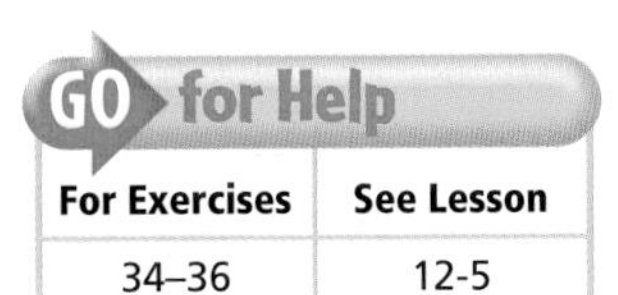

For Exercises	See Lesson
34–36	12-5

Write the number of permutations in factorial form. Then simplify.

34. D, O, G **3! = 6**
35. S, H, O, E **4! = 24**
36. D, R, U, M, S **5! = 120**

Online lesson quiz, **PHSchool.com**, **Web Code:** ara-1206

4. Assess & Reteach

Lesson Quiz

Find the number of combinations.

1. You choose 3 out of 6 people. **20**
2. Choose any two letters from the set of letters C, D, S, T, and U. **10**
3. Choose any three letters from the set of letters A, B, C, and D. **4**

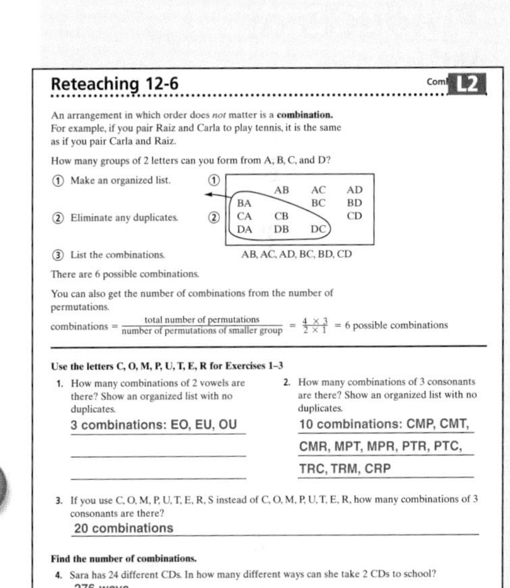

Reteaching 12-6 L2

An arrangement in which order does *not* matter is a **combination.** For example, if you pair Raiz and Carla to play tennis, it is the same as if you pair Carla and Raiz.

How many groups of 2 letters can you form from A, B, C, and D?

① Make an organized list.
② Eliminate any duplicates.

	AB	AC	AD
BA		BC	BD
CA	CB		CD
DA	DB	DC	

③ List the combinations. AB, AC, AD, BC, BD, CD

There are 6 possible combinations.

You can also get the number of combinations from the number of permutations.

$$\text{combinations} = \frac{\text{total number of permutations}}{\text{number of permutations of smaller group}} = \frac{4 \times 3}{2 \times 1} = 6 \text{ possible combinations}$$

Use the letters C, O, M, P, U, T, E, R for Exercises 1–3

1. How many combinations of 2 vowels are there? Show an organized list with no duplicates. **3 combinations: EO, EU, OU**
2. How many combinations of 3 consonants are there? Show an organized list with no duplicates. **10 combinations: CMP, CMT, CMR, MPT, MPR, PTR, PTC, TRC, TRM, CRP**
3. If you use C, O, M, P, U, T, E, R, S instead of C, O, M, P, U, T, E, R, how many combinations of 3 consonants are there? **20 combinations**

Find the number of combinations.

4. Sara has 24 different CDs. In how many different ways can she take 2 CDs to school? **276 ways**
5. Augusto has purple, green, black, red, and blue T-shirts. In how many ways can he choose 3 for his vacation? **10 ways**
6. Abdul selects three light filters from a box of ten different filters. How many different sets could he choose? **120 sets**

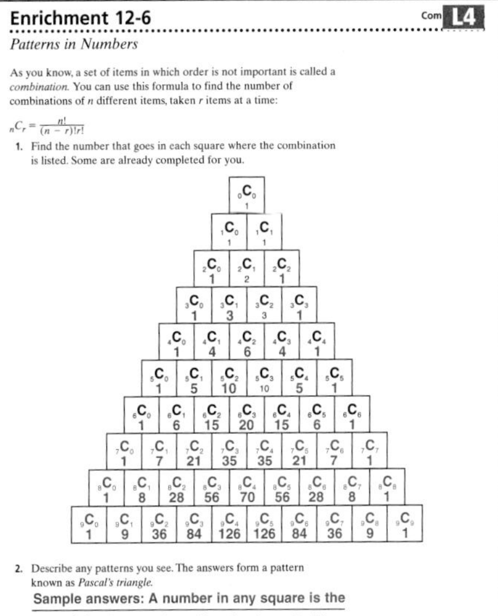

Enrichment 12-6 L4

Patterns in Numbers

As you know, a set of items in which order is not important is called a *combination*. You can use this formula to find the number of combinations of *n* different items, taken *r* items at a time:

$${}_nC_r = \frac{n!}{(n-r)!r!}$$

1. Find the number that goes in each square where the combination is listed. Some are already completed for you.

${}_0C_0$: 1
${}_1C_0$ … ${}_1C_1$: 1, 1
${}_2C_0$ … ${}_2C_2$: 1, 2, 1
${}_3C_0$ … ${}_3C_3$: 1, 3, 3, 1
${}_4C_0$ … ${}_4C_4$: 1, 4, 6, 4, 1
${}_5C_0$ … ${}_5C_5$: 1, 5, 10, 10, 5, 1
${}_6C_0$ … ${}_6C_6$: 1, 6, 15, 20, 15, 6, 1
${}_7C_0$ … ${}_7C_7$: 1, 7, 21, 35, 35, 21, 7, 1
${}_8C_0$ … ${}_8C_8$: 1, 8, 28, 56, 70, 56, 28, 8, 1
${}_9C_0$ … ${}_9C_9$: 1, 9, 36, 84, 126, 126, 84, 36, 9, 1

2. Describe any patterns you see. The answers form a pattern known as *Pascal's triangle*. **Sample answers: A number in any square is the sum of the two numbers directly above it. All numbers in the end squares of each row are 1s.**
3. Complete the next three rows of this Pascal's triangle.

Alternative Assessment

Have students work in pairs or small groups. Each student makes up a situation similar to those in Exercises 6–8. Students exchange papers and solve each others' problems. Students can challenge other groups to solve their problems.

Test Prep

Resources

For additional practice with a variety of test item formats:

- Test-Taking Strategies, p. 615
- Test Prep, p. 619
- Test-Taking Strategies with Transparencies

Checkpoint Quiz

Use this Checkpoint Quiz to check students' understanding of the skills and concepts of Lessons 12-4 through 12-6.

Resources

- All-in-One Teaching Resources Checkpoint Quiz 2
- ExamView CD-ROM
- Success Tracker™ Online Intervention

Products of Winners

In this game, students test probability. They roll two number cubes, multiply the products and get points based on whether the product is odd or even.

Guided Instruction

Before students play the game, have a volunteer read the rules. Have students predict whether Player A or Player B is more likely to win. Pair students who predict opposite letters with each other and have them test their theories by playing. When they have finished playing, discuss the results.

Resources

- two number cubes

Lessons 12-4 through 12-6

You select a letter at random from the group at the right. You replace the letter and make another selection. Find each probability.

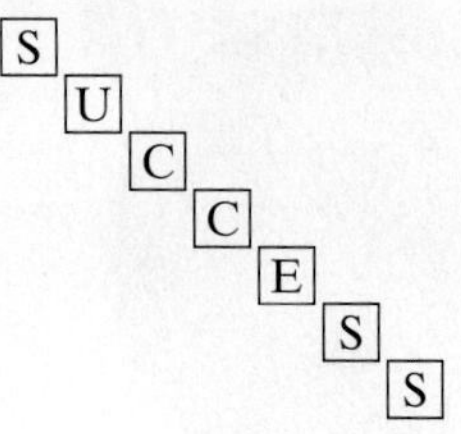

1. P(vowel, then S) $\frac{6}{49}$
2. P(vowel, then C) $\frac{4}{49}$
3. P(C, then E) $\frac{2}{49}$
4. P(T, then S) 0
5. P(U, then consonant) $\frac{5}{49}$

You have 4 blue cards, 1 red card, and 3 green cards. You select a card at random, do not replace it, and select a second card. Find each probability.

6. P(green, then red) $\frac{3}{56}$
7. P(blue, then green) $\frac{12}{56}$ or $\frac{3}{14}$
8. P(red, then blue) $\frac{4}{56}$ or $\frac{1}{14}$
9. You decide to put the pictures of six friends in a row on a bulletin board. In how many different ways can you arrange the pictures? 720 ways
10. How many different combinations of four flowers can you make from six different flowers? 15 combinations

Teams A, B, C, and D enter a cheerleading competition.

11. Give the sample space of the different orders in which the teams can perform. See margin.
12. The two best teams will advance to the next round. How many different combinations of two best teams are possible? 6 combinations

MATH GAMES — Products of Winners

What You'll Need

- two number cubes

How To Play

- Take turns rolling two number cubes. Find the product of the two numbers. If the product is even, Player A scores a point. If the product is odd, Player B scores a point.
- After 15 rolls each, the player with more points wins. Who would you rather be, Player A or Player B?

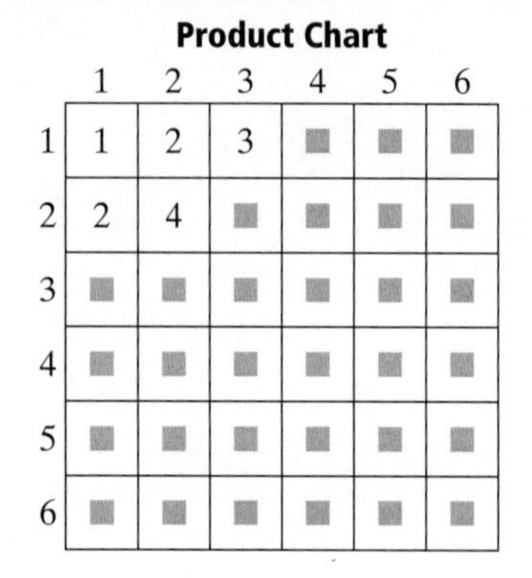

Product Chart

	1	2	3	4	5	6
1	1	2	3	■	■	■
2	2	4	■	■	■	■
3	■	■	■	■	■	■
4	■	■	■	■	■	■
5	■	■	■	■	■	■
6	■	■	■	■	■	■

11. ABCD, ABDC, ACBD, ACDB, ADBC, ADCB, BACD, BADC, BCAD, BCDA, BDAC, BDCA, CABD, CADB, CBAD, CBDA, CDAB, CDBA, DABC, DACB, DBAC, DBCA, DCAB, DCBA

Test-Taking Strategies

Eliminating Answers

Before you try to answer a multiple-choice question, you may be able to save time by eliminating some answer choices. Then you can choose your answer carefully from the remaining options.

EXAMPLE

A bag contains ten blue, six red, and four green pens. You select a pen at random. What is the probability of not choosing a green pen?

(A) $\frac{1}{5}$ (B) $\frac{4}{5}$ (C) $\frac{5}{6}$ (D) $\frac{9}{11}$

Look at the denominator of each choice. The number of possible outcomes is $10 + 6 + 4$, or 20. The denominator of the answer must be 20 or a factor of 20. Eliminate choices C and D because they have denominators of 6 and 11.

Estimate the magnitude of the answer. Since most of the pens are not green, the probability of *not* choosing a green pen is a fraction greater than $\frac{1}{2}$. Since choice A is less than $\frac{1}{2}$, you can eliminate choice A.

The correct answer is choice B.

Exercises

1. A school has 1,060 students. The results of a survey are shown.

Students Surveyed	Students Who Produced Computer Art
40	24

If the trend in the table continues, which is the best prediction of the total number of students who produced computer art? **C**

(A) 260 students (C) 640 students
(B) 480 students (D) 790 students

2. A wheel is divided evenly into three sections labeled A, B, and C. You spin it twice. Which list shows all the possible outcomes? **H**

(F) (A, B), (B, A), (A, C), (C, A), (B, C), (C, B)
(G) (A, A), (A, B), (B, A), (A, C), (C, A), (B, C), (C, C)
(H) (A, A), (A, B), (B, A), (A, C), (C, A), (B, B), (B, C), (C, B), (C, C)
(J) (A, B), (B, A), (A, C), (C, A), (B, C), (C, B), (A, A), (B, B)

Eliminating Answers

When students take multiple-choice tests, they can improve their chances of success, by the laws of probability, every time they are able to eliminate an answer choice. This feature focuses on the strategy of eliminating answer choices.

Guided Instruction

To convince students of the effectiveness of the strategy of eliminating answer choices on multiple-choice tests, present the following situation and questions:

- *You take a multiple-choice test with five answer choices. You have no idea at all which answer is correct. What is the probability that you guess the correct answer?* $\frac{1}{5}$
- *Suppose that you know one answer choice is wrong and you eliminate it. Now what is the probability that you guess the correct answer?* $\frac{1}{4}$

Resources

Test-Taking Strategies with Transparencies
- Transparency 6
- Practice Sheet, p. 24

Test-Taking Strategies with Transparencies

Test-Taking Strategies: Eliminating Answers

When solving a multiple choice problem, first try to eliminate some of the answer choices.

Be sure to cross out answers you eliminate in the test booklet—*not* on the answer sheet.

Example The original price of a coat was $65. It was discounted 20% the day you bought it. How much money did you save?
A. $78 B. $52 C. $13 D. $5

Since $78 is more than the original amount, you can immediately eliminate choice A.

By estimating you can determine that $60 times 0.20 is greater than $5, so you can eliminate choice D.

Since the question is asking for how much you saved, you can eliminate choice B because you know by estimating that you did not save more than 50% of the original price.

The answer is choice C.

Identify two answer choices you can immediately eliminate. Then solve the problem.

1. There are 32 students in Wyatt's class. At least 75% of the students have been vaccinated against chicken pox. About how many students have not had this vaccination?
A. 40 students B. 28 students
C. 16 students D. 8 students

Eliminate choice A: There are only 32 students in the class. Eliminate choice B: By estimating you know that less than half the class has not had the vaccination. The answer is choice D.

Resources

Student Edition

Extra Skills and Word Problems Practice, Ch. 12, p. 652
English/Spanish Glossary, p. 675
Formulas and Properties, p. 673
Tables, p. 670

All in One Teaching Resources

- Vocabulary and Study Skills 12F 

Differentiated Instruction

Spanish Vocabulary Workbook with Study Skills ELL
Interactive Textbook
- Audio Glossary

Online Vocabulary Quiz

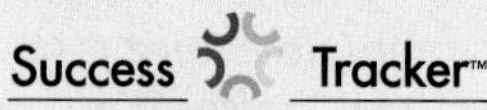

Online at PHSchool.com

Spanish Vocabulary/Study Skills ELL

Vocabulary/Study Skills L3

12F: Vocabulary Review Puzzle For use with the Chapter Review

Study Skill When using a word bank, read the words first. Then answer the questions.

Complete the crossword puzzle. Use the words from the following list.

parallelogram	conjecture	decagon	equation	mode
combination	symmetry	variable	discount	prime
independent	permutation	dependent	outcome	slope

ACROSS

2. Events are ______ if the occurrence of one event affects the probability of the occurrence of another event.
4. A figure has ______ if one side of the figure is the mirror image of the other side.
6. four-sided figure with two sets of parallel lines
8. possible result of an action
11. Events are ______ if the occurrence of one event does not affect the probability of the occurrence of another event.

DOWN

1. prediction that suggests what you expect will happen
2. difference between the original price and the sale price
3. letter that stands for a number
5. ratio that describes the steepness of a line
6. arrangement of objects in a particular order
7. number that occurs most often in a data set
9. grouping of objects in which order does not matter
10. mathematical statement with an equal sign
12. polygon with ten sides
13. whole number with only two factors, itself and the number one

Chapter 12 Review

Vocabulary Review

combination (p. 610)
complement (p. 581)
compound event (p. 598)
counting principle (p. 592)
dependent events (p. 599)
event (p. 580)
experimental probability (p. 586)
factorial (p. 606)
independent events (p. 598)
outcome (p. 580)
permutation (p. 606)
sample space (p. 591)
theoretical probability (p. 580)

Go Online PHSchool.com
For: Online vocabulary quiz
Web Code: arj-1251

Choose the correct term to complete each sentence.

1. A (combination, permutation) is a grouping of objects in which the order of the objects does not matter. combination
2. An outcome or a group of outcomes is called a(n) (event, factorial). event
3. Two events are (dependent, independent) if the occurrence of one event does not affect the probability of the occurrence of the other. independent
4. The (odds in favor of, odds against) an event is the ratio of the number of favorable outcomes to the number of unfavorable outcomes. odds in favor of
5. (Theoretical, Experimental) probability is based on observations. Experimental

Skills and Concepts

Lesson 12-1
- To find the probability and the complement of an event

You can find the **theoretical probability** of an event using this formula.

$$P(\text{event}) = \frac{\text{number of favorable outcomes}}{\text{total number of possible outcomes}}$$

You select a card at random from the cards shown at the right. Find each probability.

6. $P(\text{P})$ $\frac{1}{7}$
7. $P(\text{vowel})$ $\frac{2}{7}$
8. $P(\text{not P})$ $\frac{6}{7}$

Lesson 12-2
- To find experimental probability and to use simulations

You find the **experimental probability** of an event using this formula.

$$P(\text{event}) = \frac{\text{number of times an event occurs}}{\text{total number of trials}}$$

Games A computer game company makes random checks of its games. Of 200 games, 4 are found to be defective.

9. Find the experimental probability that a game is defective. $\frac{1}{50}$
10. If the trend continues, predict the number of defective games in a batch of 1,600. 32

11. See back of book.

Lesson 12-3
- To make and use sample spaces and to use the counting principle

The collection of all possible outcomes in a probability experiment is called a **sample space.** You can use the **counting principle** to find the number of outcomes of an event.

Use the menu below for Exercises 11–13.

Appetizers	**Soups**
Egg Rolls	Won-ton
Fried Won-tons	Sizzling Rice
Main Dishes	
Almond Chicken	
Sweet & Sour Pork	
Beef with Broccoli	

11. At the China Panda, if you order the family dinner, you choose one appetizer, one soup, and one main dish from the menu. Draw a tree diagram to show the sample space. **See margin.**
12. You ask the restaurant to choose the meal for you at random. What is the probability of getting the egg roll, won-ton soup, and almond chicken for your meal? $\frac{1}{12}$
13. Use the counting principle to find the number of possible dinners. **12 dinners**

Lesson 12-4
- To find the probability of independent and dependent events

A **compound event** consists of two or more events. Two events are **independent events** if the occurrence of one event does not affect the probabilty of the occurrence of the other. Two events are **dependent events** if the occurrence of one event affects the probability of the occurrence of the other.

A hat contains the names of eight girls and six boys. You select two names without replacing the first name. Find each probability.

14. P(boy, then boy) $\frac{30}{182}$ **or** $\frac{15}{91}$
15. P(girl, then boy) $\frac{48}{182}$ **or** $\frac{24}{91}$
16. (girl, then girl) $\frac{56}{182}$ **or** $\frac{4}{13}$
17. Two independent events A and B both have a probability of $\frac{1}{4}$. Find $P(A, \text{then } B)$. $\frac{1}{16}$

Lessons 12-5, 12-6
- To find permutations
- To find combinations

A **permutation** is an arrangement of objects in a particular order. A **combination** is a grouping of objects in which order does not matter.

18. Five students compete on a relay team. Only four of them can race at a time. How many different teams are possible? **5 teams**
19. Four students are selected for a relay team. In how many ways can they line up for the race? **24 ways**

Alternative Assessment L4

Alternative Assessment Form C

Chapter 12

INITIAL CHANGES

The lists below show the most common names given to girl and boy babies in the United States in the years 1925, 1950, and 1970.

Girls

1925	1950	1970
1. Mary	1. Linda	1. Michelle
2. Barbara	2. Mary	2. Jennifer
3. Dorothy	3. Patricia	3. Kimberly
4. Betty	4. Susan	4. Lisa
5. Ruth	5. Deborah	5. Tracy

Boys

1925	1950	1970
1. Robert	1. Robert	1. Michael
2. John	2. Michael	2. Robert
3. William	3. James	3. David
4. James	4. John	4. James
5. Charles	5. David	5. John

Show all of your work on a separate sheet of paper.

1. Does the popularity of girls' names and boys' names follow the same patterns of change? Explain your thinking with examples.
2. If you wanted to predict a name among the five most common names in 2000, would it be easier to predict a boy's name or a girl's name? Explain.
3. People use their names to identify their possessions, just as you did by putting your name on this paper. Sometimes people just put their initials on their possessions. Initials can also be used on memos, on legal documents, and on forms such as attendance sheets.
   a. How many possibilities are there for the first initial of all the English-language names in the world? Explain.
   b. How many possibilities are there for the last initial? Explain.

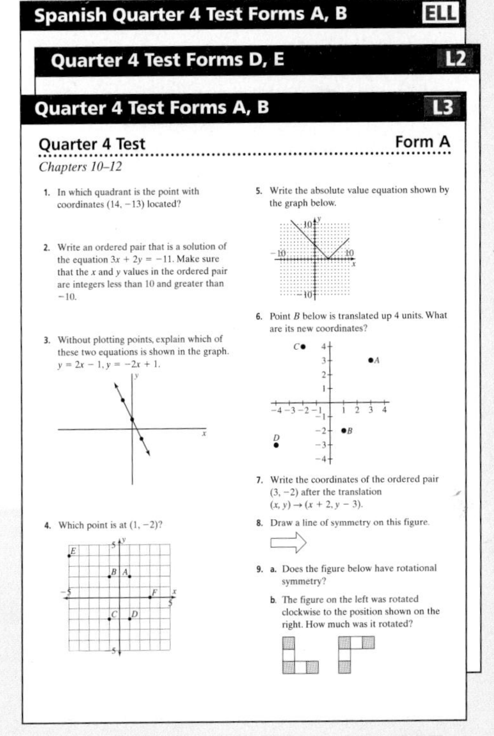

Spanish Quarter 4 Test Forms A, B ELL

Quarter 4 Test Forms D, E L2

Quarter 4 Test Forms A, B L3

Quarter 4 Test Form A

Chapters 10–12

1. In which quadrant is the point with coordinates (14, −13) located?
2. Write an ordered pair that is a solution of the equation $3x + 2y = -11$. Make sure that the x and y values in the ordered pair are integers less than 10 and greater than −10.
3. Without plotting points, explain which of these two equations is shown in the graph. $y = 2x - 1, y = -2x + 1.$
4. Which point is at (1, −2)?
5. Write the absolute value equation shown by the graph below.
6. Point B below is translated up 4 units. What are its new coordinates?
7. Write the coordinates of the ordered pair (3, −2) after the translation $(x, y) \rightarrow (x + 2, y - 3)$.
8. Draw a line of symmetry on this figure.
9. a. Does the figure below have rotational symmetry?
   b. The figure on the left was rotated clockwise to the position shown on the right. How much was it rotated?

Chapter 12 Test

Go Online PHSchool.com For: Online chapter test Web Code: ara-1252

Resources

- ExamView Assessment Suite CD-ROM
  - Ch. 12 Ready-Made Test
  - Make your own Ch. 12 test
- MindPoint Quiz Show CD-ROM
  - Chapter 12 Review

Differentiated Instruction

All in One Teaching Resources
- Below Level Chapter 12 Test L2
- Chapter 12 Test L3
- Chapter 12 Alternative Assessment L4

Spanish Assessment Resources ELL
- Below Level Chapter 12 Test L2
- Chapter 12 Test L3
- Chapter 12 Alternative Assessment L4

ExamView Assessment Suite CD-ROM
- Special Needs Test L1
- Special Needs Practice Bank L1

Online Chapter 12 Test at www.PHSchool.com L3

1. $\frac{3}{8}$; 0.375; 37.5%
2. $\frac{1}{8}$; 0.125; 12.5%

Below Level Chapter Test L2

Chapter Test L3

Chapter Test Form A

Chapter 12

Use the data below. Find the experimental probability of each event as a fraction, a decimal, and a percent.

Toss	1	2	3	4	5	6	7	8	9	10	11	12	13	14	15	16	17	18
Heads	✗		✗	✗		✗		✗	✗		✗		✗	✗	✗		✗	✗
Tails		✗			✗		✗			✗		✗				✗		

1. P(heads) $\frac{2}{3}$; $0.\overline{6}$; $66\frac{2}{3}$%
2. P(tails) $\frac{1}{3}$; $0.\overline{3}$; $33\frac{1}{3}$%

You work at a T-shirt printing business. Of the 4,700 T-shirts shipped, 564 are printed improperly.

3. What is the experimental probability that a T-shirt is printed improperly? 12%
4. Predict the number of improperly printed T-shirts in a batch of 2,000. 240 T-shirts

Use the spinner at the right to find the probability of each event.

5. P(△) $\frac{3}{8}$
6. P(□) $\frac{3}{8}$
7. P(⍙) $\frac{1}{4}$
8. P(△ or □) $\frac{3}{4}$
9. P(△ or ⍙) $\frac{5}{8}$
10. P(neither △ nor □) $\frac{1}{4}$

Suppose you have a bag that contains 4 black, 3 green, 1 red, and 2 orange marbles. Find each probability.

11. P(green) $\frac{3}{10}$
12. P(green or red) $\frac{2}{5}$
13. P(not orange) $\frac{4}{5}$
14. P(green or black) $\frac{7}{10}$
15. P(red, then orange when red is not replaced) $\frac{1}{45}$
16. P(black) $\frac{2}{5}$

The letters G E O M E T R Y are written on a set of cards. You mix the cards thoroughly. Without looking, you draw one letter, replace it, then draw another. Find each probability.

17. P(E, T) $\frac{1}{32}$
18. P(M, Y) $\frac{1}{64}$

Use the data. Find the experimental probability of each event as a fraction, a decimal, and a percent. 1–2. See margin.

Marker Color	Frequency
Purple	6
Green	2
White	3
Black	5

1. P(purple)
2. P(green)
3. P(orange) 0; 0; 0%
4. a. **Quality Control** Factory workers test 80 batteries. Four batteries are defective. What is the experimental probability that a battery is defective? $\frac{1}{20}$
   b. Assume this trend continues. Predict the number of defective batteries in a batch of 1,600. 80 defective batteries

There are six open containers arranged as shown. You toss a ball and it falls into one of the containers. Find each probability. 5. $\frac{2}{6}$ or $\frac{1}{3}$

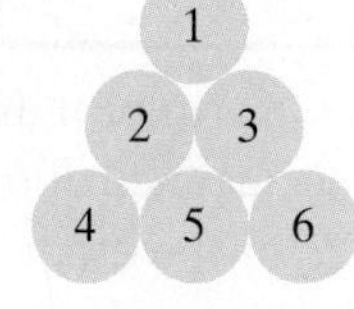

5. P(number greater than 4)
6. P(even number) $\frac{3}{6}$ or $\frac{1}{2}$
7. P(4) $\frac{1}{6}$
8. P(7) 0

You have a bag that contains 6 blue, 2 green, 3 red, and 1 white marble. You select a marble at random. Find each probability. 13. $\frac{6}{144}$ or $\frac{1}{24}$

9. P(blue) $\frac{6}{12}$ or $\frac{1}{2}$
10. P(white) $\frac{1}{12}$
11. P(not green) $\frac{10}{12}$ or $\frac{5}{6}$
12. P(red) $\frac{3}{12}$ or $\frac{1}{4}$
13. P(green, then red when green is replaced)
14. P(red, then blue when red is not replaced) $\frac{18}{132}$ or $\frac{3}{22}$
15. a. Find the number of two-letter permutations of the letters M, A, T, H. 12
    b. Find the number of two-letter combinations of the letters M, A, T, H. 6

16. Each of the letters D E T E R M I N E D is written on a card. You mix the cards thoroughly. What is the probability of selecting an E and then an M, if the first card is replaced before selecting the second card? $\frac{3}{100}$
17. **Writing in Math** Suppose you toss a coin several times and record the results. Out of 20 trials, you get tails 9 times. What is the experimental probability of getting heads? Explain why this may differ from the theoretical probability. See margin.

A car comes in the colors and models listed in the table below. Assume there is the same chance of selecting any color or model.

Colors	Models
Silver	Hatchback
Gray	Coupe
Black	Sedan

18. Give the sample space. See margin.
19. Find the probability that a car selected at random is a silver hatchback. $\frac{1}{9}$
20. Find the probability that a car selected at random is a yellow coupe. 0
21. You have the same chance of getting any one of four prizes when you buy Good Morning Cereal. You want to use a simulation to find the probability of getting all four prizes when you buy four boxes of cereal. Which statement is *not* true? B
    - (A) You can simulate the problem by using a spinner divided into four equal sections.
    - (B) A possible answer is four boxes.
    - (C) The more trials you perform, the better your results should be.
    - (D) The result of your simulation is an experimental probability.

17. $\frac{11}{20}$; although the theoretical and experimental probabilities are not always the same, you expect the experimental probability to be close to the theoretical probability.

18. (Silver, Hatchback); (Silver, Coupe); (Silver, Sedan); (Gray, Hatchback); (Gray, Coupe); (Gray, Sedan); (Black, Hatchback); (Black, Coupe); (Black, Sedan)

Multiple Choice

Read each question. Then write the letter of the correct answer on your paper.

1. Use rounding to estimate. Which sum is between 14 and 15? **B**
   - A) 13.71 + 1.5
   - B) 9.02 + 5.738
   - C) 2.69 + 12.49
   - D) 3.772 + 12.04
2. Which expression CANNOT be rewritten using the Distributive Property? **G**
   - F) 3(2 + 8)
   - G) 5(2 · 3)
   - H) (18 − 9)7
   - J) 9(14 − 6)
3. What is the solution of $-15 = m - 9$? **B**
   - A) −24
   - B) −6
   - C) 6
   - D) 24
4. Which jar of peanut butter is the best buy? **G**
   - F) an 18-oz jar for $1.69
   - G) a 30-oz jar for $2.59
   - H) a 32-oz jar for $2.89
   - J) a 24-oz jar for $2.09
5. Sarah bought a remnant of fabric $5\frac{1}{8}$ yd long to make pennants for the school contest. How many pennants can she make if $\frac{3}{4}$ yd is needed for each pennant? **B**
   - A) 5
   - B) 6
   - C) 7
   - D) 8
6. Which choice does NOT equal the others? **G**
   - F) 4% of 3,000
   - G) 40% of 30
   - H) 40% of 300
   - J) 30% of 400
7. Suppose you spin the spinners once. What is the probability that the sum of the numbers is 10? **B**

   - A) 0
   - B) $\frac{1}{4}$
   - C) $\frac{1}{2}$
   - D) $\frac{3}{4}$
8. What is the volume of a rectangular prism that has dimensions 1 in., 2 in., and 3 in.? **F**
   - F) 6 in.^{3}
   - G) 18 in.^{3}
   - H) 22 in.^{3}
   - J) 27 in.^{3}
9. How can 64 be written using exponents? **D**
   - A) 2^6
   - B) 4^3
   - C) 8^2
   - D) All of the above are correct.

Go Online PHSchool.com For: Online end-of-course test Web Code: ara-1254

10. In the diagram at the right, which two angles are adjacent angles? **F**

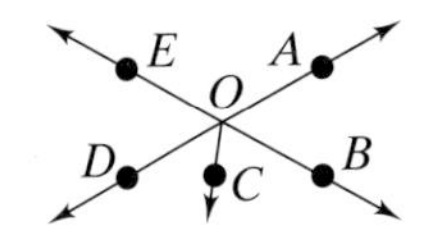

   - F) ∠EOD, ∠DOC
   - G) ∠BOC, ∠BOD
   - H) ∠AOE, ∠BOC
   - J) ∠AOB, ∠EOD
11. A circle has circumference 56.52 ft. What is its area? Use 3.14 for π. **C**
   - A) 28.26 ft^2
   - B) 56.52 ft^2
   - C) 254.34 ft^2
   - D) 1,017.36 ft^2
12. What is the solution of the inequality $-4p < 36$? **H**
   - F) $p > 9$
   - G) $p < -9$
   - H) $p > -9$
   - J) $p < 9$
13. If the area of the shaded region is 4 in.^{2}, what is the best estimate for the area of the unshaded region? **C**

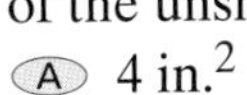

   - A) 4 in.^{2}
   - B) 8 in.^{2}
   - C) 12 in.^{2}
   - D) 16 in.^{2}
14. What percent of the letters of the alphabet are the vowels a, e, i, o, and u? **G**
   - F) about 15%
   - G) about 19%
   - H) about 30%
   - J) about 33%
15. What is the order of the numbers from least to greatest? $\frac{1}{8}, -0.18, 0.2, -\frac{2}{13}$ **C**
   - A) $-\frac{2}{13}, -0.18, \frac{1}{8}, 0.2$
   - B) $-0.18, -\frac{2}{13}, 0.2, \frac{1}{8}$
   - C) $-0.18, -\frac{2}{13}, \frac{1}{8}, 0.2$
   - D) $-\frac{2}{13}, 0.2, \frac{1}{8}, -0.18$
16. What is the solution of $\frac{x}{6} = \frac{20}{32}$? **G**
   - F) 3
   - G) 3.75
   - H) 4.8
   - J) 5

Item	1	2	3	4	5	6	7	8	9	10	11	12	13	14	15	16	17	18	19	20
Lesson	1-1	1-9	4-3	5-2	3-5	6-4	12-4	8-10	2-1	7-2	8-5	4-9	8-1	6-5	2-7	5-3	4-3	10-3	3-5	5-5

Item	21	22	23	24	25	26	27	28	29	30	31	32	33	34	35	36	37	38	39	40
Lesson	12-5	9-4	2-2	7-3	11-2	11-2	11-2	9-6	9-6	9-6	2-8	4-6	12-4	7-5	6-2	3-6	10-5	10-6	5-6	1-9

Item	41	42	43	44	45	46	47
Lesson	8-3	8-7	11-3	10-2	12-3	8-9	6-7

Test Prep

Resources

Test Prep Workbook

All in One Teaching Resources
- Cumulative Review **L3**

ExamView Assessment Suite CD-ROM
- Standardized Test Practice

Differentiated Instruction

Progress Monitoring Assessments
- Quarter 4 Test
  - Forms A & B **L3**
  - Forms D & E **L2**
- End-of-Course Test
  - Forms A & B **L3**
  - Forms D & E **L2**

Spanish Assessment Resources
- Spanish Cumulative Review **ELL**
- Spanish Quarter 4 Test
  - Forms A & B **L3**
- Spanish End-of-Course Test
  - Forms A & B **L3**

ExamView Assessment Suite CD-ROM
- Special Needs Practice Bank **L1**

Spanish Cumulative Review **ELL**

Cumulative Review **L3**

Cumulative Review

Chapters 1–12

Multiple Choice. Circle the letter of the best answer.

1. What is the area of the trapezoid below?
   9 cm, 6 cm, 12 cm
   A. 36 cm^2 B. 63 cm^2 C. 64 cm^2 D. 126 cm^2
2. What is the length of the side of a square with area 2.25 m^2?
   F. 15 m G. 1.5 m H. 0.15 m J. 0.015 m
3. A triangle has sides measuring 5 cm, 7 cm, and 13 cm. Which formula could you use to find out whether the triangle is a right triangle?
   A. $A = \frac{1}{2}bh$
   B. $C = r^2$
   C. $a^2 + b^2 = c^2$
   D. $P = s_1 + s_2 + s_3$
4. In $\triangle ABC$ below, $\overline{BD}$ is the perpendicular bisector of $\overline{AC}$. The height is 8 in. and the base is 12 in. What are the lengths of $\overline{AB}$ and $\overline{BC}$?

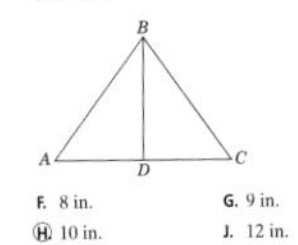

   F. 8 in. G. 9 in. H. 10 in. J. 12 in.
5. Margaret wants to paint a rectangular toy box that is 2 ft wide by 1.5 ft high by 4 ft long. One quart of paint will cover 100 ft^2. How many quarts of paint will she need to buy if she puts two coats on both the inside and outside of the toy box?
   A. 1 qt B. 2 qt C. 3 qt D. 4 qt
6. Which is between 5 and 6?
   F. $5\frac{1}{2} + 2\frac{1}{4}$ G. $4\frac{3}{5} + 2\frac{3}{4}$ H. $11\frac{2}{3} - 6\frac{7}{8}$ J. $9\frac{1}{3} - 3\frac{4}{5}$
7. Solve $x - \frac{3}{4} = 9\frac{1}{2}$.
   A. $10\frac{1}{4}$ B. 10 C. $9\frac{3}{4}$ D. $8\frac{3}{4}$
8. The length of a room is 12 ft 4 in. The width is 15 ft 6 in. What is the area of the floor of the room in square feet?
   F. $17\frac{5}{6}$ ft^2 G. 180 ft^2 H. $180\frac{1}{6}$ ft^2 J. $191\frac{1}{6}$ ft^2
9. Solve $\frac{5}{6}t = 9$.
   A. $7\frac{1}{2}$ B. 9 C. 10 D. $10\frac{4}{5}$
10. A bicycle is on sale at $\frac{1}{3}$ off the original price of $276. What is the sale price of the bike?
   F. $83 G. $92 H. $184 J. $201
11. In this figure, what is the ratio of the shaded area to the unshaded area?
   4, 4, 2, 2
   A. $\frac{1}{4}$ B. $\frac{1}{3}$ C. $\frac{3}{4}$ D. $\frac{4}{5}$

32. [2] $\frac{x}{3} + 11 = 16$
$\frac{x}{3} = 5$
$x = 15$
[1] minor computational error OR correct answer without work shown

33. [2] $\frac{2}{7} \cdot \frac{1}{6} = \frac{2}{42} = \frac{1}{21}$
[1] minor computational error OR correct answer without work shown

34. [2] $\angle A \cong \angle F$, $\angle B \cong \angle D$, $\angle C \cong \angle E$, $\overline{AB} \cong \overline{FD}$, $\overline{AC} \cong \overline{FE}$, $\overline{BC} \cong \overline{DE}$
[1] error in one or more congruency

35. [2] 0.9% = 0.009
0.009 = nine thousandths
$= \frac{9}{1{,}000}$
[1] minor computational error OR correct answer without work shown

36. [2] 1 lb = 16 oz
$\frac{1}{16} = \frac{x}{72}$
$16x = 72$
$x = 4.5$ lb
[1] minor computational error OR correct answer without work shown

37. [2] translation$(x, y) \rightarrow (x - 4, y + 4)$
[1] minor error in notation

38. [2] reflection; x-axis or $y = 0$
[1] incorrect line of symmetry

39. [2] 2 in. : 250 mi
$\frac{2}{250} = \frac{4}{x}$
$2x = 1{,}000$
$x = 500$ mi
[1] minor computational error OR correct answer without work shown

40. [2] Answers may vary. Sample:
$8(89) = 8(80 + 9)$
$= 8 \cdot 80 + 8 \cdot 9$
$= 640 + 72$
$= 712$
[1] minor computational error OR correct answer without work shown

17. Which equation has the solution 4? **B**

Ⓐ $k + 3 = -7$ Ⓒ $y - 8 = 12$
Ⓑ $5 + x = 9$ Ⓓ $1 + a = 3$

18. What is the slope of the line? **F**

Ⓕ $-\frac{3}{2}$
Ⓖ $\frac{1}{3}$
Ⓗ $\frac{1}{2}$
Ⓙ 2

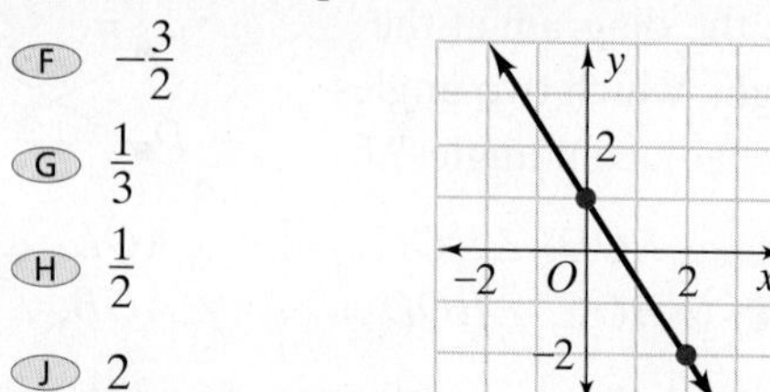

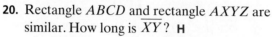

19. You plan to build a set of steps. The steps must reach a height of $8\frac{1}{2}$ ft. Each step can be no more than $7\frac{3}{4}$ in. high. What is the least number of steps you need to build? **D**

Ⓐ 11 Ⓑ 12 Ⓒ 13 Ⓓ 14

20. Rectangle *ABCD* and rectangle *AXYZ* are similar. How long is $\overline{XY}$? **H**

Ⓕ 2.5 cm Ⓗ 1.6 cm
Ⓖ 2 cm Ⓙ 1.5 cm

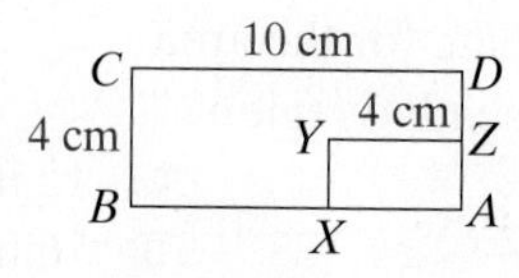

21. How many two-letter permutations of the letters C, A, R contain the letter R? **C**

Ⓐ 2 Ⓑ 3 Ⓒ 4 Ⓓ 6

22. Use the function $y = 4x - 5$. What is the value of *y* for $x = 0, 1, 2,$ and 3? **J**

Ⓕ −5, 1, 3, 2 Ⓗ −5, 9, 13, 7
Ⓖ 5, 0, 1, 2 Ⓙ −5, −1, 3, 7

23. Which expression shows the prime factorization of 120? **C**

Ⓐ $2 \cdot 3 \cdot 4 \cdot 5$ Ⓒ $2^3 \cdot 3 \cdot 5$
Ⓑ $2 \cdot 3 \cdot 20$ Ⓓ $3^2 \cdot 13$

24. What is the value of *x* in the triangle? **H**

Ⓕ 23°
Ⓖ 27°
Ⓗ 33°
Ⓙ 53°

Gridded Response

Use the bar graph below for Exercises 25–27.

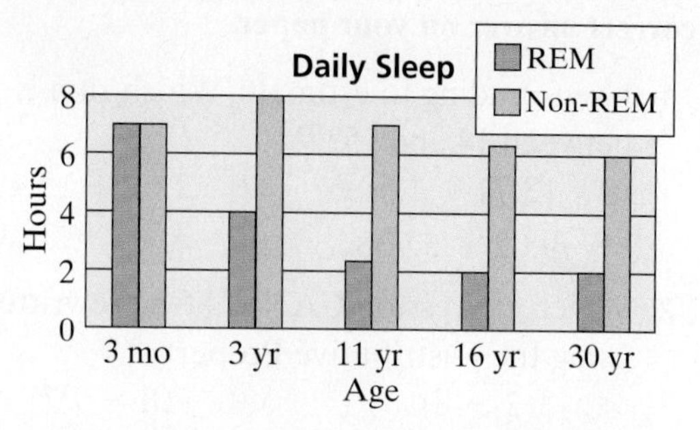

25. About how many hours per day does an 11-year-old spend in non-REM sleep? **7**

26. About how many more hours does a 3-month-old spend in REM sleep than a 30-year-old? **5**

27. About how many fewer hours does an 11-year-old spend in non-REM sleep than a 3-year-old? **1**

For Exercises 28–30, use the graph below. The graph shows the height of an object over time.

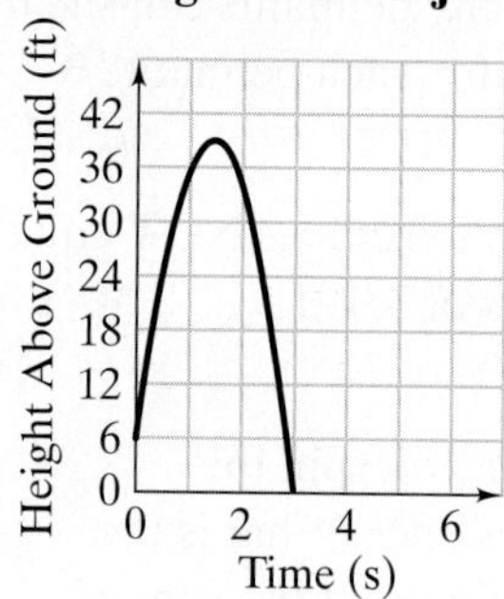

28. How many feet is the greatest height the object reaches? **39**

29. How many seconds does it take the object to hit the ground? **3**

30. How many feet is the initial height of the object? **6**

31. Write 7.8×10^2 in standard form. **780**

41a. [2] b. $A = \frac{1}{2}bh$
$= \frac{1}{2}(5.3)(5.5)$
$= \frac{1}{2}(29.15)$
$= 14.575$
$A = 14.575$
$B = 14.575 \text{ m}^2$

$A = \frac{1}{2}bh$
$= \frac{1}{2}(20)(15)$
$= (10)(15)$
$= 150$
$A = 150 \text{ in.}^2$

[1] minor computational error OR correct answer without work shown

42. [2] $a^2 + b^2 = c^2$
$70^2 + 20^2 = c^2$
$4{,}900 + 400 = c^2$
$c^2 = 5{,}300$
$c = \sqrt{5{,}300}$
$c \approx 72.8$ m
[1] minor computational error OR correct answer without work shown

Short Response 32–42. See margin.

32. Write a problem that $\frac{x}{3} + 11 = 16$ will solve.

33. Cards A through G are in a hat. You select a card at random. You select a second card without replacement. Find the probability that both cards are vowels. Show your work.

34. The triangles at the right are congruent. Write six congruencies involving corresponding parts of the triangles.

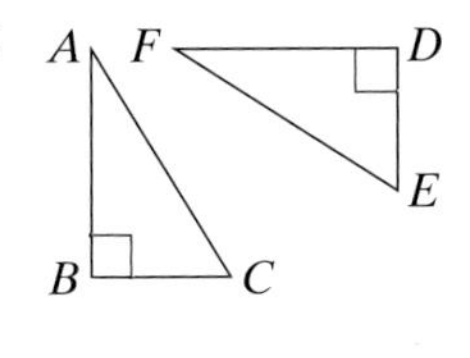

35. **a.** Write 0.9% as a decimal.
b. Write 0.9% as a fraction.

36. Find the number of pounds in 72 ounces.

Tell whether each graph shows a reflection or a translation. Name the line of symmetry or write a rule for the translation.

37.

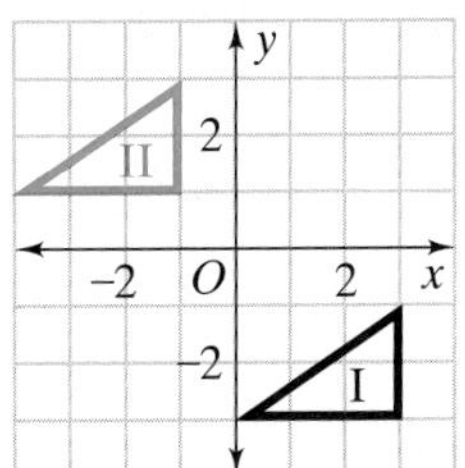

38.

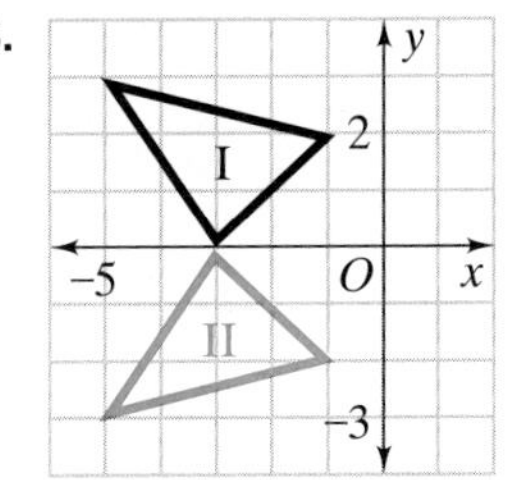

39. A map with the scale 2 in. : 250 mi shows two ponds to be 4 in. apart. How many miles apart are the ponds? Show your work.

40. Find the product 8(89) using the Distributive Property. Show your work.

41. Find the area of each triangle below.

a. **b.**

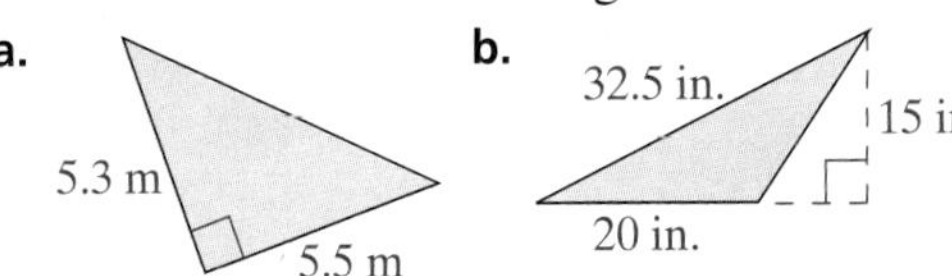

42. Find c to the nearest tenth.

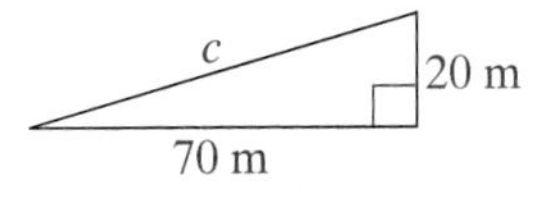

Extended Response 43–47. See margin.

43. Find the mean, median, and mode of the data in the stem-and-leaf plot. Show your work.

Stem	Leaf
7	0 0 5 8
8	1 5 6 9 9
9	4

Key: 7 | 0 means 70

44. You have $50 saved at the beginning of the month. You plan to save $25 each month. The equation $y = 25x + 50$ models your savings plan.
a. Graph this equation in the first quadrant of the coordinate plane.
b. How much will you have saved in 6 months?
c. In how many months will you have saved $150?

45. St. Francis, Torrey Pines, and Marina schools all compete for the championship in field hockey.
a. Make a table to find the sample space of possible outcomes of first, second, and third place.
b. In how many outcomes does St. Francis win with Marina in second place?
c. In how many outcomes does Torrey Pines or Marina win the championship?

46. **a.** Draw a net of the cylinder shown.

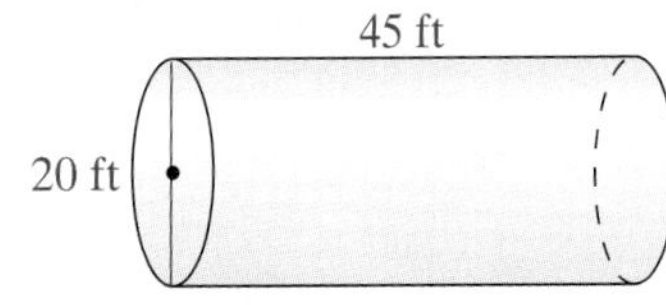

b. Find the surface area of the cylinder to the nearest tenth. Show your work.

47. You treat a friend to dinner. The cost of the food items from the menu totals $20.46. The sales tax on the food is 5%. You give a tip of 25% (before tax) for excellent service.
a. How much is the sales tax?
b. How much is the tip?
c. What is the total cost of the dinner?

43. [4] mean = $\frac{\text{sum}}{\text{num. of items}}$
70 + 70 + 75 + 78 + 81 + 85 + 86 + 89 + 89 + 94 = 817
$\frac{817}{10} = 81.7$
median = 83
mode = 70, 89

[3] minor error in one part
[2] minor error in two or more parts
[1] correct answers without work shown

44. [4] a.

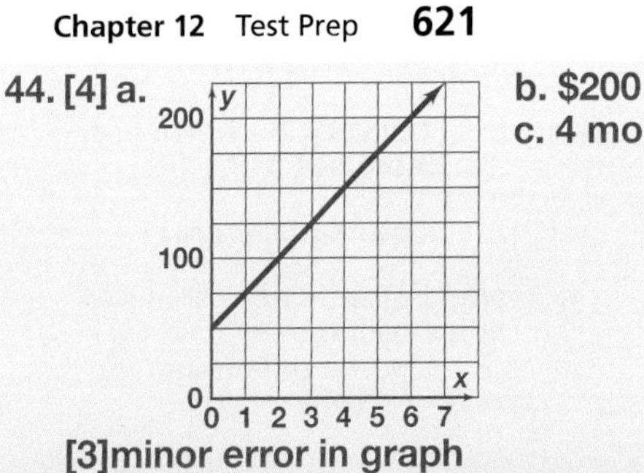

b. $200
c. 4 mo

[3] minor error in graph
[2] minor error in graph and analysis
[1] major error in graph and analysis

45. [4] a.

First Place	Second Place	Third Place
SF	TP	M
SF	M	TP
TP	SF	M
TP	M	SF
M	SF	TP
M	TP	SF

b. $\frac{1}{6}$ **c.** $\frac{2}{3}$

[3] minor error in sample space
[2] minor error in sample space and analysis
[1] major error in sample space and analysis

46. [4] a.

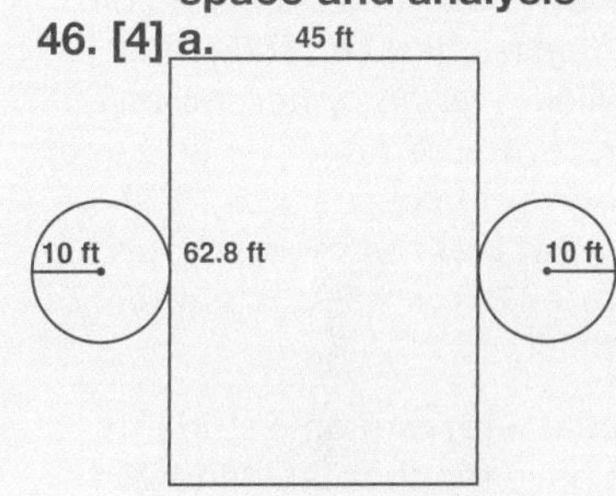

b. surface area = sum of sides
$= \pi r^2 + \pi r^2 + \ell w$
$= 2\pi r^2 + \ell w$
$= 2\pi(10)^2 + (62.8)(45)$
$= 200\pi + 2{,}826$
$= 628.3 + 2{,}826$
$= 3{,}455.8$

[3] minor error in net
[2] minor error in net and computational error in part (b)
[1] correct answer without work shown

47. [4] a. sales tax = 5% · bill
= 5% · $20.46
= 0.05 · $20.46
= $1.023
≈ $1.02

b. tip = 25% · bill
= 25% · $20.46
= 0.25 · $20.46
= $5.115
≈ $5.12

c. cost = bill + tax + tip
= $20.46 + $1.02 + $5.12
= $26.60

[3] one minor computational error
[2] two minor computational errors
[1] correct answers without work shown

Applying Probability

Students will use data from these two pages to answer the questions posed in Put It All Together.

Give one volunteer a coin and another a ball. Have the first student pick heads or tails and then flip the coin. Have the second student try to throw the ball into a wastepaper basket 5 ft away. Continue until each has taken 20 turns, recording their success rates. Ask:

- *How many times did our coin flipper guess correctly?*
- *What was his or her success rate?* **Percentage = number of correct guesses × 5**
- *What was the shooter's success rate?* **Percentage = number of baskets × 5**

Repeat with the same students, but have the shooter stand 20 ft from the basket and give the flipper a different denomination. Discuss why the success rate for the coin flip remained similar for both coins, while the probability of making a basket declined when the distance was increased.

Materials

- Two coins of different denominations
- Ball and wastepaper basket

Activating Prior Knowledge

Have students who have knowledge about basketball lead a discussion on free-throw shooting statistics.

Guided Instruction

Have a student read the opening paragraph. Arrange for a class free-throw shooting event. Students record, graph, and evaluate their results. Ask:

- *What was the highest free-throw percentage in the class?*
- *What was the class average?*
- *How does this compare to the statistics discussed earlier for professional free-throw shooting?*

Problem Solving Application

Applying Probability

Against All Odds? Your friend claims to be a great coin flipper who gets heads 50% of the time. Without doing any math, you know this is not a special skill—the results are pure chance. Suppose another friend claims to be a great free-throw shooter because of a 30% free-throw success rate. This claim is harder to evaluate. How can you tell if it's luck or skill?

Answers may vary. Samples are given for Exercises 1–3.

1a. Check students' work.

b. radius of the rim ≈ 9 in.; radius of the ball ≈ 4.75 in.; area of landing zone ≈ 594 in.^{2}

2a. Check students' work.

b. Check students' work.

c. radius of target zone ≈ 4.25 in.; area of target zone ≈ 56.7 in.^{2}

3. Probability of swish ≈ $\frac{56.7}{594}$ ≈ 0.095 = 9.5%

4. Answers may vary. Sample: A basket can be made by using the backboard, rim, or both.

Put It All Together

Materials compass, ruler, calculator

1. Suppose that your friend can hit the rim of the basket every time. Figure 1 shows the basket from above. Shots A, B and C just barely touch the rim.
   a. Copy Figure 1. Draw a circle centered on point P that connects the centers of balls A, B, and C. This is the *landing zone*. The center of the ball is within this circle for each shot.
   b. **Research** Find the radius of a men's basketball and of a basketball hoop. Calculate the area of the landing zone.
2. A "swish" shot passes through the net without touching the rim. Figure 2 shows ball D swishing through the net, falling just within the rim.
   a. Copy Figure 2. Draw two more balls (E and F) that also fall just within the rim.
   b. Draw a circle centered on point P that connects the centers of balls D, E, and F. This is the target zone. The center of the ball will be within this circle every time the ball "swishes" the net.
   c. Calculate the area of the target zone.
3. Suppose you are shooting baskets at random. Find the probability that a ball hitting the landing zone will also be in the target zone as follows:

   $$\text{Probability of "swish"} = \frac{\text{area of target zone}}{\text{area of landing zone}}$$

   Calculate this probability. Convert it to a percent.
4. **Writing in Math** A shot doesn't have to be a swish to count as a basket. Explain how else a basket can be made, and what factors affect the probability of making a basket.

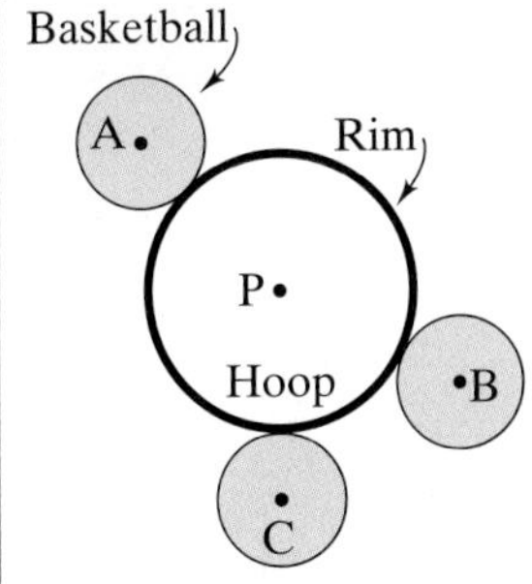

Figure 1

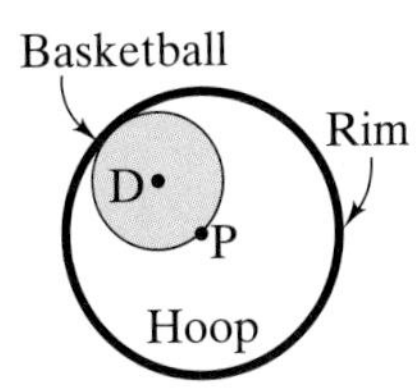

Figure 2

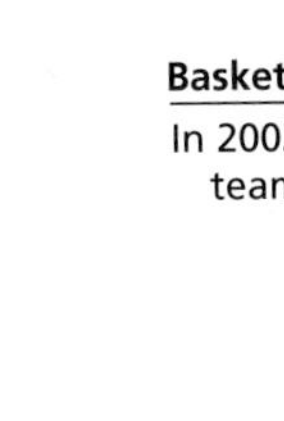

Basketball in the United States
In 2002, high school basketball teams included 540,597 boys and 456,169 girls.

For: Information about basketball
Web Code: are-1253

The skill level of the player affects the probability of making a basket.

Activity

Help students work in pairs to answer the questions.

Exercise 1b Help students recognize that the radius of the landing zone is simply the radius of a basketball plus the radius of a basketball hoop.

Differentiated Instruction

Special Needs L1

Remind students what radius is and how to find it. Also review how to find the area of a circle. Guide students to use $\frac{22}{7}$ or 3.14 as approximations for pi.

623

Chapter Projects

Board Walk

Students apply their knowledge of integers to create a board game.

Resources

All in One Teaching Resources
- Chapter 1 Project Support

Guided Instruction

Activating Prior Knowledge
Have students talk about board games that they have played. Ask them to share the strengths and weaknesses of the games. Ask:
- *What makes a board game something you would play repeatedly?* **Sample: You can make strategy choices.**

Making the Measure

Students apply their knowledge of measurement to create their own system for measuring distance.

Resources

All in One Teaching Resources
- Chapter 2 Project Support

Guided Instruction

Activating Prior Knowledge
Review common measurements of customary length. Ask students to suggest benchmarks for each. Ask:
- *What distance will you use for your ruler? Why?* **Have students justify their answers.**

Social Studies Connection
Invite students to learn more about other ancient measures, like the "cubit." Have them report on their findings.

Chapter Projects

Board Walk

Chapter 1 *Decimals and Integers*

What makes a board game so much fun? You have challenges like road blocks or false paths that make you backtrack. Then you land on a lucky square that lets you leap forward past your opponent. Best of all, you are with your friends as you play!

Create a Board Game For this chapter project, you will use integers to create a game. Then you will play your game with friends or family for a trial run. Finally you will decorate your game and bring it to class to play.

Go Online PHSchool.com
For: Information to help you complete your project
Web Code: ard-0161

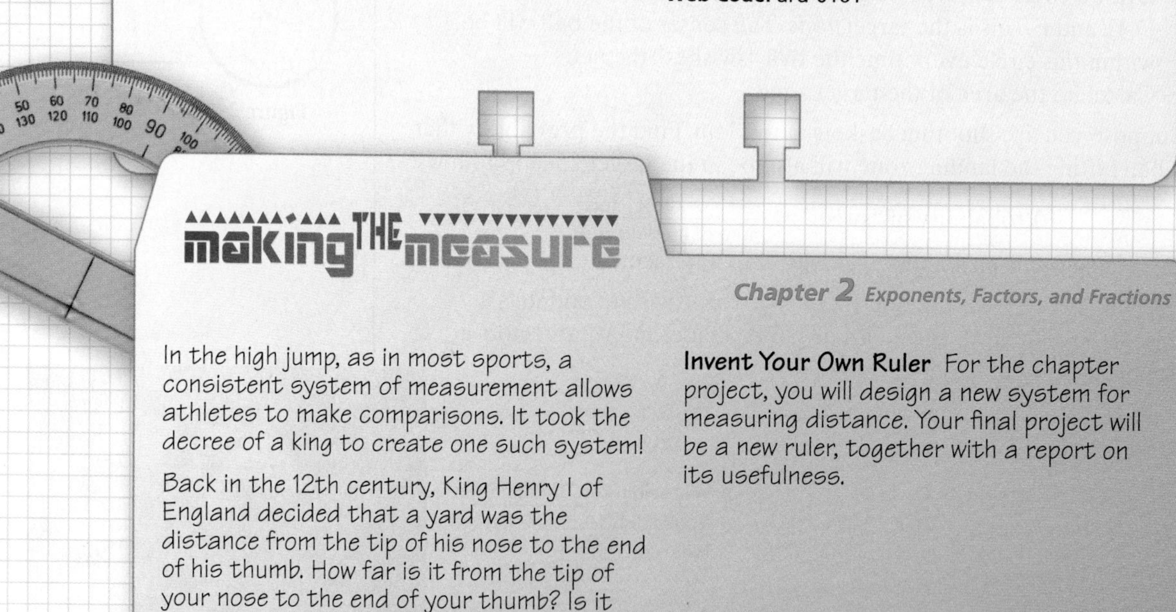

Making the Measure

Chapter 2 *Exponents, Factors, and Fractions*

In the high jump, as in most sports, a consistent system of measurement allows athletes to make comparisons. It took the decree of a king to create one such system!

Back in the 12th century, King Henry I of England decided that a yard was the distance from the tip of his nose to the end of his thumb. How far is it from the tip of your nose to the end of your thumb? Is it more than a yard or less? Is it the same distance for everyone?

Invent Your Own Ruler For the chapter project, you will design a new system for measuring distance. Your final project will be a new ruler, together with a report on its usefulness.

Go Online PHSchool.com
For: Information to help you complete your project
Web Code: ard-0261

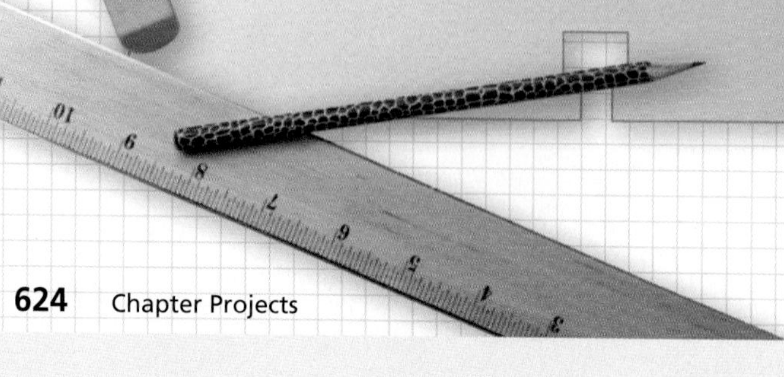

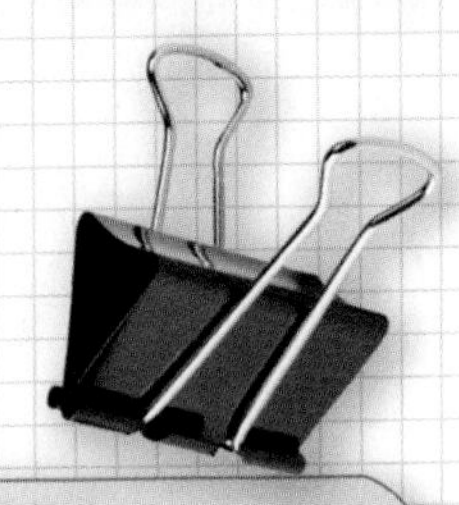

Chapter 3 Operations With Fractions

Did you ever make pancakes? The recipe can be pretty simple—an egg, some pancake mix, milk, and maybe some oil. Or forget the mix and start from scratch! Either way, you can vary the ingredients to suit your tastes. Do you want to include some wheat germ? How about some pecans, or maybe some fruit? Bananas are always in season!

Write Your Own Recipe For the chapter project, you will write your own recipe for pancakes. Your final project will be a recipe that will feed everyone in your class.

Go Online
PHSchool.com
For: Information to help you complete your project
Web Code: ard-0361

READ ALL ABOUT IT!

Chapter 4 Equations and Inequalities

Flexible hours! Great pay! Work before or after school! Newspaper deliverers needed! Suppose to earn extra money you get a job delivering newspapers in your neighborhood. You plan to save the money you make so that you can buy yourself brand new snow skiing gear.

Make a Savings Plan In this chapter, you will figure out how much time you can commit to your job, how much money you can earn per week, and how much money you need to make per week in order to reach your savings goal. As part of your final project, you will write a letter to your boss at the newspaper office describing your level of commitment as a newspaper deliverer.

Go Online
PHSchool.com
For: Information to help you complete your project
Web Code: ard-0461

Toss and Turn

Students apply their knowledge of fractions to make and then adjust a recipe.

Resources

All in One **Teaching Resources**

- Chapter 3 Project Support

Guided Instruction

Activating Prior Knowledge
Tell students that chefs constantly adjust recipes to create new dishes and to feed different numbers of diners. Ask:

- When creating a recipe, what must you think about in addition to the ingredients and how much of each to use? **Sample: Deciding where and when to buy what you need, figuring out cooking times, and so on.**

Read All About It!

Students apply their knowledge of equations and inequalities to make savings plans for themselves.

Resources

All in One **Teaching Resources**

- Chapter 4 Project Support

Guided Instruction

Activating Prior Knowledge
Ask students to share experiences they have had trying to save money over time. Ask:

- *Suppose you could save $5 each week. How could you figure out how many weeks you would have to save in order to have enough money for a $45 baseball bat and a $19 baseball?* **Sample: Write and solve an inequality: $5n \geq (45 + 19)$**

Careers
Invite students to interview bankers to find out the responsibilities, skills, and background required.

Weighty Matters

Students apply their knowledge of ratios and proportions to create a table of animal weights.

Resources

All in One **Teaching Resources**
- Chapter 5 Project Support

Guided Instruction

Activating Prior Knowledge
Brainstorm a list of common pets and their approximate weights. Then have volunteers adjust the guesses by researching typical weights for certain breeds of dogs and cats and other pets. Ask:
- *How could you find the cost of the statue of the 35-pound dog cast in gold?* **One way: Use the newspaper to find the price of gold in oz. Convert it to lb by multiplying it by 16 oz.**

Art Connection
Invite students to look through art books or other sources to find examples of sculptures of animals.

Chills and Thrills

Students apply their knowledge of percents and data collection to find out what amusement park rides people prefer.

Resources

All in One **Teaching Resources**
- Chapter 6 Project Support

Guided Instruction

Activating Prior Knowledge
Have students work together to create a list of favorite amusement park rides that they can use in their surveys. Ask:
- *How might your survey measure people's preferences?* **One way: have people use a simple point system (i.e., 3 pts for favorite, 2 pts for 2nd favorite, and so on).**

Chapter 5 Ratios, Rates, and Proportions

Have you ever loved a pet so much that you wanted a statue made of it? Imagine a statue of your pet on the front steps of your home. "Gee, what a wise way to spend hard-earned money," your admiring neighbors would say. Or maybe not. In addition to being expensive, these statues would also be heavy. For instance, a 35-lb dog cast in gold would weigh about 670 lb.

Using Specific Gravity For the chapter project, you will find the weight of different animals and the weight of different metals. Your final project will be a table of animals with their weights, the weight of their statues in different materials, and the cost of the statues.

Go Online PHSchool.com
For: Information to help you complete your project
Web Code: ard-0561

Chapter 6 Percents

Your world is spinning. You are screaming. And you are loving every minute of it! Even though you are scared, you know that you will come to a safe stop at the end of the ride.

A successful amusement park attraction must be both fun and safe. Planners of amusement parks use a lot of math to create thrills but avoid any spills.

Take a Survey For the chapter project, you will decide which rides are most likely to be most popular. Your final product will be a recommendation about which rides to include in a proposed amusement park for your town.

Go Online PHSchool.com
For: Information to help you complete your project
Web Code: ard-0661

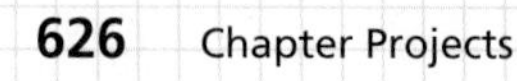

Chapter 7 Geometry

Look around you. Triangles are everywhere in construction! You see them in bridges, in buildings, in scaffolding: even in bicycle frames! This project will give you a greater appreciation of the importance of triangles in construction. You might also develop a taste for raisins!

Build a Tower For the chapter project, you will use toothpicks and raisins to build geometric shapes. Your final product will be a tower strong enough to support a baseball.

For: Information to help you complete your project
Web Code: ard-0761

Chapter 8 Measurement

Space is money! So before cargo is prepared for shipment in large containers, it is packaged in smaller containers based on its size and shape.

Cans of tuna are examples of items you buy in cylindrical containers. Would you pack two cylinders side-by-side or one above the other? One arrangement wastes cardboard! But which one?

Design Boxes for Shipping Cylinders For the chapter project, you will design boxes to hold cylindrical items. Your final product will be a model of a box that holds six cylinders.

For: Information to help you complete your project
Web Code: ard-0861

Raisin' the Roof

Students apply their knowledge of geometric shapes to build a tower.

Resources

All in One Teaching Resources
- Chapter 7 Project Support

Guided Instruction

Activating Prior Knowledge
Discuss with students that the triangle is a key structural concept that is applied to the construction of most structures. Ask:
- *How might you use the idea of triangles as supports to build your tower?* **One way: build a cube, each side of which is a square formed by four congruent right triangles.**

Shape Up and Ship Out

Students apply their knowledge of volume to design a box to hold cylinders.

Resources

All in One Teaching Resources
- Chapter 8 Project Support

Guided Instruction

Activating Prior Knowledge
Initiate a discussion of what food manufacturers consider when designing packages for their products. Ask:
- *Which is a more important design consideration, the package's size or its shape? Explain.* **Accept all reasonable responses students can justify.**

Diversity
Discuss the many different ways that people store, package, and transport goods in different cultures around the world.

Happy Landings

Students apply their knowledge of patterns and rules to create a graph showing how fast water empties from a container.

Resources

All in One Teaching Resources
- Chapter 9 Project Support

Guided Instruction

Activating Prior Knowledge
Discuss why different kinds of graphs are better suited than others to display certain kinds of data. Ask:
- *What graph do you think would work best to show the speed at which water empties from a container? Why?* **line graph; shows changes over time**

Science Connection
Have students research the speed an object falls to earth, ignoring air resistance, and report to the class.

People's Choice

Students apply their knowledge of sampling and percents to make a prediction.

Resources

All in One Teaching Resources
- Chapter 10 Project Support

Guided Instruction

Activating Prior Knowledge
Initiate a discussion in which students share their experiences with polling and sampling. Ask:
- *You wish to poll a sample of students in our school to find out how the full student body would react to a later start time. How would you constitute that sample?* **Sample: a mix of boys and girls, a mix of ages, a mix of classes**

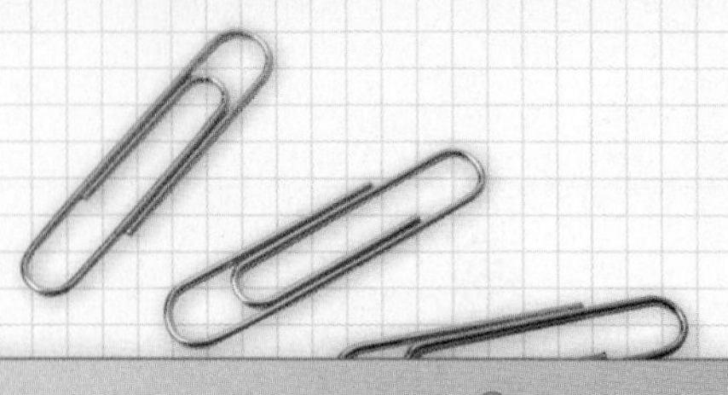

happy landings

Chapter 9 Patterns and Rules

Imagine this—you have just opened your parachute and you are floating through the air. Exciting, huh? How long it takes you to come to the ground can be predicted because the change in height versus time occurs in a predictable pattern. Many other things change in a predictable pattern, for instance, the height of a burning candle and the growth of money in a bank account.

Graphing Data For the chapter project, you will find how fast a container of water will empty if there is a hole in it. Your final project will be a graph of the data you collect.

Go Online PHSchool.com
For: Information to help you complete your project
Web Code: ard-0961

People's choice

Chapter 10 Graphing in the Coordinate Plane

You're an advertising executive, and you want to know which of three television shows is the most popular. So you plan to conduct a poll of viewers.

But how many viewers do you survey? Polling is expensive, so you don't want to poll too many. Polling too few viewers might give you the wrong information. Here's your chance to explore the process!

Finding a Sample Size Fill a container with three different kinds of beans. Use the beans to find the sample size that best predicts the percent of each kind of bean in the container.

Go Online PHSchool.com
For: Information to help you complete your project
Web Code: ard-1061

Chapter 11 Displaying and Analyzing Data

Chances are there's at least one person in a large crowd who has the same birthday as you! How many people do you think have the same favorite food? How many like the same television show? What kinds of cars do the people in the crowd have? Pollsters face questions like these all the time, and they take surveys to help answer them.

Estimate the Size of a Crowd and Take a Survey For the chapter project, you will use averages to estimate the size of a crowd. You will also take a survey and present your results in a graph.

For: Information to help you complete your project
Web Code: ard-1161

Everybody Wins

Chapter 12 Using Probability

Remember the game "Rock, Paper, Scissors"? It is an unusual game because paper wins over rock, rock wins over scissors, and scissors win over paper. You can use mathematics to create and investigate a situation with similar characteristics.

Make Three Number Cubes For this chapter project, you will design three number cubes A, B, and C, which have a surprising property: A usually beats B, B usually beats C, and C usually beats A. Your final step will be to construct your cubes.

For: Information to help you complete your project
Web Code: ard-1261

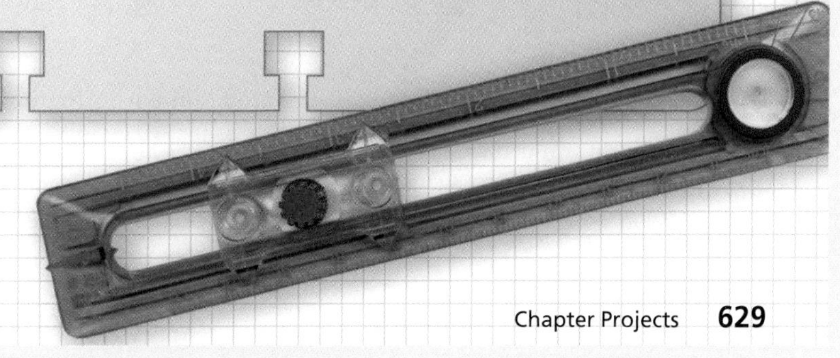

Chapter Projects

Too Many to Count

Students apply their knowledge of data analysis to estimate the size of a crowd and then graph the results.

Resources

All in One Teaching Resources
- Chapter 11 Project Support

Guided Instruction

Activating Prior Knowledge
Ask students to think about times when they were part of a large crowd, such as at a parade or at a ballgame. Ask:
- *How could you estimate the size of a large crowd?* **Sample: Use benchmarks—seating sections of a stadium, city blocks for a parade, and so on.**

Visual Learners
Have pairs of students spill a large container of counters, like coins or centimeter cubes, on a table. Ask them to find ways to divide up the counters in order to estimate how many there are.

Everybody Wins

Students apply their knowledge of probability to design a game with given attributes.

Resources

All in One Teaching Resources
- Chapter 12 Project Support

Guided Instruction

Activating Prior Knowledge
Have students talk about ways to design number cubes to meet the requirements of this project. Ask:
- *What do you think is meant by "usually"?* **Sample: That a win is more likely than not, or > 50%.**

CHAPTER 1

Exercises 1–4. Answers may vary. Samples:

1. 2
2. 3
3. 10
4. 70
5. 8.28
6. 5.23
7. 9.107
8. 2.699
9. 4.93
10. 7.917
11. 3.072
12. 46.8
13. 8.5
14. 6.3
15. 5.3
16. 23.1
17. 0.09
18. 0.0006
19. 2,700,000
20. 620
21. >
22. =
23. >
24. <
25. −100
26. 0
27. −6
28. −4
29. 0
30. −3
31. 28
32. −21
33. 45
34. −33
35. 5
36. −11

CHAPTER 1 Extra Practice

Skills

Lesson 1-1 **Use any estimation strategy to estimate.**

1. $2.7236 - 0.6512$ **2.** $2.4 + 0.86$ **3.** $106.3 \div 7.92$ **4.** 7.06×9.23

Lesson 1-2 **Find each sum or difference.**

5. $5.87 + 2.41$ **6.** $9.31 - 4.08$ **7.** $7.2 + 1.907$ **8.** $4.86 - 2.161$

Lessons 1-3 and 1-4 **Find each product or quotient.**

9. 2.9×1.7 **10.** $6.09 \cdot 1.3$ **11.** $7.68 \cdot 0.4$ **12.** $(5.2 \cdot 1.5) \cdot 6$

13. $30.6 \div 3.6$ **14.** $44.856 \div 7.12$ **15.** $17.172 \div 3.24$ **16.** $62.37 \div 2.7$

Lesson 1-5 **Write the number that makes each statement true.**

17. ■ L = 90 mL **18.** 0.6 mL = ■ L **19.** ■ mg = 2.7 kg **20.** ■ km = 620,000 m

Lesson 1-6 **Compare using <, =, or >.**

21. $|-3|$ ■ $|-2|$ **22.** $|10|$ ■ $|-10|$ **23.** $|-19|$ ■ $|9|$ **24.** $|-11|$ ■ $|-12|$

Lesson 1-7 **Find the value of each expression.**

25. $-110 + 5 - (-5)$ **26.** $3 - 6 + 3$ **27.** $(-3) + (-2) + (-1)$ **28.** $2 - 2 - 4$

29. $(-9) + 8 - (-1)$ **30.** $-7 + (12 - 8)$ **31.** $4 + 11 - (-13)$ **32.** $-14 + (-7)$

Lesson 1-8 **Find each product or quotient.**

33. $-5 \cdot (-9)$ **34.** $11 \cdot (-3)$ **35.** $-45 \div (-9)$ **36.** $\frac{-121}{11}$

Lesson 1-9 **Find the value of each expression.**

37. $2(8 - 45)$ **38.** $24 - (3 + 19)$ **39.** $3(21 + 7)$ **40.** $3(61 + 9)$

41. $(18 - 24)(7)$ **42.** $8 \cdot 6 - 47$ **43.** $(53 - 9) \div 11$ **44.** $27 \cdot (31 - 7)$

Lesson 1-10 **Find the mean, median, and mode for each situation.**

45. prices of different brands of cameras: \$150, \$100, \$240, \$220, \$195, \$225

46. number of seconds to run the 100-meter dash: 9, 13, 14, 11, 12, 12, 15, 14, 10, 13, 9, 12

Word Problems

Lesson 1-1

47. Weather In Chicago, Illinois, the average wind speed is 10.3 mi/h. In Great Falls, Montana, the average wind speed is 12.5 mi/h. What is the difference in wind speeds?

Lessons 1-2 and 1-3

48. The weight of your kitten is 4.5 kg. The weight of your friend's kitten is 2.7 kg. How much more does your kitten weigh?

49. Estimation Your new job pays $8.20 per hour. Estimate how much you will earn if you work for 4 hours.

Lesson 1-4

50. You buy four packages of shoe laces. You pay with a $10 bill and receive $3.24 in change. What is the price of each package?

Lesson 1-5

51. You are making a banner. You have a sheet of paper that is 2 m wide and want to leave a 35-cm blank margin on either side. How many centimeters can you use for text?

Lessons 1-6 through 1-8

52. Scores for a local golf tournament vary from 6 under par (-6) to 51 over par. Find the range of the scores.

53. Fitness Your pulse rate tells how fast your heart beats. Suppose your pulse after running is 160 beats per minute (bpm). After 4 min of rest, your pulse is 84 bpm. At what rate did your pulse decrease?

Lesson 1-9

54. You go with 5 friends to an art museum. The admission fee is $5.25 per person. A special exhibition costs an additional $4.75 per person. Use mental math to find the total cost.

Lesson 1-10

55. Find the mean of 2.4, 3.4, 6.1, 4.7, 2.9, 2.6, 3.3, 3.6, 2.7, and 3.2. Identify the outlier. Then find the mean without the outlier.

37. −74
38. 2
39. 84
40. 210
41. −42
42. 1
43. 4
44. 648
45. $188.\overline{3}$; 207.5; no mode
46. 12; 12; 12
47. 2.2 mi/h
48. 1.8 kg
49. $32.80
50. $1.69
51. 130 cm
52. 57
53. 19 bpm
54. $50
55. 3.49; 6.1; 3.2

Extra Skills and Word Problems

CHAPTER 2

1. 100
2. 125
3. 625
4. −36
5. 343
6. 7
7. 1
8. 4
9. 17
10. 2
11. 2
12. $\frac{7}{8}$
13. $\frac{13}{20}$
14. $\frac{1}{5}$
15. $\frac{1}{2}$
16. $\frac{2}{3}$
17. $\frac{22}{45}$
18. >
19. <
20. =
21. <
22. <
23. $\frac{63}{8}$
24. $\frac{26}{7}$
25. $\frac{13}{4}$
26. $\frac{22}{5}$
27. $\frac{61}{6}$
28. $\frac{12}{5}$
29. 0.8
30. $0.\overline{1}$
31. 0.875
32. 3.25
33. 3.5
34. $16.\overline{6}$
35. $-1.0, 0.35, \frac{3}{6}, \frac{9}{12}$
36. $-1.8, \frac{1}{4}, \frac{1}{3}, 3.5$
37. $0.375, \frac{1}{2}, 0.\overline{6}, \frac{10}{11}$
38. 5×10^3

CHAPTER 2 Extra Practice

Skills

Lesson 2-1 **Simplify. Use paper and pencil, a model, or a calculator.**

1. 100^1 2. 5^3 3. $(-5)^4$ 4. -6^2 5. $(12 - 5)^3$

Lesson 2-2 **Find the GCF of each pair of numbers.**

6. 35, 49 7. 11, 12 8. 28, 40 9. 17, 34 10. 16, 26 11. 16, 86

Lesson 2-3 **Write each fraction in simplest form.**

12. $\frac{21}{24}$ 13. $\frac{65}{100}$ 14. $\frac{15}{75}$ 15. $\frac{40}{80}$ 16. $\frac{72}{108}$ 17. $\frac{110}{225}$

Lesson 2-4 **Compare each pair of fractions. Use <, =, or >.**

18. $\frac{1}{4}$ ■ $\frac{2}{9}$ 19. $\frac{3}{7}$ ■ $\frac{1}{2}$ 20. $\frac{2}{5}$ ■ $\frac{4}{10}$ 21. $\frac{5}{6}$ ■ $\frac{7}{8}$ 22. $\frac{3}{5}$ ■ $\frac{2}{3}$

Lessons 2-5 and 2-6 **Write each mixed number as an improper fraction.**

23. $7\frac{7}{8}$ 24. $3\frac{5}{7}$ 25. $3\frac{1}{4}$ 26. $4\frac{2}{5}$ 27. $10\frac{1}{6}$ 28. $2\frac{2}{5}$

Write each fraction as a decimal.

29. $\frac{4}{5}$ 30. $\frac{1}{9}$ 31. $\frac{7}{8}$ 32. $\frac{13}{4}$ 33. $\frac{28}{8}$ 34. $\frac{100}{6}$

Lesson 2-7 **Order from least to greatest.**

35. $\frac{9}{12}, 0.35, \frac{3}{6}, -1.0$ 36. $-1.8, \frac{1}{4}, \frac{1}{3}, 3.5$ 37. $\frac{10}{11}, 0.\overline{6}, \frac{1}{2}, 0.375$

Lesson 2-8 **Write in scientific notation.**

38. 5,000 39. 160,000 40. 4,700,000 41. 7,900,000,000

Word Problems

Lesson 2-1

42. Your class is collecting pennies to give to a charity. You donate 2¢ on the first day. Each day, you double the amount you donate. How many pennies will you donate on the tenth day?

632

43. **Geometry** Let m represent the side length of one square. Let n represent the side length of another square. Find the total area of $m^2 + n^2$ for $m = 3$ and $n = 2$.

Lessons 2-2 and 2-3

44. Suppose you just fed your cat and bird. You feed your cat every 5 hours and your bird every 12 hours. In how many hours will you feed them at the same time again?

45. The shutter of a camera opens and closes quickly. For each exposure time of $\frac{1}{4}$ s, $\frac{1}{125}$ s, and $\frac{1}{250}$ s, write an equivalent fraction with a denominator of 1,000.

Lesson 2-4

46. The same number of oranges, apples, and pears are in a fruit basket. After one week, $\frac{3}{8}$ of the oranges, $\frac{1}{5}$ of the apples, and $\frac{1}{2}$ of the pears have not been eaten. Which fruit is most popular? Explain.

Lesson 2-5

47. **Baking** You can make a loaf of challah bread by braiding 6 strands of dough together. What is the number of loaves of bread you can make with 40 strands of dough? Write your answer as a mixed number.

48. The distance from your house to the mall is $18\frac{1}{2}$ miles. Write the distance as an improper fraction in fourths of a mile.

Lessons 2-6 and 2-7

49. **Surveys** When students are asked which enrichment class they prefer, 0.25 choose sign language, $\frac{10}{48}$ choose starting a business, $\frac{5}{12}$ choose robotics, and 0.125 choose origami. List their choices in decreasing order of preference.

50. A plant measures $6\frac{1}{12}$ in. Write this value as a decimal.

51. The amount of water in three different barrels is $3.\overline{6}$ gal, $\frac{16}{5}$ gal, and $3\frac{5}{6}$ gal. Which barrel contains the most water?

52. A baseball player has a batting average of .305. Write this decimal as a fraction in simplest form.

53. A stock price changes each day. The following list shows the daily price change for 7 days: 0.09, −0.70, −0.11, 0.3, 0.67, −0.28, 0.54. Order the price changes from least to greatest.

39. 1.6×10^5
40. 4.7×10^6
41. 7.9×10^9
42. $10.24
43. 13
44. 60 h
45. $\frac{250}{1,000}$ s, $\frac{8}{1,000}$ s, $\frac{4}{1,000}$ s
46. Apples; since $\frac{1}{5} < \frac{3}{8} < \frac{1}{2}$, the fruit with the least amount left is apples. So apples are most popular.
47. $6\frac{2}{3}$ loaves
48. $\frac{74}{4}$ mi
49. robotics, sign language, starting a business, origami
50. $6.08\overline{3}$ in.
51. $3\frac{5}{6}$-gal barrel
52. $\frac{61}{200}$
53. −0.70, −0.28, −0.11, 0.09, 0.3, 0.54, 0.67

CHAPTER 3

Exercises 1–8. Answers may vary. Samples:

1. about 1
2. about 2
3. about 1
4. about $1\frac{1}{2}$
5. about 18
6. about 10
7. about 65
8. about 99
9. $1\frac{1}{3}$
10. $\frac{2}{5}$
11. $\frac{1}{3}$
12. $\frac{11}{12}$
13. 7
14. $3\frac{3}{5}$
15. $7\frac{7}{8}$
16. $10\frac{13}{20}$
17. $\frac{3}{20}$
18. $\frac{1}{5}$
19. 25
20. $2\frac{3}{10}$
21. 3
22. 12
23. $2\frac{2}{9}$
24. $1\frac{7}{15}$
25. 12
26. $2\frac{1}{4}$
27. 660
28. $202\frac{2}{3}$
29. 13,000
30. $2\frac{1}{3}$
31. 25 g
32. $5\frac{1}{4}$ lb
33. 28 pt
34. 35.95 mL
35. 1,000 mg
36. 120 min

CHAPTER 3 Extra Practice

Skills

Lesson 3-1 **Use benchmarks to estimate each sum or difference.**

1. $\frac{2}{5} + \frac{7}{9}$
2. $\frac{3}{4} + \frac{5}{6}$
3. $\frac{3}{4} - \frac{1}{5}$
4. $\frac{8}{9} + \frac{7}{15}$
5. $9\frac{8}{10} + 8\frac{2}{10}$
6. $15\frac{2}{5} - 5\frac{4}{7}$
7. $71\frac{1}{5} - 5\frac{2}{3}$
8. $99\frac{9}{19} + \frac{1}{5}$

Lessons 3-2 and 3-3 **Find each sum or difference.**

9. $\frac{2}{3} + \frac{2}{3}$
10. $\frac{7}{10} - \frac{3}{10}$
11. $\frac{7}{12} - \frac{1}{4}$
12. $\frac{1}{6} + \frac{3}{4}$
13. $4\frac{3}{8} + 2\frac{5}{8}$
14. $5\frac{2}{5} - 1\frac{4}{5}$
15. $11 - 3\frac{1}{8}$
16. $7\frac{2}{5} + 3\frac{1}{4}$

Lessons 3-4 and 3-5 **Find each product or quotient.**

17. $\frac{3}{8} \cdot \frac{2}{5}$
18. $\frac{1}{4}$ of $\frac{4}{5}$
19. $\frac{5}{6}$ of 30
20. $2\frac{7}{8} \cdot \frac{4}{5}$
21. $\frac{3}{5} \div \frac{1}{5}$
22. $9 \div \frac{3}{4}$
23. $\frac{5}{6} \div \frac{3}{8}$
24. $3\frac{2}{3} \div 2\frac{1}{2}$

Lesson 3-6 **Complete.**

25. $\frac{3}{4}$ gal = ■ c
26. 4,500 lb = ■ t
27. $\frac{3}{8}$ mi = ■ yd
28. $12\frac{2}{3}$ lb = ■ oz
29. $6\frac{1}{2}$ t = ■ lb
30. 2 yd, 1 ft = ■ yd

Lesson 3-7 **Choose the more precise measurement.**

31. 25 g, 2.55 kg
32. 2 t, $5\frac{1}{4}$ lb
33. 28 pt, 15 qt
34. 7 L, 35.95 mL
35. 0.75 g, 1,000 mg
36. 120 min, 2 h

Word Problems

Lesson 3-1

37. **Estimation** Use benchmarks to estimate the total weight of two bags of cheese that weigh $\frac{1}{8}$ lb and $\frac{2}{5}$ lb.

38. **Estimation** From the following daily mileages, estimate the median number of miles a runner jogged: $9\frac{3}{5}$, $5\frac{1}{4}$, $7\frac{5}{10}$, $1\frac{7}{8}$, $6\frac{3}{4}$, $3\frac{2}{8}$.

Lessons 3-2 and 3-3

39. Write a number sentence for the model at the right.

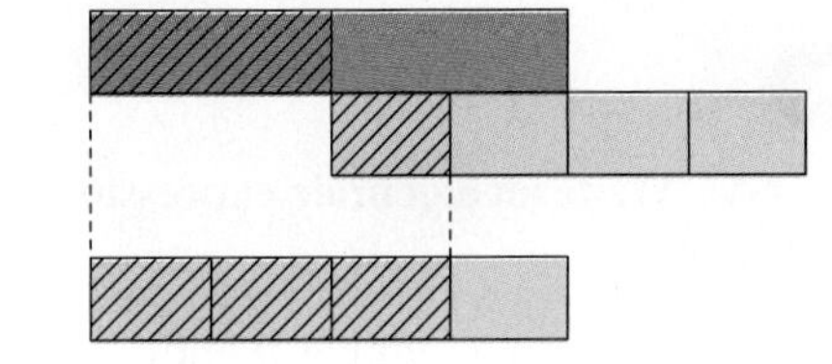

40. Geometry A side of a square is $\frac{4}{5}$ cm long. What is the perimeter of the square?

41. The Grand Canyon in Arizona has a maximum depth of $1\frac{1}{8}$ mi. The Black Canyon in Colorado has a maximum depth of $\frac{1}{2}$ mi. Find the difference between the depths.

42. You want to build a fence that has twice the perimeter of a triangular plot of land. The land plot has dimensions $1\frac{1}{2}$ yd, 2 yd, and $3\frac{1}{4}$ yd. What is the perimeter of the fence?

Lessons 3-4 and 3-5

43. At an apple orchard, you pick $10\frac{2}{5}$ lb of apples. You give $\frac{3}{8}$ of the apples to your friend. How many pounds of apples do you give your friend?

For each serving size in the table below, find the number of ounces and the number of Calories.

Nonfat Yogurt

	Servings	$\frac{1}{4}$	$\frac{1}{2}$	$\frac{3}{4}$	1	$1\frac{1}{2}$	2
44.	Ounces	a. ■	b. ■	c. ■	8	d. ■	e. ■
45.	Calories	a. ■	b. ■	c. ■	160	d. ■	e. ■

46. During a $2\frac{3}{4}$-hour assembly, each speaker spoke for $\frac{1}{4}$ of an hour. How many speakers were there?

47. Suppose you have $6\frac{4}{5}$ cantaloupes. How many fifths of a cantaloupe can you cut?

Lesson 3-6

48. A woman is 66 in. tall. What is her height in feet?

49. A marathon runner needs to run 137,280 feet to complete the race. How many miles is the race?

Lesson 3-7

50. Measurement You measure the length of a book to be 12 in. Your friend measures the same book to be 1 ft long. You know that 12 in. equals 1 ft. Does this mean that the two measurements are equally precise? Explain.

37. about $\frac{1}{2}$ lb
38. 6 mi
39. $\frac{1}{2} + \frac{1}{4} = \frac{3}{4}$
40. $3\frac{1}{5}$ cm
41. $\frac{5}{8}$ mi
42. $13\frac{1}{2}$ yd
43. $3\frac{9}{10}$ lb
44a. 2
b. 4
c. 6
d. 12
e. 16
45a. 40
b. 80
c. 120
d. 240
e. 320
46. 11 speakers
47. 34 fifths
48. $5\frac{1}{2}$ ft
49. 26 mi
50. No; an inch is a smaller unit of measure.

Extra Skills and Word Problems

CHAPTER 4

1. $n - 3$
2. $s + 17$
3. $d - 5$
4. $n \div 2 + 6$
5.

x	3(x − 1)	3x − 1	3x + 1
5	12	14	16
2	3	5	7

6. 6
7. 12
8. 23
9. 27
10. −16
11. −24
12. 47
13. 162
14. −75
15. 6
16. 40
17. −96
18. 5
19. 15
20. 5
21. 40
22. 6
23. −2
24. −4
25. 54
26.

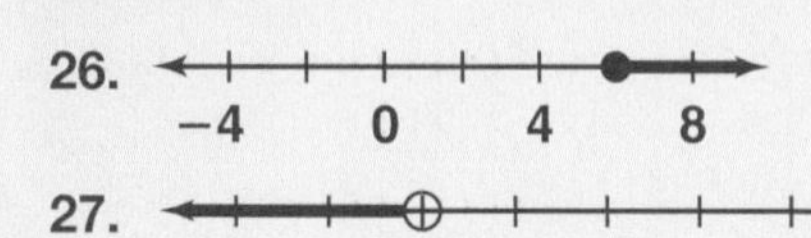

27.

28.

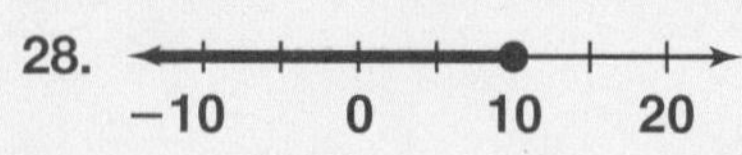

29.

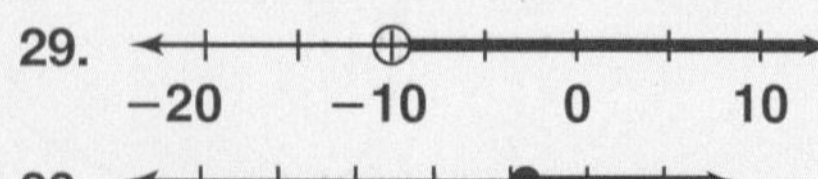

30.

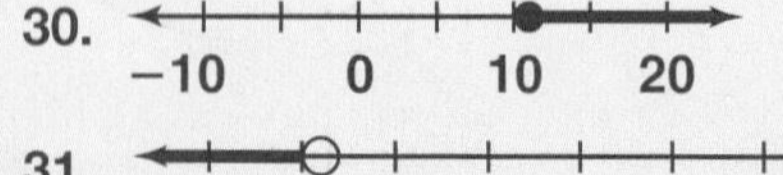

31.

32. −30 0 30 60

CHAPTER 4 Extra Practice

Skills

Lesson 4-1 Write an algebraic expression and draw a diagram for each word phrase.

1. The difference of a number n and 3
2. 17 more than s students
3. 5 fewer than d days
4. 6 more than the quotient of n and 2
5. Copy and complete the table at the right. Substitute the value on the left for the variable in the expression at the top of each column. Then evaluate.

x	3(x − 1)	3x−1	3x + 1
5	■	■	■
2	■	■	■

Lesson 4-2 Use mental math to solve each equation.

6. $5b = 30$
7. $n + 5 = 17$
8. $m - 8 = 15$
9. $\frac{z}{3} = 9$

Lesson 4-3 Solve each equation. Check your answer.

10. $t - 13 = -29$
11. $17 + d = -7$
12. $d + 112 = 159$
13. $y - 68 = 94$

Lesson 4-4 Solve each equation. Check your answer.

14. $\frac{m}{5} = -15$
15. $-7y = -42$
16. $0.4t = 16$
17. $\frac{x}{12} = -8$

Lesson 4-5 Solve each equation using number sense.

18. $7t + 5 = 40$
19. $2d - 12 = 18$
20. $5w - 18 = 7$
21. $\frac{z}{4} + 5 = 15$

Lesson 4-6 Solve each equation. Check your answer.

22. $\frac{r}{-6} + 4 = 3$
23. $12m + 24 = 0$
24. $-6g - 9 = 15$
25. $\frac{k}{-3} - 2 = -20$

Lessons 4-7 through 4-9 Solve each inequality. Graph the solution.

26. $y + 5 \geq 11$
27. $p + 7 < -3$
28. $a - 9 \leq 1$
29. $d - 3 > -13$
30. $3y \geq 33$
31. $\frac{p}{7} < -2$
32. $\frac{a}{-8} \leq -7$
33. $4d > -36$

Word Problems

Lesson 4-1

34. A dance club spends \$20 on advertising to promote its fall show. Write an algebraic expression for the amount of money left in the budget if the club starts with d dollars.

Lesson 4-2

35. Kobayashi set a world record for hot dog eating when he ate 50 hot dogs in 720 seconds. Write an equation and estimate the number of seconds it took him to eat one hot dog.

36. You mail a package at the post office. The postage costs $12.18. You pay with a $20 bill. Write an equation and estimate the amount of change you receive.

Lessons 4-3 through 4-6 **Write and solve an equation for each situation.**

37. A peregrine falcon can fly as fast as 220 mi/h. This speed is 150 mi/h faster than the maximum running speed of a cheetah. What is the speed of a cheetah?

38. A snorkeler looks at coral 4.5 feet below sea level, or at −4.5 ft. A scuba diver looks at coral located at 10 times that depth. How far below sea level does the scuba diver descend?

39. An auto rental agency offers a rate of $38 per day plus $.30/mile. After a one-day rental, Misha's bill was $74. How many miles did Misha drive?

40. A pair of running shoes costs $37 less than twice the cost of a pair of basketball sneakers. The sneakers cost $48.50. How much do the running shoes cost?

Lesson 4-7 **Write an inequality for each statement.**

41. The space shuttle can carry more than 38,000 pounds.

42. Today your break will be shorter than 15 minutes.

43. A song is less than 5 minutes long.

44. A shelf can hold at most 250 pounds.

Lessons 4-8 and 4-9 **Write and solve an inequality for each problem.**

45. A ride at an amusement park requires a rider to be at least 48 in. tall. Your little brother is 37 in. tall. How many inches must he grow in order to ride?

46. For your party, you plan a game where each player needs three spoons. You buy a box of 50 spoons. At most, how many people can play the game?

33. 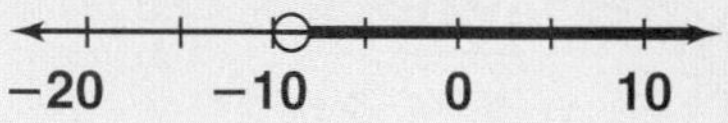

34. $d - 20$

35. about 14 s

36. about $8

37. Let s = speed of cheetah; $s + 150 = 220$; 70 mi/h.

38. Let d = depth of scuba diver; $\frac{d}{10} = -4.5$; −45 ft.

39. Let m = miles Misha drives; $38 + 0.3m = 74$; 120 mi.

40. Let r = cost of running shoes; $r = 2(48.50) - 37$; $60

41. $f > 38{,}000$

42. $b < 15$

43. $d < 5$

44. $p \leq 250$

45. $h + 37 \geq 48$; at least 11 in.

46. $3p \leq 50$; at most 16 people

Extra Skills and Word Problems

CHAPTER 5

1. 2 : 3; 2 to 3
2. $\frac{3}{5}$; 3 to 5
3. $\frac{4}{7}$; 4 : 7
4. 15 to 5; $\frac{15}{5}$
5. 25 to 50; 25 : 50
6. 48 fl oz for $3.19
7. 16 oz for $4.80
8. 6 pt for $7.14
9. 3 gal for $5.69
10. 32 oz for $4.99
11. 3 lb for $17.97
12. 5 yd for $20.30
13. 8 lb for $6.79
14. 0.5 L for $1.25
15. yes
16. no
17. yes
18. no
19. no
20. 20
21. 3
22. 3.5
23. 20
24. 2
25. $x = 8.56$, $y = 7.5$
26. $1.\overline{6}$
27. 450 ft
28. 600 ft
29. 300 ft
30. 1,200 ft
31. 465 ft
32. 37.5 ft
33. Yes; 15 out of 120 equals $\frac{1}{8}$, which is less than $\frac{3}{10}$.

Extra Practice

Skills

Lesson 5-1 Write each ratio in two other ways.

1. $\frac{2}{3}$ **2.** 3 : 5 **3.** 4 to 7 **4.** 15 : 5 **5.** $\frac{25}{50}$

Lesson 5-2 Find each unit price. Then determine the better buy.

6. soap: 32 fl oz for $2.29; 48 fl oz for $3.19
7. cereal: 12 oz for $3.95; 16 oz for $4.80
8. cider: 2 pt for $2.49; 6 pt for $7.14
9. milk: 1 gal for $1.99; 3 gal for $5.69
10. pasta: 8 oz for $1.29; 32 oz for $4.99
11. fish: 2 lb for $13.98; 3 lb for $17.97
12. fabric: 3 yd for $12.48; 5 yd for $20.30
13. rice: 2 lb for $1.89; 8 lb for $6.79
14. juice: 2 L for $5.99; 0.5 L for $1.25

Lesson 5-3 By using cross products, tell whether the ratios can form a proportion.

15. $\frac{4}{3}, \frac{12}{9}$ **16.** $\frac{8}{5}, \frac{11}{7}$ **17.** $\frac{21}{6}, \frac{7}{2}$ **18.** $\frac{6}{24}, \frac{2}{4}$ **19.** $\frac{50}{6}, \frac{3}{2}$

Lesson 5-4 Solve each proportion using cross products.

20. $\frac{12}{a} = \frac{3}{5}$ **21.** $\frac{n}{12} = \frac{4}{16}$ **22.** $\frac{7}{8} = \frac{n}{4}$ **23.** $\frac{7}{10} = \frac{14}{a}$ **24.** $\frac{7}{n} = \frac{17.5}{5}$

Lesson 5-5 Each pair of figures is similar. Find the value of each variable.

25. 6, 4, x — y, 5, 10.7

26.

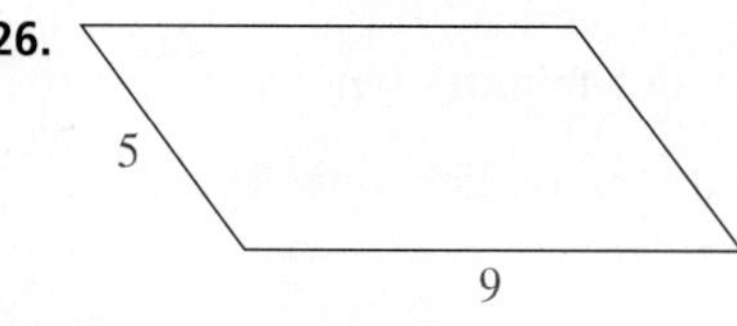

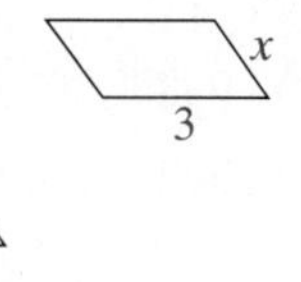

Lesson 5-6 The scale on a drawing is 0.5 in. : 15 ft. Find the actual length for each drawing length. Round to the nearest tenth, if necessary.

27. 15 in. **28.** 20 in. **29.** 10 in. **30.** 40 in. **31.** 15.5 in. **32.** 1.25 in.

Word Problems

Lesson 5-1

33. Nutrition The U.S. Department of Agriculture (USDA) recommends that no more than $\frac{3}{10}$ of your Calories come from fat. In a bowl of Tasty Crunch cereal, 15 out of 120 Calories are from fat. Is this within the USDA recommendation? Explain.

Lesson 5-2

34. Suppose you swim 500 yd in 3 min 20 s. What is your unit rate in seconds?

35. A bottle of 250 multivitamins costs $14.99. A bottle of 500 multivitamins costs $32.99. Which bottle is the better buy?

Lesson 5-3

36. For a game, you need 3 yellow marbles for every 8 red marbles. If you have 42 yellow marbles and 112 red marbles, do you have the appropriate numbers of marbles for the game? Explain.

Lesson 5-4

37. Business You sell packs of 12 pens for $3.48. At this rate, how much should you charge for a pack of 20 pens?

38. There are 385 mosquitoes in 11 ft^3 of a room. Predict the number in 15 ft^3.

39. There are 144 tulips in 8 m^2 of a garden. Predict the number in 25 m^2.

40. A recipe for fruit salad serves 4 people. It calls for 2 oranges and 16 grapes. You want to serve 10 people. How many oranges and grapes will you need?

Lesson 5-5

41. Geometry The ratio of the corresponding sides of two similar rectangles is 5 : 7. The smaller rectangle has a length of 3 cm and a width of 5 cm. Find the perimeter of the larger rectangle.

42. Indirect Measurement A student is 4 ft tall and his shadow is 3 ft long. A nearby building has a shadow 51 ft long. How tall is the building?

43. A woman stands near a telephone pole. She is 150 cm tall and her shadow is 3 m long. The shadow of the telephone pole is 30 m long. How tall is the telephone pole?

Lesson 5-6

44. The scale of a map is $\frac{1 \text{ cm}}{3.75 \text{ km}}$. Find the actual distance for a map distance of 8 cm.

45. The scale of a drawing is $\frac{1}{2}$ in. : 12 ft. Find the length of a drawing for an actual length of 84 ft.

Extra Skills and Word Problems

34. 2.5 yd/s
35. The bottle with 250 multivitamins is a better buy.
36. Yes; $\frac{3}{8}$ and $\frac{42}{112}$ are proportional because $3 \cdot 112 = 8 \cdot 42$.
37. $5.80
38. 525 mosquitoes
39. 450 tulips
40. 5 oranges and 40 grapes
41. $22\frac{2}{5}$ cm
42. 68 ft
43. 1,500 cm or 15 m
44. 30 km
45. $3\frac{1}{2}$ in.

Extra Skills and Word Problems

CHAPTER 6

Extra Practice

Skills

Lesson 6-1 Write each ratio as a percent.

1. $\frac{4}{5}$ 2. $\frac{11}{5}$ 3. $\frac{3}{25}$ 4. $\frac{19}{20}$ 5. $\frac{1}{10}$ 6. $\frac{3}{2}$

Lesson 6-2 Write each percent as a decimal.

7. 37.5% 8. 11.375% 9. 2.55% 10. 9% 11. 1.111% 12. 97.05%

Lesson 6-3 Write each percent as a fraction in simplest form.

13. 225% 14. 0.1% 15. 0.07% 16. 398% 17. 156% 18. 0.2%

Lesson 6-4 Find each answer using mental math.

19. 30% of 285 20. 51% of 326 21. 9% of 1,250 22. 49% of 88 23. 101% of 150

Lesson 6-5 Write a proportion and solve.

24. 54 is what percent of 135? 25. What percent of 48 is 2.4? 26. What percent of 200 is 120?
27. 8 is what percent of 20? 28. 32.5 is what percent of 130? 29. What percent of 150 is 27?

Lesson 6-6 Write and solve an equation to find the part of a whole.

30. 30% of 250 is what number? 31. What number is 90% of 70? 32. 45% of 200 is what number?
33. What number is 7% of 88? 34. 4% of 200 is what number? 35. What number is 22% of 1?

Lesson 6-7 Find each payment.

36. $75 with a 5% sales tax 37. $219 with a 3.5% sales tax 38. $85.65 with a 3% sales tax

Lesson 6-8 Find each percent of change. Round to the nearest tenth. State whether the change is an increase or a decrease.

39. 25 to 40 40. 95 to 45 41. 108 to 110 42. 50 to 95 43. 125 to 75
44. 8.5 to 10 45. 100 to 15 46. 63.5 to 20 47. 111 to 150 48. 25.9 to 30.2

Word Problems

Lesson 6-1

49. **Art** Draw and shade the first letter of your name on a 10-by-10 grid. Determine what percent of the grid is shaded.

CHAPTER 6

1. 80%
2. 220%
3. 12%
4. 95%
5. 10%
6. 150%
7. 0.375
8. 0.11375
9. 0.0255
10. 0.09
11. 0.01111
12. 0.9705
13. $2\frac{1}{4}$
14. $\frac{1}{1{,}000}$
15. $\frac{7}{10{,}000}$
16. $3\frac{49}{50}$
17. $1\frac{14}{25}$
18. $\frac{1}{500}$
19. 85.5
20. 166.26
21. 112.5
22. 43.12
23. 151.5
24. 40%
25. 5%
26. 60%
27. 40%
28. 25%
29. 18%
30. 75
31. 63
32. 90
33. 6.16
34. 8
35. .22
36. $78.75
37. $226.67
38. $88.22

Lesson 6-2 **The table shows the percent of the Recommended Daily Allowance for some of the nutrients in a 6-oz baked potato.**

Potato Facts

Nutrient	RDA
Magnesium	14%
Iron	34%
Vitamin B6	35%

SOURCE: National Institutes of Health

50. Write each percent as a fraction and as a decimal.

51. Nutrition Suppose you eat a 6-oz baked potato. What percent of each nutrient do you still need to meet the Recommended Daily Allowance?

Lessons 6-3 through 6-6

52. Environment A study has suggested that desert areas throughout the world could increase as much as 185% in the next 100 years due to global warming. Write the percent as a decimal and a fraction.

53. Four hundred students attended a school dance. If 35% of the students were boys, how many boys attended the dance?

54. The regular price of a backpack is \$34. A store sells the backpack at 30% off. What is the sale price?

55. An awards banquet is attended by 120 people. Ribbons are awarded for first, second, and third place in each of 25 categories. No one gets more than one ribbon. What percent of the people attending the banquet receive a ribbon?

56. A frozen yogurt shop sold 45 strawberry cones on Friday. This is 30% of the number of strawberry cones they sold that week. How many strawberry cones did they sell that week?

Lesson 6-7

57. You go to a stylist for a haircut. The cost of the haircut is \$12.50. Find the amount of a 15% tip for the stylist.

58. A salesperson earns a salary of \$2,500, plus 4% commission on sales of \$1,500. What are the salesperson's total earnings?

Lesson 6-8

59. If the cost of a dozen eggs rises from \$.99 to \$1.34, what is the percent of the increase?

60. A television is on sale for \$449.95. This is \$30 off the original price. Find the percent of the discount.

61. A bookstore pays \$4.25 for paperback books and charges \$9.95. What is the maximum percent of discount the store can offer, while making a profit of \$2 on each book? Check your answer.

39. 60% increase
40. 52.6% decrease
41. 1.9% increase
42. 90% increase
43. 40% decrease
44. 17.6% increase
45. 85% decrease
46. 68.5% decrease
47. 35.1% increase
48. 16.6% increase
49. Check students' work.
50. $\frac{7}{50}$, 0.14; $\frac{17}{50}$, 0.34; $\frac{7}{20}$, 0.35
51. Magnesium: 86%; Iron: 66%; Vitamin B6: 65%
52. 1.85; $\frac{37}{20}$
53. 140 boys
54. \$23.80
55. 62.5%
56. 150 cones
57. \$1.88
58. \$2,560
59. $35.\overline{35}$%
60. 6.25%
61. 37.2%

CHAPTER 7

1. $\overrightarrow{YZ}$
2. $\overline{SW}$
3. $\overleftrightarrow{CR}$
4. 65°; 155°
5. 75°; 165°
6. 5°; 95°
7. 44°; 134°
8. 115°
9. 90°
10. 67°
11. parallelogram; $\overline{AB}$ and $\overline{CD}$, $\overline{BC}$ and $\overline{AD}$; $\angle B$ and $\angle D$, $\angle A$ and $\angle C$
12. rectangle; $\overline{PS}$ and $\overline{QR}$, $\overline{PQ}$ and $\overline{RS}$; $\angle P$, $\angle Q$, $\angle R$, and $\angle S$
13. pentagon; $\overline{EF}$, $\overline{FG}$, $\overline{GH}$, $\overline{HJ}$, and $\overline{JE}$; $\angle E$, $\angle F$, $\angle G$, $\angle H$, and $\angle J$
14. rhombus; $\overline{LM}$, $\overline{MN}$, $\overline{NK}$, and $\overline{KL}$; $\angle L$ and $\angle N$, $\angle K$ and $\angle M$
15. $\angle C$
16. $\angle D$
17. $\overline{CD}$
18. $\overline{CE}$
19. 61.92°
20.

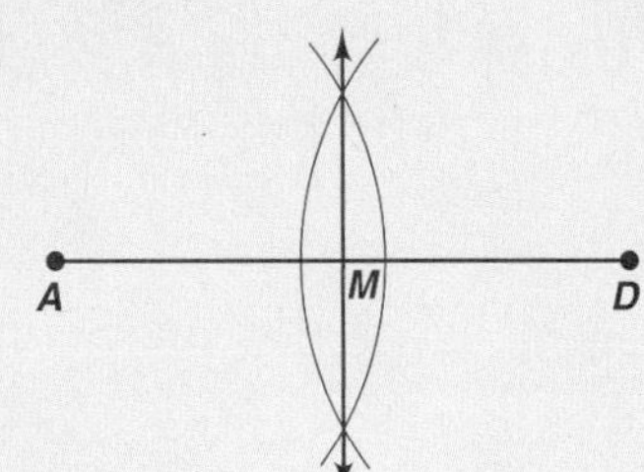

21.

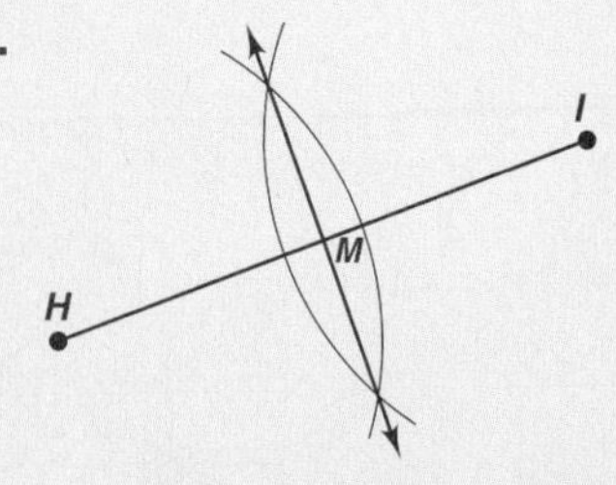

CHAPTER 7 Extra Practice

Skills

Lesson 7-1 **Name each segment, ray, or line.**

1. 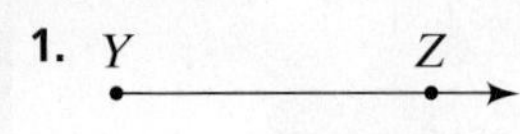

2. 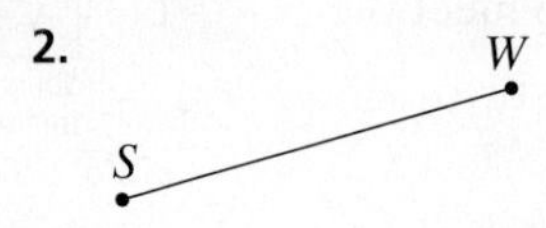

3. 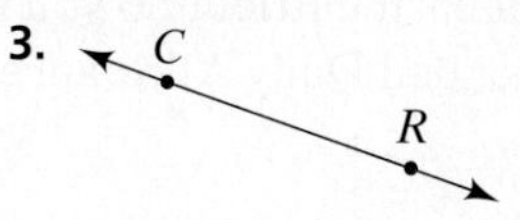

Lesson 7-2 **Find the measures of the complement and the supplement of each angle.**

4. $m\angle A = 25°$
5. $m\angle U = 15°$
6. $m\angle T = 85°$
7. $m\angle C = 46°$

Lesson 7-3 **Find x in each triangle.**

8. 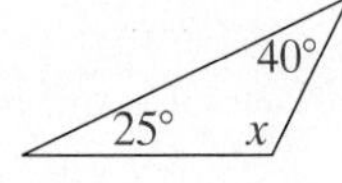

9. 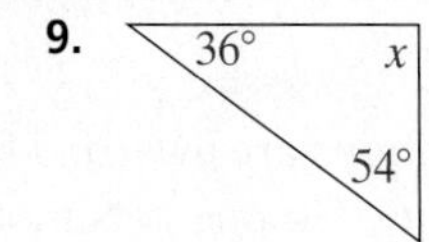

10. 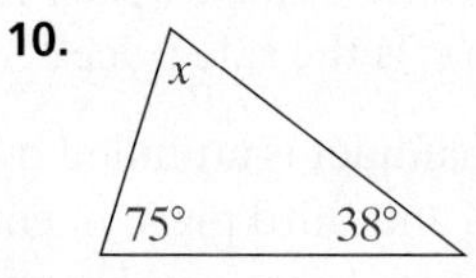

Lesson 7-4 **Classify each polygon. Then name the congruent sides and angles.**

11. 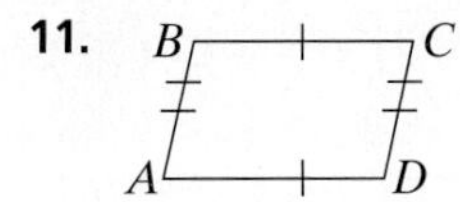

12. 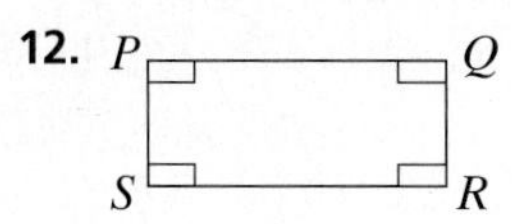

13. 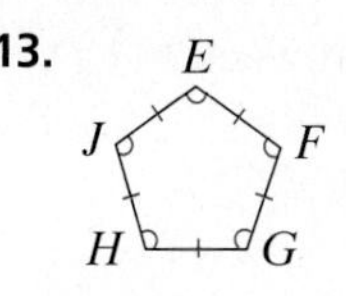

14. 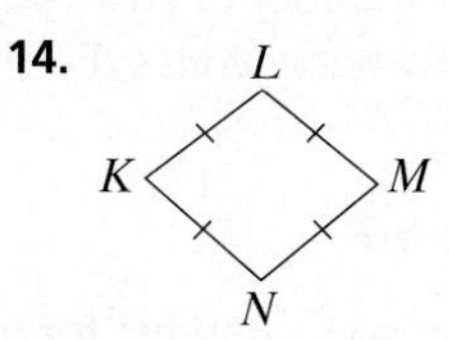

Lesson 7-5 **$\triangle ABD \cong \triangle CED$. Complete each congruence statement.**

15. $\angle A \cong$ ■
16. $\angle D \cong$ ■
17. $\overline{AD} \cong$ ■
18. $\overline{AB} \cong$ ■

Lessons 7-6 and 7-7

19. **Surveys** In a survey of 500 people, 86 preferred Brand A. Find the measure of the central angle that you would draw to represent Brand A in a circle graph.

Lesson 7-8 **Copy each segment. Then construct its perpendicular bisector.**

20. 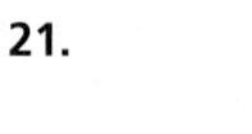

21. 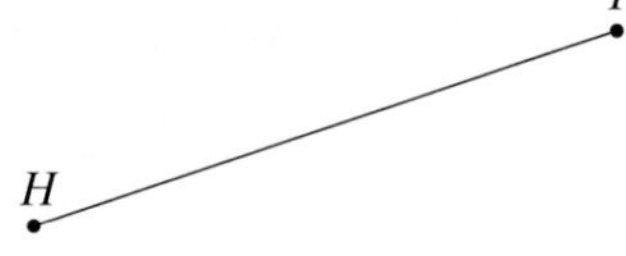

Word Problems

Lesson 7-1

22. Draw $\overline{AB}$, with $\overline{AC}$ on $\overline{AB}$.

23. Why are two lines drawn on a chalkboard *not* skew?

Lessons 7-2 and 7-3

24. Geography On a map, the measure of one of the angles formed by two intersecting roads is 79°. Find the measures of the complement and the supplement of the angle.

25. A triangular window has two angles that measure 32° and 116°. Find the measure of the third angle. Then classify the triangle.

Lesson 7-4

26. A frame for a house has two pairs of opposite sides that are parallel. What shapes can the frame be?

Lesson 7-5

27. Manufacturing Workers make sure that the same parts for a product manufactured on an assembly line are all congruent. To do this, they compare each part to a sample part. Are the triangles congruent?

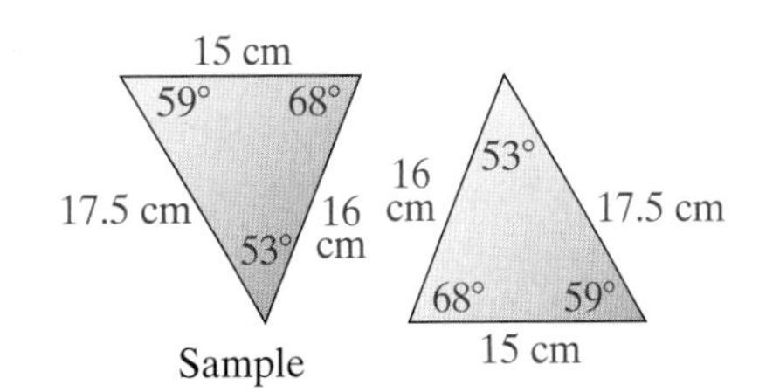

Lesson 7-6

28. Suppose your watch says 12:15 P.M. What kind of angle is the central angle formed by the two hands of your watch?

Lesson 7-7

29. Survey The data show the results of a survey on the preferred day for grocery shopping. Use the data to draw a circle graph.

Day	Percent	Day	Percent
Monday	4	Friday	17
Tuesday	5	Saturday	29
Wednesday	12	Sunday	7
Thursday	13	No preference	13

Lesson 7-8

30. Draw $\overline{CD}$ at least 3 in. long. Construct and label a segment that is one half as long as $\overline{CD}$.

Extra Skills and Word Problems

22. B, C, A

23. A chalkboard is a plane, so all lines on the chalkboard would be in the same plane.

24. 11°; 101°

25. 32°; obtuse isosceles triangle

26. parallelogram, rhombus, rectangle, square

27. yes

28. acute angle

29.

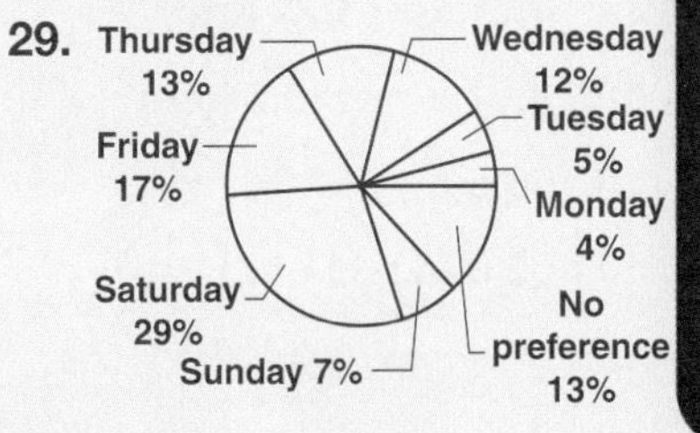

30.

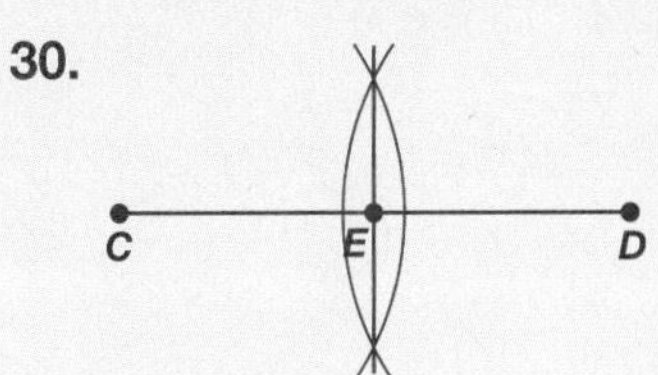

CHAPTER 8

1. 1,400 ft^2
2. 1,300 ft^2
3. 1,500 ft^2
4. 42 cm^2
5. 125 km^2
6. 180 m^2
7. 1,200 in.^{2}
8. 40 m^2
9. 240 ft^2
10. 26 mi^2
11. 4.71 km; 1.77 km^2
12. 37.7 cm; 113.04 cm^2
13. 56.5 m; 254.34 m^2
14. 62.8 in.; 314 in.^{2}
15. 1
16. 2
17. 7
18. 9
19. 30
20. 60
21. 8 m
22. 13 yd
23. 8 cm
24. triangular prism; 36 in.^{2}
25. cylinder; about 471 m^2
26. rectangular prism; 202 ft^2

CHAPTER 8 Extra Practice

Skills

Lesson 8-1 Estimate the area of each shaded region. Each square represents 100 ft^2.

1. 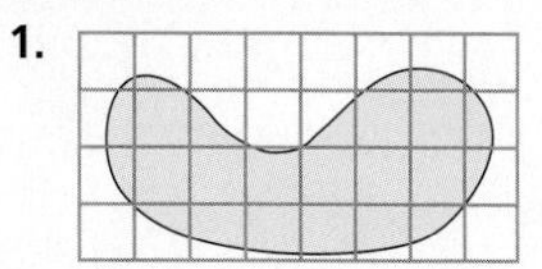

2. 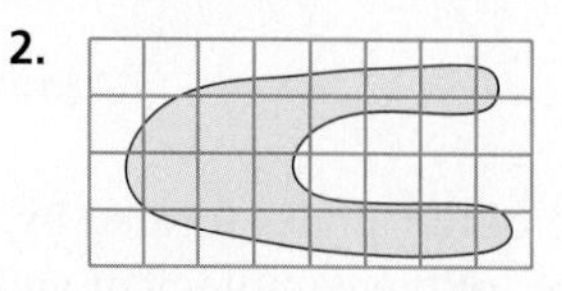

3. 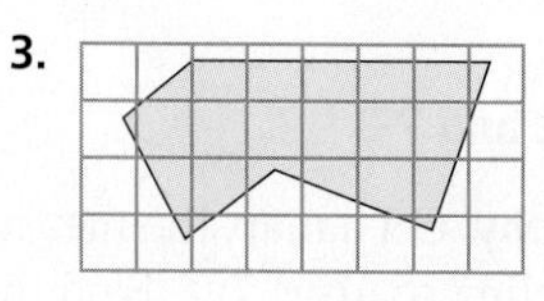

Lesson 8-2 Find each area for a triangle with base b and height h.

4. $b = 7$ cm, $h = 12$ cm
5. $b = 10$ km, $h = 25$ km
6. $b = 100$ m, $h = 3.6$ m
7. $b = 60$ in., $h = 40$ in.

Lessons 8-3 and 8-4 Find the area of each parallelogram or trapezoid.

8. 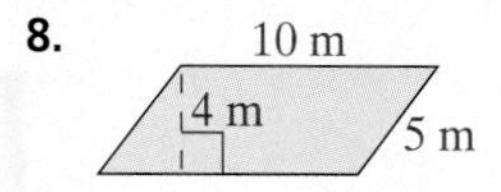

9. 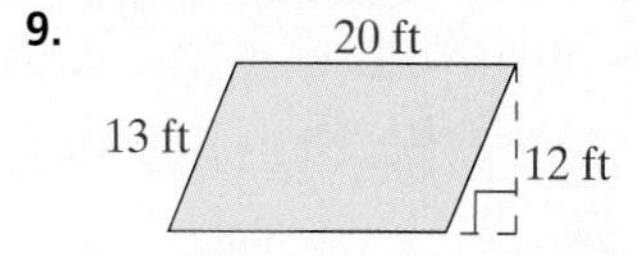

10. 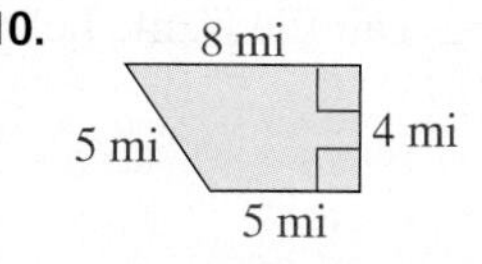

Lesson 8-5 Use $\pi \approx 3.14$ to estimate the circumference and area for each circle.

11. $d = 1.5$ km
12. $r = 6$ cm
13. $r = 9$ m
14. $d = 20$ in.

Lesson 8-6 Simplify each square root.

15. $\sqrt{1}$
16. $\sqrt{4}$
17. $\sqrt{49}$
18. $\sqrt{81}$
19. $\sqrt{900}$
20. $\sqrt{3,600}$

Lesson 8-7 Find each missing length.

21. 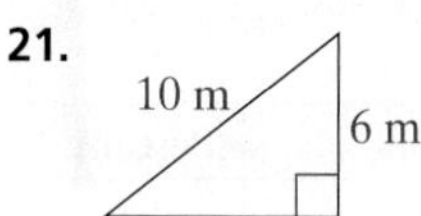

22. 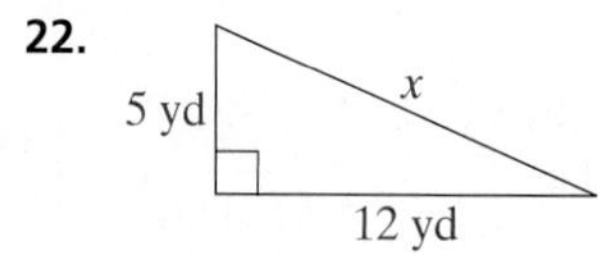

23. 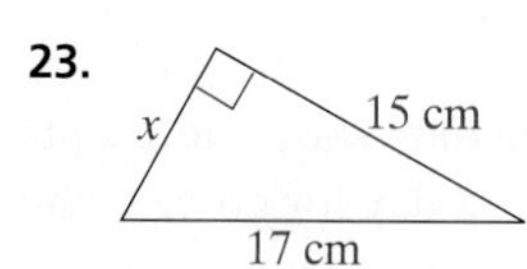

Lessons 8-8 and 8-9 Name each figure and find its surface area.

24. 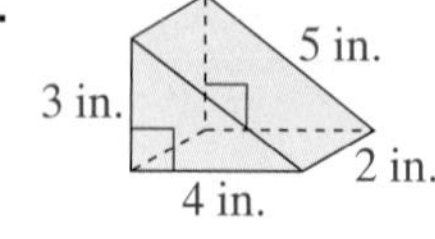

25. 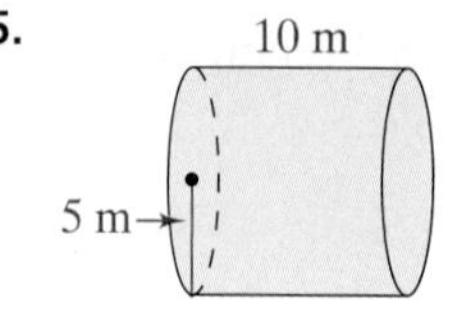

26. 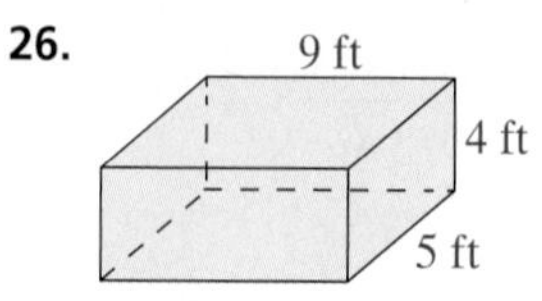

Lesson 8-10

27. Find the volume of the figures in Exercises 25 and 26.

Word Problems

Lessons 8-1 and 8-2

28. Which is a reasonable estimate for the perimeter of a school building—600 ft or 600 yd? Explain your choice.

29. **Patterns** Copy and complete the table by finding the perimeter and area of each rectangle. What happens to the perimeter and area of a rectangle when you double, triple, or quadruple the dimensions?

ℓ	w	P	A
3 in.	1 in.	■	■
6 in.	2 in.	■	■
9 in.	3 in.	■	■
12 in.	4 in.	■	■

Lessons 8-3 and 8-4

30. A triangular mirror is 12 in. wide and 37 in. tall. Find its area.

31. A park wall is designed to have the irregular shape shown. Use familiar figures to find the area of the park.

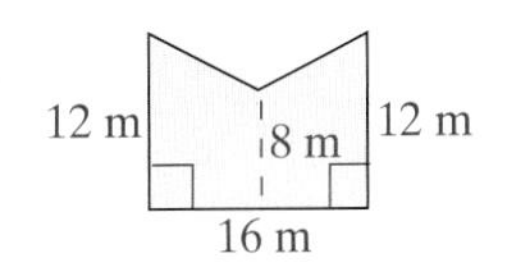

Lesson 8-5

32. A circular pool has a diameter of 20 ft. What area needs to be covered? Round to the nearest square foot.

Lessons 8-6 and 8-7

33. **Science** The distance that an object falls is given by the formula $d = 16t^2$, where d is the distance in feet and t is the time in seconds. A stone falls 1,600 ft into water. How long does it take for the stone to reach the water?

34. A ladder leans against a building. The bottom of the ladder is 5 ft from the building. The ladder reaches a window that is 17 ft from the ground. How long is the ladder? Round your answer to the nearest tenth.

Lesson 8-8

35. Use graph paper to draw a trapezoidal prism.

Lessons 8-9 and 8-10 **A jewelry box is 12 in. long, 7 in. wide, and 3 in. tall.**

36. Find the surface area of the jewelry box.

37. Find the volume of the jewelry box.

27. 785 m^3; 180 ft^3
28. Check students' work.
29. The perimeter doubles, triples, or quadruples; the area gets 4 times, 9 times, or 16 times as large.
30. 222 $in.^2$
31. 160 m^2
32. 314 ft^2
33. 10 s
34. 17.7 ft
35. Check students' work.
36. 282 $in.^2$
37. 252 $in.^3$

Extra Skills and Word Problems

CHAPTER 9

1.

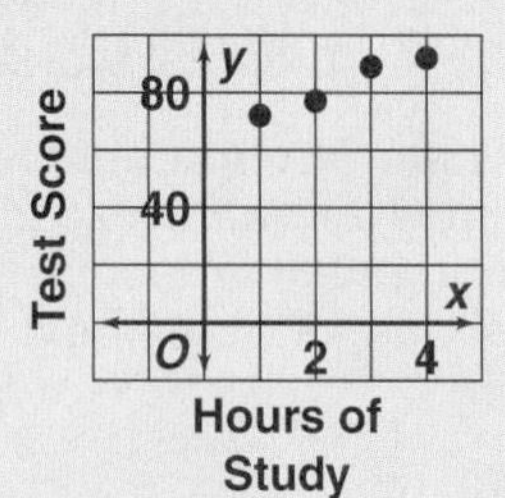

2.

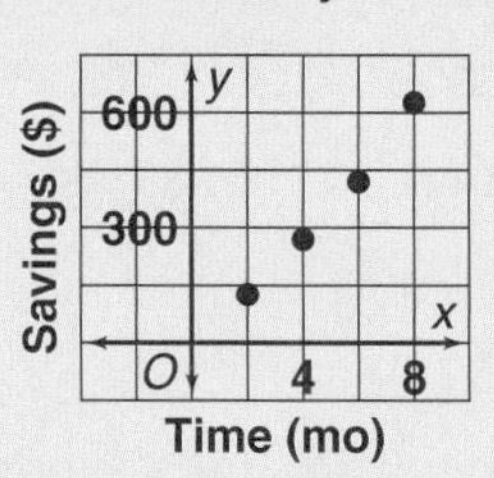

3. $q = 69$, $p = 27$;

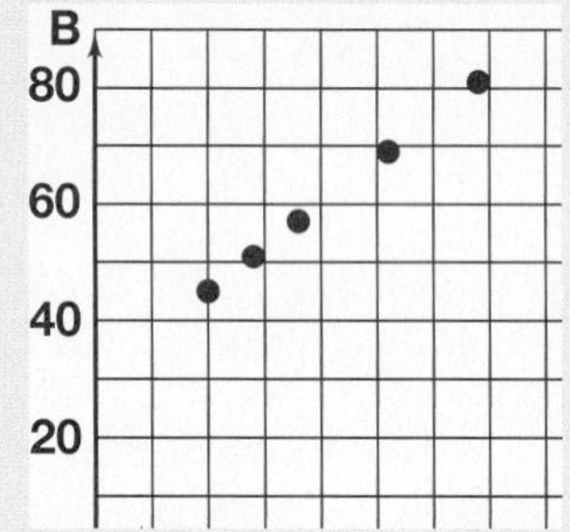

4. arithmetic
5. geometric
6. neither
7. $y = 6x$
8. $y = x - 2$
9. $y = 3x + 1$
10. $y = 10 - 2x$

11.

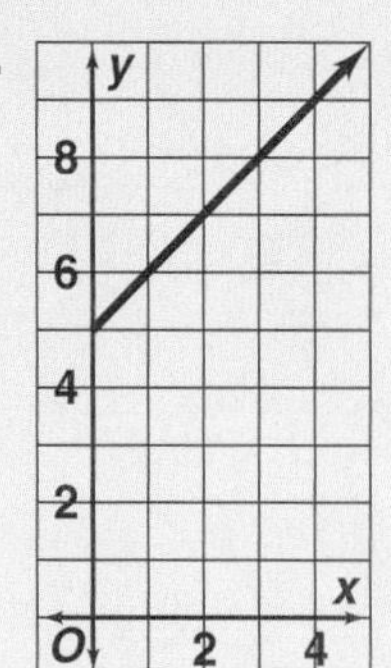

12.

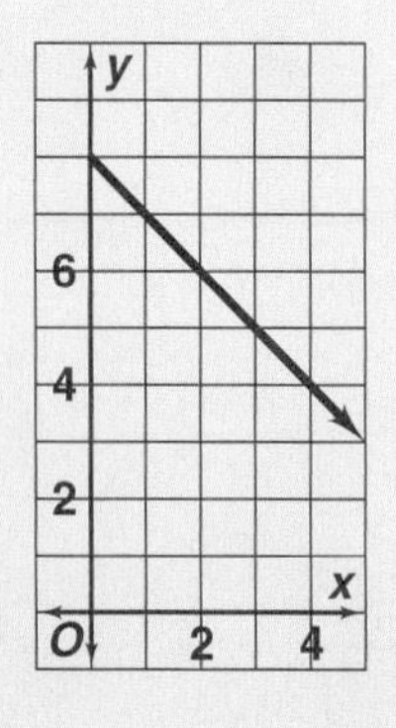

CHAPTER 9 Extra Practice

Skills

Lessons 9-1 and 9-3 **Graph the data in each table. In Exercise 3, first find the values of the variables.**

1.

Hours of Study	Science Test Score
1	72
2	77
3	89
4	92

2.

Time (mo)	Savings (dollars)
2	125
4	295
6	420
8	625

3.

A	B
15	45
17	51
19	57
23	q
p	81

Lesson 9-2 **Identify each sequence as *arithmetic, geometric,* or *neither.***

4. 7, 10, 13, 16, . . .
5. 800, 400, 200, 100, . . .
6. 50, 25, 48, 24, . . .

Lesson 9-4 **Write a rule for the function represented by each table.**

7.

x	y
0	0
1	6
2	12
3	18

8.

x	y
0	−2
1	−1
2	0
3	1

9.

x	y
0	1
1	4
2	7
3	10

10.

x	y
0	10
1	8
2	6
3	4

Lesson 9-5 **Graph each function. Use input values of 1, 2, 3, 4, and 5.**

11. $y = x + 5$
12. $y = 8 - x$
13. $y = 2x^2$

Lesson 9-6

14. On her trip to the library, Arlene walked two blocks to the bus stop in five minutes. She rode the bus for 15 min. The bus stopped three times for one minute each time. Sketch a graph to represent Arlene's trip.

Lesson 9-7 **Find the balance in each compound interest account.**

15. $1,000 principal
5% annual interest rate
4 years

16. $700 principal
4% annual interest rate
12 years

17. $1,500 principal
5.5% annual interest rate
7 years

Lesson 9-8 **Solve each equation for the variable in red.**

18. $V = \ell wh$
19. $y = mx + b$
20. $P = 2\ell + 2w$

Word Problems

Lesson 9-1

21. **Publishing** The table shows costs of printing books. Estimate the cost of printing 2,500 books and the cost of printing 7,500 books.

Number of Books	Cost ($)
5,000	175,000
10,000	290,000

Lesson 9-2

22. A boss pays new employees \$8/h the first year, \$9/h the second year, \$10/h the third year, and \$15/h the fourth year. Is the pattern *arithmetic, geometric, both,* or *neither*?

Lesson 9-3

23. There are 12 eggs in a dozen. Use this relationship to make a table that shows the number of eggs you have in 10, 20, 30, and 50 dozen eggs.

Lesson 9-4

24. Write a function rule that relates the number of miles m you travel in h hours if you drive at an average speed of 50 mi/h.

Lesson 9-5

25. **Geometry** The area of a square is a function of its side length. Write and graph a function rule to represent this concept. Describe the shape of your graph.

Lesson 9-6

26. Sketch a graph for the following situation. You run 3 blocks from the library and then walk 5 more blocks to your home. Show the distance on the vertical axis and time on the horizontal axis.

Lesson 9-7

27. **Finance** You invest \$1,000 at 5% for 9 months. How much simple interest do you earn?

Lesson 9-8

28. **Sports** The formula for a batting average a is $a = \frac{h}{n}$, where h is the number of hits and n is the number of times at bat. The highest major-league lifetime batting average was .366 by Ty Cobb, who had 4,189 hits. About how many times at bat did he have?

Extra Skills and Word Problems

13. 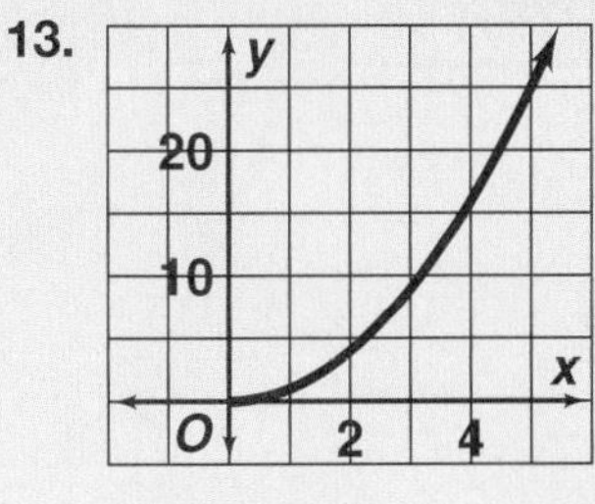

14. 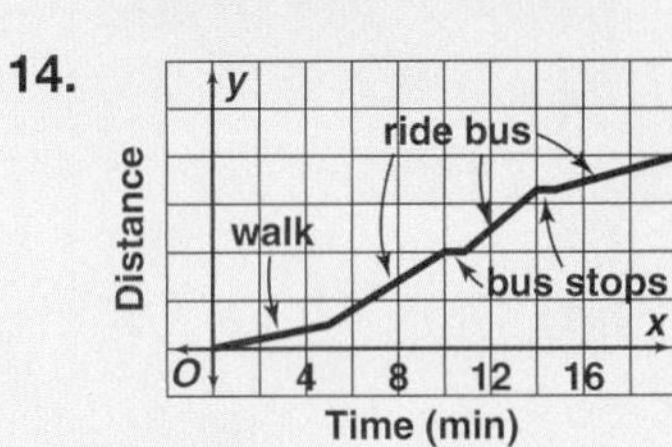

15. \$1,215.51
16. \$1,120.72
17. \$2,182.02
18. $h = \frac{V}{\ell w}$
19. $x = \frac{y - b}{m}$
20. $\ell = \frac{P - 2w}{2}$
21. about \$120,000; about \$235,000
22. neither
23.

Dozens	Number of Eggs
10	120
20	240
30	360
50	600

24. $m = 50h$
25. $A = s^2$

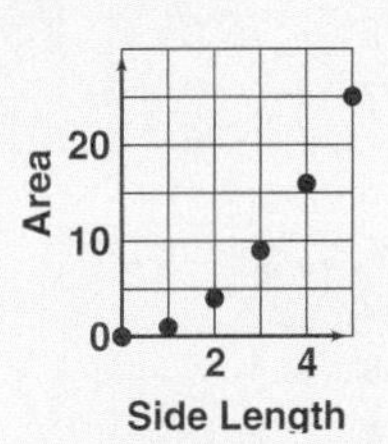

The graph is a curve that goes upward.

26. 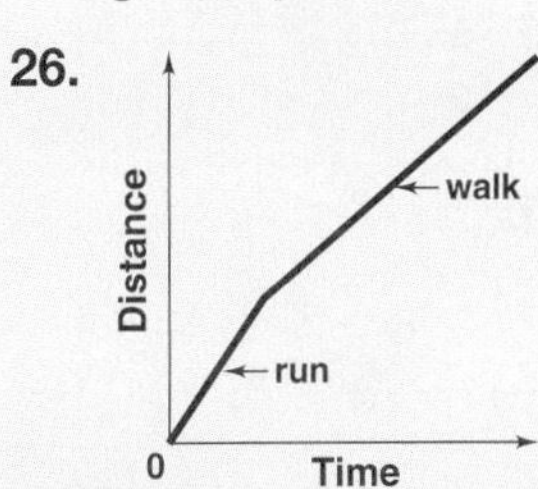

27. \$37.50
28. about 11,445 times at bat

CHAPTER 10

1. *G*
2. *E*
3. *B*
4. $-\frac{3}{4}$
5. $\frac{2}{3}$
6. −3
7. yes
8. no
9. no
10.

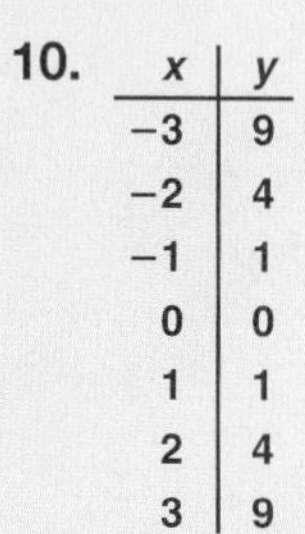

x	y
−3	9
−2	4
−1	1
0	0
1	1
2	4
3	9

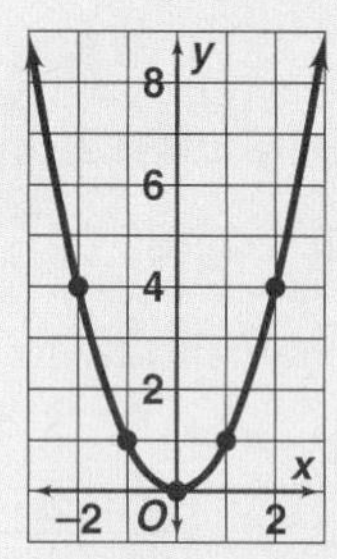

11.

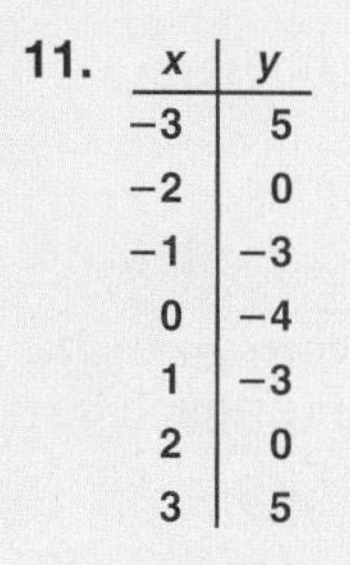

x	y
−3	5
−2	0
−1	−3
0	−4
1	−3
2	0
3	5

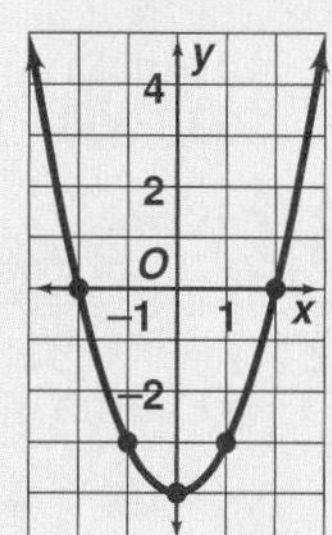

12.

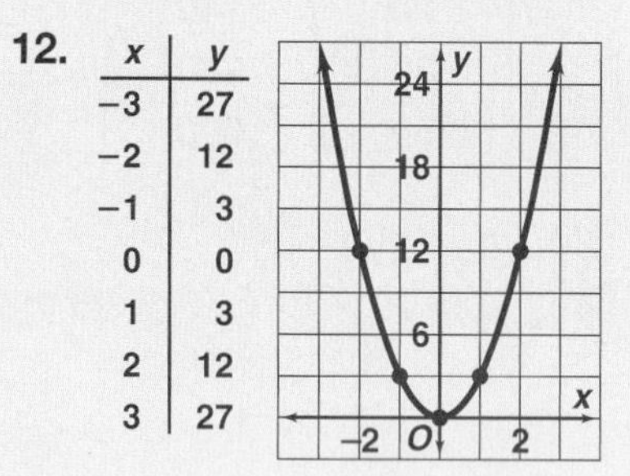

x	y
−3	27
−2	12
−1	3
0	0
1	3
2	12
3	27

13.

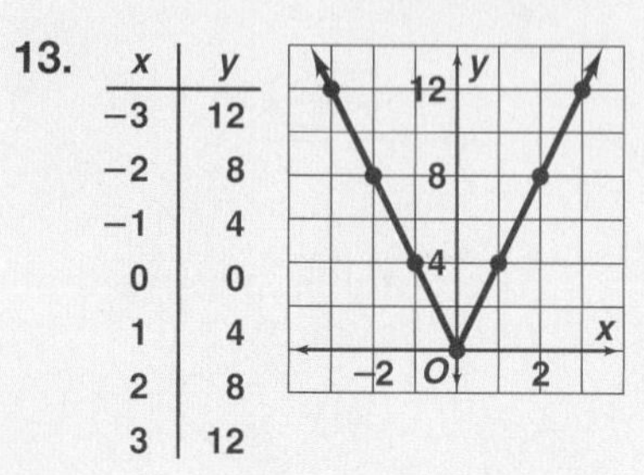

x	y
−3	12
−2	8
−1	4
0	0
1	4
2	8
3	12

14. $(x, y) \rightarrow (x - 5, y)$
15. $(x, y) \rightarrow (x + 6, y + 2)$
16. $(x, y) \rightarrow (x - 5, y + 3)$
17. reflection over the *x*-axis

CHAPTER 10 Extra Practice

Skills

Lessons 10-1 and 10-3 **Use the graph for Exercises 1–6.**

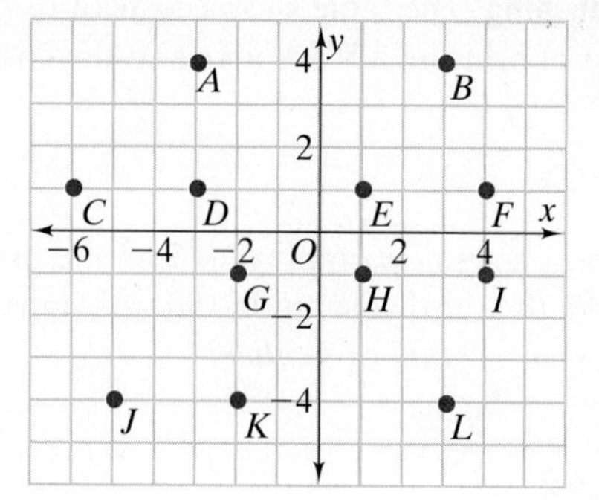

Name the point with the given coordinates.

1. $(-2, -1)$ 2. $(1, 1)$ 3. $(3, 4)$

Find the slope of the line through the points.

4. *A* and *E* 5. *G* and *E* 6. *B* and *F*

Lesson 10-2 **Tell whether each ordered pair is a solution of $y = x + 18$.**

7. $(2, 20)$ 8. $(22, 4)$ 9. $(36, -18)$

Lesson 10-4 **Make a table of solutions for each equation. Use integer values of *x* from −3 to 3. Then graph each equation.**

10. $y = x^2$ 11. $y = x^2 - 4$ 12. $y = 3x^2$ 13. $y = 4|x|$

Lesson 10-5 **Write a rule for the translation shown in each graph.**

14.

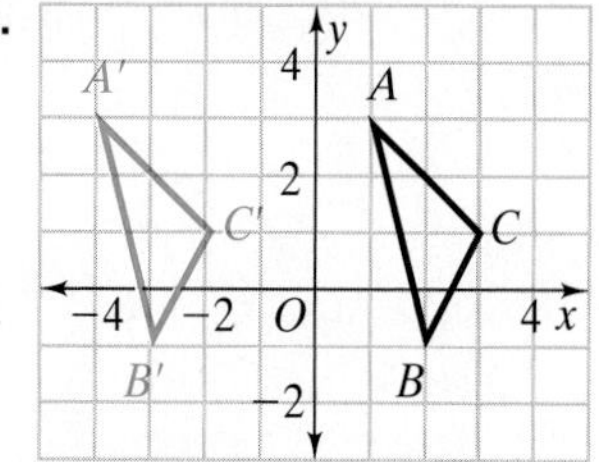

15.

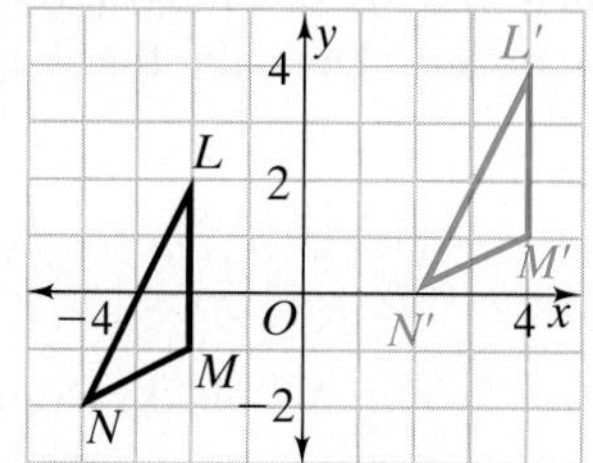

16.

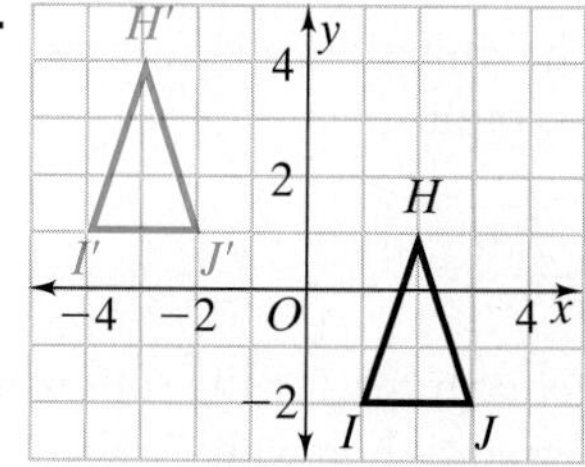

Lesson 10-6 **Use the graph from Exercises 1–6 to answer Exercise 17.**

17. Is the triangle formed by points *E, B,* and *F* a reflection, a rotation, or a translation of the image formed by points *H, L,* and *I*? Explain.

Lesson 10-7 **Graph the square *ABCD* with vertices $A(2, -3)$, $B(4, -5)$, $C(6, -3)$, and $D(4, -1)$. Then connect the vertices in order.**

18. Draw the image of *ABCD* after a rotation of 180° about point *A*.
19. Write the coordinates of the image of *ABCD*.

Word Problems

Lessons 10-1 and 10-2

20. **Geometry** Graph the points $G(1, 4)$, $H(-2, 4)$, $J(-2, -5)$, and $K(1, -5)$. Connect the points to form a polygon. Classify the polygon by its angles and its sides.

21. **Finance** You have \$35 saved. You plan to save \$5 each week from now on. The plan is modeled by $y = 5x + 35$. Graph the equation to find the amount of money you will have saved after 8 weeks.

22. **Aviation** A Learjet aircraft can travel at a speed of 400 mi/h. The equation $y = 400x$ represents the distance y traveled during x hours in flight. Graph the equation to find the distance traveled in 5 hours.

Lessons 10-3 and 10-4

23. Find the slopes of j and k in the graph at the right. Which line has a negative slope?

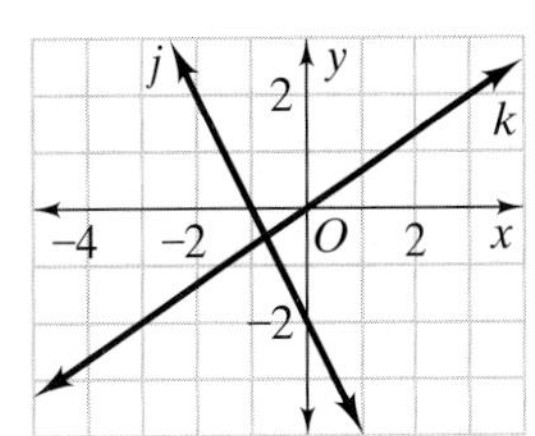

24. The slope of a piece of land is its grade. If the grade of a piece of land is 12%, what is its slope?

25. **Physics** An object falls from an initial height of 6,000 ft. The equation $h = -16t^2 + 6{,}000$ represents the height of the object, in feet, after t seconds. Make a table of values to find the height of the object after 0, 2, 4, and 10 seconds. Then graph the equation.

Lessons 10-5 through 10-7

26. You show a friend that the path you take to school can be seen as a translation on a coordinate plane. The point $M(3, 5)$ represents your house. The point $M'(2, -3)$ represents your school. Write a rule to describe your path to school.

27. Trace the figure at the right and draw the line(s) of symmetry. If there are no lines of symmetry, write *none*.

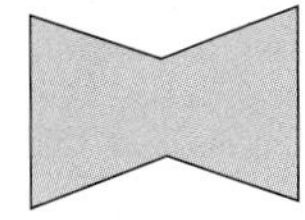

Figure II is the image of Figure I. Identify each transformation as a *translation*, a *reflection*, or a *rotation*.

28.

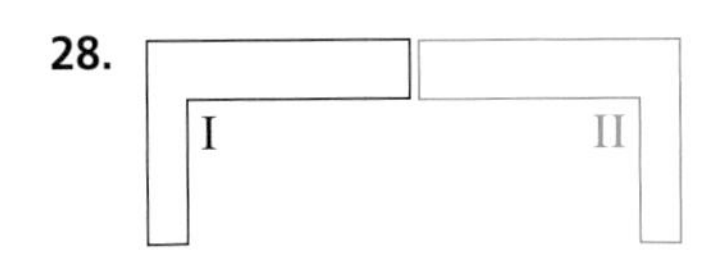

29.

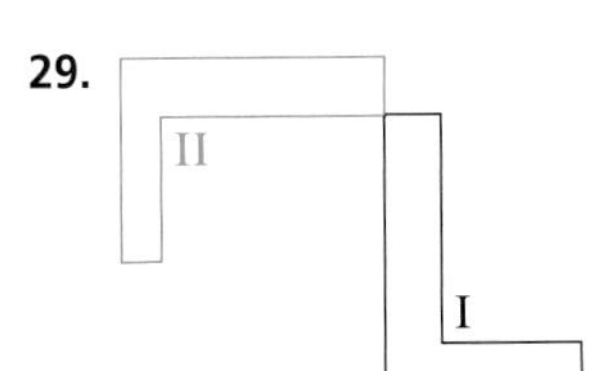

18.

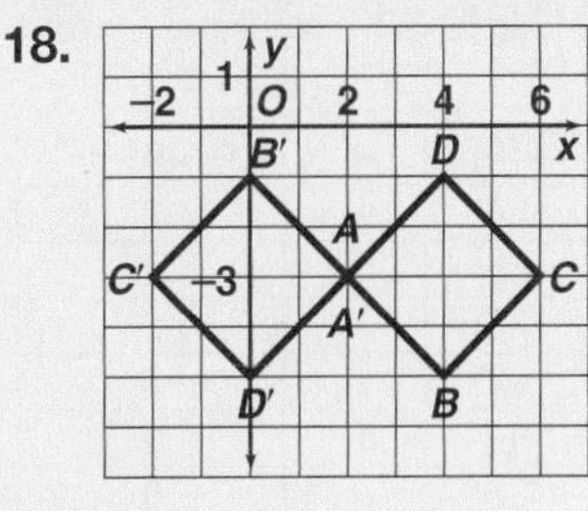

19. $A'(2, -3)$ $C'(-2, -3)$
$B'(0, -1)$ $D'(0, -5)$

20.

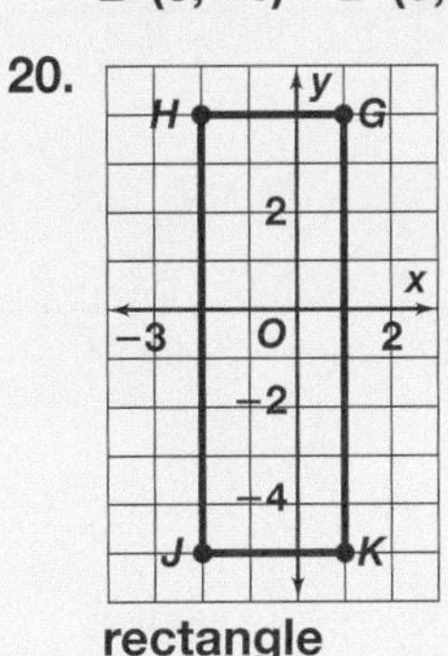

rectangle

21.

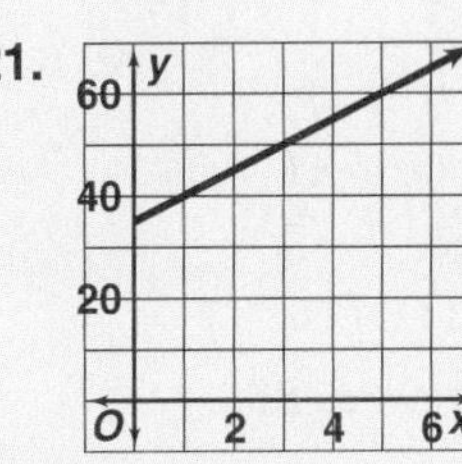

\$75

22.

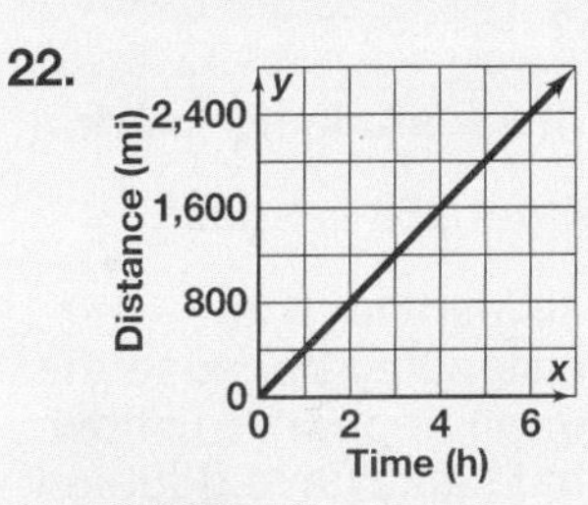

2,000 mi

23. slope of line j is -2, slope of line k is $\frac{2}{3}$, line j

24. $\frac{3}{25}$

25.

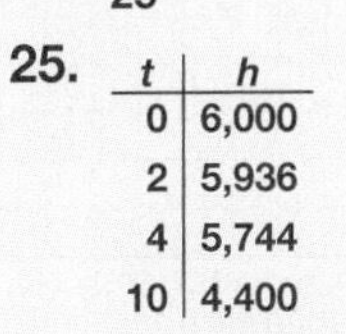

t	h
0	6,000
2	5,936
4	5,744
10	4,400

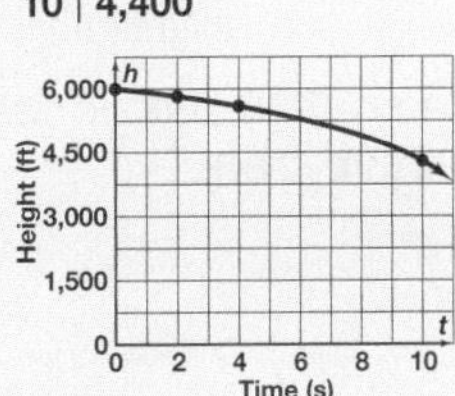

26. $(x, y) \rightarrow (x - 1, y - 8)$

27.

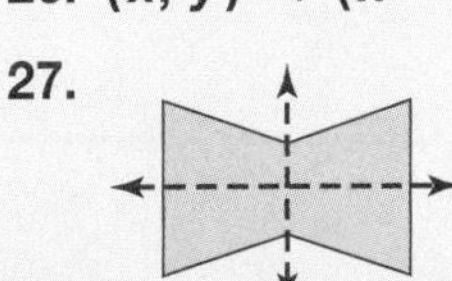

28. reflection

29. rotation

1.

Temp.	Tally	Frequency
84	II	2
85	IIII	4
86	II	2
87		0
88	III	3
89		0
90	II	2
91	I	1
92	I	1

2. 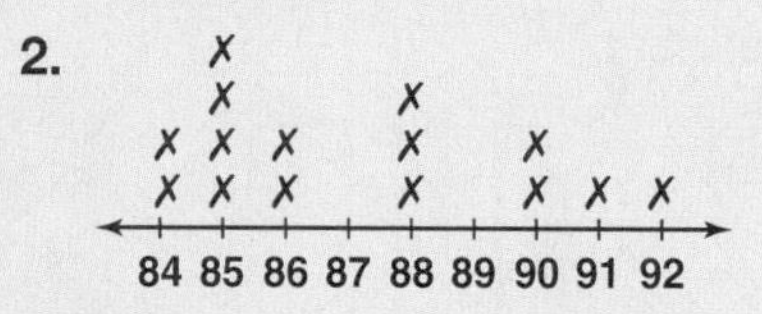

3. A

4. D

5. 18

6.

4 | 0 4 8
5 | 1 5 7 7
6 | 0 0 4
7 | 2 3 7
8 | 0 2

8 | 0 means 80

7. 42

8. Check students' work.

9. about 658 seagulls

10. Misleading; it is misleading because the graph appears to show that sales doubled from year 1 to year 5.

11. 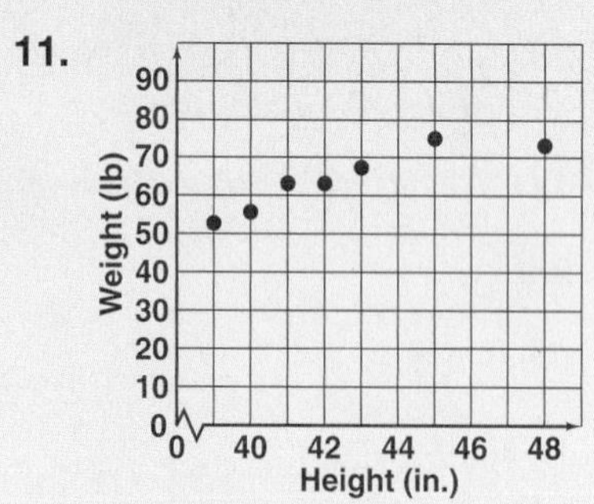

The scatter plot shows a positive trend.

CHAPTER 11 Extra Practice

Skills

Lesson 11-1 **Use the data below for Exercises 1 and 2.**
high temperatures (°F): 85, 88, 91, 84, 90, 85, 84, 90, 88, 86, 85, 92, 85, 86, 88

1. Make a frequency table.
2. Make a line plot.

Lesson 11-2 **Use the graph for Exercises 3–5.**

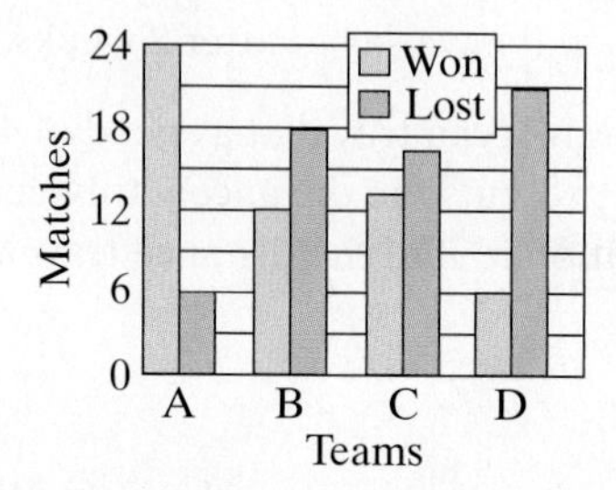

3. Which team has won the most matches?
4. Which team has won the fewest matches?
5. About how many more matches has Team A won than Team D?

Lesson 11-3 **The table shows the number of stories of some tall buildings in the United States. Use the table to answer Exercises 6 and 7.**

Number of Stories				
73	44	57	64	48
60	55	51	77	40
80	72	60	82	57

6. Draw a stem-and-leaf plot for the data.
7. What is the range of the data?

Lesson 11-4

8. Write a fair survey question and a biased survey question.

Lesson 11-5

9. Suppose 25 sea gulls were marked in a nesting area. Later, 500 sea gulls were counted, 19 of which were marked. Estimate the population.

Lesson 11-6 **The graph shows the growth of a company's DVD sales.**

Company DVD Sales

Number of DVDs Sold (millions): 0, 30, 40, 50, 60, 70, 80
Year: 1, 2, 3, 4, 5

10. Decide whether the graph appears misleading. If so, explain how the graph makes the misleading impression.

Lesson 11-7 **The data show the heights and weights of some 7-year-olds.**

Height	40 in.	39 in.	42 in.	45 in.	48 in.	43 in.	41 in.
Weight	56 lb	52 lb	62 lb	75 lb	72 lb	67 lb	62 lb

11. Make a scatter plot of the data. Describe any trend you see in the data.

Word Problems

Lesson 11-1

12. You survey students about how many times they ate at a restaurant last month. The results are shown. Make a histogram of the data.

How Many Meals Did You Eat Out Last Month?

Number of Meals	0–3	4–7	8–11	12–15
Frequency	8	12	3	7

13. You record the daily temperature (°F) for the month of June, Make a frequency table of the temperatures below.
82 81 88 88 87 92 91 85 92 83 82 84 86 89 90
87 86 84 83 84 87 86 90 91 87 86 80 84 91 90

Lesson 11-2 **Use the double bar graph.**

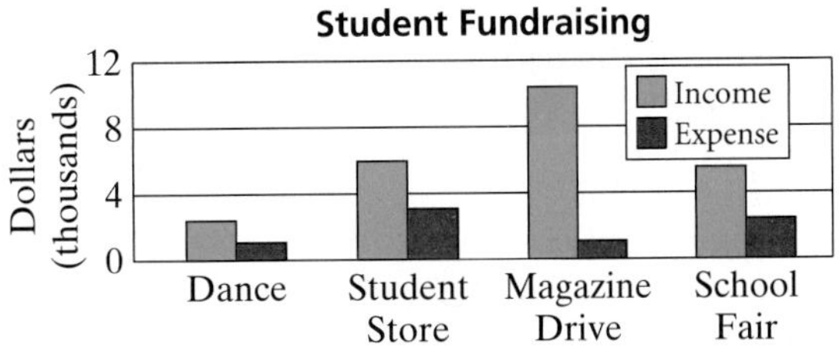

14. Which fundraisers had incomes of more than $4,000?

15. Which fundraiser made the most profit? The least profit?

Lessons 11-3 through 11-5

16. Science A botanist measures the heights of giant sunflowers in inches. Draw a stem-and-leaf plot for the height data below.
98 99 94 87 83 74 69 88 78 99 100 87 77.

17. An urban planner wants to know how road construction affects bus drivers. How can the planner survey a random sample of bus drivers?

18. Ecology Marine biologists are studying the otter population in a coastal region. There are 20 marked sea otters in the region. In a survey, the biologists count 42 sea otters, of which 12 are marked. About how many sea otters are in the area?

Lessons 11-6 and 11-7 **For Exercises 19 and 20, use the graph at right showing measles cases in the United States.**

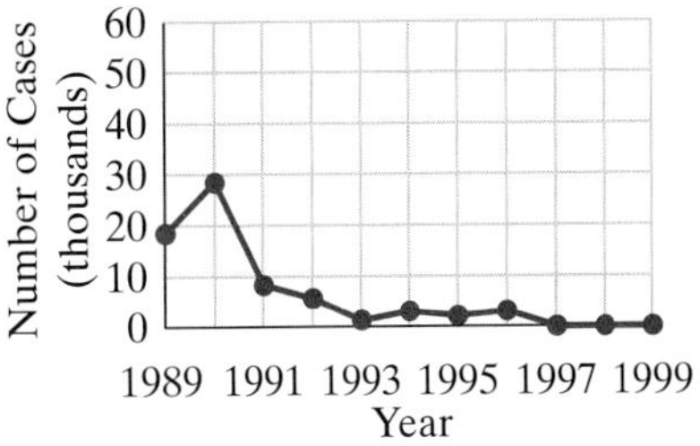

19. What does the graph suggest about the occurrence of measles in 1990?

20. Use the data to draw a graph that does not mislead.

21. What trend do you expect to see in a scatter plot in which the horizontal axis represents hours spent watching television and the vertical axis represents hours spent studying? Explain.

12.

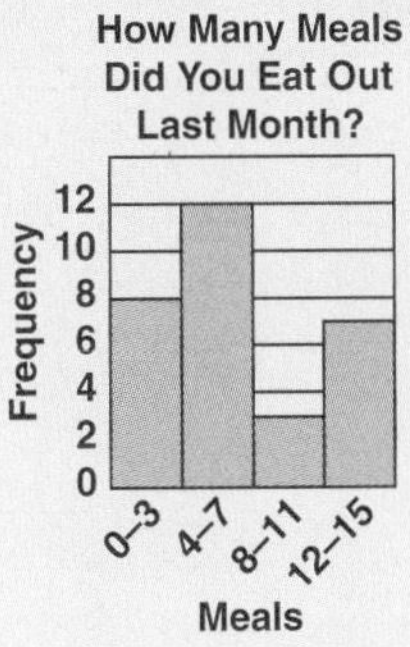

13.

Temp (F)	Tally	Frequency
80	|	1
81	|	1
82	||	2
83	||	2
84	||||	4
85	|	1
86	||||	4
87	||||	4
88	||	2
89	|	1
90	|||	3
91	|||	3
92	||	2

14. student store, magazine drive, school fair

15. magazine drive, dance

16. Heights of Giant Sunflowers (in.)

6	9
7	4 7 8
8	3 7 7 8
9	4 8 9 9
10	0

Key: 6 | 9 means 69

17. Get a list of bus drivers from each company and call every 10th driver.

18. about 70 sea otters

19. The graph suggests that the outbreak was not significant.

20.

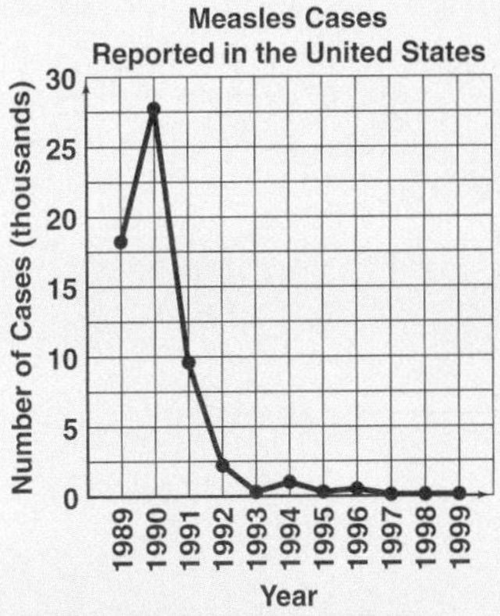

21. Negative; as the number of hours watching TV increases, the time spent studying will probably decrease.

Extra Skills and Word Problems

CHAPTER 12

1. $\frac{1}{6}$
2. $\frac{1}{3}$
3. $\frac{1}{2}$
4. $\frac{1}{20}$
5. $\frac{37}{41}$
6. $\frac{3}{28}$
7. Answers may vary. Sample: HHH, HHT, HTH, THH, TTT, TTH, THT, HTT, 8
8. Answers may vary. Sample:

	1	2	3	4	5	6
H	1H	2H	3H	4H	5H	6H
T	1T	2T	3T	4T	5T	6T

12

9. Answers may vary. Sample:

	E	F	G	H
A	EA	FA	GA	HA
B	EB	FB	GB	HB
C	EC	FC	GC	HC

12

10.

Coin 1	Coin 2	Spinner	Outcome
H	H	R	HHR
		W	HHW
		B	HHB
	T	R	HTR
		W	HTW
		B	HTB
T	H	R	THR
		W	THW
		B	THB
	T	R	TTR
		W	TTW
		B	TTB

$\frac{1}{12}$

CHAPTER 12 Extra Practice

Skills

Lesson 12-1 **You roll a number cube. Find each probability.**

1. rolling a 2
2. rolling a 3 or 5
3. rolling a 2, 4, or 6

Lesson 12-2

4. A quality control engineer at a factory inspected 300 glow sticks for quality. The engineer found 15 defective glow sticks. What is the experimental probability that a glow stick is defective?

The number of wins and losses for basketball teams are shown. Find each experimental probability.

5. Kingwood, Humble, Texas
Wins: 37, Losses: 4
Find $P(\text{Win})$.

6. Westchester, Los Angeles, Calif
Wins: 25 Losses: 3
Find $P(\text{Loss})$.

Lesson 12-3 **Make a table to show the sample space for each situation. Then find the number of outcomes.**

7. You toss three coins.
8. You spin a number 1 to 6 and toss a coin.
9. You choose one letter from each of the two sets of letters E, F, G, H and A, B, C.
10. You toss two coins and spin a spinner with three congruent sections colored red, white, and blue. Draw a tree diagram to find the sample space. Then find $P(\text{2 heads, then blue})$.

Lesson 12-4 **A bag contains 6 green marbles, 8 blue marbles, and 3 red marbles. Find *P(B)* after *A* has happened.**

11. *A*: Draw a green marble. Keep it.
B: Draw a red marble.

12. *A*: Draw a blue marble. Replace it.
B: Draw a red marble.

Lessons 12-5 and 12-6 **State whether the situation is a *permutation* or a *combination.* Then answer the question.**

13. In how many ways can a committee of 2 be chosen from 5 members?
14. In how many ways can a president and a treasurer be selected from a club of 5 members?

Word Problems

Lessons 12-1 and 12-2

15. You write the letters M, I, S, S, I, S, S, I, P, P, and I on cards and mix them in a hat. You select one card without looking. Find the probability of selecting an I.

16. Suppose you have 3 red, 3 black, and 3 blue marbles in your pocket. Does the probability of randomly selecting a black marble equal the probability of randomly selecting a blue one? Explain.

17. A quality-control inspector finds flaws in 6 of 45 tools examined. If the trend continues, what is the best prediction of the number of defective tools in a batch of 540?

18. Crispy Cereal offers one free prize in every box: a baseball card, a keychain, or a bracelet. You buy 3 boxes. Find the experimental probability of *not* getting a baseball card. Simulate the problem.

Lessons 12-3 and 12-4

19. Make a tree diagram for choosing one letter at random from each of two sets of letters: A, B, and C and W, X, Y, and Z. Then find the probability of choosing an A and a W.

20. A spinner has equal sections numbered 1 to 3. You spin the spinner 4 times. How many different outcomes are possible?

21. You drop a coin twice inside the rectangle. Estimate the probability that the coin lands in one of the circles both times. Use 3 for π.

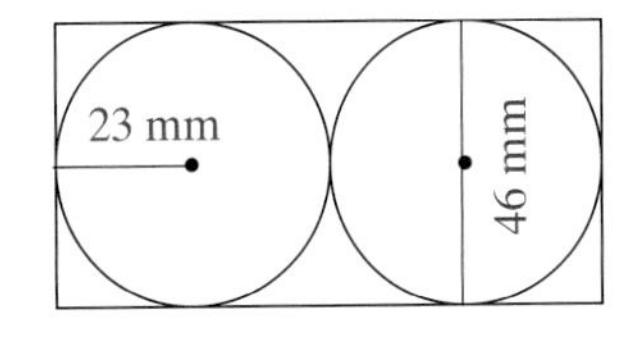

22. A volleyball team won 31 games and lost 4 games in one season. The coach has a summary sheet for each game. The coach selects two sheets at random to compare. Find P(win, then loss) without replacement.

Lessons 12-5 and 12-6

23. There are 20 teams that compete in a tournament. Find the number of different ways that two teams can finish in first and second place.

24. You scramble the letters P, A, and N. Make an organized list of the sample space. Then find the number of the groups that form real words. What are they?

25. You have six toppings to use for a pizza. How many different three-topping pizzas can you make?

26. A club of 50 people wants to select 4 members as representatives. How many different combinations of 4 people are possible?

11. $\frac{3}{16}$
12. $\frac{3}{17}$
13. combination; 10
14. permutation; 20
15. $\frac{4}{11}$
16. Yes; since you have the same number of blue marbles as black marbles, the number of favorable outcomes for each is the same.
17. about 72 defective tools
18. $\frac{8}{27}$
19. $\frac{1}{12}$
20. 81 outcomes
21. $\frac{9}{16}$
22. $\frac{62}{595}$
23. 380 ways
24. ANP, APN, NAP, NPA, PAN, PNA; 2; NAP, PAN
25. 20 pizzas
26. 230,300 combinations

Extra Skills and Word Problems

1. >
2. >
3. >
4. >
5. <
6. <
7. <
8. <
9. >
10. <
11. 3,347; 3,474; 3,734; 3,747; 3,774
12. 70,309; 70,903; 73,090; 73,909
13. 30,256,403; 30,265,403; 32,056,403; 302,056,403
14. 881,472; 881,872; 884,172; 887,142

Comparing and Ordering Whole Numbers

The numbers on a number line are in order from least to greatest.

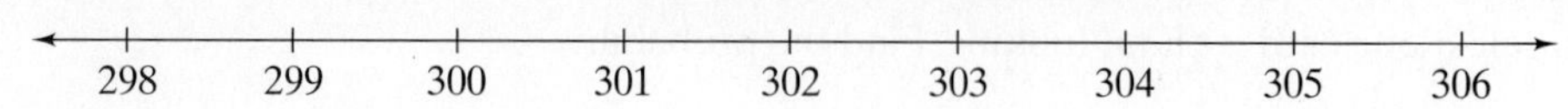

You can use a number line to compare whole numbers. Use the symbols > (is greater than) and < (is less than).

EXAMPLE

1 Use > or < to compare the whole numbers.

a. 303 ■ 299

303 is to the right of 299.

303 > 299

b. 301 ■ 305

301 is to the left of 305.

301 < 305

The value of a digit depends on its place in a number. Compare digits starting from the left.

EXAMPLE

2 Use > or < to compare the whole numbers.

a. 12,060,012,875 ■ 12,060,012,675

8 hundreds > 6 hundreds, so
12,060,012,875 > 12,060,012,675

b. 465,320 ■ 4,653,208

0 millions < 4 millions, so
465,320 < 4,653,208

Exercises

Use > or < to compare the whole numbers.

1. 3,660 ■ 360
2. 74,328 ■ 74,238
3. 88,010 ■ 8,101
4. 87,524 ■ 9,879
5. 295,286 ■ 295,826
6. 829,631 ■ 842,832
7. 932,401 ■ 932,701
8. 60,000 ■ 500,000
9. 1,609,372,002 ■ 609,172,002
10. 45,248,315,150 ■ 45,283,718,150

Order the numbers from least to greatest.

11. 3,747; 3,474; 3,774; 3,347; 3,734
12. 70,903; 70,309; 73,909; 73,090
13. 32,056,403; 302,056,403; 30,265,403; 30,256,403
14. 884,172; 881,472; 887,142; 881,872

Rounding Whole Numbers

You can use number lines to help you round numbers.

EXAMPLE

1 **a.** Round 7,510 to the nearest thousand.

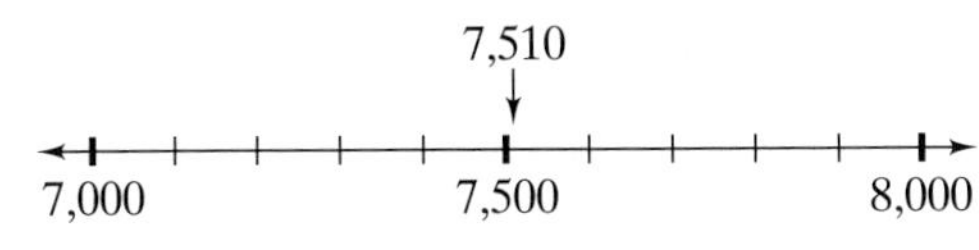

7,510 is between 7,000 and 8,000.

7,510 rounds to 8,000.

b. Round 237 to the nearest ten.

237 is between 230 and 240.

237 rounds to 240.

To round a number to a particular place, look at the digit to the right of that place. If the digit is less than 5, round down. If the digit is 5 or more, round up.

EXAMPLE

2 Round to the place of the underlined digit.

a. 3,4$\underline{6}$3,280

The digit to the right of the 6 is 3, so 3,463,280 rounds down to 3,460,000.

b. 28$\underline{9}$,543

The digit to the right of the 9 is 5, so 289,543 rounds up to 290,000.

Exercises

Round to the nearest ten.

1. 42 **2.** 89 **3.** 671 **4.** 3,482 **5.** 7,029 **6.** 661,423

Round to the nearest thousand.

7. 5,800 **8.** 3,100 **9.** 44,500 **10.** 9,936 **11.** 987 **12.** 313,591

13. 5,641 **14.** 37,896 **15.** 82,019 **16.** 808,155 **17.** 34,501 **18.** 650,828

Round to the place of the underlined digit.

19. 68,$\underline{8}$52 **20.** $\underline{4}$51,006 **21.** 3,40$\underline{6}$,781 **22.** 2$\underline{8}$,512,030 **23.** 71,2$\underline{2}$5,003

24. 96,$\underline{3}$59 **25.** 4$\underline{0}$1,223 **26.** $\underline{8}$,902 **27.** 3,6$\underline{7}$7 **28.** 2,551,$\underline{7}$50

29. 6$\underline{8}$,663 **30.** 70$\underline{1}$,803,229 **31.** 56$\underline{5}$,598 **32.** 32,$\underline{8}$10 **33.** 1,$\underline{0}$46,300

1. 40
2. 90
3. 670
4. 3,480
5. 7,030
6. 661,420
7. 6,000
8. 3,000
9. 45,000
10. 10,000
11. 1,000
12. 314,000
13. 6,000
14. 38,000
15. 82,000
16. 808,000
17. 35,000
18. 651,000
19. 68,900
20. 500,000
21. 3,407,000
22. 29,000,000
23. 71,230,000
24. 96,400
25. 400,000
26. 9,000
27. 3,680
28. 2,551,800
29. 69,000
30. 702,000,000
31. 566,000
32. 32,800
33. 1,000,000

Skills Handbook

1. 444
2. 315
3. 371
4. 640
5. 392
6. 520
7. 1,536
8. 3,663
9. 1,350
10. 1,700
11. 5,712
12. 1,809
13. 2,520
14. 2,255
15. 1,862
16. 52,287
17. 9,214
18. 30,150
19. 492
20. 2,025
21. 729
22. 822
23. 2,212
24. 5,040
25. 884
26. 2,790
27. 4,056
28. 2,838
29. 2,478
30. 2,808
31. 640
32. 2,034
33. 5,467

Multiplying Whole Numbers

When you multiply by a two-digit number, first multiply by the ones and then multiply by the tens. Add the products.

EXAMPLE

1 Multiply 62×704.

Step 1

$$\begin{array}{r} 704 \\ \times\ 62 \\ \hline 1408 \end{array}$$

Step 2

$$\begin{array}{r} 704 \\ \times\ 62 \\ \hline 1408 \\ 42240 \end{array}$$

Step 3

$$\begin{array}{r} 704 \\ \times\ 62 \\ \hline 1408 \\ +\ 42240 \\ \hline 43,648 \end{array}$$

EXAMPLE

2 Find each product.

a. 93×6

$$\begin{array}{r} 93 \\ \times\ 6 \\ \hline 558 \end{array}$$

b. 25×48

$$\begin{array}{r} 48 \\ \times\ 25 \\ \hline 240 \\ +\ 960 \\ \hline 1,200 \end{array}$$

c. 80×921

$$\begin{array}{r} 921 \\ \times\ 80 \\ \hline 73,680 \end{array}$$

Exercises

Find each product.

1. 74×6 **2.** 35×9 **3.** 53×7 **4.** 80×8 **5.** 98×4 **6.** 65×8

7. 512×3 **8.** 407×9 **9.** 225×6 **10.** 340×5 **11.** 816×7 **12.** 603×3

13. 70×36 **14.** 41×55 **15.** 38×49 **16.** 601×87 **17.** 271×34 **18.** 450×67

19. 6×82 **20.** 405×5 **21.** 81×9 **22.** 3×274 **23.** 553×4

24. 60×84 **25.** 52×17 **26.** 31×90 **27.** 78×52 **28.** 43×66

29. 826×3 **30.** 702×4 **31.** 5×128 **32.** 6×339 **33.** 781×7

Dividing Whole Numbers

First estimate the quotient by rounding the divisor, the dividend, or both. When you divide, after you bring down a digit, you must write a digit in the quotient.

EXAMPLE

Find each quotient.

a. $741 \div 8$

Estimate:
$720 \div 8 \approx 90$

$$\begin{array}{r} 92 \text{ R5} \\ 8\overline{)741} \\ -72 \\ \hline 21 \\ -16 \\ \hline 5 \end{array}$$

b. $838 \div 43$

Estimate:
$800 \div 40 \approx 20$

$$\begin{array}{r} 19 \text{ R21} \\ 43\overline{)838} \\ -43 \\ \hline 408 \\ -387 \\ \hline 21 \end{array}$$

c. $367 \div 9$

Estimate:
$360 \div 9 \approx 40$

$$\begin{array}{r} 40 \text{ R7} \\ 9\overline{)367} \\ -360 \\ \hline 7 \end{array}$$

Exercises

Divide.

1. $4\overline{)61}$	**2.** $8\overline{)53}$	**3.** $7\overline{)90}$	**4.** $3\overline{)84}$	**5.** $6\overline{)81}$
6. $6\overline{)469}$	**7.** $3\overline{)653}$	**8.** $8\overline{)645}$	**9.** $9\overline{)231}$	**10.** $4\overline{)415}$
11. $60\overline{)461}$	**12.** $40\overline{)213}$	**13.** $70\overline{)517}$	**14.** $30\overline{)432}$	**15.** $80\overline{)276}$
16. $43\overline{)273}$	**17.** $52\overline{)281}$	**18.** $69\overline{)207}$	**19.** $38\overline{)121}$	**20.** $81\overline{)433}$
21. $94\overline{)1,368}$	**22.** $62\overline{)1,147}$	**23.** $55\overline{)2,047}$	**24.** $85\overline{)1,450}$	**25.** $46\overline{)996}$
26. $94 \div 4$	**27.** $66 \div 9$	**28.** $90 \div 5$	**29.** $69 \div 6$	**30.** $58 \div 8$
31. $323 \div 5$	**32.** $849 \div 7$	**33.** $404 \div 8$	**34.** $934 \div 3$	**35.** $619 \div 6$
36. $777 \div 50$	**37.** $528 \div 20$	**38.** $443 \div 40$	**39.** $312 \div 40$	**40.** $335 \div 60$
41. $382 \div 72$	**42.** $580 \div 68$	**43.** $279 \div 43$	**44.** $232 \div 27$	**45.** $331 \div 93$
46. $614 \div 35$	**47.** $423 \div 28$	**48.** $489 \div 15$	**49.** $1,134 \div 51$	**50.** $1,103 \div 26$

1. 15 R1
2. 6 R5
3. 12 R6
4. 28
5. 13 R3
6. 78 R1
7. 217 R2
8. 80 R5
9. 25 R6
10. 103 R3
11. 7 R41
12. 5 R13
13. 7 R27
14. 14 R12
15. 3 R36
16. 6 R15
17. 5 R21
18. 3
19. 3 R7
20. 5 R28
21. 14 R52
22. 18 R31
23. 37 R12
24. 17 R5
25. 21 R30
26. 23 R2
27. 7 R3
28. 18
29. 11 R3
30. 7 R2
31. 64 R3
32. 121 R2
33. 50 R4
34. 311 R1
35. 103 R1
36. 15 R27
37. 26 R8
38. 11 R3
39. 7 R32
40. 5 R35
41. 5 R22
42. 8 R36
43. 6 R21
44. 8 R16
45. 3 R52
46. 17 R19
47. 15 R3
48. 32 R9
49. 22 R12
50. 42 R11

Skills Handbook

Place Value and Decimals

Each digit in a decimal has both a place and a value. The value of any place is one tenth the value of the place to its left. In the chart below, the digit 5 is in the hundredths place. So its value is 5 hundredths.

thousands	hundreds	tens	ones	.	tenths	hundredths	thousandths	ten-thousandths	hundred-thousandths
2	8	3	6	.	7	5	0	1	4

EXAMPLE

a. In what place is the digit 8?

hundreds

b. What is the value of the digit 8?

8 hundreds

Exercises

Use the chart above. Write the place of each digit.

1. 3 **2.** 4 **3.** 6 **4.** 7 **5.** 1 **6.** 0

Use the chart above. Write the value of each digit.

7. 3 **8.** 4 **9.** 6 **10.** 7 **11.** 1 **12.** 0

Write the value of the digit 6 in each number.

13. 0.162 **14.** 0.016 **15.** 13.672 **16.** 1,640.8 **17.** 62.135

18. 26.34 **19.** 6,025.9 **20.** 0.6003 **21.** 2,450.65 **22.** 615.28

23. 3.16125 **24.** 1.20641 **25.** 0.15361 **26.** 1.55736 **27.** 10.0563

Write the value of the underlined digit.

28. 2$\underline{4}$.0026 **29.** 14.9$\underline{3}$1 **30.** 5.78$\underline{9}$4 **31.** 0.$\underline{8}$7 **32.** 10.056$\underline{3}$

1. tens
2. hundred-thousandths
3. ones
4. tenths
5. ten-thousandths
6. thousandths
7. 3 tens
8. 4 hundred-thousandths
9. 6 ones
10. 7 tenths
11. 1 ten-thousandths
12. 0 thousandths
13. 6 hundredths
14. 6 thousandths
15. 6 tenths
16. 6 hundreds
17. 6 tens
18. 6 ones
19. 6 thousands
20. 6 tenths
21. 6 tenths
22. 6 hundreds
23. 6 hundredths
24. 6 thousandths
25. 6 ten-thousandths
26. 6 hundred-thousandths
27. 6 thousandths
28. 4 ones
29. 3 hundredths
30. 9 thousandths
31. 8 tenths
32. 3 ten-thousandths

Reading and Writing Decimals

A place value chart can help you read and write decimals. When there are no ones, write a zero before the decimal point.

billions	hundred millions	ten millions	millions	hundred thousands	ten thousands	thousands	hundreds	tens	ones	.	tenths	hundredths	thousandths	ten-thousandths	hundred-thousandths	millionths	Read
									0	.	0	7					7 hundredths
								2	3	.	0	1	4				23 and 14 thousandths
3	0	0	0	0	0	0	0	0	0	.	8						3 billion and 8 tenths
									5	.	0	0	0	1	0	2	5 and 102 millionths

EXAMPLE

a. Write thirteen ten-thousandths in numerals.

Ten-thousandths is 4 places after the decimal point. So the decimal will have 4 places after the decimal point. The number is 0.0013.

b. Write 1.025 in words.

The digit 5 is in the thousandths place. So 1.025 is one and twenty-five thousandths.

Exercises

Write a number for the given words.

1. three hundredths
2. twenty-one millions
3. six and two hundredths
4. two billion and six tenths
5. two and five hundredths
6. five thousand twelve
7. seven millionths
8. forty-one ten-thousandths
9. eleven thousandths
10. one and twenty-five millionths
11. three hundred four thousandths

Write each number in words.

12. 5,700.4
13. 3,000,000.09
14. 12.000069
15. 900.02
16. 25.00007
17. 0.00015

1. 0.03
2. 21,000,000
3. 6.02
4. 2,000,000,000.6
5. 2.05
6. 5,012
7. 0.000007
8. 0.0041
9. 0.011
10. 1.000025
11. 0.304
12. five thousand seven hundred and four tenths
13. three million and nine hundredths
14. twelve and sixty-nine millionths
15. nine hundred and two hundredths
16. twenty-five and seven hundred-thousandths
17. fifteen hundred-thousandths

Skills Handbook

1. 2.8
2. 3.8
3. 19.7
4. 401.2
5. 499.5
6. 3.9
7. 4.7
8. 20.4
9. 400.0
10. 130.0
11. 96.4
12. 125.7
13. 31.72
14. 14.87
15. 1.79
16. 0.11
17. 736.94
18. 9.61
19. 0.70
20. 4.23
21. 12.10
22. 5.77
23. 0.92
24. 4.00
25. 0.439
26. 0.065
27. 3.495
28. 8.071
29. 0.601
30. 6.007
31. 4
32. 10
33. 80
34. 105
35. 431
36. 0

Rounding Decimals

You can use number lines to help you round decimals.

EXAMPLE 1

a. Round 1.627 to the nearest tenth.

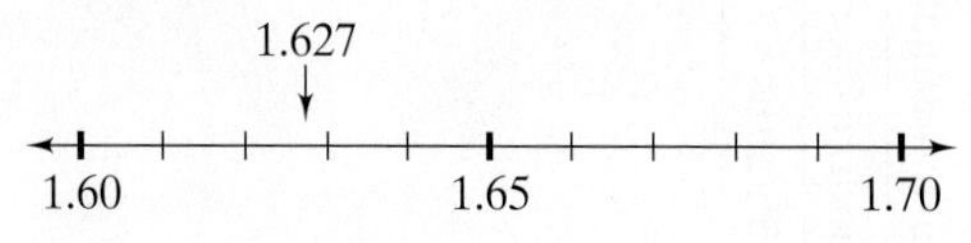

1.627 is between 1.6 and 1.7.

1.627 rounds to 1.6.

b. Round 0.248 to the nearest hundredth.

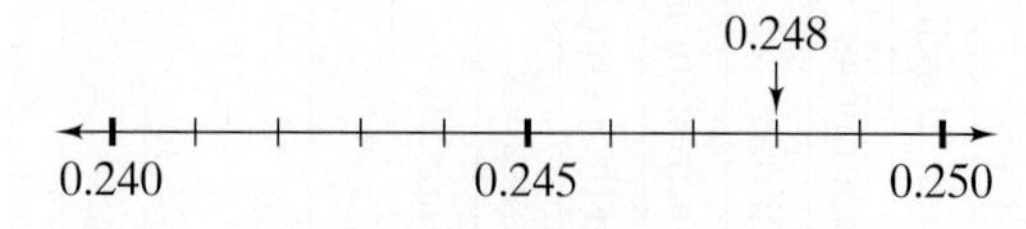

0.248 is between 0.24 and 0.25.

0.248 rounds to 0.25.

To round a number to a particular place, look at the digit to the right of that place. If the digit is less than 5, round down. If the digit is 5 or more, round up.

EXAMPLE 2

a. Round 2.4301 to the nearest whole number.

The digit to the right of 2 is 4, so 2.4301 rounds down to 2.

b. Round 0.0515 to the nearest thousandth.

The digit to the right of 1 is 5, so 0.0515 rounds up to 0.052.

Exercises

Round to the nearest tenth.

1. 2.75 **2.** 3.816 **3.** 19.72 **4.** 401.1603 **5.** 499.491 **6.** 3.949

7. 4.67522 **8.** 20.397 **9.** 399.956 **10.** 129.98 **11.** 96.4045 **12.** 125.66047

Round to the nearest hundredth.

13. 31.723 **14.** 14.869 **15.** 1.78826 **16.** 0.1119 **17.** 736.941 **18.** 9.6057

19. 0.699 **20.** 4.231 **21.** 12.09531 **22.** 5.77125 **23.** 0.9195 **24.** 4.0033

Round to the nearest thousandth.

25. 0.4387 **26.** 0.0649 **27.** 3.4953 **28.** 8.07092 **29.** 0.6008 **30.** 6.0074

Round to the nearest whole number.

31. 3.942 **32.** 10.4 **33.** 79.52 **34.** 105.3002 **35.** 431.23 **36.** 0.4962

Multiplying Decimals

When you multiply decimals, first multiply as if the factors were whole numbers. Then, count the decimal places in both factors to find how many places are needed in the product.

EXAMPLE

1 Multiply 2.5 × 1.8.

```
   1.8   ← one decimal place
 × 2.5   ← one decimal place
    90
 + 360
  4.50   ← two decimal places
```

EXAMPLE

2 Find each product.

a. 0.7 × 1.02

```
 1.02
× 0.7
0.714
```

b. 0.03 × 407

```
   407
× 0.03
 12.21
```

c. 0.62 × 2.45

```
   2.45
 × 0.62
    490
+ 14700
 1.5190
```

d. 75 × 3.06

```
   3.06
   × 75
   1530
+ 21420
 229.50
```

Exercises

Multiply.

1. 0.3 × 8
2. 5 × 0.06
3. 0.04 × 7
4. 6 × 0.8
5. 3.1 × 0.05
6. 14 × 0.2
7. 3.1 × 6
8. 0.05 × 43
9. 0.27 × 5
10. 72 × 0.6
11. 0.8 × 312
12. 4.56 × 7

13. 5 × 2.41
14. 704 × 0.3
15. 9 × 1.35
16. 1.2 × 0.3
17. 0.04 × 2.5
18. 6.6 × 0.3
19. 15.1 × 0.02
20. 0.8 × 31.3
21. 0.07 × 25.1
22. 42.2 × 0.9
23. 0.6 × 30.02
24. 0.05 × 11.8
25. 71.13 × 0.4
26. 48 × 2.1
27. 6.3 × 85
28. 0.42 × 98
29. 76 × 3.3
30. 0.77 × 51
31. 5.2 × 4.8
32. 0.12 × 6.1

1. 2.4
2. 0.3
3. 0.28
4. 4.8
5. 0.155
6. 2.8
7. 18.6
8. 2.15
9. 1.35
10. 43.2
11. 249.6
12. 31.92
13. 12.05
14. 211.2
15. 12.15
16. 0.36
17. 0.1
18. 1.98
19. 0.302
20. 25.04
21. 1.757
22. 37.98
23. 18.012
24. 0.59
25. 28.452
26. 100.8
27. 535.5
28. 41.16
29. 250.8
30. 39.27
31. 24.96
32. 0.732

Skills Handbook

1. 0.06
2. 0.08
3. 0.003
4. 0.08
5. 0.014
6. 0.012
7. 0.027
8. 0.03
9. 0.004
10. 0.07
11. 0.0025
12. 0.036
13. 0.032
14. 0.0035
15. 0.0009
16. 0.0045
17. 0.04
18. 0.042
19. 0.0007
20. 0.032
21. 0.0074
22. 0.0376
23. 0.076
24. 0.057
25. 0.085
26. 0.0248
27. 0.007
28. 0.0595
29. 0.027
30. 0.092
31. 0.0102
32. 0.072

Zeros in the Product

When you multiply with decimals, start at the right of the product to count the number of decimal places. Sometimes you need to write extra zeros to the left of a product before you can place the decimal point.

EXAMPLE

1 Multiply 0.03 × 0.51.

Step 1

0.51	← two decimal places
× 0.03	← two decimal places
153	← four decimal places

Step 2

0.51	
× 0.03	
0.0153	← Put extra zeros to the left. Then place the decimal point.

EXAMPLE

2 Find each product.

a. 0.2 × 0.3

0.3
× 0.2
0.06

b. 0.5 × 0.04

0.04
× 0.5
0.020

c. 4 × 0.02

0.02
× 4
0.08

d. 0.02 × 0.45

0.45
× 0.02
0.0090

Exercises

Multiply.

1. 0.1 × 0.6

2. 0.4 × 0.2

3. 0.05 × 0.06

4. 0.01 × 8

5. 0.7 × 0.02

6. 0.03 × 0.4

7. 0.03 × 0.9

8. 0.06 × 0.5

9. 0.2 × 0.02

10. 7 × 0.01

11. 0.05 × 0.05

12. 0.6 × 0.06

13. 0.4 × 0.08
14. 0.07 × 0.05
15. 0.03 × 0.03
16. 0.09 × 0.05
17. 0.5 × 0.08
18. 0.06 × 0.7
19. 0.07 × 0.01
20. 0.16 × 0.2
21. 0.01 × 0.74
22. 0.47 × 0.08
23. 0.76 × 0.1
24. 0.19 × 0.3
25. 0.5 × 0.17
26. 0.31 × 0.08
27. 0.14 × 0.05
28. 0.07 × 0.85
29. 0.45 × 0.06
30. 0.4 × 0.23
31. 0.17 × 0.06
32. 0.3 × 0.24

Dividing a Decimal by a Whole Number

When you divide a decimal by a whole number, first divide as if the numbers were whole numbers. Then put a decimal point in the quotient directly above the decimal point in the dividend.

EXAMPLE

1 Divide 0.256 ÷ 8.

Step 1

$$\begin{array}{r} 32 \\ 8\overline{)0.256} \\ -\ 24 \\ \hline 16 \\ -\ 16 \\ \hline 0 \end{array}$$

Step 2

$$\begin{array}{r} 0.032 \\ 8\overline{)0.256} \\ -24 \\ \hline 16 \\ -16 \\ \hline 0 \end{array}$$

← **Put extra zeros to the left. Then place the decimal point.**

EXAMPLE

2 Find each quotient.

a. 12.6 ÷ 6

$$\begin{array}{r} 2.1 \\ 6\overline{)12.6} \\ -\ 12 \\ \hline 06 \\ -\ 6 \\ \hline 0 \end{array}$$

b. 37.26 ÷ 81

$$\begin{array}{r} 0.46 \\ 81\overline{)37.26} \\ -\ 324 \\ \hline 486 \\ -\ 486 \\ \hline 0 \end{array}$$

c. 0.666 ÷ 9

$$\begin{array}{r} 0.074 \\ 9\overline{)0.666} \\ -\ 63 \\ \hline 36 \\ -36 \\ \hline 0 \end{array}$$

Exercises

Divide.

1. $4\overline{)28.56}$ **2.** $5\overline{)16.5}$ **3.** $9\overline{)6.984}$ **4.** $6\overline{)91.44}$ **5.** $4\overline{)35.16}$

6. $81\overline{)33.291}$ **7.** $22\overline{)2.42}$ **8.** $26\overline{)1{,}723.8}$ **9.** $83\overline{)15.272}$ **10.** $39\overline{)26.91}$

11. 17.52 ÷ 2 **12.** 10.53 ÷ 9 **13.** 14.49 ÷ 7 **14.** 37.14 ÷ 6

15. 0.0324 ÷ 9 **16.** 0.1352 ÷ 8 **17.** 0.0882 ÷ 6 **18.** 0.8682 ÷ 6

19. 79.599 ÷ 13 **20.** 45.918 ÷ 18 **21.** 59.7 ÷ 15 **22.** 74.664 ÷ 12

23. 12.342 ÷ 22 **24.** 29.792 ÷ 32 **25.** 22.568 ÷ 26 **26.** 11.340 ÷ 36

Skills Handbook

1. 7.14
2. 3.3
3. 0.776
4. 15.24
5. 8.79
6. 0.411
7. 0.11
8. 66.3
9. 0.184
10. 0.69
11. 8.76
12. 1.17
13. 2.07
14. 6.19
15. 0.0036
16. 0.0169
17. 0.0147
18. 0.1447
19. 6.123
20. 2.551
21. 3.98
22. 6.222
23. 0.561
24. 0.931
25. 0.868
26. 0.315

Skills Handbook

1. 0.032
2. 120
3. 0.07
4. 0.062
5. 9
6. 0.236
7. 52
8. 80
9. 5
10. 0.024
11. 1.803
12. 610
13. 823
14. 0.0004
15. 502
16. 0.165
17. 2,367
18. 4.528
19. 0.0009
20. 103
21. 426
22. 20,305
23. 0.47
24. 0.0007
25. 0.032
26. 0.0003
27. 26
28. 12,600
29. 0.0008
30. 0.067
31. 15
32. 2.35

Powers of Ten

You can use shortcuts when multiplying and dividing by powers of ten.

When you multiply by...	move the decimal point...	When you divide by...	move the decimal point...
1,000	3 places to the right.	1,000	3 places to the left.
100	2 places to the right.	100	2 places to the left.
10	1 place to the right.	10	1 place to the left.
0.1	1 place to the left.	0.1	1 place to the right.
0.01	2 places to the left.	0.01	2 places to the right.

EXAMPLE

Multiply or divide.

a. 0.3×0.01

0.00.3 ← **Move the decimal point 2 places to the left.**

$0.3 \times 0.01 = 0.003$

b. $0.18 \div 1{,}000$

0.000.18 ← **Move the decimal point 3 places to the left.**

$0.18 \div 1{,}000 = 0.00018$

Exercises

Multiply.

1. 3.2×0.01 **2.** $1{,}000 \times 0.12$ **3.** 0.7×0.1 **4.** 0.01×6.2

5. 0.09×100 **6.** 23.6×0.01 **7.** 5.2×10 **8.** $0.08 \times 1{,}000$

9. 100×0.05 **10.** 0.1×0.24 **11.** 18.03×0.1 **12.** 6.1×100

Divide.

13. $82.3 \div 0.1$ **14.** $0.4 \div 1{,}000$ **15.** $5.02 \div 0.01$ **16.** $16.5 \div 100$

17. $236.7 \div 0.1$ **18.** $45.28 \div 10$ **19.** $0.9 \div 1{,}000$ **20.** $1.03 \div 0.01$

21. $42.6 \div 0.1$ **22.** $203.05 \div 0.01$ **23.** $4.7 \div 10$ **24.** $0.07 \div 100$

Multiply or divide.

25. 0.32×0.1 **26.** $0.03 \div 100$ **27.** $2.6 \div 0.1$ **28.** $12.6 \times 1{,}000$

29. $0.8 \div 1{,}000$ **30.** 0.01×6.7 **31.** 100×0.15 **32.** $23.5 \div 10$

Zeros in Decimal Division

When you are dividing by a decimal, sometimes you need to use extra zeros in the dividend, the quotient, or both.

EXAMPLE

Find each quotient.

a. 0.14 ÷ 0.04

Multiply by 100.

$$\begin{array}{r} 3.5 \\ 0.04.\overline{)0.14.0} \\ -12 \\ \hline 20 \\ -20 \\ \hline 0 \end{array}$$

b. 0.00434 ÷ 0.07

Multiply by 100.

$$\begin{array}{r} 0.062 \\ 0.07.\overline{)0.00.434} \\ -42 \\ \hline 14 \\ -14 \\ \hline 0 \end{array}$$

c. 0.045 ÷ 3.6

Multiply by 10.

$$\begin{array}{r} 0.0125 \\ 3.6.\overline{)0.0.4500} \\ -36 \\ \hline 90 \\ -72 \\ \hline 180 \\ -180 \\ \hline 0 \end{array}$$

Exercises

Divide.

1. $0.4\overline{)0.001}$ **2.** $0.05\overline{)0.0023}$ **3.** $0.02\overline{)0.000162}$

4. $0.6\overline{)0.0015}$ **5.** $1.2\overline{)0.078}$ **6.** $0.34\overline{)0.00119}$

7. $0.12\overline{)0.009}$ **8.** $2.5\overline{)0.021}$ **9.** 0.0017 ÷ 0.02

10. 0.003 ÷ 0.6 **11.** 0.01099 ÷ 0.7 **12.** 0.104 ÷ 0.05

13. 0.0945 ÷ 0.09 **14.** 0.00045 ÷ 0.3 **15.** 0.052 ÷ 0.8

16. 0.142 ÷ 0.04 **17.** 0.034 ÷ 0.05 **18.** 0.0019 ÷ 0.2

19. 0.9 ÷ 0.2 **20.** 0.000175 ÷ 0.07 **21.** 0.0084 ÷ 1.4

22. 0.259 ÷ 3.5 **23.** 0.00468 ÷ 0.52 **24.** 0.00056 ÷ 0.16

25. 0.0612 ÷ 7.2 **26.** 0.17701 ÷ 3.1 **27.** 0.00063 ÷ 0.18

28. 0.011 ÷ 0.25 **29.** 0.3069 ÷ 9.3 **30.** 0.000924 ÷ 0.44

31. 0.03234 ÷ 0.35 **32.** 0.00123 ÷ 8.2 **33.** 0.03225 ÷ 0.75

34. 0.006 ÷ 0.75 **35.** 0.73 ÷ 0.25 **36.** 0.68 ÷ 0.002

37. 0.398 ÷ 0.05 **38.** 0.0004 ÷ 0.002 **39.** 0.125 ÷ 0.005

1. 0.0025
2. 0.046
3. 0.0081
4. 0.0025
5. 0.065
6. 0.0035
7. 0.075
8. 0.0084
9. 0.085
10. 0.005
11. 0.0157
12. 2.08
13. 1.05
14. 0.0015
15. 0.065
16. 3.55
17. 0.68
18. 0.0095
19. 4.5
20. 0.0025
21. 0.006
22. 0.074
23. 0.009
24. 0.0035
25. 0.0085
26. 0.0571
27. 0.0035
28. 0.044
29. 0.033
30. 0.0021
31. 0.0924
32. 0.00015
33. 0.043
34. 0.008
35. 2.92
36. 340
37. 7.96
38. 0.2
39. 25

Skills Handbook

1. $\frac{4}{5}$
2. $\frac{1}{6}$
3. $\frac{4}{7}$
4. 1
5. $4\frac{4}{7}$
6. $4\frac{2}{5}$
7. $\frac{5}{8}$
8. $\frac{1}{3}$
9. $\frac{3}{8}$
10. $\frac{1}{3}$
11. $\frac{3}{5}$
12. 1
13. 11
14. $5\frac{2}{5}$
15. $5\frac{2}{3}$
16. $5\frac{1}{4}$
17. $7\frac{2}{5}$
18. $11\frac{1}{2}$

Adding and Subtracting Fractions With Like Denominators

When you add or subtract fractions with the same denominator, first add or subtract the numerators. Write the answer over the denominator.

EXAMPLE

1 Add or subtract. Write the answer in simplest form.

a. $\frac{5}{16} + \frac{3}{16}$

$$\frac{5}{16} + \frac{3}{16} = \frac{8}{16} = \frac{1}{2}$$

b. $\frac{7}{8} - \frac{1}{8}$

$$\frac{7}{8} - \frac{1}{8} = \frac{6}{8} = \frac{3}{4}$$

c. $\frac{3}{5} + \frac{2}{5}$

$$\frac{3}{5} + \frac{2}{5} = \frac{5}{5} = 1$$

To add or subtract mixed numbers, add or subtract the fractions first. Then add or subtract the whole numbers.

EXAMPLE

2 Add or subtract. Write the answer in simplest form.

a. $2\frac{5}{8} + 3\frac{1}{8}$

$$2\frac{5}{8} + 3\frac{1}{8} = 5\frac{6}{8} = 5\frac{3}{4}$$

b. $4\frac{3}{4} - 1\frac{1}{4}$

$$4\frac{3}{4} - 1\frac{1}{4} = 3\frac{2}{4} = 3\frac{1}{2}$$

c. $5\frac{5}{6} + 2\frac{5}{6}$

$$5\frac{5}{6} + 2\frac{5}{6} = 7\frac{10}{6}$$
$$= 7 + 1 + \frac{4}{6}$$
$$= 8\frac{2}{3}$$

Exercises

Add or subtract. Write the answers in simplest form.

1. $\frac{2}{5} + \frac{2}{5}$ **2.** $\frac{2}{6} - \frac{1}{6}$ **3.** $\frac{2}{7} + \frac{2}{7}$ **4.** $9\frac{1}{3} - 8\frac{1}{3}$ **5.** $8\frac{6}{7} - 4\frac{2}{7}$ **6.** $3\frac{1}{10} + 1\frac{3}{10}$

7. $\frac{3}{8} + \frac{2}{8}$ **8.** $\frac{3}{6} - \frac{1}{6}$ **9.** $\frac{6}{8} - \frac{3}{8}$ **10.** $\frac{2}{9} + \frac{1}{9}$ **11.** $\frac{4}{5} - \frac{1}{5}$ **12.** $\frac{3}{4} + \frac{1}{4}$

13. $8\frac{7}{10} + 2\frac{3}{10}$ **14.** $1\frac{4}{5} + 3\frac{3}{5}$ **15.** $2\frac{2}{9} + 3\frac{4}{9}$ **16.** $8\frac{5}{8} - 3\frac{3}{8}$ **17.** $9\frac{7}{10} - 2\frac{3}{10}$ **18.** $9\frac{3}{4} + 1\frac{3}{4}$

Metric Units of Length

The basic unit of length in the metric system is the meter. All the other units are based on the meter. In the chart below, each unit is 10 times the value of the unit to its left.

Unit	Millimeter	Centimeter	Decimeter	Meter	Decameter	Hectometer	Kilometer
Symbol	mm	cm	dm	m	dam	hm	km
Value	0.001 m	0.01 m	0.1 m	1 m	10 m	100 m	1,000 m

To change a measure from one unit to another, start by using the chart to find the relationship between the two units.

EXAMPLE

Complete each equation.

a. 0.8 km = ■ m

1 km = 1,000 m

0.8 × 1,000 = 800 ← **To change km to m, multiply by 1,000.**

0.8 km = 800 m

b. 17.2 mm = ■ cm

1 mm = 0.1 cm

17.2 × 0.1 = 1.72 ← **To change mm to cm, multiply by 0.1.**

17.2 mm = 1.72 cm

c. ■ cm = 2.1 km

1 km = 100,000 cm

2.1 × 100,000 = 210,000 ← **To change km to cm, multiply by 100,000.**

210,000 cm = 2.1 km

d. ■ m = 5,200 cm

1 cm = 0.01 m

5,200 × 0.01 = 52 ← **To change cm to m, multiply by 0.01.**

52 m = 5,200 cm

Exercises

Complete each equation.

1. 1 mm = ■ cm **2.** 1 m = ■ km **3.** 1 mm = ■ m **4.** 1 cm = ■ m

5. 1.2 cm = ■ km **6.** ■ km = 45,000 mm **7.** ■ m = 30 km

8. 6.2 cm = ■ mm **9.** 3.3 km = ■ m **10.** 0.6 mm = ■ cm

11. 72 cm = ■ m **12.** 180 m = ■ mm **13.** ■ cm = 13 km

14. ■ m = 530 cm **15.** 4,900 mm = ■ m **16.** ■ cm = 24 m

17. ■ km = 106,000 cm **18.** 259,000 mm = ■ m **19.** 1,200,000 mm = ■ km

1. 0.1
2. 0.001
3. 0.001
4. 0.01
5. 0.000012
6. 0.045
7. 30,000
8. 62
9. 3,300
10. 0.06
11. 0.72
12. 180,000
13. 1,300,000
14. 5.3
15. 4.9
16. 2,400
17. 1.06
18. 259
19. 1.2

Skills Handbook

1. 1,000
2. 0.000001
3. 1,000
4. 1,000,000
5. 0.2
6. 1,300,000
7. 0.24
8. 0.0006
9. 106
10. 72,000,000
11. 1,500,000
12. 0.45
13. 4.9
14. 0.2
15. 0.008
16. 4,200
17. 57,000
18. 28,000,000
19. 9,000,000
20. 4,000,000
21. 0.87

Metric Units of Capacity

The basic unit of capacity in the metric system is the liter. All the other units are based on the liter. In the chart below, each unit is 10 times the value of the unit on the left. Note that we use a capital L as the abbreviation for *liter* to avoid confusion with the number 1.

Unit	Milliliter	Centiliter	Deciliter	Liter	Decaliter	Hectoliter	Kiloliter
Symbol	mL	cL	dL	L	daL	hL	kL
Value	0.001 L	0.01 L	0.1 L	1 L	10 L	100 L	1,000 L

To change a measure from one unit to another, start by using the chart to find the relationship between the two units.

EXAMPLE

Complete each equation.

a. 245 mL = ■ L

1 mL = 0.001 L

$245 \times 0.001 = 0.245$ ← **To change mL to L, multiply by 0.001.**

245 mL = 0.245 L

b. ■ mL = 4.5 kL

1 kL = 1,000,000 mL

$4.5 \times 1{,}000{,}000 = 4{,}500{,}000$ ← **To change kL to mL, multiply by 1,000,000.**

4,500,000 mL = 4.5 kL

Exercises

Complete each equation.

1. 1 L = ■ mL
2. 1 mL = ■ kL
3. 1 kL = ■ L
4. 1 kL = ■ mL
5. 200 L = ■ kL
6. 1.3 kL = ■ mL
7. ■ kL = 240 L
8. 0.6 mL = ■ L
9. ■ kL = 106,000 L
10. 72 kL = ■ mL
11. ■ mL = 1.5 kL
12. ■ kL = 450,000 mL
13. 4,900 L = ■ kL
14. ■ kL = 200,000 mL
15. ■ L = 8 mL
16. 4.2 L = ■ mL
17. 57,000,000 mL = ■ L
18. 28,000 kL = ■ L
19. ■ mL = 9,000 L
20. 4,000 L = ■ mL
21. 870 L = ■ kL

Metric Units of Mass

The basic unit of mass in the metric system is the gram. All the other units are based on the gram. In the chart below, each unit is 10 times the value of the unit to its left.

Unit	Milligram	Centigram	Decigram	Gram	Decagram	Hectogram	Kilogram
Symbol	mg	cg	dg	g	dag	hg	kg
Value	0.001 g	0.01 g	0.1 g	1 g	10 g	100 g	1,000 g

To change a measure from one unit to another, start by using the chart to find the relationship between the two units.

EXAMPLE

Complete each equation.

a. 2.3 kg = ■ g

1 kg = 1,000 g

2.3 × 1,000 = 2,300 ← **To change kg to g, multiply by 1,000.**

2.3 kg = 2,300 g

b. ■ g = 250 mg

1 mg = 0.001 g

250 × 0.001 = 0.25 ← **To change mg to g, multiply by 0.001.**

0.25 g = 250 mg

Exercises

Complete each equation.

1. 1 mg = ■ g
2. 1 g = ■ kg
3. 1 mg = ■ kg
4. 1 g = ■ mg
5. 1 kg = ■ g
6. 1 kg = ■ mg
7. ■ g = 8 mg
8. 1,500 mg = ■ kg
9. ■ kg = 200,000 g
10. ■ mg = 3.7 g
11. 0.6 mg = ■ g
12. 370 g = ■ kg
13. ■ kg = 300,000 mg
14. 900 g = ■ mg
15. ■ kg = 5.7 mg
16. 120 g = ■ kg
17. ■ kg = 440 g
18. ■ kg = 1,006,000 mg
19. 0.009 kg = ■ mg
20. 0.2 mg = ■ g
21. 8.6 kg = ■ g

1. 0.001
2. 0.001
3. 0.000001
4. 1,000
5. 1,000
6. 1,000,000
7. 0.008
8. 0.0015
9. 200
10. 3,700
11. 0.0006
12. 0.37
13. 0.3
14. 900,000
15. 0.0000057
16. 0.12
17. 0.44
18. 1.006
19. 9,000
20. 0.0002
21. 8,600

Skills Handbook

Tables

Table 1 Measures

Metric	Customary
Length	**Length**
10 millimeters (mm) = 1 centimeter (cm) 100 cm = 1 meter (m) 1,000 mm = 1 m 1,000 m = 1 kilometer (km)	12 inches (in.) = 1 foot (ft) 36 in. = 1 yard (yd) 3 ft = 1 yd 5,280 ft = 1 mile (mi) 1,760 yd = 1 mi
Area	**Area**
100 square millimeters (mm^2) = 1 square centimeter (cm^2) 10,000 cm^2 = 1 square meter (m^2)	144 square inches ($in.^2$) = 1 square foot (ft^2) 9 ft^2 = 1 square yard (yd^2) 4,840 yd^2 = 1 acre
Volume	**Volume**
1,000 cubic millimeters (mm^3) = 1 cubic centimeter (cm^3) 1,000,000 cm^3 = 1 cubic meter (m^3)	1,728 cubic inches ($in.^3$) = 1 cubic foot (ft^3) 27 ft^3 = 1 cubic yard (yd^3)
Mass	**Mass**
1,000 milligrams (mg) = 1 gram (g) 1,000 g = 1 kilogram (kg)	16 ounces (oz) = 1 pound (lb) 2,000 lb = 1 ton (t)
Liquid Capacity	**Liquid Capacity**
1,000 milliliters (mL) = 1 liter (L)	8 fluid ounces (fl oz) = 1 cup (c) 2 c = 1 pint (pt) 2 pt = 1 quart (qt) 4 qt = 1 gallon (gal)

Time
1 minute (min) = 60 seconds (s) 1 hour (h) = 60 min 1 day (d) = 24 h 1 year (yr) = 365 d

Table 2 Reading Math Symbols

Symbol	Meaning	Page
$\approx$	is approximately equal to	p. 4
$-$	minus (subtraction)	p. 4
$=$	is equal to	p. 4
$+$	plus (addition)	p. 5
$\div$	divide (division)	p. 5
( )	parentheses for grouping	p. 9
$\times, \cdot$	times (multiplication)	p. 15
$\lvert a \rvert$	absolute value of *a*	p. 31
$-a$	opposite of *a*	p. 31
$>$	is greater than	p. 32
$<$	is less than	p. 32
°	degrees	p. 40
[ ]	brackets for grouping	p. 48
a^n	*n*th power of *a*	p. 68
^	raise to a power (calculator key)	p. 69
...	and so on	p. 97
$\neq$	is not equal to	p. 136
$\frac{1}{a}$	reciprocal of *a*	p. 141
$\stackrel{?}{=}$	Is the statement true?	p. 174
$\geq$	is greater than or equal to	p. 205
$\leq$	is less than or equal to	p. 205
$a : b$	ratio of *a* to *b*	p. 228
$\angle A$	angle with vertex *A*	p. 251
AB	length of segment $\overline{AB}$	p. 251
$\sim$	is similar to	p. 252
$\triangle ABC$	triangle with vertices *ABC*	p. 252
%	percent	p. 274
$\overleftrightarrow{AB}$	line *AB*	p. 324
$\overrightarrow{AB}$	ray *AB*	p. 324
$\overline{AB}$	segment *AB*	p. 324
$\parallel$	is parallel to	p. 325
$\angle ABC$	angle with sides *BA* and *BC*	p. 330
$m\angle ABC$	measure of angle *ABC*	p. 330
$\cong$	is congruent to	p. 346
$\overset{\frown}{AB}$	arc *AB*	p. 351
d	diameter	p. 353
r	radius	p. 353
$\perp$	is perpendicular to	p. 362
P	perimeter	p. 375
ℓ	length	p. 375
w	width	p. 375
A	area	p. 380
b	base length	p. 380
h	height	p. 380
b_1, b_2	base lengths of a trapezoid	p. 388
C	circumference	p. 394
π	pi, an irrational number approximately equal to 3.14	p. 394
$\sqrt{x}$	nonnegative square root of *x*	p. 400
V	volume	p. 422
B	area of base	p. 422
(a, b)	ordered pair with *x*-coordinate *a* and *y*-coordinate *b*	p. 486
$F \rightarrow F'$	*F* maps onto *F′*	p. 510
A'	image of *A*, *A* prime	p. 510
P(event)	probability of the event	p. 580
$n!$	*n* factorial	p. 606

Table 3 Squares and Square Roots

Number n	Square n^2	Positive Square Root $\sqrt{n}$	Number n	Square n^2	Positive Square Root $\sqrt{n}$
1	1	1.000	**51**	2,601	7.141
2	4	1.414	**52**	2,704	7.211
3	9	1.732	**53**	2,809	7.280
4	16	2.000	**54**	2,916	7.348
5	25	2.236	**55**	3,025	7.416
6	36	2.449	**56**	3,136	7.483
7	49	2.646	**57**	3,249	7.550
8	64	2.828	**58**	3,364	7.616
9	81	3.000	**59**	3,481	7.681
10	100	3.162	**60**	3,600	7.746
11	121	3.317	**61**	3,721	7.810
12	144	3.464	**62**	3,844	7.874
13	169	3.606	**63**	3,969	7.937
14	196	3.742	**64**	4,096	8.000
15	225	3.873	**65**	4,225	8.062
16	256	4.000	**66**	4,356	8.124
17	289	4.123	**67**	4,489	8.185
18	324	4.243	**68**	4,624	8.246
19	361	4.359	**69**	4,761	8.307
20	400	4.472	**70**	4,900	8.367
21	441	4.583	**71**	5,041	8.426
22	484	4.690	**72**	5,184	8.485
23	529	4.796	**73**	5,329	8.544
24	576	4.899	**74**	5,476	8.602
25	625	5.000	**75**	5,625	8.660
26	676	5.099	**76**	5,776	8.718
27	729	5.196	**77**	5,929	8.775
28	784	5.292	**78**	6,084	8.832
29	841	5.385	**79**	6,241	8.888
30	900	5.477	**80**	6,400	8.944
31	961	5.568	**81**	6,561	9.000
32	1,024	5.657	**82**	6,724	9.055
33	1,089	5.745	**83**	6,889	9.110
34	1,156	5.831	**84**	7,056	9.165
35	1,225	5.916	**85**	7,225	9.220
36	1,296	6.000	**86**	7,396	9.274
37	1,369	6.083	**87**	7,569	9.327
38	1,444	6.164	**88**	7,744	9.381
39	1,521	6.245	**89**	7,921	9.434
40	1,600	6.325	**90**	8,100	9.487
41	1,681	6.403	**91**	8,281	9.539
42	1,764	6.481	**92**	8,464	9.592
43	1,849	6.557	**93**	8,649	9.644
44	1,936	6.633	**94**	8,836	9.695
45	2,025	6.708	**95**	9,025	9.747
46	2,116	6.782	**96**	9,216	9.798
47	2,209	6.856	**97**	9,409	9.849
48	2,304	6.928	**98**	9,604	9.899
49	2,401	7.000	**99**	9,801	9.950
50	2,500	7.071	**100**	10,000	10.000

Table 4 For Use With Problem Solving Applications

Chapter 1

Temperature Extremes and Annual Precipitation

Location	High Temperature (°F)	Low Temperature (°F)	Annual Precipitation (in.)
Anchorage, Alaska	86	−38	14.7
El Paso, Tex.	109	−8	8.0
Indianapolis, Ind.	107	−25	39.2
Miami, Fla.	100	28	59.9
Pago Pago, Amer. Samoa	98	67	193.6
San José, Costa Rica	92	49	70.8
South Pole Station	6	−107	0.1

SOURCE: *The Weather Almanac*

Chapter 9

Animal Longevity

Animal	Birth Weight (Male) (g)	Maximum Life Span (yr)
Dragonfly	1	0.1
Rat	6	3.3
Salmon	15	13
Spoonbill	45	10
Rabbit	65	13
Cat	98	28
Dog (cocker spaniel)	240	20

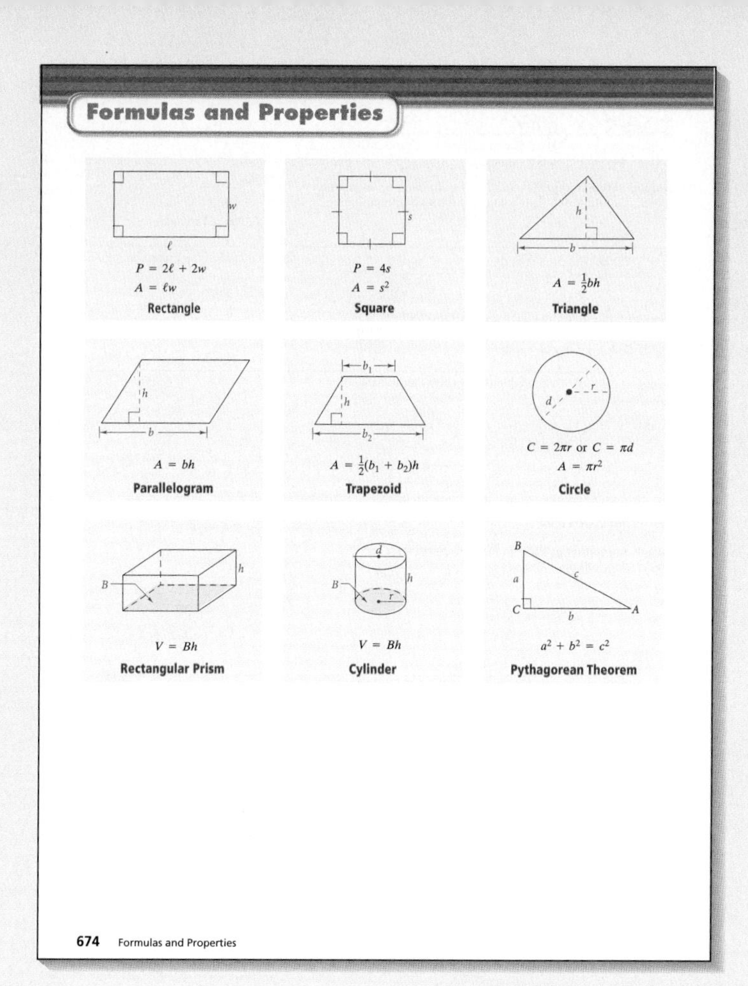

Formulas and Properties

$P = 2\ell + 2w$
$A = \ell w$
Rectangle

$P = 4s$
$A = s^2$
Square

$A = \frac{1}{2}bh$
Triangle

$A = bh$
Parallelogram

$A = \frac{1}{2}(b_1 + b_2)h$
Trapezoid

$C = 2\pi r$ or $C = \pi d$
$A = \pi r^2$
Circle

$V = Bh$
Rectangular Prism

$V = Bh$
Cylinder

$a^2 + b^2 = c^2$
Pythagorean Theorem

Properties of Real Numbers

Unless otherwise stated, the variables *a*, *b*, *c*, and *d* used in these properties can be replaced with any number represented on a number line.

Identity Properties

Addition $a + 0 = a$ and $0 + a = a$
Multiplication $a \cdot 1 = a$ and $1 \cdot a = a$

Commutative Properties

Addition $a + b = b + a$
Multiplication $a \cdot b = b \cdot a$

Associative Properties

Addition $(a + b) + c = a + (b + c)$
Multiplication $(a \cdot b) \cdot c = a \cdot (b \cdot c)$

Inverse Properties

Addition
$a + (-a) = 0$ and $-a + a = 0$
Multiplication
$a \cdot \frac{1}{a} = 1$ and $\frac{1}{a} \cdot a = 1$ $(a \neq 0)$

Distributive Properties

$a(b + c) = ab + ac$ $(b + c)a = ba + ca$
$a(b - c) = ab - ac$ $(b - c)a = ba - ca$

Properties of Equality

Addition If $a = b$, then $a + c = b + c$.
Subtraction If $a = b$, then $a - c = b - c$.
Multiplication If $a = b$, then $a \cdot c = b \cdot c$.
Division If $a = b$, and $c \neq 0$, then $\frac{a}{c} = \frac{b}{c}$.
Substitution If $a = b$, then b can replace a in any expression.
Reflexive $a = a$
Symmetric If $a = b$, then $b = a$.
Transitive If $a = b$ and $b = c$, then $a = c$.

Cross Products Property

$\frac{a}{c} = \frac{b}{d}$ is equivalent to $ad = bc$.

Zero Property

If $ab = 0$, then $a = 0$ or $b = 0$.

Closure Property

$a + b$ is a unique real number.
ab is a unique real number.

Density Property

Between any two rational numbers, there is at least one other rational number.

Properties of Inequality

Addition If $a > b$, then $a + c > b + c$.
If $a < b$, then $a + c < b + c$.
Subtraction If $a > b$, then $a - c > b - c$.
If $a < b$, then $a - c < b - c$.
Multiplication
If $a > b$ and $c > 0$, then $ac > bc$.
If $a < b$ and $c > 0$, then $ac < bc$.
If $a > b$ and $c < 0$, then $ac < bc$.
If $a < b$ and $c < 0$, then $ac > bc$.
Division
If $a > b$ and $c > 0$, then $\frac{a}{c} > \frac{b}{c}$.
If $a < b$ and $c > 0$, then $\frac{a}{c} < \frac{b}{c}$.
If $a > b$ and $c < 0$, then $\frac{a}{c} < \frac{b}{c}$.
If $a < b$ and $c < 0$, then $\frac{a}{c} > \frac{b}{c}$.
Transitive If $a > b$ and $b > c$, then $a > c$.

English/Spanish Illustrated Glossary

A

	EXAMPLES
Absolute value (p. 31) The absolute value of a number is its distance from 0 on a number line. **Valor absoluto (p. 31)** El valor absoluto de un número es su distancia del 0 en una recta numérica.	-7 is 7 units from 0, so $\lvert -7 \rvert = 7$.
Acute angle (p. 330) An acute angle is an angle with a measure between 0° and 90°. **Ángulo agudo (p. 330)** Un ángulo agudo es un ángulo que mide entre 0° y 90°.	$0° < m\angle 1 < 90°$
Acute triangle (p. 337) An acute triangle has three acute angles. **Triángulo acutángulo (p. 337)** Un triángulo acutángulo tiene tres ángulos agudos.	$\angle 1$, $\angle 2$, and $\angle 3$ are acute.
Addition Property of Equality (p. 180) The Addition Property of Equality states that if the same value is added to each side of an equation, the results are equal. **Propiedad aditiva de la igualdad (p. 180)** La propiedad aditiva de la igualdad establece que si se suma el mismo valor a cada lado de una ecuación, los resultados son iguales.	Since $\frac{20}{2} = 10$, $\frac{20}{2} + 3 = 10 + 3$. If $a = b$, then $a + c = b + c$.
Addition Property of Inequality (p. 210) The Addition Property of Inequality states that if you add the same value to each side of an inequality, the relationship between the two sides does not change. **Propiedad aditiva de la desigualdad (p. 210)** La propiedad aditiva de la desigualdad establece que si sumas el mismo valor a cada lado de una desigualdad, la relación entre los dos lados no cambia.	If $a > b$, then $a + c > b + c$. Since $4 > 2$, $4 + 11 > 2 + 11$. If $a < b$, then $a + c < b + c$. Since $4 < 9$, $4 + 11 < 9 + 11$.
Additive inverses (p. 38) Two numbers whose sum is 0 are additive inverses. **Inversos aditivo (p. 38)** Dos números cuya suma es 0 son inversos aditivos.	$(-5) + 5 = 0$

676 Glossary

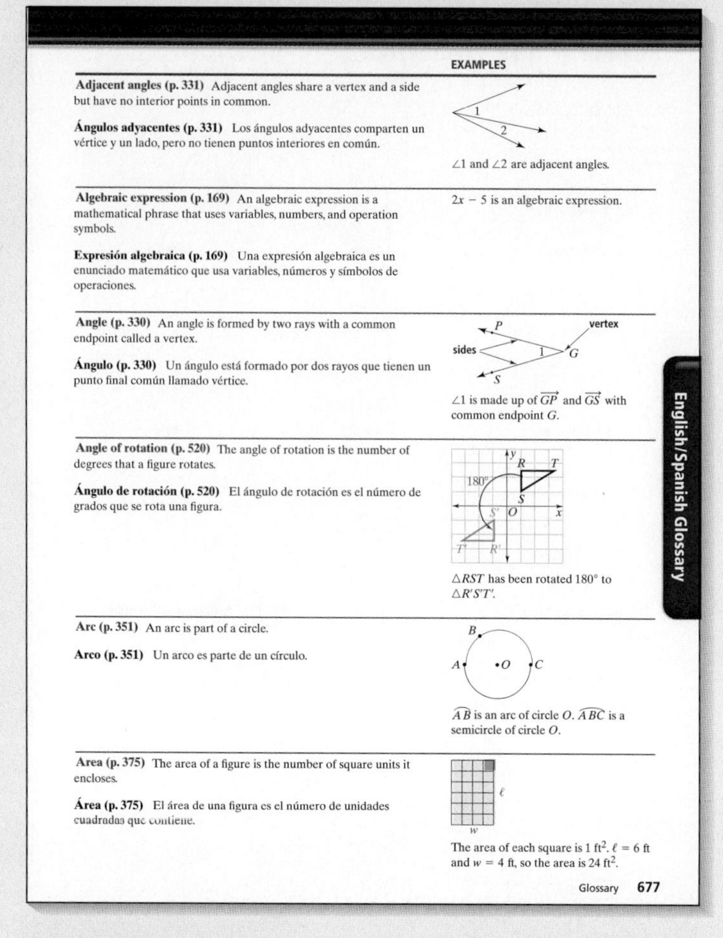

	EXAMPLES
Adjacent angles (p. 331) Adjacent angles share a vertex and a side but have no interior points in common. **Ángulos adyacentes (p. 331)** Los ángulos adyacentes comparten un vértice y un lado, pero no tienen puntos interiores en común.	$\angle 1$ and $\angle 2$ are adjacent angles.
Algebraic expression (p. 169) An algebraic expression is a mathematical phrase that uses variables, numbers, and operation symbols. **Expresión algebraica (p. 169)** Una expresión algebraica es un enunciado matemático que usa variables, números y símbolos de operaciones.	$2x - 5$ is an algebraic expression.
Angle (p. 330) An angle is formed by two rays with a common endpoint called a vertex. **Ángulo (p. 330)** Un ángulo está formado por dos rayos que tienen un punto final común llamado vértice.	$\angle 1$ is made up of $\overrightarrow{GP}$ and $\overrightarrow{GS}$ with common endpoint G.
Angle of rotation (p. 520) The angle of rotation is the number of degrees that a figure rotates. **Ángulo de rotación (p. 520)** El ángulo de rotación es el número de grados que se rota una figura.	$\triangle RST$ has been rotated 180° to $\triangle R'S'T'$.
Arc (p. 351) An arc is part of a circle. **Arco (p. 351)** Un arco es parte de un círculo.	$\overset{\frown}{AB}$ is an arc of circle O. $\overset{\frown}{ABC}$ is a semicircle of circle O.
Area (p. 375) The area of a figure is the number of square units it encloses. **Área (p. 375)** El área de una figura es el número de unidades cuadradas que contiene.	The area of each square is 1 ft². $\ell = 6$ ft and $w = 4$ ft, so the area is 24 ft².

Glossary 677

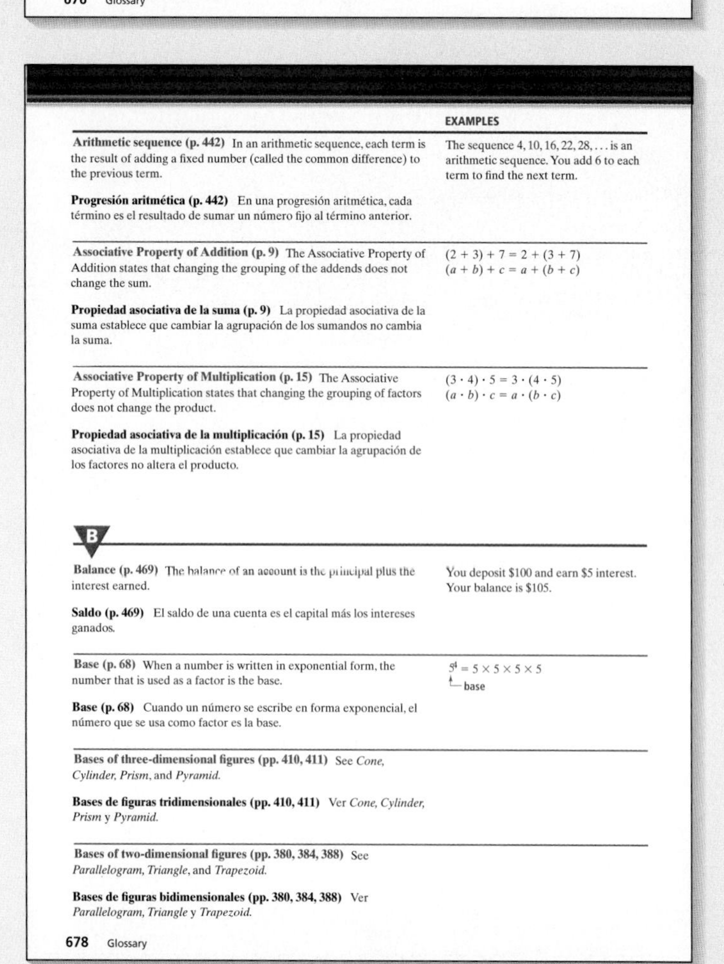

	EXAMPLES
Arithmetic sequence (p. 442) In an arithmetic sequence, each term is the result of adding a fixed number (called the common difference) to the previous term. **Progresión aritmética (p. 442)** En una progresión aritmética, cada término es el resultado de sumar un número fijo al término anterior.	The sequence 4, 10, 16, 22, 28, . . . is an arithmetic sequence. You add 6 to each term to find the next term.
Associative Property of Addition (p. 9) The Associative Property of Addition states that changing the grouping of the addends does not change the sum. **Propiedad asociativa de la suma (p. 9)** La propiedad asociativa de la suma establece que cambiar la agrupación de los sumandos no cambia la suma.	$(2 + 3) + 7 = 2 + (3 + 7)$ $(a + b) + c = a + (b + c)$
Associative Property of Multiplication (p. 15) The Associative Property of Multiplication states that changing the grouping of factors does not change the product. **Propiedad asociativa de la multiplicación (p. 15)** La propiedad asociativa de la multiplicación establece que cambiar la agrupación de los factores no altera el producto.	$(3 \cdot 4) \cdot 5 = 3 \cdot (4 \cdot 5)$ $(a \cdot b) \cdot c = a \cdot (b \cdot c)$

B

Balance (p. 469) The balance of an account is the principal plus the interest earned. **Saldo (p. 469)** El saldo de una cuenta es el capital más los intereses ganados.	You deposit $100 and earn $5 interest. Your balance is $105.
Base (p. 68) When a number is written in exponential form, the number that is used as a factor is the base. **Base (p. 68)** Cuando un número se escribe en forma exponencial, el número que se usa como factor es la base.	$5^4 = 5 \times 5 \times 5 \times 5$ base
Bases of three-dimensional figures (pp. 410, 411) See *Cone, Cylinder, Prism*, and *Pyramid*. **Bases de figuras tridimensionales (pp. 410, 411)** Ver *Cone, Cylinder, Prism* y *Pyramid*.	
Bases of two-dimensional figures (pp. 380, 384, 388) See *Parallelogram, Triangle*, and *Trapezoid*. **Bases de figuras bidimensionales (pp. 380, 384, 388)** Ver *Parallelogram, Triangle* y *Trapezoid*.	

678 Glossary

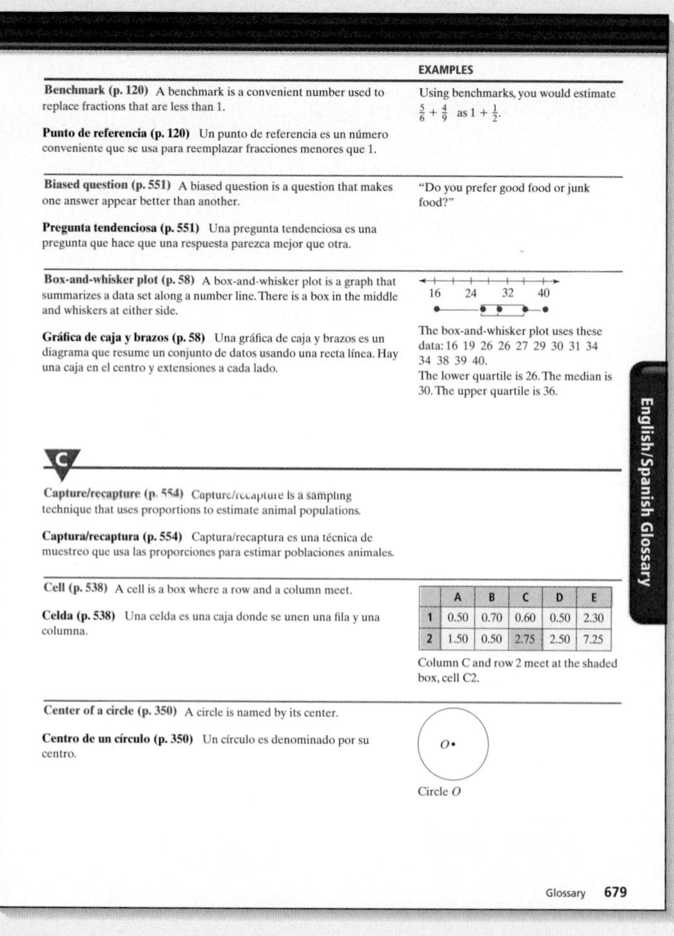

	EXAMPLES
Benchmark (p. 120) A benchmark is a convenient number used to replace fractions that are less than 1. **Punto de referencia (p. 120)** Un punto de referencia es un número conveniente que se usa para reemplazar fracciones menores que 1.	Using benchmarks, you would estimate $\frac{5}{6} + \frac{4}{9}$ as $1 + \frac{1}{2}$.
Biased question (p. 551) A biased question is a question that makes one answer appear better than another. **Pregunta tendenciosa (p. 551)** Una pregunta tendenciosa es una pregunta que hace que una respuesta parezca mejor que otra.	"Do you prefer good food or junk food?"
Box-and-whisker plot (p. 58) A box-and-whisker plot is a graph that summarizes a data set along a number line. There is a box in the middle and whiskers at either side. **Gráfica de caja y brazos (p. 58)** Una gráfica de caja y brazos es un diagrama que resume un conjunto de datos usando una recta línea. Hay una caja en el centro y extensiones a cada lado.	The box-and-whisker plot uses these data: 16 19 26 26 27 29 30 31 34 34 38 39 40. The lower quartile is 26. The median is 30. The upper quartile is 36.

C

Capture/recapture (p. 554) Capture/recapture is a sampling technique that uses proportions to estimate animal populations. **Captura/recaptura (p. 554)** Captura/recaptura es una técnica de muestreo que usa las proporciones para estimar poblaciones animales.	
Cell (p. 538) A cell is a box where a row and a column meet. **Celda (p. 538)** Una celda es una caja donde se unen una fila y una columna.	Column C and row 2 meet at the shaded box, cell C2.
Center of a circle (p. 350) A circle is named by its center. **Centro de un círculo (p. 350)** Un círculo es denominado por su centro.	Circle O

	A	B	C	D	E
1	0.50	0.70	0.60	0.50	2.30
2	1.50	0.50	2.75	2.50	7.25

Glossary 679

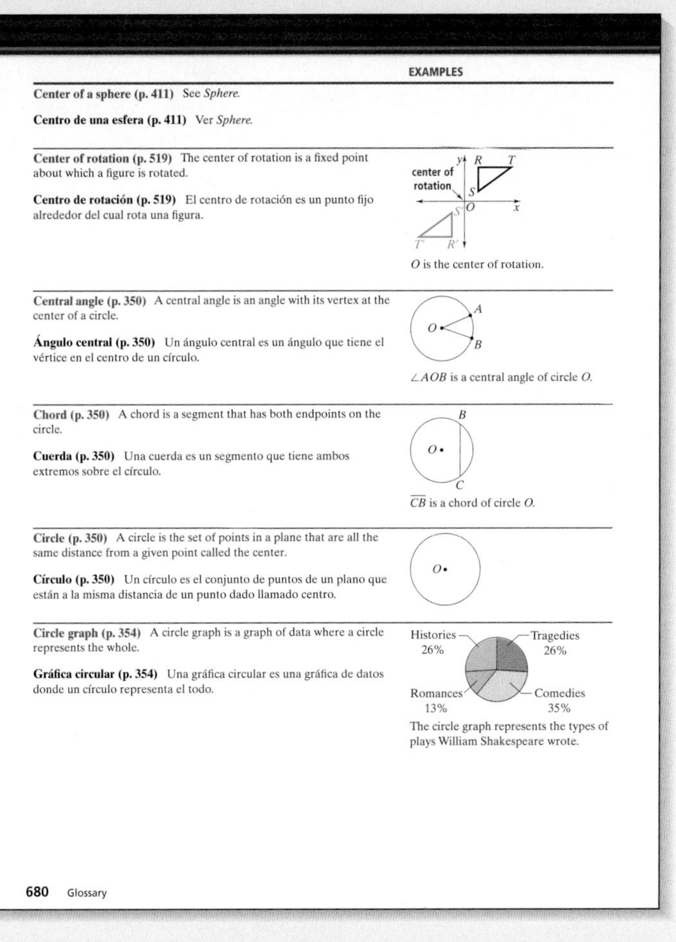

EXAMPLES

Center of a sphere (p. 411) See *Sphere.*

Centro de una esfera (p. 411) Ver *Sphere.*

Center of rotation (p. 519) The center of rotation is a fixed point about which a figure is rotated.

Centro de rotación (p. 519) El centro de rotación es un punto fijo alrededor del cual rota una figura.

O is the center of rotation.

Central angle (p. 350) A central angle is an angle with its vertex at the center of a circle.

Ángulo central (p. 350) Un ángulo central es un ángulo que tiene el vértice en el centro de un círculo.

$\angle AOB$ is a central angle of circle O.

Chord (p. 350) A chord is a segment that has both endpoints on the circle.

Cuerda (p. 350) Una cuerda es un segmento que tiene ambos extremos sobre el círculo.

$\overline{CB}$ is a chord of circle O.

Circle (p. 350) A circle is the set of points in a plane that are all the same distance from a given point called the center.

Círculo (p. 350) Un círculo es el conjunto de puntos de un plano que están a la misma distancia de un punto dado llamado centro.

Circle graph (p. 354) A circle graph is a graph of data where a circle represents the whole.

Gráfica circular (p. 354) Una gráfica circular es una gráfica de datos donde un círculo representa el todo.

The circle graph represents the types of plays William Shakespeare wrote.

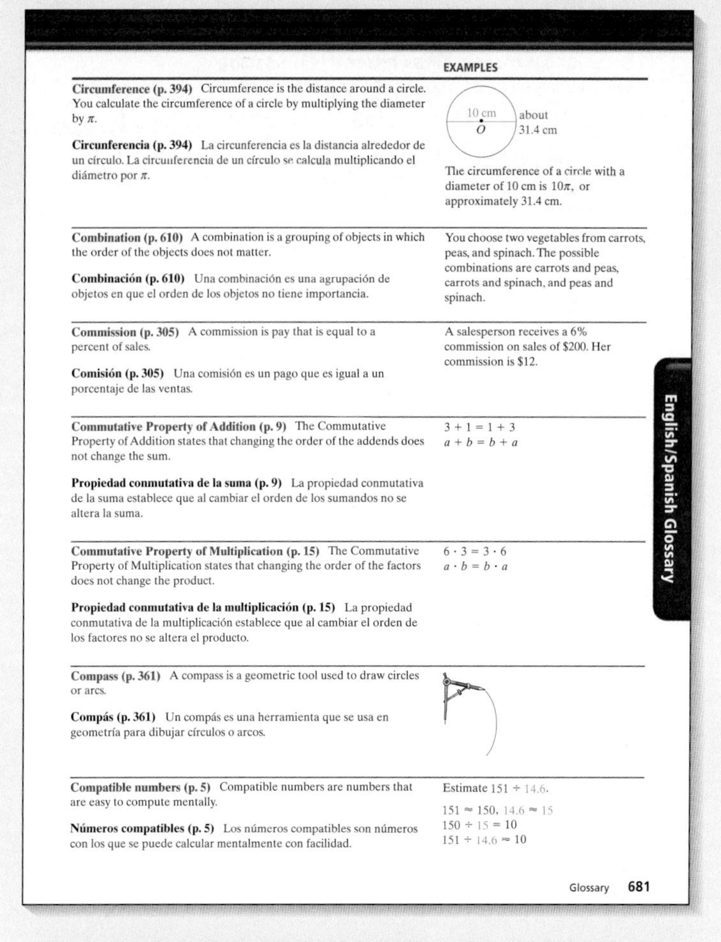

EXAMPLES

Circumference (p. 394) Circumference is the distance around a circle. You calculate the circumference of a circle by multiplying the diameter by π.

Circunferencia (p. 394) La circunferencia es la distancia alrededor de un círculo. La circunferencia de un círculo se calcula multiplicando el diámetro por π.

The circumference of a circle with a diameter of 10 cm is 10π, or approximately 31.4 cm.

Combination (p. 610) A combination is a grouping of objects in which the order of the objects does not matter.

Combinación (p. 610) Una combinación es una agrupación de objetos en que el orden de los objetos no tiene importancia.

You choose two vegetables from carrots, peas, and spinach. The possible combinations are carrots and peas, carrots and spinach, and peas and spinach.

Commission (p. 305) A commission is pay that is equal to a percent of sales.

Comisión (p. 305) Una comisión es un pago que es igual a un porcentaje de las ventas.

A salesperson receives a 6% commission on sales of \$200. Her commission is \$12.

Commutative Property of Addition (p. 9) The Commutative Property of Addition states that changing the order of the addends does not change the sum.

Propiedad conmutativa de la suma (p. 9) La propiedad conmutativa de la suma establece que al cambiar el orden de los sumandos no se altera la suma.

$3 + 1 = 1 + 3$
$a + b = b + a$

Commutative Property of Multiplication (p. 15) The Commutative Property of Multiplication states that changing the order of the factors does not change the product.

Propiedad conmutativa de la multiplicación (p. 15) La propiedad conmutativa de la multiplicación establece que al cambiar el orden de los factores no se altera el producto.

$6 \cdot 3 = 3 \cdot 6$
$a \cdot b = b \cdot a$

Compass (p. 361) A compass is a geometric tool used to draw circles or arcs.

Compás (p. 361) Un compás es una herramienta que se usa en geometría para dibujar círculos o arcos.

Compatible numbers (p. 5) Compatible numbers are numbers that are easy to compute mentally.

Números compatibles (p. 5) Los números compatibles son números con los que se puede calcular mentalmente con facilidad.

Estimate $151 \div 14.6$.
$151 \approx 150$, $14.6 \approx 15$
$150 \div 15 = 10$
$151 \div 14.6 \approx 10$

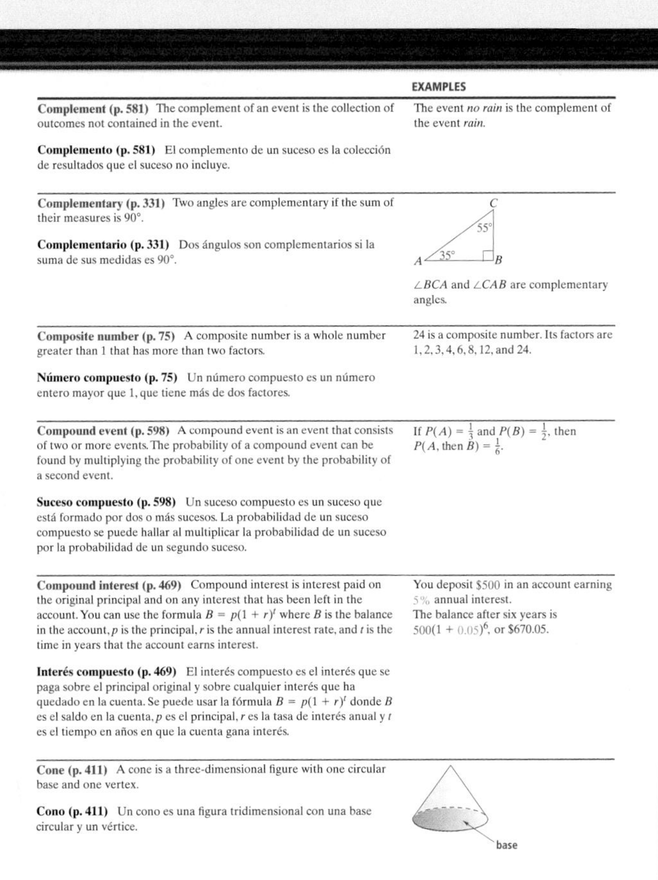

EXAMPLES

Complement (p. 581) The complement of an event is the collection of outcomes not contained in the event.

Complemento (p. 581) El complemento de un suceso es la colección de resultados que el suceso no incluye.

The event *no rain* is the complement of the event *rain.*

Complementary (p. 331) Two angles are complementary if the sum of their measures is 90°.

Complementario (p. 331) Dos ángulos son complementarios si la suma de sus medidas es 90°.

$\angle BCA$ and $\angle CAB$ are complementary angles.

Composite number (p. 75) A composite number is a whole number greater than 1 that has more than two factors.

Número compuesto (p. 75) Un número compuesto es un número entero mayor que 1, que tiene más de dos factores.

24 is a composite number. Its factors are 1, 2, 3, 4, 6, 8, 12, and 24.

Compound event (p. 598) A compound event is an event that consists of two or more events. The probability of a compound event can be found by multiplying the probability of one event by the probability of a second event.

Suceso compuesto (p. 598) Un suceso compuesto es un suceso que está formado por dos o más sucesos. La probabilidad de un suceso compuesto se puede hallar al multiplicar la probabilidad de un suceso por la probabilidad de un segundo suceso.

If $P(A) = \frac{1}{3}$ and $P(B) = \frac{1}{2}$, then $P(A, \text{then } B) = \frac{1}{6}$.

Compound interest (p. 469) Compound interest is interest paid on the original principal and on any interest that has been left in the account. You can use the formula $B = p(1 + r)^t$ where B is the balance in the account, p is the principal, r is the annual interest rate, and t is the time in years that the account earns interest.

Interés compuesto (p. 469) El interés compuesto es el interés que se paga sobre el principal original y sobre cualquier interés que ha quedado en la cuenta. Se puede usar la fórmula $B = p(1 + r)^t$ donde B es el saldo en la cuenta, p es el principal, r es la tasa de interés anual y t es el tiempo en años en que la cuenta gana interés.

You deposit \$500 in an account earning 5% annual interest. The balance after six years is $500(1 + 0.05)^6$, or \$670.05.

Cone (p. 411) A cone is a three-dimensional figure with one circular base and one vertex.

Cono (p. 411) Un cono es una figura tridimensional con una base circular y un vértice.

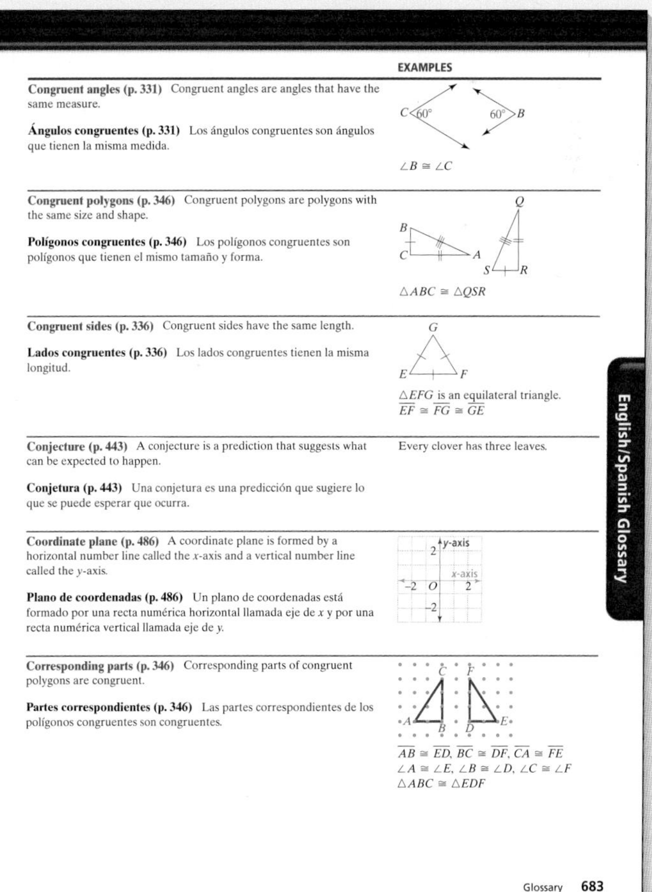

EXAMPLES

Congruent angles (p. 331) Congruent angles are angles that have the same measure.

Ángulos congruentes (p. 331) Los ángulos congruentes son ángulos que tienen la misma medida.

$\angle B \cong \angle C$

Congruent polygons (p. 346) Congruent polygons are polygons with the same size and shape.

Polígonos congruentes (p. 346) Los polígonos congruentes son polígonos que tienen el mismo tamaño y forma.

$\triangle ABC \cong \triangle QSR$

Congruent sides (p. 336) Congruent sides have the same length.

Lados congruentes (p. 336) Los lados congruentes tienen la misma longitud.

$\triangle EFG$ is an equilateral triangle.
$\overline{EF} \cong \overline{FG} \cong \overline{GE}$

Conjecture (p. 443) A conjecture is a prediction that suggests what can be expected to happen.

Conjetura (p. 443) Una conjetura es una predicción que sugiere lo que se puede esperar que ocurra.

Every clover has three leaves.

Coordinate plane (p. 486) A coordinate plane is formed by a horizontal number line called the x-axis and a vertical number line called the y-axis.

Plano de coordenadas (p. 486) Un plano de coordenadas está formado por una recta numérica horizontal llamada eje de x y por una recta numérica vertical llamada eje de y.

Corresponding parts (p. 346) Corresponding parts of congruent polygons are congruent.

Partes correspondientes (p. 346) Las partes correspondientes de los polígonos congruentes son congruentes.

$\overline{AB} \cong \overline{ED}$, $\overline{BC} \cong \overline{DF}$, $\overline{CA} \cong \overline{FE}$
$\angle A \cong \angle E$, $\angle B \cong \angle D$, $\angle C \cong \angle F$
$\triangle ABC \cong \triangle EDF$

EXAMPLES

Counting principle (p. 592) If there are m ways of making one choice from a first situation and n ways of making a choice from a second situation, then there are $m \times n$ ways to make the first choice followed by the second.

Principio de conteo (p. 592) Si hay m maneras de hacer una elección para una primera situación y n maneras de hacer una elección para una segunda situación, entonces hay $m \times n$ maneras de hacer la primera elección seguida de la segunda.

Toss a coin and roll a standard number cube. The total number of possible outcomes is $2 \times 6 = 12$.

Cross products (p. 239) For two ratios, the cross products are found by multiplying the denominator of one ratio by the numerator of the other ratio.

Productos cruzados (p. 239) En dos razones, los productos cruzados se hallan al multiplicar el denominador de una razón por el numerador de la otra razón.

In the proportion $\frac{2}{5} = \frac{10}{25}$, the cross products are $2 \cdot 25$ and $5 \cdot 10$.

Cube (p. 410) A cube is a rectangular prism whose faces are all squares.

Cubo (p. 410) Un cubo es un prisma rectangular cuyas caras son todas cuadrados.

Cubic unit (p. 421) A cubic unit is a cube whose edges are one unit long.

Unidad cúbica (p. 421) Una unidad cúbica es un cubo cuyos lados tienen una unidad de longitud.

1 cm

Cylinder (p. 410) A cylinder is a three-dimensional figure with two congruent parallel bases that are circles.

Cilindro (p. 410) Un cilindro es una figura tridimensional con dos bases congruentes paralelas que son círculos.

base
height
base

D

Decagon (p. 340) A decagon is a polygon with 10 sides.

Decágono (p. 340) Un decágono es un polígono que tiene 10 lados.

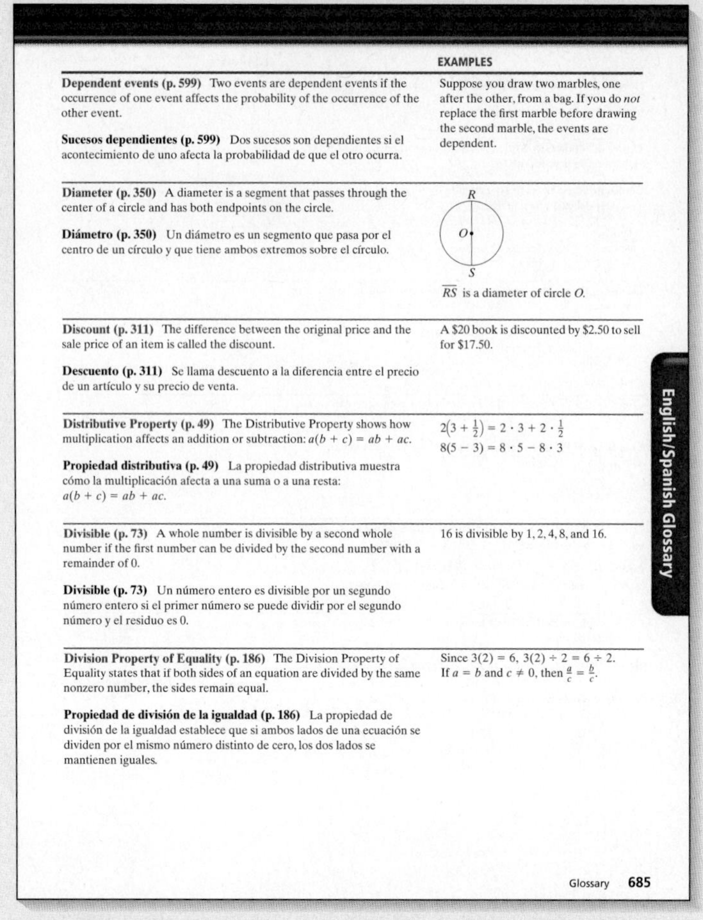

EXAMPLES

Dependent events (p. 599) Two events are dependent events if the occurrence of one event affects the probability of the occurrence of the other event.

Sucesos dependientes (p. 599) Dos sucesos son dependientes si el acontecimiento de uno afecta la probabilidad de que el otro ocurra.

Suppose you draw two marbles, one after the other, from a bag. If you do *not* replace the first marble before drawing the second marble, the events are dependent.

Diameter (p. 350) A diameter is a segment that passes through the center of a circle and has both endpoints on the circle.

Diámetro (p. 350) Un diámetro es un segmento que pasa por el centro de un círculo y que tiene ambos extremos sobre el círculo.

$\overline{RS}$ is a diameter of circle O.

Discount (p. 311) The difference between the original price and the sale price of an item is called the discount.

Descuento (p. 311) Se llama descuento a la diferencia entre el precio de un artículo y su precio de venta.

A \$20 book is discounted by \$2.50 to sell for \$17.50.

Distributive Property (p. 49) The Distributive Property shows how multiplication affects an addition or subtraction: $a(b + c) = ab + ac$.

Propiedad distributiva (p. 49) La propiedad distributiva muestra cómo la multiplicación afecta a una suma o a una resta: $a(b + c) = ab + ac$.

$2\left(3 + \frac{1}{2}\right) = 2 \cdot 3 + 2 \cdot \frac{1}{2}$
$8(5 - 3) = 8 \cdot 5 - 8 \cdot 3$

Divisible (p. 73) A whole number is divisible by a second whole number if the first number can be divided by the second number with a remainder of 0.

Divisible (p. 73) Un número entero es divisible por un segundo número entero si el primer número se puede dividir por el segundo número y el residuo es 0.

16 is divisible by 1, 2, 4, 8, and 16.

Division Property of Equality (p. 186) The Division Property of Equality states that if both sides of an equation are divided by the same nonzero number, the sides remain equal.

Propiedad de división de la igualdad (p. 186) La propiedad de división de la igualdad establece que si ambos lados de una ecuación se dividen por el mismo número distinto de cero, los dos lados se mantienen iguales.

Since $3(2) = 6$, $3(2) \div 2 = 6 \div 2$.
If $a = b$ and $c \neq 0$, then $\frac{a}{c} = \frac{b}{c}$.

English/Spanish Glossary

Glossary 685

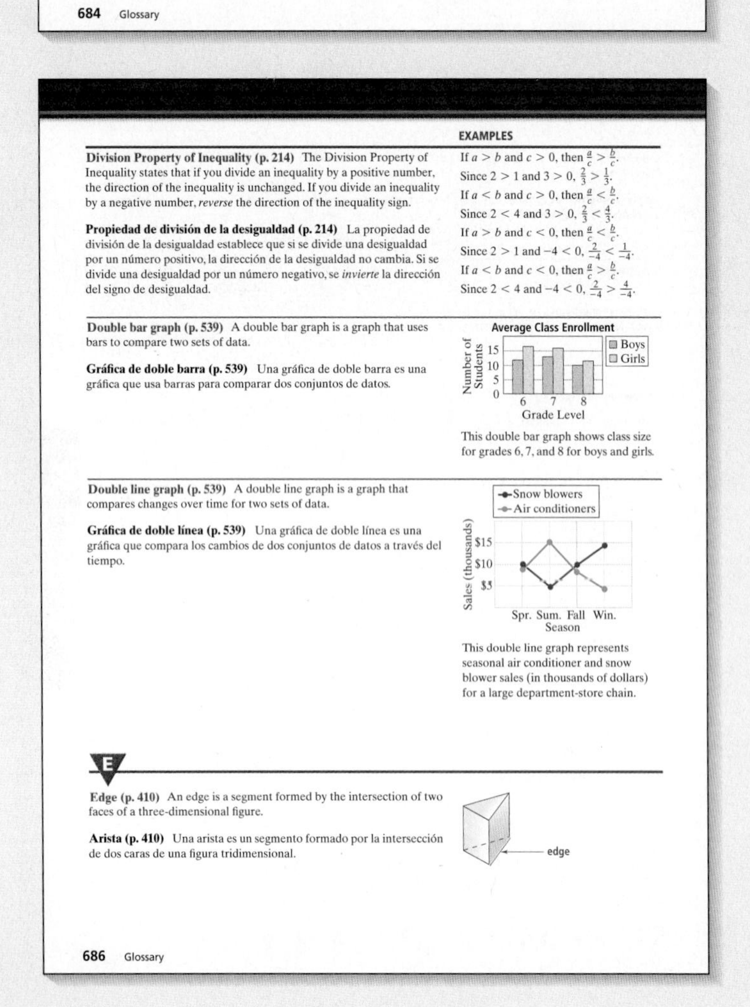

EXAMPLES

Division Property of Inequality (p. 214) The Division Property of Inequality states that if you divide an inequality by a positive number, the direction of the inequality is unchanged. If you divide an inequality by a negative number, *reverse* the direction of the inequality sign.

Propiedad de división de la desigualdad (p. 214) La propiedad de división de la desigualdad establece que si se divide una desigualdad por un número positivo, la dirección de la desigualdad no cambia. Si se divide una desigualdad por un número negativo, se *invierte* la dirección del signo de desigualdad.

If $a > b$ and $c > 0$, then $\frac{a}{c} > \frac{b}{c}$.
Since $2 > 1$ and $3 > 0$, $\frac{2}{3} > \frac{1}{3}$.
If $a < b$ and $c > 0$, then $\frac{a}{c} < \frac{b}{c}$.
Since $2 < 4$ and $3 > 0$, $\frac{2}{3} < \frac{4}{3}$.
If $a > b$ and $c < 0$, then $\frac{a}{c} < \frac{b}{c}$.
Since $2 > 1$ and $-4 < 0$, $\frac{2}{-4} < \frac{1}{-4}$.
If $a < b$ and $c < 0$, then $\frac{a}{c} > \frac{b}{c}$.
Since $2 < 4$ and $-4 < 0$, $\frac{2}{-4} > \frac{4}{-4}$.

Double bar graph (p. 539) A double bar graph is a graph that uses bars to compare two sets of data.

Gráfica de doble barra (p. 539) Una gráfica de doble barra es una gráfica que usa barras para comparar dos conjuntos de datos.

This double bar graph shows class size for grades 6, 7, and 8 for boys and girls.

Double line graph (p. 539) A double line graph is a graph that compares changes over time for two sets of data.

Gráfica de doble línea (p. 539) Una gráfica de doble línea es una gráfica que compara los cambios de dos conjuntos de datos a través del tiempo.

This double line graph represents seasonal air conditioner and snow blower sales (in thousands of dollars) for a large department-store chain.

E

Edge (p. 410) An edge is a segment formed by the intersection of two faces of a three-dimensional figure.

Arista (p. 410) Una arista es un segmento formado por la intersección de dos caras de una figura tridimensional.

686 Glossary

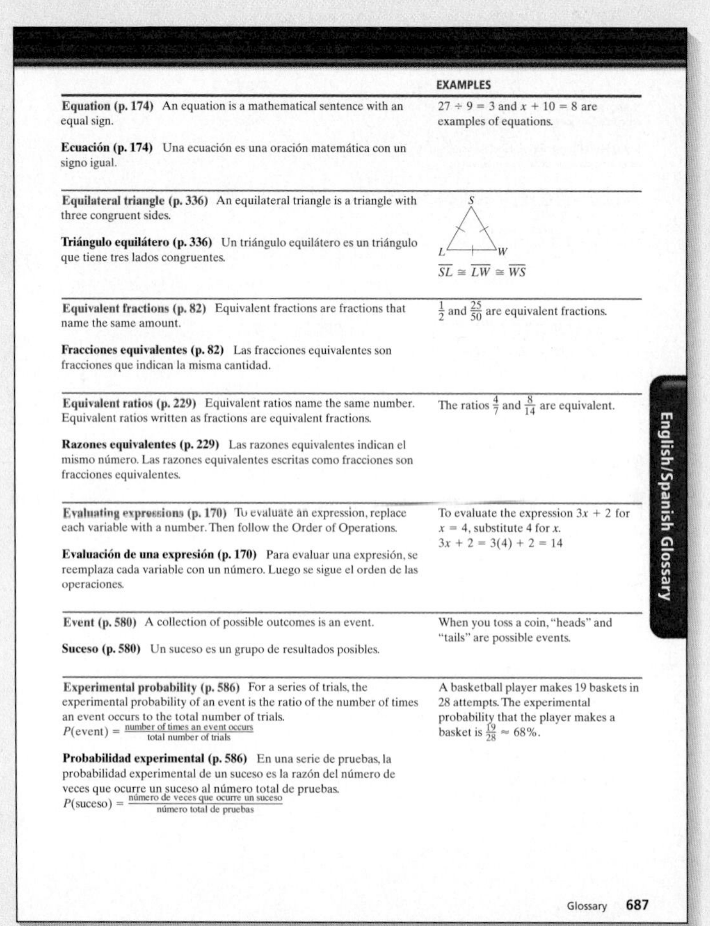

EXAMPLES

Equation (p. 174) An equation is a mathematical sentence with an equal sign.

Ecuación (p. 174) Una ecuación es una oración matemática con un signo igual.

$27 \div 9 = 3$ and $x + 10 = 8$ are examples of equations.

Equilateral triangle (p. 336) An equilateral triangle is a triangle with three congruent sides.

Triángulo equilátero (p. 336) Un triángulo equilátero es un triángulo que tiene tres lados congruentes.

$\overline{SL} \cong \overline{LW} \cong \overline{WS}$

Equivalent fractions (p. 82) Equivalent fractions are fractions that name the same amount.

Fracciones equivalentes (p. 82) Las fracciones equivalentes son fracciones que indican la misma cantidad.

$\frac{1}{2}$ and $\frac{25}{50}$ are equivalent fractions.

Equivalent ratios (p. 229) Equivalent ratios name the same number. Equivalent ratios written as fractions are equivalent fractions.

Razones equivalentes (p. 229) Las razones equivalentes indican el mismo número. Las razones equivalentes escritas como fracciones son fracciones equivalentes.

The ratios $\frac{4}{7}$ and $\frac{8}{14}$ are equivalent.

Evaluating expressions (p. 170) To evaluate an expression, replace each variable with a number. Then follow the Order of Operations.

Evaluación de una expresión (p. 170) Para evaluar una expresión, se reemplaza cada variable con un número. Luego se sigue el orden de las operaciones.

To evaluate the expression $3x + 2$ for $x = 4$, substitute 4 for x.
$3x + 2 = 3(4) + 2 = 14$

Event (p. 580) A collection of possible outcomes is an event.

Suceso (p. 580) Un suceso es un grupo de resultados posibles.

When you toss a coin, "heads" and "tails" are possible events.

Experimental probability (p. 586) For a series of trials, the experimental probability of an event is the ratio of the number of times an event occurs to the total number of trials.
$P(\text{event}) = \frac{\text{number of times an event occurs}}{\text{total number of trials}}$

Probabilidad experimental (p. 586) En una serie de pruebas, la probabilidad experimental de un suceso es la razón del número de veces que ocurre un suceso al número total de pruebas.
$P(\text{suceso}) = \frac{\text{número de veces que ocurre un suceso}}{\text{número total de pruebas}}$

A basketball player makes 19 baskets in 28 attempts. The experimental probability that the player makes a basket is $\frac{19}{28} \approx 68\%$.

English/Spanish Glossary

Glossary 687

EXAMPLES

Exponent (p. 68) An exponent tells how many times a number, or base, is used as a factor.

Exponente (p. 68) Un exponente dice cuántas veces se usa como factor un número, o base.

exponent
$3^4 = 3 \times 3 \times 3 \times 3$
Read 3^4 as *three to the fourth power.*

F

Face (p. 410) A face is a flat surface of a three-dimensional figure that is shaped like a polygon.

Cara (p. 410) Una cara es una superficie plana de una figura tridimensional que tiene la forma de un polígono.

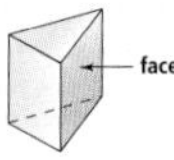

Factor (p. 75) A factor is a whole number that divides another whole number with a remainder of 0.

Divisor (p. 75) Un divisor es un número entero que divide a otro número entero y el residuo es 0.

1, 2, 3, 4, 6, 12, 18, and 36 are factors of 36.

Factorial (p. 606) A factorial is the product of all positive integers less than or equal to a number. The symbol for factorial is an exclamation point.

Factorial (p. 606) Una factorial es el producto de todos los enteros positivos menores o iguales que un número. El símbolo de factorial es un signo de cierre de exclamación.

$5! = 5 \times 4 \times 3 \times 2 \times 1 = 120$

Formula (p. 472) A formula is a rule that shows the relationship between two or more quantities.

Fórmula (p. 472) Una fórmula es una regla que muestra la relación entre dos o más cantidades.

The formula $P = 2\ell + 2w$ gives the perimeter of a rectangle in terms of its length and width.

Frequency table (p. 532) A frequency table is a table that lists each item in a data set with the number of times the item occurs.

Tabla de frecuencia (p. 532) Una tabla de frecuencia es una tabla que registra todos los elementos de un conjunto de datos y el número de veces que ocurre cada uno.

Household Telephones

Phones	Tally	Frequency
1	𝍸 |||	8
2	𝍸		6
3						4

This frequency table shows the number of household telephones for a class of students.

EXAMPLES

Function (p. 452) A function is a relationship that assigns exactly one output value for each input value.

Función (p. 452) Una función es una relación que asigna exactamente un valor resultante a cada valor inicial.

Wages s are a function of the number of hours worked w. If you earn \$6/h, then your wages can be expressed by the function $s = 6w$.

G

Geometric sequence (p. 442) In a geometric sequence, each term is the result of multiplying the previous term by a fixed number (called the common ratio).

Progresión geométrica (p. 442) En una progresión geométrica, cada término es el resultado de la multiplicación del término anterior por un número fijo llamado rayón común.

The sequence 1, 3, 9, 27, 81, . . . is a geometric sequence. You multiply each term by 3 to find the next term.

Graph of an equation (p. 492) The graph of an equation is the graph of all the points with coordinates that are solutions of the equation.

Gráfica de una ecuación (p. 492) La gráfica de una ecuación es la gráfica de todos los puntos cuyas coordenadas son soluciones a la ecuación.

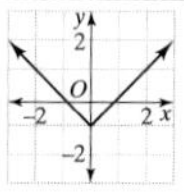

The coordinates of all the points on the graph satisfy the equation $y = |x| - 1$.

Greatest common factor (GCF) (p. 75) The greatest common factor of two or more numbers is the greatest number that is a factor of all of the numbers.

Máximo común divisor (MCD) (p. 75) El máximo común divisor de dos o más números es el mayor número que es divisor de todos los números.

The GCF of 12 and 30 is 6.

H

Height of three-dimensional figures (p. 410) See *Cylinder* and *Prism*.

Altura de figuras tridimensionales (p. 410) Ver *Cylinder* y *Prism*.

EXAMPLES

Height of two-dimensional figures (pp. 380, 384, 388) See *Parallelogram, Triangle*, and *Trapezoid.*

Altura de figuras bidimensionales (pp. 380, 384, 388) Ver *Parallelogram, Triangle* y *Trapezoid.*

Hexagon (p. 340) A hexagon is a polygon with six sides.

Hexágono (p. 340) Un hexágono es un polígono que tiene seis lados.

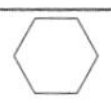

Histogram (p. 533) A histogram is a bar graph with no spaces between the bars. The height of each bar shows the frequency of data within that interval.

Histograma (p. 533) Un histograma es una gráfica de barras sin espacio entre las barras. La altura de cada barra muestra la frecuencia de los datos dentro del intervalo.

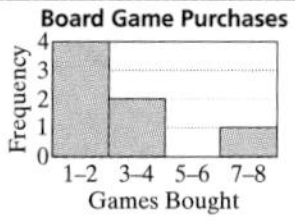

The histogram gives the frequency of board game purchases at a local toy store.

Horizontal (p. 499) Horizontal lines are parallel to the x-axis.

Horizontal (p. 499) Las rectas horizontales son paralelas al eje de x.

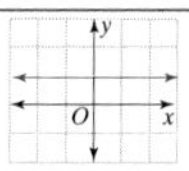

Hypotenuse (p. 405) In a right triangle, the hypotenuse is the longest side, which is opposite the right angle.

Hipotenusa (p. 405) En un triángulo rectángulo, la hipotenusa es el lado más largo, que es el lado opuesto al ángulo recto.

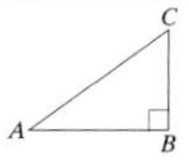

$\overline{AC}$ is the hypotenuse of $\triangle ABC$.

I

Identity Property of Addition (p. 9) The Identity Property of Addition states that the sum of 0 and a is a.

Propiedad de identidad de la suma (p. 9) La propiedad de identidad de la suma establece que la suma de cero y a es a.

$7 + 0 = 7$
$a + 0 = a$

EXAMPLES

Identity Property of Multiplication (p. 15) The Identity Property of Multiplication states that the product of 1 and a is a.

Propiedad de identidad de la multiplicación (p. 15) La propiedad de identidad de la multiplicación establece que el producto de 1 y a es a.

$7 \cdot 1 = 7$
$a \cdot 1 = a$

Image (p. 510) An image is the result of a transformation of a point, line, or figure.

Imagen (p. 510) Una imagen es el resultado de una transformación de un punto, una recta o una figura.

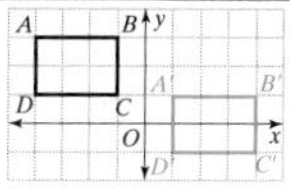

$A'B'C'D'$ is the image of $ABCD$.

Improper fraction (p. 91) An improper fraction has a numerator that is greater than or equal to its denominator.

Fracción impropia (p. 91) Una fracción impropia tiene un numerador mayor o igual que su denominador.

$\frac{24}{15}$ and $\frac{16}{16}$ are improper fractions.

Independent events (p. 598) Two events are independent events if the occurrence of one event does not affect the probability of the occurrence of the other.

Sucesos independientes (p. 598) Dos sucesos son independientes si el acontecimiento de uno no afecta la probabilidad de que el otro suceso ocurra.

Suppose you draw two marbles, one after the other, from a bag. If you replace the first marble before drawing the second marble, the events are independent.

Indirect measurement (p. 253) Indirect measurement uses proportions and similar triangles to measure distances that would be difficult to measure directly.

Medición indirecta (p. 253) La medición indirecta usa proporciones y triángulos semejantes para medir las distancias que serían difíciles de medir directamente.

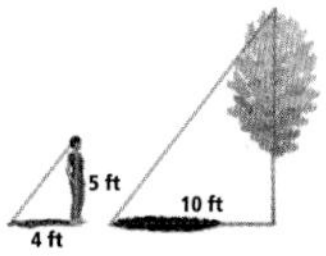

A 5-ft-tall person standing near a tree has a shadow 4 ft long. The tree has a shadow 10 ft long. The height of the tree is 12.5 ft.

Inductive reasoning (p. 443) Inductive reasoning involves looking for a pattern and writing a rule to describe the pattern in a sequence.

Razonamiento inductivo (p. 443) El razonamiento inductivo implica buscar un patrón y escribir una regla para describir el patrón en una secuencia.

By inductive reasoning, the next number in the pattern 2, 4, 6, 8, . . . is 10.

EXAMPLES

Inequality (p. 205) An inequality is a mathematical sentence that contains one of the signs <, >, ≤, ≥, or ≠.

Desigualdad (p. 205) Una desigualdad es una oración matemática que contiene uno de los signos <, >, ≤, ≥ o ≠.

$x < -5$, $x > 8$, $x \leq 1$, $x \geq -11$, $x \neq 7$

Integers (p. 31) Integers are the set of positive whole numbers, their opposites, and 0.

Enteros (p. 31) Los enteros son el conjunto de números enteros positivos, sus opuestos y el 0.

. . . −3, −2, −1, 0, 1, 2, 3, . . .

Intersecting lines (p. 325) Intersecting lines have exactly one point in common.

Rectas que se intersecan (p. 325) Las rectas que se intersecan tienen exactamente un punto en común.

Inverse operations (p. 181) Inverse operations are operations that undo each other.

Operaciones inversas (p. 181) Las operaciones inversas son las operaciones que se anulan entre ellas.

Addition and subtraction are inverse operations.

Irrational number (p. 401) An irrational number is a number that cannot be written as the ratio of two integers. In decimal form, an irrational number cannot be written as a terminating or repeating decimal.

Número irracional (p. 401) Un número irracional es un número que no se puede escribir como una razón de dos enteros. Como decimal, un número irracional no se puede escribir como decimal finito o periódico.

The numbers π and 2.41592653 . . . are irrational numbers.

Irregular polygon (p. 340) An irregular polygon is a polygon with sides that are not all congruent and/or angles that are not all congruent.

Polígono irregular (p. 340) Un polígono irregular es un polígono que tiene lados que no son todos congruentes y/o ángulos que no son todos congruentes.

KLMN is an irregular polygon.

Isosceles triangle (p. 336) An isosceles triangle is a triangle with at least two congruent sides.

Triángulo isósceles (p. 336) Un triángulo isósceles es un triángulo que tiene al menos dos lados congruentes.

$\overline{LM} \cong \overline{LB}$

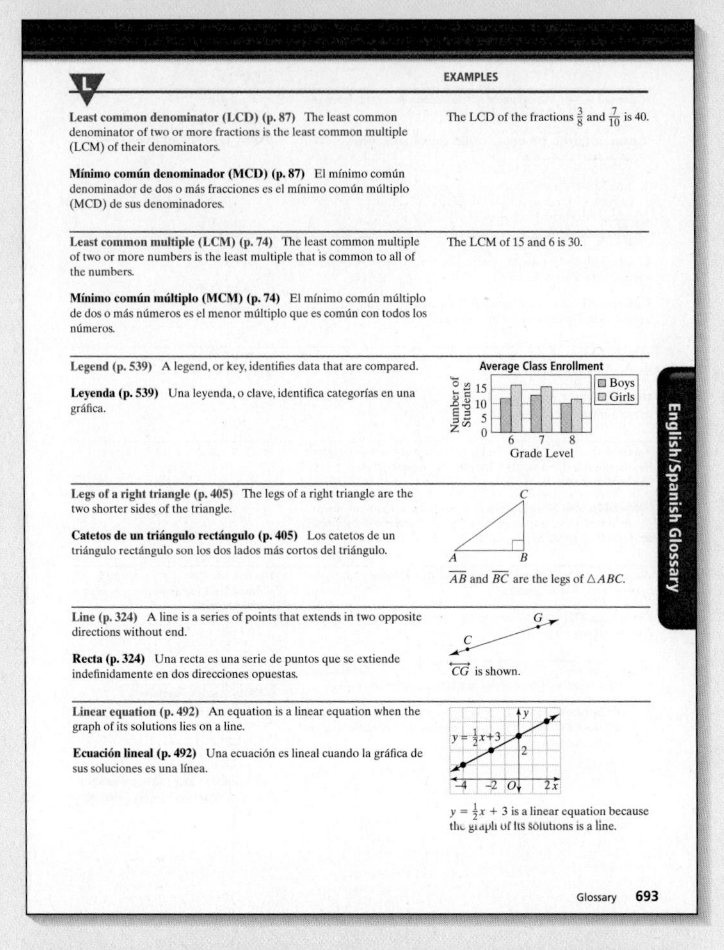

EXAMPLES

L

Least common denominator (LCD) (p. 87) The least common denominator of two or more fractions is the least common multiple (LCM) of their denominators.

Mínimo común denominador (MCD) (p. 87) El mínimo común denominador de dos o más fracciones es el mínimo común múltiplo (MCD) de sus denominadores.

The LCD of the fractions $\frac{3}{8}$ and $\frac{7}{10}$ is 40.

Least common multiple (LCM) (p. 74) The least common multiple of two or more numbers is the least multiple that is common to all of the numbers.

Mínimo común múltiplo (MCM) (p. 74) El mínimo común múltiplo de dos o más números es el menor múltiplo que es común con todos los números.

The LCM of 15 and 6 is 30.

Legend (p. 539) A legend, or key, identifies data that are compared.

Leyenda (p. 539) Una leyenda, o clave, identifica categorías en una gráfica.

Legs of a right triangle (p. 405) The legs of a right triangle are the two shorter sides of the triangle.

Catetos de un triángulo rectángulo (p. 405) Los catetos de un triángulo rectángulo son los dos lados más cortos del triángulo.

$\overline{AB}$ and $\overline{BC}$ are the legs of $\triangle ABC$.

Line (p. 324) A line is a series of points that extends in two opposite directions without end.

Recta (p. 324) Una recta es una serie de puntos que se extiende indefinidamente en dos direcciones opuestas.

$\overleftrightarrow{CG}$ is shown.

Linear equation (p. 492) An equation is a linear equation when the graph of its solutions lies on a line.

Ecuación lineal (p. 492) Una ecuación es lineal cuando la gráfica de sus soluciones es una línea.

$y = \frac{1}{2}x + 3$ is a linear equation because the graph of its solutions is a line.

English/Spanish Glossary

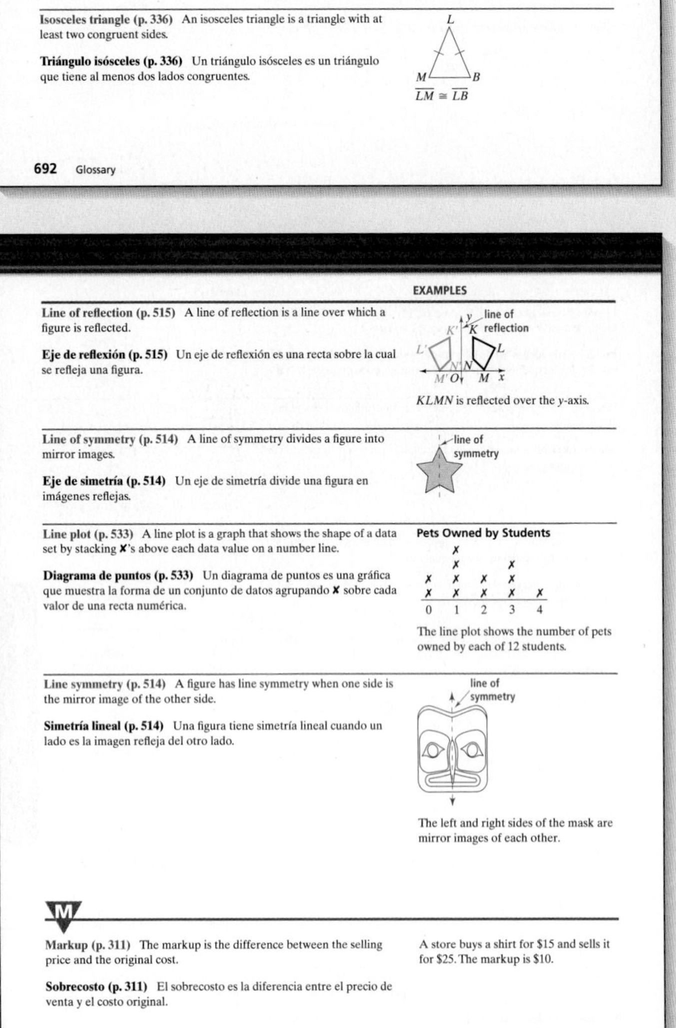

EXAMPLES

Line of reflection (p. 515) A line of reflection is a line over which a figure is reflected.

Eje de reflexión (p. 515) Un eje de reflexión es una recta sobre la cual se refleja una figura.

KLMN is reflected over the *y*-axis.

Line of symmetry (p. 514) A line of symmetry divides a figure into mirror images.

Eje de simetría (p. 514) Un eje de simetría divide una figura en imágenes reflejas.

Line plot (p. 533) A line plot is a graph that shows the shape of a data set by stacking ✗'s above each data value on a number line.

Diagrama de puntos (p. 533) Un diagrama de puntos es una gráfica que muestra la forma de un conjunto de datos agrupando ✗ sobre cada valor de una recta numérica.

Pets Owned by Students

The line plot shows the number of pets owned by each of 12 students.

Line symmetry (p. 514) A figure has line symmetry when one side is the mirror image of the other side.

Simetría lineal (p. 514) Una figura tiene simetría lineal cuando un lado es la imagen refleja del otro lado.

The left and right sides of the mask are mirror images of each other.

M

Markup (p. 311) The markup is the difference between the selling price and the original cost.

Sobrecosto (p. 311) El sobrecosto es la diferencia entre el precio de venta y el costo original.

A store buys a shirt for \$15 and sells it for \$25. The markup is \$10.

EXAMPLES

Mean (p. 53) The mean of a set of data values is the sum of the data divided by the number of data items.

Media (p. 53) La media de un conjunto de valores de datos es la suma de los datos dividida por el número de datos.

The mean temperature (°F) for the set of temperatures 44, 52, 48, 55, 61, and 67 is $\frac{44+52+48+55+61+67}{6} = 54.5$.

Median (p. 54) The median of a data set is the middle value when the data are arranged in numerical order. When there is an even number of data values, the median is the mean of the two middle values.

Mediana (p. 54) La mediana de un conjunto de datos es el valor del medio cuando los datos están organizados en orden numérico. Cuando hay un número par de valores de datos, la mediana es la media de los dos valores del medio.

Temperatures (°F) for five days arranged in order are 44, 48, 52, 55, and 58. The median temperature is 52°F because it is the middle number in the set of data.

Midpoint (p. 362) The midpoint of a segment is the point that divides the segment into two segments of equal length.

Punto medio (p. 362) El punto medio de un segmento es el punto que divide el segmento en dos segmentos de igual longitud.

$XM = YM$. M is the midpoint of $\overline{XY}$.

Mixed number (p. 91) A mixed number is the sum of a whole number and a fraction.

Número mixto (p. 91) Un número mixto es la suma de un número entero y una fracción.

$3\frac{11}{16}$ is a mixed number.
$3\frac{11}{16} = 3 + \frac{11}{16}$

Mode (p. 54) The mode of a data set is the item that occurs with the greatest frequency.

Moda (p. 54) La moda de un conjunto de datos es el dato que sucede con mayor frecuencia.

The mode of the set of prices \$2.50, \$2.75, \$3.60, \$2.75, and \$3.70 is \$2.75.

Multiple (p. 74) A multiple of a number is the product of that number and any nonzero whole number.

Múltiplo (p. 74) Un múltiplo de un número es el producto de ese número y cualquier número entero diferente de cero.

The number 39 is a multiple of 13.

Multiplication Property of Equality (p. 188) The Multiplication Property of Equality states that if each side of an equation is multiplied by the same number, the results are equal.

Propiedad multiplicativa de la igualdad (p. 188) La propiedad multiplicativa de la igualdad establece que si cada lado de una ecuación se multiplica por el mismo número, los resultados son iguales.

Since $\frac{12}{2} = 6$, $\frac{12}{2} \cdot 2 = 6 \cdot 2$.
If $a = b$, then $a \cdot c = b \cdot c$.

English/Spanish Glossary

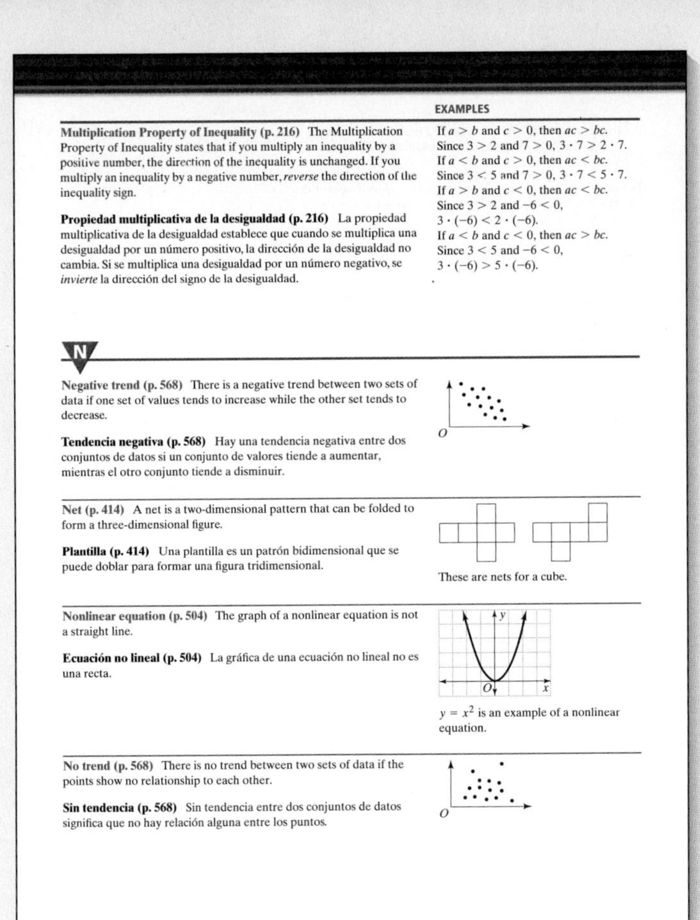

EXAMPLES

Multiplication Property of Inequality (p. 216) The Multiplication Property of Inequality states that if you multiply an inequality by a positive number, the direction of the inequality is unchanged. If you multiply an inequality by a negative number, *reverse* the direction of the inequality sign.

Propiedad multiplicativa de la desigualdad (p. 216) La propiedad multiplicativa de la desigualdad establece que cuando se multiplica una desigualdad por un número positivo, la dirección de la desigualdad no cambia. Si se multiplica una desigualdad por un número negativo, se *invierte* la dirección del signo de la desigualdad.

If $a > b$ and $c > 0$, then $ac > bc$.
Since $3 > 2$ and $7 > 0$, $3 \cdot 7 > 2 \cdot 7$.
If $a < b$ and $c > 0$, then $ac < bc$.
Since $3 < 5$ and $7 > 0$, $3 \cdot 7 < 5 \cdot 7$.
If $a > b$ and $c < 0$, then $ac < bc$.
Since $3 > 2$ and $-6 < 0$,
$3 \cdot (-6) < 2 \cdot (-6)$.
If $a < b$ and $c < 0$, then $ac > bc$.
Since $3 < 5$ and $-6 < 0$,
$3 \cdot (-6) > 5 \cdot (-6)$.

N

Negative trend (p. 568) There is a negative trend between two sets of data if one set of values tends to increase while the other set tends to decrease.

Tendencia negativa (p. 568) Hay una tendencia negativa entre dos conjuntos de datos si un conjunto de valores tiende a aumentar, mientras el otro conjunto tiende a disminuir.

Net (p. 414) A net is a two-dimensional pattern that can be folded to form a three-dimensional figure.

Plantilla (p. 414) Una plantilla es un patrón bidimensional que se puede doblar para formar una figura tridimensional.

These are nets for a cube.

Nonlinear equation (p. 504) The graph of a nonlinear equation is not a straight line.

Ecuación no lineal (p. 504) La gráfica de una ecuación no lineal no es una recta.

$y = x^2$ is an example of a nonlinear equation.

No trend (p. 568) There is no trend between two sets of data if the points show no relationship to each other.

Sin tendencia (p. 568) Sin tendencia entre dos conjuntos de datos significa que no hay relación alguna entre los puntos.

696 Glossary

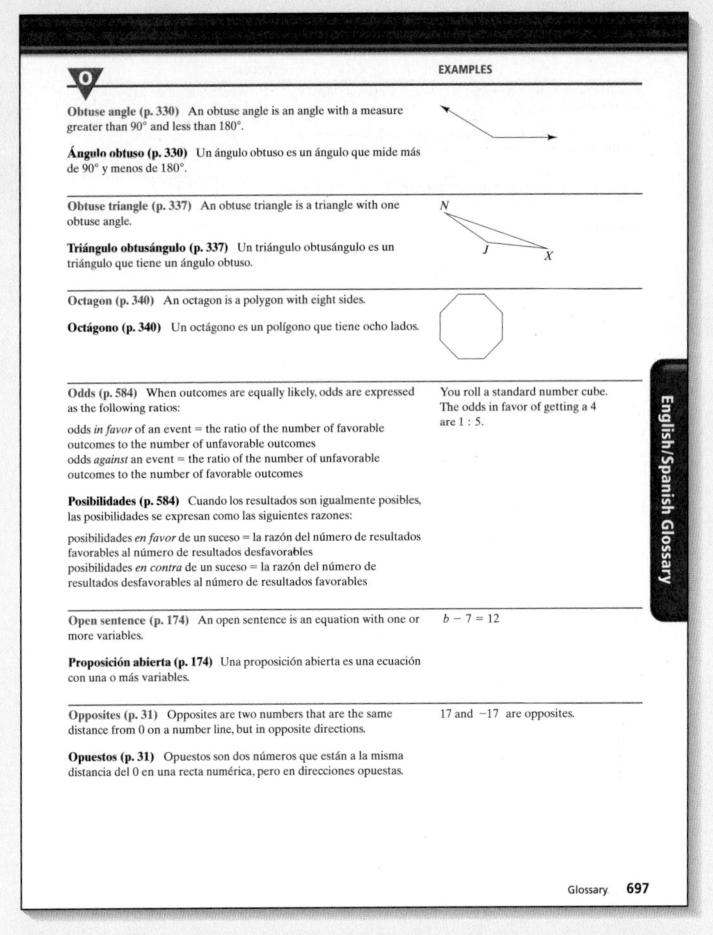

EXAMPLES

O

Obtuse angle (p. 330) An obtuse angle is an angle with a measure greater than 90° and less than 180°.

Ángulo obtuso (p. 330) Un ángulo obtuso es un ángulo que mide más de 90° y menos de 180°.

Obtuse triangle (p. 337) An obtuse triangle is a triangle with one obtuse angle.

Triángulo obtusángulo (p. 337) Un triángulo obtusángulo es un triángulo que tiene un ángulo obtuso.

Octagon (p. 340) An octagon is a polygon with eight sides.

Octágono (p. 340) Un octágono es un polígono que tiene ocho lados.

Odds (p. 584) When outcomes are equally likely, odds are expressed as the following ratios:

odds *in favor* of an event = the ratio of the number of favorable outcomes to the number of unfavorable outcomes
odds *against* an event = the ratio of the number of unfavorable outcomes to the number of favorable outcomes

Posibilidades (p. 584) Cuando los resultados son igualmente posibles, las posibilidades se expresan como las siguientes razones:

posibilidades *en favor* de un suceso = la razón del número de resultados favorables al número de resultados desfavorables
posibilidades *en contra* de un suceso = la razón del número de resultados desfavorables al número de resultados favorables

You roll a standard number cube. The odds in favor of getting a 4 are 1 : 5.

Open sentence (p. 174) An open sentence is an equation with one or more variables.

Proposición abierta (p. 174) Una proposición abierta es una ecuación con una o más variables.

$b - 7 = 12$

Opposites (p. 31) Opposites are two numbers that are the same distance from 0 on a number line, but in opposite directions.

Opuestos (p. 31) Opuestos son dos números que están a la misma distancia del 0 en una recta numérica, pero en direcciones opuestas.

17 and −17 are opposites.

Glossary 697

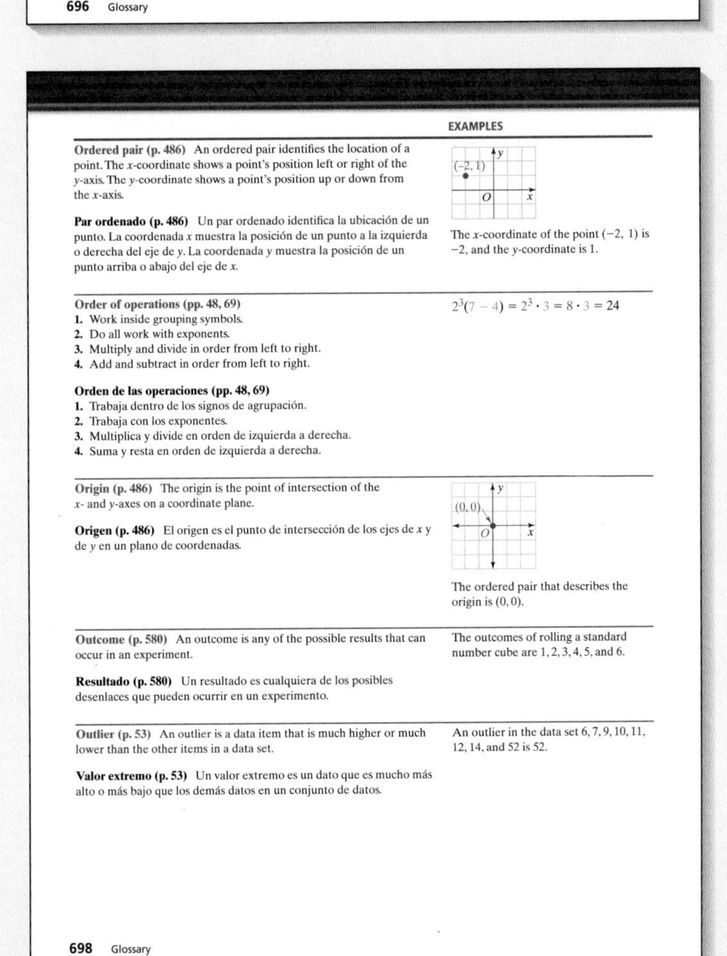

EXAMPLES

Ordered pair (p. 486) An ordered pair identifies the location of a point. The *x*-coordinate shows a point's position left or right of the *y*-axis. The *y*-coordinate shows a point's position up or down from the *x*-axis.

Par ordenado (p. 486) Un par ordenado identifica la ubicación de un punto. La coordenada *x* muestra la posición de un punto a la izquierda o derecha del eje de *y*. La coordenada *y* muestra la posición de un punto arriba o abajo del eje de *x*.

The *x*-coordinate of the point (−2, 1) is −2, and the *y*-coordinate is 1.

Order of operations (pp. 48, 69)
1. Work inside grouping symbols.
2. Do all work with exponents.
3. Multiply and divide in order from left to right.
4. Add and subtract in order from left to right.

Orden de las operaciones (pp. 48, 69)
1. Trabaja dentro de los signos de agrupación.
2. Trabaja con los exponentes.
3. Multiplica y divide en orden de izquierda a derecha.
4. Suma y resta en orden de izquierda a derecha.

$2^3(7 - 4) = 2^3 \cdot 3 = 8 \cdot 3 = 24$

Origin (p. 486) The origin is the point of intersection of the *x*- and *y*-axes on a coordinate plane.

Origen (p. 486) El origen es el punto de intersección de los ejes de *x* y de *y* en un plano de coordenadas.

The ordered pair that describes the origin is (0, 0).

Outcome (p. 580) An outcome is any of the possible results that can occur in an experiment.

Resultado (p. 580) Un resultado es cualquiera de los posibles desenlaces que pueden ocurrir en un experimento.

The outcomes of rolling a standard number cube are 1, 2, 3, 4, 5, and 6.

Outlier (p. 53) An outlier is a data item that is much higher or much lower than the other items in a data set.

Valor extremo (p. 53) Un valor extremo es un dato que es mucho más alto o más bajo que los demás datos en un conjunto de datos.

An outlier in the data set 6, 7, 9, 10, 11, 12, 14, and 52 is 52.

698 Glossary

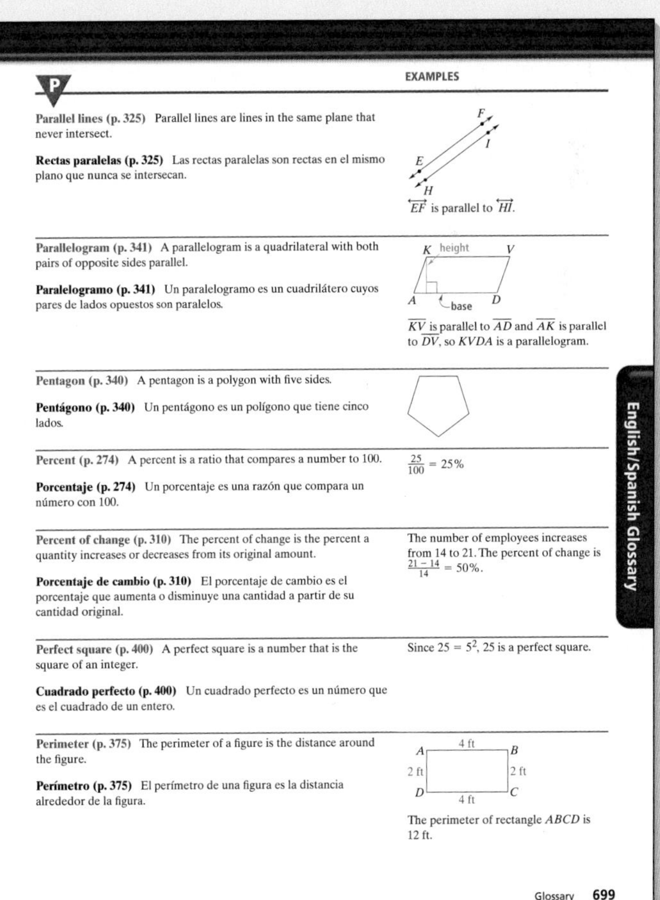

EXAMPLES

P

Parallel lines (p. 325) Parallel lines are lines in the same plane that never intersect.

Rectas paralelas (p. 325) Las rectas paralelas son rectas en el mismo plano que nunca se intersecan.

$\overleftrightarrow{EF}$ is parallel to $\overleftrightarrow{HI}$.

Parallelogram (p. 341) A parallelogram is a quadrilateral with both pairs of opposite sides parallel.

Paralelogramo (p. 341) Un paralelogramo es un cuadrilátero cuyos pares de lados opuestos son paralelos.

$\overline{KV}$ is parallel to $\overline{AD}$ and $\overline{AK}$ is parallel to $\overline{DV}$, so *KVDA* is a parallelogram.

Pentagon (p. 340) A pentagon is a polygon with five sides.

Pentágono (p. 340) Un pentágono es un polígono que tiene cinco lados.

Percent (p. 274) A percent is a ratio that compares a number to 100.

Porcentaje (p. 274) Un porcentaje es una razón que compara un número con 100.

$\frac{25}{100} = 25\%$

Percent of change (p. 310) The percent of change is the percent a quantity increases or decreases from its original amount.

Porcentaje de cambio (p. 310) El porcentaje de cambio es el porcentaje que aumenta o disminuye una cantidad a partir de su cantidad original.

The number of employees increases from 14 to 21. The percent of change is $\frac{21 - 14}{14} = 50\%$.

Perfect square (p. 400) A perfect square is a number that is the square of an integer.

Cuadrado perfecto (p. 400) Un cuadrado perfecto es un número que es el cuadrado de un entero.

Since $25 = 5^2$, 25 is a perfect square.

Perimeter (p. 375) The perimeter of a figure is the distance around the figure.

Perímetro (p. 375) El perímetro de una figura es la distancia alrededor de la figura.

The perimeter of rectangle *ABCD* is 12 ft.

Glossary 699

English/Spanish Glossary

EXAMPLES

Permutation (p. 606) A permutation is an arrangement of objects in a particular order.

Permutación (p. 606) Una permutación es un arreglo de objetos en un orden particular.

The permutations of the letters W, A, and X are WAX, WXA, AXW, AWX, XWA, and XAW.

Perpendicular bisector (p. 362) A perpendicular bisector is a segment bisector that is perpendicular to the segment.

Mediatriz (p. 362) Una mediatriz es una bisectriz de un segmento que es perpendicular a ese segmento.

K, M, A, B

$\overleftrightarrow{MK} \perp \overline{AB}$, $AM = MB$. $\overleftrightarrow{MK}$ is the perpendicular bisector of $\overline{AB}$.

Perpendicular lines (p. 362) Perpendicular lines intersect to form right angles.

Rectas perpendiculars (p. 362) Las rectas perpendiculares se intersecan para formar ángulos rectos.

R, S, D, E

$\overleftrightarrow{DE} \perp \overleftrightarrow{RS}$

Pi (p. 394) Pi (π) is the ratio of the circumference C of any circle to its diameter d.

Pi (p. 394) Pi (π) es la razón de la circunferencia C de cualquier círculo a su diámetro d.

C, d, $\pi = \frac{C}{d}$

Plane (p. 325) A plane is a flat surface that extends indefinitely in all directions.

Plano (p. 325) Un plano es una superficie plana que se extiende indefinidamente en todas las direcciones.

D, E, G, F

$DEFG$ is a plane.

Point (p. 324) A point is a location that has no size.

Punto (p. 324) Un punto es una ubicación que no tiene tamaño.

• A

A is a point.

Polygon (p. 252) A polygon is a closed figure formed by three or more line segments that do not cross.

Polígono (p. 252) Un polígono es una figura cerrada que está formada por tres o más segmentos de recta que no se cruzan.

Population (p. 550) A population is a group of objects or people about which information is wanted.

Población (p. 550) Una población es un grupo de objetos o personas sobre el que se busca información.

A class of 25 students is a sample of the population of a school.

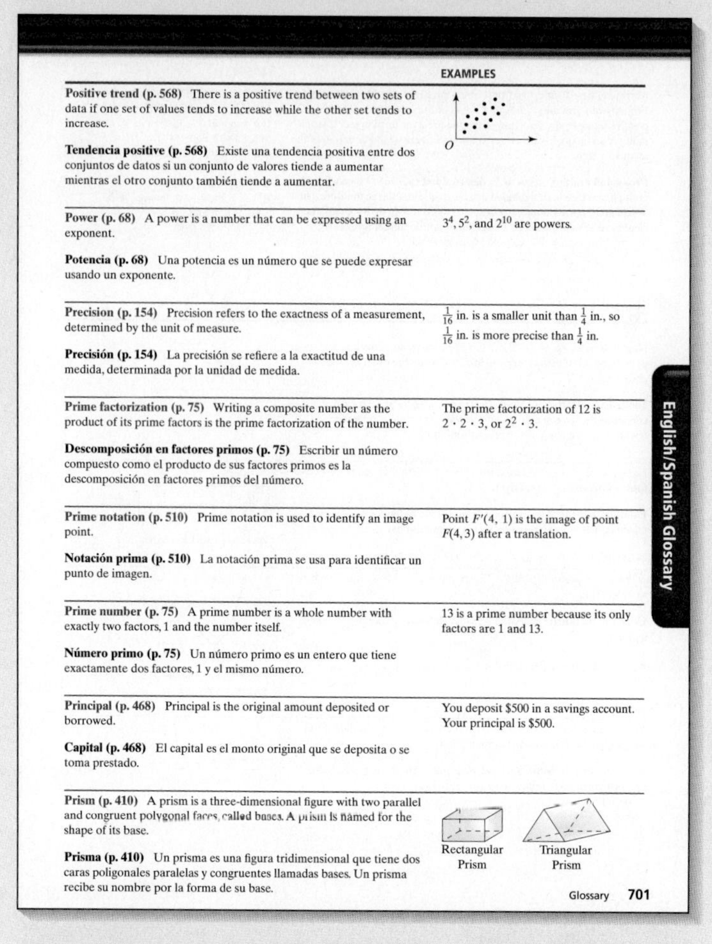

EXAMPLES

Positive trend (p. 568) There is a positive trend between two sets of data if one set of values tends to increase while the other set tends to increase.

Tendencia positive (p. 568) Existe una tendencia positiva entre dos conjuntos de datos si un conjunto de valores tiende a aumentar mientras el otro conjunto también tiende a aumentar.

Power (p. 68) A power is a number that can be expressed using an exponent.

Potencia (p. 68) Una potencia es un número que se puede expresar usando un exponente.

3^4, 5^2, and 2^{10} are powers.

Precision (p. 154) Precision refers to the exactness of a measurement, determined by the unit of measure.

Precisión (p. 154) La precisión se refiere a la exactitud de una medida, determinada por la unidad de medida.

$\frac{1}{16}$ in. is a smaller unit than $\frac{1}{4}$ in., so $\frac{1}{16}$ in. is more precise than $\frac{1}{4}$ in.

Prime factorization (p. 75) Writing a composite number as the product of its prime factors is the prime factorization of the number.

Descomposición en factores primos (p. 75) Escribir un número compuesto como el producto de sus factores primos es la descomposición en factores primos del número.

The prime factorization of 12 is $2 \cdot 2 \cdot 3$, or $2^2 \cdot 3$.

Prime notation (p. 510) Prime notation is used to identify an image point.

Notación prima (p. 510) La notación prima se usa para identificar un punto de imagen.

Point $F'(4, 1)$ is the image of point $F(4, 3)$ after a translation.

Prime number (p. 75) A prime number is a whole number with exactly two factors, 1 and the number itself.

Número primo (p. 75) Un número primo es un entero que tiene exactamente dos factores, 1 y el mismo número.

13 is a prime number because its only factors are 1 and 13.

Principal (p. 468) Principal is the original amount deposited or borrowed.

Capital (p. 468) El capital es el monto original que se deposita o se toma prestado.

You deposit $500 in a savings account. Your principal is $500.

Prism (p. 410) A prism is a three-dimensional figure with two parallel and congruent polygonal faces, called bases. A prism is named for the shape of its base.

Prisma (p. 410) Un prisma es una figura tridimensional que tiene dos caras poligonales paralelas y congruentes llamadas bases. Un prisma recibe su nombre por la forma de su base.

Rectangular Prism

Triangular Prism

English/Spanish Glossary

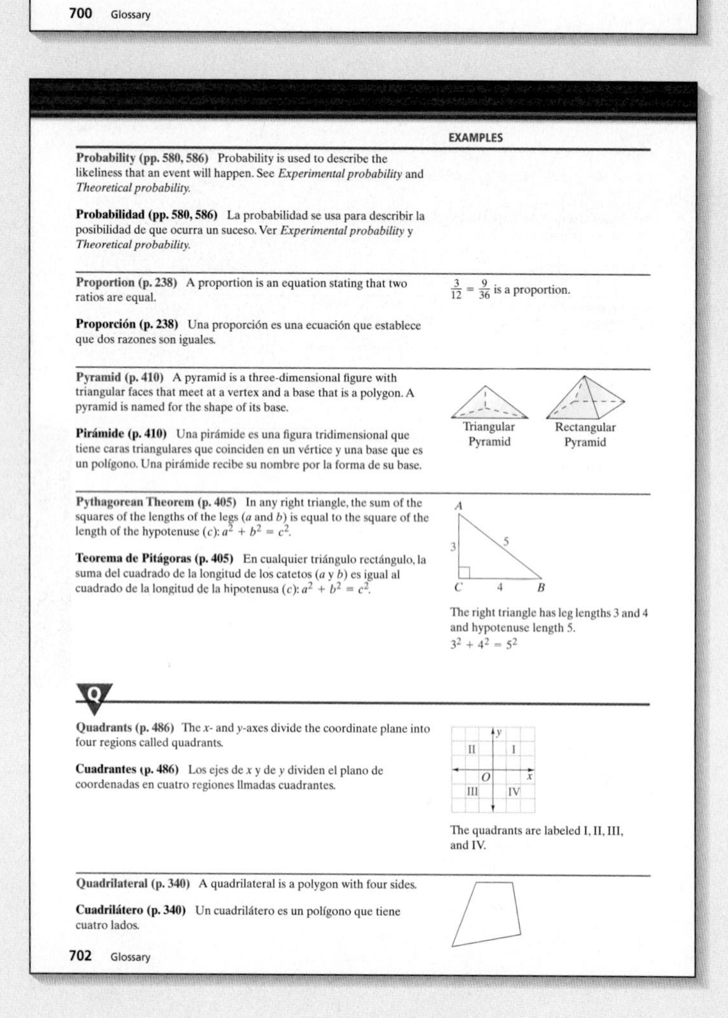

EXAMPLES

Probability (pp. 580, 586) Probability is used to describe the likeliness that an event will happen. See *Experimental probability* and *Theoretical probability*.

Probabilidad (pp. 580, 586) La probabilidad se usa para describir la posibilidad de que ocurra un suceso. Ver *Experimental probability* y *Theoretical probability*.

Proportion (p. 238) A proportion is an equation stating that two ratios are equal.

Proporción (p. 238) Una proporción es una ecuación que establece que dos razones son iguales.

$\frac{3}{12} = \frac{9}{36}$ is a proportion.

Pyramid (p. 410) A pyramid is a three-dimensional figure with triangular faces that meet at a vertex and a base that is a polygon. A pyramid is named for the shape of its base.

Pirámide (p. 410) Una pirámide es una figura tridimensional que tiene caras triangulares que coinciden en un vértice y una base que es un polígono. Una pirámide recibe su nombre por la forma de su base.

Triangular Pyramid

Rectangular Pyramid

Pythagorean Theorem (p. 405) In any right triangle, the sum of the squares of the lengths of the legs (a and b) is equal to the square of the length of the hypotenuse (c): $a^2 + b^2 = c^2$.

Teorema de Pitágoras (p. 405) En cualquier triángulo rectángulo, la suma del cuadrado de la longitud de los catetos (a y b) es igual al cuadrado de la longitud de la hipotenusa (c): $a^2 + b^2 = c^2$.

A, C, B, 3, 4, 5

The right triangle has leg lengths 3 and 4 and hypotenuse length 5.
$3^2 + 4^2 = 5^2$

Q

Quadrants (p. 486) The x- and y-axes divide the coordinate plane into four regions called quadrants.

Cuadrantes (p. 486) Los ejes de x y de y dividen el plano de coordenadas en cuatro regiones llmadas cuadrantes.

y, II, I, O, x, III, IV

The quadrants are labeled I, II, III, and IV.

Quadrilateral (p. 340) A quadrilateral is a polygon with four sides.

Cuadrilátero (p. 340) Un cuadrilátero es un polígono que tiene cuatro lados.

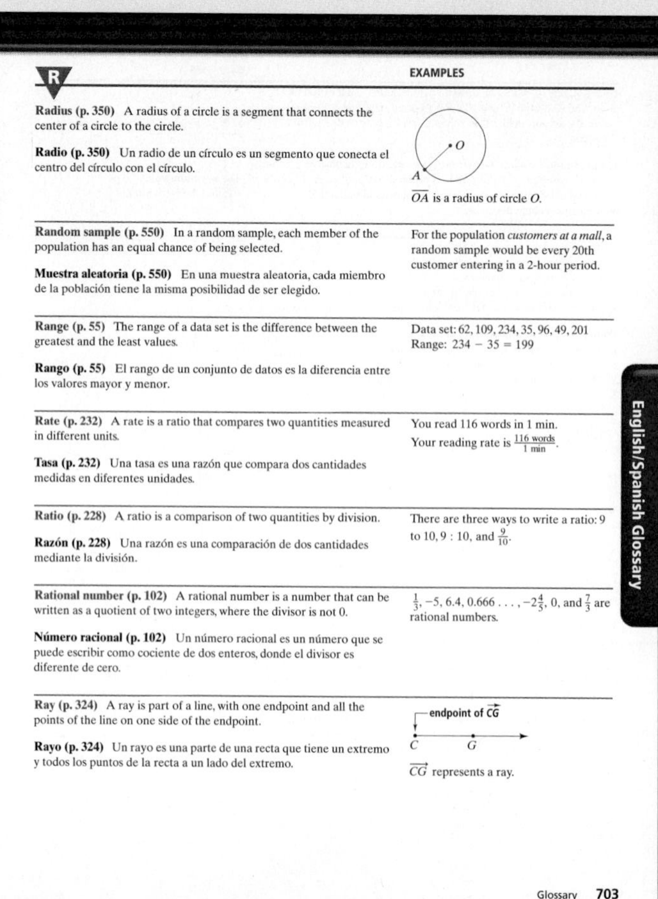

R

EXAMPLES

Radius (p. 350) A radius of a circle is a segment that connects the center of a circle to the circle.

Radio (p. 350) Un radio de un círculo es un segmento que conecta el centro del círculo con el círculo.

O, A

$\overline{OA}$ is a radius of circle O.

Random sample (p. 550) In a random sample, each member of the population has an equal chance of being selected.

Muestra aleatoria (p. 550) En una muestra aleatoria, cada miembro de la población tiene la misma posibilidad de ser elegido.

For the population *customers at a mall*, a random sample would be every 20th customer entering in a 2-hour period.

Range (p. 55) The range of a data set is the difference between the greatest and the least values.

Rango (p. 55) El rango de un conjunto de datos es la diferencia entre los valores mayor y menor.

Data set: 62, 109, 234, 35, 96, 49, 201
Range: $234 - 35 = 199$

Rate (p. 232) A rate is a ratio that compares two quantities measured in different units.

Tasa (p. 232) Una tasa es una razón que compara dos cantidades medidas en diferentes unidades.

You read 116 words in 1 min.
Your reading rate is $\frac{116 \text{ words}}{1 \text{ min}}$.

Ratio (p. 228) A ratio is a comparison of two quantities by division.

Razón (p. 228) Una razón es una comparación de dos cantidades mediante la división.

There are three ways to write a ratio: 9 to 10, 9 : 10, and $\frac{9}{10}$.

Rational number (p. 102) A rational number is a number that can be written as a quotient of two integers, where the divisor is not 0.

Número racional (p. 102) Un número racional es un número que se puede escribir como cociente de dos enteros, donde el divisor es diferente de cero.

$\frac{1}{3}$, -5, 6.4, 0.666 . . . , $-2\frac{4}{5}$, 0, and $\frac{7}{3}$ are rational numbers.

Ray (p. 324) A ray is part of a line, with one endpoint and all the points of the line on one side of the endpoint.

Rayo (p. 324) Un rayo es una parte de una recta que tiene un extremo y todos los puntos de la recta a un lado del extremo.

endpoint of $\overrightarrow{CG}$

C, G

$\overrightarrow{CG}$ represents a ray.

English/Spanish Glossary

EXAMPLES

Reciprocal (p. 141) Two numbers are reciprocals if their product is 1.

Recíproco (p. 141) Dos números son recíprocos si su producto es 1.

The numbers $\frac{4}{9}$ and $\frac{9}{4}$ are reciprocals.

Rectangle (p. 341) A rectangle is a parallelogram with four right angles.

Rectángulo (p. 341) Un rectángulo es un paralelogramo que tiene cuatro ángulos rectos.

Reflection (p. 515) A reflection, or flip, is a transformation that flips a figure over a line of reflection.

Reflexión (p. 515) Una reflexión es una transformación que voltea una figura sobre un eje de reflexión.

$K'L'M'N'$ is a reflection of $KLMN$ over the y-axis.

Regular polygon (p. 340) A regular polygon is a polygon with all sides congruent and all angles congruent.

Polígono regular (p. 340) Un polígono regular es un polígono que tiene todos los lados y todos los ángulos congruentes.

$ABDFEC$ is a regular hexagon.

Repeating decimal (p. 97) A repeating decimal is a decimal that repeats without end. The repeating block can be one or more digits.

Decimal periódico (p. 97) Un decimal periódico es un decimal que repite los mismos dígitos interminablemente. El bloque que se repite puede ser un dígito o más de un dígito.

$0.888\ldots = 0.\overline{8}$
$0.272727\ldots = 0.\overline{27}$

Rhombus (p. 341) A rhombus is a parallelogram with four congruent sides.

Rombo (p. 341) Un rombo es un paralelogramo que tiene cuatro lados congruentes.

Right angle (p. 330) A right angle is an angle with a measure of 90°.

Ángulo recto (p. 330) Un ángulo recto es un ángulo que mide 90°.

$m\angle D = 90°$

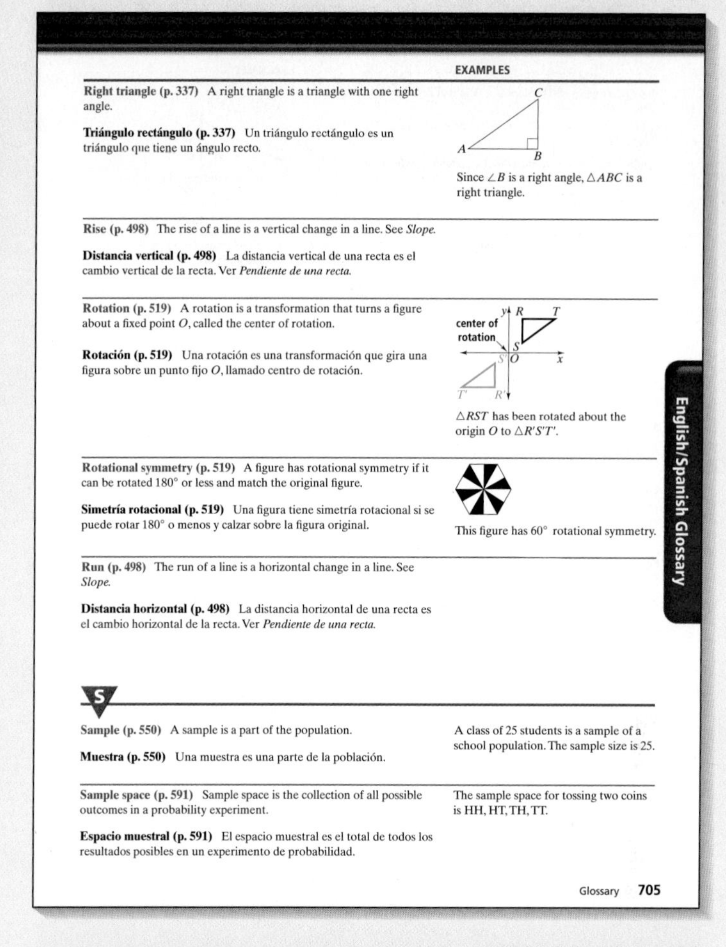

EXAMPLES

Right triangle (p. 337) A right triangle is a triangle with one right angle.

Triángulo rectángulo (p. 337) Un triángulo rectángulo es un triángulo que tiene un ángulo recto.

Since $\angle B$ is a right angle, $\triangle ABC$ is a right triangle.

Rise (p. 498) The rise of a line is a vertical change in a line. See *Slope.*

Distancia vertical (p. 498) La distancia vertical de una recta es el cambio vertical de la recta. Ver *Pendiente de una recta.*

Rotation (p. 519) A rotation is a transformation that turns a figure about a fixed point O, called the center of rotation.

Rotación (p. 519) Una rotación es una transformación que gira una figura sobre un punto fijo O, llamado centro de rotación.

$\triangle RST$ has been rotated about the origin O to $\triangle R'S'T'$.

Rotational symmetry (p. 519) A figure has rotational symmetry if it can be rotated 180° or less and match the original figure.

Simetría rotacional (p. 519) Una figura tiene simetría rotacional si se puede rotar 180° o menos y calzar sobre la figura original.

This figure has 60° rotational symmetry.

Run (p. 498) The run of a line is a horizontal change in a line. See *Slope.*

Distancia horizontal (p. 498) La distancia horizontal de una recta es el cambio horizontal de la recta. Ver *Pendiente de una recta.*

S

Sample (p. 550) A sample is a part of the population.

Muestra (p. 550) Una muestra es una parte de la población.

A class of 25 students is a sample of a school population. The sample size is 25.

Sample space (p. 591) Sample space is the collection of all possible outcomes in a probability experiment.

Espacio muestral (p. 591) El espacio muestral es el total de todos los resultados posibles en un experimento de probabilidad.

The sample space for tossing two coins is HH, HT, TH, TT.

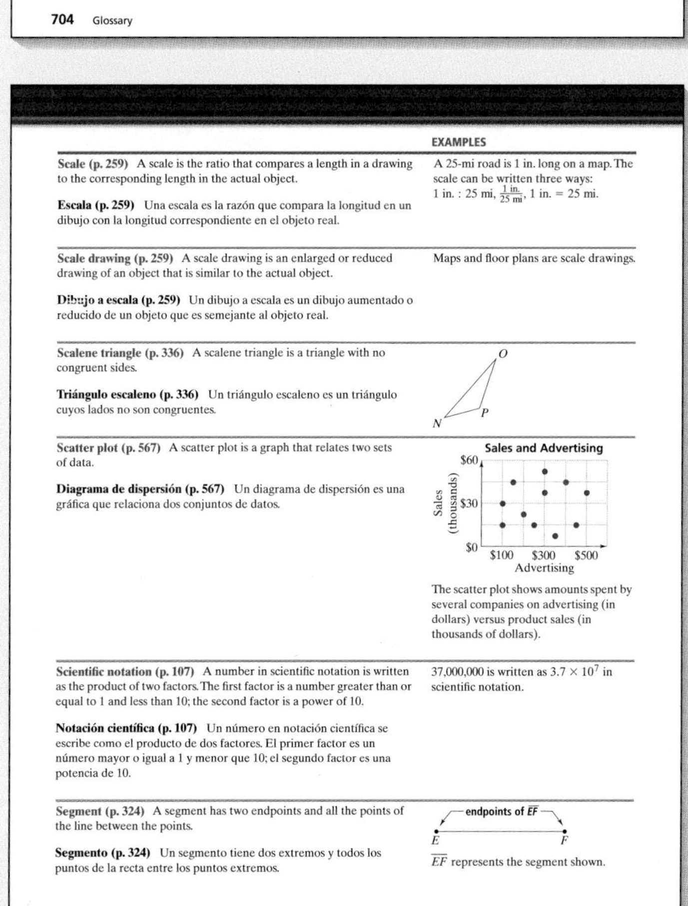

EXAMPLES

Scale (p. 259) A scale is the ratio that compares a length in a drawing to the corresponding length in the actual object.

Escala (p. 259) Una escala es la razón que compara la longitud en un dibujo con la longitud correspondiente en el objeto real.

A 25-mi road is 1 in. long on a map. The scale can be written three ways: 1 in. : 25 mi, $\frac{1 \text{ in.}}{25 \text{ mi}}$, 1 in. = 25 mi.

Scale drawing (p. 259) A scale drawing is an enlarged or reduced drawing of an object that is similar to the actual object.

Dibujo a escala (p. 259) Un dibujo a escala es un dibujo aumentado o reducido de un objeto que es semejante al objeto real.

Maps and floor plans are scale drawings.

Scalene triangle (p. 336) A scalene triangle is a triangle with no congruent sides.

Triángulo escaleno (p. 336) Un triángulo escaleno es un triángulo cuyos lados no son congruentes.

Scatter plot (p. 567) A scatter plot is a graph that relates two sets of data.

Diagrama de dispersión (p. 567) Un diagrama de dispersión es una gráfica que relaciona dos conjuntos de datos.

The scatter plot shows amounts spent by several companies on advertising (in dollars) versus product sales (in thousands of dollars).

Scientific notation (p. 107) A number in scientific notation is written as the product of two factors. The first factor is a number greater than or equal to 1 and less than 10; the second factor is a power of 10.

Notación científica (p. 107) Un número en notación científica se escribe como el producto de dos factores. El primer factor es un número mayor o igual a 1 y menor que 10; el segundo factor es una potencia de 10.

37,000,000 is written as 3.7×10^7 in scientific notation.

Segment (p. 324) A segment has two endpoints and all the points of the line between the points.

Segmento (p. 324) Un segmento tiene dos extremos y todos los puntos de la recta entre los puntos extremos.

$\overline{EF}$ represents the segment shown.

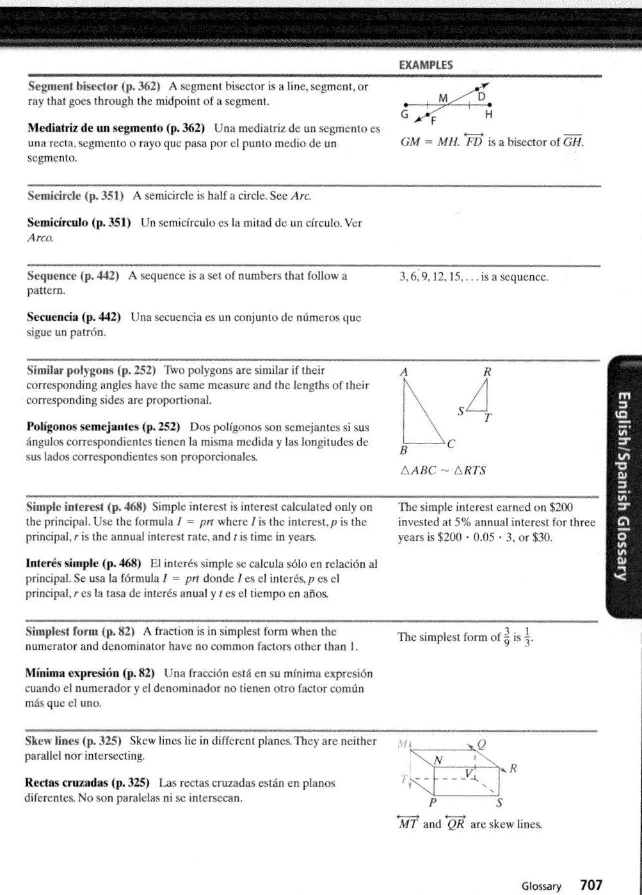

EXAMPLES

Segment bisector (p. 362) A segment bisector is a line, segment, or ray that goes through the midpoint of a segment.

Mediatriz de un segmento (p. 362) Una mediatriz de un segmento es una recta, segmento o rayo que pasa por el punto medio de un segmento.

$GM = MH$. $\overleftrightarrow{FD}$ is a bisector of $\overline{GH}$.

Semicircle (p. 351) A semicircle is half a circle. See *Arc.*

Semicírculo (p. 351) Un semicírculo es la mitad de un círculo. Ver *Arco.*

Sequence (p. 442) A sequence is a set of numbers that follow a pattern.

Secuencia (p. 442) Una secuencia es un conjunto de números que sigue un patrón.

3, 6, 9, 12, 15, . . . is a sequence.

Similar polygons (p. 252) Two polygons are similar if their corresponding angles have the same measure and the lengths of their corresponding sides are proportional.

Polígonos semejantes (p. 252) Dos polígonos son semejantes si sus ángulos correspondientes tienen la misma medida y las longitudes de sus lados correspondientes son proporcionales.

$\triangle ABC \sim \triangle RTS$

Simple interest (p. 468) Simple interest is interest calculated only on the principal. Use the formula $I = prt$ where I is the interest, p is the principal, r is the annual interest rate, and t is time in years.

Interés simple (p. 468) El interés simple se calcula sólo en relación al principal. Se usa la fórmula $I = prt$ donde I es el interés, p es el principal, r es la tasa de interés anual y t es el tiempo en años.

The simple interest earned on \$200 invested at 5% annual interest for three years is \$200 · 0.05 · 3, or \$30.

Simplest form (p. 82) A fraction is in simplest form when the numerator and denominator have no common factors other than 1.

Mínima expresión (p. 82) Una fracción está en su mínima expresión cuando el numerador y el denominador no tienen otro factor común más que el uno.

The simplest form of $\frac{3}{9}$ is $\frac{1}{3}$.

Skew lines (p. 325) Skew lines lie in different planes. They are neither parallel nor intersecting.

Rectas cruzadas (p. 325) Las rectas cruzadas están en planos diferentes. No son paralelas ni se intersecan.

$\overleftrightarrow{MT}$ and $\overleftrightarrow{QR}$ are skew lines.

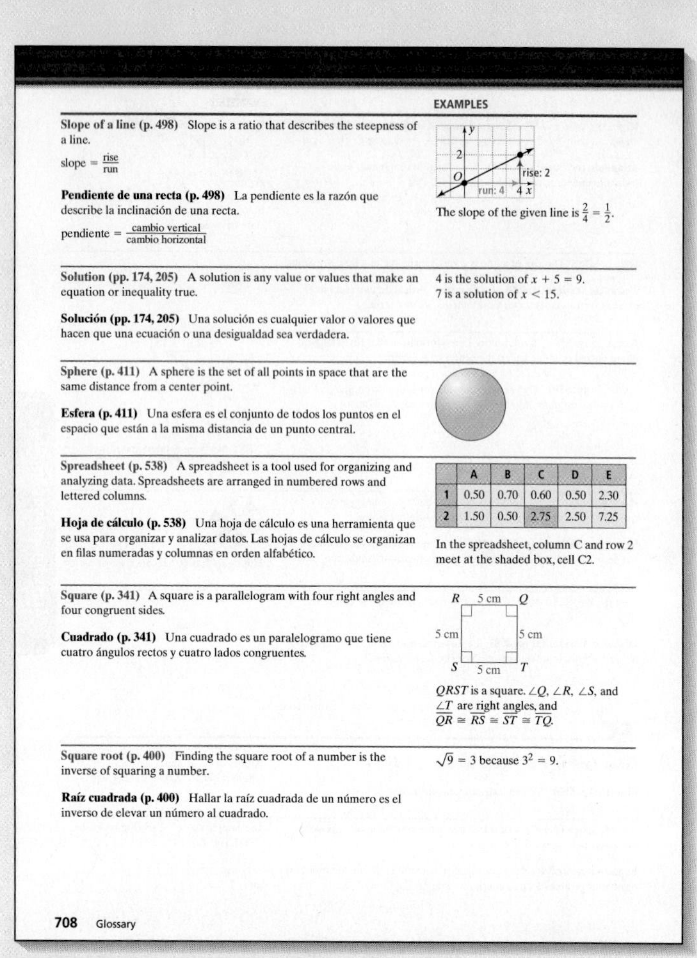

EXAMPLES

Slope of a line (p. 498) Slope is a ratio that describes the steepness of a line.

slope = $\frac{\text{rise}}{\text{run}}$

Pendiente de una recta (p. 498) La pendiente es la razón que describe la inclinación de una recta.

pendiente = $\frac{\text{cambio vertical}}{\text{cambio horizontal}}$

The slope of the given line is $\frac{2}{4} = \frac{1}{2}$.

Solution (pp. 174, 205) A solution is any value or values that make an equation or inequality true.

Solución (pp. 174, 205) Una solución es cualquier valor o valores que hacen que una ecuación o una desigualdad sea verdadera.

4 is the solution of $x + 5 = 9$.
7 is a solution of $x < 15$.

Sphere (p. 411) A sphere is the set of all points in space that are the same distance from a center point.

Esfera (p. 411) Una esfera es el conjunto de todos los puntos en el espacio que están a la misma distancia de un punto central.

Spreadsheet (p. 538) A spreadsheet is a tool used for organizing and analyzing data. Spreadsheets are arranged in numbered rows and lettered columns.

Hoja de cálculo (p. 538) Una hoja de cálculo es una herramienta que se usa para organizar y analizar datos. Las hojas de cálculo se organizan en filas numeradas y columnas en orden alfabético.

	A	B	C	D	E
1	0.50	0.70	0.60	0.50	2.30
2	1.50	0.50	2.75	2.50	7.25

In the spreadsheet, column C and row 2 meet at the shaded box, cell C2.

Square (p. 341) A square is a parallelogram with four right angles and four congruent sides.

Cuadrado (p. 341) Una cuadrado es un paralelogramo que tiene cuatro ángulos rectos y cuatro lados congruentes.

$QRST$ is a square. $\angle Q$, $\angle R$, $\angle S$, and $\angle T$ are right angles, and $\overline{QR} \cong \overline{RS} \cong \overline{ST} \cong \overline{TQ}$.

Square root (p. 400) Finding the square root of a number is the inverse of squaring a number.

Raíz cuadrada (p. 400) Hallar la raíz cuadrada de un número es el inverso de elevar un número al cuadrado.

$\sqrt{9} = 3$ because $3^2 = 9$.

708 Glossary

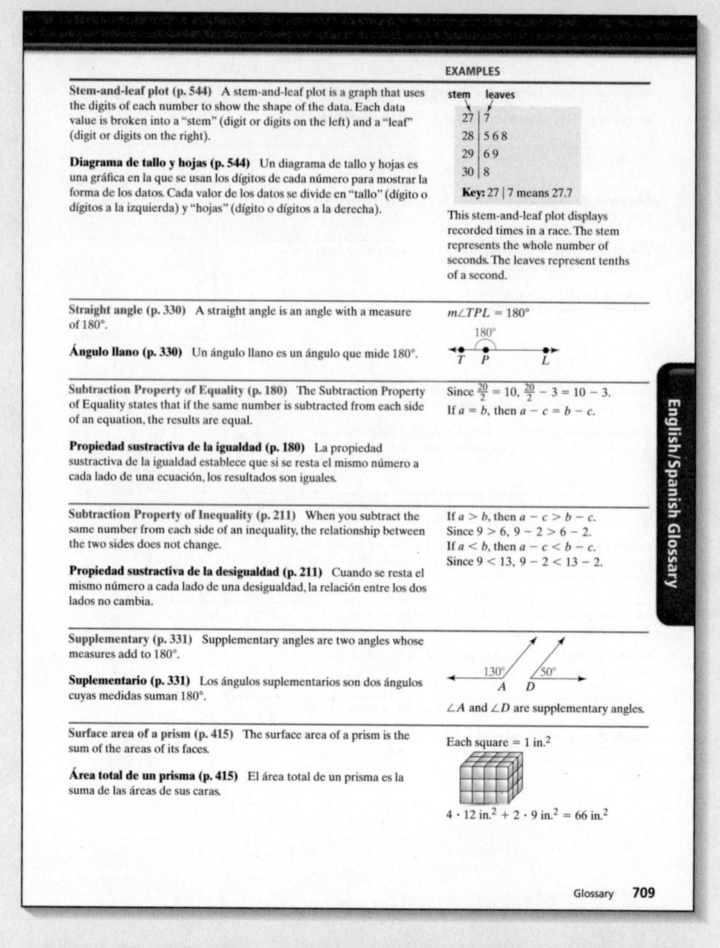

EXAMPLES

Stem-and-leaf plot (p. 544) A stem-and-leaf plot is a graph that uses the digits of each number to show the shape of the data. Each data value is broken into a "stem" (digit or digits on the left) and a "leaf" (digit or digits on the right).

Diagrama de tallo y hojas (p. 544) Un diagrama de tallo y hojas es una gráfica en la que se usan los dígitos de cada número para mostrar la forma de los datos. Cada valor de los datos se divide en "tallo" (dígito o dígitos a la izquierda) y "hojas" (dígito o dígitos a la derecha).

stem	leaves
27	7
28	5 6 8
29	6 9
30	8

Key: 27 | 7 means 27.7

This stem-and-leaf plot displays recorded times in a race. The stem represents the whole number of seconds. The leaves represent tenths of a second.

Straight angle (p. 330) A straight angle is an angle with a measure of 180°.

Ángulo llano (p. 330) Un ángulo llano es un ángulo que mide 180°.

$m\angle TPL = 180°$

Subtraction Property of Equality (p. 180) The Subtraction Property of Equality states that if the same number is subtracted from each side of an equation, the results are equal.

Propiedad sustractiva de la igualdad (p. 180) La propiedad sustractiva de la igualdad establece que si se resta el mismo número a cada lado de una ecuación, los resultados son iguales.

Since $\frac{20}{2} = 10$, $\frac{20}{2} - 3 = 10 - 3$.
If $a = b$, then $a - c = b - c$.

Subtraction Property of Inequality (p. 211) When you subtract the same number from each side of an inequality, the relationship between the two sides does not change.

Propiedad sustractiva de la desigualdad (p. 211) Cuando se resta el mismo número a cada lado de una desigualdad, la relación entre los dos lados no cambia.

If $a > b$, then $a - c > b - c$.
Since $9 > 6$, $9 - 2 > 6 - 2$.
If $a < b$, then $a - c < b - c$.
Since $9 < 13$, $9 - 2 < 13 - 2$.

Supplementary (p. 331) Supplementary angles are two angles whose measures add to 180°.

Suplementario (p. 331) Los ángulos suplementarios son dos ángulos cuyas medidas suman 180°.

$\angle A$ and $\angle D$ are supplementary angles.

Surface area of a prism (p. 415) The surface area of a prism is the sum of the areas of its faces.

Área total de un prisma (p. 415) El área total de un prisma es la suma de las áreas de sus caras.

Each square = 1 in.^{2}

$4 \cdot 12 \text{ in.}^2 + 2 \cdot 9 \text{ in.}^2 = 66 \text{ in.}^2$

English/Spanish Glossary

Glossary 709

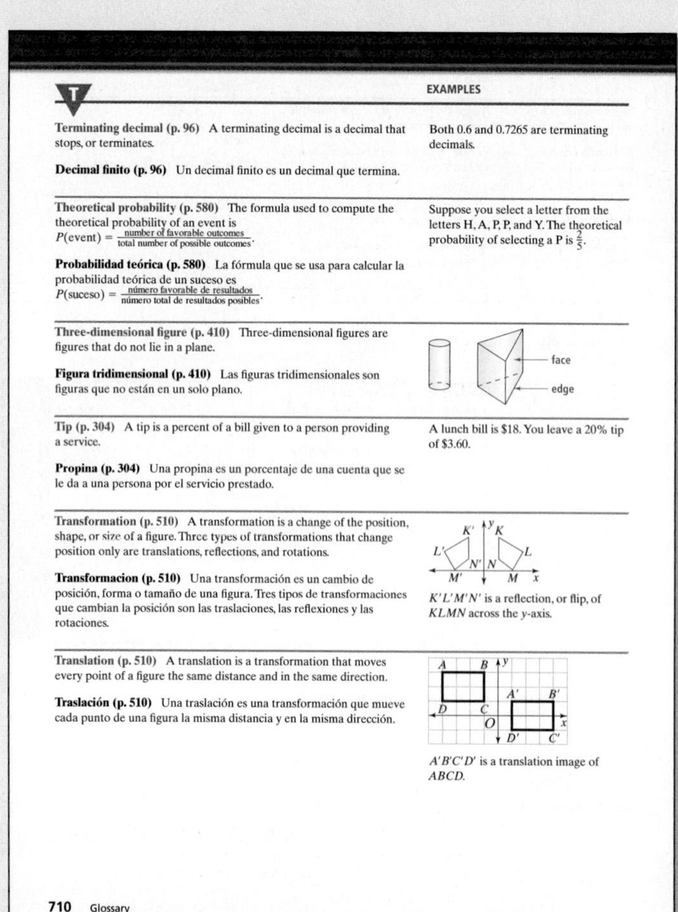

T

EXAMPLES

Terminating decimal (p. 96) A terminating decimal is a decimal that stops, or terminates.

Decimal finito (p. 96) Un decimal finito es un decimal que termina.

Both 0.6 and 0.7265 are terminating decimals.

Theoretical probability (p. 580) The formula used to compute the theoretical probability of an event is $P(\text{event}) = \frac{\text{number of favorable outcomes}}{\text{total number of possible outcomes}}$.

Probabilidad teórica (p. 580) La fórmula que se usa para calcular la probabilidad teórica de un suceso es $P(\text{suceso}) = \frac{\text{número favorable de resultados}}{\text{número total de resultados posibles}}$.

Suppose you select a letter from the letters H, A, P, P, and Y. The theoretical probability of selecting a P is $\frac{2}{5}$.

Three-dimensional figure (p. 410) Three-dimensional figures are figures that do not lie in a plane.

Figura tridimensional (p. 410) Las figuras tridimensionales son figuras que no están en un solo plano.

Tip (p. 304) A tip is a percent of a bill given to a person providing a service.

Propina (p. 304) Una propina es un porcentaje de una cuenta que se le da a una persona por el servicio prestado.

A lunch bill is \$18. You leave a 20% tip of \$3.60.

Transformation (p. 510) A transformation is a change of the position, shape, or size of a figure. Three types of transformations that change position only are translations, reflections, and rotations.

Transformacion (p. 510) Una transformación es un cambio de posición, forma o tamaño de una figura. Tres tipos de transformaciones que cambian la posición son las traslaciones, las reflexiones y las rotaciones.

$K'L'M'N'$ is a reflection, or flip, of $KLMN$ across the y-axis.

Translation (p. 510) A translation is a transformation that moves every point of a figure the same distance and in the same direction.

Traslación (p. 510) Una traslación es una transformación que mueve cada punto de una figura la misma distancia y en la misma dirección.

$A'B'C'D'$ is a translation image of $ABCD$.

710 Glossary

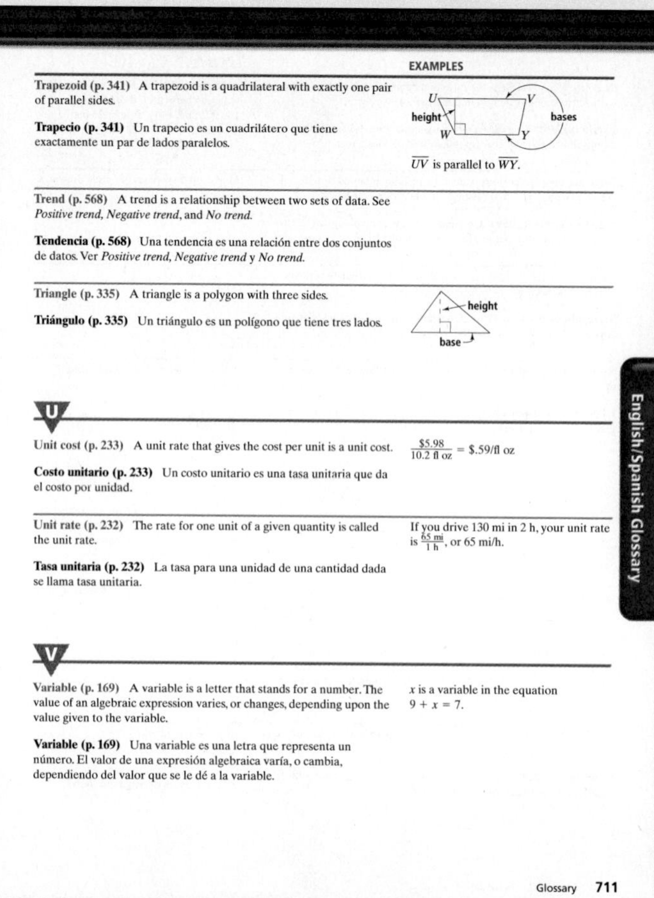

EXAMPLES

Trapezoid (p. 341) A trapezoid is a quadrilateral with exactly one pair of parallel sides.

Trapecio (p. 341) Un trapecio es un cuadrilátero que tiene exactamente un par de lados paralelos.

$\overline{UV}$ is parallel to $\overline{WY}$.

Trend (p. 568) A trend is a relationship between two sets of data. See *Positive trend*, *Negative trend*, and *No trend*.

Tendencia (p. 568) Una tendencia es una relación entre dos conjuntos de datos. Ver *Positive trend*, *Negative trend* y *No trend*.

Triangle (p. 335) A triangle is a polygon with three sides.

Triángulo (p. 335) Un triángulo es un polígono que tiene tres lados.

U

Unit cost (p. 233) A unit rate that gives the cost per unit is a unit cost.

Costo unitario (p. 233) Un costo unitario es una tasa unitaria que da el costo por unidad.

$\frac{\$5.98}{10.2 \text{ fl oz}} = \$.59/\text{fl oz}$

Unit rate (p. 232) The rate for one unit of a given quantity is called the unit rate.

Tasa unitaria (p. 232) La tasa para una unidad de una cantidad dada se llama tasa unitaria.

If you drive 130 mi in 2 h, your unit rate is $\frac{65 \text{ mi}}{1 \text{ h}}$, or 65 mi/h.

V

Variable (p. 169) A variable is a letter that stands for a number. The value of an algebraic expression varies, or changes, depending upon the value given to the variable.

Variable (p. 169) Una variable es una letra que representa un número. El valor de una expresión algebraica varía, o cambia, dependiendo del valor que se le dé a la variable.

x is a variable in the equation $9 + x = 7$.

English/Spanish Glossary

Glossary 711

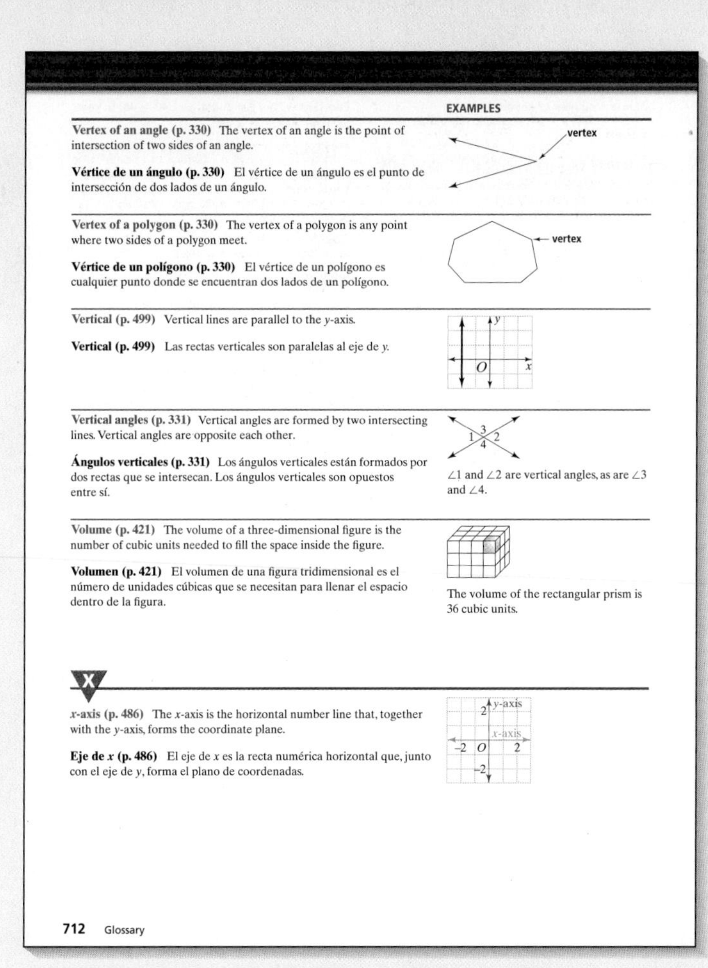

EXAMPLES

Vertex of an angle (p. 330) The vertex of an angle is the point of intersection of two sides of an angle.

Vértice de un ángulo (p. 330) El vértice de un ángulo es el punto de intersección de dos lados de un ángulo.

Vertex of a polygon (p. 330) The vertex of a polygon is any point where two sides of a polygon meet.

Vértice de un polígono (p. 330) El vértice de un polígono es cualquier punto donde se encuentran dos lados de un polígono.

Vertical (p. 499) Vertical lines are parallel to the *y*-axis.

Vertical (p. 499) Las rectas verticales son paralelas al eje de *y*.

Vertical angles (p. 331) Vertical angles are formed by two intersecting lines. Vertical angles are opposite each other.

Ángulos verticales (p. 331) Los ángulos verticales están formados por dos rectas que se intersecan. Los ángulos verticales son opuestos entre sí.

∠1 and ∠2 are vertical angles, as are ∠3 and ∠4.

Volume (p. 421) The volume of a three-dimensional figure is the number of cubic units needed to fill the space inside the figure.

Volumen (p. 421) El volumen de una figura tridimensional es el número de unidades cúbicas que se necesitan para llenar el espacio dentro de la figura.

The volume of the rectangular prism is 36 cubic units.

X

***x*-axis (p. 486)** The *x*-axis is the horizontal number line that, together with the *y*-axis, forms the coordinate plane.

Eje de *x* (p. 486) El eje de *x* es la recta numérica horizontal que, junto con el eje de *y*, forma el plano de coordenadas.

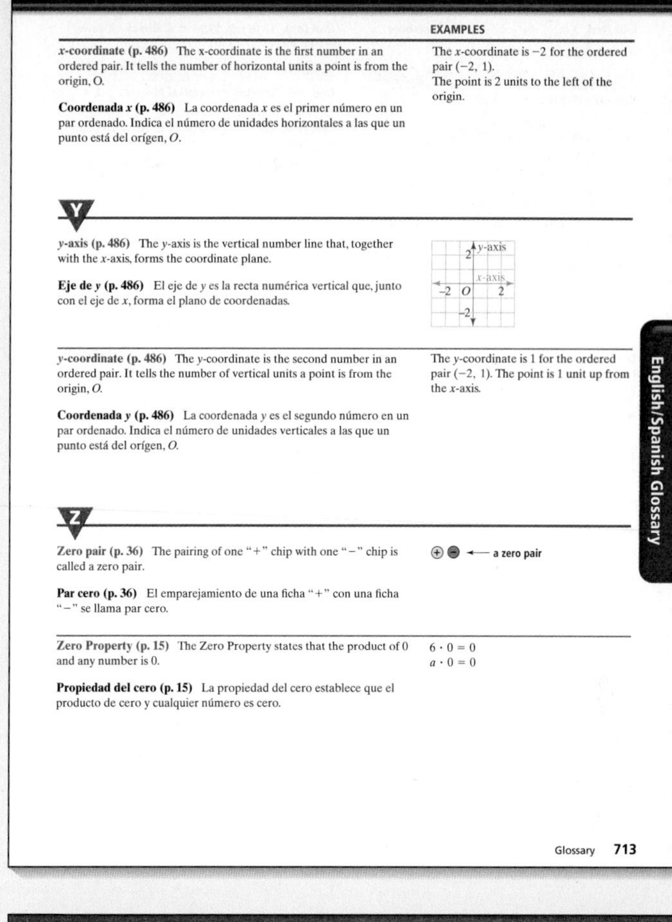

EXAMPLES

***x*-coordinate (p. 486)** The *x*-coordinate is the first number in an ordered pair. It tells the number of horizontal units a point is from the origin, O.

Coordenada *x* (p. 486) La coordenada *x* es el primer número en un par ordenado. Indica el número de unidades horizontales a las que un punto está del origen, *O*.

The *x*-coordinate is −2 for the ordered pair (−2, 1).
The point is 2 units to the left of the origin.

Y

***y*-axis (p. 486)** The *y*-axis is the vertical number line that, together with the *x*-axis, forms the coordinate plane.

Eje de *y* (p. 486) El eje de *y* es la recta numérica vertical que, junto con el eje de *x*, forma el plano de coordenadas.

***y*-coordinate (p. 486)** The *y*-coordinate is the second number in an ordered pair. It tells the number of vertical units a point is from the origin, *O*.

Coordenada *y* (p. 486) La coordenada *y* es el segundo número en un par ordenado. Indica el número de unidades verticales a las que un punto está del origen, *O*.

The *y*-coordinate is 1 for the ordered pair (−2, 1). The point is 1 unit up from the *x*-axis.

Z

Zero pair (p. 36) The pairing of one "+" chip with one "−" chip is called a zero pair.

Par cero (p. 36) El emparejamiento de una ficha "+" con una ficha "−" se llama par cero.

Zero Property (p. 15) The Zero Property states that the product of 0 and any number is 0.

Propiedad del cero (p. 15) La propiedad del cero establece que el producto de cero y cualquier número es cero.

$6 \cdot 0 = 0$
$a \cdot 0 = 0$

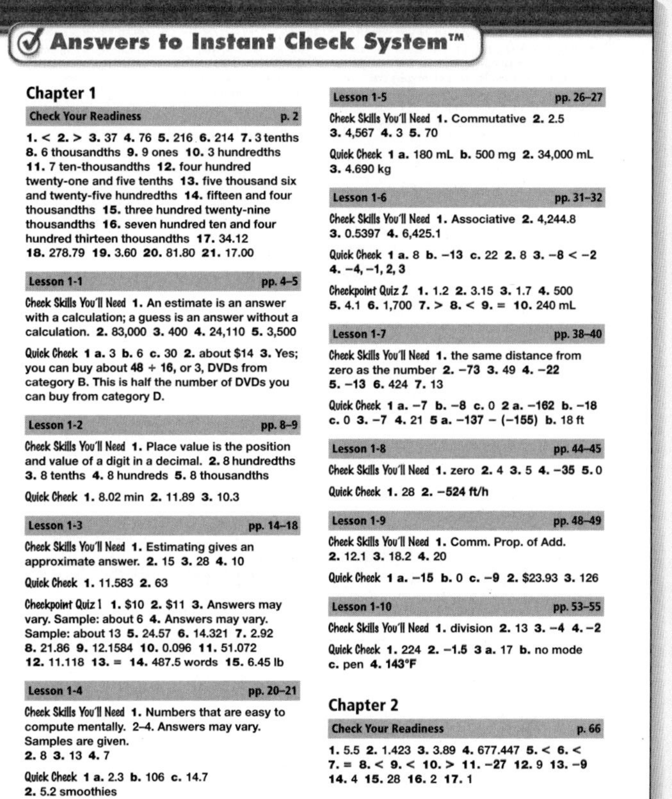

Answers to Instant Check System™

Chapter 1

Check Your Readiness p. 2

1. < **2.** > **3.** 37 **4.** 76 **5.** 216 **6.** 214 **7.** 3 tenths **8.** 6 thousandths **9.** 9 ones **10.** 3 hundredths **11.** 7 ten-thousandths **12.** four hundred twenty-one and five tenths **13.** five thousand six and twenty-five hundredths **14.** fifteen and four thousandths **15.** three hundred twenty-nine thousandths **16.** seven hundred ten and four hundred thirteen thousandths **17.** 34.12 **18.** 278.79 **19.** 3.60 **20.** 81.80 **21.** 17.00

Lesson 1-1 pp. 4–5

Check Skills You'll Need **1.** An estimate is an answer with a calculation; a guess is an answer without a calculation. **2.** 83,000 **3.** 400 **4.** 24,110 **5.** 3,500

Quick Check **1 a.** 3 **b.** 6 **c.** 30 **2.** about $14 **3.** Yes; you can buy about 48 ÷ 16, or 3, DVDs from category B. This is half the number of DVDs you can buy from category D.

Lesson 1-2 pp. 8–9

Check Skills You'll Need **1.** Place value is the position and value of a digit in a decimal. **2.** 8 hundredths **3.** 8 tenths **4.** 8 hundreds **5.** 8 thousandths

Quick Check **1.** 8.02 min **2.** 11.89 **3.** 10.3

Lesson 1-3 pp. 14–18

Check Skills You'll Need **1.** Estimating gives an approximate answer. **2.** 15 **3.** 28 **4.** 10

Quick Check **1.** 11.583 **2.** 63

Checkpoint Quiz 1 **1.** $10 **2.** $11 **3.** Answers may vary. Sample: about 6 **4.** Answers may vary. Sample: about 13 **5.** 24.57 **6.** 14.321 **7.** 2.92 **8.** 21.86 **9.** 12.1584 **10.** 0.096 **11.** 51.072 **12.** 11.118 **13.** = **14.** 487.5 words **15.** 6.45 lb

Lesson 1-4 pp. 20–21

Check Skills You'll Need **1.** Numbers that are easy to compute mentally. 2–4. Answers may vary. Samples are given.
2. 8 **3.** 13 **4.** 7

Quick Check **1 a.** 2.3 **b.** 106 **c.** 14.7
2. 5.2 smoothies

Lesson 1-5 pp. 26–27

Check Skills You'll Need **1.** Commutative **2.** 2.5 **3.** 4,567 **4.** 3 **5.** 70

Quick Check **1 a.** 180 mL **b.** 500 mg **2.** 34,000 mL **3.** 4.690 kg

Lesson 1-6 pp. 31–32

Check Skills You'll Need **1.** Associative **2.** 4,244.8 **3.** 0.5397 **4.** 6,425.1

Quick Check **1 a.** 8 **b.** −13 **c.** 22 **2.** 8 **3.** −8 < −2 **4.** −4, −1, 2, 3

Checkpoint Quiz 2 **1.** 1.2 **2.** 3.15 **3.** 1.7 **4.** 500 **5.** 4.1 **6.** 1,700 **7.** > **8.** < **9.** = **10.** 240 mL

Lesson 1-7 pp. 38–40

Check Skills You'll Need **1.** the same distance from zero as the number **2.** −73 **3.** 49 **4.** −22 **5.** −13 **6.** 424 **7.** 13

Quick Check **1 a.** −7 **b.** −8 **c.** 0 **2 a.** −162 **b.** −18 **c.** 0 **3.** −7 **4.** 21 **5 a.** −137 − (−155) **b.** 18 ft

Lesson 1-8 pp. 44–45

Check Skills You'll Need **1.** zero **2.** 4 **3.** 5 **4.** −35 **5.** 0

Quick Check **1.** 28 **2.** −524 ft/h

Lesson 1-9 pp. 48–49

Check Skills You'll Need **1.** Comm. Prop. of Add. **2.** 12.1 **3.** 18.2 **4.** 20

Quick Check **1 a.** −15 **b.** 0 **c.** −9 **2.** $23.93 **3.** 126

Lesson 1-10 pp. 53–55

Check Skills You'll Need **1.** division **2.** 13 **3.** −4 **4.** −2

Quick Check **1.** 224 **2.** −1.5 **3 a.** 17 **b.** no mode **c.** pen **4.** 143°F

Chapter 2

Check Your Readiness p. 66

1. 5.5 **2.** 1.423 **3.** 3.89 **4.** 677.447 **5.** < **6.** < **7.** = **8.** < **9.** < **10.** > **11.** −27 **12.** 9 **13.** −9 **14.** 4 **15.** 28 **16.** 2 **17.** 1

Lesson 2-1 pp. 68–69

Check Skills You'll Need **1.** before **2.** 2 **3.** 12 **4.** −10 **5.** 1

Quick Check **1 a.** 44^4 **b.** $(-2)^2$ **2 a.** 243 **b.** 1,000,000,000 **c.** 9.61 **3 a.** −27 **b.** −27 **c.** 62

Lesson 2-2 pp. 74–79

Check Skills You'll Need **1.** 0 **2.** 24 **3.** 21 **4.** 12

Quick Check **1 a.** 20 **b.** 35 **c.** 60 **2.** Composite; the factors of 15 are 1, 3, 5, and 15. Since 15 has more than two factors, 15 is composite.
3. $72 = 2^3 \cdot 3^2$ **4.** 8

Checkpoint Quiz 1 **1.** 75 **2.** 1 **3.** 54 **4.** 12 **5.** 70 **6.** 60 **7.** 1, 3, 13, 39 **8.** 1, 2, 4, 13, 26, 52 **9.** 1, 2, 5, 10, 11, 22, 55, 110 **10.** 1, 2, 4, 5, 8, 10, 20, 25, 40, 50, 100, 200 **11.** $2^5 \cdot 3$ **12.** $2 \cdot 3 \cdot 5^2$ **13.** $3^2 \cdot 5^2$ **14.** $3^2 \cdot 37$ **15.** 24 ft

Lesson 2-3 pp. 82–83

Check Skills You'll Need **1.** It is the greatest factor that divides both numbers. **2.** 2 **3.** 12 **4.** 5 **5.** 1

Quick Check **1.** Answers may vary. Sample: $\frac{8}{10}$, $\frac{12}{15}$
2. Answers may vary. Sample: $\frac{9}{15}$, $\frac{3}{5}$ **3.** $\frac{2}{3}$ **4.** $\frac{2}{5}$

Lesson 2-4 pp. 87–88

Check Skills You'll Need **1.** least common multiple **2.** 12 **3.** 20 **4.** 8 **5.** 45

Quick Check **1 a.** < **b.** < **c.** >
2. $\frac{5}{32}$ in., $\frac{1}{4}$ in., $\frac{5}{16}$ in., $\frac{3}{8}$ in.

Lesson 2-5 pp. 91–92

Check Skills You'll Need **1.** The numerator and denominator have no common factors other than 1.
2. $\frac{3}{5}$ **3.** $\frac{5}{6}$ **4.** $\frac{3}{4}$ **5.** $\frac{5}{12}$

Quick Check **1.** $\frac{21}{8}$ **2.** 3

Lesson 2-6 pp. 96–101

Check Skills You'll Need **1.** In simplest form, they are the same. 2–5. Answers may vary. Samples are given.
2. $\frac{4}{5}$, $\frac{8}{10}$ **3.** $\frac{3}{4}$, $\frac{6}{8}$ **4.** $\frac{2}{3}$, $\frac{8}{12}$ **5.** $\frac{1}{5}$, $\frac{2}{10}$

Quick Check **1.** 0.625 **2.** $0.\overline{5}$ **3 a.** $1\frac{91}{250}$ **b.** $2\frac{12}{25}$ **c.** $3\frac{3}{5}$ **4.** 1.862, $1\frac{7}{15}$, $1\frac{3}{8}$ **5.** cats, dogs, fish, birds

Checkpoint Quiz 2 **1.** $\frac{1}{2}$ **2.** $\frac{7}{10}$ **3.** $\frac{5}{8}$ **4.** > **5.** < **6.** = **7.** $4\frac{5}{6}$ **8.** $\frac{37}{9}$ **9.** $16\frac{2}{5}$ **10.** 0.00015, 0.004, 0.0112

Lesson 2-7 pp. 102–103

Check Skills You'll Need **1.** Terminating decimal; the decimal stops **2.** 0.75 **3.** $-0.\overline{7}$ **4.** $1.\overline{3}$ **5.** 0.25

Quick Check **1.** $-\frac{2}{3} < -\frac{1}{6}$ **2.** −4.2 > −4.9
3. $-6\frac{1}{4}$, −4, 6.55, $12\frac{1}{2}$

Lesson 2-8 pp. 106–107

Check Skills You'll Need **1.** factor **2.** 27 **3.** 16 **4.** 100,000 **5.** 16

Quick Check **1.** 1.69×10^5 **2.** 640,000

Chapter 3

Check Your Readiness p. 118

1. −40 **2.** −24 **3.** 100 **4.** 18 **5.** −45 **6.** 60 **7.** −3 **8.** 20 **9.** 9 **10.** 21 **11.** 12 **12.** 9 **13.** $\frac{3}{4}$ **14.** $\frac{1}{2}$ **15.** $\frac{7}{8}$ **16.** $\frac{3}{8}$ **17.** $\frac{1}{3}$ **18.** $\frac{5}{9}$ **19.** $\frac{2}{3}$ **20.** $\frac{2}{7}$

Lesson 3-1 pp. 120–121

Check Skills You'll Need **1.** When you estimate by rounding, you round a number to the nearest unit. When you use compatible numbers, you look for numbers that are easy to compute mentally. **2.** 7 **3.** 4 **4.** 8 **5.** 6 **6.** 2

Quick Check **1.** about $\frac{1}{2}$ **2.** about 6 in.
3 a. about 20 **b.** about 48 **c.** about 14
4 a. about 6 **b.** about 6 **c.** about 9

Lesson 3-2 pp. 126–127

Check Skills You'll Need **1.** equivalent **2.** 4 **3.** 8 **4.** 24

Quick Check **1 a.** $\frac{4}{5}$ **b.** $\frac{1}{4}$ **c.** 1 **2 a.** $\frac{7}{12}$ **b.** $\frac{11}{14}$ **3.** $\frac{7}{8}$ mi

Lesson 3-3 pp. 130–134

Check Skills You'll Need **1.** When two fractions have the same denominator, you can compare them by looking at the numerators. **2.** < **3.** <

Quick Check **1 a.** $4\frac{1}{3}$ **b.** $4\frac{4}{5}$ **c.** 1 **2 a.** 1 **b.** $12\frac{1}{24}$ **c.** $4\frac{3}{10}$ **3.** $1\frac{1}{2}$ ft

Checkpoint Quiz 1 **1.** about $5\frac{1}{2}$ in. **2.** about $13\frac{1}{2}$ in. **3.** about 5 in. **4.** $\frac{1}{6}$ **5.** 2 **6.** $5\frac{9}{10}$ **7.** $8\frac{19}{20}$ **8.** $\frac{2}{9}$ **9.** $11\frac{5}{8}$ **10.** $21\frac{5}{8}$ in.

Lesson 3-4 pp. 136–137

Check Skills You'll Need **1.** The numerator is greater than the denominator. **2.** $\frac{13}{5}$ **3.** $\frac{13}{2}$ **4.** $\frac{5}{3}$ **5.** $\frac{19}{4}$

Quick Check **1 a.** $\frac{3}{20}$ **b.** $\frac{2}{3}$ **c.** $\frac{8}{15}$ **2.** 63 members **3 a.** $10\frac{19}{24}$ **b.** $4\frac{17}{25}$ **c.** $15\frac{3}{32}$

Lesson 3-5 pp. 141–142

Check Skills You'll Need **1.** multiplication **2.** $\frac{2}{7}$ **3.** $\frac{5}{18}$ **4.** 45 **5.** 4

Quick Check **1 a.** $3\frac{1}{2}$ **b.** $\frac{5}{6}$ **c.** 20 **2 a.** $1\frac{25}{44}$ **b.** $\frac{3}{4}$ **c.** $1\frac{3}{16}$ **3.** $5\frac{19}{30}$ cans

Lesson 3-6 pp. 148–153

Check Skills You'll Need **1.** It consists of a whole number and a fraction. **2.** $1\frac{1}{12}$ **3.** $\frac{5}{8}$ **4.** $12\frac{1}{4}$ **5.** $4\frac{1}{4}$

Quick Check **1.** $1\frac{7}{12}$ ft **2.** $6\frac{1}{4}$ c **3.** 74 oz

Checkpoint Quiz 2 **1.** $\frac{2}{7}$ **2.** $3\frac{1}{4}$ **3.** $\frac{1}{5}$ **4.** $9\frac{11}{32}$ **5.** $\frac{11}{15}$ **6.** $11\frac{1}{9}$ yd^2 **7.** $\frac{2}{3}$ **8.** $6\frac{2}{3}$ **9.** 2 **10.** $\frac{4}{5}$ **11.** 13 **12.** $45\frac{3}{4}$ min **13.** $3\frac{1}{3}$ **14.** $2\frac{1}{4}$ **15.** $8\frac{1}{2}$

Lesson 3-7 pp. 154–155

Check Skills You'll Need **1.** meter; liter; gram **2.** 100 **3.** 0.001 **4.** 10 **5.** 1,000 **6.** 0.001

Quick Check **1.** 13 in. **2.** 12.5 g **3.** $10\frac{1}{5}$ mi **4 a.** $\frac{1}{8}$ in. **b.** $\frac{1}{16}$ in. **5.** 14.1 g **6.** 44 m

Chapter 4

Check Your Readiness p. 166

1. 4.414 **2.** 8.6 **3.** 0.79 **4.** 15.21 **5.** −36 **6.** 1 **7.** 42 **8.** −6 **9.** 20 **10.** 1 **11.** 1 **12.** 6 **13.** 7 **14.** 38 **15.** > **16.** = **17.** > **18.** <

Lesson 4-1 pp. 169–170

Check Skills You'll Need **1.** order of operations **2.** 11 **3.** 10 **4.** −10 **5.** 3

Quick Check **1.** $p - 16$ **2.** $9t$ **3.** Answers may vary. Sample: a number decreased by 50, 50 less than a number, 50 subtracted from a number **4.** 56

Lesson 4-2 pp. 174–175

Check Skills You'll Need **1.** variable **2.** $y + 4$ **3.** $v - 6$ **4.** $\frac{k}{9}$

Quick Check **1 a.** 5 **b.** 8 **2 a.** 10 **b.** −16.1 **c.** 8.8 **d.** −9 **3.** about 26 boxes

Lesson 4-3 pp. 180–182

Check Skills You'll Need **1.** equation **2.** about 0 **3.** about 18 **4.** about 25

Quick Check **1.** 168 **2.** \$7.95 **3 a.** Check students' work. **b.** 116

Lesson 4-4 pp. 186–191

Check Skills You'll Need **1.** simplest form **2.** $\frac{1}{4}$ **3.** 4 **4.** 1 **5.** a

Quick Check **1 a.** −7.2 **b.** 9 **c.** 3 **2.** $5c = 110$; \$22 **3.** −390

Checkpoint Quiz 1 **1.** $x - 4$ **2.** $3x$ **3.** $\frac{4}{x}$ **4.** $x + 9$ **5.** −4.4 **6.** 1.8 **7.** −9.1 **8.** 48 **9.** $y + 18 = 1825$; $y = 1807$ **10.** 24 years 8 months

Lesson 4-5 pp. 194–196

Check Skills You'll Need **1.** algebraic expression **2.** 1 **3.** 13 **4.** 24 **5.** −2

Quick Check **1.** Let s = son's age; $3s - 2$ **2.** 37 **3 a.** 4 **b.** 5 **c.** 11 **4.** Let b = number of 3-point baskets; $8 + 3b = 23$; 5 baskets.

Lesson 4-6 pp. 200–201

Check Skills You'll Need **1.** Multiplication Property of Equality **2.** 6 **3.** −5 **4.** −64 **5.** −18

Quick Check **1.** 1 **2.** 200 **3.** Let m = number of markers; $0.79m + 1.25 = 7.57$; 8 markers

Lesson 4-7 pp. 205–209

Check Skills You'll Need **1.** opposites **2.** > **3.** > **4.** < **5.** >

Quick Check **1.** −2, 1.4 **2.** [number line: −7 −6 −5 −4 −3 −2] **3.** $x < 4$ **4.** $t \leq 62$

Checkpoint Quiz 2 **1.** 5 **2.** 40 **3.** 2 **4.** −9 **5.** −2 **6.** −5 **7.** Let c = the cost of a skirt; $12 + 2c = 38$; \$13. **8.** [number line: −5 −4 −3 −2 −1 0] **9.** [number line: −5 −4 −3 −2 −1 0] **10.** [number line: −4 −3 −2 −1 0 1]

Lesson 4-8 pp. 210–211

Check Skills You'll Need **1.** because they "undo" each other **2.** −6 **3.** 1 **4.** 9 **5.** 7

Quick Check **1.** $y < 7$; [number line: 4 5 6 7 8 9] **2 a.** $x > -4$; [number line: −6 −5 −4 −3 −2 −1] **b.** $y < 1$; [number line: 2 1 0 1 2 3] **c.** $w \leq -9$; [number line: −11 −9 −7]

3. Let p = number of points; $p + 109 > 200$; $p > 91$; you need more than 91 points.

Lesson 4-9 pp. 214–216

Check Skills You'll Need **1.** You can add to each side of an equation or inequality without changing the relationship. **2.** $x \leq 2$ **3.** $p > 7$ **4.** $3 \geq d$ **5.** $r < -6$

Quick Check **1 a.** $p > -9$; [number line: −11 −10 −9 −8 −7 −6] **b.** $m \leq 3$; [number line: 0 1 2 3 4 5] **c.** $n > -3$; [number line: −5 −4 −3 −2 −1 0] **2.** 416 min **3.** $k > 20$; [number line: 18 20 22]

Chapter 5

Check Your Readiness p. 226

1. 14.4 **2.** 114.66 **3.** 47.7 **4.** 1.12 **5.** 2.6 **6.** 10.52 **7.** −8 **8.** 60 **9.** −2 **10.** 75 **11.** > **12.** < **13.** = **14.** > **15.** $\frac{3}{4}$ **16.** $\frac{1}{2}$ **17.** $\frac{3}{8}$ **18.** $\frac{1}{3}$ **19.** $\frac{2}{7}$

Lesson 5-1 pp. 228–229

Check Skills You'll Need **1.** Equivalent fractions are fractions that name the same amount. **2.** $\frac{1}{2}$ **3.** $\frac{7}{9}$ **4.** $\frac{4}{7}$ **5.** $\frac{1}{3}$

Quick Check **1 a.** 7 to 12, 7 : 12, $\frac{7}{12}$ **b.** 7 to 5; 7 : 5; $\frac{7}{5}$ **2.** Answers may vary. Samples: $\frac{14}{18}$, $\frac{35}{45}$ **3.** $\frac{6}{5}$ **4 a.** not equivalent **b.** not equivalent **c.** equivalent

Lesson 5-2 pp. 232–237

Check Skills You'll Need **1.** division **2.** $\frac{3}{5}$ **3.** $\frac{3}{1}$ **4.** $\frac{11}{8}$ **5.** $\frac{1}{9}$

Quick Check **1.** 70 heartbeats per min **2.** \$7.00 **3.** \$.064/fl oz, \$.056/fl oz; the 64-fl-oz bottle is the better buy.

Checkpoint Quiz 1 **1.** 7 to 52, $\frac{7}{52}$ **2.** $\frac{2}{3}$ **3.** $\frac{1}{3}$ **4.** 12 to 7 **5.** 2 : 3 **6.** 42 words/min **7.** 9 points/game **8.** \$.2633, \$.2475; the second item is the better buy. **9.** \$7.80, \$6.57; the second item is the better buy. **10.** 15 pizzas

Lesson 5-3 pp. 238–239

Check Skills You'll Need **1.** LCM **2.** > **3.** = **4.** < **5.** >

Quick Check **1.** no; $\frac{5}{6} \neq \frac{5}{7}$ **2 a.** yes; 48 = 48 **b.** yes; 36 = 36 **c.** no; 36 ≠ 40

Lesson 5-4 pp. 244–245

Check Skills You'll Need **1.** A ratio is a unit rate when you are finding the rate for one unit of something. **2.** 8 km/d **3.** 62 mi/h **4.** 25 push-ups/min **5.** 60 words/min

Quick Check **1 a.** \$6.37 **b.** \$119.51 **2 a.** 9 **b.** 2 **c.** 10 **3 a.** 16.8 **b.** 27.2 **c.** 192.5

Lesson 5-5 pp. 250–257

Check Skills You'll Need **1.** proportion **2.** 6 **3.** 12 **4.** 42 **5.** 9.5

Quick Check **1.** 20 **2.** 36 ft

Checkpoint Quiz 2 **1.** no **2.** yes **3.** no **4.** yes **5.** \$8.40 **6.** \$16.50 **7.** 120° **8.** 26° **9.** 9 **10.** 12

Lesson 5-6 pp. 259–261

Check Skills You'll Need **1.** For two ratios, the cross products are found by multiplying the denominator of each ratio by the numerator of the other ratio. **2.** 8 **3.** 15 **4.** 1 **5.** 7

Quick Check **1.** 10 m **2.** about 206 km **3.** 1 in. : 16 in. **4.** 5 in.

Chapter 6

Check Your Readiness p. 272

1. 5 **2.** 14 **3.** 5 **4.** 10 **5.** $\frac{17}{20}$ **6.** $\frac{2}{5}$ **7.** $\frac{17}{25}$ **8.** $\frac{5}{4}$ **9.** $\frac{1}{100}$ **10.** 30 **11.** 800 **12.** 8.4 **13.** 3 **14.** 80 **15.** 2.16 **16.** 200

Lesson 6-1 pp. 274–275

Check Skills You'll Need **1.** Equivalent ratios have the same value. **2–5.** Answers may vary. Samples are given. **2.** $\frac{4}{10}$, $\frac{6}{15}$ **3.** $\frac{26}{100}$, $\frac{39}{150}$ **4.** $\frac{6}{50}$, $\frac{9}{75}$ **5.** $\frac{2}{20}$, $\frac{3}{30}$

Quick Check **1.** $\frac{54}{100}$; 54% **2.**

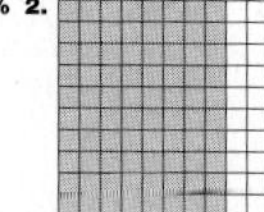

3 a. $\frac{3}{4}$; 75% **b.** $\frac{1}{2}$; 50% **c.** $\frac{7}{10}$; 70% **4.** 80%

Lesson 6-2 pp. 279–281

Check Skills You'll Need **1.** A repeating decimal is a decimal that repeats without end. **2.** 0.3125 **3.** 0.275 **4.** $0.\overline{4}$ **5.** $0.1\overline{3}$

Quick Check **1.** 60.7% **2 a.** 0.35 **b.** 0.125 **c.** 0.078 **3.** 52.5% **4.** $\frac{3}{50}$ **5 a.** 29%, $\frac{3}{10}$, $\frac{11}{25}$, 0.74 **b.** 0.08, 15%, $\frac{7}{20}$, 50%

Lesson 6-3 pp. 284–288

Check Skills You'll Need **1.** 100 **2.** 1% **3.** 98% **4.** 95% **5.** 8%

Quick Check **1.** 1.25; $\frac{5}{4}$ or $1\frac{1}{4}$ **2.** 0.0035; $\frac{7}{2,000}$ **3.** 280% **4.** 0.46%

Checkpoint Quiz 1 **1.** 0.45; $\frac{9}{20}$ **2.** 1.35; $1\frac{7}{20}$ **3.** 0.0098; $\frac{49}{5,000}$ **4.** 56% **5.** $\frac{1}{6}$, 20%, 0.245, $\frac{1}{4}$ **6.** $\frac{16}{25}$; 64% **7.**

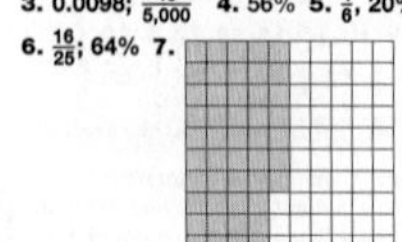

8. about 0.6% **9.** 35% **10.** 75%

Lesson 6-4 pp. 290–291

Check Skills You'll Need **1.** multiplication **2.** 1,200 **3.** 1,350 **4.** 70.2 **5.** 29.7

Quick Check **1.** 105 **2.** 960 **3 a.** about 60 **b.** about 100 **c.** about 600

Lesson 6-5 pp. 294–295

Check Skills You'll Need **1.** ratios **2.** 8 **3.** 15 **4.** 87.5

Quick Check **1.** 25% **2.** 17 **3.** 15 problems

Lesson 6-6 pp. 298–299

Check Skills You'll Need **1.** If both sides of an equation are divided by the same nonzero number, the results are equal. **2.** 17 **3.** 48

Quick Check **1.** 150 seats **2.** 16.2 **3.** about 20.5%

Lesson 6-7 pp. 304–309

Check Skills You'll Need **1.** left **2.** 0.065 **3.** 0.0425 **4.** 0.15 **5.** 0.2

Quick Check **1.** \$195.18 **2 a.** about \$8.70 **b.** about \$9.30 **c.** about \$7.50 **3.** \$192 **4.** \$849

Checkpoint Quiz 2 **1.** $58.\overline{3}$% **2.** 16 **3.** 84% **4.** 45% **5.** 68.4 **6.** 222 **7.** \$313.95 **8.** \$82.45 **9.** \$35.50 **10.** \$77

Lesson 6-8 pp. 310–311

Check Skills You'll Need **1.** 500 and $16n$ **2.** 10 **3.** 12.5 **4.** 50

Quick Check **1.** 1.9% **2.** 99.4% **3.** 40%

Chapter 7

Check Your Readiness p. 322

1. < **2.** > **3.** > **4.** < **5.** 2 **6.** 15 **7.** 12 **8.** 105 **9.** 5 **10.** 94 **11.** 80 **12.** 16 **13.** 111 **14.** 46 **15.** 108 **16.** 86.4 **17.** 16.2 **18.** 64.8

Lesson 7-1 pp. 324–325

Check Skills You'll Need **1.** <, >, ≤, ≥, ≠ **2.** [number line: −4 −2 0 2] **3.** [number line: −3 −2 −1 0 1] **4.** [number line: −2 0 2] **5.** [number line: 0 2 4 6] **6.** [number line: −1 0 1 2 3] **7.** [number line: −1 0 1 2 3 4]

Quick Check **1 a.** $\overrightarrow{PD}$ **b.** $\overline{RS}$ **c.** $\overleftrightarrow{AV}$ **2 a.** $\overline{EF}$, $\overline{HD}$ **b.** $\overline{EB}$, $\overline{CB}$, $\overline{AB}$, $\overline{GH}$, $\overline{DH}$ **c.** $\overline{FC}$, $\overline{EF}$, $\overline{BC}$, $\overline{HD}$

Lesson 7-2 pp. 330–332

Check Skills You'll Need **1.** equation **2.** 70 **3.** 140 **4.** 74 **5.** 158

Quick Check **1.** obtuse **2.** 53° **3.** 108°; 72°; 108°

Lesson 7-3 pp. 336–337

Check Skills You'll Need **1.** an angle with a measure greater than 0° and less than 90° **2.** right **3.** obtuse

Quick Check **1.** isosceles **2 a.** right **b.** acute **3.** 22°

Lesson 7-4 pp. 340–345

Check Skills You'll Need **1.** an equilateral triangle **2.** equilateral **3.** isosceles **4.** scalene

Quick Check **1 a.** The octagon is irregular because the sides are not all congruent. **b.** The triangle is regular because all sides and all angles are congruent. **c.** Multiply the side length by the number of sides the polygon has. **2 a.** hexagons, triangles **b.** squares, triangles, rhombuses **3.** Answers may vary. An example is given.

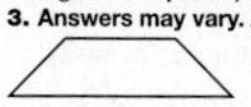

Checkpoint Quiz 1 **1.** $\overrightarrow{XZ}$ **2.** $\overleftrightarrow{NM}$ **3.** $\overline{KV}$ **4.** 78°; 168° **5.** 43°; 133° **6.** 25°; 115° **7.** 70° **8.** equilateral **9.** obtuse isosceles **10.** right scalene **11.** Equilateral $\triangle PQT$ is regular, square $QRST$ is regular, and pentagon $PQRST$ is irregular.

Lesson 7-5 pp. 346–347

Check Skills You'll Need **1.** Two polygons are similar if corresponding angles are congruent and corresponding side lengths have the same proportion. **2.** 3.15 m **3.** 56° **4.** 52°

Quick Check **1.** Congruent; the sides and angles all have the same measure. **2.** No; the sides are different lengths. **3 a.** $\overline{AB} \cong \overline{FG}$; $\overline{BC} \cong \overline{GH}$; $\overline{CD} \cong \overline{HI}$; $\overline{DA} \cong \overline{IF}$; $\angle A \cong \angle F$; $\angle B \cong \angle G$; $\angle C \cong \angle H$; $\angle D \cong \angle I$ **b.** 15.5; 139°

Lesson 7-6 pp. 350–351

Check Skills You'll Need **1.** 2 **2.** $\overline{FR}$ **3.** $\overrightarrow{ZJ}$ **4.** $\overleftrightarrow{NK}$

Quick Check **1.** $\angle AOB$, $\angle AOC$, $\angle AOD$, $\angle BOC$, $\angle BOD$, $\angle COD$ **2.** $\widehat{ZY}$, $\widehat{ZYX}$, $\widehat{YZX}$ **3.** $\widehat{CDE}$, $\overline{CE}$

Lesson 7-7 pp. 354–358

Check Skills You'll Need **1.** Every percent can be written as a decimal by moving the decimal point two places to the left. Thus 25% becomes 0.25. So taking the percent of a number is the same as multiplying by a decimal. **2.** 90 **3.** 216 **4.** 259.2

Quick Check **1.** about 470 million **2 a.** 144° **b.**

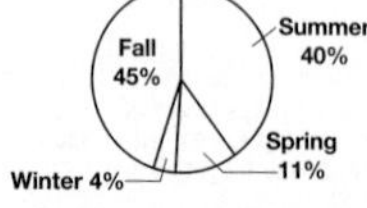

Checkpoint Quiz 2 **1.** Yes. The square root of the area is the side length. Since both squares have the same area, they have the same side length. Being squares, they both have four right angles. Congruent sides and congruent angles means the squares are congruent.

2. Town Middle School: Grade 6 25%; Grade 7 33.3%; Grade 8 41.7%

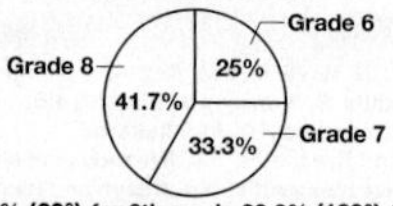

25% (90°) for 6th grade 33.3% (120°) for 7th grade, and 41.7% (150°) for 8th grade. **3.** $\overline{QP}$, $\overline{RP}$, $\overline{SP}$, $\overline{TP}$ **4.** $\overline{TR}$ **5.** $\angle RPS$, $\angle SPT$, $\angle TPQ$, and $\angle QPR$

Lesson 7-8 pp. 361–362

Check Skills You'll Need **1.** a common point **2.** $\overline{VW}$ **3.** $\overrightarrow{FG}$ **4.** $\overleftrightarrow{CL}$

Quick Check **1.** T 25 mm R; S V **2.**

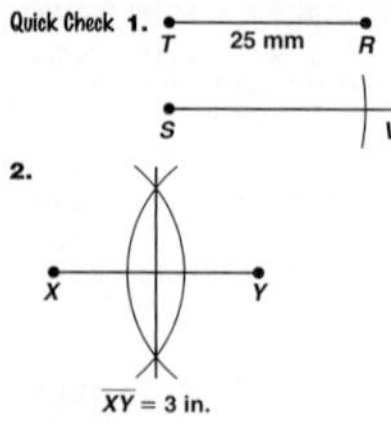

Chapter 8

Check Your Readiness p. 372

1. 3.14 **2.** 123 **3.** 108 **4.** 4 **5.** 97 **6.** 24 **7.** 144 **8.** 3 **9.** 8 **10.** 58° **11.** 63° **12.** 50°

Lesson 8-1 pp. 374–376

Check Skills You'll Need **1.** inch, foot, yard, mile **2.** 72 **3.** 4 **4.** 204

Quick Check **1.** 400 mi; driving distances are usually measured in miles. **2–4.** Answers may vary. Samples are given. **2.** about 24 units **3.** about 92 yd^2 **4.** area: about 300 ft^2; perimeter: about 100 ft

Lesson 8-2 pp. 380–381

Check Skills You'll Need **1.** Changing the order of the factors does not change the product. **2.** 450 **3.** 28.08 **4.** 57.12 **5.** 2.04

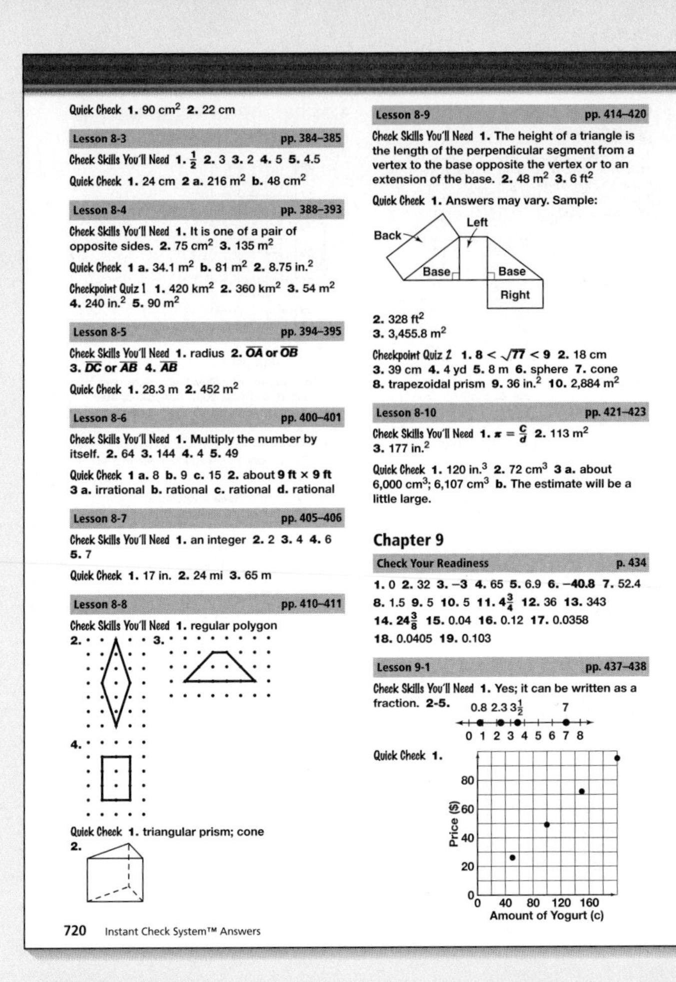

Quick Check 1. 90 cm² 2. 22 cm

Lesson 8-3 pp. 384–385

Check Skills You'll Need 1. $\frac{1}{2}$ 2. 3 3. 2 4. 5 5. 4.5

Quick Check 1. 24 cm 2 a. 216 m² b. 48 cm²

Lesson 8-4 pp. 388–393

Check Skills You'll Need 1. It is one of a pair of opposite sides. 2. 75 cm² 3. 135 m²

Quick Check 1 a. 34.1 m² b. 81 m² 2. 8.75 in.²

Checkpoint Quiz 1 1. 420 km² 2. 360 km² 3. 54 m² 4. 240 in.² 5. 90 m²

Lesson 8-5 pp. 394–395

Check Skills You'll Need 1. radius 2. $\overline{OA}$ or $\overline{OB}$ 3. $\overline{DC}$ or $\overline{AB}$ 4. $\overline{AB}$

Quick Check 1. 28.3 m 2. 452 m²

Lesson 8-6 pp. 400–401

Check Skills You'll Need 1. Multiply the number by itself. 2. 64 3. 144 4. 4 5. 49

Quick Check 1 a. 8 b. 9 c. 15 2. about 9 ft × 9 ft 3 a. irrational b. rational c. rational d. rational

Lesson 8-7 pp. 405–406

Check Skills You'll Need 1. an integer 2. 2 3. 4 4. 6 5. 7

Quick Check 1. 17 in. 2. 24 mi 3. 65 m

Lesson 8-8 pp. 410–411

Check Skills You'll Need 1. regular polygon 2. 3. 4.

Quick Check 1. triangular prism; cone 2.

Lesson 8-9 pp. 414–420

Check Skills You'll Need 1. The height of a triangle is the length of the perpendicular segment from a vertex to the base opposite the vertex or to an extension of the base. 2. 48 m² 3. 6 ft²

Quick Check 1. Answers may vary. Sample:

2. 328 ft² 3. 3,455.8 m²

Checkpoint Quiz 2 1. $8 < \sqrt{77} < 9$ 2. 18 cm 3. 39 cm 4. 4 yd 5. 8 m 6. sphere 7. cone 8. trapezoidal prism 9. 36 in.² 10. 2,884 m²

Lesson 8-10 pp. 421–423

Check Skills You'll Need 1. $\pi = \frac{C}{d}$ 2. 113 m² 3. 177 in.²

Quick Check 1. 120 in.³ 2. 72 cm³ 3 a. about 6,000 cm³; 6,107 cm³ b. The estimate will be a little large.

Chapter 9

Check Your Readiness p. 434

1. 0 2. 32 3. −3 4. 65 5. 6.9 6. −40.8 7. 52.4 8. 1.5 9. 5 10. 5 11. $4\frac{3}{4}$ 12. 36 13. 343 14. $24\frac{3}{8}$ 15. 0.04 16. 0.12 17. 0.0358 18. 0.0405 19. 0.103

Lesson 9-1 pp. 437–438

Check Skills You'll Need 1. Yes; it can be written as a fraction. 2–5.

Quick Check 1.

720 Instant Check System™ Answers

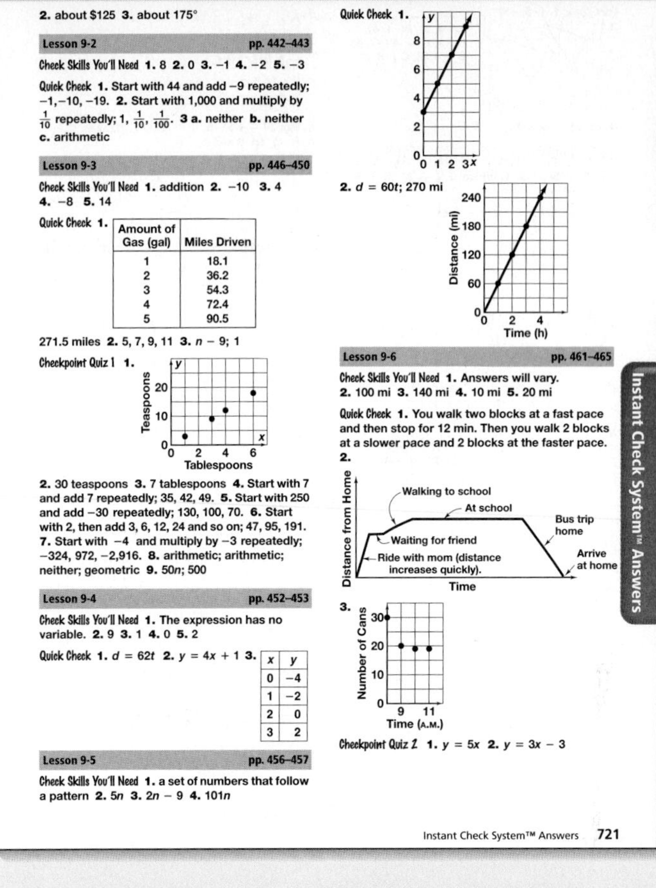

2. about $125 3. about 175°

Lesson 9-2 pp. 442–443

Check Skills You'll Need 1. 8 2. 0 3. −1 4. −2 5. −3

Quick Check 1. Start with 44 and add −9 repeatedly; −1, −10, −19. 2. Start with 1,000 and multiply by $\frac{1}{10}$ repeatedly; 1, $\frac{1}{10}$, $\frac{1}{100}$. 3 a. neither b. neither c. arithmetic

Lesson 9-3 pp. 446–450

Check Skills You'll Need 1. addition 2. −10 3. 4 4. −8 5. 14

Quick Check 1.

Amount of Gas (gal)	Miles Driven
1	18.1
2	36.2
3	54.3
4	72.4
5	90.5

271.5 miles 2. 5, 7, 9, 11 3. $n - 9$; 1

Checkpoint Quiz 1 1.

2. 30 teaspoons 3. 7 tablespoons 4. Start with 7 and add 7 repeatedly; 35, 42, 49. 5. Start with 250 and add −30 repeatedly; 130, 100, 70. 6. Start with 2, then add 3, 6, 12, 24 and so on; 47, 95, 191. 7. Start with −4 and multiply by −3 repeatedly; −324, 972, −2,916. 8. arithmetic; arithmetic; neither; geometric 9. $50n$; 500

Lesson 9-4 pp. 452–453

Check Skills You'll Need 1. The expression has no variable. 2. 9 3. 1 4. 0 5. 2

Quick Check 1. $d = 62t$ 2. $y = 4x + 1$ 3.

x	y
0	−4
1	−2
2	0
3	2

Lesson 9-5 pp. 456–457

Check Skills You'll Need 1. a set of numbers that follow a pattern 2. $5n$ 3. $2n - 9$ 4. $101n$

Quick Check 1.

2. $d = 60t$; 270 mi

Lesson 9-6 pp. 461–465

Check Skills You'll Need 1. Answers will vary. 2. 100 mi 3. 140 mi 4. 10 mi 5. 20 mi

Quick Check 1. You walk two blocks at a fast pace and then stop for 12 min. Then you walk 2 blocks at a slower pace and 2 blocks at the faster pace. 2.

3.

Checkpoint Quiz 2 1. $y = 5x$ 2. $y = 3x - 3$

Instant Check System™ Answers 721

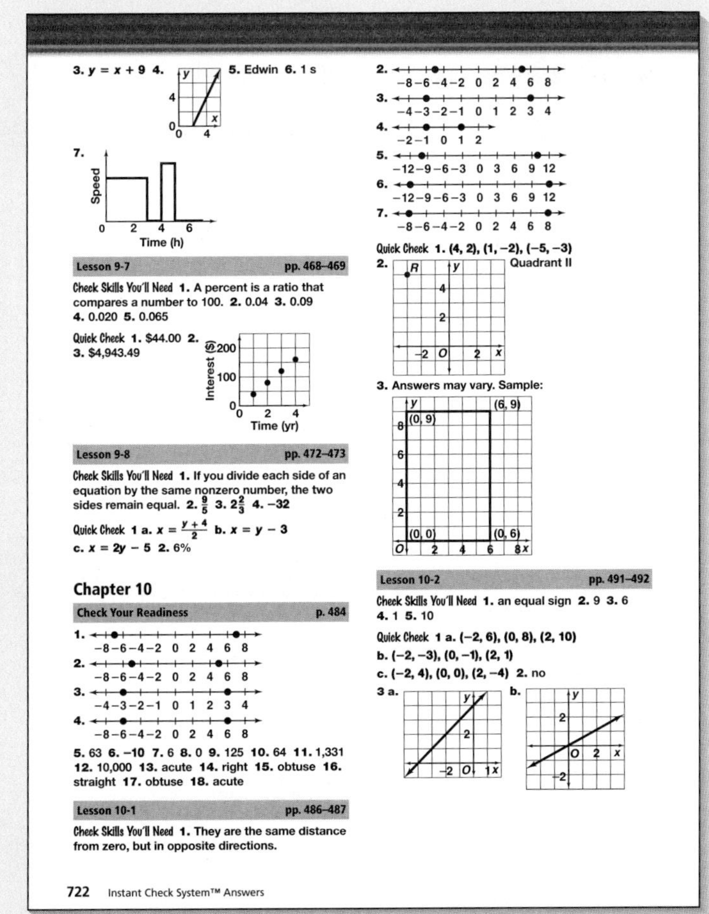

3. $y = x + 9$ 4. 5. Edwin 6. 1 s

7.

Lesson 9-7 pp. 468–469

Check Skills You'll Need 1. A percent is a ratio that compares a number to 100. 2. 0.04 3. 0.09 4. 0.020 5. 0.065

Quick Check 1. $44.00 2. 3. $4,943.49

Lesson 9-8 pp. 472–473

Check Skills You'll Need 1. If you divide each side of an equation by the same nonzero number, the two sides remain equal. 2. $\frac{9}{5}$ 3. $2\frac{2}{3}$ 4. −32

Quick Check 1 a. $x = \frac{y + 4}{2}$ b. $x = y - 3$ c. $x = 2y - 5$ 2. 6%

Chapter 10

Check Your Readiness p. 484

1. 2. 3. 4.

5. 63 6. −10 7. 6 8. 0 9. 125 10. 64 11. 1,331 12. 10,000 13. acute 14. right 15. obtuse 16. straight 17. obtuse 18. acute

Lesson 10-1 pp. 486–487

Check Skills You'll Need 1. They are the same distance from zero, but in opposite directions.

2. 3. 4. 5. 6. 7.

Quick Check 1. (4, 2), (1, −2), (−5, −3) 2. Quadrant II

3. Answers may vary. Sample:

Lesson 10-2 pp. 491–492

Check Skills You'll Need 1. an equal sign 2. 9 3. 6 4. 1 5. 10

Quick Check 1 a. (−2, 6), (0, 8), (2, 10) b. (−2, −3), (0, −1), (2, 1) c. (−2, 4), (0, 0), (2, −4) 2. no 3 a. b.

722 Instant Check System™ Answers

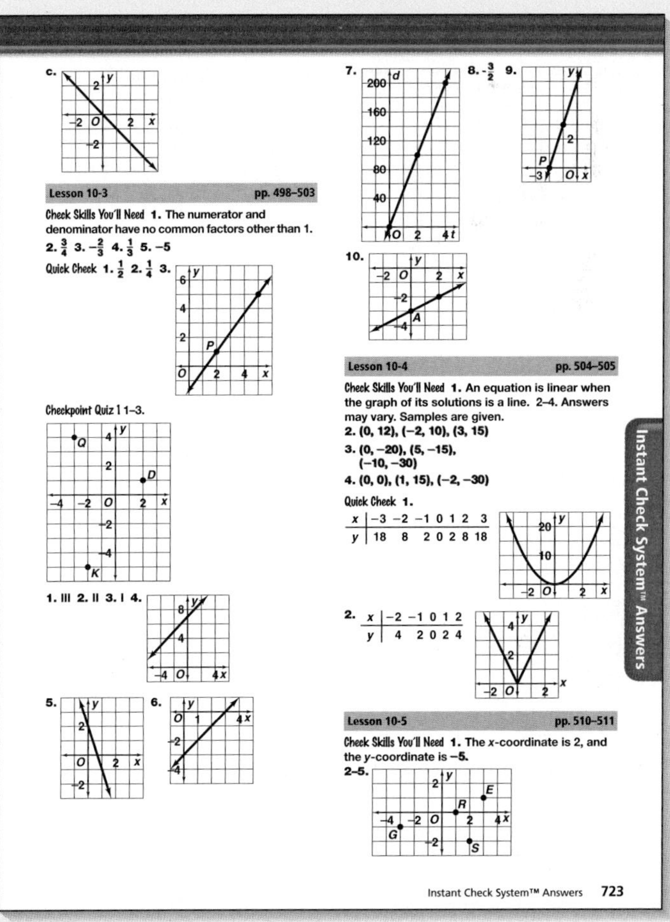

c.

Lesson 10-3 pp. 498–503

Check Skills You'll Need 1. The numerator and denominator have no common factors other than 1. 2. $\frac{3}{4}$ 3. $-\frac{2}{3}$ 4. $\frac{1}{3}$ 5. −5

Quick Check 1. $\frac{1}{2}$ 2. $\frac{1}{4}$ 3.

Checkpoint Quiz 1 1–3.

1. III 2. II 3. I 4.

5. 6.

7. 8. $-\frac{3}{2}$ 9.

10.

Lesson 10-4 pp. 504–505

Check Skills You'll Need 1. An equation is linear when the graph of its solutions is a line. 2–4. Answers may vary. Samples are given. 2. (0, 12), (−2, 10), (3, 15) 3. (0, −20), (5, −15), (−10, −30) 4. (0, 0), (1, 15), (−2, −30)

Quick Check 1.

x	−3	−2	−1	0	1	2	3
y	18	8	2	0	2	8	18

2.

x	−2	−1	0	1	2
y	4	2	0	2	4

Lesson 10-5 pp. 510–511

Check Skills You'll Need 1. The x-coordinate is 2, and the y-coordinate is −5.

2–5.

Instant Check System™ Answers 723

Instant Check System™ Answers

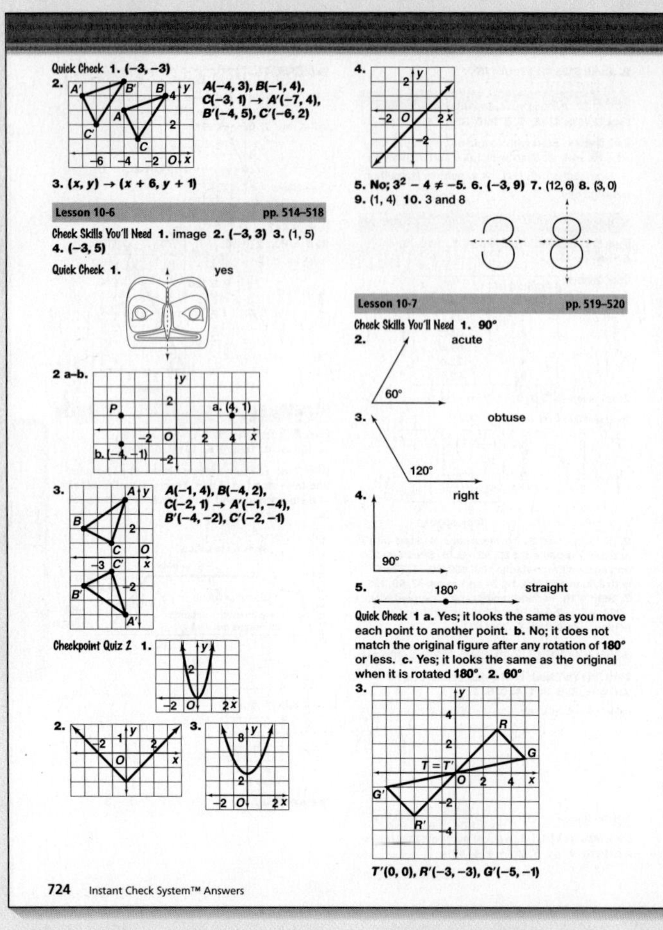

Quick Check 1. (−3, −3)
2. A(−4, 3), B(−1, 4), C(−3, 1) → A′(−7, 4), B′(−4, 5), C′(−6, 2)
3. $(x, y) \to (x + 6, y + 1)$

Lesson 10-6 pp. 514–518

Check Skills You'll Need 1. image 2. (−3, 3) 3. (1, 5) 4. (−3, 5)

Quick Check 1. yes

2 a–b. a. (4, 1) b. (−4, −1)

3. A(−1, 4), B(−4, 2), C(−2, 1) → A′(−1, −4), B′(−4, −2), C′(−2, −1)

Checkpoint Quiz 2 1.

2.

3.

4.

5. No; $3^2 - 4 \neq -5$. 6. (−3, 9) 7. (12, 6) 8. (3, 0) 9. (1, 4) 10. 3 and 8

Lesson 10-7 pp. 519–520

Check Skills You'll Need 1. 90°
2. acute
3. obtuse
4. right
5. straight

Quick Check 1 a. Yes; it looks the same as you move each point to another point. b. No; it does not match the original figure after any rotation of 180° or less. c. Yes; it looks the same as the original when it is rotated 180°. 2. 60°
3. T′(0, 0), R′(−3, −3), G′(−5, −1)

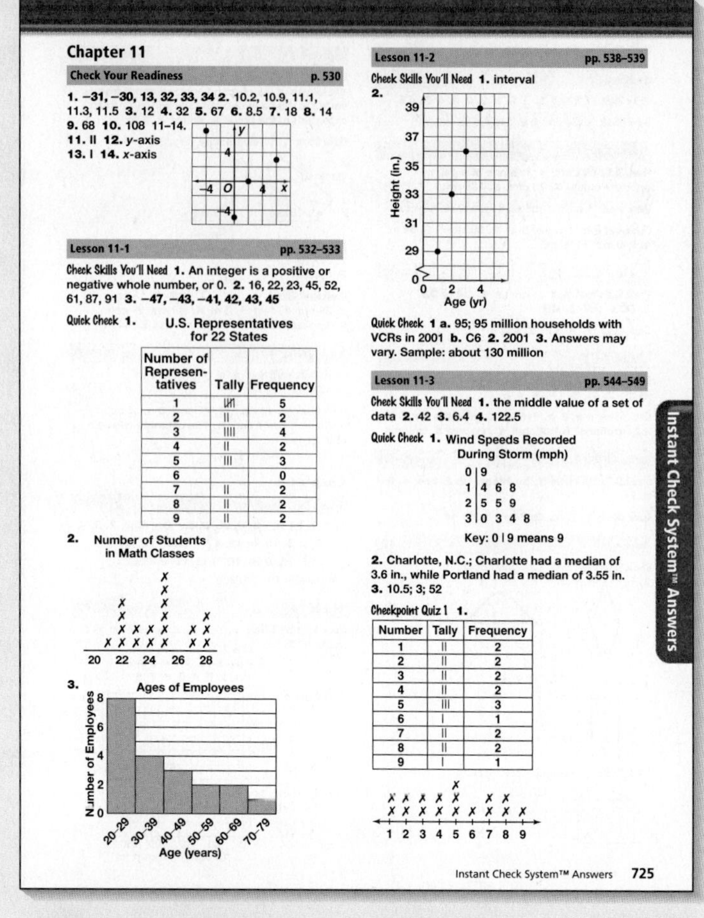

Chapter 11

Check Your Readiness p. 530

1. −31, −30, 13, 32, 33, 34 2. 10.2, 10.9, 11.1, 11.3, 11.5 3. 12 4. 32 5. 67 6. 8.5 7. 18 8. 14 9. 68 10. 108 11–14. 11. II 12. y-axis 13. I 14. x-axis

Lesson 11-1 pp. 532–533

Check Skills You'll Need 1. An integer is a positive or negative whole number, or 0. 2. 16, 22, 23, 45, 52, 61, 87, 91 3. −47, −43, −41, 42, 43, 45

Quick Check 1. U.S. Representatives for 22 States

Number of Representatives	Tally	Frequency
1	𝍸	5
2	II	2
3	IIII	4
4	II	2
5	III	3
6		0
7	II	2
8	II	2
9	II	2

2. Number of Students in Math Classes

3. Ages of Employees

Lesson 11-2 pp. 538–539

Check Skills You'll Need 1. interval
2.

Quick Check 1 a. 95; 95 million households with VCRs in 2001 b. C6 2. 2001 3. Answers may vary. Sample: about 130 million

Lesson 11-3 pp. 544–549

Check Skills You'll Need 1. the middle value of a set of data 2. 42 3. 6.4 4. 122.5

Quick Check 1. Wind Speeds Recorded During Storm (mph)

0 | 9
1 | 4 6 8
2 | 5 5 9
3 | 0 3 4 8

Key: 0 | 9 means 9

2. Charlotte, N.C.; Charlotte had a median of 3.6 in., while Portland had a median of 3.55 in. 3. 10.5; 3; 52

Checkpoint Quiz 1 1.

Number	Tally	Frequency
1	II	2
2	II	2
3	II	2
4	II	2
5	III	3
6	I	1
7	II	2
8	II	2
9	I	1

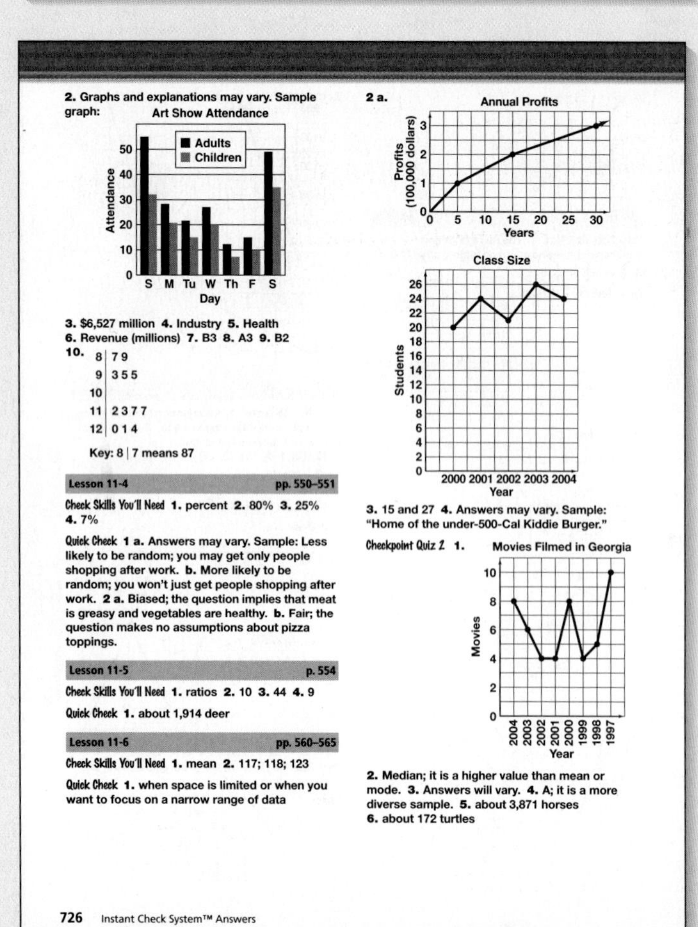

2. Graphs and explanations may vary. Sample graph: Art Show Attendance

3. $6,527 million 4. Industry 5. Health 6. Revenue (millions) 7. B3 8. A3 9. B2
10.

8 | 7 9
9 | 3 5 5
10 |
11 | 2 3 7 7
12 | 0 1 4

Key: 8 | 7 means 87

Lesson 11-4 pp. 550–551

Check Skills You'll Need 1. percent 2. 80% 3. 25% 4. 7%

Quick Check 1 a. Answers may vary. Sample: Less likely to be random; you may get only people shopping after work. b. More likely to be random; you won't just get people shopping after work. 2 a. Biased; the question implies that meat is greasy and vegetables are healthy. b. Fair; the question makes no assumptions about pizza toppings.

Lesson 11-5 p. 554

Check Skills You'll Need 1. ratios 2. 10 3. 44 4. 9

Quick Check 1. about 1,914 deer

Lesson 11-6 pp. 560–565

Check Skills You'll Need 1. mean 2. 117; 118; 123

Quick Check 1. when space is limited or when you want to focus on a narrow range of data

2 a. Annual Profits

Class Size

3. 15 and 27 4. Answers may vary. Sample: "Home of the under-500-Cal Kiddie Burger."

Checkpoint Quiz 2 1. Movies Filmed in Georgia

2. Median; it is a higher value than mean or mode. 3. Answers will vary. 4. A; it is a more diverse sample. 5. about 3,871 horses 6. about 172 turtles

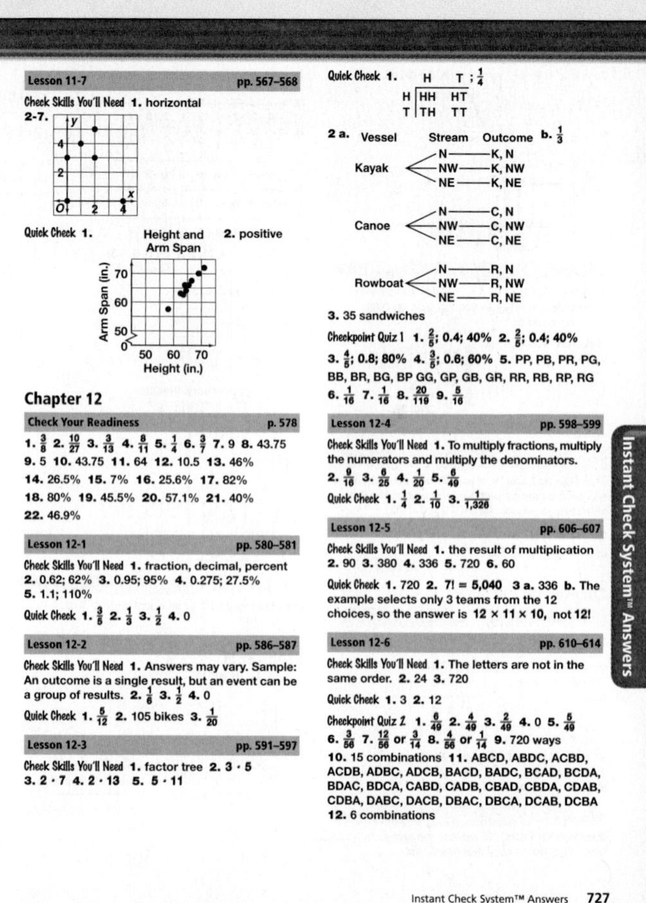

Lesson 11-7 pp. 567–568

Check Skills You'll Need 1. horizontal
2–7.

Quick Check 1. Height and Arm Span 2. positive

Chapter 12

Check Your Readiness p. 578

1. $\frac{3}{8}$ 2. $\frac{10}{27}$ 3. $\frac{3}{13}$ 4. $\frac{8}{11}$ 5. $\frac{1}{4}$ 6. $\frac{3}{7}$ 7. 9 8. 43.75 9. 5 10. 43.75 11. 64 12. 10.5 13. 46% 14. 26.5% 15. 7% 16. 25.6% 17. 82% 18. 80% 19. 45.5% 20. 57.1% 21. 40% 22. 46.9%

Lesson 12-1 pp. 580–581

Check Skills You'll Need 1. fraction, decimal, percent 2. 0.62; 62% 3. 0.95; 95% 4. 0.275; 27.5% 5. 1.1; 110%

Quick Check 1. $\frac{3}{5}$ 2. $\frac{1}{3}$ 3. $\frac{1}{2}$ 4. 0

Lesson 12-2 pp. 586–587

Check Skills You'll Need 1. Answers may vary. Sample: An outcome is a single result, but an event can be a group of results. 2. $\frac{1}{6}$ 3. $\frac{1}{2}$ 4. 0

Quick Check 1. $\frac{5}{12}$ 2. 105 bikes 3. $\frac{1}{20}$

Lesson 12-3 pp. 591–597

Check Skills You'll Need 1. factor tree 2. $3 \cdot 5$ 3. $2 \cdot 7$ 4. $2 \cdot 13$ 5. $5 \cdot 11$

Quick Check 1. T: $\frac{1}{4}$

	H	T
H	HH	HT
T	TH	TT

2 a. b. $\frac{1}{3}$

3. 35 sandwiches

Checkpoint Quiz 1 1. $\frac{2}{5}$; 0.4; 40% 2. $\frac{2}{5}$; 0.4; 40% 3. $\frac{4}{5}$; 0.8; 80% 4. $\frac{3}{5}$; 0.6; 60% 5. PP, PB, PR, PG, BB, BR, BG, BP GG, GP, GB, GR, RR, RB, RP, RG 6. $\frac{1}{16}$ 7. $\frac{1}{16}$ 8. $\frac{20}{119}$ 9. $\frac{5}{16}$

Lesson 12-4 pp. 598–599

Check Skills You'll Need 1. To multiply fractions, multiply the numerators and multiply the denominators. 2. $\frac{9}{16}$ 3. $\frac{6}{25}$ 4. $\frac{1}{20}$ 5. $\frac{6}{49}$

Quick Check 1. $\frac{1}{4}$ 2. $\frac{1}{10}$ 3. $\frac{1}{1,326}$

Lesson 12-5 pp. 606–607

Check Skills You'll Need 1. the result of multiplication 2. 90 3. 380 4. 336 5. 720 6. 60

Quick Check 1. 720 2. 7! = 5,040 3 a. 336 b. The example selects only 3 teams from the 12 choices, so the answer is $12 \times 11 \times 10$, not 12!

Lesson 12-6 pp. 610–614

Check Skills You'll Need 1. The letters are not in the same order. 2. 24 3. 720

Quick Check 1. 3 2. 12

Checkpoint Quiz 2 1. $\frac{6}{49}$ 2. $\frac{4}{49}$ 3. $\frac{2}{49}$ 4. 0 5. $\frac{5}{49}$ 6. $\frac{3}{56}$ 7. $\frac{12}{56}$ or $\frac{3}{14}$ 8. $\frac{4}{56}$ or $\frac{1}{14}$ 9. 720 ways 10. 15 combinations 11. ABCD, ABDC, ACBD, ACDB, ADBC, ADCB, BACD, BADC, BCAD, BCDA, BDAC, BDCA, CABD, CADB, CBAD, CBDA, CDAB, CDBA, DABC, DACB, DBAC, DBCA, DCAB, DCBA 12. 6 combinations

Selected Answers

Chapter 1

Lesson 1-1 pp. 6–7

EXERCISES **1.** Rounded numbers are numbers rounded to the nearest whole number, whereas compatible numbers are chosen because they are easy to compute mentally. **3.** C **5.** B **7.** 11 **9.** 90 **13.** 19 **15.** 14 **19.** 4 **21.** 10 **29.** about $2 **33.** about $26; (390 ÷ 30) × $2 **37.** 706.8, 761.8, 768.0, 768.1

Lesson 1-2 pp. 10–11

EXERCISES **1.** Assoc. Prop. of Add. **3.** 40 **5.** 19 **7.** 0 **9.** 105.8 **11.** 7.582 **19.** 5.26 **21.** 0.0645 **27.** 43.5 **29.** 39.5 **35.** > **37.** 122.4 m **39.** 4.67 in. **41.** sailfish; 0.008 mi/h **45.** 136

Lesson 1-3 pp. 16–17

EXERCISES **1.** Comm. Prop. of Mult. **3.** The product is divided by 10. **5.** 2 **7.** 0; zero property **9.** 3; commutative property **13.** 0.15 **15.** 3.915 **21.** 0 **23.** 21.5 **33.** 0.432 **39.** 13.52 million **41.** $11.04 **45.** 1.643

Lesson 1-4 pp. 22–23

EXERCISES **1.** dividend **3.** 4 **5.** 400 **7.** 6 **9.** 12 **17.** 0.75 **19.** 740 **27.** 4,490 **29.** 3.02 **33.** 14 yd **35.** The quotient is greater than the divisor; dividing by a number less than one is the same as multiplying by a whole number. **37.** 1.45 in. **41.** 6.7

Lesson 1-5 pp. 28–30

EXERCISES **1.** 1,000 **3.** 0.01 **5.** 0.064 **7.** 0.00849 **9.** 22 m **11.** 900 **13.** 58,000 **23.** C **25.** D **29.** millimeters **33.** 1,250 g **35.** 5 mugs **37.** No; a double batch needs 2.02 L. **41.** 6.35

Lesson 1-6 pp. 33–34

EXERCISES **1.** Whole numbers do not include negative numbers. **3 a.** sometimes true **b.** sometimes true **c.** sometimes true **d.** sometimes true **5.** −4 **7.** 2 **9.** 5 **11.** −12 **13.** 8 **15.** −11 **23.** 11 **25.** 1 **33.** < **35.** < **41.** −5, −2, −1, 0, 7 **45 a.** Omaha, Nebr. Bismarck, N. Dak. **47.** −3551, −3515, −3155, −3151 **49.** No; there are no integers between the integers −3 and −4. **53.** 453

Lesson 1-7 pp. 41–42

EXERCISES **1.** A; always **3.** C; never **5.** 8 **7.** −1 **9.** 1 **13.** 38 **15.** 0 **19.** 13 **21.** 25 **29.** −1 + 4 = 3 **31.** −2 **33.** 8:00 A.M. **41.** >

Lesson 1-8 pp. 46–47

EXERCISES **1.** B **3.** 4 **5.** 80 **7.** −2 **9.** −5 **11.** 36 **13.** 21 **19.** 3 **21.** −14 **31.** −1 **35.** 6 min **37 a.** 278 − 5(15) **b.** $203 **41.** 1.42

Lesson 1-9 pp. 50–51

EXERCISES **1.** multiplication **3.** 11(2) should be 11(0.2). **5.** 20; 4 **7.** 30; 2; 160 **9.** 7 **11.** 3 **17.** 145 **19.** 19.8 **27.** (4 + 4) ÷ 4 − 4 = −2 **29.** 1,840 ft² **31.** 3,351 ft **39.** −60

Lesson 1-10 pp. 55–57

EXERCISES **1.** outlier **3.** 5 **5.** Median; the outlier 45 greatly affects the mean. **7.** 3.5; 3.5; no mode; 1 **9.** 8 **13.** 8 **17.** 51 and 58 **21.** 10 **29.** Answers may vary. Samples are given. 1, 2, 3, 5, 5 **33.** 87 cartons **37.** 123, 213, 231, 312, 321

Chapter Review pp. 60–61

1. Distr. Prop. **2.** Assoc. Prop. of Add. **3.** absolute value **4.** mode **5.** order of operations

Exercises 6–9. Answers may vary. Samples are given.

6. 7; compatible numbers **7.** 52, rounding **8.** 63, rounding **9.** 6; front-end estimation **10.** 3.7 **11.** 0.567 **12.** 0.73 **13.** 16.875 **14.** 7.31 lb **15.** 0.456 **16.** 14,200 **17.** 340 **18.** 71.47 m **19.** < **20.** = **21.** > **22.** > **23.** 6 **24.** 29 **25.** −30 **26.** −25 **27.** 3.425 **28.** 13.13 **29.** 19 **30.** 162.4 **31.** 2; 2; 2; 3

Chapter 2

Lesson 2-1 pp. 70–71

EXERCISES **1.** The exponent tells you how many times the base is used as a factor. **3.** −625 **7.** B **9.** 2^6 **11.** 6^3 **15.** 81 **17.** 0.000064 **25.** 39

27. 39 **31.** 60 · 60 · 60; 60^3

33.

Power of 10	Value	Number of Zeros
10^1	10	1
10^2	100	2
10^3	1,000	3
10^4	10,000	4
10^5	100,000	5

35. 12 **37.** Positive; $(-672)^2$ is positive and larger than 192. **45.** 40

Lesson 2-2 pp. 77–78

EXERCISES **1.** A factor is a whole number that divides another whole number with a remainder of 0. A multiple of a number is the product of the number and another whole number. **3.** 2 **5.** Composite; the factors are 1, 2, 4, 7, 14 and 28. **7.** Composite; the factors are 1, 2, 13 and 26. **9.** 36 **11.** 10 **17.** 1, 2, 4, 5, 10, 20 **21.** prime **25.** $3^2 \cdot 5$ **27.** $2^2 \cdot 3 \cdot 7$ **33.** 2 **35.** 16 **43.** 1 **45.** 50 **47.** 6 boxes **49.** Yes; both can be written as $2^3 \cdot 3$. **55.** −5, −4, 3, 6

Lesson 2-3 pp. 84–85

EXERCISES **1.** No; the GCF of the numerator and denominator of a fraction in simplest form is 1. **3.** No; the student added 4 instead of multiplying by 4 **7.** $\frac{10}{12}$, $\frac{15}{18}$ **9.** $\frac{4}{18}$, $\frac{6}{27}$ **15.** $\frac{4}{12}$, $\frac{1}{3}$ **17.** $\frac{9}{27}$, $\frac{1}{3}$ **23.** $\frac{3}{4}$ **31.** Answers may vary. Sample: $\frac{5}{8}$, $\frac{10}{16}$ **33.** D **37.** Answers may vary. Sample: $\frac{2a}{2b}$, $\frac{3a}{3b}$ **41.** 17.09

Lesson 2-4 pp. 89–90

EXERCISES **1.** The LCM of the denominators of two or more fractions is the LCD. **3.** 63 **5.** 24 **7.** > **9.** < **11.** < **19.** $\frac{1}{9}$, $\frac{1}{8}$, $\frac{1}{6}$ **21.** $\frac{5}{8}$, $\frac{7}{9}$, $\frac{2}{1}$ **29.** $\frac{3}{4}$ in.; the $\frac{3}{8}$-in. nail is not long enough to go all the way through the board. **31.** = **33.** > **35.** what they say **37.** $\frac{3}{4} > \frac{7}{10}$ **45.** 72

Lesson 2-5 pp. 93–94

EXERCISES **1.** Both are larger than 1. **3.** proper fraction **5.** improper fraction **7.** 8 **9.** It is a whole number because 72 is divisible by 12. **11.** $\frac{23}{4}$ **13.** $\frac{23}{5}$ **21.** $8\frac{1}{3}$ **23.** $2\frac{7}{12}$ **33.** $\frac{1}{12}$ **35.** 1 **37.** $1\frac{5}{8}$ in., $\frac{13}{8}$ in. **39.** $4\frac{3}{8}$ mi **41.** $3\frac{2}{8}$, $3\frac{1}{4}$, $\frac{26}{8}$ **45.** −12

Lesson 2-6 pp. 99–100

EXERCISES **1.** A terminating decimal stops, whereas a repeating decimal has a block of digits that repeat without end. **3.** $1\frac{3}{8}$ **5.** $3\frac{99}{100}$ **7.** 0.5, $1\frac{1}{3}$, $1.\overline{9}$, 2 **9.** 0.8 **11.** $0.\overline{6}$ **17.** $\frac{33}{50}$ **19.** $3\frac{3}{4}$ **25.** 3.84, $3\frac{41}{50}$, 3.789, 3 **31.** < **33.** > **35.** when the numerator is not a multiple of 3 **37.** N.Y.: $\frac{4{,}572}{19{,}227}$ Tex.: $\frac{6{,}267}{22{,}490}$ Calif.: $\frac{9{,}596}{33{,}893}$ Fla.: $\frac{4{,}003}{17{,}397}$ Ohio: $\frac{2{,}779}{11{,}459}$ **39.** Fla., N.Y., Ohio, Tex., Calif. **41.** The quotients are 0.5, 5, 50, 500, 5,000, and 50,000. As the divisor gets smaller, the quotient gets larger because **divisor × quotient = dividend.** **45.** 0.36

Lesson 2-7 pp. 104–105

EXERCISES **1.** Answers may vary. Sample: A rational number is a number that can be written as a quotient of two integers. **3.** > **5.** −236, $-7\frac{1}{7}$, $-3.\overline{3}$, 0, $\frac{41}{99}$ **7.** > **9.** < **19.** −1.0, $-\frac{3}{4}$, 0.25, $\frac{3}{2}$ **25.** > **27.** > **29.** frog **31.** $\frac{3}{4}$ **35.** $\frac{1}{2}$

Lesson 2-8 pp. 108–109

EXERCISES **1.** It is written as the product of two factors, one greater than or equal to 1 and less than 10, and the other a power of 10. **3.** 3 **5.** B **7.** A **9.** 7.5×10^7 **11.** 4.4×10^4 **17.** 3,400 **27.** 1,600 lb **31.** 3.5×10^5 **33.** about 3.3×10^2 h **37.** 60

Chapter Review pp. 112–113

1. exponent **2.** prime **3.** factors **4.** equivalent fraction **5.** GCF **6.** −16 **7.** −64 **8.** 125 **9.** 60 **10.** $2^2 \cdot 3 \cdot 7$ **11.** $2 \cdot 3 \cdot 13$ **12.** $2 \cdot 3^2 \cdot 5$ **13.** $2^2 \cdot 23$ **14.** 5^3 **15.** 210 days from now **16.** $\frac{1}{6}$, $\frac{1}{4}$, $\frac{1}{3}$ **17.** $\frac{1}{4}$, $\frac{3}{8}$, $\frac{2}{5}$ **18.** $\frac{3}{8}$, $\frac{1}{2}$, $\frac{5}{6}$ **19.** $\frac{5}{9}$, $\frac{7}{12}$, $\frac{2}{3}$ **20.** store B **21.** $5\frac{3}{4}$ h **22.** 6 **23.** $2\frac{1}{4}$ **24.** $6\frac{2}{3}$ **25.** 25 **26.** 7 **27.** $0.\overline{3}$ **28.** $0.\overline{5}$ **29.** 2.5 **30.** 0.8 **31.** 0.08 **32.** $-\frac{7}{8}$, $0.\overline{3}$, $\frac{3}{4}$ **33.** $-\frac{4}{11}$, −0.3, 2.7 **34.** $-\frac{5}{8}$, −0.5, 2.2 **35.** 7.123×10^6 **36.** 906,000 **37.** 8.19×10^4 **38.** 601,500,000

Chapter 3

Lesson 3-1 pp. 122–123

EXERCISES **1.** Answers may vary. Sample: A benchmark is a convenient number used to replace a fraction. Examples are 0, $\frac{1}{2}$, and 1 for fractions between 0 and 1. **3.** 1 **5.** $\frac{1}{2}$ **7.** 1 **9.** 7 **11.** about $\frac{1}{2}$ **13.** about $\frac{1}{2}$ **21.** about 10 **23.** about 5 **29.** about 4 **35.** about 29 **37.** 2 h **39.** about $\frac{1}{2}$ t **41.** about $4\frac{1}{2}$ t **49.** $11\frac{1}{8}$

Lesson 3-2 pp. 128–129

EXERCISES **1.** Answers may vary. Sample: It doesn't make sense to add fractions with different denominators. For example, $\frac{1}{2} + \frac{1}{3} \neq \frac{2}{5}$. **3.** 16 **5.** 6 **7.** 1 **9.** $\frac{2}{3}$ **11.** $\frac{23}{40}$ **13.** $\frac{1}{18}$ **15.** $1\frac{1}{2}$ **17.** $\frac{2}{5}$ **27.** $1\frac{5}{8}$ **29.** $1\frac{3}{10}$ **37.** positive **41.** $\frac{1}{2}$ **43.** $\frac{7}{10}$ **45.** $\frac{17}{20}$ **51.** −2.4, 1.34, $\frac{7}{3}$, $\frac{25}{6}$

Lesson 3-3 pp. 132–133

EXERCISES **1.** $3\frac{3}{4}$ **3.** $2\frac{2}{5}$ **5.** $3\frac{3}{2}$ **7.** $\frac{19}{10}$ **9.** $8\frac{1}{5}$ **15.** $18\frac{5}{6}$ **19.** $11\frac{1}{3}$ **25.** $29\frac{3}{20}$ **29.** $7\frac{7}{8}$ mi **31.** $4\frac{1}{2}$ in.² **33.** $10\frac{5}{8}$ c.i. **37.** 115

Lesson 3-4 pp. 138–139

EXERCISES **3.** $2\frac{1}{4} \cdot 1\frac{2}{3} = \frac{9}{4} \cdot \frac{5}{3}$ **5.** No; you do not need to find a common denominator, since you reduce the product anyway. **7.** $\frac{1}{3}$ **9.** $\frac{5}{18}$ **15.** 12 **17.** 26 **25.** 12 **27.** $3\frac{2}{3}$ **33.** $8\frac{1}{2}$ c **35.** 0 **39.** $\frac{7}{8}$ mi **41.** 28 games **43.** 2 mi²; $6\frac{3}{5}$ mi **47.** 76

Lesson 3-5 pp. 143–145

EXERCISES **1.** $1\frac{1}{2} \div \frac{1}{2}$ **3.** 4 **5.** $\frac{2}{7}$ **7.** A **9.** 8 **11.** $\frac{5}{8}$ **21.** $\frac{6}{13}$ **23.** $1\frac{1}{2}$ **35.** Your friend found the reciprocal of $\frac{3}{8}$ without changing $5\frac{3}{8}$ to an improper fraction; $\frac{8}{43}$. **37.** 800 nails **39.** 6 costumes **41.** 1 **47.** 12 ft **53.** 2,564

Lesson 3-6 pp. 150–151

EXERCISES **1.** multiply **3.** multiply **5.** divide **7.** It takes more smaller units to equal the larger units. **9.** $4\frac{1}{2}$ **11.** $6\frac{1}{6}$ **17.** 3 **19.** $\frac{1}{2}$ **23.** 12 **25.** 5,000 **31.** 9,560 ft **33.** 2,400 **35.** $3\frac{1}{2}$ **39.** 4 ft **41.** no; quarter pound = 4 oz **47.** 3

Lesson 3-7 pp. 156–157

EXERCISES **1.** 4 in., 4 in., $3\frac{6}{8}$ in. **3.** $3\frac{12}{16}$ in. **5.** $4\frac{1}{2}$ ft **7.** $11\frac{15}{16}$ in. **9.** 56 lb **17.** 2 lb **19.** 4 ft **21.** 8.9 oz **25.** 2 cm **27.** about $4\frac{2}{8}$ in. **29.** ft **31.** cm **35.** 541 ft **41.** about 6

Chapter Review pp. 160–161

1. reciprocal **2.** benchmark **3.** precision **4.** benchmark; precision **5.** about 1 **6.** about $\frac{1}{2}$ **7.** about 1 **8.** about $\frac{1}{2}$ **9.** about 3 **10.** about 25 **11.** about 14 **12.** about 10 **13.** about 8 mi **14.** $1\frac{7}{12}$ **15.** $6\frac{13}{15}$ **16.** $15\frac{5}{12}$ **17.** $5\frac{13}{24}$ **18.** $11\frac{1}{20}$ min **19.** $\frac{1}{4}$ **20.** $\frac{9}{10}$ **21.** 9 **22.** $94\frac{23}{24}$ **23.** $39\frac{7}{12}$ m² **24.** $\frac{1}{2}$ **25.** 6 **26.** $3\frac{3}{5}$ **27.** $\frac{5}{18}$ **28.** $3\frac{2}{11}$ ft **29.** $3\frac{1}{2}$ **30.** $1\frac{1}{2}$ **31.** 5 **32.** 2 **33.** 4,000 **34.** 7 **35.** about $9\frac{1}{2}$ ft **36.** 12 c **37.** 5.5 L **38.** 8.75 m **39.** 23 h **40.** 25 g **41.** $11\frac{3}{16}$ in. **42.** 7 g **43.** 13 yd **44.** 11.6 cm **45.** 21.00 lb **46.** 5 L **47.** 5.13 m

Chapter 4

Lesson 4-1 pp. 171–172

EXERCISES **1.** An algebraic expression differs from a numerical expression because it contains at least one variable, and the value changes. **3.** addition

5–7. Answers may vary. Samples are given:

5. the product of 5 and w **7.** the quotient of w and 4 **9.** 6 **11.** 6 **13.** $\frac{p}{5}$ **17–19.** Answers may vary. Samples are given. **17.** two more than a number **19.** nine and one tenth less than a number **25.** 42 **27.** 3 **33.** $10n$ **35.** Answers may vary. Sample: Multiply 24(60) to find the number of minutes in 24 h. Then multiply the answer by 1,260, the number of beats per minute. The heart beats 1,814,400 times in 24 h. **39.** Answers may vary. Sample: The second student charges $15 to start and $3/h. **43.** 45

Lesson 4-2 pp. 176–177

EXERCISES **1.** solution **3.** yes **5.** no **7.** −3 **9.** −4 **11.** 288 **15.** −8 **17.** −4 **21.** about 19 **27.** 54 **29.** 20 **33.** $4d = 360$ **35.** 31.89 in. **37.** $4c = 67.80$; about $17; no; the total should be about 4($12.95), or $51.80. **41.** 1

Lesson 4-3 pp. 182–184

EXERCISES **1.** Inverse **3.** 12 **5.** −49 **7.** 205.4 **19.** 9 **25.** B **31.** 119 **33.** −0.9 **37.** $p - 8.45 = 21.50$; $29.95 **39.** Let r = the number of runs needed; $3 + 4 + 2 + 6 + 8 + r = 30$; **43.** 33

Lesson 4-4 pp. 188–190

EXERCISES **1.** Multiplication Property of Equality **3.** Division Property of Equality **5.** Division Property of Equality **7.** C **9.** 12 **11.** −5 **23.** −24 **25.** 100.8 **39.** 50 **41.** 85 **45.** 15 yr **47.** b is the reciprocal of a, and $b \neq 0$. **55.** $\frac{23}{28}$

Lesson 4-5 pp. 196–198

EXERCISES **1.** A one-step expression uses only one operation, while a two-step expression uses two. **3.** B **5.** C **7.** 70 **9.** Let h = your height; $6h + 1$. **13.** 1.5 **15.** −0.9 **21.** 15.2 **23.** 7 **33.** 11.5 **35.** 45 **39.** 9 min **41.** Answers may vary. Sample: You need to keep the equation "balanced." **47.** =

Lesson 4-6 pp. 202–204

EXERCISES **1.** Let m = number of miles; $2.00 + 0.50m = 5.00$; 6 mi **3.** 12 **5.** −3 **7.** 8 **31.** Let h = number of additional hours; $3.95 + 1.25h = 7.70$; 3 h. **35.** 11 **39.** 15 credits

Lesson 4-7 pp. 207–208

EXERCISES **1.** inequality **3.** C **5.** −1 **7.** −2 **11.** 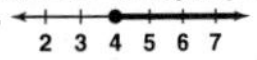 (number line: 2 3 4 5 6 7) **13.** 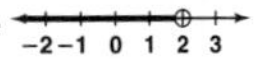 (number line: −2 −1 0 1 2 3) **19.** $x \leq 6$ **25.** $a \geq 17$; (number line: 15 16 17 18 19 20) **31.** $w \leq 3$ **33.** −3, −2, or −1 **39.** −4, −2, −1, 2, 4, 7

Lesson 4-8 pp. 212–213

EXERCISES **1.** Addition Property of Inequality **3.** C **5.** B **7.** $g \leq -6$; (number line: −9 −8 −7 −6 −5 −4) **9.** $y \geq 16$; (number line: 14 16 18) **17.** $n \geq 1$; (number line: −1 0 1 2 3 4) **19.** $p \leq -4$; (number line: −7 −6 −5 −4 −3 −2) **29.** $w > -\frac{1}{2}$ **31.** $j \geq -5$ **35.** No; the solution of $x + 5 \leq -2$ is $x \leq -7$, and the solution of $-2 \leq x + 5$ is $x \geq -7$. **39.** no more than 1,220 Cal **43.** $\frac{16}{5}$

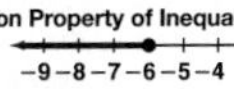

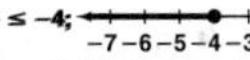

Lesson 4-9 pp. 216–218

EXERCISES **1.** The inequality symbol is reversed. **3.** > **5.** > **7.** $p > 12$; (number line: 10 11 12 13 14 15) **9.** $x < -8$; (number line: −11 −10 −9 −8 −7 −6) **19.** $p < -15$; (number line: −17 −15 −13) **21.** $w \leq -21$; (number line: −23 −21 −19) **35.** $4x \leq -44$; $x \leq -11$ **41.** $50.2 < t$ **43.** at most 3 hot dogs **45.** 4 bags of peanuts; $.50 **47.** The student should not have reversed the inequality symbol. **49.** $-3.x + 10 > 19$; $x < -3$; (number line: −6 −5 −4 −3 −2 −1) **53.** $-3.8\overline{3}$; 3.6; 2.2

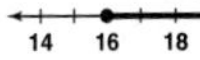

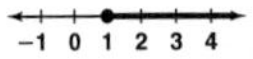

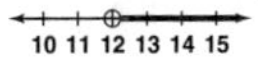

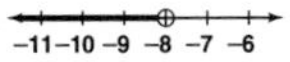

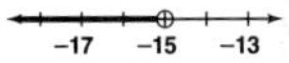

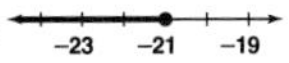

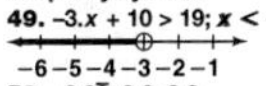

Chapter Review pp. 220–221

1. variable **2.** inequality **3.** equation **4.** algebraic expression **5.** solution of an inequality **6.** 5 **7.** 6 **8.** 13 **9.** 4 **10.** −7 **11.** 20 **12.** 4 **13.** 72 **14.** 24 **15.** 84 **16.** 99 **17.** −12 **18.** 52 **19.** 5 **20.** 43 **21.** −54 **22.** 114 tickets **23.** 1 **24.** 12 **25.** −3 **26.** $140 + 35w = 1{,}050$; 26 wk **27.** $t \leq 10$; (number line: 7 8 9 10 11 12) **28.** $r < 5$; (number line: 2 3 4 5 6 7) **29.** $h < -22$; (number line: −24 −22 −20) **30.** $p \geq -15$; (number line: −2 −1 0 1 2 3) **31.** $g > 17$; (number line: 15 17 19) **32.** $m \geq -7$; (number line: −9 −8 −7 −6 −5 −4) **33.** $x \geq 20$; (number line: 18 20 22)

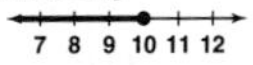

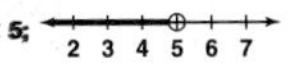

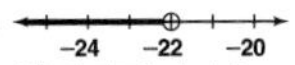

Selected Answers

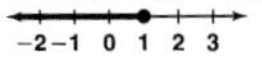

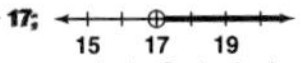

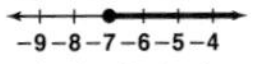

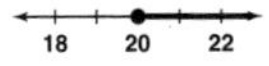

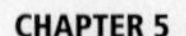

CHAPTER 5

Lesson 5-1 pp. 230–231

EXERCISES **1.** Equivalent ratios name the same amount, as do equivalent fractions.

3–5. Answers may vary. Samples are given.

3. 2 to 16 **5.** 20 to 18 **7.** $\frac{17}{34}$ **9.** equivalent **11.** equivalent **13.** 21 to 25, **21 : 25,** $\frac{21}{25}$

14–16. Answers may vary. Samples are given.

15. 12 to 14 **17.** $\frac{1}{2}$ **19.** $\frac{3}{200}$ **23.** equivalent **31 a. 8 : 4, 7.5 : 3, 3.5 : 1 b.** 10 qt antifreeze, 5 qt water **37.** 45

Lesson 5-2 pp. 234–235

EXERCISES **1.** Answers may vary. Sample: It gives the cost of one item. **3.** 147; 7 **5.** 8 g **7.** 16 points/game **11.** $30 **15.** $.93/oz **17.** $.06/fl oz, $.05/fl oz; the 50-fl-oz detergent is the better buy. **23.** $.75/pt **25.** $.60/lb **27 a.** 1 person/mi^2 **b.** No; answers may vary. Sample: Some regions of the state are more densely populated than others. **29.** about 39,421,000 times; about 108,000 times; about 75 times **33.** 29

Lesson 5-3 pp. 240–241

EXERCISES **1.** ratios **3.** 8 **5.** 9 **7.** 12 **9.** No; you need to multiply the numerator of one by the denominator of the other. They do form a proportion.

$$\frac{3}{4} \stackrel{?}{=} \frac{12}{16}$$
$$3 \cdot 16 \stackrel{?}{=} 4 \cdot 12$$
$$48 = 48$$

11. no **13.** yes **31.** no **33.** no **35.** Yes; the weight ratios $\frac{174}{29}$ and $\frac{102}{17}$ are proportional because $174 \cdot 17 = 102 \cdot 29$. **37.** yes; $\frac{26}{1} = \frac{104}{4}$

39. $\frac{4n}{3}$ and $\frac{12n}{9}$ will always form a proportion because $36n = 36n$ for all values of n. **43.** 105

Lesson 5-4 pp. 246–248

EXERCISES **1.** Answers may vary. Sample: Finding a unit rate gives you a denominator of 1, so you only need to multiply to solve the proportion. $\frac{5}{1} = \frac{x}{8}, x = 40$ **3.** 4 **5.** 12 **7.** $7.90 **9.** $59.00 **11.** 6 **13.** 6 **17.** 20 **19.** 85.75 **31.** 5 **33.** 6.9 **35.** 240 should be the numerator. $\frac{2}{3} = \frac{240}{n}$ **43.** −6, −3, −2, 1, 8

Lesson 5-5 pp. 254–255

EXERCISES **1.** Two polygons are similar if corresponding angles have the same measure and the lengths of the corresponding sides form equivalent ratios. **3.** $\overline{AB}$ **7.** 11.2 ft **15.** 6 **17.** 2.88 km **19.** 99 cm **23.** 5

Lesson 5-6 pp. 261–263

EXERCISES **1.** ratio; length **3.** No; the scale should be written as **model : actual,** or **8 ft : 6 ft. 5.** 7 in. **7.** 165 ft **9.** 495 ft **13.** about 38 mi **23.** Answers may vary. Sample: **1 cm : 1.9 m 25a. 1 in. : 4 ft b.** 0.5 in.

Chapter Review pp. 266–267

1. a scale drawing **2.** unit cost **3.** rate **4.** scale **5.** similar **6.** $\frac{3}{10}$ **7.** $\frac{16}{5}$ **8. 3 : 1 9. 15 : 4 10.** $\frac{1}{4}$ **11.** $\frac{39}{97}$, 39 to 97, 39 : 97 **12.** 6 passengers/car **13.** 75 Cal/serving **14.** 23 students/classroom **15.** $4/kg **16.** $0.28/oz, $0.31/oz; the 10-oz size is the better buy. **17.** 3 ft^2 per minute **18.** 12 **19.** 25 **20.** 3 **21.** 136 **22.** $\frac{5}{3} = \frac{x}{250{,}000}$; 417,000 board feet **23.** 7.5 ft **24.** 15 ft **25.** 45 **26.** $x = 45, y = 36$ **27.** 6,300 mi **28.** 0.75 in.

Chapter 6

Lesson 6-1 pp. 276–277

EXERCISES **1.** $\frac{32}{100}$; 32% **3.** 67% **5.** 90% **7.** $\frac{64}{100}$; 64% **11.** **15.** $\frac{4}{5}$, 80%

21. 60% **23.** 84% **25.** 55% **27.** 70% **29.** 52% **31.** about 19% **33.** about 28% **39.** −2, −3

Lesson 6-2 pp. 282–283

EXERCISES **1.** 62%, $\frac{62}{100}$, 0.62 **3.** $\frac{1}{2}$, 0.54, 55% **5.** 57% **7.** 9% **11.** 0.88 **15.** 90% **17.** 8.3% **21.** $\frac{3}{20}$ **23.** $\frac{1}{5}$ **27.** 12%, 0.25, $\frac{1}{2}$

33.

Topping	With Olives	Plain	With Onions and Green Peppers
Percent of the Pizza	37.5%	12.5%	50%
Number of Slices	3	1	4

35 a. 79%, $\frac{16}{20}$, $\frac{21}{25}$, 85%, $\frac{9}{10}$, 92% **b.** 85% **39.** $-0.9 \le x$

Lesson 6-3 pp. 286–287

EXERCISES **1.** 1.5; $\frac{3}{2}$ or $1\frac{1}{2}$ **3.** 825% **5.** 4 **7.** Answers may vary. Sample: A decimal that is less than 1% has zeros in the tenths and hundredths place, such as 0.009. A decimal that is greater than 100% has a number other than zero before the decimal point, such as 1.01. **9.** 1.3; $1\frac{3}{10}$ **11.** 3.45; $3\frac{9}{20}$ **21.** 160% **23.** 258% **35.** 270% **37.** 1,001% **39.** 1.66; $1\frac{33}{50}$

41. **43.**

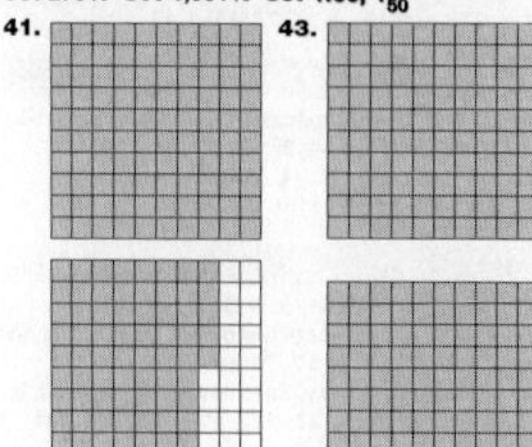

45. 0.26% **47.** No; it cannot be more than 100% fat. **49.** Yes; it is reasonable that $\frac{1}{2}$ of 1% of the seeds will not grow. **53.** $\frac{7}{25}$

Lesson 6-4 pp. 292–293

EXERCISES **1.** 29 **3.** 58.08 **5.** $\frac{1}{5} \cdot 60 = 12$ **7.** $\frac{3}{20} \cdot 40 = 6$ **9.** Answers may vary. Sample: You would probably use a fraction when the percent can be written as a fraction that is compatible with the other number. You would use a decimal for all other cases. **11.** 16 **13.** 27.6 **19.** 57.2 **21.** 474 **29.** about 30 **31.** about 80 **37.** 160.38 **39.** 42.38 **41.** 6,800 forest fires **45.** 26 students **49.** 60

Lesson 6-5 pp. 296–297

EXERCISES **1.** percent; 25% **3.** whole; 16 **5.** C **7.** B **9.** 10% **11.** 2% **15.** 5 **17.** 2.25 **21.** 18.75 **23.** 80 **27.** $\frac{90}{n} = \frac{40}{100}$; 225 **29.** $\frac{54}{144} = \frac{n}{100}$; 37.5 **31.** 37.5% **35.** The number 100 always appears as the denominator of one of the ratios, since percent means "out of 100." **41.** $\frac{1}{3}$

Lesson 6-6 pp. 300–301

EXERCISES **1.** No; "20% of 40" asks for a part, "20 is what percent of 40" asks for a percent, and "20 is 40% of what number" asks for a whole. **3.** C **5.** $625p = 500$; 80% **7.** $0.96x = 24$; 25 **13.** $x = 0.41 \cdot 800$; 328 **17.** $18 = 48x$; 37.5% **23.** 69% **27.** 49% **29.** about 190 people **31.** 45 members **37.** yes

Lesson 6-7 pp. 306–307

EXERCISES **1.** You earn 8% of the amount you sell. **3.** $3.60 **5.** $27 **7.** $.71 **9.** $77.12 **11.** about $10.35 **15.** $96 **21.** $23 **23.** 5% **25.** $6,800 **33.** $2.35/lb

Lesson 6-8 pp. 312–314

EXERCISES **1.** Answers may vary. Sample: They both involve the difference between the original price and the selling price. Percent of markup is a percent of increase and percent of discount is a percent of decrease. **3.** $\frac{15}{35}$, 43% increase **5.** $\frac{374}{748}$, 50% decrease **7.** A **9.** 13% **11.** 50% **17.** 50% **23.** 56% **29.** 25% decrease **33.** $53.30

35.

	A	B	C	D
1	Yr	Sales	Change ($)	Change (%)
2	1	200,000	—	—
3	2	240,000	40,000	20%
4	3	300,000	60,000	25%
5	4	330,000	30,000	10%

41. $471.30

Chapter Review pp. 316–317

1. C **2.** E **3.** D **4.** B **5.** A **6.** 0.65; $\frac{13}{20}$ **7.** 0.02; $\frac{1}{50}$ **8.** 0.018; $\frac{9}{500}$ **9.** 0.625; $\frac{5}{8}$ **10.** 37.5% **11.** 16% **12.** 44.82 **13.** 0.64 **14.** 97.2 **15.** 70% **16.** 47.5 **17.** 252 **18.** 12 **19.** 72 **20.** 20% **21.** $57.60 **22.** about $10.52 **23.** $268 **24.** $63 **25.** $216 **26.** $275 **27.** $225 **28.** 16.7% decrease **29.** 20% increase **30.** 15% increase **31.** 40% increase **32.** 72% decrease **33.** 16.7% decrease **34.** 31% decrease

Chapter 7

Lesson 7-1 pp. 326–327

EXERCISES **1.** Skew lines **3.** $\overleftrightarrow{LC}$ **5.** $\overrightarrow{KE}$ **7.** $\overline{BC}$, $\overline{FE}$ **9.** $\overleftrightarrow{LC}$ **11.** $\overleftrightarrow{ZV}$ **15.** $\overleftrightarrow{AE}$, $\overleftrightarrow{EF}$, $\overleftrightarrow{DH}$, $\overleftrightarrow{GH}$ **21.** yes **25.**

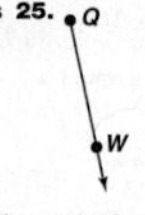

29. $\overleftrightarrow{EF}$ and $\overleftrightarrow{GH}$ do not lie in the same plane. **31.** $\overline{UV}$, $\overrightarrow{UV}$, $\overleftrightarrow{UV}$, $\overrightarrow{VU}$, $\overleftrightarrow{VW}$, $\overleftrightarrow{VY}$, $\overleftrightarrow{WY}$, $\overleftrightarrow{WX}$, $\overrightarrow{WX}$, $\overrightarrow{XW}$, $\overline{WX}$, $\overleftrightarrow{YZ}$, $\overline{YZ}$ **35.** about 9% increase

Lesson 7-2 pp. 333–334

EXERCISES **1.** Vertical angles lie opposite each other, while adjacent angles lie next to each other. **3.** acute, 45°, 135° **5.** acute, 15°, 105° **7.** obtuse **9.** 11°; 101° **11.** 78°; 168° **19.** 57° **23.** 159.8° and 69.8° **25.** The student used the wrong scale on the protractor. **27.** about 65° **29.** Answers may vary. Sample: ∠JHK and ∠BHC, ∠HCD and ∠BCL **31.** ∠HCB

Lesson 7-3 pp. 338–339

EXERCISES **1.** no **3.** obtuse scalene **5.** acute isosceles **7.** scalene **9.** obtuse **13.** 64° **17.** isosceles **21.** 44° **23.** An equilateral triangle is always isosceles because it has 3 congruent sides. An isosceles triangle is not always equilateral because it has at least 2 congruent sides. **27.** 1.16; $\frac{29}{25}$

Lesson 7-4 pp. 342–344

EXERCISES **1.** A trapezoid is a quadrilateral with only one pair of parallel sides. A parallelogram is a quadrilateral with two pairs of parallel sides. **3.** quadrilateral, parallelogram **5.** quadrilateral, parallelogram, rhombus, rectangle, square **7.** The parallelogram is irregular because the sides are not all congruent. **9.** rectangles, trapezoids **11.**

15. Trapezoid; $\overline{PQ}$ is parallel to $\overline{RS}$. **17.** JN = 5 in.; m∠J = m∠K = m∠L = m∠N = 90° **19.**

23. No; it would be a rectangle.

25.

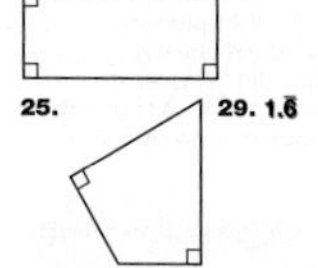

29. $1.\overline{6}$

Lesson 7-5 pp. 348–349

EXERCISES **1.** Yes. Similar polygons have congruent angles and sides in proportion. If the proportion is 1 to 1, then the polygons are not just similar, but congruent. **3.** Congruent; all corresponding sides and angles are congruent. **5.** Not congruent; not all corresponding sides and angles are congruent. **7.** $\overline{CV} \cong \overline{JP}$; $\overline{VR} \cong \overline{PX}$; $\overline{RC} \cong \overline{XJ}$; ∠C ≅ ∠J; ∠V ≅ ∠P; ∠R ≅ ∠X; VR = 9; m∠V = 83°; m∠X = 41°; m∠J = 56° **11.** △SRT **15.** $\overline{BD}$ **17.** ∠C **21.** $\overline{XY} \cong \overline{BA}$; $\overline{YZ} \cong \overline{AD}$; $\overline{ZW} \cong \overline{DC}$; $\overline{WX} \cong \overline{CB}$; ∠Y ≅ ∠A; ∠Z ≅ ∠D; ∠W ≅ ∠C; ∠X ≅ ∠B; YZ = 11; ZW = 18; m∠Y = 127°; m∠D = 83°; m∠B = 83° **25.** 25%

Lesson 7-6 pp. 352–353

EXERCISES **1.** A radius is a line from the center of a circle to a point on the circle. A diameter is a line between two points on a circle that passes through the center of the circle. **3.** M **5.** $\overline{JL}$ **7.** ∠JMK, ∠KML, ∠JML **9.** $\overline{OC}$, $\overline{OK}$, $\overline{OD}$ **13.** $\overline{TR}$, $\overline{RS}$, $\overline{ST}$ **15.** $\overline{CD}$, $\overline{DB}$, $\overline{CB}$

19–21. Answers may vary. Samples are given.

19. ∠FDG, ∠GDH **21.** △FDG **27.** 22.6 cm **31.** 400 m **35.** acute

Lesson 7-7 pp. 356–357

EXERCISES **1.** the whole circle, 360°, 100% **3.** 90° **5.** 36° **7.** sleeping **11.**

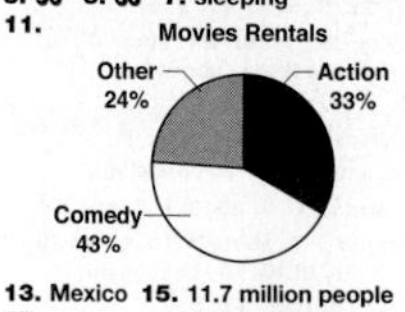

13. Mexico **15.** 11.7 million people **17.**

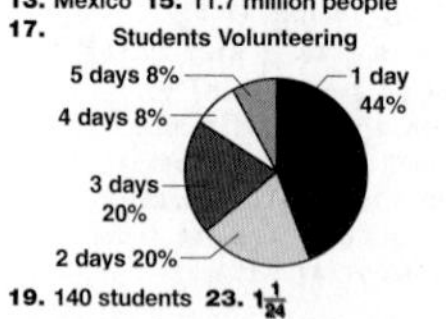

19. 140 students **23.** $1\frac{1}{24}$

Lesson 7-8 pp. 363–364

EXERCISES **1.** Answers may vary. Since Q is the midpoint of $\overline{PR}$, $\overline{PQ}$ and $\overline{QR}$ are congruent. **3.** 8 in. **5.** $3\frac{1}{2}$ m **17.**

$\overline{AB}$ = 5 in.

21. 4.2 cm **23.** 34 mm **27.**

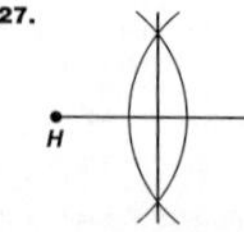

31. $\frac{5}{7}$, 5 : 7

Chapter Review pp. 366–367

1. parallel **2.** pentagon **3.** supplementary **4.** scalene **5.** acute **6.** 35°, 125° **7.** 63°, 153° **8.** 3°, 93° **9.** 78°, 168° **10.** 120°; isosceles; obtuse **11.** 60°; equilateral; acute **12.** 31°; scalene; right **13.** pentagon; regular **14.** square, regular **15.** octagon, irregular **16.** m∠M = 30°, m∠C = m∠P = 46°, m∠B = 104°, AC = 22, BC = 12, MN = 17 **17.** $\overline{TW}$, $\overline{TY}$, $\overline{TK}$, $\overline{TM}$ **18.** $\overline{MY}$ **19.** T **20.** $\overline{WY}$; $\overline{KY}$ **21.** $\overset{\frown}{WYM}$, $\overset{\frown}{YKW}$, $\overset{\frown}{KMY}$, $\overset{\frown}{MWK}$, $\overset{\frown}{WYK}$ **22.**

23.

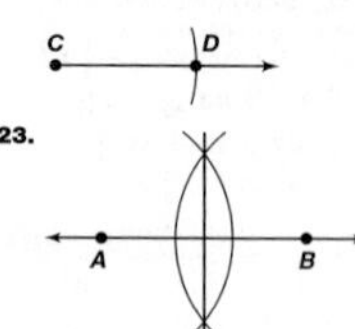

Chapter 8

Lesson 8-1 pp. 376–378

EXERCISES **1.** square **3.** C **5.** A **7.** 6 in.; it is less than a foot. **11.** about 20 ft **13.** about 875 mi^2 **15.** area: about 198 yd^2; perimeter: about 81 yd **19.** about $\frac{1}{2}$ in. **23.** Answers may vary. Sample: about 500 yd^2 **27 a.** about 7,840,000 ft^2 **b.** about 136 **31.** 150% increase

Lesson 8-2 pp. 382–383

EXERCISES **1.** right **3.** True; $A = bh$, so if the bases are equal and the heights are equal, the areas will be equal. **5.** 3 **7.** 25 m^2 **9.** 12 ft^2 **13.** 625 mi^2 **19.** 9 m; 26.4 m **21.** 10.5 cm; 21 cm^2 **23.** They have the same bases, but the height of the parallelogram must be less than the height of the rectangle because a leg is shorter than the hypotenuse. **25.** area = 161 in.^{2}; perimeter = 62 in. **29.** 2; 2 **31.** 1.52; $1\frac{13}{25}$

Lesson 8-3 — pp. 386–387

EXERCISES **1.** right **3.** 6 cm **5.** 10 cm^2 **7.** 17.5 m^2 **9.** 12.5 ft **13.** 1,440 yd^2 **15.** 72 km^2 **19.** 50,000 yd^2 **21.** 102 m^2 **23.** 10,000 km^2 **25.** 27.68 km^2 **27.** 3 ft; if $b = 4$ and $h =$ the corresponding height, then the area of the triangle is $\frac{1}{2}(4)h$; or $2h$. Since the area is 6 ft^2, $2h = 6$. So $h = 3$. **31.** 28

Lesson 8-4 — pp. 391–392

EXERCISES **1.** height **3.** $b_1 = 2.7$ in., $b_2 = 8$ in., $h = 10$ in.
5. 144 m^2 **11.** 500 km^2 **13.** 121.5 in.^{2} **15.** $5\frac{3}{4}$ ft^2 **17.** 448 in.^{2} **19.** 104 m^2; 45.6 m **23.** $480.18

Lesson 8-5 — pp. 396–397

EXERCISES **1.** No; π is nonrepeating and nonterminating. **3.** 2; 4; 4π **5.** $\frac{1}{\pi}$; $\frac{2}{\pi}$; 2 **7.** 53.4 mm **9.** 251.3 in. **13.** 314 m^2 **15.** 707 ft^2 **21.** 88 m; 616 m^2 **23.** 69.1 in.; 380.3 in.^{2} **25.** about 188 in.; about 2,827 in.^{2} **31.** $\angle O$

Lesson 8-6 — pp. 402–403

EXERCISES **1.** area **3.** yes **5.** no **7.** 10 **9.** 5 **17.** about 2 **19.** about 7 **23.** irrational **25.** rational **31.** 12 ft **35.** rational **37.** rational **39.** irrational

41.

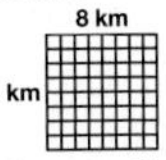

43.

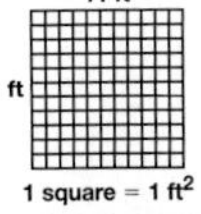

45. about 26.9 **49.** 2

Lesson 8-7 — pp. 407–408

EXERCISES **1.** hypotenuse **3.** 13.6 m **5.** 23.3 ft **7.** 17.5 m **9.** 12 ft **11.** 9 m **17.** 0.5 m **19.** 21.4 m **21.** 10 ft **23.** 7 km **25.** 3 **29.** $n = 0.60 \cdot 40$; 24

Lesson 8-8 — pp. 412–413

EXERCISES **1.** cylinder **3.** rectangle; rectangular prism **5.** pentagon; pentagonal prism **7.** cone **9.** hexagonal pyramid **13.**

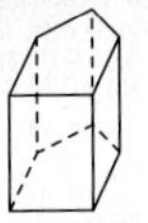

17. rectangular prism **19.** cone **21.** none **23.** 64 m^2 **25.** 7 faces; 12 edges; 7 vertices **29.** 11 m

Lesson 8-9 — pp. 416–418

EXERCISES **1.** surface area **3.** cylinder; about 5,089.4 in.^{2}

5.

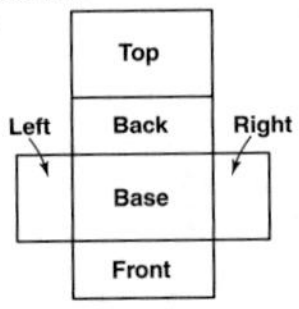

9. 264 m^2

13. 785.4 cm^2 **17.** 528 in.^{2} **19.** 1,407 ft^2 **21.** 2,890 in.^{2} **23.** 1960.4 in.^{2} **27.** 990 ft^2 **31.** rational

Lesson 8-10 — pp. 424–425

EXERCISES **1.** volume **3.** 60 in.^{3} **5.** 166.375 in.^{3} **9.** 31.5 in.^{3} **11.** about 84 in.^{3}; 88 in.^{3} **15.** 6.5 cm **17.** 18 in. **21.** about 6 million gal **23.** Yes; the height of the can is less than the height of the case. Since $3(2.5) < 7.6$, 3 cans will fit along the width of the case. Since $4(2, 5) < 11$, 4 cans will fit along the length of the case. Then the case can hold 3×4, or 12 cans. **27.** 50°; 140°

Chapter Review — pp. 428–429

1. edge **2.** hypotenuse **3.** prism **4.** circumference **5.** cone **6.** about 400 in.^{2} **7.** 7 in.; the width is less than a foot. **8.** 10.5 m^2 **9.** 38 cm^2 **10.** 160 ft^2 **11.** 25.1 in.; 50.3 in.^{2} **12.** 44 mi; 153.9 mi^2 **13.** 22 km; 38.5 km^2 **14.** 6 in. **15.** 39 yd **16.** cylinder **17.** rectangular pyramid **18.** pentagonal pyramid **19.** 22 in.^{2}; 6 in.^{3} **20.** 288 m^2; 324 m^3 **21.** 747.7 yd^2; 1,539.4 yd^3

Chapter 9

Lesson 9-1 — pp. 439–440

EXERCISES **1.** A graph shows how the data are changing. **3.** Answers may vary. Sample: 0–8; 4 intervals of 2 **5.** 0–5,000; 10 intervals of 500

7.

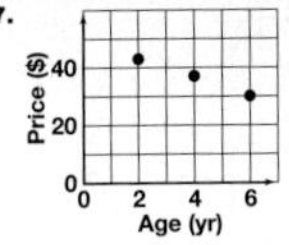

9. about 3 yr

11. about $8 **15.**

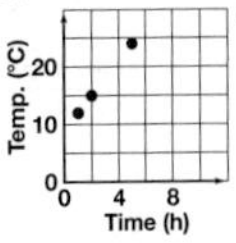

33°C

17 a.

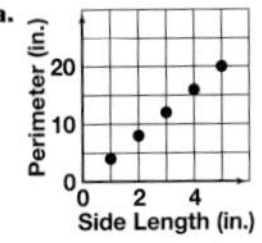

b. 2.4 in. **c.** 14 in.

d. Check students' work.

19 a.

Number of Weeks	Pay ($)
1	10
2	20
3	30
4	40
5	50

Number of Weeks	Pay ($)
1	47
2	54
3	61
4	68
5	75

b.

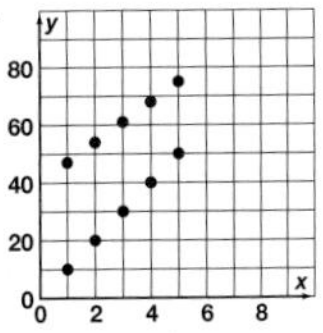

c. first neighbor: $100; second neighbor: $110; working for the second neighbor **23.** 0.05

Lesson 9-2 — pp. 444–445

EXERCISES **1.** Yes; dividing by −4 is the same as multiplying by $-\frac{1}{4}$. **3.** −6 **5.** 12.5 **7.** Start with 25 and add −4 repeatedly; 9, 5, 1. **9.** Start with 2 and multiply by −3 repeatedly; 162, −486, 1,458. **15.** neither **17.** both **21.** Answers may vary. Sample: He may be correct, but he cannot keep decreasing his time by 15 s indefinitely. **23.** 1 **25.** 1 **27 a.** 17; 16 **b.** Blue tiles: start with 1, add 0, 4, 0, 4 and so on; yellow tiles: start with 0, add 4, 0, 4, 0 and so on. **29.** 123,456 × 9 = 1,111,104 **35.** 12.5 mi

Lesson 9-3 — pp. 448–449

EXERCISES **1.** 10; 50 **3.** 324; 972; 2,916

5.

Cans of Soup	Number of Servings
3	9
4	12
5	15
6	18
7	21

7. 4, 5, 6, 7 **9.** 5, 10, 15, 20 **13.** $11n$; 110 **15.** $\frac{n}{2}$; 5 **21.**

Miles	Time (h)
10	0.4
20	0.8
30	1.2
40	1.6
50	2.0

23.

Celsius	Kelvin
0	273
10	283
20	293
40	313
80	353
120	393

25.

Side Length (in.)	Area (in.^{2})
2	4
3	9
5	25
8	64
10	100
12	144

27 a.

x	y
−4	−6
−2	−3
2	3
4	6

b. 0 **31.** 7

Lesson 9-4 — pp. 454–455

EXERCISES **1.** exactly one **3.** Each output is 5 times the input. **5.** $d = 30t$ **9.** $y = 3x + 5$

13.

x	y
0	9
1	8
2	7
3	6

15.

x	y
0	0
1	$\frac{1}{2}$
2	1
3	$\frac{3}{2}$

21. $y = (0.50)n$ **23.** $d = 48h$ **25.** Answers may vary. Sample: The y-values increase by $\frac{1}{2}$ when the x-values increase by 1. **27.** −3.875, −3.5, −0.875 **31.** 1 mm; it is less than 1 m.

Lesson 9-5 — pp. 457–459

EXERCISES **1.** a table showing the relationship between the input and output of a function

3.

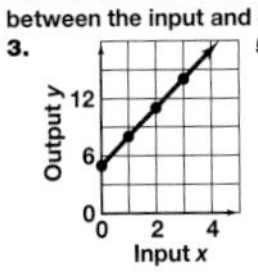

5.

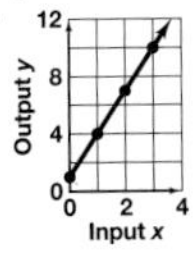

7. $d = 320t$; 3,200 mi

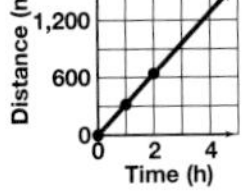

9. 750 Calories **11.** A

15.

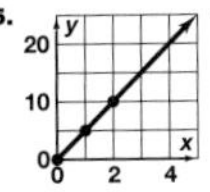

17.

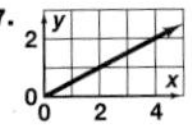

25. 15 ft^2

Lesson 9-6 — pp. 463–464

EXERCISES **1.** C **3.** B **5.** You travel at a steady pace for 2 h, stop for $\frac{1}{2}$ h, and then travel at the same steady pace for 2 h. **7.** You should not connect the points because you sell only whole-number glasses of lemonade.

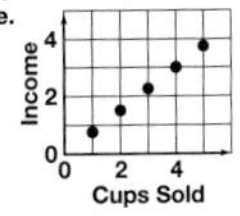

11 a. The graph shows that Josh caught up to Rafael after about 10 s. **b.** Josh **13.** The ball is 20 ft high on the way up and on the way down. It reaches that height after about 0.4 s and about 2.4 s. **15.** Answers may vary. Sample: A person threw a ball into the air. It left the person's hand 5 feet above the ground and landed on the ground about 2.75 seconds later. **19.** >

Lesson 9-7 — pp. 470–471

EXERCISES **1.** Simple interest is calculated only on the principal. Compound interest is calculated on the principal and any interest left in the account. **3.** $1,543.12 **5.** $71.40

9.

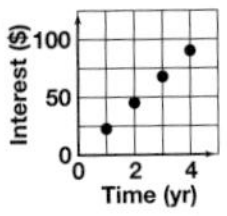

11.

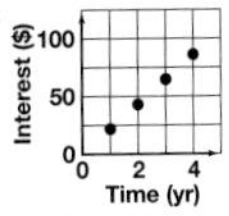

15. $1,643.94 **21.** $543.14 **25.** C **27.** E **29.** $4,370.91 **35.**

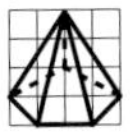

Lesson 9-8 — pp. 474–475

EXERCISES **1.** No; it cannot be used to show the relationship between two quantities. **3.** No; it cannot be used to show the relationship between two quantities. **5.** multiplication **7.** $y = \frac{x}{z}$ **9.** $r = \frac{p + 5}{3}$ **21.** $F = \frac{9}{5}C + 32$ **23.** $w = \frac{V}{\ell \cdot h}$ **25.** $r = 2t$ **27.** 7 ft **29.** 5 cm **33.** 14.4 m

Chapter Review — pp. 478–479

1. arithmetic **2.** function **3.** principal **4.** compound **5.** conjecture

6.

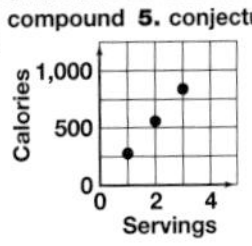

7.

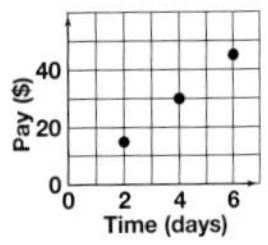

8. $82.50 **9.** arithmetic **10.** geometric **11.** neither **12.** $-6n$; −60 **13.**

x	y
0	3
2	4
4	5
6	6

14. $y = \frac{1}{2}x + 3$ **15.** $y = 2x + 2$

16.

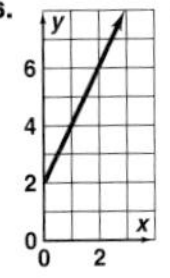

17. 7, 5, 3, 1, −1

18. Answers may vary. Sample: You travel for 2 h at a constant rate of 50 mi/h. You stop for 4 h, and then return to your starting place at a constant rate of 50 mi/h.

19.

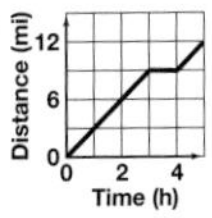

20. $450.00

21. $3,298.49 **22.** $x = \frac{z - y}{3}$ **23.** $x = -\frac{k}{4yz}$ **24.** $x = 9\left(\frac{z}{3} + 4\right)$, or $x = 3z + 36$ **25.** $9.00

Chapter 10

Lesson 10-1 — pp. 488–489

EXERCISES **1.** (0, 0) **3.** A **5.** D **7.** (−2, 2) **9.** (3, 0) **15.** IV **17.** III

23–26. Answers may vary. Samples are given.

23.

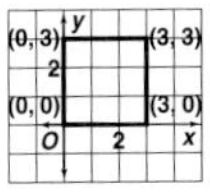

29. IV

31.

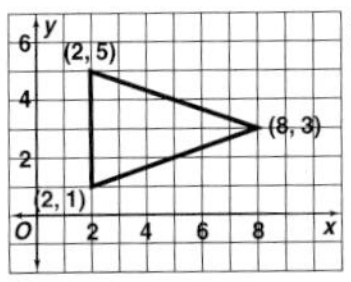

acute isoceles

33. Answers may vary. Sample: (−3, −3), (0, −3), (4, −3); if the x-coordinate is 37, the y-coordinate would be −3 because it is a straight horizontal line.

Lesson 10-2 — pp. 493–494

EXERCISES **1.** a line **3.**

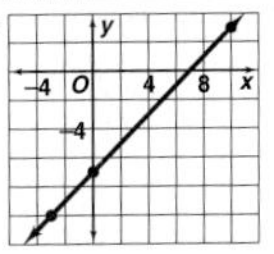

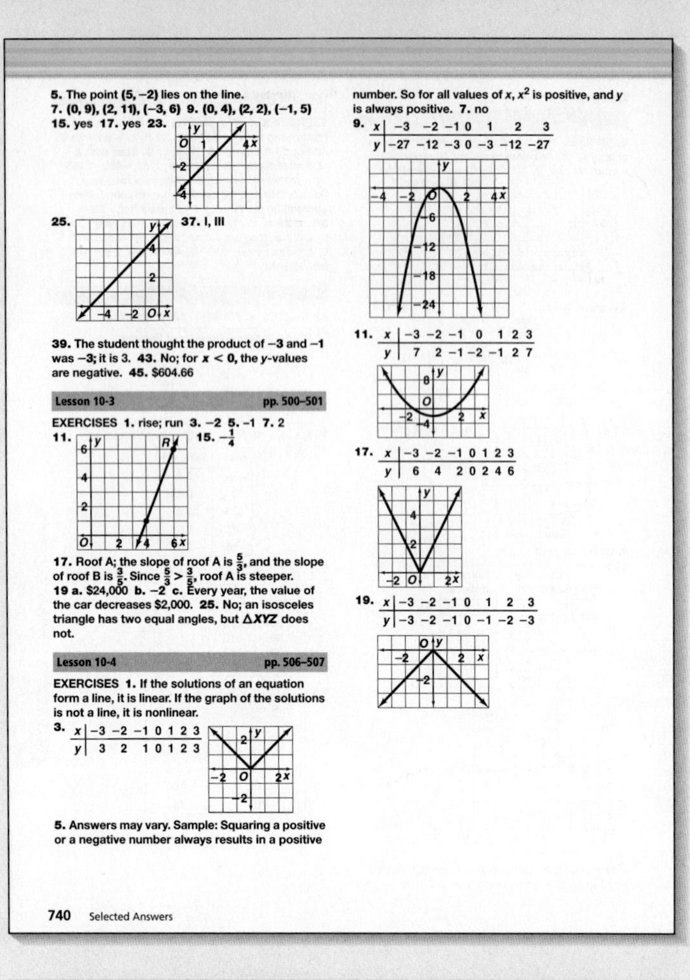

5. The point (5, −2) lies on the line. 7. (0, 9), (2, 11), (−3, 6) 9. (0, 4), (2, 2), (−1, 5) 15. yes 17. yes 23.

25.

37. I, III

39. The student thought the product of −3 and −1 was −3; it is 3. 43. No; for $x < 0$, the y-values are negative. 45. $604.66

Lesson 10-3 pp. 500–501

EXERCISES 1. rise; run 3. −2 5. −1 7. 2

11.

15. $-\frac{1}{4}$

17. Roof A; the slope of roof A is $\frac{5}{3}$, and the slope of roof B is $\frac{3}{5}$. Since $\frac{5}{3} > \frac{3}{5}$, roof A is steeper. 19 a. $24,000 b. −2 c. Every year, the value of the car decreases $2,000. 25. No; an isosceles triangle has two equal angles, but $\triangle XYZ$ does not.

Lesson 10-4 pp. 506–507

EXERCISES 1. If the solutions of an equation form a line, it is linear. If the graph of the solutions is not a line, it is nonlinear.

3.

x	−3	−2	−1	0	1	2	3
y	3	2	1	0	1	2	3

5. Answers may vary. Sample: Squaring a positive or a negative number always results in a positive number. So for all values of x, x^2 is positive, and y is always positive. 7. no

9.

x	−3	−2	−1	0	1	2	3
y	−27	−12	−3	0	−3	−12	−27

11.

x	−3	−2	−1	0	1	2	3
y	7	2	−1	−2	−1	2	7

17.

x	−3	−2	−1	0	1	2	3
y	6	4	2	0	2	4	6

19.

x	−3	−2	−1	0	1	2	3
y	−3	−2	−1	0	−1	−2	−3

740 Selected Answers

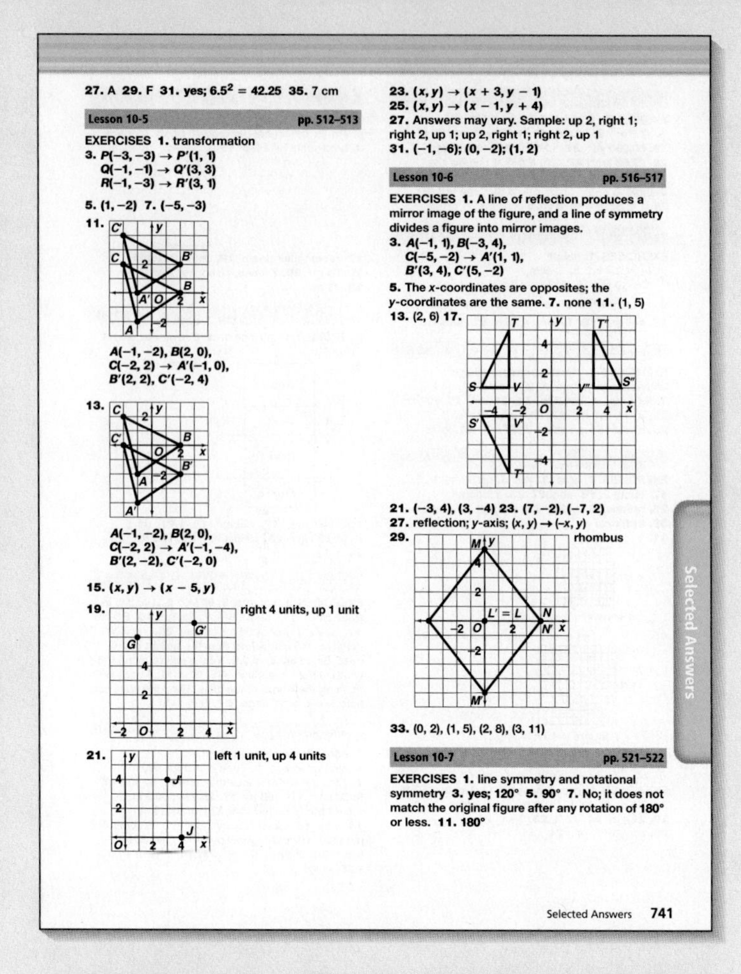

27. A 29. F 31. yes; $6.5^2 = 42.25$ 35. 7 cm

Lesson 10-5 pp. 512–513

EXERCISES 1. transformation

3. $P(-3, -3) \rightarrow P'(1, 1)$
$Q(-1, -1) \rightarrow Q'(3, 3)$
$R(-1, -3) \rightarrow R'(3, 1)$

5. (1, −2) 7. (−5, −3)

11. $A(-1, -2)$, $B(2, 0)$, $C(-2, 2) \rightarrow A'(-1, 0)$, $B'(2, 2)$, $C'(-2, 4)$

13. $A(-1, -2)$, $B(2, 0)$, $C(-2, 2) \rightarrow A'(-1, -4)$, $B'(2, -2)$, $C'(-2, 0)$

15. $(x, y) \rightarrow (x - 5, y)$

19. right 4 units, up 1 unit

21. left 1 unit, up 4 units

23. $(x, y) \rightarrow (x + 3, y - 1)$
25. $(x, y) \rightarrow (x - 1, y + 4)$
27. Answers may vary. Sample: up 2, right 1; right 2, up 1; up 2, right 1; right 2, up 1
31. (−1, −6); (0, −2); (1, 2)

Lesson 10-6 pp. 516–517

EXERCISES 1. A line of reflection produces a mirror image of the figure, and a line of symmetry divides a figure into mirror images.

3. $A(-1, 1)$, $B(-3, 4)$, $C(-5, -2) \rightarrow A'(1, 1)$, $B'(3, 4)$, $C'(5, -2)$

5. The x-coordinates are opposites; the y-coordinates are the same. 7. none 11. (1, 5) 13. (2, 6) 17.

21. (−3, 4), (3, −4) 23. (7, −2), (−7, 2)
27. reflection; y-axis; $(x, y) \rightarrow (-x, y)$
29. rhombus

33. (0, 2), (1, 5), (2, 8), (3, 11)

Lesson 10-7 pp. 521–522

EXERCISES 1. line symmetry and rotational symmetry 3. yes; 120° 5. 90° 7. No; it does not match the original figure after any rotation of 180° or less. 11. 180°

Selected Answers 741

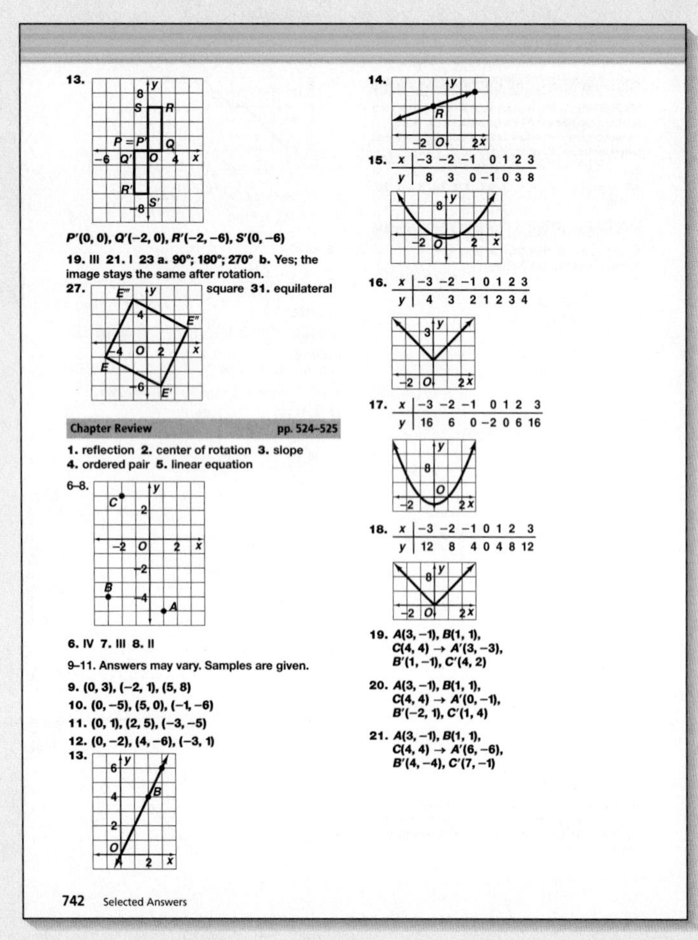

13. $P'(0, 0)$, $Q'(-2, 0)$, $R'(-2, -6)$, $S'(0, -6)$

19. III 21. I 23 a. 90°; 180°; 270° b. Yes; the image stays the same after rotation.

27. square 31. equilateral

Chapter Review pp. 524–525

1. reflection 2. center of rotation 3. slope 4. ordered pair 5. linear equation

6–8.

6. IV 7. III 8. II

9–11. Answers may vary. Samples are given.

9. (0, 3), (−2, 1), (5, 8)
10. (0, −5), (5, 0), (−1, −6)
11. (0, 1), (2, 5), (−3, −5)
12. (0, −2), (4, −6), (−3, 1)
13.

14.

15.

x	−3	−2	−1	0	1	2	3
y	8	3	0	−1	0	3	8

16.

x	−3	−2	−1	0	1	2	3
y	4	3	2	1	2	3	4

17.

x	−3	−2	−1	0	1	2	3
y	16	6	0	−2	0	6	16

18.

x	−3	−2	−1	0	1	2	3
y	12	8	4	0	4	8	12

19. $A(3, -1)$, $B(1, 1)$, $C(4, 4) \rightarrow A'(3, -3)$, $B'(1, -1)$, $C'(4, 2)$

20. $A(3, -1)$, $B(1, 1)$, $C(4, 4) \rightarrow A'(0, -1)$, $B'(-2, 1)$, $C'(1, 4)$

21. $A(3, -1)$, $B(1, 1)$, $C(4, 4) \rightarrow A'(6, -6)$, $B'(4, -4)$, $C'(7, -1)$

742 Selected Answers

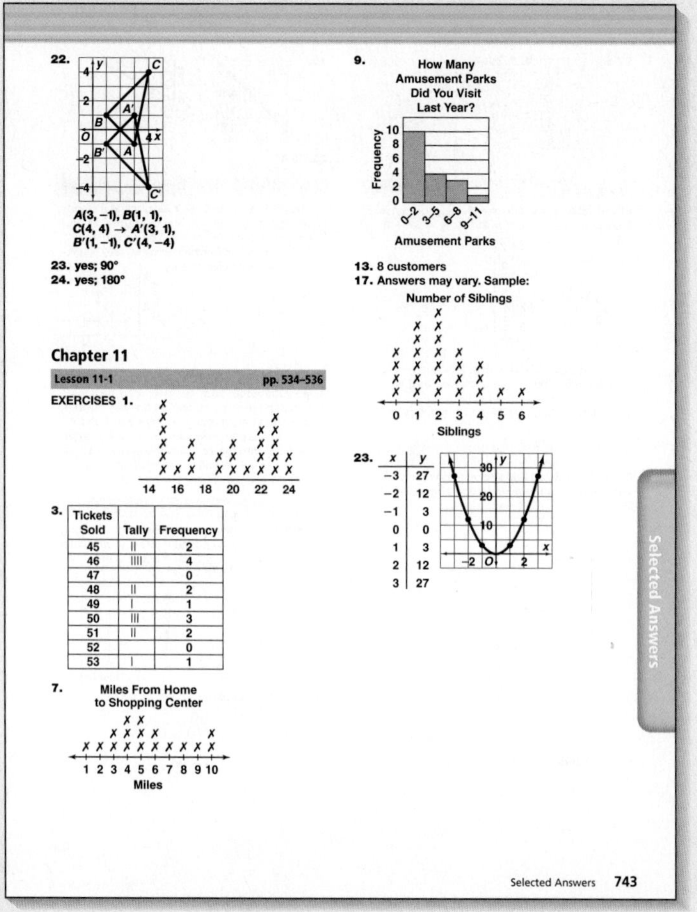

22. $A(3, -1)$, $B(1, 1)$, $C(4, 4) \rightarrow A'(3, 1)$, $B'(1, -1)$, $C'(4, -4)$

23. yes; 90°
24. yes; 180°

Chapter 11

Lesson 11-1 pp. 534–536

EXERCISES 1.

3.

Tickets Sold	Tally	Frequency
45	II	2
46	IIII	4
47		0
48	II	2
49	I	1
50	III	3
51	II	2
52		0
53	I	1

7. Miles From Home to Shopping Center

9. How Many Amusement Parks Did You Visit Last Year?

13. 8 customers
17. Answers may vary. Sample: Number of Siblings

23.

x	y
−3	27
−2	12
−1	3
0	0
1	3
2	12
3	27

Selected Answers 743

Lesson 11-2 — pp. 540–542

EXERCISES **1.** lawn mowing **3.** yes **5.** 26 **9.** A1 **11.** B3 **13.** Corpus Christi, Texas **15.** all of them **19.** 1975–2000 **25.** Double line graph; line graphs are best for showing change over time.

27 a.

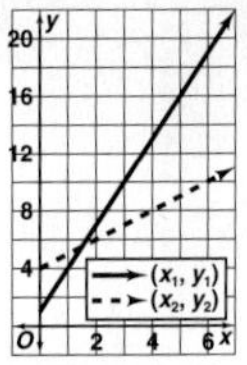

y
20
16
12
8
4
O
2
4
6
x
(x_1, y_1)
(x_2, y_2)

B3 = 4, B4 = 7, D4 = 6, D5 = 7

b. (1.5, 5.5) **31.** 1,332 m^2

Lesson 11-3 — pp. 546–547

EXERCISES **1.** The stem is the digit or digits on the left, and the leaf is the digit or digits on the right. **3.** 12 **5.**

4	3 6 7
5	1 2
6	0 1 7
7	1 2 8
8	2 6 8

Key: 8 | 2 means 82

7. High Temperatures in the Desert (°F)

9	8 9
10	0 1 3 4 8 9
11	1 2 2 3 3 8

Key: 9 | 8 means 98

9. 56 in.; 50 in. **11.** No; the median height for females is 58.5 in. **13.** Yes; 76 − 56 = 20. **15.** C **17.** 54

Lesson 11-4 — pp. 551–553

EXERCISES **1.** C **3.** B **5.** was not **7.** Part (b); the sample is more diverse. **11.** Fair; the question makes no assumptions. **13.** Biased; the question uses the terms *harsh* and *inspiring.* **19.** Yes; surveying every 10th person is random, and you should get a mix of ages and genders. **21.** Yes; you will survey people in your population without any bias. **25.** Start with 3 and add 5 repeatedly; 23, 28, 33

Lesson 11-5 — pp. 555–556

EXERCISES **1.** Researchers capture, mark, and release animals, and then capture another group of animals. The number of marked animals in the second group indicates population size. **3.** about 13,661 bass **5.** about 305 bears **7.** about 2,010 deer **9.** about 2,048 deer **17.** 63 alligators **19 a.** The proportion should be $\frac{8}{25} = \frac{38}{x}$. **b.** 119 sharks **21.** about 233 animals

Lesson 11-6 — pp. 562–564

EXERCISES **1.** no **3.** greater **5.** The vertical axis starts with 8 instead of 0. The title is also misleading since grapes are not everyone's favorite. **7.** The graph gives the impression that the largest group of students received the highest scores. There are uneven intervals on the *x*-axis.

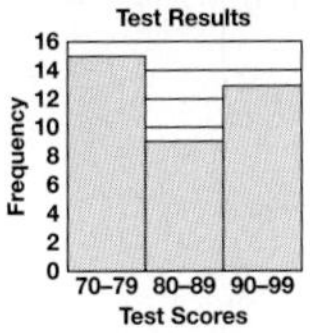

Test Results
Frequency
16
14
12
10
8
6
4
2
0
70–79
80–89
90–99
Test Scores

15. It reverses the scale on the *x*-axis. **17.** The company wants to give the impression that it is profitable even though it's not. **25.** about 45 billion

Lesson 11-7 — pp. 568–570

EXERCISES **1.** It illustrates what happens to one set of data when the other increases. **3.** yes **5.** about 47 years

7.

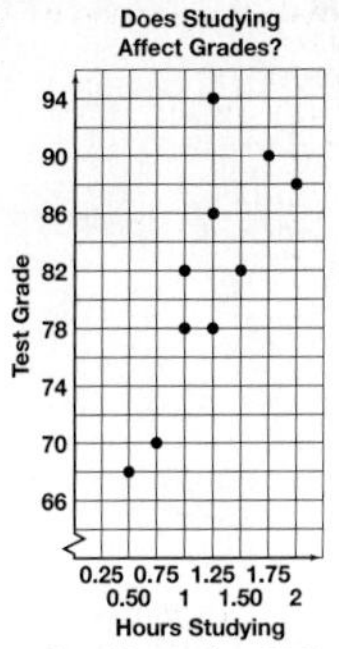

Does Studying Affect Grades?
Test Grade
94
90
86
82
78
74
70
66
0.25 0.75 1.25 1.75
0.50 1 1.50 2
Hours Studying

9. positive **11.** negative **17.** Positive; the values are both increasing. **19.** Answers may vary. Sample: 1,750 million

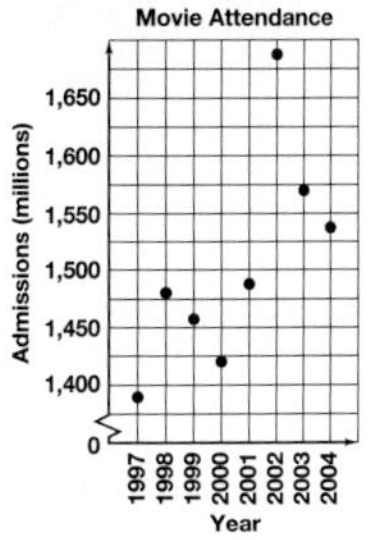

Movie Attendance
Admissions (millions)
1,650
1,600
1,550
1,500
1,450
1,400
0
1997
1998
1999
2000
2001
2002
2003
2004
Year

23. 158 seals

Chapter Review — pp. 572–573

1. stem-and-leaf **2.** negative **3.** legend **4.** line plot **5.** population

6. Number of Hours of TV (weekly)

Hours	Tally	Frequency
3	I	1
4	II	2
5	~~IIII~~	5
6	~~IIII~~	5
7	IIII	4
8	II	2
9	I	1

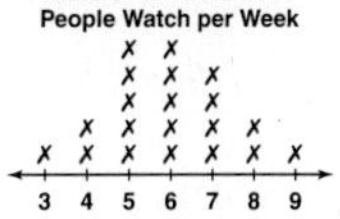

Number of Hours of TV
People Watch per Week
3 4 5 6 7 8 9

7.

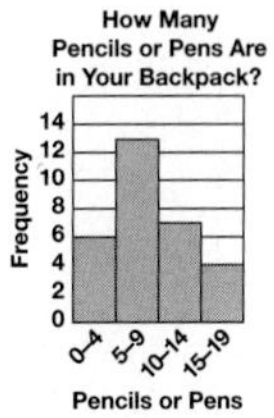

How Many
Pencils or Pens Are
in Your Backpack?
Frequency
14
12
10
8
6
4
2
0
0–4
5–9
10–14
15–19
Pencils or Pens

8. the number of times a person has flown **9.** 14 people **10.** counts to 20 **11.** counts to 20 **12.** writes name **13.**

7	0
6	0 0 0 1 6 9
5	0 0 4 5 8
4	0 0 0 5 7 7 8
3	6
2	3 7 7 9
1	6 9

Key: 1 | 6 means 16

14. Fair; the question makes no assumptions about the activity. **15.** Biased; the question assumes the ocean is calm and soothing. **16.** about 368 wolves **17.** The *y*-axis interval starts at 93, so it appears that the student who studied 6 h did twice as well as the student who studied 4 h. **18.** no trend **19.** negative

Chapter 12

Lesson 12-1 — pp. 582–583

EXERCISES **1.** Answers may vary. Sample: the result or group of results of an action **3.** B **5.** A **7.** $\frac{1}{12}$; $0.08\overline{3}$; about 8.3% **9.** $\frac{1}{2}$; 0.5; 50% **13.** 0 **15.** $\frac{3}{10}$ **25.** $\frac{3}{4}$ **27.** $\frac{2}{4}$ or $\frac{1}{2}$ **37.** $\frac{a}{b}$

Lesson 12-2 — pp. 588–589

EXERCISES **1.** Theoretical probability is computed by the formula $P(\text{event}) = \frac{\text{number of favorable outcomes}}{\text{total number of possible outcomes}}$. Experimental probability is based on experimental data or observation. **3.** 18 **5.** 5 coins; one for each baby **7.** $\frac{18}{25}$ **15.** $\frac{1}{5}$; $\frac{1}{20}$; $\frac{13}{320}$; $\frac{9}{508}$ **17.** $\frac{6}{25}$ **19.** $\frac{2}{25}$ **21.** $\frac{7}{25}$ **23.** $\frac{x}{252}$; 6 **27.** $\frac{3}{9}$ or $\frac{1}{3}$

Lesson 12-3 — pp. 593–595

EXERCISES **1.** the collection of all possible outcomes in an experiment **3.** 1 **5.** 7 **7.** 16 **9.** D **11.** 1 2 3 4 5 6; $\frac{3}{6}$ or $\frac{1}{2}$

13. 1st Spin 2nd Spin; $\frac{1}{4}$

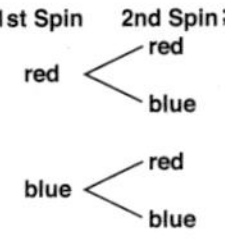

red
red
blue
blue
red
blue

15. 12 recipes **19.** $\frac{2}{16}$ or $\frac{1}{8}$ **23 a.** 16 outfits **b.** $\frac{12}{16}$ or $\frac{3}{4}$ **25.** small fruit punch, small lemonade, medium fruit punch, medium lemonade, large fruit punch, large lemonade, jumbo fruit punch, jumbo lemonade **27.** Yes; there are 24 possible orders, so the probability for each order is $\frac{1}{24}$.

Lesson 12-4 — pp. 600–602

EXERCISES **1.** Two events are independent if the occurrence of one does not affect the probability of the other occurring; two events are dependent if the occurrence of one does affect the probability of the other occurring. **3.** independent **5.** $\frac{1}{36}$ **7.** $\frac{3}{36}$ or $\frac{1}{12}$ **11.** $\frac{1}{64}$ **15.** $\frac{2}{7}$ **17.** $\frac{5}{7}$ **21.** $\frac{10}{380}$ or $\frac{1}{38}$ **25.** $\frac{1}{25}$ **27.** $\frac{24}{729}$ or $\frac{8}{243}$ **29 a.** Complementary; red, yellow, green, or broken are all the possible events of a traffic light. **b.** Mutually exclusive; since a student can only receive one grade on a test, the events have no outcomes in common. **31.** $\frac{5}{7}$ **35.**

	1	2	3
H	H1	H2	H3
T	T1	T2	T3

Lesson 12-5 — pp. 608–609

EXERCISES **1.** an arrangement of items in a particular order **3.** B **5.** C **7.** 120 **9.** 6 **13.** 3! = 6 **15.** 4! = 24 **21.** 42 **23.** 30 **27.** 24 **29.** 5,040 **31.** 3,628,800 **35.** 60 **37.** 30 **39 a.** BS, GS, RS, BD, GD, RD, BP, GP, RP **b.** 3 × 3 = 9 possibilities **43.** $\frac{1}{256}$

Lesson 12-6 — pp. 611–613

EXERCISES **1.** In a combination, only the grouping of the items matters. **3.** permutation **5.** $\frac{7 \times 6 \times 5}{3 \times 2 \times 1}$, $\frac{10 \times 9 \times 8 \times 7 \times 6}{5 \times 4 \times 3 \times 2 \times 1}$ **7.** 4 **9.** 3 **13.** 6 **19.** 60 **21.** 840 **23.** 66 ways **25.** combination; 56 sets **29.** 20 **31.** $\frac{1}{6}$ **35.** 4! = 24

Chapter Review — pp. 616–617

1. combination **2.** event **3.** independent **4.** odds in favor of **5.** Experimental **6.** $\frac{1}{7}$ **7.** $\frac{2}{7}$ **8.** $\frac{6}{7}$ **9.** $\frac{1}{50}$ **10.** 32

11. Appetizer Soup Main Dish

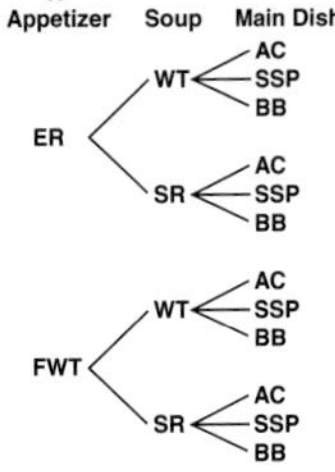

ER
WT
AC
SSP
BB
SR
AC
SSP
BB
FWT
WT
AC
SSP
BB
SR
AC
SSP
BB

12. $\frac{1}{12}$ **13.** 12 dinners **14.** $\frac{30}{182}$ or $\frac{15}{91}$ **15.** $\frac{48}{182}$ or $\frac{24}{91}$ **16.** $\frac{56}{182}$ or $\frac{4}{13}$ **17.** $\frac{1}{16}$ **18.** 5 teams **19.** 24 ways

Additional Answers

CHAPTER 2

Lesson 2-6

page 101 Activity Lab

1.

Seed Type	Fraction
A	$\frac{5}{16}$
B	$\frac{1}{4}$
C	$\frac{1}{2}$
D	$\frac{17}{35}$
E	$\frac{9}{26}$
F	$\frac{1}{3}$
G	$\frac{14}{55}$
H	$\frac{18}{35}$
I	$\frac{8}{15}$

3.

Seed Type	Number Sprouted	Number Planted
I	8	15
H	18	35
C	22	44
D	17	35
E	18	52
F	21	63
A	15	48
G	14	55
B	5	20

page 114 Chapter Test

28. Write the numerator and denominator in prime factorization form. Find the product of the common factors and divide the numerator and denominator by that product.

29. Answers may vary. Sample:

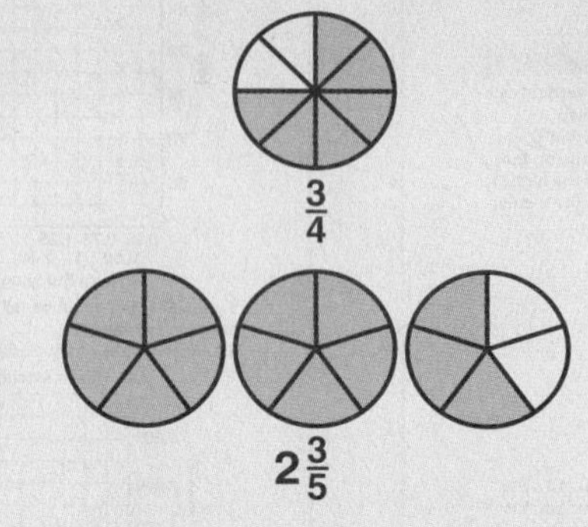

CHAPTER 3

Lesson 3-1

page 125 Activity Lab

3. $\frac{3}{4}$,

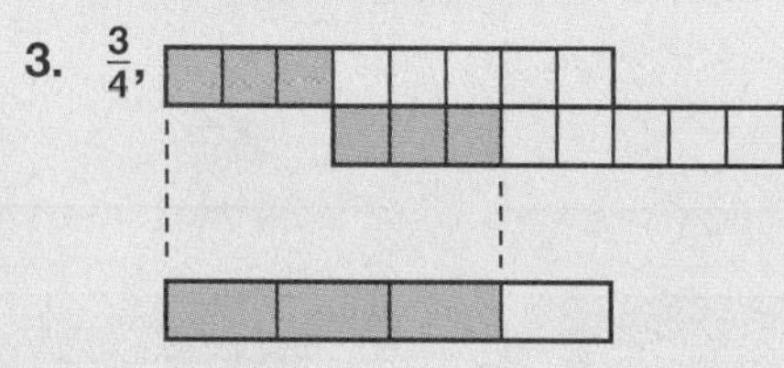

4. $\frac{2}{3}$,

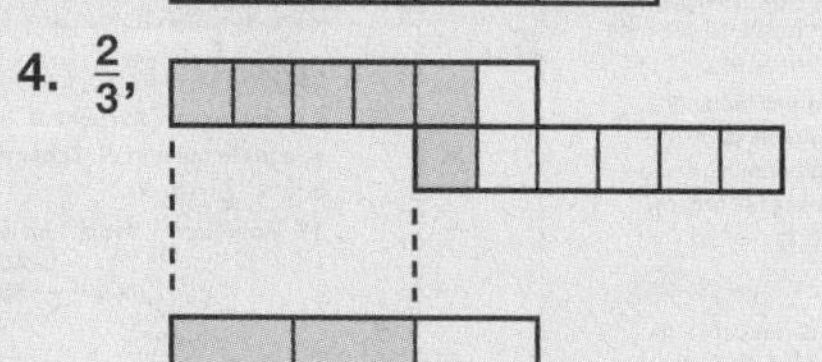

5. $\frac{9}{10}$,

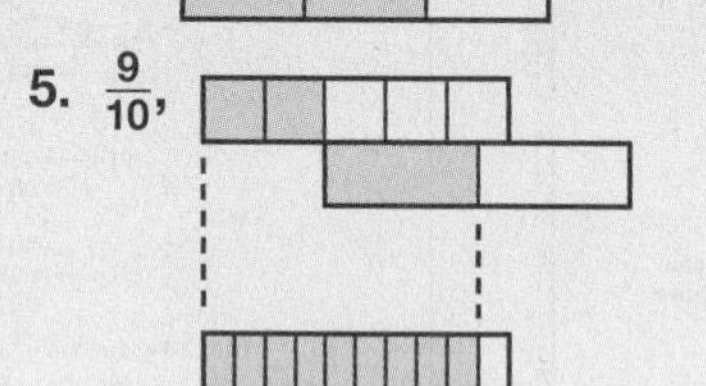

6. $\frac{1}{2}$,

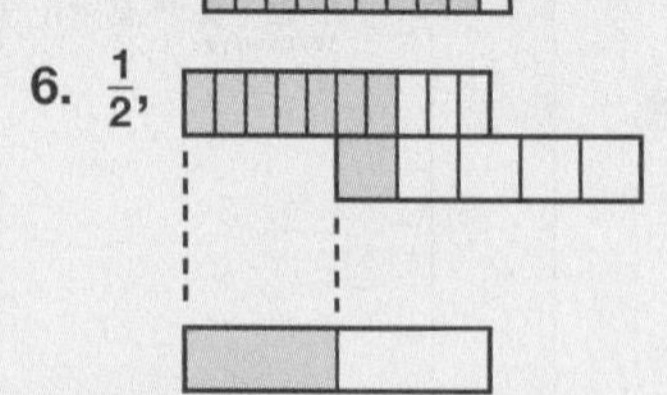

Lesson 3-3

page 135 Activity Lab

4. $\frac{2}{15}$,

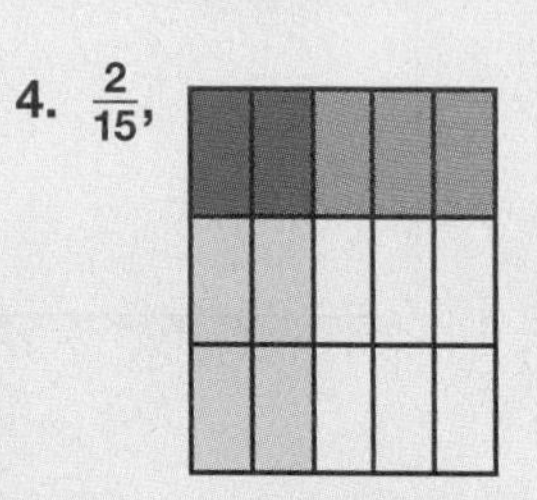

5. $\frac{5}{12}$,

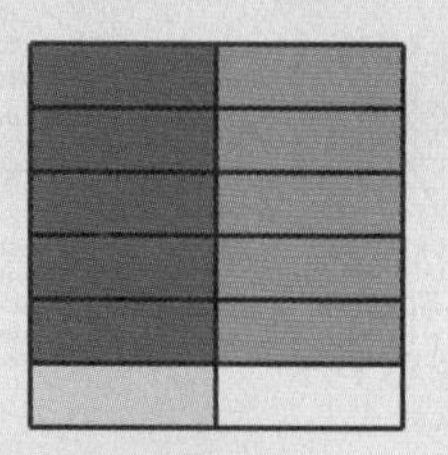

6. $\frac{1}{8}$,

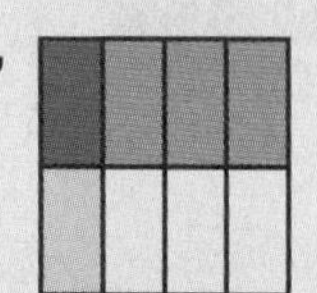

7. $\frac{6}{12}$,

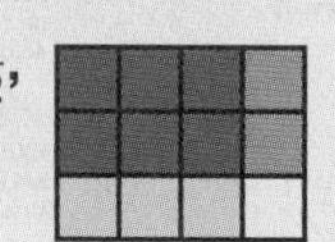

8. $\frac{3}{24}$,

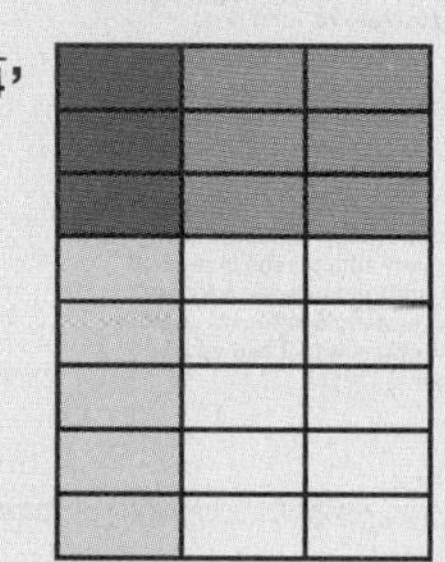

CHAPTER 4

page 168 Activity Lab

1.

A	B
1	8
2	9
5	12
10	17
30	37
50	57
n	$n + 7$

Each value in column B is 7 more than its corresponding value in column A.

2.

C	D
1	−4
2	−3
5	0
10	5
25	20
45	40
n	$n - 5$

Each value in column D is 5 less than its corresponding value in column C.

3.

E	F
1	40
2	80
5	200
10	400
100	4,000
500	20,000
n	$40n$

Each value in column F is 40 times its corresponding value in column E.

4. n represents any number in columns A, C, and E. Different operations can be applied to n to get any value in columns B, D, and F.

Lesson 4-7

pages 207–208 Homework Exercises

25. $a \geq 17$;
15 16 17 18 19 20

26. $n > 100$;
98 99 100 101 102 103

27. $p \leq 0$;
−3 −2 −1 0 1 2

28. $s \leq 65$;
50 60 70

Lesson 4-8

pages 212–214 Homework Exercises

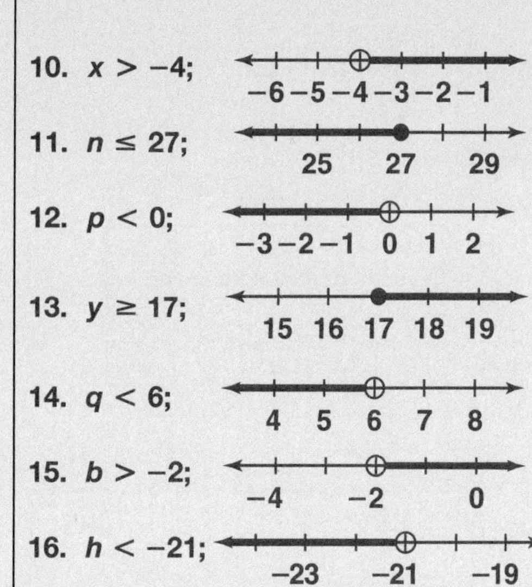

16. $h < -21$;
−23 −21 −19

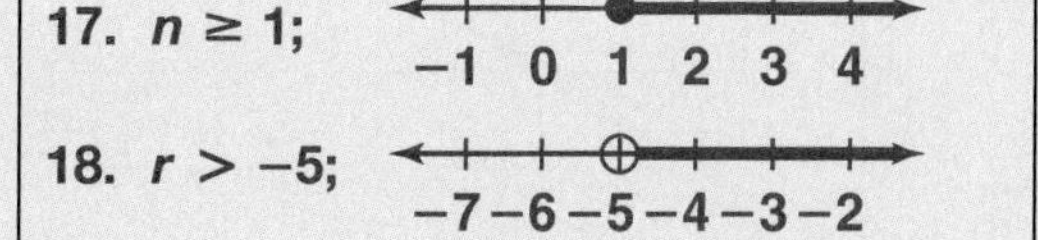

19. $p \leq -4$;
−7 −6 −5 −4 −3 −2

20. $b > -23$;
−25 −23 −21

21. $f \geq -5$;
−7 −6 −5 −4 −3 −2

22. $m > 1$;
−1 0 1 2 3

23. $x < 1$;
−1 0 1 2 3

24. $k \leq -11$;
−13 −11 −9

Lesson 4-9

pages 217–218 Homework Exercises

9. $x < -8$;
−11 −10 −9 −8 −7 −6

10. $b \leq 3$;
0 1 2 3 4 5

11. $w \geq -6$;
−8 −7 −6 −5 −4 −3

12. $d \leq 7$;
4 5 6 7 8 9

13. $t < -5$;
−8 −7 −6 −5 −4 −3

14. $g < -7$;
−9 −7 −5

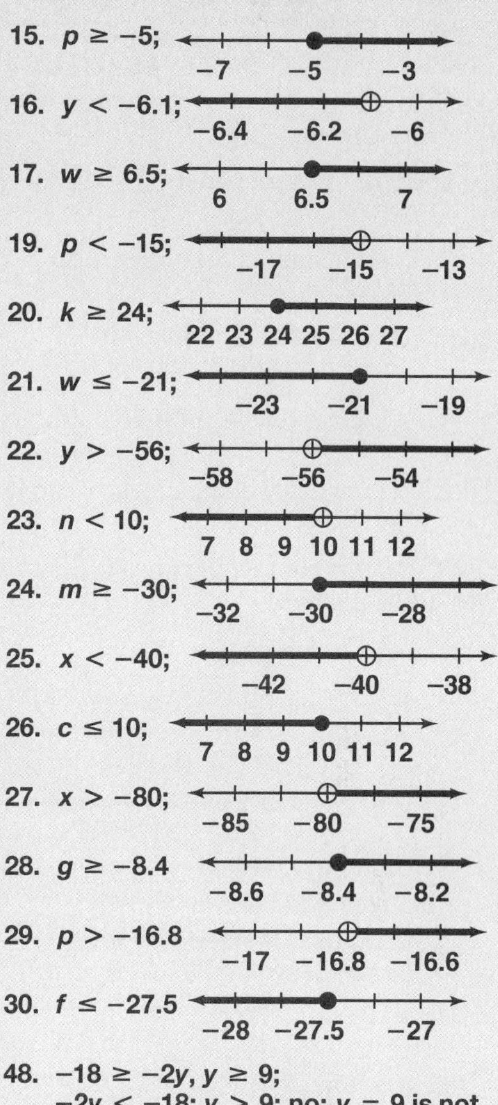

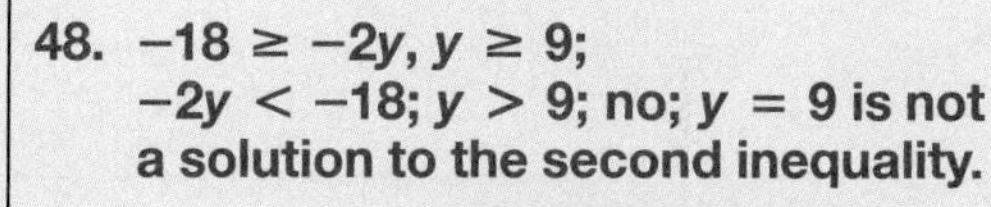

48. $-18 \geq -2y$, $y \geq 9$;
$-2y < -18$; $y > 9$; no; $y = 9$ is not a solution to the second inequality.

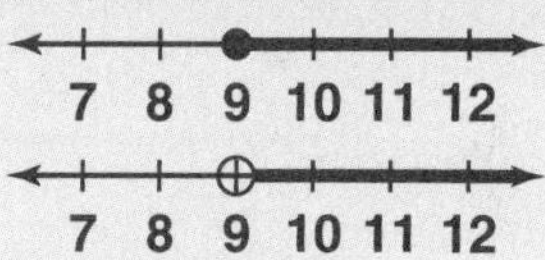

49. $-3x + 10 > 19$; $x < -3$;
−6 −5 −4 −3 −2 −1

pages 220–221 Chapter Review

27. $t \leq 10$;
7 8 9 10 11 12

28. $r < 5$;
2 3 4 5 6 7

29. $h < -22$;
−24 −22 −20

30. $p \geq -15$;
−17 −15 −13

31. $g > 17$;
15 17 19

32. $m \geq -7$; −9 −8 −7 −6 −5 −4

33. $x \geq 20$; 18 20 22

page 222 Chapter Test

19. Let b = number of bricks; $6b = 72$; 12 bricks.

20. Let a = cost per apple; $12a + 3.35 = 8.75$; \$0.45.

21. Let b = cost per button; $322b = 483$; \$1.50.

22. Let d = daily distance; $4d = 1{,}145$; 286.25 mi.

23. Let n = number of players; $\frac{n}{12} = 13$; 156 players.

24. Let c = the cost of the party; $\frac{c}{6} + 65 = 160$; \$570.

30. $n \geq 3$; 1 2 3 4 5 6

31. $y > 24$; 22 24 26

32. $m \leq -9$; −11 −9 −7

33. $w < -8$; −10 −8 −6

34. $x < -4$; −7 −6 −5 −4 −3 −2

35. $h \geq 9$; 7 8 9 10 11 12 13

36. $v < -48$; −50 −48 −46

37. $p \leq -66$; −68 −66 −64

38. $k \geq 12$; 10 12 14

39. $x > 60$; 58 60 62

40. Let m = number of additional people allowed aboard; $143 + m \leq 220$; $m \leq 77$; at most 77 people.

41. Let b = number of bottles; $1.49b \leq 12$; $b \leq 8.05$; at most 8 bottles.

42. Let m = amount Alex's mother spent; $52 + m \geq 130$; $m \geq 78$; at least \$78.

43. Answers may vary. Sample: You use inverse operations to solve both equations and inequalities. With inequalities, you sometimes have to reverse the direction of the symbol when solving.

CHAPTER 5

Lesson 5-3

page 242 Activity Lab

1.

Earnings

	1 hour	2 hours	3 hours	4 hours	5 hours
Manny	10	20	30	40	50
Ilene	13	21	29	37	45

2.

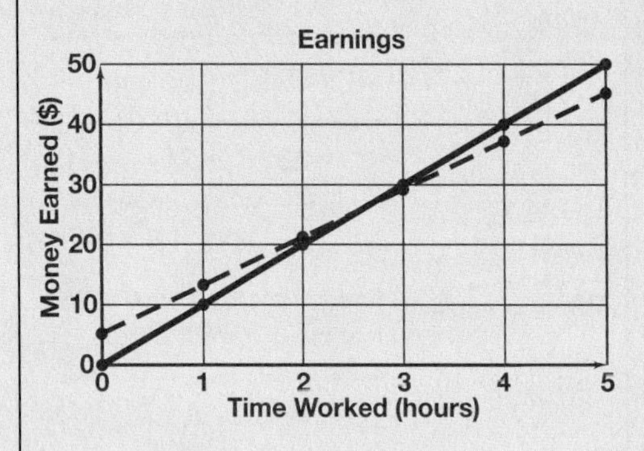

Lesson 5-4

page 251 Activity Lab

1.

Fig. 1	Angle Measure	Fig. 2	Angle Measure
∠A	90°	∠P	90°
∠B	32°	∠Q	32°
∠C	58°	∠R	58°

Angle measures are the same.

2.

Fig. 1	Side Length (cm)	Fig. 2	Side Length (cm)
AB	5	PQ	2.5
BC	6	QR	3
CA	3	RP	1.5

Lengths of Fig. 2 are all 0.5 times the lengths of Fig. 1.

3.

Fig. 3	Angle Measure	Fig. 4	Angle Measure
∠A	40°	∠P	40°
∠B	120°	∠Q	120°
∠C	20°	∠R	20°

Angle measures are the same.

Fig. 3	Side Length (cm)	Fig. 4	Side Length (cm)
AB	2	PQ	3
BC	4	QR	6
CA	5	RP	7.5

Lengths of Fig. 4 are all 1.5 times the lengths of Fig. 3.

4. Check students' work. Sample: Similar figures have the same corresponding angle measures and proportional side lengths.

Lesson 5-6

page 259 Check Skills You'll Need

1. For two ratios, the cross products are found by multiplying the denominator of each ratio by the numerator of the other ratio.

CHAPTER 6

Lesson 6-1

page 276 Homework Exercises

10.

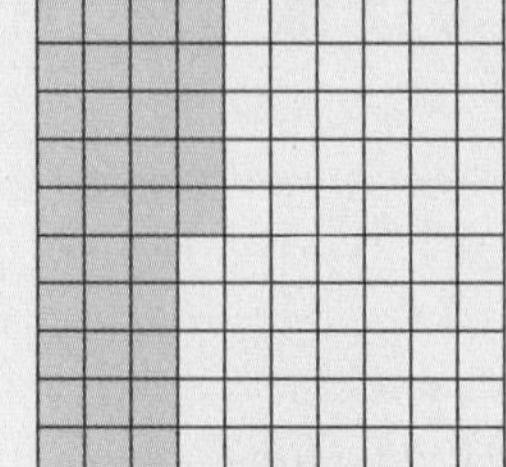

11.

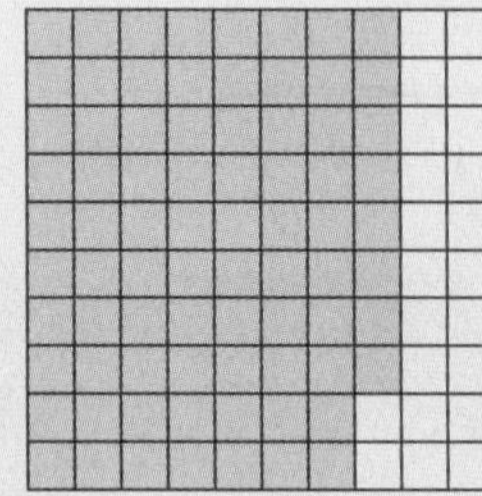

12.

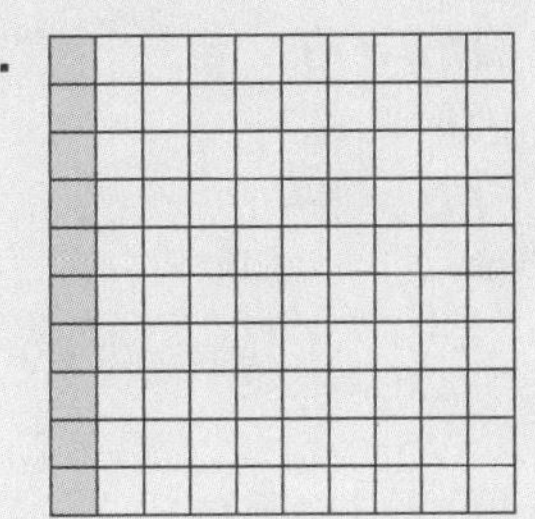

13.

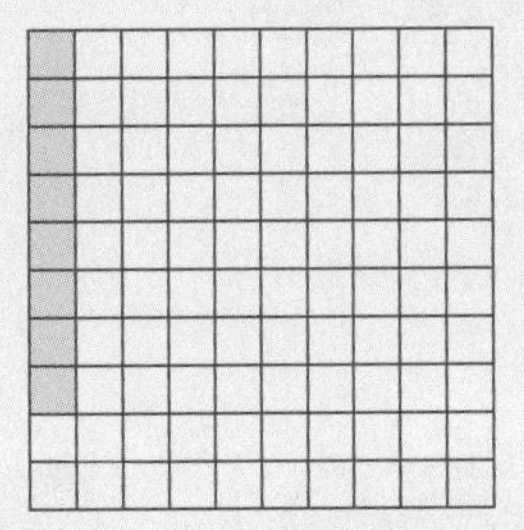

14.

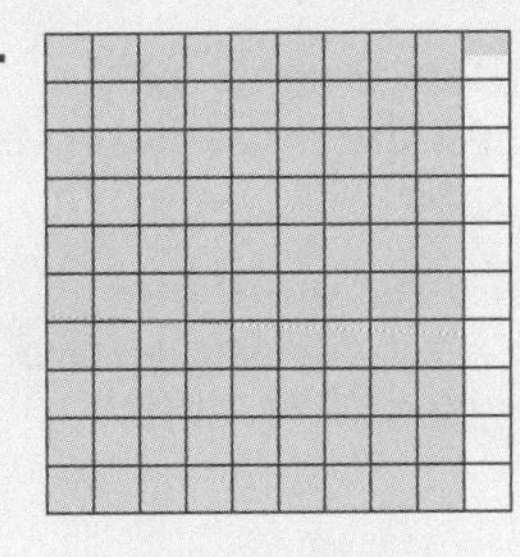

Lesson 6-2

page 283 Homework Exercises

33.

Topping	With Olives	Plain	With Onions and Green Peppers
Percent of the Pizza	37.5%	12.5%	50%
Number of Slices	3	1	4

Lesson 6-3

page 287 Homework Exercises

41.

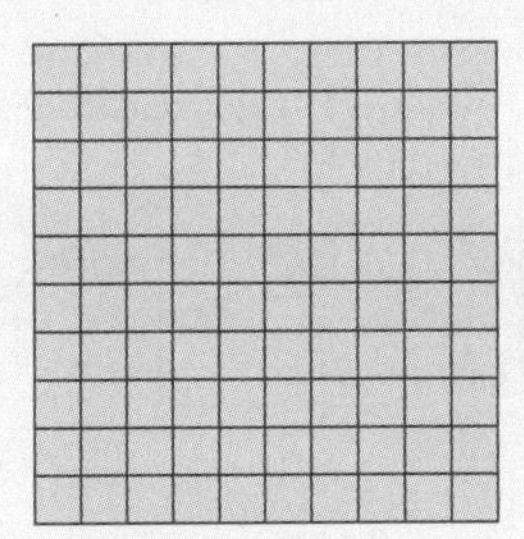

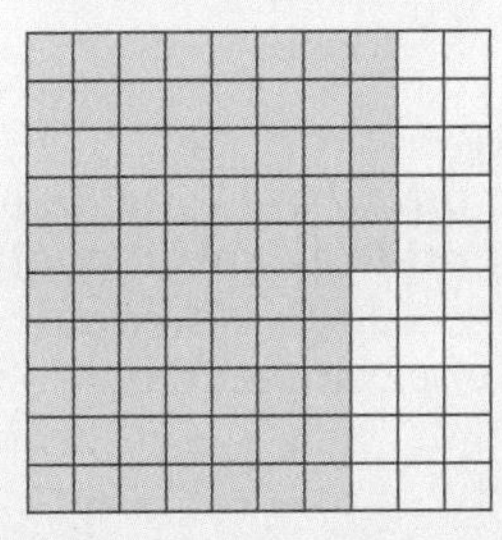

42.

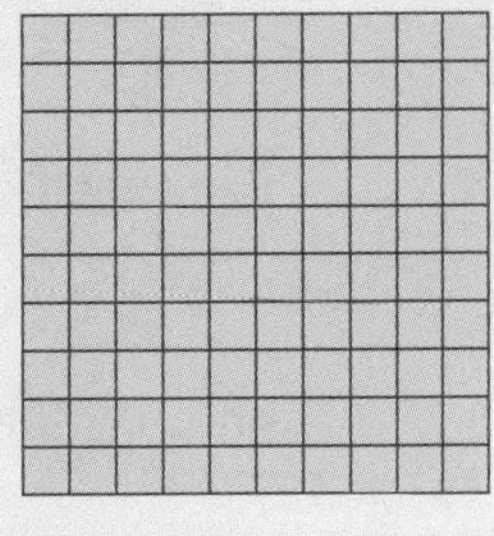

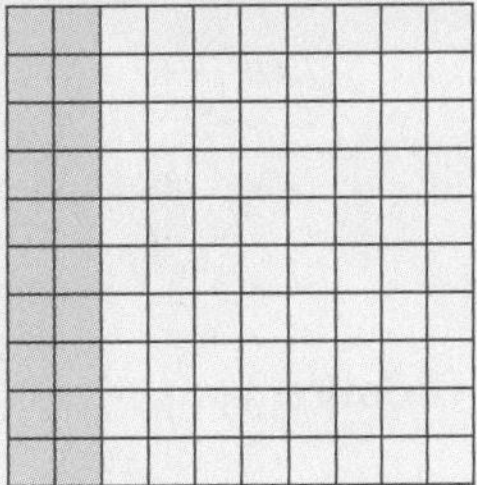

43.

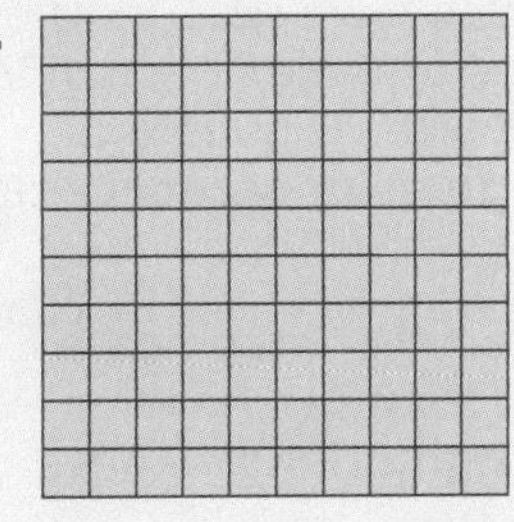

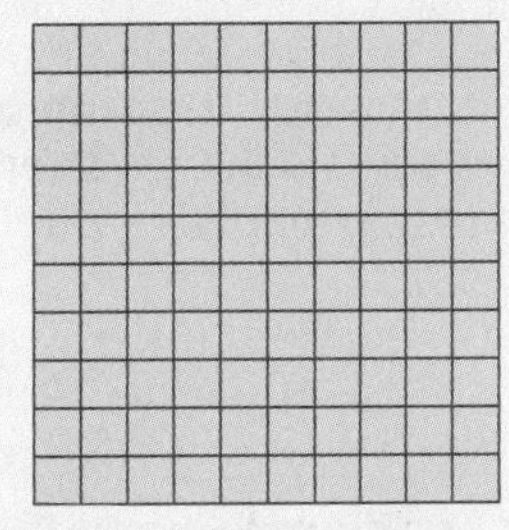

44.

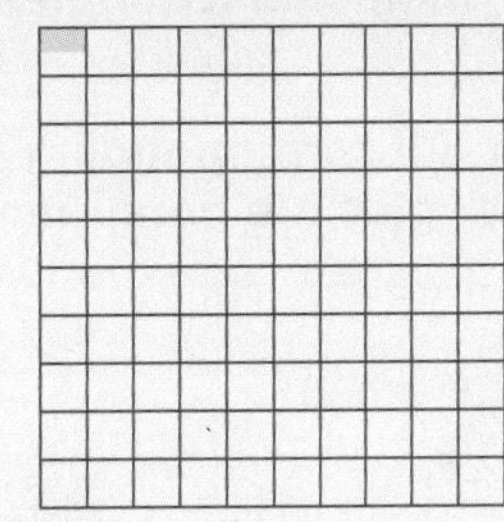

Lesson 6-4

page 289 Activity Lab

3.

Language	Percent	Number of Students
English	78%	~789
Spanish	58%	~587
Chinese	34%	~344
French	26%	~263
German	10%	~101

Lesson 6-8

page 314 Homework Exercises

35.

	A	B	C	D
1	Yr	Sales	Change ($)	Change (%)
2	1	200,000	—	—
3	2	240,000	40,000	20%
4	3	300,000	60,000	25%
5	4	330,000	30,000	10%

page 318 Chapter Test

14.

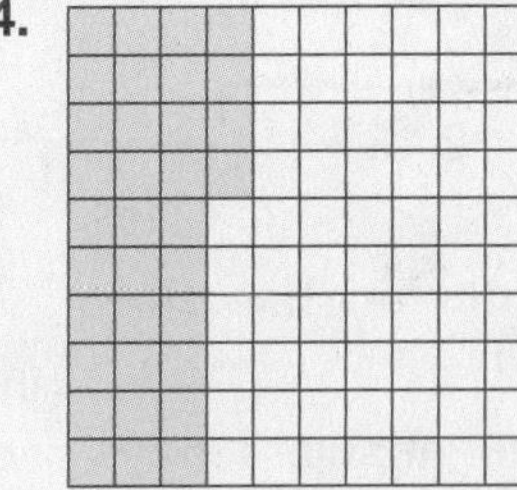

15.

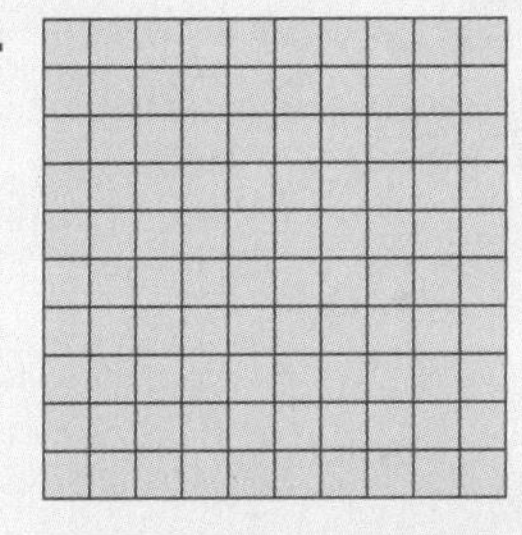

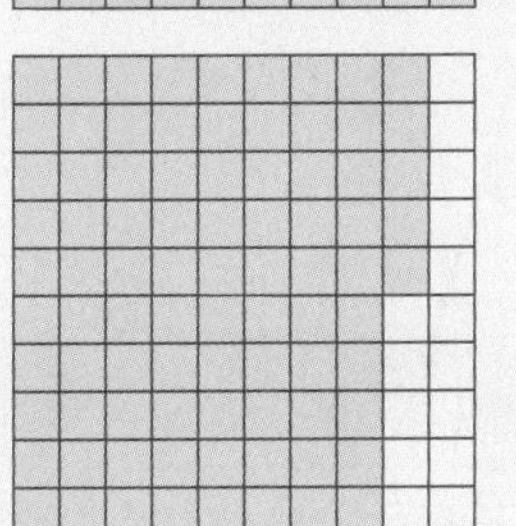

16.

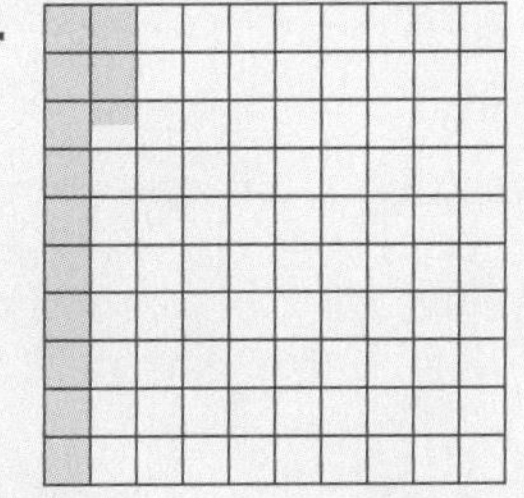

CHAPTER 7

Lesson 7-1

page 324 Check Skills You'll Need

2.

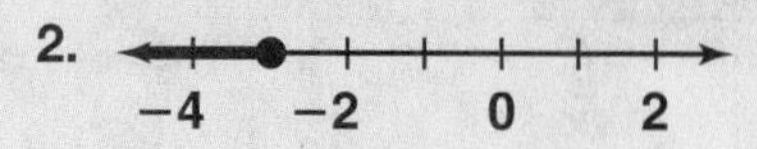

3. −3 −2 −1 0 1

4. −2 0 2

5. 0 2 4 6

6. −1 0 1 2 3

7. −1 0 1 2 3 4

page 327 Homework Exercises

Exercises 24–28. Answers may vary. Samples are given.

24.

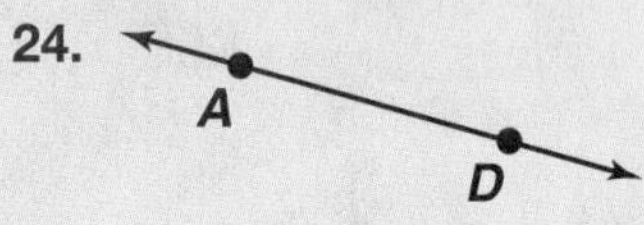

25.

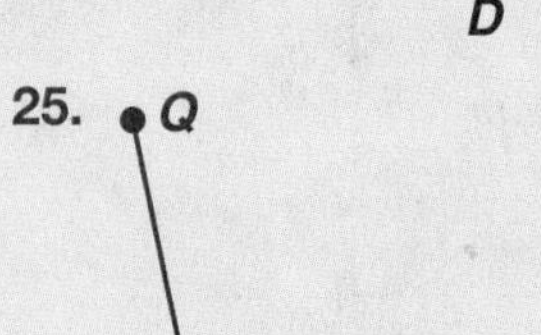

26.

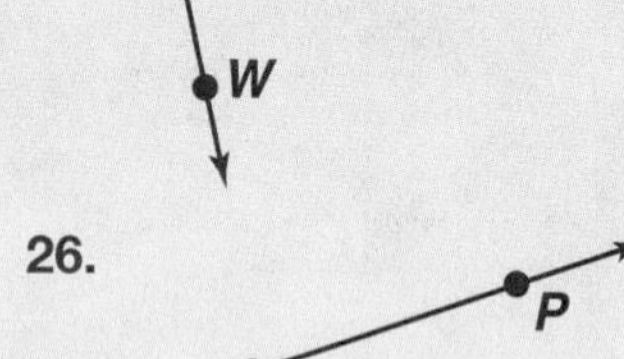

27.

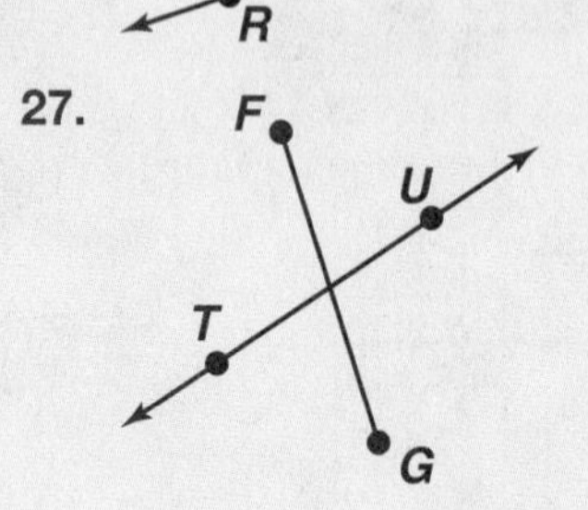

28. T B A

Lesson 7-2

page 334 Homework Exercises

32.

$m\angle A = 90 - m\angle B = 90 - 51 = 39$
$m\angle A = 39 = (3x - 12)$
$51 = 3x$
$17 = x$

Lesson 7-3

page 337 TE Closure

Scalene triangles have no congruent sides; isosceles triangles have at least two congruent sides; equilateral triangles have three congruent sides.

Acute triangles have three acute angles; obtuse triangles have one obtuse angle; right triangles have one right angle.

Write an equation using "the sum of the angle measures equals 180°." Then solve for the missing angle.

Lesson 7-5

page 347 TE Additional Examples

1. Congruent: All corresponding sides and angles are congruent.

3a. $\overline{DE} \cong \overline{OP}$,
$\overline{DF} \cong \overline{OQ}$, $\overline{EF} \cong \overline{PQ}$,
$m\angle Q \cong m\angle F$,
$m\angle P \cong m\angle E$,
$m\angle O \cong m\angle D$

3b. $OP = 24$; $m\angle Q = 53°$

page 348 Homework Exercises

7. $\overline{CV} \cong \overline{JP}$; $\overline{VR} \cong \overline{PX}$;
$\overline{RC} \cong \overline{XJ}$; $\angle C \cong \angle J$;
$\angle V \cong \angle P$; $\angle R \cong \angle X$
$VR = 9$; $m\angle V = 83°$;
$m\angle X = 41°$; $m\angle J = 56°$

8. $\overline{ED} \cong \overline{KJ}$; $\overline{DC} \cong \overline{JI}$;
$\overline{CB} \cong \overline{IH}$; $\overline{BA} \cong \overline{HG}$;
$\overline{AF} \cong \overline{GL}$; $\overline{FE} \cong \overline{LK}$;
$\angle E \cong \angle K$; $\angle D \cong \angle J$;
$\angle C \cong \angle I$; $\angle B \cong \angle H$;
$\angle A \cong \angle G$; $\angle F \cong \angle L$;
$m\angle E = 135°$;
$m\angle A = 117$; $\overline{BC} = 6$;
$LK = 11$; $m\angle J = 90°$;
$m\angle I = m\angle H = 135°$;
$m\angle L = 108°$

Lesson 7-6

page 353 Homework Exercises

29. Check students' work for diagrams. All the quadrilaterals are rectangles.

30. Check students' work for diagrams. All the triangles are scalene or isosceles right triangles.

Lesson 7-7

page 355 Quick Check

2b.

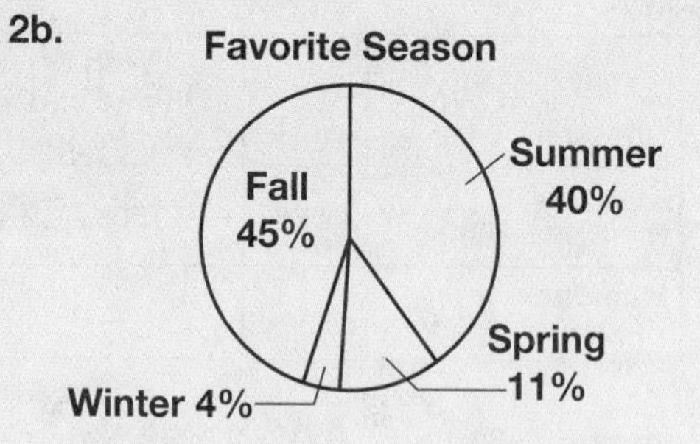

TE Additional Examples

2. Size of U.S. Households

10,854
26,724
15,309
17,152
34,666

5 or more
4 people
3 people
2 people
1 person

TE Closure

Add the parts to find the total; use a proportion to find the central angle for each part; draw a circle and mark each angle; and label each section and write a title.

pages 356–357 Homework Exercises

10. Frozen Yogurt Sales

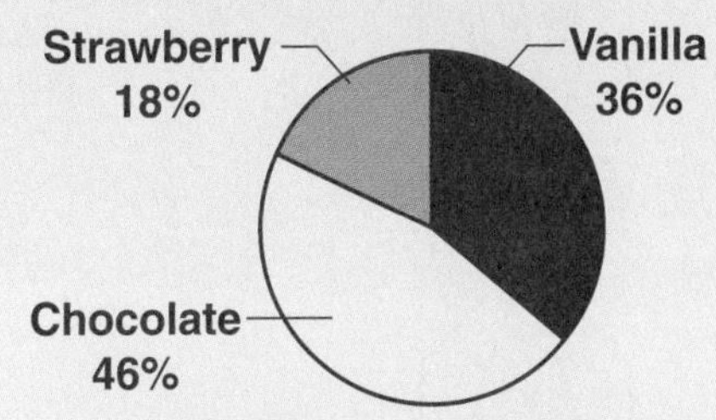

11. Movies Rentals

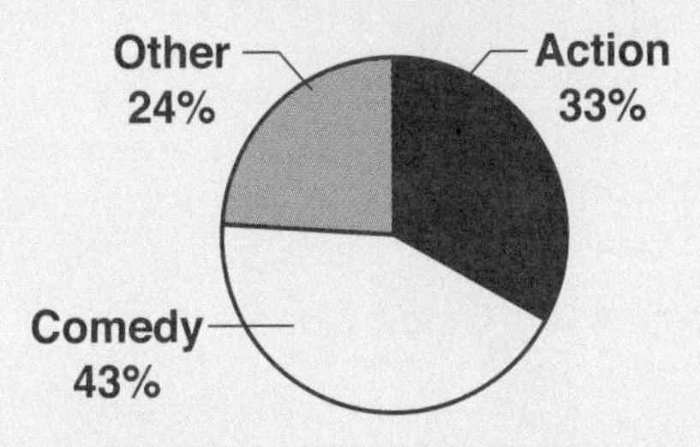

12. Mode of Transportation

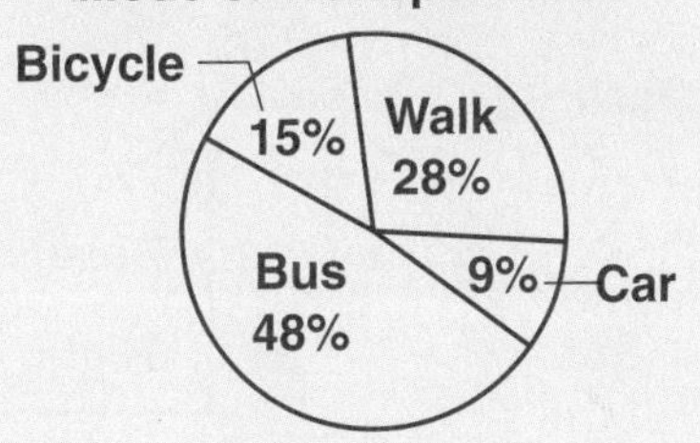

17. Students Volunteering

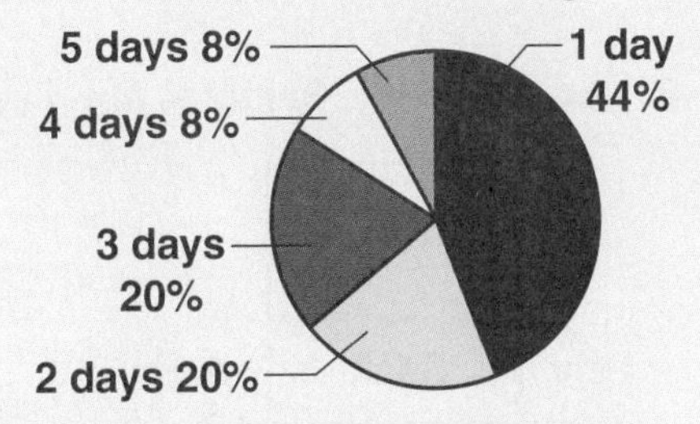

18. Answers may vary. Sample: A circle graph would be a good choice for showing an ice cream store's customers' favorite flavors among a selection of five. A circle graph would be a bad choice if you were trying to show the number of customers who like a flavor, because they can like more than one. A circle graph shows parts of a whole. When there is overlap, a Venn diagram might be more appropriate.

page 358 Checkpoint Quiz 2

2. Town Middle School

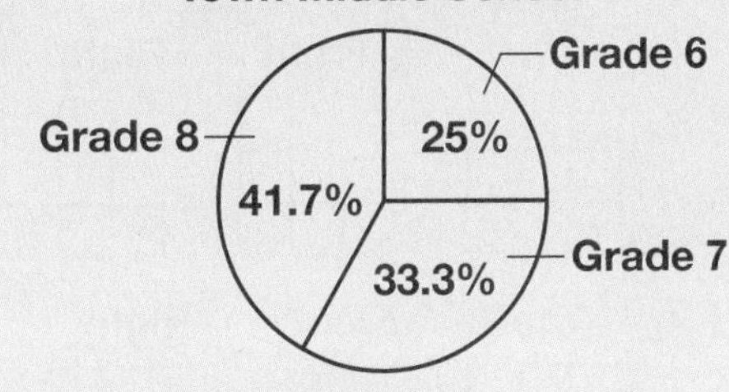

25% (90°) for 6th grade 33.3% (120°) for 7th grade, and 41.7% (150°) for 8th grade.

page 360 Guided Problem Solving

4. U.S. Blood Type Percentages

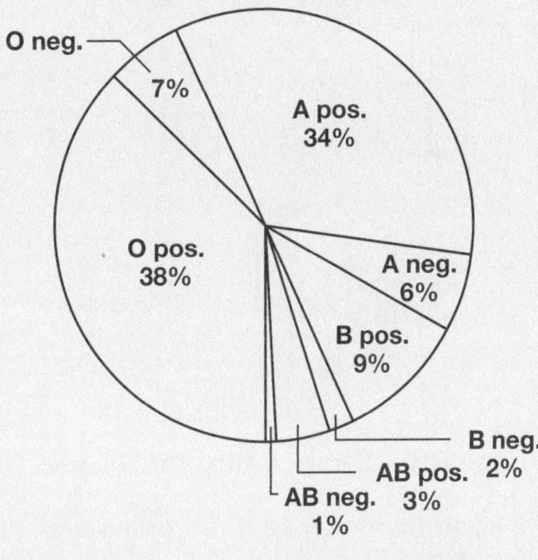

U.S. Blood Type (millions of people)

Type	Positive	Negative
O	106.8	19.7
A	95.5	16.9
B	25.3	5.6
AB	8.4	2.8

Lesson 7–8

page 363 Homework Exercises

15.

C F E D

16.

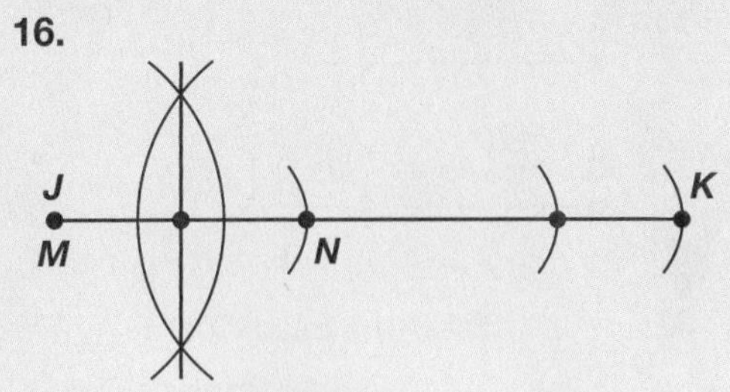

17.

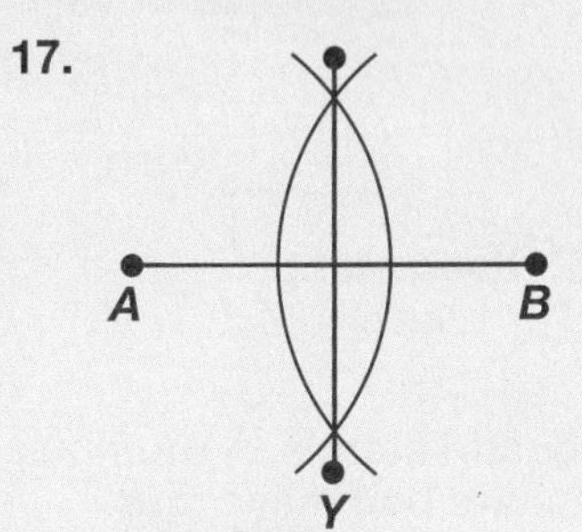

$\overline{AB}$ = 5 in.

18.

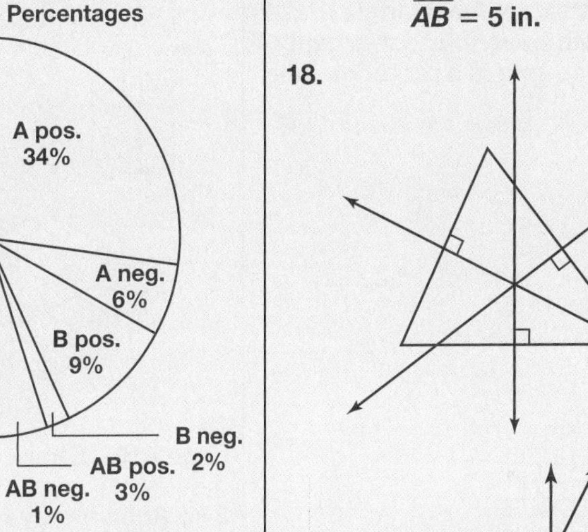

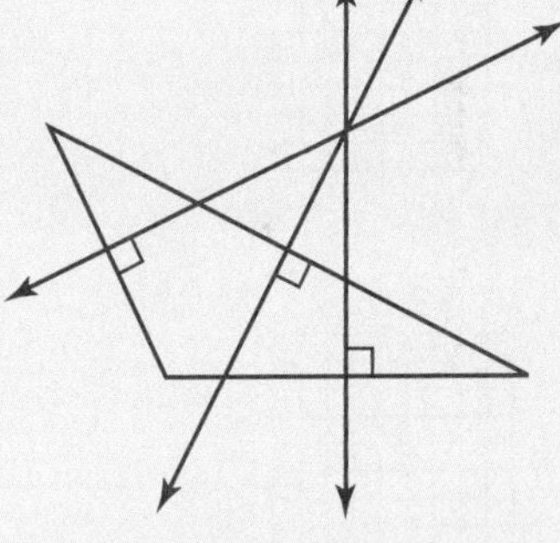

The perpendicular bisectors of the sides of any triangle will intersect in one point.

page 368 Chapter Test

12. Check students' work. An example is given.

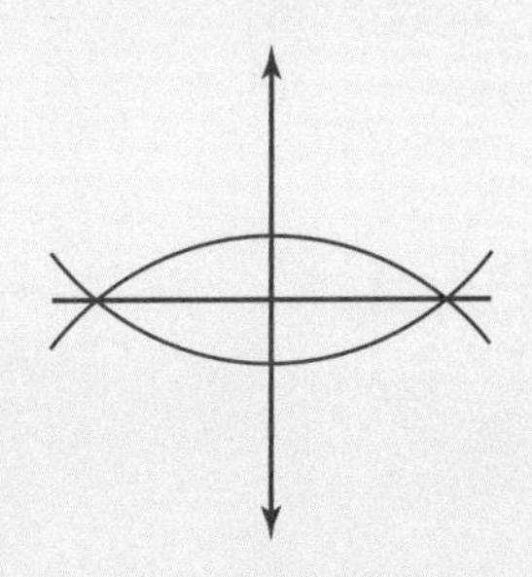

Additional Answers

13.

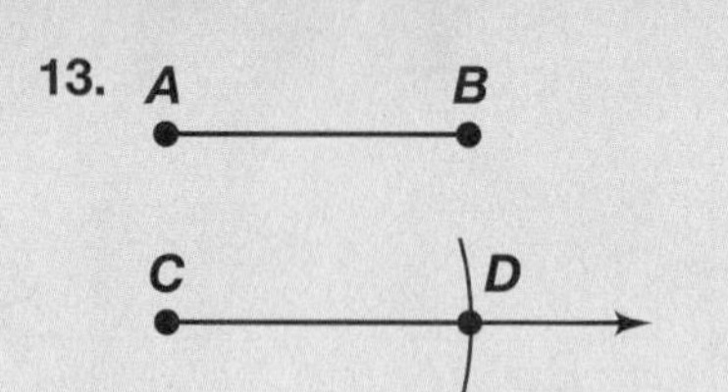

32.

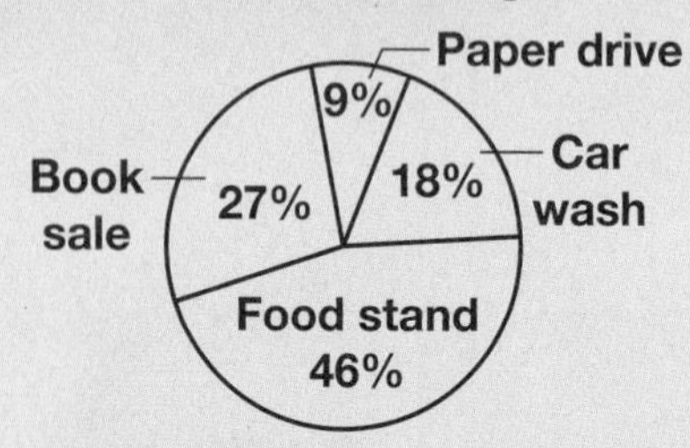

33. All are quadrilaterals. A rhombus and a square both have four congruent sides. A rectangle and a square both have four congruent angles. A square is a rectangle and a rhombus.

CHAPTER 8

Lesson 8–6

page 403 Homework Exercises

41.

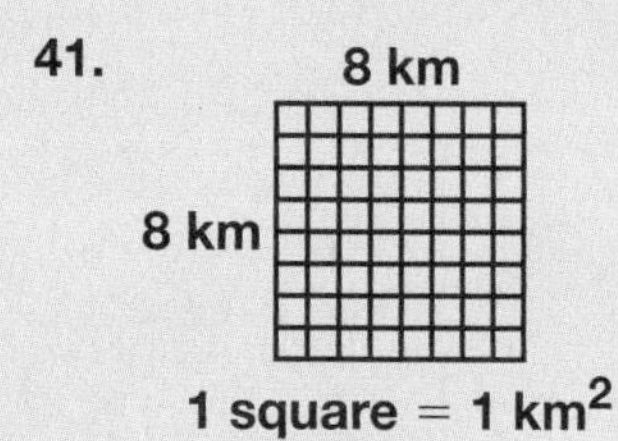

42.

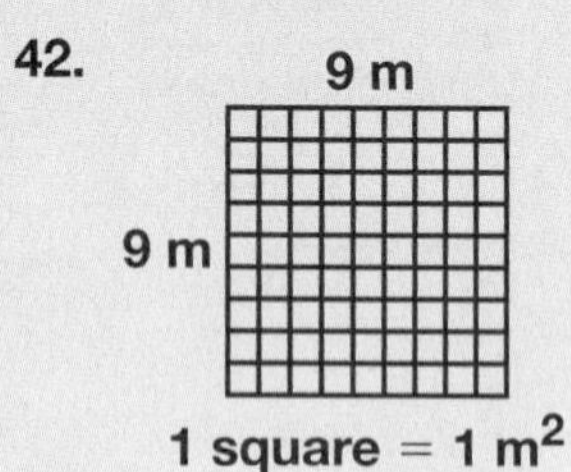

43.

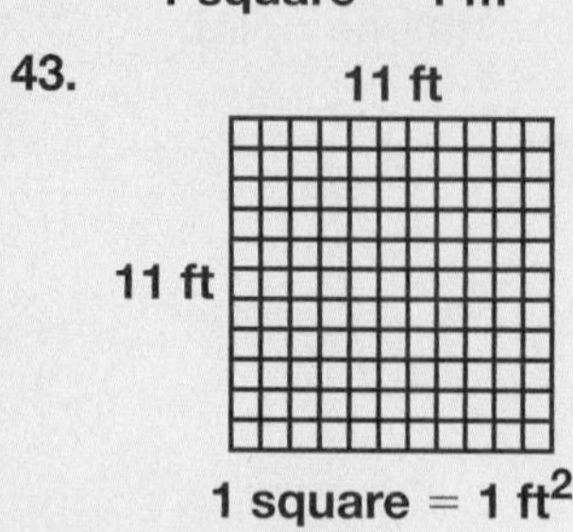

44.

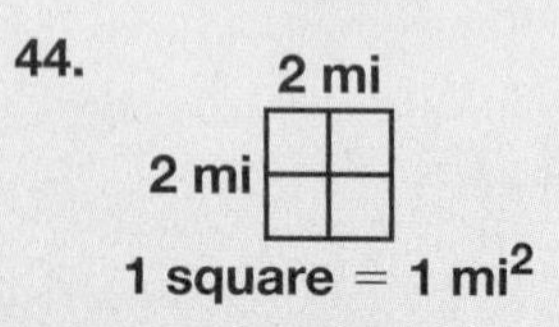

Lesson 8–8

page 409 Activity Lab

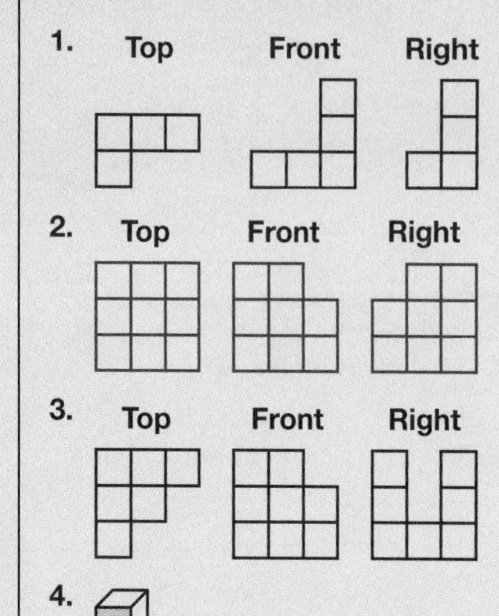

4.

5.

page 410 Check Skills You'll Need

2–4. Answers by vary. Samples are given.

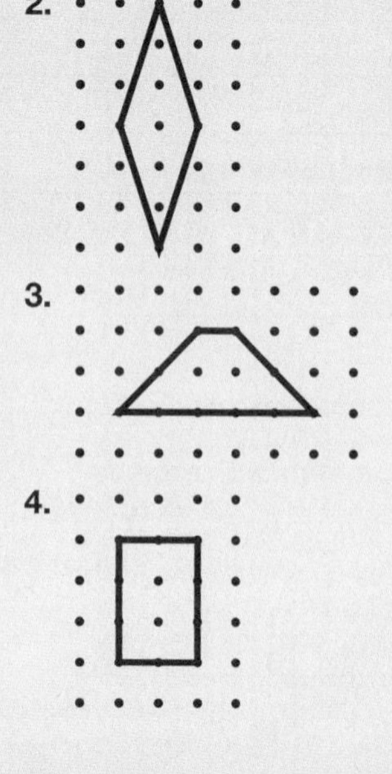

pages 412 Homework Exercises

12.

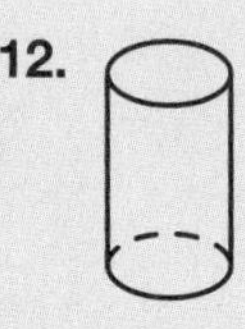

13.

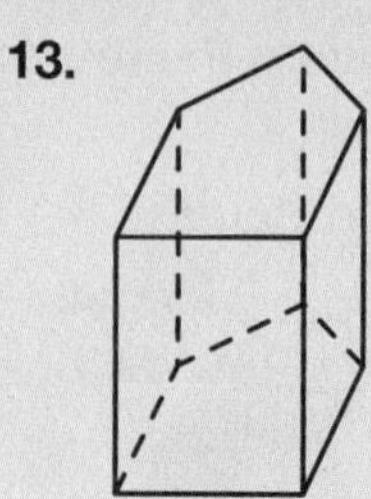

14.

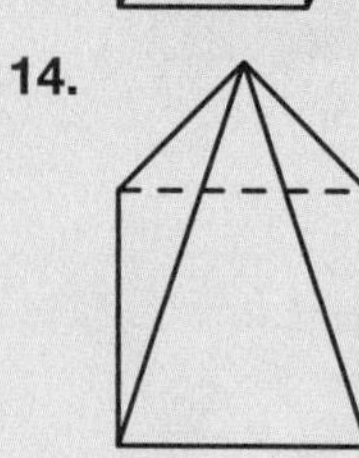

Lesson 8–9

page 414 Quick Check

1. Answers may vary. Sample:

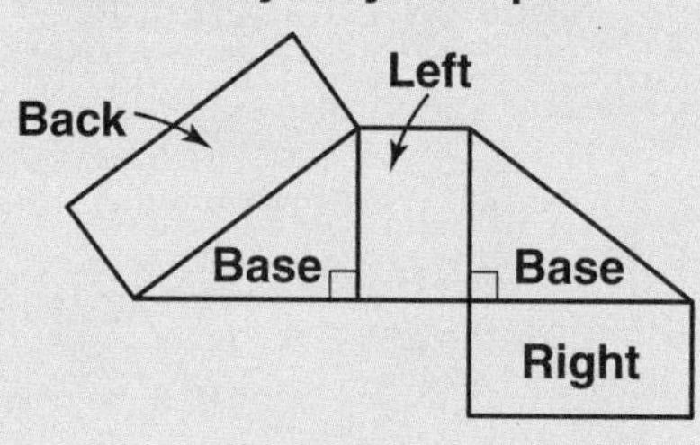

pages 417–418 Homework Exercises

5–8. Nets may vary. Samples are given.

5.

6.

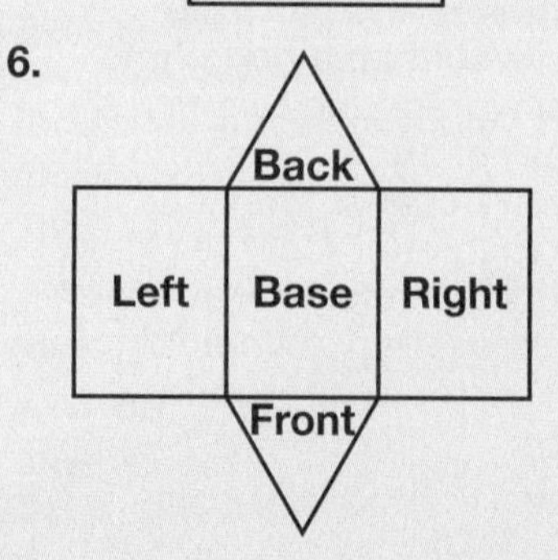

7.

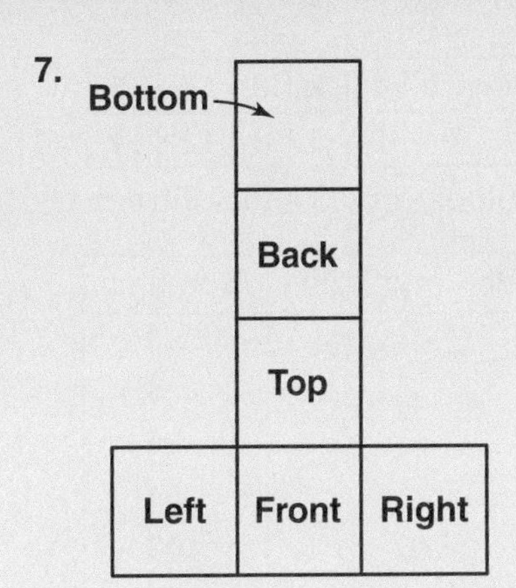

8.

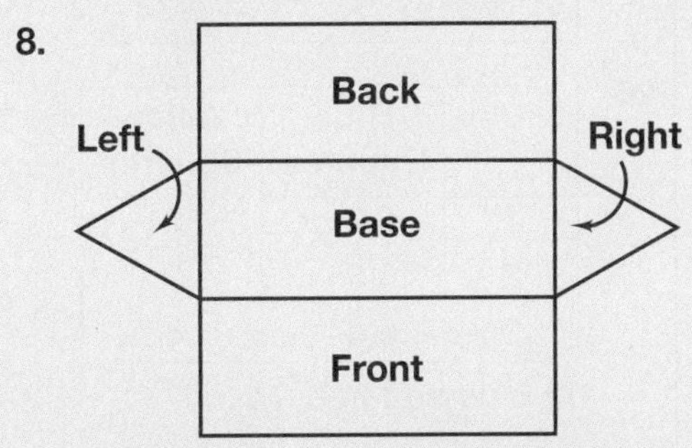

25. Doubling the radius of the base has more effect on the surface area of a cylinder because the radius is doubled and then squared.

If $r = 1$ and $h = 1$, then
$SA = 2\pi(1)^2 + 2 \cdot 1\pi \cdot 1 = 4\pi.$
If $r = 2$ and $h = 1$, then
$SA = 2\pi(2)^2 + 2 \cdot 2\pi \cdot 1 = 8\pi + 4\pi = 12\pi.$
If $r = 1$ and $h = 2$, then
$SA = 2\pi(1)^2 + 2(1)\pi(2) = 2\pi + 4\pi = 6\pi.$

26. Answers may vary. Sample:

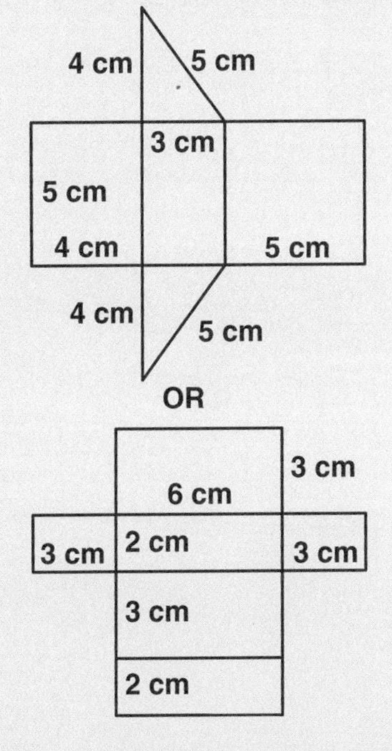

page 419 Extension

1.

Number of Unit Cubes on an Edge	Total Number of Unit Cubes	Total Number Expressed as a Power	Number of Unit Cubes With Given Number of Sides Painted			
			0	1	2	3
2	8	2^3	0	0	0	8
3	27	3^3	1	6	12	8
4	64	4^3	8	24	24	8
5	125	5^3	27	54	36	8
6	216	6^3	64	96	48	8
7	343	7^3	125	150	60	8

CHAPTER 9

page 436 Activity Lab

1b.

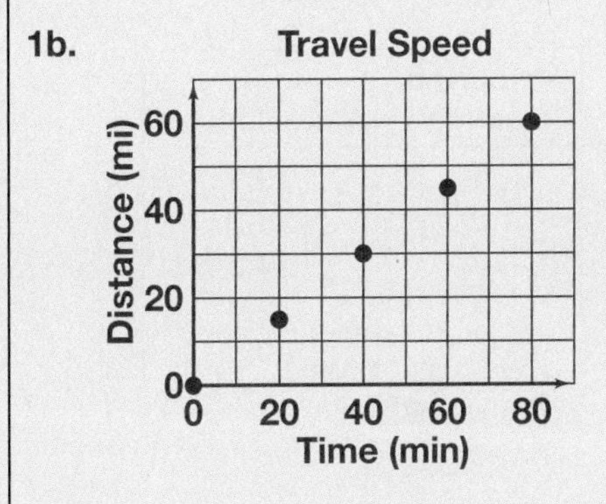

2a.

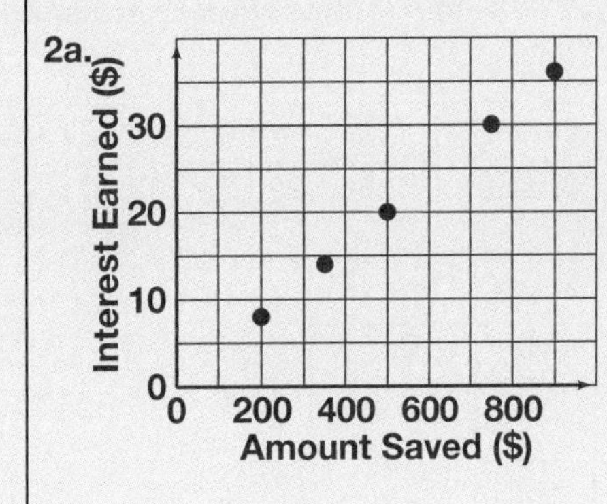

Lesson 9-1

page 437 Check Skills You'll Need

1. Yes; it can be written as a fraction.

2–5.

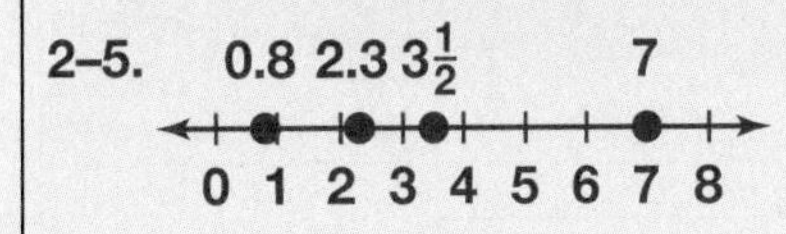

page 437 Quick Check

1.

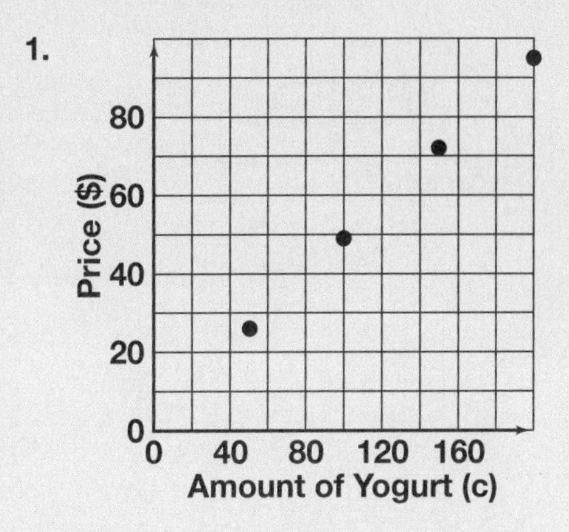

page 438 TE Additional Examples

1.

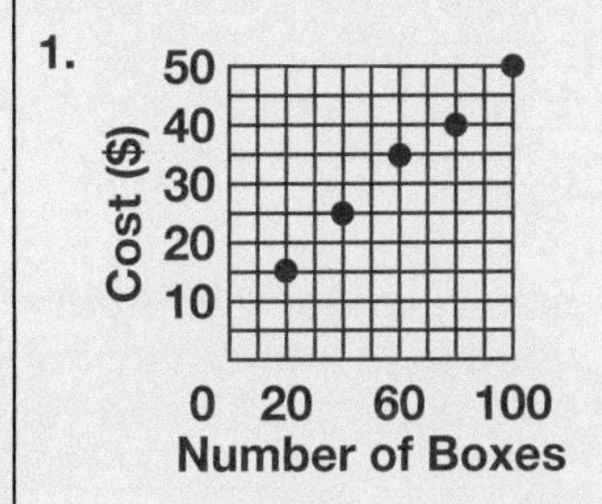

pages 439–440 Homework Exercises

14.

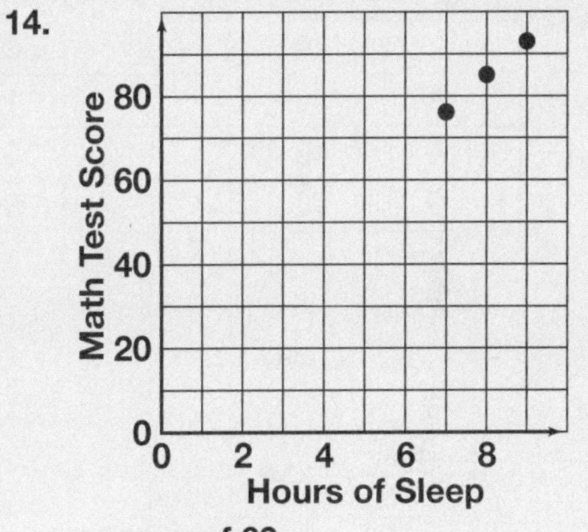

a score of 69

15.

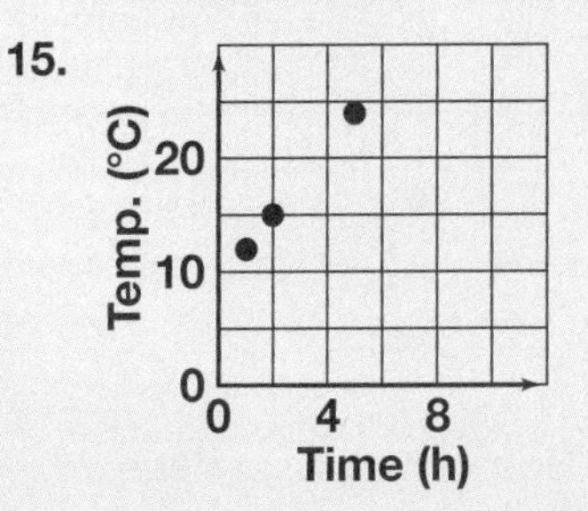

33°C

16.

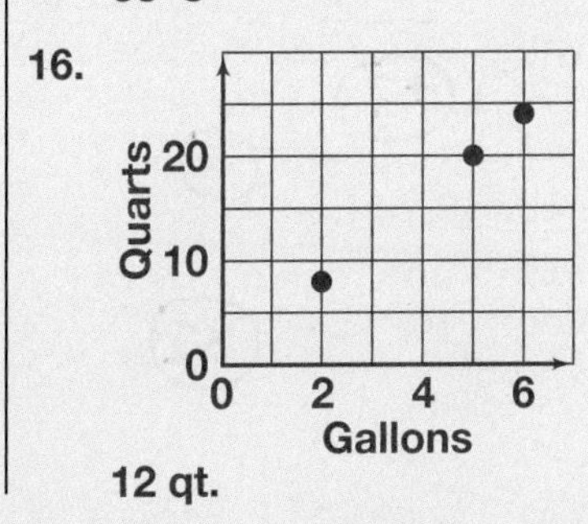

12 qt.

17a.

Perimeter (in.)

20

10

0

0 2 4

Side Length (in.)

18.

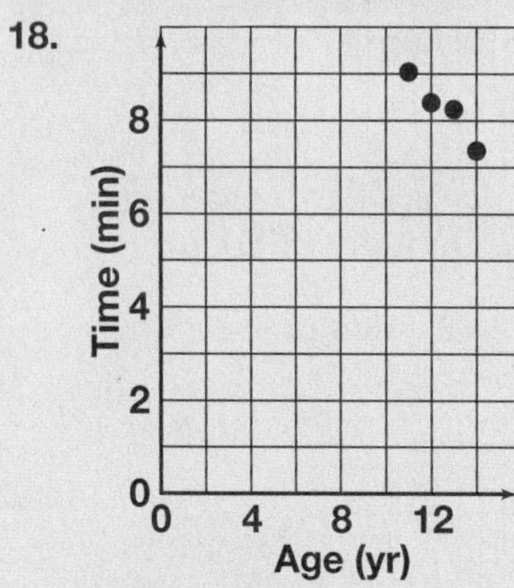

Answers may vary. Sample: About 6:25; the trend in the data may not continue between ages 14 and 18.

19a.

Number of Weeks	Pay ($)
1	10
2	20
3	30
4	40
5	50

Number of Weeks	Pay ($)
1	47
2	54
3	61
4	68
5	75

b.

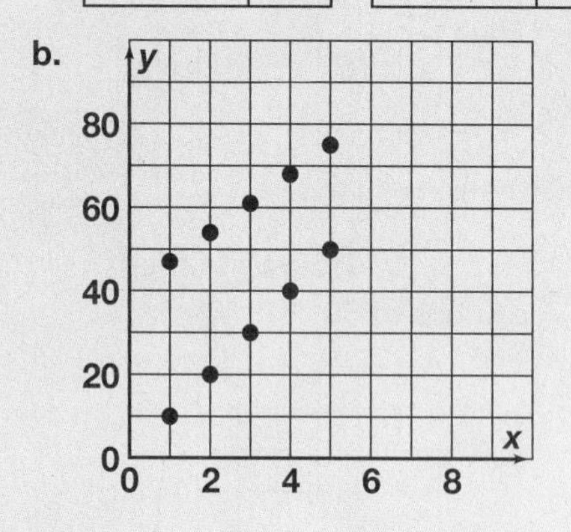

Lesson 9-2

page 444 Homework Exercises

6. Start with −8 and add 7 repeatedly; 20, 27, 34.
7. Start with 25 and add −4 repeatedly; 9, 5, 1.
8. Start with 1 and multiply by 2 repeatedly; 16, 32, 64.
9. Start with 2 and multiply by −3 repeatedly; 162, −486, 1,458.
10. Start with 600 and multiply by $-\frac{1}{2}$ repeatedly; −75, 37.5, −18.75.
11. Start with $\frac{1}{2}$ and multiply by $\frac{1}{2}$ repeatedly; $\frac{1}{32}$, $\frac{1}{64}$, $\frac{1}{128}$.
12. Start with −2 and multiply by −2 repeatedly; −32, 64, −128.
13. Start with $\frac{1}{4}$ and multiply by $\frac{1}{3}$ repeatedly; $\frac{1}{324}$, $\frac{1}{972}$, $\frac{1}{2,916}$

Lesson 9-3

page 446 Quick Check

1.

Amount of Gas (gal)	Miles Driven
1	18.1
2	36.2
3	54.3
4	72.4
5	90.5

271.5 miles

pages 448–449 Homework Exercises

20.

Time (h)	0.5	1	1.5	2
Cost ($)	12.50	25	37.50	50

21.

Miles	Time (h)
10	0.4
20	0.8
30	1.2
40	1.6
50	2.0

22.

Change in a Parking Meter ($)	Time Allowed to Park (h)
0.25	0.5
0.50	1
0.75	1.5
1.25	2.5

23.

Celsius	Kelvin
0	273
10	283
20	293
40	313
80	353
120	393

24a.

x	0	1	2	3	4	5
y	−1	2	5	8	11	14

b. Multiply x by 3 and subtract 1.

25.

Side Length (in.)	Area (in.^{2})
2	4
3	9
5	25
8	64
10	100
12	144

26.

Group	Blue Squares	Yellow Squares
1	2	2
2	3	3
3	4	4

11 blue squares

27a.

x	y
−4	−6
−2	−3
2	3
4	6

b. 0

Lesson 9-4

page 454 Homework Exercises

12.

x	y
0	2
1	3
2	4
3	5

13.

x	y
0	9
1	8
2	7
3	6

14.

x	y
0	0
1	4
2	8
3	12

15.

x	y
0	0
1	$\frac{1}{2}$
2	1
3	$\frac{3}{2}$

16.

x	y
0	0
1	−3
2	−6
3	−9

17.

x	y
0	1
1	3
2	5
3	7

18.

x	y
0	−2
1	2
2	6
3	10

19.

x	y
0	1
1	2
2	5
3	10

Lesson 9-5

page 457 Quick Check

2. $d = 60t$; 270 mi

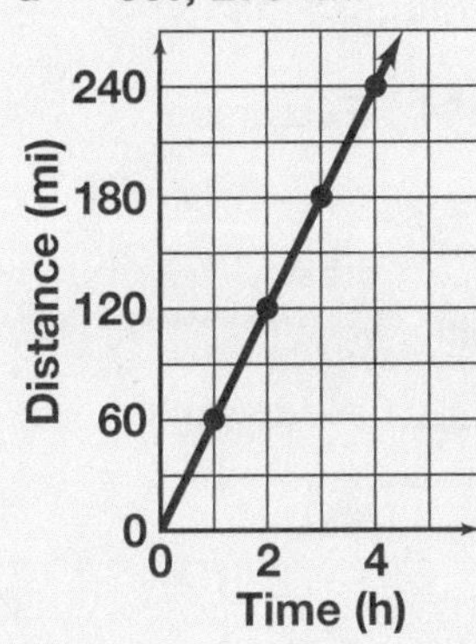

TE Additional Examples

1. Output (y) vs. Input (x)

$h = 2w + 1$; 9 in.

2.

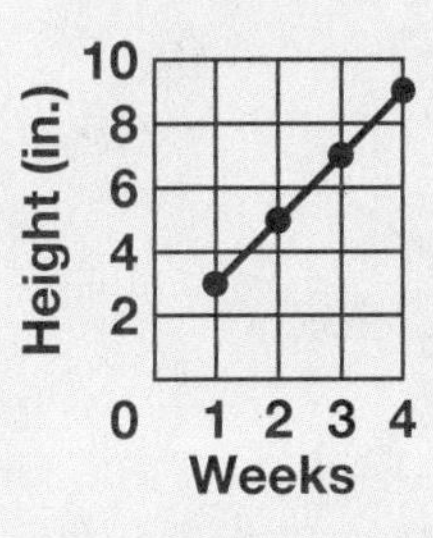

pages 458–459 Homework Exercises

4.

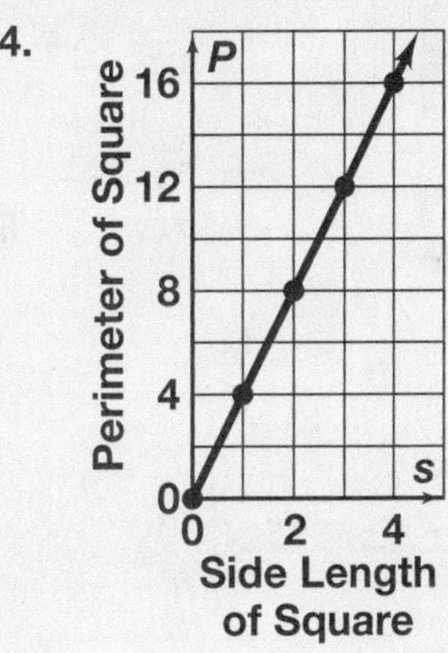

5.

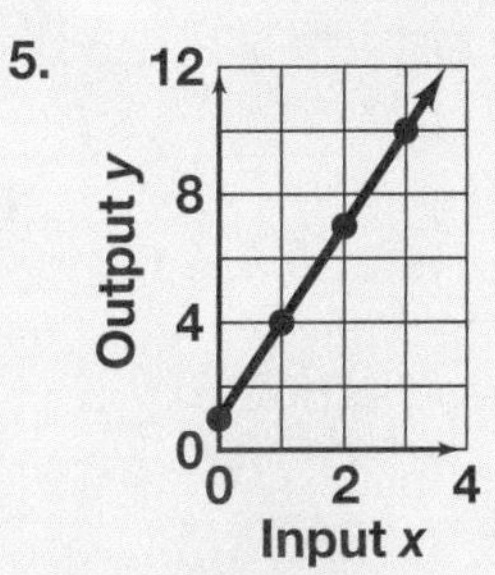

6. $S = 6t$; \$60

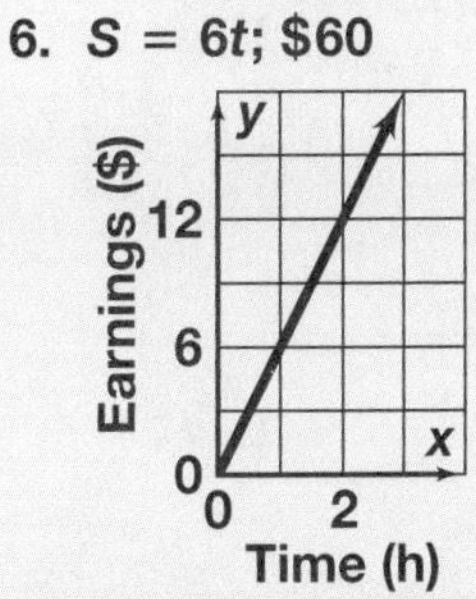

7. $d = 320t$; 3,200 mi

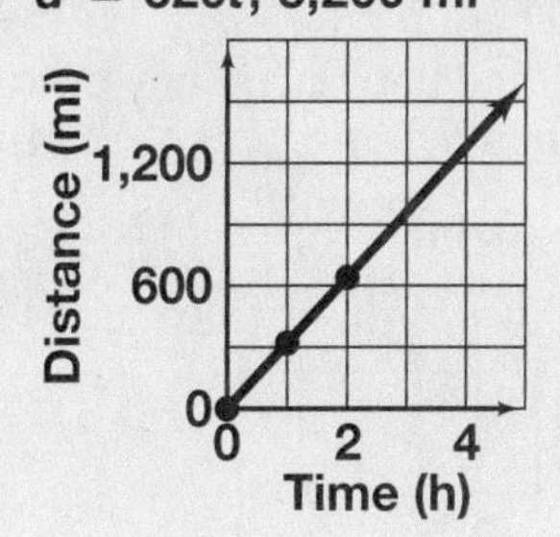

14a.

x	y
1	3
2	4
3	5
4	6

b. $y = x + 2$

15.

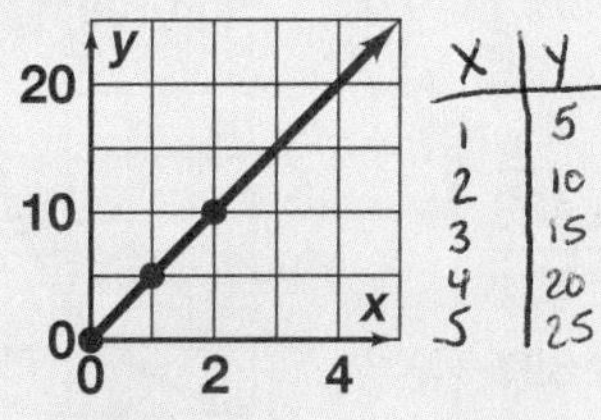

x	y
1	5
2	10
3	15
4	20
5	25

16.

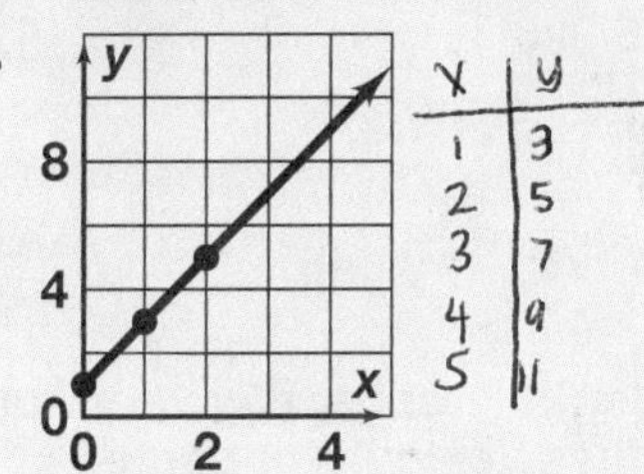

x	y
1	3
2	5
3	7
4	9
5	11

17.

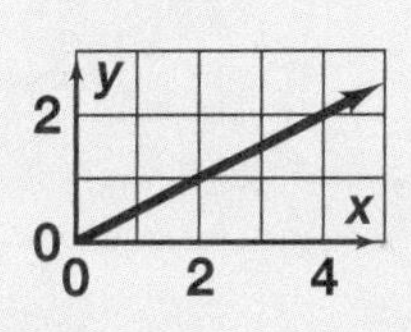

x	y
1	$\frac{1}{2}$
2	1
3	$\frac{3}{2}$
4	2
5	$\frac{5}{2}$

18.

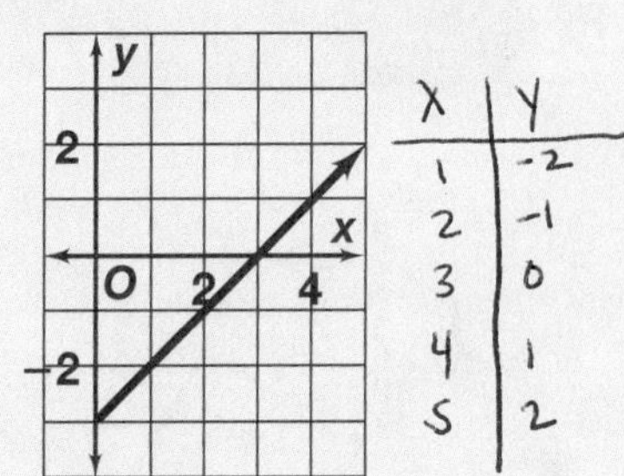

x	y
1	−2
2	−1
3	0
4	1
5	2

Additional Answers

20a. $d = 181t$

b. 181 mi/h; if you divide any of the distances by time, you get 181 mi/h.

c. about 10.4 h

d.

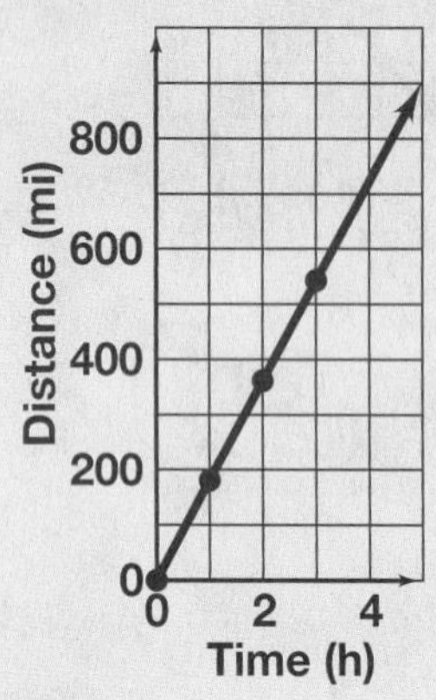

21. Answers may vary. Sample:

s	A
1	0.433
2	1.732
3	3.897
4	6.928

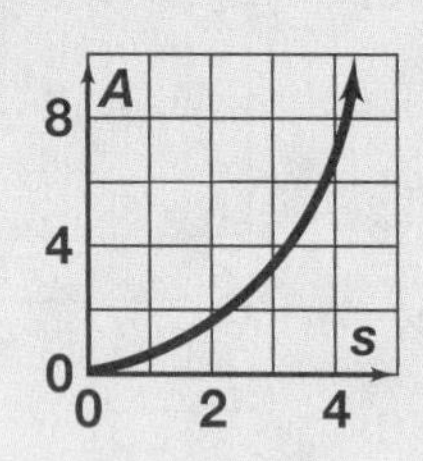

The graph is a curve.

page 460 Activity Lab

1–6. Students' graphs may vary from these calculator screens.

1.

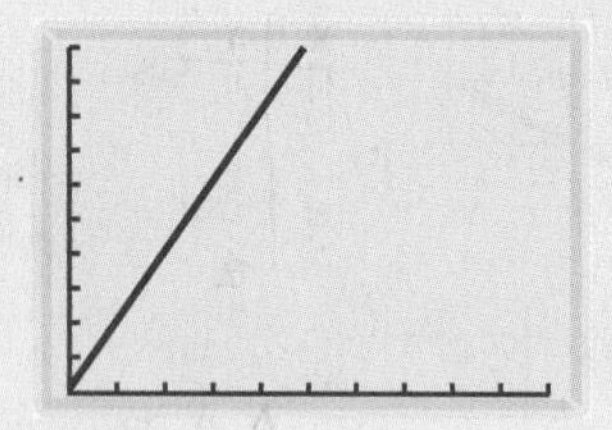

X	Y_1
0	0
1	2
2	4
3	6
4	8
5	10
6	12
X=	

2.

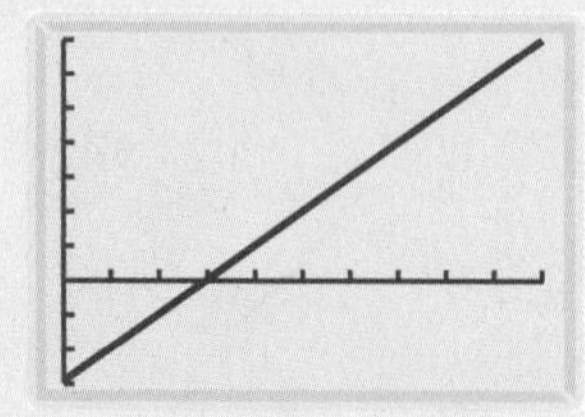

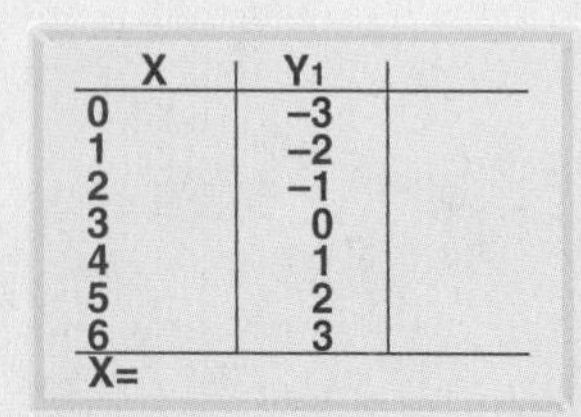

X	Y_1
0	−3
1	−2
2	−1
3	0
4	1
5	2
6	3
X=	

3.

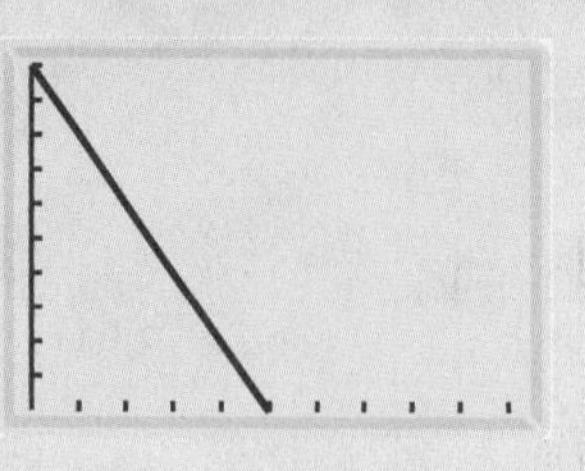

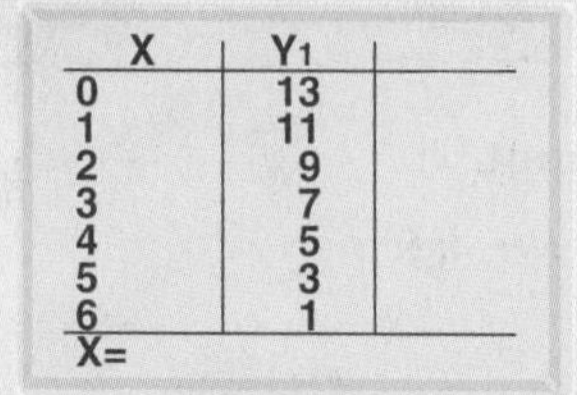

X	Y_1
0	13
1	11
2	9
3	7
4	5
5	3
6	1
X=	

4.

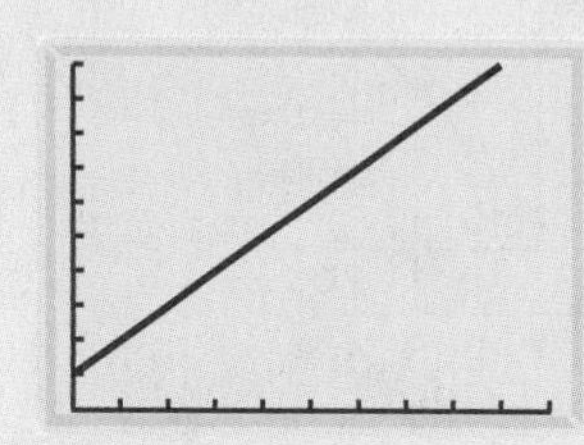

X	Y_1
0	1
1	2
2	3
3	4
4	5
5	6
6	7
X=	

5.

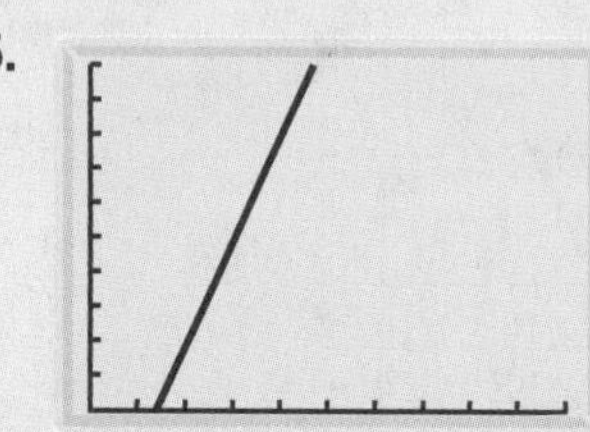

X	Y_1
0	−4
1	−1
2	2
3	5
4	8
5	11
6	14
X=	

6.

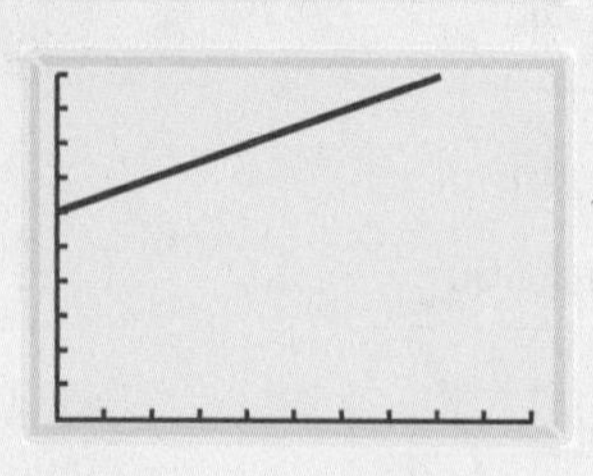

X	Y_1
0	6
1	6.5
2	7
3	7.5
4	8
5	8.5
6	9
X=	

7. Answers may vary. Sample: If you use a window of 0 to 10 for *x*, your window for *y* should go from 0 to 1,000 with increments of 100.

Lesson 9-6

page 462 Quick Check

2.

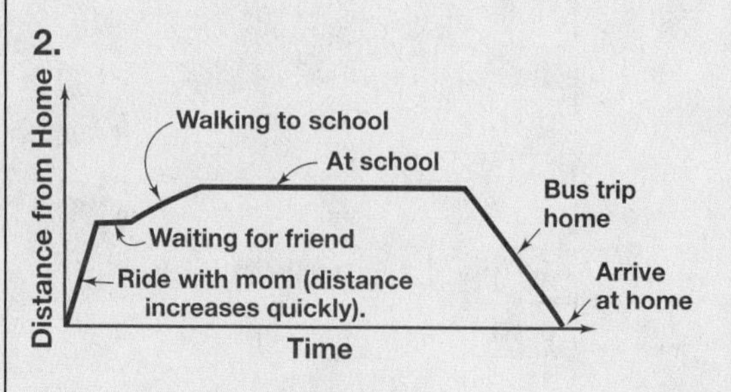

TE Additional Examples

1. Sample: Nancy walked for five min. She stopped for one min. She walked home at the same speed she started out.

2.

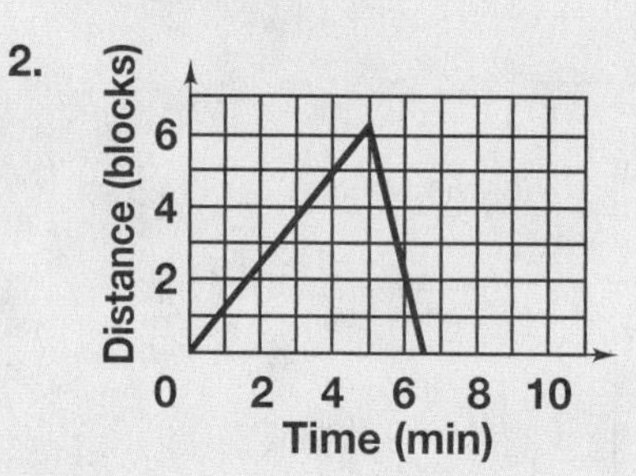

3.

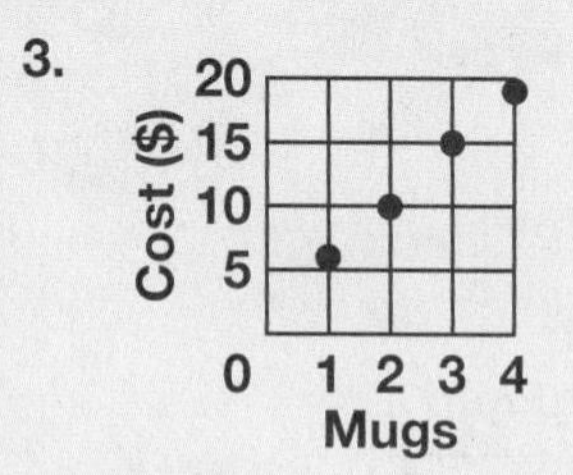

pages 463–464 Exercises

6.

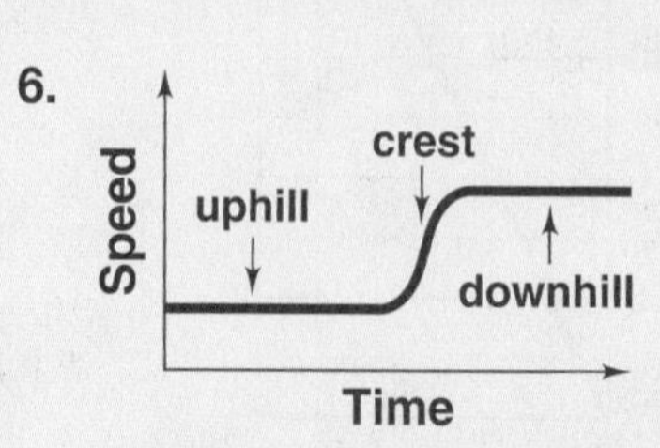

7. You should not connect the points because you sell only whole-number cups of lemonade.

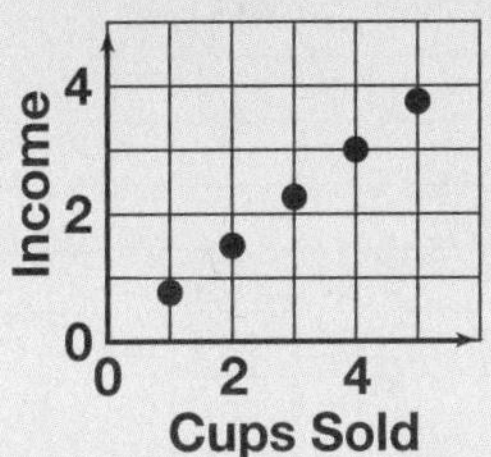

8. You can connect the points because time is continuous.

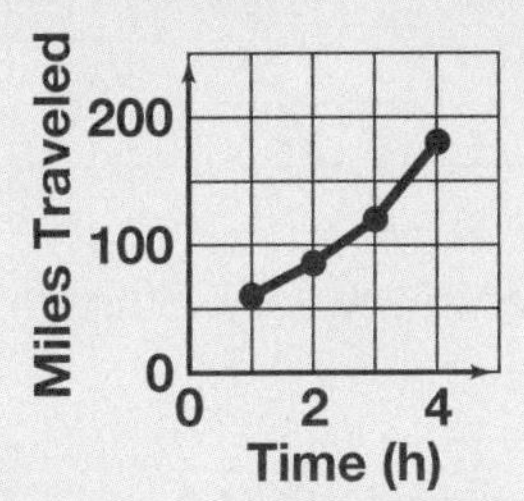

9.

10. A; the bowl has a diameter that varies with height, so the relationship between height and time will not be constant.

11a. The graph shows that Josh caught up to Rafael after about 10 s.

b. Josh

c. If the lines were parallel, then Rafael would have covered the 90 m five seconds before Josh, so Rafael would have won the race.

Lesson 9-7

pages 470–471 Homework Exercises

9.

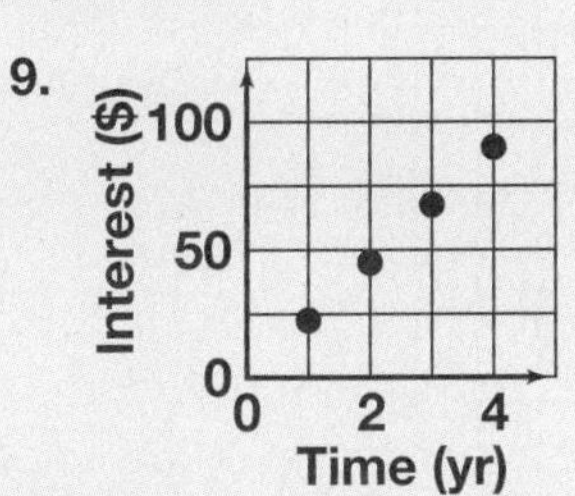

10.

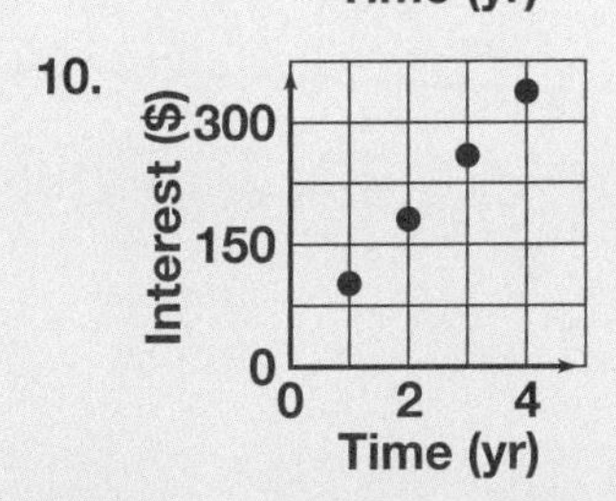

11.

12.

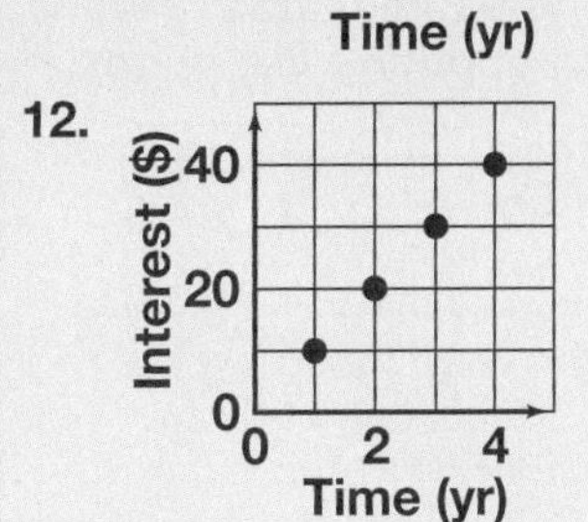

13.

14.

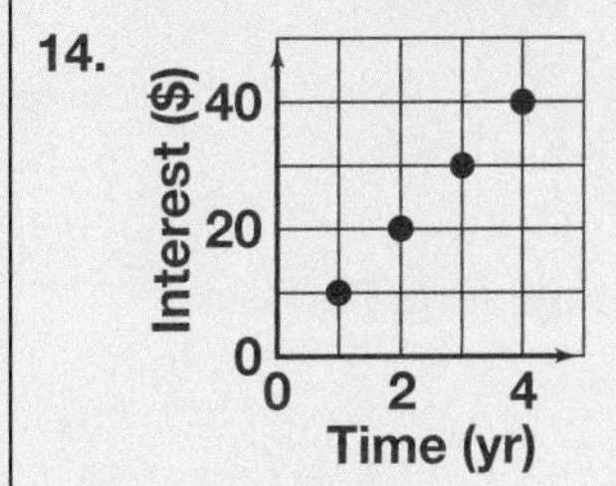

34–35. Answers may vary. Samples are given.

34.

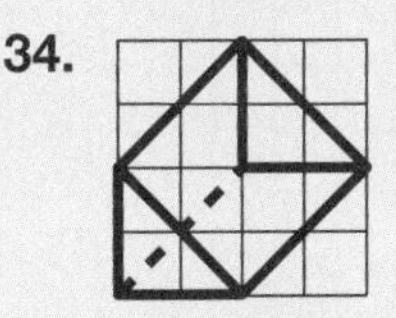

35.

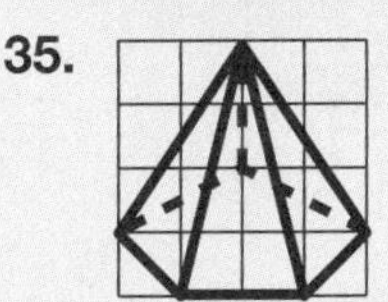

Lesson 9-8

page 476 Activity Lab

1.

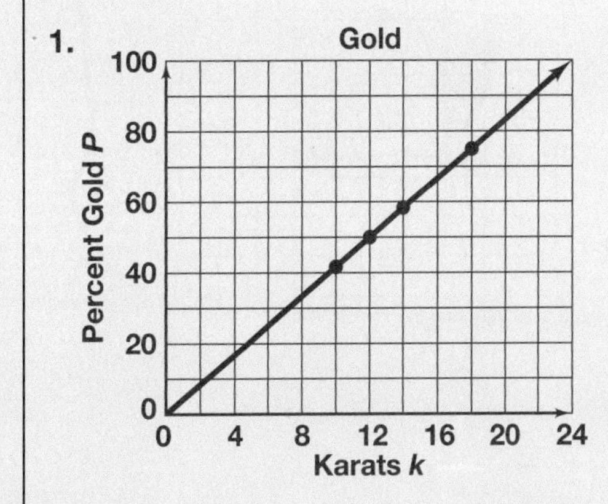

6. Slam Dunk

7. 126 inches represents 6 inches above the rim, or the height the ball needs to reach to be slam-dunked.

8. The formula accounts for the length of a person's arm above his or her head while dunking.

9. $h = \frac{4}{5}(126 - v)$

page 479 Chapter Review

13.

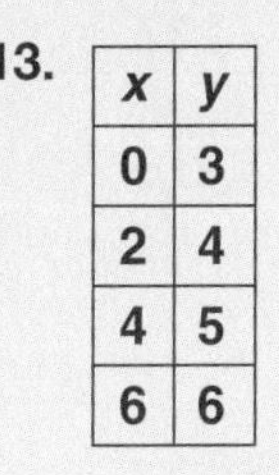

x	y
0	3
2	4
4	5
6	6

16.

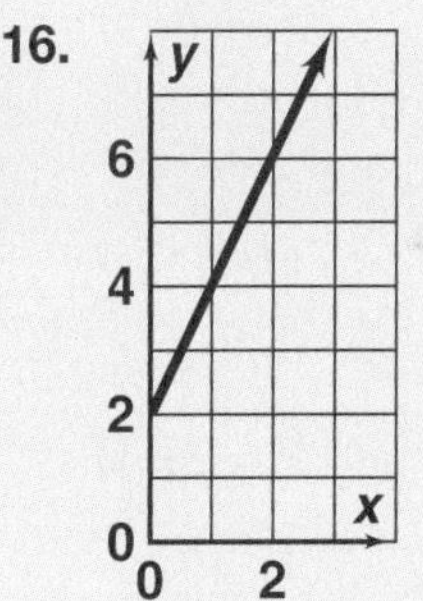

page 480 Chapter Test

1. Start with 1 and multiply by 3 repeatedly; 81, 243, 729.
2. Start with 4 and add 5 repeatedly; 24, 29, 34.
3. Start with 3, then add 1, 2, 3, 4, and so on; 13, 18, 24.
4. Start with 10 and add −2 repeatedly; 2, 0, −2.
5. Start with −23 and add 4 repeatedly; −11, −7, −3.
6. Start with 6 and multiply by 0.5 repeatedly; 0.375, 0.1875, 0.09375.
7. geometric; arithmetic; neither; arithmetic; arithmetic; geometric

8a.

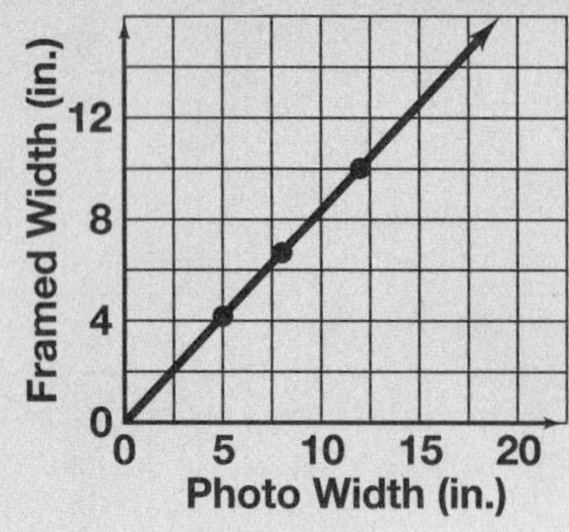

9.

Books	Cost ($)
1	2.95
2	5.90
3	8.85
4	11.80
5	14.75

10.

Side Length (in.)	Perimeter (in.)
5	20
6	24
7	28
8	32
9	36

11. $y = -5x - 2$

12. $y = 2x + 1$

13. $y = 3x$

14.

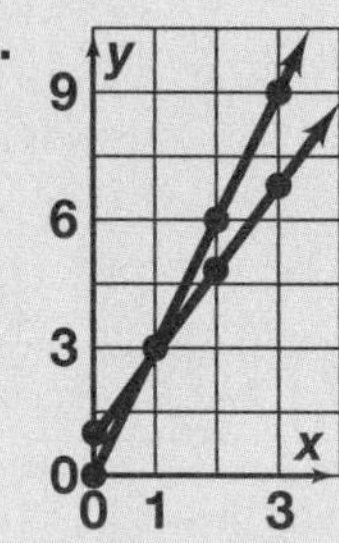

20.

Earnings ($)	Savings ($)
10	2
20	4
30	6
40	8

23.

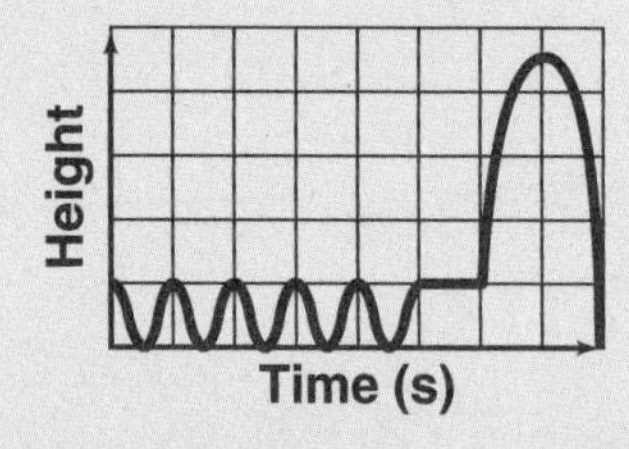

24. Answers may vary. Sample: The oven temperature is high, and then it cools down and stabilizes.

CHAPTER 10

Lesson 10-1

page 486 Check Skills You'll Need

1. They are the same distance from zero, but in opposite directions.

2.

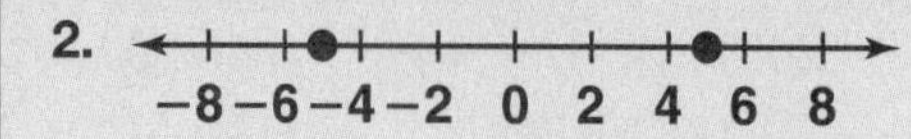

3.

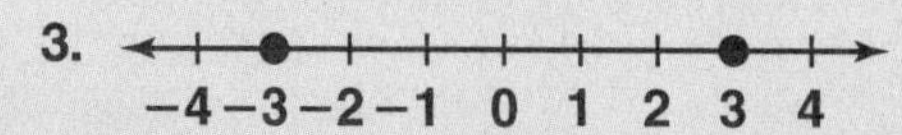

4. −2 −1 0 1 2

5. −12 −9 −6 −3 0 3 6 9 12

6.

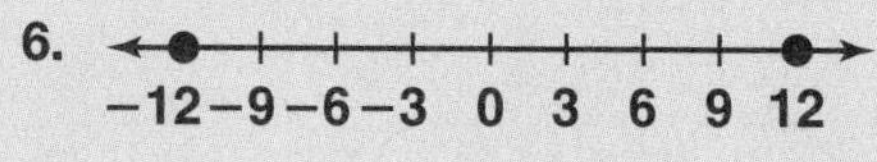

7. −8 −6 −4 −2 0 2 4 6 8

page 487 Quick Check

3. Answers may vary. Sample:

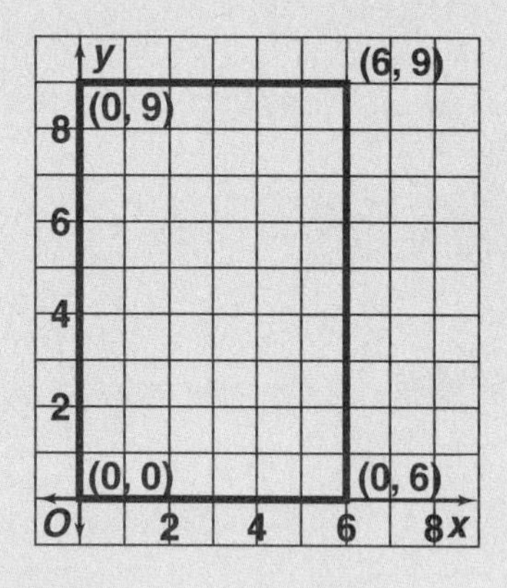

TE Additional Examples

3.

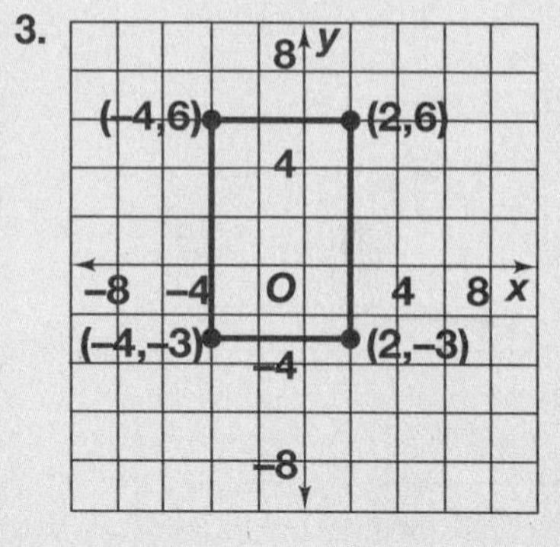

pages 488–489 Homework Exercises

15–22.

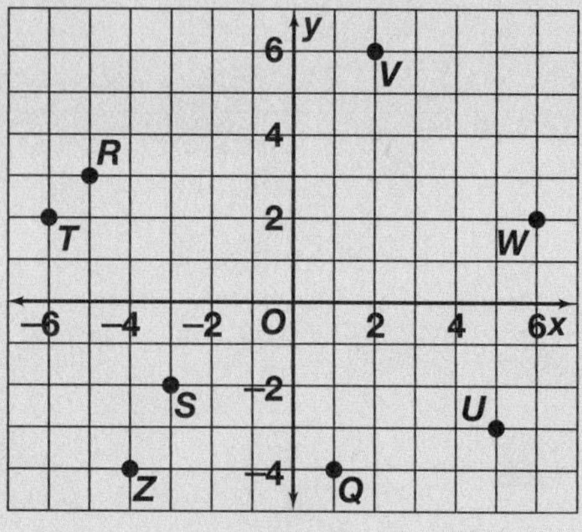

23–26. Answers may vary. Samples are given.

23.

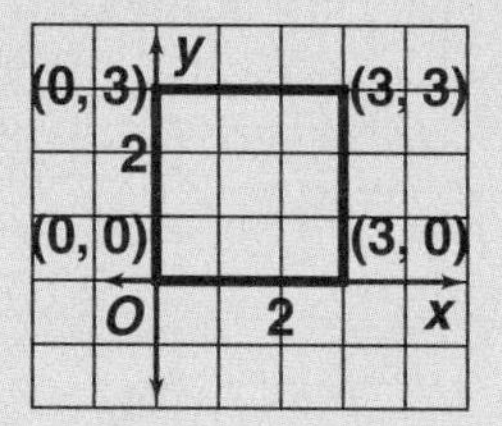

24.

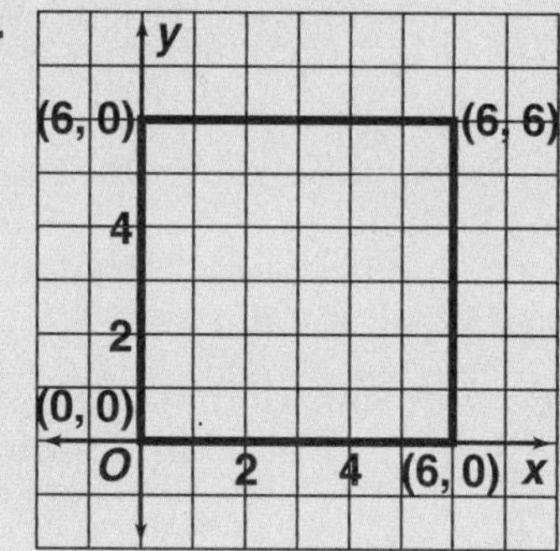

25.

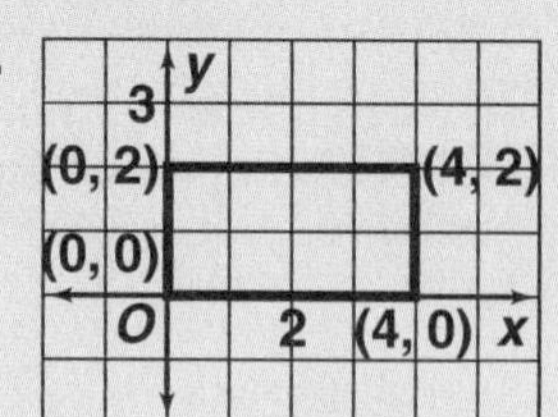

26.

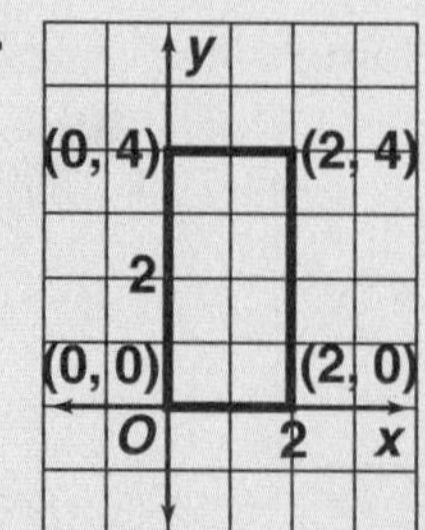

31.

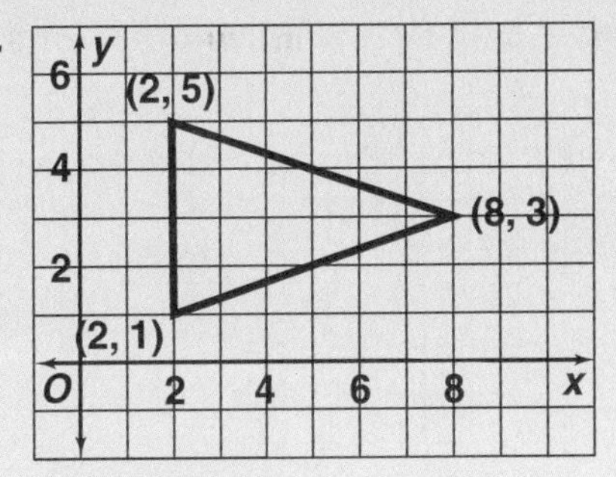

acute isoceles

32.

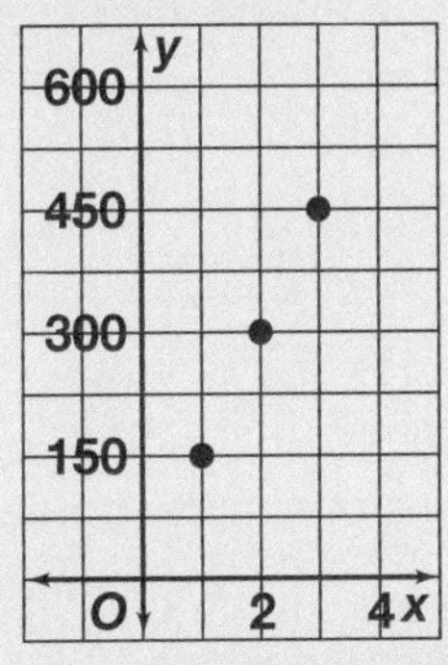

33. Answers may vary. Sample: (–3, –3), (0, –3), (4, –3)
If the *x*-coordinate is 37, the *y*-coordinate would be –3 because it is a straight horizontal line.

35. If both are positive, then the point is in quadrant I. If both are negative, then the point is in quadrant III. If the *x*-coordinate is positive and the *y*-coordinate is negative, then the point is in quadrant IV. If the *x*-coordinate is negative and the *y*-coordinate is positive, then the point is in quadrant II.

36. Answers may vary. Sample is given.

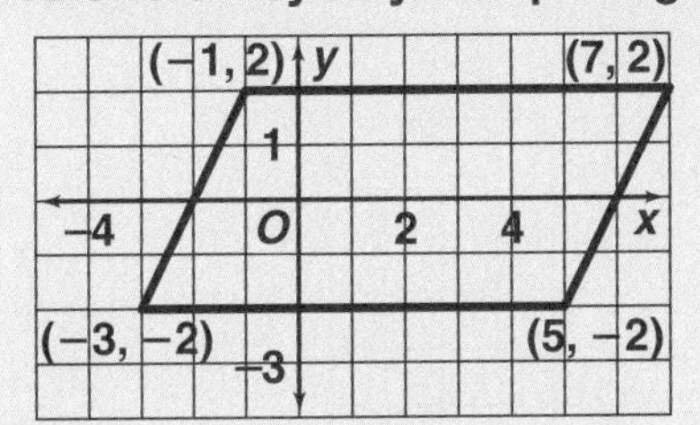

page 490 Activity Lab

2.

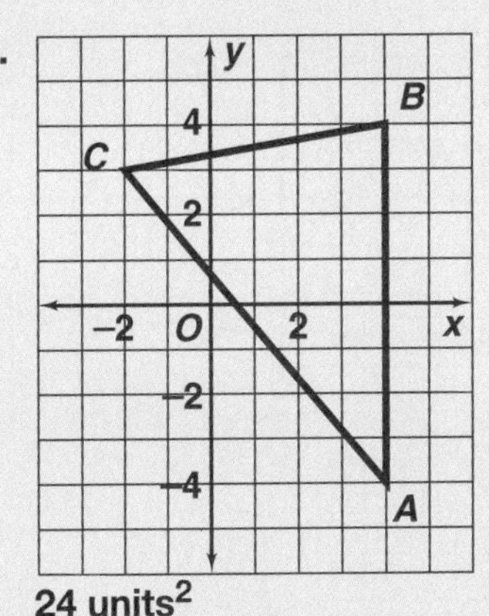

24 units^2

3.

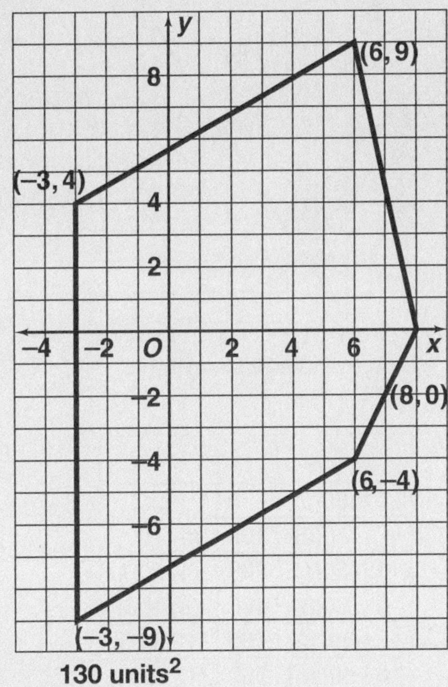

130 units^2

Lesson 10-2

page 492 Quick Check

3a.

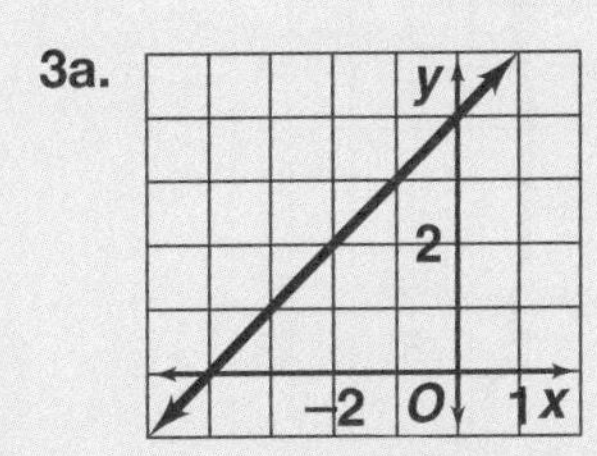

b.

c.

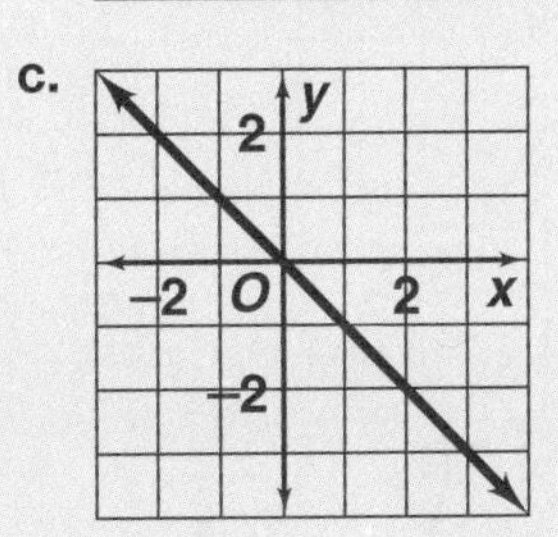

pages 493–494 Exercises

2.

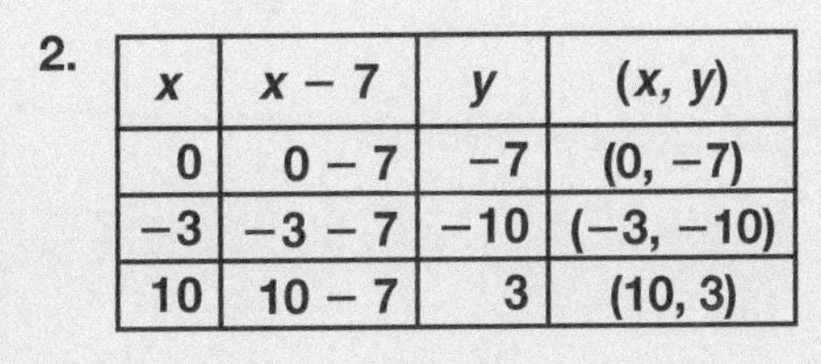

x	*x* – 7	*y*	(*x*, *y*)
0	0 – 7	–7	(0, –7)
–3	–3 – 7	–10	(–3, –10)
10	10 – 7	3	(10, 3)

3.

6–13. Answers may vary. Samples are given.

6. (0, –2), (3, 1), (–4, –6)
7. (0, 9), (2, 11), (–3, 6)
8. (0, 0), (8, 8), (–1, –1)
9. (0, 4), (2, 2), (–1, 5)
10. (0, 0), (1, 5), (–2, –10)
11. (0, 0), (1, –8), (–1, 8)
12. (0, 1), (1, 4), (–2, –5)
13. (0, –5), (1, –1), (–1, –9)

22.

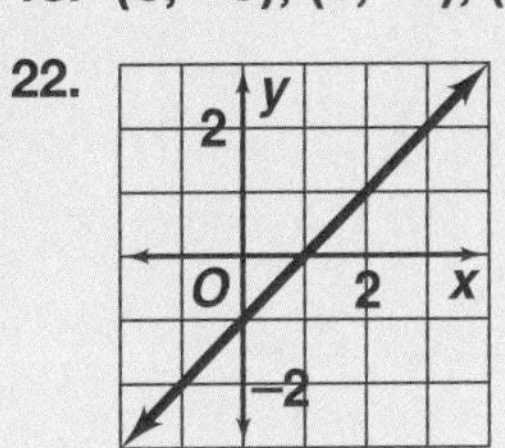

23.

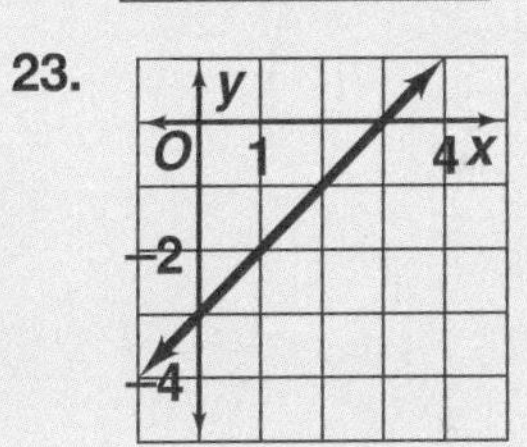

24.

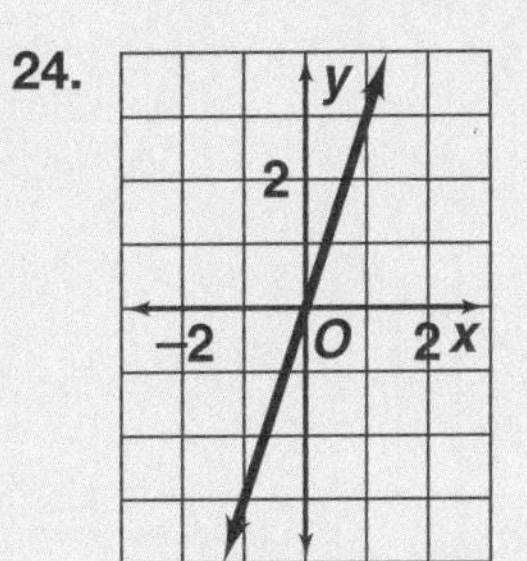

25.

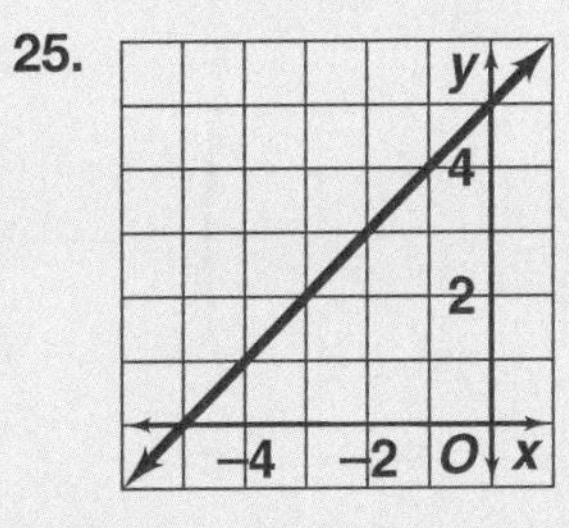

26.

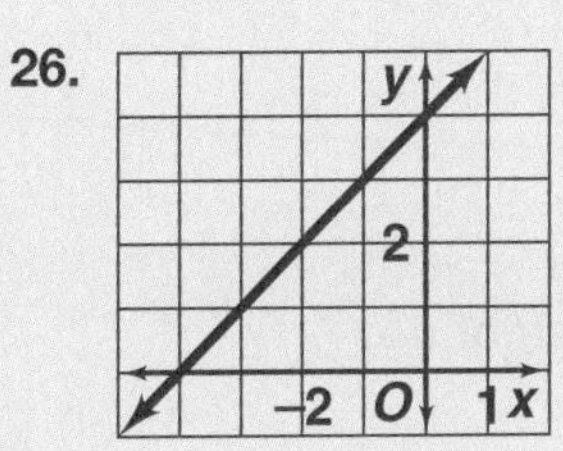

Additional Answers

27.

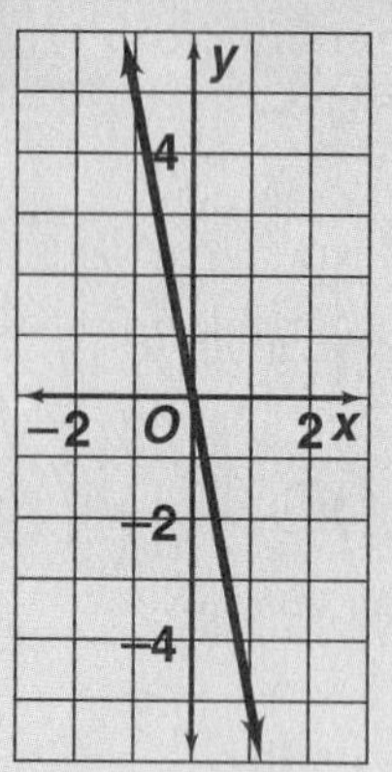

28.

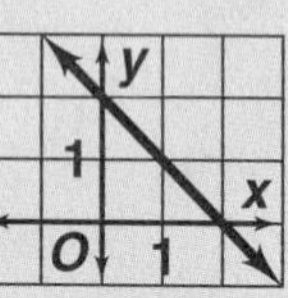

29.

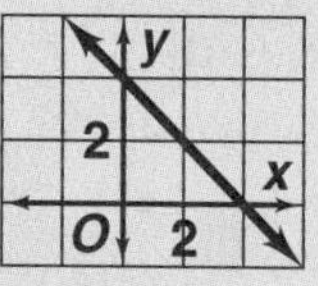

30.

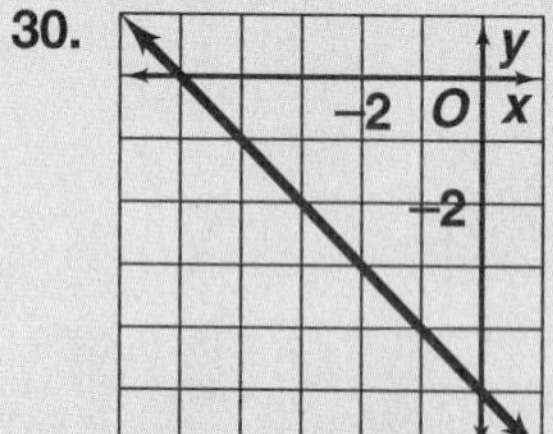

31.

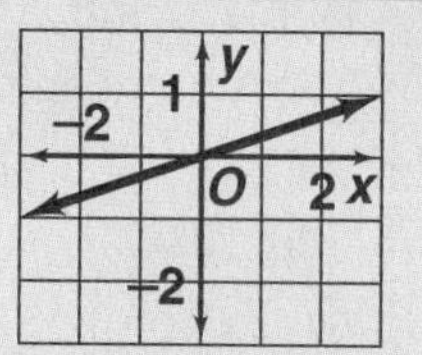

32.

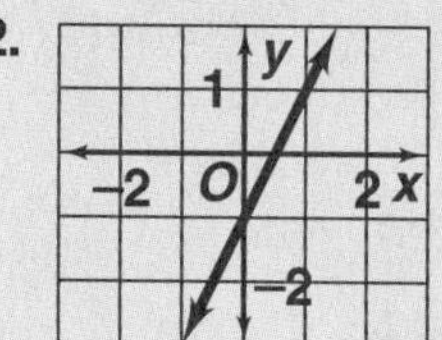

33.

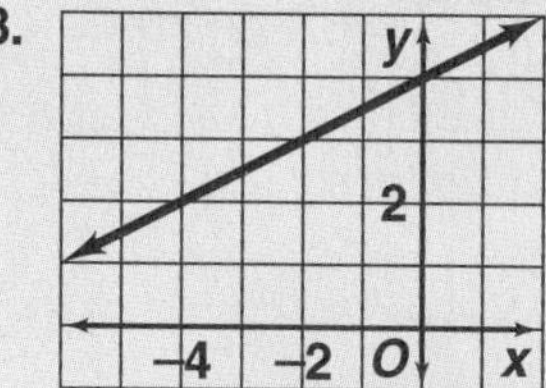

34.

Shipping Costs

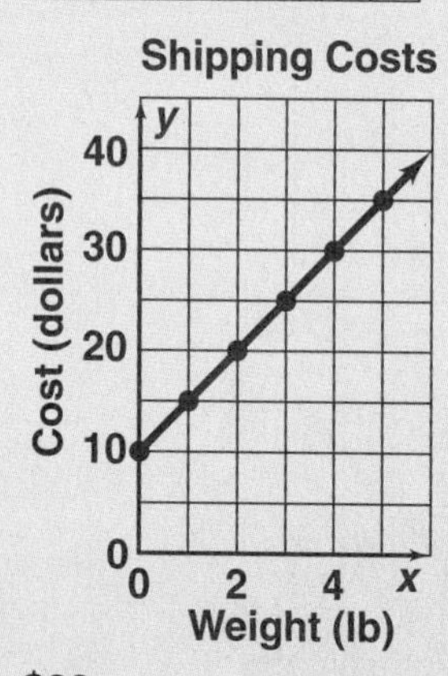

$30

35.

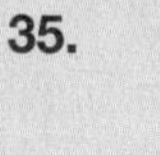

Fence Dimensions

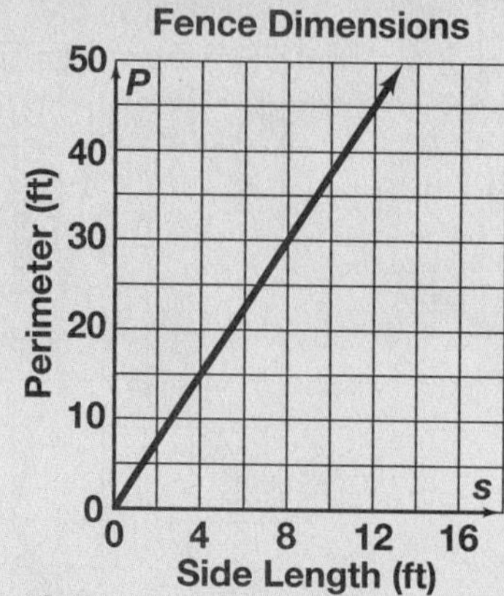

12 ft

40.

x	$50x$	y	(x, y)
0	50(0)	0	(0, 0)
2	50(2)	100	(2, 100)
4	50(4)	200	(4, 200)
8	50(8)	400	(8, 400)
12	50(12)	600	(12, 600)
16	50(16)	800	(16, 800)

yes

42.

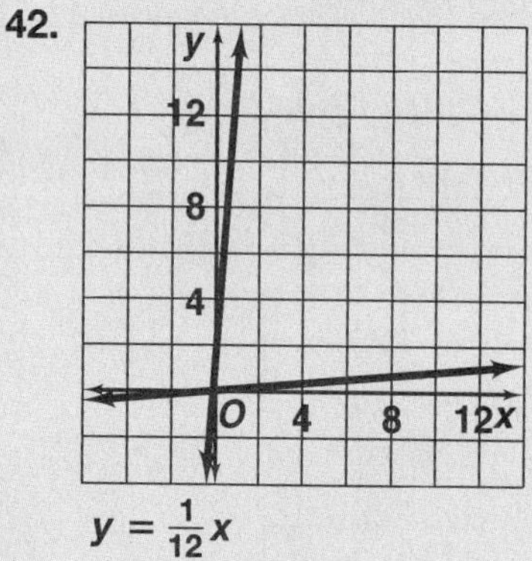

$y = \frac{1}{12}x$

Lesson 10-3

page 499 TE Additional Examples

3.

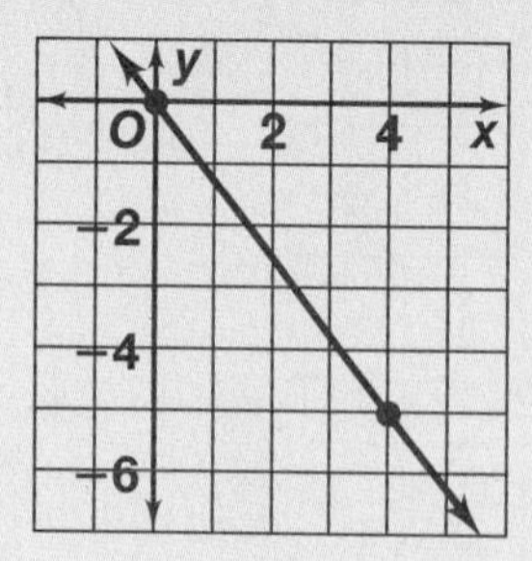

pages 500–501 Homework Exercises

13.

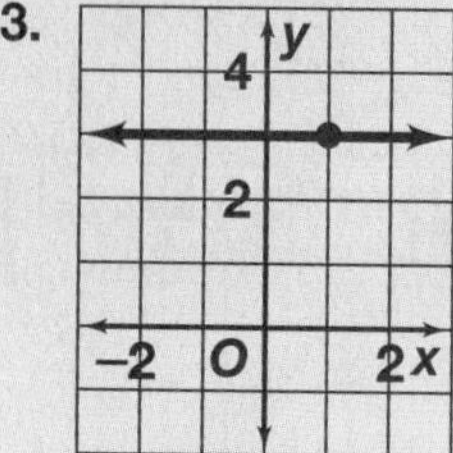

Answers may vary. Sample: Since (1, 3) and (2, 3) are both points on the line, the slope is $\frac{3-3}{2-1}$, or $\frac{0}{1}$, which is equal to 0.

17. Roof A; the slope of roof A is $\frac{5}{3}$, and the slope of roof B is $\frac{3}{5}$. Since $\frac{5}{3} > \frac{3}{5}$, roof A is steeper.

18. No; the slope of the ramp is $\frac{33}{80}$, whereas the slope in the guidelines is $\frac{1}{12}$. Since $\frac{33}{80} > \frac{1}{12}$, the ramp does not meet the guidelines.

25. No; an isoceles triangle has two equal angles, but $\triangle XYZ$ does not.

page 503 Checkpoint Quiz 1

1–3.

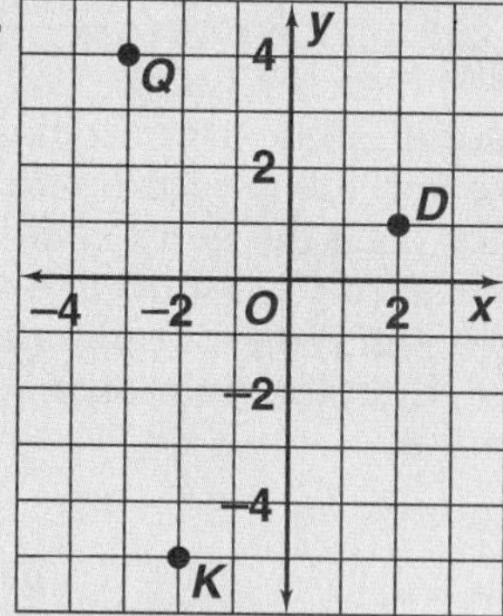

4.

5.

6.

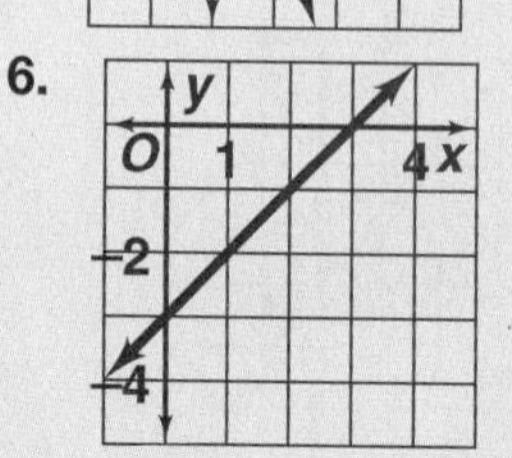

7.

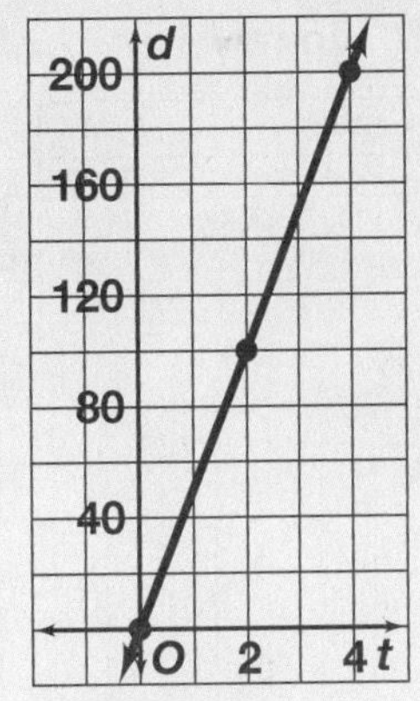

9.

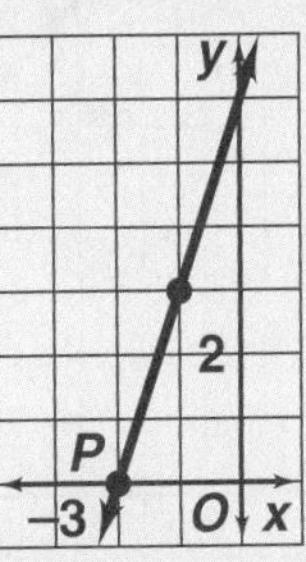

10.

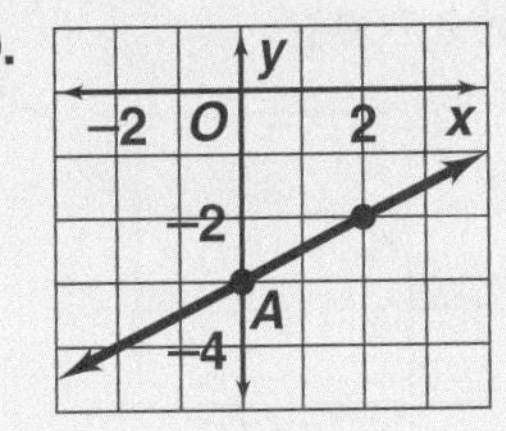

Lesson 10-4

page 504 Check Skills You'll Need

1. An equation is linear when the graph of its solutions is a line.

2–4. Answers may vary. Samples are given.

2. (0, 12), (−2, 10), (3, 15)
3. (0, −20), (5, −15), (−10, −30)
4. (0, 0), (1, 15), (−2, −30)

Quick Check

1.

x	−3	−2	−1	0	1	2	3
y	18	8	2	0	2	8	18

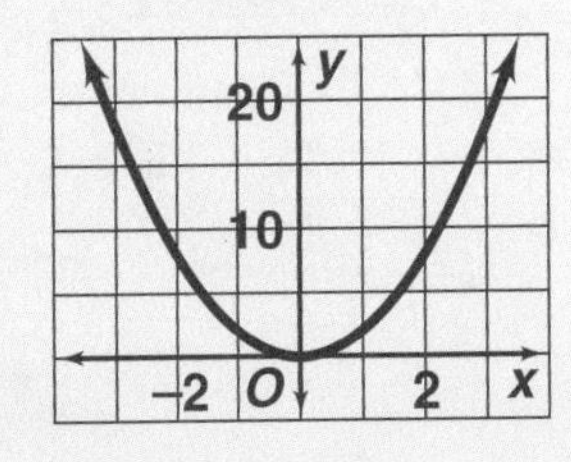

pages 506–507 Exercises

2.

x	−3	−2	−1	0	1	2	3
y	9	4	1	0	1	4	9

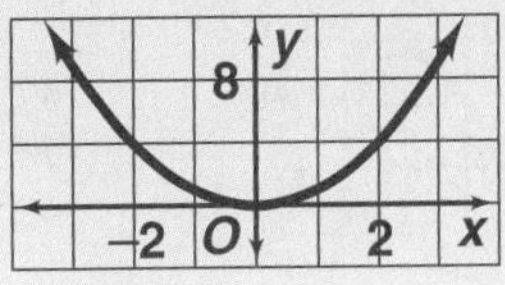

3.

x	−3	−2	−1	0	1	2	3
y	3	2	1	0	1	2	3

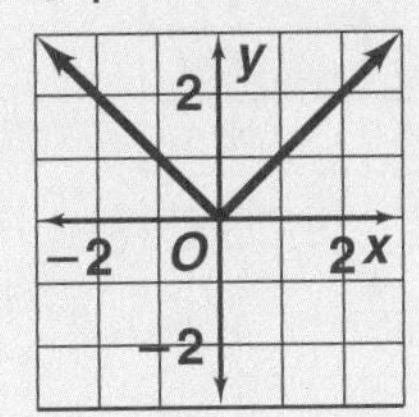

4.

x	−3	−2	−1	0	1	2	3
y	27	8	1	0	−1	−8	−27

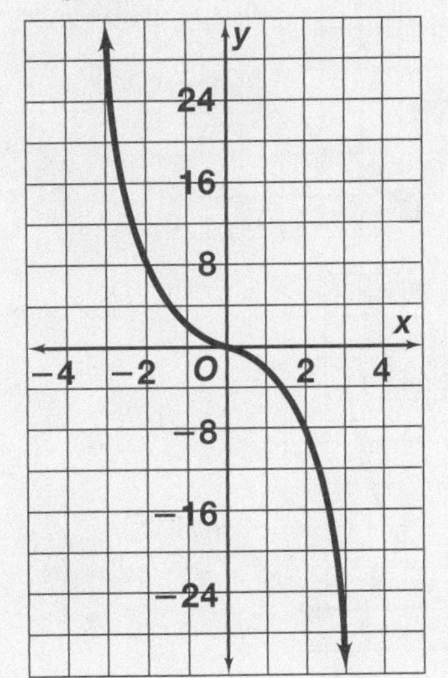

8.

x	−3	−2	−1	0	1	2	3
y	36	16	4	0	4	16	36

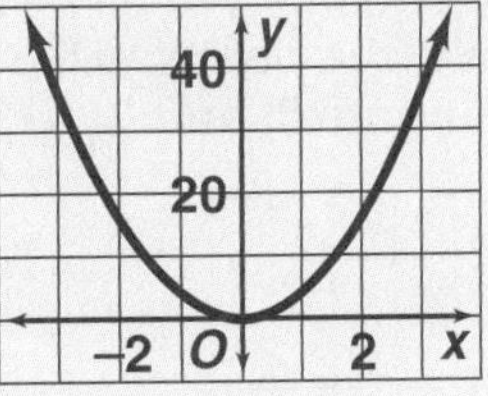

9.

x	−3	−2	−1	0	1	2	3
y	−27	−12	−3	0	−3	−12	−27

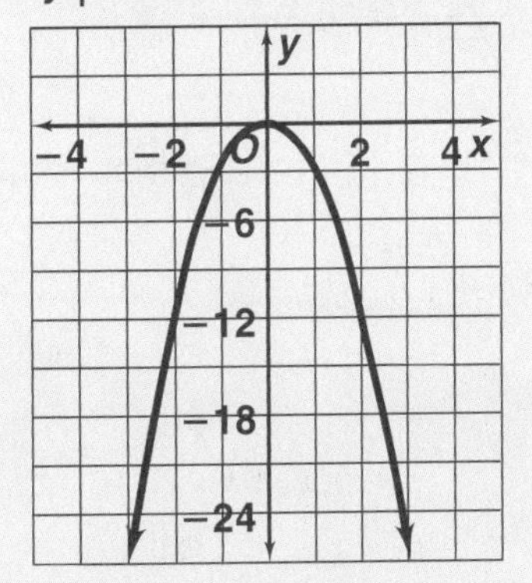

10.

x	−3	−2	−1	0	1	2	3
y	$4\frac{1}{2}$	2	$\frac{1}{2}$	0	$\frac{1}{2}$	2	$4\frac{1}{2}$

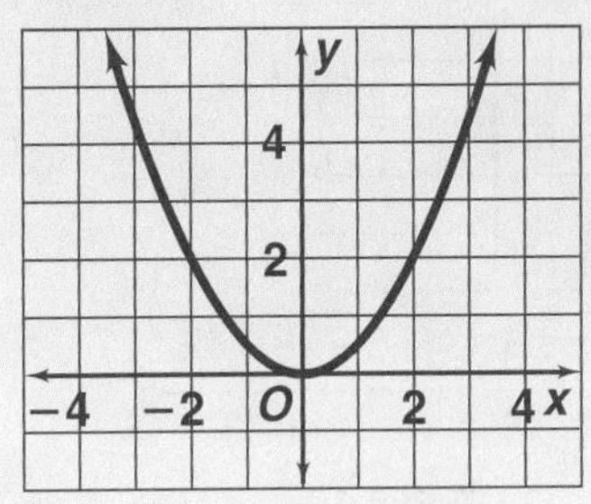

11.

x	−3	−2	−1	0	1	2	3
y	7	2	−1	−2	−1	2	7

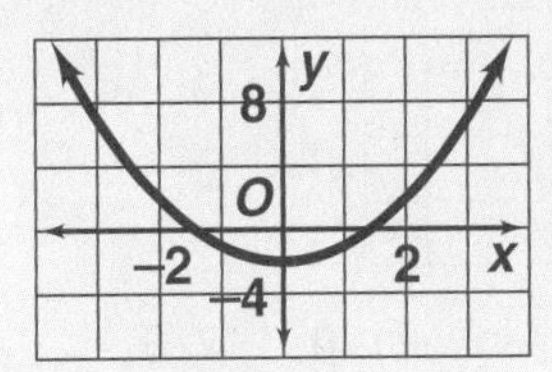

12.

x	−3	−2	−1	0	1	2	3
y	11	6	3	2	3	6	11

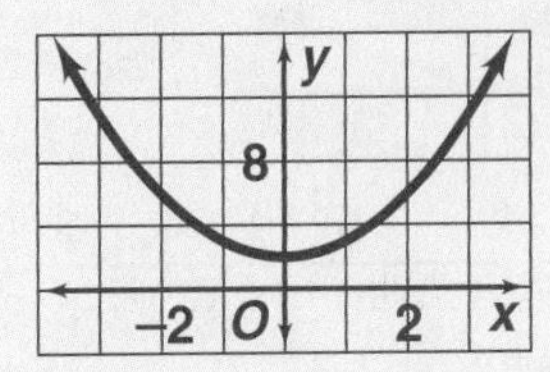

13.

x	−3	−2	−1	0	1	2	3
y	−6	−1	2	3	2	−1	−6

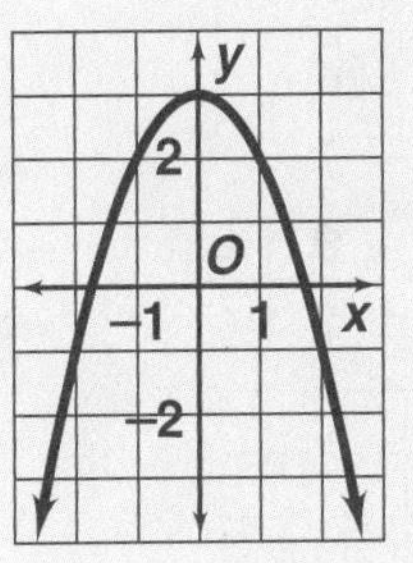

14.

x	−3	−2	−1	0	1	2	3
y	−13	−8	−5	−4	−5	−8	−13

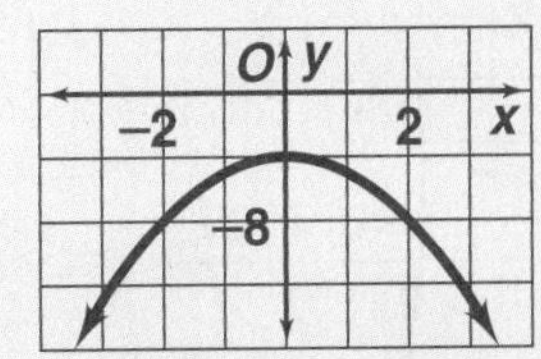

15.

x	−3	−2	−1	0	1	2	3
y	4	1	0	1	4	9	16

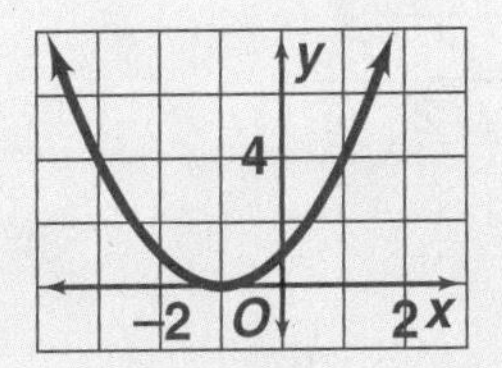

16.

x	−3	−2	−1	0	1	2	3
y	16	9	4	1	0	1	4

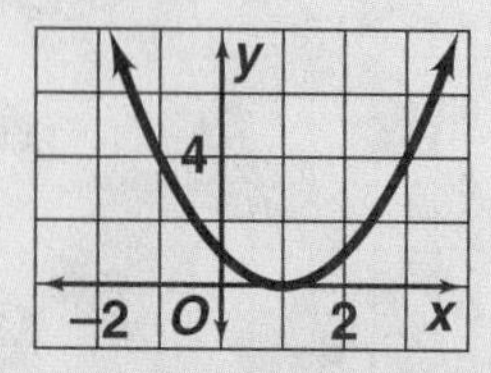

17.

x	−3	−2	−1	0	1	2	3
y	6	4	2	0	2	4	6

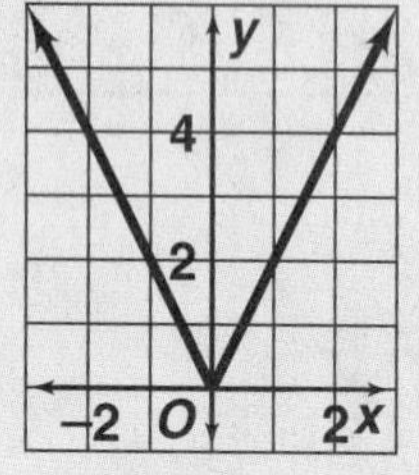

18.

x	−3	−2	−1	0	1	2	3
y	1	$\frac{2}{3}$	$\frac{1}{3}$	0	$\frac{1}{3}$	$\frac{2}{3}$	1

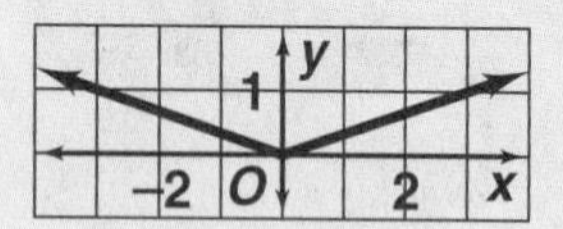

19.

x	−3	−2	−1	0	1	2	3
y	−3	−2	−1	0	−1	−2	−3

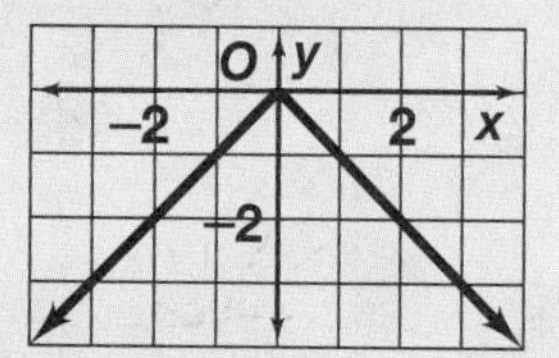

20.

x	−3	−2	−1	0	1	2	3
y	−1	0	1	2	1	0	−1

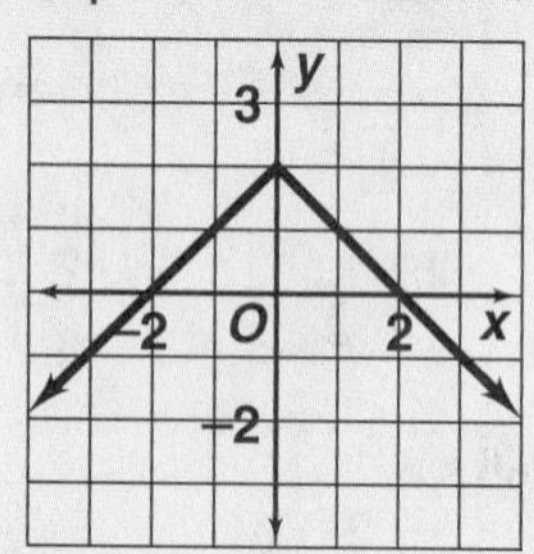

21.

x	−3	−2	−1	0	1	2	3
y	4	3	2	1	0	1	2

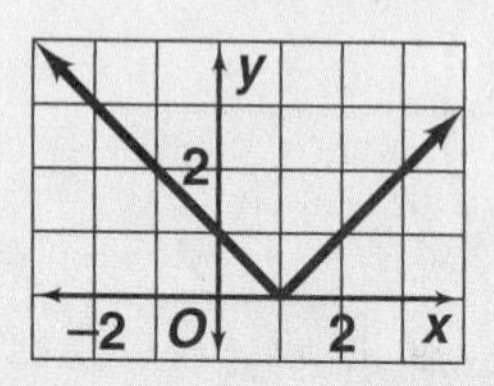

22.

x	−3	−2	−1	0	1	2	3
y	2	1	0	1	2	3	4

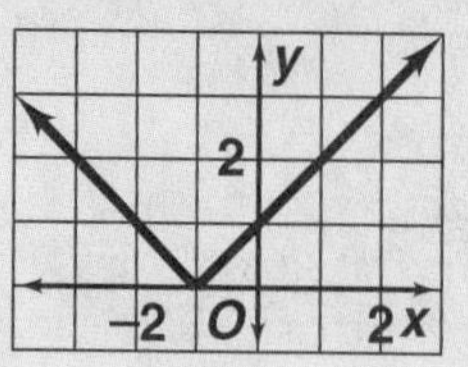

23.

t	$16t^2$	d	(t, d)
0	$16(0)^2$	0	(0, 0)
1	$16(1)^2$	16	(1, 16)
3	$16(3)^2$	144	(3, 144)
5	$16(5)^2$	400	(5, 400)

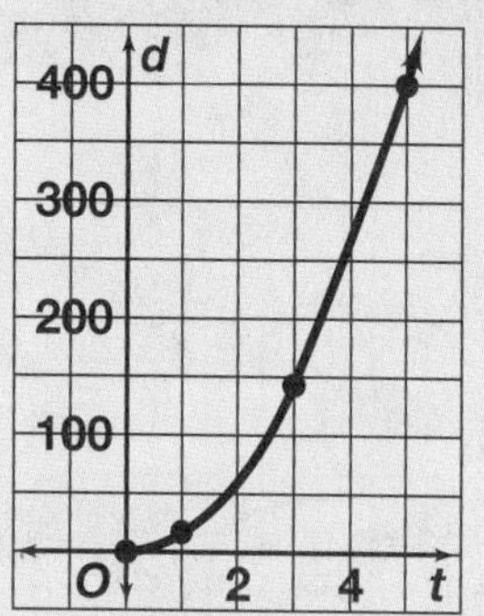

24.

25. Absolute value is always nonnegative, so the opposite of absolute value will always be nonpositive. Since y is always nonpositive, the points will always be in Quadrants III and IV.

26.

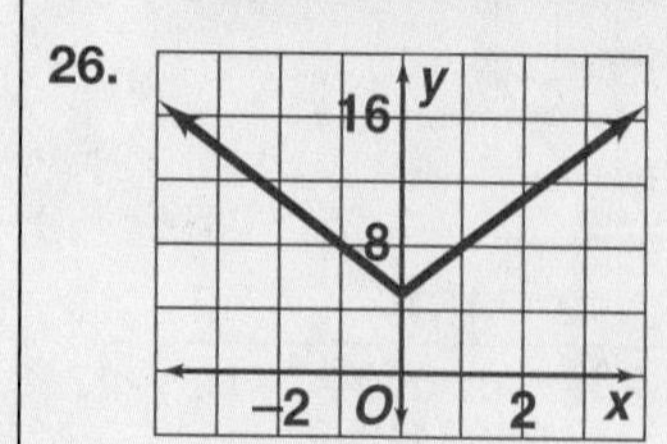

30a.

t	$-16t^2 + 12{,}000$	h	(t, h)
0	$-16(0)^2 + 12{,}000 = 0 + 12{,}000$	12,000	(0, 12,000)
5	$-16(5)^2 + 12{,}000 = -400 + 12{,}000$	11,600	(5, 11,600)
10	$-16(10)^2 + 12{,}000 = -1{,}600 + 12{,}000$	10,400	(10, 10,400)
20	$-16(20)^2 + 12{,}000 = -6{,}400 + 12{,}000$	5,600	(20, 5,600)

b.

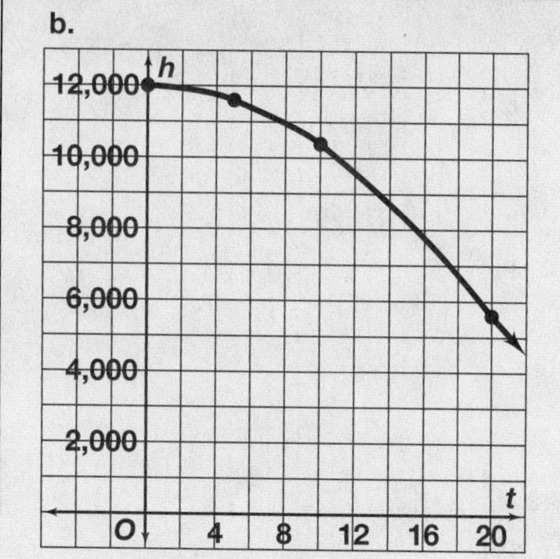

about 9,700 ft

32.

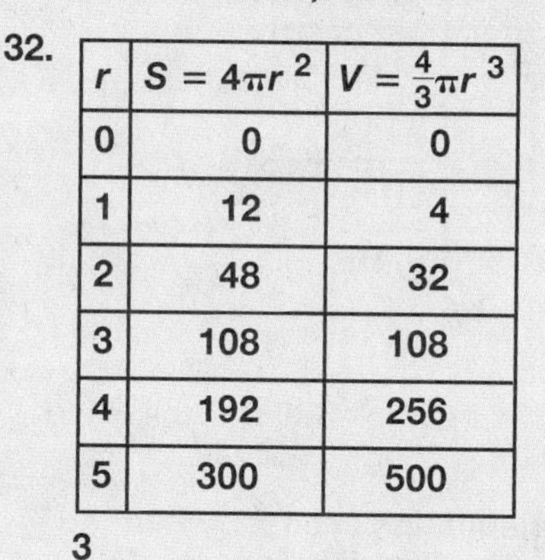

r	$S = 4\pi r^2$	$V = \frac{4}{3}\pi r^3$
0	0	0
1	12	4
2	48	32
3	108	108
4	192	256
5	300	500

3

Lesson 10-5

page 510 Check Skills You'll Need

1. The x-coordinate is 2, and the y-coordinate is −5.

2–5.

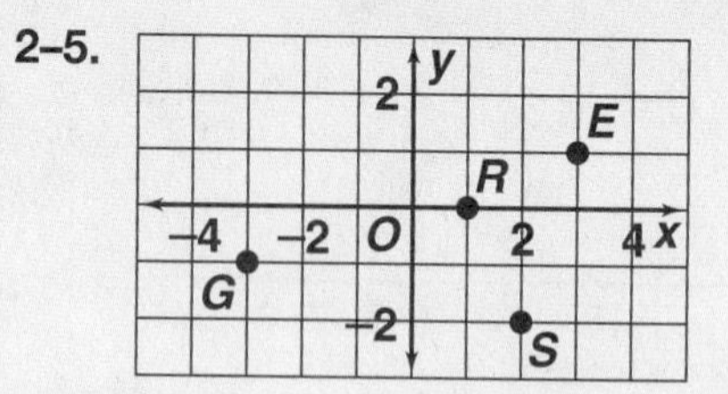

page 511 Quick Check

2\. $A(-4, 3)$, $B(-1, 4)$, $C(-3, 1) \rightarrow A'(-7, 4)$, $B'(-4, 5)$, $C'(-6, 2)$

TE Additional Examples

2\.

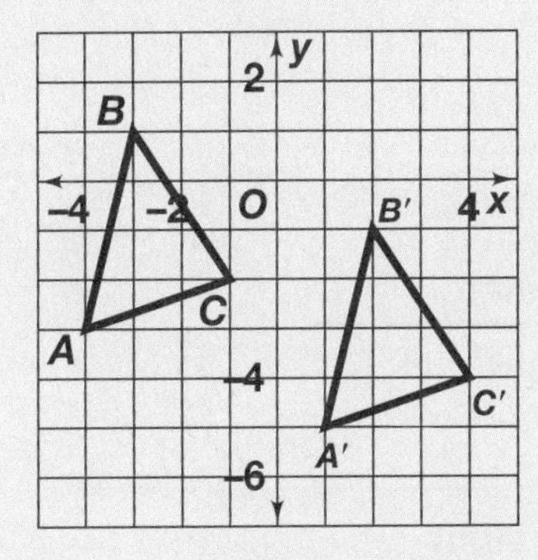

$A(-4, -3)$, $B(-3, 1)$, $C(-1, -2) \rightarrow A'(1, -5)$, $B'(2, -1)$, $C'(4, -4)$

pages 512–513 Homework Exercises

10\.

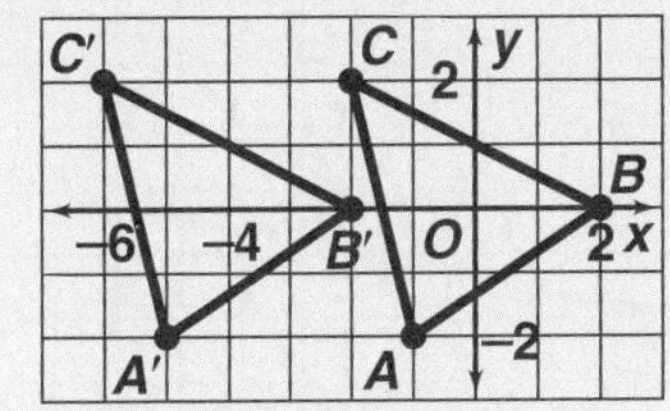

$A(-1, -2)$, $B(2, 0)$, $C(-2, 2) \rightarrow A'(-5, -2)$, $B'(-2, 0)$, $C'(-6, 2)$

11\.

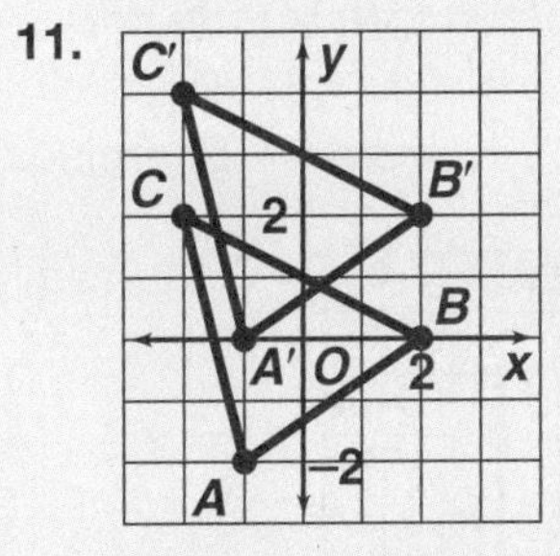

$A(-1, -2)$, $B(2, 0)$, $C(-2, 2) \rightarrow A'(-1, 0)$, $B'(2, 2)$, $C'(-2, 4)$

12\.

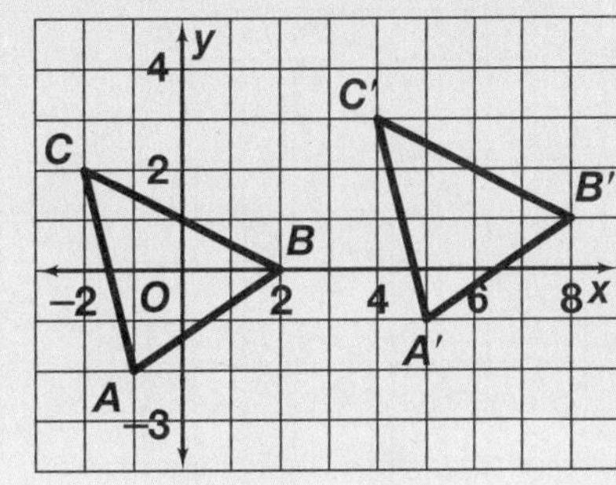

$A(-1, -2)$, $B(2, 0)$, $C(-2, 2) \rightarrow A'(5, -1)$, $B'(8, 1)$, $C'(4, 3)$

13\.

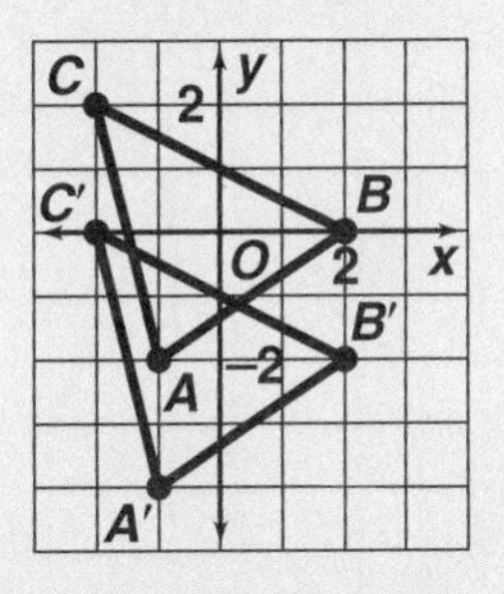

$A(-1, -2)$, $B(2, 0)$, $C(-2, 2) \rightarrow A'(-1, -4)$, $B'(2, -2)$, $C'(-2, 0)$

14\.

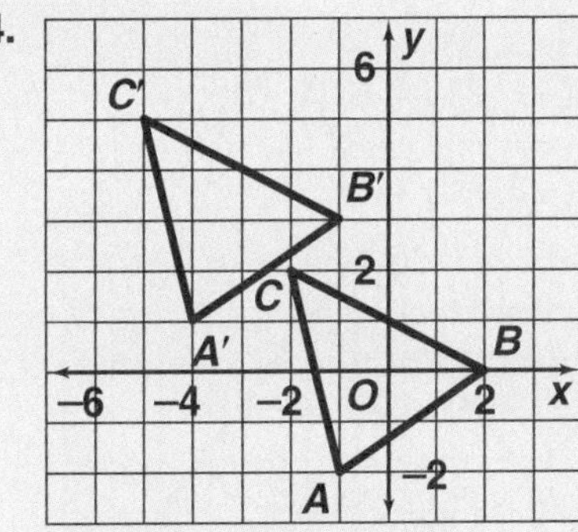

$A(-1, -2)$, $B(2, 0)$, $C(-2, 2) \rightarrow A'(-4, 1)$, $B'(-1, 3)$, $C'(-5, 5)$

18\.

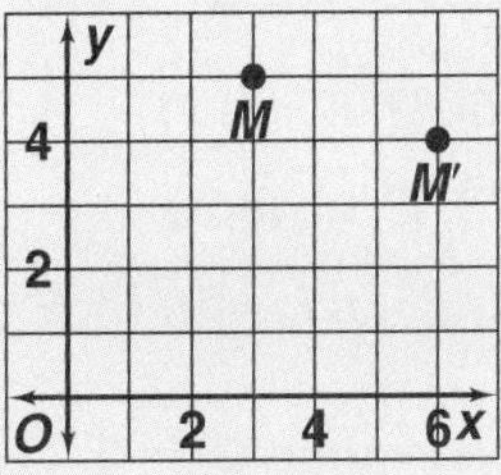

right 3 units, down 1 unit

19\.

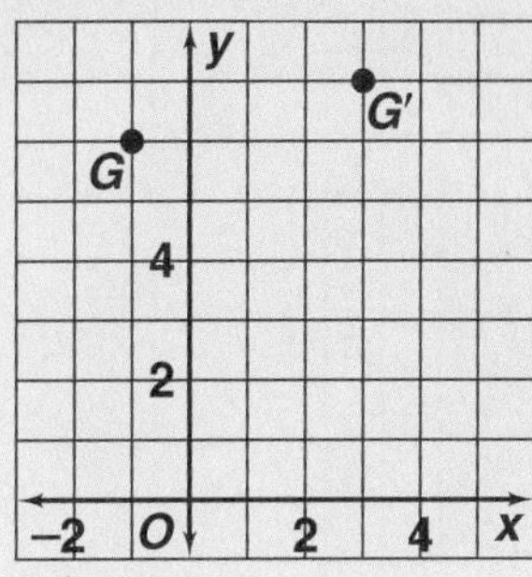

right 4 units, up 1 unit

20\.

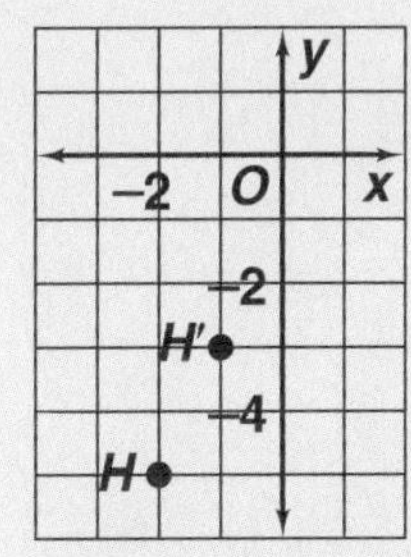

right 1 unit, up 2 units

21\.

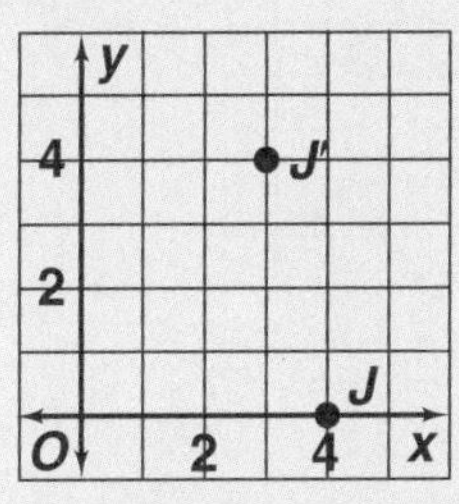

left 1 unit, up 4 units

22\. $P'(2, 4)$, $L'(-1, 4)$, $N'(2, 1)$; $(x, y) \rightarrow (x + 4, y + 3)$

26\. Answers may vary. Sample: Stating the horizontal change first fits with the format of a coordinate pair (x, y).

Lesson 10-6

pages 514–515 Quick Check

1\.

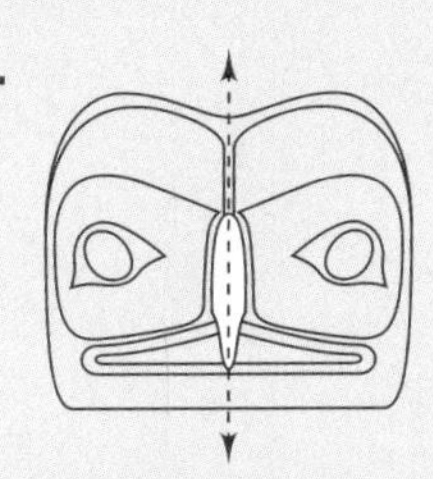

yes

2a–b.

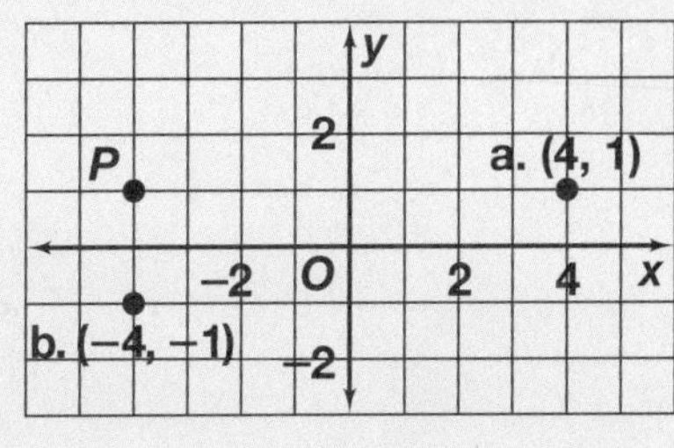

3.

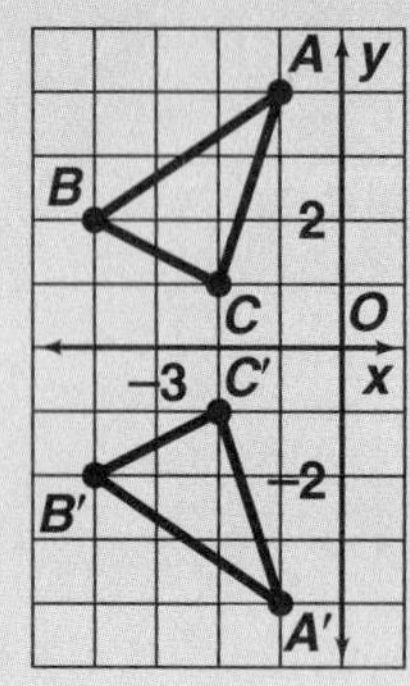

$A(-1, 4)$, $B(-4, 2)$, $C(-2, 1) \rightarrow A'(-1, -4)$, $B'(-4, -2)$, $C'(-2, -1)$

TE Additional Examples

2-3.

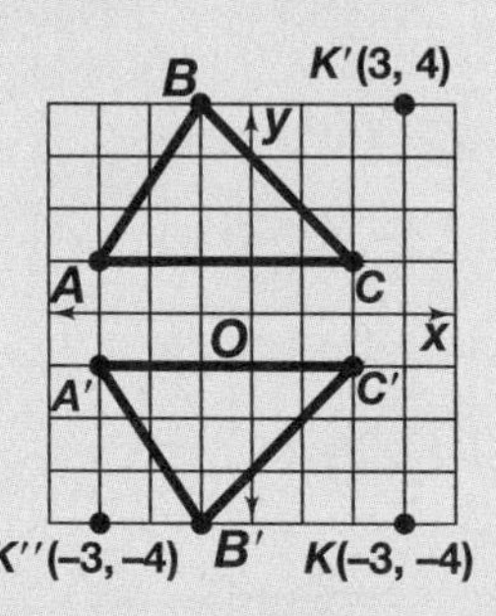

pages 516–517 Homework Exercises

6.

7. none

8.

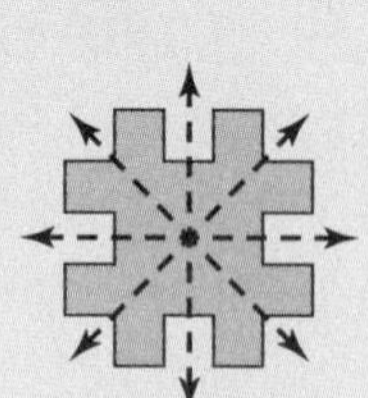

9.

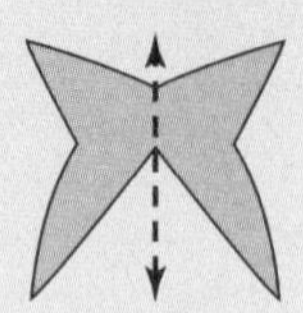

16.

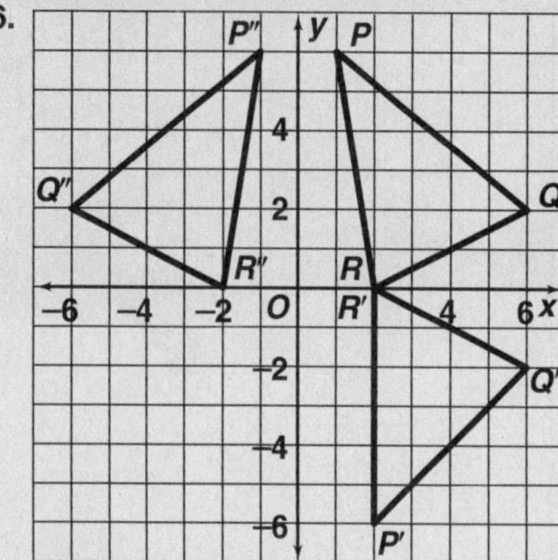

17.

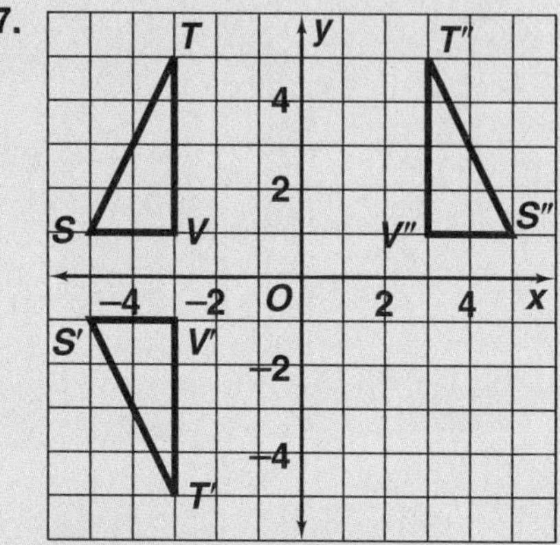

18. $A(-3, 3) \rightarrow A'(3, 3)$, $B(-1, 3) \rightarrow B'(1, 3)$, $C(-1, -1) \rightarrow C'(1, -1)$, $D(-3, -1) \rightarrow D'(3, -1)$

29.

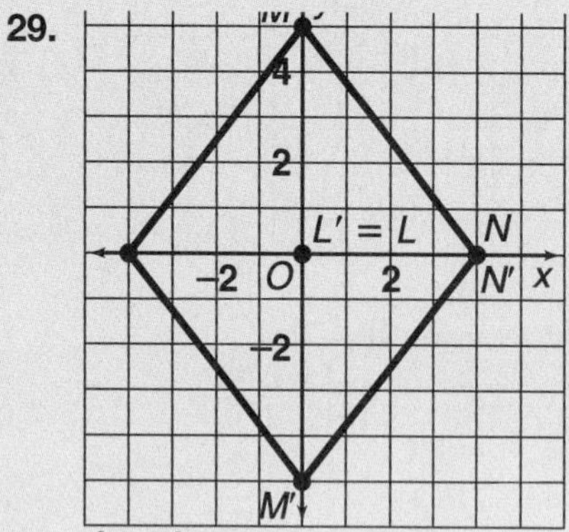

rhombus

33. (0, 2), (1, 5), (2, 8), (3, 11)

34. (0, 4), (1, 3), (2, 2), (3, 1)

35. (0, −1), (1, −3), (2, −5), (3, −7)

page 518 Checkpoint Quiz 2

10. 3 and 8

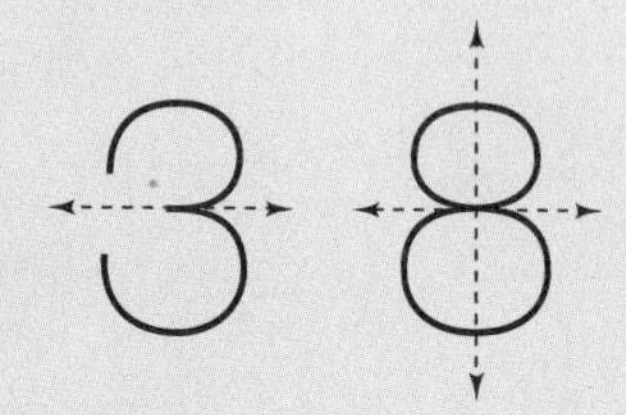

Lesson 10-7

page 519 Check Skills You'll Need

2.

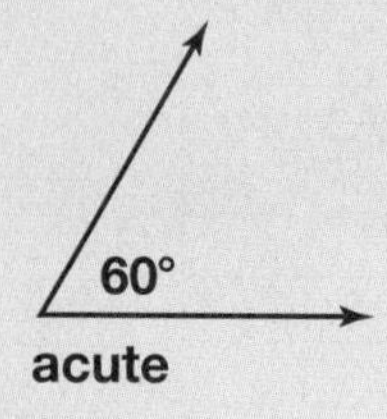

acute

3.

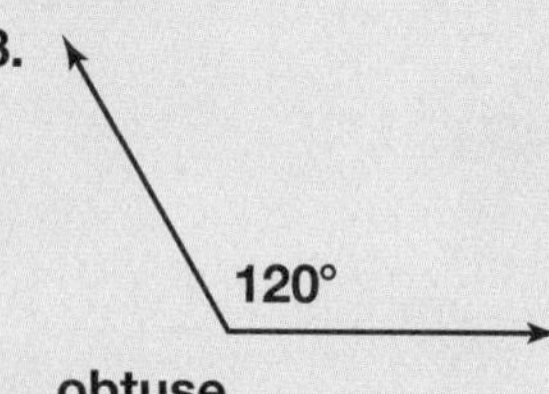

obtuse

4.

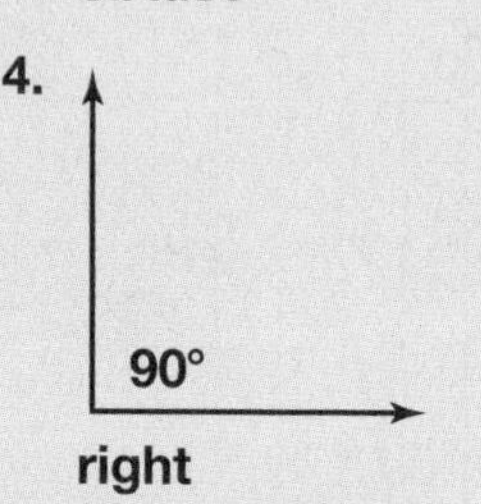

right

5. 180°

straight

page 520 Quick Check

3.

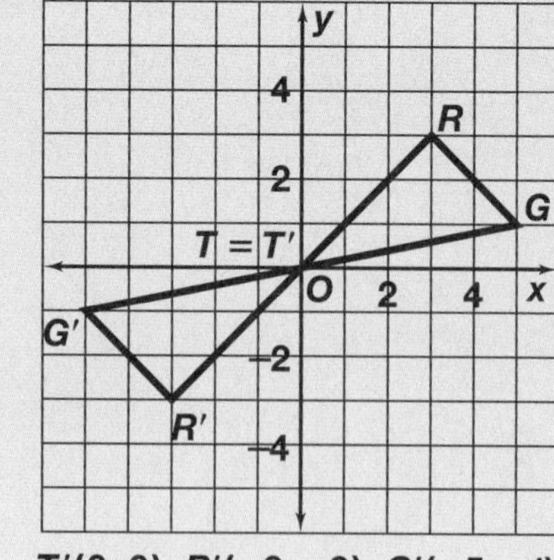

$T'(0, 0)$, $R'(-3, -3)$, $G'(-5, -1)$

pages 521–522 Homework Exercises

13.

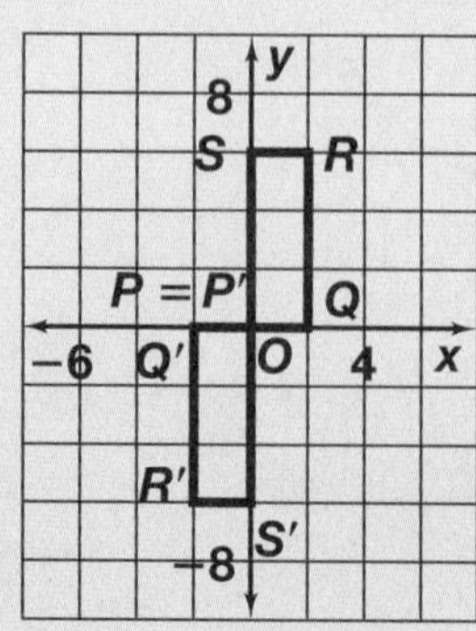

$P'(0, 0)$, $Q'(-2, 0)$, $R'(-2, -6)$, $S'(0, -6)$

14.

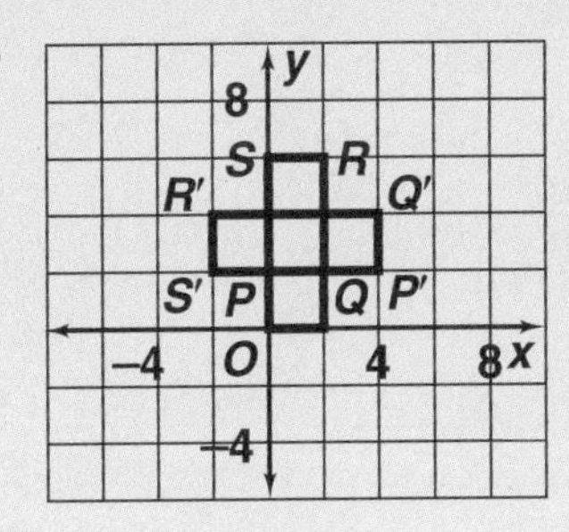

$P'(4, 2)$, $Q'(4, 4)$, $R'(-2, 4)$, $S'(-2, 2)$

15.

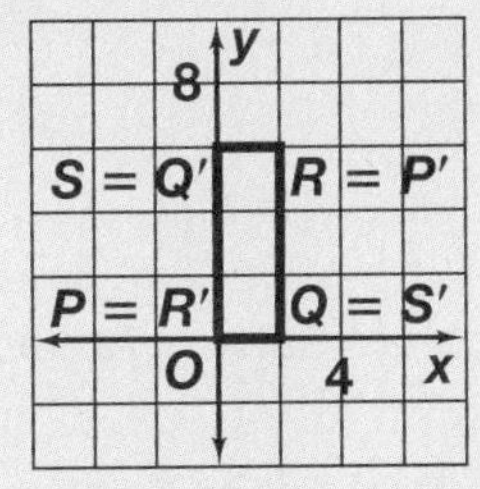

$P'(2, 6)$, $Q'(0, 6)$, $R'(0, 0)$, $S'(2, 0)$

16.

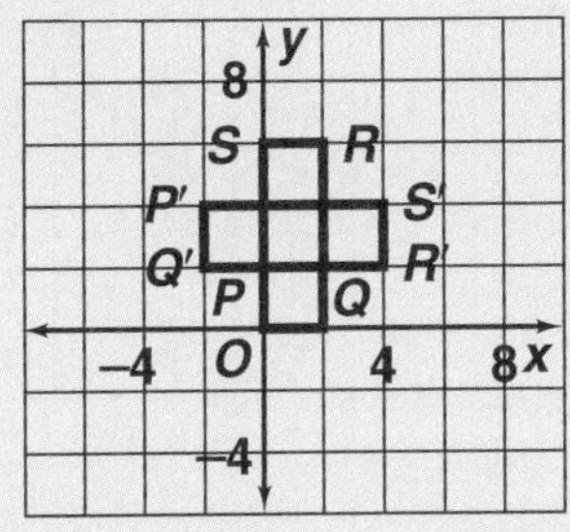

$P'(-2, 4)$, $Q'(-2, 2)$, $R'(4, 2)$, $S'(4, 4)$

27.

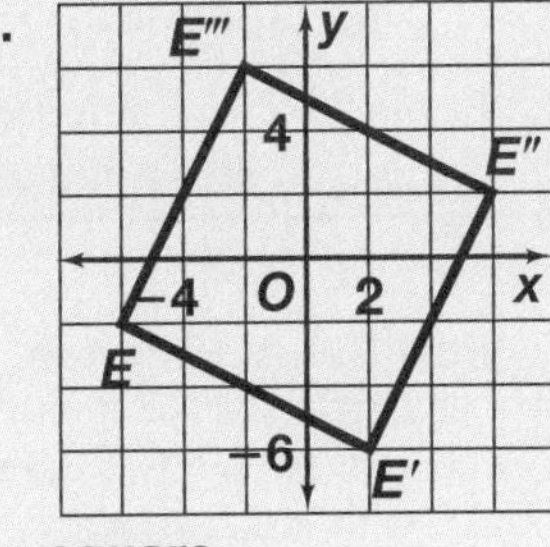

square

pages 524–525 Chapter Review

6–8.

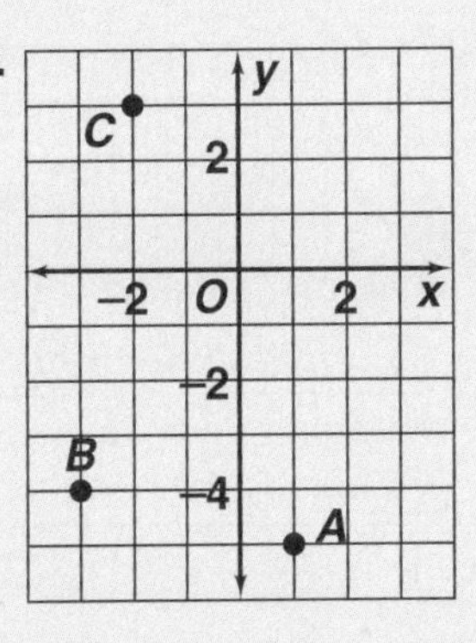

13.

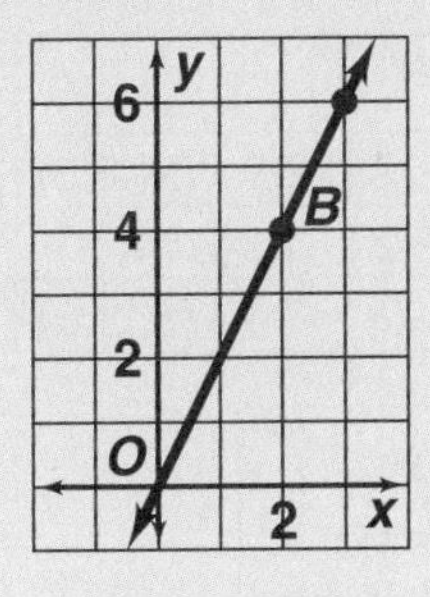

14.

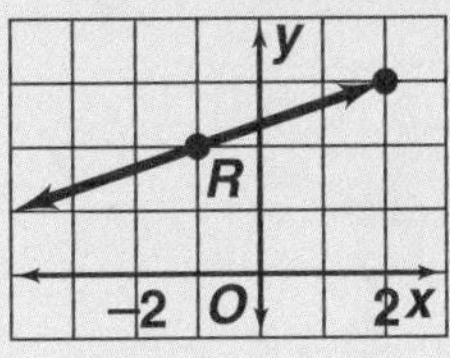

15.

x	–3	–2	–1	0	1	2	3
y	8	3	0	–1	0	3	8

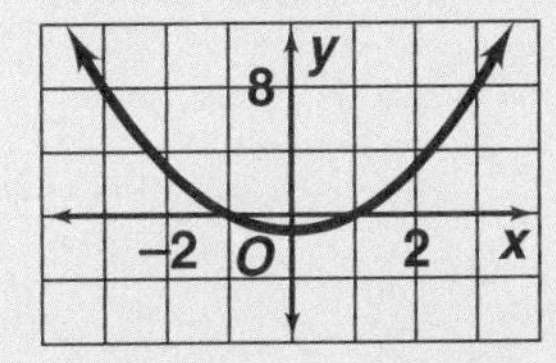

16.

x	–3	–2	–1	0	1	2	3
y	4	3	2	1	2	3	4

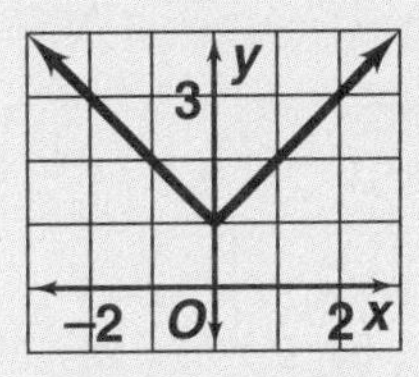

17.

x	–3	–2	–1	0	1	2	3
y	16	6	0	–2	0	6	16

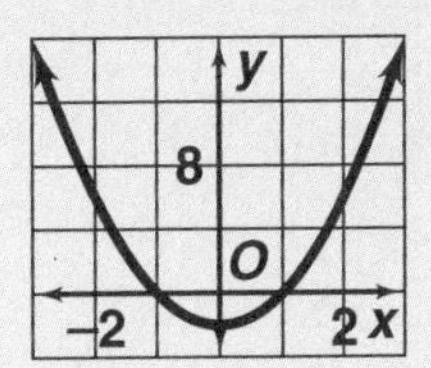

18.

x	–3	–2	–1	0	1	2	3
y	12	8	4	0	4	8	12

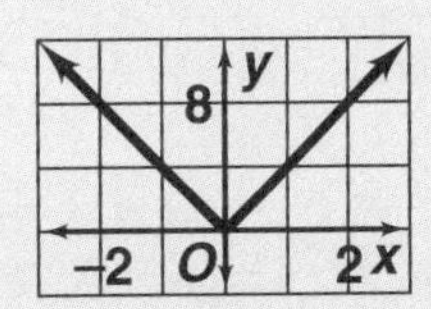

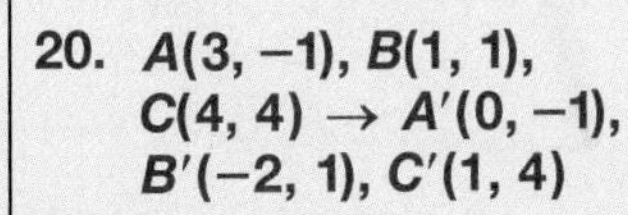

19. $A(3, -1)$, $B(1, 1)$, $C(4, 4) \rightarrow A'(3, -3)$, $B'(1, -1)$, $C'(4, 2)$

20. $A(3, -1)$, $B(1, 1)$, $C(4, 4) \rightarrow A'(0, -1)$, $B'(-2, 1)$, $C'(1, 4)$

21. $A(3, -1)$, $B(1, 1)$, $C(4, 4) \rightarrow A'(6, -6)$, $B'(4, -4)$, $C'(7, -1)$

22.

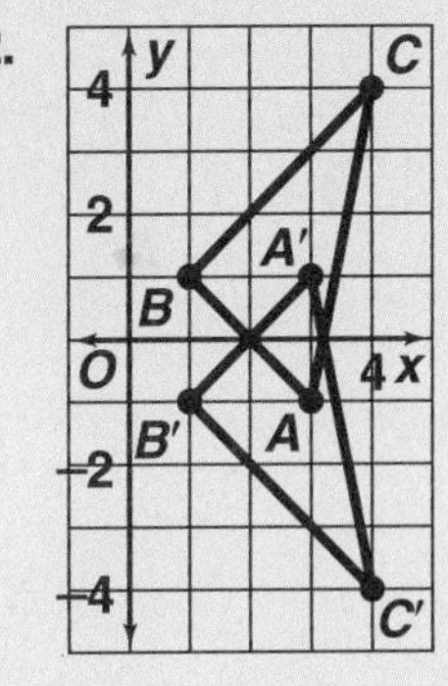

$A(3, -1)$, $B(1, 1)$, $C(4, 4) \rightarrow A'(3, 1)$, $B'(1, -1)$, $C'(4, -4)$

page 526 Chapter Test

1–4.

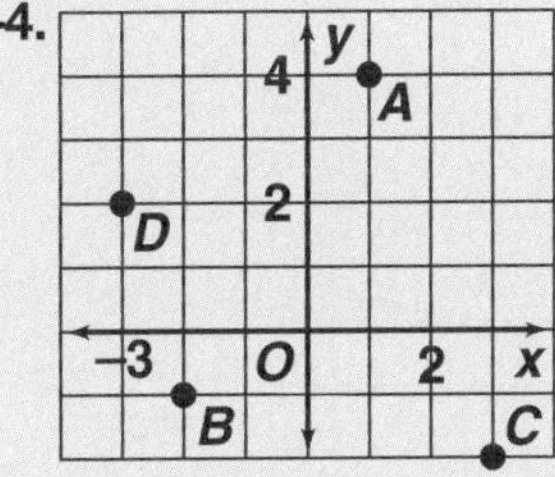

5–6. Answers may vary. Samples are given.

5.

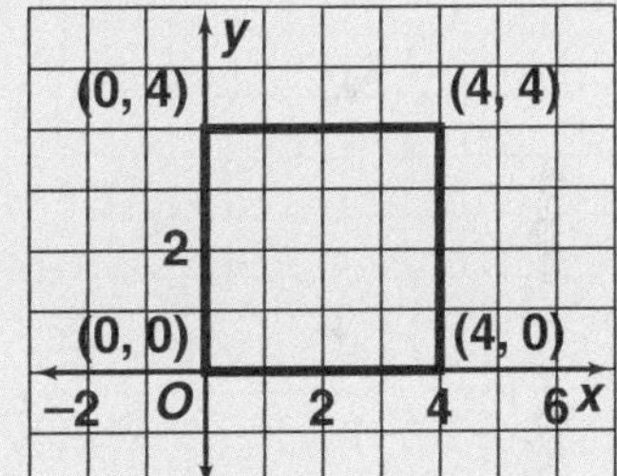

6.

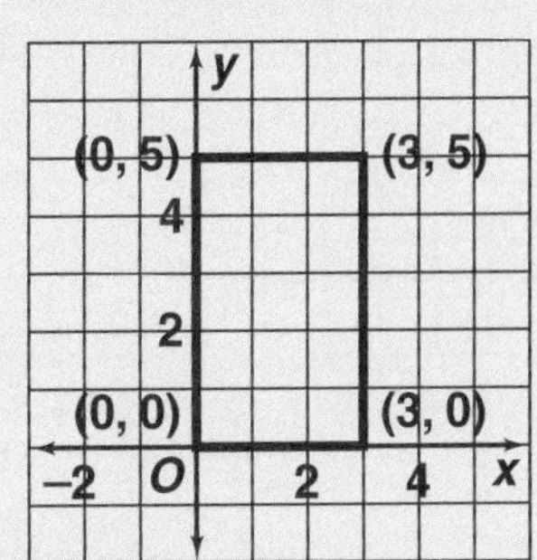

11.

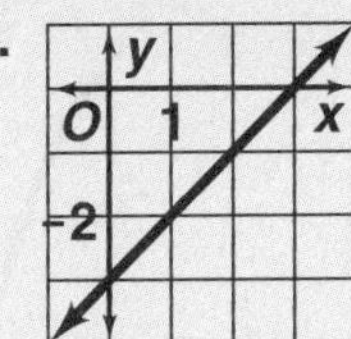

Additional Answers

12.

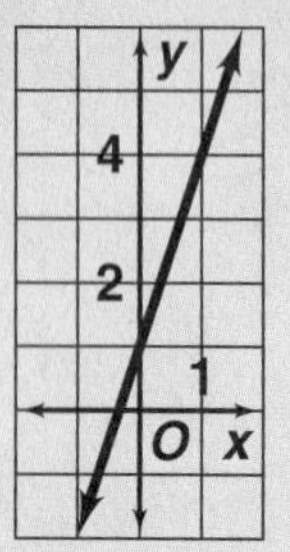

13.

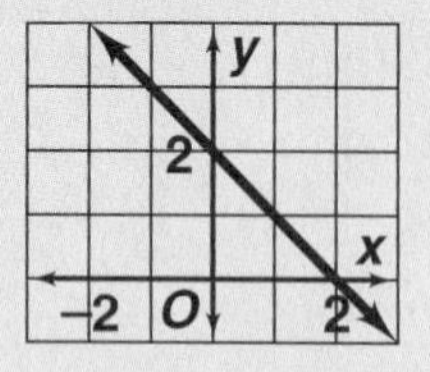

14.

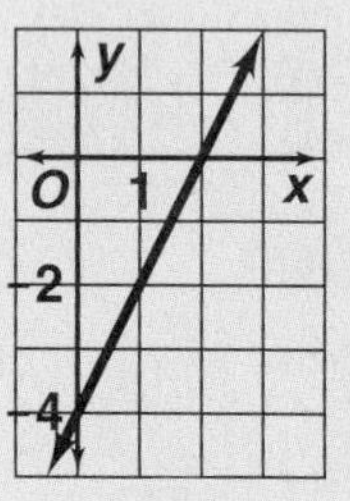

19.

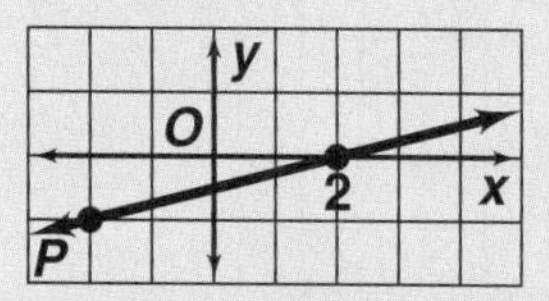

20.

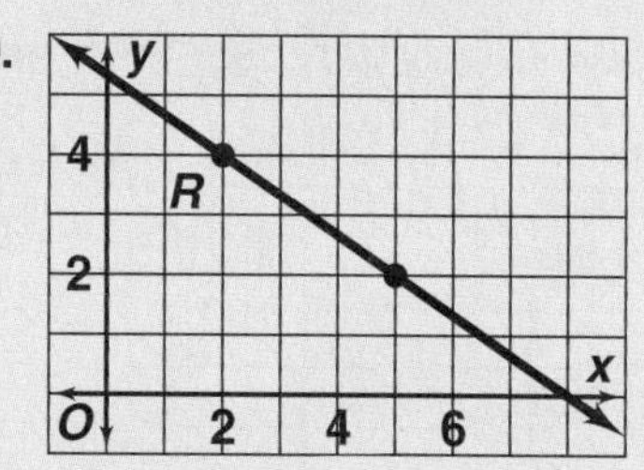

21.

x	−3	−2	−1	0	1	2	3
y	7	2	−1	−2	−1	2	7

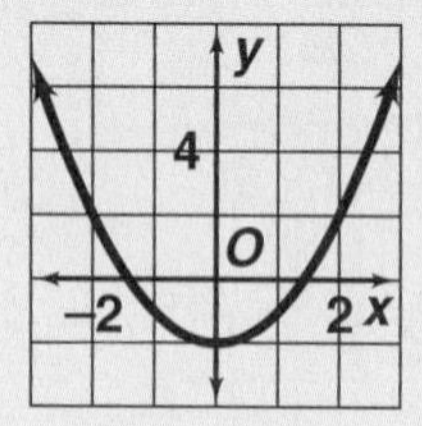

22.

x	−3	−2	−1	0	1	2	3
y	7	5	3	1	3	5	7

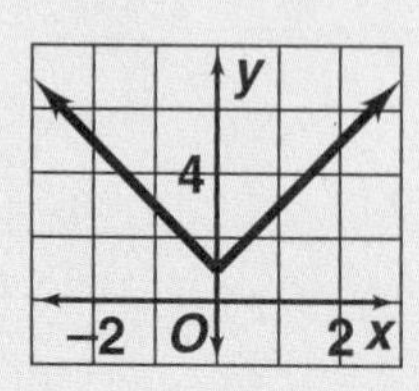

23.

x	$0.25x + 2$	y	(x, y)
2	0.25 (2) + 2	2.5	(2, 2.5)
4	0.25 (4) + 2	3	(4, 3)
6	0.25 (6) + 2	3.5	(6, 3.5)

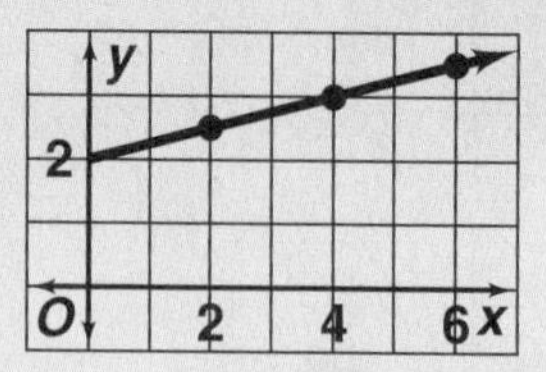

24.

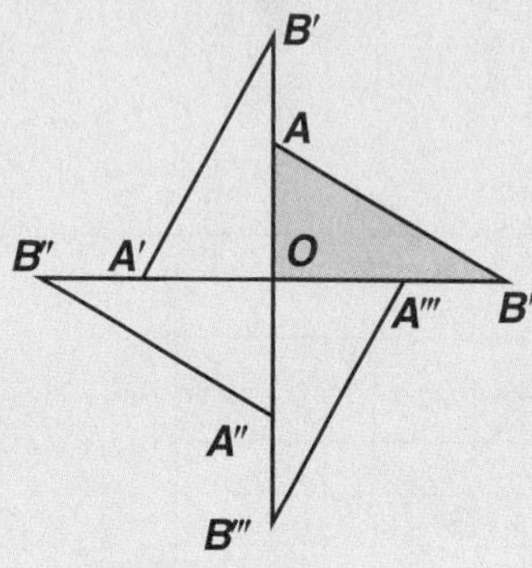

25.

$A(1, 3)$, $B(5, 8)$, $C(7, 1) \rightarrow$
$A'(2, 0)$, $B'(6, 5)$, $C'(8, -2)$

26.

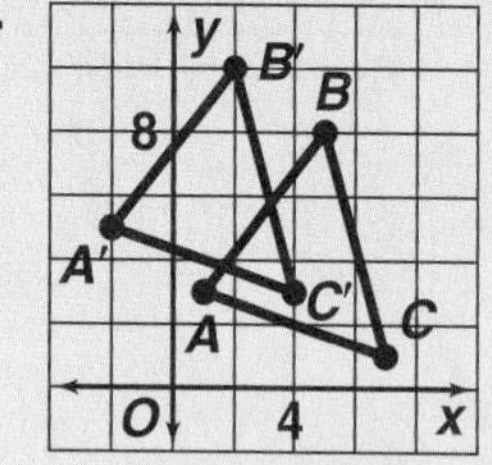

$A(1, 3)$, $B(5, 8)$, $C(7, 1) \rightarrow$
$A'(-2, 5)$, $B'(2, 10)$,
$C'(4, 3)$

27.

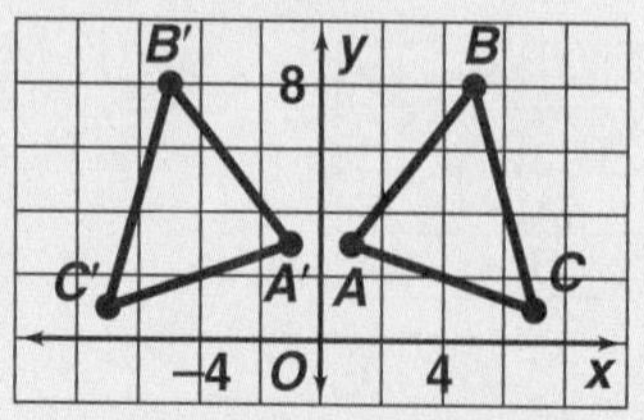

$A(1, 3)$, $B(5, 8)$, $C(7, 1) \rightarrow$
$A'(-1, 3)$, $B'(-5, 8)$,
$C'(-7, 1)$

30.

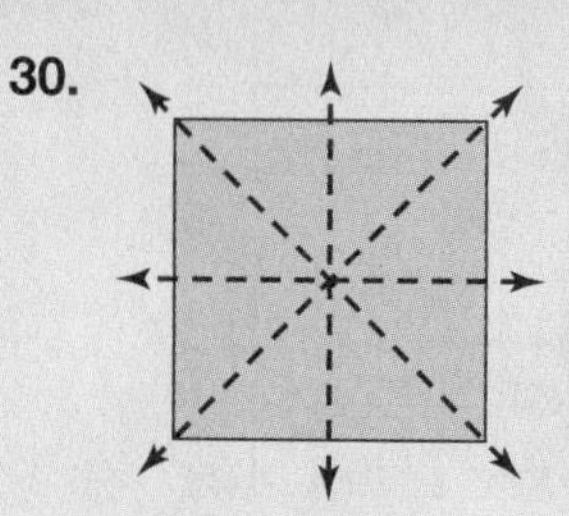

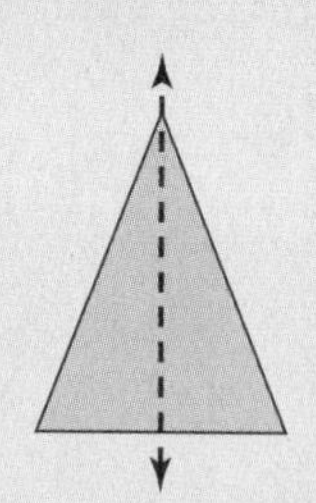

page 528 Problem Solving Application

1.

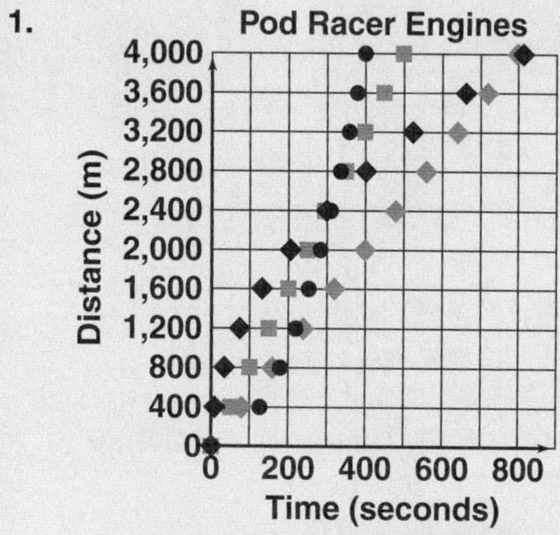

2a.

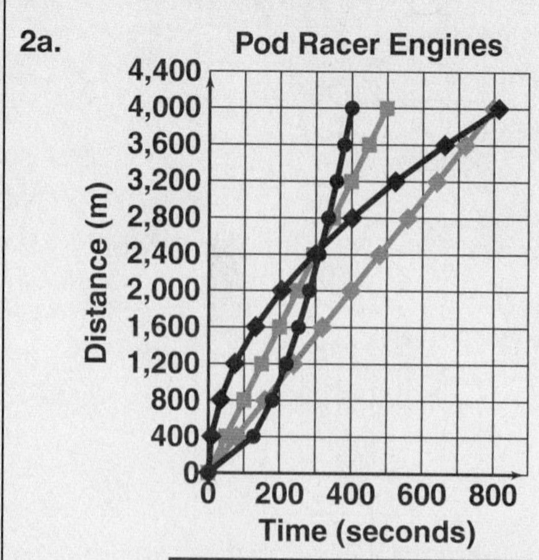

CHAPTER 11

Lesson 11-1

pages 532–533 Quick Check

1. U.S. Representatives for 22 States

Number of Representatives	Tally	Frequency
1	𝍸	5
2	II	2
3	IIII	4
4	II	2
5	III	3
6		0
7	II	2
8	II	2
9	II	2

2. Number of Students in Math Classes

20 22 24 26 28

3. Ages of Employees

Number of Employees

Age (years)

20–29 30–39 40–49 50–59 60–69 70–79

page 534 More Than One Way

Answers may vary. Sample:

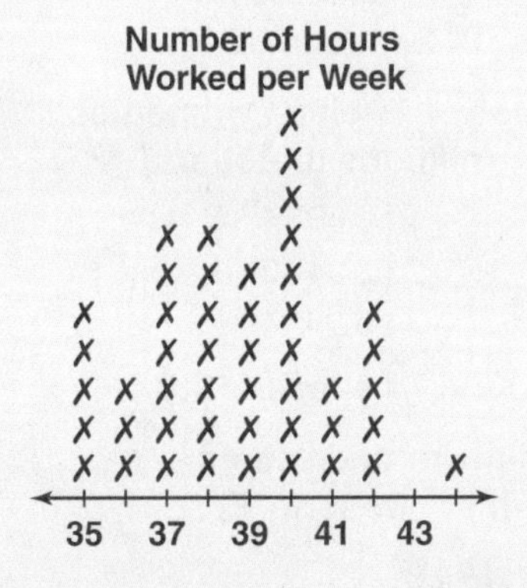

pages 534–536 Exercises

2.

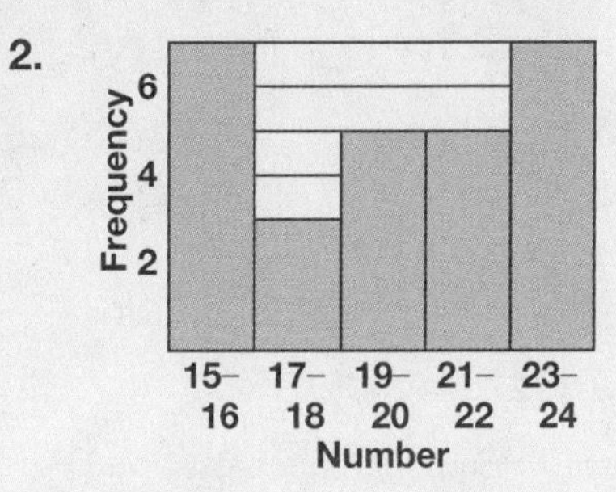

3.

Tickets Sold	Tally	Frequency
45	II	2
46	IIII	4
47		0
48	II	2
49	I	1
50	III	3
51	II	2
52		0
53	I	1

4.

Number of TVs	Tally	Frequency
1	𝍸 I	6
2	𝍸 I	6
3	𝍸	5
4	I	1

5.

Student Ages	Tally	Frequency
11	IIII	4
12	𝍸	5
13	𝍸	5
14	III	3

6. Plants Sold per Person

5 6 7 8 9 10 11 12

Plants

7.

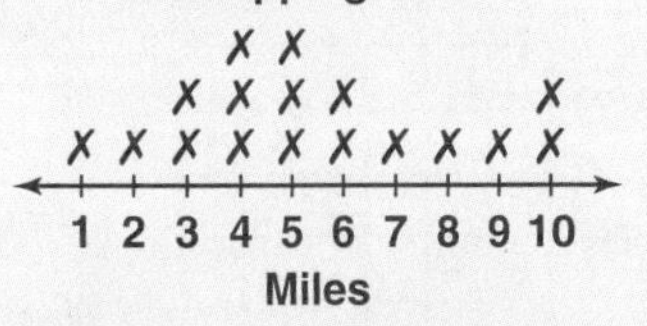

8. Blocks Walked From Home to School

1 2 3 4 5 6 7 8 9 10 11 12

Blocks

9.

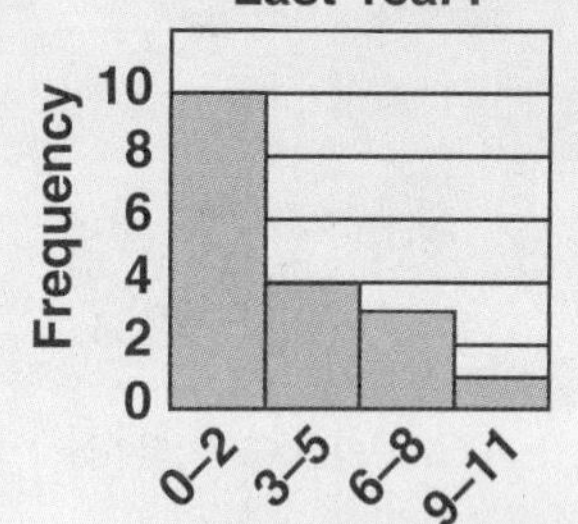

10.

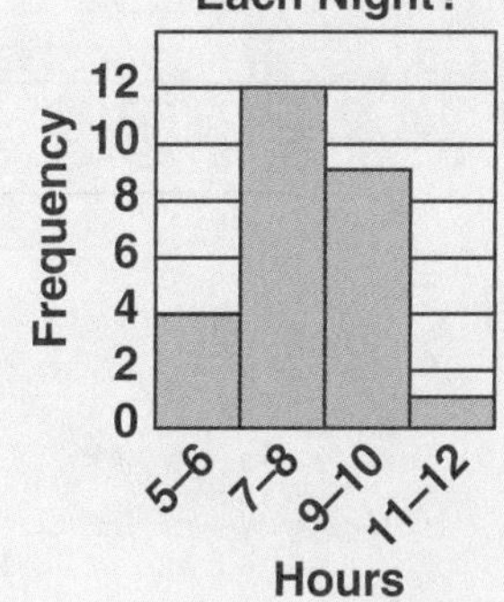

11.

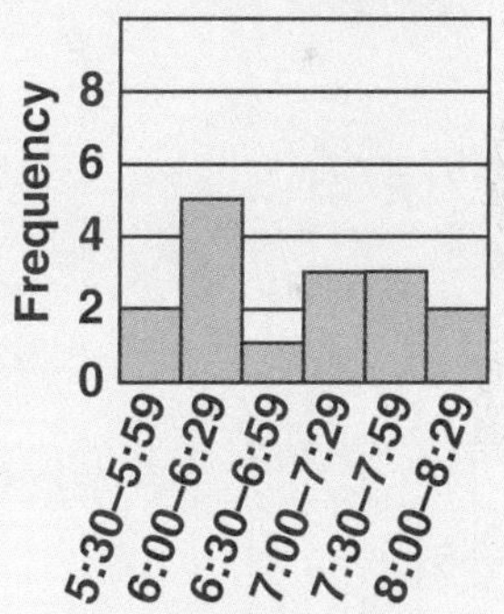

17. Answers may vary. Sample:

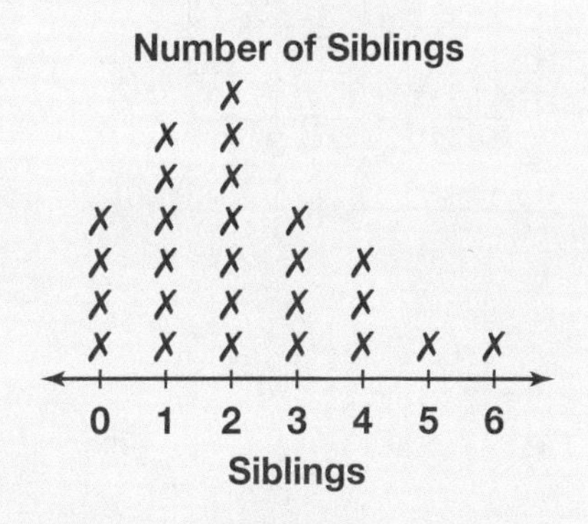

Additional Answers

18.

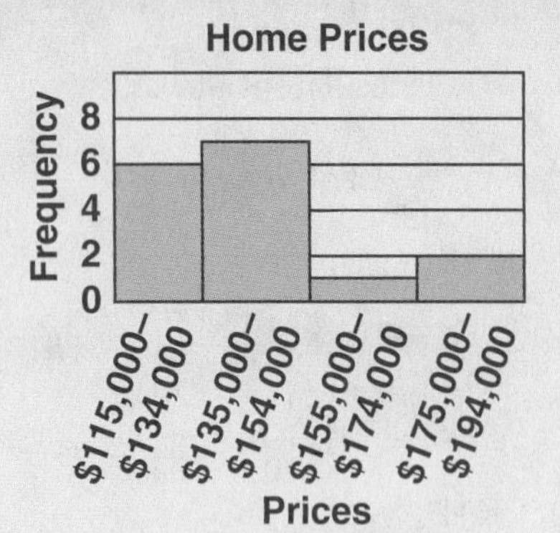

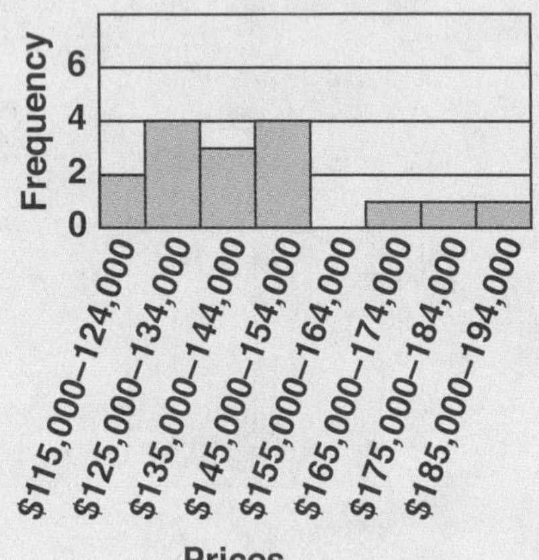

22.

x	y
−3	8
−2	3
−1	0
0	−1
1	0
2	3
3	8

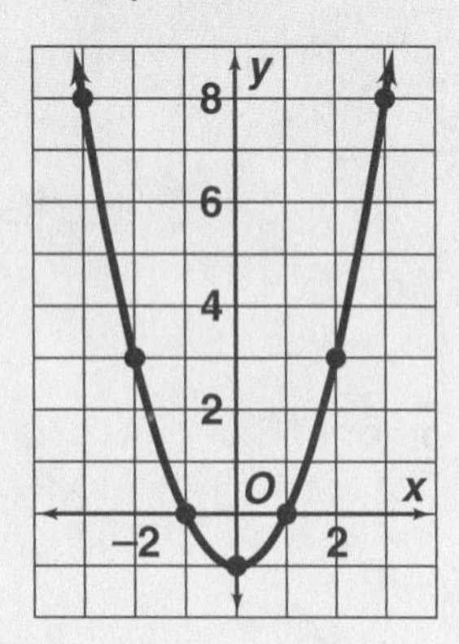

23.

x	y
−3	27
−2	12
−1	3
0	0
1	3
2	12
3	27

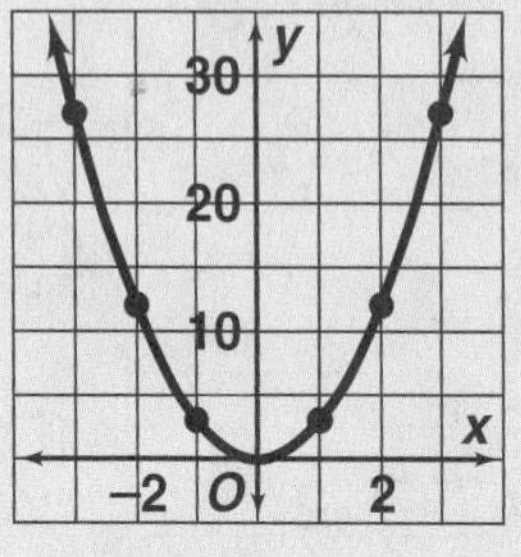

24.

x	y
−3	−27
−2	−12
−1	−3
0	0
1	−3
2	−12
3	−27

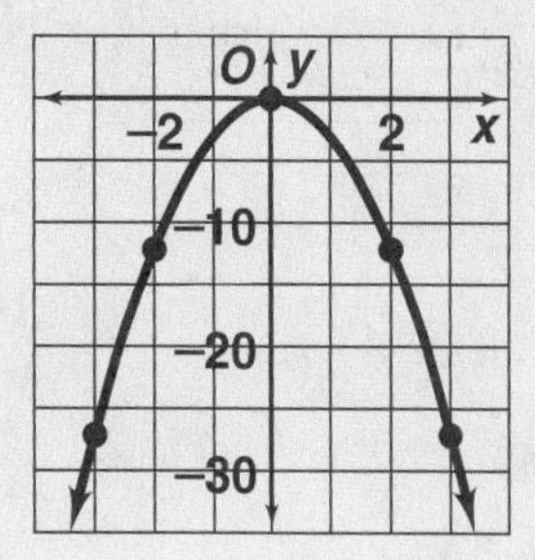

Lesson 11-2

page 538 Check Skills You'll Need

2.

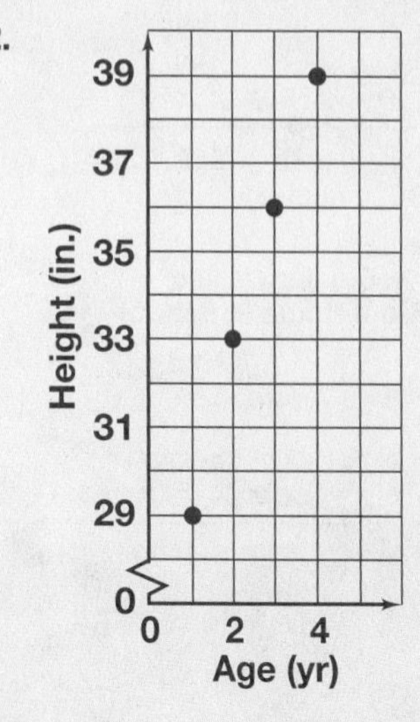

page 542 Homework Exercises

27.

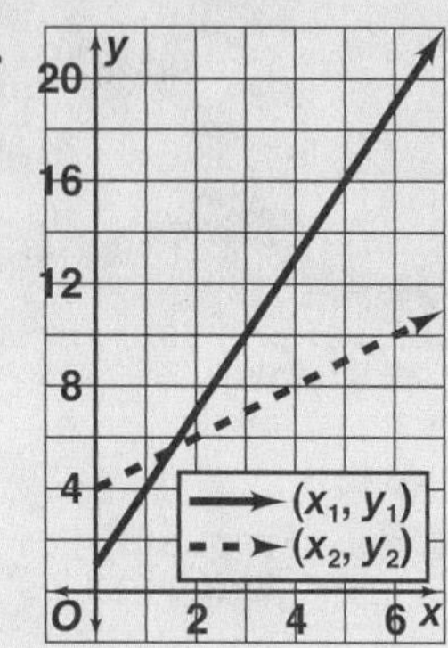

a. B3 = 4, B4 = 7, D4 = 6, D5 = 7

b. (1.5, 5.5)

page 544 Quick Check

1. Wind Speeds Recorded During Storm (mph)

0	9
1	4 6 8
2	5 5 9
3	0 3 4 8

Key: 0 | 9 means 9

page 545 Advanced Learners

1.

1	56
2	80
3	40 60 78
4	22 50
5	31
6	10
7	25

Key: 1 | 56 means $1.56

page 546 Homework Exercises

5.

4	3 6 7
5	1 2
6	0 1 7
7	1 2 8
8	2 6 8

Key: 8 | 2 means 82

6. Sales of Twelve Companies (millions of dollars)

1	3 3 4 4 4 4
2	3 4 5 5
3	6 9

Key: 1 | 3 means 1.3

7. High Temperatures in the Desert (°F)

9	8 9
10	0 1 3 4 8 9
11	1 2 2 3 3 8

Key: 9 | 8 means 98

page 549 Checkpoint Quiz 1

1.

Number	Tally	Frequency
1	||	2
2	||	2
3	||	2
4	||	2
5	|||	3
6	|	1
7	||	2
8	||	2
9	|	1

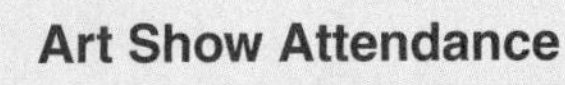

2. Graphs and explanations may vary. Sample graph:

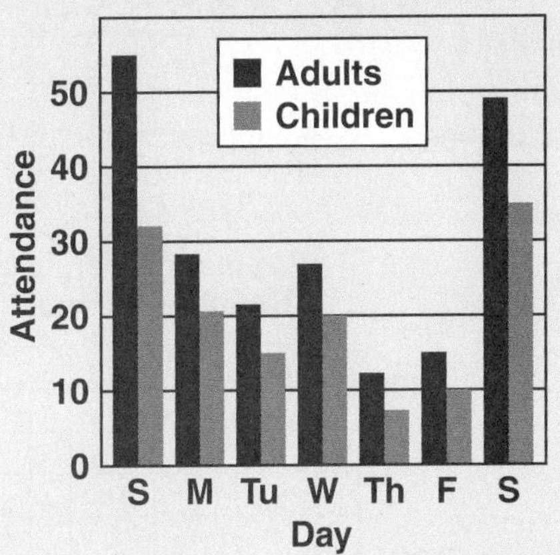

10.

8	7 9
9	3 5 5
10	
11	2 3 7 7
12	0 1 4

Key: 8 | 7 means 87

Lesson 11-4

pages 552–553 Homework Exercises

7. Part (b); the sample is more diverse.
8. Part (a); the sample is more diverse.
9. Part (b); the sample is more diverse.
10. Part (a); the sample is more diverse.

17. Survey B; survey A is biased because it uses the terms *noisy* and *overpriced;* survey B is not biased because it makes no assumptions.

19. Yes; surveying every 10th person is random, and you should get a mix of ages and genders.

20. No; you will include people outside of your population in the survey.

25. Start with 3 and add 5 repeatedly; 23, 28, 33

26. Start with −2 and add 3 repeatedly; 10, 13, 16

27. Start with 27 and subtract 11 repeatedly; −17, −28, −39

Lesson 11-5

page 555 Homework Exercises

1. Researchers capture, mark, and release animals, and then capture another group of animals. The number of marked animals in the second group indicates population size.
2. about 12,917 trout
3. about 13,661 bass

TE Closure

Use the proportion:

$$\frac{\text{number of marked animals counted}}{\text{total number of animals counted}} = \frac{\text{number of marked animals}}{\text{estimate of animal population}}.$$

pages 556 Homework Exercises

19a. The proportion should be $\frac{8}{25} = \frac{38}{x}$.

b. 119 sharks

25–29.

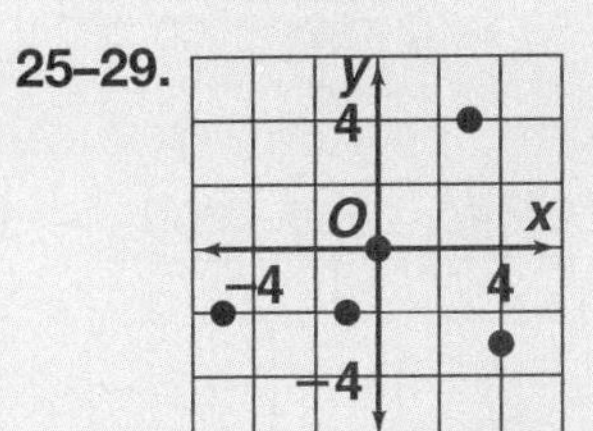

page 557 Activity Lab

1.

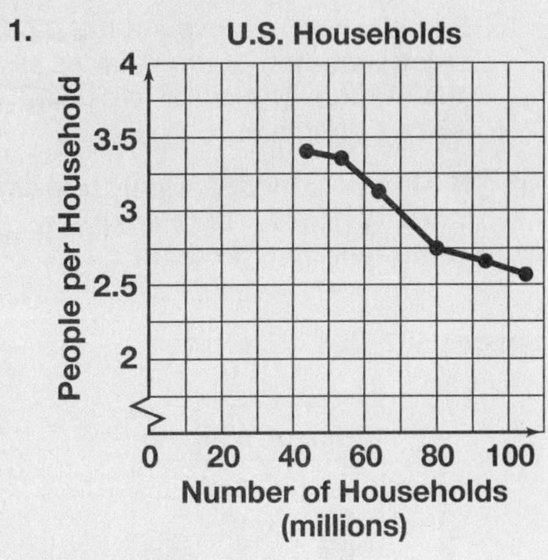

Lesson 11-6

page 561 Quick Check

2a.

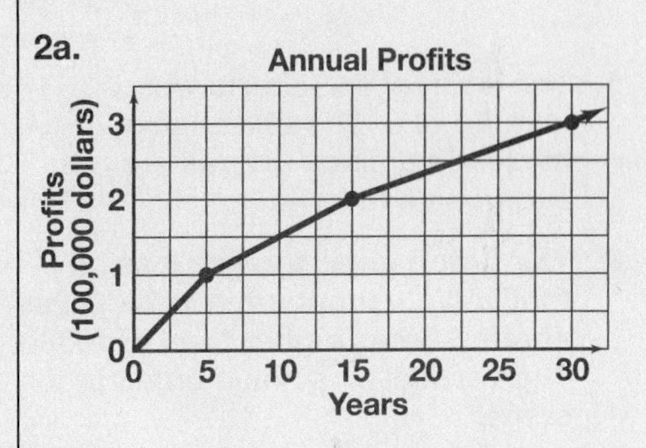

b.

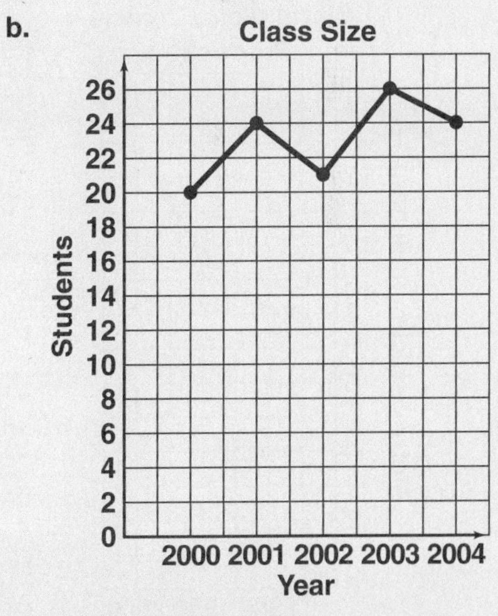

TE Additional Examples

2. The intervals on the horizontal axis are uneven. The number of viewers may not have increased as sharply as the graph illustrates.

Additional Answers

page 562 TE Additional Examples

3. Sample: The mean is not the best representation of the salaries at the station. The owner's salary skews the mean.

4. Misleading; there are only two cars under $10,000. There are many more over that amount.

pages 562–564 Exercises

4.

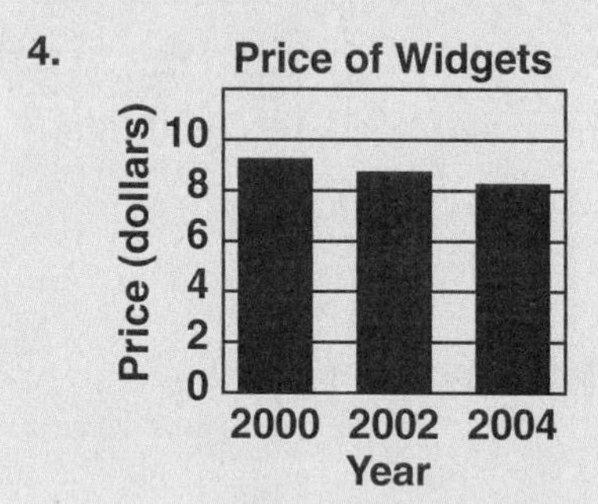

5. The vertical axis starts with 8 instead of 0. The title is also misleading since grapes are not everyone's favorite.

6. The graph gives the impression that vehicle production in 2000 was about 5 times higher than in 1980. The vertical axis has a break in the scale.

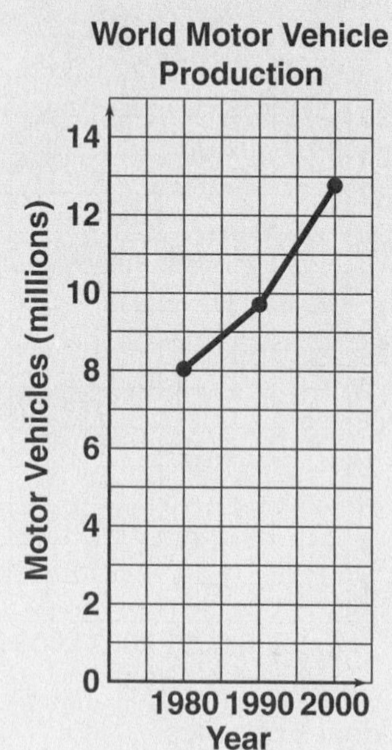

7. The graph gives the impression that the largest group of students received the highest scores. There are uneven intervals on the *x*-axis.

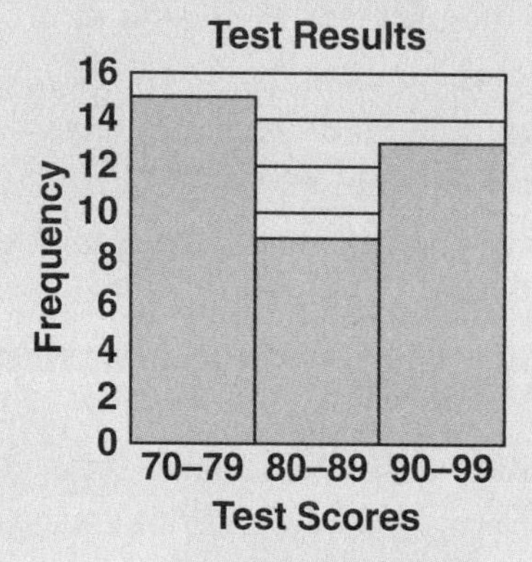

12.

Men's Olympic 100-Meter Winning Times

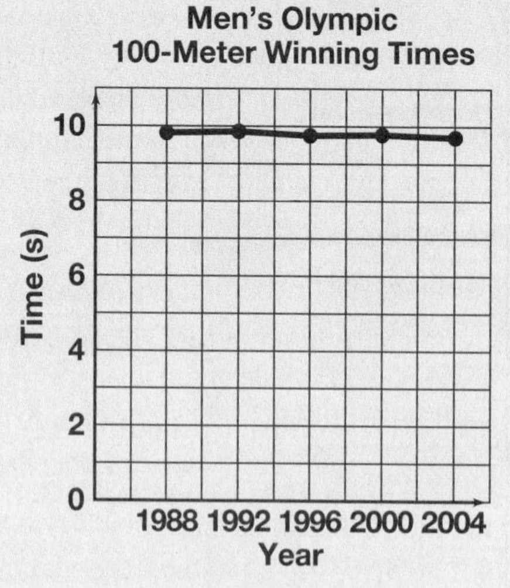

16.

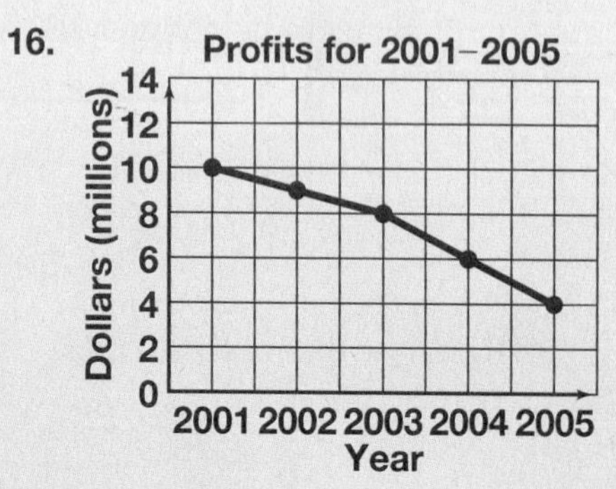

18. Answers may vary. Sample: The median; it does not reflect the lower profits.

page 565 Checkpoint Quiz 2

1. Answers may vary. Sample given.

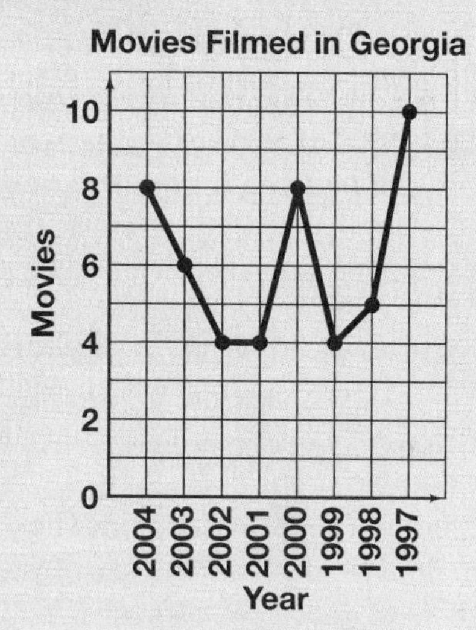

page 566 Activity Lab

3.

Moose and Wolf Populations on Isle Royale, Michigan

Lesson 11-7

page 567 Check Skills You'll Need

2–7.

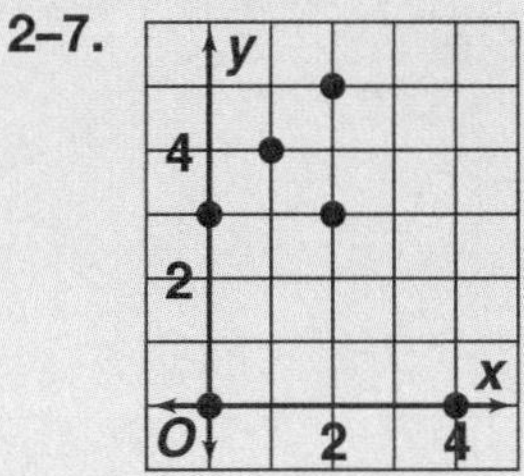

page 568 TE Additional Examples

1.

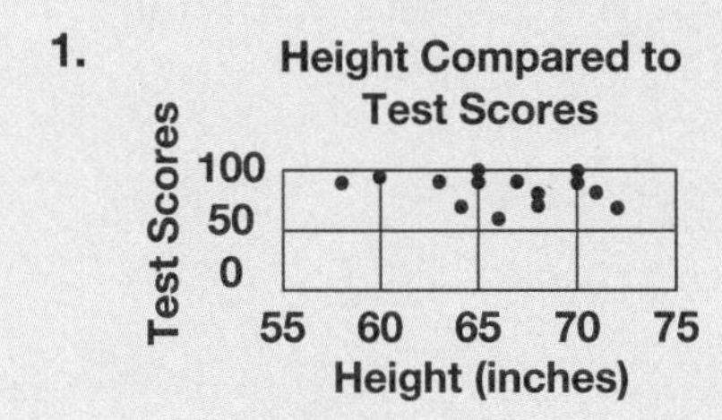

TE Closure

a graph that relates two sets of data by plotting the two sets of data as ordered pairs

A positive trend has one set of values increasing as the others increase. A negative trend has one set of values decreasing as the others increase. No trend has no relationship between the sets of data.

pages 569–570 Homework Exercises

6.

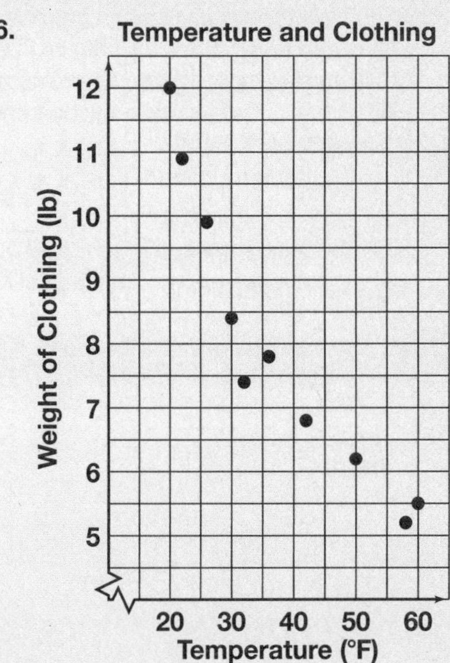

7.

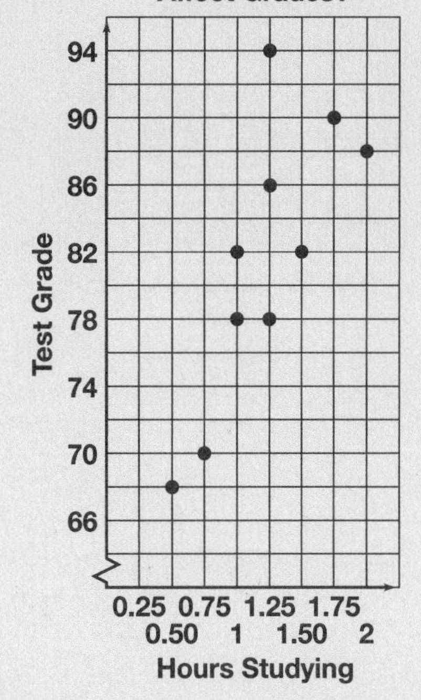

8.

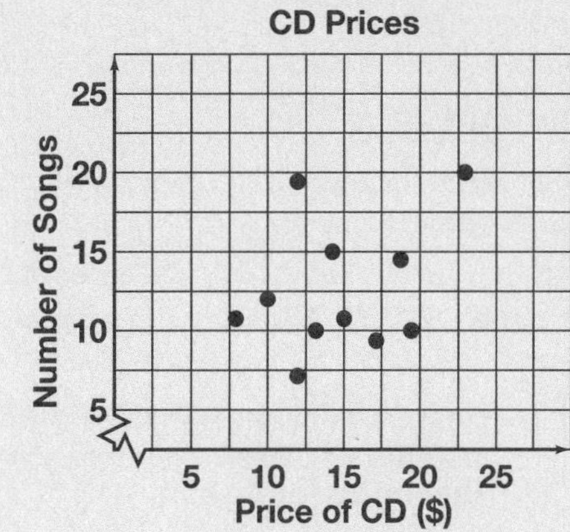

19.

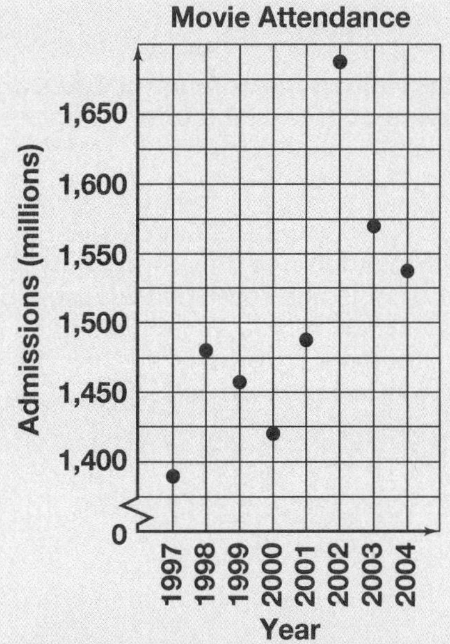

Answers may vary. Sample: 1,750 million

pages 572–573 Chapter Review

6. **Number of Hours of TV (weekly)**

Hours	Tally	Frequency
3	I	1
4	II	2
5	~~IIII~~	5
6	~~IIII~~	5
7	IIII	4
8	II	2
9	I	1

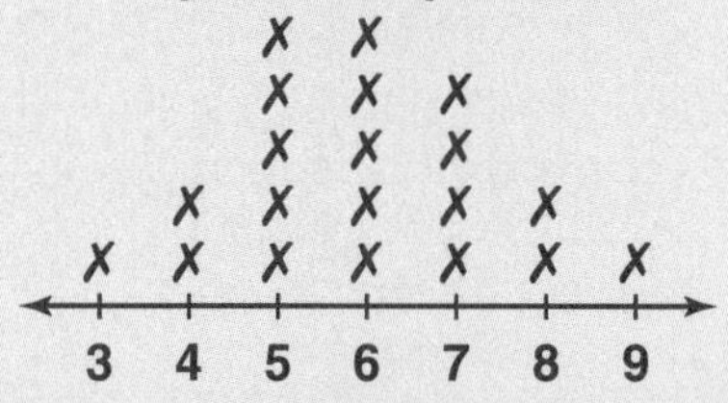

7.

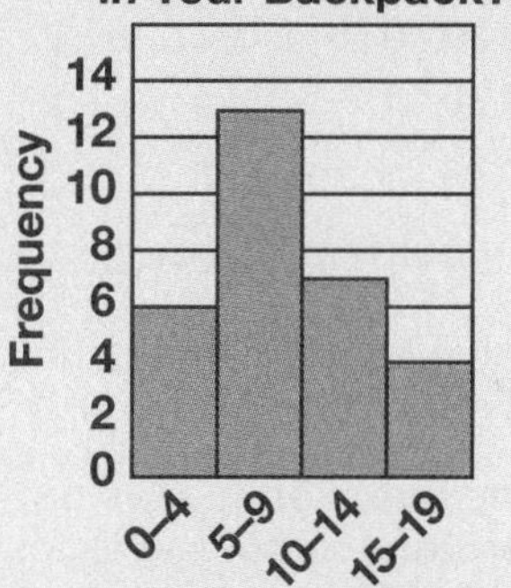

13.

```
7 | 0
6 | 0 0 0 1 6 9
5 | 0 0 4 5 8
4 | 0 0 0 5 7 7 8
3 | 6
2 | 3 7 7 9
1 | 6 9
```

Key: 1 | 6 means 16

page 574 Chapter Test

1. **Hours of Sleep**

Hours	Tally	Frequency
6	I	1
6.5	II	2
7	II	2
7.5	~~IIII~~	5
8	~~IIII~~ I	6
8.5	II	2
9	I	1
9.5	I	1

2. Hours of Sleep

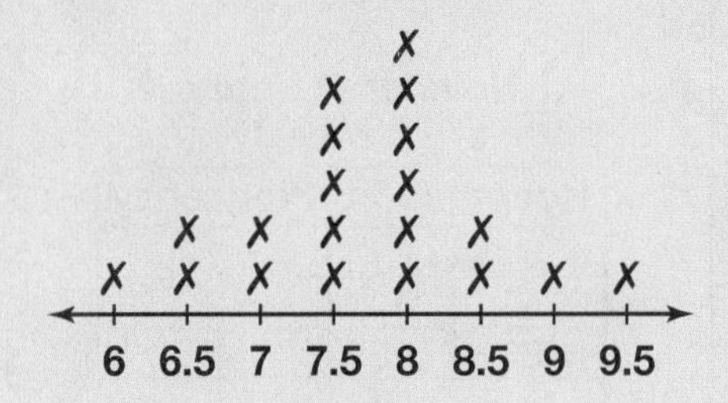

3. How Many Hours of Sleep Do You Get per Night?

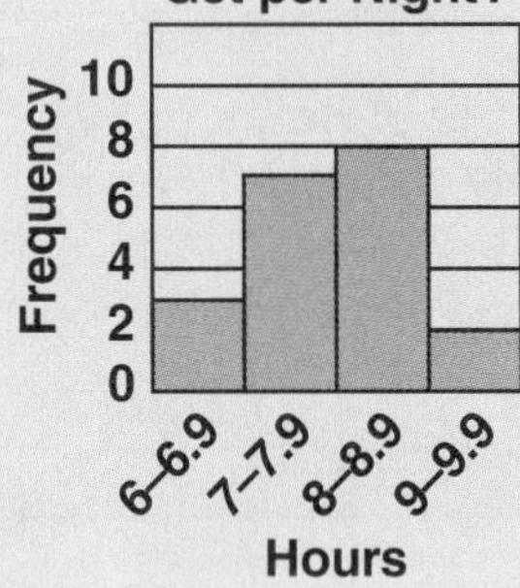

4. Hours of Sleep

Stem	Leaf
6	0 5 5
7	0 0 5 5 5 5 5
8	0 0 0 0 0 0 5 5
9	0 5

Key: 6 | 0 means 6.0

5. There would no longer be a zero after 9, and there would be three new rows ending with 12 in the stem and 0 in the leaf.

11. Fair; the question makes no assumptions about the types of shows on TV.

12. Biased; the question makes assumptions about sports being violent and documentaries being informative.

13. Biased; the question makes assumptions that summers are preferable to winters.

14. Fair; the question makes no assumptions about the seasons.

15. Answers may vary. Sample: Get a list of the people who live in the town. Call every 10th person and ask if he or she uses the library.

18.

Sales for Two Computer Companies

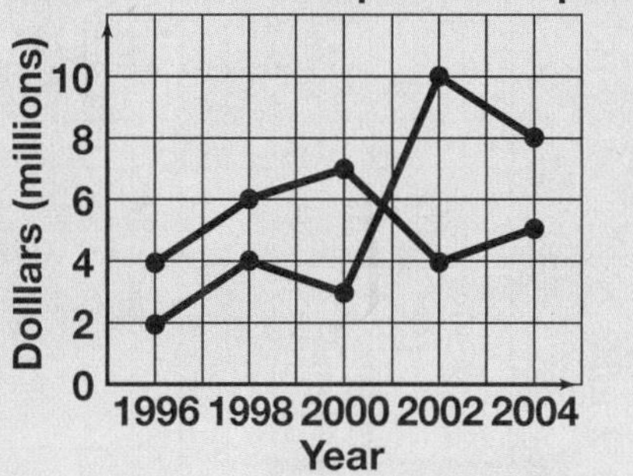

22a. Student Ages and Heights

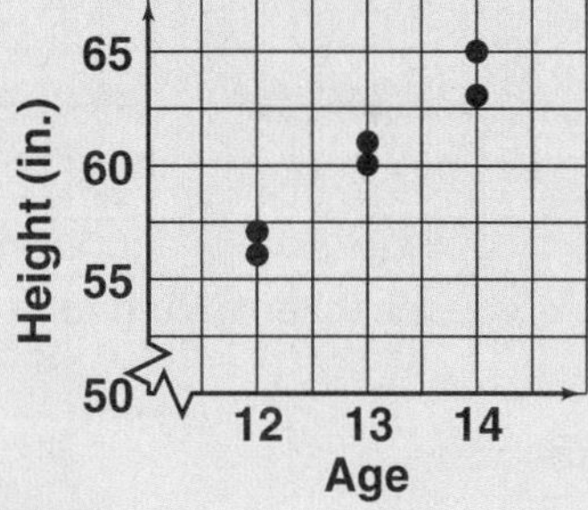

b. positive

CHAPTER 12

Lesson 12-2

page 587 TE Additional Examples

3. Results will vary, but, in a large number of trials, will approach the theoretical probability of $\frac{1}{16}$.

TE Closure

the ratio of the number of times an event occurs to the total number of trials

Sample: Decide what coins to use and what outcomes they represent; toss the coins and record each result; count how many times the successful outcome occurred; write the ratio for the experimental probability.

Lesson 12-3

page 592 Quick Check

2a.

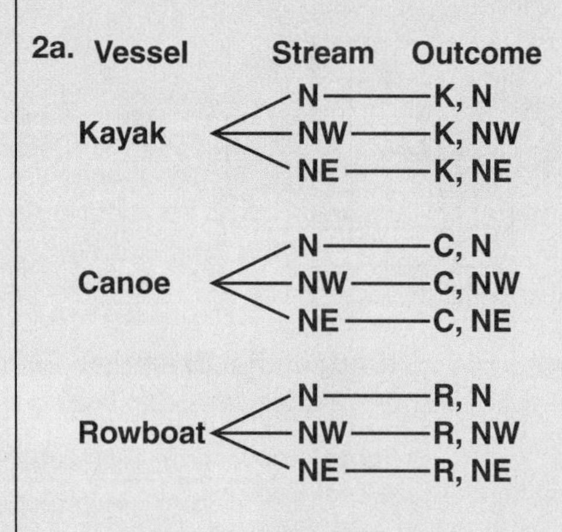

pages 593–595 Exercises

2.

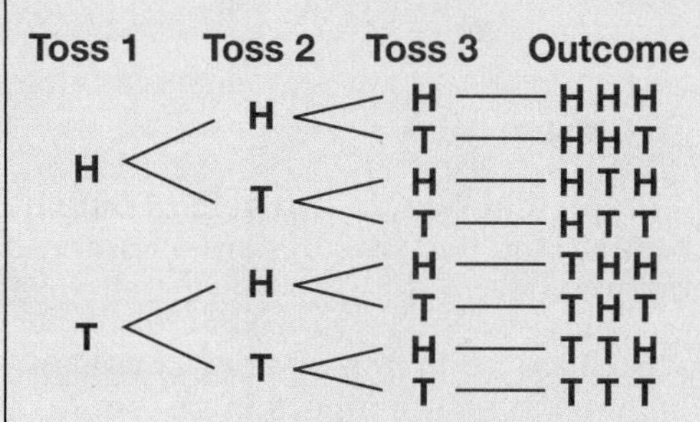

14.

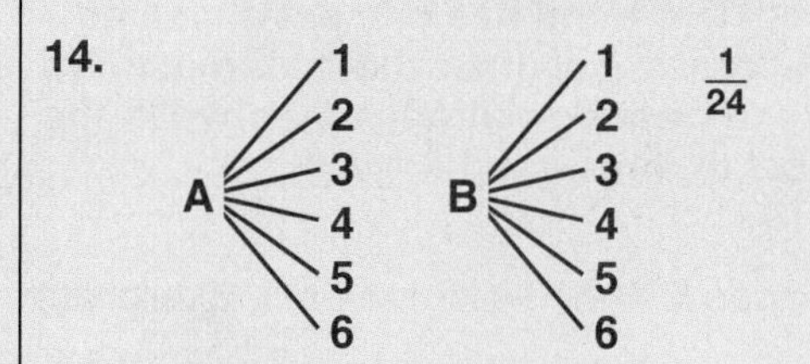

$\frac{1}{24}$

26.

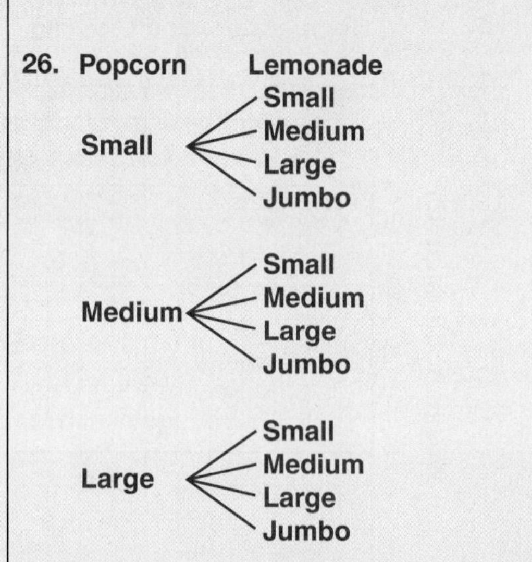

page 596 Activity Lab

4. $P(\text{match}) = \frac{12}{30}$ or $\frac{2}{5}$;

$P(\text{do not match}) = \frac{18}{30}$ or $\frac{3}{5}$

Diagrams may vary. Sample is given.

First Choice \ Second Choice	B	B	B	W	W	W
B	—	BB	BB	BW	BW	BW
B	BB	—	BB	BW	BW	BW
B	BB	BB	—	BW	BW	BW
W	WB	WB	WB	—	WW	WW
W	WB	WB	WB	WW	—	WW
W	WB	WB	WB	WW	WW	—

Lesson 12-4

page 602 Homework Exercises

29a. Complementary; red, yellow, green, or broken are all the possible events of a traffic light.

b. Mutually exclusive; since a student can only receive one grade on a test, the events have no outcomes in common.

30. Dependent; answers may vary. Sample: $P(\text{red}) = \frac{4}{9}$ and $P(\text{white}) = \frac{2}{9}$. When you pick without replacement, $P(\text{red, then white}) = \frac{4}{9} \cdot \frac{2}{8}$. The probability of the second event is affected by the occurrence of the first.

35.

	1	2	3
H	H1	H2	H3
T	T1	T2	T3

page 616 Chapter Review

11. Appetizer Soup Main Dish

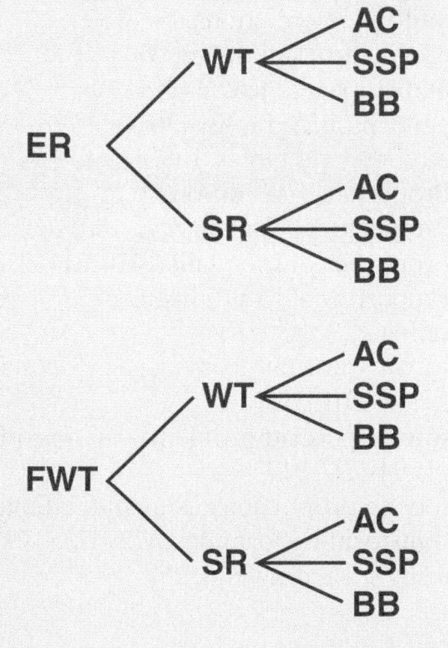

Additional Answers

Index

Teacher's Edition entries appear in blue type.

A

B

D

E

E

F

G

H

I

J

K

L

M

N

O

P

Q

R

S

T

U

V

W

X

Y

Z

Acknowledgements

Staff Credits

The people who make up the **Prentice Hall Math** team—representing design services, editorial, editorial services, educational technology, marketing, market research, photo research and art development, production services, publishing processes, and rights & permissions—are listed below. Bold type denotes core team members.

Dan Anderson, Carolyn Artin, Nick Blake, **Stephanie Bradley,** Kyla Brown, Patrick Culleton, Kathleen J. Dempsey, **Frederick Fellows, Suzanne Finn,** Paul Frisoli, Ellen Granter, **Richard Heater,** Betsy Krieble, Lisa LaVallee, Christine Lee, Kendra Lee, Cheryl Mahan, **Carolyn McGuire,** Eve Melnechuk, Terri Mitchell, Jeffrey Paulhus, Mark Roop-Kharasch, Marcy Rose, Rashid Ross, Irene Rubin, Siri Schwartzman, Vicky Shen, **Dennis Slattery,** Elaine Soares, Dan Tanguay, Tiffany Taylor, Mark Tricca, Paula Vergith, Kristin Winters, Helen Young

Additional Credits

Paul Astwood, Sarah J. Aubry, Jonathan Ashford, Peter Chipman, Patty Fagan, Tom Greene, Kevin Keane, Mary Landry, Jon Kier, Dan Pritchard, Sara Shelton, Jewel Simmons, Ted Smykal, Steve Thomas, Michael Torocsik, Maria Torti

TE Design

Susan Gerould/Perspectives

Illustration

Additional Artwork

Rich McMahon; Ted Smykal

Kenneth Batelman: **580**; Joel Dubin: **601**; John Edwards, Inc.: **184, 263, 265, 589**; Das Grup: **593, 594**; Kelly Graphics: **351, 354**; Carla Kiwior: **132, 142**; Brucie Roche: **7, 195, 235, 415, 424, 582**; John Schreiner: **218, 307, 597**; Wilkinson Studios: **423, 613**; JB Woolsey: **233, 253**; XNR Productions, Inc.: **32, 43, 55, 260, 262, 264, 383, 386, 389, 390**

Photography

Front Cover: Paul Frankian/Index Stock Imagery, Inc.
Back Cover: Corbis/Picture Quest

Title page: tl, Bob Daemmrich Photography; **tr,** Williamson Edwards/The Image Bank; **bl,** David Muench; **br,** Bob Daemmrich Photography.

Front matter Pages x, Tom Brakefield/Corbis; **xi,** Jim Craigmyle/Corbis; **xii,** Robert Tyrell; **xiii,** Joe Gemignani/Corbis; **xiv,** Getty Images, Inc; **xv,** Mitch Windham/The Image Works; **xvi,** Roger Ressmeyer/Corbis; **xvii,** Rudi Von Briel/Photo Edit; **xviii,** Jeff Greenberg/PhotoEdit; **xix,** Andersen Ross/Getty Images, Inc; **xx,** David Young-Wolff/PhotoEdit; **xxi,** Andre Jenny/Alamy; **xlviii,** Richard Haynes; **xlix,** Richard Haynes; **l,** Bryan Peterson/Getty Images, Inc.; **lii,** Cindy Charles/PhotoEdit; **liii,** Randall Hyman; **liv,** Jeff Lepore/Photo Researchers, Inc.; **lvi,** Photonica/Getty Images; **lvii,** Brian Bailey/Getty Images, Inc.; **xlviii,** Richard Haynes.

Chapter 1 Pages 3, Art Wolfe/Photo Researchers, Inc; **4**, Tom Brakefield/Corbis; **5**, Russ Lappa; **8**, Mary Kate Denny/Photo Edit; **8 ml**, Richard Haynes; **11 ml**, Gallo Images/Corbis; **11 mr**, Andrew McKim/Masterfile; **14**, Robin Nelson/Photo Edit; **16**, Robert Rathe/Getty Images; **17**, Terry Oakley/Alamy; **18**, Frank Siteman/Monkmeyer; **19**, Richard Haynes; **20**, Spencer Grant/Photo Edit; **21**, Eyewire/Getty Images, Inc. **23**, Rachel Epstein/PhotoEdit, **26 tr**, Prentice Hall photo by Jane Latta; **26 mr**, Guy Ryecart/Dorling Kindersley; **26 br**, Dorling Kindersley; **28**, Russ Poole Photography; **28 tl**, Richard Haynes; **28 mr**, Richard Haynes; **31**, Age Fotostock/SuperStock; **32**, Wayne R. Blenduke/Getty Images; **34**, Reuters NewMedia, Inc./Corbis; **35**, Richard Haynes; **37**, Richard Haynes; **41**, Keith Kent/Peter Arnold, Inc.; **43**, Richard Haynes; **44**, G. Kalt/Zefa/Corbis; **45**, Philip & Karen Smith/SuperStock; **47**, Peter Pinnock/Getty Images, Inc.; **51**, Doug Wilson/Alamy Images, **53**, Richard Haynes; **54 bl**, Clive Boursnel/Dorling Kindersley; **54 br**, Kevin Summers/Getty Images; **55**, Arthur Tilley/Getty Images, Inc.; **56**, Tim Davis/Getty Images; **57**, Gordon Clayton/Dorling Kindersley; **64 t**, Photo Franca Principe, IMSS; **64–65 b**, Digital Vision/Getty Images, Inc.

Chapter 2 Pages 67, Steve Satushek/Getty Images; **68**, Richard Haynes; **71**, Andrew Syred/Photo Researchers, Inc.; **74**, D. J. Peters/AP/Wide World Photos, **76 ml**, Richard Haynes, **76 br**, Richard Haynes; **78**, G.D.T./Getty Images, Inc.; **79 b**, Richard Haynes; **79 t**, Russ Lappa; **83**, Hans Halberstadt/Corbis; **84**, Rubberball/SuperStock; **86**, Richard Haynes; **87 mr**, A. & J. Visaoe/Peter Arnold, Inc.; **87 bl**, Richard Haynes; **88**, Jim Pickerell, **91**, Associated Press/THE HAWK EYE/AP Wide World; **92**, Tony Savino/The Image Works; **94**, The Image Works; **96**, Spencer Grant/PhotoEdit; **98**, Jeff Greenberg/PhotoEdit; **100**, Janet Foster/Masterfile; **102**, Rhoda Sidney/PhotoEdit; **105**, David Aubrey/Corbis; **106**, Jim Craigmyle/Corbis; **107**, NASA/Finley Holiday Films; **109**, *FOXTROT*©1992 Bill Amend. Reprinted with permission of Universal Press Syndicate. All rights reserved.; **116 bl**, Jim Corwin/Stock Connection/PictureQuest; **116 tl**, Dorling Kindersley; **116 tm**, Courtesy of Sony; **116 tr**, Steve Cole/Photodisc/Getty Images, Inc.; **116–117 b**, Abe Rezny/The Image Works

Chapter 3 Pages 119, Andrew Syred/SPL/Photo Researchers; **121**, Jump Run Productions/Getty Images; **123**, Peter Vanderwarker/Stock Boston; **124**, Richard Haynes; **125**, Richard Haynes; **126 mr**, David Kelly Crow/Photo Edit; **126 bl**, Richard Haynes; **127**, Addison Geary/Stock Boston; **129 ml**, Tom Stewart/Corbis; **129 mr**, John A. Rizzo/Getty Images, Inc.; **130**, Earl Carter/Southeastern Sports Photos; **131**, Marnie Burkhart/Masterfile; **133**, Tom Stewart/Corbis; **134**, Tom Stewart/Corbis; **135**, Richard Haynes; **136**, Richard Hutchings/PhotoEdit; **138**, Robert Tyrell; **139**, Kwame Zikomo/SuperStock, Inc.; **140**, Richard Haynes; **141**, Richard Haynes; **142**, Pierre Arsenault/Masterfile; **143 tl**, Richard Haynes; **143**, Richard Haynes; **145**, Douglas Faulkner/Corbis; **149**, Karl Weatherly/PhotoDisc/Getty Images, Inc.; **151**, Gallo Images/Corbis; **155 tl**, David C. Ellis/Getty Images; **155 tr**, Russ Lappa; **157**, Friedrich Von Horsten/Animals Animals/Earth Scenes; **164–165 b**, Jose Fuste Raga/Corbis; **165 mr**, Barbara Magnuson Larry Kimball/Visuals Unlimited; **165 tl**, Rainer Grosskopf/Getty Images, Inc.

Chapter 4 Pages 167, Bruce M. Herman/Photo Researchers, Inc.; **169**, Eddy Lemaistre/For Picture/Corbis; **170**, David Young-Wolff/PhotoEdit; **172**, Francois Gohier/Ardea; **174**, Keren Su/Corbis; **175**, Ariel Skelley/Corbis Stock Market; **177**, Joe Gemignani/Corbis; **180**, Richard Haynes; **181**, Lee Cohen/Getty Images; **182**, Janeart/Getty Images; **183**, Dr. Jeremy Burgess/Photo Researchers, Inc.; **185**, Richard Haynes; **187**, Richard Haynes; **189**, Will Hart; **191**, Prentice Hall; **194**, age fotostock/Superstock; **197**, Stuart Westmorland/Getty Images; **198**, David Young-Wolff/PhotoEdit; **199 t**, Richard Haynes; **200**, Ron Sachs/Corbis; **201**, Alan Thornton/Getty Images, Inc.; **202 tl**, Richard Haynes; **202 mr**, Richard Haynes; **203**, Reuters/Corbis; **204**, Patrick Clark/PhotoDisc/Getty Images, Inc.; **205**, D&J Heaton/Stock Boston; **206**, Russ Lappa; **208 tl**, SuperStock, Inc.; **208 tr**, Michael Newman/PhotoEdit; **210**, Bill Bachmann/Photo Edit **211**, Stephen Simpson/Getty Images; **213**, Kevin R. Morris/Corbis; **214**, Frances Roberts/Alamy; **215**, Jose Luis Pelaez, Inc./Corbis; **217**, Antonio Mo/Getty Images, Inc.; **224 t**, Steve Shott/Dorling Kindersley; **224–225 b**, Ron Kimball/Premium Stock/PictureQuest; **225 mr 1, 2, & 3**, Russ Lappa; **225 tl**, Steve Shott/Dorling Kindersley; **225 tr**, Dorling Kindersley

Chapter 5 Pages 227, Dave G. Houser/Corbis; **228**, Richard Haynes; **229**, AP Photo/Paul Sakuma; **231**, LWA-Dann Tardif/CORBIS; **232**, David Young-Wolff/Photo Edit; **235**, Andy Lyons/Getty Images, Inc.; **236**, BIOS/Peter Arnold, Inc.; **237 t**, Alan Levenson/Getty Images; **237 b**, Ron Kimball/Ron Kimball Stock; **238**, Richard Haynes; **241**, Johnson Space Center/NASA; **246 tl & mr**, Richard Haynes; **247**, Franklin D. Roosevelt Library; **248**, Photodisc Green/Getty Images; **251**, Richard Haynes; **252**, Alaska Stock LLC/Alamy; **255**, Brian Sytnyk/Masterfile; **257**, Richard Haynes; **258**, Richard Haynes; **259**, Getty Images, Inc.; **264**, AP/Wide World Photos; **271 tm**, www.rubberball.com

Chapter 6 Pages 273, Omni Photo Communications, Inc./Index Stock; **276**, Mitsuaki Iwago/Minden Pictures; **278**, Richard Haynes; **280**, Kelly Mooney Photography/Corbis; **281**, Peter Johnson/Corbis; **283**, Hillary Price; **284**, Robin Nelson/Photo Edit; **285**, David Young-Wolff/PhotoEdit; **287**, Tom McHugh/Photo Researchers, Inc.; **288**, Richard Haynes; **290**, Rhoda Sidney/Photo Edit; **291**, Richard Haynes; **293**, UNEP/Rougier/The Image Works; **294**, IML Image Group Ltd./Alamy; **297**, Bob Daemmrich/Stock Boston; **298 tr**, Mitch Windham/The Image Works; **298 bl**, Richard Haynes; **299**, David Young-Wolff/PhotoEdit; **301**, Bob Daemmrich/The Image Works **304**, Bernard Wolf/PhotoEdit; **305 tl**, Adamsmith Productions/Corbis; **305 bl**, Richard Haynes; **307**, Lawrence Migdale/Photo Researchers, Inc.; **310,** Mark Reinstein/The Image Works; **312 tl & mr**, Richard Haynes; **313**, Photo Researchers, Inc.; **320 bgrd, & br**, Courtesy of the U.S. Geological Survey; **321 bl, & ml**, Courtesy of the U.S. Geological Survey; **321 tl**, David Nicholls

Chapter 7 Pages 323, William A. Bake/Corbis; **325**, Rafeal Macia/Photo Researchers, Inc.; **326 mr**, Design by Jean-Charles Guillois and Ines Levy; **326 br**, James Marshall/Corbis; **328 mr**, Kevin Mallet/Dorling Kindersley; **328 tr**, Richard Haynes; **329**, Richard Haynes; **330**, Richard Pasley/Stock Boston; **331**, Richard Haynes; **332 mr & bl**, Richard Haynes; **333**, William Taufic/Corbis; **334**, Spencer Grant/PhotoEdit; **335**, Richard Haynes; **336**, Russ Lappa; **339**, Paul Hermansen/Getty Images, Inc.; **340**, 2007 Mondrian/Holtzman Trust c/o HCR International, Warrenton, VA **342**, Bill Aron/PhotoEdit; **343 tl**, Max Alexander/DK Picture Library; **343 tr**, Felicia Martinez/PhotoEdit; **343 br**, Paul Trummer/Getty Images, Inc.; **343 bl**, Cavagnaro/Visuals Unlimited; **344**, 1995 by NEA, Inc. Thaves 6–13; **345**, Peter Beck/Corbis; **346**, Roger Ressmeyer/Corbis; **347**, 2004 AFP/Getty Images; **350**, Spencer Grant/PhotoEdit; **351**, Lawrence Migdale; **353**, James L. Amos/Corbis; **354**, Richard Haynes; **355**, NASA; **357**, Spencer Grant/Photo Edit; **361**, Andreas Pollok/Getty Images; **362**, Courtesy of Wellesley College Library, Special Collections, photo by George McLean; **370 bl**, Professional Miniature Golf Association; **370 tr**, Jack Hollingsworth/Getty Images, Inc.; **370–371** Richard Hamilton Smith/Corbis; **371 tr**, SW Productions/Getty Images, Inc.

Chapter 8 Pages 373, Michael Grecco/Stock Boston; **374**, Getty Images; **375**, Jacques Descloitres, MODIS Rapid Response Team, NASA/GSFC; **379**, Richard Haynes; **380**, Janis Daemmrich/Daemmrich Photography; **380 bl**, Richard Haynes; **381**, Rudi Von Briel/Photo Edit; **384**, Bob Daemmrich/The Image Works; **387**, U.S. Coast Guard; **388**, Photo by Randy Varga; **389**, Buddy Mays/Corbis; **390 tl & br**, Richard Haynes; **392**, C Squared Studios/Getty Images, Inc.; **393**, Richard Haynes; **395**, Larry Lilac/Alamy; **397**, Lawrence Migdal/Stock Boston; **400**, Myrleen Ferguson Cate/PhotoEdit; **403**, *FOXTROT* ©1992 Bill Amend. Reprinted with permission of Universal Press Syndicate. All rights reserved. **406**, Jeff Greenberg/PhotoEdit; **408**, Michael J. Howell/Stock Boston; **409 mr & bm**, Russ Lappa; **409 tr**, Richard Haynes; **409 bl**, Russ Lappa; **410**, David Parker/Photo Researchers, Inc.; **413**, Davies & Starr/Getty Images, Inc.; **413**, John Lei/Stock Boston; **413**, Spike Mafford/Getty Images, Inc.; **416**, Richard Haynes; **418**, Amy Etra/PhotoEdit; **420**, Richard Haynes; **421 mr**, George & Monserrate Schwartz/Alamy; **421**, Richard Haynes; **432 tl**, Chris Brown/Stock Boston, Inc./PictureQuest; **432–433 m**, T.J. Florian/PictureQuest; **433 mr**, Digital Vision; **433 tm**, Phil Kember/Index Stock Imagery/PictureQuest

Chapter 9 Pages 435, Charles Bush; **438**, George Shelley/Masterfile; **440**, Bob Daemmrich/Stock Boston; **441 bl**, Russ Lappa; **441 bm**, Russ Lappa; **441 br**, Russ Lappa; **441 tr**, Richard Haynes; **442**, Hal Horwitz/Corbis; **445**, Kaz Chiba/Getty Images; **446**, SuperStock; **449 ml**, Cheryl Hogue/Visuals Unlimited; **449 mr**, Russ Lappa; **450**, Chris Marona/Photo Researchers, Inc.; **455**, Michael Newman/PhotoEdit; **456**, Royalty Free/Corbis; **457**, Mark Bernett/Stock Boston; **459**, Bettmann/Corbis; **461**, Bob Daemmrich/The Image Works; **462**, Jeff Greenberg/PhotoEdit; **464**, Lowe Art Museum, University of Miami/SuperStock, Inc.; **466**, Russ Lappa; **468 mr**, Ariel Skelley/Corbis; **468 bl**, Richard Haynes; **470**, Superstock/Alamy; **472,** AP/Wide World Photos; **473 mr & br**, Richard Haynes; **473 tl**, Richard Haynes; **475**, Richard Hutchings/Photo Researchers, Inc.; **482 ml**, Harold Wilion/Index Stock Imagery/PictureQuest; **482–483 m**, Neil Fletcher/Dorling Kindersley; **483 tm**, Ruth A. Adams/Index Stock Imagery/PictureQuest

Chapter 10 Pages 485, NASA/Masterfile; **487**, David R. Frazier/The Image Works; **489**, John R. Bracegirdle/Getty Images, Inc.; **491**, Dave Nagel/Getty Images, Inc.; **492**, Richard Haynes; **493**, Andersen Ross/Getty Images, Inc.; **494**, Lynda Richardson/Corbis; **498**, AP Wide World Photos; 499, Corbis; **500 ml**, Peter Guttman/Corbis; **500 mr**, Adam Jones/Photo Researchers, Inc.; **501**, Prentice Hall; **503**, Richard Haynes; **504**, Colorsport; **505 ml, tl, & br**, Richard Haynes; **506**, Getty Images; **507**, Steve Fitchett/Getty Images, Inc.; **508**, Richard Haynes; **510**, Victoria & Albert Museum, London/Art Resource, NY; **513**, George Hall/Corbis; **514 br**, Charles Kennard/Stock Boston; 514 bl, Jerome Wexler/Photo Researchers, Inc.; **514 mr**, Nuridsany et Perennou/Photo Researchers, Inc.; **517**, Sturgis McKeever/Photo Researchers, Inc.; **519**, Jonathan Nourok/Photo Edit; **519 br**, Foto World/Getty Images, Inc.; **528 tl**, LUCASFILM LTD. Star Wars: Episode II - Attack of the Clones © 2002 Lucasfilm Ltd. & TM. All rights reserved. Used under authorization. Unauthorized duplication is a violation of applicable law. No internet use.; **528–529 m**, LUCASFILM LTD. Star Wars: Episode II - Attack of the Clones © 2002 Lucasfilm Ltd. & TM. All rights reserved. Used under authorization. Unauthorized duplication is a violation of applicable law. No internet use.

Chapter 11 Pages 531, John Hickey/The Buffalo News; **532**, Jose Luis Pelaez, Inc./Corbis; **534 tr & ml**, Richard Haynes; **535**, Dennis MacDonald/PhotoEdit; **537,** Siede Preis/Getty Images, Inc.; **538**, Bill Aron/Photo Edit; **544**, David Young-Wolff/PhotoEdit; **546**, Jeff Greenberg/PhotoEdit; **550**, Andy King/Reuters/Landov; **552**, Anton Vengo/SuperStock, Inc.; **553**, Karen Preuss/The Image Works; **554**, Jim Brandenburg/ Minden Pictures; **556**, Amos Nachuom/Corbis; **560**, Tony Freeman/PhotoEdit; **562**, Patricia Brabant/Getty Images, Inc.; **565**, Michael Newman/PhotoEdit; **567**, Richard Haynes; **568**, David Weintraub/Photo Researchers, Inc.; **576 tr**, Randy Ury/Corbis; **576–577 br**, Jeff Maloney/Getty Images, Inc.; **577 tm**, Philip Gatward/Dorling Kindersley

Chapter 12 Pages 579, Pictor International, Ltd./PictureQuest; **580**, Jim Cummins/Getty Images; **581**, Richard Haynes; **583**, Andre Jenny/Alamy; **584**, Prentice Hall; **585**, Richard Haynes; **586**, Andy Sacks/Getty Images, Inc.; **586**, Richard Haynes; **587**, Paul Barton/Corbis; **589**, David Young-Wolff/PhotoEdit; **591**, Burke/Triolo Productions/Foodpix; **592**, Hughes Martin/Corbis; **595**, TRBPhoto/Getty Images, Inc.; **596**, Bob Daemmrich/The Image Works; **598**, Emely/Zefa/Corbis; **600 tr & ml**, Richard Haynes; **602**, Comstock/Alamy; **603**, Richard Haynes; **606**, Getty Images, Inc.; **607**, Robert Laberge/Getty Images, Inc.; **609**, Joseph Sohm; ChromoSohm Inc./Corbis; **610**, Todd Warnock/Getty Images; **611**, David Hiller/Getty Images; **613**, Russ Lappa; **622 tr**, Ken Chernus/Getty Images, Inc.; **622–623 b**, Larry Dale Gordon/Getty Images, Inc.; **623 br**, Tony Bee/photolibrary/PictureQuest

Teacher's Edition

Editorial Services: PubSmarts, LLC Pearson Education Development Group
Production Services: Argosy Publishing